e Elements

Name	Symbol	Atomic Number	Relative Atomic Mass
inium	Ac	89	(227)
minum	Al	13	26.982
ericium	Am	95	(243)
imony	Sb	51	121.76
on	Ar	18	39.948
enic	As	33	74.922
atine	At	85	(210)
ium	Ba	56	137.33
kelium	Bk	97	(247)
yllium	Be	4	9.0122
nuth	Bi	83	208.98
rium	Bh	107	(264)
on	B	5	10.81
mine	Br	35	79.904
mium	Cd	48	112.41
cium	Ca	20	40.078
ifornium	Cf	98	(251)
bon	C	6	12.011
ium	Ce	58	140.12
ium	Cs	55	132.91
orine	Cl	17	35.45
omium	Cr	24	51.996
alt	Co	27	58.933
per	Cu	29	63.546
ernicium	Cn	112	
ium	Cm	96	(247)
mstadtium	Ds	110	(271)
nium	Db	105	(262)
prosium	Dy	66	162.50
steinium	Es	99	(252)
ium	Er	68	167.26
opium	Eu	63	151.96
nium	Fm	100	(257)
orine	F	9	18.998
ovium	Fl	114	
ncium	Fr	87	(223)
olinium	Gd	64	157.25
lium	Ga	31	69.723
manium	Ge	32	72.630
d	Au	79	196.97
nium	Hf	72	178.49
sium	Hs	108	(269)
ium	He	2	4.0026
mium	Ho	67	164.93
drogen	H	1	1.008
ium	In	49	114.82
ne	I	53	126.90
ium	Ir	77	192.22
	Fe	26	55.845
pton	Kr	36	83.798
thanum	La	57	138.91
rencium	Lr	103	(262)
d	Pb	82	207.2
ium	Li	3	6.94
etium	Lu	71	174.97
rmorium	Lv	116	
gnesium	Mg	12	24.305
nganese	Mn	25	54.938
tnerium	Mt	109	(268)

Name	Symbol	Atomic Number	Relative Atomic Mass
Mendelevium	Md	101	(258)
Mercury	Hg	80	200.59
Molybdenum	Mo	42	95.96
Neodymium	Nd	[illegible]	144.24
Neon	Ne	[illegible]	20.180
Neptunium	N	[illegible]	(237)
Nickel	[illegible]	[illegible]	58.693
Niobium	[illegible]	[illegible]	92.906
Nitroge	[illegible]	[illegible]	14.007
Nob	[illegible]	[illegible]	(259)
Osmi	[illegible]	76	190.23
Oxyge	[illegible]	8	15.999
Palladiu	[illegible]	46	106.42
Phosphor	P	15	30.974
Platinum	Pt	78	195.08
Plutonium	Pu	94	(244)
Polonium	Po	84	(209)
Potassium	K	19	39.098
Praseodymium	Pr	59	140.91
Promethium	Pm	61	(145)
Protactinium	Pa	91	231.04
Radium	Ra	88	(226)
Radon	Rn	86	(222)
Rhenium	Re	75	186.21
Rhodium	Rh	45	102.91
Roentgenium	Rg	111	(272)
Rubidium	Rb	37	85.468
Ruthenium	Ru	44	101.07
Rutherfordium	Rf	104	(261)
Samarium	Sm	62	150.36
Scandium	Sc	21	44.956
Seaborgium	Sg	106	(266)
Selenium	Se	34	78.96
Silicon	Si	14	28.085
Silver	Ag	47	107.87
Sodium	Na	11	22.990
Strontium	Sr	38	87.62
Sulfur	S	16	32.06
Tantalum	Ta	73	180.95
Technetium	Tc	43	(98)
Tellurium	Te	52	127.60
Terbium	Tb	65	158.93
Thallium	Tl	81	204.38
Thorium	Th	90	232.04
Thulium	Tm	69	168.93
Tin	Sn	50	118.71
Titanium	Ti	22	47.867
Tungsten	W	74	183.84
Uranium	U	92	238.03
Vanadium	V	23	50.942
Xenon	Xe	54	131.29
Ytterbium	Yb	70	173.05
Yttrium	Y	39	88.906
Zinc	Zn	30	65.38
Zirconium	Zr	40	91.224

ic masses in this table are relative to carbon-12 and are rounded to five figures, as recommended by the International Union of Pure and ied Chemistry (IUPAC) Pure Appl. Chem., *Vol. 85, No. 5, pp. 1047–1078, 2013. For certain radioactive elements the numbers listed (in ntheses) are the mass numbers of the most stable isotopes. For H, Li, B, C, N, O, Mg, Si, S, Cl, Br, and Tl, the* conventional atomic mass, *a sentative value from its* atomic mass interval, *is provided. See page 49.*

NERAL CHEMISTRY

ICIPLES AND MODERN APPLICATIONS

ELEVENTH EDITION

RALPH H. PETRUCCI
California State University, San Bernardino

F. GEOFFREY HERRING
University of British Columbia

JEFFRY D. MADURA
Duquesne University

CAREY BISSONNETTE
University of Waterloo

PEARSON
Toronto

Editorial Director: Claudine O'Donnell
Executive Acquisitions Editor: Cathleen Sullivan
Senior Marketing Manager: Kimberly Teska
Program Manager: Darryl Kamo
Project Manager: Sarah Gallagher
Manager of Production Management: Avinash Chandra
Manager of Content Development: Suzanne Schaan
Developmental Editor: Joanne Sutherland
Media Editor: Johanna Schlaepfer
Media Developer: Shalin Banjara
Production Services: Cenveo® Publisher Services
Permissions Project Manager: Kathryn O'Handley
Photo Permissions Research: Carly Bergey, Lumina Datamatics
Text Permissions Research: Varoon Deo-Singh, MPS North America LLC.
Interior and Cover Designer: Alex Li
Cover Image: Cenveo Publisher Services

Vice-President, Cross Media and Publishing Services: Gary Bennett

Credits and acknowledgments for material borrowed from other sources and reproduced, with permission, in this t
appear on the appropriate page within the text.

---

3 16

Library and Archives Canada Cataloguing in Publication

Petrucci, Ralph H., author
General chemistry : principles and modern applications
/ Ralph H. Petrucci, F. Geoffrey Herring, Jeffrey D. Madura,
Carey Bissonnette.—Eleventh edition.

Includes index.
ISBN 978-0-13-293128-1 (bound)

1. Chemistry—Textbooks. I. Title.

QD31.3.P47 2016 540 C2015-904266-6

WARNING: Many of the compounds and chemical reactions described or pictured in this book are hazardous. Do not attempt any experiment pictured
in the text except with permission in an authorized laboratory setting and under adequate supervision.

PEARSON

ISBN 978-0-13-2

We, the authors, dedicate this edition to Ralph H. Petrucci who passed away as the final edits of this edition were being completed. The first edition of *General Chemistry: Principles and Modern Applications* was published in 1972 with Ralph as the sole author. Although the book is now in its eleventh edition, with more authors, it is still shaped by Ralph's original vision and his belief that students are very much interested in the practical applications, social significance, and historical roots of the subject areas they study, as well as their conceptual frameworks, facts, and theories. Ralph was an inspiring mentor who warmly welcomed each of us to the authoring team. We envied his clear and precise writing style and impeccable eye for detail. He was an excellent advisor to us during the preparation of the most recent editions, all of which benefited greatly from his valuable input. We will miss him dearly.

Brief Table of Contents

APPENDICES

Contents

us On Discussions on MasteringChemistry™ (www.masteringchemistry.com)

About the Authors

Ralph H. Petrucci

Ralph Petrucci received his B.S. in Chemistry from Union College, Schene NY, and his Ph.D. from the University of Wisconsin–Madison. Following ten of teaching, research, consulting, and directing the NSF Institutes for Seco School Science Teachers at Case Western Reserve University, Cleveland, Dr. Petrucci joined the planning staff of the new California State University pus at San Bernardino in 1964. There, in addition to his faculty appointme served as Chairman of the Natural Sciences Division and Dean of Acad Planning. Professor Petrucci, now retired from teaching, is also a coaut *General Chemistry* with John W. Hill, Terry W. McCreary, and Scott S. Perry.

F. Geoffrey Herring

Geoff Herring received both his B.Sc. and his Ph.D. in Physical Chem from the University of London. He is currently a Professor Emeritus i Department of Chemistry of the University of British Columbia, Vanco Dr. Herring has research interests in biophysical chemistry and has publi more than 100 papers in physical chemistry and chemical physics. Rec Dr. Herring has undertaken studies in the use of information technology interactive engagement methods in teaching general chemistry with a vi improving student comprehension and learning. Dr. Herring has ta chemistry from undergraduate to graduate levels for 30 years and has been the recipient of the Killam Prize for Excellence in Teaching.

Jeffry D. Madura, FRSC

Jeffry D. Madura is Professor and the Lambert F. Minucci Endowed Ch Computational Sciences and Engineering in the Department of Chemistry Biochemistry at Duquesne University located in Pittsburgh, PA. He ear B.A. from Thiel College in 1980 and a Ph.D. in Physical Chemistry Purdue University in 1985 under the direction of Professor Wi L. Jorgensen. The Ph.D. was followed by a postdoctoral fellowship in co tational biophysics with Professor J. Andrew McCammon at the Universi Houston. Dr. Madura's research interests are in computational chemistry biophysics. He has published more than 100 peer-reviewed papers in phy chemistry and chemical physics. Dr. Madura has taught chemistry to u graduate and graduate students for 24 years and was the recipient Dreyfus Teacher-Scholar Award. Dr. Madura was the recipient of the American Chemical Society Pittsburgh Section Award and received the E School of Natural and Environmental Sciences and the Duquesne Unive Presidential Award for Excellence in Scholarship in 2007. Dr. Madura ACS Fellow and a Fellow of the Royal Society of Chemistry. He is curr working with high school students and teachers as part of the ACS Sci Coaches program.

Carey Bissonnette

Carey Bissonnette is Continuing Lecturer in the Department of Chemist the University of Waterloo, Ontario. He received his B.Sc. from the Unive of Waterloo in 1989 and his Ph.D. in 1993 from the University of Camb in England. His research interests are in the development of method eling dynamical processes of polyatomic molecules in the gas phase. He won awards for excellence in teaching, including the University of rloo's Distinguished Teacher Award in 2005. Dr. Bissonnette has made nsive use of technology in both the classroom and the laboratory to create teractive environment for his students to learn and explore. For the past ral years, he has been actively engaged in undergraduate curriculum lopment, high-school liaison activities, and the coordination of the university's high-school chemistry contests, which are written each year by stus around the world.

Preface

"Know your audience." For this new edition, we have tried to follow this im tant advice by attending even more to the needs of students who are taking a ous journey through this material. We also know that most general chem students have career interests not in chemistry but in other areas such as bio medicine, engineering, environmental science, and agricultural sciences. An understand that general chemistry will be the only university or college chem course for some students, and thus their only opportunity to learn some pra applications of chemistry. We have designed this book for all these students.

Students of this text should have already studied some chemistry. But t with no prior background and those who could use a refresher will find tha early chapters develop fundamental concepts from the most elementary i Students who do plan to become professional chemists will also find oppor ties in the text to pursue their own special interests.

The typical student may need help identifying and applying principles visualizing their physical significance. The pedagogical features of this tex designed to provide this help. At the same time, we hope the text serv sharpen students' skills in problem solving and critical thinking. Thus, we tried to strike the proper balances between principles and applications, qualit and quantitative discussions, and rigor and simplification.

Throughout the text and on the MasteringChemistry® site (www.maste chemistry.com) we provide real-world examples to enhance the discus Examples relevant to the biological sciences, engineering, and the environm sciences are found in numerous places. This should help to bring chemistry for these students and help them understand its relevance to their career inte It also, in most cases, should help them master core concepts.

ORGANIZATION

In this edition we retain the core organization of the previous edition with notable exceptions. First, we have moved the chapter entitled Spontan Change: Entropy and Gibbs Energy forward in the text. It is now Chapter 1 moving the introduction of entropy and Gibbs energy forward in the text, w able to use these concepts in subsequent chapters. Second, we have movec chapter on chemical kinetics to Chapter 20. Consequently, the discussic chemical kinetics now appears after the chapters that rely on equilibrium thermodynamic concepts.

Like the previous edition, this edition begins with a brief overview of core cepts in Chapter 1. Then, we introduce atomic theory, including the periodic t in Chapter 2. The periodic table is an extraordinarily useful tool, and presenti early allows us to use the periodic table in different ways throughout the e chapters of the text. In Chapter 3, we introduce chemical compounds and stoichiometry. Organic compounds are included in this presentation. The e introduction of organic compounds allows us to use organic examples throug the book. Chapters 4 and 5 introduce chemical reactions. We discuss gas Chapter 6, partly because they are familiar to students (which helps them b confidence), but also because some instructors prefer to cover this material e to better integrate their lecture and lab programs. (Chapter 6 can easil deferred for coverage with the other states of matter, in Chapter 12.)

In Chapter 7, we introduce thermochemistry and discuss the energy cha that accompany physical and chemical transformations. Chapter 8 introd quantum mechanical concepts that are needed to understand the energy cha we encounter at the atomic level. This chapter includes a discussion of w ıanics, although this topic may be omitted at the instructor's discretion. ectively, Chapters 8 through 11 provide the conceptual basis for describ- he electronic structure of atoms and molecules, and the physical and ical properties of these entities. The properties of atoms and molecules hen used in Chapter 12 to rationalize the properties of liquids and solids. hapter 13 is a significant revision of Chapter 19 from the tenth edition. It duces the concept of entropy, the criteria for predicting the direction of taneous change, and the thermodynamic equilibrium condition. In pters 14–19, we apply and extend concepts introduced in Chapter 13. ever, Chapters 14–19 can be taught without explicitly covering, or refer- back to, Chapter 13.

s with previous editions, we have emphasized real-world chemistry in inal chapters that cover descriptive chemistry (Chapters 21–24), and we tried to make this material easy to bring forward into earlier parts of the Moreover, many topics in these chapters can be covered selectively, out requiring the study of entire chapters. The text ends with compre- sive chapters on nuclear chemistry (Chapter 25) and organic chemistry pters 26 and 27). Please note that an additional chapter on biochemistry pter 28) is available online.

ANGES TO THIS EDITION

have made the following important changes in specific chapters and endices:

- In Chapter 2 (Atoms and the Atomic Theory), new material is included to describe the use of atomic mass intervals and conventional atomic masses for elements such as H, Li, B, C, N, O, Mg, Si, S, Cl, Br, and Tl. Atomic mass intervals are recommended by the IUPAC because the isotopic abundances of these elements vary from one source to another, and therefore, their atomic masses cannot be considered constants of nature.
- Chapter 4 (Chemical Reactions) includes a new section that discusses the extent of reaction, and introduces a tabular approach for representing the changes in amount in terms of a single variable, representing the extent of reaction.
- In Chapter 5 (Introduction to Reactions in Aqueous Solutions), we revised Section 5-1 to differentiate between dissociation and ionization, and introduced a new figure to illustrate the dissociation of an ionic compound in water.
- Chapter 6 (Gases) makes increased use of the recommended units of pressure (e.g., Pa, kPa, and bar). Section 6-7 on the kinetic–molecular theory has been significantly revised. For example, the subsection on Derivation of Boyle's Law has been simplified and now comes after the subsections on Distribution of Molecular Speeds and The Meaning of Temperature. Section 6-8 has also been revised so that Graham's law is presented first, as an empirical law, which is then justified by using the kinetic–molecular theory.
- In Chapter 7 (Thermochemistry), we have updated the notation to ensure that we are using, for the most part, symbols that are recommended by the IUPAC. For example, standard enthalpies of reaction are represented by the symbol $\Delta_r H^\circ$ (not ΔH°) and are expressed in kJ mol^{-1} (not kJ). We have added a molecular interpretation of specific heat capacities (in Section 7-2) and an introduction to entropy (in Section 7-10).
- Chapter 8 (Electrons in Atoms) has been substantially rewritten to provide a logical introduction to the ideas leading to wave mechanics. Sections 8-2 and 8-3 of the previous edition have been combined and the

material reorganized. This chapter includes a new section that focus the energy level diagram and spectrum of the hydrogen atom. The tion entitled Interpreting and Representing the Orbitals of the Hydr Atom has been rewritten to include a discussion of the radial funct A new subsection describing the conceptual model for multielec atoms has been added to the section entitled Multielectron Atoms. sections on multielectron atoms and electron configurations have rewritten to emphasize more explicitly that the observed ground- electron configuration for an atom is the one that minimizes E_{atom} that the energies of the orbitals is only one consideration. There are new Are You Wondering? boxes in this chapter: Is the Born interpreta an idea we use to determine the final form of a wave function? and all orbital transitions allowed in atomic absorption and emission spe

- In Chapter 9 (The Periodic Table and Some Atomic Properties), a nur of sections have been rewritten to emphasize the importance of effe nuclear charge in determining atomic properties. A new section on p izability has been introduced. Several new figures have been create illustrate the variation of effective nuclear charge and atomic prope across a period or down a group (e.g., effective nuclear charges fo first 36 elements; the variation of effective nuclear charge and per screening with atomic number; the variation of average distance f the nucleus with atomic number; first ionization energies of the t row *p*-block elements; electron affinities of some of the main group ments; polarization of an atom; the variation of polarizability and atc volume with atomic number). The sections on ionization energies electron affinities have been significantly revised. Of particular note have revised the discussion of the decrease in ionization energy occurs as we move from group 2 to 13 and from group 15 to 16. Our cussion points out that various explanations have been used. The sec from the tenth edition entitled Periodic Properties of the Elements been deleted.
- Chapter 11 (Chemical Bonding II: Valence Bond and Molecular Or Theories) has been revised to include an expanded discussion of redistribution of electron density that occurs during bond formatior improved introduction to Section 11-5 Molecular Orbital Theory, an improved discussion of molecular orbital theory of the CO molecule have moved the section entitled Bonding in Metals online.
- Chapter 13 (Spontaneous Change: Entropy and Gibbs Energy) is a tot revised version of Chapter 19 from the previous edition. The cha focuses first on Boltzmann's view of entropy, which is based microstates, and then on Clausius's view, which relates entropy chang reversible heat transfer. The connection between microstates and part in-a-box model is developed to reinforce Boltzmann's view of entr Clausius's view of entropy change is used to develop expressions important and commonly encountered physical changes (e.g., pl transitions; heating or cooling at constant pressure; isothermal expan or compression of an ideal gas). These expressions are subsequently u to develop the criterion for predicting the direction of spontane change. The chapter includes a proper description of the differe between the Gibbs energy change of a system, ΔG, and the reac Gibbs energy, $\Delta_r G$. The reaction Gibbs energy ($\Delta_r G$) is used as the b for describing how the Gibbs energy of a system changes with comp tion (i.e., with respect to the extent of reaction). The derivation of equation is done in a separate section that may be used or skipped at instructor's discretion. The concepts of chemical potential and acti are also introduced.

In Chapter 14 (Solutions and Their Physical Properties), we have added a section to describe the standard thermodynamic properties of aqueous ions. We use the concepts of entropy and chemical potential in Chapter 13 to explain vapor pressure lowering and why gasoline and water don't mix.

Chapter 15 (Principles of Chemical Equilibrium) has been significantly revised to emphasize the thermodynamic basis of equilibrium and to de-emphasize aspects of kinetics. There is an increased emphasis on the thermodynamic equilibrium constant, which is expressed in terms of activities, along with an updated discussion of Le Châtelier's principle to emphasize certain limitations associated with its use (e.g., for certain reactions and initial conditions, the addition of a reactant may actually cause net change to the left). Several new worked examples are included to show how equilibrium constant expressions may be simplified and solved when the equilibrium constant is either very small or very large.

In Chapter 16 (Acids and Bases), significant changes have been made. Sections 16-1 through 16-3 have been significantly revised to provide a more logical flow and to emphasize and demonstrate that the distinction between strong and weak acids is based on the degree of ionization, which in turn depends on the magnitude of the acid ionization constant. There are two new sections, namely Sections 16-7 (Simultaneous or Consecutive Acid–Base Reactions: A General Approach) and 16-9 (Qualitative Aspects of Acid–Base Reactions). Section 16-7 focuses on writing and using material balance and charge equations. Section 16-9 focuses on predicting the equilibrium position of a general acid–base reaction. A new subsection entitled Rationalization of Acid Strengths: An Alternative Approach has been added to Section 16-10, Molecular Structure and Acid–Base Behavior. This new subsection focuses on factors that stabilize the anion formed by an acid.

In Chapter 19 (Electrochemistry), we have modified the Nernst equation to have the form $E_{cell} = E^\circ_{cell} - \frac{0.0257\ V}{z} \ln Q$. We have changed the text so that the standard hydrogen electrode is defined with respect to a pressure of 1 bar instead of 1 atm, and added a problem to the Integrative and Advanced Exercises to illustrate that this change in pressure causes only a small change in the standard reduction potentials (see Exercise 108). We have also added a section on reserve batteries.

In Appendix D, we have modified the table of Standard Electrode (Reduction) Potentials at 25 °C so that it now includes a column with the cell notation for the half-reactions.

ddition to the specific changes noted above, we have also changed much ne artwork throughout the textbook. In particular, all of the atomic and ecular orbital representations have been modified to be consistent across hapters. We have redone all of the electrostatic potential maps (EPMs) to e the same potential energy color scale unless noted in the textbook.

ERALL APPROACH

pedagogical apparatus and overall approach in this edition continue to ect contemporary thoughts on how best to teach general chemistry. We e retained the following key features of the text:

Logical approach to solving problems. All worked examples are presented consistently throughout the text by using a tripartite structure of Analyze–Solve–Assess. This presentation not only encourages students to use a logical approach in solving problems but also provides them with a way to start when they are trying to solve a problem that seem, at first, impossibly difficult. The approach is used implicit those who have had plenty of practice solving problems, but for t who are just starting out, the Analyze–Solve–Assess structure will s to remind students to (1) analyze the information and plan a stra (2) implement the strategy, and (3) check or assess their answer to er that it is a reasonable one.

- *Integrative Practice Examples and End-of-Chapter Exercises.* Users of p ous editions have given us very positive feedback about the quali the integrative examples at the end of each chapter and the variety c end-of-chapter exercises. We have added two practice examples (Pra Example A and Practice Example B) to every Integrative Example i text. Rather than replace end-of-chapter exercises with new exercise have opted to increase the number of exercises. In most chapters, at 10 new exercises have been added; and in many chapters, 20 or r exercises have been added.
- *Use of IUPAC recommendations.* We are pleased that our book serve needs of instructors and students around the globe. Because commu tion among scientists in general, and chemists in particular, is made e when we agree to use the same terms and notations, we have decid follow—with relatively few exceptions—recommendations made b International Union of Pure and Applied Chemistry (IUPAC). In partic the version of the periodic table that now appears throughout the te based on the one currently endorsed by IUPAC. The IUPAC-endorsed sion places the elements lanthanum (La) and actinium (Ac) in the thanides and actinides series, respectively, rather than in grou Interestingly, almost every other chemistry book still uses the old versi the periodic table, even though the proper placement of La and Ac has known for more than 20 years! An important change is the use of IUI recommended symbols and units for thermodynamic quantities. For e: ple, in this edition, standard enthalpies of reaction are represented b symbol $\Delta_r H^\circ$ (not ΔH°_r) and are expressed in kJ mol^{-1} (not kJ).

FEATURES OF THIS EDITION

We have made a careful effort with this edition to incorporate features that will f tate the teaching and learning of chemistry.

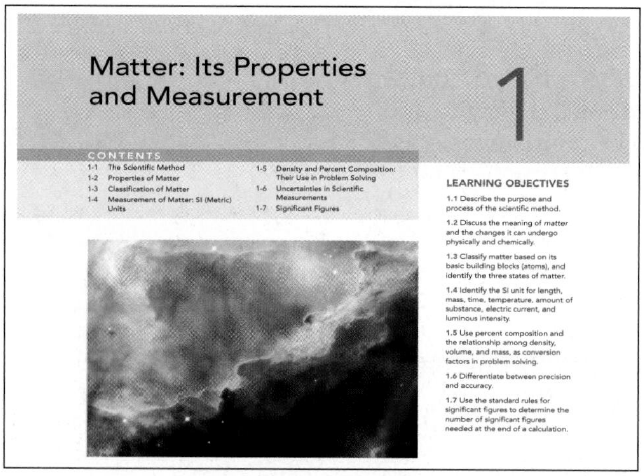

Chapter Opener

Each chapter opens with listing of the main h ings to provide a convenient overview of chapter's **Contents**. The opener also contains a of numbered **Learning Objectives** that co spond with the main sections of the chapter.

Key Terms

Key terms are boldfaced where they are defi in the text. A **Glossary** of key terms with their initions is presented in Appendix F.

The result of multiplication or division may contain only as many significant figures as the *least* precisely known quantity in the calculation.

Highlighted Boxes

Significant equations, concepts, and rules highlighted against a color background for e reference.

cept Assessment

cept Assessment questions (many of which are itative) are distributed throughout the body of chapters. They enable students to test their erstanding of basic concepts before proceeding er. Full solutions are provided in Appendix H.

> **2-4 CONCEPT ASSESSMENT**
>
> What is the single exception to the statement that all atoms comprise protons, neutrons, and electrons?

mples with Practice Examples A and B

ked-out Examples throughout the text illustrate to apply the concepts. In many instances, a ving or photograph is included to help students alize what is going on in the problem. More rtantly, all worked-out Examples now follow a rtite structure of **Analyze–Solve–Assess** to urage students to adopt a logical approach to lem solving.

vo **Practice Examples** are provided for each ked-out Example. The first, **Practice Example A**, rides immediate practice in a problem very lar to the given Example. The second, **Practice nple B**, often takes the student one step further the given Example and is similar to the end-of-pter problems in terms of level of difficulty. wers to all the Practice Examples are given in endix G.

EXAMPLE 14-5 Using Henry's Law

At 0 °C and an O_2 pressure of 1.00 atm, the aqueous solubility of $O_2(g)$ is 48.9 mL O_2 per liter. What is the molarity of O_2 in a saturated water solution when the O_2 is under its normal partial pressure in air, 0.2095 atm?

Analyze

Think of this as a two-part problem. (1) Determine the molarity of the saturated O_2 solution at 0 °C and 1 atm. (2) Use Henry's law in the manner just outlined.

Solve

Determine the molarity of O_2 at 0 °C when $P_{O_2} = 1$ atm. We are given the information that, at an O_2 pressure of 1.00 atm, a saturated solution of O_2 in water contains 48.9 mL (0.0489 L) of O_2. We also know that, at 0 °C and 1.00 atm, 1 mol O_2 occupies a volume of 22.4 L. Therefore,

$$\text{molarity} = \frac{0.0489 \text{ L } O_2 \times \dfrac{1 \text{ mol } O_2}{22.4 \text{ L } O_2}}{1 \text{ L soln}} = 2.18 \times 10^{-3} \frac{\text{mol } O_2}{\text{L soln}} = 2.18 \times 10^{-3} \text{ M}$$

Evaluate the Henry's law constant.

$$k = \frac{C}{P_{gas}} = \frac{2.18 \times 10^{-3} \text{ M}}{1.00 \text{ atm}}$$

Apply Henry's law.

$$C = k \times P_{gas} = \frac{2.18 \times 10^{-3} \text{ M}}{1.00 \text{ atm}} \times 0.2095 \text{ atm} = 4.57 \times 10^{-4} \text{ M}$$

Assess

When working problems involving gaseous solutes in a solution in which the solute is at very low concentration, use Henry's law.

PRACTICE EXAMPLE A: Use data from Example 14-5 to determine the partial pressure of O_2 above an aqueous solution at 0 °C known to contain 5.00 mg O_2 per 100.0 mL of solution.

PRACTICE EXAMPLE B: A handbook lists the solubility of carbon monoxide in water at 0 °C and 1 atm pressure as 0.0354 mL CO per milliliter of H_2O. What pressure of CO(g) must be maintained above the solution to obtain 0.0100 M CO?

rginal Notes

ginal notes help clarify important points.

> ▶ Other atomic symbols not based on English names include Cu, Ag, Sn, Sb, Au, and Hg.

ep In Mind Notes

p In Mind margin notes remind students about s introduced earlier in the text that are important to an understanding of the topic under discus-. In some instances they also warn students ut common pitfalls.

> **KEEP IN MIND**
>
> that all we know is that the second oxide is twice as rich in oxygen as the first. If the first is CO, the possibilities for the second are CO_2, C_2O_4, C_3O_6, and so on. (See also Exercise 18.)

You Wondering?

You Wondering? boxes pose and answer good stions that students often ask. Some are gned to help students avoid common misconceptions; others provide analogies or alternate lanations of a concept; and still others address arent inconsistencies in the material that the lents are learning. These topics can be assigned mitted at the instructor's discretion.

> **1-1 ARE YOU WONDERING?**
>
> **Why is it so important to attach units to a number?**
>
> In 1993, NASA started the Mars Surveyor program to conduct an ongoing series of missions to explore Mars. In 1995, two missions were scheduled that would be launched in late 1998 and early 1999. The missions were the Mars Climate Orbiter (MCO) and the Mars Polar Lander (MPL). The MCO was launched December 11, 1998, and the MPL, January 3, 1999.

cus On Discussions

erences are given near the end of each chapter a **Focus On** essay that is found on the steringChemistry® site (www.masteringmistry.com). These essays describe interesting significant applications of the chemistry dissed in the chapter. They help show the importe of chemistry in all aspects of daily life.

> MasteringCHEMISTRY www.masteringchemistry.com
>
> What is the most abundant element? This seemingly simple question does not have a simple answer. To learn more about the abundances of elements in the universe and in the Earth's crust, go to the Focus On feature for Chapter 2, entitled Occurrence and Abundances of the Elements, on the MasteringChemistry site.

Summary

2-1 Early Chemical Discoveries and the Atomic Theory—Modern chemistry began with eighteenth-century discoveries leading to the formulation of two basic laws of chemical combination, the **law of conservation of mass** and the **law of constant composition (definite proportions)**. These discoveries led to Dalton's atomic theory—that matter is composed of indestructible particles called atoms, that the atoms of an element are identical to one another but different from atoms of all other elements, and that chemical compounds are combinations of atoms of different elements. Based on this theory, Dalton proposed still another law of chemical combination, the **law of multiple proportions**.

2-2 Electrons and Other Discoveries in Atomic Physics—The first clues to the structures of atoms came through the discovery and characterization of **cathode rays (electrons)**. Key experiments were those that established the mass-to-charge ratio (Fig. 2-7) and then the charge on an electron (Fig. 2-8). Two important accidental discoveries made in the course of cathode-ray research were of X-rays and **radioactivity**. The principal types of radiation emitted by radioactive substances are **alpha (α) particles, beta (β) particles**, and **gamma (γ) rays** (Fig. 2-10).

2-3 The Nuclear Atom—Studies on the scattering of α particles by thin metal foils (Fig. 2-11) led to the concept of the nuclear atom—a tiny, but massive, positively charged nucleus surrounded by lightweight, negatively charged electrons (Fig. 2-12). A more complete description of the nucleus was made possible by the discovery of **protons** and **neutrons**. An individual atom is characterized in terms of its **atomic number (proton number) *Z*** and **mass number, *A***. The difference, $A - Z$, is the **neutron number**. The masses of individual atoms and their component parts are expressed in **atomic mass units (u)**.

Summary

A prose **Summary** is provided for each cha
The Summary is organized by the main headin
the chapter and incorporates the key terms in
faced type.

Integrative Example

For use in analytical chemistry, sodium thiosulfate solutions must be carefully prepared. In particular, the solutions must be kept from becoming acidic. In strongly acidic solutions, thiosulfate ion disproportionates into $SO_2(g)$ and $S_8(s)$.

▲ **Decomposition of thiosulfate ion**
When an aqueous solution of $Na_2S_2O_3$ is acidified, the sulfur is in the colloidal state when first formed (right).

Show that the disproportionation of $S_2O_3^{2-}(aq)$ is spontaneous for standard-state conditions in acidic solution, but not in basic solution.

Analyze

Begin by writing the half-equations and an overall equation for the disproportionation reaction. Determine E°_{cell} for the reaction and thus whether the reaction is spontaneous for standard-state conditions in acidic solution. Then make a qualitative assessment of whether the reaction is likely to be more spontaneous or less spontaneous in basic solution.

Solve

Base the overall equation on the verbal description of the reaction.

Reduction:
$$4\,S_2O_3^{2-}(aq) + 24\,H^+(aq) + 16\,e^- \longrightarrow S_8(s) + 12\,H_2O(l)$$

Oxidation:
$$4\{S_2O_3^{2-}(aq) + H_2O(l) \longrightarrow 2\,SO_2(g) + 2\,H^+(aq) + 4\,e^-\}$$

Overall:
$$8\,S_2O_3^{2-}(aq) + 16\,H^+(aq) \longrightarrow S_8(s) + 8\,SO_2(g) + 8\,H_2O(l) \quad \textbf{(22.53)}$$

To determine E°_{cell} for the reaction (22.53), use data from Figure 22-13. That figure gives an E° value for the reduction half-reaction (0.465 V) but no value for the oxidation. To obtain this missing E°, use additional data from Figure 22-13 together with the method of Example 22-1. That is, the sum of the half-equation

$$4\,SO_2(g) + 4\,H^+(aq) + 6\,e^- \longrightarrow S_4O_6^{2-}(aq) + 2\,H_2O(l)$$
$$\Delta_rG^\circ = -6FE^\circ = -6F \times 0.507\text{ V}$$

and the half-equation

$$S_4O_6^{2-}(aq) + 2\,e^- \longrightarrow 2\,S_2O_3^{2-}(aq)$$
$$\Delta_rG^\circ = -2FE^\circ = -2F \times 0.080\text{ V}$$

yields the desired new half-equation and its E° value.

$$4\,SO_2(g) + 4\,H^+(aq) + 8\,e^- \longrightarrow 2\,S_2O_3^{2-}(aq) + 2\,H_2O(l)$$
$$\Delta_rG^\circ = -F[(6 \times 0.507) + (2 \times 0.080)]\text{ V}$$
$$\Delta_rG^\circ = -8FE^\circ = -F(3.202)\text{ V}$$
$$E^\circ = (3.202/8)\text{ V} = 0.400\text{ V}$$

Now we can calculate E°_{cell} for reaction (22.53).

$$E^\circ_{cell} = E^\circ(\text{reduction}) - E^\circ(\text{oxidation}) = 0.465\text{ V} - 0.400\text{ V} = 0.065\text{ V}$$

The disproportionation is spontaneous for standard-state conditions in acidic solution.

Increasing $[OH^-]$, as would be the case in making the solution basic, means decreasing $[H^+]$. In fact, $OH^- = 1$ M corresponds to $[H^+] = 1 \times 10^{-14}$ M. Because equation (22.53) has $H^+(aq)$ on the *left* side of the equation, a decrease in $[H^+]$ favors the *reverse* reaction (by Le Châtelier's principle). At some point before the solution becomes basic, the forward reaction is no longer spontaneous.

Assess

This calculation demonstrated in a qualitative way that $S_2O_3^{2-}(aq)$ is stable in basic solutions and spontaneously disproportionates in acidic solutions. To determine the pH at which the disproportionation becomes spontaneous, one can use the Nernst equation, as seen in Exercise 100.

PRACTICE EXAMPLE A: Use information from Figure 22-17 to decide whether the nitrite anion, NO_2^-, disproportionates spontaneously in basic solution to NO_3^- and NO. Assume standard-state conditions.

PRACTICE EXAMPLE B: Does HNO_2 spontaneously disproportionate to NO_3^- and NO in acidic solution? Assume standard-state conditions. [*Hint*: Use data from Figure 22-17.]

Integrative Example

An **Integrative Example** is provided near the
of each chapter. These challenging examples s
students how to link various concepts from
chapter and earlier chapters to solve complex p
lems. Each Integrative Example is now accor
nied by a **Practice Example A** and **Pra**
Example B. Answers to these Practice Example
given in Appendix G.

Exercises

Homogeneous and Heterogeneous Mixtures

1. Which of the following do you expect to be most water soluble, and why? $C_{10}H_8(s)$, $NH_2OH(s)$, $C_6H_6(l)$, $CaCO_3(s)$.
2. Which of the following is moderately soluble both in water and in benzene [$C_6H_6(l)$], and why? **(a)** 1-butanol, $CH_3(CH_2)_2CH_2OH$; **(b)** naphthalene, $C_{10}H_8$; **(c)** hexane, C_6H_{14}; **(d)** NaCl(s).
3. Substances that dissolve in water generally do not dissolve in benzene. Some substances are moderately soluble in both solvents, however. One of the following is such a substance. Which do you think it is and why?

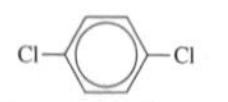

(a) *para*-Dichlorobenzene (a moth repellent)

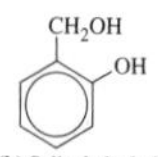

(b) Salicyl alcohol (a local anesthetic)

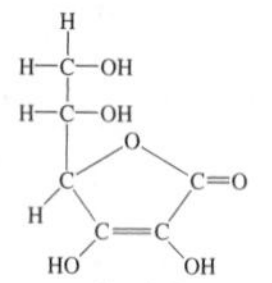

Vitamin C

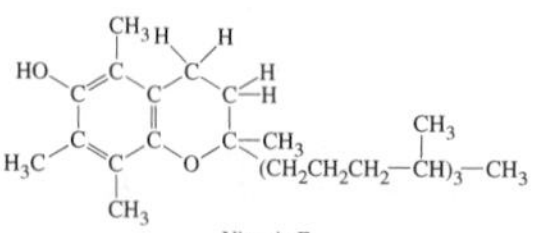

Vitamin E

End-of-Chapter Questions and Exercis

Each chapter ends with four categories of q
tions:

Exercises are organized by topic subheads and
presented in pairs. Answers to selected questions
those numbered in red) are given in Appendix G.

grative and Advanced Exercises are more anced than the preceding Exercises. They are grouped by topic or type. They integrate mater- om sections of the chapter and sometimes from tiple chapters. In some instances, they intro- e new ideas or pursue specific ideas further ı is done in the chapter. Answers to selected stions (i.e., those numbered in red) are given in endix G.

Integrative and Advanced Exercises

69. Write net ionic equations for the reactions depicted in photo **(a)** sodium metal reacts with water to produce hydrogen; photo **(b)** an excess of aqueous iron(III) chloride is added to the solution in **(a)**; and photo **(c)** the precipitate from **(b)** is collected and treated with an excess of HCl(aq).

(a) (b) (c)

70. Following are some laboratory methods occasionally used for the preparation of small quantities of chemicals. Write a balanced equation for each.
(a) preparation of $H_2S(g)$: HCl(aq) is heated with FeS(s)
(b) preparation of $Cl_2(g)$: HCl(aq) is heated with $MnO_2(s)$; $MnCl_2(aq)$ and $H_2O(l)$ are other products
(c) preparation of $N_2(g)$: Br_2 and NH_3 react in aqueous solution; NH_4Br is another product

concentration of 0.250 M Cl^-? Assume that the solution volumes are additive.

76. An unknown white solid consists of two compounds, each containing a different cation. As suggested in the illustration, the unknown is partially soluble in water. The solution is treated with NaOH(aq) and yields a white precipitate. The part of the original solid that is insoluble in water dissolves in HCl(aq) with the evolution of a gas. The resulting solution is then treated with $(NH_4)_2SO_4(aq)$ and yields a white precipitate. **(a)** Is it possible that any of the cations Mg^{2+}, Cu^{2+}, Ba^{2+}, Na^+, or NH_4^+ were present in the original unknown? Explain your reasoning. **(b)** What compounds could be in the unknown mixture (that is, what anions might be present)?

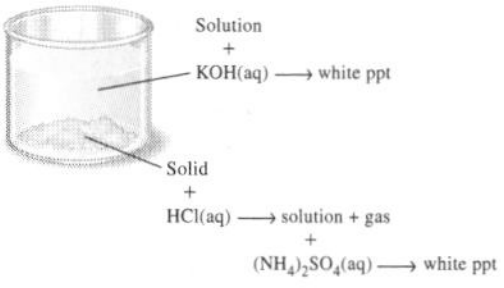

ure Problems require the highest level of skill olve. Some deal with classic experiments; some ire students to interpret data or graphs; some gest alternative techniques for problem solving; e are comprehensive in their scope; and some oduce new material. These problems are a urce that can be used in several ways: for dis- sion in class, for individually assigned home- k, or for collaborative group work. Answers to cted questions (i.e., those numbered in red) are n in Appendix G.

Feature Problems

113. Cinnamaldehyde is the chief constituent of cinnamon oil, which is obtained from the twigs and leaves of cinnamon trees grown in tropical regions. Cinnamon oil is used in the manufacture of food flavorings, perfumes, and cosmetics. The normal boiling point of cinnamaldehyde, $C_6H_5CH{=}CHCHO$, is 246.0 °C, but at this temperature it begins to decompose. As a result, cinnamaldehyde cannot be easily purified by ordinary distillation. A method that can be used instead is *steam distillation*. A heterogeneous mixture of cinnamaldehyde and water is heated until the sum of the vapor pressures of the two liquids is equal to barometric pressure. At this point, the temperature remains constant as the liquids vaporize. The mixed vapor condenses to produce two immiscible liquids; one liquid is essentially pure water and the other, pure cinnamaldehyde. The following vapor pressures of cinnamaldehyde are given: 1 mmHg at 76.1 °C; 5 mmHg at 105.8 °C; and 10 mmHg at 120.0 °C. Vapor pressures of water are given in Table 14.3.

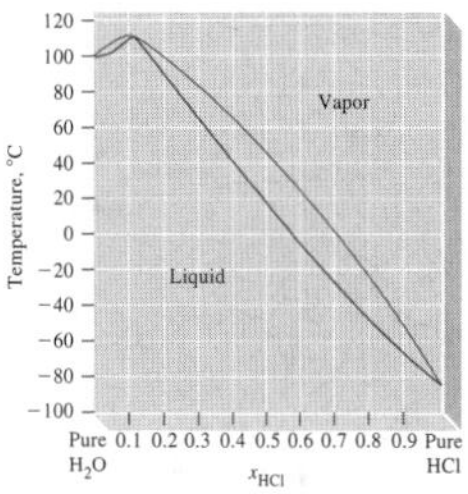

-**Assessment Exercises** are designed to help ents review and prepare for some of the types questions that often appear on quizzes and ns. Students can use these questions to decide ther they are ready to move on to the next oter or first spend more time working with the cepts in the current chapter. Answers with anations to selected questions (i.e., those num- d in red) are given in Appendix G.

Self-Assessment Exercises

77. In your own words, define the following terms or symbols: **(a)** sp^2; **(b)** σ^*_{2p}; **(c)** bond order; **(d)** π bond.

78. Briefly describe each of the following ideas: **(a)** hybridization of atomic orbitals; **(b)** σ-bond framework; **(c)** Kekulé structures of benzene, C_6H_6.

79. Explain the important distinctions between the terms in each of the following pairs: **(a)** σ and π bonds; **(b)** localized and delocalized electrons; **(c)** bonding and antibonding molecular orbitals.

80. A molecule in which sp^2 hybrid orbitals are used by the central atom in forming covalent bonds is **(a)** PCl_5; **(b)** N_2; **(c)** SO_2; **(d)** He_2.

81. The bond angle in H_2Se is best described as **(a)** between 109° and 120°; **(b)** less than in H_2S; **(c)** less than in H_2S, but not less than 90°; **(d)** less than 90°.

82. The hybridization scheme for the central atom includes a *d* orbital contribution in **(a)** I_3^-; **(b)** PCl_3; **(c)** NO_3^-; **(d)** H_2Se.

83. Of the following, the species with a bond order of 1 is **(a)** H_2^+; **(b)** Li_2; **(c)** He_2; **(d)** H_2^-.

84. The hybridization scheme for Xe in XeF_2 is **(a)** sp; **(b)** sp^3; **(c)** sp^3d; **(d)** sp^3d^2.

85. Delocalized molecular orbitals are found in **(a)** H_2; **(b)** HS^-; **(c)** CH_4; **(d)** CO_3^{2-}.

86. Explain why the molecular structure of BF_3 cannot be adequately described through overlaps involving pure *s* and *p* orbitals.

87. Why does the hybridization sp^3d not account for bonding in the molecule BrF_5? What hybridization scheme does work? Explain.

88. What is the total number of **(a)** σ bonds and **(b)** π bonds in the molecule CH_3NCO?

89. Which of the following species are paramagnetic? **(a)** B_2; **(b)** B_2^-; **(c)** B_2^+. Which species has the strongest bond?

90. Use the valence molecular orbital configuration to determine which of the following species is expected to have the lowest ionization energy: **(a)** C_2^+; **(b)** C_2; **(c)** C_2^-.

91. Use the valence molecular orbital configuration to determine which of the following species is expected to have the greatest electron affinity: **(a)** C_2^+; **(b)** Be_2; **(c)** F_2; **(d)** B_2^+.

92. Which of these diatomic molecules do you think has the greater bond energy, Li_2 or C_2? Explain.

93. For each of the following ions or molecules, decide whether the structure is best described by a single Lewis structure or by resonance structures. **(a)** $C_2O_4^{2-}$; **(b)** H_2CO; **(c)** NO_3^-.

94. Draw Lewis structures for the NO_2^- and NO_2^+ ions, and determine the likely geometry for each by using VSEPR theory. How does the hybridization of N differ in these two species?

pendices

Appendices at the back of the book provide important information:

endix A succinctly reviews of some basic **Mathematical Operations.**

endix B concisely describes **Some Basic Physical Concepts.**

endix C summarizes the conventions of **SI Units.**

endix D provides five useful **Data Tables.**

endix E provides guidelines, along with an example, for constructing **cept Maps.**

endix F consists of a **Glossary** of all the key terms in the book.

endix G provides **Answers to Practice Examples and Selected Exercises.**

endix H provides **Answers to Concept Assessment Questions.**

easy reference, the **Periodic Table of Elements** and a **Tabular Listing of nents** are presented on the inside of the front cover.

convenience, listings of **Selected Physical Constants, Some Common nversion Factors, Some Useful Geometric Formulas,** and **Location of ortant Data and Other Useful Information** are presented on the inside of back cover.

DIGITAL AND PRINT RESOURCES

For the Instructor and the Student

MasteringChemistry®
(www.masteringchemistry.com)
MasteringChemistry® is the most effective, widely used online tuto homework, and assessment system for chemistry. It helps instructors m mize class time with customizable, easy-to-assign, and automatically gra assessments that motivate students to learn outside of class and arrive pared for lecture. These assessments can easily be customized and perso ized by instructors to suit their individual teaching style. The powe gradebook provides unique insight into student and class performance e before the first test. As a result, instructors can spend class time where dents need it most.

MasteringChemistry® has always been personalized and adaptive question level by providing error-specific feedback based on actual stu responses. However, Mastering now includes new adaptive follow assignments. Content delivered to students as part of adaptive learning be automatically personalized for each student based on strengths and w nesses identified by his or her performance on Mastering Pa Assignments.

Learning Catalytics®, a "bring your own device" student engagement, ass ment, and classroom intelligence system, is also integrated MasteringChemistry®.

These resources are also available on the MasteringChemistry® site:

- A section about Bonding in Metals, to accompany Chapter 11 (Chem Bonding II: Valence Bond and Molecular Orbital Theories)
- Additional material referenced in Chapter 27 (Reactions of Org Compounds), including discussions of Organic Acids and Base Closer Look at the E2 Mechanism; and Carboxylic Acids and T Derivatives: The Addition–Elimination Mechanism
- Chapter 28 (Chemistry of the Living State)

The Pearson eText gives students access to the text whenever and wher they have access to the Internet. eText pages look exactly like the printed offering powerful new functionality for students and instructors. Users create notes, highlight text in different colors, create bookmarks, zoom, c hyperlinked words and phrases to view definitions, and view in single-p or two-page view.

For the Instructor

The Instructor Resources are available online via the Instructor Resou section of MasteringChemistry® and http://catalogue.pearsoned.ca/. following supplements are designed to facilitate lecture presentations, enc age class discussions, aid in creating tests, and foster learning:

- An **Instructor's Resource Manual**, organized by chapter, vides detailed lecture outlines, describes some common stu misconceptions, and demonstrates how to integrate the vari instructor resources into the course.

- The **Complete Solutions Manual** (978-013-292504-4) contains full solutions to all the end-of-chapter exercises and problems (including those Self-Assessment Exercises that are not discussion questions), as well as full solutions to all the Practice Examples A and B in the book. With instructor approval, arrangements can be made with the publisher to make this manual available to students.
- **Pearson's Computerized Test Bank** allows instructors to filter and select questions to create quizzes, tests, or homework. Instructors can revise questions or add their own, and may be able to choose print or online options. These questions are also available in Microsoft Word format.
- A **Test Item File** in Word provides more than 2700 questions. Many of the questions are in multiple-choice form, but there are also true/false and short-answer questions. Each question is accompanied by the correct answer, the relevant chapter section in the textbook, and a level of difficulty (i.e., 1 for Easy, 2 for Moderate, and 3 for Challenging).
- **PowerPoints Set 1** consists of all the figures and photos in the textbook in PowerPoint format.
- **PowerPoints Set 2** provides lecture outlines for each chapter of the textbook.
- **PowerPoints Set 3** provides questions for Personal Response Systems (i.e., clickers) that can be used to engage students in lectures and to obtain immediate feedback about their understanding of the concepts being presented.
- **PowerPoints Set 4** consists of the all worked examples from the textbook in PowerPoint format.
- **PowerPoints Set 5** consists of the all Practice Examples from the textbook in PowerPoint format.
- **Catalyst Laboratory Database Correlation Guide** in Excel format.
- **Focus On Discussions** consist of all the Focus On Essays referenced in the textbook which students can find on the MasteringChemistry® site (www.masteringchemistry.com).

Pearson's Learning Solutions Managers work with faculty and campus course designers to ensure that Pearson technology products, assessment tools, and online course materials are tailored to meet your specific needs. This highly qualified team is dedicated to helping schools take full advantage of a wide range of educational resources by assisting in the integration of a variety of instructional materials and media formats. Your local Pearson Education sales representative can provide you with more details on this service program.

the Student

Along with an **Access Code Card for MasteringChemistry®**, each new copy of the book is accompanied by a 12-page **Study Card** (978-013-338791-9). This card provides a convenient, concise review of some of the key concepts and topics discussed in each chapter of the textbook.

The Selected Solutions Manual (978-013-338790-2) provides full solutions to all the end-of-chapter exercises and problems that are numbered in red.

ACKNOWLEDGMENTS

We are grateful to the following instructors who provided formal review parts of the manuscript.

John Carran *Queen's University*
Chin Li Cheung *University of Nebraska, Lincoln*
Jason Clyburne *Saint Mary's University*
David Dick *College of the Rockies*
Randall S. Dumont *McMaster University*
Bryan Enderle *University of California, Davis*
David Fenske *University of the Fraser Valley*
Regina Frey *Washington University, St. Louis*
Assaf Friedler *The Hebrew University of Jerusalem*
Michael Gerken *University of Lethbridge*
Jason Grove *University of Waterloo*
Lori Jones *University of Guelph*
Muhammet Erkan Kose *North Da State University*
Masaru Kuno *University of Notre Dame*
Susan Lait *University of Lethbridge*
Jeff Landry *McMaster University*
Scott McIndoe *University of Victor*
George A. Papadantonakis *Univer of Illinois, Chicago*
Jay Shore *South Dakota State University*
Sarah West *University of Notre Dar*
Todd Whitcombe *University of Northern British Columbia*
Milton J. Wieder *Metropolitan Stat College of Denver*

We would like to thank the following instructors for technically chec selected chapters of the new edition during production.

David Dick, *College of the Rockies*
Richard A. Marta, *University of Waterloo*
Mark Quirie, *Algonquin College*
J. W. Sam Stevenson, *Marion Milit Institute*

We are most grateful to our coauthors Ralph Petrucci and Geoff Herring their guidance and mentorship over the past two editions. Their insigh comments about the various topics and revisions have been invaluabl preparing this edition, we have strived to stay true to Ralph's orig vision for this text: *Students learn best by doing; and instructors who prefe approach different from ours can adjust the order of chapters to suit their pr ences.* That is why we have added to the number of worked examples end-of-chapter exercises, and written each chapter so that it can be u independently of the others.

We would also like to acknowledge Cathleen Sullivan, Joanne Sutherl Lila Campbell, and Dawn Hunter for their encouragement and assistanc moving this edition forward.

Finally, we would like to thank our families, but especially our wi Kimberley Bissonnette and Colleen Jones, for their limitless patience enduring support.

Responding to feedback from our colleagues and students is the most im tant element in improving this book from one edition to the next. Please do hesitate to email us. Your observations and suggestions are most welcome.

CAREY BISSONNETTE
carey.bissonnette@uwaterloo.ca

JEFFRY D. MAD
madura@duq

WARNING: Many of the compounds and chemical reactions describe pictured in this book are hazardous. Do not attempt any experiment pict or implied in the text except with permission in an authorized laboratory ting and under adequate supervision.

ʼlatter: Its Properties ıd Measurement

1

NTENTS

LEARNING OBJECTIVES

1.1 Describe the purpose and process of the scientific method.

1.2 Discuss the meaning of *matter* and the changes it can undergo physically and chemically.

1.3 Classify matter based on its basic building blocks (atoms), and identify the three states of matter.

1.4 Identify the SI unit for length, mass, time, temperature, amount of substance, electric current, and luminous intensity.

1.5 Use percent composition and the relationship among density, volume, and mass as conversion factors in problem solving.

1.6 Differentiate between precision and accuracy.

1.7 Use the standard rules for significant figures to determine the number of significant figures needed at the end of a calculation.

ESA, J. Hester (ASU) /NASA

bble Space Telescope image of a cloud of hydrogen gas and dust (lower right f the image) that is part of the Swan Nebula (M17). The colors correspond to emitted by hydrogen (green), sulfur (red), and oxygen (blue). The chemical ents discussed in this text are those found on Earth and, presumably, ghout the universe.

rom the clinic that treats chemical dependency to a theatrical performance with good chemistry to the food label stating "no chemicals added," chemistry and chemicals seem an integral part of life, even if yday references to them are often misleading. A label implying the nce of chemicals in a food makes no sense. All foods consist entirely emicals, even if organically grown. In fact, all material objects—ther living or inanimate—are made up only of chemicals, and we ld begin our study with that thought clearly in mind.

y manipulating materials in their environment, people have always ticed chemistry. Among the earliest applications were glazing pottery, ting ores to produce metals, tanning hides, dyeing fabrics, and making se, wine, beer, and soap. With modern knowledge, though, chemists can decompose matter into its smallest components (atoms) and reasser those components into materials that do not exist naturally and that c exhibit unusual properties. Thus, motor fuels and thousands of chemicals in the manufacture of plastics, synthetic fabrics, pharmaceuticals, and pestic can all be made from petroleum. Modern chemical knowledge is also neede understand the processes that sustain life and to understand and con processes that are detrimental to the environment, such as the formatic smog and the destruction of stratospheric ozone. Because it relates to so m areas of human endeavor, chemistry is sometimes called the *central science*.

Early chemical knowledge consisted of the "how to" of chemistry, dis ered through trial and error. Modern chemical knowledge answers the "w as well as the "how to" of chemical change. It is grounded in principles theory, and mastering the principles of chemistry requires a system approach to the subject. Scientific progress depends on the way scientist their work—asking the right questions, designing the right experimen supply the answers, and formulating plausible explanations of their findi We begin with a closer look into the scientific method.

1-1 The Scientific Method

Science differs from other fields of study in the *method* that scientists us acquire knowledge and the special significance of this knowledge. Scien knowledge can be used to explain natural phenomena and, at times, to *pr* future events.

The ancient Greeks developed some powerful methods of acquiring kn edge, particularly in mathematics. The Greek approach was to start with tain basic assumptions or premises. Then, by the method known as *deduc* certain conclusions must logically follow. For example, if $a = b$ and b then $a = c$. Deduction alone is not enough for obtaining scientific knowle however. The Greek philosopher Aristotle *assumed* four fundamental stances: air, earth, water, and fire. All other materials, he believed, w formed by combinations of these four elements. Chemists of several centu ago (more commonly referred to as alchemists) tried, in vain, to apply four-element idea to turn lead into gold. They failed for many reasons, being that the four-element assumption is false.

The scientific method originated in the seventeenth century with such pe as Galileo, Francis Bacon, Robert Boyle, and Isaac Newton. The key to the me is to make no initial assumptions, but rather to make careful observation natural phenomena. When enough observations have been made so that a tern begins to emerge, a generalization or natural law can be formulated des ing the phenomenon. **Natural laws** are concise statements, often in mathema form, about natural phenomena. The form of reasoning in which a general s ment or natural law is inferred from a set of observations is called *induction* example, early in the sixteenth century, Polish astronomer Nicolaus Coperr (1473–1543), through careful study of astronomical observations, concluded Earth revolves around the sun in a circular orbit, although the general teac of the time, not based on scientific study, was that the sun and other heav bodies revolved around Earth. We can think of Copernicus's statement natural law. Another example of a natural law is the radioactive decay law, w dictates how long it takes for a radioactive substance to lose its radioactivity.

The success of a natural law depends on its ability to explain, or accoun observations and to predict new phenomena. Copernicus's work was a g success because he was able to predict future positions of the planets m accurately than his contemporaries. We should not think of a natural law a *absolute* truth, however. Future experiments may require us to modify the For example, Copernicus's ideas were refined a half-century later by Joha Kepler, who showed that planets travel in elliptical, not circular, orbits. To v

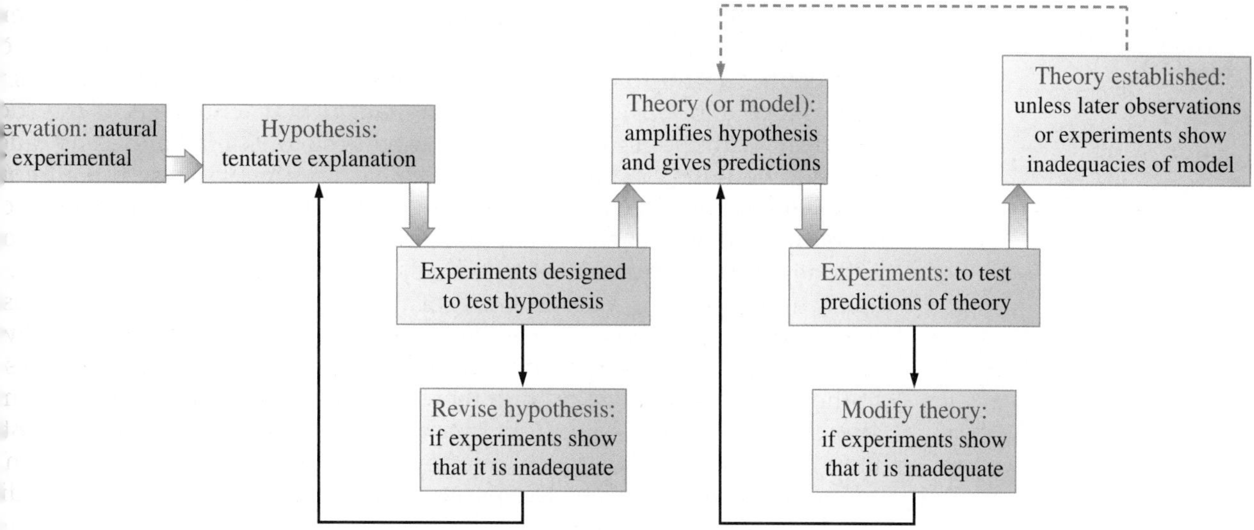

GURE 1-1
scientific method illustrated

Prof. Marvin Lang and Gary J. Shulfer/University of Wisconsin - Stevens Point

▲ Louis Pasteur (1822–1895). This great practitioner of the scientific method was the developer of the germ theory of disease, the sterilization of milk by pasteurization, and vaccination against rabies. He has been called the greatest physician of all time by some. He was, in fact, not a physician at all, but a chemist—by training and by profession.

tural law, a scientist designs *experiments* that show whether the conclusions iced from the natural law are supported by experimental results.

hypothesis is a tentative explanation of a natural law. If a hypothesis surs testing by experiments, it is often referred to as a theory. In a broader e, a **theory** is a model or way of looking at nature that can be used to explain ral laws and make further predictions about natural phenomena. When difg or conflicting theories are proposed, the one that is most successful in its ictions is generally chosen. Also, the theory that involves the smallest numof assumptions—the simplest theory—is preferred. Over time, as new evie accumulates, most scientific theories undergo modification, and some are arded.

ne **scientific method** is the combination of observation, experimentation, the formulation of laws, hypotheses, and theories. The method is illusd by the flow diagram in Figure 1-1. Scientists may develop a pattern of king about their field, known as a *paradigm*. Some paradigms may be sucful at first but then become less so. When that happens, a new paradigm be needed or, as is sometimes said, a *paradigm shift* occurs. In a way, the nod of inquiry that we call the scientific method is itself a paradigm, and e people feel that it, too, is in need of change. That is, the varied activities odern scientists are more complex than the simplified description of the ntific method presented here.* In any case, merely following a set of proces, rather like using a cookbook, will not guarantee scientific success.

nother factor in scientific discovery is chance, or serendipity. Many discoverave been made by accident. For example, in 1839, American inventor Charles dyear was searching for a treatment for natural rubber that would make it brittle when cold and less tacky when warm. During this work, he acciden spilled a rubber–sulfur mixture on a hot stove and found that the resulting uct had exactly the properties he was seeking. Other chance discoveries de X-rays, radioactivity, and penicillin. So scientists and inventors always to be alert to unexpected observations. Perhaps no one was more aware of han Louis Pasteur, who wrote, "Chance favors the prepared mind."

1-1 CONCEPT ASSESSMENT

he common saying "The exception proves the rule" a good statement of the entific method? Explain.

◀ Answers to Concept Assessment questions are given in Appendix H.

Harwood, *JCST*, **33**, 29 (2004). *JCST* is an abbreviation for *Journal of College Science Teaching*.

1-2 Properties of Matter

Dictionary definitions of chemistry usually include the terms *ma* *composition*, and *properties*, as in the statement that "chemistry is the sci that deals with the composition and properties of matter." In this and the section, we will consider some basic ideas relating to these three terr hopes of gaining a better understanding of what chemistry is all about.

Matter is anything that occupies space and displays the properties of and inertia. Every human being is a collection of matter. We all occupy sp and we describe our mass in terms of weight, a related property. (Mass weight are described in more detail in Section 1-4. Inertia is describe Appendix B.) All the objects that we see around us consist of matter. The g of the atmosphere, even though they are invisible, are matter—they occ space and have mass. Sunlight is *not* matter; rather, it is a form of ene Energy is discussed in later chapters.

Composition refers to the parts or components of a sample of ma and their relative proportions. Ordinary water is made up of two sim substances—hydrogen and oxygen—present in certain fixed proporti A chemist would say that the composition of water is 11.19% hydrogen 88.81% oxygen by mass. Hydrogen peroxide, a substance used in bleaches antiseptics, is also made up of hydrogen and oxygen, but it has a different c position. Hydrogen peroxide is 5.93% hydrogen and 94.07% oxygen by ma

Properties are those qualities or attributes that we can use to distinguish sample of matter from others; and, as we consider next, the properties of m are generally grouped into two broad categories: physical and chemical.

Physical Properties and Physical Changes

A **physical property** is one that a sample of matter displays without chan its composition. Thus, we can distinguish between the reddish brown s copper, and the yellow solid, sulfur, by the physical property of *color* (Fig.

Another physical property of copper is that it can be hammered into a sheet of foil (see Figure 1-2). Solids having this ability are said to be *malle* Sulfur is not malleable. If we strike a chunk of sulfur with a hammer, it cr bles into a powder. Sulfur is *brittle*. Another physical property of copper sulfur does not share is the ability to be drawn into a fine wire (ductility). sulfur is a far poorer conductor of heat and electricity than is copper.

Sometimes a sample of matter undergoes a change in its physical app ance. In such a **physical change**, some of the physical properties of the sar may change, but its composition remains unchanged. When liquid w freezes into solid water (ice), it certainly looks different and, in many ways, different. Yet the water remains 11.19% hydrogen and 88.81% oxygen by m

Chemical Properties and Chemical Changes

In a **chemical change**, or **chemical reaction**, one or more kinds of matte converted to new kinds of matter with different compositions. The ke

▶ FIGURE 1-2
Physical properties of sulfur and copper
A lump of sulfur (left) crumbles into a yellow powder when hammered. Copper (right) can be obtained as large lumps of native copper, formed into pellets, hammered into a thin foil, or drawn into a wire.

tifying chemical change, then, comes in observing a *change in composition.* burning of paper involves a chemical change. Paper is a complex material, its principal constituents are carbon, hydrogen, and oxygen. The chief lucts of the combustion are two gases, one consisting of carbon and oxy- (carbon dioxide) and the other consisting of hydrogen and oxygen (water, eam). The ability of paper to burn is an example of a chemical property. A **nical property** is the ability (or inability) of a sample of matter to undergo ange in composition under stated conditions.

inc reacts with hydrochloric acid solution to produce hydrogen gas and a tion of zinc chloride in water (Fig. 1-3). This reaction is one of zinc's distinc- chemical properties, just as the inability of gold to react with hydrochloric is one of gold's chemical properties. Sodium reacts not only with hydrochlo- cid but also with water. In some of their physical properties, zinc, gold, and um are similar. For example, each is malleable and a good conductor of heat electricity. In most of their chemical properties, though, zinc, gold, and um are quite different. Knowing these differences helps us to understand zinc, which does not react with water, is used in roofing nails, roof flashings, rain gutters, and sodium is not. Also, we can appreciate why gold, because s chemical inertness, is prized for jewelry and coins: It does not tarnish or In our study of chemistry, we will see why substances differ in properties how these differences determine the ways in which we use them.

Diane Hirsch/Fundamental Photographs

▲ FIGURE 1-3
A chemical property of zinc and gold: reaction with hydrochloric acid
The zinc-plated (galvanized) nail reacts with hydrochloric acid, producing the bubbles of hydrogen gas seen on its surface. The gold bracelet is unaffected by hydrochloric acid. In this photograph, the zinc plating has been consumed, exposing the underlying iron nail. The reaction of iron with hydrochloric acid imparts some color to the acid solution.

3 Classification of Matter

ter is made up of very tiny units called **atoms**. Each different type of atom is uilding block of a different chemical **element**. Presently, the International on of Pure and Applied Chemistry (IUPAC) recognizes 118 elements, but do not yet have names or symbols. The known elements range from com- substances, such as carbon, iron, and silver, to uncommon ones, such as ium and thulium. About 90 of the elements can be obtained from natural ces. The remainder do not occur naturally and have been created only in ratories. On the inside front cover you will find a complete listing of the ele- ts and also a special tabular arrangement of the elements known as the *dic table.* The periodic table is the chemist's directory of the elements. We describe it in Chapter 2 and use it throughout most of the text.

hemical **compounds** are substances comprising atoms of two or more ele- ts joined together. Scientists have identified millions of different chemical pounds. In some cases, we can isolate a molecule of a compound. A **molecule** e smallest entity having the same proportions of the constituent atoms as the compound as a whole. A molecule of water consists of three atoms: two rogen atoms joined to a single oxygen atom. A molecule of hydrogen perox- nas two hydrogen atoms and two oxygen atoms; the two oxygen atoms are d together and one hydrogen atom is attached to each oxygen atom. By con- , a molecule of the blood protein gamma globulin is made up of 19,996 atoms, hey are of just four types: carbon, hydrogen, oxygen, and nitrogen.

◀ The International Union of Pure and Applied Chemistry (IUPAC) is recognized as the world authority on chemical nomenclature, terminology, standardized methods for measurement, atomic mass, and more. Along with many other activities, IUPAC publishes journals, technical reports, and chemical databases, most of which are available at www .iupac.org.

◀ The identity of an atom is established by a feature called its atomic number (see Section 2-3). Characterizing "superheavy" elements is a daunting challenge; they are produced only a few atoms at a time and the atoms disintegrate almost instantaneously.

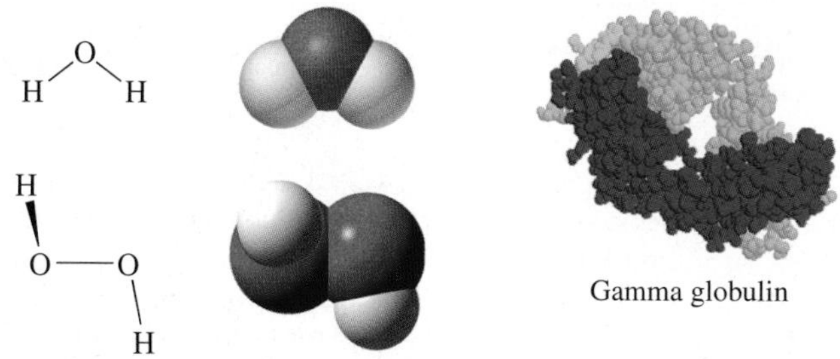

▲ Structures of water, hydrogen peroxide, and gamma globulin. Gamma globulin consists of three subunits (shown in different colors). Each subunit consists of carbon, hydrogen, oxygen, and nitrogen.

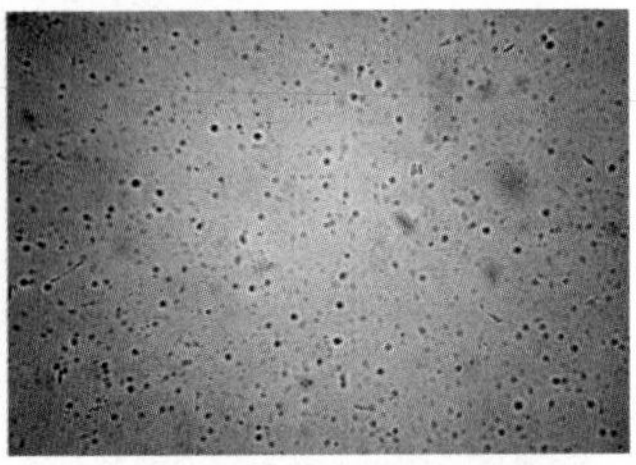

▲ Is it homogeneous or heterogeneous? When viewed through a microscope, homogenized milk is seen to consist of globules of fat dispersed in a watery medium. Homogenized milk is a *heterogeneous* mixture.

The composition and properties of an element or a compound are unif throughout a given sample and from one sample to another. Elements compounds are called **substances**. (In the chemical sense, the term *subst* should be used only for elements and compounds.) A *mixture* of substa can vary in composition and properties from one sample to another. One is uniform in composition and properties throughout is said to b **homogeneous mixture** or a *solution*. A given solution of sucrose (cane su in water is uniformly sweet throughout the solution, but the sweetnes another sucrose solution may be rather different if the sugar and water present in different proportions. Ordinary air is a homogeneous mixtu several gases, principally the *elements* nitrogen and oxygen. Seawater is a s tion of the *compounds* water, sodium chloride (salt), and a host of otl Gasoline is a homogeneous mixture or solution of dozens of compounds.

In **heterogeneous mixtures**—sand and water, for example—the com nents separate into distinct regions. Thus, the composition and physical p erties vary from one part of the mixture to another. Salad dressing, a sla concrete, and the leaf of a plant are all heterogeneous. It is usually easy to tinguish heterogeneous from homogeneous mixtures. A scheme for classif matter into elements and compounds and homogeneous and heterogene mixtures is summarized in Figure 1-4.

Separating Mixtures

▶ It is composition, particularly its variability, that helps us distinguish the several classifications of matter.

▶ Solutions can be gaseous and liquids as described here, but they can also be solids. Some alloys are examples of solid solutions.

A mixture can be separated into its components by appropriate phy means. Consider again the heterogeneous mixture of sand in water. Whe pour this mixture into a funnel lined with porous filter paper, the water pa through and sand is retained on the paper. This process of separating a s from the liquid in which it is suspended is called *filtration* (Fig. 1-5a). You probably use this procedure in the laboratory. Conversely, we cannot sepa a homogeneous mixture (solution) of copper(II) sulfate in water by filtra because all components pass through the paper. We can, however, boi solution of copper(II) sulfate and water. In the process of *distillation*, a p liquid is condensed from the vapor given off by a boiling solution. Whe

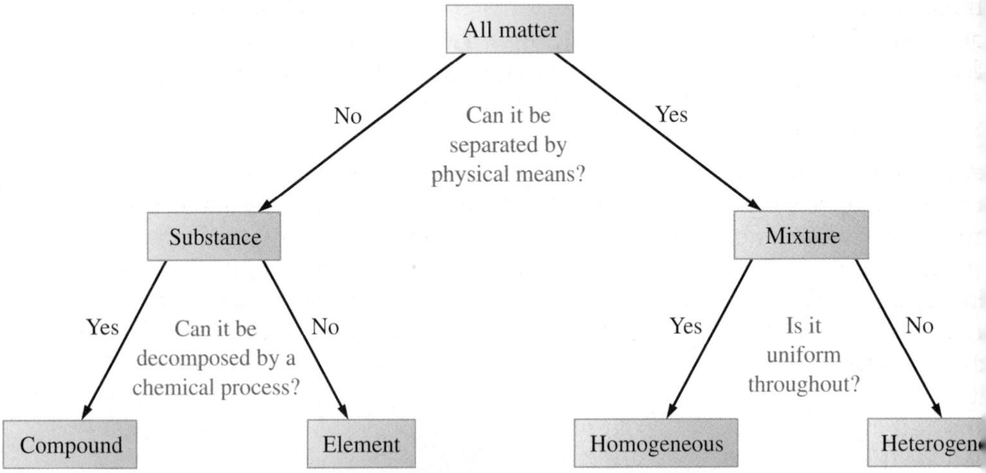

▲ FIGURE 1-4
A classification scheme for matter
Every sample of matter is either a single substance (an element or compound) or a mixture of substances. At the molecular level, an element consists of atoms of a sin type and a compound consists of two or more different types of atoms, usually join into molecules. In a homogeneous mixture, atoms or molecules are randomly mixe at the molecular level. In heterogeneous mixtures, the components are physically separated, as in a layer of octane molecules (a constituent of gasoline) floating on a layer of water molecules.

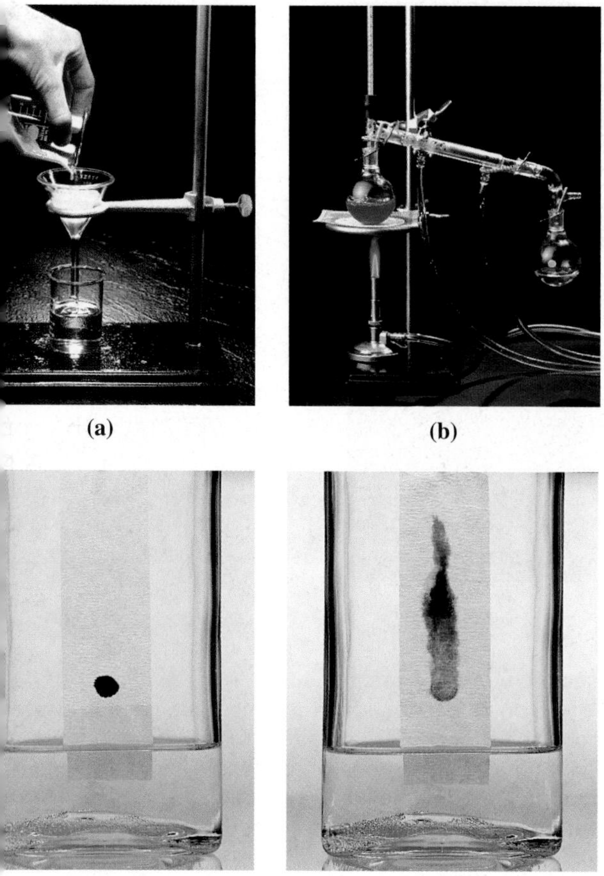

◀ FIGURE 1-5
Separating mixtures: a physical process
(a) Separation of a heterogeneous mixture by filtration: Solid copper(II) sulfate is retained on the filter paper, while liquid hexane passes through. **(b)** Separation of a homogeneous mixture by distillation: Copper(II) sulfate remains in the flask on the left as water passes to the flask on the right, by first evaporating and then condensing back to a liquid. **(c)** Separation of the components of ink by using chromatography: A dark spot of black ink can be seen just above the water line as water moves up the paper. **(d)** Water has dissolved the colored components of the ink, and these components are retained in different regions on the paper according to their differing tendencies to adhere to the paper.

(a) Carey B. Van Loon; **(b)** Carey B. Van Loon; **(c)** Richard Megna/Fundamental Photographs; **(d)** Richard Megna/Fundamental Photographs

vater has been removed by boiling a solution of copper(II) sulfate in r, solid copper(II) sulfate remains behind (Fig. 1-5b).

ıother method of separation available to modern chemists depends on iffering abilities of compounds to adhere to the surfaces of various solid ances, such as paper and starch. The technique of *chromatography* relies on rinciple. The dramatic results that can be obtained with chromatography lustrated by the separation of ink on a filter paper (Fig. 1-5c, d).

omposing Compounds

mical compound retains its identity during physical changes, but it can be nposed into its constituent elements by *chemical changes*. The decomposition npounds into their constituent elements is a more difficult matter than the physical separation of mixtures. The extraction of iron from iron oxide ores res a blast furnace. The industrial production of pure magnesium from mag-m chloride requires electricity. It is generally easier to convert a compound ther compounds by a chemical reaction than it is to separate a compound ts constituent elements. For example, when heated, ammonium dichromate nposes into the substances chromium(III) oxide, nitrogen, and water. This on, once used in movies to simulate a volcano, is illustrated in Figure 1-6.

Carey B. Van Loon

▲ FIGURE 1-6
A chemical change: decomposition of ammonium dichromate

es of Matter

er is generally found in one of three *states*: solid, liquid, or gas. In a **solid**, s or molecules are in close contact, sometimes in a highly organized gement called a *crystal*. A solid has a definite shape. In a **liquid**, the atoms olecules are usually separated by somewhat greater distances than in a . Movement of these atoms or molecules gives a liquid its most distinctive erty—the ability to flow, covering the bottom and assuming the shape of ntainer. In a **gas**, distances between atoms or molecules are much greater

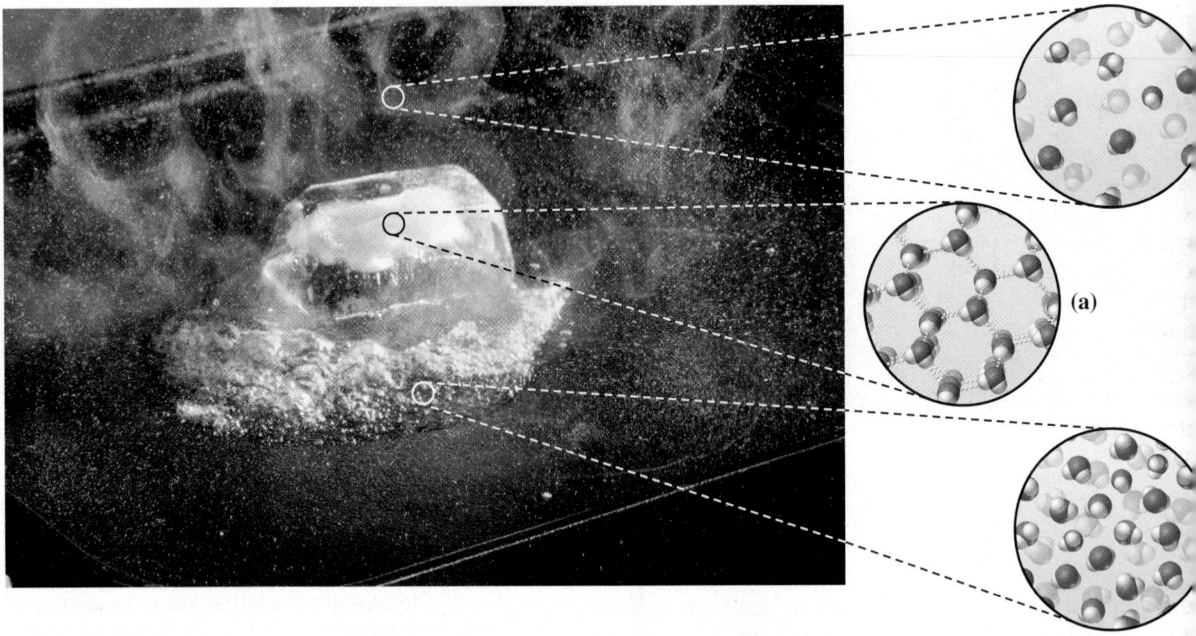

Richard Megna/Fundamental Photographs

▲ FIGURE 1-7
Macroscopic and microscopic views of matter
The picture shows a block of ice on a heated surface and the three states of water. circular insets show how chemists conceive of these states microscopically, in term molecules with two hydrogen atoms joined to one of oxygen. In ice **(a)**, the molec are arranged in a regular pattern in a rigid framework. In liquid water **(b)**, the mole are rather closely packed but move freely. In gaseous water **(c)**, the molecules are widely separated.

than in a liquid. A gas always expands to fill its container. Depending on ditions, a substance may exist in only one state of matter, or it may be prese two or three states. Thus, as the ice in a small pond begins to melt in the sp water is in two states: solid and liquid (actually, three states if we also con water vapor in the air above the pond). The three states of water are illust at two levels in Figure 1-7.

The *macroscopic level* refers to how we perceive matter with our through the outward appearance of objects. The *microscopic level* desc matter as chemists conceive of it—in terms of atoms and molecules and behavior. In this text, we will describe many macroscopic, observable pr ties of matter, but to explain these properties, we will often shift our vie the atomic or molecular level—the microscopic level.

1-4 Measurement of Matter: SI (Metric) Ur

▶ Nonnumerical information is *qualitative*, such as the color blue.

Chemistry is a *quantitative* science, which means that in many cases w measure a property of a substance and compare it with a standard hav known value of the property. We express the measurement as the produc *number* and a *unit*. The unit indicates the standard against which the meas quantity is being compared. When we say that the length of the playing fi football is 100 yd, we mean that the field is 100 times as long as a standa length called the yard (yd). In this section, we will introduce some basic of measurement that are important to chemists.

▶ The *definition of the meter*, formerly based on the atomic spectrum of ^{86}Kr, was changed to the speed of light in 1983. Effectively, the speed of light is now defined as 2.99792458×10^8 m/s.

The scientific system of measurement is called the *Système Internati d'Unités* (International System of Units) and is abbreviated **SI**. It is a mo version of the metric system, a system based on the unit of length cal *meter* (m). The meter was originally defined as 1/10,000,000 of the distance the equator to the North Pole and translated into the length of a metal bar in Paris. Unfortunately, the length of the bar is subject to change with tem ture, and it cannot be exactly reproduced. The SI system substitutes fc

BLE 1.1 SI Base Quantities

sical Quantity	Unit	Symbol
gth	meter[a]	m
s	kilogram	kg
e	second	s
perature	kelvin	K
ount of substance	mole[b]	mol
tric current[c]	ampere	A
ninous intensity[d]	candela	cd

official spelling of this unit is "metre," but we will use the American spelling.
mole is introduced in Section 2-7.
tric current is described in Appendix B and in Chapter 19.
ninous intensity is not discussed in this text.

TABLE 1.2 SI Prefixes

Multiple	Prefix
10^{18}	exa (E)
10^{15}	peta (P)
10^{12}	tera (T)
10^{9}	giga (G)
10^{6}	mega (M)
10^{3}	kilo (k)
10^{2}	hecto (h)
10^{1}	deka (da)
10^{-1}	deci (d)
10^{-2}	centi (c)
10^{-3}	milli (m)
10^{-6}	micro (μ)[a]
10^{-9}	nano (n)
10^{-12}	pico (p)
10^{-15}	femto (f)
10^{-18}	atto (a)
10^{-21}	zepto (z)
10^{-24}	yocto (y)

[a]The Greek letter μ (pronounced "mew").

dard meter bar an unchanging, reproducible quantity: 1 meter is the dis- traveled by light in a vacuum in 1/299,792,458 of a second. Length is one e seven fundamental quantities in the SI system (see Table 1.1). All other ical quantities have units that can be derived from these seven. SI is a *al* system. Quantities differing from the base unit by powers of ten are d by the use of prefixes. For example, the prefix *kilo* means "one thousand" times the base unit; it is abbreviated as k. Thus 1 *kilo*meter = 1000 meters, km = 1000 m. The SI prefixes are listed in Table 1.2.

ost measurements in chemistry are made in SI units. Sometimes we must ert between SI units, as when converting kilometers to meters. At other s we must convert measurements expressed in non-SI units into SI units, om SI units into non-SI units. In all these cases, we can use a *conversion* or a series of conversion factors in a scheme called a conversion path- Later in this chapter, we will apply conversion pathways in a method of lem solving known as *dimensional analysis*. The method itself is described me detail in Appendix A.

◀ It is a good idea to *memorize the most common SI prefixes* (such as G, M, k, d, c, m, μ, n, and p) because you can't survive in a world of science without knowing these SI prefixes.

s

describes the quantity of matter in an object. In SI the standard of mass *ilogram* (kg), which is a fairly large unit for most applications in chem- More commonly we use the unit *gram* (g).

eight is the force of gravity on an object. It is directly proportional to mass, own in the following mathematical expressions.

$$W \propto m \text{ and } W = g \times m \tag{1.1}$$

◀ The symbol $\propto$ means "proportional to." It can be replaced by an equality sign and a proportionality constant. In expression (1.1), the constant is the acceleration caused by gravity, g. (See Appendix B.)

bject has a fixed mass (m), which is independent of where or how the mass asured. Its weight (W), however, may vary because the acceleration caused avity (g) varies slightly from one point on Earth to another. Thus, an object weighs 100.0 kg in St. Petersburg, Russia, weighs only 99.6 kg in Panama it 0.4% less). The same object would weigh only about 17 kg on the moon. ugh the weight of an object varies from place to place, its mass is the same locations. The terms *weight* and *mass* are often used interchangeably, but mass is a measure of the quantity of matter. A common laboratory device for uring mass is called a balance. A balance is often called, incorrectly, a scale.

e principle used in a balance is that of counteracting the force of gravity n unknown mass with a force of equal magnitude that can be precisely ured. In older two-pan beam balances, the object whose mass is being mined is placed on one pan and counterbalancing is achieved through the of gravity acting on *weights*, objects of precisely known mass, placed on ther pan. In the type of balance most commonly seen in laboratories —the electronic balance—the counterbalancing force is a magnetic force uced by passing an electric current through an electromagnet. First, an ini- alance condition is achieved when no object is present on the balance pan.

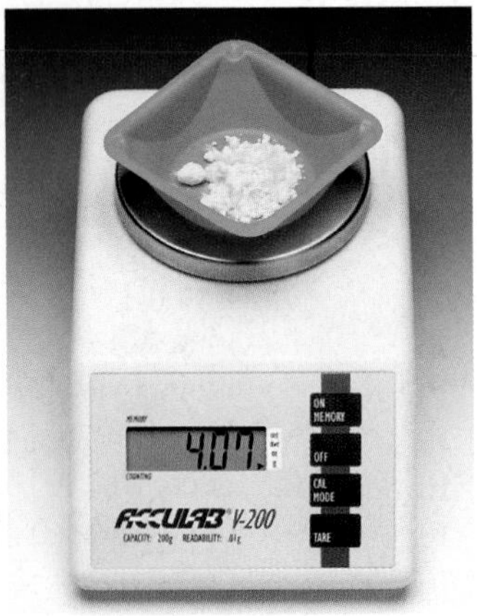

Richard Megna/Fundamental Photographs, NYC

▲ An electronic balance.

When the object to be weighed is placed on the pan, the initial balance cc tion is upset. To restore the balance condition, additional electric current r be passed through the electromagnet. The magnitude of this additional rent is proportional to the mass of the object being weighed and is transl into a mass reading that is displayed on the balance. An electronic balan shown in the margin.

1-2 CONCEPT ASSESSMENT

Would either the two-pan beam balance or the electronic balance yield the same result for the mass of an object measured on the moon as that measure for the same object on Earth? Explain.

Time

In daily use we measure time in seconds, minutes, hours, and years, dep ing on whether we are dealing with short intervals (such as the time 100 m race) or long ones (such as the time before the next appear of Halley's comet in 2062). We can use all these units in scientific v also, although in SI the standard of time is the *second* (s). A time interv 1 second is not easily established. At one time it was based on the leng a day, but this is not constant because the rate of Earth's rotation under slight variations. In 1956, the second was defined as 1/31,556,925.974 the length of the year 1900. With the advent of atomic clocks, a more pr definition became possible. The second is now defined as the duratic 9,192,631,770 cycles of a particular radiation emitted by certain atoms o element cesium (cesium-133).

▶ Electromagnetic radiation is discussed in Section 8-1.

Temperature

To establish a temperature scale, we arbitrarily set certain fixed points temperature increments called degrees. Two commonly used fixed point the temperature at which ice melts and the temperature at which water b both at standard atmospheric pressure.*

On the **Celsius** scale, the melting point of ice is 0 °C, the boiling poi water is 100 °C, and the interval between is divided into 100 equal parts ca Celsius degrees. On the **Fahrenheit** temperature scale, the melting point c is 32 °F, the boiling point of water is 212 °F, and the interval between is div into 180 equal parts called Fahrenheit degrees. Figure 1-8 compares Fahrenheit and Celsius temperature scales.

The SI temperature scale, called the **Kelvin** scale, assigns a value of ze the lowest possible temperature. The zero on the Kelvin scale is denotec and it comes at –273.15 °C. We will discuss the Kelvin temperature sca detail in Chapter 6. For now, it is enough to know the following:

- The interval on the Kelvin scale, called a *kelvin*, is the same size a Celsius degree.
- When writing a Kelvin temperature, we do not use a degree sym That is, we write 0 K or 300 K, not 0 °K or 300 °K.
- The Kelvin scale is an absolute temperature scale; there are no neg Kelvin temperatures.

▶ The SI symbol for Kelvin temperature is T and that for Celsius temperature is t but shown here as t(°C). The Fahrenheit temperature, shown here as t(°F), is not recognized in SI.

In the laboratory, temperature is most commonly measured in Ce degrees; however, these temperatures must often be converted to the K scale (in describing the behavior of gases, for example). Occasionally, par larly in some engineering applications, temperatures must be conve

*Standard atmospheric pressure is defined in Section 6-1. The effect of pressure on meltir boiling points is described in Chapter 12.

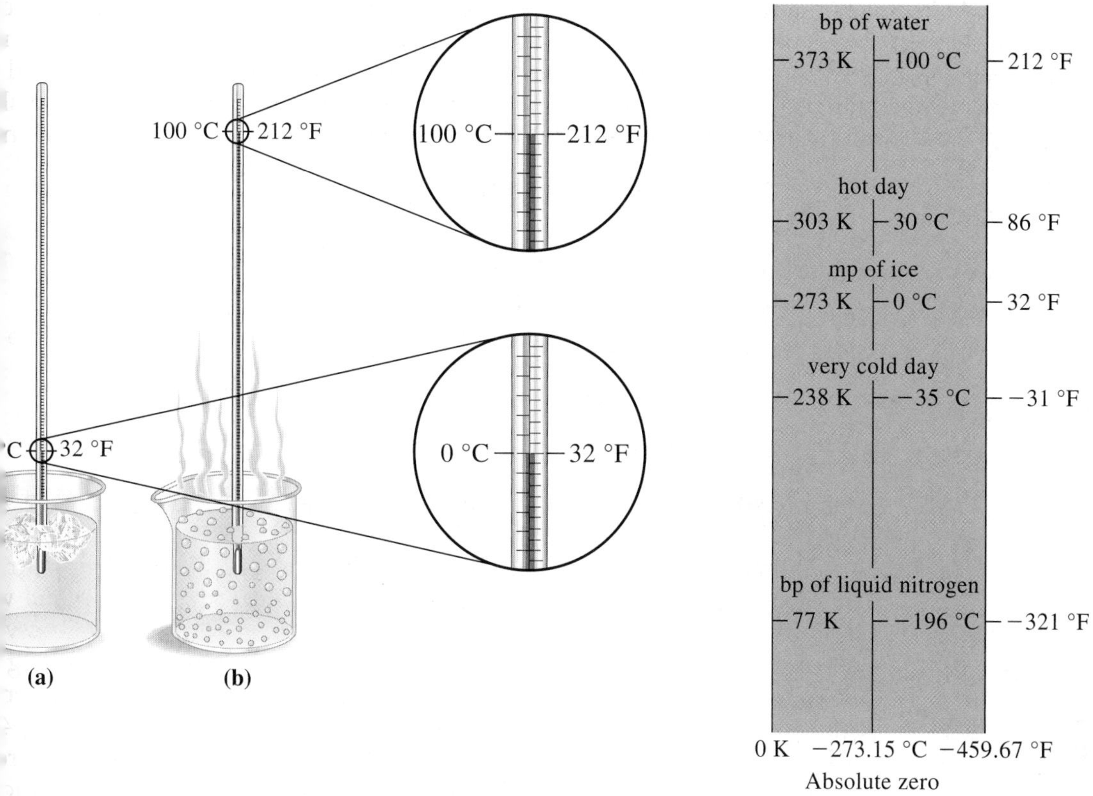

GURE 1-8
mparison of temperature scales
ne melting point (mp) of ice. **(b)** The boiling point (bp) of water.

reen the Celsius and Fahrenheit scales. Temperature conversions can be e in a straightforward way by using the algebraic equations shown below.

Kelvin from Celsius $T(\text{K}) = t(°\text{C}) + 273.15$

Fahrenheit from Celsius $t(°\text{F}) = \frac{9}{5}t(°\text{C}) + 32$

Celsius from Fahrenheit $t(°\text{C}) = \frac{5}{9}[t(°\text{F}) - 32]$

factors $\frac{9}{5}$ and $\frac{5}{9}$ arise because the Celsius scale uses 100 degrees between two chosen reference points and the Fahrenheit scale uses 180 degrees: 100 $= \frac{9}{5}$ and $100/180 = \frac{5}{9}$. The diagram in Figure 1-8 illustrates the rela-ship among the three scales for several temperatures.

XAMPLE 1-1 Converting Between Fahrenheit and Celsius Temperatures

The predicted high temperature for New Delhi, India, on a given day is 41 °C. Is this temperature higher or lower than the predicted daytime high of 103 °F for the same day in Phoenix, Arizona, reported by a newscaster?

nalyze

We are given a Celsius temperature and seek a comparison with a Fahrenheit temperature. To convert the given Celsius temperature to a Fahrenheit temperature, we use the equation given previously that expresses $t(°\text{F})$ as a function of $t(°\text{C})$.

olve

$$t(°\text{F}) = \frac{9}{5}t(°\text{C}) + 32 = \frac{9}{5}(41) + 32 = 106\ °\text{F}$$

The predicted temperature for New Delhi, 106 °F, is 3 °F higher than for Phoenix, 103 °F.

(continued)

Assess

For temperatures at which $t(°C) > -40$ °C, the Fahrenheit temperature is greater than the Celsius temper- ture. If the Celsius temperature is lower than -40 °C, then $t(°F)$ is lower than (more negative than) $t(°C$ (Fig. 1-8). Concept Assessment 1-3 asks you to think further about the relationship between $t(°C)$ and $t(°F)$.

PRACTICE EXAMPLE A: A recipe in an American cookbook calls for roasting a cut of meat at 350 °F. What is th temperature on the Celsius scale?

PRACTICE EXAMPLE B: A particular automobile engine coolant has antifreeze protection to a temperature o -22 °C. Will this coolant offer protection at temperatures as low as -15 °F?

Answers to Practice Examples are given in Appendix G.

1-3 CONCEPT ASSESSMENT

Can there be a temperature at which °C and °F have the same value? Can the be more than one such temperature? Explain.

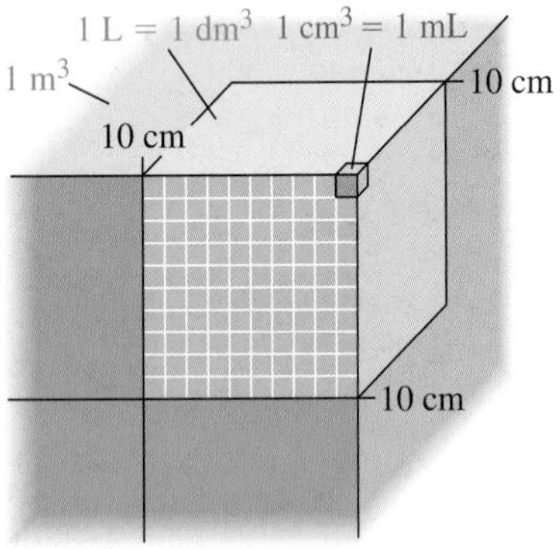

▲ FIGURE 1-9
Some metric volume units compared
The largest volume, shown in part, is the SI standard of 1 cubic meter (m^3). A cube with a length of 10 cm (1 dm) on edge (in blue) has a volume of 1000 cm^3 (1 dm^3) and is called 1 liter (1 L). The smallest cube is 1 cm on edge (red) and has a volume of 1 cm^3 = 1 mL.

▶ The official spelling is *litre*, but we will use the American spelling, *liter*.

Derived Units

The seven units listed in Table 1.1 are the SI units for the fundamental quant of length, mass, time, and so on. Many measured properties are expresse combinations of these fundamental, or base, quantities. We refer to the uni such properties as *derived units*. For example, velocity is a distance divided by time required to travel that distance. The unit of velocity is length divide time, such as m/s or m s^{-1}. Some derived units have special names. For exan the combination kg m^{-1} s^{-2} is called the *pascal* (Chapter 6) and the combina kg m^2 s^{-2} is called the *joule* (Chapter 7). Other examples are given in Append

An important measurement that uses derived units is *volume*. Volume the unit (length)^3, and the SI standard unit of volume is the *cubic meter* (More commonly used volume units are the *cubic centimeter* (cm^3) and *liter* (L). One liter is defined as a volume of 1000 cm^3, which means that *milliliter* (1 mL) is equal to 1 cm^3. The liter is also equal to one *cubic decin* (1 dm^3). Several volume units are depicted in Figure 1-9.

Non-SI Units

Although its citizens are growing more accustomed to expressing distanc kilometers and volumes in liters, the United States is one of the few coun where most units used in everyday life are still non-SI. Masses are give pounds, room dimensions in feet, and so on. In this book, we will not routi use these non-SI units, but we will occasionally introduce them in examples end-of-chapter exercises. In such cases, any necessary relationships betw non-SI and SI units will be given or can be found on the inside back cover.

1-1 ARE YOU WONDERING?

Why is it so important to attach units to a number?

In 1993, NASA started the Mars Surveyor program to conduct an ongoing ser of missions to explore Mars. In 1995, two missions were scheduled that would launched in late 1998 and early 1999. The missions were the Mars Climate Orbi (MCO) and the Mars Polar Lander (MPL). The MCO was launched December 1998, and the MPL, January 3, 1999.

Nine and a half months after launch, the MCO was to fire its main engine to hieve an elliptical orbit around Mars. The MCO engine start occurred on ptember 23, 1999, but the MCO mission was lost when the orbiter entered the artian atmosphere on a lower-than-expected trajectory. The MCO entered the v orbit because the computer on Earth used British Engineering units, whereas MCO computer used SI units.

This error in units brought the MCO 56 km above the surface of Mars instead the desired 250 km. At 250 km, the MCO would have successfully entered the sired elliptical orbit, and the $168 million orbiter would probably not have en lost.

◀ The development of science requires careful *quantitative measurement*. Theories have stood or fallen based on their agreement or otherwise with experiments in the fourth significant figure or beyond. Problem solving, the handling of units, and the use of significant figures (Section 1-7) are important in all areas of science.

Density and Percent Composition: Their Use in Problem Solving

ughout this text, we will encounter new concepts about the structure behavior of matter. One means of firming up our understanding of new concepts is to work problems that relate concepts that we already v to those we are trying to understand. In this section, we will intro- two quantities frequently required in problem solving: density and ent composition.

sity

is an old riddle: "What weighs more, a ton of bricks or a ton of cotton?" u answer that they weigh the same, you demonstrate a clear understand- f the meaning of weight and, indirectly, of the quantity of matter to which ht is proportional, that is, mass. Anyone who answers that the bricks h more than the cotton has confused the concepts of weight and density. er in a brick is more concentrated than in cotton—that is, the matter in a is confined to a smaller volume. Bricks are more dense than cotton. **sity** is the ratio of mass to volume.

$$\text{density } (d) = \frac{\text{mass } (m)}{\text{volume } (V)} \tag{1.2}$$

ass and volume are both extensive properties. An **extensive property** is *dent* on the quantity of matter observed. However, if we divide the mass substance by its volume, we obtain density, an intensive property. An **sive property** is *independent* of the amount of matter observed. Thus, the ity of pure water at 25 °C has a unique value, whether the sample fills a 1 beaker (small mass/small volume) or a swimming pool (large /large volume). Intensive properties are especially useful in chemical es because they can often be used to identify substances.

e SI base units of mass and volume are kilograms and cubic meters, ctively, but chemists generally express mass in grams and volume in centimeters or milliliters. Thus, the most commonly encountered den- nit is grams per cubic centimeter (g/cm^3) or the identical unit grams per liter (g/mL).

e mass of 1.000 L of water at 4 °C is 1.000 kg. The density of water at 4 °C is g/1000 mL, or 1.000 g/mL. At 20 °C, the density of water is 0.9982 g/mL. ity is a function of temperature because volume varies with temperature, eas mass remains constant. One reason that climate change is a concern is use as the average temperature of seawater increases, the seawater will me less dense, its volume will increase, and sea level will rise—*even if no nental ice melts.*

ke temperature, the state of matter affects the density of a substance. In ral, solids are denser than liquids and both are denser than gases, but there are notable overlaps in densities between solids and liquids. Follov are the ranges of values generally observed for densities; this informa should prove useful in solving problems.

- Solid densities: from about 0.2 g/cm^3 to 20 g/cm^3
- Liquid densities: from about 0.5 g/mL to 3–4 g/mL
- Gas densities: mostly in the range of a few grams per liter

KEEP IN MIND

that recognizing the *scale of things* is one important step in avoiding mistakes. If a solid density calculates out as 0.05 g/cm^3, or a gas density as 5.0 g/cm^3, review the work done up to that point!

In general, densities of liquids are known more precisely than thos solids (which may have imperfections in their microscopic structures). densities of elements and compounds are known more precisely than de ties of materials with variable compositions (such as wood or rubber).

The different densities of solids and liquids have several important co quences. A solid that is insoluble and floats on a liquid is *less* dense thar liquid, and it displaces a *mass* of liquid equal to its own mass. An insol solid that sinks to the bottom of a liquid is *more* dense than the liquid and places a *volume* of liquid equal to its own volume. Liquids that are immis in each other separate into distinct layers, with the most dense liquid a bottom and the least dense liquid at the top.

1-4 CONCEPT ASSESSMENT

Approximately what fraction of its volume is submerged when a 1.00 kg block of wood (d = 0.68 g/cm^3) floats on water?

KEEP IN MIND

that in a conversion pathway, all units must cancel except for the desired unit in the final result (see Appendix A-5). Also, note that the quantities given and sought are typically *extensive* properties and the conversion factor(s) are often *intensive* properties (here, density).

Density in Conversion Pathways

If we measure the mass of an object and its volume, simple division gives u density. Conversely, if we know the density of an object, we can use density conversion factor to determine the object's mass or volume. For example, a of osmium 1.000 cm on edge weighs 22.59 g. The density of osmium (the der of the elements) is 22.59 g/cm^3. What would be the mass of a cube of osr that is 1.25 inches on edge (1 in = 2.54 cm)? To solve this problem, we beg relating the volume of a cube to its length, that is, $V = l^3$. Then we can ma the *conversion pathway*:

$$\text{inches of osmium} \longrightarrow \text{cm osmium} \longrightarrow \text{cm}^3 \text{ osmium} \longrightarrow \text{g osmium}$$

(converts inches to cm) (converts cm to cm^3) (converts cm^3 to g osmium)

$$? \text{ g osmium} = \left[1.25 \text{ in} \times \frac{2.54 \text{ cm}}{1 \text{ in}}\right]^3 \times \frac{22.59 \text{ g osmium}}{1 \text{ cm}^3} = 723 \text{ g osmium}$$

At 25 °C the density of mercury, the only metal that is liquid at this tem ture, is 13.5 g/mL. Suppose we want to know the volume, in mL, of 1.0 of mercury at 25 °C. We proceed by (1) identifying the known informa 1.000 kg of mercury and d = 13.5 g/mL (at 25 °C); (2) noting what w trying to determine—a volume in milliliters (which we designate mL merc and (3) looking for the relevant conversion factors. Outlining the conve pathway will help us find these conversion factors:

$$\text{kg mercury} \longrightarrow \text{g mercury} \longrightarrow \text{mL mercury}$$

We need the factor 1000 g/kg to convert from kilograms to grams. De provides the factor to convert from mass to volume. But in this instanc need to use density in the *inverted* form. That is,

$$? \text{ mL mercury} = 1.000 \text{ kg} \times \frac{1000 \text{ g}}{1 \text{ kg}} \times \frac{1 \text{ mL mercury}}{13.5 \text{ g}} = 74.1 \text{ mL mercury}$$

amples 1-2 and 1-3 further illustrate that numerical calculations involv- lensity are generally of two types: determining density from mass and me measurements and using density as a conversion factor to relate mass volume.

AMPLE 1-2 Relating Mass, Volume, and Density

The stainless steel in the solid cylindrical rod pictured below has a density of 7.75 g/cm^3. If we want a 1.00 kg mass of this rod, how long a section must we cut off? Refer to the inside back cover for the formula to calculate the volume of a cylinder.

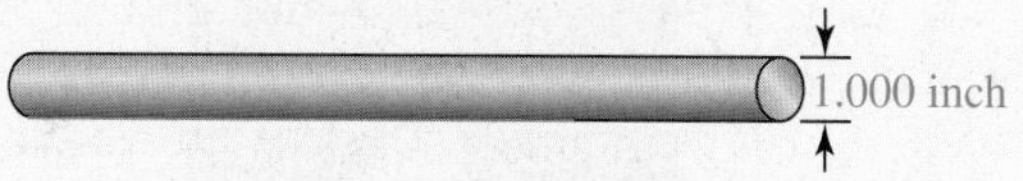

alyze

We are given the density, d, and the desired mass, m. Because $d = m/V$, we can solve for V and then use the formula for the volume of a cylinder, $V = \pi r^2 h$, to calculate h, the length of rod we seek. Two different mass units (g and kg) and two different length units (centimeters and inches) appear in the information given in this problem, so we anticipate having to make at least two unit conversions. To avoid errors, we include units in all steps.

lve

Solve equation (1.2) for V. The reciprocal of density, $1/d$, is a conversion factor for converting from mass to volume.

$$V = \frac{m}{d} = m \times \frac{1}{d}$$

Calculate the volume of the rod that will have a mass of 1.00 kg. A conversion from kg to g is required in this step.

$$V = 1.00\ \text{kg} \times \frac{1000\ \text{g}}{1\ \text{kg}} \times \frac{1\ \text{cm}^3}{7.75\ \text{g}} = 129\ \text{cm}^3$$

Solve $V = \pi r^2 h$ for h and then calculate h. We must be certain to use the radius of the rod (one-half the diameter) and to express the radius in centimeters.

$$h = \frac{V}{\pi r^2} = \frac{129\ \text{cm}^3}{3.1416 \times (0.500\ \text{in} \times 2.54\ \text{cm}/1\ \text{in})^2} = 25.5\ \text{cm}$$

sess

One way to check whether our answer is correct is to work the problem in reverse. For example, we calculate $d = 1.00 \times 10^3\ \text{g}/[3.1416 \times (1.27\ \text{cm})^2 \times 25.5\ \text{cm}] = 7.74\ \text{g/cm}^3$, which is very close to the given density. We are confident that our answer, $h = 25.5$ cm, is correct.

ACTICE EXAMPLE A: To determine the density of trichloroethylene, a liquid used to degrease electronic components, a flask is first weighed empty (108.6 g). It is then filled with 125 mL of the trichloroethylene to give a total mass of 291.4 g. What is the density of trichloroethylene in grams per milliliter?

ACTICE EXAMPLE B: Suppose that instead of using the cylindrical rod of Example 1-2 to prepare a 1.000 kg mass we were to use a solid spherical ball of copper ($d = 8.96\ \text{g/cm}^3$). What must be the radius of this ball?

AMPLE 1-3 Determining the Density of an Irregularly Shaped Solid

A chunk of coal is weighed twice while suspended from a spring scale (see Figure 1-10). When the coal is suspended in air, the scale registers 156 g; when the coal is suspended underwater at 20 °C, the scale registers 59 g. What is the density of the coal? The density of water at 20 °C is 0.9982 g cm^{-3}.

alyze

We need the ratio of mass to volume of the chunk of coal. The mass of the coal is easily obtained; it is what registers on the scale when the coal is suspended in air: 156 g. But what is the volume of this chunk of coal? The key to this calculation is the weight measurement under water. The coal weighs less than 156 g when submerged in water because the water exerts a buoyant force on the coal. The buoyant force is the difference

(continued)

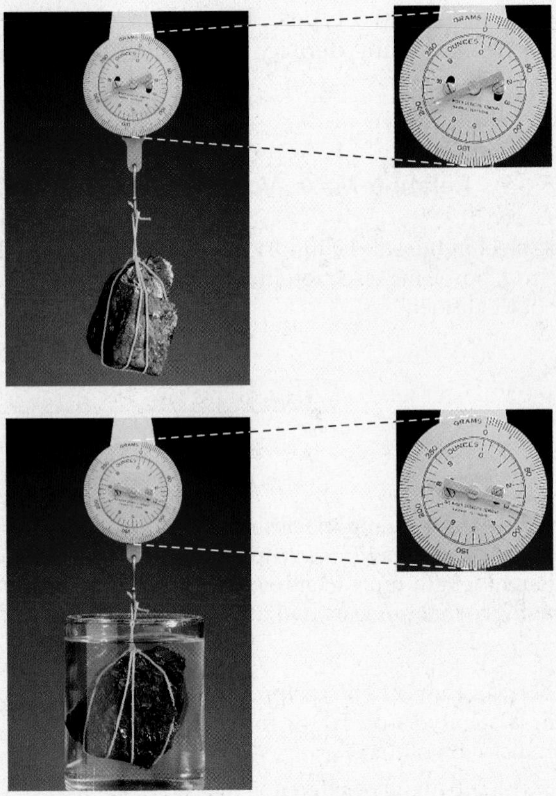

▶ FIGURE 1-10
Measuring the volume of an irregularly shaped solid
When submerged in a liquid, an irregularly shaped solid displaces a volume of liquid equal to its own volume. The necessary data can be obtained by two mass measurements of the type illustrated here; the required calculations are like those in Example 1-3.

Kristen Brochmann/Fundamental Photographs

between the two weight measurements: 156 g − 59 g = 97 g. Recall the statement on page 14 that a submerg solid displaces a volume of water equal to its own volume. We don't know this volume of water directly, but can use the mass of displaced water, 97 g, and its density, 0.9982 g/cm^3, to calculate the volume of displac water. The volume of the coal is equal to the volume of displaced water.

Solve

The mass of the chunk of coal is 156 g. If we use m_{water} to denote the mass of displaced water, then the volu of the displaced water is calculated as follows:

$$V = \frac{m_{\text{water}}}{d} = \frac{156\text{ g} - 59\text{ g}}{0.9982\text{ g/cm}^3} = 97\text{ cm}^3$$

The volume of the chunk of coal is the same as the volume of displaced water. Therefore, the density of the coal

$$d = \frac{156\text{ g}}{97\text{ cm}^3} = 1.6\text{ g/cm}^3$$

Assess

To determine the density of an object, we might think it is necessary to make measurements of both t mass and volume of the object. Example 1-3 shows that a volume measurement is not necessary. The steps our calculation can be combined to give the following expression:

(density of object)/(density of water) = (weight in water)/(weight in air − weight in water).

The expression above clearly shows that the density of an object can be det mined by making two weight measurements: one in air, and the other in a flu (such as water) of known density.

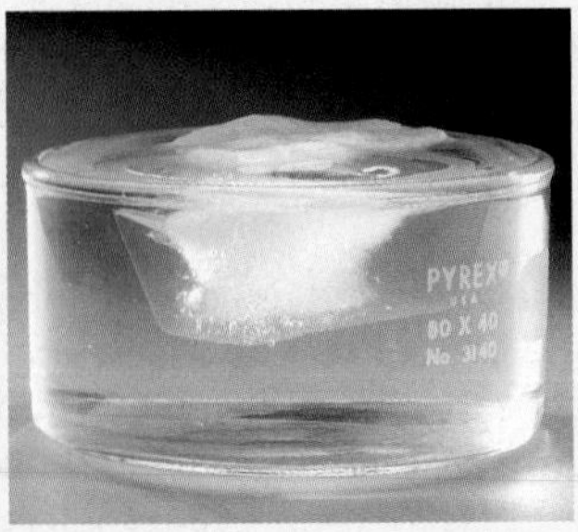

Richard Megna/Fundamental Photographs

PRACTICE EXAMPLE A: A graduated cylinder contains 33.8 mL of water. stone with a mass of 28.4 g is placed in the cylinder and the water level rises 44.1 mL. What is the density of the stone?

PRACTICE EXAMPLE B: In the situation shown in the photograph, when the cube melts completely, will the water overflow the container, will the water le in the container drop, or will the water level remain unchanged? Explain.

cent Composition as a Conversion Factor

ection 1-2, we described composition as an identifying characteristic of a ple of matter. A common way of referring to composition is through percent-. Percent (*per centum*) is the Latin for *per* (meaning "for each") and *centum* ning "100"). Thus, percent is the number of parts of a constituent in parts of the whole. To say that a seawater sample contains 3.5% sodium chlo- by mass means that there are 3.5 g of sodium chloride in every 100 g of the ater. We make the statement in terms of grams because we are talking about ent *by mass*. We can express this percent by writing the following ratios:

$$\frac{3.5 \text{ g sodium chloride}}{100 \text{ g seawater}} \quad \text{and} \quad \frac{100 \text{ g seawater}}{3.5 \text{ g sodium chloride}} \qquad \textbf{(1.3)}$$

xample 1-4, we will use one of these ratios as a conversion factor.

XAMPLE 1-4 Using Percent Composition as a Conversion Factor

A 75 g sample of sodium chloride (table salt) is to be produced by evaporating to dryness a quantity of seawater containing 3.5% sodium chloride by mass. What volume of seawater, in liters, must be taken for this purpose? Assume a density of 1.03 g/mL for seawater.

nalyze

The conversion pathway is g sodium chloride → g seawater → mL seawater → L seawater. To convert from g sodium chloride to g seawater, we need the conversion factor in expression (1.3), with g seawater in the numerator and g sodium chloride in the denominator. To convert from g seawater to mL seawater, we use the reciprocal of the density of seawater as the conversion factor. To make the final conversion, from mL seawater to L of seawater, we use the fact that 1 L = 1000 mL.

olve

Following the conversion pathway described above, we obtain

$$? \text{ L seawater} = 75 \,\cancel{\text{g sodium chloride}} \times \frac{100 \,\cancel{\text{g seawater}}}{3.5 \,\cancel{\text{g sodium chloride}}} \times \frac{1 \,\cancel{\text{mL seawater}}}{1.03 \,\cancel{\text{g seawater}}} \times \frac{1 \text{ L seawater}}{1000 \,\cancel{\text{mL seawater}}} = 2.1 \text{ L seawater}$$

ssess

In solving this problem, we set up a conversion pathway, and then we thought about the conversion factors that were required. We will make use of this approach throughout the text.

RACTICE EXAMPLE A: How many kilograms of ethanol are present in 25 L of a gasohol solution that is 90% gasoline to 10% ethanol by mass? The density of gasohol is 0.71 g/mL.

RACTICE EXAMPLE B: Common rubbing alcohol is a solution of 70.0% isopropyl alcohol by mass in water. If a 25.0 mL sample of rubbing alcohol contains 15.0 g of isopropyl alcohol, what is the density of the rubbing alcohol?

1-2 ARE YOU WONDERING?

When do we multiply and when do we divide in solving problems with percentages?

ommon way of dealing with a percentage is to convert it to decimal form (3.5% omes 0.035) and then to multiply or divide by this decimal, but students some- es can't decide which to do. Expressing percentage as a conversion factor and

(continued)

using it to produce a necessary cancellation of units gets around this difficul
Also, remember that

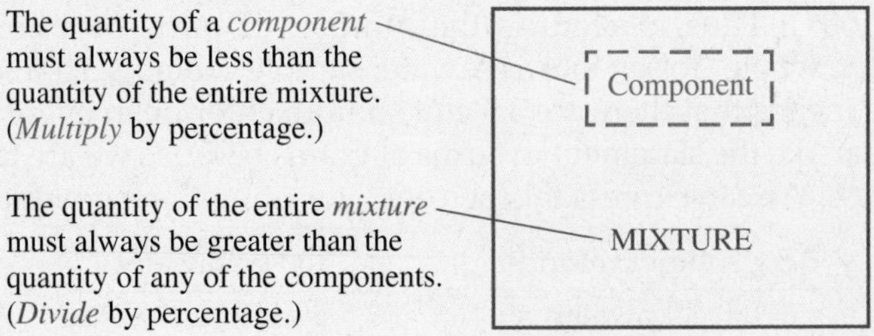

The quantity of a *component* must always be less than the quantity of the entire mixture. (*Multiply* by percentage.)
Component
The quantity of the entire *mixture* must always be greater than the quantity of any of the components. (*Divide* by percentage.)
MIXTURE

If, in Example 1-4, we had not been careful about the cancellation of units and h multiplied by percentage (3.5/100) instead of dividing by it (100/3.5), we wou have obtained the numerical answer 2.5×10^{-3}. This would be a 2.5 mL samp of seawater, weighing about 2.5 g. Clearly, a sample of seawater that *contains* 7 of sodium chloride must have a mass *greater than* 75 g.

KEEP IN MIND

that a numerical answer that defies common sense is probably wrong.

1-6 Uncertainties in Scientific Measureme

All measurements are subject to error. To some extent, measuring in ments have built-in, or inherent, errors, called **systematic errors**. (For ex ple, a kitchen scale might consistently yield results that are 25 g too hig a thermometer a reading that is 2°C too low.) Limitations in an exp menter's skill or ability to read a scientific instrument also lead to er and give results that may be either too high or too low. Such errors called **random errors**.

▶ *Random errors* are observed by scatter in the data and can be dealt with effectively by taking the average of many measurements. *Systematic errors*, conversely, are the bane of the experimental scientist. They are not readily apparent and must be avoided by carefully calibrating a method against a known sample or result. Systematic errors influence the *accuracy* of a measurement, whereas random errors are linked to the *precision* of measurements.

Precision refers to the degree of reproducibility of a measured quanti that is, the closeness of agreement when the same quantity is meas several times. The precision of a series of measurements is *high* (or g if each of a series of measurements deviates by only a small am from the average. Conversely, if there is wide deviation among the surements, the precision is *poor* (or low). **Accuracy** refers to how c a measured value is to the accepted, or actual, value. High-preci measurements are not always accurate—a large systematic error c be present. (A tight cluster of three darts near the edge of a dart boarc be considered precise but not very accurate if the intention was to strik center of the board.) Still, scientists generally strive for high precisic measurements.

To illustrate these ideas, consider measuring the mass of an ol by using the two balances shown on page 19. One of the balances is a sin pan balance that gives the mass in grams with only one decimal p The other balance is a sophisticated analytical balance that gives the in grams with four decimal places. The accompanying table gives re obtained when the object is weighed three times on each balance. Fo single-pan balance, the average of the measurements is 10.5 g, measurements ranging from 10.4 g to 10.6 g. For the analytical balance average of the measurements is 10.4978 g, with measurements ran from 10.4977 g to 10.4979 g. The scatter in the data obtained with the sin pan balance (±0.1 g) is greater than that obtained with the analy balance (±0.0001 g). Thus, the results obtained by using the single-pan ance have lower (or poorer) precision than those obtained by using the lytical balance.

Richard Megna/Fundamental Photographs

Richard Megna/Fundamental Photographs

	Pan Balance	Analytical Balance
Three measurements	10.5, 10.4, 10.6 g	10.4978, 10.4979, 10.4977 g
Their average	10.5 g	10.4978 g
Reproducibility	±0.1 g	±0.0001 g
Precision	low or poor	high or good

1-5 CONCEPT ASSESSMENT

n a set of measurements be precise without being accurate? Can the average a set of measurements be accurate and the individual measurements be precise? Explain.

Significant Figures

sider these measurements made on a low-precision balance: 10.4, 10.2, and g. The reported result is best expressed as their average, that is, 10.3 g. scientist would interpret these results to mean that the first two digits—10—nown with certainty and the last digit—3—is uncertain because it was ated. That is, the mass is known only to the nearest 0.1 g, a fact that we l also express by writing 10.3 ± 0.1 g. To a scientist, the measurement 10.3 g d to have *three* **significant figures**. If this mass is reported in kilograms r than in grams, 10.3 g = 0.0103 kg, the measurement is still expressed to significant figures even though more than three digits are shown. When ured on an analytical balance, the corresponding reported value might be 07 g—a value with *six* significant figures. The number of significant figures neasured quantity gives an indication of the capabilities of the measuring e and the precision of the measurements.

e will frequently need to determine the number of significant figures in a erical quantity. The rules for doing this, outlined in Figure 1-11, are as ws:

All nonzero digits are significant.

Zeros are also significant, but with two important *exceptions* for quantities less than one. Any zeros (1) preceding the decimal point, or (2) following the decimal point and preceding the first nonzero digit, are *not* significant.

The case of terminal zeros that precede the decimal point in quantities greater than one is *ambiguous*.

e quantity 7500 m is an example of an ambiguous case.

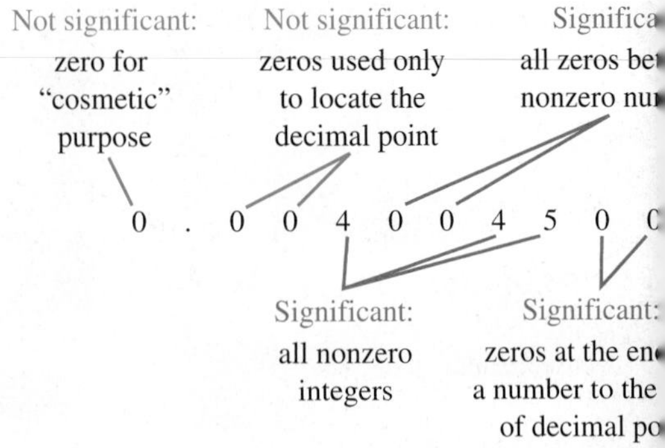

▶ FIGURE 1-11
Determining the number of significant figures in a quantity
The quantity shown here, 0.004004500, has *seven* significant figures. All nonzero digits are significant, as are the indicated zeros.

Do we mean 7500 m, measured to the nearest meter? Nearest 10 mete all the zeros are significant—if the value has *four* significant figures—we write 7500. m. That is, by writing a decimal point that is not otherwise nee we show that all zeros preceding the decimal point are significant. This nique does not help if only one of the zeros, or if neither zero, is signifi The best approach here is to use exponential notation. (Review Appendix necessary.) The coefficient establishes the number of significant figures, the power of ten locates the decimal point.

2 significant figures	*3 significant figures*	*4 significant figures*
7.5×10^3 m	7.50×10^3 m	7.500×10^3 m

Significant Figures in Numerical Calculations

Precision must neither be gained nor be lost in calculations involving sured quantities. There are several methods for determining how precise express the result of a calculation, but it is usually sufficient just to obs some simple rules involving significant figures.

▶ A more exact rule on multiplication/division is that the result should have about the same relative error—for example, expressed as parts per hundred (percent) or parts per thousand—as the least precisely known quantity. Usually the significant figure rule conforms to this requirement; occasionally, it does not (see Exercise 67).

> The result of multiplication or division may contain only as many significant figures as the *least* precisely known quantity in the calculation.

In the following chain multiplication to determine the volume of a re gular block of wood, we should round off the result to *three* significant fig Figure 1-12 may help you to understand this.

$$14.79 \text{ cm} \times 12.11 \text{ cm} \times 5.05 \text{ cm} = 904 \text{ cm}^3$$

(4 *sig. fig.*) (4 *sig. fig.*) (3 *sig. fig.*) (3 *sig. fig.*)

In adding and subtracting numbers, the applicable rule is as follows:

▶ In addition and subtraction the *absolute* error in the result can be no less than the absolute error in the least precisely known quantity. In the summation at the right, the absolute error in one quantity is ±0.1 g; in another, ±0.01 g; and in the third, ±0.001 g. The sum *must* be expressed with an absolute error of ±0.1 g.

> The result of addition or subtraction must be expressed with the same number of digits beyond the decimal point as the quantity carrying the *smallest* number of such digits.

Consider the following sum of masses.

$$\begin{array}{r} 15.02\ \text{g} \\ 9{,}986.0\ \ \text{g} \\ 3.518\ \text{g} \\ \hline 10{,}004.538\ \text{g} \end{array}$$

Carey B. Van Loon

FIGURE 1-12
gnificant figure rule in multiplication
forming the product 14.79 cm × 12.11 cm × 5.05 cm, the least precisely
own quantity is 5.05 cm. Shown on the calculators are the products of 14.79 and
.11 with 5.04, 5.05, and 5.06, respectively. In the three results, only the first two
gits, 90 . . . , are identical; variations begin in the third digit. We are certainly not
tified in carrying digits beyond the third. We express the volume as 904 cm^3.
ually, instead of a detailed analysis of the type done here, we can use a simpler
a: *The result of a multiplication may contain only as many significant figures as es the least precisely known quantity.*

sum has the same uncertainty, ±0.1 g, as does the term with the *smallest* ber of digits beyond the decimal point, 9986.0 g. Note that this calculation limited by significant figures. In fact, the sum has more significant figures than do any of the terms in the addition.

ere are two situations when a quantity appearing in a calculation may be , that is, not subject to errors in measurement. This may occur

by definition (such as 1 min = 60 s, or 1 in = 2.54 cm)

as a result of counting (such as *six* faces on a cube, or *two* hydrogen atoms in a water molecule)

◀ Later in the text, we will need to apply ideas about significant figures to logarithms. This concept is discussed in Appendix A.

Exact numbers can be considered to have an unlimited number of significant figures.

1-6 CONCEPT ASSESSMENT

ich of the following is a more precise statement of the length 1 inch: = 2.54 cm or 1 m = 39.37 in? Explain.

◀ As added practice in working with significant figures, review the calculations in Section 1-6. You will note that they conform to the significant figure rules presented here.

nding Off Numerical Results

ree significant figures, we should express 15.453 as 15.5 and 14,775 as × 10^4. If we need to drop just one digit, that is, to round off a number, the that we will follow is to increase the final digit by one unit if the digit ped is 5, 6, 7, 8, or 9 and to leave the final digit unchanged if the digit ped is 0, 1, 2, 3, or 4.* To three significant figures, 15.44 rounds off to 15.4, 5.45 rounds off to 15.5.

◀ Some people prefer the *"round 5 to even"* rule. Thus 15.55 rounds to 15.6, and 17.65 rounds to 17.6. In banking and with large data sets, rounding needs to be unbiased. With a small number of data, this is less important.

. Guare, *J. Chem. Educ.*, **68**, 818 (1991).

EXAMPLE 1-5 Applying Significant Figure Rules: Multiplication/Division

Express the result of the following calculation with the correct number of significant figures.

$$\frac{0.225 \times 0.0035}{2.16 \times 10^{-2}} = ?$$

Analyze

By inspecting the three quantities, we see that the least precisely known quantity, 0.0035, has *two* significa figures. Our result must also contain only *two* significant figures.

Solve

When we carry out the calculation above by using an electronic calculator, the result is displayed as 0.036458 In our analysis of this problem, we determined that the result must be rounded off to two significant figur and so the result is properly expressed as 0.036 or as 3.6×10^{-2}.

Assess

To check for any possible calculation error, we can estimate the correct answer through a quick mental calcu tion by using exponential numbers. The answer should be $(2 \times 10^{-1})(4 \times 10^{-3})/(2 \times 10^{-2}) \approx 4 \times 10^{-2}$, a it is. Expressing numbers in exponential notation can often help us quickly estimate what the result of a calc lation should be.

PRACTICE EXAMPLE A: Perform the following calculation, and express the result with the appropriate numb of significant figures.

$$\frac{62.356}{0.000456 \times 6.422 \times 10^{3}} = ?$$

PRACTICE EXAMPLE B: Perform the following calculation, and express the result with the appropriate numb of significant figures.

$$\frac{8.21 \times 10^{4} \times 1.3 \times 10^{-3}}{0.00236 \times 4.071 \times 10^{-2}} = ?$$

EXAMPLE 1-6 Applying Significant Figure Rules: Addition/Subtraction

Express the result of the following calculation with the correct number of significant figures.

$$(2.06 \times 10^{2}) + (1.32 \times 10^{4}) - (1.26 \times 10^{3}) = ?$$

Analyze

If the calculation is performed with an electronic calculator, the quantities can be entered just as they are writte and the answer obtained can be adjusted to the correct number of significant figures. To determine the corre number of significant figures, identify the largest quantity, and then write the other quantities with the sar power of ten as appears in the largest quantity. The answer can have no more digits beyond the decimal poi than the quantity having the smallest number of such digits.

Solve

The largest quantity is 1.32×10^{4} and thus, we write the other two quantities as 0.0206×10^{4} and 0.126×1 The result of the required calculation must be rounded off to two decimal places.

$$\begin{aligned}
&(2.06 \times 10^{2}) + (1.32 \times 10^{4}) - (1.26 \times 10^{3}) \\
&\quad = (0.0206 \times 10^{4}) + (1.32 \times 10^{4}) - (0.126 \times 10^{4}) \\
&\quad = (0.0206 + 1.32 - 0.126) \times 10^{4} \\
&\quad = 1.2146 \times 10^{4} \\
&\quad = 1.21 \times 10^{4}
\end{aligned}$$

sess

If you refer back to the margin note on page 20, you will see that there is another way to approach this problem. To determine the absolute error in the least precisely known quantity, we write the three quantities as $(2.06 \pm 0.01) \times 10^2$, $(1.32 \pm 0.01) \times 10^4$ and $(1.26 \pm 0.01) \times 10^3$. We conclude that 1.32×10^4 has the largest absolute error ($\pm 0.01 \times 10^4$) and so, the absolute error in the result of the calculation above is also $\pm 0.01 \times 10^4$. Thus, 1.2146×10^4 is rounded to 1.21×10^4.

ACTICE EXAMPLE A: Express the result of the following calculation with the appropriate number of significant figures.

$$0.236 + 128.55 - 102.1 = ?$$

ACTICE EXAMPLE B: Perform the following calculation, and express the result with the appropriate number of significant figures.

$$\frac{(1.302 \times 10^3) + 952.7}{(1.57 \times 10^2) - 12.22} = ?$$

working through the preceding examples, you likely used an electronic lator. What's nice about using electronic calculators is that we don't have to down intermediate results. In general, disregard occasional situations e intermediate rounding may be justified and store all intermediate results ur electronic calculator without regard to significant figures. Then, round the correct number of significant figures only in the final answer.

KEEP IN MIND

that addition and subtraction are governed by one significant figure rule, and multiplication and division are governed by a different rule.

asteringCHEMISTRY **www.masteringchemistry.com**

the late 1960s, scientists heatedly debated the reported discovery of a new form of water called polywater. r a discussion of the polywater debate and the importance of the scientific method in helping the scientific mmunity reach a consensus, go to the Focus On feature for Chapter 1 (The Scientific Method at Work: lywater) on the MasteringChemistry site.

Summary

The Scientific Method—The **scientific method** et of procedures used to develop explanations of al phenomena and possibly to predict additional omena. The four basic stages of the scientific method) gathering data through observations and exper- s; (2) reducing the data to simple verbal or mathe- al expressions known as **natural laws**; (3) offering isible explanation of the data through a **hypothesis**; sting the hypothesis through predictions and further imentation, leading ultimately to a conceptual model a **theory** that explains the hypothesis, often together other related hypotheses.

Properties of Matter—**Matter** is defined as any- that occupies space, possesses mass, and displays a. **Composition** refers to the component parts of a le of matter and their relative proportions. **Properties** e qualities or attributes that distinguish one sample atter from another. Properties of matter can be ed into two broad categories: **physical** and **chemical**.

Matter can undergo two types of changes: **chemical changes** or **reactions** are changes in composition; **physical changes** are changes in state or physical form and do not affect composition.

1-3 Classification of Matter—The basic building blocks of matter are called **atoms**. Matter that is composed of a collection of a single type of atom is known as an **element**. A sample of matter composed of two or more elements is known as a **compound**. A **molecule** is the smallest entity of a compound having the same proportions of the constituent atoms as does the compound as a whole. Collectively, elements and compounds compose the types of matter called **substances**. **Mixtures** of substances can be classified as **homogeneous** or **heterogeneous** (Fig. 1-4). The three states of matter are **solid**, **liquid**, and **gas**.

1-4 Measurement of Matter: SI (Metric) Units—Chemistry is a quantitative science, meaning that chemical measurements are usually expressed in terms of a number and an accompanying unit. The scientific system of measurement, called the *Système Internationale d'Unités* (abbreviated **SI**), involves seven base quantities (Table 1.1). **Mass** describes a quantity of matter. Weight measures the force of gravity on an object; weight is related to, but different from, mass. The temperature scales used by chemists are the **Celsius** and **Kelvin** scales. The **Fahrenheit** temperature scale, commonly used in daily life in the United States, is also used in some industrial settings. The three temperature scales can be related algebraically (Fig. 1-8).

1-5 Density and Percent Composition: Their Use in Problem Solving—Mass and volume are **extensive properties**; they *depend* on the amount of matter in a sample. **Density**, the ratio of the mass of a sample to its volume, is an **intensive property**, a property *independent* of the amount of matter sampled. Density is used as a conversion factor in a variety of calculations.

1-6 Uncertainties in Scientific Measureme
Measurements are subject to **systematic** and **ra** **errors**. In making a series of measurements, the deg which the measurements agree with one another is k as the **precision** of the measurement, while the deg which the measurement agrees with the actual va referred to as the **accuracy** of the measurement.

1-7 Significant Figures—The proper use of **signi figures** is important in that it prevents the suggesti a higher degree of precision in a calculated quantity is warranted by the precision of the measured qu ties used in the calculation. The precision of an ar cannot be greater than the precision of the numbers in the calculation. In addition to reporting the co number of significant figures in a calculated quantity important to know the rules for rounding off num results.

Integrative Example

Consider a 58.35 g hexagonal block of wood that is 5.00 cm on edge and 1.25 cm thick, with a 2.50 cm diameter hole d through its center. Also given are the densities of the liquids hexane (d = 0.667 g/mL) and decane (d = 0.845 g, Assume that the density of a mixture of the two liquids is a linear function of the volume percent composition of the tion. Determine the volume percent of hexane required in the solution so that the hexagonal block of wood will just b float on the solution.

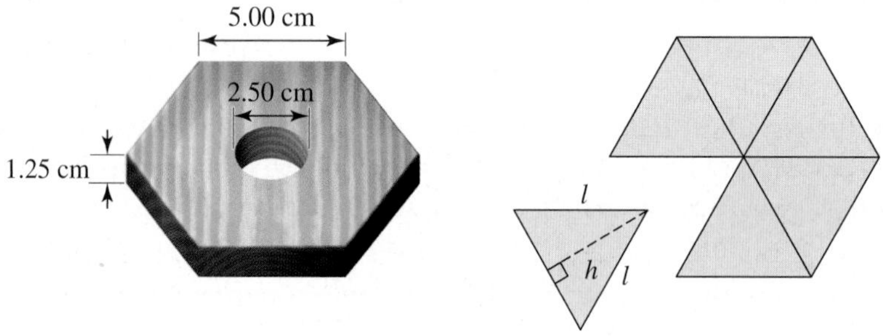

Analyze

The first task is to determine the density of the wood block, $d = m/V$. The mass is given, so the critical calculation i of the volume. The key to the volume calculation is recognizing that the volume of the block is the *difference* betwee volumes: The volume of the block if there were no hole *minus* the volume of the cylindrical hole. The second task write a simple equation relating density to the volume percent composition of the liquid solution, and then to solve equation for the volume percent hexane that yields a solution density equal to the density calculated for the wood.

Solve

The solid hexagonal block can be divided into six smaller blocks, each an equilateral triangle of length, l, and height, h. The area of a triangle is given by the formula

$$A = \frac{1}{2}(\text{base} \times \text{height}) = \frac{1}{2} \times l \times h$$

Only the base, l, is given (5.00 cm). To express h in terms of l, we use the Pythagorean theorem for the right triangle pictured, that is, $a^2 + b^2 = c^2$, rearranged to the form $a^2 = c^2 - b^2$.

$$h^2 = l^2 - \left(\frac{l}{2}\right)^2 = l^2 - \frac{l^2}{4} = \frac{3l^2}{4} \text{ and}$$

$$h = \frac{\sqrt{3}}{2} \times l$$

Now, for the area of one of the six triangles we have

$$A = \frac{1}{2} \times l \times \frac{\sqrt{3}}{2} \times l = \frac{\sqrt{3}}{4} \times l^2$$

olume of one of the triangular blocks is obtained by mul-ng the cross-sectional area A by the thickness, 1.25 cm. olume of the hexagonal block of wood, without the cylin- hole, is that of six triangular blocks.

$$V = 6 \times A \times 1.25\ \text{cm}$$
$$= 6 \times \frac{\sqrt{3}}{4} \times (5.00\ \text{cm})^2 \times 1.25\ \text{cm}$$
$$= 81.2\ \text{cm}^3$$

olume of the cylindrical hole with a radius of 1.25 cm half the 2.50 cm diameter) and a height of 1.25 cm is

$$V = \pi r^2 h = 3.1416 \times (1.25\ \text{cm})^2 \times 1.25\ \text{cm}$$
$$= 6.14\ \text{cm}^3$$

olume of the block of wood is the difference

$$V = 81.2\ \text{cm}^3 - 6.14\ \text{cm}^3 = 75.1\ \text{cm}^3$$

ensity of the wood is

$$d = \frac{m}{V} = \frac{58.35\ \text{g}}{75.1\ \text{cm}^3} = 0.777\ \text{g/cm}^3 \text{ or } 0.777\ \text{g/mL}$$

eneral formula for a linear (straight-line) relationship is

$$y = mx + b$$

e present case, let y represent the density, d, of a solu- in g/mL) and x, the volume fraction of hexane (volume nt/100). By substituting the density of pure decane into quation, we note that $x = 0$ and find that $b = 0.845$.

$$d = 0.845 = (m \times 0) + b$$

, using the density of pure hexane, $d = 0.667$, and the s $x = 1.00$ and $b = 0.845$, we obtain the value of m.

$$d = 0.667 = (m \times 1.00) + 0.845$$
$$m = 0.667 - 0.845 = -0.178$$

inal step is to find the value of x for a solution having the density as the wood: 0.777 g/mL.

$$d = 0.777 = -0.178x + 0.845$$
$$x = \frac{0.845 - 0.777}{0.178} = 0.38$$

volume fraction of hexane is 0.38, and the volume percent composition of the solution is 38% hexane and decane.

ss

e is an early point in this calculation where we can check the correctness of our work. We have two expecta- for the density of the block of wood: (1) it should be less than 1 g/cm^3 (practically all wood floats on water), (2) it must fall between 0.667 g/cm^3 and 0.845 g/cm^3. If the calculated density of the wood were outside this e, the block of wood would either float on both liquids or sink in both of them, making the rest of the calculation ssible.

nother point to notice in this calculation is that we are justified in carrying three significant figures throughout the lation up to the last step. There, because we must take the difference between two numbers of similar magnitudes, umber of significant figures drops from three to two.

CTICE EXAMPLE A: Magnalium is a solid mixture (an alloy) of aluminum metal and magnesium metal. An ularly shaped chunk of a sample of magnalium is weighed twice, once in air and once in vegetable oil, by using a g scale (see Figure 1-10). The weight in air is 211.5 g, and the weight in oil is 135.3 g. If the densities of pure inum, pure magnesium, and vegetable oil are 2.70 g/cm^3, 1.74 g/cm^3, and 0.926 g/cm^3, respectively, then what is ass percent of magnesium in this chunk of magnalium? Assume that the density of a mixture of the two metals is a r function of the mass percent composition.

CTICE EXAMPLE B: A particular sample of seawater has a density of 1.027 g/cm^3 at 10 °C and is 2.67% sodium ide by mass. Given that sodium chloride is 39.34% sodium by mass and that the mass of a single sodium atom is $\times 10^{-26}$ kg, calculate the maximum mass of sodium and the maximum number of sodium atoms that can be cted from a 1.5 L sample of this seawater.

Exercises

(see also Appendices A-1 and A-5)

The Scientific Method

1. What are the principal reasons that one theory might be adopted over a conflicting one?
2. Can one predict how many experiments are required to verify a natural law? Explain.
3. A common belief among scientists is that there exists an underlying order to nature. Einstein described this belief in the words "God is subtle, but He is not malicious." What do you think Einstein meant by this remark?
4. Describe several ways in which a scientific law d from a legislative law.
5. Describe the necessary characteristics of an ex ment that is suitable to test a theory.
6. Describe the necessary characteristics of a scie theory.

Properties and Classification of Matter

7. State whether the following properties of matter are physical or chemical.
   (a) An iron nail is attracted to a magnet.
   (b) A piece of paper spontaneously ignites when its temperature reaches 451 °F.
   (c) A bronze statue develops a green coating (patina) over time.
   (d) A block of wood floats on water.
8. State whether the following properties are physical or chemical.
   (a) A piece of sliced apple turns brown.
   (b) A slab of marble feels cool to the touch.
   (c) A sapphire is blue.
   (d) A clay pot fired in a kiln becomes hard and covered by a glaze.
9. Indicate whether each sample of matter listed is a substance or a mixture; if it is a mixture, indicate whether it is homogeneous or heterogeneous.
   (a) clean fresh air
   (b) a silver-plated spoon
   (c) garlic salt
   (d) ice
10. Indicate whether each sample of matter listed is a stance or a mixture; if it is a mixture, indicate wh it is homogeneous or heterogeneous.
    (a) a wooden beam
    (b) red ink
    (c) distilled water
    (d) freshly squeezed orange juice
11. Suggest physical changes by which the follo mixtures can be separated.
    (a) iron filings and wood chips
    (b) ground glass and sucrose (cane sugar)
    (c) water and olive oil
    (d) gold flakes and water
12. What type of change—physical or chemical—is essary to separate the following? [*Hint:* Refer to a listing of the elements.]
    (a) sugar from a sand/sugar mixture
    (b) iron from iron oxide (rust)
    (c) pure water from seawater
    (d) water from a slurry of sand in water

Exponential Arithmetic

13. Express each number in exponential notation. **(a)** 8950.; **(b)** 10,700.; **(c)** 0.0240; **(d)** 0.0047; **(e)** 938.3; **(f)** 275,482.
14. Express each number in common decimal form. **(a)** 5.12×10^{-3}; **(b)** 8.05×10^{-5}; **(c)** 291.1×10^{-4}; **(d)** 72.1×10^{-2}.
15. Express each value in exponential form. Where appropriate, include units in your answer.
    (a) speed of sound (sea level): 34,000 centimeters per second
    (b) equatorial radius of Earth: 6378 kilometers
    (c) the distance between the two hydrogen atoms in the hydrogen molecule: 74 trillionths of a meter
    (d) $\dfrac{(2.2 \times 10^3) + (4.7 \times 10^2)}{5.8 \times 10^{-3}} =$
16. Express each value in exponential form. W appropriate, include units in your answer.
    (a) solar radiation received by Earth: 173 thou trillion watts
    (b) average human cell diameter: 1 ten-milliont meter
    (c) the distance between the centers of the ato silver metal: 142 trillionths of a meter
    (d) $\dfrac{(5.07 \times 10^4) \times (1.8 \times 10^{-3})^2}{0.065 + (3.3 \times 10^{-2})} =$

ificant Figures

ndicate whether each of the following is an exact umber or a measured quantity subject to uncertainty.
a) the number of sheets of paper in a ream of paper
b) the volume of milk in a liter bottle
c) the distance between Earth and the sun
d) the distance between the centers of the two oxygen atoms in the oxygen molecule

ndicate whether each of the following is an exact umber or a measured quantity subject to uncertainty.
a) the number of pages in this text
b) the number of days in the month of January
c) the area of a city lot
d) the distance between the centers of the atoms in gold medal

xpress each of the following to *four* significant igures. **(a)** 3984.6; **(b)** 422.04; **(c)** 186,000; **(d)** 33,900; **e)** 6.321×10^4; **(f)** 5.0472×10^{-4}

Iow many significant figures are shown in each of he following? If this is indeterminate, explain why. **a)** 450; **(b)** 98.6; **(c)** 0.0033; **(d)** 902.10; **(e)** 0.02173; **f)** 7000; **(g)** 7.02; **(h)** 67,000,000

erform the following calculations; express each nswer in exponential form and with the appropriate umber of significant figures.
a) $0.406 \times 0.0023 =$
b) $0.1357 \times 16.80 \times 0.096 =$
c) $0.458 + 0.12 - 0.037 =$
d) $32.18 + 0.055 - 1.652 =$

erform the following calculations; express each umber and the answer in exponential form and with he appropriate number of significant figures.
a) $\dfrac{320 \times 24.9}{0.080} =$
b) $\dfrac{432.7 \times 6.5 \times 0.002300}{62 \times 0.103} =$
c) $\dfrac{32.44 + 4.9 - 0.304}{82.94} =$
d) $\dfrac{8.002 + 0.3040}{13.4 - 0.066 + 1.02} =$

erform the following calculations and retain the appropriate number of significant figures in each result.
a) $(38.4 \times 10^{-3}) \times (6.36 \times 10^5) =$
b) $\dfrac{(1.45 \times 10^2) \times (8.76 \times 10^{-4})}{(9.2 \times 10^{-3})^2} =$
(c) $24.6 + 18.35 - 2.98 =$
(d) $(1.646 \times 10^3) - (2.18 \times 10^2) + [(1.36 \times 10^4) \times (5.17 \times 10^{-2})] =$
(e)
$$\frac{-7.29 \times 10^{-4} + \sqrt{(7.29 \times 10^{-4})^2 + 4(1.00)(2.7 \times 10^{-5})}}{2 \times (1.00)}$$
[*Hint:* The significant figure rule for the extraction of a root is the same as for multiplication.]

24. Express the result of each of the following calculations in exponential form and with the appropriate number of significant figures.
(a)
$$(4.65 \times 10^4) \times (2.95 \times 10^{-2}) \times (6.663 \times 10^{-3}) \times 8.2 =$$
(b) $\dfrac{1912 \times (0.0077 \times 10^4) \times (3.12 \times 10^{-3})}{(4.18 \times 10^{-4})^3} =$
(c) $(3.46 \times 10^3) \times 0.087 \times 15.26 \times 1.0023 =$
(d) $\dfrac{(4.505 \times 10^{-2})^2 \times 1.080 \times 1545.9}{0.03203 \times 10^3} =$
(e)
$$\frac{(-3.61 \times 10^{-4}) + \sqrt{(3.61 \times 10^{-4})^2 + 4(1.00)(1.9 \times 10^{-5})}}{2 \times (1.00)}$$
[*Hint:* The significant figure rule for the extraction of a root is the same as for multiplication.]

25. An American press release describing the 1986 nonstop, round-the-world trip by the ultra-lightweight aircraft *Voyager* included the following data:
flight distance: 25,012 mi
flight time: 9 days, 3 minutes, 44 seconds
fuel capacity: nearly 9000 lb
fuel remaining at end of flight: 14 gal
To the maximum number of significant figures permitted, calculate
(a) the average speed of the aircraft in kilometers per hour
(b) the fuel consumption in kilometers per kilogram of fuel (assume a density of 0.70 g/mL for the fuel)

26. Use the concept of significant figures to criticize the way in which the following information was presented. "The estimated proved reserve of natural gas as of January 1, 1982, was 2,911,346 trillion cubic feet."

s of Measurement

erform the following conversions.
a) 0.127 L = ____ mL
b) 15.8 mL = ____ L
c) $981\ cm^3$ = ____ L
d) $2.65\ m^3$ = ____ cm^3

erform the following conversions.
a) 2.35 kg = ____ g
b) 792 g = ____ kg
c) 3869 mm = ____ cm
d) 0.043 cm = ____ mm

29. Perform the following conversions from non-SI to SI units. (Use information from the inside back cover, as needed.)
(a) 68.4 in = ____ cm
(b) 94 ft = ____ m
(c) 1.42 lb = ____ g
(d) 248 lb = ____ kg
(e) 1.85 gal = ____ dm^3
(f) 3.72 qt = ____ mL

30. Determine the number of the following:
(a) square meters (m^2) in 1 square kilometer (km^2)
(b) cubic centimeters (cm^3) in 1 cubic meter (m^3)
(c) square meters (m^2) in 1 square mile (mi^2) (1 mi = 5280 ft)

31. Which is the greater mass, 3245 μg or 0.00515 mg? Explain.
32. Which is the greater mass, 3257 mg or 0.000475 kg? Explain.
33. The non-SI unit, the *hand* (used by equestrians), is 4 inches. What is the height, in meters, of a horse that stands 15 hands high?
34. The unit *furlong* is used in horse racing. The units *chain* and *link* are used in surveying. There are exactly 8 furlongs in 1 mi, 10 chains in 1 furlong, and 100 links in 1 chain. To three significant figures, what is the length of 1 link in centimeters?
35. A sprinter runs the 100 yd dash in 9.3 s. At this same rate, **(a)** how long would it take the sprinter to run 100.0 m? **(b)** what is the sprinter's speed in meters per second? **(c)** how long would it take the sprinter to run a distance of 1.45 km?
36. A non-SI unit of mass used in pharmaceutical work is the *grain* (gr) (15 gr = 1.0 g). An aspirin tablet contains 5.0 gr of aspirin. A 161 lb arthritic individual takes two aspirin tablets per day.
(a) What is the quantity of aspirin in two tablets, expressed in milligrams?
(b) What is the dosage rate of aspirin, express milligrams of aspirin per kilogram of body mass
(c) At the given rate of consumption of as tablets, how many days would it take to con 1.0 kg of aspirin?
37. In SI units, land area is measured in *he* (1 hectare = 1 hm^2). The commonly used un land area in the United States is the *acre*. How acres correspond to 1 hectare? (1 mi^2 = 640 1 mi = 5280 ft, 1 ft = 12 in).
38. In an engineering reference book, you find th density of iron is 0.284 lb/in^3. What is the dens g/cm^3?
39. In a user's manual accompanying an American- automobile, a typical gauge pressure for optima formance of automobile tires is 32 lb/in^2. What pressure in grams per square centimeter and grams per square meter?
40. The volume of a red blood cell is 90.0 $\times$ 10^{-12} cm^3. Assuming that red blood cell spherical, what is the diameter of a red blood c millimeters?

Temperature Scales

41. We want to mark off a thermometer in both Celsius and Fahrenheit temperatures. On the Celsius scale, the lowest temperature mark is at −10 °C, and the highest temperature mark is at 50 °C. What are the equivalent Fahrenheit temperatures?
42. The highest and lowest temperatures on record for San Bernardino, California, are 118 °F and 17 °F, respectively. What are these temperatures on the Celsius scale?
43. The absolute zero of temperature is −273.15 °C. Should it be possible to achieve a temperature of −465 °F? Explain.
44. A family/consumer science class is given an assignment in candy-making that requires a sugar mixture to be brought to a "soft-ball" stage (234–240 °F). A student borrows a thermometer having a r from −10 °C to 110 °C from the chemistry labor to do this assignment. Will this thermometer the purpose? Explain.
45. You decide to establish a new temperature sca which the melting point of mercury (−38.9 ° 0 °M, and the boiling point of mercury (356.9 100 °M. What would be **(a)** the boiling poi water in °M; and **(b)** the temperature of abs zero in °M?
46. You decide to establish a new temperature sca which the melting point of ammonia (−77.7 is 0 °A and the boiling point of ammonia (−33.35 100 °A. What would be **(a)** the boiling point of wa °A; and **(b)** the temperature of absolute zero in °A

Density

47. A 2.18 L sample of butyric acid, a substance present in rancid butter, has a mass of 2088 g. What is the density of butyric acid in grams per milliliter?
48. A 15.2 L sample of chloroform at 20 °C has a mass of 22.54 kg. What is the density of chloroform at 20 °C, in grams per milliliter?
49. To determine the density of acetone, a 55.0 gal drum is weighed twice. The drum weighs 75.0 lb when empty and 437.5 lb when filled with acetone. What is the density of acetone expressed in grams per milliliter?
50. To determine the volume of an irregularly shaped glass vessel, the vessel is weighed empty (121.3 g) and when filled with carbon tetrachloride (283.2 g). What is the volume capacity of the vessel, in milliliters, given that the density of carbon tetrachloride is 1.59 g/mL?
51. A solution consisting of 8.50% acetone and 91.5% water by mass has a density of 0.9867 g/mL. What mass of acetone, in kilograms, is present in 7.50 the solution?
52. A solution contains 10.05% sucrose (cane suga mass. What mass of the solution, in grams, is ne for an application that requires 1.00 kg of sucrose
53. A fertilizer contains 21% nitrogen by mass. mass of this fertilizer, in kilograms, is required f application requiring 225 g of nitrogen?
54. A vinegar sample is found to have a densi 1.006 g/mL and to contain 5.4% acetic acid by How many grams of acetic acid are present in 1 of this vinegar?
55. Calculate the mass of a block of iron (d = 7.86 g/ with dimensions of 52.8 cm $\times$ 6.74 cm $\times$ 3.73 cm
56. Calculate the mass of a cylinder of stainless (d = 7.75 g/cm^3) with a height of 18.35 cm a radius of 1.88 cm.

he following densities are given at 20 °C: water, .998 g/cm^3; iron, 7.86 g/cm^3; aluminum, 2.70 g/cm^3. rrange the following items in terms of *increasing* mass.

) a rectangular bar of iron,
81.5 cm × 2.1 cm × 1.6 cm

) a sheet of aluminum foil,
12.12 m × 3.62 m × 0.003 cm

) 4.051 L of water

o determine the approximate mass of a small spher-cal shot of copper, the following experiment is erformed. When 125 pieces of the shot are counted ut and added to 8.4 mL of water in a graduated ylinder, the total volume becomes 8.9 mL. The den-ity of copper is 8.92 g/cm^3. Determine the approxi-ate mass of a single piece of shot, assuming that all f the pieces are of the same dimensions.

he density of aluminum is 2.70 g/cm^3. A square iece of aluminum foil, 22.86 cm on a side is found o weigh 2.568 g. What is the thickness of the foil, in illimeters?

he angle iron pictured here is made of steel with a ensity of 7.78 g/cm^3. What is the mass, in grams, of is object?

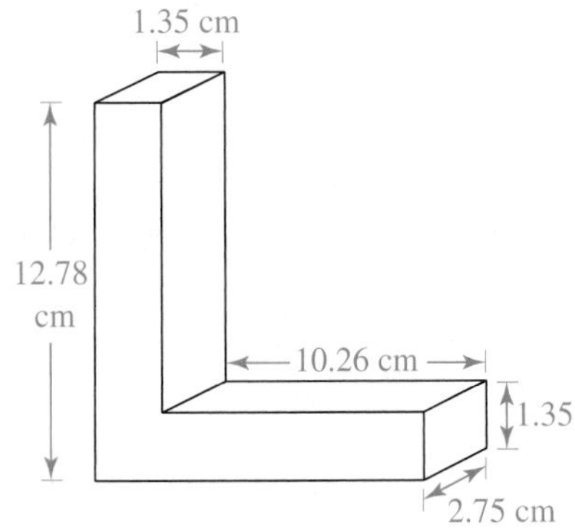

61. In normal blood, there are about 5.4 × 10^9 red blood cells per milliliter. The volume of a red blood cell is about 90.0 × 10^{-12} cm^3, and its density is 1.096 g/mL. How many liters of whole blood would be needed to collect 0.5 kg of red blood cells?

62. A technique once used by geologists to measure the density of a mineral is to mix two dense liquids in such proportions that the mineral grains just float. When a sample of the mixture in which the mineral calcite just floats is put in a special density bottle, the weight is 15.4448 g. When empty, the bottle weighs 12.4631 g, and when filled with water, it weighs 13.5441 g. What is the density of the calcite sample? (All measurements were carried out at 25 °C, and the density of water at 25 °C is 0.9970 g/mL.)

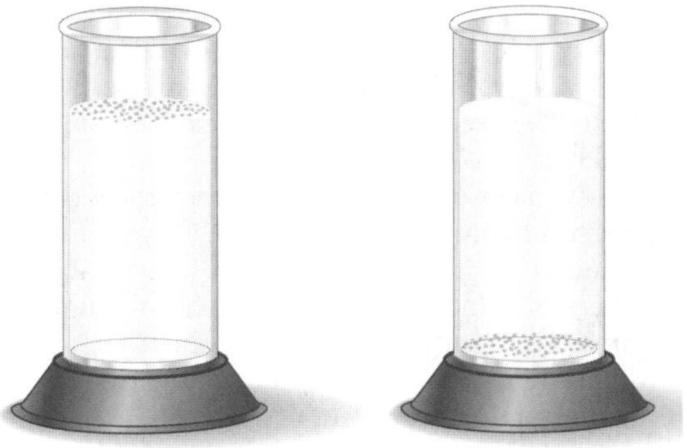

▲ At the left, grains of the mineral *calcite* float on the surface of the liquid bromoform (d = 2.890 g/mL). At the right, the grains sink to the bottom of liquid chloroform (d = 1.444 g/mL). By mixing bromoform and chloroform in just the proportions required so that the grains barely float, the density of the calcite can be determined (Exercise 62).

ent Composition

n a class of 76 students, the results of a particular xamination were 7 A's, 22 B's, 37 C's, 8 D's, 2 F's. Vhat was the percent distribution of grades, that is, % A's, % B's, and so on?

class of 84 students had a final grade distribution of 8% A's, 25% B's, 32% C's, 13% D's, 12% F's. How any students received each grade?

65. A solution of sucrose in water is 28.0% sucrose by mass and has a density of 1.118 g/mL. What mass of sucrose, in grams, is contained in 3.50 L of this solution?

66. A solution containing 12.0% sodium hydroxide by mass in water has a density of 1.131 g/mL. What volume of this solution, in liters, must be used in an application requiring 2.75 kg of sodium hydroxide?

Integrative and Advanced Exercises

According to the rules on significant figures, the product of the measured quantities 99.9 m and 1.008 m should be expressed to three significant figures—101 m^2. Yet, in this case, it would be more appropriate to express the result to *four* significant figures—100.7 m^2. Explain why.

68. For a solution containing 6.38% para-dichlorobenzene by mass in benzene, the density of the solution as a function of temperature (t) in the temperature range 15 to 65 °C is given by the equation

$$d(g/mL) = 1.5794 - 1.836 \times 10^{-3}(t - 15)$$

At what temperature will the solution have a density of 1.543 g/mL?

69. A solution used to chlorinate a home swimming pool contains 7% chlorine by mass. An ideal chlorine level for the pool is one part per million (1 ppm). (Think of 1 ppm as being 1 g chlorine per million grams of water.) If you assume densities of 1.10 g/mL for the chlorine solution and 1.00 g/mL for the swimming pool water, what volume of the chlorine solution, in liters, is required to produce a chlorine level of 1 ppm in an 18,000-gallon swimming pool?

70. A standard 1.000 kg mass is to be cut from a bar of steel having an equilateral triangular cross section with sides equal to 2.50 in. The density of the steel is 7.70 g/cm^3. How many inches long must the section of bar be?

71. The volume of seawater on Earth is about 330,000,000 mi^3. If seawater is 3.5% sodium chloride by mass and has a density of 1.03 g/mL, what is the approximate mass of sodium chloride, in tons, dissolved in the seawater on Earth (1 ton = 2000 lb)?

72. The diameter of metal wire is often referred to by its American wire-gauge number. A 16-gauge wire has a diameter of 0.05082 in. What length of wire, in meters, is found in a 1.00 lb spool of 16-gauge copper wire? The density of copper is 8.92 g/cm^3.

73. Magnesium occurs in seawater to the extent of 1.4 g magnesium per kilogram of seawater. What volume of seawater, in cubic meters, would have to be processed to produce 1.50×10^5 tons of magnesium (1 ton = 2000 lb)? Assume a density of 1.025 g/mL for seawater.

74. A typical rate of deposit of dust ("dustfall") from unpolluted air was reported as 10 tons per square mile per month. **(a)** Express this dustfall in milligrams per square meter per hour. **(b)** If the dust has an average density of 2 g/cm^3, how long would it take to accumulate a layer of dust 1 mm thick?

75. In the United States, the volume of irrigation water is usually expressed in acre-feet. One acre-foot is a volume of water sufficient to cover 1 acre of land to a depth of 1 ft (640 acres = 1 mi^2; 1 mi = 5280 ft). The principal lake in the California Water Project is Lake Oroville, whose water storage capacity is listed as 3.54×10^6 acre-feet. Express the volume of Lake Oroville in **(a)** cubic feet; **(b)** cubic meters; **(c)** U.S. gallons.

76. A Fahrenheit and a Celsius thermometer are immersed in the same medium. At what Celsius temperature will the numerical reading on the Fahrenheit thermometer be

(a) 49° less than that on the Celsius thermometer;
(b) twice that on the Celsius thermometer;
(c) one-eighth that on the Celsius thermometer;
(d) 300° more than that on the Celsius thermometer?

77. The accompanying illustration shows a 100.0 mL graduated cylinder half-filled with 8.0 g of diatomaceous earth, a material consisting mostly of silica and used as a filtering medium in swimming pools. How many milliliters of water are required to fill the cylinder to the 100.0 mL mark? The diatomaceous earth is insoluble in water and has a density of 2.2 g/cm^3.

78. The simple device pictured here, a pycnome used for precise density determinations. Fro data presented, together with the fact that the sity of water at 20 °C is 0.99821 g/mL, determi density of methanol, in grams per milliliter.

Empty 25.601 g | Filled with water at 20 °C: 35.552 g | Filled with met at 20 °C: 33.4

79. If the pycnometer of Exercise 78 is filled with et at 20 °C instead of methanol, the observed m 33.470 g. What is the density of ethanol? How cisely could you determine the composition ethanol–methanol solution by measuring its de with a pycnometer? Assume that the density solution is a linear function of the volume pe composition.

80. A pycnometer (see Exercise 78) weighs 25. empty and 35.55 g when filled with water at 2 The density of water at 20 °C is 0.9982 g/mL. 10.20 g of lead is placed in the pycnometer an pycnometer is again filled with water at 20 °C total mass is 44.83 g. What is the density of the in grams per cubic centimeter?

81. The Greater Vancouver Regional District (G chlorinates the water supply of the region at th of 1 ppm, that is, 1 kilogram of chlorine per m kilograms of water. The chlorine is introduced i form of sodium hypochlorite, which is 47.62% rine. The population of the GVRD is 1.8 millio sons. If each person uses 750 L of water per day, many kilograms of sodium hypochlorite mu added to the water supply each week to produc required chlorine level of 1 ppm?

A Boeing 767 due to fly from Montreal to Edmonton required refueling. Because the fuel gauge on the aircraft was not working, a mechanic used a dipstick to determine that 7682 L of fuel were left on the plane. The plane required 22,300 kg of fuel to make the trip. In order to determine the volume of fuel required, the pilot asked for the conversion factor needed to convert a volume of fuel to a mass of fuel. The mechanic gave the factor as 1.77. Assuming that this factor was in metric units (kg/L), the pilot calculated the volume to be added as 4916 L. This volume of fuel was added and the 767 subsequently ran out the fuel, but landed safely by gliding into Gimli Airport near Winnipeg. The error arose because the factor 1.77 was in units of pounds per liter. What volume of fuel should have been added?

The following equation can be used to relate the density of liquid water to Celsius temperature in the range from 0 °C to about 20 °C:

$$\text{cm}^3) = \frac{0.99984 + (1.6945 \times 10^{-2}t) - (7.987 \times 10^{-6}t^2)}{1 + (1.6880 \times 10^{-2}t)}$$

(a) To four significant figures, determine the density of water at 10 °C.
(b) At what temperature does water have a density of 0.99860 g/cm^3?
(c) In the following ways, show that the density passes through a maximum somewhere in the temperature range to which the equation applies.
 (i) by estimation
 (ii) by a graphical method
 (iii) by a method based on differential calculus

A piece of high-density Styrofoam measuring 24.0 cm by 36.0 cm by 5.0 cm floats when placed in a tub of water. When a 1.5 kg book is placed on top of the Styrofoam, the Styrofoam partially sinks, as illustrated in the diagram below. Assuming that the density of water is 1.00 g/mL, what is the density of Styrofoam?

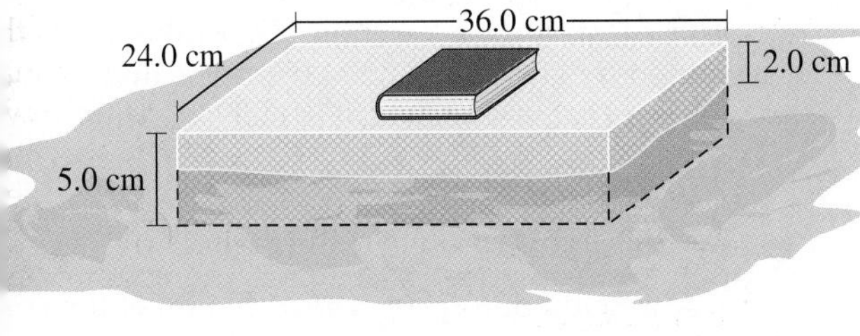

A tabulation of data lists the following equation for calculating the densities (d) of solutions of naphthalene in benzene at 30 °C as a function of the mass percent of naphthalene.

$$d(\text{g/cm}^3) = \frac{1}{1.153 - 1.82 \times 10^{-3}(\%N) + 1.08 \times 10^{-6}(\%N)^2}$$

Use the equation above to calculate **(a)** the density of pure benzene at 30 °C; **(b)** the density of pure naphthalene at 30 °C; **(c)** the density of a solution at 30 °C that is 1.15% naphthalene; **(d)** the mass percent of naphthalene in a solution that has a density of 0.952 g/cm^3 at 30 °C. [*Hint*: For (d), you need to use the quadratic formula. See Section A-3 of Appendix A.]

86. The total volume of ice in the Antarctic is about 3.01×10^7 km^3. If all the ice in the Antarctic were to melt completely, estimate the rise, h, in sea level that would result from the additional liquid water entering the oceans. The densities of ice and fresh water are 0.92 g/cm^3 and 1.0 g/cm^3, respectively. Assume that the oceans of the world cover an area, A, of about 3.62×10^8 km^2 and that the increase in volume of the oceans can be calculated as $A \times h$.

87. An empty 3.00 L bottle weighs 1.70 kg. Filled with a certain wine, it weighs 4.72 kg. The wine contains 11.5% ethyl alcohol by mass. How many grams of ethyl alcohol are there in 250.0 mL of this wine?

88. The filament in an incandescent light bulb is made from tungsten metal (d = 19.3 g/cm^3) that has been drawn into a very thin wire. The diameter of the wire is difficult to measure directly, so it is sometimes estimated by measuring the mass of a fixed length of wire. If a 0.200 m length of tungsten wire weighs 42.9 mg, then what is the diameter of the wire? Express your answer in millimeters.

89. Blood alcohol content (BAC) is sometimes reported in weight-volume percent and, when it is, a BAC of 0.10% corresponds to 0.10 g of ethyl alcohol per 100 mL of blood. In many jurisdictions, a person is considered legally intoxicated if his or her BAC is 0.10%. Suppose that a 68 kg person has a total blood volume of 5.4 L and breaks down ethyl alcohol at a rate of 10.0 grams per hour.* How many 145 mL glasses of wine, consumed over three hours, will produce a BAC of 0.10% in this 68 kg person? Assume the wine has a density of 1.01 g/mL and is 11.5% ethyl alcohol by mass. (*The rate at which ethyl alcohol is broken down varies dramatically from person to person. The value given here for the rate is a realistic, but not necessarily accurate, value.)

Feature Problems

In an attempt to determine any possible relationship between the year in which a U.S. penny was minted and its current mass (in grams), students weighed an assortment of pennies and obtained the following data.

68	1973	1977	1980	1982	1983	1985
11	3.14	3.13	3.12	3.12	2.51	2.54
)8	3.06	3.10	3.11	2.53	2.49	2.53
)9	3.07	3.06	3.08	2.54	2.47	2.53

What valid conclusion(s) might they have drawn about the relationship between the masses of the pennies within a given year and from year to year?

91. In the third century B.C., the Greek mathematician Archimedes is said to have discovered an important principle that is useful in density determinations. The story told is that King Hiero of Syracuse (in Sicily) asked Archimedes to verify that an ornate crown made for him by a goldsmith consisted of pure gold and not a gold–silver alloy. Archimedes had to do this, of course, without damaging the crown in any way. Describe how Archimedes did this, or if you don't know the rest of the story, rediscover Archimedes's principle and explain how it can be used to settle the question.

92. The Galileo thermometer shown in the photograph is based on the dependence of density on temperature. The liquid in the outer cylinder and the liquid in the partially filled floating glass balls are the same except that a colored dye has been added to the liquid in the balls. Explain how the Galileo thermometer works.

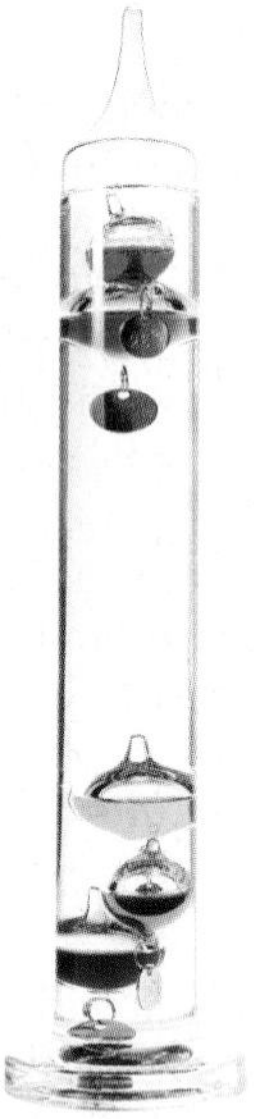

Aptyp_koK/Fotolia

93. The canoe gliding gracefully along the water in the photograph is made of concrete, which has a density of about 2.4 g/cm^3. Explain why the canoe does not sink.

Tom Leininger/University of Kansas School of Engineering

94. The accompanying sketches suggest four obs tions made on a small block of plastic materia what conclusions can be drawn from each sk and conclude by giving your best estimate c density of the plastic.

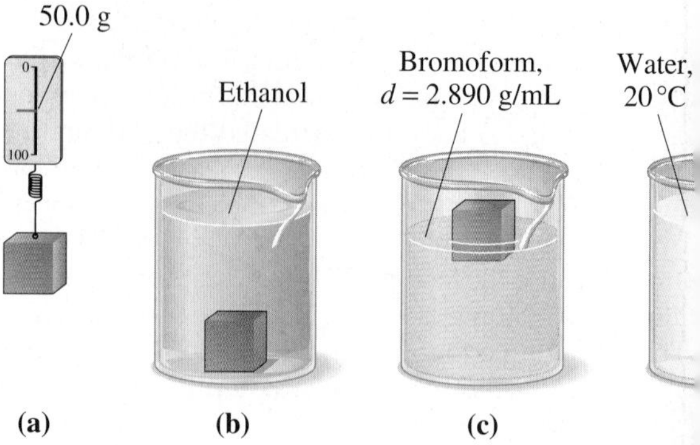

95. As mentioned on page 13, the MCO was lost becau a mix-up in the units used to calculate the force ne to correct its trajectory. Ground-based computers erated the force correction file. On Septembe 1999, it was discovered that the forces reported b ground-based computer for use in MCO navig software were low by a factor of 4.45. The error trajectory brought the MCO 56 km above the su of Mars; the correct trajectory would have bro the MCO approximately 250 km above the sur At 250 km, the MCO would have success entered the desired elliptic orbit. The data conta in the force correction file were delivered in l instead of the required SI units of newton-sec fo MCO navigation software. The newton is the SI of force and is described in Appendix B. The B Engineering (gravitational) system uses a pounc as a unit of force and ft/s^2 as a unit of acceleratic turn, the pound is defined as the pull of Earth unit of mass at a location where the accelera caused by gravity is 32.174 ft/s^2. The unit of ma this case is the slug, which is 14.59 kg. Thus,

$$\text{BE unit of force} = 1\text{ pound} = (\text{slug})(\text{ft/s}^2)$$

Use this information to confirm that

$$\text{BE unit of force} = 4.45 \times \text{SI unit of force}$$
$$1\text{ pound} = 4.45\text{ newton}$$

Self-Assessment Exercises

96. In your own words, define or explain the following terms or symbols: **(a)** mL; **(b)** % by mass; **(c)** °C; **(d)** density; **(e)** element.

97. Briefly describe each of the following ideas: **(a)** SI base units; **(b)** significant figures; **(c)** natural law; **(d)** exponential notation.

98. Explain the important distinctions between each pair of terms: **(a)** mass and weight; **(b)** intensive and extensive properties; **(c)** substance and mixture; **(d)** systematic and random errors; **(e)** hypothesis and theory.

99. A procedure designed to test the truth or the val of an explanation for many observations is c **(a)** a law; **(b)** a theory; **(c)** an experiment; (hypothesis; **(e)** none of these.

100. The fact that the volume of a fixed amount of gas fixed temperature is inversely proportional to th pressure is an example of **(a)** a hypothesis; **(b)** a th **(c)** a paradigm; **(d)** the absolute truth; **(e)** a natural

101. If a sample of matter cannot be separated by p cal means or decomposed by chemical means sample is **(a)** a homogeneous mixture; **(b)** a heterogeneous mixture; **(c)** an element; **(d)** a compound; **(e)** a pure substance.

A good example of a homogeneous mixture is
(a) a cola drink in a tightly capped bottle
(b) distilled water leaving a distillation apparatus
(c) oxygen gas in a cylinder used in welding
(d) the material produced in a kitchen blender

Compared with its mass on Earth, the mass of the same object on the moon should be **(a)** less; **(b)** more; **(c)** the same; **(d)** nearly the same, but somewhat less.

Which answer has the correct number of significant figures?
(a) $(14.7 + 24.312) \times 87.27 = 3405$
(b) $(58 + 18 + 51)/3.000 = 42$
(c) $[97.2/(114 - 37)] = 1.26$
(d) $(1.172 - 0.4963)(4.193) = 2.83$
(e) none of these

Which two of the following masses are expressed to the nearest milligram? **(a)** 32.7 g; **(b)** 0.03271 kg; **(c)** 32.7068 g; **(d)** 32.707 g; **(e)** 30.7 mg; **(f)** 3×10^3 μg.

The highest temperature of the following group is **(a)** 217 K; **(b)** 273 K; **(c)** 217 °F; **(d)** 105 °C; **(e)** 373 K.

Which of the following quantities has the greatest mass?
(a) 752 mL of water at 20 °C
(b) 1.05 L of ethanol at 20 °C (d = 0.789 g/mL)
(c) 750 g of chloroform at 20 °C (d = 1.483 g/mL)
(d) a cube of balsa wood (d = 0.11 g/cm^3) that is 19.20 cm on edge

The density of silver is 10.5 g/mL. What is the volume of 475 g of silver?
(a) 475/10.5; **(b)** 10.5/475; **(c)** $1/[10.5 \times 475]$; **(d)** 475×10.5; **(e)** none of these

The density of water is 0.9982 g/cm^3 at 20 °C. Express the density of water at 20 °C in the following units: **(a)** g/L; **(b)** kg/m^3; **(c)** kg/km^3.

Two students each made four measurements of the mass of an object. Their results are shown in the table below.

	Student A	Student B
ır measurements:	51.6, 50.8, 52.2, 50.2 g	50.1, 49.6, 51.0, 49.4 g
eir average:	51.3 g	50.0 g

The exact mass of the object is 51.0 g. Whose results are more precise, Student A's or Student B's? Whose results are more accurate?

The reported value for the volume of a rectangular piece of cardboard with the dimensions 36 cm × 20.2 cm × 9 mm should be **(a)** 6.5×10^3 cm^3; **(b)** 7×10^2 cm^3; **(c)** 655 cm^3; **(d)** 6.5×10^2 cm^3.

112. List the following in the order of increasing precision, indicating any quantities about which the precision is uncertain: **(a)** 1400 km; **(b)** 1516 kg; **(c)** 0.00304 g; **(d)** 125.34 cm; **(e)** 2000 mg.

113. *Without doing detailed calculations*, explain which of the following objects contains the greatest mass of the element iron.
(a) A 1.00 kg pile of pure iron filings.
(b) A cube of wrought iron, 5.0 cm on edge. Wrought iron contains 98.5% iron by mass and has a density of 7.7 g/cm^3.
(c) A square sheet of stainless steel 0.30 m on edge and 1.0 mm thick. The stainless steel is an alloy (mixture) containing iron, together with 18% chromium, 8% nickel, and 0.18% carbon by mass. Its density is 7.7 g/cm^3.
(d) 10.0 L of a solution characterized as follows: d = 1.295 g/mL. This solution is 70.0% water and 30.0% of a compound of iron, by mass. The iron compound consists of 34.4% iron by mass.

114. A lump of pure copper weighs 25.305 g in air and 22.486 g when submerged in water (d = 0.9982 g/mL) at 20.0 °C. Suppose the copper is then rolled into a 248 cm^2 foil of uniform thickness. What will this thickness be, in millimeters?

115. Water, a compound, is a substance. Is there any circumstance under which a sample of pure water can exist as a heterogeneous mixture? Explain.

116. In the production of ammonia, the following processes occur.
(a) *Air* is liquefied at low temperature and high *pressure*.
(b) The *temperature* of the *liquid air* is raised gradually until the oxygen boils off. Essentially pure *liquid nitrogen* remains.
(c) *Natural gas* is treated with *steam* to produce *carbon dioxide* and *hydrogen*. The proportions of the products vary with the composition of the natural gas.
(d) Hydrogen and nitrogen (both gases) are combined at high temperature and pressures. They react to form *ammonia gas*. Some unreacted hydrogen and nitrogen remain.
(e) The gases are cooled until the ammonia liquefies, and then the gaseous hydrogen and nitrogen are recirculated to react again.
For each of (a) to (e), indicate whether a physical or chemical change occurs, and briefly explain your choice. For each *italicized* word or phrase, state whether it refers to an element, a compound, a mixture, or none of these, and briefly explain your choice.

117. Appendix E describes a useful study aid known as concept mapping. Using the method presented in Appendix E, construct a concept map illustrating the different concepts presented in Sections 1-2, 1-3, and 1-4.

2 Atoms and the Atomic Theory

CONTENTS

LEARNING OBJECTIVES

2.1 Describe and distinguish between the law of conservation of mass, the law of constant composition, and the law of multiple proportions.

2.2 Discuss the discovery of electrons, and describe their basic properties, such as charge and mass.

2.3 Identify the features of the nucleus of an atom, and discuss the properties of protons, neutrons, and electrons.

2.4 Describe the meaning of *isotope*, and discuss how the masses of all the elements were deduced.

2.5 Determine the atomic mass of an element from the experimentally determined masses and percent abundances of the isotopes.

2.6 Identify different regions of the periodic table, including metals and nonmetals, noble gases, metalloids, transition metals, and main-group elements.

2.7 Identify the value of Avogadro's constant, and describe the meaning of molar mass.

2.8 Use Avogadro's constant to convert between values of mass, amount in moles, and number of atoms in a sample of element.

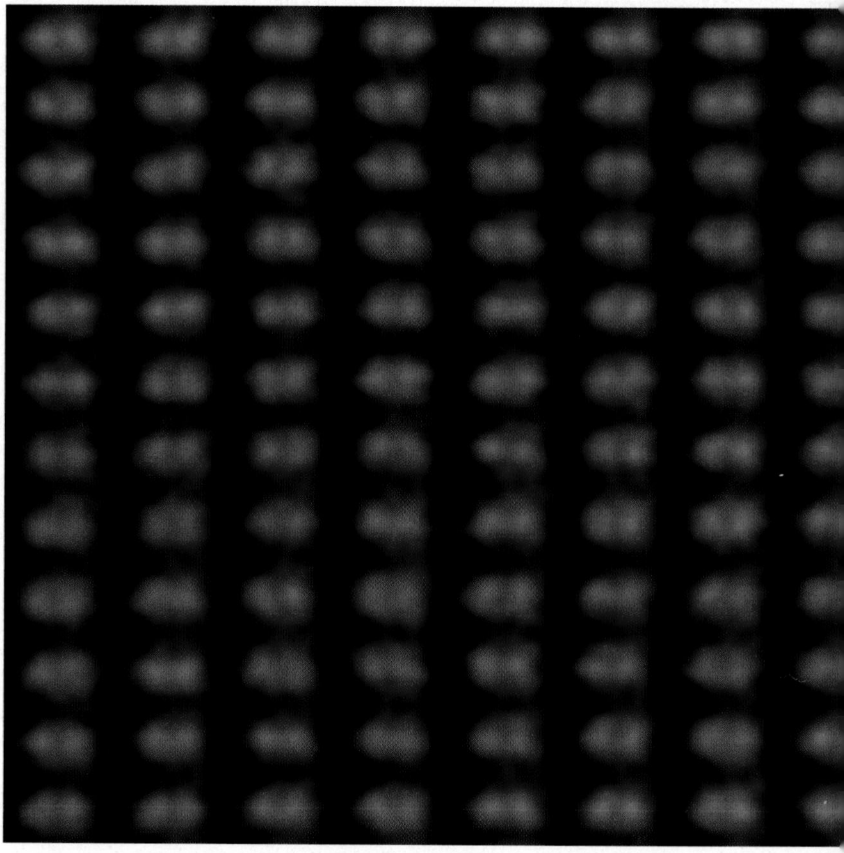

Image of silicon atoms that are only 78 pm apart; image produced by using a scann transmission electron microscope (STEM). The hypothesis that all matter is made up atoms has existed for more than 2000 years. It is only within the last few decades, however, that techniques have been developed that can render individual atoms vis

We begin this chapter with a brief survey of early chemical dis eries, culminating in Dalton's atomic theory. This is followe a description of the physical evidence leading to the mod picture of the *nuclear atom*, in which protons and neutrons are combi into a nucleus with electrons in space surrounding the nucleus. We also introduce the periodic table as the primary means of organizing ments into groups with similar properties. Finally, we will introduce ept of the mole and the Avogadro constant, which are the principal tools ounting atoms and molecules and measuring amounts of substances. We use these tools throughout the text.

Early Chemical Discoveries and the Atomic Theory

nistry has been practiced for a very long time, even if its practitioners much more interested in its applications than in its underlying princi- The blast furnace for extracting iron from iron ore appeared as early as 1300, and such important chemicals as sulfuric acid (oil of vitriol), nitric (aqua fortis), and sodium sulfate (Glauber's salt) were all well known used several hundred years ago. Before the end of the eighteenth century, principal gases of the atmosphere—nitrogen and oxygen—had been iso- l, and natural laws had been proposed describing the physical behavior of s. Yet chemistry cannot be said to have entered the modern age until the ess of combustion was explained. In this section, we explore the direct between the explanation of combustion and Dalton's atomic theory.

Carey B. Van Loon

▲ FIGURE 2-1
Two combustion reactions
The apparent product of the combustion of the match—the ash—weighs *less* than the match. The product of the combustion of the magnesium ribbon (the "smoke") weighs *more* than the ribbon. Actually, in each case, the total mass remains unchanged. To understand this, you have to know that oxygen gas enters into both combustions and that water and carbon dioxide are also products of the combustion of the match.

of Conservation of Mass

process of combustion—burning—is so familiar that it is hard to realize t a difficult riddle it posed for early scientists. Some of the difficult-to- ain observations are described in Figure 2-1.

1774, Antoine Lavoisier (1743–1794) performed an experiment in which eated a sealed glass vessel containing a sample of tin and some air. He d that the mass before heating (glass vessel + tin + air) and after heating s vessel + "tin calx" + remaining air) were the same. Through further riments, he showed that the product of the reaction, tin calx (tin oxide), isted of the original tin together with a portion of the air. Experiments like proved to Lavoisier that oxygen from air is essential to combustion and led him to formulate the **law of conservation of mass**:

> The total mass of substances present after a chemical reaction is the same as the total mass of substances before the reaction.

law is illustrated in Figure 2-2, where the reaction between silver nitrate potassium chromate to give a red solid (silver chromate) is monitored by ing the reactants on a single-pan balance—the total mass does not change. ed another way, the law of conservation of mass says that matter is neither ted nor destroyed in a chemical reaction.

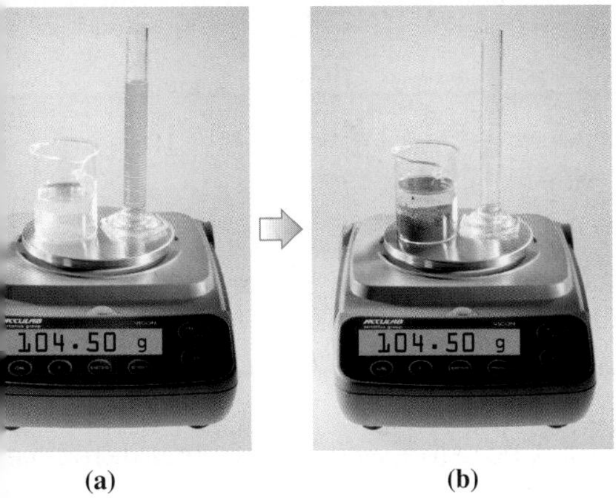

◀ FIGURE 2-2
Mass is conserved during a chemical reaction
(a) Before the reaction, a beaker with a silver nitrate solution and a graduated cylinder with a potassium chromate solution are placed on a single-pan balance, which displays their combined mass—104.50 g. **(b)** When the solutions are mixed, a chemical reaction occurs that forms silver chromate (red precipitate) in a potassium nitrate solution. Note that the total mass—104.50 g—remains unchanged.

Richard Megna/Fundamental Photographs

EXAMPLE 2-1 Applying the Law of Conservation of Mass

A 0.455 g sample of magnesium is allowed to burn in 2.315 g of oxygen gas. The sole product is magnesiu oxide. After the reaction, no magnesium remains and the mass of unreacted oxygen is 2.015 g. What mass magnesium oxide is produced?

Analyze

The total mass is unchanged. The total mass is the sum of the masses of the substances present initially. T mass of magnesium oxide is the total mass minus the mass of unreacted oxygen.

Solve

First, determine the total mass before the reaction.

mass before reaction = 0.455 g magnesium + 2.315 g oxygen
= 2.770 g mass before reaction

The total mass after the reaction is the same as before the reaction.

2.770 g mass after reaction = ? g magnesium oxide after reaction
+ 2.015 g oxygen after reaction

Solve for the mass of magnesium oxide.

? g magnesium oxide after reaction = 2.770 g mass after reaction
− 2.015 g oxygen after reaction
= 0.755 g magnesium oxide after reactic

Assess

Here is another approach. The mass of oxygen that reacted is 2.315 g − 2.015 g = 0.300 g. Thus, 0.300 g oxyg combined with 0.455 g magnesium to give 0.300 g + 0.455 g = 0.755 g magnesium oxide.

PRACTICE EXAMPLE A: A 0.382 g sample of magnesium is allowed to react with 2.652 g of nitrogen gas. The sc product is magnesium nitride. After the reaction, the mass of unreacted nitrogen is 2.505 g. What mass magnesium nitride is produced?

PRACTICE EXAMPLE B: A 7.12 g sample of magnesium is heated with 1.80 g of bromine. All the bromine is us up, and 2.07 g of magnesium bromide is the only product. What mass of magnesium remains *unreacted*?

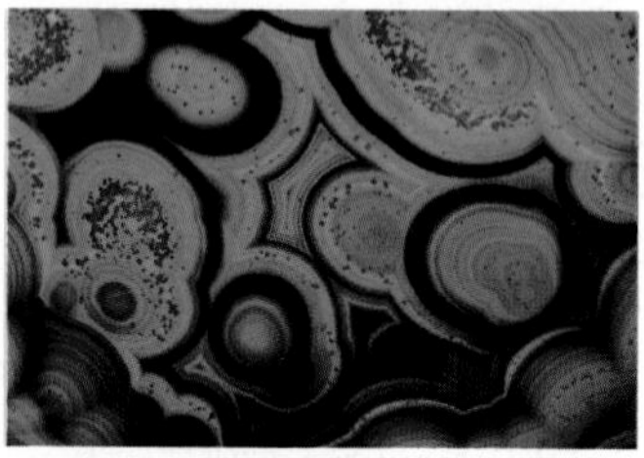

Tobkatrina/Shutterstock

(a)

Photos.com/Getty Images

(b)

▲ The mineral malachite **(a)** and the green patina on a copper roof **(b)** are both basic copper carbonate, just like the basic copper carbonate prepared by Proust in 1799.

2-1 CONCEPT ASSESSMENT

Jan Baptista van Helmont (1579–1644) weighed a young willow tree and the soil in which the tree was planted. Five years later he found that the mass of s had decreased by only 0.057 kg, while that of the tree had increased by 75 kg During that period he had added only water to the bucket in which the tree w planted. Helmont concluded that essentially all the mass gained by the tree h come from the water. Was this a valid conclusion? Explain.

Law of Constant Composition

In 1799, Joseph Proust (1754–1826) reported, "One hundred pounds of co dissolved in sulfuric or nitric acids and precipitated by the carbonate soda or potash, invariably gives 180 pounds of green carbonate."* This similar observations became the basis of the **law of constant compositio** the **law of definite proportions**:

All samples of a compound have the same composition—the same proportions by mass of the constituent elements.

To see how the law of constant composition works, consider the compc water. Water is made up of two atoms of hydrogen (H) for every atom of oxygen a fact that can be represented symbolically by a *chemical formula*, the familiar H

*The substance Proust produced is actually a more complex substance called *basic* coppe bonate. Proust's results were valid because, like all compounds, basic copper carbonate constant composition.

wo samples described below have the same proportions of the two elements, essed as percentages by mass. To determine the percent by mass of hydrogen, xample, simply divide the mass of hydrogen by the sample mass and multi- y 100%. For each sample, you will obtain the same result: 11.19% H.

Sample A and Its Composition		Sample B and Its Composition	
10.000 g		27.000 g	
1.119 g H	% H = 11.19	3.021 g H	% H = 11.19
8.881 g O	% O = 88.81	23.979 g O	% O = 88.81

XAMPLE 2-2 Using the Law of Constant Composition

In Example 2-1 we found that when 0.455 g of magnesium reacted with 2.315 g of oxygen, 0.755 g of magnesium oxide was obtained. Determine the mass of magnesium contained in a 0.500 g sample of magnesium oxide.

nalyze

We know that 0.755 g of magnesium oxide contains 0.455 g of magnesium. According to the law of constant composition, the mass ratio 0.455 g magnesium/0.755 g magnesium oxide should exist in all samples of magnesium oxide.

olve

Application of the law of constant composition gives

$$\frac{0.455\text{ g magnesium}}{0.755\text{ g magnesium oxide}} = \frac{?\text{ g magnesium}}{0.500\text{ g magnesium oxide}}$$

Solving the expression above, we obtain

$$?\text{ g magnesium} = 0.500\,\cancel{\text{g magnesium oxide}} \times \frac{0.455\text{ g magnesium}}{0.755\,\cancel{\text{g magnesium oxide}}}$$

$$= 0.301\text{ g magnesium}$$

ssess

You can also work this problem by using mass percentages. If 0.755 g of magnesium oxide contains 0.455 g of magnesium, then magnesium oxide is (0.455 g/0.755 g) × 100% = 60.3% magnesium by mass and (100% − 60.3%) = 39.7% oxygen by mass. Thus, a 0.500 g sample of magnesium oxide must contain 0.500 g × 60.3% = 0.301 g of magnesium and 0.500 g × 39.7% = 0.199 g of oxygen.

RACTICE EXAMPLE A: What masses of magnesium and oxygen must be combined to make exactly 2.000 g of magnesium oxide?

RACTICE EXAMPLE B: What substances are present, and what are their masses, after the reaction of 10.00 g of magnesium and 10.00 g of oxygen?

2-2 CONCEPT ASSESSMENT

hen 4.15 g magnesium and 82.6 g bromine react, (1) all the magnesium is ed up, (2) some bromine remains unreacted, and (3) magnesium bromide is e only product. With this information alone, is it possible to deduce the mass magnesium bromide produced? Explain.

ton's Atomic Theory

n 1803 to 1808, John Dalton, an English schoolteacher, used the two funda- tal laws of chemical combination just described as the basis of an atomic ry. His theory involved three assumptions:

Each chemical element is composed of minute, indivisible particles called atoms. Atoms can be neither created nor destroyed during a chemical change.

Sheila Terry / Science Source

▲ John Dalton (1766–1844), developer of the atomic theory. Dalton has not been considered a particularly good experimenter, perhaps because of his color blindness (a condition sometimes called daltonism). However, he did skillfully use the data of others in formulating his atomic theory. (The Granger Collection)

2. All atoms of an element are alike in mass (weight) and other prope but the atoms of one element are different from those of all other elem
3. In each of their compounds, different elements combine in a si numerical ratio, for example, one atom of A to one of B (AB), or one of A to two of B (AB_2).

If atoms of an element are indestructible (assumption 1), then the *same* a must be present after a chemical reaction as before. The total mass rem unchanged. Dalton's theory explains the law of conservation of mass. atoms of an element are alike in mass (assumption 2) and if atoms unite in numerical ratios (assumption 3), the percent composition of a compound have a unique value, regardless of the origin of the sample analyzed. Dal theory also explains the law of constant composition.

Like all good theories, Dalton's atomic theory led to a prediction—the **of multiple proportions**.

> If two elements form more than a single compound, the masses of one element combined with a fixed mass of the second are in the ratio of small whole numbers.

KEEP IN MIND

that all we know is that the second oxide is twice as rich in oxygen as the first. If the first is CO, the possibilities for the second are CO_2, C_2O_4, C_3O_6, and so on. (See also Exercise 18.)

To illustrate, consider two oxides of carbon (an oxide is a combination element with oxygen). In one oxide, 1.000 g of carbon is combined with 1.3 of oxygen, and in the other, with 2.667 g of oxygen. We see that the sec oxide is richer in oxygen; in fact, it contains twice as much oxygen as the 2.667 g/1.333 g = 2.00. We now know that the first oxide corresponds to formula CO and the second, CO_2 (Fig. 2-3).

The characteristic relative masses of the atoms of the various elem became known as atomic weights, and throughout the nineteenth cen chemists worked at establishing reliable values of relative atomic weig Mostly, however, chemists directed their attention to discovering new ments, synthesizing new compounds, developing techniques for analy materials, and in general, building up a vast body of chemical knowle Efforts to unravel the structure of the atom became the focus of physicist we see in the next several sections.

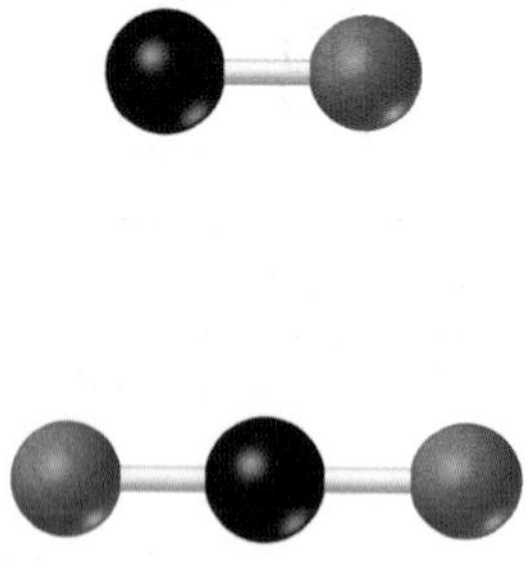

▲ FIGURE 2-3
Molecules CO and CO_2 illustrating the law of multiple proportions
The mass of carbon is the same in the two molecules, but the mass of oxygen in CO_2 is twice the mass of oxygen in CO. Thus, in accordance with the law of multiple proportions, the masses of oxygen in the two compounds, relative to a fixed mass of carbon, are in a ratio of small whole numbers, 2:1.

2-2 Electrons and Other Discoveries in Atomic Physics

Fortunately, we can acquire a qualitative understanding of atomic struc without having to retrace all the discoveries that preceded atomic phy We do, however, need a few key ideas about the interrelated phenomen electricity and magnetism, which we briefly discuss here. Electricity magnetism were used in the experiments that led to the current theor atomic structure.

Certain objects display a property called electric charge, which can be ei positive (+) or negative (−). Positive and negative charges attract each o while two positive or two negative charges repel each other. As we lear this section, all objects of matter are made up of charged particles. An ol having equal numbers of positively and negatively charged particles ca no net charge and is electrically neutral. If the number of positive cha exceeds the number of negative charges, the object has a net positive charg negative charges exceed positive charges, the object has a net negative cha Sometimes when one substance is rubbed against another, as in combing net electric charges build up on the objects, implying that rubbing separ

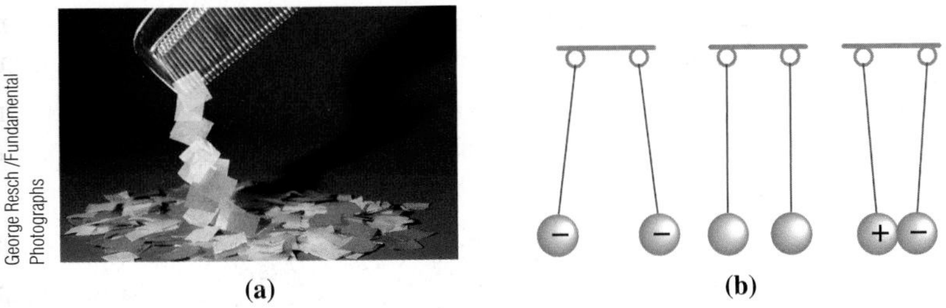

ɜURE 2-4
es between electrically charged objects
ectrostatically charged comb. If you comb your hair on a dry day, a static charge ops on the comb and causes bits of paper to be attracted to the comb. **(b)** Both ts on the left carry a negative electric charge. Objects with like charge repel each . The objects in the center lack any electric charge and exert no forces on each other. bjects on the right carry opposite charges—one positive and one negative—and t each other.

◀ We will use *electrostatics* (charge attractions and repulsions) to explain and understand many chemical properties.

e positive and negative charges (Fig. 2-4). Moreover, when a stationary ic) positive charge builds up in one place, a negative charge of equal size ears somewhere else; the charge is balanced.

gure 2-5 shows how charged particles behave when they move through field of a magnet. They are deflected from their straight-line path into a ed path in a plane perpendicular to the field. Think of the field or region nfluence of the magnet as represented by a series of invisible "lines of e" running from the north pole to the south pole of the magnet.

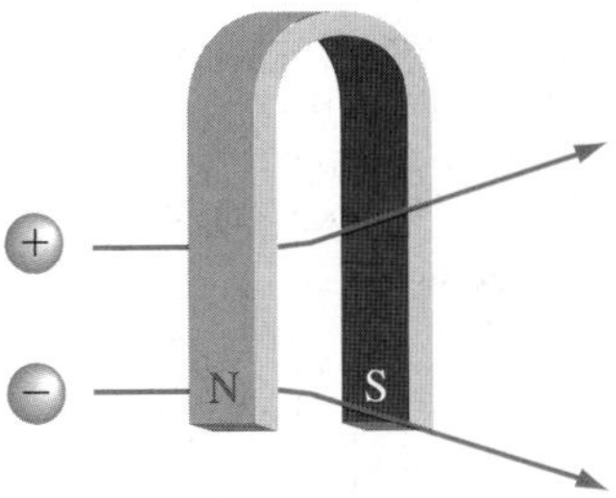

▲ FIGURE 2-5
Effect of a magnetic field on charged particles
When charged particles travel through a magnetic field so that their path is perpendicular to the field, they are deflected by the field. Negatively charged particles are deflected in one direction, and positively charged particles in the opposite direction. Several phenomena described in this section depend on this behavior.

Discovery of Electrons

, the abbreviation for cathode-ray tube, was once a familiar acronym. re liquid crystal display (LCD) was available, the CRT was the heart of puter monitors and TV sets. The first cathode-ray tube was made by hael Faraday (1791–1867) about 150 years ago. When he passed electric- through glass tubes from which most of the air had been evacuated, day discovered **cathode rays**, a type of radiation emitted by the negative inal or *cathode*. The radiation crossed the evacuated tube to the positive inal or *anode*. Later scientists found that cathode rays travel in straight s and have properties that are independent of the cathode material (that whether it is iron, platinum, and so on). The construction of a CRT is wn in Figure 2-6. The cathode rays produced in the CRT are invisible, they can be detected only by the light emitted by materials that they ke. These materials, called *phosphors*, are painted on the end of the CRT so the path of the cathode rays can be revealed. (*Fluorescence* is the term d to describe the emission of light by a phosphor when it is struck by

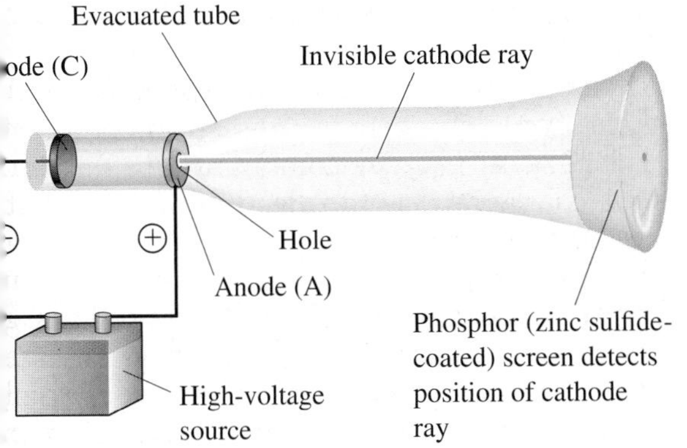

◀ FIGURE 2-6
A cathode-ray tube
The high-voltage source of electricity creates a negative charge on the electrode at the left (cathode) and a positive charge on the electrode at the right (anode). Cathode rays pass from the cathode (C) to the anode (A), which is perforated to allow the passage of a narrow beam of cathode rays. The rays are visible only through the green fluorescence that they produce on the zinc sulfide-coated screen at the end of the tube. They are invisible in other parts of the tube.

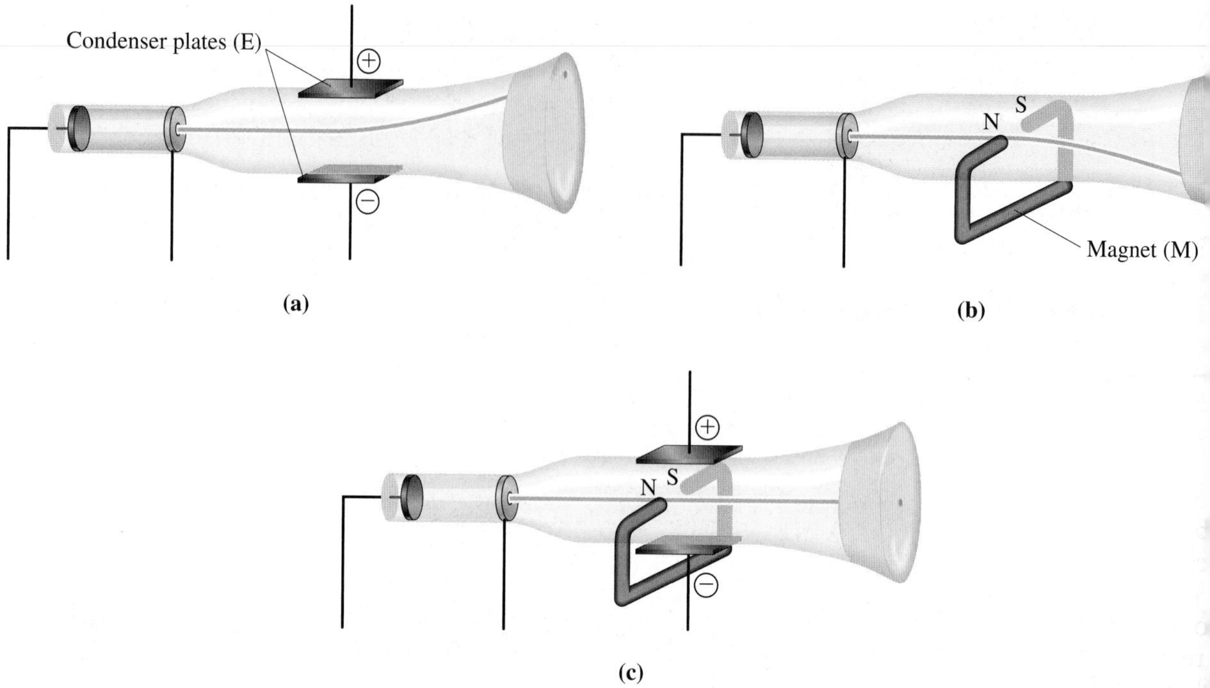

▲ FIGURE 2-7
Cathode rays and their properties
(a) Deflection of cathode rays in an electric field. The beam of cathode rays is defle as it travels from left to right in the field of the electrically charged condenser plates The deflection corresponds to that expected of negatively charged particles. **(b)** Deflection of cathode rays in a magnetic field. The beam of cathode rays is deflected as it travels from left to right in the field of the magnet (M). The deflectio corresponds to that expected of negatively charged particles. **(c)** Determining the mass-to-charge ratio, m/e, for cathode rays. The cathode-ray beam strikes the end screen undeflected if the forces exerted on it by the electric and magnetic fields are counterbalanced. By knowing the strengths of the electric and magnetic fields, together with other data, a value of m/e can be obtained. Precise measurements yi a value of -5.6857×10^{-9} g per coulomb. (Because cathode rays carry a negative charge, the sign of the mass-to-charge ratio is also negative.)

energetic radiation.) Another significant observation about cathode ray that they are deflected by electric and magnetic fields in the man expected for negatively charged particles (Fig. 2-7a, b).

In 1897, by the method outlined in Figure 2-7(c), J. J. Thomson (1856–1 established the ratio of mass (m) to electric charge (e) for cathode rays, that is, Also, Thomson concluded that cathode rays are negatively charged *fundame* particles of matter found in all atoms. (The properties of cathode rays are *inde dent* of the composition of the cathode.) Cathode rays subsequently beca known as **electrons**, a term first proposed by George Stoney in 1874.

▶ The coulomb (C) is the SI unit of electric charge (see also Appendix C).

Robert Millikan (1868–1953) determined the electronic charge e throug series of oil-drop experiments (1906–1914), described in Figure 2-8. The c rently accepted value of the electronic charge e, expressed in coulomb five significant figures, is -1.6022×10^{-19} C. By combining this value w an accurate value of the mass-to-charge ratio for an electron, we find that mass of an electron is 9.1094×10^{-28} g.

Once the electron was seen to be a fundamental particle of matter foun all atoms, atomic physicists began to speculate on how these particles w incorporated into atoms. The commonly accepted model was that proposed J. J. Thomson. Thomson thought that the positive charge necessary to coun balance the negative charges of electrons in a neutral atom was in the form

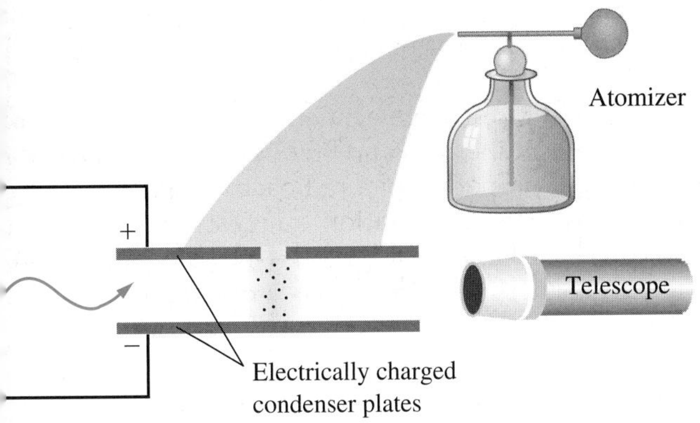

◀ FIGURE 2-8
Millikan's oil-drop experiment
Ions (charged atoms or molecules) are produced by energetic radiation, such as X-rays (X). Some of these ions become attached to oil droplets, giving them a net charge. The fall of a droplet in the electric field between the condenser plates is speeded up or slowed down, depending on the magnitude and sign of the charge on the droplet. By analyzing data from a large number of droplets, Millikan concluded that the magnitude of the charge, q, on a droplet is an *integral* multiple of the electric charge, e. That is, $q = ne$ (where $n = 1, 2, 3, \ldots$).

bulous cloud. Electrons, he suggested, floated in a diffuse cloud of posi- charge (rather like a lump of gelatin with electron "fruit" embedded in it). model became known as the plum-pudding model because of its similar- o a popular English dessert. The plum-pudding model is illustrated in re 2-9 for a neutral atom and for atomic species, called *ions*, which carry a :harge.

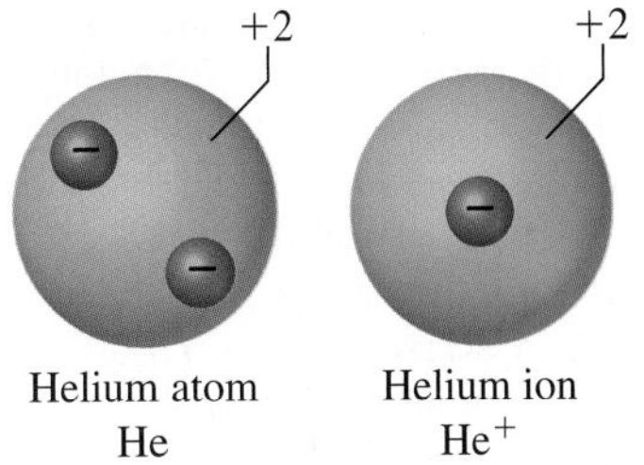

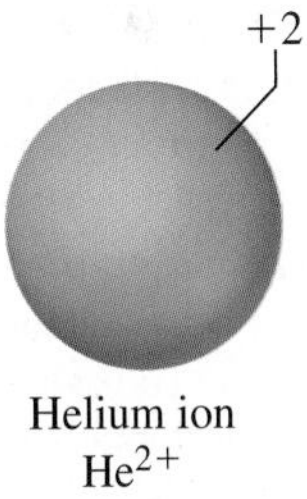

▲ FIGURE 2-9
The plum-pudding atomic model
According to this model, a helium atom would have a +2 cloud of positive charge and two electrons (−2). If a helium atom loses one electron, it becomes charged and is called an *ion*. This ion, referred to as He^+, has a net charge of 1+. If the helium atom loses both electrons, the He^{2+} ion forms.

ays and Radioactivity

ode-ray research had many important spin-offs. In particular, two natural iomena of immense theoretical and practical significance were discovered e course of other investigations.

1895, Wilhelm Roentgen (1845–1923) noticed that when cathode-ray s were operating, certain materials *outside* the tubes glowed or fluoresced. showed that this fluorescence was caused by radiation emitted by the ode-ray tubes. Because of the unknown nature of this radiation, Roentgen ed the term *X-ray*. We now recognize the X-ray as a form of high-energy tromagnetic radiation, which is discussed in Chapter 8.

ntoine Henri Becquerel (1852–1908) associated X-rays with fluorescence wondered if naturally fluorescent materials produce X-rays. To test this , he wrapped a photographic plate with black paper, placed a coin on the er, covered the coin with a uranium-containing fluorescent material, and osed the entire assembly to sunlight. When he developed the film, a clear ge of the coin could be seen. The fluorescent material had emitted radia- (presumably X-rays) that penetrated the paper and exposed the film. On occasion, because the sky was overcast, Becquerel placed the experimen- assembly inside a desk drawer for a few days while waiting for the ther to clear. On resuming the experiment, Becquerel decided to replace original photographic film, expecting that it may have become slightly osed. He developed the original film and found that instead of the ected feeble image, there was a very sharp one. The film had become ngly exposed because the uranium-containing material had emitted radia- continuously, even when it was not fluorescing. Becquerel had discovered **ioactivity**.

rnest Rutherford (1871–1937) identified two types of radiation from ioactive materials, alpha (α) and beta (β). **Alpha particles** carry two fun- nental units of positive charge and have essentially the same mass as um atoms. In fact, alpha particles are identical to He^{2+} ions. **Beta particles** negatively charged particles produced by changes occurring within the lei of radioactive atoms and have the same properties as electrons. A third n of radiation, which is not affected by electric or magnetic fields, was covered in 1900 by Paul Villard. This radiation, called **gamma rays** (γ),

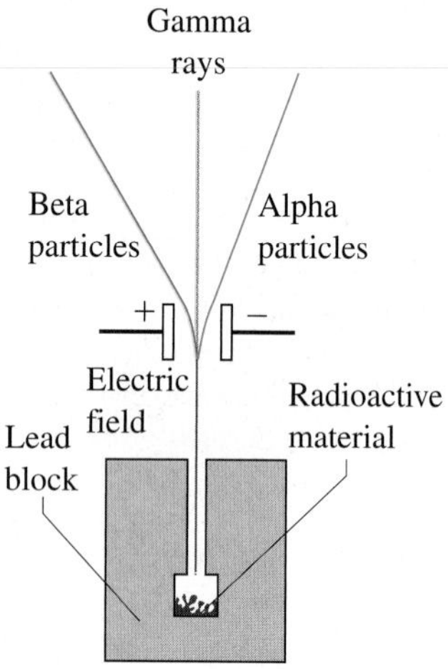

▲ FIGURE 2-10
Three types of radiation from radioactive materials
The radioactive material is enclosed in a lead block. All the radiation except that passing through the narrow opening is absorbed by the lead. When the escaping radiation is passed through an electric field, it splits into three beams. One beam is undeflected—these are gamma (γ) rays. A second beam is attracted to the negatively charged plate. These are the positively charged alpha (α) particles. The third beam, of negatively charged beta (β) particles, is deflected toward the positive plate.

is not made up of particles; it is electromagnetic radiation of extre high penetrating power. These three forms of radioactivity are illustrat Figure 2-10.

By the early 1900s, additional radioactive elements were discovered, p pally by Marie and Pierre Curie. Rutherford and Frederick Soddy made an profound finding: The chemical properties of a radioactive element *change* undergoes radioactive decay. This observation suggests that radioact involves fundamental changes at the *subatomic* level—in radioactive decay element is changed into another, a process known as *transmutation*.

2-3 The Nuclear Atom

In 1909, Rutherford, with his assistant Hans Geiger, began a line of rese using α particles as probes to study the inner structure of atoms. Base Thomson's plum-pudding model, Rutherford expected that most particl a beam of α particles would pass through thin sections of matter largely u flected, but that some α particles would be slightly scattered or deflecte they encountered electrons. By studying these scattering patterns, he hop deduce something about the distribution of electrons in atoms.

The apparatus used for these studies is pictured in Figure 2-11. Alpha p cles were detected by the flashes of light they produced when they stru zinc sulfide screen mounted on the end of a telescope. When Geiger and E Marsden, a student, bombarded very thin foils of gold with α particles, observed the following:

- The majority of α particles penetrated the foil undeflected.
- Some α particles experienced slight deflections.
- A few (about 1 in every 20,000) suffered rather serious deflections as penetrated the foil.
- A similar number did not pass through the foil at all but bounced ba the direction from which they had come.

The large-angle scattering greatly puzzled Rutherford. As he comme some years later, this observation was "about as credible as if you had fir 15-inch shell at a piece of tissue paper and it came back and hit you." By 1 though, Rutherford had an explanation. He based his explanation on a m of the atom known as the *nuclear atom* and having these features:

▶ Perhaps because he found it tedious to sit in the dark and count spots of light on a zinc sulfide screen, Geiger was motivated to develop an automatic radiation detector. The result was the well-known Geiger counter.

1. Most of the mass and all of the positive charge of an atom are centered very small region called the *nucleus*. The remainder of the atom is mo *empty space*.
2. The magnitude of the positive charge is different for different atoms ar approximately one-half the atomic weight of the element.
3. There are as many electrons outside the nucleus as there are units of p tive charge on the nucleus. The atom as a whole is electrically neutral.

▶ FIGURE 2-11
The scattering of α particles by metal foil
The telescope travels in a circular track around an evacuated chamber containing the metal foil. Most α particles pass through the metal foil undeflected, but some are deflected through large angles.

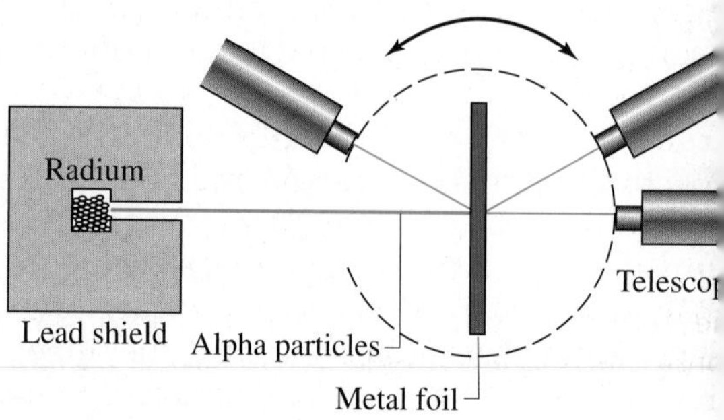

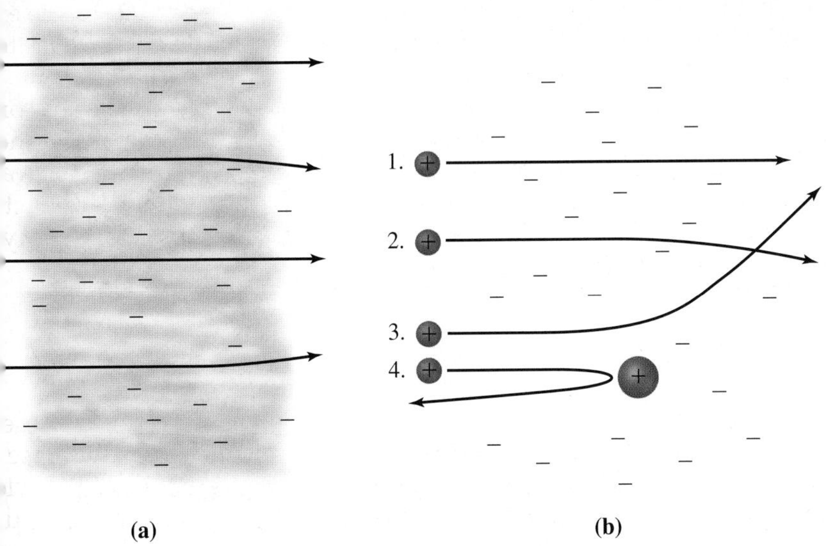

GURE 2-12
laining the results of α-particle scattering experiments
utherford's expectation was that small, positively charged α particles should pass ugh the nebulous, positively charged cloud of the Thomson plum-pudding model ly undeflected. Some would be slightly deflected by passing near electrons ent to neutralize the positive charge of the cloud). **(b)** Rutherford's explanation was d on a nuclear atom. With an atomic model having a small, dense, positively ged nucleus and extranuclear electrons, we would expect the four different types aths actually observed:

undeflected straight-line paths exhibited by most of the α particles
slight deflections of α particles passing close to electrons
severe deflections of α particles passing close to a nucleus
reflections from the foil of α particles approaching a nucleus head-on

herford's initial expectation and his explanation of the α-particle experi- ts are described in Figure 2-12.

covery of Protons and Neutrons

herford's nuclear atom suggested the existence of positively charged fun- iental particles of matter in the nuclei of atoms. Rutherford himself discov- l these particles, called **protons**, in 1919 in studies involving the scattering particles by nitrogen atoms in air. The protons were freed as a result of col- ns between α particles and the nuclei of nitrogen atoms. At about this same e, Rutherford predicted the existence in the nucleus of electrically neutral damental particles. In 1932, James Chadwick showed that a newly discov- l penetrating radiation consisted of beams of *neutral* particles. These parti- , called **neutrons**, originated from the nuclei of atoms. Thus, it has been y for about the past 100 years that we have had the atomic model suggested Figure 2-13.

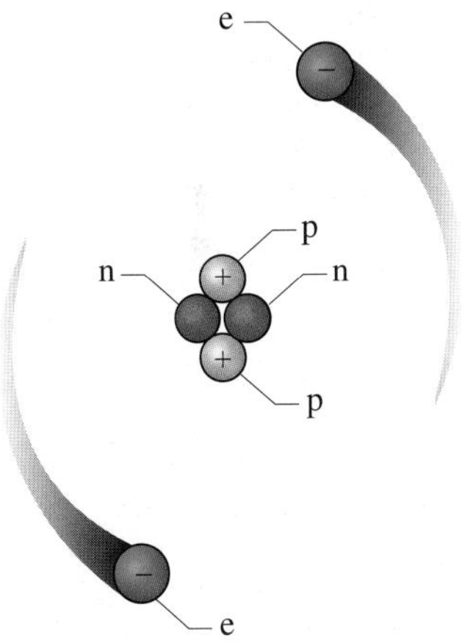

▲ FIGURE 2-13
The nuclear atom—illustrated by the helium atom
In this drawing, electrons are shown much closer to the nucleus than is the case. The actual situation is more like this: If the entire atom were represented by a room, 5 m × 5 m × 5 m, the nucleus would occupy only about as much space as the period at the end of this sentence.

2-3 CONCEPT ASSESSMENT

light of information presented to this point in the text, explain which of e three assumptions of Dalton's atomic theory (page 37) can still be nsidered correct and which cannot.

▶ The masses of the proton and neutron are different in the fourth significant figure. The charges of the proton and electron, however, are believed to be exactly equal in magnitude (but opposite in sign). The charges and masses are known much more precisely than suggested here. More precise values are given on the inside back cover.

TABLE 2.1 Properties of Three Fundamental Particles

	Electric Charge		Mass	
	SI (C)	Atomic	SI (g)	Atomic (u
Proton	$+1.6022 \times 10^{-19}$	+1	1.6726×10^{-24}	1.0073
Neutron	0	0	1.6749×10^{-24}	1.0087
Electron	-1.6022×10^{-19}	−1	9.1094×10^{-28}	0.0005485

[a]u is the SI symbol for atomic mass unit (abbreviated as amu).

Properties of Protons, Neutrons, and Electrons

The number of protons in a given atom is called the **atomic number,** or **proton number, *Z*.** The number of electrons in the atom is also equal because the atom is electrically neutral. The total number of protons and trons in an atom is called the **mass number, *A*.** The number of neutrons, **neutron number,** is $A - Z$. An electron carries an atomic unit of nega charge, a proton carries an atomic unit of positive charge, and a neutro electrically neutral. Table 2.1 presents the charges and masses of protons, trons, and electrons in two ways.

The **atomic mass unit** (described more fully on page 46) is defined as exa 1/12 of the mass of the atom known as carbon-12 (read as carbon twelve) atomic mass unit is abbreviated as amu and denoted by the symbol u. As see from Table 2.1, the proton and neutron masses are just slightly greater 1 u. By comparison, the mass of an electron is only about 1/2000th the ma the proton or neutron.

The three subatomic particles considered in this section are the only involved in the phenomena of interest to us in this text. You should be aw however, that a study of matter at its most fundamental level must cons many additional subatomic particles. The electron is believed to be a truly damental particle. However, modern particle physics now considers the n tron and proton to be composed of other, more fundamental particles.

2-4 Chemical Elements

Now that we have acquired some fundamental ideas about atomic struct we can more thoroughly discuss the concept of chemical elements.

All atoms of a particular element have the same atomic number, *Z*, a conversely, all atoms with the same number of protons are atoms of same element. The elements shown on the inside front cover have ato numbers from $Z = 1$ to $Z = 116$. Each element has a name and a distinct symbol. **Chemical symbols** are one- or two-letter abbreviations of name (usually the English name). The first (but never the second) lette the symbol is capitalized; for example, carbon, C; oxygen, O; neon, Ne; a silicon, Si. Some elements known since ancient times have symbols based their Latin names, such as Fe for iron (*ferrum*) and Pb for lead (*plumbum*). element sodium has the symbol Na, based on the Latin *natrium* for sodi carbonate. Potassium has the symbol K, based on the Latin *kalium* for po sium carbonate. The symbol for tungsten, W, is based on the German *wolfr*

▶ Just days before this text went to press, the IUPAC announced they had verified claims of the discoveries of elements 113, 115, 117, and 118. The process for determining the names and symbols of elements takes several months to complete.

▶ Other atomic symbols not based on English names include Cu, Ag, Sn, Sb, Au, and Hg.

Elements beyond uranium ($Z = 92$) do not occur naturally and must synthesized in particle accelerators (described in Chapter 25). Elements of very highest atomic numbers have been produced only on a limited num of occasions, a few atoms at a time. Inevitably, controversies have arisen ab ch research team discovered a new element and, in fact, whether a discovery made at all. However, international agreement has been reached on the first elements; each one has an official name and symbol.

:opes

epresent the composition of any particular atom, we need to specify its ıber of protons (p), neutrons (n), and electrons (e). We can do this with the bolism

$$\text{number p + number n} \rightarrow {}^{A}_{Z}\text{E} \leftarrow \text{symbol of element} \qquad \text{number p} \rightarrow Z \tag{2.1}$$

symbolism indicates that the atom is element E and that it has atomic num- Z and mass number A. For example, an atom of aluminum represented as has 13 protons and 14 neutrons in its nucleus and 13 electrons outside the eus. (Recall that an atom has the same number of electrons as protons.)

ontrary to what Dalton thought, we now know that atoms of an element ıot necessarily all have the same mass. In 1912, J. J. Thomson measured mass-to-charge ratios of positive ions formed from neon atoms. From e ratios he deduced that about 91% of the atoms had one mass and that the aining atoms were about 10% heavier. All neon atoms have 10 protons in r nuclei, and most have 10 neutrons as well. A very few neon atoms, how- , have 11 neutrons and some have 12. We can represent these three differ- types of neon atoms as

$${}^{20}_{10}\text{Ne} \qquad {}^{21}_{10}\text{Ne} \qquad {}^{22}_{10}\text{Ne}$$

toms that have the *same* atomic number (Z) but *different* mass numbers (A) called **isotopes**. Of all Ne atoms on Earth, 90.51% are ${}^{20}_{10}\text{Ne}$. The percentages ${}^{1}_{0}$Ne and ${}^{22}_{10}\text{Ne}$ are 0.27% and 9.22%, respectively. These percentages— 1%, 0.27%, 9.22%—are the **percent isotopic abundances** of the three neon opes. Sometimes the mass numbers of isotopes are incorporated into the ıes of elements, such as neon-20 (neon twenty). Percent isotopic abundances always based on *numbers*, not masses. Thus, 9051 of every 10,000 neon ns are neon-20 atoms. Some elements, as they exist in nature, consist of just a le type of atom and therefore do not have naturally occurring isotopes.* minum, for example, consists only of aluminum-27 atoms.

◀ Because neon is the only element with $Z = 10$, the symbols ^{20}Ne, ^{21}Ne, and ^{22}Ne convey the same meaning as ${}^{20}_{10}\text{Ne}$, ${}^{21}_{10}\text{Ne}$, and ${}^{22}_{10}\text{Ne}$.

◀ Odd-numbered elements tend to have fewer isotopes than do even-numbered elements. Section 25-7 will explain why.

◀ Usually all the isotopes of an element share the same name and atomic symbol. The exception is hydrogen. Isotope ${}^{2}_{1}\text{H}$ is called deuterium (symbol D), and ${}^{3}_{1}\text{H}$ is tritium (T).

s

en atoms lose or gain electrons, for example, in the course of a chemical tion, the species formed are called **ions** and carry net charges. Because an tron is negatively charged, adding electrons to an electrically neutral atom duces a negatively charged ion. Removing electrons results in a positively rged ion. The number of protons does not change when an atom becomes on. For example, $^{20}\text{Ne}^{+}$ and $^{22}\text{Ne}^{2+}$ are ions. The first one has 10 protons, eutrons, and *9* electrons. The second one also has 10 protons, but 12 neu- s and *8* electrons. The charge on an ion is equal to the number of protons *us* the number of electrons. That is

$$\text{number p + number n} \rightarrow {}^{A}_{Z}\text{E}^{\#\pm} \leftarrow \text{number p} - \text{number e} \qquad \text{number p} \rightarrow Z \tag{2.2}$$

◀ In this expression, #± indicates that the charge is written with the number (#) *before* the + or − sign. However, when the charge is 1+ or 1−, the number 1 is not included.

nother example is the $^{16}\text{O}^{2-}$ ion. In this ion, there are 8 protons (atomic nber 8), 8 neutrons (mass number − atomic number), and *10* electrons - 10 = −2).

uclide is the general term used to describe an atom with a particular atomic number and number. Although there are several elements with only one naturally occurring nuclide, it ssible to produce additional nuclides of these elements—isotopes—by artificial means ion 25-3). The artificial isotopes are radioactive, however. In all, the number of synthetic pes exceeds the number of naturally occurring ones by several fold.

EXAMPLE 2-3 Relating the Numbers of Protons, Neutrons, and Electrons in Atoms and Ions

Through an appropriate symbol, indicate the number of protons, neutrons, and electrons in **(a)** an atom o barium-135 and **(b)** the double negatively charged ion of selenium-80.

Analyze

Given the name of an element, we can find the symbol and the atomic number, Z, for that element from a li of elements or a periodic table. To determine the number of protons, neutrons, and electrons, we make use o the following relationships:

$$Z = \text{number p} \qquad A = \text{number p} + \text{number n} \qquad \text{charge} = \text{number p} - \text{number e}$$

The relationships above are summarized in expression (2.2).

Solve

(a) We are given the name (barium) and the mass number of the atom (135). From a list of the elements o a periodic table we obtain the symbol (Ba) and the atomic number ($Z = 56$), leading to the symboli representation

$$^{135}_{56}\text{Ba}$$

From this symbol one can deduce that the neutral atom has 56 protons; a neutron number o $A - Z = 135 - 56 = 79$ neutrons; and a number of electrons equal to Z, that is, 56 electrons.

(b) We are given the name (selenium) and the mass number of the ion (80). From a list of the elements or periodic table we obtain the symbol (Se) and the atomic number (34). Together with the fact that the io carries a charge of 2−, we have the data required to write the symbol

$$^{80}_{34}\text{Se}^{2-}$$

From this symbol, we can deduce that the ion has 34 protons; a neutron number of $A - Z =$ $80 - 34 = 46$ neutrons; and 36 electrons, leading to a net charge of $+34 - 36 = -2$.

Assess

When writing the symbol for a particular atom or ion, we often omit the atomic number. For example, for $^{135}_{56}\text{B}$ and $^{80}_{34}\text{Se}^{2-}$, we often use the simpler representations ^{135}Ba and $^{80}\text{Se}^{2-}$.

PRACTICE EXAMPLE A: Use the notation $^{A}_{Z}\text{E}$ to represent the isotope of silver having a neutron number of 62.

PRACTICE EXAMPLE B: Use the notation $^{A}_{Z}E$ to represent a tin ion having the same number of electrons as a atom of the isotope cadmium-112. Explain why there can be more than one answer.

2-4 CONCEPT ASSESSMENT

What is the single exception to the statement that all atoms comprise protons neutrons, and electrons?

▶ Ordinarily we expect like-charged objects (such as protons) to repel each other. The forces holding protons and neutrons together in the nucleus are very much stronger than ordinary electrical forces (Section 25-6).

▶ This definition also establishes that one atomic mass unit (1 u) is *exactly* 1/12 the mass of a carbon-12 atom.

Isotopic Masses

We cannot determine the mass of an individual atom just by adding up masses of its fundamental particles. When protons and neutrons combin form a nucleus, a very small portion of their original mass is converte energy and released. However, we cannot predict exactly how much this called nuclear binding energy will be. Determining the masses of indivic atoms, then, is something that must be done by experiment, in the follow way. By international agreement, one type of atom has been chosen a assigned a specific mass. This standard is an atom of the isotope carbon which is assigned a mass of exactly 12 atomic mass units, that is, 12 u. N the masses of other atoms relative to carbon-12 are determined with a m **spectrometer**. In this device, a beam of gaseous ions passing through elec and magnetic fields separates into components of differing masses.

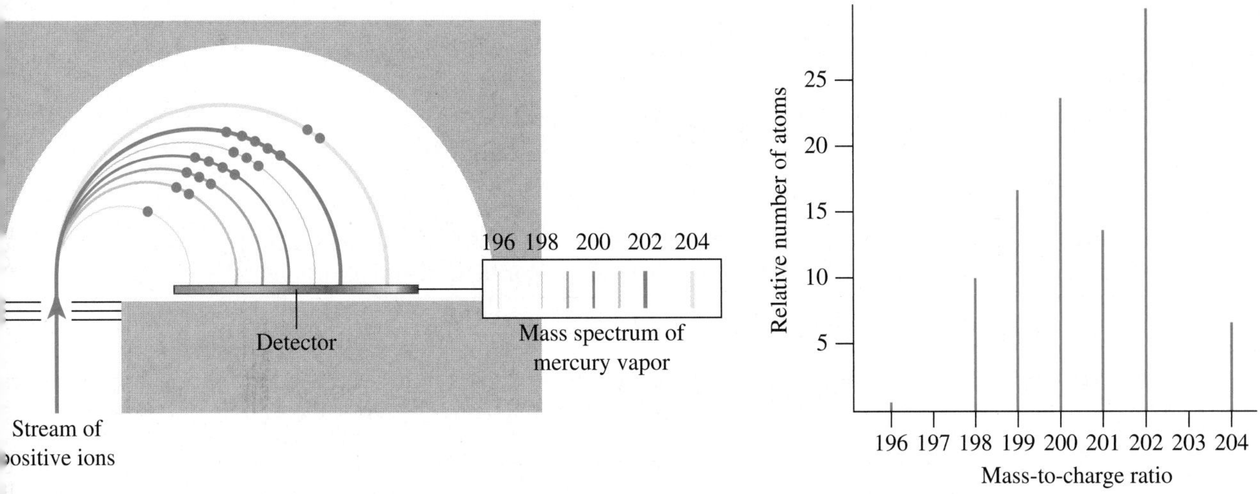

GURE 2-14
ass spectrometer and mass spectrum
is mass spectrometer, a gaseous sample is ionized by bombardment with electrons e lower part of the apparatus (not shown). The positive ions thus formed are ected to an electrical force by the electrically charged velocity selector plates and a netic force by a magnetic field perpendicular to the page. Only ions with a particular city pass through and are deflected into circular paths by the magnetic field. Ions different masses strike the detector (here a photographic plate) in different regions. more ions of a given type, the greater the response of the detector (intensity of line ne photographic plate). In the mass spectrum shown for mercury, the response of the detector (intensity of lines on photographic plate) has been converted to a scale of ive numbers of atoms. The percent isotopic abundances of the mercury isotopes are g, 0.146%; ^{198}Hg, 10.02%; ^{199}Hg, 16.84%; ^{200}Hg, 23.13%; ^{201}Hg, 13.22%; ^{202}Hg, 0%; and ^{204}Hg, 6.85%.

◀ The primary standard for atomic masses has evolved over time. For example, Dalton originally assigned H a mass of 1 u. Later, chemists took naturally occurring oxygen at 16 u to be the definition of the atomic-weight scale. Concurrently, physicists defined the oxygen-16 isotope as 16 u. This resulted in conflicting values. In 1971 the adoption of carbon-12 as the universal standard resolved this disparity.

arated ions are focused on a measuring instrument, which records their sence and amounts. Figure 2-14 illustrates mass spectrometry and a typi- mass spectrum.

lthough mass numbers are whole numbers, the actual masses of individual ns (in atomic mass units, u) are never whole numbers, except for carbon-12. vever, they are very close in value to the corresponding mass numbers, as we see for the isotope oxygen-16. From mass spectral data the ratio of the mass O to ^{12}C is found to be 1.33291. Thus, the mass of the oxygen-16 atom is

$$1.33291 \times 12\ \text{u} = 15.9949\ \text{u}$$

ch is very nearly equal to the mass number of 16.

EXAMPLE 2-4 Establishing Isotopic Masses by Mass Spectrometry

With mass spectral data, the mass of an oxygen-16 atom is found to be 1.06632 times that of a nitrogen-15 atom. Given that ^{16}O has a mass of 15.9949 u (see above), what is the mass of a nitrogen-15 atom, in u?

nalyze

Given the ratio (mass of ^{16}O)/(mass of ^{15}N) = 1.06632 and the mass of ^{16}O, 15.9949 u, we solve for the mass of ^{15}N.

olve

We know that

$$\frac{\text{mass of } ^{16}\text{O}}{\text{mass of } ^{15}\text{N}} = 1.06632$$

(continued)

We solve the preceding expression for the mass of ^{15}N and then substitute 15.9949 u for the mass of ^{16}O. W obtain the result

$$\text{mass of } ^{15}N = \frac{\text{mass of } ^{16}O}{1.06632} = \frac{15.9949 \text{ u}}{1.06632} = 15.0001 \text{ u}$$

Assess

The mass of ^{15}N is very nearly 15, as we should expect. If we had mistakenly multiplied instead of dividin by the ratio 1.06632, the result would have been slightly larger than 16 and clearly incorrect.

PRACTICE EXAMPLE A: What is the ratio of masses for $^{202}Hg/^{12}C$, if the isotopic mass for ^{202}Hg is 201.97062 u?

PRACTICE EXAMPLE B: An isotope with atomic number 64 and mass number 158 is found to have a mass rati relative to that of carbon-12 of 13.16034. What is the isotope, what is its atomic mass in u, and what is its mas relative to oxygen-16?

2-5 Atomic Mass

▶ Carbon-14, used for radiocarbon dating, is formed in the upper atmosphere. The amount of carbon-14 on Earth is too small to affect the atomic mass of carbon.

In a table of atomic masses, the value listed for carbon is about 12.01, yet atomic mass standard is *exactly* 12. Why the difference? The atomic m standard is based on a sample of carbon containing only atoms of carbon whereas naturally occurring carbon contains some carbon-13 atoms as w The existence of these two isotopes causes the observed atomic mass to greater than 12. The **atomic mass (weight)*** of an element is the averag the isotopic masses, *weighted* according to the naturally occurring ab dances of the isotopes of the element. In a weighted average, we must as greater importance—give greater weight—to the quantity that occurs m frequently. Since carbon-12 atoms are much more abundant than carbon the weighted average must lie much closer to 12 than to 13. This is the re that we get by applying the following general equation, where the rig hand side of the equation includes one term for each naturally occurr isotope.

KEEP IN MIND

that the fractional abundance is the percent abundance divided by 100%. Thus, a 98.93% abundance is a 0.9893 abundance.

$$\begin{matrix}\text{at. mass} \\ \text{of an} \\ \text{element}\end{matrix} = \left(\begin{matrix}\text{fractional} \\ \text{abundance of} \\ \text{isotope 1}\end{matrix} \times \begin{matrix}\text{mass of} \\ \text{isotope 1}\end{matrix}\right) + \left(\begin{matrix}\text{fractional} \\ \text{abundance of} \\ \text{isotope 2}\end{matrix} \times \begin{matrix}\text{mass of} \\ \text{isotope 2}\end{matrix}\right) + \dots$$

The first term on the right side of equation (2.3) represents the contribu from isotope 1; the second term represents the contribution from isotop and so on.

We will use equation (2.3), with appropriate data, in Example 2-6, but let us illustrate the ideas of fractional abundance and a weighted average different way in establishing the atomic mass of carbon. The mass spectrur a particular sample of carbon shows that 98.93% of the carbon atoms carbon-12 with a mass of exactly 12 u; the rest are carbon-13 atoms wi mass of 13.0033548378 u. Therefore:

$$\begin{aligned}\text{atomic mass of carbon} &= 0.9893 \times 12 \text{ u} + (1 - 0.9893) \times 13.0033548378 \text{ u} \\ &= 13.0033548378 \text{ u} - 0.9893 \times (13.0033548378 \text{ u} - 12 \text{ u}) \\ &= 13.0033548378 \text{ u} - 0.9893 \times (1.0033548378 \text{ u}) \\ &= 13.0033548378 \text{ u} - 0.9926 \text{ u} \\ &= 12.0108 \text{ u}\end{aligned}$$

*Since Dalton's time, atomic masses have been called atomic weights. They still are by chemists, yet what we are describing here is mass, not weight. Old habits die hard.

important to note that, in the setup just shown, 12 u and the "1" appearing ɿe factor (1 − 0.9893) are exact numbers. Thus, by applying the rules for ificant figures (see Chapter 1), the atomic mass of carbon can be reported four decimal places.

) determine the atomic mass of an element having three naturally occur- isotopes, such as potassium, we would have to include three contribu- s in the weighted average, and so on.

mic Mass Intervals and Conventional Atomic Masses

ı technological improvements in mass spectrometry, scientists can now rmine atomic masses and isotopic abundances with a very high degree of ision. This ability has led to the discovery that, for certain elements, the iso- : abundances can vary significantly from one sample to another. For exam- the highest reported value for the isotopic abundance of ^{13}C is 1.1466% (in ples of deep-sea pore water), and the lowest reported value is 0.9629% (from etene samples obtained from the ocean bottom in the North Pacific). Because ɿe variation of isotopic abundances, the experimentally determined atomic s of carbon lies within an interval that has a lower bound of 12.0096 u and an er bound of 12.0116 u. For this reason, the IUPAC has recommended that the ıic mass of carbon, and several other elements, be reported as an **atomic s interval** rather than as a single specific value (see Table 2.2). The atomic s interval for carbon is given as [12.0096, 12.0116]. Writing the standard ıic mass of carbon as [12.0096, 12.0116] indicates that its atomic mass in any nal material will be at least 12.0096 u and not more than 12.0116 u.

ı general, an atomic mass interval is expressed in the form [a, b], where a is ower bound of the interval and b is the upper bound. The interval desig- on does not imply any statistical distribution of atomic mass values veen the lower and upper bounds, nor does it represent a measure of the stical uncertainty. For example, the average of a and b is neither the most y value nor the most representative value. The difference $b - a$ does not esent the uncertainty.

or those elements with standard atomic masses given as intervals, the AC also provides **conventional atomic mass** values (Table 2.2). These con- ional values can be used when we need a specific, representative value of ıtomic mass. The values have been selected so that, for materials normally ountered, the atomic mass would be within in an interval of plus or minus in the last digit.

TABLE 2.2 Conventional Atomic Masses and Atomic Mass Intervals for Selected Elements

Atomic Number	Atomic Symbol	Atomic Mass, u: Conventional	Atomic Mass, u: Interval
1	H	1.008	[1.00784, 1.00811]
3	Li	6.94	[6.938, 6.997]
5	B	10.81	[10.806, 10.821]
6	C	12.011	[12.0096, 12.0116]
7	N	14.007	[14.00643, 14.00728]
8	O	15.999	[15.99903, 15.9997]
12	Mg	24.305	[24.304, 24.307]
14	Si	28.085	[28.084, 28.086]
16	S	32.06	[32.059, 32.076]
17	Cl	35.45	[35.446, 35.457]
35	Br	79.904	[79.901, 79.907]
81	Tl	204.38	[204.382, 204.385]

Some Representative Examples

Sometimes a qualitative understanding of the relationship between isot masses, percent isotopic abundances, and weighted-average atomic mass that we need, and no calculation is necessary, as illustrated in Example Example 2-6 and the accompanying Practice Examples provide additi applications of equation (2.3).

The table of atomic masses (inside the front cover) shows that some at masses are stated more precisely than others. For example, the atomic mass is given as 18.998 u and that of Kr is given as 83.798 u. In fact, the atomic ma fluorine is known even more precisely (18.9984032 u); the value of 18.998 u been rounded off to five significant figures. Why is the atomic mass of F kn so much more precisely than that of Kr? Only one type of fluorine atom oc naturally: fluorine-19. Determining the atomic mass of fluorine means estab ing the mass of this type of atom as precisely as possible. The atomic ma krypton is known less precisely because krypton has six naturally occurring topes. Because the percent distribution of the isotopes of krypton differs slightly from one sample to another, the weighted-average atomic mass of k ton cannot be stated with high precision.

EXAMPLE 2-5 Understanding the Meaning of a Weighted-Average Atomic Mass

The two naturally occurring isotopes of lithium, lithium-6 and lithium-7, have masses of 6.01512 u and 7.01600 respectively. Which of these two occurs in greater abundance?

Analyze

Look up the atomic mass of Li and compare it with the masses of ^{6}Li and ^{7}Li. If the atomic mass of Li is clos to that of ^{6}Li, then ^{6}Li is the more abundant isotope. If the atomic mass of Li is closer to that of ^{7}Li, then ^{7}Li the more abundant isotope.

Solve

From a table of atomic masses (inside the front cover), we see that the atomic mass of lithium is reported a an atomic mass interval [6.938, 6.997]. The conventional atomic mass value (from Table 2.2) is 6.94. Becaus the values in this range, and the conventional atomic mass value, are all much closer to 7.01600 u than 6.01512 u, lithium-7 must be the more abundant isotope.

Assess

Atomic masses of specific isotopes can be determined very precisely. The values given above for ^{6}Li and ^{7} have been rounded to five decimal places. The precise values are 6.015122795 u and 7.01600455 u.

PRACTICE EXAMPLE A: The two naturally occurring isotopes of boron, boron-10 and boron-11, have masses o 10.0129370 u and 11.0093054 u, respectively. Which of these two occurs in greater abundance?

PRACTICE EXAMPLE B: Indium has two naturally occurring isotopes and a weighted atomic mass of 114.818 One of the isotopes has a mass of 112.904058 u. Which of the following must be the second isotope: ^{111}I ^{112}In, ^{114}In, or ^{115}In? Which of the two naturally occurring isotopes must be the more abundant?

EXAMPLE 2-6 Relating the Masses and Abundances of Isotopes to the Atomic Mass of an Element

Bromine has two naturally occurring isotopes, bromine-79 and bromine-81, with masses of 78.918338 u an 80.916291 u, respectively. Bromine has an atomic mass interval of [79.901, 79.907]. Estimate the percent isotop abundances of ^{79}Br by using **(a)** the lower bound and **(b)** the upper bound of the atomic mass interval.

Analyze

We need to apply two key concepts: (1) the atomic mass of Br is a weighted average of the masses of ^{79}Br and ^{81}B and (2) the percent isotopic abundances of ^{79}Br and ^{81}Br must add up to 100%. In effect, we use equation (2. twice, setting the atomic mass of Br equal to the lower and upper bounds of the interval.

olve

The atomic mass of Br is calculated by using equation (2.3):

$$\text{atomic mass of Br} = \left(\begin{array}{c}\text{fractional} \\ \text{abundance of } {}^{79}\text{Br} \times \text{mass of } {}^{79}\text{Br}\end{array}\right) + \left(\begin{array}{c}\text{fractional} \\ \text{abundance of } {}^{81}\text{Br} \times \text{mass of } {}^{81}\text{Br}\end{array}\right)$$

The percent isotopic abundances must total 100%, and the fractional isotopic abundances must total 1. If we let f represent the fractional abundance of ^{79}Br, then the fractional abundance of ^{81}Br is $1-f$. Therefore, we may write

$$\begin{aligned}\text{atomic mass of Br} &= f \times 78.91338\text{ u} + (1-f) \times 80.916291\text{ u} \\ &= f \times (78.91338\text{ u} - 80.916291\text{ u}) + 80.916291\text{ u}\end{aligned}$$

$$f = \frac{\text{atomic mass of Br} - 80.916291\text{ u}}{78.91338\text{ u} - 80.916291\text{ u}}$$

(a) By substituting 79.901 u for the atomic mass of Br (the lower bound of the interval), we obtain the following result for the fractional abundance of ^{79}Br:

$$f = \frac{79.901\text{ u} - 80.916291\text{ u}}{78.91338\text{ u} - 80.916291\text{ u}} = 0.50817$$

An atomic mass of 79.901 u implies that the percent isotopic abundance of ^{79}Br is 50.817%. Therefore, the percent isotopic abundance of ^{81}Br is (100 − 50.817)% = 49.183%.

(b) When we set the atomic mass of Br equal to 79.907 u, the upper bound of the interval, we obtain

$$f = \frac{79.907\text{ u} - 80.916291\text{ u}}{78.91338\text{ u} - 80.916291\text{ u}} = 0.50516$$

The percent isotopic abundance of ^{79}Br is 50.516%, and the percent isotopic abundance of ^{81}Br is (100 − 50.516)% = 49.484%.

ssess

These results indicate that the percent isotopic abundance of ^{79}Br varies between 50.516% and 50.817%. Because of this variation, the atomic mass of bromine is best expressed as an atomic mass interval. When a representative value of the atomic mass of Br is required, we would use the conventional atomic mass (Table 2.2) of Br, which is 79.904 u.

RACTICE EXAMPLE A: The masses and percent isotopic abundances of the three naturally occurring isotopes of silicon are ^{28}Si, 27.9769265325 u, 92.223%; ^{29}Si, 28.976494700 u, 4.685%; ^{30}Si, 29.973377017 u, 3.092%. Calculate the weighted-average atomic mass of silicon.

RACTICE EXAMPLE B: Use data from Example 2-5 and the conventional atomic mass of Li (Table 2.2) to estimate the percent isotopic abundances of lithium-6 and lithium-7.

2-5 CONCEPT ASSESSMENT

e value listed for chromium in the table of atomic masses inside the front ver is 51.996 u. Should we conclude that naturally occurring chromium atoms all of the type $^{52}_{24}$Cr? The same table lists a value of 65.38 u for zinc. Should conclude that zinc occurs as a mixture of isotopes? Explain.

5 Introduction to the Periodic Table

ntists spend a lot of time organizing information into useful patterns. re they can organize information, however, they must possess it, and it t be correct. Botanists had enough information about plants to organize field in the eighteenth century. Because of uncertainties in atomic masses because many elements remained undiscovered, chemists were not able ganize the elements until a century later.

1 1A	2 2A	3 3B	4 4B	5 5B	6 6B	7 7B	8	9 8B	10	11 1B	12 2B	13 3A	14 4A	15 5A	16 6A	17 7A
1 H [1.007, 1.009]																
3 Li [6.938, 6.997]	4 Be 9.0122											5 B [10.80, 10.83]	6 C [12.00, 12.02]	7 N [14.00, 14.01]	8 O [15.99, 16.00]	9 F 18.998
11 Na 22.990	12 Mg [24.30, 24.31]											13 Al 26.982	14 Si [28.08, 28.09]	15 P 30.974	16 S [32.05, 32.08]	17 Cl [35.44, 35.46]
19 K 39.098	20 Ca 40.078	21 Sc 44.956	22 Ti 47.867	23 V 50.942	24 Cr 51.996	25 Mn 54.938	26 Fe 55.845	27 Co 58.933	28 Ni 58.693	29 Cu 63.546	30 Zn 65.38	31 Ga 69.723	32 Ge 72.630	33 As 74.922	34 Se 78.96	35 Br [79.90, 79.91]
37 Rb 85.468	38 Sr 87.62	39 Y 88.906	40 Zr 91.224	41 Nb 92.906	42 Mo 95.96	43 Tc (98)	44 Ru 101.07	45 Rh 102.91	46 Pd 106.42	47 Ag 107.87	48 Cd 112.41	49 In 114.82	50 Sn 118.71	51 Sb 121.76	52 Te 127.60	53 I 126.90
55 Cs 132.91	56 Ba 137.33	57–71 La–Lu	72 Hf 178.49	73 Ta 180.95	74 W 183.84	75 Re 186.21	76 Os 190.23	77 Ir 192.22	78 Pt 195.08	79 Au 196.97	80 Hg 200.59	81 Tl [204.3, 204.4]	82 Pb 207.2	83 Bi 208.98	84 Po (209)	85 At (210)
87 Fr (223)	88 Ra (226)	89–103 Ac–Lr	104 Rf (261)	105 Db (262)	106 Sg (266)	107 Bh (264)	108 Hs (277)	109 Mt (268)	110 Ds (271)	111 Rg (272)	112 Cn (285)		114 Fl (289)		116 Lv (293)	

*Lanthanide series	57 La 138.91	58 Ce 140.12	59 Pr 140.91	60 Nd 144.24	61 Pm (145)	62 Sm 150.36	63 Eu 151.96	64 Gd 157.25	65 Tb 158.93	66 Dy 162.50	67 Ho 164.93	68 Er 167.26	69 Tm 168.93	70 Yb 173.05	71 Lu 174.97
†Actinide series	89 Ac (227)	90 Th 232.04	91 Pa 231.04	92 U 238.03	93 Np (237)	94 Pu (244)	95 Am (243)	96 Cm (247)	97 Bk (247)	98 Cf (251)	99 Es (252)	100 Fm (257)	101 Md (258)	102 No (259)	103 Lr (262)

▲ FIGURE 2-15
Periodic table of the elements
Atomic masses are relative to carbon-12. For 12 elements, the atomic mass is giv as an interval (see Section 2-5). For certain radioactive elements, the numbers list parentheses are the mass numbers of the most stable isotopes. Metals are shown tan, nonmetals in blue, and metalloids in green. The noble gases (also nonmetals shown in pink. Just days before this text went to press, the IUPAC announced the had verified claims of the discovery of elements 113, 115, 117, and 118.

KEEP IN MIND

that the periodic table shown in Figure 2-15 is the one currently recommended by IUPAC. In Figure 2-15, lutetium (Lu) and lawrencium (Lr) are the last members of the lanthanide and actinide series, respectively. A strong argument* has been made for placing Lu and Lr in group 3, meaning the lanthanide series would end with ytterbium (Yb) and the actinide series would end with nobelium (Nb). To date, IUPAC has not endorsed placing Lu and Lr in group 3.

* See W. B. Jensen, *J. Chem. Educ.*, **59**, 634 (1982).

We can distinguish one element from all others by its particular se observable physical properties. For example, sodium has a low densit 0.971 g/cm^3 and a low melting point of 97.81 °C. No other element has same combination of density and melting point. Potassium, though, also a low density (0.862 g/cm^3) and low melting point (63.65 °C), much sodium. Sodium and potassium further resemble each other in that both good conductors of heat and electricity, and both react vigorously with w to liberate hydrogen gas. Gold, conversely, has a density (19.32 g/cm^3) melting point (1064 °C) that are very much higher than those of sodiur potassium, and gold does not react with water or even with ordinary acic does resemble sodium and potassium in its ability to conduct heat and e tricity, however. Chlorine is very different still from sodium, potassium, gold. It is a gas under ordinary conditions, which means that the melting p of solid chlorine (−101 °C) is far below room temperature. Also, chlorine nonconductor of heat and electricity.

Even from these very limited data, we get an inkling of a useful clas cation scheme of the elements. If the scheme is to group together elem with similar properties, then sodium and potassium should appear in same group. And if the classification scheme is in some way to disting between elements that are good conductors of heat and electricity and th that are not, chlorine should be set apart from sodium, potassium, and g The classification system we need is the one shown in Figure 2-15 (inside the front cover), known as the **periodic table** of the elements Chapter 9, we will describe how the periodic table was formulated, and will also learn its theoretical basis. For the present, we will consider on few features of the table.

▶ That elements in one group have similar properties is perhaps the most useful simplifying feature of atomic properties. Significant differences within a group do occur. The manner and reason for such differences is much of what we try to discover in studying chemistry.

Features of the Periodic Table In the periodic table, elements are listed acc ing to increasing atomic number starting at the upper left and arranged series of horizontal rows. This arrangement places similar elements in *ver* **groups** or **families**. For example, sodium and potassium are found togeth a group labeled 1 (called the *alkali metals*). We should expect other members o

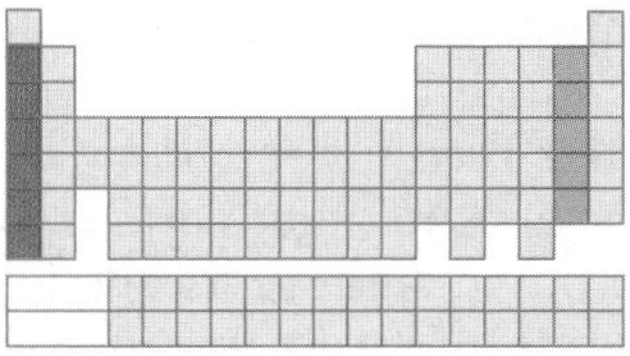

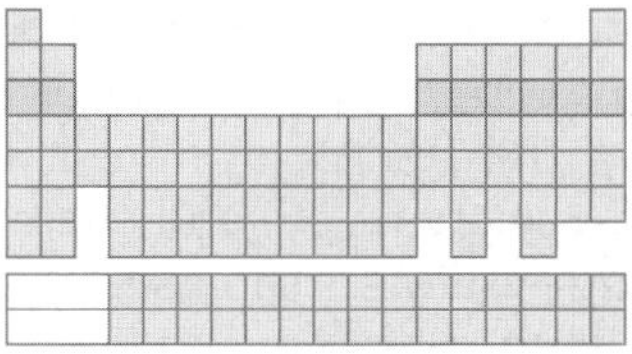

p, such as cesium and rubidium, to have properties similar to sodium and ssium. Chlorine is found at the other end of the table in a group labeled 17. e of the groups are given distinctive names, mostly related to an important erty of the elements in the group. For example, the group 17 elements are d the *halogens*, a term derived from Greek, meaning "salt former."

ach element is listed in the periodic table by placing its symbol in the mid- of a box in the table. The atomic number (*Z*) of the element is shown above symbol, and the weighted-average atomic mass of the element is shown w its symbol. Some periodic tables provide other information, such as sity and melting point, but the atomic number and atomic mass are gener- sufficient for our needs. Elements with atomic masses in parentheses, n as plutonium, Pu (244), are produced synthetically, and the number vn is the mass number of the most stable isotope.

is customary also to divide the elements into two broad categories—**metals** **nonmetals**. In Figure 2-15, colored backgrounds are used to distinguish the als (tan) from the nonmetals (blue and pink). Except for mercury, a liquid, als are solids at room temperature. They are generally malleable (capable of g flattened into thin sheets), ductile (capable of being drawn into fine wires), d conductors of heat and electricity, and have a lustrous or shiny appearance. properties of nonmetals are generally opposite those of metals; for example, metals are poor conductors of heat and electricity. Several of the nonmetals, as nitrogen, oxygen, and chlorine, are gases at room temperature. Some, as silicon and sulfur, are brittle solids. One—bromine—is a liquid.

wo other highlighted categories in Figure 2-15 are a special group of non- als known as the **noble gases** (pink), and a small group of elements, often ed **metalloids** (green), that have some metallic and some nonmetallic erties.

◀ There is lack of agreement on just which elements to label as metalloids. However, they are generally considered either to lie adjacent to the stair-step line or to be close by.

he *horizontal* rows of the table are called **periods**. (The periods are num- d at the extreme left in the periodic table inside the front cover.) The first od of the table consists of just two elements, hydrogen and helium. This is owed by two periods of eight elements each, lithium through neon and um through argon. The fourth and fifth periods contain 18 elements each, ging from potassium through krypton and from rubidium through xenon. sixth period is a long one of 32 members. To fit this period in a table that eld to a maximum width of 18 members, 15 members of the period are ed at the bottom of the periodic table. This series of 15 elements start with hanum ($Z = 57$), and these elements are called the **lanthanides**. The sev- and final period is incomplete (some members are yet to be discovered), it is known to be a long one. A 15-member series is also extracted from the nth period and placed at the bottom of the table. Because the elements in series start with actinium ($Z = 89$), they are called the **actinides**.

◀ Mendeleev's arrangement of the elements in the original periodic table was based on observed chemical and physical properties of the elements and their compounds. The arrangement of the elements in the modern periodic table is based on atomic properties—atomic number and electron configuration.

he labeling of the groups of the periodic table has been a matter of some ate among chemists. The 1 to 18 numbering system used in Figure 2-15 is one most recently adopted. Group labels previously used in the ted States consisted of a letter and a number, closely following the hod adopted by Mendeleev, the developer of the periodic table. As seen in re 2-15, the A groups 1 and 2 are separated from the remaining A groups 8) by B groups 1 through 8. The International Union of Pure and Applied mistry (IUPAC) recommended the simple 1 to 18 numbering scheme to avoid confusion between the American number and letter system and used in Europe, where some of the A and B designations were switc Currently, the IUPAC system is officially recommended by the Amer Chemical Society (ACS) and chemical societies in other nations. Because numbering systems are in use, we show both in Figure 2-15 and in the peri table inside the front cover. However, except for an occasional reminder o earlier system, we will use the IUPAC numbering system in this text.

Useful Relationships from the Periodic Table

The periodic table helps chemists describe and predict the properties of ch ical compounds and the outcomes of chemical reactions. Throughout this we will use it as an aid to understanding chemical concepts. One applica of the table worth mentioning here is how it can be used to predict li charges on simple monatomic ions.

Main-group elements are those in groups 1, 2, and 13 to 18. When m group metal atoms in groups 1 and 2 form ions, they lose the same nun of electrons as the IUPAC group number. Thus, Na atoms (group 1) lose one tron to become Na^+, and Ca atoms (group 2) lose two electrons to become C Aluminum, in group 13, loses three electrons to form Al^{3+} (here the charg "group number minus 10"). The few other metals in groups 13 and higher more than one possible ion, a matter that we deal with in Chapter 9.

When nonmetal atoms form ions, they gain electrons. The number of trons gained is normally 18 minus the IUPAC group number. Thus, an O a gains $18 - 16 = 2$ electrons to become O^{2-}, and a Cl atom gains $18 - 17$ electron to become Cl^-. The "18 minus group number" rule suggests tha atom of Ne in group 18 gains no electrons: $18 - 18 = 0$. The very limited dency of the noble gas atoms to form ions is one of several characteristic this family of elements.

EXAMPLE 2-7 Describing Relationships Based on the Periodic Table

Refer to the periodic table on the inside front cover, and indicate

(a) the element that is in group 14 and the fourth period;
(b) two elements with properties similar to those of molybdenum (Mo);
(c) the ion most likely formed from a strontium atom.

Analyze

For (a), the key concept is that the rows (periods) are numbered 1 through 7, starting from the top of the per odic table, and the groups are numbered 1 through 18, starting from the left side. For (b), the key concept is th elements in the same group have similar properties. For (c), the key concept is that main-group metal atoms groups 1 and 2 form positive ions with charges of +1 and +2, respectively.

Solve

(a) The elements in the fourth period range from K ($Z = 19$) to Kr ($Z = 36$). Those in group 14 are C, S Ge, Sn, and Pb. The only element that is common to both of these groupings is Ge ($Z = 32$).
(b) Molybdenum is in group 6. Two other members of this group that should resemble it ar chromium (Cr) and tungsten (W).
(c) Strontium (Sr) is in group 2. It should form the ion Sr^{2+}.

Assess

In Chapter 8, we will examine in greater detail reasons for the arrangement of the periodic table.

PRACTICE EXAMPLE A: Write a symbol for the ion most likely formed by an atom of each of the followin Li, S, Ra, F, I, and Al.

PRACTICE EXAMPLE B: Classify each of the following elements as a main-group or transition element. Als specify whether they are metals, metalloids, or nonmetals: Na, Re, S, I, Kr, Mg, U, Si, B, Al, As, H.

ıe elements in groups 3 to 12 are the **transition elements**, and because all ıem are metals, they are also called the **transition metals**. Like the main-ıp metals, the transition metals form positive ions, but the number of elec-s lost is not related in any simple way to the group number, mostly ıuse transition metals can form two or more ions of differing charge.

The Concept of the Mole and the Avogadro Constant

ting with Dalton, chemists have recognized the importance of relative bers of atoms, as in the statement that *two* hydrogen atoms and *one* oxygen ı combine to form *one* molecule of water. Yet it is physically impossible ount every atom in a macroscopic sample of matter. Instead, some other surement must be employed, which requires a relationship between the sured quantity, usually mass, and some known, but uncountable, number oms. Consider a practical example of mass substituting for a desired num-of items. Suppose you want to nail down new floorboards on the deck of a ntain cabin, and you have calculated how many nails you will need. If you an idea of how many nails there are in a pound, then you can buy the nails ıe pound.

ıe SI quantity that describes an amount of substance by relating it to a ber of particles of that substance is called the *mole* (abbreviated *mol*). A e is the amount of a substance that contains the same number of elemen-entities as there are atoms in exactly 12 g of pure carbon-12. The "number lementary entities (atoms, molecules, and so on)" in a mole is the **gadro constant, N_A**.

◀ Because the value of Avogadro's number depends, in part, on a measurement, the value has changed slightly over the years. The values recommended since 1986 by the Committee on Data for Science and Technology (CODATA) are listed below.

Year	Avogadro's Number
1986	6.0221367×10^{23}
1998	$6.02214199 \times 10^{23}$
2002	6.0221415×10^{23}
2006	$6.02214179 \times 10^{23}$
2010	$6.02214129 \times 10^{23}$
2014	$6.022140857 \times 10^{23}$

$$N_A = 6.022140857 \times 10^{23}\ \text{mol}^{-1} \quad \textbf{(2.4)}$$

ıe Avogadro constant consists of a number, $6.022140857 \times 10^{23}$, known as gadro's *number*, and a unit, mol^{-1}. The unit mol^{-1} signifies that the entities g counted are those present in 1 mole.

ıe value of Avogadro's number is based on both a definition and a mea-ment. A mole of carbon-12 is *defined* to be 12 g. If the mass of one carbon-12 ı is *measured* by using a mass spectrometer (see Figure 2-14), the mass ld be about 1.9926×10^{-23} g. The ratio of these two masses provides an nate of Avogadro's number. In actual fact, accurate determinations of gadro's number make use of other measurements, not the measurement of nass of a single atom of carbon-12.

ften the value of N_A is rounded off to $6.022 \times 10^{23}\ \text{mol}^{-1}$, or even to $\times 10^{23}\ \text{mol}^{-1}$.

◀ When rounding Avogadro's number or any other accurately known value, keep one more significant figure than that of the least accurate number in the calculation to avoid rounding errors.

a substance contains atoms of only a single isotope, then

$$1\ \text{mol}\ {}^{12}\text{C} = 6.02214 \times 10^{23}\ {}^{12}\text{C atoms} = 12.0000\ \text{g}$$
$$1\ \text{mol}\ {}^{16}\text{O} = 6.02214 \times 10^{23}\ {}^{16}\text{O atoms} = 15.9949\ \text{g (and so on)}$$

[ost elements are composed of mixtures of two or more isotopes so that the ıs in a sample of the element are not all of the same mass but are present in naturally occurring proportions. Thus, in one mole of carbon, most of the ıs are carbon-12, but some are carbon-13. In one mole of oxygen, most of toms are oxygen-16, but some are oxygen-17 and some are oxygen-18. As ult,

$$1\ \text{mol of C} = 6.02214 \times 10^{23}\ \text{C atoms} = 12.011\ \text{g}$$
$$1\ \text{mol of O} = 6.02214 \times 10^{23}\ \text{O atoms} = 15.999\ \text{g, and so on.}$$

ıe Avogadro constant was purposely chosen so that the mass of one mole of on-12 atoms—exactly 12 g—would have the same *numeric* value as the

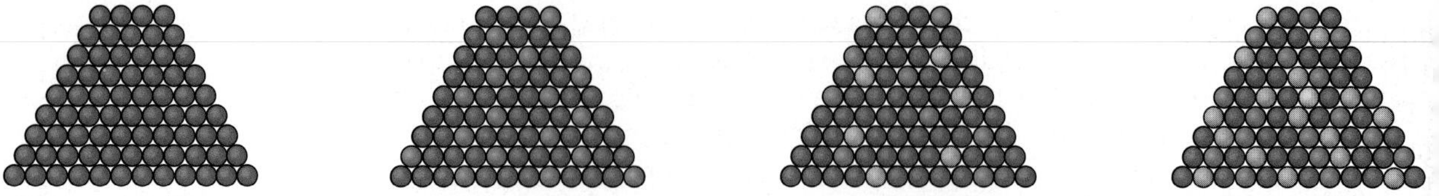

(a) 6.02214×10^{23} F atoms = 18.9984 g
(b) 6.02214×10^{23} Cl atoms = 35.45 g
(c) 6.02214×10^{23} Mg atoms = 24.305 g
(d) 6.02214×10^{23} Pb ato = 207.2 g

▲ FIGURE 2-16
Distribution of isotopes in four elements
(a) There is only one type of fluorine atom, ^{19}F (shown in red). **(b)** In chlorine, 75.77% of the atoms are ^{35}Cl (red) and the remainder are ^{37}Cl (blue). **(c)** Magnesium has one principal isotope, ^{24}Mg (red), and two minor ones, ^{25}Mg (gray) and ^{26}Mg (blue). **(d)** Lead has four naturally occurring isotopes: 1.4% ^{204}Pb (yellow), 24.1% ^{206}Pb (blue), 22.1% ^{207}Pb (gray), and 52.4% ^{208}Pb (red).

▶ The weighted-average atomic mass of carbon was calculated on page 48.

KEEP IN MIND

that molar mass has the unit g/mol.

mass of a single carbon-12 atom—exactly 12 u. As a result, for all other elem the numeric value of the mass in grams of one mole of atoms and the weigh average atomic mass in atomic mass units are equal. For example, weighted-average atomic mass of iron is 55.845 u and the mass of one mo iron atoms is 55.845 g. Thus, we can easily establish the mass of one of atoms, called the **molar mass, *M*,** from a table of atomic masses.* For exar the molar mass of iron is 55.845 g Fe/mol Fe. Figure 2-16 attempts to portra distribution of isotopes of an element, and Figure 2-17 pictures one mole ea four common elements.

2-6 CONCEPT ASSESSMENT

Dividing the molar mass of gold by the Avogadro constant yields the mass of any individual atom of naturally occurring gold. In contrast, no naturally occurring atom of silver has the mass obtained by dividing the molar mass of silver by the Avogadro constant. How can this be?

Thinking About Avogadro's Number

Avogadro's number (6.02214×10^{23}) is an enormously large number practically inconceivable in terms of ordinary experience. Suppose we counting garden peas instead of atoms. If the typical pea had a volun about 0.1 cm^3, the required pile of peas would cover the United States depth of about 6 km (4 mi). Or imagine that grains of wheat could be cou

Carey B. Van Loon

◀ FIGURE 2-17
One mole of an element
The watch glasses contain one mole of copper atoms (and one mole of sulfur atoms (right). The beaker conta one mole of mercury atoms as liquid mercury, and the balloon, of which only a small portion is visible here, contains one mole of helium atoms in the gaseous stat

*Atomic mass (atomic weight) values in tables are often written without units, especially if they are referred to as *relative* atomic masses. This simply means that the values listed are in relation to *exactly* 12 (rather than 12 u) for carbon-12. We will use the atomic mass unit (u) when referring to atomic masses (atomic weights). Most chemists do.

e rate of 100 per minute. A given individual might be able to count out t 4 billion grains in a lifetime. Even so, if all the people currently on Earth to spend their lives counting grains of wheat, they could not reach gadro's number. In fact, if all the people who ever lived on Earth had t their lifetimes counting grains of wheat, the total would still be far less Avogadro's number. (And Avogadro's number of wheat grains is far wheat than has been produced in human history.) Now consider a much efficient counting device, a modern personal computer; it is capable of ting at a rate of about 1 billion units per second. The task of counting out gadro's number would still take about 20 million years!

vogadro's number is clearly not a useful number for counting ordinary cts. However, when this inconceivably large number is used to count nceivably small objects, such as atoms and molecules, the result is a quanof material that is easily within our grasp, essentially a "handful."

▲ FIGURE 2-18 **Measurement of 7.64×10^{22} S atoms (0.127 mol S)—Example 2-8 illustrated**
The balance was set to zero (tared) when just the weighing boat was present. The sample of sulfur weighs 4.07 g.

Richard Megna/Fundamental Photographs

Using the Mole Concept in Calculations

ughout the text, the mole concept will provide conversion factors for lem-solving situations. With each new situation, we will explore how the concept applies. For now, we will deal with the relationship between numof atoms and the mole. Consider the statement: 1 mol S = 6.022×10^{23} ms = 32.06 g S. This allows us to write the conversion factors

$$\frac{1 \text{ mol S}}{6.022 \times 10^{23} \text{ S atoms}} \quad \text{and} \quad \frac{32.06 \text{ g S}}{1 \text{ mol S}}$$

calculations requiring the Avogadro constant, students often ask when to iply and when to divide by N_A. One answer is always to use the constant way that gives the proper cancellation of units. Another answer is to think rms of the expected result. In calculating a number of atoms, we expect nswer to be a very large number and certainly *never* smaller than one. The ber of moles of atoms, conversely, is generally a number of more modest and will often be less than one.

the following examples, we use atomic masses and the Avogadro constant lculations to determine the number of atoms present in a given sample. nic masses and the Avogadro constant are known rather precisely, and stus often wonder how many significant figures to carry in atomic masses or the gadro constant when performing calculations. Here is a useful rule of thumb.

ensure the maximum precision allowable, carry at least one more significant gure in well-known physical constants than in other measured quantities.

r example, in calculating the mass of 0.600 mol of sulfur, we should use the ic mass of S with *at least* four significant figures. The answer 0.600 mol S × 5 g S/mol S = 19.2 g S is a more precise response than 0.600 mol S × g S/mol S = 19.3 g S.

XAMPLE 2-8 Relating Number of Atoms, Amount in Moles, and Mass in Grams

In the sample of sulfur weighing 4.07 g pictured in Figure 2-18, **(a)** how many moles of sulfur are present, and **(b)** what is the total number of sulfur atoms in the sample?

nalyze

For (a), the conversion pathway is g S → mol S. To carry out this conversion, we multiply 4.07 g S by the conversion factor (1 mol S/32.06 g S). The conversion factor is the molar mass inverted. For (b), the conversion

(continued)

pathway is mol S → atoms S. To carry out this conversion, we multiply the quantity in moles from part (a) the conversion factor (6.022×10^{23} atoms S/1 mol S).

Solve

(a) For the conversion g S → mol S, using $(1/M)$ as a conversion factor achieves the proper cancellation units. The result of this calculation should be stored without rounding it off because it is required in part (

$$? \text{ mol S} = 4.07 \text{ g S} \times \frac{1 \text{ mol S}}{32.06 \text{ g S}} = 0.127 \text{ mol S}$$

(b) The conversion mol S → atoms S is carried out using the Avogadro constant as a conversion factor.

$$? \text{ atoms S} = 0.127 \text{ mol S} \times \frac{6.022 \times 10^{23} \text{ atoms S}}{1 \text{ mol S}} = 7.64 \times 10^{22} \text{ atoms S}$$

Assess

By including units in our calculations, we can check that proper cancellation of units occurs. Also, if our or concern is to calculate the number of sulfur atoms in the sample, the calculations carried out in parts (a) and (could be combined into a single calculation, as shown below.

$$? \text{ atoms S} = 4.07 \text{ g S} \times \frac{1 \text{ mol S}}{32.06 \text{ g S}} \times \frac{6.022 \times 10^{23} \text{ atoms S}}{1 \text{ mol S}} = 7.64 \times 10^{22} \text{ atoms S}$$

Had we rounded 4.07 g S × (1 mol S/32.06 g S) to 0.127 mol S and used the rounded result in part (b), we wou have obtained a final answer of 7.65×10^{22} atoms S. With a single line calculation, we do not have to write dov an intermediate result and we avoid round-off errors.

PRACTICE EXAMPLE A: What is the mass of 2.35×10^{24} atoms of Cu?

PRACTICE EXAMPLE B: How many lead-206 atoms are present in a 22.6 g sample of lead metal? [*Hint:* S Figure 2-16.]

Example 2-9 is perhaps the most representative use of the mole con Here it is part of a larger problem that requires other unrelated conversio tors as well. One approach is to outline a conversion pathway to get fro given to the desired information.

EXAMPLE 2-9 Combining Several Factors in a Calculation—Molar Mass, the Avogadro Constant, Percent Abundances

Potassium-40 is one of the few naturally occurring radioactive isotopes of elements of low atomic number. percent isotopic abundance among K isotopes is 0.012%. How many ^{40}K atoms are present in 225 mL of wh milk containing 1.65 mg K/mL?

Analyze

Ultimately we need to complete the conversion mL milk → atoms ^{40}K. There is no single conversion factor th allows us to complete this conversion in one step, so we anticipate having to complete several steps or conversio We are told the milk contains 1.65 mg K/mL = 1.65×10^{-3} g K/mL, and this information can be used to car out the conversion mL milk → g K. We can carry out the conversions g K → mol K → atoms K by using co version factors based on the molar mass of K and the Avogadro constant. The final conversion, atoms K - atoms ^{40}K, can be carried out by using a conversion factor based on the percent isotopic abundance of ^{40} A complete conversion pathway is shown below:

$$\text{mL milk} \rightarrow \text{mg K} \rightarrow \text{g K} \rightarrow \text{mol K} \rightarrow \text{atoms K} \rightarrow \text{atoms } ^{40}\text{K}$$

Solve

The required conversions can be carried out in a stepwise fashion, or they can be combined into a single li calculation. Let's use a stepwise approach. First, we convert from mL milk to g K.

$$? \text{ g K} = 225 \text{ mL milk} \times \frac{1.65 \text{ mg K}}{1 \text{ mL milk}} \times \frac{1 \text{ g K}}{1000 \text{ mg K}} = 0.371 \text{ g K}$$

Next, we convert from g K to mol K,

$$? \text{ mol K} = 0.371 \text{ g K} \times \frac{1 \text{ mol K}}{39.10 \text{ g K}} = 9.49 \times 10^{-3} \text{ mol K}$$

and then we convert from mol K to atoms K.

$$? \text{ atoms K} = 9.49 \times 10^{-3} \text{ mol K} \times \frac{6.022 \times 10^{23} \text{ atoms K}}{1 \text{ mol K}} = 5.71 \times 10^{21} \text{ atoms K}$$

Finally, we convert from atoms K to atoms ^{40}K.

$$? \text{ atoms } ^{40}\text{K} = 5.71 \times 10^{21} \text{ atoms K} \times \frac{0.012 \text{ atoms } ^{40}\text{K}}{100 \text{ atoms K}} = 6.9 \times 10^{17} \text{ atoms } ^{40}\text{K}$$

ssess

The final answer is rounded to two significant figures because the least precisely known quantity in the calculation, the percent isotopic abundance of ^{40}K, has two significant figures. It is possible to combine the steps above into a single line calculation.

$$? \text{ atoms } ^{40}\text{K} = 225 \text{ mL milk} \times \frac{1.65 \text{ mg K}}{1 \text{ mL milk}} \times \frac{1 \text{ g K}}{1000 \text{ mg K}} \times \frac{1 \text{ mol K}}{39.10 \text{ g K}}$$

$$\times \frac{6.022 \times 10^{23} \text{ atoms K}}{1 \text{ mol K}} \times \frac{0.012 \text{ atoms } ^{40}\text{K}}{100 \text{ atoms K}}$$

$$= 6.9 \times 10^{17} \text{ atoms } ^{40}\text{K}$$

RACTICE EXAMPLE A: How many Pb atoms are present in a small piece of lead with a volume of 0.105 cm^3? The density of Pb = 11.34 g/cm^3.

RACTICE EXAMPLE B: Rhenium-187 is a radioactive isotope that can be used to determine the age of meteorites. A 0.100 mg sample of Re contains 2.02×10^{17} atoms of ^{187}Re. What is the percent isotopic abundance of rhenium-187 in the sample?

asteringCHEMISTRY **www.masteringchemistry.com**

'hat is the most abundant element? This seemingly simple question does not have a simple answer. To arn more about the abundances of elements in the universe and in the Earth's crust, go to the Focus n feature for Chapter 2, entitled Occurrence and Abundances of the Elements, on the asteringChemistry site.

Summary

Early Chemical Discoveries and the Atomic ory—Modern chemistry began with eighteenth-ury discoveries leading to the formulation of two laws of chemical combination, the **law of conserva- of mass** and the **law of constant composition (defi- proportions)**. These discoveries led to Dalton's ic theory—that matter is composed of indestructible cles called atoms, that the atoms of an element are tical to one another but different from atoms of all r elements, and that chemical compounds are combi- ons of atoms of different elements. Based on this ry, Dalton proposed still another law of chemical bination, the **law of multiple proportions**.

Electrons and Other Discoveries in Atomic sics—The first clues to the structures of atoms came ıgh the discovery and characterization of **cathode rays trons)**. Key experiments were those that established the mass-to-charge ratio (Fig. 2-7) and then the charge on an electron (Fig. 2-8). Two important accidental discoveries made in the course of cathode-ray research were of X-rays and **radioactivity**. The principal types of radiation emitted by radioactive substances are **alpha (α) particles**, **beta (β) particles**, and **gamma (γ) rays** (Fig. 2-10).

2-3 The Nuclear Atom—Studies on the scattering of α particles by thin metal foils (Fig. 2-11) led to the concept of the nuclear atom—a tiny, but massive, positively charged nucleus surrounded by lightweight, negatively charged electrons (Fig. 2-12). A more complete description of the nucleus was made possible by the discovery of **protons** and **neutrons**. An individual atom is characterized in terms of its **atomic number (proton number) Z** and **mass number, A**. The difference, $A - Z$, is the **neutron number**. The masses of individual atoms and their component parts are expressed in **atomic mass units (u)**.

2-4 Chemical Elements—All elements from Z = 1 to Z = 116, except for elements with Z = 113 and Z = 115, have been characterized and all have been given a name and **chemical symbol**. Knowledge of the several elements following Z = 112 is more tenuous. **Nuclide** is the term used to describe an atom with a particular atomic number and a particular mass number. Atoms of the same element that differ in mass number are called **isotopes**. The **percent isotopic abundance** of an isotope and the precise mass of its atoms can be established with a **mass spectrometer** (Fig. 2-14). A special symbolism (expression 2.2) is used to represent the composition of an atom or an **ion** derived from the atom.

2-5 Atomic Mass—The **atomic mass (weight)** of an element is a weighted average based on an assigned value of exactly 12 u for the isotope carbon-12. This weighted average is calculated from the experimentally determined atomic masses and percent abundances of the naturally occurring isotopes of the element through expression (2.3). For certain elements, an **atomic mass interval** (Table 2.2) is used to indicate the range of values expected for the atomic mass because of observed variations in the isotopic abundances of these elements. For these elements, the **conventional atomic mass** can be used when a representative value of the atomic mass is required.

2-6 Introduction to the Periodic Table—The **periodic table** (Fig. 2-15) is an arrangement of the elements in horizontal rows called **periods** and ver columns called **groups** or **families**. Each group con of elements with similar physical and chemical prc ties. The elements can also be subdivided into broad egories. One categorization is that of **metals, nonme metalloids**, and **noble gases**. Another is that of **m group elements** and **transition elements (transi metals)**. Included among the transition elements ar two subcategories **lanthanides** and **actinides**. The has many uses, as will be seen throughout the Emphasis in this chapter is on the periodic table as a in writing symbols for simple ions.

2-7 The Concept of the Mole and Avogadro Constant—The **Avogadro cons** $N_A = 6.02214 \times 10^{23}\ \text{mol}^{-1}$, represents the numbe carbon-12 atoms in *exactly* 12 g of carbon-12. More ge ally, it is the number of elementary entities (for exan atoms or molecules) present in an amount known as **mole** of substance. The mass of one mole of atoms (element is called its **molar mass**, *M*.

2-8 Using the Mole Concept in Calculatio Molar mass and the Avogadro constant are used variety of calculations involving the mass, amoun moles), and number of atoms in a sample of an elen Other conversion factors may also be involved in t calculations. The mole concept is encountered in broader contexts throughout the text.

Integrative Example

A stainless steel ball bearing has a radius of 6.35 mm and a density of 7.75 g/cm^3. Iron is the principal element in s Carbon is a key minor element. The ball bearing contains 0.25% carbon, by mass. Assuming that the percent iso abundance of ^{13}C is 1.108%, how many ^{13}C atoms are present in the ball bearing?

Analyze

The goal is to determine the number of carbon-13 atoms found in a ball bearing with a particular composition. critical point in this problem is recognizing that we can relate number of atoms to mass by using molar mass Avogadro's constant. The first step is to use the radius of the ball bearing to determine its volume. The second st to determine the mass of carbon present by using the density of steel along with the percent composition. The step uses the molar mass of carbon to convert grams of carbon to moles of carbon; Avogadro's constant is then us convert moles of carbon to the number of carbon atoms. In the final step, the isotopic abundance of carbon-13 ato used to find the number of carbon-13 atoms in the total number of carbon atoms in the ball bearing.

Solve

The ball-bearing volume in cubic centimeters is found by applying the formula for the volume of a sphere, $V = 4/3\,\pi r^3$. Remember to convert the given radius from millimeters to centimeters, so that the volume will be in cubic centimeters.

$$V = \frac{4\pi}{3}\left[6.35\ \text{mm} \times \frac{1\ \text{cm}}{10\ \text{mm}}\right]^3 = 1.07\ \text{cm}^3$$

The product of the volume of the ball bearing and the density of steel equals the mass. The mass of the ball bearing multiplied by the percent carbon in the steel gives the mass of carbon present.

$$?\ \text{g C} = 1.07\ \text{cm}^3 \times \frac{7.75\ \text{g steel}}{1\ \text{cm}^3\ \text{steel}} \times \frac{0.25\ \text{g C}}{100\ \text{g steel}} = 0.021\ \text{g C}$$

The mass of carbon is first converted to moles of carbon by using the inverse of the molar mass of carbon. Avogadro's constant is then used to convert moles of carbon to atoms of carbon.

$$?\ \text{C atoms} = 0.021\ \text{g C} \times \frac{1\ \text{mol C}}{12.011\ \text{g C}} \times \frac{6.022 \times 10^{23}\ \text{C ato}}{1\ \text{mol C}}$$

$$= 1.1 \times 10^{21}\ \text{C atoms}$$

The number of ^{13}C atoms is determined by using the percent isotopic abundance of carbon-13.

$$?\ ^{13}\text{C atoms} = 1.1 \times 10^{21}\ \text{C atoms} \times \frac{1.108\ ^{13}\text{C atoms}}{100\ \text{C atoms}}$$

$$= 1.2 \times 10^{19}\ ^{13}\text{C atoms}$$

ess

number of carbon-13 atoms is smaller than the number of carbon atoms, which it should be, given that the isotopic dance of carbon-13 is just 1.108%. To avoid mistakes, every quantity should be clearly labeled with its appropriate so that units cancel properly. Two points made by this problem are, first, that the relatively small ball bearing con- a large number of carbon-13 atoms even though carbon-13's abundance is only 1.108% of all carbon atoms. Second, ize of any atom must be very small.

CTICE EXAMPLE A: Calculate the number of ^{63}Cu atoms in a cubic crystal of copper that measures exactly n on edge. The density of copper is 8.92 g/cm^3 and the percent isotopic abundance of ^{63}Cu is 69.17%.

CTICE EXAMPLE B: The United States Food and Drug Administration (USFDA) suggests a daily value of 18 mg Fe dults and for children over four years of age. The label on a particular brand of cereal states that one serving (55 g) of ereal contains 45% of the daily value of Fe. Given that the percent isotopic abundance of ^{58}Fe is 0.282%, how many full ngs of dry cereal must be eaten to consume exactly one mole of ^{58}Fe? The atomic weight of ^{58}Fe is 57.9333 u. Is it possible person to consume this much cereal in a lifetime, assuming that one full serving of cereal is eaten every day?

Exercises

of Conservation of Mass

When an iron object rusts, its mass increases. When a natch burns, its mass decreases. Do these observa- ions violate the law of conservation of mass? Explain.

Vhen a strip of magnesium metal is burned in air recall Figure 2-1), it produces a white powder that veighs more than the original metal. When a strip of nagnesium is burned in a flashbulb, the bulb weighs he same before and after it is flashed. Explain the lifference in these observations.

A 0.406 g sample of magnesium reacts with oxygen, producing 0.674 g of magnesium oxide as the only product. What mass of oxygen was consumed in the eaction?

A 1.446 g sample of potassium reacts with 8.178 g of chlorine to produce potassium chloride as the only product. After the reaction, 6.867 g of chlorine remains unreacted. What mass of potassium chloride was formed?

5. When a solid mixture consisting of 10.500 g calcium hydroxide and 11.125 g ammonium chloride is strongly heated, gaseous products are evolved and 14.336 g of a solid residue remains. The gases are passed into 62.316 g water, and the mass of the resulting solution is 69.605 g. Within the limits of experimental error, show that these data conform to the law of conservation of mass.

6. Within the limits of experimental error, show that the law of conservation of mass was obeyed in the following experiment: 10.00 g calcium carbonate (found in limestone) was dissolved in 100.0 mL hydrochloric acid (d = 1.148 g/mL). The products were 120.40 g solution (a mixture of hydrochloric acid and calcium chloride) and 2.22 L carbon dioxide gas (d = 1.9769 g/L).

of Constant Composition

n Example 2-1, we established that the mass ratio of nagnesium to magnesium oxide is 0.455 g magnesium/ 0.755 g magnesium oxide.

a) What is the ratio of oxygen to magnesium oxide, by mass?

b) What is the mass ratio of oxygen to magnesium in nagnesium oxide?

c) What is the percent by mass of magnesium in nagnesium oxide?

Samples of pure carbon weighing 3.62, 5.91, and 7.07 g vere burned in an excess of air. The masses of carbon lioxide obtained (the sole product in each case) were 13.26, 21.66, and 25.91 g, respectively.

a) Do these data establish that carbon dioxide has a ixed composition?

b) What is the composition of carbon dioxide, expressed in % C and % O, by mass?

n one experiment, 2.18 g sodium was allowed to eact with 16.12 g chlorine. All the sodium was used up, and 5.54 g sodium chloride (salt) was produced. n a second experiment, 2.10 g chlorine was allowed o react with 10.00 g sodium. All the chlorine was ised up, and 3.46 g sodium chloride was produced. Show that these results are consistent with the law of constant composition.

10. When 3.06 g hydrogen was allowed to react with an excess of oxygen, 27.35 g water was obtained. In a second experiment, a sample of water was decomposed by electrolysis, resulting in 1.45 g hydrogen and 11.51 g oxygen. Are these results consistent with the law of constant composition? Demonstrate why or why not.

11. In one experiment, the burning of 0.312 g sulfur produced 0.623 g sulfur dioxide as the sole product of the reaction. In a second experiment, 0.842 g sulfur dioxide was obtained. What mass of sulfur must have been burned in the second experiment?

12. In one experiment, the reaction of 1.00 g mercury and an excess of sulfur yielded 1.16 g of a sulfide of mercury as the sole product. In a second experiment, the same sulfide was produced in the reaction of 1.50 g mercury and 1.00 g sulfur.
    (a) What mass of the sulfide of mercury was produced in the second experiment?
    (b) What mass of which element (mercury or sulfur) remained *unreacted* in the second experiment?

Law of Multiple Proportions

13. Sulfur forms two compounds with oxygen. In the first compound, 1.000 g sulfur is combined with 0.998 g oxygen, and in the second, 1.000 g sulfur is combined with 1.497 g oxygen. Show that these results are consistent with Dalton's law of multiple proportions.

14. Phosphorus forms two compounds with chlorine. In the first compound, 1.000 g of phosphorus is combined with 3.433 g chlorine, and in the second, 2.500 g phosphorus is combined with 14.308 g chlorine. Show that these results are consistent with Dalton's law of multiple proportions.

15. The following data were obtained for compounds of nitrogen and hydrogen:

Compound	Mass of Nitrogen, g	Mass of Hydrogen, g
A	0.500	0.108
B	1.000	0.0720
C	0.750	0.108

(a) Show that these data are consistent with the law of multiple proportions.
(b) If the formula of compound B is N_2H_2, what are the formulas of compounds A and C?

16. The following data were obtained for compoun iodine and fluorine:

Compound	Mass of Iodine, g	Mass of Fluorine, g
A	1.000	0.1497
B	0.500	0.2246
C	0.750	0.5614
D	1.000	1.0480

(a) Show that these data are consistent with th of multiple proportions.
(b) If the formula for compound A is IF, what a formulas for compounds B, C, and D?

17. There are two oxides of copper. One oxide has oxygen, by mass. The second oxide has a *smalle* cent of oxygen than the first. What is the pro percent of oxygen in the second oxide?

18. The two oxides of carbon described on page 38 CO and CO_2. Another oxide of carbon has 1.10 oxygen in a 2.350 g sample. In what ratio are ca and oxygen atoms combined in molecules o third oxide? Explain.

Fundamental Charges and Mass-to-Charge Ratios

19. The following observations were made for a series of five oil drops in an experiment similar to Millikan's (see Figure 2-8). Drop 1 carried a charge of 1.28×10^{-18} C; drops 2 and 3 each carried $\frac{1}{2}$ the charge of drop 1; drop 4 carried $\frac{1}{8}$ the charge of drop 1; drop 5 had a charge four times that of drop 1. Are these data consistent with the value of the electronic charge given in the text? Could Millikan have inferred the charge on the electron from this particular series of data? Explain.

20. In an experiment similar to that described in Exercise 19, drop 1 carried a charge of 6.41×10^{-19} C; drop 2 had $\frac{1}{2}$ the charge of drop 1; drop 3 had twice the charge of drop 1; drop 4 had a charge of 1.44×10^{-18} C; and drop 5 had $\frac{1}{3}$ the charge of d Are these data consistent with the value of the tronic charge given in the text? Could Millikan inferred the charge on the electron from this pa lar series of data? Explain.

21. Use data from Table 2.1 to verify that
(a) the mass of electrons is about 1/2000 th H atoms;
(b) the mass-to-charge ratio (m/e) for positive ic considerably larger than that for electrons.

22. Determine the approximate value of m/e in gram coulomb for the ions $^{127}_{53}I^-$ and $^{32}_{16}S^{2-}$. Why are values only approximate?

Atomic Number, Mass Number, and Isotopes

23. The following radioactive isotopes have applications in medicine. Write their symbols in the form ^A_ZE. **(a)** cobalt-60; **(b)** phosphorus-32; **(c)** iron-59; **(d)** radium-226.

24. For the isotope ^{202}Hg, express the percentage of the fundamental particles in the nucleus that are neutrons.

25. Complete the following table. What mini amount of information is required to completely acterize an atom or ion?
[*Hint:* Not all rows can be completed.]

Name	Symbol	Number Protons	Number Electrons	Number Neutrons	Mass Numb
Sodium	$^{23}_{11}Na$	11	11	12	23
Silicon	—	—	—	14	—
—	—	37	—	—	85
—	^{40}K	—	—	—	—
—	—	—	33	42	—
—	$^{20}Ne^{2+}$	—	—	—	—
—	—	—	—	—	80
—	—	—	—	126	—

Arrange the following species in order of increasing (a) number of electrons; (b) number of neutrons; (c) mass.

$^{112}_{50}Sn$ $^{40}_{18}Ar$ $^{122}_{52}Te$ $^{59}_{29}Cu$ $^{120}_{48}Cd$ $^{58}_{27}Co$ $^{39}_{19}K$

For the atom ^{108}Pd with mass 107.90389 u, determine (a) the numbers of protons, neutrons, and electrons in the atom; (b) the ratio of the mass of this atom to that of an atom of $^{12}_{6}C$.

For the ion $^{228}Ra^{2+}$ with a mass of 228.030 u, determine (a) the numbers of protons, neutrons, and electrons in the ion; (b) the ratio of the mass of this ion to that of an atom of ^{16}O (refer to page 47).

An isotope of silver has a mass that is 6.68374 times that of oxygen-16. What is the mass in u of this isotope? (Refer to page 47.)

The ratio of the masses of the two naturally occurring isotopes of indium is 1.0177:1. The heavier of the two isotopes has 7.1838 times the mass of ^{16}O. What are the masses in u of the two isotopes? (Refer to page 47.)

The following data on isotopic masses are from a chemical handbook. What is the ratio of each of these masses to that of $^{12}_{6}C$? **(a)** $^{35}_{17}Cl$, 34.96885 u; **(b)** $^{26}_{12}Mg$, 25.98259 u; **(c)** $^{222}_{86}Rn$, 222.0175 u.

The following ratios of masses were obtained with a mass spectrometer: $^{19}_{9}F/^{12}_{6}C = 1.5832$; $^{35}_{17}Cl/^{19}_{9}F = 1.8406$; $^{81}_{35}Br/^{35}_{17}Cl = 2.3140$. Determine the mass of a $^{81}_{35}Br$ atom in amu.

Which of the following species has (a) equal numbers of neutrons and electrons; (b) protons, neutrons, and electrons in the ratio 9:11:8; (c) a number of neutrons equal to the number of protons plus one-half the number of electrons?

$^{24}Mg^{2+}$, ^{47}Cr, $^{60}Co^{3+}$, $^{35}Cl^{-}$, $^{124}Sn^{2+}$, ^{226}Th, ^{90}Sr

34. Given the same species as listed in Exercise 33, which has
(a) equal numbers of neutrons and protons;
(b) protons contributing more than 50% of the mass;
(c) about 50% more neutrons than protons?

35. An isotope with mass number 44 has four more neutrons than protons. This is an isotope of what element?

36. Identify the isotope X that has one more neutron than protons and a mass number equal to nine times the charge on the ion X^{3+}.

37. Iodine has many radioactive isotopes. Iodine-123 is a radioactive isotope used for obtaining images of the thyroid gland. Iodine-123 is administered to patients in the form of sodium iodide capsules that contain $^{123}I^{-}$ ions. Determine the number of neutrons, protons, and electrons in a single $^{123}I^{-}$ ion.

38. Iodine-131 is a radioactive isotope that has important medical uses. Small doses of iodine-131 are used for treating hyperthyroidism (overactive thyroid) and larger doses are used for treating thyroid cancer. Iodine-131 is administered to patients in the form of sodium iodide capsules that contain $^{131}I^{-}$ ions. Determine the number of neutrons, protons, and electrons in a single $^{131}I^{-}$ ion.

39. Americium-241 is a radioactive isotope that is used in high-precision gas and smoke detectors. How many neutrons, protons, and electrons are there in an atom of americium-241?

40. Some foods are made safer to eat by being exposed to gamma rays from radioactive isotopes, such as cobalt-60. The energy from the gamma rays kills bacteria in the food. How many neutrons, protons, and electrons are there in an atom of cobalt-60?

Atomic Mass Units, Atomic Masses

Which statement is probably true concerning the masses of *individual* chlorine atoms: *All have, some have,* or *none has* a mass of 35.45 u? Explain.

The mass of a carbon-12 atom is taken to be exactly 12 u. Are there likely to be any other atoms with an *exact* integral (whole number) mass, expressed in u? Explain.

Magnesium has three naturally occurring isotopes. Their masses are 23.985042 u, 24.985837 u, and 25.982593 u. What is the weighted-average atomic mass of magnesium in a sample for which the percent isotopic abundances of these three isotopes are 78.99%, 10.00%, and 11.01%, respectively?

There are four naturally occurring isotopes of chromium. Their masses and percent isotopic abundances are 49.9461 u, 4.35%; 51.9405 u, 83.79%; 52.9407 u, 9.50%; and 53.9389 u, 2.36%. Calculate the weighted-average atomic mass of chromium.

The two naturally occurring isotopes of silver have the following abundances: ^{107}Ag, 51.84%; ^{109}Ag, 48.16%. The mass of ^{107}Ag is 106.905092 u. What is the mass of ^{109}Ag?

46. Gallium has two naturally occurring isotopes. One of them, gallium-69, has a mass of 68.925581 u and a percent isotopic abundance of 60.11%. What must be the mass and percent isotopic abundance of the other isotope, gallium-71?

47. The three naturally occurring isotopes of potassium are ^{39}K, 38.963707 u; ^{40}K, 39.963999 u; and ^{41}K. The percent isotopic abundances of ^{39}K and ^{41}K are 93.2581% and 6.7302%, respectively. Determine the isotopic mass of ^{41}K.

48. Use the conventional atomic mass of boron to estimate the fractional isotopic abundances of the two naturally occurring isotopes, ^{10}B and ^{11}B. These isotopes have masses of 10.012937 u and 11.009305 u, respectively.

Mass Spectrometry

49. A mass spectrum of germanium displayed peaks at mass numbers 70, 72, 73, 74, and 76, with relative heights of 20.5, 27.4, 7.8, 36.5, and 7.8, respectively.
(a) In the manner of Figure 2-14, sketch this mass spectrum.
(b) Estimate the weighted-average atomic mass of germanium, and state why this result is only approximately correct.
50. Hydrogen and chlorine atoms react to form simple diatomic molecules in a 1:1 ratio, that is, HCl. The percent isotopic abundances of the chlorine isot are ^{35}Cl and ^{37}Cl are estimated to be 75.77% 24.23%, respectively. The percent isotopic a dances of ^{2}H and ^{3}H are estimated to be 0.015% less than 0.001%, respectively.
(a) How many different HCl molecules are pos and what are their mass numbers (that is, the su the mass numbers of the H and Cl atoms)?
(b) Which is the most abundant of the possible molecules? Which is the second most abundant?

The Periodic Table

51. Refer to the periodic table inside the front cover and identify
(a) the element that is in group 14 and the fourth period
(b) one element similar to and one unlike sulfur
(c) the alkali metal in the fifth period
(d) the halogen element in the sixth period
52. Refer to the periodic table inside the front cover and identify
(a) the element that is in group 11 and the sixth period
(b) an element with atomic number greater than 50 that has properties similar to the element with atomic number 18
(c) the group number of an element E that forn ion E^{2-}
(d) an element M that you would expect to forr ion M^{3+}
53. Assuming that the seventh period of the per table has 32 members, what should be the at number of (a) the noble gas following radon (b) the alkali metal following francium (Fr)?
54. Find the several pairs of elements that are "o order" in terms of increasing atomic mass and ex why the reverse order is necessary.

The Avogadro Constant and the Mole

55. What is the total number of atoms in (a) 15.8 mol Fe; (b) 0.000467 mol Ag; (c) 8.5×10^{-11} mol Na?
56. *Without doing detailed calculations*, indicate which of the following quantities contains the greatest number of atoms: 6.022×10^{23} Ni atoms, 25.0 g nitrogen, 52.0 g Cr, 10.0 cm^3 Fe ($d = 7.86$ g/cm^3). Explain your reasoning.
57. Determine
(a) the number of moles of Zn in a 415.0 g sample of zinc metal
(b) the number of Cr atoms in 147.4 kg chromium
(c) the mass of a one-trillion-atom (1.0×10^{12}) sample of metallic gold
(d) the average mass of a fluorine atom
58. Determine
(a) the number of Kr atoms in a 5.25 mg sample of krypton
(b) the molar mass, *M*, and identity of an element if the mass of a 2.80×10^{22} atom sample of the element is 2.09 g
(c) the mass of a sample of phosphorus that contains the same number of atoms as 44.75 g of magnesium
59. How many Cu atoms are present in a piece of sterling-silver jewelry weighing 33.24 g? (Sterling silver is a silver–copper alloy containing 92.5% Ag by mass.)
60. How many atoms are present in a 50.0 cm^3 sample of plumber's solder, a lead–tin alloy containing 67% Pb by mass and having a density of 9.4 g/cm^3?
61. How many ^{204}Pb atoms are present in a piece of lead weighing 215 mg? The percent isotopic abundance of ^{204}Pb is 1.4%.
62. A particular lead–cadmium alloy is 8.0% cadmiu mass. What mass of this alloy, in grams, mus weigh out to obtain a sample containing 7.25 × 10 atoms?
63. Medical experts generally believe a level of 30 μ per deciliter of blood poses a significant health (1 dL = 0.1 L). Express this level (a) in the uni Pb/L blood; (b) as the number of Pb atoms per liter of blood.
64. During a severe episode of air pollution, the centration of lead in the air was observed 3.11 μg Pb/m^3. How many Pb atoms would be pr in a 0.500 L sample of this air (the volume of ai placed in the lungs between inhaling and exhalin
65. *Without doing detailed calculations*, determine whi the following samples has the greatest numb atoms:
(a) a cube of iron with a length of 10.C ($d = 7.86$ g/cm^3)
(b) 1.00 kg of hydrogen contained in a 10,0 balloon
(c) a mound of sulfur weighing 20.0 kg
(d) a 76 lb sample of liquid mercury ($d = 13.5$ g,
66. *Without doing detailed calculations*, determine whi the following samples occupies the largest volun
(a) 25.5 mol of sodium metal ($d = 0.971$ g/cm^3)
(b) 0.725 L of liquid bromine ($d = 3.12$ g/mL)
(c) 1.25×10^{25} atoms of chromium metal ($d = 9.4$ g/cm^3)
(d) 2.15 kg of plumber's solder ($d = 9.4$ g/cm lead–tin alloy with a 2:1 atom ratio of lead to tin

Integrative and Advanced Exercises

A solution was prepared by dissolving 2.50 g potassium chlorate (a substance used in fireworks and flares) in 100.0 mL of water at 40 °C. When the solution was cooled to 20 °C, its volume was still found to be 100.0 mL, but some of the potassium chlorate had crystallized (deposited from the solution as a solid). At 40 °C, the density of water is 0.9922 g/mL, and at 20 °C, the potassium chlorate solution had a density of 1.0085 g/mL.
(a) Estimate, to two significant figures, the mass of potassium chlorate that crystallized.
(b) Why can't the answer in **(a)** be given more precisely?

William Prout (1815) proposed that all other atoms are built up of hydrogen atoms, suggesting that all elements should have integral atomic masses based on an atomic mass of one for hydrogen. This hypothesis appeared discredited by the discovery of atomic masses, such as 24.3 u for magnesium and 35.5 u for chlorine. In terms of modern knowledge, explain why Prout's hypothesis is actually quite reasonable.

Fluorine has a single atomic species, ^{19}F. Determine the atomic mass of ^{19}F by summing the masses of its protons, neutrons, and electrons, and compare your results with the value listed on the inside front cover. Explain why the agreement is poor.

Use 1×10^{-13} cm as the approximate diameter of the spherical nucleus of the hydrogen-1 atom, together with data from Table 2.1, to estimate the density of matter in a proton.

Use fundamental definitions and statements from Chapters 1 and 2 to establish the fact that 6.022×10^{23} u = 1.000 g.

In each case, identify the element in question.
(a) The mass number of an atom is 234, and the atom has 60.0% more neutrons than protons.
(b) An ion with a +2 charge has 10.0% more protons than electrons.
(c) An ion with a mass number of 110 and a 2+ charge has 25.0% more neutrons than electrons.

Determine the only possible +2 ion for which the following two conditions are both satisfied:
- The net ionic charge is *one-tenth* the nuclear charge.
- The number of neutrons is *four* more than the number of electrons.

Determine the only possible isotope (E) for which the following conditions are met:
- The mass number of E is 2.50 times its atomic number.
- The atomic number of E is equal to the mass number of another isotope (Y). In turn, isotope Y has a neutron number that is 1.33 times the atomic number of Y and equal to the neutron number of selenium-82.

Suppose we redefined the atomic mass scale by arbitrarily assigning to the naturally occurring *mixture* of chlorine isotopes an atomic mass of 35.00000 u.
(a) What would be the atomic masses of helium, sodium, and iodine on this new atomic mass scale?
(b) Why do these three elements have nearly integral (whole-number) atomic masses based on carbon-12, but not based on naturally occurring chlorine?

76. The two naturally occurring isotopes of nitrogen have masses of 14.0031 and 15.0001 u, respectively. Use the conventional atomic mass of nitrogen to estimate the percentage of ^{15}N atoms in naturally occurring nitrogen.

77. The masses of the naturally occurring mercury isotopes are ^{196}Hg, 195.9658 u; ^{198}Hg, 197.9668 u; ^{199}Hg, 198.9683 u; ^{200}Hg, 199.9683 u; ^{201}Hg, 200.9703 u; ^{202}Hg, 201.9706 u; and ^{204}Hg, 203.9735 u. Use these data, together with data from Figure 2-14, to calculate the weighted-average atomic mass of mercury.

78. Germanium has three major naturally occurring isotopes: ^{70}Ge (69.92425 u, 20.85%), ^{72}Ge (71.92208 u, 27.54%), ^{74}Ge (73.92118 u, 36.29%). There are also two minor isotopes: ^{73}Ge (72.92346 u) and ^{76}Ge (75.92140 u). Calculate the percent isotopic abundances of the two minor isotopes. Comment on the precision of these calculations.

79. From the densities of the lines in the mass spectrum of krypton gas, the following observations were made:
- Somewhat more than 50% of the atoms were krypton-84.
- The numbers of krypton-82 and krypton-83 atoms were essentially equal.
- The number of krypton-86 atoms was 1.50 times as great as the number of krypton-82 atoms.
- The number of krypton-80 atoms was 19.6% of the number of krypton-82 atoms.
- The number of krypton-78 atoms was 3.0% of the number of krypton-82 atoms.

The masses of the isotopes are

^{78}Kr, 77.9204 u	^{80}Kr, 79.9164 u	^{82}Kr, 81.9135 u
^{83}Kr, 82.9141 u	^{84}Kr, 83.9115 u	^{86}Kr, 85.9106 u

The weighted-average atomic mass of Kr is 83.80. Use these data to calculate the percent isotopic abundances of the krypton isotopes.

80. The two naturally occurring isotopes of chlorine are ^{35}Cl (34.9689 u, 75.77%) and ^{37}Cl (36.9658 u, 24.23%). The two naturally occurring isotopes of bromine are ^{79}Br (78.9183 u, 50.69%) and ^{81}Br (80.9163 u, 49.31%). Chlorine and bromine combine to form bromine monochloride, BrCl. Sketch a mass spectrum for BrCl with the relative number of molecules plotted against molecular mass (similar to Figure 2-14).

81. How many atoms are present in a 1.50 m length of 20-gauge copper wire? A 20-gauge wire has a diameter of 0.03196 in., and the density of copper is 8.92 g/cm^3.

82. Monel metal is a corrosion-resistant copper–nickel alloy used in the electronics industry. A particular alloy with a density of 8.80 g/cm^3 and containing 0.022% Si by mass is used to make a rectangular plate 15.0 cm long, 12.5 cm wide, 3.00 mm thick, and has a 2.50 cm diameter hole drilled through its center. How many silicon-30 atoms are found in this plate? The mass of a silicon-30 atom is 29.97376 u, and the percent isotopic abundance of silicon-30 is 3.10%.

83. Deuterium, ^2H (2.0140 u), is sometimes used to replace the principal hydrogen isotope ^1H in chemical studies. The percent isotopic abundance of deuterium is 0.015%. If it can be done with 100% efficiency, what mass of hydrogen gas would have to be processed to obtain a sample containing 2.50×10^{21} ^2H atoms?

84. An alloy that melts at about the boiling point of water has Bi, Pb, and Sn atoms in the ratio 10:6:5, respectively. What mass of alloy contains a total of one mole of atoms?

85. A particular silver solder (used in the electronics industry to join electrical components) is to have the *atom* ratio of 5.00 Ag/4.00 Cu/1.00 Zn. What masses of the three metals must be melted together to prepare 1.00 kg of the solder?

86. A low-melting Sn–Pb–Cd alloy called *eutectic alloy* is analyzed. The *mole* ratio of tin to lead is 2.73:1.00, and the *mass* ratio of lead to cadmium is 1.78: What is the mass percent composition of alloy?

87. In an experiment, 125 cm³ of zinc and 125 cr iodine are mixed together and the iodine is pletely converted to 164 cm³ of zinc iodide. Wha ume of zinc remains unreacted? The densities of iodine, and zinc iodide are 7.13 g/cm³, 4.93 g/ and 4.74 g/cm³, respectively.

88. Atoms are spherical and so when silver atoms together to form silver metal, they cannot fill a available space. In a sample of silver metal, app mately 26.0% of the sample is empty space. C that the density of silver metal is 10.5 g/cm³, wl the radius of a silver atom? Express your answ picometers.

Feature Problems

89. The data Lavoisier obtained in the experiment described on page 35 are as follows:

Before heating: glass vessel + tin + air

= 13 onces, 2 gros, 2.50 grains

After heating: glass vessel + tin calx + remaining air

= 13 onces, 2 gros, 5.62 grains

How closely did Lavoisier's results conform to the law of conservation of mass? (1 livre = 16 onces; 1 once = 8 gros; 1 gros = 72 grains. In modern terms, 1 livre = 30.59 g.)

90. Some of Millikan's oil-drop data are shown below. The measured quantities were not actual charges on oil drops but were proportional to these charges. Show that these data are consistent with the idea of a fundamental electronic charge.

Observation	Measured Quantity	Observation	Measured Quantity
1	19.66	8	53.91
2	24.60	9	59.12
3	29.62	10	63.68
4	34.47	11	68.65
5	39.38	12	78.34
6	44.42	13	83.22
7	49.41		

91. Before 1961, the standard for atomic masses was the isotope ^{16}O, to which physicists assigned a value of exactly 16. At the same time, chemists assigned a value of exactly 16 to the naturally occurring mixture of the isotopes ^{16}O, ^{17}O, and ^{18}O. Would you expect atomic masses listed in a 60-year-old text to be the same, generally higher, or generally lower than in this text? Explain.

92. German chemist Fritz Haber proposed paying off the reparations imposed against Germany after World War I by extracting gold from seawater. Given that (1) the amount of the reparations was \$28.8 billior lars, (2) the value of gold at the time was about \$ per troy ounce (1 troy ounce = 31.103 g), and (3) occurs in seawater to the extent of 4.67×10^{17} a per ton of seawater (1 ton = 2000 lb), how r cubic kilometers of seawater would have had processed to obtain the required amount of g Assume that the density of seawater is 1.03 g (Haber's scheme proved to be commercially in ble, and the reparations were never fully paid.)

93. Mass spectrometry is one of the most versatile powerful tools in chemical analysis because capacity to discriminate between atoms of diff masses. When a sample containing a mixtu isotopes is introduced into a mass spectromete ratio of the peaks observed reflects the ratio of the cent isotopic abundances. This ratio provides an nal standard from which the amount of a ce isotope present in a sample can be determined. T accomplished by deliberately introducing a kı quantity of a particular isotope into the sample analyzed. A comparison of the new isotope ratio first ratio allows the determination of the amount isotope present in the original sample.

An analysis was done on a rock sample to c mine its rubidium content. The rubidium conte a portion of rock weighing 0.350 g was extra and to the extracted sample was added an addit 29.45 μg of ^{87}Rb. The mass spectrum of this sp sample showed a ^{87}Rb peak that was 1.12 tim high as the peak for ^{85}Rb. Assuming that the tw topes react identically, what is the Rb content c rock (expressed in parts per million by mass) isotopic abundances and isotopic masses are sl in the table.

Isotope	% Isotopic Abundance	Atomic Mass, u
^{87}Rb	27.83	86.909
^{85}Rb	72.17	84.912

Self-Assessment Exercises

In your own words, define or explain these terms or symbols: **(a)** $^{A}_{Z}E$; **(b)** β particle; **(c)** isotope; **(d)** ^{16}O; **(e)** molar mass.

Briefly describe
(a) the law of conservation of mass
(b) Rutherford's nuclear atom
(c) weighted-average atomic mass
(d) a mass spectrum

Explain the important distinctions between each pair of terms:
(a) cathode rays and X-rays
(b) protons and neutrons
(c) nuclear charge and ionic charge
(d) periods and groups of the periodic table
(e) metal and nonmetal
(f) the Avogadro constant and the mole

A certain element contains one atom of mass 10.013 u for every four atoms of mass 11.009 u. Compute the atomic weight of the element.

When 10.0 g zinc and 8.0 g sulfur are allowed to react, all the zinc is consumed, 14.9 g zinc sulfide is produced, and the mass of unreacted sulfur remaining is
(a) 2.0 g
(b) 3.1 g
(c) 4.9 g
(d) impossible to predict from this information alone

One oxide of rubidium has 0.187 g O per gram of Rb. A possible O:Rb mass ratio for a second oxide of rubidium is **(a)** 16:85.5; **(b)** 8:42.7; **(c)** 1:2.674; **(d)** any of these.

An attempt was made to determine the atomic mass of element X. If X forms a compound with oxygen that contains 46.7% X by mass and has the formula XO, what is the atomic mass of X?

Cathode rays
(a) may be positively or negatively charged
(b) are a form of electromagnetic radiation similar to visible light
(c) have properties identical to β particles
(d) have masses that depend on the cathode that emits them

The scattering of α particles by thin metal foils established that
(a) the mass of an atom is concentrated in a positively charged nucleus
(b) electrons are fundamental particles of all matter
(c) all electrons carry the same charge
(d) atoms are electrically neutral

Which of the following have the same charge and approximately the same mass?
(a) an electron and a proton; **(b)** a proton and a neutron; **(c)** a hydrogen atom and a proton; **(d)** a neutron and a hydrogen atom; **(e)** an electron and an H^- ion

Which of the following is *not* a fundamental particle? **(a)** proton; **(b)** neutron; **(c)** beta particle; **(d)** alpha particle; **(e)** all are fundamental particles

105. Which of the following scientists did *not* contribute to determining the *structure* of the atom? **(a)** Thomson; **(b)** Rutherford; **(c)** Millikan; **(d)** Dalton; **(e)** Becquerel

106. A subatomic particle that has about the same mass as the hydrogen atom and a negative charge is called **(a)** a proton; **(b)** a neutron; **(c)** an electron; **(d)** an isotope; (e) none of these.

107. What is the correct symbol for the species that contains 18 neutrons, 17 protons, and 16 electrons?

108. The properties of magnesium will most resemble those of which of the following? **(a)** cesium; **(b)** sodium; **(c)** aluminum; **(d)** calcium; **(e)** manganese.

109. Which group in the main group of elements contains **(a)** no metals or metalloids? **(b)** only one metal or metalloid? **(c)** only one nonmetal? **(d)** only nonmetals?

110. The two species that have the same number of electrons as ^{32}S are **(a)** ^{32}Cl; **(b)** $^{34}S^+$; **(c)** $^{33}P^+$; **(d)** $^{28}Si^{2-}$; **(e)** $^{35}S^{2-}$; **(f)** $^{40}Ar^{2+}$; **(g)** $^{40}Ca^{2+}$.

111. To four significant figures, all of the following masses are possible for an individual titanium atom except one. The exception is **(a)** 45.95 u; **(b)** 46.95 u; **(c)** 47.87 u; **(d)** 47.95 u; **(e)** 48.95 u; **(f)** 49.94 u.

112. The mass of the isotope $^{84}_{36}Xe$ is 83.9115 u. If the atomic mass scale were redefined so that $^{84}_{36}Xe$ = 84 u, *exactly*, the mass of the $^{12}_{6}C$ isotope would be **(a)** 11.9115 u; **(b)** 11.9874 u; **(c)** 12 u exactly; **(d)** 12.0127 u; **(e)** 12.0885 u.

113. A 5.585-kg sample of iron (Fe) contains
(a) 10.0 mol Fe
(b) twice as many atoms as does 600.6 g C
(c) 10 times as many atoms as does 52.00 g Cr
(d) 6.022×10^{24} atoms

114. A 91.84 g sample of Ti contains **(a)** 4.175 mol of Ti; **(b)** 6.022×10^{23} Ti atoms; **(c)** 1.155×10^{24} protons; **(d)** 2.542×10^{25} electrons; **(e)** none of these.

115. There are three common iron–oxygen compounds. The one with the greatest proportion of iron has one Fe atom for every O atom and the formula FeO. A second compound has 2.327 g Fe per 1.000 g O, and the third has 2.618 g Fe per 1.000 g O. What are the formulas of these other two iron–oxygen compounds?

116. The four naturally occurring isotopes of strontium have the atomic masses 83.9134 u; 85.9093 u; 86.9089 u; and 87.9056 u. The percent isotopic abundance of the lightest isotope is 0.56% and of the heaviest, 82.58%. Estimate the percent isotopic abundances of the other two. Why is this result only a rough approximation?

117. Gold is present in seawater to the extent of 0.15 mg/ton. Assume the density of the seawater is 1.03 g/mL and determine how many Au atoms could conceivably be extracted from 0.250 L of seawater (1 ton = 2.000×10^3 lb; 1 kg = 2.205 lb).

118. Appendix E describes a useful study aid known as concept mapping. Using the method presented in Appendix E, construct a concept map illustrating the different concepts in Sections 2-7 and 2-8.

3 Chemical Compounds

CONTENTS

LEARNING OBJECTIVES

3.1 Distinguish between an empirical formula, a molecular formula, and the formula unit of an ionic compound.

3.2 Use Avogadro's constant to relate the mass and molar mass of molecules and ionic compounds to their elementary entities (atoms or ions).

3.3 Use percent composition data for a chemical compound to determine its empirical and molecular formula (and vice versa).

3.4 Use the rules for assigning oxidation states to determine the oxidation state of each element in a chemical compound.

3.5 Distinguish between organic and inorganic compounds.

3.6 Use the general rules to name simple inorganic compounds, including binary compounds, binary acids, polyatomic ions, and oxoacids.

3.7 Use the general rules to name simple organic compounds, including basic branched alkanes and alkanes with functional groups, such as alcohols and carboxylic acids.

Andrew Syred / Science Source

Scanning electron microscope image of sodium chloride crystals. Chemical compounds, their formulas, and their names are discussed in this chapter.

Water, ammonia, carbon monoxide, and carbon dioxide—all fan substances—are rather simple chemical compounds. Only slig less familiar are sucrose (cane sugar), acetylsalicylic acid (asp and ascorbic acid (vitamin C). They too are chemical compounds. In the study of chemistry is mostly about chemical compounds, and, in chapter, we will consider a number of ideas about compounds.

The common feature of all compounds is that they are composed of tv more elements. The full range of compounds can be divided into a few b categories by applying ideas from the periodic table of the elem Compounds are represented by chemical formulas, which in turn are de from the symbols of their constituent elements. In this chapter, you will how to deduce and write chemical formulas and how to use the inform rporated into chemical formulas. The chapter ends with an overview of the ionship between names and formulas—chemical nomenclature.

Types of Chemical Compounds and Their Formulas

erally speaking, two fundamental kinds of chemical bonds hold together atoms in a compound. *Covalent* bonds, which involve a sharing of electrons veen atoms, give rise to molecular compounds. *Ionic* bonds, which involve a sfer of electrons from one atom to another, give rise to ionic compounds. In section, we consider only the basic features of molecular and ionic comnds that we need as background for the early chapters of the text. Our epth discussion of chemical bonding will come in Chapters 10 and 11.

◀ In our later study of chemical bonding, we will find that the distinction between covalent and ionic bonding is not as clear-cut as these statements imply. We will consider this matter in Chapter 10.

ecular Compounds

olecular compound** is made up of discrete units called molecules, which cally consist of a small number of *nonmetal* atoms held together by covabonds. Molecular compounds are represented by **chemical formulas**, bolic representations that, at minimum, indicate

the elements present

the relative number of atoms of each element

the formula for water, the constituent elements are denoted by their sym- The relative numbers of atoms are indicated by *subscripts*. Where no subt is written, the number 1 is understood.

The two elements present

H_2O

Lack of subscript means one atom of O per molecule

Two H atoms per molecule

nother example of a chemical formula is CCl_4, which represents the comnd carbon tetrachloride. The formulas H_2O and CCl_4 both represent disentities—*molecules*. Thus, we can refer to water and carbon tetrachloride olecular compounds.

n **empirical formula** is the simplest formula for a compound; it shows the s of atoms present and their relative numbers. The subscripts in an empirformula are reduced to their simplest whole-number ratio. For example, is the empirical formula for a compound whose molecules have the fora P_4O_{10}. Generally, the empirical formula does not tell us a great deal t a compound. Acetic acid ($C_2H_4O_2$), formaldehyde (CH_2O, used to make in plastics and resins), and glucose ($C_6H_{12}O_6$, blood sugar) all have the irical formula CH_2O.

molecular formula is based on an actual molecule of a compound. In e cases, the empirical and molecular formulas are identical, such as CH_2O ormaldehyde. In other cases, the molecular formula is a multiple of the irical formula. A molecule of acetic acid, for example, consists of eight ns—two C atoms, four H atoms, and two O atoms, so the molecular fora of acetic acid is $C_2H_4O_2$. This is twice the number of atoms in the fora unit (CH_2O). Empirical and molecular formulas tell us the combining of the atoms in the compound, but they show nothing about how the ns are attached to each other. Other types of formulas, however, do convey information. Figure 3-1 shows several representations of acetic acid, the constituent that gives vinegar its sour taste.

structural formula shows the order in which atoms are bonded together in lecule and by what types of bonds. Thus, the structural formula of acetic tells us that three of the four H atoms are bonded to one of the C atoms, and

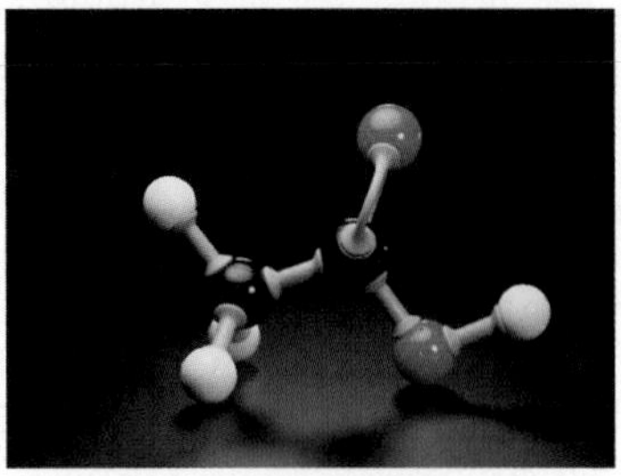

Carey B. Van Loon

Molecular model
("ball and stick")

Empirical formula: CH_2O

Molecular formula: $C_2H_4O_2$

Structural formula:

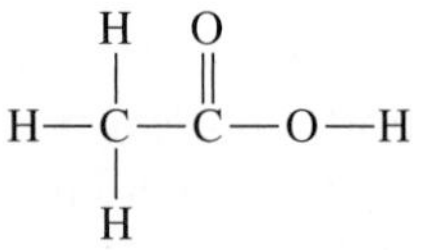

Molecular mode
("space filling"

▲ FIGURE 3-1
Several representations of the compound acetic acid
In the molecular model, the black spheres are carbon, the red are oxygen, and the white are hydrogen. To show that one H atom in the molecule is fundamentally different from the other three, the formula of acetic acid is often written as HC_2H_3C (see Section 5-3). To show that this H atom is bonded to an O atom, the formulas CH_3COOH and CH_3CO_2H are also used. For a few chemical compounds, you may different versions of chemical formulas in different sources.

the remaining H atom is bonded to an O atom. Both of the O atoms are bor to one of the C atoms, and the two C atoms are bonded to each other. The c lent bonds in the structural formula are represented by lines or dashes (—). of the bonds is represented by a double dash (=) and is called a *double* cov bond. Differences between single and double bonds are discussed later ir text. For now, just think of a double bond as being a stronger or tighter b than a single bond.

A **condensed structural formula**, which is written on a single line, i alternative, less cumbersome way of showing how the atoms of a molecul connected. Thus, the acetic acid molecule is represented as either CH_3CC or CH_3CO_2H. With this type of formula, the different ways in which th atoms are attached are still apparent.

Condensed structural formulas can also be used to show how a grou atoms is attached to another atom. Consider methylpropane, $C_4H_{1(}$ Figure 3-2(b). The structural formula shows that there is a $-CH_3$ grou atoms attached to the central carbon atom. In the condensed structural form this is indicated by enclosing the CH_3 in parentheses to the right of the ato which it is attached, thus $CH_3CH(CH_3)CH_3$. Alternatively, because the ce C atom is bonded to each of the other three C atoms, we can write the densed structural formula $CH(CH_3)_3$.

Organic compounds are made up principally of carbon and hydrogen, oxygen and/or nitrogen as important constituents in many of them. Each ca atom forms *four covalent bonds*. Organic compounds can be very complex, one way of simplifying their structural formulas is to write structures wit showing the C and H atoms explicitly. We do this by using a **line-angle forr** (also referred to as a *line structure*), in which lines represent chemical bonc carbon atom exists wherever a line ends or meets another line, and the nur of H atoms needed to complete each carbon atom's four bonds are assume be present. The symbols of other atoms or groups of atoms and the bond joining them to C atoms are written explicitly. The formula of the complex mone molecule testosterone, seen in Figure 3-2(c), is a line-angle formula.

Molecules occupy space and have a three-dimensional shape, but empi and molecular formulas do not convey any information about the sp arrangements of atoms. Structural formulas can sometimes show this, usually the only satisfactory way to represent the three-dimensional struc of molecules is with models. In a *ball-and-stick model*, atoms are represente small balls, and the bonds between atoms by sticks (see Figure 3-1). Such r els help us to visualize distances between the nuclei of atoms (bond len and the geometrical shapes of molecules. Ball-and-stick models are eas

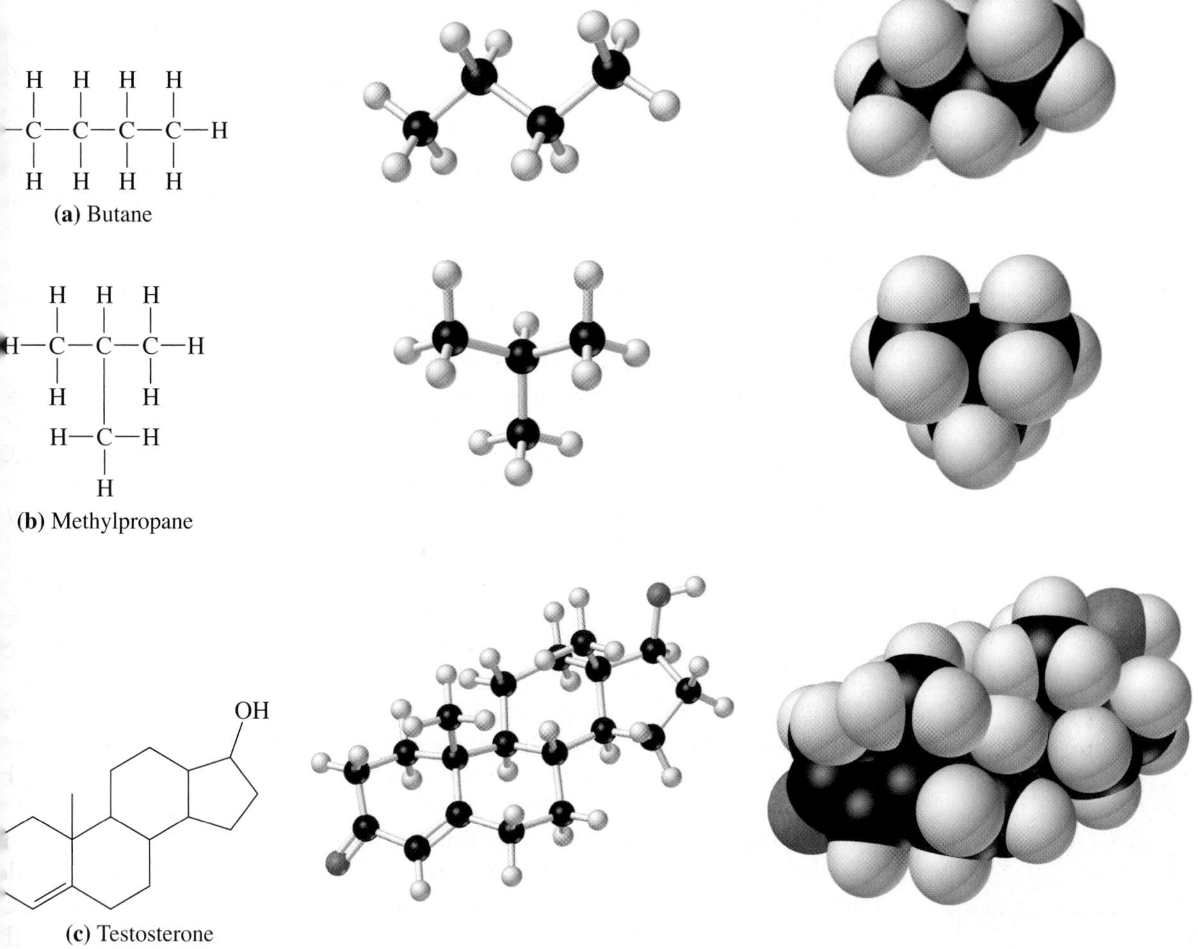

JURE 3-2
alizations of (a) butane, (b) methylpropane, and (c) testosterone

/ and interpret, but they can be somewhat misleading. Chemical bonds orces that draw atoms in a molecule into direct contact. The atoms are not apart as implied by a ball-and-stick model.

space-filling model shows that the atoms in a molecule occupy space and they are in actual contact with one another. Certain computer programs rate images of space-filling models such as those shown in Figures 3-1 3-2. A space-filling model is a more accurate representation of the size and e of a molecule because it is constructed to scale (that is, a nanometer-size cule is magnified to a millimeter or centimeter scale).

ne acetic acid molecule is made up of three types of atoms (C, H, and O) and els of the molecule reflect this fact. Different colors are used to distinguish the us types of atoms in ball-and-stick and space-filling models (see Figure 3-3). will notice that the colored spheres are of different sizes, which correspond to ize differences between the various atoms in the periodic table.

ne depictions of molecules just discussed will be used throughout this . In fact, visualization of the sizes and shapes of molecules and interpre- n of the physical and chemical properties in terms of molecular sizes and es is one of the most important aspects of modern chemistry.

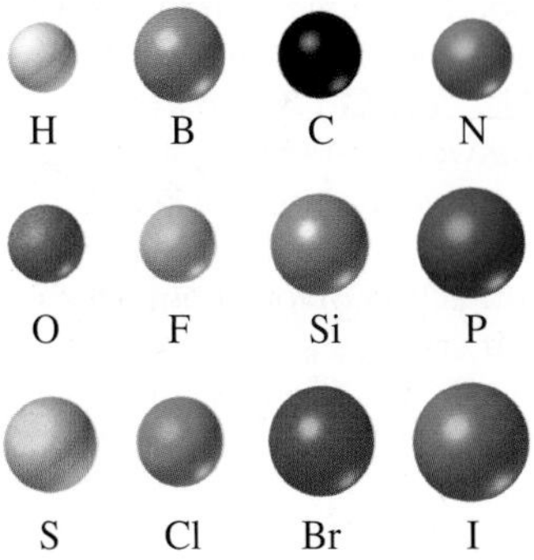

▲ FIGURE 3-3
Color scheme for use in molecular models
The sizes of atoms, reflected in the various sizes of the colored spheres, are related to the locations of the elements in the periodic table, as discussed in Section 9-3.

3-1 CONCEPT ASSESSMENT

present the succinic acid molecule, $HOOCCH_2CH_2COOH$, through empirical, lecular, structural, and line-angle formulas.

Ionic Compounds

Chemical combination of a metal and a nonmetal usually results in an i compound. An **ionic compound** is made up of positive and negative joined together by electrostatic forces of attraction (recall the attractic oppositely charged objects pictured in Figure 2-4). The atoms of metallic ments tend to lose one or more electrons when they combine with nonn atoms, and the nonmetal atoms tend to gain one or more electrons. As a resu this electron transfer, the metal atom becomes a positive ion, or **cation**, an nonmetal atom becomes a negative ion, or **anion**. We can usually deduc charge on a main-group cation or anion from the group of the periodic tab which the element belongs (recall Section 2-6). Thus, the periodic table can us to write the formulas of ionic compounds.

In the formation of sodium chloride—ordinary table salt—each sod atom gives up one electron to become a sodium ion, Na^+, and each chlo atom gains one electron to become a chloride ion, Cl^-. This fact conform the relationship between locations of the elements in the periodic table an charges on their simple ions (see page 54). For sodium chloride to be ele cally neutral, there must be one Na^+ ion for each Cl^- ion ($+1 - 1 = 0$). T the formula of sodium chloride is NaCl, and its structure, is sh in Figure 3-4.

We observe that each Na^+ ion in sodium chloride is surrounded by six ions, and vice versa, and we cannot say that any one of these six Cl^- belongs exclusively to a given Na^+ ion. Yet, the ratio of Cl^- to Na^+ ion sodium chloride is 1 : 1, and so we arbitrarily select a combination of one ion and one Cl^- ion as a formula unit. The **formula unit** of an ionic compo is the smallest electrically neutral collection of ions. The ratio of atoms (ion the formula unit is the same as in the chemical formula. Because it is buri a vast network of ions, called a crystal, a formula unit of an ionic compo does not exist as a distinct entity. Thus, it is inappropriate to call a formula of solid sodium chloride a molecule.

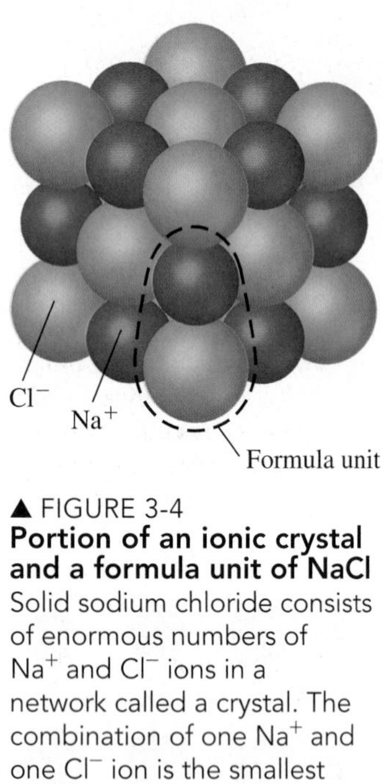

▲ FIGURE 3-4
Portion of an ionic crystal and a formula unit of NaCl
Solid sodium chloride consists of enormous numbers of Na^+ and Cl^- ions in a network called a crystal. The combination of one Na^+ and one Cl^- ion is the smallest collection of ions from which we can deduce the formula NaCl. It is a formula unit.

The situation with magnesium chloride is similar. In magnesium chlo found in trace quantities in table salt, magnesium atoms lose two electro become magnesium ions, Mg^{2+} (Mg is in group 2). To obtain an electri neutral formula unit, there must be two Cl^- ions, each with a charge of −1 every Mg^{2+} ion. The formula of magnesium chloride is $MgCl_2$.

The ions Na^+, Mg^{2+}, and Cl^- are *monatomic*, meaning that each consis a single ionized atom. By contrast, a *polyatomic* ion is made up of tw more atoms. In the nitrate ion, NO_3^-, the subscripts signify that *thr* atoms and *one* N atom are joined by covalent bonds into the single ion N Magnesium nitrate is an ionic compound made up of magnesium nitrate ions. An electrically neutral formula unit of this compound must sist of one Mg^{2+} ion and two NO_3^- ions. The formula based on this for unit is denoted by enclosing NO_3 in parentheses, followed by the subs 2; thus, $Mg(NO_3)_2$. Polyatomic ions are discussed further in Section 3-6.

3-1 ARE YOU WONDERING?

Can a compound be formed between different metal atoms?

In a metal, electrons of the atoms interact to form *metallic* bonds. The bond atoms are usually of the same element, but they may also be of different elemen giving rise to *intermetallic* compounds. The metallic bond gives metals and int metallic compounds their characteristic properties of electrical and heat cond tivity.

2 The Mole Concept and Chemical Compounds

e we know the chemical formula of a compound, we can determine its for-a mass. **Formula mass** is the mass of a *formula unit* in atomic mass units. It ways appropriate to use the term formula mass, but, for a molecular com-nd, the formula unit is an actual molecule, so we can speak of molecular s. **Molecular mass** is the mass of a molecule in atomic mass units.

ormula and molecular masses can be obtained just by adding up atomic ses (those on the inside front cover). Thus, for the molecular compound er, H_2O,

◀ The terms *formula weight* and *molecular weight* are often used in place of formula mass and molecular mass. This is similar to the situation described for atomic mass and atomic weight in the footnote on page 48.

$$\begin{aligned}\text{molecular mass } H_2O &= 2(\text{atomic mass H}) + (\text{atomic mass O})\\ &= 2(1.008\text{ u}) + 15.999\text{ u}\\ &= 18.015\text{ u}\end{aligned}$$

or the ionic compound magnesium chloride, $MgCl_2$,

◀ The terms *formula mass* and *molecular mass* have essentially the same meaning, although when referring to ionic compounds, such as NaCl and $MgCl_2$, *formula mass* is the proper term.

$$\begin{aligned}\text{formula mass } MgCl_2 &= \text{atomic mass Mg} + 2(\text{atomic mass Cl})\\ &= 24.305\text{ u} + 2(35.45\text{ u})\\ &= 95.21\text{ u}\end{aligned}$$

for the ionic compound magnesium nitrate, $Mg(NO_3)_2$,

$$\begin{aligned}\text{nula mass } Mg(NO_3)_2 &= \text{atomic mass Mg} + 2[\text{atomic mass N} + 3(\text{atomic mass O})]\\ &= 24.305\text{ u} + 2[14.007\text{ u} + 3(15.999\text{ u})]\\ &= 148.313\text{ u}\end{aligned}$$

le of a Compound

ll that in Chapter 2 a mole was defined as an amount of substance having same number of elementary entities as there are atoms in exactly 12 g of carbon-12. This definition carefully avoids saying that the entities to be nted are always atoms. As a result, we can apply the concept of a mole to quantity that we can represent by a symbol or formula—atoms, ions, for-a units, or molecules. Specifically, a *mole of compound* is an amount of com-nd containing Avogadro's number (6.02214×10^{23}) of formula units or ecules. The **molar mass** is the mass of one mole of compound—one mole olecules of a molecular compound and one mole of formula units of an c compound.

he weighted-average molecular mass of H_2O is 18.015 u, compared with ass of exactly 12 u for a carbon-12 atom. If we compare samples of water ecules and carbon atoms by using Avogadro's number of each, we get a s of 18.015 g H_2O, compared with exactly 12 g for carbon-12. The molar s of H_2O is 18.015 g H_2O/mol H_2O. If we know the formula of a com-nd, we can equate the following terms, as illustrated for H_2O, $MgCl_2$, and $NO_3)_2$.

KEEP IN MIND

that although *molecular mass* and *molar mass* sound similar and are related, they are not the same. Molecular mass is the weighted-average mass of one molecule expressed in atomic mass units, u. *Molar mass* is the mass of Avogadro's number of molecules expressed in grams per mole, g/mol. The two terms have the same numerical value but different units.

$$1\text{ mol } H_2O = 18.015\text{ g } H_2O = 6.02214 \times 10^{23}\ H_2O\text{ molecules}$$

$$1\text{ mol } MgCl_2 = 95.21\text{ g } MgCl_2 = 6.02214 \times 10^{23}\ MgCl_2\text{ formula units}$$

$$\text{ol } Mg(NO_3)_2 = 148.313\text{ g } Mg(NO_3)_2 = 6.02214 \times 10^{23}\ Mg(NO_3)_2\text{ formula units}$$

ich expressions as these provide several different types of conversion ors that can be applied in a variety of problem-solving situations. The egy that works best for a particular problem will depend, in part, on the necessary conversions are visualized. As we learned in Section 2-7, most direct link to an amount in moles is through a mass in grams, so generally the central focus of a problem is the conversion of a mass in gr to an amount in moles, or vice versa. This conversion must often be ceded or followed by other conversions involving volumes, densities, centages, and so on. As we saw in Chapter 2, one helpful tool in prob solving is to establish a conversion pathway. In Table 3.1, we summarize roles that density, molar mass, and the Avogadro constant play in a con sion pathway.

TABLE 3.1 Density, Molar Mass, and the Avogadro Constant as Conversion Factors

Density, d	converts from volume to mass
Molar mass, M	converts from mass to amount (mol)
Avogadro constant, N_A	converts from amount (mol) to elementary entities

EXAMPLE 3-1 Relating Molar Mass, the Avogadro Constant, and Formula Units of an Ionic Compound

An analytical balance can detect a mass of 0.1 mg. How many ions are present in this minimally detectab quantity of $MgCl_2$?

Analyze

The central focus is the conversion of a measured quantity, 0.1 mg $MgCl_2$, to an amount in moles. After makin the mass conversion, mg $\longrightarrow$ g, we can use the molar mass to convert from mass to amount in moles. The with the Avogadro constant as a conversion factor, we can convert from amount in moles to number o formula units. The final factor we need is based on the fact that there are *three* ions (*one* Mg^{2+} and *two* Cl^-) pe formula unit (fu) of $MgCl_2$. It is often helpful to map out a conversion pathway that starts with the informatio given and proceeds through a series of conversion factors to the information sought. For this problem, we ca begin with milligrams of $MgCl_2$ and make the following conversions:

$$\text{mg} \longrightarrow \text{g} \longrightarrow \text{mol} \longrightarrow \text{fu} \longrightarrow \text{number of ions}$$

Solve

The required conversions can be carried out in a stepwise fashion (as was done in Example 2-9), or they can b combined into a single line calculation. To avoid having to write down intermediate results and to avoi rounding errors, we'll use a single line calculation this time.

$$? \text{ ions} = 0.1 \text{ mg MgCl}_2 \times \frac{1 \text{ g MgCl}_2}{1000 \text{ mg MgCl}_2} \times \frac{1 \text{ mol MgCl}_2}{95 \text{ g MgCl}_2} \times \frac{6.0 \times 10^{23} \text{ fu MgCl}_2}{1 \text{ mol MgCl}_2} \times \frac{3 \text{ ions}}{1 \text{ fu MgCl}_2} = 2 \times 10^{18} \text{ ions}$$

Assess

The mass of the sample (0.1 mg) is given with one significant figure, and so the final answer is rounded to on significant figure. In the calculation above, the molar mass of $MgCl_2$ and the Avogadro constant are rounded t two significant figures, that is, with one more significant figure than in the measured quantity.

PRACTICE EXAMPLE A: How many grams of $MgCl_2$ would you need to obtain 5.0×10^{23} Cl^- ions?

PRACTICE EXAMPLE B: How many nitrate ions, NO_3^-, and how many oxygen atoms are present in 1.00 μg o magnesium nitrate, $Mg(NO_3)_2$?

XAMPLE 3-2 Combining Several Factors in a Calculation Involving Molar Mass

The volatile liquid ethyl mercaptan, C_2H_6S, is one of the most odoriferous substances known. It is sometimes added to natural gas to make gas leaks detectable. How many C_2H_6S molecules are contained in a 1.0 μL sample? The density of liquid ethyl mercaptan is 0.84 g/mL.

nalyze

The central focus is again the conversion of a measured quantity to an amount in moles. Because the density is given in g/mL, it will be helpful to convert the measured volume to milliliters. Then, density can be used as a conversion factor to obtain the mass in grams, and the molar mass can then be used to convert mass to amount in moles. Finally, the Avogadro constant can be used to convert the amount in moles to the number of molecules. In summary, the conversion pathway is $\mu L \rightarrow L \rightarrow g \rightarrow mol \rightarrow$ molecules.

olve

As always, the required conversions can be combined into a single line calculation. However, it is instructive to break the calculation into three steps: (1) a conversion from volume to mass, (2) a conversion from mass to amount in moles, and (3) a conversion from amount in moles to molecules. These three steps emphasize, respectively, the roles played by density, molar mass, and the Avogadro constant in the conversion pathway. (See Table 3.1.)

Convert from volume to mass.

$$? \text{ g } C_2H_6S = 1.0\ \mu L \times \frac{1 \times 10^{-6}\ L}{1\ \mu L} \times \frac{1000\ mL}{1\ L} \times \frac{0.84\ g\ C_2H_6S}{1\ mL} = 8.4 \times 10^{-4}\ g\ C_2H_6S$$

Convert from mass to amount in moles.

$$? \text{ mol } C_2H_6S = 8.4 \times 10^{-4}\ g\ C_2H_6S \times \frac{1\ mol\ C_2H_6S}{62.1\ g\ C_2H_6S} = 1.4 \times 10^{-5}\ mol\ C_2H_6S$$

Convert from moles to molecules.

$$? \text{ molecules } C_2H_6S = 1.4 \times 10^{-5}\ mol\ C_2H_6S \times \frac{6.02 \times 10^{23}\ molecules\ C_2H_6S}{1\ mol\ C_2H_6S} = 8.1 \times 10^{18}\ molecules\ C_2H_6S$$

ssess

Remember to store intermediate results in your calculator *without* rounding off. Round off at the end. The answer is rounded to two significant figures because the volume and density are given with two significant figures. Rounding errors are avoided if the required conversions are combined into a single line calculation.

$$? \text{ molecules } C_2H_6S = 1.0\ \mu L \times \frac{1 \times 10^{-6}\ L}{1\ \mu L} \times \frac{1000\ mL}{1\ L} \times \frac{0.84\ g\ C_2H_6S}{1\ mL} \times \frac{1\ mol\ C_2H_6S}{62.1\ g\ C_2H_6S} \times \frac{6.02 \times 10^{23}\ molecules\ C_2H_6S}{1\ mol\ C_2H_6S} = 8.1 \times 10^{18}\ molecules\ C_2H_6S$$

RACTICE EXAMPLE A: Gold has a density of 19.32 g/cm^3. A piece of gold foil is 2.50 cm on each side and 0.100 mm thick. How many atoms of gold are in this piece of gold foil?

RACTICE EXAMPLE B: If the 1.0 μL sample of liquid ethyl mercaptan from Example 3-2 is allowed to evaporate and distribute itself throughout a chemistry lecture room with dimensions 62 ft $\times$ 35 ft $\times$ 14 ft, will the odor of the vapor be detectable in the room? The limit of detectability is 9×10^{-4} μmol/m^3.

le of an Element—A Second Look

hapter 2, we took one mole of an element to be 6.02214×10^{23} *atoms* of the ıent. This is the only definition possible for such elements as iron, magneı, sodium, and copper, in which enormous numbers of individual spheritoms are clustered together, much like marbles in a can. But the atoms of e elements are joined together to form molecules. Bulk samples of these ıents are composed of collections of molecules. The molecules of P_4 and S_8 represented in Figure 3-5. The molecular formulas of elements that you ıld become familiar with are

$$H_2 \quad O_2 \quad N_2 \quad F_2 \quad Cl_2 \quad Br_2 \quad I_2 \quad P_4 \quad S_8$$

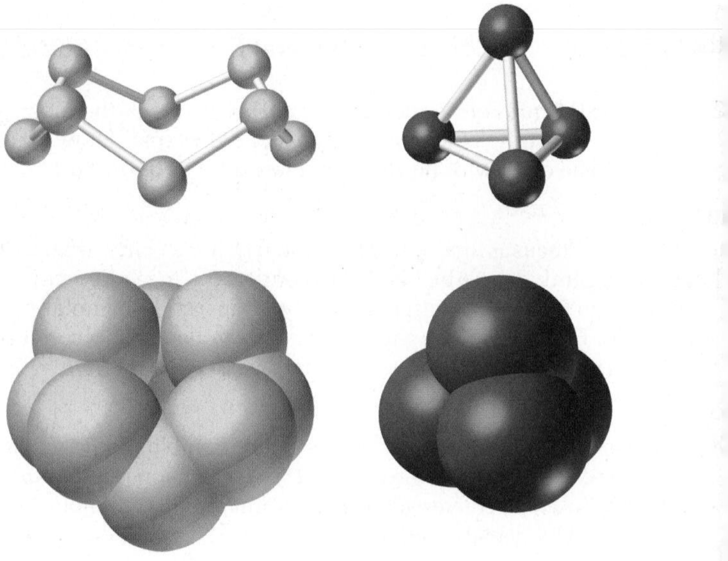

▶ FIGURE 3-5
Molecular forms of elemental sulfur and phosphorus
In a sample of solid sulfur, there are eight sulfur atoms in a sulfur molecule. In solid white phosphorus, there are four phosphorus atoms per molecule.

For these elements, we speak of an *atomic* mass or a *molecular* mass, and m mass can be expressed in two ways. Hydrogen, for example, has an atc mass of 1.008 u and a molecular mass of 2.016 u; its molar mass ca expressed as 1.008 g H/mol H or 2.016 g H_2/mol H_2.

Another phenomenon occasionally encountered is the existence o element in more than one molecular form, a situation referred to as *allot* Thus, oxygen exists in two allotropic forms, the predominantly abun diatomic oxygen, O_2, and the much less abundant allotrope *ozone*, O_3. The m mass of ordinary dioxygen is 31.998 g O_2/mol O_2, and that of ozone is 47.9 O_3/mol O_3.

3-2 CONCEPT ASSESSMENT

Without doing detailed calculations, determine which of the following quantitie has the greatest mass and which has the smallest mass: **(a)** 0.50 mol O_2; **(b)** 2.0×10^{23} Cu atoms; **(c)** 1.0×10^{24} H_2O molecules; **(d)** a 20.000 g brass weight; **(e)** 1.0 mol Ne.

3-3 Composition of Chemical Compounds

▼ Two representations of halothane.

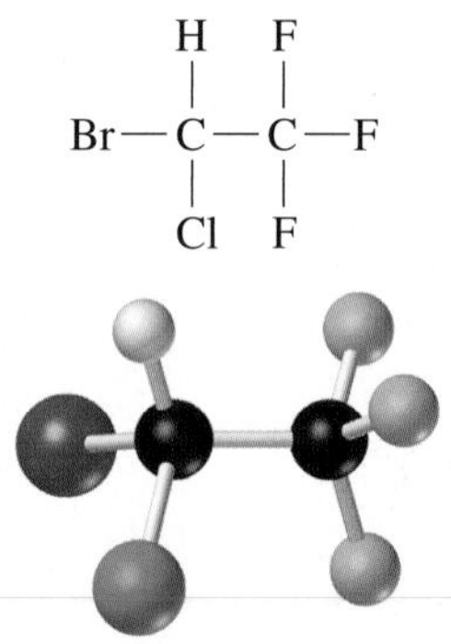

A chemical formula conveys considerable quantitative information abo compound and its constituent elements. We have already learned hov determine the molar mass of a compound, and, in this section, we cons some other types of calculations based on the chemical formula.

The colorless, volatile liquid halothane has been used as a fire extingui and also as an inhalation anesthetic. Both its empirical and molecular for las are $C_2HBrClF_3$, its molecular mass is 197.38 u, and its molar ma 197.38 g/mol, as calculated below:

$$
\begin{aligned}
M_{C_2HBrClF_3} &= 2M_C + M_H + M_{Br} + M_{Cl} + 3M_F \\
&= [(2 \times 12.011) + 1.008 + 79.904 + 35.45 + (3 \times 18.9984)]\ \text{g/m} \\
&= 197.38\ \text{g/mol}
\end{aligned}
$$

The molecular formula of $C_2HBrClF_3$ tells us that *per mole* of halothane t are two moles of C atoms, one mole each of H, Br, and Cl atoms, and t moles of F atoms. This factual statement can be turned into conversion fac swer such questions as, "How many C atoms are present per mole of hane?" In this case, the factor needed is 2 mol C/mol $C_2HBrClF_3$. That is,

$$\text{atoms} = 1.000 \text{ mol } C_2HBrClF_3 \times \frac{2 \text{ mol C}}{1 \text{ mol } C_2HBrClF_3} \times \frac{6.022 \times 10^{23} \text{ C atoms}}{1 \text{ mol C}}$$

$$= 1.204 \times 10^{24} \text{ C atoms}$$

Example 3-3, we use another conversion factor derived from the formula alothane. This factor is shown in blue in the setup, which includes other iar factors to make the conversion pathway:

$$\text{mL} \longrightarrow \text{g} \longrightarrow \text{mol } C_2HBrClF_3 \longrightarrow \text{mol F}$$

AMPLE 3-3 Using Relationships Derived from a Chemical Formula

How many moles of F atoms are in a 75.0 mL sample of halothane (d = 1.871 g/mL)?

alyze

The conversion pathway for this problem is given above. First, convert the volume of the sample to mass; this requires density as a conversion factor. Next, convert the mass of halothane to its amount in moles; this requires the inverse of the molar mass as a conversion factor. The final conversion factor is based on the formula of halothane.

lve

$$? \text{ mol F} = 75.0 \text{ mL } C_2HBrClF_3 \times \frac{1.871 \text{ g } C_2HBrClF_3}{1 \text{ mL } C_2HBrClF_3}$$

$$\times \frac{1 \text{ mol } C_2HBrClF_3}{197.4 \text{ g } C_2HBrClF_3} \times \frac{3 \text{ mol F}}{1 \text{ mol } C_2HBrClF_3}$$

$$= 2.13 \text{ mol F}$$

sess

If we had been asked for the number of moles of C instead, the final conversion factor in the calculation above would have been (2 mol C/1 mol $C_2HBrClF_3$).

ACTICE EXAMPLE A: How many grams of Br are contained in 25.00 mL of halothane (d = 1.871 g/mL)?

ACTICE EXAMPLE B: How many milliliters of halothane would contain 1.00×10^{24} Br atoms?

3-3 CONCEPT ASSESSMENT

exachlorophene, $C_{13}H_6Cl_6O_2$, a compound used in germicidal soaps, which ment contributes the greatest number of atoms and which contributes the atest mass?

culating Percent Composition from a Chemical Formula

n chemists believe that they have synthesized a new compound, a sample nerally sent to an analytical laboratory where its percent composition is rmined. This experimentally determined percent composition is then pared with the percent composition calculated from the formula of the cted compound. In this way, chemists can see if the compound obtained d be the one expected.

uation (3.1) establishes how the mass percent of an element in a comnd is calculated. In applying the equation, as in Example 3-4, think in s of the following steps.

1. **Determine the molar mass of the compound.** This is the *denominat* equation (3.1).
2. **Determine the contribution of the given element to the molar mass.** product of the formula subscript and the molar mass of the ele appears in the *numerator* of equation (3.1).
3. **Formulate the ratio of the mass of the given element to the mass o compound as a whole.** This is the ratio of the numerator from step 2 t denominator from step 1.
4. **Multiply this ratio by 100% to obtain the mass percent of the eleme**

$$\text{mass \% element} = \frac{\begin{pmatrix}\text{number of}\\ \text{atoms of element}\\ \text{per formula unit}\end{pmatrix} \times \begin{pmatrix}\text{molar mass}\\ \text{of element}\end{pmatrix}}{\text{molar mass of compound}} \times 100\%$$

The mass composition of a compound is the collection of mass percentag the individual elements in the compound.

EXAMPLE 3-4 Calculating the Mass Percent Composition of a Compound

What is the mass percent composition of halothane, $C_2HBrClF_3$?

Analyze

Apply the four-step method described above. First, determine the molar mass of $C_2HBrClF_3$. Then formul the mass ratios and convert them to mass percents. If we use molar masses that are rounded to two decin places, the calculated mass percents will be accurate to two decimal places.

Solve

The molar mass of $C_2HBrClF_3$ is 197.38 g/mol. The mass percents are

$$\%\,C = \frac{\left(2\text{ mol C} \times \frac{12.01\text{ g C}}{1\text{ mol C}}\right)}{197.38\text{ g } C_2HBrClF_3} \times 100\% = 12.17\%\,C$$

$$\%\,H = \frac{\left(1\text{ mol H} \times \frac{1.01\text{ g H}}{1\text{ mol H}}\right)}{197.38\text{ g } C_2HBrClF_3} \times 100\% = 0.51\%\,H$$

$$\%\,Br = \frac{\left(1\text{ mol Br} \times \frac{79.90\text{ g Br}}{1\text{ mol Br}}\right)}{197.38\ C_2HBrClF_3} \times 100\% = 40.48\%\,Br$$

$$\%\,Cl = \frac{\left(1\text{ mol Cl} \times \frac{35.45\text{ g Cl}}{1\text{ mol Cl}}\right)}{197.38\text{ g } C_2HBrClF_3} \times 100\% = 17.96\%\,Cl$$

$$\%\,F = \frac{\left(3\text{ mol F} \times \frac{19.00\text{ g F}}{1\text{ mol F}}\right)}{197.38\text{ g } C_2HBrClF_3} \times 100\% = 28.88\%\,F$$

Thus, the percent composition of halothane is 12.17% C, 0.51% H, 40.48% Br, 17.96% Cl, and 28.88% F.

Assess

The mass ratios appearing above are based on a sample that contains exactly *one mole* of halothane. Anoth approach is to calculate the mass of each element present in a sample that contains exactly *100* g of halothar For example, in a *100 g* sample of halothane,

$$?\text{ g C} = 100\text{ g } C_2HBrClF_3 \times \frac{1\text{ mol } C_2HBrClF_3}{197.38\text{ g } C_2HBrClF_3} \times \frac{2\text{ mol C}}{1\text{ mol } C_2HBrClF_3} \times \frac{12.01\text{ g C}}{1\text{ mol C}} = 12.17\text{ g C}$$

and so halothane is 12.17% C. The mass of carbon in a 100 g sample is numerically equal to the mass percent of carbon. If you compare the calculation of g C with that for % C given earlier, you will see that both calculations involve exactly the same factors but in a slightly different order.

RACTICE EXAMPLE A: Adenosine triphosphate (ATP) is the main energy-storage molecule in cells. Its chemical formula is $C_{10}H_{16}N_5P_3O_{13}$. What is its mass percent composition?

RACTICE EXAMPLE B: *Without doing detailed calculations*, determine which two compounds from the following list have the same percent oxygen by mass: **(a)** CO; **(b)** CH_3COOH; **(c)** C_2O_3; **(d)** N_2O; **(e)** $C_6H_{12}O_6$; **(f)** $HOCH_2CH_2OH$.

he percentages of the elements in a compound should add up to 100.00%, we can use this fact in two ways.

Check the accuracy of the computations by ensuring that the percentages total 100.00%. As applied to the results of Example 3-4:

$$12.17\% + 0.51\% + 40.48\% + 17.96\% + 28.88\% = 100.00\%$$

Determine the percentages of all the elements but one. Obtain that one by difference (subtraction). From Example 3-4:

$$\begin{aligned}\%\ \text{H} &= 100.00\% - \%\ \text{C} - \%\ \text{Br} - \%\ \text{Cl} - \%\ \text{F} \\ &= 100.00\% - 12.17\% - 40.48\% - 17.96\% - 28.88\% \\ &= 0.51\%\end{aligned}$$

ablishing Formulas from the Experimentally Determined cent Composition of Compounds

imes, a chemist isolates a chemical compound—say, from an exotic tropi- plant—and has no idea what it is. A report from an analytical laboratory on percent composition of the compound yields data needed to determine its nula.

ercent composition establishes the relative proportions of the elements in mpound on a *mass* basis. A chemical formula requires these proportions to n a *mole* basis, that is, in terms of *numbers* of atoms. Consider the following -step approach to determining a formula from the experimentally deter- ed percent composition of the compound 2-deoxyribose, a sugar that is a c constituent of DNA (deoxyribonucleic acid). The mass percent composi- of 2-deoxyribose is 44.77% C, 7.52% H, and 47.71% O.

Although we could choose any sample size, if we take one of *exactly* 100 g, the masses of the elements are numerically equal to their percentages, that is, 44.77 g C, 7.52 g H, and 47.71 g O.

Convert the masses of the elements in the 100.00 g sample to amounts in moles.

$$?\ \text{mol C} = 44.77\ \text{g C} \times \frac{1\ \text{mol C}}{12.011\ \text{g C}} = 3.727\ \text{mol C}$$

$$?\ \text{mol H} = 7.52\ \text{g H} \times \frac{1\ \text{mol H}}{1.008\ \text{g H}} = 7.46\ \text{mol H}$$

$$?\ \text{mol O} = 47.71\ \text{g O} \times \frac{1\ \text{mol O}}{15.999\ \text{g O}} = 2.982\ \text{mol O}$$

◀ Use the molar mass with one more significant figure than in the mass of the element.

Write a tentative formula based on the numbers of moles just determined.

$$C_{3.727}H_{7.46}O_{2.982}$$

4. Attempt to convert the subscripts in the tentative formula to small w numbers. This requires dividing each of the subscripts by the smallest (2.982).

$$C_{\frac{3.727}{2.982}} H_{\frac{7.46}{2.982}} O_{\frac{2.982}{2.982}} = C_{1.25} H_{2.50} O$$

If all subscripts at this point differ only slightly from whole numbe which is not the case here—round them off to whole numbers, conclu the calculation at this point.

5. If one or more subscripts is still not a whole number—which is the here—multiply all subscripts by a small whole number that will m them all integral. Thus, multiply by 4 here.

$$C_{(4\times1.25)} H_{(4\times2.50)} O_{(4\times1)} = C_5H_{10}O_4$$

The formula that we get by the method just outlined, $C_5H_{10}O_4$, is the sim possible formula—the *empirical formula*. The actual *molecular formula* ma equal to, or some multiple of, the empirical formula, such as $C_{10}H_2$ $C_{15}H_{30}O_{12}$, $C_{20}H_{40}O_{16}$, and so on. To find the multiplying factor, we must c pare the formula mass based on the empirical formula with the true molec mass of the compound. We can establish the molecular mass from a sepa experiment (by methods introduced in Chapters 6 and 14). The experimen determined molecular mass of 2-deoxyribose is 134 u. The formula mass b on the empirical formula, $C_5H_{10}O_4$, is 134.1 u. The measured molecular ma the same as the empirical formula mass. The molecular formula is also C_5H_1

We outline the five-step approach described above in the flow diag below and then apply the approach to Example 3-5, where we will find the empirical formula and the molecular formula are not the same.

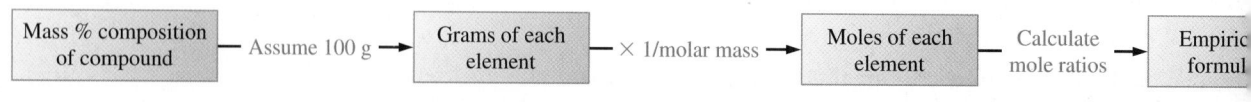

Mass % composition of compound — Assume 100 g → Grams of each element — × 1/molar mass → Moles of each element — Calculate mole ratios → Empiric formul

EXAMPLE 3-5 Determining the Empirical and Molecular Formulas of a Compound from Its Mass Percent Composition

Dibutyl succinate is an insect repellent used against household ants and roaches. Its composition is 62.58% C 9.63% H, and 27.79% O. Its experimentally determined molecular mass is 230 u. What are the empirical an molecular formulas of dibutyl succinate?

Analyze

Use the five-step approach described above.

Solve

1. Determine the mass of each element in a 100.00 g sample.

62.58 g C, 9.63 g H, 27.79 g O

2. Convert each of these masses to an amount in moles.

$$? \text{ mol C} = 62.58 \text{ g C} \times \frac{1 \text{ mol C}}{12.011 \text{ g C}} = 5.210 \text{ mol C}$$

$$? \text{ mol H} = 9.63 \text{ g H} \times \frac{1 \text{ mol H}}{1.008 \text{ g H}} = 9.55 \text{ mol H}$$

$$? \text{ mol O} = 27.79 \text{ g O} \times \frac{1 \text{ mol O}}{15.999 \text{ g O}} = 1.737 \text{ mol O}$$

3. Write a tentative formula based on these numbers of moles.

$$C_{5.210} H_{9.55} O_{1.737}$$

. Divide each of the subscripts of the tentative formula by the smallest subscript (1.737),

$$C_{\frac{5.210}{1.737}}H_{\frac{9.55}{1.737}}O_{\frac{1.737}{1.737}} = C_{2.99}H_{5.49}O$$

and round off any subscripts that differ only slightly from whole numbers; that is, round 2.99 to 3.

$$C_3H_{5.49}O$$

. Multiply all subscripts by a small whole number to make them integral (here by the factor 2), and write the empirical formula.

$$C_{2\times 3}H_{2\times 5.49}O_{2\times 1} = C_6H_{10.98}O_2$$

$$2 \times 5.49 = 10.98 \approx 11$$

Empirical formula: $C_6H_{11}O_2$

o establish the molecular formula, first determine ie empirical formula mass.

$$[(6 \times 12.0) + (11 \times 1.0) + (2 \times 16.0)]u = 115\ u$$

ince the experimentally determined formula mass 230 u) is twice the empirical formula mass, the olecular formula is twice the empirical formula.

Molecular formula: $C_{12}H_{22}O_4$

ssess

Check the result by working the problem in reverse and using numbers that are rounded off slightly. For $C_{12}H_{22}O_4$, $\%\ C \approx (12 \times 12\ u/230\ u) \times 100\% = 63\%$; $\%\ H \approx (22 \times 1\ u/230\ u) \times 100\% = 9.6\%$; and $\%\ O \approx (4 \times 16\ u/230\ u) \times 100\% = 28\%$. The calculated mass percents agree well with those given in the problem, so we can be confident that our answer is correct.

RACTICE EXAMPLE A: Sorbitol, used as a sweetener in some sugar-free foods, has a molecular mass of 182 u and a mass percent composition of 39.56% C, 7.74% H, and 52.70% O. What are the empirical and molecular formulas of sorbitol?

RACTICE EXAMPLE B: The chlorine-containing narcotic drug pentaerythritol chloral has 21.51% C, 2.22% H, and 17.64% O, by mass, as its other elements. Its molecular mass is 726 u. What are the empirical and molecular formulas of pentaerythritol chloral?

3-4 CONCEPT ASSESSMENT

plain why, if the percent composition *and* the molar mass of a compound are oth known, the molecular formula can be determined much more readily by ing the molar mass rather than a 100 g sample in the first step of the five-step ocedure in Example 3-5. In this instance, how would you then obtain the npirical formula?

3-2 ARE YOU WONDERING?

How much rounding off should you do to get integral subscripts in an empirical formula and what factors should you use to convert fractional to whole numbers?

ow much rounding off is justified depends on how precisely the elemental ialysis is done. As a result, there is no ironclad rule on the matter. For the examles in this text, if you carry all the significant figures allowable in a calculation, ou can generally round off a subscript that is within a few hundredths of a hole number (for example, 3.98 rounds off to 4). If the deviation is more than iis, you will need to adjust subscripts to integral values by multiplying by the ppropriate constant. If the appropriate constant is larger than a simple integer, ich as 2, 3, 4, or 5, you may sometimes find it easier to make the adjustment by ultiplying twice, for example, by 2 and then by 4 if the necessary constant is 8.

Combustion Analysis

Figure 3-6 illustrates an experimental method for establishing an empi formula for compounds that are easily burned, such as compounds contai carbon and hydrogen with oxygen, nitrogen, and a few other element *combustion analysis,* a weighed sample of a compound is burned in a strea oxygen gas. The water vapor and carbon dioxide gas produced in the c bustion are absorbed by appropriate substances. The increases in ma these absorbers correspond to the masses of water and carbon dioxide. We think of the matter as shown below. (The subscripts x, y, and z are inte whose values we do not know initially.)

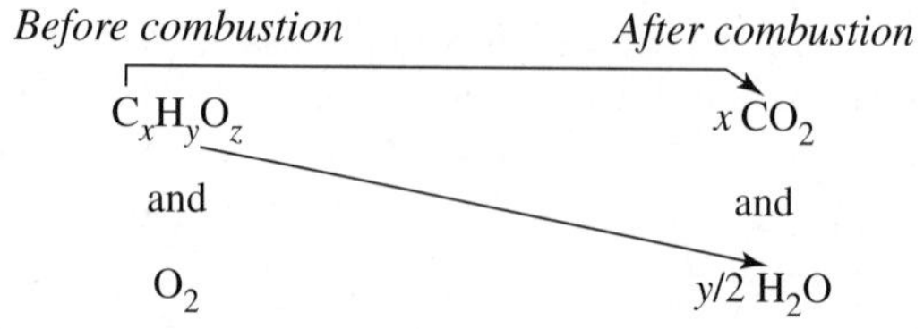

After combustion, all the carbon atoms in the sample are found in the C All the hydrogen atoms are in the H_2O. Moreover, the only source of the ca and hydrogen atoms was the sample being analyzed. Oxygen atoms in the and H_2O could have come partly from the sample and partly from the oxy gas consumed in the combustion. Thus, the quantity of oxygen in the sar has to be determined indirectly. These ideas are applied in Example 3-6.

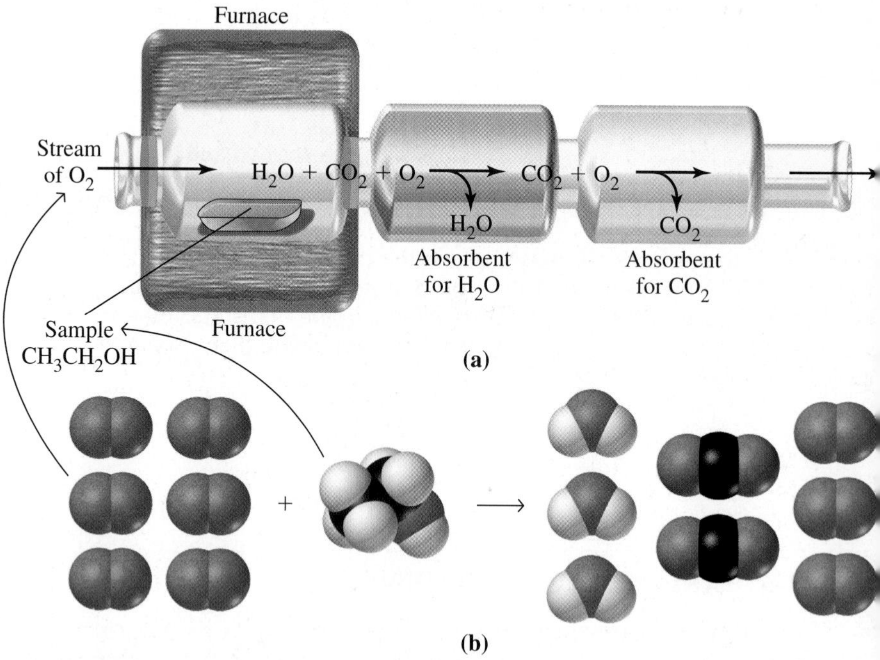

▲ FIGURE 3-6
Apparatus for combustion analysis
(a) Oxygen gas passes through the combustion tube containing the sample being analyzed. This portion of the apparatus is enclosed in a high-temperature furnace. Products of the combustion are absorbed as they leave the furnace—water vapor b magnesium perchlorate, and carbon dioxide gas by sodium hydroxide (producing sodium carbonate). The differences in mass of the absorbers, after and before the combustion, yield the masses of H_2O and CO_2 produced in the combustion reactio **(b)** A molecular picture of the combustion of ethanol. Each molecule of ethanol produces two CO_2 molecules and three H_2O molecules. Combustion takes place in an excess of oxygen, so that oxygen molecules are present at the end of the reactio Note the conservation of mass.

XAMPLE 3-6 Determining an Empirical Formula from Combustion Analysis Data

Vitamin C is essential for the prevention of scurvy. Combustion of a 0.2000 g sample of this carbon–hydrogen–oxygen compound yields 0.2998 g CO_2 and 0.0819 g H_2O. What are the percent composition and the empirical formula of vitamin C?

nalyze

After combustion, all the carbon atoms from the vitamin C sample are in CO_2 and all the hydrogen atoms are in H_2O. However, the oxygen atoms in CO_2 and H_2O come partly from the sample and partly from the oxygen gas consumed in the combustion. So, in the determination of the percent composition, we focus first on carbon and hydrogen and then on oxygen. To determine the empirical formula, we must calculate the amounts of C, H, and O in moles, and then calculate the mole ratios.

olve

Percent Composition

First, determine the mass of carbon in 0.2988 g CO_2, by converting to mol C,

$$? \text{ mol C} = 0.2998 \text{ g } CO_2 \times \frac{1 \text{ mol } CO_2}{44.009 \text{ g } CO_2} \times \frac{1 \text{ mol C}}{1 \text{ mol } CO_2} = 0.006812 \text{ mol C}$$

and then to g C.

$$? \text{ g C} = 0.006812 \text{ mol C} \times \frac{12.011 \text{ g C}}{1 \text{ mol C}} = 0.08182 \text{ g C}$$

Proceed in a similar fashion for 0.0819 g H_2O to obtain

$$? \text{ mol H} = 0.0819 \text{ g } H_2O \times \frac{1 \text{ mol } H_2O}{18.02 \text{ g } H_2O} \times \frac{2 \text{ mol H}}{1 \text{ mol } H_2O} = 0.00909 \text{ mol H}$$

and

$$? \text{ g H} = 0.00909 \text{ mol H} \times \frac{1.008 \text{ g H}}{1 \text{ mol H}} = 0.00916 \text{ g H}$$

Obtain the mass of O in the 0.2000 g sample as the difference

$$? \text{ g O} = 0.2000 \text{ g sample} - 0.08182 \text{ g C} - 0.00916 \text{ g H} = 0.1090 \text{ g O}$$

Finally, multiply the mass fractions of the three elements by 100% to obtain mass percentages.

$$\% \text{ C} = \frac{0.08182 \text{ g C}}{0.2000 \text{ g sample}} \times 100\% = 40.91\% \text{ C}$$

$$\% \text{ H} = \frac{0.00916 \text{ g H}}{0.2000 \text{ g sample}} \times 100\% = 4.58\% \text{ H}$$

$$\% \text{ O} = \frac{0.1090 \text{ g O}}{0.2000 \text{ g sample}} \times 100\% = 54.50\% \text{ O}$$

Empirical Formula

At this point we can choose either of two alternatives. The first is to obtain the empirical formula from the mass percent composition, in the same manner illustrated in Example 3-5. The second is to note that we have already determined the number of moles of C and H in the 0.2000 g sample. The number of moles of O is

$$? \text{ mol O} = 0.1090 \text{ g O} \times \frac{1 \text{ mol O}}{15.999 \text{ g O}} = 0.006813 \text{ mol O}$$

From the numbers of moles of C, H, and O in the 0.2000 g sample, we obtain the tentative empirical formula

$$C_{0.006812}H_{0.00909}O_{0.006813}$$

Next, divide each subscript by the smallest (0.006812) to obtain

$$CH_{1.33}O$$

(continued)

Finally, multiply all the subscripts by 3 to obtain

Empirical formula of vitamin C: $C_3H_4O_3$

Assess

The determination of the empirical formula does not require determining the mass percent composition as preliminary calculation. The empirical formula can be based on a sample of any size, as long as the numbers moles of the different atoms in that sample can be determined.

PRACTICE EXAMPLE A: Isobutyl propionate is the substance that flavors rum extract. Combustion of a 1.152 sample of this carbon–hydrogen–oxygen compound yields 2.726 g CO_2 and 1.116 g H_2O. What is the empiric formula of isobutyl propionate?

PRACTICE EXAMPLE B: Combustion of a 1.505 g sample of thiophene, a carbon–hydrogen–sulfur compoun yields 3.149 g CO_2, 0.645 g H_2O, and 1.146 g SO_2 as the only products of the combustion. What is the empiric formula of thiophene?

3-5 CONCEPT ASSESSMENT

When combustion is carried out in an excess of oxygen, the quantity producing the greatest mass of both CO_2 and H_2O is **(a)** 0.50 mol $C_{10}H_8$; **(b)** 1.25 mol CH **(c)** 0.500 mol C_2H_5OH; **(d)** 1.00 mol C_6H_5OH.

We have just seen how combustion reactions can be used to analyze ch cal substances, but not all samples can be easily burned. Fortunately, sev other types of reactions can be used for chemical analyses. Also, modern m ods in chemistry rely much more on physical measurements with instrum than on chemical reactions. We will cite some of these methods later in the

3-4 Oxidation States: A Useful Tool in Describing Chemical Compounds

▶ The oxidation state (O.S.) can be described as "a sometimes fictional charge." With monatomic ions, O.S. and charge are the same thing. For polyatomic ions and molecules, O.S. is fictional but equal to what the charge would be if the compounds were entirely ionic.

Most basic concepts in chemistry deal with measurable properties or phen ena. In a few instances, though, a concept has been devised more for co nience than because of any fundamental significance. This is the case with **oxidation state** (oxidation number),* which is related to the number of e trons that an atom loses, gains, or otherwise appears to use in joining w other atoms in compounds.

Consider NaCl. In this compound, an Na atom, a metal, loses one elec to a Cl atom, a nonmetal. The compound consists of the ions Na^+ and Cl^- Figure 3-4). Na^+ is in a +1 oxidation state and Cl^- is in a −1 state.

In $MgCl_2$, an Mg atom loses two electrons to become Mg^{2+}, and e Cl atom gains one electron to become Cl^-. As in NaCl, the oxidation state c is −1, but that of Mg is +2. If we take the *total* of the oxidation states of al atoms (ions) in a formula unit of $MgCl_2$, we get $+2 - 1 - 1 = 0$.

In the molecule Cl_2, the two Cl atoms are identical and should have same oxidation state. But if their total is to be zero, each oxidation state m itself be 0. Thus, the oxidation state of an atom can vary, depending on compounds in which it occurs. In the molecule H_2O, we arbitrarily assig the oxidation state of +1. Then, because the total of the oxidation states of atoms must be zero, the oxidation state of oxygen must be −2.

*Because oxidation state refers to a number, the term oxidation number is often used syr mously. We will use the two terms interchangeably.

om these examples, you can see that we need some conventions or rules ssigning oxidation states. The seven rules in Table 3.2 are sufficient to deal most cases in this text, with this understanding: *The rules must be applied e numerical order listed, and whenever two rules appear to contradict each other ch they sometimes will), follow the lower numbered rule.* Some examples are n for each rule, and the rules are applied in Example 3-7.

BLE 3.2 Rules for Assigning Oxidation States

The oxidation state (O.S.) of an individual atom in a free element (uncombined with other elements) is 0.

[*Examples:* The O.S. of an isolated Cl atom is 0; the two Cl atoms in the molecule Cl_2 both have an O.S. of 0.]

The sum of the O.S. of all the atoms in

(a) *neutral species, such as isolated atoms, molecules, and formula units, is 0;*

[*Examples:* The sum of the O.S. of all the atoms in CH_3OH and of all the ions in $MgCl_2$ is 0.]

(b) *an ion is equal to the charge on the ion.*

[*Examples:* The O.S. of Fe in Fe^{3+} is +3. The sum of the O.S. of all atoms in MnO_4^- is −1.]

In their compounds, the group 1 metals have an O.S. of +1 and the group 2 metals have an O.S. of +2.

[*Examples:* The O.S. of K is +1 in KCl and K_2CO_3; the O.S. of Mg is +2 in $MgBr_2$ and $Mg(NO_3)_2$.]

In its compounds, the O.S. of fluorine is −1.

[*Examples:* The O.S. of F is −1 in HF, ClF_3, and SF_6.]

In its compounds, hydrogen usually has an O.S. of +1.

[*Examples:* The O.S. of H is +1 in HI, H_2S, NH_3, and CH_4.]

In its compounds, oxygen usually has an O.S. of −2.

[*Examples:* The O.S. of O is −2 in H_2O, CO_2 and $KMnO_4$.]

In binary (two-element) compounds with metals, group 17 elements have an O.S. of −1; group 16 elements, −2; and group 15 elements, −3.

[*Examples:* The O.S. of Br is −1 in $MgBr_2$; the O.S. of S is −2 in Li_2S; and the O.S. of N is −3 in Li_3N.]

◀ The principal exceptions to rule 5 occur when H is bonded to metals, as in LiH, NaH, and CaH_2; exceptions to rule 6 occur in compounds with O—F bonds, such as OF_2, and in compounds where O atoms are bonded to one another, as in H_2O_2, and KO_2.

XAMPLE 3-7 Assigning Oxidation States

What is the oxidation state of the underlined element in **(a)** $\underline{P}_4$; **(b)** $\underline{Al}_2O_3$; **(c)** $\underline{Mn}O_4^-$; **(d)** $Na\underline{H}$; **(e)** $H_2\underline{O}_2$; **(f)** $\underline{Fe}_3O_4$?

nalyze

Apply the rules in Table 3.2.

olve

(a) P_4: This formula represents a molecule of elemental phosphorus. For an atom of a free element, the O.S. = 0 (rule 1). The O.S. of P in P_4 is 0.

(b) Al_2O_3: The total of the oxidation states of all the atoms in this formula unit is 0 (rule 2). The O.S. of oxygen is −2 (rule 6). The total for three O atoms is −6. The total for two Al atoms is +6. The O.S. of Al is +3.

(c) MnO_4^-: This is the formula for permanganate ion. The sum of the oxidation states of all the atoms in the ion is −1 (rule 2). The total for the four O atoms is −8. The O.S. of Mn is +7.

(d) NaH: This is a formula unit of the ionic compound sodium hydride. Rule 3 states that the O.S. of Na is +1. Rule 5 indicates that H should also have an O.S. of +1. If both atoms had an O.S. of +1, the total for the formula unit would be +2. This violates rule 2. *Rules 2 and 3 take precedence over rule 5.* Na has an O.S. of +1; the total for the formula unit is 0; and the O.S. of H must be −1.

(continued)

(e) H_2O_2: This is hydrogen peroxide. Rule 5, stating that H has an O.S. of +1, takes precedence over rul (which says that oxygen has an O.S. of −2). The sum of the oxidation states of the two H atoms is and that of the two O atoms must be −2. The O.S. of O must be −1.

(f) Fe_3O_4: The total of the oxidation states of four O atoms is −8. For three Fe atoms, the sum of the oxidati states must be +8. The O.S. per Fe atom is $\frac{8}{3}$ or $+2\frac{2}{3}$.

Assess

With practice, you should be able to do the arithmetic associated with assigning oxidation states in your hea that is, without writing down any arithmetic expressions. Also, when you have determined an oxidation sta by using arithmetic (as required by rule 2), check your result by making sure the sum of the oxidation states equal to the charge on the atom, molecule, or ion. For example, in part (c), we determined that the oxidati state of Mn in MnO_4^- is +7. We know this result is correct because the sum of the oxidation states $7 + 4(-2) = -1$, which is equal to the charge on the MnO_4^- ion.

PRACTICE EXAMPLE A: What is the oxidation state of the underlined element in each of the followin $\underline{S}_8$; $\underline{Cr}_2O_7^{2-}$; $\underline{Cl}_2O$; $K\underline{O}_2$?

PRACTICE EXAMPLE B: What is the oxidation state of the underlined element in each of the followin $\underline{S}_2O_3^{2-}$; $\underline{Hg}_2Cl_2$; $K\underline{Mn}O_4$; $H_2\underline{C}O$?

In part (f) of Example 3-7, we got the somewhat surprising answer of for the oxidation state of the iron atoms in Fe_3O_4. Before that, we saw integral values for oxidation states. How does this fractional value c about? Generally, it comes from the assumption that all the atoms of an ment have the same oxidation state in a given compound. Usually they do not always. Fe_3O_4, for example, is probably better represented as $FeO \cdot Fe$ that is, through a combination of two simpler formula units. In FeO, the of the Fe atom is +2. In Fe_2O_3, the O.S. of each of *two* Fe atoms is +3. Whe *average* the oxidation states over all three Fe atoms, we get a nonintegral va $(2 + 3 + 3)/3 = \frac{8}{3} = 2\frac{2}{3}$.

Also, we may at times need to "fragment" a formula into its constit parts before assigning oxidation states. The ionic compound NH_4NO_3 instance, consists of the ions NH_4^+ and NO_3^-. The oxidation state of NH_4^+ is −3, and in NO_3^-, +5, and we do *not* want to average them. It i more useful to know the oxidation states of the individual N atoms than it deal with an average oxidation state of +1 for the two N atoms.

Our first use of oxidation states comes in the naming of chemical c pounds in the next section.

3-6 CONCEPT ASSESSMENT

A nitrogen–hydrogen compound with molar mass 32 g/mol has N in a higher oxidation state than in NH_3. What is a plausible formula for that compound?

Carey B. Van Loon

▲ FIGURE 3-7
Two oxides of lead
These two compounds contain the same elements—lead and oxygen—but in different proportions. Their names and formulas must convey this fact: lead(IV) oxide = PbO_2 (red-brown); lead(II) oxide = PbO (yellow).

3-5 Naming Compounds: Organic and Inorganic Compounds

Throughout this chapter, we have referred to compounds mostly by their for las, but we do need to give them names. When we know the name of a c pound, we can look up its properties in a handbook, locate a chemical storeroom shelf, or discuss an experiment with a colleague. Later in the text will see cases in which different compounds have the same formula. In th instances, we will find it essential to distinguish among compounds by na We cannot give two substances the same name, yet we do want some similar in the names of similar substances (Fig. 3-7). If all compounds were referred t a common or trivial name, such as water (H_2O), ammonia (NH_3), or glu

$I_{12}O_6$), we would have to learn millions of unrelated names—an impossi-y. What we need is a systematic method of assigning names—a system of *enclature*. Several systems are used, and we will introduce each at an ropriate point in the text. Compounds formed by carbon and hydrogen or on and hydrogen together with oxygen, nitrogen, and a few other ele-ts are **organic compounds**. They are generally considered in a special ch of chemistry—*organic chemistry*—that has its own set of nomenclature s. Compounds that do not fit this description are **inorganic compounds**. branch of chemistry that concerns itself with the study of these com-nds is called *inorganic chemistry*. In the next section, we will consider the ing of inorganic compounds, and in Section 3-7, we will introduce the ing of organic compounds.

◀ *Nomenclature of Organic Chemistry* (Blue Book) and *Nomenclature of Inorganic Chemistry* (Red Book) can be found at www.iupac.org.

5 Names and Formulas of Inorganic Compounds

ary Compounds of Metals and Nonmetals

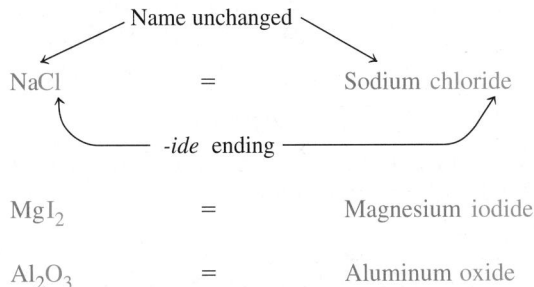

ary compounds are those formed between *two elements*. If one of the ele-ts is a metal and the other a nonmetal, the binary compound is usually le up of ions; that is, it is a binary *ionic* compound. To name a binary com-nd of a metal and a nonmetal,

write the *unmodified* name of the metal

then write the name of the nonmetal, modified to end in *-ide*

he approach is illustrated below for NaCl, MgI_2, and Al_2O_3.

Name unchanged

NaCl = Sodium chloride

-ide ending

MgI_2 = Magnesium iodide

Al_2O_3 = Aluminum oxide

onic compounds, though made up of positive and negative ions, must be *rically neutral*. The net, or total, charge of the ions in a formula unit must be). This means one Na^+ to one Cl^- in NaCl; one Mg^{2+} to two I^- in MgI_2; two to three O^{2-} in Al_2O_3; and so on. Table 3.3 lists the names and symbols of ple ions formed by metals and nonmetals. You will find this list useful en writing names and formulas of binary compounds of metals and non-als. Because some metals may form several ions, it is important to distin-sh between them in naming their compounds. The metal iron, for example, ns *two* common ions, Fe^{2+} and Fe^{3+}. The first is called the *iron(II)* ion, and second is the *iron(III)* ion. The Roman numeral immediately following the ne of the metal indicates its oxidation state or simply the charge on the ion. s, $FeCl_2$ is *iron(II)* chloride, while $FeCl_3$ is *iron(III)* chloride.

n earlier system of nomenclature that is still used to some extent applies different word endings to distinguish between two binary compounds taining the same two elements but in different proportions, such as Cu_2O CuO. In Cu_2O, the oxidation state of copper is +1, and in CuO it is +2. O is assigned the name cupr*ous* oxide, and CuO is cupr*ic* oxide. Similarly, l_2 is ferr*ous* chloride, and $FeCl_3$ is ferr*ic* chloride. The idea is to use the *-ous* ing for the lower oxidation state of the metal and *-ic* for the higher oxida-state. The *ous/ic* system has several inadequacies though, and we will not

▶ Table 3.3 simplifies considerably once you understand the pattern. All group 1 ions are +1, group 2 ions are +2, and Al^{3+}, Zn^{2+}, and Ag^+ have only the one common charge. Most other metal ions have variable charges. Monatomic halogen ions are −1 and oxygen is −2. Almost everything else can be worked out from these basics and the rule that the sum of all O.S. equals the total charge.

TABLE 3.3 Some Simple Ions

Name	Symbol	Name	Symbol
Positive ions (cations)			
Lithium ion	Li^+	Chromium(II) ion	Cr^{2+}
Sodium ion	Na^+	Chromium(III) ion	Cr^{3+}
Potassium ion	K^+	Iron(II) ion	Fe^{2+}
Rubidium ion	Rb^+	Iron(III) ion	Fe^{3+}
Cesium ion	Cs^+	Cobalt(II) ion	Co^{2+}
Magnesium ion	Mg^{2+}	Cobalt(III) ion	Co^{3+}
Calcium ion	Ca^{2+}	Copper(I) ion	Cu^+
Strontium ion	Sr^{2+}	Copper(II) ion	Cu^{2+}
Barium ion	Ba^{2+}	Mercury(I) ion	$Hg_2{}^{2+}$
Aluminum ion	Al^{3+}	Mercury(II) ion	Hg^{2+}
Zinc ion	Zn^{2+}	Tin(II) ion	Sn^{2+}
Silver ion	Ag^+	Lead(II) ion	Pb^{2+}
Negative ions (anions)			
Hydride ion	H^-	Iodide ion	I^-
Fluoride ion	F^-	Oxide ion	O^{2-}
Chloride ion	Cl^-	Sulfide ion	S^{2-}
Bromide ion	Br^-	Nitride ion	N^{3-}

use it in this text. For example, the *-ous* and *-ic* endings do not help in nam the four oxides of vanadium: VO, V_2O_3, VO_2, and V_2O_5.

Binary Compounds of Two Nonmetals

If the two elements in a binary compound are both nonmetals instead metal and a nonmetal, the compound is a molecular compound. The met of naming these compounds is similar to that just discussed. For example,

$$HCl = \text{hydrogen chloride}$$

In both the formula and the name, we write the element with the positive dation state first: HCl and *not* ClH.

EXAMPLE 3-8 Writing Formulas When Names of Compounds Are Given

Write formulas for the compounds barium oxide, calcium fluoride, and iron(III) sulfide.

Analyze

In each case, identify the cations and their charges, based on periodic table group numbers or on oxidatio states appearing as Roman numerals in names: Ba^{2+}, Ca^{2+}, and Fe^{3+}. Then identify the anions and thei charges: O^{2-}, F^-, and S^{2-}. Combine the cations and anions in the relative numbers required to produc electrically *neutral* formula units.

Solve

barium oxide: *one* Ba^{2+} and *one* O^{2-} = BaO

calcium fluoride: *one* Ca^{2+} and *two* F^- = CaF_2

iron(III) sulfide: *two* Fe^{3+} and *three* S^{2-} = Fe_2S_3

Assess

In the first case, an electrically neutral formula unit results from the combination of the charges +2 and −2 in the second case, +2 and 2 × (−1); and in the third case, 2 × (+3) and 3 × (−2).

PRACTICE EXAMPLE A: Write formulas for lithium oxide, tin(II) fluoride, and lithium nitride.

PRACTICE EXAMPLE B: Write formulas for the compounds aluminum sulfide, magnesium nitride, an vanadium(III) oxide.

XAMPLE 3-9 Naming Compounds When Their Formulas Are Given

Write acceptable names for the compounds Na_2S, AlF_3, Cu_2O.

nalyze

This task is generally easier than that of Example 3-8 because all you need to do is name the ions present. However, you must recognize that copper forms two different ions and that the cation in Cu_2O is Cu^+, copper(I).

olve

Na_2S: sodium sulfide
AlF_3: aluminum fluoride
Cu_2O: copper(I) oxide

ssess

Knowing when to use Roman numerals and when not to use them is a tricky aspect of naming ionic compounds. Because the metals of groups 1 and 2 have only one ionic form (one oxidation state), Roman numerals are not used in naming their compounds.

RACTICE EXAMPLE A: Write acceptable names for CsI, CaF_2, FeO, $CrCl_3$.

RACTICE EXAMPLE B: Write acceptable names for CaH_2, CuCl, Ag_2S, Hg_2Cl_2.

ome pairs of nonmetals form more than one binary molecular compound, we need to distinguish among them. Generally, we indicate relative num- of atoms through *prefixes*: *mono* = 1, *di* = 2, *tri* = 3, *tetra* = 4, *penta* = 5, = 6, *hepta* = 7, *octa* = 8, *nona* = 9, *deca* = 10, and so on. Thus, for the principal oxides of sulfur we write

SO_2 = sulfur *di*oxide
SO_3 = sulfur *tri*oxide

for the following boron–bromine compound

B_2Br_4 = *di*boron *tetra*bromide

dditional examples are given in Table 3.4. Note that in these examples the ix *mono-* is treated in a special way. We do not use it for the first named

TABLE 3.4 Naming Binary Molecular Compounds

Formula	Name[a]
BCl_3	Boron trichloride
CCl_4	Carbon tetrachloride
CO	Carbon monoxide
CO_2	Carbon dioxide
NO	Nitrogen monoxide
NO_2	Nitrogen dioxide
N_2O	Dinitrogen monoxide
N_2O_3	Dinitrogen trioxide
N_2O_4	Dinitrogen tetroxide
N_2O_5	Dinitrogen pentoxide
PCl_3	Phosphorus trichloride
PCl_5	Phosphorus pentachloride
SF_6	Sulfur hexafluoride

[a]When the prefix ends in *a* or *o* and the element name begins with *a* or *o*, the final vowel of the prefix is dropped for ease of pronunciation. For example, carbon *mono*xide, not carbon *mono*oxide, and dinitrogen *tetr*oxide, not dinitrogen *tetra*oxide. However, PI_3 is phosphorus *tri*iodide, not phosphorus *tri*odide.

element. Thus, NO is called nitrogen monoxide, *not mono*nitrogen *monox* Finally, several substances have common or trivial names that are so established that their systematic names are almost never used. For examp

H_2O = water (*di*hydrogen *mono*xide)

NH_3 = ammonia (H_3N = *tri*hydrogen *mono*nitride)

Binary Acids

Even though we use names like hydrogen chloride for pure binary molec compounds, we sometimes want to emphasize that their aqueous solution acids. Acids will be discussed in detail later in the text. For now, so that we recognize these substances and name them when we see their formulas, le say that an *acid* is a substance that *ionizes* or breaks down in water to proc hydrogen ions (H^+)* and anions. A *binary acid* is a two-element compoun hydrogen and a nonmetal. For example, HCl ionizes into hydrogen ions and chloride ions (Cl^-) in water; it is a binary acid. NH_3 in water is *not* an a It shows practically no tendency to produce H^+ under any conditions. I belongs to a complementary category of substances called *bases*. As we wil in Chapter 5, bases yield hydroxide ions (OH^-) in aqueous solutions.

In naming binary acids we use the prefix *hydro-* followed by the name o other nonmetal modified with an *-ic* ending. The most important binary a are listed below.

▶ The symbol (aq) signifies a substance in aqueous (water) solution.

HF(aq) = *hydro*fluor*ic* acid

HBr(aq) = *hydro*brom*ic* acid

HCl(aq) = *hydro*chlor*ic* acid

HI(aq) = *hydro*iod*ic* acid

H_2S(aq) = *hydro*sulfur*ic* acid

Polyatomic Ions

With the exception of Hg_2^{2+}, the ions listed in Table 3.3 are *monatomic*—e consists of a single atom. In **polyatomic ions**, two or more atoms are joi together by covalent bonds. These ions are common, especially among nonmetals. A number of polyatomic ions and compounds containing them listed in Table 3.5. From this table, you can see the following:

1. Polyatomic anions are more common than polyatomic cations. The m familiar polyatomic cation is the ammonium ion, NH_4^+.
2. Very few polyatomic anions carry the *-ide* ending in their names. Of th listed, only OH^- (hydroxide ion) and CN^- (cyanide ion) do. The comm endings are *-ite* and *-ate*, and some names carry prefixes, *hypo-* or *per-*.
3. An element common to many polyatomic anions is *oxygen*, usuall combination with another nonmetal. Such anions are called **oxoanions**
4. Certain nonmetals (such as Cl, N, P, and S) form a series of oxoanions c taining different numbers of oxygen atoms. Their names are related to oxidation state of the nonmetal atom to which the O atoms are bonc ranging from *hypo-* (lowest) to *per-* (highest) according to the follow scheme.

Increasing oxidation state of nonmetal →

*hypo*___*ite* ___*ite* ___*ate* *per*___*ate*

Increasing number of oxygen atoms →

*The species produced in aqueous solution is actually more complex than the simple ion The H^+ combines with an H_2O molecule to produce an ion known as the hydronium ion, H Chemists often use H^+ in place of H_3O^+, and that is what we will do until we discuss this m more fully in Chapter 5.

TABLE 3.5 Some Common Polyatomic Ions

Name	Formula	Typical Compound
Cation		
Ammonium ion	NH_4^+	NH_4Cl
Anions		
Acetate ion	CH_3COO^-	$NaCH_3COO$
Carbonate ion	CO_3^{2-}	Na_2CO_3
Hydrogen carbonate ion[a] (or bicarbonate ion)	HCO_3^-	$NaHCO_3$
Hypochlorite ion	ClO^-	$NaClO$
Chlorite ion	ClO_2^-	$NaClO_2$
Chlorate ion	ClO_3^-	$NaClO_3$
Perchlorate ion	ClO_4^-	$NaClO_4$
Chromate ion	CrO_4^{2-}	Na_2CrO_4
Dichromate ion	$Cr_2O_7^{2-}$	$Na_2Cr_2O_7$
Cyanide ion	CN^-	$NaCN$
Hydroxide ion	OH^-	$NaOH$
Nitrite ion	NO_2^-	$NaNO_2$
Nitrate ion	NO_3^-	$NaNO_3$
Oxalate ion	$C_2O_4^{2-}$	$Na_2C_2O_4$
Permanganate ion	MnO_4^-	$NaMnO_4$
Phosphate ion	PO_4^{3-}	Na_3PO_4
Hydrogen phosphate ion[a]	HPO_4^{2-}	Na_2HPO_4
Dihydrogen phosphate ion[a]	$H_2PO_4^-$	NaH_2PO_4
Sulfite ion	SO_3^{2-}	Na_2SO_3
Hydrogen sulfite ion[a] (or bisulfite ion)	HSO_3^-	$NaHSO_3$
Sulfate ion	SO_4^{2-}	Na_2SO_4
Hydrogen sulfate ion[a] (or bisulfate ion)	HSO_4^-	$NaHSO_4$
Thiosulfate ion	$S_2O_3^{2-}$	$Na_2S_2O_3$

[a]These anion names are sometimes written as a single word—for example, hydrogencarbonate, hydrogenphosphate, and so forth.

◀ Learn the most common ions first, such as NH_4^+, CO_3^{2-}, OH^-, NO_3^-, PO_4^{3-}, SO_4^{2-}, and ClO_3^-. When you understand the scheme in expression (3.3), the names of several others, such as NO_2^-, SO_3^{2-}, ClO^-, ClO_2^-, and ClO_4^- will become obvious. Over time, the rest will become more familiar to you.

All the common oxoanions of Cl, Br, and I carry a charge of −1.

Some series of oxoanions also contain various numbers of H atoms and are named accordingly. For example, HPO_4^{2-} is the *hydrogen phosphate* ion and $H_2PO_4^-$, the *dihydrogen phosphate* ion.

The prefix *thio-* signifies that a sulfur atom has been substituted for an oxygen atom. (The sulfate ion has *one* S and *four* O atoms; thiosulfate ion has *two* S and *three* O atoms.)

oacids

majority of acids are **ternary compounds**. They contain *three* different nents—hydrogen and two other nonmetals. If one of the nonmetals is oxy-, the acid is called an **oxoacid**. Think of oxoacids as combinations of hydro- ions (H^+) and oxoanions. The scheme for naming oxoacids is similar to t outlined for oxoanions, except that the ending *-ous* is used instead of *-ite* -*ic* instead of *-ate*. Several oxoacids are listed in Table 3.6. Also listed are names and formulas of compounds in which the hydrogen of the oxoacid been replaced by a metal, such as sodium. These compounds are called s; we will say much more about them in later chapters, beginning in apter 5. Acids are molecular compounds, and salts are ionic compounds.

▶ When we refer to perchloric acid, for example, we generally mean an aqueous solution of $HClO_4$.

TABLE 3.6 Nomenclature of Some Oxoacids and Their Salts

Oxidation State	Formula of Acid[a]	Name of Acid[b]	Formula of Salt	Name of Salt[b]
Cl: +1	$HClO$	*Hypo*chlor*ous* acid	$NaClO$	Sodium *hypo*chlor
Cl: +3	$HClO_2$	Chlor*ous* acid	$NaClO_2$	Sodium chlor*ite*
Cl: +5	$HClO_3$	Chlor*ic* acid	$NaClO_3$	Sodium chlor*ate*
Cl: +7	$HClO_4$	*Per*chlor*ic* acid	$NaClO_4$	Sodium *per*chlorat
N: +3	HNO_2	Nitr*ous* acid	$NaNO_2$	Sodium nitr*ite*
N: +5	HNO_3	Nitr*ic* acid	$NaNO_3$	Sodium nitr*ate*
S: +4	H_2SO_3	Sulfur*ous* acid	Na_2SO_3	Sodium sulf*ite*
S: +6	H_2SO_4	Sulfur*ic* acid	Na_2SO_4	Sodium sulf*ate*

[a]In all these acids, H atoms are bonded to O atoms, not the central nonmetal atom. Often formulas are written to reflect this fact, for instance, HOCl instead of HClO and HOClO instead of $HClO_2$.
[b]In general, the *-ic* and *-ate* names are assigned to compounds in which the central nonmeta atom has an oxidation state equal to the periodic table group number minus 10. Halogen compounds are exceptional in that the *-ic* and *-ate* names are assigned to compounds in wh the halogen has an oxidation state of +5 (even though the group number is 17).

EXAMPLE 3-10 Applying Various Rules for Naming Compounds

Name the compounds **(a)** $CuCl_2$; **(b)** ClO_2; **(c)** HIO_4; **(d)** $Ca(H_2PO_4)_2$.

Analyze

$CuCl_2$ and $Ca(H_2PO_4)_2$ are ionic compounds. To name these compounds, we must identify and name the ion ClO_2 and HIO_4 are molecular compounds. ClO_2 is a binary compound of two nonmetals, and HIO_4 an oxoacid.

Solve

(a) In this compound, the oxidation state of Cu is +2. Because Cu can also exist in the oxidation state of + we must clearly distinguish between the two possible chlorides. $CuCl_2$ is copper(II) chloride.

(b) Both Cl and O are nonmetals. ClO_2 is a binary molecular compound called chlorine dioxide.

(c) The oxidation state of I is +7. By analogy to the chlorine-containing oxoacids in Table 3.6, we shoul name this compound periodic acid (pronounced "purr-eye-oh-dic" acid).

(d) The polyatomic anion $H_2PO_4^-$ is dihydrogen phosphate ion. Two of these ions are present for ever Ca^{2+} ion in the compound calcium dihydrogen phosphate.

Assess

A fair bit of memorization is associated with naming compounds correctly. Mastery of this subject usuall requires a lot of practice.

PRACTICE EXAMPLE A: Name the compounds SF_6, HNO_2, $Ca(HCO_3)_2$, $FeSO_4$.

PRACTICE EXAMPLE B: Name the compounds NH_4NO_3, PCl_3, HBrO, $AgClO_4$, $Fe_2(SO_4)_3$.

Some Compounds of Greater Complexity

▶ In general, centered dots (·) show that a formula is a composite of two or more simpler formulas.

The copper compound that Joseph Proust used to establish the law of c stant composition (page 36) is referred to in different ways. If you look Proust's compound in a handbook of minerals, you will find it listec *malachite*, with the formula $Cu_2(OH)_2CO_3$. In a handbook that specialize pharmaceutical applications, this same compound is listed as *basic cupric bonate*, with the formula $CH_2Cu_2O_5$. In a chemistry handbook, it is listec *copper(II) carbonate dihydroxide*, with the formula $CuCO_3 \cdot Cu(OH)_2$. All need to understand at this point is that regardless of the formula you use,

XAMPLE 3-11 Applying Various Rules in Writing Formulas

Write the formula of the compound **(a)** tetranitrogen tetrasulfide; **(b)** ammonium chromate; **(c)** bromic acid; **(d)** calcium hypochlorite.

nalyze

We must apply our knowledge of prefixes (such as *tetra-*) and endings (such as *-ic*), as well as the names of common polyatomic ions (such as *ammonium*, *chromate*, and *hypochlorite*).

olve

(a) Molecules of this compound consist of *four* N atoms and *four* S atoms. The formula is N_4S_4.

(b) Two ammonium ions (NH_4^+) must be present for every chromate ion (CrO_4^{2-}). Place parentheses around NH_4^+, followed by the subscript 2. The formula is $(NH_4)_2CrO_4$. (This formula is read as "N-H-4, taken twice, C-R-O-4.")

(c) The *-ic* acid for the oxoacids of the halogens (group 17) has the halogen in the oxidation state of +5. Bromic acid is $HBrO_3$ (analogous to $HClO_3$ in Table 3.6).

(d) Here there are one Ca^{2+} and two ClO^- ions in a formula unit. This leads to the formula $Ca(ClO)_2$.

ssess

Notice that in writing formulas for compounds containing two or more polyatomic ions of the *same type*, as in (b) and (d), we put parentheses around the formula of the ion (without the charge), followed by a subscript indicating the number of ions of that type. Proper use and placement of parentheses in writing formulas is important.

RACTICE EXAMPLE A: Write formulas for the compounds **(a)** boron trifluoride, **(b)** potassium dichromate, **(c)** sulfuric acid, **(d)** calcium chloride.

RACTICE EXAMPLE B: Write formulas for the compounds **(a)** aluminum nitrate, **(b)** tetraphosphorus decoxide, **(c)** chromium(III) hydroxide, **(d)** iodic acid.

ıld obtain the same molar mass (221.116 g/mol), the same mass percent per (57.48% Cu), the same H:O mole ratio (2 mol H/5 mol O), and so on. ıort, you should be able to interpret a formula, no matter how complex its earance.

ome complex substances you are certain to encounter are known as rates. In a **hydrate**, each formula unit of the compound has associated ı it a certain number of water molecules. This does not mean that the comnds are "wet," however. The water molecules are incorporated in the solid cture of the compound. The formula shown below signifies six H_2O moles per formula unit of $CoCl_2$.

$$CoCl_2 \cdot 6\,H_2O$$

prefix for six is *hexa-* and this compound is cobalt(II) chloride *hexa*hyte. Its formula mass is that of $CoCl_2$ *plus* that associated with six H_2O: 83 u + (6 × 18.015 u) = 237.92 u. We can speak of the mass percent er in a hydrate; for $CoCl_2 \cdot 6\,H_2O$ this is

$$\%\,H_2O = \frac{(6 \times 18.015)\text{ g }H_2O}{237.92\text{ g }CoCl_2 \cdot 6\,H_2O} \times 100\% = 45.43\%$$

he water present in compounds as water of hydration can generally be oved, in part or totally, by heating. When the water is totally removed, the lting compound is said to be *anhydrous* (without water). Anhydrous comnds can be used as water absorbers, as in the use of anhydrous magnesium chlorate in combustion analysis (recall Figure 3-6). $CoCl_2$ gains and loses er quite readily and indicates this through a color change. Anhydrous Cl_2 is blue, whereas the hexahydrate is pink. This fact can be used to make a ple moisture detector (Fig. 3-8).

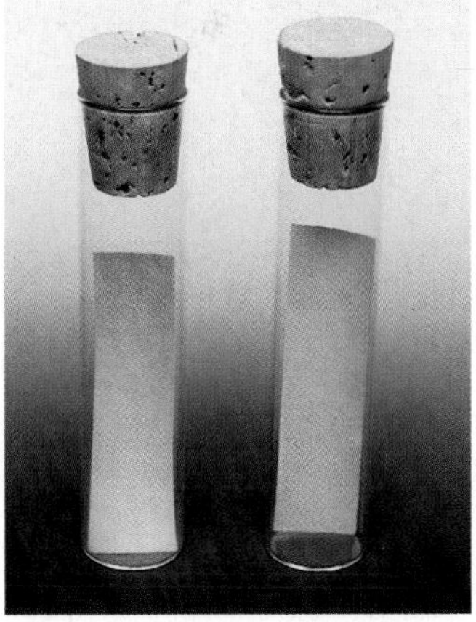

Tom Pantages

▲ FIGURE 3-8
Effect of moisture on $CoCl_2$
The piece of filter paper was soaked in a water solution of cobalt(II) chloride and then allowed to dry. When kept in dry air, the paper is blue (anhydrous $CoCl_2$). In humid air, the paper changes to pink ($CoCl_2 \cdot 6\,H_2O$).

3-7 Names and Formulas of Organic Compounds

Organic compounds abound in nature. The foods we eat are made up al exclusively of organic compounds, including not only energy-producing and carbohydrates and muscle-building proteins but also trace compou that impart color, odor, and flavor to these foods. Almost all fuels, whe used to power automobiles, trucks, trains, or airplanes, are mixtures of org compounds of a type called hydrocarbons. Most of the drugs produce pharmaceutical companies are complex organic compounds, as are com plastics. The multiplicity of organic compounds is so vast that *organic chem* exists as a distinctive field of chemistry.

The great diversity of organic compounds arises from the ability of ca atoms to combine readily with other carbon atoms and with atoms of a n ber of other elements. Carbon atoms join together to form a framewor chains or rings to which other atoms are attached. All organic compou contain carbon atoms; almost all contain hydrogen atoms; and many com ones also have oxygen, nitrogen, or sulfur atoms. These possibilities allow an almost limitless number of different organic compounds. Organic c pounds are mostly molecular; a few are ionic.

KEEP IN MIND

that each carbon atom forms four covalent bonds.

There are millions of organic compounds, many comprising highly c plex molecules. Their names are equally complicated. A systematic appr to naming these compounds is crucial, and the rules for naming inorg compounds are of little use here. The usual name, often called the commo *trivial* name, for a familiar sweetener is sucrose (sugar). The systematic n is α-D-glucopyranosyl-β-D-fructofuranoside. At this point, however, we need to recognize organic compounds and use their common names, toge with an occasional systematic name. We will look at the systematic nomen ture of organic compounds in more detail in Chapter 26.

Hydrocarbons

Compounds containing only carbon and hydrogen are called **hydrocarb** The simplest hydrocarbon contains one carbon atom and four hydro atoms—methane, CH_4 (Fig. 3-9a). As the number of carbon atoms increases number of hydrogen atoms also increases in a systematic way, depending on type of hydrocarbon. The complexity of organic chemistry arises because bon atoms can form chains and rings, and the nature of the chemical bo between the carbon atoms can vary. Hydrocarbons containing only single bo are called **alkanes**. The simplest alkane is methane, followed by ethane, C (Fig. 3-9b), and then propane, C_3H_8 (Fig. 3-9c). The fourth member of the se butane, C_4H_{10}, was shown in Figure 3-2(a). Notice that each succeeding m ber of the alkane series is formed by the addition of one C atom and tw atoms to the preceding member.

TABLE 3.7 Word Stem (or Prefix) Indicating the Number of Carbon Atoms in Simple Organic Molecules

Stem (or prefix)	Number of C Atoms
Meth-	1
Eth-	2
Prop-	3
But-	4
Pent-	5
Hex-	6
Hept-	7
Oct-	8
Non-	9
Dec-	10

The names of the alkanes are composed of two parts: a word stem (or fix) and the ending (or suffix) *-ane* indicating that the molecule is an *alk* The first four word stems in Table 3.7 reflect common names, while the indicate the number of carbon atoms in the alkane. Thus C_5H_{12} is pentane C_7H_{16} is heptane (hept = 7).

Hydrocarbon molecules with one or more double bonds between car atoms are called *alkenes*. The simplest of the alkenes is ethene (Fig. 3-9d) name consists of the stem *eth-* and the ending *-ene*. Benzene, C_6H_6 (Fig. 3- is a molecule with six carbon atoms arranged in a hexagonal ring. Molec with structures related to benzene make up a large proportion of kno organic compounds.

Refer to Figure 3-2 (page 71) and you will notice that butane and met propane have the same molecular formula, C_4H_{10}, but different structural mulas. Butane is based on a four-carbon chain, whereas in methylpropar

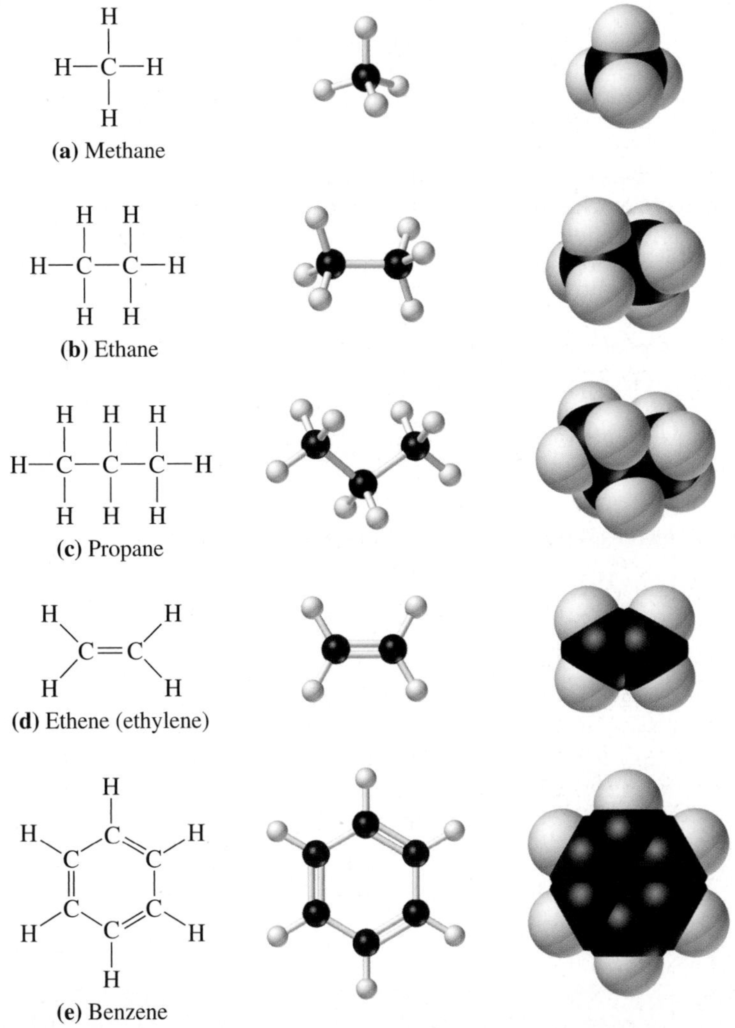

◀ FIGURE 3-9
Visualizations of some hydrocarbons

CH_3 group, called a *methyl* group, is attached to the middle carbon atom of three-carbon propane chain. Butane and methylpropane are *isomers*. **ners** are molecules that have the same molecular formula but different ngements of atoms in space. As organic molecules become more complex, possibilities for isomerism increase very rapidly.

XAMPLE 3-12 Recognizing Isomers

Are the following pairs of molecules isomers or not?

(a) $CH_3CH(CH_3)(CH_2)_3CH_3$ and $CH_3CH_2CH(CH_3)(CH_2)_3CH_3$

(b)

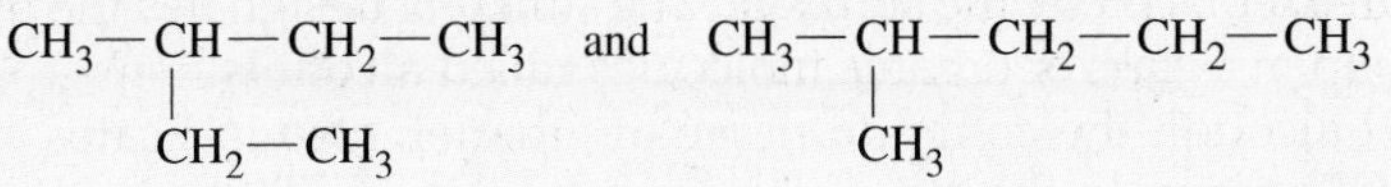

nalyze

Isomers have the same molecular formula but different structures. We first check to see if the molecular formulas are the same. If the formulas are not the same, they represent different compounds. If the formulas are the same, the compounds *may* be isomers, but only if their structures are different.

olve

(a) The molecular formula of the first compound is C_7H_{16}, while that of the second compound is C_8H_{18}. The molecules are not isomers.

(continued)

(b) These molecules have the same formula, C_6H_{14}, but they differ in structure. They are isomers. The di ference in structure is that in the first structure, a methyl side chain is on the middle carbon atom of five-carbon chain, and in the second structure it is on the second carbon atom from the end of the chai

Assess

The compounds shown in part (a) have different molecular formulas and, thus, are clearly different com pounds. We expect these compounds to have markedly different properties. The compounds shown in part (l have the same molecular formula but different molecular structures. They are different compounds and hav different properties. For example, the compound shown on the left in part (b) has a slightly higher boilin point than the compound shown on the right (63 °C versus 60 °C).

PRACTICE EXAMPLE A: Are the following pairs of molecules isomers?

(a) $CH_3C(CH_3)_2(CH_2)_3CH_3$ and $CH_3CH(CH_3)CH(CH_3)(CH_2)_3CH_3$

(b)

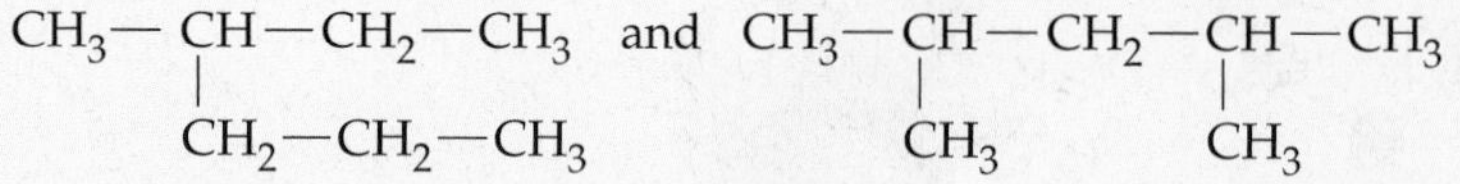

PRACTICE EXAMPLE B: Are the pairs of molecules represented by the following structural formulas isomers?

(a)

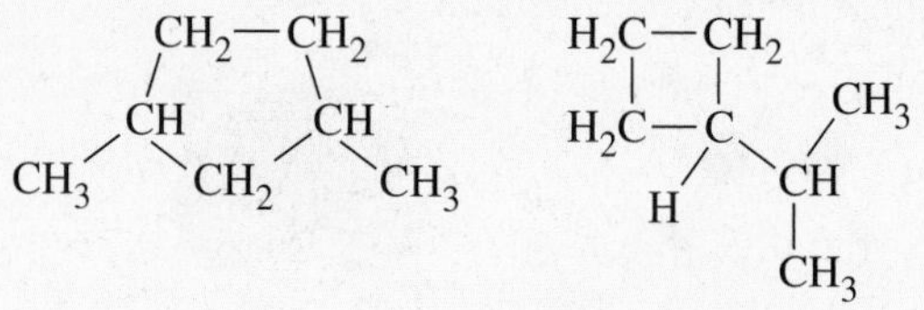

(b)

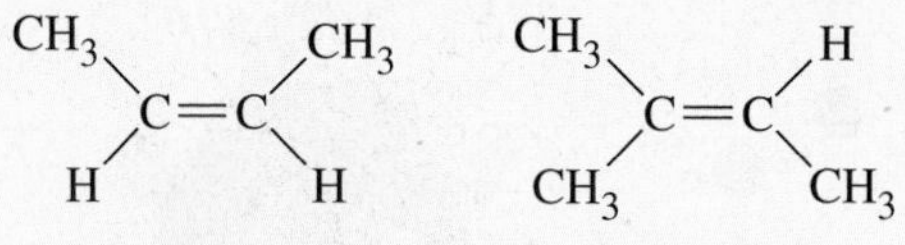

3-7 CONCEPT ASSESSMENT

In the combustion of a hydrocarbon, can the mass of H_2O produced ever exceed that of the CO_2? Explain.

Functional Groups

Carbon chains provide the framework of organic compounds; other atom groups of atoms replace one or more of the hydrogen atoms to form diffe compounds. We can illustrate this with the common alcohol molecule occurs in beer, wine, and spirits. The molecule is *ethanol*, CH_3CH_2OH, in wl one of the H atoms of ethane is replaced by an —OH group (Fig. 3-10a). systematic name ethanol is derived from the name of the alkane, ethane, the final *-e* replaced by the suffix *-ol*. The suffix *-ol* designates the presence o OH group in a class of organic molecules called **alcohols**.

Ethyl alcohol, the common name of ethanol, also indicates attachmer the —OH group to the ethane hydrocarbon chain. To name the alkane cl as a group, replace the final *-e* with *-yl*, so that ethane becomes *ethyl*, thus *e alcohol* for CH_3CH_2OH. It is often the case that the common name of one c pound, alcohol in this case, will provide the generic name for a complete c of compounds; that is, all alcohols contain at least one —OH group.

Another common alcohol is *methanol*, or wood alcohol, which has the mula CH_3OH (Fig. 3-10b). The common name for methanol is methyl alco It is interesting to note that wood alcohol is a dangerous poison, whereas grain alcohol in beer and wine is safe to consume in moderate quantities.

The —OH group in alcohols is one of the many *functional groups* foun organic compounds. **Functional groups** are individual atoms or grouping

KEEP IN MIND

that alcohols are molecular compounds in which oxygen atoms are covalently bonded to carbon atoms. They are not ionic compounds that contain hydroxide ions.

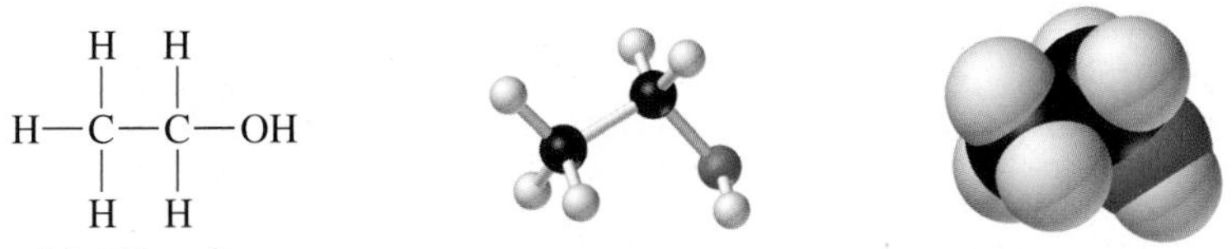

(a) Ethanol, or
Ethyl alcohol

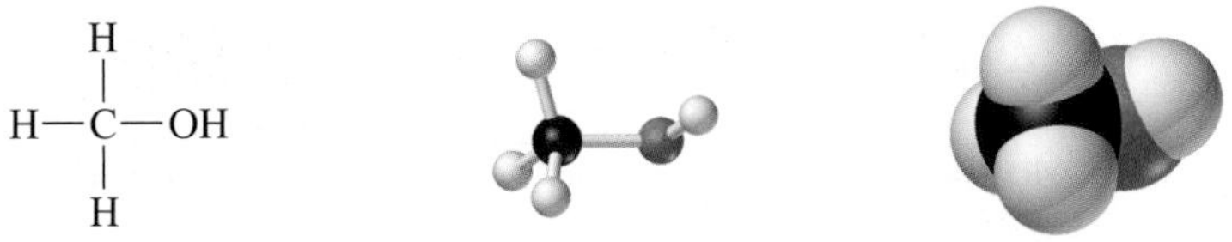

(b) Methanol, or
Methyl alcohol

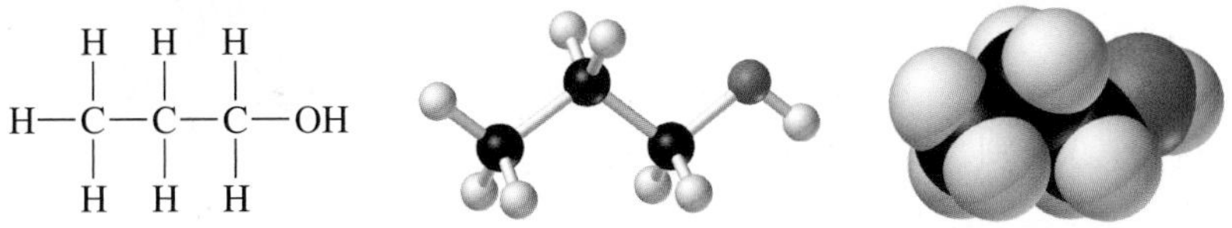

(c) Propan-1-ol, or
Propyl alcohol

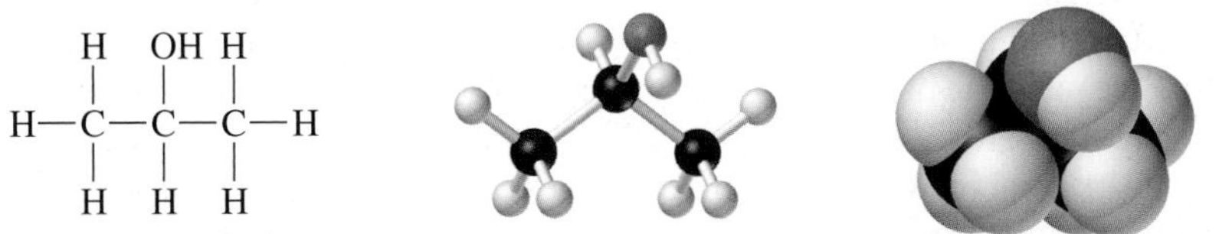

(d) Propan-2-ol, or
Isopropyl alcohol

◀ FIGURE 3-10
Visualizations of some alcohols

ns that are attached to the carbon chains or rings of organic molecules and the molecules their characteristic properties. Compounds with the same tional group generally have similar properties. The —OH group is called *hydroxyl group*. At this point only a few functional groups will be intro- ed; functional groups are discussed in more detail in Chapter 26.

he presence of functional groups also increases the possibility of isomers. example, there is only one propane molecule, C_3H_8. However, if one of the oms is replaced by a hydroxyl group, two possibilities exist for the point ttachment: at one of the end C atoms or at the middle C atom (Fig. 3-10c, This leads to two isomers. The alcohol with the —OH group attached to end carbon atom is commonly called propyl alcohol or, systematically, oan-1-ol; the 1 indicates that the —OH group is on the first or end C atom. alcohol with the —OH group attached to the middle carbon atom is com- ly called isopropyl alcohol or, systematically, propan-2-ol; the 2 indicates the —OH group is on the second C atom from the end. Notice that, in the ematic name of an alcohol, the number specifying the position of the H group is placed immediately before the suffix *-ol*.

nother important functional group is the *carboxyl* group, —COOH, or O_2H which confers acidic properties on a molecule. The C atom in the car- yl group is bound to the two O atoms in two ways. One bond is a single d to an oxygen atom that is also attached to a H atom, and the other is a ble bond to a lone O atom (Fig. 3-11a). The hydrogen attached to one of the toms in a carboxyl group is *ionizable* or *acidic*. Compounds containing the oxyl group are called **carboxylic acids**. The first carboxylic acid based on nes is *methanoic* acid, HCOOH (Fig. 3-11b). In the systematic name, the *an-* indicates one carbon atom and the *-oic* acid indicates a carboxylic acid. common name for methanoic acid is formic acid, deriving from the Latin d *formica*, meaning "ant." Formic acid is injected by an ant when it bites; leads to the burning sensation that accompanies the bite.

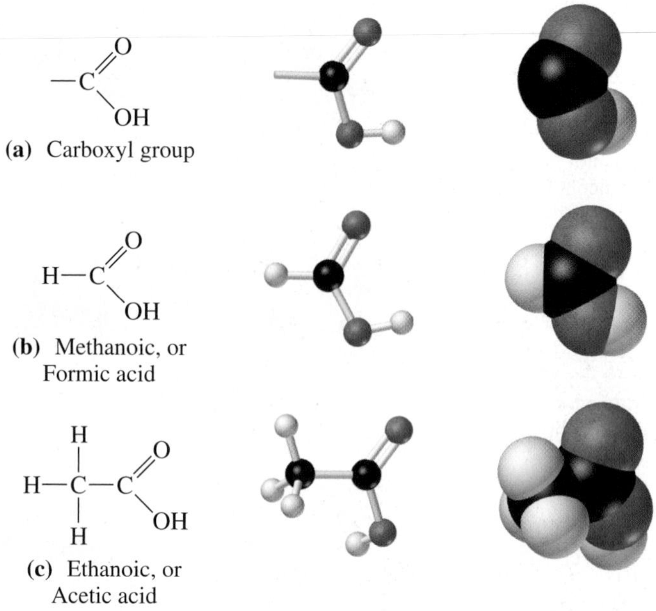

▶ FIGURE 3-11
The carboxyl group and visualizations of two carboxylic acids

The simplest carboxylic acid containing two carbon atoms is *ethanoic* more commonly known as acetic acid. The molecular formula is CH_3CO and the structure is shown in Figure 3-11(c). Vinegar is a solution of acetic in water. An additional functional group is introduced in the examples follow—a halogen atom (F, Cl, Br, I) substituting for one or more H at When present as functional groups, the halogens carry the names, *fl* *chloro-*, *bromo-*, and *iodo-*.

EXAMPLE 3-13 Recognizing Types of Organic Compounds

What type of compound is each of the following?

(a) $CH_3CH_2CH_2CH_3$ **(b)** $CH_3CHClCH_2CH_3$
(c) $CH_3CH_2CO_2H$ **(d)** $CH_3CH_2CH(OH)CH_2CH_3$

Analyze

For each compound, examine the formula to determine which functional group, if any, is present. Also co sider whether or not all the carbon–carbon bonds are single bonds.

Solve

(a) The carbon–carbon bonds are all single bonds in this hydrocarbon. This compound is an alkane.

(b) There are only single bonds in its molecules, and one H atom has been replaced by a Cl atom. This con pound is a chloroalkane.

(c) The presence of the carboxyl group, $—CO_2H$, in its molecules means that this compound is carboxylic acid.

(d) The presence of the hydroxyl group, —OH, in its molecules means that this compound is an alcohol.

Assess

Carbon–hydrogen and carbon–carbon bonds, especially carbon–carbon single bonds, are relatively unreactiv Because of this, functional groups are most important in determining the characteristic properties of organ compounds. We will encounter other functional groups and examine organic compounds in more detail i Chapters 26 and 27.

PRACTICE EXAMPLE A: What types of compounds correspond to each of the following formulas?

(a) $CH_3CH_2CH_3$ **(b)** $ClCH_2CH_2CH_3$
(c) $CH_3CH_2CH_2CO_2H$ **(d)** $CH_3CHCHCH_3$

RACTICE EXAMPLE B: What types of compounds correspond to these formulas?

(a) $CH_3CH(OH)CH_3$ **(b)** $CH_3CH(OH)CH_2CO_2H$
(c) $CH_2ClCH_2CO_2H$ **(d)** $BrCHCHCH_3$

XAMPLE 3-14 Naming Organic Compounds

Name these compounds.

(a) $CH_3CH_2CH_2CH_2CH_3$ **(b)** $CH_3CHFCH_2CH_3$
(c) $CH_3CH_2CO_2H$ **(d)** $CH_3CH_2CH(OH)CH_2CH_3$

nalyze

First, determine the type of compound. Then, count the number of carbon atoms and select the appropriate stem (or prefix) from Table 3.7 to form the name. If the compound is an alcohol, change the final *-e* in the name to *-ol;* if it is a carboxylic acid, change the final *-e* to *-oic acid*. If it is necessary to specify the position of the functional group, the carbon number is placed immediately before that part of the name to which it relates. Thus, the carbon number is placed immediately before the *-ol* suffix in the name of an alcohol and immediately before the prefix *fluoro-* in the name of a fluoroalkane.

olve

(a) The structure is that of an alkane molecule with a five-carbon chain, so the compound is pentane.

(b) The structure is that of a fluoroalkane molecule with the F atom on the *second* C atom of a four-carbon chain. The compound is called 2-fluorobutane.

(c) The carbon chain in this structure is three C atoms long, with the end C atom in a carboxyl group. The compound is propanoic acid.

(d) This structure is that of an alcohol with the hydroxyl group on the *third* C atom of a five-carbon chain. The compound is called pentan-3-ol.

ssess

In naming the compound in part (b), we stated that the F atom was bonded to the second C atom in a four-carbon chain. There is some ambiguity in that statement because, as illustrated below, we can number the C atoms in two different ways.

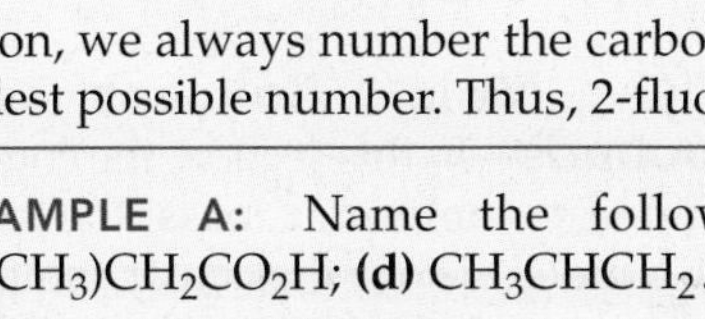

2-fluorobutane
(correct)

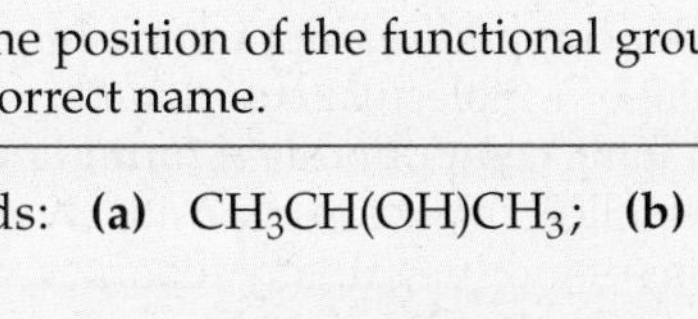

3-fluorobutane
(incorrect)

By convention, we always number the carbon atoms so that the position of the functional group is designated by the smallest possible number. Thus, 2-fluorobutane is the correct name.

RACTICE EXAMPLE A: Name the following compounds: **(a)** $CH_3CH(OH)CH_3$; **(b)** $ICH_2CH_2CH_3$; **(c)** $CH_3CH(CH_3)CH_2CO_2H$; **(d)** CH_3CHCH_2.

RACTICE EXAMPLE B: Give plausible names for the molecules that correspond to the following ball-and-stick models.

(a)

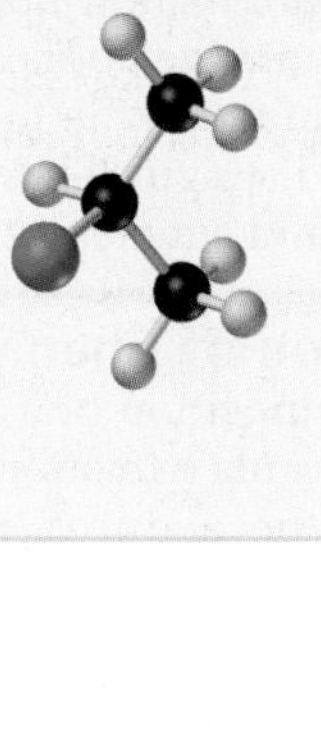

(b)

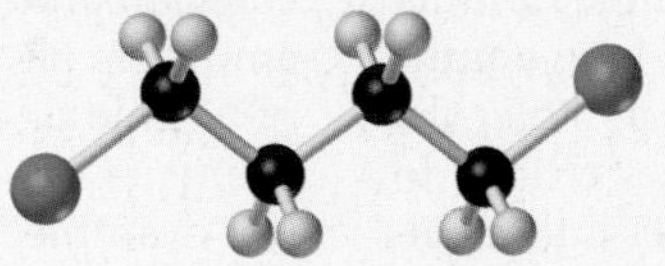

(c)

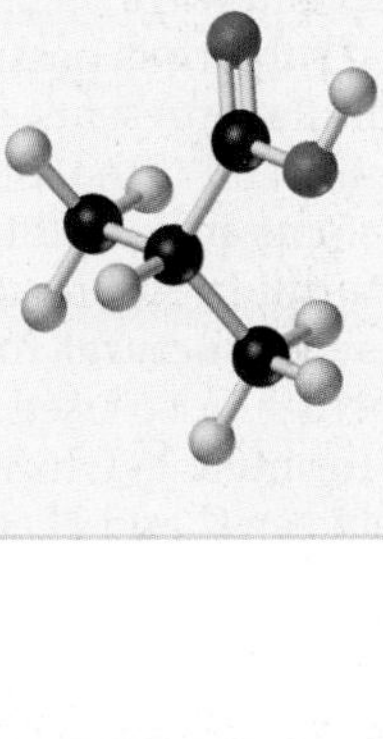

EXAMPLE 3-15 Writing Structural Formulas from the Names of Organic Compounds

Write the condensed structural formula for the organic compounds **(a)** butane; **(b)** butanoic aci **(c)** 1-chloropentane; **(d)** hexan-1-ol.

Analyze

First, identify the number of carbon atoms in the chain, and then determine the type and position of the fun tional group, if any.

Solve

(a) The word stem *but-* indicates a structure with a four-carbon chain, and the suffix *-ane* indicates an alkar No functional groups are indicated; hence, the condensed structural formula is $CH_3(CH_2)_2CH_3$.

(b) The *-oic* ending indicates that the end carbon atom of the four-carbon chain is part of a carboxylic ac group. The condensed structural formula is $CH_3(CH_2)_2CO_2H$.

(c) The prefix *chloro-* indicates the substitution of a chlorine atom for a H atom, and the 1- designates that is on the first C atom of the carbon chain. The carbon chain is five C atoms long, as signified by tl word stem *pent-*. The condensed structural formula is $CH_3(CH_2)_3CH_2Cl$.

(d) The suffix *-ol* indicates the presence of a hydroxyl group in place of a H atom, and the 1 designates th it is on the first C atom of the carbon chain. The word stem *hex-* signifies that the carbon chain is six atoms long. The condensed structural formula is $CH_3(CH_2)_4CH_2OH$.

Assess

In summary, to obtain a structural formula from the name, we split the name into its component pieces: ster prefix, and suffix. All three components provide information about the structure of the molecule.

PRACTICE EXAMPLE A: Write the condensed structural formula for the organic compounds **(a)** pentan **(b)** ethanoic acid; **(c)** 1-iodooctane (pronounced "eye-oh-dough-octane"); **(d)** pentan-1-ol.

PRACTICE EXAMPLE B: Write the line-angle formula for the organic compounds **(a)** propene; **(b)** heptan-1-(**(c)** chloroacetic acid; **(d)** hexanoic acid.

MasteringCHEMISTRY www.masteringchemistry.com

For a discussion of how chemists use mass spectrometry to establish both molecular and structural formulas, go to the Focus On feature for Chapter 3, entitled Mass Spectrometry—Determining Molecular and Structur Formulas, on the MasteringChemistry site.

Summary

3-1 Types of Chemical Compounds and Their Formulas—The two main classes of chemical compounds are **molecular compounds** and **ionic compounds**. The fundamental unit of a molecular compound is a molecule and that of an ionic compound is a **formula unit**. A formula unit is the smallest collection of positively charged ions—called **cations**—and negatively charged ions—called **anions**—that is electrically neutral overall. A **chemical formula** is a symbolic representation of a compound that can be written in several ways (Fig. 3-1). If the formula has the smallest integral subscripts possible, it is an **empirical formula**; if the formula represents an actual molecule, it is a **molecular formula**; and if the formula is written to show how individual atoms are joined together into molecules, it is a **structural formula**. Abbreviated structural formulas, called **condensed structural formulas**, are often used for organic molecules. Also used for organic molecules is the **line-angle formula**, in which all bond lines are shown except those between C and H atoms and in which the symbols C and H are mostly omitted. The relative sizes and positions of the atoms in molecules can be depicte ball-and-stick and space-filling molecular models.

3-2 The Mole Concept and Chem Compounds—In this section, the concept of atomic is extended to **molecular mass**, the mass in atomic units of a molecule of a molecular compound, and **for mass**, the mass in atomic mass units of a formula unit ionic compound. Likewise, the concept of the Avog constant and the mole is now applied to compounds, emphasis on quantitative applications involving the of a mole of compound—the **molar mass** M. For se elements, we can distinguish between a mole of mole (for example, P_4) and a mole of atoms (that is, P).

3-3 Composition of Chemical Compounds– mass percent composition of a compound can be e lished from its formula (equation 3.1). Conversely, a c ical formula can be deduced from the experimen determined percent composition of the compound nic compounds, this often involves combustion ysis (Fig. 3-6). Formulas determined from experimen- ercent composition data are empirical formulas—the lest formulas that can be written. Molecular formulas e related to empirical formulas when experimentally rmined molecular masses are available.

Oxidation States: A Useful Tool in Describing mical Compounds—The **oxidation state** of an atom ughly related to the number of electrons involved in the ation of a bond between that atom and another atom. ation states are expressed as numbers assigned accord- o a set of conventions (Table 3.2). The oxidation state ept has several uses, one of which is to aid in naming nical compounds and in writing chemical formulas.

Naming Compounds: Organic and Inorganic npounds—Assigning names to the formulas of nical compounds—nomenclature—is an important ity in chemistry. The topic is introduced in this sec- and applied to two broad categories: (1) **organic com- nds**, which are compounds formed by the elements on and hydrogen, often together with a few other ele- ts, such as oxygen and nitrogen; and (2) **inorganic pounds**, a category that includes the remaining chem- compounds. The subject of nomenclature is extended ditional contexts throughout the text.

Names and Formulas of Inorganic npounds—In this section on the nomenclature of ganic compounds, the names and formulas of two- ent or **binary compounds** are considered first. The names and symbols of some simple ions (Table 3.3) and the names of a few typical binary molecular compounds (Table 3.4) are listed. Also listed are some common ions comprising two or more atoms, **polyatomic ions**, and typical ionic compounds containing them (Table 3.5). The naming of binary acids is related to the names of the corresponding binary hydrogen compounds. The majority of acids, however, are **ternary compounds**; they consist of three elements—hydrogen and two other nonmetals. Typically, one of the elements in ternary acids is oxygen, giving rise to the name **oxoacids**. The polyatomic anions derived from oxoacids are **oxoanions**. A scheme for relating the names and formulas of oxoanions is given (expression 3.3), and the nomenclature of oxoacids and their salts is summarized (Table 3.6). Also described in this section are **hydrates**, ionic compounds having fixed numbers of water molecules associated with their formula units.

3-7 Names and Formulas of Organic Compounds—Organic compounds are based on the element carbon. **Hydrocarbons** contain only carbon and hydrogen (Fig. 3-9). **Alkane** hydrocarbon molecules contain only single bonds, and alkenes contain at least one double bond. A common phenomenon found in organic compounds is isomerism—the existence of different compounds, called **isomers**, having identical molecular formulas but different structural formulas. **Functional groups** confer distinctive properties on organic molecules when they are substituted for H atoms on carbon chains or rings. The hydroxyl group —OH is present in **alcohols** (Fig. 3-10) and the carboxyl group —COOH, in **carboxylic acids** (Fig. 3-11).

Integrative Example

ecules of a dicarboxylic acid have *two* carboxyl groups (—COOH). A 2.250 g sample of a dicarboxylic acid was ed in an excess of oxygen and yielded 4.548 g CO_2 and 1.629 g H_2O. In a separate experiment, the molecular mass of cid was found to be 174 u. From these data, what can we deduce about the structural formula of this acid?

lyze

approach will require several steps: (1) Use the combustion data to determine the percent composition of the com- nd (similar to Example 3-6). (2) Determine the empirical formula from the percent composition (similar to Example 3-5). btain the molecular formula from the empirical formula and the molecular mass. (4) Determine how the C, H, and O s represented in the molecular formula might be assembled into a dicarboxylic acid. Use molar masses with (at least) more significant figure than in the measured masses where possible; store intermediate results in your calculator with- ounding off.

e

etermine the percent composition. Calculate the mass H in 1.629 g H_2O

$$? \text{ g H} = 1.629 \text{ g H}_2\text{O} \times \frac{1 \text{ mol H}_2\text{O}}{18.015 \text{ g H}_2\text{O}} \times \frac{2 \text{ mol H}}{1 \text{ mol H}_2\text{O}} \times \frac{1.008 \text{ g H}}{1 \text{ mol H}} = 0.1823 \text{ g H}$$

d then the mass percent H in the 2.250 g sample of e dicarboxylic acid.

$$\% \text{ H} = \frac{0.1823 \text{ g H}}{2.250 \text{ g compd.}} \times 100\% = 8.102\% \text{ H}$$

so, calculate the mass of C in 4.548 g CO_2,

$$? \text{ g C} = 4.548 \text{ g CO}_2 \times \frac{1 \text{ mol CO}_2}{44.009 \text{ g CO}_2} \times \frac{1 \text{ mol C}}{1 \text{ mol CO}_2} \times \frac{12.011 \text{ g C}}{1 \text{ mol C}} = 1.241 \text{ g C}$$

followed by the mass percent C in the 2.250 g sample of dicarboxylic acid.

$$\% \text{ C} = \frac{1.241 \text{ g C}}{2.250 \text{ g compd.}} \times 100\% = 55.16\% \text{ C}$$

The mass percent O in the compound is obtained as a difference, that is,

$$\% \text{ O} = 100.00\% - 55.16\% \text{ C} - 8.102\% \text{ H} = 36.74\%$$

2. *Obtain the empirical formula from the percent composition.* The masses of the elements in 100.0 g of the compound are

$$55.16 \text{ g C} \quad 8.102 \text{ g H} \quad 36.74 \text{ g O}$$

The numbers of moles of the elements in 100.0 g of the compound are

$$55.16 \text{ g C} \times \frac{1 \text{ mol C}}{12.011 \text{ g C}} = 4.592 \text{ mol C}$$

$$8.102 \text{ g H} \times \frac{1 \text{ mol H}}{1.008 \text{ g H}} = 8.038 \text{ mol H}$$

$$36.74 \text{ g O} \times \frac{1 \text{ mol O}}{15.999 \text{ g O}} = 2.296 \text{ mol O}$$

The tentative formula based on these numbers is

$$C_{4.592}H_{8.038}O_{2.296}$$

Divide all the subscripts by 2.296 to obtain

$$C_2H_{3.50}O$$

Multiply all subscripts by two to obtain the empirical formula,

$$C_4H_7O_2$$

and then determine the empirical formula mass.

$$(4 \times 12.011 \text{ u}) + (7 \times 1.008 \text{ u}) + (2 \times 15.999 \text{ u}) = 87.098 \text{ u}$$

3. *Obtain the molecular formula.* The experimentally determined molecular mass of 174 u is twice the empirical formula mass. The molecular formula is

$$C_8H_{14}O_4$$

4. *Assemble the atoms in $C_8H_{14}O_4$ into a plausible structural formula.* The dicarboxylic acid must contain two —COOH groups. This accounts for the two C atoms, two H atoms, and all four O atoms. The remainder of the structure is based on C_6H_{12}. For example, arrange the six —CH_2 segments into a six-carbon chain and attach the —COOH groups at the ends of the chain.

$$HOOC{-}CH_2(CH_2)_4CH_2{-}COOH$$

However, there are other possibilities based on shorter chains with branches, for example

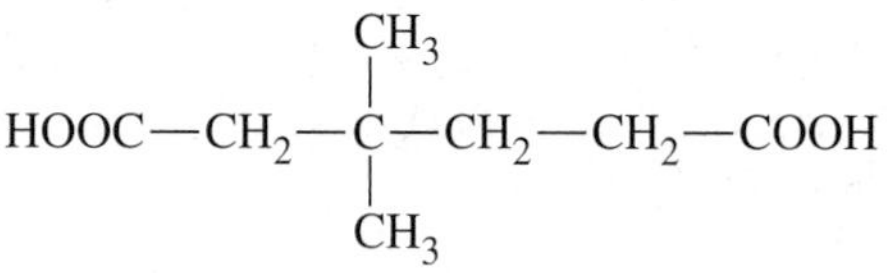

Assess

We have found a plausible structural formula, but there are many other possibilities. For example, the following three mers have a seven-carbon chain with one methyl group (—CH_3) substituted for an H atom on the chain:

$$HOOCCHCH_3(CH_2)_4COOH; \quad HOOCCH_2CHCH_3(CH_2)_3COOH; \quad HOOC(CH_2)_2CHCH_3(CH_2)_2COOH$$

In conclusion, we cannot identify a specific isomer with only the data given.

PRACTICE EXAMPLE A: A 2.4917 g sample of an unknown solid hydrate was heated to drive off all the wat hydration. The remaining solid, which weighed 1.8558 g, was analyzed and found to be 27.74% Mg, 23.57% P, and 48.69 by mass. What is the formula and name of the unknown solid hydrate?

PRACTICE EXAMPLE B: An unknown solid hydrate was analyzed and found to be 17.15% Cu, 19.14% Cl, and 60.45 by mass; the remainder was hydrogen. What are the oxidation states of copper and chlorine in this compound? What i name of this compound?

Exercises

resenting Molecules

Refer to the color scheme given in Figure 3-3, and give the molecular formulas for the molecules whose ball-and-stick models are given here.

(a)

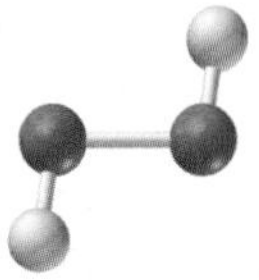

(b)

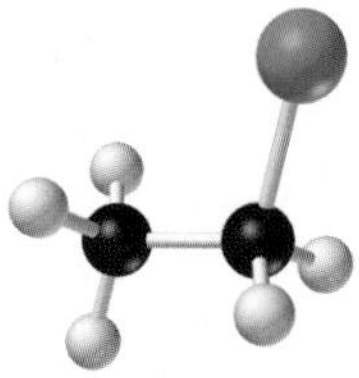

(c)

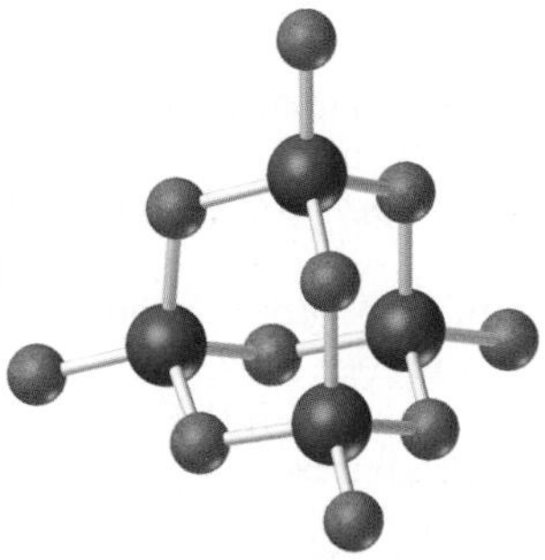

(d)

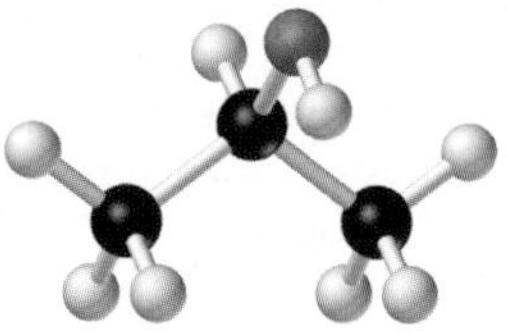

(e)

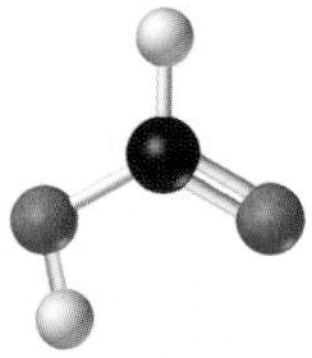

2. Give the molecular formulas for the molecules whose ball-and-stick models are given here. Refer to the color scheme in Figure 3-3.

(a)

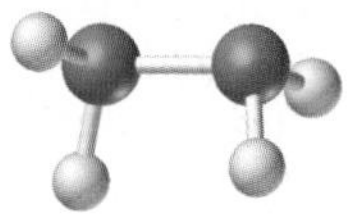

(b)

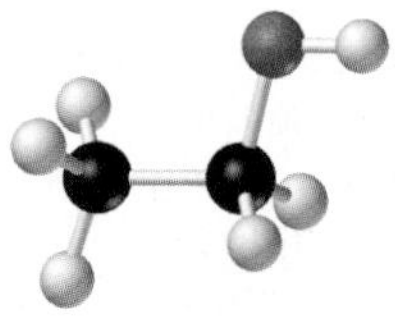

(c)

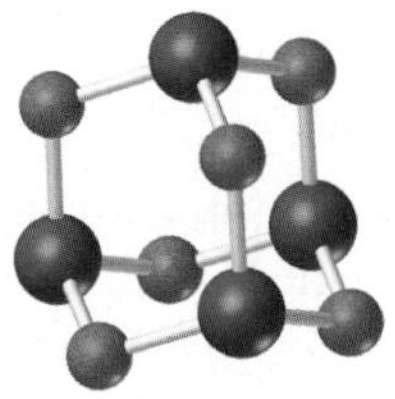

(d)

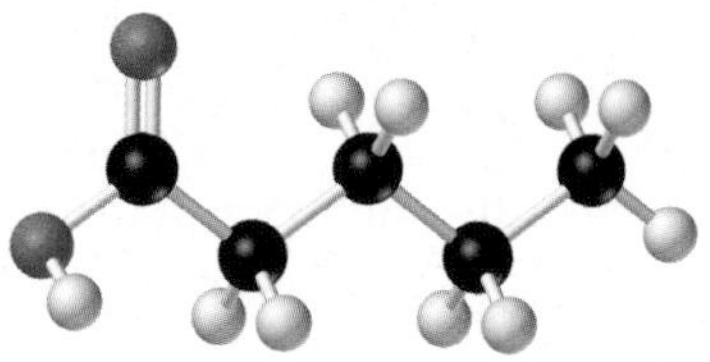

(e)

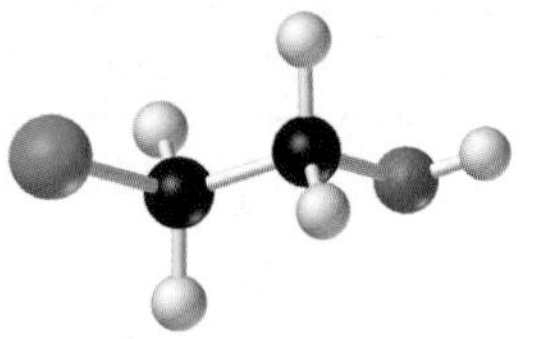

3. Give the structural formulas of the molecules shown in Exercise 1 **(b)**, **(d)**, and **(e)**.

4. Give the structural formulas of the molecules shown in Exercise 2 **(b)**, **(d)**, and **(e)**.

Avogadro Constant and the Mole

Calculate the total number of **(a)** atoms in one molecule of trinitrotoluene (TNT), $CH_3C_6H_2(NO_2)_3$; **(b)** atoms in 0.00102 mol $CH_3(CH_2)_4CH_2OH$; **(c)** F atoms in 12.15 mol $C_2HBrClF_3$.

Determine the *mass*, in grams, of

(a) 7.34 mol NO_2;

(b) 4.220×10^{25} O_2 molecules;

(c) 15.5 mol $CuSO_4 \cdot 5\,H_2O$;

(d) 2.25×10^{24} molecules of $C_2H_4(OH)_2$.

7. The amino acid methionine, which is essential in human diets, has the molecular formula $C_5H_{11}NO_2S$. Determine **(a)** its molecular mass; **(b)** the number of moles of H atoms per mole of methionine;

(c) the number of grams of C per mole of methionine;
(d) the number of C atoms in 9.07 mol methionine.

8. Determine the number of moles of Br_2 in a sample consisting of (a) 4.04×10^{22} Br_2 molecules; (b) 5.78×10^{24} Br atoms (assuming that the Br atoms react completely to form Br_2); (c) 7.82 kg bromine; (d) 3.56 L liquid bromine ($d = 3.10$ g/mL).

9. *Without doing detailed calculations*, explain which of the following has the greatest number of N atoms (a) 50.0 g N_2O; (b) 17.0 g NH_3; (c) 150 mL of liquid pyridine, C_5H_5N ($d = 0.983$ g/mL); (d) 1.0 mol N_2.

10. *Without doing detailed calculations*, determine which of the following has the greatest number of S atoms (a) 0.12 mol of solid sulfur, S_8; (b) 0.50 mol of gaseous S_2O; (c) 65 g of gaseous SO_2; (d) 75 mL of liquid thiophene, C_4H_4S ($d = 1.064$ g/mL).

11. Determine the number of *moles* of
(a) N_2O_4 in a 115 g sample
(b) N atoms in 43.5 g of $Mg(NO_3)_2$
(c) N atoms in a sample of $C_7H_5(NO_2)_3$ that has the same number of O atoms as 12.4 g $C_6H_{12}O_6$

12. Determine the *mass*, in grams, of
(a) 2.10×10^2 mol S_8
(b) 5.02×10^{22} molecules of palmitic acid, $C_{16}H$
(c) a quantity of the amino acid histidine, C_6H_9N containing 2.95 mol N atoms

13. The hemoglobin content of blood is a 15.5 g/100 mL blood. The molar mass of hemog is about 64,500 g/mol, and there are four iron atoms in a hemoglobin molecule. Approximately many Fe atoms are present in the 6 L of blood in a ical adult?

14. In white phosphorus, P atoms are joined into P_4 m cules (see Figure 3-5). White phosphorus is comm supplied in chalk-like cylindrical form. Its de is 1.823 g/cm^3. For a cylinder of white phosph 6.50 cm long and 1.22 cm in diameter, deter (a) the number of moles of P_4 present; (b) the number of P atoms.

Chemical Formulas

15. Explain which of the following statement(s) is (are) correct concerning glucose (blood sugar),$C_6H_{12}O_6$.
(a) The percentages, by mass, of C and O are the same as in CO.
(b) The ratio of C:H:O atoms is the same as in dihydroxyacetone, $(CH_2OH)_2CO$.
(c) The proportions, by mass, of C and O are equal.
(d) The highest percentage, by mass, is that of H.

16. Explain which of the following statement(s) is (are) correct for sorbic acid, $C_6H_8O_2$, an inhibitor of mold and yeast.
(a) It has a C:H:O mass ratio of 3:4:1.
(b) It has the same mass percent composition as the aquatic herbicide, acrolein, C_3H_4O.
(c) It has the same empirical formula as aspidinol, $C_{12}H_{16}O_4$, a drug used to kill parasitic worms.
(d) It has four times as many H atoms as O atoms four times as much O as H by mass.

17. For the mineral torbernite, $Cu(UO_2)_2(PO_4)_2 \cdot 8$ determine
(a) the total number of atoms in one formula un
(b) the ratio, by number, of H atoms to O atoms
(c) the ratio, by mass, of Cu to P
(d) the element present in the greatest mass perc
(e) the mass required to contain 1.00 g P

18. For the compound $Ge[S(CH_2)_4CH_3]_4$, determine
(a) the total number of atoms in one formula uni
(b) the ratio, by number, of C atoms to H atoms
(c) the ratio, by mass, of Ge to S
(d) the number of g S in 1 mol of the compound
(e) the number of C atoms in 33.10 g of the compound

Percent Composition of Compounds

19. Determine the mass percent H in the hydrocarbon decane, $C_{10}H_{22}$.

20. Determine the mass percent O in the mineral malachite, $Cu_2(OH)_2CO_3$.

21. Determine the mass percent H in the hydrocarbon isooctane, $C(CH_3)_3CH_2CH(CH_3)_2$.

22. Determine the mass percent H_2O in the hydrate $Cr(NO_3)_3 \cdot 9\,H_2O$.

23. Determine the mass percent of each of the elements in the antimalarial drug quinine, $C_{20}H_{24}N_2O_2$.

24. Determine the mass percent of each of the elements in the fungicide copper(II) oleate, $Cu(C_{18}H_{33}O_2)_2$.

25. Determine the percent, by mass, of the indicated element:
(a) Pb in tetraethyl lead, $Pb(C_2H_5)_4$, once extensively used as an additive to gasoline to prevent engine knocking
(b) Fe in Prussian blue, $Fe_4[Fe(CN)_6]_3$, a pig used in paints and printing inks
(c) Mg in chlorophyll, $C_{55}H_{72}MgN_4O_5$, the greer ment in plant cells

26. All of the following minerals are semiprecious or cious stones. Determine the mass percent of the cated element.
(a) Zr in zircon, $ZrSiO_4$
(b) Be in beryl (emerald), $Be_3Al_2Si_6O_{18}$
(c) Fe in almandine (garnet), $Fe_3Al_2Si_3O_{12}$
(d) S in lazurite (lapis lazuli), $Na_4SSi_3Al_3O_{12}$

27. *Without doing detailed calculations*, arrange the fo ing in order of increasing % Cr, by mass, and ex your reasoning: CrO, Cr_2O_3, CrO_2, CrO_3.

28. Without doing detailed calculations, explain whi the following has the greatest mass percent of su SO_2, S_2Cl_2, Na_2S, $Na_2S_2O_3$, or CH_3CH_2SH.

mical Formulas from Percent Composition

Two oxides of sulfur have nearly identical molecular masses. One oxide consists of 40.05% S. What are the simplest possible formulas for the two oxides?

An oxide of chromium used in chrome plating has a formula mass of 100.0 u and contains *four* atoms per formula unit. Establish the formula of this compound, *with a minimum of calculation.*

Diethylene glycol, used to de-ice aircraft, is a carbon–hydrogen–oxygen compound with 45.27% C and .50% H by mass. What is its empirical formula?

The food flavor enhancer monosodium glutamate (MSG) has the composition 13.6% Na, 35.5% C, .8% H, 8.3% N, 37.8% O, by mass. What is the empirical formula of MSG?

Determine the empirical formula of **(a)** the rodenticide (rat killer) warfarin, which consists of 74.01% C, .23% H, and 20.76% O, by mass; **(b)** the antibacterial agent sulfamethizole, which consists of 39.98% C, .73% H, 20.73% N, 11.84% O, and 23.72% S, by mass.

Determine the empirical formula of **(a)** benzo[*a*]pyrene, a suspected carcinogen found in cigarette smoke, consisting of 95.21% C and 4.79% H, by mass; **(b)** hexachlorophene, used in germicidal soaps, which consists of 38.37% C, 1.49% H, 52.28% Cl, and 7.86% O by mass.

A compound of carbon and hydrogen consists of 4.34% C and 5.66% H, by mass. The molecular mass of the compound is found to be 178 u. What is its molecular formula?

36. Selenium, an element used in the manufacture of photoelectric cells and solar energy devices, forms two oxides. One has 28.8% O, by mass, and the other, 37.8% O. What are the formulas of these oxides? Propose acceptable names for them.

37. Indigo, the dye for blue jeans, has a percent composition, by mass, of 73.27% C, 3.84% H, 10.68% N, and the remainder is oxygen. The molecular mass of indigo is 262.3 u. What is the molecular formula of indigo?

38. Adenine, a component of nucleic acids, has the mass percent composition: 44.45% C, 3.73% H, 51.82% N. Its molecular mass is 135.14 u. What is its molecular formula?

39. The element X forms the chloride XCl_4 containing 75.0% Cl, by mass. What is element X?

40. The element X forms the compound $XOCl_2$ containing 59.6% Cl. What is element X?

41. Chlorophyll contains 2.72% Mg by mass. Assuming one Mg atom per chlorophyll molecule, what is the molecular mass of chlorophyll?

42. Two compounds of Cl and X are found to have molecular masses and % Cl, by mass, as follows: 137 u, 77.5% Cl; 208 u, 85.1% Cl. What is element X? What is the formula for each compound?

nbustion Analysis

A 0.1888 g sample of a hydrocarbon produces 0.6260 g CO_2 and 0.1602 g H_2O in combustion analysis. Its molecular mass is found to be 106 u. For this hydrocarbon, determine its **(a)** mass percent composition; **(b)** empirical formula; **(c)** molecular formula.

Para-cresol (*p*-cresol) is used as a disinfectant and in the manufacture of herbicides. A 0.4039 g sample of this carbon–hydrogen–oxygen compound yields .1518 g CO_2 and 0.2694 g H_2O in combustion analysis. Its molecular mass is 108.1 u. For *p*-cresol, determine its **(a)** mass percent composition; **(b)** empirical formula; **(c)** molecular formula.

Dimethylhydrazine is a carbon–hydrogen–nitrogen compound used in rocket fuels. When burned in an excess of oxygen, a 0.312 g sample yields 0.458 g CO_2 and 0.374 g H_2O. The nitrogen content of a 0.486 g sample is converted to 0.226 g N_2. What is the empirical formula of dimethylhydrazine?

The organic solvent thiophene is a carbon–hydrogen–sulfur compound that yields CO_2, H_2O, and SO_2 when burned in an excess of oxygen. When subjected to combustion analysis, a 1.3020 g sample of thiophene produces 2.7224 g CO_2, 0.5575 g H_2O, and 0.9915 g SO_2. What is the empirical formula of thiophene?

47. *Without doing detailed calculations*, explain which of these compounds produces the *greatest* mass of CO_2 when 1.00 mol of the compound is burned in an excess of oxygen: CH_4, C_2H_5OH, $C_{10}H_8$, C_6H_5OH.

48. *Without doing detailed calculations*, explain which of these compounds produces the *greatest* mass of H_2O when 1.00 g of the compound is burned in an excess of oxygen: CH_4, C_2H_5OH, $C_{10}H_8$, C_6H_5OH.

49. A 1.562 g sample of the alcohol $CH_3CHOHCH_2CH_3$ is burned in an excess of oxygen. What masses of CO_2 and H_2O should be obtained?

50. Liquid ethyl mercaptan, C_2H_6S, has a density of 0.84 g/mL. Assuming that the combustion of this compound produces only CO_2, H_2O, and SO_2, what masses of each of these three products would be produced in the combustion of 3.15 mL of ethyl mercaptan?

dation States

Indicate the oxidation state of the underlined element in **(a)** $\underline{C}H_4$; **(b)** $\underline{S}F_4$; **(c)** $Na_2\underline{O}_2$; **(d)** $\underline{C}_2H_3O_2^-$; **(e)** $\underline{Fe}O_4^{2-}$.

Indicate the oxidation state of S in **(a)** SO_3^{2-}; **(b)** $S_2O_3^{2-}$; **(c)** $S_2O_8^{2-}$; **(d)** HSO_4^-; **(e)** $S_4O_6^{2-}$.

53. Chromium forms three principal oxides. Write appropriate formulas for these compounds in which the oxidation states of Cr are +3, +4, and +6, respectively.

54. Nitrogen forms five oxides in which its oxidation states are +1, +2, +3, +4, and +5, respectively. Write appropriate formulas for these compounds.

55. In many of its compounds, oxygen has an oxidation state of −2. However, there are exceptions. What is the oxidation state of oxygen in each of the following compounds? **(a)** OF_2; **(b)** O_2F_2; **(c)** CsO_2; **(d)** BaO_2.
56. Hydrogen and oxygen usually have oxidation states of +1 and −2, respectively, in their compounds. The following cases serve to remind us that ther exceptions. What are the oxidation states of the a in each of the following compounds? **(a)** M **(b)** CsO_3; **(c)** HOF; **(d)** $NaAlH_4$.

Nomenclature

57. Name these compounds: **(a)** SrO; **(b)** ZnS; **(c)** K_2CrO_4; **(d)** Cs_2SO_4; **(e)** Cr_2O_3; **(f)** $Fe_2(SO_4)_3$; **(g)** $Mg(HCO_3)_2$; **(h)** $(NH_4)_2HPO_4$; **(i)** $Ca(HSO_3)_2$; **(j)** $Cu(OH)_2$; **(k)** HNO_3; **(l)** $KClO_4$; **(m)** $HBrO_3$; **(n)** H_3PO_3.
58. Name these compounds: **(a)** $Ba(NO_3)_2$; **(b)** HNO_2; **(c)** CrO_2; **(d)** KIO_3; **(e)** LiCN; **(f)** KIO; **(g)** $Fe(OH)_2$; **(h)** $Ca(H_2PO_4)_2$; **(i)** H_3PO_4; **(j)** $NaHSO_4$; **(k)** $Na_2Cr_2O_7$; **(l)** $NH_4C_2H_3O_2$; **(m)** MgC_2O_4; **(n)** $Na_2C_2O_4$.
59. Assign suitable names to the compounds **(a)** CS_2; **(b)** SiF_4; **(c)** ClF_5; **(d)** N_2O_5; **(e)** SF_6; **(f)** I_2Cl_6.
60. Assign suitable names to the compounds **(a)** ICl; **(b)** ClF_3; **(c)** SiF_4; **(d)** PF_5; **(e)** NO_2; **(f)** S_4N_4.
61. Write formulas for the compounds: **(a)** aluminum sulfate; **(b)** ammonium dichromate; **(c)** silicon tetrafluoride; **(d)** iron(III) oxide; **(e)** tricarbon disulfide; **(f)** cobalt(II) nitrate; **(g)** strontium nitrite; **(h)** hydrobromic acid; **(i)** iodic acid; **(j)** phosphorus dichloride trifluoride.
62. Write formulas for the compounds: **(a)** magnesium perchlorate; **(b)** lead(II) acetate; **(c)** tin(IV) oxide; **(d)** hydroiodic acid; **(e)** chlorous acid; **(f)** sodium hydrogen sulfite; **(g)** calcium dihydrogen phosphate; **(h)** aluminum phosphate; **(i)** dinitrogen tetroxide; **(j)** disulfur dichloride.
63. Write a formula for **(a)** the chloride of titanium ha Ti in the O.S. +4; **(b)** the sulfate of iron havin in the O.S. +3; **(c)** an oxide of chlorine with Cl i O.S. +7; **(d)** an oxoanion of sulfur in which the ap ent O.S. of S is +7 and the ionic charge is −2.
64. Write a formula for **(a)** an oxide of nitrogen with the O.S. +5; **(b)** an oxoacid of nitrogen with N i O.S. +3; **(c)** an oxide of carbon in which the app O.S. of C is +4/3; **(d)** a sulfur-containing oxoani which the apparent O.S. of S is +2.5 and the charge is −2.
65. Name the acids: **(a)** $HClO_2$; **(b)** H_2SO_3; **(c)** **(d)** HNO_2.
66. Supply the formula for the acids: **(a)** hydrofl acid; **(b)** nitric acid; **(c)** phosphorous acid; **(d)** su acid.
67. Name the following compounds and specify w ones are best described as ionic: **(a)** OF_2; **(b)** **(c)** $CuSO_3$; **(d)** $(NH_4)_2HPO_4$.
68. Name the following compounds and specify w ones are best described as ionic: **(a)** KNO_2; **(b)** **(c)** S_2Cl_2; **(d)** $Mg(ClO)_2$; **(e)** Cl_2O.

Hydrates

69. *Without performing detailed calculations*, indicate which of the following hydrates has the greatest % H_2O by mass: $CuSO_4 \cdot 5\,H_2O$, $Cr_2(SO_4)_3 \cdot 18\,H_2O$, $MgCl_2 \cdot 6\,H_2O$, and $LiC_2H_3O_2 \cdot 2\,H_2O$.
70. *Without performing detailed calculations*, determine the hydrate of Na_2SO_3 that contains almost exactly 50% H_2O, by mass.
71. Anhydrous $CuSO_4$ can be used to dry liquids in which it is insoluble. The $CuSO_4$ is converted to $CuSO_4 \cdot 5\,H_2O$, which can be filtered off from the liquid. What is the minimum mass of anhydrous $CuSO_4$ needed to remove 12.6 g H_2O from a tankful of gasoline?
72. Anhydrous sodium sulfate, Na_2SO_4, absorbs vapor and is converted to the decahyd $Na_2SO_4 \cdot 10\,H_2O$. How much would the mass of 3 of anhydrous Na_2SO_4 increase if converted pletely to the decahydrate?
73. A certain hydrate is found to have the compos 20.3% Cu, 8.95% Si, 36.3% F, and 34.5% H_2O by What is the empirical formula of this hydrate?
74. An 8.129 g sample of $MgSO_4 \cdot x\,H_2O$ is heated un the water of hydration is driven off. The resu anhydrous compound, $MgSO_4$, weighs 3.967 g. is the formula of the hydrate?

Organic Compounds and Organic Nomenclature

75. Which of the following names is most appropriate for the molecule with the structure shown below? **(a)** butyl alcohol; **(b)** butan-2-ol; **(c)** butan-1-ol; **(d)** isopentyl alcohol.

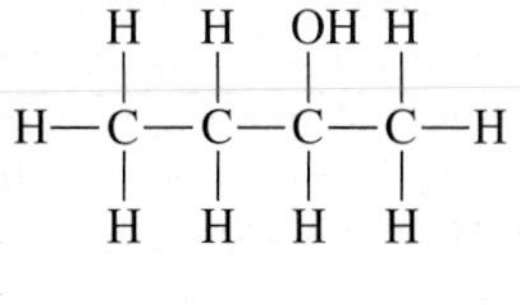

76. Which of the following names is most approp for the molecule $CH_3(CH_2)_2COOH$? **(a)** dimethy acetic acid; **(b)** propanoic acid; **(c)** butanoic **(d)** oxobutylalcohol.
77. Which of the following structures are isomers?

(a) 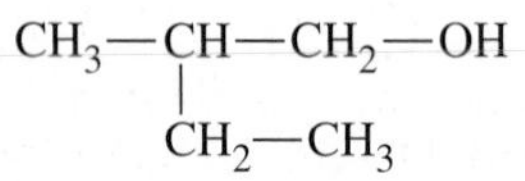

b) 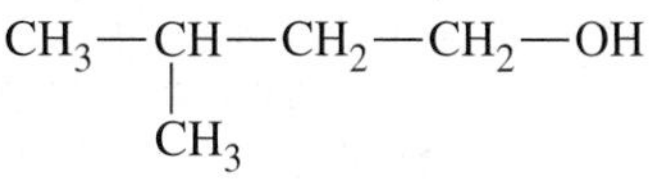

c) 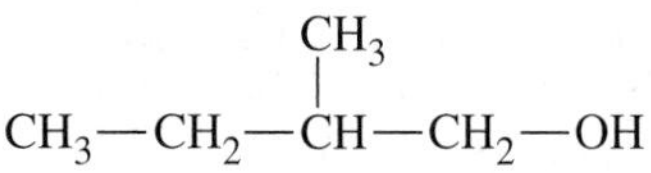

d) 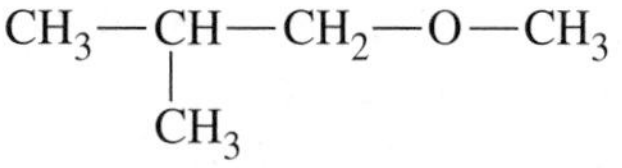

e) 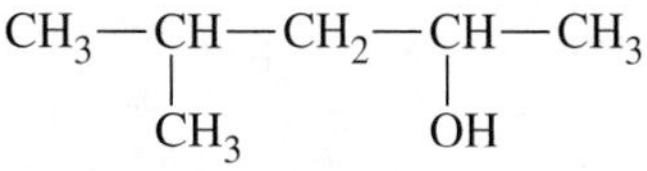

Vhich of the following structures are isomers?

a) 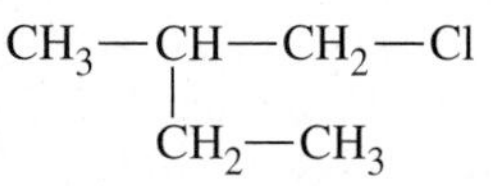

b) 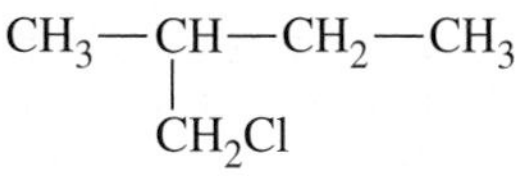

c) 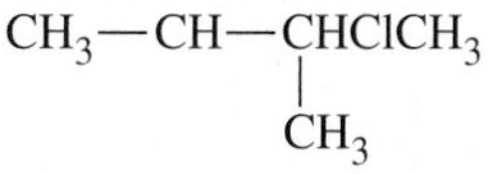

d) 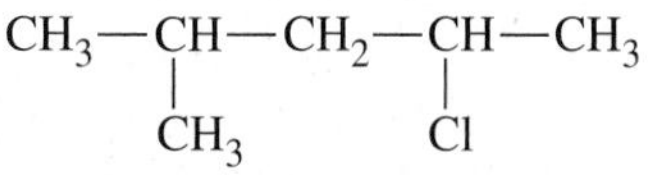

Vrite the condensed structural formulas for the organic compounds:
a) heptane (b) propanoic acid
c) 2 methylpentan-1-ol (d) fluoroethane

Vrite the condensed structural formulas for the organic compounds:
a) octane (b) heptanoic acid
c) hexan-3-ol (d) 2-chlorobutane

81. Give the name, condensed structural formula, and molecular mass of the molecule whose ball-and-stick model is shown. Refer to the color scheme in Figure 3-3.

(a) 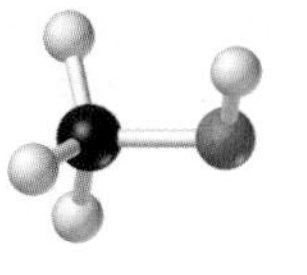 (b) 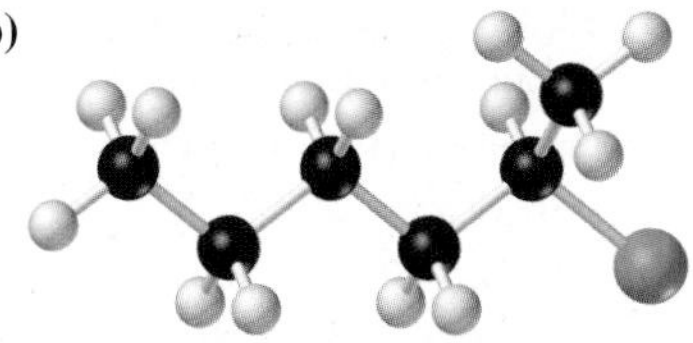

(c) 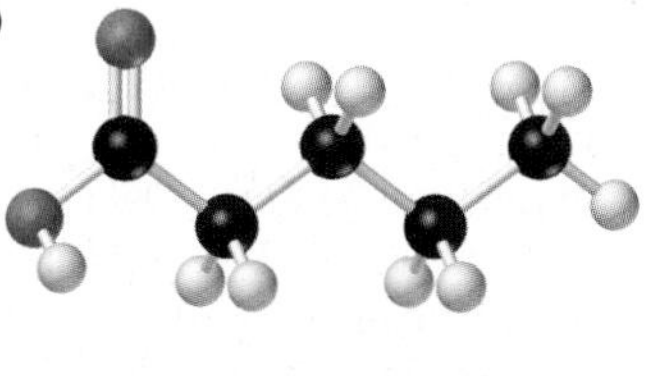 (d) 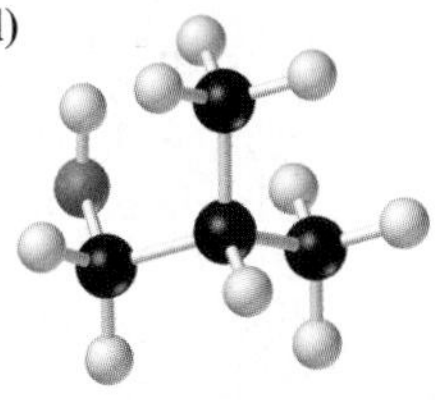

82. Give the name, condensed structural formula, and molecular mass of the molecule whose ball-and-stick model is shown. Refer to the color scheme in Figure 3-3.

(a) 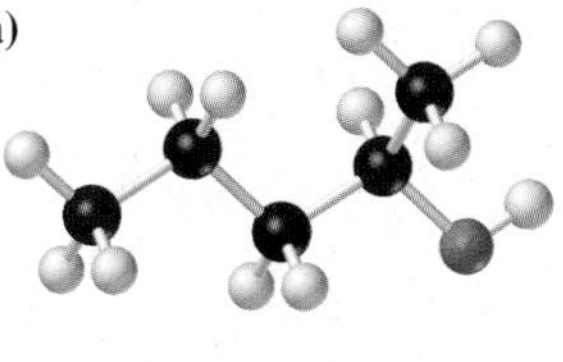 (b) 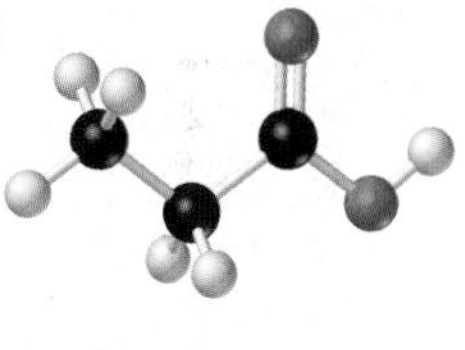

(c)  (d) 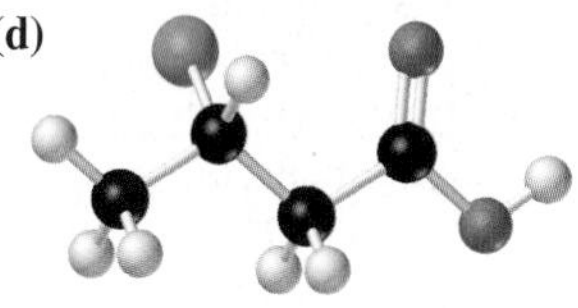

Integrative and Advanced Exercises

The mineral spodumene has the empirical formula $LiAlSi_2O_6$. Given that the percentage of lithium-6 atoms in naturally occuring lithium is 7.40%, how many lithium-6 atoms are present in a 518 g sample of spodumene?

A particular type of brass contains Cu, Sn, Pb, and Zn. A 1.1713 g sample is treated in such a way as to convert the Sn to 0.245 g SnO_2, the Pb to 0.115 g $PbSO_4$, and the Zn to 0.246 g $Zn_2P_2O_7$. What is the mass percent of each element in the sample?

A brand of lunchmeat contains 0.12% by mass of sodium benzoate, C_6H_5COONa. How many mg of Na does a person ingest by eating 3.50 oz of this meat?

The important natural sources of boron compounds are the minerals kernite, $Na_2B_4O_7 \cdot 4\,H_2O$ and borax, $Na_2B_4O_7 \cdot 10\,H_2O$. How much *additional* mass of mineral must be processed per kilogram of boron obtained if the mineral is borax rather than kernite?

87. To deposit exactly one mole of Ag from an aqueous solution containing Ag^+ requires a quantity of electricity known as one faraday (F). The electrodeposition requires that each Ag^+ ion gain one electron to become an Ag atom. Use appropriate physical constants listed on the inside back cover to obtain a precise value of the Avogadro constant, N_A.

88. By analysis, a compound was found to contain 26.58 % K and 35.45 % Cr by mass; the remainder was oxygen. What is the oxidation state of chromium in this compound? What is the name of the compound?

89. Is it possible to have a sample of S_8 that weighs 1.00×10^{-23} g? What is the smallest possible mass that a sample of S_8 can have? Express your answer to the second question in appropriate SI units so that your answer has a numerical value greater than 1. (See Table 1.2 for a list of SI prefixes.)

90. What is the molecular formula of a hydrocarbon containing n carbon atoms and only one double bond? Can such a hydrocarbon yield a greater mass of H_2O than CO_2 when burned in an excess of oxygen?

91. A hydrocarbon mixture consists of 60.0% by mass of C_3H_8 and 40.0% of C_xH_y. When 10.0 g of this mixture is burned, 29.0 g CO_2 and 18.8 g H_2O are the only products. What is the formula of the unknown hydrocarbon?

92. A 0.732 g mixture of methane, CH_4, and ethane, C_2H_6, is burned, yielding 2.064 g CO_2. What is the percent composition of this mixture **(a)** by mass; **(b)** on a mole basis?

93. The density of a mixture of H_2SO_4 and water is 1.78 g/mL. The percent composition of the mixture is to be determined by converting H_2SO_4 to $(NH_4)_2SO_4$. If 32.0 mL of the mixture gives 65.2 g $(NH_4)_2SO_4$, then what is the percent composition of the mixture?

94. In 2013, the IUPAC recommended that the atomic masses of 12 elements be expressed as an atomic mass interval rather as a single invariant value. (See Section 2-5 and Table 2.2.) For example, the IUPAC recommends that the atomic mass of Cl be given as [35.446, 35.457]. Consequently, the results of calculations involving the atomic mass of chlorine should, in principle, be reported as a range of values. Demonstrate this approach by calculating the range of values possible for the mass percent of silver in an impure sample if all the silver in a 26.39 g sample is converted to 31.56 g of silver chloride. [*Hint:* Perform two calculations, using first the lower bound and then the upper bound of the atomic mass interval of Cl.]

95. In the year 2000, the *Guinness Book of World Records* called ethyl mercaptan, C_2H_6S, the smelliest substance known. The average person can detect its presence in air at levels as low as $9 \times 10^{-4}\ \mu mol/m^3$. Express the limit of detectability of ethyl mercaptan in parts per billion (ppb). (Note: 1 ppb C_2H_6S means there is 1 g C_2H_6S per billion grams of air.) The density of air is approximately 1.2 g/L at room temperature.

96. Dry air is essentially a mixture of the following entities: N_2, O_2, Ar, and CO_2. The composition of dry air, in *mole percent*, is 78.08% N_2, 20.95% O_2, 0.93% Ar, and 0.04% CO_2. **(a)** What is the mass, in grams, of a sample of air that contains exactly one mole of the entities? **(b)** Dry air also contains other entities in much smaller amounts. For example, the mole percent of krypton (Kr) is about 1.14×10^{-4} %. Given that the density of dry air is about 1.2 g/L at room temperature, what mass of krypton could be obtained from exactly one cubic meter of dry air?

97. A public water supply was found to contain 0.8 part per billion (ppb) by mass of chloroform, $CHCl_3$. **(a)** How many $CHCl_3$ molecules would be present in a 350 mL glass of this water? **(b)** If the $CHCl_3$ in part (a) could be isolated, would this quantity be detectable on an ordinary analytical balance that measures mass with a precision of ± 0.0001 g?

98. A sample of the compound MSO_4 weighing 0.1131 g reacts with barium chloride and yields 0.2193 g $BaSO_4$. What must be the atomic mass of the metal M? [*Hint:* All the SO_4^{2-} from the MSO_4 appears in the $BaSO_4$.]

99. The metal M forms the sulfate $M_2(SO_4)_3$. A 0.7 sample of this sulfate is converted to 1.511 g Ba What is the atomic mass of M? [*Hint:* Refer to Exercise 98.]

100. A 0.622 g sample of a metal oxide with the for M_2O_3 is converted to 0.685 g of the sulfide, MS. is the atomic mass of the metal M?

101. $MgCl_2$ often occurs in table salt (NaCl) an responsible for caking of the salt. A 0.5200 g sa of table salt is found to contain 61.10% Cl, by r What is the % $MgCl_2$ in the sample? Why is the cision of this calculation so poor?

102. When 2.750 g of the oxide of lead Pb_3O_4 is stro heated, it decomposes and produces 0.0640 g of gen gas and 2.686 g of a second oxide of lead. is the empirical formula of this second oxide?

103. A 1.013 g sample of $ZnSO_4 \cdot x\,H_2O$ is dissolv water and the sulfate ion precipitated as $BaSO_4$. mass of pure, dry $BaSO_4$ obtained is 0.8223 g. is the formula of the zinc sulfate hydrate?

104. The iodide ion in a 1.552 g sample of the ionic pound MI is removed through precipitation. precipitate is found to contain 1.186 g I. What i element M?

105. An oxoacid with the formula $H_xE_yO_z$ has a for mass of 178 u, has 13 atoms in its formula unit, tains 34.80% by mass, and 15.38% by numb atoms, of the element E. What is the element E, what is the formula of this oxoacid?

106. The insecticide dieldrin contains carbon, hydr oxygen, and chlorine. When burned in an exce oxygen, a 1.510 g sample yields 2.094 g CO_2 0.286 g H_2O. The compound has a molecular ma 381 u and has half as many chlorine atoms as ca atoms. What is the molecular formula of dieldri

107. A thoroughly dried 1.271 g sample of Na_2S exposed to the atmosphere and found to gain 0. in mass. What is the percent, by mass $Na_2SO_4 \cdot 10\,H_2O$ in the resulting mixture of a drous Na_2SO_4 and the decahydrate?

108. The atomic mass of Bi is to be determined by verting the compound $Bi(C_6H_5)_3$ to Bi_2O_3. If 5.6 of $Bi(C_6H_5)_3$ yields 2.969 g Bi_2O_3, what is the at mass of Bi?

109. A piece of gold (Au) foil measuring 0.25 m 15 mm $\times$ 15 mm is treated with fluorine gas. treatment converts all the gold in the foil to 1.40 a gold fluoride. What is the formula and name fluoride? The density of gold is 19.3 g/cm^3.

110. In an experiment, 244 mL of chlorine (Cl_2, d = 2.898 g/L) combines with iodine to 1.553 g of a binary compound. In a separate ex ment, the molar mass of the compound is fou be about 467 g/mol. What is the molecular for of this compound?

111. Placing a 0.725 g copper strip in the presen iodine vapor produced a yellowish-white coatir the metal strip. The mass of the copper strip coating was 0.733 g. The coating was remove rinsing the coated metal strip in a potassium cyanate (KSCN) solution, yielding a clean co strip of mass 0.721 g. What is the empirical for of the yellowish-white compound?

Feature Problems

All-purpose fertilizers contain the essential elements nitrogen, phosphorus, and potassium. A typical fertilizer carries numbers on its label, such as "5-10-5". These numbers represent the % N, % P_2O_5, and % K_2O, respectively. The N is contained in the form of a nitrogen compound, such as $(NH_4)_2SO_4$, NH_4NO_3, or $CO(NH_2)_2$ (urea). The P is generally present as a phosphate, and the K as KCl. The expressions % P_2O_5 and % K_2O were devised in the nineteenth century, before the nature of chemical compounds was fully understood. To convert from % P_2O_5 to % P and from % K_2O to % K, the factors 2 mol P/mol P_2O_5 and 2 mol K/mol K_2O must be used, together with molar masses.

(a) Assuming three-significant-figure precision, what is the percent composition of the "5-10-5" fertilizer in % N, % P, and % K?

(b) What is the % P_2O_5 in the following compounds (both common fertilizers)? **(i)** $Ca(H_2PO_4)_2$; **(ii)** $(NH_4)_2HPO_4$.

(c) In a similar manner to the "5-10-5" fertilizer described in this exercise, how would you describe a fertilizer in which the mass ratio of $(NH_4)_2HPO_4$ to KCl is 5.00:1.00?

(d) Can a "5-10-5" fertilizer be prepared in which $(NH_4)_2HPO_4$ and KCl are the sole fertilizer components, with or without inert nonfertilizer additives? If so, what should be the proportions of the constituents of the fertilizer mixture? If this "5-10-5" fertilizer cannot be prepared, why not?

A hydrate of copper(II) sulfate, when heated, goes through the succession of changes suggested by the photograph. In this photograph, **(a)** is the original fully hydrated copper(II) sulfate; **(b)** is the product obtained by heating the original hydrate to 140 °C; **(c)** is the product obtained by further heating to 400 °C; and **(d)** is the product obtained at 1000 °C.

(a) **(b)** **(c)** **(d)**

A 2.574 g sample of $CuSO_4 \cdot x\,H_2O$ was heated to 140 °C, cooled, and reweighed. The resulting solid was reheated to 400 °C, cooled, and reweighed. Finally, this solid was heated to 1000 °C, cooled, and reweighed for the last time.

Original sample	2.574 g
After heating to 140 °C	1.833 g
After reheating to 400 °C	1.647 g
After reheating to 1000 °C	0.812 g

(a) Assuming that all the water of hydration is driven off at 400 °C, what is the formula of the original hydrate?

(b) What is the formula of the hydrate obtained when the original hydrate is heated to only 140 °C?

(c) The black residue obtained at 1000 °C is an oxide of copper. What is its percent composition and empirical formula?

114. Some substances that are only very slightly soluble in water will spread over the surface of water to produce a film that is called a *monolayer* because it is only one molecule thick. A practical use of this phenomenon is to cover ponds to reduce the loss of water by evaporation. Stearic acid forms a monolayer on water. The molecules are arranged upright and in contact with one another, rather like pencils tightly packed and standing upright in a coffee mug. The model below represents an individual stearic acid molecule in the monolayer.

(a) How many square meters of water surface would be covered by a monolayer made from 10.0 g of stearic acid?
[*Hint:* What is the formula of stearic acid?]

(b) If stearic acid has a density of 0.85 g/cm^3, estimate the length (in nanometers) of a stearic acid molecule.
[*Hint:* What is the thickness of the monolayer described in part (a)?]

(c) A very dilute solution of oleic acid in liquid pentane is prepared in the following way:

$$1.00 \text{ mL oleic acid} + 9.00 \text{ mL pentane} \longrightarrow \text{solution (1)};$$
$$1.00 \text{ mL solution (1)} + 9.00 \text{ mL pentane} \longrightarrow \text{solution (2)};$$
$$1.00 \text{ mL solution (2)} + 9.00 \text{ mL pentane} \longrightarrow \text{solution (3)};$$
$$1.00 \text{ mL solution (3)} + 9.00 \text{ mL pentane} \longrightarrow \text{solution (4)}.$$

A 0.10 mL sample of solution (4) is spread in a monolayer on water. The area covered by the monolayer is 85 cm^2. Assume that oleic acid molecules are arranged in the same way as described for stearic acid, and that the cross-sectional area of the molecule is 4.6×10^{-15} cm^2. The density of oleic acid is 0.895 g/mL. Use these data to obtain an approximate value of Avogadro's number.

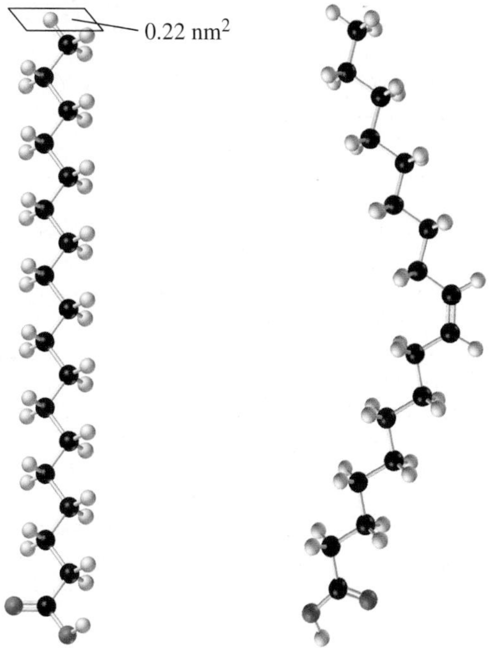

Stearic acid Oleic acid

Self-Assessment Exercises

115. In your own words, define or explain the following terms or symbols: **(a)** formula unit; **(b)** P_4; **(c)** molecular compound; **(d)** binary compound; **(e)** hydrate.

116. Briefly describe each of the following ideas or methods: **(a)** mole of a compound; **(b)** structural formula; **(c)** oxidation state; **(d)** carbon–hydrogen–oxygen determination by combustion analysis.

117. Explain the important distinctions between each pair of terms: **(a)** molecular mass and molar mass; **(b)** empirical and molecular formulas; **(c)** systematic and trivial, or common, name; **(d)** hydroxyl and carboxyl functional group.

118. Explain each term as it applies to the element nitrogen: **(a)** atomic mass; **(b)** molecular mass; **(c)** molar mass.

119. Which answer is correct? One mole of liquid bromine, Br_2, **(a)** has a mass of 79.9 g; **(b)** contains 6.022×10^{23} Br atoms; **(c)** contains the same number of atoms as in 12.01 g H_2O; **(d)** has twice the mass of 0.500 mole of gaseous Cl_2.

120. Three of the following formulas might be either an empirical or a molecular formula. The formula that must be a molecular formula is **(a)** N_2O; **(b)** N_2H_4; **(c)** NaCl; **(d)** NH_3.

121. The compound $C_7H_7NO_2$ contains **(a)** 17 atoms per mole; **(b)** equal percents by mass of C and H; **(c)** about twice the percent by mass of O as of N; **(d)** about twice the percent by mass of N as of H.

122. The greatest number of N atoms is found in **(a)** 50.0 g N_2O; **(b)** 17.0 g NH_3; **(c)** 150 mL of liquid pyridine, C_5H_5N (d = 0.983 g/mL); **(d)** 1.0 mol N_2.

123. Iron is present in red blood cells and acts to carry oxygen to the organs. Without oxygen, these organs will die. There are about 2.6×10^{13} red blood cells in the blood of an adult human, and the blood contains a total of 2.9 g of iron. How many atoms are there in each blood cell?

124. XF_3 consists of 65% F by mass. The atomic mass of the element X must be **(a)** 8 u; **(b)** 11u; **(c)** 31 u; **(d)** 35 u.

125. The oxidation state of I in the ion $H_4IO_6^-$ is **(a)** −1; **(b)** +1; **(c)** +7; **(d)** +8.

126. The oxidation state of Mn in $MgMnO_4$ is **(a)** +2; **(b)** +7; **(c)** +6; **(d)** +4; **(e)** +3.

127. The name of which compound ends with **(a)** HIO_4; **(b)** Na_2SO_3; **(c)** $KClO_2$; **(d)** HFO; **(e)** N

128. The name of $Sr(HCO_3)_2$ is **(a)** strontium ox **(b)** strontium carbonate; **(c)** sodium bicarbo **(d)** strontium bicarbonate; **(e)** none of these.

129. The formula for calcium chlorite is **(a)** Ca **(b)** $Ca(ClO_2)_2$; **(c)** $CaClO_3$; **(d)** $Ca(ClO_4)_2$.

130. Which compound has a molar mass of 51.79 g m **(a)** NaCl; **(b)** KF; **(c)** MgS; **(d)** Li_3P; **(e)** none of t

131. A formula unit of the compound $[Cu(NH_3)_4]SC$ nearly equal masses of **(a)** S and O; **(b)** N ar **(c)** H and N; **(d)** Cu and O.

132. An isomer of the compound $CH_3CH_2CHOHCH$ **(a)** $C_4H_{10}O$; **(b)** $CH_3CHOHCH_2CH_3$; **(c)** $CH_3(CH_2)_2OH$; **(d)** $CH_3CH_2OCH_2CH_3$.

133. A hydrate of Na_2SO_3 contains almost ex 50% H_2O by mass. What is the formula o hydrate?

134. Malachite is a common copper-containing mi with the formula $CuCO_3 \cdot Cu(OH)_2$. **(a)** What mass percent copper in malachite? **(b)** When chite is strongly heated, carbon dioxide and are driven off, yielding copper(II) oxide as the product. What mass of copper(II) oxide is proc per kg of malachite?

135. Acetaminophen, an analgesic and antipy drug, has a molecular mass of 151.2 u and a percent composition of 63.56% C, 6.00% H, 9 N, and 21.17% O. What is the molecular form acetaminophen?

136. Ibuprofen is a compound used in painkillers. W 2.174 g sample is burned in an excess of oxyg yields 6.029 g CO_2 and 1.709 g H_2O as the sole ucts. **(a)** What is the percent composition, by ma ibuprofen? **(b)** What is the empirical formu ibuprofen?

137. Appendix E describes a useful study aid know concept mapping. Using the method present Appendix E, construct a concept map illustratir different concepts in Sections 3-2 and 3-3.

hemical eactions

4

NTENTS

LEARNING OBJECTIVES

4.1 Write a balanced chemical equation for a chemical reaction, specifying states of matter or reaction conditions, as appropriate.

4.2 Use the methodology of converting to, between, and from moles to solve stoichiometry problems.

4.3 Determine the molarity of a solution from its measured quantities, and determine the volume of solution used in a solution dilution or a chemical reaction.

4.4 Define the terms *limiting* and *excess reactant*, and describe how to determine which reactant is the limiting one in a chemical reaction.

4.5 Determine the theoretical and percent yield of a given reaction, and distinguish between consecutive and simultaneous reactions.

4.6 Express the changes in amount for reactions that occur to a limited extent in terms of the stoichiometric numbers and the extent of reaction.

›ace shuttle *Discovery* lifts off on mission STS-26. Combustion reactions in the fuel rocket engines provide the thrust to lift the shuttle off the launch pad. In ıapter, we learn to write and use balanced chemical equations for a wide y of chemical reactions, including combustion reactions.

We are all aware that iron rusts and natural gas burns. These processes are chemical reactions. Chemical reactions are the central concern not just of this chapter but of the entire science of ıistry. In this chapter, we will establish quantitative (numerical) relahips among the substances involved in a reaction, a topic known as *on stoichiometry*. Because many chemical reactions occur in solution, ,ill also consider *solution stoichiometry* and introduce a method of ·ibing the composition of a solution called *solution molarity*.

describing chemical reactions, we often take a microscopic view and ; on the entities—atoms, ions, or molecules—that make up the sub·es involved. However, when *doing* chemistry, we often think of reac- in more macroscopic terms because, in the laboratory, we handle quantities of substances—grams or liters—that can be easily measure manipulated. In large part, reaction stoichiometry provides the relations we need to relate macroscopic amounts of substances to our microscopic of chemical reactions.

To some, stoichiometry is no more exciting than the law of conservati mass, but make no mistake—stoichiometry is important. Chemists use stoi metric principles routinely to plan experiments, analyze their results, and predictions, all of which contribute to making new discoveries and expan our knowledge of the microscopic world of atoms, molecules, and ions.

In this chapter, we will first learn to represent chemical reactions by ch cal equations, and then we will use chemical equations—and ideas from lier chapters—to establish the quantitative relationships we seek. Throug the chapter, we will discuss new aspects of problem solving and more use the mole concept.

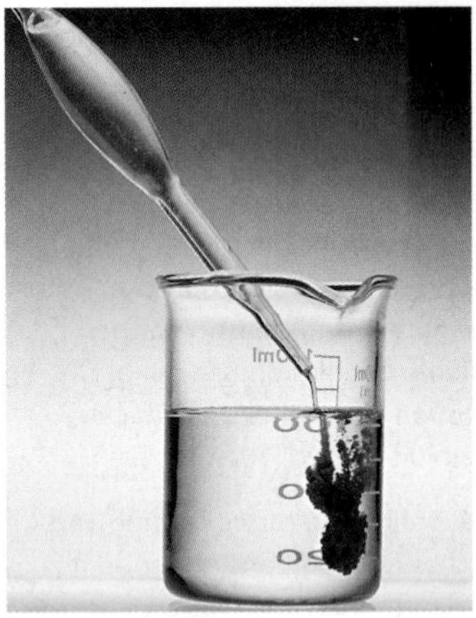

Richard Megna/Fundamental Photographs, NYC

▲ FIGURE 4-1
Precipitation of silver chromate
When aqueous solutions of silver nitrate and potassium chromate are mixed, the disappearance of the distinctive yellow color of chromate ion and the appearance of the red-brown solid, silver chromate, provide physical evidence of a reaction.

4-1 Chemical Reactions and Chemical Equations

A **chemical reaction** is a process in which one set of substances, called **react** is converted to a new set of substances, called **products**. In other words, a c ical reaction is the process by which a chemical change occurs. In many c though, nothing happens when substances are mixed; each retains its ori composition and properties. We need evidence before we can say that a rea has occurred. Some of the types of physical evidence to look for are shown

- a color change (Fig. 4-1)
- formation of a solid (precipitate) within a clear solution (Fig. 4-1)
- evolution of a gas (Fig. 4-2a)
- evolution or absorption of heat (Fig. 4-2b)

Although such observations as these usually signify that a reaction occurred, conclusive evidence still requires a detailed chemical analysis o reaction mixture to identify all the substances present. Moreover, a cher analysis may reveal that a chemical reaction has occurred even in the abs of obvious physical signs.

Just as there are symbols for elements and formulas for compounds, is a symbolic, or shorthand, way of representing a chemical reaction—the c **ical equation**. In a chemical equation, formulas for the reactants are writte the left side of the equation and formulas for the products are written o right. The two sides of the equation are joined by an arrow ($\longrightarrow$). We say the reactants *yield* the products. Consider the reaction of colorless nitr

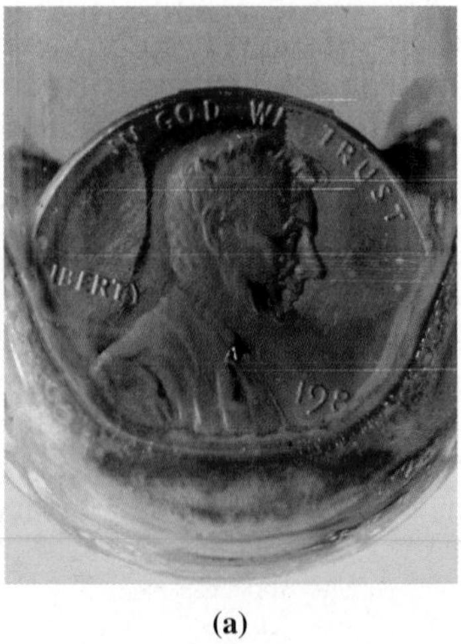

(a)

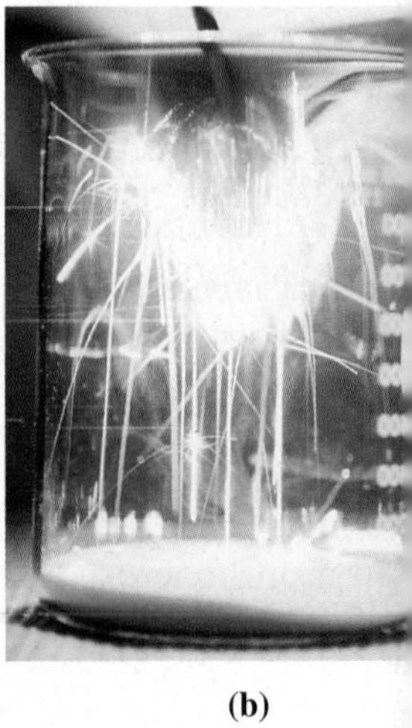

(b)

▶ FIGURE 4-2
Evidence of a chemical reaction
(a) Evolution of a gas: When a copper penny reacts with nitric acid, the red-brown gas NO_2 is evolved. **(b)** Evolution of heat: When iron gauze (steel wool) is ignited in an oxygen atmosphere, evolved heat and light provide physical evidence of a reaction.

(a) Richard Megna/Fundamental Photographs, NYC; **(b)** Tom Bochsler/Pearson Education

oxide and oxygen gases to form red-brown nitrogen dioxide gas, a reaction occurs in the manufacture of nitric acid.

$$\text{nitrogen monoxide} + \text{oxygen} \longrightarrow \text{nitrogen dioxide}$$

mplete the shorthand representation of this reaction, we must do two things:

Substitute chemical formulas for names, to obtain the following expression.

$$NO + O_2 \longrightarrow NO_2$$

n this expression, there are three O atoms on the left side (one in the molecule NO and two in the molecule O_2), but only two O atoms (in the nolecule NO_2) on the right. Because atoms are neither created nor lestroyed in a chemical reaction, this expression needs to be *balanced*.

Balance the numbers of atoms of each kind on both sides of the expression o obtain a balanced chemical equation.* In this step, the coefficient 2 is placed in front of the formulas NO and NO_2. This means that two molecules of NO are consumed and two molecules of NO_2 are produced for every molecule of O_2 consumed. In the balanced equation there are two N atoms and four O atoms on each side. In a **balanced equation**, the total number of atoms of each element present is the same on both sides of the equation. We see this below, both in the symbolic equation and in the molecular representation of the reaction.

$$2\,NO + O_2 \longrightarrow 2\,NO_2$$

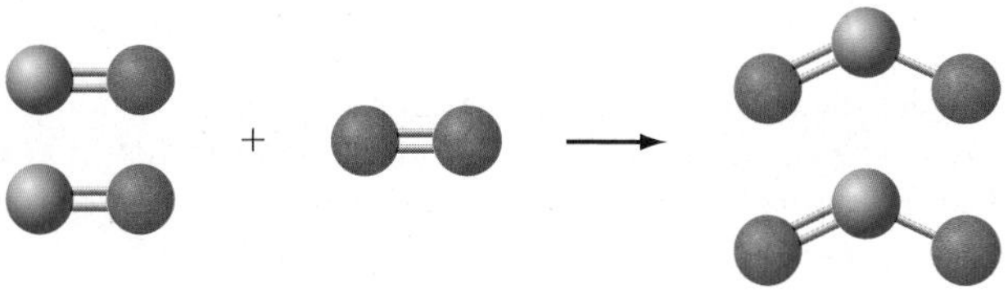

◀ Sometimes products react, partially or completely, to re-form the original reactants. Such reactions are called *reversible reactions* and are designated by a double arrow ($\rightleftharpoons$). In this chapter, we assume that any reverse reactions are negligible and that reactions go only in the forward direction. We consider reversible reactions in detail in Chapter 15.

◀ The reaction of NO and O_2 is of considerable environmental interest. Automobile engines—and all other combustion devices—make NO. One purpose of a catalytic converter is to reduce the amount of NO emitted from a tailpipe. Once in the atmosphere, NO is converted into brown NO_2 by the reaction illustrated on this page. This is why many urban atmospheres have a brown tinge.

he coefficients required to balance a chemical equation are called **stoichioric coefficients**. These coefficients are essential in relating the amounts of tants used and products formed in a chemical reaction, through a variety lculations. In balancing a chemical equation, keep the following point ind.

An equation can be balanced only by adjusting the coefficients of formulas.

he method of equation balancing described above is called *balancing by ction*. Balancing by inspection means to adjust stoichiometric coefficients ial and error until a balanced condition is found. Although the elements generally be balanced in any order, equation balancing need not be a hitiss affair. Here are some useful strategies for balancing equations.

If an element occurs in only one compound on each side of the equation, try balancing this element first.

When one of the reactants or products exists as the free element, balance this element last.

In some reactions, certain groups of atoms (for example, polyatomic ions) remain unchanged. In such cases, balance these groups as a unit.

equation—whether mathematical or chemical—must have the left and right sides equal. ould not call an expression an equation until it is balanced. The term *chemical equation* automally signifies that this balance exists. Although unnecessary, the term *balanced* is commonly when referring to a chemical equation.

▶ We will encounter a few situations in Chapters 7 and 19 where some fractional coefficients are actually required.

- It is permissible to use fractional as well as integral numbers as cc cients. At times, an equation can be balanced most easily by using or more fractional coefficients and then, if desired, clearing the fractior multiplying *all* coefficients by a common multiplier.

KEEP IN MIND

that the strategies described here work well for simple reactions. However, some reactions cannot be balanced by inspection and systematic methods must be employed. We will encounter such situations in Chapter 5.

In this chapter you should concentrate on learning to write formulas fo: reactants and products of a reaction and to balance the equation repres ing the reaction. A third task, which you will face in later chapters, is to dict the products formed when certain reactants are brought together u the appropriate conditions. Even now, however, based on concepts sented in the previous chapter, you should be able to predict the produc a *combustion reaction*. In a plentiful supply of oxygen gas, the combustic hydrocarbons and of carbon–hydrogen–oxygen compounds produces bon dioxide and water as the only products. If the compound contains su as well, sulfur dioxide will also be a product. These ideas and an equa balancing strategy are illustrated in Example 4-1.

EXAMPLE 4-1 Writing and Balancing an Equation: The Combustion of a Carbon–Hydrogen–Oxygen Compound

Liquid triethylene glycol is used as a solvent and plasticizer for vinyl and polyurethane plastics. Write a balanced chemical equation for the combustion of this compound in a plentiful supply of oxygen. A ball-and-stick model of triethylene glycol is shown here.

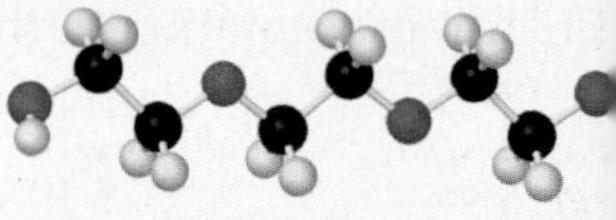

Triethylene glycol

Analyze

We deduce the formula of triethylene glycol from the molecular model. (Refer to the color scheme given on t inside back cover.) We see 6 C atoms (black), 4 O atoms (red), and 14 H atoms. The formula is $C_6H_{14}O_4$. Wh a carbon–hydrogen–oxygen compound is burned in excess oxygen, O_2, the products are CO_2 and H_2O.

Solve

Having identified the reactants and products, we write down an unbalanced chemical expression for the rea tion, showing all reactants and products, and then we balance the expression with respect to each kind of ato

Starting expression: $C_6H_{14}O_4 + O_2 \longrightarrow CO_2 + H_2O$

Balance C: $C_6H_{14}O_4 + O_2 \longrightarrow 6\,CO_2 + H_2O$

Balance H: $C_6H_{14}O_4 + O_2 \longrightarrow 6\,CO_2 + 7\,H_2O$

At this point, the right side of the expression has 19 O atoms (12 in six CO_2 molecules and 7 in seven H_2 molecules), and the left side, only 4 O atoms (in $C_6H_{14}O_4$). To obtain 15 more O atoms requires a fraction coefficient of 15/2 for O_2.

Balance O: $$C_6H_{14}O_4 + \frac{15}{2}O_2 \longrightarrow 6\,CO_2 + 7\,H_2O \quad \text{(balanced)}$$

To remove the fractional coefficient, multiply all coefficients by two:

$$2\,C_6H_{14}O_4 + 15\,O_2 \longrightarrow 12\,CO_2 + 14\,H_2O \quad \text{(balanced)}$$

Assess

To check that the equation is balanced, determine the numbers of C, H, and O atoms that appear on the each side of the equation.

Left: $(2 \times 6) = 12\,C$; $(2 \times 14) = 28\,H$; $[(2 \times 4) + (15 \times 2)] = 38\,O$

Right: $(12 \times 1) = 12\,C$; $(14 \times 2) = 28\,H$; $[(12 \times 2) + (14 \times 1)] = 38\,O$

PRACTICE EXAMPLE A: Write a balanced equation to represent the reaction of mercury(II) sulfide and calcium oxide to produce calcium sulfide, calcium sulfate, and mercury metal.

PRACTICE EXAMPLE B: Write a balanced equation for the combustion of thiosalicylic acid, $C_7H_6O_2S$, used in the manufacture of indigo dyes. A ball-and-stick model of $C_7H_6O_2S$ is shown here.

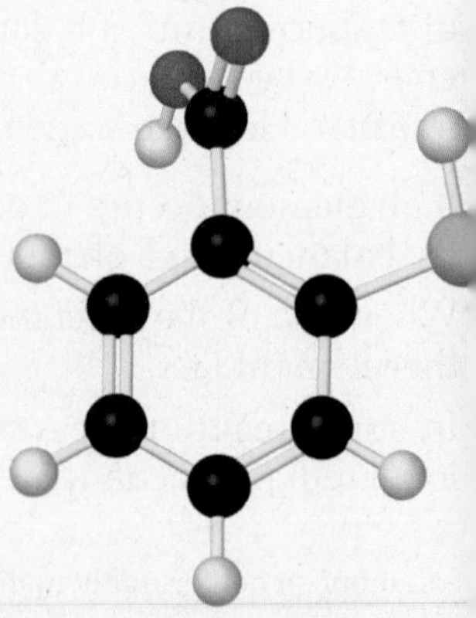

Thiosalicylic acid

es of Matter

e reaction in Example 4-1, triethylene glycol is a liquid, oxygen and carbon de are gases, and water is a liquid. Such facts as these are inconsequential if nterest is only in balancing an equation. Still, we convey a more complete sentation of the reaction by including this information, and sometimes it is tial to include such information in a chemical equation. The state of matter ysical form of reactants and products is shown by symbols in parentheses.

(g) gas (l) liquid (s) solid

, the equation for combustion of triethylene glycol can be written as

$$2\,C_6H_{14}O_4(l) + 15\,O_2(g) \longrightarrow 12\,CO_2(g) + 14\,H_2O(l)$$

◀ At the high combustion temperature, water is present as $H_2O(g)$. But when the reaction products are returned to the initial temperature, the water condenses to a liquid, $H_2O(l)$.

her commonly used symbol for reactants or products dissolved in water is

(aq) *aqueous* solution

ction Conditions

equation for a chemical reaction does not provide enough information to le you to carry out the reaction in a laboratory or chemical plant. An rtant aspect of modern chemical research involves working out the conns for a reaction. The reaction conditions are often written above or below rrow in an equation. For example, the Greek capital letter delta, Δ, means a high temperature is required—that is, the reaction mixture must be d, as in the decomposition of silver oxide.

$$2\,Ag_2O(s) \xrightarrow{\Delta} 4\,Ag(s) + O_2(g)$$

◀ In a *decomposition reaction*, a substance is broken down into simpler substances (for instance, into its elements).

n even more explicit statement of reaction conditions is shown below for the (Badische Anilin & Soda-Fabrik) process for the synthesis of methanol CO and H_2. This reaction occurs at 350 °C, under a total gas pressure that is mes as great as the normal pressure of the atmosphere, and on the surface of ture of ZnO and Cr_2O_3 acting as a *catalyst*. As we will learn later in the text, lyst is a substance that enters into a reaction in such a way that it speeds up action without itself being consumed or changed by the reaction.

◀ In a *synthesis reaction*, a new compound is formed from the reaction of two or more simpler substances, usually called the reactants or starting materials.

$$CO(g) + 2\,H_2(g) \xrightarrow[\substack{340\text{ atm}\\ ZnO,\ Cr_2O_3}]{350\text{ °C}} CH_3OH(g)$$

s important to be able to calculate how much of a particular product will oduced when certain quantities of the reactants are consumed. In the section, we will see how to use chemical equations to set up conversion rs that we can use for these and related calculations.

4-1 CONCEPT ASSESSMENT

the elements K, Cl, and O in the following equations, the requirement that number of atoms be equal on either side of the equation is met. Why is ne of them an acceptable balanced equation for the decomposition of solid assium chlorate yielding solid potassium chloride and oxygen gas?

$KClO_3(s) \longrightarrow KCl(s) + 3\,O(g)$

$KClO_3(s) \longrightarrow KCl(s) + O_2(g) + O(g)$

$KClO_3(s) \longrightarrow KClO(s) + O_2(g)$

Chemical Equations and Stoichiometry

reek, the word *stoicheion* means element. The term **stoichiometry** (stoym'-eh-tree) means, literally, to measure the elements—but from a practiandpoint, it includes all the quantitative relationships involving atomic and formula masses, chemical formulas, and chemical equations. We co ered the quantitative meaning of chemical formulas in Chapter 3, and no will explore some additional quantitative aspects of chemical equations.

The coefficients in the chemical equation

$$2\,H_2(g) + O_2(g) \longrightarrow 2\,H_2O(l)$$

mean that

$$2x \text{ molecules } H_2 + x \text{ molecules } O_2 \longrightarrow 2x \text{ molecules } H_2O$$

Suppose we let $x = 6.02214 \times 10^{23}$ (Avogadro's number). Then x mole represents *1 mole*. Thus the chemical equation also means that

$$2 \text{ mol } H_2 + 1 \text{ mol } O_2 \longrightarrow 2 \text{ mol } H_2O$$

The coefficients in the chemical equation allow us to make statements su

- *Two* moles of H_2O are *produced* for every *two* moles of H_2 *consumed.*
- *Two* moles of H_2O are *produced* for every *one* mole of O_2 *consumed.*
- *Two* moles of H_2 are *consumed* for every *one* mole of O_2 *consumed.*

Moreover, we can turn such statements into conversion factors, called stoi metric factors. A **stoichiometric factor** relates the amounts, on a mole basis, o two substances involved in a chemical reaction; thus a stoichiometric facto mole ratio. In the following examples, stoichiometric factors are printed in b

EXAMPLE 4-2 Relating the Numbers of Moles of Reactant and Product

How many moles of CO_2 are produced in the combustion of 2.72 mol of triethylene glycol, $C_6H_{14}O_4$, in excess of O_2?

Analyze

"An excess of O_2" means that there is more than enough O_2 available to permit the complete conversion of t triethylene glycol to CO_2 and H_2O. The factor for converting from moles of $C_6H_{14}O_4$ to moles of CO_2 obtained from the balanced equation for the combustion reaction.

Solve

The first step in a stoichiometric calculation is to write a balanced equation for the reaction. The balanc chemical equation for the reaction is given below.

$$2\,C_6H_{14}O_4 + 15\,O_2 \longrightarrow 12\,CO_2 + 14\,H_2O$$

Thus, 12 mol CO_2 are produced for every 2 mol $C_6H_{14}O_4$ burned. The production of 12 mol CO_2 is equivale to the consumption of 2 mol $C_6H_{14}O_4$; thus, the ratio 12 mol CO_2/2 mol $C_6H_{14}O_4$ converts from mol C_6H_{14} to mol CO_2.

$$? \text{ mol } CO_2 = 2.72 \text{ mol } C_6H_{14}O_4 \times \frac{12 \text{ mol } CO_2}{2 \text{ mol } C_6H_{14}O_4} = 16.3 \text{ mol } CO_2$$

Assess

The expression above can be written in terms of two equal ratios:

$$\frac{? \text{ mol } CO_2}{2.72 \text{ mol } C_6H_{14}O_4} = \frac{12 \text{ mol } CO_2}{2 \text{ mol } C_6H_{14}O_4}$$

You may find it easier to set up an expression in terms of ratios and then solve it for the unknown quantity.

PRACTICE EXAMPLE A: How many moles of O_2 are produced from the decomposition of 1.76 moles potassium chlorate?

$$2\,KClO_3(s) \longrightarrow 2\,KCl(s) + 3\,O_2(g)$$

PRACTICE EXAMPLE B: How many moles of Ag are produced in the decomposition of 1.00 kg of silver oxide?

$$2\,Ag_2O(s) \longrightarrow 4\,Ag(s) + O_2(g)$$

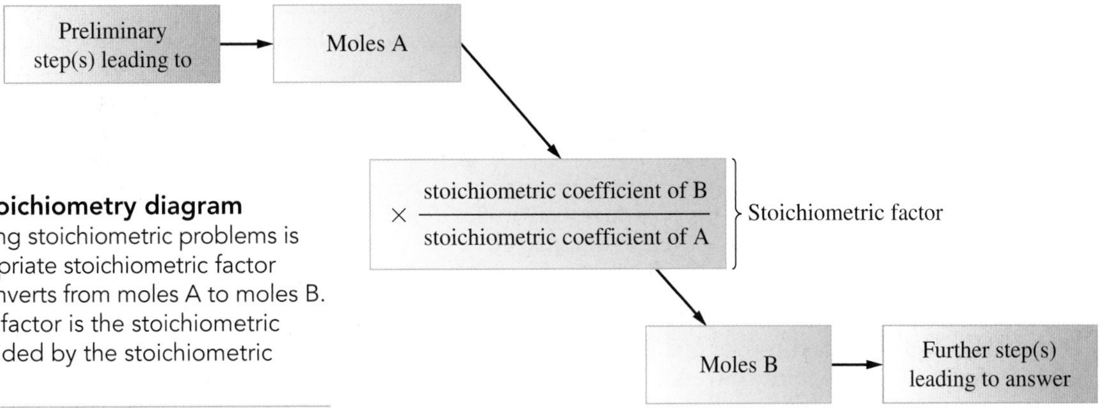

GURE 4-3
eneralized stoichiometry diagram
y step in working stoichiometric problems is ying the appropriate stoichiometric factor e ratio) that converts from moles A to moles B. stoichiometric factor is the stoichiometric ficient of B divided by the stoichiometric ficient of A.

eaction stoichiometry problems range from relatively simple to complex, all can be solved by the same general strategy. This strategy, outlined in ıre 4-3, yields information about one substance, B, from information given ıt a second substance, A. Thus, the conversion pathway is from the *known* A ıe *unknown* B. The heart of the calculation, shown in blue in Figure 4-3, is version from the number of moles of A to the number of moles of B using a or based on stoichiometric coefficients from the balanced chemical ation—that is, (stoichiometric coefficient of B)/(stoichiometric coefficient). Example 4-2 was a single-step calculation requiring only the stoichioric factor. Examples 4-3 through 4-6 require additional steps, both before and r the stoichiometric factor is employed. In short, this strategy involves the owing conversions: "to moles," "between moles," and "from moles." The versions can be done separately or they can be combined into a single-step ulation, as shown in the following examples.

KEEP IN MIND

that it is important to include units and to work from a balanced chemical equation when solving stoichiometry problems.

XAMPLE 4-3 Relating the Mass of a Reactant and a Product

What mass of CO_2 is formed in the reaction of 4.16 g triethylene glycol, $C_6H_{14}O_4$, with an excess of O_2?

nalyze

The general strategy involves the following conversions: (1) to moles, (2) between moles, and (3) from moles. In this example, the required conversions are $\text{g } C_6H_{14}O_4 \xrightarrow{1} \text{mol } C_6H_{14}O_4 \xrightarrow{2} \text{mol } CO_2 \xrightarrow{3} \text{g } CO_2$. Each numbered arrow refers to a conversion factor that changes the unit on the left to the one on the right.

olve

The conversions can be carried out by using either a stepwise approach or the conversion pathway approach. Using a *stepwise approach*, we proceed as follows.

Convert from grams of $C_6H_{14}O_4$ to moles of $C_6H_{14}O_4$ by using the molar mass of $C_6H_{14}O_4$ as a conversion factor.

$$? \text{ mol } C_6H_{14}O_4 = 4.16 \text{ g } C_6H_{14}O_4 \times \frac{1 \text{ mol } C_6H_{14}O_4}{150.2 \text{ g } C_6H_{14}O_4}$$
$$= 0.0277 \text{ mol } C_6H_{14}O_4$$

Convert from moles of $C_6H_{14}O_4$ to moles of CO_2 by using the stoichiometric factor.

$$? \text{ mol } CO_2 = 0.0277 \text{ mol } C_6H_{14}O_4 \times \frac{12 \text{ mol } CO_2}{2 \text{ mol } C_6H_{14}O_4}$$
$$= 0.166 \text{ mol } CO_2$$

Convert from moles of CO_2 to grams of CO_2 by using the molar mass of CO_2 as a conversion factor.

$$? \text{ g } CO_2 = 0.166 \text{ mol } CO_2 \times \frac{44.01 \text{ g } CO_2}{1 \text{ mol } CO_2}$$
$$= 7.31 \text{ g } CO_2$$

(continued)

In the *conversion pathway approach*, the individual steps are combined into a single line calculation, as show below.

$$? \text{ g } CO_2 = 4.16 \text{ g } C_6H_{14}O_4 \times \underbrace{\frac{1 \text{ mol } C_6H_{14}O_4}{150.2 \text{ g } C_6H_{14}O_4}}_{\text{converts to moles of } C_6H_{14}O_4} \times \underbrace{\frac{12 \text{ mol } CO_2}{2 \text{ mol } C_6H_{14}O_4}}_{\text{converts to moles of } CO_2} \times \underbrace{\frac{44.01 \text{ g } CO_2}{1 \text{ mol } CO_2}}_{\text{converts to grams of } CO_2} = 7.31 \text{ g } CO_2$$

Note the connection between the stepwise approach and the conversion pathway approach for this problem. I the conversion pathway approach, the first conversion factor converts from grams of $C_6H_{14}O_4$ to moles o $C_6H_{14}O_4$. The second conversion factor is the stoichiometric factor, and it converts from moles of $C_6H_{14}O_4$ t moles of CO_2. The third conversion factor converts from moles of CO_2 to grams of CO_2. The conversion path way approach combines all three calculations into a single line.

Assess

A quick scan of the numbers to the right of 4.16 g $C_6H_{14}O_4$ indicates that the stoichiometric factor has a value o 6; 6 × 44.01 is between 250 and 300 which, when divided by 150.2, yields a factor somewhat smaller than . The mass of CO_2 should be somewhat less than twice that of the $C_6H_{14}O_4$, and it is (compare 7.31 to 2 × 4.16 Also, note that all units cancel properly in the ultimate conversion from g $C_6H_{14}O_4$ to g CO_2.

PRACTICE EXAMPLE A: How many grams of magnesium nitride, Mg_3N_2, are produced by the reaction o 3.82 g Mg with an excess of N_2?

PRACTICE EXAMPLE B: How many grams of H_2 are required to produce 1.00 kg methanol, CH_3OH, by th reaction $CO + 2\,H_2 \longrightarrow CH_3OH$?

EXAMPLE 4-4 Relating the Masses of Two Reactants to Each Other

What mass of O_2 is consumed in the complete combustion of 6.86 g of triethylene glycol, $C_6H_{14}O_4$?

Analyze

The required conversions are $\text{g } C_6H_{14}O_4 \xrightarrow{1} \text{mol } C_6H_{14}O_4 \xrightarrow{2} \text{mol } O_2 \xrightarrow{3} \text{g } O_2$.

Solve

We will first use a stepwise approach to solve this problem.

Convert from grams of $C_6H_{14}O_4$ to moles of $C_6H_{14}O_4$ by using the molar mass of $C_6H_{14}O_4$ as a conversion factor.

$$? \text{ mol } C_6H_{14}O_4 = 6.86 \text{ g } C_6H_{14}O_4 \times \frac{1 \text{ mol } C_6H_{14}O_4}{150.2 \text{ g } C_6H_{14}O_4}$$
$$= 0.0457 \text{ mol } C_6H_{14}O_4$$

Convert from moles of $C_6H_{14}O_4$ to moles of O_2 by using the stoichiometric factor.

$$? \text{ mol } O_2 = 0.0457 \text{ mol } C_6H_{14}O_4 \times \frac{15 \text{ mol } O_2}{2 \text{ mol } C_6H_{14}O_4}$$
$$= 0.343 \text{ mol } O_2$$

Convert from moles of O_2 to grams of O_2 by using the molar mass of O_2 as a conversion factor.

$$? \text{ g } O_2 = 0.343 \text{ mol } O_2 \times \frac{32.00 \text{ g } O_2}{1 \text{ mol } O_2}$$
$$= 11.0 \text{ g } O_2$$

As in Example 4-3, the three steps can be combined into a single calculation, as shown below.

$$? \text{ g } O_2 = 6.86 \text{ g } C_6H_{14}O_4 \times \frac{1 \text{ mol } C_6H_{14}O_4}{150.2 \text{ g } C_6H_{14}O_4} \times \frac{15 \text{ mol } O_2}{2 \text{ mol } C_6H_{14}O_6} \times \frac{32.00 \text{ g } O_2}{1 \text{ mol } O_2} = 11.0 \text{ g } O_2$$

Assess

Focus on the single line calculation shown above. A quick scan of the numbers to the right of 6.86 g $C_6H_{14}O$ indicates that the stoichiometric factor has a value of 7.5; the product, 7.5 × 32.00, is about 250, which, whe divided by 150.2, yields a factor of about 250/150 = 5/3. The mass of O_2 should be about 5/3 that of $C_6H_{14}O$

and it is—that is, compare 11.0 with 5/3 of 6.86, which is about 35/3 or somewhat less than 12. As in Example 4-3, note that the proper cancellation of units occurs.

ACTICE EXAMPLE A: For the reaction $4\ NH_3 + 3\ O_3 \rightarrow 2\ N_2 + 6\ H_2O$, how many grams of NH_3 are consumed per gram of O_2?

ACTICE EXAMPLE B: In the combustion of octane, C_8H_{18}, how many grams of O_2 are consumed per gram of octane?

4-2 CONCEPT ASSESSMENT

ich statements are correct for the reaction $2\ H_2S + SO_2 \longrightarrow 3\ S + 2\ H_2O$? olain your reasoning.

- 3 mol S is produced per mole of H_2S consumed.
- 3 g S is produced for every gram of SO_2 consumed.
- 1 mol H_2O is produced per mole of H_2S consumed.
- Two-thirds of the S produced comes from H_2S.
- The number of moles of products formed equals the number of moles of reactants consumed.
- The number of moles of atoms present after the reaction is the same as the number of moles of atoms before the reaction.

hat lends great variety to stoichiometric calculations is that many other ersions may be required before and after the mol A $\longrightarrow$ mol B step at enter of the stoichiometric scheme shown in Figure 4-3. In Examples 4-3 4-4, the additional conversions involved molar mass. Other common conons may require such factors as volume, density, and percent composition. very case, however, we must always use the appropriate stoichiometric r from the chemical equation as a key conversion factor.

ne reaction between solid aluminum, Al(s), and aqueous hydrochloric , HCl(aq), can be used for preparing small volumes of hydrogen gas,), in the laboratory. A balanced chemical equation for the reaction is vn below.

$$2\ Al(s) + 6\ HCl(aq) \longrightarrow 2\ AlCl_3(aq) + 3\ H_2(g) \qquad (4.2)$$

simple laboratory setup for collecting the hydrogen gas is pictured in re 4-4. The reaction between Al(s) and HCl(aq) provides a range of possies for calculations. Examples 4-5 and 4-6 are based on this reaction.

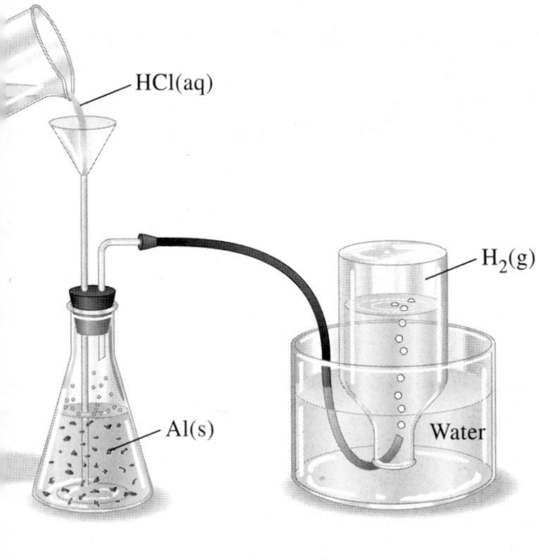

◀ FIGURE 4-4
The reaction $2\ Al(s) + 6\ HCl(aq) \longrightarrow 2\ AlCl_3(aq) + 3\ H_2(g)$
HCl(aq) is added to the flask on the left. The reaction occurs within the flask. The liberated $H_2(g)$ flows into a gas-collection apparatus, where it displaces water. Hydrogen is only very slightly soluble in water.

EXAMPLE 4-5 Additional Conversion Factors in a Stoichiometric Calculation: Volume, Density, and Percent Composition

An alloy used in aircraft structures consists of 93.7% Al and 6.3% Cu by mass. The alloy has a density 2.85 g/cm^3. A 0.691 cm^3 piece of the alloy reacts with an excess of HCl(aq). If we assume that *all* the Al but nc of the Cu reacts with HCl(aq), what is the mass of H_2 obtained? Refer to reaction (4.2).

Analyze

A simple approach to this calculation is outlined below. Each numbered arrow refers to a conversion factor th changes the unit on the left to the one on the right.

$$\text{cm}^3\text{ alloy} \xrightarrow{1} \text{g alloy} \xrightarrow{2} \text{g Al} \xrightarrow{3} \text{mol Al} \xrightarrow{4} \text{mol H}_2 \xrightarrow{5} \text{g H}_2$$

The calculation can be done in five distinct steps, or with a single setup in which the five conversions a performed in sequence.

Solve

Using a stepwise approach, we proceed as follows.

Convert from volume of alloy to grams of alloy by using the density as a conversion factor.

$$? \text{ g alloy} = 0.691 \text{ cm}^3 \text{ alloy} \times \frac{2.85 \text{ g alloy}}{1 \text{ cm}^3 \text{ alloy}} = 1.97 \text{ g alloy}$$

Convert from grams of alloy to grams of Al by using the percentage by mass of Al as a conversion factor.

$$? \text{ g Al} = 1.97 \text{ g alloy} \times \frac{93.7 \text{ g Al}}{100 \text{ g alloy}} = 1.85 \text{ g Al}$$

Convert from grams of Al to moles of Al by using the molar mass of Al as a conversion factor.

$$? \text{ mol Al} = 1.85 \text{ g Al} \times \frac{1 \text{ mol Al}}{26.98 \text{ g Al}} = 0.0684 \text{ mol Al}$$

Convert from moles of Al to moles of H_2 by using the stoichiometric factor.

$$? \text{ mol H}_2 = 0.0684 \text{ mol Al} \times \frac{3 \text{ mol H}_2}{2 \text{ mol Al}} = 0.103 \text{ mol H}_2$$

Convert from moles of H_2 to grams of H_2 using the molar mass of H_2 as a conversion factor.

$$? \text{ g H}_2 = 0.103 \text{ mol H}_2 \times \frac{2.016 \text{ g H}_2}{1 \text{ mol H}_2} = 0.207 \text{ g H}_2$$

Remember to store intermediate results without rounding off. When all of the steps are combined into a sing calculation, we do not have to write down intermediate results and we reduce the likelihood of rounding erro

$$? \text{ g H}_2 = 0.691 \text{ cm}^3 \text{ alloy} \times \frac{2.85 \text{ g alloy}}{1 \text{ cm}^3 \text{ alloy}} \times \frac{93.7 \text{ g Al}}{100 \text{ g alloy}} \times \frac{1 \text{ mol Al}}{26.98 \text{ g Al}} \times \frac{3 \text{ mol H}_2}{2 \text{ mol Al}} \times \frac{2.016 \text{ g H}_2}{1 \text{ mol H}_2} = 0.207 \text{ g H}_2$$

Assess

The units work out properly, but we must evaluate whether the answer is a reasonable one. The molar mass of Al and H_2 are approximately 27 g/mol and 2 g/mol, respectively. Equation (4.2) tells us that 1 mole of A which weighs approximately 27 g, produces 1.5 mol of H_2, which weighs $1.5 \times 2 = 3$ g. Thus, 27 g Al produc approximately 3 g H_2 and, thus, 2.7 g Al produces approximately 0.3 g H_2. In this example, we are dealing wi less than 2.7 g of Al; therefore, we expect less than 0.3 g of H_2. The answer, 0.207 g H_2, is reasonable.

PRACTICE EXAMPLE A: What volume of the aluminum-copper alloy described in Example 4-5 must be dissolve in an excess of HCl(aq) to produce 1.00 g H_2? [*Hint:* Think of this as the "inverse" of Example 4-5.]

PRACTICE EXAMPLE B: A fresh sample of the aluminum-copper alloy described in Example 4-5 yielded 1.31 g H How many grams of copper were present in the sample?

XAMPLE 4-6 Additional Conversion Factors in a Stoichiometric Calculation: Volume, Density, and Percent Composition of a Solution

A hydrochloric acid solution consists of 28.0% HCl by mass and has a density of 1.14 g/mL. What volume of this solution is required to react completely with 1.87 g Al in reaction (4.2)?

nalyze

The first challenge here is to determine where to begin. Although the problem refers to 28.0% HCl and a density of 1.14 g/mL, the appropriate starting point is with the given information—1.87 g Al. The goal of our calculation is a solution volume—mL HCl solution.

$$\text{g Al} \xrightarrow{1} \text{mol Al} \xrightarrow{2} \text{mol HCl} \xrightarrow{3} \text{g HCl} \xrightarrow{4} \text{g HCl solution} \xrightarrow{5} \text{mL HCl solution}$$

The conversion factors in the calculation involve (1) the molar mass of Al, (2) stoichiometric coefficients from equation (4.2), (3) the molar mass of HCl, (4) the percent composition of the HCl solution, and (5) the density of the HCl solution.

lve

Using a stepwise approach, we proceed as follows.

Convert from grams of Al to moles of Al by using the molar mass of Al.

$$?\ \text{mol Al} = 1.87\ \text{g Al} \times \frac{1\ \text{mol Al}}{26.98\ \text{g Al}} = 0.0693\ \text{mol Al}$$

Convert from moles of Al to moles of HCl by using the stoichiometric factor.

$$?\ \text{mol HCl} = 0.0693\ \text{mol Al} \times \frac{6\ \text{mol HCl}}{2\ \text{mol Al}} = 0.208\ \text{mol HCl}$$

Convert from moles of HCl to grams of HCl by using the molar mass of HCl.

$$?\ \text{g HCl} = 0.208\ \text{mol HCl} \times \frac{36.46\ \text{g HCl}}{1\ \text{mol HCl}} = 7.58\ \text{g HCl}$$

Convert from grams of HCl to grams of HCl solution by using the percentage by mass.

$$?\ \text{g HCl soln} = 7.58\ \text{g HCl} \times \frac{100\ \text{g HCl soln}}{28.0\ \text{g HCl}} = 27.1\ \text{g HCl soln}$$

Convert from grams of HCl solution to milliliters of HCl solution by using the density.

$$?\ \text{mL HCl soln} = 27.1\ \text{g HCl soln} \times \frac{1\ \text{mL HCl soln}}{1.14\ \text{g HCl soln}} = 23.8\ \text{mL HCl soln}$$

In the conversion pathway approach, we combine the individual steps into a single line.

$$\left(\text{g Al} \xrightarrow{1} \text{mol Al} \xrightarrow{2} \text{mol HCl} \xrightarrow{3} \text{g HCl}\right.$$

$$?\ \text{mL HCl soln} = 1.87\ \text{g Al} \times \frac{1\ \text{mol Al}}{26.98\ \text{g Al}} \times \frac{6\ \text{mol HCl}}{2\ \text{mol Al}} \times \frac{36.46\ \text{g HCl}}{1\ \text{mol HCl}}$$

$$\left.\xrightarrow{4} \text{g HCl soln} \xrightarrow{5} \text{mL HCl soln}\right)$$

$$\times \frac{100.0\ \text{g HCl soln}}{28.0\ \text{g HCl}} \times \frac{1\ \text{mL HCl soln}}{1.14\ \text{g HCl soln}}$$

$$= 23.8\ \text{mL HCl soln}$$

ssess

Let's attempt to establish whether the answer is reasonable by working the problem in reverse and using numbers that are rounded off slightly. Because the density of the solution is approximately 1 g/mL and the solution is approximately 30% HCl by mass, a 24 mL sample of the solution will contain approximately $24 \times 0.30 = 7.2$ g of HCl. Equation (4.2) tells us that 1 mol Al, or 27 g Al, reacts with 3 mol HCl, or 108 g HCl. Stated another way, 4 g HCl reacts with 1 g Al. In this example, we are using approximately 7.2 g HCl; thus, we consume approximately $7.2\ \text{g HCl} \times 1\ \text{g Al}/4\ \text{g HCl} = 1.8\ \text{g Al}$. This value is close to the actual amount of Al consumed and we conclude that our answer, 23.8 mL of HCl solution, is reasonable.

RACTICE EXAMPLE A: How many milligrams of H_2 are produced when one drop (0.05 mL) of the hydrochloric acid solution described in Example 4-6 reacts with an excess of aluminum in reaction (4.2)?

RACTICE EXAMPLE B: A particular vinegar contains 4.0% CH_3COOH by mass. It reacts with sodium carbonate to produce sodium acetate, carbon dioxide, and water. How many grams of carbon dioxide are produced by the reaction of 5.00 mL of this vinegar with an excess of sodium carbonate? The density of the vinegar is 1.01 g/mL.

4-3 CONCEPT ASSESSMENT

Without performing detailed calculations, determine which reaction produces the maximum quantity of $O_2(g)$ per gram of reactant.

(a) $2\ NH_4NO_3(s) \xrightarrow{\Delta} 2\ N_2(g) + 4\ H_2O(l) + O_2(g)$

(b) $2\ Ag_2O(s) \xrightarrow{\Delta} 4\ Ag(s) + O_2(g)$

(c) $2\ HgO(s) \xrightarrow{\Delta} 2\ Hg(l) + O_2(g)$

(d) $2\ Pb(NO_3)_2(s) \xrightarrow{\Delta} 2\ PbO(s) + 4\ NO_2(g) + O_2(g)$

4-3 Chemical Reactions in Solution

Most reactions in the general chemistry laboratory are carried out in solu This is partly because mixing the reactants in solution helps to achiev close contact between atoms, ions, or molecules necessary for a reactic occur. The stoichiometry of reactions in solutions can be described in the s way as the stoichiometry of other reactions, as we saw in Example 4-6. A new ideas that apply specifically to solution stoichiometry are also helpfu

One component of a solution, called the **solvent**, determines whethe solution exists as a solid, liquid, or gas. In this discussion we will limit selves to *aqueous solutions*—solutions in which liquid water is the solvent. other components of a solution, called **solutes**, are dissolved in the solvent use the notation NaCl(aq), for example, to describe a solution in which li water is the solvent and NaCl is the solute. The term *aqueous* does not co any information, however, about the relative proportions of NaCl and H_2 the solution. For this purpose, the property called **molarity** is commonly u

Molarity

▶ The IUPAC-preferred term for molarity is *amount concentration*. The use of the term *molarity* is still widespread.

The composition of a solution may be specified by giving its molar conce tion (or molarity), which is defined as the amount of solute, in moles, per of solution:

$$\text{molarity} = \frac{\text{amount of solute (in moles)}}{\text{volume of solution (in liters)}}$$

The expression above can be written more compactly as

$$c = \frac{n}{V} \qquad ($$

where c is the molarity in moles per liter (mol/L), n is the amount of solu moles (mol), and V is the volume of the solution in liters (L).

If 0.440 mol urea, $CO(NH_2)_2$, is dissolved in enough water to make 1.0 of solution, the solution concentration, or molarity, is

$$\frac{0.440\ \text{mol}\ CO(NH_2)_2}{1.000\ \text{L soln}} = 0.440\ \text{M}\ CO(NH_2)_2$$

The symbol M stands for the term *molar*, or mol/L. Thus, a solution that 0.440 mol $CO(NH_2)_2$/L is 0.440 M $CO(NH_2)_2$ or 0.440 molar $CO(NH_2)_2$.

Alternatively, if 0.110 mol urea is present in 250.0 mL of solution, the s tion is also 0.440 M.

$$\frac{0.110\ \text{mol}\ CO(NH_2)_2}{0.2500\ \text{L soln}} = 0.440\ \text{M}\ CO(NH_2)_2$$

When calculating concentration, we must determine the amount of solu moles from other quantities that can be readily measured, such as the ma te or the volume of a liquid solute. In Example 4-7, the mass of a liquid te is related to its volume by using density as a conversion factor. Then, ar mass is used to convert from the mass of solute to an amount of solute in es. Also, the solution volume is converted from milliliters (mL) to liters (L).

igure 4-5 illustrates a method commonly used to prepare a solution. A d sample is weighed out and dissolved in sufficient water to produce a tion of known volume—250.0 mL. To fill a beaker to a 250 mL mark is not ly precise enough as a volume measurement. As discussed in Section 1-6, e would be a systematic error because the beaker is not calibrated with icient precision. (The error in volume could be 10 to 20 mL or more.) To olve the solute in water and bring the solution level to the 250 mL mark in aduated cylinder is also not sufficiently precise. Although the graduated nder is calibrated more precisely than the beaker, the error might still be 2 mL or more. However, when filled to the calibration mark, the volumetric k pictured in Figure 4-5 contains 250.0 mL with only about 0.1 mL of error.

Vhen the molarity of a solution is known accurately, we can calculate the nber of moles of solute in a carefully measured volume of solution by g the equation below.

$$n = c \times V \qquad \textbf{(4.4)}$$

expression above is equivalent to equation (4.3), but equation (4.4) phasizes that the molarity, *c*, is a conversion factor for converting from s of solution to moles of solute. In Example 4-8, we use molarity as a conion factor in a calculation to determine the mass of solute needed to proe the solution in Figure 4-5.

KEEP IN MIND

that the volume used in molarity is that of the *solution*. It is neither that of the solvent used, nor the sum of the volumes of the solvent and solute. For example, if we add 25.0 mL ethanol to 250.0 mL water, the solution volume will be neither 250.0 mL nor 275.0 mL. We learn why in Chapter 14.

XAMPLE 4-7 Calculating Molarity from Measured Quantities

A solution is prepared by dissolving 25.0 mL ethanol, CH_3CH_2OH (d = 0.789 g/mL), in enough water to produce 250.0 mL solution. What is the molarity of ethanol in the solution?

nalyze

We must first calculate how many moles of ethanol are in a 25.0 mL sample of pure ethanol. This calculation requires the following conversions: mL ethanol $\xrightarrow{1}$ g ethanol $\xrightarrow{2}$ mol ethanol. The first conversion uses the density as a conversion factor and the second conversion uses molar mass as a conversion factor. The molarity of the solution is then calculated by using equation (4.3).

olve

The number of moles of ethanol in a 25.0 mL sample of pure ethanol is calculated below in a single line.

$$? \text{ mol } CH_3CH_2OH = 25.0 \text{ mL } CH_3CH_2OH \times \frac{0.789 \text{ g } CH_3CH_2OH}{1 \text{ mL } CH_3CH_2OH} \times \frac{1 \text{ mol } CH_3CH_2OH}{46.07 \text{ g } CH_3CH_2OH}$$
$$= 0.428 \text{ mol } CH_3CH_2OH$$

To apply the definition of molarity given in expression (4.3), note that 250.0 mL = 0.2500 L.

$$\text{molarity} = \frac{0.428 \text{ mol } CH_3CH_2OH}{0.2500 \text{ L soln}} = 1.71 \text{ M } CH_3CH_2OH$$

ssess

It is important to include the units in this calculation to ensure that we obtain the correct units for the final answer. When dealing with liquid solutes, be careful to distinguish between mL solute and mL soln.

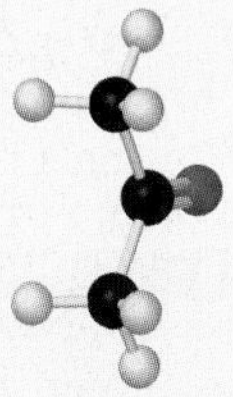
Acetone

RACTICE EXAMPLE A: A 22.3 g sample of acetone (see the model here) is dissolved in enough water to produce 1.25 L of solution. What is the molarity of acetone in this solution?

RACTICE EXAMPLE B: If 15.0 mL of acetic acid, CH_3COOH (d = 1.048 g/mL), is dissolved in enough water to produce 500.0 mL of solution, then what is the molarity of acetic acid in the solution?

(a) (b) (c)

▲ FIGURE 4-5
Preparation of 0.250 M K_2CrO_4—Example 4-8 illustrated
The solution cannot be prepared just by adding 12.1 g $K_2CrO_4(s)$ to 250.0 mL water. Instead, **(a)** the weighed quantity of $K_2CrO_4(s)$ is first added to a clean, dry 250 mL volumetric flask; **(b)** the $K_2CrO_4(s)$ is dissolved in less than 250 mL of water; and **(c)** the flask is filled to the 250.0 mL calibration mark by the careful addition (dropwise) of the remaining water.

EXAMPLE 4-8 Calculating the Mass of Solute in a Solution of Known Molarity

What mass of K_2CrO_4 is needed to prepare exactly 0.2500 L (250.0 mL) of a 0.250 M K_2CrO_4 solution in water (See Figure 4-5.)

Analyze

The conversion pathway is L soln → mol K_2CrO_4 → g K_2CrO_4. The first conversion factor is the molarit of the solution, shown below in blue, and the second conversion factor is the molar mass of K_2CrO_4.

Solve

$$? \text{ g } K_2CrO_4 = 0.2500 \text{ L soln} \times \frac{0.250 \text{ mol } K_2CrO_4}{1 \text{ L soln}} \times \frac{194.2 \text{ g } K_2CrO_4}{1 \text{ mol } K_2CrO_4}$$
$$= 12.1 \text{ g } K_2CrO_4$$

Assess

The answer has the correct units. We can check whether the answer is reasonable by working the problem i reverse and using numbers that are rounded off slightly. Because the molar mass of K_2CrO_4 is approximatel 200 g/mol and the mass of the sample is approximately 12 g, the number of moles of K_2CrO_4 in the sample i approximately $12/200 = 6/100 = 0.06$ mol. The approximate molarity is $0.06/0.250 = 0.24$ mol/L. This esti mate is close to the true molarity, and so we are confident that the answer, 12.1 g K_2CrO_4, is correct.

PRACTICE EXAMPLE A: An aqueous solution saturated with $NaNO_3$ at 25 °C is 10.8 M $NaNO_3$. What mass c $NaNO_3$ is present in 125 mL of this solution at 25 °C?

PRACTICE EXAMPLE B: What mass of $Na_2SO_4 \cdot 10\,H_2O$ is needed to prepare 355 mL of 0.445 M Na_2SO_4?

tion Dilution

mmon sight in chemistry storerooms and laboratories is rows of bottles aining solutions for use in chemical reactions. It is not practical, however, ore solutions of every possible concentration. Instead, most labs store concentrated solutions, so-called *stock solutions*, which can then be used epare more dilute solutions by adding water. The principle of dilution, h you have probably already inferred, is that the same solute that was ent in a sample of stock solution is distributed throughout the larger me of a diluted solution (see Figure 4-6).

◀ A *concentrated solution* has a relatively large amount of dissolved solute; a *dilute solution* has a relatively small amount.

hen a volume of a solution is diluted, the amount of solute *remains con-* . If we write equation (4.4) for the initial (i) undiluted solution, we obtain c_iV_i; for the final (f) diluted solution, we obtain $n_f = c_fV_f$. Because n_i is l to n_f, we obtain the following result:

$$c_iV_i = c_fV_f \tag{4.5}$$

gure 4-7 illustrates the laboratory procedure for preparing a solution by ion. Example 4-9 explains the necessary calculation.

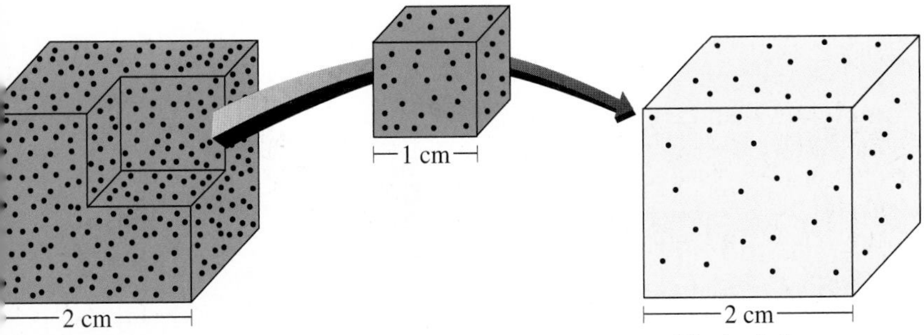

GURE 4-6
alizing the dilution of a solution
inal solution is prepared by extracting $\frac{1}{8}$ of the initial solution—1 cm^3—and diluting h water to a volume of 8 cm^3. The number of dots in the 8 cm^3 of final solution, esenting the number of solute particles, is the same as in the 1 cm^3 of initial solution.

KEEP IN MIND

that equation (4.5) applies only to dilution problems. An alternative expression uses the subscripts 1 and 2 in place of i and f: $c_1V_1 = c_2V_2$. Some students mistakenly use $c_1V_1 = c_2V_2$ to convert from moles of substance 1 to moles of substance 2 when doing stoichiometry problems. To avoid making this mistake, always use the subscripts "i" and "f" when using equation (4.5), or even better, use the subscripts "dil" (for diluted) and "conc" (for concentrated): $c_{dil}V_{dil} = c_{conc}V_{conc}$.

EXAMPLE 4-9 Preparing a Solution by Dilution

A particular analytical chemistry procedure requires 0.0100 M K_2CrO_4. What volume of 0.250 M K_2CrO_4 must be diluted with water to prepare 0.2500 L of 0.0100 M K_2CrO_4?

Analyze

First, we calculate the number of moles K_2CrO_4 that must be present in the final solution. Then, we calculate the volume of 0.250 M K_2CrO_4 that contains this amount of K_2CrO_4.

Solve

First, calculate the amount of solute that must be present in the final solution.

$$? \text{ mol } K_2CrO_4 = 0.2500 \text{ L soln} \times \frac{0.0100 \text{ mol } K_2CrO_4}{1 \text{ L soln}} = 0.00250 \text{ mol } K_2CrO_4$$

Second, calculate the volume of 0.250 M K_2CrO_4 that contains 0.00250 mol K_2CrO_4.

$$? \text{ L soln} = 0.00250 \text{ mol } K_2CrO_4 \times \frac{1 \text{ L soln}}{0.250 \text{ mol } K_2CrO_4} = 0.0100 \text{ L soln}$$

(continued)

An alternative approach is to use equation (4.5). The known factors include the volume of solution to be p pared, (V_f = 250.0 mL) and the concentrations of the final (0.0100 M) and initial (0.250 M) solutions. We m solve for the initial volume, V_i. Note that although in deriving equation (4.5) volumes were expressed in lite in applying the equation any volume unit can be used as long as we use the same unit for both V_i and V_f (m liliters in the present case). The term needed to convert volumes to liters would appear on both sides of t equation and cancel out.

$$V_i = V_f \times \frac{c_f}{c_i} = 250.0 \text{ mL} \times \frac{0.0100 \text{ M}}{0.250 \text{ M}} = 10.0 \text{ mL}$$

Assess

Let's work the problem in reverse. If we dilute 10.0 mL of 0.250 M K_2CrO_4 to 0.250 L with water, then the cc centration of the diluted solution is $0.010 \text{ L} \times 0.250 \text{ mol L}^{-1}/0.250 \text{ L} = 0.010 \text{ mol/L}$. This is the desired mol ity; therefore, the answer, 10.0 mL of 0.250 M K_2CrO_4, is correct.

PRACTICE EXAMPLE A: A 15.00 mL sample of 0.450 M K_2CrO_4 is diluted to 100.00 mL. What is the concentrati of the new solution?

PRACTICE EXAMPLE B: When left in an open beaker for a period of time, the volume of 275 mL of 0.105 M Na is found to decrease to 237 mL because of the evaporation of water. What is the new concentration of t solution?

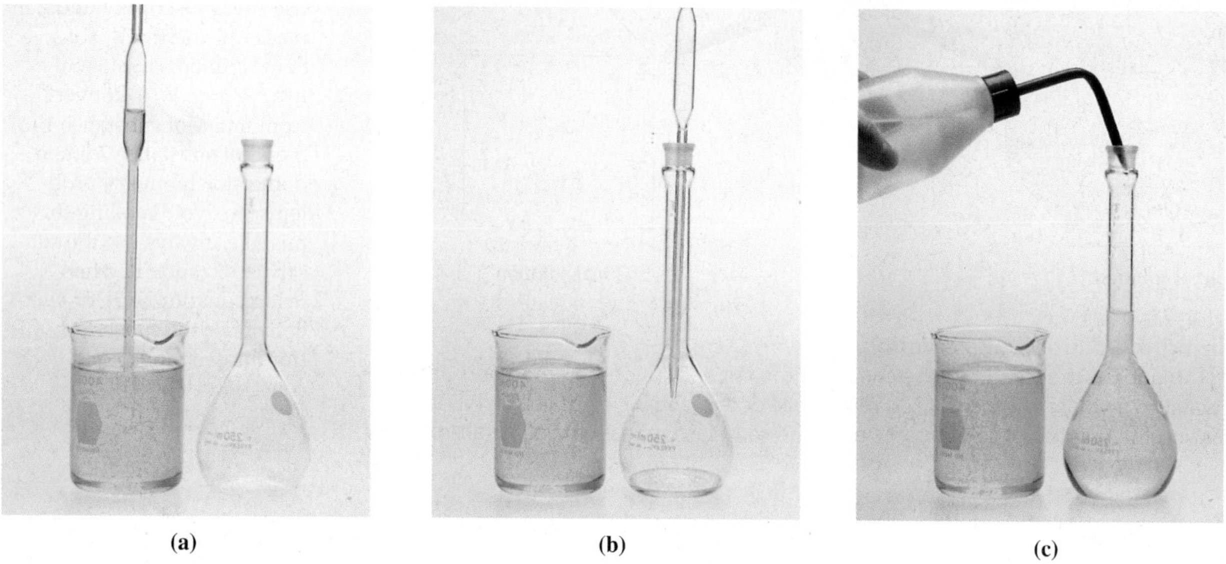

▲ FIGURE 4-7
Preparing a solution by dilution—Example 4-9 illustrated
(a) A pipet is used to withdraw a 10.0 mL sample of 0.250 M K_2CrO_4. **(b)** The pipetful of 0.250 M K_2CrO_4 is discharged into a 250.0 mL volumetric flask. **(c)** Water is then added to bring the level of the solution to the calibration mark on the neck of the flask. At this point, the solution is 0.0100 M K_2CrO_4.

4-4 CONCEPT ASSESSMENT

Without doing detailed calculations, and assuming that the volumes of solutio and water are additive, indicate the molarity of the final solution obtained as a result of

(a) adding 200.0 mL of water to 100.0 mL of 0.150 M NaCl
(b) evaporating 50.0 mL water from 250.0 mL of 0.800 M $C_{12}H_{22}O_{11}$
(c) mixing 150.0 mL 0.270 M KCl and 300.0 mL 0.135 M KCl

chiometry of Reactions in Solution

central conversion factor in Example 4-10 is the same as in previous stoi-netry problems—the appropriate stoichiometric factor. What differs from ious examples is that we use molarity as a conversion factor from solution me to number of moles of reactant in a preliminary step preceding the hiometric factor.

e will consider a number of additional examples of stoichiometric calcu-ns involving solutions in Chapter 5.

KAMPLE 4-10 Relating the Mass of a Product to the Volume and Molarity of a Reactant Solution

A 25.00 mL pipetful of 0.250 M K_2CrO_4 is added to an excess of $AgNO_3(aq)$. What mass of Ag_2CrO_4 will precipitate from the solution?

$$K_2CrO_4(aq) + 2\ AgNO_3(aq) \longrightarrow Ag_2CrO_4(s) + 2\ KNO_3(aq)$$

nalyze

The fact that an excess of $AgNO_3(aq)$ is used tells us that all of the K_2CrO_4 in the 25.00 mL sample of $K_2CrO_4(aq)$ is consumed. The calculation begins with a volume of 25.00 mL and ends with a mass of Ag_2CrO_4 expressed in grams. The conversion pathway is mL soln $\longrightarrow$ L soln $\longrightarrow$ mol K_2CrO_4 $\longrightarrow$ mol Ag_2CrO_4 $\longrightarrow$ g Ag_2CrO_4.

olve

Let's solve this problem by using a stepwise approach.

Convert the volume of $K_2CrO_4(aq)$ from milliliters to liters, and then use molarity as a conversion factor between volume of solution and moles of solute (as in Example 4-9).

$$?\ \text{mol}\ K_2CrO_4 = 25.00\ \text{mL} \times \frac{1\ \text{L}}{1000\ \text{mL}} \times \frac{0.250\ \text{mol}\ K_2CrO_4}{1\ \text{L}}$$

$$= 6.25 \times 10^{-3}\ \text{mol}\ K_2CrO_4$$

Use a stoichiometric factor from the equation to convert from moles of K_2CrO_4 to moles of Ag_2CrO_4.

$$?\ \text{mol}\ Ag_2CrO_4 = 6.25 \times 10^{-3}\ \text{mol}\ K_2CrO_4 \times \frac{1\ \text{mol}\ Ag_2CrO_4}{1\ \text{mol}\ K_2CrO_4}$$

$$= 6.25 \times 10^{-3}\ \text{mol}\ Ag_2CrO_4$$

Use the molar mass to convert from moles to grams of Ag_2CrO_4.

$$?\ \text{g}\ Ag_2CrO_4 = 6.25 \times 10^{-3}\ \text{mol}\ Ag_2CrO_4 \times \frac{331.7\ \text{g}\ Ag_2CrO_4}{1\ \text{mol}\ Ag_2CrO_4}$$

$$= 2.07\ \text{g}\ Ag_2CrO_4$$

The same final answer can be obtained more directly by combining the steps into a single line calculation.

$$?\ \text{g}\ Ag_2CrO_4 = 25.00\ \text{mL} \times \frac{1\ \text{L}}{1000\ \text{mL}} \times \frac{0.250\ \text{mol}\ K_2CrO_4}{1\ \text{L}} \times \frac{1\ \text{mol}\ Ag_2CrO_4}{1\ \text{mol}\ K_2CrO_4} \times \frac{331.7\ \text{g}\ Ag_2CrO_4}{1\ \text{mol}\ Ag_2CrO_4}$$

$$= 2.07\ \text{g}\ Ag_2CrO_4$$

ssess

The units work out properly, which is always a good sign. As we have done before, let's work the problem in reverse and use numbers that are rounded off slightly. A 2 g sample of Ag_2CrO_4 contains 2/332 = 0.006 moles of Ag_2CrO_4. The number of moles of K_2CrO_4 in the 25 mL sample of the K_2CrO_4 solution is also approximately 0.006 moles; thus, the molarity of the K_2CrO_4 solution is approximately 0.006 mol/0.025 L = 0.24 M. This result is close to the true molarity (0.250 M), and so we can be confident that our answer for the mass of Ag_2CrO_4 is correct.

RACTICE EXAMPLE A: How many milliliters of 0.250 M K_2CrO_4 must be added to excess $AgNO_3(aq)$ to produce 1.50 g Ag_2CrO_4? [*Hint:* Think of this as the inverse of Example 4-10.]

RACTICE EXAMPLE B: How many milliliters of 0.150 M $AgNO_3$ are required to react completely with 175 mL of 0.0855 M K_2CrO_4? What mass of Ag_2CrO_4 is formed?

4-4 Determining the Limiting Reactant

▶ By an *excess of a reactant,* we mean that more of the reactant is present than is consumed in the reaction. Some is left over.

▶ Chemicals are often referred to as *reagents,* and the limiting reactant in a reaction is sometimes called the *limiting reagent.*

When all the reactants are completely and simultaneously consumed in a ch cal reaction, the reactants are said to be in **stoichiometric proportions**; th they are present in the mole ratios dictated by the coefficients in the bala equation. This condition is sometimes required, for example, in certain chei analyses. At other times, as in a precipitation reaction, one of the reactants is pletely converted into products by using an excess of all the other reactants reactant that is completely consumed—the **limiting reactant**—determine quantities of products formed. In the reaction described in Example 4-10, K_2C is the limiting reactant and $AgNO_3$ is present in excess. Up to this point, we stated which reactant is present in excess and, by implication, which is the l ing reactant. In some cases, however, the limiting reactant will not be indic explicitly. If the quantities of two or more reactants are given, you must deter which is the limiting reactant, as suggested by the analogy in Figure 4- demonstrated in Example 4-11.

Sometimes our interest in a limiting reactant problem is in determi how much of an excess reactant remains, as well as how much produ formed. This additional calculation is illustrated in Example 4-12.

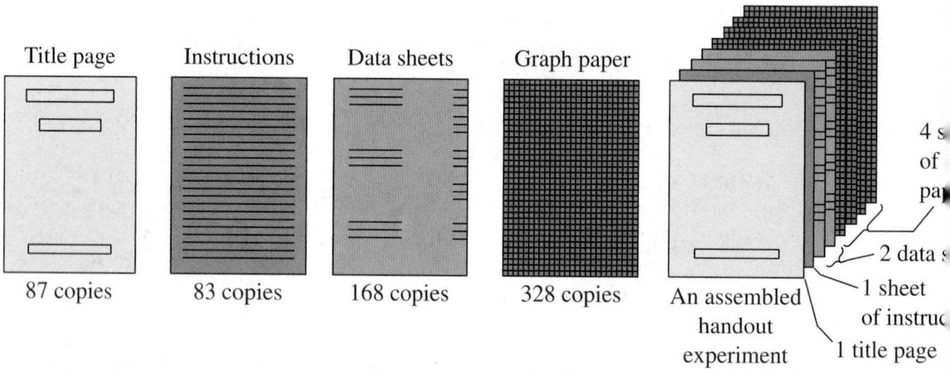

▲ FIGURE 4-8
An analogy to determining the limiting reactant in a chemical reaction—assembling a handout experiment
From the numbers of copies available and the instructions on how to assemble the handout, can you see that only 82 complete handouts are possible and that the graph paper is the limiting reactant?

EXAMPLE 4-11 Determining the Limiting Reactant in a Reaction

Phosphorus trichloride, PCl_3, is a commercially important compound used in the manufacture of pesticide gasoline additives, and a number of other products. A ball-and-stick model of PCl_3 is shown below. Liquid PC is made by the direct combination of phosphorus and chlorine.

$$P_4(s) + 6\,Cl_2(g) \longrightarrow 4\,PCl_3(l)$$

What is the maximum mass of PCl_3 that can be obtained from 125 g P_4 and 323 g Cl_2?

Analyze

The following conversions are required: mol limiting reactant → mol PCl_3 → g PCl_3. The key to solving this problem is to correctly identify the limiting reactant. One approach to this problem, outlined in Figure 4-9, is to compare the initial mole ratio of the two reactants with the ratio in which the reactants combine—6 mol Cl_2 to 1 mol P_4. If more than 6 mol Cl_2 is available per mole of P_4, chlorine is in excess and P_4 is the limiting reactant. If fewer than 6 mol Cl_2 is available per mole of P_4, chlorine is the limiting reactant.

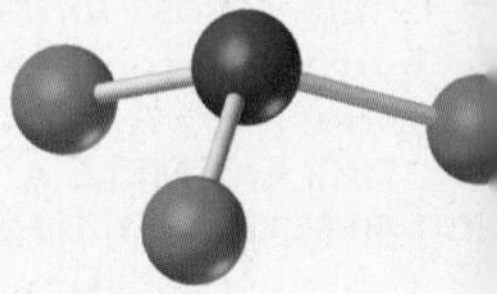

Phosphorus trichloride

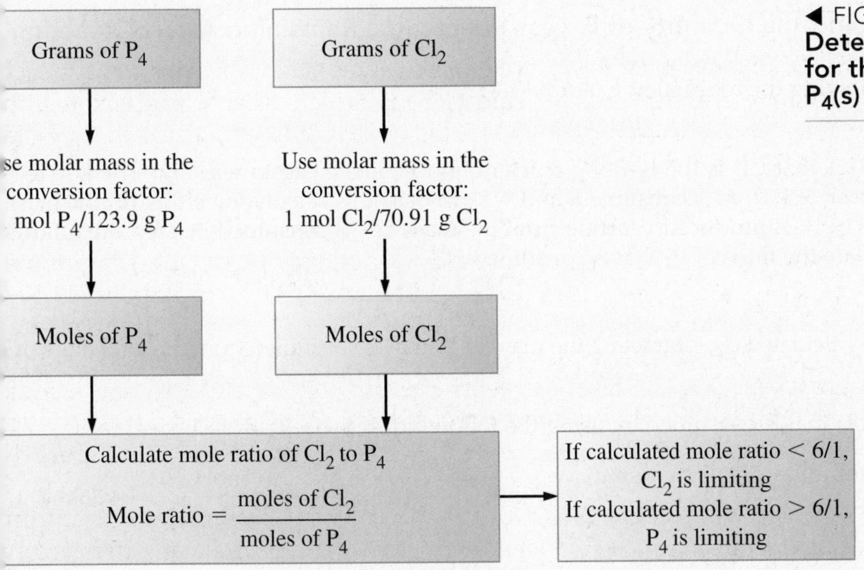

◀ FIGURE 4-9
Determining the limiting reactant for the reaction
$P_4(s) + 6\,Cl_2(g) \longrightarrow 4\,PCl_3(l)$

olve

$$? \text{ mol } Cl_2 = 323 \text{ g } Cl_2 \times \frac{1 \text{ mol } Cl_2}{70.91 \text{ g } Cl_2} = 4.56 \text{ mol } Cl_2$$

$$? \text{ mol } P_4 = 125 \text{ g } P_4 \times \frac{1 \text{ mol } P_4}{123.9 \text{ g } P_4} = 1.01 \text{ mol } P_4$$

Since there is less than 6 mol Cl_2 per mole of P_4, chlorine is the limiting reactant. The remainder of the calculation is to determine the mass of PCl_3 formed in the reaction of 323 g Cl_2 with an excess of P_4.

Having identified Cl_2 as the limiting reactant, we can complete the calculation as follows, using a stepwise approach.

Convert from grams of Cl_2 to moles of Cl_2 by using the molar mass of Cl_2.

$$? \text{ mol } Cl_2 = 323 \text{ g } Cl_2 \times \frac{1 \text{ mol } Cl_2}{70.91 \text{ g } Cl_2} = 4.56 \text{ mol } Cl_2$$

Convert from moles of Cl_2 to moles of PCl_3 by using the stoichiometric factor.

$$? \text{ mol } PCl_3 = 4.56 \text{ mol } Cl_2 \times \frac{4 \text{ mol } PCl_3}{6 \text{ mol } Cl_2} = 3.04 \text{ mol } PCl_3$$

Convert from moles of PCl_3 to grams of PCl_3 by using the molar mass of PCl_3.

$$? \text{ g } PCl_3 = 3.04 \text{ mol } PCl_3 \times \frac{137.3 \text{ g } PCl_3}{1 \text{ mol } PCl_3} = 417 \text{ g } PCl_3$$

Using the conversion pathway approach, we combine the steps into a single line calculation.

$$? \text{ g } PCl_3 = 323 \text{ g } Cl_2 \times \frac{1 \text{ mol } Cl_2}{70.91 \text{ g } Cl_2} \times \frac{4 \text{ mol } PCl_3}{6 \text{ mol } Cl_2} \times \frac{137.3 \text{ g } PCl_3}{1 \text{ mol } PCl_3} = 417 \text{ g } PCl_3$$

ssess

A different approach that leads to the same result is to do two separate calculations. First, calculate the mass of PCl_3 produced by the reaction of 323 g Cl_2 with an excess of P_4 (= 417 g PCl_3), and then calculate the mass of PCl_3 produced by the reaction of 125 g P_4 with an excess of Cl_2 (= 554 g PCl_3). Only one answer can be correct, and it must be the *smaller* of the two. An advantage of this approach is that it can be easily generalized to deal with reactions involving many reactants.

RACTICE EXAMPLE A: If 215 g P_4 is allowed to react with 725 g Cl_2 in the reaction in Example 4-11, how many grams of PCl_3 are formed?

RACTICE EXAMPLE B: If 1.00 kg each of PCl_3, Cl_2, and P_4O_{10} are allowed to react, how many kilograms of $POCl_3$ will be formed?

$$6\,PCl_3(l) + 6\,Cl_2(g) + P_4O_{10}(s) \longrightarrow 10\,POCl_3(l)$$

EXAMPLE 4-12 Determining the Quantity of Excess Reactant(s) Remaining After a Reaction

What mass of P_4 remains following the reaction in Example 4-11?

Analyze

We established in Example 4-11 that Cl_2 is the limiting reactant and P_4 is the excess reactant. The key to th problem is to calculate the mass of P_4 that is consumed, and we can base this calculation either on the mass Cl_2 consumed or on the mass of PCl_3 produced. Starting from the mass of Cl_2 consumed, we use the followi conversion pathway to calculate the mass of P_4 that is consumed.

$$\text{g } Cl_2 \longrightarrow \text{mol } Cl_2 \longrightarrow \text{mol } P_4 \longrightarrow \text{g } P_4$$

The mass of P_4 that remains is calculated by subtracting the mass of P_4 that is consumed from the total mass of I

Solve

We can calculate the mass of P_4 that is consumed by using the following stepwise approach.

Convert from grams of Cl_2 to moles of Cl_2 by using the molar mass of Cl_2.

$$? \text{ mol } Cl_2 = 323 \text{ g } Cl_2 \times \frac{1 \text{ mol } Cl_2}{70.91 \text{ g } Cl_2} = 4.56 \text{ mol Cl}$$

Convert from moles of Cl_2 to moles of P_4 by using the stoichiometric factor.

$$? \text{ mol } P_4 = 4.56 \text{ mol } Cl_2 \times \frac{1 \text{ mol } P_4}{6 \text{ mol } Cl_2} = 0.759 \text{ mol}$$

Convert from moles of P_4 to grams of P_4 by using the molar mass of P_4.

$$? \text{ g } P_4 = 0.759 \text{ mol } P_4 \times \frac{123.9 \text{ g } P_4}{1 \text{ mol } P_4} = 94.1 \text{ g } P_4$$

The single line calculation is as follows.

$$? \text{ g } P_4 = 323 \text{ g } Cl_2 \times \frac{1 \text{ mol } Cl_2}{70.91 \text{ g } Cl_2} \times \frac{1 \text{ mol } P_4}{6 \text{ mol } Cl_2} \times \frac{123.9 \text{ g } P_4}{1 \text{ mol } P_4} = 94.1 \text{ g } P_4$$

The following single line calculation shows that we obtain the same answer starting from the mass of PCl_3 th is produced.

$$? \text{ g } P_4 = 417 \text{ g } PCl_3 \times \frac{1 \text{ mol } PCl_3}{137.3 \text{ g } PCl_3} \times \frac{1 \text{ mol } P_4}{4 \text{ mol } PCl_3} \times \frac{123.9 \text{ g } P_4}{1 \text{ mol } P_4} = 94.1 \text{ g } P_4$$

The mass of P_4 remaining after the reaction is simply the difference between what was originally present ar what was consumed; that is,

$$125 \text{ g } P_4 \text{ initially} - 94.1 \text{ g } P_4 \text{ consumed} = 31 \text{ g } P_4 \text{ remaining}$$

Assess

It is possible to use the law of conservation of mass to verify that the calculated result is correct. The total ma of the reactants used in the reaction (323 g + 125 g = 448 g) must be equal to the mass of product (417 plus the mass of the P_4 that remains. Therefore, the mass of the P_4 that remains is 448 g − 417 g = 31 g. In pe forming the subtraction required in this example, the quantity with the fewest digits after the decimal poi determines how the answer should be expressed. Because there are no digits after the decimal point in 12 there can be none after the decimal point in 31.

PRACTICE EXAMPLE A: In Practice Example 4-11A, which reactant is in excess and what mass of that reacta remains after the reaction to produce PCl_3?

PRACTICE EXAMPLE B: If 12.2 g H_2 and 154 g O_2 are allowed to react, which gas and what mass of that ga remains after the reaction?

$$2 H_2(g) + O_2(g) \longrightarrow 2 H_2O(l)$$

4-5 CONCEPT ASSESSMENT

Suppose that the reaction of 1.0 mol $NH_3(g)$ and 1.0 mol $O_2(g)$ is carried to completion, producing NO(g) and $H_2O(l)$ as the only products. Which, if any, of the following statements is correct about this reaction? **(a)** 1.0 mol NO(g) is produced; **(b)** 4.0 mol NO(g) is produced; **(c)** 1.5 mol $H_2O(l)$ is produced; **(d)** al the $O_2(g)$ is consumed; **(e)** $NH_3(g)$ is the limiting reactant.

Other Practical Matters in Reaction Stoichiometry

is section we consider a few additional factors in reaction stoichiometry— in the laboratory and in the manufacturing plant. First, the calculated ome of a reaction may not be what is actually observed. Specifically, the unt of product may be, unavoidably, less than expected. Second, the route roducing a desired chemical may require several reactions carried out in ence. And third, in some cases two or more reactions may occur simulta-ısly.

oretical Yield, Actual Yield, and Percent Yield

theoretical yield of a reaction is the calculated quantity of product ected from given quantities of reactants. The quantity of product that is ally produced is called the **actual yield**. The **percent yield** is defined as

$$\text{percent yield} = \frac{\text{actual yield}}{\text{theoretical yield}} \times 100\% \qquad \textbf{(4.6)}$$

ı many reactions the actual yield almost exactly equals the theoretical d, and the reactions are said to be *quantitative*. Such reactions can be l in quantitative chemical analyses. In other reactions the actual yield ss than the theoretical yield, and the percent yield is less than 100%. reduced yield may occur for a variety of reasons. (1) The product of a tion rarely appears in a pure form, and some product may be lost dur-the necessary purification steps, which reduces the yield. (2) In many s the reactants may participate in reactions other than the one of cen-interest. These are called **side reactions**, and the unintended products called **by-products**. To the extent that side reactions occur, the yield of main product is reduced. (3) If a reverse reaction occurs, some of the ected product may react to re-form the reactants, and again the yield is than expected.

t times, the apparent yield is greater than 100%. Because we cannot get ething from nothing, this situation usually indicates an error in technique. e products are formed as a precipitate from a solvent. If the product is ghed when it is wet with solvent, the result will be a larger-than-expected s. More thorough drying of the product would give a more accurate yield rmination. Another possibility is that the product is contaminated with xcess reactant or a by-product. This makes the mass of product appear er than expected. In any case, a product must be purified before its yield is rmined.

ı Example 4-13, we establish the theoretical, actual, and percent yields of mportant industrial process.

f we know that a yield will be less than 100%, we need to adjust the unts of reactants used to produce the desired amount of product. We can-simply use the theoretical amounts of reactants; we must use more. This ıt is illustrated in Example 4-14.

nsecutive, Simultaneous, and Overall Reactions

ı in laboratory work and in manufacturing, the preferred processes are those yield a product through a single reaction. Often such processes give a higher d because there is no need to remove products from one reaction mixture for her processing in subsequent reactions. However, in many cases a multistep cess is unavoidable. **Consecutive reactions** are reactions carried out one after ther in sequence to yield a final product. In **simultaneous reactions**, two or

EXAMPLE 4-13 Determining Theoretical, Actual, and Percent Yields

Billions of kilograms of urea, $CO(NH_2)_2$, are produced annually for use as a fertilizer. A ball-and-stick model of urea is shown here. The reaction used is given below.

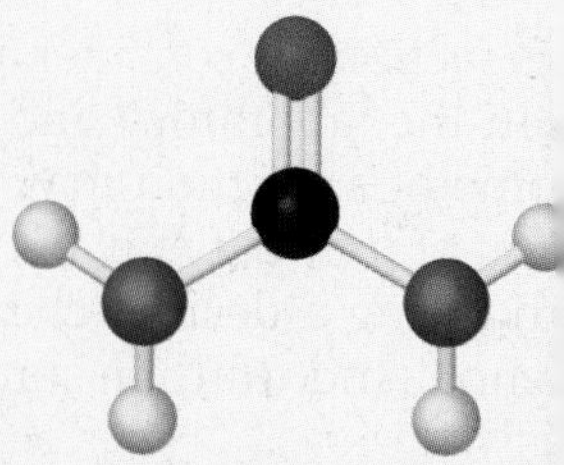

Urea

$$2\ NH_3(g) + CO_2(g) \longrightarrow CO(NH_2)_2(s) + H_2O(l)$$

The typical starting reaction mixture has a 3:1 mole ratio of NH_3 to CO_2. If 47.7 g urea forms *per mole* of CO_2 that reacts, what is the **(a)** theoretical yield; **(b)** actual yield; and **(c)** percent yield?

Analyze

The reaction mixture contains fixed amounts of NH_3 and CO_2, and so we must first determine which reacta is the limiting reactant. The stoichiometric proportions are 2 mol NH_3 : 1 mol CO_2. In the reaction mixture, tl mole ratio of NH_3 to CO_2 is 3:1. Therefore, NH_3 is the excess reactant and CO_2 is the limiting reactant. The ca culation of the theoretical yield of urea must be based on the amount of CO_2, the limiting reactant. Because tl quantity of urea is given per mole of CO_2, we should base the calculation on 1.00 mol CO_2. The following co versions are required: mol $CO_2 \rightarrow$ mol $CO(NH_2)_2 \rightarrow$ g $CO(NH_2)_2$.

Solve

(a) Let's calculate the theoretical yield by using a stepwise approach.

Convert from mol CO_2 to mol $CO(NH_2)_2$ by using the stoichiometric factor.

$$? \text{ mol } CO(NH_2)_2 = 1.00 \text{ mol } CO_2 \times \frac{1 \text{ mol } CO(NH_2)_2}{1 \text{ mol } CO_2}$$
$$= 1.00 \text{ mol } CO(NH_2)_2$$

Convert from mol $CO(NH_2)_2$ to g $CO(NH_2)_2$ by using the molar mass of $CO(NH_2)_2$.

$$? \text{ g } CO(NH_2)_2 = 1.00 \text{ mol } CO(NH_2)_2 \times \frac{60.1 \text{ g } CO(NH_2)_2}{1 \text{ mol } CO(NH_2)_2}$$
$$= 60.1 \text{ g } CO(NH_2)_2$$

Thus, 1.00 mol CO_2 is expected to yield 60.1 g $CO(NH_2)_2$, and so the theoretical yield of $CO(NH_2)_2$ is 60.1 As has been the case in all our examples, we could have combined the steps into a single line calculation.

$$\text{theoretical yield} = 1.00 \text{ mol } CO_2 \times \frac{1 \text{ mol } CO(NH_2)_2}{1 \text{ mol } CO_2} \times \frac{60.1 \text{ g } CO(NH_2)_2}{1 \text{ mol } CO(NH_2)_2} = 60.1 \text{ g } CO(NH_2)_2$$

(b) actual yield = 47.7 g $CO(NH_2)_2$

(c) $$\text{percent yield} = \frac{47.7 \text{ g } CO(NH_2)_2}{60.1 \text{ g } CO(NH_2)_2} \times 100\% = 79.4\%$$

Assess

A quick way to determine the limiting reactant is to identify the reactant that has the smallest value of (n/*coeff* where n is the number of moles of reactant available and *coeff.* is the coefficient of that reactant in the balance chemical equation. In this example, the reaction mixture contains 3 mol NH_3 and 1 mol CO_2, and so (n/*coeff.*) $3/2 = 1.5$ for NH_3 and $1/1 = 1$ for CO_2. Because $1 < 1.5$, CO_2 is the limiting reactant. To understand wh we can identify the limiting reactant by using this method, it is helpful to consider how we conve from moles of reactant to moles of product: $n \text{ mol reactant} \times \dfrac{\text{coefficient of product}}{\text{coefficient of reactant}}$. The smaller the value ((n/*coeff.*), the smaller the amount of product. The limiting reactant is the one that yields the smallest amount (product (as explained in Example 4-11); thus, the limiting reactant is the one with the smallest value of (n/*coeff.*

PRACTICE EXAMPLE A: Formaldehyde, CH_2O, can be made from methanol by the following reaction, using copper catalyst.

$$CH_3OH(g) \longrightarrow CH_2O(g) + H_2(g)$$

If 25.7 g $CH_2O(g)$ is produced per mole of methanol that reacts, what are **(a)** the theoretical yield; **(b)** the actu yield; and **(c)** the percent yield?

PRACTICE EXAMPLE B: What is the percent yield if the reaction of 25.0 g P_4 and 91.5 g Cl_2 produces 104 g PCl_3 The balanced chemical equation for the reaction is given below.

$$P_4(s) + 6\ Cl_2(g) \longrightarrow 4\ PCl_3(l)$$

EXAMPLE 4-14 Adjusting the Quantities of Reactants in Accordance with the Percent Yield of a Reaction

When heated with sulfuric or phosphoric acid, cyclohexanol, $C_6H_{11}OH$, is converted to cyclohexene, C_6H_{10}. Ball-and-stick models of cyclohexanol and cyclohexene are shown here. The balanced chemical equation for the reaction is shown below.

$$C_6H_{11}OH(l) \longrightarrow C_6H_{10}(l) + H_2O(l) \qquad \textbf{(4.7)}$$

If the percent yield is 83%, what mass of cyclohexanol must we use to obtain 25 g of cyclohexene?

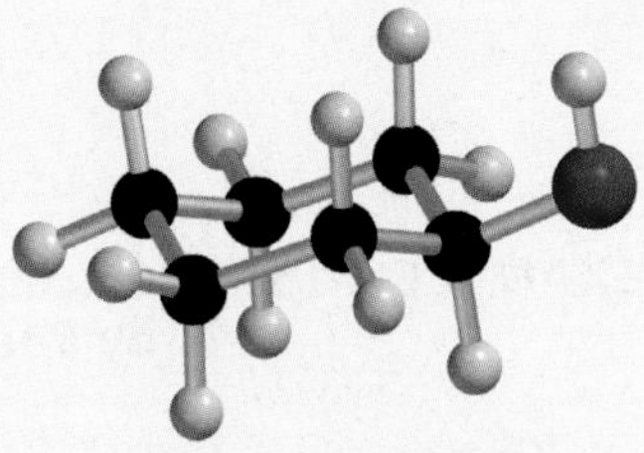

Cyclohexanol

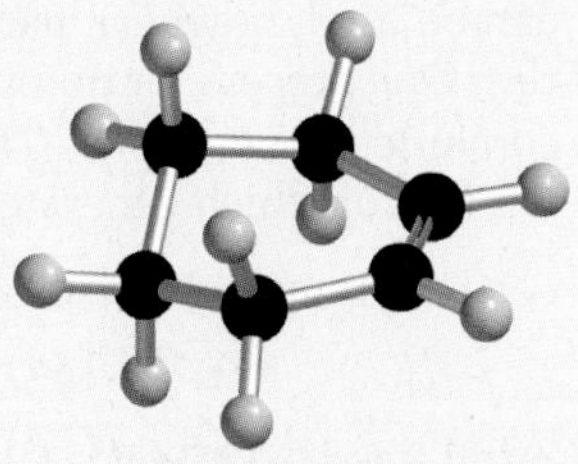

Cyclohexene

Analyze

We are given the percent yield (83%) and the actual yield of C_6H_{10} (25 g). We can use equation (4.6) to calculate the theoretical yield of C_6H_{10}. Then, we can calculate the quantity of $C_6H_{11}OH$ required to produce the theoretical yield of C_6H_{10} by using the following series of conversions: g C_6H_{10} → mol C_6H_{10} → mol $C_6H_{11}OH$ → g $C_6H_{11}OH$.

Solve

Let's use a stepwise approach.

Rearrange equation (4.6) and calculate the theoretical yield.

$$\text{theoretical yield} = \frac{25\text{ g} \times 100\%}{83\%} = 3.0 \times 10^1\text{ g}$$

The theoretical yield is expressed as 3.0×10^1 g rather than as 30 g to emphasize that there are two significant figures in the calculated result. If we expressed the result as 30 g, then the number of significant figures would be ambiguous. (See Section 1-7.) The remaining conversions are as follows.

Convert from g C_6H_{10} to mol C_6H_{10} by using the molar mass of C_6H_{10}.

$$?\text{ mol } C_6H_{10} = 3.0 \times 10^1\text{ g } C_6H_{10} \times \frac{1\text{ mol } C_6H_{10}}{82.1\text{ g } C_6H_{10}} = 0.37\text{ mol } C_6H_{10}$$

Convert from mol C_6H_{10} to mol $C_6H_{11}OH$ using the stoichiometric factor.

$$?\text{ mol } C_6H_{11}OH = 0.37\text{ mol } C_6H_{10} \times \frac{1\text{ mol } C_6H_{11}OH}{1\text{ mol } C_6H_{10}} = 0.37\text{ mol } C_6H_{11}OH$$

Convert from mol $C_6H_{11}OH$ to g $C_6H_{11}OH$ by using the molar mass of $C_6H_{11}OH$.

$$?\text{ g } C_6H_{11}OH = 0.37\text{ mol } C_6H_{11}OH \times \frac{100.2\text{ g } C_6H_{11}OH}{1\text{ mol } C_6H_{11}OH} = 37\text{ g } C_6H_{11}OH$$

We could have combined the last three steps into a single line calculation. The single line calculation is shown below, starting from a theoretical yield of 3.0×10^1 g C_6H_{10}.

$$?\text{ g } C_6H_{11}OH = 30\text{ g } C_6H_{10} \times \frac{1\text{ mol } C_6H_{10}}{82.1\text{ g } C_6H_{10}} \times \frac{1\text{ mol } C_6H_{11}OH}{1\text{ mol } C_6H_{10}} \times \frac{100.2\text{ g } C_6H_{11}OH}{1\text{ mol } C_6H_{11}OH} = 37\text{ g } C_6H_{11}OH$$

Assess

Because we have already calculated the molar masses of C_6H_{10} and $C_6H_{11}OH$, it is a simple matter to work the problem in reverse by using numbers that are slightly rounded off. A 37 g sample of $C_6H_{11}OH$ contains approximately $37/100 = 0.37$ moles of $C_6H_{11}OH$, and we expect a maximum of 0.37 moles of C_6H_{10} or $0.37 \times 82 \approx 30$ g C_6H_{10}. We obtain only 25 g C_6H_{10} and so the percent yield for the experiment is approximately $(25/30) \times 100\% = 83\%$. By working the problem in reverse, we verify that the percent yield is 83%; thus, we are confident we have solved the problem correctly.

PRACTICE EXAMPLE A: If the percent yield for the formation of urea in Example 4-13 were 87.5%, what mass of CO_2, together with an excess of NH_3, would have to be used to obtain 50.0 g $CO(NH_2)_2$?

PRACTICE EXAMPLE B: Calculate the mass of cyclohexanol ($C_6H_{11}OH$) needed to produce 45.0 g cyclohexene (C_6H_{10}) by reaction (4.7) if the reaction has a 86.2% yield and the cyclohexanol is 92.3% pure.

more substances react independently of one another in separate reactions oc ring at the same time.

Example 4-15 presents an industrial process that is carried out in two secutive reactions. The key to the calculation is to use a stoichiometric fa for each reaction. Example 4-16 deals with two reactions that occur simult ously to produce a common product, hydrogen gas.

EXAMPLE 4-15 Calculating the Quantity of a Substance Produced by Reactions Occurring Consecutively

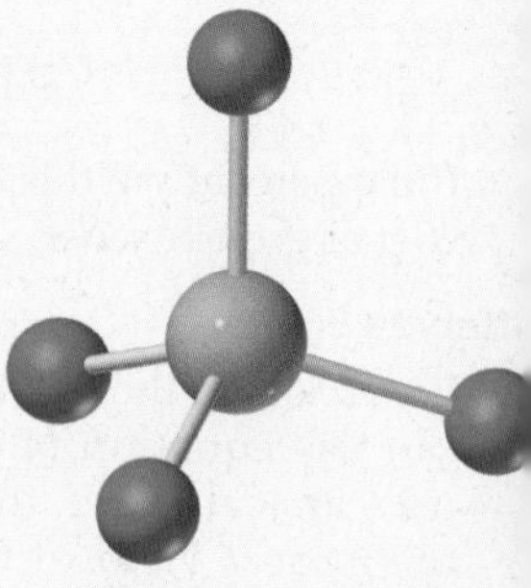

Titanium tetrachloride

Titanium dioxide, TiO_2, is the most widely used white pigment for paints, having displaced most lead-based pigments, which are environmental hazards. Before it can be used, however, naturally occurring TiO_2 must be freed of colored impurities. One process for doing this converts impure $TiO_2(s)$ to $TiCl_4(g)$, which is then converted back to pure $TiO_2(s)$. The process is based on the following reactions, the first of which generates $TiCl_4$.

$$2\,TiO_2\,(\text{impure}) + 3\,C(s) + 4\,Cl_2(g) \longrightarrow 2\,TiCl_4(g) + CO_2(g) + 2\,CO(g)$$

$$TiCl_4(g) + O_2(g) \longrightarrow TiO_2(s) + 2\,Cl_2(g)$$

What mass of carbon is consumed in producing 1.00 kg of pure $TiO_2(s)$ in this process?

Analyze

In this calculation, we begin with the product, TiO_2, and work backward to one of the reactants, C. The following conversions are required.

$$\text{kg } TiO_2 \longrightarrow \text{g } TiO_2 \longrightarrow \text{mol } TiO_2 \xrightarrow{(a)} \text{mol } TiCl_4 \xrightarrow{(b)} \text{mol C} \longrightarrow \text{g C}$$

In the conversion from mol TiO_2 to mol $TiCl_4$, labeled (a), we focus on the second reaction. In the conversion from mol $TiCl_4$ to mol C, labeled (b), we focus on the first reaction.

Solve

Using a stepwise approach, we proceed as follows.

Convert from kg TiO_2 to g TiO_2 and then to mol TiO_2 by using the molar mass of TiO_2.

$$?\text{ mol } TiO_2 = 1.00\text{ kg } TiO_2 \times \frac{1000\text{ g } TiO_2}{1\text{ kg } TiO_2} \times \frac{1\text{ mol } TiO_2}{79.88\text{ g } TiO_2} = 12.5\text{ mol } TiO_2$$

Convert from mol TiO_2 to mol $TiCl_4$ by using the stoichiometric factor from the second reaction.

$$?\text{ mol } TiCl_4 = 12.5\text{ mol } TiO_2 \times \underbrace{\frac{1\text{ mol } TiCl_4}{1\text{ mol } TiO_2}}_{(a)} = 12.5\text{ mol } TiCl_4$$

Convert from mol $TiCl_4$ to mol C by using the stoichiometric factor from the first reaction.

$$?\text{ mol C} = 12.5\text{ mol } TiCl_4 \times \underbrace{\frac{3\text{ mol C}}{2\text{ mol } TiCl_4}}_{(b)} = 18.8\text{ mol C}$$

Convert from mol C to g C by using the molar mass of C.

$$?\text{ g C} = 18.8\text{ mol C} \times \frac{12.01\text{ g C}}{1\text{ mol C}} = 226\text{ g C}$$

The conversions given above can be combined into a single line, as shown below.

$$?\text{ g C} = 1.00\text{ kg } TiO_2 \times \frac{1000\text{ g } TiO_2}{1\text{ kg } TiO_2} \times \frac{1\text{ mol } TiO_2}{79.88\text{ g } TiO_2} \times \overset{(a)}{\frac{1\text{ mol } TiCl_4}{1\text{ mol } TiO_2}} \times \overset{(b)}{\frac{3\text{ mol C}}{2\text{ mol } TiCl_4}} \times \frac{12.01\text{ g C}}{1\text{ mol C}} = 226\text{ g C}$$

Assess

To obtain the stoichiometric factor for converting from mol TiO_2 to mol C, we could have reasoned as follows From the equations for the two reactions, we see that 3 mol C will give 2 mol $TiCl_4$, which then reacts to give ar equal amount (2 mol) of TiO_2. That is, 3 mol C will give 2 mol TiO_2, or equivalently, 2 mol TiO_2 requires 3 mo C. Thus, the required stoichiometric factor is (3 mol C)/(2 mol TiO_2), the product of the two factors above labeled (a) and (b).

RACTICE EXAMPLE A: Nitric acid, HNO_3, is produced from ammonia and oxygen by the consecutive reactions

$$4\,NH_3(g) + 5\,O_2(g) \longrightarrow 4\,NO(g) + 6\,H_2O(g)$$
$$2\,NO(g) + O_2(g) \longrightarrow 2\,NO_2(g)$$
$$3\,NO_2(g) + H_2O(l) \longrightarrow 2\,HNO_3(aq) + NO(g)$$

How many grams of nitric acid can be obtained from 1.00 kg $NH_3(g)$, if NO(g) in the third reaction is *not* recycled?

RACTICE EXAMPLE B: Nitrogen gas is used to inflate automobile air bags. The gas is produced by detonation of a pellet containing NaN_3, KNO_3, and SiO_2 (all solids). When detonated, NaN_3 decomposes into $N_2(g)$ and Na(l). The Na(l) reacts almost instantly with KNO_3 producing more $N_2(g)$ as well as K_2O and Na_2O. The K_2O and Na_2O are each converted independently into harmless salts, Na_2SiO_3 and K_2SiO_3, when they react with SiO_2. Balanced chemical equations for the reactions are given below.

$$2\,NaN_3(s) \longrightarrow 3\,N_2(g) + 2\,Na(l)$$
$$10\,Na(l) + 2\,KNO_3(s) \longrightarrow N_2(s) + K_2O(s) + 5\,Na_2O(s)$$
$$K_2O(s) + SiO_2(s) \longrightarrow K_2SiO_3(s)$$
$$Na_2O(s) + SiO_2(s) \longrightarrow Na_2SiO_3(s)$$

What are the minimum masses of KNO_3 and SiO_2 that must be mixed with 95 g NaN_3 to ensure that no Na, Na_2O, or K_2O remains after the mixture is detonated?

XAMPLE 4-16 Calculating the Quantity of a Substance Produced by Reactions Occurring Simultaneously

Magnesium–aluminum alloys are widely used in aircraft construction. One particular alloy contains 70.0% Al and 30.0% Mg, by mass. How many grams of $H_2(g)$ are produced in the reaction of a 0.710 g sample of this alloy with excess HCl(aq)? Balanced chemical equations are given below for the reactions that occur.

$$2\,Al(s) + 6\,HCl(aq) \longrightarrow 2\,AlCl_3(aq) + 3\,H_2(g)$$
$$Mg(s) + 2\,HCl(aq) \longrightarrow MgCl_2(aq) + H_2(g)$$

nalyze

The two reactions given above are simultaneous reactions. Simultaneous reactions occur independently; thus, we have two conversion pathways to consider:

$$(1)\quad \text{g alloy} \longrightarrow \text{g Al} \longrightarrow \text{mol Al} \xrightarrow{(a)} \text{mol } H_2;$$
$$(2)\quad \text{g alloy} \longrightarrow \text{g Mg} \longrightarrow \text{mol Mg} \xrightarrow{(b)} \text{mol } H_2;$$

Pathways (1) and (2) are based on the first and second reactions, respectively. The total amount of H_2 produced is obtained by adding together the amounts produced by each reaction. The conversion from mol Al to mol H_2 requires a stoichiometric factor, labeled (a). The conversion from mol Mg to mol H_2 requires a different stoichiometric factor, labeled (b).

olve

Convert from g alloy to mol Al and from g alloy to mol Mg by using the mass percentages of Al and Mg and the molar masses of Al and Mg.

$$?\text{ mol Al} = 0.710\text{ g alloy} \times \frac{70.0\text{ g Al}}{100.0\text{ g alloy}} \times \frac{1\text{ mol Al}}{26.98\text{ g Al}}$$
$$= 0.0184\text{ mol Al}$$

$$?\text{ mol Mg} = 0.710\text{ g alloy} \times \frac{30.0\text{ g Mg}}{100.0\text{ g alloy}} \times \frac{1\text{ mol Mg}}{24.31\text{ g Mg}}$$
$$= 8.76 \times 10^{-3}\text{ mol Mg}$$

(continued)

Convert from mol Al to mol H_2 and from mol Mg to mol H_2 by using stoichiometric factors (a) and (b). The total number of moles of H_2 is obtained by adding together the two independent contributions.

$$? \text{ mol } H_2 = 0.0184 \text{ mol Al} \times \underbrace{\frac{3 \text{ mol } H_2}{2 \text{ mol Al}}}_{(a)} + 8.76 \times 10^{-3} \text{ mol Mg} \times \underbrace{\frac{1 \text{ mol } H_2}{1 \text{ mol Mg}}}_{(b)} = 0.0364 \text{ mol } H_2$$

Convert from mol H_2 to g H_2 by using the molar mass of H_2 as a conversion factor.

$$? \text{ g } H_2 = 0.0364 \text{ mol } H_2 \times \frac{2.016 \text{ g } H_2}{1 \text{ mol } H_2} = 0.0734 \text{ g } H_2$$

An alternative approach is to combine the steps into a single line calculation, as shown below.

$$? \text{ g } H_2 = \left(0.710 \text{ g alloy} \times \frac{70.0 \text{ g Al}}{100.0 \text{ g alloy}} \times \frac{1 \text{ mol Al}}{26.98 \text{ g Al}} \times \underbrace{\frac{3 \text{ mol } H_2}{2 \text{ mol Al}}}_{(a)} \times \frac{2.016 \text{ g } H_2}{1 \text{ mol } H_2}\right)$$

$$+ \left(0.710 \text{ g alloy} \times \frac{30.0 \text{ g Mg}}{100.0 \text{ g alloy}} \times \frac{1 \text{ mol Mg}}{24.31 \text{ g Mg}} \times \underbrace{\frac{1 \text{ mol } H_2}{1 \text{ mol Mg}}}_{(b)} \times \frac{2.016 \text{ g } H_2}{1 \text{ mol } H_2}\right) = 0.0734 \text{ g } H_2$$

Assess

In this example, the composition of the alloy is given and we solved for the amount of H_2 that is produced. T inverse problem, in which we are given the amount of H_2 produced and are asked to determine the amounts Al and Mg in the alloy, is a little harder to solve. See Practice Examples A and B below.

PRACTICE EXAMPLE A: A 1.00 g sample of a magnesium-aluminum alloy yields 0.107 g H_2 when treated w an excess of HCl(aq). What is the percentage by mass of Al in the alloy? [*Hint:* This is the inverse Example 4-16. To solve this problem, let m and $1.00 - m$ be the masses of Al and Mg, respectively, a then use these masses in the setup above to develop an equation that relates m to the total mass of obtained. Then solve for m.]

PRACTICE EXAMPLE B: A 1.500 g sample of a mixture containing only Cu_2O and CuO was treated w hydrogen to produce copper metal and water. After the water evaporated, 1.2244 g of pure copper metal w recovered. What is the percentage by mass of Cu_2O in the original mixture? Balanced chemical equations given below for the reactions involved.

$$CuO(s) + H_2(g) \longrightarrow Cu(s) + H_2O(l)$$
$$Cu_2O(s) + H_2(g) \longrightarrow 2\ Cu(s) + H_2O(l)$$

4-6 CONCEPT ASSESSMENT

Suppose that each of these reactions has a 90% yield.

$$CH_4(g) + Cl_2(g) \longrightarrow CH_3Cl(g) + HCl(g)$$
$$CH_3Cl(g) + Cl_2(g) \longrightarrow CH_2Cl_2(l) + HCl(g)$$

Starting with 50.0 g CH_4 in the first reaction and an excess of $Cl_2(g)$, the number of grams of CH_2Cl_2 formed in the second reaction is

(a) $50.0 \times 0.81 \times (85/16)$
(b) 50.0×0.90
(c) $50.0 \times 0.90 \times 0.90$
(d) $50.0 \times 0.90 \times 0.90 \times (16/50.5)(70.9/85)$

Often, we can combine a series of chemical equations for *consecu reactions* to obtain a single equation to represent the **overall reaction**. equation for this overall reaction is the **overall equation**. At times we can the overall equation for solving problems instead of working with the indi ual equations. This strategy does not work, however, if the substance of in est is not a starting material or final product but appears only in one of rmediate reactions. Any substance that is produced in one step and con-ed in another step of a multistep process is called an **intermediate**. In mple 4-15, $TiCl_4(g)$ is an intermediate.

o write an overall equation for Example 4-15, multiply the coefficients in second equation by the factor 2, add the second equation to the first, and cel any substances that appear on both sides of the overall equation.

$$2\,TiO_2\,(\text{impure}) + 3\,C(s) + 4\,Cl_2(g) \longrightarrow 2\,TiCl_4(g) + CO_2(g) + 2\,CO(g)$$
$$2\,TiCl_4(g) + 2\,O_2(g) \longrightarrow 2\,TiO_2(s) + 4\,Cl_2(g)$$
$$\textit{Overall equation:}\quad 3\,C(s) + 2\,O_2(g) \longrightarrow CO_2(g) + 2\,CO(g)$$

he result suggests that (1) we should obtain as much TiO_2 in the second tion as we started with in the first, (2) the $Cl_2(g)$ produced in the second tion can be recycled back into the first reaction, and (3) the only substances ally consumed in the overall reaction are C(s) and $O_2(g)$. You can also see we could not have used this overall equation in the calculation of mple 4-15—$TiO_2(s)$ does not appear in it.

◀ Whereas chemical equations for consecutive reactions can be combined to obtain an equation for the overall reaction, those for a set of simultaneous reactions cannot. If the two equations from Example 4-16 were added, we would obtain the equation $2\,Al + Mg + 8\,HCl \rightarrow 2\,AlCl_3 + MgCl_2 + 4\,H_2$. This equation is clearly incorrect because, for example, it suggests that to make H_2, the reaction mixture must contain both Al and Mg. However, we know that each metal can react independently with HCl to give H_2.

6 The Extent of Reaction

oughout this chapter, we have implicitly assumed that every reaction pro-ds to completion. When a reaction proceeds to completion, one (or more) of reactants is completely consumed. The reality is that, for some reactions, reactants are not completely consumed. In this section, we will introduce pproach for quantifying the extent to which a reaction proceeds from reac-s toward products.

he progress of a chemical reaction can be described by specifying how ch of a given reactant is consumed or how much of a given product is ned. More precisely, the progress can be described by specifying the *change* he amount of any reactant or product. Let's consider a specific example, ed on the following reaction:

$$2\,Li_2O_2(s) \xrightarrow{\Delta} 2\,Li_2O(s) + O_2(g) \tag{4.8}$$

◀ If the oxygen that is produced by this reaction is allowed to escape, the progress of the reaction at any point can be deduced by measuring the mass of the solids remaining. The decrease in mass tells us how much O_2 has been formed to that point.

s suppose we start with 4.18 g Li_2O_2, which is the mass of 0.0911 mol Li_2O_2, heat it until the mass of the remaining solids is 3.06 g. The difference in ss, $4.18\text{ g} - 3.06\text{ g} = 1.12\text{ g}$, represents the mass of O_2 that has been formed to point. Thus, the amount of O_2 formed by the reaction is $1.12\text{ g}/(32.0\text{ g/mol})$ .0350 mol. The amount of Li_2O formed must be twice this amount or 00 mol. The amount of Li_2O_2 consumed must also be 0.0700 mol. The results his experiment can be summarized in tabular form.

	$2\,Li_2O_2(s)$	$\xrightarrow{\Delta}$	$2\,Li_2O(s)$	+	$O_2(g)$
tial amounts:	0.0911 mol		0 mol		0 mol
anges:	-2×0.0350 mol		$+2 \times 0.0350$ mol		$+0.0350$ mol
al amounts:	$(0.0911 - 2 \times 0.0350)$ mol		2×0.0350 mol		0.0350 mol

tice the following:

- For reactants, the *changes* in amount are negative (their amounts decrease), whereas for products, the *changes* in amount are positive (their amounts increase).
- The *changes* in amount are not independent of each other. They are each a multiple of the same change. In this example, each change is a multiple of 0.0350 mol.

second point is significant because it means we can relate the final amounts he initial amounts by using a single variable. To illustrate this point, let's lace 0.0350 mol in the summary above with the symbol ξ.

◀The IUPAC recommends that the symbol ξ (Greek letter xi) be used to represent the extent of reaction.

	$2\,Li_2O_2(s)$	$\xrightarrow{\Delta}$	$2\,Li_2O(s)$	+	$O_2(g)$
initial amounts:	0.0911 mol		0 mol		0 mol
changes:	$-2\,\xi$		$+2\,\xi$		$+1\,\xi$
final amounts:	$0.0911\text{ mol} - 2\,\xi$		$2\,\xi$		$1\,\xi$

The variable ξ is a measure of how much reaction has occurred up to some p and it is called the **extent of reaction**. The extent of reaction has the unit mol can be related to the change (Δn) in the amount of any reactant or product thr the following equations:

$$\Delta n_k = \nu_k \xi \qquad (4.$$

$$\xi = \frac{\Delta n_k}{\nu_k} \qquad (4.1$$

In the expressions above, $\Delta n_k = n_{k,\text{final}} - n_{k,\text{initial}}$ is the difference betweer final and initial amounts of a particular reactant or product, k, and ν_k is the c sponding **stoichiometric number**. The stoichiometric number has the same ma tude as the stoichiometric coefficient, but it is negative for reactants and pos for products. For equation (4.8), the stoichiometric numbers are

$$\nu_{Li_2O_2} = -2 \qquad \nu_{Li_2O} = +2 \qquad \nu_{O_2} = +1$$

and the extent of reaction may be expressed in three different ways:

$$\xi = \frac{\Delta n_{Li_2O_2}}{-2} \qquad \xi = \frac{\Delta n_{Li_2O}}{+2} \qquad \xi = \frac{\Delta n_{O_2}}{+1}$$

Example 4-17 illustrates the calculation of the extent of reaction and introduc tabular approach for summarizing the changes in amounts of reactants and p ucts. This tabular approach will be particularly useful in later chapters (for ex ple, in Chapters 15 to 18), when our primary focus is on reactions that occur limited extent.

EXAMPLE 4-17 Calculating the Extent of Reaction

(a) Determine the stoichiometric numbers for the following reaction, which occurs to a limited extent at 25 °C $4\,KO_2(s) + 2\,CO_2(g) \longrightarrow 2\,K_2CO_3(s) + 3\,O_2(g)$. **(b)** In an experiment carried out at 25 °C, a reaction mixture co taining 2.478 mol KO_2 and 0.979 mol CO_2 yields 1.128 mol O_2. Calculate the extent of reaction, as well as the fin amounts of all reactants and products.

Analyze

(a) For a given reactant or product, the stoichiometric number (ν) is set equal to the stoichiometric coefficient wit a + or – sign included (a + sign for a product and a – sign for a reactant). **(b)** The extent of reaction (ξ) is define by equation (4.10) and can be used to relate the *changes* in amount to each other. The relationships among th changes in amount are conveniently summarized by constructing a table with three rows immediately below th chemical equation. The first row specifies the initial amounts. The second row specifies the *changes* in amount each of which is expressed as the product of a stoichiometric number (ν) and the extent of reaction (ξ). The thir row specifies the final amounts.

Solve

(a) The stoichiometric numbers are

$$\nu_{KO_2} = -4 \qquad \nu_{CO_2} = -2 \qquad \nu_{K_2CO_3} = +2 \qquad \nu_{O_2} = +3$$

(b) First, we construct a table with three rows below the chemical equation for the reaction. The rows specify th initial amounts, the changes in amount and the final amount, respectively.

	$4\,KO_2(s)$	+	$2\,CO_2(g)$	$\longrightarrow$	$2\,K_2CO_3(s)$	+	$3\,O_2(g)$
initial amounts:	2.478 mol		0.979 mol		0 mol		0 mol
changes:	$-4\,\xi$		$-2\,\xi$		$+2\,\xi$		$+3\,\xi$
final amounts:	$2.478 - 4\,\xi$		$0.979\text{ mol} - 2\,\xi$		$2\,\xi$		$3\,\xi$

Because the final amount of O_2 is given as 1.128 mol, we have $3\,\xi = 1.128$ mol and $\xi = \dfrac{1.128 \text{ mol}}{3} = 0.376$ mol

We can calculate the final amounts by using this value of ξ.

$$n_{KO_2,\text{final}} = 2.478 \text{ mol} - 4\xi = 2.478 \text{ mol} - 4 \times 0.376 \text{ mol} = 0.974 \text{ mol}$$

$$n_{CO_2,\text{final}} = 0.979 \text{ mol} - 2\xi = 0.979 \text{ mol} - 2 \times 0.376 \text{ mol} = 0.227 \text{ mol}$$

$$n_{K_2CO_3,\text{final}} = 2\xi = 2 \times 0.376 \text{ mol} = 0.752 \text{ mol}$$

$$n_{O_2,\text{final}} = 3\xi = 3 \times 0.376 \text{ mol} = 1.128 \text{ mol}$$

nalyze

Note that the unit of ξ is mol. We will use the extent of reaction throughout the text. For example, in Chapters 7 and 13, we will express the energy consumed or produced by a reaction in terms of the extent of reaction. In Chapters 15 to 18, a primary goal is to determine the extent of reaction for reactions that occur to a limited extent only. In these chapters, the tabular approach illustrated above will be particularly useful.

RACTICE EXAMPLE A: For an initial reaction mixture containing 0.500 mol Fe(s) and 0.500 mol H_2O(g), what is the maximum possible value for the extent of reaction, ξ? The chemical equation for the reaction is $3\,Fe(s) + 4\,H_2O(g) \longrightarrow Fe_3O_4(s) + 4\,H_2(g)$.

RACTICE EXAMPLE B: Consider the reaction $2\,N_2 + 3\,O_2 + 4\,HCl \longrightarrow 4\,NOCl + 2\,H_2O$. If the reaction of 1.00 g N_2 with excess O_2 and excess HCl yields 1.17 g NOCl, what is the extent of reaction and what percentage of the N_2 reacted?

asteringCHEMISTRY www.masteringchemistry.com

Every day, tons of chemicals are made to be used directly or as reactants in the production of other materials. Because these chemicals are sold for use in everything from food to pharmaceuticals, it is important that they made in high yield and be of high purity. To learn more about the many factors that chemists and engineers must consider when making decisions about building and operating a chemical plant, go to the Focus On feature for Chapter 4, Industrial Chemistry, on the MasteringChemistry site.

Summary

Chemical Reactions and Chemical uations—**Chemical reactions** are processes in which or more starting substances, the **reactants**, form one or re new substances, the **products**. A chemical reaction be represented symbolically in a **chemical equation**, h formulas of reactants on the left and formulas of ducts on the right; reactants and products are sepa- d by an arrow. The equation must be balanced. A **bal- ed equation** reflects the true quantitative relationships ween reactants and products. An equation is balanced placing **stoichiometric coefficients** before formulas to nify that the total number of each kind of atom is the ne on each side of the equation. The physical states of ctants and products can be indicated by symbols, such (s), (l), (g), and (aq), signifying solid, liquid, gas, and ieous solution, respectively.

2 Chemical Equations and Stoichiometry—**ichiometry** comprises quantitative relationships based atomic and molecular masses, chemical formulas, bal- ed chemical equations, and related matters. The stoi- ometry of chemical reactions makes use of conversion factors derived from balanced chemical equations and called **stoichiometric factors** (Fig. 4-3). Stoichiometric calculations usually require molar masses, densities, and percent compositions, along with other factors.

4-3 Chemical Reactions in Solution—The **molarity** of a solution is the amount of **solute** (in moles) per liter of solution (expression 4.3). Molarity can be treated as a conversion factor between solution volume and amount of solute. Molarity as a conversion factor may be applied to individual solutions, to solutions that are mixed or diluted by adding more **solvent** (Fig. 4-6; expression 4.5), and to reactions occurring in solution.

4-4 Determining the Limiting Reactant—In some reactions, one of the reactants is completely consumed, while the other reactants remain in excess. The reactant that is completely consumed determines the amount of product formed and is called the **limiting reactant**. In some reactions, the reactants are consumed simultaneously, with no reactant remaining in excess; such reactants are said to be in **stoichiometric**

proportions. In some reactions, the limiting reactant must be determined before a stoichiometric calculation can be completed.

4-5 Other Practical Matters in Reaction Stoichiometry—Stoichiometric calculations sometimes involve additional factors, including the reaction's **actual yield**, the presence of **by-products**, and how the reaction or reactions proceed. For example, some reactions yield exactly the quantity of product calculated—the **theoretical yield**. When the actual yield equals the theoretical yield, the **percent yield** is 100%. In some reactions, the actual yield is less than the theoretical, in which case the percent yield is less than 100%. Lower yields may result from the formation of by-products, substances that replace some of the desired product because of reactions other than the one of interest, called **side reactions**. Some stoichiometric calculations are complicated by the fact two or more **simultaneous reactions** may occur. In c cases, a final product may be formed through a sequ of **consecutive reactions**. An **intermediate** is any stance that is a product of one reaction and a reactant subsequent reaction. The equations for a series of cons tive reactions are often replaced with a single **ov equation** for the **overall reaction**.

4-6 The Extent of Reaction—The **extent of reac** (ξ), equation (4.10), is used to quantify and interrelate changes in amounts of reactants and products involve a reaction. The changes in amount can all be expresse a product of a **stoichiometric number** and the exter reaction. For a given reaction, the relationships among changes in amount may be summarized in tabular f (see Example 4-17).

Integrative Example

Sodium nitrite is used in the production of dyes for coloring fabrics, as a preservative in meat processing (to prevent ulism), as a bleach for fibers, and in photography. It can be prepared by passing nitrogen monoxide and oxygen g into an aqueous solution of sodium carbonate. Carbon dioxide gas is another product of the reaction.

In one experimental method, which gives a 95.0% yield, 225 mL of 1.50 M aqueous solution of sodium carbonate, 2: of nitrogen monoxide, and a large excess of oxygen gas are allowed to react. What mass of sodium nitrite is obtaine

Analyze

Five tasks must be performed in this problem: (1) Represent the reaction by a chemical equation in which the name reactants and products are replaced with formulas. (2) Balance the formula equation by inspection. (3) Determine the iting reactant. (4) Calculate the theoretical yield of sodium nitrite based on the quantity of limiting reactant. (5) expression (4.6) to calculate the actual yield of sodium nitrite.

Solve

Obtain the unbalanced equation: Substitute formulas for names.

$$Na_2CO_3(aq) + NO(g) + O_2(g) \longrightarrow NaNO_2(aq) + CO_2(g)$$

Obtain the balanced equation: Begin by noting that there must be 2 $NaNO_2$ for every Na_2CO_3.

$$Na_2CO_3(aq) + NO(g) + O_2(g) \longrightarrow 2\,NaNO_2(aq) + CO_2(g)$$

To balance the N atoms requires 2 NO on the left side:

$$Na_2CO_3(aq) + 2\,NO(g) + O_2(g) \longrightarrow 2\,NaNO_2(aq) + CO_2(g$$

To balance the O atoms requires $\frac{1}{2}$ O_2 on the left, resulting in six O atoms on each side:

$$Na_2CO_3(aq) + 2\,NO(g) + \frac{1}{2}O_2(g) \longrightarrow 2\,NaNO_2(aq) + CO_2($$

Multiply coefficients by 2 to make all of them integers in the final balanced equation.

$$2\,Na_2CO_3(aq) + 4\,NO(g) + O_2(g) \longrightarrow 4\,NaNO_2(aq) + 2\,CO$$

Determine the limiting reactant: The problem states that oxygen is in excess. To determine the limiting reactant, the number of moles of Na_2CO_3, requiring molarity as a conversion factor, must be compared with the number of moles of NO, requiring a conversion factor based on the molar mass of NO.

$$?\text{ mol } Na_2CO_3 = 0.225\text{ L soln} \times \frac{1.50\text{ mol } Na_2CO_3}{1\text{ L soln}}$$

$$= 0.338\text{ mol } Na_2CO_3$$

$$?\text{ mol NO} = 22.1\text{ g NO} \times \frac{1\text{ mol NO}}{30.01\text{ g NO}} = 0.736\text{ mol NO}$$

From the balanced equation, the stoichiometric mole ratio is

$$4\text{ mol NO} : 2\text{ mol } Na_2CO_3 = 2:1$$

The available mole ratio is

$$0.736\text{ mol NO} : 0.338\text{ mol } Na_2CO_3 = 2.18:1$$

Because the available mole ratio exceeds 2:1, NO is in excess and Na_2CO_3 is the limiting reactant.

mine the theoretical yield of the reaction: This lation is based on the amount of Na_2CO_3, miting reactant.

$$? \text{ g } NaNO_2 = 0.338 \text{ mol } Na_2CO_3 \times \frac{4 \text{ mol } NaNO_2}{2 \text{ mol } Na_2CO_3} \times \frac{69.00 \text{ g } NaNO_2}{1 \text{ mol } NaNO_2} = 46.6 \text{ g } NaNO_2$$

mine the actual yield: Use expression (4.6), h relates percent yield (95.0%), theoretical (46.6 g $NaNO_2$), and the actual yield.

$$\text{actual yield} = \frac{\text{percent yield} \times \text{theoretical yield}}{100\%} = \frac{95.0\% \times 46.6 \text{ g } NaNO_2}{100\%} = 44.3 \text{ g } NaNO_2$$

ss

solving a multistep problem, it is important to check over your work. In this example, we should first double-check he chemical equation is properly balanced. There are 4 Na's on each side, 2 C's, 12 O's, and 4 N's. We can double- that we correctly identified the limiting reactant by using a different approach. Because 0.338 mol Na_2CO_3 would 0.676 mol $NaNO_2$ (in the presence of excess NO and O_2) and 0.736 mol NO would yield 0.736 mol $NaNO_2$ (in the nce of excess Na_2CO_3 and O_2), we conclude that Na_2CO_3 must be the limiting reactant; *the limiting reactant is the one mits the amount of product obtained.* In checking the calculation of the mass of $NaNO_2$ obtained, we notice that the units out properly. To check the final step, we can use our final answer to calculate the percent yield for the experiment: /46.6) × 100% = 95%. This is the correct result for the percent yield.

CTICE EXAMPLE A: Hexamethylenediamine has the molecular formula $C_6H_{16}N_2$. It is one of the starting rials for the production of nylon. It can be prepared by the following reaction:

$$C_6H_{10}O_4(l) + NH_3(g) + H_2(g) \longrightarrow C_6H_{16}N_2(l) + H_2O(l) \text{ (not balanced)}$$

ge reaction vessel contains 4.15 kg $C_6H_{10}O_4$, 0.547 kg NH_3, and 0.172 kg H_2. If 1.46 kg $C_6H_{16}N_2$ is obtained, then what percent yield for the experiment?

CTICE EXAMPLE B: Zinc metal and aqueous hydrochloric acid, HCl(aq), react to give hydrogen gas, $H_2(g)$, aqueous zinc chloride, $ZnCl_2(aq)$. A 0.4000 g sample of *impure* zinc reacts completely when added to 750.0 mL 0179 M HCl. After the reaction, the molarity of the HCl solution is determined to be 0.00403 M. What is the ent by mass of zinc in the sample?

Exercises

ting and Balancing Chemical Equations

Balance the following equations by inspection.
(a) $SO_3 \longrightarrow SO_2 + O_2$
(b) $Cl_2O_7 + H_2O \longrightarrow HClO_4$
(c) $NO_2 + H_2O \longrightarrow HNO_3 + NO$
(d) $PCl_3 + H_2O \longrightarrow H_3PO_3 + HCl$

Balance the following equations by inspection.
(a) $P_2H_4 \longrightarrow PH_3 + P_4$
(b) $P_4 + Cl_2 \longrightarrow PCl_3$
(c) $FeCl_3 + H_2S \longrightarrow Fe_2S_3 + HCl$
(d) $Mg_3N_2 + H_2O \longrightarrow Mg(OH)_2 + NH_3$

Balance the following equations by inspection.
(a) $PbO + NH_3 \longrightarrow Pb + N_2 + H_2O$
(b) $FeSO_4 \longrightarrow Fe_2O_3 + SO_2 + O_2$
(c) $S_2Cl_2 + NH_3 \longrightarrow N_4S_4 + NH_4Cl + S_8$
(d) $C_3H_7CHOHCH(C_2H_5)CH_2OH + O_2 \longrightarrow CO_2 + H_2O$

Balance the following equations by inspection.
(a) $SO_2Cl_2 + HI \longrightarrow H_2S + H_2O + HCl + I_2$
(b) $FeTiO_3 + H_2SO_4 + H_2O \longrightarrow FeSO_4 \cdot 7\,H_2O + TiOSO_4$
(c) $Fe_3O_4 + HCl + Cl_2 \longrightarrow FeCl_3 + H_2O + O_2$
(d) $C_6H_5CH_2SSCH_2C_6H_5 + O_2 \longrightarrow CO_2 + SO_2 + H_2O$

Write balanced equations based on the information given.
(a) solid magnesium + oxygen gas ⟶ solid magnesium oxide
(b) nitrogen monoxide gas + oxygen gas ⟶ nitrogen dioxide gas
(c) gaseous ethane(C_2H_6) + oxygen gas ⟶ carbon dioxide gas + liquid water
(d) aqueous silver sulfate + aqueous barium iodide ⟶ solid barium sulfate + solid silver iodide

6. Write balanced equations based on the information given.
(a) solid magnesium + nitrogen gas ⟶ solid magnesium nitride
(b) solid potassium chlorate ⟶ solid potassium chloride + oxygen gas
(c) solid sodium hydroxide + solid ammonium chloride ⟶ solid sodium chloride + gaseous ammonia + water vapor
(d) solid sodium + liquid water ⟶ aqueous sodium hydroxide + hydrogen gas

7. Write balanced equations to represent the complete combustion of each of the following in excess oxygen: **(a)** butane, $C_4H_{10}(g)$; **(b)** isopropyl alcohol, $CH_3CH(OH)CH_3(l)$; **(c)** lactic acid, $CH_3CH(OH)COOH(s)$.

8. Write balanced equations to represent the complete combustion of each of the following in excess oxygen: **(a)** propylene, $C_3H_6(g)$; **(b)** thiobenzoic acid, $C_6H_5COSH(l)$; **(c)** glycerol, $CH_2(OH)CH(OH)CH_2OH(l)$.

9. Write balanced equations to represent:
(a) the decomposition, by heating, of solid ammonium nitrate to produce dinitrogen monoxide gas (laughing gas) and water vapor
(b) the reaction of aqueous sodium carbonate with hydrochloric acid to produce water, carbon dioxide gas, and aqueous sodium chloride
(c) the reaction of methane (CH_4), ammonia, and oxygen gases to form gaseous hydrogen cyanide (HCN) and water vapor

10. Write balanced equations to represent:
(a) the reaction of sulfur dioxide gas with oxygen gas to produce sulfur trioxide gas (one of the reactions involved in the industrial preparation of sulfuric acid)
(b) the dissolving of limestone (calcium carbonate) in water containing dissolved carbon dioxide to produce calcium hydrogen carbonate (a reaction prod temporary hardness in groundwater)
(c) the reaction of ammonia and nitrogen mon to form nitrogen gas and water vapor

11. Write a balanced chemical equation for the rea depicted below.

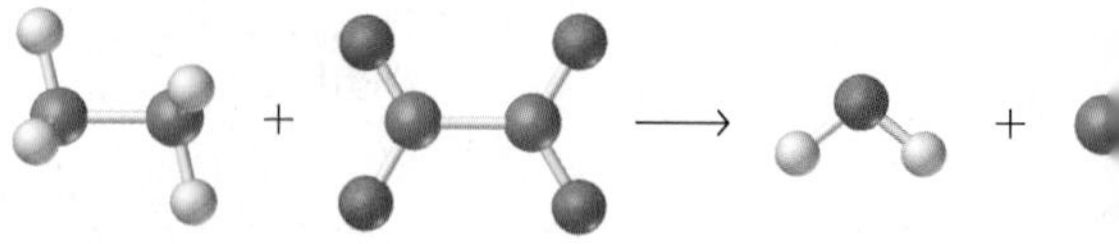

12. Write a balanced chemical equation for the rea depicted below.

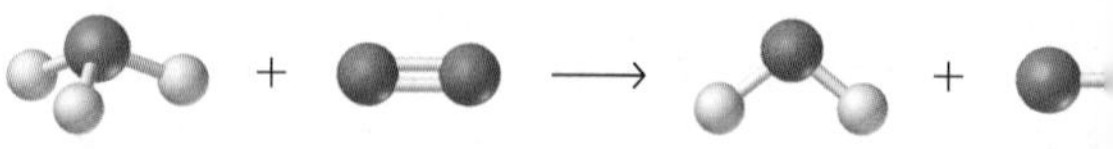

Stoichiometry of Chemical Reactions

13. In an experiment, 0.689 g Cr(s) reacts completely with 0.636 g $O_2(g)$ to form a single solid compound. Write a balanced chemical equation for the reaction.

14. A 3.104 g sample of an oxide of manganese contains 1.142 grams of oxygen. Write a balanced chemical equation for the reaction that produces the compound from Mn(s) and $O_2(g)$.

15. Iron metal reacts with chlorine gas. How many grams of $FeCl_3$ are obtained when 515 g Cl_2 reacts with excess Fe?

$$2\,Fe(s) + 3\,Cl_2(g) \longrightarrow 2\,FeCl_3(s)$$

16. If 75.8 g PCl_3 is produced by the reaction

$$6\,Cl_2(g) + P_4(s) \longrightarrow 4\,PCl_3(l)$$

how many grams each of Cl_2 and P_4 are consumed?

17. A laboratory method of preparing $O_2(g)$ involves the decomposition of $KClO_3(s)$.

$$2\,KClO_3(s) \xrightarrow{\Delta} 2\,KCl(s) + 3\,O_2(g)$$

(a) How many moles of $O_2(g)$ can be produced by the decomposition of 32.8 g $KClO_3$?
(b) How many grams of $KClO_3$ must decompose to produce 50.0 g O_2?
(c) How many grams of KCl are formed, together with 28.3 g O_2, in the decomposition of $KClO_3$?

18. A commercial method of manufacturing hydrogen involves the reaction of iron and steam.

$$3\,Fe(s) + 4\,H_2O(g) \xrightarrow{\Delta} Fe_3O_4(s) + 4\,H_2(g)$$

(a) How many grams of H_2 can be produced from 87.2 g Fe and an excess of $H_2O(g)$ (steam)?
(b) How many grams of H_2O are consumed in the conversion of 25.0 g Fe to Fe_3O_4?
(c) If 29.2 g H_2 is produced, how many grams of Fe_3O_4 must also be produced?

19. How many grams of Ag_2CO_3 are decomposed to yield 75.1 g Ag in this reaction?

$$Ag_2CO_3(s) \xrightarrow{\Delta} Ag(s) + CO_2(g) + O_2(g) \text{ (not balanced)}$$

20. How many kilograms of HNO_3 are consumed tc duce 125 kg $Ca(H_2PO_4)_2$ in this reaction?

$$Ca_3(PO_4)_2 + HNO_3 \longrightarrow Ca(H_2PO_4)_2 + Ca(NO_3)_2 \text{ (not bala}$$

21. The reaction of calcium hydride with water ca used to prepare small quantities of hydrogen gas done to fill weather-observation balloons.

$$CaH_2(s) + H_2O(l) \longrightarrow Ca(OH)_2(s) + H_2(g) \text{ (not bala}$$

(a) How many grams of $H_2(g)$ result from the tion of 127 g CaH_2 with an excess of water?
(b) How many grams of water are consumed i reaction of 56.2 g CaH_2?
(c) What mass of $CaH_2(s)$ must react with an e of water to produce 8.12×10^{24} molecules of H_2?

22. The reaction of potassium superoxide, KO_2, is us life-support systems to replace $CO_2(g)$ in expire with $O_2(g)$. The unbalanced chemical equation fc reaction is given below.

$$KO_2(s) + CO_2(g) \longrightarrow K_2CO_3(s) + O_2(g)$$

(a) How many moles of $O_2(g)$ are produced b reaction of 88.0 g $CO_2(g)$ with excess $KO_2(s)$?
(b) How many grams of $KO_2(s)$ are consume 1.000×10^3 g $CO_2(g)$ removed from expired air?
(c) How many moles of K_2CO_3 are produced per ligram of KO_2 consumed?

23. Iron ore is impure Fe_2O_3. When Fe_2O_3 is heated an excess of carbon (coke), metallic iron and ca monoxide gas are produced. From a sample o weighing 938 kg, 523 kg of pure iron is obtained. is the mass percent Fe_2O_3 in the ore sample, assu that none of the impurities contain Fe?

24. Solid silver oxide, $Ag_2O(s)$, decomposes at tem tures in excess of 300 °C, yielding metallic silver oxygen gas. A 3.13 g sample of impure silver o yields 0.187 g $O_2(g)$. What is the mass percent A in the sample? Assume that $Ag_2O(s)$ is the source of $O_2(g)$. [*Hint:* Write a balanced equatio the reaction.]

)ecaborane, $B_{10}H_{14}$, was used as a fuel for rockets in he 1950s. It reacts violently with oxygen, O_2, to pro- uce B_2O_3 and water. Calculate the percentage by nass of $B_{10}H_{14}$ in a fuel mixture designed to ensure hat $B_{10}H_{14}$ and O_2 run out at exactly the same time. Such a mixture minimizes the mass of fuel that a ocket must carry.)

'he rocket boosters of the space shuttle *Discovery*, aunched on July 26, 2005, used a fuel mixture con- aining primarily solid ammonium perchlorate, NH_4ClO_4(s), and aluminum metal. The unbalanced hemical equation for the reaction is given below.

$$Al(s) + NH_4ClO_4(s) \longrightarrow Al_2O_3(s) + AlCl_3(s) + H_2O(l) + N_2(g)$$

Vhat is the minimum mass of NH_4ClO_4 consumed, er kilogram of Al, by the reaction of NH_4ClO_4 and Al? [*Hint*: Balance the elements in the order Cl, H,), Al, N.]

27. A piece of aluminum foil measuring 10.25 cm × 5.50 cm × 0.601 mm is dissolved in excess HCl(aq). What mass of H_2(g) is produced? Use equation (4.2) and $d = 2.70$ g/cm^3 for Al.
28. An excess of aluminum foil is allowed to react with 225 mL of an aqueous solution of HCl ($d = 1.088$ g/mL) that contains 18.0% HCl by mass. What mass of H_2(g) is produced? [*Hint:* Use equation (4.2).]
29. *Without performing detailed calculations*, which of the following metals yields the greatest amount of H_2 per gram of metal reacting with HCl(aq)? **(a)** Na, **(b)** Mg, **(c)** Al, **(d)** Zn. [*Hint:* Write equations similar to (4.2).]
30. *Without performing detailed calculations*, which of the following yields the same mass of CO_2(g) per gram of compound as does ethanol, CH_3CH_2OH, when burned in excess oxygen? **(a)** H_2CO; **(b)** $HOCH_2CH_2OH$; **(c)** $HOCH_2CH(OH)CH_2OH$; **(d)** CH_3OCH_3; **(e)** C_6H_5OH.

arity

Vhat are the molarities of the following solutes when lissolved in water?
a) 2.92 mol CH_3OH in 7.16 L of solution
b) 7.69 mmol CH_3CH_2OH in 50.00 mL of solution
c) 25.2 g $CO(NH_2)_2$ in 275 mL of solution

Vhat are the molarities of the following solutes when lissolved in water?
a) 2.25×10^{-4} mol CH_3CH_2OH in 125 mL of solution
b) 57.5 g $(CH_3)_2CO$ in 525 mL of solution
c) 18.5 mL of $C_3H_5(OH)_3$ ($d = 1.26$ g/mL) in 375 mL of solution

Vhat are the molarities of the following solutes?
(a) sucrose ($C_{12}H_{22}O_{11}$) if 150.0 g is dissolved per 250.0 mL of water solution
(b) urea, $CO(NH_2)_2$, if 98.3 mg of the 97.9% pure solid is dissolved in 5.00 mL of aqueous solution
(c) methanol, CH_3OH, ($d = 0.792$ g/mL) if 125.0 mL is dissolved in enough water to make 15.0 L of solution

What are the molarities of the following solutes?
(a) aspartic acid ($H_2C_4H_5NO_4$) if 0.405 g is dissolved in enough water to make 100.0 mL of solution
(b) acetone, $(CH_3)_2CO$, ($d = 0.790$ g/mL) if 35.0 mL is dissolved in enough water to make 425 mL of solution
(c) diethyl ether, $(C_2H_5)_2O$, if 8.8 mg is dissolved in enough water to make 3.00 L of solution

How much
(a) glucose, $C_6H_{12}O_6$, in grams, must be dissolved in water to produce 75.0 mL of 0.350 M $C_6H_{12}O_6$?
(b) methanol, CH_3OH ($d = 0.792$ g/mL), in milliliters, must be dissolved in water to produce 2.25 L of 0.485 M CH_3OH?

How much
(a) ethanol, CH_3CH_2OH ($d = 0.789$ g/mL), in liters, must be dissolved in water to produce 200.0 L of 1.65 M CH_3CH_2OH?
(b) concentrated hydrochloric acid solution (36.0% HCl by mass; $d = 1.18$ g/mL), in milliliters, is required to produce 12.0 L of 0.234 M HCl?

In the United States, the concentration of glucose, $C_6H_{12}O_6$, in the blood is reported in units of milligrams per deciliter (mg/dL). In Canada, the United Kingdom, and elsewhere, the blood glucose concentration is reported in millimoles per liter (mmol/L), where 1 mmol = 1×10^{-3} mol. If a person has a blood glucose level of 85 mg/dL, then what is **(a)** the blood glucose level in mmol/L; **(b)** the molarity of glucose in the blood?

38. In many communities, water is fluoridated to prevent tooth decay. In the United States, for example, more than half of the population served by public water systems has access to water that is fluoridated at approximately 1 mg F^- per liter. **(a)** What is the molarity of F^- in water if it contains 1.2 mg F^- per liter? **(b)** How many grams of solid KF should be added to a 1.6×10^8 L water reservoir to give a fluoride concentration of 1.2 mg F^- per liter?
39. Which of the following is a 0.500 M KCl solution? **(a)** 0.500 g KCl/mL solution; **(b)** 36.0 g KCl/L solution; **(c)** 7.46 mg KCl/mL solution; **(d)** 373 g KCl in 10.00 L solution.
40. Which two solutions have the same concentration? **(a)** 55.45 g NaCl/L solution; **(b)** 5.545 g NaCl/100 g solution; **(c)** 55.45 g NaCl/kg water; **(d)** 55.45 mg NaCl/1.00 mL solution; **(e)** 5.00 mmol NaCl/5.00 mL solution.
41. Which has the higher concentration of sucrose: a 46% sucrose solution by mass ($d = 1.21$ g/mL), or 1.50 M $C_{12}H_{22}O_{11}$? Explain your reasoning.
42. Which has the greater molarity of ethanol: a white wine ($d = 0.95$ g/mL) with 11% CH_3CH_2OH by mass, or the solution described in Example 4-7? Explain your reasoning.
43. A 10.00 mL sample of 2.05 M KNO_3 is diluted to a volume of 250.0 mL. What is the concentration of the diluted solution?
44. What volume of 2.00 M $AgNO_3$ must be diluted with water to prepare 500.0 mL of 0.350 M $AgNO_3$?
45. Water is evaporated from 125 mL of 0.198 M K_2SO_4 solution until the volume becomes 105 mL. What is the molarity of K_2SO_4 in the remaining solution?

46. A 25.0 mL sample of HCl(aq) is diluted to a volume of 500.0 mL. If the concentration of the diluted solution is found to be 0.085 M HCl, what was the concentration of the original solution?

47. Given a 0.250 M K_2CrO_4 stock solution, describe how you would prepare a solution that is 0.0125 M K_2CrO_4. That is, what combination(s) of pipet and volumetric flask would you use? Typical sizes of volumetric flasks found in a general chemistry la tory are 100.0, 250.0, 500.0, and 1000.0 mL, and cal sizes of volumetric pipets are 1.00, 5.00, 25.00, and 50.00 mL.

48. Given two liters of 0.496 M KCl, describe you would use this solution to prepare 250.0 0.175 M KCl. Give sufficient details so that an student could follow your instructions.

Chemical Reactions in Solution

49. Consider the reaction below:

$$2\,AgNO_3(aq) + Na_2S(s) \longrightarrow Ag_2S(s) + 2\,NaNO_3(aq)$$

(a) How many grams of $Na_2S(s)$ are required to react completely with 27.8 mL of 0.163 M $AgNO_3$?
(b) How many grams of $Ag_2S(s)$ are obtained from the reaction in part (a)?

50. Excess $NaHCO_3$ is added to 525 mL of 0.220 M $Cu(NO_3)_2$. These substances react as follows:

$$Cu(NO_3)_2(aq) + 2\,NaHCO_3(s) \longrightarrow CuCO_3(s) + 2\,NaNO_3(aq) + H_2O(l) + CO_2(g)$$

(a) How many grams of the $NaHCO_3(s)$ will be consumed?
(b) How many grams of $CuCO_3(s)$ will be produced?

51. How many milliliters of 0.650 M K_2CrO_4 are needed to precipitate all the silver in 415 mL of 0.186 M $AgNO_3$ as $Ag_2CrO_4(s)$?

$$2\,AgNO_3(aq) + K_2CrO_4(aq) \longrightarrow Ag_2CrO_4(s) + 2\,KNO_3(aq)$$

52. Consider the reaction below.

$$Ca(OH)_2(s) + 2\,HCl(aq) \longrightarrow CaCl_2(aq) + 2\,H_2O(l)$$

(a) How many grams of $Ca(OH)_2$ are required to react completely with 415 mL of 0.477 M HCl?
(b) How many kilograms of $Ca(OH)_2$ are required to react with 324 L of a HCl solution that is 24.28% HCl by mass, and has a density of 1.12 g/mL?

53. Exactly 1.00 mL of an aqueous solution of HNO_3 is diluted to 100.0 mL. It takes 29.78 mL of 0.0142 M $Ca(OH)_2$ to convert all of the HNO_3 to $Ca(NO_3)_2$. The other product of the reaction is water. Calculate the molarity of the undiluted HNO_3 solution.

54. A 5.00 mL sample of an aqueous solution of H_3PO_4 requires 49.1 mL of 0.217 M NaOH to convert all of the H_3PO_4 to Na_2HPO_4. The other product of the reaction is water. Calculate the molarity of the H_3PO_4 solution.

55. Refer to Example 4-6 and equation (4.2). For the conditions stated in Example 4-6, determine **(a)** the number of moles of $AlCl_3$ and **(b)** the molarity of the $AlCl_3(aq)$ if the solution volume is simply the 23.8 mL calculated in the example.

56. Refer to the Integrative Example on page 140. If Na_2CO_3 in 1.42 L of aqueous solution is treated an excess of NO(g) and $O_2(g)$, what is the molar the $NaNO_2(aq)$ solution that results? (Assume the reaction goes to completion.)

57. How many grams of Ag_2CrO_4 will precipit excess $K_2CrO_4(aq)$ is added to the 415 mL of 0.1 $AgNO_3$ in Exercise 51?

58. What volume of 0.0665 M $KMnO_4$ is necessary to vert 12.5 g KI to I_2 in the reaction below? Assume H_2SO_4 is present in excess.

$$2\,KMnO_4 + 10\,KI + 8\,H_2SO_4 \longrightarrow 6\,K_2SO_4 + 2\,MnSO_4 + 5\,I_2 + 8$$

59. How many grams of sodium must react with 15 H_2O to produce a solution that is 0.175 M Na (Assume a final solution volume of 155 mL.)

$$2\,Na(s) + 2\,H_2O(l) \longrightarrow 2\,NaOH(aq) + H_2($$

60. A method of lowering the concentration of HCl(to allow the solution to react with a small quant Mg. How many milligrams of Mg must be add 250.0 mL of 1.023 M HCl to reduce the solution centration to exactly 1.000 M HCl?

$$Mg(s) + 2\,HCl(aq) \longrightarrow MgCl_2(aq) + H_2(g$$

61. A 0.3126 g sample of oxalic acid, $H_2C_2O_4$, req 26.21 mL of a particular concentration of NaOH(complete the following reaction. What is the mo of the NaOH(aq)?

$$H_2C_2O_4(s) + 2\,NaOH(aq) \longrightarrow Na_2C_2O_4(aq) + 2\,H$$

62. A 25.00 mL sample of HCl(aq) was added to a 0.1 sample of $CaCO_3$. All the $CaCO_3$ reacted, lea some excess HCl(aq).

$$CaCO_3(s) + 2\,HCl(aq) \longrightarrow CaCl_2(aq) + H_2O(l) + C($$

The excess HCl(aq) required 43.82 mL of 0.011 $Ba(OH)_2$ to complete the following reaction. W was the molarity of the original HCl(aq)?

$$2\,HCl(aq) + Ba(OH)_2(aq) \longrightarrow BaCl_2(aq) + 2\,H$$

Determining the Limiting Reactant

63. How many moles of NO(g) can be produced in the reaction of 3.00 mol $NH_3(g)$ and 4.00 mol $O_2(g)$?

$$4\,NH_3(g) + 5\,O_2(g) \xrightarrow{\Delta} 4\,NO(g) + 6\,H_2O(l)$$

64. The reaction of calcium hydride and water prod calcium hydroxide and hydrogen as products. many moles of $H_2(g)$ will be formed in the rea between 0.82 mol $CaH_2(s)$ and 1.54 mol $H_2O(l)$?

A 0.696 mol sample of Cu is added to 136 mL of 6.0 M HNO_3. Assuming the following reaction is the only one that occurs, will the Cu react completely?

$$3\,Cu(s) + 8\,HNO_3(aq) \longrightarrow 3\,Cu(NO_3)_2(aq) + 4\,H_2O(l) + 2\,NO(g)$$

How many grams of $H_2(g)$ are produced by the reaction of 1.84 g Al with 75.0 mL of 2.95 M HCl? [*Hint:* Recall equation (4.2).]

A side reaction in the manufacture of rayon from wood pulp is

$$3\,CS_2 + 6\,NaOH \longrightarrow 2\,Na_2CS_3 + Na_2CO_3 + 3\,H_2O$$

How many grams of Na_2CS_3 are produced in the reaction of 92.5 mL of liquid CS_2 (d = 1.26 g/mL) and 2.78 mol NaOH?

Lithopone is a brilliant white pigment used in water-based interior paints. It is a mixture of $BaSO_4$ and ZnS produced by the reaction

$$BaS(aq) + ZnSO_4(aq) \longrightarrow \underbrace{ZnS(s) + BaSO_4(s)}_{\text{lithopone}}$$

How many grams of lithopone are produced in the reaction of 315 mL of 0.275 M $ZnSO_4$ and 285 mL of 0.315 M BaS?

Ammonia can be generated by heating together the solids NH_4Cl and $Ca(OH)_2$. $CaCl_2$ and H_2O are also formed. **(a)** If a mixture containing 33.0 g each of NH_4Cl and $Ca(OH)_2$ is heated, how many grams of NH_3 will form? **(b)** Which reactant remains in excess, and in what mass?

70. Chlorine can be generated by heating together calcium hypochlorite and hydrochloric acid. Calcium chloride and water are also formed. **(a)** If 50.0 g $Ca(OCl)_2$ and 275 mL of 6.00 M HCl are allowed to react, how many grams of chlorine gas will form? **(b)** Which reactant, $Ca(OCl)_2$ or HCl, remains in excess, and in what mass?

71. Chromium(II) sulfate, $CrSO_4$, is a reagent that has been used in certain applications to help reduce carbon–carbon double bonds (C=C) in molecules to single bonds (C—C). The reagent can be prepared via the following reaction.

$$4\,Zn(s) + K_2Cr_2O_7(aq) + 7\,H_2SO_4(aq) \longrightarrow 4\,ZnSO_4(aq) + 2\,CrSO_4(aq) + K_2SO_4(aq) + 7\,H_2O(l)$$

What is the maximum number of grams of $CrSO_4$ that can be made from a reaction mixture containing 3.2 mol Zn, 1.7 mol $K_2Cr_2O_7$, and 5.0 mol H_2SO_4?

72. Titanium tetrachloride, $TiCl_4$, is prepared by the reaction below.

$$3\,TiO_2(s) + 4\,C(s) + 6\,Cl_2(g) \longrightarrow 3\,TiCl_4(g) + 2\,CO_2(g) + 2\,CO(g)$$

What is the maximum mass of $TiCl_4$ that can be obtained from 35 g TiO_2, 45 g Cl_2, and 11 g C?

eoretical, Actual, and Percent Yields

In the reaction of 277 g CCl_4 with an excess of HF, 187 g CCl_2F_2 is obtained. What are the **(a)** theoretical, **(b)** actual, and **(c)** percent yields of this reaction?

$$CCl_4 + 2\,HF \longrightarrow CCl_2F_2 + 2\,HCl$$

In the reaction shown, 100.0 g $C_6H_{11}OH$ yielded 64.0 g C_6H_{10}. **(a)** What is the theoretical yield of the reaction? **(b)** What is the percent yield? **(c)** What mass of $C_6H_{11}OH$ would produce 100.0 g C_6H_{10} if the percent yield is that determined in part (b)?

$$C_6H_{11}OH \longrightarrow C_6H_{10} + H_2O$$

Cryolite, Na_3AlF_6, is an important industrial reagent. It is made by the reaction below.

$$Al_2O_3(s) + 6\,NaOH(aq) + 12\,HF(g) \longrightarrow 2\,Na_3AlF_6(s) + 9\,H_2O(l)$$

In an experiment, 7.81 g Al_2O_3 and excess HF(g) were dissolved in 3.50 L of 0.141 M NaOH. If 28.2 g Na_3AlF_6 was obtained, then what is the percent yield for this experiment?

Nitrogen gas, N_2, can be prepared by passing gaseous ammonia over solid copper(II) oxide, CuO, at high temperatures. The other products of the reaction are solid copper, Cu, and water vapor. In a certain experiment, a reaction mixture containing 18.1 g NH_3 and 90.4 g CuO yields 6.63 g N_2. Calculate the percent yield for this experiment.

The reaction of 15.0 g C_4H_9OH, 22.4 g NaBr, and 32.7 g H_2SO_4 yields 17.1 g C_4H_9Br in the reaction shown. What are the **(a)** theoretical yield, **(b)** actual yield, and **(c)** percent yield of this reaction?

$$C_4H_9OH + NaBr + H_2SO_4 \longrightarrow C_4H_9Br + NaHSO_4 + H_2O$$

78. Azobenzene, an intermediate in the manufacture of dyes, can be prepared from nitrobenzene by reaction with triethylene glycol in the presence of Zn and KOH. In one reaction, 0.10 L of nitrobenzene (d = 1.20 g/mL) and 0.30 L of triethylene glycol (d = 1.12 g/mL) yields 55 g azobenzene. What are the **(a)** theoretical yield, **(b)** actual yield, and **(c)** percent yield of this reaction?

$$\underset{\text{nitrobenzene}}{2\,C_6H_5NO_2} + \underset{\text{triethylene glycol}}{4\,C_6H_{14}O_4} \xrightarrow[KOH]{Zn} \underset{\text{azobenzene}}{(C_6H_5N)_2} + 4\,C_6H_{12}O_4 + 4\,H_2O$$

79. How many grams of commercial acetic acid (97% CH_3COOH by mass) must be allowed to react with an excess of PCl_3 to produce 75 g of acetyl chloride (CH_3COCl), if the reaction has a 78.2% yield?

$$CH_3COOH + PCl_3 \longrightarrow CH_3COCl + H_3PO_3 \text{ (not balanced)}$$

80. Suppose that reactions **(a)** and **(b)** each have a 92% yield. Starting with 112 g CH_4 in reaction **(a)** and an excess of $Cl_2(g)$, how many grams of CH_2Cl_2 are formed in reaction **(b)**?

(a) $CH_4 + Cl_2 \longrightarrow CH_3Cl + HCl$

(b) $CH_3Cl + Cl_2 \longrightarrow CH_2Cl_2 + HCl$

81. An essentially 100% yield is necessary for a chemical reaction used to *analyze* a compound, but it is almost never expected for a reaction that is used to *synthesize* a compound. Explain this difference.

82. Suppose we carry out the precipitation of $Ag_2CrO_4(s)$ described in Example 4-10. If we obtain 2.058 g of precipitate, we might conclude that it is nearly p $Ag_2CrO_4(s)$, but if we obtain 2.112 g, we can be q sure that the precipitate is not pure. Explain difference.

Consecutive Reactions, Simultaneous Reactions

83. How many grams of HCl are consumed in the reaction of 425 g of a mixture containing 35.2% $MgCO_3$ and 64.8% $Mg(OH)_2$, by mass?

$$Mg(OH)_2 + 2\,HCl \longrightarrow MgCl_2 + 2\,H_2O$$
$$MgCO_3 + 2\,HCl \longrightarrow MgCl_2 + H_2O + CO_2$$

84. How many grams of CO_2 are produced in the complete combustion of 406 g of a bottled gas that consists of 72.7% propane (C_3H_8) and 27.3% butane (C_4H_{10}), by mass?

85. Dichlorodifluoromethane, once widely used as a refrigerant, can be prepared by the reactions shown. How many moles of Cl_2 must be consumed in the first reaction to produce 2.25 kg CCl_2F_2 in the second? Assume that all the CCl_4 produced in the first reaction is consumed in the second.

$$CH_4 + Cl_2 \longrightarrow CCl_4 + HCl \text{ (not balanced)}$$
$$CCl_4 + HF \longrightarrow CCl_2F_2 + HCl \text{ (not balanced)}$$

86. Carbon dioxide gas, $CO_2(g)$, produced in the combustion of a sample of ethane is absorbed in $Ba(OH)_2(aq)$, producing 0.506 g $BaCO_3(s)$. How many grams of ethane (C_2H_6) must have been burned?

$$C_2H_6(g) + O_2(g) \longrightarrow CO_2(g) + H_2O(l) \text{ (not balanced)}$$
$$CO_2(g) + Ba(OH)_2(aq) \longrightarrow BaCO_3(s) + H_2O(l)$$

87. The following process has been used to obtain iodine from oil-field brines in California. How many kilograms of silver nitrate are required in the first step for every kilogram of iodine produced in the third step?

sodium iodide + silver nitrate ⟶ silver iodide + sodium nitrate

silver iodide + iron ⟶ iron(II) iodide + silver

iron(II) iodide + chlorine gas ⟶ iron(III) chloride + solid iodine

88. Sodium bromide, used to produce silver bromide for use in photography, can be prepared as shown. How many kilograms of iron are consumed to proc 2.50×10^3 kg NaBr?

$$Fe + Br_2 \longrightarrow FeBr_2$$
$$FeBr_2 + Br_2 \longrightarrow Fe_3Br_8 \text{ (not balanced)}$$
$$Fe_3Br_8 + Na_2CO_3 \longrightarrow NaBr + CO_2 + Fe_3O_4 \text{ (not balan}$$

89. High-purity silicon is obtained using a three-s process. The first step involves heating solid sil dioxide, SiO_2, with solid carbon to give solid sil and carbon monoxide gas. In the second step, s silicon is converted into liquid silicon tetrachlor $SiCl_4$, by treating it with chlorine gas. In the last s $SiCl_4$ is treated with hydrogen gas to give ultrap solid silicon and hydrogen chloride gas.
(a) Write balanced chemical equations for the s involved in this three-step process.
(b) Calculate the masses of carbon, chlorine, hydrogen required per kilogram of silicon.

90. The following set of reactions is to be used as the b of a method for producing nitric acid, HN Calculate the minimum masses of N_2, H_2, anc required per kilogram of HNO_3.

$$N_2(g) + 3\,H_2(g) \longrightarrow 2\,NH_3(g)$$
$$4\,NH_3(g) + 5\,O_2(g) \longrightarrow 4\,NO(g) + 6\,H_2O(g)$$
$$2\,NO(g) + O_2(g) \longrightarrow 2\,NO_2(g)$$
$$3\,NO_2(g) + H_2O(l) \longrightarrow 2\,HNO_3(aq) + NO(g)$$

91. When a solid mixture of $MgCO_3$ and $CaCO_3$ is he strongly, carbon dioxide gas is given off and a solid ture of MgO and CaO is obtained. If a 24.00 g samp a mixture of $MgCO_3$ and $CaCO_3$ produces 12.00 g C then what is the percentage by mass of $MgCO_3$ ir original mixture?

92. A mixture of Fe_2O_3 and FeO was analyzed and fo to be 72.0% Fe by mass. What is the percentag mass of Fe_2O_3 in the mixture?

Integrative and Advanced Exercises

93. Write chemical equations to represent the following reactions.
(a) Limestone rock (calcium carbonate) is heated (calcined) and decomposes to calcium oxide and carbon dioxide gas.
(b) Zinc sulfide ore is heated in air (roasted) and is converted to zinc oxide and sulfur dioxide gas. (Note that oxygen gas in the air is also a reactant.)
(c) Propane gas reacts with gaseous wate produce a mixture of carbon monoxide and hy gen gases. (This mixture, called *synthesis gas*, is u to produce a variety of organic chemicals.)
(d) Sulfur dioxide gas is passed into an aque solution containing sodium sulfide and sodium bonate. The reaction products are carbon dio and an aqueous solution of sodium thiosulfate.

Write chemical equations to represent the following reactions.
(a) Calcium phosphate is heated with silicon dioxide and carbon, producing calcium silicate ($CaSiO_3$), phosphorus (P_4), and carbon monoxide. The phosphorus and chlorine react to form phosphorus trichloride, and the phosphorus trichloride and water react to form phosphorous acid.
(b) Copper metal reacts with gaseous oxygen, carbon dioxide, and water to form green basic copper carbonate, $Cu_2(OH)_2CO_3$ (a reaction responsible for the formation of the green patina, or coating, often seen on outdoor bronze statues).
(c) White phosphorus and oxygen gas react to form tetraphosphorus decoxide. The tetraphosphorus decoxide reacts with water to form an aqueous solution of phosphoric acid.
(d) Calcium dihydrogen phosphate reacts with sodium hydrogen carbonate (bicarbonate), producing calcium phosphate, sodium hydrogen phosphate, carbon dioxide, and water (the principal reaction occurring when ordinary baking powder is added to cakes, bread, and biscuits).

The three astronauts aboard *Apollo 13*, which was launched in 1970 on April 11 and returned to Earth on April 17, were kept alive during their mission, in part, because of lithium hydroxide (LiOH) canisters that were designed to remove exhaled CO_2 from the air. Solid lithium hydroxide reacts with $CO_2(g)$ to give solid Li_2CO_3 and water. With the assumption that an astronaut exhales approximately 1.00 kg CO_2 per day, what mass of LiOH was required to remove all of the CO_2 exhaled by the three-member crew on their six-day mission?

Chalkboard chalk is made from calcium carbonate and calcium sulfate, with minor impurities such as SiO_2. Only the $CaCO_3$ reacts with dilute HCl(aq). What is the mass percent $CaCO_3$ in a piece of chalk if a 3.28 g sample yields 0.981 g $CO_2(g)$?

$$CaCO_3(s) + 2\,HCl(aq) \longrightarrow CaCl_2(aq) + H_2O(l) + CO_2(g)$$

Hydrogen gas, $H_2(g)$, is passed over $Fe_2O_3(s)$ at 400 °C. Water vapor is formed together with a black residue—a compound consisting of 72.3% Fe and 27.7% O. Write a balanced equation for this reaction.

A sulfide of iron, containing 36.5% S by mass, is heated in $O_2(g)$, and the products are sulfur dioxide and an oxide of iron containing 27.6% O, by mass. Write a balanced chemical equation for this reaction.

Water and ethanol, $CH_3CH_2OH(l)$, are miscible, that is, they can be mixed in all proportions. However, when these liquids are mixed, the total volume of the resulting solution is not equal to the sum of the pure liquid volumes, and we say that the volumes are not additive. For example, when 50.0 mL of water and 50.0 mL of $CH_3CH_2OH(l)$, are mixed at 20 °C, the total volume of the solution is 96.5 mL, not 100.0 mL. (The volumes are not additive because the interactions and packing of water molecules are slightly different from the interactions and packing of CH_3CH_2OH molecules.) Calculate the molarity of CH_3CH_2OH in a solution prepared by mixing 50.0 mL of water and 50.0 mL of $CH_3CH_2OH(l)$ at 20 °C. At this temperature, the densities of water and ethanol are 0.99821 g/mL and 0.7893 g/mL, respectively.

100. When water and methanol, $CH_3OH(l)$, are mixed, the total volume of the resulting solution is not equal to the sum of the pure liquid volumes. (Refer to Exercise 99 for an explanation.) When 72.061 g H_2O and 192.25 g CH_3OH are mixed at 25 °C, the resulting solution has a density of 0.86070 g/mL. At 25 °C, the densities of water and methanol are 0.99705 g/mL and 0.78706 g/mL, respectively.
(a) Calculate the volumes of the pure liquid samples and the solution, and show that the pure liquid volumes are not additive. [*Hint*: Although the volumes are not additive, the masses are.]
(b) Calculate the molarity of CH_3OH in this solution.

101. What volume of 0.149 M HCl must be added to 1.00×10^2 mL of 0.285 M HCl so that the resulting solution has a molarity of 0.205 M? Assume that the volumes are additive.

102. What volume of 0.0175 M CH_3OH must be added to 50.0 mL of 0.0248 M CH_3OH so that the resulting solution has a molarity of exactly 0.0200 M? Assume that the volumes are additive.

103. What is the molarity of NaCl(aq) if a solution has 1.52 ppm Na? Assume that NaCl is the only source of Na and that the solution density is 1.00 g/mL. (The unit *ppm* is parts per million; here it can be taken to mean g Na per million grams of solution.)

104. How many milligrams $Ca(NO_3)_2$ must be present in 50.0 L of a solution containing 2.35 ppm Ca? [*Hint:* See also Exercise 103.]

105. A drop (0.05 mL) of 12.0 M HCl is spread over a sheet of thin aluminum foil. Assume that all the acid reacts with, and thus dissolves through, the foil. What will be the area, in cm^2, of the cylindrical hole produced? (Density of Al = 2.70 g/cm^3; foil thickness = 0.10 mm.)

$$2\,Al(s) + 6\,HCl(aq) \longrightarrow 2\,AlCl_3(aq) + 3\,H_2(g)$$

106. A small piece of zinc is dissolved in 50.00 mL of 1.035 M HCl. At the conclusion of the reaction, the concentration of the 50.00 mL sample is redetermined and found to be 0.812 M HCl. What must have been the mass of the piece of zinc that dissolved?

$$Zn(s) + 2\,HCl(aq) \longrightarrow ZnCl_2(aq) + H_2(g)$$

107. How many milliliters of 0.715 M NH_4NO_3 solution must be diluted with water to produce 1.00 L of a solution with a concentration of 2.37 mg N/mL?

108. A seawater sample has a density of 1.03 g/mL and 2.8% NaCl by mass. A saturated solution of NaCl in water is 5.45 M NaCl. How many liters of water would have to be evaporated from 1.00×10^6 L of the seawater before NaCl would begin to crystallize? (A saturated solution contains the maximum amount of dissolved solute possible.)

109. A 99.8 mL sample of a solution that is 12.0% KI by mass (d = 1.093 g/mL) is added to 96.7 mL of another solution that is 14.0% $Pb(NO_3)_2$ by mass (d = 1.134 g/mL). How many grams of PbI_2 should form?

$$Pb(NO_3)_2(aq) + 2\,KI(aq) \longrightarrow PbI_2(s) + 2\,KNO_3(aq)$$

110. Solid calcium carbonate, $CaCO_3(s)$, reacts with $HCl(aq)$ to form H_2O, $CaCl_2(aq)$, and $CO_2(g)$. If a 45.0 g sample of $CaCO_3(s)$ is added to 1.25 L of $HCl(aq)$ that is 25.7% HCl by mass ($d = 1.13$ g/mL), what will be the molarity of HCl in the solution after the reaction is completed? Assume that the solution volume remains constant.

111. A 2.05 g sample of an iron–aluminum alloy (ferroaluminum) is dissolved in excess HCl(aq) to produce 0.105 g $H_2(g)$. What is the percent composition, by mass, of the ferroaluminum?

$$Fe(s) + 2\,HCl(aq) \longrightarrow FeCl_2(aq) + H_2(g)$$
$$2\,Al(s) + 6\,HCl(aq) \longrightarrow 2\,AlCl_3(aq) + 3\,H_2(g)$$

112. A 0.155 g sample of an Al–Mg alloy reacts with an excess of HCl(aq) to produce 0.0163 g H_2. What is the percent Mg in the alloy?
[*Hint:* Write equations similar to (4.2).]

113. In a dilute nitric acid solution, copper reacts to form copper nitrate, nitrogen monoxide, and water according to the following equation:

$$3\,Cu(s) + 8\,HNO_3(aq) \longrightarrow 3\,Cu(NO_3)_2(aq) + 2\,NO(g) + 4\,H_2O(l)$$

In a vessel, 800.0 mL of 0.500 M HNO_3 is added to 3.177 g of copper, and the vessel is then sealed. After an hour, the vessel is opened and drained, and the remaining copper rinsed, dried, and weighed. The remaining mass of the copper was 0.0739 g.
(a) Calculate the extent of the reaction.
(b) Calculate the moles of NO generated by the reaction and the moles of HNO_3 remaining.

114. The following chemical equation represents the decomposition of hydrogen peroxide, H_2O_2.

$$2\,H_2O_2 \longrightarrow O_2 + 2\,H_2O$$

If the reaction started with 8.67 g of pure H_2O_2 and produced 3.74 g of O_2, what is the extent of reaction, ξ, and what percentage of the H_2O_2 reacted?

115. An organic liquid is either methyl alcohol (CH_3OH), ethyl alcohol (CH_3CH_2OH), or a mixture of the two. A 0.220 g sample of the liquid is burned in an excess of $O_2(g)$ and yields 0.352 g $CO_2(g)$. Is the liquid a pure alcohol or a mixture of the two?

116. The manufacture of ethyl alcohol, CH_3CH_2OH, yields diethyl ether, $(C_2H_5)_2O$ as a by-product. The complete combustion of a 1.005 g sample of the product of this process yields 1.963 g CO_2. What must be the mass percents of CH_3CH_2OH and of $(C_2H_5)_2O$ in this sample?

117. A mixture contains only $CuCl_2$ and $FeCl_3$. A 0.7391 g sample of the mixture is completely dissolved in water and then treated with $AgNO_3(aq)$. The following reactions occur.

$$CuCl_2(aq) + 2\,AgNO_3(aq) \longrightarrow 2\,AgCl(s) + Cu(NO_3)_2(aq)$$
$$FeCl_3(aq) + 3\,AgNO_3(aq) \longrightarrow 3\,AgCl(s) + Fe(NO_3)_3(aq)$$

If it takes 86.91 mL of 0.1463 M $AgNO_3$ solution to precipitate all the chloride as AgCl, then what is the percentage by mass of copper in the mixture?

118. Under appropriate conditions, copper sulfate, [cut off] sium chromate, and water react to form a produc[cut off] taining Cu^{2+}, CrO_4^{2-}, and OH^- ions. Analysis [cut off] compound yields 48.7% Cu^{2+}, 35.6% CrO_4^{2-} [cut off] 15.7% OH^-.
(a) Determine the empirical formula of the comp[cut off]
(b) Write a plausible equation for the reaction.

119. Write a chemical equation to represent the com[cut off] combustion of malonic acid, a compound [cut off] 34.62% C, 3.88% H, and 61.50% O, by mass.

120. Aluminum metal and iron(III) oxide react t[cut off] aluminum oxide and iron metal. What is the [cut off] mum mass of iron that can be obtained from a [cut off] tion mixture containing 2.5 g of aluminum and [cut off] of iron(III) oxide. What mass of the excess rea[cut off] remains?

121. Silver nitrate is a very expensive chemical. For [cut off] ticular experiment, you need 100.0 mL of 0.07[cut off] $AgNO_3$, but only 60 mL of 0.0500 M $AgNO_3$ is [cut off] able. You decide to pipet exactly 50.00 mL [cut off] solution into a 100.0 mL flask, add an appro[cut off] mass of $AgNO_3$, and then dilute the resulting [cut off] tion to exactly 100.0 mL. What mass of $AgNO_3$ [cut off] you use?

122. When sulfur (S_8) and chlorine are mixed in a rea[cut off] vessel, disulfur dichloride is the sole product[cut off] starting mixture below is represented by y[cut off] spheres for the S_8 molecules and green spher[cut off] the chlorine molecules.

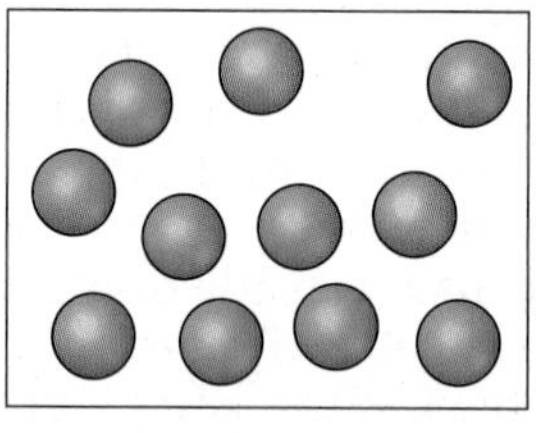

Which of the following is (are) a valid repres[cut off] tion(s) of the contents of the reaction vessel after [cut off] disulfur dichloride (represented by red spheres[cut off] formed?

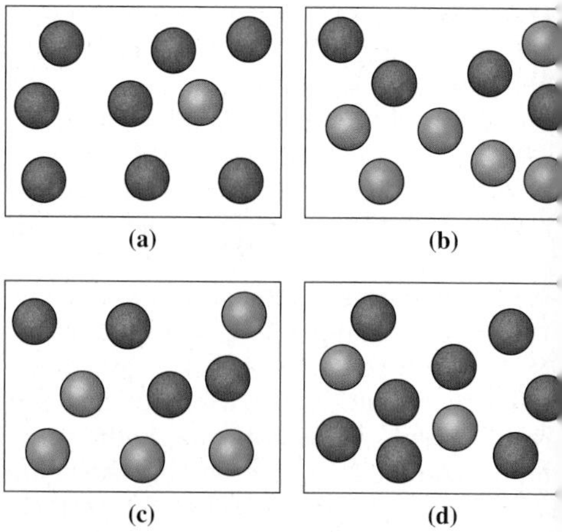

123. A method for eliminating oxides of nitrogen [cut off] NO_2) from automobile exhaust gases is to pas[cut off] exhaust gases over solid cyanuric acid, C_3N_3(O[cut off] When the hot exhaust gases come in contact [cut off] cyanuric acid, solid $C_3N_3(OH)_3$ decomposes[cut off]

isocyanic acid vapor, HNCO(g), which then reacts with NO_2 in the exhaust gases to give N_2, CO_2, and H_2O. How many grams of $C_3N_3(OH)_3$ are needed per gram of NO_2 in this method?
[*Hint*: To balance the equation for reaction between HNCO and NO_2, balance with respect to each kind of atom in this order: H, C, O, and N.]

For a specific reaction, ammonium dichromate is the only reactant and chromium(III) oxide and water are two of the three products. The third product contains only one type of atom. What is the third product and how many grams of this product are produced per kilogram of ammonium dichromate decomposed?

It is desired to produce as large a volume of 1.25 M urea [$CO(NH_2)_2$(aq)] as possible from these three sources: 345 mL of 1.29 M $CO(NH_2)_2$, 485 mL of 0.653 M $CO(NH_2)_2$, and 835 mL of 0.775 M $CO(NH_2)_2$. How can this be done? What is the maximum volume of this solution obtainable?

The mineral ilmenite, $FeTiO_3$, is an important source of titanium dioxide for use as a white pigment. In the first step in its conversion to titanium dioxide, ilmenite is treated with sulfuric acid and water to form $TiOSO_4$ and iron(II) sulfate heptahydrate. Titanium dioxide is obtained in two subsequent steps. How many kilograms of iron(II) sulfate heptahydrate are produced for every 1.00×10^3 kg of ilmenite processed?

Refer to Exercise 126. Iron(II) sulfate heptahydrate formed in the processing of ilmenite ore cannot be released into the environment. Its further treatment involves dehydration by heating to produce anhydrous iron(II) sulfate. Upon further heating, the iron(II) sulfate decomposes to iron(III) oxide, and sulfur dioxide and oxygen gases. The iron(III) oxide is used in the production of iron and steel. How many kilograms of iron(III) oxide are obtained for every 1.00×10^3 kg of iron(II) sulfate heptahydrate?

Melamine, $C_3N_3(NH_2)_3$, is used in adhesives and resins. It is manufactured in a two-step process in which urea, $CO(NH_2)_2$, is the sole starting material, isocyanic acid (HNCO) is an intermediate, and ammonia and carbon dioxide gases are by-products.
(a) Write a balanced equation for the overall reaction.
(b) What mass of melamine will be obtained from 100.0 kg of urea if the yield of the overall reaction is 84%?

Acrylonitrile is used in the production of synthetic fibers, plastics, and rubber goods. It can be prepared from propylene (propene), ammonia, and oxygen in the reaction illustrated below.
(a) Write a balanced chemical equation for this reaction.
(b) The actual yield of the reaction is 0.73 kg acrylonitrile per kilogram of propylene. What is the minimum mass of ammonia required to produce 1.00 metric ton (1000 kg) of acrylonitrile?

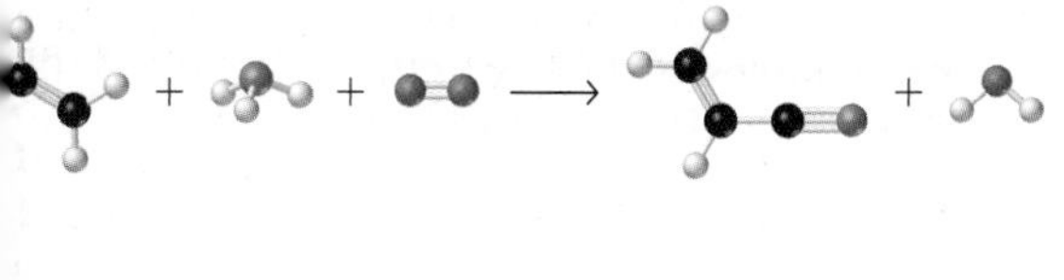

130. A fundamental principle in *green chemistry* is atom economy (AE). AE is a measure of how many atoms from the starting materials are incorporated into the desired product. For example, if a reaction incorporates all the reactant atoms into the product of interest, the reaction has a percent AE of 100%. To obtain percent AE for a reaction, we calculate the mass of the desired product that can be formed from a stoichiometric mixture of reactants, and compare this mass with the total mass of that reaction mixture. (In a stoichiometric mixture of reactants, none of the reactants are present in excess; the mole amounts are in the same ratio as the stoichiometric coefficients).

$$\%\ \text{AE} = \frac{\text{mass } (m'_{\text{P}}) \text{ of the desired product P}}{\text{total mass of a stoichiometric mixture of reactants}} \times 100$$

The prime (′) on the symbol for the mass of the desired product serves to remind us that this mass is calculated for a stoichiometric mixture of reactants. Use the definition above to calculate the percent AE for the following reactions, both of which can be used to make $C_6H_5NH_2$, the desired product.

$$C_6H_6 + (CH_3)_3SiN_3 + 2\,F_3CSO_3H + NaOH \longrightarrow C_6H_5NH_2 + N_2 + (CH_3)_3SiOSO_2CF_3 + NaF_3CSO_3 + H_2O$$

$$C_6H_6 + HNO_3 + 3\,H_2 \longrightarrow C_6H_5NH_2 + 3\,H_2O$$

131. The industrial production of hydrazine (N_2H_4) by the Raschig process is the topic of the Focus On feature for Chapter 4 on www.masteringchemistry.com. The following chemical equation represents the overall process, which actually involves three consecutive reactions.

$$2\,NH_3(aq) + Cl_2(g) + 2\,NaOH(aq) \longrightarrow N_2H_4(aq) + 2\,NaCl(aq) + 2\,H_2O(l)$$

(a) Use the definition of percent atom economy (AE) from exercise 130 to calculate, to the nearest percent, the percent AE for the Raschig process.
(b) Propose a reaction for the synthesis of N_2H_4 that has percent AE of 100%.

132. It is often difficult to determine the concentration of a species in solution, particularly if it is a biological species that takes part in complex reaction pathways. One way to do this is through a dilution experiment with labeled molecules. Instead of molecules, however, we will use fish.

An angler wants to know the number of fish in a particular pond, and so puts an indelible mark on 100 fish and adds them to the pond's existing population. After waiting for the fish to spread throughout the pond, the angler starts fishing, eventually catching 18 fish. Of these, five are marked. What is the total number of fish in the pond?

Feature Problems

133. Lead nitrate and potassium iodide react in aqueous solution to form a yellow precipitate of lead iodide. In one series of experiments, the masses of the two reactants were varied, but the *total* mass of the two was held constant at 5.000 g. The lead iodide formed was filtered from solution, washed, dried, and weighed. The table gives data for a series of reactions.

Experiment	Mass of Lead Nitrate, g	Mass of Lead Iodide, g
1	0.500	0.692
2	1.000	1.388
3	1.500	2.093
4	3.000	2.778
5	4.000	1.391

(a) Plot the data in a graph of mass of lead iodide versus mass of lead nitrate, and draw the appropriate curve(s) connecting the data points. What is the maximum mass of precipitate that can be obtained? **(b)** Explain why the maximum mass of precipitate is obtained when the reactants are in their stoichiometric proportions. What are these stoichiometric proportions expressed as a mass ratio, and as a mole ratio? **(c)** Show how the stoichiometric proportions determined in part (b) are related to the balanced equation for the reaction.

134. Baking soda, $NaHCO_3$, is made from soda ash, a common name for sodium carbonate. The soda ash is obtained in two ways. It can be manufactured in a process in which carbon dioxide, ammonia, sodium chloride, and water are the starting materials. Alternatively, it is mined as a mineral called *trona* (top photo). Whether the soda ash is mined or manufactured, it is dissolved in water and carbon dioxide is bubbled through the solution. Sodium bicarbonate precipitates from the solution.

As a chemical analyst you are presented with two samples of sodium bicarbonate—one from the manufacturing process and the other derived trona. You are asked to determine which is p and are told that the impurity is sodium carbo You decide to treat the samples with just suffi hydrochloric acid to convert all the sodium ca ate and bicarbonate to sodium chloride, ca dioxide, and water. You then precipitate silver ride in the reaction of sodium chloride with s nitrate. A 6.93 g sample of baking soda derived trona gave 11.89 g of silver chloride. A 6.78 g sa from manufactured sodium carbonate gave 11 of silver chloride. Which sample is purer, th which has the greater mass percent $NaHCO_3$?

John Cancalosi/National Geographic Society/Corbis

Trona $Na_2CO_3 \cdot NaHCO_3 \cdot 2\,H_2O$

Robert Mathena/Fundamental Photographs

Baking soda $NaHCO_3$

Self-Assessment Exercises

135. In your own words, define or explain these terms or symbols.
(a) $\xrightarrow{\Delta}$ **(b)** (aq) **(c)** stoichiometric coefficient **(d)** overall equation

136. Briefly describe **(a)** balancing a chemical equation; **(b)** preparing a solution by dilution; **(c)** determining the limiting reactant in a reaction.

137. Explain the important distinctions between **(a)** chemical formula and chemical equation; **(b)** stoichiometric coefficient and stoichiometric factor; **(c)** solute and solvent; **(d)** actual yield and percent yield; **(e)** consecutive and simultaneous reactions.

138. When the equation below is balanced, the corre of stoichiometric coefficients is **(a)** 1, 6 ⟶ 1, 3 **(b)** 1, 4 ⟶ 1, 2, 2; **(c)** 2, 6 ⟶ 2, 3, 2; **(d)** 3, 8 ⟶ 3, 4, 2.

$$? \,Cu(s) + ?\, HNO_3(aq) \longrightarrow ?\, Cu(NO_3)_2(aq) + ?\, H_2O(l) + ?\, N$$

139. A reaction mixture contains 1.0 mol $CaCN_2$ (cal cyanamide) and 1.0 mol H_2O. The maximum n ber of moles of NH_3 produced is **(a)** 3.0; **(b)** **(c)** between 1.0 and 2.0; **(d)** less than 1.0.

$$CaCN_2(s) + 3\,H_2O(l) \longrightarrow CaCO_3(s) + 2\,NH$$

Consider the chemical equation below. What is the maximum number of moles of K_2SO_4 that can be obtained from a reaction mixture containing 5.0 moles each of $KMnO_4$, KI, and H_2SO_4? **(a)** 3.0 mol; **(b)** 3.8 mol; **(c)** 5.0 mol; **(d)** 6.0 mol; **(e)** 15 mol.

$$2\ KMnO_4 + 10\ KI + 8\ H_2SO_4 \longrightarrow 6\ K_2SO_4 + 2\ MnSO_4 + 5\ I_2 + 8\ H_2O$$

In the decomposition of silver carbonate to form metallic silver, carbon dioxide gas, and oxygen gas, **(a)** one mol of oxygen gas is formed for every 2 mol of carbon dioxide gas; **(b)** 2 mol of silver metal is formed for every 1 mol of oxygen gas; **(c)** equal numbers of moles of carbon dioxide and oxygen gases are produced; **(d)** the same number of moles of silver metal are formed as of the silver carbonate decomposed.

To obtain a solution that is 1.00 M $NaNO_3$, you should prepare **(a)** 1.00 L of aqueous solution containing 100 g $NaNO_3$; **(b)** 1 kg of aqueous solution containing 85.0 g $NaNO_3$; **(c)** 5.00 L of aqueous solution containing 425 g $NaNO_3$; **(d)** an aqueous solution containing 8.5 mg $NaNO_3$/mL.

What is the volume (in mL) of 0.160 M KNO_3 that must be added to 200.0 mL of 0.240 M K_2SO_4 to produce a solution having $[K^+] = 0.400$ M?

To prepare a solution that is 0.50 M KCl starting with 100.0 mL of 0.40 M KCl, you should **(a)** add 20.0 mL of water; **(b)** add 0.075 g KCl; **(c)** add 0.10 mol KCl; **(d)** evaporate 20.0 mL of water.

An aqueous solution that is 5.30% LiBr by mass has a density of 1.040 g/mL. What is the molarity of this solution? **(a)** 0.563 M; **(b)** 0.635 M; **(c)** 0.0635 M; **(d)** 0.0563 M; **(e)** 12.0 M.

In the reaction of 2.00 mol CCl_4 with an excess of HF, 1.70 mol CCl_2F_2 is obtained.

$$CCl_4 + 2\ HF \longrightarrow CCl_2F_2 + 2\ HCl$$

(a) The theoretical yield is 1.70 mol CCl_2F_2.
(b) The theoretical yield is 1.00 mol CCl_2F_2.
(c) The theoretical yield depends on how large an excess of HF is used.
(d) The percent yield is 85%.

Consider the reaction $2\ Fe_2O_3 + 3\ C \longrightarrow 4\ Fe + 3\ CO_2$. What is the maximum mass of Fe that can be obtained from a reaction mixture containing 18.0 g Fe_2O_3 and 2.5 g C?

148. A 26.4 g sample of a mixture of NaOH and CaO contains 40.0% CaO. When the sample is treated with aqueous HCl, the following reactions occur:

$$NaOH(s) + HCl(aq) \longrightarrow NaCl(aq) + H_2O(l)$$
$$CaO(s) + 2\ HCl(aq) \longrightarrow CaCl_2(aq) + H_2O(l)$$

All of the mixture reacts, and no HCl is left over. The resulting solution is evaporated to dryness. What is the mass (in grams) of solid obtained?

149. The incomplete combustion of gasoline produces CO(g) as well as CO_2(g). Write an equation for **(a)** the complete combustion of the gasoline component octane, C_8H_{18}(l), and **(b)** incomplete combustion of octane with 25% of the carbon appearing as CO(g).

150. The minerals calcite, $CaCO_3$, magnesite, $MgCO_3$, and dolomite, $CaCO_3 \cdot MgCO_3$, decompose when strongly heated to form the corresponding metal oxide(s) and carbon dioxide gas. A 1.000 g sample known to be one of the three minerals was strongly heated and 0.477 g CO_2 was obtained. Which of the three minerals was it?

151. A 1.000 g sample of a mixture of CH_4 and C_2H_6 is analyzed by burning it completely in O_2, yielding 2.776 g CO_2. What is the percentage by mass of CH_4 in the mixture? **(a)** 93%; **(b)** 82%; **(c)** 67%; **(d)** 36%; **(e)** less than 36%.

152. Nitric acid, HNO_3, can be manufactured from ammonia, NH_3, by using the three reactions shown below.

Step 1: $4\ NH_3(g) + 5\ O_2(g) \rightarrow 4\ NO(g) + 6\ H_2O(l)$
Step 2: $2\ NO(g) + O_2(g) \rightarrow 2\ NO_2(g)$
Step 3: $3\ NO_2(g) + H_2O(l) \rightarrow 2\ HNO_3(aq) + NO(g)$

What is the maximum number of moles of HNO_3 that can be obtained from 4.00 moles of NH_3? (Assume that the NO produced in step 3 is not recycled back into step 2.) **(a)** 1.33 mol; **(b)** 2.00 mol; **(c)** 2.67 mol; **(d)** 4.00 mol; **(e)** 6.00 mol.

153. For each of the following compounds, write a balanced chemical equation for forming the compound from its elements. What is the percent atom economy in each case? **(a)** $RbBrO_4$; **(b)** H_2SO_4; **(c)** $Mg(ClO_3)_2$; **(d)** $NaNO_2$.

154. Appendix E describes a useful study aid known as concept mapping. Using the method presented in Appendix E, construct a concept map relating the topics found in Sections 4-3, 4-4, and 4-5.

5 Introduction to Reactio in Aqueous Solutions

CONTENTS

LEARNING OBJECTIVES

5.1 Distinguish between an electrolyte and a nonelectrolyte, and provide examples of each.

5.2 Use the solubility guidelines for common ionic solids to determine whether a precipitate forms in a given reaction in solution.

5.3 Identify the common strong acids and bases, and write chemical equations for reactions involving acids and bases.

5.4 Identify the oxidation states of all elements or ions in a reaction, and determine whether a redox process occurs.

5.5 Determine the balanced equation for a redox reaction by following the half-equation method in acidic and basic solutions.

5.6 Distinguish between oxidizing and reducing agents.

5.7 Use titration data to calculate the molarity of a solution.

When clear, colorless aqueous solutions of cobalt(II) chloride and sodium hydrox are mixed, a blue cloud of solid cobalt(II) hydroxide is formed. Such precipitation reactions are one of the three types of reactions considered in this chapter.

Most reactions in the general chemistry laboratory are carried o aqueous solutions—solutions for which water is the solv Aqueous solutions provide a convenient way of bringing toge accurately measured amounts of reactants, and, not surprisingly, aqu solutions feature prominently in many methods of chemical analysi this chapter, we will explore three different classes of reactions that occ aqueous solutions—precipitation, acid–base, and oxidation–reduc reactions—with the goal of understanding the nature of the substa involved, the changes that occur in these substances, and the way each r tion can be used in the laboratory for analyzing samples.

The Nature of Aqueous Solutions

try to form a mental image of an aqueous solution at the molecular level. r is the solvent in an aqueous solution, and our mental image of water might something like Figure 5-1(a). For an aqueous solution, solute particles— cules or ions—are present in much smaller number and are randomly dis- ted among the water molecules, as depicted in Figure 5-1(b).

cause we will encounter aqueous solutions of ions throughout this chapter, useful to examine the nature of such solutions in a bit more detail. An impor- characteristic of an aqueous solution of ions is that it will conduct electricity, ided the concentration of ions is not too low. An aqueous solution of ions ucts electricity because the ions move essentially independently of each r, each one carrying a certain quantity of charge. (In a metallic conductor, as copper or tungsten, electrons carry the charge.) The manner in which conduct electric current is suggested by Figure 5-2.

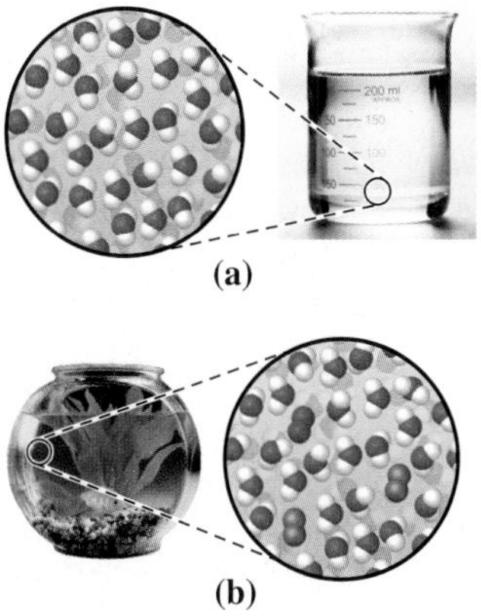

▲ FIGURE 5-1
Molecular view of water and an aqueous solution of air
(a) Water molecules (red and white) are in close proximity in liquid water. **(b)** Dissolved oxygen (red) and nitrogen (blue) molecules are far apart, separated by water molecules.

(a) Katrina Leigh/Shutterstock,
(b) C Squared Studios/Photodisc/Getty Images

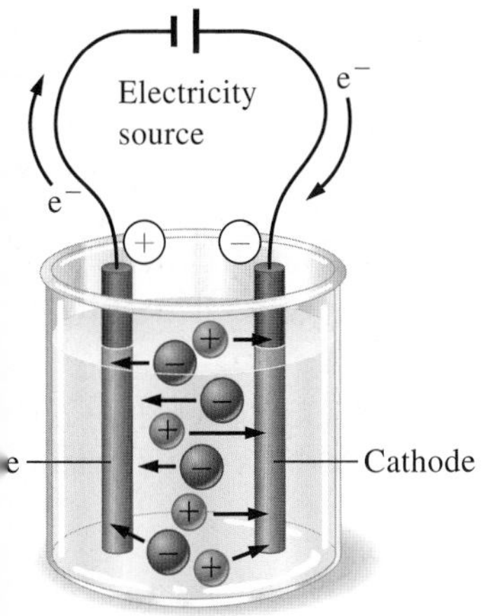

◀ FIGURE 5-2
Conduction of electricity through a solution
Two graphite rods called electrodes are placed in a solution. The external source of electricity pulls electrons from one rod and forces them onto the other, creating a positive charge on one electrode and a negative charge on the other (right). In the solution, positive ions (cations) are attracted to the negative electrode, the *cathode*; negative ions (anions) are attracted to the positive electrode, the *anode*. Thus, electric charge is carried through the solution by the migration of ions.

◀ The distinction between strong and weak electrolytes lies in their tendencies to provide ions in solution: A strong electrolyte has a strong (or high) tendency to provide ions; a weak electrolyte has a weak (or low) tendency to provide ions.

hether or not an aqueous solution is a conductor of electricity depends on nature of the solute(s). Pure water contains so few ions that it does not con- electric current. However, some solutes produce ions in solution, thereby ing the solution an electrical conductor. Solutes that provide ions when olved in water are called **electrolytes**. Solutes that that do not provide ions ater are called **nonelectrolytes**. All electrolytes provide ions in water but all electrolytes are equal in their tendencies for providing ions. A **strong trolyte** is a substance that is essentially *completely ionized* in aqueous solu- essentially all of the dissolved solute exists as ions. A **weak electrolyte** is *partially ionized* in aqueous solution: only some of the dissolved solute is erted into ions. One scheme for classifying solutes is summarized in re 5-3.

KEEP IN MIND

that the electrical conductivity of a solution depends on two factors: (1) the total concentration of the electrolyte, and (2) the extent to which the electrolyte dissociates into ions. For example, a 0.001 M HCl(aq) solution will conduct electricity, but a 1×10^{-6} M HCl(aq) solution will not, even though every HCl molecule ionizes to produce H^+ and Cl^- ions in both solutions.

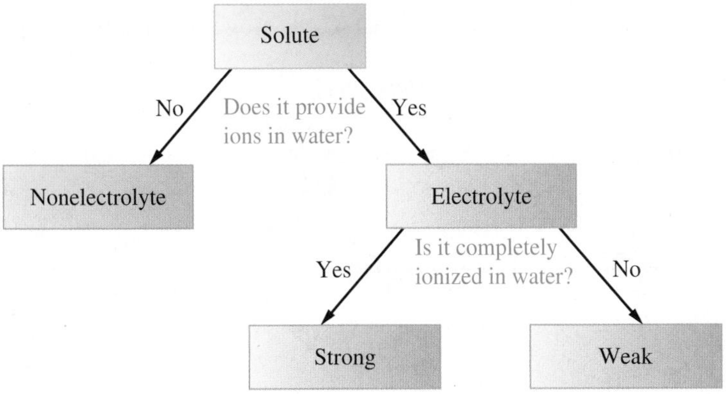

◀ FIGURE 5-3
A classification scheme for solutes

With the apparatus depicted in Figure 5-4, we can detect the presen ions in an aqueous solution by measuring how well the solution conc electricity. We can make one of three possible observations:

- *The lamp fails to light up* (Fig. 5-4a). Conclusion: no ions are preser if some are present, their concentration is extremely low). The sol is either a solution of a nonelectrolyte or a very dilute solution electrolyte. Methanol, CH_3OH, is an example of a solute that doe provide ions in water; methanol is a *nonelectrolyte*. The microsc view in Figure 5-4(a) is for an aqueous solution of methanol, ar this view we see that none of the CH_3OH molecules are ion in water.
- *The lamp lights up brightly* (Fig. 5-4b). Conclusion: the concentrati ions in solution is high. The solute is a *strong electrolyte*. Magnesium ride, $MgCl_2$, is an ionic compound that is completely ionized in w The microscopic view in Figure 5-4(b) shows that an aqueous soluti $MgCl_2$ consists of Mg^{2+} and Cl^- ions in the solvent.
- *The lamp lights up only dimly* (Fig. 5-4c). Conclusion: ions are prese solution but the concentration of ions is low. The solution could solution of a weak electrolyte, such as acetic acid (CH_3COOH), could be a dilute—but not too dilute—solution of a strong electro The microscopic view in Figure 5-4(c) is for an aqueous solutic CH_3COOH and it shows that, in water, only some of the CH_3C molecules are ionized. An aqueous solution of CH_3COOH is only a conductor of electricity.

KEEP IN MIND

that the solvent molecules are densely packed. In such diagrams as Figure 5-4, we will often show the solvent as a uniformly colored background and depict only the solute particles.

The following generalizations are helpful when deciding whether a pa ular solute in an aqueous solution is most likely to be a nonelectroly strong electrolyte, or a weak electrolyte.

- Essentially all soluble ionic compounds and only a relatively few molecula compounds are strong electrolytes.
- Most molecular compounds are either nonelectrolytes or weak electrolytes.

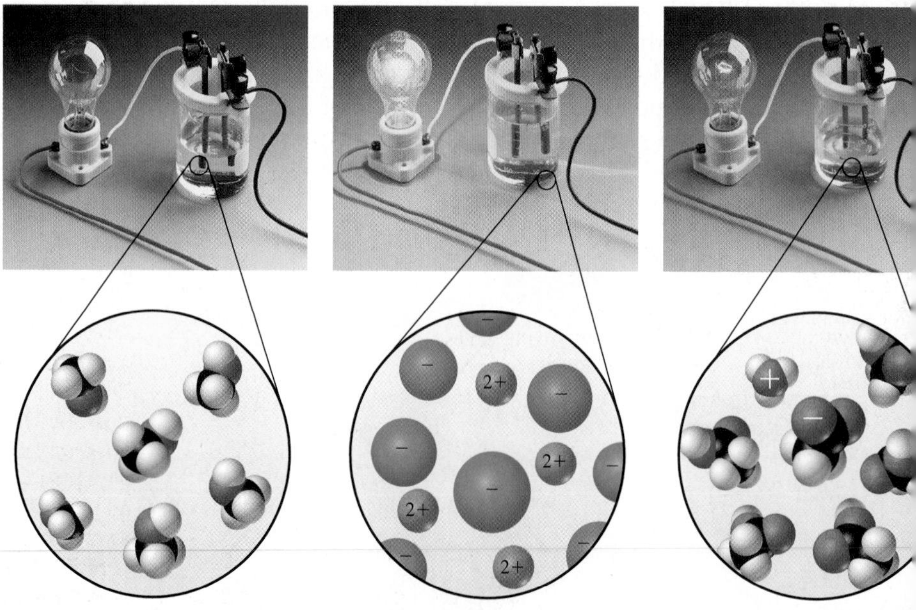

(a) (b) (c)

▶ FIGURE 5-4
Three types of electrolytes
In **(a)**, there are no ions present to speak of—only CH_3OH, molecules. Methanol (methyl alcohol), CH_3OH, is a *nonelectrolyte* in aqueous solutions. In **(b)**, the solute, $MgCl_2$, is present almost entirely as individual ions. $MgCl_2$ is a *strong electrolyte* in aqueous solutions. In **(c)**, although most of the solute is present as CH_3COOH molecules, a small fraction of the molecules ionize. CH_3COOH is a *weak electrolyte* in aqueous solution. The CH_3COOH molecules that ionize produce acetate ions CH_3COO^- and H^+ ions, and the H^+ ions attach themselves to water molecules to form hydronium ions, H_3O^+.

ociation and Ionization

is section, we investigate the processes by which electrolytes produce in solution, and by considering the extent to which these processes occur, ecide on the best way to represent the aqueous solutions depicted in re 5-4.

ı electrolyte produces ions in solution by one of two processes, namely distion or ionization. Although these two terms are often used interchangeably, do have slightly different meanings. **Dissociation** refers to the separation of tity into two or more entities, whereas **ionization** refers to the generation of r more ions. In short, an ionic compound produces ions in solution by distion. On the other hand, if a molecular compound produces ions in solution, y ionization. Let's explore the differences between these two processes in a more detail by describing the dissociation of $MgCl_2$, an ionic compound, he ionization of CH_3COOH, a molecular compound, in water to produce queous solutions depicted, respectively, in Figures 5-4(b) and 5-4(c).

ı ionic compound consists of positive and negative ions, which are typiarranged in a regular, repeating pattern as suggested by Figure 5-5. As ompound dissolves, the ions separate from one another and become surded by water molecules. An ion that is surrounded by water molecules is to be *hydrated* or *aquated*. The dissolution of an ionic solid to provide ated ions is appropriately called dissociation because the ions are initially ent in the solid, and they become separated (dissociated) from one her as the solid dissolves.

◀ We discussed some features of molecular and ionic compounds in Section 3-1.

e complete dissociation of $MgCl_2$, an ionic compound, is represented by the wing equation. The use of a right arrow ($\longrightarrow$) in this equation is appropriate use the dissociation of $MgCl_2$ proceeds essentially to completion.

$$MgCl_2(aq) \longrightarrow Mg^{2+}(aq) + 2\,Cl^-(aq) \qquad \textbf{(5.1)}$$

equation means that, in the presence of water, formula units of $MgCl_2$ letely dissociate into the separate ions. Therefore, the best representation $gCl_2(aq)$ is $Mg^{2+}(aq) + 2\,Cl^-(aq)$.

r molecular compounds, the situation is somewhat more complicated use no ions are initially present. Whenever a molecular compound provides in solution, the ions are generated by a reaction of the compound with r. The following equation represents the ionization of CH_3COOH in water:

$$CH_3COOH(aq) + H_2O(l) \rightleftharpoons H_3O^+(aq) + CH_3COO^-(aq) \qquad \textbf{(5.2)}$$

double arrow $\rightleftharpoons$ indicates that the reaction proceeds to a limited extent not to completion. For this reason, CH_3COOH is classified as a weak rolyte.

KEEP IN MIND

that a molecule containing the –COOH group is a carboxylic acid. When a carboxylic acid ionizes, it is the bond between the H and O atoms of the –COOH group that ionizes.

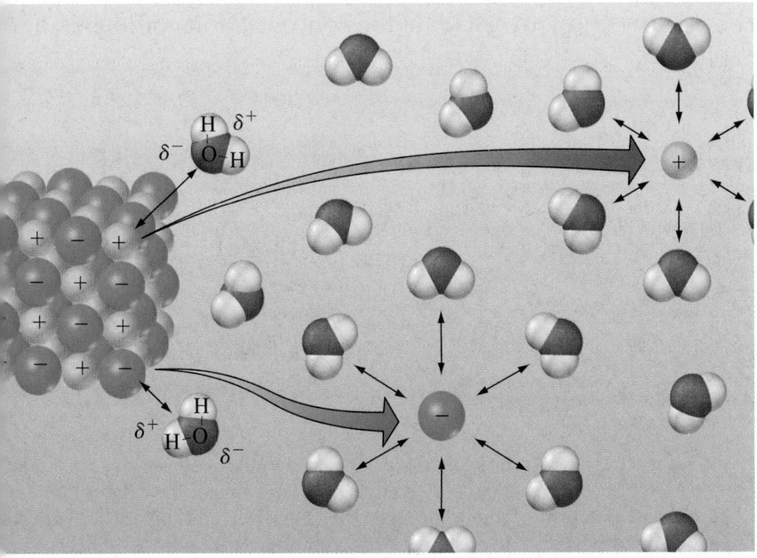

◀ FIGURE 5-5
An ionic compound dissolving in water
An ionic compound is typically a solid consisting of positive and negative ions arranged in a regular, repeating pattern. The ions are already present in the solid structure, and the dissolution of the solid involves the dissociation (separation) of the ions from one another. The clustering of water molecules around the ions and the formation of hydrated ions are key parts of the dissolution process. Water molecules are able to stabilize both positive and negative ions because the oxygen atom in a water molecule is electron rich (and thus slightly negative, δ^-) whereas the hydrogen atoms are electron poor (and thus slightly positive, δ^+).

In a solution of CH_3COOH, the relative proportions of ionized and no ized forms of CH_3COOH remain fixed but with the CH_3COOH molecul predominant species. The reason the ionization of CH_3COOH occurs to a ited extent only is that the reaction shown in equation (5.2) is highly rever some of the H_3O^+ and CH_3COO^- ions, once formed, recombine to CH_3COOH and H_2O molecules. If we could watch a particular CH_3COO g for an extended time, we would sometimes see it as the CH_3COO^- ion, but of the time it would be in a molecule, CH_3COOH. Thus, a solutic CH_3COOH is best represented as $CH_3COOH(aq)$, *not* $H_3O^+(aq) + CH_3COO$ or $H^+(aq) + CH_3COO^-(aq)$.

▶ The International Union of Pure and Applied Chemistry (IUPAC) has recommended that the H_3O^+ ion be referred to as the *oxonium* ion. However, this recommendation has not yet been universally adopted by chemists.

The H_3O^+ ion appearing in equation (5.2) is called a *hydronium ion*, a consists of a hydrogen ion H^+ (a bare proton) that is attached to a water r cule. The hydronium ion, in turn, interacts with the water molecules rounding it to form additional species, such as $H_5O_2{}^+$, $H_7O_3{}^+$, $H_9O_4{}^+$ (shov Figure 5-6), and many others. These interactions are called *hydration*. Bec the H_3O^+ ion consists of a H^+ ion attached to a water molecule, $H^+ + H_2C$ can write equation (5.2) in a simpler way by eliminating a water mol from each side of the equation:

$$CH_3COOH(aq) \rightleftharpoons H^+(aq) + CH_3COO^-(aq)$$

Equations (5.2) and (5.3) both represent the ionization of CH_3COOH in w However, the first equation is generally preferred because it emphasizes (1) the ions are generated by a reaction involving water, and (2) the H^+ i firmly attached to a water molecule.

Finally, let us consider the best way to represent a solution of CH_3C nonelectrolyte, which is depicted in Figure 5-4(a). Because a nonelectr does not ionize in solution, a solution of CH_3OH is best represente $CH_3OH(aq)$.

A Notation for Representing Concentrations of Entities in Solu

With this new information about the nature of aqueous solutions, we introduce a useful notation for solution concentrations. In a solution th 0.0050 M $MgCl_2$, we assume that the $MgCl_2$ is completely dissociated ions. Because there are two Cl^- ions for every Mg^{2+} ion, the soluti 0.0050 M Mg^{2+} but 0.0100 M Cl^-. Better still, let us introduce a special sy for the concentration of a species in solution—the bracket symbol [ ]. statement $[Mg^{2+}] = 0.0050$ M means that the concentration of the sp within the brackets—that is, Mg^{2+}—is 0.0050 mol/L. Thus,

$$\textit{in 0.0050 M MgCl}_2\text{:}\quad [Mg^{2+}] = 0.0050\text{ M}; [Cl^-] = 0.0100\text{ M}; [MgCl_2] = 0\text{ M}$$

Although we do not usually write expressions like $[MgCl_2] = 0$, it is done to emphasize that there is essentially no undissociated $MgCl_2$ in the soluti

Example 5-1 shows how to calculate the concentrations of ions in a st electrolyte solution.

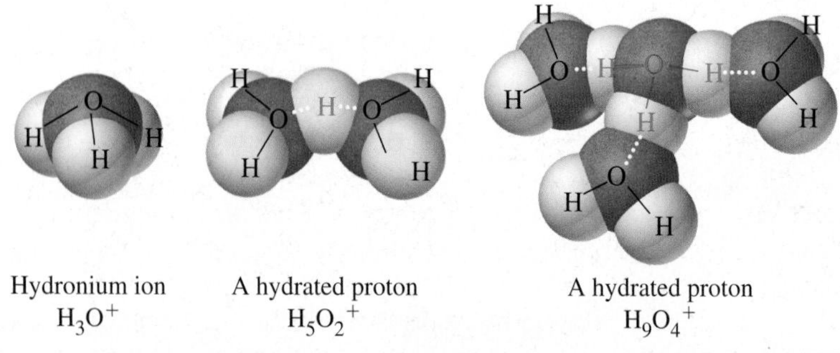

▶ FIGURE 5-6
The hydrated proton
The hydronium ion, H_3O^+, interacts with other water molecules through electrostatic attractions.

*To say that a strong electrolyte is completely dissociated into individual ions in aqueous sc is a good approximation but somewhat of an oversimplification. Some of the cations and an solution may become associated into units called *ion pairs*. Generally, though, at the low sc concentrations we will be using, assuming complete dissociation will not seriously affect our r

XAMPLE 5-1 Calculating Ion Concentrations in a Solution of a Strong Electrolyte

Aluminum sulfate, $Al_2(SO_4)_3$, is a strong electrolyte. What are the aluminum and sulfate ion concentrations in 0.0165 M $Al_2(SO_4)_3$?

nalyze

The solute is a strong electrolyte. Thus, it dissociates completely in water. First, we write a balanced chemical equation for the dissociation of $Al_2(SO_4)_3(aq)$, and then set up stoichiometric factors to relate Al^{3+} and SO_4^{2-} to the molarity of $Al_2(SO_4)_3$.

olve

The dissociation of $Al_2(SO_4)_3$ is represented by the equation below.

$$Al_2(SO_4)_3(aq) \longrightarrow 2\,Al^{3+}(aq) + 3\,SO_4^{2-}(aq)$$

The stoichiometric factors, shown in blue in the following equations, are derived from the fact that 1 mol $Al_2(SO_4)_3$ produces 2 mol Al^{3+} and 3 mol SO_4^{2-}.

$$[Al^{3+}] = \frac{0.0165\text{ mol }Al_2(SO_4)_3}{1\text{ L}} \times \frac{2\text{ mol }Al^{3+}}{1\text{ mol }Al_2(SO_4)_3} = \frac{0.0330\text{ mol }Al^{3+}}{1\text{ L}} = 0.0330\text{ M}$$

$$[SO_4^{2-}] = \frac{0.0165\text{ mol }Al_2(SO_4)_3}{1\text{ L}} \times \frac{3\text{ mol }SO_4^{2-}}{1\text{ mol }Al_2(SO_4)_3} = \frac{0.0495\text{ mol }SO_4^{2-}}{1\text{ L}} = 0.0495\text{ M}$$

ssess

For a strong electrolyte, the concentrations of the ions will always be integer multiples of the electrolyte molarity. For example, in 0.0165 M $MgCl_2$, we have $[Mg^{2+}] = 1 \times 0.0165$ M and $[Cl^-] = 2 \times 0.0165$ M.

ACTICE EXAMPLE A: Seawater is approximately 0.438 M NaCl and 0.0512 M $MgCl_2$. What is the molarity of Cl^-—that is, the total $[Cl^-]$—in seawater?

ACTICE EXAMPLE B: A water treatment plant adds fluoride ion to the water to the extent of 1.5 mg F^-/L.

(a) What is the molarity of fluoride ion in this water?

(b) If the fluoride ion in the water is supplied by calcium fluoride, what mass of calcium fluoride is present in 1.00×10^6 L of this water?

5-1 CONCEPT ASSESSMENT

Which solution is the best electrical conductor? **(a)** 0.50 M CH_3COCH_3; **(b)** 0.50 M CH_3CH_2OH; **(c)** 1.00 M $CH_2(OH)CH(OH)CH_2OH$; **(d)** 0.050 M CH_3COOH; **(e)** 0.025 M $RbNO_3$.

Which solution has the highest total molarity of ions? **(a)** 0.008 M $Ba(OH)_2$; **(b)** 0.010 M KI; **(c)** 0.011 M CH_3COOH; **(d)** 0.030 M $HOCH_2CH_2OH$; **(e)** 0.004 M $Al_2(SO_4)_3$.

Precipitation Reactions

e metal salts, such as NaCl, are quite soluble in water, while others, such gCl, are not very soluble at all. In fact, so little AgCl dissolves in water that compound is generally considered to be *insoluble*. Precipitation reactions r when certain cations and anions combine to produce an insoluble ionic called a **precipitate**. One laboratory use of precipitation reactions is in ifying the ions present in a solution, as shown in Figure 5-7. In industry, pitation reactions are used to manufacture numerous chemicals. In the ction of magnesium metal from seawater, for instance, the first step is to pitate Mg^{2+} as $Mg(OH)_2(s)$. In this section, the objective is to represent pitation reactions by chemical equations and to apply some simple rules redicting precipitation reactions.

Richard Megna/Fundamental Photographs

▲ FIGURE 5-7
Qualitative test for Cl^- in tap water
The test involves the addition of a few drops of $AgNO_3(aq)$ to tap water. The formation of a precipitate of AgCl(s) confirms the presence of Cl^-.

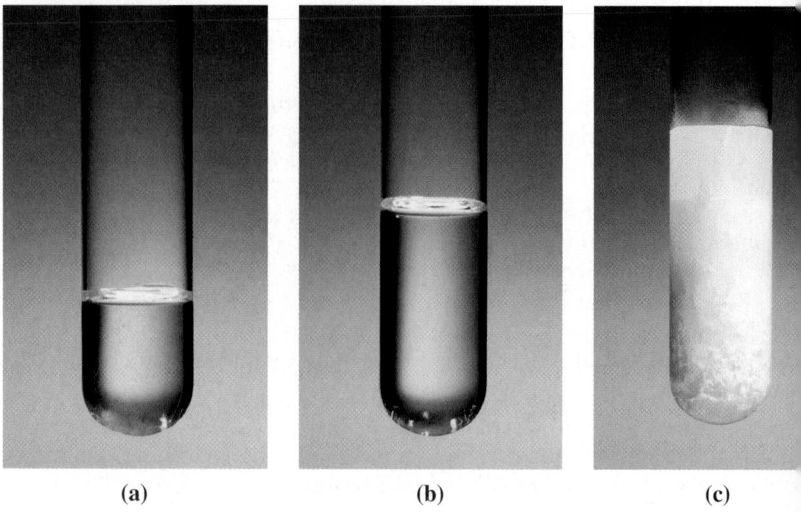

▶ FIGURE 5-8
A precipitate of silver iodide
When an aqueous solution of $AgNO_3$ **(a)** is added to one of NaI **(b)**, insoluble pale yellow or cream-colored AgI(s) precipitates from solution **(c)**.

Net Ionic Equations

The reaction of silver nitrate and sodium iodide in an aqueous solution y sodium nitrate in solution and a pale yellow or cream-colored precipita silver iodide, as shown in Figure 5-8. Applying the principles of equation ing from Chapter 4, we can write

$$AgNO_3(aq) + NaI(aq) \longrightarrow AgI(s) + NaNO_3(aq)$$

You might note a contradiction, however, between equation (5.4) and s thing we learned earlier in this chapter. In their aqueous solutions, the so ionic compounds $AgNO_3$, NaI, and $NaNO_3$—all *strong* electrolytes—sh be represented by their separate ions.

$$Ag^+(aq) + NO_3^-(aq) + Na^+(aq) + I^-(aq) \longrightarrow AgI(s) + Na^+(aq) + NO_3^-(aq)$$

We might say that equation (5.4) is the "whole formula" form of the equa whereas equation (5.5) is the "ionic" form. Notice also that in equation $Na^+(aq)$ and $NO_3^-(aq)$ appear on both sides of the equation. These ion not reactants; they go through the reaction unchanged. We call them **sp tor ions**. If we eliminate the spectator ions, all that remains is the net equation:

$$Ag^+(aq) + I^-(aq) \longrightarrow AgI(s)$$

KEEP IN MIND

that although the insoluble solid consists of ions, we don't represent ionic charges in the whole formula. That is, we write AgI(s), not $Ag^+I^-(s)$.

A **net ionic equation** is an equation that includes only the actual pa pants in a reaction, with each participant denoted by the symbol or for that best represents it. Symbols are written for individual ions, suc $Ag^+(aq)$, and whole formulas are written for insoluble solids, such as A Because net ionic equations include electrically charged species—ions— ionic equation must be balanced both for the numbers of atoms of all t and for electric charge. The same net electric charge must appear on both of the equation. Throughout the remainder of this chapter, we will repr most chemical reactions in aqueous solution by net ionic equations.

Predicting Precipitation Reactions

Suppose we are asked whether precipitation occurs when the following a ous solutions are mixed.

$$AgNO_3(aq) + KBr(aq) \longrightarrow ?$$

A good way to begin is to rewrite the expression in the ionic form.

$$Ag^+(aq) + NO_3^-(aq) + K^+(aq) + Br^-(aq) \longrightarrow ?$$

e are only two possibilities. Either some cation–anion combination leads insoluble solid—a precipitate—or no such combination is possible, and is no reaction at all.

predict what will happen without doing experiments, we need some mation about which sorts of ionic compounds are water soluble and h are water insoluble. We expect the insoluble ones to form when the opriate ions are present in solution. We don't have all-encompassing rules redicting solubilities, but a few guidelines work for the majority of com- ionic solutes. A concise form of these guidelines is presented in Table 5.1.

3LE 5.1 Solubility Guidelines for Common Ionic Solids

ow the lower-numbered guideline when two guidelines are in conflict. This ls to the correct prediction in most cases.

alts of group 1 cations (with some exceptions for Li^+) and the NH_4^+ cation are oluble.

Jitrates, acetates, and perchlorates are soluble.

alts of silver, lead, and mercury(I) are insoluble.

hlorides, bromides, and iodides are soluble.

arbonates, phosphates, sulfides, oxides, and hydroxides are insoluble sulfides of group 2 cations and hydroxides of Ca^{2+}, Sr^{2+}, and Ba^{2+} are lightly soluble).

ulfates are soluble except for those of calcium, strontium, and barium.

◀ In principle, all ionic compounds dissolve in water to some extent, though this may be very slight. For practical purposes, we consider a compound to be insoluble if the maximum amount that can dissolve is less than about 0.01 mol/L.

e guidelines from Table 5.1 are applied in the order listed, with the lower- bered guideline taking precedence in cases of a conflict. According to guidelines, AgBr(s) is insoluble in water (because rule 3 takes prece- e over rule 4) and should precipitate, whereas $KNO_3(s)$ is soluble use of rule 1). Written as an ionic equation, expression (5.7) becomes

$$^{-}(aq) + NO_3^-(aq) + K^+(aq) + Br^-(aq) \longrightarrow AgBr(s) + K^+(aq) + NO_3^-(aq)$$

he net ionic equation, we have

$$Ag^+(aq) + Br^-(aq) \longrightarrow AgBr(s) \qquad (5.8)$$

e three predictions concerning precipitation reactions made in Example 5-2 erified in Figure 5-9.

KEEP IN MIND

that when two ionic compounds form a solid precipitate, they do so by exchanging ions. In the formation of AgBr from KBr and $AgNO_3$, the following exchange takes place.

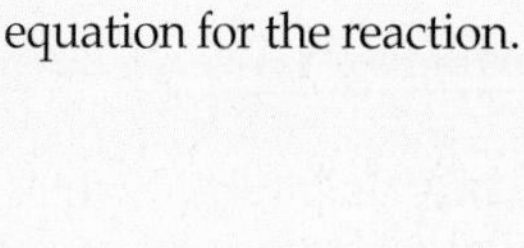

KAMPLE 5-2 Using Solubility Guidelines to Predict Precipitation Reactions

Predict whether a reaction will occur in each of the following cases. If so, write a net ionic equation for the reaction.

(a) $NaOH(aq) + MgCl_2(aq) \longrightarrow ?$
(b) $BaS(aq) + CuSO_4(aq) \longrightarrow ?$
(c) $(NH_4)_2SO_4(aq) + ZnCl_2(aq) \longrightarrow ?$

nalyze

All the compounds shown in (a), (b), and (c) are soluble and they provide ions in solution. By using the solubility guidelines in Table 5.1, determine whether the positive ions from one compound combine with the negative ions of the other to form soluble or insoluble compounds. If only soluble compounds are formed, then all ions remain in solution (no reaction). If an insoluble compound is formed, then the insoluble compound precipitates from the solution. The net ionic equation for the precipitation reaction is obtained by eliminating the spectator ions from the full ionic equation.

lve

For each of (a), (b) and (c), apply the strategy described above.

(a) In aqueous solution, we get Na^+ and OH^- from NaOH and Mg^{2+} and Cl^- from $MgCl_2$. The combination of Na^+ and Cl^- gives NaCl, a soluble compound; thus, the Na^+ and Cl^- ions remain in solution.

(continued)

However, the Mg^{2+} and OH^- ions combine to produce $Mg(OH)_2$, an insoluble compound. The fu
ionic equation is

$$2\,Na^+(aq) + 2\,OH^-(aq) + Mg^{2+}(aq) + 2\,Cl^-(aq) \longrightarrow Mg(OH)_2(s) + 2\,Na^+(aq) + 2\,Cl^-(aq)$$

With the elimination of spectator ions, we obtain

$$2\,OH^-(aq) + Mg^{2+}(aq) \longrightarrow Mg(OH)_2(s)$$

(b) In aqueous solution, we get Ba^{2+} and S^{2-} from BaS and Cu^{2+} and SO_4^{2-} from $CuSO_4$. The Ba^{2+} and SO_4^{2-} ions combine to form $BaSO_4$, an insoluble compound, and the Cu^{2+} and S^{2-} ions combine to for CuS, also an insoluble compound. The full ionic equation is

$$Ba^{2+}(aq) + S^{2-}(aq) + Cu^{2+}(aq) + SO_4^{2-}(aq) \longrightarrow BaSO_4(s) + CuS(s)$$

The equation above is also the net ionic equation because there are no spectator ions.

(c) We get NH_4^+, SO_4^{2-}, Zn^{2+}, and Cl^- ions in solution. Because all possible combinations of positive ar negative ions lead to water soluble compounds, all of the ions remain in solution. No reaction occurs.

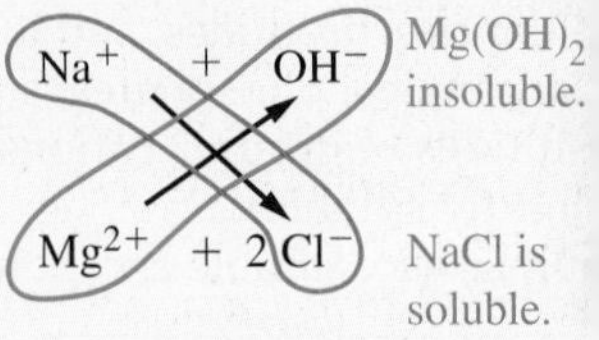

Assess

Problems of this type can also be solved by using a diagrammatic approach, which is illustrated for part (a).

As you gain experience, you should be able to go directly to a net ionic equation without first having to write an ionic equation that includes spectator ions.

PRACTICE EXAMPLE A: Indicate whether a precipitate forms by completing each equation as a net ionic equation. If no reaction occurs, so state.

(a) $AlCl_3(aq) + KOH(aq) \longrightarrow ?$
(b) $K_2SO_4(aq) + FeBr_3(aq) \longrightarrow ?$
(c) $CaI_2(aq) + Pb(NO_3)_2(aq) \longrightarrow ?$

PRACTICE EXAMPLE B: Indicate through a net ionic equation whether a precipitate forms when the following compounds in aqueous solution are mixed. If no reaction occurs, so state.

(a) sodium phosphate + aluminum chloride $\longrightarrow ?$
(b) aluminum sulfate + barium chloride $\longrightarrow ?$
(c) ammonium carbonate + lead nitrate $\longrightarrow ?$

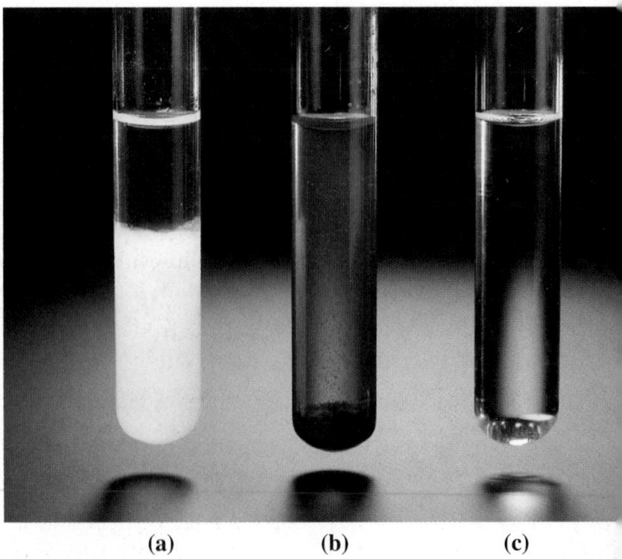

▶ FIGURE 5-9
Verifying the predictions made in Example 5-2
(a) When NaOH(aq) is added to $MgCl_2(aq)$, a white precipitate of $Mg(OH)_2(s)$ forms. **(b)** When colorless BaS(aq) is added to blue $CuSO_4(aq)$, a dark precipitate forms. The precipitate is a mixture of white $BaSO_4(s)$ and black CuS(s); a slight excess of $CuSO_4$ remains in solution. **(c)** No reaction occurs when colorless $(NH_4)_2SO_4(aq)$ is added to colorless $ZnCl_2(aq)$.

5-2 CONCEPT ASSESSMENT

ɔly the solubility guidelines in Table 5.1 to predict whether each of the owing solids is water soluble or insoluble. For which are the solubility delines inconclusive? **(a)** $Al_2(SO_4)_3$; **(b)** $Cr(OH)_3$; **(c)** K_3PO_4; **(d)** Li_2CO_3; ZnS; **(f)** $Mg(MnO_4)_2$; **(g)** $AgClO_4$; **(h)** $CaSO_4$; **(i)** PbO.

5-1 ARE YOU WONDERING?

Is an insoluble ionic compound, such as AgCl, a strong electrolyte or a weak electrolyte?

ver chloride, AgCl, is an ionic compound with very low solubility in water. ıen AgCl dissolves in water, it is 100% dissociated into Ag^+ and Cl^- ions; there no AgCl ion pairs. If we focus only on the degree of dissociation, then AgCl is rong electrolyte.

A strong electrolyte may be defined in more practical terms as a substance that, en dissolved in water, gives a solution that is a good conductor of electricity. ause AgCl has very low solubility in water, approximately 1×10^{-5} moles per r, a solution of AgCl is not a good conductor of electricity.

Some chemists would argue that AgCl is a strong electrolyte (because it is 100% sociated in aqueous solution) but some may argue that it is a weak electrolyte cause an aqueous solution of AgCl is not a good conductor of electricity).

Does it matter that chemists may not totally agree on whether AgCl should be led a strong electrolyte or a weak electrolyte? Not at all, because all chemists ee on the following facts: (1) AgCl is essentially 100% dissociated in water; only a small amount of AgCl can be dissolved in water; and (3) an aqueous ution of AgCl is not a good conductor of electricity.

Acid–Base Reactions

about acids and bases (or alkalis) date back to ancient times. The word s derived from the Latin *acidus* (sour). *Alkali* (base) comes from the Arabic *i*, referring to the ashes of certain plants from which alkaline substances e extracted. The acid–base concept is a major theme in the history of ıistry. In this section, we emphasize the view proposed by Svante enius in 1884 but also introduce a more modern theory proposed in 1923 ıomas Lowry and by Johannes Brønsted.

ls

a practical standpoint, acids can be identified by their sour taste, their y to react with a variety of metals and carbonate minerals, and the effect have on the colors of substances called *acid–base indicators*. Methyl red is id–base indicator that appears red in acidic environments and yellow wise (see Figure 5-10). From a chemist's point of view, however, an **acid** e defined as a substance that provides hydrogen ions (H^+) in aqueous ion. This definition was first proposed by Svante Arrhenius in 1884.

fferent acids exhibit different tendencies for producing H^+ ions in aque- solution. **Strong acids** have a strong tendency for producing H^+ ions. ıg acids are molecular compounds that are almost completely ionized $H^+(aq)$ and accompanying anions when in aqueous solution. Hydrogen ide, HCl, and nitric acid, HNO_3, are examples of strong acids. The ion- n of HCl in water can be represented by the following equation.

KEEP IN MIND

that $H^+(aq)$ actually represents a hydrated proton, that is, a proton attached to one H_2O molecule, as in H_3O^+, or to several H_2O molecules, as in $H_9O_4^+$.

$$HCl(aq) \longrightarrow H^+(aq) + Cl^-(aq) \quad (5.9)$$

tion (5.9) indicates that HCl is a strong electrolyte in water and that ıtially all the HCl molecules are converted into hydrated H^+ and Cl^- ions.

▶ FIGURE 5-10
An acid, a base, and an acid–base indicator
The acidic nature of lemon juice is shown by the red color of the acid–base indicator *methyl red.* The basic nature of soap is indicated by the change in color of the indicator from red to yellow.

TABLE 5.2 Common Strong Acids and Strong Bases

Acids	Bases
HCl	$LiOH$
HBr	$NaOH$
HI	KOH
$HClO_4$	$RbOH$
HNO_3	$CsOH$
H_2SO_4[a]	$Mg(OH)_2$
	$Ca(OH)_2$
	$Sr(OH)_2$
	$Ba(OH)_2$

[a]H_2SO_4 ionizes in two distinct steps. It is a strong acid only in its first ionization step (see Section 16-6).

KEEP IN MIND

that a hydrogen atom consists of one proton and one electron. Therefore, a hydrogen ion, H^+, is simply a proton.

(Recall the discussion of Section 5-1, which pointed out that, in aqueous s tion, the H^+ ion is firmly attached to a water molecule to form a H_3O^+ ior

When the strong acid HNO_3 dissolves in water, complete ionization $H^+(aq)$ and $NO_3^-(aq)$ occurs. There are so few common strong acids that make only a short list. The list of common strong acids is given in Table 5 is imperative that you memorize this list.

Weak acids are molecular compounds that have a weak tendency for ducing H^+ ions; weak acids are incompletely ionized in aqueous solution vast majority of acids are weak acids. The ionization of a weak acid is described in terms of a reversible reaction that does not go to completior described on page 156, the ionization reaction for acetic acid, CH_3COOH, be represented as

$$CH_3COOH(aq) \rightleftharpoons H^+(aq) + CH_3COO^-(aq)$$

Equation (5.10) has the following interpretation: in aqueous solution, some of the CH_3COOH molecules are converted into H^+ and CH_3COO^- The fraction of molecules that ionize can be calculated, but the calculati not simple. We will defer such calculations until Chapter 16.

Equations (5.9) and (5.10) are based on the Arrhenius theory of acids bases, and these equations might lead you to think that acids simply apart into H^+ ions and the accompanying anions when they are dissolv water. However, plenty of experimental evidence proves that this is no case. In 1923, Johannes Brønsted in Denmark and Thomas Lowry in C Britain independently proposed that the key process responsible fo properties of acids (and bases) is the transfer of an H^+ ion (a proton) one substance to another. For example, when acids dissolve in water ions are transferred from acid molecules to water molecules, as shown b for HCl and CH_3COOH.

$$HCl(aq) + H_2O(l) \longrightarrow H_3O^+(aq) + Cl^-(aq)$$

$$CH_3COOH(aq) + H_2O(l) \rightleftharpoons H_3O^+(aq) + CH_3COO^-(aq)$$

In equations (5.11) and (5.12), the acid molecules are acting as pr donors and the water molecules are acting as proton acceptors. Accordi the Brønsted–Lowry theory, an acid is a **proton donor**.

It is partly a matter of preference whether we include water as a reacta the equation for the reaction that occurs when an acid is dissolved in w Some chemists prefer to write the reaction without water as a reactant, a did in equations (5.9) and (5.10), to eliminate the clutter of "extra" water r cules. If that is also your preference, then you must remember that the H^+ i not a free proton in solution but rather is firmly bound to a water molecule exists as an H_3O^+ ion. The H_3O^+ ion is even further hydrated (see Figure Many chemists prefer to include H_2O as a reactant, as we did in equa and (5.12), to emphasize that the reactions actually involve the transfer of ns from acid molecules to water molecules.

es

a practical standpoint, we can identify bases through their bitter taste, ery feel, and effect on the colors of acid–base indicators (Fig. 5-9). The enius definition of a **base** is a substance that produces hydroxide ions) in aqueous solution. Consider a soluble ionic hydroxide, such as H. In the solid state, this compound consists of Na^+ and OH^- ions. When olid dissolves in water, the ions dissociate.

$$NaOH(aq) \longrightarrow Na^+(aq) + OH^-(aq)$$

quation above indicates that the dissociation of NaOH goes to comple- and thus NaOH is a **strong base**. Consequently, NaOH(aq) is best repre- d as Na^+(aq) plus OH^-(aq).

is true of strong acids, the number of common strong bases is small (see 5.2). They are primarily the hydroxides of group 1 and some group 2 met- Memorize the list.

rtain substances produce OH^- ions by reacting with water, not just by lving in it. Such substances, for example, ammonia, are also bases.

$$NH_3(aq) + H_2O(l) \rightleftharpoons NH_4^+(aq) + OH^-(aq) \qquad \textbf{(5.13)}$$

eaction of NH_3 with water does not go to completion; only some of the molecules ionize. For this reason, NH_3 is a called a weak electrolyte. A that is incompletely ionized in aqueous solution is a **weak base**. Most substances are weak bases.

e can also examine equation (5.13) in terms of the Brønsted–Lowry theory, h focuses on the transfer of protons from one substance to another. rding to this theory, a base is a **proton acceptor**. In equation (5.13), NH_3 ves as a proton acceptor (a Brønsted–Lowry base) and H_2O behaves as a n donor (a Brønsted–Lowry acid).

KEEP IN MIND

that NH_4^+(aq) is formed by the transfer of a proton from an H_2O to a NH_3 molecule, and NH_4^+ interacts with water in much the same way as the hydronium ion does. A ball-and-stick model of the ammonium ion is shown below.

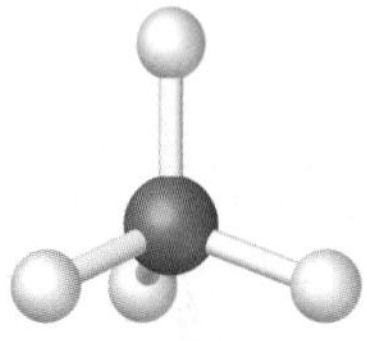

Ammonium ion

ic and Basic Solutions

ave seen that when dissolved in water, an acid produces H^+ ions and a produces OH^- ions. However, experiment shows small numbers of H^+ OH⁻ ions are present even in pure water. In pure water, the following ion occurs to a limited extent, hence the use of a double arrow ($\rightleftharpoons$) r than a single arrow ($\longrightarrow$).

$$H_2O(l) \rightleftharpoons H^+(aq) + OH^-(aq)$$

reful measurements show that $[H^+]_{water} = [OH^-]_{water} = 1.0 \times 10^{-7}$ M °C. (The subscripts on the square brackets are there to emphasize that alues given for $[H^+]$ and $[OH^-]$ are for pure water only.) Because an acid uces H^+ ions in solution, we expect that a solution of acid at 25 °C will $[H^+] > 1.0 \times 10^{-7}$ M. Such a solution is said to be *acidic*. An acidic solution greater concentration of H^+ ions than does pure water. A base produces ions, and so a solution of base will have $[OH^-] > 1.0 \times 10^{-7}$ M at 25 °C. a solution is said to be *basic*. These ideas are summarized below.

n *acidic* solution has $[H^+] > [H^+]_{water}$.
basic solution has $[OH^-] > [OH^-]_{water}$.

e statements above can be expressed another way. An acidic solution has cess of H^+ ions (compared with pure water), and a basic solution has an s of OH^- ions. We will use these ideas in Section 5-5 and encounter them in Chapter 16.

Neutralization

Perhaps the most significant property of acids and bases is the ability of to cancel or neutralize the properties of the other. In a **neutralization rea** an acid and a base react to form an aqueous solution of an ionic comp called a **salt**. As an example, consider the reaction between HCl, a strong and NaOH, a strong base:

$$\begin{array}{ccccccc} HCl(aq) & + & NaOH(aq) & \longrightarrow & NaCl(aq) & + & H_2O(l) \\ (acid) & + & (base) & \longrightarrow & (salt) & + & (water) \end{array}$$

Switching to the ionic form, we write:

$$\underbrace{H^+(aq) + Cl^-(aq)}_{(acid)} + \underbrace{Na^+(aq) + OH^-(aq)}_{(base)} \longrightarrow \underbrace{Na^+(aq) + Cl^-(aq)}_{salt} + \underbrace{H_2O(}_{water}$$

When the spectator ions are eliminated, the net ionic equation show essential nature of the neutralization of a strong acid by a strong base: H^- from the acid and OH^- ions from the base combine to form water.

$$H^+(aq) + OH^-(aq) \longrightarrow H_2O(l)$$

The situation is different when either the acid or the base in a neutraliz reaction is weak. For example, consider the neutralization reaction bet CH_3COOH, a weak acid, and NaOH:

$$\begin{array}{ccccccc} CH_3COOH(aq) & + & NaOH(aq) & \longrightarrow & NaCH_3COO(aq) & + & H_2O(l) \\ (acid) & & (base) & & (salt) & & (water) \end{array}$$

Although the reaction produces an aqueous solution of a salt, the net ionic tion for this neutralization reaction is not quite as simple as it was for the tion involving a strong acid and a strong base. As discussed earlier (page an aqueous solution of CH_3COOH is best represented as $CH_3COOH(aq)$, r $H^+(aq) + CH_3COO^-(aq)$. Also, according to rule 1 in Table 5.1, $NaCH_3CO$ soluble salt, and so $NaCH_3COO(aq)$ is best represented as $CH_3COO^-($ $Na^+(aq)$. Therefore, we can write

$$CH_3COOH(aq) + Na^+(aq) + OH^-(aq) \longrightarrow Na^+(aq) + CH_3COO^-(aq) + H_2O($$

The net ionic equation is

$$CH_3COOH(aq) + OH^-(aq) \longrightarrow CH_3COO^-(aq) + H_2O(l)$$

Thus, in the neutralization of CH_3COOH by a strong base, we can thi OH^- from the base combining directly with CH_3COOH molecules.

An analogous situation exists for a neutralization involving a weak and a strong acid. For the neutralization of the weak base NH_3 by HC example, we can think of H^+ from the acid combining directly with NH_3 ecules to form NH_4^+. The equation for the neutralization can be represent the following ionic equation

$$\underbrace{H^+(aq) + Cl^-(aq)}_{(acid)} + \underbrace{NH_3(aq)}_{(base)} \longrightarrow \underbrace{NH_4^+(aq) + Cl^-(aq)}_{(salt)}$$

or by a net ionic equation

$$H^+(aq) + NH_3(aq) \longrightarrow NH_4^+(aq)$$

All the neutralization reactions given above involve a strong acid or a s base and all of them go essentially to completion, that is, until the limiting tant is used up. Thus, we use a single arrow ($\longrightarrow$) rather than a double a ($\rightleftharpoons$) in the equations for these reactions.

▶ The formula of $NH_3(aq)$ is sometimes written as NH_4OH (ammonium hydroxide) and its ionization represented as

$$NH_4OH(aq) \rightleftharpoons NH_4^+(aq) + OH^-(aq)$$

There is, however, no hard evidence for the existence of NH_4OH, a discrete substance comprising NH_4^+ and OH^- ions. We will use only the formula $NH_3(aq)$.

ognizing Acids and Bases

s contain *ionizable* hydrogen atoms, which are generally identified by the in which the formula of an acid is written. Ionizable H atoms are sepa- d from other H atoms in the formula either by writing them first in the ecular formula or by indicating where they are found in the molecule. s, there are two ways that we can show that one H atom in the acetic acid ecule is ionizable and the other three H atoms are not.

$$\underbrace{HC_2H_3O_2 \text{ or } CH_3COOH}_{\text{acetic acid}}$$

ontrast to acetic acid, methane has four H atoms, but they are not ioniz- CH_4 is neither an acid nor a base.

substance whose formula indicates a combination of OH^- ions with ons is generally a strong base (for example, NaOH). To identify a weak , we usually need a chemical equation for the ionization reaction, as in tion (5.13). For now, NH_3 is the only weak base we will work with. Note ethanol, CH_3CH_2OH, is not a base. The OH group is not present as OH^-, er in pure ethanol or in its aqueous solutions.

e Acid–Base Reactions

$OH)_2$ is a base because it contains OH^-, but this compound is quite insol- in water. Its finely divided solid particles form a suspension in water that e familiar milk of magnesia, used as an antacid. In this suspension, $OH)_2(s)$ does dissolve very slightly, producing some OH^- in solution. If cid is added, H^+ from the acid combines with this OH^- to form water. e $Mg(OH)_2(s)$ dissolves to produce more OH^- in solution, which is ralized by more H^+, and so on. In this way, the neutralization reaction lts in the dissolving of otherwise insoluble $Mg(OH)_2(s)$. The net ionic ation for the reaction of $Mg(OH)_2(s)$ with a strong acid is

$$Mg(OH)_2(s) + 2\,H^+(aq) \longrightarrow Mg^{2+}(aq) + 2\,H_2O(l) \qquad \textbf{(5.14)}$$

$g(OH)_2(s)$ also reacts with a weak acid, such as acetic acid. In the net equation, acetic acid is written in its molecular form. But remember that e H^+ and CH_3COO^- ions are always present in an acetic acid solution. H^+ ions react with OH^- ions, as in reaction (5.14), followed by further ion- on of CH_3COOH, more neutralization, and so on. If enough acetic acid is ent, the $Mg(OH)_2$ will dissolve completely. The equation for the reaction ven below.

$$OH)_2(s) + 2\,CH_3COOH(aq) \longrightarrow Mg^{2+}(aq) + 2\,CH_3COO^-(aq) + 2\,H_2O(l) \qquad \textbf{(5.15)}$$

alcium carbonate, which is present in limestone and marble, is another er-insoluble solid that is soluble in strong and weak acids. Here the solid luces a low concentration of CO_3^{2-} ions, which combine with H^+ to form weak acid H_2CO_3. This causes more of the solid to dissolve, and so on. onic acid, H_2CO_3, is a very unstable substance that decomposes into H_2O $CO_2(g)$. The net ionic equation for the reaction of $CaCO_3$ with an acid is n below.

$$CaCO_3(s) + 2\,H^+(aq) \longrightarrow Ca^{2+}(aq) + H_2O(l) + CO_2(g) \qquad \textbf{(5.16)}$$

s, a gas is given off when $CaCO_3(s)$ reacts with an acid and dissolves. The tion represented by equation (5.16) is responsible for the erosion of marble ues by acid rain, such as the one shown in Figure 5-11. Equation (5.16) also ws that $CaCO_3(s)$ has the ability to neutralize acids. Not surprisingly, cal- n carbonate, like magnesium hydroxide, is used as an antacid.

Presselect / Alamy

▲ FIGURE 5-11
Damage caused by acid rain
This marble statue has been eroded by acid rain. Marble consists primarily of $CaCO_3$. Acids react with and dissolve marble through the reaction described in equation (5.16).

The Arrhenius definition recognizes only OH^- as a base, but wher reconsider acids and bases in more detail in Chapter 16, we will see that r ern theories identify CO_3^{2-} and many other anions, including OH^-, as b Table 5.3 lists several common anions and one cation that produce gas acid–base reactions.

TABLE 5.3 Some Common Gas-Forming Reactions

Ion	Reaction
HSO_3^-	$HSO_3^- + H^+ \longrightarrow SO_2(g) + H_2O(l)$
SO_3^{2-}	$SO_3^{2-} + 2\,H^+ \longrightarrow SO_2(g) + H_2O(l)$
HCO_3^-	$HCO_3^- + H^+ \longrightarrow CO_2(g) + H_2O(l)$
CO_3^{2-}	$CO_3^{2-} + 2\,H^+ \longrightarrow CO_2(g) + H_2O(l)$
S^{2-}	$S^{2-} + 2\,H^+ \longrightarrow H_2S(g)$
NH_4^+	$NH_4^+ + OH^- \longrightarrow NH_3(g) + H_2O(l)$

EXAMPLE 5-3 Writing Equations for Acid–Base Reactions

Write a net ionic equation to represent the reaction of **(a)** aqueous strontium hydroxide with nitric aci **(b)** solid aluminum hydroxide with hydrochloric acid.

Analyze

The reactions are neutralization reactions, which means they are of the general form acid + base → salt. Wat may also be a product. We can start with the whole formula equation, switch to the ionic equation, and the delete the spectator ions to arrive at the net ionic equation.

Solve

(a) $2\,HNO_3(aq) + Sr(OH)_2(aq) \longrightarrow Sr(NO_3)_2(aq) + 2\,H_2O(l)$

Ionic form:

$$2\,H^+(aq) + 2\,NO_3^-(aq) + Sr^{2+}(aq) + 2\,OH^-(aq) \longrightarrow Sr^{2+}(aq) + 2\,NO_3^-(aq) + 2\,H_2O(l)$$

Net ionic equation: Delete the spectator ions (Sr^{2+} and NO_3^-).

$$2\,H^+(aq) + 2\,OH^-(aq) \longrightarrow 2\,H_2O(l)$$

or, more simply,

$$H^+(aq) + OH^-(aq) \longrightarrow H_2O(l)$$

(b) $Al(OH)_3(s) + 3\,HCl(aq) \longrightarrow AlCl_3(aq) + 3\,H_2O(l)$

Ionic form:

$$Al(OH)_3(s) + 3\,H^+(aq) + 3\,Cl^-(aq) \longrightarrow Al^{3+}(aq) + 3\,Cl^-(aq) + 3\,H_2O(l)$$

Net ionic equation: Delete the spectator ion (Cl^-).

$$Al(OH)_3(s) + 3\,H^+(aq) \longrightarrow Al^{3+}(aq) + 3\,H_2O(l)$$

Assess

In part (a), the net ionic equation is $H^+(aq) + OH^-(aq) \rightarrow H_2O(l)$, as is always the case when the neutraliza tion reaction involves a soluble strong acid and a soluble strong base. In part (b), the base was not soluble; thu the net ionic equation includes a solid.

PRACTICE EXAMPLE A: Write a net ionic equation to represent the reaction of aqueous ammonia with propionic acid, CH_3CH_2COOH. Assume that the neutralization reaction goes to completion. What is the formula of the salt that results from this neutralization?

PRACTICE EXAMPLE B: Calcium carbonate is a major constituent of the hard water deposits found in teakettle and automatic coffee makers. Vinegar, which is essentially a dilute aqueous solution of acetic acid, is commonly used to remove such deposits. Write a net ionic equation for the reaction that occurs. Assume tha the neutralization reaction goes to completion.

5-3 CONCEPT ASSESSMENT

ı are given the four solids, K_2CO_3, CaO, $ZnSO_4$, and $BaCO_3$, and three solvents, O(l), HCl(aq), and H_2SO_4(aq). You are asked to prepare four solutions, each ntaining one of the four cations, that is, one with K^+(aq), one with Ca^{2+}(aq), d so on. Using water as your *first* choice, what solvent would you use to prepare ch solution? Explain your choices.

Oxidation–Reduction Reactions: Some General Principles

tical applications of oxidation–reduction reactions can be traced back sands of years to the period in human culture when metal tools were made. The metal needed to make tools was obtained by heating copper on ores, such as cuprite (Cu_2O) or hematite (Fe_2O_3), in the presence of on. Since that time, iron has become the most widely used of all metals it is produced in essentially the same way: by heating Fe_2O_3 in the pres- of carbon in a blast furnace. A simplified chemical equation for the reac- is given below.

◀ An ore is a mineral from which a metal can be extracted. Many metal ores are oxides and the metals are obtained from their oxides by the removal of oxygen.

$$Fe_2O_3(s) + 3\,CO(g) \xrightarrow{\Delta} 2\,Fe(l) + 3\,CO_2(g) \qquad (5.17)$$

◀ In a blast furnace, carbon is converted to CO, which then reacts with Fe_2O_3.

is reaction, we can think of the CO(g) as taking O atoms away from Fe_2O_3 roduce CO_2(g) and the free element iron. A commonly used term to ribe a reaction in which a substance gains O atoms is *oxidation*, and a reac- in which a substance loses O atoms is *reduction*. In reaction (5.17), CO(g) idized and Fe_2O_3(s) is reduced. Oxidation and reduction must always r together, and such a reaction is called an **oxidation–reduction**, or **redox**, **tion**. The oxygen in Fe_2O_3 can also be removed by igniting a finely led mixture of Fe_2O_3 and Al. The reaction produces a spectacular display, vn in Figure 5-12, and releases a tremendous amount of heat, which causes ron to melt. Mixtures of finely divided Fe_2O_3 and Al are used by railway kers to produce liquid iron for welding together iron railway tracks.

◀ Because it is easier to say, the term *redox* is often used instead of oxidation–reduction.

efinitions of oxidation and reduction based solely on the transfer of oms are too restrictive. By using broader definitions, many reactions can escribed as oxidation–reduction reactions, even when no oxygen is lved.

Tom Pantages

▲ FIGURE 5-12
Thermite reaction
Iron atoms of iron(III) oxide give up O atoms to Al atoms, producing Al_2O_3.

$$Fe_2O_3(s) + 2\,Al(s) \longrightarrow Al_2O_3(s) + 2\,Fe(l)$$

dation State Changes

ose we rewrite equation (5.17) and indicate the oxidation states (O.S.) of elements on both sides of the equation by using the rules listed on 85.

$$\overset{+3\ -2}{Fe_2O_3} + 3\,\overset{+2\,-2}{CO} \longrightarrow 2\,\overset{0}{Fe} + 3\,\overset{+4\,-2}{CO_2}$$

ne O.S. of oxygen is −2 everywhere it appears in this equation. That of (shown in red) changes. It *decreases* from +3 in Fe_2O_3 to 0 in the free ele- t, Fe. The O.S. of carbon (shown in blue) also changes. It *increases* from +2 O to +4 in CO_2. In terms of oxidation state changes, in an oxidation ess, the O.S. of some element increases; in a reduction process, the O.S. of e element decreases.

ven though we assess oxidation state changes by element, oxidation and ction involve the entire species in which the element is found. Thus, for reaction above, the whole compound Fe_2O_3 is reduced, not just the oms; and CO is oxidized, not just the C atom.

EXAMPLE 5-4 Identifying Oxidation–Reduction Reactions

Indicate whether each of the following is an oxidation–reduction reaction.

(a) $MnO_2(s) + 4\,H^+(aq) + 2\,Cl^-(aq) \longrightarrow Mn^{2+}(aq) + 2\,H_2O(l) + Cl_2(g)$
(b) $H_2PO_4^-(aq) + OH^-(aq) \longrightarrow HPO_4^{2-}(aq) + H_2O(l)$

Analyze

In each case, indicate the oxidation states of the elements on both sides of the equation, and look for changes

Solve

(a) The O.S. of Mn decreases from +4 in MnO_2 to +2 in Mn^{2+}. MnO_2 is reduced to Mn^{2+}. The O.S. O remains at −2 throughout the reaction, and that of H, at +1. The O.S. of Cl increases from −1 in Cl^- 0 in Cl_2. Cl^- is oxidized to Cl_2. The reaction is an oxidation–reduction reaction.

(b) The O.S. of H is +1 on both sides of the equation. Oxygen remains at O.S. −2 throughout. The O.S. phosphorus is +5 in $H_2PO_4^-$ and also +5 in HPO_4^{2-}. There are no changes in O.S. This is not a oxidation–reduction reaction. (It is, in fact, an acid–base reaction.)

Assess

Because many redox reactions involve H^+, OH^-, or insoluble ionic compounds, it is easy to confuse a red reaction with an acid–base or a precipitation reaction. It is important that you remember the defining featur of each type of reaction. Precipitation reactions involve the combination of ions in solution to produce an ins uble precipitate, acid–base reactions involve proton (H^+) transfer, and redox reactions involve changes in ox dation states.

PRACTICE EXAMPLE A: Identify whether each of the following is an oxidation–reduction reaction.

(a) $(NH_4)_2SO_4(aq) + Ba(NO_3)_2(aq) \longrightarrow BaSO_4(s) + 2\,NH_4NO_3(aq)$
(b) $2\,Pb(NO_3)_2(s) \longrightarrow 2\,PbO(s) + 4\,NO_2(g) + O_2(g)$

PRACTICE EXAMPLE B: Identify the species that is oxidized and the species that is reduced in the reaction below.

$$5\,VO^{2+}(aq) + MnO_4^-(aq) + H_2O(l) \longrightarrow 5\,VO_2^+(aq) + Mn^{2+}(aq) + 2\,H^+(aq)$$

Oxidation and Reduction Half-Reactions

The reaction illustrated in Figure 5-13 is an oxidation–reduction reaction. chemical equation for the reaction is given below.

$$Zn(s) + Cu^{2+}(aq) \longrightarrow Zn^{2+}(aq) + Cu(s)$$

We can show that the reaction is an oxidation–reduction reaction by eval ing changes in oxidation state, but there is another especially useful wa establish this. Think of the reaction as involving two **half-reactions** occur at the same time—an oxidation and a reduction. The overall reaction is sum of the two half-reactions. We can represent the half-reactions by l equations and the overall reaction by an overall equation.

Oxidation:	$Zn(s) \longrightarrow Zn^{2+}(aq) + 2\,e^-$	(
Reduction:	$Cu^{2+}(aq) + 2\,e^- \longrightarrow Cu(s)$	(
Overall:	$Zn(s) + Cu^{2+}(aq) \longrightarrow Zn^{2+}(aq) + Cu(s)$	(

In half-reaction (5.18), Zn is oxidized—its oxidation state increases from +2. This change corresponds to a loss of two electrons by each zinc atom. In reaction (5.19), Cu^{2+} is reduced—its oxidation state decreases from +2 to 0. change corresponds to the gain of two electrons by each Cu^{2+} ion. To summa

- An oxidation–reduction (redox) reaction is one in which the oxida states of the reactants change.
- **Oxidation** is a process in which the O.S. of some element *increases*. Bec an increase in O.S. may be imagined to occur by a loss of electr

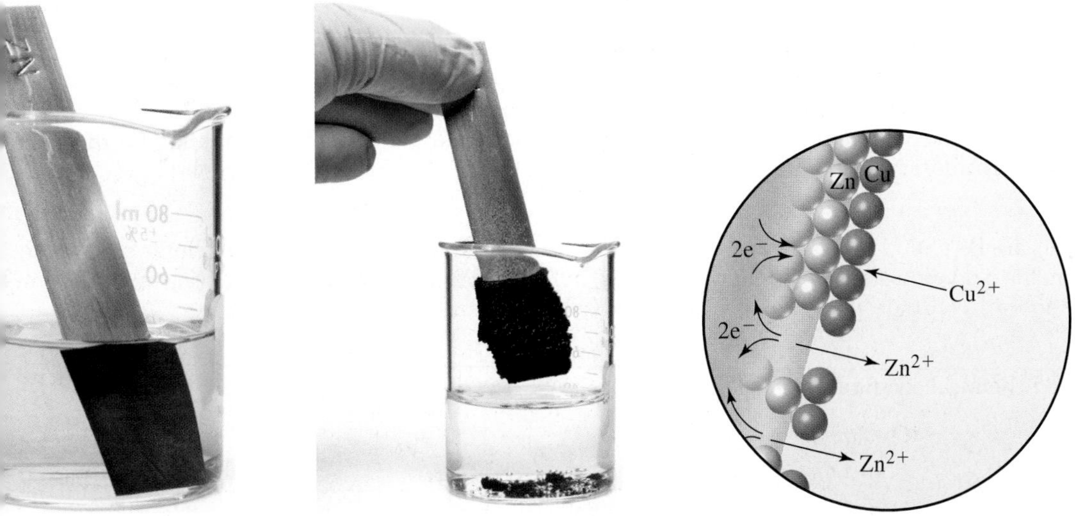

(a) (b)

GURE 5-13
xidation–reduction reaction
zinc rod above an aqueous solution of copper(II) sulfate. **(b)** Following immersion e Zn rod in the $CuSO_4(aq)$ for several hours, the blue color of $Cu^{2+}(aq)$ ppears and a deposit of copper forms on the rod. The copper metal is further zed to black CuO(s). In the microscopic view on the left, the gray spheres esent Zn atoms and the red spheres represent Cu atoms. In the reaction, Zn atoms electrons to the metal surface and enter the solution as Zn^{2+} ions. Cu^{2+} ions from olution pick up electrons and deposit on the metal surface as atoms of solid er.

ock/Science Source

electrons appear on the *right* side of the half-equation we write for an oxidation.

Reduction is a process in which the O.S. of some element *decreases*. A decrease in O.S. may be imagined to occur by a gain of electrons and so, in the half-equation we write for a reduction, electrons appear on the *left* side.

Oxidation and reduction half-reactions must always occur together, and the total number of electrons associated with the oxidation must *equal* the total number associated with the reduction.

is important to emphasize that splitting an oxidation–reduction reaction half-reactions involving electron loss and electron gain is a formalism should not be taken too literally. Although redox reactions may be imag- to involve electron transfer from one reactant to another, it is not neces- y true that electrons are actually transferred. As an example, consider the tion below.

$$NO_2^- + HClO \longrightarrow NO_3^- + H^+ + Cl^-$$

reaction is certainly a redox reaction (because the O.S. of N changes from o +5 and the O.S. of Cl changes from +1 to −1). However, the steps that r at the molecular level do not actually involve a transfer of electrons from O to NO_2^-.

igure 5-14 and Example 5-5 suggest some fundamental questions about ation–reduction. For example,

Why does Fe react with HCl(aq), displacing $H_2(g)$, whereas Cu does not?
Why does Fe form Fe^{2+} and not Fe^{3+} in this reaction?

EXAMPLE 5-5 Representing an Oxidation–Reduction Reaction Through Half-Equations and an Overall Equation

Write equations for the oxidation and reduction processes that occur and the overall equation for the reacti of iron with hydrochloric acid solution to produce $H_2(g)$ and Fe^{2+}. The reaction is shown in Figure 5-14.

Analyze

The reactants are Fe(s) and HCl(aq), and the products are $H_2(g)$ and $FeCl_2(aq)$, a soluble ionic compound. the reaction, the oxidation state of iron changes from 0 in Fe to +2 in $FeCl_2$, and the oxidation state of hydrog changes from +1 in HCl to 0 in H_2. Thus, iron is oxidized and hydrogen is reduced.

Solve

The balanced chemical equations are as follows.

$$\begin{array}{lrcl} \textit{Oxidation:} & Fe(s) & \longrightarrow & Fe^{2+}(aq) + 2\,e^- \\ \textit{Reduction:} & 2\,H^+(aq) + 2\,e^- & \longrightarrow & H_2(g) \\ \hline \textit{Overall:} & Fe(s) + 2\,H^+(aq) & \longrightarrow & Fe^{2+}(aq) + H_2(g) \end{array}$$

Assess

This example illustrates that iron dissolves in acid solution. Iron is a major component of steel and the reacti in this example contributes to the corrosion of steel that is exposed to air and moisture. For example, H^+ io from acid rain cause Fe atoms in steel to become oxidized to Fe^{2+} ions. The oxidation of iron creates small p in the steel surface and leads to corrosion.

(a) (b) (c)

▲ FIGURE 5-14
Displacement of H^+ (aq) by iron—Example 5-5 illustrated
(a) An iron nail is wrapped in a piece of copper screen. **(b)** The nail and screen are placed in HCl(aq). Hydrogen gas is evolved as the nail reacts. **(c)** The nail reacts completely and produces Fe^{2+}(aq), but the copper does not react.

PRACTICE EXAMPLE A: Represent the reaction of aluminum with hydrochloric acid to produce $AlCl_3(aq)$ and $H_2(g)$ by oxidation and reduction half-equations and an overall equation.

PRACTICE EXAMPLE B: Represent the reaction of chlorine gas with aqueous sodium bromide to produce liqui bromine and aqueous sodium chloride by oxidation and reduction half-equations and an overall equation.

Even now, you can probably see that the answers to these questions lie in relative abilities of Fe and Cu atoms to become oxidized. Fe is more ea oxidized than is Cu; also, Fe is more readily oxidized to Fe^{2+} than it is to F We can give more complete answers after we develop specific criteria desc ing oxidation and reduction in Chapter 19. For now, the information in Tabl should be helpful. The table lists some common metals that react with acic lace $H_2(g)$ and a few that do not. As noted in the table, most of the group 1 2 metals are so reactive that they will react with cold water to produce ;) and a solution of the metal hydroxide.

TABLE 5.4 Behavior of Some Common Metals with Nonoxidizing Acids[a]

React to Produce $H_2(g)$	Do Not React
Alkali metals (group 1)[b] Alkaline earth metals (group 2)[b] Al, Zn, Fe, Sn, Pb	Cu, Ag, Au, Hg

[a] A nonoxidizing acid (for example, HCl, HBr, HI) is one in which the only possible reduction half-reaction is the reduction of H^+ to H_2. Additional possibilities for metal–acid reactions are considered in Chapter 19.
[b] With the exception of Be and Mg, all group 1 and group 2 metals also react with cold water to produce $H_2(g)$; the metal hydroxide is the other product.

5-4 CONCEPT ASSESSMENT

sregarding the fact that the expressions below are not balanced, is it likely at either represents a reaction that could possibly occur? Explain.

$$MnO_4^-(aq) + Cr_2O_7^{2-}(aq) + H^+(aq) \longrightarrow MnO_2(s) + Cr^{3+}(aq) + H_2O(l)$$
$$Cl_2(g) + OH^-(aq) \longrightarrow Cl^-(aq) + ClO_3^-(aq) + H_2O(l)$$

5 Balancing Oxidation–Reduction Equations

chemical reaction, atoms are neither created nor lost; they are simply ranged. We used this idea in Chapter 4 to balance chemical equations by ection. In a redox reaction, we have additional considerations. When balng a redox reaction, we make use of a formalism in which we imagine trons are transferred from one substance to another. Using this formalism, must keep track of electrons and the charge that these electrons carry. refore, in balancing the chemical equation for a redox reaction, we focus ally on three factors: (1) the number of atoms of each type, (2) the number lectrons transferred, and (3) the total charges on reactants and products. should point out, however, that if we balance the equation with respect to number of atoms and the number of electrons transferred, then the equa- is automatically balanced with respect to the total charges.

◀ *Half-reaction* or *half-equation*, which is it? *Reaction* refers to the actual process. An *equation* is a notation we write out to indicate the formulas of the reactants and products, and their mole relationships. Many chemists refer to what we balance here as half-reactions but half-equation is more proper.

ecause it is very challenging to deal with all three of these factors simul-ously, only a small proportion of oxidation–reduction equations can be nced by simple inspection. To make this point clear, consider the follow- reactions, each of which appears to be balanced.

$$2\,MnO_4^- + H_2O_2 + 6\,H^+ \longrightarrow 2\,Mn^{2+} + 3\,O_2 + 4\,H_2O \quad (5.21)$$
$$2\,MnO_4^- + 3\,H_2O_2 + 6\,H^+ \longrightarrow 2\,Mn^{2+} + 4\,O_2 + 6\,H_2O \quad (5.22)$$
$$2\,MnO_4^- + 5\,H_2O_2 + 6\,H^+ \longrightarrow 2\,Mn^{2+} + 5\,O_2 + 8\,H_2O \quad (5.23)$$
$$2\,MnO_4^- + 7\,H_2O_2 + 6\,H^+ \longrightarrow 2\,Mn^{2+} + 6\,O_2 + 10\,H_2O \quad (5.24)$$

◀ For brevity, the physical forms of reactants and products are not included in these equations. For all ions, as well as H_2O_2, the physical form is "(aq)." Of course, the physical forms of oxygen and water are $O_2(g)$ and $H_2O(l)$.

hese are just a few of the "balanced" chemical equations we can write for the tion. However, only one, equation (5.23), is properly balanced.

o balance the chemical equation for a redox reaction, we need to make use systematic approach that considers each of the relevant factors in turn.

Although several methods are available, we emphasize one that focuses on balancing separate half-equations and then on combining the two h equations to obtain the overall chemical equation. The method is summar below. An alternative method is described in Exercise 98.

The Half-Equation Method

The basic steps in this method of balancing a redox equation are as follow

- Write and balance separate half-equations for oxidation and reductic
- Adjust coefficients in the two half-equations so that the same numbe electrons appear in each half-equation.
- Add together the two half-equations (canceling out electrons) to ob the balanced overall equation.

Balancing the half-equations involves several steps. A detailed descriptio the method is given in Table 5.5. The method is appropriate for reactions occur in an acidic solution. Because an acidic solution contains an exces H^+ ions, the method uses H^+ ions in balancing the half-equations.

TABLE 5.5 Balancing Equations for Redox Reactions in Acidic Aqueous Solutions by the Half-Equation Method: A Summary

- Write the equations for the oxidation and reduction half-reactions.
- In each half-equation
  (1) Balance atoms of all the elements except H and O
  (2) Balance oxygen by using H_2O
  (3) Balance hydrogen by using H^+
  (4) Balance charge by using electrons
- If necessary, equalize the number of electrons in the oxidation and reduction ha equations by multiplying one or both half-equations by appropriate integers.
- Add the half-equations, then cancel species common to both sides of the over equation.
- Check that both numbers of atoms and charges balance.

EXAMPLE 5-6 Balancing the Equation for a Redox Reaction in an Acidic Solution

The reaction described by expression (5.25) below is used to determine the sulfite ion concentration present i wastewater from a papermaking plant. Use the half-equation method to obtain a balanced equation for thi reaction in an acidic solution.

$$SO_3^{2-}(aq) + MnO_4^-(aq) \longrightarrow SO_4^{2-}(aq) + Mn^{2+}(aq) \qquad \text{(5.25}$$

Analyze

The reaction occurs in acidic aqueous solution. We can use the method summarized in Table 5.5 to balance it.

Solve

The O.S. of sulfur increases from +4 in SO_3^{2-} to +6 in SO_4^{2-}. The O.S. of Mn decreases from +7 in MnO_4^- to +2 in Mn^{2+}. Thus, SO_3^{2-} is oxidized and MnO_4^- is reduced.

Step 1. *Write skeleton half-equations based on the species undergoing oxidation and reduction.* The half-equations are

$$SO_3^{2-}(aq) \longrightarrow SO_4^{2-}(aq)$$
$$MnO_4^-(aq) \longrightarrow Mn^{2+}(aq)$$

Step 2. *Balance each half-equation for numbers of atoms, in this order:*

- atoms other than H and O
- O atoms, by adding H_2O with the appropriate coefficient
- H atoms, by adding H^+ with the appropriate coefficient

The other atoms (S and Mn) are already balanced in the half-equations. To balance O atoms, we add one H_2O molecule to the left side of the first half-equation and four to the right side of the second.

$$SO_3^{2-}(aq) + H_2O(l) \longrightarrow SO_4^{2-}(aq)$$
$$MnO_4^-(aq) \longrightarrow Mn^{2+}(aq) + 4\,H_2O(l)$$

To balance H atoms, we add two H^+ ions to the right side of the first half-equation and eight to the left side of the second.

$$SO_3^{2-}(aq) + H_2O(l) \longrightarrow SO_4^{2-}(aq) + 2\,H^+(aq)$$
$$MnO_4^-(aq) + 8\,H^+(aq) \longrightarrow Mn^{2+}(aq) + 4\,H_2O(l)$$

Step 3. *Balance each half-equation for electric charge.* Add the number of electrons necessary to get the same electric charge on both sides of each half-equation. By doing this, you will see that the half-equation in which electrons appear on the right side is the *oxidation half-equation.* The other half-equation, with electrons on the left side, is the *reduction half-equation.*

Oxidation: $$SO_3^{2-}(aq) + H_2O(l) \longrightarrow SO_4^{2-}(aq) + 2\,H^+(aq) + 2\,e^-$$
(net charge on each side, −2)

Reduction: $$MnO_4^-(aq) + 8\,H^+(aq) + 5\,e^- \longrightarrow Mn^{2+}(aq) + 4\,H_2O(l)$$
(net charge on each side, +2)

Step 4. *Obtain the overall redox equation by combining the half-equations.* Multiply the oxidation half-equation by 5 and the reduction half-equation by 2. This results in 10 e^- on each side of the overall equation. These terms cancel out. *Electrons must not appear in the final equation.*

$$5\,SO_3^{2-}(aq) + 5\,H_2O(l) \longrightarrow 5\,SO_4^{2-}(aq) + 10\,H^+(aq) + \cancel{10\,e^-}$$
$$2\,MnO_4^-(aq) + 16\,H^+(aq) + \cancel{10\,e^-} \longrightarrow 2\,Mn^{2+}(aq) + 8\,H_2O(l)$$

Overall: $$5\,SO_3^{2-}(aq) + 2\,MnO_4^-(aq) + 5\,H_2O(l) + 16\,H^+(aq) \longrightarrow 5\,SO_4^{2-}(aq) + 2\,Mn^{2+}(aq) + 8\,H_2O(l) + 10\,H^+(aq)$$

Step 5. *Simplify.* The overall equation should not contain the same species on both sides. Subtract 5 H_2O from each side of the equation in step 4. This leaves 3 H_2O on the right. Also subtract 10 H^+ from each side, leaving 6 H^+ on the left.

$$5\,SO_3^{2-}(aq) + 2\,MnO_4^-(aq) + 6\,H^+(aq) \longrightarrow 5\,SO_4^{2-}(aq) + 2\,Mn^{2+}(aq) + 3\,H_2O(l)$$

Step 6. *Verify.* Check the overall equation to ensure that it is balanced both for numbers of atoms and electric charge. For example, show that in the balanced equation from step 5, the net charge on each side of the equation is −6: $(5 \times -2) + (2 \times -1) + (6 \times +1) = (5 \times -2) + (2 \times +2) = -6$.

Assess

The final check completed in step 6 gives us confidence that our result is correct. This is an important step; always take the time to complete it. It is also worth pointing out that, in this example, there was only one atom per formula that was oxidized or reduced. (Refer to the skeleton half-equations given in step 1.) Many students have difficulty balancing half-equations in which more than one atom per formula is oxidized or reduced, as is the case when $Cr_2O_7^{2-}$ is reduced to Cr^{3+}. Had we used $Cr_2O_7^{2-}$ instead of MnO_4^- in equation (5.25), the balanced chemical equation for the reaction would have been $3\,SO_3^{2-} + Cr_2O_7^{2-} + 8\,H^+ \rightarrow 3\,SO_4^{2-} + 2\,Cr^{3+} + 4\,H_2O$.

PRACTICE EXAMPLE A: Balance the equation for this reaction in acidic solution.

$$Fe^{2+}(aq) + MnO_4^-(aq) \longrightarrow Fe^{3+}(aq) + Mn^{2+}(aq)$$

PRACTICE EXAMPLE B: Balance the equation for this reaction in acidic solution.

$$UO^{2+}(aq) + Cr_2O_7^{2-}(aq) \longrightarrow UO_2^{2+}(aq) + Cr^{3+}(aq)$$

Balancing Redox Equations in a Basic Solution

To balance equations for redox reactions in a basic solution, a step or must be added to the procedure used in Example 5-6. In basic solution, C not H^+, must appear in the final balanced equation. (Recall that, in b solutions, OH^- ions are present in excess.) Because both OH^- and H_2O tain H and O atoms, at times it is hard to decide on which side of the equations to put each one. One simple approach is to treat the reactio though it were occurring in an acidic solution, and balance it as in Example Then, add to each side of the overall redox equation a number of OH^- equal to the number of H^+ ions. Where H^+ and OH^- appear on the s side of the equation, combine them to produce H_2O molecules. If H_2O appears on both sides of the equation, subtract the same number of molecules from each side, leaving a remainder of H_2O on just one side. method is illustrated in Example 5-7. For ready reference, the procedu summarized in Table 5.6.

▶ As an alternative method for half-equations in basic solutions, add *two* OH^- for every O required to the O-deficient side, and add *one* H_2O to the other side (the net effect is adding one O to the O-deficient side). Then add *one* H_2O for every H required to the H-deficient side, and add *one* OH^- to the other side (the net effect is adding one H to the H-deficient side).

TABLE 5.6 Balancing Equations for Redox Reactions in Basic Aqueous Solutions by the Half-Equation Method: A Summary

- Balance the equation as if the reaction were occurring in acidic medium by using the method for acidic aqueous solutions summarized in Table 5-5.
- Add a number of OH^- ions equal to the number of H^+ ions to both sides of t overall equation.
- On the side of the overall equation containing both H^+ and OH^- ions, combi them to form H_2O molecules. If H_2O molecules now appear on both sides of the overall equation, cancel the same number from each side, leaving a remainder of H_2O on just one side.
- Check that both numbers of atoms and charges balance.

EXAMPLE 5-7 Balancing the Equation for a Redox Reaction in Basic Solution

Balance the equation for the reaction in which cyanide ion is oxidized to cyanate ion by permanganate ion in basic solution, and the permanganate is reduced to $MnO_2(s)$.

$$MnO_4^-(aq) + CN^-(aq) \longrightarrow MnO_2(s) + OCN^-(aq) \qquad (5.2$$

Analyze

The reaction occurs in basic solution. We can balance it by using the method described in Table 5.6. The hal reactions and the overall reaction are initially treated as though they were occurring in an acidic solution an finally, the overall equation is adjusted to a basic solution.

Solve

Step 1. *Write half-equations for the oxidation and reduction half-reactions, and balance them for Mn, C, and N atoms*

$$MnO_4^-(aq) \longrightarrow MnO_2(s)$$
$$CN^-(aq) \longrightarrow OCN^-(aq)$$

Step 2. *Balance the half-equations for O and H atoms. Add H_2O and/or H^+ as required.* In the MnO_4^- half-equatio there are four O's on the left and two on the right. Adding 2 H_2O balances the O's on the right. Sinc there are now four H's on the right, it is necessary to add 4 H^+ on the left side to balance them. In th CN^- half-equation, there is one O on the right but none on the left, so H_2O must be added to the le side and 2 H^+ to the right.

$$MnO_4^-(aq) + 4\,H^+(aq) \longrightarrow MnO_2(s) + 2\,H_2O(l)$$
$$CN^-(aq) + H_2O(l) \longrightarrow OCN^-(aq) + 2\,H^+(aq)$$

Step 3. *Balance the half-equations for electric charge by adding the appropriate numbers of electrons.*

Reduction: $MnO_4^-(aq) + 4H^+(aq) + 3e^- \longrightarrow MnO_2(s) + 2H_2O(l)$

Oxidation: $CN^-(aq) + H_2O(l) \longrightarrow OCN^-(aq) + 2H^+(aq) + 2e^-$

Step 4. *Combine the half-equations to obtain an overall redox equation.* Multiply the reduction half-equation by two and the oxidation half-equation by three to obtain the common multiple $6e^-$ in each half-equation. Make the appropriate cancellations of H^+ and H_2O.

$$2MnO_4^-(aq) + 8H^+(aq) + \cancel{6e^-} \longrightarrow 2MnO_2(s) + 4H_2O(l)$$
$$3CN^-(aq) + 3H_2O(l) \longrightarrow 3OCN^-(aq) + 6H^+(aq) + \cancel{6e^-}$$

Overall:
$$MnO_4^-(aq) + 3CN^-(aq) + 2H^+(aq) \longrightarrow 2MnO_2(s) + 3OCN^-(aq) + H_2O(l)$$

Step 5. *Change from an acidic to a basic medium by adding $2OH^-$ to both sides of the overall equation; combine $2H^+$ and $2OH^-$ to form $2H_2O$, and simplify.*

$$2MnO_4^-(aq) + 3CN^-(aq) + 2H^+(aq) + 2OH^-(aq) \longrightarrow 2MnO_2(s) + 3OCN^-(aq) + H_2O(l) + 2OH^-(aq)$$

$$2MnO_4^-(aq) + 3CN^-(aq) + 2H_2O(l) \longrightarrow 2MnO_2(s) + 3OCN^-(aq) + H_2O(l) + 2OH^-(aq)$$

Subtract one H_2O molecule from each side to obtain the overall balanced redox equation for reaction (5.24).

$$2MnO_4^-(aq) + 3CN^-(aq) + H_2O(l) \longrightarrow 2MnO_2(s) + 3OCN^-(aq) + 2OH^-(aq)$$

Step 6. *Verify.* Check the final overall equation to ensure that it is balanced both for number of atoms and for electric charge. For example, show that in the balanced equation from step 5, the net charge on each side of the equation is −5.

ssess

We can use the rules for assigning oxidation states (given in Table 3.2) to deduce that the O.S. of Mn changes from +7 in MnO_4^- to +4 in MnO_2. We conclude that the other substance, CN^-, is oxidized. (The rules do not allow us to assign oxidation states to C and N in CN^- or NCO^-.) Even though we cannot identify the oxidation states of C or N, we could still balance the equation for the reaction. That is one advantage of the methods we presented in Tables 5-5 and 5-6. In Chapter 10, we will discuss another method for assigning oxidation states and learn how to determine the oxidation states of C and N in such species as CN^- or NCO^-.

RACTICE EXAMPLE A: Balance the equation for this reaction in basic solution.

$$S(s) + OCl^-(aq) \longrightarrow SO_3^{2-}(aq) + Cl^-(aq)$$

RACTICE EXAMPLE B: Balance the equation for this reaction in basic solution.

$$MnO_4^-(aq) + SO_3^{2-}(aq) \longrightarrow MnO_2(s) + SO_4^{2-}(aq)$$

proportionation Reactions

ome oxidation–reduction reactions, called **disproportionation reactions,** same substance is both oxidized and reduced. An example is the decom- tion of hydrogen peroxide, H_2O_2, into H_2O and O_2:

$$2H_2O_2(aq) \longrightarrow 2H_2O(l) + O_2(g) \qquad \textbf{(5.27)}$$

eaction (5.27), the oxidation state of oxygen changes from −1 in H_2O_2 to in H_2O (a reduction) and to 0 in $O_2(g)$ (an oxidation). H_2O_2 is both oxi- ed and reduced. Reaction (5.27) produces $O_2(g)$, which bubbles out of the tion. (See Figure 5-15.)

nother example is the disproportionation of $S_2O_3^{2-}$ in acid solution:

$$S_2O_3^{2-}(aq) + 2H^+(aq) \longrightarrow S(s) + SO_2(g) + H_2O(l) \qquad \textbf{(5.28)}$$

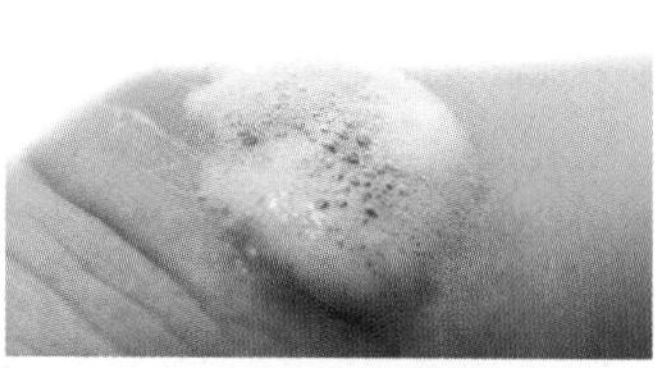

Schankz/Shutterstock

▲ FIGURE 5-15
Antiseptic action of hydrogen peroxide solution
Dilute aqueous solutions of hydrogen peroxide (usually 3% hydrogen peroxide by mass) were once commonly used to clean minor cuts and scrapes. A solution of hydrogen peroxide bubbles when poured over a cut because of the production of gaseous oxygen.

The oxidation states of S are +2 in $S_2O_3^{2-}$, 0 in S, and +4 in SO_2. Thus, S_2 is simultaneously oxidized and reduced. Solutions of sodium thiosu ($Na_2S_2O_3$) are often used in the laboratory in redox reactions, and stock s tions of $Na_2S_2O_3$ sometimes develop small deposits of sulfur, a pale ye solid, over time.

The same substance appears on the left side in each half-equation for a proportionation reaction. The balanced half-equations and overall equa for reaction (5.28) are given below.

Oxidation: $S_2O_3^{2-}(aq) + H_2O(l) \longrightarrow 2\,SO_2(g) + 2\,H^+(aq) + 4\,e^-$

Reduction: $S_2O_3^{2-}(aq) + 6\,H^+(aq) + 4\,e^- \longrightarrow 2\,S(s) + 3\,H_2O(l)$

$2\,S_2O_3^{2-}(aq) + 4\,H^+(aq) \longrightarrow 2\,S(s) + 2\,SO_2(g) + 2\,H_2O($

Overall: $S_2O_3^{2-}(aq) + 2\,H^+(aq) \longrightarrow S(s) + SO_2(g) + H_2O(l)$

5-5 CONCEPT ASSESSMENT

Is it possible for two different reactants in a redox reaction to yield a single product? Explain.

5-2 ARE YOU WONDERING?

How do we balance the equation for a redox reaction tha occurs in a medium other than an aqueous solution?

An example of such a reaction is the oxidation of $NH_3(g)$ to NO(g), the first st in the commercial production of nitric acid.

$$NH_3(g) + O_2(g) \longrightarrow NO(g) + H_2O(g)$$

Some people prefer a method called the *oxidation-state change method (see below)* f reactions of this type, but the half-equation method works just as well. All that needed is to treat the reaction as if it were occurring in an aqueous acidic solutic H^+ should appear as both a reactant and a product and cancel out in the overa equation, as seen below.

Oxidation: $4\{NH_3 + H_2O \longrightarrow NO + 5\,H^+ + 5\,e^-\}$

Reduction: $5\{O_2 + 4\,H^+ + 4\,e^- \longrightarrow 2\,H_2O\}$

Overall: $4\,NH_3 + 5\,O_2 \longrightarrow 4\,NO + 6\,H_2O$

Oxidation-state change method

In this method, changes in oxidation states are identified. That of nitrog increases from −3 in NH_3 to +2 in NO, corresponding to a "loss" of five electro per N atom. That of oxygen decreases from 0 in O_2 to −2 in NO and H_2O, corr sponding to a "gain" of two electrons per O atom. The proportion of N to O ator must be 2 N (loss of 10 e^-) to 5 O (gain of 10 e^-).

$$2\,NH_3 + \frac{5}{2}O_2 \longrightarrow 2\,NO + 3\,H_2O \quad \text{or} \quad 4\,NH_3 + 5\,O_2 \longrightarrow 4\,NO + 6\,H_2O$$

5-6 Oxidizing and Reducing Agents

Chemists frequently use the terms *oxidizing agent* and *reducing agent* to desc certain reactants in redox reactions, as in statements like "fluorine gas is a p erful oxidizing agent," or "calcium metal is a good reducing agent." Le briefly consider the meaning of these terms.

Species	O.S.
NO_3^-	+5
N_2O_4	+4
NO_2^-	+3
NO	+2
N_2O	+1
N_2	0
NH_2OH	−1
N_2H_4	−2
NH_3	−3

This species cannot be oxidized further.

This species cannot be reduced further.

◀ FIGURE 5-16
Oxidation states of nitrogen: Identifying oxidizing and reducing agents
In NO_3^- and N_2O_4, nitrogen is in one of its highest possible oxidation states (O.S.). These species are usually oxidizing agents in redox reactions. In N_2H_4 and NH_3, nitrogen is in one of its lowest oxidation states. These species are usually reducing agents.

ı a redox reaction, the substance that makes it possible for some other sub-ce to be oxidized is called the **oxidizing agent**, or **oxidant**. In doing so, the lizing agent is itself reduced. Similarly, the substance that causes some ·r substance to be reduced is called the **reducing agent**, or **reductant**. In reaction, the reducing agent is itself oxidized. Or, stated in other ways,

n oxidizing agent (oxidant)

causes another substance to be oxidized

contains an element whose oxidation state *decreases* in a redox reaction

gains electrons (electrons are found on the left side of its half-equation)

is reduced

reducing agent (reductant)

causes another substance to be reduced

contains an element whose oxidation state *increases* in a redox reaction

loses electrons (electrons are found on the right side of its half-equation)

is oxidized

ı general, a substance with an element in one of its highest possible oxida- states is an oxidizing agent. If the element is in one of its lowest possible lation states, the substance is a reducing agent.

igure 5-16 shows the range of oxidation states of nitrogen and the species to ch they correspond. The oxidation state of the nitrogen in dinitrogen tetroxide O_4) is nearly the maximum value attainable, and hence N_2O_4 is generally an lizing agent. Conversely, the nitrogen atom in hydrazine (N_2H_4) is in nearly lowest oxidation state, and hence hydrazine is generally a reducing agent. en these two liquid compounds are mixed, a vigorous reaction takes place:

$$N_2O_4(l) + 2\,N_2H_4(l) \longrightarrow 3\,N_2(g) + 4\,H_2O(g)$$

his reaction, N_2O_4 is the oxidizing agent and N_2H_4 is the reducing agent. s reaction releases so much energy that it is used in some rocket propul- ı systems.

'ertain substances in which the oxidation state of an element is between its ıest and lowest possible values may act as oxidizing agents in some instances reducing agents in others. For example, in the reaction of hydrazine with rogen to produce ammonia, hydrazine acts as an oxidizing agent.

$$N_2H_4(l) + H_2(g) \longrightarrow 2\,NH_3(g)$$

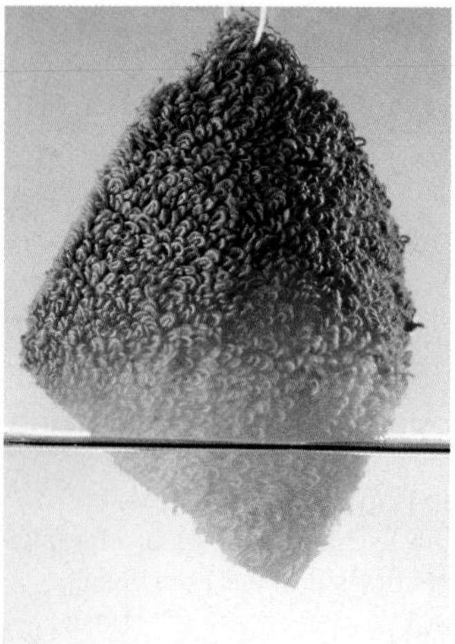

Kristen Brochmann/Fundamental Photographs

▲ FIGURE 5-17
Bleaching action of NaOCl(aq)
A red cloth becomes white when immersed in NaOCl(aq), which oxidizes the red pigment to colorless products.

Permanganate ion, MnO_4^-, is a versatile oxidizing agent that has m uses in the chemical laboratory. In the next section, we describe its use ir quantitative analysis of iron—that is, the determination of the exact (quar tive) amount of iron in an iron-containing material. Ozone, $O_3(g)$, a triat form of oxygen, is an oxidizing agent used in water purification, as in the dation of the organic compound phenol, C_6H_5OH.

$$C_6H_5OH(aq) + 14\,O_3(g) \longrightarrow 6\,CO_2(g) + 3\,H_2O(l) + 14\,O_2(g)$$

Aqueous sodium hypochlorite, NaOCl(aq), is a powerful oxidizing age is the active ingredient in many liquid chlorine bleaches. The bleaching ac of NaOCl(aq) is associated with the reduction of the OCl^- ion to Cl^-; the trons required for the reduction come from colored compounds in stains. bleaching action of NaOCl(aq) is demonstrated in Figure 5-17.

Thiosulfate ion, $S_2O_3^{2-}$, is an important reducing agent. One of its indus uses is as an *antichlor* to destroy residual chlorine from the bleaching of fib

$$S_2O_3^{2-}(aq) + 4\,Cl_2(aq) + 5\,H_2O(l) \longrightarrow 2\,HSO_4^-(aq) + 8\,H^+(aq) + 8\,Cl^-(a$$

Oxidizing and reducing agents also play important roles in biolog systems—in photosynthesis (using solar energy to synthesize gluc metabolism (oxidizing glucose), and the transport of oxygen.

EXAMPLE 5-8 Identifying Oxidizing and Reducing Agents

Hydrogen peroxide, H_2O_2, is a versatile chemical. Its uses include bleaching wood pulp and fabric and substituting for chlorine in water purification. One reason for its versatility is that it can be either a oxidizing or a reducing agent. For the following reactions, identify whether hydrogen peroxide is an oxidizin or reducing agent.

(a) $H_2O_2(aq) + 2\,Fe^{2+}(aq) + 2\,H^+(aq) \longrightarrow 2\,H_2O(l) + 2\,Fe^{3+}(aq)$

(b) $5\,H_2O_2(aq) + 2\,MnO_4^-(aq) + 6\,H^+(aq) \longrightarrow 8\,H_2O(l) + 2\,Mn^{2+}(aq) + 5\,O_2(g)$

Analyze

Before we can identify the oxidizing and reducing agents, we must first assign oxidation states, and then iden tify which substance is being oxidized and which substance is being reduced. The oxidizing agent cause another substance to be oxidized. The reducing agent causes another substance to be reduced.

Solve

(a) Fe^{2+} is oxidized to Fe^{3+} and because H_2O_2 makes this possible, it is an oxidizing agent. Viewed anothe way, we see that the oxidation state of oxygen in H_2O_2 is -1. In H_2O, it is -2. Hydrogen peroxide i reduced and thereby acts as an oxidizing agent.

(b) MnO_4^- is reduced to Mn^{2+}, and H_2O_2 makes this possible. In this situation, hydrogen peroxide is reducing agent. Or, the oxidation state of oxygen increases from -1 in H_2O_2 to 0 in O_2. Hydroge peroxide is oxidized and thereby acts as a reducing agent.

Assess

The versatility of H_2O_2 lies in its ability to act as an oxidizing agent and a reducing agent. When H_2O_2 acts a an oxidizing agent, it is reduced to H_2O, in an acidic solution, as was the case in part (a), or to OH^- in basi solution. When it acts as a reducing agent, it is oxidized to $O_2(g)$, as was the case in part (b).

PRACTICE EXAMPLE A: Is $H_2(g)$ an oxidizing or reducing agent in the reaction below? Explain.

$$2\,NO_2(g) + 7\,H_2(g) \longrightarrow 2\,NH_3(g) + 4\,H_2O(g)$$

PRACTICE EXAMPLE B: Identify the oxidizing agent and the reducing agent in the following reaction.

$$4\,Au(s) + 8\,CN^-(aq) + O_2(g) + 2\,H_2O(l) \longrightarrow 4\,[Au(CN)_2]^-(aq) + 4\,OH^-(aq)$$

5-6 CONCEPT ASSESSMENT

ıewspaper account of an accidental spill of hydrochloric acid in an area where dium hydroxide solution was also stored spoke of the potential hazardous ease of chlorine gas if the two solutions should come into contact. Was this accurate accounting of the hazard involved? Explain.

Stoichiometry of Reactions in Aqueous Solutions: Titrations

ır objective is to obtain the maximum yield of a product at the lowest cost, vould generally choose the most expensive reactant as the limiting reac- and use excess amounts of the other reactants. This is the case in most ipitation reactions. In some instances, as in determining the concentra- of a solution, we may not be interested in the products of a reaction but in the relationship between two reactants. Then we have to carry out the tion in such a way that neither reactant is in excess. A method that has been used for doing this is known as *titration*. The glassware typically l in a titration is shown in Figure 5-18.

solution of one reactant is placed in a small beaker or flask. Another tant, also in solution and commonly referred to as titrant, is in a *buret*, a ;, graduated tube equipped with a stopcock valve. The second solution is ly added to the first by manipulating the stopcock. In a **titration**, a reac- is carried out by the carefully controlled addition of one solution to ther. The trick is to stop the titration at the point where both reactants e reacted completely, a condition called the **equivalence point** of the

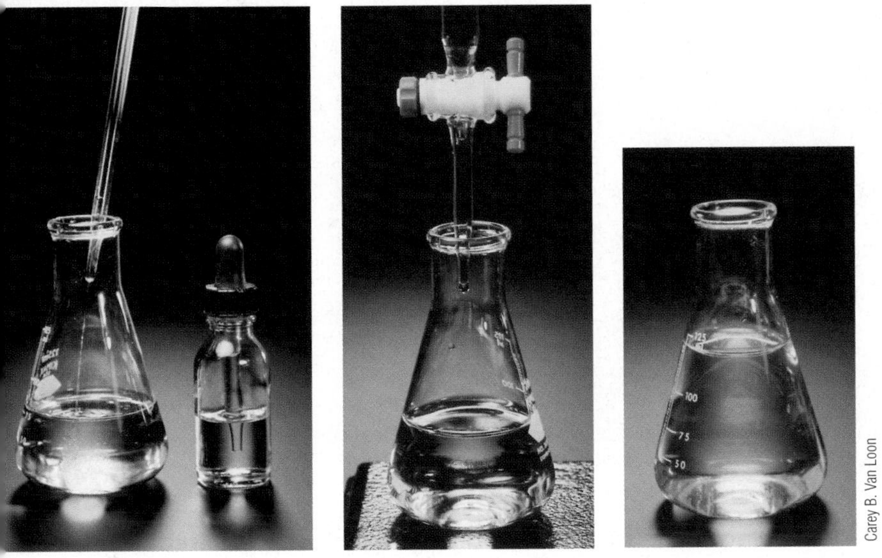

(a) (b) (c)

GURE 5-18
acid–base titration—Example 5-9 illustrated
5.00 mL sample of vinegar, a small quantity of water, and a few drops of nolphthalein indicator are added to a flask. **(b)** 0.1000 M NaOH from a previously l buret is slowly added. **(c)** As long as the acid is in excess, the solution in the flask ains colorless. When the acid has been neutralized, an additional drop of NaOH(aq) es the solution to become slightly basic. The phenolphthalein indicator turns a pink. The first lasting appearance of the pink color is taken to be the equivalence t of the titration.

titration. Key to every titration is that at the equivalence point, the two re-tants have combined in stoichiometric proportions; both have been sumed, and neither remains in excess.

In modern chemical laboratories, appropriate measuring instrument used to signal when the equivalence point is reached. However, substa called **indicators** are still widely used. A very small quantity of an approp indicator is added to the reaction mixture to produce a color change at or near the equivalence point. Figure 5-18 illustrates the neutralization of an by a base by the titration technique. Calculations that use titration data much the same as those introduced in Section 4-3, and they are also illustr in Example 5-9.

▶ The key to a successful acid–base titration is in selecting the right indicator. We learn how to do this when we consider theoretical aspects of titration in Chapter 17.

Suppose we need a $KMnO_4(aq)$ solution of exactly known molarity, c to 0.020 M. We cannot prepare this solution by weighing out the requ amount of $KMnO_4(s)$ and dissolving it in water. The solid is not pure, an actual purity (that is, the mass percent $KMnO_4$) is *not known*. Conversely can obtain iron wire in essentially pure form and allow the wire to react an acid to yield $Fe^{2+}(aq)$. $Fe^{2+}(aq)$ is oxidized to $Fe^{3+}(aq)$ by $KMnO_4(a$ an acidic solution. By determining the volume of $KMnO_4(aq)$ require oxidize a known quantity of $Fe^{2+}(aq)$, we can calculate the exact mola of the $KMnO_4(aq)$. This procedure, which determines the exact molarity solution, is called **standardization of a solution**. It is illustrated in Exar 5-10 and Figure 5-19.

▶ *Standardize* means determine the concentration of a solution, usually to three or four significant figures. It is not so important that the concentration be a round number (as 0.1000 vs. 0.1035 M), but rather that the concentration be accurately known.

(a) (b) (c)

▲ FIGURE 5-19
Standardizing a solution of an oxidizing agent through a redox titration—Example 5-10 illustrated
(a) The solution contains a known amount of Fe^{2+}, and the buret is filled with the intensely colored $KMnO_4(aq)$ to be standardized. **(b)** As it is added to the strongly acidic solution of $Fe^{2+}(aq)$, the $KMnO_4(aq)$ is immediately decolorized as a result o reaction (5.29). **(c)** When all the Fe^{2+} has been oxidized to Fe^{3+}, additional $KMnO_4($ has nothing left to oxidize and the solution turns a distinctive pink. Even a fraction of a drop of the $KMnO_4(aq)$ beyond the equivalence point is sufficient to cause thi pink coloration.

EXAMPLE 5-9 Using Titration Data to Establish the Concentrations of Acids and Bases

Vinegar is a dilute aqueous solution of acetic acid produced by the bacterial fermentation of apple cider, wine, or other carbohydrate material. The legal minimum acetic acid content of vinegar is 4% by mass. A 5.00 mL sample of a particular vinegar is titrated with 38.08 mL of 0.1000 M NaOH. Does this sample exceed the minimum limit? (Vinegar has a density of about 1.01 g/mL.)

Analyze

Acetic acid, CH_3COOH, is a weak acid and NaOH is a strong base. The reaction between CH_3COOH and NaOH is an acid–base neutralization reaction. We start by writing a balanced chemical equation for the reaction. We must convert mL NaOH to g CH_3COOH. The necessary conversions are as follows:

$$\text{mL NaOH} \longrightarrow \text{L NaOH} \longrightarrow \text{mol NaOH} \longrightarrow \text{mol } CH_3COOH \longrightarrow \text{g } CH_3COOH$$

Solve

The balanced chemical equation for the reaction is given below.

$$CH_3COOH(aq) + NaOH(aq) \longrightarrow NaCH_3COO(aq) + H_2O(l)$$

$$? \text{ g } CH_3COOH = 38.08 \text{ mL} \times \frac{1 \text{ L}}{1000 \text{ mL}} \times \frac{0.1000 \text{ mol NaOH}}{1 \text{ L}} \times \frac{1 \text{ mol } CH_3COOH}{1 \text{ mol NaOH}} \times \frac{60.05 \text{ g } CH_3COOH}{1 \text{ mol } CH_3COOH}$$

$$= 0.2287 \text{ g } CH_3COOH$$

This mass of CH_3COOH is found in 5.00 mL of vinegar of density 1.01 g/mL. The percent mass of CH_3COOH is

$$\% \, CH_3COOH = \frac{0.2287 \text{ g } CH_3COOH}{5.00 \text{ mL vinegar}} \times \frac{1 \text{ mL vinegar}}{1.01 \text{ g vinegar}} \times 100\% = 4.53\% \, CH_3COOH$$

The vinegar sample exceeds the legal minimum limit but only slightly. There is also a standard for the maximum amount of acetic acid allowed in vinegar. A vinegar producer might use this titration technique to ensure that the vinegar stays between these limits.

Assess

Such problems as this one involve many steps or conversions. Try to break the problem into simpler ones involving fewer steps or conversions. It may also help to remember that solving a stoichiometry problem involves three steps: (1) converting to moles, (2) converting between moles, and (3) converting from moles. Use molarities and molar masses to carry out volume–mole conversions and gram–mole conversions, respectively, and stoichiometric factors to carry out mole–mole conversions. The stoichiometric factors are constructed from a balanced chemical equation.

PRACTICE EXAMPLE A: A particular solution of NaOH is supposed to be approximately 0.100 M. To determine the exact molarity of the NaOH(aq), a 0.5000 g sample of $KHC_8H_4O_4$ is dissolved in water and titrated with 24.03 mL of the NaOH(aq). What is the actual molarity of the NaOH(aq)?

$$HC_8H_4O_4^-(aq) + OH^-(aq) \longrightarrow C_8H_4O_4^{2-}(aq) + H_2O(l)$$

PRACTICE EXAMPLE B: A 0.235 g sample of a solid that is 92.5% NaOH and 7.5% $Ca(OH)_2$, by mass, requires 45.6 mL of a HCl(aq) solution for its titration. What is the molarity of the HCl(aq)?

5-7 CONCEPT ASSESSMENT

10.00 mL sample of 0.311 M KOH is added to 31.10 mL of 0.100 M HCl. the resulting mixture acidic, basic, or exactly neutral? Explain.

EXAMPLE 5-10 Standardizing a Solution for Use in Redox Titrations

A piece of iron wire weighing 0.1568 g is converted to $Fe^{2+}(aq)$ and requires 26.24 mL of a $KMnO_4(aq)$ solutio
for its titration. What is the molarity of the $KMnO_4(aq)$?

$$5\,Fe^{2+}(aq) + MnO_4^-(aq) + 8\,H^+(aq) \longrightarrow 5\,Fe^{3+}(aq) + Mn^{2+}(aq) + 4\,H_2O(l) \qquad \textbf{(5.2}$$

Analyze

The key to a titration calculation is that the amounts of two reactants consumed in the titration are stoichiome rically equivalent—neither reactant is in excess. We are given a mass of Fe (0.1568 g) and must determine th number of moles of $KMnO_4$ in the 26.24 mL sample. The following conversions are required:

$$\text{g Fe} \longrightarrow \text{mol Fe} \longrightarrow \text{mol } Fe^{2+} \longrightarrow \text{mol } MnO_4^- \longrightarrow \text{mol } KMnO_4$$

The third conversion, from mol Fe^{2+} to mol MnO_4^-, requires a stoichiometric factor constructed from the coe ficients in equation (5.29).

Solve

First, determine the amount (in moles) of $KMnO_4$ consumed in the titration.

$$? \text{ mol } KMnO_4 = 0.1568 \text{ g Fe} \times \frac{1 \text{ mol Fe}}{55.847 \text{ g Fe}} \times \frac{1 \text{ mol } Fe^{2+}}{1 \text{ mol Fe}} \times \frac{1 \text{ mol } MnO_4^-}{5 \text{ mol } Fe^{2+}} \times \frac{1 \text{ mol } KMnO_4}{1 \text{ mol } MnO_4^-}$$

$$= 5.615 \times 10^{-4} \text{ mol } KMnO_4$$

The volume of solution containing the 5.615×10^{-4} mol $KMnO_4$ is 26.24 mL = 0.02624 L, which means that

$$\text{concn } KMnO_4 = \frac{5.615 \times 10^{-4} \text{ mol } KMnO_4}{0.02624 \text{ L}} = 0.02140 \text{ M } KMnO_4$$

Assess

For practical applications, such as for titrations, we use solutions with molarities that are neither very large nc very small. Typically, the molarities lie in the range 0.001 M to 0.1 M. If you calculate a molarity that is signif cantly larger than 0.1 M, or significantly smaller than 0.001 M, then you must carefully check your calculatio for possible errors.

PRACTICE EXAMPLE A: A 0.376 g sample of an iron ore is dissolved in acid, and the iron reduced to $Fe^{2+}(aq)$ and then titrated with 41.25 mL of 0.02140 M $KMnO_4$. Determine the mass percent Fe in the iron ore. [*Hint:* Use equation (5.29).]

PRACTICE EXAMPLE B: Another substance that may be used to standardize $KMnO_4(aq)$ is sodium oxalate. 0.2482 g $Na_2C_2O_4$ is dissolved in water and titrated with 23.68 mL $KMnO_4$, what is the molarity of the $KMnO_4(aq)$

$$MnO_4^-(aq) + C_2O_4^{2-}(aq) + H^+(aq) \longrightarrow Mn^{2+}(aq) + CO_2(g) + H_2O(l) \quad \text{(not balanced)}$$

MasteringCHEMISTRY www.masteringchemistry.com

Access to a plentiful supply of "pure water" is something most of us take for granted. Still, most of us would agree that purification of water is an important concern. The way in which water is purified depends on how it is to be used or on how it has been used. For a discussion of the removal or destruction of undesirable chemical substances from water, go to the Focus On feature for Chapter 5, entitled Water Treatment, on the MasteringChemistry site.

Summary

The Nature of Aqueous Solutions—Solutes in ous solution are characterized as **nonelectrolytes**, :h do not produce ions, or **electrolytes**, which pro- ions. Ionic compounds produce ions by **dissociation** :eas molecular compounds produce ions via **ioniza-** **Weak electrolytes** ionize to a limited extent, and **ıg electrolytes** dissociate or ionize almost completely ions. In addition to the molarity based on the solute as ıole, a solution's concentration can be stated in terms e molarities of the individual solute species present— :cules and ions.

Precipitation Reactions—Some reactions in ous solution involve the combination of ions to yield ter-insoluble solid—a **precipitate**. Precipitation reac- ; are generally represented by **net ionic equations**, a in which only the reacting ions and solid precipitates ;hown, and **spectator ions** are deleted. Precipitation tions usually can be predicted by using a few simple bility guidelines (Table 5.1).

Acid–Base Reactions—According to the Arrhenius ry, a substance that ionizes to produce H^+ ions in :ous solution is an **acid**. It is a **strong acid** (Table 5.2) if onization goes essentially to completion and a **weak** if the ionization is limited. Similarly, a **base** produces ¯ ions in aqueous solution and is either a **strong** : (Table 5.2) or a **weak base**, depending on the extent he ionization. According to the Brønsted–Lowry ry, in an acid–base reaction, protons are transferred ı the acid (the **proton donor**) to the base (the **proton** **ptor**). In an acid-base **neutralization reaction**, an acid a base react to form an ionic compound, a **salt**. Water ht also be a product. Some reactions in which gases evolved can also be treated as acid–base reactions le 5.3).

5-4 Oxidation–Reduction Reactions: Some General Principles—In an **oxidation–reduction (redox) reaction** certain atoms undergo an increase in oxidation state, a process called **oxidation**. Other atoms undergo a decrease in oxidation state, or **reduction**. Another useful view of redox reactions is as the combination of separate **half-reactions** for the oxidation and the reduction.

5-5 Balancing Oxidation–Reduction Equations—An effective way to balance a redox equation is to break down the reaction into separate half-reactions, write and balance half-equations for these half-reactions, and recombine the balanced half-equations into an overall balanced equation (Table 5.5). A slight variation of this method is used for a reaction that occurs in a basic aqueous solution (Table 5.6). A redox reaction in which the same substance is both oxidized and reduced is called a **disproportionation reaction**.

5-6 Oxidizing and Reducing Agents—The **oxidizing agent (oxidant)** is the key reactant in an oxidation half-reaction and is *reduced* in the redox reaction. The **reducing agent (reductant)** is the key reactant in a reduction half-reaction and is *oxidized* in the redox reaction. Some substances act only as oxidizing agents; others, only as reducing agents. Many can act as either, depending on the reaction (Fig. 5-16).

5-7 Stoichiometry of Reactions in Aqueous Solutions: Titrations—A common laboratory technique applicable to precipitation, acid–base, and redox reactions is **titration**. The key point in a titration is the **equivalence point**, which can be observed with the aid of an **indicator**. Titration data can be used to establish a solution's molarity, called **standardization of a solution**, or to provide other information about the compositions of samples being analyzed.

Integrative Example

ium dithionite, $Na_2S_2O_4$, is an important reducing agent. One interesting use is the reduction of chromate ion to luble chromium(III) hydroxide by dithionite ion, $S_2O_4^{2-}$, in basic solution. Sulfite ion is another product. The chro- e ion may be present in wastewater from a chromium-plating plant, for example. What mass of $Na_2S_2O_4$ is consumed reaction with 100.0 L of wastewater having $[CrO_4^{2-}] = 0.0148$ M?

◀ White solid sodium dithionite, $Na_2S_2O_4$, is added to a yellow solution of potassium chromate, $K_2CrO_4(aq)$ (left). A product of the reaction is gray-green chromium(III) hydroxide, $Cr(OH)_3(s)$ (right).

Analyze

The phrase "reduction of chromate" tells us that the reaction between CrO_4^{2-} and $S_2O_4^{2-}$ is a redox reaction. We obtain a balanced chemical equation for the reaction by using the method summarized in Table 5.6, and then co 100.0 L of wastewater into grams of $Na_2S_2O_4$. The necessary conversions are as follows:

$$100.0\text{ L wastewater} \longrightarrow \text{mol } CrO_4^{2-} \longrightarrow \text{mol } S_2O_4^{2-} \longrightarrow \text{mol } Na_2S_2O_4 \longrightarrow \text{g } Na_2S_2O_4$$

Solve

1. *Write an ionic expression representing the reaction.*

$$CrO_4^{2-}(aq) + S_2O_4^{2-}(aq) + OH^-(aq) \longrightarrow Cr(OH)_3(s) + SO_3^{2-}$$

2. *Balance the redox equation.* Begin by writing skeleton half-equations.

$$CrO_4^{2-} \longrightarrow Cr(OH)_3$$
$$S_2O_4^{2-} \longrightarrow SO_3^{2-}$$

Balance the half-equations for Cr, S, O, and H atoms as if the half-reactions occur in acidic solution.

$$CrO_4^{2-} + 5\,H^+ \longrightarrow Cr(OH)_3 + H_2O$$
$$S_2O_4^{2-} + 2\,H_2O \longrightarrow 2\,SO_3^{2-} + 4\,H^+$$

Balance the half-equations for charge, and label them as oxidation and reduction.

Oxidation: $S_2O_4^{2-} + 2\,H_2O \longrightarrow 2\,SO_3^{2-} + 4\,H^+ +$

Reduction: $CrO_4^{2-} + 5\,H^+ + 3\,e^- \longrightarrow Cr(OH)_3 + H_2O$

Combine the half-equations into an overall equation.

$$3 \times \{S_2O_4^{2-} + 2\,H_2O \longrightarrow 2\,SO_3^{2-} + 4\,H^+ + 2\,e^-\}$$
$$2 \times \{CrO_4^{2-} + 5\,H^+ + 3\,e^- \longrightarrow Cr(OH)_3 + H_2O\}$$
$$3\,S_2O_4^{2-} + 2\,CrO_4^{2-} + 4\,H_2O \longrightarrow 6\,SO_3^{2-} + 2\,Cr(OH)_3 + 2\,H$$

3. *Change the conditions to basic solution.* Add 2 OH^- to each side of the equation for acidic solution, and combine 2 H^+ and 2 OH^- to form 2 H_2O on the right.

$$3\,S_2O_4^{2-} + 2\,CrO_4^{2-} + 4\,H_2O + 2\,OH^- \longrightarrow 6\,SO_3^{2-} + 2\,Cr(OH)_3 + 2$$

Subtract 2 H_2O from each side of the equation to obtain the final balanced equation.

$$3\,S_2O_4^{2-}(aq) + 2\,CrO_4^{2-}(aq) + 2\,H_2O(l) + 2\,OH^-(aq) \longrightarrow 6\,SO_3^{2-}(aq) + 2\,Cr(OH$$

4. *Complete the stoichiometric calculation.* The conversion pathway is
100.0 L waste water $\longrightarrow$ mol CrO_4^{2-} $\longrightarrow$ mol $S_2O_4^{2-}$ $\longrightarrow$ mol $Na_2S_2O_4$ $\longrightarrow$ g $Na_2S_2O_4$.

$$?\text{ g } Na_2S_2O_4 = 100.0\text{ L} \times \frac{0.0148\text{ mol } CrO_4^{2-}}{1\text{ L}} \times \frac{3\text{ mol } S_2O_4^{2-}}{2\text{ mol } CrO_4^{2-}} \times \frac{1\text{ mol } Na_2S_2O_4}{1\text{ mol } S_2O_4^{2-}} \times \frac{174.1\text{ g } Na_2S_2O_4}{1\text{ mol } Na_2S_2O_4} = 387\text{ g } Na_2S_2O_4$$

Assess

In solving this problem the major effort was to balance a redox equation for a reaction under basic conditions. allowed us to find the molar relationship between dithionite and chromate ions. The remainder of the problem w stoichiometry calculation for a reaction in solution, much like Example 4-10 (page 127). A quick check of the final re involves (1) ensuring that the redox equation is balanced, and (2) noting that the number of moles of CrO_4^{2-} is a 1.5 (i.e., 100 × 0.0148), that the number of moles of $S_2O_4^{2-}$ is about 2.25 (i.e., 1.5 × 3/2), and that the mass of Na_2S_2 somewhat more than 350 (i.e., 2.25 × 175).

PRACTICE EXAMPLE A: The amount of potassium chlorate, $KClO_3$, in a 0.1432 g sample was determined as follo The sample was dissolved in 50.00 mL of 0.09101 M $Fe(NO_3)_2$ and the solution was acidified. The excess Fe^{2+} back-titrated with 12.59 mL of 0.08362 M $Ce(NO_3)_4$ solution. What is the percentage by mass of $KClO_3$ in the sam Chemical equations for the reactions involved are given below.

$$ClO_3^-(aq) + Fe^{2+}(aq) \rightarrow Cl^-(aq) + Fe^{3+}(aq) \text{ (not balanced)}$$
$$Fe^{2+}(aq) + Ce^{4+}(aq) \rightarrow Fe^{3+}(aq) + Ce^{3+}(aq)$$

PRACTICE EXAMPLE B: The amount of arsenic, As, in a 7.25 g sample was determined by converting all the ars to arsenous acid (H_3AsO_3), and then titrating H_3AsO_3 with 23.77 mL of 0.02144 M $KMnO_4$. What is the percentag mass of As in the sample? An unbalanced expression for the titration reaction is $H_3AsO_3(aq) + MnO_4^-(aq$ $H_3AsO_4(aq) + Mn^{2+}(aq)$.

Exercises

ong Electrolytes, Weak Electrolytes, and Nonelectrolytes

Using information from this chapter, indicate whether each of the following substances in aqueous solution is a nonelectrolyte, weak electrolyte, or strong electrolyte. **(a)** HC_6H_5O; **(b)** Li_2SO_4; **(c)** MgI_2; **(d)** $(CH_3CH_2)_2O$; **(e)** $Sr(OH)_2$.

Select the **(a)** best and **(b)** poorest electrical conductors from the following solutions, and explain the reason for your choices: 0.10 M NH_3; 0.10 M NaCl; 0.10 M CH_3COOH (acetic acid); 0.10 M CH_3CH_2OH (ethanol).

What response would you expect in the apparatus of Figure 5-4 if the solution tested were 1.0 M HCl? What response would you expect if the solution were both 1.0 M HCl and 1.0 M CH_3COOH?

$NH_3(aq)$ conducts electric current only weakly. The same is true for $CH_3COOH(aq)$. When these solutions are mixed, however, the resulting solution is a good conductor. How do you explain this?

Sketches **(a–c)** are molecular views of the solute in an aqueous solution. For each of the sketches, indicate whether the solute is a strong, weak, or nonelectrolyte; and which of these substances it is: sodium chloride, propionic acid, hypochlorous acid, ammonia, barium bromide, ammonium chloride, methanol.

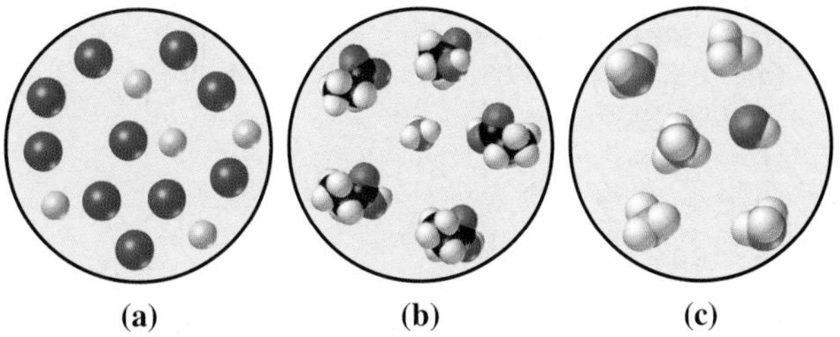

(a) (b) (c)

6. After identifying the three substances represented by the sketches in Exercise 5, sketch molecular views of aqueous solutions of the remaining four substances listed.

Concentrations

Determine the concentration of the ion indicated in each solution. **(a)** $[K^+]$ in 0.238 M KNO_3; **(b)** $[NO_3^-]$ in 0.167 M $Ca(NO_3)_2$; **(c)** $[Al^{3+}]$ in 0.083 M $Al_2(SO_4)_3$; **(d)** $[Na^+]$ in 0.209 M Na_3PO_4.

Which solution has the greatest $[SO_4^{2-}]$? **(a)** 0.075 M H_2SO_4; **(b)** 0.22 M $MgSO_4$; **(c)** 0.15 M Na_2SO_4; **(d)** 0.080 M $Al_2(SO_4)_3$; **(e)** 0.20 M $CuSO_4$.

A solution is prepared by dissolving 0.132 g $Ba(OH)_2 \cdot 8\,H_2O$ in 275 mL of water solution. What is $[OH^-]$ in this solution?

A solution is 0.126 M KCl and 0.148 M $MgCl_2$. What are $[K^+]$, $[Mg^{2+}]$, and $[Cl^-]$ in this solution?

Express the following data for cations in solution as molarities. **(a)** 14.2 mg Ca^{2+}/L; **(b)** 32.8 mg K^+/100 mL; **(c)** 225 μg Zn^{2+}/mL.

What molarity of NaF(aq) corresponds to a fluoride ion content of 0.9 mg F^-/L, the federal government's recommended limit for fluoride ion in drinking water?

13. Which of the following aqueous solutions has the highest concentration of K^+? **(a)** 0.0850 M K_2SO_4; **(b)** a solution containing 1.25 g KBr/100 mL; **(c)** a solution having 8.1 mg K^+/mL.

14. Which aqueous solution has the greatest $[H^+]$? **(a)** 0.011 M CH_3COOH; **(b)** 0.010 M HCl; **(c)** 0.010 M H_2SO_4; **(d)** 1.00 M NH_3. Explain your choice.

15. How many milligrams of MgI_2 must be added to 250.0 mL of 0.0876 M KI to produce a solution with $[I^-] = 0.1000$ M?

16. If 18.2 mL H_2O evaporates from 1.00 L of a solution containing 15.5 mg K_2SO_4/mL, what is $[K^+]$ in the solution that remains?

17. Assuming the volumes are additive, what is the $[Cl^-]$ in a solution obtained by mixing 225 mL of 0.625 M KCl and 615 mL of 0.385 M $MgCl_2$?

18. Assuming the volumes are additive, what is the $[NO_3^-]$ in a solution obtained by mixing 275 mL of 0.283 M KNO_3, 328 mL of 0.421 M $Mg(NO_3)_2$, and 784 mL of H_2O?

edicting Precipitation Reactions

Complete each of the following as a net ionic equation, indicating whether a precipitate forms. If no reaction occurs, so state.

(a) $Na^+ + Br^- + Pb^{2+} + 2\,NO_3^- \longrightarrow$

(b) $Mg^{2+} + 2\,Cl^- + Cu^{2+} + SO_4^{2-} \longrightarrow$

(c) $Fe^{3+} + 3\,NO_3^- + Na^+ + OH^- \longrightarrow$

Complete each of the following as a net ionic equation. If no reaction occurs, so state.

(a) $Ca^{2+} + 2\,I^- + 2\,Na^+ + CO_3^{2-} \longrightarrow$

(b) $Ba^{2+} + S^{2-} + 2\,Na^+ + SO_4^{2-} \longrightarrow$

(c) $2\,K^+ + S^{2-} + Ca^{2+} + 2\,Cl^- \longrightarrow$

21. Predict in each case whether a reaction is likely to occur. If so, write a net ionic equation.

(a) $HI(aq) + Zn(NO_3)_2(aq) \longrightarrow$

(b) $CuSO_4(aq) + Na_2CO_3(aq) \longrightarrow$

(c) $Cu(NO_3)_2(aq) + Na_3PO_4(aq) \longrightarrow$

22. Predict in each case whether a reaction is likely to occur. If so, write a net ionic equation.

(a) $AgNO_3(aq) + CuCl_2(aq) \longrightarrow$

(b) $Na_2S(aq) + FeCl_2(aq) \longrightarrow$

(c) $Na_2CO_3(aq) + AgNO_3(aq) \longrightarrow$

23. What reagent solution might you use to separate the cations in the following mixtures, that is, with one ion appearing in solution and the other in a precipitate? [*Hint:* Refer to Table 5.1, and consider water also to be a reagent.]
(a) $BaCl_2(s)$ and $MgCl_2(s)$
(b) $MgCO_3(s)$ and $Na_2CO_3(s)$
(c) $AgNO_3(s)$ and $Cu(NO_3)_2(s)$

24. What reagent solution might you use to separate the cations in each of the following mixtures? [*Hint:* Refer to Exercise 23.]
(a) $PbSO_4(s)$ and $Cu(NO_3)_2(s)$
(b) $Mg(OH)_2(s)$ and $BaSO_4(s)$
(c) $PbCO_3(s)$ and $CaCO_3(s)$

25. You are provided with NaOH(aq), K_2SO_4 $Mg(NO_3)_2(aq)$, $BaCl_2(aq)$, NaCl(aq), $Sr(NO_3)_2$ $AgNO_3(aq)$, and $BaSO_4(s)$. Write net ionic equa to show how you would use one or more of t reagents to obtain **(a)** $SrSO_4(s)$; **(b)** Mg(OH) **(c)** KCl(aq).

26. Write net ionic equations to show how you would one or more of the reagents in Exercise 25 to ob **(a)** $BaSO_4(s)$; **(b)** AgCl(s); **(c)** $KNO_3(aq)$.

Acid–Base Reactions

27. Complete each of the following as a *net ionic equation.* If no reaction occurs, so state.
(a) $Ba^{2+} + 2\,OH^- + CH_3COOH \longrightarrow$
(b) $H^+ + Cl^- + CH_3CH_2COOH \longrightarrow$
(c) $FeS(s) + H^+ + I^- \longrightarrow$
(d) $K^+ + HCO_3^- + H^+ + NO_3^- \longrightarrow$
(e) $Mg(s) + H^+ \longrightarrow$

28. Every antacid contains one or more ingredients capable of reacting with excess stomach acid (HCl). The essential neutralization products are CO_2 and/or H_2O. Write net ionic equations to represent the neutralizing action of the following popular antacids.
(a) Alka-Seltzer (sodium bicarbonate)
(b) Tums (calcium carbonate)
(c) milk of magnesia (magnesium hydroxide)
(d) Maalox (magnesium hydroxide, aluminum hydroxide)
(e) Rolaids [$NaAl(OH)_2CO_3$]

29. In this chapter, we described an acid as a subst capable of producing H^+ and a salt as the ionic pound formed by the neutralization of an acid base. Write ionic equations to show that sod hydrogen sulfate has the characteristics of both a and an acid (sometimes called an *acid salt*).

30. A neutralization reaction between an acid and a is a common method of preparing useful salts. net ionic equations showing how the following could be prepared in this way: **(a)** $(NH_4)_2H$ **(b)** NH_4NO_3; and **(c)** $(NH_4)_2SO_4$.

31. Which solutions would you use to precipitate M from an aqueous solution of $MgCl_2$? Explain choice. **(a)** $KNO_3(aq)$; **(b)** $NH_3(aq)$; **(c)** $H_2SO_4($ **(d)** $HC_2H_3O_2(aq)$.

32. Determine which of the following react(s) with HC to produce a gas, and write a net ionic equation(s the reaction(s). **(a)** Na_2SO_4; **(b)** $KHSO_3$; **(c)** Zn(O **(d)** $CaCl_2$.

Oxidation–Reduction (Redox) Equations

33. Assign oxidation states to the elements involved in the following reactions. Indicate which are redox reactions and which are not.
(a) $MgCO_3(s) + 2\,H^+(aq) \longrightarrow Mg^{2+}(aq) + H_2O(l) + CO_2(g)$
(b) $Cl_2(aq) + 2\,Br^-(aq) \longrightarrow 2\,Cl^-(aq) + Br_2(aq)$
(c) $Ag(s) + 2\,H^+(aq) + NO_3^-(aq) \longrightarrow Ag^+(aq) + H_2O(l) + NO_2(g)$
(d) $2\,Ag^+(aq) + CrO_4^{2-}(aq) \longrightarrow Ag_2CrO_4(s)$

34. Explain why these reactions cannot occur as written.
(a) $Fe^{3+}(aq) + MnO_4^-(aq) + H^+(aq) \longrightarrow Mn^{2+}(aq) + Fe^{2+}(aq) + H_2O(l)$
(b) $H_2O_2(aq) + Cl_2(aq) \longrightarrow ClO^-(aq) + O_2(g) + H^+(aq)$

35. Complete and balance these half-equations.
(a) $SO_3^{2-} \longrightarrow S_2O_3^{2-}$ (acidic solution)
(b) $HNO_3 \longrightarrow N_2O(g)$ (acidic solution)
(c) $Al(s) \longrightarrow Al(OH)_4^-$ (basic solution)
Indicate whether oxidation or reduction is involved.

36. Complete and balance these half-equations.
(a) $C_2O_4^{2-} \longrightarrow CO_2$ (acidic solution)
(b) $Cr_2O_7^{2-} \longrightarrow Cr^{3+}$ (acidic solution)
(c) $MnO_4^- \longrightarrow MnO_2$ (basic solution)
Indicate whether oxidation or reduction is involve

37. Balance these equations for redox reactions occur in acidic solution.
(a) $MnO_4^- + I^- \longrightarrow Mn^{2+} + I_2(s)$
(b) $BrO_3^- + N_2H_4 \longrightarrow Br^- + N_2(g)$
(c) $VO_4^{3-} + Fe^{2+} \longrightarrow VO^{2+} + Fe^{3+}$
(d) $UO^{2+} + NO_3^- \longrightarrow UO_2^{2+} + NO(g)$

38. Balance these equations for redox reactions occurr in acidic solution.
(a) $P_4(s) + NO_3^- \longrightarrow H_2PO_4^- + NO(g)$
(b) $S_2O_3^{2-} + MnO_4^- \longrightarrow SO_4^{2-} + Mn^{2+}$
(c) $HS^- + HSO_3^- \longrightarrow S_2O_3^{2-}$
(d) $Fe^{3+} + NH_3OH^+ \longrightarrow Fe^{2+} + N_2O(g)$

39. Balance these equations for redox reactions in b solution.
(a) $MnO_2(s) + ClO_3^- \longrightarrow MnO_4^- + Cl^-$
(b) $Fe(OH)_3(s) + OCl^- \longrightarrow FeO_4^{2-} + Cl^-$
(c) $ClO_2 \longrightarrow ClO_3^- + Cl^-$
(d) $Ag(s) + CrO_4^{2-} \longrightarrow Ag^+ + Cr(OH)_3(s)$

Balance these equations for redox reactions occurring in basic solution.
(a) $CrO_4^{2-} + S_2O_4^{2-} \longrightarrow Cr(OH)_3(s) + SO_3^{2-}$
(b) $[Fe(CN)_6]^{3-} + N_2H_4 \longrightarrow [Fe(CN)_6]^{4-} + N_2(g)$
(c) $Fe(OH)_2(s) + O_2(g) \longrightarrow Fe(OH)_3(s)$
(d) $CH_3CH_2OH + MnO_4^- \longrightarrow CH_3COO^- + MnO_2(s)$

Balance these equations for disproportionation reactions.
(a) $Cl_2(g) \longrightarrow Cl^- + ClO_3^-$ (basic solution)
(b) $S_2O_4^{2-} \longrightarrow S_2O_3^{2-} + HSO_3^-$ (acidic solution)

Balance these equations for disproportionation reactions.
(a) $MnO_4^{2-} \longrightarrow MnO_2(s) + MnO_4^-$ (basic solution)
(b) $P_4(s) \longrightarrow H_2PO_2^- + PH_3(g)$ (basic solution)
(c) $S_8(s) \longrightarrow S^{2-} + S_2O_3^{2-}$ (basic solution)
(d) $As_2S_3 + H_2O_2 \longrightarrow AsO_4^{3-} + SO_4^{2-}$

Write a balanced equation for these redox reactions.
(a) The oxidation of nitrite ion to nitrate ion by permanganate ion, MnO_4^-, in acidic solution (MnO_4^- ion is reduced to Mn^{2+}).
(b) The reaction of manganese(II) ion and permanganate ion in basic solution to form solid manganese dioxide.
(c) The oxidation of ethanol by dichromate ion in acidic solution, producing chromium(III) ion, acetaldehyde (CH_3CHO), and water as products.

44. Write a balanced equation for the redox reactions.
(a) The reaction of aluminum metal with hydroiodic acid.
(b) The reduction of vanadyl ion (VO^{2+}) to vanadic ion (V^{3+}) in acidic solution with zinc metal as the reducing agent.
(c) The oxidation of methanol by chlorate ion in acidic solution, producing carbon dioxide gas, water, and chlorine dioxide gas as products.

45. The following reactions do not occur in aqueous solutions. Balance their equations by the half-equation method, as suggested in Are You Wondering 5-2.
(a) $CH_4(g) + NO(g) \longrightarrow CO_2(g) + N_2(g) + H_2O(g)$
(b) $H_2S(g) + SO_2(g) \longrightarrow S_8(s) + H_2O(g)$
(c) $Cl_2O(g) + NH_3(g) \longrightarrow N_2(g) + NH_4Cl(s) + H_2O(l)$

46. The following reactions do not occur in aqueous solutions. Balance their equations by the half-equation method, as suggested in Are You Wondering 5-2.
(a) $CH_4(g) + NH_3(g) + O_2(g) \longrightarrow HCN(g) + H_2O(g)$
(b) $NO(g) + H_2(g) \longrightarrow NH_3(g) + H_2O(g)$
(c) $Fe(s) + H_2O(l) + O_2(g) \longrightarrow Fe(OH)_3(s)$

idizing and Reducing Agents

What are the oxidizing and reducing agents in the following redox reactions?
(a) $5\,SO_3^{2-} + 2\,MnO_4^- + 6\,H^+ \longrightarrow 5\,SO_4^{2-} + 2\,Mn^{2+} + 3\,H_2O$
(b) $2\,NO_2(g) + 7\,H_2(g) \longrightarrow 2\,NH_3(g) + 4\,H_2O(g)$
(c) $2\,[Fe(CN)_6]^{4-} + H_2O_2 + 2\,H^+ \longrightarrow 2\,[Fe(CN)_6]^{3-} + 2\,H_2O$

48. Thiosulfate ion, $S_2O_3^{2-}$, is a reducing agent that can be oxidized to different products, depending on the strength of the oxidizing agent and other conditions. By adding H^+, H_2O, and/or OH^- as necessary, write redox equations to show the oxidation of $S_2O_3^{2-}$ to
(a) $S_4O_6^{2-}$ by I_2 (iodide ion is another product)
(b) HSO_4^- by Cl_2 (chloride ion is another product)
(c) SO_4^{2-} by OCl^- in basic solution (chloride ion is another product)

utralization and Acid–Base Titrations

What volume of 0.0962 M NaOH is required to exactly neutralize 10.00 mL of 0.128 M HCl?

The exact neutralization of 10.00 mL of 0.1012 M $H_2SO_4(aq)$ requires 23.31 mL of NaOH. What must be the molarity of the NaOH(aq)?

$$H_2SO_4(aq) + 2\,NaOH(aq) \longrightarrow Na_2SO_4(aq) + 2\,H_2O(l)$$

How many milliliters of 2.155 M KOH are required to titrate 25.00 mL of 0.3057 M CH_3CH_2COOH (propionic acid)?

How many milliliters of 0.0750 M $Ba(OH)_2$ are required to titrate 200.0 mL of 0.0165 M HNO_3?

An NaOH(aq) solution cannot be made up to an exact concentration simply by weighing out the required mass of NaOH, because the NaOH is not pure. Also, water vapor condenses on the solid as it is being weighed. The solution must be standardized by titration. For this purpose, a 25.00 mL sample of an NaOH(aq) solution requires 28.34 mL of 0.1085 M HCl. What is the molarity of the NaOH(aq)?

$$HCl(aq) + NaOH(aq) \longrightarrow NaCl(aq) + H_2O(l)$$

54. Household ammonia, used as a window cleaner and for other cleaning purposes, is $NH_3(aq)$. The NH_3 present in a 5.00 mL sample is neutralized by 28.72 mL of 1.021 M HCl. The net ionic equation for the neutralization is

$$NH_3(aq) + H^+(aq) \longrightarrow NH_4^+(aq)$$

What is the molarity of NH_3 in the sample?

55. We want to determine the acetylsalicyclic acid content of a series of aspirin tablets by titration with NaOH(aq). Each of the tablets is expected to contain about 0.32 g of $HC_9H_7O_4$. What molarity of NaOH(aq) should we use for titration volumes of about 23 mL? (This procedure ensures good precision and allows the titration of two samples with the contents of a 50 mL buret.)

$$HC_9H_7O_4(aq) + OH^-(aq) \longrightarrow C_9H_7O_4^-(aq) + H_2O(l)$$

56. For use in titrations, we want to prepare 20 L of HCl(aq) with a concentration known to four significant figures. This is a two-step procedure beginning with the preparation of a solution of about 0.10 M HCl. A sample of this dilute HCl(aq) is titrated with a NaOH(aq) solution of known concentration.
(a) How many milliliters of concentrated HCl(aq) (d = 1.19 g/mL; 38% HCl, by mass) must be diluted with water to 20.0 L to prepare 0.10 M HCl?
(b) A 25.00 mL sample of the approximately 0.10 M HCl prepared in part (a) requires 20.93 mL of 0.1186 M NaOH for its titration. What is the molarity of the HCl(aq)?
(c) Why is a titration necessary? That is, why not prepare a standard solution of 0.1000 M HCl simply by an appropriate dilution of the concentrated HCl(aq)?

57. A 25.00 mL sample of 0.132 M HNO_3 is mixed with 10.00 mL of 0.318 M KOH. Is the resulting solution acidic, basic, or exactly neutralized?

58. A 7.55 g sample of $Na_2CO_3(s)$ is added to 125 mL of a vinegar that is 0.762 M CH_3COOH. Will the resulting solution still be acidic? Explain.

59. Refer to Example 5-9. Suppose the analysis of all vinegar samples uses 5.00 mL of the vinegar and 0.1000 M NaOH for the titration. What volume of the 0.1000 M NaOH would represent the legal minimum 4.0%, by mass, acetic acid content of the vinegar? That is, calculate the volume of 0.1000 M NaOH so that if a titration requires more than this volume, the legal minimum limit is met (less than this volume, and the limit is not met).

60. The electrolyte in a lead storage battery must have a concentration between 4.8 and 5.3 M H_2SO_4 if the battery is to be most effective. A 5.00 mL sample of a battery acid requires 49.74 mL of 0.935 M NaOH for its complete reaction (neutralization). Does the concentration of the battery acid fall within the desired range? [*Hint:* Keep in mind that the H_2SO_4 produces tw ions per formula unit.]

61. Which of the following points in a titration is re sented by the molecular view shown in the sketch

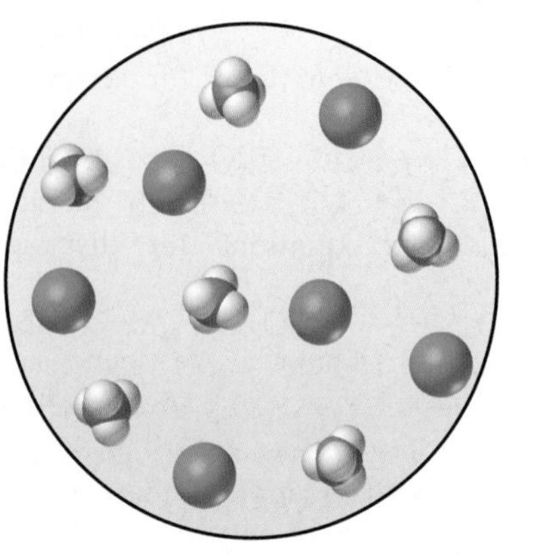

(a) 20% of the necessary titrant added in the titra of $NH_4Cl(aq)$ with HCl(aq)
(b) 20% of the necessary titrant added in the titra of $NH_3(aq)$ with HCl(aq)
(c) the equivalence point in the titration of NH_3(with HCl(aq)
(d) 120% of the necessary titrant added in the titra of $NH_3(aq)$ with HCl(aq)

62. Using the sketch in Exercise 61 as a guide, sketch molecular view of a solution in which
(a) HCl(aq) is titrated to the equivalence point w KOH(aq)
(b) $CH_3COOH(aq)$ is titrated halfway to the equ lence point with NaOH(aq).

Stoichiometry of Oxidation–Reduction Reactions

63. A $KMnO_4(aq)$ solution is to be standardized by titration against $As_2O_3(s)$. A 0.1078 g sample of As_2O_3 requires 22.15 mL of the $KMnO_4(aq)$ for its titration. What is the molarity of the $KMnO_4(aq)$?

$$5\,As_2O_3 + 4\,MnO_4^- + 9\,H_2O + 12\,H^+ \longrightarrow 10\,H_3AsO_4 + 4\,Mn^{2+}$$

64. Refer to Example 5-6. Assume that the only reducing agent present in a particular wastewater is SO_3^{2-}. If a 25.00 mL sample of this wastewater requires 31.46 mL of 0.02237 M $KMnO_4$ for its titration, what is the molarity of SO_3^{2-} in the wastewater?

65. An iron ore sample weighing 0.9132 g is dissolved in HCl(aq), and the iron is obtained as $Fe^{2+}(aq)$. This solution is then titrated with 28.72 mL of 0.05051 M $K_2Cr_2O_7$. What is the mass percent Fe in the ore sample?

$$6\,Fe^{2+} + 14\,H^+ + Cr_2O_7^{2-} \longrightarrow 6\,Fe^{3+} + 2\,Cr^{3+} + 7\,H_2O$$

66. The concentration of $Mn^{2+}(aq)$ can be determine titration with $MnO_4^-(aq)$ in basic solution 50.00 mL sample of $Mn^{2+}(aq)$ requires 78.42 m 0.04997 M $KMnO_4$ for its titration. What is [Mn^{2+} the sample?

$$Mn^{2+} + MnO_4^- \longrightarrow MnO_2(s) \quad \text{(not balance}$$

67. The titration of 5.00 mL of a saturated solutio sodium oxalate, $Na_2C_2O_4$, at 25 °C requires 25.8 of 0.02140 M $KMnO_4$ in acidic solution. What mas $Na_2C_2O_4$ in grams would be present in 1.00 L of saturated solution?

$$C_2O_4^{2-} + MnO_4^- \longrightarrow Mn^{2+} + CO_2(g) \quad \text{(not balanc}$$

68. Refer to the Integrative Example. In the treatmen 1.00×10^2 L of a wastewater solution that is 0.012 CrO_4^{2-}, how many grams of **(a)** $Cr(OH)_3(s)$ wo precipitate; **(b)** $Na_2S_2O_4$ would be consumed?

Integrative and Advanced Exercises

Write net ionic equations for the reactions depicted in photo **(a)** sodium metal reacts with water to produce hydrogen; photo **(b)** an excess of aqueous iron(III) chloride is added to the solution in **(a)**; and photo **(c)** the precipitate from **(b)** is collected and treated with an excess of HCl(aq).

(a) (b) (c)

& **(b)** William H. Breazeale; **(c)** Tom Pantages

Following are some laboratory methods occasionally used for the preparation of small quantities of chemicals. Write a balanced equation for each.
(a) preparation of $H_2S(g)$: HCl(aq) is heated with FeS(s)
(b) preparation of $Cl_2(g)$: HCl(aq) is heated with $MnO_2(s)$; $MnCl_2(aq)$ and $H_2O(l)$ are other products
(c) preparation of $N_2(g)$: Br_2 and NH_3 react in aqueous solution; NH_4Br is another product
(d) preparation of chlorous acid: an aqueous suspension of solid barium chlorite is treated with dilute $H_2SO_4(aq)$

When concentrated $CaCl_2(aq)$ is added to $Na_2HPO_4(aq)$, a white precipitate forms that is 38.7% Ca by mass. Write a net ionic equation representing the probable reaction that occurs.

You have a solution that is 0.0250 M $Ba(OH)_2$ and the following pieces of equipment: 1.00, 5.00, 10.00, 25.00, and 50.00 mL pipets and 100.0, 250.0, 500.0, and 1000.0 mL volumetric flasks. Describe how you would use this equipment to produce a solution in which $[OH^-]$ is 0.0100 M.

Sodium hydroxide used to make standard NaOH(aq) solutions for acid–base titrations is invariably contaminated with some sodium carbonate. **(a)** Explain why, except in the most precise work, the presence of this sodium carbonate generally does not seriously affect the results obtained, for example, when NaOH(aq) is used to titrate HCl(aq). **(b)** Conversely, show that if Na_2CO_3 comprises more than 1% to 2% of the solute in NaOH(aq), the titration results are affected.

A 110.520 g sample of mineral water is analyzed for its magnesium content. The Mg^{2+} in the sample is first precipitated as $MgNH_4PO_4$, and this precipitate is then converted to $Mg_2P_2O_7$, which is found to weigh 0.0549 g. Express the quantity of magnesium in the sample in parts per million (that is, in grams of Mg per million grams of H_2O).

What volume of 0.248 M $CaCl_2$ must be added to 335 mL of 0.186 M KCl to produce a solution with a concentration of 0.250 M Cl^-? Assume that the solution volumes are additive.

76. An unknown white solid consists of two compounds, each containing a different cation. As suggested in the illustration, the unknown is partially soluble in water. The solution is treated with NaOH(aq) and yields a white precipitate. The part of the original solid that is insoluble in water dissolves in HCl(aq) with the evolution of a gas. The resulting solution is then treated with $(NH_4)_2SO_4(aq)$ and yields a white precipitate. **(a)** Is it possible that any of the cations Mg^{2+}, Cu^{2+}, Ba^{2+}, Na^+, or NH_4^+ were present in the original unknown? Explain your reasoning. **(b)** What compounds could be in the unknown mixture (that is, what anions might be present)?

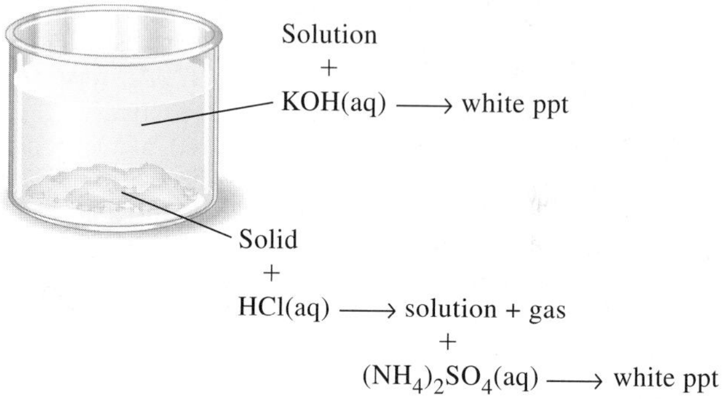

77. Balance these equations for reactions in acidic solution.
(a) $IBr + BrO_3^- + H^+ \longrightarrow IO_3^- + Br^- + H_2O$
(b) $C_2H_5NO_3 + Sn \longrightarrow NH_2OH + CH_3CH_2OH + Sn^{2+}$
(c) $As_2S_3 + NO_3^- \longrightarrow H_3AsO_4 + S + NO$
(d) $H_5IO_6 + I_2 \longrightarrow IO_3^- + H^+ + H_2O$
(e) $S_2F_2 + H_2O \longrightarrow S_8 + H_2S_4O_6 + HF$

78. Balance these equations for reactions in basic solution.
(a) $Fe_2S_3 + H_2O + O_2 \longrightarrow Fe(OH)_3 + S$
(b) $O_2^- + H_2O \longrightarrow OH^- + O_2$
(c) $CrI_3 + H_2O_2 \longrightarrow CrO_4^{2-} + IO_4^-$
(d) $Ag + CN^- + O_2 + OH^- \longrightarrow [Ag(CN)_2]^- + H_2O$
(e) $B_2Cl_4 + OH^- \longrightarrow BO_2^- + Cl^- + H_2O + H_2$

79. A method of producing phosphine, PH_3, from elemental phosphorus, P_4, involves heating the P_4 with H_2O. An additional product is phosphoric acid, H_3PO_4. Write a balanced equation for this reaction.

80. Iron (Fe) is obtained from rock that is extracted from open pit mines and then crushed. The process used to obtain the pure metal from the crushed rock produces solid waste, called *tailings*, which are stored in disposal areas near the mines. The tailings pose a serious environmental risk because they contain sulfides, such as pyrite (FeS_2), which oxidize in air to produce metal ions and H^+ ions that can enter into surface water or ground water. The oxidation of FeS_2 to Fe^{3+} is described by the unbalanced chemical expression below.

$$FeS_2(s) + O_2(g) + H_2O(l) \longrightarrow Fe^{3+}(aq) + SO_4^{2-}(aq) + H^+(aq) \quad \text{(not balanced)}$$

Thus, the oxidation of pyrite produces Fe^{3+} and H^+ ions that can leach into surface or ground water. The leaching of H^+ ions causes the water to become very acidic. To prevent acidification of nearby ground or surface water, limestone ($CaCO_3$) is added to the tailings to neutralize the H^+ ions:

$$CaCO_3(s) + 2\,H^+(aq) \longrightarrow Ca^{2+}(aq) + H_2O(l) + CO_2(g)$$

(a) Balance the equation above for the reaction of FeS_2 and O_2. [*Hint:* Start with the half-equations $FeS_2(s) \rightarrow Fe^{3+}(aq) + SO_4^{2-}(aq)$ and $O_2(g) \rightarrow H_2O(l)$.] **(b)** What is the minimum amount of $CaCO_3(s)$ required, per kilogram of tailings, to prevent contamination if the tailings contain 3% S by mass? Assume that all the sulfur in the tailings is in the form FeS_2.

81. A sample of battery acid is to be analyzed for its sulfuric acid content. A 1.00 mL sample weighs 1.303 g. This 1.00 mL sample is diluted to 250.0 mL, and 10.00 mL of this diluted acid requires 34.12 mL of 0.00498 M $Ba(OH)_2$ for its titration. What is the mass percent of H_2SO_4 in the battery acid? (Assume that complete neutralization of the H_2SO_4 occurs.)

82. A piece of marble (assume it is pure $CaCO_3$) reacts with 2.00 L of 2.52 M HCl. After dissolution of the marble, a 10.00 mL sample of the resulting solution is withdrawn, added to some water, and titrated with 24.87 mL of 0.9987 M NaOH. What must have been the mass of the piece of marble? Comment on the precision of this method; that is, how many significant figures are justified in the result?

83. The reaction below can be used as a laboratory method of preparing small quantities of $Cl_2(g)$. If a 62.6 g sample that is 98.5% $K_2Cr_2O_7$ by mass is allowed to react with 325 mL of HCl(aq) with a density of 1.15 g/mL and 30.1% HCl by mass, how many grams of $Cl_2(g)$ are produced?

$$Cr_2O_7^{2-} + H^+ + Cl^- \longrightarrow Cr^{3+} + H_2O + Cl_2(g) \quad \text{(not balanced)}$$

84. Refer to Example 5-10. Suppose that the $KMnO_4(aq)$ were standardized by reaction with As_2O_3 instead of iron wire. If a 0.1304 g sample that is 99.96% As_2O_3 by mass had been used in the titration, how many milliliters of the $KMnO_4(aq)$ would have been required?

$$As_2O_3 + MnO_4^- + H^+ + H_2O \longrightarrow H_3AsO_4 + Mn^{2+} \quad \text{(not balanced)}$$

85. A new method under development for water treatment uses chlorine dioxide rather than chlorine. One method of producing ClO_2 involves passing $Cl_2(g)$ into a concentrated solution of sodium chlorite. $Cl_2(g)$ and sodium chlorite are the sole reactants, and NaCl(aq) and $ClO_2(g)$ are the sole products. If the reaction has a 97% yield, what mass of ClO_2 is produced per gallon of 2.0 M $NaClO_2(aq)$ treated in this way?

86. The active component in one type of calcium dietary supplement is calcium carbonate. A 1.2450 g tablet of the supplement is added to 65.00 mL of 0.4984 M HCl and allowed to react. After completion of the reaction, the excess HCl(aq) requires 38.45 mL of 0.2257 M NaOH for its titration to the equivalence point. What is the calcium content of the tablet, expressed in ligrams of Ca^{2+}?

87. A 0.4324 g sample of a potassium hydroxide–lith hydroxide mixture requires 28.28 mL of 0.3520 M for its titration to the equivalence point. What is mass percent lithium hydroxide in this mixture?

88. Chile saltpeter is a natural source of $NaNO_3$; it contains $NaIO_3$. The $NaIO_3$ can be used as a sourc iodine. Iodine is produced from sodium iodate two-step process occurring under acidic conditior

$$IO_3^-(aq) + HSO_3^-(aq) \longrightarrow I^-(aq) + SO_4^{2-}(aq) \quad \text{(not balan}$$

$$I^-(aq) + IO_3^-(aq) \longrightarrow I_2(s) + H_2O(l) \quad \text{(not balan}$$

In the illustration, a 5.00 L sample of a $NaIO_3$ solution containing 5.80 g $NaIO_3$/L is treated witł stoichiometric quantity of $NaHSO_3$ (no exces either reactant). Then, a further quantity of the ir $NaIO_3(aq)$ is added to the reaction mixture to b about the second reaction. **(a)** How many gram $NaHSO_3$ are required in the first step? **(b)** What a tional volume of the starting solution must be ac in the second step?

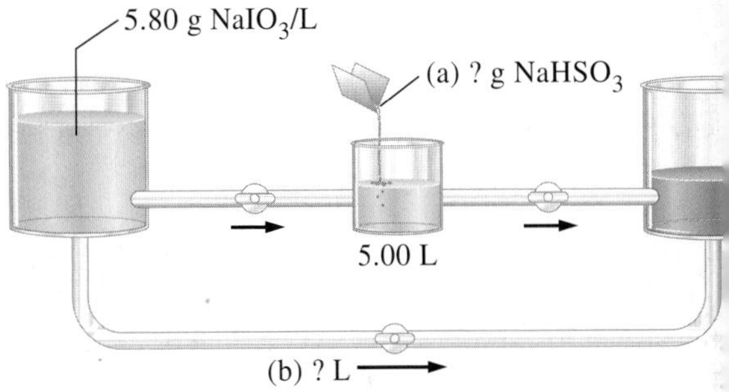

89. The active ingredients in a particular antacid t are aluminum hydroxide, $Al(OH)_3$, and magnes hydroxide, $Mg(OH)_2$. A 5.00×10^2 mg sampl the active ingredients was dissolved in 50.0 m 0.500 M HCl. The resulting solution, which was acidic, required 16.5 mL of 0.377 M NaOH for neu ization. What are the mass percentages of Al(C and $Mg(OH)_2$ in the sample?

90. A compound contains only Fe and O. A 0.2729 g ple of the compound was dissolved in 50 mL of centrated acid solution, reducing all the iron to ions. The resulting solution was diluted to 10C and then titrated with a 0.01621 M $KMnO_4$ solu The unbalanced chemical expression for reac between Fe^{2+} and MnO_4^- is given below.

$$MnO_4^-(aq) + Fe^{2+}(aq) \longrightarrow Mn^{2+}(aq) + Fe^{3+}(aq) \quad \text{(not balan}$$

The titration required 42.17 mL of the $KMnO_4$ s tion to reach the pink endpoint. What is the emp formula of the compound?

91. Warfarin, $C_{19}H_{16}O_4$, is the active ingredient use some anticoagulant medications. The amour warfarin in a particular sample was determ

as follows. A 13.96 g sample was first treated with an alkaline I_2 solution to convert $C_{19}H_{16}O_4$ to CHI_3. This treatment gives one mole of CHI_3 for every mole of $C_{19}H_{16}O_4$ that was initially present in the sample. The iodine in CHI_3 is then precipitated as AgI(s) by treatment with excess $AgNO_3(aq)$:

$$CHI_3(aq) + 3\,AgNO_3(aq) + H_2O(l) \longrightarrow 3\,AgI(s) + 3\,HNO_3(aq) + CO(g)$$

If 0.1386 g solid AgI were obtained, then what is the percentage by mass of warfarin in the sample analyzed?

Copper refining traditionally involves "roasting" insoluble sulfide ores (CuS) with oxygen. Unfortunately, the process produces large quantities of $SO_2(g)$, which is a major contributor to pollution and acid rain. An alternative process involves treating the sulfide ore with $HNO_3(aq)$, which dissolves the CuS without generating any SO_2. The unbalanced chemical expression for the reaction is given below.

$$CuS(s) + NO_3^-(aq) \longrightarrow Cu^{2+}(aq) + NO(g) + HSO_4^-(aq) \quad \text{(not balanced)}$$

What volume of concentrated nitric acid solution is required per kilogram of CuS? Assume that the concentrated nitric acid solution is 70% HNO_3 by mass and has a density of 1.40 g/mL.

93. Phosphorus is essential for plant growth, but an excess of phosphorus can be catastrophic in aqueous ecosystems. Too much phosphorus can cause algae to grow at an explosive rate and this robs the rest of the ecosystem of oxygen. Effluent from sewage treatment plants must be treated before it can be released into lakes or streams because the effluent contains significant amounts of $H_2PO_4^-$ and HPO_4^{2-}. (Detergents are a major contributor to phosphorus levels in domestic sewage because many detergents contain Na_2HPO_4.) A simple way to remove $H_2PO_4^-$ and HPO_4^{2-} from the effluent is to treat it with lime, CaO, which produces Ca^{2+} and OH^- ions in water. The OH^- ions convert $H_2PO_4^-$ and HPO_4^{2-} ions into PO_4^{3-} ions and, finally, Ca^{2+}, OH^-, and PO_4^{3-} ions combine to form a precipitate of $Ca_5(PO_4)_3OH(s)$.

(a) Write balanced chemical equations for the four reactions described above.

[*Hint*: The reactants are CaO and H_2O; $H_2PO_4^-$ and OH^-; HPO_4^{2-} and OH^-; Ca^{2+}, PO_4^{3-}, and OH^-.]

(b) How many kilograms of lime are required to remove the phosphorus from a 1.00×10^4 L holding tank filled with contaminated water, if the water contains 10.0 mg of phosphorus per liter?

Feature Problems

Sodium cyclopentadienide, NaC_5H_5, is a common reducing agent in the chemical laboratory, but there is a problem in using it: NaC_5H_5 is contaminated with tetrahydrofuran (THF), C_4H_8O, a solvent used in its preparation. The THF is present as $NaC_5H_5 \cdot (THF)_x$, and it is generally necessary to know exactly how much of this $NaC_5H_5 \cdot (THF)_x$ is present. This is accomplished by allowing a small amount of the $NaC_5H_5 \cdot (THF)_x$ to react with water,

$$NaC_5H_5 \cdot (C_4H_8O)_x + H_2O \longrightarrow NaOH(aq) + C_5H_5{-}H + x\,C_4H_8O$$

followed by titration of the NaOH(aq) with a standard acid. From the sample data tabulated below, determine the value of x in the formula $NaC_5H_5 \cdot (THF)_x$.

	Trial 1	Trial 2
ss of $NaC_5H_5 \cdot (THF)_x$	0.242 g	0.199 g
ume of 0.1001 M HCl equired to titrate NaOH(aq)	14.92 mL	11.99 mL

Manganese is derived from pyrolusite ore, an impure manganese dioxide. In the procedure used to analyze a pyrolusite ore for its MnO_2 content, a 0.533 g sample is treated with 1.651 g oxalic acid ($H_2C_2O_4 \cdot 2\,H_2O$) in an acidic medium. Following this reaction, the excess oxalic acid is titrated with 0.1000 M $KMnO_4$, 30.06 mL being required. What is the mass percent MnO_2 in the ore?

$$H_2C_2O_4 + MnO_2 + H^+ \longrightarrow Mn^{2+} + H_2O + CO_2 \quad \text{(not balanced)}$$

$$H_2C_2O_4 + MnO_4^- + H^+ \longrightarrow Mn^{2+} + H_2O + CO_2 \quad \text{(not balanced)}$$

96. The Kjeldahl method is used in agricultural chemistry to determine the percent protein in natural products. The method is based on converting all the protein nitrogen to ammonia and then determining the amount of ammonia by titration. The percent nitrogen in the sample under analysis can be calculated from the quantity of ammonia produced. Interestingly, the majority of protein molecules in living matter contain just about 16% nitrogen.

A 1.250 g sample of meat is heated with concentrated sulfuric acid and a catalyst to convert all the nitrogen in the meat to $(NH_4)_2SO_4$. Then excess NaOH(aq) is added to the mixture, which is heated to expel $NH_3(g)$. All the nitrogen from the sample is found in the $NH_3(g)$, which is then absorbed in and neutralized by 50.00 mL of dilute $H_2SO_4(aq)$. The excess $H_2SO_4(aq)$ requires 32.24 mL of 0.4498 M NaOH for its titration. A separate 25.00 mL sample of the dilute $H_2SO_4(aq)$ requires 22.24 mL of 0.4498 M NaOH for its titration. What is the percent protein in the meat?

97. Blood alcohol content (BAC) is often reported in weight–volume percent (w/v%). For example, a BAC of 0.10% corresponds to 0.10 g CH_3CH_2OH per 100 mL

A Method for Balancing Equations for Oxidation–Reduction Reactions That Occur in an Acidic or a Basic Aqueous Solution

1. Assign oxidation states to each element in the reaction and identify the species being oxidized and reduced.
2. Write separate, unbalanced equations for the oxidation and reduction half-reactions.
3. Balance the separate half-equations, in this order:
   - first with respect to the element being oxidized or reduced
   - then by adding electrons to one side or the other to account for the number of electrons produced (oxidation) or consumed (reduction)
4. Combine the half-reactions algebraically so that the total number of electrons cancels out.
5. Balance the net charge by either adding OH^- (for basic solutions) or H^+ (for acidic solutions).
6. Balance the O and H atoms by adding H_2O.
7. Check that the final equation is balanced with respect to each type of atom and with respect to charge.

of blood. Estimates of BAC can be obtained from breath samples by using a number of commercially available instruments, including the Breathalyzer for which a patent was issued to R. F. Borkenstein in 1958. The chemistry behind the Breathalyzer is described by the oxidation–reduction reaction below, which occurs in acidic solution:

$$\underset{\text{ethyl alcohol}}{CH_3CH_2OH(g)} + \underset{\text{(yellow-orange)}}{Cr_2O_7^{2-}(aq)} \longrightarrow$$

$$CH_3COOH(aq) + \underset{\text{(green)}}{Cr^{3+}(aq)} \quad \text{(not balanced)}$$

A Breathalyzer instrument contains two ampules, each of which contains 0.75 mg $K_2Cr_2O_7$ dissolved in 3 mL of 9 M $H_2SO_4(aq)$. One of the ampules is used as reference. When a person exhales into the tube of the Breathalyzer, the breath is directed into one of the ampules, and ethyl alcohol in the breath converts $Cr_2O_7^{2-}$ into Cr^{3+}. The instrument compares the colors of the solutions in the two ampules to determine the breath alcohol content (BrAC), and then converts this into an estimate of BAC. The conversion of BrAC into BAC rests on the assumption that 2100 mL of air exhaled from the lungs contains the same amount of alcohol as 1 mL of blood. With the theory and assumptions described in this problem, calculate the molarity of $K_2Cr_2O_7$ in the ampules before and after a breath test in which a person with a BAC of 0.05% exhales 0.500 L of his breath into a Breathalyzer instrument.

98. In this problem, we describe an alternative met for balancing equations for oxidation-reduction r tions. The method is similar to the method given viously in Tables 5.5 and 5.6, but it places m emphasis on the assignment of oxidation states. (method summarized in Tables 5.5 and 5.6 does require you to assign oxidation states.) An emph on oxidation states is warranted because oxida states are useful not only for keeping track of e trons but also for predicting chemical properties. method is summarized in the table above.

The method offers a couple of advantages. First method applies to both acidic and basic environm because we balance charges by using either H^+ acidic environments) or OH^- (for basic envi ments). Second, the method is somewhat more cient than the method we described previo because, in the method described here, we bala only once for charge and only once for hydrogen oxygen. In the other method, we focus on the h equations separately and must balance twice charge and twice for hydrogen and oxygen.

Use the alternative method described above to ance the following oxidation-reduction equations

(a) $Cr_2O_7^{2-}(aq) + Cl^-(aq) \longrightarrow Cr^{3+}(aq) + Cl_2(g)$ (acidic solut

(b) $C_2O_4^{2-}(aq) + MnO_4^-(aq) \longrightarrow CO_3^{2-}(aq) + MnO_2(s)$ (basic solut

Self-Assessment Exercises

99. In your own words, define or explain the terms or symbols **(a)** $\rightleftharpoons$ **(b)** [ ]; **(c)** spectator ion; **(d)** weak acid.

100. Briefly describe **(a)** half-equation method of balancing redox equations; **(b)** disproportionation reaction; **(c)** titration; **(d)** standardization of a solution.

101. Explain the important distinctions between **(a)** a strong electrolyte and strong acid; **(b)** an oxidizing agent and reducing agent; **(c)** precipitation reactions and neutralization reactions; **(d)** half-reaction and overall reaction.

102. The number of moles of hydroxide ion in 0.300 0.0050 M $Ba(OH)_2$ is **(a)** 0.0015; **(b)** 0.0030; **(c)** 0.0 **(d)** 0.010.

103. The highest $[H^+]$ will be found in an aqueous s tion that is **(a)** 0.10 M HCl; **(b)** 0.10 M NH_3; **(c)** 0.1 CH_3COOH; **(d)** 0.10 M H_2SO_4.

104. To precipitate Zn^{2+} from $Zn(NO_3)_2(aq)$, **(a)** NH_4Cl; **(b)** $MgBr_2$; **(c)** K_2CO_3; **(d)** $(NH_4)_2SO_4$

105. When treated with dilute HCl(aq), the solid reacts to produce a gas is **(a)** $BaSO_3$; **(b)** Z **(c)** NaBr; **(d)** Na_2SO_4.

An unknown solid compound dissolves readily when added to water, forming a solution that conducts electricity. A precipitate forms when $Ba(NO_3)_2(aq)$ is added to a solution of this compound, but not when $Cu(NO_3)_2(aq)$ is added. When the unknown solid is added to $CH_3COOH(aq)$, no gas is produced. When the solid is added to NaOH (aq), a gas with the pungent odor of ammonia is produced. What is the possible identity of the unknown solid?

What is the net ionic equation for the reaction that occurs when an aqueous solution of KI is added to an aqueous solution of $Pb(NO_3)_2$?

When aqueous sodium carbonate, Na_2CO_3, is treated with dilute hydrochloric acid, HCl, the products are sodium chloride, water, and carbon dioxide gas. What is the net ionic equation for this reaction?

Describe the synthesis of each of the following ionic compounds, starting from solutions of sodium and nitrate salts. Then write the net ionic equation for each synthesis.
(a) $Zn_3(PO_4)_2$;
(b) $Cu(OH)_2$;
(c) $NiCO_3$.

Consider the following redox reaction:

$$4\,NO(g) + 3\,O_2(g) + 2\,H_2O(l) \longrightarrow 4\,NO_3^-(aq) + 4\,H^+(aq)$$

(a) Which species is oxidized?
(b) Which species is reduced?
(c) Which species is the oxidizing agent?
(d) Which species is the reducing agent?
(e) Which species gains electrons?
(f) Which species loses electrons?

Balance the following oxidation–reduction equations.
(a) $Cl_2(aq) \longrightarrow Cl^-(aq) + ClO^-(aq)$ (basic solution)
(b) $C_2O_4^{2-}(aq) + MnO_4^-(aq) \longrightarrow Mn^{2+}(aq) + CO_2(g)$ (acidic solution)

In the equation

$$?\,Fe^{2+}(aq) + O_2(g) + 4\,H^+(aq) \longrightarrow ?\,Fe^{3+}(aq) + 2\,H_2O(l)$$

the missing coefficients **(a)** are each 2; **(b)** are each 4; **(c)** can have any values as long as they are the same; **(d)** must be determined by experiment.

What is the simplest ratio a:b when the equation below is properly balanced?

$$ClO^-(aq) + b\,I_2(aq) \xrightarrow[\text{solution}]{\text{acidic}} c\,Cl^-(aq) + d\,IO_3^-(aq)$$

(a) 2:5; **(b)** 5:2; **(c)** 1:5; **(d)** 5:1; **(e)** 2:3.

114. In the half-reaction in which NpO_2^+ is converted to Np^{4+}, the number of electrons appearing in the half-equation is **(a)** 1; **(b)** 2; **(c)** 3; **(d)** 4.

115. Which list of compounds contains a nonelectrolyte, a weak electrolyte, and a strong electrolyte? **(a)** CO_2, NaCl, $MnSO_4$; **(b)** H_2SO_4, CH_3COOH, CuCl; **(c)** SO_2, HF, $FeSO_4$; **(d)** $Ba(ClO_3)_2$, $K_2S_2O_3$, $NaMnO_4$; **(e)** none of these.

116. Which list of compounds contains a weak acid, a weak base, and a salt? **(a)** HCl, NH_3, Na_2SO_4; **(b)** HNO_2, NH_3, NH_4NO_2; **(c)** HCl, $Ca(OH)_2$, $CaSO_4$; **(d)** HNO_2, KOH, Cs_2CrO_4; **(e)** none of these.

117. Which list of compounds contains two soluble compounds and an insoluble one?
(a) $HgBr_2$, $MnSO_4$, $Na_2C_2O_4$;
(b) $Na_2S_2O_3$, NH_4Cl, CoI_2;
(c) MnS, $Cu(OH)_2$, Al_2O_3;
(d) $Pb(ClO_4)_2$, $Ca(NO_3)_2$, Hg_2SO_4;
(e) none of these.

118. Classify each of the following statements as true or false.
(a) Barium chloride, $BaCl_2$, is a weak electrolyte in aqueous solution.
(b) In the reaction $H^-(aq) + H_2O(l) \rightarrow H_2(g) + OH^-(aq)$, water acts as both an acid and an oxidizing agent.
(c) A precipitate forms when aqueous sodium carbonate, $Na_2CO_3(aq)$, is treated with excess aqueous hydrochloric acid, HCl(aq).
(d) Hydrofluoric acid, HF, is a strong acid in water.
(e) Compared with a 0.010 M solution of $NaNO_3$, a 0.010 M solution of $Mg(NO_3)_2$ is a better conductor of electricity.

119. Which of the following reactions are oxidation–reduction reactions?
(a) $H_2CO_3(aq) \longrightarrow H_2O(l) + CO_2(g)$
(b) $2\,Li(s) + 2\,H_2O(l) \longrightarrow 2\,LiOH(aq) + H_2(g)$
(c) $4\,Ag(s) + PtCl_4(aq) \longrightarrow 4\,AgCl(s) + Pt(s)$
(d) $2\,HClO_4(aq) + Ca(OH)_2(aq) \longrightarrow 2\,H_2O(l) + Ca(ClO_4)_2(aq)$

120. Similar to Figure 5-4(c), but using the formulas HAc, Ac^-, and H_3O^+, give a more accurate representation of $CH_3COOH(aq)$ in which ionization is 5% complete.

121. Appendix E describes a useful study aid known as concept mapping. Using the method presented in Appendix E, construct a concept map illustrating the different concepts introduced in Sections 5-4, 5-5, and 5-6.

6 Gases

CONTENTS

LEARNING OBJECTIVES

6.1 Describe the principle of measuring the pressure of a gas by using an open-end manometer.

6.2 Identify the relationships between the volume, temperature, and pressure of a gas by using Boyle's, Charles's, and Avogadro's laws.

6.3 Use the ideal gas equation and the general gas equation to determine the relationship between *V*, *T*, and *P* of a gas in two states.

6.4 Describe how to find the relationship between the state properties (*P*, *V*, *T*) of a gas and its molar mass or density.

6.5 Discuss how the law of combining volumes can be used to simplify a problem involving the reaction of gases.

6.6 Discuss Dalton's law of partial pressures, and use it to determine the pressure of a gas collected over water.

6.7 Describe how mass and temperature affect the distribution of molecular speeds of gases.

6.8 Use Graham's law to determine the molar mass of an unknown gas when comparing its rate of diffusion with a known gas.

6.9 Identify conditions under which gases behave ideally versus nonideally, and describe how molecular size and intermolecular forces cause deviations from ideal gas behavior.

Carlos Caetano/Shutterstock

Hot-air balloons have intrigued people from the time the simple gas laws fundamental to their operation came to be understood more than 200 years ago

You shouldn't overinflate a bicycle tire, or discard an aerosol can i incinerator, or search for a gas leak with an open flame. In each there is a danger of explosion. These and many other observat concerning gases can be explained by concepts considered in this cha The behaviors of the bicycle tire and the aerosol can are based on relat ships among pressure, temperature, volume, and amount of gas. O examples of the behavior of gases can be seen in a balloon filled with he or hot air rising in air and carbon dioxide gas vaporizing from a block of ice and sinking to the floor. An understanding of the lifting power of lig than-air balloons comes in large part from knowledge of gas densities

◀ FIGURE 6-1
The gaseous states of three halogens (group 17)
The greenish yellow gas is $Cl_2(g)$; the brownish red gas is $Br_2(g)$ above a small pool of liquid bromine; the violet gas is $I_2(g)$ in contact with grayish-black solid iodine. Most other common gases, such as H_2, O_2, N_2, CO, and CO_2, are colorless.

dependence on molar mass, temperature, and pressure. Predicting how far how fast gas molecules migrate through air requires knowing something t the phenomenon of diffusion.

r a quantitative description of the behavior of gases, we will employ some le gas laws and a more general expression called the *ideal gas equation*. e laws will be explained by the kinetic–molecular theory of gases. The s covered in this chapter extend the discussion of reaction stoichiometry the previous two chapters and lay some groundwork for use in the follow- hapter on thermochemistry. The relationships between gases and the other s of matter—liquids and solids—are discussed in Chapter 12.

Properties of Gases: Gas Pressure

e characteristics of gases are familiar to everyone. Gases expand to fill containers and assume the shapes of their containers. They diffuse into nother and mix in all proportions. We cannot see individual particles of a although we can see the bulk gas if it is colored (Fig. 6-1). Some gases, as hydrogen and methane, are combustible; whereas others, such as m and neon, are chemically unreactive.

ur properties determine the physical behavior of a gas: the amount of the in moles) and the volume, temperature, and pressure of the gas. If we v any three of these, we can usually calculate the value of the remaining by using a mathematical equation called an *equation of state* (such as the gas equation, given on page 206). To some extent we have already dis- ed the properties of amount, volume, and temperature, but we need to ider the idea of pressure.

Kristen Brochmann/Fundamental Photographs

▲ FIGURE 6-2
Illustrating the pressure exerted by a solid
The two cylinders have the same mass and exert the same force on the supporting surface ($F = g \times m$). The tall, thin one has a smaller area of contact, however, and exerts a greater pressure ($P = F/A$).

Concept of Pressure

lloon expands when it is inflated with air, but what keeps the balloon in istended shape? A plausible hypothesis is that molecules of a gas are in tant motion, frequently colliding with one another and with the walls of container. In their collisions, the gas molecules exert a force on the con- r walls. This force keeps the balloon distended. It is not easy, however, to sure the total force exerted by a gas. Instead of focusing on this total force, onsider instead the gas pressure. **Pressure** is defined as a force per unit that is, a force divided by the area over which the force is distributed. re 6-2 illustrates the idea of pressure exerted by a solid.

SI, the unit of force is a *newton* (N), which is the force, *F*, required to pro- an acceleration of one meter per second per second (1 m s^{-2}) in a one- ram mass (1 kg), that is, $1 \text{ N} = 1 \text{ kg m s}^{-2}$. The corresponding force per

▶ If you have not taken a physics course, consult Appendix B for a brief discussion of force and work.

unit area—pressure—is expressed in the unit N/m^2. A pressure of one ne per square meter is defined as one **pascal (Pa)**. Thus, a pressure in pascal

$$P\,(\text{Pa}) = \frac{F\,(\text{N})}{A\,(\text{m}^2)}$$

A pascal is a rather small pressure unit, so the **kilopascal (kPa)** is more monly used. The pascal honors Blaise Pascal (1623–1662), who studied sure and its transmission through fluids—the basis of modern hydraulics

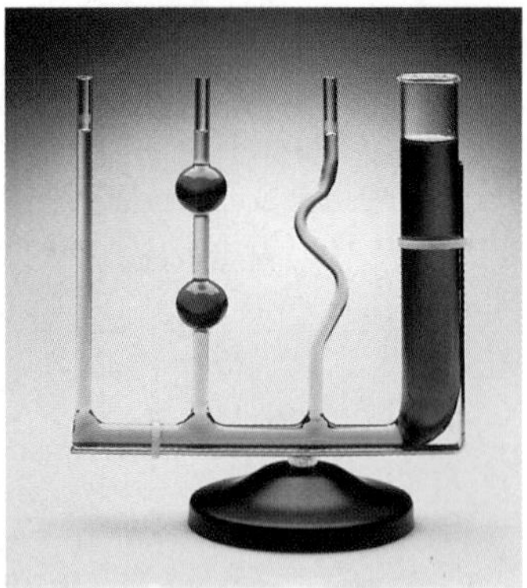

Richard Megna/Fundamental Photographs

▲ FIGURE 6-3
The concept of liquid pressure
All the interconnected vessels fill to the same height. As a result, the liquid pressures are the same despite the different shapes and volumes of the containers.

Liquid Pressure

Because it is difficult to measure the total force exerted by gas molecules also difficult to apply equation (6.1) to gases. The pressure of a gas is us measured indirectly, by comparing it with a liquid pressure. Figure 6-3 trates the concept of liquid pressure and suggests that the pressure of a li depends only on the height of the liquid column and the density of the li To confirm this statement, consider a liquid with density d, contained cylinder with cross-sectional area A, filled to a height h.

Now recall that (1) weight is a force, and weight (W) and mass (m) are portional: $W = g \times m$. (2) The mass of a liquid is the product of its volume density: $m = V \times d$. (3) The volume of a cylinder is the product of its h and cross-sectional area: $V = h \times A$. We use these facts to derive the equat

$$P = \frac{F}{A} = \frac{W}{A} = \frac{g \times m}{A} = \frac{g \times V \times d}{A} = \frac{g \times h \times \cancel{A} \times d}{\cancel{A}} = g \times h \times d$$

Thus, because g is a constant, *liquid pressure is directly proportional to the l density and the height of the liquid column.*

Barometric Pressure

In 1643, Evangelista Torricelli constructed the device pictured in Figure 6 measure the pressure exerted by the atmosphere. This device is call **barometer**.

If a glass tube that is open at both ends stands upright in a container of cury (Fig. 6-4a), the mercury levels inside and outside the tube are the sam create the situation in Figure 6-4(b), we seal one end of a long glass tube, pletely fill the tube with Hg(l), cover the open end, and invert the tube a container of Hg(l). Then we reopen the end that is submerged in the cury. The mercury level in the tube falls to a certain height and stays t Something keeps the mercury at a greater height inside the tube than out Some tried to ascribe this phenomenon to forces within the tube, but Torr understood that the forces involved originated outside the tube.

In the open-end tube (Fig. 6-4a), the atmosphere exerts the same pres on the surface of the mercury both inside and outside the tube, and the li levels are equal. Inside the closed-end tube (Fig. 6-4b), there is no air abov mercury (only a trace of mercury vapor). The atmosphere exerts a force o surface of the mercury in the outside container. This force is transm through the liquid, holding up the mercury column within the tube. The umn exerts a downward pressure that depends on its height and the de of Hg(l). When the pressure at the bottom of the mercury column is equ the pressure of the atmosphere, the column height is maintained.

The height of mercury in a barometer provides a measure of **barometric sure**. Barometric pressures may be expressed in a unit called **millimet mercury (mmHg)**, defined as the pressure exerted by a column of mer that is exactly 1 mm in height when the density of mercury is e to 13.5951 g/cm^3(0 °C) and the acceleration due to gravity, g, is equ 9.80655 m/s^2. Notice that this unit of pressure assumes specific values fo density of mercury and the acceleration due to gravity. This is because the de

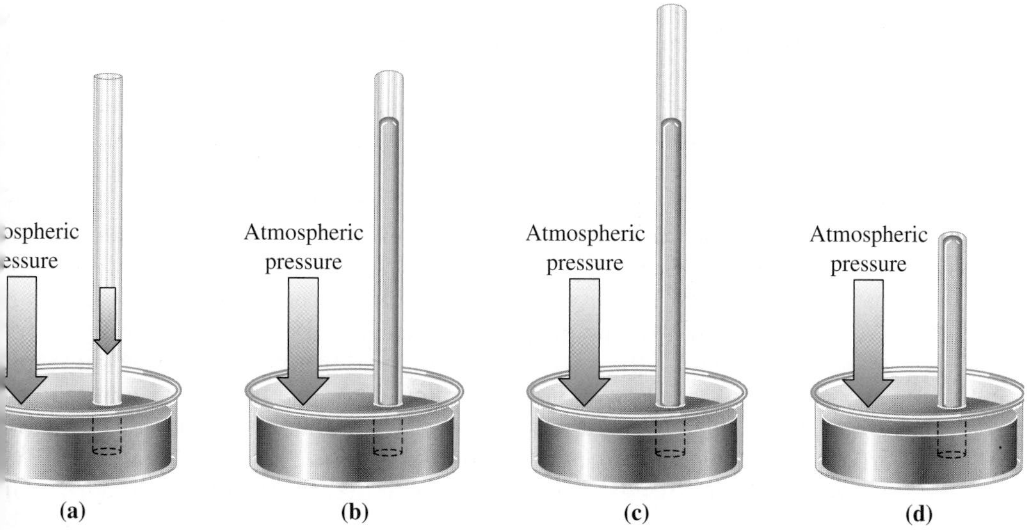

GURE 6-4
asurement of atmospheric pressure with a mercury barometer
ws represent the pressure exerted by the atmosphere. **(a)** The liquid mercury s are equal inside and outside the open-end tube. **(b)** A column of mercury mm high is maintained in the closed-end tube, regardless of the overall height of tube **(c)** as long as it exceeds 760 mm. **(d)** A column of mercury fills a closed-end that is shorter than 760 mm. In the closed-end tubes in (b) and (c), the region ve the mercury column is devoid of air and contains only a trace of mercury vapor.

g(l) depends on temperature and g depends on the specific location on Earth all p. 9). Typically, the pressure exerted by the atmosphere can support a col- n of mercury that is about 760 mm high and thus, atmospheric pressure is cally about 760 mmHg. Let's use equation (6.2) to calculate the pressure ted by a column of mercury that is exactly 760 mm high when the density of cury is $d = 13.5951\ \text{g/cm}^3 = 1.35951 \times 10^4\ \text{kg/m}^3$ and $g = 9.80655\ \text{m/s}^2$.

$$P = (9.80665\ \text{m s}^{-2})(0.760000\ \text{m})(1.35951 \times 10^4\ \text{kg m}^{-3})$$
$$= 1.01325 \times 10^5\ \text{kg m}^{-1}\ \text{s}^{-2}$$

◀ In this calculation, we have written the units of d and g in the form kg m^{-3} and m s^{-2}, respectively, rather than as kg/m^3 and m/s^2. You must become equally comfortable with using either negative exponents or a slash (/) when working with derived units. For example, the unit $\text{kg m}^{-1}\ \text{s}^{-2}$ may also be written as $\text{kg/(m s}^2)$.

unit that arises in this calculation, $\text{kg m}^{-1}\ \text{s}^{-2}$, is the SI unit for pressure , as mentioned earlier, is called the pascal (Pa). A pressure of exactly 325 Pa or 101.325 kPa, has special significance because in the SI system of s, **one standard atmosphere (atm)** is defined to be exactly equal to 325 Pa, or 101.325 kPa. Another pressure unit that is sometimes encoun- d is a unit called a **torr** and denoted by the symbol **Torr**. This unit honors icelli and is defined as exactly 1/760 of a standard atmosphere. The fol- ing expression shows the relationships among these units.

$$1\ \text{atm} = 760\ \text{Torr} \approx 760\ \text{mmHg} \qquad \textbf{(6.3)}$$

ndicated above, the units torr and millimeters of mercury are not strictly al. This is because 760 Torr is *exactly* equal to 101,325 Pa but 760 mmHg is *approximately* equal to 101,325 Pa (that is, to about six or seven significant res). The difference between a torr and a millimeter of mercury is too ll to worry about, except in highly accurate work. Thus, in this text, we use the pressure units of Torr and mmHg interchangeably.

Mercury is a relatively rare, expensive, and poisonous liquid. Why u rather than water in a barometer? As we will see in Example 6-1 extreme height required for a water barometer is a distinct disadvan Whereas atmospheric pressure can be measured with a mercury baron less than 1 m high, a water barometer would have to be as tall as a th story building.

When you use a drinking straw, you reduce the air pressure abov liquid inside the straw by inhaling. Atmospheric pressure on the liquid side the straw then pushes the liquid up the straw and into your moutl old-fashioned hand suction pump for pumping water (once common in areas) works by the same principle. The result of Example 6-1 indicates, ever, that even if all the air could be removed from inside a pipe, atmosp pressure outside the pipe could not raise water to a height of more than a 10 m. Thus, a suction pump works only for shallow wells. To pump v from a deep well, a mechanical pump is required. The mechanical p pushes the water upward by using a force that is greater than the force c atmosphere pushing the water down.

EXAMPLE 6-1 Comparing Liquid Pressures

What is the height of a column of water that exerts the same pressure as a column of mercury 76.0 cm (760 mm) high?

Analyze

Equation (6.2) shows that, for a given liquid pressure, the column height is inversely proportional to the liquid density. The lower the liquid density, the greater the height of the liquid column. Mercury is 13.6 times as dense as water (13.6 g/cm^3 versus 1.00 g/cm^3). If columns of water and mercury exert the same pressure, then the column of water is 13.6 times as high as the column of mercury.

Solve

Although we have already reasoned out the answer, we can arrive at the same conclusion by applying equation (6.2) twice, and then setting the two pressures equal to each other. Equation (6.2) can be used to describe the pressure of the mercury column of known height and the pressure of the water column of unknown height. Then we can set the two pressures equal to each other.

$$\text{pressure of Hg column} = g \times h_{Hg} \times d_{Hg} = g \times 76.0\text{ cm} \times 13.6\text{ g/cm}^3$$

$$\text{pressure of H}_2\text{O column} = g \times h_{H_2O} \times d_{H_2O} = g \times h_{H_2O} \times 1.00\text{ g/cm}^3$$

$$\cancel{g} \times h_{H_2O} \times 1.00\text{ g/cm}^3 = \cancel{g} \times 76.0\text{ cm} \times 13.6\text{ g/cm}^3$$

$$h_{H_2O} = 76.0\text{ cm} \times \frac{13.6\text{ g/cm}^3}{1.00\text{ g/cm}^3} = 1.03 \times 10^3\text{ cm} = 10.3\text{ m}$$

Assess

We can think about equation (6.2) in another way. For a column of liquid of fixed height, the greater the density of the liquid, the greater the pressure exerted by the liquid column. A column of mercury that is 760 mm high will exert a pressure 13.6 times as great as a column of water that is 760 mm high.

PRACTICE EXAMPLE A: A barometer is filled with diethylene glycol ($d = 1.118\text{ g/cm}^3$). The liquid height is found to be 9.25 m. What is the barometric (atmospheric) pressure expressed in millimeters of mercury?

PRACTICE EXAMPLE B: A barometer is filled with triethylene glycol. The liquid height is found to be 9.14 m when the atmospheric pressure is 757 mmHg. What is the density of triethylene glycol?

6-1 CONCEPT ASSESSMENT

Explain how the action of a water siphon is related to that of a suction pump.

nometers

ough a mercury barometer is indispensable for measuring the pressure of tmosphere, it is rarely used alone to measure other gas pressures. The dif- ty with a barometer is in placing it inside the container of gas whose pres- is to be measured. However, the pressure of the gas to be measured can ompared with barometric pressure by using a **manometer**. Figure 6-5 trates the principle of an open-end manometer. When the gas pressure g measured and the prevailing atmospheric (barometric) pressure are l, the heights of the mercury columns in the two arms of the manometer qual. A difference in height of the two arms signifies a difference between gas pressure and barometric pressure.

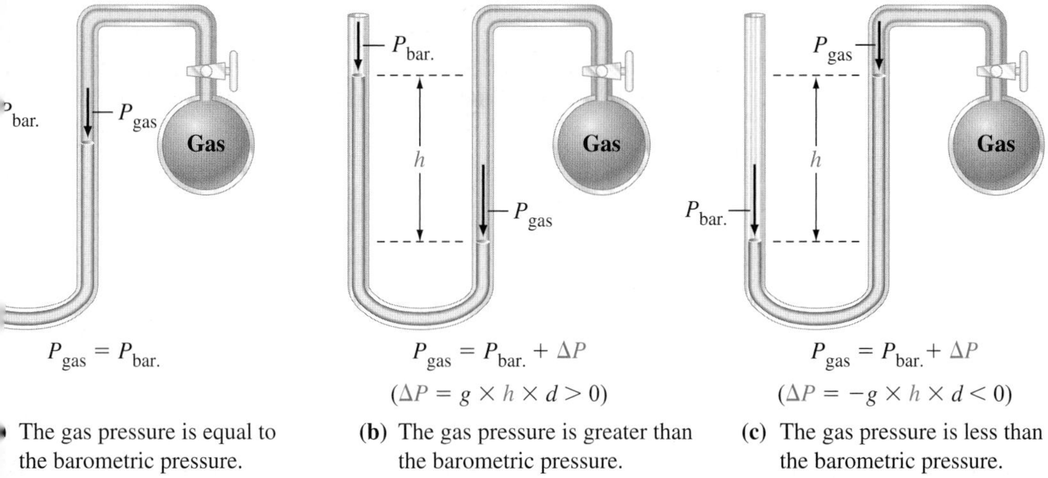

GURE 6-5
surement of gas pressure with an open-end manometer
ossible relationships between barometric pressure and a gas pressure under urement are pictured here and described in Example 6-2. If P_{gas} and $P_{bar.}$ are essed in mmHg, then ΔP is numerically equal to the height h expressed in millimeters.

XAMPLE 6-2 Using a Manometer to Measure Gas Pressure

When the manometer in Figure 6-5(c) is filled with liquid mercury ($d = 13.6\ g/cm^3$), the barometric pressure is 748.2 mmHg, and the difference in mercury levels is 8.6 mmHg. What is the gas pressure P_{gas}?

nalyze

We must first establish which is greater: the barometric pressure or the gas pressure. In Figure 6-5(c), the barometric pressure forces liquid mercury down the tube toward the gas sample. The barometric pressure is greater than the gas pressure. Thus, $\Delta P = P_{gas} - P_{bar.} < 0$.

olve

The gas pressure is less than the barometric pressure. Therefore, we subtract 8.6 mmHg from the barometric pressure to obtain the gas pressure.

$$P_{gas} = P_{bar.} + \Delta P = 748.2\ \text{mmHg} - 8.6\ \text{mmHg} = 739.6\ \text{mmHg}$$

ssess

Because all pressures are expressed in millimeters of mercury, the pressure difference (ΔP) is numerically equal to the difference in mercury levels. Thus, the density of mercury does not enter into the calculation.

RACTICE EXAMPLE A: Suppose that the mercury level in Example 6-2 is 7.8 mm higher in the arm open to the atmosphere than in the closed arm. What would be the value of P_{gas}?

RACTICE EXAMPLE B: Suppose $P_{bar.}$ and P_{gas} are those described in Example 6-2, but the manometer is filled with liquid glycerol ($d = 1.26\ g/cm^3$) instead of mercury. What would be the difference in the two levels of the liquid?

Units of Pressure: A Summary

Table 6.1 lists several different units used to express pressure. The u shown in red are used frequently by chemists, even though they are not of the SI system. The atmosphere is a useful unit because volumes of g are often measured at the prevailing atmospheric pressure. Typic the atmospheric pressure is close to 1 atm, or 760 Torr. The units shown in are those preferred in the SI system.

The units shown in black in Table 6.1 are based on the unit **bar**. One bar i times as large as a kilopascal. Atmospheric pressure is typically close to 1 bar unit *millibar* is commonly used by meteorologists.

Although we can generally choose freely among the pressure units in Tabl when doing calculations involving gases, we will encounter situations require SI units. This is the case in Example 6-3.

TABLE 6.1 Some Common Pressure Units

Atmosphere	atm	
Millimeter of mercury	mmHg	1 atm ≈ 760 mmHg
Torr	Torr	= 760 Torr
Pascal	Pa	= 101,325 Pa
Kilopascal	kPa	= 101.325 kPa
Bar	bar	= 1.01325 bar
Millibar	mbar	= 1013.25 mbar

EXAMPLE 6-3 Using SI Units of Pressure

The 1.000 kg red cylinder in Figure 6-2 has a diameter of 4.10 cm. What pressure, expressed in Torr, does th cylinder exert on the surface beneath it?

Analyze

We must apply equation (6.2). It is best to use SI units and obtain a pressure in SI units (Pa), and then conve the pressure to the required units (Torr).

Solve

Expression (6.1) defines pressure as force divided by area.

$$P = \frac{F}{A}$$

The force exerted by the cylinder is its weight.

$$F = W = m \times g$$

The mass is 1.000 kg, and g (the acceleration due to gravity) is $9.81\ \text{m s}^{-2}$. The product of these two terms is the force in newtons.

$$F = m \times g = 1.000\ \text{kg} \times 9.81\ \text{m s}^{-2} = 9.81\ \text{N}$$

The force is exerted on the area of contact between the cylinder and the underlying surface. This circular area is calculated by using the radius of the cylinder—one-half the 4.10 cm diameter, expressed in meters.

$$A = \pi r^2 = 3.1416 \times \left(2.05\ \text{cm} \times \frac{1\ \text{m}}{100\ \text{cm}}\right)^2 = 1.32 \times 10^{-3}\ \text{n}$$

The force divided by the area (in square meters) gives the pressure in pascals.

$$P = \frac{F}{A} = \frac{9.81\ \text{N}}{1.32 \times 10^{-3}\ \text{m}^2} = 7.43 \times 10^3\ \text{Pa}$$

The relationship between the units Torr and pascal (Table 6.1) is used for the final conversion.

$$P = 7.43 \times 10^3\ \text{Pa} \times \frac{760\ \text{Torr}}{101{,}325\ \text{Pa}} = 55.7\ \text{Torr}$$

Assess

It is difficult to tell at a glance whether this is a reasonable result. To check our result, let us focus instead or cylindrical column of mercury that is 55.7 mm high and has a diameter of 4.10 cm. This column of mercury al exerts a pressure of 55.7 mmHg = 55.7 Torr. The volume of mercury in this column is $V = \pi r^2 h = \pi \times (2.05\ \text{cm}$ $\times\ 5.57\ \text{cm} = 73.5\ \text{cm}^3$. The density of mercury is about $13.6\ \text{g/cm}^3$ (page 194) and thus, the mass of the mercu column is $73.5\ \text{cm}^3 \times 13.6\ \text{g/cm}^3 = 1.00 \times 10^3\ \text{g} = 1.00\ \text{kg}$. This is exactly the mass of the steel cylinder.

RACTICE EXAMPLE A: The 1.000 kg green cylinder in Figure 6-2 has a diameter of 2.60 cm. What pressure, expressed in Torr, does this cylinder exert on the surface beneath it?

RACTICE EXAMPLE B: We want to increase the pressure exerted by the 1.000 kg red cylinder in Example 6-3 to 100.0 mbar by placing a weight on top of it. What must be the mass of this weight? Must the added weight have the same cross-sectional area as the cylinder? Explain.

2 The Simple Gas Laws

is section, we consider relationships involving the pressure, volume, temper-e, and amount of a gas. Specifically, we will see how one variable depends on her, as the remaining two are held fixed. Collectively, these relationships are red to as the simple gas laws. You can use these laws in problem solving, but will probably prefer the equation developed in the next section—the ideal gas tion. You may find that the greatest use of the simple gas laws is in solidify-our qualitative understanding of the behavior of gases.

le's Law

62, working with air, Robert Boyle discovered the first of the simple gas , now known as **Boyle's law**.

a fixed amount of gas at a constant temperature, the gas volume is ersely proportional to the gas pressure. **(6.4)**

onsider the gas in Figure 6-6. It is confined in a cylinder closed off by a freely ing "weightless" piston. The pressure of the gas depends on the total weight ed on top of the piston. This weight (a force) divided by the area of the piston s the gas pressure. If the weight on the piston is doubled, the pressure dou-and the gas volume decreases to one-half its original value. If the pressure of as is tripled, the volume decreases to one-third. Conversely, if the pressure is ced by one-half, the gas volume doubles, and so on. Mathematically, the rse relationship between gas pressure and volume is expressed as

$$P \propto \frac{1}{V} \quad \text{or} \quad PV = a \text{ (a constant)} \qquad \textbf{(6.5)}$$

n the proportionality sign ($\propto$) is replaced with an equal sign and a pro-ionality constant, the product of the pressure and volume of a fixed unt of gas at a given temperature is seen to be a constant (a). The value of ends on the amount of gas and the temperature. The graph in Figure 6-6 at of $PV = a$. It is called a hyperbola.

e equation $PV = a$ can be used to derive another equation that is useful ituations in which a gas undergoes a change at constant temperature. If vrite equation (6.5) for the initial state (i) and for the final state (f), we get

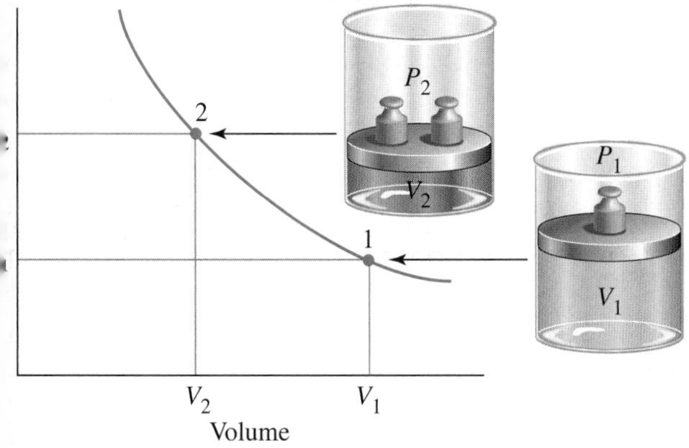

◀ FIGURE 6-6
Relationship between gas volume and pressure—Boyle's law
When the temperature and amount of gas are held constant, gas volume is inversely proportional to the pressure: A doubling of the pressure causes the volume to decrease to one-half its original value.

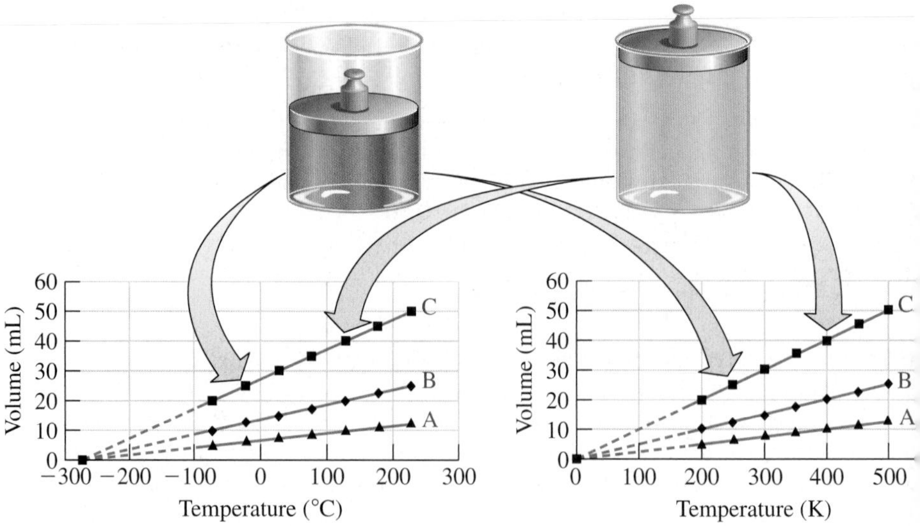

▲ FIGURE 6-7
Gas volume as a function of temperature
Volume is plotted against temperature on two different scales—Celsius and Kelvin. volumes of three different gases (A, B, and C) are measured at 1 atm and 500 K. As temperature is lowered, the volume decreases as predicted by Charles's law. Thus, 250 K (−23 °C), for example, the volume of gas C has become 25 mL, one-half of t original 50 mL. Although the relationship between volume and temperature is line both the Celsius and Kelvin temperature scales, the volume is directly proportional to the absolute temperature. That is, the volume must be zero at a temperature of zero. Only the Kelvin scale meets this requirement.

▶ It is probably fair to say that the absolute zero of temperature was "discovered" by noting that a plot of V (or P) versus. T for any gas extrapolates to −273 °C.

$P_i V_i = a$ and $P_f V_f = a$. Because both PV products are equal to the same v of a, we obtain the result

$$P_i V_i = P_f V_f \qquad (n \text{ constant}, T \text{ constant})$$

The equation above is often used to relate pressure and volume changes.

6-2 CONCEPT ASSESSMENT

A 50.0 L cylinder contains nitrogen gas at a pressure of 21.5 atm. The content of the cylinder are emptied into an evacuated tank of unknown volume. If the final pressure in the tank is 1.55 atm, then what is the volume of the tank? **(a)** (21.5/1.55) × 50.0 L; **(b)** (1.55/21.5) × 50.0 L; **(c)** 21.5/(1.55 × 50) L; **(d)** 1.55/(21.5 × 50) L.

Charles's Law

The relationship between the volume of a gas and temperature was dis ered by the French physicists Jacques Charles in 1787 and, independentl Joseph Louis Gay-Lussac, who published it in 1802.

Figure 6-7 pictures a fixed amount of gas confined in a cylinder. The pressu held constant at 1 atm while the temperature is varied. The volume of increases as the temperature is raised and decreases as the temperature is low The relationship is linear. Figure 6-7 shows the linear dependence of volum temperature for three gases at three different initial conditions. One point in mon to the three lines is their intersection with the temperature axis. Altho they differ at every other temperature, the gas volumes all reach a value of ze the same temperature. The temperature at which the volume of a hypothe

*All gases condense to liquids or solids before the temperature approaches absolute zero. when we speak of the volume of a gas, we mean the free volume among the gas molecule the volume of the molecules themselves. Thus, the gas we refer to here is *hypothetical*. It is whose molecules have mass but no volume and that does not condense to a liquid or solid

ecomes zero is the absolute zero of temperature: −273.15 °C on the Celsius or 0 K on the **absolute**, or **Kelvin**, scale. The relationship between the Kelvin erature, T, and the Celsius temperature, t, is shown below in equation (6.6).

$$T(\text{K}) = t(^\circ\text{C}) + 273.15 \tag{6.6}$$

graph on the right hand side of Figure 6-7 shows that, from a volume of at 0 K, the gas volume is directly proportional to the Kelvin temperature. statement of **Charles's law** given below in (6.7) summarizes the relation- between volume and temperature.

e volume of a fixed amount of gas at constant pressure is directly portional to the Kelvin (absolute) temperature. **(6.7)**

athematical terms, Charles's law is

$$V \propto T \quad \text{or} \quad V = bT \text{ (where } b \text{ is a constant)} \tag{6.8}$$

value of the constant b depends on the amount of gas and the pressure. es *not* depend on the identity of the gas.

om either expression (6.7) or (6.8), we see that doubling the Kelvin tem- ture of a gas causes its volume to double. Reducing the Kelvin temperature ne-half (say, from 300 to 150 K) causes the volume to decrease to one-half, so on.

quation (6.8) can be used to derive an equation that is useful for situations hich a gas undergoes a change at constant pressure. If we apply equation twice, once for the initial state (i) and once for the final state (f), we get $_i) = b$ and $(V_f/T_f) = b$. Because both (V/T) quotients are equal to the e value of b, we obtain the result

$$\frac{V_i}{T_i} = \frac{V_f}{T_f} \quad (n \text{ constant}, P \text{ constant})$$

equation above is often used to relate volume and temperature changes.

6-3 CONCEPT ASSESSMENT

alloon is inflated to a volume of 2.50 L inside a house that is kept at 24 °C. en it is taken outside on a very cold winter day. If the temperature outside is 5 °C, what will be the volume of the balloon when it is taken outside? Assume t the quantity of air in the balloon and its pressure both remain constant. (248/297) × 2.50 L; **(b)** (297/248) × 2.50 L; **(c)** 248/(297 × 2.50) L; 297/(248 × 2.50) L.

6-4 CONCEPT ASSESSMENT

ubling a gas temperature from 100 K to 200 K causes a gas volume to uble. Would you expect a similar doubling of the gas volume when a gas is ated from 100 °C to 200 °C? Explain.

ndard Conditions of Temperature and Pressure

use gas properties depend on temperature and pressure, it is useful to a set of standard conditions of temperature and pressure that can be used omparing different gases. The standard temperature for gases is taken to be = 273.15 K and standard pressure, 1 bar = 100 kPa = 10^5 Pa. **Standard itions of temperature and pressure** are usually abbreviated as **STP**. It is rtant to emphasize that STP was defined differently in the past, and some

◀ When converting from °C to K, apply the addition and subtraction significant figure rule: the two-significant-figure 25 °C, when added to 273.15, becomes the three-significant-figure 298 K. Similarly, 25.0 °C becomes 298.2 K (that is, 25.0 + 273.15 = 298.15, which rounds to 298.2).

◀ Charles's ideas about the effect of temperature on the volume of a gas were probably influenced by his passion for hot-air balloons, a popular craze of the late eighteenth century.

▲ Charles experimented with the first hydrogen-filled balloons, much like the one shown here, though smaller. He also invented most of the features of modern ballooning, including the suspended basket and the valve to release gas.

◀ Although IUPAC has recommended that one *bar* should replace the *atmosphere* as the standard state condition for gas law and thermodynamic data, use of the atmosphere persists.

▲ **Amedeo Avogadro (1776–1856)—a scientist ahead of his time**
Avogadro's hypothesis and its ramifications were not understood by his contemporaries but were effectively communicated by Stanislao Cannizzaro (1826–1910) about 50 years later.

Stamp from the Private Collection of Proffeser C.M.Lang. Photography by Gary J. Shulfer, University of Wisconsin, Stevens Point."1956, Italy(Scott#714)"; Scott Standared Postage Stamp Catalogue, Scott Pub. Co., Sidney, Ohio

▶ Avogadro's hypothesis and statements derived from it apply only to gases. There is no similar relationship for liquids or solids.

texts and chemists still use the old definition. The old definition, which based on a standard pressure of 1 atm, is discouraged. In this text, we use definition recommended by the International Union of Pure and App Chemistry (IUPAC):

Standard Temperature and Pressure (STP): 0 °C and 1 bar = 10^5 Pa

Avogadro's Law

In 1808, Gay-Lussac reported that gases react by volumes in the ratio of s whole numbers. One proposed explanation was that equal volumes of gas the same temperature and pressure contain equal numbers of atoms. Da did not agree with this proposition, however. If Gay-Lussac's proposi were true, then the reaction of hydrogen and oxygen to form water woul $H(g) + O(g) \longrightarrow HO(g)$, with combining volumes of 1 : 1 : 1, rather tha 2 : 1 : 2 that was observed.

In 1811, Amedeo Avogadro resolved this dilemma by proposing not onl "equal volumes–equal numbers" hypothesis, but also that molecules of a may break up into half-molecules when they react. Using modern terminol we would say that O_2 molecules split into atoms, which then combine molecules of H_2 to form H_2O molecules. In this way, the volume of ox needed is only one-half that of hydrogen. Avogadro's reasoning is outline Figure 6-8.

Avogadro's equal volumes–equal numbers hypothesis can be state either of two ways.

1. Equal volumes of different gases compared at the same temperature pressure contain equal numbers of molecules.
2. Equal numbers of molecules of different gases compared at the same perature and pressure occupy equal volumes.

A relationship that follows from *Avogadro's hypothesis*, often ca **Avogadro's law**, is as follows.

At a fixed temperature and pressure, the volume of a gas is directly proportional to the amount of gas. **(6**

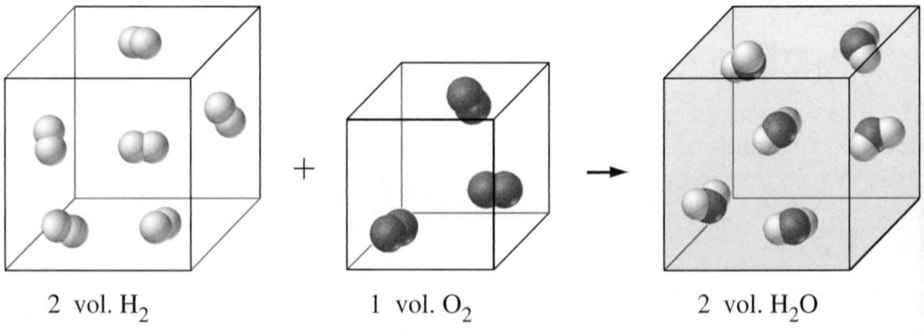

▲ FIGURE 6-8
Formation of water—actual observation and Avogadro's hypothesis
In the reaction $2\,H_2(g) + 1\,O_2(g) \longrightarrow 2\,H_2O(g)$, only one-half as many O_2 molecu are required as are H_2 molecules. If equal volumes of gases contain equal number molecules, this means the volume of $O_2(g)$ is one-half that of $H_2(g)$. The combining ratio by volume is 2 : 1 : 2.

LE 6.2 Densities and Molar Volumes of Various Gases

	Molar Mass, g mol^{-1}	Density (at STP), g L^{-1}	Molar Volume,[a] L mol^{-1} (at STP)	(at 0 °C, 1 atm)
	2.01588	8.87104×10^{-2}	22.724	22.427
	4.00260	0.17615	22.722	22.425
l gas	–	–	22.711	22.414
	28.0134	1.23404	22.701	22.404
	28.0101	1.23375	22.696	22.399
	31.9988	1.41034	22.689	22.392
	16.0425	0.70808	22.656	22.360
	71.0019	3.14234	22.595	22.300
	44.0095	1.95096	22.558	22.263
)	44.0128	1.95201	22.550	22.255
-6	30.0690	1.33740	22.483	22.189
3	17.0352	0.76139	22.374	22.081
	146.0554	6.52800	22.374	22.081
-8	44.0956	1.98318	22.235	21.944
	64.064	2.89190	22.153	21.863

ce: All data are from the National Institute of Standards and Technology (NIST) *iistry WebBook*, available online at http://webbook.nist.gov/chemistry/.

molar volume is equal to the molar mass divided by the density. The molar volume at and 1 atm is obtained by dividing the molar volume at STP by 1.01325.

number of moles of gas (n) is doubled, the volume doubles, and so on. thematical statement of this fact is

$$V \propto n \qquad \text{and} \qquad V = k \times n$$

e constant k, which is equal to V/n, is the volume per mole of gas, a quan- ve call the molar volume. Molar volumes of gases vary with temperature ressure but experiment reveals that, for given values of T and P, the molar nes of all gases are approximately equal. (In Section 6-9, we will discuss the molar volumes of gases are not exactly equal.) The data in Table 6.2 that the molar volume of a gas is approximately 22.414 L at 0 °C and 1 atm 2.711 L at STP. The following statement summarizes these observations.

◀ In general, relating the amount of gas and its volume is best done with the ideal gas equation (Section 6-3).

$$1 \text{ mol gas} = 22.414 \text{ L (at 0 °C, 1 atm)} = 22.711 \text{ L (at STP)} \qquad \textbf{(6.10)}$$

gure 6-9 should help you to visualize 22.414 L of a gas.

◀ FIGURE 6-9
Molar volume of a gas visualized
The wooden cube is 28.2 cm on edge and has approximately the same volume as one mole of gas at 1 atm and 0 °C: 22.414 L. By contrast, the volume of the basketball is 7.5 L; the soccer ball, 6.0 L; and the football, 4.4 L.

▶ The ideal gas equation will probably not be supplied on exams and should be memorized. Values of R given here, however, will be available. There is really only one R but like many properties and constants, its value can be expressed in a variety of units.

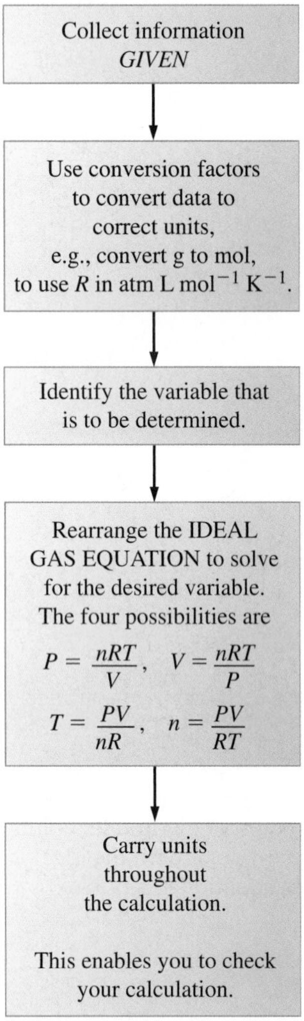

▲ **Applying the ideal gas equation**

TABLE 6.3 Five Common Values* of R
0.082057338 atm L mol^{-1} K^{-1}
0.083144598 bar L mol^{-1} K^{-1}
8.3144598 kPa L mol^{-1} K^{-1}
8.3144598 Pa m^3 mol^{-1} K^{-1}
8.3144598 J mol^{-1} K^{-1}

* Based on the CODATA recommended 2014 values of the fundamental physical constants (www.physics.nist.gov/cuu/Constants/).

6-3 Combining the Gas Laws: The Ideal Gas Equation and the General Gas Equation

Each of the three simple gas laws describes the effect that changes ir variable have on the gas volume when the other two variables are constant.

1. Boyle's law describes the effect of pressure, $V \propto 1/P$.
2. Charles's law describes the effect of temperature, $V \propto T$.
3. Avogadro's law describes the effect of the amount of gas, $V \propto n$.

These three laws can be combined into a single equation—the **idea equation**—that includes all four gas variables: volume, pressure, tem ture, and amount of gas.

The Ideal Gas Equation

In accord with the three simple gas laws, the volume of a gas is *directly* portional to the amount of gas, *directly* proportional to the Kelvin tempera and *inversely* proportional to pressure. That is,

$$V \propto \frac{nT}{P} \quad \text{and} \quad V = \frac{RnT}{P}$$

$$PV = nRT \qquad (6$$

A gas whose behavior conforms to the ideal gas equation is called an **ide perfect, gas**. Before we can apply equation (6.11), we need a value for the stant R, called the **gas constant**. One way to obtain this is to substitute equation (6.11) the molar volume of an ideal gas at 0 °C and 1 atm. How the value of R will then depend on which units are used to express pre and volume. With a molar volume of 22.4140 L and pressure in atmosph

$$R = \frac{PV}{nT} = \frac{1\text{ atm} \times 22.4140\text{ L}}{1\text{ mol} \times 273.15\text{ K}} = 0.082057\text{ atm L mol}^{-1}\text{ K}^{-1}$$

Using the SI units of m^3 for volume and Pa for pressure gives

$$R = \frac{PV}{nT} = \frac{101{,}325\text{ Pa} \times 2.24140 \times 10^{-2}\text{ m}^3}{1\text{ mol} \times 273.15\text{ K}} = 8.3145\text{ Pa m}^3\text{ mol}^{-1}\text{ K}^{-1}$$

The units Pa m^3 mol^{-1} K^{-1} also have another significance. The pascal has kg m^{-1} s^{-2}, so the units m^3 Pa become kg m^2 s^{-2}, which is the SI ur energy—the joule. Thus R also has the value

$$R = 8.3145\text{ J mol}^{-1}\text{ K}^{-1}$$

We will use this value of R when we consider the energy involved i expansion and compression.

Common values of the gas constant are listed in Table 6.3, and you have a chance to use all of them in the Practice Examples and end-of-ch exercises in this chapter. A general strategy for applying the ideal gas equ is illustrated in the diagram in the margin. This strategy is used in Exampl and 6-5.

XAMPLE 6-4 Calculating a Gas Volume with the Ideal Gas Equation

What is the volume occupied by 13.7 g $Cl_2(g)$ at 45 °C and 98.4 kPa?

nalyze

This is a relatively straightforward application of the ideal gas equation. We are given an amount of gas (in grams), a pressure (in kPa), and a temperature (in °C). Before using the ideal gas equation, we must express the amount in moles and the temperature in Kelvin. Include units throughout the calculation to ensure that the final result has acceptable units.

olve

$$P = 98.4\ \text{kPa}$$
$$V = ?$$
$$n = 13.7\ \text{g Cl}_2 \times \frac{1\ \text{mol Cl}_2}{70.90\ \text{g Cl}_2} = 0.193\ \text{mol Cl}_2$$
$$R = 8.314\ \text{kPa L mol}^{-1}\ \text{K}^{-1}$$
$$T = 45\ °\text{C} + 273 = 318\ \text{K}$$

Divide both sides of the ideal gas equation by P to solve for V.

$$\frac{PV}{P} = \frac{nRT}{P} \quad \text{and} \quad V = \frac{nRT}{P}$$

$$V = \frac{nRT}{P} = \frac{0.193\ \text{mol} \times 8.314\ \text{kPa L mol}^{-1}\ \text{K}^{-1} \times 318\ \text{K}}{98.4\ \text{kPa}} = 5.19\ \text{L}$$

ssess

A useful check of the calculated result is to make certain the units cancel properly. In the setup above, all units cancel except for L, a unit of volume. Keep in mind that when canceling units, such a unit as mol^{-1} is the same as 1/mol. Thus, $\text{mol} \times \text{mol}^{-1} = 1$ and $\text{K} \times \text{K}^{-1} = 1$.

RACTICE EXAMPLE A: What is the volume occupied by 20.2 g $NH_3(g)$ at −25° C and 752 mmHg?

RACTICE EXAMPLE B: At what temperature will a 13.7 g Cl_2 sample exert a pressure of 0.993 bar when confined in a 7.50 L container?

XAMPLE 6-5 Calculating a Gas Pressure with the Ideal Gas Equation

What is the pressure, in kilopascals, exerted by 1.00×10^{20} molecules of N_2 in a 305 mL flask at 175 °C?

nalyze

We are given an amount of gas (in molecules), a volume (in mL), and a temperature (in °C). Before using these quantities in the ideal gas equation, we must express the amount in moles, the volume in liters, and the temperature in Kelvin. Include units throughout the calculation to ensure that the final result has acceptable units.

olve

Because we seek a pressure in kilopascals, let us use the form of the ideal gas equation having

$$R = 8.3145\ \text{kPa L mol}^{-1}\ \text{K}^{-1}$$

The first step is to convert from molecules to moles of a gas, n.

$$n = 1.00 \times 10^{20}\ \text{molecules N}_2 \times \frac{1\ \text{mol N}_2}{6.022 \times 10^{23}\ \text{molecules N}_2}$$
$$= 0.000166\ \text{mol N}_2$$

(continued)

Convert from milliliters to liters and then to cubic meters (recall Figure 1-9).

$$V = 305\text{ mL} \times \frac{1\text{ L}}{1000\text{ mL}} = 0.305\text{ L}$$

Express gas temperature on the Kelvin scale.

$$T = (175 + 273)\text{ K} = 448\text{ K}$$

Rearrange the ideal gas equation to the form $P = nRT/V$, and substitute the above data.

$$P = \frac{nRT}{V} = \frac{0.000166\text{ mol} \times 8.3145\text{ kPa L mol}^{-1}\text{ K}^{-1} \times 448\text{ K}}{0.305\text{ L}}$$
$$= 2.03\text{ kPa}$$

Assess

Again, we see from the cancellation of units above that only the desired unit—a pressure unit—remains.

PRACTICE EXAMPLE A: How many moles of He(g) are in a 5.00 L storage tank filled with helium at 10.5 at pressure at 30.0 °C?

PRACTICE EXAMPLE B: How many molecules of $N_2(g)$ remain in an ultrahigh vacuum chamber of 3.45 r volume when the pressure is reduced to 6.67×10^{-7} Pa at 25 °C?

The General Gas Equation

Sometimes a gas is described under two different sets of conditions. In situations, the ideal gas equation must be applied twice—to an initial cc tion and a final condition.

Initial condition (i)	*Final condition* (f)
$P_iV_i = n_iRT_i$	$P_fV_f = n_fRT_f$
$R = \frac{P_iV_i}{n_iT_i}$	$R = \frac{P_fV_f}{n_fT_f}$

The above expressions are equal to each other because each is equal to R.

$$\frac{P_iV_i}{n_iT_i} = \frac{P_fV_f}{n_fT_f} \qquad (6.$$

Expression (6.12) is called the **general gas equation**. It is often applied in in which one or two of the gas properties are held constant, and the equ can be simplified by eliminating these constants. For example, if a cons mass of gas is subject to changes in temperature, pressure, and volume, n_i n_f cancel because they are equal (constant moles); thus we have

$$\frac{P_iV_i}{T_i} = \frac{P_fV_f}{T_f} \quad (n\text{ constant})$$

This equation is sometimes referred to as the *combined gas law*. In Exampl both volume and mass are constant, and this establishes the simple relatior between gas pressure and temperature known as *Amontons's law*: The pressu a fixed amount of gas confined to a fixed volume is directly proportional t Kelvin temperature.

Using the Gas Laws

When confronted with a problem involving gases, students sometimes der which gas equation to use. Gas law problems can often be thought more than one way. When a problem involves a comparison of two gas two states (initial and final) of a single gas, use the general gas equation (after eliminating any term (n, P, T, V) that remains constant. Otherwise the ideal gas equation (6.11).

EXAMPLE 6-6 Applying the General Gas Equation

The situation pictured in Figure 6-10(a) is changed to that in Figure 6-10(b). What is the gas pressure in Figure 6-10(b)?

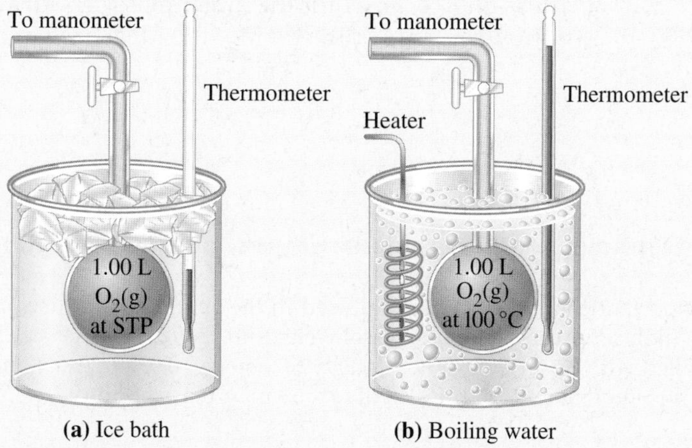

▲ FIGURE 6-10
Pressure of a gas as a function of temperature—Example 6-6 visualized
The amount of gas and volume are held constant.
(a) 1.00 L $O_2(g)$ at STP; **(b)** 1.00 L $O_2(g)$ at 100 °C.

nalyze

Identify the quantities in the general gas equation that remain constant. Cancel out these quantities and solve the equation that remains.

olve

In this case, the amount of O_2 is constant ($n_i = n_f$) and the volume is constant ($V_i = V_f$).

$$\frac{P_i \cancel{V_i}}{\cancel{n_i} T_i} = \frac{P_f \cancel{V_f}}{\cancel{n_f} T_f} \quad \text{and} \quad \frac{P_i}{T_i} = \frac{P_f}{T_f} \quad \text{and} \quad P_f = P_i \times \frac{T_f}{T_i}$$

Since $P_i = 1.00$ bar, $T_i = 273$ K, and $T_f = 373$ K, then

$$P_f = 1.00\ \text{bar} \times \frac{373\ \text{K}}{273\ \text{K}} = 1.37\ \text{bar}$$

ssess

We can base our check on a qualitative, intuitive understanding of what happens when a gas is heated in a closed container. Its pressure increases (possibly to the extent that the container bursts). If, by error, we had used the ratio of temperatures 273 K/373 K, the final pressure would have been less than 1.00 bar—an impossible result.

RACTICE EXAMPLE A: A 1.00 mL sample of $N_2(g)$ at 36.2 °C and 2.14 atm is heated to 37.8 °C, and the pressure changed to 1.02 atm. What volume does the gas occupy at this final temperature and pressure?

RACTICE EXAMPLE B: Suppose that in Figure 6-10 we want the pressure to remain at 1.00 bar when the $O_2(g)$ is heated to 100 °C. What mass of $O_2(g)$ must we release from the flask?

Applications of the Ideal Gas Equation

ough the ideal gas equation can always be used as it was presented in tion (6.11), it is useful to recast it into slightly different forms for some ications. We will consider two such applications in this section: determi-n of molar masses and gas densities.

Molar Mass Determination

If we know the volume of a gas at a fixed temperature and pressure, we solve the ideal gas equation for the amount of the gas in moles. Because number of moles of gas (n) is equal to the mass (m) of gas divided by the m mass (M), if we know the mass and number of moles of gas, we can solve expression $n = m/M$ for the molar mass, M. An alternative is to make the stitution $n = m/M$ directly into the ideal gas equation.

$$PV = \frac{mRT}{M}$$

EXAMPLE 6-7 Determining a Molar Mass with the Ideal Gas Equation

Propylene is an important commercial chemical used in the synthesis of other organic chemicals and in pr duction of plastics (polypropylene). A glass vessel weighs 40.1305 g when clean, dry, and evacuated; it weig 138.2410 g when filled with water at 25.0 °C (density of water = 0.9970 g/mL) and 40.2959 g when filled wi propylene gas at 740.3 mmHg and 24.0 °C. What is the molar mass of propylene?

Analyze

We are given a pressure (in mmHg), a temperature (in °C), and information that will enable us to determi the amount of gas (in grams) and the volume of the vessel. If we express these quantities in atmospheres, Kelvi moles, and liters, respectively, then we can use equation (6.13), with $R = 0.082057 \text{ atm L mol}^{-1} \text{ K}^{-1}$, to calcula the molar mass of the gas.

Solve

First determine the mass of water required to fill the vessel.

$$\text{mass of water to fill vessel} = 138.2410 \text{ g} - 40.1305 \text{ g} = 98.1105 \text{ g}$$

Use the density of water in a conversion factor to obtain the volume of water (and hence, the volume of the glass vessel).

$$\text{volume of water (volume of vessel)} = 98.1105 \text{ g } H_2O \times \frac{1 \text{ mL } H_2O}{0.9970 \text{ g } H_2O} = 98.41 \text{ mL} = 0.09841 \text{ L}$$

The mass of the gas is the difference between the weight of the vessel filled with propylene gas and the weight of the empty vessel.

$$\text{mass of gas} = 40.2959 \text{ g} - 40.1305 \text{ g} = 0.1654 \text{ g}$$

The values of temperature and pressure are given.

$$T = (24.0 + 273.15) \text{ K} = 297.2 \text{ K}$$

$$P = 740.3 \text{ mmHg} \times \frac{1 \text{ atm}}{760 \text{ mmHg}} = 0.9741 \text{ atm}$$

Substitute data into the rearranged version of equation (6.13).

$$M = \frac{mRT}{PV} = \frac{0.1654 \text{ g} \times 0.082057 \text{ atm L mol}^{-1} \text{ K}^{-1} \times 297.2 \text{ K}}{0.9741 \text{ atm} \times 0.09841 \text{ L}}$$

$$= 42.08 \text{ g mol}^{-1}$$

Assess

Cancellations leave the units g and mol^{-1}. The unit g mol^{-1} or g/mol is that for molar mass, the quantity we a seeking. We can use another approach to solving this problem. We can substitute the pressure (0.9741 atm temperature (297.2 K), and volume (0.09841 L) into the ideal gas equation to calculate the number of moles the gas sample (0.003931 mol). Because the sample contains 0.003931 mol and has a mass of 0.165 g, the mol mass is 0.165 g/0.003931 mol = 42.0 g mol^{-1}. The advantage of this alternative approach is that it makes use only the ideal gas equation; you do not have to memorize or derive equation (6.13) for the cases when yo might need it.

PRACTICE EXAMPLE A: The same glass vessel used in Example 6-7 is filled with an unknown gas at 772 mmH and 22.4 °C. The gas-filled vessel weighs 40.4868 g. What is the molar mass of the gas?

PRACTICE EXAMPLE B: A 1.27 g sample of an oxide of nitrogen, believed to be either NO or N_2O, occupies volume of 1.07 L at 25 °C and 0.982 bar. Which oxide is it?

uation (6.13) is used to determine a molar mass in Example 6-7, but note he equation can also be used when the molar mass of a gas is known and nass of a particular sample of the gas is sought.

ppose that we want to determine the formula of an unknown hydrocar- With combustion analysis we can establish the mass percent composi- and from this, we can determine the empirical formula. The method of nple 6-7 gives us a molar mass, in g mol^{-1}, which is numerically equal to the cular mass, in u. This is all the information we need to establish the true cular formula of the hydrocarbon (see Exercise 96).

Densities

etermine the density of a gas, we can start with the density equation, n/V. Then we can express the mass of gas as the product of the number of s of gas and the molar mass: $m = n \times M$. This leads to

$$d = \frac{m}{V} = \frac{n \times M}{V} = \frac{n}{V} \times M$$

, with the ideal gas equation, we can replace n/V by its equivalent, P/RT, tain

$$d = \frac{m}{V} = \frac{MP}{RT} \qquad \textbf{(6.14)}$$

e density of a gas at STP can easily be calculated by dividing its molar by the molar volume (22.7 L/mol). For $O_2(g)$ at STP, for example, the ity is 32.0 g/22.7 L = 1.41 g/L. Equation (6.14) can be used for other con- ns of temperature and pressure.

KEEP IN MIND

that gas densities are typically much smaller than those of liquids and solids. Gas densities are usually expressed in grams per *liter* rather than grams per *milliliter*.

KAMPLE 6-8 Using the Ideal Gas Equation to Calculate a Gas Density

What is the density of oxygen gas (O_2) at 298 K and 0.987 bar?

nalyze

The gas is identified, and therefore the molar mass can be calculated. We are given a temperature in Kelvin and a pressure in bars, so we can use equation (6.14) directly with $R = 0.08314 \text{ bar L mol}^{-1} \text{K}^{-1}$.

olve

The molar mass of O_2 is 32.0 g mol^{-1}. Now, use equation (6.14).

$$d = \frac{m}{V} = \frac{MP}{RT} = \frac{32.00 \text{ g mol}^{-1} \times 0.987 \text{ bar}}{0.08314 \text{ bar L mol}^{-1} \text{K}^{-1} \times 298 \text{ K}} = 1.27 \text{ g/L}$$

ssess

We can solve this problem in another way. To calculate the density of a gas at a certain temperature and pressure, use a 1.00 L sample of the gas. The mass of a 1.00 L sample is equal to the density in grams per liter. To calculate the mass of a 1.00 L sample, first use the ideal gas equation to calculate the number of moles in the sample, and then convert the amount in moles to an amount in grams by using the molar mass as a conversion factor. In the present case, the amount of $O_2(g)$ in a 1.00 L sample at 0.987 bar and 298 K is 0.0398 mol O_2, or 1.27 g O_2. Because a 1.00 L sample of O_2 at this temperature and pressure has a mass of 1.27 g, the density is 1.27 g/L.

ACTICE EXAMPLE A: What is the density of helium gas at 298 K and 0.987 bar? Based on your answer, explain why we can say that helium is "lighter than air." [*Hint:* The average molar mass of the molecules in air is 28.8 g mol^{-1}.]

ACTICE EXAMPLE B: The density of a sample of gas is 1.00 g/L at 745 mmHg and 109 °C. What is the molar mass of the gas?

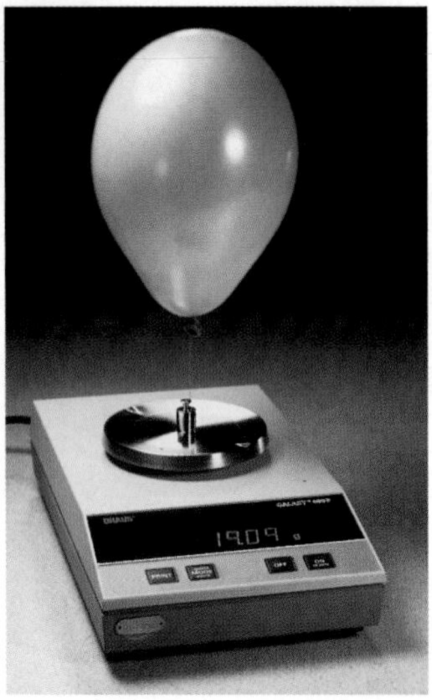

Carey B. Van Loon

▲ FIGURE 6-11
The helium-filled balloon exerts a lifting force on the 20.00 g weight, so that the balloon and the weight together weigh only 19.09 g.

The density of gases differs from that of solids and liquids in two impo ways.

1. Gas densities depend strongly on pressure and temperature, increasi the gas pressure increases and decreasing as the temperature incre Densities of liquids and solids also depend somewhat on temperature they depend far less on pressure.
2. The density of a gas is directly proportional to its molar mass. No si relationship exists between density and molar mass for liquids and sc

An important application of gas densities is in establishing condition lighter-than-air balloons. A gas-filled balloon will rise in the atmosphere if the density of the gas is less than that of the surrounding air. Becaus densities are directly proportional to molar masses, the lower the molar of the gas, the greater its lifting power. The lowest molar mass is th hydrogen, but hydrogen is flammable and forms explosive mixtures wit The explosion of the dirigible *Hindenburg* in 1937 spelled the en transoceanic travel by hydrogen-filled airships. Now, airships such a Goodyear blimps use helium, which has a molar mass only twice th hydrogen (Fig. 6-11) but is inert. Hydrogen is still used for weather and observational balloons.

Another alternative is to fill a balloon with *hot* air. Equation (6.14) indi that the density of a gas is *inversely* proportional to temperature. Hot air i dense than cold air. However, because the density of air decreases ra with altitude, there is a limit to how high a hot-air balloon or any gas- balloon can rise.

6-5 Gases in Chemical Reactions

Reactions involving gases as reactants or products (or both) are no strang us. We now have a new tool to apply to reaction stoichiometry calculat the ideal gas equation. Specifically, we can now handle information a gases in terms of volumes, temperatures, and pressures, as well as by and amount in moles. A practical application is the nitrogen-forming rea in an automobile air-bag safety system, which utilizes the rapid decom tion of sodium azide.

$$2\,NaN_3(s) \xrightarrow{\Delta} 2\,Na(l) + 3\,N_2(g)$$

The essential components of the system are an ignition device and a pellet taining sodium azide and appropriate additives. When activated, the sy inflates an air bag in 20 to 60 ms and converts Na(l) to a harmless solid res

For reactions involving gases, we can (1) use stoichiometric factors to r the amount of a gas to amounts of other reactants or products, and (2) us ideal gas equation to relate the amount of gas to volume, temperature, pressure. In Example 6-9, we use this approach to determine the volun $N_2(g)$ produced in a typical air-bag system.

Law of Combining Volumes

If the reactants and products involved in a stoichiometric calculatio gases, sometimes we can use a particularly simple approach. Conside reaction.

$$2\,NO(g) + O_2(g) \longrightarrow 2\,NO_2(g)$$
$$2\text{ mol }NO(g) + 1\text{ mol }O_2(g) \longrightarrow 2\text{ mol }NO_2(g)$$

XAMPLE 6-9 Using the Ideal Gas Equation in Reaction Stoichiometry Calculations

What volume of N_2, measured at 98.0 kPa and 26 °C, is produced when 75.0 g NaN_3 is decomposed?

$$2\,NaN_3(s) \xrightarrow{\Delta} 2\,Na(l) + 3\,N_2(g)$$

nalyze

The following conversions are required.

$$g\,NaN_3 \longrightarrow mol\,NaN_3 \longrightarrow mol\,N_2 \longrightarrow L\,N_2$$

The molar mass of NaN_3 is used for the first conversion. The second conversion makes use of a stoichiometric factor constructed from the coefficients of the chemical equation. The ideal gas equation is used to complete the final conversion.

olve

$$?\,mol\,N_2 = 75.0\,g\,NaN_3 \times \frac{1\,mol\,NaN_3}{65.01\,g\,NaN_3} \times \frac{3\,mol\,N_2}{2\,mol\,NaN_3} = 1.73\,mol\,N_2$$

$$P = 98.0\,kPa$$
$$V = ?$$
$$n = 1.73\,mol$$
$$R = 8.314\,kPa\,L\,mol^{-1}\,K^{-1}$$
$$T = (26 + 273)\,K = 299\,K$$
$$V = \frac{nRT}{P} = \frac{1.73\,mol \times 8.314\,kPa\,L\,mol^{-1}\,K^{-1} \times 299\,K}{98.0\,kPa} = 43.9\,L$$

ssess

75.0 g NaN_3 is slightly more than one mole ($M \approx 65$ g/mol). From this amount of NaN_3 we should expect a little more than 1.5 mol $N_2(g)$. At 0 °C and 100 kPa, 1.5 mol $N_2(g)$ would occupy a volume of $1.5 \times 22.7 = 34$ L. Because the temperature is higher than 0 °C and the pressure is lower than 100 kPa, the sample should have a volume somewhat greater than 34 L.

RACTICE EXAMPLE A: How many grams of NaN_3 are needed to produce 20.0 L of $N_2(g)$ at 30.0 °C and 776 mmHg?

RACTICE EXAMPLE B: How many grams of Na(l) are produced per liter of $N_2(g)$ formed in the decomposition of sodium azide if the gas is collected at 25 °C and 1.00 bar?

ose the gases are compared at the same T and P. Under these conditions, mole of gas occupies a particular volume, call it V liters; two moles of gas py $2V$ liters; and so on.

$$2\,V\,L\,NO(g) + V\,L\,O_2(g) \longrightarrow 2\,V\,L\,NO_2(g)$$

e divide each coefficient by V, we get the following result:

$$2\,L\,NO(g) + 1\,L\,O_2(g) \longrightarrow 2\,L\,NO_2(g)$$

s, the volume ratio of the gases consumed and produced in a chemical is ame as the mole ratio, provided the volumes are all measured at the same perature and pressure.

hat we have just done is to develop, in modern terms, Gay-Lussac's **law ombining volumes**. We previewed this law on page 204 by suggesting the volumes of gases involved in a reaction are in the ratio of small whole bers. The small whole numbers are simply the stoichiometric coefficients e balanced equation. We apply this law in Example 6-10.

EXAMPLE 6-10 Applying the Law of Combining Volumes

Zinc blende, ZnS, is the most important zinc ore. Roasting (strong heating) of ZnS in oxygen is the first step i the commercial production of zinc.

$$2\,ZnS(s) + 3\,O_2(g) \xrightarrow{\Delta} 2\,ZnO(s) + 2\,SO_2(g)$$

What volume of $SO_2(g)$ can be obtained from 1.00 L $O_2(g)$ and excess ZnS(s)? Both gases are measured a 25 °C and 98.0 kPa.

Analyze

The reactant and product being compared are both gases, and both are at the same temperature and pressur Therefore, we can use the law of combining volumes and treat the coefficients in the balanced chemical equa tion as if they had units of liters.

Solve

The stoichiometric factor (shown below in blue) converts from L $O_2(g)$ to L $SO_2(g)$.

$$?\,L\,SO_2(g) = 1.00\,L\,O_2(g) \times \frac{2\,L\,SO_2(g)}{3\,L\,O_2(g)} = 0.667\,L\,SO_2(g)$$

Assess

Some students would solve this problem by using the following sequence of conversions: L O_2 — mol O_2 $\longrightarrow$ mol SO_2 $\longrightarrow$ L SO_2. This approach is acceptable but not as simple as the approach we used.

PRACTICE EXAMPLE A: The first step in making nitric acid is to convert ammonia to nitrogen monoxide. This done under conditions of high temperature and in the presence of a platinum catalyst. What volume of O_2(g is consumed per liter of NO(g) formed?

$$4\,NH_3(g) + 5\,O_2(g) \xrightarrow[850\,°C]{Pt} 4\,NO(g) + 6\,H_2O(g)$$

PRACTICE EXAMPLE B: If all gases are measured at the same temperature and pressure, what volume of NH_3(g is produced when 225 L $H_2(g)$ are consumed in the reaction $N_2(g) + H_2(g) \longrightarrow NH_3(g)$ (not balanced)?

6-5 CONCEPT ASSESSMENT

Would the answer in Example 6-10 be greater than, less than, or equal to 0.667 L if **(a)** both gases were measured at STP; **(b)** if the O_2 were measured a STP and the SO_2 were measured at 25 °C and 98.0 kPa?

6-6 Mixtures of Gases

The simple gas laws, such as Boyle's and Charles's laws, were based or behavior of air—a mixture of gases. So, the simple gas laws and the idea equation apply to a *mixture* of nonreactive gases as well as to individual g Where possible, the simplest approach to working with gaseous mixtures is to use for the value of n the *total* number of moles of the gaseous mixture (

As a specific example, consider a mixture of gases in a vessel of fixed vol V at temperature T. The total pressure of the mixture is determined by the number of moles:

$$P_{tot} = \frac{n_{tot}RT}{V} \qquad (T \text{ constant}, V \text{ constant})$$

For fixed values of T and P, the total volume of a mixture of gases is also d mined by the total number of moles:

$$V_{tot} = \frac{n_{tot}RT}{P} \qquad (T \text{ constant}, P \text{ constant})$$

XAMPLE 6-11 Applying the Ideal Gas Equation to a Mixture of Gases

What is the pressure, in bar, exerted by a mixture of 1.0 g H_2 and 5.00 g He when the mixture is confined to a volume of 5.0 L at 20 °C?

nalyze

For fixed T and V, the total pressure of a mixture of gases is determined by the total number of moles of gas: $P_{tot} = n_{tot}RT/V$.

olve

$$n_{tot} = \left(1.0\text{ g H}_2 \times \frac{1\text{ mol H}_2}{2.02\text{ g H}_2}\right) + \left(5.00\text{ g He} \times \frac{1\text{ mol He}}{4.003\text{ g He}}\right)$$

$$= 0.50\text{ mol H}_2 + 1.25\text{ mol He} = 1.75\text{ mol gas}$$

$$P_{tot} = \frac{1.75\text{ mol} \times 0.0831\text{ bar L mol}^{-1}\text{ K}^{-1} \times (20 + 273)\text{ K}}{5.0\text{ L}} = 8.5\text{ bar}$$

ssess

It is also possible to solve this problem by starting from equation (6.12). Because 1 mol of ideal gas occupies 22.7 L at 273 K and 1 bar, the pressure exerted by 1.75 mol of gas in a 5.0 L vessel at 293 K is $(1.75\text{ mol}/1.00\text{ mol}) \times (293\text{ K}/273\text{ K}) \times (22.7\text{ L}/5.0\text{ L}) \times 1.0\text{ bar} = 8.5\text{ bar}$.

RACTICE EXAMPLE A: What will be the total gas pressure, in bar, if 12.5 g Ne is added to the mixture of gases described in Example 6-11 and the temperature is then raised to 55 °C? [*Hint:* What is the new number of moles of gas? What effect does raising the temperature have on the pressure of a gas at constant volume?]

RACTICE EXAMPLE B: 2.0 L of $O_2(g)$ and 8.0 L of $N_2(g)$, each at 0.00 °C and 1.00 atm, are mixed together. The nonreactive gaseous mixture is compressed to occupy 2.0 L at 298 K. What is the pressure exerted by this mixture?

hn Dalton made an important contribution to the study of gaseous mixtures. roposed that in a mixture, each gas expands to fill the container and exerts ame pressure (called its **partial pressure**) that it would if it were alone in the ainer. **Dalton's law of partial pressures** states that the total pressure of a ure of gases is the sum of the partial pressures of the components of the mix- as shown in Figure 6-12. For a mixture of gases, A, B, and so on,

$$P_{tot} = P_A + P_B + \cdots \qquad (6.16)$$

KEEP IN MIND

that when this expression is used, $V_A = V_B = \cdots = V_{tot}$

a gaseous mixture of n_A moles of A, n_B moles of B, and so on, the volume gas would individually occupy at a pressure equal to P_{tot} is

$$V_A = n_ART/P_{tot}; V_B = n_BRT/P_{tot}; \cdots$$

total volume of the gaseous mixture is

$$V_{tot} = V_A + V_B \cdots$$

the commonly used expression *percent by volume* is

$$\text{volume \% A} = \frac{V_A}{V_{tot}} \times 100\%; \text{volume \% B} = \frac{V_B}{V_{tot}} \times 100\%; \cdots$$

KEEP IN MIND

that when using this expression, $P_A = P_B = \cdots = P_{tot}$

an derive a particularly useful expression from the following ratios,

$$\frac{P_A}{P_{tot}} = \frac{n_A(\cancel{RT/V_{tot}})}{n_{tot}(\cancel{RT/V_{tot}})} = \frac{n_A}{n_{tot}} \quad \text{and} \quad \frac{V_A}{V_{tot}} = \frac{n_A(\cancel{RT/P_{tot}})}{n_{tot}(\cancel{RT/P_{tot}})} = \frac{n_A}{n_{tot}}$$

ch means that

$$\frac{n_A}{n_{tot}} = \frac{P_A}{P_{tot}} = \frac{V_A}{V_{tot}} = x_A \qquad (6.17)$$

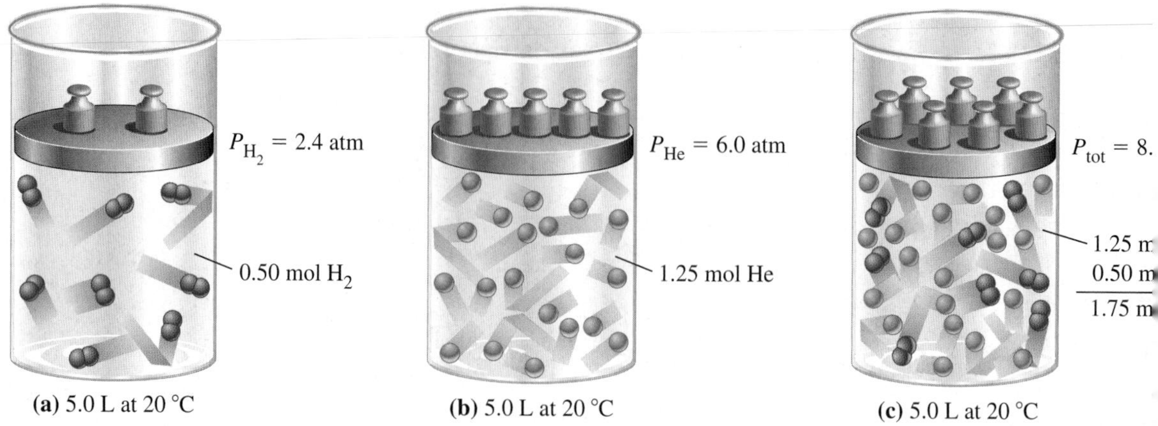

▲ FIGURE 6-12
Dalton's law of partial pressures illustrated
The pressure of each gas is proportional to the number of moles of gas. The total pressure is the sum of the partial pressures of the individual gases.

The term n_A/n_{tot} is given a special name, the mole fraction of A, x_A. The **fraction** of a component in a mixture is the fraction of all the molecules in mixture contributed by that component. The sum of all the mole fractions mixture is one.

As illustrated in Example 6-12, we can often think about mixtures of g in more than one way.

EXAMPLE 6-12 Calculating the Partial Pressures in a Gaseous Mixture

What are the partial pressures of H_2 and He in the gaseous mixture described in Example 6-11?

Analyze

The ideal gas equation can be applied to each gas individually to obtain the partial pressure of each gas.

Solve

One approach involves a direct application of Dalton's law in which we calculate the pressure that each ga would exert if it were alone in the container.

$$P_{H_2} = \frac{n_{H_2} \times RT}{V} = \frac{0.50 \text{ mol} \times 0.0831 \text{ bar L mol}^{-1} \text{K}^{-1} \times 293 \text{ K}}{5.0 \text{ L}} = 2.4 \text{ bar}$$

$$P_{He} = \frac{n_{He} \times RT}{V} = \frac{1.25 \text{ mol} \times 0.0831 \text{ bar L mol}^{-1} \text{K}^{-1} \times 293 \text{ K}}{5.0 \text{ L}} = 6.1 \text{ bar}$$

Expression (6.17) gives us a simpler way to answer the question because we already know the number of moles of each gas and the total pressure from Example 6-11 (P_{tot} = 8.4 atm).

$$P_{H_2} = \frac{n_{H_2}}{n_{tot}} \times P_{tot} = \frac{0.50}{1.75} \times 8.5 \text{ bar} = 2.4 \text{ bar}$$

$$P_{He} = \frac{n_{He}}{n_{tot}} \times P_{tot} = \frac{1.25}{1.75} \times 8.5 \text{ bar} = 6.1 \text{ bar}$$

Assess

An effective way of checking an answer is to obtain the same answer when the problem is done in differen ways, as was the case here.

PRACTICE EXAMPLE A: A mixture of 0.197 mol CO_2(g) and 0.00278 mol H_2O(g) is held at 30.0 °C and 2.50 atm What is the partial pressure of each gas?

PRACTICE EXAMPLE B: The percent composition of air by volume is 78.08% N_2, 20.95% O_2, 0.93% Ar, and 0.036 CO_2. What are the partial pressures of these four gases in a sample of air at a barometric pressure of 748 mmHg

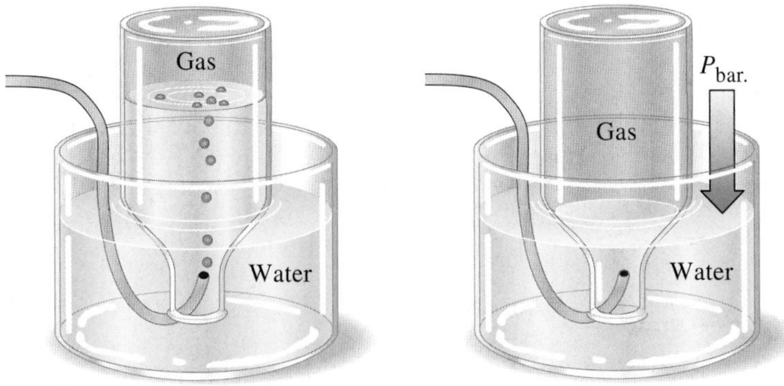

GURE 6-13
ecting a gas over water
bottle is filled with water and its open end is held below the water level in the ainer. Gas from a gas-generating apparatus is directed into the bottle. As gas mulates in the bottle, water is displaced from the bottle into the container. To make total gas pressure in the bottle equal to barometric pressure, the position of the bottle t be adjusted so that the water levels inside and outside the bottle are the same.

◀ Some early experimenters used mercury instead of water in this device so as to be able to collect gases soluble in water.

6-6 CONCEPT ASSESSMENT

ithout doing a detailed calculation, state the outcome(s) you would expect Figure 6-12(c) if an additional 0.50 mol H_2 were added to the cylinder in gure 6-12(a): **(a)** the mole fraction of H_2 would double; **(b)** the partial pressure He would remain the same; **(c)** the mole fraction of He would remain the me; **(d)** the total gas pressure would increase by 50%; **(e)** the total mass gas would increase by 1.0 g.

he device pictured in Figure 6-13, a *pneumatic trough*, played a crucial role solating gases in the early days of chemistry. The method works, of course, y for gases that are insoluble in and do not react with the liquid being dis- ced. Many important gases meet these criteria. For example, H_2, O_2, and are all essentially insoluble in and unreactive with water.

gas collected in a pneumatic trough filled with water is said to be *collected water* and is "wet." It is a mixture of two gases—the desired gas and water or. The gas being collected expands to fill the container and exerts its partial ssure, P_{gas}. Water vapor, formed by the evaporation of liquid water, also fills container and exerts a partial pressure, P_{H_2O}. The pressure of the water or depends only on the temperature of the water, as shown in Table 6.4.

According to Dalton's law, the total pressure of the wet gas is the sum of the partial pressures. The total pressure can be made equal to the prevailing ssure of the atmosphere (barometric pressure) by adjusting the position of bottle; thus we can write

$$P_{tot} = P_{bar.} = P_{gas} + P_{H_2O} \quad \text{or} \quad P_{gas} = P_{bar.} - P_{H_2O}$$

ce P_{gas} has been established, it can be used in stoichiometric calculations as strated in Example 6-13.

TABLE 6.4 Vapor Pressure of Water at Various Temperatures

	Vapor Pressure	
t, °C	kPa	mmHg
15.0	1.706	12.79
17.0	1.938	14.54
19.0	2.198	16.49
21.0	2.448	18.66
23.0	2.811	21.08
25.0	3.170	23.78
30.0	4.247	31.86
50.0	12.35	92.65

EXAMPLE 6-13 Collecting a Gas over a Liquid (Water)

In the following reaction, 81.2 mL of $O_2(g)$ is collected over water at 23 °C and barometric pressure 751 mmHg. What mass of $Ag_2O(s)$ decomposed? (The vapor pressure of water at 23 °C is 21.1 mmHg.)

$$2\,Ag_2O(s) \longrightarrow 4\,Ag(s) + O_2(g)$$

(continued)

Analyze

The key concept is that the gas collected is wet, that is, a *mixture* of $O_2(g)$ and water vapor. Us $P_{bar.} = P_{O_2} + P_{H_2O}$ to calculate P_{O_2}, and then use the ideal gas equation to calculate the number of moles of O The following conversions are used to complete the calculation: mol $O_2 \longrightarrow$ mol $Ag_2O \longrightarrow$ g Ag_2O.

Solve

$$P_{O_2} = P_{bar.} - P_{H_2O} = 751\ \text{mmHg} - 21.1\ \text{mmHg} = 730\ \text{mmHg}$$

$$P_{O_2} = 730\ \text{mmHg} \times \frac{1\ \text{atm}}{760\ \text{mmHg}} = 0.961\ \text{atm}$$

$$V = 81.2\ \text{mL} = 0.0812\ \text{L}$$

$$n = ?$$

$$R = 0.08206\ \text{atm L mol}^{-1}\ \text{K}^{-1}$$

$$T = (23 + 273)\ \text{K} = 296\ \text{K}$$

$$n = \frac{PV}{RT} = \frac{0.961\ \text{atm} \times 0.0812\ \text{L}}{0.08206\ \text{atm L mol}^{-1}\ \text{K}^{-1} \times 296\ \text{K}} = 0.00321\ \text{mol}$$

From the chemical equation we obtain a factor to convert from moles of O_2 to moles of Ag_2O. The molar mas of Ag_2O provides the final factor.

$$?\ \text{g}\ Ag_2O = 0.00321\ \text{mol}\ O_2 \times \frac{2\ \text{mol}\ Ag_2O}{1\ \text{mol}\ O_2} \times \frac{231.7\ \text{g}\ Ag_2O}{1\ \text{mol}\ Ag_2O} = 1.49\ \text{g}\ Ag_2O$$

Assess

The determination of the number of moles of O_2 in the sample is the key calculation. We can quickly estimate th number of moles of O_2 in the sample by using the fact that for typical conditions ($T \approx 298$ K, $P \approx 760$ mmHg) the molar volume of an ideal gas is about 24 L. The number of moles of gas (mostly O_2) in the sample is approxi mately 0.08 L/24 L $\text{mol}^{-1} \approx 0.003$ mol. This estimate is quite close to the value calculated above.

PRACTICE EXAMPLE A: The reaction of aluminum with hydrochloric acid produces hydrogen gas. The balance chemical equation for the reaction is given below.

$$2\ Al(s) + 6\ HCl(aq) \longrightarrow 2\ AlCl_3(aq) + 3\ H_2(g)$$

If 35.5 mL of $H_2(g)$ is collected over water at 26 °C and a barometric pressure of 755 mmHg, how many mole of HCl must have been consumed? The vapor pressure of water at 26 °C is 25.2 mmHg.

PRACTICE EXAMPLE B: An 8.07 g sample of impure Ag_2O decomposes into solid silver and $O_2(g)$. If 395 mL O_2(g is collected over water at 25 °C and 749.2 mmHg barometric pressure, then what is the percent by mass o Ag_2O in the sample? The vapor pressure of water at 25 °C is 23.8 mmHg.

6-7 Kinetic–Molecular Theory of Gases

Let us apply some terminology that we introduced in Section 1-1 on the sci tific method: The simple gas laws and the ideal gas equation are usec predict gas behavior. They are *natural laws*. To explain the gas laws, we nee *theory*. One theory developed during the mid-nineteenth century is called **kinetic–molecular theory of gases**. It is based on the *model* illustratec Figure 6-14 and outlined as follows.

- A gas is composed of a very large number of extremely small particles (n ecules or, in some cases, atoms) in constant, random, straight-line motio
- Molecules of a gas are separated by great distances. The gas is mo empty space. (The molecules are treated as so-called point masses though they have mass but no volume.)
- Molecules collide only fleetingly with one another and with the wall their container, and most of the time molecules are not colliding.
- There are assumed to be no forces between molecules except very bri during collisions. That is, each molecule acts independently of all others and is unaffected by their presence, except during collisions.

Individual molecules may gain or lose energy as a result of collisions. In a collection of molecules at constant temperature, however, *the total energy remains constant.*

ne validity of this model can be ascertained only by comparing predictions d on the model with experimental facts. As we will see, the predictions based nis model are consistent with several observed macroscopic properties.

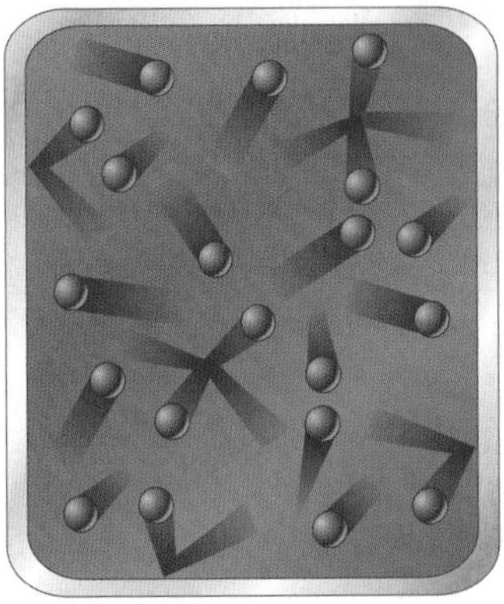

▲ FIGURE 6-14
Visualizing molecular motion
Molecules of a gas are in constant, random, straight-line motion and exhibit a distribution of speeds. They undergo many collisions with one another and with the container wall.

tribution of Molecular Speeds

all the molecules in a gas sample travel at the same speed. Because of the number of molecules, we cannot know the speed of each molecule, but an make a statistical prediction of how many molecules have a particular d, u. To make such a prediction, we can use the following equation, which ribes the distribution of speeds in a gas sample. Roughly speaking, for a ified value of u, in m/s, the corresponding value of F represents the frac- of molecules having speed between u and $(u + 1)$ m/s.

$$F(u) = 4\pi\left(\frac{M}{2\pi RT}\right)^{3/2} u^2 e^{-(Mu^2/2RT)} \qquad \textbf{(6.18)}$$

equation above is called the Maxwell–Boltzmann distribution of speeds, onor of James Clerk Maxwell, who first derived it in 1860, and Ludwig zmann who later provided significant insights into the physical origins of equation. We will not attempt to derive the distribution because to do so uires complex mathematics. Instead, we will accept it as valid and use it to us understand how the distribution of speeds depends on molar mass, nd temperature, T.

◀ Maxwell derived the equation in 1860, but it took until 1955 for direct experimental verification of this equation to be made. The experimental determination of the molecular speed distribution is described on page 221.

plot of $F(u)$ versus u is shown in Figure 6-15 for a sample of $H_2(g)$ at 0 °C. shape of the distribution is easily justified. The distribution depends on the duct of two opposing factors: a factor that is proportional to u^2 and an onential factor, $e^{-Mu^2/2RT}$. As u increases, the u^2 factor increases from a value ero, while the exponential factor decreases from a value of one. The u^2 fac- avors the presence of molecules with high speeds and is responsible for e being few molecules with speeds near zero. The exponential factor ors low speeds and limits the number of molecules that can have high eds. Because F is the product of these two opposing factors, F increases from lue of zero, reaches a maximum, and then decreases as u increases. Notice the distribution is not symmetrical about its maximum.

n Figure 6-16, we show how the distribution of molecular speeds depends emperature and molar mass. When we compare the distributions for g) at 273 K and 1000 K, we see that the range of speeds broadens as the perature increases and that the distribution shifts toward higher speeds.

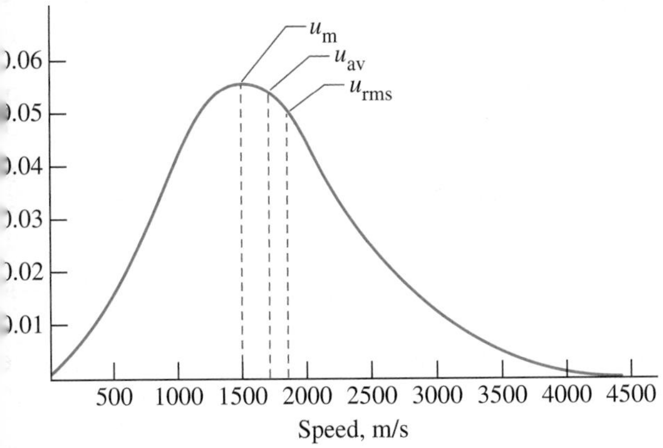

◀ FIGURE 6-15
Distribution of molecular speeds—hydrogen gas at 0 °C
The percentages of molecules with a certain speed are plotted as a function of the speed. Three different speeds are noted on the graph. The most probable speed, u_m, is approximately 1500 m/s; the average speed, u_{av}, is approximately 1700 m/s; and the root-mean-square speed, u_{rms}, is approximately 1800 m/s. Notice that $u_m < u_{av} < u_{rms}$.

▶ The area under each curve is the same. Each curve is for the same total number of molecules. Raising T (or lowering M) is like stretching a rubber graph. The same number of molecules are spread out over a wider range of speeds.

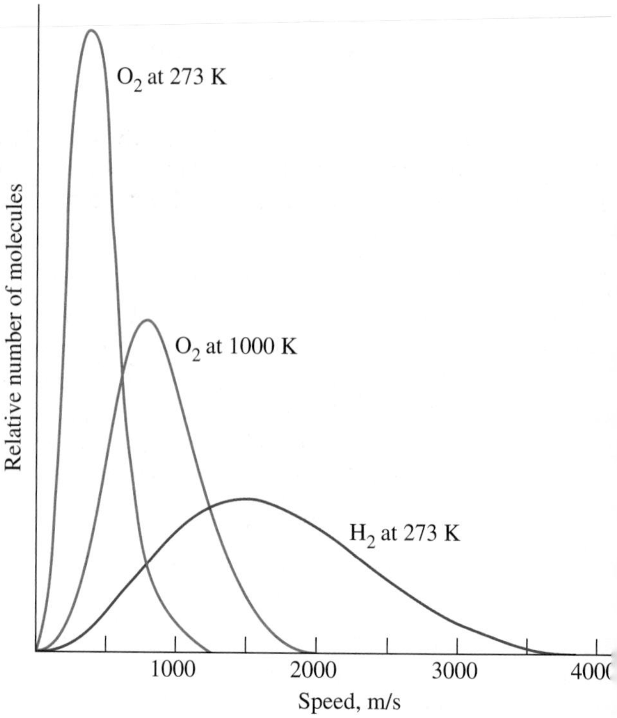

▶ FIGURE 6-16
Distribution of molecular speeds—the effect of mass and temperature
The relative numbers of molecules with a certain speed are plotted as a function of the speed. Note the effect of temperature on the distribution for oxygen molecules and the effect of mass—oxygen must be heated to a very high temperature to have the same distribution of speeds as does hydrogen at 273 K.

The distributions for $O_2(g)$ at 273 K and $H_2(g)$ at 273 K reveal that the lig the gas, the broader the range of speeds.

Let's focus now on the three characteristic speeds identified in Figure 6 The most probable speed, or modal speed, is denoted by u_m; more molec have this speed than any other speed. The average speed is denoted by The **root-mean-square speed, u_{rms},** is obtained from the average of u^2, w is denoted by $\overline{u^2}$. More specifically, $u_{rms} = \sqrt{\overline{u^2}}$. (In this context, the ove reminds us that we are referring to the average value of u^2.) These charact tic speeds can be calculated by using the following expressions:

▶ These equations are obtained by solving expressions involving the Maxwell–Boltzmann distribution function, $F(u)$. The derivation of these equations involves the use of calculus. The equation for u_{rms} is developed in a simpler way on page 225.

$$u_m = \sqrt{\frac{2RT}{M}} \qquad u_{av} = \sqrt{\frac{8RT}{\pi M}} \qquad u_{rms} = \sqrt{\frac{3RT}{M}} \qquad (\text{6}$$

When using any of expressions (6.19), we must express the gas constant as

$$R = 8.3145 \text{ J K}^{-1} \text{ mol}^{-1}$$

and the molar mass, M, in *kilograms* per mole, as we show in Example 6-1

EXAMPLE 6-14 Calculating a Root-Mean-Square Speed

Which is the greater speed, that of a bullet fired from a high-powered M-16 rifle (2180 mi/h) or th root-mean-square speed of H_2 molecules at 25 °C?

Analyze

This is a straightforward application of equation (6.19). We must use SI units: $R = 8.3145 \text{ J K}^{-1} \text{ mol}^{-1}$ and $M = 2.016 \times 10^{-3} \text{ kg mol}^{-1}$. Recall that $1 \text{ J} = 1 \text{ kg m}^2 \text{ s}^{-2}$.

olve

Determine u_{rms} of H_2 with equation (6.19).

$$u_{rms} = \sqrt{\frac{3 \times 8.3145\,\text{kg m}^2\,\text{s}^{-2}\,\text{mol}^{-1}\,\text{K}^{-1} \times 298\,\text{K}}{2.016 \times 10^{-3}\,\text{kg mol}^{-1}}}$$

$$= \sqrt{3.69 \times 10^6\,\text{m}^2/\text{s}^2} = 1.92 \times 10^3\,\text{m/s}$$

The remainder of the problem requires us either to convert 1.92×10^3 m/s to a speed in miles per hour, or 2180 mi/h to meters per second. Then we can compare the two speeds. When we do this, we find that 1.92×10^3 m/s corresponds to 4.29×10^3 mi/h. The root-mean-square speed of H_2 molecules at 25 °C is greater than the speed of the high-powered rifle bullet.

ssess

The cancellation of units yields a result for u_{rms} with the correct units (m/s). Also, Figure 6-15 shows that u_{rms} for H_2 is a bit greater than 1500 m/s at 273 K. At 298 K, u_{rms} should be slightly greater than it is at 273 K.

RACTICE EXAMPLE A: Which has the greater root-mean-square speed at 25 °C, NH_3(g) or HCl(g)? Calculate u_{rms} for the one with the greater speed.

RACTICE EXAMPLE B: At what temperature are u_{rms} of H_2 and the speed of the M-16 rifle bullet given in Example 6-14 the same?

6-7 CONCEPT ASSESSMENT

ithout performing an actual calculation, indicate which has the greater ns, He(g) at 1000 K or H_2(g) at 250 K.

6-1 ARE YOU WONDERING?

How can the distribution of molecular speeds be demonstrated experimentally?

is can be done with the apparatus shown in Figure 6-17. An oven and tached evacuated chamber are separated by a wall with a small hole in it. Gas olecules are heated in the oven, emerge through the hole, and pass through a ries of slits, called *collimators*, that herd the molecules into a beam. The numr of molecules in the beam is kept low so that collisions between them will not sturb the beam.

The molecular beam passes through a series of rotating disks. Each disk has a it cut in it. The slits on successive disks are offset from each other by a certain gle. A molecule passing through the first rotating disk will pass through the cond disk only if its velocity is such that it arrives at the disk at the exact oment that the second slit appears. Thus, for a given rotation speed, only ose molecules with the appropriate velocity can pass through the entire series disks.

The number of molecules that pass through the disks and arrive at the detecr is recorded for each chosen speed of rotation. The number of molecules for ch speed of rotation is then plotted against the rotation speed. From the mensions of the apparatus, the rotation speeds of the disks can be converted molecular speeds, and a plot similar to the one in Figure 6-15 can be tained.

(continued)

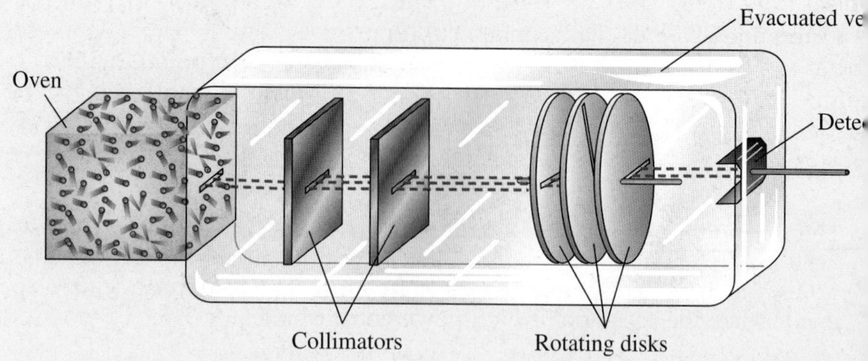

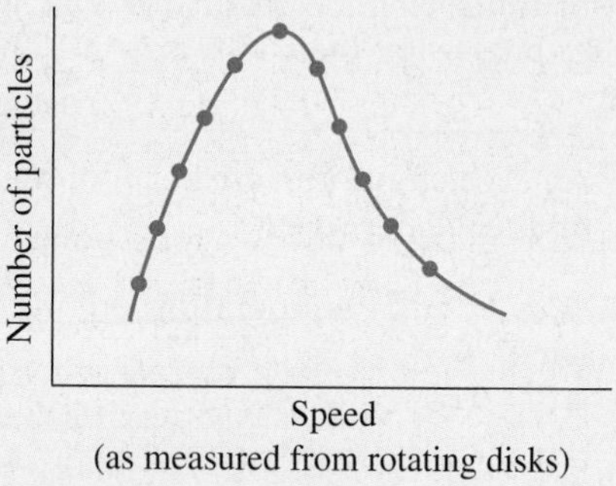

▲ FIGURE 6-17
Distribution of molecular speeds—an experimental determination
Only those molecules with the correct speed to pass through all rotating disks will reach the detector, where they can be counted. By changing the rate of rotation of the disks, the complete distribution of molecular speeds can be determined.

The Meaning of Temperature

We can use the equation above for the root-mean-square speed to develo expression for the average kinetic energy, $\overline{E_k}$, of a collection of molecules an interesting new idea about temperature. The average kinetic energy collection of molecules, each having a mass $m = M/N_A$, is the average of $\frac{1}{2}$ Therefore,

$$\overline{E_k} = \frac{1}{2}m\overline{u^2} = \frac{1}{2} \times \frac{M}{N_A} \times (u_{rms})^2$$

Since $u_{rms} = \sqrt{3RT/M}$ and $u_{rms}^2 = 3RT/M$, the expression above simplifie

$$\overline{E_k} = \frac{3}{2}\left(\frac{RT}{N_A}\right) \quad ($$

Because R and N_A are constants, equation (6.20) states that $\overline{E_k}$ = constant In other words:

> The Kelvin temperature (T) of a gas is directly proportional to the average translational kinetic energy ($\overline{E_k}$) of its molecules: $T \propto \overline{E_k}$. **(6.**

The idea expressed in expression (6.21) also helps us to understand wh happening at the molecular level when objects with different temperat come into contact with each other. Molecules in the hotter object have average, higher kinetic energies than do the molecules in the colder ob en the two objects come into contact with each other, molecules in the hot- object give up some of their kinetic energy through collisions with mole- s in the colder object. The transfer of energy continues until the average etic energies of the molecules in the two objects become equal, that is, until temperatures equalize. Finally, the idea expressed in equation (6.21) pro- es a new way of looking at the absolute zero of temperature: *It is the tem- ture at which translational molecular motion should cease.*

6-2 ARE YOU WONDERING?

What's the lowest temperature we can reach?

harles's law suggests that there is an absolute zero of temperature, 0 K, but can we tain this temperature? The answer is no, but we can come mighty close. Current tempts have resulted in temperatures lower than 1 nanokelvin! However, it is not mply a matter of putting some hot atoms in a refrigerator operating near 0 K. We ave seen that the temperature of a sample of gas is proportional to the kinetic nergy of the gas molecules; therefore, to cool atoms down we have to remove their netic energy. Simple cooling will not do the trick, because the refrigerator would ways have to be at a lower temperature than the atoms being cooled.

Extremely cold atoms can be produced by removing their kinetic energy by opping them in their tracks. This has been accomplished by using a technique alled laser cooling, in which a laser light is directed at a beam of atoms, hitting em head-on and dramatically slowing them down. Once the atoms are cooled, tersecting beams of six lasers are used to reduce their energies still further. The ample of cold atoms is then trapped by a magnetic field for about 1 s. In 1995, a am at the University of Colorado successfully cooled a beam of rubidium Rb) atoms to 1.7×10^{-7} K using this procedure. The coldest human-made tem- erature, 450 picokelvin, was reported by scientists from the Massachusetts nstitute of Technology in 2003.*

*A. E. Leanhardt, T. A. Pasquini, M. Saba, A. Schirotzek, Y. Shin, D. Kielpinski, D. E. ritchard, and W. Ketterle, *Science*, **301**, 1513 (2003).

rivation of Boyle's Law

his section, we will demonstrate that the kinetic–molecular theory of gases vides a satisfactory explanation of Boyle's law. Boyle's law was stated thematically in equation (6.5):

$$PV = a$$

value of a depends on the number, N, of molecules in the sample and the perature, T. Because pressure is force divided by area, the key to deriving ation (6.5) is in assessing the forces associated with molecules hitting the lls of the container. Here, we present a simplified approach for justifying ation (6.5); a rigorous derivation is quite complicated. In our simplified roach, we focus first on a single molecule moving toward a wall perpen- ular to its path to obtain an expression for the pressure it exerts on the wall. en, we modify the expression to take account of the fact that a sample of gas tains many molecules with a distribution of speeds and moving in random ections.

To begin, let's focus on a molecule traveling along the x direction toward a l perpendicular to its path, as suggested by Figure 6-18. The speed of the lecule is denoted by u_x. The force exerted on the wall by the molecule may expressed as the product of two factors: the *momentum transfer*, or impulse, collision and the number of collisions per unit time—the *collision frequency*.

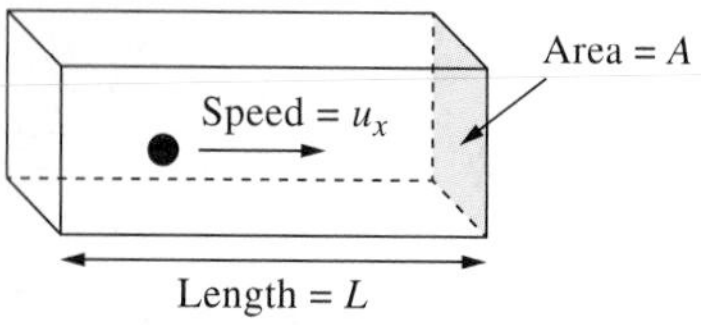

▲ FIGURE 6-18
A model for calculating the pressure exerted by a single molecule
A particle moving with speed u_x toward the end wall of a cubical box has initial momentum mu_x. After deflecting off the wall, it has final momentum $-mu_x$. The change in momentum is $-mu_x - mu_x = -2mu_x$.

Momentum is, by definition, the product of mass and *velocity*. For the m ecule in Figure 6-18, the initial velocity is $+u_x$; therefore, the momentum be the collision is $+mu_x$. When the molecule hits the wall, it is reflected witl losing any energy and travels in the opposite direction. The velocity momentum of the particle after the collision are, respectively, $-u_x$ and − The momentum change of the molecule is the final momentum minus initial momentum: $-mu_x - (+mu_x) = -2mu_x$. The momentum transfe the wall is $+2mu_x$. This represents the momentum transfer per collision. must multiply this quantity by the number of collisions per unit time molecule makes with the end wall. The time between collisions is equa the time it takes for the molecule to travel *twice* the length of the box, $2L$. a particle moving with speed u_x, the time it takes to travel a distance 2 $t = 2L/u_x$. Because $2L/u_x$ is the time between collisions, the reciprocal of quantity is the number of collisions per unit time or the collision freque Notice that the collision frequency, $u_x/2L$, is directly proportional to the m ecular speed.

The force exerted by the molecule on the wall is the product of the mon tum transfer and the collision frequency: $(+2mu_x)(u_x/2L) = mu_x^2/L$. Using fact that pressure is force divided by area, we obtain the following result the pressure caused by the repeated collisions of a single molecule:

$$P = \frac{\text{force}}{\text{area}} = \frac{mu_x^2/L}{A} = \frac{mu_x^2}{AL} = \frac{mu_x^2}{V} \qquad \text{(one molec}$$

▶ For motion in one dimension, velocity is the speed with a + or − sign included to indicate the direction of travel. For example, for a particle moving with constant speed u_x, the velocity is $+u_x$ if the x-coordinate of the particle is increasing but $-u_x$ if the x-coordinate is decreasing.

In the last step, we made use of the fact that the volume, V, of the cubical is equal to the area, A, of the end wall times the length, L.

Of course, a sample of gas contains many molecules with a distributio speeds and moving in random directions. To deduce how the express above must be modified, we focus first on a sample containing N molecu each traveling in the x direction but with different speeds. The express above is easily modified: we multiply by N and replace u_x^2 with $\overline{u_x^2}$, the aver value of u_x^2. Thus, the expression for pressure becomes

$$P = \frac{Nm\overline{u_x^2}}{V} \qquad \text{(}N\text{ molecu}$$

To better understand the meaning of $\overline{u_x^2}$, consider five molecules with u_x v ues of 400, 450, 525, 585, and 600 m/s. We find $\overline{u_x^2}$ by squaring the u_x valu adding the squares, and dividing by the number of particles, in this case fi

$$\overline{u^2} = \frac{(400\text{ m/s})^2 + (450\text{ m/s})^2 + (525\text{ m/s})^2 + (585\text{ m/s})^2 + (600\text{ m/s})^2}{5}$$
$$= 2.68 \times 10^5\text{ m}^2/\text{s}^2$$

We have one final factor to consider. There is absolutely nothing special ab the x direction, so we should expect that $\overline{u_x^2} = \overline{u_y^2} = \overline{u_z^2} = \frac{1}{3}\overline{u^2}$, where the quan $\overline{u^2} = \overline{u_x^2} + \overline{u_y^2} + \overline{u_z^2}$ is the average of u^2, taking into account all the molecu not just those moving in the x direction. The quantity $\overline{u^2}$ is called the *me square speed*. When we substitute $\frac{1}{3}\overline{u^2}$ for $\overline{u_x^2}$ in our previous expression fo we obtain the following result:

$$P = \frac{1}{3}\frac{Nm\overline{u^2}}{V} \qquad \text{(6}$$

KEEP IN MIND

that *mean-square speed* is not the same as *square mean speed*. The mean, or average, speed is $(400 + 450 + 525 + 585 + 600)/5 = 512$ m/s. The square of this average, the square mean speed, is $(512\text{ m/s})^2 = 2.62 \times 10^5\text{ m}^2/\text{s}^2$. Mean-square speed is the quantity that must be used in equation (6.22).

This is the basic equation of the kinetic–molecular theory of gases. We c rewrite equation (6.22) as $PV = \frac{1}{3}Nm\overline{u^2}$. If it is true that $\overline{u^2}$ depends only temperature, then equation (6.22) is a mathematical statement of Boyle's la

Ve can write $PV = \frac{1}{3}N m\overline{u^2}$ in another form by expressing the number, N, of lecules as nN_A where n is the amount in moles and $N_A = 6.022 \times 10^{23}\ \text{mol}^{-1}$. s gives $PV = \frac{1}{3}nN_A m\overline{u^2}$. But $N_A m$ is just the molar mass, M. Therefore,

$$PV = \frac{1}{3}nM\overline{u^2}$$

comparing this result to the ideal gas equation, $PV = nRT$, we conclude that $= \frac{1}{3}M\overline{u^2}$ and, thus, $\overline{u^2} = 3RT/M$. Since the root-mean-square speed, u_{rms}, efined as $u_{\text{rms}} = \sqrt{\overline{u^2}}$, we can write

$$u_{\text{rms}} = \sqrt{\overline{u^2}} = \sqrt{\frac{3RT}{M}}$$

ıs, we have now also justified the result given earlier for u_{rms} (see expres- ı 6.19).

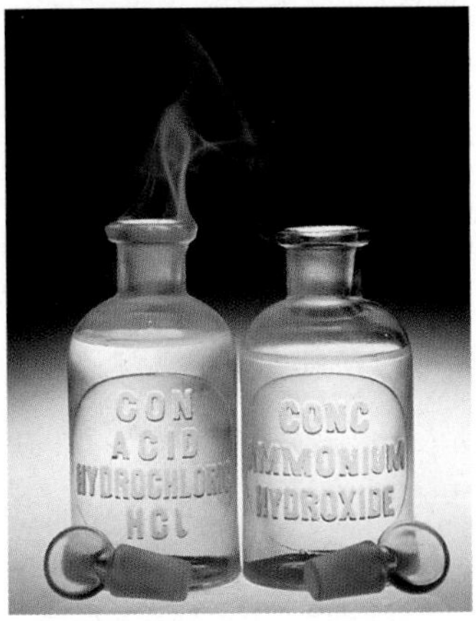

Richard Megna/Fundamental Photographs, NYC.

▲ FIGURE 6-19
Diffusion of $NH_3(g)$ and $HCl(g)$
$NH_3(g)$ escapes from $NH_3(aq)$ (but labeled ammonium hydroxide in this photograph), and $HCl(g)$ escapes from $HCl(aq)$. The gases diffuse toward each other, and, where they meet, a white cloud of ammonium chloride forms as a result of the following reaction:

$$NH_3(g) + HCl(g) \longrightarrow NH_4Cl(s)$$

Because of their greater average speed, NH_3 molecules diffuse faster than HCl. As a result, the cloud forms close to the mouth of the HCl(aq) container.

8 Gas Properties Relating to the Kinetic–Molecular Theory

nolecular speed of 1500 m/s corresponds to 5400 km/h or about 3400 mi/h. m this, it might seem that a given gas molecule could travel very long dis- ces over a very short time, but this is not quite the case. Every gas molecule lergoes collisions with other gas molecules and, as a result, keeps changing ection. Gas molecules follow a tortuous path, which slows them down in get- ; from one point to another. Still, the net rate at which gas molecules move in articular direction does depend on their average speeds, which in turn ends on the temperature and molar mass of the gas (expression 6.19). In this tion, we will focus on two processes involving gases, with the goal of devel- ng an understanding of the factors affecting the rates at which they occur.

Diffusion is the migration of molecules as a result of random molecular tion. Figure 6-19 pictures a common diffusion seen in a chemistry labora- y. The diffusion of two or more gases results in an intermingling of the mol- les and, in a closed container, soon produces a homogeneous mixture, as own in Figure 6-20(a). A related phenomenon, **effusion**, is the escape of gas lecules from their container through a tiny orifice or pinhole. The effusion a hypothetical mixture of two gases is suggested by Figure 6-20(b).

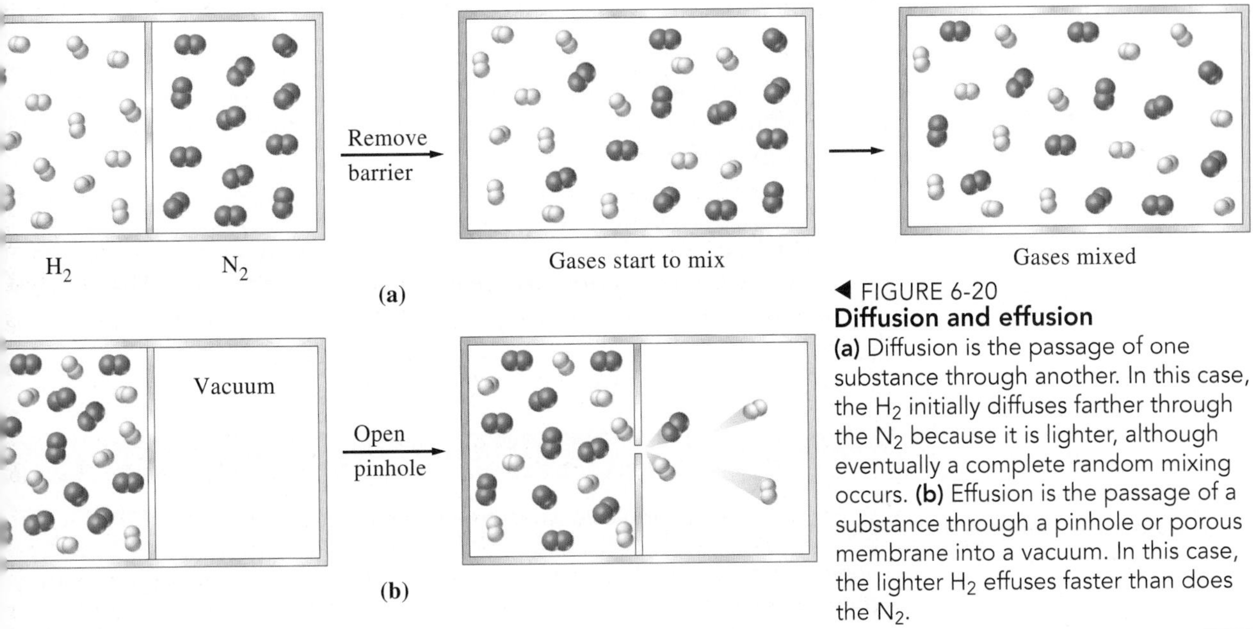

◀ FIGURE 6-20
Diffusion and effusion
(a) Diffusion is the passage of one substance through another. In this case, the H_2 initially diffuses farther through the N_2 because it is lighter, although eventually a complete random mixing occurs. **(b)** Effusion is the passage of a substance through a pinhole or porous membrane into a vacuum. In this case, the lighter H_2 effuses faster than does the N_2.

Both the rate of diffusion and the rate of effusion depend on a numbe factors. However, for effusion at least, the following statement describes fa accurately how the rate of effusion depends on the molar mass of the gas.

> The rate of effusion of a gas is inversely proportional to the square root of it molar mass. (6.2

Expression (6.23) is known as **Graham's law**, originally proposed by Bri chemist Thomas Graham around 1830. Graham's law has serious limitati that you should be aware of. It can be used to describe effusion only for ga at very low pressures so that molecules escape through an orifice individua not as a jet of gas. Also, the orifice must be tiny so that no collisions occu molecules pass through. Graham proposed his law initially to describe the fusion of gases, but the law actually does not apply to diffusion. Molecule a diffusing gas undergo collisions with each other and with the gas into wh they are diffusing. Some even move in the opposite direction to the net fl Nevertheless, diffusion does occur, and gases of low molar mass do diff faster than those of higher molar mass. We cannot, however, use Graha law to make quantitative predictions about rates of diffusion.

▶ The reason for specifying the condition that the gases must be at the same temperature and pressure will become apparent in the next subsection, where we use the kinetic–molecular theory to explain the basis of Graham's law.

Starting from Graham's law, we can develop an expression for compar the relative rates of effusion of two different gases, A and B, at the same te perature and pressure. Since, by Graham's law, we can write

$$\text{rate of effusion of A} \propto 1/\sqrt{M_A} \qquad \text{and} \qquad \text{rate of effusion of B} \propto 1/\sqrt{M_B}$$

then

$$\frac{\text{rate of effusion of A}}{\text{rate of effusion of B}} = \frac{1/\sqrt{M_A}}{1/\sqrt{M_B}} = \sqrt{\frac{M_B}{M_A}} \qquad (6$$

Equation (6.24) can be used in a variety of ways. For example, it can be us to determine which of two gases effuses faster, which does so in a shorter ti which travels farther in a given time, and so on. An effective way to do t is to note that in every case, a ratio of effusion rates, times, distances, a so on, is equal to the square root of a ratio of molar masses, as indicated equation (6.25).

$$\text{ratio of} \begin{cases} \text{molecular speeds} \\ \text{effusion rates} \\ \text{effusion times} \\ \text{distances traveled by molecules} \\ \text{amounts of gas effused} \end{cases} = \sqrt{\text{ratio of two molar masses}} \qquad (6$$

When using equation (6.25), first reason qualitatively and work out wheth the ratio of properties should be greater than or less than one. Then set up ratio of molar masses accordingly. Examples 6-15 and 6-16 illustrate this l of reasoning.

6-8 CONCEPT ASSESSMENT

Which answer(s) is (are) true when comparing 1.0 mol $H_2(g)$ at STP and 0.50 mol He(g) at STP? The two gases have equal **(a)** average molecular kinetic energies; **(b)** root-mean-square speeds; **(c)** masses; **(d)** volumes; **(e)** densities; **(f)** effusion rates through the same orifice.

EXAMPLE 6-15 Comparing Amounts of Gases Effusing Through an Orifice

If 2.2×10^{-4} mol $N_2(g)$ effuses through a tiny hole in 105 s, then how much $H_2(g)$ would effuse through the same orifice in 105 s?

Analyze

Let us reason qualitatively: H_2 molecules are lighter than N_2 molecules, so $H_2(g)$ should effuse faster than $N_2(g)$ when the gases are compared at the same temperature. Before we set the ratio

$$\frac{\text{mol } H_2 \text{ effused}}{\text{mol } N_2 \text{ effused}}$$

equal to $\sqrt{\text{ratio of molar masses}}$, we must ensure that the ratio of molar masses is *greater than* 1.

Solve

$$\frac{?\ \text{mol } H_2}{2.2 \times 10^{-4}\ \text{mol } N_2} = \sqrt{\frac{M_{N_2}}{M_{H_2}}} = \sqrt{\frac{28.014}{2.016}} = 3.728$$

$$?\ \text{mol } H_2 = 3.728 \times 2.2 \times 10^{-4} = 8.2 \times 10^{-4}\ \text{mol } H_2$$

Assess

We could have estimated the result before calculating it. Because the ratio of molar masses is approximately 14, the ratio of effusion rates is approximately $\sqrt{14}$, which is slightly smaller than 4. Therefore, H_2 will effuse almost 4 times as fast as N_2 and almost 4 times as much H_2 will effuse in the same period.

PRACTICE EXAMPLE A: In Example 6-15, how much $O_2(g)$ would effuse through the same orifice in 105 s?

PRACTICE EXAMPLE B: In Example 6-15, how long would it take for 2.2×10^{-4} mol H_2 to effuse through the same orifice as the 2.2×10^{-4} mol N_2?

EXAMPLE 6-16 Relating Effusion Times and Molar Masses

A sample of Kr(g) escapes through a tiny hole in 87.3 s. The same amount of an unknown gas escapes in 42.9 s under identical conditions. What is the molar mass of the unknown gas?

Analyze

Because the unknown gas effuses faster, it must have a smaller molar mass than Kr. Before we set the ratio

$$\frac{\text{effusion time for unknown}}{\text{effusion time for Kr}}$$

equal to $\sqrt{\text{ratio of two molar masses}}$, we must make sure the ratio of molar masses is smaller than one. Thus, the ratio of molar masses must be written with the molar mass of the lighter gas (the unknown gas) in the numerator.

Solve

$$\frac{\text{effusion time for unknown}}{\text{effusion time for Kr}} = \frac{42.9\ \text{s}}{87.3\ \text{s}} = \sqrt{\frac{M_{unk}}{M_{Kr}}} = 0.491$$

$$M_{unk} = (0.491)^2 \times M_{Kr} = (0.491)^2 \times 83.80 = 20.2\ \text{g/mol}$$

Assess

Use the final result and work backward. The molar mass of the unknown is about 4 times as small as that of Kr. Because *effusion rate* $\propto 1/\sqrt{M}$, the unknown will effuse about $\sqrt{4} = 2$ times as fast as Kr. The effusion times show that the unknown does indeed effuse 2 times as fast as Kr.

PRACTICE EXAMPLE A: Under the same conditions as in Example 6-16, another unknown gas requires 131.3 s to escape. What is the molar mass of this unknown gas?

PRACTICE EXAMPLE B: Given all the same conditions as in Example 6-16, how long would it take for a sample of ethane gas, C_2H_6, to effuse?

Derivation of Graham's Law

The derivation of Graham's law begins with the recognition that the ra effusion is directly proportional to (1) the area, A, of the pinhole thrc which the gas escapes and (2) the rate at which gas molecules hit (or co with) the pinhole. The rate at which gas molecules hit the pinhole is, in directly proportional to the average speed, u_{av}, of the molecules and the n ber of molecules per unit volume, N/V (or, alternatively, the number of m per unit volume, n/V). In other words, the faster the molecules are mo and the greater their concentration, the greater the rate of effusion. T we expect

$$\text{rate of effusion} = \text{constant} \times A \times u_{av} \times \frac{n}{V}$$

According to the kinetic–molecular theory, $u_{av} = \sqrt{8RT/\pi M}$. Also, $PV =$ and thus, $n/V = P/(RT)$. Consequently,

$$\text{rate of effusion} = \text{constant} \times A \times \sqrt{\frac{8RT}{\pi M}} \times \frac{P}{RT}$$

For a fixed set of experimental conditions, the quantities A, P, and T ca considered constants and we may write

$$\text{rate of effusion} \propto 1/\sqrt{M}$$

This result is a mathematical statement of Graham's law, expression (6 which is what we set out to prove.

When high-pressure $UF_6(g)$ is forced through a barrier having millior submicroscopic holes per square centimeter, molecules containing the iso ^{235}U pass through the barrier slightly faster than those containing ^{238}U, ju expected from expression (6.24), and therefore $UF_6(g)$ contains a slig higher ratio of ^{235}U to ^{238}U than it did previously. The gas has bec enriched in ^{235}U. Carrying this process through several thousand pa yields a product highly enriched in ^{235}U.

Applications of Diffusion

Up to this point, we have focused more on effusion than on diffus However, diffusion of gases into one another has many practical applicati Natural gas and liquefied petroleum gas (LPG) are odorless; for comme use, a small quantity of a gaseous organic sulfur compound, methyl merc tan, CH_3SH, is added to them. The mercaptan has an odor that can be dete in parts per billion (ppb) or less. When a leak occurs, which can lead to asph iation or an explosion, we rely on the diffusion of this odorous compound f warning.

During World War II, the Manhattan Project (the secret, U.S. governm run program for developing the atomic bomb) used a method cal gaseous diffusion to separate the desired isotope ^{235}U from the predo nant ^{238}U. The method is based on the fact that uranium hexafluoride is of the few compounds of uranium that can be obtained as a gas at mode temperatures.

6-9 Nonideal (Real) Gases

The data in Table 6.2 provide us with clear evidence that real gases are "ideal." We should comment briefly on the conditions under which a real ga ideal or nearly so and what to do when the conditions lead to nonideal behav A useful measure of how much a gas deviates from ideal gas behavior is foun its compressibility factor. The *compressibility factor* of a gas is the ratio PV/n From the ideal gas equation we see that for an ideal gas, $PV/nRT = 1$. For a gas, the compressibility factor can have values that are significantly diffe ı 1. Values of the compressibility factor are given in Table 6.5 for a variety of ·s at 300 K and 10 bar. The data in Table 6.5 show that the deviations from ıl gas behavior can be small or large, depending on the gas. At 300 K 10 bar, He, H_2, CO, N_2, and O_2 behave almost ideally ($PV/nRT \approx 1$) but $_3$ and SF_6 do not ($PV/nRT \approx 0.88$). In Figure 6-21, the compressibility factor otted as a function of pressure for three different gases. The principal con-ion from this plot is that all gases behave ideally at sufficiently low pres-·s, say, below 1 atm, but that deviations set in at increased pressures. At very ı pressures, the compressibility factor is always greater than one.

Ionideal gas behavior can be described as follows: Boyle's law predicts that ery high pressures, a gas volume becomes extremely small and approaches). This cannot be, however, because the molecules themselves occupy space are practically incompressible, as suggested in Figure 6-22. Because of the te size of the molecules, the PV product at high pressures is larger than dicted for an ideal gas, and the compressibility factor is greater than one. ther consideration is that intermolecular forces exist in gases. Figure 6-23 ws that because of attractive forces between the molecules, the force of the isions of gas molecules with the container walls is less than expected for an ıl gas. Intermolecular forces of attraction account for compressibility factors ess than one. These forces become increasingly important at low tempera-·s, where translational molecular motion slows down. To summarize:

Gases tend to behave *ideally* at *high temperatures* and *low pressures*.

Gases tend to behave *nonideally* at *low temperatures* and *high pressures*.

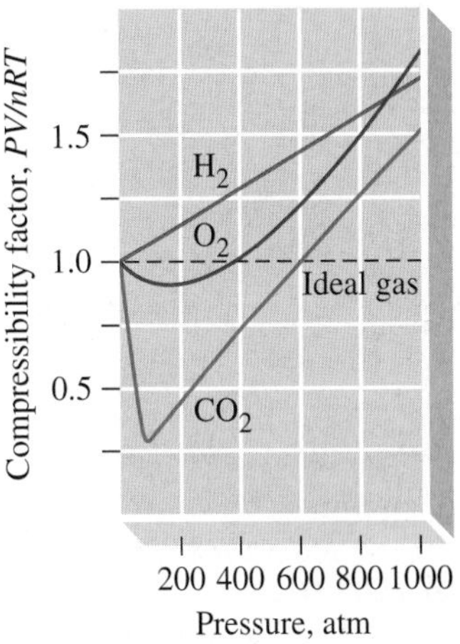

▲ FIGURE 6-21
The behavior of real gases—compressibility factor as a function of pressure at 0 °C
Values of the compressibility factor less than one signify that intermolecular forces of attraction are largely responsible for deviations from ideal gas behavior. Values greater than one are found when the volume of the gas molecules themselves is a significant fraction of the total gas volume.

e van der Waals Equation

umber of equations can be used for real gases, equations that apply over a er range of temperatures and pressures than the ideal gas equation. Such ations are not as general as the ideal gas equation. They contain terms that e specific, but different, values for different gases. Such equations must ect for the volume associated with the molecules themselves and for inter-lecular forces of attraction. Of all the equations that chemists use for model-the behavior of real gases, the **van der Waals equation**, equation (6.26), is the plest to use and interpret.

$$\left(P + \frac{n^2a}{V^2}\right)(V - nb) = nRT \tag{6.26}$$

◀ The van der Waals equation reproduces the observed behavior of gases with moderate accuracy. It is most accurate for gases comprising approximately spherical molecules that have small dipole moments. We will discuss molecular shapes and dipole moments in Chapter 10.

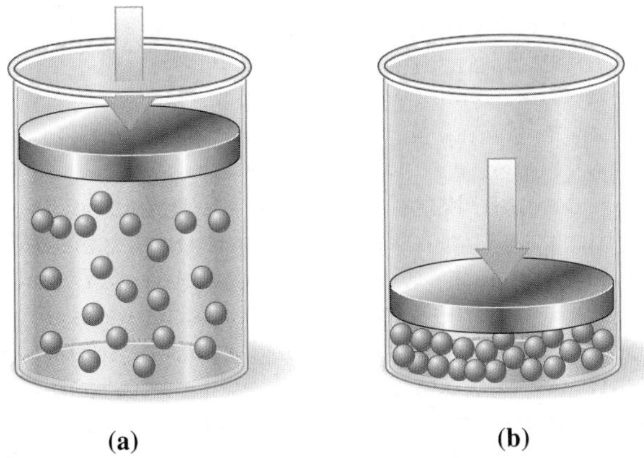

IGURE 6-22
e effect of finite molecular size
ı), a significant fraction of the container is empty space and the gas can still be ıpressed to a smaller volume. In **(b)**, the molecules occupy most of the available space. volume of the system is only slightly greater than the total volume of the molecules.

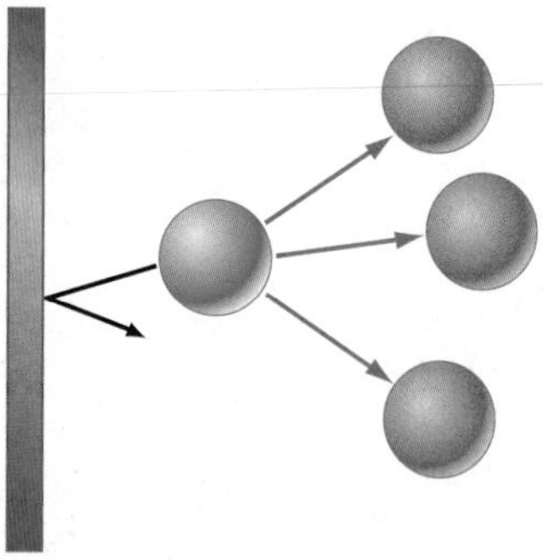

▲ FIGURE 6-23
Intermolecular forces of attraction
Attractive forces of the red molecules for the green molecule cause the green molecule to exert less force when it collides with the wall than if these attractions did not exist.

TABLE 6.5 van der Waals Constants and Compressibility Factors (at 10 bar and 300 K) for Various Gases

Gas	van der Waals Constants		Compressibility Fact
	a, bar L^2 mol^{-2}	b, L mol^{-1}	
H_2	0.2452	0.0265	1.006
He	0.0346	0.0238	1.005
Ideal gas	0	0	1
N_2	1.370	0.0387	0.998
CO	1.472	0.0395	0.997
O_2	1.382	0.0319	0.994
CH_4	2.303	0.0431	0.983
NF_3	3.58	0.0545	0.965
CO_2	3.658	0.0429	0.950
N_2O	3.852	0.0444	0.945
C_2H_6	5.580	0.0651	0.922
NH_3	4.225	0.0371	0.887
SF_6	5.580	0.0651	0.880
C_3H_8	9.39	0.0905	a
SO_2	7.857	0.0879	a

Source: van der Waals constants are from the *CRC Handbook of Chemistry and Physics*, 95th e David R. Lide (ed.), Boca Raton, FL: Taylor & Francis Group, 2015. Compressibility factor are calculated by using data from the National Institute of Standards and Technology (NIS Chemistry WebBook, available online at http://webbook.nist.gov/chemistry/.
[a]At 10 bar and 300 K, C_3H_8 and SO_2 are liquids.

The equation incorporates two molecular parameters, a and b, whose va vary from molecule to molecule, as shown in Table 6.5.

The van der Waals equation and the ideal gas equation both have the fo *pressure factor* × *volume factor* = nRT. The van der Waals equation use modified pressure factor, $P + n^2a/V^2$, in place of P and a modified volu factor, $V - nb$, in place of V. In the modified volume factor, the term accounts for the volume of the molecules themselves. The parameter called the *excluded volume per mole,* and, to a rough approximation, it is the ume that one mole of gas occupies when it condenses to a liquid. Because molecules are not point masses, the volume of the container must be smaller than nb, and the volume available for molecular motion is $V - nb$. suggested in Figure 6-22(b), the volume available for molecular motio quite small at high pressures.

To explain the significance of the term n^2a/V^2 in the modified press factor, it is helpful to solve equation (6.26) for P:

$$P = \frac{nRT}{V - nb} - \frac{n^2a}{V^2}$$

Provided V is not too small, the first term in the equation above is appro mately equal to the pressure exerted by an ideal gas: $nRT/(V - nb$ $nRT/V = P_{\text{ideal}}$. The equation given above for P predicts that the press exerted by a real gas will be less than that of an ideal gas. Figure 6-23 il trates why. Because of attractive forces, molecules near the container w are attracted toward the molecules behind them; as a result, the gas ex less force on the container walls. The term n^2a/V^2 takes into account decrease in pressure caused by intermolecular attractions. In 1873, the Du physicist Johannes van der Waals reasoned that the decrease in press caused by intermolecular attractions should be proportional to the squar the concentration, and so the decrease in pressure is represented in the fo n^2a/V^2. The proportionality constant, a, provides a measure of how stron the molecules attract each other.

close examination of Table 6.5 shows that the values of both a and b ease as the sizes of the molecules increase. The smaller the values of a and ne more closely the gas resembles an ideal gas. Deviations from ideality, as asured by the compressibility factor, become more pronounced as the ues of a and b increase. In Example 6-17 we calculate the pressure of a real by using the van der Waals equation. Solving the equation for either n or V nore difficult, however (see Exercise 121).

◀ In Chapter 12, we will examine intermolecular forces of attraction in greater detail and establish why the strength of the attractive intermolecular forces increases as the sizes of the molecules increase.

EXAMPLE 6-17 Using the van der Waals Equation to Calculate the Pressure of a Gas

Use the van der Waals equation to calculate the pressure exerted by 1.00 mol $Cl_2(g)$ confined to a volume of 2.00 L at 273 K. For Cl_2, $a = 6.34$ bar L^2 mol^{-2} and $b = 0.0542$ L mol^{-1}.

Analyze

This is a straightforward application of equation (6.26). It is important to include units to make sure the units cancel out properly.

Solve

Solve equation (6.26) for P.

$$P = \frac{nRT}{V - nb} - \frac{n^2 a}{V^2}$$

Then substitute the following values into the equation.

$$n = 1.00 \text{ mol};\ V = 2.00 \text{ L};\ T = 273 \text{ K};\ R = 0.08314 \text{ bar L mol}^{-1}\text{ K}^{-1}$$

$$n^2 a = (1.00)^2 \text{ mol}^2 \times 6.34 \frac{\text{bar L}^2}{\text{mol}^2} = 6.34 \text{ bar L}^2$$

$$nb = 1.00 \text{ mol} \times 0.0542 \text{ L mol}^{-1} = 0.0542 \text{ L}$$

$$P = \frac{1.00 \text{ mol} \times 0.08314 \text{ bar L mol}^{-1}\text{ K}^{-1} \times 273 \text{ K}}{(2.00 - 0.0542) \text{ L}} - \frac{6.34 \text{ bar L}^2}{(2.00)^2 \text{ L}^2}$$

$$P = 11.7 \text{ bar} - 1.6 \text{ bar} = 10.1 \text{ bar}$$

Assess

The pressure calculated with the ideal gas equation is 11.3 bar. By including only the b term in the van der Waals equation, we get a value of 11.7 bar. Including the a term reduces the calculated pressure by about 1.6 bar. Under the conditions of this problem, intermolecular forces of attraction are the main cause of the departure from ideal behavior. Although the deviation from ideality here is rather large, in problem-solving situations, you can generally assume that the ideal gas equation will give satisfactory results.

PRACTICE EXAMPLE A: Substitute $CO_2(g)$ for $Cl_2(g)$ in Example 6-17, given the values $a = 3.66$ bar L^2 mol^{-2} and $b = 0.0429$ L mol^{-1}. Which gas, CO_2 or Cl_2, shows the greater departure from ideal gas behavior? [*Hint:* For which gas do you find the greater difference in calculated pressures, first using the ideal gas equation and then the van der Waals equation?]

PRACTICE EXAMPLE B: Substitute $CO(g)$ for $Cl_2(g)$ in Example 6-17, given the values $a = 1.47$ bar L^2 mol^{-2} and $b = 0.0395$ L mol^{-1}. Including CO_2 from Practice Example 6-17A, which of the three gases—Cl_2, CO_2, or CO—shows the greatest departure from ideal gas behavior?

6-9 CONCEPT ASSESSMENT

Following are the measured densities at 20.0 °C and 1 atm pressure of three gases: O_2, 1.331 g/L; OF_2, 2.26 g/L; NO, 1.249 g/L. Arrange them in the order of increasing adherence to the ideal gas equation. [*Hint:* What property can you calculate and compare with a known value?]

The blanket of gases surrounding Earth forms our atmosphere. It not only protects us from harmful radiation but also plays an essential role in moving water, essential to life, from the oceans to the land. For a discussion of the regions and composition of Earth's atmosphere, go to the Focus On feature for Chapter 6 on the MasteringChemistry site.

Summary

6-1 Properties of Gases: Gas Pressure—A gas is described in terms of its **pressure**, temperature, volume, and amount. Gas pressure is most readily measured by comparing it with the pressure exerted by a liquid column, usually mercury (equation 6.2). The pressure exerted by a column of mercury in a **barometer** and called the **barometric pressure** is equal to the prevailing pressure of the atmosphere (Fig. 6-4). Other gas pressures can be measured with a **manometer** (Fig. 6-5). Pressure can be expressed in a variety of units (Table 6.1), including the SI units **pascal (Pa)** and **kilopascal (kPa)**. Also commonly used are **bar; millimeter of mercury (mmHg); torr (Torr)**, where 1 Torr $\approx$ 1 mmHg; and **atmosphere (atm)**, where 1 atm = 760 Torr $\approx$ 760 mmHg.

6-2 The Simple Gas Laws—The most common simple gas laws are **Boyle's law** relating gas pressure and volume (equation 6.5, Fig. 6-6); **Charles's law** relating gas volume and temperature (equation 6.8, Fig. 6-7); and **Avogadro's law**, relating volume and amount of gas. Some important ideas that originate from the simple gas laws are the **Kelvin (absolute)** scale of temperature (equation 6.6), the **standard conditions of temperature and pressure (STP)**, and the molar volume of a gas at STP—22.7 L/mol (expression 6.10).

6-3 Combining the Gas Laws: The Ideal Gas Equation and the General Gas Equation—The simple gas laws can be combined into the **ideal gas equation**, $PV = nRT$ (equation 6.11), where R is called the **gas constant**. A gas whose behavior can be predicted with this equation is known as an **ideal**, or **perfect, gas**. The ideal gas equation can be solved for any one of the variables when all the others are known. The **general gas equation** (equation 6.12) is a useful variant of the ideal gas equation for describing the behavior of a gas when certain variables are held constant and others are allowed to change.

6-4 Applications of the Ideal Gas Equation—An important application of the ideal gas equation is its use in determining molecular masses (equation 6.13) and gas densities (equation 6.14).

6-5 Gases in Chemical Reactions—Because it relates the volume of a gas at a given temperature and pressure to the amount of gas, the ideal gas equation o enters into stoichiometric calculations for reacti involving gases. In calculations based on the volume two gaseous reactants and/or products measured at same temperature and pressure, the **law of combin volumes** is generally applicable.

6-6 Mixtures of Gases—The ideal gas equa applies to mixtures of ideal gases as well as to pure ga The enabling principle, known as **Dalton's law of pa pressures**, is that each gas expands to fill the contai exerting the same pressure as if it were alone in the c tainer (Fig. 6-12). The total pressure is the sum of th **partial pressures** (equation 6.16). A useful concept in d ing with mixtures of gases is that of **mole fraction**, fraction of the molecules in a mixture contributed by e component (equation 6.17). In the common procedur collecting a gas over water (Fig. 6-13), the particular being isolated is mixed with water vapor.

6-7 Kinetic–Molecular Theory of Gases—' **kinetic–molecular theory of gases** yields basic exp sions (equations 6.20, 6.21, and 6.22) that show how temperature and pressure of a gas are related to the av age kinetic energy or the **root-mean-square speed** (u_{rms} molecules. An important aspect of the kinetic–molecu theory is the concept of a distribution of molecular spe (Figs. 6-15 and 6-16).

6-8 Gas Properties Relating to the Kinet Molecular Theory—The **diffusion** and **effusion** gases (Fig. 6-20) can be described by the kinetic–molecu theory. Using an approximation known as **Graham's l** molar masses can be determined by measuring rates effusion (equation 6.24).

6-9 Nonideal (Real) Gases—Because of fir molecular size and intermolecular forces of attract (Figs. 6-22 and 6-23), real gases generally behave idea only at high temperatures and low pressures. Ot equations of state, such as the **van der Waals equati** (equation 6.26), take into account the factors causi nonideal behavior and often work when the ideal g equation fails.

Integrative Example

Combustion of 1.110 g of a gaseous hydrocarbon yields 3.613 g CO_2 and 1.109 g H_2O, and no other products. A 0.28 sample of the hydrocarbon occupies a volume of 132 mL at 24.8 °C and 1.00 bar. Write a plausible structural formula a hydrocarbon corresponding to these data.

lyze

the combustion data for the 1.110 g sample of hydrocarbon and the method of Example 3-6 on page 83 to determine mpirical formula. Use the *P–V–T* data in equation (6.13) for the 0.288 g sample to determine the molar mass and molar mass of the hydrocarbon. By comparing the empirical formula mass and the molecular mass, establish the molecformula. Now write a structural formula consistent with the molecular formula, keeping in mind the point made on 70: each carbon atom forms four bonds.

e

ulate the number of moles of C and H in .110 g sample of hydrocarbon based on nasses of CO_2 and H_2O obtained in its bustion.

$$? \text{ mol C} = 3.613 \text{ g } CO_2 \times \frac{1 \text{ mol } CO_2}{44.01 \text{ g } CO_2} \times \frac{1 \text{ mol C}}{1 \text{ mol } CO_2} = 0.08209 \text{ mol C}$$

$$? \text{ mol H} = 1.109 \text{ g } H_2O \times \frac{1 \text{ mol } H_2O}{18.02 \text{ g } H_2O} \times \frac{2 \text{ mol H}}{1 \text{ mol } H_2O} = 0.1231 \text{ mol H}$$

these numbers of moles as the provisional cripts in the formula.

$$C_{0.08209}H_{0.1231}$$

de each provisional subscript by the ller of the two to obtain the empirical nula.

$$C_{\frac{0.08209}{0.08209}}H_{\frac{0.1231}{0.08209}}$$

$$CH_{1.500} = C_2H_3$$

etermine the molar mass, use a modified of equation (6.13).

$$M = \frac{mRT}{PV} = \frac{0.288 \text{ g} \times 0.08314 \text{ bar L mol}^{-1} \text{K}^{-1} \times (24.8 + 273.2) \text{ K}}{1.00 \text{ bar} \times 0.132 \text{ L}}$$

$$= 54.1 \text{ g mol}^{-1}$$

empirical formula mass is

$$\left(2 \text{ C atoms} \times \frac{12.0 \text{ u}}{1 \text{ C atom}}\right) + \left(3 \text{ H atoms} \times \frac{1.01 \text{ u}}{1 \text{ H atom}}\right) = 27.0 \text{ u}$$

molar mass based on the empirical nula, 27.0 g mol^{-1}, is almost exactly onethe observed molar mass of 54.1 g mol^{-1}. molecular formula of the hydrocarbon is

$$C_{2\times2}H_{2\times3} = C_4H_6$$

four-carbon alkane is butane, C_4H_{10}. noval of 4 H atoms to obtain the formula $_6$ is achieved by inserting two C-to-C ble bonds.

$H_2C{=}CH{-}CH{=}CH_2$ or $H_2C{=}C{=}CH{-}CH_3$

other possibilities involve the presence C-to-C triple bond.

$H_3C{-}C{\equiv}C{-}CH_3$ or $H_3C{-}CH_2{-}C{\equiv}CH$

ess

combination of combustion data and gas-law data yields a molecular formula with certainty. However, because of nerism the exact structural formula cannot be pinpointed. All that we can say is that the hydrocarbon might have any of the four structures shown, but it might be still another structure, for example, based on a ring of C atoms rather n a straight chain.

ACTICE EXAMPLE A: When a 0.5120 g sample of a gaseous hydrocarbon was burned in excess oxygen, 1.687 g CO_2 0.4605 g H_2O were obtained. The density of the compound, in its vapor form, is 1.637 g/L at 25 °C and 101.3 kPa. ermine the molecular formula of the hydrocarbon, and draw a plausible structural formula for the molecule.

ACTICE EXAMPLE B: An organic compound contains only C, H, N, and O. When the compound is burned in gen, with appropriate catalysts, nitrogen gas (N_2), carbon dioxide (CO_2), and water vapor (H_2O) are produced. A)23 g sample of the compound yielded 151.2 mg CO_2, 69.62 mg H_2O, and 9.62 mL of $N_2(g)$ at 0.00 °C and 1.00 atm. The sity of the compound, in its vapor form, was found to be 3.57 g L^{-1} at 127 °C and 748 mmHg. What is the molecular mula of the compound?

Exercises

Pressure and Its Measurement

1. Convert each pressure to an equivalent pressure in atmospheres. **(a)** 736 mmHg; **(b)** 0.776 bar; **(c)** 892 Torr; **(d)** 225 kPa.
2. Calculate the height of a mercury column required to produce a pressure **(a)** of 0.984 atm; **(b)** of 928 Torr; **(c)** equal to that of a column of water 142 ft high.
3. Calculate the height of a column of liquid benzene ($d = 0.879\ g/cm^3$), in meters, required to exert a pressure of 0.970 atm.
4. Calculate the height of a column of liquid glycerol ($d = 1.26\ g/cm^3$), in meters, required to exert the same pressure as 3.02 m of $CCl_4(l)$ ($d = 1.59\ g/cm^3$).
5. What is the pressure (in mmHg) of the gas inside the apparatus below if $P_{bar.} = 740$ mmHg, $h_1 = 30$ mm, and $h_2 = 50$ mm?

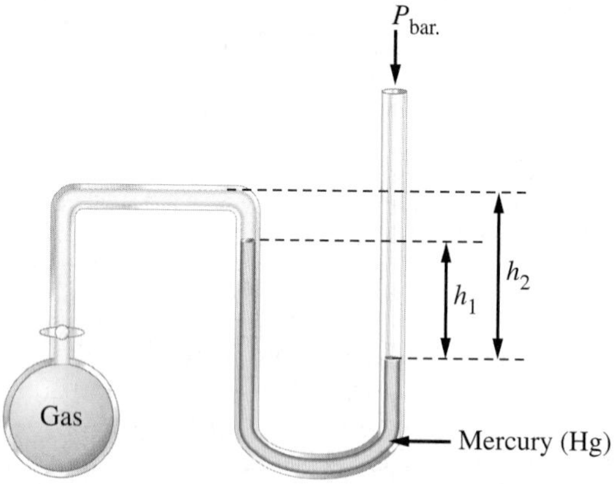

6. What is the pressure (in mmHg) of the gas insid apparatus below if $P_{bar.} = 740$ mmHg, $h_1 = 30$ and $h_2 = 40$ mm?

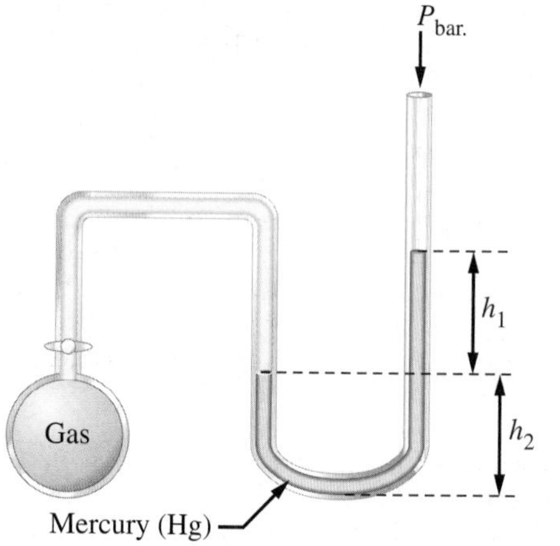

7. At times, a pressure is stated in units of *mass* per area rather than *force* per unit area. Express $P = 1$ in the unit kg/cm^2.
[*Hint:* How is a mass in kilograms related to a for
8. Express $P = 1$ atm in pounds per square inch (ps
[*Hint:* Refer to Exercise 7.]

The Simple Gas Laws

9. A sample of $O_2(g)$ has a volume of 26.7 L at 762 Torr. What is the new volume if, with the temperature and amount of gas held constant, the pressure is **(a)** lowered to 385 Torr; **(b)** increased to 3.68 atm?
10. An 886 mL sample of Ne(g) is at 752 mmHg and 26 °C. What will be the new volume if, with the pressure and amount of gas held constant, the temperature is **(a)** increased to 98 °C; **(b)** lowered to −20 °C?
11. If 3.0 L of oxygen gas at 177 °C is cooled at constant pressure until the volume becomes 1.50 L, then what is the final temperature?
12. We want to change the volume of a fixed amount of gas from 725 mL to 2.25 L while holding the temperature constant. To what value must we change the pressure if the initial pressure is 105 kPa?
13. A 35.8 L cylinder of Ar(g) is connected to an evacuated 1875 L tank. If the temperature is held constant and the final pressure is 721 mmHg, what must have been the original gas pressure in the cylinder, in atmospheres?
14. A sample of $N_2(g)$ occupies a volume of 42.0 mL under the existing barometric pressure. Increasing the pressure by 85 mmHg reduces the volume to 37.7 mL. What is the prevailing barometric pressure, in limeters of mercury?
15. A weather balloon filled with He gas has a volum $2.00 \times 10^3\ m^3$ at ground level, where the atmosph pressure is 1.000 atm and the temperature 27 °C. A the balloon rises high above Earth to a point wl the atmospheric pressure is 0.340 atm, its vol increases to $5.00 \times 10^3\ m^3$. What is the temperatu the atmosphere at this altitude?
16. The photographs show the contraction of an argon-fi balloon when it is cooled by liquid nitrogen. To w

approximate fraction of its original volume will the balloon shrink when it is cooled from a room temperature of 22 °C to a final temperature of about −22 °C?

What is the mass of argon gas in a 75.0 mL volume at STP?

What volume of gaseous chlorine at STP would you need to obtain a 250.0 g sample of gas?

A 27.6 mL sample of $PH_3(g)$ (used in the manufacture of flame-retardant chemicals) is obtained at STP.
(a) What is the mass of this sample, in milligrams?
(b) How many molecules of PH_3 are present?

A 5.0×10^{17} atom sample of radon gas is obtained.
(a) What is the mass of this sample, in micrograms?
(b) What is the volume of this sample at STP, in microliters?

21. You purchase a bag of potato chips at an ocean beach to take on a picnic in the mountains. At the picnic, you notice that the bag has become inflated, almost to the point of bursting. Use your knowledge of gas behavior to explain this phenomenon.

22. Scuba divers know that they must not ascend quickly from deep underwater because of a condition known as the *bends*, discussed in Chapter 14. Another concern is that they must constantly exhale during their ascent to prevent damage to the lungs and blood vessels. Describe what would happen to the lungs of a diver who inhaled compressed air at a depth of 30 m and held her breath while rising to the surface.

neral Gas Equation

A sample of gas has a volume of 4.25 L at 25.6 °C and 748 mmHg. What will be the volume of this gas at 26.8 °C and 742 mmHg?

A 10.0 g sample of a gas has a volume of 5.25 L at 25 °C and 102 kPa. If 2.5 g of the same gas is added to this *constant* 5.25 L volume and the temperature raised to 62 °C, what is the new gas pressure?

25. A constant-volume vessel contains 12.5 g of a gas at 21 °C. If the pressure of the gas is to remain constant as the temperature is raised to 210 °C, how many grams of gas must be released?

26. A 34.0 L cylinder contains 305 g $O_2(g)$ at 22 °C. How many grams of $O_2(g)$ must be released to reduce the pressure in the cylinder to 1.15 atm if the temperature remains constant?

al Gas Equation

What is the volume, in liters, occupied by 89.2 g $CO_2(g)$ at 37 °C and 98.3 kPa?

A 12.8 L cylinder contains 35.8 g O_2 at 46 °C. What is the pressure of this gas, in kilopascals?

Kr(g) in a 18.5 L cylinder exerts a pressure of 11.2 atm at 28.2 °C. How many grams of gas are present?

A 72.8 L constant-volume cylinder containing 7.41 g He is heated until the pressure reaches 3.50 atm. What is the final temperature in degrees Celsius?

A laboratory high vacuum system is capable of evacuating a vessel to the point that the amount of gas remaining is 5.0×10^9 molecules per cubic meter. What is the residual pressure in pascals?

32. What is the pressure, in pascals, exerted by 1242 g $CO_2(g)$ when confined at −25 °C to a cylindrical tank 25.0 cm in diameter and 1.75 m high?

33. What is the molar volume of an ideal gas at **(a)** 25 °C and 1.00 atm; **(b)** 100 °C and 748 Torr?

34. At what temperature is the molar volume of an ideal gas equal to 22.4 L, if the pressure of the gas is 2.5 atm?

termining Molar Mass

A 0.418 g sample of gas has a volume of 115 mL at 66.3 °C and 99.0 kPa. What is the molar mass of this gas?

What is the molar mass of a gas found to have a density of 0.841 g/L at 415 K and 96.7 kPa?

What is the molecular formula of a gaseous fluoride of sulfur containing 70.4% F and having a density of approximately 4.5 g/L at 20 °C and 1.0 atm?

A 2.650 g sample of a gaseous compound occupies 428 mL at 24.3 °C and 742 mmHg. The compound consists of 15.5% C, 23.0% Cl, and 61.5% F, by mass. What is its molecular formula?

39. A gaseous hydrocarbon weighing 0.231 g occupies a volume of 102 mL at 23 °C and 749 mmHg. What is the molar mass of this compound? What conclusion can you draw about its molecular formula?

40. A 132.10 mL glass vessel weighs 56.1035 g when evacuated and 56.2445 g when filled with the gaseous hydrocarbon acetylene at 749.3 mmHg and 20.02 °C. What is the molar mass of acetylene? What conclusion can you draw about its molecular formula?

s Densities

A particular application calls for $N_2(g)$ with a density of 1.80 g/L at 32 °C. What must be the pressure of the $N_2(g)$ in millimeters of mercury? What is the molar volume under these conditions?

42. Monochloroethylene gas is used to make polyvinylchloride (PVC). It has a density of 2.56 g/L at 22.8 °C and 101 kPa. What is the molar mass of monochloroethylene? What is the molar volume under these conditions?

43. In order for a gas-filled balloon to rise in air, the density of the gas in the balloon must be less than that of air.
(a) Consider air to have a molar mass of 28.96 g/mol; determine the density of air at 25 °C and 1.00 atm, in g/L.
(b) Show by calculation that a balloon filled with carbon dioxide at 25 °C and 1 atm could not be expected to rise in air at 25 °C.

44. Refer to Exercise 43, and determine the minimum temperature to which the balloon described in part (b) would have to be heated before it could begin to in air. (Ignore the mass of the balloon itself.)

45. The density of phosphorus vapor is 2.64 g/L at 31 and 1.03 bar. What is the molecular formula of phosphorus under these conditions?

46. A particular gaseous hydrocarbon that is 82.7% C 17.3% H by mass has a density of 2.33 g/L at 2 and 746 mmHg. What is the molecular formula of hydrocarbon?

Gases in Chemical Reactions

47. What volume of $O_2(g)$ is consumed in the combustion of 75.6 L $C_3H_8(g)$ if both gases are measured at STP?

48. How many liters of $H_2(g)$ at STP are produced per gram of Al(s) consumed in the following reaction?

$$2\,Al(s) + 6\,HCl(aq) \longrightarrow 2\,AlCl_3(aq) + 3\,H_2(g)$$

49. A particular coal sample contains 3.28% S by mass. When the coal is burned, the sulfur is converted to $SO_2(g)$. What volume of $SO_2(g)$, measured at 23 °C and 738 mmHg, is produced by burning 1.2×10^6 kg of this coal?

50. One method of removing $CO_2(g)$ from a spacecraft is to allow the CO_2 to react with LiOH. How many liters of $CO_2(g)$ at 25.9 °C and 1.00 bar can be removed per kilogram of LiOH consumed?

$$2\,LiOH(s) + CO_2(g) \longrightarrow Li_2CO_3(s) + H_2O(l)$$

51. A 3.57 g sample of a KCl–$KClO_3$ mixture is decomposed by heating and produces 119 mL $O_2(g)$, measured at 22.4 °C and 98.3 kPa. What is the mass percent of $KClO_3$ in the mixture?

$$2\,KClO_3(s) \longrightarrow 2\,KCl(s) + 3\,O_2(g)$$

52. Hydrogen peroxide, H_2O_2, is used to disinfect co lenses. How many milliliters of $O_2(g)$ at 22 °C 1.00 bar can be liberated from 10.0 mL of an aqu solution containing 3.00% H_2O_2 by mass? The de of the aqueous solution of H_2O_2 is 1.01 g/mL.

$$2\,H_2O_2(aq) \longrightarrow 2\,H_2O(l) + O_2(g)$$

53. Calculate the volume of $H_2(g)$, measured at 2 and 751 Torr, required to react with 28.5 L CO measured at 0 °C and 760 Torr, in this reaction.

$$3\,CO(g) + 7\,H_2(g) \longrightarrow C_3H_8(g) + 3\,H_2O(l$$

54. The Haber process is the principal method for fi nitrogen (converting N_2 to nitrogen compounds).

$$N_2(g) + 3\,H_2(g) \longrightarrow 2\,NH_3(g)$$

Assume that the reactant gases are completely c verted to $NH_3(g)$ and that the gases behave ideal
(a) What volume of $NH_3(g)$ can be produced f 152 L $N_2(g)$ and 313 L of $H_2(g)$ if the gases are m sured at 315 °C and 5.25 atm?
(b) What volume of $NH_3(g)$, measured at 25 °C 727 mmHg, can be produced from 152 L $N_2(g)$ 313 L $H_2(g)$, measured at 315 °C and 5.25 atm?

Mixtures of Gases

55. What is the volume, in liters, occupied by a mixture of 15.2 g Ne(g) and 34.8 g Ar(g) at 7.24 bar pressure and 26.7 °C?

56. A balloon filled with $H_2(g)$ at 0.0 °C and 1.00 atm has a volume of 2.24 L. What is the final gas volume if 0.10 mol He(g) is added to the balloon and the temperature is then raised to 100 °C while the pressure and amount of gas are held constant?

57. A gas cylinder of 53.7 L volume contains $N_2(g)$ at a pressure of 28.2 atm and 26 °C. How many grams of Ne(g) must we add to this same cylinder to raise the total pressure to 75.0 atm?

58. A 2.35 L container of $H_2(g)$ at 762 mmHg and 24 °C is connected to a 3.17 L container of He(g) at 728 mmHg and 24 °C. After mixing, what is the total gas pressure, in millimeters of mercury, with the temperature remaining at 24 °C?

59. Which actions would you take to establish a pressure of 2.00 atm in a 2.24 L cylinder containing 1.60 g $O_2(g)$ at 0 °C? **(a)** add 1.60 g O_2; **(b)** release 0.80 g O_2; **(c)** add 2.00 g He; **(d)** add 0.60 g He.

60. A mixture of 4.0 g $H_2(g)$ and 10.0 g He(g) in a 5. flask is maintained at 0 °C.
(a) What is the total pressure in the container?
(b) What is the partial pressure of each gas?

61. A 2.00 L container is filled with Ar(g) at 752 mm and 35 °C. A 0.728 g sample of C_6H_6 vapor is t added.
(a) What is the total pressure in the container?
(b) What is the partial pressure of Ar and of C_6H_6

62. The chemical composition of air that is exha (expired) is different from ordinary air. A typical an sis of expired air at 37 °C and 1.00 atm, expresse percent by volume, is 74.2% N_2, 15.2% O_2, 3.8% C 5.9% H_2O, and 0.9% Ar. The composition of ordin air is given in Practice Example 6-12B.
(a) What is the ratio of the partial pressure of CO_2 in expired air to that in ordinary air?
(b) Would you expect the density of expired air to greater or less than that of ordinary air at the sa temperature and pressure? Explain.

(c) Confirm your expectation by calculating the densities of ordinary air and expired air at 37 °C and 1.00 atm.

In the drawing below, 1.00 g $H_2(g)$ is maintained at 1 atm pressure in a cylinder closed off by a freely moving piston. Which sketch, **(a)**, **(b)**, or **(c)**, best represents the mixture obtained when 1.00 g He(g) is added? Explain.

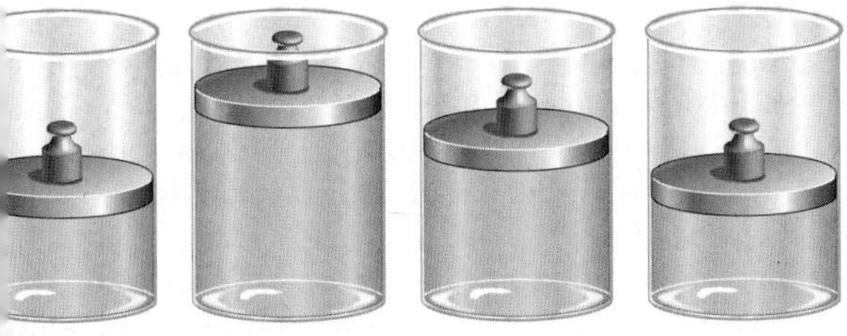

In the drawing above, 1.00 g $H_2(g)$ at 300 K is maintained at 1 atm pressure in a cylinder closed off by a freely moving piston. Which sketch, **(a)**, **(b)**, or **(c)**, best represents the mixture obtained when 0.50 g $H_2(g)$ is added and the temperature is reduced to 275 K? Explain your answer.

65. A 4.0 L sample of O_2 gas has a pressure of 1.0 bar. A 2.0 L sample of N_2 gas has a pressure of 2.0 bar. If these two samples are mixed and then compressed in a 2.0 L vessel, what is the final pressure of the mixture? Assume that the temperature remains unchanged.

66. The following figure shows the contents and pressures of three vessels of gas that are joined by a connecting tube.

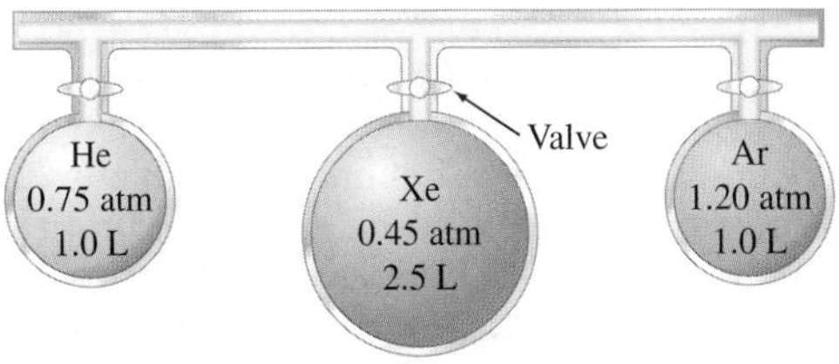

After the valves on the vessels are opened, the final pressure is measured and found to be 0.675 atm. What is the total volume of the connecting tube? Assume that the temperature remains constant.

llecting Gases over Liquids

A 1.65 g sample of Al reacts with excess HCl, and the liberated H_2 is collected over water at 25 °C at a barometric pressure of 744 mmHg. What volume of gaseous mixture, in liters, is collected?

$$2\,Al(s) + 6\,HCl(aq) \longrightarrow 2\,AlCl_3(aq) + 3\,H_2(g)$$

An 89.3 mL sample of wet $O_2(g)$ is collected over water at 21.3 °C at a barometric pressure of 756 mmHg (vapor pressure of water at 21.3 °C = 19 mmHg). **(a)** What is the partial pressure of $O_2(g)$ in the sample collected, in millimeters of mercury? **(b)** What is the volume percent O_2 in the gas collected? **(c)** How many grams of O_2 are present in the sample?

A sample of $O_2(g)$ is collected over water at 24 °C. The volume of gas is 1.16 L. In a subsequent experiment, it is determined that the mass of O_2 present is 1.46 g. What must have been the barometric pressure at the time the gas was collected? (The vapor pressure of water at 24 °C is 22.4 Torr.)

A 1.072 g sample of He(g) is found to occupy a volume of 8.446 L when collected over hexane at 25.0 °C and 738.6 mmHg barometric pressure. Use these data to determine the vapor pressure of hexane at 25.0 °C.

71. At elevated temperatures, solid sodium chlorate ($NaClO_3$) decomposes to produce sodium chloride, NaCl, and O_2 gas. A 0.8765 g sample of impure sodium chlorate was heated until the production of oxygen ceased. The oxygen gas was collected over water and occupied a volume of 57.2 mL at 23.0 °C and 734 Torr. Calculate the mass percentage of $NaClO_3$ in the original sample. Assume that none of the impurities produce oxygen on heating. The vapor pressure of water is 21.07 Torr at 23.0 °C.

72. When solid $KClO_3$ is heated strongly, it decomposes to form solid potassium chloride, KCl, and O_2 gas. A 0.415 g sample of impure $KClO_3$ is heated strongly and the O_2 gas produced by the decomposition is collected over water. When the wet O_2 gas is cooled back to 26 °C, the total volume is 229 mL and the total pressure is 323 Torr. What is the mass percentage of $KClO_3$ in the original sample? Assume that none of the impurities produce oxygen on heating. The vapor pressure of water is 25.22 Torr at 26 °C.

etic–Molecular Theory

Calculate u_{rms}, in meters per second, for $Cl_2(g)$ molecules at 30 °C.

The u_{rms} of H_2 molecules at 273 K is 1.84×10^3 m/s. At what temperature is u_{rms} for H_2 twice this value?

Refer to Example 6-14. What must be the molecular mass of a gas if its molecules are to have a root-mean-square speed at 25 °C equal to the speed of the M-16 rifle bullet?

Refer to Example 6-14. Noble gases (group 18) exist as atoms, not molecules (they are monatomic). Cite one noble gas whose u_{rms} at 25 °C is higher than the speed of the rifle bullet and one whose u_{rms} is lower.

77. At what temperature will u_{rms} for Ne(g) be the same as u_{rms} for He at 300 K?

78. Determine u_m, $\overline{u}$, and u_{rms} for a group of ten automobiles clocked by radar at speeds of 38, 44, 45, 48, 50, 55, 55, 57, 58, and 60 mi/h, respectively.

79. Calculate the average kinetic energy, $\overline{E_k}$, for $O_2(g)$ at 298 K and 1.00 atm.

80. Calculate the total kinetic energy, in joules, of 155 g $N_2(g)$ at 25 °C and 1.00 atm. [*Hint*: First calculate the average kinetic energy, $\overline{E_k}$.]

Diffusion and Effusion of Gases

81. If 0.00484 mol $N_2O(g)$ effuses through an orifice in a certain period of time, how much $NO_2(g)$ would effuse in the same time under the same conditions?
82. A sample of $N_2(g)$ effuses through a tiny hole in 38 s. What must be the molar mass of a gas that requires 64 s to effuse under identical conditions?
83. What are the ratios of the diffusion rates for the pairs of gases **(a)** N_2 and O_2; **(b)** H_2O and D_2O (D = deuterium, i.e., ${}^2_1H$); **(c)** ${}^{14}CO_2$ and ${}^{12}CO_2$; **(d)** ${}^{235}UF_6$ and ${}^{238}UF_6$?
84. Which of the following visualizations best represents the distribution of O_2 and SO_2 molecules near an orifice some time after effusion occurs in the direction indicated by the arrows? The initial condition was one of equal numbers of O_2 molecules (●) and SO_2 molecules (●) on the left side of the orifice. Explain.

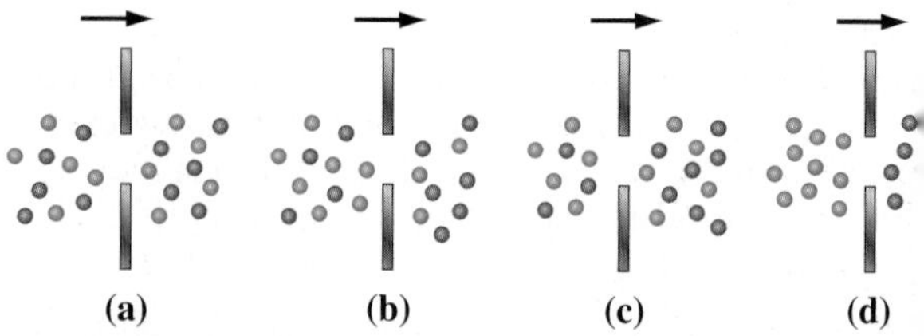

85. It takes 22 hours for a neon-filled balloon to shrir half its original volume at STP. If the same ball had been filled with helium, then how long wou have taken for the balloon to shrink to half its orig volume at STP?
86. The molar mass of radon gas was first estimatec comparing its diffusion rate with that of merc vapor, Hg(g). What is the molar mass of radon if cury vapor diffuses 1.082 times as fast as radon Assume that Graham's law holds for diffusion.

Nonideal Gases

87. Refer to Example 6-17. Recalculate the pressure of $Cl_2(g)$ by using both the ideal gas equation and the van der Waals equation at the temperatures **(a)** 100 °C; **(b)** 200 °C; **(c)** 400 °C. From the results, confirm the statement that a gas tends to be more ideal at high temperatures than at low temperatures.
88. Use both the ideal gas equation and the van der Waals equation to calculate the pressure exerted by 1.50 mol of $SO_2(g)$ when it is confined at 298 K to a volume of **(a)** 100.0 L; **(b)** 50.0 L; **(c)** 20.0 L; **(d)** 10.0 L. Under which of these conditions is the pressure calculated with the ideal gas equation within a few percent of that calculated with the van der Waals equation? Use values of *a* and *b* from Table 6.5.
89. Use the value of the van der Waals constant *b* He(g), given in Table 6.5, to estimate the radiu of a single helium atom. Give your answer in p meters. [*Hint*: The volume of a sphere of radius $4\pi r^3/3$.]
90. **(a)** Use the value of the van der Waals constant *b* $CH_4(g)$, given in Table 6.5, to estimate the radiu the CH_4 molecule. (See Exercise 89.) How does y estimate of the radius compare with the v $r = 228$ pm, obtained experimentally from an an sis of the structure of solid methane? **(b)** The der of $CH_4(g)$ is 66.02 g mL^{-1} at 100 bar and 325 K. W is the value of the compressibility factor at this perature and pressure?

Integrative and Advanced Exercises

91. Explain why it is necessary to include the density of Hg(l) and the value of the acceleration due to gravity, *g*, in a precise definition of a millimeter of mercury (page 196).
92. Assume the following initial conditions for the graphs labeled A, B, and C in Figure 6-7. (A) 10.0 mL at 400 K; (B) 20.0 mL at 400 K; (C) 40.0 mL at 400 K. Use Charles's law to calculate the volume of each gas at 0, −100, −200, −250, and −270 °C. Show that the volume of each gas becomes zero at −273.15 °C.
93. Consider the diagram to the right. The "initial" sketch illustrates, both at the macroscopic and molecular levels, an initial condition: 1 mol of a gas at 273 K and 1.00 bar. With as much detail as possible, illustrate the final condition after each of the following changes.
(a) The pressure is changed to 250 mmHg while standard temperature is maintained.
(b) The temperature is changed to 140 K while standard pressure is maintained.
(c) The pressure is changed to 0.5 bar while the temperature is changed to 550 K.
(d) An additional 0.5 mol of gas is introduced intc cylinder, the temperature is changed to 135 °C, the pressure is changed to 2.25 bar.

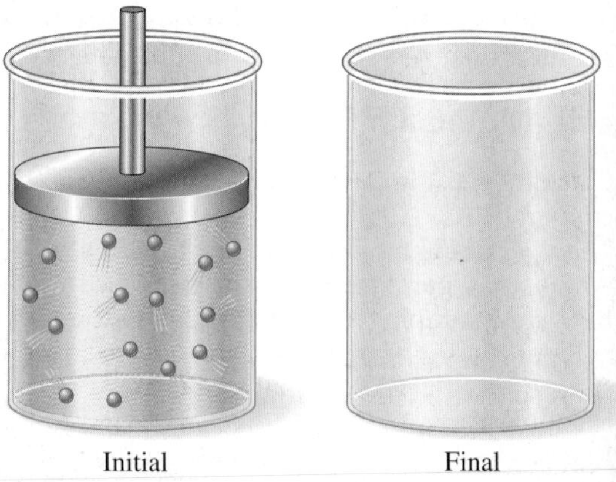

Two evacuated bulbs of equal volume are connected by a tube of negligible volume. One of the bulbs is placed in a constant-temperature bath at 225 K and the other bulb is placed in a constant-temperature bath at 350 K. Exactly 1 mol of an ideal gas is injected into the system. Calculate the final number of moles of gas in each bulb.

A compound is 85.6% carbon by mass. The rest is hydrogen. When 10.0 g of the compound is evaporated at 50.0 °C, the vapor occupies 6.30 L at 1.00 atm pressure. What is the molecular formula of the compound?

A 0.7178 g sample of a hydrocarbon occupies a volume of 390.7 mL at 65.0 °C and 99.2 kPa. When the sample is burned in excess oxygen, 2.4267 g CO_2 and 0.4967 g H_2O are obtained. What is the molecular formula of the hydrocarbon? Write a plausible structural formula for the molecule.

A 3.05 g sample of $NH_4NO_3(s)$ is introduced into an evacuated 2.18 L flask and then heated to 250 °C. What is the total gas pressure, in atmospheres, in the flask at 250 °C when the NH_4NO_3 has completely decomposed?

$$NH_4NO_3(s) \longrightarrow N_2O(g) + 2\,H_2O(g)$$

Ammonium nitrite, NH_4NO_2, decomposes according to the chemical equation below.

$$NH_4NO_2(s) \longrightarrow N_2(g) + 2\,H_2O(g)$$

What is the total volume of products obtained when 128 g NH_4NO_2 decomposes at 819 °C and 101 kPa?

A mixture of 1.00 g H_2 and 8.60 g O_2 is introduced into a 1.500 L flask at 25 °C. When the mixture is ignited, an explosive reaction occurs in which water is the only product. What is the total gas pressure when the flask is returned to 25 °C? (The vapor pressure of water at 25 °C is 23.8 mmHg.)

In the reaction of $CO_2(g)$ and solid sodium peroxide (Na_2O_2), solid sodium carbonate (Na_2CO_3) and oxygen gas are formed. This reaction is used in submarines and space vehicles to remove expired $CO_2(g)$ and to generate some of the $O_2(g)$ required for breathing. Assume that the volume of gases exchanged in the lungs equals 4.0 L/min, the CO_2 content of expired air is 3.8% CO_2 by volume, and the gases are at 25 °C and 735 mmHg. If the $CO_2(g)$ and $O_2(g)$ in the above reaction are measured at the same temperature and pressure, **(a)** how many milliliters of $O_2(g)$ are produced per minute and **(b)** at what rate is the $Na_2O_2(s)$ consumed, in grams per hour?

What is the partial pressure of $Cl_2(g)$, in millimeters of mercury, at 0.00 °C and 1.00 atm in a gaseous mixture that consists of 46.5% N_2, 12.7% Ne, and 40.8% Cl_2, by mass?

A gaseous mixture of He and O_2 has a density of 0.518 g/L at 25 °C and 721 mmHg. What is the mass percent He in the mixture?

When working with a mixture of gases, it is sometimes convenient to use an *apparent molar mass* (a weighted-average molar mass). Think in terms of replacing the mixture with a hypothetical single gas. What is the apparent molar mass of air, given that air is 78.08% N_2, 20.95% O_2, 0.93% Ar, and 0.036% CO_2, by volume?

104. A mixture of $N_2O(g)$ and $O_2(g)$ can be used as an anesthetic. In a particular mixture, the partial pressures of N_2O and O_2 are 612 Torr and 154 Torr, respectively. Calculate **(a)** the mass percentage of N_2O in this mixture, and **(b)** the apparent molar mass of this anesthetic. [*Hint*: For part (b), refer to Exercise 103.]

105. Gas cylinder A has a volume of 48.2 L and contains $N_2(g)$ at 8.35 atm at 25 °C. Gas cylinder B, of unknown volume, contains He(g) at 9.50 atm and 25 °C. When the two cylinders are connected and the gases mixed, the pressure in each cylinder becomes 8.71 atm. What is the volume of cylinder B?

106. The accompanying sketch is that of a closed-end manometer. Describe how the gas pressure is measured. Why is a measurement of $P_{bar.}$ not necessary when using this manometer? Explain why the closed-end manometer is more suitable for measuring low pressures and the open-end manometer more suitable for measuring pressures nearer atmospheric pressure.

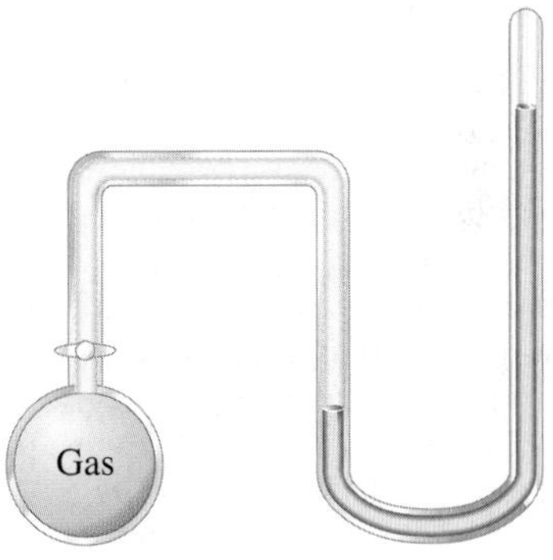

107. Producer gas is a type of fuel gas made by passing air or steam through a bed of hot coal or coke. A typical producer gas has the following composition in percent by volume: 8.0% CO_2, 23.2% CO, 17.7% H_2, 1.1% CH_4, and 50.0% N_2.
(a) What is the density of this gas at 23 °C and 763 mmHg, in grams per liter?
(b) What is the partial pressure of CO in this mixture at 0.00 °C and 1 atm?
(c) What volume of air, measured at 23 °C and 741 Torr, is required for the complete combustion of 1.00×10^3 L of this producer gas, also measured at 23 °C and 741 Torr?
[*Hint:* Which three of the constituent gases are combustible?]

108. What volume of air, measured at 298 K and 101 kPa, is required to burn 2.00 kg C_8H_{18}? Air is approximately 78.1% N_2 and 20.9% O_2, by volume. Other gases make up the remaining 1.0%.

109. A mixture of $H_2(g)$ and $O_2(g)$ is prepared by electrolyzing 1.32 g water, and the mixture of gases is collected over water at 30 °C and 748 mmHg. The volume of "wet" gas obtained is 2.90 L. What must be the vapor pressure of water at 30 °C?

$$2\,H_2O(l) \xrightarrow{\text{electrolysis}} 2\,H_2(g) + O_2(g)$$

110. Aluminum (Al) and iron (Fe) each react with hydrochloric acid solution (HCl) to produce a chloride salt and hydrogen gas, $H_2(g)$. A 0.1924 g sample of a mixture of Al and Fe is treated with excess HCl solution. A volume of 159 mL of H_2 gas is collected over water at 19.0 °C and 841 Torr. What is the percent (by mass) of Fe in the mixture? The vapor pressure of water at 19.0 °C is 16.5 Torr.

111. A 0.168 L sample of $O_2(g)$ is collected over water at 26 °C and a barometric pressure of 737 mmHg. In the gas that is collected, what is the percent water vapor **(a)** by volume; **(b)** by number of molecules; **(c)** by mass? (Vapor pressure of water at 26 °C = 25.2 mmHg.)

112. A breathing mixture is prepared in which He is substituted for N_2. The gas is 79% He and 21% O_2, by volume. **(a)** What is the density of this mixture in grams per liter at 25 °C and 1.00 atm? **(b)** At what pressure would the He–O_2 mixture have the same density as that of air at 25 °C and 1.00 atm? See Exercise 103 for the composition of air.

113. Chlorine dioxide, ClO_2, is sometimes used as a chlorinating agent for water treatment. It can be prepared from the reaction below:

$$Cl_2(g) + 4\,NaClO(aq) \longrightarrow 4\,NaCl(aq) + 2\,ClO_2(g)$$

In an experiment, 1.0 L $Cl_2(g)$, measured at 10.0 °C and 4.66 atm, is dissolved in 0.750 L of 2.00 M $NaClO(aq)$. If 25.9 g of pure ClO_2 is obtained, then what is the percent yield for this experiment?

114. The amount of ozone, O_3, in a mixture of gases can be determined by passing the mixture through a solution of excess potassium iodide, KI. Ozone reacts with the iodide ion as follows:

$$O_3(g) + 3\,I^-(aq) + H_2O(l) \longrightarrow O_2(g) + I_3^-(aq) + 2\,OH^-(aq)$$

The amount of I_3^- produced is determined by titrating with thiosulfate ion, $S_2O_3^{2-}$:

$$I_3^-(aq) + 2\,S_2O_3^{2-}(aq) \longrightarrow 3\,I^-(aq) + S_4O_6^{2-}(aq)$$

A mixture of gases occupies a volume of 53.2 L at 18 °C and 0.993 atm. The mixture is passed slowly through a solution containing an excess of KI to ensure that all the ozone reacts. The resulting solution requires 26.2 mL of 0.1359 M $Na_2S_2O_3$ to titrate to the end point. Calculate the mole fraction of ozone in the original mixture.

115. A 0.1052 g sample of $H_2O(l)$ in an 8.050 L sample of dry air at 30.1 °C evaporates completely. To what temperature must the air be cooled to give a relative humidity of 80.0%? Vapor pressures of water: 20 °C, 17.54 mmHg; 19 °C, 16.49 mmHg; 18 °C, 15.48 mmHg; 17 °C, 14.54 mmHg; 16 °C, 13.63 mmHg; 15 °C, 12.79 mmHg. [*Hint*: Go to Focus On feature for Chapter 6 on the MasteringChemistry site, www.masteringchemistry.com, for a discussion of relative humidity.]

116. An alternative to Figure 6-6 is to plot P against $1/V$. The resulting graph is a straight line passing through the origin. Use Boyle's data from Feature Problem 125 to draw such a straight-line graph. What factors would affect the *slope* of this straight line? Explain.

117. We have noted that atmospheric pressure depends on altitude. Atmospheric pressure as a function of altitude can be calculated with an equation kn as the barometric formula:

$$P = P_0 \times 10^{-Mgh/2.303RT}$$

In this equation, P and P_0 can be in any pres units, for example, Torr. P_0 is the pressure at level, generally taken to be 1.00 atm or its equiva The units in the exponential term must be SI u however. Use the barometric formula to

(a) estimate the barometric pressure at the top o Whitney in California (altitude: 14,494 ft; assu temperature of 10 °C);

(b) show that barometric pressure decreases by thirtieth in value for every 900-ft increase in altit

118. Consider a sample of $O_2(g)$ at 298 K and 1.0 Calculate **(a)** u_{rms} and **(b)** the fraction of molec that have speed equal to u_{rms}.

119. A nitrogen molecule (N_2) having the average ki energy at 300 K is released from Earth's surfa travel upward. If the molecule could move upv without colliding with other molecules, then how would it go before coming to rest? Give your ans in kilometers. [*Hint*: When the molecule comes to the potential energy of the molecule will be where m is the molecular mass in kilogr $g = 9.81\ m\ s^{-2}$ is the acceleration due to gravity, a is the height, in meters, above Earth's surface.]

120. For $H_2(g)$ at 0 °C and 1 atm, calculate the centage of molecules that have speed **(a)** 0 m **(b)** 500 m s^{-1}; **(c)** 1000 m s^{-1}; **(d)** 1500 m s^{-1}; **(e)** 2000 m s^{-1}; **(f)** 2500 m s^{-1}; **(g)** 3500 m s^{-1}. Gi your results to obtain your own version of Figure

121. If the van der Waals equation is solved for volun cubic equation is obtained.

(a) Derive the equation below by rearranging e tion (6.26).

$$V^3 - n\left(\frac{RT + bP}{P}\right)V^2 + \left(\frac{n^2a}{P}\right)V - \frac{n^3ab}{P} = 0$$

(b) What is the volume, in liters, occupied by 1 $CO_2(g)$ at a pressure of 12.5 atm and 286 K? For CO_2 $a = 3.61\ atm\ L^2\ mol^{-2}$ and $b = 0.0429\ L\ mol^{-1}$. [*Hint*: Use the ideal gas equation to obtain an mate of the volume. Then refine your estim either by trial and error, or using the method of cessive approximations. See Appendix A, p A5–A6, for a description of the method of succes approximations.]

122. According to the *CRC Handbook of Chemistry Physics* (95th ed.), the molar volume of O_2(g 0.2168 L mol^{-1} at 280 K and 10 MPa. (Note: 1 MI 1×10^6 Pa.)

(a) Use the van der Waals equation to calculate pressure of one mole of $O_2(g)$ at 280 K if volume is 0.2168 L. What is the % error in the culated pressure? The van der Waals constant $a = 1.382\ bar\ L^2\ mol^{-2}$ and $b = 0.0319$ L mol

(b) Use the ideal gas equation to calculate the ume of one mole of $O_2(g)$ at 280 K and 10 M What is the % error in the calculated volume

123. A particular equation of state for $O_2(g)$ has the f

$$P\overline{V} = RT\left(1 + \frac{B}{\overline{V}} + \frac{C}{\overline{V}^2}\right)$$

where $\overline{V}$ is the molar volume, $B = -21.89\ cm^3/mol$ and $C = 1230\ cm^6/mol^2$.

(a) Use the equation to calculate the pressure exerted by 1 mol $O_2(g)$ confined to a volume of $500\ cm^3$ at 273 K.

(b) Is the result calculated in part (a) consistent with that suggested for $O_2(g)$ by Figure 6-21? Explain.

A 0.156 g sample of a magnesium–aluminum alloy dissolves completely in an excess of HCl(aq). The liberated $H_2(g)$ is collected over water at 5 °C when the barometric pressure is 752 Torr. After the gas is collected, the water and gas gradually warm to the prevailing room temperature of 23 °C. The pressure of the collected gas is again equalized against the barometric pressure of 752 Torr, and its volume is found to be 202 mL. What is the percent composition of the magnesium–aluminum alloy? (Vapor pressure of water: 6.54 mmHg at 5 °C and 21.07 mmHg at 23 °C).

Feature Problems

Shown below is a diagram of Boyle's original apparatus. At the start of the experiment, the length of the

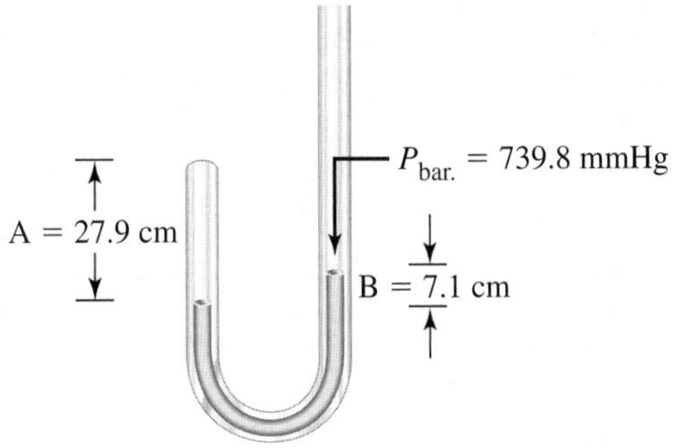

air column (A) on the left was 30.5 cm and the heights of mercury in the arms of the tube were equal. When mercury was added to the right arm of the tube, a difference in mercury levels (B) was produced, and the entrapped air on the left was compressed into a shorter length of the tube (smaller volume) as shown in the illustration for A = 27.9 cm and B = 7.1 cm. Boyle's values of A and B, in centimeters, are listed as follows:

A:	30.5	27.9	25.4	22.9	20.3
B:	0.0	7.1	15.7	25.7	38.3
A:	17.8	15.2	12.7	10.2	7.6
B:	53.8	75.4	105.6	147.6	224.6

Barometric pressure at the time of the experiment was 739.8 mmHg. Assuming that the length of the air column (A) is proportional to the volume of air, show that these data conform reasonably well to Boyle's law.

126. In 1860, Stanislao Cannizzaro showed how Avogadro's hypothesis could be used to establish the atomic masses of elements in gaseous compounds. Cannizzaro took the atomic mass of hydrogen to be exactly one and assumed that hydrogen exists as H_2 molecules (molecular mass = 2). Next, he determined the volume of $H_2(g)$ at 0.00 °C and 1.00 atm that has a mass of exactly 2 g. This volume is 22.4 L. Then he assumed that 22.4 L of any other gas would have the same number of molecules as in 22.4 L of $H_2(g)$. (Here is where Avogadro's hypothesis entered in.) Finally, he reasoned that the ratio of the mass of 22.4 L of any other gas to the mass of 22.4 L of $H_2(g)$ should be the same as the ratio of their molecular masses. The sketch below illustrates Cannizzaro's reasoning in establishing the atomic weight of oxygen as 16. The gases in the table all contain the element X. Their molecular masses were determined by Cannizzaro's method. Use the percent composition data to deduce the atomic mass of X, the number of atoms of X in each of the gas molecules, and the identity of X.

Compound	Molecular Mass, u	Mass Percent X, %
Nitryl fluoride	65.01	49.4
Nitrosyl fluoride	49.01	32.7
Thionyl fluoride	86.07	18.6
Sulfuryl fluoride	102.07	31.4

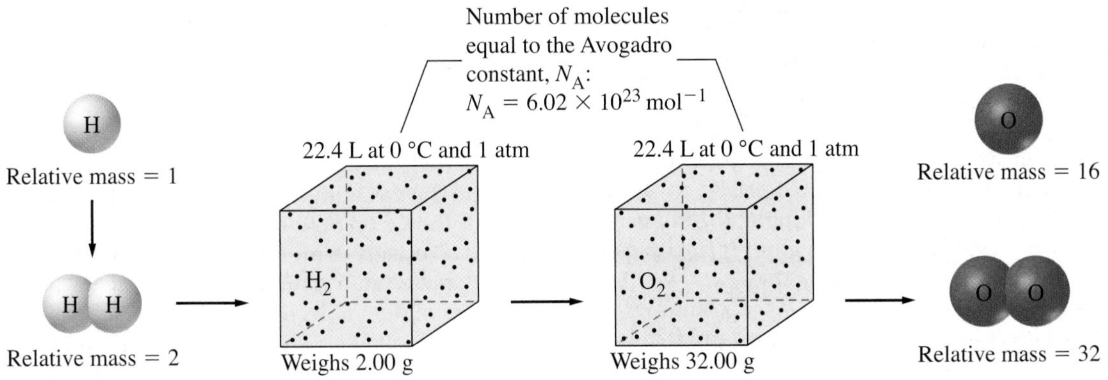

127. In research that required the careful measurement of gas densities, John Rayleigh, a physicist, found that the density of $O_2(g)$ had the same value whether the gas was obtained from air or derived from one of its compounds. The situation with $N_2(g)$ was different, however. The density of $N_2(g)$ had the same value when the $N_2(g)$ was derived from any of various compounds, but a *different* value if the $N_2(g)$ was extracted from air. In 1894, Rayleigh enlisted the aid of William Ramsay, a chemist, to solve this apparent mystery; in the course of their work they discovered the noble gases.
(a) Why do you suppose that the $N_2(g)$ extracted from liquid air did *not* have the same density as $N_2(g)$ obtained from its compounds?
(b) Which gas do you suppose had the greater density: $N_2(g)$ extracted from air or $N_2(g)$ prepared from nitrogen compounds? Explain.
(c) The way in which Ramsay proved that nitrogen gas extracted from air was itself a mixture of gases involved allowing this nitrogen to react with magnesium metal to form magnesium nitride. Explain the significance of this experiment.
(d) Calculate the *percent difference* in the densities at 0.00 °C and 1.00 atm of Rayleigh's $N_2(g)$ extracted from air and $N_2(g)$ derived from nitrogen compounds. [The volume percentages of the major components of air are 78.084% N_2, 20.946% O_2, 0.934% Ar, and 0.0379% CO_2.]

128. The equation $d/P = M/RT$, which can be derived from equation (6.14), suggests that the ratio of the density (d) to pressure (P) of a gas at constant temperature should be a constant. The gas density data at the end of this question were obtained for $O_2(g)$ at various pressures at 273.15 K.
(a) Calculate values of d/P, and with a graph other means determine the ideal value of the d/P for $O_2(g)$ at 273.15 K.
[*Hint:* The ideal value is that associated with a fect (ideal) gas.]
(b) Use the value of d/P from part (a) to calculate cise value for the atomic mass of oxygen, and con this value with that listed on the inside front cove

P, mmHg:	760.00	570.00	380.00	190
d, g/L:	1.428962	1.071485	0.714154	0.3

129. A sounding balloon is a rubber bag filled with H and carrying a set of instruments (the payl Because this combination of bag, gas, and pay has a smaller mass than a corresponding volu air, the balloon rises. As the balloon rises, it exp From the table below, estimate the maximum h to which a spherical balloon can rise given the of balloon, 1200 g; payload, 1700 g: quantity of H in balloon, 120 ft^3 at 0.00 °C and 1.00 atm; dia of balloon at maximum height, 25 ft. Air pre and temperature as functions of altitude are:

Altitude, km	Pressure, mbar	Temperatur
0	1.0×10^3	288
5	5.4×10^2	256
10	2.7×10^2	223
20	5.5×10^1	217
30	1.2×10^1	230
40	2.9×10^0	250
50	8.1×10^{-1}	250
60	2.3×10^{-1}	256

Self-Assessment Exercises

130. In your own words, define or explain each term or symbol. **(a)** atm; **(b)** STP; **(c)** R; **(d)** partial pressure; **(e)** u_{rms}.

131. Briefly describe each concept or process: **(a)** absolute zero of temperature; **(b)** collection of a gas over water; **(c)** effusion of a gas; **(d)** law of combining volumes.

132. Explain the important distinctions between **(a)** barometer and manometer; **(b)** Celsius and Kelvin temperature; **(c)** ideal gas equation and general gas equation; **(d)** ideal gas and real gas.

133. Which exerts the greatest pressure, **(a)** a 75.0 cm column of Hg(l) (d = 13.6 g/mL); **(b)** a column of air 10 mi high; **(c)** a 5.0 m column of $CCl_4(l)$ (d = 1.59 g/mL); **(d)** 10.0 g $H_2(g)$ at STP?

134. For a fixed amount of gas at a fixed pressure, changing the temperature from 100.0 °C to 200 K causes the gas volume to **(a)** double; **(b)** increase, but not to twice its original value; **(c)** decrease; **(d)** stay the same.

135. Two gases were mixed into a 5.000 L container at 291.0 K. Gas A was originally confined in 14.20 L at 1.081 bar and 303.1 K. Gas B was originally confined in 1.251 L at 26.77 bar and 327. **(a)** What is the final total pressure in the 5.0 container? **(b)** What is the partial pressure o A? **(c)** What is the partial pressure of gas B?

136. A fragile glass vessel will break if the internal sure equals or exceeds 2.0 bar. If the vessel is se at 0 °C and 1.0 bar, then at what temperature the vessel break? Assume that the vessel doe expand when heated.

137. Which of the following choices represents molar volume of an ideal gas at 25 °C and 1.5
(a) $(298 \times 1.5/273) \times 22.4$ L;
(b) 22.4 L;
(c) $(273 \times 1.5/298) \times 22.4$ L;
(d) $[298/(273 \times 1.5)] \times 22.4$ L;
(e) $[273/(298 \times 1.5)] \times 22.4$ L.

138. The gas with the greatest density at STP is **(a)** **(b)** Kr; **(c)** SO_3; **(d)** Cl_2.

139. Precisely 1 mol of helium and 1 mol of neo mixed in a container. **(a)** Which gas has the gr average molecular speed? **(b)** Which type of m cule strikes the wall of the container more quently? **(c)** Which gas exerts the larger pressu

. If the Kelvin temperature of a gas doubles, then which of the following also doubles? **(a)** the average molecular speed; **(b)** the speed of every molecule; **(c)** the kinetic energy of every molecule; **(d)** the average kinetic energy of the molecules; **(e)** none of these.

. The postulates of the kinetic molecular theory of gases include all those that follow *except* **(a)** no forces exist between molecules; **(b)** molecules are point masses; **(c)** molecules are repelled by the wall of the container; **(d)** molecules are in constant random motion; **(e)** all are postulates.

. Consider the statements (a) to (e) below. Assume that $H_2(g)$ and $O_2(g)$ behave ideally. State whether each of the following statements is true or false. For each false statement, explain how you would change it to make it a true statement.
(a) Under the same conditions of temperature and pressure, the average kinetic energy of O_2 molecules is less than that of H_2 molecules.
(b) Under the same conditions of temperature and pressure, H_2 molecules move faster, on average, than O_2 molecules.
(c) The volume of 1.00 mol of $H_2(g)$ at 25.0 °C, 1.00 atm is 22.4 L.
(d) The volume of 2.0 g $H_2(g)$ is equal to the volume of 32.0 g $O_2(g)$, at the same temperature and pressure.
(e) In a mixture of H_2 and O_2 gases, with partial pressures P_{H_2} and P_{O_2}, respectively, the total pressure is the larger of P_{H_2} and P_{O_2}.

A sample of $O_2(g)$ is collected over water at 23 °C and a barometric pressure of 751 Torr. The vapor pressure of water at 23 °C is 21 mmHg. The partial pressure of $O_2(g)$ in the sample collected is **(a)** 21 mmHg; **(b)** 751 Torr; **(c)** 0.96 atm; **(d)** 1.02 atm.

At 0 °C and 0.500 atm, 4.48 L of gaseous NH_3 **(a)** contains 6.02×10^{22} molecules; **(b)** has a mass of 17.0 g; **(c)** contains 0.200 mol NH_3; **(d)** has a mass of 3.40 g.

To establish a pressure of 2.00 atm in a 2.24 L cylinder containing 1.60 g $O_2(g)$ at 0 °C, **(a)** add 1.60 g O_2; **(b)** add 0.60 g He(g); **(c)** add 2.00 g He(g); **(d)** release 0.80 g $O_2(g)$.

146. Carbon monoxide, CO, and hydrogen react according to the equation below.

$$3\,CO(g) + 7\,H_2(g) \longrightarrow C_3H_8(g) + 3\,H_2O(g)$$

What volume of which reactant gas remains if 12.0 L CO(g) and 25.0 L $H_2(g)$ are allowed to react? Assume that the volumes of both gases are measured at the same temperature and pressure.

147. A mixture of 5.0×10^{-5} mol $H_2(g)$ and 5.0×10^{-5} mol $SO_2(g)$ is placed in a 10.0 L container at 25 °C. The container has a pinhole leak. After a period of time, the partial pressure of $H_2(g)$ in the container **(a)** is less than that of the $SO_2(g)$; **(b)** is equal to that of the $SO_2(g)$; **(c)** exceeds that of the $SO_2(g)$; **(d)** is the same as in the original mixture.

148. Under which conditions is Cl_2 most likely to behave like an ideal gas? Explain. **(a)** 100 °C and 10.0 atm; **(b)** 0 °C and 0.50 atm; **(c)** 200 °C and 0.50 atm; **(d)** 400 °C and 10.0 atm.

149. Without referring to Table 6.5, state which species in each of the following pairs has the greater value for the van der Waals constant *a*, and which one has the greater value for the van der Waals constant *b*. **(a)** He or Ne; **(b)** CH_4 or C_3H_8; **(c)** H_2 or Cl_2.

150. Explain why the height of the mercury column in a barometer is independent of the diameter of the barometer tube.

151. A gaseous hydrocarbon that is 82.7% C and 17.3% H by mass has a density of 2.35 g/L at 25 °C and 752 Torr. What is the molecular formula of this hydrocarbon?

152. Draw a box to represent a sample of air containing N_2 molecules (represented as squares) and O_2 molecules (represented as circles) in their correct proportions. How many squares and circles would you need to draw to also represent the $CO_2(g)$ in air through a single mark? What else should you add to the box for this more complete representation of air? [*Hint:* See Exercise 103.]

153. Appendix E describes a useful study aid known as concept mapping. Using the method presented in Appendix E, construct a concept map illustrating the different concepts to show the relationships among all the gas laws described in this chapter.

7 Thermochemistry

CONTENTS

LEARNING OBJECTIVES

7.1 Differentiate between open, closed, and isolated systems and between kinetic energy, potential energy, heat, and work.

7.2 Quantify the amount of heat exchanged between a system and its surroundings by using the mass and specific heat capacity of a substance.

7.3 Determine the heat of reaction from data obtained by using a calorimeter.

7.4 Use the correct equation for pressure–volume work to quantify an energy transfer occurring as a result of compression or expansion of gases.

7.5 Describe the first law of thermodynamics as a function of heat and work, and explain what is meant by the term *state function*.

7.6 Distinguish between the constant volume heat of reaction (q_V) and the constant-pressure heat of reaction (q_P), and identify the relationships among q_V, q_P, the internal energy change (ΔU) and enthalpy change (ΔH).

7.7 Use Hess's law to determine an unknown heat of reaction.

7.8 Use standard enthalpies of formation to determine an unknown heat of reaction.

7.9 Use thermochemical data to assess fuels as energy sources, including fossil fuels and other alternatives.

7.10 Explain the difference between a spontaneous and nonspontaneous process, and explain why the enthalpy change is not a reliable criterion for predicting the direction of spontaneous change.

Potassium reacts with water, liberating sufficient heat to ignite the hydrogen evol The transfer of heat between substances in chemical reactions is an important as of thermochemistry.

Natural gas consists mostly of methane, CH_4. As we learne Chapter 4, the combustion of a hydrocarbon, such as meth yields carbon dioxide and water as products. More important, h ever, is another "product" of this reaction, which we have not previo mentioned: heat. This heat can be used to produce hot water in a w heater, to heat a house, or to cook food.

Thermochemistry is the branch of chemistry concerned with the effects that accompany chemical reactions. To understand the relation veen heat and chemical and physical changes, we must start with some c definitions. We will then explore the concept of heat and the methods l to measure the transfer of energy across boundaries. Another form of gy transfer is work, and, in combination with heat, we will define the first of thermodynamics. At this point, we will establish the relationship veen heats of reaction and changes in internal energy and enthalpy. We see that the tabulation of the change in internal energy and change in alpy can be used to calculate, directly or indirectly, energy changes during nical and physical changes. Finally, concepts introduced in this chapter answer a host of practical questions, such as why natural gas is a better than coal and why the energy value of fats is greater than that of carbohyes and proteins.

◀ Thermochemistry is a subfield of a larger discipline called *thermodynamics*. The broader aspects of thermodynamics are considered in Chapters 13 and 14.

Getting Started: Some Terminology

nis section, we introduce and define some very basic terms. Most are dised in greater detail in later sections, and your understanding of these ns should grow as you proceed through the chapter.

et us think of the universe as being comprised of a system and its surround-. A **system** is the part of the universe chosen for study, and it can be as large as ne oceans on Earth or as small as the contents of a beaker. Most of the systems will examine will be small and we will look, particularly, at the transfer of gy (as heat and work) and *matter* between the system and its surroundings. **surroundings** are that part of the universe outside the system with which the em interacts. Figure 7-1 pictures three common systems: first, as we see them then, in an abstract form that chemists commonly use. An **open system** freely anges energy and matter with its surroundings (Fig. 7-1a). A **closed system**

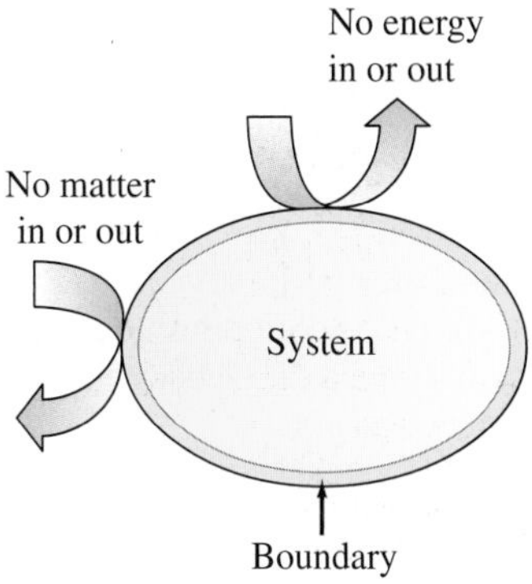

▲ **Isolated system**
Neither energy nor matter is transferred between the system and its surroundings.

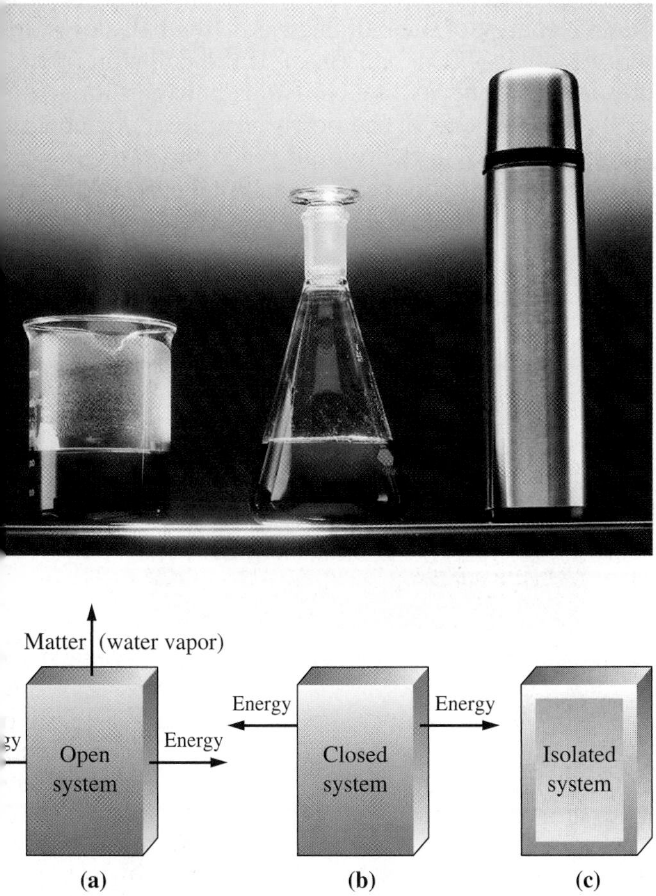

◀ FIGURE 7-1
Systems and their surroundings
(a) *Open system*. The beaker of hot coffee transfers energy to the surroundings—it loses heat as it cools. Matter is also transferred in the form of water vapor. **(b)** *Closed system*. The flask of hot coffee transfers energy (heat) to the surroundings as it cools. Because the flask is stoppered, no water vapor escapes and no matter is transferred. **(c)** *Isolated system*. Hot coffee in an insulated container approximates an isolated system. No water vapor escapes, and, for a time at least, little heat is transferred to the surroundings. (Eventually, though, the coffee in the container cools to room temperature.)

can exchange energy, but not matter, with its surroundings (Fig. 7-1b). An **isol** **system** does not interact with its surroundings (approximated in Figure 7-1c)

The remainder of this section says more, in a general way, about energy its relationship to work. Like many other scientific terms, *energy* is deri from Greek. It means "work within." **Energy** is the capacity to do work. **W** is done when a force acts through a distance. Moving objects do work w they slow down or are stopped. Thus, when one billiard ball strikes ano and sets it in motion, work is done. The energy of a moving object is ca **kinetic energy** (the word *kinetic* means "motion" in Greek). We can the relationship between work and energy by comparing the units for th two quantities. The kinetic energy (E_k) of an object is based on its mass and velocity (u) through the first equation below; work (w) is related to f [mass (m) × acceleration (a)] and distance (d) by the second equation.

▶ As discussed in Appendix B-1, the SI unit for acceleration is m s^{-2}. We encountered this unit previously (page 196)—the acceleration due to gravity was given as g = 9.80665 m s^{-2}.

$$E_k = \frac{1}{2}mu^2$$

$$w = m \times a \times d$$

When mass, speed, acceleration, and distance are expressed in SI units, units of both kinetic energy and work will be kg m^2 s^{-2}, which is the SI of energy—the *joule* (J). That is, 1 J = 1 kg m^2 s^{-2}.

▶ The joule is a unit of work, heat, and energy, but work and heat are not forms of energy but *processes* by which the energy of a system is changed.

The bouncing ball in Figure 7-2 suggests something about the natu energy and work. First, to lift the ball to the starting position, we have to a a force through a distance (to overcome the force of gravity). The work we "stored" in the ball as energy. This stored energy has the potential to do w when released and is therefore called potential energy. **Potential energ** energy resulting from condition, position, or composition; it is an energy a ciated with forces of attraction or repulsion between objects.

When we release the ball, it is pulled toward Earth's center by the forc gravity—it falls. Potential energy is converted to kinetic energy during fall. The kinetic energy reaches its maximum just as the ball strikes the sur On its rebound, the kinetic energy of the ball decreases (the ball slows do and its potential energy increases (the ball rises). If the collision of the with the surface were perfectly *elastic*, like collisions between molecule the kinetic–molecular theory, the sum of the potential and kinetic energie the ball would remain constant. The ball would reach the same maxim height on each rebound, and it would bounce forever. But we know this doe

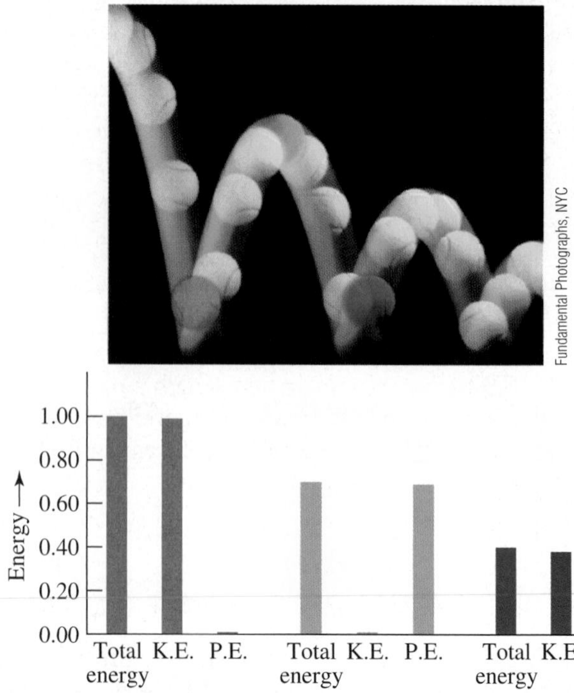

▶ FIGURE 7-2
Potential energy (P.E.) and kinetic energy (K.E.)
The energy of the bouncing tennis ball changes continuously from potential to kinetic energy and back again. The maximum potential energy is at the top of each bounce, and the maximum kinetic energy occurs at the moment of impact. The sum of P.E. and K.E. decreases with each bounce as the thermal energies of the ball and the surroundings increase. The ball soon comes to rest. The bar graph below the bouncing balls illustrates the relative contributions that the kinetic and potential energy make to the total energy for each ball position. The red bars correspond to the red ball, green bars correspond to the green ball, and the blue bars correspond to the blue ball.

pen—the bouncing ball soon comes to rest. All the energy originally sted in the ball as potential energy (by raising it to its initial position) tually appears as additional kinetic energy of the atoms and molecules make up the ball, the surface, and the surrounding air. This kinetic energy ciated with random molecular motion is called **thermal energy**.

general, thermal energy is proportional to the temperature of a system, as sug- ed by the kinetic theory of gases (see page 222). The more vigorous the motion e molecules in the system, the higher the temperature and the greater is its nal energy. However, the thermal energy of a system also depends on the ber of particles present, so that a small sample at a high temperature (for exam- cup of coffee at 75 °C) may have less thermal energy than a larger sample at ver temperature (for example, a swimming pool at 30 °C). Thus, temperature thermal energy must be carefully distinguished. Equally important, we need to nguish between energy changes produced by the action of forces through nces—*work*—and those involving the transfer of thermal energy—*heat*.

7-1 CONCEPT ASSESSMENT

nsider the following situations: a stick of dynamite exploding deep within a ountain cavern, the titration of an acid with base in a laboratory, and a cylinder a steam engine with all of its valves closed. To what type of thermodynamic tems do these situations correspond?

2 Heat

t is energy transferred between a system and its surroundings as a result of nperature difference. Energy that passes from a warmer body (with a higher perature) to a colder body (with a lower temperature) is transferred as heat. e molecular level, molecules of the warmer body, through collisions, lose tic energy to those of the colder body. Thermal energy is transferred—"heat s"—until the average molecular kinetic energies of the two bodies become same, until the temperatures become equal. Heat, like work, describes gy in transit between a system and its surroundings.

ot only can heat transfer cause a change in temperature but, in some nces, it can also change a state of matter. For example, when a solid is ed, the molecules, atoms, or ions of the solid move with greater vigor and ntually break free from their neighbors by overcoming the attractive s between them. Energy is required to overcome these attractive forces. ing the process of melting, for example, the temperature remains constant thermal energy transfer (heat) is used to overcome the forces holding the l together. A process occurring at a constant temperature is said to be *ermal*. Once a solid has melted completely, any further heat flow will the temperature of the resulting liquid.

lthough we commonly use expressions like "heat is lost," "heat is gained," t flows," and "the system loses heat to the surroundings," you should not these statements to mean that a system contains heat. It does not. The gy content of a system, as we shall see in Section 7-5, is a quantity called *nternal energy*. Heat is simply a form in which a quantity of energy may be *ferred* across a boundary between a system and its surroundings.

is reasonable to expect that the quantity of heat, q, required to change the perature of a substance depends on

how much the temperature is to be changed

the quantity of substance

the nature of the substance (type of atoms or molecules)

e will soon learn that the quantity of heat required to achieve a certain perature change also depends on the conditions under which the substance

Bilwissedition Ltd. & Co. KG / Alamy

▲ James Joule (1818–1889)—an amateur scientist
Joule's primary occupation was running a brewery, but he also conducted scientific research in a home laboratory. His precise measurements of quantities of heat formed the basis of the law of conservation of energy.

is heated (for example, whether the substance is heated under the conditic constant volume or constant pressure). Historically, the quantity of required to change the temperature of one gram of water by one degree Ce has been called the **calorie** (**cal**). The calorie is a small unit of energy, anc unit *kilocalorie* (kcal) has also been widely used. The SI unit for heat is simpl basic SI energy unit, the joule (J).

$$1 \text{ cal} = 4.184 \text{ J} \qquad (7$$

Although the joule is used almost exclusively in this text, the calorie is wi encountered in older scientific literature. In the United States, the kilocalo commonly used for measuring the energy content of foods (see the Focus On ture for Chapter 7 on www.masteringchemistry.com).

Heat Capacity

The quantity of heat required to change the temperature of a system by degree is called the **heat capacity** of the system. Heat capacity is represente the symbol C. To obtain the heat capacity of a system (for example, a reaction sel or 10 grams of water), we deliver a known quantity of heat, q, and mea the temperature change produced, ΔT. The heat capacity, C, is then calculat

▶ The Greek letter delta, Δ, indicates a *change* in some quantity. For example, $\Delta T = T_f - T_i$, where T_f is the final temperature and T_i is the initial temperature.

$$C = \frac{q}{\Delta T} \qquad (7$$

In equation (7.3), the temperature change is $\Delta T = T_f - T_i$, where T_f is final temperature and T_i is the initial temperature. Since the unit of q is jc (J) and that of ΔT is °C or K, heat capacity has units of J °C^{-1} or J K^{-1}. The u J °C^{-1} and J K^{-1} are equivalent and interchangeable: a temperature *change* i is equal to a temperature *change* in K. For example, when temperature of a tem changes from 0 °C (or 273.15 K) to 20 °C (or 293.15 K), the tempera change is 20 °C or 20 K.

Once we know the heat capacity of a system, we can use it to conve observed temperature change into a quantity of heat (or vice versa) by sol equation (7.3) for q or ΔT. For example,

$$q = C\Delta T \qquad (7$$

For reasons made clear in the next section, heat capacity depends on whe the system is heated at constant pressure or at constant volume. To disting between the heat capacities for heating at constant pressure and at constant ume, the symbols C_p and C_V are sometimes used. Most of the processes we consider occur at constant pressure, so we will normally make use of cons pressure heat capacities and often omit the subscript p on C.

For pure substances, the heat capacity is often expressed per mole o gram of substance. The **molar heat capacity** at constant pressure (symbol (is the quantity of heat required to raise the temperature of one mole of a stance by one degree under the condition of constant pressure. The cons pressure **specific heat capacity** (symbol, c_p), sometimes called the *specific h* is the quantity of heat required to change the temperature of one gram of a stance by one degree at constant pressure. Here are the values for water at 2

▶ Heat capacities and specific heat capacities are somewhat temperature dependent. For example, the specific heat capacity of water is 4.1813 J g^{-1} °C^{-1} at 25 °C and 4.1794 J g^{-1} °C^{-1} at 40 °C.

$$\text{For water at 25 °C:} \quad C_{p,m} = \frac{75.326 \text{ J}}{\text{mol °C}} \qquad c_p = \frac{4.1813 \text{ J}}{\text{g °C}}$$

Furthermore, for a sample containing a known amount (n) or mass (m) single (pure) substance, the constant-pressure heat capacity of the samp $C = nC_{p,m} = mc_p$. Therefore, equation (7.4) takes the following form.

$$q = C_p\Delta T = nC_{p,m}\,\Delta T = mc_p\Delta T \qquad (7$$

*The original meaning of specific heat was that of a *ratio*: the quantity of heat req to change the temperature of a mass of substance divided by the quantity of heat requi produce the same temperature change in the same mass of water—this definition would specific heat capacity dimensionless. The meaning given here is more commonly used.

Example 7-1, the objective is to calculate a quantity of heat based on the unt of a substance, the specific heat capacity of that substance, and its perature change. In solving the problem, we use the specific heat capacity conversion factor to convert a temperature change, in °C, to a quantity of in J.

XAMPLE 7-1 Calculating a Quantity of Heat

How much heat is required to raise the temperature of 100.0 mL of water (approximately 100.0 g) from room temperature, typically 21.0 °C, to body temperature, typically 37.0 °C? (Assume the specific heat capacity of water is $4.18\ \text{J g}^{-1}\ {}^\circ\text{C}^{-1}$ throughout this temperature range.)

nalyze

To answer this question, we begin by multiplying the specific heat capacity by the mass of water to obtain the heat capacity of the system. To find the amount of heat required to produce the desired temperature change we multiply the heat capacity by the temperature difference.

olve

The specific heat capacity is the heat capacity of 1.00 g water:

$$\frac{4.18\ \text{J}}{\text{g water }^\circ\text{C}}$$

The heat capacity of the system (100.0 g water) is

$$100.0\ \cancel{\text{g water}} \times \frac{4.18\ \text{J}}{\cancel{\text{g water}}\ ^\circ\text{C}} = 418\frac{\text{J}}{^\circ\text{C}}$$

The required temperature change in the system is

$$(37.0 - 21.0)\ ^\circ\text{C} = 16.0\ ^\circ\text{C}$$

The heat required to produce this temperature change is

$$418\frac{\text{J}}{\cancel{^\circ\text{C}}} \times 16.0\ \cancel{^\circ\text{C}} = 6.69 \times 10^3\ \text{J}$$

ssess

Remember that specific heat capacity is the heat capacity *per gram* of substance. Also note that the change in temperature is determined by subtracting the initial temperature from the final temperature. This will be important in determining the sign on the value you determine for heat, as will become apparent in the next section.

RACTICE EXAMPLE A: How much heat, in kilojoules (kJ), is required to raise the temperature of 237 g of water from 4.0 to 37.0 °C?

RACTICE EXAMPLE B: How much heat, in kilojoules (kJ), is required to raise the temperature of 2.50 kg Hg(l) from −20.0 to −6.0 °C? Assume a density of 13.6 g/mL and a molar heat capacity of $28.0\ \text{J mol}^{-1}\ {}^\circ\text{C}^{-1}$ for Hg(l).

uation (7.5), the temperature change is expressed as $\Delta T = T_f - T_i$, where the final temperature and T_i is the initial temperature. When the temperaof a system increases ($T_f > T_i$), ΔT and q are *positive*. A positive value of q fies that heat is absorbed or *gained* by the system. When the temperature of tem decreases ($T_f < T_i$), ΔT and q are *negative*. A negative value of q signihat heat is evolved or *lost* by the system.

◀ The symbol > means "greater than," and < means "less than."

nother idea that enters into calculations of quantities of heat is the **law of ervation of energy**: In interactions between a system and its surroundings, otal energy remains *constant*—energy is neither created nor destroyed. lied to the exchange of heat, this means that

$$q_{\text{system}} + q_{\text{surroundings}} = 0 \qquad \textbf{(7.6)}$$

Thus, heat *gained* by a system is *lost* by its surroundings, and vice versa.

$$q_{system} = -q_{surroundings}$$

Experimental Determination of Specific Heat Capacities

Let us consider how the law of conservation of energy is used in the experiment outlined in Figure 7-3. The object is to determine the specific heat capacity of lead. The transfer of energy, as heat, from the lead to the cold water causes the temperature of the lead to decrease and that of the water to increase, until the lead and water are at the same temperature. Either the lead or the water can be considered the system. If we consider lead to be the system, we can write $q_{lead} = q_{system}$. Furthermore, if the lead and water are maintained in a thermally insulated enclosure, we can assume that $q_{water} = q_{surroundings}$. Then, applying equation (7.7), we have

$$q_{lead} = -q_{water}$$

We complete the calculation in Example 7-2.

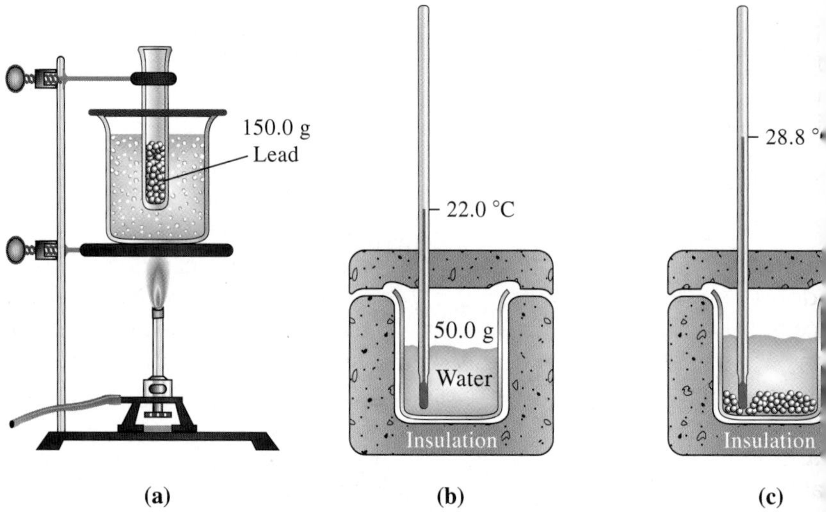

▶ FIGURE 7-3
Determining the specific heat capacity of lead—Example 7-2 illustrated
(a) A 150.0 g sample of lead is heated to the temperature of boiling water (100.0 °C). **(b)** A 50.0 g sample of water is added to a thermally insulated beaker, and its temperature is found to be 22.0 °C. **(c)** The hot lead is dumped into the cold water, and the temperature of the final lead–water mixture is 28.8 °C.

EXAMPLE 7-2 Determining a Specific Heat Capacity from Experimental Data

Use data presented in Figure 7-3 to calculate the specific heat capacity of lead.

Analyze

Keep in mind that if we know any four of the five quantities—q, m, specific heat capacity, T_f, T_i—we can solve equation (7.5) for the remaining one. We know from Figure 7-3 that a known quantity of lead is heated and then dumped into a known amount of water at a known initial temperature. Once the system comes to equilibrium, the water temperature is also the final temperature of the lead. In this type of question, we will use equation (7.5).

Solve

First, use equation (7.5) to calculate q_{water}.

$$q_{water} = 50.0\ \text{g water} \times \frac{4.18\ \text{J}}{\text{g water}\ ^\circ\text{C}} \times (28.8 - 22.0)\ ^\circ\text{C} = 1.4 \times 10^3\ \text{J}$$

From equation (7.8) we can write

$$q_{lead} = -q_{water} = -1.4 \times 10^3\ \text{J}$$

Now, from equation (7.5) again, we obtain

$$q_{lead} = 150.0\ \text{g lead} \times \text{specific heat capacity of lead} \times (28.8 - 100.0)\ ^\circ\text{C} = -1.4 \times 10^3\ \text{J}$$

$$\text{specific heat capacity of lead} = \frac{-1.4 \times 10^3\ \text{J}}{150.0\ \text{g} \times (28.8 - 100.0)\ ^\circ\text{C}} = \frac{-1.4 \times 10^3\ \text{J}}{150.0\ \text{g} \times -71.2\ ^\circ\text{C}} = 0.13\ \text{J g}^{-1}\ ^\circ\text{C}^{-1}$$

ssess

The key concept to recognize is that energy, in the form of heat, flowed from the lead, which is our system, to the water, which is part of the surroundings. A quick way to make sure that we have done the problem correctly is to check the sign on the final answer. For specific heat capacity, the sign should always be positive and have the units of $J\ g^{-1}\ °C^{-1}$.

RACTICE EXAMPLE A: When 1.00 kg lead (specific heat capacity $= 0.13\ J\ g^{-1}\ °C^{-1}$) at 100.0 °C is added to a quantity of water at 28.5 °C, the final temperature of the lead–water mixture is 35.2 °C. What is the mass of water present?

RACTICE EXAMPLE B: A 100.0 g copper sample (specific heat capacity $= 0.385\ J\ g^{-1}\ °C^{-1}$) at 100.0 °C is added to 50.0 g water at 26.5 °C. What is the final temperature of the copper–water mixture?

7-2 CONCEPT ASSESSMENT

th a minimum of calculation, estimate the final temperature reached when 0.0 mL of water at 10.00 °C is added to 200.0 mL of water at 70.00 °C. What sic principle did you use and what assumptions did you make in arriving at s estimate?

cific Heat Capacities of Some Substances

oted, specific heat capacity is a measure of how much energy (heat) must dded to 1 g of substance to raise the temperature by 1 °C. When a subce is heated, the added energy must be absorbed by the atoms, molecules, ns of a system. In general, the greater the number of entities in 1 g and the ter the number of ways these entities can absorb energy, the greater the e of the specific heat capacity. (We discuss the different ways molecules absorb added energy on page 259.)

ble 7.1 lists specific heat capacities of some familiar substances. For many tances, the specific heat capacity is less than $1\ J\ g^{-1}\ °C^{-1}$. A few substances— (l), in particular—have specific heat capacities that are substantially larger these. Why does liquid water have a high specific heat capacity? The fact water molecules form hydrogen bonds, which we discuss in Chapter 12, is contributing factor but so too is the low molar mass of water. The number of ecules in 1 g of water is much greater than the number of atoms in, for exam- 1 g of lead. Thus, when the same quantity of heat is added to 1 g of water to 1 g of lead, the added energy is spread over a larger number of entities in water than in the lead. Thus, the average energy of water molecules eases by a smaller amount than that of lead atoms. The smaller increase in age energy manifests itself as a smaller increase in temperature because, as established in Chapter 6, temperature is a measure of the average (translal) energy of the particles in the system. Lead and mercury have low specific capacities because their molar masses are quite large. One-gram samples of e substances have far fewer atoms to absorb added energy.

he observation that $H_2O(s)$, $H_2O(l)$, and $H_2O(g)$ have different specific capacities indicates that the physical state (solid, liquid, or gas) also plays e. In the solid phase, the motions and arrangements of H_2O molecules are erent than they are in the liquid and gas phases. Because of these differs, the solid, liquid, and gas have different specific heat capacities.

nother consideration is the structural complexity of the molecules themes. Carbon dioxide, CO_2, and propane, C_3H_8, have molar masses of /mol, yet the specific heat capacity of $C_3H_8(g)$ is substantially larger than that $O_2(g)$. The reason is that C_3H_8 molecules are structurally more complex than molecules, and C_3H_8 molecules have more ways to absorb added energy.

TABLE 7.1 Some Specific Heat Capacities, $J\ g^{-1}\ °C^{-1}$

Solids	
$Pb(s)$	0.130
$Cu(s)$	0.385
$Fe(s)$	0.449
$S_8(s)$	0.708
$P_4(s)$	0.769
$Al(s)$	0.897
$Mg(s)$	1.023
$H_2O(s)$	2.11
Liquids	
$Hg(l)$	0.140
$Br_2(l)$	0.474
$CCl_4(l)$	0.850
$CH_3COOH(l)$	2.15
$CH_3CH_2OH(l)$	2.44
$H_2O(l)$	4.18
Gases	
$CO_2(g)$	0.843
$N_2(g)$	1.040
$C_3H_8(g)$	1.67
$NH_3(g)$	2.06
$H_2O(g)$	2.08
$H_2(g)$	14.3

Source: CRC Handbook of Chemistry and Physics, 90th ed., David R. Lide (ed.), Boca Raton, FL: Taylor & Francis Group, 2010.

7-3 CONCEPT ASSESSMENT

Two objects of the same mass absorb the same amount of heat when placed in a flame, but the temperature of one object increases more than the temperature o the other. Which object has the greater specific heat capacity?

7-3 Heats of Reaction and Calorimetry

In Section 7-1, we introduced the notion of *thermal energy*—kinetic energy ass ated with random molecular motion. Another type of energy that contribut the internal energy of a system is **chemical energy**. This is energy associated chemical bonds and intermolecular attractions. If we think of a chemical reac as a process in which some chemical bonds are broken and others are forn then, in general, we expect the chemical energy of a system to change as a resu a reaction. Furthermore, we might expect some of this energy change to appe heat. A **heat of reaction**, q_{rxn}, is the quantity of heat exchanged between a sys and its surroundings when a chemical reaction occurs within the syster *constant temperature*. One of the most common reactions studied is the combus reaction. This is such a common reaction that we often refer to the *heat of com tion* when describing the heat released by a combustion reaction.

If a reaction occurs in an *isolated* system, that is, one that exchanges no ter or energy with its surroundings, the reaction produces a change in the t mal energy of the system—the temperature either increases or decrea Imagine that the previously isolated system is allowed to interact with its roundings. The heat of reaction is the quantity of heat exchanged between system and its surroundings as the system is restored to its initial tempera (Fig. 7-4). In actual practice, we do not physically restore the system to its in temperature. Instead, we calculate the quantity of heat that *would be* exchar in this restoration. To do this, a probe (thermometer) is placed within the tem to record the temperature change produced by the reaction. Then, we the temperature change and other system data to calculate the heat of reac that would have occurred at constant temperature.

Two widely used terms related to heats of reaction are exothermic endothermic reactions. An **exothermic reaction** is one that produces a temp ture increase in an isolated system or, in a nonisolated system, gives off heat tc surroundings. For an exothermic reaction, the heat of reaction is a negative q tity ($q_{rxn} < 0$). In an **endothermic reaction**, the corresponding situation is a perature decrease in an isolated system or a gain of heat from the surround

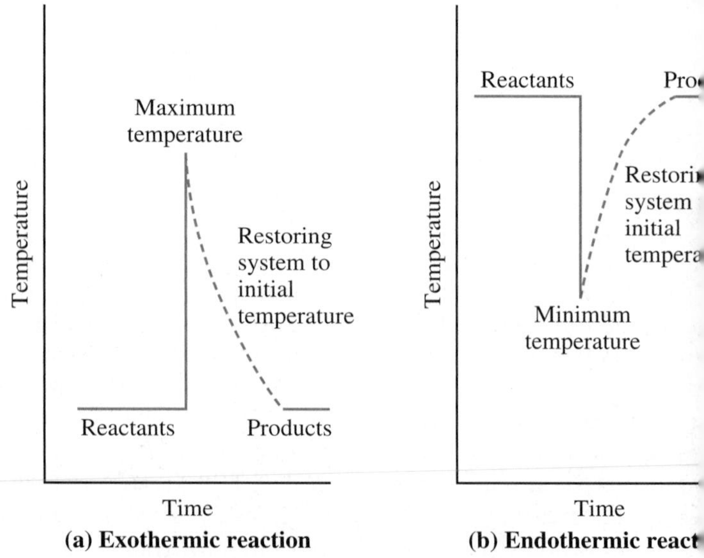

▶ FIGURE 7-4
Conceptualizing a heat of reaction at constant temperature
The solid lines indicate the initial temperature and the **(a)** maximum and **(b)** minimum temperature reached in an isolated system, in an exothermic and an endothermic reaction, respectively. The broken lines represent pathways to restoring the system to the initial temperature. The heat of reaction is the heat lost or gained by the system in this restoration.

(a)

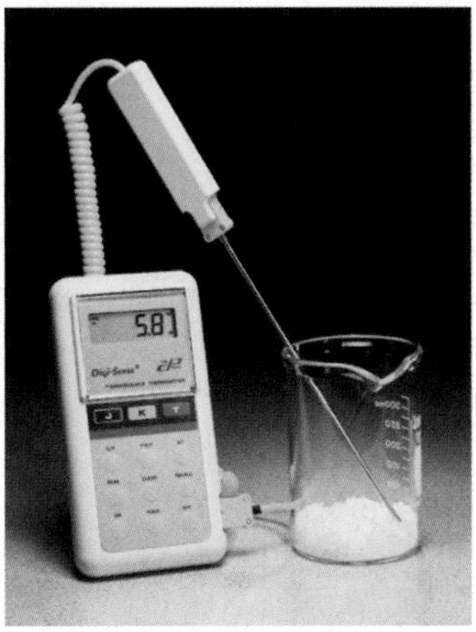

(b)

◀ Exothermic and endothermic reactions
(a) An exothermic reaction. Slaked lime, $Ca(OH)_2$, is produced by the action of water on quicklime, (CaO). The reactants are mixed at room temperature, but the temperature of the mixture rises to 40.5 °C.

$$CaO(s) + H_2O(l) \longrightarrow Ca(OH)_2(s)$$

(b) An endothermic reaction. $Ba(OH)_2 \cdot 8\,H_2O(s)$ and $NH_4Cl(s)$ are mixed at room temperature, and the temperature falls to 5.8 °C in the reaction.

$$Ba(OH)_2 \cdot 8\,H_2O(s) + 2\,NH_4Cl(s) \longrightarrow BaCl_2 \cdot 2\,H_2O(s) + 2\,NH_3(aq) + 8\,H_2O(l)$$

(a) & **(b)** Carey B. Van Loon

nonisolated system. In this case, the heat of reaction is a positive quantity > 0). Heats of reaction are experimentally determined in a **calorimeter**, a ce for measuring quantities of heat. We will consider two types of calorime- in this section, and we will treat both of them as *isolated* systems.

nb Calorimetry

ıre 7-5 shows a **bomb calorimeter**, which is ideally suited for measuring heat evolved in a combustion reaction. The system is everything within the ble-walled outer jacket of the calorimeter. This includes the bomb and its ents, the water in which the bomb is immersed, the thermometer, the stirrer, so on. The system is *isolated* from its surroundings. When the combustion tion occurs, chemical energy is converted to thermal energy, and the temper- e of the system rises. The heat of reaction, as described earlier, is the quantity eat that the system would have to *lose* to its surroundings to be restored to its al temperature. This quantity of heat, in turn, is just the *negative* of the ther- energy gained by the calorimeter and its contents ($q_{calorim}$).

$$q_{rxn} = -q_{calorim} \text{ (where } q_{calorim} = q_{bomb} + q_{water} \cdots) \qquad \textbf{(7.9)}$$

KEEP IN MIND
that the temperature of a reaction mixture usually changes during a reaction, so the mixture must be returned to the initial temperature (actually or hypothetically) before we assess how much heat is exchanged with the surroundings.

the calorimeter is assembled in exactly the same way each time we use it— is, use the same bomb, the same quantity of water, and so on—we can define *at capacity of the calorimeter*. This is the quantity of heat required to raise the perature of the calorimeter assembly by one degree Celsius. When this heat city is multiplied by the observed temperature change, we get $q_{calorim}$.

◀ The heat capacity of a bomb calorimeter must be determined by experiment.

$$q_{calorim} = \text{heat capacity of calorim} \times \Delta T \qquad \textbf{(7.10)}$$

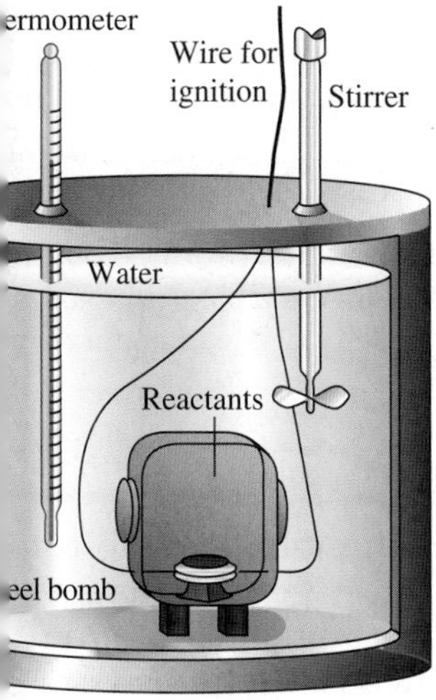

◀ FIGURE 7-5
A bomb calorimeter assembly
An iron wire is embedded in the sample in the lower half of the bomb. The bomb is assembled and filled with $O_2(g)$ at high pressure. The assembled bomb is immersed in water in the calorimeter, and the initial temperature is measured. A short pulse of electric current heats the sample, causing it to ignite. The final temperature of the calorimeter assembly is determined after the combustion. Because the bomb confines the reaction mixture to a fixed volume, the reaction is said to occur at *constant volume*. The significance of this fact is discussed in Section 7-6.

And from $q_{calorim}$, we then establish q_{rxn}, as in Example 7-3, where we de mine the heat of combustion of sucrose (table sugar).

EXAMPLE 7-3 Using Bomb Calorimetry Data to Determine a Heat of Reaction

The combustion of 1.010 g sucrose, $C_{12}H_{22}O_{11}$, in a bomb calorimeter causes the temperature to rise from 24.9 to 28.33 °C. The heat capacity of the calorimeter assembly is 4.90 kJ/°C. **(a)** What is the heat of combustion o sucrose expressed in kilojoules per mole of $C_{12}H_{22}O_{11}$? **(b)** Verify the claim of sugar producers that one tea spoon of sugar (about 4.8 g) contains only 19 Calories.

Analyze

We are given a specific heat capacity and two temperatures, the initial and the final, which indicate that we ar to use equation (7.5). In these kinds of experiments one obtains the amount of heat generated by the reactio by measuring the temperature change in the surroundings. This means that $q_{rxn} = -q_{calorim}$.

Solve

(a) Calculate $q_{calorim}$ with equation (7.10).

$$q_{calorim} = 4.90\ \text{kJ/°C} \times (28.33 - 24.92)\ \text{°C} = (4.90 \times 3.41)\ \text{kJ} = 16.7\ \text{kJ}$$

Now, using equation (7.9), we get

$$q_{rxn} = -q_{calorim} = -16.7\ \text{kJ}$$

This is the heat of combustion of the 1.010 g sample.
Per gram $C_{12}H_{22}O_{11}$:

$$q_{rxn} = \frac{-16.7\ \text{kJ}}{1.010\ \text{g}\ C_{12}H_{22}O_{11}} = -16.5\ \text{kJ/g}\ C_{12}H_{22}O_{11}$$

Per mole $C_{12}H_{22}O_{11}$:

$$q_{rxn} = \frac{-16.5\ \text{kJ}}{\text{g}\ C_{12}H_{22}O_{11}} \times \frac{342.3\ \text{g}\ C_{12}H_{22}O_{11}}{1\ \text{mol}\ C_{12}H_{22}O_{11}} = -5.65 \times 10^3\ \text{kJ/mol}\ C_{12}H_{22}O_{11}$$

(b) To determine the caloric content of sucrose, we can use the heat of combustion per gram of sucros determined in part (a), together with a factor to convert from kilojoules to kilocalories. (Becaus 1 cal = 4.184 J, 1 kcal = 4.184 kJ.)

$$?\ \text{kcal} = \frac{4.8\ \text{g}\ C_{12}H_{22}O_{11}}{1\ \text{tsp}} \times \frac{-16.5\ \text{kJ}}{1\ \text{g}\ C_{12}H_{22}O_{11}} \times \frac{1\ \text{kcal}}{4.184\ \text{kJ}} = -19\ \text{kcal/tsp}$$

1 food Calorie (1 Calorie with a capital C) is actually 1000 cal, or 1 kcal. Therefore, 19 kcal = 19 Calories. Th claim is justified.

Assess

A combustion reaction is an exothermic reaction, which means that energy flows, in the form of heat, from th reaction system to the surroundings. Therefore, the *q* for a combustion reaction is negative.

PRACTICE EXAMPLE A: Vanillin is a natural constituent of vanilla. It is also manufactured for use in artifici vanilla flavoring. The combustion of 1.013 g of vanillin, $C_8H_8O_3$, in the same bomb calorimeter as i Example 7-3 causes the temperature to rise from 24.89 to 30.09 °C. What is the heat of combustion of vanilli expressed in kilojoules per mole?

PRACTICE EXAMPLE B: The heat of combustion of benzoic acid, $C_6H_5COOH(s)$, is −26.42 kJ/g. One method o obtaining the heat capacity of a bomb calorimeter is to measure the temperature change produced by th combustion of a given mass of benzoic acid. If the combustion of a 1.176 g sample of benzoic acid causes temperature *increase* of 4.96 °C in a bomb calorimeter assembly, what is the heat capacity of the assembly?

The Coffee-Cup Calorimeter

In the general chemistry laboratory you are much more likely to run the simple calorimeter pictured in Figure 7-6 (on page 256) than a bo calorimeter. We mix the reactants (generally in aqueous solution) Styrofoam cup and measure the temperature change. Styrofoam is a g insulator, so there is very little heat transfer between the cup and the ounding air during the experiment. We treat the system—the cup and its ents—as an *isolated* system.

s with the bomb calorimeter, the heat of reaction is defined as the quantity eat that would be exchanged with the surroundings in restoring the imeter to its initial temperature. But, again, the calorimeter is not physi- restored to its initial conditions. We simply take the heat of reaction to be *egative* of the quantity of heat producing the temperature change in the imeter. That is, we use equation (7.9): $q_{rxn} = -q_{calorim}$.

Example 7-4, we make certain assumptions to simplify the calculation, or more precise measurements, these assumptions would not be made Exercise 25).

KAMPLE 7-4 Determining a Heat of Reaction from Calorimetric Data

In the neutralization of a strong acid with a strong base, the essential reaction is the combination of $H^+(aq)$ and $OH^-(aq)$ to form water (recall page 164).

$$H^+(aq) + OH^-(aq) \longrightarrow H_2O(l)$$

Two solutions, 25.00 mL of 2.50 M HCl(aq) and 25.00 mL of 2.50 M NaOH(aq), both initially at 21.1 °C, are added to a Styrofoam-cup calorimeter and allowed to react. The temperature rises to 37.8 °C. Determine the heat of the neutralization reaction, expressed per mole of H_2O formed. Is the reaction endothermic or exothermic?

nalyze

In addition to assuming that the calorimeter is an isolated system, assume that all there is in the system to absorb heat is 50.00 mL of water. This assumption ignores the fact that 0.0625 mol each of NaCl and H_2O are formed in the reaction, that the density of the resulting NaCl(aq) is not exactly 1.00 g/mL, and that its specific heat capacity is not exactly $4.18\ J\ g^{-1}\ °C^{-1}$. Also, ignore the small heat capacity of the Styrofoam cup itself.

Because the reaction is a neutralization reaction, let us call the heat of reaction q_{neutr}. Now, according to equation (7.9), $q_{neutr} = -q_{calorim}$, and if we make the assumptions described above, we can solve the problem.

olve

We begin with

$$q_{calorim} = 50.00\ mL \times \frac{1.00\ g}{1\ mL} \times 4.18 \frac{J}{g\ °C} \times (37.8 - 21.1)\ °C = 3.49 \times 10^3\ J$$

$$q_{neutr} = -q_{calorim} = -3.49 \times 10^3\ J = -3.49\ kJ$$

In 25.00 mL of 2.50 M HCl, the amount of H^+ is

$$?\ mol\ H^+ = 25.00\ mL \times \frac{1\ L}{1000\ mL} \times \frac{2.50\ mol}{1\ L} \times \frac{1\ mol\ H^+}{1\ mol\ HCl} = 0.0625\ mol\ H^+$$

Similarly, in 25.00 mL of 2.50 M NaOH there is 0.0625 mol OH^-. Thus, the H^+ and the OH^- combine to form 0.0625 mol H_2O. (The two reactants are in *stoichiometric* proportions; neither is in excess.)

The amount of heat produced per mole of H_2O is

$$q_{neutr} = \frac{-3.49\ kJ}{0.0625\ mol} = -55.8\ kJ/mol$$

Because q_{neutr} is a *negative* quantity, the neutralization reaction is *exothermic*.

ssess

Even though, in this example, we considered a specific reaction, the result $q_{neutr} = -55.8\ kJ/mol$ is more general. We will obtain the same value of q_{neutr} by considering any strong acid-strong base reaction because the net ionic equation is the same for all strong acid–strong base reactions.

ACTICE EXAMPLE A: Two solutions, 100.0 mL of 1.00 M $AgNO_3(aq)$ and 100.0 mL of 1.00 M NaCl(aq), both initially at 22.4 °C, are added to a Styrofoam-cup calorimeter and allowed to react. The temperature rises to 30.2 °C. Determine q_{rxn} per mole of AgCl(s) in the reaction.

$$Ag^+(aq) + Cl^-(aq) \longrightarrow AgCl(s)$$

(continued)

PRACTICE EXAMPLE B: Two solutions, 100.0 mL of 1.020 M HCl and 50.0 mL of 1.988 M NaOH, both initially 24.52 °C, are mixed in a Styrofoam-cup calorimeter. What will be the final temperature of the mixture? Ma the same assumptions, and use the heat of neutralization established in Example 7-4. [*Hint:* Which is t limiting reactant?]

The bomb and coffee-cup calorimeters are just two examples of calorim used in experiments. Another type of calorimeter is the ice calorimeter, i duced in exercise 105.

▲ FIGURE 7-6
A Styrofoam "coffee-cup" calorimeter
The reaction mixture is in the inner cup. The outer cup provides additional thermal insulation from the surrounding air. The cup is closed off with a cork stopper through which a thermometer and a stirrer are inserted and immersed into the reaction mixture. The reaction in the calorimeter occurs under the *constant pressure* of the atmosphere. We consider the difference between constant-volume and constant-pressure reactions in Section 7-6.

7-4 CONCEPT ASSESSMENT

How do we determine the heat capacity of the solution calorimeter (coffee-cu calorimeter)?

7-4 Work

We have just learned that heat effects generally accompany chemical reactio some reactions, work is also involved—that is, the system may do work c surroundings or vice versa. Consider the decomposition of potassium chlor potassium chloride and oxygen. Suppose that this decomposition is carrie in the strange vessel pictured in Figure 7-7. The walls of the container resist ing under the pressure of the expanding $O_2(g)$ except for the piston that c off the cylindrical top of the vessel. The pressure of the $O_2(g)$ exceeds the a spheric pressure and the piston is lifted—the system does work on the surro ings. Can you see that even if the piston were removed, work still would be as the expanding $O_2(g)$ pushed aside other atmospheric gases? Work invc in the expansion or compression of gases is called **pressure–volume w** Pressure–volume, or *P–V*, work is the type of work performed by explosive by the gases formed in the combustion of gasoline in an automobile engine.

Now let us switch to a somewhat simpler situation to see how to calcul quantity of *P–V* work.

In the hypothetical apparatus pictured in Figure 7-8(a), a weightless pist attached to a weightless wire support, to which a weightless pan is attache the pan are two identical weights just sufficient to stop the gas from expan The gas is confined by the cylinder walls and piston, and the space abov piston is a vacuum. The cylinder is contained in a constant-temperature bath, which keeps the temperature of the gas constant. Now imagine that o the two weights is removed, leaving half the original mass on the pan. L call this remaining mass *M*. The gas will expand and the remaining weigh move against gravity, the situation represented by Figure 7-8(b). Afte

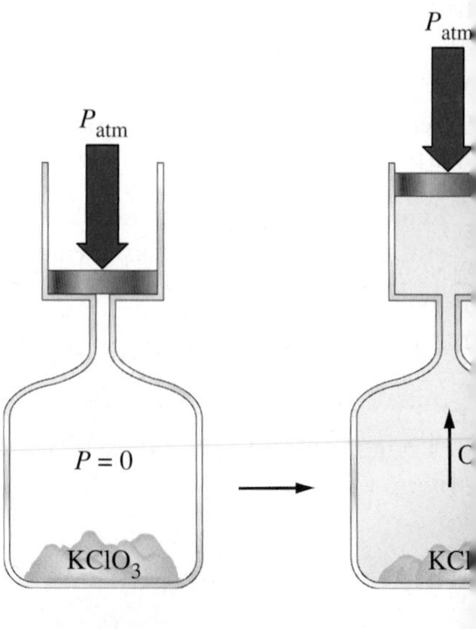

▶ FIGURE 7-7
Illustrating work (expansion) during the chemical reaction
$2\ KClO_3(s) \longrightarrow 2\ KCl(s) + 3\ O_2(g)$
The oxygen gas that is formed pushes back the weight and, in doing so, does work on the surroundings.

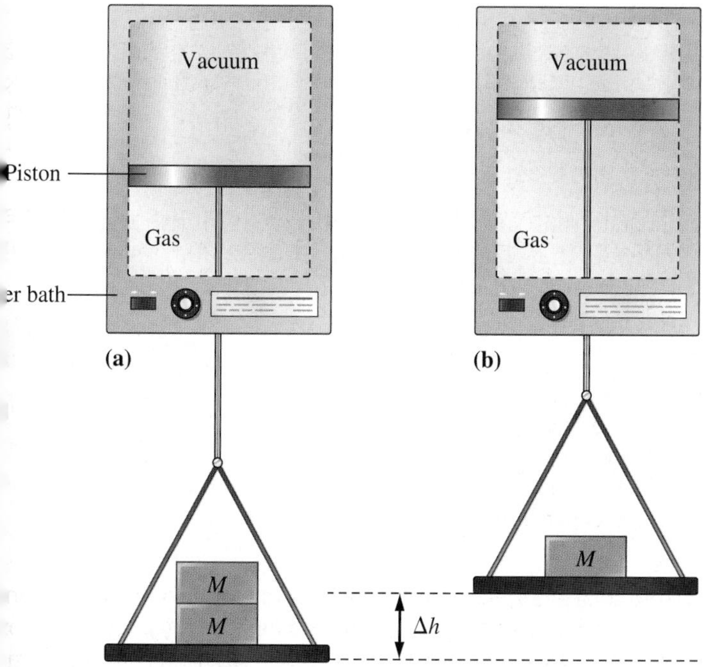

◀ FIGURE 7-8
Pressure–volume work
(a) In this hypothetical apparatus, a gas is confined by a massless piston of area A. A massless wire is attached to the piston and the gas is held back by two weights with a combined mass of $2M$ resting on the massless pan. The cylinder is immersed in a large water bath in order to keep the gas temperature constant. The initial state of the gas is $P_i = 2\,Mg/A$ with a volume V_i at temperature, T. **(b)** When the external pressure on the confined gas is suddenly lowered by removing one of the weights, the gas expands, pushing the piston up by the distance, Δh. The increase in volume of the gas (ΔV) is the product of the cross-sectional area of the cylinder (A) and the distance (Δh). The final state of the gas is $P_f = Mg/A$, V_f, and T.

ansion, we find that the piston has risen through a vertical distance, Δh; that volume of gas has doubled; and that the pressure of the gas has decreased. Jow let us see how pressure and volume enter into calculating how much sure–*volume* work the expanding gas does. First we can calculate the work e by the gas in moving the weight of mass M through a displacement Δh. all from equation (7.1) that the work can be calculated by

$$\text{work}\ (w) = \text{force}\ (M \times g) \times \text{distance}\ (\Delta h) = -M \times g \times \Delta h$$

magnitude of the force exerted by the weight is $M \times g$, where g is the leration due to gravity. The negative sign appears because the force is actin a direction opposite to the piston's direction of motion.

Jow recall equation (6.1)—pressure = force $(M \times g)$/area (A)—so that if expression for work is multiplied by A/A we get

$$w = -\frac{M \times g}{A} \times \Delta h \times A = -P_{ext}\Delta V \qquad \textbf{(7.11)}$$

he "pressure" part of the pressure–volume work is seen to be the external sure (P_{ext}) on the gas, which in our thought experiment is equal to the weight ing down on the piston and is given by Mg/A. Note that the product of area (A) and height (Δh) is equal to a volume—the volume change, ΔV, luced by the expansion.

wo significant features to note in equation (7.11) are the *negative* sign and the r P_{ext}. The negative sign is necessary to conform to sign conventions that we introduce in the next section. When a gas expands, ΔV is positive and w is tive, signifying that energy leaves the system as work. When a gas is comsed, ΔV is negative and w is positive, signifying that energy (as work) enters system. P_{ext} is the *external* pressure—the pressure against which a system nds or the applied pressure that compresses a system. In some instances the nal pressure in a system will be essentially equal to the external pressure, in h case the pressure in equation (7.11) is expressed simply as P.

◀ Work is negative when energy is transferred out of the system and is positive when energy is transferred into the system. This is consistent with the signs associated with the heat of a reaction (q) during exothermic and endothermic processes.

pressure is stated in bars or atmospheres and volume in liters, the unit of k is bar L or atm L. However, the SI unit of work is the joule. To convert bar L to J, or from atm L to J, we use one of the following relationships, of which are *exact*.

◀ The unit atm L, often written as L atm, is the liter-atmosphere. The use of this unit still persists.

$$1\ \text{bar L} = 100\ \text{J} \qquad 1\ \text{atm L} = 101.325\ \text{J}$$

These relationships are easily established by comparing values of the constant, R, given in Table 6.3. For example, because $R = 8.3145\ \text{J mol}^{-1}\ \text{K}^{-}$ 0.083145 bar L mol^{-1} K^{-1}, we have

$$\frac{8.3145\ \text{J}\ \cancel{\text{mol}^{-1}}\ \cancel{\text{K}^{-1}}}{0.083145\ \text{bar L}\ \cancel{\text{mol}^{-1}}\ \cancel{\text{K}^{-1}}} = 100\ \frac{\text{J}}{\text{bar L}}$$

This result confirms that 1 bar L = 100 J. How do we establish that 1 atm exactly 101.325 J? Recall that 1 atm is exactly 1.01325 bar (see Table Thus, 1 atm L = 1.01325 bar L = 1.01325 × 100 J = 101.325 J.

EXAMPLE 7-5 Calculating Pressure–Volume Work

Suppose the gas in Figure 7-8 is 0.100 mol He at 298 K, the two weights correspond to an external pressure 2.40 atm in Figure 7-8(a), and the single weight in Figure 7-8(b) corresponds to an external pressure of 1.20 atm How much work, in joules, is associated with the gas expansion at constant temperature?

Analyze

We are given enough data to calculate the initial and final gas volumes (note that the identity of the gas does n enter into the calculations because we are assuming ideal gas behavior). With these volumes, we can obtain Δ The external pressure in the pressure–volume work is the *final* pressure: 1.20 atm. The product $-P_{ext} \times \Delta$ must be multiplied by a factor to convert work in liter-atmospheres to work in joules.

Solve

First calculate the initial and final volumes.

$$V_{\text{initial}} = \frac{nRT}{P_i} = \frac{0.100\ \text{mol} \times 0.0821\ \text{atm L mol}^{-1}\ \text{K}^{-1} \times 298\ \text{K}}{2.40\ \text{atm}} = 1.02\ \text{L}$$

$$V_{\text{final}} = \frac{nRT}{P_f} = \frac{0.100\ \text{mol} \times 0.0821\ \text{atm L mol}^{-1}\ \text{K}^{-1} \times 298\ \text{K}}{1.20\ \text{atm}} = 2.04\ \text{L}$$

$$\Delta V = V_f - V_i = 2.04\ \text{L} - 1.02\ \text{L} = 1.02\ \text{L}$$

$$w = -P_{ext} \times \Delta V = -1.20\ \text{atm} \times 1.02\ \text{L} \times \frac{101\ \text{J}}{1\ \text{atm L}} = -1.24 \times 10^2\ \text{J}$$

Assess

The negative value signifies that the expanding gas (i.e., the system) does work on its surroundings. Keep i mind that the ideal gas equation embodies Boyle's law: The volume of a fixed amount of gas at a fixed ten perature is inversely proportional to the pressure. Thus, in Example 7-5 we could simply write that

$$V_f = 1.02\ \text{L} \times \frac{2.40\ \text{atm}}{1.20\ \text{atm}}$$

$$V_f = 2.04\ \text{L}$$

PRACTICE EXAMPLE A: How much work, in joules, is involved when 0.225 mol N_2 at a constant temperature 23 °C is allowed to expand by 1.50 L in volume against an external pressure of 0.750 atm? [*Hint:* How much this information is required?]

PRACTICE EXAMPLE B: How much work is done, in joules, when an external pressure of 2.50 atm is applied, at constant temperature of 20.0 °C, to 50.0 g $N_2(g)$ in a 75.0 L cylinder? The cylinder is like that shown in Figure 7-

7-5 CONCEPT ASSESSMENT

A gas in a 1.0 L closed cylinder has an initial pressure of 10.0 bar. It has a final pressure of 5.0 bar. The volume of the cylinder remained constant during this time. What form of energy was transferred across the boundary to cause this change? In which direction did the energy flow?

5 The First Law of Thermodynamics

absorption or evolution of heat and the performance of work require nges in the energy of a system and its surroundings. When considering the gy of a system, we use the concept of internal energy and how heat and k are related to it.

nternal energy, ***U***, is the total energy (both kinetic and potential) in a sys-, including *translational kinetic energy* of molecules, the energy associated molecular rotations and vibrations, the energy stored in chemical bonds intermolecular attractions, and the energy associated with electrons in ns. Some of these forms of internal energy are illustrated in Figure 7-9. rnal energy also includes energy associated with the interactions of pro- and neutrons in atomic nuclei, although this component is unchanged in nical reactions. A system contains *only* internal energy. A system does not ain energy in the form of heat or work. Heat and work are the means by ch a system exchanges energy with its surroundings. *Heat and work exist during a change in the system.* The relationship between heat (q), work (w), changes in internal energy (ΔU) is dictated by the law of conservation of gy, expressed in the form known as the **first law of thermodynamics**.

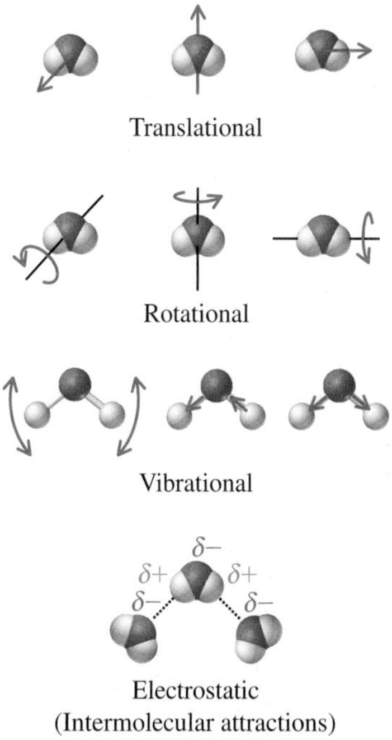

▲ FIGURE 7-9
Some contributions to the internal energy of a system
The models represent water molecules, and the arrows represent the types of motion they can undergo. In the intermolecular attractions between water molecules, the symbols $\delta+$ and $\delta-$ signify a separation of charge, producing centers of positive and negative charge that are smaller than ionic charges. These intermolecular attractions are discussed in Chapter 12.

$$\Delta U = q + w \tag{7.12}$$

solated system is unable to exchange either heat or work with its sur-ndings, so that $\Delta U_{\text{isolated system}} = 0$, and we can say

The energy of an isolated system is constant.

/hen using equation (7.12) we must keep these important points in mind.

Any energy *entering* the system carries a *positive* sign. Thus, if heat is *absorbed* by the system, $q > 0$. If work is done *on* the system, $w > 0$.

Any energy *leaving* the system carries a *negative* sign. Thus, if heat is *given off* by the system, $q < 0$. If work is done *by* the system, $w < 0$.

In general, the internal energy of a system changes as a result of energy entering or leaving the system as heat and/or work. If, on balance, more energy enters the system than leaves, ΔU is *positive*. If more energy leaves than enters, ΔU is *negative*.

A consequence of $\Delta U_{\text{isolated system}} = 0$ is that $\Delta U_{\text{system}} = -\Delta U_{\text{surroundings}}$; that is, energy is conserved.

se ideas are summarized in Figure 7-10 and illustrated in Example 7-6.

KEEP IN MIND

that heat is the disordered flow of energy and work is the ordered flow of energy.

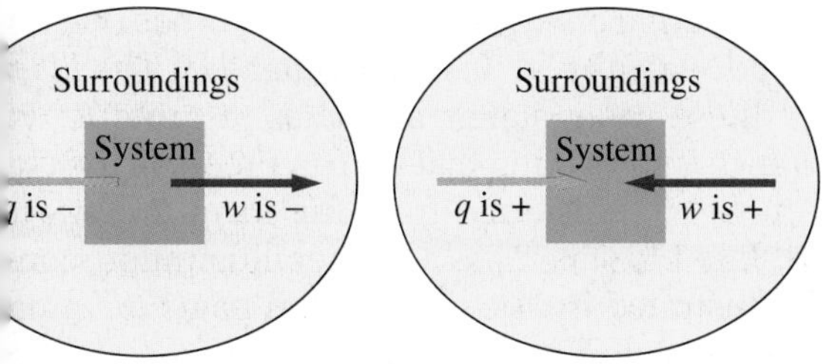

◀ FIGURE 7-10
Illustration of sign conventions used in thermodynamics
Arrows represent the direction of heat flow (⟶) and work (⟶). In the left diagram, the minus (−) signs signify energy leaving the system and entering the surroundings. In the right diagram the plus (+) signs refer to energy entering the system from the surroundings. These sign conventions are consistent with the expression $\Delta U = q + w$.

EXAMPLE 7-6 Relating ΔU, q, and w Through the First Law of Thermodynamics

A gas, while expanding, absorbs 25 J of heat and does 243 J of work. What is ΔU for the gas?

Analyze

The key to problems of this type lies in assigning the correct signs to the quantities of heat and work. Becaus heat is absorbed by (enters) the system, q is *positive*. Because work done *by* the system represents energy *leavin* the system, w is *negative*. You may find it useful to represent the values of q and w, with their correct sign within parentheses. Then complete the algebra.

Solve

$$\Delta U = q + w = (+25\,\text{J}) + (-243\,\text{J}) = 25\,\text{J} - 243\,\text{J} = -218\,\text{J}$$

Assess

The negative sign for the change in internal energy, ΔU, signifies that the system, in this case the gas, has lost energ

PRACTICE EXAMPLE A: In compressing a gas, 355 J of work is done on the system. At the same time, 185 J of he escapes from the system. What is ΔU for the system?

PRACTICE EXAMPLE B: If the internal energy of a system *decreases* by 125 J at the same time that the syster *absorbs* 54 J of heat, does the system do work or have work done on it? How much?

7-6 CONCEPT ASSESSMENT

When water is injected into a balloon filled with ammonia gas, the balloon shrinks and feels warm. What are the sources of heat and work, and what are the signs of q and w in this process?

Functions of State

To describe a system completely, we must indicate its temperature, its press and the kinds and amounts of substances present. When we have done this have specified the *state* of the system. Any property that has a unique valu a specified state of a system is said to be a **function of state**, or a **state funct** For example, a sample of pure water at 20 °C (293.15 K) and under a pressu 100 kPa is in a specified state. The density of water in this state is 0.99820 g/ We can establish that this density is a unique value—a function of state—ir following way: Obtain three different samples of water—one purified by ex sive distillation of groundwater; one synthesized by burning pure H_2(g pure O_2(g); and one prepared by driving off the water of hydration f $CuSO_4 \cdot 5H_2O$ and condensing the gaseous water to a liquid. The densiti the three different samples for the state that we specified will all be the s 0.99820 g/mL. Thus, the value of a function of state depends on the state o system, and not on how that state was established.

KEEP IN MIND

that a pressure of 100 kPa is equal to 1 bar.

The internal energy of a system is a function of state, although there is no ple measurement or calculation that we can use to establish its value. That is cannot write down a value of U for a system in the same way that we can w $d = 0.99820$ g/mL for the density of water at 20 °C. Fortunately, we don't nee know actual values of U. Consider, for example, heating 10.0 g of ice at 0 ° a final temperature of 50 °C. The internal energy of the ice at 0 °C has one un value, U_1, while that of the liquid water at 50 °C has another, U_2. The *diffe* in internal energy between these two states also has a unique va $\Delta U = U_2 - U_1$, and this difference *is* something that we can precisely measu is the quantity of energy that must be transferred from the surroundings to system during the change from state 1 to state 2. As a further illustration, cons the scheme outlined here and illustrated by the diagram on page 261. Ima that a system changes from state 1 to state 2 and then back to state 1.

$$\text{State 1 } (U_1) \xrightarrow{\Delta U} \text{State 2 } (U_2) \xrightarrow{-\Delta U} \text{State 1 } (U_1)$$

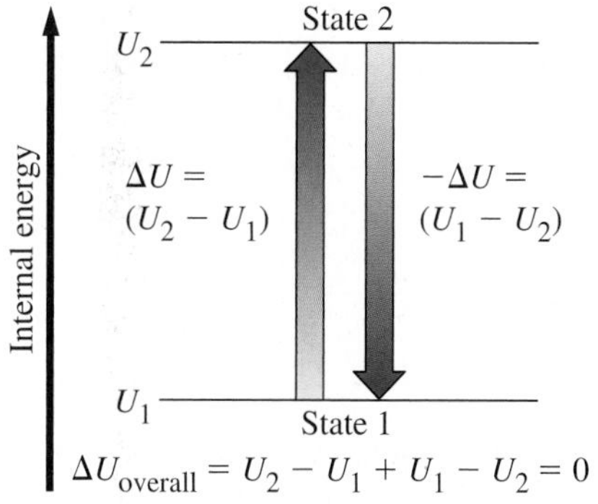

ause U has a unique value in each state, ΔU also has a unique value; it is $- U_1$. The change in internal energy when the system is returned from state state 1 is $-\Delta U = U_1 - U_2$. Thus, the *overall* change in internal energy is

$$\Delta U + (-\Delta U) = (U_2 - U_1) + (U_1 - U_2) = 0$$

s means that the internal energy returns to its initial value of U_1, which it st do, since it is a function of state. It is important to note here that when we erse the direction of change, we change the sign of ΔU.

:h-Dependent Functions

ike internal energy and changes in internal energy, heat (q) and work (w) *not* functions of state. Their values depend on the path followed when a :em undergoes a change. We can see why this is so by considering again process described by Figure 7-8 and Example 7-5. Think of the 0.100 mol of at 298 K and under a pressure of 2.40 atm as *state 1*, and under a pressure .20 atm as *state 2*. The change from state 1 to state 2 occurred in a single). Suppose that in another instance, we allowed the expansion to occur ɔugh an intermediate state pictured in Figure 7-11. That is, suppose the ·rnal pressure on the gas was first reduced from 2.40 atm to 1.80 atm (at ıch point, the gas volume would be 1.36 L). Then, in a second stage, ıced from 1.80 atm to 1.20 atm, thereby arriving at state 2.

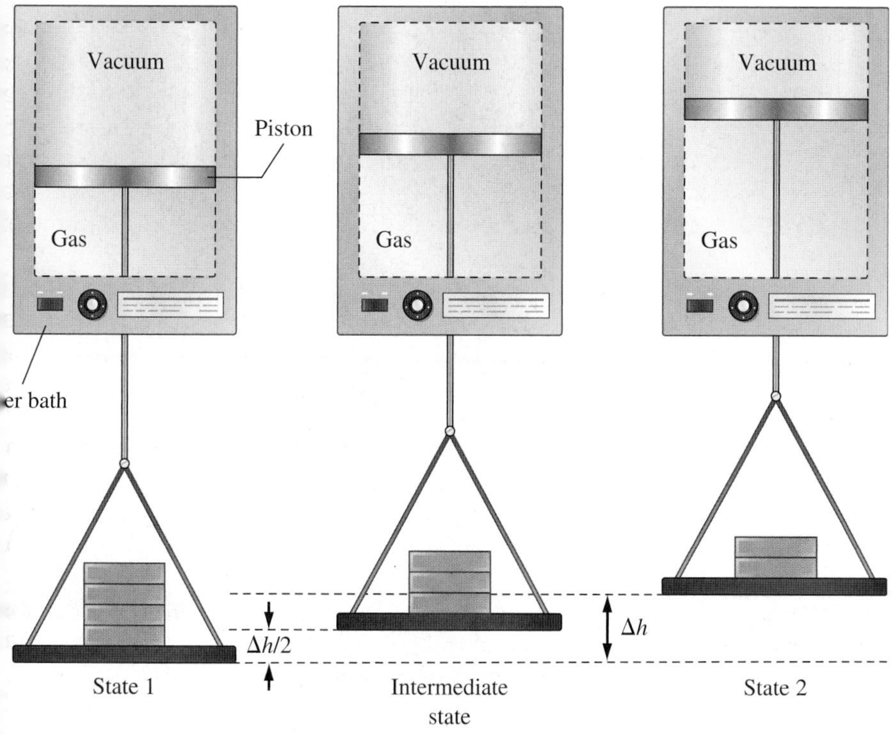

◀ FIGURE 7-11
A two-step expansion for the gas shown in Figure 7-8
In the initial state there are four weights of mass $M/2$ holding the gas back. In the intermediate state one of these weights has been removed and in the final state a second weight of mass $M/2$ has been removed. The initial and final states in this figure are the same as in Figure 7-8. This two-step expansion helps us to establish that the work of expansion depends on the path taken.

We calculated the amount of work done by the gas in a single-stage expan in Example 7-5; it was $w = -1.24 \times 10^2$ J. The amount of work done in the stage process is the sum of two terms: the pressure–volume work for each s of the expansion.

$$
\begin{aligned}
w &= -1.80 \text{ atm} \times (1.36 \text{ L} - 1.02 \text{ L}) - 1.20 \text{ atm} \times (2.04 \text{ L} - 1.36 \text{ L}) \\
&= -0.61 \text{ atm L} - 0.82 \text{ atm L} \\
&= -1.43 \text{ atm L} \times \frac{101 \text{ J}}{1 \text{ atm L}} = -1.44 \times 10^2 \text{ J}
\end{aligned}
$$

KEEP IN MIND

that if w differs in the two expansion processes, q must also differ, and in such a way that $q + w = \Delta U$ has a unique value, as required by the first law of thermodynamics.

The value of ΔU is the same for the single- and two-stage expansion proce because internal energy is a function of state. However, we see that slig more work is done in the two-stage expansion. Work is not a function of s it is path dependent. In the next section, we will stress that heat is also dependent.

Now consider a different way to carry out the expansion from state 1 to st (see Figure 7-12). The weights in Figures 7-8 and 7-11 have now been repl by an equivalent amount of sand so that the gas is in state 1. Imagine sar removed very slowly from this pile—say, one grain at a time. When exactly the sand has been removed, the gas will have reached state 2. This very s expansion proceeds in a nearly reversible fashion. From a thermodynamic spective, a process is reversible if the changes produced in the system and surroundings (and therefore, the universe) can be completely undone reversing the steps. A **reversible process** involves an infinite number of ir mediate states, with the system variables changing from their initial value their final values by infinitesimally small amounts. The process illustrate Figure 7-12 is not quite reversible because grains of sand have more tha infinitesimal mass. In this approximately reversible process we have ma very large number of intermediate expansions. This process provides m work than when the gas expands directly from state 1 to state 2.

The important difference between the expansion in a finite number of s and the reversible expansion is that the gas in the reversible process is alwa equilibrium with its surroundings whereas in a stepwise process this is n the case. For processes involving a finite number of steps, changes produce the system or surroundings cannot be totally undone by reversing the st Such processes are said to be **irreversible**.

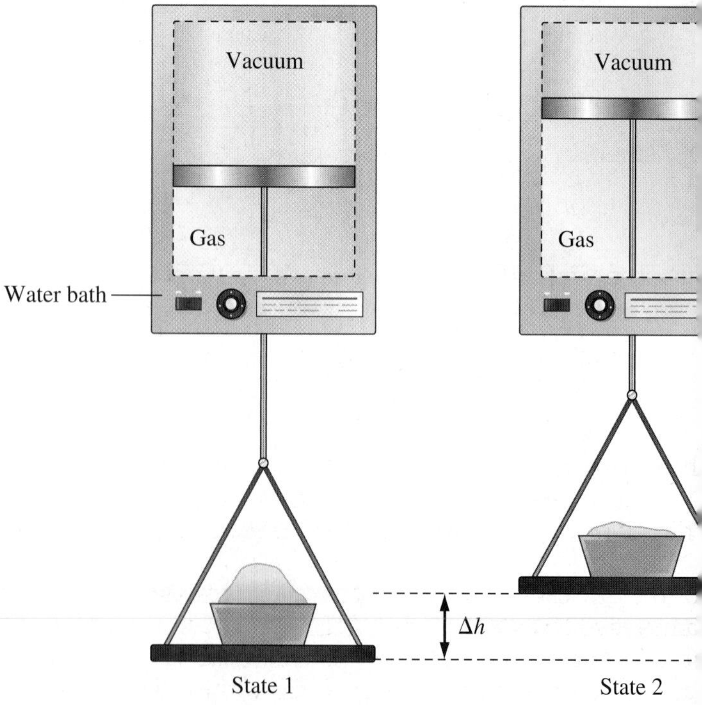

▶ FIGURE 7-12
A different method of achieving the expansion of a gas
In this expansion process, the weights in Figures 7-8 and 7-11 have been replaced by a pan containing sand, which has a mass of $2M$ equivalent to that of the weights in the initial state. In the final state the mass has been reduced to M.

comparing the quantity of work done in the two different expansions ;. 7-8 and 7-11), we found them to be different, thereby proving that work is state function. Additionally, the quantity of work performed is greater in wo-step expansion (Fig. 7-11) than in the single-step expansion (Fig. 7-8). leave it to the interested student to demonstrate, through Feature lem 125, that the maximum possible work is that done in a reversible nsion (Fig. 7-12).

gure 7-13 shows three different paths leading to the same internal energy ıge in a system and illustrates qualitatively that q and w are not state tions.

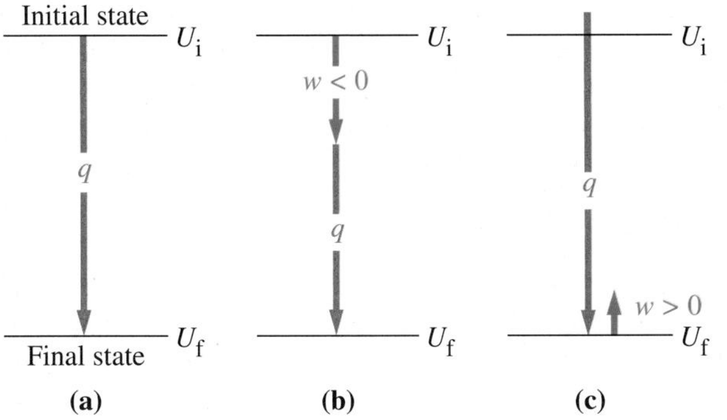

◀ FIGURE 7-13
Three different paths leading to the same internal energy change in a system
In path **(a)**, no work is done and $\Delta U = q$. In path **(b)**, work is done by the system ($w < 0$). In path **(c)**, the surroundings do work on the system ($w > 0$).

7-7 CONCEPT ASSESSMENT

sample can be heated very slowly or very rapidly. The darker shading in the stration indicates a higher temperature. Which of the two sets of diagrams you think corresponds to reversible heating and which to spontaneous, or eversible, heating?

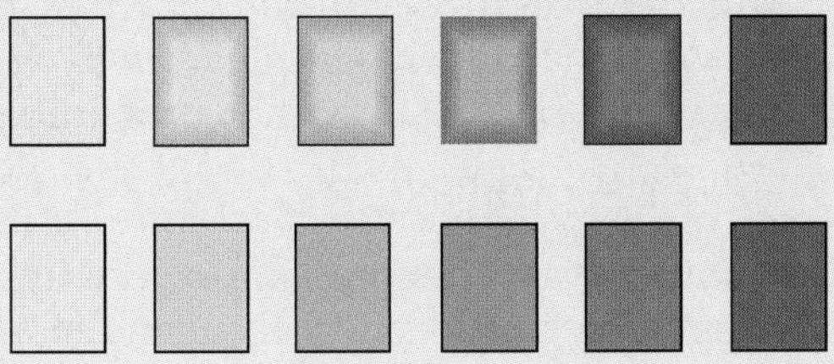

◀ Although no perfectly reversible process exists, the melting and freezing of a substance at its transition temperature is an example of a process that is nearly reversible: pump in heat (melts), take out heat (freezes).

Application of the First Law to Chemical and Physical Changes

ıis section, we apply the first law of thermodynamics to a system that ergoes a chemical change. Let's represent the chemical change as follows:

$$\underset{\substack{\text{(initial state)} \\ U_i}}{a \text{ mol A} + b \text{ mol B} + \cdots} \longrightarrow \underset{\substack{\text{(final state)} \\ U_f}}{c \text{ mol C} + d \text{ mol D} + \cdots}$$

he uppercase letters A, B, C, D, and so on, represent different substances. lowercase letters a, b, c, d, and so on, represent the stoichiometric ficients in the balanced chemical equation for the reaction: $+ b\text{B} + \cdots \rightarrow c\text{C} + d\text{D} + \cdots$

or the process above, $\Delta U = U_f - U_i$. According to the first law of thermo- amics, we can also say that $\Delta U = q + w$. If the process is carried out in a b calorimeter (see Figure 7-5), then the initial and final volumes of the sys- are the same (the system is confined within the bomb). Because the volume is constant, $\Delta V = 0$, no work is done, $w = -P_{ext}\Delta V = 0$. If the heat tr ferred for this constant volume process is denoted as q_V, then we see that

$$\Delta U = q_V \quad (7$$

We arrive at the following conclusion: For a constant-volume process, as a reaction in a bomb calorimeter, the heat transferred is equal to the inte energy change, ΔU, of the system.

Chemical reactions are not ordinarily carried out at constant volum bomb calorimeters. More often, they are carried out in beakers, flasks, other containers open to the atmosphere and under the *constant pressure* o atmosphere. In many reactions carried out at constant pressure, a s amount of pressure–volume work is done as the system expands or contr (recall Figure 7-7). Let's assume the volume of the system changes from an tial volume, V_i, to a final volume, V_f, under a constant external pressure, we represent the heat transferred in this constant-pressure process by q_P the work done by $w = -P_{ext}\Delta V$, then, by the first law, we have

$$\Delta U = q_P - P_{ext}\Delta V = q_P - P_{ext}(V_f - V_i) = q_P - P_{ext}V_f + P_{ext}V_i$$

Now, we substitute $U_f - U_i$ for ΔU; P_iV_i for $P_{ext}V_i$, and P_fV_f for $P_{ext}V_f$. last two substitutions are possible because the system pressure is equal to in both the initial and final states: $P_i = P_f = P_{ext}$. By making these subs tions, we obtain

$$U_f - U_i = q_P - P_fV_f + P_iV_i$$

which can be rearranged to give

$$(U_f + P_fV_f) - (U_i + P_iV_i) = q_P$$

The left side of this expression is the change in the quantity $U +$ The quantities U, P, and V are all state functions, so $U + PV$ must also state function. This state function is called the **enthalpy,** H, and is the sur the internal energy and the pressure–volume product: $H = U + PV$. expression above can be written in a very simple form by replacing $U_f +$ with H_f, the final enthalpy, and $U_i + P_iV_i$ with H_i, the initial enthalpy.

$$\Delta H = H_f - H_i = q_P \quad (7$$

Equation (7.14) is a simple way of expressing a very important idea: f constant-pressure process, such as a reaction occurring in a container ope the atmosphere, the heat transferred is equal to the enthalpy change, ΔH the system. Equation (7.14) is nothing more than a statement of the first for a constant-pressure process.

Let's explore further the relationship between ΔH and ΔU. For a const pressure process, we can write $q = q_P = \Delta H$ and $w = -P\Delta V$. By ma these substitutions into the first law, $\Delta U = q + w$, we obtain

$$\Delta U = \Delta H - P\Delta V \quad ($$

▶ ΔU in equation (7.15) is, strictly speaking, the internal energy change at constant pressure. Setting $\Delta U = q_V$ in this equation is an approximation but one that is usually valid.

The last term in this expression is the energy (work) associated with change in volume of the system under a constant external pressure. To as just how significant pressure–volume work is, let's consider the follow chemical change, which is also illustrated in Figure 7-14.

$$2 \text{ mol CO(g)} + 1 \text{ mol } O_2\text{(g)} \longrightarrow 2 \text{ mol } CO_2\text{(g)}$$

7-8 CONCEPT ASSESSMENT

Suppose a system is subjected to the following changes: a 40 kJ quantity of heat is added and the system does 15 kJ of work; then the system is returned its original state by cooling and compression. What is the value of ΔH?

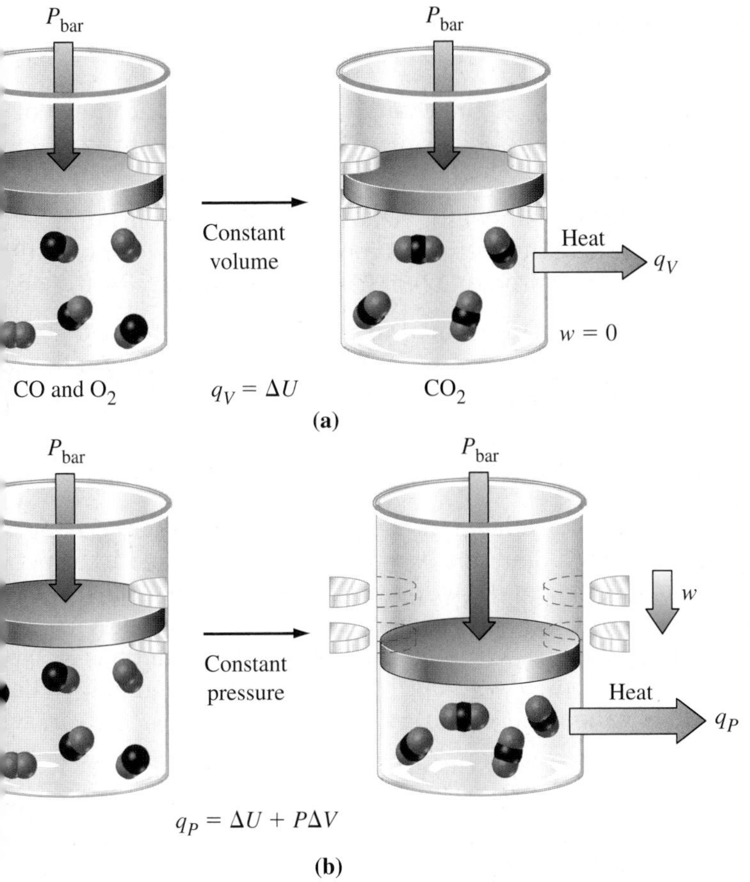

◀ FIGURE 7-14
Comparing q_V and q_P for the process $2\ mol\ CO(g) + 1\ mol\ O_2(g) \longrightarrow 2\ mol\ CO_2(g)$
(a) No work is performed at constant volume because the piston cannot move because of the stops placed through the cylinder walls; $q_V = \Delta U = -563.5$ kJ. **(b)** When the process is carried out at constant pressure, the stops are removed. This allows the piston to move and the surroundings do work on the system, causing it to shrink into a smaller volume. More heat is evolved than in the constant-volume process; $q_P = \Delta H = -566.0$ kJ.

ıe heat transferred in this process is measured under constant-pressure ditions at a constant temperature of 298 K, we get −566.0 kJ, indicating 566.0 kJ of energy has left the system as heat: $\Delta H = -566.0$ kJ. To evalu- the pressure–volume work, we begin by writing

$$P\Delta V = P(V_f - V_i)$$

n we can use the ideal gas equation to write this alternative expression.

$$P\Delta V = RT(n_{f,\,gas} - n_{i,\,gas}) = \Delta n_{gas}\,RT$$

'e, $n_{f,gas}$ is the number of moles of gas in the products ($2\ mol\ CO_2$) and ıs is the number of moles of gas in the reactants ($2\ mol\ CO + 1\ mol\ O_2$). ıs,

$$P\Delta V = 0.0083145\ kJ\ mol^{-1}\ K^{-1} \times 298\ K \times [2 - (2 + 1)]\ mol = -2.5\ kJ$$

change in internal energy is

$$\begin{aligned}\Delta U &= \Delta H - P\Delta V \\ &= -566.0\ kJ - (-2.5\ kJ) \\ &= -563.5\ kJ\end{aligned}$$

s calculation shows that the $P\Delta V$ term is quite small compared to ΔH and t ΔU and ΔH are almost the same. An additional interesting fact here is that volume of the system decreases as a consequence of the work done on the tem by the surroundings.

'he result obtained above can be expressed more generally as

$$\Delta H = \Delta U + \Delta n_{gas}RT \qquad (7.16)$$

or, since $\Delta H = q_P$ and $\Delta U = q_V$, as

$$q_P = q_V + \Delta n_{gas}RT \quad (7$$

In these expressions, Δn_{gas} is the change in the number of moles of ; Using equations (7.16) and (7.17), it is relatively straightforward to estab the following points:

- For a given chemical change, the heat transferred at constant press (q_P) is equal to the heat transferred at constant volume (q_V) only if the no net consumption or production of gas (i.e., when $\Delta n_{gas} = 0$).
- Because RT is approximately equal to 2.5 kJ mol^{-1} at 298 K, the ma tude of the difference between q_P and q_V is typically only a few k joules. For example, in a process for which $\Delta n_{gas} = +1$ mol, $q_P - q$ about 2.5 kJ at 298 K.

Enthalpy of Reaction: $\Delta_r H$

Up to this point, we have considered internal energy and enthalpy changes a system in which specified amounts of reactants are converted into speci amounts of products. For example, we saw that when 2 mol CO(g) and 1 O_2(g) react to give 2 mol CO_2(g) at constant pressure, the enthalpy chang the system is −566 kJ. Another way of providing this information is to exp the enthalpy change *per mole of reaction*. The corresponding enthalpy chai denoted by $\Delta_r H$, is called the **enthalpy of reaction**. Thus, we can wi

▶ Strictly speaking, the enthalpy of reaction, $\Delta_r H$, is the rate of change of H with respect to the extent of reaction, ξ. Thus, $\Delta_r H = -566$ kJ mol^{-1} indicates that the enthalpy of the system decreases by 566 kJ *per mole of reaction*. On the other hand, ΔH is the enthalpy change of the system, expressed in J or kJ. To determine the value of ΔH, we must know the extent of reaction. $\Delta_r H$ and ΔH also differ in that $\Delta_r H$ is an *intensive* quantity whereas ΔH is an *extensive* quantity.

$$2\,CO(g) + O_2(g) \longrightarrow 2\,CO_2(g) \qquad \Delta_r H = -566 \text{ kJ mol}^{-1}$$

As established in Chapter 4, *one mole of reaction* refers to the situatio which the extent of reaction is equal to one mole; that is, the conversio 2 mol CO and 1 mol O_2 to 2 mol CO_2.

Now, consider the enthalpy of reaction for the combustion of sucrose.

$$C_{12}H_{22}O_{11}(s) + 12\,O_2(g) \longrightarrow 12\,CO_2(g) + 11\,H_2O(l) \quad (7$$
$$\Delta_r H = -5.65 \times 10^3 \text{ kJ mol}^{-1}$$

Thus, the amount of heat evolved is -5.65×10^3 kJ *per mole of react* Consequently, when the extent of reaction is equal to one mole ($\xi = 1$ m 1 mol $C_{12}H_{22}O_{11}$(s) reacts with 12 mol O_2(g) to produce 12 mol CO_2(g) 11 mol H_2O(l) and 5.65×10^3 kJ of heat. Interestingly, because $\Delta n_{gas} = 0$ this process, the amount of heat evolved is the same whether the reactio carried out a constant pressure or at constant volume.

EXAMPLE 7-7 Stoichiometric Calculations Involving Quantities of Heat

The enthalpy of reaction for the combustion of sucrose, $C_{12}H_{22}O_{11}$(s), is $\Delta_r H = -5.65 \times 10^3$ kJ/mol. How much heat is associated with the complete combustion of 1.00 kg of sucrose?

Analyze

The first step is to determine the number of moles in 1.00 kg of sucrose, and then use that value and the $\Delta_r H$ value for the reaction to calculate the quantity of heat produced.

Solve

Express the quantity of sucrose in moles.

$$? \text{ mol} = 1.00 \text{ kg } C_{12}H_{22}O_{11} \times \frac{1000 \text{ g } C_{12}H_{22}O_{11}}{1 \text{ kg } C_{12}H_{22}O_{11}} \times \frac{1 \text{ mol } C_{12}H_{22}O_{11}}{342.3 \text{ g } C_{12}H_{22}O_{11}} = 2.92 \text{ mol } C_{12}H_{22}O_{11}$$

Use the $\Delta_r H$ value to formulate a conversion factor (shown in blue) -5.65×10^3 kJ to convert from mol $C_{12}H_{22}O_{11}$ to kJ of heat.

$$? \text{ kJ} = 2.92 \text{ mol } C_{12}H_{22}O_{11} \times \frac{-5.65 \times 10^3 \text{ kJ}}{1 \text{ mol } C_{12}H_{22}O_{11}} = -1.65 \times 10^4 \text{ kJ}$$

The negative sign denotes that heat is given off in the combustion.

ssess

As discussed on page 252, the heat produced by a combustion reaction is not immediately transferred to the surroundings. You can use data from Table 7.1 to show that the heat released by this reaction is *more* than that required to raise the temperature of the products to 100 °C.

RACTICE EXAMPLE A: What mass of sucrose must be burned to produce 1.00×10^3 kJ of heat?

RACTICE EXAMPLE B: A 25.0 mL sample of 0.1045 M HCl(aq) was neutralized by NaOH(aq). Use the result of Example 7-4 to determine the heat evolved in this neutralization.

Physical Significance of Enthalpy Change

he preceding discussion, we used the first law of thermodynamics, $= q + w$, to show that $\Delta H = q_P$ (see equation 7.14). $\Delta H = q_P$ is just ther form of the first law, one that is particularly convenient for constant-sure processes.

/hat is the physical significance or meaning of ΔH? The answer to this stion is remarkably simple: ΔH represents the heat transferred under stant-pressure conditions. In other words, we use two different symbols, and q_P, to represent the same thing. One of these symbols refers to a prop- (H) of the system and the other to something we can measure (q).

oes the property H have a simple physical or molecular interpretation? answer to this question is, perhaps disappointingly, no. By definition, $U + PV$. Each of U, P, and V are easily interpreted or explained. The rnal energy, U, represents the total energy of a system, which is distributed ng the various molecular motions and interactions. The pressure, P, of a em is the force per unit area exerted by the molecules of the system and V st the volume occupied by the system. However, the combination $U + PV$ s not have any simple physical meaning or molecular interpretation. This bination of quantities is introduced for convenience only. Had we not oduced the definition $H = U + PV$, equation (7.14) would be expressed $(U + PV) = q_P$ or $\Delta U + P\Delta V = q_P$.

n summary, because most processes are carried out at constant pressure, most often measure the heat transferred as q_P, a quantity that may also be esented as ΔH.

halpy Change Accompanying a Change tate of Matter

en a liquid is in contact with the atmosphere, energetic molecules at the ace of the liquid can overcome forces of attraction to their neighbors and s into the gaseous, or vapor, state. We say that the liquid *vaporizes*. If the perature of the liquid is to remain constant, the liquid must absorb heat n its surroundings to replace the energy carried off by the vaporizing mol-les. The heat required to vaporize a fixed quantity of liquid is called the halpy (or heat) of vaporization. Usually the fixed quantity of liquid chosen ne mole, and we can call this quantity the *molar enthalpy of vaporization*. For mple,

$$H_2O(l) \longrightarrow H_2O(g) \qquad \Delta_{vap}H = 44.0 \text{ kJ mol}^{-1} \text{ at } 298.15 \text{ K}$$

described the melting of a solid in a similar fashion (page 247). The energy uirement in this case is called the enthalpy (or heat) of fusion. For the melting ne mole of ice, we can write

$$H_2O(s) \longrightarrow H_2O(l) \qquad \Delta_{fus}H = 6.01 \text{ kJ mol}^{-1} \text{ at } 273.15 \text{ K}$$

◀ According to the International Union of Pure and Applied Chemistry (IUPAC), the subscript used to denote a chemical process should be used as a subscript on the Δ symbol.

We can use the data represented in these equations, together with other appropriate data, to answer questions like those posed in Example 7-8 and its accompanying Practice Examples.

EXAMPLE 7-8 Enthalpy Changes Accompanying Changes in States of Matter

Calculate ΔH for the process in which 50.0 g of water is converted from liquid at 10.0 °C to vapor at 25.0 °C.

Analyze

The key to this calculation is to view the process as proceeding in two steps: first raising the temperature of liquid water from 10.0 to 25.0 °C, and then completely vaporizing the liquid at 25.0 °C. The total enthalpy change is the sum of the changes in the two steps.

Solve

HEATING WATER FROM 10.0 TO 25.0 °C

This heat requirement can be determined by the method shown in Example 7-1; that is, we apply equation (7.5

$$? \text{ kJ} = 50.0 \text{ g } H_2O \times \frac{4.18 \text{ J}}{\text{g } H_2O \text{ °C}} \times (25.0 - 10.0) \text{ °C} \times \frac{1 \text{ kJ}}{1000 \text{ J}} = 3.14 \text{ kJ}$$

VAPORIZING WATER AT 25.0 °C

For this part of the calculation, the quantity of water must be expressed in moles so that we can then use the molar enthalpy of vaporization at 25 °C: $\Delta_{vap}H = 44.0 \text{ kJ/mol}$.

$$? \text{ kJ} = 50.0 \text{ g } H_2O \times \frac{1 \text{ mol } H_2O}{18.02 \text{ g } H_2O} \times \frac{44.0 \text{ kJ}}{1 \text{ mol } H_2O} = 122 \text{ kJ}$$

TOTAL ENTHALPY CHANGE

$$\Delta H = 3.14 \text{ kJ} + 122 \text{ kJ} = 125 \text{ kJ}$$

Assess

Note that the enthalpy change is positive, which reflects that the system (i.e., the water) gains energy. The reverse would be true for condensation of water at 25.0 °C and cooling it to 10.0 °C.

PRACTICE EXAMPLE A: What is the enthalpy change when a cube of ice 2.00 cm on edge is brought from -10.0 °C to a final temperature of 23.2 °C? For ice, use a density of 0.917 g/cm^3, a specific heat capacity of 2.01 J g^{-1} °C^{-1}, and an enthalpy of fusion of 6.01 kJ/mol.

PRACTICE EXAMPLE B: What is the maximum mass of ice at -15.0 °C that can be completely converted to water vapor at 25.0 °C if the available heat for this transition is 5.00×10^3 kJ?

Standard States and Standard Enthalpies of Reaction

The measured enthalpy change for a reaction has a unique value *only* if the initial state (reactants) and final state (products) are precisely described. If we define a particular state as *standard* for the reactants and products, we can then say that the standard enthalpy change is the enthalpy change in a reaction in which the reactants and products are in their standard states. This so-called **standard enthalpy of reaction** is denoted with a degree symbol, $\Delta_r H°$.

The **standard state** of a solid or liquid substance is the pure element or compound at a pressure of *1 bar* (10^5 Pa)* and at the temperature of interest. For a gas, the standard state is the pure gas behaving as an (hypothetical) ideal gas at a pressure of 1 bar and the temperature of interest. Although temperature is not part of the definition of a standard state, it still must be specified in tabulated values of $\Delta_r H°$, because $\Delta_r H°$ depends on temperature. The values given in this text are all for 298.15 K (25 °C) unless otherwise stated.

*The International Union of Pure and Applied Chemistry (IUPAC) recommended that the standard-state pressure be changed from 1 atm to 1 bar more than 30 years ago, but some tables are still based on the 1 atm standard. Fortunately, the differences in values resulting from this change in standard-state pressure are very small—almost always small enough to be ignored.

the rest of this chapter, we will mostly use standard enthalpy changes. will explore the details of nonstandard conditions in Chapter 13.

halpy Diagrams

negative sign of $\Delta_r H$ in equation (7.18) means that the enthalpy of the lucts is lower than that of the reactants. This *decrease* in enthalpy appears eat evolved to the surroundings. The combustion of sucrose is an exother- reaction. In the reaction

$$N_2(g) + O_2(g) \longrightarrow 2\,NO(g) \qquad \Delta_r H° = 180.50 \text{ kJ mol}^{-1}$$

products have a *higher* enthalpy than the reactants; $\Delta_r H$ is positive. To luce this increase in enthalpy, heat is absorbed from the surroundings. The tion is endothermic. An **enthalpy diagram** is a diagrammatic representa- of enthalpy changes in a process. Figure 7-15 shows how exothermic and othermic reactions can be represented through such diagrams.

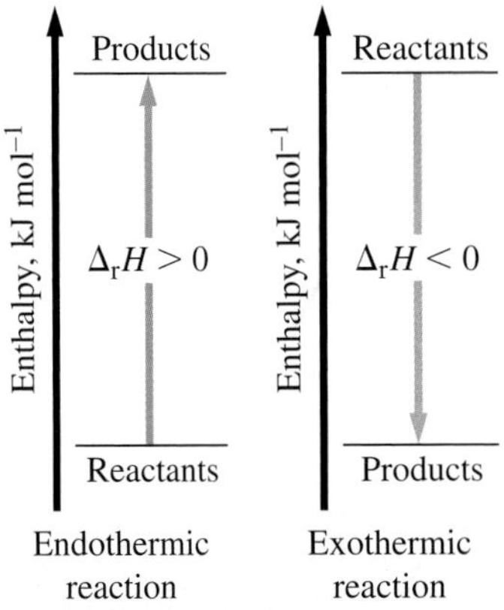

▲ FIGURE 7-15
Enthalpy diagrams
Horizontal lines represent absolute values of enthalpy. The higher a horizontal line, the greater the value of *H* that it represents. Vertical lines or arrows represent changes in enthalpy ($\Delta_r H$). Arrows pointing up signify increases in enthalpy—endothermic reactions. Arrows pointing down signify decreases in enthalpy—exothermic reactions.

7-1 ARE YOU WONDERING?

Why does $\Delta_r H$ depend on temperature?

ne difference in $\Delta_r H°$ for a reaction at two different temperatures is determined by e amount of heat involved in changing the reactants and products from one tem- erature to the other under constant pressure. These quantities of heat can be calcu- ted with the help of equation (7.5): q_P = heat capacity × temperature change = ΔT. We write an expression of this type for each reactant and product and com- ne these expressions with the measured $\Delta_r H°$ value at one temperature to obtain e value of $\Delta_r H°$ at another. This method is illustrated in Figure 7-16 and applied Exercise 118.

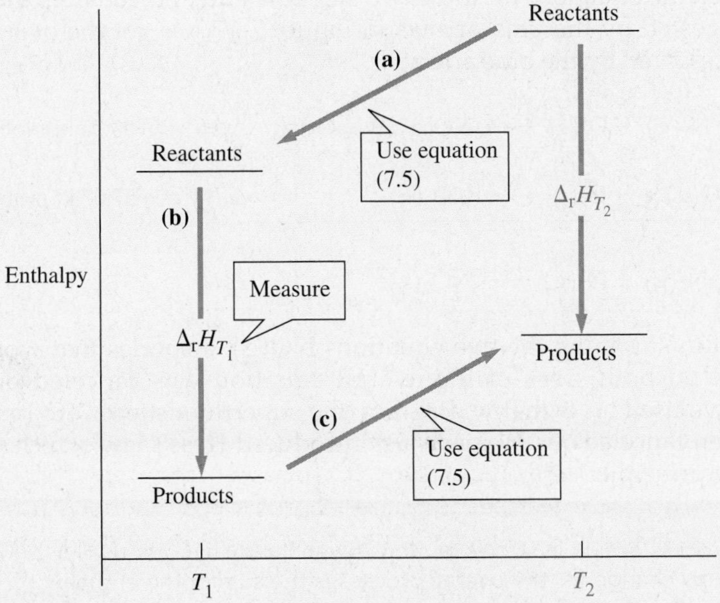

▲ FIGURE 7-16
Conceptualizing $\Delta_r H$ as a function of temperature
n the three-step process outlined here, **(a)** the reactants are cooled from the emperature T_2 to T_1. **(b)** The reaction is carried out at T_1, and **(c)** the products re warmed from T_1 to T_2. When the quantities of heat associated with each tep are combined, the result is the same as if the reaction had been carried out at T_2, that is, $\Delta_r H_{T_2}$.

7-7 Indirect Determination of $\Delta_r H$: Hess's La

One of the reasons that the enthalpy concept is so useful is that a large num of heats of reaction can be calculated from a small number of measureme The following features of enthalpy change ($\Delta_r H$) make this possible.

- **$\Delta_r H°$ Depends on the Way the Reaction is Written.** Consider the s dard enthalpy change in the formation of NO(g) from its elemen 25 °C.

$$N_2(g) + O_2(g) \longrightarrow 2\,NO(g) \qquad \Delta_r H° = 180.50\ \text{kJ mol}^{-1}$$

To express the enthalpy change in terms of *one mole* of NO(g), we di all coefficients *and the $\Delta_r H$ value* by *two*.

$$\frac{1}{2}N_2(g) + \frac{1}{2}O_2(g) \longrightarrow NO(g) \qquad \Delta_r H° = \frac{1}{2} \times 180.50 = 90.25\ \text{kJ mol}^{-1}$$

▶ Although we have avoided fractional coefficients previously, we need them here. The coefficient of NO(g) must be one.

- **$\Delta_r H°$ Changes Sign When a Process Is Reversed.** As we learne page 261, if a process is reversed, the change in a function of s reverses sign. Thus, $\Delta_r H°$ for the *decomposition* of one mole of NO($-\Delta_r H°$ for the *formation* of one mole of NO(g).

$$NO(g) \longrightarrow \frac{1}{2}N_2(g) + \frac{1}{2}O_2(g) \qquad \Delta_r H° = -90.25\ \text{kJ mol}^{-1}$$

- **Hess's Law of Constant Heat Summation.** To describe the stanc enthalpy change for the formation of $NO_2(g)$ from $N_2(g)$ and $O_2(g)$,

$$\frac{1}{2}N_2(g) + O_2(g) \longrightarrow NO_2(g) \qquad \Delta_r H° = ?$$

we can think of the reaction as proceeding in two steps: First we f NO(g) from $N_2(g)$ and $O_2(g)$, and then $NO_2(g)$ from NO(g) and O_2 When the equations for these two steps are added together in the mar suggested by the gray arrows in Figure 7-17, we get the overall re represented by the blue arrow.

$$\frac{1}{2}N_2(g) + O_2(g) \longrightarrow \cancel{NO(g)} + \cancel{\frac{1}{2}O_2(g)} \qquad \Delta_r H° = +90.25\ \text{kJ mol}^{-1}$$

$$\cancel{NO(g)} + \cancel{\frac{1}{2}O_2(g)} \longrightarrow NO_2(g) \qquad \Delta_r H° = -57.07\ \text{kJ mol}^{-1}$$

$$\frac{1}{2}N_2(g) + O_2(g) \longrightarrow NO_2(g) \qquad \Delta_r H° = +33.18\ \text{kJ mol}^{-1}$$

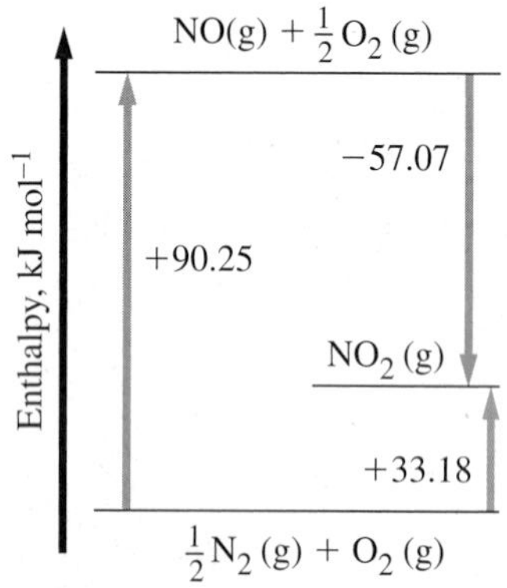

▲ FIGURE 7-17
An enthalpy diagram illustrating Hess's law
The numerical values are $\Delta_r H°$ values in kJ mol^{-1}. Whether the reaction occurs through a single step (blue arrow) or in two steps (gray arrows), $\Delta_r H° = 33.18$ kJ mol^{-1} for the overall reaction $\frac{1}{2}N_2(g) + O_2(g) \longrightarrow NO_2(g)$.

Note that in summing the two equations NO(g), a species that would h appeared on both sides of the overall equation was canceled out. A because we used an enthalpy diagram, the superfluous term ½$O_2(g)$ ente in and then canceled out. We have just introduced **Hess's law**, which states following principle:

> If a process occurs in stages or steps (even if only hypothetically), the enthalpy change for the overall process is the sum of the enthalpy changes for the individual steps.

Hess's law is simply a consequence of the state function property enthalpy. Regardless of the path taken in going from the initial state to final state, $\Delta_r H$ (or $\Delta_r H°$ if the process is carried out under standard co tions) has the same value.

Suppose we want the standard enthalpy change for the reaction

$$3\,C(\text{graphite}) + 4\,H_2(g) \longrightarrow C_3H_8(g) \qquad \Delta_r H° = ? \qquad (7$$

v should we proceed? If we try to get graphite and hydrogen to react, a slight tion will occur, but it will not go to completion. Furthermore, the product not be limited to propane (C_3H_8); several other hydrocarbons will form as . The fact is that we cannot directly measure $\Delta_r H°$ for reaction (7.19). Instead, nust resort to an *indirect calculation* from $\Delta_r H°$ values that can be established xperiment. Here is where Hess's law is of greatest value. It permits us to calte enthalpy changes that we cannot measure directly. In Example 7-9, we use dard enthalpies of combustion to calculate $\Delta_r H°$ for a reaction.

XAMPLE 7-9 Applying Hess's Law

The standard enthalpies of combustion of C(graphite), $H_2(g)$ and $C_3H_8(g)$ are –393.5, –285.8, and –2219.9 kJ mol^{-1}, respectively. Use these values to calculate $\Delta_r H°$ for reaction (7.19).

$$3\,C(\text{graphite}) + 4\,H_2(g) \longrightarrow C_3H_8(g) \qquad \Delta_r H° = ?$$

nalyze

To determine an enthalpy change with Hess's law, we need to combine the appropriate chemical equations. A good starting point is to write chemical equations for the given combustion reactions based on *one mole* of the indicated reactant. Recall (see page 114) that the products of the combustion of carbon–hydrogen–oxygen compounds are $CO_2(g)$ and $H_2O(l)$.

olve

Begin by writing the following equations

$$\textbf{(a)}\ C_3H_8(g) + 5\,O_2(g) \longrightarrow 3\,CO_2(g) + 4\,H_2O(l) \qquad \Delta_r H° = -2219.9\text{ kJ mol}^{-1}$$

$$\textbf{(b)}\ C(\text{graphite}) + O_2(g) \longrightarrow CO_2(g) \qquad \Delta_r H° = -393.5\text{ kJ mol}^{-1}$$

$$\textbf{(c)}\ H_2(g) + \frac{1}{2}O_2(g) \longrightarrow H_2O(l) \qquad \Delta_r H° = -285.8\text{ kJ mol}^{-1}$$

Because our objective in reaction (7.19) is to *produce* $C_3H_8(g)$, the next step is to find a reaction in which $C_3H_8(g)$ is formed—the *reverse* of reaction **(a)**.

$$-\textbf{(a)}:\ 3\,CO_2(g) + 4\,H_2O(l) \longrightarrow C_3H_8(g) + 5\,O_2(g) \quad \Delta_r H° = -(-2219.9\text{ kJ mol}^{-1}) = +2219.9\text{ kJ mol}^{-1}$$

Now, we turn our attention to the reactants, C(graphite) and $H_2(g)$. To get the proper number of moles of each, we must multiply equation **(b)** by three and equation **(c)** by four.

$$3 \times \textbf{(b)}:\ 3\,C(\text{graphite}) + 3\,O_2(g) \longrightarrow 3\,CO_2(g) \qquad \Delta_r H° = 3(-393.5\text{ kJ mol}^{-1}) = -1181\text{ kJ mol}^{-1}$$

$$4 \times \textbf{(c)}:\ 4\,H_2(g) + 2\,O_2(g) \longrightarrow 4\,H_2O(l) \qquad \Delta_r H° = 4(-285.8\text{ kJ mol}^{-1}) = -1143\text{ kJ mol}^{-1}$$

Here is the overall change we have described: 3 mol C(graphite) and 4 mol $H_2(g)$ have been consumed, and 1 mol $C_3H_8(g)$ has been produced. This is exactly what is required in equation (7.19). We can now combine the three modified equations.

$$\begin{array}{llll} -\textbf{(a)}: & \cancel{3\,CO_2(g)} + \cancel{4\,H_2O(l)} \longrightarrow C_3H_8(g) + \cancel{5\,O_2(g)} & \Delta_r H° = +2219.9\text{ kJ mol}^{-1} \\ 3 \times \textbf{(b)}: & 3\,C(\text{graphite}) + \cancel{3\,O_2(g)} \longrightarrow \cancel{3\,CO_2(g)} & \Delta_r H° = -1181\text{ kJ mol}^{-1} \\ 4 \times \textbf{(c)}: & 4\,H_2(g) + \cancel{2\,O_2(g)} \longrightarrow \cancel{4\,H_2O(l)} & \Delta_r H° = -1143\text{ kJ mol}^{-1} \\ \hline & 3\,C(\text{graphite}) + 4\,H_2(g) \longrightarrow C_3H_8(g) & \Delta_r H° = -104\text{ kJ mol}^{-1} \end{array}$$

ssess

Hess's law is a powerful technique to determine the enthalpy of reaction by using a series of unrelated reactions, along with their enthalpies of reaction. In this example, we took three unrelated combustion reactions and were able to determine the enthalpy of reaction of another reaction.

RACTICE EXAMPLE A: The standard heat of combustion of propene, $C_3H_6(g)$, is −2058 kJ/mol. Use this value and other data from this example to determine $\Delta_r H°$ for the hydrogenation of propene to propane.

$$CH_3CH{=}CH_2(g) + H_2(g) \longrightarrow CH_3CH_2CH_3(g) \qquad \Delta_r H° = ?$$

RACTICE EXAMPLE B: From the data in Practice Example 7-9A and the following equation, determine the standard enthalpy of combustion of one mole of $CH_3CH(OH)CH_3(l)$.

$$CH_3CH{=}CH_2(g) + H_2O(l) \longrightarrow CH_3CH(OH)CH_3(l) \qquad \Delta_r H° = -52.3\text{ kJ mol}^{-1}$$

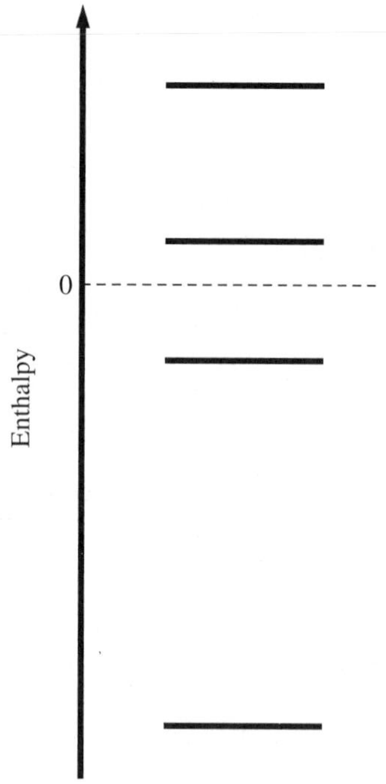

7-9 CONCEPT ASSESSMENT

The heat of reaction between carbon (graphite) and the corresponding stoichiometric amounts of hydrogen gas to form $C_2H_2(g)$, $C_2H_4(g)$, and $C_2H_6(g)$ are 226.7, 52.3 and −84.7 kJ mol^{-1}, respectively. Relate these values to the enthalpy diagram shown in the margin. Indicate on the diagram the standard enthalpy of reaction for $C_2H_2(g) + 2\,H_2(g) \longrightarrow C_2H_6(g)$.

7-8 Standard Enthalpies of Formation

In the enthalpy diagrams we have drawn, we have not written any numer values on the enthalpy axis. This is because we cannot determine *absolute* ues of enthalpy, H. However, enthalpy *is* a function of state, so *change* enthalpy, ΔH, have unique values. We can deal just with these chan Nevertheless, as with many other properties, it is still useful to have a star point, a zero value.

Consider a map-making analogy: What do we list as the height of a mo tain? Do we mean by this the vertical distance between the mountaintop the center of Earth? Between the mountaintop and the deepest trench in ocean? No. By agreement, we mean the vertical distance between the mo taintop and mean sea level. We *arbitrarily* assign to mean sea level an eleva of zero, and all other points on Earth are relative to this zero elevation. The vation of Mt. Everest is +8848 m; that of Badwater, Death Valley, Californi −86 m. We do something similar with enthalpies. We relate our zero to enthalpies of certain forms of the elements and determine the enthalpie other substances relative to this zero.

The **standard enthalpy of formation** ($\Delta_f H°$) of a substance is the entha *change* that occurs in the formation of one mole of the substance in the stand state from the *reference* forms of the elements in their standard states. The re ence forms of the elements in all but a few cases are the most stable forms of elements at one bar and the given temperature. The degree symbol denotes the enthalpy change is a standard enthalpy change, and the subscript f signi that the reaction is one in which a substance is formed from its elements. formation of the most stable form of an element from itself is no change at Therefore:

KEEP IN MIND

that we use the expression "standard enthalpy of formation" even though what we are describing is actually a standard enthalpy *change*.

The standard enthalpy of formation is 0 for a pure element in its reference form.

Listed here are the most stable forms of several elements at 298.15 K, temperature at which thermochemical data are commonly tabulated.

$$Na(s) \quad H_2(g) \quad N_2(g) \quad O_2(g) \quad C(graphite) \quad Br_2(l)$$

The situation with carbon is an interesting one. In addition to graphite, carb also exists naturally in the form of diamond. However, because there is a meas able enthalpy difference between them, they cannot both be assigned $\Delta_f H° =$

$$C(graphite) \longrightarrow C(diamond) \qquad \Delta_r H° = 1.9 \text{ kJ mol}^{-1}$$

We choose as the reference form the more stable form, the one with lower enthalpy. Thus, we assign $\Delta_f H°(graphite) = 0$, and $\Delta_f H°(diamond$ 1.9 kJ/mol.

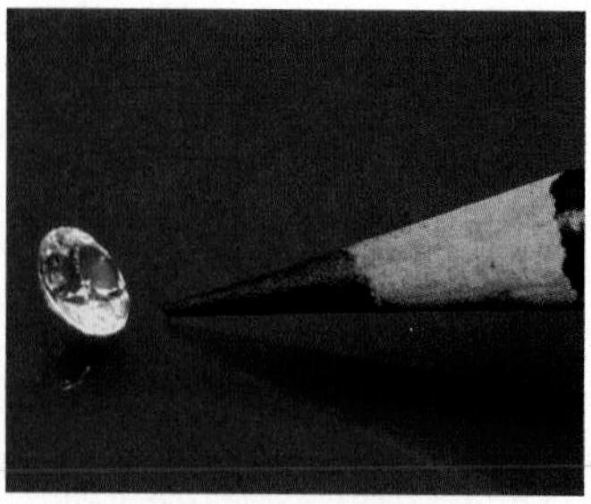
Tom Pantages

▲ Diamond and graphite.

Although we can obtain bromine in either the gaseous or liquid state at .15 K, $Br_2(l)$ is the most stable form. $Br_2(g)$, if obtained at 298.15 K and ar pressure, immediately condenses to $Br_2(l)$.

$$Br_2(l) \longrightarrow Br_2(g) \qquad \Delta_f H° = 30.91 \text{ kJ mol}^{-1}$$

e enthalpies of formation are $\Delta_f H°[Br_2(l)] = 0$ and $\Delta_f H°[Br_2(g)] =$ 91 kJ/mol.

A rare case in which the reference form is not the most stable form is the ele- nt phosphorus. Although over time it converts to solid red phosphorus, id white phosphorus has been chosen as the reference form.

$$P(s, white) \longrightarrow P(s, red) \qquad \Delta_f H° = -17.6 \text{ kJ mol}^{-1}$$

e standard enthalpies of formation are $\Delta_f H°[P(s, white)] = 0$ and $H°[P(s, red)] = -17.6$ kJ/mol.

Standard enthalpies of formation of some common substances are pre- ted in Table 7.2. Figure 7-18 emphasizes that both positive and negative ndard enthalpies of formation are possible. It also suggests that standard halpies of formation are related to molecular structure.

We will use standard enthalpies of formation in a variety of calculations. en, the first thing we must do is write the chemical equation to which a $H°$ value applies, as in Example 7-10.

▲ Liquid bromine vaporizing.

ABLE 7.2 Some Standard Molar Enthalpies of Formation, $_f H°$ at 298.15 K

ıbstance	kJ/mol[a]	Substance	kJ/mol[a]
O(g)	−110.5	HBr(g)	−36.40
$O_2(g)$	−393.5	HI(g)	26.48
$H_4(g)$	−74.81	$H_2O(g)$	−241.8
$_2H_2(g)$	226.7	$H_2O(l)$	−285.8
$_2H_4(g)$	52.26	$H_2S(g)$	−20.63
$_2H_6(g)$	−84.68	$NH_3(g)$	−46.11
$_3H_8(g)$	−103.8	NO(g)	90.25
$H_{10}(g)$	−125.6	$N_2O(g)$	82.05
$H_3OH(l)$	−238.7	$NO_2(g)$	33.18
$_2H_5OH(l)$	−277.7	$N_2O_4(g)$	9.16
F(g)	−271.1	$SO_2(g)$	−296.8
Cl(g)	−92.31	$SO_3(g)$	−395.7

'alues are for reactions in which one mole of substance is formed. Most of the data have en rounded off to four significant figures.

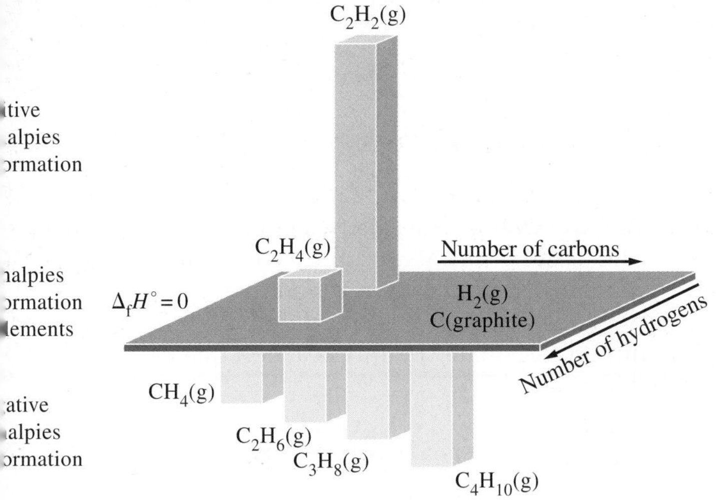

◀ FIGURE 7-18
Some standard enthalpies of formation at 298.15 K
Standard enthalpies of formation of elements are shown in the central plane, with $\Delta_f H° = 0$. Substances with positive enthalpies of formation are above the plane, while those with negative enthalpies of formation are below the plane.

EXAMPLE 7-10 Relating a Standard Enthalpy of Formation to a Chemical Equation

The enthalpy of formation of formaldehyde is $\Delta_f H° = -108.6$ kJ/mol at 298.15 K. Write the chemical equation to which this value applies.

Analyze

The equation must be written for the formation of one mole of gaseous HCHO. The most stable forms of the elements at 298.15 K and 1 bar are gaseous H_2 and O_2 and solid carbon in the form of graphite (Fig. 7-19). Note that we need one fractional coefficient in this equation.

Solve

$$H_2(g) + \frac{1}{2}O_2(g) + C(\text{graphite}) \longrightarrow HCHO(g) \qquad \Delta_f H° = -108.6 \text{ kJ mol}^{-1}$$

Assess

When answering these types of problems, we must remember to use the elements in their most stable form under the given conditions. In this example, the stated conditions were 298.15 K and 1 bar.

PRACTICE EXAMPLE A: The standard enthalpy of formation for the amino acid leucine, $C_6H_{13}O_2N(s)$, is -637.3 kJ/mol. Write the chemical equation to which this value applies.

PRACTICE EXAMPLE B: How is $\Delta_r H°$ for the following reaction related to the standard enthalpy of formation of $NH_3(g)$ listed in Table 7.2? What is the value of $\Delta_r H° = ?$

$$2\,NH_3(g) \longrightarrow N_2(g) + 3\,H_2(g) \qquad \Delta_r H° = ?$$

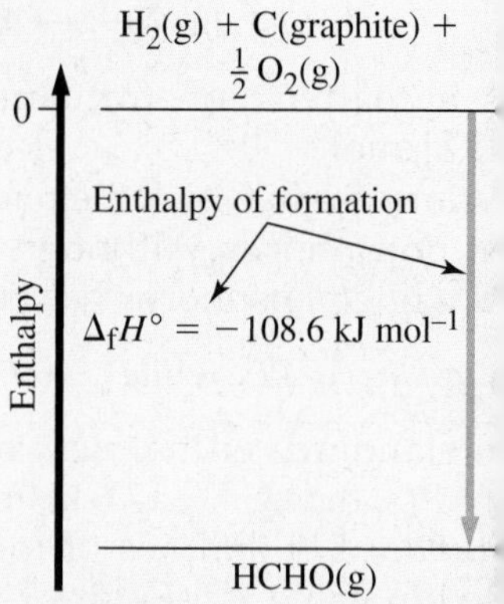

▲ FIGURE 7-19 **Standard enthalpy of formation of formaldehyde, HCHO(g)** The formation of HCHO(g) from its elements in their standard states is an exothermic reaction. The hea[t] evolved per mole of HCHO(g[)] formed is the standard enthalpy (heat) of formation.

7-2 ARE YOU WONDERING?

What is the significance of the sign of a $\Delta_f H°$ value?

A compound having a positive value of $\Delta_f H°$ is formed from its elements by a[n] endothermic reaction. If the reaction is reversed, the compound decomposes int[o] its elements in an exothermic reaction. We sometimes say that the compound [is] unstable with respect to its elements. This does not mean that the compound car[n]not be made, but it does suggest a tendency for the compound to enter into chem[m]ical reactions yielding products with lower enthalpies of formation.

When no other criteria are available, chemists sometimes use enthalp[y] change as a rough indicator of the likelihood of a chemical reaction occurring— exothermic reactions generally being more likely to occur unassisted tha[n] endothermic ones. We'll present much better criteria later in the text.

Standard Enthalpies of Reaction

We have learned that if the reactants and products of a reaction are in th[eir] standard states, the enthalpy change is a *standard* enthalpy change, which [we] can denote as $\Delta_r H°$. One of the primary uses of standard enthalpies of for[ma]tion is in calculating standard enthalpies of reaction.

Let us use Hess's law to calculate the standard enthalpy of reaction for [the] decomposition of sodium bicarbonate, a minor reaction that occurs when b[ak]ing soda is used in baking.

$$2\,NaHCO_3(s) \longrightarrow Na_2CO_3(s) + H_2O(l) + CO_2(g) \qquad \Delta_r H° = ? \qquad (7$$

n Hess's law, we see that the following four equations yield equation (7.20) en added together.

$$2\,NaHCO_3(s) \longrightarrow 2\,Na(s) + H_2(g) + 2\,C(graphite) + 3\,O_2(g) \qquad \Delta_r H^\circ = -2 \times \Delta_f H^\circ[NaHCO_3(s)]$$

$$2\,Na(s) + C(graphite) + \frac{3}{2}\,O_2(g) \longrightarrow Na_2CO_3(s) \qquad \Delta_r H^\circ = \Delta_f H^\circ[Na_2CO_3(s)]$$

$$H_2(g) + \frac{1}{2}\,O_2(g) \longrightarrow H_2O(l) \qquad \Delta_r H^\circ = \Delta_f H^\circ[H_2O(l)]$$

$$C(graphite) + O_2(g) \longrightarrow CO_2(g) \qquad \Delta_r H^\circ = \Delta_f H^\circ[CO_2(g)]$$

$$2\,NaHCO_3(s) \longrightarrow Na_2CO_3(s) + H_2O(l) + CO_2(g) \qquad \Delta_r H^\circ = ?$$

quation (a) is the *reverse* of the equation representing the formation of moles of $NaHCO_3(s)$ from its elements. This means that $\Delta_r H^\circ$ for reac- (a) is the *negative* of twice $\Delta_f H^\circ[NaHCO_3(s)]$. Equations (b), (c), and (d) resent the formation of *one* mole each of $Na_2CO_3(s)$, $CO_2(g)$, and $H_2O(l)$. ıs, we can express the value of $\Delta_r H^\circ$ for the decomposition reaction as

$$H^\circ = \Delta_f H^\circ[Na_2CO_3(s)] + \Delta_f H^\circ[H_2O(l)] + \Delta_f H^\circ[CO_2(g)] - 2 \times \Delta_f H^\circ[NaHCO_3(s)] \quad \textbf{(7.21)}$$

Ve can use the enthalpy diagram in Figure 7-20 to visualize the Hess's law cedure and to show how the state function property of enthalpy enables us to ve at equation (7.21). Imagine the decomposition of sodium bicarbonate tak- place in two steps. In the first step, suppose a vessel contains 2 mol $NaHCO_3$, ich is allowed to decompose into 2 mol Na(s), 2 mol C(graphite), 1 mol $H_2(g)$, 3 mol $O_2(g)$, as in equation (a) above. In the second step, recombine the ol Na(s), 2 mol C(graphite), 1 mol $H_2(g)$, and 3 mol $O_2(g)$ to form the prod- s according to equations (b), (c), and (d) above.

The pathway shown in Figure 7-20 *is not* how the reaction actually occurs. s does not matter, though, because enthalpy is a state function and the nge of any state function is independent of the path chosen. The enthalpy nge for the overall reaction is the sum of the standard enthalpy changes of individual steps.

$$\Delta_r H^\circ = \Delta_r H^\circ_{decomposition} + \Delta_r H^\circ_{recombination}$$

$$\Delta_r H^\circ_{decomposition} = -2 \times \Delta_f H^\circ[NaHCO_3(s)]$$

$$\Delta_r H^\circ_{recombination} = \Delta_f H^\circ[Na_2CO_3(s)] + \Delta_f H^\circ[H_2O(l)] + \Delta_f H^\circ[CO_2(g)]$$

hat

$$H^\circ = \Delta_f H^\circ[Na_2CO_3(s)] + \Delta_f H^\circ[H_2O(l)] + \Delta_f H^\circ[CO_2(g)] - 2 \times \Delta_f H^\circ[NaHCO_3(s)]$$

Equation (7.21) is a specific application of a general relationship for a stan- d enthalpy of reaction.

Consider the following general equation for a reaction involving the sub- nces A, B, C, D, and so on.

$$a\text{A} + b\text{B} + \ldots \longrightarrow c\text{C} + d\text{D} + \ldots$$

The lowercase letters *a*, *b*, *c*, *d*, and so on, represent the stoichiometric coeffi- nts required to balance the equation. The standard enthalpy of reaction, H°, is obtained using the following equation:

$$H^\circ = \underbrace{[c \times \Delta_f H^\circ_C + d \times \Delta_f H^\circ_D + \ldots]}_{\text{weighted sum of } \Delta_f H^\circ \text{ values for the products}} - \underbrace{[a \times \Delta_f H^\circ_A + b \times \Delta_f H^\circ_B + \ldots]}_{\text{weighted sum of } \Delta_f H^\circ \text{ values for the reactants}} \quad \textbf{(7.22)}$$

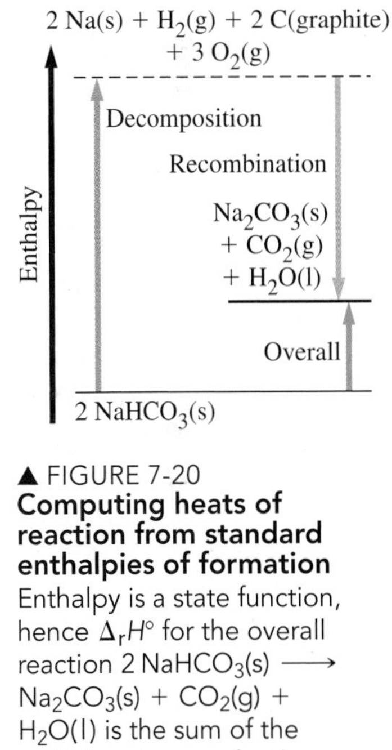

▲ FIGURE 7-20
Computing heats of reaction from standard enthalpies of formation
Enthalpy is a state function, hence $\Delta_r H^\circ$ for the overall reaction $2\,NaHCO_3(s) \longrightarrow Na_2CO_3(s) + CO_2(g) + H_2O(l)$ is the sum of the enthalpy changes for the two steps shown. The thin line represents the enthalpy for the reactants, the dashed horizontal line represents the enthalpy of the elements, and the thick, darker horizontal line represents the enthalpy of the products.

▶ FIGURE 7-21
Diagrammatic representation of equation (7.22)
The horizontal lines represent the standard enthalpies for the reactants, elements, and products.

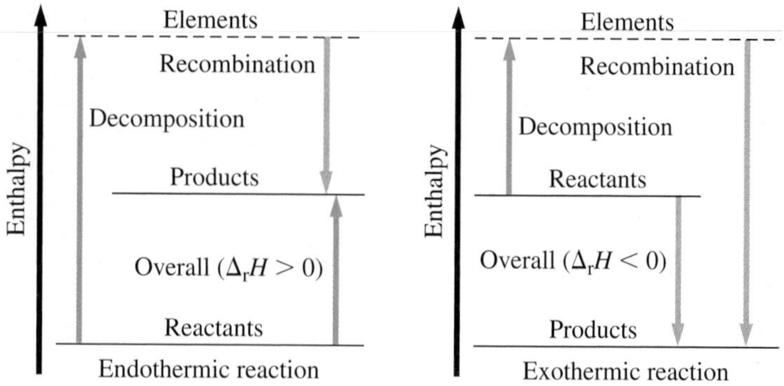

Equation (7.22) can be also be expressed as a single weighted sum by w ing it in terms of the stoichiometric numbers (ν), which were introduced i Chapter 4—see page 138.

$$\Delta_r H° = \nu_A \times \Delta_f H°_A + \nu_B \times \Delta_f H°_B + \nu_C \times \Delta_f H°_C + \nu_D \times \Delta_f H°_D + \cdot$$
$$= \sum \nu \times \Delta_f H° \quad (7$$

▶ The symbol Σ (Greek, sigma) means "the sum of."

Recall that the stoichiometric number (ν) for a given reactant or produ just the stoichiometric coefficient with a $+$ or $-$ sign included: a + sign f product and a $-$ sign for a reactant. Thus, for the generalized reaction we considering, the stoichiometric numbers are

$$\nu_A = -a \qquad \nu_B = -b \qquad \nu_C = +c \qquad \nu_D = +d$$

If we substitute these values into equation (7.23), we get equation (7. thereby proving that equation (7.23) is just another way of expressing ec tion (7.22).

In the weighted sums in the equations above for $\Delta_r H°$, the terms are forr by multiplying the standard molar enthalpies of formation by the correspc ing stoichiometric coefficients or stoichiometric numbers, both of wl are simply numbers (without units). As a result, $\Delta_r H°$ has units of kJ mo The basis for these equations is shown in Figure 7-21 and is appliec Example 7-11. At this point, you should also recognize that $\Delta_r H°$ is enthalpy change *per mole of reaction* for the following process.

pure, unmixed reactants (each in its standard state) ⟶ pure, unmixed products (each in its standard stat

The process involves the complete conversion of stoichiometric amou of pure, unmixed reactants in their standard states to stoichiometric amount pure, unmixed products in their standard states. This concept will be furt explored in Chapter 13, when we learn about other thermodynamic quantiti

EXAMPLE 7-11 Calculating $\Delta_r H°$ from Tabulated Values of $\Delta_f H°$

Let us apply equation (7.22) to calculate the standard enthalpy of combustion of ethane, $C_2H_6(g)$, a componen of natural gas.

Analyze

This type of problem is a straightforward application of equation (7.22). Appendix D has a table of thermody namic data which includes the standard enthalpy of formation for a number of compounds.

Solve

The reaction is

$$C_2H_6(g) + \frac{7}{2} O_2(g) \longrightarrow 2\,CO_2(g) + 3\,H_2O(l)$$

The relationship we need is equation (7.22). The data we substitute into the relationship are from Table 7.2.

$$\Delta_r H° = \{2 \times \Delta_f H°[CO_2(g)] + 3 \times \Delta_f H°[H_2O(l)]\}$$
$$- \{1 \times \Delta_f H°[C_2H_6(g)] + \frac{7}{2} \times \Delta_f H°[O_2(g)]\}$$
$$= 2 \times (-393.5\ kJ/mol) + 3 \times (-285.8\ kJ/mol)$$
$$- 1 \times (-84.7\ kJ/mol) - \frac{7}{2} \times (0\ kJ/mol)$$
$$= -787.0\ kJ\ mol^{-1} - 857.4\ kJ\ mol^{-1} + 84.7\ kJ\ mol^{-1} = -1559.7\ kJ\ mol^{-1}$$

Assess

In these types of problems, we must make sure to subtract the sum of the products' standard enthalpies of formation from the sum of the reactants' standard enthalpies of formation. We must also keep in mind that the standard enthalpy of formation of an element in its reference form is zero. Thus, we can drop the term involving $\Delta_f H°[O_2(g)]$ at any time in the calculation.

PRACTICE EXAMPLE A: Use data from Table 7.2 to calculate the standard enthalpy of combustion of ethanol, $C_2H_5OH(l)$, at 298.15 K.

PRACTICE EXAMPLE B: Calculate the standard enthalpy of combustion at 298.15 K *per mole* of a gaseous fuel that contains C_3H_8 and C_4H_{10} in the mole fractions 0.62 and 0.38, respectively.

A type of calculation as important as the one illustrated in Example 7-11 is determination of an unknown $\Delta_f H°$ value from a set of known $\Delta_f H°$ values and a known standard enthalpy of reaction, $\Delta_r H°$. As shown in Example 7-12, the essential step is to rearrange equation (7.22) to isolate the unknown $\Delta_f H°$ on one side of the equation. Also shown is a way of organizing data that you may find helpful.

EXAMPLE 7-12 Calculating an Unknown $\Delta_f H°$ Value

Use the data here and in Table 7.2 to calculate $\Delta_f H°$ of benzene, $C_6H_6(l)$.

$$2\,C_6H_6(l) + 15\,O_2(g) \longrightarrow 12\,CO_2(g) + 6\,H_2O(l) \qquad \Delta_r H° = -6535\ kJ\ mol^{-1}$$

Analyze

We have a chemical equation and know the standard enthalpy of reaction. We are asked to determine a standard enthalpy of formation. Equation (7.22) relates a standard enthalpy of reaction to standard enthalpy of formations for reactants and products. To begin, we organize the data needed in the calculation by writing the chemical equation for the reaction with $\Delta_f H°$ data listed under the chemical formulas.

Solve

	$2\,C_6H_6(l)$	$+ 15\,O_2(g)$	$\longrightarrow 12\,CO_2(g)$	$+ 6\,H_2O(l)$	$\Delta_r H° = -6535\ kJ\ mol^{-1}$
$\Delta_f H°$, kJ/mol	?	0	−393.5	−285.8	

Now, we can substitute known data into expression (7.22) and rearrange the equation to obtain a lone term on the left: $\Delta_f H°[C_6H_6(l)]$. The remainder of the problem simply involves numerical calculations.

$$\Delta_r H° = \{12 \times (-393.5\ kJ/mol) + 6 \times (-285.8\ kJ/mol)\} - 2 \times \Delta_f H°[C_6H_6(l)]$$
$$= -6535\ kJ\ mol^{-1}$$
$$\Delta_f H°[C_6H_6(l)] = \frac{\{-4722\ kJ\ mol^{-1} - 1715\ kJ\ mol^{-1}\} + 6535\ kJ\ mol^{-1}}{2} = 49\ kJ/mol$$

Assess

By organizing the data as shown, we were able to identify what is unknown and see how to use equation (7.22). To obtain the correct answer, we also needed to use the correct states for the compounds. In combustion reactions, the water in the product is always liquid. If we had used the standard enthalpy of formation for gaseous water, we would have obtained the wrong answer.

(continued)

PRACTICE EXAMPLE A: The overall reaction that occurs in photosynthesis in plants is

$$6\,CO_2(g) + 6\,H_2O(l) \longrightarrow C_6H_{12}O_6(s) + 6\,O_2(g) \qquad \Delta_rH° = 2803\ kJ\ mol^{-1}$$

Determine the standard enthalpy of formation of glucose, $C_6H_{12}O_6(s)$, at 298.15 K.

PRACTICE EXAMPLE B: A handbook lists the standard enthalpy of combustion of gaseous dimethyl ethe $(CH_3)_2O(g)$, at 298.15 K as −31.70 kJ/g. What is the standard molar enthalpy of formation of dimethyl ether a 298.15 K?

Reactions Involving Ions in Solution

Many chemical reactions in aqueous solution are best thought of as reacti between ions and best represented by net ionic equations. Consider the n tralization of a strong acid by a strong base. Using the enthalpy of neutral tion that we obtained in Example 7-4, we can write

$$H^+(aq) + OH^-(aq) \longrightarrow H_2O(l) \qquad \Delta_rH° = -55.8\ kJ\ mol^{-1} \qquad (7$$

We should also be able to calculate this enthalpy of neutralization by us enthalpy of formation data in expression (7.22), but this requires us have enthalpy of formation data for individual ions. And there is a sli problem in getting these. We cannot create ions of a single type in a chem reaction. We always produce cations and anions simultaneously, as in the re tion of sodium and chlorine to produce Na^+ and Cl^- in NaCl. We must cho a particular ion to which we assign an enthalpy of formation of *zero* in aqueous solutions. We then compare the enthalpies of formation of other i to this reference ion. The ion we arbitrarily choose for our zero is H^+(a Now let us see how we can use expression (7.22) and data from equation (7 to determine the enthalpy of formation of $OH^-(aq)$.

$$\Delta_rH° = 1 \times \Delta_fH°[H_2O(l)] - \{1 \times \Delta_fH°[H^+(aq)]$$
$$+ 1 \times \Delta_fH°[OH^-(aq)]\} = -55.8\ kJ\ m$$

$$\Delta_fH°[OH^-(aq)] = 55.8\ kJ\ mol^{-1} + \{1 \times \Delta_fH°[H_2O(l)]\} - \{1 \times \Delta_fH°[H^+(aq)]\}$$

$$\Delta_fH°[OH^-(aq)] = 55.8\ kJ\ mol^{-1} - 285.8\ kJ\ mol^{-1} - 0\ kJ\ mol^{-1} = -230.0\ kJ/mol$$

Table 7.3 lists data for several common ions in aqueous solution. Enthalp of formation in solution depend on the solute concentration. These data representative for *dilute* aqueous solutions (about 1 M), the type of solut that we normally deal with. Some of these data are used in Example 7-13.

TABLE 7.3 Some Standard Molar Enthalpies of Formation, $\Delta_fH°$ of Ions in Aqueous Solution at 298.15 K

Ion	kJ/mol	Ion	kJ/mo
H^+	0	OH^-	−230.
Li^+	−278.5	Cl^-	−167.
Na^+	−240.1	Br^-	−121.
K^+	−252.4	I^-	−55.
NH_4^+	−132.5	NO_3^-	−205.
Ag^+	105.6	CO_3^{2-}	−677.
Mg^{2+}	−466.9	S^{2-}	33.
Ca^{2+}	−542.8	SO_4^{2-}	−909.
Ba^{2+}	−537.6	$S_2O_3^{2-}$	−648.
Cu^{2+}	64.77	PO_4^{3-}	−1277
Al^{3+}	−531		

XAMPLE 7-13 Calculating the Enthalpy Change in an Ionic Reaction

Given that $\Delta_f H°[BaSO_4(s)] = -1473$ kJ/mol, what is the standard enthalpy change for the precipitation of barium sulfate?

nalyze

First, write the net ionic equation for the reaction and introduce the relevant data. Then make use of equation (7.22).

olve

Start by organizing the data in a table.

	$Ba^{2+}(aq)$ +	$SO_4^{2-}(aq)$ ⟶	$BaSO_4(s)$	$\Delta_r H° = ?$
$\Delta_f H°$, kJ/mol	−537.6	−909.3	−1473	

Then substitute data into equation (7.22).

$$\begin{aligned}\Delta_r H° &= 1 \times \Delta_f H°[BaSO_4(s)] - 1 \times \Delta_f H°[Ba^{2+}(aq)] - 1 \times \Delta_f H°[SO_4^{2-}(aq)] \\ &= 1 \times (-1473\ \text{kJ/mol}) - 1 \times (-537.6\ \text{kJ/mol}) - 1 \times (-909.3\ \text{kJ/mol}) \\ &= -1473\ \text{kJ mol}^{-1} + 537.6\ \text{kJ mol}^{-1} + 909.3\ \text{kJ mol}^{-1} = -26\ \text{kJ mol}^{-1}\end{aligned}$$

ssess

The standard enthalpy of reaction determined here is the heat given off by the system (i.e., the ionic reaction).

RACTICE EXAMPLE A: Given that $\Delta_f H°[AgI(s)] = -61.84$ kJ/mol, what is the standard enthalpy change for the precipitation of silver iodide?

RACTICE EXAMPLE B: The standard enthalpy change for the precipitation of $Ag_2CO_3(s)$ is −39.9 kJ per mole of $Ag_2CO_3(s)$ formed. What is $\Delta_f H°[Ag_2CO_3(s)]$?

7-10 CONCEPT ASSESSMENT

; it possible to calculate a heat of reaction at 373.15 K by using standard nthalpies of formation at 298.15 K? If so, explain how you would do this, and ndicate any additional data you might need.

9 Fuels as Sources of Energy

e of the most important uses of thermochemical measurements and calcula-ns is in assessing materials as energy sources. For the most part, these mate-ls, called fuels, liberate heat through the process of combustion. We will efly survey some common fuels, emphasizing matters that a thermochemi- background helps us to understand.

◀ Although the formula $C_m(H_2O)_n$ suggests a "hydrate" of carbon, in carbohydrates, there are no H_2O units, as there are in hydrates, such as $CuSO_4 \cdot 5\,H_2O$. H and O atoms are simply found in the same numerical ratio as in H_2O.

ssil Fuels

e bulk of current energy needs are met by petroleum, natural gas, and l—so-called fossil fuels. These fuels are derived from plant and animal life millions of years ago. The original source of the energy locked into these ls is solar energy. In the process of *photosynthesis*, CO_2 and H_2O, in the pres-ce of enzymes, the pigment chlorophyll, and sunlight, are converted to *bohydrates*. These are compounds with formulas $C_m(H_2O)_n$, where m and re positive integers. For example, in the sugar glucose $m = n = 6$, that is, $(H_2O)_6 = C_6H_{12}O_6$. Its formation through photosynthesis is an *endothermic* ocess, represented as

$$O_2(g) + 6\,H_2O(l) \xrightarrow[\text{sunlight}]{\text{chlorophyll}} C_6H_{12}O_6(s) + 6\,O_2(g) \quad \Delta_r H° = +2.8 \times 10^3\ \text{kJ mol}^{-1} \tag{7.25}$$

◀ Glucose is what we call an energy-carrying molecule. Energy-carrying molecules store energy from one source (e.g., light), and use it elsewhere (e.g., as heat). Examples of biological energy-carrying molecules are ATP and NADH.

When reaction (7.25) is reversed, as in the combustion of glucose, he evolved. The combustion reaction is *exothermic*.

The complex carbohydrate cellulose, with molecular masses ranging u 500,000 u, is the principal structural material of plants. When plant life dec poses in the presence of bacteria and out of contact with air, O and H at are removed and the approximate carbon content of the residue increase the progression

Peat ⟶ lignite (32% C) ⟶ sub-bituminous coal (40% C) ⟶
bituminous coal (60% C) ⟶ anthracite coal (80

It may take about 300 million years for this process to progress all the wa anthracite coal. Coal, then, is a combustible organic rock consisting of carl hydrogen, and oxygen, together with small quantities of nitrogen, sulfur, mineral matter (ash). (One proposed formula for a "molecule" of bitumin coal is $C_{153}H_{115}N_3O_{13}S_2$.)

Petroleum and natural gas formed in a different way. The remains of pl and animals living in ancient seas fell to the ocean floor, where they w decomposed by bacteria and covered with sand and mud. Over time, the s and mud were converted to sandstone by the weight of overlying layers of s and mud. The high pressures and temperatures resulting from this overly sandstone rock formation transformed the original organic matter into pe leum and natural gas. The ages of these deposits range from about 250 millio 500 million years.

A typical natural gas consists of about 85% methane (CH_4), 10% eth (C_2H_6), 3% propane (C_3H_8), and small quantities of other combustible noncombustible gases. A typical petroleum consists of several hundred dif ent hydrocarbons that range in complexity from C_1 molecules (CH_4) to C_4 higher (such as $C_{40}H_{82}$).

One way to compare different fuels is through their heats of combusti In general, *the higher the heat of combustion, the better the fuel*. Table 7.4 l approximate heats of combustion for the fossil fuels. These data show t *biomass* (living matter or materials derived from it—wood, alcoh municipal waste) is a viable fuel, but that fossil fuels yield more energy unit mass.

TABLE 7.4 Approximate Heats of Combustion of Some Fuels

Fuel	Heat of Combustion kJ/g
Municipal waste	−12.7
Cellulose	−17.5
Pinewood	−21.2
Methanol	−22.7
Peat	−20.8
Bituminous coal	−28.3
Isooctane (a component of gasoline)	−47.8
Natural gas	−49.5

Problems Posed by Fossil Fuel Use There are two fundamental proble with the use of fossil fuels. First, fossil fuels are essentially *nonrenewable* ene sources. The world consumption of fossil fuels is expected to increase for foreseeable future (Fig. 7-22), but when will Earth's supply of these fuels out? There is currently a debate about whether oil production has peaked n and is about to decline, or whether it will peak more toward the middle of this century. The second problem with fossil fuels is their environmental eff Sulfur impurities in fuels produce oxides of sulfur. The high temperatures as ciated with combustion cause the reaction of N_2 and O_2 in air to form oxide nitrogen. Oxides of sulfur and nitrogen are implicated in air pollution and important contributors to the environmental problem known as acid ra Another inevitable product of the combustion of fossil fuels is carbon dioxi one of the "greenhouse" gases leading to *global warming* and potential chan in Earth's climate.

▶ Environmental issues associated with oxides of sulfur and nitrogen are discussed more fully in later chapters.

Global Warming—An Environmental Issue Involving Carbon Dioxide do not normally think of CO_2 as an air pollutant because it is essentially n toxic and is a natural and necessary component of air. Its ultimate effect on environment, however, could be very significant. A buildup of $CO_2(g)$ in atmosphere may disturb the energy balance on Earth.

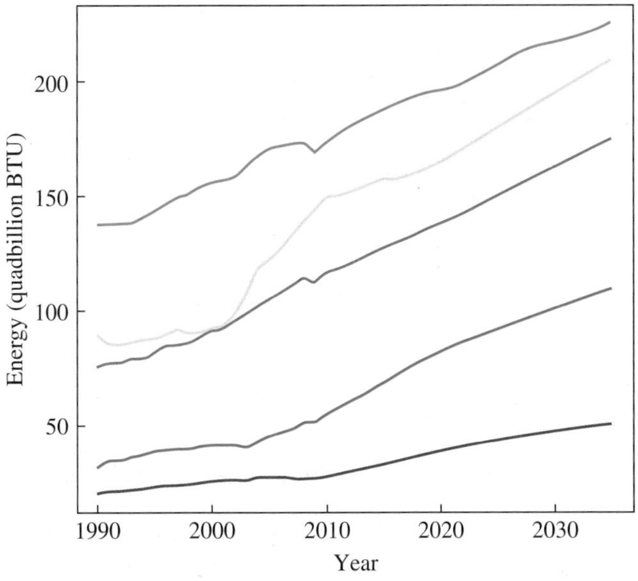

FIGURE 7-22
rld primary energy consumption by energy source
se graphs show the history of energy consumption since 1990, with predictions to 35. Petroleum (blue line) is seen to be the major source of energy for the eseeable future, followed by coal (yellow) and natural gas (pink), which are about the ne. Other sources of energy included are renewable power (green) and nuclear wer (purple). The unit Btu is a measure of energy and stands for British thermal unit e Exercise 95). [Source: U.S. Energy Information Administration International Energy tlook 2011 DOE/EIA-0484(2011).]

Earth's atmosphere, discussed in the Focus On feature for Chapter 6 on the steringChemistry website, is largely transparent to visible and UV radia-n from the sun. This radiation is absorbed at Earth's surface, which is rmed by it. Some of this absorbed energy is reradiated as infrared radia-n. Certain atmospheric gases, primarily CO_2, methane, and water vapor, sorb some of this infrared radiation, and the energy thus retained in the nosphere produces a warming effect. This process, outlined in Figure 7-23, often compared to the retention of thermal energy in a greenhouse and is led the "greenhouse effect."* The natural greenhouse effect is essential in intaining the proper temperature for life on Earth. Without it, Earth would permanently covered with ice.

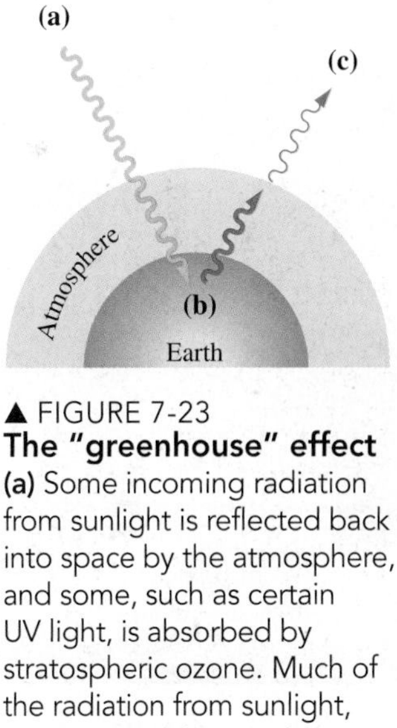

▲ FIGURE 7-23
The "greenhouse" effect
(a) Some incoming radiation from sunlight is reflected back into space by the atmosphere, and some, such as certain UV light, is absorbed by stratospheric ozone. Much of the radiation from sunlight, however, reaches Earth's surface. **(b)** Earth's surface re-emits some of this energy as infrared radiation. **(c)** Some of the infrared radiation leaving the Earth's surface is absorbed by CO_2 and other greenhouse gases and is redirected back towards the Earth's surface. The redirected infrared radiation warms the atmosphere.

Over the past 400,000 years, the atmospheric carbon dioxide concentration s varied from 180 to 300 parts per million with the preindustrial-age con-ntration at about 285 ppm. By 2011, the level had increased to about) ppm and is still rising (Fig. 7-24). Increasing atmospheric carbon dioxide ncentrations result from the burning of carbon-containing fuels such as od, coal, natural gas, and gasoline and from the deforestation of tropical gions (plants, through photosynthesis, consume CO_2 from the atmos-ere). The expected effect of a CO_2 buildup is an increase in Earth's average nperature, a **global warming**. Some estimates are that a doubling of the $)_2$ concentration over that of preindustrial times could occur before the end the present century and that this doubling could produce an average bal temperature increase of 1.5 to 4.5 °C.

Predicting the probable effects of a CO_2 buildup in the atmosphere is done gely through computer models, and it is very difficult to know all the factors

*lass, like CO_2, is transparent to visible and some UV light but absorbs infrared radiation. glass in a greenhouse, though, acts primarily to prevent the bulk flow of warm air out of the enhouse.

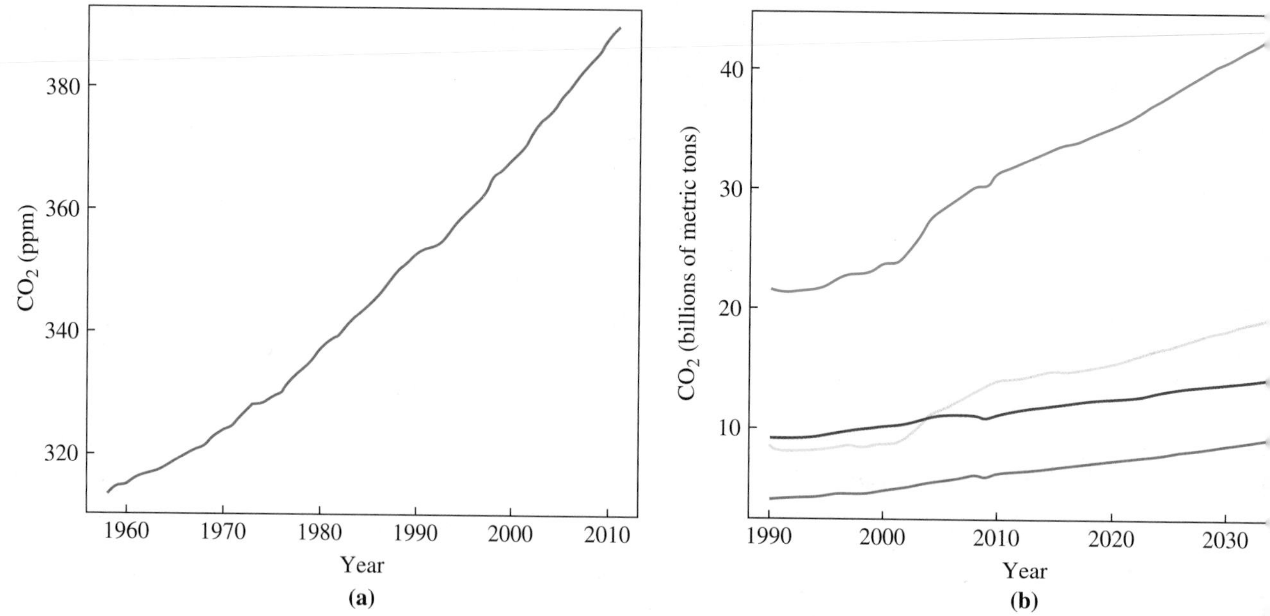

▲ FIGURE 7-24
Increasing carbon dioxide concentration of the atmosphere
(a) The global average atmospheric carbon dioxide level over a 53-year span, express in parts per million by volume, as measured by a worldwide cooperative sampling network. **(b)** The actual and predicted CO_2 emissions for a 45-year span due to the combustion of natural gas (pink line), coal (yellow), and petroleum (dark blue), togeth with the total of all CO_2 emissions (light blue). The CO_2 concentration of the atmosph continues to increase, from approximately 375 ppm in 2003 to 390 ppm in 2011.

that should be included in these models and the relative importance of th factors. For example, global warming could lead to the increased evaporat of water and increased cloud formation. In turn, an increased cloud co could reduce the amount of solar radiation reaching Earth's surface and some extent, offset global warming.

Some of the significant possible effects of global warming are

- local temperature changes. The average annual temperature for Ala and Northern Canada has increased by 1.9 °C over the past 50 ye Alaskan winter temperatures have increased by an average of 3.5 over this same time period.
- a rise in sea level caused by the thermal expansion of seawater a increased melting of continental ice caps. A potential increase in sea le of up to 1 m by 2100 would displace tens of millions of inhabitants Bangladesh alone.
- the migration of plant and animal species. Vegetation now characteristi certain areas of the globe could migrate into regions several hundred k meters closer to the poles. The areas in which diseases, such as malaria, endemic could also expand.

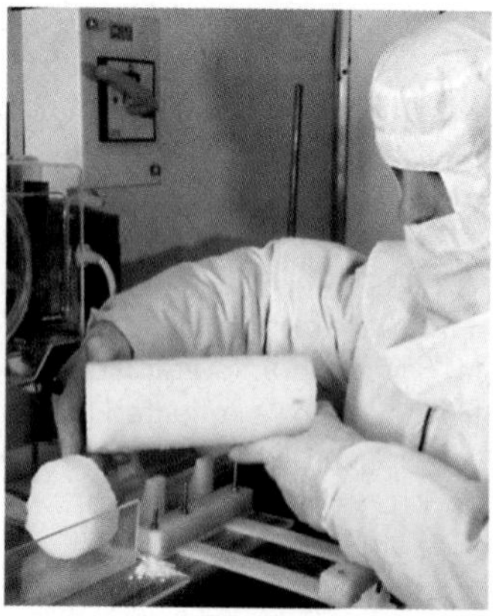

Simon Fraser/Science Photo Library

▲ An ice core from the ice sheet in Antarctica is cut into sections in a refrigerated clean room. The ice core is then analyzed to determine the amount and type of trapped gases and trace elements it contains. These data provide information regarding past changes in climate and current trends in the pollution of the atmosphere.

Although some of the current thinking involves speculation, a growing body evidence supports the likelihood of global warming, which contributes sign cantly to climate change. For example, analyses of tiny air bubbles trapped in Antarctic ice cap show a strong correlation between the atmospheric CO_2 c centration and temperature for the past 160,000 years—low temperatures dur periods of low CO_2 levels and higher temperatures with higher levels of CO_2

CO_2 is not the only greenhouse gas. Several gases are even stronger infra absorbers—specifically, methane (CH_4), ozone (O_3), nitrous oxide (N_2O), a chlorofluorocarbons (CFCs). Furthermore, atmospheric concentrations some of these gases have been growing at a faster rate than that of CO_2.

ategies beyond curtailing the use of chlorofluorocarbons and fossil fuels ve emerged for countering a possible global warming. Like several other jor environmental issues, some aspects of climate change are not well lerstood, and research, debate, and action are all likely to occur simultane- sly for a long time to come.

al and Other Energy Sources

he United States, reserves of coal far exceed those of petroleum and natural s. Despite this relative abundance, however, the use of coal has not reased significantly in recent years. In addition to the environmental effects d above, the expense and hazards involved in the deep mining of coal are nsiderable. Surface mining, which is less hazardous and expensive than ep mining, is also more damaging to the environment. One promising pos- ility for using coal reserves is to convert coal to gaseous or liquid fuels, ner in surface installations or while the coal is still underground.

sification of Coal Before cheap natural gas became available in the 1940s, s produced from coal (variously called producer gas, town gas, or city gas) s widely used in the United States. This gas was manufactured by passing am and air through heated coal and involved such reactions as

$$\text{(graphite)} + H_2O(g) \longrightarrow CO(g) + H_2(g) \qquad \Delta_r H° = +131.3 \text{ kJ mol}^{-1} \qquad (7.26)$$

$$CO(g) + H_2O(g) \longrightarrow CO_2(g) + H_2(g) \qquad \Delta_r H° = -41.2 \text{ kJ mol}^{-1} \qquad (7.27)$$

$$C(\text{graphite}) + O_2(g) \longrightarrow 2\,CO(g) \qquad \Delta_r H° = -221.0 \text{ kJ mol}^{-1} \qquad (7.28)$$

$$\text{(graphite)} + 2\,H_2(g) \longrightarrow CH_4(g) \qquad \Delta_r H° = -74.8 \text{ kJ mol}^{-1} \qquad (7.29)$$

e principal gasification reaction (7.26) is highly endothermic. The heat uirements for this reaction are met by the carefully controlled partial burn- of coal (reaction 7.28).

A typical producer gas consists of about 23% CO, 18% H_2, 8% CO_2, and CH_4 by volume. It also contains about 50% N_2 because air is used in its oduction. Because the N_2 and CO_2 are noncombustible, producer gas has ly about 10% to 15% of the heat value of natural gas. Modern gasification ocesses include several features:

. They use $O_2(g)$ instead of air, thereby eliminating $N_2(g)$ in the product.
. They provide for the removal of noncombustible $CO_2(g)$ and sulfur impu- rities. For example,

$$CaO(s) + CO_2(g) \longrightarrow CaCO_3(s)$$
$$2\,H_2S(g) + SO_2(g) \longrightarrow 3\,S(s) + 2\,H_2O(g)$$

. They include a step (called *methanation*) to convert CO and H_2, in the pres- ence of a catalyst, to CH_4.

$$CO(g) + 3\,H_2(g) \xrightarrow{\text{catalyst}} CH_4(g) + H_2O(l)$$

e product is called *substitute natural gas* (SNG), a gaseous mixture with com- sition and heat value similar to that of natural gas.

quefaction of Coal The first step in obtaining liquid fuels from coal gener- y involves gasification of coal, as in reaction (7.26). This step is followed by talytic reactions in which liquid hydrocarbons are formed.

$$n\,CO + (2n + 1)H_2 \longrightarrow C_nH_{2n+2} + n\,H_2O$$

still another process, liquid methanol is formed.

$$CO(g) + 2\,H_2(g) \longrightarrow CH_3OH(l) \qquad (7.30)$$

In 1942, about 32 million gallons of aviation fuel were made from coa Germany. In South Africa, the Sasol process for coal liquefaction has be major source of gasoline and a variety of other petroleum products and ch icals for more than 50 years.

Methanol

Methanol, CH_3OH, can be obtained from coal by reaction (7.30). It can als produced by thermal decomposition (pyrolysis) of wood, manure, sewag municipal waste. The heat of combustion of methanol is only about one- that of a typical gasoline on a mass basis, but methanol has a high oct number (106) compared with 100 for the gasoline hydrocarbon isooctane about 92 for premium gasoline. Methanol has been tested and used as a in internal combustion engines and is cleaner burning than gasol Methanol can also be used for space heating, electric power generation, cells, and as a reactant to make a variety of other organic compounds.

Ethanol

Ethanol, CH_3CH_2OH, is produced mostly from ethylene, C_2H_4, which in t is derived from petroleum. Current interest centers on the productior ethanol by the fermentation of organic matter, a process known through recorded history. Ethanol production by fermentation is probably m advanced in Brazil, where sugarcane and cassava (manioc) are the plant n ter (biomass) used. In the United States, corn-based ethanol is used chiefl an additive to gasoline to improve its octane rating and reduce air polluti Also, a 90% gasoline–10% ethanol mixture is used as an automotive f under the name *gasohol*.

Biofuels

Biofuels are renewable energy sources that are similar to fossil fuels. Biofu are fuels derived from dead biological material, most commonly plants. Fo fuels are derived from biological material that has been dead for a very lo time. The use of biofuels is not new; several car inventors had envisioned th vehicles running on such fuels as peanut oil, hemp-derived fuel, and etha Bioethanol is derived from the fermentation of carbohydrates found in m sugar or starch crops, such as sugarcane or corn, respectively. Reacting v etable oil with a base–alcohol mixture produces a compound commonly cal a biodiesel. A typical petro–diesel compound is the hydrocarbon ceta ($C_{16}H_{34}$), and the typical biodiesel compound contains oxygen atoms, as ill trated in the figure below. The standard enthalpies of combustion of petro–diesel and the biodiesel are very similar.

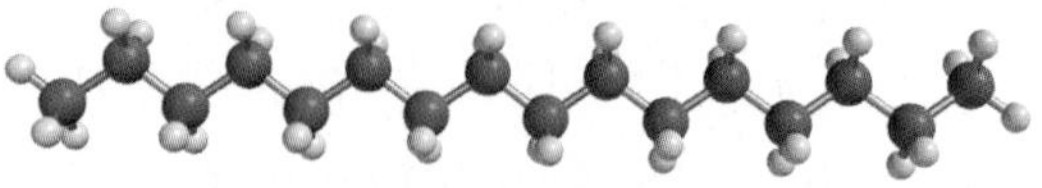

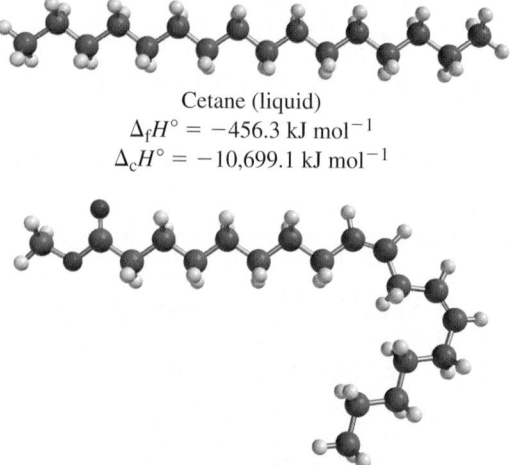

Cetane (liquid)
$\Delta_f H° = -456.3 \text{ kJ mol}^{-1}$
$\Delta_c H° = -10{,}699.1 \text{ kJ mol}^{-1}$

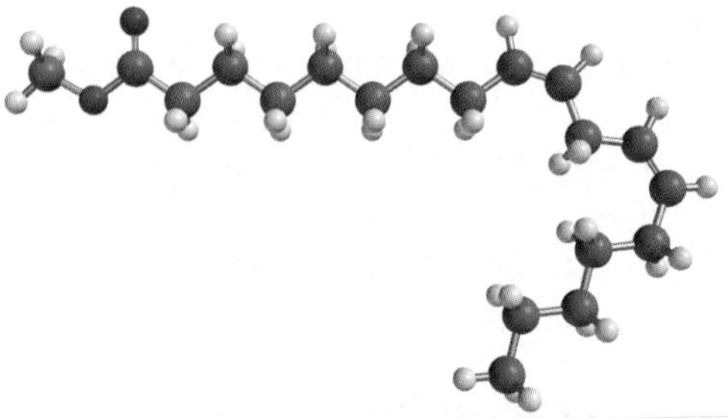

Linoleic acid methyl ester (liquid)
$\Delta_f H° = -604.88 \text{ kJ mol}^{-1}$
$\Delta_c H° = -11{,}690.1 \text{ kJ mol}^{-1}$

Although biofuels are appealing replacements for fossil fuels, their wide-
read adoption has several potential drawbacks. One major concern is the
d-versus-fuel issue. Typical plants used for food (e.g., sugarcane) are
irces of biofuels, which drives up the cost of food. A positive aspect of biofu-
is that they are carbon neutral; that is, the $CO_2(g)$ produced by the burning
a biofuel is then used by plants for new growth, resulting in no net gain of
bon in the atmosphere. Biofuels and their use have many other advantages
d disadvantages. Importantly, chemical knowledge of these compounds is
eded to address these issues.

drogen

other fuel with great potential is hydrogen. Its most attractive features are that

- on a per gram basis, its heat of combustion is more than twice that of methane and about three times that of gasoline;
- the product of its combustion is H_2O, not CO and CO_2 as with gasoline.

rrently, the bulk of hydrogen used commercially is made from petroleum
d natural gas, but for hydrogen to be a significant fuel of the future, efficient
thods must be perfected for obtaining hydrogen from other sources, espe-
lly water. Alternative methods of producing hydrogen and the prospects of
eloping an economy based on hydrogen are discussed later in the text.

ternative Energy Sources

mbustion reactions are only one means of extracting useful energy from
terials. We have also seen two other alternative energy sources, bioethanol
d biodiesel. Yet another alternative is to carry out reactions that yield the
ne products as combustion reactions in electrochemical cells called *fuel cells*.
e energy is released as electricity rather than as heat (see Section 19-5). Solar
rgy can be used directly, without recourse to photosynthesis. Nuclear
cesses can be used in place of chemical reactions (Chapter 25). Other alter-
ive sources in various stages of development and use include hydroelectric
rgy, geothermal energy, and tidal and wind power.

10 Spontaneous and Nonspontaneous Processes: An Introduction

applying the concepts introduced in this chapter, we can quantify the
rgy changes involved in a variety of physical and chemical processes.
rgy changes are certainly important. When thinking of the feasibility of a
cess or a reaction, designing a new synthesis, or comparing sources of
rgy, a chemist is aided by knowing or being able to calculate the enthalpy
nge(s) involved because, as established in this chapter, the enthalpy
nge for a process or a reaction at constant pressure represents the heat
rmal energy) that we must supply or can expect to obtain.

An interesting aspect of all physical and chemical changes that we have not
emphasized is that they all occur naturally or *spontaneously* in one direc-
n. For example, a shiny iron (Fe) pipe will eventually acquire a coating of
t (Fe_2O_3). An equation representing the formation of rust is

$$4\,Fe(s) + 3\,O_2(g) \longrightarrow 2\,Fe_2O_3(s)$$

Once the rust begins to form, the formation of rust continues without any
litional action or external influence. The reverse process, the conversion of
rust coating to give iron metal and oxygen, does not happen naturally or
ntaneously under ambient conditions. Before we consider possible expla-
ions for these observations, we should explore the scientific meaning of the
ns *spontaneous* and *nonspontaneous*.

A **spontaneous process** is a process that occurs in a system left to it once started, no action from outside the system (no external action) is ne sary to make the process continue. Conversely, a **nonspontaneous pro** will not occur *unless* some external action is continuously applied.

Some of our everyday experiences might compel us to believe that there natural tendency for systems to proceed toward states of lower energy. example, a tightly wound spring naturally unwinds to release its stc energy, a ball rolls downhill, and water flows to a lower level. A common ture of these processes is that the potential energy of the system decrease the energy change associated with a process (or the enthalpy change f constant-pressure process) the important consideration for establisł whether the process is spontaneous or nonspontaneous?

In the 1870s, Pierre Marcellin Berthelot and Julius Thomsen independe proposed that the direction of spontaneous change is the direction in wl the enthalpy of a system decreases. In a system in which enthalpy decrea heat is given off by the system to the surroundings. Berthelot and Thor developed the hypothesis that exothermic reactions should be spontane Let us return to the example of the rusting of iron, and a few other examp to test this hypothesis.

The spontaneous rusting of iron supports the hypothesis because, a turns out, the following reaction is exothermic:

$$4\,Fe(s) + 3\,O_2(g) \longrightarrow 2\,Fe_2O_3(s) \qquad \Delta_rH^\circ = -1648.4 \text{ kJ mol}^{-1}$$

Although the rusting of iron occurs slowly, it does so continuously. A result, the amount of iron decreases and the amount of rust increases un final state of equilibrium is reached in which essentially all the iron has b converted to iron(III) oxide. The reverse situation, the extraction of pure from iron(III) oxide, is not impossible, but it is certainly nonspontaneous fact, this nonspontaneous reverse process is involved in the manufactur iron from iron ore.

Now consider the following reaction, which we introduced on page 253:

$$Ba(OH)_2 \cdot 8\,H_2O(s) + 2\,NH_4Cl(s) \longrightarrow BaCl_2(s) + 10\,H_2O(l) + 2NH_3(g)$$

$$\Delta_rH^\circ = +165 \text{ kJ m}$$

This reaction is endothermic, yet it is *spontaneous* as written, which is a c tradiction to the Berthelot–Thomsen hypothesis.

Let us now briefly consider the melting of ice above 0 °C and the freezin water below 0 °C, examples that reinforce the notion that spontane processes can be exothermic or endothermic. The melting of ice is endotl mic yet spontaneous at temperatures above 0 °C.

$$H_2O(s) \longrightarrow H_2O(l) \qquad \Delta_rH^\circ = 6.010 \text{ kJ mol}^{-1}$$

The reverse process, freezing, is exothermic and spontaneous below 0 °C

$$H_2O(l) \longrightarrow H_2O(s) \qquad \Delta_rH^\circ = -6.010 \text{ kJ mol}^{-1}$$

Clearly, the enthalpy change for a process is not a reliable criterion for p dicting the direction of spontaneous change. Still, we have established the lowing important points:

- **If a process is spontaneous, the reverse process is nonspontaneous.** example, the rusting of an iron pipe is spontaneous under ambient c ditions, whereas the conversion of rust to iron and oxygen is not. At te peratures above 0 °C, the melting of ice is spontaneous but the freez of water is not.
- **Both spontaneous and nonspontaneous processes are possible, only spontaneous processes will occur *without intervent*** Nonspontaneous processes require the system to be acted on by an ex nal agent.

- **Some spontaneous processes occur very slowly, and others occur rather rapidly.** For example, the melting of an ice cube that has been dropped into cold water is a spontaneous process that occurs slowly. The melting of an ice cube that has been dropped into hot water is a spontaneous process that occurs rapidly. The main point is that spontaneous does not mean fast.
- **Some spontaneous processes are exothermic, and others are endothermic.** The formation of rust and the freezing of water (at temperatures below 0 °C) are examples of exothermic processes that occur spontaneously. The reaction of $Ba(OH)_2 \cdot 8\,H_2O(s)$ and $2\,NH_4Cl(s)$ and the melting of ice (at temperatures above 0 °C) are examples of endothermic processes that occur spontaneously.

\s we will see in Chapter 13, the enthalpy change for a process is one thing we need to ›w to be able to decide whether the process will occur spontaneously, that is, on its own hout any external influence. In that chapter, we will see that the criterion for the direc- 1 of spontaneous change is based not only on energy or enthalpy changes but also on nges in another system property called *entropy*. Simply stated, entropy measures the dis- sal of energy and, more specifically, the number of ways a given quantity of energy can dispersed, or distributed, among the particles of the system. The concepts discussed in chapter (for example, describing and quantifying the energy changes occurring in a ious processes) will help us understand and use entropy changes for predicting the ection of spontaneous change.

lasteringCHEMISTRY www.masteringchemistry.com

Have you ever considered where you get the energy to carry out your daily activities? The Focus On eature for Chapter 7 on the MasteringChemistry site, entitled Fats, Carbohydrates, and Energy torage, provides some insight into the body's energy storage system.

Summary

Getting Started: Some Terminology—The ect of a thermochemical study is broken down into the **em** of interest and the portions of the universe with ch the system may interact, the **surroundings**. An **n system** can exchange both energy and matter with urroundings. A **closed system** can exchange only gy and not matter. An **isolated system** can exchange her energy nor matter with its surroundings (Fig. 7-1). **rgy** is the capacity to do work, and **work** is performed n a force acts through a distance. Energy can be her characterized (Fig. 7-2) as **kinetic energy** (energy ciated with matter in motion) or **potential energy** rgy resulting from the position or composition of ter). Kinetic energy associated with random molecular ion is sometimes called **thermal energy**.

Heat—**Heat** is energy transferred between a system its surroundings as a result of a temperature difference veen the two. In some cases, heat can be transferred at stant temperature, as in a change in state of matter in system. A quantity of heat is the product of the heat acity of the system and the temperature change (equa- 7.4). In turn, **heat capacity** is the product of mass and **cific heat capacity**, the amount of heat required to ige the temperature of one gram of substance by one ree Celsius. Alternatively, heat capacity is the product nount (in moles) and **molar heat capacity**, the quantity eat required to increase the temperature of one mole of substance by 1 °C. Historically, the unit for measuring heat has been the **calorie** (**cal**), but the SI unit of heat is the joule, the same as for other forms of energy (equation 7.2). Energy transfers between a system and its surroundings must conform to the **law of conservation of energy**, meaning that all heat lost by a system is gained by its surroundings (equation 7.6).

7-3 Heats of Reaction and Calorimetry—In a chemical reaction, a change in the **chemical energy** associated with the reactants and products may appear as heat. The **heat of reaction** is the quantity of heat exchanged between a system and its surroundings when the reaction occurs at a constant temperature. In an **exothermic reaction**, heat is given off by the system; in an **endothermic reaction** the system absorbs heat. Heats of reaction are determined in a **calorimeter**, a device for measuring quantities of heat (equation 7.10). Exothermic combustion reactions are usually studied in a **bomb calorimeter** (Fig. 7-5). A common type of calorimeter used in the general chemistry laboratory is constructed from ordinary Styrofoam cups (Fig. 7-6).

7-4 Work—In some reactions an energy transfer between a system and its surroundings occurs as work. This is commonly the work involved in the expansion or compression of gases (Fig. 7-8) and is called **pressure–volume work** (equation 7.11).

7-5 The First Law of Thermodynamics—**Internal energy** (U) is the total energy (both kinetic and potential) in a system. The **first law of thermodynamics** relates changes in the internal energy of a system (ΔU) to the quantities of heat (q) and work (w) exchanged between the system and its surroundings. The relationship is $\Delta U = q + w$ (equation 7.12) and requires that a set of sign conventions be consistently followed. A **function of state (state function)** has a value that depends only on the exact condition or state in which a system is found and not on how that state was reached. Internal energy is a state function. A path-dependent function, such as heat or work, depends on how a change in a system is achieved. A change that is accomplished through an infinite number of infinitesimal steps is a **reversible process** (Fig. 7-12), whereas a change accomplished in one step or a finite series of steps is **irreversible**.

7-6 Application of the First Law to Chemical and Physical Changes—With work limited to pressure–volume work, the heat transferred in a constant-volume process is equal to the internal energy change (equation 7.13). For constant-pressure processes, a more useful function is **enthalpy** (H), defined as the internal energy (U) of a system plus the pressure–volume product (PV). The heat transferred in a constant pressure process is equal to the enthalpy change (ΔH) of the system (equation 7.14). Most often, the heat of a reaction is reported as a $\Delta_r H$ value, the enthalpy change per mole of reaction. A substance under a pressure of 1 bar (10^5 Pa) and at the temperature of interest is said to be in its **standard state**. If the reactants and products of a reaction are in their standard states, the enthalpy change in a reaction is called the **standard enthalpy of reaction** and designated as $\Delta_r H°$. Enthalpy changes can be represented schematically through **enthalpy diagrams** (Fig. 7-15).

7-7 Indirect Determination of $\Delta_r H$: Hess's La
Often an unknown $\Delta_r H$ value can be established i rectly through **Hess's law**, which states that an ov enthalpy change is the sum of the enthalpy changes o individual steps leading to the overall process (Fig. 7-

7-8 Standard Enthalpies of Formation—By trarily assigning an enthalpy of zero to the reference f of the elements in their standard states, the enthalpy ch in the formation of a compound from its elements bec a **standard enthalpy of formation** ($\Delta_f H°$). Using tabu standard enthalpies of formation (Table 7.2), it is possi calculate standard enthalpies of reactions without havi perform additional experiments (equation 7.22).

7-9 Fuels as Sources of Energy—One of the applications of thermochemistry is in the study o combustion of fuels as energy sources. Currently, the p cipal fuels are the fossil fuels, but potential altern fuels are also mentioned in this chapter and discuss more depth later in the text. One of the problems wit use of fossil fuels is the potential for **global warming.**

7-10 Spontaneous and Nonspontane Processes: An Introduction—A process that pro without external intervention is said to be a **spontan process**. A **nonspontaneous process** cannot occur wit external intervention. If a process is spontaneous in direction, then it is nonspontaneous in the reverse dire Some spontaneous processes are exothermic, and other endothermic, so the criterion for spontaneous change not be based on enthalpy changes alone. The directi spontaneous change involves changes in another pro called entropy. Entropy provides a measure of the nu of ways a given quantity of energy can be dispersed, or tributed, among the particles of the system.

Integrative Example

When charcoal is burned in a limited supply of oxygen in the presence of H_2O, a mixture of CO, H_2, and other nonc bustible gases (mostly CO_2) is obtained. Such a mixture is called *synthesis gas*. This gas can be used to synthesize org compounds, or it can be burned as a fuel. A typical synthesis gas consists of 55.0% CO(g), 33.0% H_2(g), and 1 noncombustible gases (mostly CO_2), *by volume*. To what temperature can 25.0 kg water at 25.0 °C be heated with the liberated by the combustion of 0.205 m^3 of this typical synthesis gas, measured at 25.0 °C and 102.6 kPa press

Analyze

First, use the ideal gas equation to calculate the total number of moles of gas, and then use equation (6.17) to establis number of moles of each combustible gas. Next, write an equation for the combustion of each gas. Use these equat and enthalpy of formation data to calculate the total amount of heat released by the combustion. Finally, use equa (7.5) to calculate the temperature increase when this quantity of heat is absorbed by the 25.0 kg of water. The final w temperature is then easily established.

Solve

Substitute the applicable data into the ideal gas equation using SI units, with $R = 8.3145$ Pa m^3 mol^{-1} K^{-1}. Solve for n.

$$n = \frac{PV}{RT} = \frac{102.6\ \text{kPa} \times 1000\ \text{Pa}/1\ \text{kPa} \times 0.205\ \text{m}^3}{8.3145\ \text{Pa m}^3\ \text{mol}^{-1}\ \text{K}^{-1} \times 298.2\ \text{K}}$$
$$= 8.48\ \text{mol gas}$$

Now, calculate the amounts of CO and H_2, converting the volume percents to mole fractions and using equation (6.17).

$$n_{CO} = n_{tot} \times x_{CO} = 8.48\ \text{mol} \times 0.550 = 4.66\ \text{mol CO}$$
$$n_{H_2} = n_{tot} \times x_{H_2} = 8.48\ \text{mol} \times 0.330 = 2.80\ \text{mol } H_2$$

e an equation for the combus- of CO(g), list $\Delta_f H°$ data eath the equation, and deter- e $\Delta_c H°$ per mole of CO(g).

$$\mathrm{CO(g)} + \frac{1}{2}\mathrm{O_2(g)} \longrightarrow \mathrm{CO_2(g)}$$

$$\Delta_f H°: \quad -110.5\ \text{kJ/mol} \qquad 0\ \text{kJ/mol} \qquad -393.5\ \text{kJ/mol}$$

$$\Delta_c H° = 1 \times (-393.5\ \text{kJ/mol}) - 1 \times (-110.5\ \text{kJ/mol}) = -283.0\ \text{kJ mol}^{-1}$$

e another equation for the bustion of $H_2(g)$ again listing ° data beneath the equation, determining $\Delta_c H°$ per mole 2(g).

$$\mathrm{H_2(g)} + \frac{1}{2}\mathrm{O_2(g)} \longrightarrow \mathrm{H_2O(l)}$$

$$\Delta_f H°: \quad 0\ \text{kJ/mol} \qquad 0\ \text{kJ/mol} \qquad -285.8\ \text{kJ/mol}$$

$$\Delta_c H° = 1 \times (-285.8\ \text{kJ/mol}) = -285.8\ \text{kJ mol}^{-1}$$

rmine the total heat released in combustion of the amounts of nd H_2 in the 0.205 m^3 of gas.

$$4.66\ \text{mol} \times (-283.0\ \text{kJ/mol}) + 2.80\ \text{mol} \times (-285.8\ \text{kJ/mol}) = -2.12 \times 10^3\ \text{kJ}$$

quantity of heat absorbed by 25.0 kg of water is

$$q_{\text{water}} = -q_{\text{comb}} = -\left(-2.12 \times 10^3\ \text{kJ} \times \frac{1000\ \text{J}}{1\ \text{kJ}}\right) = 2.12 \times 10^6\ \text{J}$$

range equation (7.5) to solve he temperature change in the $\times 10^4$ g (25.0 kg) of water.

$$\Delta T = \frac{q_{\text{water}}}{m_{\text{water}} \times c_{p,\text{water}}}$$

$$\Delta T = \frac{2.12 \times 10^6\ \text{J}}{\left(2.50 \times 10^4\ \text{g} \times 4.18\ \frac{\text{J}}{\text{g °C}}\right)} = 20.3\ \text{°C}$$

n the initial temperature and temperature change, deter- e the final temperature.

$$T_f = T_i + \Delta T = 25.0\ \text{°C} + 20.3\ \text{°C} = 45.3\ \text{°C}$$

ess

assumption that the gas sample obeys the ideal gas law is probably valid since the temperature of the gas 0 °C) is not particularly low and the gas pressure, about 1 atm, is not particularly high. However, the implicit mption that all the heat of combustion could be transferred to the water was probably not valid. If the transfer e to occur in an ordinary gas-fired water heater, some of the heat would undoubtedly be lost through the exhaust t. Thus, our calculation was of the highest temperature that could possibly be attained. Note that in using the l gas equation the simplest approach was to work with SI units because those were the units of the data that were n.

CTICE EXAMPLE A: The enthalpy of combustion for hexadec-1-ene, $C_{16}H_{32}$, is $-10{,}539.0\ \text{kJ mol}^{-1}$, and of hexadecane, $C_{16}H_{34}$, is $-10{,}699.1\ \text{kJ mol}^{-1}$. What is the enthalpy of hydrogenation of hexadec-1-ene to adecane?

CTICE EXAMPLE B: A chemist mixes 56 g CaO, powdered lime, with 0.10 L of water at 25 °C. After the comple- of the reaction, $\mathrm{CaO(s)} + \mathrm{H_2O(l)} \rightarrow \mathrm{Ca(OH)_2(s)}$, what are the contents of the reaction vessel? Use data from this oter and Appendix D and the fact that $\Delta_{\text{vap}}H°[\mathrm{H_2O(l)}] = 40.6\ \text{kJ mol}^{-1}$ at 100 °C. [*Hint:* Assume that none of the $OH)_2$ dissolves in the water that remains, and the heat released by the reaction is absorbed by the water and the solid $OH)_2$.]

Exercises

Heat Capacity (Specific Heat Capacity)

1. Calculate the quantity of heat, in kilojoules, **(a)** required to raise the temperature of 9.25 L of water from 22.0 to 29.4 °C; **(b)** associated with a 33.5 °C decrease in temperature in a 5.85 kg aluminum bar (specific heat capacity of aluminum = $0.903\ \mathrm{J\ g^{-1}\ °C^{-1}}$).
2. Calculate the final temperature that results when **(a)** a 12.6 g sample of water at 22.9 °C absorbs 875 J of heat; **(b)** a 1.59 kg sample of platinum at 78.2 °C gives off 1.05 kcal of heat ($c_p = 0.032\ \mathrm{cal\ g^{-1}\ °C^{-1}}$).
3. Refer to Example 7-2. The experiment is repeated with several different metals substituting for the lead. The masses of metal and water and the initial temperatures of the metal and water are the same as in Figure 7-3. The final temperatures are **(a)** Zn, 38.9 °C; **(b)** Pt, 28.8 °C; **(c)** Al, 52.7 °C. What is the specific heat capacity of each metal, expressed in $\mathrm{J\ g^{-1}\ °C^{-1}}$?
4. A 75.0 g piece of Ag metal is heated to 80.0 °C and dropped into 50.0 g of water at 23.2 °C. The final temperature of the Ag–H_2O mixture is 27.6 °C. What is the specific heat capacity of silver?
5. A 465 g chunk of iron is removed from an oven and plunged into 375 g water in an insulated container. The temperature of the water increases from 26 to 87 °C. If the specific heat capacity of iron is $0.449\ \mathrm{J\ g^{-1}\ °C^{-1}}$, what must have been the original temperature of the iron?
6. A piece of stainless steel ($c_p = 0.50\ \mathrm{J\ g^{-1}\ °C^{-1}}$) is transferred from an oven at 201 °C into 150 mL of water at 23.2 °C. The water temperature rises to 55.4 °C. What is the mass of the steel? How precise is this method of mass determination? Explain.
7. A 1.00 kg sample of magnesium at 40.0 °C is a to 1.00 L of water maintained at 20.0 °C i insulated container. What will be the final tem ture of the Mg–H_2O mixture (specific heat cap of Mg = $1.024\ \mathrm{J\ g^{-1}\ °C^{-1}}$)?
8. Brass has a density of $8.40\ \mathrm{g/cm^3}$ and a specific capacity of $0.385\ \mathrm{J\ g^{-1}\ °C^{-1}}$. A 15.2 $\mathrm{cm^3}$ piece of at an initial temperature of 163 °C is dropped in insulated container with 150.0 g water initial 22.4 °C. What will be the final temperature o brass–water mixture?
9. A 74.8 g sample of copper at 143.2 °C is added insulated vessel containing 165 mL of glyc $C_3H_8O_3(l)$ ($d = 1.26$ g/mL), at 24.8 °C. The final perature is 31.1 °C. The specific heat capacity of per is $0.385\ \mathrm{J\ g^{-1}\ °C^{-1}}$. What is the heat capaci glycerol in $\mathrm{J\ mol^{-1}\ °C^{-1}}$?
10. A 69.0 g sample of gold at 127.1 °C is added to an lated vessel containing 543.0 mL of water at 25. The final temperature is 25.4 °C. What is the sp heat capacity of gold in $\mathrm{J\ g^{-1}\ °C^{-1}}$? The specific capacity of water is $4.18\ \mathrm{J\ g^{-1}\ °C^{-1}}$, and its densit 25.0 °C) is $0.997\ \mathrm{g\ mL^{-1}}$.
11. In the form of heat, 6.052 J of energy is transferre 1.0 L sample of air ($d = 1.204\ \mathrm{mg/cm^3}$) at 20. The final temperature of the air is 25.0 °C. What i heat capacity of air in J/K?
12. What is the final temperature (in °C) of 1.24 water with an initial temperature of 20.0 °C after 6. of heat is added to it?

Heats of Reaction

13. How much heat, in kilojoules, is associated with the production of 283 kg of slaked lime, $Ca(OH)_2$?

$$CaO(s) + H_2O(l) \longrightarrow Ca(OH)_2(s) \qquad \Delta_rH° = -65.2\ \mathrm{kJ\ mol^{-1}}$$

14. The standard enthalpy of reaction for the combustion of octane is $\Delta_rH° = -5.48 \times 10^3$ kJ/mol $C_8H_{18}(l)$. How much heat, in kilojoules, is liberated *per gallon* of octane burned? (Density of octane = 0.703 g/mL; 1 gal = 3.785 L.)
15. How much heat, in kilojoules, is evolved in the complete combustion of **(a)** 1.325 g $C_4H_{10}(g)$ at 25 °C and 1 atm; **(b)** 28.4 L $C_4H_{10}(g)$ at STP; **(c)** 12.6 L $C_4H_{10}(g)$ at 23.6 °C and 738 mmHg? Assume that the enthalpy of reaction does not change significantly with temperature or pressure. The complete combustion of butane, $C_4H_{10}(g)$, is represented by the equation

$$C_4H_{10}(g) + \frac{13}{2}O_2(g) \longrightarrow 4\,CO_2(g) + 5\,H_2O(l) \qquad \Delta_rH° = -2877\ \mathrm{kJ\ mol^{-1}}$$

16. Upon complete combustion, the indicated substa evolve the given quantities of heat. Write a bala equation for the combustion of 1.00 mol of substance, including the enthalpy of reaction, for the reaction.
    (a) 0.584 g of propane, $C_3H_8(g)$, yields 29.4 kJ
    (b) 0.136 g of camphor, $C_{10}H_{16}O(s)$, yields 5.27 k
    (c) 2.35 mL of acetone, $(CH_3)_2CO(l)$ (0.791 g/mL), yields 58.3 kJ
17. The combustion of methane gas, the principal stituent of natural gas, is represented by the equa

$$CH_4(g) + 2\,O_2(g) \longrightarrow CO_2(g) + 2\,H_2O(l) \qquad \Delta_rH° = -890.3\ \mathrm{kJ\ m}$$

(a) What mass of methane, in kilograms, mus burned to liberate 2.80×10^7 kJ of heat?
(b) What quantity of heat, in kilojoules, is liberate the complete combustion of 1.65×10^4 L of CH measured at 18.6 °C and 768 mmHg?
(c) If the quantity of heat calculated in part (b) c be transferred with 100% efficiency to water, v volume of water, in liters, could be heated from 8 60.0 °C as a result?

Refer to the Integrative Example. What volume of the synthesis gas, measured at STP and burned in an open flame (constant-pressure process), is required to heat 40.0 gal of water from 15.2 to 65.0 °C? (1 gal = 3.785 L.)

The combustion of hydrogen–oxygen mixtures is used to produce very high temperatures (approximately 2500 °C) needed for certain types of welding operations. Consider the reaction to be

$$\text{g)} + \frac{1}{2}\,O_2(g) \longrightarrow H_2O(g) \quad \Delta_r H^\circ = -241.8\ \text{kJ mol}^{-1}$$

What is the quantity of heat evolved, in kilojoules, when a 180 g mixture containing equal parts of H_2 and O_2 by mass is burned?

Thermite mixtures are used for certain types of welding, and the thermite reaction is highly exothermic.

$$Fe_2O_3(s) + 2\,Al(s) \longrightarrow Al_2O_3(s) + 2\,Fe(s) \qquad \Delta_r H^\circ = -852\ \text{kJ mol}^{-1}$$

1.00 mol of granular Fe_2O_3 and 2.00 mol of granular Al are mixed at room temperature (25 °C), and a reaction is initiated. The liberated heat is retained within the products, whose combined specific heat capacity over a broad temperature range is about $0.8\ \text{J g}^{-1}\,°\text{C}^{-1}$. (The melting point of iron is 1530 °C.) Show that the quantity of heat liberated is more than sufficient to raise the temperature of the products to the melting point of iron.

A 0.205 g pellet of potassium hydroxide, KOH, is added to 55.9 g water in a Styrofoam coffee cup. The water temperature rises from 23.5 to 24.4 °C. [Assume that the specific heat capacity of dilute KOH(aq) is the same as that of water.]
(a) What is the approximate heat of solution of KOH, expressed as kilojoules per mole of KOH?
(b) How could the precision of this measurement be improved *without* modifying the apparatus?

The heat of solution of KI(s) in water is +20.3 kJ/mol KI. If a quantity of KI is added to sufficient water at 24.3 °C in a Styrofoam cup to produce 175.0 mL of 2.50 M KI, what will be the final temperature? (Assume a density of 1.30 g/mL and a specific heat capacity of $2.7\ \text{J g}^{-1}\,°\text{C}^{-1}$ for 2.50 M KI.)

23. You are planning a lecture demonstration to illustrate an endothermic process. You want to lower the temperature of 1400 mL water in an insulated container from 25 to 10 °C. Approximately what mass of $NH_4Cl(s)$ should you dissolve in the water to achieve this result? The heat of solution of NH_4Cl is +14.7 kJ/mol NH_4Cl.

24. Care must be taken in preparing solutions of solutes that liberate heat on dissolving. The heat of solution of NaOH is −44.5 kJ/mol NaOH. To what maximum temperature may a sample of water, originally at 24 °C, be raised in the preparation of 500 mL of 4.0 M NaOH? Assume the solution has a density of 1.08 g/mL and specific heat capacity of $4.00\ \text{J g}^{-1}\,°\text{C}^{-1}$.

25. Refer to Example 7-4. The product of the neutralization is 0.500 M NaCl. For this solution, assume a density of 1.02 g/mL and a specific heat capacity of $4.02\ \text{J g}^{-1}\,°\text{C}^{-1}$. Also, assume a heat capacity for the Styrofoam cup of 10 J/°C, and recalculate the heat of neutralization.

26. The heat of neutralization of HCl(aq) by NaOH(aq) is −55.84 kJ/mol H_2O produced. If 50.00 mL of 1.05 M NaOH is added to 25.00 mL of 1.86 M HCl, with both solutions originally at 24.72 °C, what will be the final solution temperature? (Assume that no heat is lost to the surrounding air and that the solution produced in the neutralization reaction has a density of 1.02 g/mL and a specific heat capacity of $3.98\ \text{J g}^{-1}\,°\text{C}^{-1}$.)

27. Acetylene (C_2H_2) torches are used in welding. How much heat (in kJ) evolves when 5.0 L of C_2H_2 ($d = 1.0967\ \text{kg/m}^3$) is mixed with a stoichiometric amount of oxygen gas? The combustion reaction is

$$C_2H_2(g) + \frac{5}{2}\,O_2(g) \longrightarrow 2\,CO_2(g) + H_2O(l) \qquad \Delta_r H^\circ = -1299.5\ \text{kJ mol}^{-1}$$

28. Propane (C_3H_8) gas ($d = 1.83\ \text{kg/m}^3$) is used in most gas grills. What volume (in liters) of propane is needed to generate 273.8 kJ of heat?

$$C_3H_8(g) + 5\,O_2(g) \longrightarrow 3\,CO_2(g) + 4\,H_2O(l) \qquad \Delta_r H^\circ = -2219.9\ \text{kJ mol}^{-1}$$

thalpy Changes and States of Matter

What mass of ice can be melted with the same quantity of heat as required to raise the temperature of 3.50 mol $H_2O(l)$ by 50.0 °C? [$\Delta_{fus}H^\circ = 6.01$ kJ/mol $H_2O(s)$]

What will be the final temperature of the water in an insulated container as the result of passing 5.00 g of steam, $H_2O(g)$, at 100.0 °C into 100.0 g of water at 25.0 °C? ($\Delta_{vap}H^\circ = 40.6$ kJ/mol H_2O)

A 125 g stainless steel ball bearing ($c_p = 0.50\ \text{J g}^{-1}\,°\text{C}^{-1}$) at 525 °C is dropped into 75.0 mL of water at 28.5 °C in an open Styrofoam cup. As a result, the water is brought to a boil when the temperature reaches 100.0 °C. What mass of water vaporizes while the boiling continues? ($\Delta_{vap}H^\circ = 40.6$ kJ/mol H_2O)

32. If the ball bearing described in Exercise 31 is dropped onto a large block of ice at 0 °C, what mass of liquid water will form? ($\Delta_{fus}H^\circ = 6.01$ kJ/mol H_2O)

33. The enthalpy of sublimation (solid → gas) for dry ice (i.e., CO_2) is $\Delta_{sub}H^\circ = 571$ kJ/kg at −78.5 °C. If 125.0 J of heat is transferred to a block of dry ice that is −78.5 °C, what volume of CO_2 gas ($d = 1.98$ g/L) will be generated?

34. The enthalpy of vaporization for $N_2(l)$ is 5.56 kJ/mol. How much heat (in J) is required to produce 1.0 L of $N_2(g)$ at 77.36 K and 1.0 atm?

Calorimetry

35. A sample gives off 5228 cal when burned in a bomb calorimeter. The temperature of the calorimeter assembly increases by 4.39 °C. Calculate the heat capacity of the calorimeter, in kilojoules per degree Celsius.
36. The following substances undergo complete combustion in a bomb calorimeter. The calorimeter assembly has a heat capacity of 5.136 kJ/°C. In each case, what is the final temperature if the initial water temperature is 22.43 °C? **(a)** 0.3268 g caffeine, $C_8H_{10}O_2N_4$ (heat of combustion = −1014.2 kcal/mol caffeine); **(b)** 1.35 mL of methyl ethyl ketone, $C_4H_8O(l)$, d = 0.805 g/mL (heat of combustion = −2444 kJ/mol methyl ethyl ketone).
37. A bomb calorimetry experiment is performed with xylose, $C_5H_{10}O_5(s)$, as the combustible substance. The data obtained are

mass of xylose burned:	1.183 g
heat capacity of calorimeter:	4.728 kJ/°C
initial calorimeter temperature:	23.29 °C
final calorimeter temperature:	27.19 °C

(a) What is the heat of combustion of xylose, in kilojoules per mole? **(b)** Write the chemical equation for the complete combustion of xylose, and represent the value of $\Delta_r H$ in this equation. (Assume for this reaction that $\Delta U \approx \Delta_r H$.)

38. A coffee-cup calorimeter contains 100.0 mL of 0.300 M HCl at 20.3 °C. When 1.82 g Zn(s) is added, the temperature rises to 30.5 °C. What is the heat of reaction per mol Zn? Make the same assumptions as in Example 7-4, and also assume that there is no heat lost with the $H_2(g)$ that escapes.

$$Zn(s) + 2H^+(aq) \longrightarrow Zn^{2+}(aq) + H_2(g)$$

39. A 0.75 g sample of KCl is added to 35.0 g H_2O in a Styrofoam cup and stirred until it dissolves. The temperature of the solution drops from 24.8 to 23.6 °C. **(a)** Is the process endothermic or exothermic? **(b)** What is the heat of solution of KCl expressed in kilojoules per mole of KCl?
40. The heat of solution of potassium acetate in w is −15.3 kJ/mol KCH_3COO. What will be the temperature when 0.241 mol KCH_3COO is disso in 815 mL water that is initially at 25.1 °C?
41. A 1.620 g sample of naphthalene, $C_{10}H_8(s)$, is c pletely burned in a bomb calorimeter assembly a temperature increase of 8.44 °C is noted. If the he combustion of naphthalene is −5156 kJ/mol C what is the heat capacity of the bomb calorimeter
42. Salicylic acid, $C_7H_6O_3$, has been suggested calorimetric standard. Its heat of combustio -3.023×10^3 kJ/mol $C_7H_6O_3$. From the follov data determine the heat capacity of a bomb calorin assembly (that is, the bomb, water, stirrer, thermo ter, wires, and so forth).

mass of salicylic acid burned:	1.201 g
initial calorimeter temperature:	23.68 °C
final calorimeter temperature:	29.82 °C

43. Refer to Example 7-3. Based on the heat of combus of sucrose established in the example, what shoul the temperature change (ΔT) produced by the c bustion of 1.227 g $C_{12}H_{22}O_{11}$ in a bomb calorim assembly with a heat capacity of 3.87 kJ/°C?
44. A 1.397 g sample of thymol, $C_{10}H_{14}O(s)$ (a preserva and a mold and mildew preventative), is burned bomb calorimeter assembly. The temperature incr is 11.23 °C, and the heat capacity of the bomb calor ter is 4.68 kJ/°C. What is the heat of combustion of mol, expressed in kilojoules per mole of $C_{10}H_{14}O$?
45. A 5.0 g sample of NaCl is added to a Styrofoam cu water, and the change in water temperature is 5. The heat of solution of NaCl is 3.76 kJ/mol. Wh the mass (in g) of water in the Styrofoam cup?
46. We can determine the purity of solid material using calorimetry. A gold ring (for pure gold, spe heat capacity = 0.1291 J g^{-1} K^{-1}) with mass of 1 is heated to 78.3 °C and immersed in 50.0 g of 23. water in a constant-pressure calorimeter. The temperature of the water is 31.0 °C. Is this a pure s ple of gold?

Pressure–Volume Work

47. Calculate the quantity of work associated with a 3.5 L expansion of a gas (ΔV) against a pressure of 748 mmHg in the units **(a)** atm L; **(b)** joules (J); **(c)** calories (cal).
48. Calculate the quantity of work, in joules, associated with the compression of a gas from 5.62 L to 3.37 L by a constant pressure of 1.23 atm.
49. A 1.00 g sample of Ne(g) at 1 atm pressure and 27 °C is allowed to expand into an *evacuated* vessel of 2.50 L volume. Does the gas do work? Explain.
50. Compressed air in aerosol cans is used to free electronic equipment of dust. Does the air do any work as it escapes from the can?
51. In each of the following processes, is any work done when the reaction is carried out at constant pressure in a vessel open to the atmosphere? If so, is work don the reacting system or on it? **(a)** Neutralizatio $Ba(OH)_2(aq)$ by HCl(aq); **(b)** conversion of gas nitrogen dioxide to gaseous dinitrogen tetro **(c)** decomposition of calcium carbonate to calc oxide and carbon dioxide gas.
52. In each of the following processes, is any work c when the reaction is carried out at constant pressu a vessel open to the atmosphere? If so, is work don the reacting system or on it? **(a)** Reaction of nitr monoxide and oxygen gases to form gaseous nitr dioxide; **(b)** precipitation of magnesium hydro by the reaction of aqueous solutions of NaOH $MgCl_2$; **(c)** reaction of copper(II) sulfate and w vapor to form copper(II) sulfate pentahydrate.

If 325 J of work is done by a system at a pressure of 1.0 atm and 298 K, what is the change in the volume of the system?

54. A movable piston in a cylinder holding 5.0 L $N_2(g)$ is used to lift a 2.41 kg object to a height of 2.6 meters. How much work (in J) was done by the gas?

st Law of Thermodynamics

What is the change in internal energy of a system if the system **(a)** absorbs 58 J of heat and does 58 J of work; **(b)** absorbs 125 J of heat and does 687 J of work; **(c)** evolves 280 cal of heat and has 1.25 kJ of work done on it?

What is the change in internal energy of a system if the *surroundings* **(a)** transfer 235 J of heat and 128 J of work to the system; **(b)** absorb 145 J of heat from the system while doing 98 J of work on the system; **(c)** exchange no heat, but receive 1.07 kJ of work from the system?

The internal energy of a fixed quantity of an ideal gas depends only on its temperature. A sample of an ideal gas is allowed to expand at a constant temperature (isothermal expansion). **(a)** Does the gas do work? **(b)** Does the gas exchange heat with its surroundings? **(c)** What happens to the temperature of the gas? **(d)** What is ΔU for the gas?

In an *adiabatic* process, a system is thermally insulated—there is no exchange of heat between system and surroundings. For the adiabatic expansion of an ideal gas **(a)** does the gas do work? **(b)** Does the internal energy of the gas increase, decrease, or remain constant? **(c)** What happens to the temperature of the gas? [*Hint*: Refer to Exercise 57.]

59. Do you think the following observation is in any way possible? An ideal gas is expanded isothermally and is observed to do twice as much work as the heat absorbed from its surroundings. Explain your answer. [*Hint:* Refer to Exercises 57 and 58.]

60. Do you think the following observation is any way possible? A gas absorbs heat from its surroundings while being compressed. Explain your answer. [*Hint:* Refer to Exercises 55 and 56.]

61. There are other forms of work besides P–V work. For example, electrical work is defined as the potential × change in charge, $w = \phi\,\Delta q$. If a charge in a system is changed from 10 C to 5 C in a potential of 100 V and 45 J of heat is liberated, what is the change in the internal energy? (Note: 1 V = 1 J/C)

62. Another form of work is extension, defined as the tension × change in length, $w = f\,\Delta l$. A piece of DNA has an approximate tension of f = 10 pN. What is the change in the internal energy of the adiabatic stretching of DNA by 10 pm?

ating ΔH and ΔU

Only one of the following quantities is equal to the heat of a chemical reaction, *regardless of how the reaction is carried out*. Which one and why? **(a)** q_V; **(b)** q_P; **(c)** $\Delta U - w$; **(d)** ΔU; **(e)** $\Delta_r H$.

Determine whether ΔH is equal to, greater than, or less than ΔU for the following processes. Keep in mind that "greater than" means more positive or less negative, and "less than" means less positive or more negative. Assume that the only significant change in volume during a constant pressure process is that associated with changes in the amounts of gases.
(a) The complete combustion of one mole of butan-1-ol.
(b) The complete combustion of one mole of glucose, $C_6H_{12}O_6(s)$.
(c) The decomposition of solid ammonium nitrate to produce liquid water and gaseous dinitrogen monoxide.

65. The heat of combustion of propan-2-ol at 298.15 K, determined in a bomb calorimeter, is −33.41 kJ/g. For the combustion of one mole of propan-2-ol, determine **(a)** ΔU, and **(b)** $\Delta_r H$.

66. Write an equation to represent the combustion of thymol referred to in Exercise 44. Include in this equation the values for ΔU and ΔH.

ss's Law

The standard enthalpy of formation of $NH_3(g)$ is −46.11 kJ/mol. What is $\Delta_r H°$ for the following reaction?

$$\frac{2}{3}NH_3(g) \longrightarrow \frac{1}{3}N_2(g) + H_2(g) \qquad \Delta_r H° =$$

Use Hess's law to determine $\Delta_r H°$ for the reaction $CO(g) + \frac{1}{2}O_2(g) \longrightarrow CO_2(g)$, given that

$$C(\text{graphite}) + \frac{1}{2}O_2(g) \longrightarrow CO(g) \qquad \Delta_r H° = -110.54\ \text{kJ mol}^{-1}$$

$$C(\text{graphite}) + O_2(g) \longrightarrow CO_2(g) \qquad \Delta_r H° = -393.51\ \text{kJ mol}^{-1}$$

69. Use Hess's law to determine $\Delta_r H°$ for the reaction $C_3H_4(g) + 2\,H_2(g) \longrightarrow C_3H_8(g)$, given that

$$H_2(g) + \frac{1}{2}O_2(g) \longrightarrow H_2O(l) \qquad \Delta_r H° = -285.8\ \text{kJ mol}^{-1}$$

$$C_3H_4(g) + 4\,O_2(g) \longrightarrow 3\,CO_2(g) + 2\,H_2O(l) \qquad \Delta_r H° = -1937\ \text{kJ mol}^{-1}$$

$$C_3H_8(g) + 5\,O_2(g) \longrightarrow 3\,CO_2(g) + 4\,H_2O(l) \qquad \Delta_r H° = -2219.1\ \text{kJ mol}^{-1}$$

70. Given the following information:

$$\frac{1}{2}N_2(g) + \frac{3}{2}H_2(g) \longrightarrow NH_3(g) \qquad \Delta_rH_1^\circ$$

$$NH_3(g) + \frac{5}{4}O_2(g) \longrightarrow NO(g) + \frac{3}{2}H_2O(l) \qquad \Delta_rH_2^\circ$$

$$H_2(g) + \frac{1}{2}O_2(g) \longrightarrow H_2O(l) \qquad \Delta_rH_3^\circ$$

Determine Δ_rH° for the following reaction, expressed in terms of $\Delta_rH_1^\circ$, $\Delta_rH_2^\circ$, and $\Delta_rH_3^\circ$.

$$N_2(g) + O_2(g) \longrightarrow 2\,NO(g) \qquad \Delta_rH^\circ = ?$$

71. For the reaction $C_2H_4(g) + Cl_2(g) \longrightarrow C_2H_4Cl_2(l)$, determine Δ_rH°, given that

$$4\,HCl(g) + O_2(g) \longrightarrow 2\,Cl_2(g) + 2\,H_2O(l) \qquad \Delta_rH^\circ = -202.4\text{ kJ mol}^{-1}$$

$$2\,HCl(g) + C_2H_4(g) + \frac{1}{2}O_2(g) \longrightarrow C_2H_4Cl_2(l) + H_2O(l) \qquad \Delta_rH^\circ = -318.7\text{ kJ mol}^{-1}$$

72. Determine Δ_rH° for this reaction from the data below.

$$N_2H_4(l) + 2\,H_2O_2(l) \longrightarrow N_2(g) + 4\,H_2O(l)$$

$$N_2H_4(l) + O_2(g) \longrightarrow N_2(g) + 2\,H_2O(l) \qquad \Delta_rH^\circ = -622.2\text{ kJ mol}^{-1}$$

$$H_2(g) + \frac{1}{2}O_2(g) \longrightarrow H_2O(l) \qquad \Delta_rH^\circ = -285.8\text{ kJ mol}^{-1}$$

$$H_2(g) + O_2(g) \longrightarrow H_2O_2(l) \qquad \Delta_rH^\circ = -187.8\text{ kJ mol}^{-1}$$

73. Substitute natural gas (SNG) is a gaseous mixture containing $CH_4(g)$ that can be used as a fuel. One reaction for the production of SNG is

$$4\,CO(g) + 8\,H_2(g) \longrightarrow 3\,CH_4(g) + CO_2(g) + 2\,H_2O(l) \qquad \Delta_rH^\circ = ?$$

Use appropriate data from the following list to determine Δ_rH° for this SNG reaction.

$$C(\text{graphite}) + \frac{1}{2}O_2(g) \longrightarrow CO(g) \qquad \Delta_rH^\circ = -110.5\text{ kJ mol}^{-1}$$

$$CO(g) + \frac{1}{2}O_2(g) \longrightarrow CO_2(g) \qquad \Delta_rH^\circ = -283.0\text{ kJ mol}^{-1}$$

$$H_2(g) + \frac{1}{2}O_2(g) \longrightarrow H_2O(l) \qquad \Delta_rH^\circ = -285.8\text{ kJ mol}^{-1}$$

$$C(\text{graphite}) + 2\,H_2(g) \longrightarrow CH_4(g) \qquad \Delta_rH^\circ = -74.81\text{ kJ mol}^{-1}$$

$$CH_4(g) + 2\,O_2(g) \longrightarrow CO_2(g) + 2\,H_2O(l) \qquad \Delta_rH^\circ = -890.3\text{ kJ mol}^{-1}$$

74. CCl_4, an important commercial solvent, is prepared by the reaction of $Cl_2(g)$ with a carbon compound. Determine Δ_rH° for the reaction

$$CS_2(l) + 3\,Cl_2(g) \longrightarrow CCl_4(l) + S_2Cl_2(l)$$

Use appropriate data from the following listing.

$$CS_2(l) + 3\,O_2(g) \longrightarrow CO_2(g) + 2\,SO_2(g) \qquad \Delta_rH^\circ = -1077\text{ kJ mo}$$

$$2\,S(s) + Cl_2(g) \longrightarrow S_2Cl_2(l) \qquad \Delta_rH^\circ = -58.2\text{ kJ mo}$$

$$C(s) + 2\,Cl_2(g) \longrightarrow CCl_4(l) \qquad \Delta_rH^\circ = -135.4\text{ kJ mo}$$

$$S(s) + O_2(g) \longrightarrow SO_2(g) \qquad \Delta_rH^\circ = -296.8\text{ kJ mo}$$

$$SO_2(g) + Cl_2(g) \longrightarrow SO_2Cl_2(l) \qquad \Delta_rH^\circ = +97.3\text{ k mo}$$

$$C(s) + O_2(g) \longrightarrow CO_2(g) \qquad \Delta_rH^\circ = -393.5\text{ kJ mo}$$

$$CCl_4(l) + O_2(g) \longrightarrow COCl_2(g) + Cl_2O(g) \qquad \Delta_rH^\circ = -5.2\text{ kJ mo}$$

75. Use Hess's law and the following data

$$CH_4(g) + 2\,O_2(g) \longrightarrow CO_2(g) + 2\,H_2O(g) \qquad \Delta_rH^\circ = -802\text{ kJ mo}$$

$$CH_4(g) + CO_2(g) \longrightarrow 2\,CO(g) + 2\,H_2(g) \qquad \Delta_rH^\circ = +247\text{ kJ mo}$$

$$CH_4(g) + H_2O(g) \longrightarrow CO(g) + 3\,H_2(g) \qquad \Delta_rH^\circ = +206\text{ kJ mo}$$

to determine Δ_rH° for the following reaction, important source of hydrogen gas

$$CH_4(g) + \frac{1}{2}O_2(g) \longrightarrow CO(g) + 2\,H_2(g)$$

76. The standard heats of combustion (Δ_rH°) buta-1,3-diene, $C_4H_6(g)$; butane, $C_4H_{10}(g)$; and H are −2540.2, −2877.6, and −285.8 kJ mol^{-1}, res tively. Use these data to calculate the heat of hy genation of buta-1,3-diene to butane.

$$C_4H_6(g) + 2\,H_2(g) \longrightarrow C_4H_{10}(g) \qquad \Delta_rH^\circ =$$

[*Hint:* Write equations for the combustion reacti In each combustion, the products are $CO_2(g)$ $H_2O(l)$.]

77. One glucose molecule, $C_6H_{12}O_6(s)$, is converte two lactic acid molecules, $CH_3CH(OH)COOH$ during glycolysis. Given the combustion reaction glucose and lactic acid, determine the stanc enthalpy for glycolysis.

$$C_6H_{12}O_6(s) + 6\,O_2(g) \longrightarrow 6\,CO_2(g) + 6\,H_2O(l) \qquad \Delta_rH^\circ = -2808\text{ kJ m}$$

$$CH_3CH(OH)COOH(s) + 3\,O_2(g) \longrightarrow 3\,CO_2(g) + 3\,H_2O(l) \qquad \Delta_rH^\circ = -1344\text{ kJ m}$$

78. The standard enthalpy of fermentation of glucos ethanol is

$$C_6H_{12}O_6(s) \longrightarrow 2\,CH_3CH_2OH(l) + 2\,CO_2(g) \qquad \Delta_rH^\circ = -72\text{ kJ m}$$

Use the standard enthalpy of combustion for cose to calculate the enthalpy of combustion ethanol.

ndard Enthalpies of Formation

Use standard enthalpies of formation from Table 7.2 and equation (7.22) to determine the standard enthalpy of reaction in the following reactions.
(a) $C_3H_8(g) + H_2(g) \longrightarrow C_2H_6(g) + CH_4(g)$;
(b) $2\,H_2S(g) + 3\,O_2(g) \longrightarrow 2\,SO_2(g) + 2\,H_2O(l)$.

Use standard enthalpies of formation from Tables 7.2 and 7.3 and equation (7.22) to determine the standard enthalpy of reaction in the following reaction.

$$NH_4^+(aq) + OH^-(aq) \longrightarrow H_2O(l) + NH_3(g).$$

Use the information given here, data from Appendix D, and equation (7.22) to calculate the standard enthalpy of formation per mole of ZnS(s).

$$2\,ZnS(s) + 3\,O_2(g) \longrightarrow 2\,ZnO(s) + 2\,SO_2(g) \qquad \Delta_rH° = -878.2\ \text{kJ mol}^{-1}$$

Use the data in Figure 7-18 and information from Section 3-7 to establish possible relationships between the molecular structure of the hydrocarbons and their standard enthalpies of formation.

Use standard enthalpies of formation from Table 7.2 to determine $\Delta_rH°$ at 25 °C for the following reaction.

$$2\,Cl_2(g) + 2\,H_2O(l) \longrightarrow 4\,HCl(g) + O_2(g) \qquad \Delta_rH° = ?$$

Use data from Appendix D to calculate $\Delta_rH°$ for the following reaction at 25 °C.

$$Fe_2O_3(s) + 3\,CO(g) \longrightarrow 2\,Fe(s) + 3\,CO_2(g) \qquad \Delta_rH° = ?$$

Use data from Table 7.2 to determine the standard heat of combustion of $C_2H_5OH(l)$, if reactants and products are maintained at 25 °C and 1 bar.

Use data from Table 7.2, together with the fact that $\Delta_rH° = -3509\ \text{kJ mol}^{-1}$ for the complete combustion of pentane, $C_5H_{12}(l)$, to calculate $\Delta_rH°$ for the reaction below.

$$5\,CO(g) + 11\,H_2(g) \longrightarrow C_5H_{12}(l) + 5\,H_2O(l) \qquad \Delta_rH° = ?$$

87. Use data from Table 7.2 and $\Delta_rH°$ for the following reaction to determine the standard enthalpy of formation of $CCl_4(g)$ at 25 °C and 1 bar.

$$CH_4(g) + 4\,Cl_2(g) \longrightarrow CCl_4(g) + 4\,HCl(g) \qquad \Delta_rH° = -397.3\ \text{kJ mol}^{-1}$$

88. Use data from Table 7.2 and $\Delta_rH°$ for the following reaction to determine the standard enthalpy of formation of hexane, $C_6H_{14}(l)$, at 25 °C and 1 bar.

$$2\,C_6H_{14}(l) + 19\,O_2(g) \longrightarrow 12\,CO_2(g) + 14\,H_2O(l) \qquad \Delta_rH° = -8326\ \text{kJ mol}^{-1}$$

89. Use data from Table 7.3 and Appendix D to determine the standard enthalpy change in the following reaction.

$$Al^{3+}(aq) + 3\,OH^-(aq) \longrightarrow Al(OH)_3(s) \qquad \Delta_rH° = ?$$

90. Use data from Table 7.3 and Appendix D to determine $\Delta_rH°$ the following reaction.

$$Mg(OH)_2(s) + 2\,NH_4^+(aq) \longrightarrow Mg^{2+}(aq) + 2\,H_2O(l) + 2\,NH_3(g) \qquad \Delta_rH° = ?$$

91. The decomposition of limestone, $CaCO_3(s)$, into quicklime, CaO(s), and $CO_2(g)$ is carried out in a gas-fired kiln. Use data from Appendix D to determine how much heat is required to decompose 1.35×10^3 kg $CaCO_3(s)$. (Assume that heats of reaction are the same as at 25 °C and 1 bar.)

92. Use data from Table 7.2 to calculate the volume of butane, $C_4H_{10}(g)$, measured at 24.6 °C and 756 mmHg, that must be burned to liberate 5.00×10^4 kJ of heat.

93. Ants release formic acid (HCOOH) when they bite. Use the data in Table 7.2 and the standard enthalpy of combustion for formic acid ($\Delta_rH° = -255$ kJ/mol) to calculate the standard enthalpy of formation for formic acid.

94. Calculate the enthalpy of combustion for lactic acid by using the data in Table 7.2 and the standard enthalpy of formation for lactic acid [$CH_3CH(OH)COOH(s)$]: $\Delta_fH° = -694.0$ kJ/mol.

Integrative and Advanced Exercises

A British thermal unit (Btu) is defined as the quantity of heat required to change the temperature of 1 lb of water by 1 °F. Assume the specific heat capacity of water to be independent of temperature. How much heat is required to raise the temperature of the water in a 40 gal water heater from 48 to 145 °F in **(a)** Btu; **(b)** kcal; **(c)** kJ?

What volume of 18.5 °C water must be added, together with a 1.23 kg piece of iron at 68.5 °C, so that the temperature of the water in the insulated container shown in the figure remains constant at 25.6 °C?

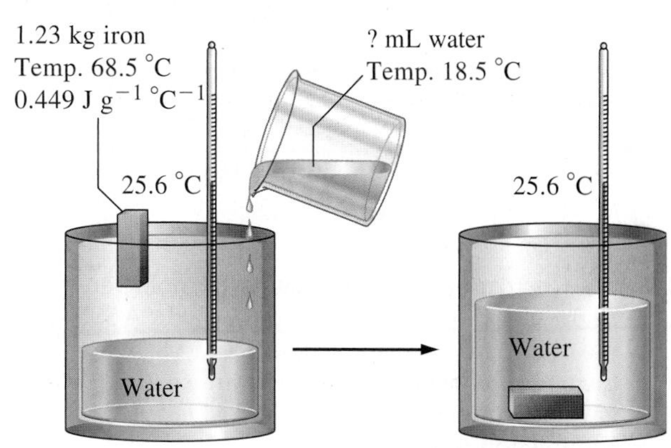

97. A 7.26 kg shot (as used in the sporting event, the shot put) is dropped from the top of a building 168 m high. What is the maximum temperature increase that could occur in the shot? Assume a specific heat capacity of $0.47\ J\ g^{-1}\ °C^{-1}$ for the shot. Why would the actual measured temperature increase likely be less than the calculated value?

98. An alternative approach to bomb calorimetry is to establish the heat capacity of the calorimeter, *exclusive* of the water it contains. The heat absorbed by the water and by the rest of the calorimeter must be calculated separately and then added together. A bomb calorimeter assembly containing 983.5 g water is calibrated by the combustion of 1.354 g anthracene. The temperature of the calorimeter rises from 24.87 to 35.63 °C. When 1.053 g citric acid is burned in the same assembly, but with 968.6 g water, the temperature increases from 25.01 to 27.19 °C. The heat of combustion of anthracene, $C_{14}H_{10}(s)$, is −7067 kJ/mol $C_{14}H_{10}$. What is the heat of combustion of citric acid, $C_6H_8O_7$, expressed in kJ/mol?

99. The method of Exercise 98 is used in some bomb calorimetry experiments. A 1.148 g sample of benzoic acid is burned in excess $O_2(g)$ in a bomb immersed in 1181 g of water. The temperature of the water rises from 24.96 to 30.25 °C. The heat of combustion of benzoic acid is −26.42 kJ/g. In a second experiment, a 0.895 g powdered coal sample is burned in the same calorimeter assembly. The temperature of 1162 g of water rises from 24.98 to 29.81 °C. How many metric tons (1 metric ton = 1000 kg) of this coal would have to be burned to release 2.15×10^9 kJ of heat?

100. A handbook lists two different values for the heat of combustion of hydrogen: 33.88 kcal/g if $H_2O(l)$ is formed, and 28.67 kcal/g if $H_2O(g)$ is formed. Explain why these two values are different, and indicate what property this difference represents. Devise a means of verifying your conclusions.

101. Determine the missing values of $\Delta_r H°$ in the diagram shown below.

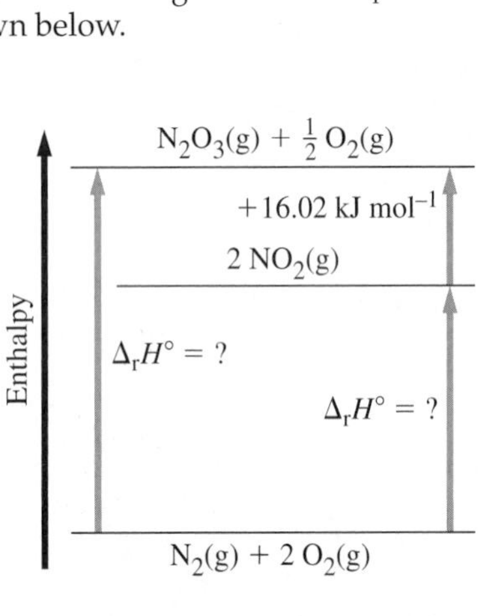

102. A particular natural gas consists, in mole percents, of 83.0% CH_4, 11.2% C_2H_6, and 5.8% C_3H_8. A 385 L sample of this gas, measured at 22.6 °C and 739 mmHg, is burned at constant pressure in an excess of oxygen gas. How much heat, in kilojoules, is evolved in the combustion reaction?

103. An overall reaction for a coal gasification proces

$$2\,C(\text{graphite}) + 2\,H_2O(g) \longrightarrow CH_4(g) + CO$$

Show that this overall equation can be establi by an appropriate combination of equations Section 7-9.

104. Which of the following gases has the greater value on a per liter (STP) basis? That is, which the greater heat of combustion? [*Hint:* The only bustible gases are CH_4, C_3H_8, CO, and H_2.]
(a) coal gas: 49.7% H_2, 29.9% CH_4, 8.2% N_2, 6.9% 3.1% C_3H_8, 1.7% CO_2, and 0.5% O_2, by volume.
(b) sewage gas, 66.0% CH_4, 30.0% CO_2, and N_2, by volume.

105. A calorimeter that measures an exothermic he reaction by the quantity of ice that can be melt called an ice calorimeter. Now consider that 0.1 of methane gas, $CH_4(g)$, at 25.0 °C and 744 mm is burned at constant pressure in air. The heat l ated is captured and used to melt 9.53 g ice at ($\Delta_{fus}H$ of ice = 6.01 kJ/mol).
(a) Write an equation for the complete combus of CH_4, and show that combustion is incomple this case.
(b) Assume that CO(g) is produced in the incom combustion of CH_4, and represent the combustio best you can through a single equation with s whole numbers as coefficients. ($H_2O(l)$ is anc product of the combustion.)

106. For the reaction

$$C_2H_4(g) + 3\,O_2(g) \longrightarrow 2\,CO_2(g) + 2\,H_2O(l) \quad \Delta_r H° = -1410.9\ \text{kJ m}$$

if the H_2O were obtained as a gas rather than a uid, **(a)** would the heat of reaction be greater (negative) or smaller (less negative) than that cated in the equation? **(b)** Explain your ans **(c)** Calculate the value of $\Delta_r H°$ in this case.

107. Some of the butane, $C_4H_{10}(g)$, in a 200.0 L cylind 26.0 °C is withdrawn and burned at a constant sure in an excess of air. As a result, the pressure c gas in the cylinder falls from 2.35 atm to 1.10 The liberated heat is used to raise the temperatu 132.5 L of water in a heater from 26.0 to 62. Assume that the combustion products are CC and $H_2O(l)$ exclusively, and determine the effici of the water heater. (That is, what percent of the of combustion was absorbed by the water?)

108. The metabolism of glucose, $C_6H_{12}O_6$, yields CC and $H_2O(l)$ as products. Heat released in the pro is converted to useful work with about 70% ciency. Calculate the mass of glucose metabolize a 58.0 kg person in climbing a mountain with ar vation gain of 1450 m. Assume that the work formed in the climb is about four times that req to simply lift 58.0 kg by 1450 m. ($\Delta_f H°$ of $C_6H_{12}C$ is −1273.3 kJ/mol.)

109. An alkane hydrocarbon has the formula C_nH The enthalpies of formation of the alkanes dec (become more negative) as the number of C a increases. Starting with butane, $C_4H_{10}(g)$, for additional CH_2 group in the formula, the enthal

formation, $\Delta_f H°$, changes by about -21 kJ/mol. Use this fact and data from Table 7.2 to estimate the heat of combustion of heptane, $C_7H_{16}(l)$.

Upon complete combustion, a 1.00 L sample (at STP) of a natural gas gives off 43.6 kJ of heat. If the gas is a mixture of $CH_4(g)$ and $C_2H_6(g)$, what is its percent composition, *by volume*?

Under the entry H_2SO_4, a reference source lists many values for the standard enthalpy of formation. For example, for pure $H_2SO_4(l)$, $\Delta_f H° = -814.0$ kJ/mol; for a solution with 1 mol H_2O per mole of H_2SO_4, -841.8; with 10 mol H_2O, -880.5; with 50 mol H_2O, -886.8; with 100 mol H_2O, -887.7; with 500 mol H_2O, -890.5; with 1000 mol H_2O, -892.3; with 10,000 mol H_2O, -900.8; and with 100,000 mol H_2O, -907.3.
(a) Explain why these values are not all the same.
(b) The value of $\Delta_f H°[H_2SO_4(aq)]$ in an infinitely dilute solution is -909.3 kJ/mol. What data from this chapter can you cite to confirm this value? Explain.
(c) If 500.0 mL of 1.00 M $H_2SO_4(aq)$ is prepared from pure $H_2SO_4(l)$, what is the approximate change in temperature that should be observed? Assume that the $H_2SO_4(l)$ and $H_2O(l)$ are at the same temperature initially and that the specific heat capacity of the $H_2SO_4(aq)$ is about 4.2 J g^{-1} °C^{-1}.

Refer to the discussion of the gasification of coal (page 283), and show that some of the heat required in the gasification reactions (equations 7.26 and 7.27) can be supplied by the *methanation* reaction. This fact contributes to the success of modern processes that produce *synthetic natural gas* (SNG).

A 1.103 g sample of a gaseous carbon–hydrogen–oxygen compound that occupies a volume of 582 mL at 765.5 Torr and 25.00 °C is burned in an excess of $O_2(g)$ in a bomb calorimeter. The products of the combustion are 2.108 g $CO_2(g)$, 1.294 g $H_2O(l)$, and enough heat to raise the temperature of the calorimeter assembly from 25.00 to 31.94 °C. The heat capacity of the calorimeter is 5.015 kJ/°C. Write an equation for the combustion reaction, and indicate $\Delta_r H°$ for this reaction at 25.00 °C.

Several factors are involved in determining the cooking times required for foods in a microwave oven. One of these factors is specific heat capacity. Determine the approximate time required to warm 250 mL of chicken broth from 4 °C (a typical refrigerator temperature) to 50 °C in a 700 W microwave oven. Assume that the density of chicken broth is about 1 g/mL and that its specific heat capacity is approximately 4.2 J g^{-1} °C^{-1}.

Suppose you have a setup similar to the one depicted in Figure 7-8 except that there are two different weights rather than two equal weights. One weight is a steel cylinder 10.00 cm in diameter and 25 cm long, the other weight produces a pressure of 745 Torr. The temperature of the gas in the cylinder in which the expansion takes place is 25.0 °C. The piston restraining the gas has a diameter of 12.00 cm, and the height of the piston above the base of the gas expansion cylinder is 8.10 cm. The density of the steel is 7.75 g/cm^3. How much work is done when the steel cylinder is suddenly removed from the piston?

116. When one mole of sodium carbonate decahydrate (washing soda) is gently warmed, 155.3 kJ of heat is absorbed, water vapor is formed, and sodium carbonate heptahydrate remains. On more vigorous heating, the heptahydrate absorbs 320.1 kJ of heat and loses more water vapor to give the monohydrate. Continued heating gives the anhydrous salt (soda ash) while 57.3 kJ of heat is absorbed. Calculate ΔH for the conversion of one mole of washing soda into soda ash. Estimate ΔU for this process. Why is the value of ΔU only an estimate?

117. The oxidation of $NH_3(g)$ to NO(g) in the Ostwald process must be very carefully controlled in terms of temperature, pressure, and contact time with the catalyst. This is because the oxidation of $NH_3(g)$ can yield any one of the products $N_2(g)$, $N_2O(g)$, NO(g), and $NO_2(g)$, depending on conditions. Show that oxidation of $NH_3(g)$ to $N_2(g)$ is the most exothermic of the four possible reactions.

118. In the Are You Wondering 7-1 box, the temperature variation of enthalpy is discussed, and the equation q_P = heat capacity × temperature change = $C_p \times \Delta T$ was introduced to show how enthalpy changes with temperature for a constant-pressure process. Strictly speaking, the heat capacity of a substance at constant pressure is the slope of the line representing the variation of enthalpy (H) with temperature, that is

$$C_p = \frac{dH}{dT} \qquad \text{(at constant pressure)}$$

where C_p is the heat capacity of the substance in question. Heat capacity is an extensive quantity and heat capacities are usually quoted as molar heat capacities $C_{p,\mathrm{m}}$, the heat capacity of one mole of substance, which is an intensive property. The heat capacity at constant pressure is used to estimate the change in enthalpy due to a change in temperature. For infinitesimal changes in temperature,

$$dH = C_p dT \qquad \text{(at constant pressure)}$$

To evaluate the change in enthalpy for a particular temperature change, from T_1 to T_2, we write

$$\int_{H(T_1)}^{H(T_2)} dH = H(T_2) - H(T_1) = \int_{T_1}^{T_2} C_p dT$$

If we assume that C_p is independent of temperature, then we recover equation (7.5)

$$q_p = \Delta H = C_p \Delta T$$

On the other hand, we often find that the heat capacity is a function of temperature; a convenient empirical expression is

$$C_{p,\mathrm{m}} = a + bT + \frac{c}{T^2}$$

What is the change in molar enthalpy of N_2 when it is heated from 25.0 °C to 100.0 °C? The molar heat capacity of nitrogen is given by

$$C_{p,\mathrm{m}} = \left(28.58 + 3.77 \times 10^{-3}\,T - \frac{0.5 \times 10^5}{T^2}\right) \mathrm{J\,mol^{-1}\,K^{-1}}$$

119. How much heat is required to convert 10.0 g of ice at −5.0 °C to steam at 100.0 °C? The temperature-dependent constant-pressure specific heat capacity of ice is $c_p(T)/(\text{kJ kg}^{-1}\text{K}^{-1}) = 1.0187T - 1.49 \times 10^{-2}$. The temperature-dependent constant-pressure specific heat for water is $c_p(T)/(\text{kJ kg}^{-1}\text{K}^{-1}) = -1.0 \times 10^{-7}T^3 + 1.0 \times 10^{-4}T^2 - 3.92 \times 10^{-2}T + 8.7854$.

120. The standard enthalpy of formation of gaseous H_2O at 298.15 K is −241.82 kJ mol^{-1}. Using the ideas contained in Figure 7-16, estimate its value at 100.0 °C given the following values of the molar heat capacities at constant pressure: $H_2O(g)$: 33.58 J K^{-1} mol^{-1}; $H_2(g)$: 28.84 J K^{-1} mol^{-1}; $O_2(g)$: 29.37 J K^{-1} mol^{-1}. Assume the heat capacities are independent of temperature.

121. Cetane, $C_{16}H_{34}$, is a typical petrodiesel with a standard enthalpy of combustion of −10,699.1 kJ mol^{-1}. Methyl linoleate, $C_{19}H_{34}O_2$, is a biodiesel with a standard enthalpy of combustion of −11,690.1 kJ mol^{-1}. V volume of methyl linoleate provides the same energ one liter of cetane? The densities of cetane and me linoleate are 0.773 and 0.885 g mL^{-1}, respectively.

122. Carbon dioxide emissions have been implicated major factor in climate change. Which of the follo liquid fuels, when burned completely in oxyge 25 °C, generates the smallest amount of CO_2 *per kilo* of energy output?
Methanol, CH_3OH ($\Delta_f H° = -238.7$ kJ mol^{-1});
cetane, $C_{16}H_{34}$ ($\Delta_f H° = -456.3$ kJ mol^{-1});
methyl linoleate, $C_{19}H_{34}O_2$ ($\Delta_f H° = -604.9$ kJ mo
octane, C_8H_{18} ($\Delta_f H° = -250.1$ kJ mol^{-1}).
The $\Delta_f H°$ values for $CO_2(g)$ and $H_2O(l)$ are −393.5 −285.8 kJ mol^{-1}, respectively.

Feature Problems

123. James Joule published his definitive work related to the first law of thermodynamics in 1850. He stated that "the quantity of heat capable of increasing the temperature of one pound of water by 1 °F requires for its evolution the expenditure of a mechanical force represented by the fall of 772 lb through the space of one foot." Validate this statement by relating it to information given in this text.

124. Based on specific heat capacity measurements, Pierre Dulong and Alexis Petit proposed in 1818 that the specific heat capacity of an element is inversely related to its atomic weight (atomic mass). Thus, by measuring the specific heat capacity of a new element, its atomic weight could be readily established.
(a) Use data from Table 7.1 and inside the front cover to plot a *straight-line* graph relating atomic mass and specific heat capacity. Write the equation for this straight line.
(b) Use the measured specific heat capacity of 0.23 J g^{-1} $°\text{C}^{-1}$ and the equation derived in part (a) to obtain an approximate value of the atomic mass of cadmium, an element discovered in 1817.
(c) To raise the temperature of 75.0 g of a particular metal by 15 °C requires 450 J of heat. What might this metal be?

125. We can use the heat liberated by a neutralization reaction as a means of establishing the stoichiometry of the reaction. The data in the table are for the reaction of 1.00 M NaOH with 1.00 M citric acid, $C_6H_8O_7$, in a total solution volume of 60.0 mL.

mL 1.00 M NaOH Used	mL 1.00 M Citric Acid Used	ΔT, °C
20.0	40.0	4.7
30.0	30.0	6.3
40.0	20.0	8.2
50.0	10.0	6.7
55.0	5.0	2.7

(a) Plot ΔT versus mL 1.00 M NaOH, and ide the exact stoichiometric proportions of NaOH citric acid at the equivalence point of the neutra tion reaction.
(b) Why is the temperature change in the neu ization greatest when the reactants are in their stoichiometric proportions? That is, why not us excess of one of the reactants to ensure that the tralization has gone to completion to achieve maximum temperature increase?
(c) Rewrite the formula of citric acid to re more precisely its acidic properties. Then wr balanced net ionic equation for the neutraliza reaction.

126. In a student experiment to confirm Hess's law reaction

$$NH_3(\text{concd aq}) + HCl(aq) \longrightarrow NH_4Cl(aq$$

was carried out in two different ways. First, 8.0 of concentrated $NH_3(aq)$ was added to 100.0 m 1.00 M HCl in a calorimeter. (The $NH_3(aq)$ slightly in excess.) The reactants were initial 23.8 °C, and the final temperature after neutra tion was 35.8 °C. In the second experiment, air bubbled through 100.0 mL of concentr $NH_3(aq)$, sweeping out $NH_3(g)$ (see sketch). $NH_3(g)$ was neutralized in 100.0 mL of 1.00 M The temperature of the concentrated $NH_3(aq$ from 19.3 to 13.2 °C. At the same time, the temp ture of the 1.00 M HCl rose from 23.8 to 42.9 °C was neutralized by $NH_3(g)$. Assume that all s tions have densities of 1.00 g/mL and specific capacities of 4.18 J g^{-1} $°\text{C}^{-1}$.
(a) Write the two equations and $\Delta_r H$ values fo processes occurring in the second experim Show that the sum of these two equations is same as the equation for the reaction in the experiment.

(b) Show that, within the limits of experimental error, $\Delta_r H$ for the overall reaction is the same in the two experiments, thereby confirming Hess's law.

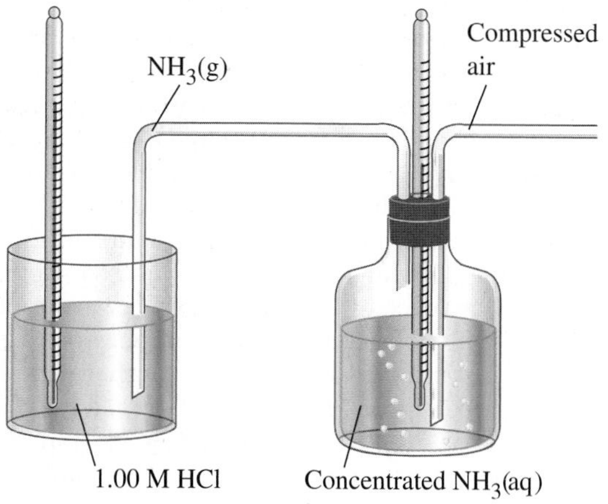

When an ideal gas is heated, the change in internal energy is limited to increasing the average translational kinetic energy of the gas molecules. Thus, there is a simple relationship between ΔU of the gas and the change in temperature that occurs. Derive this relationship with the help of ideas about the kinetic–molecular theory of gases developed in Chapter 6. After doing so, obtain numerical values (in $J\ mol^{-1}\ K^{-1}$) for the following molar heat capacities.

(a) the heat capacity, C_V, for one mole of gas under constant-volume conditions

(b) the heat capacity, C_p, for one mole of gas under constant-pressure conditions

128. Refer to Example 7-5 dealing with the work done by 0.100 mol He at 298 K in expanding in a single step from 2.40 to 1.20 atm. Review also the two-step expansion (2.40 atm $\longrightarrow$ 1.80 atm $\longrightarrow$ 1.20 atm) described on page 261 (see Figure 7-11).

(a) Determine the total work that would be done if the He expanded in a series of steps, at 0.10 atm intervals, from 2.40 to 1.20 atm.

(b) Represent this total work on the graph below, in which the quantity of work done in the two-step expansion is represented by the sum of the colored rectangles.

(c) Show that the maximum amount of work would occur if the expansion occurred in an infinite number of steps. To do this, express each infinitesimal quantity of work as $dw = -P\,dV$ and use the methods of integral calculus (integration) to sum these quantities. Assume ideal behavior for the gas.

(d) Imagine reversing the process, that is, compressing the He from 1.20 to 2.40 atm. What are the maximum and minimum amounts of work required to produce this compression? Explain.

(e) In the isothermal compression described in part (d), what is the change in internal energy assuming ideal gas behavior? What is the value of q?

(f) Using the formula for the work derived in part (c), obtain an expression for q/T. Is this new function a state function? Explain.

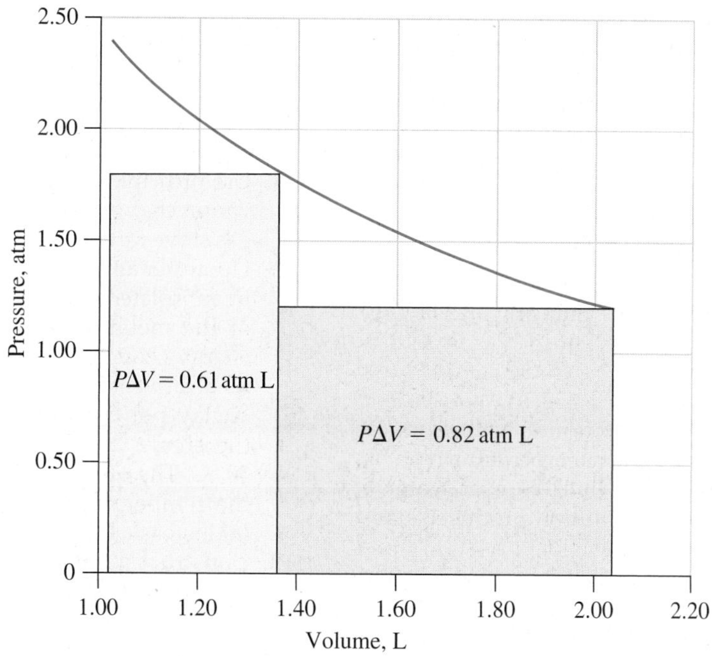

Look up the specific heat capacity of several elements, and plot the products of the specific heat capacities and atomic masses as a function of the atomic masses. Based on the plot, develop a hypothesis to explain the data. How could you test your hypothesis?

Self-Assessment Exercises

130. In your own words, define or explain the following terms or symbols: **(a)** $\Delta_r H$; **(b)** $-P\Delta V$; **(c)** $\Delta_f H°$; **(d)** standard state; **(e)** fossil fuel.

131. Briefly describe each of the following ideas or methods: **(a)** law of conservation of energy; **(b)** bomb calorimetry; **(c)** function of state; **(d)** enthalpy diagram; **(e)** Hess's law.

132. Explain the important distinctions between each pair of terms: **(a)** system and surroundings; **(b)** heat and work; **(c)** specific heat capacity and heat capacity; **(d)** endothermic and exothermic; **(e)** constant-volume process and constant-pressure process.

133. The temperature increase of 225 mL of water at 25 °C contained in a Styrofoam cup is noted when a 125 g sample of a metal at 75 °C is added. With reference to Table 7.1, the greatest temperature increase will be noted if the metal is **(a)** lead; **(b)** aluminum; **(c)** iron; **(d)** copper.

134. A plausible final temperature when 75.0 mL of water at 80.0 °C is added to 100.0 mL of water at 20 °C is **(a)** 28 °C; **(b)** 40 °C; **(c)** 46 °C; **(d)** 50 °C.

135. $\Delta U = 100$ J for a system that gives off 100 J of heat and **(a)** does no work; **(b)** does 200 J of work; **(c)** has 100 J of work done on it; **(d)** has 200 J of work done on it.

136. The heat of solution of NaOH(s) in water is −41.6 kJ/mol NaOH. When NaOH(s) is dissolved in water the solution temperature **(a)** increases; **(b)** decreases; **(c)** remains constant; **(d)** either increases or decreases, depending on how much NaOH is dissolved.

137. The standard molar enthalpy of formation of CO_2(g) is equal to **(a)** 0; **(b)** the standard molar heat of combustion of graphite; **(c)** the sum of the standard molar enthalpies of formation of CO(g) and O_2(g); **(d)** the standard molar heat of combustion of CO(g).

138. Write the formation reaction for each of the following compounds: **(a)** $SnCl_2$(s); **(b)** C_6H_5COOH(s); **(c)** $COCl_2$(g).

139. Compute $\Delta_r H°$ for the following reactions. The value of $\Delta_f H°$ in kJ mol^{-1} is given for each substance below its formula.

(a) SiO_2 (s) + 4 HF(g) ⟶ SiF_4(g) + 2 H_2O(g)
  −910.9   −271.1   −1615.0   −241.8

(b) 2 CuS(s) + 3 O_2(g) ⟶ 2 CuO(s) + 2 SO_2(g)
  −53.1   0.0   −157.3   −296.8

140. When dissolved in water, 1.00 mol LiCl produces 37.12 kJ of heat. What is the final temperature in (in °C) when 5.00 g LiCl dissolves in 110.0 g of water at 20.00 °C? Assume that the solution produced has a specific heat capacity of 4.00 J g^{-1} °C^{-1}.

141. When an element is involved in a formation reaction, it does *not* have to be **(a)** pure; **(b)** at 1.00 M concentration; **(c)** at 1.00 bar pressure; **(d)** in its most stable form; **(e)** none of these.

142. The standard state of a substance is **(a)** the pure form at 1 bar; **(b)** the most stable form at 25 °C and 1 bar; **(c)** the most stable form at 0 °C; **(d)** the pure gas form at 25 °C; **(e)** none of these.

143. Which two of the following statements are fa **(a)** $q_V = q_P$ for the reaction N_2(g) + O_2(g) 2 NO(g); **(b)** $\Delta_r H > 0$ for an endothermic reac **(c)** By convention, the most stable form of an eler must always be chosen as the reference form assigned the value $\Delta_f H° = 0$; **(d)** ΔU and $\Delta_r H$ reaction can never have the same value; **(e)** $\Delta_r H$ for the neutralization of a strong acid by a strong

144. A 1.22 kg piece of iron at 126.5 °C is dropped 981 g water at 22.1 °C. The temperature rise 34.4 °C. What will be the final temperature if same piece of iron at 99.8 °C is dropped into 32 of glycerol, $HOCH_2CH(OH)CH_2OH$(l) at 26.2 For glycerol, $d = 1.26$ g/mL; $C_p = 219$ J mol^{-1} K

145. Write the balanced chemical equations for react that have the following as their standard enth changes.

(a) $\Delta_f H° = +82.05$ kJ/mol N_2O(g)
(b) $\Delta_f H° = -394.1$ kJ/mol SO_2Cl_2(l)
(c) $\Delta_c H° = -1527$ kJ/mol CH_3CH_2COOH(l)

146. The standard molar heats of combustio C(graphite) and CO(g) are −393.5 and −283 kJ/ respectively. Use those data and that for the fol ing reaction

$$CO(g) + Cl_2(g) \longrightarrow COCl_2(g)$$
$$\Delta_r H° = -108 \text{ kJ m}$$

to calculate the standard molar enthalpy of fo tion of $COCl_2$(g).

147. Can a chemical compound have a standard enth of formation of zero? If so, how likely is this to oc Explain.

148. Is it possible for a chemical process to have ΔU and $\Delta H > 0$? Explain.

149. Use principles from this chapter to explain the ol vation that professional chefs prefer to cook w gas stove rather than an electric stove.

150. Hot water and a piece of cold metal come into co in an isolated container. When the final tempera of the metal and water are identical, is the energy change in this process **(a)** zero; **(b)** nega **(c)** positive; **(d)** not enough information.

151. A clay pot containing water at 25 °C is place the shade on a day in which the temperatu 30 °C. The outside of the clay pot is kept moist. the temperature of the water inside the clay **(a)** increase; **(b)** decrease; **(c)** remain the same?

152. Construct a concept map encompassing the i behind the first law of thermodynamics.

153. Construct a concept map to show the use of enth for chemical reactions.

154. Construct a concept map to show the i relationships between path-dependent and p independent quantities in thermodynamics.

8

lectrons in Atoms

ONTENTS

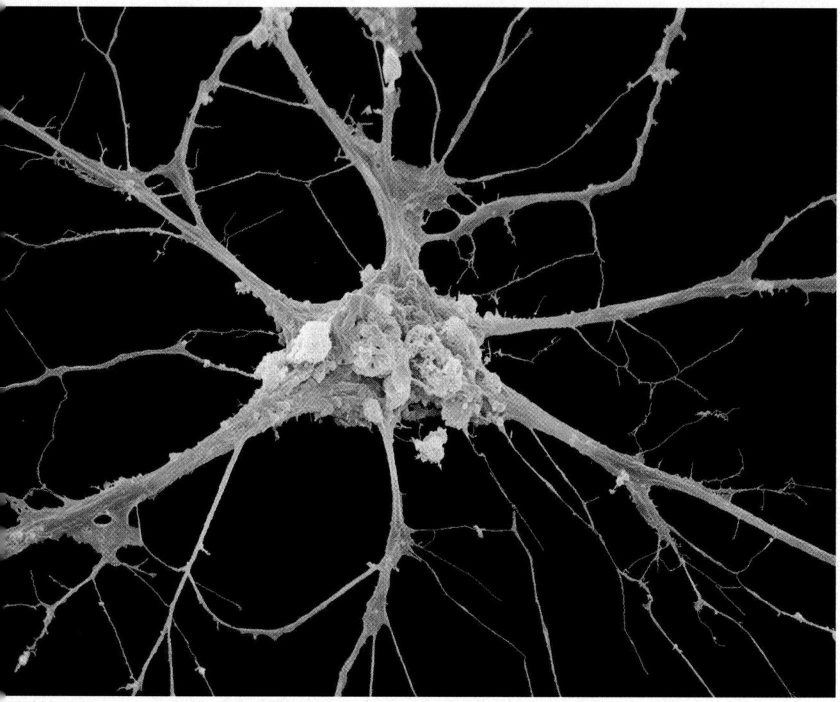

image of two neurons (gray objects) is produced by an electron microscope that on the wave properties of electrons discussed in this chapter.

At the end of the nineteenth century, some observers of the scientific scene believed that it was nearly time to close the books on the field of physics. They thought that with the accumulated knowledge of previous two or three centuries, the main work left to be done was to y this body of physics—classical physics—to such fields as chemistry biology.

nly a few fundamental problems remained, including an explanation rtain details of light emission and a phenomenon known as the photo-tric effect. But the solution to these problems, rather than marking an in the study of physics, spelled the beginning of a new golden age of sics. These problems were solved through a bold new proposal—the

LEARNING OBJECTIVES

8.1 Describe the amplitude, frequency, and wavelength of a wave and the relationships among them. Identify the various types of electromagnetic waves and their order within the electromagnetic series.

8.2 Discuss how the observation of blackbody radiation, the photoelectric effect, and atomic line spectra contributed to the development of quantum theory.

8.3 Construct an energy-level diagram for the hydrogen atom, and use it to explain why the spectrum of the hydrogen atom contains a limited number of wavelength components.

8.4 Describe the two revolutionary ideas by de Broglie and Heisenberg that led to the development of quantum mechanics.

8.5 Discuss the energy levels and wave functions of a particle in a one-dimensional box.

8.6 Explain the organization of hydrogen atom orbitals into shells and subshells.

8.7 Describe the shape, nodes (angular and radial), and orientations in three-dimensional space of the *s*, *p*, and *d* orbitals.

8.8 Identify the quantum numbers used to characterize electron spin.

8.9 Explain why, in multielectron atoms, orbitals with different values of ℓ within a principal shell have different energies.

8.10 Use the aufbau process to predict ground-state electron configurations of atoms.

8.11 Use the position of an element in the periodic table to predict the ground-state electron configuration of its atoms.

quantum theory—a scientific breakthrough of epic proportions. In this chap we will see that to explain phenomena at the atomic and molecular level, cl sical physics is inadequate—only the quantum theory will do.

The aspect of quantum mechanics emphasized in this chapter is h electrons are described through features known as quantum numbers a electron orbitals. The model of atomic structure developed here will expl many of the topics discussed in the next several chapters: periodic trends the physical and chemical properties of the elements, chemical bonding, a intermolecular forces.

Our understanding of the electronic structures of atoms will be gained studying the interactions of electromagnetic radiation and matter. The chap begins with background information about electromagnetic radiation, a then turns to connections between electromagnetic radiation and ato structure. The best approach to learning material in this chapter is to conc trate on the basic ideas relating to atomic structure, many of which are ill trated through the in-text examples. At the same time, pursue further det of interest in some of the Are You Wondering features and portions Sections 8-5, 8-7, and 8-9.

▶ Water waves, sound waves, and seismic waves (which produce earthquakes) are unlike electromagnetic radiation. They require a material medium for their transmission.

8-1 Electromagnetic Radiation

Electromagnetic radiation is a form of energy transmission in which elec and magnetic fields are propagated as waves through empty space (a vacu or through a medium, such as glass. A **wave** is a disturbance that transm energy through space or a material medium. Anyone who has sat in a small b on a large body of water has experienced wave motion. The wave moves ac the surface of the water, and the disturbance alternately lifts the boat and all it to drop. Although water waves may be more familiar, let us use a sim example to illustrate some important ideas and terminology about waves traveling wave in a rope.

Imagine tying one end of a long rope to a post and holding the other en your hand (Fig. 8-1). Imagine also that you have marked one small segmer the rope with red ink. As you move your hand up and down, you set u wave motion in the rope. The wave travels along the rope toward the dis post, but the colored segment simply moves up and down. In relation to center line (the broken line in Figure 8-1), the wave consists of *crests*, or h points, where the rope is at its greatest height above the center line, *troughs*, or low points, where the rope is at its greatest depth below the ce line. The maximum height of the wave above the center line or the maxim depth below is called the **amplitude**. The distance between the tops of successive crests (or the bottoms of two troughs) is called the **wavelen** designated by the Greek letter lambda, λ.

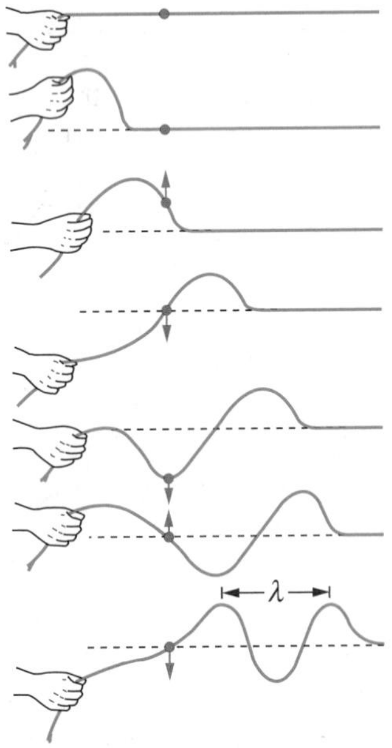

▲ FIGURE 8-1
The simplest wave motion—traveling wave in a rope
As a result of the up-and-down hand motion (top to bottom), waves pass along the long rope from left to right. This one-dimensional moving wave is called a traveling wave. The wavelength of the wave, λ—the distance between two successive crests—is identified.

Wavelength is one important characteristic of a wave. Another feature, **quency**, designated by the Greek letter nu, ν, is the number of crest troughs that pass through a given point per unit of time. Frequency has unit, time^{-1}, usually s^{-1} (per second), meaning the number of events or cy per second. The product of the length of a wave (λ) and the frequency shows how far the wave front travels in a unit of time. This is the speed o wave. Thus, if the wavelength in Figure 8-1 were 0.5 m and the frequency, (meaning three complete up-and-down hand motions per second), the sp of the wave would be $0.5\ \text{m} \times 3\ \text{s}^{-1} = 1.5\ \text{m/s}$.

We cannot actually see an electromagnetic wave as we do the trave wave in a rope, but we can try to represent it as in Figure 8-2. As the fig shows, the magnetic field component lies in a plane perpendicular to electric field component. An electric field is the region around an electric charged particle. The presence of an electric field can be detected by mea ing the force on an electrically charged object when it is brought into the fi

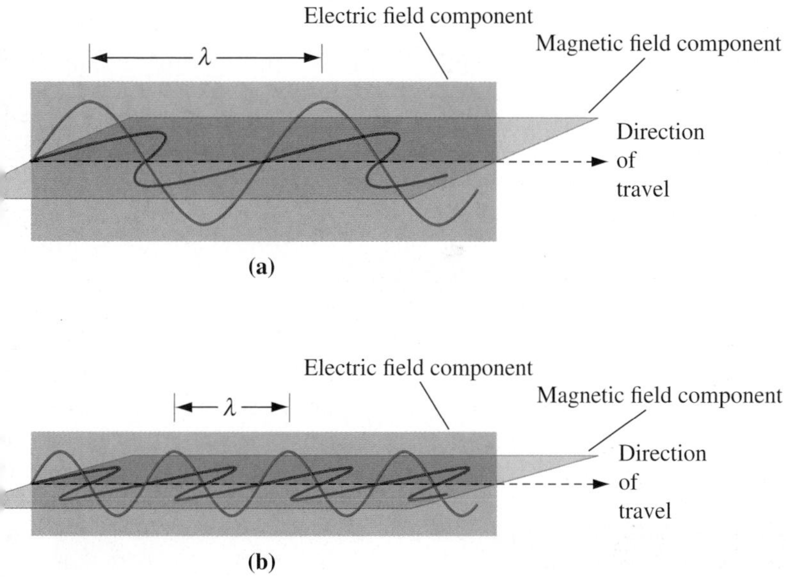

◀ FIGURE 8-2
Electromagnetic waves
This sketch of two different electromagnetic waves shows the propagation of mutually perpendicular oscillating electric and magnetic fields. For a given wave, the wavelengths, frequencies, and amplitudes of the electric and magnetic field components are identical. If these views are of the same instant of time, we would say that **(a)** has the longer wavelength and lower frequency, and **(b)** has the shorter wavelength and higher frequency.

agnetic field is found in the region surrounding a magnet. According to a ry proposed by James Clerk Maxwell (1831–1879) in 1865, electromag- c radiation—a propagation of electric and magnetic fields—is produced n accelerating electrically charged particle (a charged particle whose city changes). Radio waves, for example, are a form of electromagnetic ation produced by causing oscillations (fluctuations) of the electric cur- in a specially designed electrical circuit. With visible light, another n of electromagnetic radiation, the accelerating charged particles are the trons in atoms or molecules.

◀ Electromagnetic waves are *transverse* waves—the electric and magnetic fields are *perpendicular* to the perceived direction of motion. To a first approximation, water waves are also transverse waves. Sound waves, by contrast, are *longitudinal*. This effect is the result of small pulses of pressure that move in the *same* direction as the sound travels.

quency, Wavelength, and Speed Electromagnetic Radiation

SI unit for frequency, s^{-1}, is the **hertz (Hz)**, and the basic SI wavelength is the meter (m). Because many types of electromagnetic radiation have short wavelengths, however, smaller units, including those listed below, also used. The angstrom, named for the Swedish physicist Anders ström (1814–1874), is not an SI unit.

1 centimeter (cm) $= 1 \times 10^{-2}$ m

1 millimeter (mm) $= 1 \times 10^{-3}$ m

1 micrometer (μm) $= 1 \times 10^{-6}$ m

1 nanometer (nm) $= 1 \times 10^{-9}$ m $= 1 \times 10^{-7}$ cm $= 10$ Å

1 angstrom (Å) $= 1 \times 10^{-10}$ m $= 1 \times 10^{-8}$ cm $= 100$ pm

1 picometer (pm) $= 1 \times 10^{-12}$ m $= 1 \times 10^{-10}$ cm $= 10^{-2}$ Å

distinctive feature of electromagnetic radiation is its *constant* speed of 792458 $\times 10^8$ m s^{-1} in a vacuum, often referred to as the **speed of light**. speed of light is represented by the symbol *c*, and the relationship between speed and the frequency and wavelength of electromagnetic radiation is

◀ The speed of light is commonly rounded off to 3.00×10^8 m s^{-1}.

$$c = \nu \times \lambda \qquad (8.1)$$

re 8-3 indicates the wide range of possible wavelengths and frequencies some common types of electromagnetic radiation and illustrates this ortant fact: The wavelength of electromagnetic radiation is shorter for frequencies and longer for low frequencies. Example 8-1 illustrates the of equation (8.1).

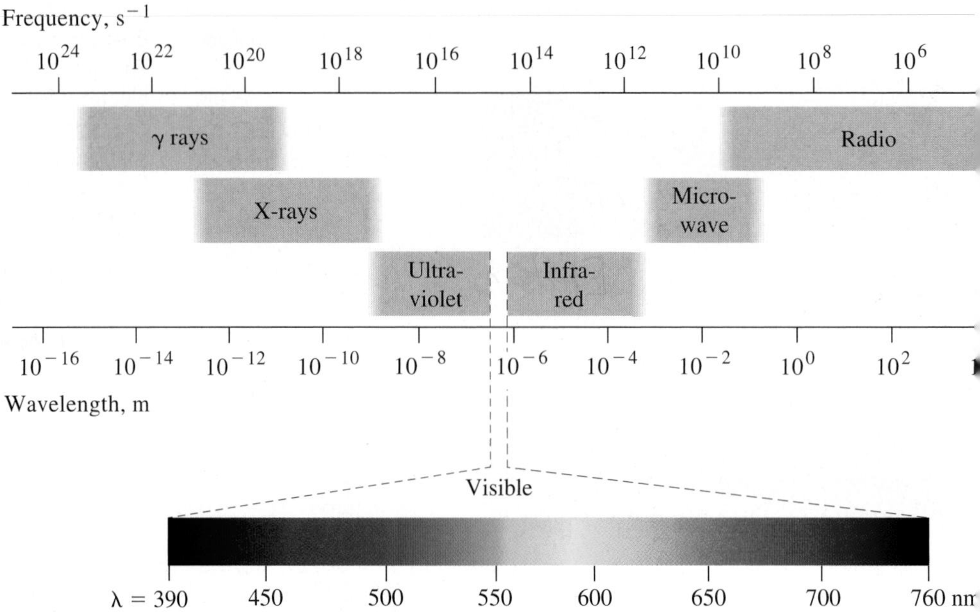

▲ FIGURE 8-3
The electromagnetic spectrum
The visible region, which extends from violet at the shortest wavelength to red at the longest wavelength, is only a small portion of the entire spectrum. The approximate wavelength and frequency ranges of some other forms of electromagnetic radiation are also indicated.

EXAMPLE 8-1 Relating Frequency and Wavelength of Electromagnetic Radiation

Most of the light from a sodium vapor lamp has a wavelength of 589 nm. What is the frequency of this radiatio

Analyze

To use equation (8.1), we first convert the wavelength of the light from nanometers to meters, since the spee of light is in m s^{-1}. Then, we rearrange it to the form $\nu = c/\lambda$ and solve for ν.

Solve

Change the units of λ from nanometers to meters.

$$\lambda = 589\ \text{nm} \times \frac{1 \times 10^{-9}\ \text{m}}{1\ \text{nm}} = 5.89 \times 10^{-7}\ \text{m}$$

$$c = 2.998 \times 10^{8}\ \text{m s}^{-1}$$

$$\nu = ?$$

Rearrange equation (8.1) to the form $\nu = c/\lambda$, and solve for ν.

$$\nu = \frac{c}{\lambda} = \frac{2.998 \times 10^{8}\ \text{m s}^{-1}}{5.89 \times 10^{-7}\ \text{m}} = 5.09 \times 10^{14}\ \text{s}^{-1} = 5.09 \times 10^{14}\ \text{Hz}$$

Assess

The essential element here is to recognize the need to change the units of λ. This change is often needed whe converting wavelength to frequency and vice versa.

PRACTICE EXAMPLE A: The light from red LEDs (light-emitting diodes) is commonly seen in many electron devices. A typical LED produces 690 nm light. What is the frequency of this light?

PRACTICE EXAMPLE B: An FM radio station broadcasts on a frequency of 91.5 megahertz (MHz). What is th wavelength of these radio waves in meters?

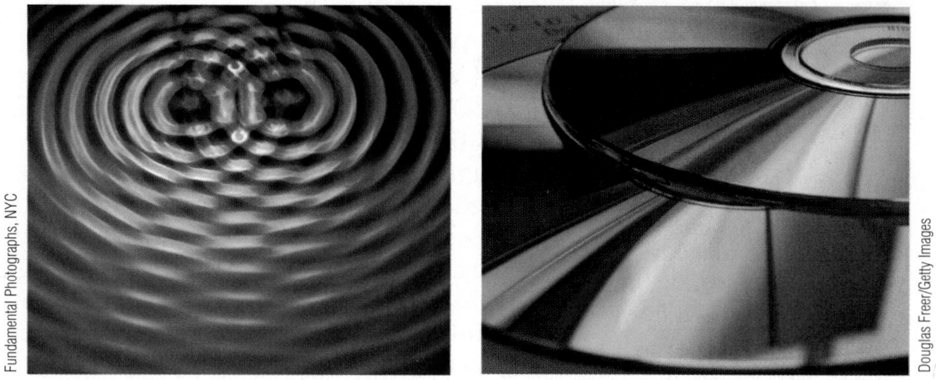

(a) (b)

GURE 8-4
nples of interference
tones and ripples.
D reflection.

An Important Characteristic of Electromagnetic Waves

The properties of electromagnetic radiation that we will use most extensively are those just introduced—amplitude, wavelength, frequency, and speed. Another essential characteristic of electromagnetic radiation, which will underpin our discussion of atomic structure later in the chapter, is described next.

If two pebbles are dropped close together into a pond, ripples (waves) emerge from the points of impact of the two stones. The two sets of waves intersect, and there are places where the waves disappear and places where the waves persist, creating a crisscross pattern (Fig. 8-4a). Where the waves are "in step" upon meeting, their crests coincide, as do their troughs. The waves combine to produce the highest crests and deepest troughs in the water. The waves are said to be *in phase*, and the addition of the waves is called *constructive interference* (Fig. 8-5a). Where the waves meet in such a way that the peak of one wave occurs at the trough of another, the waves cancel and the water is flat (Fig. 8-5b). These out-of-step waves are said to be *out of phase*, and the cancellation of the waves is called *destructive interference*.

An everyday illustration of interference involving electromagnetic waves is seen in the rainbow of colors that shine from the surface of a compact disc

EP IN MIND

t destructive interference urs when waves are of phase by one-half velength. If waves are out phase by more or less than , but also not completely phase, then only partial tructive interference urs.

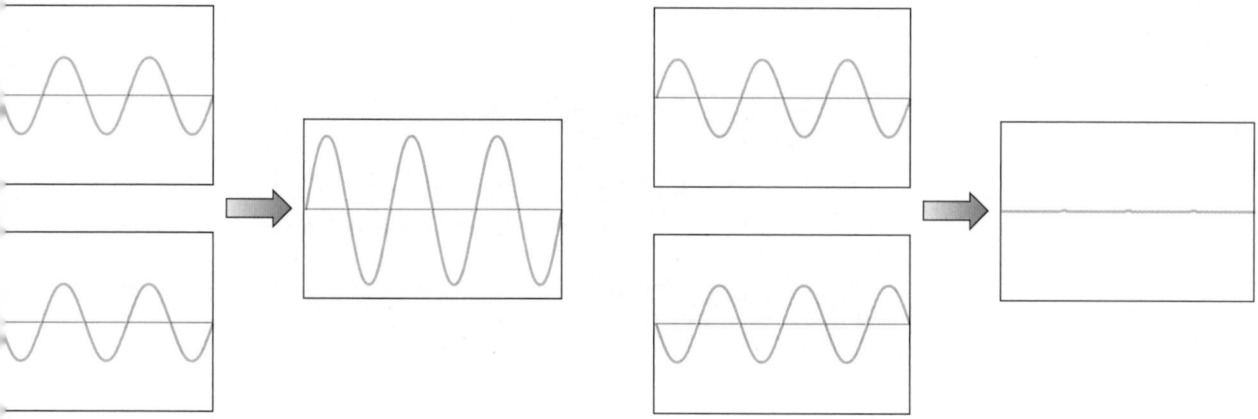

(a) (b)

▲ FIGURE 8-5
Interference in two overlapping light waves
(a) In constructive interference, the troughs and crests are in step (in phase), leading to addition of the two waves. **(b)** In destructive interference, the troughs and crests are out of step (out of phase), leading to cancellation of the two waves.

▲ FIGURE 8-6
Refraction of light
Light is refracted (bent) as it passes from air into the glass prism, and again as it emerges from the prism into air.

(Fig. 8-4b). White light, such as sunlight, contains all the colors of the r bow. The colors differ in wavelength (and frequency), and when these ferent wavelength components are reflected off the tightly spaced groove the CD, they travel slightly different distances. This creates phase dif ences that depend on the angle at which we hold the CD to the light sou The light waves in the beam interfere with each other, and, for a given a between the incoming and reflected light, all colors cancel except one. L waves of that color interfere constructively and reinforce one another. T as we change the angle of the CD to the light source, we see different co The dispersion of different wavelength components of a light beam thro the interference produced by reflection from a grooved surface is ca **diffraction**.

Diffraction is a phenomenon that can be explained only as a propert waves. Both the physical picture and mathematics of interference and diff tion are the same for water waves and electromagnetic waves.

▶ The *wave nature of light* is demonstrated by its ability to be dispersed by diffraction and refraction.

The Visible Spectrum

The speed of light is lower in any medium than it is in a vacuum. Also, speed is different in different media. As a consequence, light is refracte bent, when it passes from one medium to another (Fig. 8-6). Moreover, altho electromagnetic waves all have the same speed in a vacuum, waves of diffe wavelengths have slightly different speeds in air and other media. Thus, wh beam of white light is passed through a transparent medium, the wavele components are refracted differently. The light is dispersed into a band of co a *spectrum*. In Figure 8-7(a), a beam of white light (for example, sunlight) is persed by a glass prism into a continuous band of colors corresponding to the wavelength components from red to violet. This is the visible spect shown in Figure 8-3 and also seen in a rainbow, where the medium disperses the sunlight is droplets of water (Fig. 8-7b).

8-1 CONCEPT ASSESSMENT

Red laser light is passed through a device called a frequency doubler. What is the approximate color of the light that exits the frequency doubler? How are the wavelengths of the original light and the frequency-doubled light related?

▶ The *importance of light* to chemistry is that light is a form of energy and that by studying light–matter interactions we can detect energy changes in atoms and molecules. Another means of monitoring the energy of a system is through observations of heat transfer. Light can be more closely controlled and thus gives us more detailed information than can be obtained with heat measurements.

(a) (b)

▲ FIGURE 8-7
The spectrum of "white" light
(a) Dispersion of light through a prism. Red light is refracted the least and violet lig the most when "white" light is passed through a glass prism. The other colors of th visible spectrum are found between red and violet. **(b)** Rainbow near a waterfall. Here, water droplets are the dispersion medium.

(a) Andrea Danti/Shutterstock; **(b)** Photos.com/Jupiterimages

2 Prelude to Quantum Theory

development of quantum theory was driven by several experiments, each ɔlving the interaction of light and matter. To explain the results obtained ɪese experiments, scientists had to reformulate the physical laws that govern ɔehavior of particles at the atomic scale. In this section, we focus on a few of e experiments and discuss how they contributed to the development of ortant new ideas and undoubtedly the biggest scientific revolution of the 100 years.

▲ Light emission by molten iron.

ckbody Radiation

are aware that hot objects emit light of different colors, from the dull red of lectric-stove heating element to the bright white of a light bulb filament or ten iron. Light emitted by a hot radiating object can be dispersed by a prism roduce a continuous color spectrum. As seen in Figure 8-8, the light intensity es smoothly with wavelength, peaking at a wavelength fixed by the source perature. Classical physics could not provide a complete explanation of t emission by heated solids, a process known as *blackbody radiation*. Classical ry predicts that the intensity of the radiation emitted would increase indefi-ly as λ decreases (or as ν increases), as indicated by the dashed lines in re 8-8. In 1900, to explain the fact that the intensity does not increase indefi-ly, Max Planck (1858–1947) made a revolutionary proposal: *Energy, like mat-s discontinuous*. Here, then, is the essential difference between the classical sics of Planck's time and the new quantum theory that he proposed: sical physics places no limitations on the amount of energy a system may ess, whereas quantum theory limits this energy to a set of specific values. *difference* between any two allowed energies of a system also has a specific e, called a **quantum** of energy. This means that when the energy increases ı one allowed value to another, it increases by a tiny jump, or quantum. e is a way of thinking about a quantum of energy: It bears a similar relation- to the total energy of a system as a single atom does to an entire sample atter.

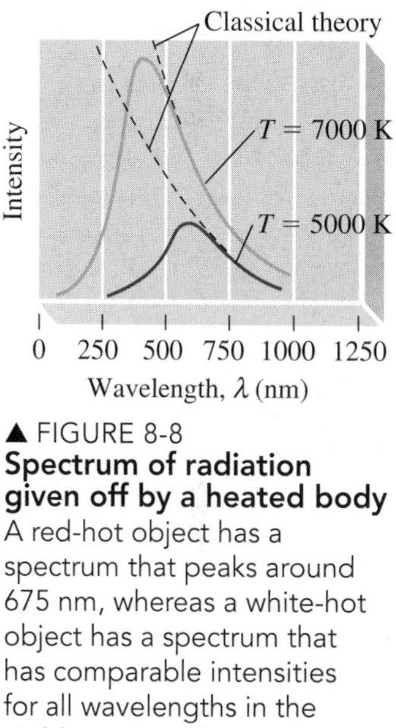

▲ FIGURE 8-8
Spectrum of radiation given off by a heated body
A red-hot object has a spectrum that peaks around 675 nm, whereas a white-hot object has a spectrum that has comparable intensities for all wavelengths in the visible region. The sun has a blackbody temperature of about 5750 K. Objects emit radiation at *all* temperatures, not just at high temperatures. For example, night-vision goggles make infrared radiation emitted by objects visible in the dark.

he model Planck used for the emission of electromagnetic radiation was of a group of atoms on the surface of the heated object oscillating together ı the same frequency. Planck's approach was equivalent to assuming that group of atoms, the oscillator, must have an energy corresponding to the ation

$$\epsilon = nh\nu$$

re ϵ is the energy, n is a positive integer, ν is the oscillator frequency, and a constant that had to be determined by experiment. By using his theory experimental data for the distribution of frequencies with temperature, ıck established the following value for the constant h. We now call it **ıck's constant**, and it has the value

$$h = 6.62607 \times 10^{-34}\ \mathrm{J\ s}$$

ıck's postulate can be rephrased in this more general way: The energy of a ntum of electromagnetic radiation is proportional to the frequency of the ation—the higher the frequency, the greater the energy. This is summa-d by what we now call Planck's equation.

◀ Planck's equation can be used to develop relationships among frequency, wave-length, and energy. By using this information, the relative energies of radiation on the electromagnetic spectrum can be compared.

$$E = h\nu \qquad (8.2)$$

Akg-images/Newscom

▲ **Max Planck (1858–1947)**
The results obtained by Planck in his analysis of blackbody radiation were developed within about eight weeks. Recounting this period many years later, Planck remarked, "After a few weeks of the most strenuous work of my life, the darkness lifted and an unexpected vista began to appear."

8-1 ARE YOU WONDERING?

How do Planck's ideas account for the fact that the intensity of blackbody radiation drops off at higher frequencies?

Planck was aware of the work of Ludwig Boltzmann, who, with James Maxwe had derived an equation to account for the distribution of molecular speed Boltzmann had shown that the relative chance of finding a molecule with a pa ticular speed was related to its kinetic energy by the following expression.

$$\text{relative chance} \propto e^{\left(-\frac{\text{kinetic energy}}{k_B T}\right)}$$

where k_B is the Boltzmann constant, and T is the Kelvin temperature. You will al notice that the curve of intensity versus wavelength in Figure 8-8 bears a stro resemblance to the distribution of molecular speeds in Figure 6-15. Plan assumed that the energies of the groups of atoms oscillating to emit blackbo radiation were distributed according to the Boltzmann distribution law. That the relative chance of an oscillator having the energy $nh\nu$ is proportional $e^{-nh\nu/k_BT}$, where n is an integer, 1, 2, 3, and so on. So this expression shows that t chance of an oscillator having a high frequency is lower than for oscillators ha ing lower frequencies because as n increases, $e^{-nh\nu/k_BT}$, decreases. The assumptic that the energy of the oscillators in the light-emitting source cannot have contin ous values leads to excellent agreement between theory and experiment.

At the time Planck made his quantum hypothesis, scientists had ha previous experience with macroscopic physical systems that required existence of separate energy levels and that energy may only be emitte absorbed in specific quanta. Their experience was that there were no t retical limits on the energy of a system and that the transfer of energy continuous. Thus it is not surprising that scientists, including Planck l self, were initially skeptical of the quantum hypothesis. It had b designed to explain radiation from heated bodies and certainly could be accepted as a general principle until it had been tested on o applications.

Only after the quantum hypothesis was successfully applied to phen ena other than blackbody radiation did it acquire status as a great new sc tific theory. The first of these successes came in 1905 with Albert Einste quantum explanation of the photoelectric effect.

The Photoelectric Effect

In 1888, Heinrich Hertz discovered that when light strikes the surface of tain metals, electrons are ejected. This phenomenon is called the **photoele effect** and the electrons emitted through this process are called photo trons. The salient feature of the photoelectric effect is that electron emis only occurs when the frequency of the incident light exceeds a partic threshold value (ν_0). When this condition is met,

- the number of electrons emitted depends on the intensity of the inci light, but
- the kinetic energies of the emitted electrons depend on the frequen the light.

These observations, especially the dependency on frequency, could n explained by classical wave theory. However, Albert Einstein showed that are exactly what would be expected with a particle interpretation of radia In 1905, Einstein proposed that electromagnetic radiation has particle lities and that "particles" of light, subsequently called **photons** by J. Lewis, have a characteristic energy given by

$$E_{\text{photon}} = h\nu \quad (8.3)$$

n the particle model, a photon of energy $h\nu$ strikes a bound electron, which orbs the photon energy. If the photon energy, $h\nu$, is greater than the energy ding the electron to the surface (a quantity known as the *work function*), a toelectron is liberated. Thus, the lowest frequency light producing the toelectric effect is the threshold frequency, and any energy in excess of the k function appears as kinetic energy in the emitted photoelectrons. The k function is represented by the symbol Φ and is, by definition, the minim energy needed to extract an electron from a metal's surface.

◀ Light–matter interactions usually involve *one photon per atom or electron*. Thus, to escape from a photoelectric surface, an electron must do so with the energy from a single photon collision. The electron cannot accumulate the energy from several hits by photons.

n the discussion that follows, based on the experimental setup shown in ure 8-9, we will see how the threshold frequency and work function are luated. Also, we will see that the photoelectric effect provides an independent evaluation of Planck's constant, h.

n Figure 8-9, light (designated $h\nu$) is allowed to shine on a piece of metal in evacuated chamber. The electrons emitted by the metal (photoelectrons) el to the upper plate and complete an electric circuit set up to measure the toelectric current through an ammeter. Figure 8-9(b) illustrates the variation he photoelectric current, I_p, detected by the ammeter as the frequency (ν) intensity of the incident light is increased. We see that no matter how nse the light, no current flows if the frequency is below the threshold frency, ν_0, and no photoelectric current is produced. In addition no matter v weak the light, there is a photoelectric current if $\nu > \nu_0$. The magnitude he photoelectric current is, as shown in Figure 8-9(b), directly proportional he intensity of the light, so that the number of photoelectrons increases

◀ With the advent of lasers we have observed the simultaneous absorption of two photons by one electron. Instances of two adjacent molecules cooperatively absorbing one photon are also known. Such occurrences are exceptions to the more normal one photon/one electron phenomena.

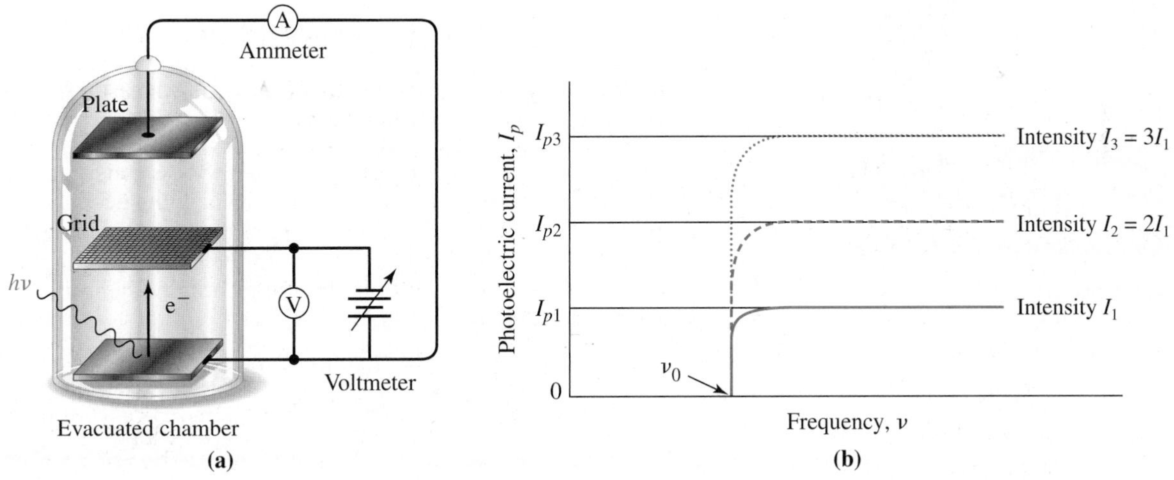

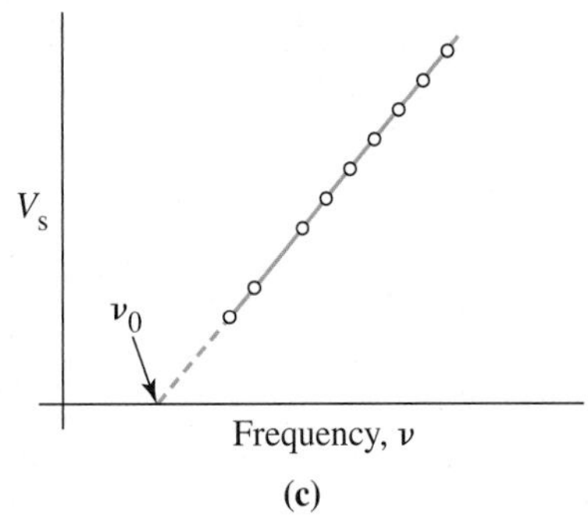

GURE 8-9
photoelectric effect
Schematic diagram of the apparatus for photoelectric effect measurements. The photoelectric current, I_p, measured as a function of frequency, ν, for e different intensities of light, I. The photoelectric current appears only if greater than the threshold value, ν_0. For $\nu > \nu_0$, the photoelectric current eases proportionally with the intensity of the light. For example, when the nsity of light is increased by a factor of two, from I_1 to $I_2 = 2I_1$, the toelectric current increases by a factor of two, from I_{p_1} to $I_{p_2} = 2I_{p_1}$. topping voltage of photoelectrons as a function of frequency of incident ation. The stopping voltage (V_s) is plotted against the frequency of the dent radiation. The threshold frequency (ν_0) of the metal is found by apolation.

with the intensity of the incident light. Therefore, we can associate light in sity with the number of photons arriving at a point per unit time.

A second circuit is set up to measure the velocity of the photoelectrons, hence their kinetic energy. In this circuit, a potential difference (voltag maintained between the photoelectric metal and an open-grid electr placed below the upper plate. For electric current to flow, electrons must through the openings in the grid and onto the upper plate. The nega potential on the grid acts to slow down the approaching electrons. the potential difference between the grid and the emitting metal is increase point is reached at which the photoelectrons are stopped at the grid and current ceases to flow through the ammeter. The potential difference at point is called the *stopping voltage*, V_s. At the stopping voltage, the kin energy of the photoelectrons has been converted to potential ene expressed through the following equation (in which m, u, and e are the m speed, and charge of an electron, respectively).

$$\frac{1}{2}mu^2 = eV_s$$

As a result of experiments of the type just described, we find that V_s is portional to the frequency of the incident light but independent of the l intensity. Also, as shown in Figure 8-9, if the frequency, ν, is below the *thres frequency*, ν_0, no photoelectric current is produced. At frequencies greater t ν_0, the empirical equation for the stopping voltage is

$$V_s = k(\nu - \nu_0)$$

The constant k is independent of the metal used, but ν_0 varies from one m to another. Although there is no relation between V_s and the light intensity, photoelectric current, I_p, is proportional to the intensity of the light as il trated in Figure 8-9(b).

Bettmann/Corbis

▲ Albert Einstein (1879–1955)
Albert Einstein received a Nobel Prize for his work on the photoelectric effect. However, he is better known for his development of the theory of relativity, and $E = mc^2$.

8-2 ARE YOU WONDERING?

In what ways is a photon the same as, or different from, other more familiar particles?

To explain the photoelectric effect, light of frequency, ν, is considered a stream particle-like entities (photons), each of which travels with speed c and carries energy given by equation (8.3). Emission of a photoelectron is imagined to occ as the result of a collision between a photon of the incident light and an electr in the target. As a result of the collision, the energy and momentum of the phot are transferred to the electron.

Classically, we think of a particle as having a mass m and speed u. Because of i motion, the particle possesses kinetic energy $E_k = \left(\frac{1}{2}\right)mu^2$ and momentu $p = mu$. A photon is like a particle in that it is a carrier of both energy ar momentum, but it is unlike a "regular" particle in that it has no mass. How is that a photon, with zero mass, possesses momentum? The answer lies Einstein's theory of special relativity. Einstein derived the following expressio which relates the energy and momentum of a particle.

$$E^2 = (pc)^2 + (m_0c^2)^2$$

In the expression above, m_0 is the rest mass, or intrinsic mass, of the particle. By de inition, it is the measured mass of the particle when it is at rest with respect to t person or detector making the measurement. For a photon, $m_0 = 0$, and so t expression above reduces to $E = pc$. Because $E = h\nu$ for a photon, we can write

$$p = \frac{h\nu}{c} = \frac{h}{\lambda}$$

om this expression for p, we see that the wave and particle models of light are inti-
ately connected. We will see in Section 8-4 that the expression $p = h/\lambda$ applies to
particles, not just photons.
The equation $p = h/\lambda$ also helps us understand the effect of a transfer of momen-
m in a collision of a photon with another particle, such as an electron. If a photon
ansfers some of its momentum to another particle, then the momentum, p, of the
oton decreases and, as a consequence, its wavelength, λ, increases. The change in
avelength that occurs when light is scattered by electrons in atoms in a crystal (the
ompton effect) was first observed in 1923. The Compton effect provides additional
nfirmation that light consists of particle-like entities that can transfer momentum
other particles through collisions.

he work function, Φ, for a given metal represents the minimum quantity
ork—and hence, the minimum quantity of energy—needed to extract an
tron from a metal's surface. According to Einstein's model, light of frency ν_0 consists of photons with just enough energy to liberate electrons.
s, the work function may be expressed as the product of Planck's constant
the threshold frequency $\Phi = h\nu_0$, and as the product of the charge on the
tron, e, and the potential, V_0, that has to be overcome, $\Phi = eV_0$. Therefore,
$h\nu_0 = eV_0$. Thus, the threshold frequency for the photoelectric effect is
n by the expression

$$\nu_0 = \frac{\Phi}{h} = \frac{eV_0}{h}$$

e the work function is a characteristic of the metal used in the experiment, ν_0
so a characteristic of the metal, as confirmed by experiment.
Vhen a photon of energy $h\nu$ strikes an electron in the metal's surface, some
e energy is used to do the work of freeing the electron, and the rest is used
npart kinetic energy to the liberated electron. Thus, by conservation of
gy, we have

$$E_{\text{photon}} = \Phi + \frac{1}{2}mu^2$$

ince $E_{\text{photon}} = h\nu$ and $\Phi = eV_0$, we can also write

$$\frac{1}{2}mu^2 + eV_0 = h\nu$$

ch gives

$$eV_s = \frac{1}{2}mu^2 = h\nu - eV_0$$

ch is identical to the empirically determined equation for V_s with $k = h/e$
$h\nu_0 = eV_0$. Careful experiments showed that the constant h had the same
e as determined by Planck for blackbody radiation. The additional fact that
number of photoelectrons increases with the intensity of light indicates that
should associate light intensity with the number of photons arriving at a
nt per unit time.

8-2 CONCEPT ASSESSMENT

e wavelength of light needed to eject electrons from hydrogen atoms is
.2 nm. When light of 80.0 nm is shone on a sample of hydrogen atoms,
ectrons are emitted from the hydrogen gas. If, in a different experiment, the
avelength of the light is changed to 70.0 nm, what is the effect compared to
e use of 80.0 nm light? Are more electrons emitted? If not, what happens?

EXAMPLE 8-2 Using Planck's Equation to Calculate the Energy of Photons of Light

For radiation of wavelength 242.4 nm, the longest wavelength that will bring about the photodissociation O_2, what is the energy of **(a)** one photon, and **(b)** a mole of photons of this light?

Analyze

To use Planck's equation, we need the frequency of the radiation. We can get this from equation (8.1) after fir expressing the wavelength in meters. Planck's equation is written for one photon of light. We emphasize th by including the unit in the value of h. Once we have the energy per photon, we can multiply it by th Avogadro constant to convert to a per-mole basis.

Solve

(a) First, calculate the frequency of the radiation.

$$\nu = \frac{c}{\lambda} = \frac{2.998 \times 10^{8}\,\mathrm{m\,s^{-1}}}{242.4 \times 10^{-9}\,\mathrm{m}} = 1.237 \times 10^{15}\,\mathrm{s^{-1}}$$

Then, calculate the energy of a single photon.

$$E = h\nu = 6.626 \times 10^{-34}\,\frac{\mathrm{J\,s}}{\mathrm{photon}} \times 1.237 \times 10^{15}\,\mathrm{s^{-1}}$$
$$= 8.196 \times 10^{-19}\,\mathrm{J/photon}$$

(b) Calculate the energy of a mole of photons.

$$E = 8.196 \times 10^{-19}\,\mathrm{J/photon} \times 6.022 \times 10^{23}\,\mathrm{photons/mol}$$
$$= 4.936 \times 10^{5}\,\mathrm{J/mol}$$

Assess

We can see from this example that when the energy of a single photon is expressed in SI units, the energy rather small and perhaps difficult to interpret. However, the amount of energy carried by a *mole* of photons something we can easily relate to. As shown above, light with a wavelength of 242.4 nm has an energy conte of 493.6 kJ/mol, which is similar in magnitude to the internal energy and enthalpy changes of chemical rea tions (see Chapter 7).

PRACTICE EXAMPLE A: The protective action of ozone in the atmosphere comes through ozone's absorption UV radiation in the 230 to 290 nm wavelength range. What is the energy, in kilojoules per mole, associate with radiation in this wavelength range?

PRACTICE EXAMPLE B: Chlorophyll absorbs light at energies of 3.056×10^{-19} J/photon and 4.414×10^{-19} J/photo To what color and frequency do these absorptions correspond?

Atomic Emission Spectra

The visible spectrum in Figure 8-7 is said to be a *continuous spectrum* bec the light being dispersed consists of many wavelength components. If source of a spectrum produces light having only a relatively small numb wavelength components, then a *discontinuous spectrum* is observed. For ex ple, if the light source is an electric discharge passing through a gas, only tain colors are seen in the spectrum (Fig. 8-10a, b). Or if the light source is a flame into which an ionic compound has been introduced, the flame acquire a distinctive color indicative of the metal ion present (Fig. 8-10c–e each case, the emitted light produces a spectrum consisting of only a lim number of discrete wavelength components, observed as colored lines dark spaces between them. These discontinuous spectra are called **atomi line, spectra**. The production of the line spectrum of helium is illustrate Figure 8-11. The light source is a lamp containing helium gas at a low press When an electric discharge is passed through the lamp, helium atoms ab energy, which they then emit as light. The light is passed through a narrow and then dispersed by a prism. The colored components of the light

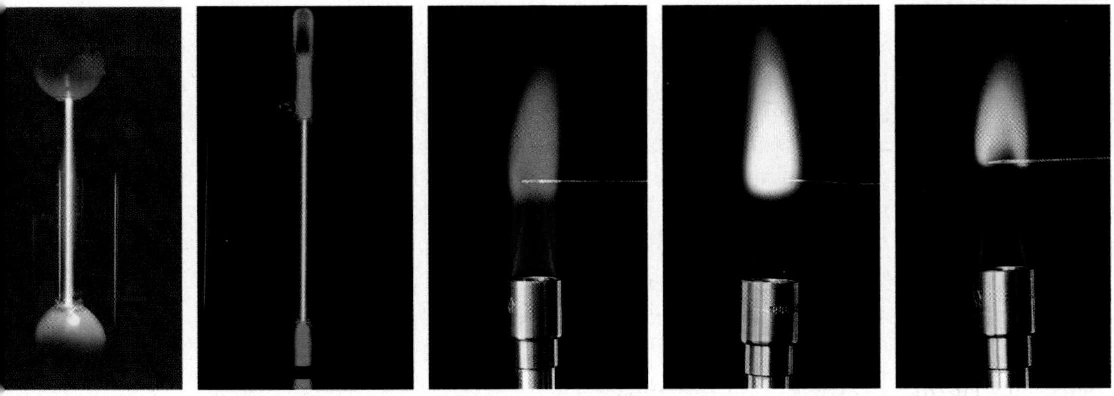

GURE 8-10
rces for light emission
t emitted by an electric discharge through **(a)** hydrogen gas and **(b)** neon gas. Light
ted when compounds of the alkali metals are excited in the gas flames: **(c)** lithium,
odium, and **(e)** potassium.

y B. Van Loon; **(b) to (e)** Tom Pantages

ected and recorded on photographic film. Each wavelength component ears as an image of the slit: a thin line. In all, five lines in the spectrum of um can be seen with the unaided eye.

ach element has its own distinctive line spectrum—a kind of atomic finger- t. Robert Bunsen (1811–1899) and Gustav Kirchhoff (1824–1887) developed first spectroscope and used it to identify elements. In 1860, they discovered w element and named it cesium (Latin, *caesius*, sky blue) because of the dis- tive blue lines in its spectrum. They discovered rubidium in 1861 in a simi- way (Latin, *rubidius*, deepest red). Still another element characterized by its que spectrum is helium (Greek, *helios*, the sun). Its spectrum was observed ing the solar eclipse of 1868, but helium was not isolated on Earth for ther 27 years.

◀ Bunsen designed a special gas burner for his spectroscopic studies. This burner, the common laboratory Bunsen burner, produces very little background radiation to interfere with spectral observations.

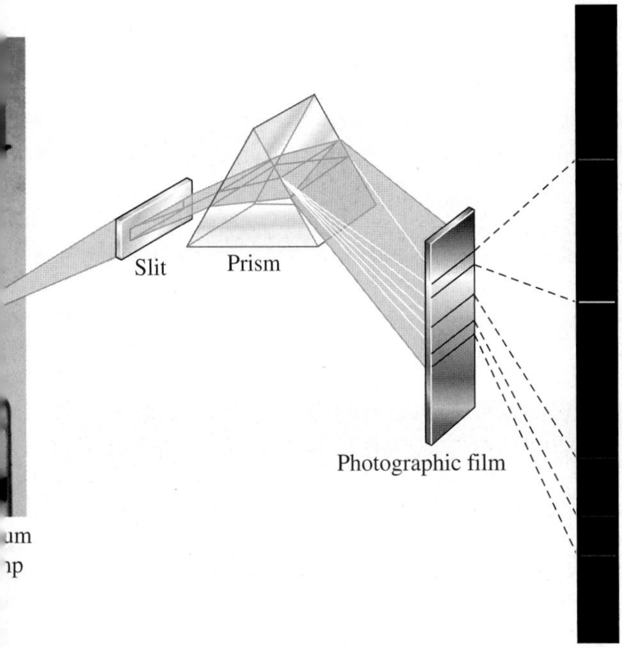

◀ FIGURE 8-11
The atomic, or line, spectrum of helium
The apparatus pictured here, in which the spectral lines are photographed, is called a *spectrograph*. If the observations are made by visual sighting alone, the device is called a *spectroscope*. If the positions and brightness of the lines are measured and recorded by other than visual or photographic means, the term generally used is *spectrometer*.

Among the most extensively studied atomic spectra has been the hydro spectrum. Light from a hydrogen lamp appears to the eye as a reddish-pu (Fig. 8-10a). The principal wavelength component of this light is red ligh wavelength 656.3 nm. Three other lines appear in the visible spectrun atomic hydrogen, however: a greenish-blue line at 486.1 nm, a violet lin 434.0 nm, and another violet line at 410.1 nm. The visible atomic spectrun hydrogen is shown in Figure 8-12. In 1885, Johann Balmer, apparently thro trial and error, deduced the following formula for the wavelengths of th spectral lines:

$$\lambda = \frac{Bm^2}{m^2 - n^2}$$

In this equation, B is a constant having the value 364.6 nm, and m and n resent integers. When n is set equal to 2 and m is set equal to 3, the wavelen of the red line is obtained. With $n = 2$ and $m = 4$, the wavelength of the green blue line is obtained. The remaining two lines in the visible spectrum obtained by using $n = 2$ with $m = 5$ and $m = 6$. An important use of Balm formula was the identification of spectral lines of hydrogen in other region the electromagnetic spectrum, such as those corresponding to $n = 2$ and m to $m = 11$, found in the ultraviolet spectra of white stars seen by astronom years earlier. The series of lines having $n = 2$ is now known as the Bal series. Balmer also speculated that if other values of n were used, then ot series in the infrared and ultraviolet regions could be generated. We will that this is indeed the case.

Balmer's equation was later found to be a special case of the Rydberg mula devised by Johannes Rydberg in 1888.

$$\frac{1}{\lambda} = \frac{4}{B}\left(\frac{1}{n^2} - \frac{1}{m^2}\right) = R_H\left(\frac{1}{n^2} - \frac{1}{m^2}\right) \quad (8$$

R_H is the Rydberg constant for the hydrogen atom, the value of wh is $1.09678 \times 10^7\ m^{-1}$. The wavelengths of the lines in the Balmer series obtained by using $n = 2$ and $m > n$ in equation (8.4).

The fact that the atomic emission spectra consist of only limited number well-defined wavelength lines suggests that only a limited number of ene values are available to excited gaseous atoms. Why is the energy of an at restricted to a limited number of energy values? The search for an answe this question not only provided scientists with a great opportunity to le about the structures of atoms but also led them to one of the greatest bre throughs of modern science, namely, quantum theory.

8-3 CONCEPT ASSESSMENT

When comet Shoemaker–Levy 9 crashed into Jupiter's surface, scientists viewe the event with spectrographs. What did they hope to discover?

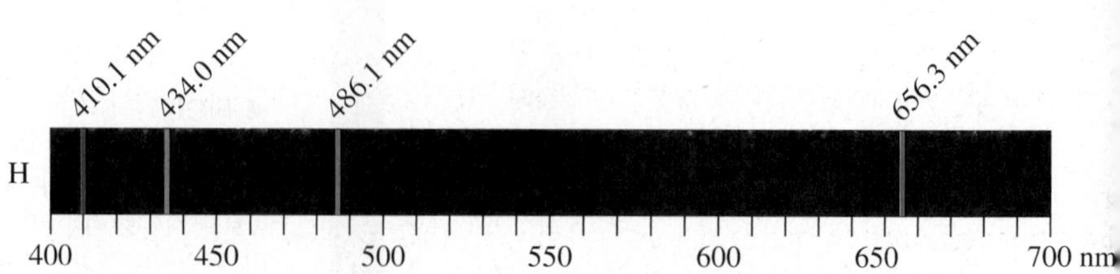

▲ FIGURE 8-12
The Balmer series for hydrogen atoms—a line spectrum
The four lines shown are the only ones visible to the unaided eye. Additional, closel spaced lines lie in the ultraviolet (UV) region.

Early Attempt to Understand Atomic Emission Spectra: ɘ Bohr Model

ording to the Rutherford model of the nuclear atom (Section 2-3), the elec- ıs in an atom are arranged outside the nucleus of an atom. But how are the trons arranged and how do they behave? If, for example, the negatively rged electrons were stationary, then they would be pulled into the posi- ly charged nucleus. Therefore, the electrons in an atom must be in motion. is assumed that electrons move around the nucleus like the planets orbit- the sun, a problem arises. According to classical physics, orbiting electrons constantly accelerating and, thus, radiating energy. By losing energy, the trons would be drawn ever closer to the nucleus and soon spiral into it. In 3, Niels Bohr (1885–1962) tried to resolve this problem by postulating the owing for a hydrogen atom:

The electron moves about the nucleus with speed u in one of a fixed set of circular orbits; as long as the electron remains in a given orbit, its energy is constant and no energy is emitted. Thus, each orbit is characterized by a fixed radius, r, and a fixed energy, E.

The electron's angular momentum, $\ell = mur$, is an integer multiple of $(h/2\pi)$, that is $\ell = n \times (h/2\pi)$, with $n = 1, 2, 3$, and so on.

An atom emits energy as a photon when the electron falls from an orbit of higher energy and larger radius to an orbit of lower energy and smaller radius.

▲ **Niels Bohr (1885–1962)** In addition to his work on the hydrogen atom, Bohr headed the Institute of Theoretical Physics in Copenhagen, which became a mecca for theoretical physicists in the 1920s and 1930s.

Stamp from the private collection of Professor C. M. Lang. Photography by Gary J. Shulfer, University of Wisconsin, Stevens Point. "1963, Denmark (Scott #409)"; Scott Standard Postage Stamp Catalogue, Scott Pub. Co., Sidney, Ohio.

condition in point 2 introduces an integer, n, to restrict the angular momen- ı, ℓ, to specific values: $h/(2\pi)$, h/π, $3h/(2\pi)$, $2h/\pi$, and so on. The integer n ılled a *quantum number* and the condition on ℓ is an example of a *quantization lition*. Bohr could not provide a physical justification for this quantization dition. He deduced it by working backward from equation (8.4). Thus, by ıg classical theory and imposing a quantization condition, Bohr was able to ive equations for the energies and radii of the allowed orbits. Exercise 120 ıses on the derivation of these equations, the most important of which is the owing equation for the energy:

$$E_n = -\frac{R_H}{n^2} \qquad n = 1, 2, 3, \ldots \qquad \textbf{(8.5)}$$

is a numerical constant, called the Rydberg constant, with a value of $= 2.17868 \times 10^{-18}$ J.

According to equation (8.5):

- **The energy of the hydrogen atom is quantized.** By this we mean the energy is restricted to specific values: $-R_H$, $-(R_H/4)$, $-(R_H/9)$, $-(R_H/16)$, and so on.
- **All the allowed energy values are negative.** The theory that leads to equation (8.5) employs the convention that *the energy of the electron is defined to be zero when the electron is free of the nucleus*, that is, when it is infinitely far away from the nucleus. Physically, $n = \infty$ corresponds to the situation in which the electron is free of the nucleus.

n the next section, we use equation (8.5) to explain certain features of the ission spectrum of the hydrogen atom. In this regard, the Bohr model is ıarkably successful. However, the model is highly problematic. From a ctical standpoint, it cannot be generalized to explain the emission spectra ıtoms or ions with more than one electron. From a fundamental stand- nt, the model is an uneasy mixture of classical physics and unjustifiable ıntization conditions. Bohr himself described his model simply as a way to represent several experimental facts, none of which could be explaine using only classical physics.

Modern quantum theory, also called quantum physics or quantum mech ics, replaced Bohr's theory in 1926. Quantization arises naturally by using qu tum mechanics. It is not assumed or imposed beforehand as a condition, as done by Bohr. As we will soon see, the circular orbits that are so prominen Bohr's model of the hydrogen atom are absent in the model based on quan mechanics. Despite the fact that Bohr's model of the hydrogen atom is wron was an important scientific development because it prompted a parad shift—the quantum leap—from classical physics to the new quantum physi

EXAMPLE 8-3 Understanding the Meaning of Quantization of Energy

Is there an energy level for the hydrogen atom having $E = -1.00 \times 10^{-20}$ J?

Analyze

Rearrange equation (8.5) for n^2 and solve for n. If the value of n is an integer, then the given energy correspond to an energy level for the hydrogen atom.

Solve

Let us rearrange equation (8.5), solve for n^2, and then for n.

$$n^2 = \frac{-R_H}{E_n}$$

$$= \frac{-2.179 \times 10^{-18}\ \text{J}}{-1.00 \times 10^{-20}\ \text{J}} = 2.179 \times 10^2 = 217.9$$

$$n = \sqrt{217.9} = 14.76$$

Because the value of n is not an integer, this is not an allowed energy level for the hydrogen atom.

Assess

Equation (8.5) places a severe restriction on the energies allowed for a hydrogen atom.

PRACTICE EXAMPLE A: Is there an energy level for the hydrogen atom, $E_n = -2.69 \times 10^{-20}$ J?

PRACTICE EXAMPLE B: The energy of an electron in a hydrogen atom is -4.45×10^{-20} J. What level does it occupy

8-3 Energy Levels, Spectrum, and Ionizatio Energy of the Hydrogen Atom

With equation (8.5), we can calculate the energies of the allowed energy sta or *energy levels*, of the hydrogen atom. These levels are represented schem cally in Figure 8-13. This representation is called an energy-level diagram. S a diagram shows the order of the allowed energy levels and helps us visua the energy differences between these levels. These energy differences ar particular interest because, as we will soon see, the *energy difference* betwe pair of energy levels is something that can be measured. We will reinforce idea in this section by using the energy-level diagram of the hydrogen ato interpret not only the atomic line spectrum, such as that shown in Figure 8 but also the concept of *ionization energy*, which is the energy required to rem an electron from an atom.

Spectroscopy and Atomic Line Spectra

Normally, the electron in a hydrogen atom is found in the lowest energy le that is, with $n = 1$. This lowest energy level is known as the **ground state**. W the electron gains a quantum of energy, it moves to a higher level (n = 2, 3,

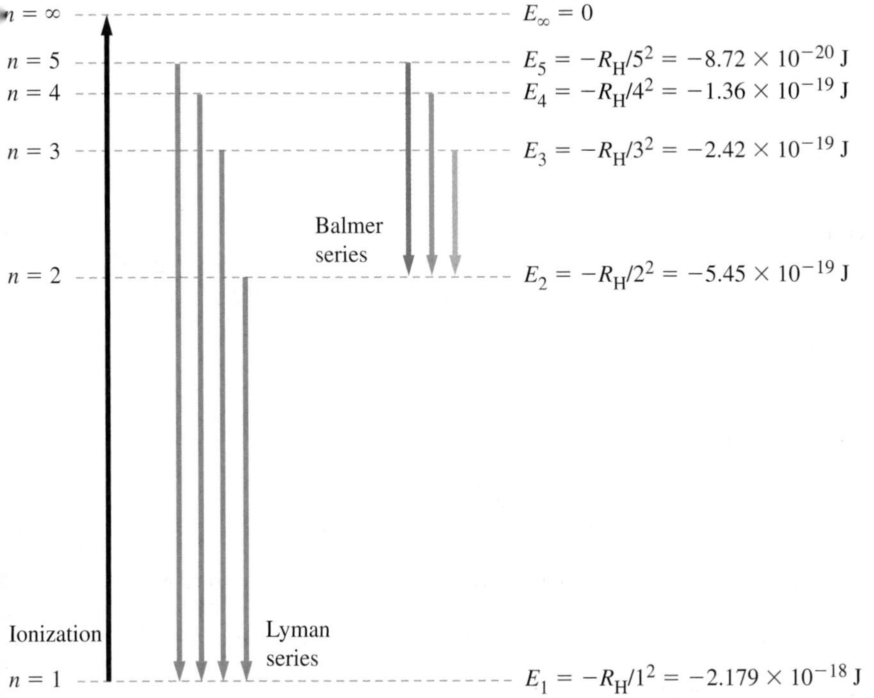

◀ FIGURE 8-13
Energy-level diagram for the hydrogen atom
Two of the series in the emission spectrum of the hydrogen atom are identified by downward-pointing arrows. The Balmer series arises from transitions in which electrons in excited atoms fall from higher energy levels to the $n = 2$ energy level. The first three lines in the Balmer series are shown here (in color). The Lyman series (gray lines) arises from transitions in which electrons in excited atoms fall from higher energy levels to the ground state ($n = 1$). These lines are in the ultraviolet. The black line indicates the situation when the electron in a hydrogen atom has acquired sufficient energy (2.179×10^{-18} J) to become free of the nucleus. In such a situation, the hydrogen atom is ionized.

n), and the atom is in an **excited state**. When the electron drops from a ier energy level to a lower energy level, a unique quantity of energy is tted—the difference in energy between the two levels. Equation (8.5) can be d to derive an expression for the difference in energy between two levels, re n_f is the final level and n_i is the initial one:

$$\Delta E = E_f - E_i = \frac{-R_H}{n_f^2} - \frac{-R_H}{n_i^2} = -R_H\left(\frac{1}{n_f^2} - \frac{1}{n_i^2}\right) \tag{8.6}$$

represents the energy change for the atom. The energy of the photon, ton, either absorbed or emitted, is equal to the *magnitude* of this energy difnce, $|\Delta E|$. Because $E_{photon} = h\nu$ and $E_{photon} = |\Delta E|$, we can write

$$E_{photon} = h\nu = |\Delta E| = |E_f - E_i| \tag{8.7}$$

ch emphasizes that the energy of a photon is always positive. (Think of ioton as a certain quantity of energy that can be absorbed or emitted by tom.)

xample 8-4 uses equations (8.6) and (8.7) as a basis for calculating the wave-th of a line in the emission spectrum of the hydrogen atom. Because the difnces between energy levels are limited in number, so too are the energies of emitted photons. Therefore, only certain wavelengths (or frequencies) are rved for the spectral lines.

XAMPLE 8-4 Calculating the Wavelength of a Line in the Hydrogen Spectrum

Determine the wavelength of the line in the Balmer series of hydrogen corresponding to the transition from $n = 5$ to $n = 2$.

nalyze

The transition is from a higher to a lower energy level, so energy (a photon) is emitted by the atom. According to equation (8.7), the energy of the emitted photon $E_{photon} = h\nu$ is equal to $|\Delta E|$, the magnitude of the energy difference between the two levels involved. First, we use equation (8.6) to calculate the energy difference, ΔE. Then, we obtain E_{photon} and ν by applying equation (8.7). Finally, by rearranging equation (8.1), we calculate $\lambda = c/\nu$.

(continued)

Solve

The specific data for equation (8.6) are $n_i = 5$ and $n_f = 2$.

$$\Delta E = -2.179 \times 10^{-18}\,\mathrm{J}\left(\frac{1}{2^2} - \frac{1}{5^2}\right)$$
$$= -2.179 \times 10^{-18} \times (0.25000 - 0.04000)$$
$$= -4.576 \times 10^{-19}\,\mathrm{J}$$

Rearranging $E_{photon} = |\Delta E| = h\nu$ gives the frequency

$$\nu = \frac{E_{photon}}{h} = \frac{4.576 \times 10^{-19}\,\mathrm{J}}{6.626 \times 10^{-34}\,\mathrm{J\,s}} = 6.906 \times 10^{14}\,\mathrm{s}^{-1}$$

Rearranging $c = \lambda\nu$ for the wavelength gives the following result:

$$\lambda = \frac{c}{\nu} = \frac{2.998 \times 10^{8}\,\mathrm{m\,s}^{-1}}{6.906 \times 10^{14}\,\mathrm{s}^{-1}} = 4.341 \times 10^{-7}\,\mathrm{m} = 434.1\,\mathrm{nm}$$

Assess

Note the good agreement between this result and the data in Figure 8-12. The color of the spectral line is deter mined by the energy difference, ΔE, while the intensity is determined by the number of hydrogen atom undergoing this transition. The greater the number of atoms undergoing the same transition, the greater th number of emitted photons, resulting in greater intensity. As a final point, notice that the energy difference, Δ the energy change for the atom, is negative: The energy of the atom decreases because of the transition tha occurs. A common mistake made by students is to calculate a negative frequency (ν) or (λ) because they forg to use the magnitude, or absolute value, of ΔE. Negative values for ν or λ are not appropriate; frequency (th number of cycles per second) and wavelength (the distance between successive maxima) of electromagneti are, by definition, positive quantities.

PRACTICE EXAMPLE A: Determine the wavelength of light absorbed in an electron transition from $n = 2$ t $n = 4$ in a hydrogen atom.

PRACTICE EXAMPLE B: Refer to Figure 8-13 and determine which transition produces the longest wavelengt line in the Lyman series of the hydrogen spectrum. What is the wavelength of this line in nanometers and i angstroms?

As shown in Example 8-4, quantization of energy provides the basis understanding the emission spectra of atoms. An emission spectru obtained when individual atoms in a collection of atoms (roughly 10² them) are excited to the various possible excited states of the atom. The at then relax to states of lower energy by emitting photons of various frequ cies. These ideas are summarized schematically in Figure 8-14(a).

Figure 8-14(b) illustrates an alternative technique in which we pass elec magnetic radiation, such as white light, through a sample of atoms in t ground states and then pass the emerging light through a prism. Now observe which frequencies of light the atoms *absorb*. This form of spectrosc is called *absorption spectroscopy*.

Figure 8-14 can help us understand how to relate the frequency of the l emitted or absorbed by an atom to the energy levels involved. In the cas emission (Fig. 8-14a), we have $E_f = E_i - h\nu$ and so $h\nu = E_i - E_f$. For abs tion (Fig. 8-14b), we have $E_f = E_i + h\nu$ and $h\nu = E_f - E_i$. Thus,

$$\nu = \frac{E_i - E_f}{h} \qquad (8$$

when a photon is emitted, and

$$\nu = \frac{E_f - E_i}{h} \qquad (8.$$

when a photon is absorbed.

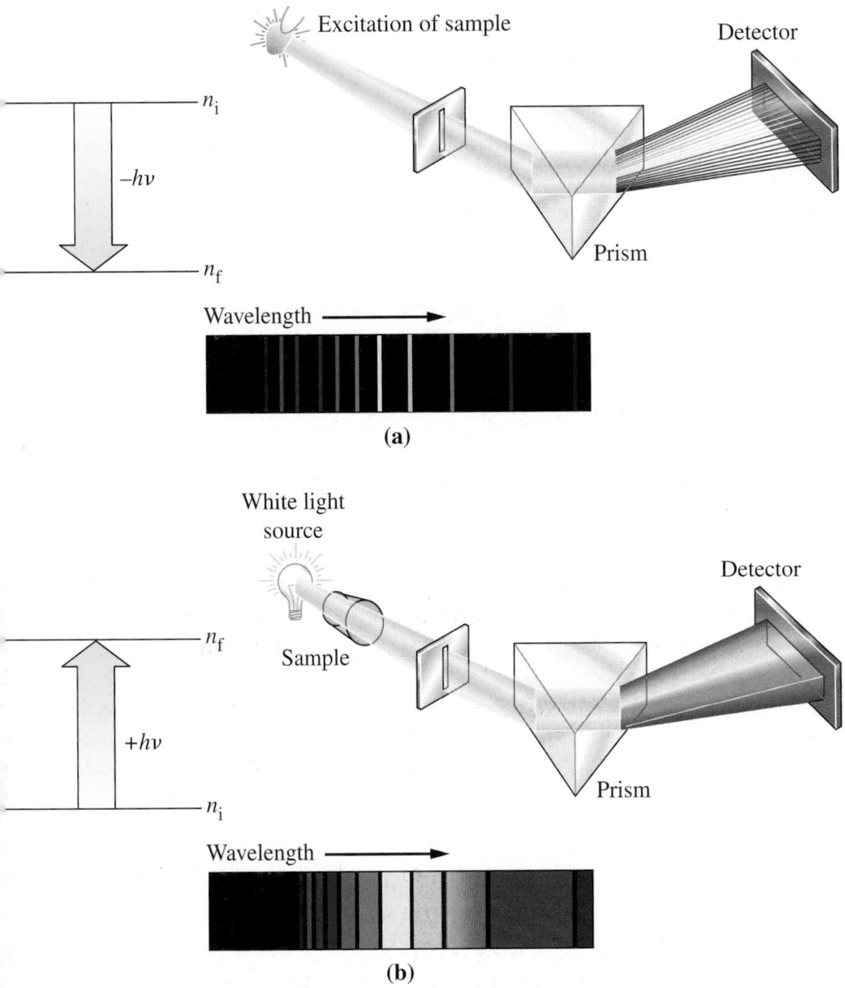

◀ FIGURE 8-14
Emission and absorption spectroscopy
(a) Emission spectroscopy. Bright lines are observed on a dark background of the photographic plate. **(b)** Absorption spectroscopy. Dark lines are observed on a bright background on the photographic plate.

n principle, we obtain exactly the same information about the quantized rgy levels of a system by using either emission spectroscopy or absorp- n spectroscopy. The choice of which technique to use is influenced by er considerations. If the sample contains a relatively small number of ms, emission spectroscopy might be the preferred technique because it a higher sensitivity. (It is easier to detect a very dim line on a dark back- und than to detect a faint dark line on a bright background.) If sensitiv- is not a concern, then perhaps absorption spectroscopy might be the ferred technique. Absorption spectra are often less complicated than ission spectra. An excited sample will contain atoms in a variety of tes, each being able to drop down to any of several lower states. An orbing sample generally is cool and transitions are possible only from ground state. The Balmer series is not seen, for example, in absorption m cold hydrogen atoms.

nization Energy of Hydrogen and Hydrogen-Like Ions

can use ideas from the preceding sections to calculate the energy required emove the electron from the ground state ($n = 1$) of a hydrogen atom. Let's this by considering the special case in which the energy of a photon orbed by a hydrogen atom is just enough to remove the electron from the = 1 level. The electron is freed, the atom is ionized, and the energy of the e electron is zero. Using $E_i = E_1$ and $E_f = 0$ in equation (8.8 b), and rearrang- the expression, we obtain

$$E_{photon} = h\nu = E_f - E_i = 0 - E_1 = -E_1$$

We define the *ionization energy*, $E_i(H)$ of the hydrogen atom, as

▶ Although the IUPAC recommends using the symbol E_i for ionization energy, other symbols are commonly used (e.g., I and IE).

$$E_i(H) = -E_1 = -\left(-\frac{R_H}{1^2}\right) = R_H$$

$E_i(H)$ represents the energy required to remove the electron from the grou[nd] state of the hydrogen atom.

The ideas just developed about the ionization of atoms are applied [in] Example 8-5 to *hydrogen-like species*, such as Li^{2+} or Be^{3+}, which have only o[ne] electron. In these species, the electron interacts with a nucleus of charge + [Z,] where Z is the atomic number. The corresponding energy-level expression is

$$E_n = \frac{-Z^2 R_H}{n^2} \qquad (8\text{.9})$$

The dependence of the energy on Z^2 can be rationalized as follows. The ener[gy] depends on both the magnitude of the charges and the separation betw[een] them. Since a greater value of Z affects both of these factors, the overall dep[en]dence is Z^2.

EXAMPLE 8-5 Applying Conservation of Energy to the Ionization of a Hydrogen-Like Ion

Determine the kinetic energy of the electron ionized from a Li^{2+} ion in its ground state, using a photon of frequency $5.000 \times 10^{16}\ s^{-1}$.

Analyze

When a photon of a given energy ionizes a species, any excess energy is transferred as kinetic energy to the electron; that is, $E_{photon} = E_i(Li^{2+}) + E_k(\text{electron})$. The energy of the electron in the Li^{2+} ion is calculated by using equation (8.9), and the energy of the photon is calculated by using Planck's relationship. The difference is the kinetic energy of the electron.

Solve

$$E_1 = \frac{-3^2 \times 2.179 \times 10^{-18}\ J}{1^2} = -1.961 \times 10^{-17}\ J$$

The energy of a photon of frequency $5.000 \times 10^{16}\ s^{-1}$ is

$$E_{proton} = h\nu = 6.626 \times 10^{-34}\ J\ s \times 5.000 \times 10^{16}\ s^{-1} = 3.313 \times 10^{-17}\ J\ photon^{-1}$$

The kinetic energy of the electron is given by $E_k(\text{electron}) = E_{photon} - E_i(Li^{2+})$; that is,

$$\text{kinetic energy} = 3.313 \times 10^{-17}\ J - 1.961 \times 10^{-17}\ J = 1.352 \times 10^{-17}\ J$$

Assess

Notice the similarity between the energy conservation expression used in solving this problem ($E_{photon} = E_i(Li^{2+}) + E_k(\text{electron})$) and the one used in explaining the photoelectric effect ($E_{photon} = F + E_k(\text{electron})$).

PRACTICE EXAMPLE A: Determine the wavelength of light emitted in an electron transition from $n = 5$ to $n = 3$ in a Be^{3+} ion.

PRACTICE EXAMPLE B: The frequency of the $n = 3$ to $n = 2$ transition for an unknown hydrogen-like ion occurs at a frequency 16 times that of the hydrogen atom. What is the identity of the ion?

8-4 CONCEPT ASSESSMENT

Which of the following electronic transitions in a hydrogen atom will lead to the emission of a photon with the shortest wavelength, $n = 1$ to $n = 4$, $n = 4$ to $n = 2$, $n = 3$ to $n = 2$?

·4 Two Ideas Leading to Quantum Mechanics

the previous section, we pointed out that the interpretation of atomic line ectra posed a difficult problem for classical physics and that Bohr had some ccess in explaining the emission spectrum for the hydrogen atom. However, cause his model was not correct, he was unable to explain all features of the drogen emission spectrum and could not explain the spectra of multielectron ms at all. A decade or so after Bohr's work on hydrogen, two landmark ideas nulated a new approach to quantum mechanics. Those ideas are considered this section and the new quantum mechanics—wave mechanics—in the next.

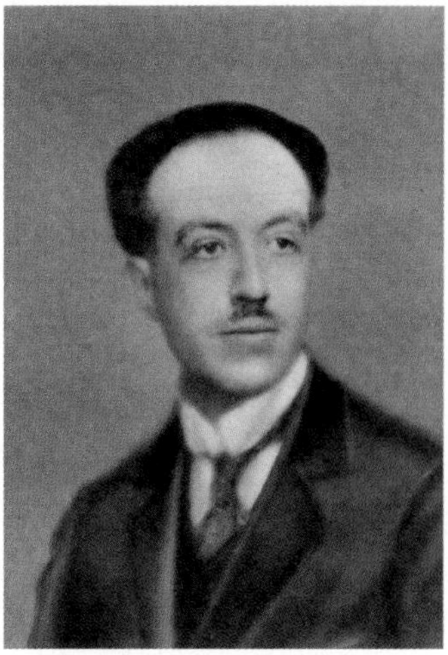

Mary Evans Picture Library / Alamy

▲ **Louis de Broglie (1892–1987)**
De Broglie conceived of the wave–particle duality of small particles while working on his doctorate degree. He was awarded the Nobel Prize in physics 1929 for this work.

ave–Particle Duality

explain the photoelectric effect, Einstein suggested that light has particle-like operties, which are displayed through photons. Other phenomena, however, ch as the dispersion of light into a spectrum by a prism, are best understood in ms of the wave theory of light. Light, then, appears to have a *dual* nature.

In 1924, Louis de Broglie, considering the nature of light and matter, offered tartling proposition: *Small particles of matter may at times display wave-like perties.*

De Broglie argued that the relationship $p = h/\lambda$, derived by Einstein for the mentum of a photon (see Are You Wondering 8-2), should also apply to ticles of matter. For a particle of mass m moving with speed u, the momen-n is $p = mu$, and so the relationship $p = h/\lambda$ can be written in the form

$$\lambda = \frac{h}{p} = \frac{h}{mu} \tag{8.10}$$

KEEP IN MIND

that in equation (8.10), wavelength is in meters, mass is in kilograms, and velocity is in meters per second. Planck's constant must also be expressed in units of mass, length, and time. This requires replacing the joule by the equivalent units kg m^2 s^{-2}.

Equation (8.10) is de Broglie's famous relationship for the wavelength of at he called a *phase wave*. Although de Broglie had no doubt about the physi- reality of the phase wave, he was reluctant to commit to a physical interpre- ion of it. In the concluding sentences of his doctoral thesis, de Broglie plained that his definition of the phase wave was left purposefully vague; he ferred instead to let his work stand as "a formal scheme whose physical con- t is not yet fully determined." Today, de Broglie's phase waves are called *mat- waves*. If matter waves exist for small particles, then beams of particles, such as ctrons, should exhibit the characteristic properties of waves, namely diffraction call page 306). If the distance between the objects that the waves scatter from is out the same as the wavelength of the radiation, diffraction occurs and an inter- ence pattern is observed. For example, X-rays are highly energetic photons with associated wavelength of about 1 Å (100 pm). X-rays are scattered by the regular ay of atoms in the metal aluminum, where the atoms are about 2 Å (200 pm) art, producing the diffraction pattern shown in Figure 8-15(a).

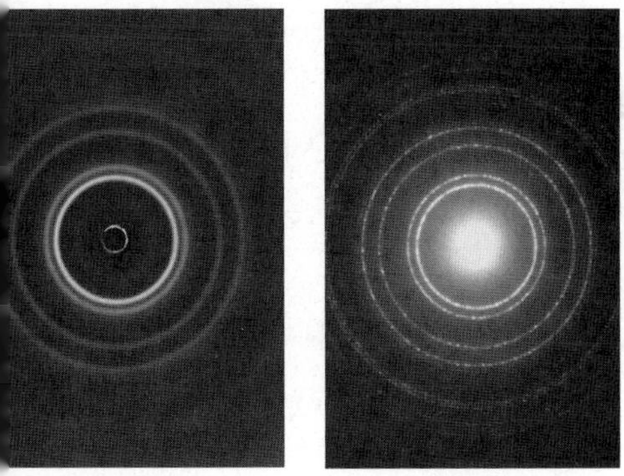

(a) (b)

◀ FIGURE 8-15
Wave properties of electrons demonstrated
(a) Diffraction of X-rays by metal foil.
(b) Diffraction of electrons by metal foil, confirming the wave-like properties of electrons.

In 1927, C. J. Davisson and L. H. Germer of the United States showed t a beam of slow electrons is diffracted by a crystal of nickel. In a sim experiment in that same year, G. P. Thomson of Scotland directed a bear electrons at a thin metal foil. He obtained the same pattern for the diff tion of electrons by aluminum foil as with X-rays of the same wavelen (Fig. 8-15b).

Thomson and Davisson shared the 1937 Nobel Prize in physics for t electron diffraction experiments. George P. Thomson was the son of Thomson, who had won the Nobel Prize in physics in 1906 for his discov of the electron. It is interesting to note that Thomson the father showed the electron is a particle, and Thomson the son showed that the electron wave. Father and son together demonstrated the **wave–particle dualit** electrons.

The wavelength calculated in Example 8-6, 24.2 pm, is significant w compared to, for example, the distance between neighboring atoms in a minum metal (200 pm). It is only when wavelengths are comparable atomic or nuclear dimensions that wave–particle duality is important. concept has little meaning when applied to large (macroscopic) objects, s as baseballs and automobiles, because their wavelengths are too smal measure. For these macroscopic objects, the laws of classical physics quite adequate.

The Uncertainty Principle

The laws of classical physics permit us to make precise predictions. For exa ple, we can calculate the exact point at which a rocket will land after it is fi The more precisely we measure the variables that affect the rocket's traject (path), the more accurate our calculation (prediction) will be. In effect, ther no limit to the accuracy we can achieve. In classical physics, nothing is lef chance—physical behavior can be predicted with certainty.

EXAMPLE 8-6 Calculating the Wavelength Associated with a Beam of Particles

What is the wavelength associated with electrons traveling at one-tenth the speed of light?

Analyze

To calculate the wavelength, we use equation (8.10). To use it, we have to collect the electron mass, the electron velocity, and Planck's constant, and then adjust the units so that they are expressed in terms of kg, m, and s.

Solve

The electron mass, expressed in kilograms, is 9.109×10^{-31} kg (recall Table 2.1).
The electron velocity is $u = 0.100 \times c = 0.100 \times 3.00 \times 10^{8}\ \mathrm{m\,s^{-1}} = 3.00 \times 10^{7}\ \mathrm{m\,s^{-1}}$.
Planck's constant $h = 6.626 \times 10^{-34}\ \mathrm{J\,s} = 6.626 \times 10^{-34}\ \mathrm{kg\,m^2\,s^{-2}\,s} = 6.626 \times 10^{-34}\ \mathrm{kg\,m^2\,s^{-1}}$.
Substituting these data into equation (8.10), we obtain

$$\lambda = \frac{6.626 \times 10^{-34}\ \mathrm{kg\,m^2\,s^{-1}}}{(9.109 \times 10^{-31}\ \mathrm{kg})(3.00 \times 10^{7}\ \mathrm{m\,s^{-1}})}$$
$$= 2.42 \times 10^{-11}\ \mathrm{m} = 24.2\ \mathrm{pm}$$

Assess

By converting the unit J to $\mathrm{kg\,m^2\,s^{-2}}$, we are able to obtain the wavelength in meters.

PRACTICE EXAMPLE A: Assuming Superman has a mass of 91 kg, what is the wavelength associated with him if he is traveling at one-fifth the speed of light?

PRACTICE EXAMPLE B: To what velocity (speed) must a beam of protons be accelerated to display a de Broglie wavelength of 10.0 pm? Obtain the proton mass from Table 2.1.

uring the 1920s, Niels Bohr and Werner Heisenberg considered hypothet- experiments to establish just how precisely the behavior of subatomic icles can be determined. The two variables that must be measured are the tion of the particle (x) and its momentum ($p = mu$). The conclusion they hed is that there must *always* be uncertainties in measurement such that product of the uncertainty in position, Δx, and the uncertainty in momen-, Δp, is

$$\Delta x \Delta p \geq \frac{h}{4\pi} \quad (8.11)$$

he significance of this expression, called the **Heisenberg uncertainty prin-e**, is that we cannot measure position and momentum with great precision ultaneously. An experiment designed to locate the position of a particle great precision cannot also measure the momentum of the particle pre-ly, and vice versa. In simpler terms, if we know precisely where a particle ve cannot also know precisely where it has come from or where it is going. e know precisely how a particle is moving, we cannot also know precisely ere it is. In the subatomic world, things must always be "fuzzy." Why uld this be so?

he Heisenberg uncertainty principle (expression 8.11) implies that for a y precise measurement of position, x, many values of momentum, p, are sible. One way to rationalize this result is to conceive of a highly localized ticle as a superposition of many matter waves of different de Broglie wave-gths, as suggested by Figure 8-16. The superposition of many waves of dif-nt wavelengths produces an interference pattern, which tends to localize resultant wave, and the particle it describes, to a region of space. Each con-uting wavelength corresponds to a different value of the momentum

▲ **Werner Heisenberg (1901–1976)**
In addition to his enunciation of the uncertainty principle, for which he won the Nobel Prize in physics in 1932, Heisenberg also developed a mathematical description of the hydrogen atom that gave the same results as Schrödinger's equation (page 332). Heisenberg (left) is shown here dining with Niels Bohr.

◀ The uncertainty principle is not easy for most people to accept. Einstein spent a good deal of time from the middle 1920s until his death in 1955 attempting, unsuccessfully, to disprove it.

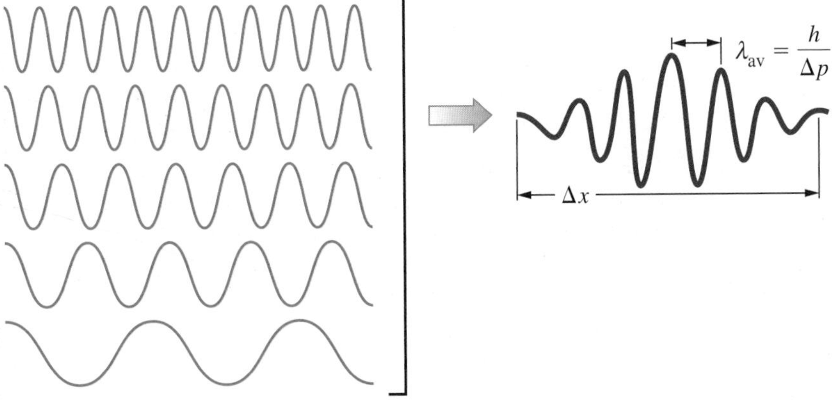

IGURE 8-16
uncertainty principle interpreted graphically
ollection of waves with varying wavelengths (left) can combine into a "wave packet" nt). The superposition of the different wavelengths yields an average wavelength) and causes the wave packet to be more localized (Δx) than the individual es. The greater the number of wavelengths that combine, the more precisely an ociated particle can be located, that is, the smaller Δx. However, because each of wavelengths corresponds to a different value of momentum according to the de glie relationship, the greater is the uncertainty in the resultant momentum.

(because $p = h/\lambda$). In general, the more localized the resultant wave (whic also called a *wave packet*), the greater the range of momentum values that tribute to it. On the other hand, if the momentum is very precisely kno then a very small range of wavelengths contributes to the wave packet. superposition of waves of similar wavelengths gives a wave packet that is highly localized. Thus, the more precisely we know the momentum, the localized the wave packet and the more uncertain we are about the positio the particle.

The concept of wave-particle duality and the Heisenberg uncertainty pri ple have a profound influence on how we should conceive of, and describe electron. An electron is neither a particle nor a wave but somehow both. A the more certain we are about some aspect of an electron's behavior, the less tain we are about other aspects. With these ideas in mind, we turn our atten to a modern description of electrons in atoms.

One electron-volt (1 eV) is the energy acquired by an electron as it falls through an electric potential difference of 1 volt. ▼

EXAMPLE 8-7 Calculating the Uncertainty in the Position of an Electron

A 12 eV electron can be shown to have a speed of 2.05×10^6 m/s. Assuming that the precision (uncertainty) c this value is 1.5%, with what maximum precision can we simultaneously measure the position of the electron

Analyze

When given an uncertainty as a percentage, we have to convert it to a fraction by dividing by 100%. The uncer tainty of the velocity is then obtained by multiplying this number by the actual velocity.

Solve

The uncertainty in the electron speed is

$$\Delta u = 0.015 \times 2.05 \times 10^6 \text{ m s}^{-1} = 3.1 \times 10^4 \text{ m s}^{-1}$$

The electron mass, 9.109×10^{-31} kg (recall Table 2.1), is known much more precisely than the electron speed which means that

$$\Delta p = m\Delta u = 9.109 \times 10^{-31} \text{ kg} \times 3.1 \times 10^4 \text{ m s}^{-1}$$
$$= 2.8 \times 10^{-26} \text{ kg m s}^{-1}$$

From expression (8.11), the uncertainty in the electron's position is

$$\Delta x = \frac{h}{4\pi\Delta p} = \frac{6.63 \times 10^{-34} \text{ kg m}^2 \text{ s}^{-1}}{4 \times 3.14 \times 2.8 \times 10^{-26} \text{ kg m s}^{-1}} = 1.9 \times 10^{-9} \text{ m} = 1.9 \times 10^3 \text{ pm}$$

Assess

The uncertainty in the electron's position is about 10 atomic diameters. Given the uncertainty in its speed there is no way to pin down the electron's position with any greater precision.

PRACTICE EXAMPLE A: Superman has a mass of 91 kg and is traveling at one-fifth the speed of light. If the speec at which Superman travels is known with a precision of 1.5%, what is the uncertainty in his position?

PRACTICE EXAMPLE B: What is the uncertainty in the speed of a beam of protons whose position is known witl the uncertainty of 24 nm?

8-5 CONCEPT ASSESSMENT

An electron has a mass approximately 1/2000th of the mass of a proton. Assuming that a proton and an electron have similar wavelengths, how would their speeds compare?

5 Wave Mechanics

Broglie's relationship suggests that electrons are matter waves and thus uld display wavelike properties. A consequence of this wave–particle duals the limited precision in determining an electron's position and momentum osed by the Heisenberg uncertainty principle. How then are we to view trons in atoms? To answer this question, we must begin by identifying two es of waves.

nding Waves

an ocean, the wind produces waves on the surface whose crests and ighs travel great distances. These are called *traveling waves*. In the traveling ve shown in Figure 8-1, every portion of a very long rope goes through an ntical up-and-down motion. The wave transmits energy along the entire gth of the rope. An alternative form of a wave is seen in the vibrations in a cked guitar string, suggested by Figure 8-17.

egments of the string experience up-and-down displacements with time, l they oscillate or vibrate between the limits set by the blue curves. The ortant aspect of these waves is that the crests and troughs of the wave occur ixed positions and the amplitude of the wave at the fixed ends is zero. Of spe- interest is the fact that the magnitudes of the oscillations differ from point to nt along the wave, including certain points, called *nodes*, that undergo no dis-cement at all. A wave with these characteristics is called a **standing wave.**

Ve might say that the permitted wavelengths of standing waves are quan-ed. They are equal to twice the path length (L) divided by a whole number , that is,

$$\lambda = \frac{2L}{n} \quad \text{where } n = 1, 2, 3, \ldots \text{ and the total number of nodes} = n + 1 \qquad \textbf{(8.12)}$$

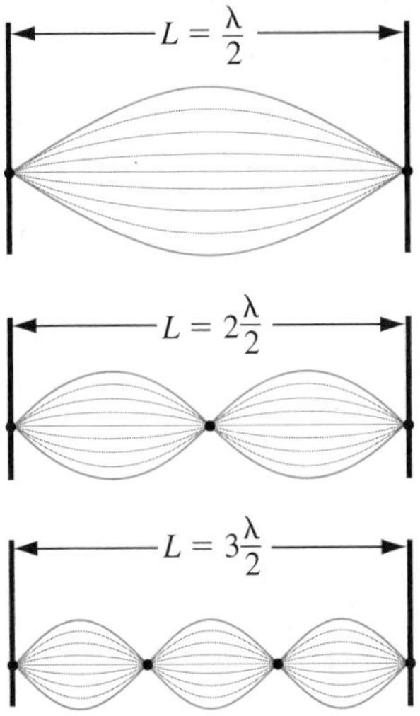

▲ FIGURE 8-17
Standing waves in a string
The string can be set into motion by plucking it. The blue boundaries outline the range of displacements at each point for each standing wave. The relationships between the wavelength, string length, and the number of nodes—points that are not displaced—are given by equation (8.12). The nodes are marked by bold dots.

The plucked guitar string can be represented by a one-dimensional standing ve. In an analogous fashion, an electron in a circular orbit might also be repre-ted by a standing wave, one having an integral number of wavelengths that fit ctly the circumference of the orbit, as suggested in Figure 8-18. Although such nodel combines both the particle and the wave nature of the electron, it is not propriate for describing the electron in a hydrogen atom. As we will see, the rect model for the hydrogen atom is based on a three-dimensional treatment.

◀ Beating a drum produces a two-dimensional standing wave, and ringing a spherical bell produces a three-dimensional standing wave.

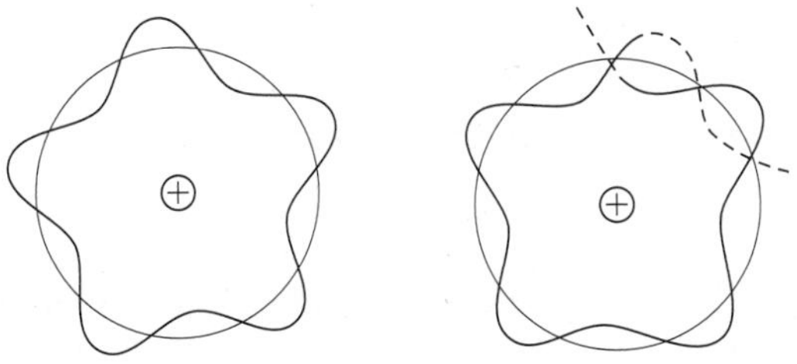

IGURE 8-18
e electron as a matter wave
se patterns are two-dimensional cross-sections of a much more complicated e-dimensional wave. The wave pattern in **(a)**, a *standing wave*, is an acceptable resentation. It has an integral number of wavelengths (five) about the nucleus; cessive waves reinforce one another. The pattern in **(b)** is unacceptable. The number vavelengths is nonintegral, and successive waves tend to cancel each other; that is, crest in one part of the wave overlaps a trough in another part of the wave, and re is no resultant wave at all.

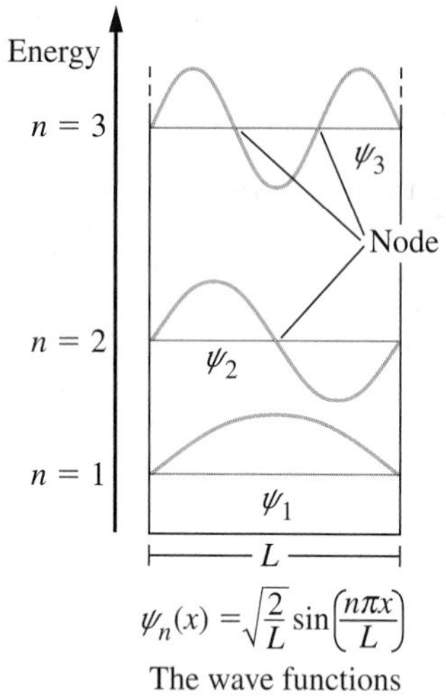

▲ FIGURE 8-19
The standing waves of a particle in a one-dimensional box
The first three wave functions and their energies are shown in relation to the position of the particle within the box. The wave function changes sign at the nodes.

Particle in a Box: Standing Waves, Quantum Particles, and Wave Functions

In 1927, Erwin Schrödinger, an expert on the theory of vibrations and stand waves, suggested that an electron (or any other particle) exhibiting wave properties should be describable by a mathematical equation called a **w function**. The wave function, denoted by the Greek letter psi, ψ, should co spond to a standing wave within the boundary of the system being describ The simplest system for which we can write a wave function is another o dimensional system, that of a quantum particle confined to move in a o dimensional box, a line. The wave function for this so-called "particle i box" looks like those of a guitar string (Fig. 8-17), but now it represents matter waves of a particle. Since the particle is constrained to be in the box, waves also must be in the box, as illustrated in Figure 8-19.

If the length of the box is L and the particle moves along the x directi then the equation for the standing wave is

$$\psi_n(x) = \sqrt{\frac{2}{L}}\sin\left(\frac{n\pi x}{L}\right),\ n = 1, 2, 3, \ldots \qquad \text{(8}$$

where the quantum number, n, labels the wave function.

This wave function is a sine function. To illustrate, consider the c where $n = 2$.

When

$x = 0,$	$\sin 2\pi(0)/L = \sin 0 = 0,$	$\psi_n(x) = 0$
$x = L/4,$	$\sin 2\pi(L/4)/L = \sin \pi/2 = 1,$	$\psi_n(x) = (2/L)^{1/2}$
$x = L/2,$	$\sin 2\pi(L/2)/L = \sin \pi = 0,$	$\psi_n(x) = 0$
$x = 3L/4,$	$\sin 2\pi(3L/4)/L = \sin 3\pi/2 = -1,$	$\psi_n(x) = -(2/L)^{1/2}$
$x = L,$	$\sin 2\pi(L)/L = \sin 2\pi = 0,$	$\psi_n(x) = 0$

At one end of the box ($x = 0$), both the sine function and the wave function zero. At one-fourth the length of the box ($x = L/4$), the sine function and wave function both reach their maximum values. At the midpoint of the b both are again zero; the wave function has a node. At three-fourths the l length, both functions reach their minimum values (negative quantities), anc the farther end of the box, both functions are again zero.

8-3 ARE YOU WONDERING?

How did we arrive at Equation (8.13)?

The answer to how we arrived at equation (8.13) lies in the equation that gives the form of the wave function and the boundaries within which the quantum mechanical particle is confined. If you are familiar with differential calculus, you will recognize the equation below as a differential equation. Specifically, it describes a one-dimensional standing wave for the simple system of a particle in a box. The solution to this equation is the wave function for the system.

$$\frac{d^2\psi}{dx^2} = -\left(\frac{2\pi}{\lambda}\right)^2\psi$$

Notice the form of the wave equation: By differentiating the wave function twice, we obtain the wave function times a constant. Many functions satisfy this requirement. For example, two trigonometric functions that have this property are the sine and cosine functions. First, let us consider the function $\psi = A\cos(ax)$

where A and a are constants having nonzero values. If we differentiate ψ twice with respect to x, we obtain

$$\frac{d\psi}{dx} = -aA\sin(ax) \qquad \frac{d^2\psi}{dx^2} = -a^2A\cos(ax) = -a^2\psi$$

By comparing the two expressions we have for $d^2\psi/dx^2$, we can identify $a = (2\pi/\lambda)$. Second, for the function $\psi = A\sin(ax)$, we have

$$\frac{d\psi}{dx} = aA\cos(ax) \qquad \frac{d^2\psi}{dx^2} = -a^2A\sin(ax) = -a^2\psi$$

Again, we identify $a = (2\pi/\lambda)$. Which of the two functions, $\psi = A\cos(2\pi x/\lambda)$ or $\psi = A\sin(2\pi x/\lambda)$, can be used to describe a standing wave in a box that extends from $x = 0$ to $x = L$? We must have $\psi = 0$ when $x = 0$ and when $x = L$. When $x = 0$, we have $A\cos(2\pi x/\lambda) = A\cos 0 = A \neq 0$ and $A\sin(2\pi x/\lambda) = A\sin 0 = 0$. Thus, the condition at $x = 0$ establishes that $\psi = A\sin(2\pi x/\lambda)$ is the correct function to use. In carrying out this procedure, we have used a boundary condition of the system to help choose the correct form of the wave function. This is a common procedure when solving quantum mechanical problems. The other boundary condition, that $\psi = 0$ at $x = L$, is achieved by applying the standing wave requirement in equation (8.12). Substituting $\lambda = (2L/n)$ into the expression $\psi = A\sin(2\pi x/\lambda)$, we obtain

$$\psi_n = A\sin\left(\frac{n\pi x}{L}\right)$$

where n is identified as a quantum number, $n = 1, 2, 3, 4, \ldots$

The determination of A is all that remains. However, to determine A, we need to know how to interpret the wave function. We will return to the determination of A in Are You Wondering 8-4.

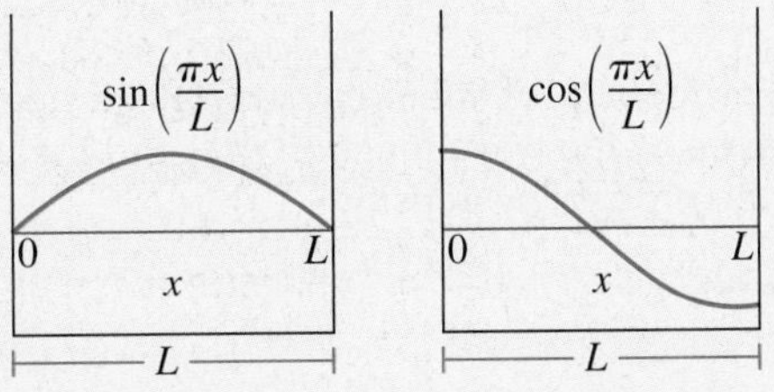

▲ Illustration of why $\cos(\pi x/L)$ is an unacceptable solution for the particle in a box. The function $\sin(\pi x/L)$ correctly goes to zero at the edges of the box, but $\cos(\pi x/L)$ does not.

What sense can we make of the wave function and the quantum number? First, consider the quantum number, n. What can we relate it to? The particle that we are considering is freely moving (not acted upon by any outside forces) with a kinetic energy given by the expression

$$E_k = \frac{1}{2}mu^2 = \frac{m^2u^2}{2m} = \frac{p^2}{2m} \tag{8.14}$$

Now, to associate this kinetic energy with a wave, we can use de Broglie's relationship ($\lambda = h/p$) to get

$$E_k = \frac{p^2}{2m} = \frac{h^2}{2m\lambda^2}$$

The wavelengths of the matter wave have to fit the standing wave conditi described earlier for the standing waves of a guitar string (equation 8. Substituting the wavelength of the matter wave from equation (8.12) into equation for the energy of the wave yields

$$E_k = \frac{h^2}{2m\lambda^2} = \frac{h^2}{2m(2L/n)^2} = \frac{n^2h^2}{8mL^2}$$

So we see that the standing wave condition naturally gives rise to qua zation of the wave's energy, with the allowable values determined by value of n. Note also that as we decrease the size of the box, the kin energy of the particle increases, and according to the uncertainty princi our knowledge of the momentum must decrease. A final noteworthy poir that the energy of the particle *cannot be zero*. The lowest possible energy, responding to $n = 1$, is called the **zero-point energy**. Because the zero-pc energy is not zero, the particle cannot be at rest. This observation is con tent with the uncertainty principle because the position and moment both must be uncertain, and there is nothing uncertain about a parti at rest.

The Born Interpretation of the Wave Function

The particle-in-a-box model helps us see the origin of the quantizatior energy, but how are we to interpret the wave function, ψ? The answe this question was provided in 1926 by German physicist Max Bo According to Born's view, wave mechanics does not answer the questi "What is the precise position of a particle?" but rather, "What is the pro bility of finding a particle within a specified volume of space?" Moreo Born argued that it is the value of ψ^2, not the value of ψ itself, that determi the probability.

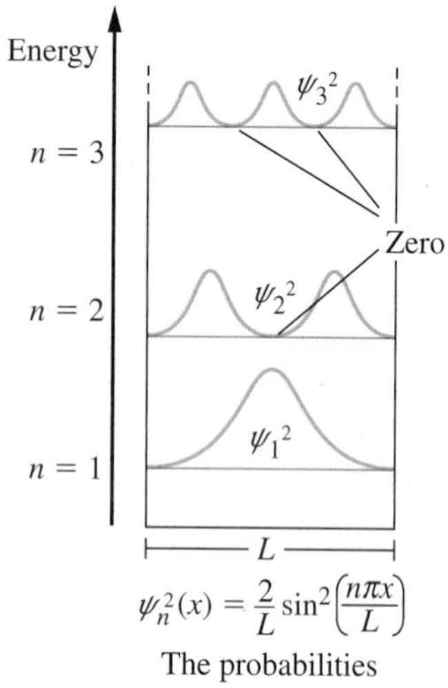

▲ FIGURE 8-20
The probabilities of a particle in a one-dimensional box
The squares of the first three wave functions and their energies are shown in relation to the position of the particle within the box. There is no chance of finding the particle at the points where $\psi^2 = 0$.

> The total probability of finding a particle in a small volume of space is the produc of the square of the wave function, ψ^2, and the volume of interest. The factor ψ is called the probability density.

Now let us return to a particle constrained to a one-dimensional path i box and look at the probabilities for the wave functions. These are shown Figure 8-20. First, notice that even where the wave function is negative, probability density is positive, as it should be in all cases. Next, look at probability density for the wave function corresponding to $n = 1$. The hig est value of ψ^2 is at the center of the box; that is, the particle is most likely be found there. The probability density for the state with $n = 2$ indica that the particle is most likely to be found between the center of the box a the walls.

A final consideration of the particle-in-a-box model concerns its extensi to a three-dimensional box. In this case, the particle can move in all th directions—x, y, and z—and the quantization of energy is described by t following expression,

$$E_{n_x n_y n_z} = \frac{h^2}{8m}\left[\frac{n_x^2}{L_x^2} + \frac{n_y^2}{L_y^2} + \frac{n_z^2}{L_z^2}\right]$$

where there is one quantum number for each dimension. Thus, a thre dimensional system needs three quantum numbers. With these particle-in box ideas, we can now discuss solving the quantum mechanical problem of t hydrogen atom.

8-4 ARE YOU WONDERING?

Is the Born interpretation an idea we use to determine the final form of a wave function?

The answer is yes. Let's illustrate this by considering the wave function for the lowest energy level of a particle in a box. In Are You Wondering 8-3, we established that $\psi_n(x) = A \sin(n\pi x/L)$. Therefore, the wave function for the lowest energy level, $n = 1$, is

$$\psi_1(x) = A \sin\left(\frac{\pi x}{L}\right)$$

We know that the particle must be somewhere between $x = 0$ and $x = L$. According to the Born interpretation, $\psi_1^2(x)$ is the probability per unit length of finding the electron, and $\psi_1^2(x)dx$ is the probability of finding the particle between the two points x and $x + dx$. The total probability of finding the particle between $x = 0$ and $x = L$ is the sum (integral) of all these probabilities, and it must be equal to 1. Mathematically, we represent this idea as

$$\int_0^L \psi_1^2(x)dx = A^2 \int_0^L \sin^2\left(\frac{\pi x}{L}\right)dx = 1$$

The integral $\int_0^L \sin^2(\pi x/L)\,dx$ has the value $L/2$, so that

$$A^2\left(\frac{L}{2}\right) = 1 \quad \text{and} \quad A = \sqrt{\frac{2}{L}}$$

By using the Born interpretation, we have completed the derivation of equation (8.13). The procedure that we performed to ensure that the total probability is equal to 1 is called *normalization*.

EXAMPLE 8-8 Using the Wave Functions of a Particle in a One-Dimensional Box

What is the fraction, as a percentage, of the total probability of finding, between points at 0 pm and 30 pm, an electron in the $n = 5$ level of a one-dimensional box 150 pm long?

Analyze

If an electron is in the $n = 5$ level, then we have a 100% chance of finding it in that level. The $n = 5$ wave function has 4 nodes 30 pm apart, and there are five maxima in ψ^2 at 15 pm, 45 pm, 75 pm, 105 pm, and 135 pm for a one-dimensional box 150 pm long.

Solve

The position at 30 pm corresponds to a node in the wave function, and there are four of these. The endpoints are not nodes because, strictly speaking, at a node, the wave function has to pass *through* zero, that is, change sign. The total area between 0 and 30 pm of ψ^2 represents 20% of the total probability because there are five peaks in the ψ^2 function. Therefore, between 0 pm and 30 pm, we have a 20% probability of finding the particle.

Assess

We must remember that the particle we are considering exhibits wave–particle duality, making it inappropriate to ask a question about how it gets from one side of the node to the other (but that is an appropriate question for a *classical* particle). All we know is that in the $n = 5$ state, for example, the particle is in the box somewhere. When we make a measurement, we'll find the particle on one side of a node or the other. Between 0 and 30 pm, we have a 20% chance of finding the particle, and the maximum chance occurs at 15 pm.

(continued)

PRACTICE EXAMPLE A: What is the fraction, as a percentage of the total probability of finding, between points 50 pm and 75 pm, an electron in the $n = 6$ level of a one-dimensional box 150 pm long?

PRACTICE EXAMPLE B: A particle is confined to a one-dimensional box 300 pm long. For the sta having $n = 3$, at what points (not counting the ends of the box) does the particle have zero probability of beir found?

EXAMPLE 8-9 Calculating Transition Energy and Photon Wavelength for the Particle in a Bo

What is the energy difference between the ground state and the first excited state of an electron contained in one-dimensional box 1.00×10^2 pm long? Calculate the wavelength of the photon that could excite the ele tron from the ground state to the first excited state.

Analyze

The energy of an electron (E_n) in level n is

$$E_n = \frac{n^2h^2}{8mL^2}$$

We can write expressions for E_n and E_{n+1}, subtract them, and then substitute the values for h, m, and L. Th ground state corresponds to $n = 1$, and the first excited state corresponds to $n = 2$. Finally, we can calculat the wavelength of the photon from the Planck relationship and $c = \lambda\nu$.

Solve

The energies for the states $n = 1$ and $n = 2$ are

$$E_{\text{ground state}} = E_1 = \frac{h^2}{8mL^2}(1^2)$$

$$E_{\text{first excited state}} = E_2 = \frac{h^2}{8mL^2}(2^2)$$

The energy difference is

$$\Delta E = E_2 - E_1 = \frac{3h^2}{8mL^2}$$

The electron mass is 9.109×10^{-31} kg, Planck's constant $h = 6.626 \times 10^{-34}$ J s, and the length of the box i 1.00×10^{-10} m. (Recall: 1 pm $= 10^{-12}$ m.) Substituting these data into the equation, we obtain

$$\Delta E = \frac{3(6.626 \times 10^{-34}\,\text{J s})^2}{8(9.109 \times 10^{-31}\,\text{kg})(1.00 \times 10^{-10}\,\text{m})^2}$$
$$= 1.81 \times 10^{-17}\,\text{J}$$

By using Planck's constant and this value as the energy of a photon, we can calculate the frequency of th photon and then the wavelength. Combining these steps,

$$\lambda = \frac{hc}{E_{\text{photon}}} = \frac{hc}{\Delta E} = \frac{6.626 \times 10^{-34}\,\text{J s} \times 3.00 \times 10^8\,\text{m s}^{-1}}{1.81 \times 10^{-17}\,\text{J}} = 11.0 \times 10^{-9}\,\text{m} = 11.0\,\text{nm}$$

Assess

If we needed the energy of the photon in kJ mol^{-1}, we would have had to multiply 1.8×10^{-17} J by 10^{-3} kJ/J and $N_A = 6.022 \times 10^{23}$ mol^{-1}.

PRACTICE EXAMPLE A: Calculate the wavelength of the photon emitted when an electron in a box 5.0×10^1 pm long falls from the $n = 5$ level to the $n = 3$ level.

PRACTICE EXAMPLE B: A photon of wavelength 24.9 nm excites an electron in a one-dimensional box from the ground state to the first excited state. Estimate the length of the box.

8-6 CONCEPT ASSESSMENT

or a particle in a one-dimensional box, in which state (value of n) is the
reatest probability of finding the particle at one-quarter the length of the box
om either end?

6 Quantum Theory of the Hydrogen Atom

will now use ideas from Section 8-5 to develop a conceptual model for
lerstanding the hydrogen atom, a simple system consisting of a single elec-
n interacting with just one nucleus. This simple model system is arguably
of the most important models in chemistry because it provides the basis
understanding multielectron atoms, the organization of elements in
periodic table, and, ultimately, the physical and chemical properties of the
ments and their compounds. As we explore this model, we will introduce
cepts and terminology that are used throughout chemistry.

3efore we begin, let's summarize a few key ideas from Section 8-5. We
rned that if a particle is confined to a one-dimensional box, the energy of
particle is quantized. That is, the particle can possess only certain quantities
energy. In addition, we learned that the state of the particle, or the matter
ve associated with it, can be characterized by a quantum number, n, and
cribed by a wave function, ψ_n, that can be analyzed to reveal certain gen-
l features. For the particle in a box, not only do we find that ψ_n has $n - 1$
les, but also we discover an interesting correlation between the energy of
h state and the number of nodes in the associated wave function: The
rgy of the particle increases with the number of nodes.

How does the system of a particle in a box help us understand the hydro-
n atom? The electron in a hydrogen atom is also confined, not literally by
penetrable walls but in principle because of its attraction to the nucleus. If
accept the basic idea that the electron in a hydrogen atom is "confined" by
attraction to the nucleus, then it should come as no surprise that the
rgy of the hydrogen atom is also quantized. The allowed energies will not
the same as for the particle in a box, but the energies will be restricted to
tain values nonetheless. We should also expect that the state of the electron
l be characterized by quantum numbers and described by a wave function
t can be analyzed to reveal certain important features. By the end of the
xt section, we will see that all these assertions are true.

e Schrödinger Equation

1927, Erwin Schrödinger proposed an equation for the hydrogen atom
it incorporated both the particle and the wave nature of the electron (see
e You Wondering 8-5). The **Schrödinger equation** is a wave equation that
ist be solved to obtain the energy levels and wave functions needed to
scribe a quantum mechanical system. Solving the Schrödinger equation is
complicated process. We will not go into the details of solving it but
stead describe and interpret the solutions by using ideas introduced in ear-
r sections.

Solving the Schrödinger equation for the hydrogen atom gives the
me expression for the energy levels, equation (8.5), that we encountered
eviously:

$$E_n = -\frac{R_H}{n^2}$$

8-5 ARE YOU WONDERING?

What is the Schrödinger equation for the hydrogen atom?

The Schrödinger equation is accepted as a basic postulate of quantum mechanics. It cannot be derived from other equations. However, the form of the Schrödinger equation can be justified as follows. We start with the equation for a standing wave in one dimension:

$$\frac{d^2\psi}{dx^2} = -\left(\frac{2\pi}{\lambda}\right)^2\psi$$

The next step is to substitute de Broglie's relationship for the wavelength of a matter wave.

$$\frac{d^2\psi}{dx^2} = -\left(\frac{2\pi}{h}p\right)^2\psi$$

Finally, we use the relationship between momentum and kinetic energy, equation (8.14), to obtain

$$-\frac{h^2}{8\pi^2 m}\frac{d^2\psi}{dx^2} = E_k\psi$$

This is the Schrödinger equation of a free particle moving in one dimension. Suppose instead that the particle is subjected to force, the strength of which varies as the particle moves from point to point. For such a situation, we write the expression above in a different way, replacing E_k by $E - V(x)$, where E is the total energy (a constant) and $V(x)$ is the potential energy. The function $V(x)$ takes into account the possibility that the potential energy of the particle changes with its position, as is the case when the particle is subjected to a force. We obtain

$$-\frac{h^2}{8\pi^2 m}\frac{d^2\psi}{dx^2} + V(x)\psi = E\psi$$

Extending this treatment to three dimensions, we obtain the Schrödinger equation for the hydrogen atom or hydrogen-like ion, where we understand $V(r)$ to be $(-e)(Ze)/4\pi\epsilon_0 r$), the potential energy associated with the interaction of the electron (charge $= -e$), and the nucleus of the one electron atom or ion (charge $= Ze$). (See Appendix B.)

$$-\frac{h^2}{8\pi^2 m_e}\left(\frac{\partial^2\psi}{\partial x^2} + \frac{\partial^2\psi}{\partial y^2} + \frac{\partial^2\psi}{\partial z^2}\right) - \frac{Ze^2}{4\pi\epsilon_0 r}\psi = E\psi$$

This is the equation that Schrödinger obtained. In the equation above, $\partial^2\psi/\partial x^2$ means that we differentiate ψ twice with respect to x, treating the other variables (y and z) as constants. The notation $\partial^2\psi/\partial x^2$ is used instead of $d^2\psi/dx^2$ because ψ depends on more than one variable.

Following a suggestion by Eugene Wigner, Schrödinger used spherical polar coordinates to solve the equation above rather than the Cartesian coordinates x, y, and z. That is, he substituted the values of x, y, and z in terms of spherical polar coordinates given in the caption for Figure 8-21 and performed the necessary lengthy algebra to collect the variables r, θ, and ϕ. The equation he obtained is

$$-\frac{h^2}{8\pi^2\mu r^2}\left[\frac{\partial}{\partial r}\left(r^2\frac{\partial\psi}{\partial r}\right) + \frac{1}{\sin\theta}\frac{\partial}{\partial\theta}\left(\sin\theta\frac{\partial\psi}{\partial\theta}\right) + \frac{1}{\sin^2\theta}\frac{\partial^2\psi}{\partial\phi^2}\right] - \frac{Ze^2}{4\pi\epsilon_0 r}\psi = E\psi \quad (8.15)$$

where the mass of the electron has been replaced by the more correct reduced mass of the atom, μ, given by

$$\frac{1}{\mu} = \frac{1}{m_e} + \frac{1}{m_{nucleus}} \quad \text{or} \quad \mu = \frac{m_e\, m_{nucleus}}{m_e + m_{nucleus}}$$

This is the Schrödinger equation in spherical polar coordinates for a hydrogen-like ion of atomic number Z or the hydrogen atom if $Z = 1$. The solutions are shown in Table 8.2 on page 338.

ı the equation $E_n = -R_H/n^2$, n is the *principal quantum number* and takes ɔnly nonzero integer values: $n = 1, 2, 3, \ldots, \infty$. As noted earlier, R_H is a ıerical constant, called the Rydberg constant, the value of which is ıined from the following expression of fundamental constants, all of which ear in the Schrödinger equation:

$$R_H = \frac{\mu e^4}{8\epsilon_0^2 h^2} = 2.17869 \times 10^{-18}\ \text{J} \qquad \textbf{(8.16)}$$

constants appearing in equation (8.16) are defined in Table 8.1. The value R_H calculated from these fundamental constants agrees with value ıined by Rydberg from Balmer's empirical equation (8.4), once the for- .a above has been divided by hc to convert from J to m^{-1}. This agreement not only a scientific triumph for Schrödinger but also significant in estab- ing quantum theory as one of the most significant advances in science.

olutions to the Schrödinger equation for the hydrogen atom give not only rgy levels but also wave functions. These wave functions are called **itals** to distinguish them from the orbits of the incorrect Bohr theory. The hematical form of these orbitals is more complex than for the particle in a , but nonetheless they can be interpreted in a straightforward way.

Vave functions are most easily analyzed in terms of the three variables ıired to define a point with respect to the nucleus. In the usual Cartesian rdinate system, these three variables are the x, y, and z dimensions. In the erical polar coordinate system, they are r, the distance of the point from nucleus, and the angles θ (theta) and ϕ (phi), which describe the orienta- of the distance line, r, with respect to the x, y, and z axes (Fig. 8-21). ıer coordinate system could be used in solving the Schrödinger equation. Iowever, in the spherical polar system, the orbitals can be expressed as a duct of two separate factors: a *radial factor*, R, that depends only on r, and *ngular factor*, Y, that depends on θ (theta) and ϕ (phi) That is,

$$\psi(r, \theta, \phi) = R(r)Y(\theta, \phi)$$

'he radial factor $R(r)$ is also called the **radial wave function**, and the ular factor $Y(\theta, \phi)$ is also called the **angular wave function**. Each orbital, ıas three quantum numbers to define it since the hydrogen atom is a e-dimensional system. The particular set of quantum numbers confers ticular functional forms to $R(r)$ and $Y(\theta, \phi)$, which are most conveniently resented in graphical form. In Section 8-8, we will use various graphical resentations of orbitals to deepen our understanding of the description of trons in atoms.

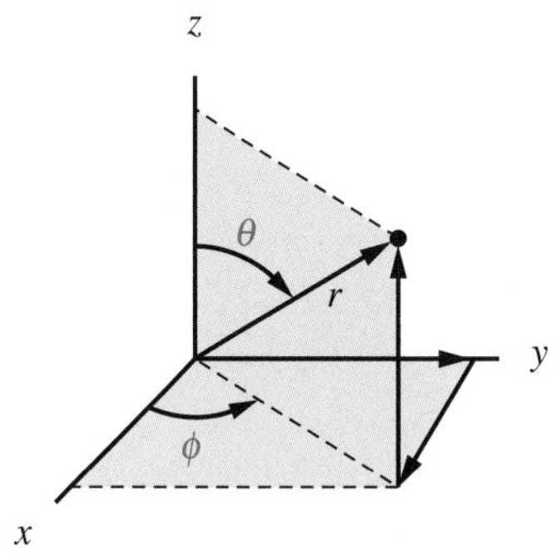

▲ FIGURE 8-21
The relationship between spherical polar coordinates and Cartesian coordinates
The coordinates x, y, and z are expressed in terms of the distance r and the angles θ and ϕ.

BLE 8.1 Values[a] of the Fundamental Constants Used in the lculation of the Rydberg Constant, R_H

duced mass	$\mu = m_e m_p/(m_e + m_p)$
ctron mass	$m_e = 9.10938356 \times 10^{-31}$ kg
ɔton mass	$m_p = 1.672621898 \times 10^{-27}$ kg
mentary charge	$e = 1.6021766208 \times 10^{-19}$ C
nck's constant	$h = 6.626070040 \times 10^{-34}$ J s
rmittivity of vacuum	$\epsilon_0 = 8.854187817 \times 10^{-12}\ C^2\ J^{-1}\ m^{-1}$

ıese are the 2014 CODATA recommended values (http://physics.nist.gov/cuu/ nstants/index.html), which became available in 2015 and replace the previous values m 2010. CODATA is the Committee on Data for Science and Technology.

In the remainder of this section, we will explore the combinations of qu tum numbers that produce different orbitals and discover that the orbital the hydrogen atom (in fact, the orbitals of all atoms) are organized i groups and subgroups (called *shells* and *subshells*).

Assigning Quantum Numbers

The following relationships involving the three quantum numbers arise fr the solution of the Schrödinger wave equation for the hydrogen atom. In solution the values of the quantum numbers are fixed *in the order listed.*

The first number to be fixed is the *principal quantum number, n,* which r have only a positive, nonzero integral value.

$$n = 1, 2, 3, 4, \ldots \quad (8$$

Second is the *orbital angular momentum quantum number, ℓ,* which may be z or a positive integer, but not larger than $n - 1$ (where n is the principal qu tum number).

$$\ell = 0, 1, 2, 3, \ldots, n - 1 \quad (8$$

Third is the *magnetic quantum number,* m_ℓ, which may be a negative or posi integer, including zero, and ranging from $-\ell$ to $+\ell$ (where ℓ is the orb angular momentum quantum number).

$$m_\ell = -\ell, (-\ell + 1), \ldots, -2, -1, 0, 1, 2 \ldots, (\ell - 1), +\ell \quad (8$$

EXAMPLE 8-10 Applying Relationships Among Quantum Numbers

Can an orbital have the quantum numbers $n = 2, \ell = 2$, and $m_\ell = 2$?

Analyze

We must determine whether the given set of quantum numbers is allowed by the rules expressed in equations (8.17), (8.18), and (8.19).

Solve

No. The ℓ quantum number cannot be greater than $n - 1$. Thus, if $n = 2$, ℓ can be only 0 or 1. And if ℓ can be only 0 or 1, m_ℓ cannot be 2; m_ℓ must be 0 if $\ell = 0$ and may be $-1, 0$, or $+1$ if $\ell = 1$.

Assess

It is important that we remember the physical significance of the various quantum numbers, as well as the rules interrelating their values. We will soon see that the quantum number n determines the radial distribution and the average *distance* of the electron and, thus, is most important in determining the *energy* of an electron. Quantum number ℓ determines the angular distribution or *shape* of an orbital. The relationships among the quantum numbers impart a logical organization of orbitals into shells and subshells.

PRACTICE EXAMPLE A: Can an orbital have the quantum numbers $n = 3, \ell = 0$, and $m_\ell = 0$?

PRACTICE EXAMPLE B: For an orbital with $n = 3$ and $m_\ell = 1$, what is (are) the possible value(s) of ℓ?

Principal Shells and Subshells

All orbitals with the same value of n are in the same **principal electronic sh** or **principal level**, and all orbitals with the same n and ℓ values are in same **subshell**, or **sublevel**.

Principal electronic shells are numbered according to the value of n. The f principal shell consists of orbitals with $n = 1$; the second principal shell orbitals with $n = 2$; and so on. The value of n relates to the energy and m probable distance of an electron from the nucleus. The higher the value o the greater the electron energy and the farther, on average, the electron is fr the nucleus. The principal quantum number, therefore, has a physical sign

:e, as do the other quantum numbers. The quantum number ℓ determines angular distribution, or *shape*, of an orbital and m_ℓ determines the *orienta-* of the orbital.

he number of subshells in a principal electronic shell is the same as the nber of allowed values of the orbital angular momentum quantum num- ℓ. In the first principal shell, with $n = 1$, the only allowed value of ℓ is 0, there is a single subshell. The second principal shell ($n = 2$), with the wed ℓ values of 0 and 1, consists of two subshells; the third principal shell = 3) has three subshells ($\ell = 0, 1$, and 2); and so on. Or, to put the matter in ther way, because there are n possible values of the ℓ quantum number, is, $0, 1, 2, \ldots (n - 1)$, the number of subshells in a principal shell is equal ne principal quantum number. As a result, there is one subshell in the prin- l shell with $n = 1$, two subshells in the principal shell with $n = 2$, and so The name given to a subshell, regardless of the principal shell in which it is nd, depends on the value of the ℓ quantum number. The first four sub- lls are

s subshell	*p subshell*	*d subshell*	*f subshell*
$\ell = 0$	$\ell = 1$	$\ell = 2$	$\ell = 3$

he number of orbitals in a subshell is the same as the number of allowed ıes of m_ℓ for the particular value of ℓ. Recall that the allowed values of m_ℓ $0, \pm1, \pm2, \ldots, \pm\ell$, and thus the total number of orbitals in a subshell is + 1. The names of the orbitals are the same as the names of the subshells in ch they appear.

s orbitals	*p orbitals*	*d orbitals*	*f orbitals*
$\ell = 0$	$\ell = 1$	$\ell = 2$	$\ell = 3$
$m_\ell = 0$	$m_\ell = 0, \pm1$	$m_\ell = 0, \pm1, \pm2$	$m_\ell = 0, \pm1, \pm2, \pm3$
one *s* orbital in an *s* subshell	three *p* orbitals in a *p* subshell	five *d* orbitals in a *d* subshell	seven *f* orbitals in an *f* subshell

'o designate the particular principal shell in which a given subshell or ital is found, we use a combination of a number and a letter. For example, symbol 2*p* is used to designate both the *p* subshell of the second principal ll and any of the three *p* orbitals in that subshell. Some of the points dis- sed here are illustrated in Example 8-11.

EXAMPLE 8-11 Relating Orbital Designations and Quantum Numbers

Write an orbital designation corresponding to the quantum numbers $n = 4, \ell = 2, m_\ell = 0$.

Analyze

To write orbital designations you need to recall the conventions associated with the quantum numbers n and ℓ. For the quantum number n we use only the number while for the quantum number ℓ we use the following letters: $\ell = 0, s; \ell = 1, p; \ell = 2, d$; and so on.

Solve

The magnetic quantum number, m_ℓ, is not reflected in the orbital designation. The type of orbital is determined by the ℓ quantum number. Because $\ell = 2$, the orbital is of the *d* type. Because $n = 4$, the orbital designation is 4*d*.

Assess

This is another type of problem in which we need to have memorized the quantum number rules and their designations. This information will be important in the later chapters.

PRACTICE EXAMPLE A: Write an orbital designation corresponding to the quantum numbers $n = 3, \ell = 1$, and $m_\ell = 1$.

PRACTICE EXAMPLE B: Write all the combinations of quantum numbers that define hydrogen-atom orbitals with the same energy as the 3*s* orbital.

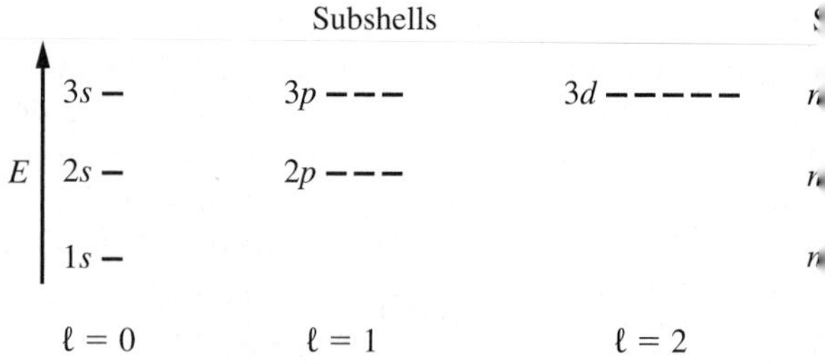

Each subshell is made up of $(2\ell + 1)$ orbitals.

▶ FIGURE 8-22
Shells and subshells of a hydrogen atom
The hydrogen atom orbitals are organized into shells and subshells.

The Energies of Principal Shells and Subshells in One-Electron Species

▶ In Section 8-10 and in Chapter 24, we will see that orbital energies of multielectron atoms also depend on the quantum numbers ℓ and m_ℓ.

As we saw in Section 8-3, the energy levels of the hydrogen atom c hydrogen-like species are given by equation (8.9):

$$E_n = \frac{-Z^2 R_H}{n^2}$$

For a given value of Z, the energies depend only on the principal quant number, n. This means that all the subshells within a principal electronic s have the same energy, as do all the orbitals within a subshell. Levels with same energy are said to be **degenerate**. Figure 8-22 shows an energy-level gram and the arrangement of shells and subshells for a hydrogen atom. energy-level diagram for other one-electron species, such as He^+, Li^{2+}, B B^{4+}, C^{5+}, and so on, is similar to that shown in Figure 8-22.

8-6 ARE YOU WONDERING?

Are all orbital transitions allowed in atomic absorption and emission spectra?

The short answer to this question is no. Let's suppose that the state of the electron in a hydrogen atom changes from some initial state $(n_i, \ell_i, m_{\ell,i})$ to some final state $(n_f, \ell_f, m_{\ell,f})$ as a result of the atom absorbing or emitting a photon. From the discussion in Section 8-6, we know that the photon energy, $E_{photon} = h\nu$, must be equal to $|\Delta E|$, the magnitude of the energy difference between the initial and fina states. However, this is not the only rule that must be obeyed. Other rules, callec selection rules, must also be obeyed. The selection rules are summarized below.

Selection Rule	Comment
Δn = any integer	The transition must also obey $\lvert\Delta E\rvert = E_{photon}$.
$\Delta\ell = -1$ or $+1$	The allowed orbital transitions include $s \rightleftarrows p$, $p \rightleftarrows d$, etc., but not $s \rightleftarrows s$, $p \rightleftarrows p$, etc., or $s \rightleftarrows d$.
$\Delta m_\ell = -1, 0$ or $+1$	The restriction for Δm_ℓ applies only if the spectrum is measured in the presence of an applied magnetic field.

We will not attempt to justify these selection rules except to say that the selection rule for $\Delta\ell$ arises from the fact that a photon carries not only a certain quantity of energy but also one unit of angular momentum. Therefore, when an atom absorbs or emits a photon, not only does the energy of the atom change, but the angular momentum of the atom also increases or decreases by one unit. Because of the selection rule for $\Delta\ell$, an electron in an s orbital, for example, cannot ndergo a transition to a d orbital. For such a transition, $\Delta\ell = 2 - 0 = 2$, which ; not allowed by the selection rules. On the other hand, a transition from an s rbital to a p orbital is allowed ($\Delta\ell = 1 - 0 = 1$), as is a transition from a p rbital to a d orbital ($\Delta\ell = 2 - 1 = 1$). Transitions between orbitals having the ame value of ℓ, such as $s \rightarrow s$, $p \rightarrow p$, and so on, are not allowed because for ıese transitions, $\Delta\ell = 0$.

As we will see in Section 8-8, a fourth quantum number, m_s, is needed to com- letely describe an electron. The selection rule for m_s is $\Delta m_s = 0$, indicating that ıe value of m_s does not change when a photon is absorbed or emitted.

7 Interpreting and Representing the Orbitals of the Hydrogen Atom

major undertaking in this section will be to describe the three-dimensional ›ability density distributions obtained for the various orbitals in the hydro- atom. Through the Born interpretation of wave functions (page 328), we represent the probability densities of the orbitals of the hydrogen atom as aces that encompass most of the electron probability. We will see that the ›ability density for each type of orbital has its own distinctive shape. In lying this section, it is important for you to remember that, even though we offer some additional quantitative information about orbitals, your pri- y concern should be to acquire a broad qualitative understanding. It is this litative understanding that you can apply in our later discussion of how tals enter into a description of chemical bonding.

hroughout this discussion, recall that orbitals are wave functions, mathe- ical solutions of the Schrödinger wave equation. The wave function itself no physical significance. However, the square of the wave function, ψ^2, is a ntity that is related to probabilities. Probability density distributions based ψ^2 are three-dimensional, and it is these three-dimensional regions that we ın when we refer to the shape of an orbital.

◀ In Chapter 11, we will discover important uses of the wave function, ψ, itself as a basis for discussing bonding between atoms.

'he forms of the radial wave function $R(r)$ and the angular wave function , ϕ) for a one-electron, hydrogen-like atom are shown in Table 8.2. The first ıg to note is that the angular part of the wave function for an s orbital, $\frac{}{\pi}\Big)^{1/2}$, is always the same, regardless of the principal quantum number. Next, e that the angular parts of the p and d orbitals are also independent of the ntum number n. Therefore all orbitals of a given type (s, p, d, f) have the ıe angular behavior. It is also worth noting that the names given to the angu- ›arts are related to their functional forms in Cartesian coordinates. Also note the equations in Table 8.2 are in a general form where the atomic number Z ıcluded. This means that the equations apply to any one-electron atom, that o a hydrogen atom or a hydrogen-like ion. Finally, note that the term σ earing throughout the table is equal to $2Zr/na_0$. The quantity a_0 is called the ır **radius**, the value of which can be related to other constants appearing in Schrödinger equation:

$$a_0 = \frac{\epsilon_0 h^2}{\pi m_e e^2} = 5.29177 \times 10^{-11}\,\text{m} = 52.9177\,\text{pm}$$

s distance is the radius of the lowest energy orbit in Bohr's model. The name en to this quantity commemorates the pioneering work of Niels Bohr.

'o obtain the wave function for a particular state, we simply multiply the ial part by the angular part. We begin, however, by looking separately at radial and angular parts of the wave functions for $n = 1$, 2, and 3.

TABLE 8.2 The Angular and Radial Parts of the Wave Functions for a Hydrogen-Like Atom

Angular Part $Y(\theta, \phi)$ Cartesian	Spherical Polar	Radial Part $R_{n\ell}(r)$ $\sigma = \frac{2Zr}{na_0}$
$Y(s) = \left(\frac{1}{4\pi}\right)^{1/2}$	(same as Cartesian)	$R_{1s} = 2\left(\frac{Z}{a_0}\right)^{3/2} e^{-\sigma/2}$
		$R_{2s} = \frac{1}{2\sqrt{2}}\left(\frac{Z}{a_0}\right)^{3/2}(2 - \sigma)e^{-\sigma/2}$
		$R_{3s} = \frac{1}{9\sqrt{3}}\left(\frac{Z}{a_0}\right)^{3/2}(6 - 6\sigma + \sigma^2)e^{-\sigma/2}$
$Y(p_x) = \left(\frac{3}{4\pi}\right)^{1/2}\frac{x}{r}$	$= \left(\frac{3}{4\pi}\right)^{1/2}\sin\theta\cos\phi$	$R_{2p} = \frac{1}{2\sqrt{6}}\left(\frac{Z}{a_0}\right)^{3/2}\sigma e^{-\sigma/2}$
$Y(p_y) = \left(\frac{3}{4\pi}\right)^{1/2}\frac{y}{r}$	$= \left(\frac{3}{4\pi}\right)^{1/2}\sin\theta\sin\phi$	$R_{3p} = \frac{1}{9\sqrt{6}}\left(\frac{Z}{a_0}\right)^{3/2}(4 - \sigma)\sigma e^{-\sigma/2}$
$Y(p_z) = \left(\frac{3}{4\pi}\right)^{1/2}\frac{z}{r}$	$= \left(\frac{3}{4\pi}\right)^{1/2}\cos\theta$	
$Y(d_{z^2}) = \left(\frac{5}{16\pi}\right)^{1/2}\frac{3z^2 - r^2}{r^2}$	$= \left(\frac{5}{16\pi}\right)^{1/2}(3\cos^2\theta - 1)$	$R_{3d} = \frac{1}{9\sqrt{30}}\left(\frac{Z}{a_0}\right)^{3/2}\sigma^2 e^{-\sigma/2}$
$Y(d_{x^2-y^2}) = \left(\frac{15}{16\pi}\right)^{1/2}\frac{x^2 - y^2}{r^2}$	$= \left(\frac{15}{16\pi}\right)^{1/2}\sin^2\theta\cos 2\phi$	
$Y(d_{xy}) = \left(\frac{15}{4\pi}\right)^{1/2}\frac{xy}{r^2}$	$= \left(\frac{15}{16\pi}\right)^{1/2}\sin^2\theta\sin 2\phi$	
$Y(d_{xz}) = \left(\frac{15}{4\pi}\right)^{1/2}\frac{xz}{r^2}$	$= \left(\frac{15}{4\pi}\right)^{1/2}\sin\theta\cos\theta\cos\phi$	
$Y(d_{yz}) = \left(\frac{15}{4\pi}\right)^{1/2}\frac{yz}{r^2}$	$= \left(\frac{15}{4\pi}\right)^{1/2}\sin\theta\cos\theta\sin\phi$	

$$\psi(1s) = R(r) \times = \frac{2e^{-r/a_0}}{a_0^{3/2}} \times = \frac{e^{-r/a_0}}{\sqrt{(\pi a_0^3)}}$$

The Radial Functions

Ultimately, the radial functions determine how the probability density f particular state (orbital) changes with the distance, r, from the nucleus; t they provide the information we need to compare the sizes of different orbi

The radial functions for $n = 1$, 2, and 3 with the appropriate values of ℓ illustrated in Figure 8-23. Notice that each radial function decays exponenti with increasing r and some cross the horizontal axis one or more times be finally decaying to zero. You may also notice that some of the radial functi have a nonzero value at $r = 0$, whereas others have a value of zero at $r = 0$. following points summarize the main features of the radial functions and be verified by referring to Figure 8-23:

- The radial function decays exponentially to a value of zero as r increa consequently, we can think of each orbital as having a certain size. In g eral, the larger the value of n, the larger the orbital. This can be seer comparing, for example, the radial functions for the 1s and 3s orbitals. radial function for the 1s orbital decays to a value of nearly zero by $r =$ for the 3s orbital, the radial function decays to a value of nearly zer approximately $r = 20a_0$. It is for this reason that we say, for example, a 3s orbital is "larger" than a 1s orbital.
- The radial function crosses the horizontal axis $n - \ell - 1$ times be finally decaying to a value of zero. The point at which the radial func crosses the horizontal axis corresponds to a **radial node**. Thus, the ra

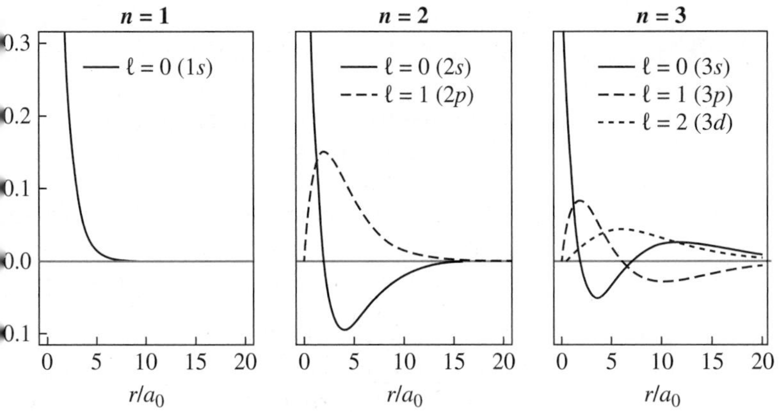

GURE 8-23
ial functions of hydrogen orbitals
radial functions, $R(r)$, for orbitals of the hydrogen atom having $n = 1$, 2, or 3. As ıssed in the text, the number of radial nodes for a given orbital is equal to the ber of times $R(r)$ crosses the horizontal axis. In general, the number of radial nodes ual to $n - \ell - 1$. For s orbitals, $R(r)$ has its maximum value at $r = 0$, whereas for r orbitals (p, d, f, and so on), $R(r) = 0$ at $r = 0$.

function has $n - \ell - 1$ radial nodes. For example, a 1s orbital has $1 - 0 - 1 = 0$ radial nodes and a 3p orbital has $3 - 1 - 1 = 1$ radial node.

For s orbitals ($\ell = 0$), the radial factor has a nonzero value at the nucleus. For $\ell \neq 0$ (that is, for p, d, f, etc. orbitals), the radial function has a value of zero at the nucleus. For example, the radial functions for the 1s, 2s, and 3s orbitals all have a cusp at $r = 0$ (the nucleus) whereas the radial functions for the 2p, 3p, and 3d orbitals are all zero at the nucleus. Consequently, the probability density for an s orbital has its maximum value at $r = 0$ whereas for other orbitals (p, d, f, etc.), the probability density is zero at $r = 0$.

Iot surprisingly, the main features summarized above arise from the math- tical forms of the radial functions (Table 8.2). For example, the radial func- s decay exponentially to a value of zero because the exponential factor $e^{-\sigma/2}$ ears in all of them, where $\sigma = 2Zr/na_0$. The number of radial nodes and the e of the radial function at the nucleus are, however, determined by the fac- multiplying $e^{-\sigma/2}$. By examining Table 8.2 carefully, you will discover that, ach radial function, the exponential factor is multiplied by $\sigma^{\ell} \times f(\sigma)$, re f represents a polynomial in σ of degree $n - \ell - 1$. For example, in the al function for the 3p orbital, the exponential factor is multiplied by $\sigma) \times \sigma$, with $(4 - \sigma)$ a polynomial of order $3 - 1 - 1 = 1$. A polynomial of er $n - \ell - 1$ will cross the horizontal axis up to $n - \ell - 1$ times, and so ı radial function crosses the horizontal axis this number of times. The factor s also important because it affects the behavior of the radial function at). Because $\sigma = 2Zr/na_0$, the factor σ^{ℓ} has a value of zero at $r = 0$, except n $\ell = 0$. When $\ell = 0$, σ^{ℓ} always has a value of 1.

Ve have rationalized main features of the radial functions by considering the general form of these functions. To go further, we must consider the ise forms of these functions. Consider for example the radial function for 2s orbital of hydrogen ($Z = 1$), which can be written in the following form:

$$R(r) = \frac{1}{2\sqrt{2}}\frac{1}{a_0^{3/2}}\left(2 - \frac{r}{a_0}\right)e^{-r/2a_0}$$

values of r less than $2a_0$, R is positive, and for $r > 2a_0$, it is negative. Thus, radial function for the 2s orbital has a radial node at $r = 2a_0$. The 3s orbital two radial nodes: one at $r = 1.9\,a_0$ and another at $r = 7.1\,a_0$. (See Exercise 107.)

The Angular Functions

To view the angular functions, we will plot them in the form of polar gra Figure 8-24. In a polar graph, the magnitude of the function at a partic value of the angles is given as the distance from the origin. The graph Figure 8-24 are cross-sections of the complete three-dimensional graphs a as a consequence, show the behavior of $Y(\theta, \phi)$ in only a single plane, function of either θ or ϕ alone. The planes selected for the figure are those most clearly show the shapes of the particular angular functions. Let us r examine the shapes of these angular wave functions in a bit more detail.

s orbitals For s orbitals ($\ell = 0$), the angular function is $Y = (1/4\pi)^{1/2}$. function has no angular dependence, and so it has the same value for all ues of θ and ϕ. For example, it has the value $(1/4\pi)^{1/2}$ when $\theta = \pi/2$ $\phi = 0$ and also when $\theta = 0$ and $\phi = \pi/2$. The polar graph of this function sphere. For this reason, s orbitals have a spherical shape.

p orbitals For p orbitals ($\ell = 1$), there are three angular functi Although the mathematical forms of these functions (Table 8.2) are differ their polar graphs reveal that they are identical in shape but oriented dif ently in space. The three p orbitals are labeled p_x, p_y, or p_z, to signify that t are oriented along the x, y, or z axes. In contrast to what we saw for s orbi the angular functions for the p orbitals do not have a constant value. They functions of θ and ϕ and, therefore, p orbitals do not have spherical sha We can explain the shapes of the p orbitals by focusing on the angular fu tion for the p_z orbital. By referring to Table 8.2, we find that the ang function for the p_z orbital is proportional to $\cos\theta$. Thus, the p_z wave func has an angular maximum along the positive z axis, for there $\theta = 0$ $\cos(0) = +1$. Along the negative z axis, the p_z wave function has its m negative value, for there $\theta = \pi$ and $\cos(\pi) = -1$. The designation p_z help remember that this angular function has its maximum magnitude along z axis. Everywhere in the xy plane $\theta = \pi/2$ and $\cos\theta = 0$, so the xy plar a node. Because this node arises in the angular function, it is called an *an lar node*. A similar analysis of the p_x and p_y orbitals shows that they are si lar to the p_z orbital, but with angular nodes in the yz and xz planes, res tively. In three dimensions, the polar graph for each p orbital consists of spheres tangent to the origin, as shown in Figure 8-24. The phase (positiv negative) is included in these graphs to indicate where Y has positive or r ative values. We will see in Chapter 11 that the phase of the orbital is important consideration when developing models for describing chem bonding.

***d* orbitals** The angular functions with $\ell = 2$ are more complicated, as ca seen from their mathematical forms (Table 8.2). It turns out that the ang functions for d orbitals ($\ell = 2$) possess two angular nodes, whereas p orbi ($\ell = 1$) possess one angular node and s orbitals ($\ell = 0$) have no ang nodes. In general, the number of angular nodes is equal to the value of ℓ.

Let's illustrate some of these ideas by considering the angular function the $d_{x^2-y^2}$ orbital. The angular function for this orbital is proportiona $\sin^2\theta\cos 2\phi$. How should we visualize this function? We can proceed setting $\theta = \pi/2$ and plotting the function $\cos 2\phi$ as a polar graph. Exan Figure 8-21, and you will see that the angle $\theta = \pi/2$ corresponds to the plane. By setting $\theta = \pi/2$, we obtain the cross-section shown in Figure 8 The angular function consists of four lobes oriented along the x and y a The phase (sign) of Y in various regions is indicated by the red and blue li Notice that the phase is positive for two of the lobes and negative for the o two. Also take note of the alternation in phase as we move either clockwis counterclockwise from one lobe to another.

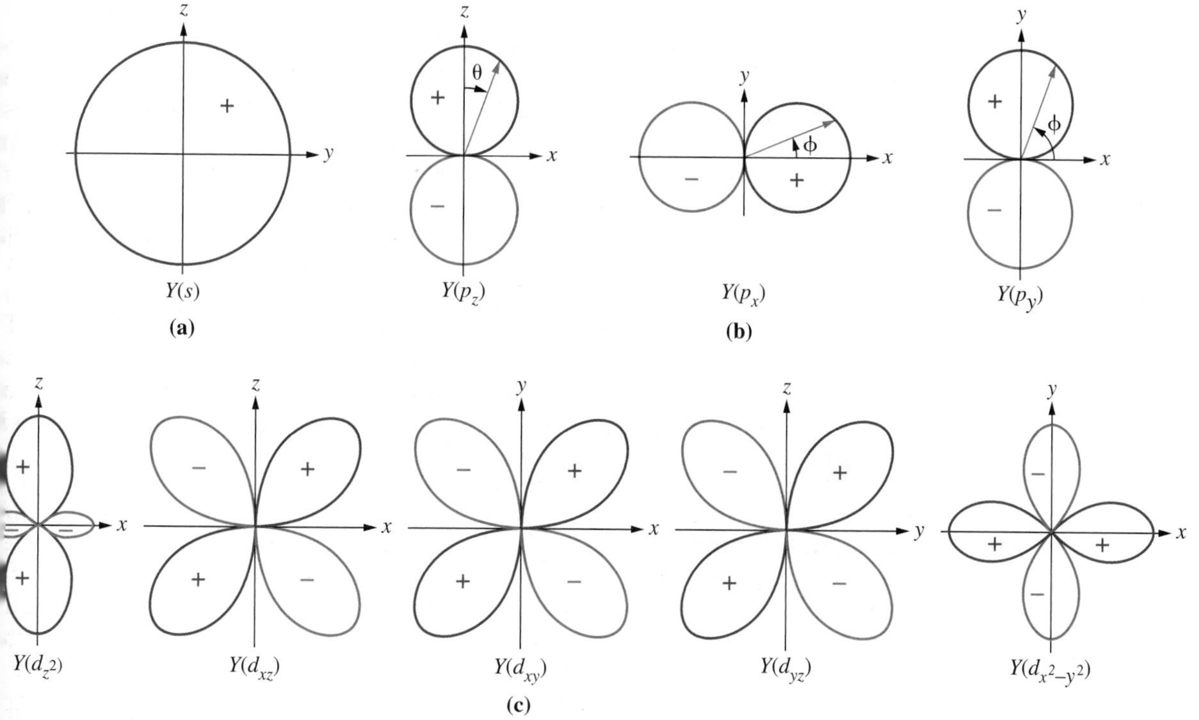

FIGURE 8-24
gular functions of the s, p, and d orbitals
e angular functions Y for s, p, and d orbitals are shown in **(a)**, **(b)**, and **(c)**, pectively. As illustrated in part (b), the distance from the origin to a point on the ve (red arrow) gives the magnitude of the angular function for a given value of θ or where θ is the angle measured from the z axis and ϕ is the angle measured in from x axis in the xy plane. See also Figure 8-21. The colors blue and red are used to icate whether the angular function has a positive value (blue) or a negative value d) in that region.

Cross-sections of the angular functions for the d_{xy}, d_{xz}, d_{yz}, and d_{z^2} are also played in Figure 8-24. We observe that four of them have the same shape as $-y^2$, but they are oriented differently with respect to the axes. As is the case the $d_{x^2-y^2}$ orbital, the d_{xy}, d_{xz}, and d_{z^2} orbitals each have two nodal planes. e d_{z^2} orbital has quite a different shape but also has two angular nodes. The gular nodes for the d_{z^2} orbital are conical surfaces.

The angular functions for f, g, h, and so on, orbitals have rather complicated apes because of the larger number of angular nodes. These orbitals are not en encountered, and so we will not consider their shapes at all.

e Wave Functions and the Shapes of the Orbitals

mentioned at the start of this section, the complete wave function is given the product of a radial function and an angular function. To construct the nplete wave function for one orbital of a hydrogen atom, we use expresns from Table 8.2 with $Z = 1$. Let us construct the wave function for the 1s ital by combining the appropriate radial and angular functions. In the ression below, the radial function is shown in red; the angular function is wn in blue.

$$\psi(1s) = R(r) \times Y(\theta, \phi) = \frac{2e^{-r/a_0}}{a_0^{3/2}} \times \frac{1}{\sqrt{4\pi}} = \frac{e^{-r/a_0}}{\sqrt{(\pi a_0^3)}}$$

How can we represent $\psi(1s)$ using a graph? One way is to pass a pl through the nucleus (for example, the *xy* plane) and plot a graph of the val of $\psi(1s)$ as perpendicular distances above or below the many points in plane at which the electron might be found. The resultant graph, showr Figure 8-25(a), looks like a symmetrical, cone-shaped "hill" (think of a v cano), with its peak directly above the nucleus. As we do in topographi maps of Earth's surface, we can project the three-dimensional surface ont two-dimensional contour map. The contour map is shown below the surf in Figure 8-25(a) and separately in Figure 8-25(b). The circular contour li join points for which $\psi(1s)$ has the same value. For contours close to nucleus, $\psi(1s)$ has a large (positive) value. For contours farther away, $\psi(1s)$ a lower value.

Another way of representing $\psi(1s)$ is as an *isosurface* (Fig. 8-25c). $\psi(1s)$ the same value at all points on this isosurface (a sphere). Because the isos face of $\psi(1s)$ is spherical, we can say that a 1*s* orbital is spherical.

Still another way of representing $\psi(1s)$ is shown in Figure 8-25(d). In su a graph, the density of points is highest where $\psi(1s)$ has its largest valu This representation shows that a 1*s* orbital is spherical but also conveys

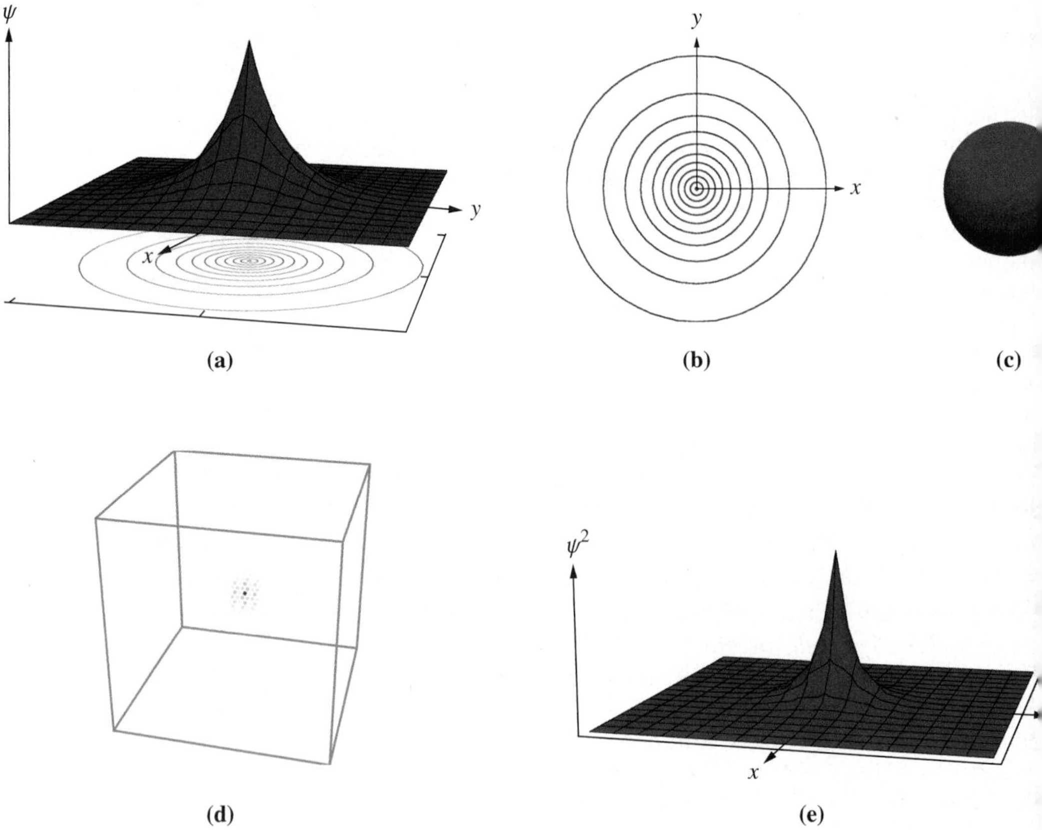

▲ FIGURE 8-25
Representations of the wave function and electron probability density of t 1*s* orbital
(a) In this diagram, the value of ψ is represented by the height above the *xy* plane (t *xy* plane is an arbitrary choice; any plane could have been chosen). **(b)** A contour ma of the wave function for the 1*s* orbital in the *xy* plane. **(c)** A reduced-scale three-dimensional representation of the 1*s* orbital. ψ has the same value at all points on th surface; thus, the surface represents points of constant value. For this reason, the surface is called an isosurface. **(d)** A "foggy" plot of the 1*s* orbital. The density of points, or their opacity, is highest where the magnitude of ψ has its largest values. **(e)** In this diagram, the electron probability density, ψ^2, is represented by the height above the *xy* plane.

ial distribution of the probability density. Consequently, Figure 8-25(d) much better representation of a 1*s* orbital than the isosurface shown in ıre 8-25(c).

ınally, we turn our attention to the graphical representation of $\psi^2(1s)$, the ıbility density. As we established in Section 8-5, the probability of finding electron in a small volume of space in the vicinity of a given point is given he values of ψ^2. (Recall the Born interpretation.) For a 1*s* orbital we have

◀ The probability density is also sometimes called the *electron density*.

$$\psi^2(1s) = \frac{1}{\pi}\left(\frac{1}{a_0}\right)^3 e^{-2r/a_0} \tag{8.20}$$

ı Figure 8-25(e), $\psi^2(1s)$ is represented as a surface: the perpendicular ht from a point in the *xy* plane to a point on the surface is equal to the ıe of $\psi^2(1s)$ at that point. Thus, the surface shows the variation of probabil- lensity from point to point. The probability density is highest near the eus and decreases with increasing distance from the nucleus.

low let's look at the wave function of the 2*s* orbital. Again, the radial func- is shown in red and the angular function in blue:

$$) = R(r) \times Y(\theta,\phi) = \frac{1}{2\sqrt{2}}\frac{1}{a_0^{3/2}}\left(2 - \frac{r}{a_0}\right)e^{-r/2a_0} \times \frac{1}{\sqrt{4\pi}} = \frac{1}{4}\left(\frac{1}{2\pi a_0^3}\right)^{1/2}\left(2 - \frac{r}{a_0}\right)e^{-r/2a_0}$$

wave function for the 2*s* orbital possesses a radial node at $r = 2a_0$ because factor $(2 - r/a_0)$ changes sign at that distance. The electron probability sity for the 2*s* orbital is given by

$$\psi^2(2s) = \frac{1}{8\pi}\left(\frac{1}{a_0}\right)^3\left(2 - \frac{r}{a_0}\right)^2 e^{-r/a_0} \tag{8.21}$$

ıparing expressions (8.19) and (8.20), we see that the exponential function changed from e^{-2r/a_0} for the 1*s* orbital to e^{-r/a_0} for the 2*s* orbital. As a result, wave function for the 2*s* orbital decays more slowly than that of the 1*s* tal and extends farther from the nucleus.

he fact that the wave function for the 2*s* orbital extends farther from the leus than that of the 1*s* orbital, together with the presence of the radial node, ıns that the 2*s* orbital is bigger than a 1*s* orbital and contains a radial node. se features are illustrated in Figure 8-26, which compares the 1*s*, 2*s*, and 3*s* tals. Note that the 3*s* orbital exhibits two radial nodes and is larger than both 1*s* and the 2*s* orbitals. The fact that the number of nodes increases as the ·gy is increased is characteristic of high-energy standing waves. To highlight change in phase of an orbital in progressing outward from the nucleus, we e adopted the modern usage of different colors to indicate regions where ψ a positive value (blue) or a negative value (red). Thus, in Figure 8-26 the 1*s* tal is blue throughout; the 2*s* orbital starts out blue and then switches to red; , finally, the 3*s* orbital starts out blue, changes to red, and then changes back lue, reflecting the presence of two radial nodes.

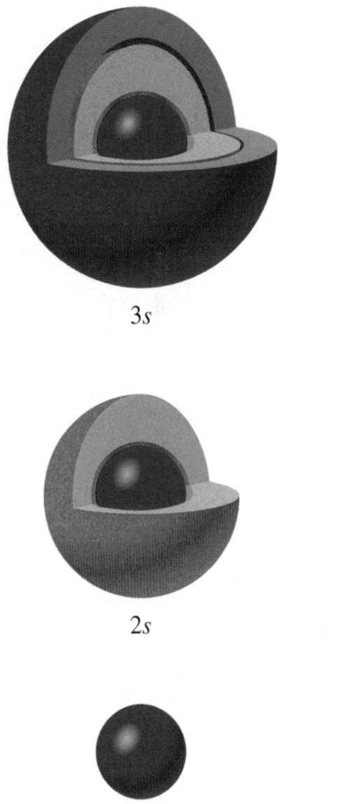

▲ FIGURE 8-26
Three-dimensional representations of the 1*s*, 2*s*, and 3*s* orbitals
The first three *s* orbitals of the hydrogen atom. Note the increasing size of the orbital in proceeding from 1*s* to 2*s* and on to 3*s*.

low let's look at the wave function of the $2p_x$ orbital. Combining the radial angular parts we get

$$) = R(r) \times Y(\theta,\phi) = \frac{1}{2\sqrt{6}}\frac{1}{a_0^{3/2}}e^{-r/2a_0} \times \left(\frac{3}{4\pi}\right)^{1/2}\sin(\theta)\cos(\phi) = \frac{1}{4}\left(\frac{1}{2\pi a_0^3}\right)^{1/2} e^{-r/2a_0}\sin(\theta)\cos(\phi)$$

.s discussed previously, a 2*p* orbital has no radial nodes. In contrast to *s* tals, which are nonzero at $r = 0$, *p* orbitals vanish at $r = 0$. This difference have an important consequence when we consider multielectron atoms.

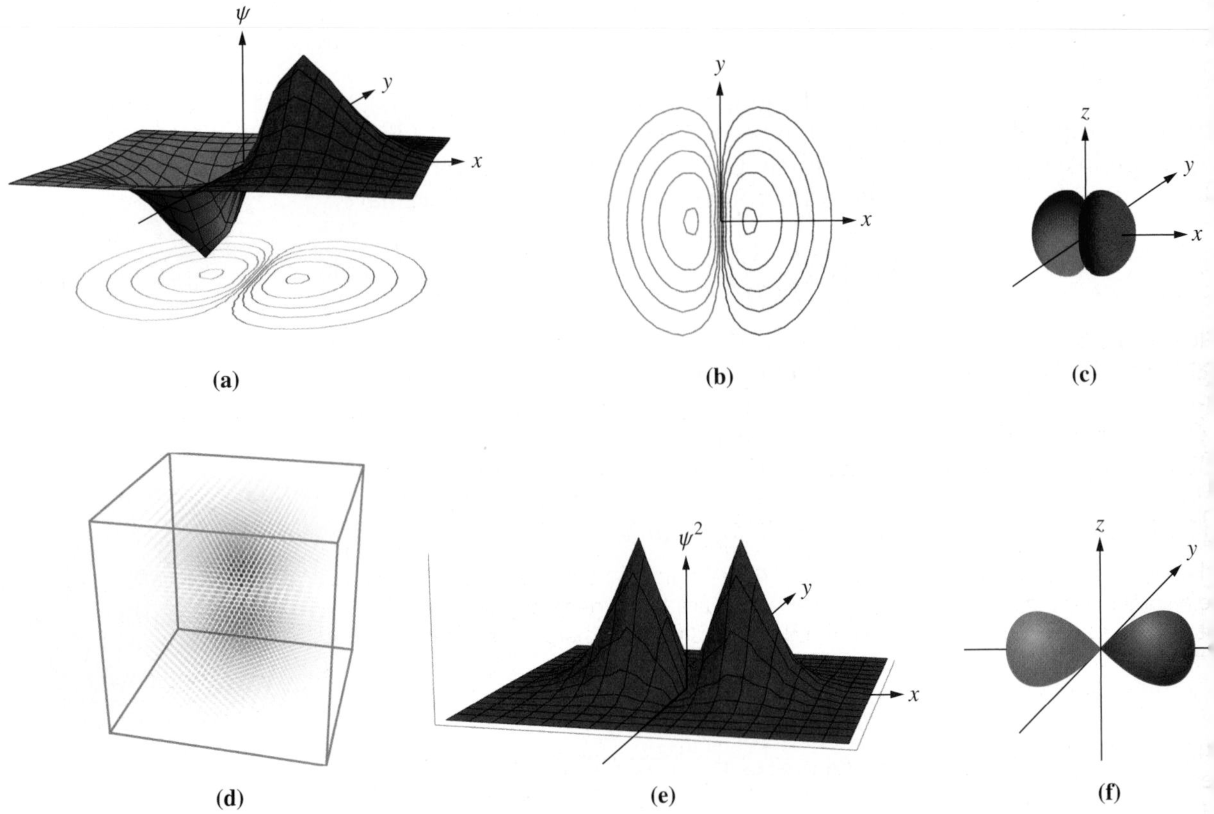

▲ FIGURE 8-27
Representations of the wave function and electron probability density of $2p_x$ orbital
(a) The wave function, ψ, for the $2p_x$ orbital of the hydrogen atom. The value of ψ is plotted as a distance above or below the *xy* plane. The nucleus is imagined to be a the origin, at $x = 0$ and $y = 0$ in this diagram. The colors are used to indicate regions which ψ has either a positive (blue) or negative (red) value. **(b)** A contour map of the wave function for the $2p_x$ orbital in the *xy* plane. **(c)** A three-dimensional representa of the $2p_x$ orbital. ψ has the same magnitude at all points on this surface. **(d)** In this "foggy" plot for the $2p_x$ orbital, the density of points is highest where the magnituc of ψ has its largest values. **(e)** In this diagram, the electron probability density, ψ^2, is represented by the height above the *xy* plane. **(f)** Simplified representation of a $2p_x$ orbital used throughout this text.

We have displayed the wave function, contour map, isosurface, "fog plot, and square of the wave function in Figure 8-27 for the $2p_x$ orbital Figures 8-27(a) and 8-27(e), the height above the xy plane represents the va of the wave function (Fig. 8-27a) or the square (Fig. 8-27e) of the wave fu tion. In Figure 8-27(a), the alternation in phase is readily apparent, wherea Figure 8-27(e), the change of phase is not apparent. In Figure 8-27(f), we sh a simplified representation of the $2p_x$ orbital, which we will use through the remainder of the text.

All three of the p orbitals are shown in Figure 8-28 and are seen tc directed along the three perpendicular axes of the Cartesian system. Ag we have used different colors to represent the phase alternation in th orbitals. However, we must remember that these refer only to the phase the original wave function, *not* to ψ^2.

We will not construct wave functions for the d orbitals but simply sh them. The wave function, contour maps, and isosurfaces for the $3d_{xy}$ and orbitals are shown in Figure 8-29; these graphs are realistic representation the shapes of these orbitals. Simplified representations of the all five d orb are shown in Figure 8-30. Two of the d orbitals ($d_{x^2-y^2}$ and d_{z^2}) are dire

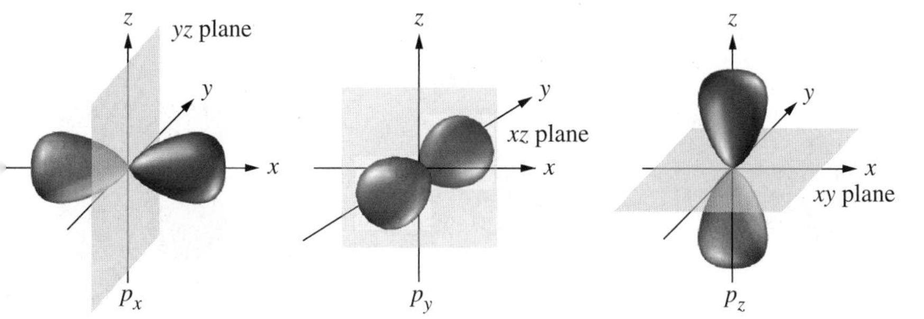

GURE 8-28
plified representations of the three 2*p* orbitals
se simplified representations are used throughout the text to show schematically that 2*p* orbitals have one angular node (one nodal plane). The *p* orbitals are usually esented as directed along the perpendicular *x*, *y*, and *z* axes, and the symbols p_x, p_y, p_z are often used. The p_z orbital has $m_\ell = 0$. The situation with p_x and p_y is more plex, however. Each of these orbitals has contributions from both $m_\ell = 1$ and $= -1$. Our main concern is just to recognize that *p* orbitals occur in sets of three and be represented in the orientations shown here. In higher-numbered shells, *p* orbitals a somewhat different appearance, but we will use these general shapes for all *p* tals. The colors of the lobes signify the different phases of the original wave function.

ng the three perpendicular axes of the Cartesian system, and the remaining e (d_{xy}, d_{xz}, d_{yz}) point between these Cartesian axes. A key feature of the *d* tals is the presence of two angular nodes (nodal surfaces). The *d* orbitals important in understanding the chemistry of the transition elements, as we see in Chapter 23.

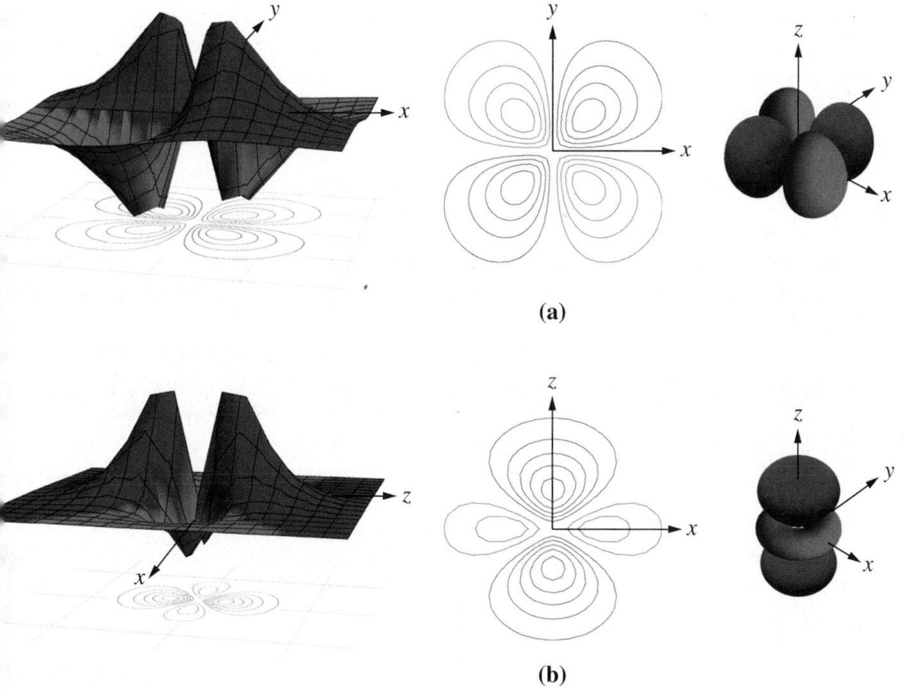

GURE 8-29
resentations of the $3d_{xy}$ and $3d_{z^2}$ orbitals of the hydrogen atom
wave function, contour map, and isosurface are shown in **(a)** for the $3d_{xy}$ orbital in **(b)** for the $3d_{z^2}$ orbital.

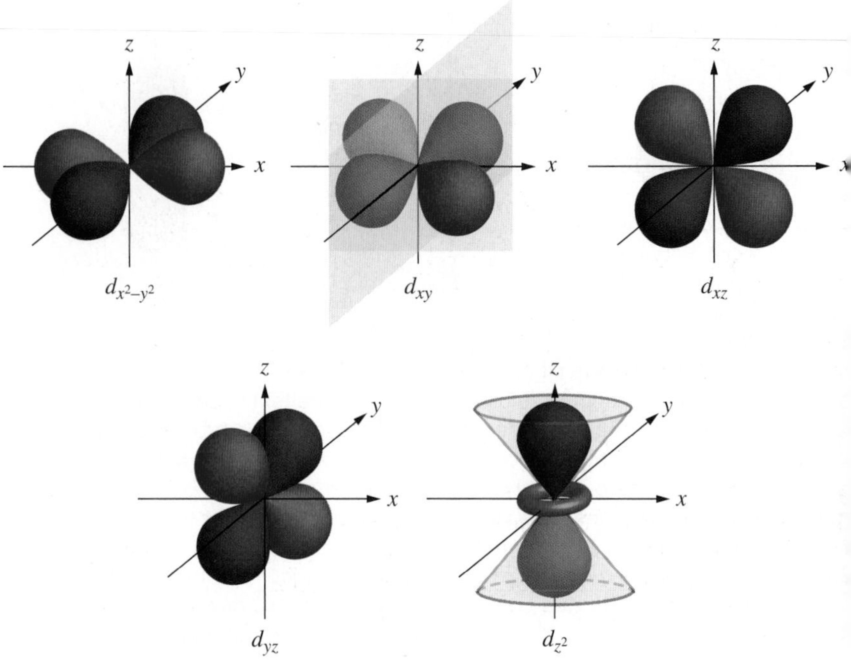

▲ FIGURE 8-30
Representations of the five *d* orbitals
The designations *xy*, *xz*, *yz*, and so on, are related to the values of the quantum number m_ℓ, but this is a detail that we will not pursue in the text. The number of nc surfaces for an orbital is equal to the ℓ quantum number. For *d* orbitals, there are tv such surfaces. The nodal planes for the d_{xy} orbital are shown here. (The nodal surfa for the d_{z^2} orbital are actually cone-shaped.)

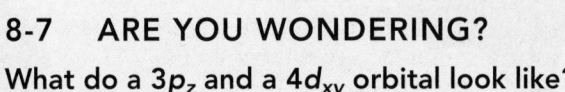

8-7 ARE YOU WONDERING?

What do a $3p_z$ and a $4d_{xy}$ orbital look like?

When considering the shapes of the atomic orbitals with higher principal quantu numbers, we can draw on what has already been discussed. For example, the 3 orbital has $3 - 1 - 1 = 1$ radial node and 1 angular node, for a total of 2 node Figure 8-31 shows a contour plot of the value of the $3p_z$ wave function in the plane in the manner of Figure 8-27(b). We notice that the $3p_z$ orbital has the san general shape as a $2p_z$ orbital because of the angular node, but the radial node h appeared as a circle (dashed in Figure 8-31). The appearance of the $3p_z$ orbit is that of a smaller *p* orbital inside a larger one. Similarly the $4d_{xy}$ orbital appears a smaller d_{xy} inside a larger one. However, it must be emphasized that each pl represents a single orbital, not one orbital nested inside another. In Figure 8-31 tl radial node is indicated by the dashed circle and the presence of the node is inc cated by the alternation in color. This idea can be extended to enable us to sketc orbitals of increasing principal quantum number.

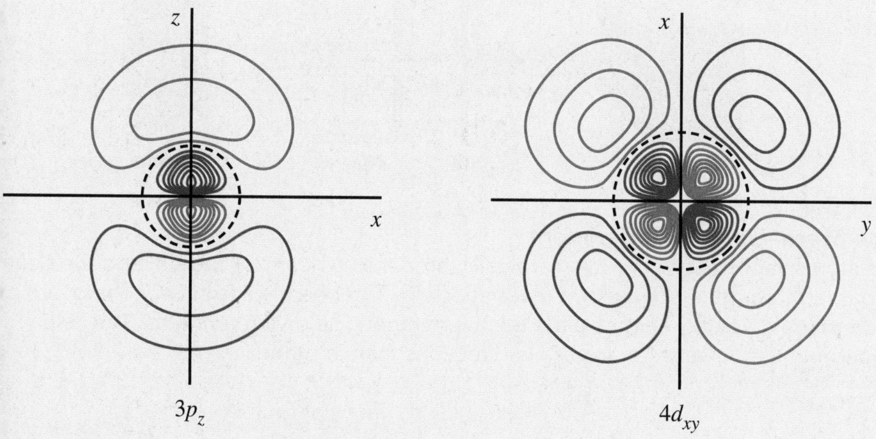

▲ FIGURE 8-31
Contour plots for the 3p_z and 4d_{xy} orbitals
The relative phases in these orbitals are shown by the colors red and blue. The radial nodes are represented by the dashed circles.

8-7 CONCEPT ASSESSMENT

Vhat type of orbital has three angular nodes and one radial node?

8 Electron Spin: A Fourth Quantum Number

ve mechanics provides three quantum numbers with which we can velop a description of electron orbitals. However, in 1925, George lenbeck and Samuel Goudsmit proposed that some unexplained features of hydrogen spectrum could be understood by assuming that an electron acts if it spins, much as Earth spins on its axis. As suggested by Figure 8-32, re are two possibilities for **electron spin**. Thus, these two possibilities uire a fourth quantum number, the electron spin quantum number m_s. e electron spin quantum number may have a value of $+\frac{1}{2}$ (also denoted by arrow ↑) or $-\frac{1}{2}$ (denoted by the arrow ↓); the value of m_s does not depend any of the other three quantum numbers.

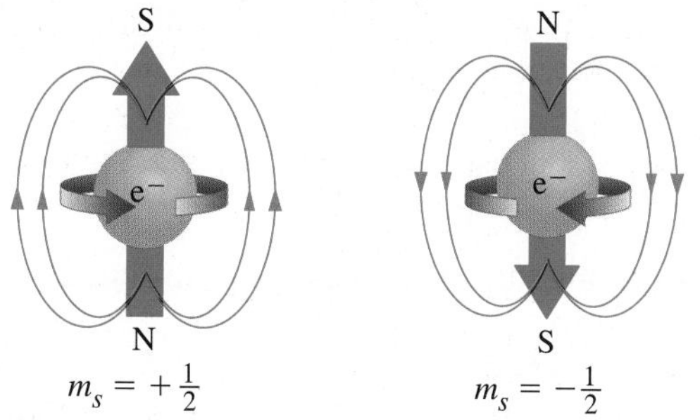

IGURE 8-32
ctron spin visualized
possibilities for electron spin are shown with their associated magnetic fields. electrons with opposing spins have opposing magnetic fields that cancel, ing no net magnetic field for the pair.

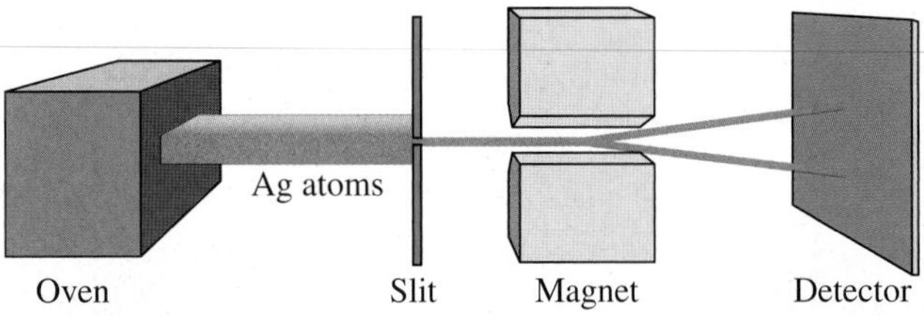

▲ FIGURE 8-33
The Stern–Gerlach experiment
Ag atoms vaporized in the oven are collimated into a beam by the slit, and the bear is passed through a nonuniform magnetic field. The beam splits in two. The beam c atoms would not experience a force if the magnetic field were uniform. The field strength must be stronger in certain directions than in others.

Actually, electron spin is characterized by using two quantum number and m_s. The s quantum number determines the *magnitude* of the magnetic fi produced and m_s, the *orientation* of this field. For an electron, s is always eq to $\frac{1}{2}$, and we say that an electron is a "spin $\frac{1}{2}$" particle. For other particles, s have other values. For example, $s = 1$ for a photon. For a given value of s, allowed values of m_s are $-s, -s+1, -s+2, \ldots, s$. For $s = \frac{1}{2}$, the possible val for m_s are $-\frac{1}{2}$ and $\frac{1}{2}$. As long as we keep in mind that $s = \frac{1}{2}$ for all electro we can safely omit explicit reference to the quantum number s when cha terizing an electron's spin.

What is the evidence that the phenomenon of electron spin exis An experiment by Otto Stern and Walter Gerlach in 1920, though desig for another purpose, seems to yield this proof (Fig. 8-33). Silver was vap ized in an oven, and a beam of silver atoms was passed through a nonu form magnetic field, where the beam split in two. Here is a simplif explanation.

1. An electron, because of its spin, generates a magnetic field.
2. A pair of electrons with opposing spins has no net magnetic field.
3. A silver atom in its lowest energy state has only one unpaired elect The direction of the net magnetic field produced depends only on the s of the unpaired electron.
4. In a beam of a large number of silver atoms there is an equal chance that unpaired electron will have a spin of $+\frac{1}{2}$ or $-\frac{1}{2}$. The magnetic field indu by the silver atoms interacts with the nonuniform field, and the bean silver atoms splits into two beams.

Electronic Structure of the H Atom: Representing the Four Quantum Numbers

Now that we have described the four quantum numbers, we are in a positio bring them together into a description of the electronic structure of hydrogen atom. The electron in a ground-state hydrogen atom is found at lowest energy level. This corresponds to the principal quantum number n = and because the first principal shell consists only of an s orbital, the orb quantum number $\ell = 0$. The only possible value of the magnetic quant number is $m_\ell = 0$. Either spin state is possible for the electron, and we do know which it is unless we do an experiment like that of Uhlenbeck Goudsmit's. Thus,

$$n = 1 \qquad \ell = 0 \qquad m_\ell = 0 \qquad m_s = +\frac{1}{2} \text{ or } -\frac{1}{2}$$

emists often say that the electron in the ground-state hydrogen atom is in 1s orbital, or that it is a 1s electron, and they represent this by the notation

$$1s^1$$

ere the superscript 1 indicates one electron in the 1s orbital. Either spin e is allowed, but we do not designate the spin state in this notation.

n the excited states of the hydrogen atom, the electron occupies orbitals ı higher values of n. Thus, when excited to the level with $n = 2$, the elec- can occupy either the 2s or one of the 2p orbitals; all have the same energy. ause the probability density extends farther from the nucleus in the 2s and rbitals than in the 1s orbital, the excited-state atom is larger than is the ınd-state atom. The excited states just described can be represented as

$$2s^1 \text{ or } 2p^1$$

ı the remaining sections of the chapter this discussion will be extended to the tronic structures of atoms having more than one electron—*multielectron* atoms.

KEEP IN MIND

that orbitals are mathematical functions and not themselves physical regions in space. However, it is customary to refer to an electron that is described by a particular orbital as being "in the orbital."

XAMPLE 8-12 Choosing an Appropriate Combination of the Four Quantum Numbers: n, ℓ, m_ℓ, and m_s

From the following sets of quantum numbers (n, ℓ, m_ℓ, m_s), identify the set that is correct, and state the orbital designation for those quantum numbers:

$$(2,1,0,0) \quad \left(2,0,1,\frac{1}{2}\right) \quad \left(2,2,0,\frac{1}{2}\right) \quad \left(2,-1,0,\frac{1}{2}\right) \quad \left(2,1,0,-\frac{1}{2}\right)$$

nalyze

We know that if $n = 2$, ℓ has two possible values: 0 or 1. The range of values for m_ℓ is given by equation (8.19), and $m_s = \pm\frac{1}{2}$. By using this information, we can judge which combination is correct.

olve

(n, ℓ, m_ℓ, m_s)	**Comment**
$(2,1,0,0)$	The value of m_s is incorrect.
$\left(2,0,1,\frac{1}{2}\right)$	The value of m_ℓ is incorrect.
$\left(2,2,0,\frac{1}{2}\right)$	The value of ℓ is incorrect.
$\left(2,-1,0,\frac{1}{2}\right)$	The value of ℓ is incorrect.
$\left(2,1,0,-\frac{1}{2}\right)$	All the quantum numbers are correct.

The correct combination of quantum numbers has $n = 2$, $\ell = 1$, $m_\ell = 0$, and $m_s = -\frac{1}{2}$, which corresponds to a 2p orbital.

ssess

The combination of quantum numbers identified above for an electron in a 2p orbital is one of six possible combinations. The other five combinations for an electron in a 2p orbital are $\left(2,1,0,\frac{1}{2}\right)$, $\left(2,1,-1,-\frac{1}{2}\right)$, $\left(2,1,-1,\frac{1}{2}\right)$, $\left(2,1,1,-\frac{1}{2}\right)$, and $\left(2,1,1,\frac{1}{2}\right)$.

(continued)

PRACTICE EXAMPLE A: Determine which set of the following quantum numbers (n, ℓ, m_ℓ, m_s) is wrong an indicate why:

$$(3, 2, -2, 1) \quad \left(3, 1, -2, \frac{1}{2}\right) \quad \left(3, 0, 0, \frac{1}{2}\right) \quad \left(2, 3, 0, \frac{1}{2}\right) \quad \left(1, 0, 0, -\frac{1}{2}\right) \quad \left(2, -1, -1, \frac{1}{2}\right)$$

PRACTICE EXAMPLE B: Identify the error in each set of quantum numbers below:

$$(2, 1, 1, 0) \quad \left(1, 1, 0, \frac{1}{2}\right) \quad \left(3, -1, 1, -\frac{1}{2}\right) \quad \left(0, 0, 0, -\frac{1}{2}\right) \quad \left(2, 1, 2, \frac{1}{2}\right)$$

8-9 Multielectron Atoms

Schrödinger developed his wave equation for the hydrogen atom—an a containing just one electron. For multielectron atoms, a new factor ari mutual repulsion between electrons. The repulsion between the elect means that the electrons in a multielectron atom tend to stay away from another, and their motions become inextricably entangled. The approxir approach taken to solve this many-particle problem is to consider the e trons, one by one, in the environment established by the nucleus and the o electrons. When this is done, the electron orbitals obtained are of the s types as those obtained for the hydrogen atom; they are called *hydrogen* orbitals. Compared with the hydrogen atom, the angular parts of the orb of a multielectron atom are unchanged, but the radial parts are different.

A Conceptual Model for Multielectron Atoms

As suggested above, the results obtained for the hydrogen atom provide basis of a very useful conceptual model for describing electrons in a multie tron atom. However, we must anticipate that adjustments will need t made because, in a multielectron atom, we have interactions of electrons only with the nucleus but with other electrons. A wealth of evidence, exp mental and theoretical, supports the validity of a conceptual model base the following points.*

1. The quantum mechanical wave function for a multielectron atom ca approximated as a superposition of orbitals, each bearing some res blance to those describing the quantum states of the hydrogen atom. I orbital in a multielectron atom describes how a single electron behave the field of a nucleus under the average influence of all the other electr
2. The total energy of an atom with N electrons has the general f $E_{\text{atom}} = F - G$, where F represents a sum of orbital energ $F = \epsilon_1 + \epsilon_2 + \epsilon_3 + \cdots + \epsilon_N$, and G takes account of electron–elec repulsions. The orbital energy ϵ_1, for example, is the energy of electron a particular orbital interacting with the nucleus *under the average influ of all the other electrons*. In general, the orbital energies increase increases and, for equal values of n, increase as ℓ increases, as sugge by Figure 8-34.
3. The order in which we assign electrons to specific orbitals is base minimizing E_{atom}. Orbitals that minimize the value of F may not neces ily minimize E_{atom}. Therefore, we must be careful not to place too m emphasis on the energies of the orbitals themselves.

The form of the equation $E_{\text{atom}} = F - G$ might look, at first, a little odd bec it seems to suggest that the energy of an atom is lowest when electron–elec

* See, for example, F. Pilar, *J. Chem. Educ.*, **55**, 2 (1978).

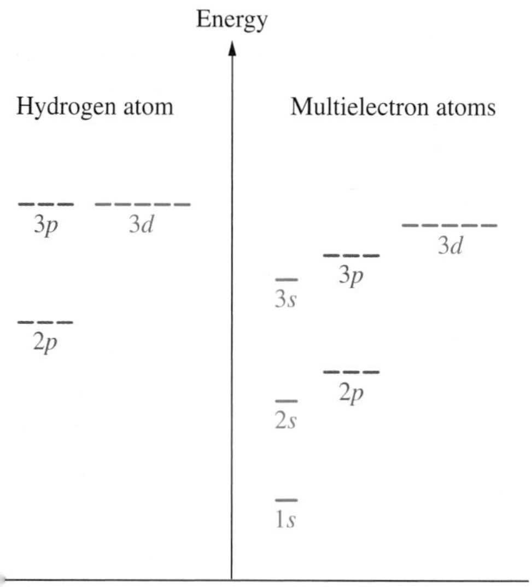

◀ FIGURE 8-34
Orbital energy-level diagrams for the hydrogen atom and a multielectron atom
This is a schematic diagram showing the relative energies of orbitals for the $n = 1$, 2, and 3 shells of the hydrogen atom and a multielectron atom. For the hydrogen atom, orbitals within a principal shell—for example, 3*s*, 3*p*, and 3*d*—have the same energy and are said to be energetically degenerate. However, in a multielectron atom, orbitals within a principal shell have different energies. In general, for a multielectron atom, orbital energies increase with the value of *n* and for a fixed value of *n*, with the value of ℓ. The diagram also illustrates that the energy of a given orbital (e.g., 1*s*) decreases as the atomic number, *Z*, increases. It is important not to try to rationalize the orbital filling order by using a diagram of this type because, as described in the text, the orbital filling order cannot be explained in terms of the orbital energies alone. It is not unusual to find that a lower total energy for the atom can be obtained by placing an electron in an orbital of higher energy.

ılsions, *G*, are greatest. The situation is not quite that simple because each tal energy (and therefore *F*) already includes the effects of electron–electron ılsions. In fact, electron–electron repulsions are double counted by *F*. For nple, ϵ_1 includes the repulsion between electrons 1 and 2 but so too does ϵ_2. use *F* double counts the effects of electron–electron repulsions, we must sub- *G* from *F* to obtain the correct value for E_{atom}.

summary, each electron in a multielectron atom is described by (or upies") an orbital that is qualitatively similar to a hydrogen-like orbital 2*s*, 2*p*, 3*s*, etc.). We can imagine building up an atom electron by electron ssigning electrons to the various orbitals in a way that gives the lowest ible value to E_{atom}.

efore examining the rules for assigning electrons to the various orbitals, vill first discuss the concepts of penetration and shielding. These concepts help us explain why, in a multielectron atom, orbitals with different es of ℓ within a principal shell have different energies.

etration and Shielding

k about the attractive force of the atomic nucleus for one particular elec- some distance from the nucleus. Electrons in orbitals closer to the nucleus *n* or *shield* the nucleus from electrons farther away. In effect, the screening rons reduce the effectiveness of the nucleus in attracting the particular e-distant electron. They effectively reduce the nuclear charge felt by nore-distant electron.

he magnitude of the reduction of the nuclear charge depends on the types rbitals the inner electrons are in and the type of orbital that the screened ron is in. We have seen that *s* orbitals have a high probability density at nucleus, whereas *p* and *d* orbitals have zero probability densities at the eus. Thus, electrons in *s* orbitals are more effective at screening the eus from outer electrons than are electrons in *p* or *d* orbitals. This ability of rons in *s* orbitals that allows them to get close to the nucleus is called *pen- ion*. An electron in an orbital with good penetration is better at screening one with low penetration.

KEEP IN MIND

that orbital-wave functions extend farther out from the nucleus as *n* increases. Thus, an electron in a 3*s* or 3*p* orbital has a higher probability of being farther from the nucleus than does an electron in a 1*s* orbital.

'e must consider a different kind of probability distribution to describe penetration to the nucleus by orbital electrons. Rather than considering probability at a point, which we did to ascribe three-dimensional shapes rbitals, we need to consider the probability of finding the electron any- re in a spherical shell of radius *r* and an infinitesimal thickness *dr*. This of probability is expressed in terms of the *radial distribution function*, ch is defined as $r^2R^2(r)$. That we must consider the product $r^2R^2(r)$ and

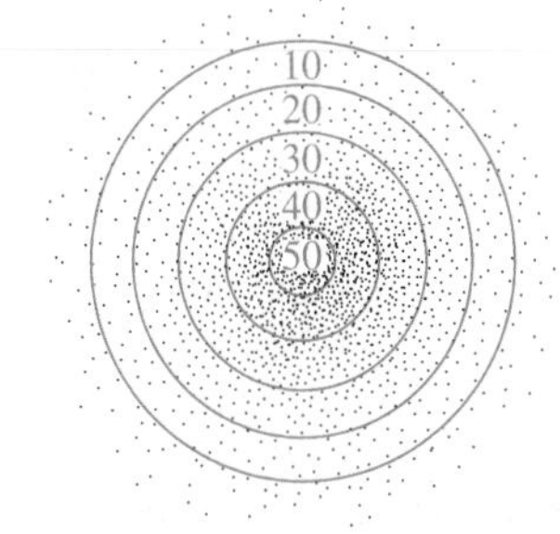

▲ FIGURE 8-35
Dartboard analogy to a 1s orbital
Imagine that a single dart (electron) is thrown at a dartboard 1500 times. The board contains 95% of all the holes; it is analogous to the 1s orbital. Where is a thrown dart most likely to hit? The number of holes per unit area is greatest in the "50" region—that is, the 50 region has the greatest probability density. The most likely score is "30," however, because the most probable area hit is in the 30 ring and not the 50 ring, which is smaller than the 30 ring. The 30 *ring* on the dartboard is analogous to a spherical *shell* of 53 pm radius within the larger sphere representing the 1s orbital.

not just $R^2(r)$ can be justified fairly easily by considering the special cas an electron in the 1*s* orbital. In such a situation, the probability density *particular point* is

$$\psi^2(1s) = (1/4\pi)R^2(1s)$$

The probability of finding the electron between r and $r + dr$ is

$$\psi^2(1s)\,dV_{\text{shell}} = \left(\frac{1}{4\pi}\right)R^2(1s) \times dV_{\text{shell}}$$

where dV_{shell} is the volume of a thin spherical shell of thickness dr. It ca shown (see Exercise 110) that dV_{shell} is equal to $4\pi r^2 dr$, and so the probab of finding the electron between r and $r + dr$ is

$$\left(\frac{1}{4\pi}\right)R^2(1s) \times 4\pi r^2 dr = r^2R^2(1s)\,dr$$

That is, the probability is proportional to $r^2R^2(r)$, not to $R^2(r)$. Althoug obtained this result by considering an electron in a 1*s* orbital, the result is, in completely general. Figure 8-35 offers a dartboard analogy to clarify the dis tion between probability at a point and probability in a region of space.

Radial distribution functions for some hydrogenic (hydrogen-like) orbital plotted in Figure 8-36. The radial probability density, $R^2(r)$, for a 1*s* orbital dicts that the maximum probability for a 1*s* electron is *at* the nucleus. How because the volume of this region is vanishingly small ($r = 0$), the radial pr bility distribution $[r^2R^2(r)]$ is zero at the nucleus. The electron in a hydrogen a is most likely to be found 53 pm from the nucleus; this is where the radial pr bility distribution reaches a maximum. This distance is exactly equal to the radius, a_0. The boundary surface within which there is a 95% probability of ing an electron is a much larger sphere, one with a radius of about 141 pm comparing the radial probability curves for the 1*s*, 2*s*, and 3*s* orbitals, we find a 1*s* electron has a greater probability of being close to the nucleus than a 2*s* tron does, which in turn has a greater probability than does a 3*s* electron. In c paring 2*s* and 2*p* orbitals, a 2*s* electron has a greater chance of being close to

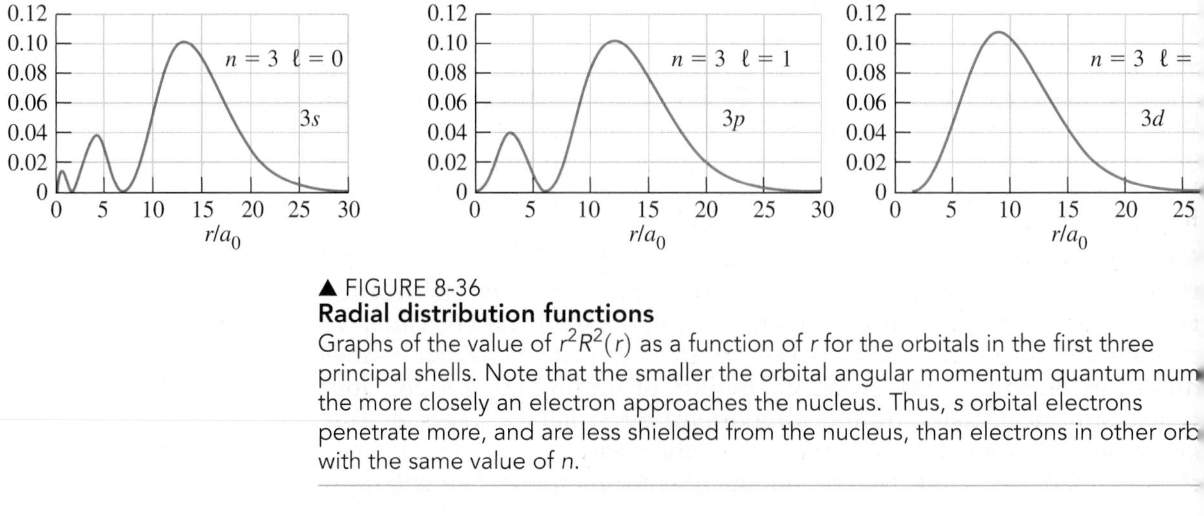

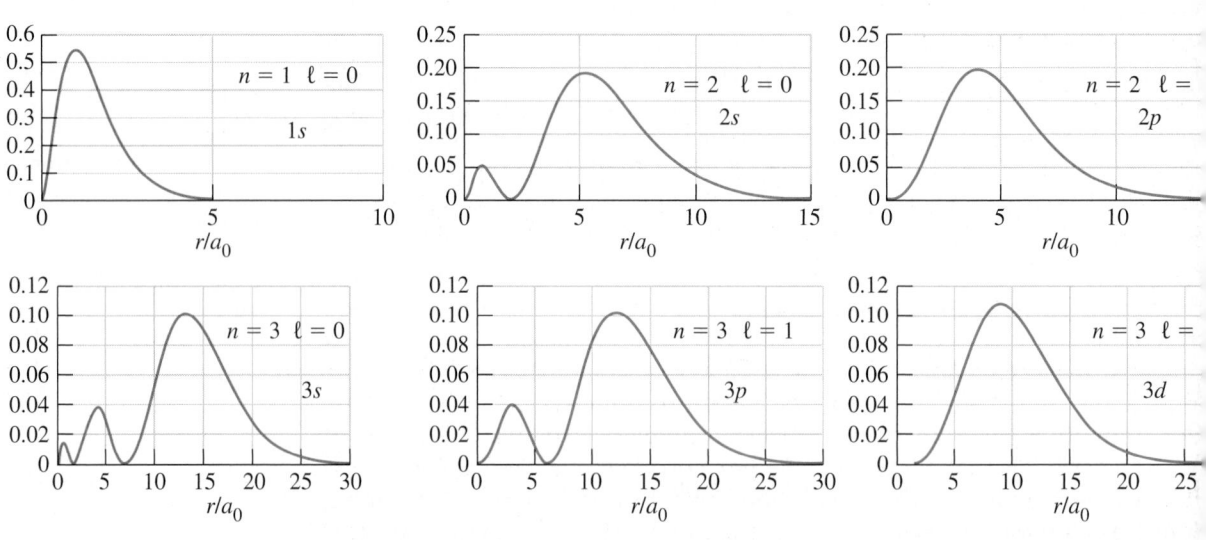

▲ FIGURE 8-36
Radial distribution functions
Graphs of the value of $r^2R^2(r)$ as a function of r for the orbitals in the first three principal shells. Note that the smaller the orbital angular momentum quantum num the more closely an electron approaches the nucleus. Thus, s orbital electrons penetrate more, and are less shielded from the nucleus, than electrons in other orb with the same value of *n*.

leus than a $2p$ electron does. The $2s$ electron exhibits greater penetration than 2p electron. Electrons having a high degree of penetration effectively "block the v" of an electron in an outer orbital "looking" for the nucleus.

'he nuclear charge that an electron would experience if there were no intering electrons is Z, the atomic number. The nuclear charge that an electron ually experiences is reduced by intervening electrons to a value of Z_{eff}, ed the **effective nuclear charge**. The less of the nuclear charge that an outer tron "sees" (that is, the smaller the value of Z_{eff}), the smaller is the attrac- of the electron to the nucleus, and hence the *higher* is the energy of the tal in which the electron is found.

o summarize, compared with a p electron in the same principal shell, an s tron is more penetrating and not as well screened. The s electron experies a higher Z_{eff}, is held more tightly, and is at a lower energy than a p elec-. Similarly, the p electron is at a lower energy than a d electron in the same ncipal shell. Thus, the energy ordering of subshells is $ns < np < nd$, as strated in Figure 8-34. Orbitals within a given subshell have the same rgy because all the orbitals in the subshell have the same radial characters and thereby experience the same effective nuclear charge, Z_{eff}. As a ult, all three p orbitals of a principal shell have the same energy; all five d tals have the same energy; and so on.

KEEP IN MIND

that, similar to the situation in equation (8.9), the energy of an orbital (E_n) is given by the proportionality

$$E_n \propto -\frac{Z_{eff}^2}{n^2}.$$

10 Electron Configurations

electron configuration of an atom is a designation of how electrons are ributed among various orbitals in principal shells and subshells. In later pters, we will find that many of the physical and chemical properties of nents can be correlated with electron configurations. In this section, we see how the results of wave mechanics, expressed as a set of rules, can us to write probable electron configurations for the elements.

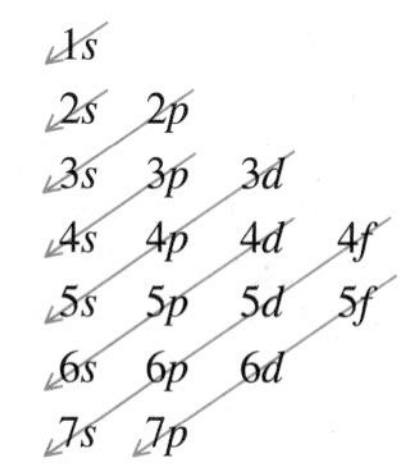

▲ FIGURE 8-37
The order of filling of electronic subshells
Beginning with the top line, follow the arrows, and the order obtained is the same as in expression (8.22).

es for Assigning Electrons to Orbitals

Electrons occupy orbitals in a way that minimizes the energy of the atom. As explained on page 350, the total energy of an atom depends not only on the orbital energies but also on the electronic repulsions that arise from placing electrons in particular orbitals. That is, the orbital filling order cannot be reliably predicted by consideration of orbital energies alone. The exact order of filling of orbitals has been established by experiment, principally through spectroscopy and magnetic studies, and it is this order based on experiment that we must follow in assigning electron configurations to the elements. With only a few exceptions, the order in which orbitals fill is

$$1s, 2s, 2p, 3s, 3p, 4s, 3d, 4p, 5s, 4d, 5p, 6s, 4f, 5d, 6p, 7s, 5f, 6d, 7p \qquad \textbf{(8.22)}$$

◀ This order of filling corresponds roughly to the order of increasing orbital energy, but the overriding principle governing the order of filling of orbitals is that the energy of the atom as a whole be kept at a minimum.

It is equally important to remember that, for the reasons described above, this filling order does not represent the relative energy ordering of the orbitals. Some students find the diagram pictured in Figure 8-37 a useful way to remember this order, but the best method of establishing the order of filling of orbitals is based on the periodic table, as we will see in Section 8-11.

Only two electrons may occupy the same orbital, and these electrons must have opposite spins. In 1926, Wolfgang Pauli explained complex features of emission spectra associated with atoms in magnetic fields by proposing that no two electrons in an atom can have the same set of quantum numbers—the **Pauli exclusion principle**. If two electrons (labeled 1 and 2) occupy the same orbital, then $n_1 = n_2$, $\ell_1 = \ell_2$, and $m_{\ell_1} = m_{\ell_2}$. By applying the Pauli exclusion principle, we see that the two electrons must have different values of m_s, the spin quantum number.

Because of this limit of two electrons per orbital, the capacity
subshell for electrons can be obtained by doubling the number of orbi
in the subshell. Thus, the *s* subshell consists of *one* orbital with a capa
of *two* electrons; the *p* subshell consists of *three* orbitals with a total ca
ity of *six* electrons; and so on.

3. **When orbitals of identical energy (degenerate orbitals) are available, e trons initially occupy these orbitals singly and with parallel spins.** rule means that we must place electrons singly in each orbital with par spins before pairing them to ensure that **Hund's rule** is followed. A sim fied statement of Hund's rule is that, for a given configuration, the arra ment having the maximum number of parallel spins is lower in energy t any other arrangement arising from the same configuration. This beha can be rationalized as follows. Because electrons all carry the same ele charge, if the available orbitals all have the same energy, then by pla them in different orbitals the electrons are spatially as far apart as possi Why is the atom's energy lower when the electrons' spins are parallel? answer to this question may seem odd: Electrons with parallel spins r each other more, and thus shield each other less, than if their spins w opposite. Thus, with the spins parallel, the attraction of each electron to nucleus is greater than if the electrons had opposite spins. The overall e is that, for a set of degenerate orbitals, having electrons in different orb with their spins parallel lowers the total energy of the atom.*

Representing Electron Configurations

Before we assign electron configurations to atoms of the different eleme we need to introduce methods of representing these configurations. The e tron configuration of a carbon atom is shown in three different ways:

spdf notation (condensed):	C	$1s^2 2s^2 2p^2$
spdf notation (expanded):	C	$1s^2 2s^2 2p_x^1 2p_y^1$
orbital diagram:	C	[↑↓] [↑↓] [↑][↑][ ]
		1*s* 2*s* 2*p*

In each of these methods we assign six electrons because the atomic nur of carbon is 6. Two of these electrons are in the 1*s* subshell, two in the 2*s*, two in the 2*p*. The condensed ***spdf*** **notation** denotes only the total numbe electrons in each subshell; it does not show how electrons are distribu among orbitals of equal energy. In the expanded *spdf* notation, Hund's ru reflected in the assignment of electrons to the 2*p* subshell—two 2*p* orbitals singly occupied and one remains empty. The **orbital diagram** breaks d each subshell into individual orbitals (drawn as boxes).

Electrons in orbitals are shown as arrows. An arrow pointing up correspo to one type of spin $(+\frac{1}{2})$, and an arrow pointing down to the other (Electrons in the same orbital with opposing (opposite) spins are said to be *p* (↑↓). The electrons in the 1*s* and 2*s* orbitals of the carbon atom are pai Electrons in different, singly occupied orbitals of the same subshell have same, or *parallel*, spins (arrows pointing in the same direction). This is veyed in the orbital diagram for carbon, where we write [↑][↑][ ] rather t [↑][↓][ ] for the 2*p* subshell. Both experiment and theory confirm that an e tron configuration in which electrons in singly occupied orbitals have par spins is a better representation of the lowest energy state of an atom than other electron configuration that we can write. The configuration represente the orbital diagram [↑][↓][ ] is, in fact, an excited state of carbon; any orbital gram with unpaired spins that are not parallel constitutes an excited state.

The most stable or the most energetically favorable configurations for isol atoms, those discussed here, are called *ground-state electron configurations*. Lat

▶ When listed in tables, as in Appendix D, electron configurations are usually written in the condensed *spdf* notation.

* See R. Boyd, *Nature*, **310**, 480 (1984).

text we will briefly mention some electron configurations that are not the ›t stable. Atoms with such configurations are said to be in an *excited state*.

ə Aufbau Process

write electron configurations we will use the **aufbau process**. *Aufbau* is a man word that means "building up," and what we do is assign electron figurations to the elements in order of increasing atomic number. To pro-d from one atom to the next, we add a proton and some neutrons to the leus and then describe the orbital into which the added electron goes.

Z = 1, H. The lowest energy state for the electron is the 1*s* orbital. The electron configuration is $1s^1$.

Z = 2, He. A second electron goes into the 1*s* orbital, and the two electrons have opposing spins, $1s^2$.

Z = 3, Li. The third electron cannot be accommodated in the 1*s* orbital (Pauli exclusion principle). It goes into the lowest energy orbital available, 2*s*. The electron configuration is $1s^2 2s^1$.

Z = 4, Be. The configuration is $1s^2 2s^2$.

Z = 5, B. Now the 2*p* subshell begins to fill: $1s^2 2s^2 2p^1$.

Z = 6, C. A second electron also goes into the 2*p* subshell, but into one of the remaining empty *p* orbitals (Hund's rule) with a spin parallel to the first 2*p* electron. (See figure to the right.)

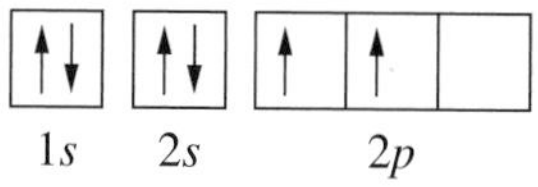

Z = 7–10, N through Ne. In this series of four elements, the filling of the subshell is completed. The number of unpaired electrons reaches a maximum (three) with nitrogen and then decreases to zero with neon.

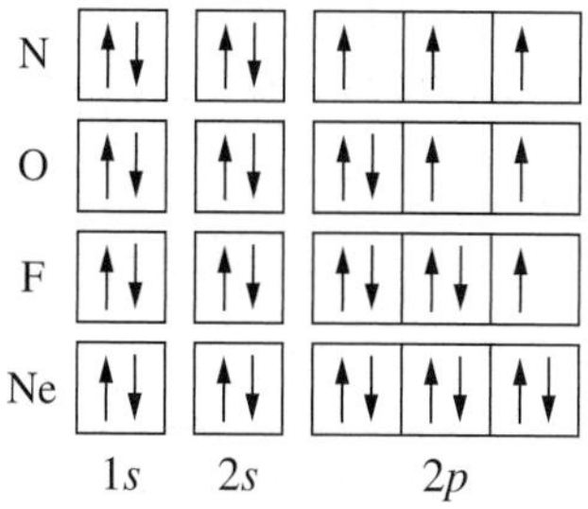

Z = 11–18, Na through Ar. The filling of orbitals for this series of eight elements closely parallels the eight elements from Li through Ne, except that electrons go into 3*s* and 3*p* subshells. Each element has the 1*s*, 2*s*, and 2*p* subshells filled. Because the configuration $1s^2 2s^2 2p^6$ is that of neon, we will call this the neon core, represent it as [Ne], and concentrate on the electrons beyond the core. Electrons that are added to the electronic shell of highest principal quantum number (the outermost, or *valence shell*) are called **valence electrons**. The electron configuration of Na is written below in a form called a *noble-gas-core-abbreviated electron configuration*, consisting of [Ne] as the noble gas core and $3s^1$ as the configuration of the valence electron. For the other third-period elements, only the valence-shell electron configurations are shown.

Na	Mg	Al	Si	P	S	Cl	Ar
$[Ne]3s^1$	$3s^2$	$3s^2 3p^1$	$3s^2 3p^2$	$3s^2 3p^3$	$3s^2 3p^4$	$3s^2 3p^5$	$3s^2 3p^6$

Z = 19 and 20, K and Ca. After argon, instead of 3*d*, the next subshell to fill s 4*s*. Using the symbol [Ar] to represent the noble gas core, $1s^2 2s^2 2p^6 3s^2 3p^6$, ve get the electron configurations shown below for K and Ca.

$$\text{K: [Ar]}4s^1 \quad \text{and} \quad \text{Ca: [Ar]}4s^2$$

Z = 21–30, Sc through Zn. In this next series of elements, electrons fill he *d* orbitals of the third shell. The *d* subshell has a total capacity of ten electrons—ten elements are involved. There are two possible way write the electron configuration of scandium.

$$\text{(a) Sc: [Ar]}3d^{1}4s^{2} \quad \text{or} \quad \text{(b) Sc: [Ar]}4s^{2}3d^{1}$$

Both methods are commonly used. Method (a) groups together all subshells of a principal shell and places subshells of the highest princ quantum level last. Method (b) lists orbitals in the apparent orde which they fill. In this text, we will use method (a).

▶ Although method (b) conforms better to the order in which orbitals fill, method (a) better represents the order in which electrons are lost on ionization, as we will see in the next chapter.

The electron configurations of this series of ten elements are li below in both the orbital diagram and the *spdf* notation.

		3d					4s	
Sc:	[Ar]	↑					↑↓	[Ar]$3d^{1}4s^{2}$
Ti:	[Ar]	↑	↑				↑↓	[Ar]$3d^{2}4s^{2}$
V:	[Ar]	↑	↑	↑			↑↓	[Ar]$3d^{3}4s^{2}$
Cr:	[Ar]	↑	↑	↑	↑	↑	↑	[Ar]$3d^{5}4s^{1}$
Mn:	[Ar]	↑	↑	↑	↑	↑	↑↓	[Ar]$3d^{5}4s^{2}$
Fe:	[Ar]	↑↓	↑	↑	↑	↑	↑↓	[Ar]$3d^{6}4s^{2}$
Co:	[Ar]	↑↓	↑↓	↑	↑	↑	↑↓	[Ar]$3d^{7}4s^{2}$
Ni:	[Ar]	↑↓	↑↓	↑↓	↑	↑	↑↓	[Ar]$3d^{8}4s^{2}$
Cu:	[Ar]	↑↓	↑↓	↑↓	↑↓	↑↓	↑	[Ar]$3d^{10}4s^{1}$
Zn:	[Ar]	↑↓	↑↓	↑↓	↑↓	↑↓	↑↓	[Ar]$3d^{10}4s^{2}$

The *d* orbitals fill in a fairly regular fashion in this series, but there are exceptions: chromium (Cr) and copper (Cu). These exceptions involve subshell that is either half-filled with electrons, as with Cr ($3d^{5}$), or c pletely filled, as with Cu ($3d^{10}$).

Z = 31–36, Ga through Kr. In this series of six elements, the 4*p* subs is filled, ending with krypton.

$$\text{Kr: [Ar]}3d^{10}4s^{2}4p^{6}$$

Z = 37–54, Rb to Xe. In this series of 18 elements, the subshells fill in order 5*s*, 4*d*, and 5*p*, ending with the configuration of xenon.

$$\text{Xe: [Kr]}4d^{10}5s^{2}5p^{6}$$

Z = 55–86, Cs to Rn. In this series of 32 elements, with a few excepti the subshells fill in the order 6*s*, 4*f*, 5*d*, 6*p*. The configuration of radon

$$\text{Rn: [Xe]}4f^{14}5d^{10}6s^{2}6p^{6}$$

Z = 87–?, Fr to ? Francium starts a series of elements in which the shells that fill are 7*s*, 5*f*, 6*d*, and presumably 7*p*, although atoms in w filling of the 7*p* subshell is expected have only recently been discove and are not yet characterized.

Appendix D gives a complete listing of ground-state electron configurat for all atoms.

8-8 ARE YOU WONDERING?

Why does the orbital filling order given by expression (8.22) fail for chromium and copper?

:hromium (Cr) and copper (Cu) are the first two elements for which the orbital lling order given in expression (8.22) fails to give the correct prediction for the round-state electron configuration.

Element	Predicted Configuration	Observed Configuration
Cr (Z = 24)	[Ar] $3d^4 4s^2$	[Ar] $3d^5 4s^1$
Cu (Z = 29)	[Ar] $3d^9 4s^2$	[Ar] $3d^{10} 4s^1$

The observed ground-state configurations for both Cr and Cu involve half-lled subshells or filled subshells. Thus, the supposed "special stability" of half-lled and filled subshells is sometimes used as an explanation for why Cr and Cu ave the observed configurations. Such an explanation raises the question, "What the origin of this special stability?" If this special stability exists, then all the toms below Cr in group 6 and below Cu in group 11 should also have half-filled r filled subshells. However, experiment reveals that this is not always the case. 1ost notably, for tungsten (W), the ground-state configuration is the predicted ne, [Xe] $4f^{14}5d^4 6s^2$, not [Xe] $4f^{14}5d^5 6s^1$.

The following statements summarize what you should take away from this iscussion.

1. The observed ground-state electron configuration is always the one that gives the lowest total energy for the atom. As discussed in the text, electron motions in a multielectron atom are highly correlated; consequently, the total energy of an atom is, in some cases, a very delicate balance between electron–nuclear attractions and electron–electron repulsions.
2. Nearly all the exceptions to the predicted filling order (so-called *anomalous configurations*) involve either filled or half-filled subshells. Explaining these exceptions is not only rather complicated but also probably best done case by case.

As a final comment, it should not be too surprising that some atoms have anomalous" electron configurations. Given that the total energy of an atom epends on the correlated motions of many electrons, it might be surprising that single filling order, expression (8.22), works as often as it does.

:XAMPLE 8-13 Recognizing Correct and Incorrect Ground State and Excited State Atomic Orbital Diagrams

Which of the following orbital diagrams is incorrect? Explain. Which of the correct diagrams corresponds to an excited state and which to the ground state of the neutral atom?

nalyze

When faced with a set of orbital diagrams, the best strategy is to investigate each one and apply Hund's rule and the Pauli exclusion principle, the former to decide on ground or excited states, and the latter for the correctness of the diagram.

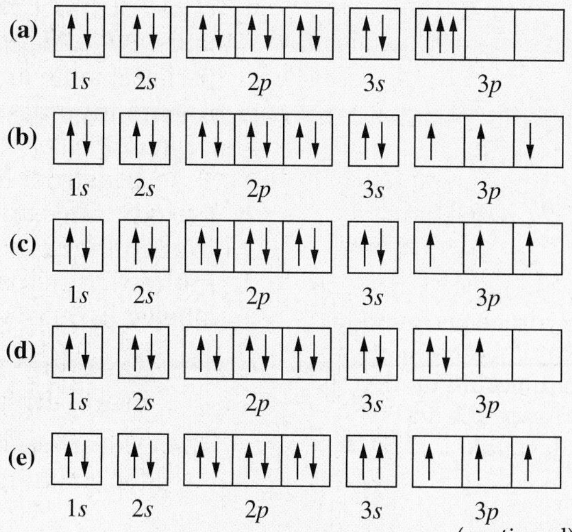

olve

(a) By scanning diagram (a), we see that all the orbitals 1*s*, 2*s*, 2*p*, and 3*s* are filled with two electrons of opposite spin, conforming to the Pauli exclusion principle. However, the 3*p* orbital contains three electrons, which violates this principle.

(b) In diagram (b), the orbitals 1*s*, 2*s*, 2*p*, and 3*s* are filled with two electrons of opposite spin, which is correct. The 3*p* level contains three electrons in

(continued)

separate orbitals, conforming to Hund's rule, but two of them have opposite spin to the other; cons quently, this is an excited state of the element.

(c) When we compare diagram (c) with diagram (b), we see that all the three electrons in the $3p$ subsh have the same spin, and so this is the ground state.

(d) When we compare diagram (d) with diagram (b), we see that of the three electrons in the $3p$ subshe two are paired and one is not. Again, this is an excited state.

(e) By scanning diagram (e), we see that all the orbitals $1s, 2s$, and $2p$ are filled with two electrons opposite spin. However, the $3s$ orbital contains two electrons with the same spin, which violates t Pauli principle. This diagram is incorrect.

Assess

Orbital diagrams are a useful way to display electronic configurations, but we must take care to obey Hund rule and the Pauli exclusion principle.

PRACTICE EXAMPLE A: Which two of the following orbital diagrams are equivalent?

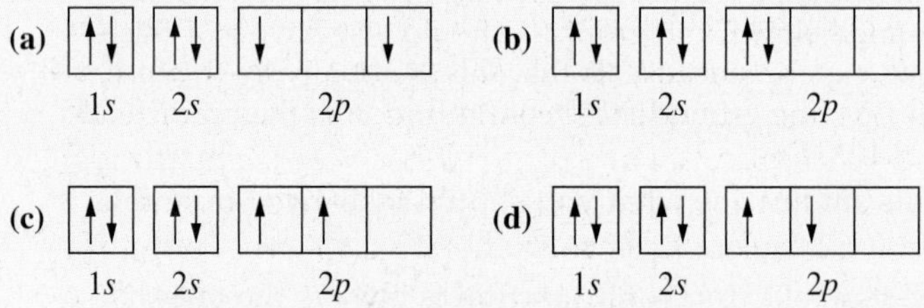

PRACTICE EXAMPLE B: Does the following orbital diagram for a neutral species correspond to the ground sta or an excited state?

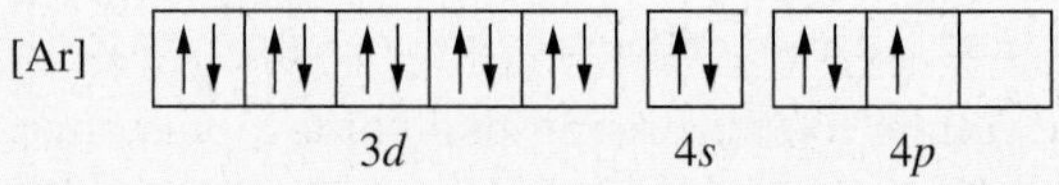

8-11 Electron Configurations and the Periodic Table

We have just described the aufbau process of making probable assignme electrons to the orbitals in atoms. Although electron configurations may s rather abstract, they actually lead us to a better understanding of the peri table. Around 1920, Niels Bohr began to promote the connection betwee periodic table and quantum theory. The chief link, he pointed out, is in tron configurations. *Elements in the same group of the table have similar ele configurations.*

To construct Table 8.3, we have taken three groups of elements from periodic table and written their electron configurations. The similarity in tron configuration within each group is readily apparent. If the shell o highest principal quantum number—the outermost, or valence, shel labeled n, then

- The group 1 atoms (alkali metals) have *one* outer-shell (valence) ele in an s orbital, that is, ns^1.
- The group 17 atoms (halogens) have *seven* outer-shell (valence) electro the configuration ns^2np^5.

▶ Hydrogen is found in group 1 because of its electron configuration, $1s^1$. However, it is not an alkali metal.

The group 18 atoms (noble gases)—with the exception of helium, which has only two electrons—have outermost shells with *eight* electrons, in the configuration ns^2np^6.

nough it is not correct in all details, Figure 8-38 relates the aufbau process he periodic table by dividing the table into the following four blocks of nents according to the subshells being filled:

***s* block.** The *s* orbital of highest principal quantum number (n) fills. The *s* block consists of groups 1 and 2 (plus He in group 18).

***p* block.** The *p* orbitals of highest quantum number (n) fill. The *p* block consists of groups 13, 14, 15, 16, 17, and 18 (except He).

***d* block.** The *d* orbitals of the electronic shell $n - 1$ (the next to outermost) fill. The *d* block includes groups 3, 4, 5, 6, 7, 8, 9, 10, 11, and 12.

***f* block.** The *f* orbitals of the electronic shell $n - 2$ fill. The *f*-block elements are the lanthanides and the actinides.

nother point to notice from Table 8.3 is that the electron configuration sists of a noble-gas core corresponding to the noble gas from the previous iod plus the additional electrons required to satisfy the atomic number. ognizing this and dividing the periodic table into blocks can simplify the of assigning electron configurations. For example, strontium is in group 2, second *s*-block group, so that its valence-shell configuration is $5s^2$ since it is he fifth period. The remaining electrons are in the krypton core configura- n (the noble gas in the previous period); thus the electron configuration r is

$$\text{Sr: [Kr]}5s^2$$

or the *p*-block elements in groups 13 to 18, the number of valence electrons om 1 to 6. For example, aluminum is in period 3 and group 13, its valence- ll electron configuration is $3s^23p^1$. We use $n = 3$ since Al is in the third iod and we have to accommodate three electrons after the neon core, which tains 10 electrons. Thus the electron configuration of Al is

$$\text{Al: [Ne]}3s^23p^1$$

BLE 8.3 Electron Configurations of Some Groups of Elements

oup	Element	Configuration
	H	$1s^1$
	Li	$[\text{He}]2s^1$
	Na	$[\text{Ne}]3s^1$
	K	$[\text{Ar}]4s^1$
	Rb	$[\text{Kr}]5s^1$
	Cs	$[\text{Xe}]6s^1$
	Fr	$[\text{Rn}]7s^1$
	F	$[\text{He}]2s^22p^5$
	Cl	$[\text{Ne}]3s^23p^5$
	Br	$[\text{Ar}]3d^{10}4s^24p^5$
	I	$[\text{Kr}]4d^{10}5s^25p^5$
	At	$[\text{Xe}]4f^{14}5d^{10}6s^26p^5$
	He	$1s^2$
	Ne	$[\text{He}]2s^22p^6$
	Ar	$[\text{Ne}]3s^23p^6$
	Kr	$[\text{Ar}]3d^{10}4s^24p^6$
	Xe	$[\text{Kr}]4d^{10}5s^25p^6$
	Rn	$[\text{Xe}]4f^{14}5d^{10}6s^26p^6$

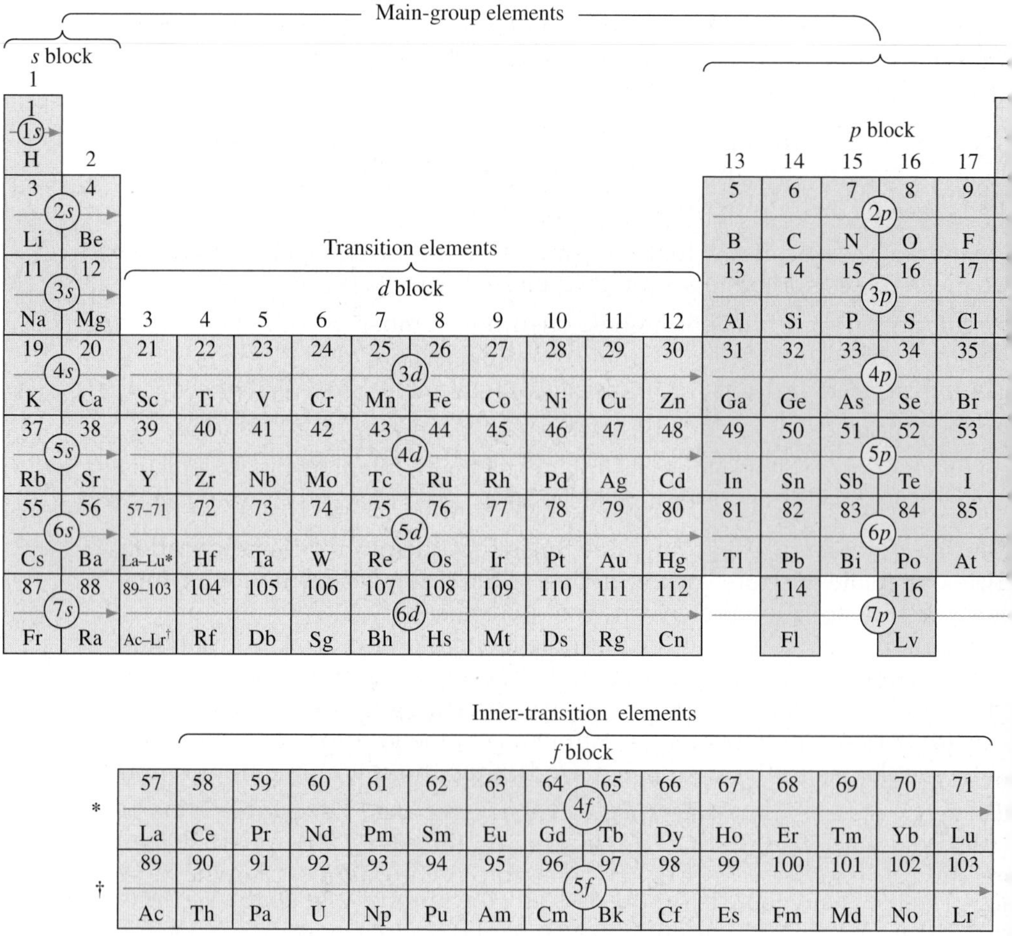

▲ FIGURE 8-38
Electron configurations and the periodic table
To use this figure as a guide to the aufbau process, locate the position of an element the table. Subshells listed ahead of this position are filled. For example, germanium ($Z = 32$) is located in group 14 of the blue $4p$ row. The filled subshells are $1s^2$, $2s^2$, $2p^6$, $3s^2$, $3p^6$, $4s^2$, and $3d^{10}$. At ($Z = 32$), a second electron has entered the $4p$ subsh The electron configuration of Ge is $[Ar]3d^{10}4s^24p^2$. Exceptions to the orderly filling c subshells suggested here are found among a few of the *d*-block and some of the *f*-blo elements.

Gallium is also in group 13, but in period 4. Its valence-shell electron c figuration is $4s^24p^1$. To write the electron configuration of Ga, we can s with the electron configuration of the noble gas that closes the third per argon, and we add to it the subshells that fill in the fourth period: $4s$, $3d$, $4p$. The $3d$ subshell must fill with 10 electrons before the $4p$ subshell begin fill. Consequently, the electron configuration of gallium must be

$$\text{Ga: } [\text{Ar}]3d^{10}4s^24p^1$$

Thallium is in group 13 and period 6. Its valence-shell electron configu tion is $6s^26p^1$. Again, we indicate the electron configuration of the noble that closes the fifth period as a core, and add the subshells that fill in the si period: $6s$, $4f$, $5d$, and $6p$.

$$\text{Tl: } [\text{Xe}]4f^{14}5d^{10}6s^26p^1$$

The elements in group 13 have the common valence configuration ns^2 again illustrating the repeating pattern of valence electron configurations dc a group, which is the basis of the similar chemical properties of the eleme within a group of the periodic table.

The transition elements correspond to the d block, and their electron configu-
tions are established in a similar manner. To write the electron configuration a transition element, start with the electron configuration of the noble gas that oses the prior period and add the subshells that fill in the period of the transi- on element being considered. The s subshell fills immediately after the preced- g noble gas; most transition metal atoms have two electrons in the s subshell the valence shell, but some have only one. Thus, vanadium ($Z = 23$), which s two valence electrons in the 4s subshell and core electrons in the configura- on of the noble gas argon, must have *three* 3d electrons ($2 + 18 + 3 = 23$).

$$\text{V: [Ar]}3d^34s^2$$

Chromium ($Z = 24$), as we have seen before, has only one valence electron the 4s subshell and core electrons in the argon configuration. Consequently must have *five* 3d electrons ($1 + 18 + 5 = 24$).

$$\text{Cr: [Ar]}3d^54s^1$$

◀ The electron configurations for the *lower d- and f-block* elements contain many exceptions that need not be memorized. Few people know all of them. Anyone needing any of these configurations can look them up when needed in tables, such as in Appendix D.

Copper ($Z = 29$) also has only one valence electron in the 4s subshell in dition to its argon core, so the copper atom must have *ten* 3d electrons $+ 18 + 10 = 29$).

$$\text{Cu: [Ar]}3d^{10}4s^1$$

Chromium and copper are two exceptions to the straightforward filling of omic subshells in the first d-block row. An examination of the electron con- gurations of the heavier elements (Appendix D) will reveal that there are her special cases that are not easily explained—for example, gadolinium has e configuration $[\text{Xe}]4f^76d^16s^2$. Examples 8-14 through 8-16 provide several ore illustrations of the assignment of electron configurations using the ideas esented here.

8-8 CONCEPT ASSESSMENT

The following orbital diagram represents an excited state of an atom. Identify the atom and give the orbital diagram corresponding to its ground state orbital diagram.

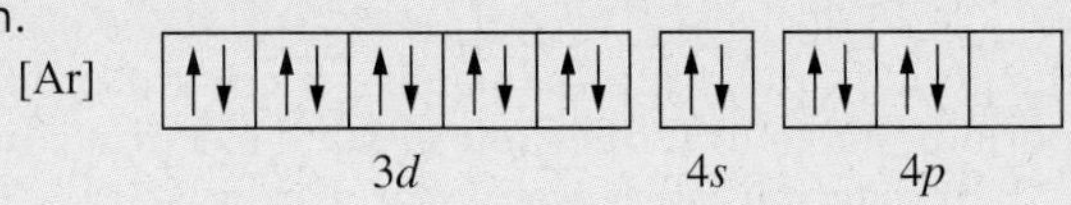

EXAMPLE 8-14 Using *spdf* Notation for an Electron Configuration

(a) Identify the element having the electron configuration

$$1s^22s^22p^63s^23p^5$$

(b) Write the electron configuration of arsenic.

Analyze

The total number of electrons in a neutral atomic species is equal to the atomic number of the element. All electrons must be accounted for in an electron configuration.

Solve

(a) Add the superscript numerals ($2 + 2 + 6 + 2 + 5$) to obtain the atomic number 17. The element with this atomic number is chlorine.

(b) Arsenic ($Z = 33$) is in period 4 and group 15. Its valence-shell electron configuration is $4s^24p^3$. The noble gas that closes the third period is Ar ($Z = 18$), and the subshells that fill in the fourth period are 4s, 3d, and 4p, in that order. Note that we account for 33 electrons in the configuration

$$\text{As: [Ar]}3d^{10}4s^24p^3$$

(continued)

Assess

As long as we count the number of electrons accurately and know the order of the orbitals, we should be able to interpret or write the correct electronic configuration.

PRACTICE EXAMPLE A: Identify the element having the electron configuration $1s^22s^22p^63s^23p^63d^24s^2$.

PRACTICE EXAMPLE B: Use *spdf* notation to show the electron configuration of iodine. How many electrons does the I atom have in its 3*d* subshell? How many unpaired electrons are there in an I atom?

EXAMPLE 8-15 Representing Electron Configurations

Write **(a)** the electron configuration of mercury, and **(b)** an orbital diagram for the electron configuration of tin.

Analyze

To write the electronic configuration, we locate the element on the periodic table and then ascertain which subshells are filled. We must be careful, with high-atomic-number elements, to take into account the lanthanide and actinide elements.

Solve

(a) Mercury, in period 6 and group 12, is the transition element at the end of the third transition series, in which the 5*d* subshell fills ($5d^{10}$). The noble gas that closes period 5 is xenon, and the lanthanide series intervenes between xenon and mercury, in which the 4*f* subshell fills ($4f^{14}$). When we put all these facts together, we conclude that the electron configuration of mercury is

$$[\text{Xe}]4f^{14}5d^{10}6s^2$$

(b) Tin is in period 5 and group 14. Its valence-shell electron configuration is $5s^25p^2$. The noble gas that closes the fourth period is Kr ($Z = 36$), and the subshells that fill in the fifth period are 5*s*, 4*d*, and 5*p*. Note that all subshells are filled in the orbital diagram except for 5*p*. Two of the 5*p* orbitals are occupied by single electrons with parallel spins; one 5*p* orbital remains empty.

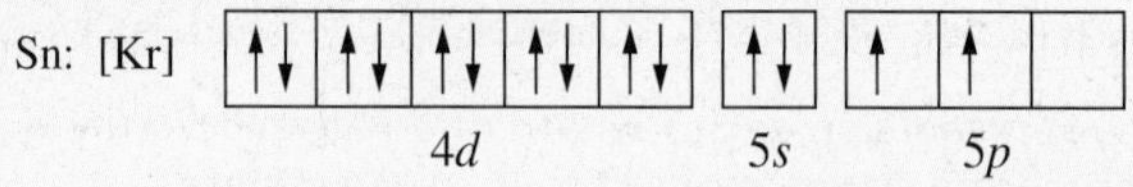

Assess

As illustrated by Figure 8-38, the structure of the periodic table approximately reflects the orbital filling order given by expression (8.22). Learn to use the periodic table to write ground-state electron configurations of atoms quickly, rather than using expression (8.22) or diagrams such as given in Figure 8-37.

PRACTICE EXAMPLE A: Represent the electron configuration of iron with an orbital diagram.

PRACTICE EXAMPLE B: Represent the electron configuration of bismuth with an orbital diagram.

EXAMPLE 8-16 Relating Electron Configurations to the Periodic Table

Indicate the number of **(a)** valence electrons in an atom of bromine; **(b)** 5*p* electrons in an atom of tellurium; **(c)** unpaired electrons in an atom of indium; **(d)** 3*d* and 4*d* electrons in a silver atom.

Analyze

Determine the atomic number and the location of each element in the periodic table. Then, explain the significance of its location.

Solve

(a) Bromine ($Z = 35$) is in group 17. There are seven outer-shell, or valence, electrons in all atoms in this group.

(b) Tellurium ($Z = 52$) is in period 5 and group 16. There are six outer-shell electrons, two of them are *s*, and the other four are *p*. The valence-shell electron configuration of tellurium is $5s^25p^4$; the tellurium atom has four 5*p* electrons.

(c) Indium ($Z = 49$) is in period 5 and group 13. The electron configuration of its inner shells is $[Kr]4d^{10}$. All the electrons in this inner-shell configuration are paired. The valence-shell electron configuration is $5s^2 5p^1$. The two 5*s* electrons are paired, and the 5*p* electron is unpaired. The In atom has one unpaired electron.

(d) Ag ($Z = 47$) is in period 5 and group 11. The noble gas that closes period 4 is krypton ($Z = 36$). By using the aufbau process to assign the 11 outer-shell electrons of silver to the 5*d* and 4*d* orbitals, we predict the valence-shell configuration of silver is $4d^9 5s^2$. We have good reason to believe that the actual valence-shell configuration is probably $4d^{10}5s^1$ not $4d^9 5s^2$. (Ag is immediately below Cu and the valence-shell configuration of Cu is $3d^{10}4s^1$ not $3d^9 4s^2$.) Appendix D confirms that the valence-shell configuration of silver is, in fact, $4d^{10}5s^1$. Thus, a silver atom has ten 3*d* electrons and ten 4*d* electrons.

Assess

By considering the position of an atom in the periodic table, we can quickly determine the electron configuration, the number of valence electrons, the number of electrons in a particular subshell, or the number of unpaired electrons. Part (d) of this problem serves as a reminder that, for the lower *d*- and *f*-block elements, the actual electron configurations may be different from those predicted by using the aufbau process.

PRACTICE EXAMPLE A: For an atom of Sn, indicate the number of **(a)** electronic shells that are either filled or partially filled; **(b)** 3*p* electrons; **(c)** 5*d* electrons; and **(d)** unpaired electrons.

PRACTICE EXAMPLE B: Indicate the number of **(a)** 3*d* electrons in Y atoms; **(b)** 4*p* electrons in Ge atoms; and **(c)** unpaired electrons in Au atoms.

Mastering**CHEMISTRY** www.masteringchemistry.com

Laser devices are in use everywhere—in compact disc players, bar-code scanners, laboratory instruments, and in cosmetic, dental, and surgical procedures. Lasers produce light with highly desirable properties by a process called stimulated emission. For a discussion of how lasers work, go to the Focus On feature for Chapter 8, Helium–Neon Lasers, on the MasteringChemistry site.

Summary

Electromagnetic Radiation—**Electromagnetic radiation** is a type of energy transmission in the form of a wave. The waves of electromagnetic radiation are characterized by an **amplitude**, the maximum height of wave crests and maximum depth of wave troughs, a **wavelength**, the distance between wave crests and **frequency**, ν, which signifies how often the fluctuations occur. Frequency is measured in **hertz, Hz** (cycles per second). Wavelength and frequency are related by the equation (8.1): $c = \lambda\nu$, where c is the **speed of light**. The wave character of electromagnetic radiation means that the waves can be dispersed into individual components of different wavelengths, a diffraction pattern, by striking a closely grooved surface (Fig. 8-4).

Prelude to Quantum Theory—The study of electromagnetic radiation emitted from hot objects led to Planck's theory, which postulates that quantities of energy can have only certain values, with the smallest unit of energy being that of a **quantum**. The energy of a quantum is given by equation (8.2): $E = h\nu$, where h is **Planck's constant**. Einstein's interpretation of the **photoelectric effect**—the ability of light to eject electrons when striking certain surfaces (Fig. 8-9)—led to a new interpretation of electromagnetic radiation: Light has a particle-like nature in addition to its wave-like properties. Light particles are called **photons**. The energy of a photon is related to the frequency of the radiation by $E_{photon} = h\nu$.

Light emitted from excited atoms and ions consists of a limited number of wavelength components, which can be dispersed by a prism to produce **atomic** or **line spectra** (Fig. 8-11). The first attempt to explain atomic (line) spectra was made by Niels Bohr who postulated that the electron in a hydrogen atom exists in a circular orbit designated by a quantum number, n, that restricts the energy of the electron to certain values (equation 8.5).

8-3 Energy Levels, Spectrum, and Ionization Energy of the Hydrogen Atom—The energy levels of the hydrogen atom depend on a quantum number, n, which can take on the values $n = 1, 2, 3$, and so on. The lowest energy state, with $n = 1$, is called the **ground state**.

Levels with $n > 1$ are called **excited states**. The state with quantum number $n = \infty$ corresponds to an ionized hydrogen atom. The allowed energy levels can be represented using an **energy-level diagram**. Transitions between the various levels are accompanied by either the absorption or the emission of photons, the energies of which match the magnitude of the energy difference, $|\Delta E|$, between the two levels (equation 8.7). Atomic *absorption spectroscopy* and *emission spectroscopy* are experimental techniques for the detection of photons absorbed or emitted by a sample of gas atoms.

8-4 Two Ideas Leading to Quantum Mechanics—Louis de Broglie postulated **wave–particle duality** in which particles of matter such as protons and electrons would at times display wave-like properties (equation 8.10). Because of an inherent uncertainty of the position and momentum of a wave-like particle, Heisenberg postulated that we cannot simultaneously know a subatomic particle's precise momentum and its position, a proposition referred to as the **Heisenberg uncertainty principle** (expression 8.11).

8-5 Wave Mechanics—The application of the concept of wave–particle duality requires that we view the electron in a system in terms of a **wave function** that corresponds to a **standing wave** within the boundary of the system (Figs. 8-17 and 8-18). Application of these ideas to a particle in a one-dimensional box shows that at the lowest energy level the energy of the particle is nonzero that is, the system has a **zero-point energy**.

8-6 Quantum Theory of the Hydrogen Atom—The solution of the **Schrödinger equation** for the hydrogen atom provides wave functions called **orbitals**, which are the product of a **radial wave function**, $R(r)$, and an **angular wave function**, $Y(\theta,\phi)$. The three quantum numbers arising from the Schrödinger wave equation are the *principal quantum number*, n, the *orbital angular momentum quantum number*, ℓ, and the *magnetic quantum number*, m_ℓ. All orbitals with the same value of n are in the same **principal electronic shell (principal level)**, and all orbitals with the same values of n and ℓ are in the same **subshell (sublevel)**. The orbitals with different values of ℓ (0, 1, 2, 3, and so on) are designated s, p, d, f (Fig. 8-22). Orbitals in the same subshell of a hydrogen-like species have the same energy and are said to be **degenerate**.

8-7 Interpreting and Representing the Orbitals of the Hydrogen Atom—Interpreting the solutions to the Schrödinger equation for the hydrogen atom leads to a description of the shapes of the electron probability distributions for electrons in the s, p, and d orbitals. The nu… of nodes ($n - 1$) in an orbital increases as n incr… Nodes are where the wave function changes sign… total number of nodes is equal to the number of … **nodes**, $n - \ell - 1$, plus the number of angular nodes…

8-8 Electron Spin: A Fourth Quantum Numb…
Stern and Gerlach demonstrated that electrons pos… quality called **electron spin** (Figs. 8-32 and 8-33). The… tron spin quantum number, m_s, takes the value $+\frac{1}{2}$ or…

8-9 Multielectron Atoms—The wave functio… multielectron atom can be approximated as a superpo… of orbitals, each of which is qualitatively similar to a h… gen-like orbital. In multielectron atoms, orbitals with … ent values of ℓ are not degenerate (Fig. 8-34). The l… degeneracy within a principal shell is explained in ter… the different **effective nuclear charge, Z_{eff}**, experienc… electrons in different subshells.

8-10 Electron Configurations—**Electron confi… tion** describes how the electrons are distributed amon… various orbitals in principal shells and subshells … atom. Electrons fill orbitals in a way (expression 8.22… minimizes the total energy of an atom. The **Pauli excl… principle** states that a maximum of two elec… may occupy an orbital. **Hund's rule** says that when d… erate orbitals are available, electrons initially occupy … orbitals singly with parallel spins. Electron configura… are represented by either expanded or condensed … **notation** or an **orbital diagram** (page 354). The **au… process** is used to assign electron configurations to th… ments of the periodic table. Electrons added to the sh… highest quantum number in the aufbau process are … **valence electrons**.

8-11 Electron Configurations and the Peri… Table—Elements in the same group of the periodic … have similar electron configurations. Groups 1 and … respond to the **s block** with filled or partially … valence-shell s orbitals. Groups 13 through 18 corres… to the **p block** with filled or partially filled valence… p orbitals. The ***d* block** corresponds to groups 3 throu… as the $n - 1$ energy level is being filled—that is, ha… filled or partially filled d orbitals. In the ***f*-block** elem… also called the lanthanides and actinides, the $n - 2$ … fills with electrons; that is, they have filled or par… filled f orbitals.

Integrative Example

Microwave ovens are not just being put to use in the kitchen—they are also useful in the chemical laboratory, par… larly in drying samples for chemical analysis. A typical microwave oven uses microwave radiation with a waveleng… 12.2 cm.

Are there any electronic transitions between consecutive levels in the hydrogen atom that could *conceivably* pro… microwave radiation of wavelength 12.2 cm? Estimate the principal quantum levels between which the transition oc…

Analyze

Use the wavelength of microwaves to calculate the frequency of the radiation. Calculate the energy of the photon tha… this frequency. Then, estimate the values of the quantum numbers involved by using equations (8.6) and (8.7).

ve

:alculate the frequency of the microwave radiaion. Microwaves are a form of electromagnetic adiation and thus travel at the speed of light, .998 × 10^8 m s^{-1}. Convert the wavelength to neters, and then use the equation

$$\nu = c/\lambda$$

$$\nu = \frac{2.998 \times 10^8\ \text{m s}^{-1}}{12.2\ \text{cm} \times 1\ \text{m}/100\ \text{cm}} = 2.46 \times 10^9\ \text{Hz}$$

:alculate the energy associated with one hoton of the microwave radiation by using quation (8.3).

$$E_{\text{photon}} = h\nu = 6.626 \times 10^{-34}\ \text{J s} \times 2.46 \times 10^9\ \text{s}^{-1}$$
$$= 1.63 \times 10^{-24}\ \text{J}$$

)etermine whether the expression E_{photon} = 1.63 × 10^{-24} J = $|\Delta E|$ can be satisfied for a ransition between consecutive levels by first leveloping an expression for ΔE when the atom nakes a transition from $n_i = n + 1$ to $n_f = n$. To btain the expression, we substitute these values or n_f and n_i into equation (8.6).

$$\Delta E = 2.179 \times 10^{-18}\ \text{J}\left(\frac{1}{(n+1)^2} - \frac{1}{n^2}\right)$$
$$= 2.179 \times 10^{-18}\ \text{J}\left(\frac{n^2 - (n+1)^2}{n^2(n+1)^2}\right)$$
$$= -2.179 \times 10^{-18}\ \text{J}\left(\frac{2n+1}{n^2(n+1)^2}\right)$$

ubstitute the values for E_{photon} and ΔE into quation (8.7) to obtain an expression that we nust solve for n.

$$E_{\text{photon}} = 1.63 \times 10^{-24}\ \text{J} = 2.179 \times 10^{-18}\ \text{J}\left(\frac{2n+1}{n^2(n+1)^2}\right)$$
$$\frac{1.63 \times 10^{-24}\ \text{J}}{2.179 \times 10^{-18}\ \text{J}} = 7.48 \times 10^{-7} = \left(\frac{2n+1}{n^2(n+1)^2}\right)$$

.ook at Figure 8-13, the simplified energy-level liagram for the hydrogen atom. Energy differnces between the low-lying levels are of the rder 10^{-19} to 10^{-20} J. These are orders of magniude (10^4 to 10^5 times) greater than the energy per hoton of 1.63 × 10^{-24} J from part 2. Note, iowever, that the energy differences become rogressively smaller as n increases. As n pproaches ∞, the energy differences approach ero, and some transitions between these high n evels should correspond to microwave radiation. Thus we expect n to be large, so that to a good pproximation we can neglect one with respect to n nd write

$$7.48 \times 10^{-7} = \left(\frac{2n+1}{n^2(n+1)^2}\right) \approx \left(\frac{2n}{n^2n^2}\right) \approx \left(\frac{2}{n^3}\right)$$

;olving for n

$$n \approx \left(\frac{2}{7.48 \times 10^{-7}}\right)^{1/3} \approx 138.8$$

Ve can check this result by substituting this value of n = 139 into the exact expression.

$$7.48 \times 10^{-7} = \frac{2n+1}{n^2(n+1)^2} = \frac{2(139)+1}{139^2(139+1)^2} = 7.37 \times 10^{-7}$$

The agreement is not very good, so let's try n = 138.

$$7.48 \times 10^{-7} = \frac{2n+1}{n^2(n+1)^2} = \frac{2(138)+1}{138^2(138+1)^2} = 7.53 \times 10^{-7}$$

This provides closer agreement. The value of the rincipal quantum number is n = 138.

e emission of a photon for the deexcitation of an electron from n = 139 to n = 138 produces a wavelength for that oton in the microwave region.

sess

might question whether the n = 139 state is still a bound state or whether the energy required to create this state ises ionization (see Exercise 106).

ACTICE EXAMPLE A: Calculate the de Broglie wavelength of a helium atom at 25 °C and moving at the rootan-square velocity. At what temperature would the average helium atom be moving fast enough to have a de Broglie velength comparable to that of the size of a typical atom, about 300 pm?

PRACTICE EXAMPLE B: By using a two-photon process (that is, two sequential excitations), a chemist is abl excite the electron in a hydrogen atom to the 5*d* level. Not all excitations are possible; they are governed by selection r (see Are You Wondering 8-6). Use the selection rules to identify the possible intermediate levels (more than one possible) involved, and calculate the frequencies of the two photons involved in each process. Identify the transit allowed when a sample of hydrogen atoms excited to the 5*d* level exhibits an emission spectrum. When a sampl gaseous sodium atoms is similarly excited to the 5*d* level, what would be the difference in the emission spect observed?

Exercises

Electromagnetic Radiation

1. A hypothetical electromagnetic wave is pictured here. What is the wavelength of this radiation?

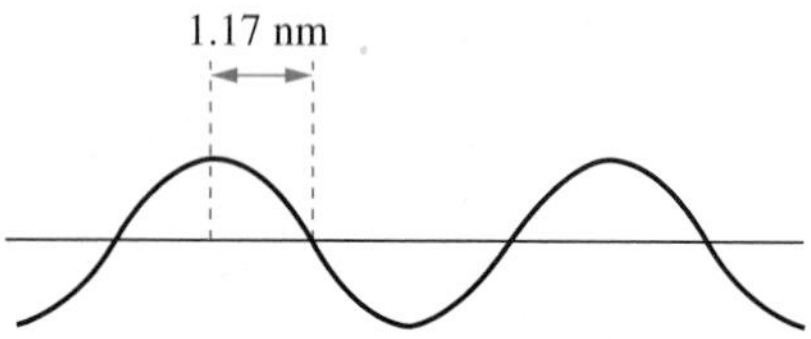

2. For the electromagnetic wave described in Exercise 1, what are **(a)** the frequency, in hertz, and **(b)** the energy, in joules per photon?
3. The magnesium spectrum has a line at 266.8 nm. Which of these statements about this radiation is (are) correct? Explain.
   (a) It has a higher frequency than radiation with wavelength 402 nm.
   (b) It is visible to the eye.
   (c) It has a greater speed in a vacuum than does red light of wavelength 652 nm.
   (d) Its wavelength is longer than that of X-rays.
4. The most intense line in the cerium spectrum i 418.7 nm.
   (a) Determine the frequency of the radiation pro ing this line.
   (b) In what part of the electromagnetic spectrum this line occur?
   (c) Is it visible to the eye? If so, what color is it? If is this line at higher or lower energy than visible li
5. *Without doing detailed calculations,* determine whic the following wavelengths represents light of highest frequency: **(a)** 6.7×10^{-4} cm; **(b)** 1.23 **(c)** 80 nm; **(d)** 6.72 μm.
6. *Without doing detailed calculations,* arrange the fol ing electromagnetic radiation sources in orde increasing frequency: **(a)** a red traffic light;
   (b) a 91.9 MHz radio transmitter;
   (c) light with a frequency of $3.0 \times 10^{14}\ s^{-1}$;
   (d) light with a wavelength of 49 nm.
7. How long does it take light from the sun, 93 mi miles away, to reach Earth?
8. In astronomy, distances are measured in *light-y* the distance that light travels in one year. What is distance of one light-year expressed in kilometers

Photons and the Photoelectric Effect

9. Determine
   (a) the energy, in joules per photon, of radiation of frequency $7.39 \times 10^{15}\ s^{-1}$;
   (b) the energy, in kilojoules per mole, of radiation of frequency $1.97 \times 10^{14}\ s^{-1}$.
10. Determine
   (a) the frequency, in hertz, of radiation having an energy of 8.62×10^{-21} J/photon;
   (b) the wavelength, in nanometers, of radiation with 360 kJ/mol of energy.
11. A certain radiation has a wavelength of 574 nm. What is the energy, in joules, of **(a)** one photon; **(b)** a mole of photons of this radiation?
12. What is the wavelength, in nanometers, of light with an energy content of 2112 kJ/mol? In what portion of the electromagnetic spectrum is this light?
13. *Without doing detailed calculations,* indicate which of following electromagnetic radiations has the grea energy per photon and which has the least: **(a)** 662 **(b)** 2.1×10^{-5} cm; **(c)** 3.58 μm; **(d)** 4.1×10^{-6} m.
14. *Without doing detailed calculations,* arrange the foll ing forms of electromagnetic radiation in *increa* order of energy per mole of photons: **(a)** radia with $\nu = 3.0 \times 10^{15}\ s^{-1}$; **(b)** an infrared heat la **(c)** radiation having $\lambda = 7000$ Å; **(d)** dental X-ray
15. In what region of the electromagnetic spectr would you expect to find radiation having an ene per photon 100 times that associated with 988 radiation?
16. High-pressure sodium vapor lamps are used in st lighting. The two brightest lines in the sodium s trum are at 589.00 and 589.59 nm. What is the differe in energy per photon of the radiations correspond to these two lines?

The lowest-frequency light that will produce the photoelectric effect is called the *threshold frequency*.
(a) The threshold frequency for indium is $9.96 \times 10^{14}\ s^{-1}$. What is the energy, in joules, of a photon of this radiation?
(b) Will indium display the photoelectric effect with UV light? With infrared light? Explain.

18. The minimum energy required to cause the photoelectric effect in potassium metal is 3.69×10^{-19} J. Will photoelectrons be produced when visible light shines on the surface of potassium? If 520 nm radiation is shone on potassium, what is the velocity of the ejected electrons?

omic Spectra

Use the Balmer equation (8.4) to determine
(a) the frequency, in s^{-1}, of the radiation corresponding to $n = 5$;
(b) the wavelength, in nanometers, of the line in the Balmer series corresponding to $n = 7$;
(c) the value of n corresponding to the Balmer series line at 380 nm.

How would the Balmer equation (8.4) have to be modified to predict lines in the infrared spectrum of hydrogen? [*Hint:* Compare equations (8.4) and (8.6).]

What is ΔE for the transition of an electron from $n = 6$ to $n = 3$ in a hydrogen atom? What is the frequency of the spectral line produced?

What is ΔE for the transition of an electron from $n = 5$ to $n = 2$ in a hydrogen atom? What is the frequency of the spectral line produced?

To what value of n in equation (8.4) does the line in the Balmer series at 389 nm correspond?

24. The Lyman series of the hydrogen spectrum can be represented by the equation

$$\nu = 3.2881 \times 10^{15}\ s^{-1}\left(\frac{1}{1^2} - \frac{1}{n^2}\right) \ (\text{where } n = 2, 3, \ldots)$$

(a) Calculate the maximum and minimum wavelength lines, in nanometers, in this series.
(b) What value of n corresponds to a spectral line at 95.0 nm?
(c) Is there a line at 108.5 nm? Explain.

25. Calculate the wavelengths, in nanometers, of the first four lines of the Balmer series of the hydrogen spectrum, starting with the *longest* wavelength component.

26. A line is detected in the hydrogen spectrum at 1880 nm. Is this line in the Balmer series? Explain.

ergy Levels and Spectrum of the Hydrogen Atom

Calculate the energy, in joules, of a hydrogen atom when the electron is in the sixth energy level.

Calculate the increase in energy, in joules, when an electron in the hydrogen atom is excited from the first to the third energy level.

What are the **(a)** frequency, in s^{-1}, and **(b)** wavelength, in nanometers, of the light emitted when the electron in a hydrogen atom drops from the energy level $n = 7$ to $n = 4$? **(c)** In what portion of the electromagnetic spectrum is this light?

Without doing detailed calculations, indicate which of the following electron transitions requires the greatest amount of energy to be *absorbed* by a hydrogen atom: from **(a)** $n = 1$ to $n = 2$; **(b)** $n = 2$ to $n = 4$; **(c)** $n = 3$ to $n = 9$; **(d)** $n = 10$ to $n = 1$.

For a hydrogen atom, determine
(a) the energy level corresponding to $n = 8$;
(b) whether there is an energy level at -2.500×10^{-19} J;
(c) the ionization energy, if the electron is initially in the $n = 6$ level.

Without doing detailed calculations, indicate which of the following electron transitions in the hydrogen atom results in the emission of light of the longest wavelength. **(a)** $n = 4$ to $n = 3$; **(b)** $n = 1$ to $n = 2$; **(c)** $n = 1$ to $n = 6$; **(d)** $n = 3$ to $n = 2$.

What electron transition in a hydrogen atom, starting from $n = 7$, will produce light of wavelength 410 nm?

34. What electron transition in a hydrogen atom, ending in $n = 3$, will produce light of wavelength 1090 nm?

35. The emission spectrum below for a one-electron (hydrogen-like) species in the gas phase shows all the lines, before they merge together, resulting from transitions to the ground state from higher energy states. Line A has a wavelength of 103 nm.

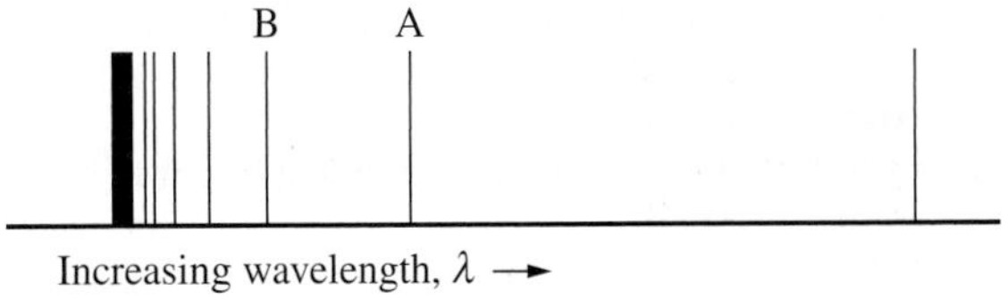

(a) What are the upper and lower principal quantum numbers corresponding to the lines labeled A and B?
(b) Identify the one-electron species that exhibits the spectrum.

36. The emission spectrum below for a one-electron (hydrogen-like) species in the gas phase shows all the lines, before they merge together, resulting from transitions to the first excited state from higher energy states. Line A has a wavelength of 434 nm.

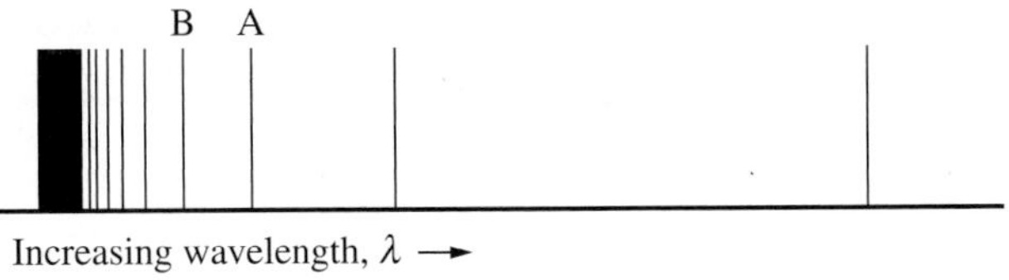

(a) What are the upper and lower principal quantum numbers corresponding to the lines labeled A and B?
(b) Identify the one-electron species that exhibits the spectrum.

37. The emission spectrum below for a one-electron (hydrogen-like) species in the gas phase shows all the lines, before they merge together, resulting from transitions to the first excited state from higher energy states. Line A has a wavelength of 27.1 nm.

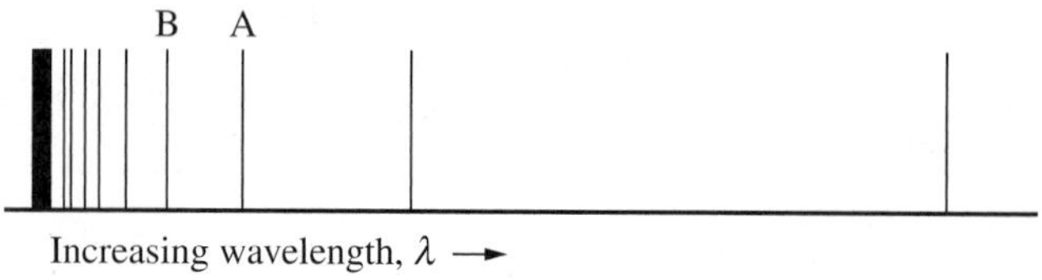

(a) What are the upper and lower principal quantum numbers corresponding to the lines labeled A and B?
(b) Identify the one-electron species that exhibits spectrum.

38. The emission spectrum below for a one-elect (hydrogen-like) species in the gas phase shows all lines, before they merge together, resulting from t sitions to the ground state from higher energy sta Line A has a wavelength of 10.8 nm.

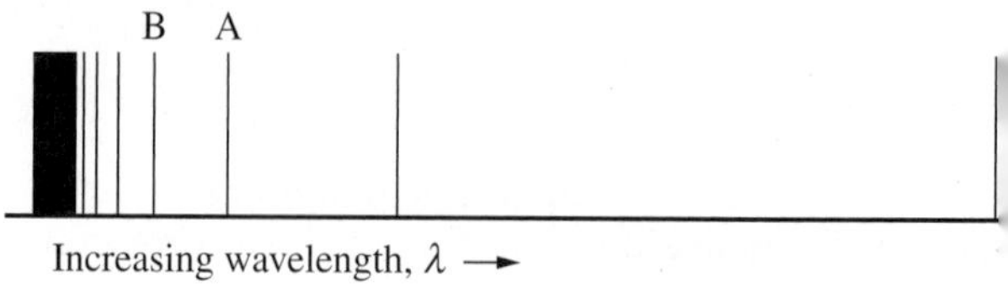

(a) What are the upper and lower principal quant numbers corresponding to the lines labeled A and
(b) Identify the one-electron species that exhibits spectrum.

Wave–Particle Duality

39. Which must possess a greater velocity to produce matter waves of the same wavelength (such as 1 nm), protons or electrons? Explain your reasoning.

40. What must be the velocity, in meters per second, of a beam of electrons if they are to display a de Broglie wavelength of 850 nm?

41. Calculate the de Broglie wavelength, in nanometers, associated with a 145 g baseball traveling at a speed of 168 km/h. How does this wavelength compare v typical nuclear or atomic dimensions?

42. What is the wavelength, in nanometers, associa with a 9.7 g bullet with a muzzle velocity of 887 m that is, considering the bullet to be a matter wa Comment on the feasibility of an experimental m surement of this wavelength.

The Heisenberg Uncertainty Principle

43. The uncertainty relation $\Delta x \Delta p \geq h/(4\pi)$, expression (8.11), is valid for motion in any direction. For circular motion, the relation may be expressed as $\Delta r \Delta p \geq h/(4\pi)$, where Δr is the uncertainty in radial position and Δp is the uncertainty in the momentum along the *radial* direction. Describe how Bohr's model of the hydrogen atom violates the uncertainty relation expressed in the form $\Delta r \Delta p \geq h/(4\pi)$.

44. Although Einstein made some early contributions to quantum theory, he was never able to accept the Heisenberg uncertainty principle. He stated, "God does not play dice with the Universe." What do you suppose Einstein meant by this remark? In reply to Einstein's remark, Niels Bohr is supposed to have said, "Albert, stop telling God what to do." What do you suppose Bohr meant by this remark?

45. A proton is accelerated to one-tenth the velocity light, and this velocity can be measured with a pr sion of 1%. What is the uncertainty in the positio this proton?

46. Show that the uncertainty principle is not signific when applied to large objects such as automobi Assume that m is precisely known; assign a reas able value to either the uncertainty in position or uncertainty in velocity, and estimate a value of other.

47. What must be the velocity of electrons if their ass ated wavelength is to equal the Bohr radius, a_0?

48. What must be the velocity of electrons if their ass ated wavelength is to equal the *longest* wavelen line in the Lyman series? [*Hint:* Refer to Figure 8-1

Wave Mechanics

49. A standing wave in a string 42 cm long has a total of six nodes (including those at the ends). What is the wavelength, in centimeters, of this standing wave?

50. What is the length of a string that has a standing wave with four nodes (including those at the ends) and $\lambda = 17$ cm?

51. Calculate the wavelength of the electromagnetic ra ation required to excite an electron from the grou state to the level with $n = 4$ in a one-dimensional b 5.0×10^1 pm long.

52. An electron in a one-dimensional box requires a wa length of 618 nm to excite an electron from the n = level to the $n = 4$ level. Calculate the length of the b

. An electron in a 20.0 nm box is excited from the ground state into a higher energy state by absorbing a photon of wavelength 8.60×10^{-5} m. Determine the final energy state.

. Calculate the wavelength of the electromagnetic radiation required to excite a proton from the ground state to the level with $n = 4$ in a one-dimensional box 5.0×10^1 pm long.

55. Describe some of the differences between the orbits of the Bohr atom and the orbitals of the wave mechanical atom. Are there any similarities?

56. The greatest probability of finding the electron in a small-volume element of the 1*s* orbital of the hydrogen atom is at the nucleus. Yet the most probable distance of the electron from the nucleus is 53 pm. How can you reconcile these two statements?

uantum Numbers and Electron Orbitals

. Select the correct answer and explain your reasoning. An electron having $n = 3$ and $m_\ell = 0$ **(a)** must have $m_s = +\frac{1}{2}$; **(b)** must have $\ell = 1$; **(c)** may have $\ell = 0, 1,$ or 2; **(d)** must have $\ell = 2$.

. Write an acceptable value for each of the missing quantum numbers.
(a) $n = 3, \ell = ?, m_\ell = 2, m_s = +\frac{1}{2}$
(b) $n = ?, \ell = 2, m_\ell = 1, m_s = -\frac{1}{2}$
(c) $n = 4, \ell = 2, m_\ell = 0, m_s = ?$
(d) $n = ?, \ell = 0, m_\ell = ?, m_s = ?$

. What type of orbital (i.e., 3*s*, 4*p*, ...) is designated by these quantum numbers?
(a) $n = 5, \ell = 1, m_\ell = 0$
(b) $n = 4, \ell = 2, m_\ell = -2$
(c) $n = 2, \ell = 0, m_\ell = 0$

. Which of the following statements is (are) correct for an electron with $n = 4$ and $m_\ell = 2$? Explain.
(a) The electron is in the fourth principal shell.
(b) The electron may be in a *d* orbital.
(c) The electron may be in a *p* orbital.
(d) The electron must have $m_s = +\frac{1}{2}$.

61. Concerning the electrons in the shells, subshells, and orbitals of an atom, how many can have
(a) $n = 4, \ell = 2, m_\ell = 1$, and $m_s = +\frac{1}{2}$?
(b) $n = 4, \ell = 2$, and $m_\ell = 1$?
(c) $n = 4$ and $\ell = 2$?
(d) $n = 4$?
(e) $n = 4, \ell = 2$, and $m_s = +\frac{1}{2}$?

62. Concerning the concept of subshells and orbitals,
(a) How many subshells are found in the $n = 3$ level?
(b) What are the names of the subshells in the $n = 3$ level?
(c) How many orbitals have the values $n = 4$ and $\ell = 3$?
(d) How many orbitals have the values $n = 3, \ell = 2$, and $m_\ell = -2$?
(e) What is the total number of orbitals in the $n = 4$ level?

he Shapes of Orbitals and Radial Probabilities

. Calculate the finite value of r, in terms of a_0, at which the node occurs in the wave function of the 2*s* orbital of a hydrogen atom.

. Calculate the finite value of r, in terms of a_0, at which the node occurs in the wave function of the 2*s* orbital of a Li^{2+} ion.

. Show that the probability of finding a $2p_y$ electron in the *xz* plane is zero.

. Show that the probability of finding a $3d_{xz}$ electron in the *xy* plane is zero.

. Prepare a two-dimensional plot of $Y(\theta, \phi)$ for the p_y orbital in the *xy* plane.

. Prepare a two-dimensional plot of $Y^2(\theta, \phi)$ for the p_y orbital in the *xy* plane.

. Using a graphical method, show that in a hydrogen atom the radius at which there is a maximum probability of finding an electron is a_0 (53 pm).

70. Use a graphical method or some other means to show that in a Li^{2+} ion, the radius at which there is a maximum probability of finding an electron is $\frac{a_0}{3}$ (18 pm).

71. Identify the orbital that has **(a)** one radial node and one angular node; **(b)** no radial nodes and two angular nodes; **(c)** two radial nodes and three angular nodes.

72. Identify the orbital that has **(a)** two radial nodes and one angular node; **(b)** five radial nodes and zero angular nodes; **(c)** one radial node and four angular nodes.

73. A contour map for an atomic orbital of hydrogen is shown at the top of page 370 for the *xy* and *xz* planes. Identify the orbital.

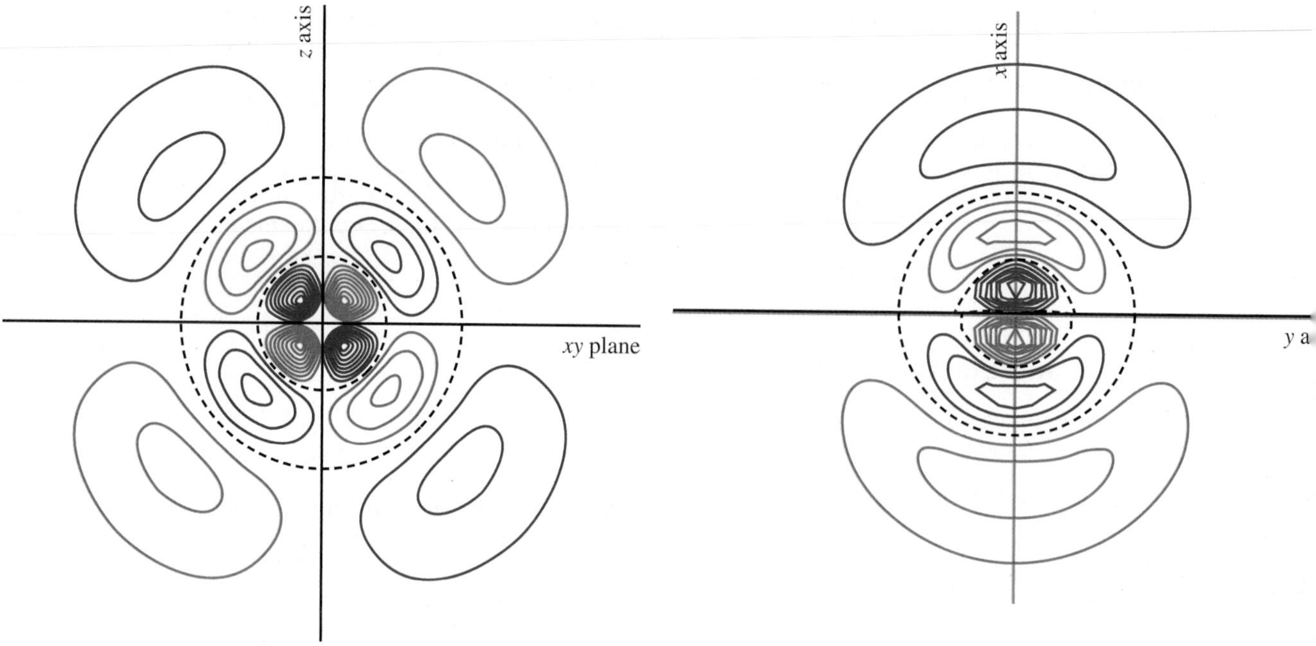

74. A contour map for an atomic orbital of hydrogen is shown below for the xy and xz planes. Identify the type ($s, p, d, f, g \ldots$) of orbital.

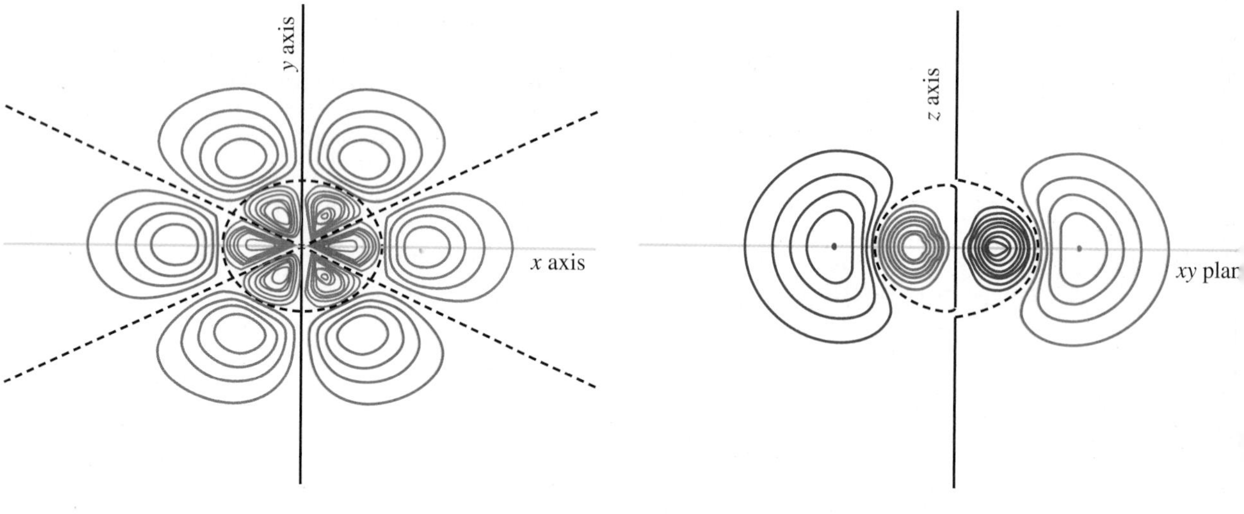

Electron Configurations

75. On the basis of the periodic table and rules for electron configurations, indicate the number of **(a)** $2p$ electrons in N; **(b)** $4s$ electrons in Rb; **(c)** $4d$ electrons in As; **(d)** $4f$ electrons in Au; **(e)** unpaired electrons in Pb; **(f)** elements in group 14 of the periodic table; **(g)** elements in the sixth period of the periodic table.

76. Based on the relationship between electron configurations and the periodic table, give the number of **(a)** outer-shell electrons in an atom of Sb; **(b)** electrons in the fourth principal electronic shell of Pt; **(c)** elements whose atoms have six outer-shell electrons; **(d)** unpaired electrons in an atom of Te; **(e)** transition elements in the sixth period.

77. Which of the following is the correct orbital diagr for the ground-state electron configuration of phosp rus? Explain what is wrong with each of the others.

(a) [Ne] 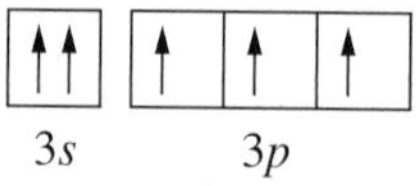

(b) [Ne] 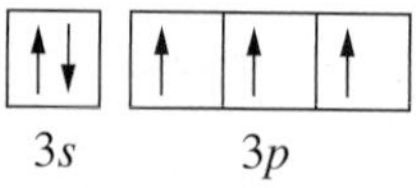

(c) [Ne] 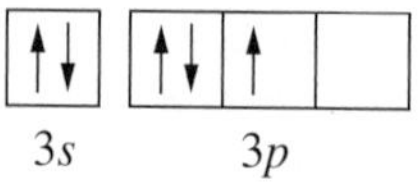

(d) [Ne] 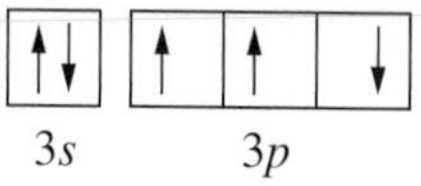

Which of the following is the correct orbital diagram for the ground-state electron configuration of molybdenum? Explain what is wrong with each of the others.

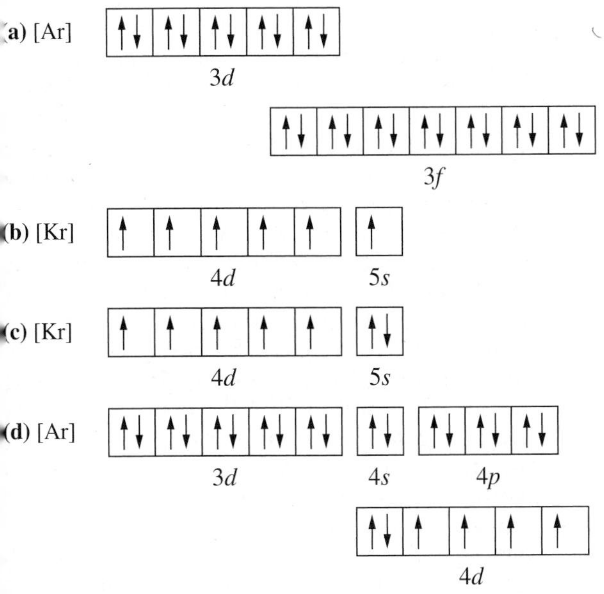

Use the basic rules for electron configurations to indicate the number of **(a)** unpaired electrons in an atom of P; **(b)** 3*d* electrons in an atom of Br; **(c)** 4*p* electrons in an atom of Ge; **(d)** 6*s* electrons in an atom of Ba; **(e)** 4*f* electrons in an atom of Au.

Use orbital diagrams to show the distribution of electrons among the orbitals in **(a)** the 4*p* subshell of Br; **(b)** the 3*d* subshell of Co^{2+}, given that the two electrons lost are 4*s*; **(c)** the 5*d* subshell of Pb.

The recently discovered element 114, Flerovium, should most closely resemble Pb.
(a) Write the electron configuration of Pb.
(b) Propose a plausible electron configuration for element 114.

Without referring to any tables or listings in the text, mark an appropriate location in the blank periodic table provided for each of the following: **(a)** the fifth-period noble gas; **(b)** a sixth-period element whose atoms have three unpaired *p* electrons; **(c)** a *d*-block element having one 4*s* electron; **(d)** a *p*-block element that is a metal.

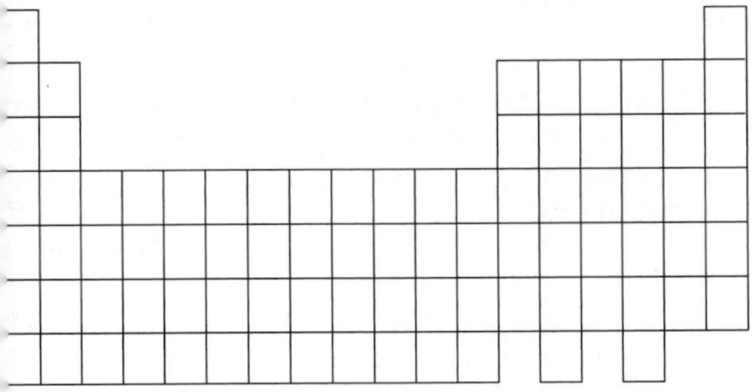

83. Which of the following electron configurations corresponds to the ground state and which to an excited state?

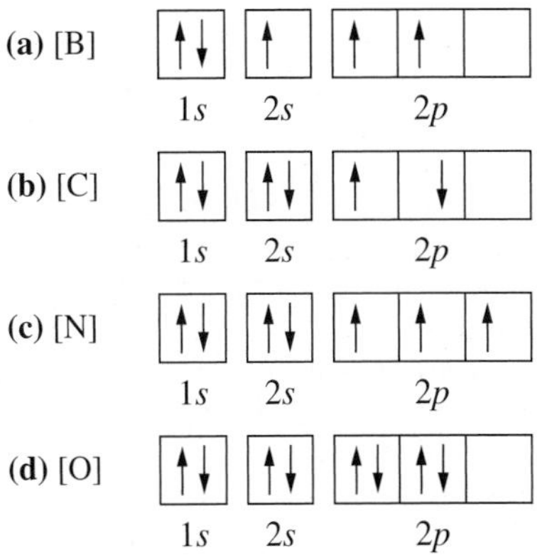

84. To what neutral atom do the following valence-shell configurations correspond? Indicate whether the configuration corresponds to the ground state or an excited state.

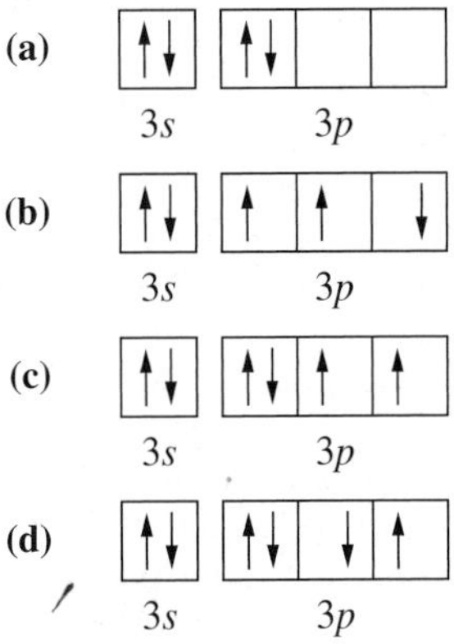

85. What is the expected ground-state electron configuration for each of the following elements? **(a)** mercury; **(b)** calcium; **(c)** polonium; **(d)** tin; **(e)** tantalum; **(f)** iodine.

86. What is the expected ground-state electron configuration for each of the following elements? **(a)** tellurium; **(b)** cesium; **(c)** selenium; **(d)** platinum; **(e)** osmium; **(f)** chromium.

87. The following electron configurations correspond to the ground states of certain elements. Name each element. **(a)** $[Rn]6d^27s^2$; **(b)** $[He]2s^22p^2$; **(c)** $[Ar]3d^34s^2$; **(d)** $[Kr]4d^{10}5s^25p^4$; **(e)** $[Xe]4f^26s^26p^1$.

88. The following electron configurations correspond to the ground states of certain elements. Name each element. **(a)** $[Ar]3d^{10}4s^24p^3$; **(b)** $[Ne]3s^23p^4$; **(c)** $[Ar]3d^14s^2$; **(d)** $[Kr]4d^65s^2$; **(e)** $[Xe]4f^{12}6s^2$.

Integrative and Advanced Exercises

89. Derive the Balmer and Rydberg equations from equation (8.6).

90. Electromagnetic radiation can be transmitted through a vacuum or empty space. Can heat be similarly transferred? Explain.

91. The *work function* is the energy that must be supplied to cause the release of an electron from a photoelectric material. The corresponding photon frequency is the threshold frequency. The higher the energy of the incident light, the more kinetic energy the electrons have in moving away from the surface. The work function for mercury is equivalent to 435 kJ/mol photons.
(a) Can the photoelectric effect be obtained with mercury by using visible light? Explain.
(b) What is the kinetic energy, in joules, of the ejected electrons when light of 215 nm strikes a mercury surface?
(c) What is the velocity, in meters per second, of the ejected electrons in part (b)?

92. Infrared lamps are used in cafeterias to keep food warm. How many photons per second are produced by an infrared lamp that consumes energy at the rate of 95 W and is 14% efficient in converting this energy to infrared radiation? Assume that the radiation has a wavelength of 1525 nm.

93. In 5.0 s, a 75 watt light source emits 9.91×10^{20} photons of a monochromatic (single wavelength) radiation. What is the color of the emitted light?

94. Determine the de Broglie wavelength of the electron ionized from a He^+ ion in its ground state using light of wavelength 208 nm.

95. The Pfund series of the hydrogen spectrum has as its *longest* wavelength component a line at 7400 nm. Describe the electron transitions that produce this series. That is, give a quantum number that is common to this series.

96. Between which two levels of the hydrogen atom must an electron fall to produce light of wavelength 1876 nm?

97. Use appropriate relationships from the chapter to determine the wavelength of the line in the emission spectrum of He^+ produced by an electron transition from $n = 5$ to $n = 2$.

98. Draw an energy-level diagram that represents all the possible lines in the emission spectrum of hydrogen atoms produced by electron transitions, in one or more steps, from $n = 5$ to $n = 1$.

99. An atom in which just one of the outer-shell electrons is excited to a very high quantum level n is called a "high Rydberg" atom. In some ways, all these atoms resemble a hydrogen atom with its electron in a high n level. Explain why you might expect this to be the case.

100. If all other rules governing electron configurations were valid, what would be the electron configuration of cesium if **(a)** there were *three* possibilities for electron spin; **(b)** the quantum number ℓ could have the value n?

101. Ozone, O_3, absorbs ultraviolet radiation and dissociates into O_2 molecules and O atoms: $O_3 + h\nu \longrightarrow O_2 + O$. A 1.00 L sample of air at 22 °C and 748 mmHg contains 0.25 ppm of O_3. How much energy, in jou must be absorbed if all the O_3 molecules in the s ple of air are to dissociate? Assume that each pho absorbed causes one O_3 molecule to dissociate, that the wavelength of the radiation is 254 nm.

102. Radio signals from *Voyager 1* in the 1970s w broadcast at a frequency of 8.4 GHz. On Earth, radiation was received by an antenna able to de signals as weak as 4×10^{-21} W. How many phot per second does this detection limit represent?

103. Certain metal compounds impart colors to flame sodium compounds, yellow; lithium, red; bari green—and flame tests can be used to detect th elements. **(a)** At a flame temperature of 800 °C, collisions between gaseous atoms with aver kinetic energies supply the energies required for emission of visible light? **(b)** If not, how do y account for the excitation energy?

104. The angular momentum of an electron in the B hydrogen atom is mur, where m is the mass of the e tron, u, its velocity, and r, the radius of the Bohr or The angular momentum can have only the val $nh/2\pi$, where n is an integer (the number of the B orbit). Show that the *circumferences* of the various B orbits are integral multiples of the de Broglie wa lengths of the electron treated as a matter wave.

105. A molecule of chlorine can be dissociated into ato by absorbing a photon of sufficiently high energy. A excess energy is translated into kinetic energy as atoms recoil from one another. If a molecule of chlor at rest absorbs a photon of 300 nm wavelength, w will be the velocity of the two recoiling atoms? Assu that the excess energy is equally divided between two atoms. The bond energy of Cl_2 is 242.6 kJ mol^{-}

106. Refer to the Integrative Example. Determine whet or not $n = 138$ is a bound state. If it is, what sort state is it? What is the radius of the orbit and h many revolutions per second does the electron m about the nucleus?

107. Using the relationships given in Table 8.2, find the fir values of r, in terms of a_0, of the nodes for a 3*s* orbita

108. Use a graphical method or some other means determine the radius at which the probability finding a 2*s* orbital is maximum.

109. Using the relationships in Table 8.2, prepare a ske of the 95% probability surface of a $4p_x$ orbital.

110. Given that the volume of a sphere is $V = (4/3)\pi$ show that the volume, dV, of a thin spherical shell radius r and thickness dr is $4\pi r^2 dr$. [Hint: This exerc can be done easily and elegantly by using calculus can also be done without using calculus by expressi the volume of a thin spherical shell as a volume diff ence, $(4/3)\pi(r + dr)^3 - (4/3)\pi r^3$, and simplifying expression. To obtain the correct result by using the ter approach, you must remember that dr represent very small distance.]

111. In the ground state of a hydrogen atom, what is t probability of finding an electron anywhere in sphere of radius **(a)** a_0, or **(b)** $2a_0$? [*Hint:* This exerc requires calculus.]

When atoms in excited states collide with unexcited atoms they can transfer their excitation energy to those atoms. The most efficient energy transfer occurs when the excitation energy matches the energy of an excited state in the unexcited atom. Assuming that we have a collection of excited hydrogen atoms in the $2s^1$ excited state, are there any transitions of He^+ that could be most efficiently excited by the hydrogen atoms?

Feature Problems

Principal Spectral Lines of Some Period 4 Transition Elements (nm)

V	306.64	309.31	318.40	318.54	327.11	437.92	438.47	439.00
Cr	357.87	359.35	360.53	361.56	425.44	427.48	428.97	520.45
Mn	257.61	259.37	279.48	279.83	403.08	403.31	403.45	
Fe	344.06	358.12	372.00	373.49	385.99			
Ni	341.48	344.63	345.85	346.17	349.30	351.51	352.45	361.94

We have noted that an emission spectrum is a kind of "atomic fingerprint." The various steels are alloys of iron and carbon, usually containing one or more other metals. Based on the principal lines of their atomic spectra, which of the metals in the table above are likely to be present in a steel sample whose hypothetical emission spectrum is pictured? Is it likely that still other metals are present in the sample? Explain.

ypothetical emission spectrum
real spectrum, the photographic images of the spectral s would differ in depth and thickness depending on the ıgths of the emissions producing them. Some of the tral lines would not be seen because of their faintness.

Balmer seems to have deduced his formula for the visible spectrum of hydrogen just by manipulating numbers. A more common scientific procedure is to graph experimental data and then find a mathematical equation to describe the graph. Show that equation (8.4) describes a straight line. Indicate which variables must be plotted, and determine the numerical values of the slope and intercept of this line. Use data from Figure 8-12 to confirm that the four lines in the visible spectrum of hydrogen fall on the straight-line graph.

The Rydberg–Ritz combination principle is an empirical relationship proposed by Walter Ritz in 1908 to explain the relationship among spectral lines of the hydrogen atom. The principle states that the spectral lines of the hydrogen atom include frequencies that are either the sum or the difference of the frequencies of two other lines. This principle is obvious to us, because we now know that spectra arise from transitions between energy levels, and the energy of a transition is proportional to the frequency.

The frequencies of the first ten lines of an emission spectrum of hydrogen are given in the table at the bottom of this page. In this problem, use ideas from this chapter to identify the transitions involved, and apply the Rydberg–Ritz combination principle to calculate the frequencies of other lines in the spectrum of hydrogen.

(a) Use Balmer's original equation, $\lambda = Bm^2/(m^2 - n^2)$, with $B = 346.6$ nm, to develop an expression for the frequency $\nu_{m,n}$ of a line involving a transition from level m to level n, where $m > n$.

(b) Use the expression you derived in (a) to calculate the expected ratio of the frequencies of the first two lines in each of the Lyman, Balmer, and Paschen series: $\nu_{2,1}/\nu_{3,1}$ (for the Lyman series); $\nu_{3,2}/\nu_{3,1}$ (for the Balmer series); and $\nu_{4,3}/\nu_{5,3}$ (for the Paschen series). Compare your calculated ratios to the observed ratio $2.465263/2.921793 = 0.843750$ to identify the series as the Lyman, Balmer, or Paschen series. For each line in the series, specify the transition (quantum numbers) involved. Use a diagram, such as that given in Figure 8-13, to summarize your results.

(c) Without performing any calculations, and starting from the Rydberg formula, equation (8.4), show that $\nu_{2,1} + \nu_{3,2} = \nu_{3,1}$, and thus, $\nu_{3,1} - \nu_{2,1} = \nu_{3,2}$. This is an illustration of the Rydberg–Ritz combination principle: the frequency of a spectral line is equal to the sum or difference of frequencies of other lines.

(d) Use the Rydberg–Ritz combination principle to determine, if possible, the frequencies for the other two series named in (b). [*Hint*: The diagram you drew in part (b) might help you identify the appropriate combinations of frequencies.]

(e) Identify the transition associated with a line of frequency 2.422405×10^{13} s^{-1}, one line in a series of lines discovered in 1953 by C. J. Humphreys.

equencies ($\times 10^{15}$ s^{-1}) of the First Ten Lines in an Emission Spectrum of Hydrogen

465263	2.921793	3.081578	3.155536	3.195711	3.219935	3.235657	3.246436	3.254147	3.259851

116. Emission and absorption spectra of the hydrogen atom exhibit line spectra characteristic of quantized systems. In an absorption experiment, a sample of hydrogen atoms is irradiated with light with wavelengths ranging from 100 to 1000 nm. In an emission spectrum experiment, the hydrogen atoms are excited through an energy source that provides a range of energies from 1230 to 1240 kJ mol^{-1} to the atoms. Assume that the absorption spectrum is obtained at room temperature, when all atoms are in the ground state.
(a) Calculate the position of the lines in the absorption spectrum.
(b) Calculate the position of the lines in the emission spectrum.
(c) Compare the line spectra observed in the two experiments. In particular, will the number of lines observed be the same?

117. Diffraction of radiation takes place when the distance between the scattering centers is comparable to the wavelength of the radiation.
(a) What velocity must helium atoms possess to be diffracted by a film of silver atoms in which the spacing is 100 pm?
(b) Electrons accelerated through a certain potential are diffracted by a thin film of gold. Would you expect a beam of protons accelerated through the same potential to be diffracted when it strikes the film of gold? If not, what would you expect to see instead?

118. The emission spectrum below is for hydrogen atoms in the gas phase. The spectrum is of the first few emission lines from principal quantum number 6 down to all possible lower levels.

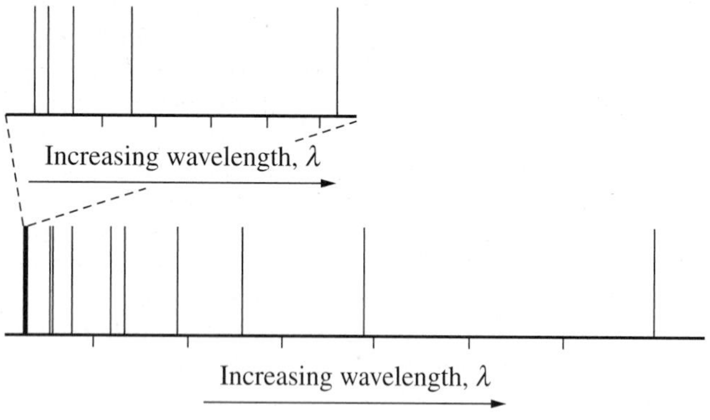

As discussed in Are You Wondering 8-6, not all possible de-excitations are possible; the transitions are governed by *selection rules*. Using the selection rules from Are You Wondering 8-6, identify the transitions, in terms of the types of orbital (s, p, d, f), involved, that are observed in the spectrum shown above.

In the presence of a magnetic field, the lines split into more lines according to the magnetic quantum number. Using the selection rule for m_ℓ, identify the line(s) in the spectrum that split(s) into the greatest number of lines.

119. (This exercise requires calculus.) In this exercise, ideas from this chapter to develop the solutio the particle-in-a-box problem. We begin by wri the Schrödinger equation for a particle of ma moving in one dimension:

$$-\left(\frac{h^2}{8\pi^2 m}\right)\frac{d^2\psi}{dx^2} + V(x)\psi = E\psi$$

The equation above is the one-dimensional ver of equation (8.15). For a particle in a box, there ar forces acting on the particle (except at the bounda of the box), and so the potential energy, V, of the ticle is constant. Without loss of generality, we assume that the value of V is zero in the box.
(a) Show that, for a particle in a box, the equa above can be written in the form $d^2\psi/dx^2 = -a$ where $a^2 = 8\pi^2 mE/h^2$.
(b) Show that $\psi = A\sin(ax)$ is a solution to equation $d^2\psi/dx^2 = -a^2\psi^2$, by differentiating ψ tv with respect to x.
(c) Following the same approach you used in show that $\psi = A\cos(ax)$ is also a solution to equation $d^2\psi/dx^2 = -a^2\psi^2$.
(d) For a particle in a box, the probability density, must be zero at $x = 0$. To ensure that this is so, must have $\psi = 0$ at $x = 0$. This requirement is ca a *boundary condition*. Use this boundary conditio establish that the wave function for a particle in box must be of the form $\psi = A\sin(ax)$, $\psi = A\cos(ax)$.
(e) Using the result from (d), show that the bou ary condition $\psi = 0$ at $x = L$ requires that $aL =$ so that the wave function may be written $\psi = A\sin(n\pi x/L)$. [*Hint*: $\sin z = 0$ when z is an i ger multiple of π.]
(f) Using the result $aL = n\pi$ from (e) and the that $a^2 = 8\pi^2 mE/h^2$, as established in (a), show $E = n^2h^2/(8mL^2)$.
(g) We know for sure (the probability is 1) that particle must be somewhere between $x = 0$ and x Mathematically, we express this condition $\int_0^L \psi^2 dx = 1$. It is called a *normalization condit* Using the result $\psi = A\sin(n\pi x/L)$ from (e), sh that the normalization condition requires t $A = \sqrt{2/L}$. [*Hint*: The integral $\int_0^L \sin^2(n\pi x/L)dx$ the value $L/2$.]

Working through this problem will walk through the basic procedure for solving a quant mechanical problem: Writing down the Schrödin equation for the system of interest (part a); establish the general form of the solutions (parts b and c); using appropriate boundary conditions and a norm ization condition to determine not only the spec form of ψ but also the allowed values for E (parts d-

. In 1913, Danish physicist Neils Bohr proposed a theory for the hydrogen atom in which the electron is imagined to be moving around a stationary nucleus in one of many possible circular orbits, each of which has a fixed energy and radius. By using classical physics and imposing a quantization condition, Bohr derived equations for the energies and radii of these orbits. Derive Bohr's equations by using the following steps. Note: Steps (a), (b), and (d) are based on fundamental ideas from classical physics. Step (c) introduces a new idea, a quantization condition, that causes the energies and radii of the orbits to take on certain well-defined values.
(a) Write down an expression for the total energy, E, of the electron (mass m_e) moving in a circular orbit of radius r with speed u. [*Hint*: See Appendix B, specifically equations (B.12) and (B.14).]
(b) Use the condition that the force of attraction between the electron and proton has the same magnitude as the centrifugal force, $m_e u^2/r$, to show that the total energy of the electron is $E = -e^2/(8\pi\epsilon_0 r)$. [*Hint*: See equation (B.13) in Appendix B.]
(c) Use the information from (b), along with the quantization condition that the orbital angular momentum, $\ell = mur$, of the electron in the nth orbit ($n = 1, 2, 3$, etc.) is $n \times h/(2\pi)$ to show that the energy and radius of the nth orbit are, respectively, $E_n = -R_\infty/n^2$ and $r_n = a_0 \times n^2$, with $R_\infty = m_e e^4/(8\epsilon_0^2 h^2) = 2.17987 \times 10^{-18}$ J and $a_0 = h^2\epsilon_0/(\pi m_e e^2)$. $= 5.29177 \times 10^{-11}$ m. [*Hint*: Use the conditions given in (b) and (c) to eliminate both u and r from the expression given in (b) for E.]
(d) Convert $R_\infty = 2.17987 \times 10^{-18}$ J to $R_H = 2.17869 \times 10^{-18}$ J by replacing m_e in the expression for R_∞ with the so-called reduced mass $\mu = m_e m_p/(m_e + m_p)$, where $m_p = 1.67262 \times 10^{-27}$ kg is the mass of the proton. The conversion of R_∞ to R_H corrects for the fact that, because the proton is not infinitely massive compared to the electron, the nucleus is not actually stationary.

Self-Assessment Exercises

1. In your own words, define the following terms or symbols: **(a)** λ; **(b)** ν; **(c)** h; **(d)** ψ; **(e)** principal quantum number, n.

2. Briefly describe each of the following ideas or phenomena: **(a)** atomic (line) spectrum; **(b)** photoelectric effect; **(c)** matter wave; **(d)** Heisenberg uncertainty principle; **(e)** electron spin; **(f)** Pauli exclusion principle; **(g)** Hund's rule; **(h)** orbital diagram; **(i)** electron charge density; **(j)** radial electron density.

3. Explain the important distinctions between each pair of terms: **(a)** frequency and wavelength; **(b)** ultraviolet and infrared light; **(c)** continuous and discontinuous spectra; **(d)** traveling and standing waves; **(e)** quantum number and orbital; **(f)** *spdf* notation and orbital diagram; **(g)** *s* block and *p* block; **(h)** main group and transition element; **(i)** the ground state and excited state of a hydrogen atom.

4. Describe two ways in which the orbitals of multielectron atoms resemble hydrogen orbitals and two ways in which they differ from hydrogen orbitals.

125. Explain the phrase *effective nuclear charge*. How is this related to the shielding effect?

126. With the help of sketches, explain the difference between a p_x, p_y, and p_z orbital.

127. With the help of sketches, explain the difference between a $2p_z$ and $3p_z$ orbital.

128. If traveling at equal speeds, which of the following matter waves has the longest wavelength? Explain. **(a)** electron; **(b)** proton; **(c)** neutron; **(d)** α particle (He^{2+}).

129. For electromagnetic radiation transmitted through a vacuum, state whether each of the following properties is directly proportional to, inversely proportional to, or independent of the frequency: **(a)** velocity; **(b)** wavelength; **(c)** energy per mole. Explain.

130. Construct a concept map representing the ideas of quantum mechanics.

131. Construct a concept map representing the atomic orbitals of hydrogen and their properties.

132. Construct a concept map for the configurations of multielectron atoms.

9 The Periodic Table and Some Atomic Properties

CONTENTS

LEARNING OBJECTIVES

9.1 State the periodic law, and discuss the contributions made by Meyer, Mendeleev, Ramsay, and Mosely in establishing the form of the periodic table.

9.2 Discuss the relative tendencies of metals, nonmetals, and metalloids to acquire or lose electrons in terms of their ground-state electron configurations.

9.3 Identify trends in atomic radii across a period and down a group, and interpret these trends in terms of effective nuclear charge. Predict whether an ion is larger or smaller than the neutral atom from which it is derived.

9.4 Describe and explain the variation of ionization energy across a period and down a group.

9.5 Describe and explain the variation of electron affinity across a period and across a group.

9.6 Distinguish between paramagnetic and diamagnetic atoms and ions.

9.7 Discuss the relationship between polarizability and atomic volume. Describe the variation of polarizability across a period and down a group.

A scanning tunneling microscope image of 48 iron atoms adsorbed onto a surface copper atoms. The iron atoms were moved into position with the tip of the scannin tunneling microscope in order to create a barrier that forced some electrons of the copper atoms into a quantum state seen here as circular rings of electron density. The colors are from the computer rendering of the image. In this chapter we discus the periodic table and the properties of atoms and ions.

The periodic table unifies so much of what is known about the el ments, in terms of their physical and chemical properties, that w continue to use it even though there are aspects of it we don't y fully understand. Our understanding of the rationale underlying the pe odic table took a great leap forward about 50 years after the table was cr ated, in large part because of the insights provided by quantum mechani

The basis of the periodic table is the electron configurations of the el ments, a topic we studied in Chapter 8. In this chapter, we will use th table as a backdrop for a discussion of some properties of element including atomic radii, ionization energies, electron affinities, and polari abilities. These atomic properties also arise in the discussion of chemic bonding in the following two chapters, and the periodic table itself will b our indispensable guide throughout much of the remainder of the text.

1 Classifying the Elements: The Periodic Law and the Periodic Table

869, Dmitri Mendeleev and Lothar Meyer independently proposed the **iodic law**:

Vhen the elements are arranged in order of increasing atomic mass, certain sets f properties recur periodically.

Ieyer based his periodic law on the property called atomic volume—the nic mass of an element divided by the density of its solid form. We now this property *molar volume.*

$$\text{atomic (molar) volume (cm}^3\text{/mol)} = \text{molar mass (g/mol)} \times 1/d\ (\text{cm}^3\text{/g}) \quad \textbf{(9.1)}$$

yer presented his results as a graph of atomic volume against atomic mass. w it is customary to plot his results as molar volume against atomic num- as seen in Figure 9-1. Notice how high atomic volumes recur periodically the alkali metals Li, Na, K, Rb, and Cs. Later, Meyer examined other physi- properties of the elements and their compounds, such as hardness, com- ssibility, and boiling points, and found that these also vary periodically.

ndeleev's Periodic Table

previously described, the periodic table is a tabular arrangement of the ele- nts that groups similar elements together. Mendeleev's work attracted re attention than Meyer's for two reasons: He left blank spaces in his table undiscovered elements, and he corrected some atomic mass values. The nks in his table came at atomic masses 44, 68, 72, and 100 for the elements now know as scandium, gallium, germanium, and technetium. Two of the mic mass values he corrected were those of indium and uranium.

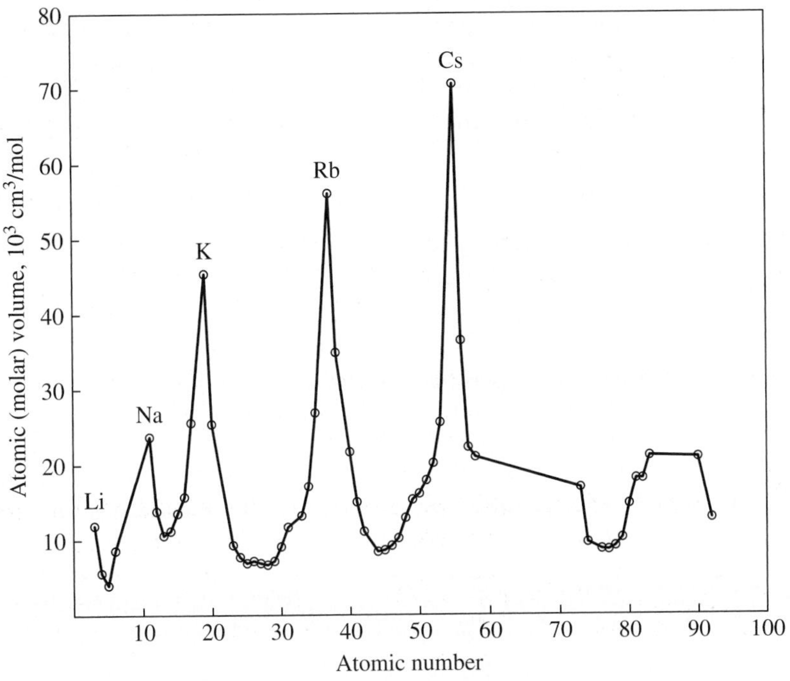

FIGURE 9-1
illustration of the periodic law—variation of atomic volume with omic number

s adaptation of Meyer's graph plots atomic volumes against atomic numbers. The ta are from an English translation of Meyer's 1883 book *Moderen Therien*. The graph ows peaks at the alkali metals (Li, Na, K, Rb, and Cs). Gaseous elements are cluded because the molar volume of a gas is a measure of the volume of empty ace, not the volume of a mole of atoms.

Reihen	Gruppe I. — R^2O	Gruppe II. — RO	Gruppe III. — R^2O^3	Gruppe IV. RH^4 RO^2	Gruppe V. RH^3 R^2O^5	Gruppe VI. RH^2 RO^3	Gruppe VII. RH R^2O^7	Gruppe VIII. — RO^4
1	H = 1							
2	Li = 7	Be = 9,4	B = 11	C = 12	N = 14	O = 16	F = 19	
3	Na = 23	Mg = 24	Al = 27,3	Si = 28	P = 31	S = 32	Cl = 35,5	
4	K = 39	Ca = 40	— = 44	Ti = 48	V = 51	Cr = 52	Mn = 55	Fe = 56, Co = 59, Ni = 59, Cu = 63.
5	(Cu = 63)	Zn = 65	— = 68	— = 72	As = 75	Se = 78	Br = 80	
6	Rb = 85	Sr = 87	?Yt = 88	Zr = 90	Nb = 94	Mo = 96	— = 100	Ru = 104, Rh = 104, Pd = 106, Ag = 108
7	(Ag = 108)	Cd = 112	In = 113	Sn = 118	Sb = 122	Te = 125	J = 127	
8	Cs = 133	Ba = 137	?Di = 138	?Ce = 140	–	–	–	– – – –
9	(–)	–	–	–	–	–	–	
10	–	–	?Er = 178	?La = 180	Ta = 182	W = 184	–	Os = 195, Ir = 197, Pt = 198, Au = 199
11	(Au = 199)	Hg = 200	Tl = 204	Pb = 207	Bi = 208			
12	–	–	–	Th = 231	–	U = 24		

Stamp from the private collection of Professor C. M. Lang. Photography by Gary J. Shulf University of Wisconsin, Stevens Point. "1957, Russia (Scott #1906) and 1969, Russia (#3607)"; Scott Standard Postage Stamp Catalogue, Scott Pub. Co., Sidney, Ohio

▲ Dmitri Mendeleev (1834–1907)
Mendeleev's discovery of the periodic table came from his attempts to systematize properties of the elements for presentation in a chemistry textbook. His highly influential book went through eight editions in his lifetime and five more after his death.

In his periodic table Mendeleev arranged the elements into eight groups (Gruppe) and twelve rows (Reihen). The formulas are written as Mendeleev wrote them. R^2O, RO, and so on, are formulas of the element oxides (such as Li_2O, MgO, ...); RH^4, RH^3, and so forth, are formulas of the element hydrides (such as CH_4, NH_3, ...).

▶ Other properties of the alkali metals are discussed in Section 9-7.

In Mendeleev's table, similar elements fall in vertical groups, and the pro erties of the elements change gradually from top to bottom in the group. As example, we have seen that the alkali metals (Mendeleev's group I) have hi molar volumes (Fig. 9-1). They also have low melting points, which decrea in the order

$$\text{Li } (174\ ^\circ\text{C}) > \text{Na } (97.8\ ^\circ\text{C}) > \text{K } (63.7\ ^\circ\text{C}) > \text{Rb } (38.9\ ^\circ\text{C}) > \text{Cs } (28.5\ ^\circ\text{C})$$

In their compounds, the alkali metals exhibit the oxidation state +1, formi ionic compounds, such as $NaCl$, KBr, CsI, Li_2O, and so on.

Discovery of New Elements

Three elements predicted by Mendeleev were discovered shortly after the appe ance of his 1871 periodic table (gallium, 1875; scandium, 1879; germanium, 188 Table 9.1 illustrates how closely Mendeleev's predictions for eka-silicon ag with the observed properties of the element germanium, discovered in 18 Often, new ideas in science take hold slowly, but the success of Mendelee predictions stimulated chemists to adopt his table fairly quickly.

▶ The term *eka* is derived from Sanskrit and means "first." That is, eka-silicon means, literally, "first comes silicon" (and then comes the unknown element).

TABLE 9.1 Properties of Germanium: Predicted and Observed

Property	Predicted Eka-silicon (1871)	Observed Germanium (1886)
Atomic mass	72	72.6
Density, g/cm^3	5.5	5.47
Color	dirty gray	grayish white
Density of oxide, g/cm^3	EsO_2: 4.7	GeO_2: 4.703
Boiling point of chloride	$EsCl_4$: below 100 °C	$GeCl_4$: 86 °C
Density of chloride, g/cm^3	$EsCl_4$: 1.9	$GeCl_4$: 1.887

ne group of elements that Mendeleev did not anticipate was the noble es. He left no blanks for them. William Ramsay, their discoverer, proposed ing them in a separate group of the table. Because argon, the first noble discovered (1894), had an atomic mass greater than that of chlorine and parable to that of potassium, Ramsay placed the new group, which alled group 0, between the halogen elements (group VII) and the alkali als (group I).

mic Number as the Basis for the Periodic Law

ndeleev placed certain elements out of the order of increasing atomic s to get them into the proper groups of his periodic table. He assumed was because of errors in atomic masses. With improved methods of ermining atomic masses and with the discovery of argon (group 0, atomic s 39.9), which was placed ahead of potassium (group I, atomic mass), it became clear that a few elements might always remain "out of er." At the time, these out-of-order placements were justified by chemical lence. Elements were placed in the groups that their chemical behavior ated. There was no theoretical explanation for this reordering. Matters nged in 1913 as a result of some research by Henry G. J. Moseley on the ay spectra of the elements.

Bettmann/Corbis

▲ **Henry G. J. Moseley (1887–1915)**
Moseley was one of a group of brilliant scientists whose careers were launched under Ernest Rutherford. Moseley was tragically killed at Gallipoli, in Turkey, during World War I.

s we learned in Chapter 2, X-rays are a high-frequency form of electro-gnetic radiation produced when a cathode-ray (electron) beam strikes the de of a cathode-ray tube (see Figure 9-2a). The anode is called the *target*. ay emission can be explained in the following way. If the bombarding ctrons have sufficient energy, they can eject electrons from the inner itals of target metal atoms. Electrons from higher orbitals then drop vn to fill the vacancies, emitting X-ray photons with energies correspond- to the difference in energy between the originating level and the vacancy el. Moseley reasoned that because the energy required to eject an electron ends on the nuclear charge, the frequencies of emitted X-rays should end on the nuclear charges of atoms in the target. Using techniques vly developed by the father–son team of W. Henry Bragg and Lawrence Bragg, Moseley obtained photographic images of X-ray spectra l assigned frequencies to the spectral lines. His spectra for the elements m Ca to Zn are reproduced in Figure 9-2(b).

Moseley was able to correlate X-ray frequencies to numbers equal to the clear charges and corresponding to the positions of elements in ndeleev's periodic table. For example, aluminum, the thirteenth element in table, was assigned an *atomic number* of 13. Moseley's equation is $= A(Z - b)^2$, where ν is the X-ray frequency, Z is the atomic number, and nd b are constants. Moseley used this relationship to predict three new ele-nts ($Z = 43, 61$, and 75), which were discovered in 1937, 1945, and 1925, pectively. Also, he proved that in the portion of the periodic table with ich he worked (from $Z = 13$ to $Z = 79$), there could be no additional new ments beyond those three. All available atomic numbers had been igned. From the standpoint of Moseley's work, then, we should restate the iodic law.

Similar properties recur periodically when elements are arranged according to increasing atomic number.

scription of a Modern Periodic Table: The Long Form

ndeleev's periodic table consisted of 8 groups, but most modern periodic les are arranged in 18 groups of elements. Let us briefly review the descrip-n of the periodic table given in Section 2-6.

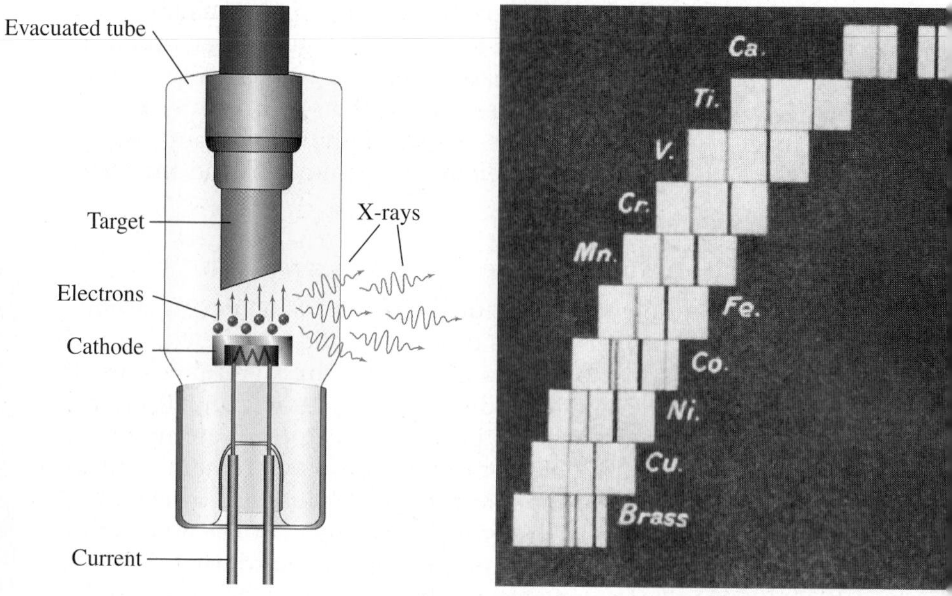

(a) (b)

▲ FIGURE 9-2
Schematic of an X-ray tube and Moseley's X-ray spectra of several elemen
(a) A heated filament emits electrons by a process called thermionic emission. The electrons are accelerated by a high voltage and collide with the metal target. The hig energetic electrons ionize electrons from the inner shells of the metal atoms of the tar Subsequently, electrons from higher orbitals drop down to occupy the vacancies and doing so, emit X-ray photons that correspond to the energy difference between the tw orbitals. **(b)** In this photograph from Moseley's 1913 paper, you can see two lines for e element, beginning with Ca at the top. With each successive element, the lines are displaced to the left, the direction of increasing X-ray frequency in these experiments. Where more than two lines appear, the sample contained one or more other elements or impurities. Notice, for example, that one line in the Co spectrum matches a line in Fe spectrum, and another matches a line in the Ni spectrum. Brass, which is an alloy o copper and zinc, shows two lines for Cu and two for Zn.

In the periodic table (see inside front cover), the vertical groups br together elements with similar properties. The horizontal periods of the ta are arranged in order of increasing atomic number from left to right. T groups are numbered at the top, and the periods at the extreme left. The f two groups—the *s* block—and the last six groups—the *p* block—together c stitute the *main-group elements*. Because they come between the *s* block and *p* block, the *d* block elements are known as the *transition elements*. The *f* bl elements, sometimes called the *inner transition elements*, would extend table to a width of 32 members if incorporated in the main body of the tab The table would generally be too wide to fit on a printed page, and so *f* block elements are extracted from the table and placed at the botto The 15 elements following barium ($Z = 56$) are called the *lanthanides*, and 15 following radon ($Z = 88$) are called the *actinides*.

9-2 Metals and Nonmetals and Their Ions

In Section 2-6, we established two categories of elements, *metals* and *nonmet* and described some of their physical properties. Most metals are good co ductors of heat and electricity, are malleable and ductile, and have moder ıigh melting points. In general, nonmetals are nonconductors of heat and :tricity and are nonmalleable (brittle) solids, though a number of non- :als are gases at room temperature.

'hrough the color scheme of the periodic table on the inside front cover, see that the majority of the elements are metals (tan) and that nonmetals ıe) are confined to the right side of the table. The noble gases (pink) are ıted as a special group of nonmetals. Metals and nonmetals are often sep- ted by a stairstep diagonal line, and several elements near this line are ːn called metalloids (green). **Metalloids** are elements that look like metals l in some ways behave like metals but also have some nonmetallic proper- .

◀ Metalloids (such as silicon) are semiconductors and materials composed of metalloids play an important role in microcomputer technology.

n Mendeleev's version of the periodic table, the positions of the elements re based on readily observable physical and chemical properties. The ılting arrangement placed elements with similar properties in the same tical group. What causes the elements in the same group to have similar perties? In Chapter 8, we learned that the *atoms of elements in the same group e similar electron configurations*. Thus, it may be that similarities in the phys- l and chemical properties of the elements arise from similarities in the :tron configurations of their constituent atoms.

.et's now briefly explore a few of the links between the electron configura- ıs of atoms and some observations about the elements, starting with the ıle gases.

KEEP IN MIND

that an atom can have, in principle, one of many different electron configurations. In most contexts, the electron configuration of an atom usually refers to the configuration of lowest energy, that is, the ground-state electron configuration. For temperatures typically encountered, most atoms are usually in their ground electronic states.

ıble Gases

heir lowest energy state, atoms of the noble gases have the maximum num- of electrons permitted in the valence shell of an atom: two in helium ($1s^2$) l eight in the other noble gas atoms (ns^2np^6). These electron configurations very difficult to alter and seem to confer a high degree of chemical inert- ;s to the noble gases. It is interesting to note, then, that the *s*-block metals, ether with Al in group 13, tend to lose enough electrons to acquire the elec- n configurations of the noble gases. Conversely, nonmetals tend to gain ıugh electrons to achieve the same configurations.

◀ Compounds containing radon, xenon, and krypton have been prepared. Recently, compounds containing argon have also been prepared.

ain-Group Metal Ions

e atoms of elements of groups 1 and 2—the most active metals—have elec- n configurations that differ from those of the noble gas of the preceding iod by only one and two electrons in the *s* orbital of a new electron shell. If atom is stripped of its outer-shell electron, it becomes the *positive ion* K^+ h the electron configuration [Ar]. A Ca atom acquires the [Ar] configura- n following the removal of two electrons.

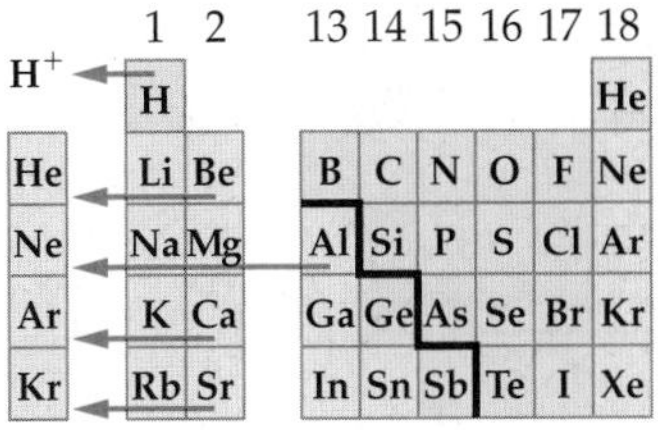

▲ Metals tend to lose electrons to attain noble gas electron configurations.

$$K([Ar]4s^1) \longrightarrow K^+([Ar]) + e^-$$
$$Ca([Ar]4s^2) \longrightarrow Ca^{2+}([Ar]) + 2e^-$$

e energy required to bring about ionization is often provided by other ıcesses occurring at the same time (such as an attraction between positive and ;ative ions). Aluminum is the only *p*-block metal that forms an ion with a ble gas electron configuration—Al^{3+}. This is because all other *p*-block ele- nts would have to remove 10 *d* electrons to attain the electron configuration the previous noble gas. The electron configurations of the other *p*-block tal ions are summarized in Table 9.2.

ain-Group Nonmetal Ions

e atoms of groups 17 and 16—the most active nonmetals—have one and o electrons fewer than the noble gas at the end of the period. Groups 17 and

TABLE 9.2 Electron Configurations of Some Metal Ions[a]

"Noble Gas"		"Pseudo-Noble Gas"[b]	"18 + 2"[c]	Other
Li^+	Be^{2+}	Ga^{3+}	In^+	Cr^{2+}, Cr^{3+}
Na^+	Mg^{2+}	Tl^{3+}	Tl^+	Mn^{2+}, Fe^{2+}
K^+	Ca^{2+}	Cu^+	Sn^{2+}	Fe^{3+}, Co^{2+}
Rb^+	Sr^{2+}	Ag^+, Au^+	Pb^{2+}	Ni^{2+}, Cu^{2+}
Cs^+	Ba^{2+}	Zn^{2+}	Sb^{3+}	
Fr^+	Ra^{2+}		Bi^{3+}	
Al^{3+}				

[a]Main-group metal ions are printed in black and transition metal ions in blue.
[b]In the configuration labeled "pseudo-noble gas," all electrons of the outermost shell have been lost. The next-to-outermost electron shell of the atom becomes the outermost shell of the ion and contains 18 electrons, for example, Ga^{3+}: $[Ne]3s^23p^63d^{10}$.
[c]In the configuration labeled "18 + 2" all outer-shell electrons except the two *s* electrons ar lost, producing an ion with 18 electrons in the next-to-outermost shell and 2 electrons in th outermost, for example, Sn^{2+}: $[Ar]3d^{10}4s^24p^64d^{10}5s^2$.

16 atoms can acquire the electron configurations of noble gas atoms by *gain* the appropriate numbers of electrons.

$$Cl\ ([Ne]3s^23p^5) + e^- \longrightarrow Cl^-\ ([Ar])$$
$$S\ ([Ne]3s^23p^4) + 2\,e^- \longrightarrow S^{2-}\ ([Ar])$$

In most cases, the energy of a nonmetal atom is lowered by the addition an electron. However, energy is always required to add more than one el tron. The necessary energy is often supplied by other processes that oc simultaneously (such as an attraction between positive and negative ior Nonmetal ions with a charge of −3 are rare. However, some metal nitric containing the nitride ion, N^{3-}, and some metal phosphides containing phosphide ion, P^{3-}, are known.

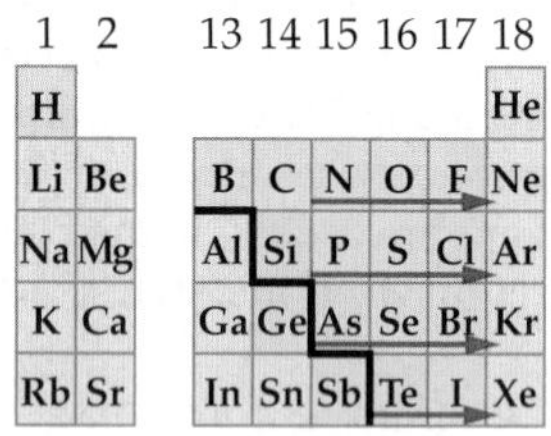

1 2 13 14 15 16 17 18
H He
Li Be B C N O F Ne
Na Mg Al Si P S Cl Ar
K Ca Ga Ge As Se Br Kr
Rb Sr In Sn Sb Te I Xe

▲ Nonmetals tend to gain electrons to attain noble-gas electron configurations.

Transition Metal Ions

In Section 8-11, we established that most transition metal atoms, in th ground electronic states, have two electrons in the *ns* orbital and one or m electrons in the $(n - 1)\,d$ orbitals. For example, the ground-state electron co figuration of Ti is $[Ar]3d^24s^2$ and that of Mn is $[Ar]3d^54s^2$. Experiments ha established that when these atoms ionize, the 4*s* electrons are removed fi not the 3*d* electrons.

When a transition metal atom ionizes, electrons from the *ns* orbital are removed first.

▶ A useful mnemonic is that the electron configuration of a cation can be obtained from the electron configuration of the parent atom by removing those electrons in orbitals with the *highest* quantum number first.

Thus, without exception, transition metal ions with charges of +2 or higl have all their valence electrons in the $(n - 1)\,d$ orbitals. For example, t ground-state electron configurations of Ti^{2+} and Mn^{2+} are, respectively, [Ar] and $[Ar]3d^5$. In the formation of transition metal ions with charges of +3 higher, one or more $(n - 1)\,d$ electrons might be lost together with the *ns* el trons. For example, the ground-state configurations of Ti^{4+} and Mn^{3+} are [and $[Ar]3d^4$, respectively.

A few transition metal atoms acquire noble-gas electron configurations wh forming cations, as do Sc in Sc^{3+} and Ti in Ti^{4+}, but most transition me ns do not (see Table 9.2). An iron atom does not acquire a noble-gas elec- configuration when it loses its $4s^2$ electrons to form the ion Fe^{2+},

$$\text{Fe}\,([\text{Ar}]3d^6 4s^2) \longrightarrow \text{Fe}^{2+}\,([\text{Ar}]3d^6) + 2\,\text{e}^-$$

does it with the loss of an additional $3d$ electron to form the ion Fe^{3+}.

$$\text{Fe}\,([\text{Ar}]3d^6 4s^2) \longrightarrow \text{Fe}^{3+}\,([\text{Ar}]3d^5) + 3\,\text{e}^-$$

drogen

ough all the other elements have a definite place in the periodic table, rogen does not. Because the ground-state electron configuration of the tom ($1s^1$) resembles that of the group 1 metals (ns^1), hydrogen is often ed in group 1, even though we classify it as a nonmetal. Hydrogen does ear to become metallic when subjected to pressures of about 2 million bar, these are hardly ordinary laboratory conditions. Because hydrogen, like the gens, is one electron short of having a noble-gas electron configuration, it is etimes placed in group 17; however, hydrogen does not resemble the halo- s very much. For example, F_2 and Cl_2 are excellent oxidizing agents, but H_2 reducing agent. Still another alternative places hydrogen by itself at the top he periodic table and near the center.

3 Sizes of Atoms and Ions

arlier chapters, we discovered the importance of atomic masses in matters ting to stoichiometry. To understand certain physical and chemical pro- ties, we need to know something about atomic sizes. In this section we cribe atomic radius, the first of a group of atomic properties that we will nine in this chapter.

omic Radius

ortunately, atomic radius is hard to define. We have seen that atomic tals extend, in principle, to infinity. Although the probability of finding an tron decreases with increasing distance from the nucleus, there is always zero probability of finding an electron at very large distances from the leus. Thus, an atom has no precise outer boundary. We might describe an *tive* atomic radius as, say, the distance from the nucleus within which 95% ne electron charge density is found, but this distance cannot be measured erimentally.

rom an experimental standpoint, we cannot make a measurement of the ius of a single, isolated atom. We can, however, obtain a measure of the (radius) of an atom when it is combined with other atoms. For this rea- , we define atomic radius in terms of internuclear distance.

Ve will emphasize an atomic radius based on the distance between the lei of two atoms joined by a chemical bond. The **covalent radius** is one-half distance between the nuclei of two identical atoms joined by a single cova- bond. The **ionic radius** is based on the distance between the nuclei of ions ed by an ionic bond. Because the ions are not identical in size, this distance st be properly apportioned between the cation and anion. One way to ortion the electron density between the ions is to define the radius of one and then infer the radius of the other ion. The convention we have chosen se is to assign O^{2-} an ionic radius of 140 pm. An alternative apportioning eme is to use F^- as the reference ionic radius. When using ionic radii data, fully note which convention is used and do not mix radii from the different ventions. Starting with a radius of 140 pm for O^{2-}, the radius of Mg^{2+} can obtained from the internuclear distance in MgO, the radius of Cl^- from the rnuclear distance in $MgCl_2$, and the radius of Na^+ from the internuclear

▶ Despite the SI convention, the angstrom unit, Å, is still widely used by X-ray crystallographers and others who work with atomic and molecular dimensions.

distance in NaCl. For metals, we define a **metallic radius** as one-half the tance between the nuclei of two atoms in contact in the crystalline solid m Similarly, in a solid sample of a noble gas the distance between the center neighboring atoms is called the **van der Waals radius**. There is much del about the values of the atomic radii of noble gases because the experime determination of the van der Waals radii is difficult; consequently, the ato radii of noble gases are left out of the discussion of trends in atomic radii.

The angstrom unit, Å, has long been used for atomic dimensi (1 Å = 10^{-10} m). The angstrom, however, is not a recognized SI unit. Th units are the nanometer (nm) and picometer (pm).

$$1\,\text{nm} = 1 \times 10^{-9}\,\text{m};\; 1\,\text{pm} = 1 \times 10^{-12}\,\text{m};\; 1\,\text{nm} = 1000\,\text{pm} \qquad (9$$

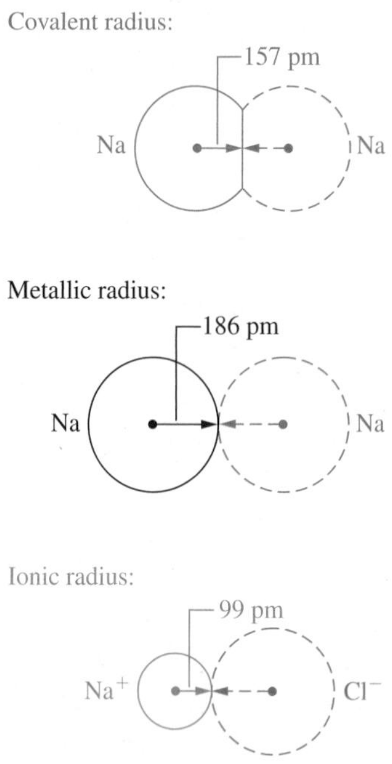

▲ FIGURE 9-3
Covalent, metallic, and ionic radii compared
Atomic radii are represented by the solid arrows. The covalent radius is based on the diatomic molecule $Na_2(g)$, found only in gaseous sodium. The metallic radius is based on adjacent atoms in solid sodium, Na(s). The value of the ionic radius of Na^+ is obtained by the comparative method described in the text.

Figure 9-3 illustrates the definitions of covalent, ionic, and metallic radi comparing these three radii for sodium. Figure 9-4 is a plot of atomic rac against atomic number for a large number of elements. In this plot meta radii are used for metals and covalent radii for nonmetals. Figure 9-4 sugg certain trends in atomic radii, for example, large radii for group 1, decreas across the periods to smaller radii for group 17. A careful examinatio Figure 9-4 reveals the following:

> Atomic radius decreases from left to right through a period of elements and increases from top to bottom through a group.

Before we interpret these trends, let us return to a topic introducec Chapter 8.

Screening and Penetration

In Chapter 8, we compared radial distribution functions for various orbi (Fig. 8-36) and established that *s* electrons penetrate better than *p* electrc which in turn penetrate better than *d* electrons. We can reinforce this idea, extend it further, by considering the radial distribution functions plottec

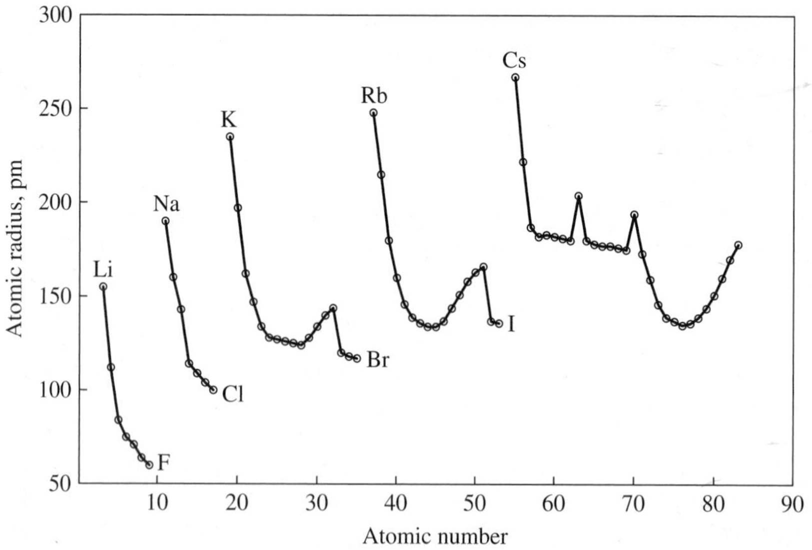

▲ FIGURE 9-4
Atomic radii
The values plotted are metallic radii for metals and covalent radii for nonmetals. Dat for the noble gases are not included because of the difficulty of measuring covalent radii for these elements (only Kr and Xe compounds are known). The explanations usually given for the several small peaks in the middle of some periods are beyond the scope of this discussion.

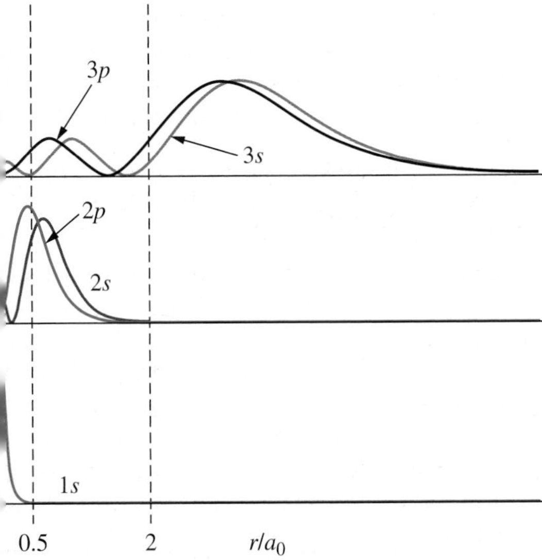

◀ FIGURE 9-5
Radial distribution functions for aluminum
Graphs of $r^2R(r)^2$ as a function of r for the 1s, 2s, 2p, 3s, and 3p orbitals of aluminum. These graphs were obtained by using the radial functions from Table 8.2 with $Z_{eff} = 12.591$ (for 1s), $Z_{eff} = 8.214$ (for 2s), $Z_{eff} = 8.963$ (for 2p), $Z_{eff} = 4.117$ (for 3s), and $Z_{eff} = 4.066$ (for 3p). The functions for the $n = 2$ and $n = 3$ orbitals were multiplied by factors of three and ten, respectively, and shifted vertically by different amounts to ensure that all the graphs are equally visible. The dashed vertical lines indicate distances within which electrons in the $n = 1$ and $n = 2$ shells are typically found: $0.5a_0 \approx 26$ pm for $n = 1$ and $2a_0 \approx 106$ pm for $n = 2$.

re 9-5 for several orbitals of aluminum. Keep in mind the physical mean- of these plots: The area under the curve between any two distances, say, r_1 r_2, represents the probability of finding the electron between r_1 and r_2. ce that there is essentially 100% probability of finding the $n = 1$ electrons in 26 pm of the nucleus and the $n = 2$ electrons within 106 pm of the eus. On the other hand, the probability is high (greater than 85%) that the 3 electrons will be found at distances greater than 106 pm. Therefore, we say the inner (core) electrons of the aluminum atom (electrons having er n values) *screen* or *shield* the outer (valence) electrons from experiencing full nuclear charge, as suggested by Figure 9-6.

he effect of screening is to reduce the magnitude of the nuclear charge felt given electron. Thus, let us define the **effective nuclear charge, Z_{eff}**, felt given electron as

$$Z_{eff} = Z - S \quad (9.3)$$

n this expression, Z is the actual nuclear charge and S is the amount of ge that is screened out by all the other electrons. Think of S as the average nber of electrons between the nucleus and the electron of interest. trons in different orbitals can be assigned different values of Z_{eff}, the value ending on the extent to which a given electron is screened or shielded by r electrons. (These ideas are explored in more detail in Exercise 69.)

o gain an appreciation of the possible magnitude of S, consider again the alu- um atom, for which $Z = 13$. The ground-state electron configuration of Al is

$$\text{Al: } 1s^22s^22p^63s^23p^1$$

e assume that the inner core electrons ($1s^22s^22p^6$) screen the nuclear charge ectly and that the outer 3s and 3p electrons do not screen each other, then 3s and 3p electrons would each experience a nuclear charge of - 10 = +3. However, neither assumption—full screening by inner elec- s and no screening by outer electrons—is correct. These assumptions re the fact that the electrons, both inner and outer, occupy orbitals with dif- nt radial distributions and different degrees of penetration.

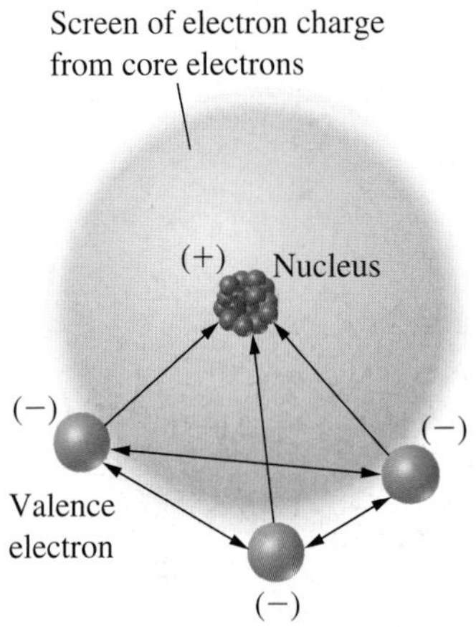

▲ FIGURE 9-6
The shielding effect and effective nuclear charge, Z_{eff}
Three valence electrons (blue) are attracted to the nucleus of a Al atom. The atom's +13 nuclear charge is screened by the 10 core electrons (gray), but not perfectly. The valence electrons also screen each other somewhat. The result is an *effective* nuclear charge, Z_{eff}, closer to +4 than to +3.

9-1 CONCEPT ASSESSMENT

stimate Z_{eff} for a 3p electron in Si by assuming the inner (core) electrons screen uter (valence) electrons perfectly and the outer electrons do not screen each her.

Realistic values of Z_{eff} can be obtained from an analysis of the wave f tions of multielectron atoms, as described in Are You Wondering 9-1. Valu Z_{eff} for the valence electrons for the first 36 elements are shown in Figure The following points can be established by careful examination of these val

- **For all atoms except H, Z_{eff} for a valence electron is significantly than Z, the actual nuclear charge.** The effective nuclear charge is than the actual nuclear charge because of screening.
- **Valence electrons are screened significantly by inner (core) elect but only slightly by each other.** The screening of valence electron core electrons is best studied by focusing on Z_{eff} for the valence elec of an alkali metal atom. For example, Z_{eff} for the 3*s* electron in a Na a is about +2.5, which is significantly less than the actual nuclear charg +11. Thus, for the valence electron in a Na atom, about 8.5 units of ch are canceled out by the core electrons. Typically, core electrons are 80 100% effective in screening the valence electrons. The mutual scree of valence electrons is best studied by focusing on the variation of Z a given electron across a period. Consider, for example, the third- atoms Na through Ar. With the assumption that the valence electron not screen each other, Z_{eff} for a 3*s* electron would increase by one uni each unit increase in Z as we move across the period: +2.5 for Na, +3. Mg, +4.5 for Al, and so on. However, we observe that Z_{eff} increase something less than one unit each time (for example, by about +0.8 u from Na to Mg and from Mg to Al). Roughly speaking, valence elect are no more than one-third (33%) effective in screening each other.
- **For electrons in the same shell, Z_{eff} decreases as ℓ increases.** Cons Z_{eff} for the 3*s* and 3*p* electrons of the third-row atoms Al through

	1	2	13	14	15	16	17	1
	1 H							2 H
1*s*	1.000							1.6
	3 Li	4 Be	5 B	6 C	7 N	8 O	9 F	1 N
2*s*	1.279	1.912	2.576	3.217	3.847	4.492	5.128	5.7
2*p*			2.421	3.136	3.834	4.453	5.100	5.7
	11 Na	12 Mg	13 Al	14 Si	15 P	16 S	17 Cl	1 A
3*s*	2.507	3.308	4.117	4.903	5.642	6.637	7.068	7.7
3*p*			4.066	4.285	4.886	5.482	6.116	6.7
	19 K	20 Ca	31 Ga	32 Ge	33 As	34 Se	35 Br	3 K
4*s*	3.495	4.398	7.067	8.044	8.944	9.758	10.553	11.
4*p*			6.222	6.780	7.449	8.287	9.028	9.7

	21 Sc	22 Ti	23 V	24 Cr	25 Mn	26 Fe	27 Co	28 Ni	29 Cu	3 Z
4*s*	4.632	4.817	4.981	5.133	5.283	5.434	5.576	5.711	5.858	5.9
3*d*	7.120	8.141	8.983	9.757	10.528	11.180	11.855	12.350	13.201	13.8

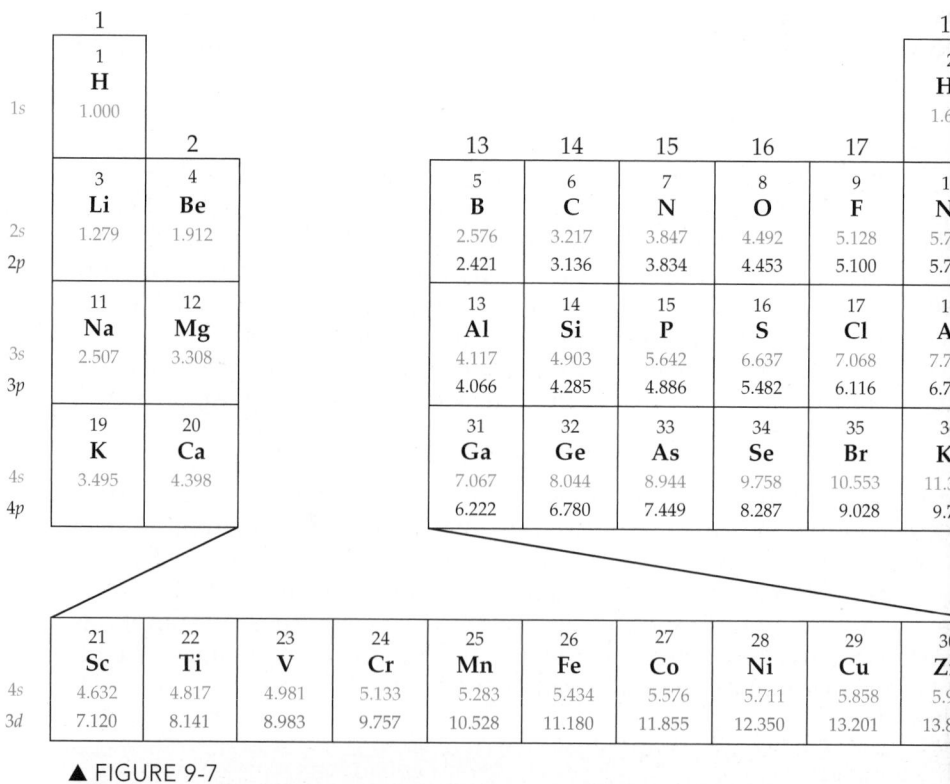

▲ FIGURE 9-7
Effective nuclear charges, Z_{eff}, of valence electrons
Values in blue are for *ns* electrons; values in black are for *np* electrons; and values ir red are for 3*d* electrons. The values are from *Inorganic Chemistry, Principles of Structure and Reactivity*, Fourth Edition, J. E. Huheey, E. A. Keiter, R. L. Keiter, Harp Collins, New York (1993).

Because a 3*s* electron penetrates to the nucleus better than a 3*p* electron, Z_{eff} of a 3*s* electron is greater than that of a 3*p* electron.

Z_{eff} increases from left to right across a period and from top to bottom down a group. Within a period, the increase in Z_{eff} is comparable to the increase in Z itself because the valence electrons are not very effective at screening each other. However, within a group, the increase in Z_{eff} for a valence electron is much smaller than the increase in Z because core electrons are very effective at screening valence electrons.

◀ To a reasonable approximation, Z_{eff} increases linearly with Z across a period and linearly with *n* down a group.

lthough we can assign a Z_{eff} value to each electron in an atom, we can simy our discussion considerably by focusing on Z_{eff} for the least strongly nd electron in an atom. The least strongly bound electron is, on average, farthest from the nucleus and the most easily removed. Trends in atomic ii and ionization energies can, for the most part, be explained by focusing Z_{eff} for the least strongly bound electron. (We will define and discuss ionion energies in Section 9-4.)

he variation of Z_{eff} across a period and down a group for the least strongly nd electron in an atom is illustrated in Figure 9-8(a). Generally speaking, increases from left to right across a period and from top to bottom down a up. We observe that the slope of the line joining atoms in the same period pe = 0.64) is greater than that of the line joining atoms in the same group pe = 0.086 or 0.15). In other words, Z_{eff} increases much more significantly oss a period than it does down a group. For atoms of the second period, for mple, Z_{eff} increases by about 0.64 units for every unit increase in Z whereas atoms of groups 1 or 18, the increase in Z_{eff} is only 0.086 or 0.15 units of rge for each unit increase in Z. The observation that Z_{eff} increases more sigcantly with Z across a period than down a group justifies our earlier asser1 that the mutual screening of electrons within a shell is not as significant or ctive as the screening of valence electrons by core electrons.

n Figure 9-8(b), we focus instead on the percentage of the nuclear charge eened out by other electrons.

$$\text{percent screening} = \frac{S}{Z} \times 100 \qquad (9.4)$$

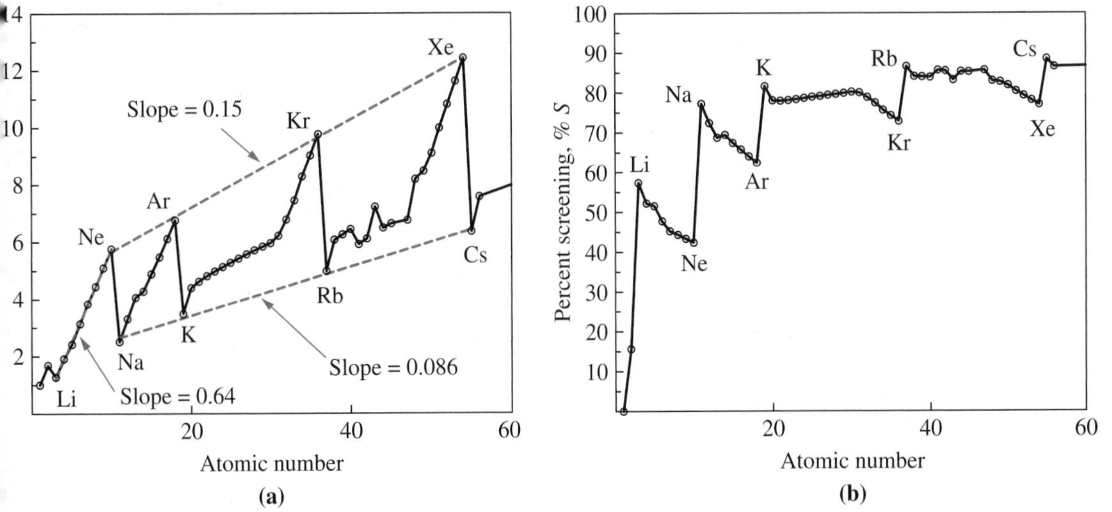

IGURE 9-8
iation of effective nuclear charge and percent screening
The effective nuclear charge for the least strongly bound electron in an atom eases significantly from left to right across a period but much less significantly down oup. **(b)** The percentage of the nuclear charge that is screened out by other ctrons decreases from left to right across a period and increases down a group. ctrons in the same shell (same value of *n*) cancel out a much smaller fraction of the lear charge than do electrons in inner shells (smaller values of *n*).

At the start of a period, the percent screening is relatively large beca according to the aufbau procedure, the last electron added goes into a hig shell (an orbital with a higher n value). This electron is significantly scree by electrons in inner shells. For example, at the start of each period, about (to 85% of the nuclear charge felt by the ns electron is screened out by electr in inner shells.

Figure 9-8(b) also shows that for atoms of the main group (the s an blocks), percent screening decreases across a period and is approximately c stant across the d block.

The Effects of Penetration and Screening

The following expression allows us to estimate, for an electron in a gi orbital, the average value of the distance of the electron from the nucleus:

▶ For ns and np electrons, the factor in curly brackets has a value between 1.25 and 1.5, and may be considered approximately constant.

$$\bar{r}_{n\ell} = \frac{n^2 a_0}{Z_{\text{eff}}}\left\{1 + \frac{1}{2}\left[1 - \frac{\ell(\ell + 1)}{n^2}\right]\right\} \quad (\text{}$$

In the expression above, Z_{eff} is the effective nuclear charge for the electror interest and a_0 is the Bohr radius. For each atom, we expect the least stron bound electron to be, on average, farthest from the nucleus. Let's use Z_{eff} that electron in equation (9.5) to obtain an estimate of the radius of each at The results are shown in Figure 9-9. Notice that the variation of $\bar{r}_{n\ell}$ with very similar to that of atomic radius (Fig. 9-4), and thus it seems reasonabl use equation (9.5) to rationalize the variation of atomic radius across a per and down a group.

1. **Variation of atomic radii within a period of the periodic table.** Equat (9.5) suggests that the radius of an atom is approximately proportiona n^2/Z_{eff}. For the main group elements, as we move from left to right acro period, n^2 is a constant and, as shown in Figure 9-8(a), Z_{eff} increases rat significantly. Thus, the decrease in radius across the period is attribute

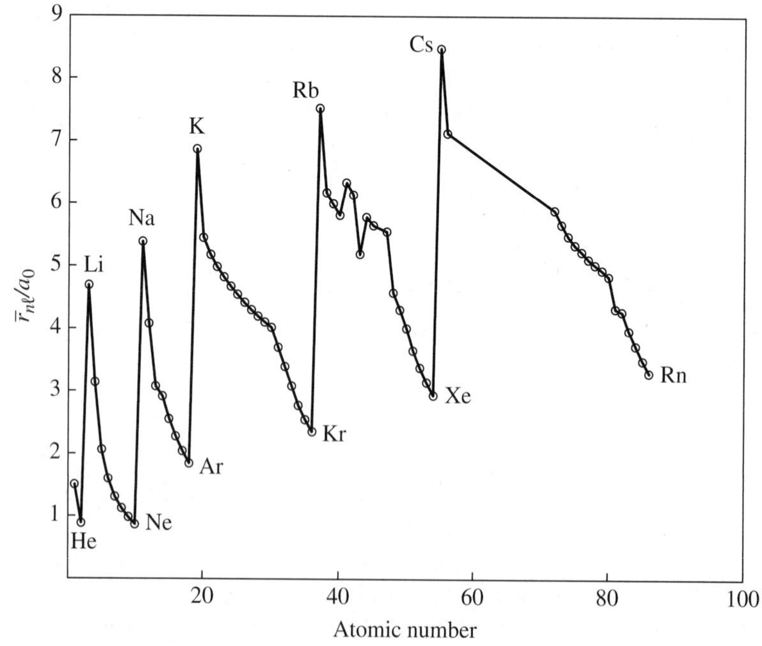

▲ FIGURE 9-9
The average distance from the nucleus for the least strongly bound electr
The average distance, $\bar{r}_{n\ell}$, for the least strongly bound electron of decreases from le to right across a period and increases down a group.

the increase in the effective nuclear charge felt by the least strongly bound electron.

Variation of atomic radii within a group of the periodic table. We have already established that electrons in the outermost shell of an atom are significantly screened by those in inner shells. Thus, for atoms closer to the bottom of the group, the outer electrons occupy orbitals that extend over much larger distances, and we expect the radius of an atom to increase from top to bottom within a group.

It is not immediately obvious that equation (9.5) correctly predicts that atomic radii increase from top to bottom in a group. As we move from top to bottom within a group, the values of both n^2 and Z_{eff} increase. However, the value of n^2 increases more significantly than Z_{eff} as we proceed from top to bottom in a group. For example, in group 17, the value of n^2 increases by a factor of 6.25 as we move down the group (from $2^2 = 4$ to $5^2 = 25$) but the value of Z_{eff} for the np electrons increases by a factor of only 1.77 (from 5.100 to 9.028). Because the proportional increase in the value of n^2 is much greater than that of Z_{eff}, the value of n^2/Z_{eff} increases and so too does the value of $\bar{r}_{n\ell}$.

Variation in atomic radius within a transition series. With the transition elements, the situation is a little different from that described above. In Figure 9-4, it is apparent that the atomic radii of transition elements tend to be about the same across a period but with a few unusual peaks. It is beyond the scope of this text to explain the exceptions; however, the general trend is not difficult to understand. In a series of transition elements, additional electrons go into an *inner* electron shell, where they participate in shielding outer-shell electrons from the nucleus. At the same time, the number of electrons in the *outer* shell tends to remain constant. Thus, the outer-shell electrons experience a roughly comparable force of attraction to the nucleus throughout a transition series. Consider Fe, Co, and Ni. Fe has 26 protons in the nucleus and 24 inner-shell electrons. In Co ($Z = 27$), there are 25 inner-shell electrons, and in Ni ($Z = 28$), there are 26. In each case, the two outer-shell electrons are under the influence of about the same net charge (about +2). That is, Z_{eff} for the $4s$ electrons of the first transition series is approximately constant. Thus, atomic radii do not change very much for this series of three elements, namely, 124 pm for Fe and 125 pm for Co and Ni.

KEEP IN MIND

that the *inner* shell referred to here is the $3d$ shell.

9-1 ARE YOU WONDERING?

Where do estimates of the screening by electrons come from?

nese estimates come from an analysis of the wave functions of multielectron atoms. n exact solution of the Schrödinger equation can be obtained for the H atom, but r multielectron atoms, only approximate solutions are possible. The principle of e calculation is to assume each electron in the atom occupies an orbital much like ose of the hydrogen atom. However, the functional form of the orbital is based on nother assumption: that the electron moves in an effective or average field dictated y all the other electrons. With this assumption, the complicated multielectron hrödinger equation is converted into a set of simultaneous equations—one for ch electron. Each equation contains the unknown effective field and the unknown nctional form of the orbital for the electron. The approach to solving such a set of juations is to guess at the functional forms of the orbitals, calculate an average

(continued)

potential for each electron to move in, and then solve for a new set of orbitals—o
for each electron. The expectation is that the new orbitals are better than the initi
guess. The new orbitals are then used to calculate a new effective field for th
electrons, and the whole process is repeated until the calculated orbitals do n
change much.

This iterative procedure, called the *self-consistent field* (*SCF*) *method*, wa
devised by Douglas Hartree in 1936, before the advent of computers. Current
the wave functions of atoms and molecules are obtained by implementing th
SCF procedures on computers. The use of computers to calculate molecula
properties from a wave function by the SCF procedure has lead to the term *mol*
cular modeling. Molecular modeling has become a tool in modern chemic
research.

The atomic orbitals obtained from SCF calculations closely resemble th
atomic orbitals of the hydrogen atom in many ways. The angular dependence
the orbitals is identical, so that we can identify s, p, d, f orbitals by their characte
istic shapes. The radial functions of the orbitals are different because the effecti
field is different from the one in the hydrogen atom, but the principal quantu
number can still be defined. Thus, each electron in a multielectron atom has ass
ciated with it the four quantum numbers n, ℓ, m_ℓ, and m_s. Estimates of screenin
constants are based on an analysis of the radial functions obtained from SCF ca
culations.

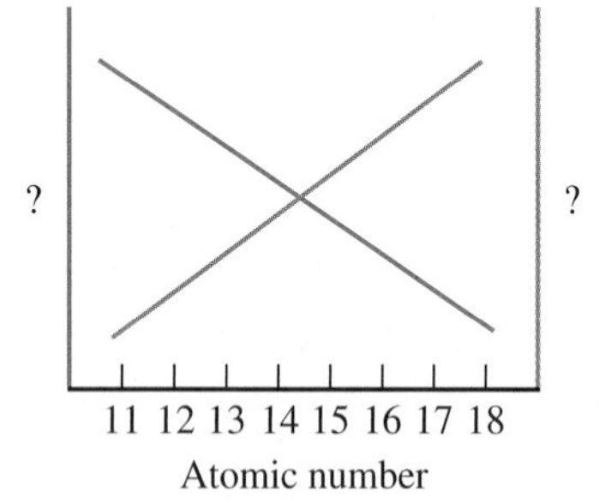

9-2 CONCEPT ASSESSMENT

The graph in the margin represents the variation of Z_{eff} and atomic radius with
atomic number. Which axis and correspondingly colored line corresponds to Z_e
and which to atomic radius?

EXAMPLE 9-1 Relating Atomic Size to Position in the Periodic Table

Refer only to the periodic table on the inside front cover, and determine which is the largest atom: Sc, Ba, or Se

Analyze

We first find the element in the periodic table and decide whether or not the elements are in the same perio
and whether they are on the right or left of the periodic table. We can then use the rules noted above to decid
on the relative sizes of atoms (or ions).

Solve

Sc and Se are both in the fourth period, and we would expect Sc to be larger than Se because atomic size
decrease from left to right in a period. Ba is in the sixth period and so has more electronic shells than eithe
Sc or Se. Furthermore, it lies even closer to the left side of the table (group 2) than does Sc (group 3). We ca
say with confidence that the Ba atom should be the largest of the three.

Assess

The positions of the atoms in the periodic table help us determine their relative sizes. In this case we hav
shown that $r_{Ba} > r_{Sc} > r_{Se}$. The actual atomic radii are Se, 117 pm; Sc, 161 pm; and Ba, 217 pm.

PRACTICE EXAMPLE A: Use the periodic table on the inside front cover to predict which is the smallest atom
As, I, or S.

PRACTICE EXAMPLE B: Which of the following atoms do you think is closest in size to the Na atom: Br, Ca, K, o
Al? Explain your reasoning, and do not use any tabulated data from the chapter in reaching your conclusion.

c Radius

n a metal atom loses one or more electrons to form a positive ion, the tive nuclear charge exceeds the negative charge of the electrons in the lting cation. The nucleus draws the electrons in closer, and, as a conse- ce, the following holds true.

Cations are smaller than the atoms from which they are formed.

gure 9-10 compares four species: the atoms Na and Mg and the ions Na^+ and $^+$. As expected, the Mg atom is smaller than the Na atom, and the cations maller than the corresponding atoms. Na^+ and Mg^{2+} are **isoelectronic**— have equal numbers of electrons (10) in identical configurations, $^2 2p^6$. Mg^{2+} is smaller than Na^+ because its nuclear charge is larger (+12, pared with +11 for Na).

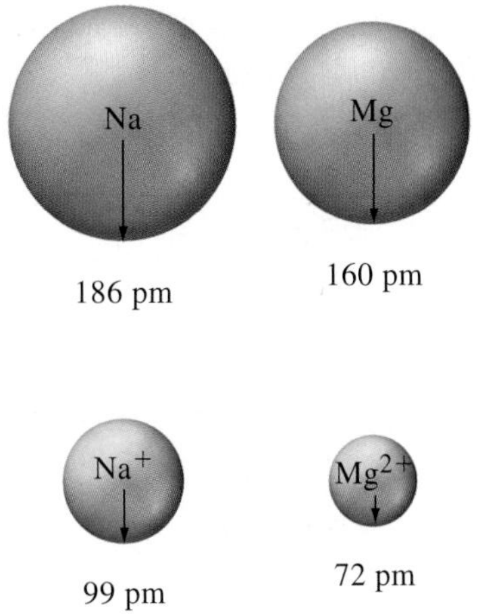

▲ FIGURE 9-10
A comparison of atomic and ionic sizes
Metallic radii are shown for Na and Mg and ionic radii for Na^+ and Mg^{2+}.

For isoelectronic cations, the more positive the ionic charge, the smaller the ionic radius.

'hen a nonmetal atom gains one or more electrons to form a negative ion on), the nuclear charge remains constant, but Z_{eff} is reduced because of the tional electron(s). The electrons are not held as tightly and repulsions ng the electrons increase. The electrons of the anion are more spread out they are in the atom, and thus, the radius of the anion is greater than that e atom. This generalization can be expressed in more quantitative terms onsidering a specific example: The covalent radius of Cl is 100 pm reas the ionic radius of Cl^- is 181 pm (Figure 9-11).

Anions are larger than the atoms from which they are formed. For isoelec- tronic anions, the more negative the charge, the larger the ionic radius.

nowledge of atomic and ionic radii can be used to modify the physical erties of certain materials. One example concerns strengthening of glass. mal window glass contains Na^+ and Ca^{2+} ions. The glass is brittle and ters easily when struck by a hard blow. One way to strengthen the glass is place the Na^+ ions at the surface with K^+ ions. The K^+ ions are larger and ıp the surface sites, leaving less opportunity for cracking than with the ller Na^+ ions. The result is a shatter-resistant glass.
nother example is the striking result when Cr^{3+} ions replace about 1% of Al^{3+} ions in aluminum oxide, Al_2O_3. This substitution is possible use Cr^{3+} ions are only slightly larger (by 9 pm) than Al^{3+} ions. Pure alu- um oxide is colorless, but with this small amount of chromium(III) ion, it beautiful red color. This impure Al_2O_3 is the gem known as a ruby. ies and other gemstones can be made artificially and are used as jewelry in devices such as lasers. The color of the ruby is further discussed in pter 24.
igure 9-11, arranged in the format of the periodic table, shows relative s of typical atoms and ions, and summarizes the generalizations described is section.

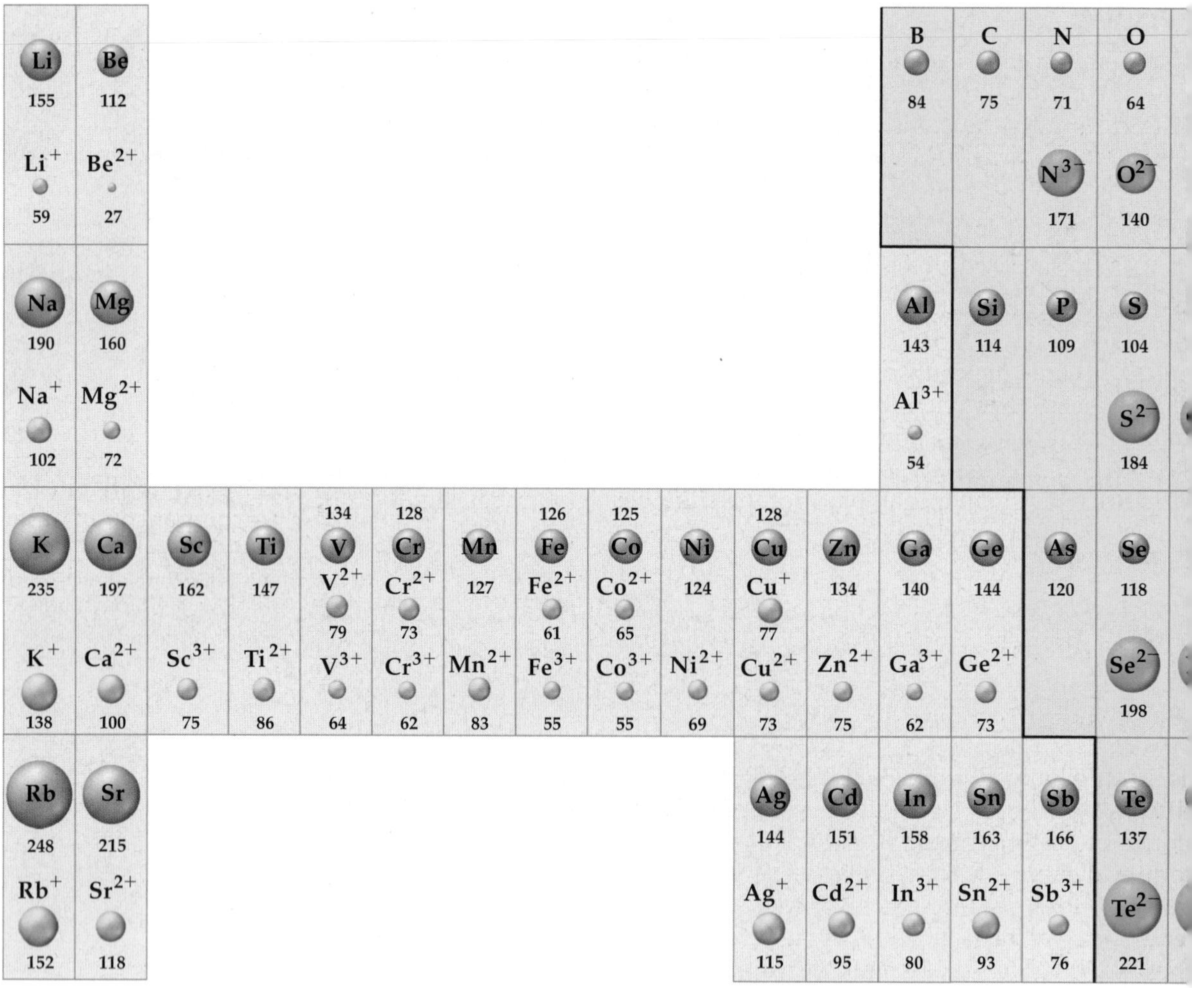

▲ FIGURE 9-11
A comparison of some atomic and ionic radii
The values given, in picometers (pm), are metallic radii for metals, single covalent radii for nonmetals, and ionic radii for the ions indicated. Gold spheres represent neutral atoms; blue spheres represent cations; and green spheres represent anions.

EXAMPLE 9-2 Comparing the Sizes of Cations and Anions

Refer only to the periodic table on the inside front cover, and arrange the following species in order of increasing size: K^{+}, Cl^{-}, S^{2-}, and Ca^{2+}.

Analyze

The key lies in recognizing that the four species are *isoelectronic*, having the electron configuration of argon: $1s^22s^22p^63s^23p^6$. When considering isoelectronic cations, the higher the charge on the ion, the smaller the ion.

Solve

The larger charge on the calcium ion means that Ca^{2+} is smaller than K^{+}. Because K^{+} has a higher nuclear charge than Cl^{-} ($Z = 19$, compared with $Z = 17$), it is smaller than Cl^{-}. For isoelectronic anions, the higher the charge, the larger the ion. S^{2-} is larger than Cl^{-}. The order of increasing size is

$$Ca^{2+} < K^{+} < Cl^{-} < S^{2-}$$

Assess

We can summarize the generalizations about isoelectronic atoms and ions into a single statement: Among isoelectronic species, the greater the atomic number, the smaller the size.

PRACTICE EXAMPLE A: Refer only to the periodic table on the inside front cover, and arrange the following species in order of increasing size: Ti^{2+}, V^{3+}, Ca^{2+}, Br^-, and Sr^{2+}.

PRACTICE EXAMPLE B: Refer only to the periodic table on the inside front cover, and determine which species is in the *middle* position when the following five are ranked according to size: the atoms N, Cs, and As and the ions Mg^{2+} and Br^-.

9-3 CONCEPT ASSESSMENT

ı the blank periodic table in the margin, locate the following:

) The smallest group 13 atom

) The smallest period 3 atom

:) The largest anion of a nonmetal in period 3

) The largest group 13 cation

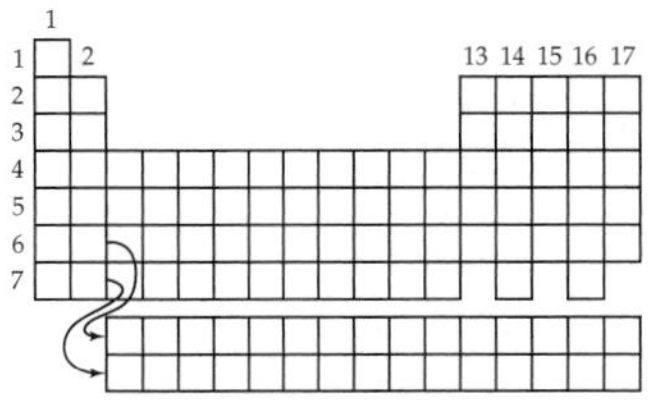

4 Ionization Energy

iscussing metals, we talked about metal atoms losing electrons and eby altering their electron configurations. But atoms do not eject electrons ıtaneously. Electrons are attracted to the positive charge on the nucleus of tom, and energy is needed to overcome that attraction. The more easily its trons are lost, the more metallic an atom is considered to be. The **ioniza- energy, E_i**, is the quantity of energy a *gaseous* atom must absorb to be able κpel an electron. The electron that is lost is the one that is highest in energy, therefore, is most loosely held.

ɔnization energies are usually measured through experiments in which ɔous atoms at low pressures are bombarded with photons of sufficient gy to eject an electron from the atom. Here are two typical values.

$$Mg(g) \longrightarrow Mg^+(g) + e^- \qquad E_i(Mg) = 738\ kJ/mol$$
$$Mg^+(g) \longrightarrow Mg^{2+}(g) + e^- \qquad E_i(Mg^+) = 1451\ kJ/mol$$

he symbol $E_i(Mg)$ represents the *first* ionization energy of Mg—the energy ıired to strip one electron from a neutral gaseous atom.* $E_i(Mg^+)$ repre- s the ionization energy of Mg^+ and thus, the *second* ionization energy Mg. Further ionization energies are denoted by $E_i(Mg^{2+})$, $E_i(Mg^{3+})$, so on. Each succeeding ionization energy is invariably larger than the eding one due to the progressively increasing Z_{eff}. In the case of magne- n, for example, in the second ionization, the electron, once freed, has to e away from an ion with a charge of +2 (Mg^{2+}). More energy must be sted than for a freed electron to move away from an ion with a charge of Mg^+). This is a direct consequence of Coulomb's law, which states, in , that the force of attraction between oppositely charged particles is ctly proportional to the magnitudes of the charges.

◀ A distinction between valence electrons and core electrons can be made based on the ionization energies for removing electrons one by one. The ionization energies of valence electrons are much smaller and show a big jump when the first core electron is removed.

ıization energies are sometimes expressed in the unit electron-volt (eV). One electron-volt is nergy acquired by an electron as it falls through an electric potential difference of 1 volt. It is a small energy unit, especially suited to describing processes involving individual atoms. When ation is based on a *mole* of atoms, kJ/mol is the preferred unit (1 eV/atom = 96.49 kJ/mol). ɛtimes the term *ionization potential* is used instead of ionization energy.

▶ FIGURE 9-12
First ionization energies as a function of atomic number
Ionization energies generally increase from left to right across a period (black lines) and decrease from top to bottom within a group (red dashed lines). More energy is required to ionize noble gas atoms than to ionize atoms immediately preceding or following them. The alkali metal atoms are the most easily ionized.

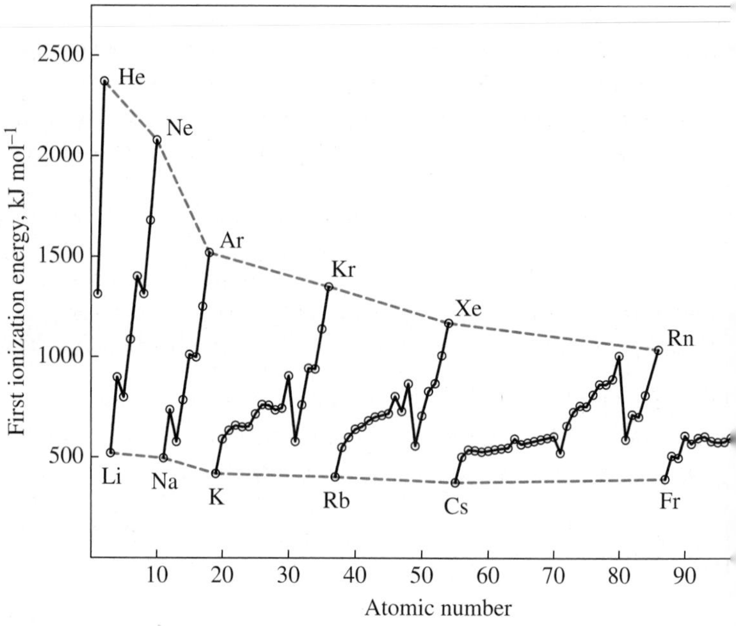

First ionization energies (E_i) for many atoms are plotted in Figure 9
Notice the following:

With relatively few exceptions, ionization energies increase from left to right across a period and decrease from top to bottom within a group.

The variation of ionization energies across a period or within a grou essentially the opposite of that observed for atomic radii. This observa leads us to the following conclusion:

Ionization energies decrease as atomic radii increase.

This statement above is easily rationalized: the farther an electron is f the nucleus, the more easily it can be removed. The decrease in ioniza energy that accompanies an increase in atomic radius is evident wher compare the ionization energies and atomic radii of the alkali metal at (Table 9.3).

Table 9.4 lists ionization energies for the third-period elements. With m exceptions, the trend in moving across a period (follow the colored strip that atomic radii decrease, ionization energies increase, and the elem

TABLE 9.3 Atomic Radii and First Ionization Energies of the Alkali Metal (Group 1) Elements

	Atomic Radius, pm	Ionization Energy, kJ mol^{-1}
Li	155	520.2
Na	190	495.8
K	235	418.8
Rb	248	403.0
Cs	267	375.7

TABLE 9.4 Ionization Energies of the Third-Period Elements (in kJ mol^{-}

	Na	Mg	Al	Si	P	S	Cl	Ar
First	495.8	737.7	577.6	786.5	1,012	999.6	1,251.1	1,5
Second	4,562	1,451	1,817	1,577	1,903	2,251	2,297	2,6
Third		7,733	2,745	3,232	2,912	3,361	3,822	3,9
Fourth			11,580	4,356	4,957	4,564	5,158	5,7
Fifth				16,090	6,274	7,013	6,542	7,2
Sixth					21,270	8,496	9,362	8,7
Seventh						27,110	11,020	12,0

ome less metallic, or more nonmetallic, in character. Table 9.4 lists stepwise zation energies. Note particularly the large breaks that occur along the ag diagonal line. Consider magnesium as an example.

he removal of two electrons from the $n = 3$ shell of Mg gives an ion with configuration [He]$2s^2 2p^6$. To remove a third electron requires taking an tron from the lower energy $n = 2$ shell. Consequently, the third ionization rgy of Mg is much larger than the second ionization energy—so much er that Mg^{3+} is not produced in ordinary chemical processes. Similarly, we not encounter Na^{2+} or Al^{4+} in ordinary chemical processes.

low let us turn our attention to the exceptions to the regular trend in first zation energies. For example, why is the first ionization energy of Al smaller n that of Mg and the first ionization energy of S smaller than that of P?

ecause Al is immediately to the right of Mg, we might expect the first ion-ion energy of Al to be larger than that of Mg, yet it is not. To understand y, we must consider the particular electrons lost. As illustrated by the ital diagrams in Figure 9-13, Mg loses a 3*s* electron whereas Al loses a 3*p* tron. Although Z_{eff} for the 3*p* electron of Al is greater than that of the 3*s* trons in Mg, the 3*p* electron of Al is less penetrating, of higher energy, and e easily removed.

Vhy is the first ionization energy of S lower than that of P? As suggested by ure 9-14, not only is it lower, but the first ionization energies of S, Cl, and are all lower than expected and considerably lower than the values pre-ed by extrapolating the first ionization energies of the elements immedi-y preceding them. We will consider two explanations,* both of which focus he fact that the ionization of S, Cl, or Ar involves the removal of a paired tron whereas the ionization of Al, Si, or P involves the removal of an aired electron. (See Figure 9-13.)

According to the first explanation, electron–electron repulsions are the key sideration. Paired electrons occupy the same orbital and are, on average, er together than electrons in separate orbitals. Thus, they experience extra ulsion and are more easily removed. Although the electron–electron repul-s increase with the number of electrons in the 3*p* orbitals, there is a signif-nt increase in the electron repulsions, and a corresponding decrease in the t ionization energy (approximately 226 kJ mol^{-1}), once orbital sharing ins, that is, as we proceed from P([Ne]$3s^2 3p^3$) to S([Ne]$3s^2 3p^4$).

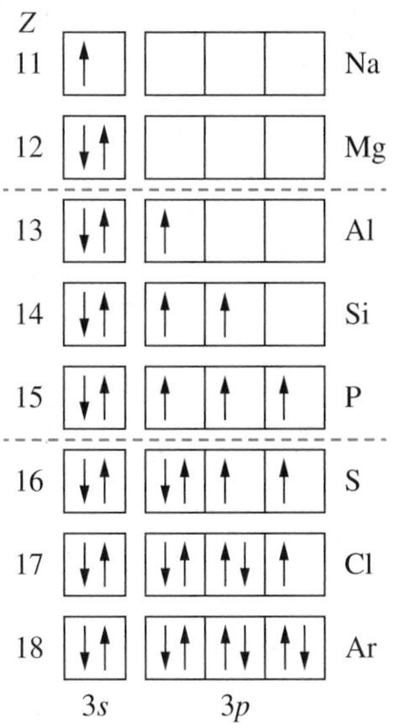

▲ FIGURE 9-13
Orbital diagrams showing the valence electron configurations for atoms of the third row elements
The dashed lines divide these elements into three sets, according to the type of electron that is removed on ionization. For Na and Mg, an electron is removed from a 3*s* orbital. For Al, Si, and P, an unpaired 3*p* electron is removed whereas for S, Cl, and Ar, a paired 3*p* electron is removed. When the first ionization energies of these atoms are plotted as a function of atomic number, *Z*, the corresponding graph (Figure 9-14) is piecewise linear, with a significant drop observed as we move from one set to the next.

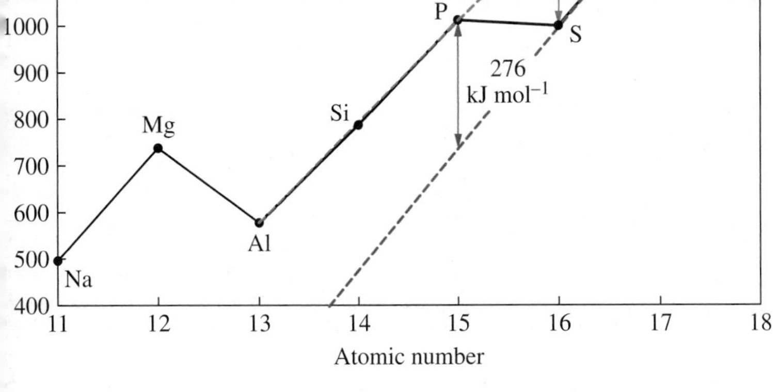

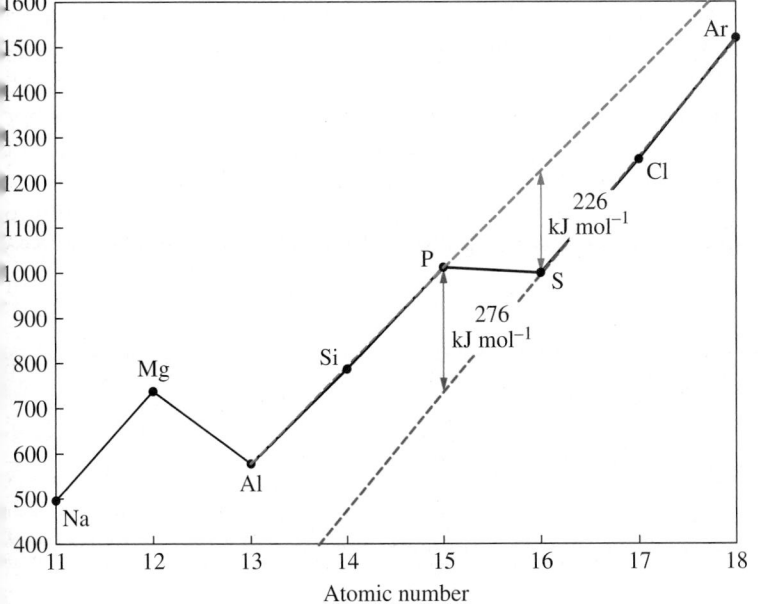

◀ FIGURE 9-14
First ionization energies of the third row *p*-block elements
The variation of the first ionization energy with *Z* is piecewise linear, with significant decreases occurring between Mg and Al and between P and S. Compared with the elements immediately preceding them, S, Cl, and Ar have ionization energies that are lower than expected by extrapolation (red dashed line). Similarly, compared with the elements immediately following them, Al, Si, and P have ionization energies that are higher than expected by back-extrapolation (blue dashed line).

ne following discussion is based on arguments made by various authors, including P. Cann, *em. Educ.*, **77,** 1056 (2000) and R. J. Boyd, *Nature*, **310,** 489 (1984).

The second explanation focuses instead on the extent of screening and strength of the electron–nucleus attractions. Unpaired electrons with par spins tend to avoid each other more, screen each other less, experien higher effective nuclear charge, interact more strongly with the nucleus, are harder to remove. According to this line of reasoning, the e electron–nuclear attraction causes the first ionization energy of P([Ne]$3s^2$ to be greater by approximately 276 kJ mol^{-1} than the value expected by backward-extrapolation of the first ionization energies of S, Cl, and Conversely, paired electrons screen each other to a greater extent, interact strongly with the nucleus, and are more easily removed.

Is the observed dip in the first ionization energy that occurs as we m from group 15 to 16 caused by increased electron–electron repulsions o decreased electron–nucleus attractions? It is difficult to answer this q tion with complete certainty, but it is clear that the dip occurs once orb sharing begins. The difficulty of providing an unambiguous explana for the observed dip is not totally unexpected. As we pointed ou Chapter 8, the energy of an atom is a delicate balance of electron–elec repulsions and electron–nucleus attractions. The rationalization of ion tion energies is further complicated by the fact that, for example, the ionization energy of P depends on the energies of both P([Ne]$3s^23p^3$) P^+([Ne]$3s^23p^2$). Similarly, the first ionization energy of S depends on energies of both S([Ne]$3s^23p^4$) and S^+([Ne]$3s^23p^3$). Thus, the decreas ionization energy that occurs as we move from P to S depends on a deli balance of electron–electron repulsions and electron–nucleus attractior four different species.

EXAMPLE 9-3 Relating Ionization Energies

Refer to the periodic table on the inside front cover, and arrange the following in the expected order of increas ing first ionization energy: As, Sn, Br, Sr.

Analyze

Ionization energies decrease as atomic radii increase. Thus, if we arrange these four atoms according t decreasing radius, we will likely have arranged them according to increasing ionization energy. The larges atoms are to the left and the bottom of the periodic table. The smallest atoms are to the right and toward th top of the periodic table.

Solve

Of the four atoms, the one that best fits the large-atom category is Sr. Although none of the four atoms is par ticularly close to the top of the table, Br is the farthest to the right. This fixes the two extremes: Sr with the low est ionization energy and Br with the highest. A tin atom should be larger than an arsenic atom, and thus S should have a lower ionization energy than As. The expected order of *increasing* ionization energies i $Sr < Sn < As < Br$.

Assess

The generalization that ionization energies decrease as atomic radii increase ignores the exceptions that occu when making comparisons between atoms of groups 2 and 13, as well as atoms of groups 15 and 16.

PRACTICE EXAMPLE A: Refer to the periodic table on the inside front cover, and arrange the following in th expected order of increasing first ionization energy: Cl, K, Mg, S.

PRACTICE EXAMPLE B: Refer to the periodic table on the inside front cover, and determine which element i most likely in the *middle* position when the following five elements are arranged according to first ionizatio energy: Rb, As, Sb, Br, Sr.

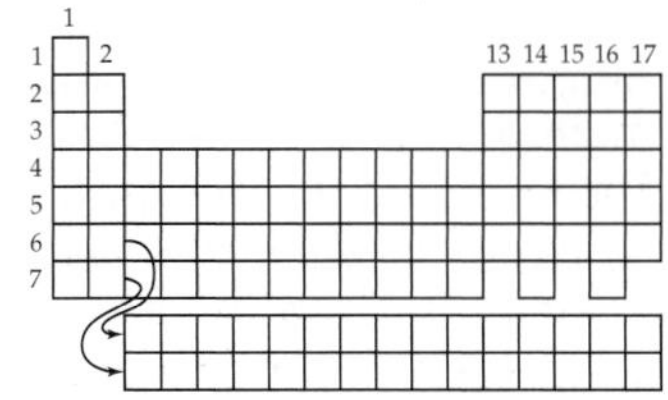

9-4 CONCEPT ASSESSMENT

n the blank periodic table in the margin, locate the following:

a) The group 14 element with the highest first ionization energy

) The element with the greatest first ionization energy in period 4

c) A *p*-block element in period 4 that has a lower first ionization energy than the element immediately preceding it and the element directly following it

5 Electron Affinity

ization energy is the energy change for the removal of an electron. Let's sider the energy change associated with the addition of an electron. The rmochemical equation for the addition of an electron to a fluorine atom is

$$F(g) + e^- \longrightarrow F^-(g) \qquad \Delta_{ea}H = -328 \text{ kJ mol}^{-1}$$

tice that the process above is *exothermic*, meaning that energy is given off en an F atom gains an electron. **Electron affinity, E_{ea},** can be defined as the halpy change, $\Delta_{ea}H$, that occurs when an atom in the gas phase gains an ctron. According to this definition, the electron affinity of fluorine is a neg-e quantity.

Ve have defined electron affinity to reflect the tendency for a neutral atom ain an electron. An alternative definition refers to the energy change in the cess: $X^-(g) \longrightarrow X(g) + e^-$; that is, reflecting the tendency of an anion to an electron. This alternative definition leads to the opposite signs for E_{ea} ues from those written in this text. You should be prepared to see electron nities expressed in both ways in the chemical literature.

ome representative values of $\Delta_{ea}H$ are listed in Figure 9-15 and plotted in ure 9-16. To interpret these values, we have to consider the type of orbital vhich the incoming electron has to be accommodated and the effect of the oming electron on the electron–electron repulsions and electron–nucleus actions. For many atoms, $\Delta_{ea}H$ is negative, an indication that there is gen-lly net attraction between an atom and an incoming electron. This net

1	2	13	14	15	16	17	18
H -72.8							He >0
Li -59.6	Be >0	B −26.7	C −121.8	N +7	O −141.0	F −328.0	Ne >0
Na -52.9	Mg >0	Al −42.5	Si −133.6	P −72	S −200.4	Cl −349.0	Ar >0
K -48.4	Ca −2.37	Ga −28.9	Ge −119.0	As −78	Se −195.0	Br −324.6	Kr >0
Rb -46.9	Sr −5.03	In −28.9	Sn −107.3	Sb −103.2	Te −190.2	I −295.2	Xe >0
Cs -45.5	Ba −13.95	Tl −19.2	Pb −35.1	Bi −91.2	Po −186	At −270	Rn >0

◀ FIGURE 9-15 **Electron affinities of main-group elements** Values are in kilojoules per mole for the process $X(g) + e^- \longrightarrow X^-(g)$.

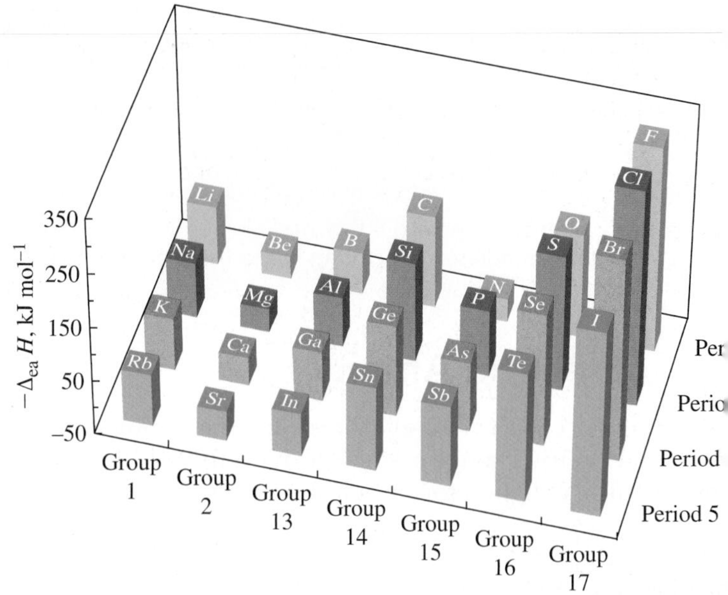

▶ FIGURE 9-16
Electron affinities of some of the main group elements
The electron affinity of species X is represented here by the value of $-\Delta_{ea}H$, where $\Delta_{ea}H$ is the enthalpy change for the process $X(g) + e^- \rightarrow X^-(g)$. The more exothermic this process, the greater the electron affinity of X. Within a period, the elements in groups 16 and 17 have the highest electron affinities. Nitrogen (in group 15) and the elements in group 12 have positive (or only slightly negative) values of $\Delta_{ea}H$ and have little or no tendency to gain an electron.

attraction arises because the electrons of the neutral atom do not comple shield an incoming electron from the nuclear charge.

Let's focus first on the variation of $\Delta_{ea}H$ across a period. Ignoring for moment the atoms of groups 2, 15, and 18, we see from Figures 9-15 and 9-16 as we move from left to right across a period, the addition of an electron beco more favorable, with $\Delta_{ea}H$ taking on increasingly negative values. The pro of adding an electron becomes more exothermic* because Z_{eff} for the added tron increases as we proceed from left to right across a given period.

Let's now consider the variation of $\Delta_{ea}H$ within a group, again igno groups 2, 15, and 18. Typically, the value of $\Delta_{ea}H$ is most negative for the a of the third period and less negative as we move down the group. For ex ple, in group 17, the Cl atom has the most negative $\Delta_{ea}H$ value. Progres toward the bottom of that group, $\Delta_{ea}H$ becomes less negative because incoming electron is accommodated in larger and larger orbitals and is farther from (and less attracted to) the nucleus.

For most groups, the atom of the second row has a lower electron affi ($\Delta_{ea}H$ is less negative) than the atom of the third row. Why is this? The ato orbitals of the second row atoms are much smaller (more compact) than th of third row atoms. Consequently, when an electron is added to a second atom, it is likely that the incoming electron encounters strong repulsive fo from other electrons in the atom and is not as tightly bound as we might erwise expect.

For some atoms, $\Delta_{ea}H$ is positive (or has a very small negative val These atoms have no tendency (or little tendency) to gain an electron. Th the case with the noble gases (group 18), where an added electron would h to enter the *s* orbital of the next electronic shell. Other cases include atom groups 2 and 12, for which the added electron would have to enter an orb of the next subshell.

It is interesting to note that the nitrogen atom also shows little tendenc gain an electron. The positive value for $\Delta_{ea}H$ indicates that N^-([He]$2s^2 2p$ of slightly higher energy (slightly less stable) than N([He]$2s^2 2p^3$). Presuma N^- is less stable than N because of the increase in electron–electron repuls or the decrease in electron–nucleus attractions caused by the pairing of e trons that accompanies the addition of an electron to the nitrogen atom.

*It is somewhat awkward to speak of larger and smaller with the term *electron affinity*. A st tendency to gain an electron, which implies a high "affinity" for an electron, as with F and reflected through a *low* value of E_{ea}—a large *negative* value.

ussed in Section 9-4, the pairing of electrons affects both the electron–
ron repulsions and electron–nucleus attractions.)
. considering the gain of a second electron by a nonmetal atom, we unter positive electron affinities. Here the electron to be added is oaching not a neutral atom, but a negative ion. There is a strong repulsive between the electron and the ion, and the energy of the system increases. s, for an element like oxygen, the first electron affinity is negative and the nd is positive.

$$O(g) + e^- \longrightarrow O^-(g) \qquad E_{ea,1} = \Delta_{ea} H[O(g)] = -141.0\ kJ/mol$$

$$O^-(g) + e^- \longrightarrow O^{2-}(g) \qquad E_{ea,2} = \Delta_{ea} H[O^-(g)] = +744\ kJ/mol$$

high positive value of $E_{ea,2}$ makes the formation of *gaseous* O^{2-} seem unlikely. The ion O^{2-} can exist, however, in ionic compounds, such as)(s), where formation of the ion is accompanied by other energetically rable processes.

9-5 CONCEPT ASSESSMENT

the blank periodic table in the margin locate the group expected to have
the most negative electron affinities in each period
the least negative electron affinities in each period
all positive electron affinities in each period

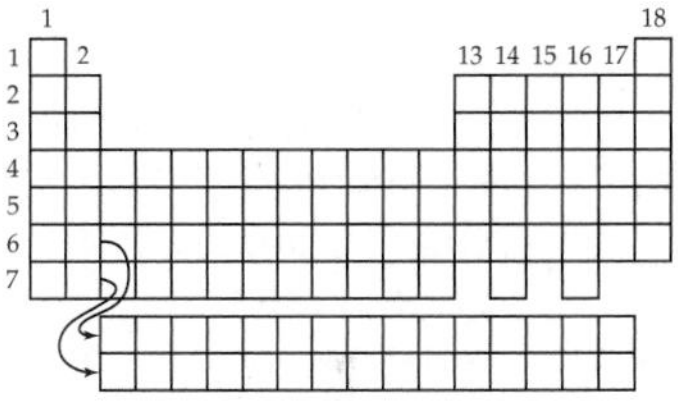

Magnetic Properties

mportant property related to the electron configurations of atoms and is their behavior in a magnetic field. As discussed on page 347, an elec- generates a magnetic field because of its spin. In a **diamagnetic** atom or all electrons are paired and the individual magnetic effects cancel out. A nagnetic species is weakly repelled by a magnetic field. A **paramagnetic** or ion has unpaired electrons, and the individual magnetic effects do not el out. The unpaired electrons possess a magnetic moment that causes the or ion to be attracted to an external magnetic field. The more unpaired rons present, the stronger is this attraction.
anganese has a paramagnetism corresponding to five unpaired electrons, ch is consistent with the electron configuration

Mn: [Ar] 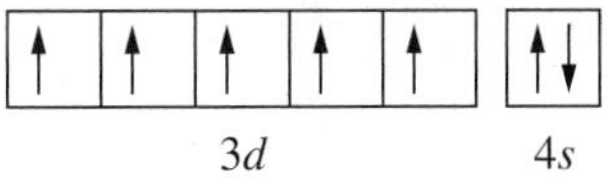

n a manganese atom loses two electrons, it becomes the ion Mn^{2+}, which ramagnetic, and the strength of its paramagnetism corresponds to five aired electrons.

Mn^{2+}: [Ar] 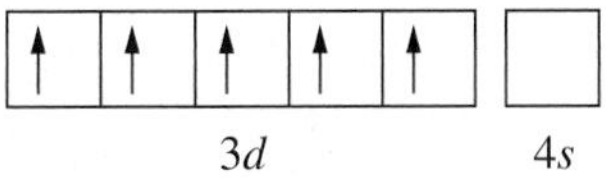

n a third electron is lost to produce Mn^{3+}, the ion has a paramagnetism esponding to four unpaired electrons. The third electron lost is one of the aired 3*d* electrons.

Mn^{3+}: [Ar] 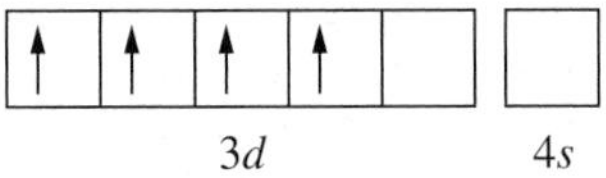

EXAMPLE 9-4 Determining the Magnetic Properties of an Atom or Ion

Which of the following would you expect to be diamagnetic and which paramagnetic?

(a) Na atom **(b)** Mg atom **(c)** Cl^- ion **(d)** Ag atom

Analyze

To determine whether or not an atom or ion is paramagnetic, we need to determine whether there are ar unpaired electrons. If there is at least one unpaired electron, the species is paramagnetic.

Solve

(a) Paramagnetic. The Na atom has a single 3*s* electron outside the Ne core. This electron is unpaired.

(b) Diamagnetic. The Mg atom has *two* 3*s* electrons outside the Ne core. They must be paired, as are all other electrons in the Ne core.

(c) Diamagnetic. Cl^- is isoelectronic with Ar and has all electrons paired $(1s^2 2s^2 2p^6 3s^2 3p^6)$.

(d) Paramagnetic. We do not need to work out the exact electron configuration of Ag. Because the atom 47 electrons—an odd number—at least one of the electrons must be unpaired (recall the Stern–Gerl experiment, page 348).

Assess

If the total number of electrons is odd, there must be at least one unpaired electron and so the species is par magnetic. If the total number of electrons is even, the species may be paramagnetic or diamagnetic, but t chances are good that the species is paramagnetic. Of all the known atoms, only 18 are diamagnetic in the ground electronic states: those from groups 2, 12, and 18, as well as Yb and No. For an ion to be diamagnetic, must be isoelectronic with one of these 18 atoms.

PRACTICE EXAMPLE A: Which of the following are paramagnetic and which are diamagnetic: Zn, Cl, K^+, O^2 and Al?

PRACTICE EXAMPLE B: Which has the greater number of unpaired electrons, Cr^{2+} or Cr^{3+}? Explain.

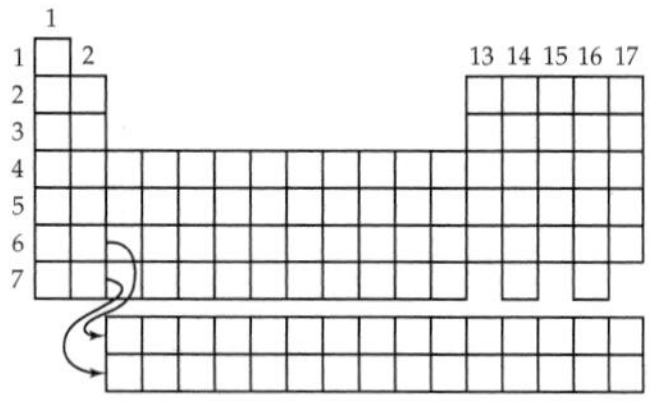

9-6 CONCEPT ASSESSMENT

On the blank periodic table in the margin locate the following:

(a) The period 4 transition element having a cation in the +3 oxidation state that is diamagnetic

(b) The period 5 element existing in the −2 oxidation state as an anion that is diamagnetic

(c) The period 4 transition element having a +2 cation that is paramagnetic and has a half-filled *d* subshell

9-7 Polarizability

For an isolated atom, the average distribution of electronic charge abou nucleus is spherical (Fig. 9-17a). This is not the case for an atom in the vic of another atom, molecule, or ion or in an externally applied electric field example, when an atom is placed in the electric field between two oppos charged parallel plates (Fig. 9-17b), the position of the much heavier nucle left essentially unchanged but the electron cloud is distorted (shifted tov the positively charged plate). Such an atom is said to be polarized. For a p ized atom, the centers of positive and negative charges are displaced f each other. The magnitude of this displacement depends on how easily electron cloud of the atom can be distorted. The **polarizability** of an atom vides a measure of the extent to which its electron cloud can be distorted example, by the application of an externally applied electric field or by

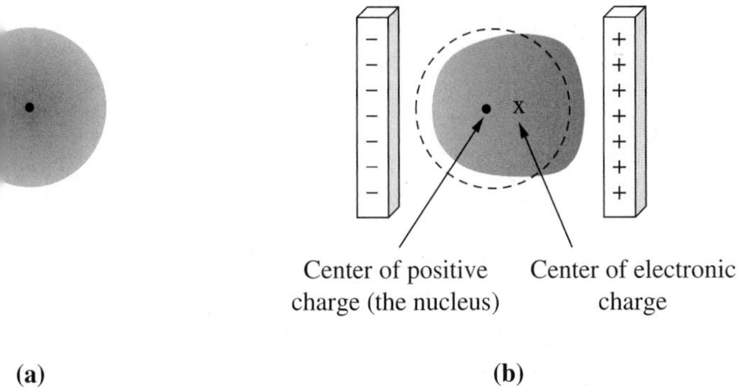

◀ FIGURE 9-17
Polarization of an atom
(a) For an isolated atom, the distribution of electronic charge about the nucleus is spherical. **(b)** In an electric field, the distribution of electronic charge is nonspherical, and the centers of positive and negative charge no longer coincide. The atom is said to be polarized.

roach of another atom, molecule, or ion. It is often expressed in units of ume. The polarizability of an atom depends on how diffuse or spread out lectron cloud is, and in general,

Polarizability increases with the size of the atom.

hus, polarizability decreases from left to right across a period and eases from top to bottom within a group. The polarizability of an atom is ilar in magnitude to the atomic volume calculated from atomic radii, as gested by Figure 9-18.

)o all the electrons in an atom contribute equally to the polarizability? No. ntum mechanical calculations on atoms reveal that the loosely bound nce electrons contribute more to the polarizability than the tightly bound er electrons. This result is not totally unexpected because, as we have ady learned, the valence electrons are, on average, farther from the leus and experience a smaller effective nuclear charge than do the inner trons. Thus, the valence electrons experience a greater shift in position the inner electrons when an atom is placed an externally applied electric d or is approached by another atom, molecule, or ion.

◀ We will see later that the polarizabilities of atoms, molecules, and ions enter into discussions of, for example, chemical bonding, intermolecular forces, phase changes, solvation, and chemical reactivity.

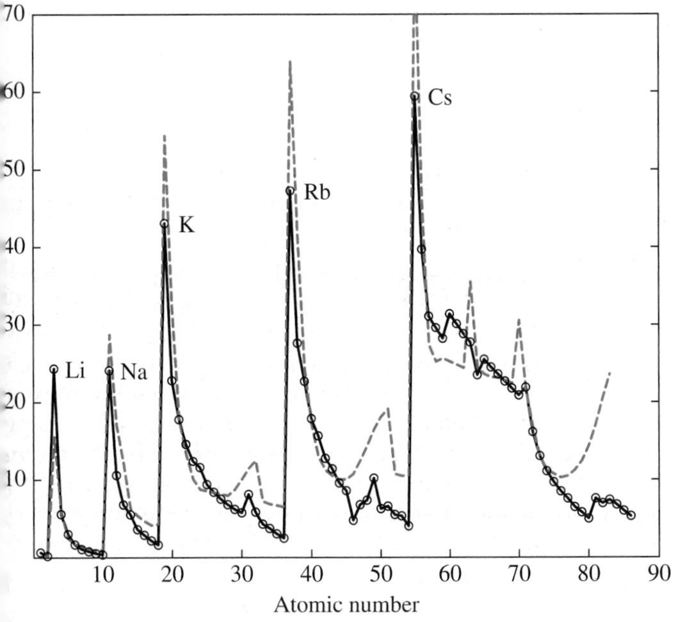

◀ FIGURE 9-18
Polarizabilities and atomic volumes
The variation of polarizability with atomic number (solid black line) closely resembles that of atomic volume (dashed red line). The atomic volume is calculated as $(4/3)\pi r^3$, where r is the atomic radius as defined in Figure 9-4. Both polarizability and atomic volume decrease from left to right across a period and increase from top to bottom in a group.

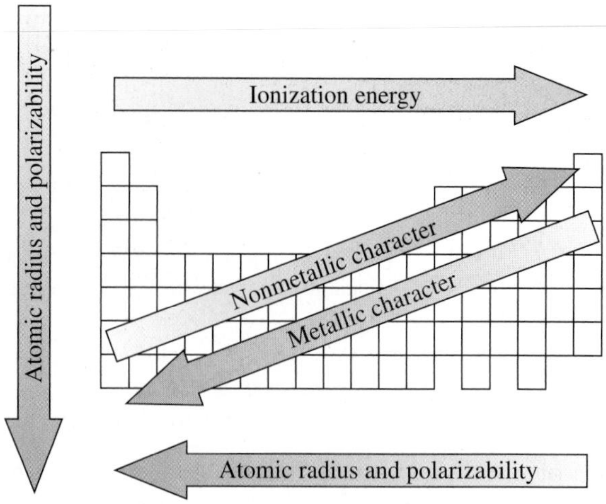

▶ FIGURE 9-19
Atomic properties and the periodic table—a summary
Atomic radius refers to metallic radius for metals and covalent radius for nonmetals. Ionization energies refer to first ionization energy. Metallic character relates generally to the ability to lose electrons, and nonmetallic character to the ability to gain electrons.

In the next few chapters, we will use our knowledge of the variation of ionization energy, electron affinity, and polarizability to explain various asp of bonding. Figure 9-19 provides a useful summary.

9-7 CONCEPT ASSESSMENT

The Al atom and Al^{3+} ion have essentially the same mass but very different polarizabilities. Which has the greater polarizability, Al or Al^{3+}?

MasteringCHEMISTRY **www.masteringchemistry.com**

Based on its position in the periodic table, mercury should be a solid with a melting point well over 300 °C. Yet, it is a liquid at room temperature. For a discussion of why mercury is a liquid, unlike other metallic elements, go to the Focus On feature for Chapter 9, The Periodic Law and Mercury, on the MasteringChemistry site.

Summary

9-1 Classifying the Elements: The Periodic Law and the Periodic Table—The experimental basis of the periodic table of the elements is the **periodic law**: Certain properties recur periodically when the elements are arranged by increasing atomic number. The theoretical basis is that the properties of an element are related to the electron configuration of its atoms, and elements in the same group of the periodic table have similar electron configurations.

9-2 Metals and Nonmetals and Their Ions—The three classes of elements of the periodic table are the *nonmetals*, *metals*, and *metalloids*. **Metalloids** have some properties characteristic of metals and some characteristics of nonmetals. The nonmetals are further divided into the noble gases and the remainder of the main-group metals, while the metals include the main-group m and transition elements.

9-3 Sizes of Atoms and Ions—Types of atomic include **covalent radii**, **metallic radii**, and **van der W radii** (Fig. 9-3). In general, atomic radii decrease acr period and increase down a group of the periodic (Figs. 9-4 and 9-11), mirroring the variation in **effe nuclear charge**, Z_{eff}, (Figs. 9-7 and 9-8a) across a pe and down a group. The **ionic radii** of positive ions smaller than the neutral atom, whereas negative ion larger than the parent atom (Figs. 9-10 and 9-11). I radii exhibit adherence to the periodic law similar to atomic radii. When atoms or ions have the same nber of electrons, they are said to be **isoelectronic.** en the radii of isoelectronic species are compared, the e negative the charge, the larger the radius of the ion tom.

Ionization Energy—A study of **ionization ener-** , E_i, shows that the periodic relationship observed is erned by the variation of Z_{eff}—that is, the ionization rgy decreases down a group and increases across a od (Fig. 9-12, Tables 9.3 and 9.4).

Electron Affinity—**Electron affinity**, E_{ea}, is the rgy change when an electron is added to a gaseous n. Interpreting trends is somewhat more difficult for tron affinity than for ionization energy because of the additional electron repulsions that arise when an electron is added to an atom.

9-6 Magnetic Properties—The magnetic properties of an atom or ion stem from the presence or absence of unpaired electrons. **Paramagnetic** atoms and ions have one or more unpaired electrons. In **diamagnetic** atoms and ions, all electrons are paired.

9-7 Polarizability—The **polarizability** of an atom provides a measure of the extent to which its electron cloud can be distorted (polarized) by an electric field or the approach of another atom, molecule, or ion (Fig. 9-17). Polarizability increases as the size of the atom increases (Fig. 9-18).

Integrative Example

en the ionization energies of a series of isoelectronic atoms and ions are compared, an interesting relationship bserved for some of them. In particular, if the square root of the ionization energy (in kJ mol^{-1}) for the series Be^{+}, B^{2+}, C^{3+}, N^{4+}, O^{5+}, and F^{6+} is plotted against the atomic number (Z) of the species, a linear relationship is ined. The corresponding graph for the series Na, Mg^{+}, Al^{2+}, Si^{3+}, P^{4+}, S^{5+}, and Cl^{6+}, is also linear. The graph is shown w.

he equations for the two lines joining the points are

$$\text{Second-row elements:} \quad \sqrt{E_i} = 18.4Z - 32.0 \tag{9.6}$$

$$\text{Third-row elements:} \quad \sqrt{E_i} = 13.7Z - 127 \tag{9.7}$$

lain the origin of these relationships and the differences in the numerical coefficients.

alyze

first notice that the electron configuration of the second-row atoms and ions is $1s^22s^1$, that is, a single electron ($2s^1$) ond the helium core ($1s^2$). Similarly, for the third-row atoms and ions the electron configuration is $1s^22s^22p^63s^1$, that single electron ($3s^1$) beyond the neon core ($1s^22s^22p^6$).

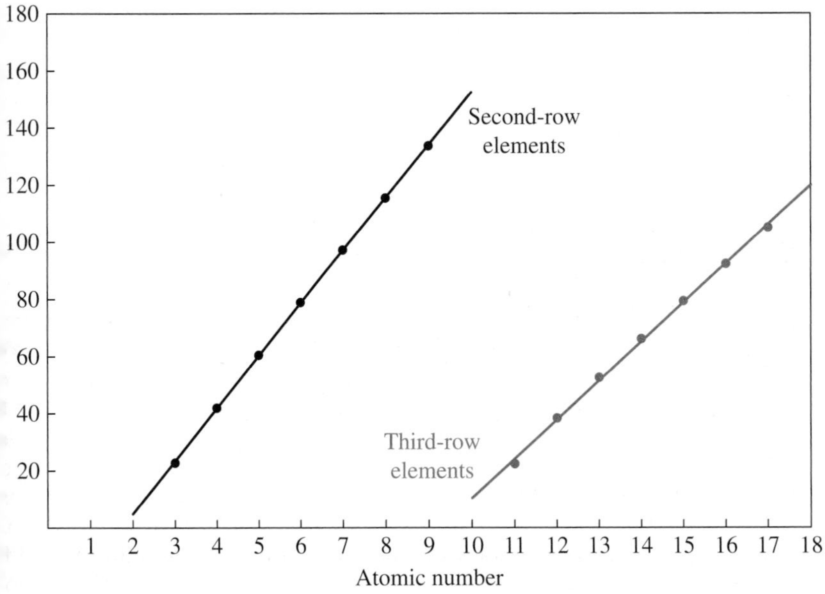

◀ Variation of $\sqrt{E_i}$ with atomic number for the second-row elements (black line) and the third-row elements (red line).

In both series of atoms and ions, the inner-core electrons screen the single valence-shell electron from the nuc
These species are all reminiscent of the Bohr atom so that as an approximation we can use the expression for the en levels for hydrogen-like atoms or hydrogen-like ions (equation 8.9). Specifically, we should be able to use equation (to derive equations for the energy required to remove the electron from the valence shells of a hydrogen-like species, is, the ionization energy (E_i). Once we have these equations, we can compare them with the equations for the straight-line graphs.

Solve

Equation (8.9), with the substitutions $Z = Z_{eff}$, and $R_H = 2.179 \times 10^{-18}$ J, gives

$$E_n = -2.179 \times 10^{-18}\,\text{J} \times \left(\frac{Z_{eff}^2}{n^2}\right)$$

E_n has the unit J atom^{-1}, which must be converted to kJ mol^{-1}—the unit of E_i in the two straight-line equations given to us.

$$E_n = -2.179 \times 10^{-18}\frac{\text{J}}{\text{atom}} \times 6.022 \times 10^{23}\frac{\text{atom}}{\text{mol}} \times \frac{Z}{n}$$

$$= -1.3122 \times 10^6 \frac{Z_{eff}^2}{n^2}\,\text{J mol}^{-1} \times \frac{1\,\text{kJ}}{1000\,\text{J}}$$

The energy required to remove an electron from an orbital with principal quantum number n in a hydrogen-like species—the ionization energy—is the *negative* of that shown above, that is,

$$E_i = -E_n = 1312.2 \times \frac{Z_{eff}^2}{n^2}\,\text{kJ mol}^{-1}$$

Equation (9.8) shows that the ionization energy (E_i) is a linear function of $(Z_{eff})^2$ and the straight-line graphs (equations 9.6 and 9.7) show that $\sqrt{E_i}$ is a linear function of Z. We are on the right track. However, we must now take into account that we are considering one-electron systems with a nucleus shielded by a closed shell.

First consider the second-row series (Li, Be^+, B^{2+}, C^{3+}, N^{4+}, O^{5+}, and F^{6+}), all members of which have the electron configuration $1s^22s^1$. If we assume that the closed shell $1s^2$ perfectly screens the outer electron $2s^1$, the value of Z_{eff} for this series will be $Z - 2$. Thus we should substitute $Z_{eff} = Z - 2$ into equation (9.8), and also $n = 2$ since ionization occurs from the 2s orbital, to obtain

$$E_i = 1312.2 \times \frac{(Z-2)^2}{n^2} = 1312.2 \times \frac{(Z-2)^2}{2^2}$$

Taking the square root of both sides and clearing the fraction gives

$$\sqrt{E_i} = 36.22 \times \left(\frac{Z-2}{2}\right) = 18.11Z - 36.22 \quad (9$$

Let us now look at the third-row series. In this case the configuration of the isoelectronic series is $1s^22s^22p^63s^1$ so that if we assume perfect screening of the $3s^1$ electron by the ten inner-core electrons we have

$$Z_{eff} = Z - 10$$

Proceeding as before and remembering that ionization occurs from the $n = 3$ level, we obtain

$$E_i = 1312.2 \times \frac{(Z-10)^2}{n^2} = 1312.2 \times \frac{(Z-10)^2}{3^2}$$

and

$$\sqrt{E_i} = 36.22 \times \left(\frac{Z-10}{3}\right) = 12.07Z - 120.7 \quad (9$$

ess

omparing equations (9.9) and (9.10), we find that the difference in the slope (coefficient for Z) is due to the differ- in the principal quantum number of the orbital from which the ionization occurs. The difference in the intercepts is both to the principal quantum number from which the ionization occurs and to the number of electrons screening the nce-shell electron.

quation (9.9) for the second-row elements is in remarkable agreement with the empirically observed equation (9.6) at eginning of this example, especially considering we started from an equation for the energy levels of a one-electron ies.

lthough the general form of equation (9.10) is correct for the third-row series, the agreement between the numerical tants is not as good. This is to be expected because we have assumed perfect screening by the core electrons, which pletely ignores the different characteristics of the electrons composing the inner core. The intricate, correlated ions of the electrons in the core leads to a complicated combination of screening and penetration that cannot be unted for with our simple model.

CTICE EXAMPLE A: Francium ($Z = 87$) is an extremely rare radioactive element formed when actinium = 89) undergoes alpha-particle emission. Francium occurs in natural uranium minerals, but estimates are that little e than 15 g of francium exists in the top 1 km of Earth's crust. Few of francium's properties have been measured, but e can be inferred from its position in the periodic table. Estimate the melting point, density, and atomic (metallic) us of francium. [*Hint:* Plot each property versus atomic number, Z, and extrapolate to $Z = 87$.]

CTICE EXAMPLE B: Discuss the likelihood that element 168, should it ever be synthesized in sufficient quantity, ld be a "noble liquid" at 298 K and 1 bar. Could element 168 be a "noble solid" at 298 K and 1 bar? Use *spdf* notation ow the electron configuration you would expect for element 168. [*Hint:* Prepare graphs of boiling point versus ic number and melting point versus atomic number. Extrapolate to $Z = 168$.]

Element	mp, K	bp, K
Argon	83.95	87.45
Helium		4.25
Krypton	116.5	120.9
Neon	24.48	27.3
Radon	202	211.4
Xenon	161.3	166.1

Exercises

e Periodic Law

Use data from Figure 9-1 and equation (9.1) to estimate the density of the recently discovered element 114.

Suppose that lanthanum ($Z = 57$) were a newly discovered element having a density of 6.145 g/cm^3. Estimate its molar mass.

The following densities, in grams per cubic centimeter, are for the listed elements in their standard states at 298 K. Show that density is a periodic property of these elements: Al, 2.699; Ar, 0.0018; As, 5.778; Br, 3.100; Ca, 1.550; Cl, 0.0032; Ga, 5.904; Ge, 5.323; Kr, 0.0037; Mg, 1.738; P, 1.823; K, 0.856; Se, 4.285; Si, 2.336; Na, 0.968; S, 2.069.

4. The following melting points are in degrees Celsius. Show that melting point is a periodic property of these elements: Al, 660; Ar, −189; Be, 1278; B, 2300; C, 3350; Cl, −101; F, −220; Li, 179; Mg, 651; Ne, −249; N, −210; O, −218; P, 590; Si, 1410; Na, 98; S, 119.

e Periodic Table

Mendeleev's periodic table did not preclude the possibility of a new group of elements that would fit within the existing table, as was the case with the noble gases. Moseley's work did preclude this possibility. Explain this difference.

Explain why the several periods in the periodic table do not all have the same number of members.

Assuming that the seventh period is 32 members long, what should be the atomic number of the noble gas following radon (Rn)? Of the alkali metal following francium (Fr)? What would you expect their approximate atomic masses to be?

8. Concerning the incomplete seventh period of the periodic table, what should be the atomic number of the element **(a)** for which the filling of the 6*d* subshell is completed; **(b)** that should most closely resemble bismuth; **(c)** that should be a noble gas?

Atomic Radii and Ionic Radii

9. For each of the following pairs, indicate the atom that has the *larger* size: **(a)** Te or Br; **(b)** K or Ca; **(c)** Ca or Cs; **(d)** N or O; **(e)** O or P; **(f)** Al or Au.
10. Indicate the *smallest* and the *largest* species (atom or ion) in the following group: Al atom, F atom, As atom, Cs^+ ion, I^- ion, N atom.
11. Explain why the radii of atoms do not simply increase uniformly with increasing atomic number.
12. The masses of individual atoms can be determined with great precision, yet there is considerable uncertainty about the exact size of an atom. Explain why this is the case.
13. Which is **(a)** the smallest atom in group 13; **(b)** the smallest of the following atoms: Te, In, Sr, Po, Sb? Why?
14. How would you expect the sizes of the hydrogen ion, H^+, and the hydride ion, H^-, to compare with that of the H atom and the He atom? Explain.
15. Arrange the following in expected order of increasing radius: Br, Li^+, Se, I^-. Explain your answer.
16. Explain why the generalizations presente Figure 9-19 cannot be used to answer the ques Which is larger, an Al atom or an I atom?
17. Among the following ions, several pairs are *iso tronic*. Identify these pairs. Fe^{2+}, Sc^{3+}, Ca^{2+}, F^-, C Co^{3+}, Sr^{2+}, Cu^+, Zn^{2+}, Al^{3+}.
18. The following species are isoelectronic with the n gas krypton. Arrange them in order of increa radius and comment on the principles involve doing so: Rb^+, Y^{3+}, Br^-, Sr^{2+}, Se^{2-}.
19. All the isoelectronic species illustrated in the text the electron configurations of noble gases. Can ions be isoelectronic *without* having noble-gas e tron configurations? Explain.
20. Is it possible for two different atoms to be isoe tronic? two different cations? two different anior cation and an anion? Explain.

Ionization Energies; Electron Affinities

21. Use principles established in this chapter to arrange the following atoms in order of *increasing* value of the first ionization energy: Sr, Cs, S, F, As.
22. Are there any atoms for which the second ionization energy is smaller than the first? Explain.
23. Some electron affinities are negative quantities, and some are zero or positive. Why is this not also the case with ionization energies?
24. How much energy, in joules, must be absorbed to convert to Na^+ all the atoms present in 1.00 mg of *gaseous* Na? The first ionization energy of Na is 495.8 kJ/mol.
25. How much energy, in kilojoules, is required to remove all the third-shell electrons in a mole of gaseous silicon atoms?
26. What is the maximum number of Cs^+ ions that can be produced per joule of energy absorbed by a sample of gaseous Cs atoms?
27. The production of gaseous bromide ions from bromine molecules can be considered a two-step process in which the first step is

$$Br_2(g) \longrightarrow 2\,Br(g) \qquad \Delta_r H = +193\ \text{kJ mol}^{-1}$$

Is the formation of $Br^-(g)$ from $Br_2(g)$ an endothermic or exothermic process?
28. Use ionization energies and electron affinities listed in the text to determine whether the following reaction is endothermic or exothermic.

$$Mg(g) + 2\,F(g) \longrightarrow Mg^{2+}(g) + 2\,F^-(g)$$

29. The Na^+ ion and the Ne atom are isoelectronic. ease of loss of an electron by a gaseous Ne atom, ionization energy, has a value of 2081 kJ/mol. ease of loss of an electron from a gaseous Na^+ second ionization energy, has a value of 4562 kJ/ Why are these values not the same?
30. From the data in Figure 9-12, the formation gaseous anion Li^- appears energetically favora That is, energy is given off when gaseous Li at accept electrons. Comment on the likelihood of fc ing a stable compound containing the Li^- ion, suc Li^+Li^- or Na^+Li^-.
31. Compare the elements Al, Si, S, and Cl.
    (a) Place the elements in order of increasing ior tion energy.
    (b) Place the elements in order of increasing elec affinity.
    (c) Place the elements in order of increasing pc izability.
32. Compare the elements Na, Mg, O, and P.
    (a) Place the elements in order of increasing ion tion energy.
    (b) Place the elements in order of increasing e tron affinity.
    (c) Place the elements in order of increasing pol ability.

Magnetic Properties

33. Unpaired electrons are found in only one of the following species. Indicate which one, and explain why: F^-, Ca^{2+}, Fe^{2+}, S^{2-}.
34. Which of the following species has the greatest number of unpaired electrons **(a)** Ge; **(b)** Cl; **(c)** Cr^{3+}; **(d)** Br^-?
35. Which of the following species would you expe be diamagnetic and which paramagnetic? **(a)** **(b)** Cr^{3+}; **(c)** Zn^{2+}; **(d)** Cd; **(e)** Co^{3+}; **(f)** Sn^{2+}; **(g)** Br.
36. Write electron configurations consistent with the lowing data on numbers of unpaired electr Ni^{2+}, 2; Cu^{2+}, 1; Cr^{3+}, 3.

Must all atoms with an odd atomic number be paramagnetic? Must all atoms with an even atomic number be diamagnetic? Explain.

38. Neither Co^{2+} nor Co^{3+} has 4*s* electrons in its electron configuration. How many unpaired electrons would you expect to find in each of these ions? Explain.

edictions Based on the Periodic Table

Use ideas presented in this chapter to indicate **(a)** three metals that you would expect to exhibit the photoelectric effect with visible light and three that you would not; **(b)** the noble gas element that should have the highest density in the liquid state; **(c)** the approximate first ionization energy E_i of fermium ($Z = 100$); **(d)** the approximate density of solid radium ($Z = 88$).

Arrange the following atoms in order of increasing polarizability: F, Na, P, As, Br.

Arrange the following species in order of increasing polarizability: N, S, Be, K, O.

For the following groups of elements, select the one that has the property noted:
(a) the largest atom: Mg, Mn, Mo, Ba, Bi, Br
(b) the lowest first ionization energy: B, Sr, Al, Br, Mg, Pb
(c) the most negative electron affinity: As, B, Cl, K, Mg, S
(d) the largest number of unpaired electrons: F, N, S^{2-}, Mg^{2+}, Sc^{3+}, Ti^{3+}

Of the species Cl, Cl^+, and Cl^-, which has the highest polarizability? Which has lowest polarizability?

Of the species Na^+, Na, F, and F^-, which has the highest polarizability? Which has lowest polarizability?

45. Match each of the lettered items on the left with an appropriate numbered item on the right. All the numbered items should be used, and some more than once.

(a) $Z = 32$	**1.** two unpaired *p* electrons
(b) $Z = 8$	**2.** diamagnetic
(c) $Z = 53$	**3.** more negative electron affinity than elements on either side of it in the same period
(d) $Z = 38$	
(e) $Z = 48$	
(f) $Z = 20$	**4.** first ionization energy lower than that of Ca but greater than that of Cs

46. Match each of the lettered items in the column on the left with the most appropriate numbered item(s) in the column on the right. Some of the numbered items may be used more than once and some not at all.

(a) Tl	**1.** an alkaline earth metal
(b) $Z = 70$	**2.** element in period 5 and group 15
(c) Ni	
(d) $[Ar]4s^2$	**3.** largest atomic radius of all the elements
(e) a metalloid	
(f) a nonmetal	**4.** an element in period 4 and group 16
	5. $3d^8$
	6. one *p* electron in the shell of highest *n*
	7. lowest ionization energy of all the elements
	8. an *f*-block element

47. Which of the following ions are unlikely to be found in chemical compounds: K^+, Ga^{4+}, Fe^{6+}, S^{2-}, Ge^{5+}, or Br^-? Explain briefly.

48. Which of the following ions are likely to be found in chemical compounds: Na^{2+}, Li^+, Al^{4+}, F^{2-}, or Te^{2-}? Explain briefly.

Integrative and Advanced Exercises

Four atoms and/or ions are sketched below in accordance with their relative atomic and/or ionic radii.

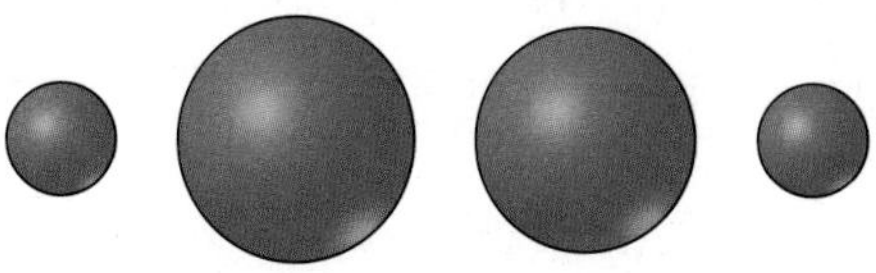

Which of the following sets of species are compatible with the sketch? Explain. **(a)** C, Ca^{2+}, Cl^-, Br^-; **(b)** Sr, Cl, Br^-, Na^+; **(c)** Y, K, Ca, Na^+; **(d)** Al, Ra^{2+}, Zr^{2+}, Mg^{2+}; **(e)** Fe, Rb, Co, Cs.

Sketch a periodic table that would include *all* the elements in the main body of the table. How many "numbers" wide would the table be?

In Mendeleev's time, indium oxide, which is 82.5% In by mass, was thought to be InO. If this were the case, in which group of Mendeleev's table (page 378) should indium be placed?

52. Instead of accepting the atomic mass of indium implied by the data in Exercise 51, Mendeleev proposed that the formula of indium oxide is In_2O_3. Show that this assumption places indium in the proper group of Mendeleev's periodic table on page 378.

53. Refer to Figure 9-11 and explain why the difference between the ionic radii of the −1 and −2 anions does not remain constant from top to bottom of the periodic table.

54. Explain why the third ionization energy of Li(g) is an easier quantity to calculate than either the first or second ionization energies. Calculate the third ionization energy for Li, and express the result in kJ/mol.

55. Two elements, A and B, have the electron configurations shown.

$$A = [Ar]4s^1 \quad B = [Ar]3d^{10}4s^24p^3$$

(a) Which element is a metal?
(b) Which element has the greater ionization energy?
(c) Which element has the larger atomic radius?
(d) Which element has the greater electron affinity?

56. Two elements, A and B, have the electron configurations shown.

$$A = [Kr]4s^2 \quad B = [Ar]3d^{10}4s^24p^5$$

(a) Which element is a metal?
(b) Which element has the greater ionization energy?
(c) Which element has the larger atomic radius?
(d) Which element has the greater electron affinity?

57. Studies done in 1880 showed that a chloride of uranium had 37.34% Cl by mass and an approximate formula mass of 382 u. Other data indicated the specific heat of uranium to be 0.0276 cal $g^{-1}\,°C^{-1}$. Are these data in agreement with the atomic mass of uranium assigned by Mendeleev, 240 u? [*Hint:* Refer to Feature Problem 124 of Chapter 7.]

58. Assume that atoms are hard spheres, and use the metallic radius of 186 pm for Na to estimate the volumes of one Na atom and of one mole of Na atoms. How does your result compare with the atomic volume found in Figure 9-1? Why is there so much disagreement between the two values?

59. When sodium chloride is strongly heated in a flame, the flame takes on the yellow color associated with the emission spectrum of sodium atoms. The reaction that occurs in the *gaseous* state is

$$Na^+(g) + Cl^-(g) \longrightarrow Na(g) + Cl(g).$$

Calculate $\Delta_r H$ for this reaction.

60. Use information from Chapters 8 and 9 to calculate the *second* ionization energy for the He atom. Compare your result with the tabulated value of 5251 kJ mol^{-1}.

61. Refer only to the periodic table on the inside front cover, and arrange the following ionization energies in order of increasing value: the first ionization energy of F; the second ionization energy of Ba; the third ionization energy of Sc; the second ionization energy of Na; the third ionization energy of Mg. Explain the basis of any uncertainties.

62. Refer to the footnote on page 393. Then use values of basic physical constants and other data from the appendices to show that 1 eV/atom = 96.49 kJ mol^{-1}.

63. The ionization energies of Li, Be^+, B^{2+}, and C^{3+} respectively, 520, 1757, 3659, and 6221 kJ mol^{-1}. ionization energies Na, Mg^+, Al^{2+}, and Si^{3+} are (f Table 9.4) 495.8, 1451, 2745, and 4356 kJ mol^{-1}. Pl graph of the square roots of the ionization ener versus the nuclear charge for these two series. Exp the observed relationship with the aid of Bo expression for the binding energy of an electron one-electron atom.

64. Elements 114–116 have recently been reported t synthesized. Using data given below and the peri law, fill in the missing data for these elements.

Sn	50	Sb	51	Te	52
$5s^25p^2$	118.7	$5s^25p^3$	121.8	$5s^25p^4$	127
2	145	3	145	1	140
16.29	107.3	18.19	103.2	20.46	190
708.6	7.31	834	6.69	869.3	6.2
Pb	82	Bi	83	Po	84
$6s^26p^2$	207.2	$6s^26p^3$	208.9	$6s^26p^4$	209
2	180	3	160	1	190
18.26	35.1	21.31	91.2	22.97	183
715.6	11.35	703	9.75	812.1	9.3
Fl	114	Unp	115	Lv	116
?	?	?	?	?	?
?	?	?	?	?	?
?	?	?	?	?	?
?	?	?	?	?	?

The entries for each element are organized as follows:

Atomic symbol	Z
Valence configuration	Atomic mass
No. of unpaired electrons	Atomic radius (p
Molar volume (cm^3)	Electron affinity (kJ mol^{-1})
First ionization energy (kJ mol^{-1})	Density (g cm^{-3})

Feature Problems

65. The work functions for a number of metals are given in the following table. How do the work functions vary
(a) down a group?
(b) across a period?
(c) Estimate the work function for potassium and compare it with a published value.
(d) What periodic property is the work function most like?

Metal	Work Function, J × 10^{19}
Al	6.86
Cs	3.45
Li	4.6
Mg	5.86
Na	4.40
Rb	3.46

. The following are a few elements and their characteristic X-ray wavelengths:

Element	X-ray Wavelength, pm
Mg	987
S	536
Ca	333
Cr	229
Zn	143
Rb	93

Use these data to determine the constants A and b in Moseley's relationship (page 379). Compare your value of A with the value obtained from Bohr's theory for the frequencies emitted by one-electron atoms. Suggest a reasonable interpretation of the quantity b.

. Gaseous sodium atoms absorb quanta with the energies shown in the table below.

Energy of Quanta, kJ mol^{-1}	Electron Configuration
0	$[Ne]3s^1$
203	$[Ne]3p^1$
308	$[Ne]4s^1$
349	$[Ne]3d^1$
362	$[Ne]4p^1$

(a) The ionization energy of the ground state is 496 kJ mol^{-1}. Calculate the ionization energies for each of the states given in the table.
(b) Calculate Z_{eff} for each state.
(c) Calculate $\bar{r}_{n\ell}$ for each state.
(d) Interpret the results obtained from parts (b) and (c) in terms of penetration and screening.

. A method for estimating electron affinities is to extrapolate Z_{eff} values for atoms and ions that contain the same number of electrons as the negative ion of interest. Use the data in the table to answer the questions that follow.

Atom or Ion: E_i(kJ mol^{-1})	Atom or Ion: E_i(kJ mol^{-1})	Atom or Ion: E_i(kJ mol^{-1})
Ne: 2080	F: 1681	O: 1314
Na^+: 4565	Ne^+: 3963	F^+: 3375
Mg^{2+}: 7732	Na^{2+}: 6912	Ne^{2+}: 6276
Al^{3+}: 11,577	Mg^{3+}: 10,548	Na^{3+}: 9540

(a) Estimate the electron affinity of F, and compare it with the experimental value.
(b) Estimate the electron affinities of O and N.
(c) Examine your results in terms of penetration and screening.

69. We have seen that the wave functions of hydrogen-like atoms contain the nuclear charge Z for hydrogen-like atoms and ions, but modified through equation (9.3) to account for the phenomenon of shielding or screening. In 1930, John C. Slater devised the following set of empirical rules to calculate a shielding constant for a designated electron in the orbital ns or np:

(i) Write the electron configuration of the element, and group the subshells as follows: $(1s)$, $(2s, 2p)$, $(3s, 3p)$, $(3d)$, $(4s, 4p)$, $(4d)$, $(4f)$, $(5s, 5p)$, etc.
(ii) Electrons in groups to the right of the (ns, np) group contribute nothing to the shielding constant for the designated electron.
(iii) All the other electrons in the (ns, np) group shield the designated electron to the extent of 0.35 each.
(iv) All electrons in the $n - 1$ shell shield to the extent of 0.85 each.
(v) All electrons in the $n - 2$ shell, or lower, shield completely—their contributions to the shielding constant are 1.00 each.

When the designated electron being shielded is in an nd or nf group, rules (ii) and (iii) remain the same but rules (iv) and (v) are replaced by

(vi) Each electron in a group lying to the left of the nd or nf group contributes 1.00 to the shielding constant.

These rules are a simplified generalization based on the average behavior of different types of electrons. Use these rules to do the following:
(a) Calculate Z_{eff} for a valence electron of oxygen.
(b) Calculate Z_{eff} for the $4s$ electron in Cu.
(c) Calculate Z_{eff} for a $3d$ electron in Cu.
(d) Evaluate the Z_{eff} for the valence electrons in the group 1 elements (including H), and show that the ionization energies observed for this group are accounted for by using the Slater rules. [*Hint:* Do not overlook the effect of n on the orbital energy.]
(e) Evaluate Z_{eff} for a valence electron in the elements Li through Ne, and use the results to explain the observed trend in first ionization energies for these elements.
(f) Using the radial functions given in Table 8.2 and Z_{eff} estimated with the Slater rules, compare plots of the radial probability for the $3s$, $3p$, and $3d$ orbitals for the H atom and the Na atom. What do you observe from these plots regarding the effect of shielding on radial probability distributions?

Self-Assessment Exercises

In your own words, define the following terms: **(a)** isoelectronic; **(b)** valence-shell electrons; **(c)** metal; **(d)** nonmetal; **(e)** metalloid.

Briefly describe each of the following ideas or phenomena: **(a)** the periodic law; **(b)** ionization energy; **(c)** electron affinity; **(d)** paramagnetism.

72. Explain the important distinctions between each pair of terms: **(a)** actinide and lanthanide element; **(b)** covalent and metallic radius; **(c)** atomic number and effective nuclear charge; **(d)** ionization energy and electron affinity; **(e)** paramagnetic and diamagnetic.

73. The element whose atoms have the electron configuration $[Kr]4d^{10}5s^25p^3$ **(a)** is in group 13 of the periodic table; **(b)** bears a similarity to the element Bi; **(c)** is similar to the element Te; **(d)** is a transition element.
74. The fourth-period element with the largest atom is **(a)** K; **(b)** Br; **(c)** Pb; **(d)** Kr.
75. Which of the following has the largest radius **(a)** an Ar atom; **(b)** a K^+ ion; **(c)** a Ca^{2+} ion; **(d)** a Cl^- ion?
76. The highest first ionization energy of the following is that of **(a)** Cs; **(b)** Cl; **(c)** I; **(d)** Li.
77. The most negative electron affinity of the following elements is that of **(a)** Br; **(b)** Sn; **(c)** Ba; **(d)** Li.
78. An ion that is isoelectronic with Se^{2-} is **(a)** S^{2-}; **(b)** I^-; **(c)** Xe; **(d)** Sr^{2+}.
79. Write electron configurations to show the first two ionizations for Cs. Explain why the second ionization energy is much greater than the first.
80. Explain why the first ionization energy of Mg is greater that of Na, whereas the second ionization of Na is greater than that of Mg.
81. Answer each of the following questions:
(a) Which of the elements P, As, and S has the largest atomic radius?
(b) Which of the following has the smallest radius: Xe, O^{2-}, N^{3-}, or F^-?
(c) Which should have the largest difference between the first and second ionization energy: Al, Si, P, or Cl?
(d) Which has the largest ionization energy: C, Si, or Sn?
(e) Which has the largest electron affinity: Na, B, Al, or C?
82. The first ionization energies of Si, P, S, and Cl are given in Table 9.4. Briefly provide an explanation for this trend.
83. Find three pairs of elements that are out of order in the periodic table in terms of their atomic masses. Why is it necessary to invert their order in the table?
84. For the atom $^{119}_{50}Sn$, indicate the number of **(a)** protons in the nucleus; **(b)** neutrons in the nucleus; **(c)** $4d$ electrons; **(d)** $3s$ electrons; **(e)** $5p$ electrons; **(f)** electrons in the valence shell.
85. Refer to the periodic table on the inside front cover and indicate **(a)** the most nonmetallic element; **(b)** the transition metal with lowest atomic number; **(c)** a metalloid whose atomic number is exactly midway between those of two noble gas elements.
86. Give the symbol of the element **(a)** in group 14 that has the smallest atoms; **(b)** in period 5 that has the largest atoms; **(c)** in group 17 that has the lowest first ionization energy.
87. Refer only to the periodic table on the inside front cover and indicate which of the atoms, Bi, S, Ba, As, and Ca, **(a)** is most metallic; **(b)** is most nonmetallic; **(c)** has the intermediate value when the five are arranged in order of increasing first ionization energy.
88. Arrange the following elements in order of decreasing metallic character: Sc, Fe, Rb, Br, O, Ca, F, Te.
89. In multielectron atoms many of the periodic tr[...] can be explained in terms of Z_{eff}. Consider the [fol]lowing statements and discuss whether or no[t] statement is true or false.
(a) Electrons in a p orbital are more effective [...] electrons in the s orbitals in shielding other elect[...] from the nuclear charge.
(b) Z_{eff} for an electron in an s orbital is lower [...] that for an electron in a p orbital in the same shel[...]
(c) Z_{eff} is usually less than Z.
(d) Electrons in orbitals having $\ell = 1$ penetrate b[...] than those with $\ell = 2$.
(e) Z_{eff} for the orbitals of the elements Na [...] Mg($3s$), Al($3p$), P($3p$), and S($3p$) are in the o[...] $Z_{eff}(Na) < Z_{eff}(Mg) > Z_{eff}(Al) < Z_{eff}(P) > Z_{ef}$[...]
90. Consider a nitrogen atom in the ground state [...] comment on whether the following statement[s] true or false.
(a) Z_{eff} for an electron in a $2s$ orbital is greater [...] that for the $1s$ orbital.
(b) The Z_{eff} for the $2p$ and $2s$ orbitals is the same.
(c) More energy is required to remove an elec[...] from a $2s$ orbital than from the $2p$ orbital.
(d) The $2s$ electron is less shielded than th[e] [...] electron.
91. Describe how the ionization energies of [...] ions He^-, Li^-, Be^-, B^-, C^-, N^-, O^-, and F^- vary [...] atomic number.
92. Describe how the ionization energies of [...] ions Be^+, B^+, C^+, N^+, O^+, F^+, Ne^+, and Na^+ vary [...] atomic number.
93. Which element Na or Mg is likely to have Δ[...] greater than zero?
94. Why, in general, is the addition of an electron t[o] [...] atom an exothermic process?
95. When compared to a nonmetal of the same peri[...] metal will have a larger **(a)** atomic radius; **(b)** io[...] tion energy; **(c)** electron affinity; **(d)** atomic nu[...] **(e)** none of these.
96. Which of the following is an example of a metall[...] **(a)** S; **(b)** Zn; **(c)** Ge; **(d)** Re; **(e)** none of these.
97. Which of the following has a smaller radius th[...] neon atom? **(a)** Mg^{2+}; **(b)** F^-; **(c)** O^{2-}; **(d)** K^+; **(e)** [...] of these.
98. Which electron is lost when an atom ionizes? **(a)** [...] electron with the highest principal quantum num[...] **(b)** the electron with lowest principal quantum n[...]ber; **(c)** an outer-shell electron with the highest v[...] of the orbital angular momentum quantum num[...] **(d)** the electron with highest orbital angular mom[en]tum quantum number; **(e)** none of these.
99. The electrons lost when Fe ionizes to Fe^{2+} are **(a**[...] **(b)** $3d$; **(c)** $4s$; **(d)** $3p$; **(e)** none of these.
100. Construct a concept map (see Appendix E) conn[...]ing the ideas that govern the periodic law and [...] periodic variation of atomic properties.

:hemical Bonding I:
asic Concepts

10

ONTENTS

LEARNING OBJECTIVES

10.1 Distinguish between a Lewis symbol and a Lewis structure.

10.2 Describe the use of the octet rule for writing a Lewis structure.

10.3 Differentiate between the use of electronegativities and electrostatic potential maps for describing the distribution of electron density in a molecule.

10.4 Describe the strategy for writing Lewis structures, and the use of formal charges for determining the plausibility of a given Lewis structure.

10.5 Describe the concept of resonance and the difference between equivalent and nonequivalent resonance structures.

10.6 Identify three commonly encountered exceptions to the octet rule.

10.7 Use VSEPR theory to determine the shapes of molecules.

10.8 Describe the relationship between bond order and bond length between two atoms.

10.9 Use bond-dissociation energies to estimate the enthalpy change for a gas-phase reaction.

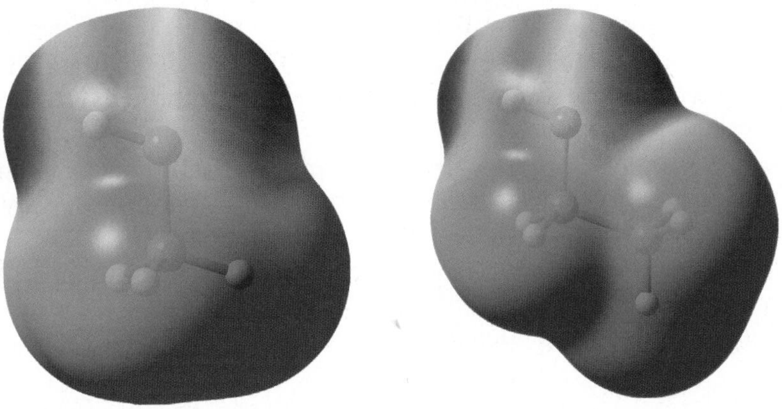

nputer-generated electrostatic potential maps of methanol (CH_3OH) and anol (CH_3CH_2OH). The surface encompassing each molecule shows the extent lectron charge density while the colors show the distribution of charge in the ecule. In this chapter, we study ideas that enable us to predict the geometric oes and polarity of molecules.

onsider all that we already know about chemical compounds. We can determine their compositions and write their formulas. We can represent the reactions of compounds by chemical equa-ns and perform stoichiometric and thermochemical calculations based these equations. And we can do all this without really having to con-er the ultimate structure of matter—the structure of atoms and mole-es. Yet the shape of a molecule—that is, the arrangement of its atoms space and their connectivities to one another—often defines its chem-y. If water had a different shape, its properties would be significantly ferent, and life as we know it would not be possible.

In this chapter, we will describe the interactions between atoms called *mical bonds*. Most of the discussion centers on the Lewis theory, which ovides one of the simplest methods of representing chemical bonding. will also explore another relatively simple theory, one for predicting obable molecular shapes. Throughout the chapter, we will try to relate these theories to what is known about molecular structures from experime measurements. In Chapter 11 we will examine the subject of chemical bon in greater depth, and in Chapter 12 we will describe intermolecular forc forces between molecules—and explore further the relationship between m cular shape and the properties of substances.

10-1 Lewis Theory: An Overview

▶ Since 1962, a number of compounds of Xe and Kr have been synthesized. As we will see in this chapter, a focus on noble-gas electron configurations can still be useful, even if the idea that they confer complete inertness is invalid.

In the period from 1916 to 1919, two Americans, Gilbert N. Lewis and Irv Langmuir, and a German, Walther Kossel, advanced an important prop about chemical bonding: Something unique in the electron configuration noble gas atoms accounts for their inertness, and atoms of other elem combine with one another to acquire electron configurations like thos noble gas atoms. The theory that grew out of this model has been most clo associated with Gilbert N. Lewis and is called the **Lewis theory**. Some fu mental ideas associated with Lewis's theory follow:

1. Electrons, especially those of the outermost (valence) electronic shell, p a fundamental role in chemical bonding.
2. In some cases, electrons are *transferred* from one atom to another. Posi and negative ions are formed and attract each other through electrost forces called **ionic bonds**.
3. In other cases, one or more pairs of electrons are *shared* between ato A bond formed by the sharing of electrons between atoms is calle **covalent bond**.
4. Electrons are transferred or shared in such a way that each atom acqu an especially stable electron configuration. Usually this is a noble gas c figuration, one with eight outer-shell electrons, or an **octet**.

▶ The term *covalent* was introduced by Irving Langmuir.

Lewis Symbols and Lewis Structures

Lewis developed a special set of symbols for his theory. A **Lewis symbol** c sists of a chemical symbol to represent the nucleus and *core* (inner-sh *electrons* of an atom, together with dots placed around the symbol to repres the *valence* (outer-shell) *electrons*. Thus, the Lewis symbol for silicon, which the electron configuration $[Ne]3s^23p^2$, is

$$\cdot\dot{\underset{\cdot}{\mathrm{Si}}}\cdot$$

Bettmann/Corbis

▲ **Gilbert Newton Lewis (1875–1946)**
Lewis's contribution to the study of chemical bonding is evident throughout this text. Equally important, however, was his pioneering introduction of thermodynamics into chemistry.

Electron spin had not yet been proposed when Lewis framed his the and so he did not show that two of the valence electrons ($3s^2$) are paired two ($3p^2$) are unpaired. We will write Lewis symbols in the way Lewis c We will place single dots on the sides of the symbol, up to a maximum of f Then we will pair up dots until we reach an octet. Lewis symbols are c monly written for main-group elements but much less often for transit elements. Lewis symbols for several main-group elements are writter Example 10-1.

A **Lewis structure** is a combination of Lewis symbols that represents eit the transfer or the sharing of electrons in a chemical bond.

Ionic bonding (transfer of electrons):

$$\underset{\text{Lewis symbols}}{\mathrm{Na}\times + \cdot\ddot{\underset{\cdot\cdot}{\mathrm{Cl}}}\mathbf{:}} \longrightarrow \underset{\text{Lewis structure}}{[\mathrm{Na}]^+[{}^{\times}_{\bullet}\ddot{\underset{\cdot\cdot}{\mathrm{Cl}}}\mathbf{:}]^-} \quad (1$$

Covalent bonding (sharing of electrons):

$$\underset{\text{Lewis symbols}}{\mathrm{H}\times + \cdot\ddot{\underset{\cdot\cdot}{\mathrm{Cl}}}\mathbf{:}} \longrightarrow \underset{\text{Lewis structure}}{\mathrm{H}\overset{\cdot}{\times}\ddot{\underset{\cdot\cdot}{\mathrm{Cl}}}\mathbf{:}} \quad (1$$

ıese two examples, we designated the electrons involved in bond formation ·rently—(×) from one atom and (•) from the other. This helps to empha- that an electron is transferred in ionic bonding and that a pair of electrons is ·ed in covalent bonding. Of course, it is impossible to distinguish between trons, and henceforth we will use only dots (•) to represent electrons in ·is structures. In Lewis theory, we use square brackets to identify ions, as we in equation (10.1). The charge on the ion is given as a superscript.

ewis's work dealt mostly with covalent bonding, which we will empha- throughout this chapter. However, Lewis's ideas also apply to ionic bond- and we briefly describe this application next.

XAMPLE 10-1 Writing Lewis Symbols

Write Lewis symbols for the following elements: **(a)** N, P, As, Sb, Bi; **(b)** Al, I, Se, Ar.

nalyze

The position of the element in the periodic table determines the number of valence electrons in the Lewis symbol. For main-group elements, the number of valence electrons, and hence the number of dots appearing in a Lewis symbol, is equal to the group number for the *s*-block elements and to the group number minus 10 for the *p*-block elements.

olve

(a) These are group 15 elements, and their atoms all have five valence electrons (ns^2np^3). The Lewis symbols all have five dots.

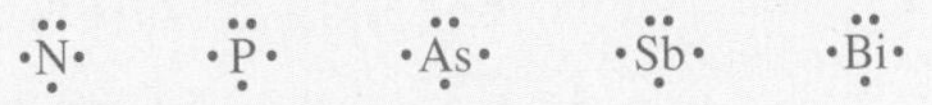

(b) Al is in group 13; I, in group 17; Se, in group 16; Ar, in group 18.

ssess

This example, although very straightforward, is very important. The accurate counting of valence electrons is essential for many aspects of chemical bonding.

RACTICE EXAMPLE A: Write Lewis symbols for Mg, Ge, K, and Ne.

RACTICE EXAMPLE B: Write the Lewis symbols expected for Sn, Br^-, Tl^+, and S^{2-}.

wis Structures for Ionic Compounds

ection 3-2, we learned that the formula unit of an ionic compound is the ıplest electrically neutral collection of cations and anions from which the ·mical formula of the compound can be established. The Lewis structure of ium chloride (equation 10.1) represents its formula unit. For an ionic com- ınd of a main-group element, (1) the Lewis symbol of the metal ion has no :s if all the valence electrons are lost, and (2) the ionic charges of both cations l anions are shown. These ideas are further illustrated through Example 10-2.

◀ No bond is 100% ionic. All ionic bonds have some covalent character.

EXAMPLE 10-2 Writing Lewis Structures of Ionic Compounds

Write Lewis structures for the following compounds: **(a)** BaO; **(b)** $MgCl_2$; **(c)** aluminum oxide.

Analyze

Our approach here is to write the Lewis symbol and determine how many electrons each atom must gain or lose to acquire a noble-gas-electron configuration.

(continued)

Solve

(a) Ba loses two electrons, and O gains two. In the equation below, we use curved red arrows, each with half an arrowhead, to show the movement of single electrons.

$$Ba\cdot + \cdot\overset{..}{\underset{..}{O}}: \longrightarrow [Ba]^{2+} [:\overset{..}{\underset{..}{O}}:]^{2-}$$

Lewis structure

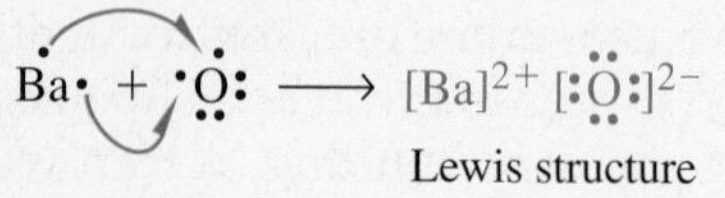

(b) A Cl atom can accept only one electron because it already has seven valence electrons. One more electron will give it a complete octet. Conversely, a Mg atom must lose two electrons to have the electron configuration of the preceding noble gas neon. So two Cl atoms are required for each Mg atom.

$$Mg\cdot + \begin{matrix} \cdot\overset{..}{\underset{..}{Cl}}: \\ \\ \cdot\overset{..}{\underset{..}{Cl}}: \end{matrix} \longrightarrow [Mg]^{2+} 2[:\overset{..}{\underset{..}{Cl}}:]^-$$

Lewis structure

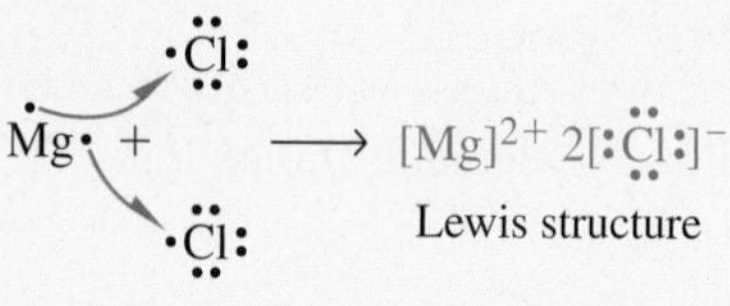

(c) The formula of aluminum oxide follows directly from the Lewis structure. The combination of one Al atom, which loses three electrons, and one O atom, which gains two, leaves an excess of one lost electron. To match the numbers of electrons lost and gained, the formula unit must be based on *two* Al atoms and *three* O atoms.

$$\begin{matrix} Al\cdot & \cdot\overset{..}{\underset{..}{O}}: \\ & \cdot\overset{..}{\underset{..}{O}}: & \longrightarrow & 2[Al]^{3+} \ 3[:\overset{..}{\underset{..}{O}}:]^{2-} \\ Al\cdot & \cdot\overset{..}{\underset{..}{O}}: \end{matrix}$$

Lewis structure

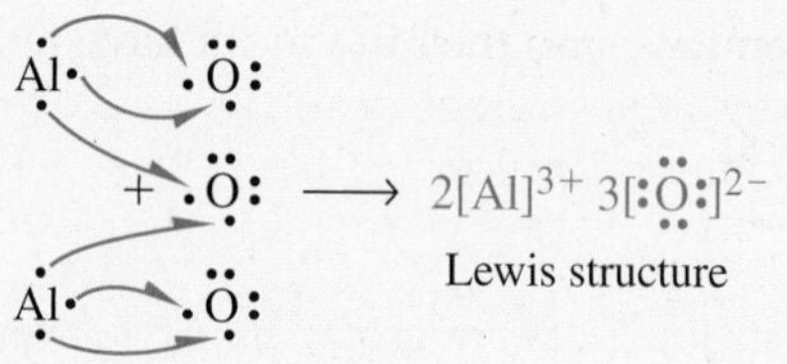

Assess

We almost never write Lewis structures for ionic compounds, except when we want to emphasize the ratio in which the ions combine. The structures of ionic compounds are much more complicated than is suggested by the Lewis structure. See, for example, the structure of NaCl shown in Figure 10-1.

PRACTICE EXAMPLE A: Write plausible Lewis structures for (a) Na_2S and (b) Mg_3N_2.

PRACTICE EXAMPLE B: Write plausible Lewis structures for (a) calcium iodide; (b) barium sulfide; (c) lithium oxide.

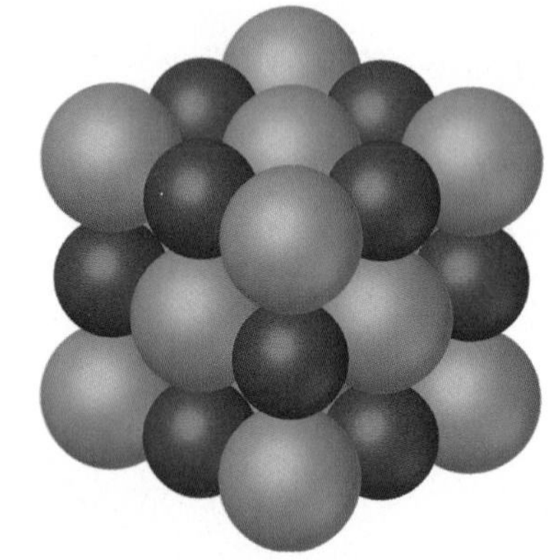

The compounds described in Example 10-2 are *binary ionic compounds* consisting of monatomic cations and monatomic anions. Commonly encountered *ternary ionic compounds* consist of monatomic and polyatomic ions. Bonds between atoms within the polyatomic ions is covalent. Some ternary ionic compounds are considered later in the chapter.

With the exception of ion pairs, such as (Na^+Cl^-), that may be found in the *gaseous* state, formula units of solid ionic compounds do not exist as separate entities. Instead, each cation is surrounded by anions and each anion by cations. These very large numbers of ions are arranged in an orderly network of alternating cations and anions called an *ionic crystal* (Fig. 10-1). Ionic crystal structures and the energy changes accompanying the formation of ionic crystals are described in Chapter 12.

▲ FIGURE 10-1
Portion of an ionic crystal
This structure of alternating Na^+ and Cl^- ions extends in all directions and involves countless numbers of ions.

🔍 **10-1   CONCEPT ASSESSMENT**

How many valence electrons do the Lewis symbols for the elements in group 16 have? Which of the following are correct Lewis symbols for sulfur?

$$:\overset{.}{\underset{}{S}}: \quad :\overset{.}{\underset{..}{S}}: \quad \cdot\overset{.}{\underset{.}{S}}\cdot \quad \cdot\overset{.}{\underset{.}{S}}:$$

-2 Covalent Bonding: An Introduction

hlorine atom shows a tendency to gain an electron, as indicated by its elec- affinity (-349 kJ mol^{-1}). From which atom, sodium or hydrogen, can the tron most readily be extracted? Neither atom gives up an electron freely, the energy required to extract an electron from Na ($E_i = 496$ kJ mol^{-1}) is ch smaller than that for H ($E_i = 1312$ kJ mol^{-1}). In Chapter 9 we learned t the lower its ionization energy, the more metallic an element is; sodium is ch more metallic than hydrogen (recall Figure 9-19). In fact, hydrogen is sidered to be a nonmetal. A hydrogen atom in the gaseous state does not e up an electron to another nonmetal atom. Bonding between a hydrogen m and a chlorine atom involves the sharing of electrons, which leads to a *alent bond.*

To emphasize the sharing of electrons, let us think of the Lewis structure of l in this manner.

e broken circles represent the outermost electron shells of the bonded atoms. e number of dots lying on or within each circle represents the effective num- of electrons in each valence shell. The H atom has two dots, as in the electron figuration of He. The Cl atom has eight dots, corresponding to the outer- ll configuration of Ar. Note that we counted the two electrons between H l Cl (:) *twice*. These two electrons are shared by the H and Cl atoms. This red pair of electrons constitutes the covalent bond. Written below are two litional Lewis structures of simple molecules.

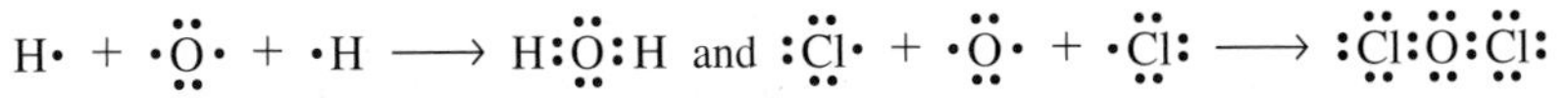

Water — Dichlorine monoxide

◀ The Lewis structures for H_2O and Cl_2O suggest that these molecules have a linear shape. They do not. Lewis theory by itself does not address the question of molecular shape (see Section 10-7).

As was the case for Cl in HCl, the O atom in the Lewis structure of H_2O and Cl_2O is surrounded by eight electrons (when the bond-pair electrons are uble counted). In attaining these eight electrons, the O atom conforms to the **et rule**—a requirement of eight valence-shell electrons for the atoms in a wis structure. Note, however, that the H atom is an exception to this rule. e H atom can accommodate only two valence-shell electrons.

Lewis theory helps us to understand why elemental hydrogen and chlorine st as diatomic molecules, H_2 and Cl_2. In each case, a pair of electrons is red between the two atoms. The sharing of a single pair of electrons ween bonded atoms produces a **single covalent bond**. To underscore the portance of electron pairs in the Lewis theory the term **bond pair** applies to air of electrons in a covalent bond, while **lone pair** applies to electron pairs t are not involved in bonding. Also, in writing Lewis structures it is cus- nary to replace bond pairs with lines (—). These features are shown in the lowing Lewis structures.

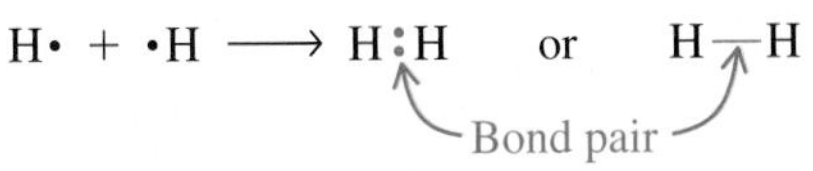

(10.3)

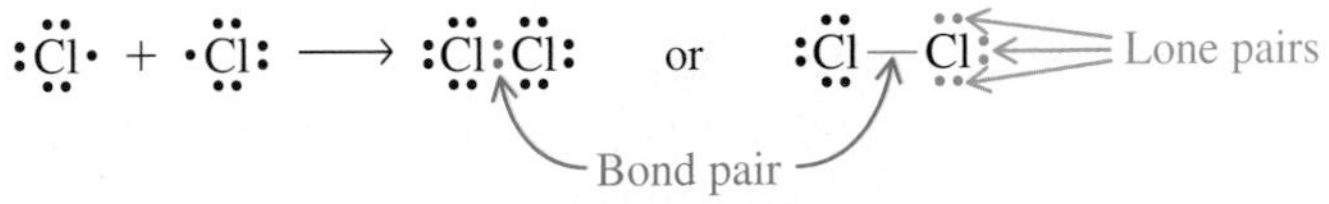

(10.4)

ordinate Covalent Bonds

e Lewis theory of bonding describes a covalent bond as the sharing of a pair electrons, but this does not necessarily mean that each atom contributes an ctron to the bond. A covalent bond in which a single atom contributes both the electrons to a shared pair is called a **coordinate covalent bond**.

EXAMPLE 10-3 **Writing Simple Lewis Structures**

Write a Lewis structure for the ammonia molecule, NH_3.

Analyze

To write a Lewis structure we must know the number of valence electrons associated with each atom.

Solve

The valence electrons can then be represented in the Lewis symbols, as shown here.

$$H\cdot \quad H\cdot \quad H\cdot \quad \cdot \overset{\cdot\cdot}{\underset{\cdot\cdot}{N}}:$$

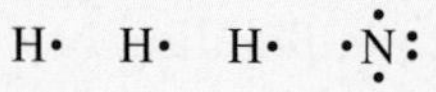

Now we can assemble one N and three H atoms into a structure that gives the N atom a valence-shell octet and each of the H atoms two valence electrons (producing the electron configuration of He).

$$\begin{array}{c} H \\ H:\overset{\cdot\cdot}{\underset{\cdot\cdot}{N}}: \\ H \end{array}$$

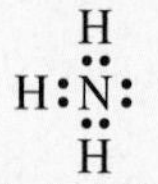

Assess

The application of the octet rule has led us to the correct Lewis structure for ammonia, but, as we will see later in this text, many molecules do not obey the octet rule.

PRACTICE EXAMPLE A: Write Lewis structures for Br_2, CH_4, and $HOCl$.

PRACTICE EXAMPLE B: Write Lewis structures for NI_3, N_2H_4, and C_2H_6.

If we attempt to attach a fourth H atom to the Lewis structure of NH_3 show in Example 10-3, we encounter a difficulty. The electron brought by the fourth atom would raise the total number of valence electrons around the N atom *nine*, so there would no longer be an octet. The *molecule* NH_4 does not form, the *ammonium ion*, $NH_4{}^+$, does, as suggested in Figure 10-2. That is, the lo pair of electrons on a NH_3 molecule extracts an H atom from a HCl molecule and the electrons in the H—Cl bond remain on the Cl atom. The result is equ alent to a H^+ ion joining with the NH_3 molecule to form the $NH_4{}^+$ ion,

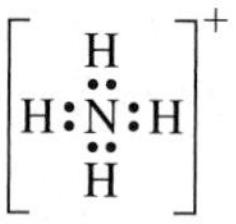

$$\left[\begin{array}{c} H \\ H:\overset{\cdot\cdot}{\underset{\cdot\cdot}{N}}:H \\ H \end{array} \right]^+ \tag{1}$$

As shown in Figure 10-2, the electron pair from the H—Cl bond remains the Cl atom, converting it to a Cl^- ion.

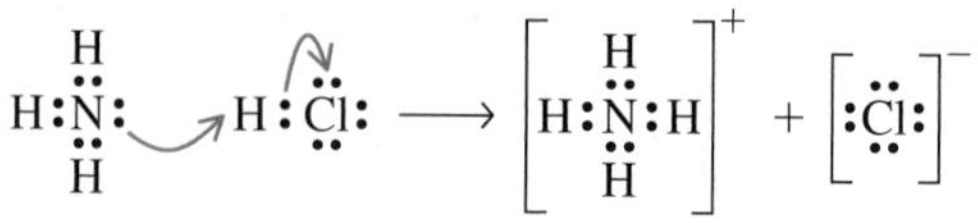

▶ FIGURE 10-2
Formation of the ammonium ion, $NH_4{}^+$
The H atom of HCl leaves its electron with the Cl atom and, as H^+, attaches itself to the NH_3 molecule through the lone-pair electrons on the N atom. The ions $NH_4{}^+$ and Cl^- are formed.

$$H:\overset{\cdot\cdot}{\underset{H}{N}}: \quad \overset{\curvearrowleft}{H}:\overset{\cdot\cdot}{\underset{\cdot\cdot}{Cl}}: \longrightarrow \left[\begin{array}{c} H \\ H:\overset{\cdot\cdot}{\underset{\cdot\cdot}{N}}:H \\ H \end{array} \right]^+ + \left[:\overset{\cdot\cdot}{\underset{\cdot\cdot}{Cl}}: \right]^-$$

The bond formed between the N atom of NH_3 and the H^+ ion in struct (10.5) is a *coordinate covalent bond*. It is important to note, however, that on the bond has formed, it is impossible to say which of the four N—H bond the coordinate covalent bond. Thus, a coordinate covalent bond is indist guishable from a regular covalent bond.

Another example of coordinate covalent bonding is found in the famil hydronium ion.

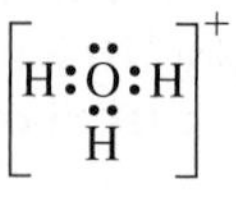

$$\left[\begin{array}{c} H:\overset{\cdot\cdot}{\underset{\cdot\cdot}{O}}:H \\ H \end{array} \right]^+ \tag{1}$$

10-2 CONCEPT ASSESSMENT

'hat types of bonds can be used to describe the chemical bonds in BF_4^-?

ıltiple Covalent Bonds

he preceding description of the Lewis model for covalent chemical bonding, have used a single pair of electrons between two atoms to describe a single alent bond. Often, however, more than one pair of electrons must be shared n atom is to attain an octet (noble gas electron configuration). CO_2 and N_2 two molecules in which atoms share more than one pair of electrons.

'irst, let's apply the ideas about Lewis structures to CO_2. From the Lewis nbols, we see that the C atom can share a valence electron with each O m, thus forming two carbon-to-oxygen single bonds.

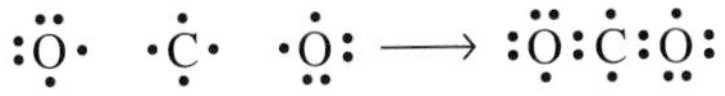

: this leaves the C atom and both O atoms still shy of an octet. The problem olved by shifting the unpaired electrons into the region of the bond, as icated by the red arrows.

(10.7)

Lewis structure (10.7), the bonded atoms are seen to share *two* pairs of elecns (a total of four electrons) between them—a **double covalent bond** (=).

Now let's try our hand at writing a Lewis structure for the N_2 molecule. r first attempt might again involve a single covalent bond and the incorrect ıcture shown below.

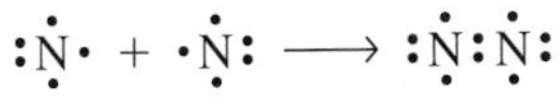

(*Incorrect*)

ch N atom appears to have only six outer-shell electrons, not the expected ht. The situation can be corrected by bringing the four unpaired electrons o the region between the N atoms and using them for additional bond pairs. all, we now show the sharing of *three* pairs of electrons between the N atoms. e bond between the N atoms in N_2 is a **triple covalent bond** (≡). Double d triple covalent bonds are known as **multiple covalent bonds**.

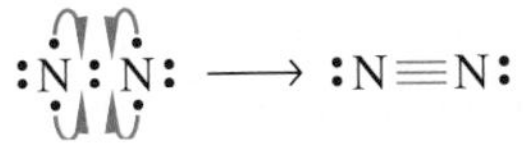

(10.8)

The triple covalent bond in N_2 is a very strong bond that is difficult to break a chemical reaction. The unusual strength of this bond makes $N_2(g)$ quite ert. As a result, $N_2(g)$ coexists with $O_2(g)$ in the atmosphere and forms ides of nitrogen only in trace amounts at high temperatures. The lack of ctivity of N_2 with O_2 is an essential condition for life on Earth. The inertss of $N_2(g)$ also makes it difficult to synthesize nitrogen compounds.

Another molecule whose Lewis structure features a multiple bond is O_2, ıich has a double bond.

(10.9)

e blue question mark suggests that there is some doubt about the validity of ucture (10.9), and the source of the doubt is illustrated in Figure 10-3. The strucre fails to account for the *paramagnetism* of oxygen—the O_2 molecule must have *paired* electrons. Unfortunately, no completely satisfactory Lewis structure is ssible for O_2, but in Chapter 11, bonding in the O_2 molecule is described in a ıy that accounts for both the double bond and the observed paramagnetism.

◀ Throughout this chapter, we use curved red arrows to help us visualize the movement of electrons. IUPAC recommends the use of an arrow with a half arrowhead when a single electron is moved and an arrow with a full arrowhead when a pair of electrons is moved.

Richard Megna/Fundamental Photographs

▲ FIGURE 10-3
Paramagnetism of oxygen
Liquid oxygen is attracted into the magnetic field of a large magnet.

KEEP IN MIND

that merely being able to write a plausible Lewis structure does not prove that it is the correct electronic structure. Proof can come only through confirming experimental evidence.

We could continue applying ideas introduced in this section, but our abi to write plausible Lewis structures will be greatly aided by a couple of ı ideas that we introduce in Section 10-3.

10-3 CONCEPT ASSESSMENT

In which groups of the periodic table are the elements most likely to use multiple bonds?

10-3 Polar Covalent Bonds and Electrostatic Potential Maps

We have introduced ionic and covalent bonds as though they are of two d tinctly different types: ionic bonds involving a *complete transfer* of electr and covalent bonds involving an *equal sharing* of electron pairs. Such is the case, however, and most chemical bonds fall between the two extreme 100% ionic and 100% covalent. A covalent bond in which electrons are shared equally between two atoms is called a **polar covalent bond**. In suc bond, electrons are displaced toward the more nonmetallic element. T unequal sharing of the electrons leads to a partial negative charge on more nonmetallic element, signified by $\delta-$, and a corresponding partial po tive charge on the more metallic element, designated by $\delta+$. Thus we can r resent the polar bond in HCl by a Lewis structure in which the partial char $\delta+$ and $\delta-$ indicate that the bond pair of electrons lies closer to the Cl th to the H.

$$^{\delta+}\mathrm{H}\ :\underset{\cdot\cdot}{\overset{\cdot\cdot}{\mathrm{Cl}}}:^{\delta-}$$

The advent of inexpensive, fast computers has allowed chemists to deve methods for displaying the electron distribution within molecules. This dis bution is obtained, in principle, by solving the Schrödinger equation for a m ecule. Although the solution can be obtained only by using approxim methods, these methods provide an **electrostatic potential map**, a way visualize the charge distribution within a molecule.

Before discussing these maps let us first review the notion of electron de sity, or charge density, introduced in Chapter 8. There we saw that the beha ior of electrons in atoms can be described by mathematical functions call orbitals. The probability of finding an electron at some point in the thre dimensional region associated with an orbital is related to the square of atomic orbital function. Typically, we refer to the region encompassing 95% the probability of finding the electron as the shape of the orbital. In a simi way we can map the total electron density throughout a molecule, that is, r just the density of a single orbital. The electron density surface that enco passes 95% of the charge density in ammonia is depicted in Figure 10-4.

The *electrostatic potential* is the work done in moving a unit of positi charge at a constant speed from one region of a molecule to another. The ele trostatic potential map is obtained by hypothetically probing an electron de sity surface with a positive point charge. The positive point charge will attracted to an electron-rich region—a region of excess negative charge wh all the charges of the nuclei and electrons have been taken into account—a the electrostatic potential will be *negative*. Conversely, if the point charge placed in an electron-poor region, a region of excess positive charge, the po tive point charge will be repelled, and the electrostatic potential will *positive*. The procedure for making an electrostatic potential map is illustrat in Figure 10-4, which shows the distribution of electron density in ammoni

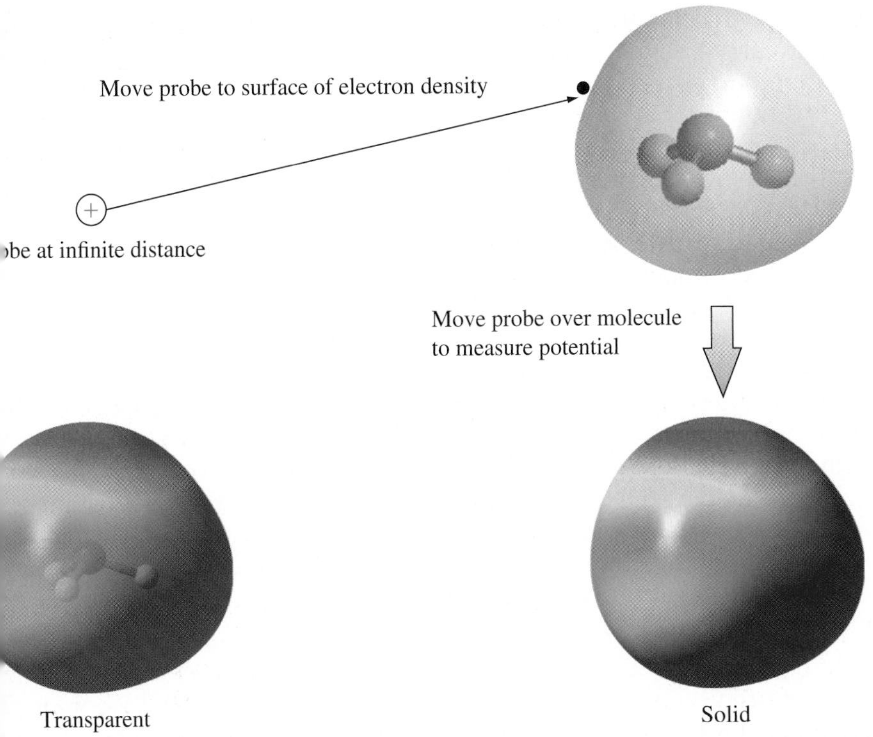

◀ FIGURE 10-4
Determination of the electrostatic potential map for ammonia
The electrostatic potential at any point on the charge density surface of a molecule is defined as the change in energy that occurs when a unit positive charge is brought to this point, starting from another point that is infinitely far removed from the molecule. The surface encompassing the ammonia molecule is analogous to the 95% surface of electron charge density for atomic orbitals discussed in Chapter 8. The electrostatic potential map gives information about the distribution of electron charge within this surface.

\n electrostatic potential map gives information about the distribution of :tron charge in a molecule. For example, in a neutral molecule, if the poten- at a point is positive, it is likely that an atom at this point carries a net posi- › charge. An arbitrary "rainbow" color scheme is adopted in the display of an :trostatic potential map. Red, the low-energy end of the spectrum, is used for ions of the most negative electrostatic potential, and blue is used to color ions of the most positive electrostatic potential. Intermediate colors represent ›rmediate values of the electrostatic potential. Thus, the potential increases m red through yellow to blue, as seen in the scale in Figure 10-5. For example,

◀ Electrostatic potential is the work done in moving a unit of positive charge at a constant speed from one region of a molecule to another.

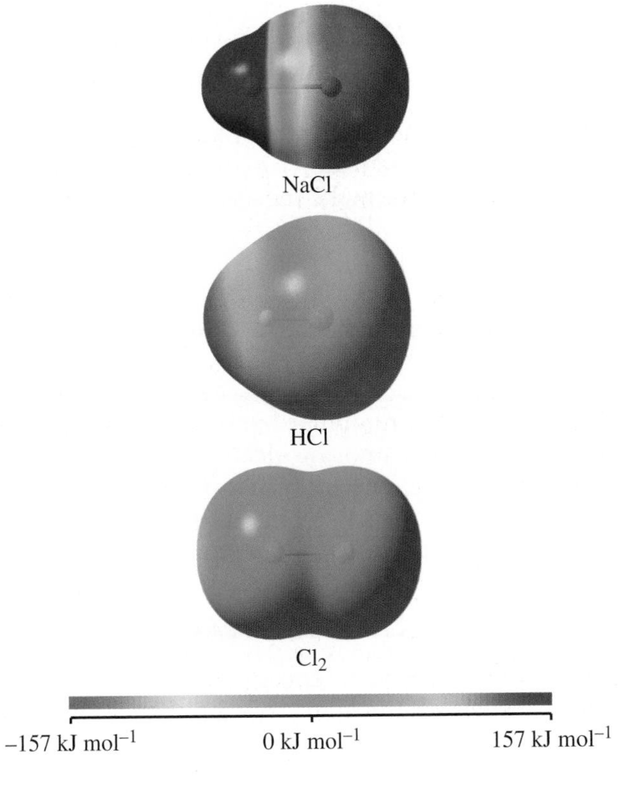

◀ FIGURE 10-5
The electrostatic potential maps for sodium chloride, hydrogen chloride, and chlorine
The dark red and dark blue on the electrostatic potential map correspond to the extremes of the electrostatic potential, negative to positive, for the particular molecule for which the map is calculated. To get a reliable comparison of different molecules, the values of the extremes in electrostatic potential (in $kJ\ mol^{-1}$) must be the same for all of the molecules compared. In the maps shown here the range is -157 to $157\ kJ\ mol^{-1}$.

the blue-green color surrounding the hydrogen atoms in Figure 10-4 sugg that they carry a slight positive charge. The nitrogen atom, being closest to red region, carries a net negative charge.

Let us now look at the computed electrostatic potential maps for NaCl, and HCl (Fig. 10-5). We see that Cl_2 has a uniform distribution of electron ch density as depicted by the uniform color distribution in the electrostatic pote map. This is typical for a nonpolar covalent bond and occurs in all diatomic ecules containing identical atoms. The sodium chloride molecule, convers exhibits a highly nonuniform distribution of electron charge density. The sod atom is almost exclusively in the blue extreme of positive charge and the chlo in the red extreme of negative charge. This electrostatic potential map is typ of an ionic bond, yet it is clear from the map that the transfer of electron den from the sodium atom to the chlorine atom is not complete. That is, the N bond is not completely ionic. Experiments show that the bond is about 80% io The molecule HCl also has an unsymmetrical distribution of electron charge sity, as indicated by the gradation of color in the electrostatic potential map. hydrogen atom has a partial positive charge, as indicated by the pale b Correspondingly, the chlorine atom has a partial negative charge, as indicate the yellow-green color. The electrostatic potential map clearly depicts the p nature of the bond in HCl.

Electronegativity

We expect the H—Cl bond to be polar because the Cl atom has a gre affinity for electrons than does the H atom. Electron affinity is an ato property, however, and more meaningful predictions about bond polari are those based on a molecular property, one that relates to the ability atoms to lose or gain electrons when they are part of a molecule rather t isolated from other atoms.

Electronegativity (EN) describes an atom's ability to compete for electr with other atoms to which it is bonded. As such, electronegativity is relate ionization energy (E_i) and electron affinity (E_{ea}). To see how they are rela consider the reaction between two hypothetical elements, A and B, wh could give the products A^+B^- or A^-B^+. We represent these two reactions by expressions

$$A + B \longrightarrow A^+B^- \qquad \Delta E_1 = E_i(A) + E_{ea}(B) \qquad (10$$

$$A + B \longrightarrow A^-B^+ \qquad \Delta E_2 = E_i(B) + E_{ea}(A) \qquad (10$$

If the bonding electrons are shared approximately equally in these hypoth cal structures, we would expect that $\Delta E_1 = \Delta E_2$ because neither extre (A^+B^- or A^-B^+) is favored. If we make the assumption that the resultant bo is nonpolar, then

$$E_i(A) + E_{ea}(B) = E_i(B) + E_{ea}(A)$$

which gives, after collecting terms for each atom,

$$E_i(A) - E_{ea}(A) = E_i(B) - E_{ea}(B) \qquad (10.$$

Equation (10.12) tells us that a nonpolar bond will result when the differe between the ionization energy and the electron affinity is the same for bo atoms involved in the bond. The quantity ($E_i - E_{ea}$) provides a meas of the ability of an atom to attract electrons (or electron charge density) itself relative to some other atom. Thus it is related to the electronegativity the atom.

$$EN_A \propto E_i(A) - E_{ea}(A)$$

An element with a high ionization energy and an electron affinity that large and negative, such as fluorine, will have a large electronegativ

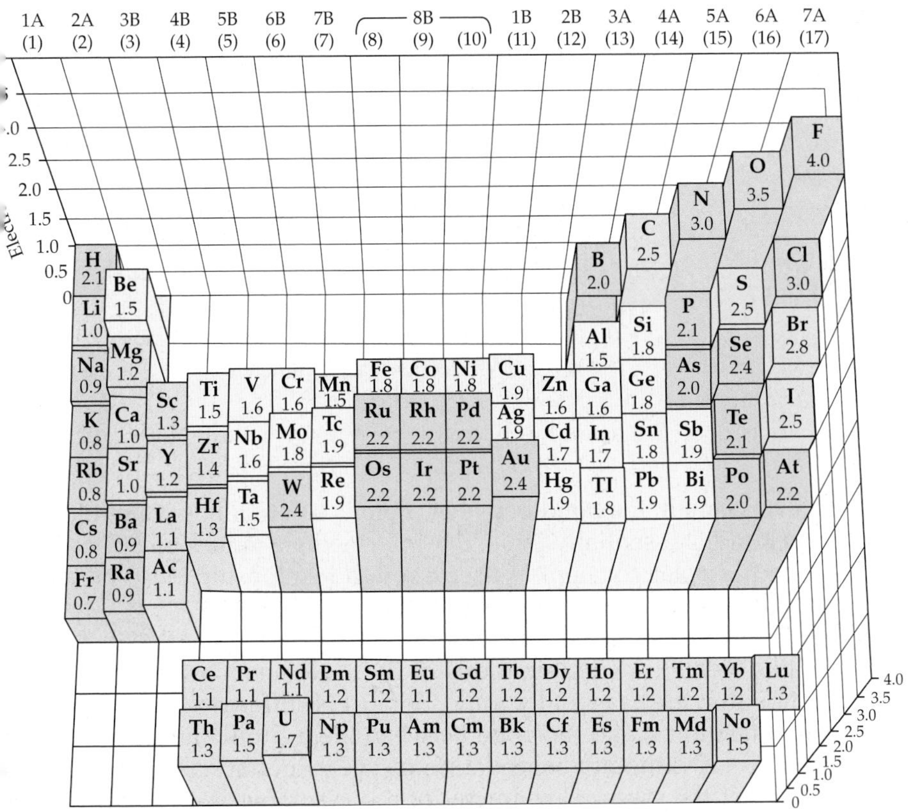

◀ FIGURE 10-6
Electronegativities of the elements
As a general rule, electronegativities *decrease* from *top to bottom* in a group and *increase* from *left to right* in a period of elements. The values are from L. Pauling, *The Nature of the Chemical Bond*, 3rd ed., Cornell University, Ithaca, NY, 1960, page 93. Values may be somewhat different when based on other electronegativity scales.

tive to an atom with a low ionization energy and a small electron affinity, h as sodium.

There are several methods for converting qualitative comparisons to actual nerical values of the electronegativities of the elements. Some of these meth- are described and compared in Exercise 128. One widely used electronega- ty scale, with values given in Figure 10-6, is that devised by Linus Pauling 01–1994). Pauling's EN values range from about 0.7 to 4.0. In general, the ver its EN, the more metallic the element is, and the higher the EN, the more nmetallic it is. From Figure 10-6 we also see that electronegativity decreases m top to bottom in a group and increases from left to right in a period of the iodic table. These are the expected trends when we interpret electronegativity erms of the quantity ($E_i - E_{ea}$). That is, as the ionization energy (E_i) increases oss the period we expect the electronegativity to increase. The distinction ween electron affinity and electronegativity is clearly seen when we consider electron affinities of fluorine (-328 kJ mol^{-1}) and chlorine (-349 kJ mol^{-1}): hough the electron affinity of Cl (-349 kJ mol^{-1}) is somewhat more negative n that of F (-328 kJ mol^{-1}), the EN of Cl (3.0) is significantly lower than that F (4.0) because of the decreased ionization energy of Cl (1251 kJ mol^{-1}) rela- e to F (1681 kJ mol^{-1}).

◀ Pauling's electronegativity scale is based on bond energies (Section 10-9). The bond energy for the A–B bond represents the energy change for the process AB(g) → A(g) + B(g). Pauling observed that the bond energy, D_{A-B}, for the A–B bond is greater than the average of the A–A and B–B bond energies, $\frac{1}{2}(D_{A-A} + D_{B-B})$. His first set of electronegativity values were calculated by assuming the magnitude of $EN_A - EN_B$ is directly proportional to the difference between D_{A-B} and $\frac{1}{2}(D_{A-A} + D_{B-B})$.

10-4 CONCEPT ASSESSMENT

Vith the aid of only a periodic table, decide which is the most electronegative tom of each of the following sets of elements: **(a)** As, Se, Br, I; **(b)** Li, Be, Rb, Sr; **c)** Ge, As, P, Sn.

Electronegativity values allow an insight into the amount of polar character a covalent bond based on **electronegativity difference, ΔEN**—the absolute lue of the difference in EN values of the bonded atoms. If ΔEN for two

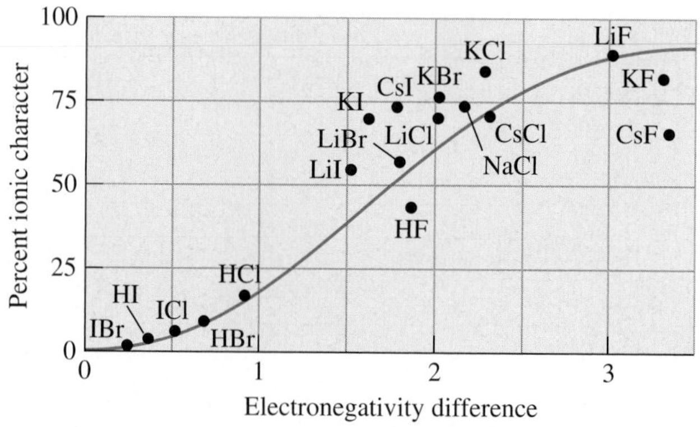

▶ FIGURE 10-7
Percent ionic character of a chemical bond as a function of electronegativity difference
The red curve represents the equation:
$\%$ ionic character $= 100 \times [1 - e^{-0.25(EN_A - EN_B)^2}]$.

▶ Although Figure 10-7 suggests that the bond between two identical metal atoms should be covalent [as it is in $Li_2(g)$, for example], in *solid* metals, where bonding extends throughout a network of many, many atoms, the bonding is of a type called *metallic* (explored in the next chapter).

atoms is very small, the bond between them is essentially covalent. If ΔEN large, the bond is essentially ionic. For intermediate values of ΔEN, the bo is described as polar covalent. A useful rough relationship between ΔEN a percent ionic character of a bond is presented in Figure 10-7.

Large EN differences are found between the more metallic and the m nonmetallic elements. Combinations of these elements are expected to p duce bonds that are essentially ionic. Small EN differences are expected two nonmetal atoms, and the bond between them should be essentially co lent. Thus, even without a compilation of EN values at hand, you should able to predict the essential character of a bond between two atoms. Sim assess the metallic/nonmetallic characters of the bonded elements from periodic table (recall Figure 9-19).

EXAMPLE 10-4 Assessing Electronegativity Differences and the Polarity of Bonds

(a) Which bond is more polar, H—Cl or H—O?
(b) What is the percent ionic character of each of these bonds?

Analyze

To decide which bond is more polar, look up EN values for H, Cl, and O in Figure 10-6, and then compute electronegativity differences, ΔEN, for H—Cl and H—O bonds. The greater the electronegativity difference, the more polar the bond. To determine the percentage ionic character, we use the curve in Figure 10-7.

Solve

(a) $EN_H = 2.1$; $EN_{Cl} = 3.0$; $EN_O = 3.5$. For the H—Cl bond, $\Delta EN = 3.0 - 2.1 = 0.9$. For the H—O bond, $\Delta EN = 3.5 - 2.1 = 1.4$. Because its ΔEN is somewhat greater, we expect the H—O bond to be the more polar bond.
(b) Determine the percent ionic character from Figure 10-7.

H—Cl bond: $\Delta EN = 0.9$ ≈18% ionic

H—O bond: $\Delta EN = 1.4$ ≈39% ionic

Assess

In this example, we used EN values to decide which of two bonds is more polar. EN values can also be used to decide which end of a given bond will be slightly negative. For example, because EN_{Cl} is greater than EN_H, we conclude that the Cl end of the H—Cl bond will be slightly negative; thus, the H end will be slightly positive.

PRACTICE EXAMPLE A: Which of the following bonds are the most polar, that is, have the greatest ionic character: H—Br, N—H, N—O, P—Cl?

PRACTICE EXAMPLE B: Which is the most polar bond: C—S, C—P, P—O, or O—F?

illustrate the variation of bond polarity with electronegativity using elec-
static potential maps, consider the electrostatic potential maps for HCl,
r, and HI displayed below.

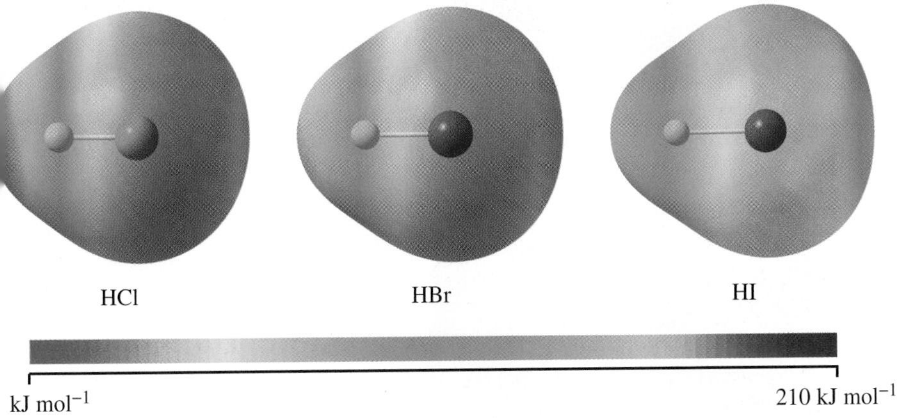

◀ The electrostatic potential map shown here for HCl has a different range of colors than the one shown in Figure 10-5. In Fig. 10-5, the electrostatic potential map for NaCl shows the greatest range of colors because, of the molecules shown there, NaCl has the greatest ionic character. Here, the electrostatic potential map for HCl shows the greatest range of colors because, compared to HBr and HI, HCl has the greatest ionic character. (See Figure 10-7.)

n these electrostatic potential maps, the color on the H atom ranges from k blue in HCl to pale blue in HI, which is consistent with a decreasing pos- e charge on the H atom. Correspondingly, the halogen atom becomes less —signifying a decreasing negative charge in going from chlorine to iodine. Electrostatic potential maps are a powerful way of displaying the variation polarity within a group of related molecules. We will use computed elec- static potential maps later in this chapter and in subsequent chapters, enever charge separation within a molecule contributes significantly to derstanding the topic at hand.

EXAMPLE 10-5 Identifying a Molecular Structure Using Electronegativity and Electrostatic Potential Maps

Two electrostatic potential maps are shown below. One corresponds to NaF and the other to NaH. Which map corresponds to which molecule?

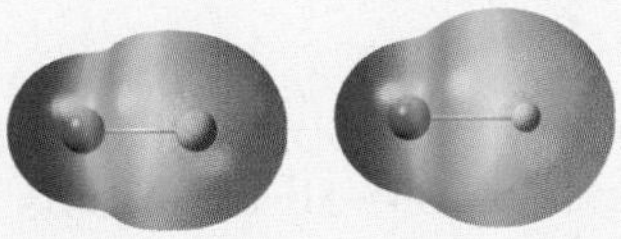

Analyze

Look up EN values for H, F, and Na in Figure 10-6, and then compute electronegativity differences for NaF and NaH bonds. The bond with the greatest electronegativity difference will be more polar, and its electrostatic potential map will show a greater range of colors.

Solve

$EN_H = 2.1$; $EN_{Na} = 0.9$; $EN_F = 4.0$. For the H—Na bond, $\Delta EN = 2.1 - 0.9 = 1.2$. For the F—Na bond, $\Delta EN = 4.0 - 0.9 = 3.1$. Because ΔEN for NaF is greater, we expect the F—Na bond to be the more polar bond. We conclude that the electrostatic potential map on the left represents NaF.

Assess

It may seem surprising that, in the electrostatic potential maps shown for NaF and NaH, the charge density surface around the H "atom" in NaH appears larger than the F "atom" in NaF. Bear in mind that the bonds in both molecules have significant ionic character, and so, when comparing the electrostatic potentials maps for these two molecules, it is more appropriate to think in terms of F^- and H^- ions. Various studies suggest that the NaH bond is probably between 50% and 80% ionic and that the NaF bond is about 90% ionic. Studies on solid NaH suggest that the radius of a H^- ion is somewhere between that of F^- (133 pm) and that of Cl^- (181 pm).

(continued)

PRACTICE EXAMPLE A: Which of the following electrostatic potential maps corresponds to IF, and which to IBr

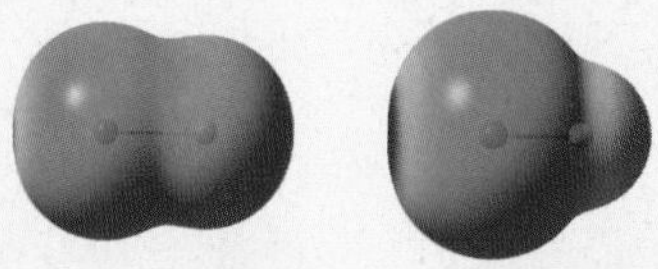

PRACTICE EXAMPLE B: Which of the following electrostatic potential maps corresponds to CH_3OH, and which to CH_3SH?

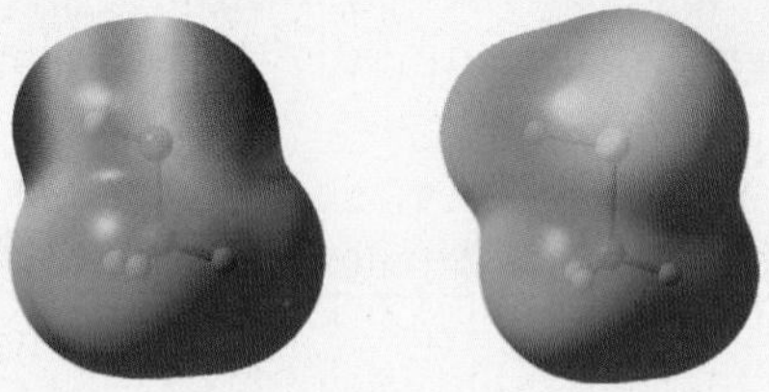

10-4 Writing Lewis Structures

In this section we combine the ideas introduced in the preceding three sectic with a few new concepts to write a variety of Lewis structures. Let us be with a reminder of some of the essential features of Lewis structures that have already encountered.

- *All* the valence electrons of the atoms in a Lewis structure must appear the structure.
- *Usually*, all the electrons in a Lewis structure are paired.
- *Usually*, each atom acquires an outer-shell octet of electrons. Hydrog however, is limited to two outer-shell electrons.
- *Sometimes*, multiple covalent bonds (double or triple bonds) are need Multiple covalent bonds are formed most readily by C, N, O, P, and S ator

Skeletal Structures

The usual starting point in writing a Lewis structure is to designate t **skeletal structure**—all the atoms in the structure arranged in the order which they are bonded to one another. In a skeletal structure with more th two atoms, we generally need to distinguish between central and termi atoms. A **central atom** is bonded to two or more atoms, and a **terminal atom** bonded to just one other atom. As an example, consider ethanol, CH_3CH_2O Its skeletal structure is the same as the following structural formula. In th structure, the *central atoms*—both C atoms and the O atom—are printed in re The *terminal atoms*—all six H atoms—are printed in blue.

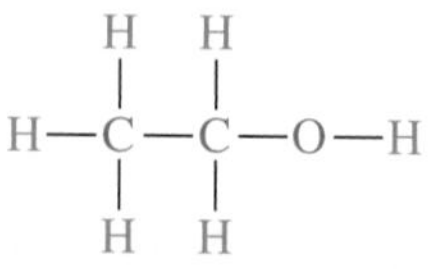

(10.

Here are a few additional facts about central atoms, terminal atoms, and skele structures.

- *Hydrogen atoms are always terminal atoms.* This is because an H atom c accommodate only two electrons in its valence shell, so it can form on one bond to another atom. (An interesting and rare exception occurs some boron–hydrogen compounds.)

Central atoms are generally those with the lowest electronegativity. In the skeletal structure (10.13), the atoms of lowest electronegativity (EN = 2.1) happen to be H atoms, but as noted above, H atoms can be only terminal atoms. Next lowest in electronegativity (EN = 2.5) are the C atoms, and these are central atoms. The O atom has the highest electronegativity (3.5) but nevertheless is also a central atom. For O to be a terminal atom in structure (10.13) would require it to exchange places with an H atom, but this would make the H atom a central atom and that is not possible. The chief cases where O atoms are central atoms are in structures with a *peroxo* linkage (—O—O—) or a *hydroxy* group (—O—H). Otherwise, expect an O atom to be a *terminal* atom.

Carbon atoms are always central atoms. This is a useful fact to keep in mind when writing Lewis structures of organic molecules.

- Except for the very large number of chain-like organic molecules, *molecules and polyatomic ions generally have compact, symmetrical structures.* Thus, of the two skeletal structures below, the more compact structure on the right is the one actually observed for phosphoric acid, H_3PO_4.

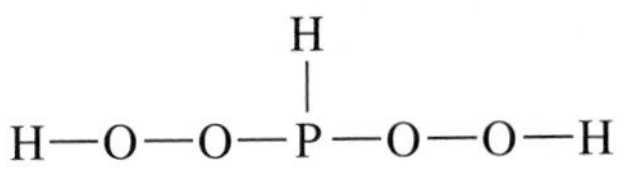

(*Incorrect*)

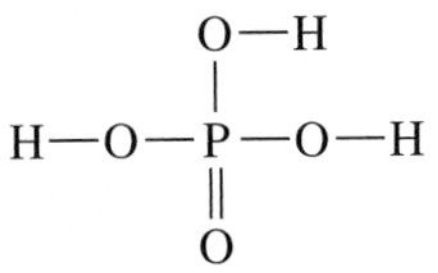

(*Correct*)

Strategy for Writing Lewis Structures

this point, let us incorporate a number of the ideas that we have considered far into a specific approach to writing Lewis structures. This strategy is signed to give you a place to begin, as well as consecutive steps to follow to hieve a plausible Lewis structure.

Determine the total number of valence electrons that must appear in the structure.

Examples: In the *molecule* CH_3CH_2OH, there are 4 valence electrons for each C atom, or 8 for the two C atoms; 1 for each H atom, or 6 for the six H atoms; and 6 for the lone O atom. The total number of valence electrons in the Lewis structure of CH_3CH_2OH is

$$8 + 6 + 6 = 20$$

In the *polyatomic ion* PO_4^{3-}, there are 5 valence electrons for the P atom and 6 for each O atom, or 24 for all four O atoms. To produce the charge of −3, an additional 3 valence electrons must be brought into the structure. The total number of valence electrons in the Lewis structure of PO_4^{3-} is

$$5 + 24 + 3 = 32$$

In the *polyatomic ion* NH_4^+, there are 5 valence electrons for the N atom and 1 for each H atom, or 4 for all four H atoms. To account for the charge of +1, one of the electrons must be *lost*. The total number of valence electrons in NH_4^+ is

$$5 + 4 - 1 = 8$$

Identify the central atoms(s) and terminal atoms.

Write a plausible skeletal structure. Join the atoms in the skeletal structure by *single* covalent bonds (single dashes, representing two electrons each).

For each bond in the skeletal structure, subtract *two* from the total number of valence electrons.

With the valence electrons remaining, *first* complete the octets of the terminal atoms. *Then*, to the extent possible, complete the octets of the central atom(s). If there are just enough valence electrons to complete octets for the atoms, the structure at this point is a satisfactory Lewis structure.

6. If one or more central atoms are left with an incomplete octet after step move lone-pair electrons from one or more terminal atoms to form *multi* covalent bonds to central atoms. Do this to the extent necessary to give atoms complete octets, thereby producing a plausible Lewis structure.

Figure 10-8 summarizes this procedure for writing Lewis structures.

▶ It requires a lot of practice to become proficient at writing Lewis structures. Begin by writing structures of molecules that have only one central atom before trying to write the structures of more complicated molecules.

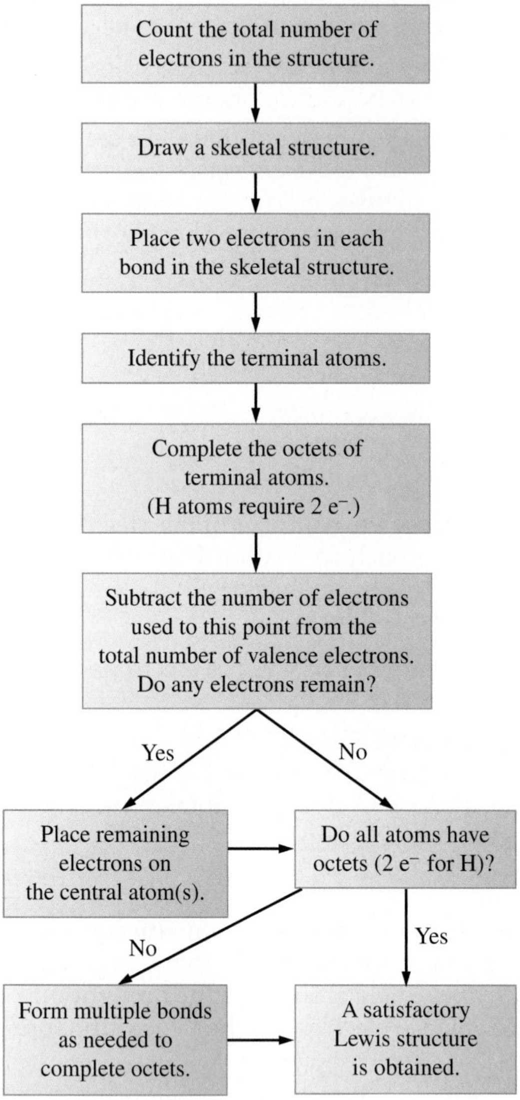

▶ FIGURE 10-8
Summary scheme for drawing Lewis structures

EXAMPLE 10-6 Applying the General Strategy for Writing Lewis Structures

Write a plausible Lewis structure for *cyanogen*, C_2N_2, a poisonous gas used as a fumigant and rocket propellant.

Analyze

Here, we apply the scheme for constructing Lewis structures (Fig. 10-8).

Solve

Step 1. Determine the total number of valence electrons. Each of the two C atoms (group 14) has *four* valence electrons, and each of the two N atoms (group 15) has *five*. The total number of valence electrons is $4 + 4 + 5 + 5 = 18$.

Step 2. Identify the central atom(s) and terminal atoms. Because the C atoms have a lower electronegativity (2.5) than do the N atoms (3.0), C atoms are central atoms, and N atoms are terminal atoms.

Step 3. Write a plausible skeletal structure by joining atoms through *single* covalent bonds.

N—C—C—N

Step 4. Subtract *two* electrons for each bond in the skeletal structure. The three bonds in this structure account for 6 of the 18 valence electrons. This leaves 12 valence electrons to be assigned.

Step 5. Complete octets for the terminal N atoms, and to the extent possible, the central C atoms. The remaining 12 valence electrons are sufficient only to complete the octets of the N atoms.

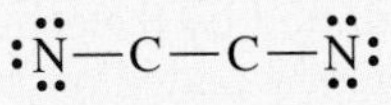

Step 6. Move lone pairs of electrons from the terminal N atoms to form multiple bonds to the central C atoms. Each C atom has only four electrons in its valence shell and needs four more to complete an octet. Thus, each C atom requires two additional pairs of electrons, which it acquires if we move two lone pairs from each N atom into its bond with a C atom, as shown below.

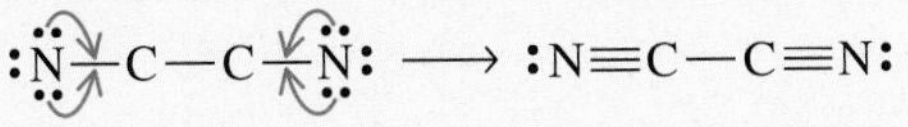

Assess

The construction of correct Lewis structures is an important skill that all chemists have to master. It is imperative to be able to apply the scheme without referring to the steps on page 425 or in Figure 10-8.

PRACTICE EXAMPLE A: Write plausible Lewis structures for **(a)** CS_2, **(b)** HCN, and **(c)** $COCl_2$.

PRACTICE EXAMPLE B: Write plausible Lewis structures for **(a)** formic acid, HCOOH, and **(b)** acetaldehyde, CH_3CHO.

EXAMPLE 10-7 Writing a Lewis Structure for a Polyatomic Ion

Write the Lewis structure for the *nitronium* ion, NO_2^+.

Analyze

Again, we use the strategy illustrated in Figure 10-8.

Solve

Step 1. Determine the total number of valence electrons. The N atom (group 15) has *five* valence electrons, and each of the two O atoms (group 16) has *six*. However, *one* valence electron must be removed to produce the charge of +1. The total number of valence electrons is $5 + 6 + 6 - 1 = 16$

Step 2. Identify the central atom(s) and terminal atoms. The N atom has a lower electronegativity (3.0) than the O atoms (3.5). N is the central atom, and the O atoms are the terminal atoms.

Step 3. Write a plausible skeletal structure by joining atoms through *single* covalent bonds.

O—N—O

Step 4. Subtract *two* electrons for each bond in the skeletal structure. The two bonds in this structure account for 4 of the 16 valence electrons. This leaves 12 valence electrons to be assigned.

Step 5. Complete octets for the terminal O atoms, and to the extent possible, the central N atom. The remaining 12 valence electrons are sufficient only to complete the octets of the O atoms.

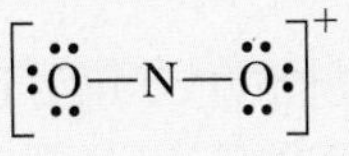

(continued)

Step 6. Move lone pairs of electrons from the terminal O atoms to form multiple bonds to the central N atom The N atom has only four electrons in its valence shell and needs four more to complete an octet. Thus, the N atom requires two additional pairs of electrons, which it acquires if we move one lone pair from each O atom into its bond with the N atom, as shown below.

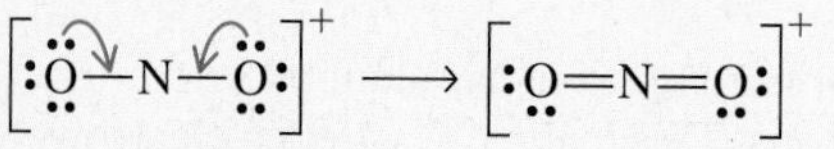

(10.14

Assess

After drawing a Lewis structure, and before moving on to the next step of a problem or to the next exercise check the structure. Each atom is surrounded by 8 electrons (each atom has an octet), and the structure has a total valence of 16 (we have not inadvertently added or dropped electrons). In assessing the structure, we must remember that each line represents *two* electrons (a bonding pair).

PRACTICE EXAMPLE A: Write plausible Lewis structures for the following ions: **(a)** NO^+; **(b)** $N_2H_5^+$; **(c)** O_2^{2-}.

PRACTICE EXAMPLE B: Write plausible Lewis structures for the following ions: **(a)** BF_4^-; **(b)** NH_3OH^+; **(c)** NCO^-

Formal Charge

Instead of writing Lewis structure (10.14) for the nitronium ion in Example 1(we might have written the following structure.

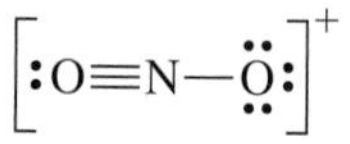

(*Improbable*) **(10.**

Despite the fact that this structure satisfies the usual requirements—the c rect number of valence electrons and an octet for each atom—we have mark it improbable because it fails in one additional requirement. Have you notic that in our strategy for writing Lewis structures, once the total number valence electrons has been determined, there is no need to keep track of wh electrons came from which atoms? Nevertheless, after we have a plausi Lewis structure, we can go back and assess where each electron apparen came from, and in this way we can evaluate formal charges. **Formal char (FC)** are apparent charges on certain atoms in a Lewis structure that ar when atoms have not contributed equal numbers of electrons to the coval bonds joining them. In cases where more than one Lewis structure seems p sible, formal charges are used to ascertain which sequence of atoms a arrangement of bonds is most satisfactory.

The formal charge on an atom in a Lewis structure is the number of valer electrons in the free (uncombined) atom minus the number of electro assigned to that atom in the Lewis structure, with the electrons assigned in t following way.

- Count *lone-pair electrons* as belonging entirely to the atom on which th are found.
- Divide *bond-pair electrons* equally between the bonded atoms.

Assigning electrons (e^-) in this way is equivalent to writing that

$$e^- \text{ assigned to a bonded atom in a Lewis structure}$$
$$= \text{number lone-pair } e^- + \frac{1}{2}\text{number bond-pair}$$

Because formal charge is the difference between the assignment of valen electrons to a free (uncombined) atom and to the atom in a Lewis structure, can be expressed as

$$\text{C} = \text{number valence e}^- \text{ in free atom} - \text{number lone-pair e}^- - \frac{1}{2}\text{number bond-pair e}^- \quad (10.16)$$

Jow, let us assign formal charges to the atoms in structure (10.15), proceed-from left to right.

:O≡ FC = 6 valence e^- in O − 2 lone-pair e^- $-\frac{1}{2}$ (6 bond-pair e^-) = 6 − 2 − 3 = +1

≡N— FC = 5 valence e^- in N − 0 lone-pair e^- $-\frac{1}{2}$ (8 bond-pair e^-) = 5 − 0 − 4 = +1

-Ö: FC = 6 valence e^- in O − 6 lone-pair e^- $-\frac{1}{2}$ (2 bond-pair e^-) = 6 − 6 − 1 = −1

'ormal charges in a Lewis structure can be shown by using small numbers.

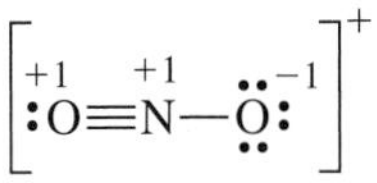

(10.17)

'he following are general rules that can help to determine the plausibility Lewis structure based on its formal charges.

- The sum of the formal charges in a Lewis structure must equal *zero* for a neutral molecule and must equal the magnitude of the charge for a polyatomic ion. [Thus for structure (10.17), this sum is +1 + 1 − 1 = +1.]
- Where formal charges are required, they should be as small as possible.
- Negative formal charges usually appear on the most electronegative atoms; positive formal charges, on the least electronegative atoms.
- Structures having formal charges of the same sign on adjacent atoms are unlikely.

◀ We will see some exceptions to the idea that formal charges should be kept to a minimum in Section 10-6.

.ewis structure (10.17) conforms to the first two rules, but is not in good ordance with the third rule. Despite the fact that O is the most electronega- element in the structure, one of the O atoms has a positive formal charge. greatest failing, though, is in the fourth rule. Both the O atom on the left l the N atom adjacent to it have positive formal charges. Structure (10.17) is the most satisfactory Lewis structure. By contrast, the Lewis structure of $_2^+$ derived in Example 10-7 has only one formal charge, +1, on the central tom. It conforms to the rules completely and is the most satisfactory Lewis icture.

EXAMPLE 10-8 Using Formal Charges in Writing Lewis Structures

Write the most plausible Lewis structure of nitrosyl chloride, NOCl, one of the oxidizing agents present in *aqua regia*, a mixture of concentrated nitric and hydrochloric acids capable of dissolving gold.

Analyze

Although the formula is written as NOCl, we can reject the skeletal structure N—O—Cl because it places the most electronegative atom as the central atom. (We are asked to consider N—O—Cl in Practice Example A.) Having ruled out N—O—Cl as a possible skeletal structure, we are left with the following as possibilities:

O—Cl—N and O—N—Cl

(continued)

To determine the best structure, we must first complete the skeletal structures and then assign formal charges. The best structure will have the fewest and smallest formal charges.

Solve

Regardless of the skeletal structure chosen, the number of valence electrons (dots) and bonds that must appear in the final Lewis structure is

$$5 \text{ from N} + 6 \text{ from O} + 7 \text{ from Cl} = 18$$

When we apply the four steps listed below to the two possible skeletal structures, we obtain a total of four Lewis structures—two for each skeletal structure. This doubling occurs because in step 4, there are two ways to complete the octets of the central atoms. The final Lewis structures obtained are labeled (a_1), (a_2), (b_1), and (b_2).

(a)

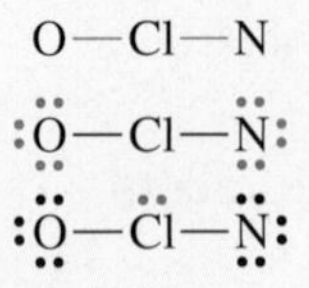

1. Assign four electrons.
2. Assign twelve more electrons.
3. Assign the last two electrons.

(b)

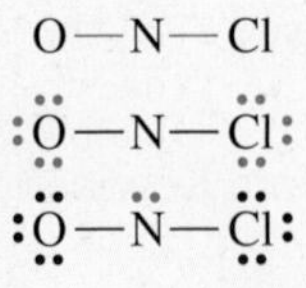

4. Complete the octet on the central atom.

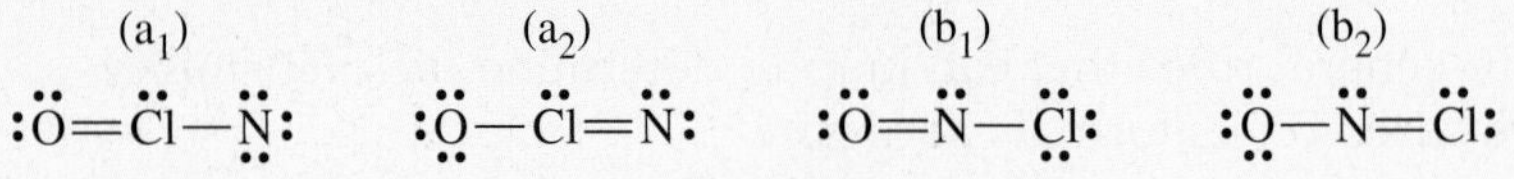

Evaluate formal charges by using equation (10.16). In structure (a_1),

for the N atom,

$$\text{FC} = 5 - 6 - \frac{1}{2}(2) = -2$$

for the O atom,

$$\text{FC} = 6 - 4 - \frac{1}{2}(4) = 0$$

for the Cl atom,

$$\text{FC} = 7 - 2 - \frac{1}{2}(6) = +2$$

Proceed in a similar manner for the other three structures. Summarize the formal charges for the four structures.

	(a_1)	(a_2)	(b_1)	(b_2)
N:	−2	−1	0	0
O:	0	−1	0	−1
Cl:	+2	+2	0	+1

Select the best Lewis structure in terms of the formal-charge rules. First, note that all four structures obey the requirement that formal charges of a neutral molecule add up to zero. In structure (a_1), the formal charges are large (+2 on Cl and −2 on N) and the negative formal charge is not on the most electronegative atom. Structure (a_2) has formal charges on all atoms, one of them large (+2 on Cl). Structure (b_1) is the ideal we seek—no formal charges. In structure (b_2), we again have formal charges. The best Lewis structure of nitrosyl chloride is

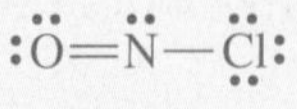

Assess

Based on structure (b_1), ONCl is a better way to write the formula of nitrosyl chloride.

PRACTICE EXAMPLE A: Write a Lewis structure for nitrosyl chloride based on the skeletal structure N—O—Cl, and show that this structure is not as plausible as the one obtained in Example 10-8.

PRACTICE EXAMPLE B: Write two Lewis structures for cyanamide, NH_2CN, an important chemical of the fertilizer and plastics industries. Use the formal charge concept to choose the more plausible structure.

10-5 CONCEPT ASSESSMENT

›r molecules, the most satisfactory Lewis structure may have no formal charges C = 0) in some cases and formal charges in others. For polyatomic ions, inimally the most satisfactory Lewis structure has a formal charge on at least e atom. Explain the basis of these observations.

10-1 ARE YOU WONDERING?

Do formal charges represent actual charges on the atoms?

›rmal charges are not actual charges, which can be seen from a comparison of e electrostatic potential map of the HCN molecule and the formal charges erived from the Lewis structure.

$$\underset{0}{H}-\underset{0}{C}\equiv\underset{0}{N}\text{:}$$

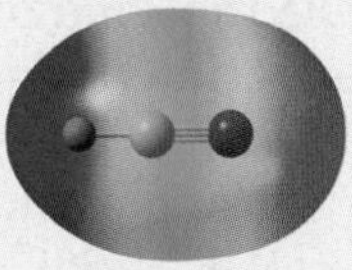

Although the formal charges are all zero, the electrostatic potential map shows at the H atom in the HCN molecule is slightly positive (blue) and that the nitro- en atom is slightly negative (red). In molecules, the true charges on atoms are sually, but not always, between +1 and −1. For example, in the HCl molecule, e charge on H is about +0.17 and that on Cl is about −0.17. (See page 447.)
The method used for assigning formal charges is really just a form of "electron ›okkeeping." In this text, we have now discussed two different concepts— xidation states and formal charges—that are used for electron bookkeeping. xidation states and formal charges are both very useful. They are compared in e table below.

	Interpretation	Comments
Oxidation state	The charge an atom would have if the bonding electrons in each bond were *transferred* to the more electronegative atom.	• The oxidation state concept tends to exaggerate the ionic character of the bonding between atoms. • Oxidation states are used to predict and rationalize chemical properties of compounds.
Formal charge	The charge an atom would have if the bonding electrons in each bond were *divided equally* between the two atoms involved.	• The formal charge concept tends to exaggerate the covalent charac- ter of the bonding between atoms. • Formal charges are used to assess which Lewis structure is the most satisfactory representation of the true structure.

For many molecules, the bonding is closer to being "pure covalent" than it is to eing "pure ionic" and so the formal charge on an atom is often—but not lways—numerically closer to the true charge. That's why we focus on formal harges when assessing the relative importance of different Lewis structures. That eing said, it is important to emphasize that chemists still question and debate vhether it is true that the best structure is the one having the fewest and smallest ›rmal charges.

10-5 Resonance

The ideas presented in the previous section allow us to write many Le structures, but some structures still present problems. We describe these pr lems in the next two sections.

Although we usually think of the formula of oxygen as O_2, there are a ally two different oxygen molecules. Familiar oxygen is *dioxygen*, O_2; the o molecule is *trioxygen*—ozone, O_3. The term used to describe the existenc two or more forms of an element that differ in their bonding and molec structure is *allotropy*—O_2 and O_3 are allotropes of oxygen. Ozone is found urally in the stratosphere and is also produced in the lower atmosphere constituent of smog.

When we apply the usual rules for Lewis structures for ozone, we come with these *two* possibilities.

▶ Bond lengths are discussed more fully in Section 10-8.

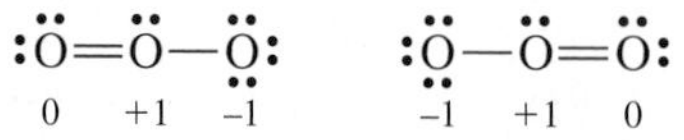

Each structure suggests that one oxygen-to-oxygen bond is single and other is double. Yet experimental evidence indicates that the two oxyg to-oxygen bonds are the same; each has a length of 127.8 pm. This bond len is shorter than the O—O single-bond length of 147.5 pm in hydrogen pe ide, H—O—O—H, but it is longer than the double-bond length of 120.74 pn diatomic oxygen, :O=O:. The bonds in ozone are intermediate between a gle and a double bond. The difficulty is resolved if we say that the true Le structure of O_3 is *neither* of the previously proposed structures but a comp ite, or *hybrid*, of the two, a fact that we can represent as

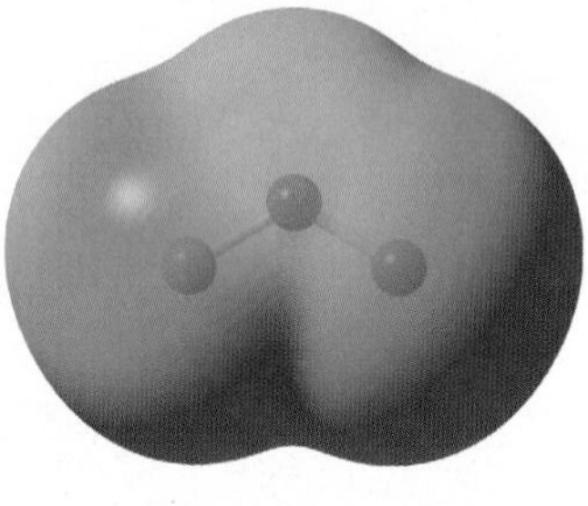

▲ **Electrostatic potential map of ozone**
All the atoms in an ozone molecule have the same electronegativity and yet the distribution of electron density is nonuniform. The reason will become apparent when we describe in a more sophisticated way the bonding in this molecule.

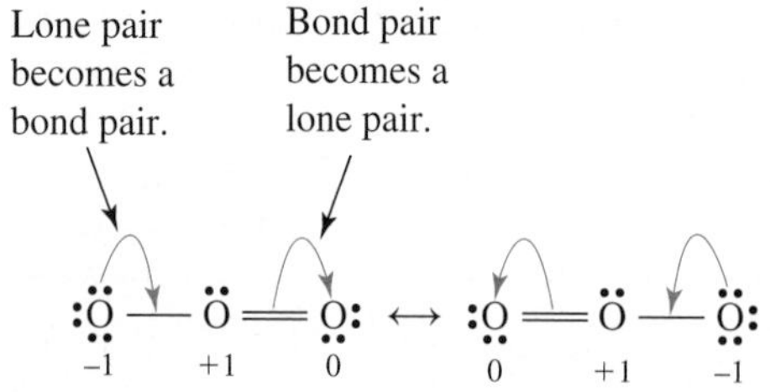

(10

The situation in which two or more plausible Lewis structures contrib to the "correct" structure is called **resonance**. The true structure i *resonance hybrid* of plausible contributing structures. Acceptable contribut structures to a resonance hybrid must all have the same skeletal structure (atomic positions cannot change); they can differ only in how electrons distributed within the structure. In expression (10.18), the two contribut structures are joined by a double-headed arrow. The arrow does *not* m that the molecule has one structure part of the time and the other struct the rest of the time. *It has the same structure all the time.* By averaging the sin bond in one structure with the double bond in the other, we might say t the oxygen-to-oxygen bonds in ozone are halfway between a single a double bond, that is, 1.5 bonds. The fact that the electrons in ozone are tributed over the whole molecule so as to produce two equivalent bond readily seen in the electrostatic potential map of ozone, shown in the marg

The two resonance structures in expression (10.18) are equivalent; that they contribute equally to the structure of the resonance hybrid. In ma cases, several contributing resonance structures do not contribute equally. example, consider the azide anion, N_3^-, for which three resonance structu are given below.

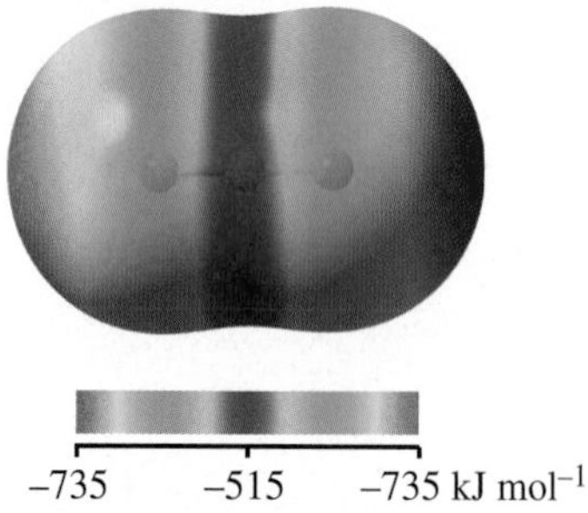

▲ FIGURE 10-9
Electrostatic potential map of the azide ion
The electrostatic potential near the terminal N atoms is more negative than it is for the central N atom. This observation indicates that the central structure (at the bottom of this page) is the most important resonance structure for the N_3^- ion.

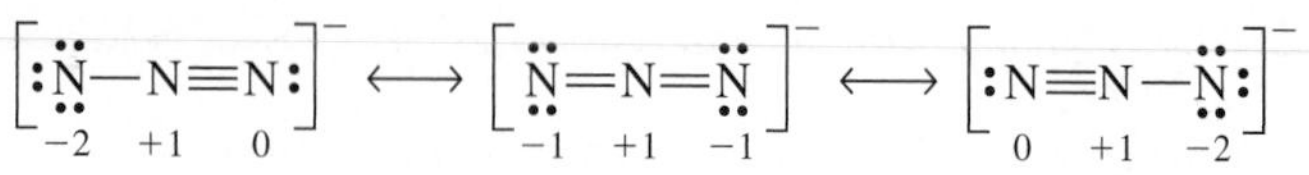

can decide which resonance structure likely contributes most to the hybrid pplying the general rules for formal charges (page 429). The central reso-ce structure avoids the unlikely large formal charge of -2 found on an tom in the other two structures. Consequently, we expect that the structure $=\overset{+}{N}=\overset{..}{N}{}^{-}$ contributes most to the resonance hybrid of the azide anion. choice is supported by the electrostatic potential map shown in Figure 10-9.

EXAMPLE 10-9   Representing the Lewis Structure of a Resonance Hybrid

Write the Lewis structure of the acetate ion, CH_3COO^-.

Analyze

A key concept is that resonance structures differ only in how electrons are distributed within the structure. We cannot change the positions of the atoms. First, we draw a skeletal structure (see the electrostatic potential map below), and then we complete it by using the strategy we've used previously. Finally, we generate additional structures (resonance structures) by moving electron pairs.

Solve

The skeletal structure has the three H atoms as terminal atoms bonded to a C atom as a central atom. The second C atom is also a central atom bonded to the first. The two O atoms are terminal atoms bonded to the second C atom.

$$\begin{array}{ccc} H & & O \\ | & & | \\ H-C&-&C-O \\ | & & \\ H & & \end{array}$$

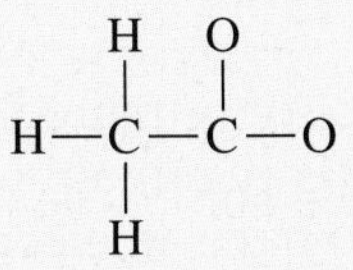

The number of valence electrons (dots) that must appear in the Lewis structure is

$$(3 \times 1) + (2 \times 4) + (2 \times 6) + 1 = 3 + 8 + 12 + 1 = 24$$

$\quad$ ↑

From H $\quad$ From C $\quad$ From O $\quad$ To establish charge of $1-$

Twelve of the valence electrons are used in the bonds in the skeletal structure, and the remaining twelve are distributed as lone-pair electrons on the two O atoms.

$$\left[\begin{array}{ccc} H & \overset{..}{:}\overset{-1}{\overset{..}{O}}\!: & \\ | & | & \\ H-C&-&\underset{+1}{C}-\overset{..}{\underset{..}{\underset{-1}{O}}}\!: \\ | & & \\ H & & \end{array}\right]^{-}$$

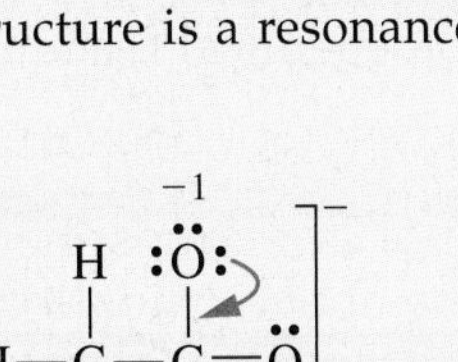

In completing the octet of the C atom on the right, we discover that we can write two completely equivalent Lewis structures, depending on which of the two O atoms furnishes the lone pair of electrons to form a carbon-to-oxygen double bond. The true Lewis structure is a resonance hybrid of the following two contributing structures.

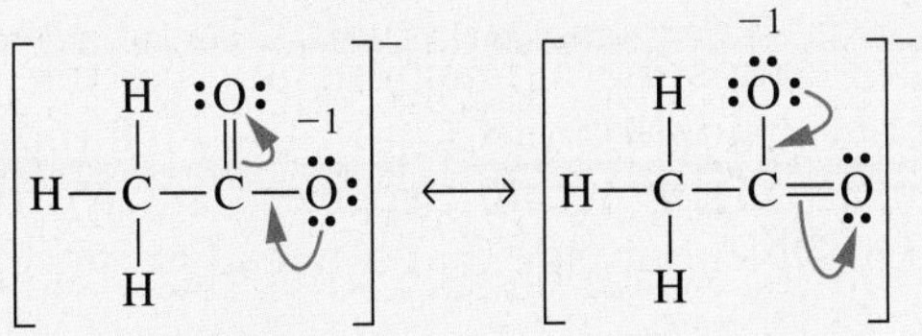

$$\left[\begin{array}{ccc} H & :\overset{..}{O}\!: & \\ | & \parallel & \\ H-C&-&C{-}\overset{..}{\underset{..}{O}}\!:^{-1} \\ | & & \\ H & & \end{array}\right]^{-} \longleftrightarrow \left[\begin{array}{ccc} H & :\overset{..}{\overset{-1}{O}}: & \\ | & | & \\ H-C&-&C=\overset{..}{\underset{..}{O}} \\ | & & \\ H & & \end{array}\right]^{-}$$ (10.19)

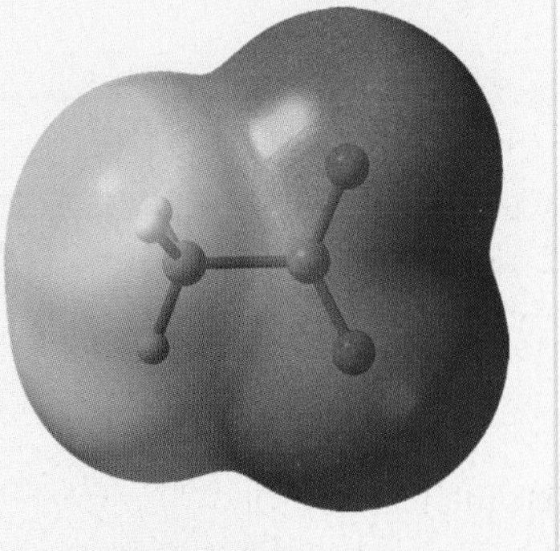

▲ **Acetate anion**

Assess

Even though the formal process of converting one resonance structure to another moves electrons, resonance is not meant to indicate the motion of electrons. The acetate anion has a structure that is a composite of the two resonance forms that we have constructed.

PRACTICE EXAMPLE A: Draw Lewis structures to represent the resonance hybrid for the SO_2 molecule.

PRACTICE EXAMPLE B: Draw Lewis structures to represent the resonance hybrid for the nitrate ion.

10-6 CONCEPT ASSESSMENT

Is resonance possible in the acetic acid (CH_3CO_2H) molecule? Explain.

10-6 Exceptions to the Octet Rule

The octet rule has been our mainstay in writing Lewis structures, and it continue to be one. Yet at times, we must depart from the octet rule, as we see in this section.

Odd-Electron Species

The molecule NO has 11 valence electrons, an odd number. If the numbe valence electrons in a Lewis structure is odd, there must be an unpaired e tron somewhere in the structure. Lewis theory deals with electron pairs does not tell us where to put the unpaired electron; it could be on either th or the O atom. To obtain a structure free of formal charges, however, we put the unpaired electron on the N atom.

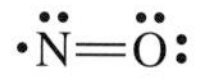

The presence of unpaired electrons causes odd-electron species to be p magnetic. NO is paramagnetic. Molecules with an even number of electr are expected to have all electrons paired and to be diamagnetic. An impor exception is seen in the case of O_2, which is paramagnetic despite hav 12 valence electrons. Lewis theory does not provide a good electronic st ture for O_2, but the molecular orbital theory that we will consider in the n chapter is much more successful.

▶ Experimental evidence for the paramagnetism of O_2 is shown in Figure 10-3.

The number of stable odd-electron molecules is quite limited. More comm are **free radicals**, or simply *radicals*, highly reactive molecular fragments w one or more unpaired electrons. The formulas of free radicals are usually w ten with a dot to emphasize the presence of an unpaired electron, such as in *methyl* radical, $\cdot CH_3$, and the *hydroxyl* radical, $\cdot OH$. The Lewis structure these two free radicals are

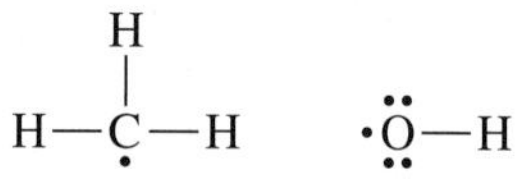

Both of these free radicals are commonly encountered as transitory spe in flames. In addition, $\cdot OH$ is formed in the atmosphere in trace amounts a result of photochemical reactions.

$$\cdot OH + CO \longrightarrow CO_2 + \cdot H$$

Many important atmospheric reactions involve free radicals as reacta such as in the above oxidation of CO to CO_2. Free radicals, because of th unpaired electron, are highly reactive species. The hydroxyl radical, for exa ple, is implicated in DNA damage that can lead to cancer.

Incomplete Octets

Our initial attempt to write the Lewis structure of boron trifluoride leads t structure in which the B atom has only *six* electrons in its valence shell— *incomplete octet*.

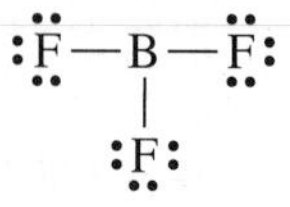

(10

have learned to complete the octets of central atoms by shifting lone-pair trons from terminal atoms to form multiple bonds. One of three equivalent ctures with a boron-to-fluorine double bond is shown below.

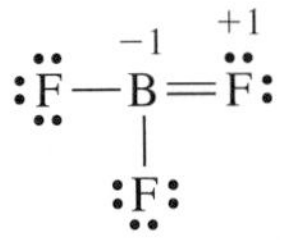

(10.21)

n observation in support of structure (10.21) is that the B—F bond length F_3 (130 pm) is less than expected for a single bond. A shorter bond sugts that more than two electrons are present, that is, that there is multipled character in the bond. However, the placement of formal charges in cture (10.18) breaks an important rule—negative formal charge should be nd on the more electronegative atom in the bond. In this structure, the pose formal charge is on the most electronegative of all atoms—F.

he high electronegativity of fluorine (4.0) and the much lower one of on (2.0) suggest an appreciable ionic character to the boron-to-fluorine d (see Figure 10-6). This suggests the possibility of such ionic structures as following.

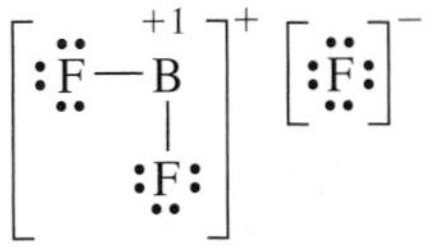

(10.22)

iew of its molecular properties and chemical behavior, the best represenon of BF_3 appears to be a resonance hybrid of structures (10.20, 10.21, and 2), with perhaps the most important contribution made by the structure h an incomplete octet (10.20). Whichever BF_3 structure we choose to phasize, an important characteristic of BF_3 is its strong tendency to form a rdinate covalent bond with a species capable of donating an electron pair he B atom. This can be seen in the formation of the BF_4^- ion.

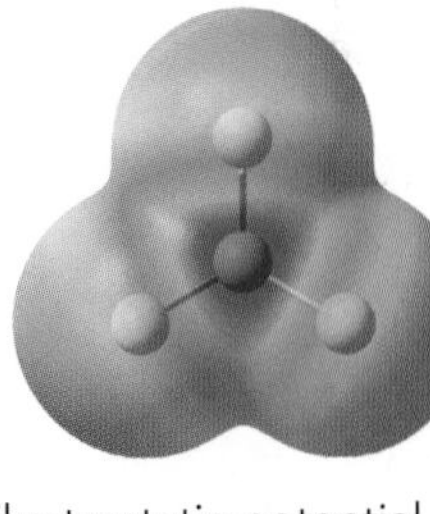

▲ Electrostatic potential map of the BF_3 molecule

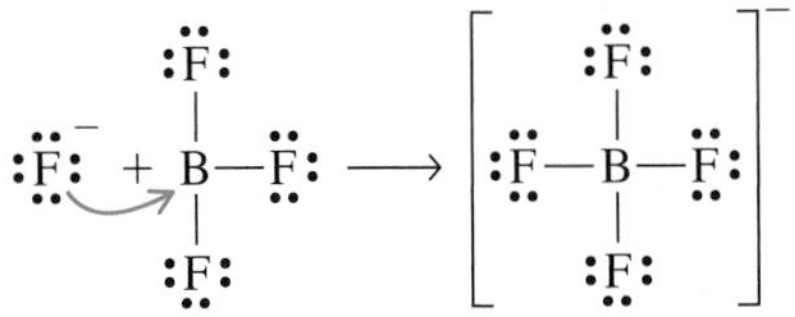

F_4^-, the bonds are single bonds and the bond length is 145 pm.

he number of species with incomplete octets is limited to some beryllium, on, and aluminum compounds. Perhaps the best examples are the boron rides. Bonding in the boron hydrides will be discussed in Chapter 22.

panded Valence Shells

have consistently tried to write Lewis structures in which all atoms except ave a complete octet, that is, in which each atom has eight valence elecns. There are a few Lewis structures that break this rule by having 10 or n 12 valence electrons around the central atom, creating what is called an **anded valence shell**. Describing bonding in these structures is an area of ive interest among chemists.

Molecules with expanded valence shells typically involve nonmetal atoms the third period and beyond that are bonded to highly electronegative ms. For example, phosphorus forms two chlorides, PCl_3 and PCl_5. We can te a Lewis structure for PCl_3 with the octet rule. In PCl_5, with five Cl atoms ded directly to the central P atom, the outer shell of the P atom appears to e *ten* electrons. We might say that the valence shell has expanded to ten ctrons. In the SF_6 molecule, the valence shell appears to expand to 12.

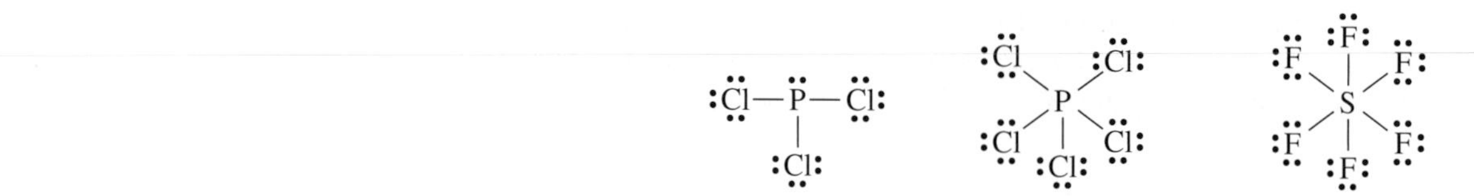

Octet — Expanded valence shell — Expanded valence shell

Expanded valence shells have also been used in cases where they appea
give a better Lewis structure than strict adherence to the octet rule, as s
gested by the two Lewis structures for the sulfate ion that follow.

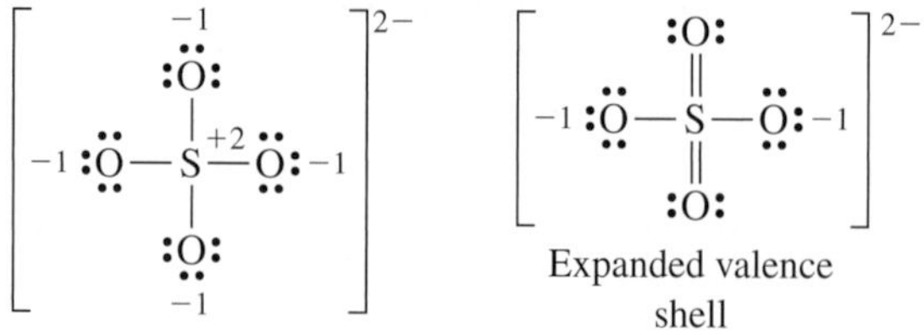

Normal octet — Expanded valence shell

The argument for including the expanded valence-shell structure is tha
reduces formal charges. Also, the experimentally determined sulfur-to-oxy
bond lengths in SO_4^{2-} and H_2SO_4 are in agreement with this idea.
experimental results for H_2SO_4, summarized in structure (10.23), indicate
the S—O bond with O as a central atom and with an attached H atom is lor
than the S—O bond with O as a terminal atom.

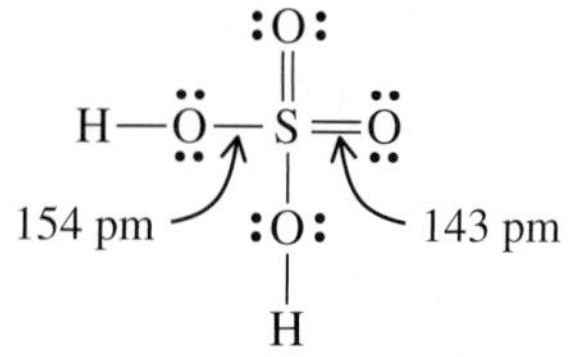

(1

Experimental evidence appears to support using an expanded valence s
in the Lewis structure of sulfuric acid. The experimentally determined S–
bond length in the sulfate anion—149 pm—lies between the two S—O b
lengths found in sulfuric acid, suggesting a partial double-bond character.
expanded valence-shell structure is suggestive of this partial double-bo
character, whereas the octet structure is not. For the sulfate anion, best ag
ment with the observed S—O bond lengths is found in a resonance hyb
having strong contributions from a series of resonance structures (10.24) ba
on expanded valence shells.

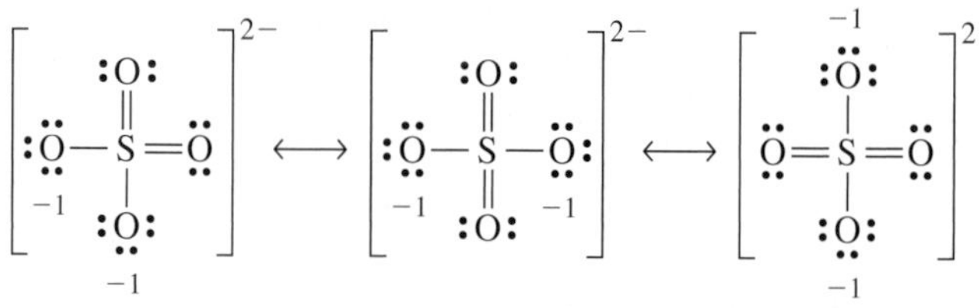

(10

The problem with expanded valence-shell structures is, of course,
explain where the "extra" electrons go. This expansion has been rationali
by assuming that after the 3*s* and 3*p* subshells of the central atom fill to cap
ity (eight electrons), extra electrons go into the empty 3*d* subshell. If
assume that the energy difference between the 3*p* and 3*d* levels is not v
large, the valence-shell expansion scheme seems reasonable. But is this a va
assumption? The use of the 3*d* orbitals for valence-shell expansion is a ma
of scientific dispute.* Although unresolved questions about the expanc

*L. Suidan et al., *J. Chem. Educ.*, **72**, 583 (1995); G. H. Purser, *J. Chem. Educ.*, **78**, 981 (2001).

ence-shell concept may be unsettling, the point to keep in mind is that the nodified octet rule works perfectly well for most uses of Lewis structures. will return to this topic, together with several other unsettled issues, in the cluding section of Chapter 11.

10-7 CONCEPT ASSESSMENT

n page 432, we reasoned that because of resonance, oxygen–oxygen bonds in $_3$ were halfway between single and double bonds, that is, 1.5 bonds. Do you xpect the sulfur–oxygen bonds in SO_2 to be single, double, or 1.5 bonds? Explain our answer, bearing in mind the ability of sulfur to expand its valence shell.

-7 Shapes of Molecules

Lewis structure for water gives the impression that the constituent atoms arranged in a straight line.

$$\mathrm{H{-}\ddot{\underset{\cdot\cdot}{O}}{-}H}$$

wever, the experimentally determined shape of the molecule is not linear. molecule is *bent*, as shown in Figure 10-10. Does it really matter that the O molecule is bent rather than linear? The answer is, decidedly, yes. In apter 14, we will find that it also accounts for the ability of liquid water to solve so many different substances.

What we seek in this section is a simple model for predicting the approxi-te shape of a molecule. Unfortunately, Lewis theory tells us nothing about shapes of molecules, but it is an excellent place to begin. The next step is to an idea based on repulsions between valence-shell electron pairs. We will cuss this idea after defining a few terms.

By molecular shape, we mean the geometric figure we get when joining nuclei of bonded atoms by straight lines. Figure 10-10 depicts the *tri-mic* (three-atom) water molecule using a ball-and-stick model. The balls resent the three atoms in the molecule, and the straight lines (sticks), bonds between atoms. In reality, the atoms in the molecule are in close ntact, but for clarity we show only the centers of the atoms. To have a nplete description of the shape of a molecule, we need to know not only bond lengths, the distances between the nuclei of bonded atoms, but o the **bond angles**, the angles between adjacent lines representing nds. We will concentrate on bond angles in this section and bond lengths Section 10-8.

A diatomic molecule has only one bond and no bond angle. Because the ometric shape determined by two points is a straight line, *all diatomic mol-les are linear*. A triatomic molecule has two bonds and one bond angle. If bond angle is 180°, the three atoms lie on a straight line, and the mole-e is *linear*. For any other bond angle, a triatomic molecule is said to be *gular, bent*, or *V-shaped*. Some polyatomic molecules with more than three ms have planar or even linear shapes. More commonly, however, the nters of the atoms in these molecules define a three-dimensional geomet-figure.

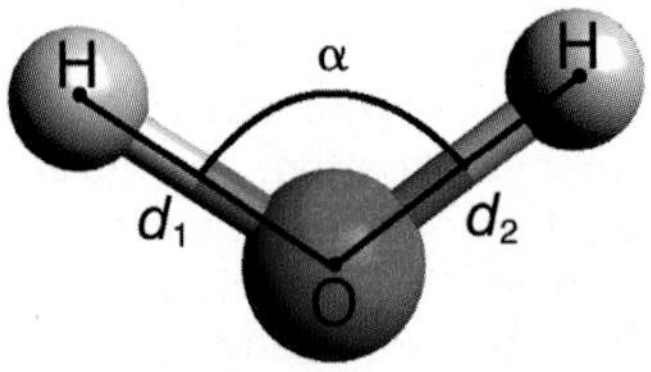

▲ FIGURE 10-10
Geometric shape of a molecule
To establish the shape of the triatomic H_2O molecule shown here, we need to determine the distances between the nuclei of the bonded atoms and the angle between adjacent bonds. In H_2O, the bond lengths $d_1 = d_2 = 95.8$ pm and the bond angle $\alpha = 104.45°$.

ence-Shell Electron-Pair Repulsion (VSEPR) Theory

shape of a molecule is established by experiment or by a quantum mechan- calculation confirmed by experiment. The results of these experiments d calculations are generally in good agreement with the **valence-shell ctron-pair repulsion theory (VSEPR theory)**. In VSEPR theory, we focus on s of electrons in the *valence* electron shell of a central atom in a structure.

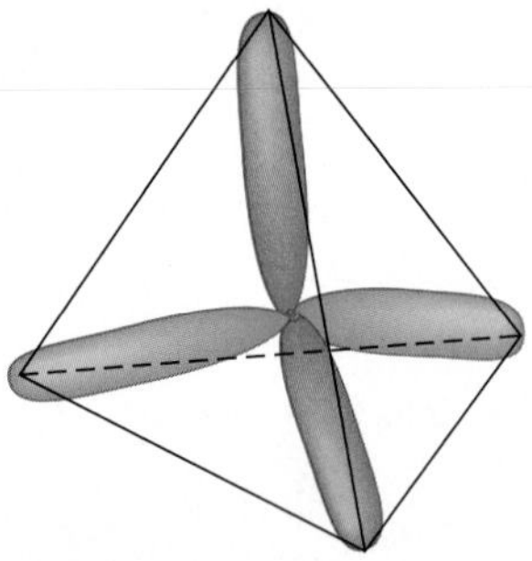

▲ FIGURE 10-11
Balloon analogy to valence-shell electron-pair repulsion
When two elongated balloons are twisted together, they separate into four lobes. To minimize interferences, the lobes spread out into a tetrahedral pattern. (A regular tetrahedron has four faces, each an equilateral triangle.) The lobes are analogous to valence-shell electron pairs.

> Electron pairs repel each other, whether they are in chemical bonds (bond pairs) or unshared (lone pairs). Electron pairs assume orientations about an atom to minimize repulsions.

This, in turn, results in particular geometric shapes for molecules.

Another aspect of VSEPR theory is a focus not just on electron pairs bu electron *groups*. A group of electrons can be a pair, either a lone pair or a b pair, or it can be a single unpaired electron on an atom with an incomp octet, as in NO. A group can also be a double or triple bond between atoms. Thus in the $\ddot{\underset{..}{O}}=C=\ddot{\underset{..}{O}}$ molecule, the central C atom has *two* elect groups in its valence shell. Each of the double bonds with its two elect pairs is treated as *one* electron group.

Consider the methane molecule, CH_4, in which the central C atom has acqu the electron configuration of Ne by forming covalent bonds with four H atom

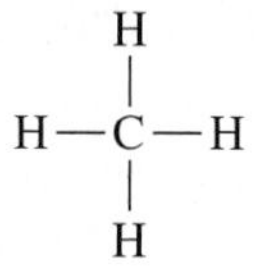
H
H—C—H
H

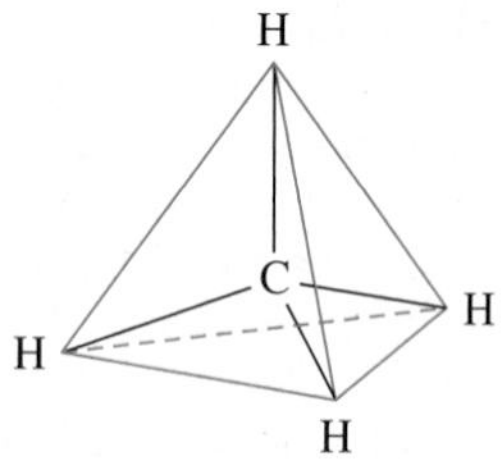
H
C
H
H
H

What orientation will the four electron groups (bond pairs) assume? The loon analogy of Figure 10-11 suggests that electron-group repulsions will f the groups as far apart as possible—to the corners of a tetrahedron having C atom at its center. The VSEPR method predicts, correctly, that CH_4 *tetrahedral* molecule.

Having established that the molecular shape of methane is tetrahedral, following question arises: How can we represent the three-dimensional shap molecule on a sheet of paper? In the diagram in the margin, we have enclose methane molecule in a tetrahedron (red lines).

We see that the C—H bonds point to the vertices of the tetrahedron. two of the C—H bonds define a plane, and in the figures below we cho two C—H bonds (shown in blue) to lie in the plane of the page.

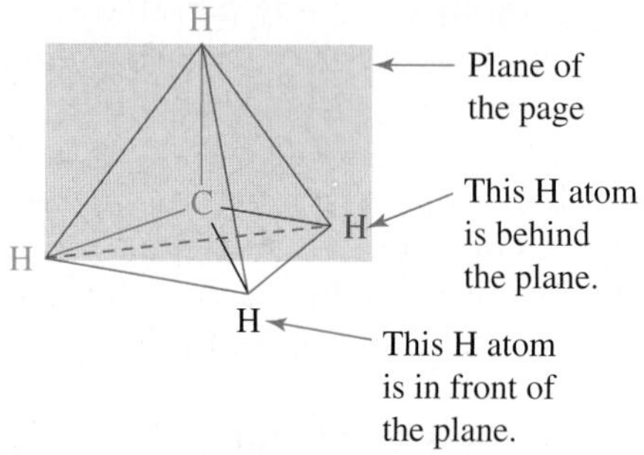
H
Plane of the page
C
This H atom is behind the plane.
H
H
H
This H atom is in front of the plane.

When examining the figure above, we can see that the other two C– bonds do not lie in the plane of the page. One of the C—H bonds points ou the plane of the page (toward us) and the other points behind (away from us the figure in the lower margin, we use a solid wedge to represent the bond t points toward us and a dashed wedge to represent the bond that points awa

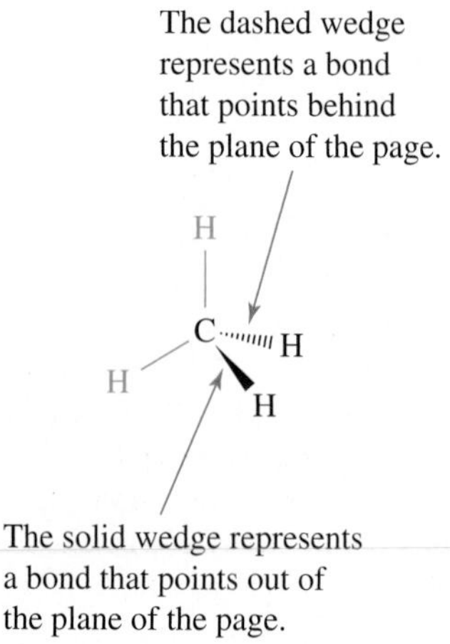
The dashed wedge represents a bond that points behind the plane of the page.
H
C
H
H
H
The solid wedge represents a bond that points out of the plane of the page.

The *dash and wedge* symbolism described above is routinely used chemists to represent three-dimensional structures of molecules. The symb ism is based on the following conventions:

- Ordinary lines are used to represent bonds that lie in the plane of the pa
- Solid wedges are used to represent bonds that point toward the view that is, in front of the plane of the paper.
- Dashed wedges are used to represent bonds that point away from viewer, that is, behind the plane of the paper. The current conventio to place the narrow end of the wedge next to the atom in the plane of paper and the wide end near the atom that points away from the view

ı NH_3 and H_2O, the central atom is also surrounded by four groups of trons in a tetrahedral arrangement, but these molecules *do not* have a tetraral shape.

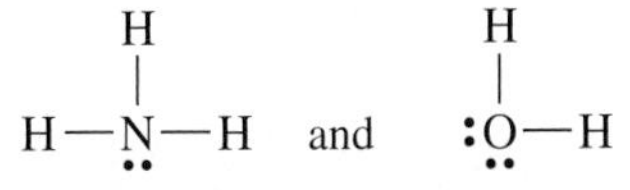

e is the situation: VSEPR theory predicts the distribution of electron ıps, and in these molecules, electron groups are arranged tetrahedrally ıt the central atom. The shape of a molecule, however, is determined by the tion of the atomic nuclei. To avoid confusion, we will call the geometric disıtion of electron groups the **electron-group geometry** and the geometric ıngement of the atomic nuclei—the actual determinant of the molecular pe—the **molecular geometry**.

ı the NH_3 molecule, only three of the electron groups are bond pairs; the th is a lone pair. Joining the N nucleus to the H nuclei by straight lines outs a pyramid with the N atom at the apex and the three H atoms at the base; called a *trigonal pyramid*. We say that the electron-group geometry is tetraral and the molecular geometry is trigonal-pyramidal.

ı the H_2O molecule, two of the four electron groups are bond pairs and are lone pairs. The molecular shape is obtained by joining the two H lei to the O nucleus with straight lines. For H_2O, the electron-group geomis tetrahedral and the molecular geometry is V-shaped, or bent. In the diaı below, the Lewis structure for water is drawn in two ways.

ıe first diagram, which is how we usually draw the structure, all the atoms lie he plane of the paper. In the second structure, we use dash and wedge sym; to indicate that one of the bonds points toward us and the other points away. 'he geometric shapes of CH_4, NH_3, and H_2O are summarized in ıre 10-12. In the VSEPR notation used in Figure 10-12, A is the central atom, X

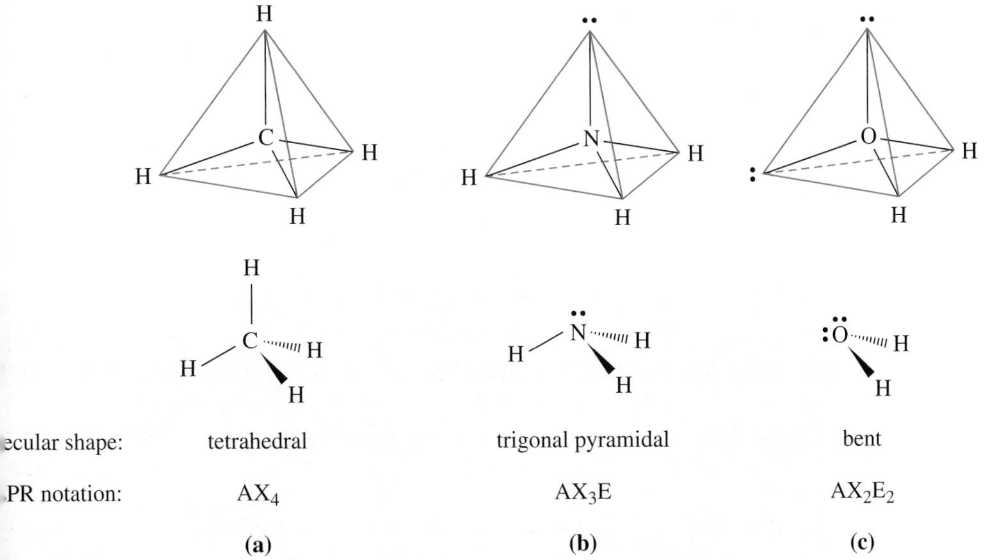

GURE 10-12
ecular shapes based on tetrahedral electron-group geometry of CH_4, NH_3, and H_2O
hree molecules have a tetrahedral arrangement of electron groups around the central atom. However, molecular shapes nolecular geometries) are established by focusing only on the positions of the atoms bonded to a common center. end that lone pairs are invisible when visualizing the molecular shapes. In **(a)**, there are no lone pairs and the C atom sits e center of a tetrahedron; the molecular shape is tetrahedral. Pretending the lone pair on the N atom in **(b)** is invisible, the om is the vertex in a pyramid having a triangular base; the molecular shape is called trigonal pyramidal. In **(c)**, the O and oms form a V-shape and the molecular shape is V-shaped or bent.

is a terminal atom or group of atoms bonded to the central atom, and E lone pair of electrons. Thus, the symbol AX_2E_2 signifies that *two* atom groups (X) are bonded to the central atom (A). The central atom also *two* lone pairs of electrons (E). H_2O is an example of a molecule of AX_2E_2 type.

For tetrahedral electron-group geometry, we expect bond angles of 10 known as the *tetrahedral* bond angle. In the CH_4 molecule, the measu bond angles are, in fact, 109.5 °. The bond angles in NH_3 and H_2O slightly smaller: 107 ° for the H—N—H bond angle and 104.5 ° for H—O—H bond angle. We can explain these less-than-tetrahedral b angles by the fact that the charge cloud of the lone-pair electrons spre out. This forces the bond-pair electrons closer together and reduces bond angles.

VSEPR theory works best for second-period elements. The predicted b angle of 109.5 ° for H_2O is close to the measured angle of 104.5 °. For however, the predicted value of 109.5 ° is not in good agreement with observed value, 92 °. Even though VSEPR theory does not give an accu prediction for the angle in H_2S, it does provide an indication that the mole is bent.

Possibilities for Electron-Group Distributions

The most common situations are those in which central atoms have two, th four, five, or six electron groups distributed around them.

Electron-group geometries

- two electron groups: linear
- three electron groups: trigonal planar
- four electron groups: tetrahedral
- five electron groups: trigonal bipyramidal
- six electron groups: octahedral

Figure 10-13 extends the balloon analogy to these cases. The cases for five- six-electron groups are typified by PCl_5 and SF_6, molecules with expan valence shells.

The molecular geometry is the same as the electron-group geometry when all electron groups are bond pairs. These are for the VSEPR notation (that is, AX_2, AX_3, AX_4, and so on). In Table 10.1, the AX_n cases are illustrate

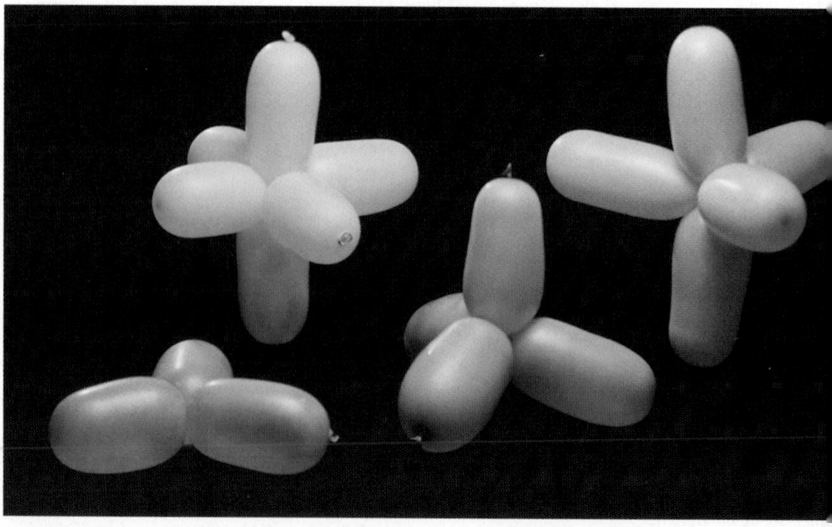

▶ FIGURE 10-13
Several electron-group geometries illustrated
The electron-group geometries pictured are trigonal-planar (orange), tetrahedral (gray), trigonal-bipyramidal (pink), and octahedral (yellow). The atoms at the ends of the balloons are not shown and are not important in this model.

BLE 10.1 Molecular Geometry as a Function of Electron-Group Geometry

mber ctron oups	Electron-Group Geometry	Number of Lone Pairs	VSEPR Notation	Molecular Geometry	Ideal Bond Angles	Example	
	linear	0	AX_2	 (linear)	180°	$BeCl_2$	 Carey B. Van Loon
	trigonal planar	0	AX_3	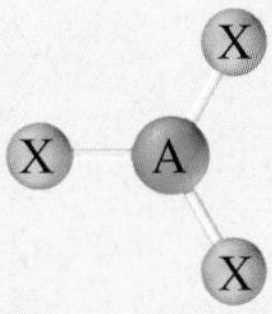 (trigonal planar)	120°	BF_3	 Carey B. Van Loon
	trigonal planar	1	AX_2E	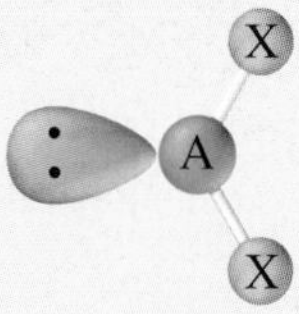 (bent)	120°	SO_2[a]	
	tetrahedral	0	AX_4	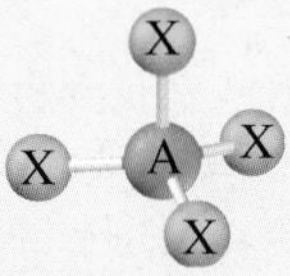 (tetrahedral)	109.5°	CH_4	 Carey B. Van Loon
	tetrahedral	1	AX_3E	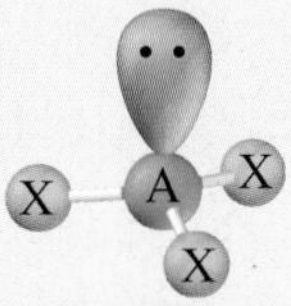 (trigonal pyramidal)	109.5°	NH_3	
	tetrahedral	2	AX_2E_2	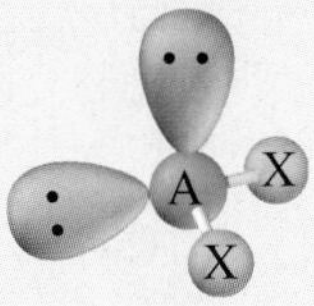 (bent)	109.5°	OH_2	
	trigonal bipyramidal	0	AX_5	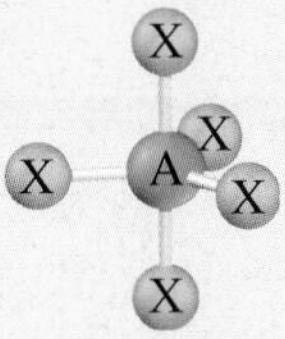 (trigonal bipyramidal)	90°, 120°	PCl_5	 Carey B. Van Loon

(continued)

TABLE 10.1 Molecular Geometry as a Function of Electron-Group Geometry (Continued)

Number of Electron Groups	Electron-Group Geometry	Number of Lone Pairs	VSEPR Notation	Molecular Geometry	Ideal Bond Angles	Example
	trigonal bipyramidal	1	AX_4E[b]	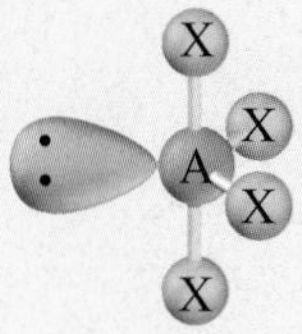 (seesaw)	90°, 120°	SF_4
	trigonal bipyramidal	2	AX_3E_2	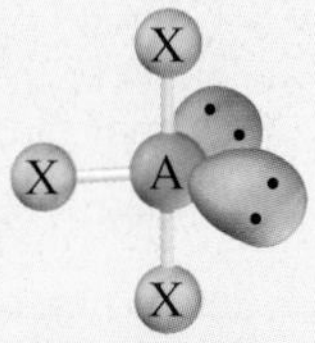 (T-shaped)	90°	ClF_3
	trigonal bipyramidal	3	AX_2E_3	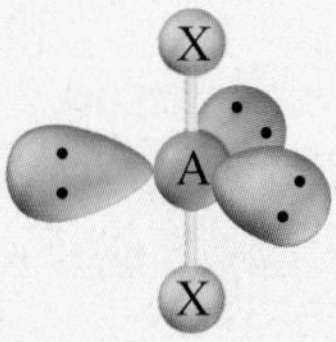 (linear)	180°	XeF_2
6	octahedral	0	AX_6	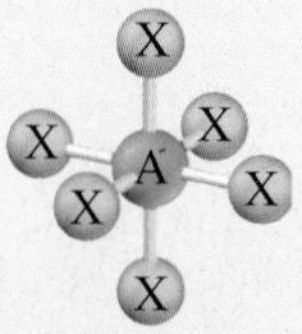 (octahedral)	90°	SF_6
	octahedral	1	AX_5E	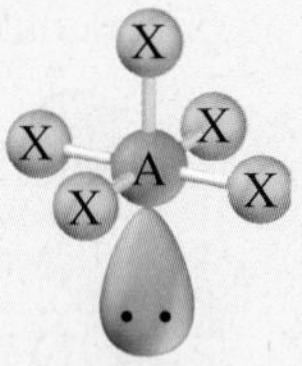 (square pyramidal)	90°	BrF_5
	octahedral	2	AX_4E_2	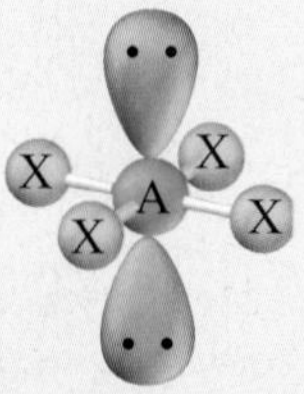 (square planar)	90°	XeF_4

Carey B. Van Loon

[a]For a discussion of the structure of SO_2, see page 444.
[b]For a discussion of the placement of the lone-pair electrons in this structure, see page 443.

tographs of ball-and-stick models. If one or more electron groups are lone s, the molecular geometry is different from the electron-group geometry, ough still derived from it. The relationship between electron-group geome- nd molecular geometry is summarized in Table 10.1. To understand all the s in Table 10.1, we need two more ideas.

The closer together two groups of electrons are forced, the stronger the repulsion between them. The repulsion between two electron groups is much stronger at an angle of 90° than at 120° or 180°.

Lone-pair electrons spread out more than do bond-pair electrons. As a result, the repulsion of one lone pair of electrons for another lone pair is greater than, say, between two bond pairs. The order of repulsive forces, from strongest to weakest, is

lone pair–lone pair > lone pair–bond pair > bond pair–bond pair

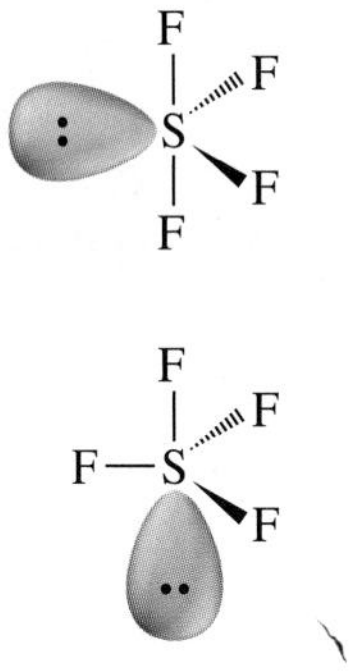

onsider SF_4, with the VSEPR notation AX_4E. Two possibilities for its structure presented in the margin, but only one is correct. The correct structure (top)

EXAMPLE 10-10 Using VSEPR Theory to Predict a Geometric Shape

Predict the molecular geometry of the polyatomic anion ICl_4^-.

nalyze

To solve this problem, apply the four steps outlined on page 444.

olve

Step 1. Write the Lewis structure. The number of valence electrons is

From I	+	From Cl	+	To establish ionic charge of −1	
(1 × 7)		(4 × 7)		1	= 36

To join 4 Cl atoms to the central I atom and to provide octets for all the atoms, we need 32 electrons. In order to account for all 36 valence electrons, we need to place an *additional* four electrons around the I atom as lone pairs. That is, we are forced to expand the valence shell of the I atom to accommodate all the electrons required in the Lewis structure.

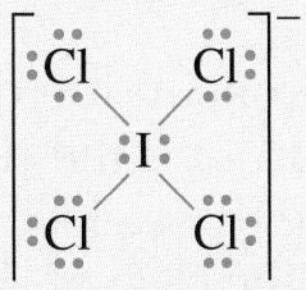

Step 2. There are six electron groups around the I atom, four *bond pairs* and two *lone pairs*.

Step 3. The electron-group geometry (the orientation of six electron groups) is *octahedral*.

Step 4. The ICl_4^- anion is of the type AX_4E_2, which according to Table 10.1, leads to a molecular geometry that is square planar.

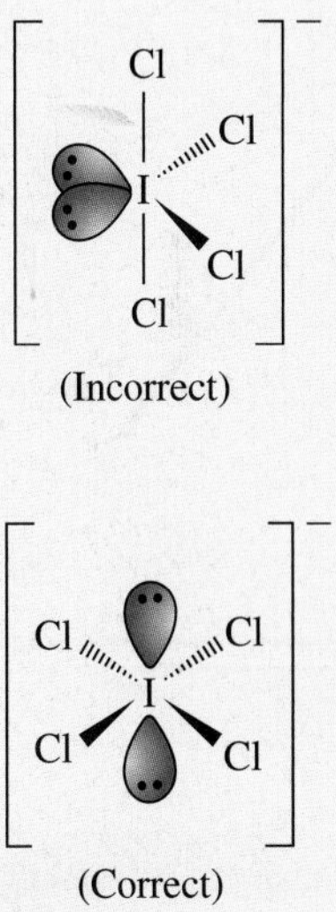

▲ FIGURE 10-14
Example 10-10 illustrated
The observed structure of ICl_4^- is square planar. The iodine atom has a formal charge of −1 (not shown in the structures above).

ssess

Figure 10-14 suggests two possibilities for distributing bond pairs and lone pairs in ICl_4^-. The square planar structure is correct because the lone pair–lone pair interaction is kept at 180°. In the incorrect structure, this interaction is at 90°, which results in a strong repulsion.

RACTICE EXAMPLE A: Predict the molecular geometry of nitrogen trichloride.

RACTICE EXAMPLE B: Predict the molecular geometry of phosphoryl chloride, $POCl_3$, an important chemical in the manufacture of gasoline additives, hydraulic fluids, and fire retardants.

KEEP IN MIND

that we need to expand a valence shell only if more than eight electrons must be accommodated by the central atom in a Lewis structure. This leads to structures based on five or six electron groups. Otherwise, octet-based Lewis structures are perfectly satisfactory when applying the VSEPR theory.

places a lone pair of electrons in the equatorial plane of the bipyramid. As a re *two* lone pair–bond pair interactions are 90°. In the incorrect structure (bottom) lone pair of electrons is at the bottom axial position of the bipyramid and resul *three* lone pair–bond pair interactions of 90°. This is a less favorable arrangeme

Applying VSEPR Theory

The following four-step strategy can be used for predicting the shape molecules.

1. Draw a plausible Lewis structure of the species (molecule or polyatomic i
2. Determine the number of electron groups around the central atom, and id tify them as being either *bond-pair* electron groups or *lone pairs* of electro
3. Establish the electron-group geometry around the central atom—lir trigonal planar, tetrahedral, trigonal bipyramidal, or octahedral.
4. Determine the molecular geometry from the positions of the at bonded directly to the central atom. (Refer to Table 10.1.)

10-8 CONCEPT ASSESSMENT

The ions ICl_2^- and ICl_2^+ differ by only two electrons. Would you expect them t have the same geometric shape? Explain.

Structures with Multiple Covalent Bonds

In a multiple covalent bond, all electrons in the bond are confined to region between the bonded atoms, and together constitute one group of e trons. Let us test this idea by predicting the molecular geometry of su dioxide. S is the central atom, and the total number of valence electror $3 \times 6 = 18$. The Lewis structure is the resonance hybrid of the three c tributing structures shown below.

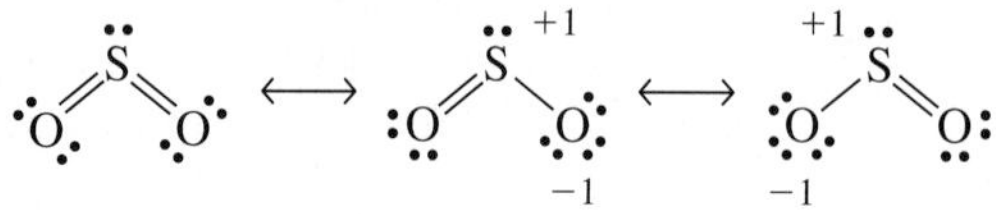

Because we count the electrons in the double covalent bond as one gro the electron-group geometry around the central S atom is that of *three* elec groups—*trigonal-planar*. Of the three electron groups, two are bonding gro and one is a lone pair. This is the case of AX_2E (see Table 10.1). The molec shape is *angular*, or *bent*, with an expected bond angle of 120°. (The measu bond angle in SO_2 is 119°.)

EXAMPLE 10-11 Using VSEPR Theory to Predict the Shape of a Molecule with a Multiple Covalent Bond

Predict the molecular geometry of formaldehyde, H_2CO, used to make a number of polymers, such as melamine resins. The Lewis structure of the H_2CO molecule is shown here.

Analyze

We see from the Lewis structure that the carbon–oxygen bond is a double bond. When considering molecules with double or triple bonds, we treat a double or triple bond as a single electron group.

Solve

There are three electron groups around the C atom, two groups in the carbon-to-hydrogen single bonds an the third group in the carbon-to-oxygen double bond. The electron-group geometry for three electron group is *trigonal-planar*. Because all the electron groups are involved in bonding, the VSEPR notation for this molecule is AX_3. The molecular geometry is also trigonal-planar.

Assess

Because the geometry is trigonal planar, we expect the angle between the two H—C bonds to be close to 120°.

PRACTICE EXAMPLE A: Predict the shape of the COS molecule.

PRACTICE EXAMPLE B: Nitrous oxide, N_2O, is the familiar laughing gas used as an anesthetic in dentistry. Predict the shape of the N_2O molecule.

olecules with More Than One Central Atom

hough many of the structures of interest to us have only one central atom, EPR theory can also be applied to molecules or polyatomic anions with more n one central atom. In such cases, the geometric distribution of terminal atoms und *each* central atom must be determined and the results then combined into ngle description of the molecular shape. We use this idea in Example 10-12.

EXAMPLE 10-12 Applying VSEPR Theory to a Molecule with More Than One Central Atom

Methyl isocyanate, CH_3NCO, is used in the manufacture of insecticides, such as carbaryl (Sevin). In the CH_3NCO molecule, the three H atoms and the O atom are terminal atoms and the two C and one N atom are central atoms. Draw the structure of this molecule, using dash and wedge symbols, and indicate the various bond angles.

Analyze

We must first draw a plausible Lewis structure. Then, we determine the electron group geometry around each atom and estimate the angles between pairs of bonds.

Solve

The number of valence electrons in the structure is

$$\begin{array}{ccccccc} \text{From C} & & \text{From N} & & \text{From O} & & \text{From H} \\ (2 \times 4) & + & (1 \times 5) & + & (1 \times 6) & + & (3 \times 1) = 22 \end{array}$$

In drawing the skeletal structure and assigning valence electrons, we first obtain a structure with incomplete octets. By shifting the indicated electrons, we can give each atom an octet.

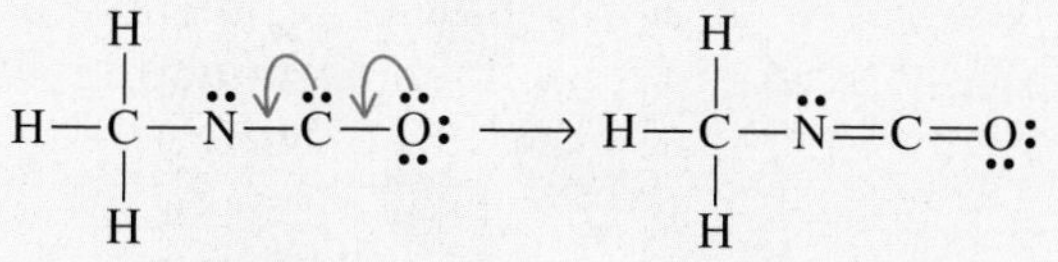

The C atom on the left has four electron groups around it—all bond pairs. The shape of this end of the molecule is *tetrahedral*. The C atom to the right, by forming two double bonds, is treated as having two groups of electrons around it. This distribution is *linear*. For the N atom, three groups of electrons are distributed in a *trigonal planar* manner. The C—N—C bond angle should be about 120°.

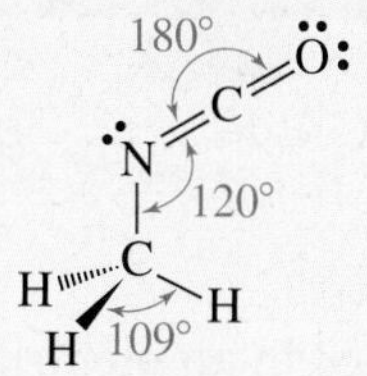

Assess

The strategy outlined above can be applied to molecules of varying complexity.

(continued)

PRACTICE EXAMPLE A: Sketch, by using dash and wedge symbols, the methanol molecule, CH_3OH. Indicate th bond angles in this molecule.

PRACTICE EXAMPLE B: Glycine, an amino acid, has the formula H_2NCH_2COOH. Sketch, by using dash an wedge symbols, the glycine molecule, and indicate the various bond angles.

10-9 CONCEPT ASSESSMENT

Methyl isocyanate, CH_3NCO, can be represented as a hybrid of three Lewis structures. The most satisfactory structure is the one given above in Example 10-12. Draw the other two structures. On the basis of formal charges, which of the structures is least satisfactory?

Molecular Shapes and Dipole Moments

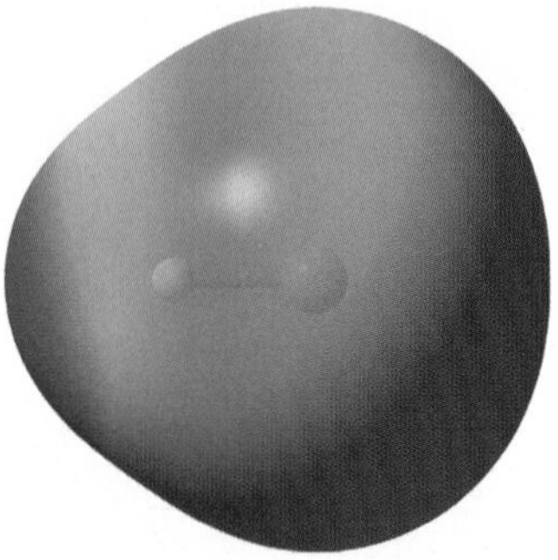

▲ Electrostatic potential map of HCl

Let us recall some facts that we learned about polar covalent bond Section 10-3. In the HCl molecule, the Cl atom is more electronegative t the H atom. Electrons are displaced toward the Cl atom, as shown in electrostatic potential map in the margin. The HCl molecule is a **polar molec** In the representation below, we use a cross-base arrow ($\leftarrow\!\!\!\vdash$) that point the atom that attracts electrons more strongly.

$$^{\delta+}\text{H} \;\mapsto\; \text{Cl}^{\delta-}$$

The extent of the charge displacement in a polar covalent bond is give the **dipole moment**, μ. The dipole moment is the product of a partial ch (δ) and distance (d).

$$\mu = \delta \times d \qquad \textbf{(1}$$

If the product, $\delta \times d$, has a value of 3.34×10^{-30} coulomb · meter (C · the dipole moment, μ, has a value called 1 *debye*, D (pronounced duh-b One experimental method of determining dipole moments is based on behavior of polar molecules in an electric field, suggested in Figure 10-15.

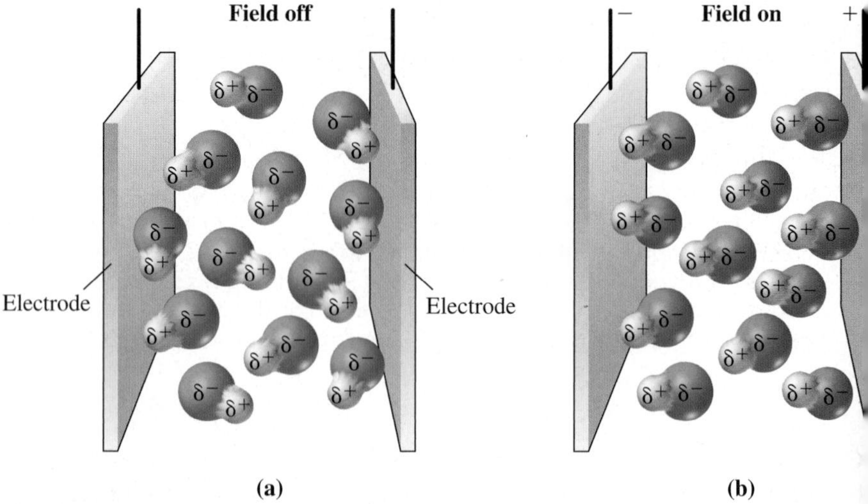

▲ FIGURE 10-15
Polar molecules in an electric field
The device pictured is called an electrical condenser (or capacitor). It consists of a pai electrodes separated by a medium that does not conduct electricity but consists of p molecules. **(a)** When the field is off, the molecules orient randomly. **(b)** When the elec field is turned on, the polar molecules orient in the field between the charged plates that the negative ends of the molecules are toward the positive plate and vice versa.

he polarity of the H—Cl bond, as demonstrated on page 418, involves a : of the electron charge density toward the Cl atom, and this produces a sep- ion of the centers of positive and negative charge. Suppose, instead of a shift ectron charge density, we think of an equivalent situation—the transfer of a *ion* of the charge of an electron from the H atom to the Cl atom through the re internuclear distance. Let us determine the magnitude of this partial ge, δ. To do this, we need the measured dipole moment, 1.03 D; the H—Cl d length, 127.4 pm; and equation (10.25) rearranged to

$$\delta = \frac{\mu}{d} = \frac{1.03\ \text{D} \times 3.34 \times 10^{-30}\ \text{C}\cdot\text{m/D}}{127.4 \times 10^{-12}\ \text{m}} = 2.70 \times 10^{-20}\ \text{C}$$

charge is about 17% of the charge on an electron (1.602×10^{-19} C) and sug- s that HCl is about 17% ionic. This assessment of the percent ionic character he H—Cl bond agrees well with the 20% we made based on electro- ativity differences (recall Example 10-4).

CO_2. Carbon dioxide molecules are *nonpolar*. To understand this observa- tion, we need to distinguish between the displacement of electron charge density in a particular bond and in the molecule as a whole. The elec- tronegativity difference between C and O causes a displacement of elec- tron charge density toward the O atom in each carbon-to-oxygen bond and gives rise to a *bond dipole*. However, because the two bond dipoles are equal in magnitude and point in opposite directions, they cancel each other and lead to a *resultant* dipole moment of zero for the molecule. The symmetrical nature of the electron charge density is clear in the electrosta- tic potential map for CO_2 shown in the margin.

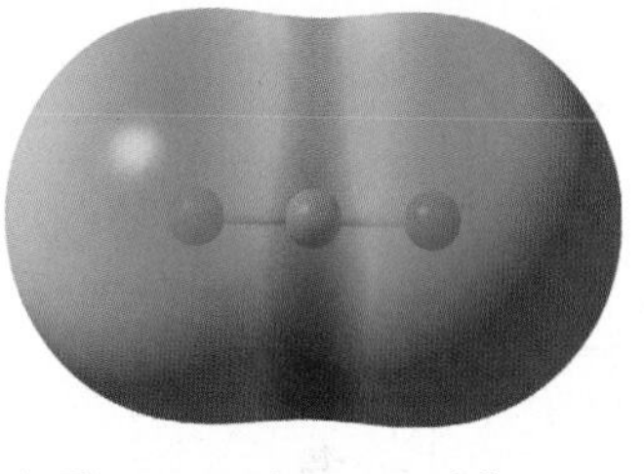
▲ Electrostatic potential map of carbon dioxide

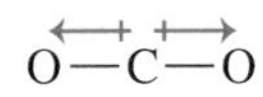

$\mu = 0$

The fact that CO_2 is nonpolar is experimental proof that the three atoms in the molecule lie along a straight line in the order O—C—O. Of course, we can also predict that CO_2 is a linear molecule with the VSEPR theory, based on the Lewis structure

$$:\ddot{O}=C=\ddot{O}:$$

H_2O. Water molecules are *polar*. They have bond dipoles because of the electronegativity difference between H and O, and the bond dipoles com- bine to produce a resultant dipole moment of 1.84 D. The electrostatic potential map for water provides visual evidence of a net dipole moment on the water molecule. The molecule cannot be linear, for this would lead to a cancellation of bond dipoles, just as with CO_2. We have predicted with the VSEPR theory that the H_2O molecule is bent, and the observation that it is a polar molecule simply confirms the prediction.

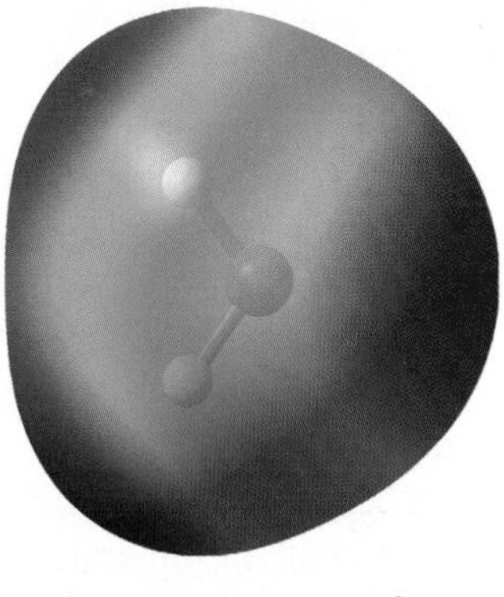
▲ Electrostatic potential map of water

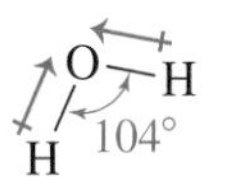

CCl_4. Carbon tetrachloride molecules are *nonpolar*. Based on the elec- tronegativity difference between Cl and C, we expect a bond dipole for the C—Cl bond. The fact that the resultant dipole moment is *zero* means that the bond dipoles must be oriented in such a way that they cancel. The tetra- hedral molecular geometry of CCl_4 provides the symmetrical distribution of bond dipoles that leads to this cancellation, as shown in Figure 10-16(a). Can you see that the molecule will be polar if one of the Cl atoms is replaced by an atom with a different electronegativity, say H? In the mole- cule, $CHCl_3$, there is a resultant dipole moment (Fig. 10-16b).

KEEP IN MIND

that the lack of a molecular dipole moment cannot distinguish between the two possible molecular geometries: tetrahedral and square planar. To do this, other experimental evidence, such as X-ray diffraction, is required.

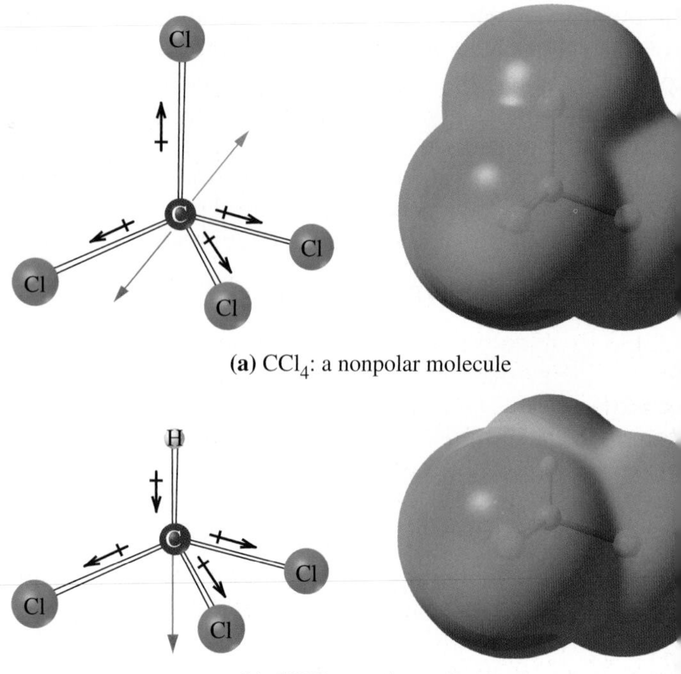

(a) CCl_4: a nonpolar molecule

(b) $CHCl_3$: a polar molecule

▶ FIGURE 10-16
Molecular shapes and dipole moments
(a) The resultant of two of the C—Cl bond dipoles is shown as a red arrow, and that of the other two, as a blue arrow. The red and blue arrows point in opposite directions and cancel. The CCl_4 molecule is nonpolar. The balance of the charge distribution in CCl_4 is clearly seen in the electrostatic potential map. **(b)** The individual bond dipoles do combine to yield a resultant dipole moment (red arrow) of 1.04 D. The electrostatic potential map indicates that the hydrogen atom has a partial positive charge.

EXAMPLE 10-13 Determining the Relationship Between Geometric Shapes and the Resultar Dipole Moments of Molecules

Which of these molecules would you expect to be polar: Cl_2, ICl, BF_3, NO, SO_2?

Analyze

We will use the methods described above to determine the shape of the molecule, and then ascertain wheth or not bond dipoles, if present, produce a net permanent dipole moment.

Solve

Polar: ICl, NO, SO_2. ICl and NO are diatomic molecules with an electronegativity difference between th bonded atoms. SO_2 is a bent molecule with an electronegativity difference between the S and O atoms.

Nonpolar: Cl_2 and BF_3. Cl_2 is a diatomic molecule of identical atoms; hence no electronegativity difference. Fo BF_3, refer to Table 10.1. BF_3 is a symmetrical planar molecule (120° bond angles). The B—F bond dipoles canc each other.

Assess

Bond dipoles are vector quantities. When adding them together, we must add them as vectors, that is, "heac to-tail," as illustrated below.

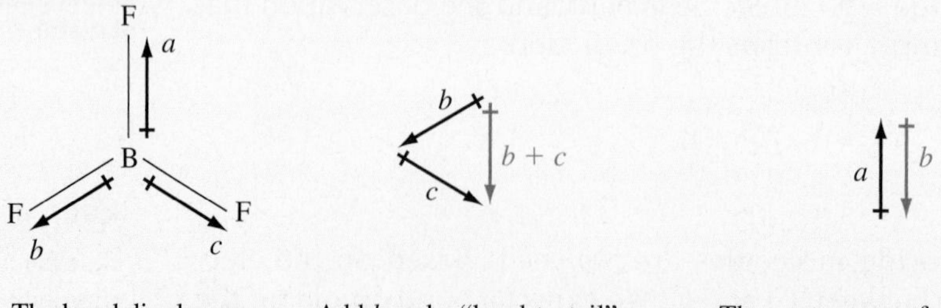

The bond dipoles are labeled *a*, *b*, and *c*.

Add *b* and *c* "head-to-tail" to form the vector sum *b* + *c*.

The vector sum of *a* and *b* + *c* is a vector of zero length (no dipole).

PRACTICE EXAMPLE A: Only one of the following molecules is polar. Which is it, and why? SF_6, H_2O_2, C_2H_4.

PRACTICE EXAMPLE B: Only one of the following molecules is *nonpolar*. Which is it, and why? Cl_3CCH_3, PCl CH_2Cl_2, NH_3.

10-10 CONCEPT ASSESSMENT

e molecule NH_3 has the dipole moment μ = 1.47 D, whereas for the similar
olecule NF_3, μ = 0.24 D. Why do you suppose there is such a large difference
these two values? [*Hint:* What is the effect of the lone-pair electrons on the
atom?]

-8 Bond Order and Bond Lengths

term **bond order** describes whether a covalent bond is *single* (bond er = 1), *double* (bond order = 2), or *triple* (bond order = 3). Think of elecs as the "glue" that binds atoms together in covalent bonds. The higher bond order—that is, the more electrons present—the more glue and the e tightly the atoms are held together.

ond length is the distance between the centers of two atoms joined by a alent bond. A double bond between atoms is shorter than a single bond, and ple bond is shorter still. You can see this relationship clearly in Table 10.2 by paring the three different bond lengths for the nitrogen-to-nitrogen bond. example, the measured length of the nitrogen-to-nitrogen triple bond in N_2 09.8 pm, whereas the nitrogen-to-nitrogen single bond in hydrazine, J—NH_2, is 147 pm.

erhaps you can now also better understand the meaning of covalent radius we introduced in Section 9-3. The single-bond covalent radius is one-half the ance between the centers of identical atoms joined by a *single* covalent bond. s, the single-bond covalent radius of chlorine in Figure 9-11 (100 pm) is onethe bond length given in Table 10.2, that is, $\frac{1}{2} \times 199$ pm. A rough generalon is:

The length of the covalent bond between two atoms can be approximated as the sum of the covalent radii of the two atoms.

ome of these ideas about bond length are applied in Example 10-14.

ABLE 10.2 Some Average Bond Lengths[a]

ond	Bond Length, pm	Bond	Bond Length, pm	Bond	Bond Length, pm
—H	74.14	C—C	154	N—N	145
—C	110	C=C	134	N=N	123
—N	100	C≡C	120	N≡N	109.8
—O	97	C—N	147	N—O	136
—S	132	C=N	128	N=O	120
—F	91.7	C≡N	116	O—O	145
—Cl	127.4	C—O	143	O=O	121
—Br	141.4	C=O	120	F—F	143
—I	160.9	C—Cl	178	Cl—Cl	199
				Br—Br	228
				I—I	266

Most values (C—H, N—H, C—H, and so on) are averaged over a number of species ntaining the indicated bond and may vary by a few picometers. Where a diatomic olecule exists, the value given is the actual bond length in that molecule (H_2, N_2, IF, and so on) and is known more precisely.

EXAMPLE 10-14 Estimating Bond Lengths

Provide the best estimate you can of these bond lengths for the **(a)** the nitrogen-to-hydrogen bonds in NH
(b) the bromine-to-chlorine bond in BrCl.

Analyze

If no bond length is listed for a particular bond, A—B, then look up the bond lengths for A—A and B—B. Th
A—B bond length can then be estimated as one-half the A—A bond length plus one-half the B—B bond lengt

Solve

(a) The Lewis structure of ammonia (page 416) shows the nitrogen-to-hydrogen bonds as single bond The value listed in Table 10.2 for the N—H bond is 100 pm, so this is the value we would predict. (Th measured N—H bond length in NH_3 is 101.7 pm.)

(b) There is no bromine-to-chlorine bond length in Table 10.2, so we need to calculate an approximate bon length using the relationship between bond length and covalent radii. BrCl contains a Br—Cl *sing* bond [imagine substituting one Br atom for one Cl atom in structure (10.4)]. The length of the Br—C bond is one-half the Cl—Cl bond length *plus* one-half the Br—Br bond length: $\left(\frac{1}{2} \times 199\ \text{pm}\right)$ $\left(\frac{1}{2} \times 228\ \text{pm}\right) = 214\ \text{pm}$. (The measured bond length is 213.8 pm.)

Assess

The data in Table 10.2 can be used to make estimates of bond lengths in a variety of molecules.

PRACTICE EXAMPLE A: Estimate the bond lengths of the carbon-to-hydrogen bonds and the carbon-to-bromin bond in CH_3Br.

PRACTICE EXAMPLE B: In the thiocyanate ion, SCN^-, the length of the carbon-to-nitrogen bond is 115 pm. Writ a plausible Lewis structure for this ion and describe its geometric shape.

An interesting situation arises for molecules in which resonance is pres In such molecules, fractional bond orders are possible. Consider, for exam the carbonate anion, CO_3^{2-}, shown below.

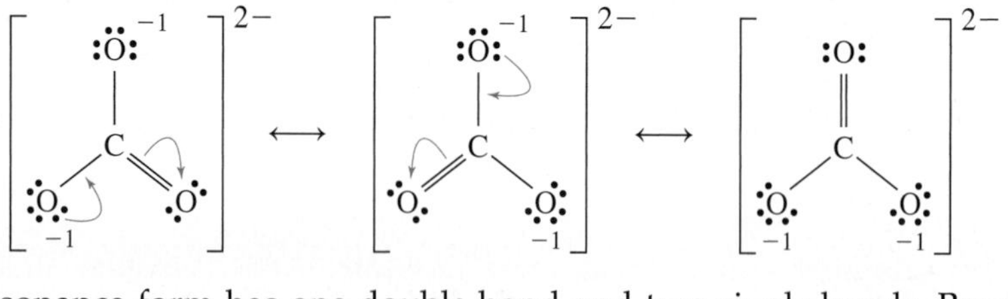

Each resonance form has one double bond and two single bonds. Because actual structure of the carbonate anion is an average of these three structu the average bond order is $\frac{1}{3}(1 + 1 + 2) = 1\frac{1}{3}$. The CO bond distance in carbonate anion is 129 pm, which is intermediate between a C—O single b (143 pm) and a C=O double bond (120 pm), as we might expect for a f tional bond order.

10-11 CONCEPT ASSESSMENT

NO_2^- and NO_2^+ are made up of the same atoms. How would you expect the nitrogen-to-oxygen bond lengths in these two ions to compare?

10-9 Bond Energies

Together with bond lengths, bond energies can be used to assess the suita ity of a proposed Lewis structure. Bond energy, bond length, and bond or are interrelated properties in this sense: the higher the bond order, the sho the bond between two atoms and the greater the bond energy.

nergy is *released* when isolated atoms join to form a covalent bond, and rgy must be *absorbed* to break apart covalently bonded atoms. **Bond- ociation energy**, *D*, is the quantity of energy required to break one mole of alent bonds in a *gaseous* species. The SI units are kilojoules per mole of ds (kJ mol^{-1}).

Ve can think of bond-dissociation energy as an enthalpy change or a heat of tion, as discussed in Chapter 7. For example,

ond breakage: $H_2(g) \longrightarrow 2\,H(g) \qquad \Delta_r H = D(H{-}H) = +435.93\ kJ\ mol^{-1}$

ond formation: $2\,H(g) \longrightarrow H_2(g) \qquad \Delta_r H = -D(H{-}H) = -435.93\ kJ\ mol^{-1}$

: is not hard to picture the meaning of bond energy for a diatomic mole- , because there is only one bond in the molecule. It is also not difficult to that the bond-dissociation energy of a diatomic molecule can be expressed er precisely, as is that of $H_2(g)$. With a polyatomic molecule, such as H_2O, situation is different (Fig. 10-17). The energy needed to dissociate one mole I atoms by breaking one O—H bond per H_2O molecule,

$$H{-}OH(g) \longrightarrow H(g) + OH(g) \qquad \Delta_r H = D(H{-}OH) = +498.7\ kJ\ mol^{-1}$$

ifferent from the energy required to dissociate one mole of H atoms by ıking the bonds in OH(g):

$$O{-}H(g) \longrightarrow H(g) + O(g) \qquad \Delta_r H = D(O{-}H) = +428.0\ kJ\ mol^{-1}$$

'he two O—H bonds in H_2O are identical; therefore, they should have ıtical energies. This energy, which we can call the O—H bond energy in), is the *average* of the two values listed above: 463.4 kJ mol^{-1}. The O—H d energy in other molecules containing the OH group will be somewhat erent from that in H—O—H. For example, in methanol, CH_3OH, the -H bond-dissociation energy, which we can represent as $D(H{-}OCH_3)$, is 8 kJ mol^{-1}. The usual method of tabulating bond energies (Table 10.3) is *verages*. An **average bond energy** is the average of bond-dissociation ener- ; for a number of different species containing the particular bond. lerstandably, average bond energies cannot be stated as precisely as spe- : bond-dissociation energies.

s you can see from Table 10.3, double bonds have higher bond energies ı do single bonds between the same atoms, but they are *not* twice as large. ıle bonds are stronger still, but their bond energies are *not* three times as

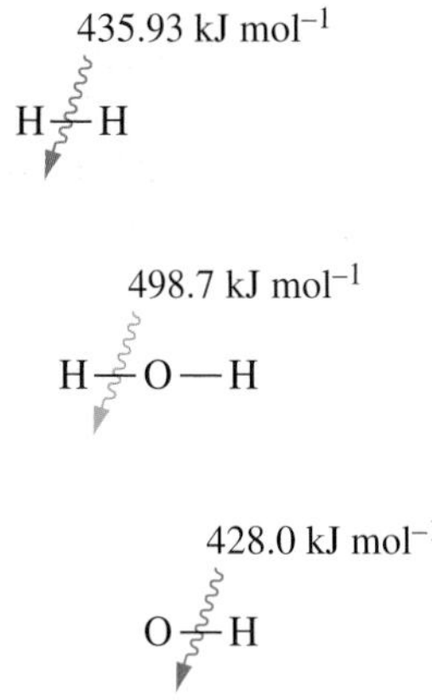

▲ FIGURE 10-17
Some bond energies compared
The same quantity of energy, 435.93 kJ mol^{-1}, is required to break all H—H bonds. In H_2O, more energy is required to break the first bond (498.7 kJ mol^{-1}) than to break the second (428.0 kJ mol^{-1}). The second bond broken is that in the OH radical. The O—H bond energy in H_2O is the average of the two values: 463.4 kJ mol^{-1}.

KEEP IN MIND

that tabulated bond energies are for isolated molecules in the gaseous state. They do not apply to molecules in close contact in liquids and solids.

BLE 10.3 Some Average Bond Energies[a]

nd	Bond Energy, kJ mol^{-1}	Bond	Bond Energy, kJ mol^{-1}	Bond	Bond Energy, kJ mol^{-1}
—H	436	C—C	347	N—N	163
—C	414	C=C	611	N=N	418
—N	389	C≡C	837	N≡N	946
—O	464	C—N	305	N—O	222
—S	368	C=N	615	N=O	590
—F	565	C≡N	891	O—O	142
—Cl	431	C—O	360	O=O	498
—Br	364	C=O	736[b]	F—F	159
—I	297	C—Cl	339	Cl—Cl	243
				Br—Br	193
				I—I	151

lthough all data are listed with about the same precision (three significant figures), some ues are actually known more precisely. Specifically, the values for the diatomic mole- les H_2, HF, HCl, HBr, HI, N_2 (N≡N), O_2 (O=O), F_2, Cl_2, Br_2, and I_2 are actually nd-dissociation energies, rather than average bond energies.

ne value for the C=O bonds in CO_2 is 799 kJ mol^{-1}.

large as single bonds between the same atoms. This observation about b order and bond energy will seem quite reasonable after multiple bonds more fully described in the next chapter.

Bond energies also have some interesting uses in thermochemistry. F reaction involving *gases*, visualize the process

$$\text{gaseous reactants} \longrightarrow \text{gaseous atoms} \longrightarrow \text{gaseous products}$$

In this hypothetical process, we first break all the bonds in reactant m cules and form gaseous atoms. For this step, the enthalpy change is (bond breakage) = ΣBE (reactants), where BE stands for bond energy. N we allow the gaseous atoms to recombine into product molecules. In step, bonds are formed and $\Delta_r H$ (bond formation) = −ΣBE (products). enthalpy change of the reaction, then, is

KEEP IN MIND

that the difference in equation (10.26) is calculated as reactants minus products and not products minus reactants.

$$\Delta H = \Delta_r H\,(\text{bond breakage}) + \Delta_r H\,(\text{bond formation})$$
$$\approx \Sigma\text{BE}\,(\text{reactants}) - \Sigma\text{BE}\,(\text{products}) \qquad \text{(1}$$

The approximately equal sign (≈) in expression (10.26) signifies that s of the bond energies used are likely to be *average* bond energies rather true bond-dissociation energies. Also, a number of terms often cancel because some of the same types of bonds appear in the products as in the r tants. We can base the calculation of $\Delta_r H$ just on the *net* number and type bonds broken and formed, as illustrated in Example 10-15.

EXAMPLE 10-15 Calculating an Enthalpy of Reaction from Bond Energies

The reaction of methane (CH_4) and chlorine produces a mixture of products called chloromethanes. One o these is monochloromethane, CH_3Cl, used in the preparation of silicones. Calculate $\Delta_r H$ for the reaction

$$CH_4(g) + Cl_2(g) \longrightarrow CH_3Cl(g) + HCl(g)$$

Analyze

To identify which bonds are broken and formed, it helps to draw structural formulas (or Lewis structures), a in Figure 10-18. To apply expression (10.26) literally, we would break *four* C—H bonds and *one* Cl—Cl bon and form *three* C—H bonds, *one* C—Cl bond, and *one* H—Cl bond. The *net* change, however, is the breakin of *one* C—H bond and *one* Cl—Cl bond, followed by the formation of *one* C—Cl bond and *one* H—Cl bond

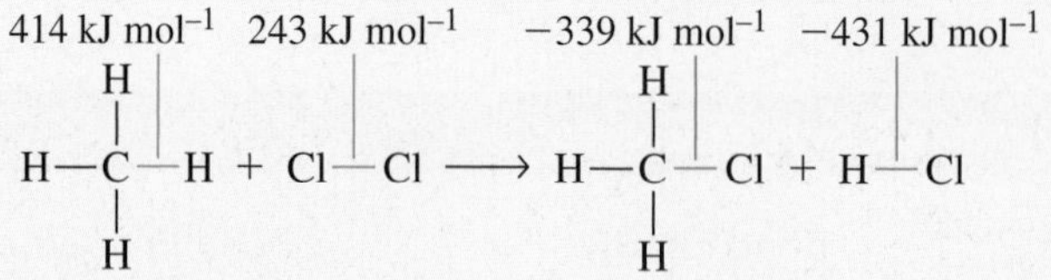

▲ FIGURE 10-18
Net bond breakage and formation in a chemical reaction—Example 10-15 illustrated
Bonds that are broken are shown in red and bonds that are formed, in blue. Bonds that remain unchanged are black. The net change is that *one* C—H and *one* Cl—Cl bond break and *one* C—Cl and *one* H—Cl bond form.

Solve

$\Delta_r H$ *for net bond breakage:*	C—H bonds	$1 \times +414$ kJ mol^{-1}
	Cl—Cl bonds	$1 \times +243$ kJ mol^{-1}
	sum:	$+657$ kJ mol^{-1}
$\Delta_r H$ *for net bond formation:*	C—Cl bonds	1×-339 kJ mol^{-1}
	H—Cl bonds	1×-431 kJ mol^{-1}
	sum:	-770 kJ mol^{-1}

Enthalpy of reaction: $\Delta_r H = (657 - 770)\ \text{kJ mol}^{-1} = -113\ \text{kJ mol}^{-1}$

ssess

A number of terms cancel out because some of the same types of bonds appear in both reactants and products. Such a situation is not uncommon.

RACTICE EXAMPLE A: Use bond energies to estimate the enthalpy change for the reaction

$$2\,H_2(g) + O_2(g) \longrightarrow 2\,H_2O(g)$$

RACTICE EXAMPLE B: Use bond energies to estimate the enthalpy of formation of $NH_3(g)$.

n Chapter 7, we learned how to calculate $\Delta_r H$ for any reaction using $\Delta_f H°$ ues—see equation (7.21). Equation (10.26) gives us another way to calculate I for a gas-phase reaction. There is no advantage to using bond energies over nalpy-of-formation data. Enthalpies of formation are generally known rather cisely, whereas bond energies are only average values. But when enthalpy-ormation data are lacking, bond energies can prove particularly useful.

nother way to use bond energies is in predicting whether a reaction will *ndothermic* or *exothermic*. In general, if

$$\underset{\text{(reactants)}}{\text{weak bonds}} \longrightarrow \underset{\text{(products)}}{\text{strong bonds}} \qquad \Delta_r H < 0 \qquad \text{(exothermic)}$$

$$\underset{\text{(reactants)}}{\text{strong bonds}} \longrightarrow \underset{\text{(products)}}{\text{weak bonds}} \qquad \Delta_r H > 0 \qquad \text{(endothermic)}$$

mple 10-16 applies this idea to a reaction involving highly reactive, unsta-species for which enthalpies of formation are not normally listed.

XAMPLE 10-16 Using Bond Energies to Predict Exothermic and Endothermic Reactions

One of the steps in the formation of monochloromethane (Example 10-15) is the reaction of a gaseous chlorine *atom* (a chlorine radical) with a molecule of methane. The products are an unstable methyl radical and HCl(g). Is this reaction endothermic or exothermic?

$$CH_4 + \cdot Cl(g) \longrightarrow \cdot CH_3(g) + HCl(g)$$

nalyze

In the reaction, one C—H bond is broken for every H—Cl bond formed. Thus, we must compare the bond energies for the C—H and H—Cl bonds to decide whether the reaction is endothermic or exothermic.

olve

For every molecule of CH_4 that reacts, *one* C—H bond *breaks*, requiring 414 kJ per mole of bonds; and *one* H—Cl bond *forms*, releasing 431 kJ per mole of bonds. Because more energy is released in forming new bonds than is absorbed in breaking old ones, we predict that the reaction is exothermic.

ssess

In the example above, we had to break only C—H bonds and form only H—Cl bonds. Most reactions involve breaking and forming several types of bonds, and so it is usually not obvious whether the reaction will be exothermic or endothermic. In such cases, we must calculate $\Delta_r H$, by using equation (10.26), to see whether $\Delta_r H > 0$ or $\Delta_r H < 0$.

RACTICE EXAMPLE A: Is the following reaction endothermic or exothermic?

$$CH_3COCH_3(g) + H_2(g) \longrightarrow (CH_3)_2CH(OH)(g)$$

RACTICE EXAMPLE B: Predict whether the following reaction should be exothermic or endothermic:

$$H_2O(g) + Cl_2(g) \longrightarrow \frac{1}{2}O_2(g) + 2\,HCl(g).$$

MasteringCHEMISTRY www.masteringchemistry.com

It is almost a certainty that you are familiar with microwave radiation—have you ever cooked or heated fooc in a microwave oven? Microwave radiation plays a prominent role in a branch of science called rotational (or microwave) spectroscopy. For a discussion of rotational spectroscopy and how it is used not only to mak precise measurements of bond lengths but also to detect molecules in interstellar space, go to the Focus On feature for Chapter 10, Molecules in Space: Measuring Bond Lengths, on the MasteringChemistry site.

Summary

10-1 Lewis Theory: An Overview—A **Lewis symbol** represents the valence electrons of an atom by using dots placed around the chemical symbol. A **Lewis structure** is a combination of Lewis symbols used to represent chemical bonding. Normally, all the electrons in a Lewis structure are paired, and each atom in the structure acquires an **octet**—that is, there are eight electrons in the valence shell. In **Lewis theory**, chemical bonds are classified as **ionic bonds**, which are formed by electron transfer between atoms, or **covalent bonds**, which are formed by electrons shared between atoms. Most bonds, however, have partial ionic and partial covalent characteristics.

10-2 Covalent Bonding: An Introduction—Atoms in molecules are often surrounded by eight valence-shell electrons (an octet) and thus conform to the **octet rule**. In covalent bonds, pairs of electrons shared between two atoms to form the bonds are called **bond pairs**, whereas pairs of electrons not shared in a chemical bond are called **lone pairs**. A single pair of electrons shared between two atoms constitutes a **single covalent bond**. When both electrons in a covalent bond between two atoms are provided by only one of the atoms, the bond is called a **coordinate covalent bond**. To construct a Lewis structure for a molecule in which all atoms obey the octet rule, it is often necessary for atoms to share more than one pair of electrons, thus forming **multiple covalent bonds**. Two shared electron pairs constitute a **double covalent bond** and three shared pairs a **triple covalent bond**.

10-3 Polar Covalent Bonds and Electrostatic Potential Maps—A covalent bond in which the electron pair is not shared equally by the bonded atoms is called a **polar covalent bond**. Whether a bond is polar or not can be predicted by comparing the **electronegativity (EN)** of the atoms involved (Fig. 10-6). The greater the **electronegativity difference** (Δ**EN**) between two atoms in a chemical bond, the more polar the bond and the more ionic its character. The electron charge distribution in a molecule can be visualized by computing an **electrostatic potential map** (Figs. 10-4 and 10-5). The variation of charge in the molecule is represented by a color spectrum in which red is the most negative and blue the most positive. Electrostatic potential maps are a powerful way of representing electron charge distribution in both polar and nonpolar molecules.

10-4 Writing Lewis Structures—To draw the Lewis structure of a covalent molecule, one needs to know the **skeletal structure**—that is, which is the **central atom** and what atoms are bonded to it. Atoms that are bonded tc one other atom are called **terminal atoms** (structure 1C Typically, the atom with the lowest electronegativity central atom. At times, the concept of **formal ch** (expression 10.16) is useful in selecting a skeletal struc and assessing the plausibility of a Lewis structure.

10-5 Resonance—Often, more than one plaus Lewis structure can be written for a species; this situa is called **resonance**. In these cases the true structure resonance hybrid of two or more contributing structu

10-6 Exceptions to the Octet Rule—There are c exceptions to the octet rule. (1) Odd-electron species, s as NO, have an unpaired electron and are paramagn Many of these species are reactive molecular fragme such as OH, called **free radicals**. (2) A few molecules h incomplete octets in their Lewis structures, that is, enough electrons to provide an octet for every at (3) **Expanded valence shells** occur in some compounc nonmetals of the third period and beyond. In these, valence shell of the central atom must be expanded to 1 12 electrons in order to write a Lewis structure.

10-7 Shapes of Molecules—A powerful methoc predicting the **molecular geometry**, or molecular shape, species is the **valence-shell electron-pair repulsion th** (**VSEPR theory**). The shape of a molecule or polyatomic depends on the geometric distribution of valence-shell tron groups—the **electron-group geometry**—and whe these groups contain bonding electrons or lone p (Fig. 10-11, Table 10.1). The angles between the elec groups provide a method for predicting **bond angles** molecule. An important use of information about the sh of molecules is in establishing whether bonds in a mole combine to produce a resultant **dipole moment** (Fig. 10 Molecules with a resultant dipole moment are **polar m cules**; those with no resultant dipole moment are nonpol

10-8 Bond Order and Bond Lengths—Single, c ble, and triple covalent bonds are said to have a **b order** of 1, 2, and 3, respectively. **Bond length** is the tance between the centers of two atoms joined by a co lent bond. The greater the bond order, the shorter bond length (Table 10.2).

10-9 Bond Energies—**Bond-dissociation energy** is the quantity of energy required to break one mol covalent bonds in a gaseous molecule. **Average b energies** (Table 10.3) can be used to estimate entha changes for reactions involving gases.

Integrative Example

yl fluoride is a reactive gas useful in rocket propellants. Its mass percent composition is 21.55% N, 49.23% O, and 3% F. Its density is 2.7 g/L at 20 °C and 1.00 atm pressure. Describe the nitryl fluoride molecule as completely as ible—that is, its formula, Lewis structure, molecular shape, and polarity.

lyze

, determine the empirical formula of nitryl fluoride from the composition data and the molar mass based on that for- a. Next, determine the true molar mass from the vapor density data. Now the two results can be compared to establish nolecular formula. Then, write a plausible Lewis structure based on the molecular formula, and apply VSEPR theory e Lewis structure to predict the molecular shape. Finally, assess the polarity of the molecule from the molecular shape electronegativity values.

ve

etermine the empirical formula, use the method of nple 3-5 (page 80). In 100.0 g of the compound,

$$\text{mol N} = 21.55\ \text{g N} \times \frac{1\ \text{mol N}}{14.007\ \text{g N}} = 1.539\ \text{mol N}$$

$$\text{mol O} = 49.23\ \text{g O} \times \frac{1\ \text{mol O}}{15.999\ \text{g O}} = 3.077\ \text{mol O}$$

$$\text{mol F} = 29.23\ \text{g F} \times \frac{1\ \text{mol F}}{18.998\ \text{g F}} = 1.539\ \text{mol F}$$

empirical formula is

$$N_{1.539}O_{3.077}F_{1.539} = NO_2F$$

molar mass based on this formula is

$$14 + 32 + 19 = 65\ \text{g/mol}$$

the method of Example 6-7 (page 210) to determine nolar mass of the gas.

$$\text{molar mass} = \frac{mRT}{PV} = \frac{dRT}{P}$$

$$= \frac{2.7\ \text{g/L} \times 0.0821\ \text{L atm mol}^{-1}\ \text{K}^{-1} \times 293\ \text{K}}{1.00\ \text{atm}}$$

$$= 65\ \text{g/mol}$$

use the two molar mass results are the same, the ecular and empirical formulas are the same: NO_2F.

use N has the lowest electronegativity, it should ne lone central atom in the Lewis structure. There are equivalent contributing structures to a resonance rid.

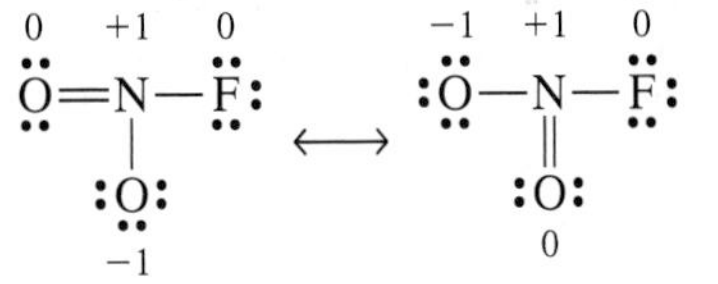

ee electron groups around the N atom produce a trig- planar electron-group geometry. All the electron ıps participate in bonding, so the molecular geome- s also trigonal planar. The predicted bond angles are . (The experimentally determined F—N—O bond e is 118°.)

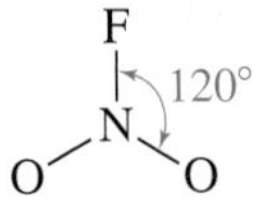

molecule has a symmetrical shape, but because the tronegativity of F is different from that of O, should expect the electron charge distribution in the ecule to be nonsymmetrical, leading to a small resul- dipole moment. NO_2F is a polar molecule.

ess

he contributing structures to the resonance hybrid vis structure, the F atom has a formal charge of zero the two O atoms have an average formal charge of As a result, we might expect the oxygen-atom region he molecule to be the most negative in electron rge density and the fluorine region to be more neu- . This conclusion is confirmed by the electrostatic ential map for NO_2F. Notice also the positive charge

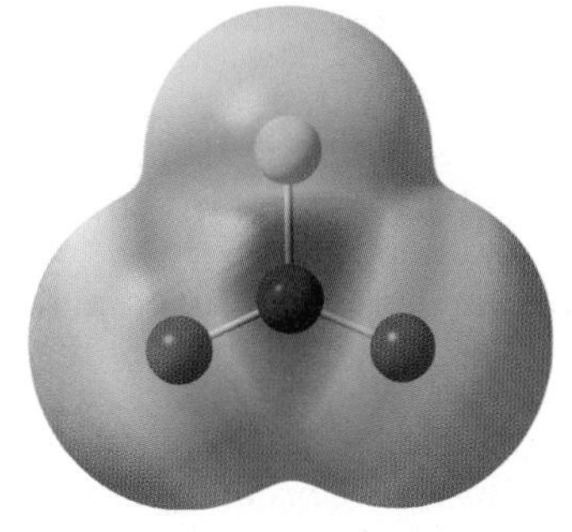

density around the nitrogen nucleus, which is in accord with the fact that the nitrogen is the least electronegative o three elements.

PRACTICE EXAMPLE A: Phosphorus pentachloride, $PCl_5(g)$, can be made from the reaction of $PCl_3(g)$ and Cl By using only data from Appendix D, estimate the average bond energy for the P—Cl bond. Assume that the P- bond energies are the same in PCl_3 and PCl_5. With reference to the geometries of PCl_3 and PCl_5, explain why the P- bond energies are probably not the same in these two molecules.

PRACTICE EXAMPLE B: The condensed structural formulas of formamide and formaldoxime are H_2NCHO H_2CNOH, respectively. One of these molecules is much more stable than the other. **(a)** Sketch the structures of t molecules, using dash and wedge symbols, indicate the various bond angles, and use bond energies to determine w molecule is more stable. **(b)** Experiment shows that the H—N—H angle in formamide is 119° and the N—C—O a is 124°. Draw a Lewis structure for formamide that is consistent with this structural information.

Exercises

Lewis Theory

1. Write Lewis symbols for the following atoms. **(a)** Kr; **(b)** Ge; **(c)** N; **(d)** Ga; **(e)** As; **(f)** Rb.
2. Write Lewis symbols for the following ions. **(a)** H^-; **(b)** Sn^{2+}; **(c)** K^+; **(d)** Br^-; **(e)** Se^{2-}; **(f)** Sc^{3+}.
3. Write plausible Lewis structures for the following molecules that contain only single covalent bonds. **(a)** FCl; **(b)** I_2; **(c)** SF_2; **(d)** NF_3; **(e)** H_2Te.
4. Each of the following molecules contains at least one multiple (double or triple) covalent bond. Give a plausible Lewis structure for **(a)** HCN; **(b)** $SC(NH_2)_2$; **(c)** F_2CO; **(d)** Cl_2SO; **(e)** C_2H_2; **(f)** SO_2.
5. By means of Lewis structures, represent bonding between the following pairs of elements: **(a)** Cs and Br; **(b)** H and Sb; **(c)** B and Cl; **(d)** Cs and Cl; **(e)** Li and O; **(f)** Cl and I. Your structures should show whether the bonding is essentially ionic or covalent.
6. Which of the following have Lewis structures that *do not* obey the octet rule: NF_3, $B(OH)_3$, $SiF_6{}^{2-}$, SO_3, $PH_4{}^+$, $PO_4{}^{3-}$, ClO_2, C_2H_4, $SO(CH_3)$.
7. Give several examples for which the following statement proves to be incorrect. "All atoms in a Lewis structure have an octet of electrons in their valence shells."
8. Suggest reasons why the following do not exist as stable molecules: **(a)** H_3; **(b)** HHe; **(c)** He_2; **(d)** H_3O.
9. Describe what is wrong with each of the following Lewis structures.

   (a) H—H—N̈—Ö—H

   (b) Ca—Ö:

10. Describe what is wrong with each of the follov Lewis structures.

    (a) :Ö—C̈l—Ö:

    (b) [·Ċ=N̈:]$^-$

11. Only one of the following Lewis structures is cor Select that one and indicate the errors in the other

    (a) cyanate ion [:Ö—C=N̈:]$^-$

    (b) carbide ion [C≡C:]$^{2-}$

    (c) hypochlorite ion [:C̈l—Ö:]$^-$

    (d) nitrogen(II) oxide :N̈=Ö:

12. Indicate what is wrong with each of the follow Lewis structures. Replace each one with a m acceptable structure.

    (a) Mg :Ö:

    (b) [:Ö—N̈=Ö:]$^+$

    (c) [:C̈l]$^+$[:Ö:]$^{2-}$[C̈l:]$^+$

    (d) [:S̈—C=N̈:]$^-$

Ionic Bonding

13. Write Lewis structures for the following ionic compounds: **(a)** calcium chloride; **(b)** barium sulfide; **(c)** lithium oxide; **(d)** sodium fluoride.
14. Under appropriate conditions, both hydrogen and nitrogen can form monatomic anions. What are the Lewis symbols for these ions? What are the Lewis structures of the compounds **(a)** lithium hydride; **(b)** calcium hydride; **(c)** magnesium nitride?
15. Derive the correct formulas for the following ic compounds by writing Lewis structures. **(a)** lithi sulfide; **(b)** sodium fluoride; **(c)** calcium iod **(d)** scandium chloride.
16. Each of the following ionic compounds consists combination of monatomic and polyatomic ic Represent these compounds with Lewis structu **(a)** $Al(OH)_3$; **(b)** $Ca(CN)_2$; **(c)** NH_4F; **(d)** KC **(e)** $Ba_3(PO_4)_2$.

rmal Charge

Assign formal charges to each of the atoms in the following structures.

(a) $[H—C≡C:]^-$

(b) 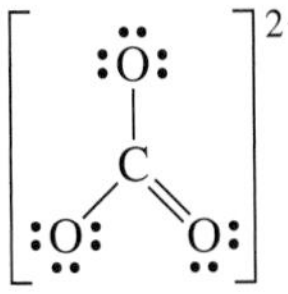

(c) $[CH_3—CH—CH_3]^+$

Assign formal charges to each of the atoms in the following structures.

(a) 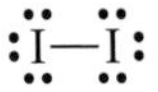

(b) 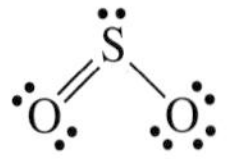

(c) 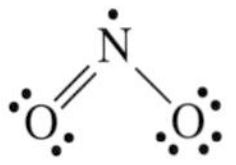

Both oxidation state and formal charge involve conventions for assigning valence electrons to bonded atoms in compounds, but clearly they are not the same. Describe several ways in which these concepts differ.

Although the notion that a Lewis structure in which formal charges are zero or held to a minimum seems to apply in most instances, describe several significant situations in which this appears not to be the case.

21. What is the formal charge of the indicated atom in each of the following structures?
(a) the central O atom in O_3
(b) Al in AlH_4^-
(c) Cl in ClO_3^-
(d) Si in SiF_6^{2-}
(e) Cl in ClF_3

22. Assign formal charges to the atoms in the following species, and then select the more likely skeletal structure.
(a) H_2NOH or H_2ONH
(b) SCS or CSS
(c) NFO or FNO
(d) $SOCl_2$ or $OSCl_2$ or OCl_2S
(e) F_3SN and F_3NS

23. The concept of formal charge helped us to choose the more plausible of the Lewis structures for NO_2^+ given in expressions (10.14) and (10.15). Can it similarly help us to choose a single Lewis structure as most plausible for CO_2H^+? Explain.

24. Show that the idea of minimizing the formal charges in a structure is at times in conflict with the observation that compact, symmetrical structures are more commonly observed than elongated ones with many central atoms. Use ClO_4^- as an illustrative example.

wis Structures

Write acceptable Lewis structures for the following molecules: **(a)** H_2NNH_2; **(b)** HOClO; **(c)** $(HO)_2SO$; **(d)** HOOH; **(e)** SO_4^{2-}.

Two molecules that have the same formulas but different structures are said to be isomers. (In isomers, the same atoms are present but linked together in different ways.) Draw acceptable Lewis structures for *two* isomers of C_2O_4. [*Hint:* The C atoms and two O atoms form a square.]

The following polyatomic anions involve covalent bonds between O atoms and the central nonmetal atom. Propose an acceptable Lewis structure for each. **(a)** SO_3^{2-}; **(b)** NO_2^-; **(c)** CO_3^{2-}; **(d)** HO_2^-.

Represent the following ionic compounds by Lewis structures: **(a)** barium hydroxide; **(b)** sodium nitrite; **(c)** magnesium iodate; **(d)** aluminum sulfate.

Write a plausible Lewis structure for crotonaldehyde, $CH_3CHCHCHO$, a substance used in tear gas and insecticides.

Write a plausible Lewis structure for C_3O_2, a substance known as carbon suboxide.

31. Write Lewis structures for the molecules represented by the following molecular models.

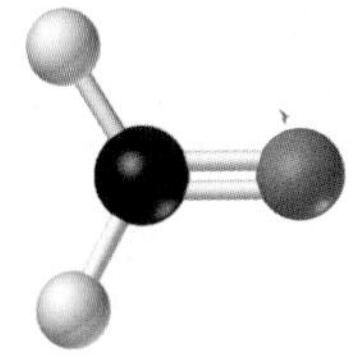

(a)

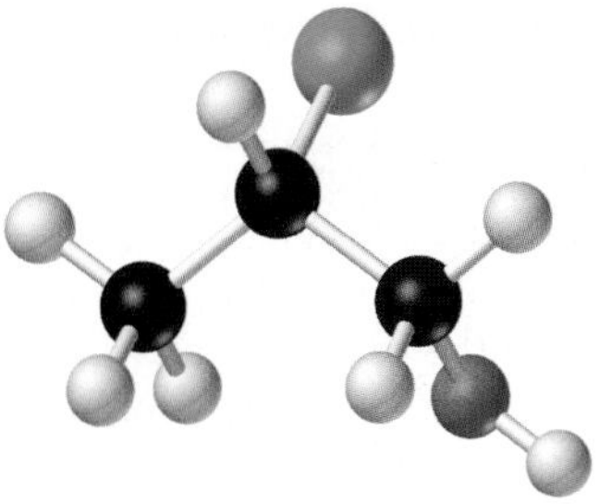

(b)

32. Write Lewis structures for the molecules represented by the following molecular models.

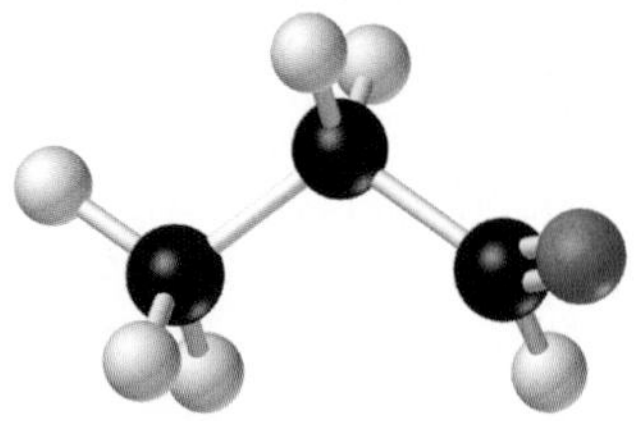

(a)

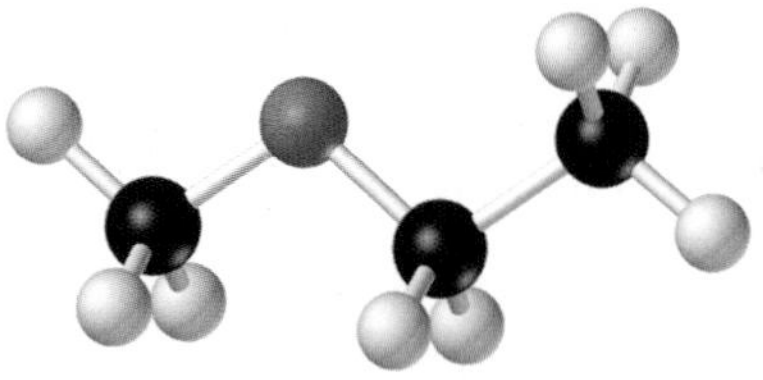

(b)

33. Write Lewis structures for the molecules represented by the following line-angle formulas. [*Hint:* Recall page 70 and Figure 3-2.]

(a) 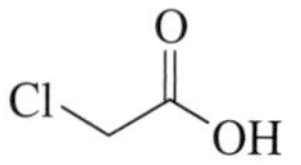

(b) 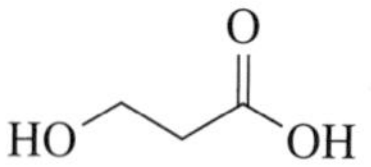

34. Write Lewis structures for the molecules represented by the following line-angle formulas. [*Hint:* Recall page 70 and Figure 3-2.]

(a) 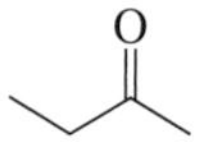

(b) 

35. Identify the main group that the element X belong in each of the following Lewis structures. For types of molecule shown, give an example that ex

(a) 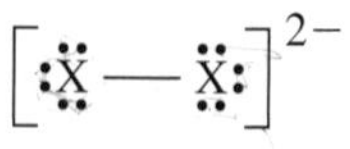

(b) 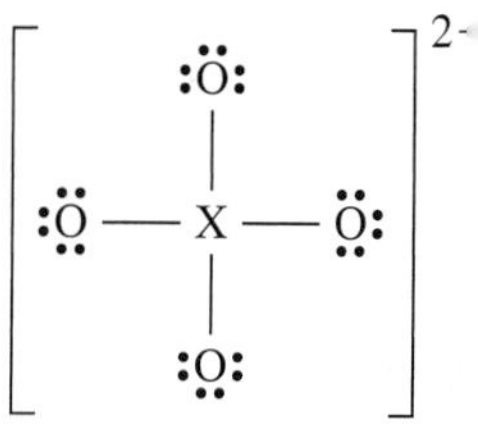

(c) 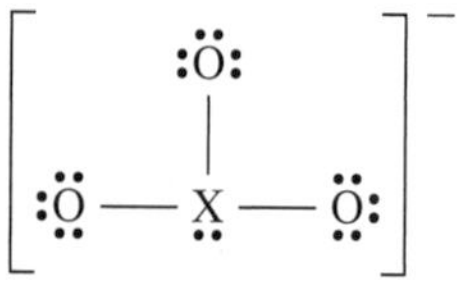

(d) 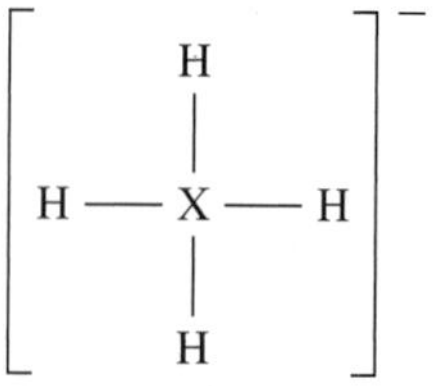

36. Identify the main group that the element X belong in each of the following Lewis structures. For types of molecule shown, give an example that ex

(a) 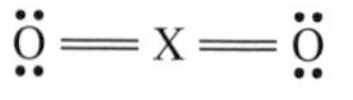

(b) 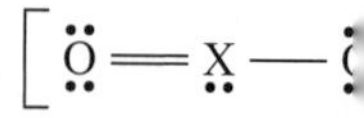

(c) 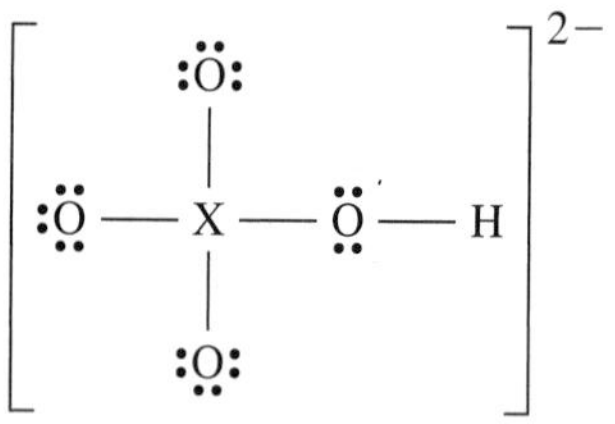

(d) 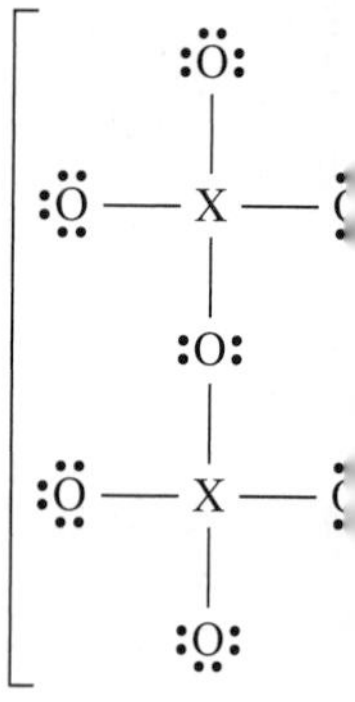

Polar Covalent Bonds and Electrostatic Potential Maps

37. Use your knowledge of electronegativities, but *do not* refer to tables or figures in the text, to arrange the following bonds in terms of *increasing* ionic character: C—H, F—H, Na—Cl, Br—H, K—F.

38. Which of the following molecules would you expect to have a resultant dipole moment (μ)? Explain. **(a)** F_2; **(b)** NO_2; **(c)** BF_3; **(d)** HBr; **(e)** H_2CCl_2; **(f)** SiF_4; **(g)** OCS.

39. What is the percent ionic character of each of the following bonds? **(a)** S—H; **(b)** O—Cl; **(c)** Al—O; **(d)** As—O.

40. Plot the data of Figure 10-6 as a function of atomic number. Does the property of electronegativity conform to the periodic law? Do you think it should?

41. Use a cross-base arrow ($\leftrightarrow$) to represent the polarity of the bond in each of the following diatomic molecules. Then use the data below to calculate, in the manner described on page 447, the partial charges (δ) on the atoms in each molecule. Express the partial charges as a decimal fraction of the elementary charge, e $= 1.602 \times 10^{-19}$ C, for example $\delta = +0.17$e or $\delta = -0.17$e.

	Bond Length, pm	Dipole Moment, D
ClF	162.8	0.8881
RbF	227.0	8.547
SnO	183.3	4.3210
BaO	194.0	7.954

42. Use a cross-base arrow ($\leftrightarrow$) to represent the pola of the bond in each of the following diatomic mo cules. Then use the data below to calculate the par charges (δ) on the atoms in each molecule. Express partial charges in the manner described in Exercise

	Bond Length, pm	Dipole Moment, D
OH	98.0	1.66
CH	131.1	1.46
CN	117.5	1.45
CS	194.4	1.96

Which electrostatic potential map corresponds to $F_2C{=}O$, and which to $H_2C{=}O$?

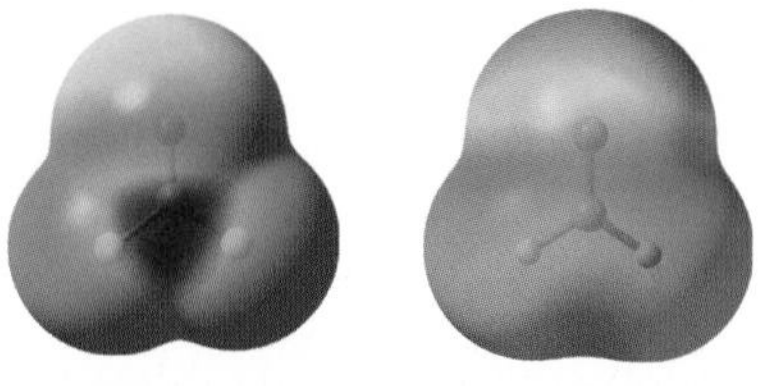

Match the correct electrostatic potential map corresponding to HOCl, FOCl, and HOF.

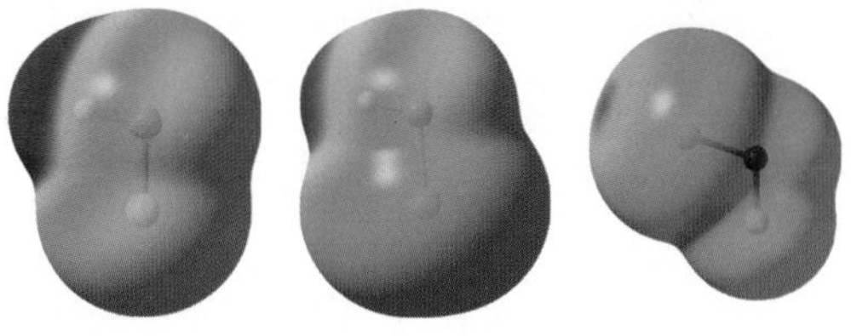

45. Two electrostatic potential maps are shown, one corresponding to a molecule containing only S and F, the other Si and F. Match them. What are the molecular formulas of the compounds?

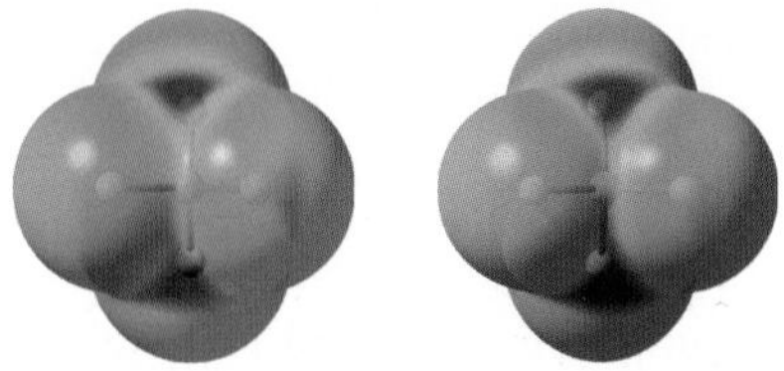

46. Two electrostatic potential maps are shown, one corresponding to a molecule containing only Cl and F, the other P and F. Match them. What are the molecular formulas of the compounds?

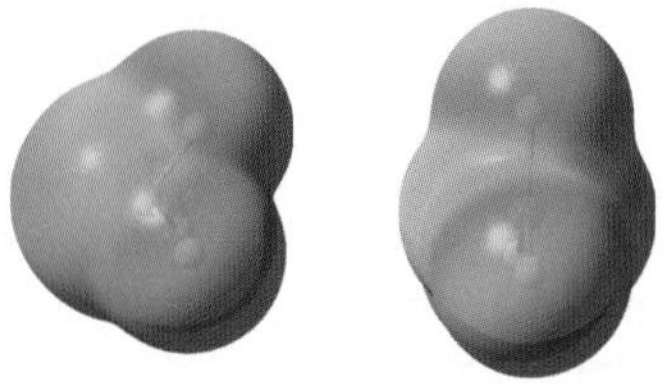

sonance

Through appropriate Lewis structures, show that the phenomenon of resonance is involved in the nitrite ion.

Which of the following species requires a resonance hybrid for its Lewis structure? Explain. **(a)** CO_2; **(b)** OCl^-; **(c)** CO_3^{2-}; **(d)** OH^-.

Dinitrogen oxide (nitrous oxide, or "laughing gas") is sometimes used as an anesthetic. Here are some data about the N_2O molecule: N—N bond length = 113 pm; N—O bond length = 119 pm. Use these data and other information from the chapter to comment on the plausibility of each of the following Lewis structures shown. Are they all valid? Which ones do you think contribute most to the resonance hybrid?

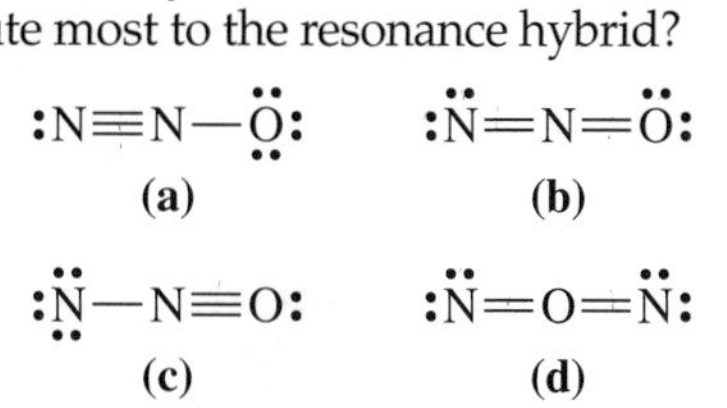

The Lewis structure of nitric acid, $HONO_2$, is a resonance hybrid. How important do you think the contribution of the following structure is to the resonance hybrid? Explain.

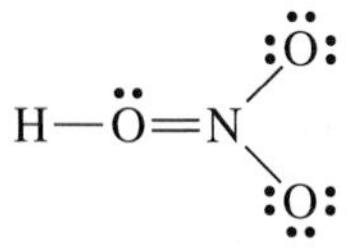

51. Draw Lewis structures for the following species, indicating formal charges and resonance where applicable:
(a) HCO_2^-
(b) HCO_3^-
(c) FSO_3^-
(d) $N_2O_3^{2-}$ (the nitrogen atoms are joined centrally with one oxygen atom on one N and two on the other)

52. Draw Lewis structures for the following species, indicating formal charges and resonance where applicable:
(a) $HOSO_3^-$
(b) H_2NCN
(c) FCO_2^-
(d) S_2N_2 (a cyclic structure with S and N alternating)

dd-Electron Species

Write plausible Lewis structures for the following odd-electron species: **(a)** CH_3; **(b)** ClO_2; **(c)** NO_3.

Write plausible Lewis structures for the following free radicals: **(a)** $\cdot C_2H_5$; **(b)** $HO_2\cdot$; **(c)** $ClO\cdot$.

Which of the following species would you expect to be diamagnetic and which paramagnetic? **(a)** OH^-; **(b)** OH; **(c)** NO_3; **(d)** SO_3; **(e)** SO_3^{2-}; **(f)** HO_2.

56. Write a plausible Lewis structure for NO_2, and indicate whether the molecule is diamagnetic or paramagnetic. Two NO_2 molecules can join together (*dimerize*) to form N_2O_4. Write a plausible Lewis structure for N_2O_4, and comment on the magnetic properties of the molecule.

Expanded Valence Shells

57. In which of the following species is it *necessary* to employ an expanded valence shell to represent the Lewis structure: PO_4^{3-}, PI_3, ICl_3, $OSCl_2$, SF_4, ClO_4^-? Explain your choices.

58. Describe the carbon-to-sulfur bond in H_2CSF_4. T is, is it most likely a single, double, or triple bond

Molecular Shapes

59. Use VSEPR theory to predict the geometric shapes of the following molecules and ions: **(a)** N_2; **(b)** HCN; **(c)** NH_4^+; **(d)** NO_3^-; **(e)** NSF.
60. Use VSEPR theory to predict the geometric shapes of the following molecules and ions: **(a)** PCl_3; **(b)** SO_4^{2-}; **(c)** $SOCl_2$; **(d)** SO_3; **(e)** BrF_4^+.
61. Each of the following is either linear, angular (bent), planar, tetrahedral, or octahedral. Indicate the correct shape of **(a)** H_2S; **(b)** N_2O_4; **(c)** HCN; **(d)** $SbCl_6^-$; **(e)** BF_4^-.
62. Predict the geometric shapes of **(a)** CO; **(b)** $SiCl_4$; **(c)** PH_3; **(d)** ICl_3; **(e)** $SbCl_5$; **(f)** SO_2; **(g)** AlF_6^{3-}.
63. One of the following ions has a *trigonal-planar* shape: SO_3^{2-}; PO_4^{3-}; PF_6^-; CO_3^{2-}. Which ion is it? Explain.
64. Two of the following have the same shape. Which two, and what is their shape? What are the shapes of the other two? NI_3, HCN, SO_3^{2-}, NO_3^-.
65. Each of the following molecules contains one or more multiple covalent bonds. Draw plausible Lewis structures to represent this fact, and predict the shape of each molecule. **(a)** CO_2; **(b)** Cl_2CO; **(c)** $ClNO_2$.
66. Sketch the probable geometric shape of a molecule of **(a)** N_2O_4 (O_2NNO_2); **(b)** C_2N_2 (NCCN); **(c)** C_2H_6 (H_3CCH_3); **(d)** C_2H_6O (H_3COCH_3).
67. Use the VSEPR theory to predict the shapes of the anions **(a)** ClO_4^-; **(b)** $S_2O_3^{2-}$ (that is, SSO_3^{2-}); **(c)** PF_6^-; **(d)** I_3^-.
68. Use the VSEPR theory to predict the shape of **(a)** molecule OSF_2; **(b)** the molecule O_2SF_2; **(c)** the SF_5^-; **(d)** the ion ClO_4^-; **(e)** the ion ClO_3^-.
69. The molecular shape of BF_3 is planar (see Table 1 If a fluoride ion is attached to the B atom of through a coordinate covalent bond, the ion results. What is the shape of this ion?
70. Explain why it is not necessary to find the Le structure with the smallest formal charges to mak successful prediction of molecular geometry in VSEPR theory. For example, write Lewis struct for SO_2 having different formal charges, and pre the molecular geometry based on these structures
71. Comment on the similarities and differences in molecular structure of the following triatomic spe CO_2, NO_2^-, O_3, and ClO_2^-.
72. Comment on the similarities and differences in molecular structure of the following four-at species: NO_3^-, CO_3^{2-}, SO_3^{2-}, and ClO_3^-.
73. Draw a plausible Lewis structure for the follow series of molecules and ions: **(a)** ClF_2^-; **(b)** C **(c)** ClF_4^-; **(d)** ClF_5. Describe the electron group geo try and molecular structure of these species.
74. Draw a plausible Lewis structure for the follow series of molecules and ions: **(a)** SiF_6^{2-}; **(b)** P **(c)** $AsCl_5$; **(d)** ClF_3; **(e)** XeF_4. Describe the electron gr geometry and molecular structure of these species

Shapes of Molecules with More Than One Central Atom

75. Sketch the propyne molecule, $CH_3C{\equiv}CH$. Indicate the bond angles in this molecule. What is the maximum number of atoms that can be in the same plane?
76. Sketch the propene molecule, $CH_3CH{=}CH_2$. Indicate the bond angles in this molecule. What is the maximum number of atoms that can be in the same plane?
77. Lactic acid has the formula $CH_3CH(OH)COOH$. Sketch the lactic acid molecule, and indicate the various bond angles.
78. Levulinic acid has the formula $CH_3(CO)CH_2C$ COOH. Sketch the levulinic acid molecule, and in cate the various bond angles.
79. Sketch, by using the dash and wedge symbolism, H_2NCH_2CHO molecule, and indicate the vari bond angles.
80. One of the isomers of chloromethanol has the form $ClCH_2OH$. Sketch, by using the dash and wedge sy bolism, this isomer of chloromethanol, and indic the various bond angles.

Polar Molecules

81. Predict the shapes of the following molecules, and then predict which would have resultant dipole moments: **(a)** SO_2; **(b)** NH_3; **(c)** H_2S; **(d)** C_2H_4; **(e)** SF_6; **(f)** CH_2Cl_2.
82. Which of the following molecules would you expect to be polar? **(a)** HCN; **(b)** SO_3; **(c)** CS_2; **(d)** OCS; **(e)** $SOCl_2$; **(f)** SiF_4; **(g)** POF_3. Give reasons for your conclusions.
83. The molecule H_2O_2 has a resultant dipole moment of 2.2 D. Can this molecule be linear? If not, describe a shape that might account for this dipole moment.
84. Refer to the Integrative Example. A compound rela to nitryl fluoride is nitrosyl fluoride, FNO. For t molecule, indicate **(a)** a plausible Lewis structure a **(b)** the geometric shape. **(c)** Explain why the m sured resultant dipole moment for FNO is larger th the value for FNO_2.

nd Lengths

Without referring to tables in the text, indicate which of the following bonds you would expect to have the greatest bond length, and give your reasons. **(a)** O_2; **(b)** N_2; **(c)** Br_2; **(d)** BrCl.

Estimate the lengths of the following bonds and indicate whether your estimate is likely to be too high or too low: **(a)** I—Cl; **(b)** C—F.

A relationship between bond lengths and single-bond covalent radii of atoms is given on page 449. Use this relationship together with appropriate data from Table 10.2 to estimate these single-bond lengths. **(a)** I—Cl; **(b)** O—Cl; **(c)** C—F; **(d)** C—Br.

88. In which of the following molecules would you expect the oxygen-to-oxygen bond to be the *shortest*? Explain. **(a)** H_2O_2, **(b)** O_2, **(c)** O_3.

89. Refer to the Integrative Example. Use data from the chapter to estimate the length of the N—F bond in FNO_2.

90. Write a Lewis structure of the hydroxylamine molecule, H_2NOH. Then, with data from Table 10.2, determine all the bond lengths.

nd Energies

A reaction involved in the formation of ozone in the upper atmosphere is $O_2 \longrightarrow 2\,O$. *Without* referring to Table 10.3, indicate whether this reaction is endothermic or exothermic. Explain.

Use data from Table 10.3, but *without performing detailed calculations*, determine whether each of the following reactions is exothermic or endothermic.

(a) $CH_4(g) + I(g) \longrightarrow \cdot CH_3(g) + HI(g)$

(b) $H_2(g) + I_2(g) \longrightarrow 2\,HI(g)$

Use data from Table 10.3 to estimate the enthalpy change ($\Delta_r H$) for the following reaction.

$$C_2H_6(g) + Cl_2(g) \longrightarrow C_2H_5Cl(g) + HCl(g) \qquad \Delta_r H = ?$$

One of the chemical reactions that occurs in the formation of photochemical smog is $O_3 + NO \longrightarrow NO_2 + O_2$. Estimate $\Delta_r H$ for this reaction by using appropriate Lewis structures and data from Table 10.3.

Estimate the standard enthalpies of formation at 25 °C and 1 bar of **(a)** OH(g); **(b)** $N_2H_4(g)$. Write Lewis structures and use data from Table 10.3, as necessary.

Use $\Delta_r H$ for the reaction in Example 10-15 and other data from Appendix D to estimate $\Delta_f H°$ [$CH_3Cl(g)$].

Use bond energies from Table 10.3 to estimate $\Delta_r H$ for the following reaction.

$$C_2H_2(g) + H_2(g) \longrightarrow C_2H_4(g) \qquad \Delta_r H = ?$$

Equations (1) and (2) can be combined to yield the equation for the formation of $CH_4(g)$ from its elements.

(1)	C(s)	$\longrightarrow$	C(g)	$\Delta_r H = 717\ \text{kJ mol}^{-1}$
(2)	$C(g) + 2\,H_2(g)$	$\longrightarrow$	$CH_4(g)$	$\Delta_r H = ?$
Overall:	$C(s) + 2\,H_2(g)$	$\longrightarrow$	$CH_4(g)$	$\Delta_f H° = -75\ \text{kJ mol}^{-1}$

Use the preceding data and a bond energy of 436 kJ mol^{-1} for H_2 to estimate the C—H bond energy. Compare your result with the value listed in Table 10.3.

99. One reaction involved in the sequence of reactions leading to the destruction of ozone is

$$NO_2(g) + O(g) \longrightarrow NO(g) + O_2(g)$$

Calculate $\Delta_r H°$ for this reaction by using the thermodynamic data in Appendix D. Use your $\Delta_r H°$ value, plus data from Table 10.3, to estimate the nitrogen–oxygen bond energy in NO_2. [*Hint:* The structure of nitrogen dioxide, NO_2, is best represented as a resonance hybrid of two equivalent Lewis structures.]

100. A reaction involved in the sequence of reactions leading to the destruction of ozone is

$$O_3(g) + O(g) \longrightarrow 2\,O_2(g) \qquad \Delta_r H° = -394\ \text{kJ mol}^{-1}$$

Estimate the oxygen-oxygen bond energy in ozone by using the oxygen–oxygen bond energy in dioxygen from Table 10.3. Compare this value with the O—O and O=O bond energies in Table 10.3. How could you explain any differences?

Integrative and Advanced Exercises

1. Given the bond-dissociation energies: nitrogen-to-oxygen bond in NO, 631 kJ mol^{-1}; H—H in H_2, 436 kJ mol^{-1}; N—H in NH_3, 389 kJ mol^{-1}; O—H in H_2O, 463 kJ mol^{-1}; calculate $\Delta_r H$ for the reaction below.

$$2\,NO(g) + 5\,H_2(g) \longrightarrow 2\,NH_3(g) + 2\,H_2O(g)$$

2. The following statements are not made as carefully as they might be. Criticize each one.

(a) Lewis structures with formal charges are incorrect.

(b) Triatomic molecules have a planar shape.

(c) Molecules in which there is an electronegativity difference between the bonded atoms are polar.

103. A compound consists of 42.44% N and 57.56% F, by mass. Write a plausible Lewis structure based on the empirical formula of this compound.

104. A 0.325 g sample of a gaseous hydrocarbon occupies a volume of 193 mL at 749 mmHg and 26.1 °C. Determine the molecular mass, and write a plausible condensed structural formula for this hydrocarbon.

105. A 1.24 g sample of a hydrocarbon, when completely burned in an excess of $O_2(g)$, yields 4.04 g CO_2 and 1.24 g H_2O. Draw a plausible structural formula for the hydrocarbon molecule. [*Hint:* There is more than one possible arrangement of the C and H atoms.]

106. Draw Lewis structures for two different molecules with the formula C_3H_4. Is either of these molecules linear? Explain.

107. Sodium azide, NaN_3, is the nitrogen gas-forming substance used in automobile air-bag systems. It is an ionic compound containing the azide ion, N_3^-. In this ion, the two nitrogen-to-nitrogen bond lengths are 116 pm. Describe the resonance hybrid Lewis structure of this ion.

108. Use the bond-dissociation energies of $N_2(g)$ and $O_2(g)$ in Table 10.3, together with data from Appendix D, to estimate the bond-dissociation energy of NO(g).

109. Hydrogen azide, HN_3, is a liquid that explodes violently when subjected to physical shock. In the HN_3 molecule, one nitrogen-to-nitrogen bond length is 113 pm, and the other is 124 pm. The H—N—N bond angle is 112°. Draw Lewis structures and a sketch of the molecule consistent with these facts.

110. A few years ago the synthesis of a salt containing the N_5^+ ion was reported. What is the likely shape of this ion—linear, bent, zigzag, tetrahedral, seesaw, or square-planar? Explain your choice.

111. Carbon suboxide has the formula C_3O_2. The carbon-to-carbon bond lengths are 130 pm and carbon-to-oxygen, 120 pm. Propose a plausible Lewis structure to account for these bond lengths, and predict the shape of the molecule.

112. In certain polar solvents, PCl_5 undergoes an ionization reaction in which a Cl^- ion leaves one PCl_5 molecule and attaches itself to another. The products of the ionization are PCl_4^+ and PCl_6^-. Draw a sketch showing the changes in geometric shapes that occur in this ionization (that is, give the shapes of PCl_5, PCl_4^+, and PCl_6^-).

$$2\,PCl_5 \rightleftharpoons PCl_4^+ + PCl_6^-$$

113. Estimate the enthalpy of formation of HCN using bond energies from Table 10.3, data from elsewhere in the text, and the reaction scheme outlined as follows.

$$\text{(1) } C(s) \rightarrow C(g) \quad \Delta_r H° = ?$$
$$\text{(2) } C(g) + \tfrac{1}{2}N_2(g) + \tfrac{1}{2}H_2(g) \rightarrow HCN(g) \quad \Delta_r H° = ?$$
$$\textit{Overall: } C(g) + \tfrac{1}{2}N_2(g) + \tfrac{1}{2}H_2(g) \rightarrow HCN(g) \quad \Delta_f H° = ?$$

114. The standard enthalpy of formation of $H_2O_2(g)$ is -136 kJ mol^{-1}. Use this value, with other appropriate data from the text, to estimate the oxygen-to-oxygen single-bond energy. Compare your result with the value listed in Table 10.3.

115. Use the VSEPR theory to predict a probable shape of the molecule F_4SCH_2, and explain the source of any ambiguities in your prediction.

116. The standard enthalpy of formation of methanet $CH_3SH(g)$, is -22.9 kJ mol^{-1}. Methanethiol ca synthesized by the reaction of gaseous methanol $H_2S(g)$. Water vapor is another product. Use this ir mation and data from elsewhere in the text to estir the carbon-to-sulfur bond energy in methanethiol.

117. For LiBr, the dipole moment (measured in the phase) and the bond length (measured in the s state) are 7.268 D and 217 pm, respectively. For N the corresponding values are 9.001 D and 236.1 **(a)** Calculate the percent ionic character for bond. **(b)** Compare these values with the expe ionic character based on differences in electroneg ity (see Figure 10-7). **(c)** Account for any difference the values obtained in these two different ways.

118. One possibility for the electron-group geometry *seven* electron groups is pentagonal-bipyramida found in the IF_7 molecule. Write the VSEPR nota for this molecule. Sketch the structure of the m cule, labeling all the bond angles.

119. The extent to which an acid (HA) ionizes in w depends upon the stability of the anion (A^-); more stable the anion, the more extensive is the sociation of the acid. The anion is most stable w the negative charge is distributed over the wh anion rather than localized at one particular at Consider the following acids: acetic a fluoroacetic acid, cyanoacetic acid, and nitroac acid. Draw Lewis structures for their anions, inc ing contributing resonance structures. Rank acids in order of increasing extent of ionizat Electrostatic potential maps for the four anions provided on the next page. Identify which map co sponds to which anion, and discuss whether the m confirm conclusions based on Lewis structures.

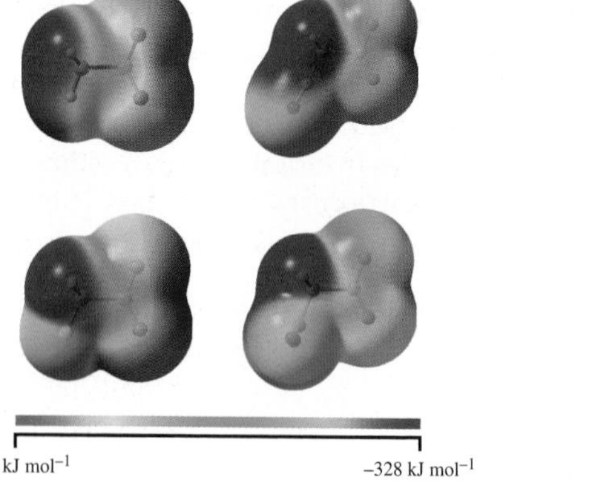

120. R. S. Mulliken proposed that the electronegativ (EN) of an atom is given by

$$EN = k \times (E_i - E_{ea})$$

where E_i and E_{ea} are the ionization energy and el tron affinity of the atom, respectively. Using electron affinities and ionization energy val for the halogen atoms up to iodine, estimate the va of k by employing the electronegativity values Figure 10-6. Estimate the electron affinity of At.

121. When molten sulfur reacts with chlorine gas, a v smelling orange liquid forms. When analyzed,

liquid compound has the empirical formula SCl. Several possible Lewis structures are shown below. Criticize these structures and choose the best one.

(a) 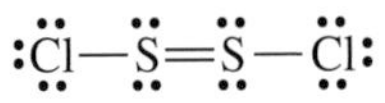

(b) 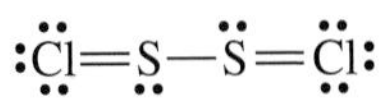

(c) 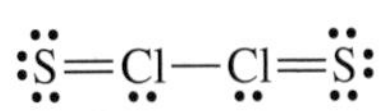

(d) 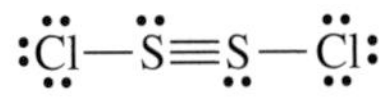

(e) 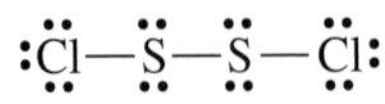

. Hydrogen azide, HN_3, can exist in two forms. One form has the three nitrogen atoms connected in a line; and the nitrogen atoms form a triangle in the other. Construct Lewis structures for these isomers and describe their shapes. Other interesting derivatives are nitrosyl azide (N_4O) and trifluoromethyl azide (CF_3N_3). Describe the shapes of these molecules based on a line of nitrogen atoms.

123. A pair of isoelectronic species for C and N exist with the formula X_2O_4 in which there is an X—X bond. A corresponding fluoride of boron also exists. Draw Lewis structures for these species and describe their shapes.

124. Acetone $(CH_3)_2C{=}O$, a ketone, will react with a strong base (A^-) to produce the enolate anion, $CH_3(C{=}O)CH_2^-$. Draw the Lewis structure of the enolate anion, and describe the relative contributions of any resonance structures.

125. The species PBr_4^- has been synthesized and has been described as a tetrahedral anion. Comment on this description.

126. One of the allotropes of sulfur is a ring of eight sulfur atoms. Draw the Lewis structure for the S_8 ring. Is the ring likely to be planar? The S_8 ring can be oxidized to produce S_8O. In S_8O, the oxygen atom is bonded to one of the S atoms and the S_8 ring is still intact. Draw the Lewis structure for S_8O.

127. One of the allotropes of phosphorus consists of four phosphorus atoms at the corners of a tetrahedron. Draw a Lewis structure for this allotrope that satisfies the octet rule. The P_4 molecule can be oxidized to P_4O_6, where the oxygen atoms insert between the phosphorus atoms. Draw the Lewis structure of this oxide. Are the P—O—P bonds linear?

Feature Problems

. In this problem, we examine the basis of three different electronegativity scales and work through the same types of calculations as those performed by the people who initially suggested these scales. The scale developed by Robert Mulliken employs ionization energies (E_i) and electron affinities (E_{ea}) whereas the scale developed by Linus Pauling is based on bond dissociation energies (D). The scale developed by A. Louis Allred and Eugene G. Rochow employs effective nuclear charges (Z_{eff}) and covalent radii (r_{cov}). The key equations for each scale are given below.

ectronegativity cale	Defining Equation
auling[a]	$EN_A - EN_B = \sqrt{\dfrac{D_{A-B} - \frac{1}{2}(D_{A-A} + D_{B-B})}{1\ \text{eV}}}$
ulliken[b, c]	$EN = 0.336 \times \left(\dfrac{E_i + E_{ea}}{2\ \text{eV}}\right) - 0.165$
llred-Rochow[c]	$EN = \dfrac{3590\ Z_{eff}}{(r_{cov}/1\ \text{pm})^2} + 0.744$

riginally, Pauling defined EN_H to be 2.1, the value chosen to e the elements C to F electronegativity values from 2.5 to 4.0.

trictly speaking, the E_i and E_{ea} values used in this expression not the experimentally observed values for an isolated atom those calculated for an atom as it exists in a molecule. Also, in this formula is the energy change for $X^-(g) \rightarrow X(g) + e^-$.
e constants in these equations (0.336 and –0.165 or 3590 and 44) are chosen to ensure the EN values span essentially the ne range as Pauling's values.

In devising his scale, Pauling observed that the bond energy, D_{A-B}, for the A–B bond is greater than the average of the A–A and B–B bond energies, $\frac{1}{2}(D_{A-A} + D_{B-B})$, and he attributed the increase in bond strength to the partial ionic character of the A–B bond.

Mulliken argued that the ionization energy (E_i) and electron affinity (E_{ea}) are of equal importance for the electronegativity of an atom. Therefore, he suggested that the average of these two quantities, $(E_i + E_{ea})/2$, be used to define the electronegativity of an atom.

Allred and Rochow focused on the attractive Coulombic force between an electron near the "surface" of an atom and the nucleus of that atom. They argued that the magnitude of this force is proportional to $(e)(Z_{eff}e)/r_{cov}^2 = e^2Z_{eff}/r_{cov}^2$, where $e = 1.602 \times 10^{-19}$ C is the magnitude of the charge of an electron, $Z_{eff}e$ is the nuclear charge experienced by an electron near the atom's surface, and r_{cov} is the covalent radius and a realistic measure of the size of an atom.

The values of D, Z_{eff}, and r_{cov} given below are the actual values used by Pauling and by Allred and Rochow in their original papers. Use the data below and the equations above to calculate the electronegativities of F, Cl, Br, and I. Summarize your results in a table having four columns: Atom, EN(Pauling), EN(Mulliken), EN(Allred-Rochow). [*Hint:* The Pauling values you calculate will not be exactly equal to those in Figure 10-6. The values in Figure 10-6 are based on bond dissociation energies from a wider range of molecules than we are considering in this problem.]

	Atom, X				
	H	F	Cl	Br	I
E_i, eV	13.5985	17.423	12.9677	11.8139	10.4513
E_{ea}, eV	0.7542	3.399	3.617	3.365	3.059
D(H—X), eV	4.44	6.39	4.38	3.74	3.07
D(X—X), eV	4.44	2.80	2.468	1.962	1.535
Z_{eff}	1	4.86	5.75	7.25	7.25
r_{cov}, pm	–	71.5	99.4	114.2	133.4

129. On page 447, the bond angle in the H_2O molecule is given as 104° and the resultant dipole moment as $\mu = 1.84$ D.
(a) By an appropriate geometric calculation, determine the value of the H—O bond dipole in H_2O.
(b) Use the same method as in part (a) to estimate the bond angle in H_2S, given that the H—S bond dipole is 0.67 D and that the resultant dipole moment is $\mu = 0.93$ D.
(c) Refer to Figure 10-16. Given the bond dipoles 1.87 D for the C—Cl bond and 0.30 D for the C—H bond, together with $\mu = 1.04$ D, estimate the H—C—Cl bond angle in $CHCl_3$.

130. Alternative strategies to the one used in this chapter have been proposed for applying the VSEPR theory to molecules or ions with a single central atom. In general, these strategies do not require writing Lewis structures. In one strategy, we write

(1) the total number of electron pairs = [(numbe valence electrons) ± (electrons required for ic charge)]/2
(2) the number of bonding electron pairs = (num of atoms) − 1
(3) the number of electron pairs around cen atom = total number of electron pairs − 3 × [nu ber of terminal atoms (excluding H)]
(4) the number of lone-pair electrons = numbe central atom pairs − number of bonding pairs

After evaluating items 2, 3, and 4, establish the VSE notation and determine the molecular shape. Use method to predict the geometrical shapes of the lowing: **(a)** PCl_5; **(b)** NH_3; **(c)** ClF_3; **(d)** SO_2; **(e)** C **(f)** PCl_4^+. Justify each of the steps in the strategy, explain why it yields the same results as the VSE method based on Lewis structures. How does strategy deal with multiple bonds?

Self-Assessment Exercises

131. In your own words, define the following terms: **(a)** valence electrons; **(b)** electronegativity; **(c)** bond-dissociation energy; **(d)** double covalent bond; **(e)** coordinate covalent bond.

132. Briefly describe each of the following ideas: **(a)** formal charge; **(b)** resonance; **(c)** expanded valence shell; **(d)** bond energy.

133. Explain the important distinctions between **(a)** ionic and covalent bonds; **(b)** lone-pair and bond-pair electrons; **(c)** molecular geometry and electron-group geometry; **(d)** bond dipole and resultant dipole moment; **(e)** polar molecule and nonpolar molecule.

134. Of the following species, the one with a triple covalent bond is **(a)** NO_3^-; **(b)** CN^-; **(c)** CO_2; **(d)** $AlCl_3$.

135. The formal charges on the O atoms in the ion $[ONO]^+$ is **(a)** −2; **(b)** −1; **(c)** 0; **(d)** +1.

136. Which molecule is nonlinear? **(a)** SO_2; **(b)** CO_2; **(c)** HCN; **(d)** NO.

137. Which molecule is nonpolar? **(a)** SO_3; **(b)** CH_2Cl_2; **(c)** NH_3; **(d)** FNO.

138. The highest bond-dissociation energy is found in **(a)** O_2; **(b)** N_2; **(c)** Cl_2; **(d)** I_2.

139. The greatest bond length is found in **(a)** O_2; **(b)** I **(c)** Br_2; **(d)** BrCl.

140. Draw plausible Lewis structures for the follow species; use expanded valence shells where nec sary. **(a)** Cl_2O; **(b)** PF_3; **(c)** CO_3^{2-}; **(d)** BrF_5.

141. Predict the shapes of the following sulfur-contain species. **(a)** SO_2; **(b)** SO_3^{2-}; **(c)** SO_4^{2-}.

142. Which of the following ionic compounds composed of only nonmetal atoms? **(a)** NH_4N **(b)** $Al_2(SO_4)_3$; **(c)** Na_2SO_3; **(d)** $AlCl_3$; **(e)** none of the

143. Which of the following molecules does not obey octet rule? **(a)** HCN; **(b)** PF_3; **(c)** CS_2; **(d)** NO; **(e)** nc of these.

144. Which of the following molecules has no po bonds? **(a)** H_2CO; **(b)** CCl_4; **(c)** OF_2; **(d)** N_2O; **(e)** nc of these.

145. The electron-group geometry of H_2O is **(a)** tetral dral; **(b)** trigonal planar; **(c)** bent; **(d)** linear; **(e)** no of these.

146. For each of the following compounds, give t names of the electron-group geometry and the mo cular shape. Sketch the molecule and indicate on t sketch the direction of the dipole moment, if any. F the sketches, use the wedge-and-dash notation. **(a)** SiF_4; **(b)** NF_3 **(c)** SF_4; **(d)** IF_5.

. Use bond enthalpies from Table 10.3 to determine whether $CH_4(g)$, $CH_3OH(g)$, $H_2CO(g)$, or $HCOOH(g)$ produces the most energy *per gram* when burned completely in $O_2(g)$ to give $CO_2(g)$ and $H_2O(g)$. Is there any relationship between the oxidation state of carbon and the heat of combustion (in kJ kg^{-1} or kJ mol^{-1})?

. Without referring to tables or figures in the text other than the periodic table, indicate which of the following atoms, Bi, S, Ba, As, or Mg, has the intermediate value when they are arranged in order of increasing electronegativity.

. Use data from Tables 10.2 and 10.3 to determine for each bond in this following structure **(a)** the bond length and **(b)** the bond energy.

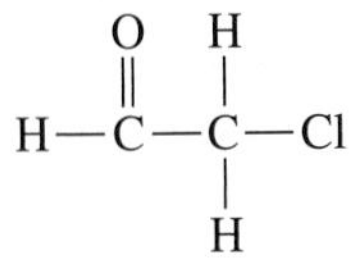

150. What is the VSEPR theory? On what physical basis is the VSEPR theory founded?
151. Use the NH_3 molecule as an example to explain the difference between molecular geometry and electron-group geometry.
152. If you have four electron pairs around a central atom, under what circumstances can you have a pyramidal molecule? Similarly, how can you have a bent molecule? What are the expected bond angles in each case?
153. Draw three resonance structures for the sulfine molecule, H_2CSO. Do not consider ring structures.
154. Construct a concept map illustrating the connections between Lewis dot structures, the shapes of molecules, and polarity.

11 Chemical Bonding II: Valence Bond and Molecular Orbital Theories

CONTENTS

LEARNING OBJECTIVES

11.1 Describe the redistribution of electron density and the variation of energy as a function of internuclear distance that accompany bond formation.

11.2 Use valence bond theory to describe bond formation in terms of the overlap of atomic orbitals.

11.3 Discuss the concept of orbital hybridization and use appropriate hybridization schemes to describe bonding in molecules.

11.4 Distinguish between sigma and pi bonds.

11.5 Distinguish between a bonding and an antibonding orbital, and describe how they each affect the bond order of a diatomic molecule.

11.6 Use molecular orbital theory to describe delocalized *p* electrons.

11.7 Describe the location and significance of the bond critical point in an electron density map of a molecule.

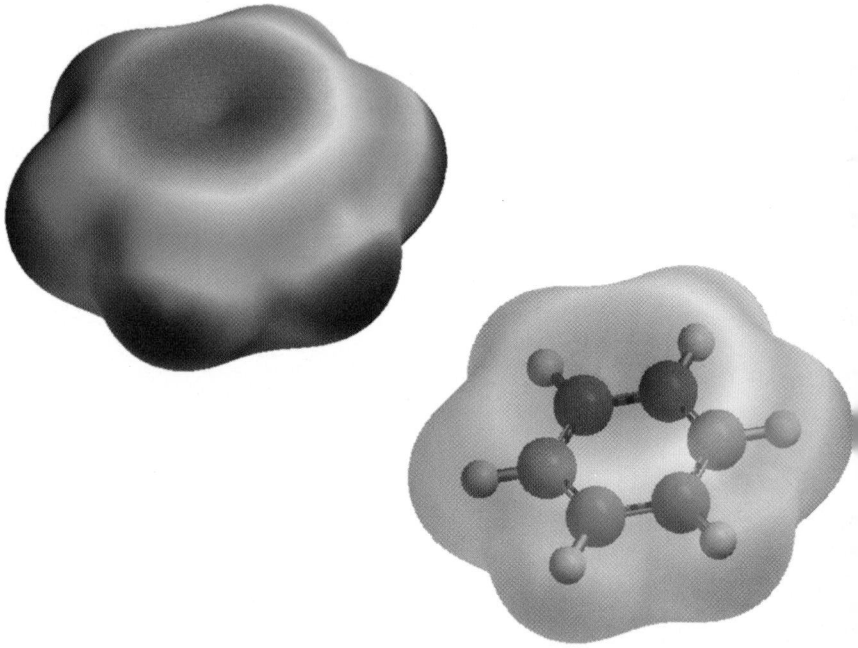

Electrostatic potential maps of benzene (one solid and one transparent) showing the negative charge density caused by the π molecular orbitals of benzene.

Although the Lewis theory has been useful in our discussion chemical bonding, some cases require more sophisticat approaches. One such approach involves the familiar *s*, *p*, and atomic orbitals, or mixed-orbital types called *hybrid orbitals*. A secor approach involves the creation of a set of orbitals that belongs to a mol cule as a whole. Electrons are then assigned to these *molecular orbitals*.

Our purpose in this chapter is not to try to master theories of covale bonding in all their details. We want simply to discover how these theori provide models that yield deeper insights into the nature of chemic bonding than do Lewis structures alone.

-1 What a Bonding Theory Should Do

hydrogen molecule is a simple model for discussing bonding theories. gine bringing together two H atoms that are initially very far apart. When H atoms are infinitely far apart, the two H atoms do not interact with each er, and by convention the net energy of interaction between the H atoms is). As the two H atoms approach each other, three types of interactions occur: each electron is attracted to the other nucleus (illustrated by a red dashed line ig. 11-1); (2) the electrons repel each other (illustrated by a blue dashed in Fig. 11-1); and (3) the two nuclei repel each other (illustrated by a black hed line in Fig. 11-1).

Ve can plot the net energy of interaction of the two H atoms as a function of distance between the atomic nuclei as illustrated in Figure 11-2. The inter-on energy is equal to zero when the atoms are very far apart (condition a). At rmediate distances (condition b), attractive interactions dominate, leading to es that draw the atoms closer together. At very small internuclear distances ndition d), repulsive interactions dominate, yielding forces that push the ms apart. At one particular internuclear distance (74 pm, condition c) energy reaches its lowest value (−436 kJ/mol). This is the condition in ich the two H atoms combine into a H_2 molecule through a covalent bond. nuclei continuously move back and forth; that is, the molecule vibrates, the average internuclear distance is about 74 pm. This internuclear distance responds to the *bond length*. The energy corresponds to the negative of the *d-dissociation energy*.

\ theory of covalent bonding should help us understand why a given mol-le has its particular set of observed properties—bond-dissociation ener-s, bond lengths, bond angles, and so on. There are several approaches to lerstanding bonding. The approach used depends on the situation because erent methods have different strengths and weaknesses. The strength of Lewis theory is in the ease with which it can be applied; a Lewis structure be written rather quickly. VSEPR theory makes it possible to propose mol-lar shapes that are generally in good agreement with experimental results.

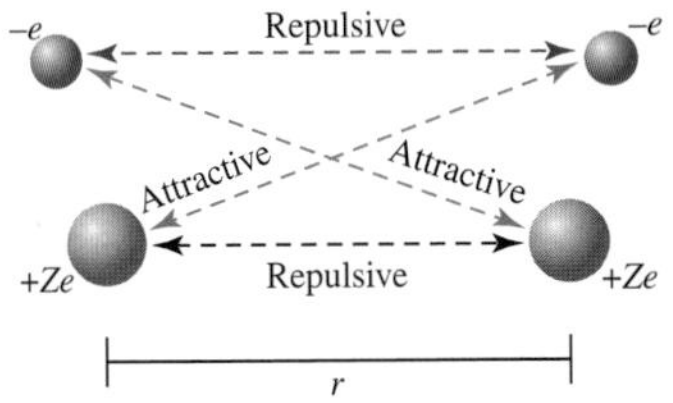

▲ FIGURE 11-1
Type of interactions between two hydrogen atoms
The types of interactions that occur as two hydrogen atoms, infinitely separated, approach each other. The dashed lines represent the types of interactions (red for attractive interactions and blue and black for repulsive interactions). The solid black line represents the internuclear distance, r, between the two hydrogen atoms.

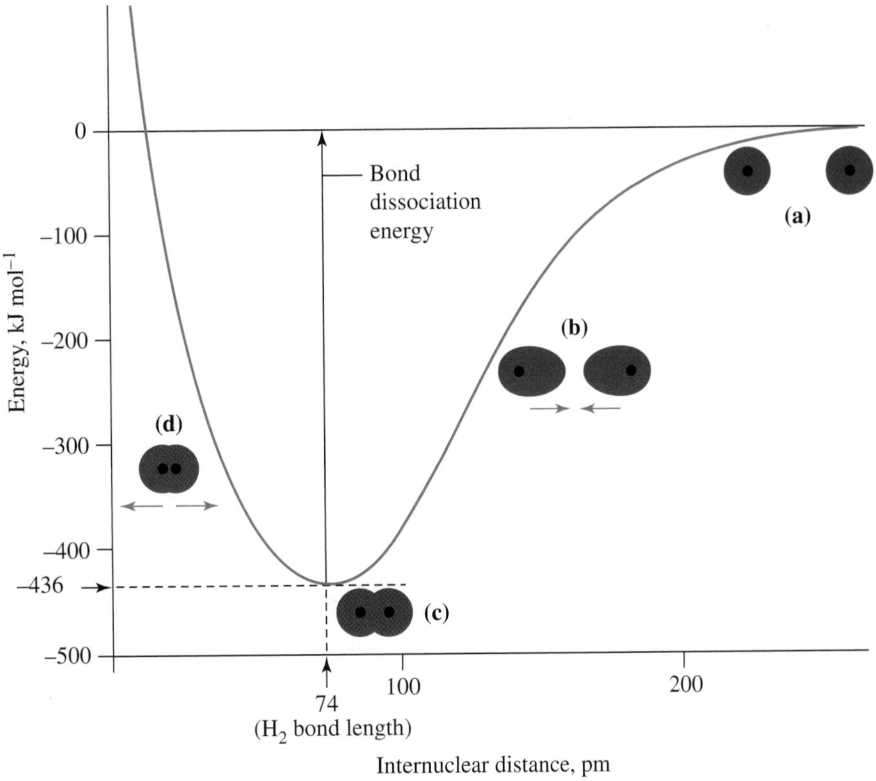

IGURE 11-2
ergy of interaction of two drogen atoms plotted for ernuclear separations from o to infinity
The energy is defined as zero en the two H atoms are infinitely arated. **(b)** Where the curve oes downward from right to left, net interaction is attractive. en the interaction is attractive, energy decreases as the ernuclear distance, r, decreases. The H_2 molecule has its lowest ergy, and is most stable, for an ernuclear distance of 74 pm. Where the curve slopes upward m right to left, the interaction epulsive. When the interaction is ulsive, the energy increases as internuclear distance, r, creases.

However, neither method yields quantitative information about bond er gies and bond lengths, and the Lewis theory has problems with odd-elect species and situations in which it is not possible to represent a molec through a single structure (resonance).

Before examining theories of chemical bonding, it will be helpful to c sider why these theories are all firmly rooted in quantum theory.

The Covalent Bond: A Quantum Mechanical Concept

In previous chapters, we associated a chemical bond with the sharing of or more electron pairs. It is tempting, and perhaps too convenient, to descr the stability of a chemical bond in terms of simple electrostatics in whic negatively charged electron pair between a pair of positively charged nu serves as the "glue" that holds the nuclei together. Such a description, ho ever, is incomplete. We learned in Chapter 8 that the structure and stability an atom cannot be explained without, at some point, invoking quant mechanical principles. Similarly, the stability of a molecule, and the descr tion of its bonds, requires principles from quantum theory. A covalent bonc simply put, a quantum mechanical concept.

Let us restrict ourselves momentarily to a purely classical description c covalent bond between two atoms, in which an electron from each aton moved from a point close to its own nucleus to a point between the two nuc Initially, each electron is close to one nucleus, and the two electrons are rather apart. In other words, each electron is rather strongly attracted to its o nucleus, and because the electrons are rather far apart from each other, the el tron–electron repulsion is relatively low. The movement of the electrons t point between the two nuclei reduces the electron–nucleus attractions a increases the electron–electron repulsions. As a result, the movement of electro from near their respective nuclei to the internuclear region involves an incre in (potential) energy. Thus, by focusing only on the electrostatic interactions, i difficult to rationalize how the sharing of electrons has a stabilizing effect.

Explaining the stability of a chemical bond requires us to consider, fror quantum mechanical point of view, the redistribution of electron density fro the nuclei to the internuclear region. The redistribution of electron density th occurs in the formation of a bond between two H atoms is illustrated schem ically in Figure 11-3(a), which compares the electron density of the H_2 mo cule with that of the nonbonded H atoms.

From Figure 11-3(a), we see that bond formation in H_2 involves the trans of electron density from regions outside both nuclei to regions near to a between both nuclei. Consequently, the electrons in H_2 spend time not only the internuclear region but also nearer to one of the nuclei than they would a nonbonded atom. Overall, each electron spends a significant fraction of time in regions where the net force exerted by the electron on the nuclei ten to draw the nuclei together (Fig. 11-3b) and much less time in regions whe the net force tends to draw the nuclei apart (Fig 11-3c).

Figure 11-3(a) likely exaggerates how much of the electron density is tra ferred into the internuclear region on bond formation. For H_2, the net trans of electron density into the internuclear region is only about 16%.* With suc small percentage of the electron density transferred into the internucle region, we might ask: Does the transfer of electron density into the intern clear region contribute most to bond formation? Somewhat surprisingly, t answer is no. It turns out that the contraction of the electron density pe toward the individual atomic nuclei (Fig. 11-3a) contributes more to t energy lowering of the system than does the transfer of electron density ir the internuclear region. The contraction leads to a significant decrease potential energy because each electron is closer, on average, to one of t nuclei than the electron in an isolated H atom.

* F. Rioux, *Chem. Educ.*, **8**, 10 (2003).

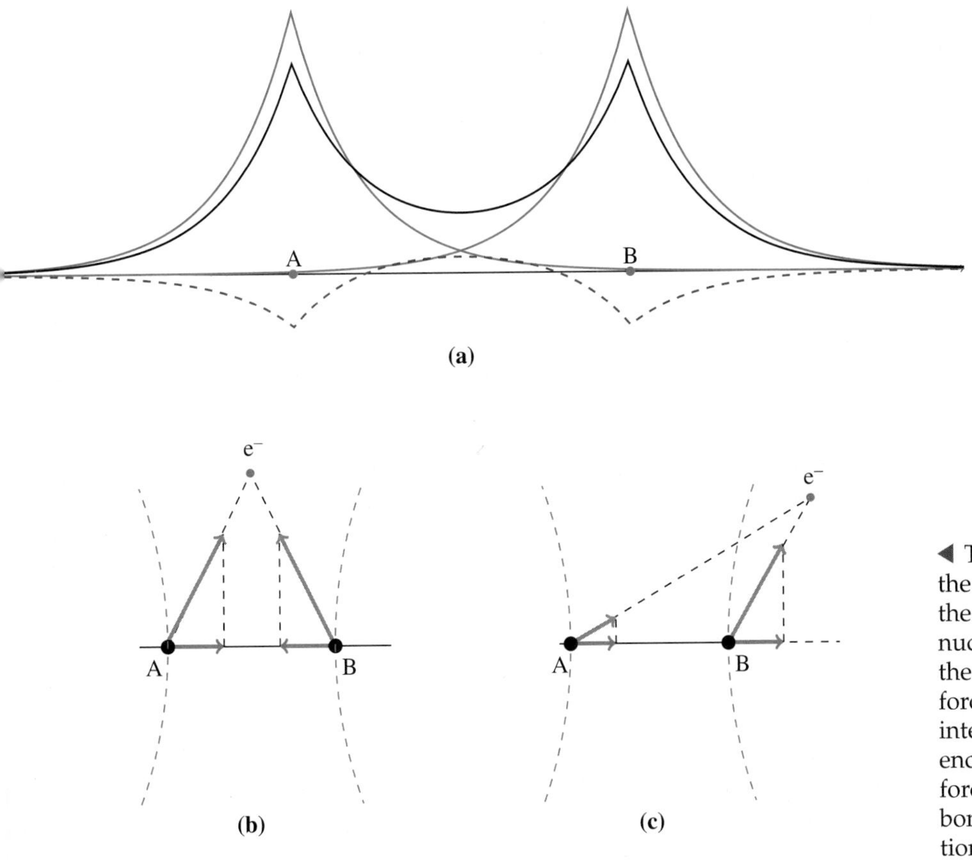

◀ The gray arrows represent the attractive forces between the electron and each nucleus. The red arrows are the components of these forces that are parallel to the internuclear axis. The differences between the component forces on A and B lead to bonding, as in (b) or separation, as in (c).

IGURE 11-3
listribution of electron density for bond formation in H_2
he black line represents the electron density in H_2. The red line represents the ctron densities for the two nonbonded H atoms. The dashed blue line represents difference between these electron densities: the electron density in H_2 minus the ctron densities for the two nonbonded atoms. Bond formation involves a *traction* of electron density toward the nuclei (represented by the black line inside red line at the left of A and the right of B) and a *redistribution* of electron density the internuclear region (represented by the black line above the red line). Overall, d formation involves a transfer of electron density from regions outside both nuclei he regions near to and between the two nuclei. **(b)** For an electron between the ndaries shown (red dashed lines), the net force exerted by an electron on the nuclei ds to draw the nuclei together. **(c)** Outside the boundaries shown, the net force rted by an electron on the nuclei tends to separate them.

The main point of the preceding discussion is that, ultimately, it is a large crease in potential energy caused by the contraction of electron density to ions nearer to the nuclei that contributes most to the stability of a chemical nd. We can justify this assertion, at least qualitatively, by considering the ative importance of other factors: the kinetic energy of the electrons, the ctron–electron repulsion, and the nuclear–nuclear repulsion.

Let's focus first on the total kinetic energy of the electrons. The total kinetic ergy of the electrons increases on bond formation, but there are two comting effects. We can unravel these effects by first developing the idea that kinetic energy of an electron decreases as the volume of space available to ncreases. Using the de Broglie relation, $\lambda = h/p$, where λ is the de Broglie velength and $p = mv$, we can write the kinetic energy of an electron as $= \frac{1}{2}mv^2 = h^2/(2m\lambda^2)$. The movement of an electron from a confined space ear one nucleus) to a larger region (between two nuclei) corresponds to an crease in the de Broglie wavelength, λ, and thus to a decrease in kinetic energy. Therefore, the kinetic energy of an electron will be lowered whe moves within a larger region encompassing both nuclei than when it mc only in the vicinity of one nucleus. However, we know that the net transfe electron density into the internuclear region is generally quite small and is set by a net contraction of the electron density toward the individual nu The net contraction of electron density toward the nuclei causes an increas kinetic energy. The overall effect is that the formation of a chemical b involves a small increase in the kinetic energy of the electrons.

There are three contributions to the total potential energy of the electron H_2: electron–nuclear attractions, electron–electron repulsion, and nucle nuclear repulsion. The net contraction of the electron density toward the ir vidual nuclei causes a significant increase in the electron–nuclei attractic Although electron–electron repulsion is destabilizing, its contribution is ty cally smaller in magnitude than the electron–nuclear attractions. The cor bution from nuclear–nuclear repulsion is typically much smaller than tha electron–nuclear attractions for two reasons: (1) the distance between nuclei is typically much larger than that between the electrons and the nu and (2) the transfer of electron density into the internuclear region causes nuclear–nuclear repulsion to be reduced even further.

In summary, bond formation involves a relatively small increase in kinetic energy of the electrons and a relatively large decrease in the poten energy of the electrons. The decrease in (potential) energy arises prima from increased electron–nuclear attractions brought about by the contract of electron density toward the nuclei.

We will now deepen our understanding of chemical bonding by empha ing a quantum mechanical perspective and describing bond formatior terms of orbitals.

11-2 Introduction to the Valence Bond Method

▶ What we are calling "overlap" is actually an interpenetration of two orbitals.

Recall the region of high electron probability in a H atom that we describec Chapter 8 through the mathematical function called a 1*s* orbital (page 341). the two H atoms pictured in Figure 11-2 approach each other, these regi begin to interpenetrate. We say that the two orbitals overlap. When the t atoms are close enough that their atomic orbitals overlap, a covalent bond be formed. Bond formation is imagined to occur through a redistributior electron probability density, as illustrated in Figure 11-3(a). It involves increase in electron probability density between the two positively charg nuclei. In the process, the energy of the system is lowered.

A description of covalent bond formation in terms of atomic orbital over is called the **valence bond method**. The creation of a covalent bond in valence bond method is normally based on the overlap of half-filled orbit but sometimes such an overlap involves a filled orbital on one atom and empty orbital on another. The valence bond method gives a *localized* elect model of bonding: Core electrons and lone-pair valence electrons retain same orbital locations as in the separated atoms, but the bonding electrons not. Instead, they are described by an electron probability density tl includes the region of orbital overlap and both nuclei.

Figure 11-4 shows the imagined overlap of atomic orbitals in the formati of hydrogen-to-sulfur bonds in hydrogen sulfide. Note especially that maxim overlap between the 1*s* orbital of a H atom and a 3*p* orbital of a S atom occ along a line joining the centers of the H and S atoms. The two half-filled sul 3*p* orbitals that overlap in H_2S are perpendicular to each other, and the valer bond method suggests a H—S—H bond angle of 90°. This is in good agr ment with the observed angle of 92°.

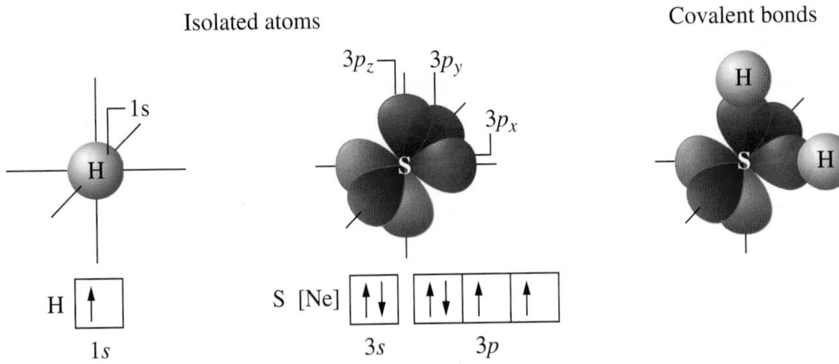

FIGURE 11-4
nding in H_2S represented by atomic orbital overlap
S, only 3*p* orbitals are shown. The phases of the lobes of the sulfur 3*p* orbitals are
own in blue and red for positive and negative, respectively. The choice of which lobe
ositive is arbitrary. However, once this choice is made, the other lobe is necessarily
gative. Bond formation can be represented diagrammatically by the overlap of
itals that are in phase (same color), although the hydrogen 1*s* orbital is colored
ow here, not blue, for clarity.

KEEP IN MIND

the actual orbital shapes are pointed out on pages 343–346. The orbitals here, and throughout the rest of the book, are schematic representations.

EXAMPLE 11-1 Using the Valence Bond Method to Describe a Molecular Structure

Describe the phosphine molecule, PH_3. by the valence bond method.

Analyze

We use four steps when applying the valence bond method. First, we identify the valence orbitals of the central atom. Second, we sketch the valence orbitals. Third, we bring in the atoms to be bonded to the central atom and sketch the orbital overlap. Finally, we describe the resulting structure. These steps are illustrated in Figure 11-5.

Solve

Step 1. Draw valence-shell orbital diagrams for the separate atoms.

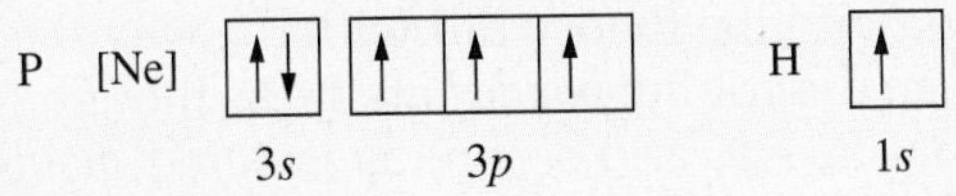

Step 2. Sketch the orbitals of the central atom (P) that are involved in the overlap. These are the half-filled 3*p* orbitals (Fig. 11-5).

Step 3. Complete the structure by bringing together the bonded atoms and representing the orbital overlap.

Step 4. Describe the structure. PH_3 is a *trigonal-pyramidal* molecule. The three H atoms lie in the same plane. The P atom is situated at the top of the pyramid above the plane of the H atoms, and the three H—P—H bond angles are 90°.

Assess

The predicted H—P—H bond angle is 90°, and the experimentally measured bond angles are 93° to 94°. These are in good agreement.

PRACTICE EXAMPLE A: Use the valence bond method to describe bonding and the expected molecular geometry in nitrogen triiodide, NI_3.

PRACTICE EXAMPLE B: Describe the molecular geometry of NH_3, first using the VSEPR method and then using the valence bond method described above. How do your answers differ? Which method seems to be more appropriate in this case? Explain.

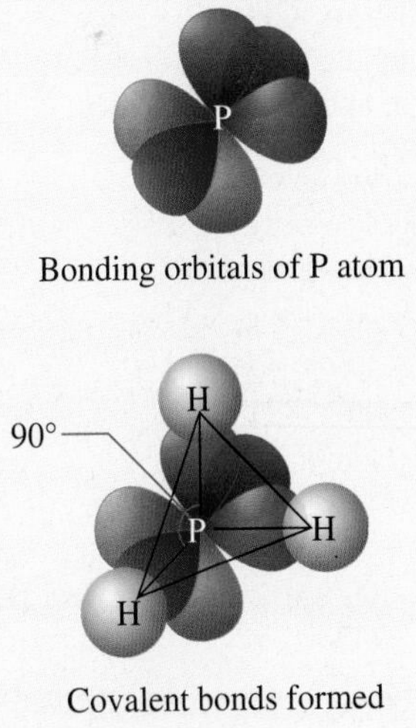

▲ FIGURE 11-5
Bonding and structure of the PH_3 molecule—Example 11-1 illustrated
Only bonding orbitals are shown. The 1*s* orbitals (yellow) of three H atoms overlap with the three 3*p* orbitals of the P atom.

11-3 Hybridization of Atomic Orbitals

If we try to extend the unmodified valence bond method of Section 11-2 t greater number of molecules, we are quickly disappointed. In most cases, descriptions of molecular geometry based on the simple overlap of unmc fied atomic orbitals do not conform to observed measurements. For examp based on the *ground-state* electron configuration of the valence shell of carb

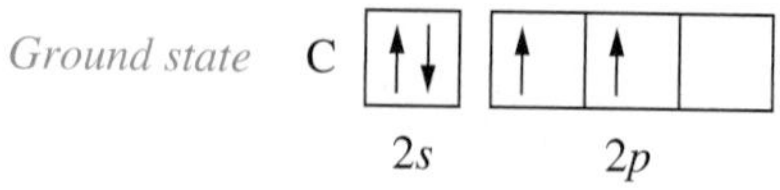

and employing only half-filled orbitals, we expect the existence of a molec with the formula CH_2 and a bond angle of 90°. The CH_2 molecule is a higl reactive molecule observed only under specially designed circumstances.

The simplest hydrocarbon observed under normal laboratory condition methane, CH_4. This is a stable, unreactive molecule with a molecular form consistent with the octet rule of the Lewis theory. To obtain this molecular f mula by the valence bond method, we need an orbital diagram for carbon which there are four unpaired electrons so that orbital overlap leads to fc C—H bonds. To get such a diagram, imagine that one of the 2*s* electrons i ground-state C atom absorbs energy and is promoted to the empty 2*p* orbit The resulting electron configuration is that of an *excited state.*

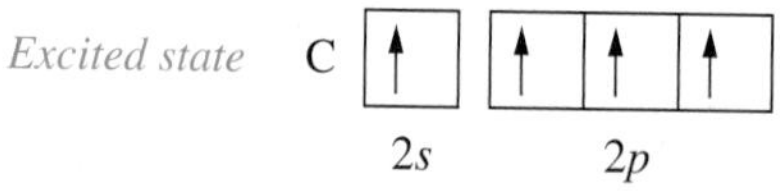

▲ FIGURE 11-6
Ball-and-stick model of methane, CH_4
The molecule has a tetrahedral structure, and the H—C—H bond angles are 109.5°.

The electron configuration of this excited state suggests a molecule w three mutually perpendicular C—H bonds based on the 2*p* orbitals of the atom (90° bond angles). The fourth bond would be directed to whatever po tion in the molecule could accommodate the fourth H atom. This descriptic however, does not agree with the experimentally determined H—C— bond angles, all four of which are found to be 109.5°, the same as predicted VSEPR theory (Fig. 11-6). A bonding scheme based on the excited-state ele tron configuration does a poor job of explaining the bond angles in CH_4.

The problem is not with the theory but with the way the situation has be defined. We have been describing *bonded* atoms as though they have the sar kinds of orbitals (that is, *s*, *p*, and so on) as isolated, *nonbonded* atoms. Th assumption worked rather well for H_2S and PH_3, but we have no reason expect these unmodified pure atomic orbitals to work equally well in all case

Given that the CH_4 molecule has a tetrahedral geometry, a fact establishe by experiment and supported by VSEPR theory, let us imagine that as t four tetrahedrally arranged H atoms are brought toward the central carbo the valence orbitals of C are transformed into a new set of orbitals. The "nev orbitals are more appropriate for bonding in a tetrahedral arrangement. Th imagined transformation of the orbitals of the C atom is illustrated below.

Valence orbitals of an isolated C atom

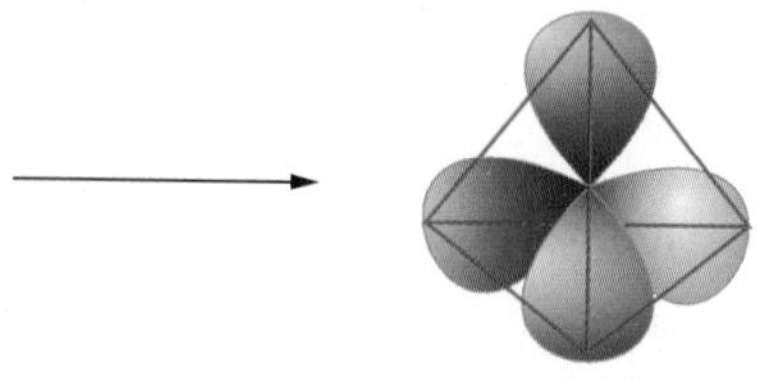

Valence orbitals of the C atom in CH_4

▶ The algebraic combination of wave functions is, in fact, a linear combination of atomic orbitals; that is, they are simply added or subtracted. The resultant linear combinations are solutions to the Schrödinger equation of the molecule.

The transformation described involves replacing four atomic orbitals wi four "new" orbitals, each of which points toward the vertex of a tetrahedro Because atomic orbitals are mathematical expressions, this transformation ca be represented mathematically by taking appropriate algebraic combinations the wave functions representing the 2*s* and three 2*p* orbitals. To obtain four ne

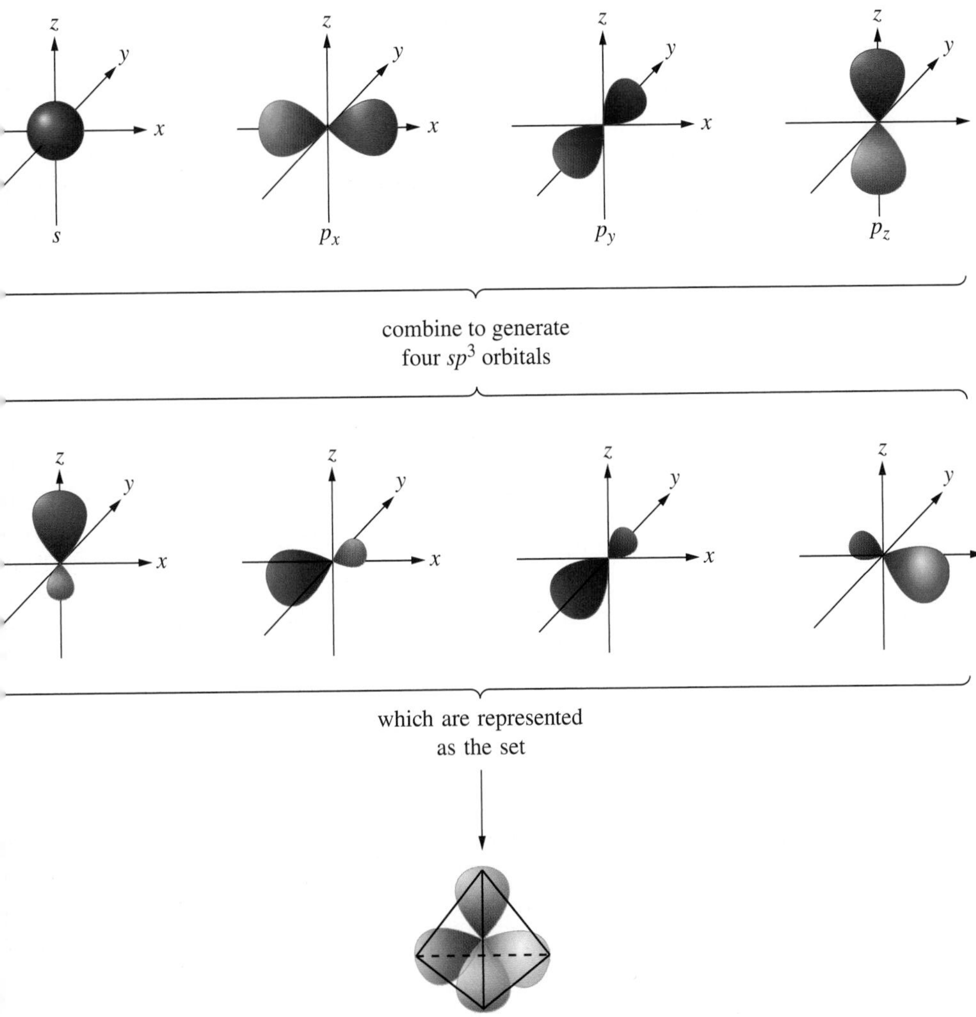

FIGURE 11-7
e sp^3 hybridization scheme
:his diagram, the *s* orbital is shown entirely in blue, suggesting that the corresponding ve function has only positive values. This is a simplified representation because, as we rned in Chapter 8, a 2*s* orbital possesses a radial node, and the wave function has a ferent sign on either side of that radial node. By showing the *s* orbital entirely in blue, are defining the wave function for the orbital to be such that it takes on positive ues at large values of *r*. We may focus on the sign of the wave function at large values *r* because, in discussions of bonding, the outermost portions of the atomic orbitals ntribute the most to the orbital overlaps involved.

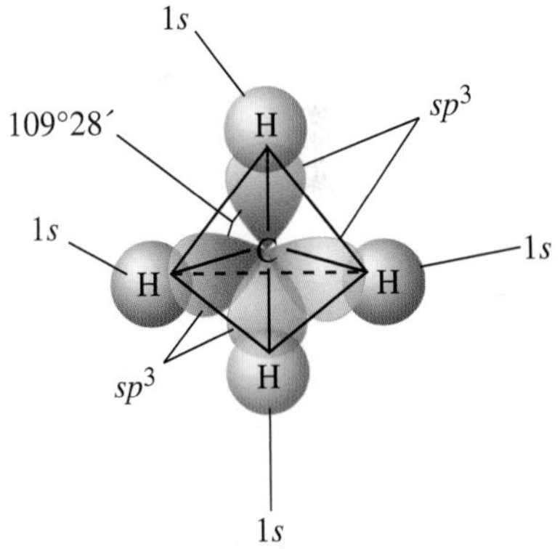

▲ FIGURE 11-8
Bonding and structure of CH_4
The four carbon orbitals are sp^3 hybrid orbitals (purple). Those of the hydrogen atoms (yellow) are 1*s*. The structure is tetrahedral, with H—C—H bond angles of 109.5° (more precisely, 109.471°). Remember that the hydrogen orbitals and the carbon hybrid orbitals have the same phase, but we have colored the hydrogen orbitals yellow for clarity.

bitals, four different algebraic combinations are required, each combination presenting one of the new orbitals. We do not need the mathematical expres- ons representing these combinations. Instead, we will focus only on the graph- al representations of these new orbitals and discuss some of their features.

The mathematical process of transforming pure atomic orbitals into formulated atomic orbitals for bonded atoms is called **hybridization**, and e new orbitals are called **hybrid orbitals**. These newly formed hybrid bitals are still atomic orbitals. Figure 11-7 illustrates the hybridization of ne *s* and three *p* orbitals into a new set of four **sp^3 hybrid orbitals**. The mbol sp^3 signifies that one *s* and three *p* orbitals are involved. Each sp^3 brid orbital has 25% *s* character and 75% *p* character and thus has energy at is intermediate between those of the 2*s* and 2*p* orbitals. For CH_4, each of e sp^3 orbitals on the C atom is used to form a bond with a H atom, as sug- ested by Figure 11-8.

A useful representation of sp^3 hybridization of the valence-shell orbital carbon is given below.

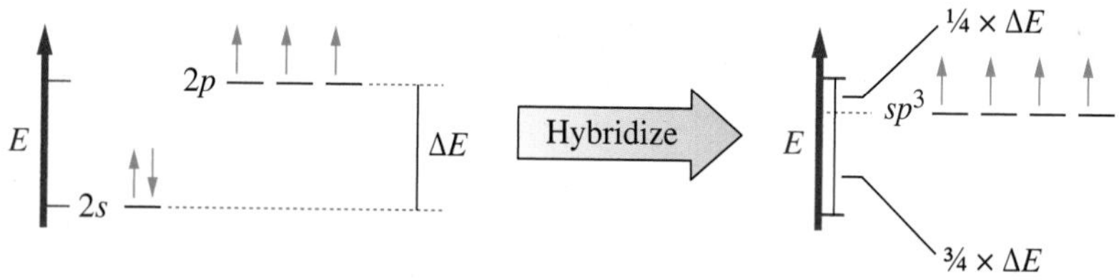

Let us denote the energies of the 2s and 2p orbitals by E_{2s} a $E_{2p} = E_{2s} + \Delta E$, where $\Delta E = E_{2p} - E_{2s}$. Because the sp^3 orbitals are 25% and 75% 2p, the energy of the sp^3 orbitals is given by

$$E_{sp^3} = \frac{1}{4} \times E_{2s} + \frac{3}{4} \times E_{2p} = E_{2s} + \frac{3}{4} \times \Delta E = E_{2p} - \frac{1}{4} \times \Delta E$$

The last two expressions indicate that an sp^3 orbital is, compared wit 2s orbital, shifted upward in energy by $\frac{3}{4}$ of the energy difference ΔE a compared with a 2p orbital, shifted downwards in energy by $\frac{1}{4}$ of the ene difference.

Up to this point, we have discussed only the sp^3 hybridization scheme a used it to describe the bonding in the CH_4 molecule. Many other hybridi tion schemes have been devised as a way to explain bonding in molecules practically any shape. Before using the concept of hybridization or introd ing other hybridization schemes to describe bonding in other molecules will be helpful to emphasize the following points.

1. For a given hybridization scheme, *the number of hybrid orbitals equals total number of atomic orbitals that are combined*. Furthermore, both hybridization scheme and the resulting hybrid orbitals are represented a symbol that identifies the numbers and kinds of orbital involved. will soon discuss the *sp* and sp^2 hybridization schemes and discover th these hybridization schemes involve one *s* orbital and either one or tw orbitals.
2. The objective of a hybridization scheme is an after-the-fact rationalizati of the experimentally observed shape of a molecule. Hybridization is n an actual physical phenomenon. We cannot observe electron density d tributions changing from those of pure orbitals to those of hybrid orbita
3. For some molecules, describing bond formation in terms of hybrid orbit is not appropriate. Nevertheless, the concept of hybridization works ve well for many molecules, especially for carbon-containing molecules. Th the concept of hybridization is used a great deal in organic chemistry.

Bonding H_2O in and NH_3

Applied to H_2O and NH_3, VSEPR theory describes a tetrahedral electro group geometry for *four* electron groups. This, in turn, requires an s hybridization scheme for the central atoms in H_2O and NH_3. This scheme su gests angles of 109.5° for the H—O—H bond in water and the H—N— bonds in NH_3. These angles are in reasonably good agreement with the expe mentally observed bond angles of 104.5° in water and 107° in NH_3. Bondi in NH_3, for example, can be described in terms of the following valence-sh orbital diagram for nitrogen.

▶ Notice that hybrid orbitals can accommodate lone-pair electrons as well as bonding electrons.

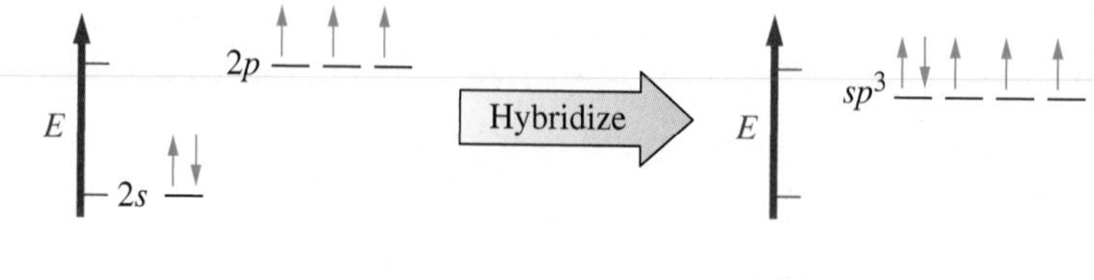

ecause one of the sp^3 orbitals is occupied by a lone pair of electrons, only three half-filled sp^3 orbitals are involved in bond formation. This suggests trigonal-pyramidal molecular geometry depicted in Figure 11-9, just as s VSEPR theory.

ven though the sp^3 hybridization scheme seems to work quite well for H_2O NH_3, both theoretical and experimental (spectroscopic) evidence favors a cription based on *unhybridized p* orbitals of the central atoms. The -O—H and H—N—H bond angle expected for 1*s* and 2*p* atomic orbital rlaps is 90°, which does not conform to the observed bond angles. One pos- e explanation is that because O—H and N—H bonds have considerable ic character, repulsions between the positive partial charges associated with H atoms force the H—O—H and H—N—H bonds to "open up" to val- greater than 90°. The issue of how best to describe the bonding orbitals in) and NH_3 is still unsettled and underscores the occasional difficulty of ding a single theory that is consistent with all the available evidence.

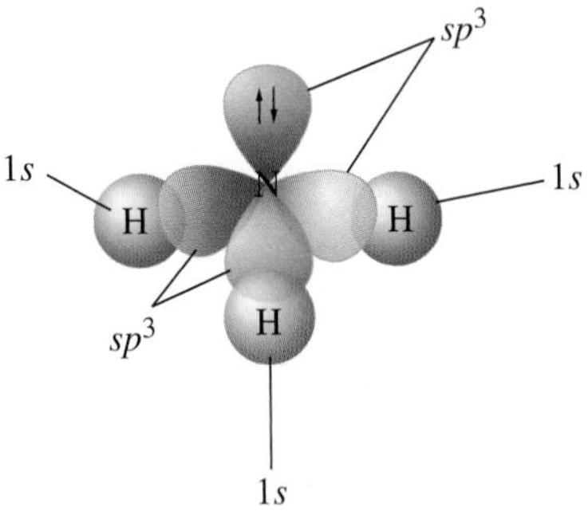

▲ FIGURE 11-9
sp^3 hybrid orbitals and bonding in NH_3
An sp^3 hybridization scheme conforms to a molecular geometry in close agreement with experimental observations. Excluding the orbital occupied by a lone pair of electrons, the centers of the atoms form a trigonal pyramid. The hydrogen orbitals are colored yellow for clarity, but they have the same phase as the nitrogen hybrid orbitals.

Hybrid Orbitals

rbon's group 13 neighbor, boron, has *four* orbitals but only *three* electrons in valence shell. For most boron compounds, the appropriate hybridization eme combines the 2*s* and two 2*p* orbitals into *three* **sp^2 hybrid orbitals** and ves one *p* orbital unhybridized. Valence-shell orbital diagrams for this ridization scheme for boron are shown here, and the scheme is further tlined in Figure 11-10.

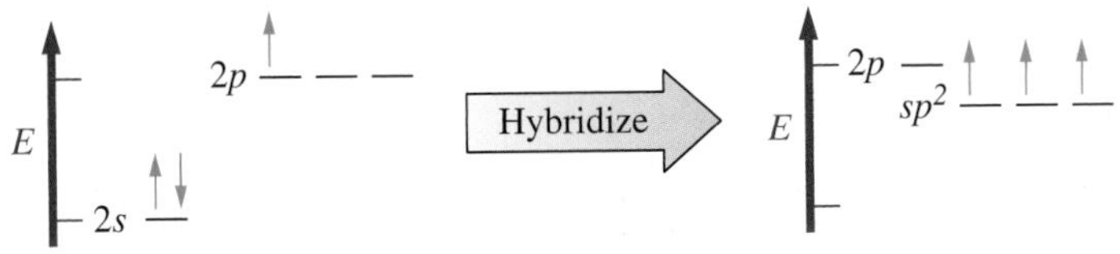

The sp^2 hybridization scheme corresponds to trigonal-planar electron- oup geometry and 120° bond, as in BF_3. Note again that in the hybridization emes of valence bond theory, the number of orbitals is conserved; that is, in sp^2 hybridized atom there are still four orbitals: three sp^2 hybrids and an hybridized *p* orbital.

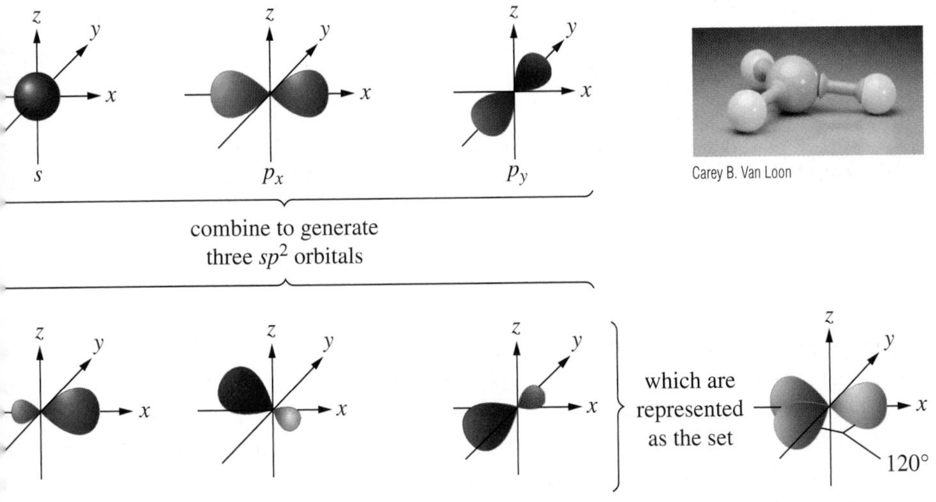

◀ FIGURE 11-10
The sp^2 hybridization scheme
The sp^2 hybrid orbitals point toward the corners of an equilateral triangle and are appropriate for describing bonding in a trigonal planar arrangement of atoms. Keep in mind that when this hybridization scheme is used to describe bonds formed by atoms other than boron, there is also an unhybridized *p* orbital (not shown) that is oriented along the *z* axis.

Hybrid Orbitals

ron's group 2 neighbor, beryllium, has *four* orbitals and only *two* elec- ons in its valence shell. In the hybridization scheme that best describes rtain *gaseous* beryllium compounds, the 2*s* and one 2*p* orbital of Be are bridized into *two* **sp hybrid orbitals**, and the remaining two 2*p* orbitals

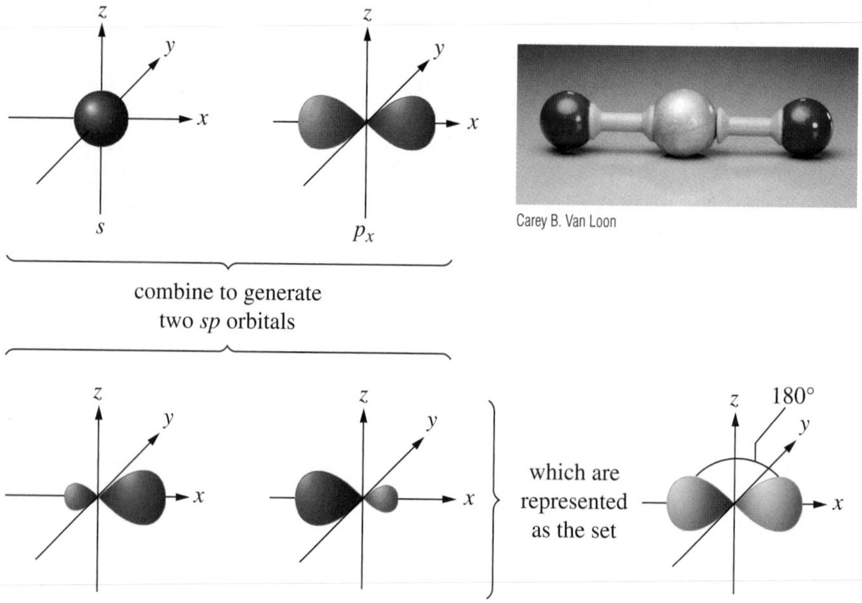

▲ FIGURE 11-11
The *sp* hybridization scheme
The *sp* hybrid orbitals are appropriate for describing bonding in a linear arrangemen
of atoms. Keep in mind that when this hybridization scheme is used to describe bon
formed by atoms other than beryllium, there are also two unhybridized *p* orbitals (nc
shown) that are oriented along the *y* and *z* axis.

▶ In hybridization, molecular orbital, and valence bond theory, not only is energy conserved, but also the number of orbitals is conserved. For example, for an sp^2 hybridized atom, there is still one *p* orbital left over and for an *sp* hybridized atom, there are two unhybridized *p* orbitals left over. Carbon readily uses the leftover *p* orbitals to form π bonds (see page 482). In contrast, silicon, the element one below carbon, does not use the *p* orbitals as readily to form π bonds. As we will see in Section 11-4, the formation of a π bond involves the side-to-side overlap of unhybridized *p* orbitals. An unhybridized 3*p* orbital of silicon does not project out far enough to form π bonds.

are left unhybridized. Valence-shell orbital diagrams of beryllium in t
hybridization scheme are shown here, and the scheme is further outlined
Figure 11-11.

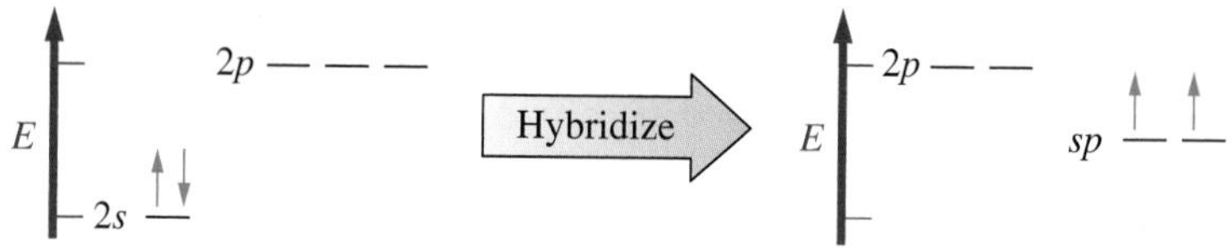

The *sp* hybridization scheme corresponds to a linear electron-group geomet
and a 180° bond angle, as in $BeCl_2(g)$.

11-1 CONCEPT ASSESSMENT

Criticize the following statement: The hybridization of the C atom in CH_3^+ and CH_3^- are both expected to be the same as in CH_4.

11-1 ARE YOU WONDERING?

What do we mean when we say that atomic orbitals mix to form a hybrid orbital?

As pointed out in the text, orbital hybridization or orbital mixing is not an actual physical phenomenon—it is a mathematical process of transforming pure atomic orbitals for isolated atoms into new atomic orbitals for bonded atoms. In particular, a hybrid atomic orbital is the result of a mathematical combination (algebraic addition and subtraction) of the wave functions describing two or more atomic orbitals. When the algebraic functions that represent *s* and *p* orbitals are added, a new function is produced; this is an *sp* hybrid. When the same algebraic functions are subtracted, another new function is produced; this is a second *sp* hybrid. The hybridization process is shown in Figure 11-12, where we see the consequence of the phase of the *p* orbital when we add the *s* and *p* orbitals: The negative phase of the *p* orbital cancels part of the positive *s* orbital. This leads to the teardrop-shaped orbital pointing in the direction of the positive lobe of the *p* orbital. As shown in Figure 11-12, subtraction of the two orbitals reverses this situation. The two ways of combining an *s* and a *p* orbital generate the two equivalent *sp* hybrid orbitals, each having its greatest amplitude (or electron density if we square the amplitude) in a direction 180° from the other. A similar procedure is used to construct the three sp^2 and the four sp^3 hybrid orbitals, although the combinations of orbitals are slightly more complicated.

p^3d and sp^3d^2 Hybrid Orbitals

describe hybridization schemes that correspond to the 5- and 6-electron-oup geometries of VSEPR theory, we need to go beyond the *s* and *p* subells of the valence shell, and traditionally this has meant including *d*-orbital ntributions. We can achieve the *five* half-filled orbitals of phosphorus to count for the five P—Cl bonds in PCl_5 and its trigonal-bipyramidal molecar geometry through the hybridization of the *s*, three *p*, and one *d* orbital of e valence shell into *five* **sp^3d hybrid orbitals**.

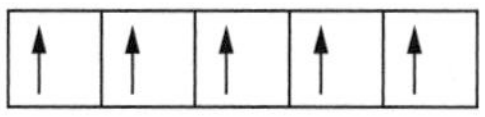

sp^3d

We can achieve the *six* half-filled orbitals of sulfur to account for the six —F bonds in SF_6 and its octahedral molecular geometry through the bridization of the *s*, three *p*, and two *d* orbitals of the valence shell into *six* **$^3d^2$ hybrid orbitals**.

sp^3d^2

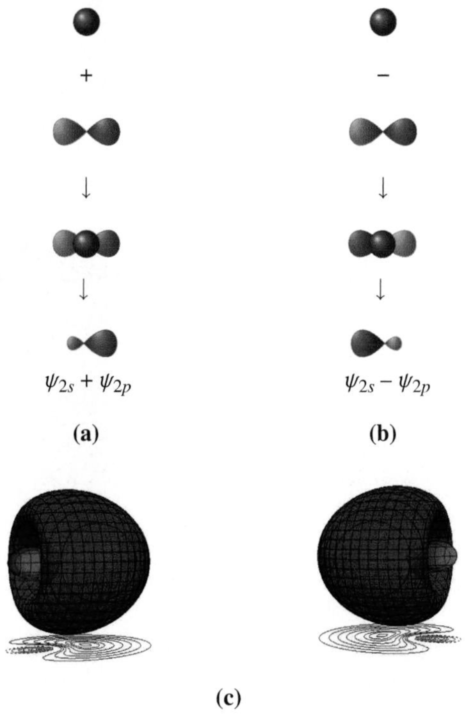

◀ FIGURE 11-12
The construction of *sp* hybrid orbitals
The linear combinations of the *s* and *p* orbitals used to form *sp* hybrid orbitals are depicted. The combination $\psi_{2s} + \psi_{2p}$ is shown in **(a)**, and the combination $\psi_{2s} - \psi_{2p}$ is shown in **(b)**. In **(c)**, the three-dimensional forms of the *sp* hybrids are shown above their contour maps.

The sp^3d and sp^3d^2 hybrid orbitals and two of the molecular geometrie which they can occur are featured in Figure 11-13.

We have previously stated that hybridization is not a real phenomenon, an after-the-fact rationalization of an experimentally determined res Perhaps there is no better illustration of this point than the issue of the s and sp^3d^2 hybrid orbitals. In discussing the concept of the expanded vale shell in Chapter 10, we noted that valence-shell expansion would seen require *d* electrons in bonding schemes, but theoretical considerations c serious doubt on *d*-electron participation. The same doubt, of course, exte to the use of *d* orbitals in hybridization schemes.

Despite the difficulty posed by hybridization schemes involving *d* orbit the *sp*, sp^2, and sp^3 hybridization schemes are well established and very cc monly encountered, particularly among the second-period elements.

11-2 CONCEPT ASSESSMENT

Give the formula of a compound or ion composed of arsenic and fluorine in which the arsenic atom has a sp^3d^2 hybridization state.

Hybrid Orbitals and the Valence-Shell Electron-Pair Repulsion (VSEPR) Theory

In the previous section, we used either the experimental geometry or geometry predicted by VSEPR theory to help us decide on the appropri hybridization scheme for the central atom. The concept of hybridizati arose before the formulation of the VSEPR theory as we used it in Chapter

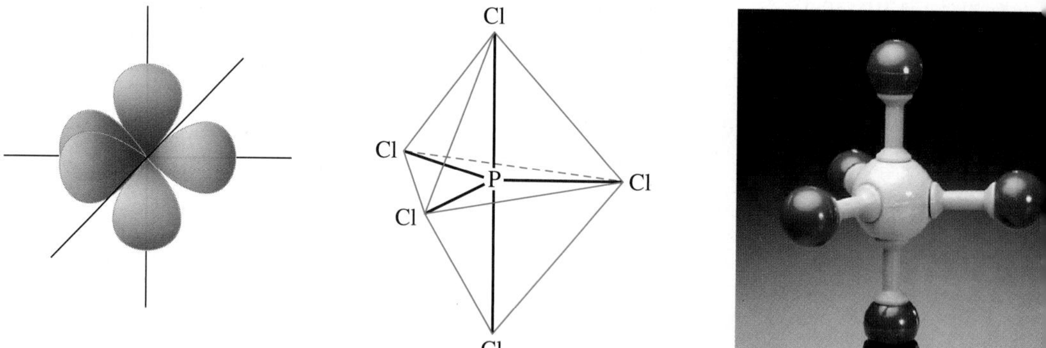

(a) sp^3d orbitals Trigonal-bipyramidal structure

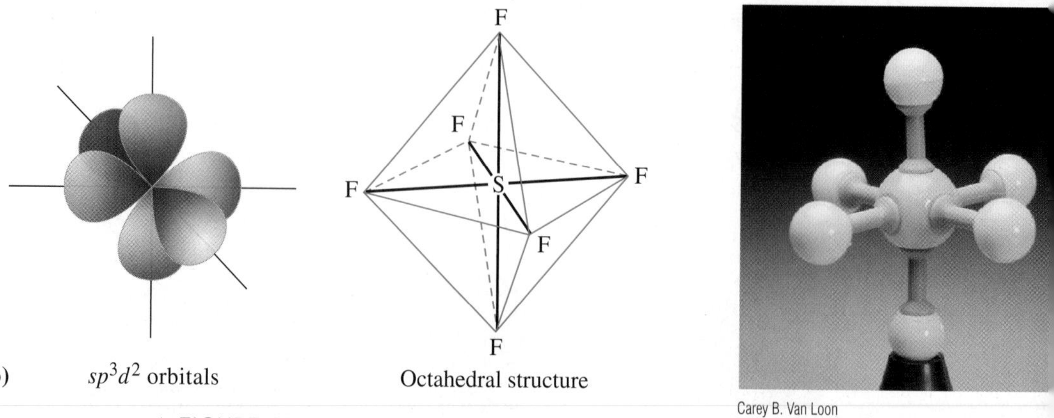

(b) sp^3d^2 orbitals Octahedral structure

▲ FIGURE 11-13
sp^3d and sp^3d^2 hybrid orbitals

1931, Linus Pauling introduced the concept of hybridization of orbitals to ount for the known geometries of CH_4, H_2O, and NH_3. It was first sug- ted by Nevil Vincent Sidgwick and Herbert Marcus Powell in 1940 that lecular geometry was determined by the arrangement of electron pairs in valence shell, and this suggestion was subsequently developed into the of rules known as VSEPR by Ronald Gillespie and Ronald Nyholm in 7. The advantage of VSEPR is that it has a predictive capability based on wis structures, whereas hybridization schemes, as described here, require a or knowledge of the molecular geometry. So how should we proceed to scribe the bonding in molecules? We can choose the likely hybridization eme for a central atom in a structure in the valence bond method by

- writing a plausible Lewis structure for the species of interest
- using VSEPR theory to predict the probable electron-group geometry of the central atom
- selecting the hybridization scheme corresponding to the electron-group geometry

e procedure outlined above is illustrated in Figure 11-14 using the molecule 4 as an example.

As suggested by Table 11.1, the hybridization scheme adopted for a central m should be the one producing the same number of hybrid orbitals as there valence-shell electron groups, and in the same geometric orientation. us, an sp^3 hybridization scheme for the central atom predicts that four brid orbitals are distributed in a tetrahedral fashion. This results in molecu- structures that are tetrahedral, trigonal-pyramidal, or angular, depending how many hybrid orbitals are involved in orbital overlap and how many ntain lone-pair electrons, corresponding to the VSEPR notations AX_4, AX_3E, d AX_2E_2 respectively.

The *s* and *p* orbital hybridization schemes are especially important in ganic compounds, whose principal elements are C, O, and N, in addition to H. will consider some important applications to organic chemistry in the next tion.

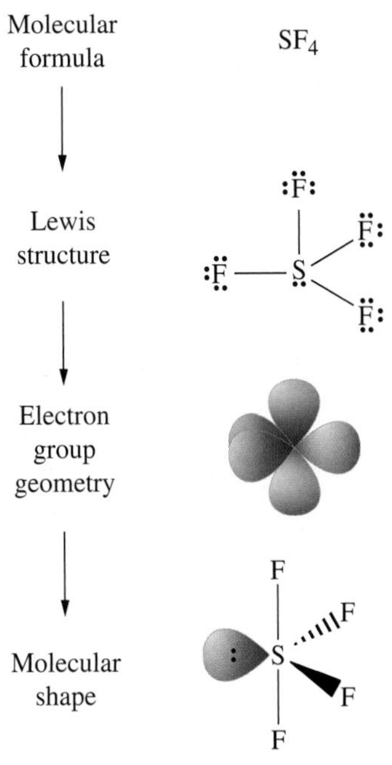

▲ FIGURE 11-14
Using electron-group geometry to determine hybrid orbitals

TABLE 11.1 Some Hybrid Orbitals and Their Geometric Orientations

Hybrid Orbitals	Geometric Orientation	Example
sp	Linear	$BeCl_2$
sp^2	Trigonal-planar	BF_3
sp^3	Tetrahedral	CH_4
sp^3d	Trigonal-bipyramidal	PCl_5
sp^3d^2	Octahedral	SF_6

EXAMPLE 11-2 Proposing a Hybridization Scheme to Account for the Shape of a Molecule

Predict the shape of the XeF_4 molecule and a hybridization scheme consistent with this prediction.

Analyze

Follow the procedure outlined in Figure 11-14.

Solve

1. **Write a plausible Lewis structure.** The Lewis structure we write must account for 36 valence electrons—*eight* from the Xe atom and *seven* each from the *four* F atoms. To place this many electrons in the

(continued)

Lewis structure, we must expand the valence shell of the Xe atom to accommodate 12 electrons. The Lewis structure is

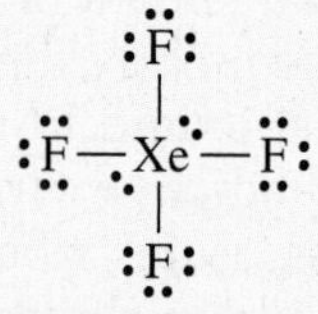

2. **Use the VSEPR theory to establish the electron-group geometry of the central atom.** From the Lewis structure, we see that there are *six* electron groups around the Xe atom. Four electron groups are bond pairs and two are lone pairs. The electron-group geometry for *six* electron groups is *octahedral.*
3. **Describe the molecular geometry.** The VSEPR notation for XeF_4 is AX_4E_2, and the molecular geometry is *square-planar* (see Table 10.1). The four pairs of bond electrons are directed to the corners of a square, and the lone pairs of electrons are found above and below the plane of the Xe and F atoms, as shown here.

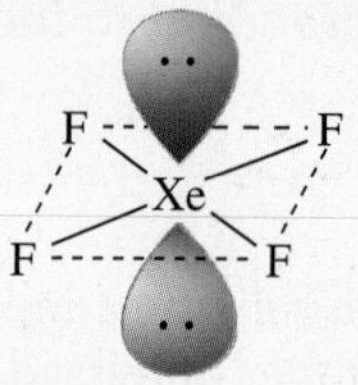

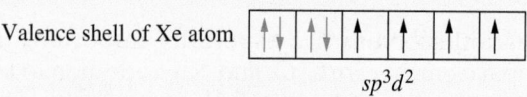

4. **Select a hybridization scheme that corresponds to the VSEPR prediction.** The only hybridization scheme consistent with an octahedral distribution of *six* electron groups is sp^3d^2. The orbital diagram for this scheme shows clearly that *four* of the eight valence electrons of the central Xe atom singly occupy four of the sp^3d^2 orbitals. The remaining *four* valence electrons of the atom occupy the remaining two sp^3d^2 orbitals as lone pairs. These are the lone pairs of electrons situated above and below the plane of the Xe and F atoms in the above sketch.

Valence shell of Xe atom
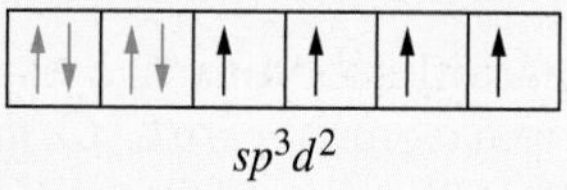

Assess

Note that the number of electron pairs in the electron-group geometry dictates how many orbitals are used in the hybridization scheme. We see that a combination of VSEPR and hybridization theory is an appealing way to describe the shape of and bonding in a molecule. However, we should emphasize a point made earlier on page 478: The inclusion of *d* orbitals in hybridization schemes is questionable. In other words, the use of the sp^3d^2 hybridization scheme for xenon is probably not the best way to describe the bonding in XeF_4.

PRACTICE EXAMPLE A: Describe the molecular geometry and propose a plausible hybridization scheme for the central atom in the ion Cl_2F^+.

PRACTICE EXAMPLE B: Describe the molecular geometry and propose a plausible hybridization scheme for the central atom in the ion BrF_4^+.

11-2 ARE YOU WONDERING?

Should we use the VSEPR method (Section 10-7) or the valence bond method in rationalizing the geometric shape of a molecule?

There is no "correct" method for describing molecular structures. The only correct information is the experimental evidence from which the structure is established. Once this experimental evidence is in hand, you may find it easier to rationalize this evidence by one method or another. For H_2S, the valence bond method, which suggests a bond angle of 90°, seems to do a better job of explaining the observed 92° bond angle than does the VSEPR theory. For the Lewis structure,

—S—H, VSEPR theory predicts a tetrahedral electron-group geometry, which turn suggests a tetrahedral bond angle—that is, 109.5°. However, by modifying is initial VSEPR prediction to accommodate lone-pair–lone-pair and lone-air–bond-pair repulsions (see page 443), the predicted bond angle is less than)9.5°.

VSEPR theory gives reasonably good results in the majority of cases. Unless ou have specific information to suggest otherwise, describing a molecular nape with the VSEPR theory is a good bet. It is important to remember that both e VSEPR and valence bond methods are simply models we use to rationalize e shapes and bonding of polyatomic molecules and as such, should be viewed ith a critical eye, always keeping experimental results in sight.

KEEP IN MIND

that the VSEPR method uses empirical data to give an approximate molecular geometry, whereas the valence bond method relates to the orbitals used in bonding based on a given geometry.

11-3 CONCEPT ASSESSMENT

'hat hybridization do you expect for the central atom in a molecule that has a quare-pyramidal geometry?

-4 Multiple Covalent Bonds

o different types of orbital overlap occur when multiple bonds are cribed by the valence bond method. In our discussion we will use as spe-c examples the carbon-to-carbon double bond in ethylene, C_2H_4, and the bon-to-carbon triple bond in acetylene, C_2H_2.

nding in C_2H_4

ylene has a carbon-to-carbon double bond in its Lewis structure.

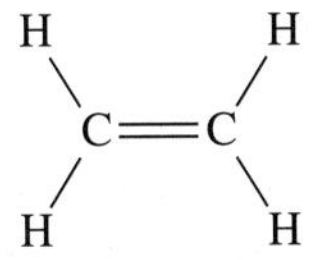

ylene is a planar molecule. The H—C—H and H—C—C bond angles close to 120°. VSEPR theory treats each C atom as being surrounded by *e* electron groups in a trigonal-planar arrangement. VSEPR theory does *not* tate that the two —CH_2 groups be coplanar, but as we will see, valence nd theory does.

The hybridization scheme that produces a set of hybrid orbitals with a gonal-planar orientation is sp^2. The valence-shell orbital diagrams of carbon this scheme are

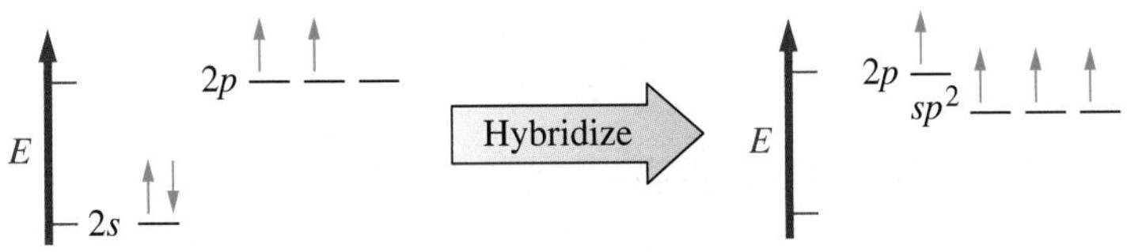

The $sp^2 + p$ orbital set is pictured in Figure 11-15. One of the bonds ween the carbon atoms results from the overlap of sp^2 hybrid orbitals from h atom. This overlap occurs along the line joining the nuclei of the two oms. Orbitals that overlap in this end-to-end fashion produce a **sigma nd**, designated **σ bond**. Figure 11-15 shows that a second bond between C atoms results from the overlap of the unhybridized *p* orbitals. In this nd, there is a region of high electron density above and below the plane of carbon and hydrogen atoms. The bond produced by this side-to-side erlap of two parallel orbitals is called a **pi bond**, designated **π bond**.

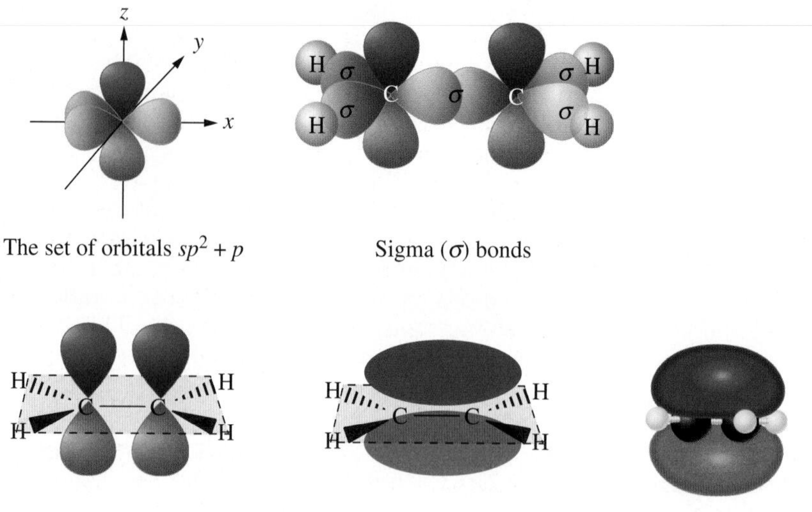

The set of orbitals $sp^2 + p$

Sigma (σ) bonds

Overlap of p orbitals leading to a pi (π) bond

Computed π orbital of ethylene

▶ FIGURE 11-15
Sigma (σ) and pi (π) bonding in C_2H_4
The purple orbitals are sp^2 hybrid orbitals; the red and blue orbitals are 2p, with the colors indicating their phase. The sp^2 hybrid orbitals overlap along the line joining the bonded atoms—a σ bond. The 2p orbitals overlap in a side-to-side fashion and form a π bond. Notice that the phase of the p orbitals is retained.

An alternative definition for σ and π bonds is based on the number of nc planes that are parallel to the bond axis: a σ bond has no nodal planes pa lel to the bond axis and a π bond has 1 such nodal plane.

The ball-and-stick model in Figure 11-16 illustrates bonding in ethylen helps to show that

- the shape of a molecule is determined only by the orbitals form σ bonds (the σ-bond framework).
- rotation about the double bond is severely restricted. To twist one —C group out of the plane of the other would reduce the amount of overla the p orbitals and weaken the π bond. The double bond is rigid, and C_2H_4 molecule is planar.

KEEP IN MIND

that only one of the bonds in a multiple bond is a σ bond; the others are π bonds—one π bond in a double bond and two in a triple bond.

Additionally, in carbon-to-carbon multiple bonds, the σ bond invol more extensive overlap than does the π bond. As a result, a carbon-to-carl double bond (σ + π) is stronger than a single bond (σ), but not twice strong (from Table 10.3, C—C, 347 kJ/mol; C═C, 611 kJ/mol; C≡ 837 kJ/mol).

▲ FIGURE 11-16
Ethylene, C_2H_4
The H—C—H and H—C—C bond angles are 120°. The model also distinguishes between the σ bond between the C atoms (the solid line connecting the two carbon atoms) and the π bond (the blue lobe above and the red lobe below the plane of the molecule).

Bonding in C_2H_2

Bonding in acetylene, C_2H_2, is similar to that in C_2H_4, but with these dif ences: The Lewis structure of C_2H_2 features a triple covalent bo H—C≡C—H. The molecule is *linear*, as found by experiment and as expec from VSEPR theory. A hybridization scheme to produce hybrid orbitals in a ear orientation is *sp*. The valence-shell orbital diagrams representing hybridization are

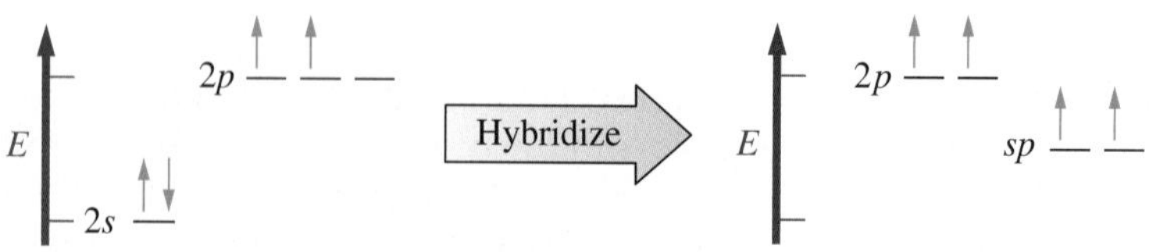

In the triple bond in C_2H_2, one of the carbon-to-carbon bonds is a σ bond a *two* are π bonds, as suggested in Figure 11-17.

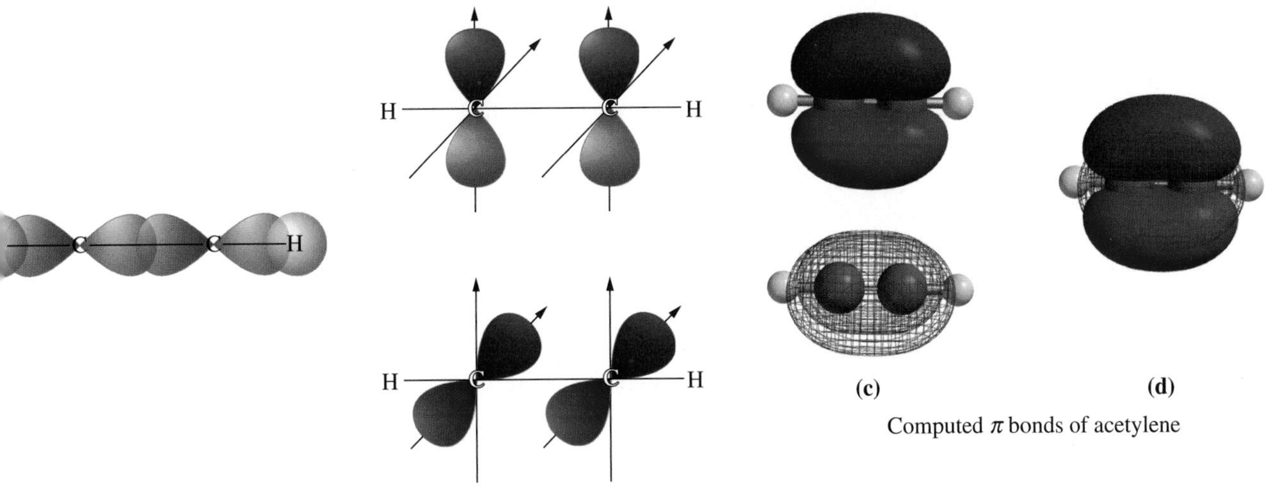

(a) Formation of σ bonds

(b) Formation of π bonds

(c) (d) Computed π bonds of acetylene

IGURE 11-17
nd π bonding in C_2H_2
The σ-bond framework that joins the atoms H—C—C—H through 1s orbitals of H atoms and *sp* orbitals of the C atoms is illustrated. **(b)** Two π bonds are formed. e π bond (upper part of diagram) is formed by a pair of *p* orbitals, which are ented along an axis in the plane of the page. The other π bond (lower part of gram) is formed by a pair of *p* orbitals oriented along an axis perpendicular to the ne of the page. **(c)** These computed models of the π bonds described in (b) provide ore realistic representation of their shapes. One of the π orbitals is represented by ansparent (mesh) surface to distinguish it from the other π bond. **(d)** The two nputed π bonds are shown together.

EXAMPLE 11-3 Proposing Hybridization Schemes Involving σ and π Bonds

Formaldehyde gas, H_2CO, is used in the manufacture of plastics; in aqueous solution, it is the familiar biological preservative called formalin. Describe the molecular geometry and a bonding scheme for the H_2CO molecule.

Analyze

The number of electron groups around a central atom dictates the number of atomic orbitals that undergo hybridization. The carbon atom in formaldehyde contains three electron groups (remember, a double bond is counted as one group), which means that three atomic orbitals are hybridized to form three sp^2 orbitals. We will follow the procedure shown in Figure 11-14.

Solve

1. **Write the Lewis structure.** C is the central atom, and H and O are terminal atoms. The total number of valence electrons is 12. Note that this structure requires a carbon-to-oxygen double bond.

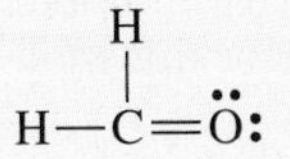

2. **Determine the electron-group geometry of the central C atom.** The σ-bond framework is based on three electron groups around the central C atom. VSEPR theory, based on the distribution of three electron groups, suggests a trigonal-planar molecule with 120° bond angles.
3. **Identify the hybridization scheme that conforms to the electron-group geometry.** A trigonal-planar orientation of orbitals is associated with sp^2 hybrid orbitals.

(continued)

4. **Identify the orbitals of the central atom that are involved in orbital overlap.** The C atom is hybridized to produce the orbital set $sp^2 + p$, as in C_2H_4. Two of the sp^2 hybrid orbitals are used to form σ bonds with the H atoms. The remaining sp^2 hybrid orbital is used to form a σ bond with oxygen. The unhybridized p orbital of the C atom is used to form a π bond with O.

Valence shell of the C atom: 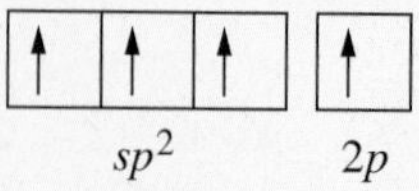

5. **Sketch the bonding orbitals of the central and terminal atoms.** The bonding orbitals of the central C atom described above are pictured in Figure 11-18(a). The sp^2 hybrid orbitals are shown in purple, and the pure p orbitals, in blue and red. The H atoms have only $1s$ orbitals available for bonding. For oxygen, the half-filled $2p$ orbital can be used for end-to-end overlap in the σ bond to carbon, and a half-filled $2p$ orbital can participate in the side-to-side overlap leading to a π bond. Thus, the valence-shell orbital diagram we can use for oxygen is

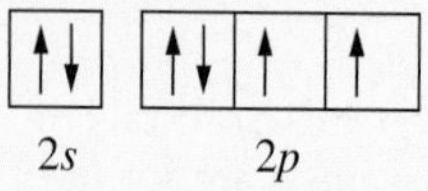

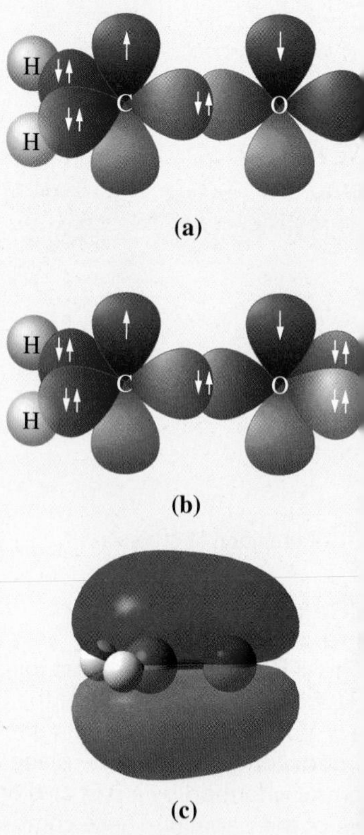

▲ FIGURE 11-18
Bonding and structure of the H_2CO molecule—Example 11-3 illustrated
(a) The orbital set $sp^2 + p$ is used for the C atom, $1s$ orbitals for H, and two half-filled $2p$ orbitals for O. For simplicity, only bonding orbitals of the valence shells are shown. **(b)** An alternative description of the C=O bond uses the orbital set $sp^2 + p$ for the O atom. **(c)** Computed π orbital for the CO group.

Assess

The bonding and structure of the H_2CO molecule are suggested by the three-dimensional sketch in Figure 11-18(c). A Lewis structure is given below (Fig. 11-19).

One of the main purposes of invoking the use of hybrid orbitals is to describe molecular geometry (for example, bond angles), and so we generally apply hybridization schemes only to central atoms, although hybridization of terminal atoms (except H) may also be invoked.

Thus, an alternative description of the C=O bond in H_2CO is obtained by assuming that the O atom is also sp^2 hybridized. Since the O atom is surrounded by three groups, a bonded C atom and two lone pairs, sp^2 hybridization is the appropriate hybridization scheme. Assuming that the O atom is sp^2 hybridized, the carbon–oxygen bond consists of a σ bond, arising from the overlap of an sp^2 orbital on the C atom with an sp^2 orbital on the O atom, and a π bond, arising from the side-to-side overlap of $2p$ orbitals, as illustrated in Figure 11-18(b). The other two sp^2 orbitals on the O atom are used to accommodate the two lone pairs.

The H_2CO molecule has four bonds around carbon (three σ bonds and one π bond). These four bonds are used to bond three groups: two H atoms and one O atom. In determining the electron-group geometry and bond angles around a particular atom, we focus on the number of σ bonds and lone pairs, if any.

PRACTICE EXAMPLE A: Describe a plausible bonding scheme and the molecular geometry of dimethyl ether, CH_3OCH_3.

PRACTICE EXAMPLE B: Acetic acid, the acidic component of vinegar, has the formula CH_3COOH. Describe the molecular geometry and a bonding scheme for this molecule.

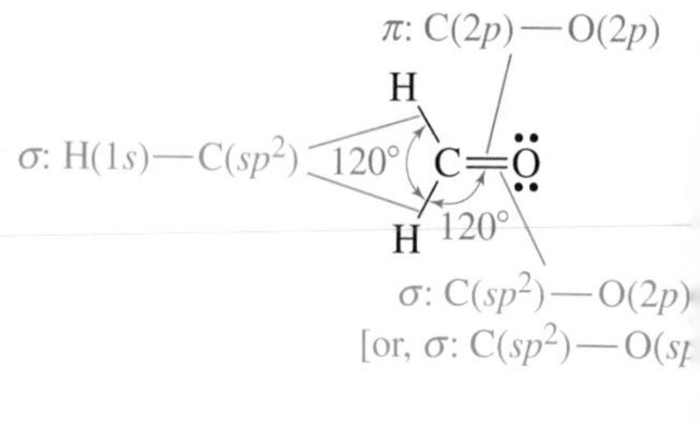

▶ FIGURE 11-19
Bonding in H_2CO—a schematic representation

)rawing three-dimensional sketches to show orbital overlaps, as in ıre 11-18(c), is not easy. A simpler, two-dimensional representation of the ding scheme for formaldehyde is shown at the bottom of the previous e in Figure 11-19. Bonds between atoms are drawn as straight lines. They labeled σ or π, and the orbitals that overlap are indicated.

Ve have stressed a Lewis structure as the first step in describing a bonding eme. Sometimes the starting point is a description of the species obtained xperiment. Example 11-4 illustrates such a case.

EXAMPLE 11-4 Using Experimental Data to Assist in Selecting a Hybridization and Bonding Scheme

Formic acid, HCOOH, is an irritating substance released by ants when they bite (*formica* is Latin, meaning "ant"). A structural formula with bond angles is given here. Propose a hybridization and bonding scheme consistent with this structure.

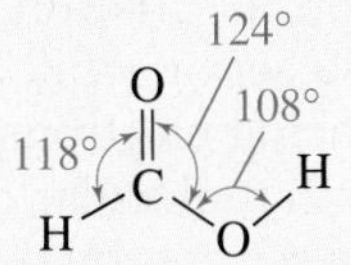

Analyze

When we are given the bond angles for a molecule, we know the hybridization scheme is dictated by the bond angles of that atom; for example, if the bond angle is close to 109.5°, we expect sp^3 hybridization.

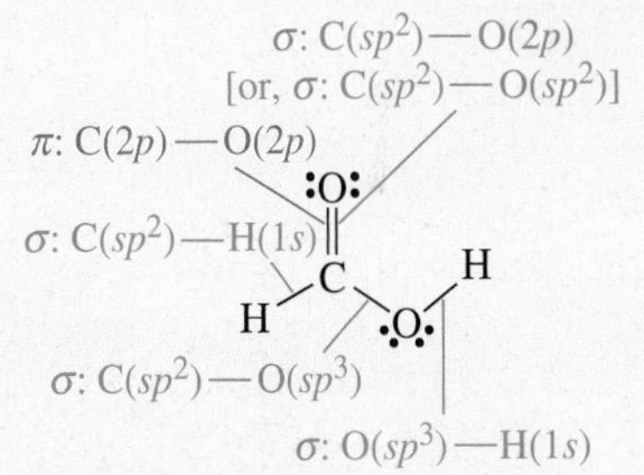

▲ FIGURE 11-20
Bonding and structure of HCOOH—Example 11-4 illustrated

Solve

The 118° H—C—O bond angle on the left is very nearly the 120° angle for a trigonal-planar distribution of three groups of electrons. This requires an sp^2 hybridization scheme for the C atom. The 124° O—C—O bond angle is also close to the 120° expected for sp^2 hybridization. The C—O—H bond angle of 108° is close to the tetrahedral angle—109.5°. The O atom on the right employs an sp^3 hybridization scheme. The four σ and one π bonds and the orbital overlaps producing them are indicated in Figure 11-20.

Assess

The structure originally given does not show the lone pair electrons; they are there implicitly. We can infer the presence of the lone pairs from the observed bond angles. We would have deduced the sp^3 hybridization at the oxygen atom by using a Lewis structure and VSEPR theory.

As discussed in Example 11-3, an alternative description of the C=O bond is obtained by assuming that this O atom is also sp^2 hybridized. In other words, we may also label the C=O bond as σ: C(sp^2)—O(sp^2) plus π: C(2p)—O(2p).

PRACTICE EXAMPLE A: Acetonitrile is an industrial solvent. Propose a hybridization and bonding scheme consistent with its structure.

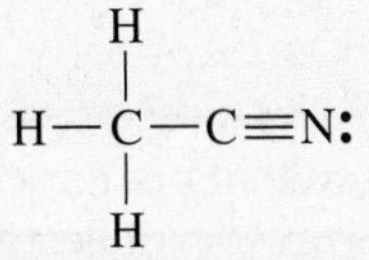

PRACTICE EXAMPLE B: A reference source on molecular structures lists the following data for dinitrogen monoxide (nitrous oxide), N_2O: Bond lengths: N—N = 113 pm; N—O = 119 pm; bond angle = 180°. Show that the Lewis structure of N_2O is a resonance hybrid of two contributing structures, and describe a plausible hybridization and bonding scheme for each.

11-4 CONCEPT ASSESSMENT

he molecule diazine has the molecular formula N_2H_2. What is the hybridization f the nitrogen and does the molecule contain a double or triple bond?

11-5 Molecular Orbital Theory

Lewis structures, VSEPR theory, and the valence bond method make a po combination for describing covalent bonding and molecular structures. T are satisfactory for most of our purposes. Sometimes, however, chemists n a greater understanding of molecular structures and properties than th methods provide. None of these methods, for instance, provides an expla tion of the electronic spectra of molecules, why oxygen is paramagnetic why H_2^+ is a stable species. To address these questions, we need a differ method of describing chemical bonding.

This method, called **molecular orbital theory**, starts with a simple pictur molecules, but it quickly becomes complex in its details. We will provide c an overview and then focus primarily on the application of molecular orb theory to diatomic molecules. We will begin our overview by comparing m cular orbital theory with the conceptual model we introduced in Chapter 8 multielectron atoms.

As discussed in Chapter 8 (see page 350), we imagine that an atom available to it a set of orbitals (1*s*, 2*s*, 2*p*, etc.) and can be built up electron electron by placing electrons into these orbitals in a specified order. In a si lar manner, with molecular orbital theory, we imagine that each molecule available to it a set of orbitals, called molecular orbitals (MOs) and can be b up electron by electron by placing electrons into these orbitals in a specif order. Unlike atomic orbitals, which are centered on a single nucleus, mole lar orbitals are defined with respect to all the nuclei.

Let's explore molecular orbital theory more deeply by considering results of a molecular orbital calculation for F_2. Figure 11-21 shows the m ecular orbitals (MOs) for F_2. These MOs are obtained from an analysis of total wave function describing the F_2 molecule in its lowest energy st The total wave function itself is obtained by solving the Schrödinger eq tion for the F_2 molecule. We see that there are ten MOs, and these MOs not associated with just one F nucleus but are distributed around b nuclei. We can easily rationalize the number and the shapes of the mole lar orbitals by making use of a method known as the *linear combinatio atomic orbitals*, which is usually abbreviated as LCAO. The LCAO metho based on the idea that because a molecule is composed of atoms, the ato orbitals (AOs) of those atoms can be used as the basis of a method describing how each electron in a molecule interacts with all nuclei simu neously. That is, the atomic orbitals can be used as a basis for describing MOs of a molecule. In practice, the LCAO method starts from the prem that each MO can be represented mathematically as a (linear) combinat of all AOs.

The LCAO method incorporates the following ideas.

- **Each MO can be represented mathematically as a (linear) combinati of AOs.** Conceptually, we imagine that the AOs are transformed i (replaced by) a set of molecular orbitals when the atoms combine to fo the molecule. This conceptualization does not correspond to any real physical process; it is a way of applying what we know about electrons atoms and extending that description to molecules. A difficult aspect applying the approach is in deducing the appropriate combinations AOs to use. For diatomic molecules, however, the appropriate combi tions are rather simple.
- **The total number of MOs is equal to the total number of AOs.** example, the electrons in a ground state F atom are described by f atomic orbitals: 1*s*, 2*s*, $2p_x$, $2p_y$, and $2p_z$. Thus, the electrons for two i lated (nonbonded) F atoms are described by a total of ten orbitals (f centered on one F atom and five centered on the other). The ten AOs c be combined mathematically to form ten MOs that are appropriate describing electrons in the F_2 molecule.

▶ Molecular orbital theory states that the number of molecular orbitals formed is equal to the number of atomic orbitals combined.

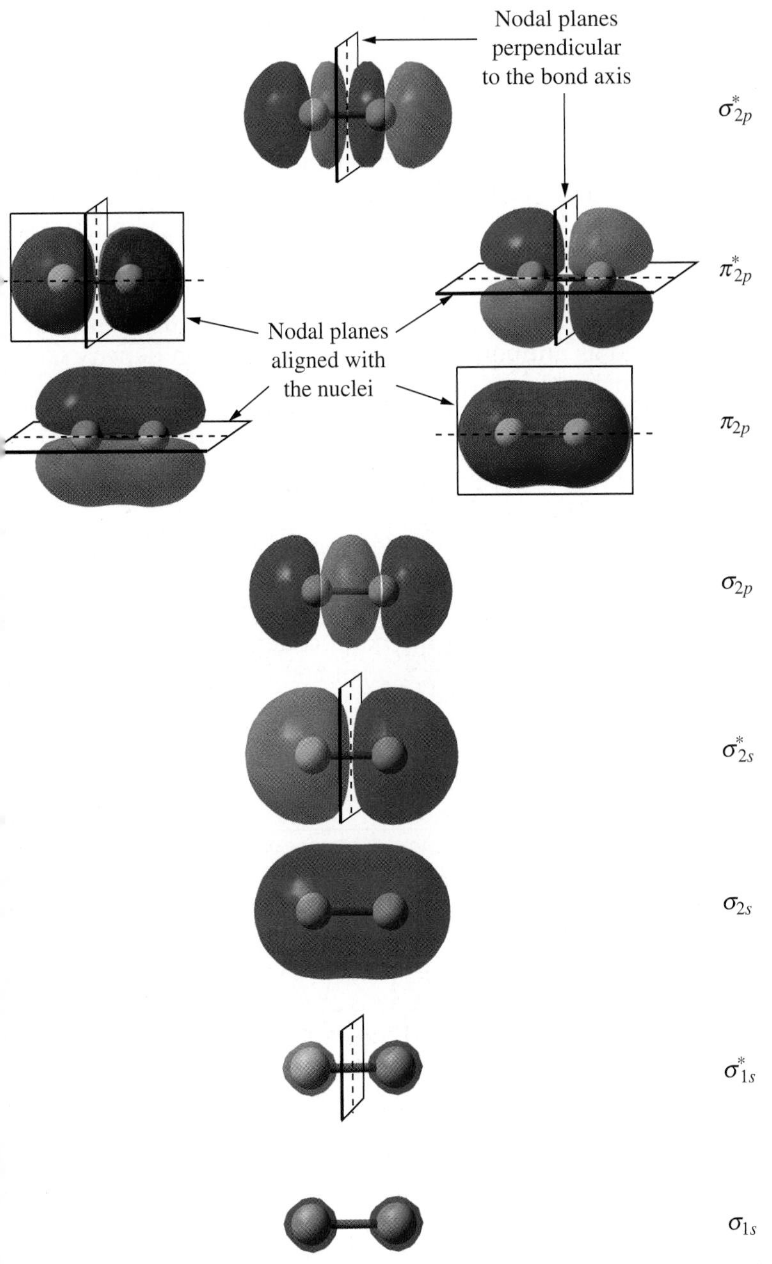

◀ FIGURE 11-21
Computed molecular orbitals of F_2
The molecular orbitals of F_2 can be classified as σ or π and as bonding or antibonding. The classification depends on the presence or absence of certain types of nodal planes. The symbol π indicates the presence of a nodal plane aligned with the two nuclei whereas the symbol σ indicates the absence of such a nodal plane. An antibonding orbital is labeled with an asterisk (*) and is characterized by the presence of a nodal plane perpendicular to and between the two nuclei. In a bonding orbital, there is no such nodal plane. The subscript on σ or π is a reference to the atomic orbitals (one on each atom) that are combined mathematically in the LCAO method to form a particular molecular orbital. For example, the σ_{2p} orbital is represented mathematically as a linear combination of 2*p* orbitals, one on each F atom. The formation of bonding and antibonding orbitals from 1*s* and 2*p* orbitals is illustrated in Figs. 11-22 and 11-24.

- **MOs can be classified as bonding, antibonding, or nonbonding.** This classification is made by considering the electron density, which is the square of the wave function representing a MO, in the internuclear region. Bonding orbitals will have electron density between the nuclei (e.g., σ_{2s} in Figure 11-21). Antibonding orbitals will have a node between the nuclei (e.g., σ_{2s}* in Figure 11-21). *Nonbonding orbitals* will typically consist of AOs isolated on one or more atoms and neither add to nor detract from bond formation. For example, in Figure 11-21 we observe that the nonbonding σ_{1s} and σ_{1s}* MOs are composed of 1*s* AOs on each F atom. For most molecules, MOs composed of core AOs are nonbonding.
- **The electron configuration for the molecule is obtained by placing the electrons into the MOs in a particular order.** This is analogous to the conceptual model we used for multielectron atoms. The only difference here is that we are filling MOs, not AOs. Not surprisingly, rules we used for atoms, such as the Pauli exclusion principle and Hund's rule, also apply to molecules. That is, the maximum number of electrons per

orbital is two, and if there are two electrons in one MO, they must h opposite spins.

KEEP IN MIND

that $1s_A + 1s_B$ is a shorthand way of writing $\psi_{1s_A} + \psi_{1s_B}$. Thus, in this context, $1s_A$ and ψ_{1s_A} both refer to the wave function describing a 1*s* orbital centered on atom A. Analogous statements can be made concerning $1s_B$ and ψ_{1s_B}.

Figure 11-22 summarizes the LCAO method through the combination of 1*s* orbitals of atomic H into two molecular orbitals in a H_2 molecule. The r ecular orbital resulting from the addition of the 1*s* orbitals ($1s_A + 1s_B$) **bonding molecular orbital**. For this orbital, we see that there is electron c sity between the nuclei and that the molecular orbital has a cylindrical sh This bonding molecular orbital is given the label σ_{1s} and is lower in ene than the 1*s* atomic orbitals. The molecular orbital resulting from subtraction of the 1*s* orbitals ($1s_A - 1s_B$) is an **antibonding molecular orb** The orbital is described as antibonding because the electron density is zer a point between the nuclei. (More precisely, the wave function has a node point between the two nuclei.) An antibonding orbital is distinguished fro bonding orbital by a superscript asterisk (*). The antibonding orbital resul from the subtraction of two 1s orbital is thus labeled as σ_{1s}*. The σ_{1s}* is hig in energy than the 1*s* atomic orbital.

A stable molecular species has more electrons in bonding orbitals tha antibonding orbitals. For example, if the excess of bonding over antibonc electrons is *two*, this corresponds to a *single* covalent bond in Lewis theory molecular orbital theory, we say that the bond order is 1. **Bond order** is c half the difference between the number (no.) of bonding and antibonding e trons (e^-), that is,

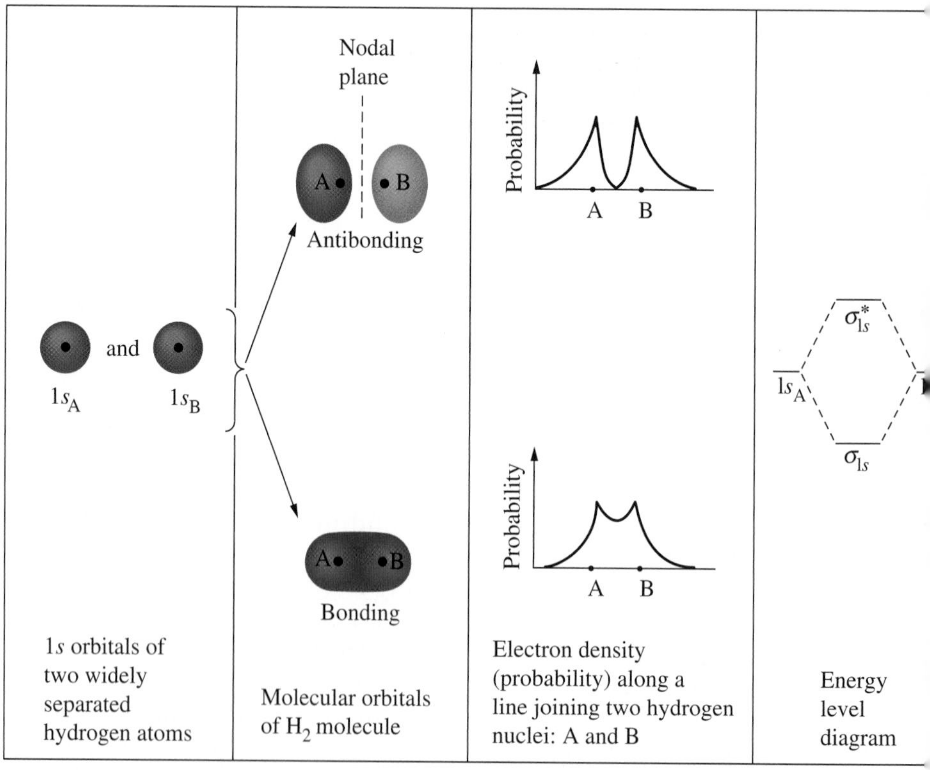

▲ FIGURE 11-22
The interaction of two hydrogen atoms according to molecular orbital theory
The energy of the bonding σ_{1s} molecular orbital is lower and that of the antibonding σ^*_{1s} molecular orbital is higher than the energies of the 1s atomic orbitals. In the bonding molecular orbital, electron density is concentrated in regions near to and between the t nuclei. In the antibonding orbital, electron density is low in the internuclear region and high in the regions outside the two nuclei. A node between the two nuclei is a defining feature of an antibonding orbital.

$$\text{bond order} = \frac{\text{no. of } e^- \text{ in bonding MOs} - \text{no. of } e^- \text{ in antibonding MOs}}{2} \quad \textbf{(11.1)}$$

KEEP IN MIND

that equation (11.1) is only valid for diatomic molecules.

atomic Molecules of the First-Period Elements

's use the ideas just outlined to describe some molecular species of the first-riod elements, H and He (Fig. 11-23).

H_2^+ This species has a single electron. It enters the σ_{1s} orbital, a bonding molecular orbital. Using equation (11.1), we see that the bond order is $(1 - 0)/2 = \frac{1}{2}$. This is equivalent to a one-electron, or *half*, bond, a bond type that is not easily described by the Lewis theory.

H_2 This molecule has two electrons, both in the σ_{1s} orbital. The bond order is $(2 - 0)/2 = 1$. With Lewis theory and the valence bond method, we describe the bond in H_2 as single covalent.

He_2^+ This ion has three electrons. Two electrons are in the σ_{1s} orbital, and one is in the σ^*_{1s} orbital. This species exists as a stable ion with a bond order of $(2 - 1)/2 = \frac{1}{2}$.

He_2 Two electrons are in the σ_{1s} orbital, and two are in the σ^*_{1s}. The bond order is $(2 - 2)/2 = 0$. No bond is produced—He_2 is not a stable species.

EXAMPLE 11-5 Relating Bond Energy and Bond Order

The bond energy of H_2 is 436 kJ/mol. Estimate the bond energies of H_2^+ and He_2^+.

Analyze

The strength of a bond is directly proportional to its bond order. If we double the bond order, we double the strength (approximately).

Solve

The bond order in H_2 is 1, equivalent to a single bond. In both H_2^+ and He_2^+, the bond order is $\frac{1}{2}$. We should expect the bonds in these two species to be only about half as strong as in H_2—about 220 kJ/mol.

Assess

The actual bond energies are 255 kJ/mol and 251 kJ/mol for H_2^+ and He_2^+, respectively, showing that our approximation is reasonable.

PRACTICE EXAMPLE A: The bond energy of Li_2 is 106 kJ/mol. Estimate the bond energy of Li_2^+.

PRACTICE EXAMPLE B: Do you think the ion H_2^- is stable? Explain.

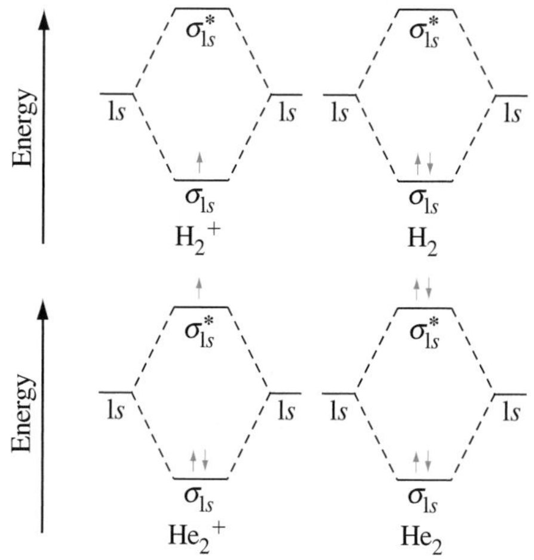

◀ FIGURE 11-23
Molecular orbital diagrams for the diatomic molecules and ions of the first-period elements
The 1s energy levels of the isolated atoms are shown to the left and right of each diagram. The line segments in the middle represent the molecular orbital energy levels—lower than the 1s levels for σ_{1s} and higher than the 1s levels for σ^*_{1s}.

11-5 CONCEPT ASSESSMENT

A ground state H_2 molecule can absorb electromagnetic radiation to form the ion H_2^+ or an excited state with an electron promoted to the σ^*_{1s} orbital. Which process requires the greater amount of energy? Which species is most stable?

Molecular Orbitals of the Second-Period Elements

For diatomic molecules and ions of H and He, we had to combine onl orbitals. In the second period, the situation is more interesting because must work with both 2*s* and 2*p* orbitals. This results in *eight* molecular orbi Let's see how this comes about.

The molecular orbitals formed by combining 2*s* atomic orbitals are simila those from 1*s* atomic orbitals, except they are at a higher energy. The situa for combining 2*p* atomic orbitals, however, is different. Two possible ways fo atomic orbitals to combine into molecular orbitals are shown in Figure 11 end-to-end and side-to-side. The best overlap for *p* orbitals is along a stra line (that is, end-to-end). This combination produces σ-type molecular orbi σ_{2p} and σ^*_{2p}. In forming the bonding and antibonding combinations along internuclear axis, we must take into account the phase of the 2*p* orbitals. We up the atomic orbitals as shown in Figure 11-24(a), with the positive (blue) l of each function pointing to the internuclear region. Then, since the wave fu tions are in phase, the addition of the two wave functions leads to an increas electron density in the internuclear region and produces a σ_{2p} orbital. When two atomic orbitals are set up as shown in Figure 11-24(b), with lobes of op site phase pointing into the internuclear region, a nodal plane midway betw the nuclei is formed, leading to an antibonding σ^*_{2p} orbital.

▶ Note that assigning blue to the positive lobe is quite arbitrary. The important aspect is that the orbitals are in phase to create a bonding orbital.

Only one pair of *p* orbitals can combine in an end-to-end fashion. The ot two pairs must combine in a parallel or side-to-side fashion to produce π-t molecular orbitals: π_{2p} and π^*_{2p}. The two possible ways for the side-to-side co bination of a pair of 2*p* orbitals are shown in Figure 11-24(c–f). The π_{2p} bo ing orbital (Fig. 11-24c) is formed by adding the *p* orbital on one nucleus to orbital on the other nucleus, in such a way that the positive and negative lo of one orbital are in phase with the positive and negative lobes of the oth orbital on the other nucleus. This produces additional electron den between the nuclei, but in a much less direct way than in the σ orbital beca the additional electron density is not found along the internuclear axis. π^*_{2p} antibonding orbital is formed by subtracting the two *p* orbitals perpend lar to the internuclear axes, as shown in Figure 11-24(d). Now, in addition to nodal plane that contains the nuclei, a node is formed between the nuclei, this is a characteristic of antibonding character. There are actually *four* π-t molecular orbitals (two bonding and two antibonding) because there are pairs of 2*p* atomic orbitals arranged in a parallel fashion.

The energy-level diagram for the molecular orbitals formed from ato orbitals of the second principal electronic shell is related to the atomic orb energy levels. For example, molecular orbitals formed from 2*s* orbitals are lower energy than those formed from 2*p* orbitals—the same relationship between the 2*s* and 2*p* atomic orbitals. Another expectation is that σ-t bonding orbitals should have lower energies than π-type because end-to- overlap of 2*p* orbitals should be more extensive than side-to-side overl resulting in a lower energy. This ordering is shown in Figure 11-25(a). In c structing this energy-level diagram, we have made the assumption th orbitals mix only with *s* orbitals and *p* orbitals mix only with *p* orbit However, if we use this assumption for some diatomic molecules, we make predictions that do not match experimental results.

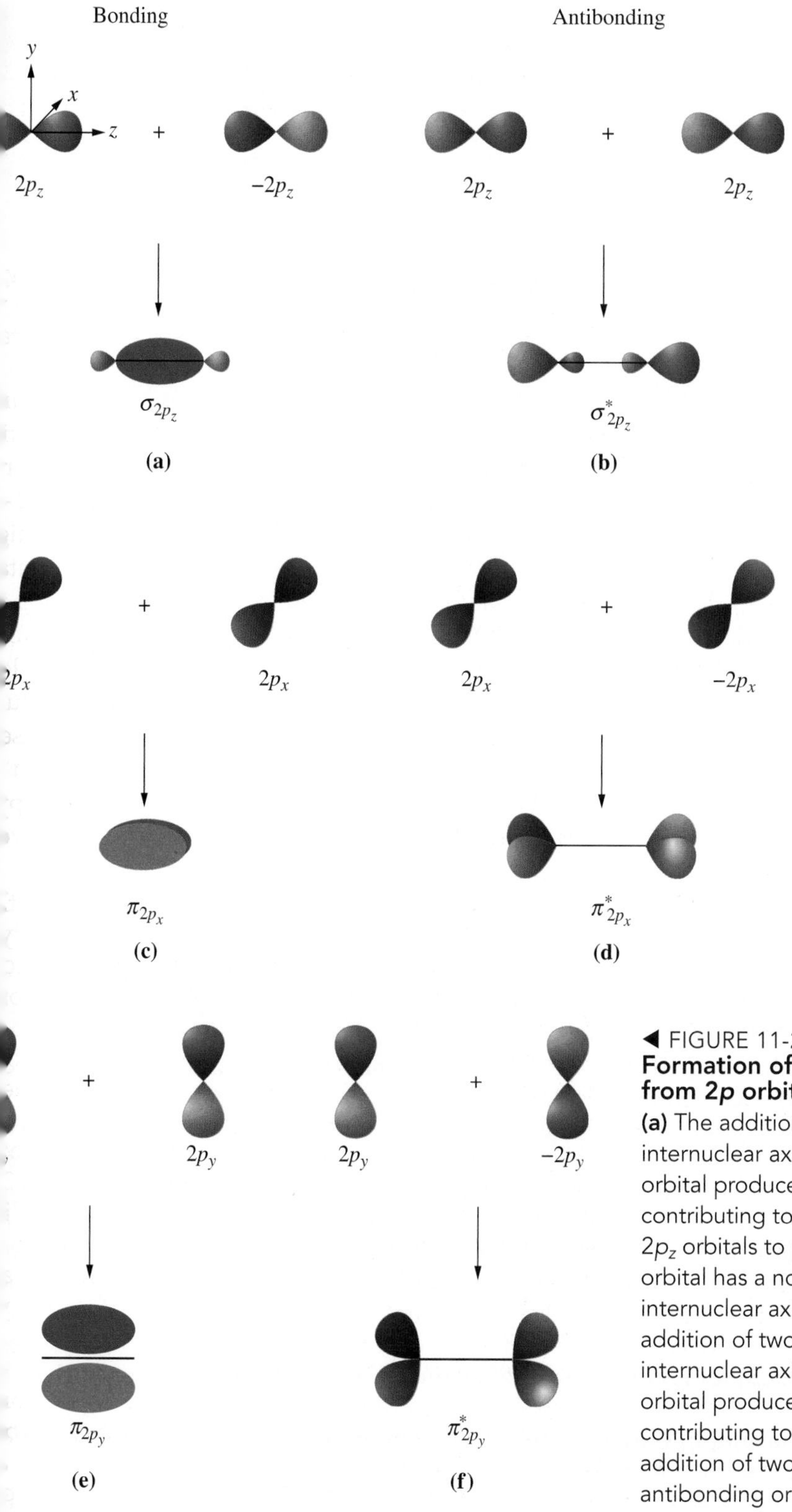

KEEP IN MIND

that the different colors of the orbitals depicted in these various figures represent the phases of the orbitals.

KEEP IN MIND

that subtracting two wave functions that are in phase is equivalent to *adding* the same functions when they are out of phase.

◀ FIGURE 11-24
Formation of bonding and antibonding orbitals from 2*p* orbitals
(a) The addition of two $2p_z$ orbitals along the internuclear axis to form a σ_{2p} molecular orbital. This orbital produces electron density between the nuclei, contributing to a chemical bond. **(b)** The addition of two $2p_z$ orbitals to produce an antibonding π^*_{2p} orbital. This orbital has a nodal plane perpendicular to the internuclear axis, as do all antibonding orbitals. **(c)** The addition of two $2p_x$ orbitals perpendicular to the internuclear axis to form a π_{2p} molecular orbital. This orbital produces electron density between the nuclei, contributing to a multiple chemical bond. **(d)** The addition of two $2p_x$ orbitals to produce a π^*_{2p} antibonding orbital with a nodal plane. **(e)** The addition of two $2p_y$ orbitals perpendicular to the internuclear axis to form another π_{2p} bonding molecular orbital. **(f)** The addition of two $2p_y$ orbitals perpendicular to the internuclear axis to form another π^*_{2p} antibonding molecular orbital.

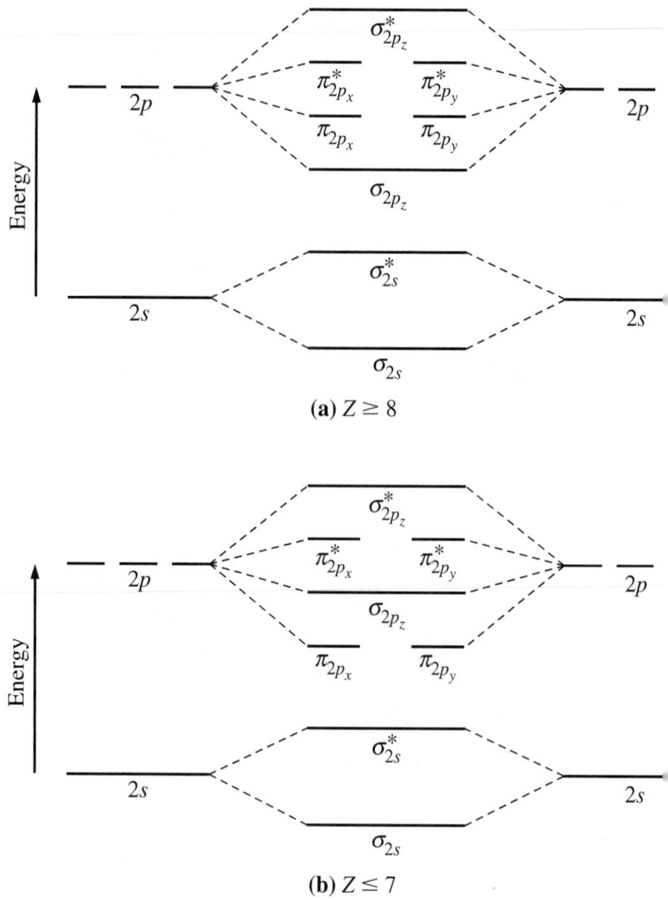

▶ FIGURE 11-25
The two possible molecular orbital energy-level schemes for diatomic molecules of the second-period elements
(a) The expected ordering when σ_{2p} lies below the π_{2p}. This is the ordering for elements with $Z \geq 8$. **(b)** The modified ordering due to *s* and *p* orbital mixing when σ_{2p} lies above the π_{2p} is the ordering for elements with $Z \leq 7$. In this figure we have assumed that the *z*-axis is the internuclear axis.

We need to take into account the fact that both the 2*s* and 2*p* orbitals fc molecular orbitals (σ_{2s} and σ_{2p}) that produce electron density in the sa region between the nuclei. These two σ orbitals are of such similar energy shape that they themselves mix to form modified σ orbitals. The modifie orbitals each contain a fraction of the original σ_{2s} and σ_{2p}. The modified (with some σ_{2p} mixed in) goes down in energy, and the modified σ_{2p} (w some σ_{2s} mixed in) goes up in energy, producing a different ordering energy levels. The important aspect of this mixing is that the modified σ_2 pushed up in energy above the π_{2p} orbitals (Fig. 11-25b).

For the molecular orbitals in O_2 and F_2, the situation is as expected beca the energy difference between the 2*s* and 2*p* orbitals is large, and little *s* an mixing takes place; that is, the σ_{2s} and σ_{2p} orbitals are *not* modified described above. For other diatomic molecules of the second-period eleme (for example, C_2 and N_2), the π_{2p} orbitals are at a lower energy than because the energy difference between the 2*s* and 2*p* orbitals is smaller, 2*s*–2*p* orbital interactions affect the way in which atomic orbitals combi This leads to the modified σ_{2s} and σ_{2p} orbitals described above.

Here is how we assign electrons to the molecular orbitals of the diato molecules of the second-period elements: We start with the σ_{1s} and σ^*_{1s} orbi filled. Then we add electrons, in order of increasing energy, to the availa molecular orbitals of the second principal shell. Figure 11-26 shows the el tron assignments for the homonuclear diatomic molecules of the seco period elements. Some molecular properties are also listed in the figure.

▶ A *homonuclear* diatomic molecule (X_2) is one in which both atoms are of the same kind. A *heteronuclear* diatomic molecule (XY) is one in which the two atoms are different.

Just as we might arrange the valence-shell atomic orbitals of an atom, can arrange the second-shell molecular orbitals of a diatomic molecule in order of increasing energy. Then we can assign electrons to these orbit thereby obtaining a *molecular orbital occupancy diagram*. If we assign the ei

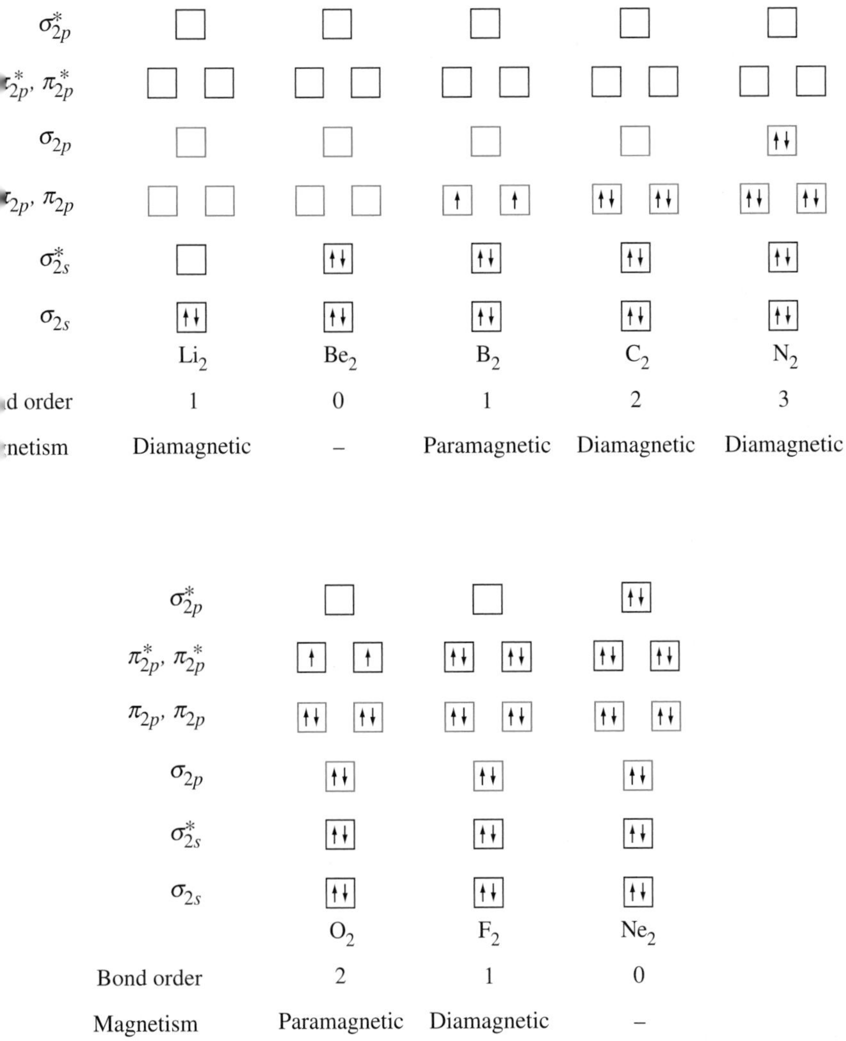

◀ FIGURE 11-26
Molecular orbital occupancy diagrams for the homonuclear diatomic molecules of the second-period elements
For O_2, F_2, and Ne_2, the σ_{2p} orbital (green box) is lower in energy than the π_{2p} orbitals (red boxes). In all cases, the σ_{1s} and σ^*_{1s} molecular orbitals are filled but not shown.

ence electrons of the molecule C_2 to the diagram in Figure 11-25(a), we ain

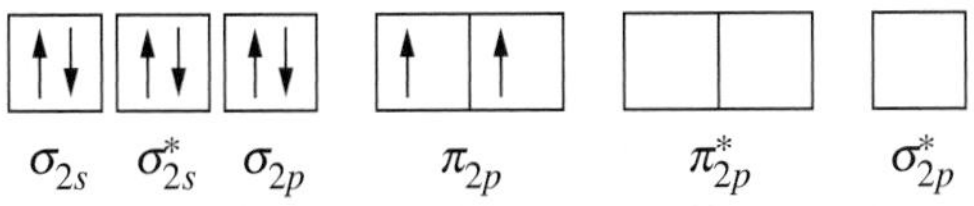

KEEP IN MIND

that Hund's rule applies to molecules as well as atoms.

Experiment shows that the C_2 molecule is diamagnetic, not paramagnetic, l the configuration just described is *incorrect*. So here we see the importance he modified energy-level diagram in Figure 11-25(b). Assignment of eight ctrons to the following molecular orbital occupancy diagram is consistent h the observation that C_2 is diamagnetic.

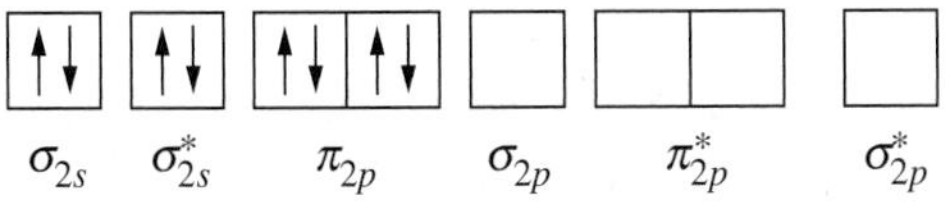

is modified energy-level diagram is used for homonuclear diatomic mole-es involving elements with atomic numbers from three through seven.

◀ Instead of writing the molecular orbital occupancy diagram, we can write the ground state configuration of O_2 as $\sigma^2_{2s}\sigma^{*2}_{2s}\sigma^2_{2p}\pi^4_{2p}\pi^{*2}_{2p}$. When writing the electron configuration for a molecule, we follow the convention we used when writing the electron configuration for an atom. That is, we write the orbital designation by placing the number of electrons as a superscript.

Special Look at O_2

lecular orbital theory helps us understand some of the previously unex-ined features of the O_2 molecule. Each O atom brings six valence electrons the diatomic molecule, O_2. In the molecular orbital occupancy diagram below, we see that when 12 valence electrons are assigned to molecul orbitals, the molecule has *two unpaired electrons*. This explains the paramagn ism of O_2 (see page 417).

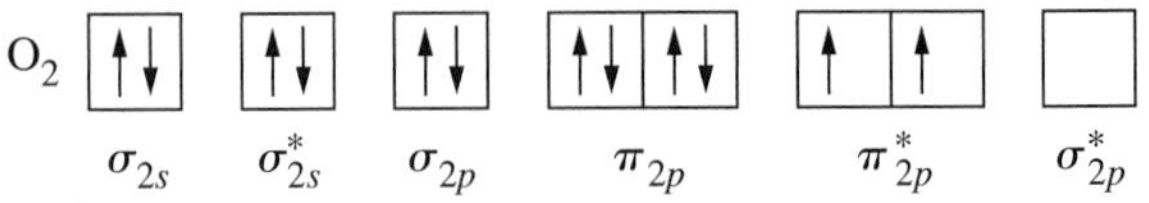

(11

There are eight valence electrons in bonding orbitals and four in antibondi orbitals so the bond order is two. A bond order of two corresponds to a co lent double bond.

EXAMPLE 11-6 Writing a Molecular Orbital Occupancy Diagram and Determining Bond Order

Represent bonding in O_2^+ with a molecular orbital occupancy diagram, and determine the bond order in this ion.

Analyze

The O_2^+ ion has 11 valence electrons. We can assign these to the available molecular orbitals in accordance with the ideas stated on page 488. Alternatively, we can remove one electron from an orbital in the molecular orbital diagram given in equation (11.2).

Solve

In the following diagram, there is an excess of *five* bonding electrons over antibonding ones. The bond order is 2.5.

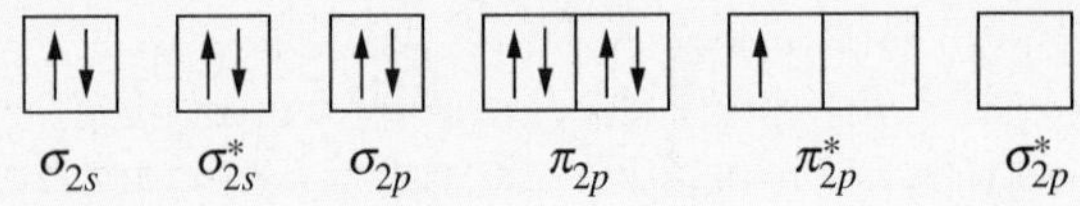

Assess

The electronic configuration of O_2^+ is $\sigma_{2s}^2\sigma_{2s}^{*2}\sigma_{2p}^2\pi_{2p}^4\pi_{2p}^{*1}$. Note that in this diagram the molecule, the molecular orbital occupancy diagram without $2s$–$2p$ mixing is used, because the $2s$–$2p$ separation in oxygen is large.

PRACTICE EXAMPLE A: Refer to Figure 11-26. Write a molecular orbital occupancy diagram, determine the bond order, and write the electronic configurations of **(a)** N_2^+; **(b)** Ne_2^+; **(c)** C_2^{2-}.

PRACTICE EXAMPLE B: The bond lengths for O_2^+, O_2, O_2^-, and O_2^{2-} are 112, 121, 128, and 149 pm, respectively. Are these bond lengths consistent with the bond order determined from the molecular orbital occupancy diagram? Explain.

11-6 CONCEPT ASSESSMENT

In valence bond theory, π bonds are always accompanied by a σ bond. Can the same also be said of the molecular orbital theory for diatomic molecules?

A Look at Heteronuclear Diatomic Molecules

The ideas that we developed for homonuclear diatomic species can extended, with some care, to give us an idea of the bonding in heteronucl diatomic species. To illustrate how to construct a molecular orbital diagr for a heteronuclear diatomic species, let us consider the molecule carb monoxide, CO. Our first consideration is the relative placement of the $2s$ a $2p$ orbitals of C and O. As suggested by Figure 11-27, the energy of a gi orbital decreases as the atomic number increases. Therefore, the $2s$ orbita the O atom must be placed at a lower energy than the $2s$ orbital of Similarly, the $2p$ orbitals of O must be placed at a lower energy than thos C. Our next consideration is to decide whether or not the σ molecular orbi

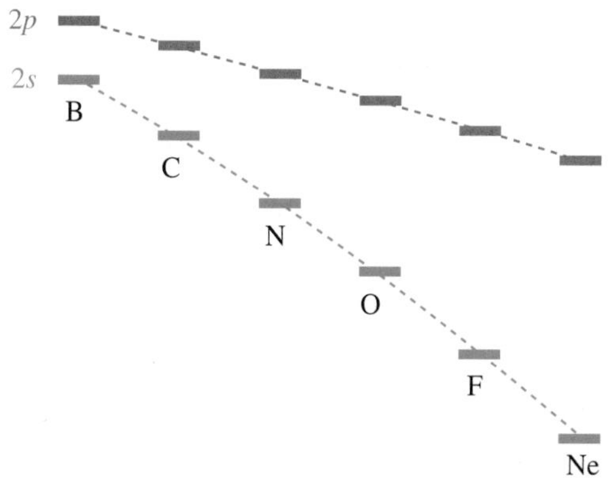

◀ FIGURE 11-27
The relative energies of 2*s* and 2*p* orbitals of some second-period elements
The 2*s* and 2*p* orbitals decrease in energy from left to right across a period because of an increase in the effective nuclear charge. The energy separation between the 2*s* and 2*p* orbitals also increases from left to right across the period.

CO should be formulated by considering only the combinations of C(2*s*) th O(2*s*) and C(2*p*) with O(2*p*). Figure 11-27 can help us decide: In terms of ergy, the 2*p* orbital of O is approximately halfway between the 2*s* and 2*p* itals on C. Thus, the combination of O(2*p*) with C(2*s*) is probably as impor- t as the combination of O(2*p*) with C(2*p*). In other words, we expect the σ itals of CO to be mixtures of 2*s* and 2*p* orbitals from both C and O and, nsequently, the energy ordering of the molecular orbitals of CO will be sim- r to that shown in Figure 11-25(b). (Recall that the energy ordering in ure 11-25(b) is appropriate whenever the 2*s* orbital of one atom is close in ergy to the 2*p* orbital of the other.)

A molecular orbital diagram for the CO molecule is shown in Figure 11-28. e molecular orbitals are labeled by using a simplified notation in which the nbol σ or π is preceded by an integer to indicate the relative energy ordering he orbitals of each type. Thus, the σ orbital of lowest energy is labeled 1σ, the t lowest 2σ, and so on. Similarly, the π orbitals of lowest energy are labeled the next lowest 2π, and so on. The energy ordering of the molecular orbitals, $< 2\sigma < 3\sigma < 4\sigma < 1\pi < 3\sigma < 2\pi < 4\sigma$, is not totally unexpected; it is

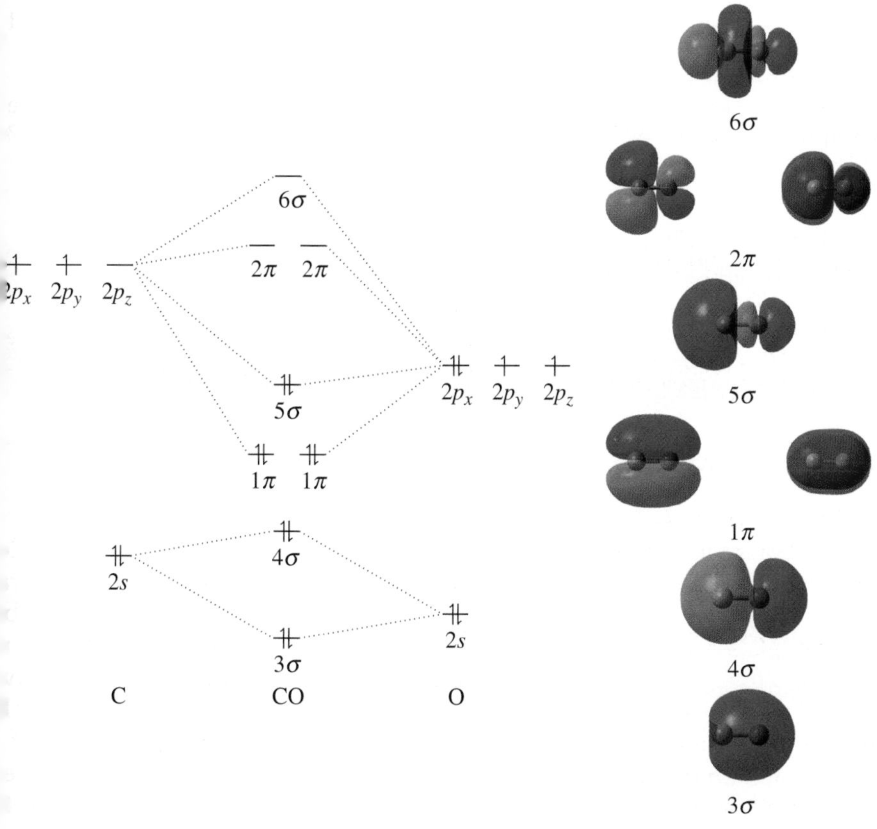

◀ FIGURE 11-28
The molecular orbital diagram of CO
Notice that the CO molecular orbitals are ordered in the same manner as the molecular orbitals in Figure 11-25(b). The 1π orbitals (bonding) and 2π orbitals (antibonding) are mixtures of 2*p* orbitals from C and O. The 3σ (bonding), 4σ (nonbonding), 5σ (nonbonding), and 6σ (antibonding) orbitals are mixtures of 2*s* and 2*p* orbitals from both C and O. The 1σ and 2σ orbitals which are mixtures of the 1*s* orbitals from C and O are not shown.

consistent with the order presented earlier (Fig. 11-25b) when the 2*s* orbit one atom is close in energy to the 2*p* orbital of the other.

According to Figure 11-28, the configuration of CO is $3\sigma^2\,4\sigma^2\,1\pi^4\,5\sigma^4$, v the 5σ orbital the *highest occupied molecular orbital* (HOMO). Before we can culate the bond order, we must first classify the occupied molecular orbital bonding, antibonding, or nonbonding. The 1π orbitals are bonding whe the 2π orbitals are antibonding. This is established by recalling that th orbitals are mixtures of 2*p* orbitals from C and O, and observing that the and 2π orbitals are, respectively, lower and higher in energy than th orbitals of C and O. The classification of the σ orbitals is a little trickier. As have already established, all the σ orbitals are mixtures of 2*s* and 2*p* from b C and O. Because the 3σ and 6σ orbitals are, respectively, lower and highe energy than the 2*s* and 2*p* orbitals of C and O, we may tentatively classify as bonding and 6σ as antibonding. What about the 4σ and 5σ orbitals? observe that the energy of the 4σ orbital places it approximately half between the 2*s* and 2*p* orbitals on O, whereas the 5σ orbital is approxima halfway between the 2*s* and 2*p* orbitals on C. It seems reasonable to classify as a nonbonding 2*s*–2*p* hybrid orbital localized on O and 5σ as a nonbonc 2*s*–2*p* hybrid orbital localized on C.

The tentative classification of the σ orbitals given above is supportec both quantum mechanical calculations* on the CO molecule and experime data. With the assumption that the 4σ and 5σ orbitals are nonbonding, bonding in CO comes from the two electrons in the 3σ orbital and the two e tron pairs in the degenerate 1π orbitals, producing an effective bond order

▶ For a discussion of the orbital structure of CO, see the Focus On feature for Chapter 11, Photoelectron Spectroscopy, on the MasteringChemistry website, www.masteringchemistry.com.

Now, let us use the molecular orbital diagram for CO to predict the b order for the free radical, NO, which has just one more electron than When adding electrons to the molecular orbitals of Figure 11-28, the last e tron of the NO molecule goes into a 2π orbital. Thus, the configuration is

$$3\sigma^2 4\sigma^2 1\pi^4 5\sigma^2 2\pi^1$$

The bond order becomes 2.5. We predict the bond energy in NO to be than in CO, as is observed experimentally.

EXAMPLE 11-7 Writing Electron Configurations for Heteronuclear Diatomic Species

Write the ground state electron configuration for the cyanide ion, CN^-, and determine the bond order for thi ion. Clearly state any assumptions.

Analyze

The ion is isoelectronic with CO. Because both C and N have a small 2*s*−2*p* energy gap, we assume tha 2*s*−2*p* mixing will be important for CN^- and that the 5σ orbital is higher in energy than the 1π orbitals.

Solve

The number of valence electrons to be assigned to the molecular orbitals is $4 + 5 + 1 = 10$. Because of th 2*s*−2*p* mixing, we use the modified order of molecular orbitals and write the configuration of CN^- as

$$3\sigma^2 4\sigma^2 1\pi^4 5\sigma^2$$

With the assumption that 4σ and 5σ are nonbonding, as they are for CO, the bond order is $(2 + 4)/2 = 3$.

Assess

As expected the bond order in CN^- is 3, as it is in the isoelectronic molecule CO. In addition the Lewis struc ture also gives a triple bond.

PRACTICE EXAMPLE A: Write the electron configuration for CN^+, and determine the bond order.

PRACTICE EXAMPLE B: Write the electron configuration for BN, and determine the bond order.

*The 2σ orbital is best described as a weakly bonding orbital localized on O whereas 3σ weakly antibonding orbital localized on C. See Y. Liu et. al., *J. Chem. Ed.*, **89**, 355 (2012).

11-7 CONCEPT ASSESSMENT

'ould you expect NeO to be a stable molecule?

-6 Delocalized Electrons: An Explanation Based on Molecular Orbital Theory

ection 11-4, we discussed localized π bonds, such as those in ethylene, C_2H_4. ne molecules, such as benzene (C_6H_6) and substances related to it—aromatic npounds—have a network of π bonds that extends over several nuclei. In section, we will combine ideas from the bonding theories we have studied s far to consider bonding in benzene and a few other polyatomic molecules. will describe the bonding in these molecules by using valence bond theory the σ bonds and molecular orbital theory for the π bonds.

n general, the application of molecular orbital theory to polyatomic mole- :s is rather complicated. This is because the molecular orbitals (more pre- ly, the combinations of atomic orbitals contributing to the various molecu- orbitals) are not easily deduced. Thus, we will not attempt to deduce them. tead, we will simply present the molecular orbitals as needed and use them ielp us understand the bonding in each case.

◀ The term *aromatic* relates to the fragrant aromas associated with some (but by no means all) of these compounds.

▲ Dame Kathleen Lonsdale first determined the X-ray crystal structure of benzene. Her experiment demonstrated that the benzene molecule is flat, as predicted by theorists.

nding in Benzene

865, Friedrich Kekulé advanced the first good proposal for the structure of zene. He suggested that the C_6H_6 molecule consists of a flat, hexagonal ring six carbon atoms joined by alternating single and double covalent bonds. h C atom is joined to two other C atoms and to one H atom. To explain the that the carbon-to-carbon bonds are all alike, Kekulé suggested that the sin- and double bonds continually oscillate from one position to the other. Today, say that the two possible Kekulé structures are actually contributing struc- :s to a resonance hybrid. This view is suggested by Figure 11-29.

Ve can gain a more thorough understanding of bonding in the benzene lecule by combining the valence bond and molecular orbital methods. A ond framework for the observed planar structure can be constructed with ° bond angles by using sp^2 hybridization at each carbon atom. End-to-end rlap of the sp^2 orbitals produces σ bonds. The six remaining $2p$ orbitals are d to construct the delocalized π bonds. Figure 11-30 gives a valence bond ory representation of bonding in C_6H_6.

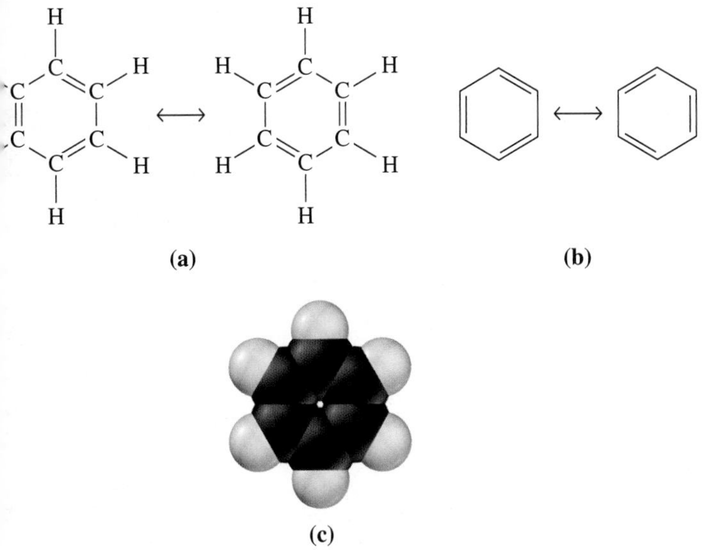

◀ FIGURE 11-29
Resonance in the benzene molecule and the Kekulé structures
(a) Two equivalent Kekulé structures for benzene, C_6H_6, each showing alternate carbon-to-carbon single and double bonds. These two resonance structures contribute equally to the true structure. **(b)** A simplified representation of the two Kekulé structures. A carbon atom is at each corner of the hexagonal structure, and a hydrogen atom is bonded to each carbon atom. (The symbols for carbon and hydrogen, as well as the C—H bonds, are customarily omitted in these structures.) **(c)** A space-filling model.

▶ FIGURE 11-30
Bonding in benzene, C_6H_6, by the valence-bond method
(a) Carbon atoms use sp^2 and p orbitals. Each carbon atom forms three σ bonds, two with neighboring C atoms in the hexagonal ring and a third with a H atom. **(b)** The unhybridized $2p$ orbitals on the carbon atoms of benzene, which produce the delocalized π bonds in benzene. **(c)** Because the π bonding is delocalized around the benzene ring, the molecule is often represented by a hexagon with an inscribed circle.

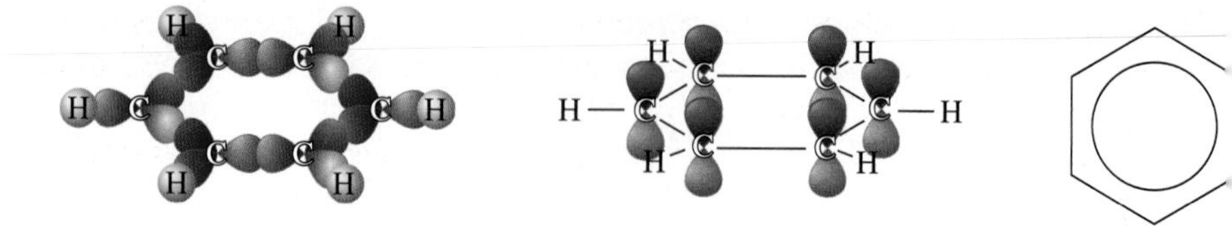

(a) σ bond framework **(b)** Carbon $2p$ orbitals to be used in π bonding **(c)** Symboli representatic

We do not need to think in terms of an oscillation between two structu (Kekulé) or of a resonance hybrid for benzene. The π bonds are not locali between specific carbon atoms but are spread out around the six-membe ring. To represent this *delocalized* π bonding, the symbol for benzene is o written as a hexagon with an inscribed circle (Fig. 11-30c).

We can best understand the delocalized π bonds through molecular orb theory, in which we imagine replacing the six $2p$ atomic orbitals of C atoms (see Figure 11-30b) with six molecular orbitals of the π type. combinations of atomic orbitals, and the resulting molecular orbitals, shown in Figure 11-31. Three of the π molecular orbitals are bonding, three are antibonding. The π orbital of lowest in energy has all six $2p$ orbi in phase (all the blue lobes are on the same side of the σ framework) and c sequently, there are no nodes between adjacent carbon atoms. The next π-bonding molecular orbitals have the same energy; that is, they are ener ically degenerate (Fig. 11-31b). They each have one node, which is rep sented by a dashed line in Figure 11-31(a). That these two orbitals are hig in energy than the one having no nodes should not come as a surp because, as we have already seen, the energy of an orbital increases with number of nodes. The next pair of orbitals, which are antibonding π orbi have two nodes, and the final orbital has three nodes (Fig. 11-31a). The co puted π molecular orbitals of benzene are also included in Figure 11-31(c).

The three bonding orbitals fill with six electrons (one $2p$ electron from eac atom), and the three antibonding orbitals remain empty. The bond order ass ated with the six electrons in π-bonding molecular orbitals is $(6 - 0)/2 =$

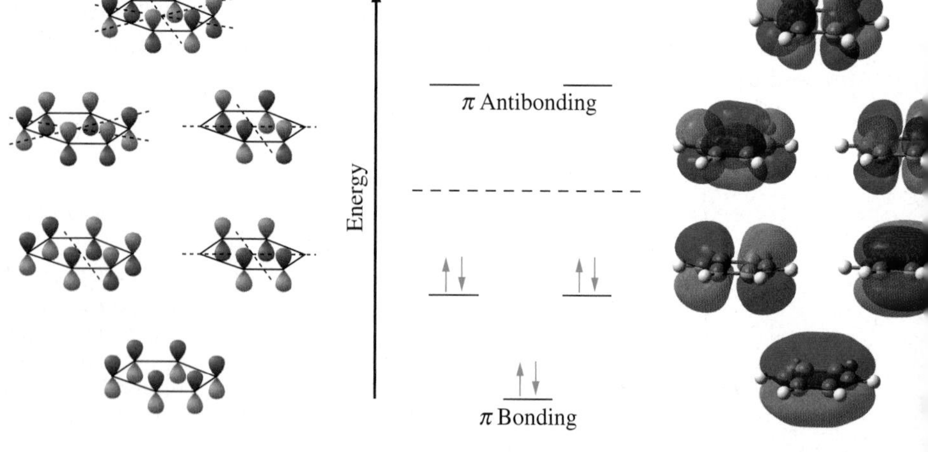

π orbitals of benzene Computed π orbitals

(a) (b) (c)

▲ FIGURE 11-31
π molecular orbital diagram for C_6H_6
Of the six π molecular orbitals, three are bonding orbitals and each of these is filled with an electron pair. The three antibonding molecular orbitals at higher energy remain empty.

three bonds are distributed among the six C atoms, which amounts to 3/6, half-bond, between each pair of C atoms. Add to this the σ bonds in the ond framework, and we have a bond order of 1.5 for each carbon-to-carbon d. This is exactly what we also get by averaging the two Kekulé structures of ıre 11-29.

'he π molecular orbitals shown in Figure 11-31 are spread out among all six toms instead of being shared between pairs of C atoms. For this reason, se molecular orbitals are called **delocalized molecular orbitals**. Electrons upying the three bonding π molecular orbitals create two donut-shaped ions of electron density in C_6H_6: one above and one below the plane of the nd H atoms.

her Structures with Delocalized Molecular Orbitals

using delocalized bonding schemes, we can avoid writing two or more conuting structures to a resonance hybrid, as is so often required in the Lewis ory. Consider the ozone molecule, O_3, that we used to introduce the cept of resonance in Section 10-5. In place of the resonance hybrid based on se contributing structures,

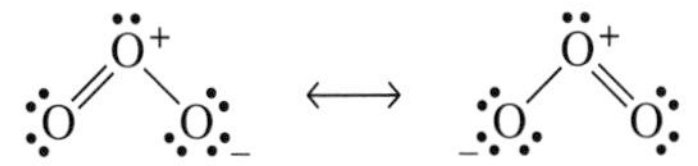

can write the single structure shown in Figure 11-32. Here are the ideas t lead to Figure 11-32.

With VSEPR theory, we predict a trigonal-planar electron-group geometry (the measured bond angle is 117°). The hybridization scheme chosen for the central O atom is sp^2, and although we normally do not need to invoke hybridization for terminal atoms, this case is simplified if we assume sp^2 hybridization for the terminal O atoms as well. Thus, each O atom uses the orbital set $sp^2 + p$.

Of the 18 valence electrons in O_3, assign 14 to the sp^2 hybrid orbitals of the σ-bond framework. Four of these are bonding electrons (red) and ten are lone-pair electrons (blue).

The three unhybridized 2*p* orbitals are replaced by *three* molecular orbitals of the π type (Fig. 11-33). One of these orbitals is a bonding

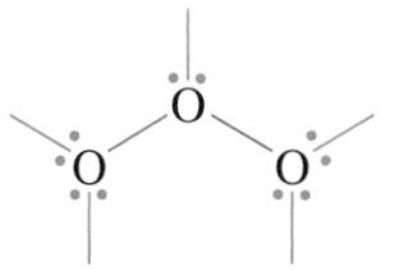

(a) σ bond framework

(b) Delocalized π molecular orbital

▲ FIGURE 11-32
Structure of the ozone molecule, O_3
(a) The σ-bond framework and the assignment of bond-pair (|) and lone-pair (:) electrons to sp^2 hybrid orbitals are discussed in points 1 and 2. **(b)** The π molecular orbitals and assignments of electrons to them are discussed in points 3 and 4.

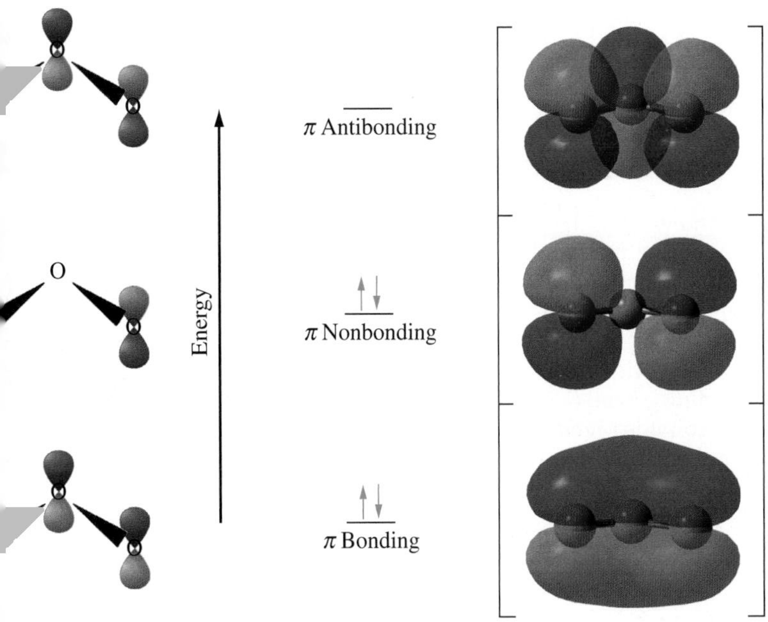

◀ FIGURE 11-33
π bonding orbitals of the ozone molecule O_3
The π-bonding molecular orbital has all the 2*p* orbitals in phase. The π-antibonding molecular orbital has all the three 2*p* orbitals out of phase. The π-nonbonding molecular orbital has a single node and makes zero contribution to the wave function from the central atom. This orbital is called *nonbonding* because there is no region of electron density between the central atom and its neighbors. The nonbonding orbital is at the same energy as the original 2*p* orbitals on the oxygen atoms, whereas the π-bonding molecular orbital is stabilized with respect to the original orbitals and the π-antibonding molecular orbital is destabilized by an equal amount with respect to the original 2*p* orbitals. The energy level diagram is shown in the middle of this figure, with the computed π molecular orbitals shown on the right.

molecular orbital, and the second is antibonding. The third is a *nonbonı molecular orbital*. A nonbonding molecular orbital has the same energy as atomic orbitals from which it is formed, and it neither adds to nor detr from bond formation.

4. The remaining four valence electrons are assigned to the π molecu orbitals. Two go into the bonding orbital and two into the nonbond orbital. The antibonding orbital remains empty.
5. The bond order associated with the π molecular orbitals is $(2 - 0)/2$ = This π bond is distributed between the two O—O bonds and amount one-half of a π bond for each.

The points listed here lead to a total bond order of 1.5 for the O—O bonds in This is equivalent to averaging the two Lewis structures. The O—O bo length suggested by this method was described in Section 10-5.

EXAMPLE 11-8 Using Delocalized Molecular Orbitals to Describe Bonding in NO_3^-

For the NO_3^- ion, one molecular orbital in the π system is bonding and one is antibonding. How many non bonding π orbitals are there? Which of the π molecular orbitals are occupied? What is the predicted bonc order for the nitrogen-to-oxygen bonds?

Analyze

First, we focus on electrons associated with the σ bond framework and determine hybridization schemes fo the atoms. Then, we determine the number of electrons in the π system, and the number of 2*p* orbitals involved forming the π molecular orbitals. The key to solving this problem is reasoning out the number of each type of molecular orbital (bonding, antibonding, nonbonding) in the π system. The final step is to assign electrons in the π system to the appropriate molecular orbitals.

Solve

The total number of electrons in the NO_3^- ion is $(3 \times 6 + 5 + 1) = 24$. Resonance structures for the NO_3^- ion and the corresponding σ bond framework are shown below. Because the N atom is bonded to three atoms, it i sp^2-hybridized. Each O atom may also be considered to be sp^2-hybridized, with one of its sp^2 hybrid orbital used to form a σ bond with N and other two accommodating the lone pairs.

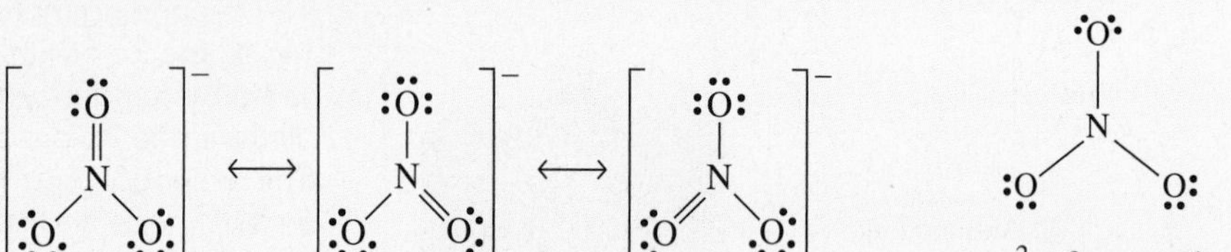

sp^2 σ framework

The number of electrons in the σ bond framework is 18, and so the number of electrons in the π system is $24 - 18 = 6$. These 6 electrons must be assigned to π molecular orbitals obtained from combining four 2*p* orbitals (one on each atom). We must now reason out the number of nonbonding molecular orbitals there are ir the π system.

Starting from four 2*p* orbitals, we must obtain a total of four π molecular orbitals. From the informatior given in the problem statement, we know that one of the molecular orbitals is a bonding orbital (all the 2*p* orbitals in phase) and another is an antibonding orbital (all the 2*p* orbitals out of phase). The π-bonding orbita is lower in energy and the π-antibonding orbital is higher in energy than the 2*p* orbitals. Thus, there must be $4 - 2 = 2$ *nonbonding* molecular orbitals in the π system. These nonbonding orbitals are equal in energy to the

$2p$ orbitals, and thus higher in energy than the bonding molecular orbital but lower in energy than the antibonding molecular orbital. The molecular orbital occupancy diagram below illustrates how the six π electrons are assigned to the π molecular orbitals.

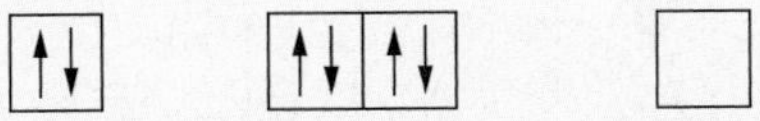

π Bonding π Nonbonding π Antibonding

Delocalized π bonding molecular orbital of NO_3^-

Thus, the overall π bond order in the NO_3^- ion is one, and as suggested by the diagram, the π bond is spread equally over three nitrogen–oxygen bonds. Each nitrogen-to-oxygen bond consists of one σ bond and one-third of a π bond, and so the bond order of each nitrogen-to-oxygen bond is $1\frac{1}{3} = 1.33$.

Assess

To solve this problem, we used the following important properties of molecular orbitals: (1) the number of molecular orbitals in the π network equals the number of $2p$ atomic orbitals contributing to it; and (2) for a set of molecular orbitals arising from the same set of atomic orbitals, the energy ordering of the orbitals is typically (from lowest to highest energy): bonding < nonbonding < antibonding. Are You Wondering 11-3 provides schematic representations of the π molecular orbitals of NO_3^-.

PRACTICE EXAMPLE A: Represent chemical bonding in the molecule SO_3 by using a combination of localized and delocalized orbitals.

PRACTICE EXAMPLE B: Represent chemical bonding in the ion NO_2^- by using a combination of localized and delocalized orbitals.

11-8 CONCEPT ASSESSMENT

Would you expect the delocalized π-bonding framework in HCO_2^- to be similar to that in ozone or to that in the nitrate anion?

11-3 ARE YOU WONDERING?

What do the π molecular orbitals of NO_3^- look like?

The construction of these orbitals is shown in Figure 11-34. The π-bonding molecular orbital has the four $2p$ orbitals (one on the N atom and one each on the three oxygen atoms) all in phase. The corresponding antibonding π orbital has two nodes and is at the highest energy. Since we are combining four $2p$ orbitals we expect to get four molecular orbitals. The remaining two molecular orbitals will each have one node. The only way to create a molecular orbital with one node in NO_3^- is for the molecular orbital to have a node at the nitrogen atom. Such a molecular orbital must be nonbonding with respect to nitrogen. That the nitrogen does not contribute to the degenerate nonbonding orbitals can be clearly seen in the computed molecular orbitals depicted in Figure 11-34.

(continued)

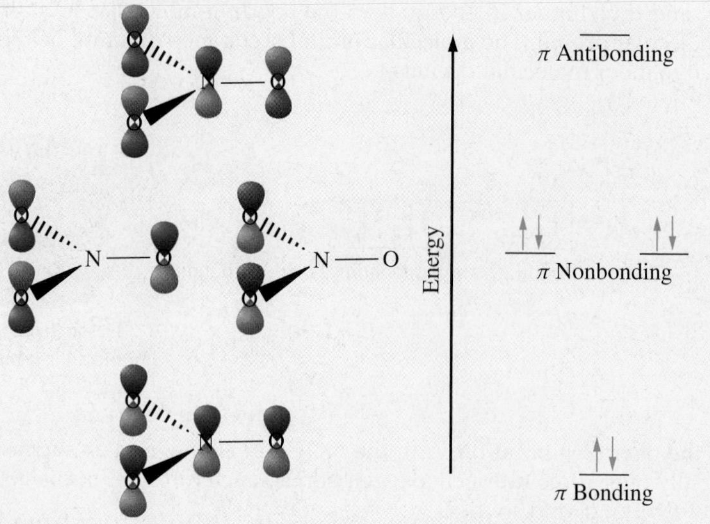

▲ FIGURE 11-34
π bonding orbitals of the nitrate anion, NO_3^-
The π-bonding molecular orbital has all the *p* orbitals in phase, whereas the π-antibonding orbital has all the *p* orbitals out of phase. Each of the nonbonding orbitals has a node at the nitrogen atom.

To illustrate the importance of molecular orbital theory, we will see how it is used to explain the colors of plants. Two pigment molecules typically isolated from vegetables are β-carotene

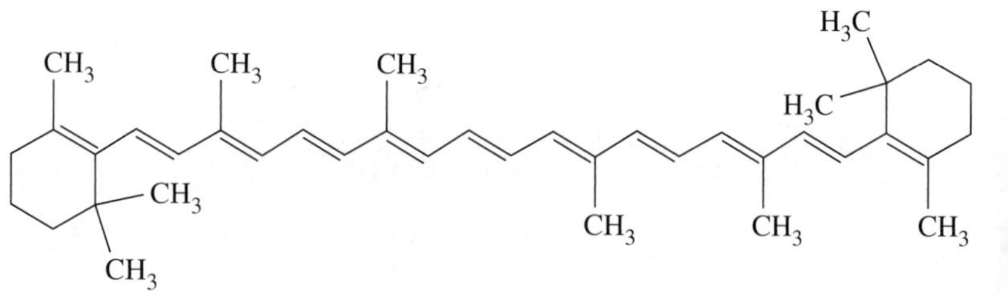

β-Carotene, $C_{40}H_{56}$

found in carrots and leaves, and lycopene

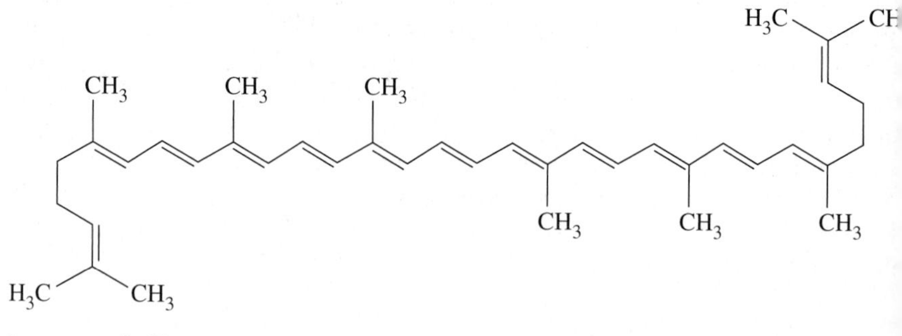

Lycopene, $C_{40}H_{58}$

which is present in tomatoes. The common feature of these molecules is the contiguous π system. The many *p* orbitals on the trigonal-planar carbon atoms contribute to many π molecular orbitals. As a consequence, the molecules have many π molecular energy levels that become very closely spaced (see Figure 11-35).

In molecules with long extended π systems, the *highest occupied molecular orbital* (HOMO) is very close in energy to the *lowest unoccupied molecular orbital*

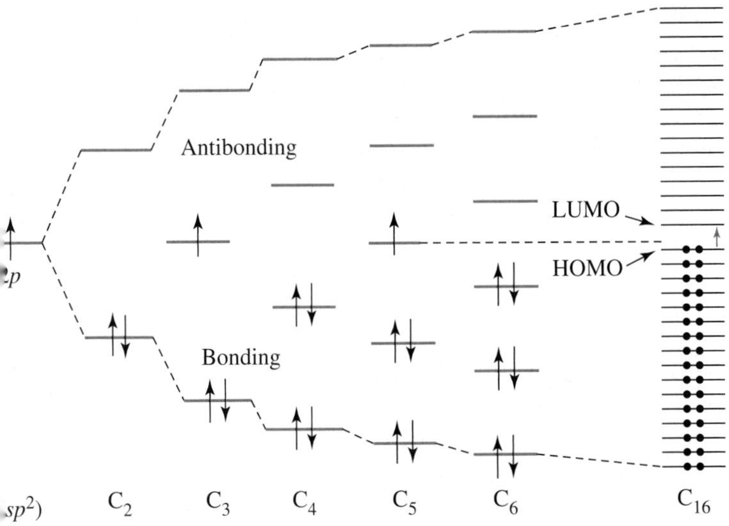

◀ FIGURE 11-35
The formation of π molecular orbitals in a long-chain polyene
The formation of such an extended π system requires an alternation of double and single bonds, as occurs in such molecules as carotene. There are many closely spaced energy levels, and the HOMO–LUMO gap is quite small. Molecules with such an extended π system are often colored because photons of visible light can excite electrons from the HOMO to the LUMO.

'MO). As a result it takes very little energy to excite an electron from the MO to the LUMO (Fig. 11-35). Photons of visible light have enough energy to te the electrons across the energy gap between the HOMO and the LUMO, the absorption of these photons is responsible for the colors that we see.

Ideas introduced in this chapter can be used to describe bonding in metals and to account for some of their properties. See the Appendix to Chapter 11, Bonding in Metals, on the MasteringChemistry site (www.masteringchemistry.com).

MasteringCHEMISTRY

-7 Some Unresolved Issues: Can Electron Density Plots Help?

Chapters 10 and 11 we have presented a wide range of views of chemical ding, from simple Lewis theory to the more advanced valence bond and mo- ılar orbital approaches. We must emphasize, however, that each of these mod- has its deficiencies, and their uncritical use can lead to incorrect conclusions. Here are some of the unresolved issues we have encountered: Employing anded valence shells in Lewis structures created the quandary of where to ommodate the extra valence electrons in such molecules as SF_4 and SF_6 ge 435), specifically, are *d*-orbitals used in the bonding description? A ated issue is whether to use expanded valence shells to minimize formal rges in anions such as SO_4^{2-} (page 436). Still another issue is whether EPR theory or hybridization schemes of the valence bond method gives the re fundamental view of molecular shapes. Finally, we might wonder how valence bond and molecular orbital theories are related. In this section, we l attempt to provide answers to these questions—stressing the significance electron density calculations.

nding in the Molecule SF_6

st let's see if it is possible to describe the bonding in SF_6 while maintaining the et rule. One proposal has been to introduce resonance structures of the form

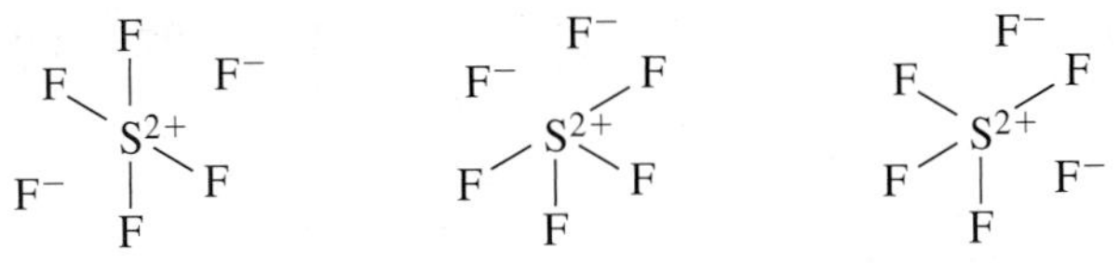

with the actual electronic structure being a combination of these resona structures. The resonance structures illustrate the concept of *hyperconjuga* a situation in which the number of electron pairs used to bond other atom a central atom is less than the number of bonds formed. In the case of SF_6, electron pairs bond six F atoms to a central S atom and the octet rule is served. When resonance structures are written in this way, the assumptio that the bond lines represent covalent bonds, whereas the other bonds fully ionic. The molecule SF_6 has equivalent bonds and consequently wc require a total of 15 structures of the type shown. This description impli charge of +2 on the sulfur and a charge of −1/3 on each F atom, correspc ing to a collective charge of −2 on the six F atoms. This description of bond then, has the appeal of describing the polarity of the bonds while get around the problem that arises when the Lewis structure is written as shc below, namely, where do the "extra" electrons go on the S atom?

Should we use hyperconjugation to describe the bonding in SF_6? (answer is to compare the suggested charges on the S and F atoms with th obtained from a quantum-mechanical calculation. The calculation giv charge of +3.17 on sulfur and −0.53 on each fluorine. To describe bond through hyperconjugation that is in better agreement with the quant mechanical calculation, we would have to use additional resonance structu with higher charges and fewer covalent bonds. Such an approach is cle cumbersome, and adoption of this large number of structures is not justi just to satisfy the octet rule.

The problem in describing molecules with expanded valence shells, so-ca *hypervalent* molecules, is that there is no generally accepted way of denot polar bonds in a structure. Furthermore, we must remember that Lewis devi his "rule of eight" in an era when only a few molecules, such as PCl_5 and were known, and he did not consider these exceptions to be of any great sig icance. Why was that? Lewis viewed the "rule of two" to be of greater fun mental importance than the rule of eight; that is, the electron density caused the electron pair is paramount in understanding bonding. Bonding in molec with expanded valence shells is not a consequence of a special type of bondi Bonds in these molecules are similar to those in other molecules and can v from predominantly covalent to predominantly ionic.

If we do not use hyperconjugation, how are we to describe where the "ex electrons go? This question arises because we seem implicitly to think ab bonding in terms of hybridization of atomic orbitals. Thus, in order to desc bonding in the methane molecule consistent with its tetrahedral geometry introduced the concept of sp^3 hybrid orbitals. That is, the hybridization introduced *because* of the geometry. The geometry of a molecule can be de mined only experimentally or estimated by using VSEPR theory.

We extended the concept of hybridization of orbitals to molecules w expanded valence shells to include *d* orbitals, that is, sp^3d and sp^3d^2 hyb orbitals to accommodate five- and six-electron pairs, respectively. Although is an appealing idea, it has come into question because quantum-mechan calculations have shown that the wave functions contain very little contribut from *d* orbitals. Thus, it appears that to describe bonding in SF_6 we should av using *d* orbitals in hybridization schemes.

How are we to proceed? Where are the electrons in the SF_6 molecule? Re that we have already employed the results of quantum-mechanical calculati in constructing electrostatic potential maps (page 418). Let's turn again to results of such calculations to improve our understanding of the bonding SF_6 and similar molecules.

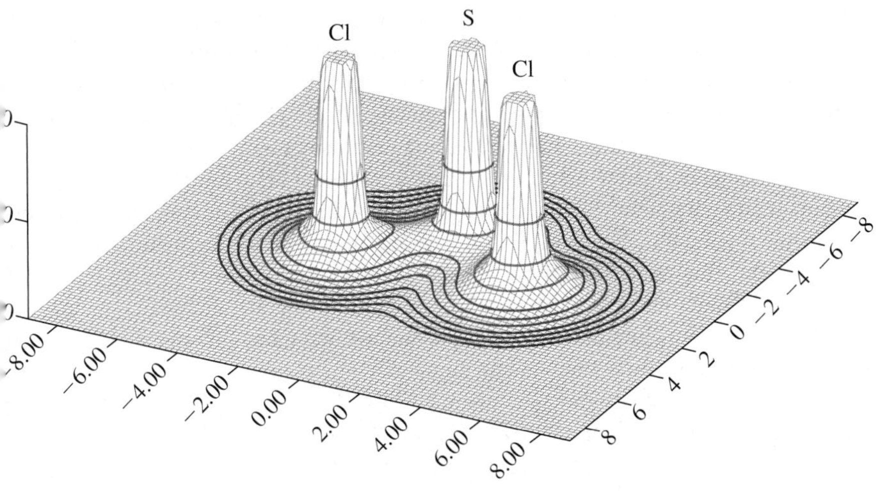

GURE 11-36
ensity contour map of the electron density of SCl_2 in the plane of nuclei
isodensity contour lines, in atomic units (au), are shown in color, in the order 0.001, 2, 0.004, 0.008 (four outermost contours in blue); 0.2, 0.4, 0.8 (next three). sities are truncated at 2.00 au (innermost red contour). The atomic unit of electron sity = $e/a_0^3 = 1.081 \times 10^{12}$ C m^3, where a_0 is the Bohr radius (adapted from Matta Gillespie, *J. Chem. Ed.*, **79**, 1141, 2002).

KEEP IN MIND

that a contour map represents changes in topology of a surface; similar maps are used by mountaineers to plan their ascent of a mountain.

nding in the Molecule SCl_2

continue our discussion consider first a simpler molecule—SCl_2. Figure 11-36 ws how the electron density (ρ) varies in the plane that contains the sulfur m and the two chlorine atoms. The most striking feature of this diagram is t the electron density is very high at each nucleus; in fact, we have truncated very large maxima in order to show other features in the diagram. An espe-ly significant feature is the small ridge of electron density between the sulfur each of the chlorine atoms. This ridge of electron density represents a trans- of electron density from the atomic orbitals to the internuclear region and, pite its modest height, contributes to bond formation.

An alternative representation of the electron density distribution is a con-r map. Such a contour map for SCl_2 is shown in Figure 11-37. The lines wn between the S atom and each Cl atom in Figure 11-37 represent the es along the top of the small ridge of electron density between these atoms

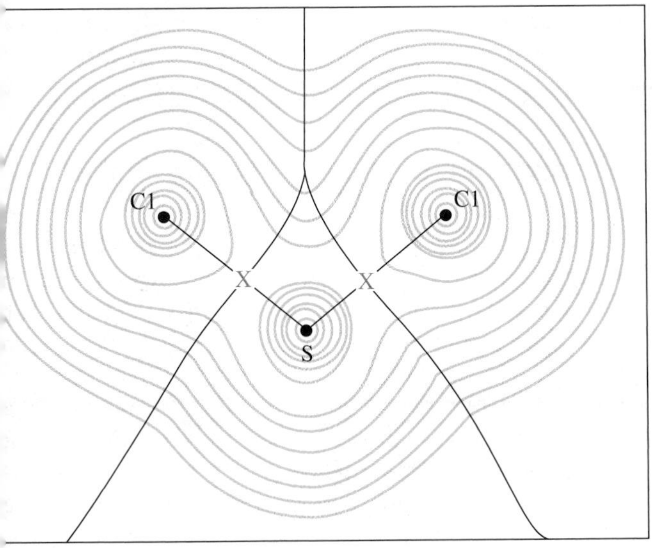

◀ FIGURE 11-37
Contour map of the electron density in SCl_2
The electron density increases from the outermost isodensity contour at 0.001 au in incremental steps of 2×10^{-3} au, 4×10^{-2} au, 8×10^{-1} au, 16×10^{0} au, and so on. The lines connecting the nuclei are the bond paths. The bond critical points are depicted by red Xs (adapted from Matta and Gillespie, *J. Chem. Ed.*, **79**, 1141, 2002).

that we saw in Figure 11-36. The lines between the chlorine atoms and the fur atom are called the *bond paths* and represent the chemical bond betw the S and one Cl atom as we would normally draw it in a Lewis structure Figure 11-37, notice the vertical line above the S atom that splits in two before the sulfur atom is reached, with each segment intersecting the b paths for the two sulfur-chlorine bonds between the S and a Cl atom. bifurcated line represents a path tracing the minimum in the electron den analogous to a path along the valley floor between the "mountains" of e tron density. The point where this line intersects each bond line is called *bond critical point*. The electron density at the bond critical point can be use describe the type of bond connecting a pair of atoms in a molecule. The gre the electron density, the higher the bond order is.

Bonding in the Molecule H_2SO_4 and the Anion SO_4^{2-}

To decide whether or not to use expanded valence shells to minimize for charges, we will consider the sulfuric acid molecule and the sulfate an Figure 11-38 is a three-dimensional representation of the electron density di bution surfaces at a value of 0.002 atomic units (au); they correspond to the face encompassing about 98% of the electron density in H_2SO_4 and SO_4^{2-}. If choose a surface with a higher electron density value, then we include les the electron density distribution. Below the 98% surfaces are two electron d sity plots at increasing densities for the surfaces being calculated, also show Figure 11-38. What do these tell us? When we reach a density for the calcula surface just greater than the density at the bond critical point, the electron d sity in that bond disappears and we have established the amount of elect

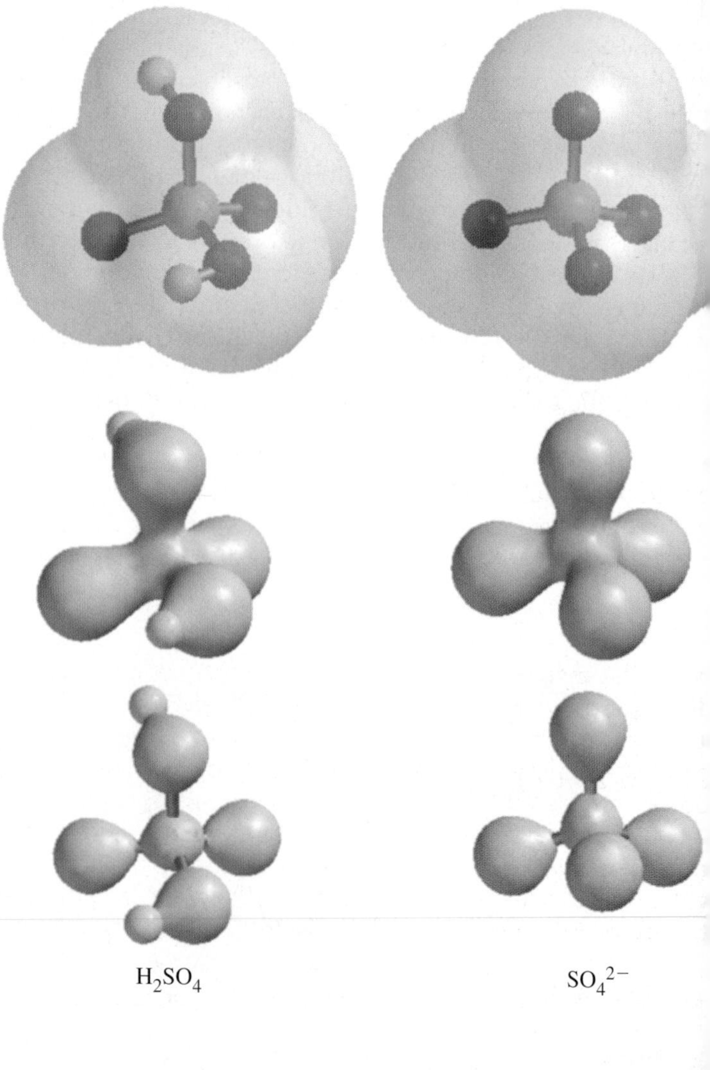

▶ FIGURE 11-38
Three-dimensional plots of the electron density in H_2SO_4 and SO_4^{2-}
The values for the outer isodensity envelope are set at 0.002 au, 0.22 au, and 0.28 au in the three figures for both species.

sity in that bond. If there are bonds in the molecule that have more elec- ı density at their bond critical points, then electron density will still appear he three-dimensional representation. To illustrate this point, observe that .22 au of electron density, the electron density between the sulfur atom the two oxygen atoms attached to hydrogen atoms has disappeared (the k from the ball-and-stick model is apparent), but between the sulfur and two oxygen atoms that do not have a hydrogen atom, it is still visible. The tron density between the sulfur atom and the nonprotonated oxygen (the gen atom that does not have a hydrogen attached) atoms does not disap- r until the density of the calculated surface is increased to 0.33 au. We con- le that there is more electron density in the bonds between the sulfur atom the nonprotonated oxygen atoms than in the bonds between sulfur and gen attached to a proton. This extra electron density can be represented in ewis structure that places a double bond between the central S atom and two terminal O atoms, thereby reducing the formal charges seen in the et Lewis structure of H_2SO_4.

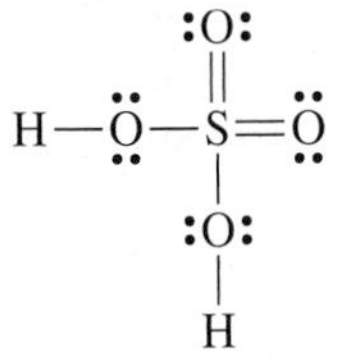
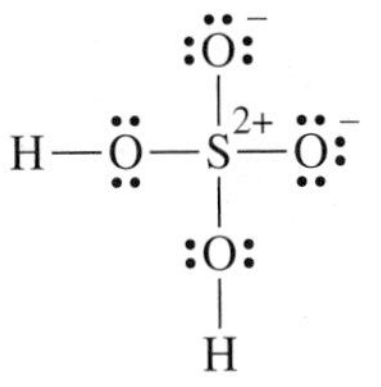

considering the electron density surface for the sulfate anion, we see that bond critical density is 0.28 au for the four bonds between the sulfur atom l the oxygen atoms. This is similar to that for the bond between sulfur and nonprotonated oxygen atoms and greater than that for the sulfur–oxygen ıds attached to protonated oxygen atoms. The presence of this higher bond ical point in the sulfate anion suggests that maybe the Lewis structure that ıimizes formal charges is the best. All four sulfur–oxygen bonds have the ne bond critical point, which corresponds to the possible resonance struc- es that can be written.

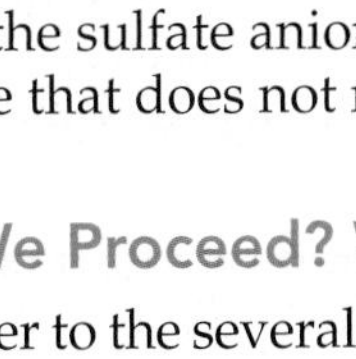
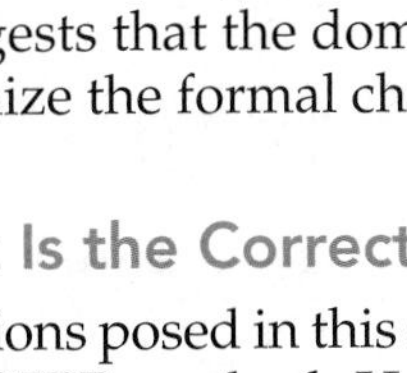

have reached a point at which minimizing the formal charges by using oanded valence shells seems best. However, one detailed analysis of the vefunction of the sulfate anion suggests that the dominant form is the sim- octet structure that does not minimize the formal charge.*

ow Should We Proceed? What Is the Correct Formulation?

rhaps the answer to the several questions posed in this section lies in the work R. J. Gillespie, a coauthor of the VSEPR method. He proposes that Lewis uctures be written as Lewis would have written them—that is, with no oanded valence shells. In cases of highly polar bonds, there can be, simulta- ously, considerable electron density between the atoms (a significantly cova- t bond) *and* a large charge separation between the atoms (a significantly ionic nd). These two factors provide a better understanding of the strength of polar valent bonds in molecules such as BF_3 (with a B—F bond dissociation thalpy of 613 kJ mol^{-1}) or SiF_4 (with a Si—F bond dissociation enthalpy

. Suidan, J. K. Badenhoop, E. D. Glendenning, and F. Weinhold, *J. Chem. Educ.*, **72,** 583 (1995).

of 567 kJ mol^{-1}). By contrast, the nonpolar covalent C—C bond dissocia enthalpy is only 345 kJ mol^{-1}. An analysis of the electron densities in BF_3 SiF_4 shows that they do have a combination of highly covalent and ionic b characteristics, leading to very strong bonds.

The controversy as to how best to write Lewis structures will no doubt c tinue in the chemical literature, but you should not be too dismayed by situation. Our approach to depicting the electronic structure of a molecul based on the simplest Lewis structure and its concomitant use in determin the shape of a molecule through VSEPR theory. In order to probe more dee into the nature of a chemical bond—for example, to understand experime results, such as bond enthalpy values—we must analyze a computed elect density map for that molecule rather than rely just on the Lewis structure.

MasteringCHEMISTRY www.masteringchemistry.com

The orbital structures of molecules are studied using photoelectron spectroscopy. The method involves passin high-energy photons through a gaseous sample of molecules, and measuring the kinetic energies of the ejected electrons. For a discussion of Photoelectron Spectroscopy, go to the Focus On feature for Chapter 11 on the MasteringChemistry site.

Summary

11-1 What a Bonding Theory Should Do—A basic requirement of a bonding theory is that it provide a better description of the electronic structure of molecules than the simple ideas of the Lewis model. The transfer of electron density from regions outside the nuclei to regions nearer and between the nuclei produces a significant decrease in the potential energy of the electrons and contributes substantially to the stability of a chemical bond.

11-2 Introduction to the Valence Bond Method—**Valence bond method** considers a covalent bond in terms of the overlap of atomic orbitals of the bonded atoms.

11-3 Hybridization of Atomic Orbitals—Some molecules can be described in terms of the overlap of simple orbitals, but often orbitals that are a composite of simple orbitals—**hybrid orbitals**—are needed. The **hybridization** scheme chosen is the one that produces an orientation of hybrid orbitals to match the electron-group geometry predicted by the VSEPR theory (Fig. 11-14). ***sp* hybrid orbitals** (Fig. 11-11) are associated with linear electron-group geometries; sp^2 **hybrid orbitals** (Fig. 11-10) with trigonal planar geometries; sp^3 **hybrid orbitals** (Fig. 11-7) with tetrahedral geometries; sp^3d **hybrid orbitals** with trigonal bipyramidal geometries; and sp^3d^2 **hybrid orbitals** (Fig. 11-13) with octahedral geometries.

11-4 Multiple Covalent Bonds—End-to-end overlap of orbitals produces **σ (sigma) bonds**. Side-to-side overlap of two *p* orbitals produces a **π (pi) bond**. Single covalent bonds are σ bonds. A double bond consists of one σ bond and one π bond (Fig. 11-15). A triple bond consists of one σ bond and two π bonds (Fig. 11-17). The geometric shape of a species determines the σ-bond framework, and π bonds are added as required to complete the bonding description.

11-5 Molecular Orbital Theory—In **molecular orb theory**, electrons are assigned to molecular orbitals. numbers and kinds of molecular orbitals are related to atomic orbitals used to generate them. Electron den between atoms is high in **bonding molecular orbitals** very low in **antibonding orbitals** (Fig. 11-22). **Bond ord** one-half the difference between the numbers of electron bonding molecular orbitals and in antibonding molecu orbitals (equation 11.1). Molecular orbital energy-level grams and an aufbau process can be used to describe electronic structure of a molecule; this is similar to what done for atomic electron configurations in Chapter 8.

11-6 Delocalized Electrons: An Explanati Based on Molecular Orbital Theory—Bonding in benzene molecule, C_6H_6, is partly based on the concep **delocalized molecular orbitals**. These are regions of h electron density that extend over several atoms in a m cule (Fig. 11-31). Delocalized molecular orbitals also prov an alternative to the concept of resonance in other molecu and ions.

11-7 Some Unresolved Issues: Can Electr Density Plots Help?—Electron density plots can used as a guide in understanding the bonding in m cules that don't necessarily have simple Lewis structu In molecules such as SF_6, we employed the concept hyperconjugation. For SCl_2 the electron density was a lyzed in terms of bond paths and bond critical points. H_2SO_4 and SO_4^{2-}, bond critical points derived fr charge density point out that drawing Lewis structu with the lowest formal charge are not necessarily the b way to represent the bonding in these compounds. In g eral one should write Lewis structures without expand valence shells.

Integrative Example

'drogen azide, HN_3, and its salts (metal azides) are unstable substances used in detonators for high explosives. Sodium .de, NaN_3, is used in air-bag safety systems in automobiles (see page 212). A reference source lists the following data HN_3. (The subscripts a, b, and c distinguish the three N atoms from one another.) Bond lengths: $N_a—N_b$ = 124 pm; $—N_c$ = 113 pm. Bond angles: $H—N_a—N_b$ = 112.7°; $N_a—N_b—N_c$ = 180°.

Write two contributing structures to the resonance hybrid for HN_3, and describe a plausible hybridization and bond- ; scheme for each structure.

alyze

: can use data from Table 10.2 to estimate the bond order for the two nitrogen-to-nitrogen bonds. From this information can write plausible Lewis structures, and by applying VSEPR theory to the Lewis structures, we can predict a likely)metric shape of the molecule. Finally, with this information we can propose hybridization schemes for the central ›ms and an overall bonding scheme for the molecule.

lve

)m Table 10.2, the average bond lengths for N-to-N ıds are 145 pm for a single bond, 123 pm for a double ıd, and 110 pm for a triple bond. Thus it is likely that the $—N_b$ bond (124 pm) has a considerable double-bond ıracter, and the $N_b—N_c$ bond (113 pm) has a consider- e triple-bond character.

The HN_3 molecule has a total of 16 valence electrons in lectron pairs. The plausible Lewis structures have N_a and atoms as central atoms, the N_c and H atoms as terminal ms, and bonds reflecting the observed bond lengths.

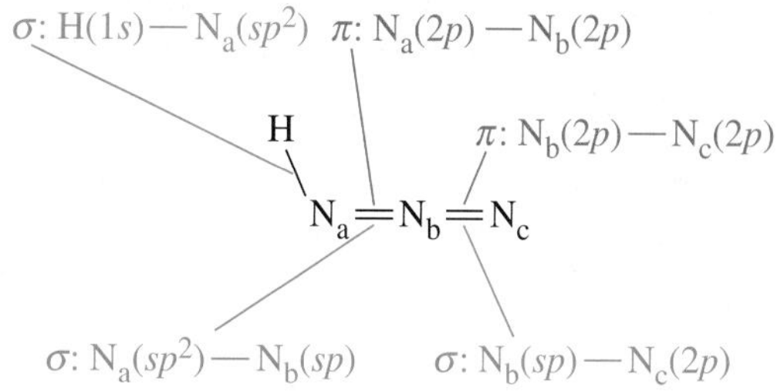

Structure (I)

) $H—\ddot{N}_a=\overset{+}{N}_b=\ddot{N}_c^{-}$ (II) $H—\ddot{N}_a^{-}—\overset{+}{N}_b\equiv N_c$:

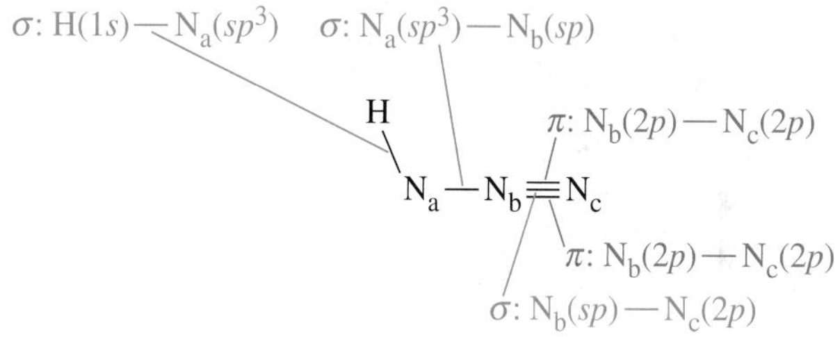

Structure (II)

According to VSEPR theory, in both structures (I) and (II) electron-group geometry around N_b is linear. This corre- ›nds to *sp* hybridization. In structure (I), the electron-group)metry around N_a is trigonal-planar, corresponding to hybridization; in structure (II), the electron-group geome- around N_a is tetrahedral, corresponding to sp^3 hybridiza- ı. These hybridization schemes, the orbital overlaps, and geometric structures of the two resonance structures are icated on the right.

sess

these two resonance structures, which is the most favor- e? We could try to use formal charges, but we find that structure (I), the formal charges on N_a, N_b, and N_c, are 0, , and −1, respectively; correspondingly in structure (II), formal charges are −1, +1, and 0, respectively. Because formal charges are so similar, we cannot make any initive conclusion as to which structure is favored. The est way to decide is to compare the observed molecular ıcture with that suggested by the two hybridization- emes given above. In structure (I), N_a is sp^2 hybridized :hat a $H—N_a—N_b$ angle is expected to be close to 120°; ereas in structure (II), the hybridization on N_a is sp^3 so t the $H—N_a—N_b$ angle is expected to be close to 109°.)erimentally it is found that the $H—N_a—N_b$ angle is °, so that the hybridization scheme in structure (II) is to preferred. The electrostatic potential map for HN_3 is ›wn on the right, and we observe that N_a is relatively ;atively charged as compared to N_b and N_c—in accord h our preferred structure (II).

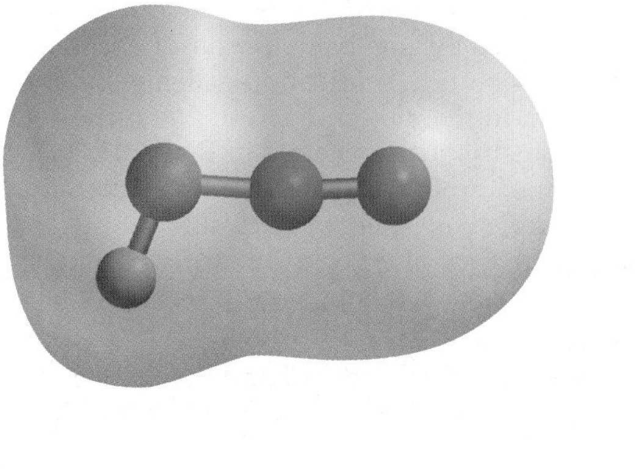

PRACTICE EXAMPLE A: Melamine is a carbon–hydrogen–nitrogen compound used in the manufacture adhesives, protective coatings, and textile finishing (such as in wrinkle-free, wash-and-wear fabrics). Its mass perc composition is 28.57% C, 4.80% H, and 66.64% N. The melamine molecule features a six-member ring with alternati carbon and nitrogen atoms. Half the nitrogen atoms and all the H atoms are outside the ring. For melamine, **(a)** writ plausible Lewis structure, **(b)** describe bonding in the molecule by the valence bond method, and **(c)** describe bonding the ring system through molecular orbital theory.

PRACTICE EXAMPLE B: Dimethylglyoxime (DMG) is a carbon–hydrogen–nitrogen–oxygen compound witl molecular mass of 116.12 u. In a combustion analysis, a 2.464 g sample of DMG yields 3.735 g CO_2 and 1.530 g H_2O. I separate experiment, the nitrogen in a 1.868 g sample of DMG is converted to $NH_3(g)$ and the NH_3 is neutralized passing it into 50.00 mL of 0.3600 M $H_2SO_4(aq)$. After neutralization of the NH_3 the excess $H_2SO_4(aq)$ requires 18.63 of 0.2050 M NaOH(aq) for its neutralization. Using these data, determine for dimethylglyoxime **(a)** the most plausi Lewis structure, and **(b)** in the manner of Figure 11-19, a plausible bonding scheme.

Exercises

Valence Bond Method

1. Indicate several ways in which the valence bond method is superior to Lewis structures in describing covalent bonds.
2. Explain why it is necessary to hybridize atomic orbitals when applying the valence bond method—that is, why are there so few molecules that can be described by the overlap of pure atomic orbitals only?
3. Describe the molecular geometry of H_2O suggested by each of the following methods: **(a)** Lewis theory; **(b)** valence bond method using simple atomic orbitals; **(c)** VSEPR theory; **(d)** valence bond method using hybridized atomic orbitals.
4. Describe the molecular geometry of NH_3 suggested by each of the following methods: **(a)** Lewis theory; **(b)** valence bond method using simple atomic orbitals; **(c)** VSEPR theory; **(d)** valence bond method using hybridized atomic orbitals.
5. In which of the following, CO_3^{2-}, SO_2, CCl_4, CO, NO_2^-, would you expect to find sp^2 hybridization of the central atom? Explain.
6. In the manner of Example 11-1, describe the probable structure and bonding in **(a)** HI; **(b)** BrCl; **(c)** H_2Se; **(d)** OCl_2.
7. For each of the following species, identify the central atom(s) and propose a hybridization scheme for those atom(s): **(a)** CO_2; **(b)** $HONO_2$; **(c)** ClO_3^-; **(d)** BF_4^-.
8. Propose a plausible Lewis structure, geometric structure, and hybridization scheme for the ONF molecule.
9. Describe a hybridization scheme for the central Cl atom in the molecule ClF_3 that is consistent with the geometric shape pictured in Table 10.1. Which orbitals of the Cl atom are involved in overlaps, and which are occupied by lone-pair electrons?
10. Describe a hybridization scheme for the central S atom in the molecule SF_4 that is consistent with the geometric shape pictured in Table 10.1. Which orbitals of the S atom are involved in overlaps, and which are occupied by lone-pair electrons?
11. Match each of the following species with one of these hybridization schemes: sp, sp^2, sp^3, sp^3d, sp^3d^2. **(a)** PF_6^-; **(b)** COS; **(c)** $SiCl_4$; **(d)** NO_3^-; **(e)** AsF_5.
12. Propose a hybridization scheme to account for bonds formed by the central carbon atom in each of the following molecules: **(a)** hydrogen cyanide, HCN; **(b)** methyl alcohol, CH_3OH; **(c)** acetone, $(CH_3)_2C$ **(d)** carbamic acid,

O
‖
H_2NCOH

13. Indicate which of the following molecules and ions linear, which are planar, and which are neither. Th propose hybridization schemes for the central ato **(a)** $Cl_2C{=}CCl_2$; **(b)** $N{\equiv}C{-}C{\equiv}N$; **(c)** $F_3C{-}C{\equiv}$ **(d)** $[S{-}C{\equiv}N]^-$.
14. In the manner of Figure 11-18, indicate the structu of the following molecules in terms of the overlap simple atomic orbitals and hybrid orbitals: **(a)** CH_2C **(b)** OCN^-; **(c)** BF_3.
15. Write Lewis structures for the following molecul and then label each σ and π bond. **(a)** HCN; **(b)** C_2 **(c)** $CH_3CHCHCCl_3$; **(d)** HONO.
16. Represent bonding in the carbon dioxide molecu CO_2, by **(a)** a Lewis structure and **(b)** the valen bond method. Identify σ and π bonds, the necess hybridization scheme, and orbital overlap.
17. Use the method of Figure 11-19 to represent bondi in each of the following molecules: **(a)** CCl_4; **(b)** ON **(c)** HONO; **(d)** $COCl_2$.
18. Use the method of Figure 11-19 to represent bonding each of the following ions: **(a)** NO_2^-; **(b)** I_3^-; **(c)** C_2O_4 **(d)** HCO_3^-.
19. The molecular model below represents citric acid, acidic component of citrus juices. Represent bonding the citric acid molecule using the method of Figure 11 to indicate hybridization schemes and orbital overlap

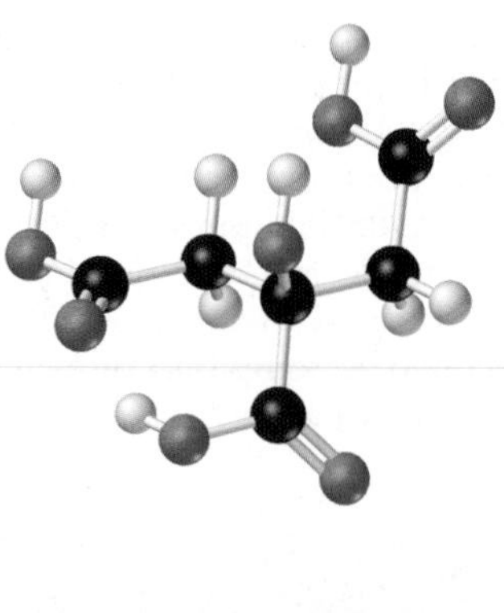

Malic acid is a common organic acid found in unripe apples and other fruit. With the help of the molecular model shown below, represent bonding in the malic acid molecule, using the method of Figure 11-19 to indicate hybridization schemes and orbital overlaps.

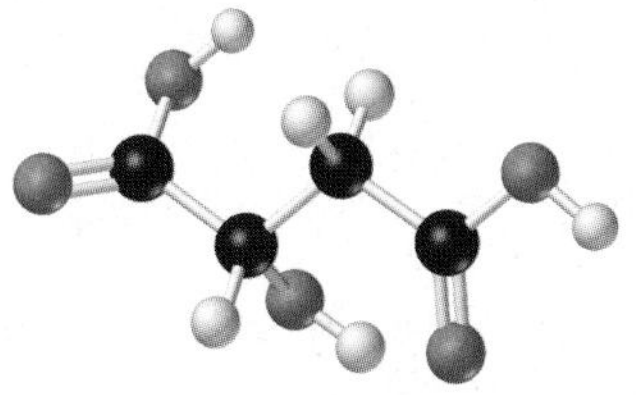

Shown below are ball-and-stick models. Describe hybridization and orbital-overlap schemes consistent with these structures.

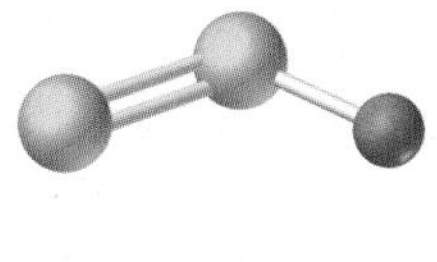

(a) S_2O

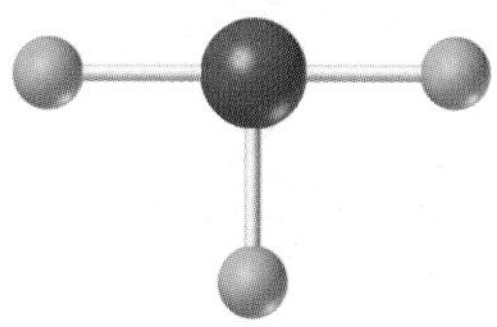

(b) BrF_3

Shown below are ball-and-stick models. Describe hybridization and orbital-overlap schemes consistent with these structures.

(a) XeF_2

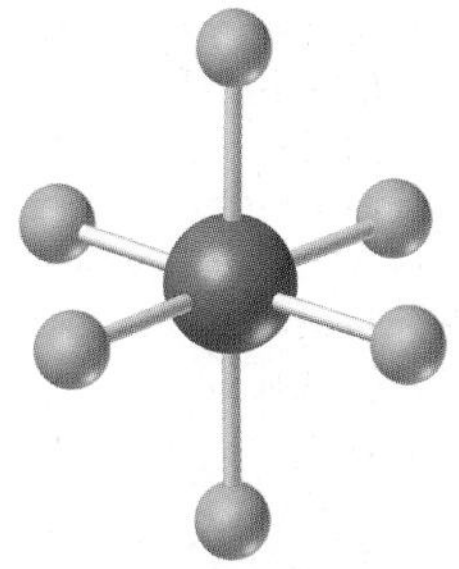

(b) IF_6^+

Propose a bonding scheme that is consistent with the structure for propynal. [*Hint:* Consult Table 10.2 to assess the multiple-bond character in some of the bonds.]

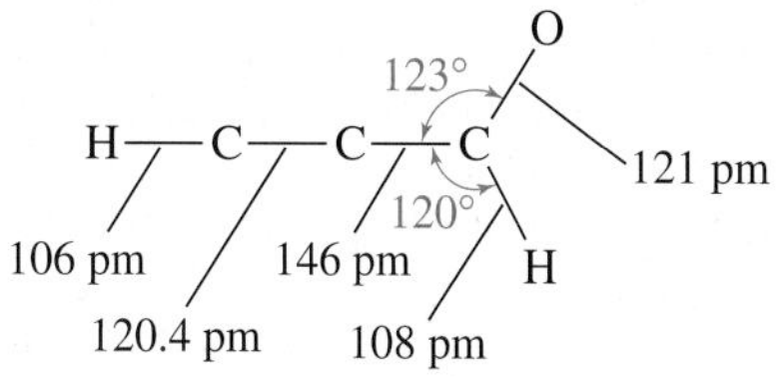

24. The structure of the molecule allene, CH_2CCH_2, is shown here. Propose hybridization schemes for the C atoms in this molecule.

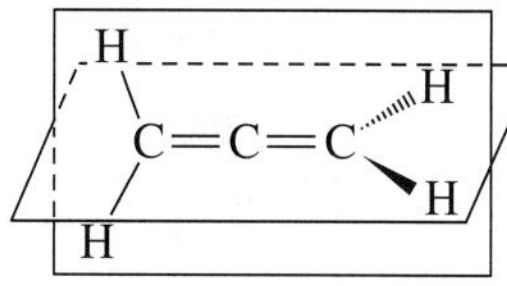

25. Angelic acid, shown below, occurs in sumbol root, a herb used as a stimulant.

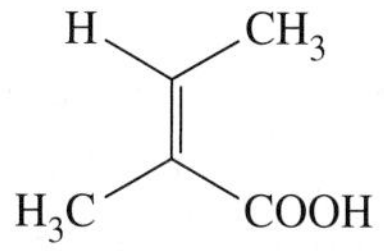

Represent the bonding in the angelic acid molecule by using the method in Figure 11-19 to indicate hybridization schemes and orbital overlaps. What is the maximum number of atoms that can lie in the same plane?

26. Dimethylolpropionic acid, shown below, is used in the preparation of resins.

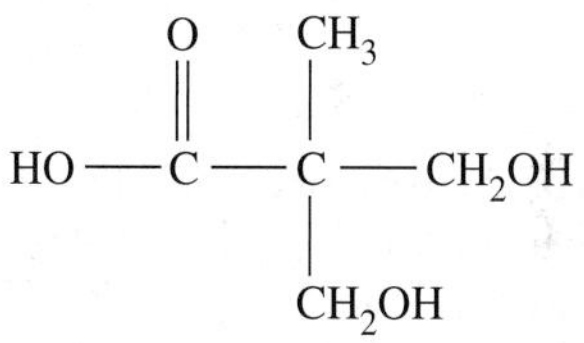

Represent the bonding in the dimethylolpropionic acid molecule by using the method in Figure 11-19 to indicate hybridization schemes and orbital overlaps. What is the maximum number of atoms that can lie in the same plane?

olecular Orbital Theory

Explain the essential difference in how the valence-bond method and molecular orbital theory describe a covalent bond.

Describe the bond order of diatomic carbon, C_2, with Lewis theory and molecular orbital theory, and explain why the results are different.

$N_2(g)$ has an exceptionally high bond energy. Would you expect either N_2^- or N_2^{2-} to be a stable diatomic species in the gaseous state? Explain.

The paramagnetism of gaseous B_2 has been established. Explain how this observation confirms that the π_{2p} orbitals are at a lower energy than the σ_{2p} orbital for B_2.

31. In our discussion of bonding, we have not encountered a bond order higher than triple. Use the energy-level diagrams of Figure 11-26 to show why this is to be expected.

32. Is it correct to say that when a diatomic molecule loses an electron, the bond energy always decreases (that is, that the bond is always weakened)? Explain.

33. For the following pairs of molecular orbitals, indicate the one you expect to have the lower energy, and state the reason for your choice. **(a)** σ_{1s} or σ_{1s}^*; **(b)** σ_{2s} or σ_{2p}; **(c)** σ_{1s}^* or σ_{2s}; **(d)** σ_{2p} or σ_{2p}^*.

34. For each of the species C_2^+, O_2^-, F_2^+, and NO^+,
(a) Write the molecular orbital occupancy diagram (as in Example 11-6).
(b) Determine the bond order, and state whether you expect the species to be stable or unstable.
(c) Determine if the species is diamagnetic or paramagnetic; and if paramagnetic, indicate the number of unpaired electrons.

35. Write plausible molecular orbital occupancy diagrams for the following *heteronuclear* diatomic species: **(a)** NO; **(b)** NO^+; **(c)** CO; **(d)** CN; **(e)** CN^-; **(f)** CN^+; **(g)** BN.

36. We have used the term "isoelectronic" to refer to atoms with identical electron configurations. In molecular orbital theory, this term can be applied to molecular species as well. Which of the species in Exercise 35 are isoelectronic?

37. Consider the molecules NO^+ and N_2^+ and use molecular orbital theory to answer the following:
(a) Write the molecular orbital configuration of each ion (ignore the 1*s* electrons).
(b) Predict the bond order of each ion.
(c) Which of these ions is paramagnetic? Whic diamagnetic?
(d) Which of these ions do you think has the gre bond length? Explain.

38. Consider the molecules CO^+ and CN^- and use m ular orbital theory to answer the following:
(a) Write the molecular orbital configuration of ion (ignore the 1*s* electrons).
(b) Predict the bond order of each ion.
(c) Which of these ions is paramagnetic? Whic diamagnetic?
(d) Which of these ions do you think has the gre bond length? Explain.

39. Construct the molecular orbital diagram for Would you expect the bond length of CF^+ to be lo or shorter than that of CF?

40. Construct the molecular orbital diagram for S Would you expect the bond length of $SrCl^+$ t longer or shorter than that of SrCl?

Delocalized Molecular Orbitals

41. Explain why the concept of delocalized molecular orbitals is essential to an understanding of bonding in the benzene molecule, C_6H_6.

42. Explain how it is possible to avoid the concept of resonance by using molecular orbital theory.

43. In which of the following molecules would you ex to find delocalized molecular orbitals? Exp **(a)** C_2H_4; **(b)** SO_2; **(c)** H_2CO.

44. In which of the following ions would you expect to delocalized molecular orbitals? Explain. **(a)** HC **(b)** CO_3^{2-}; **(c)** CH_3^+.

Integrative and Advanced Exercises

45. The Lewis structure of N_2 indicates that the nitrogen-to-nitrogen bond is a triple covalent bond. Other evidence suggests that the σ bond in this molecule involves the overlap of *sp* hybrid orbitals.
(a) Draw orbital diagrams for the N atoms to describe bonding in N_2.
(b) Can this bonding be described by either sp^2 or sp^3 hybridization of the N atoms? Can bonding in N_2 be described in terms of unhybridized orbitals? Explain.

46. Show that both the valence bond method and molecular orbital theory provide an explanation for the existence of the covalent molecule Na_2 in the gaseous state. Would you predict Na_2 by the Lewis theory?

47. A group of spectroscopists believe that they have detected one of the following species: NeF, NeF^+, or NeF^-. Assume that the energy-level diagrams of Figure 11-25 apply, and describe bonding in these species. Which of these species would you expect the spectroscopists to have observed?

48. Lewis theory is satisfactory to explain bonding in the ionic compound K_2O, but it does not readily explain formation of the ionic compounds potassium superoxide, KO_2, and potassium peroxide, K_2O_2.
(a) Show that molecular orbital theory can provide this explanation.
(b) Write Lewis structures consistent with the mc ular orbital explanation.

49. The compound potassium sesquoxide has the emp cal formula K_2O_3. Show that this compound car described by an appropriate combination of po sium, peroxide, and superoxide ions. Write a Le structure for a formula unit of the compound.

50. Draw a Lewis structure for the urea molec $CO(NH_2)_2$, and predict its geometric shape with VSEPR theory. Then revise your assessment of molecule, given the fact that all the atoms lie in same plane, and all the bond angles are 120°. Prop a hybridization and bonding scheme consistent v these experimental observations.

51. Methyl nitrate, CH_3NO_3, is used as a rocket propell The skeletal structure of the molecule is CH_3ONO_2. N and three O atoms all lie in the same plane, but CH_3 group is not in the same plane as the NO_3 gro The bond angle C—O—N is 105°, and the bond a O—N—O is 125°. One nitrogen-to-oxygen bo length is 136 pm, and the other two are 126 pm.
(a) Draw a sketch of the molecule showing its g metric shape.
(b) Label all the bonds in the molecule as σ or π, a indicate the probable orbital overlaps involved.
(c) Explain why all three nitrogen-to-oxygen bo lengths are not the same.

Fluorine nitrate, $FONO_2$, is an oxidizing agent used as a rocket propellant. A reference source lists the following data for FO_aNO_2. (The subscript "a" shows that this O atom is different from the other two.)

Bond lengths: N—O = 129 pm;
N—O_a = 139 pm; O_a—F = 142 pm
Bond angles: O—N—O = 125°;
F—O_a—N = 105°
NO_aF plane is perpendicular to the O_2NO_a plane

Use these data to construct a Lewis structure(s), a three-dimensional sketch of the molecule, and a plausible bonding scheme showing hybridization and orbital overlaps.

Draw a Lewis structure(s) for the nitrite ion, NO_2^-. Then propose a bonding scheme to describe the σ and π bonding in this ion. What conclusion can you reach about the number and types of π molecular orbitals in this ion? Explain.

Think of the reaction shown here as involving the transfer of a fluoride ion from ClF_3 to AsF_5 to form the ions ClF_2^+ and AsF_6^-. As a result, the hybridization scheme of each central atom must change. For each reactant molecule and product ion, indicate **(a)** its geometric structure and **(b)** the hybridization scheme for its central atom.

$$ClF_3 + AsF_5 \longrightarrow (ClF_2^+)(AsF_6^-)$$

In the gaseous state, HNO_3 molecules have two nitrogen-to-oxygen bond distances of 121 pm and one of 140 pm. Draw a plausible Lewis structure(s) to represent this fact, and propose a bonding scheme in the manner of Figure 11-19.

He_2 does not exist as a stable molecule, but there is evidence that such a molecule can be formed between electronically excited He atoms. Write an electron configuration for He_2 to account for this.

The molecule formamide, $HCONH_2$, has the approximate bond angles H—C—O, 123°; H—C—N, 113°; N—C—O, 124°; C—N—H, 119°; H—N—H, 119°. The C—N bond length is 138 pm. Two Lewis structures can be written for this molecule, with the true structure being a resonance hybrid of the two. Propose a hybridization and bonding scheme for each structure.

Pyridine, C_5H_5N, is used in the synthesis of vitamins and drugs. The molecule can be thought of in terms of replacing one CH unit in benzene with a N atom. Draw orbital diagrams to show the orbitals of the C and N atoms involved in the σ and π bonding in pyridine. How many bonding and antibonding π-type molecular orbitals are present? How many delocalized electrons are present?

One of the characteristics of antibonding molecular orbitals is the presence of a nodal plane. Which of the bonding molecular orbitals considered in this chapter have nodal planes? Explain how a molecular orbital can have a nodal plane and still be a bonding molecular orbital.

The ion F_2Cl^- is linear, but the ion F_2Cl^+ is bent. Describe hybridization schemes for the central Cl atom consistent with this difference in structure.

Ethyl cyanoacetate, a chemical used in the synthesis of dyes and pharmaceuticals, has the mass percent composition: 53.09% C, 6.24% H, 12.39% N, and 28.29% O. In the manner of Figure 11-19, show a bonding scheme for this substance. The scheme should designate orbital overlaps, σ and π bonds, and expected bond angles.

62. A certain monomer used in the production of polymers has one nitrogen atom and the mass composition 67.90% C, 5.70% H, and 26.40% N. Sketch the probable geometric structure of this molecule, labeling all the expected bond lengths and bond angles.

63. A solar cell that is 15% efficient in converting solar to electric energy produces an energy flow of 1.00 kW/m^2 when exposed to full sunlight.
(a) If the cell has an area of 40.0 cm^2, what is the power output of the cell, in watts?
(b) If the power calculated in part (a) is produced at 0.45 V, how much current does the cell deliver?

64. Toluene-2,4-diisocyanate is used in the manufacture of polyurethane foam. An incomplete structure is shown below. Describe the hybridization scheme for the atoms marked with an asterisk, and indicate the values of the bond angles marked α and β.

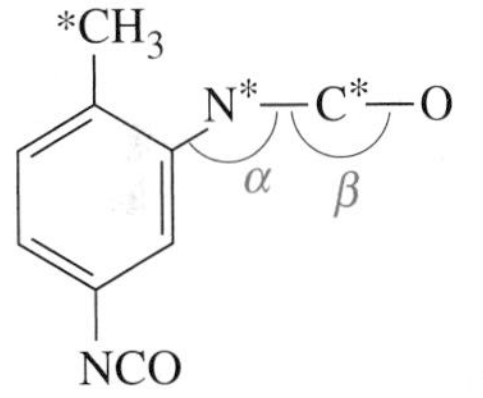

65. Histidine, an essential amino acid, serves as a part of the active center in many enzymes. It is the precursor to histamine, a neurotransmitter and a component of the body's immune response. The structure of histidine is shown below.

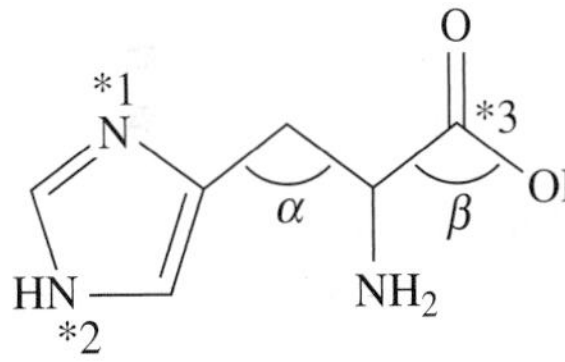

Identify the hybridization scheme for the atoms marked with an asterisk, and indicate the values of the bond angles marked α and β.

66. The anion I_4^{2-} is linear, and the anion I_5^- is V-shaped, with a 95° angle between the two arms of the V. For the central atoms in these ions, propose hybridization schemes that are consistent with these observations.

67. Pentadiene, C_5H_8, has three isomers, depending on the position of the two double bonds. Determine the shape of these isomers by using VSEPR theory. Describe the bonding in these molecules by using the valence bond method. Do the shapes agree in the two theories? Use molecular orbital theory to decide which of these molecules has a delocalized π system. Sketch the molecular orbital and an energy-level diagram.

68. A conjugated hydrocarbon has an alternation of double and single bonds. Draw the molecular orbitals of the π system of 1,3,5-hexatriene. If the energy required to excite an electron from the HOMO to the LUMO corresponds to a wavelength of 256 nm, do you expect the wavelength for the corresponding excitation in 1,3,5,7-octatetraene to be a longer or shorter wavelength? [*Hint:* Refer to Figure 11-34.]

69. An elusive intermediate of atmospheric reactions of HONO may be nitrosyl O-hydroxide, HOON. Electronic structure calculations seem to indicate that HOON is best represented by a combination of three resonance structures, with major contribution fro radical-pair structure (involving HO and NO radic significant contribution from a molecular struct and a small contribution from an ion-pair struc (involving HO^- and NO^+ ions). Use the informatio represent the structure of HOON in terms of th three resonance structures.

Feature Problems

70. Resonance energy is the difference in energy between a real molecule—a resonance hybrid—and its most important contributing structure. To determine the resonance energy for benzene, we can determine an energy change for benzene and the corresponding change for one of the Kekulé structures. The resonance energy is the difference between these two quantities.
(a) Use data from Appendix D to determine the enthalpy of hydrogenation of liquid benzene to liquid cyclohexane.
(b) Use data from Appendix D to determine the enthalpy of hydrogenation of liquid cyclohexene to liquid cyclohexane.

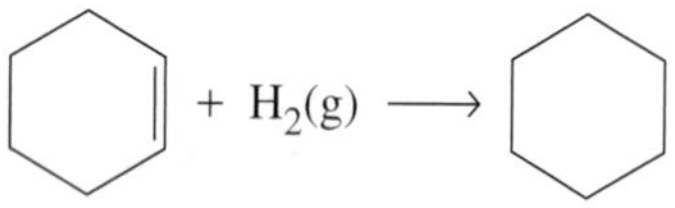

For the enthalpy of formation of liquid cyclohexene, use $\Delta_f H^\circ = -38.5$ kJ/mol.
(c) Assume that the enthalpy of hydrogenation of 1,3,5-cyclohexatriene is three times as great as that of cyclohexene, and calculate the resonance energy of benzene.
(d) Another way to assess resonance energy is through bond energies. Use bond energies from Table 10.3 (page 451) to determine the total enthalpy change required to break all the bonds in a Kekulé structure of benzene. Next, determine the enthalpy change for the dissociation of $C_6H_6(g)$ into its gaseous atoms by using data from Table 10.3 and Appendix D. Then calculate the resonance energy of benzene.

71. Furan, C_4H_4O, is a substance derivable from oat hulls, corn cobs, and other cellulosic waste. It is a starting material for the synthesis of other chemicals used as pharmaceuticals and herbicides. The furan molecule is planar and the C and O atoms are bonded into a five-membered pentagonal ring. The H atoms are attached to the C atoms. The chemical behavior of the molecule suggests that it is a resonance hybrid of several contributing structures. These structures show that the double bond character is associated with the entire ring in the form of a π electron cloud.
(a) Draw Lewis structures for the several contributing structures to the resonance hybrid mentioned above.
(b) Draw orbital diagrams to show the orbitals that are involved in the σ and π bonding in furan. [*Hint:* You need use only one of the contribut structures, such as the one with no formal charg
(c) How many π electrons are there in the furan m ecule? Show that this number of π electrons is same, regardless of the contributing structure you for this assessment.

72. As discussed in Are You Wondering 11-1, the hybrid orbitals are algebraic combinations of th and p orbitals. The required combinations of $2s$ an orbitals are

$$\psi_1(sp) = \frac{1}{\sqrt{2}}[\psi(2s) + \psi(2p_z)]$$

$$\psi_2(sp) = \frac{1}{\sqrt{2}}[\psi(2s) - \psi(2p_z)]$$

(a) By combining the appropriate functions give Table 8.2, construct a polar plot in the manne Figure 8-24 for each of the above functions in the plane. In a polar plot, the value of r/a_0 is set a fixed value (for example, 1). Describe the sha and phases of the different portions of the hyb orbitals, and compare them with those shown Figure 11-12.
(b) Convince yourself that the combinations empl ing the $2p_x$ or $2p_y$ orbital also give similar hyb orbitals but pointing in different directions.
(c) The combinations for the sp^2 hybrids in the plane are

$$\psi_1(sp^2) = \frac{1}{\sqrt{3}}\psi(2s) + \frac{\sqrt{2}}{\sqrt{3}}\psi(2p_x)$$

$$\psi_2(sp^2) = \frac{1}{\sqrt{3}}\psi(2s) - \frac{1}{\sqrt{6}}\psi(2p_x) + \frac{1}{\sqrt{2}}\psi(2p_y)$$

$$\psi_3(sp^2) = \frac{1}{\sqrt{3}}\psi(2s) - \frac{1}{\sqrt{6}}\psi(2p_x) - \frac{1}{\sqrt{2}}\psi(2p_y)$$

By constructing polar plots (in the xy plane), sh that these functions correspond to the sp^2 hybr depicted in Figure 11-10.

73. In Chapter 10, we saw that electronegativity differen determine whether bond dipoles exist in a molecule a that molecular shape determines whether bond dipo cancel (nonpolar molecules) or combine to produ a resultant dipole moment (polar molecules). Th the ozone molecule, O_3, has no bond dipoles because all the atoms are alike. Yet, O_3 *does* have a resultant dipole moment: $\mu = 0.534$ D. The electrostatic potential map for ozone is shown below. Use the electrostatic potential map to decide the direction of the dipole. Using the ideas of delocalized bonding in molecules, can you rationalize this electrostatic potential map?

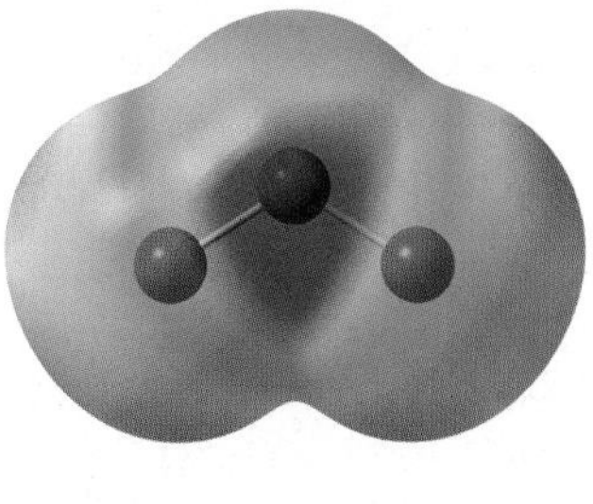

Borazine, $B_3N_3H_6$ is often referred to as inorganic benzene because of its similar structure. Like benzene, borazine has a delocalized π system. Describe the molecular orbitals of the π system. Identify the highest occupied molecular orbital (HOMO) and the lowest unoccupied molecular orbital (LUMO). How many nodes does the LUMO possess?

Which of the following combinations of orbitals give rise to bonding molecular orbitals? For those combinations that do, label the resulting bonding molecular orbital as σ or π.

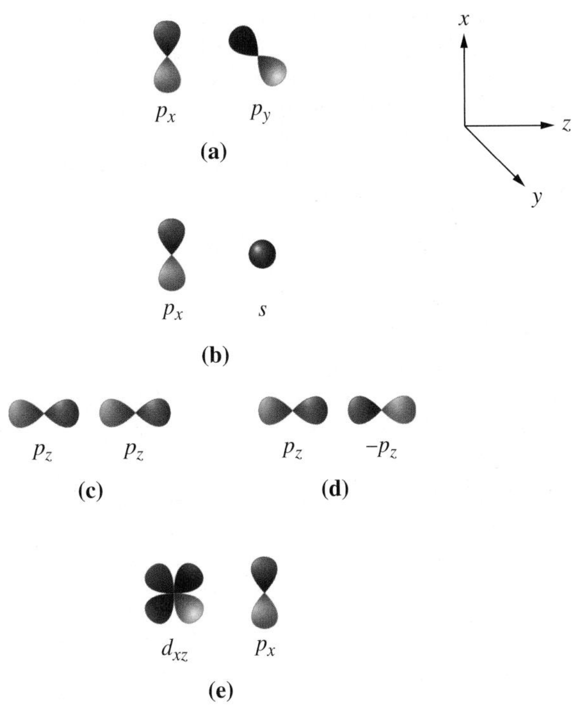

76. Construct a molecular orbital diagram for HF, and label the molecular orbitals as bonding, antibonding, or nonbonding.

Self-Assessment Exercises

In your own words, define the following terms or symbols: **(a)** sp^2; **(b)** σ^*_{2p}; **(c)** bond order; **(d)** π bond.

Briefly describe each of the following ideas: **(a)** hybridization of atomic orbitals; **(b)** σ-bond framework; **(c)** Kekulé structures of benzene, C_6H_6.

Explain the important distinctions between the terms in each of the following pairs: **(a)** σ and π bonds; **(b)** localized and delocalized electrons; **(c)** bonding and antibonding molecular orbitals.

A molecule in which sp^2 hybrid orbitals are used by the central atom in forming covalent bonds is **(a)** PCl_5; **(b)** N_2; **(c)** SO_2; **(d)** He_2.

The bond angle in H_2Se is best described as **(a)** between 109° and 120°; **(b)** less than in H_2S; **(c)** less than in H_2S, but not less than 90°; **(d)** less than 90°.

The hybridization scheme for the central atom includes a *d* orbital contribution in **(a)** I_3^-; **(b)** PCl_3; **(c)** NO_3^-; **(d)** H_2Se.

Of the following, the species with a bond order of 1 is **(a)** H_2^+; **(b)** Li_2; **(c)** He_2; **(d)** H_2^-.

The hybridization scheme for Xe in XeF_2 is **(a)** *sp*; **(b)** sp^3; **(c)** sp^3d; **(d)** sp^3d^2.

Delocalized molecular orbitals are found in **(a)** H_2; **(b)** HS^-; **(c)** CH_4; **(d)** CO_3^{2-}.

Explain why the molecular structure of BF_3 cannot be adequately described through overlaps involving pure *s* and *p* orbitals.

87. Why does the hybridization sp^3d not account for bonding in the molecule BrF_5? What hybridization scheme does work? Explain.

88. What is the total number of **(a)** σ bonds and **(b)** π bonds in the molecule CH_3NCO?

89. Which of the following species are paramagnetic? **(a)** B_2; **(b)** B_2^-; **(c)** B_2^+. Which species has the strongest bond?

90. Use the valence molecular orbital configuration to determine which of the following species is expected to have the lowest ionization energy: **(a)** C_2^+; **(b)** C_2; **(c)** C_2^-.

91. Use the valence molecular orbital configuration to determine which of the following species is expected to have the greatest electron affinity: **(a)** C_2^+; **(b)** Be_2; **(c)** F_2; **(d)** B_2^+.

92. Which of these diatomic molecules do you think has the greater bond energy, Li_2 or C_2? Explain.

93. For each of the following ions or molecules, decide whether the structure is best described by a single Lewis structure or by resonance structures. **(a)** $C_2O_4^{2-}$; **(b)** H_2CO; **(c)** NO_3^-.

94. Draw Lewis structures for the NO_2^- and NO_2^+ ions, and determine the likely geometry for each by using VSEPR theory. How does the hybridization of N differ in these two species?

95. In which of the following is the central atom *sp* hybridized? **(a)** $BeCl_2$; **(b)** BCl_3; **(c)** CCl_4; **(d)** NCl_3; **(e)** none of these.

96. Which of the following can be used to explain why all bond distances and angles in methane, CH_4, are the same? **(a)** resonance; **(b)** delocalization of electrons; **(c)** bond polarities; **(d)** electronegativity; **(e)** orbital hybridization.

97. According to molecular orbital theory, the O_2^{2-} ion has which of the following? **(a)** two unpaired electrons; **(b)** a bond order of two; **(c)** its highest energy electron in a σ^* orbital; **(d)** no 2*s* electrons; **(e)** all of these.

98. What is the angle between the hybrid orbitals obtained by combining the 2*s* and two 2*p* orbitals of an atom? **(a)** 90°; **(b)** 120°; **(c)** 180°; **(d)** 109.5°; **(e)** none of these.

99. Consider the molecule with the Lewis structure given below.

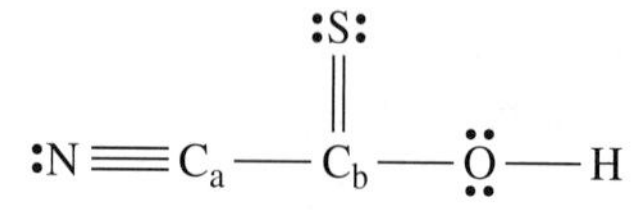

(a) How many σ and π bonds are there?
(b) What is the appropriate hybridization scheme each of C_a, C_b, and O?
(c) In which orbitals are the lone pairs located?
(d) What are the (ideal) values of the following b angles?

$N{-}C_a{-}C_b$ $C_a{-}C_b{=}S$ $C_a{-}C_b{-}O$ $C_b{-}O{-}$

100. Construct a concept map that embodies the idea valence bond theory.

101. Construct a concept map that connects the idea molecular orbital theory.

102. Construct a concept map that describes the in connection between valence bond theory and mo ular orbital theory in the description of resona structures.

ıtermolecular Forces: iquids and Solids

12

ONTENTS

LEARNING OBJECTIVES

12.1 Discuss the different types of intermolecular forces, such as London dispersion, dipole–dipole, and hydrogen bonding forces.

12.2 Describe how the properties of liquids are related to intermolecular forces.

12.3 Discuss how temperature changes affect the state of a substance (i.e., solid, liquid, and gas).

12.4 Identify the triple, melting, boiling, and critical points on a phase diagram.

12.5 Differentiate between a network covalent solid, an ionic solid, a molecular solid, a metallic solid, and give one example of each.

12.6 Describe the packing of spheres for simple cubic, body-centered cubic, and face-centered cubic structures.

12.7 Describe the Born–Fajans–Haber cycle and how it can be used.

this scene from Antarctica, water exists in three states of matter—solid in the ice, uid in the sea, and gas in the atmosphere. Solids, liquids, and gases were mpared at the macroscopic and microscopic levels in Chapter 1 (Fig. 1-7).

When we make ice cubes by placing water in a tray in a freezer, energy is removed from the water molecules, which gradually slow down. Attractive (intermolecular) forces between the mol- ules take over, and the water solidifies into ice. When an ice cube melts, ergy from the surroundings is absorbed by the water molecules, which ercome the intermolecular forces within the ice cube and enter the liquid ate. In our study of gases, we intentionally sought conditions in which e intermolecular forces were negligible. This approach allowed us to scribe gases with the ideal gas equation and to explain their behavior ith the kinetic–molecular theory of gases. To describe the other states of atter—liquids and solids—we must first be able to identify the various termolecular forces. We then consider some interesting properties of uids and solids related to the strengths of these forces.

◀ Two of the many natural phenomena described in this chapter include the more ordered structure of the solid compared with the liquid state and the variation of density with the state of matter.

12-1 Intermolecular Forces

In our study of gases, we noted that at high pressures and low temperatu intermolecular forces cause gas behavior to depart from ideality. When th forces are sufficiently strong compared with the thermal energy, a gas c denses to a liquid. That is, the intermolecular forces keep the molecules in s close proximity that they are confined to a definite volume, as expected for liquid state.

In this section, we will examine the types of intermolecular forces known lectively as **van der Waals forces**. The intermolecular forces contributing to term $a(n/V)^2$ in the van der Waals equation for nonideal gases (equation 6. are of this type.

Two molecular properties—the dipole moment (see Section 10-7) and pol izability (see Section 9-7)—are essential for describing the physical basis attractive intermolecular forces. These properties are used to describe the tribution of electron density within a molecule. Before discussing differ types of intermolecular interactions, we'll review some of the points we ma earlier about these two molecular properties.

Dipole Moment and Polarizability

▶ A bond dipole results from a difference in electronegativities, which causes the electrons between a pair of bonded atoms to be pulled toward the more electronegative atom.

A dipole moment, μ, exists in a molecule when the centers of positive and ne tive charge do not coincide. One way to establish whether a molecule posses a dipole moment is by forming a summation of *bond dipoles*, taking into acco their magnitudes and direction, as illustrated in Figure 10-16 for the molecu CCl_4 and $CHCl_3$. The CCl_4 molecule does not possess a permanent dip moment—the bond dipoles exactly cancel—and is said to be nonpolar. On other hand, the $CHCl_3$ molecule possesses a dipole moment—the bond dipo do not cancel—and is said to be polar. It is important to be able to distingu between polar and nonpolar molecules because polar molecules interact w each other in more ways than do nonpolar molecules.

▶ As pointed out in Section 9-7, the electrons in a molecule do not contribute equally to its polarizability. Electrons farther from the nuclei are less firmly held, are more easily displaced, and contribute more to the polarizability than do the electrons closer to the nuclei.

The polarizability, α, of a molecule provides a measure of the extent to wh its electron cloud can be distorted from its "normal" or "average" shape, example, by the application of an externally applied electric field or by t approach of another molecule. The polarizability of a molecule depends on h diffuse or "spread out" its electron cloud is. Polarizability is often expressed units of volume, which suggests that the polarizability of a molecule is related the volume of its charge cloud.

We now turn our attention to different types of interactions that contribute the attractions between molecules and account for differences in the physi properties of compounds.

Dipole–Dipole Interactions

The first type of intermolecular interaction we will discuss is the interacti between a pair of *polar* molecules. A polar molecule has a permanent dip moment, so polar molecules tend to line up with the positive end of one dip directed toward the negative ends of neighboring dipoles (Figure 12-1). Th partial ordering of molecules can cause a substance to persist as a solid or liqu at temperatures higher than otherwise expected.

As an example of the influence of dipole–dipole interactions on physic properties, consider Figure 12-2, which depicts the molecules CF_4 and CH These two molecules have very similar polarizabilities, but CHF_3 is polar a CF_4 is not. This seemingly simple difference has a significant effect: T observed boiling point of CHF_3 is more than 40 °C higher than that of CF_4.

There are two reasons that the polar CHF_3 molecules interact more strong with each other than do the nonpolar CF_4 molecules:

- *dipole–dipole interactions,* which refer to the head-to-tail interactions of t CHF_3 dipole moments, as already discussed and illustrated in Figure 12

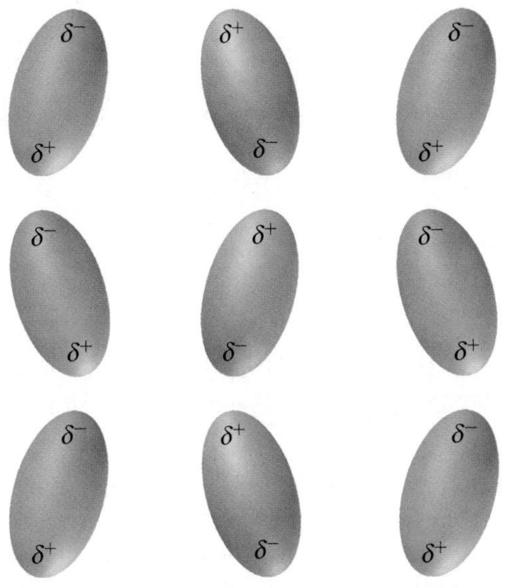

◀ FIGURE 12-1
Dipole–dipole interactions
Dipoles tend to arrange themselves with the positive end of one dipole pointed toward the negative end of a neighboring dipole. Ordinarily, thermal motion upsets this orderly array. Nevertheless, this tendency for dipoles to align themselves can affect physical properties, such as the melting points of solids and the boiling points of liquids.

- *dipole–induced dipole interactions,* which arise because the dipole moment on one molecule induces a change in the dipole moment of a neighboring molecule

pole–induced dipole interactions result from the fact that all molecules are larizable. The electronic charge cloud of a molecule (polar or nonpolar) will ways be distorted to some extent by the approach of a polar molecule. The wly formed dipole—which is different from the original dipole—is called an *luced dipole.* The interaction between a dipole and an induced dipole causes e attraction between a pair of polar molecules to be slightly greater than that edicted by interaction of the unmodified dipole moments. (See Exercise 121.)

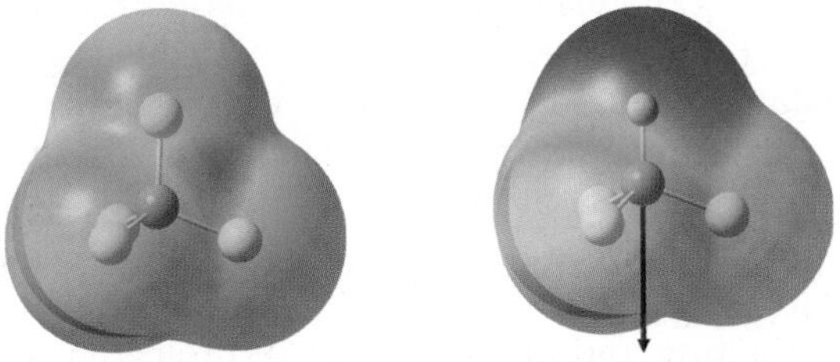

CF_4
$\mu = 0$ (nonpolar)
$\alpha = 3.8 \times 10^{-24}$ cm^3

CHF_3
$\mu = 1.65$ D (polar)
$\alpha = 3.6 \times 10^{-24}$ cm^3

FIGURE 12-2
ectrostatic potential maps and properties of CF_4 and CHF_3
e CF_4 and CHF_3 molecules have very similar polarizabilities, but CHF_3 is polar and CF_4 not. The dipole moment vector of CHF_3 is represented by the blue arrow. The ference in the dipole moments manifests itself as a dramatic difference in boiling point. discussed, the intermolecular attractions between CHF_3 molecules include additional ntributions—namely, dipole–dipole and dipole–induced dipole interactions—that are t present between pairs of CF_4 molecules.

◀ The SI unit for dipole moment is C m. However, the non-SI unit, debye (D), is often used instead. One debye is approximately 3.34×10^{-30} C m.

ispersion Forces

though the intermolecular interactions described in the previous section are portant for understanding the interactions between polar molecules, these teractions are usually not the most important. Another type of interaction ust exist between molecules because, for example, even nonpolar sub-ances—such as He, N_2, CF_4, and so on—will liquefy if the temperature is wered sufficiently. To understand the nature of this other type of interaction,

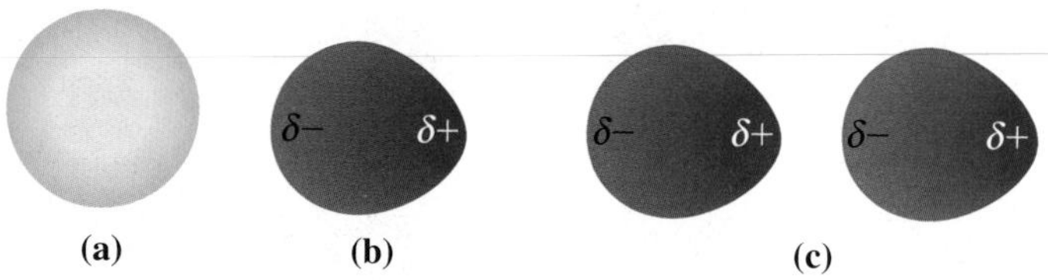

▲ FIGURE 12-3
Instantaneous and induced dipoles
(a) In the *normal condition*, a nonpolar molecule has a symmetrical charge distributi
(b) In the *instantaneous condition*, a displacement of the electronic charge produce
instantaneous dipole with a charge separation represented as δ+ and δ−. **(c)** In an *induced dipole*, the instantaneous dipole on the left induces a charge separation in molecule on the right. The result is an instantaneous dipole–induced dipole attracti

which always contributes to the attraction between molecules—polar nonpolar—we need to expand on what we already know about polarizabil

There is an important characteristic of polarizability that we have not mentioned, namely, its dynamic (time-varying) nature. Because the electr in a molecule are in constant motion, it is possible that at some particu instant—purely by chance—electrons are concentrated in one region of a m ecule. This displacement of electrons causes, for example, a normally non lar species to become momentarily polar. An *instantaneous dipole* is form That is, the molecule has an instantaneous dipole moment. After this, el trons in a neighboring molecule may be displaced to produce a dipole— induced dipole. Taken together, these two events lead to an intermolecu force of attraction (Fig. 12-3). We can call this interaction an instantanec dipole–induced dipole attraction, but the names more commonly used **dispersion force** and **London force**, the latter in honor of Fritz London w in 1928, offered a theoretical explanation of these forces.

Generally speaking, dispersion forces become stronger (more attractive) polarizability increases. Therefore, substances made up of larger, more pol izable molecules tend to have higher boiling points, as suggested by the d in Table 12.1. The data in Table 12.1 also show that, roughly speaking, polar ability increases with molecular mass. This correlation is not totally un pected. A molecule with many atoms is not only massive but also has ma electrons. Therefore, a massive molecule generally has a large, polarizal charge cloud. Because of the correlation between polarizability and mass, i fair to say that the melting and boiling points of molecular substances tend increase with increasing molecular mass. For instance, helium (atomic mass, 4 has a boiling point of 4 K, whereas radon (atomic mass, 222 u) has a boili

▶ Recall from Chapter 3, a molecular substance is made up of molecules. The molecules interact with each other through relatively weak intermolecular forces. The atoms of a given molecule are held together by relatively strong covalent bonds.

TABLE 12.1 Some Properties of Selected Nonpolar Compounds

Compound	Molar Mass, u	Polarizability,* 10^{-25} cm^3	Boiling Point, K
H_2	2.016	8.04	20.35
O_2	32.00	15.7	90.19
N_2	28.01	17.4	77.35
CH_4	16.04	25.9	109.15
CH_3CH_3	30.07	44.7	184.55
Cl_2	70.90	46.1	238.25
$CH_3CH_2CH_3$	44.10	62.9	231.05
CCl_4	153.81	112	349.95

*Sometimes polarizability is referred to as polarizability volume. Note that the units of polarizability given above have the units of volume. Thus, polarizability provides a measur of the atomic or molecular volume. Polarizability values are from the *CRC Handbook of Chemistry and Physics*, 93rd edition.

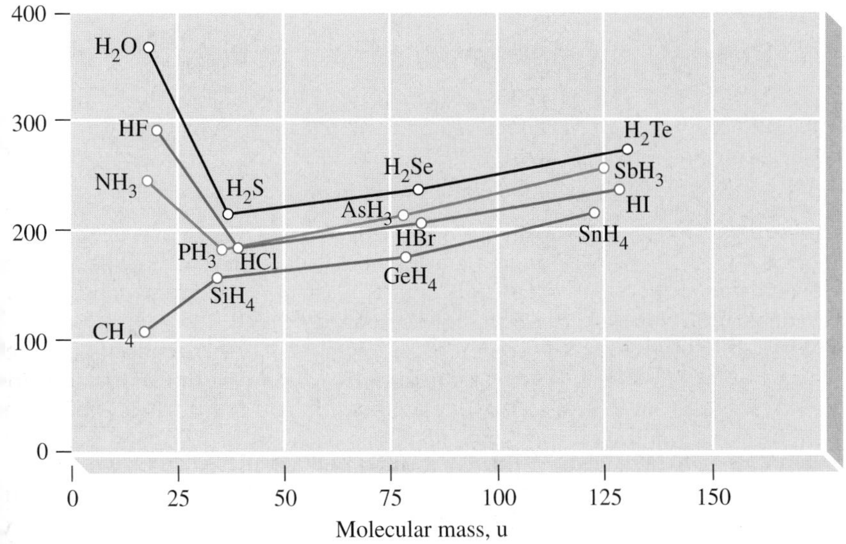

◀ FIGURE 12-4
Comparison of boiling points of some hydrides of the elements of groups 14, 15, 16, and 17
The values for NH_3, H_2O, and HF are unusually high compared with those of other members of their groups.

int of 211 K. However, it is important to remember the underlying reason r the observation that boiling points tend to increase with molecular mass: *increasing polarizability of the atoms or molecules.*

ydrogen Bonding

gure 12-4, in which the boiling points of a series of similar compounds are plot-d as a function of molecular mass, demonstrates some features that we cannot plain by the types of intermolecular forces considered to this point. The hydro-n compounds (hydrides) of the group 14 elements display normal behavior;

EXAMPLE 12-1 Comparing Physical Properties of Polar and Nonpolar Substances

Which would you expect to have the higher boiling point, the hydrocarbon fuel butane, C_4H_{10}, or the organic solvent acetone, $(CH_3)_2CO$?

Analyze

Because the two substances have the same molecular mass (58 u), we expect that the polarizabilities of C_4H_{10} and $(CH_3)_2CO$ molecules will be approximately equal. Thus, we have to look elsewhere for a factor on which to base our prediction.

The next consideration is the polarity of the molecules. The electronegativity difference between C and H is so small that we generally expect hydrocarbons, such as butane, to be nonpolar. However, we notice that one of the molecules contains a carbon–oxygen bond, and thus, a strong carbon-to-oxygen dipole. At times, it is helpful to sketch the structure of a molecule to see whether symmetrical features cause bond dipoles to cancel. It is not necessary to sketch the structure of the acetone molecule to deduce that it is a polar molecule. The C=O bond dipole in acetone cannot be offset by other bond dipoles. Thus, acetone is polar.

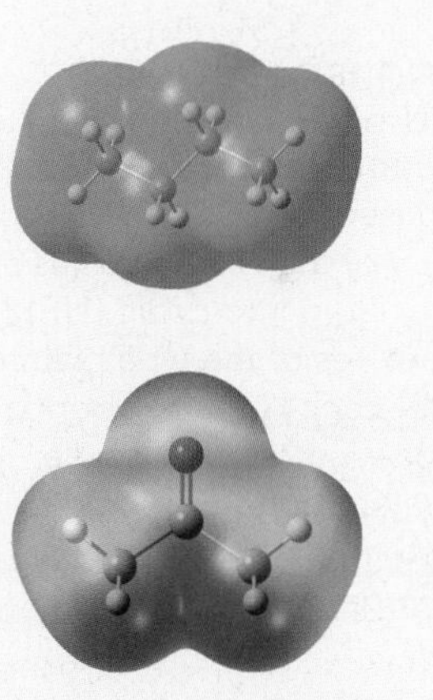

▲ **Butane and acetone**
The diagrams show electrostatic potential maps for butane and acetone. The red color indicates regions of high negative electrostatic potential.

Solve

Given two substances with the same molecular mass, one polar and one non-polar, we expect the polar substance—acetone—to have the higher boiling point. (The measured boiling points are butane, −0.5 °C; acetone, 56.2 °C.)

Assess

In general, when comparing the properties of different substances, we must consider the various types of intermolecular forces and the factors that affect the strength of each type of force. Although it wasn't important here, the three-dimensional shape of a molecule is usually a very important consideration and it is usually necessary to sketch the molecular structure to see how molecular shape plays a role.

(continued)

PRACTICE EXAMPLE A: Which of the following substances would you expect to have the highest boiling point C_3H_8, CO_2, CH_3CN? Explain.

PRACTICE EXAMPLE B: Arrange the following in the expected order of increasing boiling point: C_8H_{18} $CH_3CH_2CH_2CH_3$, $(CH_3)_3CH$, C_6H_5CHO (octane, butane, isobutane, and benzaldehyde respectively).

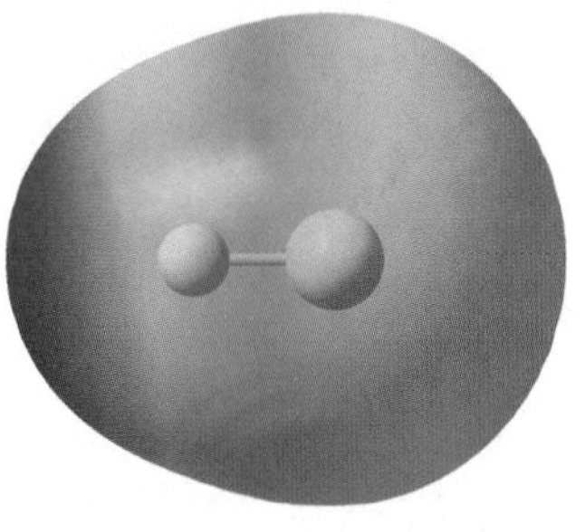

▲ **Electrostatic potential map of HF**

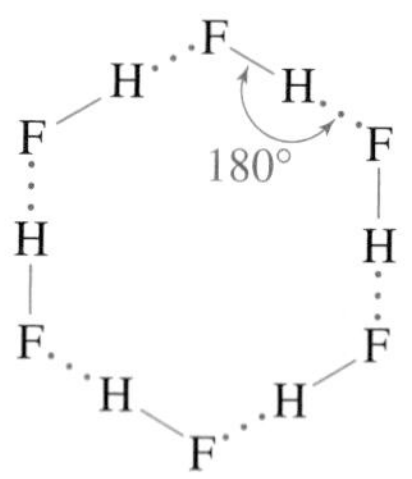

▲ FIGURE 12-5
Hydrogen bonding in gaseous hydrogen fluoride
In gaseous hydrogen fluoride, many of the HF molecules are associated into cyclic $(HF)_6$ structures of the type pictured here. Each H atom is bonded to one F atom by a single covalent bond (—) and to another F atom through a hydrogen bond (···).

that is, the boiling points increase regularly as the molecular mass increases. there are three striking exceptions in groups 15, 16, and 17. The boiling point NH_3, H_2O, and HF are as high or higher than those of any other hydride in th group—not lowest, as we might expect. A special type of intermolecular fo causes this exceptional behavior, as we see for hydrogen fluoride in Figure 1 The main points established in the figure are outlined below.

- The alignment of HF dipoles places an H atom between two F ato Because of the very small size of the H atom, the dipoles come cl together and produce strong *dipole–dipole* attractions.
- Although an H atom is covalently bonded to one F atom, it is also wea bonded to the F atom of a nearby HF molecule. This occurs throug lone pair of electrons on the F atom. Each H atom acts as a bric between two F atoms.
- The bond angle between two F atoms bridged by an H atom (that is, angle F—HF···F) is about 180°.

The type of intermolecular force just described is called a **hydrogen bo** although it is simply an electrostatic attraction and not an actual chemi bond like a covalent bond. In a hydrogen bond an H atom is covalen bonded to a highly electronegative atom, which attracts electron density aw from the H nucleus. This in turn allows the H nucleus, a proton, to be simul neously attracted to a lone pair of electrons on a highly electronegative at in a neighboring molecule.

In general, a hydrogen bond is depicted as X—H··· Y—, where the th dots denote the hydrogen bond. The X—H fragment is typically referred as the hydrogen bond donor since the fragment X—H provides the hyd gen as part of the hydrogen bond. The fragment Y— is known as the hyd gen bond acceptor since it accepts the hydrogen as part of the hydrog bond. Figure 12-6 illustrates the definition of a hydrogen bond along wi several examples. Compared with other intermolecular forces, hydrog bonds are relatively strong, having energies of the order of 15 to 40 kJ mol By contrast, single covalent bonds (also known as intramolecular bonds) a much stronger still—greater than 150 kJ mol^{-1}. (See Table 10.3 for furth comparisons.)

Hydrogen Bonding in Water

Ordinary water is certainly the most common substance in which hydrog bonding occurs. Figure 12-7 shows how one water molecule is held to fo neighbors in a tetrahedral arrangement by hydrogen bonds. In ice, hydrog bonds hold the water molecules in a rigid but rather open structure. As i melts, only a fraction of the hydrogen bonds are broken. One indication of th is the relatively low heat of fusion of ice (6.01 kJ mol^{-1}). It is much less th we would expect if all the hydrogen bonds were to break during melting.

The open structure of ice shown in Figure 12-7(b) gives ice a low densi When ice melts, some of the hydrogen bonds are broken. This allows the wat molecules to be more compactly arranged, accounting for the increase in de sity when ice melts. That is, the number of H_2O molecules per unit volume greater in the liquid than in the solid.

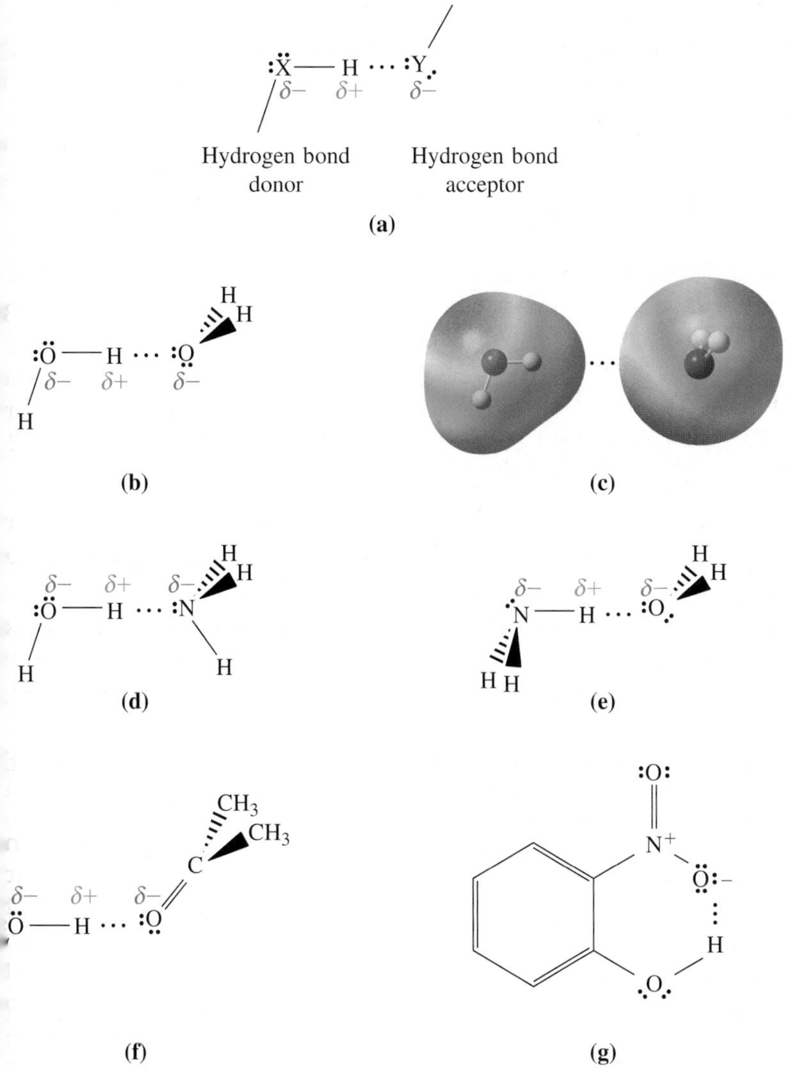

◀ FIGURE 12-6
The hydrogen bond illustrated
(a) A hydrogen bond involves a hydrogen bond donor (shown on the left) and a hydrogen bond acceptor (shown on the right). **(b)** Hydrogen bonding between H_2O molecules represented by using Lewis structures. **(c)** The electrostatic potential maps for the H_2O molecules forming a hydrogen bond show that the electron-rich (red) region of one water molecule interacts with the electron-deficient (blue) region of the other water molecule. **(d)** In this hydrogen bond between water and ammonia molecules, the water molecule is the donor and the ammonia molecule is the acceptor. **(e)** For this hydrogen bond, the ammonia molecule is the donor and the water molecule is the acceptor. **(f)** Hydrogen bonding between water and acetone molecules. **(g)** An intramolecular hydrogen bond.

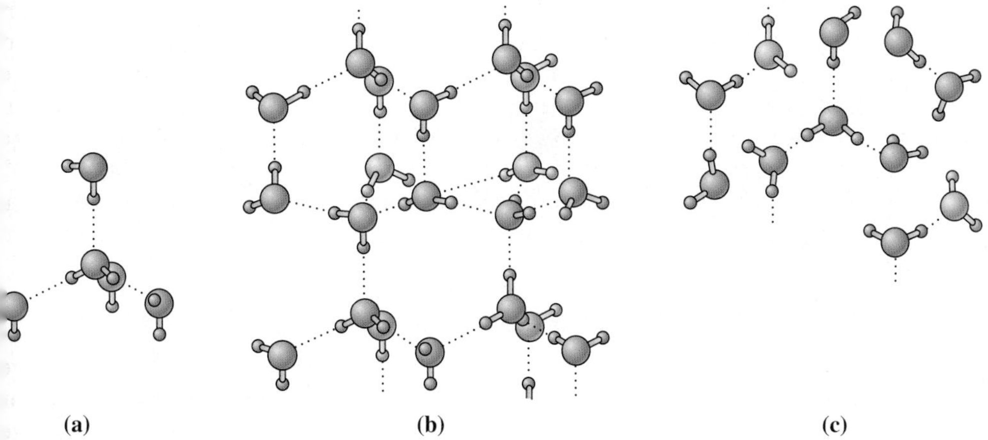

FIGURE 12-7
ydrogen bonding in water
) Each water molecule is linked to four others through hydrogen bonds. The arrangement is trahedral. Each H atom is situated along a line joining two O atoms, but closer to one O atom (100 pm) an to the other (180 pm). **(b)** For the crystal structure of ice, H atoms lie between pairs of O atoms, ɟain closer to one O atom than to the other. (Molecules behind the plane of the page are light blue.) atoms are arranged in bent hexagonal rings arranged in layers. This characteristic pattern is similar to e hexagonal shapes of snowflakes. **(c)** In the liquid, water molecules have hydrogen bonds to only some their neighbors. This allows the water molecules to pack more densely in the liquid than in the solid.

Richard Megna/Fundamental Photographs, NYC

▲ FIGURE 12-8
Solid and liquid densities compared
The sight of ice cubes floating on liquid water (left) is a familiar one; ice is less dense than liquid water. The more common situation, however, is that of paraffin wax (right). Solid paraffin is denser than the liquid and sinks to the bottom of the beaker.

As liquid water is heated above the melting point, hydrogen bonds c tinue to break. The molecules become even more closely packed, and density of the liquid water continues to increase. Liquid water attains maximum density at 3.98 °C. Above this temperature, the water behave a "normal" fashion: Its density decreases as temperature increases. The unus freezing-point behavior of water explains why a freshwater lake freezes fr the top down. When the water temperature falls below 4 °C, the denser wa sinks to the bottom of the lake and the colder surface water freezes. The over the top of the lake then tends to insulate the water below from furt heat loss. This allows fish to survive the winter in a lake that has been fro over. Without hydrogen bonding, all lakes would freeze from the bottom and fish, small bottom-feeding animals, and aquatic plants would survive the winter. The density relationship between liquid water a ice is compared in Figure 12-8 with the more common liquid–solid dens relationship.

Other Properties Affected by Hydrogen Bonding

Water is one example of a substance whose properties are affected by hyd gen bonding. There are numerous others. In acetic acid, CH_3COOH, pairs molecules tend to join together into *dimers* (double molecules), both in t liquid and in the vapor states (Fig. 12-9). Not all the hydrogen bonds disrupted when liquid acetic acid vaporizes, and, as a result, the heat vaporization is abnormally low.

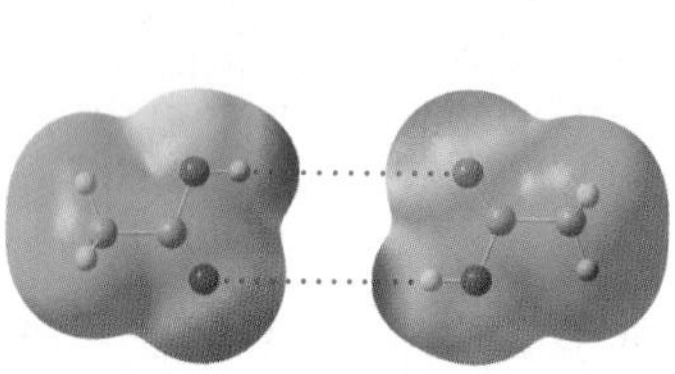

▲ FIGURE 12-9
An acetic acid dimer
Electrostatic potential maps showing hydrogen bonding.

Certain trends in viscosity can also be explained by hydrogen bonding. alcohols, the H atom in a —OH group in one molecule can form a hydrog bond to the O atom in a neighboring alcohol molecule. An alcohol molec with two —OH groups (a *diol*) has more possibilities for hydrogen-bo formation than a comparable alcohol with a single —OH group. Havi stronger intermolecular forces, we expect the diol to flow more slowly, that to have a greater viscosity, than the simple alcohol. When still more —C groups are present (*polyols*), we expect a further increase in viscosity. The comparisons are illustrated by the three common alcohols below. (The unit is a centipoise. The SI unit of viscosity is $1\ \mathrm{N\,s\,m^{-2}} = 10\ \mathrm{P} = 1000\ \mathrm{cP}$. T Greek letter eta, η, is typically used as a symbol for viscosity.)

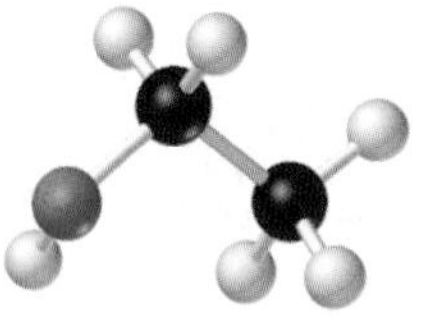

Ethyl alcohol
(ethanol)
at 20 °C: $\eta = 1.20$ cP

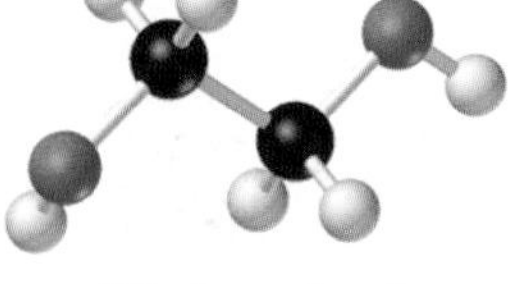

Ethylene glycol
(1,2-ethanediol)
at 20 °C: $\eta = 19.9$ cP

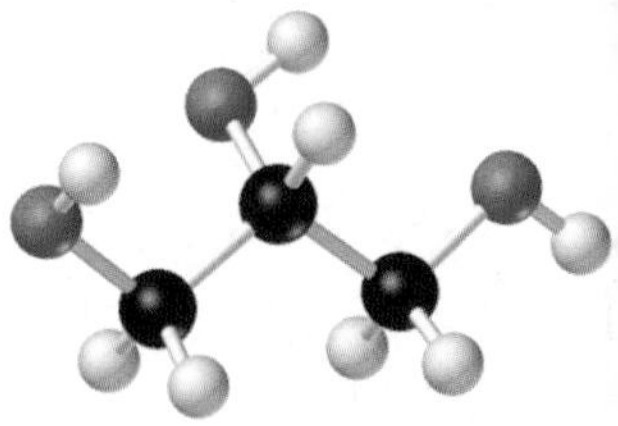

Glycerol
(1,2,3-propanetriol)
at 20 °C: $\eta = 1490$ cP

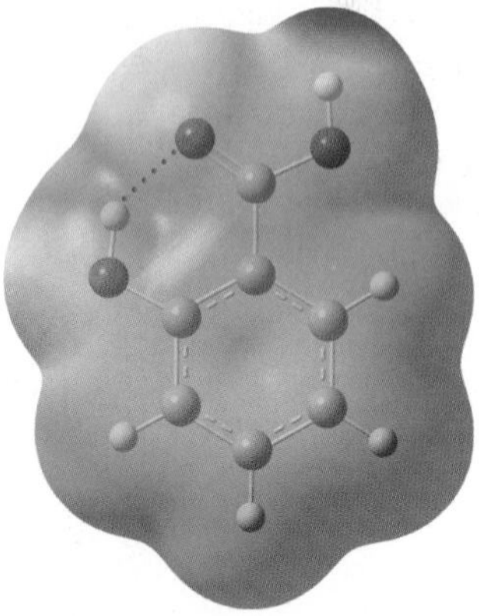

▲ Electrostatic potential map of salicylic acid showing intramolecular hydrogen bonding.

*Inter*molecular and *Intra*molecular Hydrogen Bonding

All the examples of hydrogen bonding presented to this point have involve an intermolecular force *between two molecules*, and this is called an *intermol cular hydrogen bond*. Another possibility occurs in molecules with an H ato covalently bonded to one highly electronegative atom (for example, O or N and with another highly electronegative atom nearby in the same molecul This type of hydrogen bonding *within a molecule* is called *intramolecul hydrogen bonding*. As shown in the molecular model of salicylic acid on th facing page, an intramolecular hydrogen bond (represented by a dotted lin joins the —OH group to the doubly bonded oxygen atom of the —COO up on the same molecule. To underscore the importance of molecular ›metry in establishing the conditions necessary for intramolecular hydro- bonding, we need only turn to an isomer of salicylic acid called *para-* łroxybenzoic acid. In this molecule, the H atom of the —COOH group is close to the doubly bonded O atom of the same group to form a hydro- bond, and the H atom of the —OH group on the opposite side of the lecule is too far away. Intramolecular hydrogen bonding does not occur his situation.

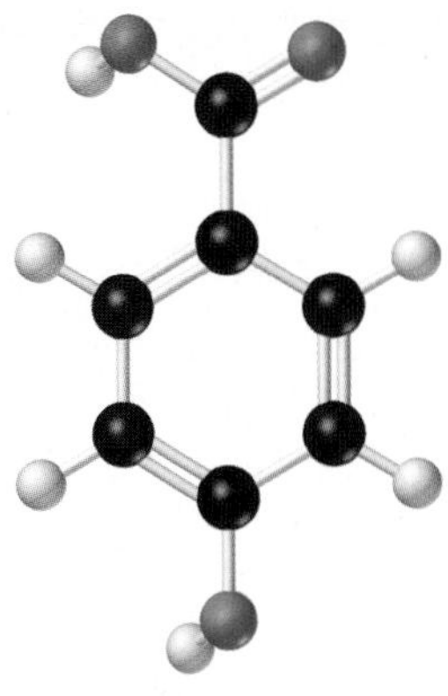

▲ There is no intramolecular hydrogen bonding in *para*-hydroxybenzoic acid.

·drogen Bonding In Living Matter

ne chemical reactions in living matter involve complex structures, such as ›teins and DNA, and in these reactions certain bonds must be easily broken d re-formed. Hydrogen bonding is a type of bonding with energies of just right magnitude to allow this. We will also find that both intra- and inter- lecular hydrogen bonding is involved in these complex structures.

Hydrogen bonding seems to provide an answer to the puzzle of how ne trees are able to grow to great heights. In Chapter 6 (page 198), we rned that atmospheric pressure is capable of pushing a column of water a maximum height of only about 10 m—other factors must be involved in nsporting water to the tops of redwood trees up to 100 m tall. Hydrogen nding seems to be a factor in transporting water in trees. Thin columns of ter (in xylem, a plant tissue) extend from the roots to the leaves in the ry tops of trees. In these columns, the water molecules are hydrogen- nded to one another, with each water molecule acting like a link in a 1esive chain. When one water molecule evaporates from a leaf, another olecule in the chain moves to take its place and all the other molecules are lled up the chain. Ultimately, a new water molecule joins the chain in the ›t system. In the next chapter, we will learn about another factor in trans- rting water in trees: osmotic pressure and its ability to force water ough a membrane.

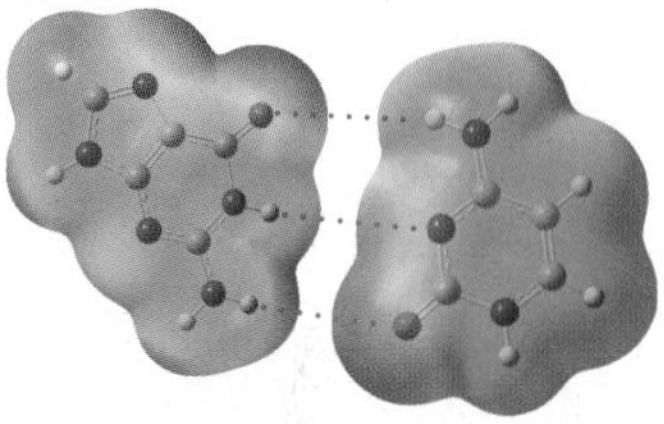

▲ **Hydrogen bonding between guanine (left) and cytosine (right) in DNA**

Alyona Burchette/Shutterstock

▲ **Sequoia trees**
The mystery of how these trees can bring water to leaves that are hundreds of feet up may be explained by hydrogen bonding.

12-1 CONCEPT ASSESSMENT

Vhat are the types of intermolecular interactions in $CH_3CH_2NH_2(l)$, and which is he strongest?

mmary of van der Waals Forces

hen assessing the importance of van der Waals forces, consider the follow- ; statements.

- *Dispersion (London) forces exist between all molecules.* They involve displacements of all the electrons in molecules, and they increase in strength with increasing molecular size. The forces also depend on molecular shapes.
- *Forces associated with permanent dipoles* involve displacements of electron pairs in bonds rather than in molecules as a whole. These forces are found only in substances with resultant dipole moments (polar molecules). Their existence *adds* to the effect of dispersion forces also present.
- *When comparing substances of roughly comparable molecular sizes,* dipole forces can produce significant differences in properties such as melting point, boiling point, and enthalpy of vaporization.
- *When comparing substances of widely different molecular sizes,* dispersion forces are usually more significant than dipole forces.

Let's see how these statements relate to the data in Table 12.2, which cludes a rough breakdown of van der Waals forces into dispersion forces

TABLE 12.2 Intermolecular Forces and Properties of Selected Substances

	Molecular Mass, u	Polarizability, 10^{-25} cm^3	Dipole Moment, D	Van der Waals Forces		$\Delta_{vap}H$ kJ mol^{-1}	Boiling Point, K
				% Dispersion	% Dipole		
F_2	38.00	13.8	0	100	0	6.86	85.01
HCl	36.46	26.3	1.08	77.4	22.6	16.15	188.11
HBr	80.91	36.1	0.82	93.6	6.4	17.61	206.43
HI	127.91	54.4	0.44	99.2	0.8	19.77	237.80

▶ In our discussion of London forces, we use molecular mass only as a guide to the number of electrons present in a molecule; the forces holding a molecule in a liquid are not gravitational in nature.

and forces caused by dipoles. HCl and F_2 have comparable molecular siz but because HCl is polar, it has a significantly larger $\Delta_{vap}H$ and a higher b ing point. Within the series HCl, HBr, and HI, molecular size increases shar and $\Delta_{vap}H$ and boiling points increase in the order HCl < HBr < The more polar nature of HCl and HBr relative to HI is not sufficient to reve the trends produced by the increasing molecular sizes—dispersion forces are predominant intermolecular forces.

12-1 ARE YOU WONDERING?

Are there other types of intermolecular forces?

Table 12.2 summarizes some of the more commonly encountered interaction involving atoms, molecules, and ions, along with their typical strengths.

In Table 12.3 one of the strongest forces between ions or molecules arises fron the electrostatic attraction of opposite charge (ion–ion). This type of interaction is what gives rise to the high melting points of ionic solids, along with their brittle nature. In such molecules as biomolecules, interactions between differen charged groups, also known as salt bridges, increase their stability.

As we discovered in Chapter 11, molecules have dipole moments because o electronegativity differences between atoms. The dipole moment in a molecule will interact with a charged ion to form an ion–dipole interaction. Ion–dipole interactions are important in understanding the dissolution of salts.

In the absence of charges and dipole moments, other higher-order moments (e.g., quadrupole moments) become dominant. An example of quadrupole-quadrupole interactions would be between two CO_2 molecules.

Although CO_2 does not possess a dipole moment, the polarity of the carbon–oxygen bonds gives rise to a quadrupole moment, as illustrated below. Whereas a dipole moment is represented by two partial charges, $\delta+$ and $\delta-$, a quadrupole moment is represented by four partial charges.

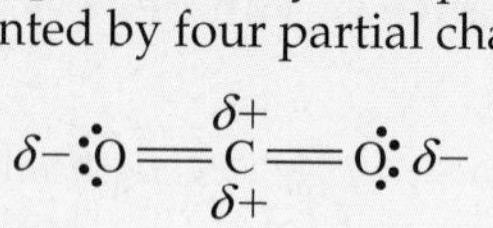

Carey B. Van Loon

▲ FIGURE 12-10
An effect of surface tension illustrated
Despite being denser than water, the needle is supported on the surface of the water. The property of surface tension accounts for this unexpected behavior.

12-2 Some Properties of Liquids

Surface Tension

The observation of a needle floating on water, as pictured in Figure 12-10, puzzling. Steel is much denser than water and should not float. Somethi must overcome the force of gravity on the needle, allowing it to remain su pended on the surface of the water. What is this special quality associated wi the surface of liquid water?

Figure 12-11 suggests an important difference in the forces experienced molecules within the bulk of a liquid and by those at the surface. Interi

TABLE 12.3 Summary of Noncovalent Interactions

Force	Energy,[a] kJ/mol	Example	Model
Intermolecular			
London dispersion	0.05–40	$CH_4 \cdots CH_4$	
Dipole–induced dipole	2–10	$CH_3(CO)CH_3 \cdots C_5H_{12}$	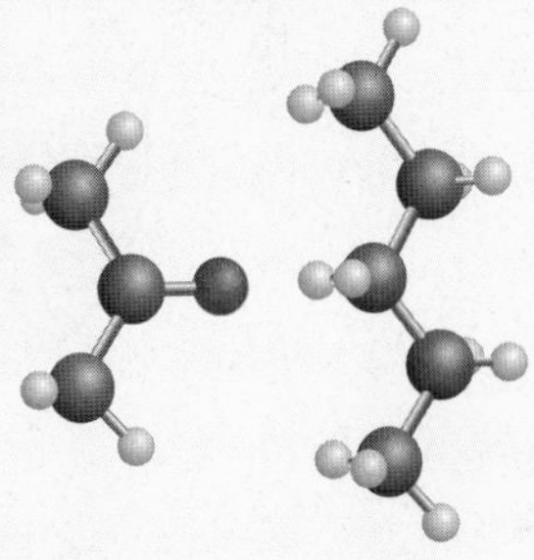
Ion–induced dipole	3–15	$Li^+ \cdots C_5H_{12}$	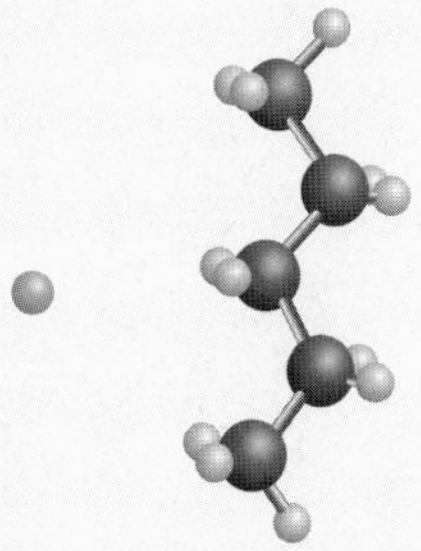
Dipole–dipole	5–25	$H_2O \cdots CO$	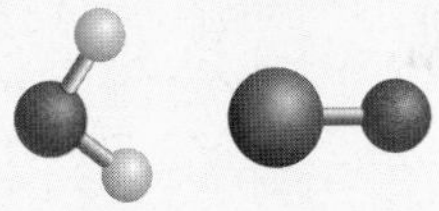
Hydrogen bond	10–40	$CH_3OH \cdots H_2O$	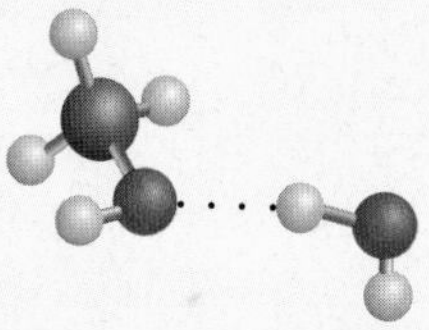
Ion–dipole	40–600	$K^+ \cdots H_2O$	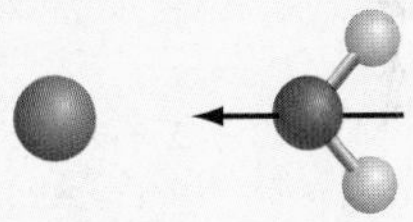
Ion–ion	400–4000	$Lys^+ \cdots Glu^-$	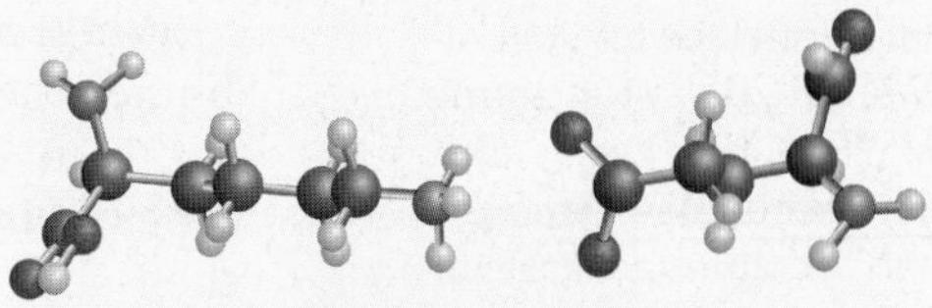
Interatomic			
London dispersion	0.05–40	$Ar \cdots Ar$	
Ion–ion	400–4000	$Na^+ \cdots Cl^-$	 
Metallic	100–1000	$Ag \cdots Ag$	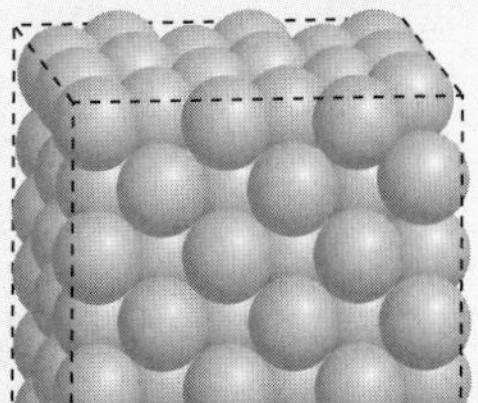

[a] These are gas phase values.

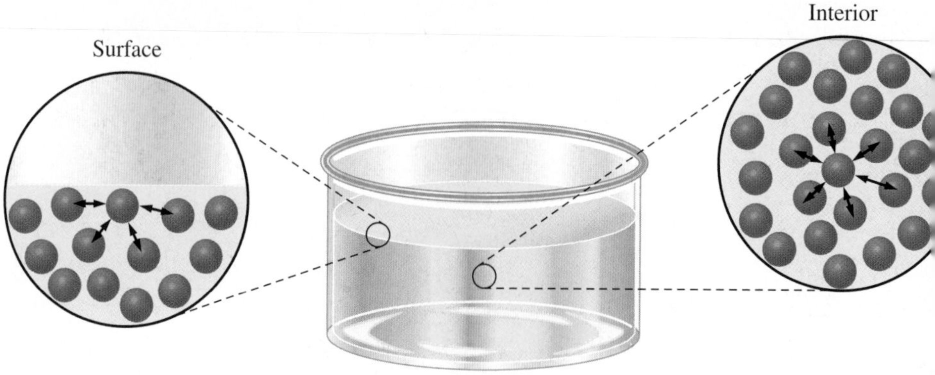

▲ FIGURE 12-11
Intermolecular forces in a liquid
Molecules at the surface are attracted only by other surface molecules and by molecules below the surface. Molecules in the interior experience forces from neighboring molecules in all directions.

molecules have more neighbors and experience more attractive intermole lar interactions than surface molecules. The increased number of attractic by neighboring molecules places an interior molecule in a more stable en ronment (lower energy state) than a surface molecule. Consequently, as ma molecules as possible tend to enter the bulk of a liquid, while as few as pos ble remain at the surface. Thus, liquids tend to maintain a minimum surfa area. To increase the surface area of a liquid requires that molecules be mov from the interior to the surface of a liquid, and this requires that work done. The steel needle of Figure 12-10 remains suspended on the surface the water because energy is required to spread the surface of the water ov the top of the needle.

Surface tension is the energy, or work, required to increase the surface area a liquid. Surface tension is often represented by the Greek letter gamma (γ) a has the units of energy per unit area, typically joules per square meter ($J\ m^{-2}$). the temperature—and hence the intensity of molecular motion—increases, int molecular forces become less effective. Less work is required to extend the su face of a liquid, meaning that surface tension *decreases* with *increased* temperatu

▶ The surface tension of water at 20 °C, for example, is $7.28 \times 10^{-2}\ J\ m^{-2}$, and that of mercury is more than six times as large, at $47.2 \times 10^{-2}\ J\ m^{-2}$.

When a drop of liquid spreads into a film across a surface, we say that t liquid *wets* the surface. Whether a drop of liquid wets a surface or retains spherical shape and stands on the surface depends on the strengths of tw types of intermolecular forces. The forces exerted between molecules holdi them together in the drop are **cohesive forces**, and the forces between liqu molecules and the surface are **adhesive forces**. If cohesive forces are stro compared with adhesive forces, a drop maintains its shape. If adhesive forc are strong enough, the energy requirement for spreading the drop into a fil is met through the work done by the collapsing drop.

Water wets many surfaces, such as glass and certain fabrics. This characte istic is essential to its use as a cleaning agent. If glass is coated with a film of or grease, water no longer wets the surface and water droplets stand on th glass, as shown in Figure 12-12. When we clean glassware in the laborato we have done a good job if water forms a uniform thin film on the glass. Wh we wax a car, we have done a good job if water uniformly beads up all alo the surface.

Adding a detergent to water has two effects: The detergent solution di solves grease to expose a clean surface, and the detergent lowers the surfa tension of water. Lowering the surface tension means lowering the energ required to spread drops into a film. Substances that reduce the surface te sion of water and allow it to spread more easily are known as *wetting agen*

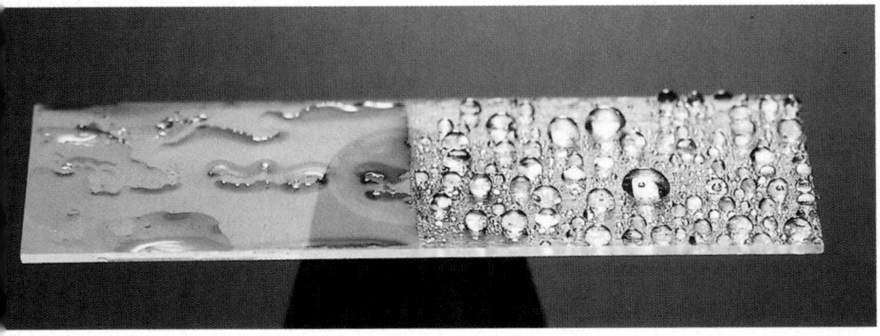

'IGURE 12-12
tting of a surface
ter spreads into a thin film on a clean glass surface (left). If the glass is coated with or grease, the adhesive forces between the water and oil are not strong enough to ead the water, and droplets stand on the surface (right).

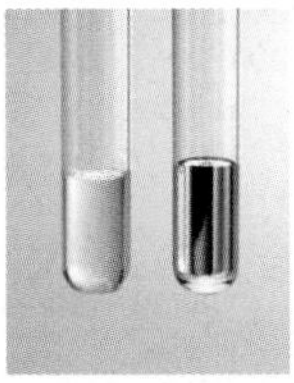

▲ FIGURE 12-13
Meniscus formation
Water wets glass (left). The meniscus is concave—the bottom of the meniscus is below the level of the water–glass contact line. Mercury does not wet glass. The meniscus is convex—the top of the meniscus is above the mercury–glass contact line.

ey are used in applications ranging from dish washing to industrial)cesses.

Figure 12-13 illustrates another familiar observation. If the liquid in the ıss tube is water, the water is drawn slightly up the walls of the tube by hesive forces between water and glass. The interface between the water and : air above it, called a *meniscus,* is concave, or curved in. With liquid mer-ry, the meniscus is convex, or curved out. Cohesive forces in mercury, con-ting of metallic bonds between Hg atoms, are strong; mercury does not wet ss. The effect of meniscus formation is greatly magnified in tubes of small ımeter, called *capillary tubes.* In the *capillary action* shown in Figure 12-14, the ter level inside the capillary tube is noticeably higher than outside. The ıking action of a sponge depends on the rise of water into capillaries of a rous material, such as cellulose. The penetration of water into soils also pends in part on capillary action. Conversely, mercury—with its strong ıesive forces and weaker adhesive forces—does not show a capillary rise. ther, mercury in a glass capillary tube will have a lower level than the mer-ry outside the capillary.

▲ FIGURE 12-14
Capillary action
A thin film of water spreads up the inside walls of the capillary because of strong adhesive forces between water and glass (water wets glass). The pressure below the meniscus falls slightly. Atmospheric pressure then pushes a column of water up the tube to eliminate the pressure difference. The *smaller* the diameter of the capillary, the *higher* the liquid rises. Because its magnitude is also directly proportional to surface tension, capillary rise provides a simple experimental method of determining surface tension, described in Exercise 122.

scosity

ıother property at least partly related to intermolecular forces is **viscosity**— iquid's resistance to flow. The stronger the intermolecular forces of attrac-n, the greater the viscosity. When a liquid flows, one portion of the liquid)ves with respect to neighboring portions. Cohesive forces within the liquid ːate an internal friction, which reduces the rate of flow. In liquids of low vis-sity, such as ethyl alcohol and water, the effect is weak, and they flow easily. quids such as honey and heavy motor oil flow much more sluggishly. We y that they are *viscous.* One method of measuring viscosity is to time the fall a steel ball through a certain depth of liquid (Fig. 12-15). The greater the vis-sity of the liquid, the longer it takes for the ball to fall. Because intermolecular ːces of attraction can be offset by higher molecular kinetic energies, viscosity nerally *decreases* with *increased* temperature for liquids.

◀ For liquids, viscosity decreases with increasing temperature, but for gases, the viscosity increases with increasing temperature.

12-2 CONCEPT ASSESSMENT

The viscosity of automotive motor oil is designated by its SAE number, such as 40 W. When compared in a ball viscometer (Fig. 12-15), the ball drops much faster through 10 W oil than through 40 W oil. Which of these two oils provides better winter service in the Arctic region of Canada? Which is best suited for summer use in the American Southwest? Which oil has the stronger intermolecular forces of attraction?

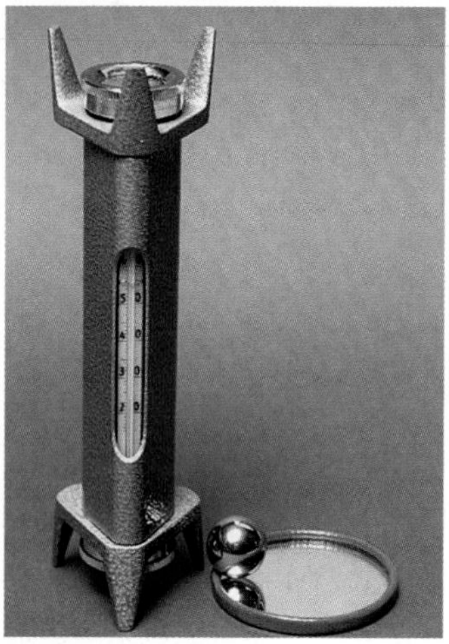

Tom Pantages

▲ FIGURE 12-15
Measuring viscocity
By measuring the velocity of a ball dropping through a liquid, a measure of the liquid viscosity can be obtained.

Enthalpy of Vaporization

In our study of the kinetic–molecular theory (Section 6-7), we saw that the spe and kinetic energies of molecules vary over a wide range at any given tempe ture (Fig. 6-16). Then, in Chapter 7, we learned that molecules having kinetic en gies sufficiently above the average value are able to overcome intermolecu forces of attraction and escape from the surface of the liquid into the gaseous st This passage of molecules from the surface of a liquid into the gaseous, or va state is called **vaporization** or **evaporation**. Vaporization occurs more readily w

- *increased temperature*—more molecules have sufficient kinetic energy overcome intermolecular forces of attraction in the liquid.
- *increased surface area* of the liquid—a greater proportion of the liquid m ecules are at the surface.
- *decreased strength of intermolecular forces*—the kinetic energy needed overcome intermolecular forces of attraction is less, and more molecu have enough energy to escape.

Because the molecules lost through evaporation are much more energetic th average, the average kinetic energy of the remaining molecules decreases. T temperature of the evaporating liquid falls. This accounts for the cooling sensati you feel when a volatile liquid, such as ethyl alcohol, evaporates on your skin.

The quantity of heat that must be absorbed to vaporize one mole of liquid constant temperature and constant pressure is called the *enthalpy of vapori tion,* $\Delta_{vap}H$. Values of $\Delta_{vap}H$ are often expressed in the unit kJ mol^{-1}, as seen Table 12.4, and thus refer to a process in which one mole of liquid is vaporize Because vaporization is an endothermic process, $\Delta_{vap}H$ is always positive.

The differences in the $\Delta_{vap}H$ values in Table 12.4 reflect the differences in intermolecular forces contributing to the attraction between molecules in ea liquid. The data in Table 12.4 illustrate the significant influence of hydrog bonding. Molecules of dimethyl ether, $(CH_3)_2O$, do not form hydrogen bon with each other, and neither do diethyl ether molecules, $(CH_3CH_2)_2O$. These t liquids have the lowest $\Delta_{vap}H$ values and boiling points of the liquids listed Table 12.4 even though, for example, $(CH_3CH_2)_2O$ molecules are, by far, the m polarizable. For the other liquids—CH_3OH, CH_3CH_2OH, and H_2O—hydrog bonds are formed between molecules but to varying degrees. Water has t largest $\Delta_{vap}H$ value and highest boiling point because each water molecule linked to four others through hydrogen bonds, as illustrated in Figure 12-7(The CH_3OH and CH_3CH_2OH molecules form fewer hydrogen bonds, a fact th can be rationalized by noting that the replacement of a hydrogen atom in t H_2O molecule with a CH_3 group or a CH_3CH_2 group corresponds to the loss o hydrogen bond donor. Consequently, both $CH_3OH(l)$ and $CH_3CH_2OH(l)$ have smaller $\Delta_{vap}H$ value and a lower boiling point than $H_2O(l)$.

TABLE 12.4 Intermolecular Forces and Enthalpies of Vaporization at 298 K[a]

Liquid	μ, Debye	α, 10^{-25} cm^3	% Dispersion	% Dipole	$\Delta_{vap}H$, kJ mol^{-1}	Boiling Point, °C
Dimethyl ether, $(CH_3)_2O$	1.30	52.9	84.6	15.4	18.5[b]	−24.8[c]
Diethyl ether, $(CH_3CH_2)_2O$	1.10	102	96.3	3.7	27.1	34.5
Methanol, CH_3OH	1.70	32.9	47.6	42.4	37.4	64.6
Ethanol, CH_3CH_2OH	1.69	54.1	69.8	30.2	42.3	78.3
Water, H_2O	1.85	14.5	13.8	86.2	44.0	100

[a]All data are from the *CRC Handbook of Chemistry and Physics*, 95th edition.
[b]$\Delta_{vap}H$ values are temperature and pressure dependent (see Exercise 93). The $\Delta_{vap}H$ values for these substances at their respectiv normal boiling points are 21.5 kJ mol^{-1} for $(CH_3)_2O$; 26.5 kJ mol^{-1} for $(CH_3CH_2)_2O$; 35.2 kJ mol^{-1} for CH_3OH; 38.6 kJ mol^{-1} for CH_3CH_2OH; and 40.7 kJ mol^{-1} for H_2O.
[c]These are the normal boiling points, measured at 1 atm.

In summary, when hydrogen bonds are present, the enthalpy of vaporization ıd the boiling point increase with the number of hydrogen bonds; and when ey are not, $\Delta_{vap}H$ and the boiling point tend to increase with the molecular ılarizability.

The conversion of a gas or vapor to a liquid is called **condensation**. ›ndensation is the reverse of vaporization. Thus,

$$\Delta_{cond}H = -\Delta_{vap}H$$

·cause $\Delta_{vap}H$ is always positive, $\Delta_{cond}H$ is always negative. Condensation is *exothermic* process. This explains why burns produced by a given mass of eam (vaporized water) are much more severe than burns produced by the same ass of hot water. Hot water causes burns by releasing heat as it cools. Steam leases a large quantity of heat when it condenses to liquid water, followed by e further release of heat as the hot water cools.

EXAMPLE 12-2 Estimating the Heat Evolved in the Condensation of Steam

A 0.750 L sample of steam obtained at the normal boiling point of water was allowed to condense on a slightly cooler surface. Using data from Table 12.4, estimate the quantity of heat evolved. Why is the result only an estimate?

Analyze

First, let us describe the steam sample a bit more precisely. Steam is water vapor, $H_2O(g)$; and when in equilibrium with liquid water at its normal boiling point, the steam is at 1.000 atm pressure.

Solve

We are given a volume of gas (0.750 L) at a fixed temperature (100.00 °C) and pressure (1.000 atm), and so we use the ideal gas equation to calculate the number of moles of $H_2O(g)$. That is,

$$n_{H_2O} = \frac{PV}{RT} = \frac{1.000 \text{ atm} \times 0.750 \text{ L}}{0.08206 \text{ L atm mol}^{-1} \text{ K}^{-1} \times (273.15 + 100.00) \text{ K}}$$
$$= 0.0245 \text{ mol}$$

On a molar basis, estimate the enthalpy of condensation of the $H_2O(g)$ to be the *negative* of the value of $\Delta_{vap}H$ of H_2O given in Table 12.4, that is, $-44.0 \text{ kJ mol}^{-1}$. For the 0.0245 mol sample of steam,

$$\Delta_{cond}H = 0.0245 \text{ mol} \times (-44.0 \text{ kJ mol}^{-1}) = -1.08 \text{ kJ}$$

Assess

This result is only an estimate for two reasons: (1) $\Delta_{cond}H$, like $\Delta_{vap}H$, is temperature dependent. The value used was for 298 K, whereas it should have been for 373 K, a value that was not given. (2) The condensed liquid water was at a temperature lower than 373 K ("slightly cooler surface"). An additional small quantity of heat was liberated as the condensed steam cooled to that lower temperature.

PRACTICE EXAMPLE A: How much heat is required to vaporize a 2.35 g sample of diethyl ether at 298 K?

PRACTICE EXAMPLE B: Calculate a more accurate answer to Example 12-2 by using $\Delta_{vap}H = 40.7 \text{ kJ mol}^{-1}$ for water at 100 °C, 85.0 °C as the temperature of the surface on which the steam condenses, and $4.21 \text{ J g}^{-1} \text{ °C}^{-1}$ as the average specific heat of $H_2O(l)$ in the temperature range 85 to 100 °C.

ɑpor Pressure

ater left in an open beaker evaporates completely. A different condition sults if the beaker with the water is placed in a closed container. As shown Figure 12-16, in a container with both liquid and vapor present, vaporization and condensation occur simultaneously. If sufficient liquid is present, eventually a condition is reached in which the amount of vapor remains constant. This ndition is one of *dynamic equilibrium*. Dynamic equilibrium always implies at two opposing processes are occurring simultaneously and at equal rates.

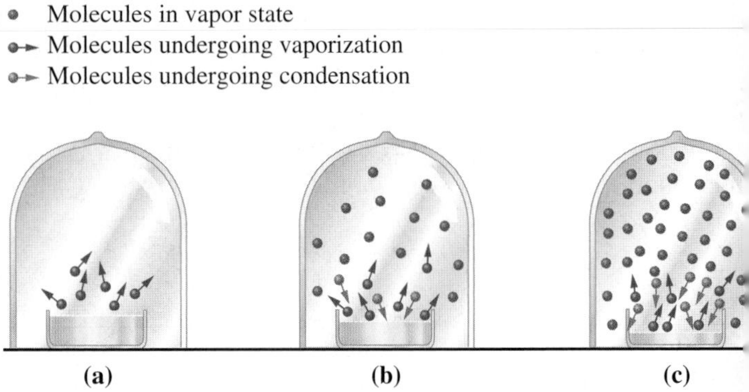

▶ FIGURE 12-16
Establishing liquid–vapor equilibrium
(a) A liquid is allowed to evaporate into a closed container. Initially, only vaporization occurs. **(b)** Condensation begins. The rate at which molecules evaporate is greater than the rate at which they condense, and the number of molecules in the vapor state continues to increase. **(c)** The rate of condensation is equal to the rate of vaporization. The number of vapor molecules remains constant over time, as does the pressure exerted by this vapor.

As a result, there is no net change with time once equilibrium has been esta lished. A symbolic representation of the liquid–vapor equilibrium is shov below.

$$\text{liquid} \underset{\text{condensation}}{\overset{\text{vaporization}}{\rightleftharpoons}} \text{vapor}$$

▶ Gasoline is a mixture of volatile hydrocarbons and is an important precursor of smog, whether it is vaporized from oil refineries, filling-station operations, automobile gas tanks, or gas-powered lawn mowers.

The pressure exerted by a vapor in dynamic equilibrium with its liquid called the **vapor pressure**. Liquids with high vapor pressures at room ten perature are said to be *volatile*, and those with very low vapor pressures a *nonvolatile*. Whether a liquid is volatile or not is determined primarily the strengths of its intermolecular forces—the weaker these forces, the mo volatile the liquid (the higher its vapor pressure). Diethyl ether and aceto are volatile liquids; at 25 °C their vapor pressures are 534 and 231 mmH respectively. Water at ordinary temperatures is a moderately volatile liqui at 25 °C, its vapor pressure is 23.8 mmHg. Mercury is essentially a no volatile liquid; at 25 °C, its vapor pressure is 0.0018 mmHg.

As an excellent first approximation, the vapor pressure of a liquid depen only on the particular liquid and its temperature. Vapor pressure depends neither the amount of liquid nor the amount of vapor, as long as some of ea is present at equilibrium. These statements are illustrated in Figure 12-17.

A graph of vapor pressure as a function of temperature is known as a **vap pressure curve**. Vapor pressure curves always have the appearance of those Figure 12-18: *Vapor pressure increases with temperature.* Vapor pressures of wa at different temperatures are presented in Table 12.5.

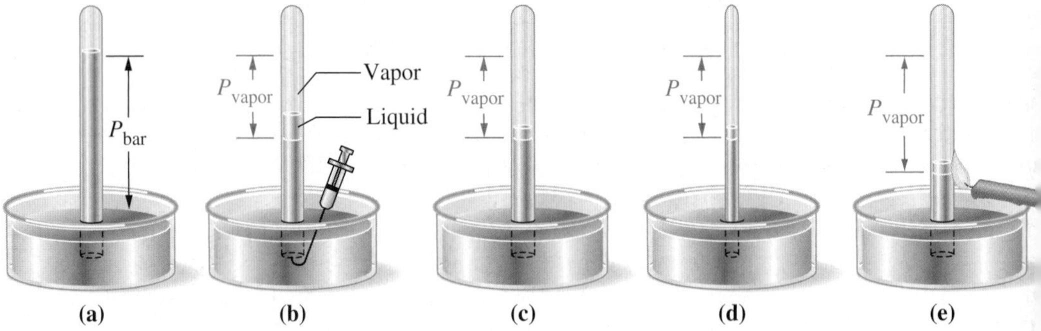

▲ FIGURE 12-17
Vapor pressure illustrated
(a) A mercury barometer. **(b)** The pressure exerted by the vapor in equilibrium with a liqu injected to the top of the mercury column depresses the mercury level. **(c)** Compared w (b), the vapor pressure is independent of the volume of liquid injected. **(d)** Compared w (c), the vapor pressure is independent of the volume of vapor present. **(e)** Vapor pressur increases with an increase in temperature.

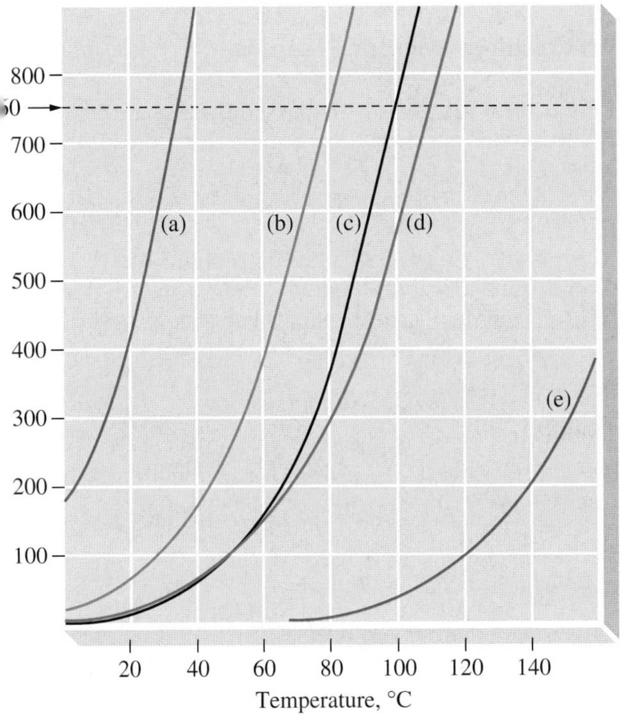

◀ FIGURE 12-18
Vapor pressure curves of several liquids
(a) Diethyl ether, $(CH_3CH_2)_2O$; **(b)** benzene, C_6H_6; **(c)** water, H_2O; **(d)** toluene, $C_6H_5CH_3$; **(e)** aniline, $C_6H_5NH_2$. The normal boiling points are the temperatures at the intersection of the dashed line at P = 760 mmHg with the vapor pressure curves.

ABLE 12.5 Vapor Pressure of Water at Various Temperatures

mperature, C	Pressure, mmHg	Temperature, °C	Pressure, mmHg	Temperature, °C	Pressure, mmHg
).0	4.6	29.0	30.0	93.0	588.6
).0	9.2	30.0	31.8	94.0	610.9
).0	17.5	40.0	55.3	95.0	633.9
.0	18.7	50.0	92.5	96.0	657.6
.0	19.8	60.0	149.4	97.0	682.1
3.0	21.1	70.0	233.7	98.0	707.3
.0	22.4	80.0	355.1	99.0	733.2
5.0	23.8	90.0	525.8	100.0	760.0
5.0	25.2	91.0	546.0	110.0	1074.6
7.0	26.7	92.0	567.1	120.0	1489.1
3.0	28.3				

easuring Vapor Pressure

gure 12-17 suggests one method of determining vapor pressure—inject a all sample of the target liquid at the top of a mercury barometer, and mea- re the depression of the mercury level. The method does not give very pre- e results, however, and it is not useful for measuring vapor pressures that either very low or quite high. Better results are obtained with methods in ich the pressure above a liquid is continuously varied and measured, and liquid–vapor equilibrium temperature is recorded. In short, the boiling int of the liquid changes in accordance with the change in the pressure ove the liquid, and the vapor pressure curve of the liquid can be traced. The essure measurements are made with either a closed-end or open-end anometer (page 199). A method that is useful for determining very low por pressures is based on the rate of effusion of a gas through a tiny orifice. this method, equations from the kinetic–molecular theory (Section 6-7) are plied. Example 12-3 illustrates a method (called the transpiration method) which an inert gas is saturated with the vapor under study. Then the ideal s equation is used to calculate the vapor pressure.

EXAMPLE 12-3 Using the Ideal Gas Equation to Calculate a Vapor Pressure

A sample of 113 L of helium gas at 1360 °C and prevailing barometric pressure is passed through molten silver at the same temperature. The gas becomes saturated with silver vapor, and the liquid silver loses 0.120 g in mass. What is the vapor pressure of liquid silver at 1360 °C?

Analyze

Let's assume that after the gas has become saturated with silver vapor, its volume remains at 113 L. This assumption will be valid if the vapor pressure of the silver is quite low compared with the barometric pressure. According to Dalton's law of partial pressures (page 215) we can deal with the silver vapor as if it were a single gas occupying a volume of 113 L.

Solve

The data required in the ideal gas equation are listed below.

$P = ?$ $\qquad V = 113\ \text{L}$

$R = 0.08206\ \text{L atm mol}^{-1}\ \text{K}^{-1}$ $\qquad T = 1360 + 273.15 = 1633\ \text{K}$

$$n = 0.120\ \text{g Ag} \times \frac{1\ \text{mol Ag}}{107.9\ \text{g Ag}} = 0.00111\ \text{mol Ag}$$

$$P = \frac{nRT}{V}$$

$$P = \frac{0.00111\ \text{mol} \times 0.08206\ \text{L atm mol}^{-1}\ \text{K}^{-1} \times 1633\ \text{K}}{113\ \text{L}}$$

$$= 1.32 \times 10^{-3}\ \text{atm (1.00 Torr)}$$

Assess

The assumption we made appears to be valid, because the experimental vapor pressure of liquid silver at 1360 °C is 1 mmHg or 1.32×10^{-3} atm.

PRACTICE EXAMPLE A: Equilibrium is established between liquid hexane, C_6H_{14}, and its vapor at 25.0 °C. A sample of the vapor is found to have a density of 0.701 g/L. Calculate the vapor pressure of hexane at 25.0 °C, expressed in Torr.

PRACTICE EXAMPLE B: With the help of Figure 12-18, estimate the density of the vapor in equilibrium with liquid diethyl ether at 20.0 °C.

Using Vapor Pressure Data

One use of vapor pressure data is in calculations dealing with the collection gases over liquids, particularly water (Section 6-6). Another use, illustrated Example 12-4, is in predicting whether a substance exists solely as a gas (vapo or as a liquid and vapor in equilibrium.

EXAMPLE 12-4 Making Predictions with Vapor Pressure Data

As a result of a chemical reaction, 0.132 g H_2O is produced and maintained at a temperature of 50.0 °C in a closed flask of 525 mL volume. Will the water be present as liquid only, vapor only, or liquid and vapor in equilibrium (Fig. 12-19)?

Analyze

Let's consider each of the three possibilities in the order that they are given.

Solve

LIQUID ONLY

With a density of about 1 g/mL, a 0.132 g sample of H_2O has a volume of only about 0.13 mL. There is no way that the sample could completely fill a 525 mL flask. The condition of liquid only is *impossible*.

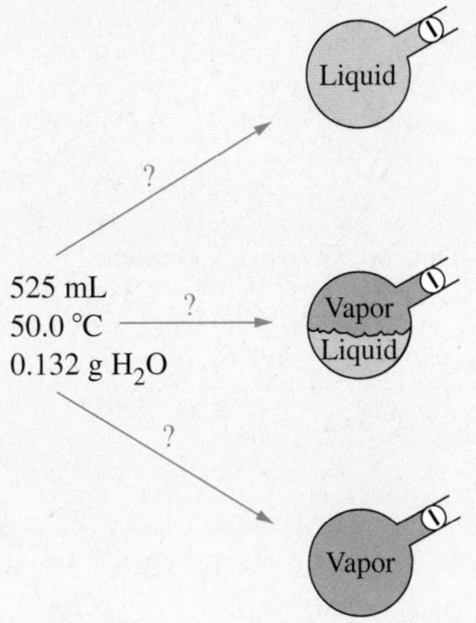

◀ FIGURE 12-19
Predicting states of matter—Example 12.4 illustrated
For the conditions given on the left, which of the final conditions pictured on the right will result?

VAPOR ONLY

The portion of the flask that is not occupied by liquid water must be filled with something (it cannot remain a vacuum). That something is water vapor. The question is, will the sample vaporize completely, leaving no liquid? Let's use the ideal gas equation to calculate the pressure that would be exerted if the entire 0.132 g H_2O were present in the gaseous state.

$$P = \frac{nRT}{V}$$

$$= \frac{0.132 \text{ g } H_2O \times \dfrac{1 \text{ mol } H_2O}{18.02 \text{ g } H_2O} \times 0.08206 \text{ L atm mol}^{-1} \text{ K}^{-1} \times 323.2 \text{ K}}{0.525 \text{ L}}$$

$$= 0.370 \text{ atm} \times \frac{760 \text{ mmHg}}{1 \text{ atm}} = 281 \text{ mmHg}$$

Now compare this calculated pressure with the vapor pressure of water at 50.0 °C (Table 12.5). The calculated pressure—281 mmHg—greatly exceeds the vapor pressure—92.5 mmHg. Water formed in the reaction as $H_2O(g)$ condenses to $H_2O(l)$ when the gas pressure reaches 92.5 mmHg, for this is the pressure at which the liquid and vapor are in equilibrium at 50.0 °C. The condition of vapor only is *impossible*.

LIQUID AND VAPOR

This is the only possibility for the final condition in the flask. Liquid water and water vapor coexist in equilibrium at 50.0 °C and 92.5 mmHg.

Assess

We found the solution to this problem through the application of the ideal gas equation and our understanding of vapor pressure. Note that in the first two steps, we considered the two extremes, with the first being just liquid water and the second all vapor.

PRACTICE EXAMPLE A: If the reaction described in this example resulted in H_2O produced and maintained at 80.0 °C, would the water be present as vapor only or as liquid and vapor in equilibrium? Explain.

PRACTICE EXAMPLE B: For the situation described in Example 12.4, what mass of water is present as liquid and what mass as vapor?

n Equation for Expressing Vapor Pressure Data

you look for vapor pressure data on a liquid in a handbook or in data tables, u are unlikely to find graphs like Figure 12-18. Also, with the exception of a v liquids, such as water and mercury, you are unlikely to find data tables like ble 12.5. What you will find, instead, are mathematical equations relating por pressures and temperatures. Such equations can summarize in one line ta that might otherwise take a full page. Equation (12.1) is a particularly comon form of vapor pressure equation. It expresses the natural logarithm (ln)

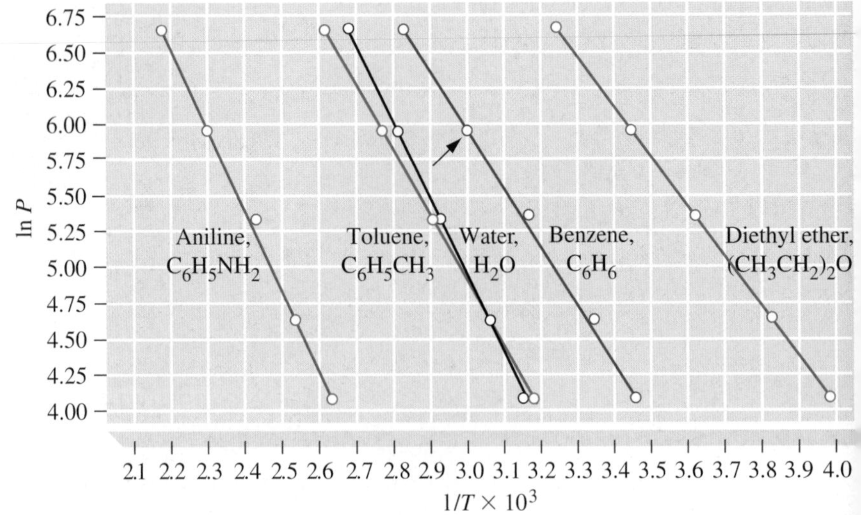

▶ FIGURE 12-20
Vapor pressure data plotted as ln *P* versus 1/*T*
Pressures are in millimeters of mercury, and temperatures are in Kelvin. Data from Figure 12-18 have been recalculated and replotted as in the following example: For benzene at 60 °C, the vapor pressure is 400 mmHg; ln P = ln 400 = 5.99. $T = 60 + 273 = 333$ K; $1/T = 1/333 = 0.00300 = 3.00 \times 10^{-3}$; $1/T \times 10^3 = 3.00 \times 10^{-3} \times 10^3 = 3.00$. The point corresponding to (3.00, 5.99) is marked by the black arrow.

of vapor pressure as a function of the reciprocal of the Kelvin temperatu (1/*T*). The relationship is that of a straight line, and the straight-line plots f the liquids featured in Figure 12-18 are drawn in Figure 12-20.

▶ Refer to Appendix A to see how the constant *B* is eliminated to give equation (12.2).

$$\underbrace{\ln P}_{} = \underbrace{-A}_{}\underbrace{\left(\frac{1}{T}\right)}_{} + \underbrace{B}_{} \qquad (12.$$

Equation of straight line: $y = m \times x + b$

To use equation (12.1), we need values for the two constants, *A* and *B*. The co stant *A* is related to the enthalpy of vaporization of the liquid: $A = \Delta_{vap}H/$ where $\Delta_{vap}H$ is expressed in the unit J mol^{-1} and the value used for *R* 8.3145 J mol^{-1} K^{-1}. It is customary to eliminate *B* by rewriting equation (12.1) f two different temperatures, in a form called the Clausius–Clapeyron equation.

KEEP IN MIND

that the heat of vaporization in this equation cannot be $\Delta_{vap}H°$, since in general the pressure is not 1 bar.

$$\ln\left(\frac{P_2}{P_1}\right) = -\frac{\Delta_{vap}H}{R}\left(\frac{1}{T_2} - \frac{1}{T_1}\right) \qquad (12.2)$$

We apply equation (12.2) in Example 12-5.

EXAMPLE 12-5 Applying the Clausius–Clapeyron Equation

Calculate the vapor pressure of water at 35.0 °C using data from Tables 12.4 and 12.5.

Analyze

Starting with the Clausius-Clapeyron equation, we recognize that we need four pieces of data to solve for the fifth. Since we are asked to calculate a vapor pressure, we will need two temperatures, a pressure, and the enthalpy of vaporization.

Solve

Designate the unknown vapor pressure as P_1 at the temperature T_1. That is,

$$P_1 = ? \qquad T_1 = (35.0 + 273.15)\,\text{K} = 308.2\,\text{K}$$

For P_2 and T_2 choose known data for a temperature close to 35.0 °C, for example, 40.0 °C.

$$P_2 = 55.3\,\text{mmHg} \qquad T_2 = (40.0 + 273.15)\,\text{K} = 313.2\,\text{K}$$

For $\Delta_{vap}H$, let's assume that the value given in Table 12.4 applies throughout the temperature range from 30.0 °C to 40.0 °C.

$$\Delta_{vap}H = 44.0\,\text{kJ mol}^{-1} \times \frac{1000\,\text{J}}{1\,\text{kJ}} = 44.0 \times 10^3\,\text{J mol}^{-1}$$

Now substitute these values into equation (12.2) to obtain

$$\ln\left(\frac{55.3\ \text{mmHg}}{P_1}\right) = -\frac{44.0 \times 10^3\ \text{J mol}^{-1}}{8.3145\ \text{J mol}^{-1}\ \text{K}^{-1}}\left(\frac{1}{313.2} - \frac{1}{308.2}\right)\text{K}^{-1}$$

$$= -5.29 \times 10^3(0.003193 - 0.003245) = 0.28$$

Next, determine that $e^{0.28} = 1.32$ (see Appendix A). Thus,

$$\frac{55.3\ \text{mmHg}}{P_1} = e^{0.28} = 1.32$$

$$P_1 = 55.3\ \text{mmHg}/1.32 = 41.9\ \text{mmHg}$$

Assess

Here, P_1 must be *smaller than* P_2 because $T_1 < T_2$. Thus, regardless of how we write equation (12.2)—different formulations are possible—or choose (T_1, P_1) and (T_2, P_2), we are guided by the fact that vapor pressure always increases with temperature. One way to check our answer is to repeat the calculation by using this pressure as the known pressure to see whether we obtain the pressure given in the problem. We can also check this against the experimentally determined vapor pressure of water at 35.0 °C, which is 42.175 mmHg.

PRACTICE EXAMPLE A: A handbook lists the vapor pressure of methyl alcohol as 100 mmHg at 21.2 °C. What is its vapor pressure at 25.0 °C?

PRACTICE EXAMPLE B: A handbook lists the normal boiling point of isooctane, a gasoline component, as 99.2 °C and its enthalpy of vaporization ($\Delta_{vap}H$) as 35.76 kJ mol^{-1} C_8H_{18}. Calculate the vapor pressure of isooctane at 25 °C.

iling and the Boiling Point

en a liquid is heated in a container *open to the atmosphere*, there is a particu- temperature at which vaporization occurs throughout the liquid rather than ply at the surface. Vapor bubbles form within the bulk of the liquid, rise to surface, and escape. The pressure exerted by escaping molecules equals t exerted by molecules of the atmosphere, and **boiling** is said to occur. ring boiling, energy absorbed as heat is used only to convert molecules of uid to vapor. The temperature remains constant until all the liquid has led away, as is dramatically illustrated in Figure 12-21. The temperature at ich the vapor pressure of a liquid is equal to standard atmospheric pressure tm = 760 mmHg) is the **normal boiling point**. In other words, the normal ling point is the boiling point of a liquid at 1 atm pressure. The normal boil- points of several liquids can be determined from the intersection of the hed line in Figure 12-18 with the vapor pressure curves for the liquids.

◀ When a pan of water is put on the stove to boil, small bubbles are usually observed as the water begins to warm. These are bubbles of dissolved air being expelled. Once the water boils, however, all the dissolved air is expelled and the bubbles consist only of water vapor.

◀ FIGURE 12-21
Boiling water in a paper cup
An empty paper cup heated over a Bunsen burner quickly bursts into flame. If a paper cup is filled with water, it can be heated for an extended time as the water boils. This is possible for three reasons: (1) Because of the high heat capacity of water, the temperature of the water and cup is kept well below that required for the combustion of paper to occur. (2) As the water boils, large quantities of heat ($\Delta_{vap}H$) are required to convert the liquid to its vapor. (3) The temperature of the cup does not rise above the boiling point of water as long as liquid water remains. The boiling point of 99.9 °C instead of 100.0 °C suggests that the prevailing barometric pressure was slightly below 1 atm.

▶ A more extreme case is that on the summit of Mt. Everest, where a climber would barely be able to heat a cup of tea to 70 °C.

Figure 12-18 also helps us see that the boiling point of a liquid va significantly with barometric pressure. Shifting the dashed line show $P = 760$ mmHg to higher or lower pressures, and the new points of intersec with the vapor pressure curves come at different temperatures. Barometric p sures below 1 atm are commonly encountered at high altitudes. At an altitud 1609 m (that of Denver, Colorado), barometric pressure is about 630 mmHg. boiling point of water at this pressure is 95 °C (203 °F). It takes longer to c foods under conditions of lower boiling-point temperatures. A three-mir boiled egg takes longer than three minutes to cook. We can counteract the ef of high altitudes by using a pressure cooker. In a pressure cooker, the cook water is maintained under higher-than-atmospheric pressure and its boi temperature increases, for example, to about 120 °C at 2 atm pressure.

Richard Megna/Fundamental Photographs, NYC

▲ A liquid boils at low pressure
Water boils when its vapor pressure equals the pressure on its surface. Bubbles form throughout the liquid.

12-3 CONCEPT ASSESSMENT

Why does a three-minute boiled egg take longer than three minutes to cook ir Switzerland and not on Manhattan Island in New York City?

The Critical Point

In describing boiling, we made an important qualification: Boiling occurs " container open to the atmosphere." If a liquid is heated in a *sealed* contai boiling does not occur. Instead, the temperature and vapor pressure rise c tinuously. Pressures many times atmospheric pressure may be attained. If the right quantity of liquid is sealed in a glass tube and the tube is heated, a Figure 12-22, the following phenomena can be observed:

▶ Although the term *gas* can be used exclusively, sometimes the term *vapor* is used for the gaseous state at temperatures *below* T_c and gas at temperatures *above* T_c.

- The density of the liquid decreases, that of the vapor increases, and ev tually the two densities become equal.
- The surface tension of the liquid approaches zero. The interface betw the liquid and vapor becomes less distinct and eventually disappears

The **critical point** is the point at which these conditions are reached and liquid and vapor become indistinguishable. The temperature at the crit point is the critical temperature, T_c, and the pressure is the critical press P_c. The critical point is the highest point on a vapor pressure curve and rep sents the highest temperature at which the liquid can exist. Several crit temperatures and pressures are listed in Table 12.6.

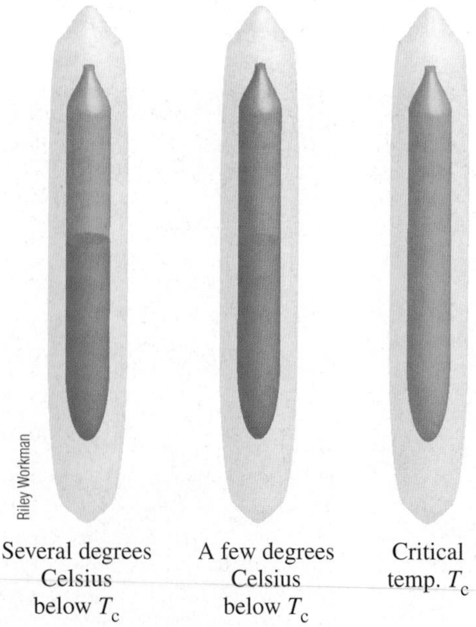

▶ FIGURE 12-22
Attainment of the critical point for chlorine, Cl_2
In a sealed container, the meniscus separating a liquid from its vapor is visible at temperatures below the critical point but becomes barely visible at the instant the critical point is reached. At the critical point—the liquid and vapor become indistinguishable.

ABLE 12.6 Some Critical Temperatures, T_c, and Critical Pressures, P_c

'ermanent" gases[a]	T_c, K	P_c, bar	"Nonpermanent" gases[b]	T_c, K	P_c, bar
$_2$	33.3	12.8	CO_2	304.2	72.9
$_2$	126.2	33.5	HCl	324.6	82.1
$_2$	154.8	50.1	NH_3	405.7	112.5
H_4	191.1	45.8	SO_2	431.0	77.7
			H_2O	647.3	218.3

ermanent gases cannot be liquefied at 25 °C (298 K).
Ionpermanent gases can be liquefied at 25 °C.

A gas can be liquefied only at temperatures *below* its critical temperature, If room temperature is *below* T_c, this liquefaction can be accomplished just applying sufficient pressure. If room temperature is *above* T_c, however, led pressure *and* a lowering of temperature to a value below T_c are uired. We will comment further on the liquefaction of gases on page 542.

12-4 CONCEPT ASSESSMENT

ompare the critical temperatures of NH_3 and N_2 (Table 12.6). Which gas has ne stronger intermolecular forces?

EXAMPLE 12-6 Relating Intermolecular Forces and Physical Properties

Arrange the following substances in the order in which you would expect their boiling points to increase: $CCl_4, Cl_2, ClNO, N_2$.

Analyze

Recall that boiling point trends are related to intermolecular forces. We should begin by identifying the types and strengths of intermolecular forces at work.

Solve

Three of the substances are nonpolar. For these, the strengths of dispersion forces, and hence the boiling points, should increase with increasing molecular mass, that is, $N_2 < Cl_2 < CCl_4$. ClNO has a molecular mass (65.5 u) comparable to that of Cl_2 (70.9 u), but the ClNO molecule is polar (bond angle $\approx$ 120°). This suggests stronger intermolecular forces and a higher boiling point for ClNO than for Cl_2. We should not expect the boiling point of ClNO to be higher than that of CCl_4, however, because of the large difference in their molecular masses (65.5 u compared with 154 u). The expected order is $N_2 < Cl_2 < ClNO < CCl_4$. (The observed boiling points are 77.3, 239.1, 266.7, and 349.9 K, respectively.)

Assess

Even though one molecule (ClNO) is polar, it does not have the highest boiling point, indicating that dispersion forces can be stronger than dipole–dipole forces.

PRACTICE EXAMPLE A: Arrange the following in the expected order of increasing boiling point: Ne, He, $Cl_2, (CH_3)_2CO, O_2, O_3$.

PRACTICE EXAMPLE B: Following are some values of $\Delta_{vap}H$ for several liquids at their normal boiling points: H_2, 0.92 kJ mol^{-1}; CH_4, 8.16 kJ mol^{-1}; C_6H_6, 31.0 kJ mol^{-1}; CH_3NO_2, 34.0 kJ mol^{-1}. Explain the differences among these values.

12-5 CONCEPT ASSESSMENT

xplain why CCl_4 has a higher boiling point than CH_3Cl, despite the polarity of H_3Cl.

12-3 Some Properties of Solids

We mentioned some properties of solids (for example, malleability, ductili at the beginning of this text, and we will continue to consider additional pro erties. For now, we will comment on some properties that allow us to think solids in relation to the other states of matter—liquids and gases.

Melting, Melting Point, and Heat of Fusion

In a crystalline solid, there is a regular, ordered arrangement of particles. T particles may be atoms, ions, or molecules. As a crystalline solid is heated, atoms, ions, or molecules vibrate more vigorously. Eventually a temperatur reached at which these vibrations disrupt the ordered crystalline structu The atoms, ions, or molecules can slip past one another, and the solid loses definite shape and is converted to a liquid. This process is called **melting**, fusion, and the temperature at which it occurs is the **melting point**. T reverse process, the conversion of a liquid to a solid, is called **freezing**, solidification, and the temperature at which it occurs is the **freezing poi** The melting point of a solid and the freezing point of its liquid are identical. this temperature, solid and liquid coexist in equilibrium.

If we add heat uniformly to a solid–liquid mixture at equilibrium, the te perature remains constant while the solid melts. Only when all the solid h melted does the temperature begin to rise. Conversely, if we remove heat u formly from a solid–liquid mixture at equilibrium, the liquid freezes at a co stant temperature. The quantity of heat required to melt a solid is the *entha of fusion*, $\Delta_{fus}H$. Some typical enthalpies of fusion, expressed in kilojoules p mole, are listed in Table 12.7. Perhaps the most familiar example of a melti (and freezing) point is that of water, 0 °C. This is the temperature at which l uid and solid water, in contact with air and under standard atmospheric pr sure, are in equilibrium. The enthalpy of fusion of water is 6.01 kJ mol which we can express as

$$H_2O(s) \longrightarrow H_2O(l) \qquad \Delta_{fus}H = +6.01 \text{ kJ mol}^{-1} \qquad (1$$

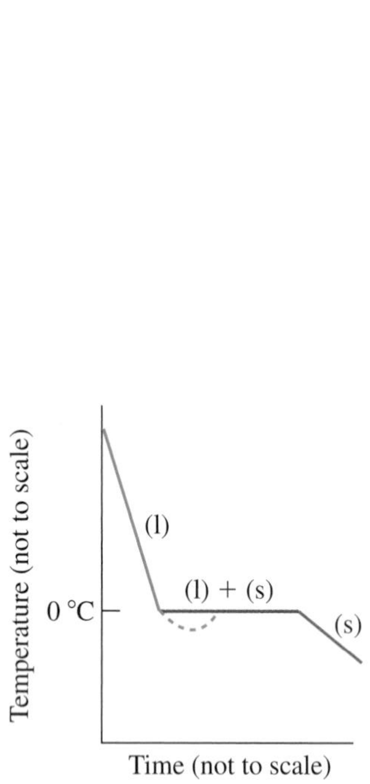

▲ FIGURE 12-23
Cooling curve for water
The broken-line portion represents the condition of supercooling that occasionally occurs. (l) = liquid; (s) = solid.

Here is an easy way to determine the freezing point of a liquid. Allow the l uid to cool, and measure the liquid temperature as it falls with time. Wh freezing begins, the temperature *remains constant* until all the liquid has froz Then the temperature is again free to fall as the solid cools. If we plot tempe tures against time, we get a graph known as a *cooling curve*. Figure 12-23 i cooling curve for water. We can also run this process backward, that is, by sta ing with the solid and adding heat. Now the temperature remains consta while melting occurs. This temperature–time plot is called a *heating cur* Generally speaking, the appearance of the heating curve is that of a cooli curve that has been flipped from left to right. A heating curve for water sketched in Figure 12-24.

Often, an experimentally determined cooling curve does not look quite l the solid–line plot in Figure 12-23. The temperature may drop below t freezing point without any solid appearing. This condition is known

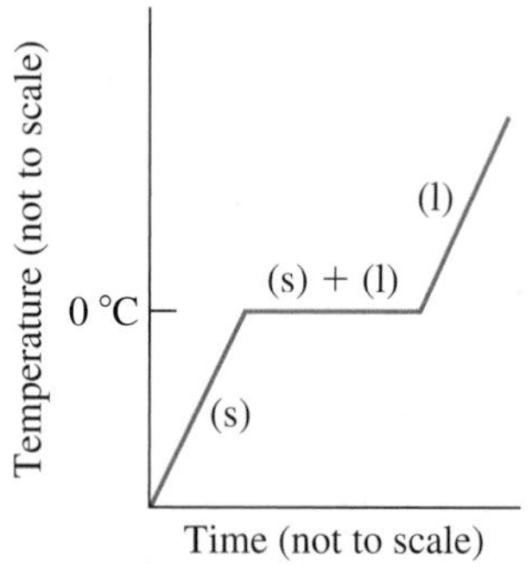

▲ FIGURE 12-24
Heating curve for water
This curve traces the changes that occur as ice is heated from below the melting point to produce liquid water somewhat above the melting point.

TABLE 12.7 Some Enthalpies of Fusion

Substance	Melting Point, °C	$\Delta_{fus}H$, kJ mol^{-1}
Mercury, Hg	−38.9	2.30
Sodium, Na	97.8	2.60
Methyl alcohol, CH_3OH	−97.7	3.21
Ethyl alcohol, CH_3CH_2OH	−114	5.01
Water, H_2O	0.0	6.01
Benzoic acid, C_6H_5COOH	122.4	18.08
Naphthalene, $C_{10}H_8$	80.2	18.98

ercooling. In order for a crystalline solid to start forming from a liquid, a :leation event must occur. A nucleation event can be classified as either nogeneous or heterogeneous. Homogeneous nucleation is when molecules small region of the liquid form a small crystal. Heterogeneous nucleation olves the formation of a small crystal on the surface of an inert particle (for mple, a suspended dust particle) or on scratches etched in the container. If liquid contains very few of these small crystals, the liquid may supercool a time before freezing. When a supercooled liquid does begin to freeze, vever, the temperature rises back to the normal freezing point while freez- is completed. We can always recognize supercooling through a slight dip cooling curve just before the horizontal portion.

◀ Examples of supercooled substances are water droplets in the sky. They remain liquid at temperatures well below the freezing point. When they find a bit of dust on which they can nucleate, the droplets spontaneously turn to ice.

blimation

e liquids, solids can also give off vapors, although because of the stronger rmolecular forces present, solids are generally not as volatile as liquids at a en temperature. The direct passage of molecules from the solid to the vapor e is called **sublimation**. The reverse process, the passage of molecules from vapor to the solid state, is called **deposition**. When sublimation and depo- on occur at equal rates, a dynamic equilibrium exists between a solid and vapor. The vapor exerts a characteristic pressure called the *sublimation pres- :*. A plot of sublimation pressure as a function of temperature is called a *limation curve*. The *enthalpy of sublimation* ($\Delta_{sub}H$) is the quantity of heat ded to convert a solid to vapor. At the sublimation point, sublimation lid $\longrightarrow$ vapor) is equivalent to melting (solid $\longrightarrow$ liquid) followed by orization (liquid $\longrightarrow$ vapor). This suggests the following relationship ong $\Delta_{fus}H$, $\Delta_{vap}H$, and $\Delta_{sub}H$ at the melting point.

$$\Delta_{sub}H = \Delta_{fus}H + \Delta_{vap}H \quad \textbf{(12.4)}$$

: value of $\Delta_{sub}H$ obtained with equation (12.4) can replace the enthalpy of orization in the Clausius–Clapeyron equation (12.2), so that sublimation ssures can be calculated as a function of temperature.

Two familiar solids with significant sublimation pressures are ice and dry (solid carbon dioxide). If you live in a cold climate, you are aware that w may disappear from the ground even though the temperature may fail ise above 0 °C. Under these conditions, the snow does not melt; it sub- es. The sublimation pressure of ice at 0 °C is 4.58 mmHg. That is, the solid has a vapor pressure of 4.58 mmHg at 0 °C. If the air is not already satu- :d with water vapor, the ice will sublime. The sublimation and deposition odine are pictured in Figure 12-25.

Carey B. Van Loon

▲ FIGURE 12-25
Sublimation of iodine
Even at temperatures well below its melting point of 114 °C, solid iodine exhibits an appreciable sublimation pressure. Here, purple iodine vapor is produced at about 70 °C. Deposition of the vapor to solid iodine occurs on the colder walls of the flask.

12-6 CONCEPT ASSESSMENT

ecall the discussion of dew and frost formation (see the Focus On feature for hapter 6, *Earth's Atmosphere*, at www.masteringchemistry.com). Do the urroundings absorb or lose heat when water vapor condenses to dew or frost? the quantity of heat per gram of $H_2O(g)$ condensed the same whether the ondensate is dew or frost? Explain.

2-4 Phase Diagrams

agine constructing a pressure–temperature graph in which each point on graph represents a condition under which a substance might be found. At temperatures and high pressures, such as the red points in Figure 12-26, expect the atoms, ions, or molecules of a substance to be in a close orderly angement—a solid. At high temperatures and low pressures—the brown

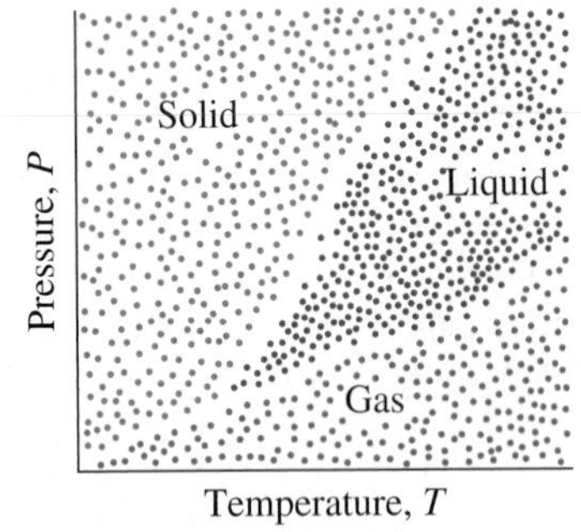

▲ FIGURE 12-26
Temperature, pressure, and states of matter
The outline of a phase diagram is suggested by the distribution of points. The red points identify the temperatures and pressures at which solid is the stable phase; the blue points identify the temperatures and pressures at which liquid is the stable phase; and the brown points represent the temperatures and pressures at which gas is the stable phase. (See also Figures 12-27 and 12-28.)

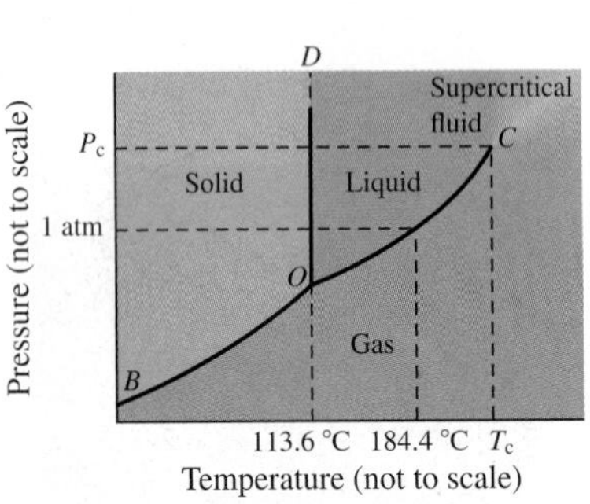

▲ FIGURE 12-27
Phase diagram for iodine
Note that the melting point and triple point temperatures for iodine are essentially the same. Generally, large pressure increases are required to produce even small changes in solid–liquid equilibrium temperatures. The pressure and temperature axes on a phase diagram are generally not drawn to scale so that the significant features of the diagram can be more readily emphasized.

points in Figure 12-26—we expect the gaseous state; and at intermediate te peratures and pressures, we expect a liquid (blue points in Figure 12-26).

Figure 12-26 is a **phase diagram**, a graphical representation of the conditio of temperature and pressure at which solids, liquids, and gases (vapors) exi either as single phases, or states, of matter or as two or more phases in equili rium with one another. The different regions of the diagram correspond to si gle phases, or states, of matter. Straight or curved lines where single-pha regions adjoin represent two phases in equilibrium.

Iodine

One of the simplest phase diagrams is that of iodine shown in Figure 12- The curve OC is the vapor pressure curve of liquid iodine, and C is the criti point. OB is the sublimation curve of solid iodine. The nearly vertical line C represents the effect of pressure on the melting point of iodine; it is called t *fusion curve.* The point O has a special significance. It defines the *unique* te perature and pressure at which the *three* states of matter, solid, liquid, a gas, coexist in equilibrium. It is called a **triple point**. For iodine, the trip point is at 113.6 °C and 91.6 mmHg. The normal melting point (113.6 °C) a the boiling point (184.4 °C) are the temperatures at which a line at $P = 1$ a intersects the fusion and vapor pressure curves, respectively. Melting essentially unaffected by pressure in the limited range from 91.6 mmHg 1 atm, and the normal melting point and the triple point are at almost t same temperature.

The sublimation curve for iodine in Figure 12-27 appears to be a continuati of the vapor pressure curve, but if the data are plotted to scale, a discontinuity seen at the triple point O. Moreover, this must *always* be the case. If these t curves were continuous, then the lines representing the variation of ln P w $1/T$ (Fig. 12-20) would have the same slope—but this is not possible. The va of $\Delta_{vap}H$ determines the slope of the vapor pressure line (recall equation 12 whereas $\Delta_{sub}H$ determines the slope of the sublimation line. However, these t enthalpy changes can never be the same, because $\Delta_{sub}H = \Delta_{vap}H + \Delta_{fus}H$.

The extreme range of temperatures and pressures required for the ent phase diagram precludes plotting it to scale. This is why the axes are labe "not to scale."

Carbon Dioxide

The case of carbon dioxide, shown in Figure 12-28, differs from that of iodine one important respect—the pressure at the triple point O is greater than 1 atm line at $P = 1$ atm intersects the *sublimation curve,* not the vapor pressure curve solid CO_2 is heated in an open container, it sublimes away at a constant temp ature of −78.5 °C. It *does not melt* at atmospheric pressure (and so is called "c ice"). Because it maintains a low temperature and does not produce a liquid melting, dry ice is widely used in freezing and preserving foods.

Liquid CO_2 can be obtained at pressures above 5.1 atm and it is most f quently encountered in CO_2 fire extinguishers. All three states of matter involved in the action of these fire extinguishers. When the liquid CO_2 released, most of it quickly vaporizes. The heat required for this vaporizatio extracted from the remaining $CO_2(l)$, which has its temperature lowered to point that it freezes and falls as a $CO_2(s)$ "snow." In turn, the $CO_2(s)$ quic sublimes to $CO_2(g)$. All of this helps to quench a fire by displacing the around the fire with a "blanket" of $CO_2(g)$ and by cooling the area somewha

Supercritical Fluids

Because the liquid and gaseous states become identical and indistinguishabl the critical point, it is difficult to know what to call the state of matter at temp atures and pressures above the critical point. For example, this state of ma has the high density of a liquid and the low viscosity of a gas. The term tha

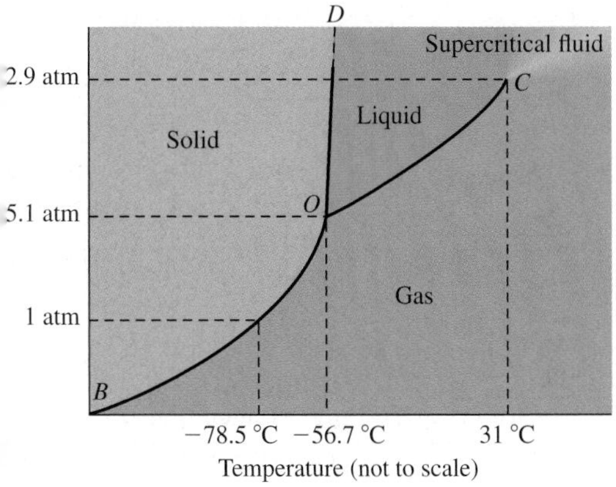

IGURE 12-28
ase diagram for carbon dioxide
eral aspects of this diagram are scribed in the text. An additional ture not shown here is the curvature he fusion curve *OD* to the right at y high pressures, ultimately reaching nperatures above the critical nperature.

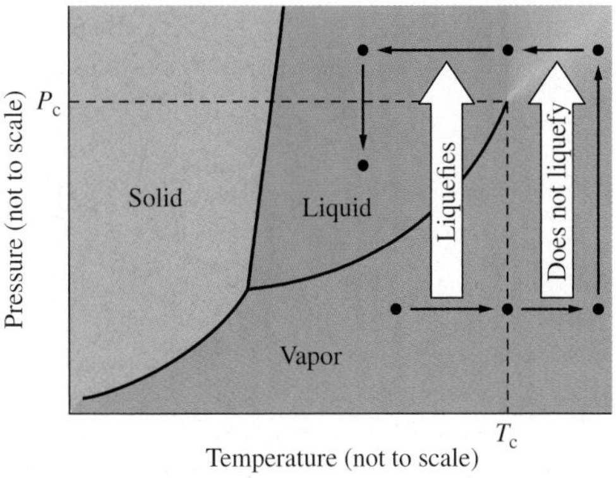

▲ FIGURE 12-29
Critical point
Applying pressure to a gas at temperatures below the critical point, T_c, causes a liquid to form with the appearance of a meniscus, a discontinuous phase change. Applying pressure above the critical point simply increases the density of the supercritical fluid. In a path traced by the small arrows, gas changes to liquid without exhibiting a discontinuous phase transition.

w commonly used is *supercritical fluid* (SCF). Above the critical temperature, amount of pressure can liquefy a supercritical fluid. Consider the generic ase diagram in Figure 12-29. The path of dots starting with a vapor below the tical temperature takes us to a low-density gas above the critical temperature. nen the pressure is greatly increased, we produce a supercritical fluid of much ater density. If, while the pressure exceeds the critical pressure, P_c, the temrature is reduced below the critical temperature, we obtain a liquid. Even th further reduction of pressure, the sample remains in the liquid phase. In lowing the path described, we have gone from a gas to a liquid without serving a liquid–gas interface. The only way to observe the liquid–vapor erface is to cross the phase boundary below the critical temperature. Note t in the present case we could observe the liquid–vapor interface by lowering pressure on the liquid to a point on the vapor pressure curve.

Although we do not ordinarily think of liquids or solids as being soluble gases, *volatile* liquids and solids are. The mole fraction solubility is simply ratio of the vapor pressure (or the sublimation pressure) to the total gas essure. And liquids and solids become much more soluble in a gas that is ove its critical pressure and temperature, mostly because the density of SCF is high and approaches that of a liquid. Molecules in supercritical ids, being in much closer proximity than in ordinary gases, can exert ong attractive forces on the molecules of a liquid or solid solute. SCFs disy solvent properties similar to ordinary liquid solvents. To vary the presre of an SCF means to vary its density and also its solvent properties. us, a given SCF, such as carbon dioxide, can be made to behave like many ferent solvents.

Until recently, the principal method of decaffeinating coffee was to extract caffeine with a solvent, such as methylene chloride (CH_2Cl_2). This solvent objectionable because it is hazardous in the workplace and difficult to comtely remove from the coffee. Now, supercritical fluid CO_2 is used. In one ocess, green coffee beans are brought into contact with CO_2 at about 90 °C d 160 to 220 atm. The caffeine content of the coffee is reduced from its rmal 1% to 3% to about 0.02%. When the temperature and pressure of the $)_2$ are reduced, the caffeine precipitates and the CO_2 is recycled.

▲ **Decaffeinated coffee**
"Naturally" decaffeinated coffee is made through a process that uses supercritical fluid CO_2 as a solvent to dissolve the caffeine in green coffee beans. Afterward, the beans are roasted and sold to consumers.

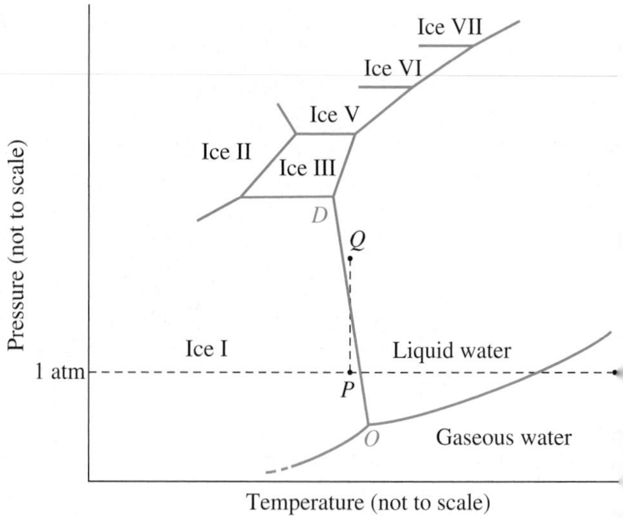

▶ FIGURE 12-30
Phase diagram for water
Point *O*, the triple point, is at 0.0098 °C and 4.58 mmHg. (The normal melting point is at exactly 0 °C and 760 mmHg.) The critical point, *C*, is at 374.1 °C and 218.2 atm. At point *D* the temperature is −22.0 °C and the pressure is 2045 atm. The negative slope of the fusion curve, *OD* (greatly exaggerated here), and the significance of the broken straight lines are discussed in the text.

Water

▶ Since the fusion curve in the phase diagram of water is negative, that means that ice floats. If ice did not float, then the polar seas would fill with ice that would never melt. Most solids are more dense than their liquid state. Water is, fortuitously for us, an exception.

▶ An increase in pressure to 125 atm lowers the freezing point of water by only about 1 °C.

The phase diagram of water (Fig. 12-30) presents several new features. On that the fusion curve *OD* has a *negative* slope; that is, it slopes toward pressure axis. The melting point of ice *decreases* with an increase in pressu and this is rather unusual behavior for a solid (bismuth and antimony a behave in this way). However, because large changes in pressure are requi to produce even small decreases in the melting point, we do not commo observe this melting behavior of ice. One example that has been given co from ice-skating. Presumably, the pressure of the skate blades melts the i and the skater skims along on a thin lubricating film of liquid water. T explanation is unlikely, however, because the pressure of the blades does produce a significant lowering of the melting point and certainly can explain the ability to skate on ice at temperatures much below the freezi point. (Recent experimental evidence suggests that molecules in a very t surface layer on ice are mobile in the same way as in liquid water, and t mobility persists even at very low temperatures.)

Another feature illustrated in the phase diagram of water is **polymorphis** the existence of a solid substance in more than one form. Ordinary ice, cal ice I, exists under ordinary pressures. The other forms exist only at high pr sures. Polymorphism is more the rule than the exception among solids. Wher occurs, a phase diagram has triple points in addition to the usual solid-liqu vapor triple point. For example at point *D* in Figure 12-30, ice I, ice III, and liq H_2O are in equilibrium at −22.0 °C and 2045 atm. Note that the fusion cur for the forms of ice other than ice I have *positive* slopes. Thus, the triple po with ice VI, ice VII, and liquid water is at 81.6 °C and 21,700 atm.

Phases and Phase Transitions

What's the difference between a phase and a state of matter? These terms te to be used synonymously, but there is a small distinction between them. As have already noted, there are just *three* states of matter: solid, liquid, and gas *phase* is any sample of matter with definite composition and uniform propert that is distinguishable from other phases with which it is in contact. Thus, can describe liquid water in equilibrium with its vapor as a two-phase mixtu The liquid is one phase and the gas, or vapor, is the other. In this case, the pha (liquid and gas) are the same as the states of matter present (liquid and gas).

We can describe the equilibrium mixture at the triple point *D* in Figure 12 as a *three-phase* mixture, even though only *two* states of matter are present (so and liquid). Two of the phases are in the solid state—the polymorphic for ice I and ice III. For mixtures of two or more components, different phases m ist in the liquid state as well as in the solid state. For example, most mixtures triethylamine, $N(CH_2CH_3)_3$, and water at 25 °C separate into two physi-lly distinct liquid phases. One is a saturated solution of triethylamine in ater and the other, a saturated solution of water in triethylamine. Because the essure–temperature diagrams we have been describing can accommodate all e phases in a system, we call them *phase* diagrams. We call the crossing of a o-phase curve in a phase diagram a *phase transition*.

Listed below are six common names assigned to phase transitions.

melting(s $\longrightarrow$ l)	freezing (l $\longrightarrow$ s)
vaporization (l $\longrightarrow$ g)	condensation (g $\longrightarrow$ l)
sublimation (s $\longrightarrow$ g)	deposition (g $\longrightarrow$ s)

Following are two useful generalizations about the changes that occur hen crossing a two-phase equilibrium curve in a phase diagram.

- From lower to higher temperatures along a *constant-pressure* line (an isobar), enthalpy *increases*. (Heat is absorbed.)
- From lower to higher pressures along a *constant-temperature* line (an isotherm), volume *decreases*. (The phase at the higher pressure has the higher density.)

e second generalization helps us to understand why a fusion curve usually s a positive slope. Typical behavior is for a solid to have a greater density an the corresponding liquid. Example 12-7 illustrates how we can use a ase diagram to describe the phase transitions that a substance can undergo.

◀ Because ice I is less dense than $H_2O(l)$, the fusion curve *OD* in Figure 12-30 has a negative slope.

EXAMPLE 12-7 Interpreting a Phase Diagram

A sample of ice is maintained at 1 atm and at a temperature represented by point *P* in Figure 12-30. Describe what happens when **(a)** the temperature is raised, at constant pressure, to point *R*, and **(b)** the pressure is raised, at constant temperature, to point *Q*. The sketches Figure 12-31 suggest the conditions at points *P*, *Q*, and *R*.

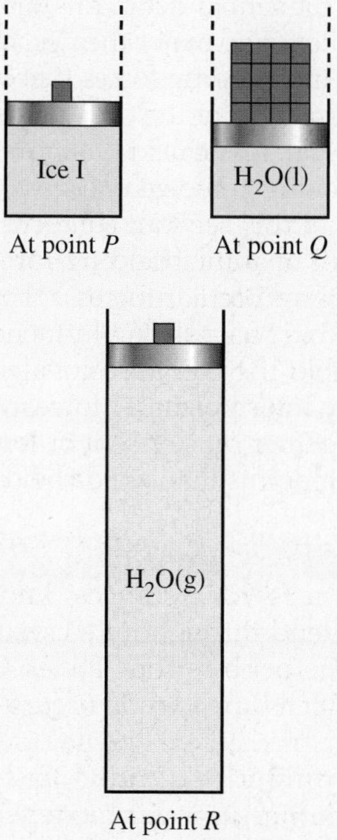

◀ FIGURE 12-31 **Example 12-7 illustrated** A sample of pure water is confined in a cylinder by a freely moving piston surmounted by weights to establish the confining pressure. Sketched here are conditions at the points labeled *P*, *Q*, and *R* in Figure 12-30. The transition from point *P* to *Q* is accomplished by changing the pressure at constant temperature (isothermal). The transition from point *P* to *R* is accomplished by changing the temperature at constant pressure (isobaric).

Analyze

Recall that the lines separating the different phases represent coexistence lines. At these coexistence lines, the system is a mixture of both phases. On either side of those lines, the system is in that particular phase. Also recall that as the system moves from one phase to another at the coexistence lines, the temperature remains constant until all of one phase is converted to another.

Solve

(a) When the temperature reaches a point on the fusion curve *OD* (0 °C), ice begins to melt. The temperature remains constant as ice is converted to liquid. When melting is complete, the temperature again increases. No vapor appears in the cylinder until the temperature reaches 100 °C, at which point the vapor pressure is 1 atm. When all the liquid has vaporized, the temperature is again free to rise to a final value of *R*.

(continued)

(b) Because solids are not very compressible, very little change occurs until the pressure reaches the point of intersection of the constant-temperature line *PQ* with the fusion curve *OD*. Here melting begins. A significant *decrease* in volume occurs (about 10%) as ice is converted to liquid water. After melting, additional pressure produces very little change because liquids are not very compressible.

Assess

Phase diagrams are very useful for understanding the conditions needed to observe the different phases of matter. We should now be able to use the phase diagram in Figure 12-30 to determine the pressure required to observe sublimation instead of melting.

PRACTICE EXAMPLE A: With as much detail as possible, describe the phase changes that would occur if a sample of water represented by point *R* in Figure 12-30 were brought first to point *P* and then to point *Q*.

PRACTICE EXAMPLE B: Draw a sketch showing the condition prevailing along the line *PR* when 1.00 mol of water has been brought to the point where exactly one-half of it has vaporized. Compare this to the condition at point *R* in Figure 12-31, assuming that this is also based on 1.00 mol of water. For example, is the volume of the system the same as that in Figure 12-31? If not, is it larger or smaller, and by how much? Assume that the temperature at point *R* is the same as the critical temperature of water and that water vapor behaves as an ideal gas.

12-7 CONCEPT ASSESSMENT

One method of restoring water-damaged books after a fire is extinguished in a library is by "freeze drying" them in evacuated chambers. Describe how this method might work.

12-5 The Nature of Bonding in Solids

In the remainder of this chapter, we focus on solids. Our discussion of solids w follow two major themes. In this section, we discuss solids in terms of the natu of the bonding forces that hold the solid together. These forces contribute to t formation of the orderly packing arrangements that are observed in crystalli solids. In the next section, we will describe the structures of crystalline solids looking at the geometries of some common packing arrangements.

In this section, we focus primarily on solids that are held together by co lent or ionic bonding forces. However, solids may also be held together intermolecular forces or by metallic bonding forces. The nature of the bondi forces has a strong influence on the physical properties of a solid, as shown Table 12.8. Because covalent and ionic bonding forces are much stronger th the intermolecular forces we have been discussing in this chapter, solids he together by covalent or ionic bonding forces are harder and have higher me ing points than solids held together by intermolecular forces.

Network Covalent Solids

In a few substances, known as **network covalent solids**, covalent bor extend throughout a crystalline solid. In these cases, the entire crystal is h together by strong forces. Consider, for example, two of the allotropic forms which pure carbon occurs—diamond and graphite.

Diamond Figure 12-32 shows one way that carbon atoms can bond one another in a very extensive array or crystal. The two-dimensional Le structure (Fig. 12-32a) is useful only in suggesting that the bonding sche involves ever-increasing numbers of C atoms leading to a giant molecule does not give any insight into the three-dimensional structure of molecule. For this, we need the portion of the crystal shown in Figure 12-32(Each atom is bonded to four others. Atoms 1, 2, and 3 lie in a plane, w

ABLE 12.8 Characteristics of Crystalline Solids

pes	Structural Particles	Strongest Contributing Forces[a]	Typical Properties	Examples
etallic	Cations and delocalized electrons	Metallic bonds	Hardness varies from soft to very hard; melting point varies from low to very high; lustrous; ductile; malleable; very good conductors of heat and electricity	Na, Mg, Al, Fe, Sn, Cu, Ag, W
nic	Cations and anions	Electrostatic attractions	Hard; moderate to very high melting points; nonconductors as solids, but good electric conductors as liquids; many are soluble in polar solvents such as water	NaCl, MgO, $NaNO_3$
etwork valent	Atoms	Covalent bonds	Most are very hard and either sublime or melt at very high temperatures; most are nonconductors of electricity	C(diamond), C(graphite), SIC, AlN, SiO_2
olecular				
Nonpolar	Atoms or nonpolar molecules	Dispersion forces	Soft; extremely low to moderate melting points (depending on molar mass); sublime in some cases; soluble in some nonpolar solvents	He, Ar, H_2, CO_2, CCl_4, CH_4, I_2
Polar	Polar molecules	Dispersion forces and dipole–dipole attractions	Low to moderate melting points; soluble in some polar and some nonpolar solvents	$(CH_3)_2O$, $CHCl_3$, HCl
Hydrogen-Bonded	Molecules with H bonded to N, O, or F	Hydrogen bonds	Low to moderate melting points; soluble in some hydrogen-bonded solvents and some polar solvents	H_2O, NH_3

[a]Generally speaking, more than one type of force contributes. For each case above, we have listed only the strongest of all the ntributing forces.

m 4 above the plane. Atoms 1, 2, 3, and 5 define a tetrahedron with atom 4 ts center. When viewed from a particular direction, a nonplanar hexagonal angement of carbon atoms (gray) is also seen.

f silicon atoms are substituted for half the carbon atoms in this structure, resulting structure is that of silicon carbide (carborundum). Both diamond d silicon carbide are extremely hard, and this accounts for their extensive as abrasives. In fact, diamond is the hardest substance known. To scratch oreak diamond or silicon carbide crystals, covalent bonds must be broken. ese two materials are also nonconductors of electricity and do not melt or blime except at very high temperatures. SiC sublimes at 2700 °C, and diand melts above 3500 °C.

◀ Another silicon-containing network covalent solid is ordinary silica—silicon dioxide, SiO_2.

aphite Carbon atoms can bond together to produce a solid with properties y different from those of diamond. In graphite, bonding involves the orbital $sp^2 + p$. The three sp^2 orbitals are directed in a plane at angles of 120 °C, and p orbitals overlap in the manner described for carbon atoms in benzene, C_6H_6

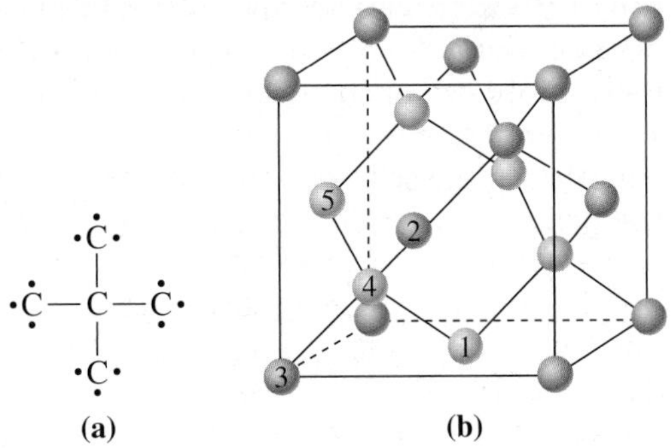

(a) (b)

FIGURE 12-32
e diamond structure
A portion of the Lewis structure. The crystal structure shows each oon atom bonded to four others tetrahedral fashion. The segment he entire crystal shown here is ed a unit cell.

KEEP IN MIND

that four bonds directed from a central atom to the corners of a tetrahedron correspond to the sp^3 hybridization scheme.

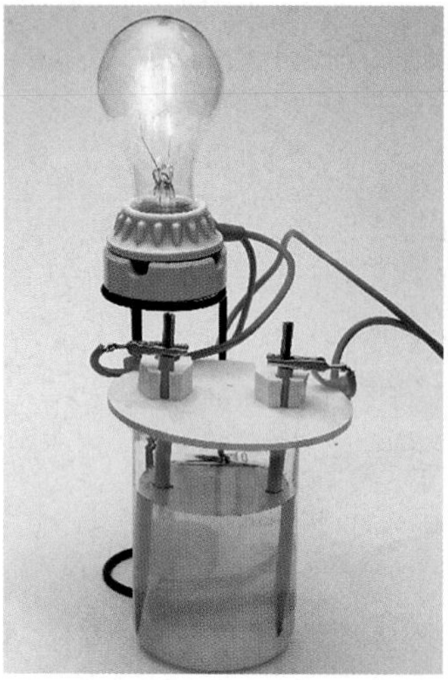

Tom Pantages

▲ **Graphite conducts electricity** Delocalized electrons in graphite allow the conduction of electricity. In the photo, pencil "lead,"a mixture of graphite and clay, is used as electodes to complete the circuit. The beaker contains a solution of ions that carry the current between the pencil electrodes.

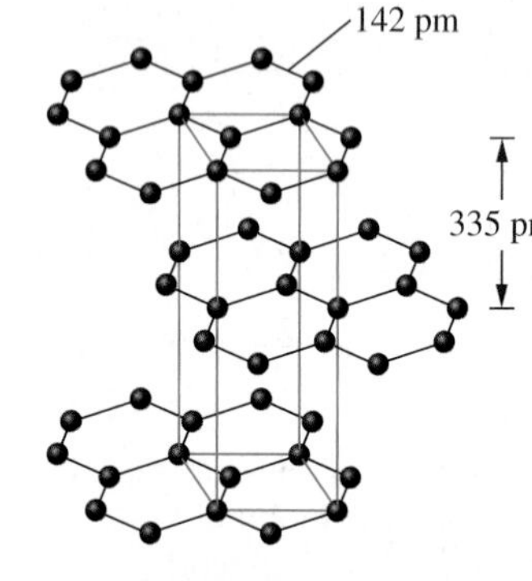

▲ FIGURE 12-33
The graphite structure

(see Figure 11-30). Thus, the *p* electrons are delocalized—that is, not restricte the region between two C atoms but shared among many C atoms within a pl of C atoms. This type of bonding produces the crystal structure show Figure 12-33. Each carbon atom forms strong covalent bonds with three ne boring carbon atoms in the same plane, giving rise to layers of carbon atoms hexagonal arrangement. Bonding within layers is strong, but the intermolec forces between layers are the much weaker van der Waals forces. We can see through bond distances. The C—C bond distance within a layer is 142 pm (c pared with 139 pm in benzene); between layers, it is 335 pm.

Its unique crystal structure gives graphite some distinctive propert Because bonding between layers is weak, the layers can glide over one ano rather easily. As a result, graphite is a good lubricant, either in dry form or i oil suspension.* If a mild pressure is applied to a piece of graphite, layers of graphite flake off; this is what happens when we use a graphite pencil. A because the *p* electrons are delocalized, they migrate through the planes of bon atoms when an electric field is applied; graphite conducts electricity. important use of graphite is as electrodes in batteries and in industrial elect ysis. Diamond is not an electrical conductor because all its valence electrons localized or permanently fixed into single covalent bonds.

Other Allotropes of Carbon

In 1985, the first of what is now known to be an extensive series of allotro of carbon was discovered. In experiments designed to mimic conditi found near red-giant stars, a number of carbon-containing molecules w discovered and characterized through mass spectroscopy. The strongest p in the mass spectrum came at 720 u, corresponding to the molecule C_{60}. F time, proposing a plausible structure for this molecule proved to be a c lenge. Neither diamond- nor graphite-type structures could account f molecule with 60 carbon atoms because "dangling" bonds would remai the edges of the structures. The structure that was finally proposed, and c firmed by X-ray crystallography, is that of a *truncated icosahedron*—a th dimensional figure composed of 12 pentagonal and 20 hexagonal faces, w a carbon atom at each of its 60 vertices (Fig. 12-34c). This figure resembl soccer ball and also certain geodesic domes. In fact, the resemblance to geodesic dome led first to the proposed name "buckminsterfullerenes," t simply, *fullerenes*, and finally the colloquial expression "buckyballs." (geodesic dome is an architectural form pioneered by R. Buckminster Full Since 1985, many other fullerenes have been discovered, includ C_{70}, C_{74}, and C_{82}. Fullerenes can also form compounds, some by attach atoms or groups of atoms to their surfaces, others by encasing an atom ins the fullerene structure. To date, several thousand fullerene compounds h been prepared.

Research on fullerenes has led to the discovery of a related type of car allotrope—*nanotubes*. A nanotube can be thought of as a two-dimensio array of hexagonal rings of carbon atoms, which is called a graphene sheet. analogous macroscopic structure is a sheet of chicken wire. Now imag rolling the graphene sheet into a cylinder (something that a section cut fro roll of chicken wire seems to do naturally). Finally, cap each end of the cy drical graphene sheet with half of a fullerene (Fig. 12-35). The diameter these tubes are of the order of a few nanometers (hence the name *nano*tu Their lengths can vary from several nanometers to a micrometer or m Nanotubes possess unusual electronic and mechanical properties that o the promise of some applications in the macroscopic world and proba

*The lubricating properties of graphite also appear to depend on the presence of molec between the layers of carbon atoms. When graphite is strongly heated in a vacuum it becom much poorer lubricant.

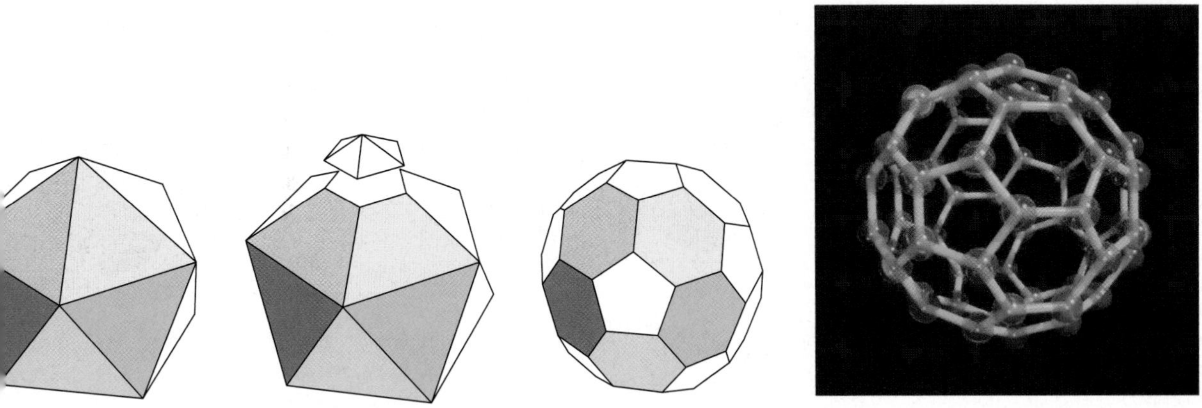

(a) (b) (c) (d)

Tom Pantages

GURE 12-34
erenes
ın icosahedron, a shape formed by 20 equilateral triangles. Five triangles meet at ı of the 12 vertices. **(b)** Truncating or cutting off a vertex reveals a new pentagonal . **(c)** The truncated icosahedron. Twelve pentagons have replaced the original ertices, and the 20 equilateral triangles have been replaced by 20 hexagons. ʼhe C_{60} molecule.

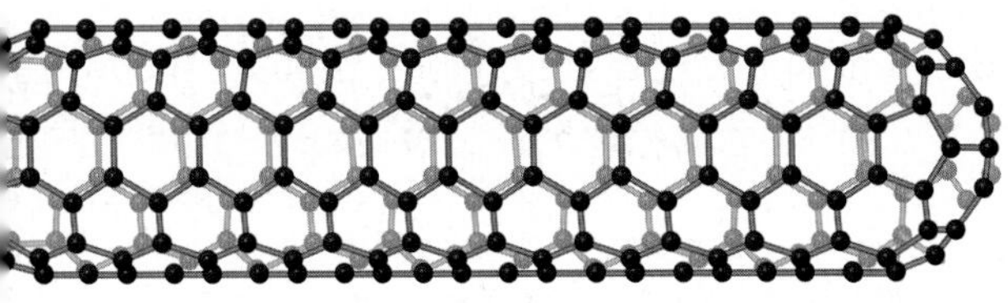

(a)

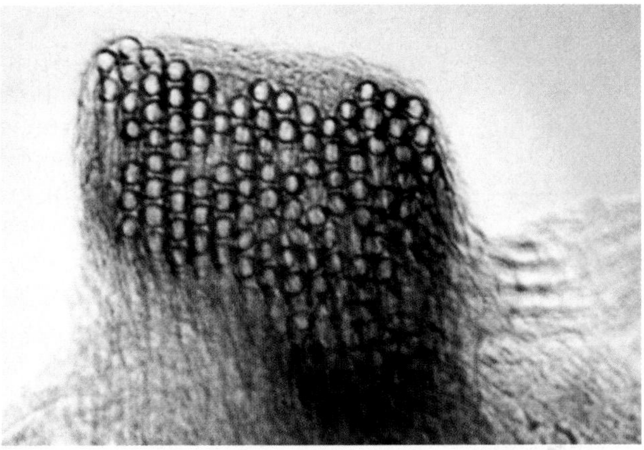

(b)

Boris I. Yakobson and Richard E. Smalley, Rice University

IGURE 12-35
ıotubes
3all-and-stick model of a small nanotube. **(b)** A bundle ngle-wall nanotubes.

ıy more in the submicroscopic world of *nanotechnology*. For example, nan-bes might one day be used to form the molecular wires of nanoscale elec-ıic devices.

ic Solids

en predicting properties of an **ionic solid**, we often face this question: How icult is it to break up an ionic crystal and separate its ions? This question is ressed by the *lattice energy* of a crystal. Let us define **lattice energy** as the rgy given off when separated *gaseous* ions—positive and negative—come ether to form *one mole* of a *solid* ionic compound. Lattice energies can be useful naking predictions about the melting points and water solubilities of ionic ıpounds. We will examine how to calculate lattice energies in Section 12-7. times, however, we need to make only *qualitative* comparisons of *interionic es*, and the following generalization works quite well.

The attractive force between a pair of oppositely charged ions increases with increased charge on the ions and with decreased ionic sizes.

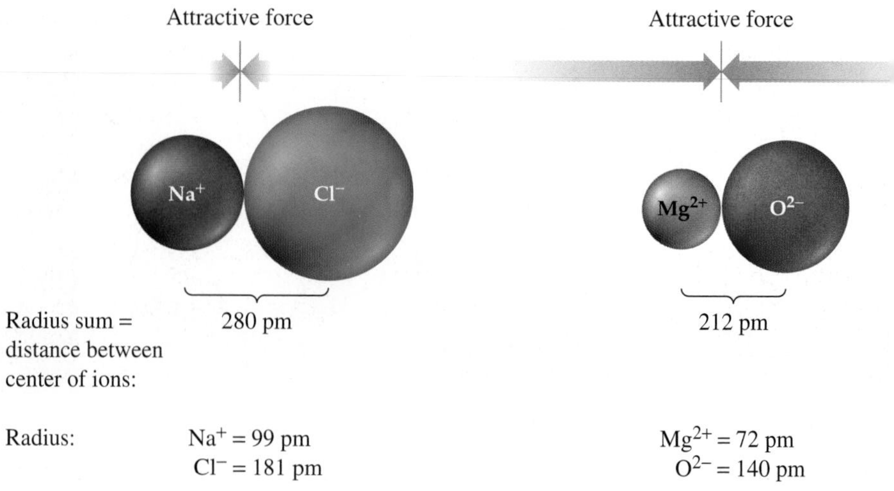

▶ FIGURE 12-36
Interionic forces of attraction
Because of the higher charges on the ions and the closer proximity of their centers, the interionic attractive force between Mg^{2+} and O^{2-} is about seven times as great as between Na^+ and Cl^-.

This idea is based on Coulomb's law (Appendix B) and illustrated in Figure 12

For most ionic compounds, lattice energies are great enough that ions not readily detach themselves from the crystal and pass into the gaseous st Ionic solids do not sublime at ordinary temperatures. We can melt ionic so by supplying enough thermal energy to disrupt the crystalline lattice. In g eral, the higher the lattice energy of an ionic compound, the higher is its m ing point.

The energy required to break up an ionic crystal when it dissolves res from the interaction of ions in the crystal with molecules of the solvent. extent to which an ionic solid dissolves in a solvent, however, depends in p on the lattice energy of the ionic solid. As a rough rule, though, the lower lattice energy, the greater the quantity of an ionic solid that can be dissolve a given quantity of solvent.

EXAMPLE 12-8 Predicting Physical Properties of Ionic Compounds

Which has the higher melting point, KI or CaO?

Analyze

Trends in melting points, just as we observed for boiling points, depend on intermolecular forces. For ioni compounds the melting points are dependent on the interionic forces, which are related to Coulomb's law.

Solve

Ca^{2+} and O^{2-} are more highly charged than K^+ and I^-. Also, Ca^{2+} is smaller than K^+, and O^{2-} is smaller tha I^-. We would expect the interionic forces in crystalline CaO to be much larger than in KI. CaO should have th higher melting point. (The observed melting points are 677 °C for KI and 2590 °C for CaO.)

Assess

We see that two factors contribute to the interionic forces in this problem. The first is the charge and the secon is the radius of each ion.

PRACTICE EXAMPLE A: Cite one ionic compound that you would expect to have a lower melting point than K and one with a higher melting point than CaO.

PRACTICE EXAMPLE B: Which would you expect to have the greater solubility in water, NaI or $MgCl_2$? Explain

Molecular Solids

Molecular solids are made up of discrete molecules that interact via intern ecular forces. Several examples are given in Table 12.8. At very low temp tures, the noble gases form solids that are held together by London forces, a thus, they are also considered molecular solids.

o picture a crystalline solid having molecules as its structural units, con- r solid methane, CH_4. Each CH_4 molecule occupies a volume equivalent to here with a radius of 228 pm, and the molecules are organized into a regu- repeating pattern. For CH_4, the repeating pattern is represented by a cube has a CH_4 molecule at each corner and at the center of each face, as shown ne diagram in the margin. Because the CH_4 molecule is nonpolar and inca- le of forming hydrogen bonds, the CH_4 molecules interact via London es only.

he situation is similar for other molecular solids, except that the details of packing arrangement may be different (e.g., the repeating pattern may not ased on a cubic arrangement of molecules), and the forces holding the solid ether may also include dipole–dipole or hydrogen bonding interactions.

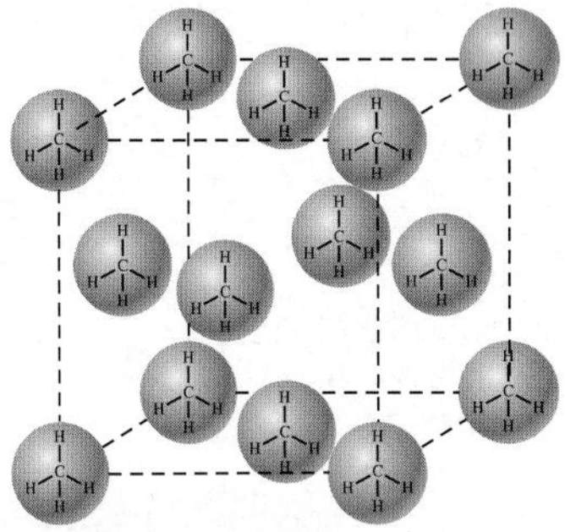

▲ An example of a molecular solid
In solid methane, CH_4, the molecules are organized in a repeating pattern that is represented by a cube having a CH_4 molecule at each corner and at the center of each face. We will see in the next section that this arrangement is described as face-centered cubic (fcc).

tallic Solids

ding in metals can be explained as a network of positive ions immersed in a of electrons. That is, the electrons in the valence shell of the metal atoms highly delocalized. For this reason, metals are very good conductors of tricity. Metallic bonding forces are strong in comparison with those arising n intermolecular forces (Table 12.3), and thus, **metallic solids** have consid- ly higher melting points than molecular solids.

-6 Crystal Structures

stals are solid structures that have plane surfaces, sharp edges, and regular metric shapes. They have aroused interest from earliest times, whether as ice, salt, quartz, or gemstones. Only in relatively recent times have we come to a damental understanding of the crystalline state. This understanding started h the invention of the optical microscope and was greatly expanded follow- the discovery of X-rays. The key idea, now supported by countless experi- nts, is that the regularity observed in crystals at the macroscopic level is due to nderlying regular pattern in the arrangement of atoms, ions, or molecules.

stal Lattices

can probably think of a number of situations in which you have had to l with repeating patterns in one or two dimensions. These might include jects like stringing beads to make a necklace, wallpapering a room, or cre- g a design with floor tiles. The structures of crystals, however, must be cribed through *three*-dimensional patterns. These patterns are outlined inst a framework called a *lattice*, comprising the intersections of three sets arallel planes. Figure 12-37 shows the special case for a lattice in which the nes are equidistant and mutually perpendicular (intersecting at 90° les). This is called a *cubic* lattice, and it can be used to describe a number of stals. For other crystals, the appropriate lattice may involve planes that are equidistant or that intersect at angles other than 90°. In all, there are seven sibilities for crystal lattices, but we will emphasize only the cubic lattice.

attice planes intersect to produce three-dimensional figures having six es arranged in three sets of parallel planes. These figures are called *paral- pipeds*. In Figure 12-37, the parallelepipeds are cubes. A parallelepiped ed the **unit cell** can be used to generate the entire lattice by replicating it ng the three perpendicular directions. (Consider a two-dimensional anal- : A single floor tile is like a unit cell and, wherever the first tile is placed, entire floor can be covered by adding identical tiles in the two perpendic- r directions from the initial one.) Where possible, we arrange the three- nensional space lattice so that the centers of the structural particles of the stal (atoms, ions, or molecules) are situated at lattice points. If a unit cell structural particles only at its corners, it is called a *primitive* or simple unit , because it is the simplest unit cell we can consider. But some unit cells

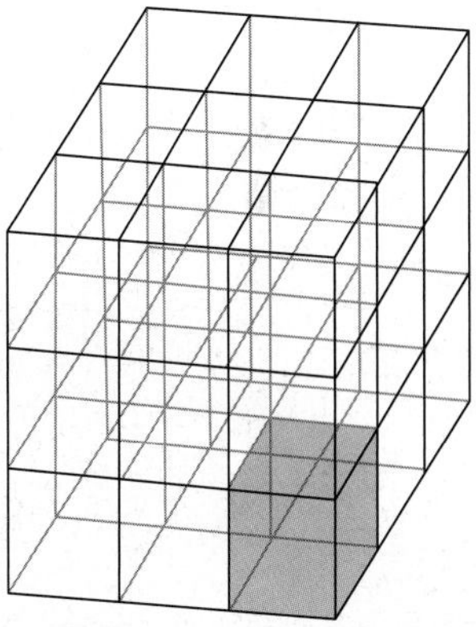

▲ FIGURE 12-37
The cubic lattice
One parallelepiped formed by the intersection of mutually perpendicular planes is shaded in green—it is a cube. An endless lattice can be generated by simple displacements of the green cube in the three perpendicular directions (that is, left and right, up and down, and forward and backward).

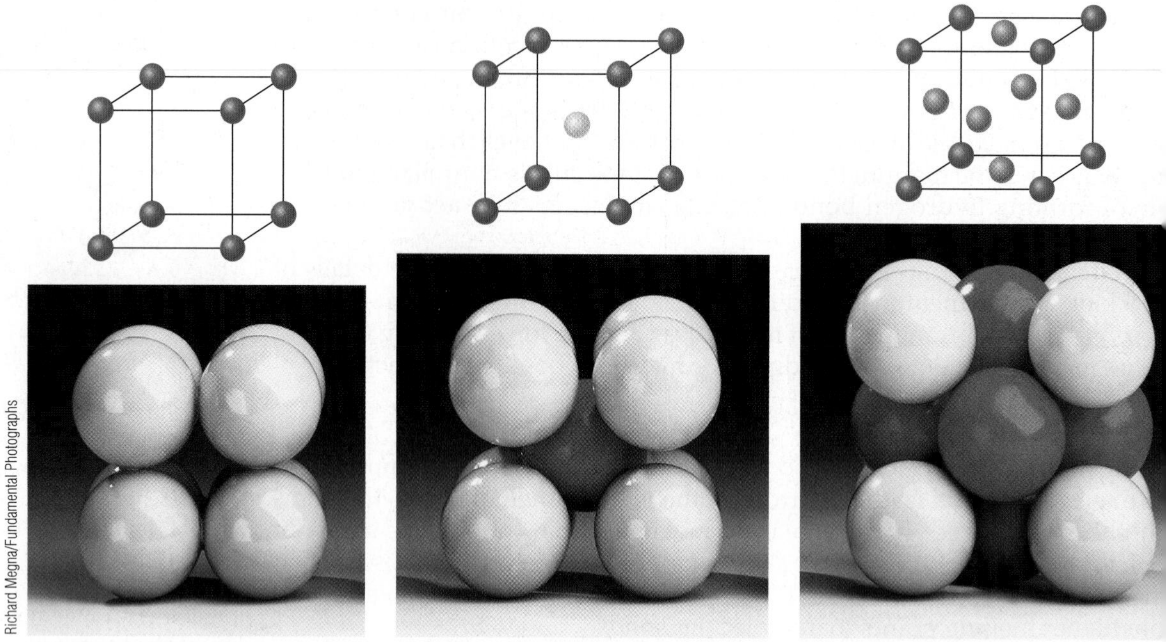

Simple cubic Body-centered cubic Face-centered cubic

▲ FIGURE 12-38
Unit cells in the cubic crystal system
In the line-and-ball drawings in the top row, only the centers of spheres (atoms) are shown at their respective positions in the unit cells. The space-filling models in the bottom row show contacts between spheres (atoms). In the simple cubic cell, spher come into contact along each edge. In the body-centered cubic (bcc) cell, contact o the spheres is along the cube diagonal. In the face-centered cubic (fcc) cell, contact along the diagonal of each face. The spheres shown here are identical atoms; color is used only for emphasis.

have more structural particles than those found at its corners. In the **bo centered cubic (bcc)** structure, there is a structural particle at the center of cube as well as at each corner of the unit cell. In a **face-centered cubic (** structure, there is a structural particle at the center of each face as well a each corner. These unit cells are shown in Figure 12-38.

▲ A closest packed pyramid of cannonballs. Oranges at a fruit stand are often packed in cubic closest packed pyramids so that they will not slip.

Closest Packed Structures

Unlike boxes, which can be stacked to fill all space, when spheres are stac together, there must always be some unfilled space. In some arrangeme however, the spheres come into as close contact as possible, and the volum the holes, or voids, is at a minimum. These arrangements are known as clos packed structures and are the basis of a number of crystal structures.

To analyze the closest packed structures in Figure 12-39(a), imagine layer of spheres, layer A (red), in which each sphere is in contact with six oth arranged in a hexagonal fashion around it. Among the spheres, there are ho The hole between three spheres resembling a triangle is called a *trigonal* (Fig. 12-39b). Once the first sphere is placed in the next layer, layer B (yello the entire pattern for that layer is fixed. Again, there are holes in layer B, but holes are of two different types. *Tetrahedral holes* fall directly over *sphere* layer A, and *octahedral holes* fall directly over *holes* in layer A (Fig. 12-39b).

There are two possibilities for C, the third layer (Fig. 12-39a). In one arran ment, called **hexagonal closest packed (hcp)**, all the tetrahedral holes are c ered. Layer C is identical to layer A, and the structure begins to repeat itself the other arrangement, called **cubic closest packed**, all the octahedral holes covered. The spheres in layer C (blue) are out of line with those in layer A. O when the fourth layer is added does the structure begin to repeat itself.

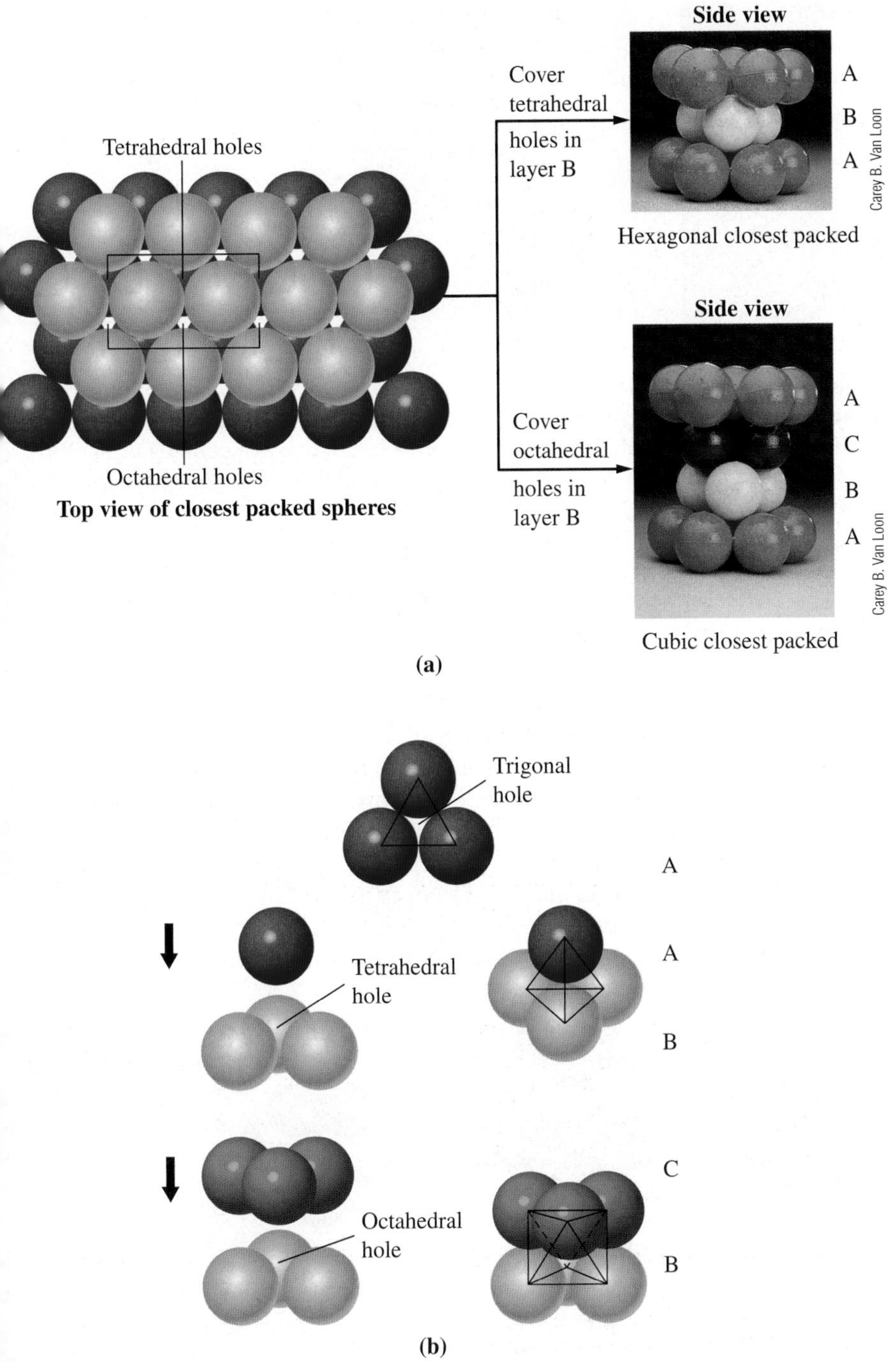

FIGURE 12-39
osest packed structures
Spheres in layer A are red. Those in layer B are yellow, and in layer C, blue. **(b)** The es in closest packed structures. The trigonal hole is formed by three spheres in one he layers. The tetrahedral hole is formed when a sphere in the upper layer sits in dimple of the lower layer. The octahedral hole is formed between two groups hree spheres in two layers.

Study Figure 12-40 and you will see that the cubic closest packed structure s a face-centered cubic unit cell. The unit cell of the hexagonal closest cked structure is shown in Figure 12-41. In both the hcp and fcc structures, les account for only 25.96% of the total volume. For the simple cubic and bcc it cells, holes account for a higher percentage of the total volume. In the bcc it cell, holes account for 31.98% of the total volume. The best examples of

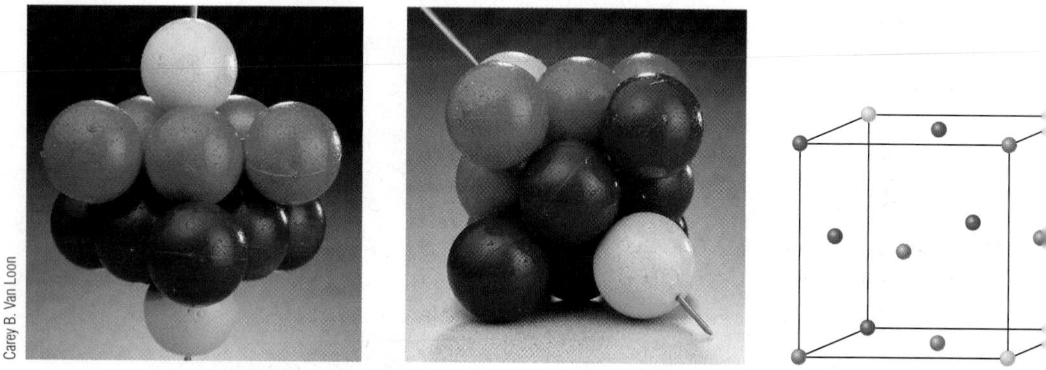

▲ FIGURE 12-40
A face-centered cubic unit cell for the cubic closest packing of spheres
The 14 spheres on the left are extracted from a larger array of spheres in a cubic closest packed structure. The two middle layers each have six atoms; the top and bottom layers, one. Rotation of the group of 14 spheres reveals the fcc unit cell (rig

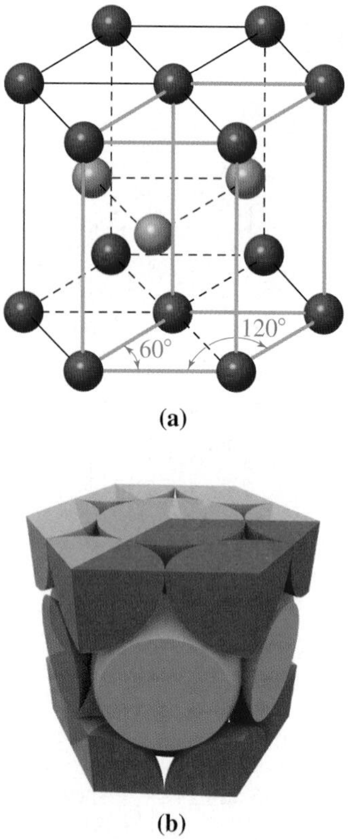

▲ FIGURE 12-41
The hexagonal closest packed (hcp) crystal structure
(a) A unit cell is highlighted in heavy green lines. The atoms that are part of that cell are joined in solid lines. Note that the unit cell is a parallelepiped but not a cube. Three adjoining unit cells are depicted. The highlighted unit cell and broken-line regions together show the layering (ABA) described in Figure 12-39. **(b)** The hexagonal prism showing parts of the shared spheres at the corners and the single sphere at the center of the unit cell.

crystal structures based on the packing of spheres are found among the n als. Some examples are listed in Table 12.9.

Coordination Number and Number of Atoms per Unit Cell

In crystals with atoms as their structural units, each atom is in contact w several others. For example, can you see in Figure 12-38 that the center aton the bcc unit cell is in contact with each corner atom? The number of atc with which a given atom is in contact is called its *coordination number*. For bcc structure, this is 8; for the fcc and hcp structures, the coordination num is 12. The easiest way to visualize the coordination number 12 is from the l ering of spheres described in Figure 12-39. Each sphere is in contact with others in the same layer, *three* in the layer above, and *three* in the layer belo

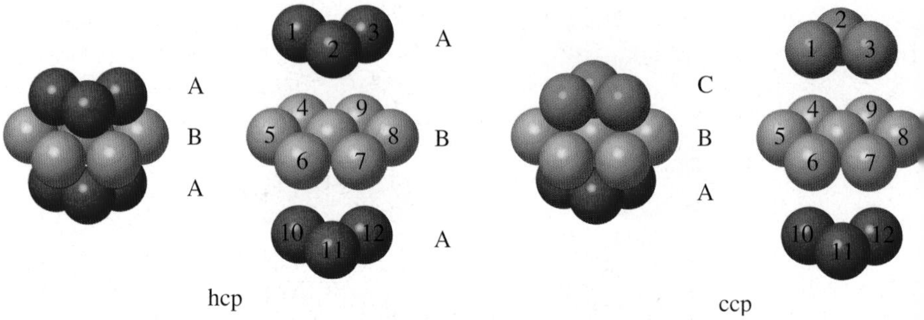

▲ **Illustrating the coordination number for the hcp and ccp structures**

TABLE 12.9 Four Ways of Packing Identical Spheres

Unit Cell	Coordination Number	Atoms per Unit Cell	Volume Occupied, %	Example
Simple cubic	6	1	52	Po
Body-centered cubic (bcc)	8	2	68	Fe, Na, K, W
Hexagonal closest packed (hcp)	12	2	74	Cd, Mg, Ti, Zn
Face-centered cubic (fcc)	12	4	74	Ag, Cu, Pb

Note: Face-centered cubic (fcc) is equivalent to cubic closest packed (ccp).

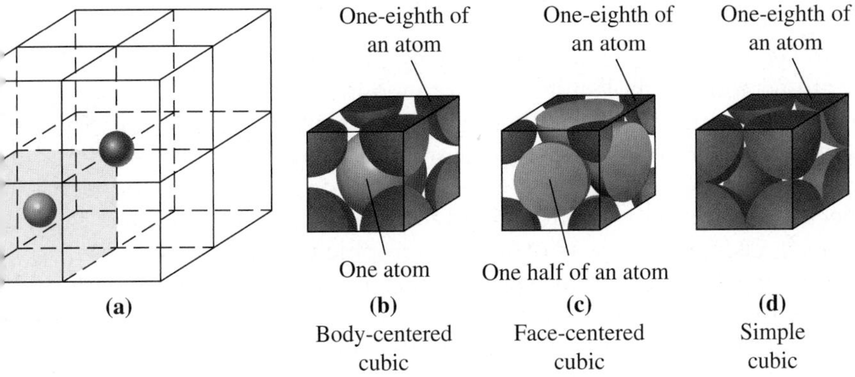

IGURE 12-42
portioning atoms among cubic unit cells
Eight unit cells are outlined. Attention is directed to the blue-shaded unit cell. For ty, only the centers of two atoms are pictured. The atom in the center of the blue cell ongs entirely to that cell. The corner atom is seen to be shared by all eight unit cells. The shared spheres of a body-centered unit cell. **(c)** The shared spheres of a face-tered unit cell. **(d)** The shared spheres of a simple cubic unit cell. The spheres wn here are identical atoms; color is used only for emphasis.

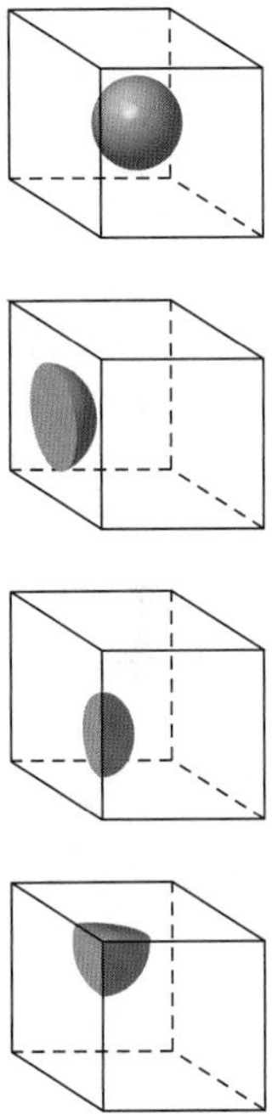

▲ **How spheres are shared between or among unit cells**
For a sphere in the middle of the unit cell, there is no sharing; on a face 1/2 of the sphere is in the unit cell; at an edge only 1/4 of the sphere is in the unit cell; and in a corner only 1/8 is contained within the unit cell.

Although it takes nine atoms to draw the bcc unit cell, it is wrong to conclude t the unit cell consists of nine atoms. As shown in Figure 12-42(a), only the ter atom belongs *entirely* to the bcc unit cell. The other atoms are shared with er unit cells. The corner atoms are shared among eight adjoining unit cells. ly one-eighth of each corner atom should be thought of as belonging entirely a given unit cell (Fig. 12-42b). Thus, the eight corner atoms collectively con-ute the equivalent of *one* atom to the unit cell. The total number of atoms in a unit cell, then, is *two* [that is, $1 + (8 \times \frac{1}{8})$]. For the hcp unit cell of Figure 12-41, also get *two* atoms per unit cell if we use the correct counting procedure. e corner atoms account for $\frac{1}{8} \times 8 = 1$ atom, and the central atom belongs irely to the unit cell. In the fcc unit cell, the corner atoms account for $8 = 1$ atom, and those in the center of the faces for $\frac{1}{2} \times 6 = 3$ atoms. The unit cell contains *four* atoms (Fig. 12-42c). The simple cubic unit cell con-s only *one* atom per unit cell (Fig. 12-42d).

12-2 ARE YOU WONDERING?

How do we calculate the volume of the holes in a structure?

o illustrate this, consider the bcc structure. The ratio of the occupied volume to he unit cell volume is

$$f_V = \frac{\text{volume of spheres in unit cell}}{\text{volume of unit cell}}$$

f the radius of the atom is r, the volume of a sphere is $(4/3)\pi r^3$, and as shown in igure 12-45 on page 557 the cube edge l is $r(4/\sqrt{3})$. Based on two complete pheres in the unit cell, we have

$$f_V = \frac{2 \times (4/3)\pi r^3}{[r(4/\sqrt{3})]^3} = 0.6802$$

Thus, 68.02% of the unit cell is occupied and 31.98% of the unit cell is empty. Note also hat this percentage of empty space is the same regardless of the radius of the sphere.

X-Ray Diffraction

We can see macroscopic objects by using visible light and our eyes. To “s
how atoms, ions, or molecules are arranged in a crystal requires light of m
shorter wavelength. When a beam of X-rays encounters atoms, the X-r
interact with electrons in the atoms and the original beam is scattered in
directions. The pattern of this scattered radiation is related to the distribut
of electronic charge in the atoms and/or molecules. The scattered X-rays
produce a visible pattern, as on a photographic film, and it is then possibl
infer the microscopic structure of the substance from this visible pattern. H
successful we are in making inferences depends on the amount of the sc
tered radiation recovered, that is, on how much “information” is gather
The power of the X-ray diffraction method has been greatly increased by
use of high-speed computers to process vast amounts of X-ray data.

▶ The X-ray diffraction method was originated by Max von Laue (Nobel Prize, 1914) but carried further by the Braggs. William Lawrence Bragg was only 25 years old when he and his father, William Henry Bragg, won the Nobel Prize in 1915.

Figure 12-43 suggests a method of scattering X-rays from a crystal. X-
data can be explained by a geometric analysis proposed by W. H. Bragg a
W. L. Bragg in 1912 and illustrated in Figure 12-44. The figure shows two r
in a monochromatic (single-wavelength) X-ray beam, labeled *a* and *b*. Wav
is diffracted, or scattered, by one plane of atoms or ions in a crystal and wa
b from the next plane below. Wave *b* travels a greater distance than wav
The additional distance is $2d \sin \theta$. The intensity of the scattered radiation w
be greatest if waves *a* and *b* reinforce each other, that is, if their crests a
troughs line up. To satisfy this requirement, the additional distance trave
by wave *b* must be an integral multiple of the wavelength of the X-rays.

$$n\lambda = 2d \sin \theta \qquad \textbf{(1}$$

From the measured angle θ yielding the maxmum intensity for the scatte
X-rays, and the X-ray wavelength (λ), we can calulate the spacing (d) betw

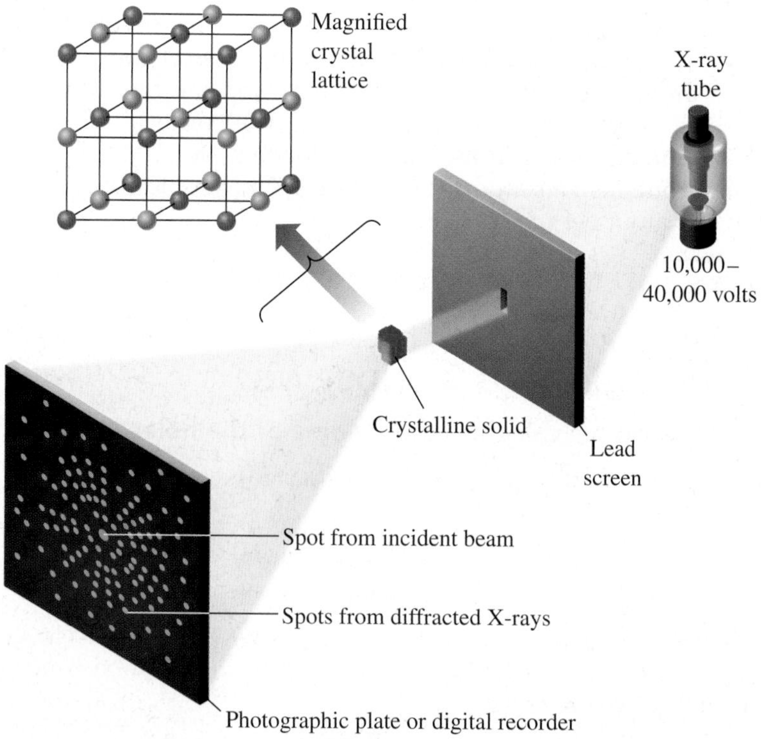

▲ FIGURE 12-43
Diffraction of X-rays by a crystal
In X-ray diffraction, the scattering is usually from no more than 20 planes deep in a crystal. The size of the single crystal is to have enough of the surface available for diffraction, yet the diffraction is dominated by a few of the surface planes.

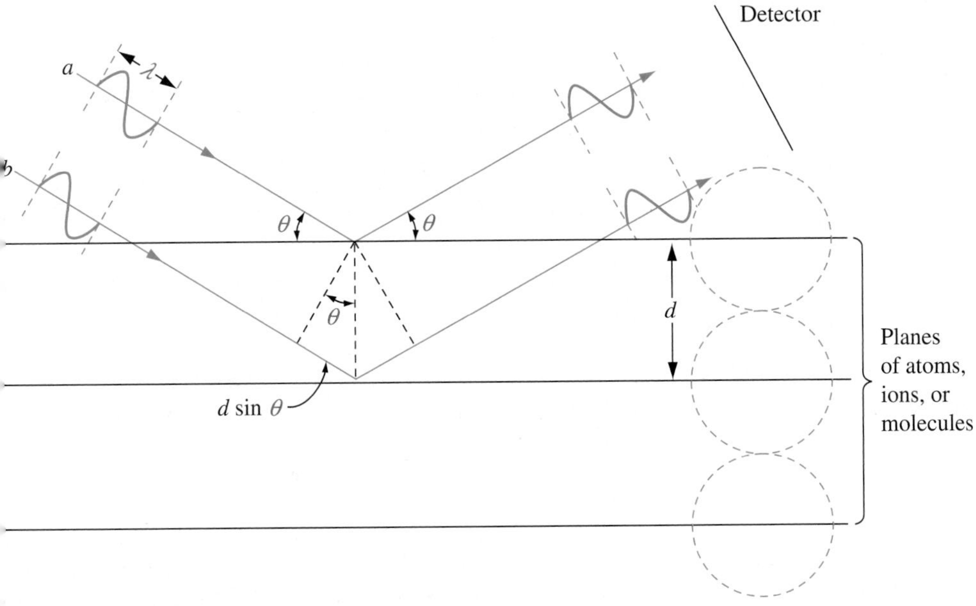

IGURE 12-44
termination of crystal structure by X-ray diffraction
two triangles outlined by dashed lines are identical. The hypotenuse of each ngle is equal to the interatomic distance, d. The side opposite the angle θ thus a length of $d \sin \theta$. Wave b travels farther than wave a by the distance $2d \sin \theta$.

mic planes. With different orientations of the crystal, we can determine mic spacings and electron densities for different directions through the stal, in short, the crystal structure.

Once a crystal structure is known, certain other properties can be deter- ned by calculation. Example 12-9 shows the determination of a metallic

EXAMPLE 12-9 Using X-Ray Data to Determine an Atomic Radius

At room temperature, iron crystallizes in a bcc structure. By X-ray diffraction, the edge of the cubic cell corresponding to Figure 12-45 is found to be 287 pm. What is the radius of an iron atom?

Analyze

Nine atoms are associated with a bcc unit cell. One atom is located at each of the eight corners of the cube and one at the center. The three atoms along a diagonal through the cube are in contact. The length of the cube diagonal (the distance from the farthest upper-right corner to the nearest lower-left corner) is four times the atomic radius. Also shown in Figure 12-45 is the fact the diagonal of a cube is equal to $\sqrt{3} \times l$. The length of an edge, l, is what is given.

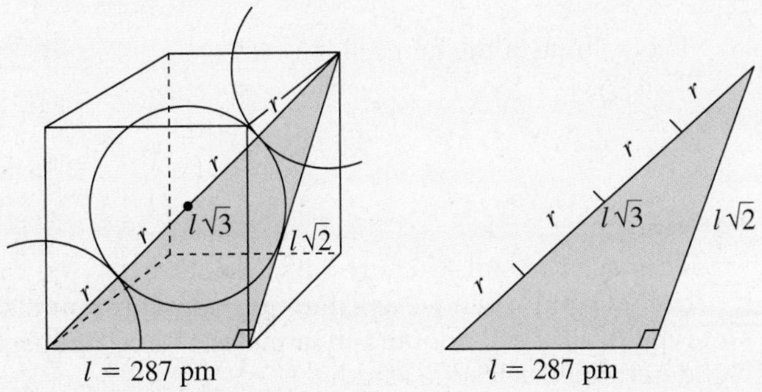

▲ FIGURE 12-45
Determination of the atomic radius of iron—Example 12-9 illustrated
The right triangle must conform to the Pythagorean formula $a^2 + b^2 = c^2$. That is, with l as an edge of the cube, $(l)^2 + (l\sqrt{2})^2 = (l\sqrt{3})^2$, or $(l)^2 + 2(l)^2 = 3(l)^2$.

Solve

Setting the length of the cube diagonal, in terms of atomic radii, equal to the expression relating the diagonal to the cube, we have

$$4r = l\sqrt{3}$$

(continued)

which is used to solve for the atomic radius of an iron atom:

$$r = \frac{\sqrt{3} \times 287\ \text{pm}}{4} = \frac{1.732 \times 287\ \text{pm}}{4} = 124\ \text{pm}$$

Assess

We see that it is important to know the atomic arrangement for each basic unit cell and to know that atoms at the corners are shared between unit cells.

PRACTICE EXAMPLE A: Potassium crystallizes in the bcc structure. What is the length of the unit cell in this structure? Use the metallic radius of potassium given in Figure 9-11.

PRACTICE EXAMPLE B: Aluminum crystallizes in an fcc structure. Given that the atomic radius of Al is 143.1 pm, what is the volume of a unit cell?

radius, and Example 12-10 estimates the density of a crystalline solid. F both of these calculations, we need to sketch, or in some way visualize unit cell of the crystal. In particular, we need to see which atoms are direct contact.

EXAMPLE 12-10 Relating Density to Crystal Structure Data

Use data from Example 12-9, together with the molar mass of Fe and the Avogadro constant, to calculate the density of iron.

Analyze

To calculate the density, we need the mass of the unit cell (in grams) and its volume (in cm^3). From Table 12.9, we find that there are *two* Fe atoms per bcc unit cell, which we can use to calculate the mass of the unit cell. In Example 12-9, we saw that the length of a unit cell is $l = 287\ \text{pm} = 287 \times 10^{-12}\ \text{m} = 2.87 \times 10^{-8}\ \text{cm}$, which we can use to find the density of iron.

Solve

We need the mass of these two atoms, and the key to getting this is a conversion factor based on the fact that $1\ \text{mol Fe} = 6.022 \times 10^{23}\ \text{Fe atoms} = 55.85\ \text{g Fe}$.

$$m = 2\ \text{Fe atoms} \times \frac{55.85\ \text{g Fe}}{6.022 \times 10^{23}\ \text{Fe atoms}} = 1.855 \times 10^{-22}\ \text{g Fe}$$

The volume of the unit cell is $V = l^3 = (2.87 \times 10^{-8})^3\ \text{cm}^3$. Density is the ratio of mass to volume.

$$\text{density of Fe} = \frac{m}{V} = \frac{1.855 \times 10^{-22}\ \text{g Fe}}{2.36 \times 10^{-23}\ \text{cm}^3} = 7.86\ \text{g Fe cm}^{-3}$$

Assess

The use of crystal structure data is another way we can determine the density of materials. This method is especially useful when we are studying new compounds or materials of which we have only a small quantity. The type of unit cell adopted by a metal can be deduced by using the experimental density and experimentally determined cell edge length.

PRACTICE EXAMPLE A: Use the result of Practice Example 12-9A, the molar mass of K, and the Avogadro constant to calculate the density of potassium.

PRACTICE EXAMPLE B: Use the result of Practice Example 12-9B, the molar mass of Al, and its density ($2.6984\ \text{g cm}^{-3}$) to evaluate the Avogadro constant, N_A.
[*Hint:* From the volume of a unit cell and the density of Al, you can determine the mass of a unit cell. Knowing the number of Al atoms in the fcc unit cell, you can determine the mass per Al atom.]

12-8 CONCEPT ASSESSMENT

uppose a first-order diffraction is observed at an angle θ for a particular plane f atoms by using X-rays of wavelength λ. In order to observe a second-order iffraction from the same plane of atoms at the same angle θ, the wavelength f the X-rays must be a multiple of λ. What is that multiple?

ic Crystal Structures

ve try to apply the packing-of-spheres model to an ionic crystal, we run into complications: (1) Some of the ions are positively charged and some are gatively charged, and (2) the cations and anions are of different sizes. What can expect, however, is that oppositely charged ions will come into close ximity. Generally we think of them as being in contact. Like-charged ions, ause of mutual repulsions, are not in direct contact. We can think of some ic crystals as a fairly closely packed arrangement of ions of one type with les filled by ions of the opposite charge. The relative sizes of cations and ons are important in establishing a particular packing arrangement.

A common arrangement in binary ionic solids is the face-centered cubic angement. Very often one of the ions, usually the anion, can be viewed as opting the face-centered cubic structure while the cation occupies one of the les between the closest-packed spheres. The three types of holes of the cubic sest packed structure—trigonal, tetrahedral, and octahedral—are shown in ure 12-46. The size of the holes is related to the radius, R, of the anions used form the structure. Figure 12-47 shows a cross section through an octahe- al hole. The radius of the cation, r, that can just fit into the hole can be found using the Pythagorean formula, as follows.

$$(2R)^2 + (2R)^2 = (2R + 2r)^2$$
$$2\sqrt{2}R = 2R + 2r$$
$$(2\sqrt{2} - 2)R = 2r$$
$$(\sqrt{2} - 1)R = r$$
$$r = 0.414\,R$$

nilar calculations can be used for tetrahedral and trigonal holes, for which = 0.225 R and $r = 0.155\,R$, respectively. These calculations show that in the bic closest packed structure, the octahedral hole is bigger than the tetra- dral hole.

Another arrangement adopted by binary ionic solids is the simple cubic angement. The simple cubic arrangement is not a closest packed structure d has larger holes than the cubic closest packed arrangement. The simple bic structure has a cubic hole at the center of the unit cell. The size of the cubic le is $r = 0.732\,R$; of the cubic unit cells considered here, this is the largest hole.

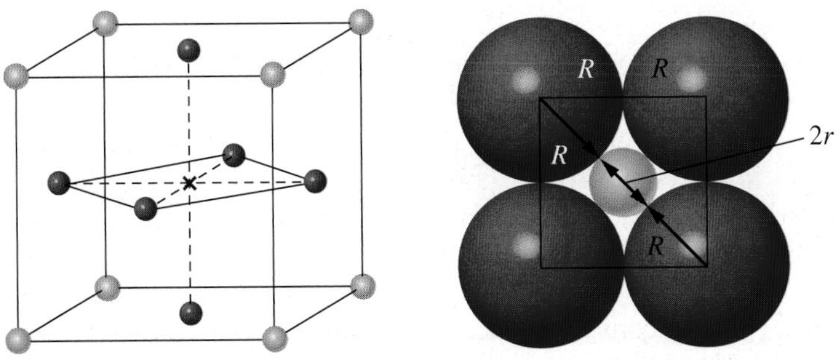

FIGURE 12-47
oss section of an octahedral hole

KEEP IN MIND

that cations almost always have smaller radii than anions. This was discussed in Chapter 9.

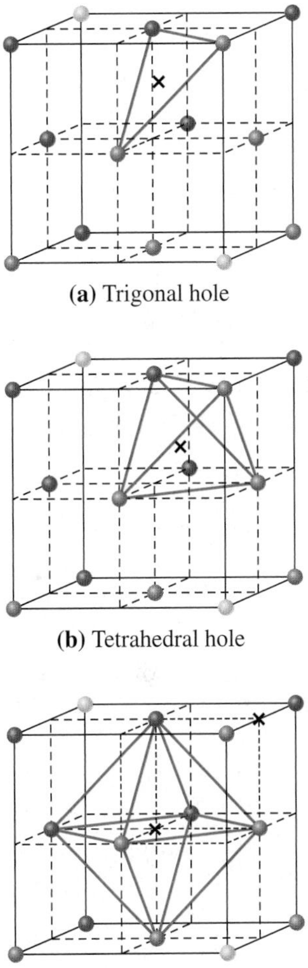

(a) Trigonal hole

(b) Tetrahedral hole

(c) Octahedral hole

▲ FIGURE 12-46
Holes in a face-centered cubic unit cell
(a) The trigonal hole is formed by two face-centered spheres and one corner sphere. **(b)** The tetrahedral hole is formed by three face-centered spheres and one corner sphere. **(c)** The octahedral hole is formed by all six face-centered spheres in the cube.

Which hole does the cation occupy in a closest-packed array of anions? ' cation occupies a hole that maximizes the attractions between the cation a anion and minimizes the repulsions between the anions. This can be acco plished by accommodating cations into holes that are slightly smaller than actual size of the ion. This pushes the anions of the closest packed array sligh apart, reducing repulsions, while the anion and cation are in contact, maximiz attractions. Therefore, if a cation is to occupy a tetrahedral hole, the ion should bigger than the tetrahedral hole but smaller than the octahedral hole; that is,

$$0.225\, R_{\text{anion}} < r_{\text{cation}} < 0.414\, R_{\text{anion}}$$

or in terms of the *radius ratio* of the cation (r) to the anion (R)

$$0.225 < (r_{\text{cation}}/R_{\text{anion}}) < 0.414$$

Similarly, if a cation is to occupy an octahedral hole, the radii will be goverr by the radius ratio inequality

$$0.414 < (r_{\text{cation}}/R_{\text{anion}}) < 0.732$$

where the upper limit of the inequality corresponds to the hole in a sim cubic lattice. When the cation is too large, that is, bigger than 0.732 R, anions adopt a simple cubic structure, which allows the cation to be acco modated in the cubic hole of the lattice.

To summarize, if

$0.225 < (r_{\text{cation}}/R_{\text{anion}}) < 0.414$	tetrahedral hole of fcc array of anions occupied by the cation
$0.414 < (r_{\text{cation}}/R_{\text{anion}}) < 0.732$	octahedral hole of fcc array of anions occupied by the cation
$0.732 < (r_{\text{cation}}/R_{\text{anion}})$	cubic hole of simple cubic array of anion occupied by the cation

The criteria given here provide a useful way of rationalizing the structures binary ionic solids. However, as with all simplified models, we must be aw of the limitations of the model. In developing the criteria given, we ha assumed that there are no interactions other than coulombic attractions betwe the ions. The criteria will fail if this is not the case. Nonetheless, we will find criteria to be very useful.

In defining a unit cell of an ionic crystal we must choose a unit cell that

- by translation in three dimensions generates the entire crystal
- is consistent with the formula of the compound
- indicates the coordination numbers of the ions

Unit cells of crystalline NaCl and CsCl are pictured in Figures 12-48 and 12- respectively. We can investigate these structures for their consistency with t formula of the compound and the type of hole the cation occupies.

The radius ratio for NaCl is

$$\frac{r_{Na^+}}{R_{Cl^-}} = \frac{99\text{ pm}}{181\text{ pm}} = 0.55$$

KEEP IN MIND

that a cubic closest packed array of spheres produces a face-centered cubic unit cell.

We expect Na^+ ions to occupy the octahedral holes of the cubic closest pack arrays of Cl^- ions. The sodium chloride unit cell is shown in Figure 12-4 To establish the formula of the compound, we must apportion the 27 ions Figure 12-48 among the unit cell and its neighboring unit cells. Recall fro Figure 12-42 on how this apportioning is done. Each Cl^- ion in a corner positi is shared by *eight* unit cells, and each Cl^- in the center of a face is shared by *t* unit cells. This leads to a total number of Cl^- ions in the unit cell $(8 \times \frac{1}{8}) + (6 \times \frac{1}{2}) = 1 + 3 = 4$. There are Na^+ ions along the edges of the u cell, and each edge is shared by *four* unit cells. The Na^+ ion in the very center of t

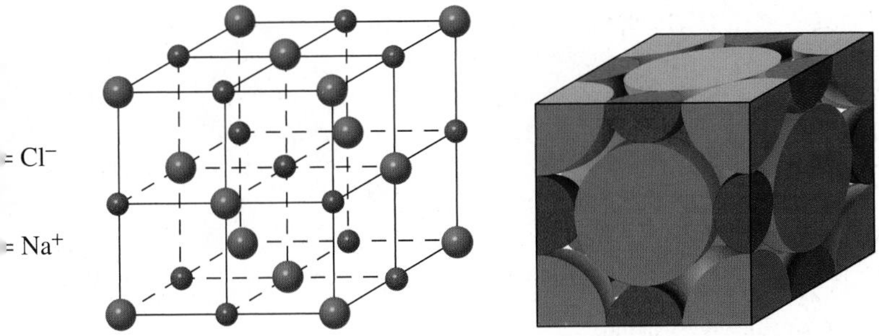

◀ FIGURE 12-48
The sodium chloride unit cell
For clarity, only the centers of the ions are shown. Oppositely charged ions are actually in contact. We can think of this structure as an fcc lattice of Cl^- ions, with Na^+ ions filling the octahedral holes.

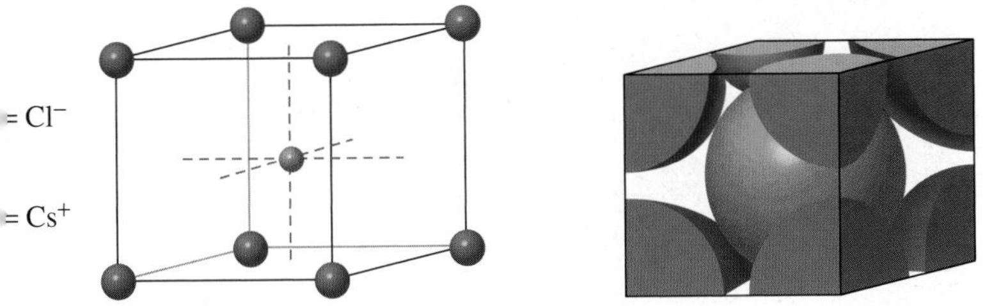

◀ FIGURE 12-49
The cesium chloride unit cell
The Cs^+ ion is in the center of the cube, with Cl^- ions at the corners. In reality, each Cl^- is in contact with the Cs^+ ion. An alternative unit cell has Cl^- at the center and Cs^+ at the corners.

t cell belongs entirely to that cell. Thus, the total number of Na^+ ions in a unit is $(12 \times \frac{1}{4}) + (1 \times 1) = 3 + 1 = 4$. The unit cell has the equivalent of 4 Na^+ 1 4 Cl^- ions. The ratio of Na^+ to Cl^- is $4:4 = 1:1$, corresponding to the formula Cl.

To establish the coordination number in an ionic crystal, count the number nearest neighbor ions of opposite charge to any given ion in the crystal. In Cl, each Na^+ is surrounded by *six* Cl^- ions. The coordination numbers of h Na^+ and Cl^- are *six*. By contrast, the coordination numbers of Cs^+ and Cl^- Figure 12-49 are *eight*. The difference in the structure of CsCl from that of Cl can be accounted for in terms of the radius ratio for this compound.

$$\frac{r_{Cs^+}}{R_{Cl^-}} = \frac{169 \text{ pm}}{181 \text{ pm}} = 0.934$$

EXAMPLE 12-11 Relating Ionic Radii and the Dimensions of a Unit Cell of an Ionic Crystal

The ionic radii of Na^+ and Cl^- in NaCl are 99 and 181 pm, respectively. What is the edge length of the unit cell?

Analyze

Again, the key to solving this problem lies in understanding geometric relationships in the unit cell. Along each edge of the unit cell (see Figure 12-48), two Cl^- ions are in contact with one Na^+. The edge length is equal to the radius of one Cl^- plus the diameter of Na^+, plus the radius of another Cl^-.

Solve

The solution is,

$$\begin{aligned} \text{Length} &= (r_{Cl^-}) + (r_{Na^+}) + (r_{Na^+}) + (r_{Cl^-}) \\ &= 2(r_{Na^+}) + 2(r_{Cl^-}) \\ &= (2 \times 99) + (2 \times 181) = 560 \text{ pm} \end{aligned}$$

Assess

As we saw earlier, we need to remember the atomic arrangement of the unit cell to determine its the geometric relationships.

PRACTICE EXAMPLE A: The ionic radius of Cs^+ is 167 pm. Use Figure 12-49 and information in Examples 12-9 and 12-11 to determine the edge length of the unit cell of CsCl.

PRACTICE EXAMPLE B: Use the length of the unit cell of NaCl obtained in Example 12-11, together with the molar mass of NaCl and the Avogadro constant, to estimate the density of NaCl.

▶ FIGURE 12-50
Some unit cells of greater complexity
(a) The zinc blende structure is an fcc lattice of anions, with cations filling half the tetrahedral holes. There are four cations and four anions per cell.
(b) The fluorite structure is an fcc lattice of cations, with anions filling all the tetrahedral holes. There are four cations and eight anions per cell.
(c) In the ball-and-stick model of TiO_2, two of the O^{2-} ions are shown to lie entirely within the boundaries of the unit cell. In actuality, these two O^{2-} ions extend beyond the boundaries of the unit cell and penetrate the boundaries of adjacent unit cells, as illustrated by the space-filling model. Likewise, O^{2-} ions from adjacent unit cells penetrate the boundaries of this unit cell. When the fractions of spheres representing the ions are added, there are two Ti^{4+} ions and four O^{2-} ions per cell.

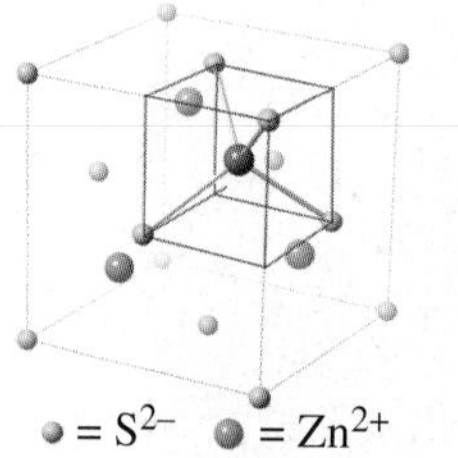

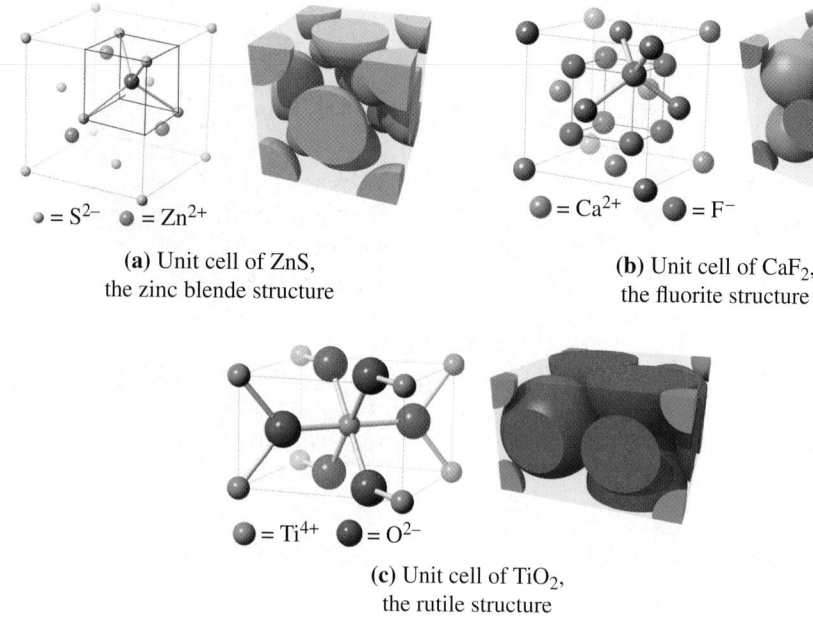

(a) Unit cell of ZnS, the zinc blende structure

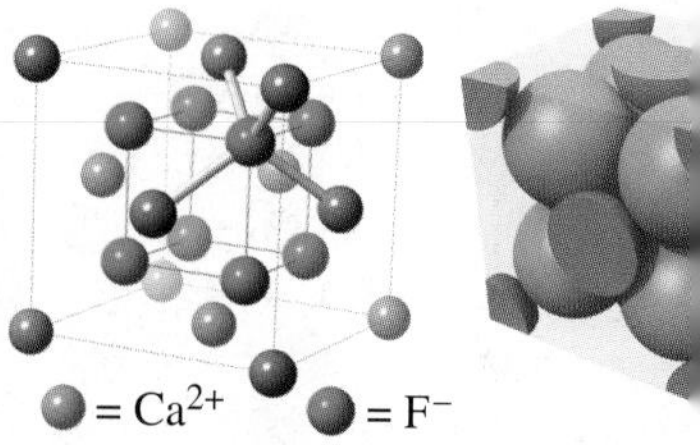

(b) Unit cell of CaF_2, the fluorite structure

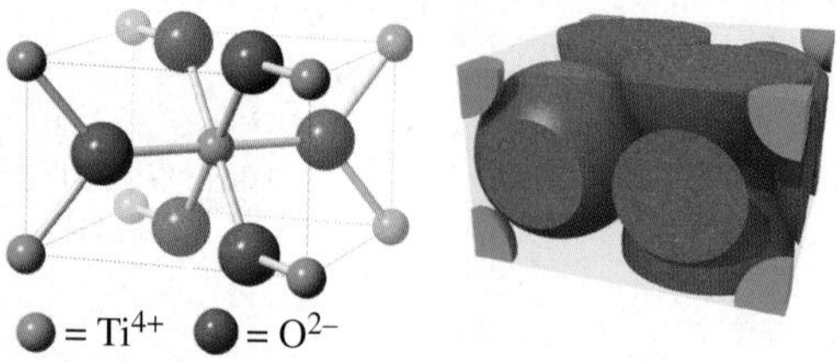

(c) Unit cell of TiO_2, the rutile structure

We expect from the radius ratio inequalities that the Cs^+ ion will occupy a cu hole in a simple cubic lattice of Cl^- ions. This unit cell is in accord with the o to-one ratio of Cs^+ to Cl^- ions since there is one Cs^+ in the center of the unit and $8 \times (\frac{1}{8})$ Cl^- ions at the corners.

Ionic compounds of the type $M^{2+}X^{2-}$ (for example, MgO, BaS, CaO) n form crystals of the NaCl type. However, if the cation is small enough, as in case of Zn^{2+}, it can occupy the tetrahedral holes. The radius ratio for ZnS is 0. so to satisfy the stoichiometry only half of the tetrahedral holes (there are ei of them) are occupied, to correspond to the four S^{2-} forming the face-cente cubic array (Fig. 12-50a). For substances with the formulas MX_2 or M_2X, crystal structures are more complex. Because the cations and anions occur unequal numbers, the crystals have *two* coordination numbers, one for cation and another for the anion.

CaF_2 (the fluorite structure) has twice as many fluoride ions as calcium ic (Fig. 12-50b). The coordination number of Ca^{2+} is *eight,* and that of F^- is *fo* This is easiest to see by looking at the Ca^{2+} ion in the middle of a face. There four F^- ions within the unit cell that are nearest neighbors. In addition, the fo F^- ions in the next unit cell (the one that shares the face-centered Ca^{2+} ion) also nearest neighbors. This gives a coordination number of eight for the Ca The F^- ions each have one corner Ca^{2+} ion and three face-centered Ca^{2+} ions nearest neighbors, giving a coordination number of four. In TiO_2 (the rut structure, Figure 12-50c), Ti^{4+} has a coordination number of *six* and O^{2-}, *thr* In this structure, two of the O^{2-} ions are within the interior of the cell, two in the top face, and two in the bottom face of the cell. Ti^{4+} ions are at the corn and the center of the cell.

12-9 CONCEPT ASSESSMENT

Buckminsterfullerene C_{60} crystallizes in a face-centered cubic array. If potassium atoms fill all the tetrahedral and octahedral holes, what is the formula of the resulting compound?

2-7 Energy Changes in the Formation of Ionic Crystals

e concept of lattice energy, which we introduced qualitatively in Section 12-5, nost useful when stated in quantitative terms. It is difficult, however, to cal-ate a lattice energy directly. The problem is that oppositely charged ions act one another and like-charged ions repel one another, and these interac-ns must be considered at the same time. More commonly, lattice energy is ermined *indirectly* through an application of Hess's law known as the n–Fajans–Haber cycle, named after its originators Max Born, Kasimir Fajans, d Fritz Haber. The crux of the method is to design a sequence of steps in ich enthalpy changes are known for all the steps but one—the step in which rystal lattice is formed from gaseous ions.

Figure 12-51 illustrates a five-step method for finding the lattice energy NaCl.

◀ NaCl has a lattice energy of about $\frac{1}{4}$ that of MgO. This is because the coulombic attraction between the ions is proportional to $(-1)(+1)$ in the former and $(-2)(+2)$ in the latter.

Step 1. Sublime solid Na.
Step 2. Dissociate $\frac{1}{2}$ $Cl_2(g)$ into Cl(g).
Step 3. Ionize Na(g) to $Na^+(g)$.
Step 4. Convert Cl(g) to $Cl^-(g)$.
Step 5. Combine $Na^+(g)$ and $Cl^-(g)$ to form NaCl(s).

KEEP IN MIND

that the unit kJ mol^{-1} refers to kilojoules *per mole of reaction*.

e overall change in these five steps is the same as the reaction in which NaCl(s) is med from its elements in their standard states—that is, $\Delta_r H_{overall} =$ $H°[NaCl(s)]$. From Appendix D, we see that $\Delta_f H°[NaCl(s)] = -411$ kJ mol^{-1}, in the following setup, the lattice energy of NaCl is the only unknown.

$$\text{Na(s)} \longrightarrow \cancel{\text{Na(g)}} \qquad \Delta_r H_1 = \Delta_{sub} H = +107 \text{ kJ mol}^{-1}$$

$$\frac{1}{2}\text{Cl}_2\text{(g)} \longrightarrow \cancel{\text{Cl(g)}} \qquad \Delta_r H_2 = \frac{1}{2} \text{ Cl—Cl bond energy} = +122 \text{ kJ mol}^{-1}$$

$$\cancel{\text{Na(g)}} \longrightarrow \cancel{\text{Na}^+\text{(g)}} + \cancel{e^-} \qquad \Delta_r H_3 = \text{1st ioniz. energy} = +496 \text{ kJ mol}^{-1}$$

$$\cancel{\text{Cl(g)}} + \cancel{e^-} \longrightarrow \cancel{\text{Cl}^-\text{(g)}} \qquad \Delta_r H_4 = \text{electron affinity of Cl} = -349 \text{ kJ mol}^{-1}$$

$$\cancel{\text{Na}^+\text{(g)}} + \cancel{\text{Cl}^-\text{(g)}} \longrightarrow \text{NaCl(s)} \qquad \Delta_r H_5 = \text{lattice energy of NaCl} = ?$$

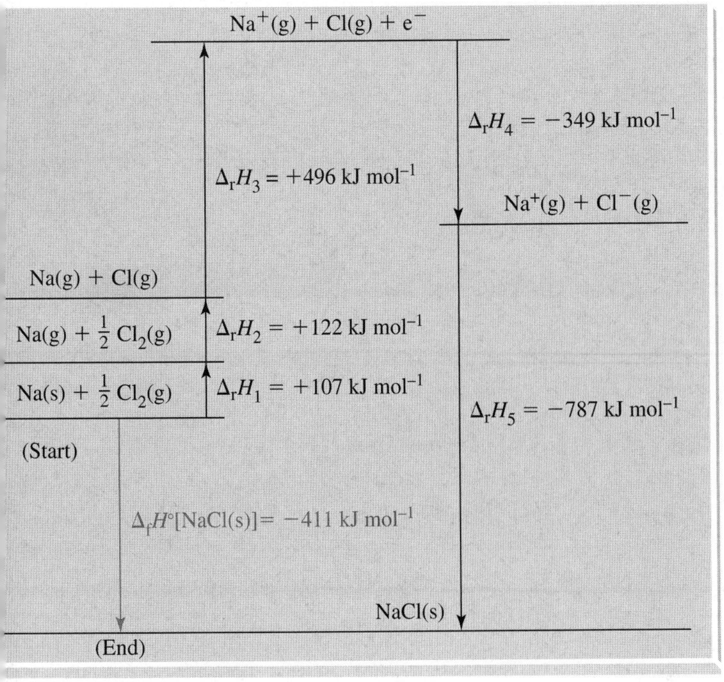

◀ FIGURE 12-51
Enthalpy diagram for the formation of an ionic crystal
Shown here is a five-step sequence for the formation of NaCl(s) from its elements in their standard states. The sum of the five enthalpy changes gives $\Delta_f H°[NaCl(s)]$. The equivalent one-step reaction for the formation of NaCl(s) directly from Na(s) and $Cl_2(g)$ is shown in color. (The vertical arrows representing $\Delta_r H$ values are not to scale.)

$$\text{overall:} \qquad Na(s) + \frac{1}{2}Cl_2(g) \longrightarrow NaCl(s)$$

$$\Delta_r H_{overall} = -411\text{ kJ mol}^{-1} = \Delta_r H_1 + \Delta_r H_2 + \Delta_r H_3 + \Delta_r H_4 + \Delta_r H_5$$
$$-411\text{ kJ mol}^{-1} = (107 + 122 + 496 - 349)\text{ kJ mol}^{-1} + \Delta_r H_5$$
$$\Delta_r H_5 = \text{lattice energy} = (-411 - 107 - 122 - 496 + 349)\text{ kJ mol}^{-1} = -787\text{ kJ m}$$

▶ A commonly used convention defines lattice energy in terms of the breakup of a crystal rather than its formation. By that convention, all lattice energies are positive quantities and $\Delta_r H_5 = -(\text{lattice energy})$.

One way to use the concept of lattice energy is in making predictions abo the possibility of synthesizing ionic compounds. In Example 12-12, we calc late the enthalpy of formation of MgCl(s). Then, we predict the likelihood obtaining this compound.

EXAMPLE 12-12 Relating Enthalpy of Formation, Lattice Energy, and Other Energy Quantities

With the following data, calculate $\Delta_f H°$ of MgCl(s): Enthalpy of sublimation of Mg(s): +146 kJ mol^{-1}; enthalpy of dissociation of Cl_2(g): +244 kJ mol^{-1}; first ionization energy of Mg(g): +738 kJ mol^{-1}; electron affinity of Cl(g): −349 kJ mol^{-1}; lattice energy of MgCl(s): −676 kJ mol^{-1}.

Analyze

We begin this problem by drawing an enthalpy diagram for the formation of an ionic solid, MgCl(s). From the diagram we find that the lattice energy ($\Delta_r H_5$) is known, and the unknown is $\Delta_r H_{overall}$, which is the enthalpy of formation of MgCl(s).

Solve

The various reactions are given as part of the example:

$$Mg(s) \longrightarrow \cancel{Mg(g)} \qquad \Delta_r H_1 = +146\text{ kJ mol}^{-1}$$
$$\frac{1}{2}Cl_2(g) \longrightarrow \cancel{Cl(g)} \qquad \Delta_r H_2 = \frac{1}{2}(+244)\text{ kJ mol}^{-1}$$
$$\cancel{Mg(g)} \longrightarrow \cancel{Mg^+(g)} + \cancel{e^-} \qquad \Delta_r H_3 = +738\text{ kJ mol}^{-1}$$
$$\cancel{Cl(g)} + \cancel{e^-} \longrightarrow \cancel{Cl^-(g)} \qquad \Delta_r H_4 = -349\text{ kJ mol}^{-1}$$
$$\cancel{Mg^+(g)} + \cancel{Cl^-(g)} \longrightarrow MgCl(s) \qquad \Delta_r H_5 = -676\text{ kJ mol}^{-1}$$

$$\text{overall:} \qquad Mg(s) + \frac{1}{2}Cl_2(g) \longrightarrow MgCl(s)$$

$$\Delta_r H_{overall} = \Delta_f H°[MgCl(s)] = \Delta_r H_1 + \Delta_r H_2 + \Delta_r H_3 + \Delta_r H_4 + \Delta_r H_5$$
$$= (146 + 122 + 738 - 349 - 676)\text{ kJ mol}^{-1} = -19\text{ kJ mol}^{-1}$$

Assess

These types of problems are another form of Hess's law. The enthalpy diagram is just a way for us to keep the equations straight.

PRACTICE EXAMPLE A: The enthalpy of sublimation of cesium is 78.2 kJ mol^{-1}, and $\Delta_f H°[CsCl(s)] =$ −442.8 kJ mol^{-1}. Use these values, together with other data from the text, to calculate the lattice energy of CsCl(s).

PRACTICE EXAMPLE B: Given the following data, together with data included in Example 12-12, calculate $\Delta_f H°$ of $CaCl_2$(s): Enthalpy of sublimation of Ca(s), +178.2 kJ mol^{-1}; first ionization energy of Ca(g), +590 kJ mol^{-1}; second ionization energy of Ca(g), +1145 kJ mol^{-1}; lattice energy of $CaCl_2$(s), −2223 kJ mol^{-1}.

Example 12-12 suggests that we can obtain MgCl(s) as a stable compound it has a slightly negative enthalpy of formation. Why have we been writi $MgCl_2$ all this time instead of MgCl? You might think that because MgCl ha Mg-to-Cl ratio of 1:1 and $MgCl_2$ has a ratio of 1:2, MgCl should form if Mg reacts with a limited amount of Cl_2(g). But this is not the case. No matter hc limited the amount of Cl_2(g) available, the only compound that forms $MgCl_2$. To understand this, repeat the calculation of Example 12-12 for the fc mation of $MgCl_2$(s), and you will obtain an enthalpy of formation that y much more negative than that for MgCl(s) (see Exercise 87). Even though energy requirement to produce Mg^{2+} is larger than to produce Mg^{+}, the late energy is very much greater for $MgCl_2(s)$ than for MgCl(s). This because the *doubly* charged Mg^{2+} ions exert a much stronger force on Cl^{-} s than do *singly* charged Mg^{+} ions. The reaction between magnesium and orine does not stop at MgCl but continues on to the more stable $MgCl_2$.

Is $NaCl_2$ a stable compound? Here, the answer is no. The additional lattice ergy associated with $NaCl_2$ over NaCl is not nearly enough to compensate the very high second ionization energy of sodium (see Exercise 114).

Mastering CHEMISTRY www.masteringchemistry.com

The acronym LCD, which stands for liquid crystal display, needs little introduction. Using the word liquid to describe crystals appears contradictory based upon the material in this chapter. For a discussion of the properties and uses of materials that we call liquid crystals, go to the Focus On feature for Chapter 12, Liquid Crystals, on the MasteringChemistry website.

Summary

-1 Intermolecular Forces—The intermolecular ces that occur between molecules are collectively known van der Waals forces. The most common intermolecular ces of attraction are those between instantaneous and luced dipoles (**dispersion forces**, or **London forces**). The gnitudes of dispersion forces depend on how easily elecn displacements within molecules cause a temporary balance of electron charge distribution, that is, on the larizability of the molecule. In polar substances, there are o dipole–dipole forces. Some hydrogen-containing subnces exhibit significant intermolecular attractions called **drogen bonds,** in which H atoms bonded to highly elecnegative atoms—N, O, or F—in a molecule are simultaously attracted to other highly electronegative atoms in e same molecule or in different molecules. Hydrogen nding has a profound effect on physical properties, such boiling points (Fig. 12-4) and is a vital intermolecular ce in living systems.

-2 Some Properties of Liquids—**Surface tension,** e energy required to extend the surface of a liquid, and **scosity,** a liquid's resistance to flow, are properties ated to intermolecular forces. Familiar phenomena such drop shape, meniscus formation, and capillary action pend on surface tension. Specifically, these phenomena e influenced by the balance between **cohesive forces,** termolecular forces between molecules in a liquid, and **hesive forces,** intermolecular forces between liquid olecules and a surface. **Vapor pressure,** the pressure erted by a vapor in equilibrium with a liquid, is a meare of the volatility of a liquid and is related to the ength of intermolecular forces. The conversion of a liqd to a vapor is called **vaporization** or **evaporation**; the verse process is called **condensation**. The dependence of por pressure on temperature is represented by a **vapor essure curve** (Fig. 12-18) and can be expressed in the Clausius–Clapeyron equation (equation 12.2). When the pressure exerted by the escaping molecules from the surface of the liquid equals the pressure exerted by the molecules in the atmosphere, **boiling** is said to occur. The **normal boiling point** is the temperature at which the vapor pressure of the liquid equals 1 atm. The **critical point** is the condition of temperature and pressure at which a liquid and its vapor become indistinguishable (Fig. 12-22).

12-3 Some Properties of Solids—When crystalline solids are heated, a temperature is reached where the solid state is converted to a liquid—**melting** occurs. The temperature at which this occurs is the **melting point.** When liquids are cooled, the crystalline material will form during the process of **freezing**, and the temperature at which this occurs is the **freezing point.** Under certain conditions solids can directly convert into vapor by the process of **sublimation**. The reverse process is called **deposition**. Among the properties of a solid affected by intermolecular forces are its sublimation (vapor) pressure and its melting point.

12-4 Phase Diagrams—A **phase diagram** (Figs. 12-26 to 12-30) is a graphical plot of conditions under which solids, liquids, and gases (vapors) exist, as single phases or as two or more phases in equilibrium with one another. Significant points on a phase diagram are the **triple point** (where all three phases coexist), melting point, boiling point, and critical point, beyond which a supercritical fluid is possible. Some substances can exist in different forms in the solid state; such behavior is called **polymorphism** (Fig. 12-30).

12-5 The Nature of Bonding in Solids—Solids can be classified according to the nature of bonding. In **network covalent solids** chemical bonds extend throughout a crystalline structure. For these substances the chemical bonds are the forces holding the atoms in place. **Ionic solids** are composed of ions held in place through interionic forces of attraction. Solids that are composed of discrete molecules are known as **molecular solids**. These molecules are held in place through the different intermolecular forces. Metal atoms form **metallic solids** through the delocalization of electrons. The delocalized electrons freely move throughout the solid, lending to it various properties, such as conductivity. **Lattice energy** is the energy released when separated gaseous ions come together to form one mole of an ionic solid.

12-6 Crystal Structures—Some crystal structures can be described in terms of the packing of spheres. Depending on the way in which the spheres are packed, different **unit cells** are obtained (Fig. 12-37). The hexagonal unit cell is obtained with **hexagonal closest packed (hcp)** spheres; a **face-centered cubic (fcc)** unit cell is obtair with **cubic closest packed** spheres (Figs. 12-38 and 39). A **body-centered cubic (bcc)** (Fig. 12-38) unit cel found in some cases where spheres are not packed closely as in the hcp and fcc structures. The dimensi of the unit cell can be determined by X-ray diffracti and these dimensions can be used in calculating ato radii and densities. An important consideration w ionic crystals is that the ions are not all of the same s or charge. Ionic crystals can often be viewed as an ar of anions with the cations fitting in the holes within anion array.

12-7 Energy Changes in the Formation of Io Crystals—Lattice energies of ionic crystals can be rela to certain atomic and thermodynamic properties means of the Born–Fajans–Haber cycle (Fig. 12-51).

Integrative Example

Use data from the table of physical properties of hydrazine, N_2H_4, to calculate the partial pressure of $N_2H_4(g)$ whe container filled with an equilibrium mixture of $N_2H_4(g)$ and $N_2H_4(l)$ at 25.0 °C is cooled to the temperature of ice–water bath.

Property	Value
Freezing point	2.0 °C
Boiling point	113.5 °C
Critical temperature	380 °C
Critical pressure	145.4 atm
Enthalpy of fusion	12.66 kJ mol^{-1}
Heat capacity of liquid	98.84 J mol^{-1} °C^{-1}
Density of liquid at 25.0 °C	1.0036 g mL^{-1}
Vapor pressure at 25.0 °C	14.4 Torr

Analyze

At a temperature below its freezing point of 2.0 °C, the hydrazine will be present as a solid in equilibrium with its vap We are seeking the sublimation pressure of $N_2H_4(s)$ at the melting point of ice, 0 °C. At its freezing point of 2.0 °C, t hydrazine coexists in three phases—liquid, solid, and vapor. We must first determine the vapor pressure of hydrazine 2.0 °C. Then we can then use the Clausius–Clapeyron equation (12.2) to calculate the vapor (sublimation) pressure at 0 Our principal task will be to identify the data needed to apply the Clausius–Clapeyron equation, three times in all, detailed in the stepwise solution to the problem.

Solve

To determine a value of $\Delta_{vap}H$, choose the vapor pressure data at 25.0 °C for T_1 and P_1 and at the normal boiling point for T_2 and P_2 for substitution into equation (12.2).

$$\ln\left(\frac{P_2}{P_1}\right) = -\frac{\Delta_{vap}H}{R}\left(\frac{1}{T_2} - \frac{1}{T_1}\right)$$

$$\ln\left(\frac{760\text{ Torr}}{14.4\text{ Torr}}\right) = -\frac{\Delta_{vap}H}{8.3145\text{ J mol}^{-1}\text{ K}^{-1}} \times \left(\frac{1}{386.7\text{ K}} - \frac{1}{298.2\text{ K}}\right) = 3.967$$

$$\Delta_{vap}H = -\frac{8.3145\text{ J mol}^{-1}\text{ K}^{-1} \times 3.967}{(0.002586 - 0.003353)\text{ K}^{-1}} = 4.30 \times 10^4\text{ J mol}^{-1}$$

Return to the Clausius–Clapeyron equation (12.2) by using the value of $\Delta_{vap}H$ just obtained. Also use the same data as in the first calculation for T_1 and P_1, but now with $T_2 = 2.0 + 273.15 = 275.2$ K and P_2 as an unknown. Solve the equation for P_2.

$$\ln\left(\frac{P_2}{14.4}\right) = -\frac{4.30 \times 10^4\text{ J mol}^{-1}}{8.3145\text{ J mol}^{-1}\text{ K}^{-1}} \times \left(\frac{1}{275.2\text{ K}} - \frac{1}{298.2\text{ K}}\right)$$

$$= -5.17 \times 10^3 \times (2.80 \times 10^{-4}) = -1.45$$

$$P_2/14.4 = e^{-1.45} = 0.235$$

$$P_2 = 0.235 \times 14.4 = 3.38\text{ Torr}$$

now have the triple point data for lrazine. The triple point temperature :.0 °C (275.2 K) and the triple point ssure is 3.38 Torr. In our final appli- on of equation (12.2), we use those a as T_2 and P_2. The temperature T_1 is C (273.2 K) and the unknown subli- tion pressure is P_1. The enthalpy nge needed in this final calculation st be the enthalpy of sublimation, ich is

$$\Delta_{sub}H = \Delta_{fus}H + \Delta_{vap}H$$
$$= \left(12.66\frac{kJ}{mol} \times \frac{1000\ J}{1\ kJ}\right) + 4.30 \times 10^4\ J\ mol^{-1} = 5.57 \times 10^4\ J\ mol^{-1}$$

ally, we substitute these data into ıation (12.2) and solve for P_1, the ›limation pressure of hydrazine at C.

$$\ln\left(\frac{3.38\ Torr}{P_1}\right) = -\frac{5.57 \times 10^4\ J\ mol^{-1}}{8.3145\ J\ mol^{-1}K^{-1}} \times \left(\frac{1}{275.2} - \frac{1}{273.2}\right)K^{-1}$$
$$\ln\left(\frac{3.38\ Torr}{P_1}\right) = -6.70 \times 10^3 \times (-2.66 \times 10^{-5}) = 0.178$$
$$\frac{3.38\ Torr}{P_1} = e^{0.178} = 1.19$$
$$P_1 = \frac{3.38\ Torr}{1.19} = 2.84\ Torr$$

sess

serve that, compared with the vapor pressure at 25 °C (14.4 Torr), the calculated triple point pressure (3.38 Torr) is aller; and the sublimation pressure at 0 °C (2.84 Torr) is smaller still. This is certainly the trend expected for the three ues. In the three situations in which equation (12.2) is used, the first one is the most subject to error because the differ- :e between T_2 and T_1 is 89 °C, while in the other two it is 23 °C and 2 °C, respectively. Both $\Delta_{vap}H$ and $\Delta_{sub}H$ are doubtedly temperature-dependent.

ACTICE EXAMPLE A: The normal boiling point of isooctane (a gasoline component with a high octane rating) 99.2 °C, and its $\Delta_{vap}H$ is 35.76 kJ mol^{-1}. Because isooctane and water have nearly identical boiling points, will ·y have nearly equal vapor pressures at room temperature? If not, which would you expect to be more volatile? plain.

ACTICE EXAMPLE B: The *second* electron affinity of oxygen is, by definition, the energy change for adding an ctron to O^- to form O^{2-}:

$$O^-(g) + e^- \longrightarrow O^{2-}(g) \quad E_{ea,2} = ?$$

e second electron affinity cannot be measured directly, but it can be obtained indirectly by using the Born–Haber cycle an ionic compound containing the O^{2-} ion.

(a) Show that $E_{ea,2}$ can be calculated from the enthalpy of formation and lattice energy of MgO(s), the enthalpy of sublimation of Mg(s), the ionization energies of Mg, the bond energy of O_2, and the $E_{ea,1}$ for O(g).

(b) The lattice energy of MgO is −3925 kJ mol^{-1}. Combine this with other values in the text to estimate $E_{ea,2}$ for oxygen.

Exercises

termolecular Forces

. For each of the following substances describe the importance of dispersion (London) forces, dipole–dipole interactions, and hydrogen bonding: **(a)** HCl; **(b)** Br_2; **(c)** ICl; **(d)** HF; **(e)** CH_4.

. When another atom or group of atoms is substituted for one of the hydrogen atoms in benzene, C_6H_6, the boiling point changes. Explain the order of the following boiling points: C_6H_6, 80 °C; C_6H_5Cl, 132 °C; C_6H_5Br, 156 °C; C_6H_5OH, 182 °C.

3. Arrange the liquids represented by the following molecular models in the expected order of increasing viscosity at 25 °C.

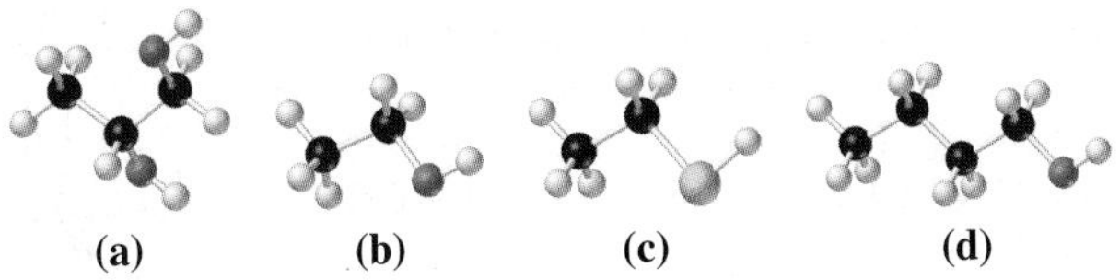

4. Arrange the liquids represented by the following molecular models in the expected order of increasing normal boiling point.

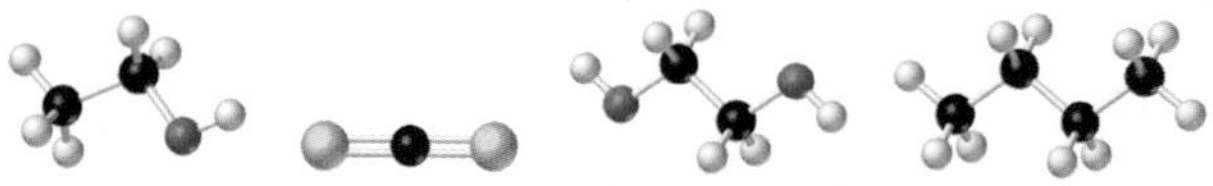

(a) (b) (c) (d)

5. One of the following substances is a liquid at room temperature and the others are gaseous: CH_3OH; C_3H_8; N_2; N_2O. Which do you think is the liquid? Explain.
6. In which of the following compounds might intramolecular hydrogen bonding be an important factor? Explain. **(a)** $CH_3CH_2COCH_3$; **(b)** $CH_3NH_2CH_2CH_2COOH$; **(c)** $CH_3CH_2CHFCH_2OH$; **(d)** *ortho*-phthalic acid.

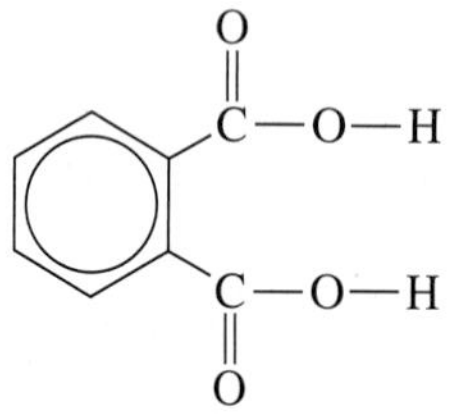

ortho-Phthalic acid

7. How many water molecules can hydrogen bonc methanol?
8. What is the maximum number of hydrogen bo that can form between two acetic acid molecules?
9. In DNA the nucleic acid bases form hydrogen bo between them, which are responsible for the forı tion of the double-stranded helix. Arrange the ba guanine and cytosine to give the maximum num of hydrogen bonds.

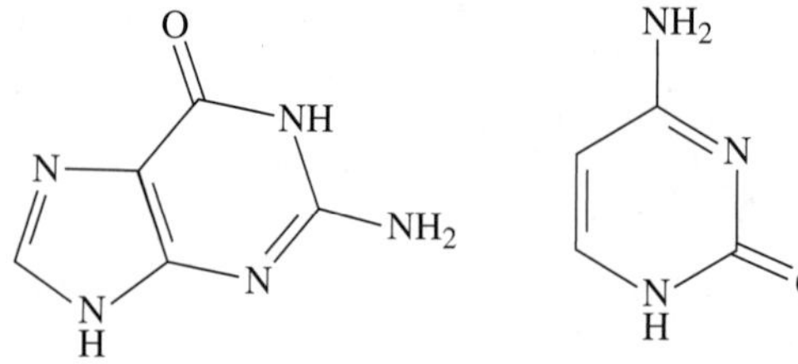

Guanine Cytosine

10. Water molecules will form small, stable cluste Draw one possible water cluster by using six wa molecules and maximizing the number of hydrog bonds for each water molecule.

Surface Tension and Viscosity

11. Silicone oils, such as $H_3C[SiO(CH_3)_2]_n Si(CH_3)$, are used in water repellents for treating tents, hiking boots, and similar items. Explain how silicone oils function.
12. Surface tension, viscosity, and vapor pressure are all related to intermolecular forces. Why do surface tension and viscosity decrease with temperature, whereas vapor pressure increases with temperature?
13. Is there any scientific basis for the colloquial expression "slower than molasses in January"? Explain.
14. A television commercial claims that a product makes water "wetter." Can there be any basis to this claim? Explain.
15. Rank the following in order of increasing surfa tension (at room temperature): **(a)** CH_3O **(b)** $HOCH_2CH_2OH$; **(c)** $CH_3CH_2OCH_2CH_3$.
16. Would you predict the surface tension of t-butyl al hol, $(CH_3)_3COH$, to be greater than or less than th of *n*-butyl alcohol, $CH_3CH_2CH_2CH_2OH$? Explain.
17. Butanol and pentane have approximately t same mass, however, the viscosity (at 20 °C) butanol is η = 2.948 cP, and the viscosity of pentane η = 0.240 cP. Explain this difference.
18. Carbon tetrachloride (CCl_4) and mercury have sir lar viscosities at 20 °C. Explain.

Vaporization

19. As a liquid evaporated from an open container, its temperature was observed to remain roughly constant. When the same liquid evaporated from a thermally insulated container (a vacuum bottle or Dewar flask), its temperature was observed to drop. How would you account for this difference?
20. Explain why vaporization occurs only at the surface of a liquid until the boiling point temperature is reached. That is, why does vapor not form throughout the liquid at all temperatures?
21. The enthalpy of vaporization of benzene, $C_6H_6(l)$, is 33.9 kJ mol^{-1} at 298 K. How many liters of $C_6H_6(g)$, measured at 298 K and 95.1 mmHg, are formed when 1.54 kJ of heat is absorbed by $C_6H_6(l)$ at a constant temperature of 298 K?
22. A vapor volume of 1.17 L forms when a sample of li uid acetonitrile, CH_3CN, absorbs 1.00 kJ of heat at normal boiling point (81.6 °C and 1 atm). What $\Delta_{vap}H$ in kilojoules per mole of CH_3CN?
23. Use data from the Integrative Example (page 566) determine how much heat is required to conve 25.00 mL of liquid hydrazine at 25.0 °C to hydrazi vapor at its normal boiling point.
24. How much heat is required to raise the temperature 215 g $CH_3OH(l)$ from 20.0 to 30.0 °C and then vapo ize it at 30.0 °C? Use data from Table 12.4 and a mol heat capacity of $CH_3OH(l)$ of 81.1 J mol^{-1}K^{-1}.
25. How many liters of $CH_4(g)$, measured at 23.4 °C ar 768 mmHg, must be burned to provide the heat needed vaporize 3.78 L of water at 100 °C? For CH_4, $\Delta_{comb}H$

-8.90×10^2 kJ mol^{-1}. For $H_2O(l)$ at 100 °C, $d = 0.958$ g cm^{-3}, and $\Delta_{vap}H = 40.7$ kJ mol^{-1}.

A 50.0 g piece of iron at 152 °C is dropped into 20.0 g $H_2O(l)$ at 89 °C in an open, thermally insulated container. How much water would you expect to vaporize, assuming no water splashes out? The specific heats of iron and water are 0.45 and 4.21 J g^{-1} °C^{-1}, respectively, and $\Delta_{vap}H = 40.7$ kJ mol^{-1} H_2O.

por Pressure and Boiling Point

From Figure 12-18, estimate **(a)** the vapor pressure of $C_6H_5NH_2$ at 100 °C; **(b)** the normal boiling point of $C_6H_5CH_3$.

Use data in Figure 12-20 to estimate **(a)** the normal boiling point of aniline; **(b)** the vapor pressure of diethyl ether at 25 °C.

Equilibrium is established between $Br_2(l)$ and $Br_2(g)$ at 25.0 °C. A 250.0 mL sample of the vapor weighs 0.486 g. What is the vapor pressure of bromine at 25.0 °C, in millimeters of mercury?

The density of acetone vapor in equilibrium with liquid acetone, $(CH_3)_2CO$, at 32 °C is 0.876 g L^{-1}. What is the vapor pressure of acetone at 32 °C, expressed in kilopascals?

A double boiler is used when a careful control of temperature is required in cooking. Water is boiled in an outside container to produce steam, and the steam condenses on the outside walls of an inner container in which cooking occurs. (A related laboratory device is called a steam bath.) **(a)** How is heat energy conveyed to the food to be cooked in a double boiler? **(b)** What is the maximum temperature that can be reached in the inside container?

One popular demonstration in chemistry labs is performed by boiling a small quantity of water in a metal can (such as a used soda can), picking up the can with tongs and quickly submerging it upside down in cold water. The can collapses with a loud and satisfying pop. Give an explanation of this crushing of the can. (Note: If you try this demonstration, do not heat the can over an open flame.)

Pressure cookers achieve a high cooking temperature to speed the cooking process by heating a small amount of water under a constant pressure. If the pressure is set at 2 atm, what is the boiling point of the water? Use information from Table 12.5.

34. Use data from Table 12.5 to estimate **(a)** the boiling point of water in Santa Fe, New Mexico, if the prevailing atmospheric pressure is 640 mmHg; **(b)** the prevailing atmospheric pressure at Lake Arrowhead, California, if the observed boiling point of water is 94 °C.

35. A 25.0 L volume of He(g) at 30.0 °C is passed through 6.220 g of liquid aniline ($C_6H_5NH_2$) at 30.0 °C. The liquid remaining after the experiment weighs 6.108 g. Assume that the He(g) becomes saturated with aniline vapor and that the total gas volume and temperature remain constant. What is the vapor pressure of aniline at 30.0 °C?

36. A 7.53 L sample of $N_2(g)$ at 742 mmHg and 45.0 °C is bubbled through $CCl_4(l)$ at 45.0 °C. Assuming the gas becomes saturated with $CCl_4(g)$, what is the volume of the resulting gaseous mixture if the total pressure remains at 742 mmHg and the temperature remains at 45 °C? The vapor pressure of CCl_4 at 45 °C is 261 mmHg.

37. Some vapor pressure data for Freon-12, CCl_2F_2, once a common refrigerant, are −12.2 °C, 2.0 atm; 16.1 °C, 5.0 atm; 42.4 °C, 10.0 atm; 74.0 °C, 20.0 atm. Also, bp = −29.8 °C, T_c = 111.5 °C, P_c = 39.6 atm. Use these data to plot the vapor pressure curve of Freon-12. What approximate pressure would be required in the compressor of a refrigeration system to convert Freon-12 vapor to liquid at 25.0 °C?

38. A 10.0 g sample of liquid water is sealed in a 1515 mL flask and allowed to come to equilibrium with its vapor at 27 °C. What is the mass of $H_2O(g)$ present when equilibrium is established? Use vapor pressure data from Table 12.5.

e Clausius–Clapeyron Equation

Cyclohexanol has a vapor pressure of 10.0 mmHg at 56.0 °C and 100.0 mmHg at 103.7 °C. Calculate its enthalpy of vaporization, $\Delta_{vap}H$.

The vapor pressure of methyl alcohol is 40.0 mmHg at 5.0 °C. Use this value and other information from the text to estimate the normal boiling point of methyl alcohol.

The normal boiling point of acetone, an important laboratory and industrial solvent, is 56.2 °C and its $\Delta_{vap}H$ is 25.5 kJ mol^{-1}. At what temperature does acetone have a vapor pressure of 375 mmHg?

42. The vapor pressure of trichloromethane (chloroform) is 40.0 Torr at −7.1 °C. Its enthalpy of vaporization is 29.2 kJ mol^{-1}. Calculate its normal boiling point.

43. Benzaldehyde, C_6H_5CHO, has a normal boiling point of 179.0 °C and a critical point at 422 °C and 45.9 atm. Estimate its vapor pressure at 100.0 °C.

44. With reference to Figure 12-20, which is the more volatile liquid, benzene or toluene? At approximately what temperature does the less volatile liquid have the same vapor pressure as the more volatile one at 65 °C?

ritical Point

Which substances listed in Table 12.6 can exist as liquids at room temperature (about 20.0 °C)? Explain.

46. Can SO_2 be maintained as a liquid under a pressure of 100 atm at 0 °C? Can liquid methane be obtained under the same conditions?

Melting and Freezing

47. The normal melting point of copper is 1357 K, and $\Delta_{fus}H$ of Cu is 13.05 kJ mol^{-1}. **(a)** How much heat, in kilojoules, is evolved when a 3.78 kg sample of molten Cu freezes? **(b)** How much heat, in kilojoules, must be absorbed at 1357 K to melt a bar of copper that is 75 cm × 15 cm × 12 cm? (Assume d = 8.92 g/cm^3 for Cu.)

48. An ice calorimeter measures quantities of heat by quantity of ice melted. How many grams of ice wo be melted by the heat released in the complete cc bustion of 1.60 L of propane gas, $C_3H_8(g)$, measu at 20.0 °C and 735 mmHg? [*Hint*: What is the stand molar enthalpy of combustion of $C_3H_8(g)$?]

States of Matter and Phase Diagrams

49. An 80.0 g piece of dry ice, $CO_2(s)$, is placed in a 0.500 L container, and the container is sealed. If this container is held at 25 °C, what state(s) of matter must be present? [*Hint*: Refer to Table 12.6 and Figure 12-28.]

50. Sketch a plausible phase diagram for hydrazine (N_2H_4) from the following data: triple point (2.0 °C and 3.4 mmHg), the normal melting point (2 °C), the normal boiling point (113.5 °C), and the critical point (380 °C and 145 atm). The density of the liquid is less than that of the solid. Label significant data points on this diagram. Are there any features of the diagram that remain uncertain? Explain.

51. Shown here is a portion of the phase diagram for phosphorus.
(a) Indicate the phases present in the regions labeled with a question mark.
(b) A sample of solid red phosphorus cannot be melted by heating in a container open to the atmosphere. Explain why this is so.
(c) Trace the phase changes that occur when the pressure on a sample is reduced from point A to B, at constant temperature.

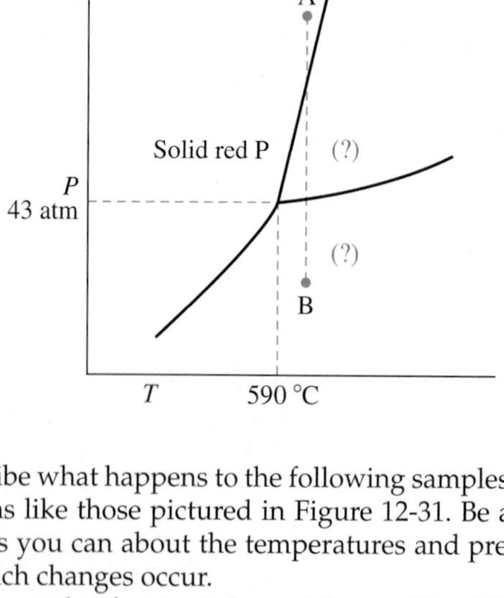

52. Describe what happens to the following samples in situations like those pictured in Figure 12-31. Be as specific as you can about the temperatures and pressures at which changes occur.
(a) A sample of water is heated from −20 to 200 °C at a constant pressure of 600 Torr.
(b) The pressure on a sample of iodine is increased from 90 mmHg to 100 atm at a constant temperature of 114 °C.
(c) A sample of carbon dioxide at 35 °C is cooled −100 °C at a constant pressure of 50 atm. [*Hint*: Re also to Table 12.6.]

53. A 0.240 g sample of $H_2O(l)$ is sealed into an eva ated 3.20 L flask. What is the pressure of the vapo the flask if the temperature is **(a)** 30.0 °C; **(b)** 50.0 **(c)** 70.0 °C?

54. A 2.50 g sample of $H_2O(l)$ is sealed in a 5.00 L flask 120.0 °C.
(a) Show that the sample exists completely as vap
(b) Estimate the temperature to which the flask m be cooled before liquid water condenses.

55. Use appropriate phase diagrams and data fr Table 12.6 to determine whether any of the followi is likely to occur naturally at or near Earth's surf anywhere on Earth. Explain. **(a)** $CO_2(s)$; **(b)** $CH_4($ **(c)** $SO_2(g)$; **(d)** $I_2(l)$; **(e)** $O_2(l)$.

56. Trace the phase changes that occur as a sample $H_2O(g)$, originally at 1.00 mmHg and −0.10 °C, compressed at constant temperature until the pr sure reaches 100 atm.

57. To an insulated container with 100.0 g $H_2O(l)$ 20.0 °C, 175 g steam at 100.0 °C and 1.65 kg of ice 0.0 °C are added.
(a) What mass of ice remains unmelted after equil rium is established?
(b) What *additional* mass of steam should be int duced into the insulated container to just melt all the ice?

58. A 54 cm^3 ice cube at −25.0 °C is added to a therma insulated container with 400.0 mL $H_2O(l)$ at 32.0 What will be the final temperature in the contair and what state(s) of matter will be present? (Speci heats: $H_2O(s)$, 2.01 J g^{-1} °C^{-1}; $H_2O(l)$, 4.18 J g °C Densities: $H_2O(s)$, 0.917 g/cm^3; $H_2O(l)$, 0.998 g/cr Also, $\Delta_{fus}H$ of ice = 6.01 kJ mol^{-1}.)

59. You decide to cool a can of soda pop quickly in t freezer compartment of a refrigerator. When you ta out the can, the soda pop is still liquid; but when y open the can, the soda pop immediately freez Explain why this happens.

60. Why is the triple point of water (ice–liquid–vapor better fixed point for establishing a thermomet scale than either the melting point of ice or the boili point of water?

Network Covalent Solids

61. Based on data presented in the text, would you expect diamond or graphite to have the greater density? Explain.

62. Diamond is often used as a cutting medium in gla cutters. What property of diamond makes this pos ble? Could graphite function as well?

Silicon carbide, SiC, crystallizes in a form similar to diamond, whereas boron nitride, BN, crystallizes in a form similar to graphite.
(a) Sketch the SiC structure as in Figure 12-32(b).
(b) Propose a bonding scheme for BN.

64. Are the fullerenes network covalent solids? What makes them different from diamond and graphite? It has been shown that carbon can form chains in which every other carbon atom is bonded to the next carbon atom by a triple bond. Is this allotrope of carbon a network covalent solid? Explain.

...nic Bonding and Properties

The melting points of NaF, NaCl, NaBr, and NaI are 988, 801, 755, and 651 °C, respectively. Are these data consistent with ideas developed in Section 12-5? Explain.

Use Coulomb's law (see Appendix B) to verify the conclusion concerning the relative strengths of the attractive forces in the ion pairs $Na^{+}Cl^{-}$ and $Mg^{2+}O^{2-}$ presented in Figure 12-36.

67. The hardness of crystals is rated based on Mohs hardness values. The higher the Mohs value, the harder the material is to scratch. Which crystal will have the highest Mohs value: NaF, NaCl, or KCl?

68. Will the mineral villaumite (NaF) or periclase (MgO) have a higher Mohs hardness value (see Exercise 67)?

...ystal Structures

Explain why there are *two* arrangements for the closest packing of spheres rather than a single one.

Argon, copper, sodium chloride, and carbon dioxide all crystallize in the fcc structure. How can this be when their physical properties are so different?

Consider the two-dimensional lattice shown here.

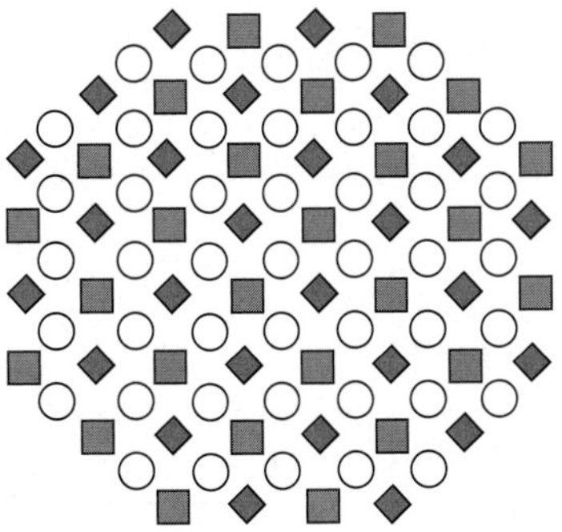

(a) Identify a unit cell.
(b) How many of each of the following elements are in the unit cell: ◆, ■, and ○?
(c) Indicate some simpler units than the unit cell, and explain why they cannot function as a unit cell.

. As we saw in Section 12-6, stacking spheres always leaves open space. Consider the corresponding situation in two dimensions: Squares can be arranged to cover all the area, but circles cannot. For the arrangement of circles pictured here, what percentage of the area remains uncovered?

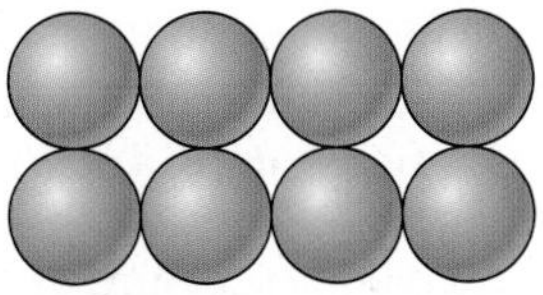

. Tungsten has a body-centered cubic crystal structure. Using a metallic radius of 139 pm for the W atom, calculate the density of tungsten.

. Magnesium crystallizes in the hcp arrangement shown in Figure 12-41. The dimensions of the unit cell are height, 520 pm; length on an edge, 320 pm. Calculate the density of Mg(s), and compare with the measured value of 1.738 g/cm^3.

75. Polonium (Po) is the only element known to take on the simple cubic crystal system. The distance between nearest neighbor Po atoms in this structure is 335 pm.
(a) What is the diameter of a Po atom?
(b) What is the density of Po metal?
(c) At what angle (in degrees) to the parallel faces of the Po unit cells would first-order diffraction be observed when using X-rays of wavelength 1.785×10^{-10} m?

76. Germanium has a cubic unit cell with a side edge of 565 pm. The density of germanium is 5.36 g/cm^3. What is the crystal system adopted by germanium?

77. Silicon tetrafluoride molecules are arranged in a body-centered cubic unit cell. How many silicon atoms are in the unit cell?

78. Two views, a top and side view, for the unit cell for rutile (TiO_2) are shown here. **(a)** How many titanium atoms (blue) are in this unit cell? **(b)** How many oxygen atoms (red) are in this unit cell?

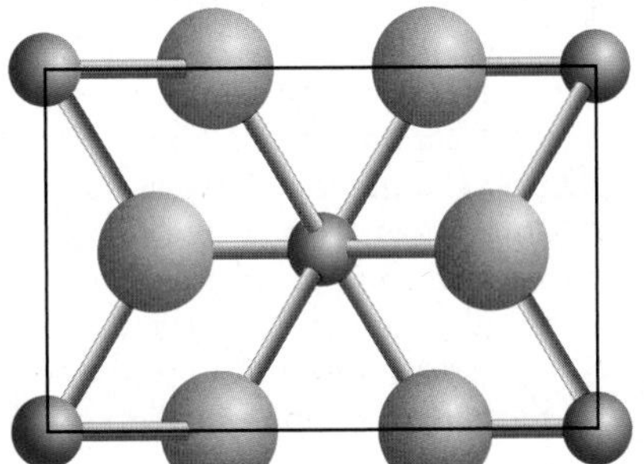

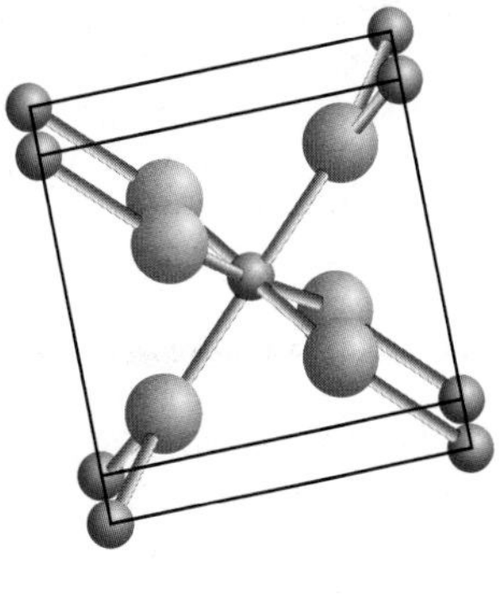

Ionic Crystal Structures

79. Show that the unit cells for CaF_2 and TiO_2 in Figure 12-50 are consistent with their formulas.
80. Using methods similar to Examples 12-10 and 12-11, calculate the density of CsCl. Use 169 pm as the radius of Cs^+.
81. The crystal structure of magnesium oxide, MgO, is of the NaCl type (Fig. 12-48). Use this fact, together with ionic radii from Figure 9-11, to establish the following.
 (a) the coordination numbers of Mg^{2+} and O^{2-};
 (b) the number of formula units in the unit cell;
 (c) the length and volume of a unit cell;
 (d) the density of MgO.
82. Potassium chloride has the same crystal structure NaCl. Careful measurement of the internuclear dista between K^+ and Cl^- ions gave a value of 314.54 pm. T density of KCl is 1.9893 g/cm^3. Use these data to eva ate the Avogadro constant, N_A.
83. Use data from Figure 9-11 to predict the type of cu unit cell adopted by **(a)** CaO; **(b)** CuCl; **(c)** LiO_2 (radius of the O_2^- ion is 128 pm).
84. Use data from Figure 9-9 to predict the type of cu unit cell adopted by **(a)** BaO; **(b)** CuI; **(c)** LiS_2. (The ra of Ba^{2+} and S_2^- ions are 135 and 198 pm, respectivel

Lattice Energy

85. *Without doing calculations*, indicate how you would expect the lattice energies of LiCl(s), KCl(s), RbCl(s), and CsCl(s) to compare with the value of $-787\ \text{kJ mol}^{-1}$ determined for NaCl(s) on page 563. [*Hint:* Assume that the enthalpies of sublimation of the alkali metals are comparable in value. What atomic properties from Chapter 9 should you compare?]
86. Determine the lattice energy of KF(s) from the following data: $\Delta_f H°[\text{KF(s)}] = -567.3\ \text{kJ mol}^{-1}$; enthalpy of sublimation of K(s), $89.24\ \text{kJ mol}^{-1}$; enthalpy of dissociation of $F_2(g)$, $159\ \text{kJ mol}^{-1}\ F_2$; E_i for K(g), $418.9\ \text{kJ mol}^{-1}$; E_{ea} for F(g), -328kJ mol^{-1}.
87. Refer to Example 12-12. Together with data given there, use the data here to calculate $\Delta_f H°$ for 1 mol $MgCl_2(s)$. Explain why you would expect $MgCl_2$ be a much more stable compound than MgCl. (T second ionization energy of Mg is $1451\ \text{kJ mol}^{-1}$; lattice energy of $MgCl_2(s)$ is $-2526\ \text{kJ mol}^{-1}\ MgCl_2$.
88. In ionic compounds with certain metals, hydrog exists as the hydride ion, H^-. Determine the el tron affinity of hydrogen; that is, $\Delta_r H$ for t process $H(g) + e^- \rightarrow H^-(g)$. To do so, use data fro Section 12-7; the bond energy of $H_2(g)$ fro Table 10.3; $-812\ \text{kJ mol}^{-1}$ for the lattice energy NaH(s); and $-57\ \text{kJ mol}^{-1}$ NaH for the enthalpy formation of NaH(s).

Integrative and Advanced Exercises

89. When a wax candle is burned, the fuel consists of *gaseous* hydrocarbons appearing at the end of the candle wick. Describe the phase changes and processes by which the solid wax is ultimately consumed.
90. The normal boiling point of water is 100.00 °C and the enthalpy of vaporization at this temperature is $\Delta_{vap}H = 40.657\ \text{kJ mol}^{-1}$. What would be the boiling point of water if it were based on a pressure of 1 bar instead of the standard atm?
91. A supplier of cylinder gases warns customers to determine how much gas remains in a cylinder by weighing the cylinder and comparing this mass to the original mass of the full cylinder. In particular, the customer is told not to try to estimate the mass of gas available from the measured gas pressure. Explain the basis of this warning. Are there cases where a measurement of the gas pressure *can* be used as a measure of the remaining available gas? If so, what are they?
92. Use the following data and data from Appendix D to determine the quantity of heat needed to convert 15.0 g of solid mercury at −50.0 °C to mercury vapor at 25 °C. Specific heats: Hg(s), $24.3\ \text{J mol}^{-1}\ \text{K}^{-1}$; Hg(l), $28.0\ \text{J mol}^{-1}\ \text{K}^{-1}$. Melting point of Hg(s), −38.87 °C. Heat of fusion, $2.33\ \text{kJ mol}^{-1}$.
93. To vaporize 1.000 g water at 20 °C requires 2447 J of heat. At 100 °C, 10.00 kJ of heat will convert 4.430 g $H_2O(l)$ to $H_2O(g)$. Do these observations conform to your expectations? Explain.
94. Estimate how much heat is absorbed when 1.00 g Instant Car Kooler vaporizes. Comment on the effe tiveness of this spray in cooling the interior of a c Assume the spray is 10% $C_2H_5OH(aq)$ by ma the temperature is 55 °C, the heat capacity of air $29\ \text{J mol}^{-1}\ \text{K}^{-1}$, and use $\Delta_{vap}H$ data from Table 12.
95. Because solid *p*-dichlorobenzene, $C_6H_4Cl_2$, su limes rather easily, it has been used as a moth rep lent. From the data given, estimate the sublimati pressure of $C_6H_4Cl_2(s)$ at 25 °C. For $C_6H_4Cl_2$; mp 53.1 °C; vapor pressure of $C_6H_4Cl_2(l)$ at 54.8 is 10.0 mmHg; $\Delta_{fus}H = 17.88\ \text{kJ mol}^{-1}$; $\Delta_{vap}H$ $72.22\ \text{kJ mol}^{-1}$.
96. A 1.05 mol sample of $H_2O(g)$ is compressed in a 2.61 L flask at 30.0 °C. Describe the point(s) Figure 12-30 representing the final condition.
97. One handbook lists the sublimation pressure of so benzene as a function of *Kelvin* temperature, *T*, $\log P\ (\text{mmHg}) = 9.846 - 2309/T$. Another han book lists the vapor pressure of *liquid* benzene as function of *Celsius* temperature, *t*, as $\log P\ (\text{mmHg})$ $6.90565 - 1211.033/(220.790 + t)$. Use these equ tions to estimate the normal melting point of benzer and compare your result with the listed val of 5.5 °C.
98. By the method used to graph Figure 12-20, plot ln versus $1/T$ for liquid white phosphorus, and estima

(a) its normal boiling point and **(b)** its enthalpy of vaporization, $\Delta_{vap}H$, in $kJ\ mol^{-1}$. Vapor pressure data: 76.6 °C, 1 mmHg; 128.0 °C, 10 mmHg; 166.7 °C, 40 mmHg; 197.3 °C, 100 mmHg; 251.0 °C, 400 mmHg.

9. Assume that a skater has a mass of 80 kg and that his skates make contact with 2.5 cm^2 of ice. **(a)** Calculate the pressure in atm exerted by the skates on the ice. **(b)** If the melting point of ice decreases by 1.0 °C for every 125 atm of pressure, what would be the melting point of the ice under the skates?

0. Estimate the boiling point of water in Leadville, Colorado, elevation 3170 m. To do this, use the barometric formula relating pressure and altitude: $P = P_0 \times 10^{-Mgh/2.303\,RT}$ (where P = pressure in atm; $P_0 = 1$ atm; acceleration due to gravity, $g = 9.81\ m\ s^{-2}$; molar mass of air, $M = 0.02896\ kg\ mol^{-1}$; $R = 8.3145\ J\ mol^{-1}\ K^{-1}$; and T is the Kelvin temperature). Assume the air temperature is 10.0 °C and that $\Delta_{vap}H = 41\ kJ\ mol^{-1}\ H_2O$.

1. Inspection of the straight-line graphs in Figure 12-20 suggests that the graphs for benzene and water intersect at a point that falls off the page. At this point, the two liquids have the same vapor pressure. Estimate the temperature and the vapor pressure at this point by a calculation based on data obtainable from the graphs.

2. A cylinder containing 151 lb Cl_2 has an inside diameter of 10 in. and a height of 45 in. The gas pressure is 100 psi (1 atm = 14.7 psi) at 20 °C. Cl_2 melts at −103 °C, boils at −35 °C, and has its critical point at 144 °C and 76 atm. In what state(s) of matter does the Cl_2 exist in the cylinder?

3. In acetic acid vapor, some molecules exist as monomers and some as dimers (see Figure 12-9). If the density of the vapor at 350 K and 1 atm is 3.23 g/L, what percentage of the molecules must exist as dimers? Would you expect this percent to increase or decrease with temperature?

4. A 685 mL sample of Hg(l) at 20 °C is added to a large quantity of liquid N_2 kept at its boiling point in a thermally insulated container. What mass of N_2(l) is vaporized as the Hg is brought to the temperature of the liquid N_2? For the specific heat of Hg(l) from 20 to −39 °C use $0.138\ J\ g^{-1}\ °C^{-1}$, and for Hg(s) from −39 to −196 °C, $0.126\ J\ g^{-1}\ °C^{-1}$. The density of Hg(l) is 13.6 g/mL, its melting point is −39 °C, and its enthalpy of fusion is $2.30\ kJ\ mol^{-1}$. The boiling point of N_2(l) is −196 °C, and its $\Delta_{vap}H$ is $5.58\ kJ\ mol^{-1}$.

5. Sketched here are two hypothetical phase diagrams for a substance, but neither of these diagrams is possible. Indicate what is wrong with each of them.

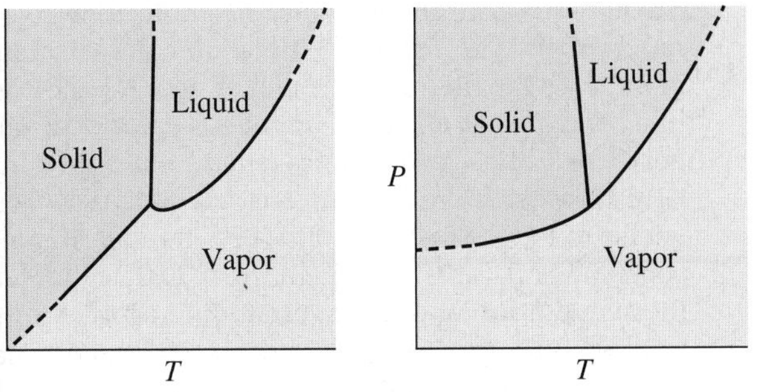

106. A chemistry handbook lists the following equation for the vapor pressure of NH_3(l) as a function of temperature. What is the normal boiling point of NH_3(l)?

$$\log_{10} P(\text{mmHg}) = 9.95028 - 0.003863T - \frac{1473.17}{T}$$

107. The triple point temperature of bismuth is 544.5 K and the normal boiling point is 1832 K. Imagine that a 1.00 mol sample of bismuth is heated at a constant rate of $1.00\ kJ\ min^{-1}$ in an apparatus in which the sample is maintained under a constant pressure of 1 atm. In the manner shown in Figure 12-24 and as much to scale as possible, that is in terms of times and temperatures, sketch the heating curve that would be obtained in heating the sample from 300 K to 2000 K. Use the following data. $\Delta_{fus}H = 10.9\ kJ\ mol^{-1}$ for Bi(s); $\Delta_{vap}H = 151.5\ kJ\ mol^{-1}$ for Bi(l); average molar heat capacities, in $J\ mol^{-1}\ K^{-1}$, 28 for Bi(s), 31 for Bi(l), and 21 for Bi(g). [*Hint:* Under the conditions described, no vapor appears until the normal boiling point is reached.]

108. The crystal structure of lithium sulfide (Li_2S), is pictured here. The length of the unit cell is 5.88×10^2 pm. For this structure, determine

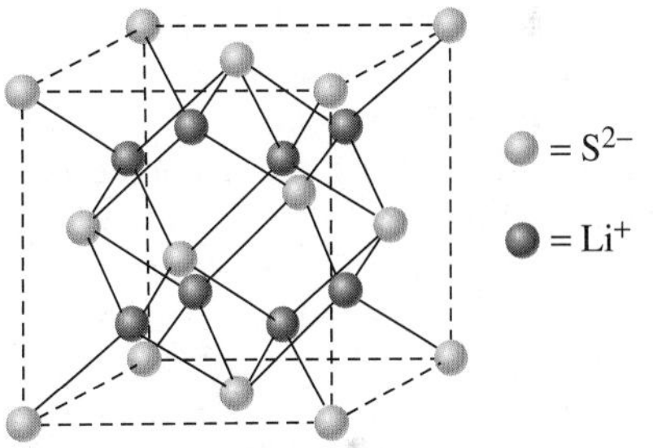

(a) the coordination numbers of Li^+ and S^{2-};
(b) the number of formula units in the unit cell;
(c) the density of Li_2S.

109. Refer to Figure 12-44 and Figure 12-48. Suppose that the two planes of ions pictured in Figure 12-44 correspond to the top and middle planes of ions in the NaCl unit cell in Figure 12-48. If the X-rays used have a wavelength of 154.1 pm, at what angle θ would the diffracted beam have its greatest intensity? [*Hint:* Use $n = 1$ in equation (12.5).]

110. Use the analyses of a bcc structure on page 555 and the fcc structure in Exercise 146 to determine the percent voids in the packing-of-spheres arrangement found in the fcc crystal structure.

111. One way to describe ionic crystal structures is in terms of cations filling voids among closely packed anions. Show that in order for cations to fill the tetrahedral voids in a close packed arrangement of anions, the radius ratio of cation, r_c, to anion, r_a, must fall between the following limits $0.225 < r_c/r_a < 0.414$.

112. Use the unit cell of diamond in Figure 12-32(b) and a carbon-to-carbon bond length of 154.45 pm, together with other relevant data from the text, to calculate the density of diamond.

113. The enthalpy of formation of NaI(s) is $-288\ kJ\ mol^{-1}$. Use this value, together with other data in the text, to calculate the lattice energy of NaI(s). [*Hint:* Use data from Appendix D also.]

114. Show that the formation of $NaCl_2$(s) is very unfavorable; that is, $\Delta_f H°[NaCl_2(s)]$ is a large *positive* quantity. To do this, use data from Section 12-7 and assume that the lattice energy for $NaCl_2$ would be about the same as that of $MgCl_2$, -2.5×10^3 kJ mol^{-1}.

115. A crystalline solid contains three types of ions, Na^+, O^{2-}, and Cl^-. The solid is made up of cubic unit cells that have O^{2-} ions at each corner, Na^+ ions at the center of each face, and Cl^- ions at the center of the cells. What is the chemical formula of the compound? What are the coordination numbers for the O^{2-} and Cl^- ions? If the length of one edge of the unit cell is a, what is the shortest distance from the center of a Na^+ ion to the center of an O^{2-} ion? Similarly, what is the shortest distance from the center of a Cl^- ion to the center of an O^{2-} ion?

116. A certain mineral has a cubic unit cell with calcium at each corner, oxygen at the center of each face, and titanium at its body center. What is the formula of the mineral? An alternate way of drawing the unit cell has calcium at the center of each cubic unit cell. What are the positions of titanium and oxygen in such a representation of the unit cell? How many oxygen atoms surround a particular titanium atom in either representation?

117. Calculate the radius ratio (r_+/r_-) for CaF_2. Suggest an alternative structure to that shown in Figure 12-50(b) that better conforms to the radius ratio you compute.

118. In some barbecue grills the electric lighter consists of a small hammer-like device striking a small crystal, which generates voltage and causes a spark between wires that are attached to opposite surfaces of the crystal. The phenomenon of causing an electric potential through mechanical stress is known as the piezoelectric effect. One type of crystal that exhibits the piezoelectric effect is lead zirconate titanate. In this perovskite crystal structure, a titanium(IV) ion sits in the middle of a tetragonal unit cell with dimensions 0.403 nm $\times$ 0.398 nm $\times$ 0.398 nm. At each corner i lead(II) ion, and at the center of each face is an oxyg anion. Some of the Ti(IV) are replaced by Zr(IV). T substitution, along with Pb(II), results in the pie electic behavior.

(a) How many oxygen ions are in the unit cell?

(b) How many lead(II) ions are in the unit cell?

(c) How many titanium(IV) ions are in the unit ce

(d) What is the density of the unit cell?

119. Ionic liquids (ILs) are salts that are in the liquid sta At a given temperature, ILs have lower vapor pr sures than molecular compounds in the liquid st because the forces of attraction between opposite charged ions are much stronger than intermolecu forces. Thus, ILs tend to be much less volatile a less flammable than many other liquids. ILs are interest because of their potential role as "safer" a "greener" solvents. Two examples of ionic liqui are 1-butyl-3-methylimidazolium tetrafluorobora [Bmim][BF_4], and 1-allyl-3-methylimidazolium ch ride, [Amim]Cl, both of which consist of a relative large organic cation and an inorganic anion.

(a) Look up and then draw the structures of the io making up these two ILs.

(b) Find the melting points for these two ILs and NaCl.

(c) Explain why the melting points of these two I are much lower than that of NaCl.

120. In a 1999 study of cobalt nanocrystals, D. P. Dine and M. G. Bawendi discovered that cobalt forms interesting cubic structure unlike any of the cul structures described in this chapter. They called th new form ϵ-cobalt to distinguish it from the more co monly encountered hcp and fcc forms of cobalt. F ϵ-cobalt, the unit cell has an edge length of 609.7 p and contains 20 atoms. The density of ϵ-cobalt $\rho = 8.635$ g cm^{-3}. Use these data to estimate t number of cobalt atoms in a spherical nanocrystal ϵ-cobalt if the diameter of the nanocrystal is 2 nm.

Feature Problems

121. Intermolecular forces play vital and varied roles in nature. For example, these forces enable gecko lizards to climb walls and hang upside down from ceilings, seemingly defying gravity. Intermolecular forces—more specifically, hydrogen bonds—are the reason that DNA molecules, carriers of the genetic code for most living organisms, exist as a double helix. The helical structure of proteins, the molecules that catalyze biochemical reactions occurring in our bodies and regulate metabolic processes, is also the result of hydrogen bonding. In Section 12-1, we learned about the physical basis of different types of intermolecular forces, such as dipole–dipole, dipole–induced dipole, and instantaneous dipole–induced dipole (dispersion) interactions. We also discussed the relative strengths of these different types of interactions and the percent contributions they make to the attraction between molecules. This problem focuses on doing calculations to verify the claims made in Section 12-1.

For two *identical* molecules separated from each oth by a distance much greater than their own dime sions, the average potential energy of interaction, E, approximately

$$E = -\frac{1}{r^6}\left[\frac{2\mu^4}{3k_B T}\cdot\frac{1}{(4\pi\epsilon_0)^2} + 2\mu^2\alpha\cdot\frac{1}{(4\pi\epsilon_0)} + \frac{3}{4}\alpha^2 E_i\right]$$

In the equation above, μ is the molecular dipo moment in C m, α is the molecular polarizabili in m^3, E_i is the first ionization energy of the molecu in J, and r is the center of mass separation in between the two molecules. In additic $\epsilon_0 = 8.854 \times 10^{-12}$ C^{2}J^{-1}m^{-1} is the permittivity vacuum, $k_B = 1.3807 \times 10^{-23}$ J K^{-1} is the Boltzmar constant, and T is the temperature in K. The first ter in the equation above represents the dipole–dipo interaction, the second term represents t dipole–induced dipole interaction, and the third ter represents the dispersion interaction.

Use the equation above and data from the table that follows to answer the questions below. Assume the center of mass separation, r, between molecules is exactly 400 pm and the temperature is 298 K.

(a) For each substance in the table, calculate E, as well as the contributions to E from dipole–dipole, dipole–induced dipole, and dispersion interactions. Express E and the various contributions to E in kJ mol^{-1}.

(b) Use your results from (a) to calculate, for each substance, the percent contribution made by each type of interaction.

(c) What is the range of values of E calculated in (a)? Briefly comment on how these values compare in magnitude with the (covalent) bond energies, D, given in Table 10.3.

(d) Use your results from (a) to prepare three separate graphs of $\Delta_{vap}H$ versus $-E$, one graph for *each* class of compounds shown in the table (halides, alcohols, and hydrocarbons). What do these graphs illustrate?

(e) The formula for E indicates that the contribution from dipole–dipole interactions decreases as temperature increases. Explain.

Substance[a]	μ, D	α, 10^{-25} cm^3	E_i, kJ mol^{-1}	$\Delta_{vap}H$,[b] kJ mol^{-1}
Halides				
HF	1.826	8.0	1548	7.49[c]
HCl	1.1086	26.3	1230	16.15
HBr	0.8272	36.1	1125	17.61[d]
HI	0.448	54.4	1002	19.76
Water and Alcohols				
H_2O	1.8546	14.5	1218	40.65
CH_3OH	1.70	32.9	1047	35.21
CH_3CH_2OH	1.69	54.1	1006	38.56
$CH_3(CH_2)_2OH$	1.55	67.4	982	41.44
$CH_3(CH_2)_3OH$	1.66	88.8	964	43.29
Hydrocarbons				
CH_4	0.0	25.93	1217	8.19
CH_3CH_3	0.0	44.7	1115	14.69
$CH_3CH_2CH_3$	0.0	62.9	1057	19.04
$(CH_3)_3CH$	0.132	81.4	1020	21.30
$CH_3(CH_2)_2CH_3$	0.0	82.0	1016	22.44
$CH_3(CH_2)_3CH_3$	0.0	99.9	992	25.70

[a] All values in this table are from the *CRC Handbook*, 95th edition, except where noted.
[b] The $\Delta_{vap}H$ values are measured at the substance's normal boiling point.
[c] This value is from the *Handbook of Inorganic Compounds*, 2nd edition, by Dale L. Perry.
[d] This value is from the *Air Liquide Gas Encyclopedia*, http://encyclopedia.airliquide.com.

2. In a capillary rise experiment, the height (h) to which a liquid rises depends on the density (d) and surface tension (γ) of the liquid and the radius of the capillary (r). The equation relating these quantities and the acceleration due to gravity (g) is $h = 2\gamma/dgr$. The sketch provides data obtained with ethanol. What is the surface tension of ethanol?

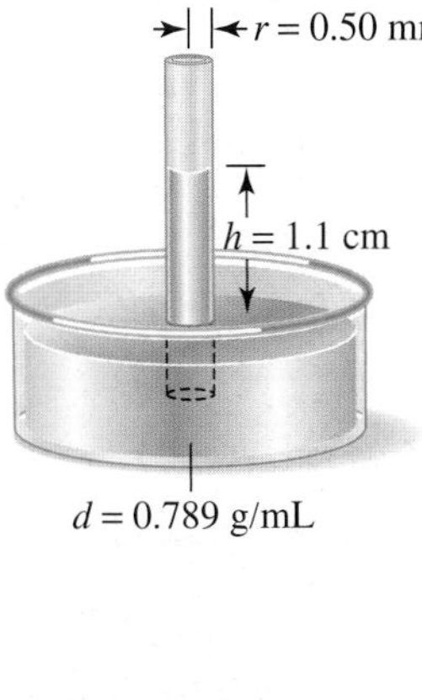

123. We have learned that the enthalpy of vaporization of a liquid is generally a function of temperature. If we wish to take this temperature variation into account, we cannot use the Clausius–Clapeyron equation in the form given in the text (that is, equation 12.2). Instead, we must go back to the differential equation upon which the Clausius–Clapeyron equation is based and reintegrate it into a new expression. Our starting point is the following equation describing the rate of change of vapor pressure with temperature in terms of the enthalpy of vaporization, the difference in molar volumes of the vapor (V_g), and liquid (V_l), and the temperature.

$$\frac{dP}{dT} = \frac{\Delta_{vap}H}{T(V_g - V_l)}$$

Because in most cases the volume of one mole of vapor greatly exceeds the molar volume of liquid,

we can treat the V_l term as if it were zero. Also, unless the vapor pressure is unusually high, we can treat the vapor as if it were an ideal gas; that is, for one mole of vapor, $PV = RT$. Make appropriate substitutions into the above expression, and separate the P and dP terms from the T and dT terms. The appropriate substitution for $\Delta_{vap}H$ means expressing it as a function of temperature. Finally, integrate the two sides of the equation between the limits P_1 and P_2 on one side and T_1 and T_2 on the other.
(a) Derive an equation for the vapor pressure of $C_2H_4(l)$ as a function of temperature, if $\Delta_{vap}H = 15{,}971 + 14.55\,T - 0.160\,T^2$ (in J mol^{-1}).
(b) Use the equation derived in (a), together with the fact that the vapor pressure of $C_2H_4(l)$ at 120 K is 10.16 Torr, to determine the normal boiling point of ethylene.

124. All solids contain defects or imperfections of structure or composition. Defects are important because they influence properties, such as mechanical strength. Two common types of defects are a missing ion in an otherwise perfect lattice, and the slipping of an ion from its normal site to a hole in the lattice. The holes discussed in this chapter are often called *interstitial sites*, since the holes are in fact interstices in the array of spheres. The two types of defects described here are called *point defects* because they occur within specific sites. In the 1930s, two solid-state physicists, W. Schottky and J. Fraenkel, studied the two types of point defects: A Schottky defect corresponds to a missing ion in a lattice, while a Fraenkel defect corresponds to an ion that is displaced into an interstitial site.
(a) An example of a Schottky defect is the absence of a Na^+ ion in the NaCl structure. The absence of a Na^+ ion means that a Cl^- ion must also be absent to preserve electrical neutrality. If one NaCl unit is missing per unit cell, does the overall stoichiometry change, and what is the change in density?
(b) An example of a Fraenkel defect is the movement of a Ag^+ ion to a tetrahedral interstitial site from its normal octahedral site in AgCl, which has a structure like NaCl. Does the overall stoichiometry of the compound change, and do you expect the density to change?
(c) Titanium monoxide (TiO) has a sodium chloride-like structure. X-ray diffraction data show that the edge length of the unit cell is 418 pm. The density of the crystal is 4.92 g/cm^3. Do the data indicate the presence of vacancies? If so, what type of vacancies?

125. In an ionic crystal lattice each cation will be attracted by anions next to it and repulsed by cations near it. Consequently the coulomb potential leading to the lattice energy depends on the type of crystal. To get the total lattice energy you must sum all of the electrostatic interactions on a given ion. The general form of the electrostatic potential is

$$V = \frac{Q_1 Q_2 e^2}{d_{12}}$$

where Q_1 and Q_2 are the charges on ions 1 and 2, d_{12} is the distance between them in the crystal lattice. and e is the charge on the electron.
(a) Consider the linear "crystal" shown below.

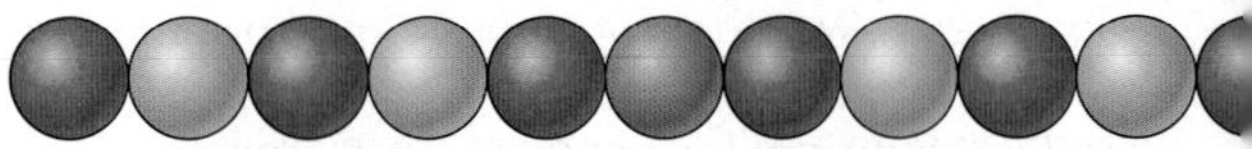

The distance between the centers of adjace spheres is R. Assume that the blue sphere and tl green spheres are cations and that the red spher are anions. Show that the total electrostatic energy

$$V = -\frac{Q^2e^2}{d} \times \ln 2$$

(b) In general, the electrostatic potential in a crys can be written as

$$V = -k_M \frac{Q^2e^2}{R}$$

where k_M is a geometric constant, called tl Madelung constant, for a particular crystal syste under consideration. Now consider the NaCl cry tal structure and let R be the distance between tl centers of sodium and chloride ions. Show that considering three layers of nearest neighbors to central chloride ion, k_M is given by

$$k_M = \left(6 - \frac{12}{\sqrt{2}} + \frac{8}{\sqrt{3}} - \frac{6}{\sqrt{4}} + \cdots\right)$$

(c) Carry out the same calculation for the Cs structure. Are the Madelung constants the same?

126. Plot the following data first as boiling point vers polarizability, and then as boiling point vers molecular mass. What conclusions can you dra from these plots?

Compound	Polarizability, 10^{-25} cm^3	Mass, u	Boiling Point, K
H_2	7.90	2.016	20.35
N_2	17.6	28.01	77.35
O_2	16.0	32.00	90.188
Cl_2	46.1	70.90	238.25
HF	24.6	20.01	292.69
HCl	26.3	36.46	188.25
HBr	36.1	80.91	206.15
HI	54.4	127.91	237.77
N_2O	30.0	44.01	184.65
CO	19.5	28.01	81.65
SO_2	37.2	64.06	263.15
H_2S	37.8	34.08	212.45
CS_2	87.4	76.13	319.45
NH_3	22.6	17.03	239.8
HCN	25.9	27.03	299.15
CH_4	26.0	16.04	109.15
C_2H_6	44.7	30.07	184.55
$CH_2{=}CH_2$	42.6	28.05	169.45
$CH{\equiv}CH$	33.3	26.04	189.15
C_3H_8	62.9	44.01	231.05
C_6H_6	103	78.11	353.25
CH_3Cl	45.6	50.49	248.95
CH_2Cl_2	64.8	84.93	313.15
$CHCl_3$	82.3	119.37	334.85
CCl_4	105	153.81	349.95
CH_3OH	32.3	32.04	338.15

7. The Born–Fajans–Haber cycle uses thermodynamic cycles to determine lattice energy. An alternative to the Born–Fajans–Haber method is one based on fundamental principles. Because the dominant interactions in an ionic crystal are Coulomb interactions, we can use the theory of electrostatics to calculate the lattice energy. Kapustinskii used these ideas and proposed the following equation:

$$U = \frac{120{,}250\, \nu\, Z^+ Z^-}{r_0}\left(1 - \frac{34.5}{r_0}\right)(\text{kJ mol}^{-1})$$

where the number of ions per formula unit is given by ν and r_0 is equal to the sum of the ionic radii, $r_+ + r_-$ (pm). Use the equation to complete the following table:

Compound	Lattice Energy, kJ mol^{-1}	r_-, pm	r_+, pm
NaCl		181	99
LaF_3		133	117
Na_2SO_4	−3389		99

Self-Assessment Exercises

8. In your own words, define or explain the following terms or symbols: **(a)** $\Delta_{vap}H$; **(b)** T_c; **(c)** instantaneous dipole; **(d)** coordination number; **(e)** unit cell.

9. Briefly describe each of the following phenomena or methods: **(a)** capillary action; **(b)** polymorphism; **(c)** sublimation; **(d)** supercooling; **(e)** determining the freezing point of a liquid from a cooling curve.

0. Explain the important distinctions between each pair of terms: **(a)** adhesive and cohesive forces; **(b)** vaporization and condensation; **(c)** triple point and critical point; **(d)** face-centered and body-centered cubic unit cell; **(e)** tetrahedral and octahedral hole.

1. Which of the following liquid properties depends on the strength of intermolecular attractions? **(a)** surface tension; **(b)** boiling point; **(c)** vapor pressure; **(d)** heat of vaporization; **(e)** all of these.

2. A liquid is in equilibrium with its vapor in a closed container. The lid of the container is removed briefly, allowing some of the vapor to escape, and then replaced. What is the immediate result of the vapor escaping? **(a)** vaporization rate decreases; **(b)** condensation rate decreases; **(c)** vaporization rate increases; **(d)** condensation rate increases; **(e)** none of these.

3. The magnitude of one of the following properties must always increase with temperature; that one is **(a)** surface tension; **(b)** density; **(c)** vapor pressure; **(d)** $\Delta_{vap}H$.

4. Of the compounds HF, CH_4, CH_3OH, N_2H_4, and $CHCl_3$, hydrogen bonding is an important intermolecular force in **(a)** none of these; **(b)** two of these; **(c)** three of these; **(d)** all but one of these; **(e)** all of these.

5. In the responses below, the vapor pressure of trichloroethene is listed for a given temperature. In which response does the given temperature correspond to the normal boiling point? **(a)** 40 Torr at 40.1 °C; **(b)** 100 Torr at 61.3 °C; **(c)** 400 Torr at 100.0 °C; **(d)** 760 Torr at 120.8 °C; **(e)** none of these.

6. The normal boiling point of acetone is 56.2 °C, and the molar heat of vaporization is 32.0 kJ mol^{-1}. What is the boiling temperature of acetone under a pressure of 50.0 mmHg?

7. A metal that crystallizes in the body-centered cubic (bcc) structure has a crystal coordination number of **(a)** 6; **(b)** 8; **(c)** 12; **(d)** any even number between 4 and 12.

8. A unit cell of an ionic crystal **(a)** shares some ions with other unit cells; **(b)** is the same as the formula unit; **(c)** is any portion of the crystal that has a cubic shape; **(d)** must contain the same number of cations and anions.

139. If the triple point pressure of a substance is greater than 1 atm, which two of the following conclusions are valid?
(a) The solid and liquid states of the substance cannot coexist at equilibrium.
(b) The melting point and boiling point of the substance are identical.
(c) The liquid state of the substance cannot exist.
(d) The liquid state cannot be maintained in a beaker open to air at 1 atm pressure.
(e) The melting point of the solid must be greater than 0 °C.
(f) The gaseous state at 1 atm pressure cannot be condensed to the solid at the triple point temperature.

140. In each of the following pairs, which would you expect to have the higher boiling point? **(a)** C_7H_{16} or $C_{10}H_{22}$; **(b)** C_3H_8 or $(CH_3)_2O$; **(c)** CH_3CH_2SH or CH_3CH_2OH.

141. One of the substances is out of order in the following list based on increasing boiling point. Identify it, and put it in its proper place: N_2, O_3, F_2, Ar, Cl_2. Explain your reasoning.

142. Arrange the following substances in the expected order of increasing melting point: KI, Ne, K_2SO_4, C_3H_8, CH_3CH_2OH, MgO, $CH_2OHCHOHCH_2OH$.

143. Is it possible to obtain a sample of ice from liquid water without ever putting the water in a freezer or other enclosure at a temperature below 0 °C? If so, how might this be done?

144. The phenomena in Figure 12-22 will be seen at the critical temperature only if the proper amount of liquid is placed in the sealed tube initially. Why should this be the case? What would you expect to see if too little liquid was present initially? If too much liquid was present?

145. The following data are given for CCl_4. Normal melting point, −23 °C; normal boiling point, 77 °C; density of liquid 1.59 g/mL; $\Delta_{fus}H$ = 3.28 kJ mol^{-1}; vapor pressure at 25 °C, 110 Torr.
(a) What phases—solid, liquid, and/or gas—are present if 3.50 g CCl_4 is placed in a closed 8.21 L container at 25 °C?

(b) How much heat is required to vaporize 2.00 L of $CCl_4(l)$ at its normal boiling point?

146. The fcc unit cell is a cube with atoms at each of the corners and in the center of each face, as shown here. Copper has the fcc crystal structure. Assume an atomic radius of 128 pm for a Cu atom.

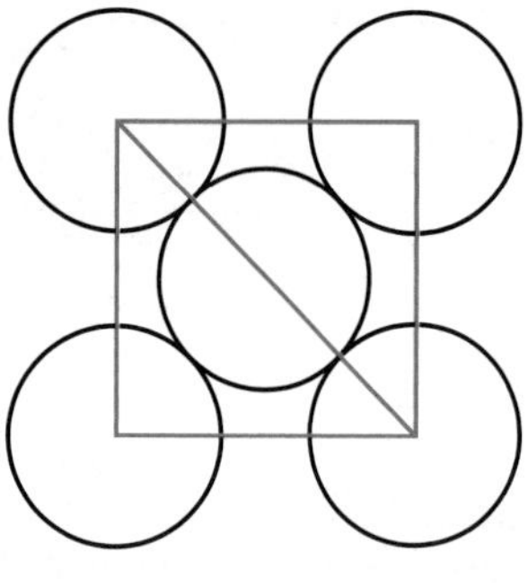

(a) What is the length of the unit cell of Cu?
(b) What is the volume of the unit cell?
(c) How many atoms belong to the unit cell?
(d) What percentage of the volume of the unit cell is occupied?
(e) What is the mass of a unit cell of copper?
(f) Calculate the density of copper.

147. Of the following liquids at 20 °C, which has the sm est surface tension? **(a)** CH_3OH; **(b)** CH_3CH_2C **(c)** $CH_3CH_2CH_2OH$; **(d)** $CH_3CH_2CH_2CH_2OH$.

148. Of the following liquids at 20 °C, which has the sm est viscosity? **(a)** dodecane, $C_{12}H_{26}$; **(b)** *n*-nona C_9H_{20}; **(c)** *n*-heptane, C_7H_{16}; **(d)** *n*-pentane, C_5H_{12}

149. Would you expect an ionic solid or a network co lent solid to have the higher melting point?

150. Consider the following ions: Na^+, K^+, Ca^{2+}, Mg^{2+}, Br^-, O^{2-}, and S^{2-}. Which cation and which anion you expect to combine to form the highest melti compound? Carefully explain your choice.

151. In the lithium iodide crystal, the Li–I distance is 3.02 Calculate the iodide radius, assuming that the iod ions are in contact.

152. Which of the following phase transitions is m likely to occur when the pressure on a metallic so increases? **(a)** bcc to sc; **(b)** fcc to sc; **(c)** bcc to **(d)** fcc to sc.

153. Construct a concept map representing the differ types of intermolecular forces and their origin.

154. Construct a concept map using the ideas of packing spheres and the structure of metal and ionic crysta

155. Construct a concept map showing the ideas co tained in a phase diagram.

›pontaneous Change: :ntropy and Gibbs :nergy

13

ONTENTS

LEARNING OBJECTIVES

13.1 Describe the concepts of microstate and entropy, and discuss how they are related. Identify situations in which entropy generally increases, and describe them in terms of the microstates involved.

13.2 Describe how Clausius's equation can be used to obtain equations for calculating entropy changes for simple physical changes, including phase changes, constant pressure heating/cooling, and isothermal expansion/compression. Apply the resulting equations to calculate entropy changes.

13.3 Explain how the standard molar entropy of a substance is obtained. Use the standard molar entropies of reactants and products to determine the entropy change for a chemical reaction.

13.4 State the second law of thermodynamics, and identify the relationship between Gibbs energy, enthalpy, and entropy.

13.5 Predict the direction of spontaneous chemical change by using values of the standard Gibbs energy of reaction ($\Delta_r G°$) and the thermodynamic reaction quotient (Q).

13.6 Use the van't Hoff equation to calculate the equilibrium constant as a function of temperature.

13.7 Describe how the coupling of chemical reactions may make a nonspontaneous process become a spontaneous one.

13.8 Discuss the relationships among chemical potential, activity and the Gibbs energy of a mixture.

ermodynamics originated in the early nineteenth century with attempts to orove the efficiency of steam engines. However, the laws of thermodynamics are lely useful throughout the field of chemistry and in biology and physics, as we cover in this chapter.

Our everyday experiences have conditioned us to accept that certain things happen naturally in one direction only. For example, a bouncing ball eventually comes to rest on the floor, but a ball at rest ll not begin to bounce. An ice cube placed in hot water eventually melts, t a glass of water will not produce an ice cube and hot water. A shiny iron il rusts in air, but a rusty nail will not naturally shed its rusty exterior to oduce a shiny nail. In this chapter, we explore concepts needed to underind why change happens naturally in one direction only.

At the end of Chapter 7, we noted some chemical and physical processes at proceeded in a certain direction without external influence, that is,

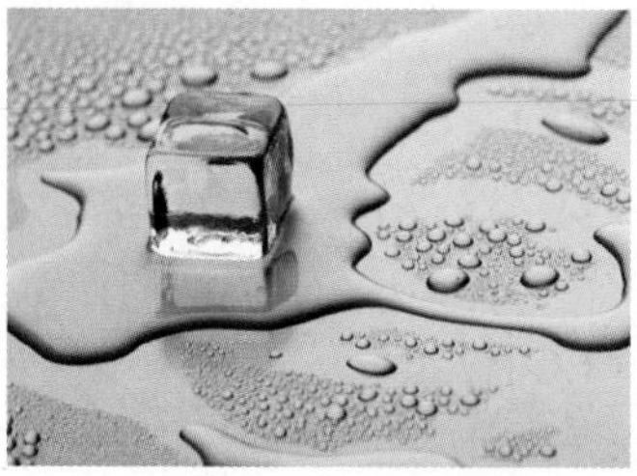

▲ The melting of an ice cube occurs spontaneously at temperatures above 0 °C.

spontaneously (Section 7-10). Among those examples, we saw situations which the enthalpy, H, of the system increased, decreased, or stayed the sa Clearly, the enthalpy change, ΔH, is not a reliable criterion for decid whether or not a particular change will occur spontaneously.

In 1850, Clausius introduced the concept of entropy to explain the direct of spontaneous change. Twenty-seven years later Ludwig Boltzmann p posed an alternative view of entropy based on probability theory. Not surp ingly, Clausius's and Boltzmann's definitions of entropy were eventua shown to be equivalent. So what is entropy and why is it important? Sim stated, entropy measures the dispersal of energy. It is an important conc because a great deal of experimental evidence supports the notion that ene spontaneously "spreads out" or "disperses" if it is not hindered from do so. Entropy is the yardstick for measuring the dispersal of energy.

In this chapter, we will continue to interpret observations about mac scopic systems by using a microscopic point of view. We will develop a c ceptual model for understanding entropy and learn how to evaluate entro changes for a variety of physical and chemical processes. Most importan we will define the criterion for spontaneous change and discover tha considers not only the entropy change for the system but also that of the s roundings. Finally, we will also learn about another important thermo namic quantity, called Gibbs energy, which can also be used for understandi the direction of spontaneous change.

13-1 Entropy: Boltzmann's View

▶ Spontaneous: "proceeding from natural feeling or native tendency without external constraint . . . ; developing without apparent external influence, force, cause, or treatment" (*Merriam-Webster's Collegiate Dictionary*, online, 2000).

We will soon see that the criterion for spontaneous change can be expressed terms of a thermodynamic quantity called entropy. Let's first focus our att tion on developing a conceptual model for understanding entropy. Then, will be able to use entropy, more specifically entropy changes, to explain w certain processes are spontaneous and others are not.

Microstates

The modern interpretation of entropy is firmly rooted in the idea that a mac scopic system is made up of many particles (often 10^{23} or more). Consider, example, a fixed amount, n, of an ideal gas at temperature T in a container volume V. The pressure of the gas is $P = nRT/V$. At the macroscopic level, state of the gas is easily characterized by giving the values of n, T, V, and The state of the gas won't change without some external influence (e.g., adding more gas, increasing the temperature, compressing the gas). Howev on the microscopic level, the state of the system is not so easily characteriz The molecules are in continuous random motion, experiencing collisions w each other or the walls of the container. The positions, velocities, and energ of individual molecules change from one instant to the next. The main poin that for a given macroscopic state, characterized by n, T, P, and V, there many possible microscopic configurations (or microstates), each of whi might be characterized by giving the position, velocity, and energy of eve molecule in the gas. Stated another way, the macroscopic properties of the g such as its temperature, pressure, and volume, could be described by any o of a very large number of microscopic configurations.

▶ A simple model for an ideal gas is obtained by treating the gas as a collection of noninteracting particles confined to a three-dimensional box. As we saw in Chapter 8, for a particle confined to a box, the kinetic energy (translational energy) is quantized. Each microstate corresponds to a particular way of distributing the molecules among the available translational energy levels.

In this discussion, we suggested that the microstate of an ideal gas could described by giving the position, velocity, and energy of every molecule in t gas. However, such a description is not consistent with quantum mechan because, according to the Heisenberg uncertainty principle (page 323), ex values for the position and velocity (or momentum) of a particle cannot simultaneously specified. To be consistent with quantum mechanics, microstate is characterized by specifying the quantum state (quantum numb and energy) of every particle or by specifying how the particles are distribu among the quantized energy levels. Thus, according to quantum mechanics

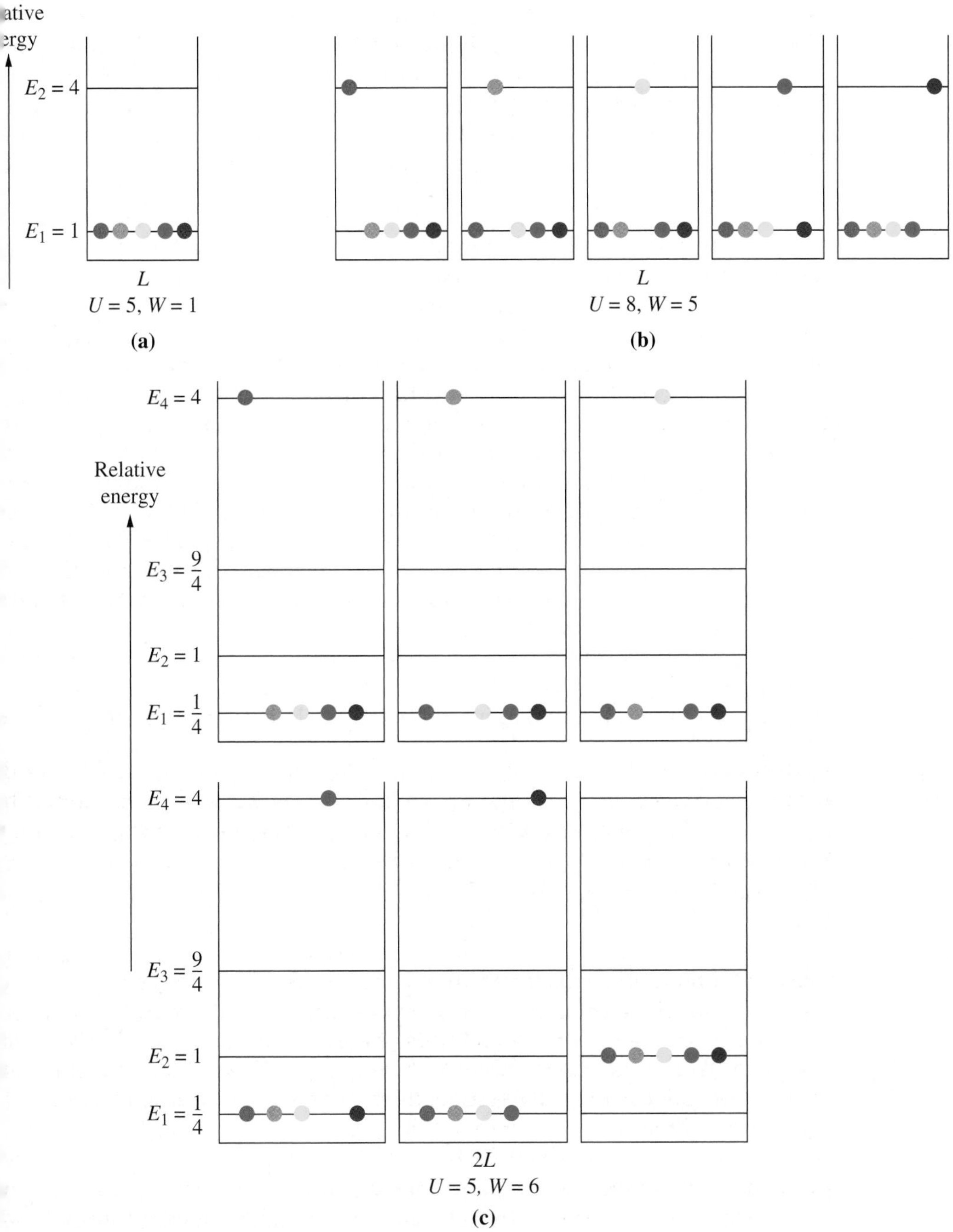

FIGURE 13-1
umeration of microstates
e distribution of five different particles among among the particle-in-a-box energy els. The energies are expressed as multiples of $h^2/8mL^2$. **(a)** For a box of length L, re is only one possible microstate when $U = 5 \times (h^2/8mL^2)$. **(b)** When the internal ergy is increased to $8 \times (h^2/8mL^2)$, the number of microstates increases to five cause more energy levels are accessible. **(c)** When length of the box is increased to the number of microstates for $U = 5 \times (h^2/8mL^2)$ increases to six. More energy els are accessible because the energy levels are shifted to lower values and are re closely spaced.

icrostate is a specific microscopic configuration describing how the particles a system are distributed among the available energy levels.

Let's explore the concept of a microstate by considering a system of five rticles confined to a one-dimensional box of length L. (We discussed the odel of a particle in a box on page 326.) To start, we use the energy level expression $E_n = n^2h^2/8mL^2$ to calculate a few energy levels. Representati energy levels, expressed as multiples of $h^2/8mL^2$, are shown in the diagrams Figure 13-1. Let's place the five particles among these energy levels with t constraint that the total energy, U, of the system must be $5 \times (h^2/8mL^2)$. T only possible arrangement (Fig. 13-1a) has all the particles in the $n = 1$ lev The total energy of this microstate is obtained by adding the particle energi

$$U = (1 + 1 + 1 + 1 + 1) \times (h^2/8mL^2) = 5 \times (h^2/8mL^2)$$

Notice that, for the situation just discussed, the state of the system can described in two ways. At the macroscopic level, the state of the system described by specifying the total energy, U, and the length, L, of the box. At t molecular level, the state of the system is described in terms of a microst having all particles in the $n = 1$ level. If we use the symbol W to represent t number of microstates, we have for this case $W = 1$. Notice that for this to energy, one energy level is accessible to the particles, namely, the $n = 1$ level

Let's suppose we increase the total energy of the system to $8 \times (h^2/8m$ without changing the length of the box. Such an increase in U can be achieve for example, by raising the temperature of the system. Figure 13-1(b) sho that, for $U = 8 \times (h^2/8mL^2)$, there are five possible microstates ($W = 5$) and increase in the number of energy levels that are accessible to the particles. see that as the total energy (or temperature) of a system increases, so too the number of microstates and the number of accessible energy levels.

Now suppose we increase the length of the box from L to $2L$ but keep the to energy fixed at a value of $5 \times (h^2/8mL^2)$. The increase in L may be considered "expansion" of the system. Figure 13-1(c) shows that, for this total energy, th are six possible microstates ($W = 6$) and a greater number of energy levels a accessible to the particles. We observe that, for a fixed total energy, the numb of microstates increases as the box length increases, that is, as the syste expands. The number of microstates increases when the box length increas because, for the larger box, the various levels are not only lower in energy b also more closely spaced. Thus, the number of energy levels that are accessib by the particles increases.

The point of this discussion was to illustrate not only the enumeration microstates through the distribution of particles among the available energy le els but also that W, the number of microstates, increases with both the total ener and total space available to the particles of the system. The number of accessib energy levels also increases with the total energy and total space available. No we make the connection between the number of microstates, W, and entropy.

Central Library for Physics, Vienna, Austria

▲ A bust marking Ludwig Boltzmann's tomb in Vienna
Boltzmann's famous equation is inscribed on the tomb. At the time of Boltzmann's death, the term *log* was used for both natural logarithms and logarithms to the base ten; the symbol ln had not yet been adopted.

The Boltzmann Equation for Entropy

Entropy, *S*, is a thermodynamic property that is related to the way in whi the energy of a system is distributed among the available energy leve Ludwig Boltzmann made this important conceptual breakthrough when associated the number of energy levels in the system with the number of wa of arranging the particles (atoms, ions, or molecules) in these energy leve Boltzmann derived the relationship

$$S = k_B \ln W \qquad \textbf{(13}$$

where S is the entropy, k_B is the Boltzmann constant, and W is the number microstates. The Boltzmann constant is related to the gas constant R a Avogadro's number, N_A, by the expression $k_B = R/N_A$. We can think of k_B as t gas constant per molecule. (Although we did not specifically introduce k_B our discussion of kinetic–molecular theory, R/N_A appears in equation (6.19 Using $R = 8.3145\ \text{J mol}^{-1}\ \text{K}^{-1}$ and $N_A = 6.0221 \times 10^{23}\ \text{mol}^{-1}$, we obta $k_B = 1.3807 \times 10^{-23}\ \text{J K}^{-1}$. The constant k_B also appears in the followi important relationship derived by Boltzmann for the probability, p, that a sy tem has energy E.

$$p \propto e^{-E/k_B T}$$

ar in mind that the concept of quantization of energy was not developed ring Boltzmann's lifetime (1844–1906), so he did not express his probability v in terms of the quantized energy levels of the particles in the system. A odern form of this law is

$$N_i \propto e^{-E_i/k_B T}$$

ıere N_i is the number of particles in the system having energy E_i. The expres-n above holds only when the number of particles is extremely large.

Equation (13.1) justifies our earlier assertion that entropy provides a easure of the dispersal of energy: the greater the value of W, the greater the mber of ways of distributing the total energy of a system among the energy els, and the greater the entropy, S.

The Boltzmann equation, equation (13.1), is deceptively simple. It suggests ıt S can be calculated simply by enumerating the possible microstates. As monstrated earlier, the enumeration of microstates is straightforward when system contains only five particles. For a system of containing a mole of rticles, the number of microstates defies comprehension (see Exercise 3):

$$W \approx 10^{(10^{23})} = 10^{100,000,000,000,000,000,000,000}$$

◀ In this expression, the power of ten is a 1 followed by 23 zeros. The value of W is a 1 followed by 10^{23} zeros. If you could write down these digits on a piece of paper at a rate of one million digits every second, it would take you 10^{17} seconds or 3 billion years to complete the task.

is number is not only incomprehensible but also uncountable. Fortunately, don't have to count the microstates. We must simply accept that, for a given stem, a very large number of microstates are possible.

The simplicity of equation (13.1) contrasts the enormity of Boltzmann's ıievement. To obtain this result, he and others (particularly James Clerk ıxwell and J. Willard Gibbs) developed a new branch of physics for describ-; systems containing a large number of particles. This branch of physics is w called *statistical physics* or *statistical mechanics*. A key idea in statistical chanics is that from the multitude of possible microstates, not all are equally ely. Think of flipping a coin a million times. Is it possible that you would flip ads each time? Yes. Is it likely? No. We can be quite certain that if we were to a coin a million times, the outcome will almost always be something very se to 50% heads and 50% tails. Boltzmann and others realized that, among all the possible microstates, the most probable ones contribute the most to the croscopic state. By focusing on the microstates of highest probability, ltzmann was able to derive expressions for evaluating W for a given macro-pic state. With such expressions in hand, the impossibility of counting the crostates is circumvented.

Fortunately, we will not encounter or use the complicated expressions for that are obtained by applying statistical mechanics. Our goal is to develop understanding of how the number of microstates and entropy of a system ange when, for example, the total energy, temperature, or volume changes. have already established the following key ideas.

- When the space available to the particles of a system is fixed, W and S increase as the total energy, U, increases or as T increases.
- When the total energy of a system is fixed, W and S increase as the space available to the particles increases.

Let's reinforce these ideas by using equation (13.1) to help us understand, m a microscopic point of view, how W and S change for a few simple ocesses.

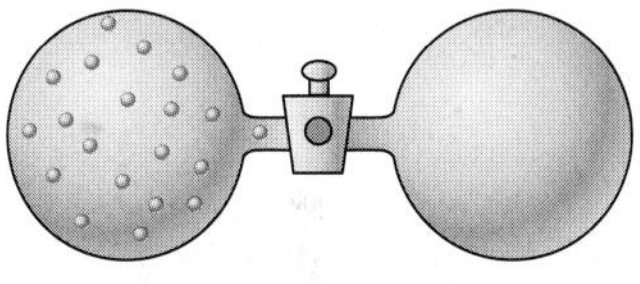

(a) Initial condition

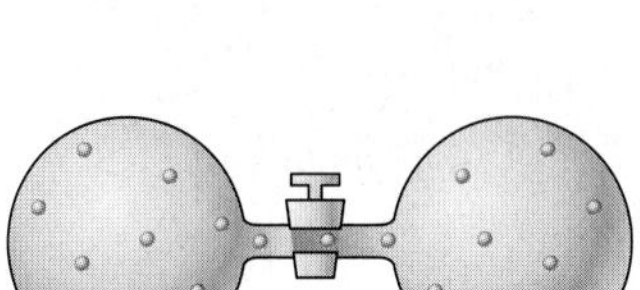

(b) After expansion into vacuum

▲ FIGURE 13-2
Expansion of an ideal gas into a vacuum
(a) Initially, an ideal gas is confined to the bulb on the left at 1.00 bar pressure.
(b) When the stopcock is opened, the gas expands into the identical bulb on the right. The final condition is one in which the gas is equally distributed between the two bulbs at a pressure of 0.50 bar.

icroscopic Interpretation of Entropy Change

the start of this chapter, we remarked that energy spontaneously disperses less it is prevented from doing so. To decide whether or not the energy of a stem has become "more dispersed" in some process, we must calculate anges in entropy.

Let's consider the isothermal expansion of an ideal gas as illustrated in ure 13-2. This figure depicts two identical glass bulbs joined by a stopcock.

Initially, the bulb on the left contains an ideal gas at 1.00 bar pressure, and t bulb on the right is evacuated. When the valve is opened, the gas spon neously expands into the evacuated bulb. After this expansion, the molecu are dispersed throughout the apparatus, with essentially equal numbers molecules in both bulbs. The final pressure is 0.50 bar. In this spontaneo process, it is obvious that the volume of the gas changes. But what about t energy or entropy of the gas? Do either of these quantities change? To answ this question, we must calculate ΔU and ΔS.

One of the characteristics of an ideal gas is that its internal energy (does not depend on the gas pressure but only on the temperature. For exa ple, the internal energy of a monatomic ideal gas, such as He or Ne, $U = \frac{3}{2}RT$ (see Section 6–7). Therefore, for the isothermal expansion of ideal gas, $\Delta U = 0$. Also, the enthalpy change is zero: $\Delta H = 0$. This mea that the expansion is not caused by the system dropping to a lower ener state. A convenient mental image to explain the expansion is that the g molecules tend to spread out into the larger volume available to them at t reduced pressure. A more fundamental description of the underlying cau is that, for the same total energy, in the expanded volume there are mo available translational energy levels among which the gas molecules can distributed. That is, the number of microstates, W, is greater for the syste in the larger volume than in the smaller volume. We can justify this idea using Figure 13-1, which shows that the number of microstates increas when the space available to the particles increases (compare Figs. 13-1a a 13-1c). Thus, when a system expands, we have not only $V_f > V_i$ but al $W_f > W_i$ and $\ln W_f > \ln W_i$. Using Boltzmann's equation, equation (13. we obtain

▶ By definition, $H = U + PV$, so $\Delta H = \Delta U + \Delta(PV)$. For a fixed amount of an ideal gas, we can write $\Delta(PV) = \Delta(nRT) = nR\Delta T$ and so, $\Delta H = \Delta U + nR\Delta T$. For an isothermal process involving an ideal gas, we have $\Delta U = 0$ and $\Delta T = 0$. Thus, $\Delta H = 0$.

$$\Delta S = k_B \ln W_f - k_B \ln W_i = k_B(\ln W_f - \ln W_i) > 0$$

Therefore, entropy increases when a gas expands at constant temperature in a larger volume. Is the increase in entropy somehow connected to the spo taneity of this process? Let's explore this idea further by examining anoth spontaneous process.

Consider the situation depicted in Figure 13-3, which describes the spon neous mixing of ideal gases. We can represent this process symbolically as

$$A(g) + B(g) \longrightarrow \text{mixture of } A(g) \text{ and } B(g)$$

The final state is the mixed state, and the initial state is the un-mixed sta Let's demonstrate that $\Delta U = 0$ and $\Delta S > 0$ for this process. Because the mo cules do not interact (they are ideal gases), the mixing is really just two spo taneous expansions occurring simultaneously. Let ΔU_A, ΔS_A, ΔU_B, and Δ represent the internal energy and entropy changes for gases A and B. described above, for the spontaneous expansion of a gas, the internal energy constant and entropy increases: $\Delta U_A = \Delta U_B = 0$ and $\Delta S_A > 0$ and $\Delta S_B >$ Thus, $\Delta U = \Delta U_A + \Delta U_B = 0$ and $\Delta S = \Delta S_A + \Delta S_B > 0$.

So, for both the spontaneous expansion of an ideal gas and the spon neous mixing of ideal gases, there is no change in internal energy (enthalpy) but an increase in entropy. It seems possible that *increases in entro underlie spontaneous processes.* We will soon see that the characteristic featu of a spontaneous process is that it causes the entropy of the universe increase.

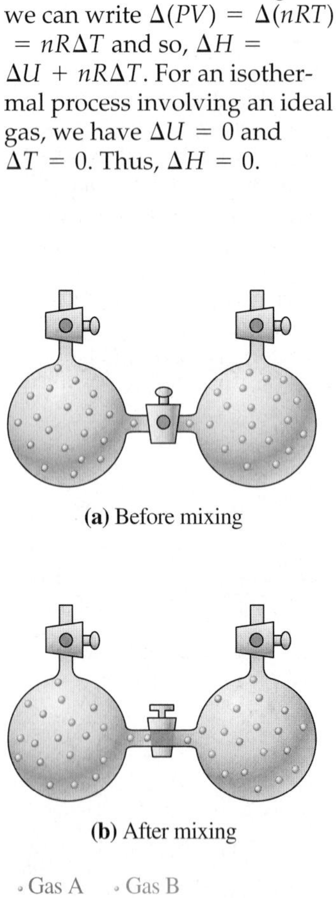

▲ FIGURE 13-3
The mixing of ideal gases
The total volume of the system and the total gas pressure remain fixed. The net change is that **(a)** before mixing, each gas is confined to half the total volume (a single bulb) at a pressure of 1.00 bar, and **(b)** after mixing, each gas has expanded into the total volume (both bulbs) and exerts a partial pressure of 0.50 bar.

13-1 CONCEPT ASSESSMENT

Expansion of a gas into a vacuum is not only spontaneous but also instantaneous. Is it generally true that a spontaneous process is also instantaneous? Explain.

Let us explore the connection between the number of microstates (W) and tropy (S) a little further by using a small one-dimensional crystal of four rous oxide molecules, NO. Let's suppose that, at $T = 0$ K, the four NO mol- iles are aligned as [NO···NO···NO···NO]. This arrangement represents nicrostate, and there are no other microstates like this one. Therefore, $W = 1$ d $S = 0$. Now consider what happens when the temperature is raised just ough to allow a single NO molecule to rotate. Such an increase in tempera- e corresponds to an increase in the internal energy of the system. Now four crostates are possible: [NO···NO···NO···ON], [NO···NO···ON···NO], O···ON···NO···NO], and [ON···NO···NO···NO]. W equals 4. By ng Boltzmann's equation, we find that $S = k_B \ln 4 = (1.3807 \times 10^{-23}\ \mathrm{J\,k^{-1}})$ $4 = 1.9141 \times 10^{-23}\ \mathrm{J\,K^{-1}}$. This is yet another example illustrating that as energy of the system increases, the number of microstates increases; there- e, entropy of the system will increase.

EXAMPLE 13-1 Relating Changes in the Number of Microstates to a Volume Change

A system containing four neon atoms is confined to a one-dimensional box. The system undergoes an expansion from 905 pm to 1810 pm at a fixed total energy of 14.0×10^{-24} J. **(a)** Determine the number of microstates for both the initial and final states of the system by illustrating, in the manner of Figure 13-1, the various possibilities for distributing the atoms among the particle-in-a-box energy levels. Use a different color for each atom. **(b)** Calculate the entropy change for the system.

Analyze

(a) We want to construct diagrams similar to those shown in Figure 13-1. To construct such diagrams, we use the particle-in-a-box energy level expression, $E_n = n^2h^2/(8mL^2)$, to calculate the energies of the levels for the initial ($L = 905$ pm) and final ($L = 1810$ pm) states. Then, we determine the number of different ways that the atoms can be placed in these energy levels, keeping in mind that for both the initial and final states, the total energy of the system must be equal to 14.0×10^{-24} J. **(b)** To calculate the entropy change $\Delta S = S_f - S_i$, we use the number of microstates for the initial and final states in Boltzmann's equation, $S = k_B \ln W$.

Solve

(a) Before using the particle-in-a-box energy level expression, we need to express the mass of a neon atom in the SI unit of kilograms.

$$m = 20.18\,\frac{\mathrm{g}}{\mathrm{mol}} \times \frac{10^{-3}\ \mathrm{kg}}{1\ \mathrm{g}} \times \frac{1\ \mathrm{mol}}{6.022 \times 10^{23}\ \mathrm{atoms}} = 3.351 \times 10^{-26}\,\frac{\mathrm{kg}}{\mathrm{atom}}$$

The energy levels for the 905 pm and 1810 pm boxes are given by

$$L = 905\ \mathrm{pm}: \quad E_n = \frac{(6.626 \times 10^{-34}\ \mathrm{J})^2}{(8)(3.351 \times 10^{-26}\ \mathrm{kg})(905 \times 10^{-12}\ \mathrm{m})^2} \times n^2 = (2.00 \times 10^{-24}\ \mathrm{J}) \times n^2$$

$$L = 1810\ \mathrm{pm}: \quad E_n = \frac{(6.626 \times 10^{-34}\ \mathrm{J})^2}{(8)(3.351 \times 10^{-26}\ \mathrm{kg})(1810 \times 10^{-12}\ \mathrm{m})^2} \times n^2 = (0.500 \times 10^{-24}\ \mathrm{J}) \times n^2$$

The energies of the first four levels are

	$E_n/10^{-24}$ J	
n	$L = 905$ pm	$L = 1810$ pm
1	2.00	0.500
2	8.00	2.00
3	18.0	4.50
4	32.0	8.00

As shown in the diagrams that follow, there are 4 microstates for the system in the 905 pm box, and 8 microstates for the system in the 1810 pm box, each having a total energy of 14.0×10^{-24} J.

(continued)

For the 905 pm box: 4 microstates

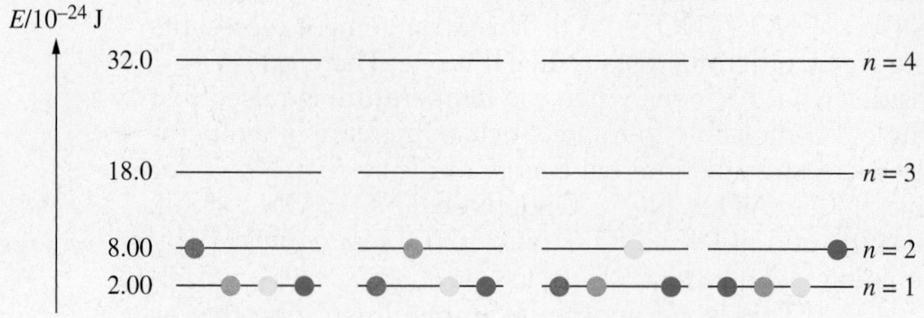

For the 1810 pm box: 8 microstates

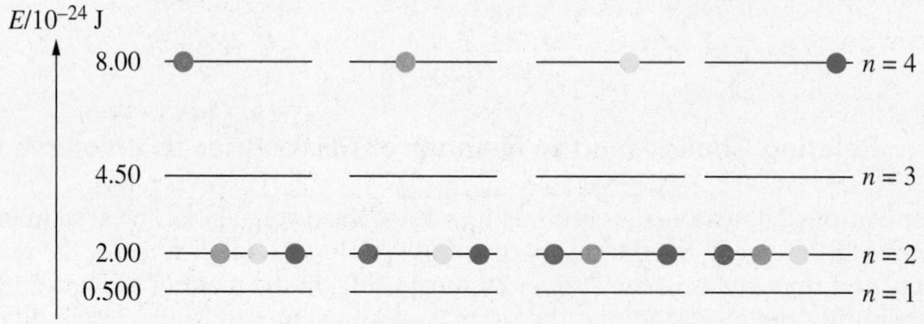

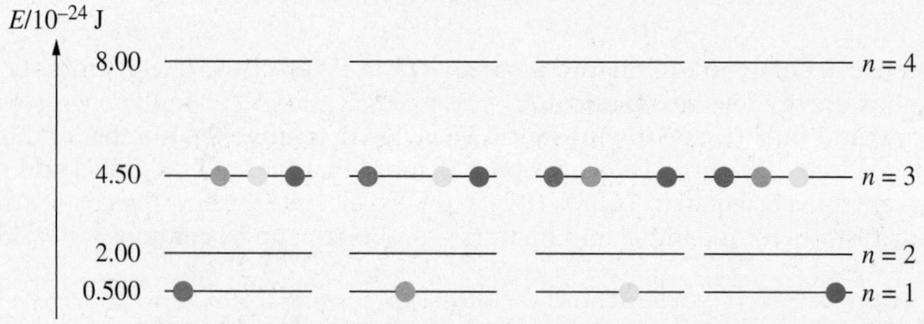

(b) The entropy change for the system is

$$\Delta S = k_B \ln 8 - k_B \ln 4 = k_B(\ln 8 - \ln 4) = k_B \ln (8/4) = k_B \ln 2$$
$$= 1.3807 \times 10^{-23}\ \text{J/K} \times 0.693 = 9.57 \times 10^{-24}\ \text{J/K}$$

Assess

This example illustrates that for a fixed total energy, both the number of microstates and the entropy increase as volume increases.

PRACTICE PROBLEM A: How many microstates are there for four neon atoms confined to a 905 pm box if the total energy is 20.0×10^{-24} J?

PRACTICE PROBLEM B: Which of the following, **(a)** or **(b)**, represents a change in volume?

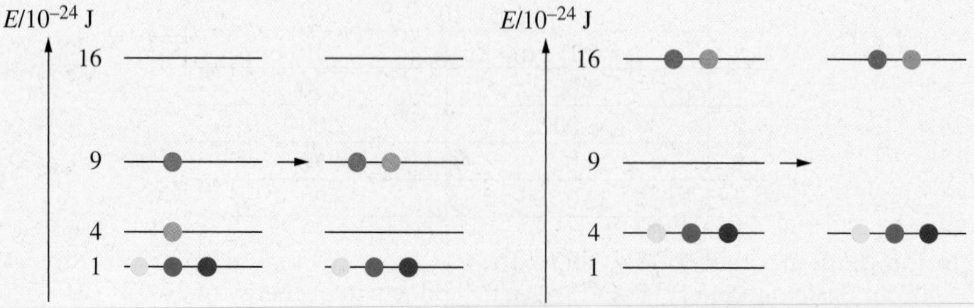

(a) **(b)**

escribing Entropy Changes for Some Simple Processes

 can use ideas from earlier sections to construct mental pictures for under- nding how the entropy of a system changes during a process. We begin by ting that, in many cases, an increase or a decrease in the number of crostates (or the number of accessible energy levels) parallels an increase or crease in the *number* of microscopic particles and the *space* available to them. a consequence, we can often make qualitative predictions about entropy ange by focusing on those two factors. Let's test this idea by considering the ee spontaneous endothermic processes of melting, vaporization, and disso- ion, as illustrated in Figure 13-4.

In the melting of ice, a crystalline solid is replaced by a less structured liquid. lecules that were relatively fixed in position in the solid, being limited to rational motion, are now free to move about a bit. The molecules have gained ne translational and rotational motion. The number of accessible microscopic ergy levels has increased and so has the entropy. In the vaporization process, iquid is replaced by an even less structured gas. Molecules in the gaseous te, because they can move within a large free volume, have many more acces- le energy levels than do those in the liquid state. In the gas, energy can be read over a much greater number of microscopic energy levels than in the liq- l. The entropy of the gaseous state is much higher than that of the liquid state.

In the dissolving of ammonium nitrate in water, for example, a crystalline id and a pure liquid are replaced by a mixture of ions and water molecules the liquid (solution) state. This situation is somewhat more involved than first two because some decrease in entropy is associated with the clustering water molecules around the ions because of ion–dipole forces. The increase entropy that accompanies the destruction of the solid's crystalline lattice pre- minates, however, and for the overall dissolution process, $\Delta S > 0$.

In each of the three spontaneous endothermic processes discussed here, increase in entropy ($\Delta S > 0$) outweighs the fact that heat must be sorbed ($\Delta H > 0$), and each process is spontaneous.

In summary, four situations generally produce an *increase* in entropy:

- Pure liquids or liquid solutions are formed from solids.
- Gases are formed from either solids or liquids.
- The number of molecules of gas increases as a result of a chemical reaction.
- The temperature of a substance increases. (Increased temperature means an increased number of accessible energy levels for the increased molecular motion, whether it be vibrational motion of atoms or ions in a solid, or translational and rotational motion of molecules in a liquid or gas.)

 apply these generalizations in Example 13-2.

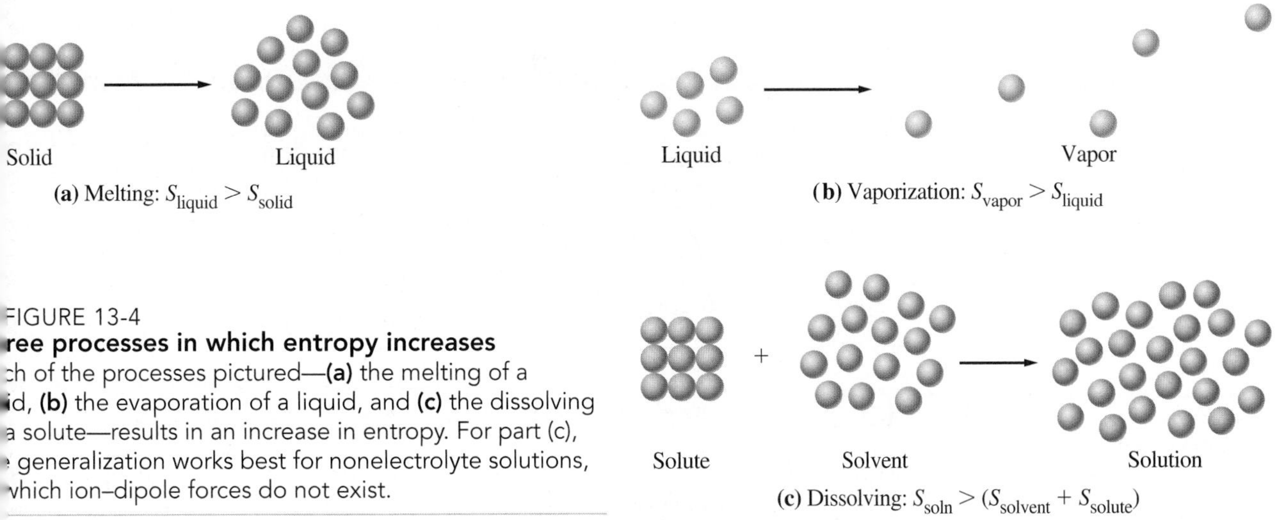

FIGURE 13-4
ree processes in which entropy increases
ch of the processes pictured—**(a)** the melting of a id, **(b)** the evaporation of a liquid, and **(c)** the dissolving a solute—results in an increase in entropy. For part (c), generalization works best for nonelectrolyte solutions, which ion–dipole forces do not exist.

EXAMPLE 13-2 Making Qualitative Predictions of Entropy Changes in Physical and Chemical Processes

Predict whether each of the following processes involves an increase or a decrease in entropy or whether the outcome is uncertain.

(a) The decomposition of ammonium nitrate (a fertilizer and a highly explosive compound): $2\,NH_4NO_3(s) \longrightarrow 2\,N_2(g) + 4\,H_2O(g) + O_2(g)$.

(b) The conversion of SO_2 to SO_3 (a key step in the manufacture of sulfuric acid): $2\,SO_2(g) + O_2(g) \longrightarrow 2\,SO_3(g)$.

(c) The extraction of sucrose from cane sugar juice: $C_{12}H_{22}O_{11}(aq) \longrightarrow C_{12}H_{22}O_{11}(s)$.

(d) The "water gas shift" reaction (involved in the gasification of coal): $CO(g) + H_2O(g) \longrightarrow CO_2(g) + H_2(g)$.

Analyze

Apply the generalizations summarized on the previous page. Three of the processes are chemical reactions, and for those processes, we should first consider whether the number of molecules of gas increases or decreases.

Solve

(a) Here, a solid yields a large quantity of gas. Entropy increases.

(b) Three moles of gaseous reactants produce two moles of gaseous products. The loss of one mole of gas indicates a loss of volume available to a smaller number of gas molecules. This loss reduces the number of possible configurations for the molecules in the system and the number of accessible microscopic energy levels. Entropy decreases.

(c) The sucrose molecules are reduced in mobility and in the number of forms in which their energy can be stored when they leave the solution and arrange themselves into a crystalline state. Entropy decreases.

(d) The entropies of the four gases are likely to be different because their molecular structures are different. The number of moles of gases is the same on both sides of the equation, however, so the entropy change is likely to be small if the temperature is constant. On the basis of just the generalizations listed above, we cannot determine whether entropy increases or decreases.

Assess

As we will soon see, the ability to predict an increase or a decrease in entropy will help us to understand when a process will proceed spontaneously in the forward direction.

PRACTICE EXAMPLE A: Predict whether entropy increases or decreases in each of the following reactions. **(a)** The Claus process for removing H_2S from natural gas: $2\,H_2S(g) + SO_2(g) \longrightarrow 3\,S(s) + 2\,H_2O(g)$; **(b)** the decomposition of mercury(II) oxide: $2\,HgO(s) \longrightarrow 2\,Hg(l) + O_2(g)$.

PRACTICE EXAMPLE B: Predict whether entropy increases or decreases or whether the outcome is uncertain in each of the following reactions. **(a)** $Zn(s) + Ag_2O(s) \longrightarrow ZnO(s) + 2\,Ag(s)$; **(b)** the chlor-alkali process, $2\,Cl^-(aq) + 2\,H_2O(l) \xrightarrow{\text{electrolysis}} 2\,OH^-(aq) + H_2(g) + Cl_2(g)$.

13-2 CONCEPT ASSESSMENT

Figure 13-2 illustrates a spontaneous process through the expansion of an ideal gas into an evacuated bulb. Use the one-dimensional particle-in-a-box model to represent the initial condition of Figure 13-2. Use a second particle-in-a-box model to represent the system after expansion into the vacuum. Use these models to explain on a microscopic basis why this expansion is spontaneous. [*Hint:* Assume that the volume of the bulbs is analogous to the length of the box.]

13-2 Entropy Change: Clausius's View

▶ The word *entropy* is derived from the Greek words *en* (meaning "in") and *trope* (meaning "change or transformation").

Rudolf Clausius, a German physicist, introduced the term entropy in 18 many years before Boltzmann presented his now famous equation. Claus made significant contributions to thermodynamics, including providin

:ailed analysis of an important thermodynamic cycle, called the Carnot :le, on which the so-called ideal heat engine is based. A heat engine is signed to convert heat into work, and the ideal heat engine achieves the ximum possible efficiency for this conversion. During his analysis of the rnot cycle, Clausius discovered a new thermodynamic property, or state ıction, S, that he called entropy. Although Clausius was not able to provide interpretation of what S represents, he showed not only that S exists but ɔ how it behaves.

The period between Clausius's discovery of entropy and Boltzmann's inter- tation of it was remarkable in that scientists could not give a good answer :he question "What is entropy?" but they could answer very practical ques- ns like "How does entropy change when a system expands at constant pres- e?" or "How does entropy change when the temperature of a system reases at constant volume?" By asking and answering these very practical l precise questions, scientists discovered that S is a property that behaves in ertain way for all natural changes and exactly the opposite way for all natural changes.

Clausius proved that changes in S could be related to heat transfer, pro- led heat was transferred in a "reversible" way. As discussed in Chapter 7 ge 262), a reversible process involves changing the system variables by initesimal amounts. For example, to raise the temperature of a system from o T_f in a reversible way means that the temperature is increased by infini- imal amounts dT, as suggested below, until finally the temperature T_f is ched.

Initial state — An infinite number of intermediate states — Final state

$$T_i \xrightarrow{\delta q_{rev}} T_i + dT \xrightarrow{\delta q_{rev}} \cdots \xrightarrow{\delta q_{rev}} \boxed{T \xrightarrow{\delta q_{rev}} T + dT} \xrightarrow{\delta q_{rev}} \cdots \xrightarrow{\delta q_{rev}} T_f$$

e change in state is imagined to proceed through an infinite number of ermediate states by delivering infinitesimally small amounts of heat, δq_{rev}, :h time. The subscript "rev" on δq emphasizes that we are considering a ersible process. In the reversible process illustrated above, the step for ich the temperature changes from T to $T + dT$ (red box) represents any one the infinite number of intermediate states. Clausius suggested that in each p of this reversible process, the entropy changes by a small amount, dS, en by

$$dS = \frac{\delta q_{rev}}{T} \qquad \textbf{(13.2)}$$

ere T is the temperature just before delivering the small quantity of heat, ev. The total entropy change, ΔS, for the system is obtained by adding the ues of $dS = \delta q_{rev}/T$ for each step. Mathematically, the summation of all se infinitesimal quantities is obtained by using the calculus technique of egration. See Are You Wondering 13-1.

Let's try to rationalize equation (13.2) by using ideas from Section 13-1. We that dS is directly proportional to the quantity of heat transferred. This ms reasonable because the more energy added to a system (as heat), the ater the number of energy levels available to the microscopic particles. ising the temperature also increases the availability of energy levels, but for iven quantity of heat the proportional increase in the number of energy lev- is greatest at low temperatures. It seems reasonable then that dS should be ersely proportional to the Kelvin temperature. Notice that, because dS is portional to δq_{rev} and inversely proportional to T, the unit of entropy nge is J/K or J K^{-1}.

13-1 ARE YOU WONDERING?

How is equation (13.2) used to calculate a finite entropy change, ΔS?

The infinitesimal change in entropy, dS, that accompanies an infinitesima reversible heat flow, δq_{rev}, is $dS = \delta q_{\text{rev}}/T$. Now imagine the change in a systen from state 1 to state 2 is carried out in a series of such infinitesimal reversibl steps. Summation of all these infinitesimal quantities through the calculus tech nique of integration yields ΔS.

$$\Delta S = \int \frac{\delta q_{\text{rev}}}{T}$$

If the change of state is isothermal (carried out at constant temperature), we ca write

$$\Delta S = \int \frac{\delta q_{\text{rev}}}{T} = \frac{1}{T}\int \delta q_{\text{rev}} = \frac{q_{\text{rev}}}{T}$$

Starting from appropriate expressions for δq_{rev}, ΔS can be related to other sys tem properties. For the isothermal, reversible expansion of an ideal ga $\delta q_{\text{rev}} = -\delta w_{\text{rev}}$, leading to equation (13.6), which describes ΔS in terms of gas vol umes. For a reversible change in temperature at constant pressure, $\delta q_{\text{rev}} = C_p\,dT$ which leads to the entropy change for a change in temperature (see Exercise 95).

Notice that an infinitesimal entropy change is represented by dS whereas a infinitesimal quantity of heat is represented by δq. The difference in notatio arises because the results obtained by adding dS values or by adding δq value have fundamentally different interpretations. The dS values add to give a numbe that represents a *change* in a property of the system (in this case S). The δq value add to give q, the total quantity of heat transferred. Heat is energy in transitio and not a system property, so notation such as dq or Δq is not appropriate.

The significance of equation (13.2) is that it relates S, more specificall small change in S, to quantities (heat and temperature) that are easily int preted and measured. However, to calculate the entropy change for a syst using this expression, we must devise a way to accomplish a given change a reversible way. This is not always easy, but as we see below, it can be dc fairly easily for certain types of changes.

Phase Transitions

We can use the expression $dS = \delta q_{\text{rev}}/T$ to obtain an equation for ΔS whe pure substance undergoes a phase change at constant temperature, T. Let's c sider the constant pressure vaporization of a liquid at its vaporization tempe ture, T_{vap}. The process below shows that to carry out the vaporization i reversible way, heat must be delivered in infinitesimal amounts. This is acco plished by placing the system in contact with another system having a tempe ture that is only infinitesimally greater than T_{vap}. (The vaporization process can be reversed at any time by lowering the temperature infinitesimally below T_{v}

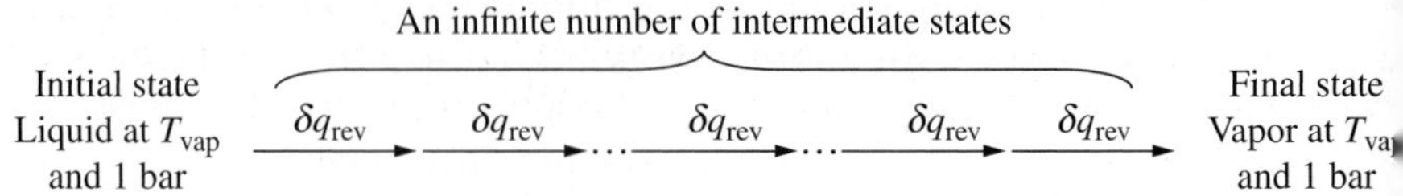

Each infinitesimal quantity of heat is used to convert a small amount of liq to an equivalent amount of vapor at a constant temperature. That is, the hea used to overcome the intermolecular forces of attraction, not to cause an incre temperature. The total entropy change for the vaporization is obtained by ding the values of $dS = \delta q_{rev}/T_{vap}$. Because the temperature is constant, we accomplish this by first adding the values of δq_{rev} and then dividing the ult by T_{vap}. The total heat transferred is $q_{vap} = \Delta_{vap}H°$ and so, we can write

$$\Delta_{vap}S = \frac{q_{vap}}{T_{vap}} = \frac{\Delta_{vap}H°}{T_{vap}}$$

e equation above can be written more generally as

$$\Delta_{tr}S = \frac{\Delta_{tr}H°}{T_{tr}} \tag{13.3}$$

ere the symbol tr represents a transition. When applying equation (13.3), are specific about which transition is involved by replacing tr with, for mple, fus for the melting of solid or vap for the vaporization of liquid. so, equation (13.3) applies only if the phase transition occurs at the usual or rmal transition temperature. The reason is that a phase transition can be ried out reversibly only at the normal transition temperature.

Consider the reversible melting (fusion) of ice at its normal melting point, which $\Delta_{fus}H° = 6.02\ \text{kJ mol}^{-1}$.

$$H_2O(s, 1\ \text{atm}, 273.15\ \text{K}) \rightleftharpoons H_2O(l, 1\ \text{atm}, 273.15\ \text{K})$$

e are justified in using $\Delta_{fus}H°$ here because 1 atm ≈ 1 bar.) The standard ropy change is

$$\Delta_{fus}S° = \frac{\Delta_{fus}H°}{T_{fus}} = \frac{6.02 \times 10^3\ \text{J mol}^{-1}}{273.15\ \text{K}} = 22.0\ \text{J mol}^{-1}\ \text{K}^{-1}$$

tropy changes depend on the quantities of substances involved and are netimes expressed on a per-mole basis.

A useful generalization known as Trouton's rule states that for many liqls at their normal boiling points, the standard molar entropy of vaporizan has a value of about 87 J mol^{-1} K^{-1}.

KEEP IN MIND

that the normal melting point and normal boiling point are determined at 1 atm pressure. The difference between 1 atm and the standard state pressure of 1 bar is so small that we can usually ignore it.

$$\Delta_{vap}S° = \frac{\Delta_{vap}H°}{T_{vap}} \approx 87\ \text{J mol}^{-1}\ \text{K}^{-1} \tag{13.4}$$

For instance, the values of $\Delta_{vap}S°$ for benzene (C_6H_6) and octane (C_8H_{18}) 87.1 and 86.2 J mol^{-1} K^{-1}, respectively. If the increased accessibility of croscopic energy levels produced in transferring one mole of molecules m liquid to vapor at 1 bar is roughly comparable for different liquids, then should expect similar values of $\Delta_{vap}S°$.

Instances in which Trouton's rule fails are also understandable. In water d in ethanol, for example, hydrogen bonding among molecules produces a ver entropy than would otherwise be expected in the liquid state. nsequently, the entropy increase in the vaporization process is greater than rmal, and so $\Delta_{vap}S° > 87\ \text{J mol}^{-1}\ \text{K}^{-1}$.

EXAMPLE 13-3 Determining the Entropy Change for a Phase Transition

What is the standard molar entropy change for the vaporization of water at 373 K given that the standard molar enthalpy of vaporization is 40.7 kJ mol^{-1} at this temperature?

Analyze

This is an example of a phase transition, which means we can make use of $\Delta_{tr}S° = \dfrac{\Delta_{tr}H°}{T_{tr}}$.

(continued)

Solve

Although a chemical equation is not necessary, writing one can help us see the process we use to find the value of $\Delta_{vap}S°$.

$$H_2O(l, 1\text{ atm}) \rightleftharpoons H_2O(g, 1\text{ atm}) \qquad \Delta_{vap}H° = 40.7\text{ kJ mol}^{-1}$$
$$\Delta_{vap}S° = ?$$

$$\Delta_{vap}S° = \frac{\Delta_{vap}H°}{T_{vap}} = \frac{40.7\text{ kJ mol}^{-1}}{373\text{ K}} = 0.109\text{ kJ mol}^{-1}\text{ K}^{-1}$$
$$= 109\text{ J mol}^{-1}\text{ K}^{-1}$$

Assess

When solving this type of problem, we should check the sign of ΔS. Here, we expect an increase in entropy (ΔS is positive) because, as discussed on page 587, the entropy of a gas is much higher than that of a liquid.

PRACTICE EXAMPLE A: What is the standard molar entropy of vaporization, $\Delta_{vap}S°$, for CCl_2F_2, a chlorofluorocarbon that once was heavily used in refrigeration systems? Its normal boiling point is −29.79 °C, and $\Delta_{vap}H° = 20.2\text{ kJ mol}^{-1}$.

PRACTICE EXAMPLE B: The entropy change for the transition from solid rhombic sulfur to solid monoclinic sulfur at 95.5 °C is $\Delta_{tr}S° = 1.09\text{ J mol}^{-1}\text{ K}^{-1}$. What is the standard molar enthalpy change, $\Delta_{tr}H°$, for this transition?

Heating or Cooling at Constant Pressure

Equation (13.2) can also be used to obtain an expression for the entro change, ΔS, for a substance that is heated (or cooled) at constant pressur without a phase change—from an initial temperature T_i to a final temperat T_f. The heating of a substance from T_i to T_f can be carried out in a reversi way by making sure that, at each stage of the heating process, the substanc in contact with a heat source having a temperature that is only infinitesima greater than the temperature, T, of the substance itself. At each stage of heating process, the substance absorbs a quantity of heat equal to δq_{rev} and temperature of the system increases from T to $T + dT$. The quantity of h absorbed is, by applying equation (7.5), $\delta q_{rev} = C_p\, dT$, where C_p is the co stant pressure heat capacity of the substance being heated. Thus, the cor sponding entropy change is $dS = (C_p/T)dT$. By adding these infinitesin entropy changes, for the infinite number of intermediate states between T_i a T_f, we obtain (using the calculus technique of integration) the result

▶ If you are familiar with the calculus technique of integration, then you will know that $\int_a^b (1/x)dx = \ln(b/a)$. Thus, $\int_{T_i}^{T_f} (C_p/T)dT = C_p \ln(T_f/T_i)$ provided that C_p has the same (constant) value between T_i and T_f.

$$\Delta S \approx C_p \ln\left(\frac{T_f}{T_i}\right) \qquad \text{constant pressure heating or cooling} \qquad (1$$

This result rests on the assumption that the heat capacity, C_p, has the sa value for all temperatures between T_i and T_f. Using equation (13.5) with $T_f >$ we obtain $\Delta S > 0$. In other words, the entropy of a substance increases wh the temperature increases. Using equation (13.5) with $T_f < T_i$, we obt $\Delta S < 0$. That is, the entropy decreases when the temperature decreases.

Changes in State for an Ideal Gas

Later in this chapter, we will discuss the thermodynamics of chemical re tions. Somewhat surprisingly, a reaction involving ideal gases is the starti point for applying thermodynamic principles to chemical reactions, and wh such a reaction is assumed to occur at constant temperature, the amounts a the partial pressures of the gases will change. Therefore, we will want to kn how, for example, the entropy of an ideal gas changes with pressure. We c establish some key ideas by using equation (13.2) to obtain an expression the entropy change, ΔS, for an ideal gas that is expanded or compress

thermally from an initial volume V_i to a final volume V_f. During the expan-
n or compression, the pressure of the gas will change from P_i to P_f.

To expand or compress a gas in a reversible way, we must ensure that the ernal pressure is different from the gas pressure by only an infinitesimal ount. (See Figure 7-12 for a method of expanding a gas in a nearly ersible fashion.) The reversible expansion of a gas can be represented by following process.

Initial state — An infinite number of intermediate states — Final state

$$\begin{pmatrix} V_i \\ P_i \end{pmatrix} \xrightarrow[P_{ext} = P_i - dP]{\delta q_{rev},\ \delta w_{rev}} \begin{pmatrix} V_i + dV \\ P_i - dP \end{pmatrix} \rightarrow \cdots \rightarrow \boxed{\begin{pmatrix} V \\ P \end{pmatrix} \xrightarrow[P_{ext} = P_i - dP]{\delta q_{rev},\ \delta w_{rev}} \begin{pmatrix} V + dV \\ P - dP \end{pmatrix}} \rightarrow \cdots \rightarrow \begin{pmatrix} V_f \\ P_f \end{pmatrix}$$

At each stage, the external pressure, P_{ext}, is adjusted so that it is only infinitesi-
lly smaller than the gas pressure, P. Because the gas pressure is slightly greater n the external pressure, the gas will expand but only by an infinitesimal ount dV. Let's focus on the step in which the volume changes from V to $V + dV$ l the pressure changes from P to $P - dP$ (red box). This step represents any one he infinite number of intermediate states. For this step, the work done is

$$\delta w_{rev} = -P_{ext}dV = -(P - dP)dV = -PdV + dPdV \approx -PdV$$

are justified in neglecting the term $dPdV$ because it is the product of two y small quantities whereas $-PdV$ is a product involving only one small antity. For the isothermal expansion of an ideal gas, the change in internal ergy is zero. (See page 584.) Thus, for this step, we must have $= 0 = \delta q_{rev} + \delta w_{rev}$, and

$$\delta q_{rev} = -\delta w_{rev} = PdV = \left(\frac{nRT}{V}\right)dV$$

ing this result in equation (13.2), we get

$$dS = \frac{\delta q_{rev}}{T} = \left(\frac{nR}{V}\right)dV$$

adding these infinitesimal entropy changes, for the infinite number of ermediate states between V_i and V_f, we obtain (using the calculus technique ntegration)

$$\Delta S = nR \ln\left(\frac{V_f}{V_i}\right) \quad \text{isothermal volume change for an ideal gas} \qquad \textbf{(13.6)}$$

ce the temperature is constant, we can write $P_f V_f = P_i V_i$ or $V_f/V_i = P_i/P_f$. erefore, we can also write

$$S = nR \ln\left(\frac{P_f}{P_i}\right) = -nR \ln\left(\frac{P_f}{P_i}\right) \quad \text{isothermal pressure change for an ideal gas} \qquad \textbf{(13.7)}$$

Using equations (13.6) and (13.7), we can establish that the entropy of an al gas increases ($\Delta S > 0$) if the volume increases ($V_f > V_i$) or if the pres-
e decreases ($P_f < P_i$). These results are not unexpected because, as we saw lier, the number of microstates and entropy increase as the space available the particles increases.

Equation (13.7) is, in fact, a special case of the following general expression calculating the entropy change of an ideal gas.

$$\Delta S = C_p \ln\left(\frac{T_f}{T_i}\right) - nR \ln\left(\frac{P_f}{P_i}\right) \qquad \textbf{(13.8)}$$

me important equations for calculating entropy changes are presented in ble 13.1, along with the conditions under which they may be used.

◀ To illustrate that we are justified in neglecting the term $dPdV$, let's use $P = 1$ bar, a typical value, and approximate the infinitesimal quantities dP and dV as 1×10^{-10} bar and 1×10^{-10} L, respectively. We have $dPdV \approx 1 \times 10^{-20}$ bar L and $-PdV \approx -1 \times 10^{-10}$ bar L. Therefore, $-PdV + dPdV$
$= -1 \times 10^{-10}$ bar L
$+ 1 \times 10^{-20}$ bar L
$\approx -1 \times 10^{-10}$ bar L
$= -PdV$.

◀ To derive equation (13.8), we must make use of the first and second laws, the relationship between internal energy and enthalpy, certain characteristics of ideal gases, and a considerable amount of calculus. The derivation of this equation is typically presented in more advanced physical chemistry courses.

TABLE 13.1 Some Equations for Calculating Enthalpy and Entrop Changes

Process	Enthalpy Change	Entropy Change
Phase transition at the usual (normal) transition temperature	$\Delta_{tr}H°$	$\Delta_{tr}S = \dfrac{\Delta_{tr}H°}{T_{tr}}$
Heating or cooling at constant pressure (no phase transition; C_p has constant value)	$\Delta H = C_p\Delta T$	$\Delta S = C_p \ln\left(\dfrac{T_f}{T_i}\right)$
Change in state from (T_i, P_i) to (T_f, P_f) for an ideal gas*	$\Delta H = C_p\Delta T$	$\Delta S = C_p \ln\left(\dfrac{T_f}{T_i}\right) - nR \ln\left(\dfrac{P_f}{P_i}\right.$

*The enthalpy H of an ideal gas depends only on temperature, T. Consequently, ΔH depends only on ΔT.

Remember that each equation for ΔS was obtained by applying equat (13.2) to the appropriate process. We will find occasion to use these expr sions for ΔS throughout this chapter.

13-2 ARE YOU WONDERING?

Is there a microscopic approach to obtaining equation (13.6)?

To do this, we use the ideas of Ludwig Boltzmann. Consider an ideal gas at an ini tial volume V_i and allow the gas to expand isothermally to a final volume V_f. B using the Boltzmann equation, we find that for the change in entropy,

$$\Delta S = S_f - S_i = k_B \ln W_f - k_B \ln W_i$$

$$\Delta S = k_B \ln \frac{W_f}{W_i}$$

where k_B is the Boltzmann constant, S_i and S_f are the initial and final entropies respectively, and W_i and W_f are the number of microstates for the initial and fina macroscopic states of the gas, respectively. We must now obtain a value for th ratio W_f/W_i. To do that, suppose that there is only a single gas molecule in a con tainer. The number of microstates available to this single molecule should b proportional to the number of positions where the molecule can be and, hence, t the volume of the container. That is also true for each molecule in a system c $n \times N_A$ particles (n is the amount in moles and N_A is Avogadro's number). Th number of microstates available to the whole system is

$$W_{total} = W_{particle\ 1} \times W_{particle\ 2} \times W_{particle\ 3} \times \cdots$$

Because the number of microstates for each particle is proportional to th volume V of the container, the number of microstates for $n \times N_A$ ideal gas mol ecules is

$$W \propto V^{nN_A}$$

Thus, the ratio of the microstates for isothermal expansion is

$$\frac{W_f}{W_i} = \left(\frac{V_f}{V_i}\right)^{nN_A}$$

We can now calculate ΔS as follows:

$$\Delta S = k_B \ln \frac{W_f}{W_i} = k_B \ln\left(\frac{V_f}{V_i}\right)^{nN_A} = nN_A k_B \ln\left(\frac{V_f}{V_i}\right) = nR \ln\left(\frac{V_f}{V_i}\right)$$

where R is the gas constant. This equation, which gives the entropy change for th isothermal expansion of an ideal gas, is equation (13.6).

EXAMPLE 13-4 Calculating the Entropy Change for Heating Ice Under Constant Pressure

Calculate the entropy change for the following constant pressure process.

$$H_2O(s, 100 \text{ g}, -10\ °C, 1 \text{ bar}) \longrightarrow H_2O(l, 100 \text{ g}, 10\ °C, 1 \text{ bar})$$

The molar heat capacities of ice and water are, respectively, $C_{p,m} = 37.12\ J\ mol^{-1}\ K^{-1}$ and $C_{p,m} = 75.3\ J\ mol^{-1}\ K^{-1}$. The enthalpy of fusion for ice is $\Delta_{fus}H° = 6.01\ kJ\ mol^{-1}$ at 0 °C.

Analyze

We must think of a way to carry out this process reversibly so that we may use the equations given in Table 13.1. Reversible, constant pressure heating is imagined to occur as the result of placing the system (100 g of H_2O) in contact with a heat source, the temperature of which is gradually increased from –10 to 10 °C and always just slightly greater than that of the H_2O. During this reversible heating process, several changes occur. First, the ice is heated reversibly from –10 °C to 0 °C and then melted reversibly at 0 °C. Finally, the water is heated reversibly from 0 °C to 10 °C. The entropy changes for these steps can be calculated by using equations from Table 13.1 and added together to give the total entropy change.

Solve

The amount of H_2O is $n = (100\ g/18.02\ g\ mol^{-1}) = 5.55$ mol. For the reversible heating of ice from –10 °C to 0 °C:

$$\Delta S_1^\circ = C_p(\text{ice}) \ln\left(\frac{T_f}{T_i}\right) = (5.55\ \text{mol})(37.12\ \text{J mol}^{-1}\ \text{K}^{-1}) \ln\left(\frac{273\ \text{K}}{263\ \text{K}}\right) = 7.69\ \text{J K}^{-1}$$

For the reversible melting of ice at 0 °C:

$$\Delta S_2^\circ = \frac{\Delta_{fus}H°}{T_{fus}} = \frac{(5.55\ \text{mol})(6010\ \text{J mol}^{-1})}{273\ \text{K}} = 122.18\ \text{J K}^{-1}$$

For the reversible heating of water from 0 °C to 10 °C:

$$\Delta S_3^\circ = C_p(\text{water}) \ln\left(\frac{T_f}{T_i}\right) = (5.55\ \text{mol})(75.3\ \text{J mol}^{-1}\ \text{K}^{-1}) \ln\left(\frac{283\ \text{K}}{273\ \text{K}}\right) = 15.03\ \text{J K}^{-1}$$

The total entropy change is the sum of the individual steps:

$$\Delta S_{total}^\circ = \Delta S_1^\circ + \Delta S_2^\circ + \Delta S_3^\circ = (7.69 + 122.18 + 15.03)\ \text{J K}^{-1} = 144.90\ \text{J K}^{-1}$$

Assess

The overall entropy change is positive, which is what we expect from an increase in temperature. More importantly, H_2O is converted from a solid—a state of lower entropy—to a liquid, a state of higher entropy. The phase change makes by far the largest contribution ($+122.18\ J\ K^{-1}$) to the total entropy change. The contributions from the heating of ice ($+7.69\ J\ K^{-1}$) and the heating of water ($+15.03\ J\ K^{-1}$) are much smaller. The entropy change is greater (almost double) for the heating of water than for the heating of ice even though the temperature increase is the same for both (+10 °C) because, as discussed in Section 7-2, liquid water has many more ways to disperse the energy entering the system than ice does.

PRACTICE EXAMPLE A: One mole of neon gas, initially at 300 K and 1.00 bar, expands adiabatically (i.e., with no heat lost to the surroundings) against a constant external pressure of 0.50 bar until the gas pressure is also 0.50 bar. The final temperature of the gas is 240 K. What is ΔS for the gas? The molar heat capacity of Ne(g) is $20.8\ J\ mol^{-1}\ K^{-1}$.

PRACTICE EXAMPLE B: One mole of neon gas, initially at 300 K and 1.00 bar, expands isothermally against a constant external pressure of 0.50 bar until the gas pressure is also 0.50 bar. What is ΔS for the gas?

13-3 Combining Boltzmann's and Clausius's Ideas: Absolute Entropies

From the equation $S = k_B \ln W$, we see that if $W = 1$, then we must have $S = 0$. Is there any reason to expect that the macroscopic state of a system might be described by a single microstate? It turns out that the answer to this question is yes. Imagine lowering the temperature of a system to smaller and sma values. As we lower the temperature, the total energy available to the partic of the system decreases. As a result, the particles are forced to occupy levels lower and lower energy until finally all the particles in the system are in lowest energy level. When all the particles are in the lowest energy level, have $W = 1$ and so, $S = 0$. This idea is embodied in the following stateme which is one way of stating the **third law of thermodynamics**.

> The entropy of a pure perfect crystal at 0 K is zero.

We justified the statement above by using Boltzmann's equation for However, before Boltzmann presented his famous equation, scientists h already deduced the third law of thermodynamics by studying isotherr processes, such as phase transitions or reactions, at very low temperatu They found that for a wide variety of isothermal processes, ΔS approache value of zero as the temperature approached 0 K. This general observat suggests that the entropies of all substances must be the same at 0 K. Wh Boltzmann presented his equation for S, it was clear that $S = 0$ is the corr value for entropy at 0 K.

As we saw in the previous section, Clausius's approach to entropy provi equations for calculating entropy changes of substances that are heated cooled, or that undergo a phase transition. It turns out that we can use th equations, together with the fact that $S = 0$ at $T = 0$ K, to assign a spec value to the entropy of any substance at 298.15 K. Consider the followi process for a hypothetical substance X, which is assumed to be a solid at (and 1 bar and a gas at 298.15 K and 1 bar.

$$\begin{array}{ccc} \text{X(s, 0 K, 1 bar)} & \longrightarrow & \text{X(g, 298.15 K, 1 bar)} \\ S_i^\circ = 0 & & S_f^\circ = ? \end{array}$$

This constant pressure process involves the following sequence of chang where T_{low} represents a very low temperature. (Recall that 0 K is unattainab

▶ The expressions for ΔS_1°, ΔS_2°, and ΔS_3° are approximations because the heat capacity of each phase changes slightly over the large temperature ranges involved. Thus, the assumption that C_p is constant is not valid. More accurate results are obtained by taking into account the temperature variation of the various heat capacities. See Exercise 95.

1. Heating the solid from T_{low} to its melting point, T_{fus}.

$$\text{X(s, } T_{low}\text{, 1 bar)} \longrightarrow \text{X(s, } T_{fus}\text{, 1 bar)} \qquad \Delta S_1^\circ \approx C_p(\text{solid}) \ln(T_{fus}/T_{low})$$

2. Melting the solid at T_{fus}.

$$\text{X(s, } T_{fus}\text{, 1 bar)} \longrightarrow \text{X(l, } T_{fus}\text{, 1 bar)} \qquad \Delta S_2^\circ = \Delta_{fus}H^\circ/T_{fus}$$

3. Heating the liquid from T_{fus} to its boiling point, T_{vap}.

$$\text{X(l, } T_{fus}\text{, 1 bar)} \longrightarrow \text{X(l, } T_{vap}\text{, 1 bar)} \qquad \Delta S_3^\circ \approx C_p(\text{liquid}) \ln(T_{vap}/T_{fus})$$

4. Vaporizing the liquid at T_{vap}.

$$\text{X(l, } T_{vap}\text{, 1 bar)} \longrightarrow \text{X(g, } T_{vap}\text{, 1 bar)} \qquad \Delta S_4^\circ = \Delta_{vap}H^\circ/T_{vap}$$

5. Heating the vapor from T_{vap} to 298.15 K.

$$\text{X(g, } T_{vap}\text{, 1 bar)} \longrightarrow \text{X(g, 298.15 K, 1 bar)} \quad \Delta S_5^\circ \approx C_p(\text{gas}) \ln(298.15\text{ K}/T_{fus})$$

The entropy changes for the phase changes can be calculated usi equation (13.3), provided the corresponding enthalpy changes and transit temperatures are known. The entropy changes for the heating of solid, liqu and gas can also be calculated provided the heat capacity of each phase known. By adding the entropy changes for these steps, we obtain the to entropy change ΔS. Remembering that $S_i^\circ = 0$ and treating $\Delta S = S_f^\circ - S_i^\circ$ a known quantity, we have $S_f^\circ = \Delta S$. In this manner, for any substance, can assign a specific value to S° at 298.15 K. Figure 13-5 illustrates the meth for CH_3Cl.

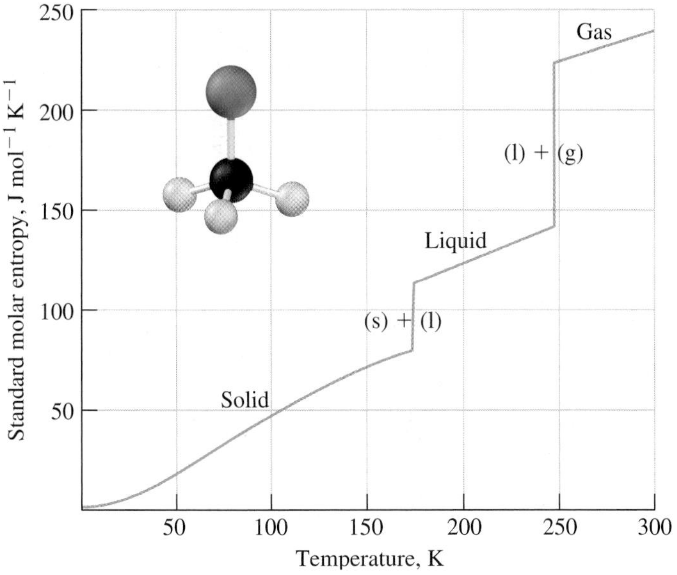

IGURE 13-5
lar entropy as a function of temperature
standard molar entropy of methyl chloride, CH_3Cl, is plotted at various temperatures n 0 to 298.15 K, with the phases noted. The vertical segment between the solid and id phases corresponds to $\Delta_{fus}S$; the other vertical segment, to $\Delta_{vap}S$. By the third of thermodynamics, an entropy of *zero* is expected at 0 K. Experimental methods not be carried to that temperature, however, so an extrapolation is required.

The absolute entropy of one mole of substance in its standard state is called **standard molar entropy**, $S°$. Standard molar entropies of a number of sub- nces at 298.15 K are tabulated in Appendix D. These values may be used to culate the **standard reaction entropy,** $\Delta_r S°$, for a reaction. Consider the fol- ving general equation for a reaction.

$$a\,\mathrm{A} + b\,\mathrm{B} + \ldots \longrightarrow c\,\mathrm{C} + d\,\mathrm{D} + \ldots$$

he equation above, the uppercase letters A, B, C, D, etc., represent different stances and the lowercase letters a, b, c, d, etc., represent the coefficients uired to balance the equation. $\Delta_r S°$ for the reaction is obtained using the owing equation, which has a familiar form (recall equation 7.22).

$$\Delta_r S° = \underbrace{[c\,S°_C + d\,S°_D + \ldots]}_{\text{Weighted sum of } S° \text{ values for products}} - \underbrace{[a\,S°_A + b\,S°_B + \ldots]}_{\text{Weighted sum of } S° \text{ values for reactants}} \qquad \textbf{(13.9)}$$

◀ As established on page 276 for $\Delta_r H°$, the equation for $\Delta_r S°$ can also be expressed in the form

$$\Delta_r S° = \sum_{\text{all substances}} \nu \times S°$$
$$= \sum_{\text{products}} \nu \times S° - \sum_{\text{reactants}} |\nu| \times S°$$

n these weighted sums, the terms are formed by multiplying the standard lar entropy by the corresponding stoichiometric coefficient. The coefficients simply numbers (without units) and so $\Delta_r S°$ has the same unit as $S°$, that is $ol^{-1}\,K^{-1}$. Example 13–5 shows how to use this equation.

EXAMPLE 13-5 Calculating the Standard Entropy of Reaction

Use data from Appendix D to calculate the standard reaction entropy at 298.15 K for the conversion of nitrogen monoxide to nitrogen dioxide (a step in the manufacture of nitric acid).

$$2\,NO(g) + O_2(g) \longrightarrow 2\,NO_2(g) \qquad \Delta_r S° = ?$$

(continued)

Analyze

The standard reaction entropy, $\Delta_r S°$, is calculated from standard molar entropies by applying equation (13.5).

Solve

Equation (13.5) takes the form

$$\Delta_r S° = 2S°_{NO_2(g)} - 2S°_{NO(g)} - S°_{O_2(g)}$$
$$= [(2 \times 240.1) - (2 \times 210.8) - 205.1]\ \text{J mol}^{-1}\,\text{K}^{-1} = -146.5\ \text{J mol}^{-1}\,\text{K}^{-1}$$

Assess

Some qualitative reasoning can be applied as a useful check on this calculation. If three moles of gaseous reactants are completely converted it two moles of gaseous products (a net decrease in the number of moles of gas), then the entropy of the system should decrease. That is, we should expect $\Delta_r S° < 0$.

PRACTICE EXAMPLE A: Use data from Appendix D to calculate the standard reaction entropy at 298.15 K for the synthesis of ammonia from its elements.

$$N_2(g) + 3\,H_2(g) \longrightarrow 2\,NH_3(g) \qquad \Delta_r S° = ?$$

PRACTICE EXAMPLE B: N_2O_3 is an unstable oxide that readily decomposes. The standard reaction entropy for the decomposition of N_2O_3 to nitrogen monoxide and nitrogen dioxide at 25 °C is $\Delta_r S° = 138.5\ \text{J mol}^{-1}\,\text{K}^{-1}$. What is the standard molar entropy of $N_2O_3(g)$ at 25 °C?

Example 13-5 uses the standard molar entropies of $NO_2(g)$ and NO(Why is the value for $NO_2(g)$, 240.1 J $\text{mol}^{-1}\,\text{K}^{-1}$, greater than that of NO(210.8 J $\text{mol}^{-1}\,\text{K}^{-1}$? The simple answer is that because the NO_2 molecule ha greater number of atoms and a somewhat more complex structure than d the NO molecule, a system of NO_2 molecules has a greater number of ways make use of a fixed amount of energy than does a system of NO molecul For example, when a gas absorbs energy (e.g., heat), some of the energy g into raising the average translational energy of the molecules. However, th are other ways for energy to be used. One possibility, shown in Figure 13-6 that the vibrational energies of molecules can be increased. In the *diato* molecule NO, only one type of vibration is possible; in the *triatomic* molec NO_2, three types are possible. Because there are more ways of distributi energy among NO_2 molecules than among NO molecules, $NO_2(g)$ ha higher molar entropy than does NO(g) at the same temperature. In gene we can say that

▶ In general, at low temperatures, because the quanta of energy are so small, translational energies are most important in establishing the entropy of gaseous molecules. As the temperature increases and the quanta of energy become larger, first rotational energies become important, and finally, at still higher temperatures, vibrational modes of motion start to contribute to the entropy.

> Standard molar entropy, $S°$, increases as molecular complexity increases (i.e., as the number of atoms per molecule increases).

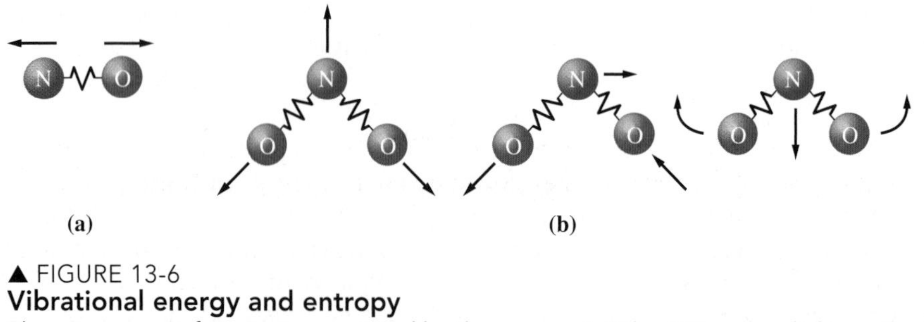

▲ FIGURE 13-6
Vibrational energy and entropy
The movement of atoms is suggested by the arrows. **(a)** The NO molecule has only on type of vibrational motion, whereas **(b)** the NO_2 molecule has three. This difference he account for the fact that the molar entropy of $NO_2(g)$ is greater than that of NO(g).

The idea that standard molar entropy increases with the number of atoms a molecule is illustrated in Figure 13-7.

13-3 ARE YOU WONDERING?

Is the standard molar entropy related to the amount of energy stored in a substance?

'he answer to this question is yes, provided we focus only on substances that are olids at 298.15 K, and we interpret the energy stored in a substance to be the mount of heat required to raise the temperature of the solid from 0 K to 298.15 K t a constant pressure of 1 bar. The following figure is a plot of standard molar ntropy, $S°$, versus $\Delta H°$, for several monatomic solids. Here, $\Delta H°$ represents the nthalpy change per mole of solid when it is heated from 0 to 298.15 K at a contant pressure equal to 1 bar.

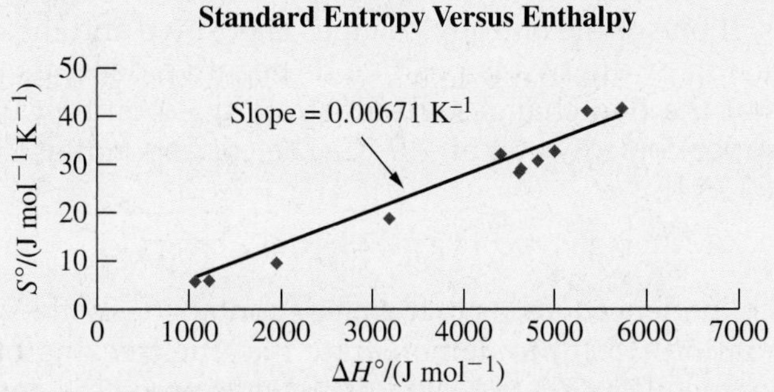

From the graph, we see that, for monatomic solids,

$$S°_{298.15} \approx \Delta H° \times 0.00671\ K^{-1}$$

The result indicates that the standard entropy value, $S°$, at 298.15 K is proporional to the heat absorbed by that solid to get it from 0 to 298.15 K with the proortionality constant 0.00671 K^{-1}. The heat that is absorbed by the solid in this rocess is dispersed through the various energy levels of the solid. The simple elationship clearly shows that the greater the amount of heat (energy) that is bsorbed when one mole of the solid is heated from 0 to 298.15 K, the greater its tandard molar entropy. Thus, the standard molar entropy of a solid is a direct neasure of the amount of energy stored in the solid.

The proportionality constant 0.00671 K^{-1} may be expressed as a temperature y taking the reciprocal of this value: $1/(0.00671\ K^{-1}) = 149$ K. This temperature s exactly half way between 0 K and 298.15 K. We explore reasons for this in Exercise 94.

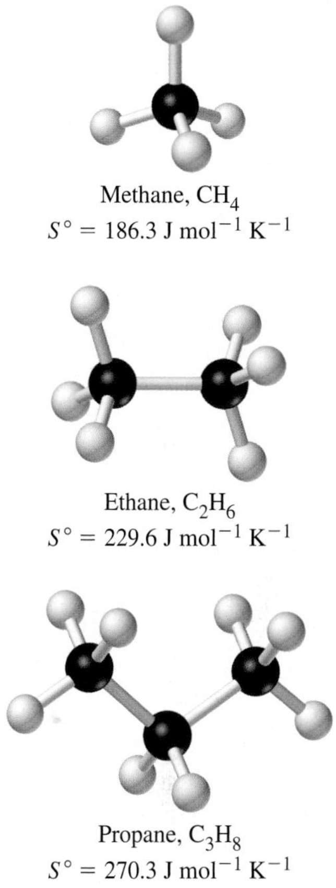

▲ FIGURE 13-7
Standard molar entropies of some hydrocarbons

3-4 Criterion for Spontaneous Change: The Second Law of Thermodynamics

Section 13-1, we came to the tentative conclusion that processes in which the ropy increases should be spontaneous and that processes in which entropy creases should be nonspontaneous. But this statement can present difficul-, for example, of how to explain the spontaneous freezing of liquid water at) °C. Because crystalline ice has lower molar entropy than does liquid water, freezing of water is a process for which entropy *decreases*. The way out of s dilemma is to recognize that *two* entropy changes must always be considd simultaneously: the entropy change of the system itself, $\Delta S_{sys} = \Delta S$, and entropy change of the surroundings, ΔS_{surr}. The criterion for spontaneous

▶ We will usually use ΔS to represent the entropy change for the system. However, in some instances, we use the symbol ΔS_{sys} for extra emphasis.

change must be based on the sum of the two, called the entropy change of universe, ΔS_{univ}. Although it is beyond the scope of this discussion to ver the following expression, or explain how scientists came to recognize its val ity, the expression provides the basic criterion for spontaneous change.

$$\Delta S_{univ} = \Delta S_{sys} + \Delta S_{surr} > 0 \qquad \textbf{(13.}$$

Equation (13.10) is one way of stating the **second law of thermodynami** Another way is through the following statement.

> All spontaneous processes produce an increase in the entropy of the universe.

According to expression (13.10), if a process produces positive entro changes in both the system and its surroundings, the process is surely spon neous. If both entropy changes are negative, the process is just as surely n spontaneous. If one of the entropy changes is positive and the other negati whether the sum of the two is positive or negative depends on the relat magnitudes of the two changes. To illustrate this point let us consider freezing of super-cooled water at –10 °C. The corresponding Kelvin tempe ture is $T = 263.15$ K.

$$H_2O(l, 263.15\ K) \longrightarrow H_2O(s, 263.15\ K)$$

Everyday experience tells us that water spontaneously freezes at 263.15 Let's use equation (13.10) to demonstrate that the freezing of super-cool water is spontaneous at 263.15 K, and the reverse process is nonspontaneou

Before we attempt to calculate the entropy changes for the system and surroundings, we must think about how to carry out the desired change i reversible way. Because the water is below its normal freezing point, we ca not carry out this process in a reversible way by keeping the temperature 263.15 K. To carry out the process in a reversible way, we must first carefu raise the temperature of the water to 273.15 K. At this temperature, wa freezes reversibly. Once all the water has frozen, we slowly lower the temp ature back to 263.15 K. The following thermodynamic cycle summari how super-cooled water can be converted to ice at 263.15 K in a reversil way. The overall process is shown in black and the reversible path represe ing the overall process is shown in blue.

$$H_2O(l, 263.15\ K) \xrightarrow{\Delta S = ?,\ \Delta H = ?} H_2O(s, 263.15\ K)$$

$$\Delta S_1, \Delta H_1 \downarrow \qquad\qquad \uparrow \Delta S_3, \Delta H_3$$

$$H_2O(l, 273.15\ K) \xrightarrow{\Delta S_2,\ \Delta H_2} H_2O(s, 273.15\ K)$$

The enthalpy and entropy changes for the three steps shown in blue can calculated using equations from Table 13.1, assuming the system conta exactly 1 mole. (The constant pressure molar heat capacities of ice and wa are 37.3 and 75.3 J mol^{-1} K^{-1}, respectively. The enthalpy of fusion of ice 6.01 kJ mol^{-1} K^{-1} at 273.15 K.)

1. Reversible heating of water from 263.15 K to 273.15 K

$$\Delta H_1 = C_p(\text{water})\Delta T$$
$$= (1\ \text{mol})(75.3\ \text{J mol}^{-1}\text{K}^{-1})(+10\ \text{K}) = +753\ \text{J}$$

$$\Delta S_1 = C_p(\text{water}) \ln (T_f/T_i)$$
$$= (1\ \text{mol})(75.3\ \text{J mol}^{-1}\text{K}^{-1}) \ln (273.15\ \text{K}/263.15\ \text{K}) = +2.81\ \text{J/K}$$

Reversible freezing of water at 273.15 K

$$\Delta H_2 = -(1\text{ mol})\Delta_{\text{fus}}H° = -(1\text{ mol})(6.01 \times 10^3\text{ J mol}^{-1}) = -6010\text{ J}$$

$$\Delta S_2 = -(1\text{ mol})\Delta_{\text{fus}}H°/T_{\text{fus}}$$
$$= -(1\text{ mol})(6.01 \times 10^3\text{ J mol}^{-1}/273.15\text{ K}) = -22.0\text{ J mol}^{-1}$$

Reversible cooling of ice from 273.15 K to 263.15 K

$$\Delta H_3 = C_p(\text{ice})\Delta T$$
$$= (1\text{ mol})(37.3\text{ J mol}^{-1}\text{ K}^{-1})(-10\text{ K}) = -373\text{ J}$$

$$\Delta S_3 = C_p(\text{ice}) \ln (T_f/T_i)$$
$$= (1\text{ mol})(37.3\text{ J mol}^{-1}\text{ K}^{-1}) \ln (263.15\text{ K}/273.15\text{ K}) = -1.39\text{ J mol}^{-1}$$

Notice that the enthalpy and entropy changes are both positive for the heat- process, but they are both negative for the freezing and cooling processes. e values of ΔS and ΔH for the overall process are obtained by adding the responding values for the individual steps.

$$H = \Delta H_1 + \Delta H_2 + \Delta H_3 = 753\text{ J} - 6010\text{ J} - 373\text{ J} = -5630\text{ J}$$

$$S = \Delta S_1 + \Delta S_2 + \Delta S_3 = 2.81\text{ J mol}^{-1} - 22.0\text{ J mol}^{-1} - 1.39\text{ J mol}^{-1} = -20.6\text{ J mol}^{-1}$$

The entropy of the system decreases because, in the solid, the molecules are re restricted in their motions; therefore, there are fewer ways to distribute total energy of the system among the molecules. So, why is the freezing of er-cooled water spontaneous? It must be because the entropy change for surroundings is greater than 20.6 J mol^{-1}. Let's calculate ΔS_{surr} to verify t this is indeed the case.

The entropy change for the surroundings is determined by the total amount heat that flows into the surroundings and the temperature of the surround- s. Let's assume the surroundings are, in general, large enough that any antity of heat released to (or absorbed from) the surroundings causes no re than an infinitesimal change in the temperature of the surroundings. erefore, we can calculate the entropy change for the surroundings as

$$\Delta S_{\text{surr}} = \frac{-q_{\text{sys}}}{T_{\text{surr}}} \tag{13.11}$$

ere q_{sys} represents the actual amount of heat that is absorbed or released by system. Following the usual sign conventions, established in Chapter 7 e page 249), $q_{\text{sys}} > 0$ when the system absorbs heat and $q_{\text{sys}} < 0$ when the tem releases heat.

For the overall process described above, we have $q_{\text{sys}} = \Delta H = -5630$ kJ, ich means that 5630 kJ of heat flows into the surroundings. The entropy nge for the surroundings is

$$\Delta S_{\text{surr}} = \frac{+5630\text{ J}}{263.15\text{ K}} = +21.4\text{ J mol}^{-1}$$

d that of the universe is

$$\Delta S_{\text{univ}} = \Delta S + \Delta S_{\text{surr}} = -20.6\text{ J mol}^{-1} + 21.4\text{ J mol}^{-1} = +0.8\text{ J mol}^{-1} > 0$$

ce $\Delta S_{\text{univ}} > 0$, the freezing of super-cooled water at 263.15 K is spontaneous.

What about the reverse process, the conversion of $H_2O(s)$ to $H_2O(l)$ at 3.15 K? For this process, we would have $\Delta H = +5630$ J, $\Delta S = +20.6$ J mol^{-1}, $_{\text{surr}} = -21.4$ J mol^{-1}, and $\Delta S_{\text{univ}} = -0.8$ J mol^{-1}. The second law requires t $\Delta S_{\text{univ}} > 0$, so we conclude that the melting of ice at 263.15 K is nonspon- eous.

The ideas above can be summarized as follows.

- If $\Delta S_{\text{univ}} > 0$, the process is *spontaneous*.
- If $\Delta S_{\text{univ}} < 0$, the process is *nonspontaneous*.
- If $\Delta S_{\text{univ}} = 0$, the process is *reversible*.

13-3 CONCEPT ASSESSMENT

Using ideas from this section, explain why highly exothermic processes tend to be spontaneous. Under what condition(s) will a highly exothermic process be nonspontaneous?

Bettmann/Corbis

▲ **J. Willard Gibbs (1839–1903)—a great "unknown" scientist**
Gibbs, a Yale University professor of mathematical physics, spent most of his career without recognition, partly because his work was abstract and partly because his important publications were in little-read journals. Yet today, Gibbs's ideas serve as the basis of most of chemical thermodynamics.

Gibbs Energy and Gibbs Energy Change

We could use expression (13.10) as the basic criterion for spontaneous chan but it would be preferable to have a criterion that could be applied to *the s tem itself*, without having to worry about changes in the surroundings. Suc criterion can be developed in a straightforward manner for processes such phase changes and chemical reactions that occur at constant temperature a constant pressure.

To develop this new criterion, let us explore a hypothetical process that occ at constant temperature T and constant pressure P, with work limited to pr sure-volume work. We will assume that the surroundings also have temperat T. Because pressure is constant, we have $\Delta H_{sys} = q_p$ and by equation (13.11), entropy change in the surroundings is $\Delta S_{surr} = -\Delta H_{sys}T$.* Now substitute t value for ΔS_{surr} into equation (13.10) and multiply by T. We obtain

$$T\Delta S_{univ} = T\Delta S_{sys} - \Delta H_{sys} = -(\Delta H_{sys} - T\Delta S_{sys})$$

Finally, multiply by –1 (change signs).

$$-T\Delta S_{univ} = \Delta H_{sys} - T\Delta S_{sys}$$

The right side of this equation has terms involving *only the system*. On the side appears the term ΔS_{univ}, which embodies the criterion for spontane change, that for a spontaneous process, $\Delta S_{univ} > 0$.

The equation above is generally cast in a somewhat different form by int ducing a new thermodynamic function, called the **Gibbs energy, *G*.** T Gibbs energy for a system is defined by the equation

$$G = H - TS \quad \textbf{(13.}$$

The **Gibbs energy change, ΔG,** for a process at constant T is

$$\Delta G = \Delta H - T\Delta S \quad \textbf{(13.}$$

In equation (13.13), all the terms refer to *changes for the system*. All reference the surroundings has been eliminated. Also, when we compare the expr sions for $-T\Delta S_{univ}$ and ΔG given above, we get

$$\Delta G = -T\Delta S_{univ}$$

Now, by noting that ΔG is *negative* when ΔS_{univ} is *positive*, we have our fi criterion for spontaneous change based on properties of only the system its

For a process occurring at constant T and P, the following statements h true.

- If $\Delta G < 0$ (*negative*), the process is *spontaneous*.
- If $\Delta G > 0$ (*positive*), the process is *nonspontaneous*.
- If $\Delta G = 0$ (*zero*), the process is *reversible* and the system has reached eq librium.

Table 13.2 provides a summary of the criteria we can use for determini whether or not a particular process is spontaneous, nonspontaneous, reversible.

*We cannot similarly substitute $\Delta H_{sys}/T$ for ΔS_{sys}. A process that occurs spontaneousl generally far removed from an equilibrium condition and is therefore *irreversible*. We can substitute δq for an irreversible process into equation (13.2).

ABLE 13.2 Criteria for Spontaneous Change

	Criteria	
’pe of Change	General Conditions	Constant T and P
pontaneous	$\Delta S_{univ} > 0$	$\Delta G < 0$
onspontaneous	$\Delta S_{univ} < 0$	$\Delta G > 0$
eversible	$\Delta S_{univ} = 0$	$\Delta G = 0$

pplying the Gibbs Energy Criteria for Spontaneous Change

e can use ideas from Table 13.2 together with equation (13.13) to make some alitative predictions. Altogether there are four possibilities for ΔG on the sis of the signs of ΔH and ΔS. These possibilities are outlined in Table 13.3 d demonstrated in Example 13-6.

If ΔH is *negative* and ΔS is *positive*, the expression $\Delta G = \Delta H - T\Delta S$ is nega- e at all temperatures. The process is spontaneous at all temperatures. This rresponds to the situation noted previously in which both ΔS_{sys} and ΔS_{surr} positive and ΔS_{univ} is also positive.

Unquestionably, if a process is accompanied by an *increase* in enthalpy (heat absorbed) and a *decrease* in entropy, ΔG is positive at all temperatures and process is nonspontaneous. This corresponds to a situation in which both $_{sys}$ and ΔS_{surr} are negative and ΔS_{univ} is also negative.

The questionable cases are those in which the entropy and enthalpy anges work in opposition—that is, with ΔH and ΔS *both* negative or *both* sitive. In these cases, whether a reaction is spontaneous or not (that is, hether ΔG is negative or positive) depends on temperature. In general, if a ction has negative values for both ΔH and ΔS, it is spontaneous at *lower* nperatures, whereas if ΔH and ΔS are both positive, the reaction is sponta- ous at *higher* temperatures.

◀ For cases 2 and 3, there is a particular temperature at which a process switches from being spontaneous to being nonspontaneous. Section 13-8 explains how to determine such a temperature.

ABLE 13.3 Applying the Criteria for Spontaneous Change: $\Delta G - \Delta H - T\Delta S$

ase	ΔH	ΔS	ΔG	Result	Example
	–	+	–	spontaneous at all temp.	$2\,N_2O(g) \longrightarrow 2\,N_2(g) + O_2(g)$
	–	–	{– / +}	spontaneous at low temp.; nonspontaneous at high temp.	$H_2O(l) \longrightarrow H_2O(s)$
	+	+	{+ / –}	nonspontaneous at low temp.; spontaneous at high temp.	$2\,NH_3(g) \longrightarrow N_2(g) + 3\,H_2(g)$
	+	–	+	nonspontaneous at all temp.	$3\,O_2(g) \longrightarrow 2\,O_3(g)$

13-4 CONCEPT ASSESSMENT

The normal boiling point of water is 100 °C. At 120 °C and 1 atm, is ΔH or $T\Delta S$ greater for the vaporization of water?

EXAMPLE 13-6 Using Enthalpy and Entropy Changes to Predict the Direction of Spontaneous Change

Under what temperature conditions would the following reactions occur spontaneously?

(a) $2\,NH_4NO_3(s) \longrightarrow 2\,N_2(g) + 4\,H_2O(g) + O_2(g) \qquad \Delta_r H° = -236.0\ kJ\ mol^{-1}$

(b) $I_2(g) \longrightarrow 2\,I(g)$

(continued)

Analyze

(a) The reaction is exothermic, and in Example 13-2(a) we concluded that $\Delta S > 0$ because large quantities of gases are produced.

(b) Because one mole of gaseous reactant produces two moles of gaseous product, we expect entropy to increase. But what is the sign of ΔH? We could calculate ΔH from enthalpy of formation data, but there is no need to. In the reaction, covalent bonds in $I_2(g)$ are broken and no new bonds are formed. Because energy is absorbed to break bonds, ΔH must be positive. With $\Delta H > 0$ and $\Delta S > 0$, case 3 in Table 13.3 applies.

Solve

(a) With $\Delta H < 0$ and $\Delta S > 0$, this reaction should be spontaneous at all temperatures (case 1 in Table 13.3). $NH_4NO_3(s)$ exists only because the decomposition occurs very slowly. (We will investigate the factors affecting the rates of chemical reactions in Chapter 20.)

(b) ΔH is larger than $T\Delta S$ at low temperatures, and the reaction is nonspontaneous. At high temperatures, the $T\Delta S$ term becomes larger than ΔH, ΔG becomes negative, and the reaction is spontaneous.

Assess

We observe that reaction spontaneity depends on a balance of enthalpy, entropy, and temperature. Table 13.3 is a good summary of the conditions in which reactions will be spontaneous or nonspontaneous.

PRACTICE EXAMPLE A: Which of the four cases in Table 13.3 would apply to each of the following reactions?

(a) $N_2(g) + 3\,H_2(g) \longrightarrow 2\,NH_3(g)$, $\quad \Delta_r H° = -92.22 \text{ kJ mol}^{-1}$

(b) $2\,C(\text{graphite}) + 2\,H_2(g) \longrightarrow C_2H_4(g)$, $\quad \Delta_r H° = 52.26 \text{ kJ mol}^{-1}$

PRACTICE EXAMPLE B: Under what temperature conditions would the following reactions occur spontaneously? **(a)** The decomposition of calcium carbonate into calcium oxide and carbon dioxide. The reaction is endothermic. **(b)** The "roasting" of zinc sulfide in oxygen to form zinc oxide and sulfur dioxide. This exothermic reaction releases 439.1 kJ for every mole of zinc sulfide that reacts.

Example 13-6(b) illustrates why there is an upper temperature limit for t stabilities of chemical compounds. No matter how positive the value of Δ for dissociation of a molecule into its atoms, the term $T\Delta S$ will eventua exceed ΔH in magnitude for sufficiently large values of T. For those tempe tures, dissociation will be spontaneous. Known temperatures range from ne absolute zero to the interior temperatures of stars (about 3×10^7 K). Molecu exist only at limited temperatures (up to about 1×10^4 K or about 0.0 of this total temperature range).

▶ A related observation is that only a small fraction of the mass of the universe is in molecular form.

Gibbs Energy Change and Work

Up to this point, we have made use of the fact that, for a process occurring constant T and constant P, the Gibbs energy change is less than or equal zero: $\Delta G \leq 0$. This result was developed on page 602 by focusing on a proce in which work is limited to pressure-volume work. How is this result chang if other types of work are possible? Before answering this question, let us fi mention some examples of non-PV work that are relevant to chemist Examples include the work associated with forcing electrons through a circ (electrical work), the work associated with raising a column of liquid, t work of stretching an elastic material (e.g., a muscle), or the work of sustai ing nerve activity. It turns out that, if other types of work are possible, then t second law guarantees that $\Delta G \leq w_{\text{non-}PV}$, where $w_{\text{non-}PV}$ represents the su total of all other forms of work. Clearly, if $w_{\text{non-}PV} = 0$, then we get back t usual condition that $\Delta G \leq 0$.

According to the sign conventions established for work in Chapter 7 (w positive when work is done on the system; w is negative when work is do by the system), we can write $w_{\text{non-}PV} = -|w_{\text{non-}PV}|$ when the system do non-PV work on the surroundings, where $|w_{\text{non-}PV}|$ is the magnitude of t ıount of non-*PV* work we can obtain from the process. Thus, we may write $\leq -|w_{\text{non-}PV}|$ and from this we can say

$$|w_{\text{non-}PV}| \leq -\Delta G$$

hat does this result imply? First, we get nonzero, non-*PV* work only if ΔG the process is negative. Second, $-\Delta G$ represents the maximum amount of ergy available to do non-*PV* work. Thus, for example, if ΔG for a process is 00 kJ, then the maximum amount of energy available for non-*PV* work is 0 kJ. Finally, if the process is carried out reversibly, then the maximum ıount of energy available for non-*PV* work is $|\Delta G|$, otherwise it is less than $G|$. Because $|\Delta G|$ determines how much energy is available to do non-*PV* ›rk, *G* was once called the Gibbs free energy or simply free energy by most emists. In this context, "free" refers to the availability of energy for doing n-*PV* work and is not meant to imply that the actual financial cost of obtain- this energy is zero. (Without a doubt, there are always nonzero financial sts associated with obtaining energy from a process.) In Chapter 19, we will how the Gibbs energy change for a system can be converted into electrical ›rk.

As discussed above, ΔG for a process reflects the amount of energy that is ailable to perform non-*PV* work. An important application is to use the ergy generated by a spontaneous process to do the work of making another nspontaneous process occur. For example, the biochemical oxidation of one ›le of glucose, $C_6H_{12}O(s)$, at 37 °C gives $\Delta G = -37$ kJ. The energy obtained ›m this process is used in living systems to drive the formation of adenosine phosphate (ATP) from adenosine diphosphate (ADP) and phosphate. ATP needed to sustain reactions involved in muscle expansion and contraction.

3-5 Gibbs Energy Change of a System of Variable Composition: $\Delta_r G°$ and $\Delta_r G$

established in the preceding section, the Gibbs energy change of a system is key quantity for predicting the spontaneity of change under the condition constant *T* and constant *P*, a condition that is frequently encountered g., with phase transitions and chemical reactions). In such situations, the bbs energy change arises not because of changes in *T* or *P*, but from changes the amounts of substances, that is, from a change in composition of the sys- n. For example, in a phase transition occurring at constant *T* and constant *P*, composition of the system changes because a certain amount of a substance converted from one phase to another. In a chemical reaction, the composition the system changes because certain amounts of reactants are converted into oducts. In this section, we introduce the concepts needed to evaluate the ange in Gibbs energy that arises from a change in composition.

andard Gibbs Energy of Reaction, $\Delta_r G°$

e **standard Gibbs energy of reaction**, $\Delta_r G°$, is defined as the Gibbs energy ange per mole of reaction for the following process.

Pure, unmixed reactants (each in its standard state)	$\longrightarrow$	Pure, unmixed products (each in its standard state)

e process involves the complete conversion of stoichiometric amounts of ıre, unmixed reactants in their standard states to stoichiometric amounts pure, unmixed products in their standard states. The Gibbs energy change this process can be obtained from tabulated thermochemical data of pure bstances in their standard states (see Appendix D). To see how this is done, nsider the following general equation representing the process above.

$$aA + bB + \ldots \longrightarrow cC + dD + \ldots$$

Following the same approach that was used in Chapter 7 to establish the res given in equation (7.22) for $\Delta_r H°$, it is possible to show that

$$\Delta_r G° = \underbrace{[c\,\Delta_f G°_C + d\,\Delta_f G°_D + \ldots]}_{\text{weighted sum of } \Delta_f G° \text{ values for the products}} - \underbrace{[a\,\Delta_f G°_A + b\,\Delta_f G°_B + \ldots]}_{\text{weighted sum of } \Delta_f G° \text{ values for the reactants}} \quad (13.1$$

where $\Delta_f G°$, the **standard Gibbs energy of formation**, is the Gibbs ener change *per mole* for a reaction in which a substance in its standard state is form from its elements in their reference forms in their standard states. As was the ca for standard enthalpies of formation, we have $\Delta_f G° = 0$ for an element in its r erence form at a pressure of 1 bar. Standard Gibbs energies of formation fo variety of substances are tabulated in Appendix D. The value of $\Delta_r G°$ calcula using equation (13.14) has the unit J mol^{-1} because $\Delta_f G°$ values have the u J mol^{-1} and the coefficients *a*, *b*, *c*, *d*, etc., are simply numbers with no units.

Some additional relationships involving $\Delta_r G°$ are similar to those present for enthalpy in Section 7-7: (1) $\Delta_r G°$ changes sign when a process is revers (2) $\Delta_r G°$ for an overall reaction can be obtained by adding together the Δ_r values for the individual steps; and (3) $\Delta_r G°$ is equal to $\Delta_r H° - T\Delta_r S°$. T last relationship is helpful in situations for which $\Delta_r H°$ and $\Delta_r S°$ values a known or more easily obtained, as illustrated in Example 13-7.

EXAMPLE 13-7 Calculating $\Delta_r G°$ for a Reaction

Determine $\Delta_r G°$ at 298.15 K for the reaction

$$2\,NO(g) + O_2(g) \longrightarrow 2\,NO_2(g) \quad (\text{at } 298.15\text{ K}) \qquad \Delta_r H° = -114.1\text{ kJ mol}^{-1}$$
$$\Delta_r S° = -146.5\text{ J K}^{-1}\text{ mol}^{-1}$$

Analyze

Because we have values of $\Delta_r H°$ and $\Delta_r S°$, the most direct method of calculating $\Delta_r G°$ is to use the expression $\Delta_r G° = \Delta_r H° - T\Delta_r S°$.

Solve

Note that the unit for the standard molar enthalpy is kJ mol^{-1}, and for the standard molar entropy it is J mol^{-1} K^{-1}. Before combining the $\Delta_r H°$ and $\Delta_r S°$ values to obtain a $\Delta_r G°$ value, the $\Delta_r H°$ and $\Delta_r S°$ values must be expressed in the same energy unit (for instance, kJ).

$$\Delta_r G° = -114.1\text{ kJ mol}^{-1} - (298.15\text{ K} \times -0.1465\text{ kJ mol}^{-1}\text{K}^{-1})$$
$$= -114.1\text{ kJ mol}^{-1} + 43.68\text{ kJ mol}^{-1}$$
$$= -70.4\text{ kJ mol}^{-1}$$

Assess

This example says that all the reactants and products are maintained at 25 °C and 1 bar pressure. Under these conditions, the Gibbs energy change is −70.4 kJ for oxidizing two moles of NO to two moles of NO_2. To do this, it is necessary to replenish the reactants so as to maintain the standard conditions.

PRACTICE EXAMPLE A: Determine $\Delta_r G°$ at 298.15 K for the reaction $4\,Fe(s) + 3\,O_2(g) \longrightarrow 2\,Fe_2O_3(s)$. $\Delta_r H° = -1648$ kJ mol^{-1} and $\Delta_r S° = -549.3$ J mol^{-1} K^{-1}.

PRACTICE EXAMPLE B: Determine $\Delta_r G°$ for the reaction in Example 13-7 by using $\Delta_f G°$ values from Appendix D. Compare the two results.

13-5 CONCEPT ASSESSMENT

For the reaction below, $\Delta_r G° = 326.4$ kJ mol^{-1}:

$$3\,O_2(g) \longrightarrow 2\,O_3(g)$$

What is the Gibbs energy change for the system when 1.75 mol $O_2(g)$ at 1 bar reacts completely to give $O_3(g)$ at 1 bar?

bbs Energy of Reaction for Nonstandard Conditions, $\Delta_r G$

Section 13-4, we established that for a spontaneous process at constant T and nstant P, the Gibbs energy of the system always decreases: $(\Delta G)_{T,P} < 0$. nsequently, to predict the direction of spontaneous chemical change, we need equation that relates a change in G to a change in composition. With such an uation, we can decide whether G will increase or decrease when the composi-n of a system changes by some amount. With that knowledge, we can state equivocally whether reaction to the left or to the right is spontaneous, for a en initial condition. In this section we present the equation, without derivation, d focus on interpreting and using it properly. We defer the derivation of this uation to Section 13-8. The motivation for this approach is quite simply stated: r most applications, knowing the meaning of the equation and how to use it is imately more helpful than knowing or remembering how it can be derived.

To place this discussion on firmer ground, let's consider a system contain- $N_2(g)$, $H_2(g)$, and $NH_3(g)$, treated as ideal gases, and let n_{N_2}, n_{H_2}, and n_{NH_3} present the initial amounts, in moles, of each gas. Suppose that the composi-n of the system changes, at constant T and constant P, because the following action advances by an infinitesimal amount, $d\xi$. The changes occurring in e system are summarized below.

$$N_2(g) + 3\,H_2(g) \underset{d\xi < 0}{\overset{d\xi > 0}{\rightleftharpoons}} 2\,NH_3(g)$$

	$N_2(g)$	$3\,H_2(g)$	$2\,NH_3(g)$	
Initial state:	n_{N_2}	n_{H_2}	n_{NH_3}	G_i
Change:	$-\,d\xi$	$-\,3\,d\xi$	$+\,2\,d\xi$	$+\,dG$
Final state:	$n_{N_2} - d\xi$	$n_{H_2} - 3\,d\xi$	$n_{NH_3} + 2\,d\xi$	$G_f = G_i + dG$

e have arrows pointing to the left and to the right because we want to con-ler the effect on G of a change in composition arising from reaction to the ht ($\longrightarrow$) or to the left ($\longleftarrow$). The sign of $d\xi$ determines the direction of action. A positive value ($d\xi > 0$) describes reaction from left to right, nereas a negative value ($d\xi < 0$) describes reaction from right to left. As sug-sted by the summary above, an infinitesimally small change in composition, presented by $d\xi$, causes an infinitesimally small change, dG, in the Gibbs ergy of the system.

To decide whether the reaction to the right or to the left is spontaneous, we ed to know how G changes with ξ, the extent of reaction, and more specifi-lly, the rate of change of G with respect to ξ. The rate of change of G with spect to the extent of reaction is represented by the symbol $\Delta_r G$ and it is lled the **Gibbs energy of reaction**. As we will demonstrate in Section 13-8, e equation for $\Delta_r G$ has the following remarkably simple form.

$$\Delta_r G = \Delta_r G° + RT \ln Q \qquad \textbf{(13.15)}$$

We have already discussed the meaning of the term $\Delta_r G°$. In the term RT Q, the quantity Q is called the **reaction quotient**. Because equation (13.15) lates the value of $\Delta_r G$ for arbitrary nonstandard conditions to its value under ndard conditions, we anticipate that the term $RT \ln Q$ includes the effects of inging the system from standard conditions (pure, unmixed substances, each $P° = 1$ bar) to the actual nonstandard conditions (a mixture of substances th a total pressure P). For the present example, the reaction quotient takes e following form, a result that will be established in Section 13-8.

$$Q = \frac{(P_{NH_3}/P°)^2}{(P_{N_2}/P°)\,(P_{H_2}/P°)^3}$$

In general, the reaction quotient depends on the composition of the sys-m, in this case through the partial pressures of the gases. So, for a specified mposition (e.g., for given values of the partial pressures), we can calculate

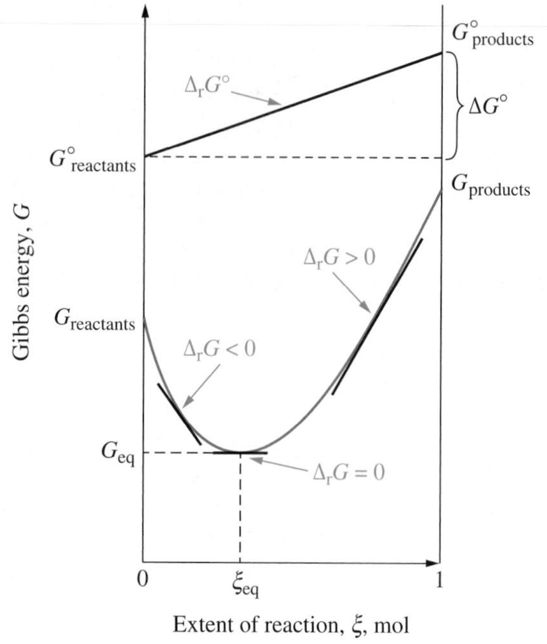

▲ FIGURE 13-8
Variation of G at constant *T* and constant *P*
At $\xi = 0$ mol, the system contains stoichiometric amounts of reactants only and at $\xi = 1$ mol, it contains stoichiometric amounts of products only. The quantities $G^\circ_{reactants}$ and $G^\circ_{products}$ are the Gibbs energies for stoichiometric amounts of pure, unmixed reactants and of pure, unmixed products in their standard states. $\Delta G^\circ = G^\circ_{products} - G^\circ_{reactants}$ is the Gibbs energy change for converting stoichiome amounts of pure, unmixed reactants in their standard states to stoichiometric amoun of pure, unmixed products in their standard states, whereas the standard Gibbs ener of reaction, $\Delta_r G^\circ = \Delta G^\circ/(1\ \text{mol})$, is the slope of the line joining the points marked by $G^\circ_{products}$ and $G^\circ_{reactants}$. The quantities $G_{reactants}$ and $G_{products}$ represent the Gibbs energies of reactants and of products after stoichiometric amounts of them are broug to a total pressure *P*. The Gibbs energy of reaction, $\Delta_r G$, is the rate of change of *G* w respect to the extent of reaction. Its value at any point is equal to the slope of the tangent to the curve at that point. At equilibrium, *G* reaches its minimum value and $\Delta_r G = 0$. The extent of reaction and the Gibbs energy at equilibrium are denoted by ξ_{eq} and G_{eq}, respectively. When the extent of reaction is less than ξ_{eq}, the slope (Δ_rC is negative and, therefore, *G* decreases as ξ increases. When the extent of reaction is greater than ξ_{eq}, the slope ($\Delta_r G$) is positive and *G* increases as ξ increases.

a value of *Q* and, by applying equation (13.15), a value of $\Delta_r G$. Figure 1: can help us understand the meaning of the value of $\Delta_r G$ and also how decide whether the reaction to the left or to the right is spontaneous for t conditions specified. Notice from Figure 13-8 that $\Delta_r G$ is the slope of the ta gent at a particular point in a graph of *G* versus the extent of the reactic With the help of Figure 13-8, we can establish the following important idea

1. **If $\Delta_r G < 0$, then *G* decreases as ξ increases.** When $\Delta_r G < 0$, the reacti proceeds spontaneously from left to right to achieve a decrease in t value of *G*. The reverse reaction, from right to left, causes the value of *G* increase and is therefore nonspontaneous.
2. **If $\Delta_r G > 0$, then *G* increases as ξ increases.** To achieve a decrease in t value of *G*, the value of ξ must decrease. This means the reaction must proce in the reverse direction, from right to left. Thus, when $\Delta_r G > 0$, reaction fr right to left is spontaneous and reaction from left to right is nonspontaneou
3. **When $\Delta_r G = 0$, the system has attained the minimum possible value of** We say that the system has reached equilibrium. When $\Delta_r G = 0$, reacti to the right (increasing ξ) or to the left (decreasing ξ) is accompanied by increase in *G*.

These results are summarized in Table 13.4.

TABLE 13.4 Predicting the Direction of Spontaneous Chemical Change

$\Delta_r G$	Spontaneous Reaction
< 0	Left to right (⟶)
> 0	Right to left (⟵)
$= 0$	Equilibrium (⇌)

An alternative explanation of Figure 13-8 is illustrated in the following dia-
m for the generalized reaction $aA(g) + bB(g) \longrightarrow cC(g) + dD(g)$.

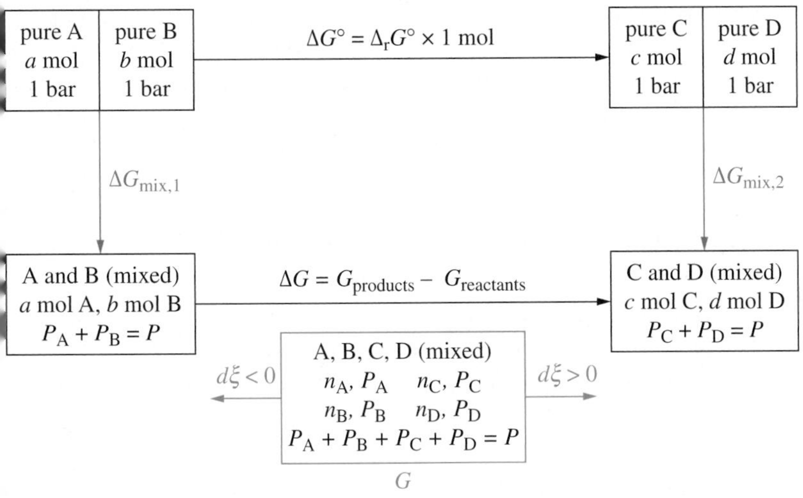

The top portion of this diagram highlights that $\Delta G°$ is the change in Gibbs ergy for the system when stoichiometric amounts of unmixed reactants in eir standard states are converted completely into stoichiometric amounts of mixed products in their standard states. Although this conversion does not rrespond to any real situation (it is a hypothetical process), we observe that $G°$ is equal to $\Delta_r G°$ times one mole of reaction. The bottom portion of the agram describes a more realistic situation, in which a mixture initially conning only A and B is converted to a mixture containing only C and D. The nversion is carried out at a constant pressure P. The Gibbs energy change for e complete conversion of a mixture of A and B to a mixture of C and D is ΔG. e diagram shows that the hypothetical and realistic processes can be related two mixing processes (red arrows), one involving the mixing of reactants and e other involving the mixing of products. The mixing processes also involve inging the mixture to a total pressure P. The Gibbs energy changes for these o mixing processes are denoted by $\Delta G_{mix,1}$ and $\Delta G_{mix,2}$. From the diagram, e can establish the result $\Delta G = \Delta G° - \Delta G_{mix,1} + \Delta G_{mix,2}$.

Now, let's turn our focus to the box in blue in the bottom portion of the diaam. This box represents the system containing arbitrary amounts of A, B, C, d D at total pressure P, somewhere between a mixture of stoichiometric nounts of reactants and a mixture of stoichiometric amounts of products. e Gibbs energy of the system at that particular point is G. Depending on the lues of n_A, n_B, n_C, and n_D, the reaction goes either to the left ($d\xi < 0$) or to the ght ($d\xi > 0$), the corresponding change in G given by $dG = \Delta_r G \times d\xi$. is the sign of $\Delta_r G$ that indicates whether the reaction to the left or right is ontaneous.

Equation (13.15) is the relationship we need to predict whether the reaction the right or to the left is spontaneous under any conditions of composition, ovided that the temperature and pressure at which we observe the reaction e constant. Because of the importance of this equation, it is imperative that e are perfectly clear about the meaning of the quantities appearing in it.

- $\Delta_r G°$ is the standard Gibbs energy of reaction in J mol^{-1}. As discussed previously, it is the Gibbs energy change per mole of reaction when stoichiometric amounts of pure unmixed reactants in their standard states are completely converted into stoichiometric amounts of pure unmixed products in their standard states. The Gibbs energy change for this process is $\Delta_r G° \times 1$ mol. The value of $\Delta_r G°$ is often calculated using thermochemical data for pure substances in their standard states (equation 13.14).

- Q represents the reaction quotient, and its value depends on the compo tion of the system. Therefore, the effect of composition is contained in t term $RT \ln Q$. We will say more about Q in the discussion that follows.
- $\Delta_r G$ represents the rate of change of Gibbs energy with respect to t extent of reaction for a system under the specified conditions of comp sition. It has the unit J mol^{-1}. Its value is easily calculated once the valu of $\Delta_r G°$ and Q have been determined. As shown in Table 13.4, it is t sign of $\Delta_r G$ that matters most. The value of $\Delta_r G$ can be used to the es mate the change in G that is caused by a given change in compositic For example, when the composition of a system changes because o reaction that advances by an infinitesimal amount $d\xi$, the correspondi change in Gibbs energy is $dG = \Delta_r G \times d\xi$.

In Example 13-8, we apply these concepts to a system containing specifi amounts of N_2(g), H_2(g), and NH_3(g). Following this example, we discuss t general form for the reaction quotient, Q, so that we can apply equation (13.1 to any reaction.

EXAMPLE 13-8 Calculating $\Delta_r G$ from Thermodynamic Data and Specified Reaction Conditions

A system contains H_2, N_2, and NH_3 gases, each with a partial pressure of 0.100 bar. The temperature is held constant at 298.15 K. Calculate the Gibbs energy of reaction, $\Delta_r G$, for the formation of NH_3(g) from N_2(g) and H_2(g) under these conditions, and predict whether formation or consumption of NH_3 is spontaneous.

Analyze

We are asked to calculate the Gibbs energy of reaction for a specified reaction and an initial condition. To calculate $\Delta_r G$, we can use equation (13.15), which means we need values for the standard Gibbs energy of reaction, $\Delta_r G°$, and the reaction quotient Q.

Solve

The first step is to write a balanced equation for the reaction.

$$N_2(g) + 3\,H_2(g) \longrightarrow 2\,NH_3(g)$$

The expression to be evaluated is equation (13.15).

$$\Delta_r G = \Delta_r G° + RT \ln Q$$

Next, determine $\Delta_r G°$ by using equation (13.14) and $\Delta_f G°$ values from Appendix D.

$$\begin{aligned}\Delta_r G° &= 2\Delta_f G°[NH_3(g)] - \Delta_f G°[N_2(g)] - 3\Delta_f G°[H_2(g)]\\ &= [2(-16.45) - 1(0) - 3(0)]\ \text{kJ mol}^{-1}\\ &= -32.90\ \text{kJ mol}^{-1}\end{aligned}$$

The value of Q is calculated from the data given in the problem. We make use of the fact that $P° = 1$ bar.

$$Q = \frac{(P_{NH_3}/P°)^2}{(P_{N_2}/P°)(P_{H_2}/P°)^3} = \frac{(0.10)^2}{(0.10)(0.10)^3} = 100$$

Now we use equation (13.15) to obtain the value of $\Delta_r G$.

$$\begin{aligned}\Delta_r G &= \Delta_r G° + RT \ln Q\\ &= -32.90\ \text{kJ mol}^{-1} + (8.314 \times 10^{-3}\ \text{kJ mol}^{-1}\ \text{K}^{-1})(298.15\ \text{K}) \ln 100\\ &= -21.5\ \text{kJ mol}^{-1}\end{aligned}$$

The value of –21.7 kJ mol^{-1} for $\Delta_r G$ represents the rate of change of Gibbs energy with respect to the extent of reaction for the system under the specified conditions of composition. The sign of $\Delta_r G$ indicates that the rate of change is negative (G decreases as the reaction proceeds in the forward direction). Therefore, for the given initial conditions, the reaction proceeds spontaneously from left to right, leading to the formation of more NH_3.

Assess

For the given initial conditions, reaction to the right (formation of more NH_3) is spontaneous. However, different initial conditions might lead to spontaneous consumption of NH_3. For example, if the initial partial pressures were $P_{NH_3} = 10$ bar, and $P_{N_2} = P_{H_2} = 1.0 \times 10^{-2}$ bar, then $Q = 1 \times 10^{10}$ and $\Delta_r G = 24$ kJ mol$^{-1} > 0$.

In this case, the forward reaction is nonspontaneous and the reverse reaction (consumption of NH_3) is spontaneous.

PRACTICE EXAMPLE A: What is the partial pressure of NH_3 in the ammonia synthesis reaction if the Gibbs energy of reaction is –82.00 kJ mol^{-1} and the partial pressures of hydrogen and nitrogen are each 0.500 bar? The temperature is 298.15 K.

PRACTICE EXAMPLE B: What is the minimum value of Q required to make the reverse reaction, conversion of NH_3 to N_2 and H_2, spontaneous at 298.15 K?

e Thermodynamic Reaction Quotient, Q

e saw earlier that the reaction quotient for the reaction

$$N_2(g) + 3\,H_2(g) \longrightarrow 2\,NH_3(g)$$

given by

$$Q = \frac{(P_{NH_3}/P^\circ)^2}{(P_{N_2}/P^\circ)(P_{H_2}/P^\circ)^3}$$

e can rewrite this expression in a slightly different but more general way by fining $a = P/P^\circ$ as the **activity** of an ideal gas. Thus, for an ideal gas, the activ- is simply the partial pressure of the gas divided by $P^\circ = 1$ bar, the standard te pressure. Although we will have more to say about activity in Section 13-8, now we need only say that ultimately, the activity of a substance depends not ly on the amount of substance but also on the form in which it appears in the stem. The following rules summarize how the activity of various substances is fined (see also Table 13.5). It is beyond the scope of this discussion to explain e reasons for defining activities in these ways, so we will simply accept these finitions and use them. However, it is important to note that the activity of a bstance is defined with respect to a specific reference state.

- **For solids and liquids:** The activity $a = 1$. The reference state is the pure solid or liquid.
- **For gases:** With ideal gas behavior assumed, the activity is replaced by the numerical value of the gas pressure in bar. The reference state is an ideal gas at 1 bar at the temperature of interest. Thus, the activity of a gas at 0.50 bar pressure is $a = (0.50\ \text{bar})/(1\ \text{bar}) = 0.50$. (Recall also that 1 bar of pressure is almost identical to 1 atm.)
- **For solutes in aqueous solution:** With ideal solution behavior assumed (for example, no interionic interactions), the activity is replaced by the numerical value of the molarity. The reference state is an ideal solution having a concentration of 1 M at the temperature of interest. Thus, the activity of the solute in a 0.25 M solution is $a = (0.25\ \text{M})/(1\ \text{M}) = 0.25$.

TABLE 13.5 Activities of Solids, Liquids, Gases, and Solutes*

Form of Substance	Activity
X(s)	$a = 1$
X(l)	$a = 1$
X(g)	$a = P_X/P^\circ$
X(aq)	$a = [X]/c^\circ$

*In these relationships, P° = 1 bar and $c^\circ = 1\ \text{mol L}^{-1}$.

terms of activities, the expression above for Q may be written as

$$Q = \frac{(a_{NH_3})^2}{(a_{N_2})(a_{H_2})^3}$$

though this expression for Q is for a very specific reaction, it reveals that) Q is a quotient formed by writing the activities of the products in the numer- or and the activities of the reactants in the denominator; and (2) the activity of ch reactant or product is raised to a power equal to the corresponding coeffi- ent from the balanced equation for the reaction. To illustrate the general proce- ire for writing the reaction quotient, consider again the following general uation for a chemical reaction.

$$a\,A + b\,B + \ldots \longrightarrow c\,C + d\,D + \ldots$$

To write the reaction quotient for a reaction, we proceed as follows.

1. **1.** Form a quotient in which the activities of products appear in the nume tor and those of reactants appear in the denominator. In both the nume tor and the denominator, the activities are combined by multiplying the (The resulting expression is not yet the reaction quotient.)

$$\frac{a_C \times a_D \times \cdots}{a_A \times a_B \times \cdots}$$

2. **2.** To obtain the reaction quotient, Q, raise each activity to a power equal the corresponding coefficient from the balanced chemical equation.

$$Q = \frac{a_C^c \times a_D^d \times \cdots}{a_A^a \times a_B^b \times \cdots} \qquad \textbf{(13.}$$

The value of the reaction quotient depends on the composition of the syste because, as described in the preceding discussion, the activities typica depend on the pressures or concentrations of the substances involved.

Relationship of $\Delta_r G°$ to the Equilibrium Constant K

We encounter an interesting situation when we apply equation (13.15) to reaction at equilibrium. We have learned that at equilibrium, $\Delta_r G = 0$ and we can write

$$0 = \Delta_r G° + RT \ln Q_{eq}$$

The subscript eq on Q emphasizes that the equation, as written, appl only if the system has reached equilibrium. If we rearrange the expressi above for $\ln Q_{eq}$, we get $\ln Q_{eq} = -\Delta_r G°/RT$. For a given reaction at a partic lar temperature, $\Delta_r G°$ has a specific value, as demonstrated by Example 13 Therefore, $\ln Q_{eq}$ and Q_{eq} also have certain fixed values once the reacti and the temperature are specified. Let's use the symbol K to represent t value of Q_{eq} and call it the **equilibrium constant**. So, by definition, the eq librium constant, K, represents the value of the reaction quotient, Q, equilibrium. By replacing Q_{eq} with K, we can write the equation above in t following form.

$$\Delta_r G° = -RT \ln K \qquad \textbf{(13.1}$$

If we have a value of $\Delta_r G°$ for a reaction at a given temperature, we can u equation (13.17) to calculate an equilibrium constant K. This means that t tabulation of thermodynamic data in Appendix D can serve as a direct sour of countless equilibrium constant values at 298.15 K. Knowing the value of for a reaction turns out to be extremely useful because, as we will see Chapter 15, the value of K and a set of initial reaction conditions are all that v need to be able to predict the equilibrium composition of a system.

We can use equation (13.17) to rewrite equation (13.15) as

$$\Delta_r G = \Delta_r G° + RT \ln Q = -RT \ln K + RT \ln Q$$

By combining the logarithmic terms, we obtain

$$\Delta_r G = RT \ln (Q/K) \qquad \textbf{(13.1}$$

This equation can be used to establish a method for predicting the direction spontaneous change. The method involves comparing the values of Q and Again, there are three specific cases to consider. They are described below ar summarized in Table 13.6.

1. **$Q < K$:** In this case, $Q/K < 1$ and $\Delta_r G = RT \ln (Q/K) < 0$. As alrea established, when $\Delta_r G < 0$, the forward reaction is spontaneous. In th situation, the reaction proceeds spontaneously in the direction that caus the value of Q to increase.

ABLE 13.6 Using Q and K to Predict the Direction of Spontaneous Chemical Change

ondition	Spontaneous Reaction
$< K$	Left to right ($\longrightarrow$)
$> K$	Right to left ($\longleftarrow$)
$= K$	Equilibrium ($\rightleftharpoons$)

. **$Q > K$:** Here we have $Q/K > 1$ and $\Delta_r G = RT \ln (Q/K) > 0$. Therefore, the reverse reaction is spontaneous. The reaction proceeds spontaneously in the direction that causes the value of Q to decrease.

. **$Q = K$:** When the reaction quotient is equal to the equilibrium constant, $\Delta_r G = RT \ln (1) = 0$ and the system has reached equilibrium.

fferent Forms of the Equilibrium Constant

hen the reaction quotient for a reaction is written in terms of activities, the rresponding equilibrium constant is called the **thermodynamic equilib-m constant**. Activities are dimensionless (unitless) quantities and there-re, the thermodynamic equilibrium constant is also a dimensionless antity. The thermodynamic equilibrium constant is appropriate for use in uation (13.15).

Consider again the general equation for a reaction that has reached uilibrium.

$$a\,A + b\,B + \ldots \rightleftharpoons c\,C + d\,D + \ldots$$

e may write the equilibrium condition $Q_{eq} = K$ for this reaction as follows.

$$K = \frac{(a_{C,eq})^c\,(a_{D,eq})^d \cdots}{(a_{A,eq})^a\,(a_{B,eq})^b \cdots} \tag{13.19}$$

KEEP IN MIND

that by writing an equation for a reaction with double arrows, each with a half arrowhead ($\rightleftharpoons$), we are emphasizing that the reaction has reached equilibrium.

ne subscript eq on the activities emphasizes that equilibrium values for these antities must be used. Although the notation used above is clear, few emists use it because the numerous subscripts and parentheses make the sulting expression appear rather cluttered. Many chemists would instead rite the expression above in the abbreviated form

$$K = \frac{a_C^c\,a_D^d}{a_A^a\,a_B^b}$$

ne expression above is much simpler, but it hides the true meaning of what is tended and can be easily misinterpreted. It seems to imply that the value of changes as the values of the activities change. However, that is not the tended meaning. To the left of the equal sign, we have K, the equilibrium nstant. To the right of the equal sign, we have the reaction quotient, Q, pressed in terms of the activities. By setting these two things equal, we ean that the reaction quotient has the value K, which is of course only true at uilibrium. Thus, when writing or using the simplified expressions above for equilibrium values of the activities are implied. This seems an obvious, and rhaps even a trivial, point but it is important to remember that, as the com-sition of the system changes, the value of Q changes, not the value of K.

Because the activities are expressed in different ways for different types of bstances, the equilibrium condition—equation (13.19)—is also expressed in fferent ways, depending on the types of substances involved. Let us explore is issue in more detail.

Reactions Involving Gases. For a reaction that involves only gases, we m write equation (13.19) as (assuming ideal behavior)

$$K = \frac{(P_{\text{C,eq}}/P°)^c\,(P_{\text{D,eq}}/P°)^d\cdots}{(P_{\text{A,eq}}/P°)^a\,(P_{\text{B,eq}}/P°)^b\cdots} = \frac{P^c_{\text{C,eq}}\,P^d_{\text{D,eq}}\cdots}{P^a_{\text{A,eq}}\,P^b_{\text{B,eq}}\cdots} \times \left(\frac{1}{P°}\right)^{\Delta\nu}$$

where $P° = 1$ bar and $\Delta\nu = (c + d + \cdots) - (a + b + \cdots)$ is the sum of co ficients for products minus the sum of coefficients for reactants. Let us defi

$$K_p = \frac{P^c_{\text{C,eq}}\,P^d_{\text{D,eq}}}{P^a_{\text{A,eq}}\,P^b_{\text{B,eq}}} \qquad (13.$$

Notice that K_p has the same form as K, equation (13.19), except that part pressures have taken the place of activities. Now, we can express K as

$$K = K_p \times (1/P°)^{\Delta\nu} \qquad (13.$$

In principle, when using equation (13.21), we can express the pressures in a pressure unit, but using a unit other than bar requires an extra calculation, th is, the evaluation of the factor $(1/P°)^{\Delta\nu}$. If we express pressures in bar, then th factor has a numerical value of 1. However, if we choose instead to expre pressures in atmospheres (atm), then $P° = 1\text{ bar} = 1/1.01325$ atm a $(1/P°)^{\Delta\nu}$ has a value of $(1/1.01325\text{ atm})^{\Delta\nu}$. This complication can be avoid entirely if we simply agree to express all pressures in bar and substitute th values *without units* into the expression for K_p. By doing so, we obtain the co rect value of K without having to worry about units.

Reactions in Aqueous Solution. For a reaction that occurs in aqueous sol tion, the activities are expressed in terms of concentrations (page 611). F example, the activity of A(aq) is expressed as $a_{\text{A}} = [\text{A}]/c°$, where $c°$ 1 mol/L. We can write similar expressions for B(aq), C(aq), D(aq), etc. For reaction in aqueous solution, equation (13.19) takes the form

$$K = K_c \times (1/c°)^{\Delta\nu} \qquad (13.$$

where, as before, $\Delta\nu = (c + d + \cdots) - (a + b + \cdots)$ and

$$K_c = \frac{[\text{C}]^c_{\text{eq}}\,[\text{D}]^d_{\text{eq}}\cdots}{[\text{A}]^a_{\text{eq}}\,[\text{B}]^b_{\text{eq}}\cdots} \qquad (13.$$

The factor $(1/c°)^{\Delta\nu}$ appearing in equation (13.22) has the unit $(\text{mol/L})^{-\Delta\nu}$. discussed above for reactions involving gases, the issue of units can avoided and the use of equation (13.23) simplified if we choose judiciously express concentrations in mol/L and substitute their values *without units* in the expression for K_c.

Reactions in a Heterogeneous System. In the previous cases, the substanc involved in the reaction were either all gases or all dissolved in aqueous sol tion. Those systems are *homogeneous* because all the substances are in the san phase and constitute a homogeneous mixture. As established in the previo discussion, for a reaction in a homogeneous system, the thermodynamic equ librium constant, K, may expressed in terms of either K_p or K_c. This is not ne essarily the case for a heterogeneous system, where the substances exist different phases. To make this point clear, let us consider the following rea tion which involves substances in a variety of different forms.

$$2\,\text{Al(s)} + 6\,\text{H}^+\text{(aq)} \rightleftharpoons 2\,\text{Al}^{3+}\text{(aq)} + 3\,\text{H}_2\text{(g)}$$

or this reaction, the equilibrium condition is

$$K = \frac{(a_{Al^{3+},eq})^2 (a_{H_2,eq})^3}{(a_{Al,eq})^2 (a_{H^+,eq})^6} \approx \frac{([Al^{3+}]_{eq}/c°)^2 (P_{H_2,eq}/P°)^3}{(1)^2 ([H^+]_{eq}/c°)^6}$$

$$= \frac{[Al^{3+}]^2_{eq} P^3_{H_2,eq}}{[H^+]^6_{eq}} \times (c°)^4 \times \left(\frac{1}{P°}\right)^3$$

hat is, for reactions in heterogeneous systems, the thermodynamic equilibrium constant might involve both pressures and concentrations. In such cases, is not appropriate to associate K with either K_p or K_c.

EXAMPLE 13-9 Writing Thermodynamic Equilibrium Constant Expressions

For the following reversible reactions, write thermodynamic equilibrium constant expressions, making appropriate substitutions for activities. Then relate K to K_c or K_p, where this can be done.

(a) The water gas reaction

$$C(s) + H_2O(g) \rightleftharpoons CO(g) + H_2(g)$$

(b) Formation of a saturated aqueous solution of lead(II) iodide, a very slightly soluble solute

$$PbI_2(s) \rightleftharpoons Pb^{2+}(aq) + 2\,I^-(aq)$$

(c) Oxidation of sulfide ion by oxygen gas (used in removing sulfides from wastewater, as in pulp and paper mills)

$$O_2(g) + 2\,S^{2-}(aq) + 2\,H_2O(l) \rightleftharpoons 4\,OH^-(aq) + 2\,S(s)$$

Analyze

In each case, once we have made the appropriate substitutions for activities, if all factors in our expression are molarities, the thermodynamic equilibrium constant is easily related to K_c. If all factors are partial pressures, K is easily related to K_p. If both molarities *and* partial pressures appear in the expression, however, K is not related simply to K_c or K_p.

Solve

(a) The activity of solid carbon is 1. Activities of the gases are expressed in terms of partial pressures.

$$K = \frac{a_{CO(g)} a_{H_2(g)}}{a_{C(s)} a_{H_2O(g)}} = \frac{(P_{CO}/P°)(P_{H_2}/P°)}{(P_{H_2O}/P°)} = \frac{(P_{CO})(P_{H_2})}{(P_{H_2O})} \times \frac{1}{P°} = K_p \times \frac{1}{P°}$$

(b) The activity of solid lead(II) iodide is 1. Activities of the aqueous ions are expressed in terms of molarities.

$$K = \frac{a_{Pb^{2+}(aq)} \left(a_{I^-(aq)}\right)^2}{a_{PbI_2(s)}} = \left(\frac{[Pb^{2+}]}{c°}\right)\left(\frac{[I^-]}{c°}\right)^2 = [Pb^{2+}][I^-]^2 \times \left(\frac{1}{c°}\right)^3 = K_c \times \left(\frac{1}{c°}\right)^3$$

(c) The activity of both the solid sulfur and the liquid water is 1. The activities of $OH^-(aq)$ and $S^{2-}(aq)$ are expressed in terms of molarities and the activity of $O_2(g)$ is expressed in terms of partial pressure. Thus, the resulting K is not related to K_c or K_p.

$$K = \frac{\left(a_{OH^-(aq)}\right)^4 \left(a_{S(s)}\right)^2}{a_{O_2(g)} \left(a_{S^{2-}(aq)}\right)^2 \left(a_{H_2O(l)}\right)^2} = \frac{([OH^-]/c°)^4 (1)^2}{(P_{O_2}/P°)([S^{2-}]/c°)^2 (1)^2} = \frac{[OH^-]^4}{P_{O_2}[S^{2-}]^2} \times \left(\frac{1}{c°}\right)^2 \times P°$$

Assess

For each of the expressions given above, equilibrium values of activities, partial pressures, and molarities are implied. Also, these are thermodynamic equilibrium expressions since they are written in terms of their activities. The values of the thermodynamic equilibrium constant will be dimensionless. All the expressions given above for K include factors involving powers of $c°$ or $P°$. As described in the text (page 611), we can avoid the complication of having to evaluate these factors by judiciously choosing to express concentrations in mol/L and pressures in bar.

PRACTICE EXAMPLE A: Write thermodynamic equilibrium constant expressions for each of the following reactions. Relate these to K_c or K_p where appropriate.

(a) $Si(s) + 2\,Cl_2(g) \rightleftharpoons SiCl_4(g)$

(b) $Cl_2(g) + H_2O(l) \rightleftharpoons HOCl(aq) + H^+(aq) + Cl^-(aq)$

(continued)

PRACTICE EXAMPLE B: Write a thermodynamic equilibrium constant expression to represent the reaction of solid lead(II) sulfide with aqueous nitric acid to produce solid sulfur, a solution of lead(II) nitrate, and nitrogen monoxide gas. Base the expression on the balanced net ionic equation for the reaction.

We have now acquired all the tools with which to perform one of the m practical calculations of chemical thermodynamics: *determining the equilibri constant for a reaction from tabulated data.* Example 13-10, which demonstra this application, uses thermodynamic properties of ions in aqueous soluti as well as of compounds. An important idea to note about the thermodynar properties of ions is that they are relative to $H^+(aq)$, which, by convention assigned values of *zero* for $\Delta_f H°$, $\Delta_f G°$, and $S°$. This means that entropies lis for ions are not absolute entropies, as they are for compounds. Negative v ues of $S°$ simply denote an entropy less than that of $H^+(aq)$.

EXAMPLE 13-10 Calculating the Equilibrium Constant of a Reaction from the Standard Gibbs Energy of Reaction

Determine the equilibrium constants K and K_c at 298.15 K for the dissolution of magnesium hydroxide in an acidic solution.

$$Mg(OH)_2(s) + 2\,H^+(aq) \rightleftharpoons Mg^{2+}(aq) + 2\,H_2O(l)$$

Analyze

The key to solving this problem is to find a value of $\Delta_r G°$ and then to use the expression $\Delta_r G° = -RT \ln K$.

Solve

We can obtain $\Delta_r G°$ from standard Gibbs energies of formation listed in Appendix D. Note that because its value is zero, the term $\Delta_f G°[H^+(aq)]$ is not included.

$$\Delta_r G° = 2\Delta_r G°[H_2O(l)] + \Delta_r G°[Mg^{2+}(aq)] - \Delta_r G°[Mg(OH)_2(s)]$$
$$= 2(-237.1\ \text{kJ mol}^{-1}) + (-454.8\ \text{kJ mol}^{-1}) - (-833.5\ \text{kJ mol}^{-1})$$

$$\Delta_r G° = -RT \ln K = -95.5\ \text{kJ mol}^{-1} = -95.5 \times 10^3\ \text{J mol}^{-1}$$

Now solve for ln K and K.

$$\ln K = \frac{-\Delta_r G°}{RT} = \frac{-(-95.5 \times 10^3\ \text{J mol}^{-1})}{8.3145\ \text{J mol}^{-1}\ \text{K}^{-1} \times 298.15\ \text{K}} = 38.5$$

$$K = e^{38.5} = 5 \times 10^{16}$$

The value of K obtained above is the thermodynamic equilibrium constant. Because the activities of both $Mg(OH)_2(s)$ and $H_2O(l)$ are 1, the expression for K may be written as

$$K = \frac{a_{Mg^{2+}(aq)}\left(a_{H_2O(l)}\right)^2}{a_{Mg(OH)_2(s)}\left(a_{H^+(aq)}\right)^2} = \frac{([Mg^{2+}]/c°)}{([H^+]/c°)^2} = \frac{[Mg^{2+}]}{[H^+]^2} \times c° = K_c \times c°$$

Therefore,

$$K_c = K \times \left(\frac{1}{c°}\right) = 5 \times 10^{16} \times \left(\frac{1}{1\,M}\right) = 5 \times 10^{16}\ M^{-1}$$

Assess

Notice that the thermodynamic equilibrium constant is dimensionless (no units) and that although K_c has the same value as K, it has the unit M^{-1}. Finally, because the activities of both $Mg(OH)_2(s)$ and $H_2O(l)$ are 1 and the activities of $Mg^{2+}(aq)$ and $H^+(aq)$ can be expressed in terms of molarities, we could have obtained the relationship between K and K_c directly by applying equation (13.22):

$$K = K_c \times \left(\frac{1}{c°}\right)^{\Delta\nu} = K_c \times \left(\frac{1}{c°}\right)^{1-2} = K_c \times \left(\frac{1}{c°}\right)^{-1} = K_c \times c°$$

PRACTICE EXAMPLE A: Determine the thermodynamic equilibrium constant at 298.15 K for $AgI(s) \rightleftharpoons Ag^+(aq) + I^-(aq)$.

PRACTICE EXAMPLE B: What is the value of the thermodynamic equilibrium constant at 298.15 K for the reaction of solid manganese dioxide with HCl(aq) to give manganese(II) ion in solution and chlorine gas?

TABLE 13.7 Significance of the Magnitude of $\Delta_r G°$ and K (at 298 K)

$\Delta_r G°$, kJ mol^{-1}	K	Significance
+200	9.1×10^{-36}	Equilibrium favors reactants
+100	3.0×10^{-18}	
+50	1.7×10^{-9}	
+10	1.8×10^{-2}	
+1.0	6.7×10^{-1}	Equilibrium calculation is necessary
0	1.0	
−1.0	1.5	
−10	5.6×10^{1}	
−50	5.8×10^{8}	Equilibrium favors products
−100	3.3×10^{17}	
−200	1.1×10^{35}	

terpreting the Values of $\Delta_r G°$ and K

solving equation (13.17) for K, we see that the value of K is determined by e value of $\Delta_r G°$.

$$K = e^{-\Delta_r G°/RT}$$

ne expression above can be used to calculate values of K at 298 K for different lues of $\Delta_r G°$. Several values are presented in Table 13.7. Notice that positive lues of $\Delta_r G°$ give values of K that are less than one, and the more positive e $\Delta_r G°$ value, the smaller the value of K. Negative values of $\Delta_r G°$ give values K that are greater than one, and the more negative the $\Delta_r G°$ value, the eater the value of K.

If we think of $\Delta_r G°$ as providing a measure of the thermodynamic stability products relative to reactants, then $\Delta_r G° < 0$ means that products in their andard states are thermodynamically more stable than reactants in eir standard states. Conversely, $\Delta_r G° > 0$ means that products in their stan-rd states are thermodynamically less stable than reactants in their standard ates. Because the values of $\Delta_r G°$ and K are related, K may also be considered measure of the thermodynamic stability of products relative to reactants. A rge value of K implies that products are thermodynamically more stable an reactants and, thus, equilibrium favors products. A small value of K nplies that products are thermodynamically less stable than reactants and uilibrium favors reactants.

Let us explore in more detail how the Gibbs energy of a system changes ith composition, with the intention of developing an understanding of how e position of equilibrium correlates with the magnitude of the equilibrium nstant K. The graphs in Figure 13-9 are plots of G versus the extent of reac-on, ξ, for three different situations: an intermediate value of K (Fig. 13-9a), a ery small value of K (Fig. 13-9b), and a very large value of K (Fig. 13-9c). To terpret the graphs, it will be helpful to focus on the meaning of the magni-de of K. Without loss of generality, we can focus on the following gas-phase action

$$a\,\text{A(g)} + b\,\text{B(g)} + \ldots \longrightarrow c\,\text{C(g)} + d\,\text{D(g)} + \ldots$$

r which

$$K = \frac{P^c_{\text{C,eq}}\,P^d_{\text{D,eq}}\cdots}{P^a_{\text{A,eq}}\,P^b_{\text{B,eq}}\cdots} \times \left(\frac{1}{P°}\right)^{\Delta\nu}$$

small value of K suggests that the values of $P_{\text{C,eq}}$ and $P_{\text{D,eq}}$ are small com-ared with those of $P_{\text{A,eq}}$ and $P_{\text{B,eq}}$. If the values of $P_{\text{C,eq}}$ and $P_{\text{D,eq}}$ are small,

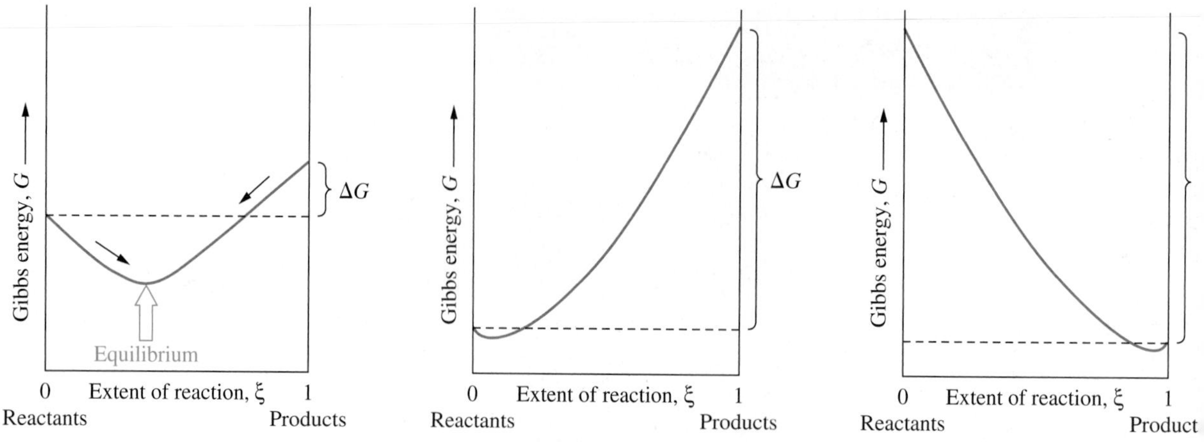

▲ FIGURE 13-9
The position of equilibrium for small, large, and intermediate values of K
Gibbs energy is plotted against the extent of reaction for a reaction $a\,A + b\,B \longrightarrow c\,C + d\,D$. For $\xi = 0$ mol, the system contains stoichiometric amounts of reactants (a mol A and b mol B) at a total pressure P. For $\xi = 1$ mol, the system contains stoichiometric amounts of products (c mol C and d mol D) at a total pressure P. As described in the text, $\Delta G = -1 \text{ mol} \times RT \ln K + \text{constant}$. **(a)** When $K \approx 1$, the value of ΔG is small and the equilibrium position corresponds to having appreciable amounts of both reactants and products. **(b)** When $K \ll 1$, ΔG has a large, positive value and the equilibrium position corresponds to an equilibrium position closer to reactants (a small value of ξ). **(c)** When $K \gg 1$, ΔG has a large negative value and the equilibrium position corresponds to an equilibrium position closer to products (a larger value of ξ).

it is because the conversion of A and B to C and D has not occurred to appreciable extent. Thus, a small value of K corresponds to an equilibriu position that is closer to reactants than to products, as suggested Figure 13-9(b). Conversely, a large value of K corresponds to an equilibriu position that is closer to products (Fig. 13-9c). Between these two extrem the equilibrium position corresponds to a situation in which apprecia amounts of both reactants and products are present (Fig. 13-9a). Figure 13 also illustrates that, from a theoretical standpoint, no chemical reaction go totally to completion.

Figure 13-9 can also help us understand the relationship between ΔG ar the value of K. When K is not too large or too small, ΔG is also relatively sm (Fig. 13-9a). However, if K is very small, complete conversion of reactants products would be accompanied by a very large increase in Gibbs ener (Fig. 13-9b). Finally, when K is very large, complete conversion of reactants products would give a very large decrease in Gibbs energy.

Equation (13.17) shows that the value of K is related in a simple way to t value of $\Delta_r G°$. But what is the relationship between ΔG and K? To obtain answer to this question, we refer to the diagram on page 609, which helped establish the result

$$\Delta G = \Delta_r G° \times 1 \text{ mol} - \Delta G_{\text{mix},1} + \Delta G_{\text{mix},2}$$

In the expression above, $\Delta G_{\text{mix},1}$ represents the Gibbs energy changes for co verting pure reactants, each at 1 bar, to a mixture of reactants at pressure Similarly, $\Delta G_{\text{mix},2}$ represents the Gibbs energy changes for converting pu products, each at 1 bar, to a mixture of products at pressure P. From equatic (13.17), we have $\Delta_r G° = -RT \ln K$ and thus, we may write the expressic above as

$$\Delta G = -RT \ln K - \Delta G_{\text{mix},1} + \Delta G_{\text{mix},2} = -RT \ln K + C$$

ıere $C = -\Delta G_{mix,1} + \Delta G_{mix,2}$ represents a constant, the value of which cludes the effects of bringing pure reactants each at a pressure of 1 bar to a xture of reactants with total pressure P, and of bringing pure products each a pressure of 1 bar to a mixture of products with total pressure P. The pression above for ΔG suggests that (1) the larger the value of K, the more gative the value of ΔG and thus, the greater the decrease in G; and (2) the ıaller the value of K, the more positive the value of ΔG and the greater the crease in G.

summary,

- A large K value ($K \gg 1$) indicates that products in their standard states are thermodynamically more stable than reactants in their standard states and the equilibrium position is closer to products. Complete conversion of reactants to products would be accompanied by a large decrease in G.
- A small K value ($K \ll 1$) indicates that products in their standard states are thermodynamically less stable than reactants in their standard states and the equilibrium position is closer to reactants. Complete conversion of reactants to products would be accompanied by a large increase in G.
- When $K \approx 1$, the equilibrium position corresponds to having appreciable amounts of both reactants and products. Complete conversion of reactants to products would be accompanied by a relatively small change in G.

3-6 $\Delta_r G°$ and K as Functions of Temperature

this section, we will describe how to use the value of $\Delta_r G°$ or K at one temerature to obtain their values at another temperature. In the method illusated in Example 13-11, we assume that $\Delta_r H°$ and $\Delta_r S°$ are independent of mperature. Even with this assumption, $\Delta_r G° = \Delta_r H° - T\Delta_r S°$ is strongly mperature-dependent because the temperature factor, T, multiplies $\Delta_r S°$.

◀ The assumption that $\Delta_r H°$ and $\Delta_r S°$ are independent of temperature is valid provided that the heat capacity of the reacting system does not change substantially when stoichiometric amounts of pure reactants are converted completely into stoichiometric amounts of pure products.

EXAMPLE 13-11 Determining the Relationship Between an Equilibrium Constant and Temperature by Using Equations for Gibbs Energy of Reaction

At what temperature will the equilibrium constant for the formation of NOCl(g) be $K = 1.00 \times 10^3$? Data for this reaction at 25 °C are

$$2\,NO(g) + Cl_2(g) \rightleftharpoons 2\,NOCl(g) \quad \Delta_r G° = -40.9\ \text{kJ mol}^{-1} \quad \Delta_r H = -77.1\ \text{kJ mol}^{-1} \quad \Delta_r S° = -121.3\ \text{J mol}^{-1}\ \text{K}^{-1}$$

Analyze

To determine an unknown temperature from a known equilibrium constant, we need an equation in which both of these terms appear. The required equation is $\Delta_r G° = -RT \ln K$. However, to solve for the unknown temperature, we need the value of $\Delta_r G°$ at that temperature. We know the value of $\Delta_r G°$ at 25 °C (-40.9 kJ mol^{-1}), but we also know that this value will be different at other temperatures. We can assume, however, that the values of $\Delta_r H°$ and $\Delta_r S°$ will not change much with temperature. This means that we can obtain a value of $\Delta_r G°$ from the equation $\Delta_r G° = \Delta_r H° - T\Delta_r S°$, where T is the *unknown* temperature and the values of $\Delta_r H°$ and $\Delta_r S°$ are those at 25 °C. Now we have two equations that we can set equal to each other.

Solve

That is,

$$\Delta_r G° = \Delta_r H° - T\Delta_r S° = -RT \ln K$$

(continued)

We can gather the terms with T on the right,

$$\Delta_r H° = T\Delta_r S° - RT \ln K = T(\Delta_r S° - R \ln K)$$

and solve for T.

$$T = \frac{\Delta_r H°}{\Delta_r S° - R \ln K}$$

Now substitute values for $\Delta_r H°$, $\Delta_r S°$, R, and $\ln K$.

$$T = \frac{-77.1 \times 10^3 \text{ J mol}^{-1}}{-121.3 \text{ J mol}^{-1} \text{ K}^{-1} - [8.3145 \text{ J mol}^{-1} \text{ K}^{-1} \times \ln (1.00 \times 10^3)]}$$

$$= \frac{-77.1 \times 10^3 \text{ J mol}^{-1}}{-121.3 \text{ J mol}^{-1} \text{ K}^{-1} - (8.3145 \times 6.908) \text{ J mol}^{-1} \text{ K}^{-1}}$$

$$= \frac{-77.1 \times 10^3 \text{ J mol}^{-1}}{-178.7 \text{ J mol}^{-1} \text{ K}^{-1}} = 431 \text{ K}$$

Assess

Although the answer shows three significant figures, the final result should probably be rounded to just two significant figures. The assumption we made about the constancy of $\Delta_r H°$ and $\Delta_r S°$ is probably no more valid than that.

PRACTICE EXAMPLE A: At what temperature will the formation of $NO_2(g)$ from $NO(g)$ and $O_2(g)$ have $K_p = 1.50 \times 10^2$? For the reaction $2\,NO(g) + O_2(g) \rightleftharpoons 2\,NO_2(g)$ at 25 °C, $\Delta_r H° = -114.1 \text{ kJ mol}^{-1}$ and $\Delta_r S° = -146.5 \text{ J mol}^{-1}\text{K}^{-1}$.

PRACTICE EXAMPLE B: For the reaction $2\,NO(g) + Cl_2(g) \rightleftharpoons 2\,NOCl(g)$, what is the value of K at **(a)** 25 °C? **(b)** 75 °C? Use data from Example 13-11.

An alternative to the method outlined in Example 13-11 is to relate the equilibrium constant and temperature directly, without specific reference to a Gibbs energy change. We start with the same two expressions as in Example 13-11

$$-RT \ln K = \Delta_r G° = \Delta_r H° - T\Delta_r S°$$

and divide by $-RT$

$$\ln K = \frac{-\Delta_r H°}{RT} + \frac{\Delta_r S°}{R} \quad (13.24)$$

If we assume that $\Delta_r H°$ and $\Delta_r S°$ are constant, equation (13.24) implies that a plot of $\ln K$ versus $1/T$ is a straight line with a slope of $-\Delta_r H°/R$ and a y-intercept of $\Delta_r S°/R$. Table 13.8 lists equilibrium constants as a function of th

TABLE 13.8 Equilibrium Constants, K, for the Reaction $2\,SO_2(g) + O_2(g) \rightleftharpoons 2\,SO_3(g)$ at Several Temperatures

T, K	$1/T$, K^{-1}	K	ln K
800	12.5×10^{-4}	9.1×10^2	6.81
850	11.8×10^{-4}	1.7×10^2	5.14
900	11.1×10^{-4}	4.2×10^1	3.74
950	10.5×10^{-4}	1.0×10^1	2.30
1000	10.0×10^{-4}	3.2×10^0	1.16
1050	9.52×10^{-4}	1.0×10^0	0.00
1100	9.09×10^{-4}	3.9×10^{-1}	−0.94
1170	8.5×10^{-4}	1.2×10^{-1}	−2.12

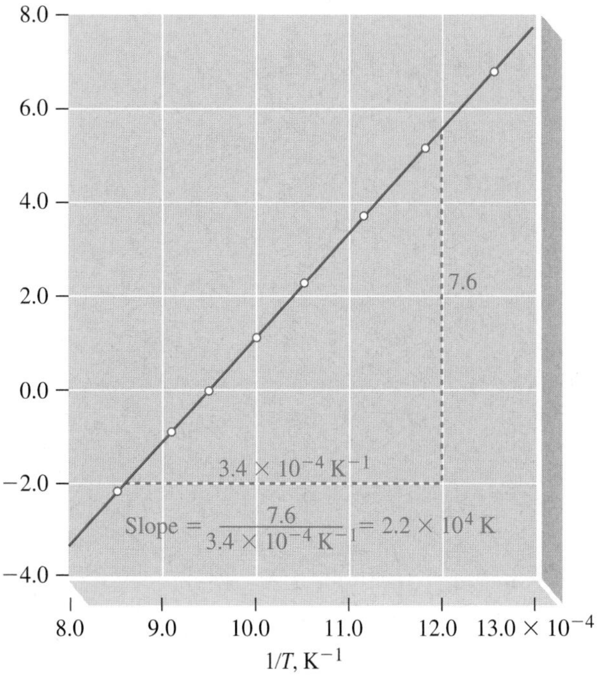

◀ FIGURE 13-10
Temperature dependence of the equilibrium constant K for the reaction

$$2\,SO_2(g) + O_2(g) \rightleftharpoons 2\,SO_3(g)$$

This graph can be used to establish the enthalpy of reaction, $\Delta_r H°$ (see equation 13.25).

$$\text{slope} = -\Delta_r H°/R = 2.2 \times 10^4\ K$$
$$\Delta_r H° = -8.3145\ J\ mol^{-1}\ K^{-1} \times 2.2 \times 10^4\ K$$
$$= -1.8 \times 10^5\ J\ mol^{-1}$$
$$= -1.8 \times 10^2\ kJ\ mol^{-1}$$

ciprocal of Kelvin temperature for the reaction of $SO_2(g)$ and $O_2(g)$ that rms $SO_3(g)$. The ln K and $1/T$ data from Table 13.8 are plotted in Figure 13-10 nd yield the expected straight line.

Now we can follow the procedure used in Appendix A-4 to derive the lausius–Clapeyron equation. We can write equation (13.24) twice, for two ifferent temperatures and with the corresponding equilibrium constants. hen, if we subtract one equation from the other, we obtain the result shown ere,

$$\ln\frac{K_2}{K_1} = -\frac{\Delta_r H°}{R}\left(\frac{1}{T_2} - \frac{1}{T_1}\right) \qquad \textbf{(13.25)}$$

here T_2 and T_1 are two Kelvin temperatures; K_2 and K_1 are the equilibrium nstants at those temperatures; $\Delta_r H°$ is the enthalpy of reaction, expressed in mol^{-1}; and R is the gas constant, expressed as 8.3145 J mol^{-1} K^{-1}. Jacobus n't Hoff (1852–1911) derived equation (13.25), which is often referred to as *e van't Hoff equation.*

◀ For endothermic reactions, K increases as T increases whereas for exothermic reactions, K decreases as T increases. You can verify this through equation (13.25) by setting $T_1 = 100$ K, $T_2 = 1000$ K, and $K_1 = 1$ and solving for K_2, first with $\Delta_r H° = +1.0 \times 10^5$ J mol^{-1} and then with $\Delta_r H° = -1.0 \times 10^5$ J mol^{-1}.

KEEP IN MIND

that the Clausius–Clapeyron equation (12.2) is just a special case of equation (13.25) in which the equilibrium constants are equilibrium vapor pressures and $\Delta_r H° = \Delta_{vap} H°$.

EXAMPLE 13-12 Relating Equilibrium Constants and Temperature Through the van't Hoff Equation

Use data from Table 13.8 and Figure 13-10 to estimate the temperature at which $K = 1.0 \times 10^6$ for the reaction

$$2\,SO_2(g) + O_2(g) \rightleftharpoons 2\,SO_3(g)$$

Analyze

By consulting Table 13.8, we see that $K = 9.1 \times 10^2$ at 800 K for this reaction and the value of K increases as T decreases. Thus, the value of K will be equal to 1.0×10^6 at a temperature lower than 800 K. To find this temperature, we can use equation (13.25) with $T_1 = ?$ K, $K_1 = 1.0 \times 10^6$, $T_2 = 800$ K, $K_2 = 9.1 \times 10^2$, and $\Delta_r H° = -1.8 \times 10^5$ J mol^{-1} (from Figure 13-10). We expect $T_1 < T_2$.

(continued)

Solve

We substitute $K_1 = 1.0 \times 10^6$, $T_2 = 800$ K, $K_2 = 9.1 \times 10^2$, and $\Delta_r H° = -1.8 \times 10^5$ J mol^{-1} into equation (13.25) and solve for T_1.

$$\ln\left(\frac{K_2}{K_1}\right) = -\frac{\Delta_r H°}{R}\left(\frac{1}{T_2} - \frac{1}{T_1}\right)$$

$$\ln\left(\frac{9.1 \times 10^2}{1 \times 10^6}\right) = -\frac{(-1.8 \times 10^5 \text{ J mol}^{-1})}{8.3145 \text{ J mol}^{-1}\text{ K}^{-1}}\left(\frac{1}{800 \text{ K}} - \frac{1}{T_1}\right)$$

$$-7.00 = 2.2 \times 10^4 \text{ K}\left(\frac{1}{800 \text{ K}} - \frac{1}{T_1}\right)$$

$$\frac{-7.00}{2.2 \times 10^4 \text{ K}} = \frac{1}{800 \text{ K}} - \frac{1}{T_1}$$

$$\frac{1}{T_1} = \frac{1}{800 \text{ K}} + \frac{-7.00}{2.2 \times 10^4 \text{ K}}$$

$$\frac{1}{T_1} = (1.25 \times 10^{-5} \text{ K}^{-1}) + (3.2 \times 10^{-4} \text{ K}^{-1}) = 1.57 \times 10^{-3} \text{ K}^{-1}$$

$$T_1 = \frac{1}{1.57 \times 10^{-3} \text{ K}^{-1}} = 637 \text{ K}$$

Assess

A common error in this type of problem is the use of incorrect temperature units. Express T in Kelvin (K).

PRACTICE EXAMPLE A: Estimate the temperature at which $K = 5.8 \times 10^{-2}$ for the reaction in Example 13-12. Use data from Table 13.8 and Figure 13-10.

PRACTICE EXAMPLE B: What is the value of K_p for the reaction $2\ SO_2(g) + O_2(g) \rightleftharpoons 2\ SO_3(g)$ at 235 °C? Use data from Table 13.8, Figure 13-10, and the van't Hoff equation (13.25).

13-7 Coupled Reactions

We have seen two ways to obtain product from a nonspontaneous reactic (1) change the reaction conditions to ones that make the reaction spontaneo (mostly by changing the temperature), and (2) combine a pair of reactions, o with a positive $\Delta_r G$ and one with a negative $\Delta_r G$, to obtain a spontaneo overall reaction. Such paired reactions are called **coupled reactions**. Consid the extraction of a metal from its oxide.

When copper(I) oxide is heated to 673 K, no copper metal is obtained. T decomposition of Cu_2O to form products in their standard states (for instanc $P_{O_2} = 1.00$ bar) is nonspontaneous at 673 K.

▶ In general chemistry, simple examples of reactions are generally used. In fact, in almost all interesting cases, one reaction is coupled to another, and so forth. No better example exists than the complex cycles of coupled chemical reactions in biological processes.

$$Cu_2O(s) \xrightarrow{\Delta} 2\ Cu(s) + \frac{1}{2} O_2(g) \qquad \Delta_r G°_{673\text{ K}} = +125 \text{ kJ mol}^{-1} \qquad \textbf{(13.2}$$

Suppose this nonspontaneous decomposition reaction is coupled with the pa tial oxidation of carbon to carbon monoxide—a spontaneous reaction. T overall reaction (13.27) is spontaneous when reactants and products are their standard states because $\Delta_r G°$ has a negative value.

$$Cu_2O(s) \longrightarrow 2\ Cu(s) + \cancel{\tfrac{1}{2} O_2(g)} \qquad \Delta_r G°_{673\text{ K}} = +125 \text{ kJ mol}^{-1}$$

$$C(s) + \cancel{\tfrac{1}{2} O_2(g)} \longrightarrow CO(g) \qquad \Delta_r G°_{673\text{ K}} = -175 \text{ kJ mol}^{-1}$$

$$Cu_2O(s) + C(s) \longrightarrow 2\ Cu(s) + CO(g) \qquad \Delta_r G°_{673\text{ K}} = -50 \text{ kJ mol}^{-1} \qquad \textbf{(13.2}$$

Note that reactions (13.26) and (13.27) are not the same, even though each s Cu(s) as a product. The purpose of coupled reactions, then, is to produce spontaneous overall reaction by combining two other processes: one non- ontaneous and one spontaneous. Many metallurgical processes employ upled reactions, especially those that use carbon or hydrogen as reducing ents.

To sustain life, organisms must synthesize complex molecules from simpler es. If carried out as single-step reactions, these syntheses would generally accompanied by increases in enthalpy, decreases in entropy, and increases Gibbs energy—in short, they would be nonspontaneous and would not cur. In living organisms, changes in temperature and electrolysis are not able options for dealing with nonspontaneous processes. Here, coupled actions are crucial. See Focus On 13-1 on the MasteringChemistry website r an example.

3-8 Chemical Potential and Thermodynamics of Spontaneous Chemical Change

this section, we focus on deriving equation (13.15), the equation that is used predict the direction of spontaneous change in a system that undergoes a ange in composition at constant T and constant P. Perhaps surprisingly, the ncepts we need to derive equation (13.15) find their origin in an equation that scribes how the Gibbs energy of an ideal gas depends on temperature, pres- re, and the amount, n. The equation obtained for an ideal gas can be general- ed to all substances by introducing the concepts of chemical potential and tivity, arguably two of the most important concepts in chemical thermody- mics. As we will see, the concepts of chemical potential and activity are inter- vined; it is nearly impossible to speak of one without reference to the other. owever, they are precisely what we need to describe the Gibbs energy change r a system undergoing a change in composition at constant T and constant P.

ibbs Energy of an Ideal Gas

begin, consider the following the isothermal process for an ideal gas, X.

$$X(g, T, P^\circ = 1 \text{ bar}) \longrightarrow X(g, T, P)$$

this process, the pressure of the gas is changed isothermally from 1 bar to a nal pressure P. Because T is constant, we can write $\Delta G = \Delta H - T\Delta S$ for this ocess. Recall that $\Delta H = 0$ for an isothermal process involving an ideal gas age 584) and so we have $\Delta G = -T\Delta S$. Using equation (13.7) for ΔS with $= P$ and $P_i = P^\circ$, we obtain

$$\Delta G = nRT \ln\left(\frac{P}{P^\circ}\right)$$

we use the symbols G° and G to represent the Gibbs energy of the gas in the itial and final states, respectively, we have $\Delta G = G - G^\circ$ and we can express e equation above in the form

$$G = G^\circ + nRT \ln\left(\frac{P}{P^\circ}\right) \quad \textbf{(13.28)}$$

This equation shows how the Gibbs energy of an ideal gas changes with ressure at constant temperature. If we divide all terms by n, we obtain

$$G_m = G_m^\circ + RT \ln\left(\frac{P}{P^\circ}\right) \quad \textbf{(13.29)}$$

here $G_m = G/n$ and $G_m^\circ = G^\circ/n$ are the molar Gibbs energies (Gibbs energy er mole) for, respectively, the gas at T and P and the gas at T and 1 bar. Both uantities have the unit J mol^{-1}.

Gibbs Energy of an Ideal Gas Mixture

Let's now consider a mixture containing several ideal gases A(g), B(g), C(etc. If the amounts of these gases are denoted by n_A, n_B, n_C, etc., then the Gib energy of the mixture is

$$G = n_A G_{m,A} + n_B G_{m,B} + n_C G_{m,C} + \cdots \quad \text{(13.}$$

where $G_{m,A}$, $G_{m,B}$, $G_{m,C}$, etc. are the molar Gibbs energies of the gases, each which has the form given by equation (13.29) with P replaced by the approp ate partial pressure P_A, P_B, P_C, etc. We now ask, what is the change in Gib energy for the system if the amount of A(g) is changed by an amount Δ without changing the amounts of the other gases? The Gibbs energy change

$$\begin{aligned}\Delta G &= G_f - G_i \\ &= [(n_A + \Delta n_A)G_{m,A} + n_B G_{m,B} + n_C G_{m,C} + \cdots] \\ &\quad - [n_A G_{m,A} + n_B G_{m,B} + n_C G_{m,C} + \cdots] \\ &= G_{m,A}\Delta n_A\end{aligned}$$

Notice that ΔG for the system is directly proportional to Δn_A, with $G_{m,A}$ pr viding the link between these two quantities. Thus, for an ideal gas mixtu the molar Gibbs energy $G_{m,A}$ allows us to relate the change in Gibbs energy a change in the amount of gas A.

Chemical Potential and Activity

Gilbert N. Lewis, a leader in the development of thermodynamics, recogniz that equation (13.29) could be generalized so that it applies to any substan (solid, liquid, gas, dissolved solute, etc.) by writing

$$\mu = \mu^\circ + RT \ln a \quad \text{(13.3}$$

Through this expression, Lewis introduced not only the concept of chemic potential, represented by the symbol μ (Greek letter mu), but also the conce of activity, represented by the symbol a. The quantity μ is the chemical pote tial of the substance for the given conditions of T and P, whereas μ° is t chemical potential of the substance in a well-defined reference state at tempe ature T and a pressure of 1 bar. By comparing equations (13.29) and (13.31), w see that, for an ideal gas, $\mu = G_m$, $\mu^\circ = G^\circ_m$, and $a = P/P^\circ$. That is, for a ideal gas, the chemical potential is simply the molar Gibbs energy and t activity is P/P°.

Equation (13.30) can also be expressed more generally by writing

$$G = n_A\mu_A + n_B\mu_B + n_C\mu_C + \cdots \quad \text{(13.3}$$

This equation relates the Gibbs energy of a mixture of substances to t amounts and chemical potentials of those substances. As argued above f $G_{m,A}$, the chemical potential μ_A relates the change in G to a change in t amount of substance A.

Let us explore the concept of chemical potential a little further. The **chem cal potential, μ,** of a substance refers to its ability or potential to change t Gibbs energy of the system. If a substance has a high chemical potential, means that a small change in the amount of that substance will cause a rel tively large change in the Gibbs energy of the system. On the other hand, if t chemical potential is low, then even a large change in amount produces only small change in the Gibbs energy. More precisely, when the amount of su stance A in a system is changed by an infinitesimally small amount dn_A, t corresponding change in Gibbs energy is

KEEP IN MIND

that dn_A represents an infinitesimally small change in the amount of substance A.

$$\begin{aligned}dG &= G_f - G_i \\ &= [(n_A + dn_A)\mu_A + n_B\mu_B + n_C\mu_C + \cdots] - [n_A\mu_A + n_B\mu_B + n_C\mu_C + \cdots]\end{aligned}$$

hich simplifies to

$$dG = \mu_A dn_A \qquad (13.33)$$

juation (13.33) shows that the change in Gibbs energy is directly propor-
nal to the change in amount of A; the constant of proportionality is μ_A, the emical potential of A. We may also think of μ_A as the rate of change of G ith respect to the amount of substance A. For example, if the chemical tential of a substance A in a system is constant and equal to $_A = 10\ \text{J mol}^{-1}$, then increasing the amount of that substance by 0.01 mol uses the Gibbs energy of the system to increase by approximately J $\text{mol}^{-1} \times 0.01\ \text{mol} = 0.1$ J. (This calculation is approximate because the ange in amount, 0.01 mol, is not infinitesimally small.)

With the chemical potential, μ, of a substance expressed in the form of juation (13.31), $\mu = \mu^\circ + RT \ln a$, we can see that as the activity a creases, then so too does the chemical potential. This, in essence, is the eaning of activity: a direct measure of the chemical potential of a substance id, therefore, of the ability or potential of a substance to change the Gibbs ergy of a system. The greater the activity, the greater the chemical potenil and the greater the ability of that substance to change the Gibbs energy the system.

We must emphasize that the concept of activity was introduced by Lewis r a rather simple reason: to ensure that the chemical potential of a substance ways has the form given by equation (13.31). To ensure that this is so, activis of various substances must be defined as described on page 611.

Can we say more about what the activity of a substance might represent? ot really, but we can say a little about what it does not represent. The activity a substance is sometimes described as an "effective pressure" or an "effecve concentration," but such vague descriptions are not at all illuminating id are somewhat misleading because these terms incorrectly suggest that) activity is designed to provide a different or better measure of pressure or ncentration, and (2) activities have the unit of pressure or concentration hen in fact they are dimensionless quantities. So, it is not particularly helpful think of activity as an effective pressure or an effective concentration. It is uch better to think of pressure and concentration as providing a measure of tivity.

The significance of the concept of chemical potential cannot be overstated. or example, a recurring theme in thermodynamics is that a substance moves ontaneously from a region of high chemical potential to a region of low emical potential. Typically, the spontaneity of physical and chemical ianges may be explained with reference to not only entropy changes but also fferences in chemical potentials.

Our interest in chemical potentials (and activities) arises from the fact that e Gibbs energy of a mixture can be expressed in terms of the amounts of the ibstances and their chemical potentials, as shown by equations (13.31) and 3.32). These two equations, and the second law of thermodynamics, can be sed to develop the criterion for predicting the direction of spontaneous iemical change.

xpressing $\Delta_r G^\circ$ in Terms of Chemical Potentials

lthough the standard Gibbs energy of reaction, $\Delta_r G^\circ$, is usually expressed in rms of $\Delta_f G^\circ$ values, as shown by equation (13.14), it is also possible to press it in terms of the chemical potentials of the substances involved. To see ow this is done, consider the following change in composition.

<u>Initial State</u>	<u>Final State</u>
Pure A at 1 bar, $n_A = a$ mol	Pure C at 1 bar, $n_C = c$ mol
Pure B at 1 bar, $n_B = b$ mol	Pure D at 1 bar, $n_D = d$ mol
$G_i = n_A\mu_A^\circ + n_B\mu_B^\circ$	$G_f = n_C\mu_C^\circ + n_D\mu_D^\circ$

The change in Gibbs energy is, by applying equation (13.32),

$$\begin{aligned}\Delta G &= G_f - G_i \\ &= [(c\text{ mol})\mu_C^\circ + (d\text{ mol})\mu_D^\circ + \cdots] - [(a\text{ mol})\mu_A^\circ + (b\text{ mol})\mu_B^\circ + \cdots] \\ &= \{[c\,\mu_C^\circ + d\,\mu_D^\circ + \cdots] - [a\,\mu_A^\circ + b\,\mu_B^\circ + \cdots]\} \times 1\text{ mol}\end{aligned}$$

Since $\Delta_r G^\circ$ is the Gibbs energy change *per mole*, we have $\Delta_r G^\circ = \Delta G/(1$ m
and thus,

$$\Delta_r G^\circ = \underbrace{[c\,\mu_C^\circ + d\,\mu_D^\circ + \cdots]}_{\text{Weighted sum of } \mu^\circ \text{ values for products}} - \underbrace{[a\,\mu_A^\circ + b\,\mu_B^\circ + \cdots]}_{\text{Weighted sum of } \mu^\circ \text{ values for reactants}} \quad \textbf{(13.}$$

The chemical potentials have the unit J mol^{-1} and the coefficients a, b, c, anc
are simply numbers (no units). Again, we see that $\Delta_r G^\circ$ has the unit J mol
We will encounter this last expression for $\Delta_r G^\circ$ again.

Criterion for Predicting the Direction of Spontaneous Chemical Change

In Section 13-4, we established that, for a spontaneous process at constant
and constant P, the Gibbs energy of the system always decreases: $(\Delta G)_{T,P} <$
This idea can also be expressed as $(dG)_{T,P} < 0$, for a system that undergoe
spontaneous change that is infinitesimally small. We will find this seco
expression the most useful for developing an equation that can be used f
predicting the direction of spontaneous change in a system in which a chen
cal reaction occurs.

Consider a system containing substances A, B, C, and D, and let n_A, n_B, r
and n_D represent the initial amounts, in moles, of each substance. Suppose th
the composition of the system changes, at constant T and constant P, becau
the following reaction advances by an infinitesimal amount, $d\xi$. The chang
occurring in the system are summarized below.

$$a\,\text{A} + b\,\text{B} \underset{d\xi < 0}{\overset{d\xi > 0}{\rightleftarrows}} c\,\text{C} + d\,\text{D}$$

	a A	b B	c C	d D	
Initial:	n_A	n_B	n_C	n_D	G_i
Change:	$-a\,d\xi$	$-b\,d\xi$	$+c\,d\xi$	$+d\,d\xi$	$+dG$
Final:	$n_A - a\,d\xi$	$n_B - b\,d\xi$	$n_C + c\,d\xi$	$n_D + d\,d\xi$	$G_f = G_i + dG$

We have arrows pointing to the left and to the right because we want to co
sider the effect on G of a change in composition arising from reaction to t
right ($\longrightarrow$) or to the left ($\longleftarrow$). The sign of $d\xi$ determines the direction
reaction: A positive value ($d\xi > 0$) describes reaction from left to rig
whereas a negative value ($d\xi < 0$) describes reaction from right to left. As su
gested by the summary above, an infinitesimally small change in compositio
represented by $d\xi$, causes an infinitesimally small change, dG, in the Gib
energy of the system. We can use equation (13.30) to write expressions for
and G_i, and thus relate the change, dG, in Gibbs energy to the chemical pote
tials of A, B, C, and D:

$$\begin{aligned}dG &= G_f - G_i \\ &= [(n_A - ad\xi)\mu_A + (n_B - bd\xi)\mu_B + (n_C + cd\xi)\mu_C + (n_D + dd\xi)\mu_D] \\ &\quad - [n_A\mu_A + n_B\mu_B + n_C\mu_C + n_D\mu_D] \\ &= -a\mu_A d\xi - b\mu_B d\xi + c\mu_C d\xi + d\mu_D d\xi \\ &= [-a\mu_A - b\mu_B + c\mu_C + d\mu_D]d\xi \\ &= [(c\mu_C + d\mu_D) - (a\mu_A + b\mu_B)]d\xi\end{aligned}$$

ABLE 13.9 Using the Expression $dG = \Delta_r G \times d\xi \leq 0$

$_rG$	$d\xi$	dG	Conclusion
	+	+	The forward reaction is nonspontaneous.
	–	–	The reverse reaction is spontaneous.
	+	–	The forward reaction is spontaneous.
	–	+	The reverse reaction is nonspontaneous.
	+ or –	0	The system has reached equilibrium.

e quantity in square brackets has a form similar to that given in uation (13.34) for $\Delta_r G°$, except that it refers to nonstandard conditions. erefore, let's represent the quantity in square brackets by the symbol $\Delta_r G$. us, we have

$$dG = \Delta_r G \times d\xi \qquad \textbf{(13.35)}$$

ıere

$$\Delta_r G = (c\mu_C + d\mu_D) - (a\mu_A + b\mu_B) \qquad \textbf{(13.36)}$$

e see immediately that $\Delta_r G$ has the unit J mol^{-1}. (The unit of $\Delta_r G$ can be ablished by considering the units of dG and $d\xi$, or by remembering that the emical potentials themselves have the unit J mol^{-1}.)

Equation (13.35) is the key to predicting the direction of spontaneous ange, so we must be certain to interpret it properly. It shows clearly that if $G < 0$, then we must have reaction occurring in the forward direction $\xi > 0$) to ensure that G decreases ($dG < 0$), as required by the second law. ı the other hand, if $\Delta_r G > 0$, then we must have reaction occurring in the verse direction ($d\xi < 0$) to ensure that G decreases. Table 13.9 summarizes e various possibilities.

The task that remains is to obtain an expression for $\Delta_r G$, and we do this by iting the chemical potentials for A, B, C, and D in the form of equation (13.31). r example, the chemical potential of A is

$$\mu_A = \mu_A^\circ + RT \ln a_A$$

nilar expressions hold for μ_B, μ_C, and μ_D. Thus, we may write equation (13.36)

$$\begin{aligned}
G &= (c\mu_C + d\mu_D) - (a\mu_A + b\mu_B) \\
&= c(\mu_C^\circ + RT \ln a_C) + d(\mu_D^\circ + RT \ln a_D) - [a(\mu_A^\circ + RT \ln a_A) + b(\mu_B^\circ + RT \ln a_B)] \\
&= c\mu_C^\circ + d\mu_D^\circ - (a\mu_A^\circ + b\mu_B^\circ) + cRT \ln a_C + dRT \ln a_D - aRT \ln a_A - bRT \ln a_B \\
&= c\mu_C^\circ + d\mu_D^\circ - (a\mu_A^\circ + b\mu_B^\circ) + RT \ln a_C^c + RT \ln a_D^d - RT \ln a_A^a - RT \ln a_B^b \\
&= c\mu_C^\circ + d\mu_D^\circ - (a\mu_A^\circ + b\mu_B^\circ) + RT \ln\left(\frac{a_C^c\, a_D^d}{a_A^a\, a_B^b}\right) \\
&= \Delta_r G° + RT \ln\left(\frac{a_C^c\, a_D^d}{a_A^a\, a_B^b}\right)
\end{aligned}$$

the last step, we have made use of the fact that $G° = c\mu_C^\circ + d\mu_D^\circ - (a\mu_A^\circ + b\mu_B^\circ)$ because, shown by equation (13.34), $\Delta_r G°$ n be expressed in terms of standard chemical potentials. The quotient in the garithmic term is the general form of the thermodynamic reaction quotient, Thus, we can write the equation above in its most general form as

$$\Delta_r G = \Delta_r G° + RT \ln Q$$

ıe relationship above is equation (13.15), which is what we set out to derive.

MasteringCHEMISTRY www.masteringchemistry.com

Adenosine triphosphate (ATP) is the energy currency used in nature to drive reactions. The conversion of adenosine diphosphate (ADP) to ATP is a nonspontaneous process. For a discussion of how nature is able to convert ADP to ATP, go to the Focus On feature for Chapter 13, Coupled Reactions in Biological Systems, on the MasteringChemistry site.

Summary

13-1 Entropy: Boltzmann's View—**Entropy,** S, is a thermodynamic property that is related to the way in which the energy of a system is distributed among the available energy levels. A **microstate** is a specific microscopic configuration describing how the particles of a system are distributed among the available energy levels. Boltzmann's formula for entropy (equation 13.1) indicates that entropy is proportional to ln W, where W is the number of microstates and, therefore, Boltzmann's formula provides the basis for a microscopic view of entropy. For example, the spontaneous expansion of ideal gases involves an increase in entropy that can be rationalized in terms of an increase in the number of accessible energy levels and an increase in the number of microstates.

13-2 Entropy Change: Clausius's View—Clausius, using macroscopic observations, proposed that the entropy change for a process can be related to the amount of heat transferred divided by the temperature (equation 13.2), provided the process is imagined to occur in a reversible way. Consequently, an entropy change, ΔS, has the unit $J\,K^{-1}$. Clausius's formula can be used to obtain expressions for calculating entropy changes for a variety of physical changes, including phase transitions, constant pressure heating or cooling, or the isothermal expansion or compression of an ideal gas. Table 13.1 summarizes formulas for calculating entropy changes for these different processes.

13-3 Combining Boltzmann's and Clausius's Ideas: Absolute Entropies—A statement of the **third law of thermodynamics** is that the entropy of a pure perfect crystal at 0 K is zero. Thus, we can assign a specific (absolute) value to the entropy of a pure substance. This is in marked contrast to the situation for other thermodynamic properties, such internal energy and enthalpy, for which absolute values cannot be assigned. The absolute entropy of one mole of substance in its standard state is called the **standard molar entropy,** $S°$. Standard molar entropies of reactants and products can be used to calculate the **standard reaction entropy,** $\Delta_r S°$ (equation 13.9).

13-4 Criterion for Spontaneous Change: The Second Law of Thermodynamics—The basic criterion for spontaneous change is that the entropy change of the universe, which is the sum of the entropy change of the system plus that of the surroundings, must be greater than zero (equation 13.10). This statement is known as the **second law of thermodynamics.** An equivalent criterion applied to the system alone is based on a thermodynamic function known as the **Gibbs energy,** G. For an isothermal process, the **Gibbs energy change,** $\Delta_r G$, is the enthalpy change for the system (ΔH) minus the product of the temperat and entropy change for the system ($T\Delta S$) (equation 13. Table 13.3 summarizes the criteria for spontaneous cha based on Gibbs energy change.

13-5 Gibbs Energy of a System of Variab Composition: $\Delta_r G°$ and $\Delta_r G$—The **standard Gib energy of reaction,** $\Delta_r G°$, is based on the conversion of s ichiometric amounts of reactants in their standard state stoichiometric amounts of products in their standard sta Values of $\Delta_r G°$ are often calculated from tabulated value **standard Gibbs energies of formation,** $\Delta_f G°$ at 298.1 (equation 13.14). For nonstandard conditions, the **Gib energy of reaction,** $\Delta_r G$, is equal to $\Delta_r G°$ plus RT ln (equation 13.15) where Q, the **reaction quotient,** ta account of the initial (nonstandard) conditions. The valu $\Delta_r G$ can be used to predict the direction of spontane chemical change as described in Table 13.4. The relations between the standard Gibbs energy change and the **eq librium constant** for a reaction is $\Delta_r G° = -RT$ ln (equation 13.17). The constant, K, is the value of the react quotient Q at equilibrium and is called the **thermodynan equilibrium constant** (equation 13.19). Both Q and K expressed in terms of the activities of reactants and pr ucts (equations 13.16 and 13.19). The activities can related to molarities and gas partial pressures by means c few simple conventions (Table 13.5). The direction of sp taneous chemical change can also be predicted by comp ing the values of Q and K (Table 13.6).

13-6 $\Delta_r G°$ and K as Functions of Temperature By starting with the relationship between standard Gib energy change and the equilibrium constant, the va Hoff equation—relating the equilibrium constant a temperature—can be written (equation 13.25). With t equation, tabulated data at 25 °C can be used to determi equilibrium constants not just at 25 °C but at other te peratures as well.

13-7 Coupled Reactions—Nonspontaneous p cesses can be made spontaneous by coupling them w spontaneous reactions and by taking advantage of t state function property of G. **Coupled reactions,** that paired reactions that yield a spontaneous overall reacti occur in metallurgical processes and in biochemical tra formations.

13-8 Chemical Potential and Thermodynami of Spontaneous Chemical Change—G. N. Lev introduced the concept of chemical potential, represent by the symbol μ and the concept of activity, represent by the symbol a. The **chemical potential,** μ, of a substan ers to its ability or potential to change the Gibbs energy the system. The activity is defined in a way that ensures t the chemical potential has a specific form (equation 31). The criterion for predicting the direction of sponta- neous chemical change can be obtained by considering the Gibbs energy of a mixture, which can be expressed in terms of the chemical potentials of the substances involved (equation 13.32).

Integrative Example

e synthesis of methanol is of great importance because methanol can be used directly as a motor fuel, mixed with gaso- e for fuel use, or converted to other organic compounds. The synthesis reaction, carried out at about 500 K, is

$$CO(g) + 2\,H_2(g) \rightleftharpoons CH_3OH(g)$$

nat are the values of K and K_p at 500 K?

alyze

r approach to this problem begins with determining $\Delta_r G°$ from Gibbs energy of formation data and using $\Delta_r G°$ to d K at 298 K. The next step is to calculate $\Delta_r H°$ from enthalpy of formation data and use this value together with K at 3 K in expression (13.25) to find K at 500 K. Finally, we obtain K_p by using equation (13.21).

lve

ite the equation for methanol synthesis; place Gibbs ergy of formation data from Appendix D under formu- in the equation, and use these data to calculate $\Delta_r G°$ 298 K.

$$\begin{array}{lccc} & CO(g) & + 2\,H_2(g) \rightleftharpoons & CH_3OH(g) \\ \Delta_f G°,\ \text{kJ mol}^{-1} & -137.2 & 0 & -162.0 \end{array}$$

$$\Delta_r G° = 1 \times (-162.0\ \text{kJ mol}^{-1}) - 1 \times (-137.2\ \text{kJ mol}^{-1}) = -24.8\ \text{kJ mol}^{-1}$$

calculate K_p at 298 K, use $\Delta_r G°$ at 298 K, written as 4.8 × 10^3 J mol^{-1}, in the expression $\Delta_r G° = -RT \ln K$.

$$\ln K = -\Delta_r G°/RT = \frac{-(-24.8 \times 10^3\ \text{J mol}^{-1})}{8.3145\ \text{J mol}^{-1}\ \text{K}^{-1} \times 298\ \text{K}} = 10.0$$

$$K = e^{10.0} = 2.2 \times 10^4$$

determine $\Delta_r H°$ at 298 K, use standard enthalpy of mation data from Appendix D, applied in the same nner as was previously used for $\Delta_r G°$.

$$\begin{array}{lccc} & CO(g) & + 2\,H_2(g) \rightleftharpoons & CH_3OH(g) \\ \Delta_f H°,\ \text{kJ mol}^{-1} & -110.5 & 0 & -200.7 \end{array}$$

$$\Delta_r H° = 1 \times (-200.7\ \text{kJ mol}^{-1}) - 1 \times (-110.5\ \text{kJ mol}^{-1}) = -90.2\ \text{kJ mol}^{-1}$$

e the van't Hoff equation with $K = 2.2 \times 10^4$ at 298 K d $\Delta_r H° = -90.2 \times 10^3\ \text{J mol}^{-1}$. Solve for K at 500 K.

$$\ln \frac{K}{2.2 \times 10^4} = \frac{-90.2 \times 10^3\ \text{J mol}^{-1}}{8.3145\ \text{J mol}^{-1}\ \text{K}^{-1}}\left(\frac{1}{500\ \text{K}} - \frac{1}{298\ \text{K}}\right) = -14.7$$

obtain K_p, we use equation (13.21) with = 1 − (1 + 2) = −2

$$\frac{K}{2.2 \times 10^4} = e^{-14.7} = 4 \times 10^{-7} \qquad K = 9 \times 10^{-3}$$

$$K = K_p \times (1/P°)^{-2}$$

$$K_p = K \times (1/P°)^2 = 9 \times 10^{-3}\ \text{bar}^{-2}$$

sess

ere are two important points to make. First, notice that the thermodynamic equilibrium constant, K, has no units and at K_p has the unit $(1/\text{bar})^2$ in this case. However, their numerical values are equal. Sometimes, chemists give K_p and K_c lues without units, but they do so in the knowledge that the appropriate power of bar or mol L^{-1} can be easily luded again when explicitly required. Second, notice that, for this exothermic reaction ($\Delta_r H° < 0$), an increase in tem- rature causes the value of K to decrease and thus, causes the equilibrium position to shift from products towards reac- nts. For an endothermic reaction ($\Delta_r H° > 0$), the opposite is true: Increasing the temperature causes the value of K to rease and the equilibrium position to shift from reactants toward products.

RACTICE EXAMPLE A: Dinitrogen pentoxide, N_2O_5, is a solid with a high vapor pressure. Its vapor pressure at °C is 100 mmHg, and the solid sublimes at a pressure of 1.00 atm at 32.4 °C. What is $\Delta_r G°$ 25 °C for the reaction $O_5(s) \longrightarrow N_2O_5(g)$?

RACTICE EXAMPLE B: A plausible reaction for the production of ethylene glycol (used as antifreeze) is

$$2\,CO(g) + 3H_2(g) \longrightarrow CH_2(OH)CH_2OH(l)$$

(continued)

The following thermodynamic properties of $CH_2(OH)CH_2OH(l)$ at 25 °C are given: $\Delta_f H° = -454.8\ kJ\ mol^{-1}$ a $\Delta_f G° = -323.1\ kJ\ mol^{-1}$. Use these data, together with values from Appendix D, to obtain a value of $S°$, the stand molar entropy of $CH_2(OH)CH_2OH(l)$ at 25 °C.

Exercises

Entropy and Spontaneous Change

1. Consider a system of five distinguishable particles confined to a one-dimensional box of length L. Describe how the following actions affect the number of accessible microstates and the entropy of the system:
(a) increasing the length of the box to $2L$ for fixed total energy
(b) increasing the total energy for constant length L

2. Consider a sample of ideal gas initially in a volume V at temperature T and pressure P. Does the entropy of this system increase, decrease, or stay the same in the following processes?
(a) The gas expands isothermally.
(b) The pressure is increased at constant temperature.
(c) The gas is heated at constant pressure.

3. The standard molar entropy of $H_2(g)$ is $S° = 130.7\ J\ mol^{-1}\ K^{-1}$ at 298 K. Use this value, together with Boltzmann's equation, to determine the number of microstates, W, for one mole of $H_2(g)$ at 298 K and 1 bar. Reflect on the magnitude of your calculated value by describing how to write the number in decimal form. [*Hint:* To write the number 1×10^{15}, for example, in decimal form, we would write down a 1 followed by 15 zeros. The relationship $\ln W = 2.303 \log W$ might also be useful.]

4. In a 1985 paper in the *Journal of Chemical Thermodynamics*, the standard molar entropy of chalcopyrite, $CuFeS_2$, is given as $0.012\ J\ mol^{-1}\ K^{-1}$ at 5 K. [See R. A. Robie et. al., *J. Chem. Thermodyn.*, **17**, 481 (1985).] Use this value to estimate the number of microstates for one picogram (1×10^{-12} g) of $CuFeS_2$ at 5 K and 1 bar. Report your answer in the manner described in Exercise 3.

5. Indicate whether each of the following changes represents an increase or a decrease in entropy in a system, and explain your reasoning: **(a)** the freezing of ethanol; **(b)** the sublimation of dry ice; **(c)** the burning of a rocket fuel.

6. Arrange the entropy changes of the following processes, all at 25 °C, in the expected order of increasing ΔS, and explain your reasoning:
(a) $H_2O(l, 1\ bar) \longrightarrow H_2O(g, 1\ bar)$
(b) $CO_2(s, 1\ bar) \longrightarrow CO_2(g, 0.01\ bar)$
(c) $H_2O(l, 1\ bar) \longrightarrow H_2O(g, 0.01\ bar)$

7. Use ideas from this chapter to explain this famous remark attributed to Rudolf Clausius (1865): "Die Energie der Welt ist konstant; die Entropie der Welt strebt einem Maximum zu." ("The energy of the world is constant; the entropy of the world increases toward a maximum.")

8. Comment on the difficulties of solving environmental pollution problems from the standpoint of entropy changes associated with the formation of pollutants and with their removal from the environment.

9. Indicate whether entropy increases or decreases each of the following reactions. If you cannot be c tain simply by inspecting the equation, explain wh
(a) $CCl_4(l) \longrightarrow CCl_4(g)$
(b) $CuSO_4 \cdot 3\,H_2O(s) + 2\,H_2O(g) \longrightarrow CuSO_4 \cdot 5\,H_2O$
(c) $SO_3(g) + H_2(g) \longrightarrow SO_2(g) + H_2O(g)$
(d) $H_2S(g) + O_2(g) \longrightarrow H_2O(g) + SO_2(g)$ (not balanc

10. Which substance in each of the following pairs wo have the greater entropy? Explain.
(a) at 75 °C and 1 bar: 1 mol $H_2O(l)$ or 1 mol $H_2O($
(b) at 5 °C and 1 bar: 50.0 g Fe(s) or 0.80 mol Fe(s)
(c) 1 mol Br_2 (l, 1 bar, 8 °C) or 1 mol Br_2 (s, 1 bar, −8
(d) 0.312 mol SO_2 (g, 0.110 bar, 32.5 °C) or 0.284 n O_2 (g, 15.0 bar, 22.3 °C)

11. *Without performing any calculations* or using data fr Appendix D, predict whether $\Delta_r S°$ for each of the f lowing reactions is positive or negative. If it is i possible to determine the sign of $\Delta_r S°$ from the inf mation given, indicate why.
(a) $CaO(s) + H_2O(l) \longrightarrow Ca(OH)_2(s)$
(b) $2\,HgO(s) \longrightarrow 2\,Hg(l) + O_2(g)$
(c) $2\,NaCl(l) \longrightarrow 2\,Na(l) + Cl_2(g)$
(d) $Fe_2O_3(s) + 3\,CO(g) \longrightarrow 2\,Fe(s) + 3\,CO_2(g)$
(e) $Si(s) + 2\,Cl_2(g) \longrightarrow SiCl_4(g)$

12. By analogy to $\Delta_f H°$ and $\Delta_f G°$ how would you defi standard entropy of formation? Which would ha the largest standard entropy of formation: $CH_4($ $CH_3CH_2OH(l)$, or $CS_2(l)$? First make a qualitati prediction; then test your prediction with data fr Appendix D.

13. Calculate the entropy change, ΔS, for the followi processes. If necessary, look up required data Appendix D.
(a) A mole of He(g) undergoes an expansion from to $2V$ at 298 K.
(b) The temperature of one mole of $CH_4(g)$ increased from 298 K to 325 K at a constant pressure 1 bar.

14. Calculate the entropy change, ΔS, for the followi processes. If necessary, look up required data Appendix D.
(a) The pressure of one mole of $O_2(g)$ is increas from P to $2P$ at 298 K.
(b) The temperature of one mole of $CO_2(g)$ increased from 298 K to 355 K at a constant volume 20.0 L.

15. In Example 13-3, we dealt with $\Delta_{vap}H°$ and $\Delta_{vap}S°$ water at 100 °C.
(a) Use data from Appendix D to determine valu for these two quantities at 25 °C.

(b) From your knowledge of the structure of liquid water, explain the differences in $\Delta_{vap}H°$ values and in $\Delta_{vap}S°$ values between 25 °C and 100 °C.

Pentane is one of the most volatile of the hydrocarbons in gasoline. At 298.15 K, the following enthalpies of formation are given for pentane: $\Delta_f H°[C_5H_{12}(l)] = -173.5$ kJ mol^{-1}; $\Delta_f H°[C_5H_{12}(g)] = -146.9$ kJ mol^{-1}.
(a) Estimate the normal boiling point of pentane.
(b) Estimate $\Delta_{vap}G°$ for pentane at 298 K.
(c) Comment on the significance of the sign of $\Delta_{vap}G°$ for pentane at 298 K.

Which of the following substances would obey Trouton's rule most closely: HF, $C_6H_5CH_3$ (toluene), or CH_3OH (methanol)? Explain your reasoning.

18. Estimate the normal boiling point of bromine, Br_2, in the following way: Determine $\Delta_{vap}H°$ for Br_2 from data in Appendix D. Assume that $\Delta_{vap}H°$ remains constant and that Trouton's rule is obeyed.

19. In what temperature range can the following equilibrium be established? Explain.

$$H_2O(l, 0.50\text{ bar}) \rightleftharpoons H_2O(g, 0.50\text{ bar})$$

20. Refer to Figure 12-28 and equation (13.13). Which has the lowest Gibbs energy at 1 atm and −60 °C: solid, liquid, or gaseous carbon dioxide? Explain.

bbs Energy and Spontaneous Change

Which of the following changes in a thermodynamic property would you expect to find for the reaction $Br_2(g) \longrightarrow 2\,Br(g)$ *at all temperatures?* Explain. **(a)** $\Delta H < 0$; **(b)** $\Delta S > 0$; **(c)** $\Delta G < 0$; **(d)** $\Delta S < 0$.

If a reaction can be carried out only because of an external influence, such as the use of an external source of power, which of the following changes in a thermodynamic property *must* apply? Explain. **(a)** $\Delta H > 0$; **(b)** $\Delta S > 0$; **(c)** $\Delta G = \Delta H$; **(d)** $\Delta G > 0$.

Indicate which of the four cases in Table 13.3 applies to each of the following reactions. If you are unable to decide from only the information given, state why.

(a) $PCl_3(g) + Cl_2(g) \longrightarrow PCl_5(g)$ $\quad \Delta_r H° = -87.9$ kJ mol^{-1}

(b) $CO_2(g) + H_2(g) \longrightarrow CO(g) + H_2O(g)$ $\quad \Delta_r H° = +41.2$ kJ mol^{-1}

(c) $NH_4CO_2NH_2(s) \longrightarrow 2\,NH_3(g) + CO_2(g)$ $\quad \Delta_r H° = +159.2$ kJ mol^{-1}

Indicate which of the four cases in Table 13.3 applies to each of the following reactions. If you are unable to decide from only the information given, state why.

(a) $H_2O(g) + \frac{1}{2}O_2(g) \longrightarrow H_2O_2(g)$ $\quad \Delta_r H° = +105.5$ kJ mol^{-1}

(b) $C_6H_6(l) + \frac{15}{2}O_2(g) \longrightarrow 6\,CO_2(g) + 3\,H_2O(g)$ $\quad \Delta_r H° = -3135$ kJ mol^{-1}

(c) $NO(g) + \frac{1}{2}Cl_2(g) \longrightarrow NOCl(g)$ $\quad \Delta_r H° = -38.54$ kJ mol^{-1}

25. For the mixing of ideal gases (see Figure 13-3), explain whether a positive, negative, or zero value is expected for ΔH, ΔS, and ΔG.

26. In Chapter 14, we will see that, for the formation of an ideal solution of liquid components, $\Delta H = 0$. What would you expect for the values of ΔS and ΔG? (Is each value positive, negative, or zero?)

27. Explain why **(a)** some exothermic reactions do not occur spontaneously, and **(b)** some reactions in which the entropy of the system increases do not occur spontaneously.

28. Explain why you would expect a reaction of the type $AB(g) \longrightarrow A(g) + B(g)$ always to be spontaneous at *high* rather than at low temperatures.

andard Gibbs Energy of Reaction, $\Delta_r G°$

From the data given in the following table, determine $\Delta_r S°$ for the reaction $NH_3(g) + HCl(g) \longrightarrow NH_4Cl(s)$. All data are at 298 K.

	$\Delta_f H°$, kJ mol^{-1}	$\Delta_f G°$, kJ mol^{-1}
$NH_3(g)$	−46.11	−16.48
$HCl(g)$	−92.31	−95.30
$NH_4Cl(s)$	−314.4	−202.9

Use data from Appendix D to determine values of $\Delta_r G°$ for the following reactions at 25 °C.
(a) $C_2H_2(g) + 2\,H_2(g) \longrightarrow C_2H_6(g)$
(b) $2\,SO_3(g) \longrightarrow 2\,SO_2(g) + O_2(g)$
(c) $Fe_3O_4(s) + 4\,H_2(g) \longrightarrow 3\,Fe(s) + 4\,H_2O(g)$
(d) $2\,Al(s) + 6\,H^+(aq) \longrightarrow 2\,Al^{3+}(aq) + 3\,H_2(g)$

At 298 K, for the reaction $2\,PCl_3(g) + O_2(g) \longrightarrow 2\,POCl_3(l)$, $\Delta_r H° = -620.2$ kJ mol^{-1} and the standard molar entropies, in J mol^{-1} K^{-1}, are $PCl_3(g)$, 311.8; $O_2(g)$, 205.1; and $POCl_3(l)$, 222.4. Determine **(a)** $\Delta_r G°$ at 298 K and **(b)** whether the reaction proceeds spontaneously in the forward or the reverse direction when reactants and products are in their standard states.

32. At 298 K, for the reaction $2\,H^+(aq) + 2\,Br^-(aq) + 2\,NO_2(g) \longrightarrow Br_2(l) + 2\,HNO_2(aq)$, $\Delta_r H° = -61.6$ kJ mol^{-1} and the standard molar entropies are $H^+(aq)$, 0 J mol^{-1} K^{-1}; $Br^-(aq)$, 82.4 J mol^{-1} K^{-1}; $NO_2(g)$, 240.1 J mol^{-1} K^{-1}; $Br_2(l)$, 152.2 J mol^{-1} K^{-1}; $HNO_2(aq)$, 135.6 J mol^{-1} K^{-1}. Determine **(a)** $\Delta_r G°$ at 298 K and **(b)** whether the reaction proceeds spontaneously in the forward or the reverse direction when reactants and products are in their standard states.

33. The following $\Delta_r G°$ values are given for 25 °C.
(1) $N_2(g) + 3\,H_2(g) \longrightarrow 2\,NH_3(g)$ $\quad \Delta_r G° = -33.0$ kJ mol^{-1}

(2) $4\,NH_3(g) + 5\,O_2(g) \longrightarrow 4\,NO(g) + 6\,H_2O(l)$
$\Delta_rG° = -1010.5$ kJ mol^{-1}
(3) $N_2(g) + O_2(g) \longrightarrow 2\,NO(g)$
$\Delta_rG° = +173.1$ kJ mol^{-1}
(4) $N_2(g) + 2\,O_2(g) \longrightarrow 2\,NO_2(g)$
$\Delta_rG° = +102.6$ kJ mol^{-1}
(5) $2\,N_2(g) + O_2(g) \longrightarrow 2\,N_2O(g)$
$\Delta_rG° = +208.4$ kJ mol^{-1}

Combine the preceding equations, as necessary, to obtain $\Delta_rG°$ values for each of the following reactions.

(a) $N_2O(g) + \frac{3}{2}O_2(g) \longrightarrow 2\,NO_2(g)$ $\quad\Delta_rG° = ?$
(b) $2\,H_2(g) + O_2(g) \longrightarrow 2\,H_2O(l)$ $\quad\Delta_rG° = ?$
(c) $2\,NH_3(g) + 2\,O_2(g) \longrightarrow N_2O(g) + 3\,H_2O(l)$ $\quad\Delta_rG° = ?$

Of reactions (a), (b), and (c), which would tend to go to completion at 25 °C, and which would reach an equilibrium condition with significant amounts of all reactants and products present?

34. The following $\Delta_rG°$ values are given for 25 °C.
(1) $SO_2(g) + 3\,CO(g) \longrightarrow COS(g) + 2\,CO_2(g)$
$\Delta_rG° = -246.4$ kJ mol^{-1}
(2) $CS_2(g) + H_2O(g) \longrightarrow COS(g) + H_2S(g)$
$\Delta_rG° = -41.5$ kJ mol^{-1}
(3) $CO(g) + H_2S(g) \longrightarrow COS(g) + H_2(g)$
$\Delta_rG° = +1.4$ kJ mol^{-1}
(4) $CO(g) + H_2O(g) \longrightarrow CO_2(g) + H_2(g)$
$\Delta_rG° = -28.6$ kJ mol^{-1}

Combine the preceding equations, as necessary, to obtain $\Delta_rG°$ values for the following reactions.
(a) $COS(g) + 2\,H_2O(g) \longrightarrow SO_2(g) + CO(g) + 2\,H_2(g)$ $\quad\Delta_rG° = ?$
(b) $COS(g) + 3\,H_2O(g) \longrightarrow SO_2(g) + CO_2(g) + 3\,H_2(g)$ $\quad\Delta_rG° = ?$
(c) $COS(g) + H_2O(g) \longrightarrow CO_2(g) + H_2S(g)$
$\Delta_rG°$ =

Of reactions (a), (b), and (c), which is spontaneou[s in] the forward direction when reactants and produ[cts] are present in their standard states?

35. Write an equation for the combustion of one mole [of] benzene, $C_6H_6(l)$, and use data from Appendix D [to] determine $\Delta_rG°$ at 298 K if the products of the co[m]bustion are **(a)** $CO_2(g)$ and $H_2O(l)$, and **(b)** $CO_2(g)$ and $H_2O(g)$. Describe how you might determine [the] *difference* between the values obtained in (a) and [(b)] without having either to write the combustion equ[a]tion or to determine $\Delta_rG°$ values for the combust[ion] reactions.

36. Use molar entropies from Appendix D, together w[ith] the following data, to estimate the bond-dissociat[ion] energy of the F_2 molecule.

$$F_2(g) \longrightarrow 2\,F(g) \quad \Delta_rG° = 123.9 \text{ kJ mol}^{-1}$$

Compare your result with the value listed in Table 1[0.3].

37. Assess the feasibility of the reaction

$$N_2H_4(g) + 2\,OF_2(g) \longrightarrow N_2F_4(g) + 2\,H_2O(g)$$

by determining each of the following quantities [for] this reaction at 25 °C.
(a) $\Delta_rS°$ (The standard molar entropy of $N_2F_4(g)$ [is] 301.2 J mol^{-1} K^{-1}.)
(b) $\Delta_rH°$ (Use data from Table 10.3 and F—[F] and N—F bond energies of 222 and 301 kJ mol[$^{-1}$], respectively.)
(c) $\Delta_rG°$
Is the reaction feasible? If so, is it favored at high [or] low temperatures?

38. Solid ammonium nitrate can decompose to dinitrog[en] oxide gas and liquid water. What is $\Delta_rG°$ at 298 K? [Is] the decomposition reaction favored at temperatu[res] above or below 298 K?

The Thermodynamic Equilibrium Constant

39. For each of the following reactions, write down the relationship between K and either K_p or K_c, as appropriate.
(a) $2\,SO_2(g) + O_2(g) \rightleftharpoons 2\,SO_3(g)$
(b) $HI(g) \rightleftharpoons \frac{1}{2}H_2(g) + \frac{1}{2}I_2(g)$
(c) $NH_4HCO_3(s) \rightleftharpoons NH_3(g) + CO_2(g) + H_2O(l)$

40. $H_2(g)$ can be prepared by passing steam over hot iron: $3\,Fe(s) + 4\,H_2O(g) \rightleftharpoons Fe_3O_4(s) + 4\,H_2(g)$.
(a) Write an expression for the thermodynamic equilibrium constant for this reaction.
(b) Explain why the partial pressure of $H_2(g)$ is independent of the amounts of Fe(s) and $Fe_3O_4(s)$ present.
(c) Can we conclude that the production of $H_2(g)$ from $H_2O(g)$ could be accomplished regardless of the proportions of Fe(s) and $Fe_3O_4(s)$ prese[nt]? Explain.

41. In the synthesis of gaseous methanol from carb[on] monoxide gas and hydrogen gas, the followi[ng] equilibrium concentrations were determined [at] 483 K: $[CO(g)] = 0.0911$ M, $[H_2(g)] = 0.0822$ [M,] and $[CH_3OH(g)] = 0.00892$ M. Calculate the eq[ui]librium constant and $\Delta_rG°$ for the reacti[on] $CO(g) + 2\,H_2(g) \longrightarrow CH_3OH(g)$.

42. Calculate the equilibrium constant and $\Delta_rG°$ for t[he] reaction $CO(g) + 2\,H_2(g) \longrightarrow CH_3OH(g)$ at 48[3 K] by using the data tables from Appendix D. [*Hint:* T[he] value you obtain for K will be slightly different fr[om] the value you calculated in Exercise 41.]

Relationships Involving Δ_rG, $\Delta_rG°$, Q, and K

43. Use data from Appendix D to determine K at 298 K for the reaction $N_2O(g) + \frac{1}{2}O_2(g) \rightleftharpoons 2\,NO(g)$.

44. Use data from Appendix D to establish for the reaction $2\,N_2O_4(g) + O_2(g) \rightleftharpoons 2\,N_2O_5(g)$:
(a) $\Delta_rG°$ at 298 K for the reaction as written;
(b) K at 298 K.

45. Use data from Appendix D to determine values at 29[8 K] of $\Delta_rG°$ and K for the following reactions. (*Note:* T[he] equations are not balanced.)
(a) $HCl(g) + O_2(g) \rightleftharpoons H_2O(g) + Cl_2(g)$
(b) $Fe_2O_3(s) + H_2(g) \rightleftharpoons Fe_3O_4(s) + H_2O(g)$
(c) $Ag^+(aq) + SO_4^{2-}(aq) \rightleftharpoons Ag_2SO_4(s)$

In Example 13-2, we were unable to conclude by inspection whether $\Delta_r S°$ for the reaction $CO(g) + H_2O(g) \longrightarrow CO_2(g) + H_2(g)$ should be positive or negative. Use data from Appendix D to obtain $\Delta_r S°$ at 298 K.

Use thermodynamic data at 298 K to decide in which direction the reaction

$$2\,SO_2(g) + O_2(g) \rightleftharpoons 2\,SO_3(g)$$

is spontaneous when the partial pressures of SO_2, O_2, and SO_3 are 1.0×10^{-4}, 0.20, and 0.10 bar, respectively.

Use thermodynamic data at 298 K to decide in which direction the reaction

$$H_2(g) + Cl_2(g) \rightleftharpoons 2\,HCl(g)$$

is spontaneous when the partial pressures of H_2, Cl_2, and HCl are all 0.5 bar.

For the reaction below, $\Delta_r G° = 27.07\ \text{kJ mol}^{-1}$ at 298 K.

$$CH_3CO_2H(aq) + H_2O(l) \rightleftharpoons CH_3CO_2^-(aq) + H_3O^+(aq)$$

Use this thermodynamic quantity to decide in which direction the reaction is spontaneous when the concentrations of $CH_3CO_2H(aq)$, $CH_3CO_2^-(aq)$, and $H_3O^+(aq)$ are 0.10 M, 1.0×10^{-3} M, and 1.0×10^{-3} M, respectively.

For the reaction below, $\Delta_r G° = 29.05\ \text{kJ mol}^{-1}$ at 298 K.

$$NH_3(aq) + H_2O(l) \rightleftharpoons NH_4^+(aq) + OH^-(aq)$$

Use this thermodynamic quantity to decide in which direction the reaction is spontaneous when the concentrations of $NH_3(aq)$, $NH_4^+(aq)$, and $OH^-(aq)$ are 0.10 M, 1.0×10^{-3} M, and 1.0×10^{-3} M, respectively.

For the reaction $2\,NO(g) + O_2(g) \longrightarrow 2\,NO_2(g)$ all but one of the following equations is correct. Which is *incorrect*, and why? **(a)** $K = K_p$; **(b)** $\Delta_r S° = (\Delta_r H° - \Delta_r G°)/T$; **(c)** $K = e^{-\Delta_r G°/RT}$; **(d)** $\Delta_r G = \Delta_r G° + RT \ln Q$.

Why is $\Delta_r G°$ such an important property of a chemical reaction, even though the reaction is generally carried out under *nonstandard* conditions?

At 1000 K, an equilibrium mixture in the reaction $CO_2(g) + H_2(g) \rightleftharpoons CO(g) + H_2O(g)$ contains 0.276 mol H_2, 0.276 mol CO_2, 0.224 mol CO, and 0.224 mol H_2O.
(a) What is K at 1000 K?
(b) Calculate $\Delta_r G°$ at 1000 K.
(c) In which direction would a spontaneous reaction occur if the following were brought together at 1000 K: 0.0750 mol CO_2, 0.095 mol H_2, 0.0340 mol CO, and 0.0650 mol H_2O?

For the reaction $2\,SO_2(g) + O_2(g) \rightleftharpoons 2\,SO_3(g)$, $K_c = 2.8 \times 10^2\ \text{M}^{-1}$ at 1000 K.
(a) What is $\Delta_r G°$ at 1000 K? [*Hint:* What is *K*?]
(b) If 0.40 mol SO_2, 0.18 mol O_2, and 0.72 mol SO_3 are mixed in a 2.50 L flask at 1000 K, in what direction will a net reaction occur?

55. For the following equilibrium reactions, calculate $\Delta_r G°$ at the indicated temperature.
(a) $H_2(g) + I_2(g) \rightleftharpoons 2\,HI(g)$ $\quad K = 50.2$ at 445 °C
(b) $N_2O(g) + \frac{1}{2}O_2(g) \rightleftharpoons 2\,NO(g)$ $\quad K = 8.5 \times 10^{-13}$ at 25 °C
(c) $N_2O_4(g) \rightleftharpoons 2\,NO_2(g)$ $\quad K = 0.114$ at 25 °C
(d) $2\,Fe^{3+}(aq) + Hg_2^{2+}(aq) \rightleftharpoons 2\,Fe^{2+}(aq) + 2\,Hg^{2+}(aq)$ $\quad K = 9.14 \times 10^{-6}$ at 25 °C

56. Two equations can be written for the dissolution of $Mg(OH)_2(s)$ in acidic solution.

$$Mg(OH)_2(s) + 2\,H^+(aq) \rightleftharpoons Mg^{2+}(aq) + 2\,H_2O(l) \qquad \Delta_r G° = -95.5\ \text{kJ mol}^{-1}$$

$$\frac{1}{2}Mg(OH)_2(s) + H^+(aq) \rightleftharpoons \frac{1}{2}Mg^{2+}(aq) + H_2O(l) \qquad \Delta_r G° = -47.8\ \text{kJ mol}^{-1}$$

(a) Explain why these two equations have different $\Delta_r G°$ values.
(b) Will K for these two equations be the same or different? Explain.

57. At 298 K, $\Delta_f G°[CO(g)] = -137.2\ \text{kJ mol}^{-1}$ and $K = 6.5 \times 10^{11}$ for the reaction $CO(g) + Cl_2(g) \rightleftharpoons COCl_2(g)$. Use these data to determine $\Delta_f G°[COCl_2(g)]$, and compare your result with the value in Appendix D.

58. Use thermodynamic data from Appendix D to calculate values of K for the reaction describing the dissolution of the following sparingly soluble solutes: **(a)** AgBr; **(b)** $CaSO_4$; **(c)** $Fe(OH)_3$. [*Hint:* The dissolution process is the reverse of the precipitation reaction. Precipitation reactions were discussed in Chapter 5.]

59. To establish the law of conservation of mass, Lavoisier carefully studied the decomposition of mercury(II) oxide:

$$HgO(s) \longrightarrow Hg(l) + \frac{1}{2}O_2(g)$$

At 25 °C, $\Delta_r H° = +90.83\ \text{kJ mol}^{-1}$ and $\Delta_r G° = +58.54\ \text{kJ mol}^{-1}$.
(a) Show that the partial pressure of $O_2(g)$ in equilibrium with HgO(s) and Hg(l) at 25 °C is extremely low.
(b) What conditions do you suppose Lavoisier used to obtain significant quantities of oxygen?

60. Currently, CO_2 is being studied as a source of carbon atoms for synthesizing organic compounds. One possible reaction involves the conversion of CO_2 to methanol, CH_3OH.

$$CO_2(g) + 3\,H_2(g) \longrightarrow CH_3OH(g) + H_2O(g)$$

With the aid of data from Appendix D, determine
(a) if this reaction proceeds to any significant extent at 25 °C;
(b) if the production of $CH_3OH(g)$ is favored by raising or lowering the temperature from 25 °C;
(c) K for this reaction at 500 K.

$\Delta_r G°$ and K as Functions of Temperature

61. Use data from Appendix D to establish at 298 K for the reaction:

$$2\,NaHCO_3(s) \longrightarrow Na_2CO_3(s) + H_2O(l) + CO_2(g)$$

(a) $\Delta_r S°$; **(b)** $\Delta_r H°$; **(c)** $\Delta_r G°$; **(d)** K.

62. A possible reaction for converting methanol to ethanol is

$$CO(g) + 2\,H_2(g) + CH_3OH(g) \longrightarrow C_2H_5OH(g) + H_2O(g)$$

(a) Use data from Appendix D to calculate $\Delta_r H°$ $\Delta_r S°$, and $\Delta_r G°$ for this reaction at 25 °C.
(b) Is this reaction thermodynamically favored at high or low temperatures? At high or low pressures? Explain.
(c) Estimate K for the reaction at 750 K.

63. What must be the temperature if the following reaction has $\Delta_r G° = -45.5$ kJ mol^{-1}, $\Delta_r H° = -24.8$ kJ mol^{-1}, and $\Delta_r S° = 15.2$ J mol^{-1} K^{-1}?

$$Fe_2O_3(s) + 3\,CO(g) \longrightarrow 2\,Fe(s) + 3\,CO_2(g)$$

64. Estimate K at 100 °C for the reaction $2\,SO_2(g) + O_2(g) \rightleftharpoons 2\,SO_3(g)$. Use data from Table 13.8 and Figure 13-10.

65. The synthesis of ammonia by the Haber process occurs by the reaction $N_2(g) + 3\,H_2(g) \rightleftharpoons 2\,NH_3(g)$ at 400 °C. Using data from Appendix D and assuming that $\Delta_r H°$ and $\Delta_r S°$ are essentially unchanged in the temperature interval from 25 to 400 °C, estimate K at 400 °C.

66. Use data from Appendix D to determine **(a)** $\Delta_r H°$, $\Delta_r S°$, and $\Delta_r G°$ at 298 K and **(b)** K at 875 K for the water gas shift reaction, used commercially to produce $H_2(g)$: $CO(g) + H_2O(g) \rightleftharpoons CO_2(g) + H_2(g)$. [*Hint:* Assume that $\Delta_r H°$ and $\Delta_r S°$ are essentially unchanged in this temperature interval.]

67. In Example 13-12, we used the van't Hoff equation to determine the temperature at which $K = 1.0 \times 10^6$ for the reaction $2\,SO_2(g) + O_2(g) \rightleftharpoons 2\,SO_3(g)$. Obtain another estimate of this temperature with data fr Appendix D and equations (13.13) and (13.17). Comp your result with that obtained in Example 13-12.

68. The following equilibrium constants have been de mined for the reaction $H_2(g) + I_2(g) \rightleftharpoons 2\,HI($ $K = 50.0$ at 448 °C and 66.9 at 350 °C. Use these d to estimate $\Delta_r H°$ for the reaction.

69. For the reaction $N_2O_4(g) \rightleftharpoons 2\,NO_2(g)$, $\Delta_r H° = +57.2$ kJ mol^{-1} and $K = 0.113$ at 298 K.
(a) What is K at 0 °C?
(b) At what temperature will $K = 1.00$?

70. Use data from Appendix D and the van't Hoff equati (13.25) to estimate a value of K at 100 °C for the re tion $2\,NO(g) + O_2(g) \rightleftharpoons 2\,NO_2(g)$. [*Hint:* Fi determine K at 25 °C. What is $\Delta_r H°$ for the reaction?

71. For the reaction

$$CO(g) + 3\,H_2(g) \rightleftharpoons CH_4(g) + H_2O(g),$$
$$K = 2.15 \times 10^{11} \text{ at } 200$$
$$K = 4.56 \times 10^{8} \text{ at } 260$$

determine $\Delta_r H°$ by using the van't Hoff equati (13.25) and by using tabulated data in Appendix Compare the two results, and comment on how go the assumption is that $\Delta_r H°$ is essentially indep dent of temperature in this case.

72. Sodium carbonate, an important chemical used in production of glass, is made from sodium hydrog carbonate by the reaction

$$2\,NaHCO_3(s) \rightleftharpoons Na_2CO_3(s) + CO_2(g) + H_2O($$

Data for the temperature variation of K for t reaction are $K = 1.66 \times 10^{-5}$ at 30 °C; 3.90×10^{-4} 50 °C; 6.27×10^{-3} at 70 °C; and 2.31×10^{-1} at 100
(a) Plot a graph similar to Figure 13-10, and det mine $\Delta_r H°$for the reaction.
(b) Calculate the temperature at which the total pressure above a mixture of $NaHCO_3(s)$ a $Na_2CO_3(s)$ is 2.00 bar.

Coupled Reactions

73. Titanium is obtained by the reduction of $TiCl_4(l)$, which in turn is produced from the mineral rutile (TiO_2).
(a) With data from Appendix D, determine $\Delta_r G°$ at 298 K for this reaction.

$$TiO_2(s) + 2\,Cl_2(g) \longrightarrow TiCl_4(l) + O_2(g)$$

(b) Show that the conversion of $TiO_2(s)$ to $TiCl_4(l)$, with reactants and products in their standard states, is spontaneous at 298 K if the reaction in (a) is coupled with the reaction

$$2\,CO(g) + O_2(g) \longrightarrow 2\,CO_2(g)$$

74. Following are some standard Gibbs energies of formation, $\Delta_f G°$, at 1000 K: NiO(s), −115 kJ mol^{-1}; MnO(s), −280 kJ mol^{-1}; TiO_2(s), −630 kJ mol^{-1}. The standard Gibbs energy of formation of CO(g) at 1000 K is −250 kJ mol^{-1}. Use the method of coupled reactions (page 622) to determine which of these metal oxides can be reduced to the metal by a spontaneous reaction with carbon at 1000 K and with all reacta and products in their standard states.

75. In biochemical reactions* the phosphorylation amino acids is an important step. Consider the f lowing two reactions and determine whether t phosphorylation of arginine with ATP is spontaneo

$$ATP + H_2O \longrightarrow ADP + P \quad \Delta_r G°' = -31.5 \text{ kJ mo}$$
$$\text{arginine} + P \longrightarrow \text{phosphorarginine} + H_2O$$
$$\Delta_r G°' = 33.2 \text{ kJ mo}$$

76. The synthesis of glutamine from glutamic acid is giv by $Glu^- + NH_4^+ \longrightarrow Gln + H_2O$. The standa Gibbs energy of reaction* at pH = 7 and $T = 310$ K $\Delta_r G°' = 14.8$ kJ mol^{-1}. Will this reaction be spon neous if coupled with the hydrolysis of ATP?

$$ATP + H_2O \longrightarrow ADP + P \quad \Delta_r G°' = -31.5 \text{ kJ mol}$$

*$\Delta_r G°'$ denotes the Gibbs energy of reaction for the *biolog standard state,* which has [H+] equal to 1.0×10^{-7} M, not 1 Refer to Focus On 13-1: Coupled Reactions in Biologi Systems.

Integrative and Advanced Exercises

In an *adiabatic* process, there is no exchange of heat between the system and its surroundings. Of the quantities ΔS, ΔS_{surr}, and ΔS_{univ}, which one(s) must always be equal to zero for a spontaneous adiabatic process? Under what condition(s) will an adiabatic process have $\Delta S = \Delta S_{surr} = \Delta S_{univ} = 0$?

Calculate ΔS and ΔS_{univ} when 1.00 mol of $H_2O(l)$, initially at 10 °C, is converted to $H_2O(g)$ at 125 °C at a constant pressure of 1.00 bar. The molar heat capacities of $H_2O(l)$ and $H_2O(g)$ are, respectively, 75.3 J mol^{-1} K^{-1} and 33.6 J mol^{-1} K^{-1}. The standard enthalpy of vaporization of water is 40.66 kJ mol^{-1} at 100 °C.

Consider the following hypothetical process in which heat flows from a low to a high temperature. For copper, the molar heat capacity at constant pressure is 0.385 J mol^{-1} K^{-1}. (For simplicity, you may assume that no heat is lost to the surroundings and that the volume changes are negligible.)

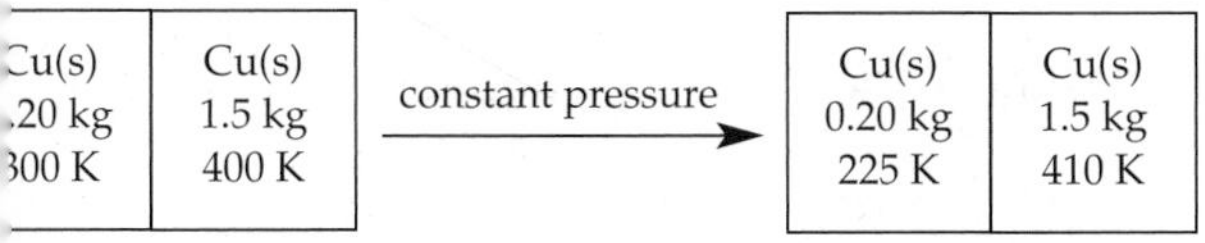

(a) Show that the process conserves energy (i.e., show that the heat absorbed by warmer block of metal is equal to the heat released from the colder block).
(b) Calculate ΔS_{univ} for the process to show that the process is nonspontaneous.

One mole of argon gas, Ar(g), undergoes a change in state from 25.6 L and 0.877 bar to 15.1 L and 2.42 bar. What are ΔH and ΔS for the argon gas? For Ar(g), the molar heat capacity at constant pressure is 20.79 J mol^{-1} K^{-1}.

Use data from Appendix D to estimate **(a)** the normal boiling point of mercury and **(b)** the vapor pressure of mercury at 25 °C.

Consider the vaporization of water: $H_2O(l) \longrightarrow H_2O(g)$ at 100 °C, with $H_2O(l)$ in its standard state, but with the partial pressure of $H_2O(g)$ at 2.0 atm. Which of the following statements about this vaporization at 100 °C are true? Explain. **(a)** $\Delta_r G° = 0$, **(b)** $\Delta_r G = 0$, **(c)** $\Delta_r G° > 0$, **(d)** $\Delta_r G > 0$.

At 298 K, 1.00 mol BrCl(g) is introduced into a 10.0 L vessel, and equilibrium is established in the reaction $BrCl(g) \rightleftharpoons \frac{1}{2} Br_2(g) + \frac{1}{2} Cl_2(g)$. Calculate the amounts of each of the three gases present when equilibrium is established. [*Hint:* First, use data from Appendix D, as necessary, to calculate the equilibrium constant K for the reaction. Then, relate the equilibrium amounts of the gases to their initial amounts through the extent of reaction, ξ. You might find it helpful to use a tabular approach, as was done on page 626. Finally, use the equilibrium condition $Q = K$ to solve for ξ.]

Use data from Appendix D and other information from this chapter to estimate the temperature at which the dissociation of $I_2(g)$ becomes appreciable [for example, with the $I_2(g)$ 50% dissociated into I(g) at 1 atm total pressure].

85. The following table shows the enthalpies and Gibbs energies of formation of three metal oxides at 25 °C. **(a)** Which of these oxides can be most readily decomposed to the free metal and $O_2(g)$? **(b)** For the oxide that is most easily decomposed, to what temperature must it be heated to produce $O_2(g)$ at 1.00 atm pressure?

	$\Delta_f H°$, kJ mol^{-1}	$\Delta_f G°$, kJ mol^{-1}
PbO(red)	−219.0	−188.9
Ag_2O	−31.05	−11.20
ZnO	−348.3	−318.3

86. The following data are given for the two solid forms of HgI_2 at 298 K.

	$\Delta_f H°$, kJ mol^{-1}	$\Delta_f G°$, kJ mol^{-1}	$S°$, J mol^{-1} K^{-1}
HgI_2(red)	−105.4	−101.7	180
HgI_2(yellow)	−102.9	(?)	(?)

Estimate values for the two missing entries. To do this, assume that for the transition $HgI_2(red) \longrightarrow HgI_2(yellow)$, the values of $\Delta_r H°$ and $\Delta_r S°$ at 25 °C have the same values that they do at the equilibrium temperature of 127 °C.

87. Oxides of nitrogen are produced in high-temperature combustion processes. The essential reaction is $N_2(g) + O_2(g) \rightleftharpoons 2\,NO(g)$. At what approximate temperature will an *equimolar* mixture of $N_2(g)$ and $O_2(g)$ be 1.0% converted to NO(g)? [*Hint:* Use data from Appendix D as necessary.]

88. Use the following data together with other data from the text to determine the temperature at which the equilibrium pressure of water vapor above the two solids in the following reaction is 75 Torr.

$$CuSO_4 \cdot 3\,H_2O(s) \rightleftharpoons CuSO_4 \cdot H_2O(s) + 2\,H_2O(g)$$

	$\Delta_f H°$, kJ mol^{-1}	$\Delta_f G°$, kJ mol^{-1}	$S°$, J mol^{-1} K^{-1}
$CuSO_4 \cdot 3\,H_2O(s)$	−1684.3	−1400.0	221.3
$CuSO_4 \cdot H_2O(s)$	−1085.8	−918.1	146.0

89. From the data given in Exercise 72, estimate a value of $\Delta_r S°$ at 298 K for the reaction

$$2\,NaHCO_3(s) \longrightarrow Na_2CO_3(s) + H_2O(g) + CO_2(g)$$

90. The normal boiling point of cyclohexane, C_6H_{12}, is 80.7 °C. Estimate the temperature at which the vapor pressure of cyclohexane is 100.0 mmHg.

91. The term *thermodynamic stability* refers to the sign of $\Delta_f G°$. If $\Delta_f G°$ is negative, the compound is stable with respect to decomposition into its elements. Use the data in Appendix D to determine whether $Ag_2O(s)$ is thermodynamically stable at **(a)** 25 °C and **(b)** 200 °C.

92. At 0 °C, ice has a density of 0.917 g mL^{-1} and an absolute entropy of 37.95 J mol^{-1} K^{-1}. At this temperature, liquid water has a density of 1.000 g mL^{-1} and an absolute entropy of 59.94 J mol^{-1} K^{-1}. The pressure corresponding to these values is 1 bar. Calculate ΔG, ΔS, and ΔH for the melting of two moles of ice at its normal melting point.

93. The decomposition of the poisonous gas phosgene is represented by the equation $COCl_2(g) \rightleftharpoons CO(g) + Cl_2(g)$. Values of K for this reaction are $K = 6.7 \times 10^{-9}$ at 99.8 °C and $K = 4.44 \times 10^{-2}$ at 395 °C. At what temperature is $COCl_2$ 15% dissociated when the total gas pressure is maintained at 1.00 atm?

94. Assume that the constant pressure heat capacity, C_p, of a solid is a linear function of temperature of the form $C_p = aT$, where a is a constant. Starting from the expressions below for $S°$ and $\Delta H°$, show that $S°/\Delta H° = 2/(298.15\ \text{K}) = 0.00671\ \text{K}^{-1}$, a claim made in Are You Wondering 13-3.

$$S° = \int_{0\,\text{K}}^{298.15\,\text{K}} \frac{C_p}{T}\,dT \qquad \Delta H° = \int_{0\,\text{K}}^{298.15\,\text{K}} C_p dT$$

95. The standard molar entropy of solid hydrazine at its melting point of 1.53 °C is 67.15 J mol^{-1} K^{-1}. The enthalpy of fusion is 12.66 kJ mol^{-1}. For $N_2H_4(l)$ in the interval from 1.53 °C to 298.15 K, the molar heat capacity at constant pressure is given by the expression $C_{p,m} = 97.78 + 0.0586(T - 280)$. Determine the standard molar entropy of $N_2H_4(l)$ at 298.15 K. [*Hint:* The heat absorbed to produce an infinitesimal change in the temperature of a substance is $\delta q = C_p\,dT$.]

96. Use the following data to estimate, $S°[C_6H_6(g, 1\ \text{atm})]$ at 298.15 K. For $C_6H_6(s, 1\ \text{atm})$ at its melting point of 5.53 °C, $S°$ is 128.82 J mol^{-1} K^{-1}. The enthalpy of fusion is 9.866 kJ mol^{-1}. From the melting point to 298.15 K, the average heat capacity of liquid benzene is 134.0 J mol^{-1} K^{-1}. The enthalpy of vaporization of $C_6H_6(l)$ at 298.15 K is 33.85 kJ mol^{-1}, and in the vaporization, $C_6H_6(g)$ is produced at a pressure of 95.13 Torr. Imagine that this vapor could be compressed to 1 atm pressure without condensing and while behaving as an ideal gas. [*Hint:* Refer to the preced[ing] exercise, and note the following: For infinitesimal qu[anti]ties, $dS = \delta q/dT$; for the isothermal compression of [an] ideal gas, $\delta q = -dw$; and for pressure–volume w[ork] $\delta w = -P\,dV$.

97. Consider a system that has four energy levels, w[ith] energy $\varepsilon = 0, 1, 2,$ and 3 energy units, and three pa[rti]cles labeled A, B, and C. The total energy of the s[ys]tem, in energy units, is 3. How many microstates c[an] be generated?

98. In Figure 13-5 the temperature dependence of [the] standard molar entropy for chloroform is plott[ed]. **(a)** Explain why the slope for the standard mo[lar] entropy of the solid is greater than the slope for [the] standard molar entropy of the liquid, which [is] greater than the slope for the standard molar entro[py] of the gas. **(b)** Explain why the change in the sta[n]dard molar entropy from solid to liquid is smal[ler] than that for the liquid to gas.

99. The following data are from a laboratory experim[ent] that examines the relationship between solubility a[nd] thermodynamics. In this experiment $KNO_3(s)$ [is] placed in a test tube containing some water. The so[lu]tion is heated until all the $KNO_3(s)$ is dissolved a[nd] then allowed to cool. The temperature at which cr[ys]tals appear is then measured. From this experim[ent] we can determine the equilibrium constant, Gib[bs] energy, enthalpy, and entropy for the reaction. Use [the] following data to calculate $\Delta_r G$, $\Delta_r H$, and $\Delta_r S$ for [the] dissolution of $KNO_3(s)$. (Assume the initial mass [of] $KNO_3(s)$ was 20.2 g.)

Total Volume, mL	Temperature Crystals Formed, K
25.0	340
29.2	329
33.4	320
37.6	313
41.8	310
46.0	306
51.0	303

Data reported by J. O. Schreck, *J. Chem. Educ.*, **73**, 426 (1996).

Feature Problems

100. A tabulation of more precise thermodynamic data than are presented in Appendix D lists the following values for $H_2O(l)$ and $H_2O(g)$ at 298.15 K, at a standard state pressure of 1 bar.

	$\Delta_f H°$, kJ mol^{-1}	$\Delta_f G°$, kJ mol^{-1}	$S°$, J mol^{-1} K^{-1}
$H_2O(l)$	−285.830	−237.129	69.91
$H_2O(g)$	−241.818	−228.572	188.825

(a) Use these data to determine, in two different ways, $\Delta_r G°$ at 298.15 K for the vaporization: $H_2O\ (l, 1\ \text{bar}) \rightleftharpoons H_2O(g, 1\ \text{bar})$.

(b) Use the result of part (a) to obtain the value o[f] for this vaporization and, hence, the vapor pressure [of] water at 298.15 K.

(c) The vapor pressure in part (b) is in the unit b[ar]. Convert the pressure to millimeters of mercury.

(d) Start with the value $\Delta_r G° = 8.590$ kJ mol^{-1}, a[nd] calculate the vapor pressure of water at 298.15 K i[n a] fashion similar to that in parts (b) and (c). In this w[ay] demonstrate that the results obtained in a thermo[dy]namic calculation do not depend on the conventi[on] we choose for the standard state pressure, as long [as] we use standard state thermodynamic data consist[ent] with that choice.

. The graph shows how $\Delta_r G°$ varies with temperature for three different oxidation reactions: the oxidations of C(graphite), Zn, and Mg to CO, ZnO, and MgO, respectively. Such graphs as these can be used to show the temperatures at which carbon is an effective reducing agent to reduce metal oxides to the free metals. As a result, such graphs are important to metallurgists. Use these graphs to answer the following questions.

(a) Why can Mg be used to reduce ZnO to Zn at all temperatures, but Zn cannot be used to reduce MgO to Mg at any temperature?

(b) Why can C be used to reduce ZnO to Zn at some temperatures but not at others? At what temperatures can carbon be used to reduce zinc oxide?

(c) Is it possible to produce Zn from ZnO by its direct decomposition without requiring a coupled reaction? If so, at what approximate temperatures might this occur?

(d) Is it possible to decompose CO to C and O_2 in a spontaneous reaction? Explain.

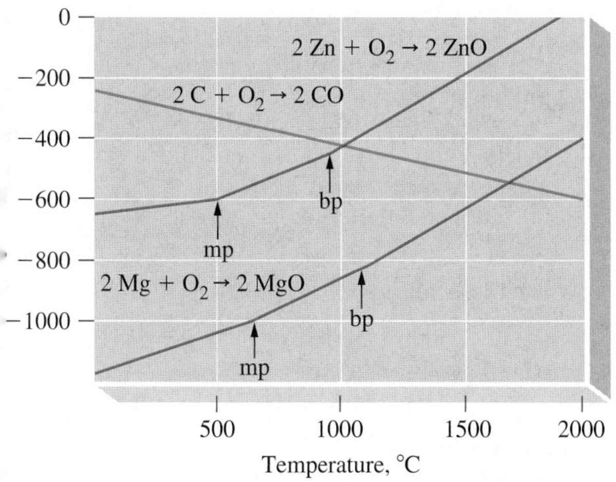

$\Delta_r G°$ for three reactions as a function of temperature. The ctions are indicated by the equations written above the phs. The points noted by arrows are the melting points) and boiling points (bp) of zinc and magnesium.

(e) To the set of graphs, add straight lines representing the reactions

$$C(graphite) + O_2(g) \longrightarrow CO_2(g)$$
$$2\,CO(g) + O_2(g) \longrightarrow 2\,CO_2(g)$$

given that the three lines representing the formation of oxides of carbon intersect at about 800 °C. [*Hint:* At what other temperature can you relate $\Delta_r G°$ and temperature?]

The slopes of the three lines described above differ sharply. Explain why this is so—that is, explain the slope of each line in terms of principles governing Gibbs energy change.

(f) The graphs for the formation of oxides of other metals are similar to the ones shown for Zn and Mg; that is, they all have positive slopes. Explain why carbon is such a good reducing agent for the reduction of metal oxides.

. In a heat engine, heat (q_h) is absorbed by a working substance (such as water) at a high temperature (T_h). Part of this heat is converted to work (w), and the rest (q_l) is released to the surroundings at the lower temperature (T_l). The *efficiency* of a heat engine is the ratio w/q_h. The second law of thermodynamics establishes the following equation for the maximum efficiency of a heat engine, expressed on a percentage basis.

$$\text{efficiency} = \frac{w}{q_h} \times 100\% = \frac{T_h - T_l}{T_h} \times 100\%$$

In a particular electric power plant, the steam leaving a steam turbine is condensed to liquid water at 41 °C (T_l) and the water is returned to the boiler to be regenerated as steam. If the system operates at 36% efficiency,

(a) What is the minimum temperature of the steam [$H_2O(g)$] used in the plant?

(b) Why is the actual steam temperature probably higher than that calculated in part (a)?

(c) Assume that at T_h the $H_2O(g)$ is in equilibrium with $H_2O(l)$. Estimate the steam pressure at the temperature calculated in part (a).

(d) Is it possible to devise a heat engine with greater than 100 percent efficiency? With 100 percent efficiency? Explain.

103. Refer to Focus On 13-1: Coupled Reactions in Biological Systems. The Gibbs energy available from the complete combustion of 1 mol of glucose to carbon dioxide and water is

$$C_6H_{12}O_6(aq) + 6\,O_2(g) \longrightarrow 6\,CO_2(g) + 6\,H_2O(l)$$
$$\Delta_r G° = -2870 \text{ kJ mol}^{-1}$$

(a) Under biological standard conditions, compute the maximum number of moles of ATP that could form from ADP and phosphate if all the energy of combustion of 1 mol of glucose could be utilized.

(b) The actual number of moles of ATP formed by a cell under aerobic conditions (that is, in the presence of oxygen) is about 38. Calculate the efficiency of energy conversion of the cell.

(c) Consider these typical physiological conditions.

$$P_{CO_2} = 0.050 \text{ bar}; P_{O_2} = 0.132 \text{ bar};$$
$$[\text{glucose}] = 1.0 \text{ mg/mL}; \text{pH} = 7.0;$$
$$[\text{ATP}] = [\text{ADP}] = [P_i] = 0.00010 \text{ M}.$$

Calculate $\Delta_r G°$ for the conversion of 1 mol ADP to ATP and $\Delta_r G°$ for the oxidation of 1 mol glucose under these conditions.

(d) Calculate the efficiency of energy conversion for the cell under the conditions given in part (c). Compare this efficiency with that of a diesel engine that attains 78% of the theoretical efficiency operating with T_h = 1923 K and T_l = 873 K. Suggest a reason for your result. [*Hint:* See the preceding exercise.]

104. The entropy of materials at T = 0 K should be zero; however, for some substances, this is not true. The difference between the measured value and expected value of zero is known as residual entropy. This residual entropy arises because the molecules can have a number of different orientations in the crystal and can

be estimated by applying Boltzmann's formula (equation 13.1) with an appropriate value for W.
(a) Given that a CO molecule can have one of two possible orientations in a crystal (CO or OC), calculate the residual entropy for a crystal consisting of one mole of CO at $T = 0$ K.
(b) Ice consists of water molecules arranged in a tetrahedral pattern. In this configuration the question is whether a given hydrogen atom is halfway between two oxygen atoms or closer to one oxygen atom than the other. Starting with a series of basic assumptions, Linus Pauling used the agreement of his calculated residual entropy and the experimental value to determine that a given hydrogen at is closer to one oxygen atom. An approach us by Pauling to calculate the residual entropy can summarized as follows.
i) Place four oxygen atoms in a tetrahedral arran ment around a central water molecule.
ii) Determine the number of orientations of the c tral water molecule, assuming that it forms t hydrogen bonds with the surrounding oxygen ato
iii) Using this approach, plus the additional obs vation made by Pauling that only one-quarter of orientations identified in (ii) are accessible, calcul the residual entropy of one mole of H_2O.

Self-Assessment Exercises

105. In your own words, define the following symbols: **(a)** ΔS_{univ}; **(b)** $\Delta_f G°$; **(c)** K; **(d)** $S°$; **(e)** $\Delta_r G°$.

106. Briefly describe each of the following ideas, methods, or phenomena: **(a)** absolute molar entropy; **(b)** coupled reactions; **(c)** Trouton's rule; **(d)** evaluation of an equilibrium constant from tabulated thermodynamic data.

107. Explain the important distinctions between each of the following pairs: **(a)** spontaneous and nonspontaneous processes; **(b)** the second and third laws of thermodynamics; **(c)** $\Delta_r G°$ and $\Delta_r G°$; **(d)** ΔG and $\Delta_r G$.

108. For a process to occur spontaneously, **(a)** the entropy of the system must increase; **(b)** the entropy of the surroundings must increase; **(c)** both the entropy of the system and the entropy of the surroundings must increase; **(d)** the net change in entropy of the system and surroundings considered together must be a positive quantity; **(e)** the entropy of the universe must remain constant.

109. The Gibbs energy of a reaction can be used to assess which of the following? **(a)** how much heat is absorbed from the surroundings; **(b)** how much work the system does on the surroundings; **(c)** the net direction in which the reaction occurs to reach equilibrium; **(d)** the proportion of the heat evolved in an exothermic reaction that can be converted to various forms of work.

110. The reaction, $2\ Cl_2O(g) \longrightarrow 2\ Cl_2(g) + O_2(g)$ $\Delta_r H° = -161\ \text{kJ mol}^{-1}$, is expected to be **(a)** spontaneous at all temperatures; **(b)** spontaneous at low temperatures, but nonspontaneous at high temperatures; **(c)** nonspontaneous at all temperatures; **(d)** spontaneous at high temperatures only.

111. If $\Delta_r G° = 0$ for a reaction, it must also be true that **(a)** $K = 0$; **(b)** $K = 1$; **(c)** $\Delta_r H° = 0$; **(d)** $\Delta_r S° = 0$; **(e)** the equilibrium activities of the reactants and products do not depend on the initial conditions.

112. Two correct statements about the reversible reaction $N_2(g) + O_2(g) \rightleftharpoons 2\ NO(g)$ are **(a)** $K = K_p$; **(b)** the equilibrium amount of NO increases with an increased total gas pressure; **(c)** the equilibri amount of NO increases if an equilibrium mixt is transferred from a 10.0 L container to a 20. container; **(d)** $K = K_c$; **(e)** the composition of equilibrium mixture of the gases is independent the temperature.

113. *Without performing detailed calculations,* indic whether any of the following reactions would occu a measurable extent at 298 K.
(a) Conversion of dioxygen to ozone:
$$3\ O_2(g) \longrightarrow 2\ O_3(g)$$
(b) Dissociation of N_2O_4 to NO_2:
$$N_2O_4(g) \longrightarrow 2\ NO_2(g)$$
(c) Formation of BrCl:
$$Br_2(l) + Cl_2(g) \longrightarrow 2\ BrCl(g)$$

114. Explain briefly why
(a) the change in entropy in a system is not alway suitable criterion for spontaneous change;
(b) $\Delta_r G°$ is so important in dealing with the quest of spontaneous change, even though the conditic employed in a reaction are very often nonstandar

115. A handbook lists the following standa enthalpies of formation at 298 K for cyclopenta C_5H_{10}: $\Delta_f H°[C_5H_{10}(l)] = -105.9\ \text{kJ mol}^{-1}$ a $\Delta_f H°[C_5H_{10}(g)] = -77.2\ \text{kJ mol}^{-1}$.
(a) Estimate the normal boiling point of cyclopenta
(b) Estimate $\Delta_r G°$ for the vaporization of cyclop tane at 298 K.
(c) Comment on the significance of the sign of Δ at 298 K.

116. Consider the reaction $NH_4NO_3(s) \longrightarrow N_2O(g)$ $2\ H_2O(l)$ at 298 K.
(a) Is the forward reaction endothermic exothermic?
(b) What is the value of $\Delta_r G°$ at 298 K?
(c) What is the value of K at 298 K?
(d) Does the reaction tend to occur spontaneou at temperatures above 298 K, below 298 K, both, neither?

7. Which of the following graphs of Gibbs energy versus the extent of reaction represents an equilibrium constant closest to 1?

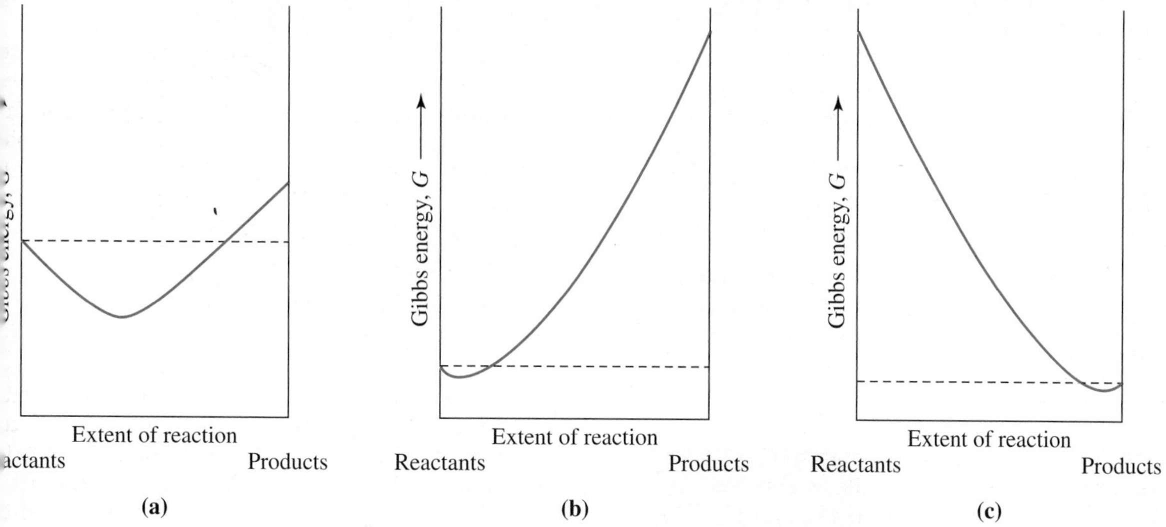

118. At room temperature and normal atmospheric pressure, is the entropy change of the universe positive, negative, or zero for the transition of carbon dioxide solid to liquid?

14 Solutions and Their Physical Properties

CONTENTS

LEARNING OBJECTIVES

14.1 Describe the meaning of solvent and solute and their relative amounts in dilute or concentrated solutions.

14.2 Differentiate between molarity and molality, and describe the effect of temperature on each.

14.3 Identify the three kinds of interactions that determine whether or not a solution is ideal.

14.4 Differentiate between saturated, unsaturated, and supersaturated solutions. Explain how to prepare a supersaturated solution.

14.5 Describe how the solubility of a gas is affected by temperature and by pressure.

14.6 Explain how a boiling-point diagram (a plot of boiling point versus mole fraction) differs for an ideal solution versus an azeotrope.

14.7 Describe what is meant by osmosis, osmotic pressure, and reverse osmosis.

14.8 Describe the processes and applications of freezing-point depression and boiling-point elevation of nonelectrolyte solutions.

14.9 Explain why colligative properties of electrolyte solutions are more difficult to calculate than for nonelectrolyte solutions.

14.10 Describe a colloidal mixture and the Tyndall effect.

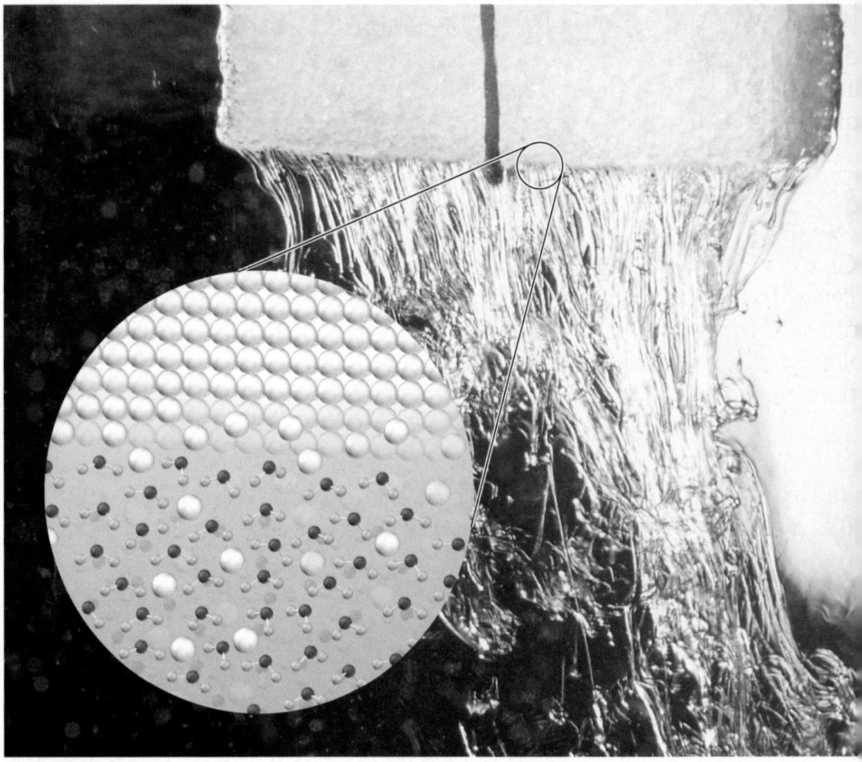

The dissolving of a cube of sugar (sucrose) is seen here as swirls of a higher-density sucrose solution falling through the lower-density water. The microscopic view shows the sucrose molecules, represented by white spheres, leaving the lattice and dispersir among the water molecules, represented by the red-and-white balls and sticks.

Residents of cold climate regions know they must add antifreeze the water in the cooling system of an automobile in the winter. T antifreeze–water mixture has a much lower freezing point than dc pure water. In this chapter we will learn why.

To restore body fluids to a dehydrated individual by intravenous inj tion, pure water cannot be used. A solution with just the right value o physical property known as osmotic pressure is necessary, and tl requires a solution of a particular concentration of solutes. Again, in tl chapter we will learn why.

Altogether, we will explore several solution properties whose values ɔend on solution concentration. Our emphasis will be on describing solu- n phenomena and their applications and explaining these phenomena at molecular level.

4-1 Types of Solutions: Some Terminology

Chapters 1 and 4 we learned that a solution is a *homogeneous mixture*. It is *nogeneous* because its composition and properties are uniform, and it is a *cture* because it contains two or more substances in proportions that can be ied. Recall that a solution is composed of a solvent and one or more solutes. e **solvent** is the component that is present in the greatest quantity or that ermines the state of matter in which the solution exists. A **solute** is said to dissolved in the solvent. A *concentrated* solution has a relatively large quan- of dissolved solute(s), and a *dilute* solution has only a small quantity. nsider solutions containing sucrose (cane sugar) as one of the solutes in the vent water: Pancake syrup is a concentrated solution, whereas a sweetened ɔ of coffee is much more dilute.

Although liquid solutions are most common, solutions can exist in gaseous d solid states as well. For instance, the U.S. five-cent coin, the nickel, is a id solution of 75% Cu and 25% Ni. Solid solutions with a metal as the sol- nt are also called *alloys*.* Table 14.1 lists a few common solutions.

4-2 Solution Concentration

Chapters 4 and 5 we learned that to describe a solution fully we must know *concentration*—a measure of the quantity of solute in a given quantity of sol- nt (or solution). The concentration unit we stressed in those chapters was larity. In this section, we describe several methods of expressing concentra- n, each of which serves a different purpose.

ass Percent, Volume Percent, and Mass/Volume Percent

we dissolve 5.00 g NaCl in 95.0 g H_2O, we get 100.0 g of a solution that is 0% NaCl, by *mass*. Mass percent is widely used in industrial chemistry. us, we might read that the action of 78% H_2SO_4(aq) on phosphate rock $Ca_3(PO_4)_2 \cdot CaF_2$] produces 46% H_3PO_4(aq).

ABLE 14.1 Some Common Solutions

olution	Components
aseous solutions	
ir	N_2, O_2, and several others
atural gas	CH_4, C_2H_6, and several others
quid solutions	
eawater	H_2O, NaCl, and many others
inegar	H_2O, CH_3COOH (acetic acid)
oda pop	H_2O, CO_2, $C_{12}H_{22}O_{11}$ (sucrose), and several others
olid solutions	
ellow brass	Cu, Zn
alladium–Hydrogen	Pd, H_2

e term *alloy* can also apply to certain heterogeneous mixtures, such as the common two-phase id mixture of lead and tin known as *solder*, or to intermetallic compounds, such as the er–tin compound Ag_3Sn that, in the past, was mixed with mercury in *dental amalgam*.

Because liquid volumes are so easily measured, some solutions are p pared on a *volume* percent basis. For example, a handbook lists a freezi point of −15.6 °C for a methyl alcohol–water antifreeze solution that is 25. CH_3OH, by volume. Such a solution could be prepared by dissolving 25.0 $CH_3OH(l)$ with water until the total solution volume is 100.0 mL.

Another possibility is to express the *mass* of solute and *volume* of soluti An aqueous solution with 0.9 g NaCl in 100.0 mL of solution is said to be 0. NaCl (*mass/volume*). Mass/volume percent is extensively used in the medi and pharmaceutical fields.

Parts per Million, Parts per Billion, and Parts per Trillion

In solutions where the mass or volume percent of a component is very low, often switch to other units to describe solution concentration. For examp 1 mg solute/L solution amounts to only 0.001 g/L. A solution that is this dil will have the same density as water, approximately 1 g/mL; therefc the solution concentration is 0.001 g solute/1000 g solution, which is same as 1g solute/1,000,000 g solution. We can describe the solute concent tion more succinctly as 1 *part per million* **(ppm)**. For a solution with o 1 μg solute/L solution, the situation is 1×10^{-6} g solute/1000 g solution, 1.0 g solute/1×10^9 g solution. Here, the solute concentration is 1 *part per lion* **(ppb)**. If the solute concentration is only 1 ng solute/L solution, the co centration is 1 *part per trillion* **(ppt)**.

▶ A part per billion: The average head of hair has 10,000 hairs. Therefore, 1 part per billion is one hair out of 100,000 people.

KEEP IN MIND

that 1 ppm = 1 mg/L, 1 ppb = 1 μg/L, and 1 ppt = 1 ng/L.

Because these terms are widely used in environmental reporting, they m be more familiar than other units that chemists use. For example, a consun in California might read in an annual water quality report from the munici water department that the maximum contaminant level allowed for nitr ion is 45 ppm and for carbon tetrachloride, 0.5 ppb.

Mole Fraction and Mole Percent

To relate certain physical properties (such as vapor pressure) to solution co centration, we need a unit in which all solution components are expressed a mole basis. We can do this with the mole fraction. The *mole fraction* of co ponent *i*, designated x_i, is the fraction of all the molecules in a solution that of type *i*. The mole fraction of component *j* is x_j, and so on. The mole fracti of a solution component is defined as

$$x_i = \frac{\text{amount of component } i \text{ (in moles)}}{\text{total amount of all solution components (in moles)}}$$

The sum of the mole fractions of all the solution components is 1.

$$x_i + x_j + x_k + \cdots = 1$$

The *mole percent* of a solution component is the percent of all the molecu in solution that are of a given type. Mole percents are mole fractions mu plied by 100%.

14-1 CONCEPT ASSESSMENT

In one mole of a solution with a mole fraction of 0.5 water, how many water molecules would there be?

Molarity

In Chapters 4 and 5 we introduced molarity to provide a conversion fac relating the amount of solute and the volume of solution. We used it in varic stoichiometric calculations. As we learned at that time,

$$\text{molarity } (c) = \frac{\text{amount of solute (in moles)}}{\text{volume of solution (in liters)}}$$

olality

ppose we prepare a solution at 20 °C by using a volumetric flask calibrated at °C. Then suppose we warm this solution to 25 °C. As the temperature creases from 20 to 25 °C, the amount of solute remains constant, but the solu-n volume increases slightly (by about 0.1%). The number of moles of solute r liter—the molarity—*decreases* slightly (by about 0.1%). This temperature pendence of molarity can be a problem in experiments demanding a high ecision. That is, the solution might be used at a temperature different from the e at which it was prepared, and so its molarity is not exactly the one written the label. A concentration unit that is *independent* of temperature, and also oportional to mole fraction in dilute solutions, is **molality** (m)—the number moles of solute per kilogram of *solvent* (not of solution). A solution in which)0 mol of urea, $CO(NH_2)_2$, is dissolved in 1.00 kg of water is described as a)0 molal solution and designated as 1.00 m $CO(NH_2)_2$. Molality is defined as

$$\text{molality } (m) = \frac{\text{amount of solute (in moles)}}{\text{mass of solvent (in kilograms)}}$$

◀ Molal units, being the number of moles per kg of solvent, are independent of temperature in contrast to molar units, being the number of moles per liter.

The concentration of a solution is expressed in several different ways in ample 14-1. The calculation in Example 14-2 is perhaps more typical: A con-ntration is converted from one unit (molarity) to another (mole fraction).

EXAMPLE 14-1 Expressing a Solution Concentration in Various Units

An ethanol–water solution is prepared by dissolving 10.00 mL of ethanol, CH_3CH_2OH (d = 0.789 g/mL), in a sufficient volume of water to produce 100.0 mL of a solution with a density of 0.982 g/mL (Fig. 14-1). What is the concentration of ethanol in this solution expressed as **(a)** volume percent; **(b)** mass percent; **(c)** mass/volume percent; **(d)** mole fraction; **(e)** mole percent; **(f)** molarity; **(g)** molality?

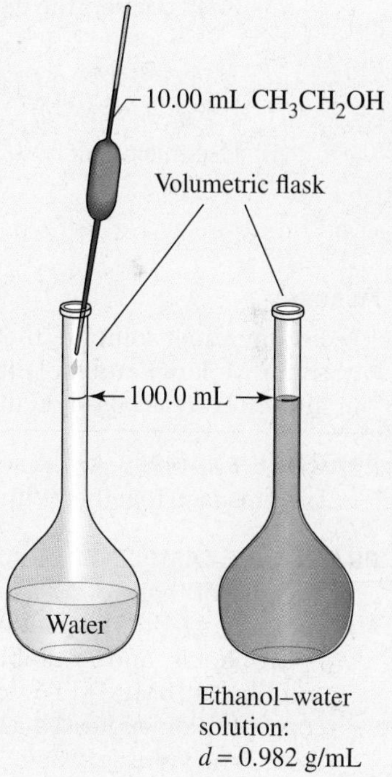

▲ FIGURE 14-1 **Preparation of an ethanol–water solution—Example 14-1 illustrated** A 10.00 mL sample of CH_3CH_2OH is added to some water in the volumetric flask. The solution is mixed, and more water is added to bring the total volume to 100.0 mL.

Analyze

Each part of this problem uses an equation presented in the text. Expressing concentrations in these different units will illustrate the similarities and differences among volume percent, mass percent, mass/volume percent, mole fraction, mole percent, molarity, and molality.

Solve

(a) Volume percent ethanol

$$\text{volume percent ethanol} = \frac{10.00\text{ mL ethanol}}{100.0\text{ mL solution}} \times 100\% = 10.00\%$$

(b) Mass percent ethanol

$$\text{mass ethanol} = 10.00\text{ mL ethanol} \times \frac{0.789\text{ g ethanol}}{1.00\text{ mL ethanol}}$$

$$= 7.89\text{ g ethanol}$$

$$\text{mass soln} = 100.0\text{ mL soln} \times \frac{0.982\text{ g soln}}{1.0\text{ mL solution}} = 98.2\text{ g soln}$$

$$\text{mass percent ethanol} = \frac{7.89\text{ g ethanol}}{98.2\text{ g solution}} \times 100\% = 8.03\%$$

(c) Mass/volume percent ethanol

$$\text{mass/volume percent ethanol} = \frac{7.89\text{ g ethanol}}{100.0\text{ mL solution}} \times 100\% = 7.89\%$$

(continued)

(d) Mole fraction of ethanol
Convert the mass of ethanol from part (b) to an amount in moles.

$$? \text{ mol } CH_3CH_2OH = 7.89 \text{ g } CH_3CH_2OH \times \frac{1 \text{ mol } CH_3CH_2OH}{46.07 \text{ g } CH_3CH_2OH}$$

$$= 0.171 \text{ mol } CH_3CH_2OH$$

Determine the mass of water present in 100.0 mL of solution.

$$98.2 \text{ g soln} - 7.89 \text{ g ethanol} = 90.3 \text{ g water}$$

Convert the mass of water to the number of moles present.

$$? \text{ mol } H_2O = 90.3 \text{ g } H_2O \times \frac{1 \text{ mol } H_2O}{18.02 \text{ g } H_2O} = 5.01 \text{ mol } H_2O$$

$$x_{CH_3CH_2OH} = \frac{0.171 \text{ mol } CH_3CH_2OH}{0.171 \text{ mol } CH_3CH_2OH + 5.01 \text{ mol } H_2O} = \frac{0.171}{5.18} = 0.0330$$

(e) Mole percent ethanol

$$\text{mole percent } CH_3CH_2OH = x_{CH_3CH_2OH} \times 100\% = 0.0330 \times 100\% = 3.30\%$$

(f) Molarity of ethanol
Divide the number of moles of ethanol from part (d) by the solution volume, 100.0 mL = 0.1000 L.

$$\text{molarity} = \frac{0.171 \text{ mol } CH_3CH_2OH}{0.1000 \text{ L soln}} = 1.71 \text{ M}$$

(g) Molality of ethanol
First, convert the mass of water present in 100.0 mL of solution [from part (d)] to the unit kg.

$$? \text{ kg } H_2O = 90.3 \text{ g } H_2O \times \frac{1 \text{ kg } H_2O}{1000 \text{ g } H_2O} = 0.0903 \text{ kg } H_2O$$

Use this result and the number of moles of CH_3CH_2OH from part (d) to establish the molality.

$$\text{molality} = \frac{0.171 \text{ mol } CH_3CH_2OH}{0.0903 \text{ kg } H_2O} = 1.89 \text{ mol kg}^{-1}$$

Assess

For the same solution, the volume percent, mass percent, and mass/volume percent are not necessarily the same. Molarity and molality are also not the same values, because molarity is based on the volume of solution and molality is based on the mass of the solvent.

PRACTICE EXAMPLE A: A solution that is 20.0% ethanol, by volume, is found to have a density of 0.977 g/mL. Use this fact, together with data from Example 14-1, to determine the mass percent ethanol in the solution.

PRACTICE EXAMPLE B: Ionic liquids (ILs) are salts with relatively low melting points and vapor pressures. Because of their low volatilities, chemists continue to investigate the potential of ILs as safer and "greener" solvents. 1-Butyl-3-methylimidazolium hexafluorophosphate, [BMIM][PF6], is a viscous, colorless, hydrophobic, and insoluble ionic liquid with a molar mass of 284.1 g/mol and a density of 1.38 g/mL. In a solution of [BMIM][PF6] and carbon dioxide at 8 MPa, the mole fraction of CO_2 is 0.60. What is the solution concentration expressed as the **(a)** molarity of carbon dioxide and **(b)** molality of carbon dioxide? Assume that there is no volume change when CO_2 is added to [BMIM][PF6].

EXAMPLE 14-2 Converting Molarity to Mole Fraction

Laboratory ammonia is 14.8 M NH_3(aq) with a density of 0.8980 g/mL. What is x_{NH_3} in this solution?

Analyze

In this problem we note that no volume of solution is stated, suggesting that our calculation can be based on any fixed volume of our choice. A convenient volume to work with is one liter. We need to determine the number of moles of NH_3 and of H_2O in one liter of the solution.

Solve

Find the number of moles of NH_3 by using the definition of molarity.

$$\text{moles of } NH_3 = 1.00 \text{ L} \times \frac{14.8 \text{ mol } NH_3}{1 \text{ L}} = 14.8 \text{ mol } NH_3$$

For moles of H_2O, first find mass of the solution by using solution density.

$$\text{mass of soln} = 1000.0 \text{ mL soln} \times \frac{0.8980 \text{ g soln}}{1.0 \text{ mL solution}} = 898.0 \text{ g soln}$$

Then use moles of NH_3 and molar mass to find the mass of NH_3.

$$\text{mass of } NH_3 = 14.8 \text{ mol } NH_3 \times \frac{17.03 \text{ g } NH_3}{1 \text{ mol } NH_3} = 252 \text{ g } NH_3$$

Find the mass of H_2O by subtracting the mass of NH_3 from the solution mass.

$$\text{mass of } H_2O = 898.0 \text{ g soln} - 252 \text{ g } NH_3 = 646 \text{ g } H_2O$$

Find moles of H_2O by multiplying by the inverse of the molar mass for H_2O.

$$\text{moles of } H_2O = 646 \text{ g } H_2O \times \frac{1 \text{ mol } H_2O}{18.02 \text{ g } H_2O} = 35.8 \text{ mol } H_2O$$

Find the mole fraction of ammonia x_{NH_3} by dividing moles NH_3 by the total number of moles of NH_3 and H_2O in the solution.

$$x_{NH_3} = \frac{14.8 \text{ mol } NH_3}{14.8 \text{ mol } NH_3 + 35.8 \text{ mol } H_2O} = 0.292$$

Assess

By using the solution concentration definitions, we were able to convert from one concentration unit to another. This skill is used frequently by chemists.

PRACTICE EXAMPLE A: A 16.00% aqueous solution of glycerol, $HOCH_2CH(OH)CH_2OH$, by mass, has a density of 1.037 g/mL. What is the mole fraction of glycerol in this solution?

PRACTICE EXAMPLE B: A 10.00% aqueous solution of sucrose, $C_{12}H_{22}O_{11}$, by mass, has a density of 1.040 g/mL. What is **(a)** the molarity; **(b)** the molality; and **(c)** the mole fraction of $C_{12}H_{22}O_{11}$, in this solution?

14-2 CONCEPT ASSESSMENT

Which of the several concentration units described in Section 14-2 are temperature-dependent and which are not? Explain.

14-3 Intermolecular Forces and the Solution Process

We can often understand a process if we analyze its energy requirements; this approach can help us to explain why some substances mix to form solutions and others do not. In this section, we focus on the behavior of molecules in solution, specifically on intermolecular forces and their contribution to the energy required for the dissolution process.

Enthalpy of Solution

In the formation of some solutions, heat is given off to the surroundings; in other cases, heat is absorbed. An enthalpy of solution, $\Delta_{soln}H$, can be rather easily measured—for example, in the coffee-cup calorimeter of Figure 7-6—but why should some solution processes be exothermic, whereas others are endothermic?

Let's think in terms of a three-step approach to $\Delta_{soln}H$. First, solvent molecules must be separated from one another to make room for the solute molecules. Some energy is required to overcome the forces of attraction between solvent molecules. As a result, this step should be an endothermic one: $\Delta H > 0$. Second, the solute molecules must be separated from one another. This step, too, will consume energy and should be endothermic. Finally, we imagine that the solvent and solute molecules are attracted to one another. Therefore, when the separated solvent and solute molecules combine to form a solution, we expect energy to be released. This is an exothermic

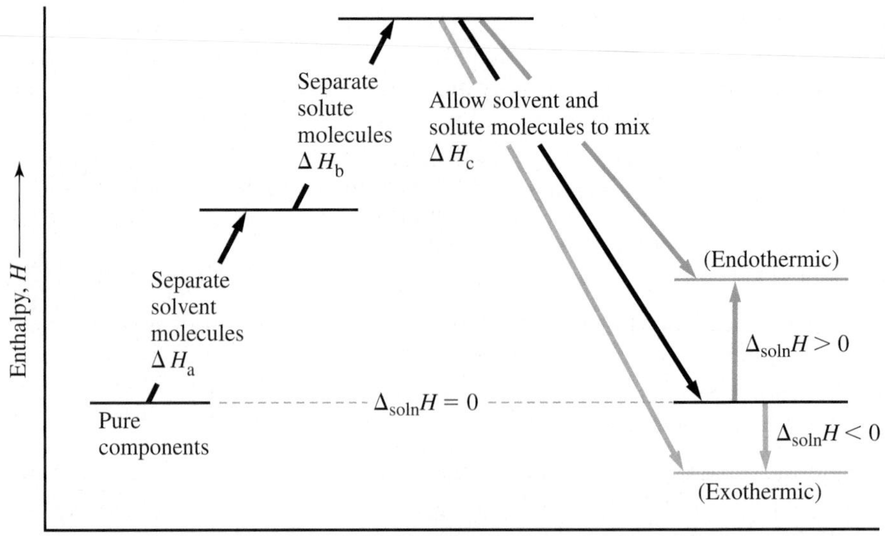

▲ FIGURE 14-2
Enthalpy diagram for solution formation
The solution process can be endothermic (blue arrow), exothermic (red arrow), or has $\Delta_{soln}H = 0$ (black arrow), depending on the magnitude of the enthalpy change in the mixing step.

step: $\Delta H < 0$. The enthalpy of solution is the sum of the three enthal changes just described, and depending on their relative values, $\Delta_{soln}H$ either positive (endothermic) or negative (exothermic). This three-step proc is summarized by equation (14.1) and Figure 14-2.

(a)	pure solvent	$\longrightarrow$	separated solvent molecules	ΔH_a >
(b)	pure solute	$\longrightarrow$	separated solute molecules	ΔH_b >
(c)	separated solvent and solute molecules	$\longrightarrow$	solution	ΔH_c <
Overall:	pure solvent + pure solute	$\longrightarrow$	solution	

$$\Delta_{soln}H = \Delta H_a + \Delta H_b + \Delta H_c \quad \textbf{(14}$$

The enthalpy changes ΔH_a and ΔH_b can sometimes be identified with oth more familiar, enthalpy changes. For example, when the solute and solvent both liquids (a common situation), ΔH_a is equal to $\Delta_{vap}H$ for the solvent a ΔH_b is equal to $\Delta_{vap}H$ for the solute. Similarly, if the solute and solvent atomic or molecular solids, then ΔH_a is equal to $\Delta_{sub}H$ for the solvent a ΔH_b is equal to $\Delta_{sub}H$ for the solute.

The enthalpy change ΔH_c includes two contributions: the enthalpy chan for the mixing of solvent and solute in the gas phase plus the enthal change for the mixture to undergo a phase change from gas to either the so or the liquid state.

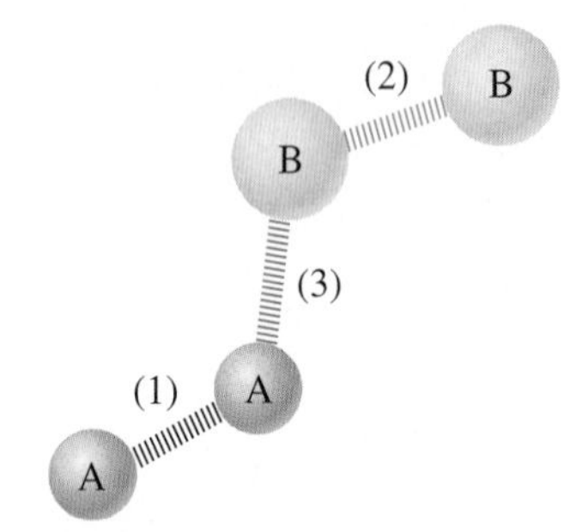

▲ FIGURE 14-3
Intermolecular forces in a solution
The intermolecular forces of attraction, represented here by dashed lines, are between: (1) solvent molecules, A–A; (2) solute molecules, B–B; and (3) solvent and solute molecules, A–B.

Intermolecular Forces in Mixtures

We see from equation (14.1) that the magnitude and sign of $\Delta_{soln}H$ depen on the values of the three terms ΔH_a, ΔH_b, and ΔH_c. These, in turn, depe on the strengths of the *three* kinds of intermolecular forces of attraction rep sented in Figure 14-3. Four possibilities for the relative strengths of these int molecular forces are described in the discussion that follows.

1. If the intermolecular forces of attraction shown in Figure 14-3 are of t same type or of equal strength, the solute and solvent molecules m

randomly. A homogeneous mixture or solution results. Because properties of solutions of this type can generally be predicted from the properties of the pure components, they are called **ideal solutions**. There is no overall enthalpy change in the formation of an ideal solution from its components, and $\Delta_{soln}H = 0$. This means that ΔH_c in equation (14.1) is equal in magnitude and opposite in sign to the sum of ΔH_a and ΔH_b. Many mixtures of liquid hydrocarbons fit this description, or very nearly so (Fig. 14-4).

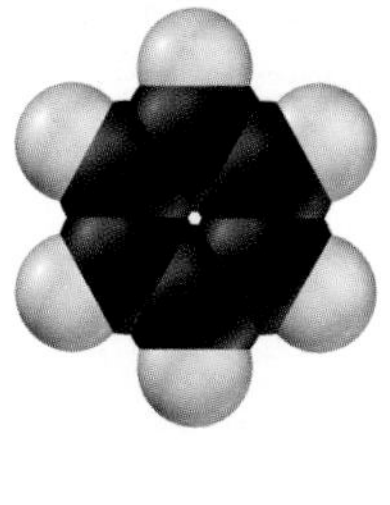

(a)

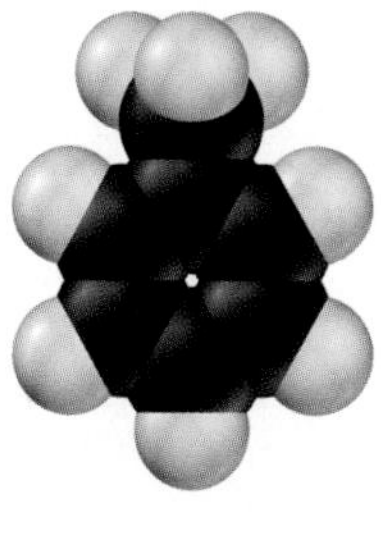

(b)

▲ FIGURE 14-4
Two components of a nearly ideal solution
Think of the $-CH_3$ group in toluene **(b)** as a small "bump" on the planar benzene ring **(a)**. Substances with similar molecular structures have similar intermolecular forces of attraction.

If the forces between unlike molecules are more attractive than those between like molecules, a solution also forms. The properties of such solutions generally cannot be predicted, however, and they are called *nonideal solutions*. Interactions between solute and solvent molecules (ΔH_c) release more heat than the heat absorbed to separate the solvent and solute molecules ($\Delta H_a + \Delta H_b$). The solution process is exothermic ($\Delta_{soln}H < 0$). Solutions of acetone and chloroform fit this type. As suggested by Figure 14-5, weak hydrogen bonding occurs between the two kinds of molecules, but the conditions for hydrogen bonding are not met in either of the pure liquids alone.*

If the forces between solute and solvent are *somewhat weaker* than between molecules of the same kind, complete mixing may still occur, the solution formed is *nonideal*. The solution has a higher enthalpy than the pure components, and the solution process is endothermic. This type of behavior is observed in mixtures of carbon disulfide (CS_2), a nonpolar liquid, and acetone, a polar liquid. In these mixtures, the acetone molecules are attracted to other acetone molecules by dipole–dipole interactions and hence show a preference for other acetone molecules as neighbors. A possible explanation of how a solution process can be endothermic and still occur is found on page 649.

Finally, if forces of attraction between unlike molecules are *much weaker* than those between like molecules, the components remain segregated in a *heterogeneous mixture*. Dissolution does not occur to any significant extent.

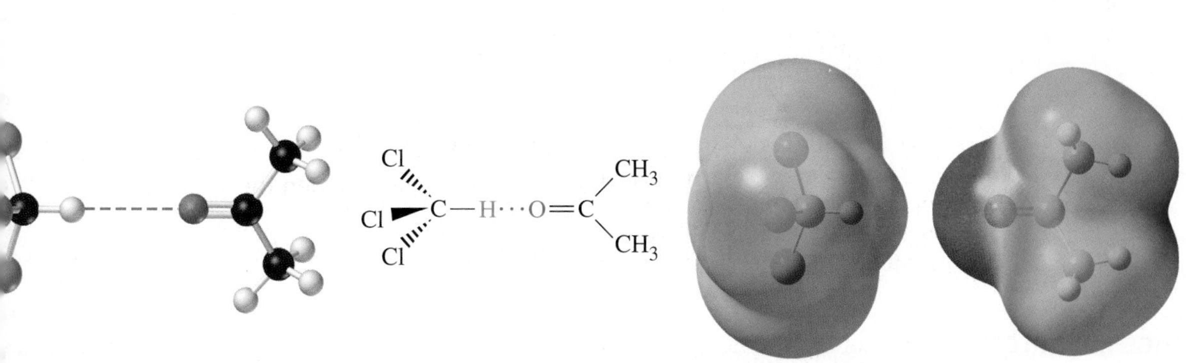

FIGURE 14-5
ermolecular force between unlike molecules leading to a nonideal ution
e interaction between these molecules is illustrated by using three different presentations: ball and stick, line representation, and electrostatic potential maps. drogen bonding between $CHCl_3$ (chloroform) and $(CH_3)_2CO$ (acetone) molecules oduces forces of attraction between unlike molecules that exceed those between e molecules.

most cases, H atoms bonded to C atoms cannot participate in hydrogen bonding. In a mole- e like $CHCl_3$, however, the three Cl atoms have a strong electron-withdrawing effect on elec- ns in the C—H bond ($\mu = 1.01$ D). The H atom is then attracted to a lone pair of electrons on O atom of $(CH_3)_2CO$ (but not to Cl atoms in other $CHCl_3$ molecules).

▶ Remember that "like dissolves like."

As an oversimplified summary of the four cases described in the precedir paragraphs, we can say that "like dissolves like." That is, substances with sin lar molecular structures are likely to exhibit similar intermolecular forces attraction and to be soluble in one another. Substances with dissimilar stru tures are likely not to form solutions. Of course, in many cases, parts of t structures may be similar and parts may be dissimilar. Then it is a matter of tr ing to establish which are the more important parts, a matter we explore Example 14-3.

EXAMPLE 14-3 Using Intermolecular Forces to Predict Solution Formation

Predict whether or not a solution will form in each of the following mixtures and whether the solution is likely to be ideal: **(a)** ethyl alcohol, CH_3CH_2OH, and water; **(b)** the hydrocarbons hexane, $CH_3(CH_2)_4CH_3$, and octane, $CH_3(CH_2)_6CH_3$; **(c)** octanol, $CH_3(CH_2)_6CH_2OH$, and water.

Analyze

Keep in mind that ideal or nearly ideal solutions are not too common. They require the solvent and solute(s) to be quite similar in structure.

(a) If we think of water as H—OH, ethyl alcohol is similar to water. (Just substitute the group CH_3CH_2— for one of the H atoms in water.) Both molecules meet the requirements of hydrogen bonding as an important intermolecular force. The strengths of the hydrogen bonds between like molecules and between unlike molecules are likely to differ, however.

(b) In hexane, the carbon chain is six atoms long, and in octane it is eight. Both substances are virtually nonpolar, and intermolecular attractive forces (of the dispersion type) should be quite similar both in the pure liquids and in the solution.

(c) At first sight, this case may seem similar to (a), with the substitution of a hydrocarbon group for a H atom in H—OH. Here, however, the carbon chain is *eight* members long. This long carbon chain is much more important than the terminal —OH group in establishing the physical properties of octanol. Viewed from this perspective, octanol and water are quite *dissimilar*.

Solve

(a) We expect ethyl alcohol and water to form *nonideal* solutions.

(b) We expect a solution to form, and it should be nearly *ideal*.

(c) We do not expect a solution to form.

Assess

In these types of problems a strong understanding of both molecular structure and intermolecular forces is required. Keep in mind the statement "like dissolves like."

In our answer to part (c), we observed that octanol does not form a solution with water; however, alcohols, such as butyl alcohol, $CH_3CH_2CH_2CH_2OH$, have a limited solubility in water (9 grams per 100 grams of water). The aqueous solubilities of alcohols fall off fairly rapidly as the hydrocarbon chain length increases beyond four.

PRACTICE EXAMPLE A: Which of the following organic compounds do you think is most readily soluble in water? Explain.

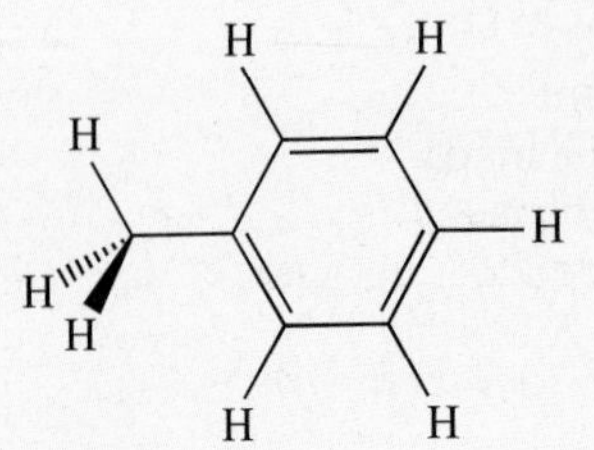

(a) Toluene

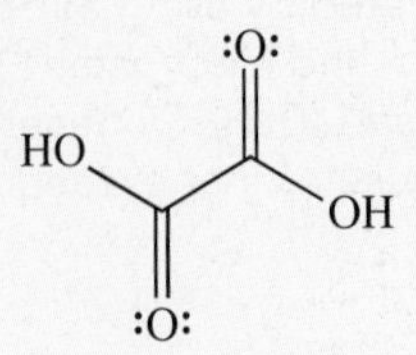

(b) Oxalic acid

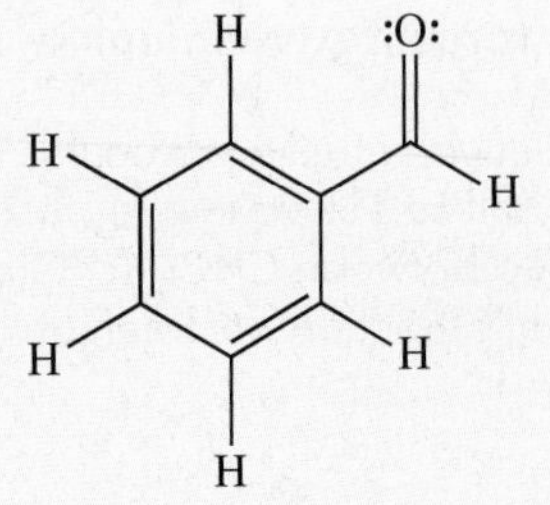

(c) Benzaldehyde

PRACTICE EXAMPLE B: In which solvent is solid iodine likely to be more soluble, water or carbon tetrachloride? Explain.

The approach described for determining whether the solute and solvent ll mix to form a homogeneous liquid–liquid mixture focuses only on the ture of the intermolecular forces. From a thermodynamic perspective, such approach considers only the energetics of the solution process ., only the enthalpy changes involved) and neglects entropy effects. Thus, edictions made by using the approach described above are not always cor- t. A more sophisticated approach would be to consider the Gibbs energy of ution, $\Delta_{soln}G = \Delta_{soln}H - T\Delta_{soln}S$. As we established in Chapter 13, the n of $\Delta_{soln}G$ will indicate whether a process is spontaneous or not: If $_{ln}G < 0$ the solution process is spontaneous, and if $\Delta_{soln}G > 0$, it is not ontaneous. By focusing on $\Delta_{soln}G$, the formation of a solution can be inves- ated from both an enthalpic and an entropic perspective.

14-1 ARE YOU WONDERING?

What is the nature of the intermolecular forces in a mixture of carbon disulfide and acetone?

Carbon disulfide is a nonpolar molecule, and so in the pure substance the only ntermolecular forces are weak London dispersion forces; carbon disulfide is a volatile liquid. Acetone is a polar molecule, and in the pure substance lipole–dipole forces are strong. Acetone is somewhat less volatile than carbon lisulfide. In a solution of acetone in carbon disulfide (case 3 on page 647), the lipoles of acetone molecules polarize carbon disulfide molecules, giving rise to *lipole–induced dipole* interactions.

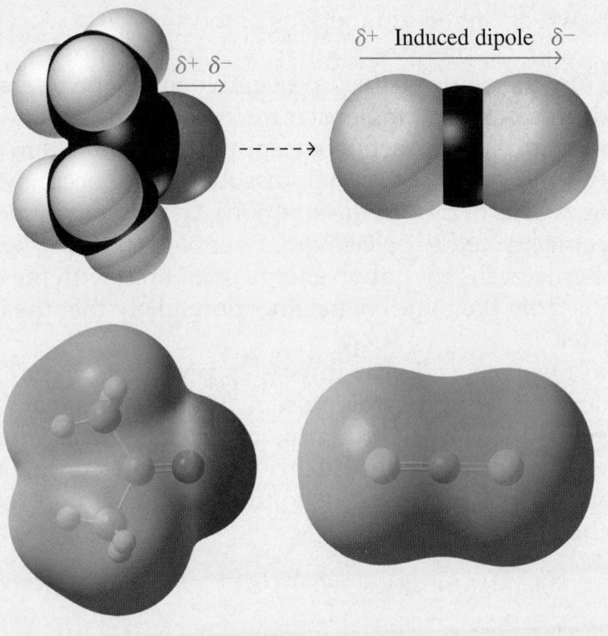

The dipole–induced dipole forces between acetone and carbon disulfide mole- cules are weaker than the dipole–dipole interactions among acetone molecules, causing the acetone molecules to be relatively less stable in their solutions with carbon disulfide than they are in pure acetone. As a result, acetone–carbon disul- fide mixtures are nonideal solutions.

ormation of Ionic Solutions

assess the energy requirements for the formation of aqueous solutions of nic compounds, we turn to the process pictured in Figure 14-6. Water

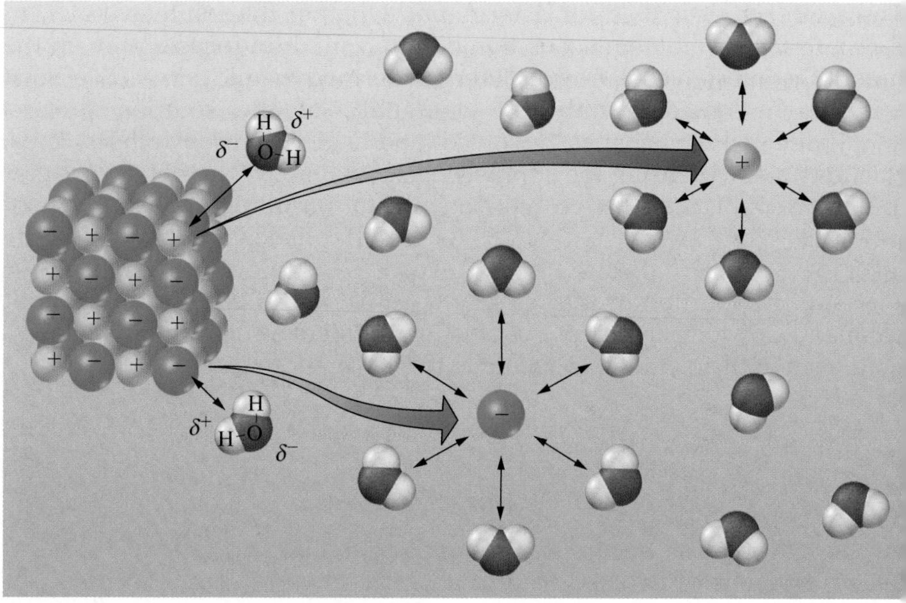

▲ FIGURE 14-6
An ionic crystal dissolving in water
Clustering of water dipoles around the surface of the ionic crystal and the formation of hydrated ions in solution are the key factors in the dissolution process.

dipoles are shown clustered around ions at the surface of a crystal. The neg tive ends of water dipoles are pointed toward the positive ions, and t positive ends of water dipoles toward negative ions. The interaction betwe an ion and a dipole is an intermolecular force known as an ion–dipole force these ion–dipole forces of attraction are strong enough to overcome the inte onic forces of attraction in the crystal, dissolving will occur. Moreover, the ion–dipole forces also persist in the solution. An ion surrounded by a clus of water molecules is said to be *hydrated*. Energy is *released* when ions becor hydrated. The greater the hydration energy compared with the energy need to separate ions from the ionic crystal, the more likely that the ionic solid w dissolve in water.

▶ We discussed lattice energy in Section 12-7.

We can again use a *hypothetical* three-step process to describe the dissol tion of an ionic solid. The energy requirement to dissociate an ionic solid ir separated gaseous ions, an endothermic process, is the *negative of* the latti energy. Energy is released in the next two steps—hydration of the gaseo cations and anions. The enthalpy of solution is the sum of these three Δ_r values, described below for NaCl.

$$NaCl(s) \longrightarrow Na^+(g) + Cl^-(g) \qquad \Delta_r H_1 = (-\text{lattice energy of NaCl}) > 0$$

$$Na^+(g) \xrightarrow{H_2O} Na^+(aq) \qquad \Delta_r H_2 = (\text{hydration energy of } Na^+) < 0$$

$$Cl^-(g) \xrightarrow{H_2O} Cl^-(aq) \qquad \Delta_r H_3 = (\text{hydration energy of } Cl^-) < 0$$

$$NaCl(s) \xrightarrow{H_2O} Na^+(aq) + Cl^-(aq) \; \Delta_{soln}H = \Delta_r H_1 + \Delta_r H_2 + \Delta_r H_3 \approx +5\text{ kJ mo}$$

The dissolution of sodium chloride in water is *endothermic*, and this also the case for the vast majority (about 95%) of soluble ionic compounc Why does NaCl dissolve in water if the process is endothermic? It mig appear that an endothermic process would not occur because of t increase in enthalpy. Because NaCl does actually dissolve in water, the must be another factor involved. In fact, *two* factors must be considered

▶ A common misconception is that an endothermic process cannot be spontaneous.

termining whether a process will occur spontaneously. Enthalpy change only one of them. The other factor, called *entropy* (see page 580), concerns e natural tendency for microscopic particles—atoms, ions, or molecules— spread themselves out in the space available to them. The dispersed con-tion of the microscopic particles in NaCl(aq) compared with pure NaCl(s) d H_2O(l) offsets the +5 kJ mol^{-1} increase in enthalpy in the solution ocess. In summary, if the hypothetical three-step process for solution for-ation is *exothermic*, we expect dissolution to occur; but we also expect a lution to form for an endothermic solution process, as long as $\Delta_{soln}H$ is t too large.

andard Thermodynamic Properties of Aqueous Ions

Chapter 7, we learned how to combine thermodynamic properties of dif-rent substances to calculate the heat absorbed or released by a reaction. In apter 13, we learned how to combine thermodynamic properties of fferent substances to calculate equilibrium constants for reactions. Many emical reactions occur in aqueous solution and involve ions. To be able to lculate heats of reaction or equilibrium constants for such reactions, we ed values for the thermodynamic properties of the aqueous ions involved. this section, we discuss the standard thermodynamic properties of ions in lution, particularly with respect to how their values are established and terpreted.

Values of $\Delta_f H°$, $\Delta_f G°$, $S°$, and C_p for a selection of aqueous ions are given in ble 14.2. Notice that the thermodynamic properties of H^+(aq) are all equal zero. These values are established by convention.

$$\Delta_f H°[H^+(aq)] = 0 \quad S°[H^+(aq)] = 0 \quad \Delta_f G°[H^+(aq)] = 0 \quad C_p[H^+(aq)] = 0$$

ABLE 14.2 Standard Thermodynamic Properties of Some Aqueous Ions[a]

Ion	$\Delta_f H°$, kJ mol^{-1}	$\Delta_f G°$, kJ mol^{-1}	$S°$, J mol^{-1} K^{-1}	C_p, J mol^{-1} K^{-1}
H^+(aq)	0	0	0	0
Li^+(aq)	−278.5	−293.3	13.4	62
Na^+(aq)	−240.1	−261.9	59.0	42
K^+(aq)	−252.4	−283.3	102.5	12
Rb^+(aq)	−251.2	−284.0	121.5	−9
Cs^+(aq)	−258.3	−292.0	133.1	−23
NH_4^+(aq)	−132.5	−79.31	113.4	69
Be^{2+}(aq)	−382.8	−397.7	−129.7	
Mg^{2+}(aq)	−466.9	−454.8	−138.1	−16
Ca^{2+}(aq)	−542.8	−553.6	−53.1	−27
Sr^{2+}(aq)	−545.8	−559.5	−32.6	−37
Ba^{2+}(aq)	−537.6	−560.8	9.6	−48
Al^{3+}(aq)	−531.0	−485.0	−321.7	−119
F^-(aq)	−332.6	−278.8	−13.8	−116
Cl^-(aq)	−167.2	−131.2	56.5	−126
Br^-(aq)	−121.6	−104.0	82.4	−132
I^-(aq)	−55.19	−51.57	111.3	−121
OH^-(aq)	−230.0	−157.2	−10.75	−140
NO_3^-(aq)	−207.4	−111.3	146.4	−71
CH_3COO^-(aq)	−486.0	−369.3	86.6	+26
SO_4^{2-}(aq)	−909.3	−744.5	20.1	−276
PO_4^{3-}(aq)	−1277.4	−1018.7	−220.5	−495

[a] All data are from the CRC *Handbook of Chemistry and Physics*, 95th edition, except for the heat capacity values, which are from L. G. Hepler and J. K. Hovey, *Can. J. Chem.*, **74**, 639 (1996).

This convention is adopted because it is impossible to prepare a solution just one type of ion and, therefore, to measure independently the properties $H^+(aq)$ or any other aqueous ion. An aqueous solution of ions contains at lea two different types of ions (e.g., a cation and an anion); consequently, *any* the modynamic property of such a solution includes contributions from all t ions. For example, the thermodynamic properties of a solution of HCl inclu contributions from $H^+(aq)$ and $Cl^-(aq)$. By defining all the thermodynam properties of $H^+(aq)$ to be zero, the thermodynamic properties of $Cl^-(aq)$ c be set equal to those of HCl(aq). With the thermodynamic properties $Cl^-(aq)$ now established, the thermodynamic properties of $Na^+(aq)$, for exa ple, can be obtained by making measurements on solutions of NaCl(aq). W can continue in this manner to obtain values for the thermodynamic prope ties of other aqueous ions.

The values of $\Delta_f H°$ and $\Delta_f G°$ for an aqueous ion are based on a formati reaction in which the aqueous ion is formed from its elements, each in its sta dard state. Formation reactions for $H^+(aq)$, $Na^+(aq)$, and $Cl^-(aq)$ are giv below, along with the corresponding $\Delta_f H°$ values. Notice that the formati reaction for an aqueous ion also involves the production or consumption electrons.

$$\frac{1}{2}H_2(g, 1\text{ bar}) \xrightarrow{H_2O} H^+(aq, 1\text{ molal}) + e^- \qquad \Delta_f H° = 0\ (\text{by definition})$$

$$Na(s, 1\text{ bar}) \xrightarrow{H_2O} Na^+(aq, 1\text{ molal}) + e^- \qquad \Delta_f H° = -240.1\text{ kJ mol}^{-1}$$

$$\frac{1}{2}Cl_2(g, 1\text{ bar}) + e^- \xrightarrow{H_2O} Cl^-(aq, 1\text{ molal}) \qquad \Delta_f H° = -167.2\text{ kJ mol}^{-1}$$

As mentioned above, the values given in Table 14.2 are obtained by defini the values for $H^+(aq)$ as zero. Consequently, the values of $\Delta_f H°$ and $\Delta_f G°$ pr vide only a *relative* measure of the enthalpy and Gibbs energy changes for t formation of an aqueous ion. For example, $\Delta_f H°[Na^+(aq)] = -240.1$ kJ mo indicates that the formation of $Na^+(aq)$ is 240.1 kJ mol^{-1} more exothermic th the formation of $H^+(aq)$.

From Table 14.2, we see that the values of $S°$ or C_p for some aqueous io are negative, which is a strong indication that we must interpret the values $S°$ and C_p differently for aqueous ions than for pure substances. (Recall: T standard molar entropies and molar heat capacities of pure substances a always positive.) These values provide a relative measure of the entropy an heat capacity *changes* that occur in a solution *per mole* of ion added, assumi that the ion is added at constant T and constant P and without any appreci ble change in the (overall) composition of the solution. For $F^-(aq)$, the valu are $S° = -13.8\text{ J mol}^{-1}\text{ K}^{-1}$ and $C_p = -106.7\text{ J mol}^{-1}\text{ K}^{-1}$. The negative va ues of $S°$ and C_p for $F^-(aq)$ indicate that the entropy and heat capaci changes produced by adding F^- to water are more negative (or less positiv than the changes produced by adding H^+ to water. Ions with negative valu for $S°$ may be considered "entropy lowering" because they have a strong tendency than H^+ to orient nearby water molecules. Ions with negative va ues for C_p may be considered "heat capacity lowering" because they have stronger tendency than H^+ to disrupt the hydrogen bonding network th exists in pure water. Consequently, less heat is required to produce a giv temperature change. Table 14.2 shows that, for ions of elements in the san group (for example, F^-, Cl^-, Br^-, and I^-, or Li^+, Na^+, K^+, Rb^+, and Cs^+), t $S°$ values tend to increase down the group whereas the C_p values tend decrease.

▶ In more advanced treatments of thermodynamics, $S°$ and C_p for an aqueous ion are called partial molar quantities. They represent the rates of change of S and C_p of the solution with respect to the amount of that ion. These rates of change are represented mathematically as (partial) derivatives of S and C_p with respect to the amount, n, of a particular ion.

We can use data for $H_2O(l)$, from Appendix D, as well as the data f $Na^+(aq)$ and $Cl^-(aq)$ from Table 14.2, to estimate the heat capacity of 1 mol NaCl(aq). In a sample that contains exactly 1000 g of H_2O, we also have 1 m

22.99 g) of Na^+ and 1 mol (or 35.45 g) of Cl^-. Therefore, the total mass of sample is 1058.44 g, and its heat capacity is

$$C_p = n_{H_2O}C_p[H_2O(l)] + n_{Na^+}C_p[Na^+(aq)] + n_{Cl^-}C_p[Cl^-(aq)]$$
$$= \frac{1000\ g}{18.015\ g\ mol^{-1}} \times 75.3\ J\ mol^{-1}\ K^{-1} + 1\ mol \times 46.4\ J\ mol^{-1}\ K^{-1}$$
$$+ 1\ mol \times (-136.4\ J\ mol^{-1}\ K^{-1})$$
$$= 4179.9\ J\ K^{-1} + 46.4\ J\ K^{-1} - 136.4\ J\ K^{-1}$$
$$= 4089.9\ J\ K^{-1}$$

◀ This calculation is approximate because the C_p values vary with molality. We have ignored this complication.

The heat capacity calculated above is for a sample weighing 1058.44 g. We n express the heat capacity *per gram of solution* (in other words, as a specific at capacity) by dividing the heat capacity above by 1058.44 g. We obtain

$$\text{ecific heat capacity of 1 molal NaCl(aq)} = \frac{4089.9\ J\ K^{-1}}{1058.44\ g} = 3.86\ J\ K^{-1}\ g^{-1}$$

The specific heat capacity of water is 4.18 $J\ K^{-1}\ g^{-1}$ (see Table 7.1), so we ve demonstrated by calculation that 1 molal NaCl(aq) has a lower specific at capacity than water. This means that it is easier (less heat is required) to se the temperature of 1 g of 1 molal NaCl(aq) by 1 K than to raise the temrature of 1 g of water by 1 K. Stated another way, a given quantity of heat ll cause a greater temperature change in 1 g of 1 molal NaCl(aq) than in 1 g water. In general, the specific heat capacity of a salt solution will always be wer than that of water. See Figure 14-7. The reason is that when an ion is

KEEP IN MIND

that ion–dipole forces are stronger than hydrogen bonding forces. See Table 12.3.

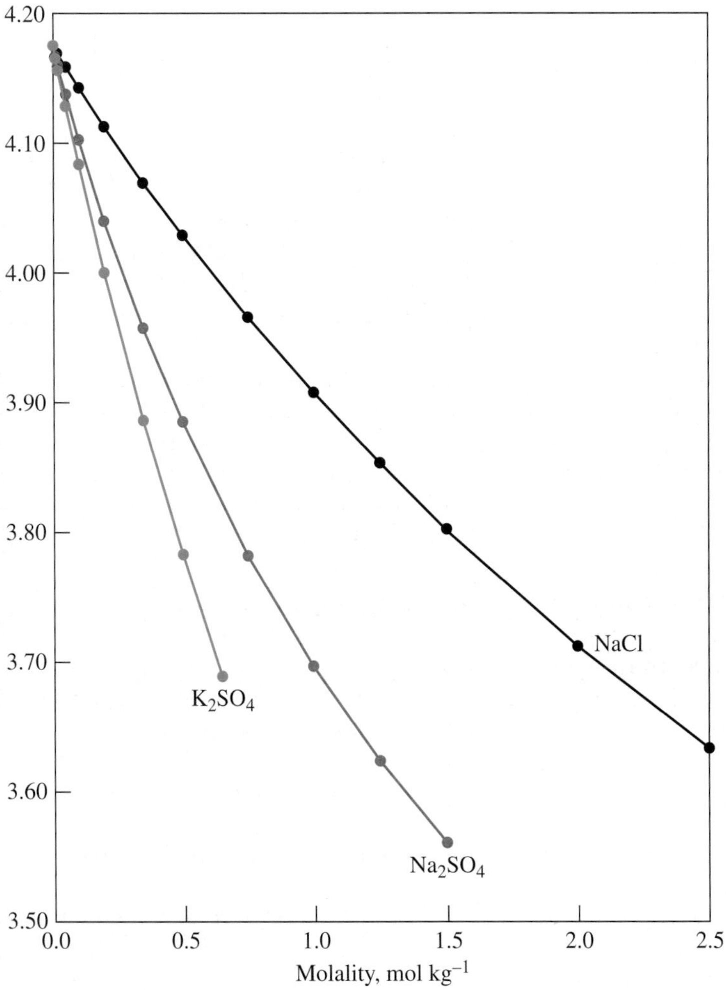

◀ FIGURE 14-7
Trends in specific heat for aqueous salt solutions
The variation of specific heat with concentration is illustrated for aqueous solutions of sodium chloride (black curve), sodium sulfate (red curve), and potassium sulfate (blue curve). In general, as the concentration of the salt increases, the specific heat of the solution decreases. The curve for Na_2SO_4(aq) lies below that of NaCl(aq) because water molecules interact more strongly with a sulfate ion (SO_4^{2-}) than with a chloride ion (Cl^-). The specific heats of Na_2SO_4(aq) and K_2SO_4(aq) are different because the cation–water interactions are different: $Na^+ \cdots H_2O$ versus $K^+ \cdots H_2O$.

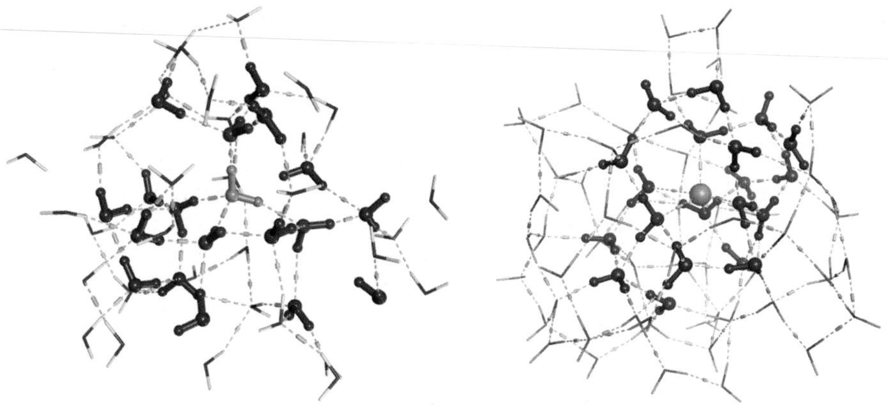

(a) (b)

▲ FIGURE 14-8
Intermolecular interactions in pure water and for a sodium ion in water
Dashed lines represent intermolecular interactions. The length of the cylinder in the middle of the dashed line provides a measure of the strength of the interaction: The longer the cylinder, the stronger the interaction. In **(a)**, a central water molecule (orange) forms hydrogen bonds with four other water molecules (blue). These inner-layer water molecules form hydrogen bonds with other water molecules (green). The water molecules in this second layer in turn form hydrogen bonds to other water molecules (red-and-white sticks). In **(b)**, a sodium ion (orange) is surrounded by six water molecules (blue). These six water molecules constitute the first hydration shell. A second hydration shell (green) is also shown. Water molecules beyond the second hydration shell are also shown (red-and-white sticks). They are relatively far from the sodium ion and are considered part of the bulk solution.

added to water, the hydrogen bonding network that exists in pure water disrupted. Water molecules nearest the ion tend to be oriented differently th they are in pure water, because of relatively strong ion–dipole forces, and the water molecules nearest the ion do not hydrogen bond as effectively wi nearby water molecules (Fig. 14-8). The weakening of the interactions betwe water molecules means that less energy (heat) is required to raise the tempe ature of the solution.

Before leaving this section, let us see if we can rationalize differences in t specific heats of different aqueous salt solutions. Figure 14-7 is a graph of sp cific heat versus concentration for three aqueous salt solutions. In all thr cases, we observe a decrease in specific heat as the salt concentrati increases. We also observe that the order of specific heats is K_2SO_4(a $<$ Na_2SO_4(aq) $<$ NaCl(aq). The specific heat of Na_2SO_4(aq) is lower than that NaCl(aq) primarily because the $SO_4{}^{2-}$ ion, with a charge of -2, interacts mo strongly with water molecules than does the singly charged Cl^- ion ar causes a greater disruption in the hydrogen bonding network of water. Th specific heat of K_2SO_4(aq) is lower than that of Na_2SO_4(aq) because, eve though the Na^+–H_2O interactions are stronger than the K^+–H_2O intera tions, the larger K^+ ion disrupts the hydrogen bond network in water mo than the smaller Na^+ ion does.

14-4 Solution Formation and Equilibrium

In the previous section, we described what happens at the molecular (micr scopic) level when solutions form. In this section, we will describe soluti formation in terms of phenomena that we can actually observe, that is, macroscopic view.

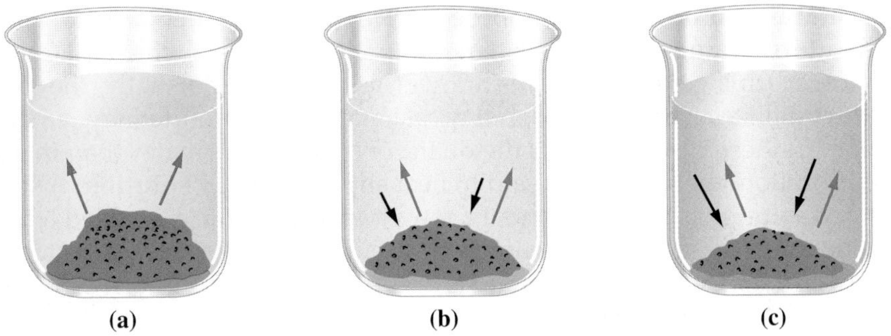

=IGURE 14-9
rmation of a saturated solution
e lengths of the arrows represent the rate of dissolution (↑) and the rate of stallization (↓). **(a)** When solute is first placed in the solvent, only dissolution occurs. After a time, the rate of crystallization becomes significant. **(c)** The solution is urated when the rates of dissolution and crystallization become equal.

Figure 14-9 suggests what happens when a solid solute and liquid solvent e mixed. At first, only dissolution occurs, but soon the reverse process of ystallization becomes increasingly important; and some dissolved atoms, ns, or molecules return to the undissolved state. When dissolution and crys- lization occur at the same rate, the solution is in a state of dynamic equilib- um. The quantity of dissolved solute remains constant with time, and the ution is said to be a **saturated solution**. The concentration of the saturated lution is called the **solubility** of the solute in the given solvent. Solubility ries with temperature, and a solubility–temperature graph is called a *solu- ity curve*. Some typical solubility curves are shown in Figure 14-10.

If, in preparing a solution, we start with less solute than would be present the saturated solution, the solute completely dissolves, and the solution is

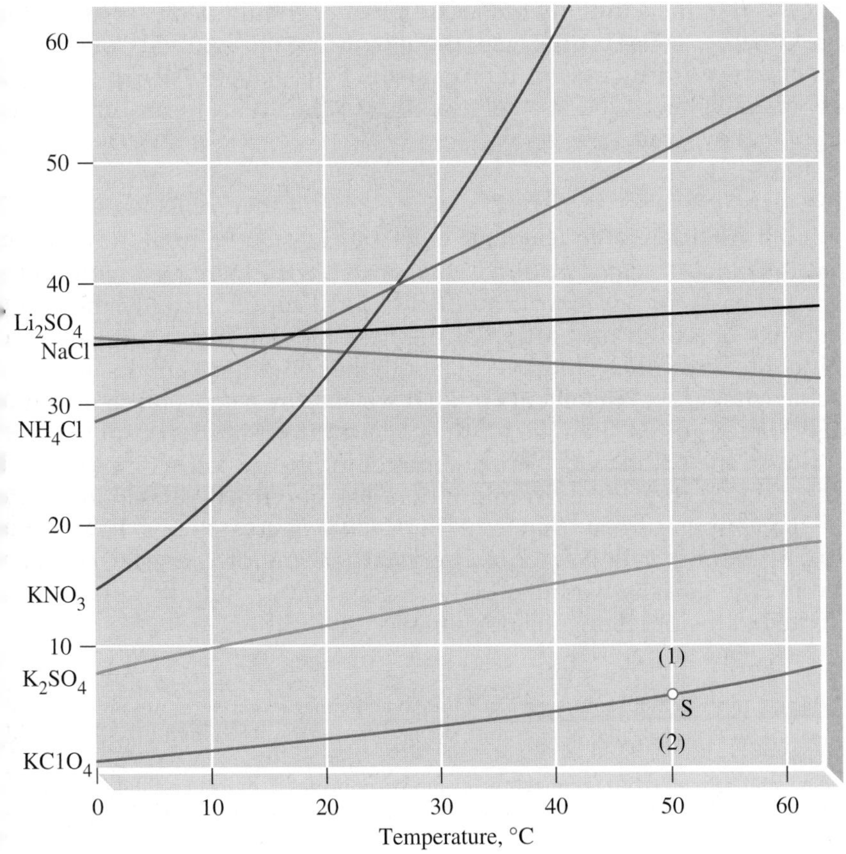

◀ FIGURE 14-10
Aqueous solubility of several salts as a function of temperature
Solubilities can be expressed in many ways: molarities, mass percent, or, as in this figure, grams of solute per 100 g H_2O. For each solubility curve (as shown here for $KClO_4$), points on the curve (S) represent saturated solutions. Regions above the curve (1) correspond to supersaturated solutions and below the curve (2), to unsaturated solutions.

an **unsaturated solution**. But suppose we prepare a saturated solution at c temperature and then change the temperature to a value at which the solubl ity is lower (this generally means a lower temperature). Usually, the exc solute crystallizes from solution, but occasionally all the solute may remain solution. In these cases, because the quantity of solute is greater than in a s urated solution, the solution is said to be a **supersaturated solution**. A sup saturated solution is unstable, and if a few crystals of solute are added to se as particles on which crystallization can occur, the excess solute crystalliz Figure 14-10 shows how unsaturated and supersaturated solutions can be re resented with a solubility curve.

Solubility as a Function of Temperature

As a general observation, the solubilities of ionic substances (about 95% them) *increase* with increasing temperature. Exceptions to this generalizati tend to be found among compounds containing the anions SO_3^{2-}, SO_4 AsO_4^{3-}, and PO_4^{3-}.

In Chapter 15 we will learn to predict how an equilibrium conditi changes with such variables as temperature and pressure by using an id known as *Le Châtelier's principle*. One statement of the principle is that h added to a system at equilibrium stimulates the heat-absorbing, or endoth mic, reaction. This suggests that when $\Delta_{soln}H > 0$, raising the temperatu stimulates dissolving and *increases* the solubility of the solute. Conversely $\Delta_{soln}H < 0$ (exothermic), the solubility *decreases* with increasing temperatu In this case, crystallization—being endothermic—is favored over dissolving

We must be careful in applying the relationship we just described. The p ticular value of $\Delta_{soln}H$ that establishes whether solubility increases decreases with increased temperature is that associated with dissolving small quantity of solute in a solution that is already very nearly saturated. some cases, this heat effect is altogether different from what is observed wh a solute is added to the pure solvent. For example, when NaOH is dissolved water, there is a sharp increase in temperature—*an exothermic* process. Tl fact suggests that the solubility of NaOH in water should decrease as the te perature is raised. What is observed, though, is that the solubility of NaOH water *increases* with increased temperature. This is because when a sm quantity of NaOH is added to a solution that is already nearly saturated, he is *absorbed*, not evolved.*

Richard Megna/Fundamental Photographs

▲ FIGURE 14-11
Recrystallization of KNO_3
Colorless crystals of KNO_3 separated from an aqueous solution of KNO_3 and $CuSO_4$ (an impurity). The pale blue color of the solution is produced by Cu^{2+}, which remains in solution.

Fractional Crystallization

Compounds synthesized in chemical reactions are generally impure, but t fact that the solubilities of most solids increase with increased temperatu provides the basis for one simple method of purification. Usually, the impu solid consists of a high proportion of the desired compound and lesser pr portions of the impurities. Suppose that both the compound and its impuriti are soluble in a particular solvent and that we prepare a concentrated soluti at a high temperature. Then we let the concentrated solution cool. At low temperatures, the solution becomes saturated in the desired compound. T excess compound crystallizes from solution. The impurities remain in soluti because the temperature is still too high for these to crystallize.† This meth of purifying a solid, called **fractional crystallization**, or **recrystallization**, pictured in Figure 14-11. Example 14-4 illustrates how solubility curves can used to predict the outcome of a fractional crystallization.

*The solid in equilibrium with saturated NaOH(aq) over a range of temperatures around 25 °C $NaOH \cdot H_2O(s)$. It is actually the temperature dependence of the solubility of this hydrate th we have been discussing.

†This is the usual behavior, but at times, one or more impurities may form a solid solution w the compound being recrystallized. In these cases simple recrystallization does not work a method of purification.

EXAMPLE 14-4 Applying Solubility Data in Fractional Crystallization

A solution is prepared by dissolving 95 g NH_4Cl in 200.0 g H_2O at 60 °C. **(a)** What mass of NH_4Cl will recrystallize when the solution is cooled to 20 °C? **(b)** How might we improve the yield of NH_4Cl?

Analyze

We need to know the solubility of NH_4Cl at 20 °C and at 95 °C. We obtain the required data from Figure 14-10, which shows the solubility of several salts as a function of temperature.

Solve

(a) Using Figure 14-10, we estimate that the solubility of NH_4Cl at 20 °C is 37 g NH_4Cl/100 g H_2O. The quantity of NH_4Cl in the saturated solution at 20 °C is

$$200.0 \text{ g } H_2O \times \frac{37 \text{ g } NH_4Cl}{100 \text{ g } H_2O} = 74 \text{ g } NH_4Cl$$

The mass of NH_4Cl recrystallized is 95 − 74 = 21 g.

(b) The yield of NH_4Cl in (a) is rather poor—21 g out of 95 g, or 22%. We can do better: (1) The solution at 60 °C, although concentrated, is not saturated. Using Figure 14-10, we estimate that a saturated solution at 60 °C has 55 g NH_4Cl/100 g H_2O. Thus, the 95 g NH_4Cl requires less than 200.0 g H_2O to make a saturated solution. At 20 °C, a smaller quantity of saturated solution would contain less NH_4Cl than in (a), and the yield of recrystallized NH_4Cl would be greater. (2) Instead of cooling the solution to 20 °C, we might cool it to 0 °C. Here the solubility of NH_4Cl is less than at 20 °C, and more solid would recrystallize. (3) Still another possibility is to start with a solution at a temperature higher than 60 °C, say closer to 100 °C. The mass of water needed for the saturated solution would be less than at 60 °C. Note that options (1) and (3) both require changing the conditions by using a different amount of water from that originally specified.

Assess

The amount of dissolved salt can be increased by increasing the volume of solvent or by increasing the temperature. Keep in mind that fractional crystallization works best when the quantities of impurities are small and the solubility curve of the desired solute rises steeply with temperature.

PRACTICE EXAMPLE A: Calculate the quantity of NH_4Cl that would be obtained if suggestions (1) and (2) in Example 14-4(b) were followed. [*Hint*: Use data from Figure 14-10. What mass of water is needed to produce a saturated solution containing 95 g NH_4Cl at 60 °C?]

PRACTICE EXAMPLE B: Use Figure 14-10 to examine the solubility curves for the three potassium salts: $KClO_4$, K_2SO_4, and KNO_3. If saturated solutions of these salts at 40 °C are cooled to 20 °C, rank the salts in order of highest percent yield for the recrystallization.

4-5 Solubilities of Gases

'hy does a freshly opened can of soda pop fizz, and why does the soda go flat ter a time? To answer questions like these requires an understanding of the lubilities of gases. As discussed in this section, the effect of temperature on e solubility of gases is generally different from that on solid solutes. dditionally, the pressure of a gas strongly affects its solubility.

ffect of Temperature

'e cannot make an all-inclusive generalization about the effect of temperature on e solubilities of gases in solvents. It is certainly true, though, that the solubilities most gases in water *decrease* with an increase in temperature. This is true of $_2$(g) and O_2(g)—the major components of air—and of air itself (Fig. 14-12). his fact helps to explain why many types of fish can survive only in cold water. here is not enough dissolved air (oxygen) in warm water to sustain them.

For solutions of gases in organic solvents, the situation is often the reverse of at just described; that is, gases may become more soluble at higher temperares. The solubility behavior of the noble gases in water is more complex. The lubility of each gas decreases with an increase in temperature, reaching a

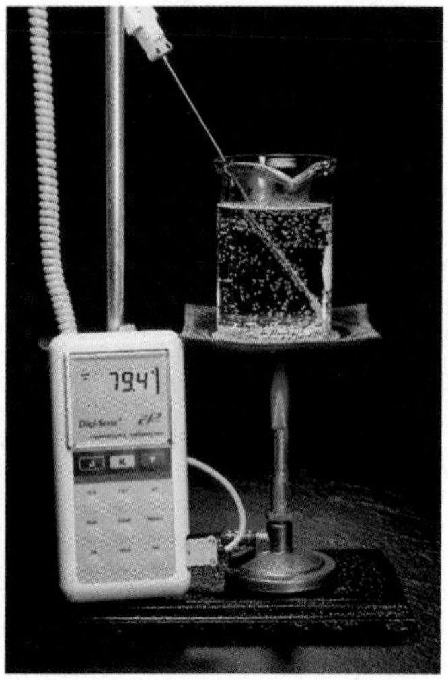

Carey B. Van Loon

▲ FIGURE 14-12
Effect of temperature on the solubilities of gases
Dissolved air is released as water is heated, even at temperatures well below the boiling point.

minimum at a certain temperature; then the solubility trend reverses directic with the gas becoming more soluble with an increase in temperature. For exa ple, for helium at 1 atm pressure, this minimum solubility in water comes at 35 °

Effect of Pressure

Pressure affects the solubility of a gas in a liquid much more than does tempe ture. The English chemist William Henry (1775–1836) found that *the solubility c gas increases with increasing pressure.* A mathematical statement of **Henry's law**

$$C = k \times P_{gas} \quad \textbf{(14.2)}$$

In this equation, C represents the solubility of a gas in a particular solvent a fixed temperature, P_{gas} is the partial pressure of the gas above the solutic and k is a proportionality constant. To evaluate the proportionality constant we need to have one measurement of the solubility of the gas at a known pre sure and temperature. For example, the aqueous solubility of $N_2(g)$ at 0 °C ar 1.00 atm is 23.54 mL N_2 per liter. The Henry's law constant, k, is

$$k = \frac{C}{P_{gas}} = \frac{23.54 \text{ mL N}_2\text{/L}}{1.00 \text{ atm}}$$

Suppose we want to increase the solubility of the $N_2(g)$ to a value 100.0 mL N_2 per liter. Equation (14.2) suggests that to do so, we must increa the pressure of $N_2(g)$ above the solution. That is,

$$P_{N_2} = \frac{C}{k} = \frac{100.0 \text{ mL N}_2\text{/L}}{(23.54 \text{ mL N}_2\text{/L})/1.00 \text{ atm}} = 4.25 \text{ atm}$$

At times, we are required to change the units used to express a gas solub ity at the same time that the pressure is changed. This variation is illustrated Example 14-5.

We can rationalize Henry's law as follows: In a saturated solution, the ra of evaporation of gas molecules from solution and the rate of condensation gas molecules into the solution are equal. Both of these rates depend on th number of molecules per unit volume. With increasing pressure on the sy tem, the number of molecules per unit volume in the gaseous state increas (through an increase in the gas pressure), and the number of molecules p unit volume must also increase in the solution (through an increase in conce tration). Figure 14-13 illustrates this rationalization.

We see a practical application of Henry's law in carbonated beverages. Th dissolved gas is carbon dioxide, and the higher the gas pressure maintaine above the soda pop, the more CO_2 that dissolves. When a bottle of soda opened, some gas is released. As the gas pressure above the solution drops, di solved CO_2 is expelled, usually fast enough to cause fizzing. In sparkling wine the dissolved CO_2 is also under pressure, but rather than being added artificial as in soda pop, the CO_2 is produced by a fermentation process within the bottl

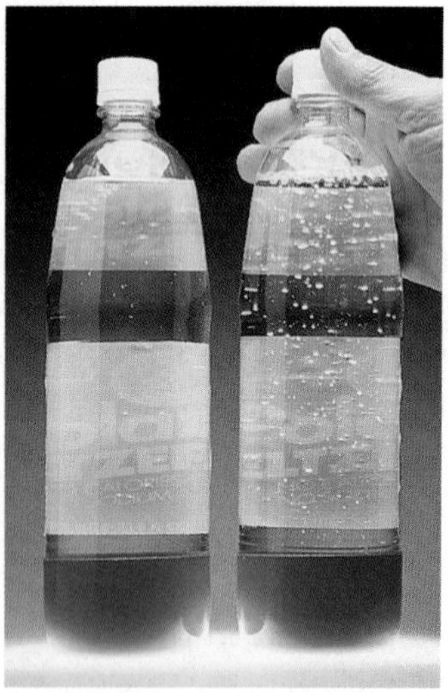

Charles D. Winters / Science Source

▲ The unopened bottle of soda water is under a high pressure of $CO_2(g)$. When a similar bottle is opened, the pressure quickly drops and some of the $CO_2(g)$ is released from solution (bubbles).

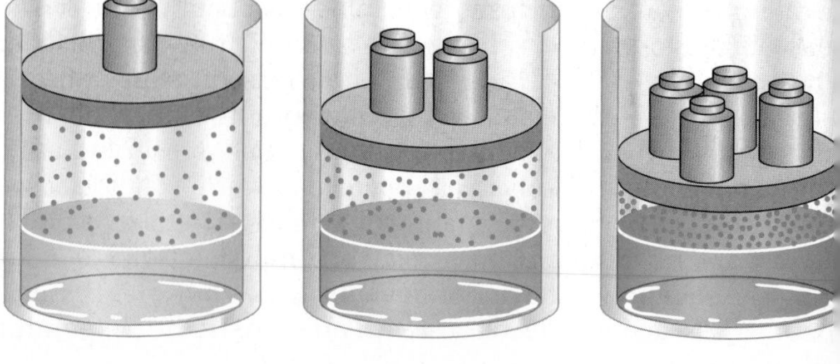

▶ FIGURE 14-13
Effect of pressure on the solubility of a gas
The concentration of dissolved gas (suggested by the depth of color) is proportional to the pressure of the gas above the solution (suggested by the density of the dots).

EXAMPLE 14-5 Using Henry's Law

At 0 °C and an O_2 pressure of 1.00 atm, the aqueous solubility of $O_2(g)$ is 48.9 mL O_2 per liter. What is the molarity of O_2 in a saturated water solution when the O_2 is under its normal partial pressure in air, 0.2095 atm?

Analyze

Think of this as a two-part problem. (1) Determine the molarity of the saturated O_2 solution at 0 °C and 1 atm. (2) Use Henry's law in the manner just outlined.

Solve

Determine the molarity of O_2 at 0 °C when $P_{O_2} = 1$ atm. We are given the information that, at an O_2 pressure of 1.00 atm, a saturated solution of O_2 in water contains 48.9 mL (0.0489 L) of O_2. We also know that, at 0 °C and 1.00 atm, 1 mol O_2 occupies a volume of 22.4 L. Therefore,

$$\text{molarity} = \frac{0.0489 \text{ L} \times \dfrac{1 \text{ mol}}{22.4 \text{ L}}}{1 \text{ L soln}} = 2.18 \times 10^{-3} \frac{\text{mol}}{\text{L soln}} = 2.18 \times 10^{-3} \text{ M}$$

Evaluate the Henry's law constant.

$$k = \frac{C}{P_{gas}} = \frac{2.18 \times 10^{-3} \text{ M}}{1.00 \text{ atm}}$$

Apply Henry's law.

$$C = k \times P_{gas} = \frac{2.18 \times 10^{-3} \text{ M}}{1.00 \text{ atm}} \times 0.2095 \text{ atm} = 4.57 \times 10^{-4} \text{ M}$$

Assess

When working problems involving gaseous solutes in a solution in which the solute is at very low concentration, use Henry's law.

PRACTICE EXAMPLE A: Use data from Example 14-5 to determine the partial pressure of O_2 above an aqueous solution at 0 °C known to contain 5.00 mg O_2 per 100.0 mL of solution.

PRACTICE EXAMPLE B: A handbook lists the solubility of carbon monoxide in water at 0 °C and 1 atm pressure as 0.0354 mL CO per milliliter of H_2O. What pressure of CO(g) must be maintained above the solution to obtain 0.0100 M CO?

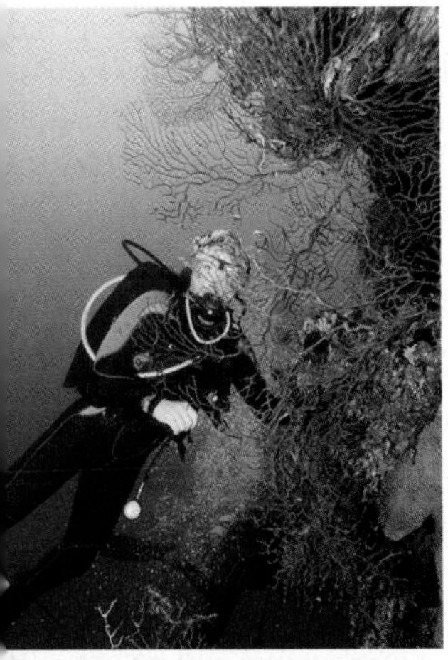

To avoid the painful and ɑngerous condition of the ɛnds, divers must not surface ɔo quickly from great depths.

Deep-sea diving provides us with still another example of Henry's law. Divers must carry a supply of air to breathe while underwater. If they are to stay submerged for any period of time, they must breathe compressed air. High-pressure air, however, is much more soluble in the blood and other body fluids than is air at normal pressures. When a diver returns to the surface, excess dissolved $N_2(g)$ is released as tiny bubbles from body fluids. When the ascent to the surface is made too quickly, N_2 diffuses out of the blood too quickly, causing severe pain in the limbs and joints, probably by interfering with the nervous system. This dangerous condition, known as "the bends," can be avoided if the diver ascends very slowly or spends time in a decompression chamber. Another effective method is to substitute a helium–oxygen mixture for compressed air. Helium is less soluble in blood than is nitrogen.

Henry's law (equation 14.2) fails for gases at high pressures; it also fails if the gas ionizes in water or reacts with water. For example, at 20 °C and with $P_{HCl} = 1$ atm, a saturated solution of HCl(aq) is about 20 M. But to prepare 10 M HCl, we do not need to maintain $P_{HCl} = 0.5$ atm above the solution, nor is $P_{HCl} = 0.05$ atm above 1 M HCl. In fact, for 1 M HCl, the odor of HCl(g) is undetectable. The reason we cannot detect any HCl(g) is that HCl ionizes in aqueous solutions, and in dilute solutions there are almost no molecules of HCl.

$$\text{HCl(g)} \xrightarrow{H_2O} \text{H}^+\text{(aq)} + \text{Cl}^-\text{(aq)}$$

Henry's law applies only to equilibrium between molecules of a gas and t same *molecules* in solution.

14-3 CONCEPT ASSESSMENT

Do you think that Henry's law works better for solutions of HCl(g) in benzene, C_6H_6, than it does for solutions of HCl(g) in water? If so, why?

14-6 Vapor Pressures of Solutions

Separating compounds from one another is a task that chemists commor face. If the compounds are volatile liquids, this separation often can achieved by *distillation*. To understand how distillation works, we need know something about the vapor pressures of solutions. This knowledge w also enable us to deal with other important solution properties, such as boili points, freezing points, and osmotic pressures.

To simplify the following discussion we will consider only solutions with tw components, solvent A and solute B. In the 1880s, the French chem F. M. Raoult found that a dissolved solute *lowers* the vapor pressure of the so vent. **Raoult's law** states that the partial pressure exerted by solvent vapor abo an ideal solution, P_A, is the product of the mole fraction of solvent in the sol tion, x_A, and the vapor pressure of the pure solvent at the given temperature, P

$$P_A = x_A P_A^* \qquad (14.3)$$

Equation (14.3) relates to Raoult's observation that a dissolved solute low ers the vapor pressure of the solvent because if $x_A + x_B = 1.00$, x_A must less than 1.00, and P_A must be smaller than P_A^*. Strictly speaking, Raoult's la applies only to ideal solutions and to all volatile components of the solutior However, even in nonideal solutions, the law often works reasonably well for t *solvent* in *dilute* solutions, for example, solutions in which $x_{solv} > 0.98$.

You may be wondering: Why does the solvent vapor pressure alwa decrease when solute is added? To explain, let us consider the addition of nonvolatile solute to a pure liquid solvent. We begin with the fact that wh the gas phase and liquid phase are in equilibrium with each other, the chem cal potential of the solvent molecules, μ_A, is the same in the two phase (Recall that the subscript A refers to the solvent.)

▶ Chemical potential is defined and discussed in Section 13-8.

$$\mu_A(g) = \mu_A(l)$$

The chemical potential of the solvent molecules in the liquid phase can represented in the form given in equation (13.31), with the activity, *a*, of t solvent set equal to its mole fraction, x_A.

$$\mu_A(l) = \mu_A^\circ(l) + RT \ln x_A$$

In the expression above, $\mu_A^\circ(l)$ is the chemical potential of the pure liqui Figure 14-14 shows how the chemical potential of the solvent in a solutic varies with its mole fraction.

When solute is added to the solvent, the mole fraction, x_A, of the solvent w be less than one. Since $\ln x_A$ is negative for $x_A < 1$, the chemical potential of t solvent in the liquid solution, $\mu_A(l)$, will be less than the chemical potential the pure liquid solvent, $\mu_A^\circ(l)$. Since the chemical potentials are unequal, sc vent molecules will move from the region of high chemical potential (in th case, the gas phase) to the region of low chemical potential (the liquid phas until the chemical potentials are again equal. The movement of solvent mol cules from the gas phase to the liquid phase results in a lower vapor pressure

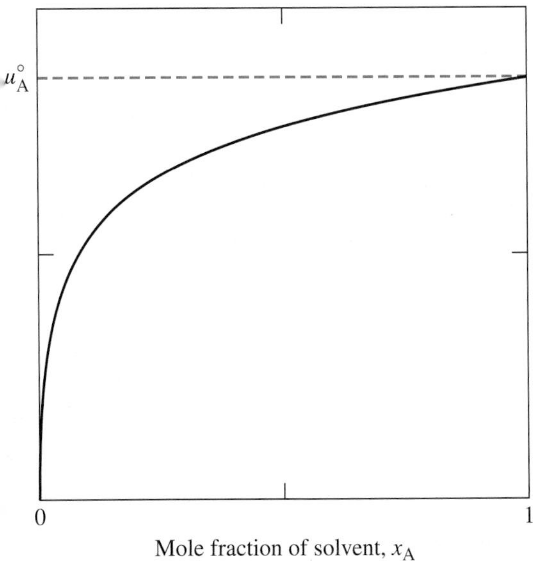

◀ FIGURE 14-14
Chemical potential of the solvent as a function of mole fraction
The chemical potential of the solvent in a solution is always less than or equal to that of the pure solvent.

Another way to rationalize this result is by focusing on entropies rather than chemical potentials. As illustrated in Figure 14-15, the entropy of an ideal ution is higher than that of pure solvent, a fact that is readily explained by ng Boltzmann's equation for entropy, equation (13.1). When a nonvolatile ute is added to the solvent, there is an increase in the number of microstates cause of the multitude of ways of distributing the solute molecules among the vent molecules. Therefore, the entropy of the solution (S_{soln}) is greater than e entropy of the pure liquid (S_{liq}). In an ideal solution, the intermolecular ces of attraction are the same as for the pure liquid solvent. Therefore, we pect the enthalpy of vaporization, $\Delta_{vap}H$, to be the same whether vaporiza-n of the solvent occurs from the pure solvent or from the solution. nsequently, the entropy of vaporization, $\Delta_{vap}S = \Delta_{vap}H/T$, is also the same. nce the entropy of the solution is initially greater than that of the pure solvent, e entropy of the vapor produced by the vaporization of solvent from the ideal ution must be greater than the entropy of the vapor obtained from the pure vent: $S_{soln} + \Delta_{vap}S$ is greater than $S_{liq} + \Delta_{vap}S$. For the vapor above the ution to have the higher entropy, its volume must be greater and its pressure ver than that of the vapor above the pure solvent. (Recall: In Chapter 13, we ablished that the entropy of an ideal gas increases as the pressure decreases as the volume increases. See the discussion following equation (13.7).)

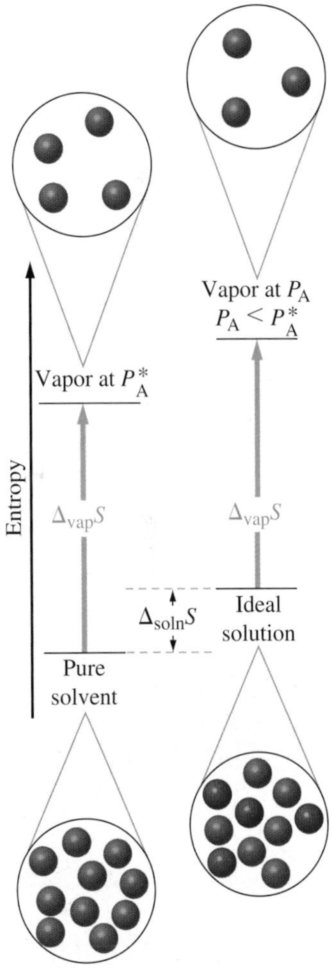

▲ FIGURE 14-15
An entropy-based rationale of Raoult's law
If $\Delta_{vap}S$ has the same value for vaporization from the pure solvent and from an ideal solution, the pressure of the vapor obtained from the solution is less than that of the vapor obtained from the pure solvent: $P_A < P_A^*$.

14-4 CONCEPT ASSESSMENT

An alternative statement of Raoult's law is that the fractional lowering of the vapor pressure of the solvent, $(P_A^* - P_A)/P_A^*$, is equal to the mole fraction of solute(s), x_B. Show that this statement is equivalent to equation (14.3).

quid–Vapor Equilibrium: Ideal Solutions

e results of Examples 14-6 and 14-7, together with similar data for other nzene–toluene solutions, are plotted in Figure 14-16. This figure consists of four es—three straight and one curved—spanning the entire concentration range.

The red line shows how the vapor pressure of benzene varies with the solution mposition. Because benzene in benzene–toluene solutions obeys Raoult's law, e red line has the equation $P_{benz} = x_{benz}P^*_{benz}$. The blue line shows how the por pressure of toluene varies with solution composition and indicates that luene also obeys Raoult's law. The dashed black line shows how the *total* vapor essure varies with the solution composition. Can you see that each pressure this black line is the sum of the pressures on the two straight lines that lie below it? Thus, point 3 represents the total vapor pressure (point 1 + point 2) of a benzene–toluene solution in which $x_{benz} = 0.500$ (see Example 14-6).

EXAMPLE 14-6 Predicting Vapor Pressures of Ideal Solutions

The vapor pressures of pure benzene and pure toluene at 25 °C are 95.1 and 28.4 mmHg, respectively. A solution is prepared in which the mole fractions of benzene and toluene are both 0.500. What are the partial pressures of the benzene and toluene above this solution? What is the total vapor pressure?

Analyze

We saw in Figure 14-4 that benzene–toluene solutions should be ideal. We expect Raoult's law to apply to both solution components.

Solve

$$P_{benz} = x_{benz}P^*_{benz} = 0.500 \times 95.1 \text{ mmHg} = 47.6 \text{ mmHg}$$
$$P_{tol} = x_{tol}P^*_{tol} = 0.500 \times 28.4 \text{ mmHg} = 14.2 \text{ mmHg}$$
$$P_{total} = P_{benz} + P_{tol} = 47.6 \text{ mmHg} + 14.2 \text{ mmHg} = 61.8 \text{ mmHg}$$

Assess

In this example we assumed these to be ideal solutions, which allowed us to use Raoult's law. We observe that the vapor pressure of each component is lowered because of the presence of the other component.

PRACTICE EXAMPLE A: The vapor pressure of pure hexane and pentane at 25 °C are 149.1 mmHg and 508.5 mmHg, respectively. If a hexane–pentane solution has a mole fraction of hexane of 0.750, what are the vapor pressures of hexane and pentane above the solution? What is the total vapor pressure?

PRACTICE EXAMPLE B: Calculate the vapor pressures of benzene, C_6H_6, and toluene, C_7H_8, and the total pressure at 25 °C above a solution with equal *masses* of the two liquids. Use the vapor pressure data given in Example 14-6.

EXAMPLE 14-7 Calculating the Composition of Vapor in Equilibrium with a Liquid Solution

What is the composition of the vapor in equilibrium with the benzene–toluene solution of Example 14-6?

Analyze

We are being asked to find the mole fraction of benzene and of toluene in the vapor. From Example 14-6 we know the vapor pressure of pure benzene and pure toluene. We have already calculated the partial vapor pressures; now we need to apply the definition of mole fraction.

Solve

The ratio of each partial pressure to the total pressure is the mole fraction of that component in the vapor. (This is another application of equation 6.17.) The mole-fraction composition of the vapor is

$$x_{benz} = \frac{P_{benz}}{P_{total}} = \frac{47.6 \text{ mmHg}}{61.8 \text{ mmHg}} = 0.770$$
$$x_{tol} = \frac{P_{tol}}{P_{total}} = \frac{14.2 \text{ mmHg}}{61.8 \text{ mmHg}} = 0.230$$

Assess

The mole fraction of benzene in the vapor is 0.770, whereas in the liquid the mole fraction of benzene is 0.5. For toluene the mole fraction in the vapor is 0.230, whereas in the liquid the mole fraction of toluene is 0.5. This difference in mole-fraction vapor composition caused by the difference in vapor pressures of the two components is the central concept of fractional distillation, which is discussed next.

PRACTICE EXAMPLE A: What is the composition of the vapor in equilibrium with the hexane–pentane solution described in Practice Example 14-6A?

PRACTICE EXAMPLE B: What is the composition of the vapor in equilibrium with the benzene–toluene solution described in Practice Example 14-6B?

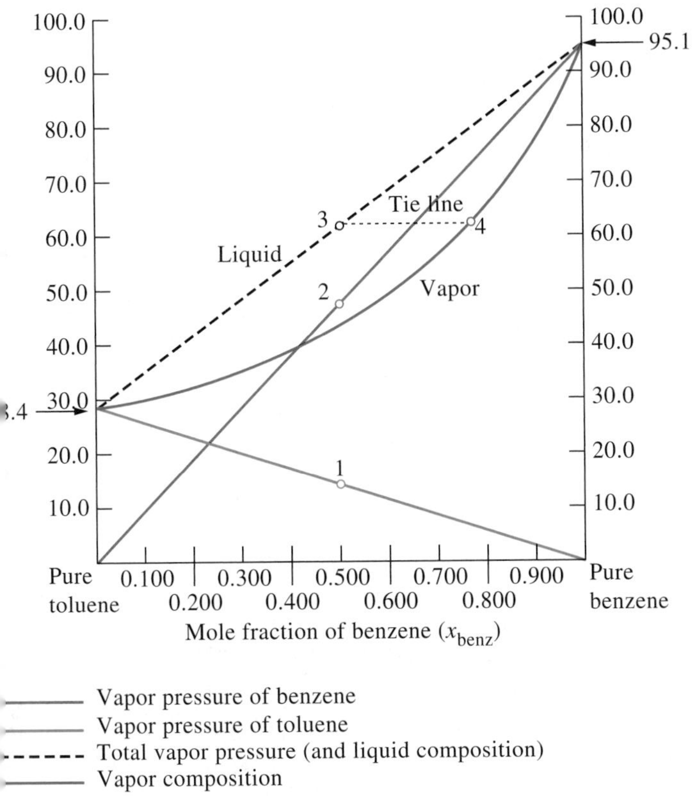

◀ FIGURE 14-16
Liquid–vapor equilibrium for benzene–toluene mixtures at 25 °C
In this diagram, partial pressures and the total pressure of the vapor are plotted as a function of the solution and vapor compositions.

As we calculated in Example 14-7, the vapor in equilibrium with a solution which $x_{benz} = 0.500$ is richer still in benzene. The vapor has $x_{benz} = 0.770$ oint 4). The line joining points 3 and 4 is called a *tie line*. Imagine establish-g a series of tie lines throughout the composition range. The vapor ends of ese tie lines can be joined to form the green curve in Figure 14-16. From the ative placement of the liquid and vapor curves, we see that for ideal solu-ons of two components, *the vapor phase is richer in the more volatile component an is the liquid phase.*

14-5 CONCEPT ASSESSMENT

Describe a case in which the liquid and vapor curves in a diagram such as Figure 14-16 would converge into a single curve. Is such a case likely to exist?

ractional Distillation

et's look at liquid–vapor equilibrium in benzene–toluene mixtures in a some-hat different way. Instead of plotting vapor pressures as a function of the solu-on and vapor compositions, let's plot normal boiling temperature—the mperature at which the *total* vapor pressure of the solution is 1 atm. The result-g graph is shown in Figure 14-17. This graph is useful in explaining **fractional istillation**, a procedure for separating volatile liquids from one another.

Notice that the graph starts at a high temperature (110.6 °C), the boiling point toluene—and ends at a lower temperature (80.0 °C), the boiling point of ben-ne. This is the reverse of the situation in Figure 14-16. Also, the vapor curve lies ove the liquid curve in Figure 14-17, not below, as is the case in Figure 14-16.

Figure 14-17 indicates that a benzene–toluene solution with $x_{benz} = 0.30$ oils at a temperature of 98.6 °C and is in equilibrium with a vapor in which $x_{enz} = 0.51$. Imagine extracting some of this vapor and cooling it to the point here it condenses to a liquid. The new liquid will have $x_{benz} = 0.51$ and rep-esents the conclusion of stage 1 in Figure 14-17. Now imagine repeating the

KEEP IN MIND

that the placement of the two curves in liquid–vapor equilibrium diagrams is such that the vapor is richer in the more volatile component than is the liquid. The more volatile component is the one with the higher vapor pressure or lower boiling point.

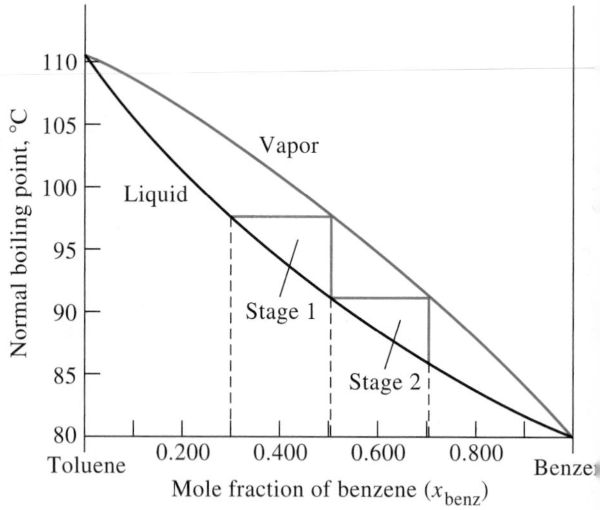

▶ FIGURE 14-17
Liquid–vapor equilibrium for benzene–toluene mixtures at 1 atm
In this diagram, the normal boiling points of solutions are plotted as a function of solution and vapor compositions.

winhorse/E+/Getty Images

▲ Fractional distillation is used in many industrial processes.

process, that is, vaporizing a solution with $x_{benz} = 0.51$ and condensing th vapor. The new liquid at the end of stage 2 has $x_{benz} = 0.71$. By repeating th cycle, the vapor becomes progressively richer in benzene. As pictured i Figure 14-18, boiling solutions in equilibrium with vapor can be spread ou over a long column, called a *fractionating column*, in which the equilibriu temperatures range from lowest at the top of the column to highest at the bo tom. The most volatile component in the solution emerges from the top of th column as a vapor that is condensed to a liquid and removed. The lea volatile component concentrates in the pot at the bottom of the colum Fractional distillation of a solution of many volatile components, such a petroleum, can be carried out in such a way that the components are wit drawn from the top of the column and condensed, one by one.

Liquid–Vapor Equilibrium: Nonideal Solutions

We cannot construct a liquid–vapor equilibrium diagram for nonideal solu tions in the simple manner illustrated in Figure 14-16. For example, vapo pressures in acetone–chloroform solutions are *lower* than we would predi for ideal solutions and boiling temperatures are correspondingly *higher*. I acetone–carbon disulfide solutions, conversely, vapor pressures are high

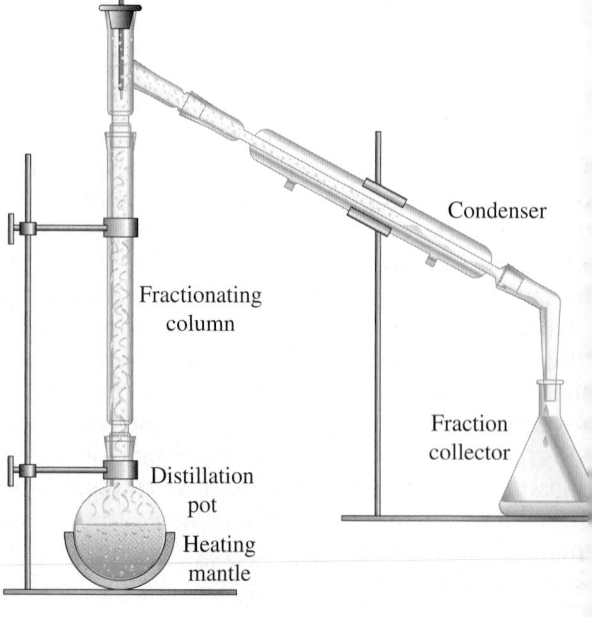

▶ FIGURE 14-18
Fractional distillation
The fractionating column is packed with glass beads or stainless steel turnings. Initially, as vapor rises from the pot and encounters these cooler objects, it condenses to a liquid. As the beads or turnings heat up, the liquid–vapor equilibrium front moves progressively up the column. Soon, liquid–vapor equilibrium occurs throughout the column, but with the equilibrium temperature changing continuously from the hottest regions at the bottom of the column to the coolest at the top. The vapor emerging from the top of the column is condensed to a liquid in the water-cooled condenser. The first fraction collected contains the most volatile component (lowest boiling point). Later fractions are less volatile liquids. The least volatile (highest boiling point) components remain as a residue in the distillation pot.

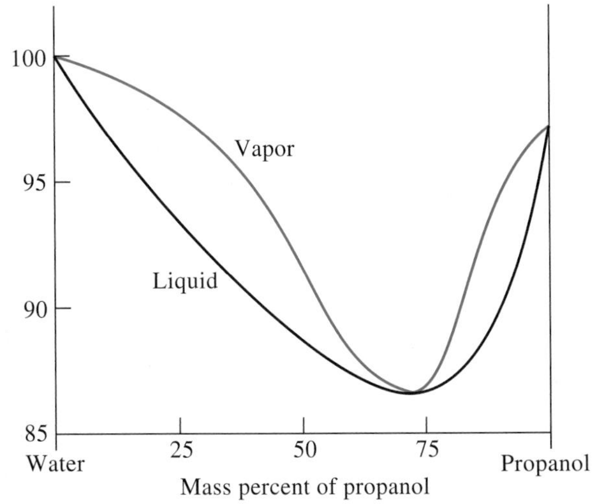

◀ FIGURE 14-19
A minimum boiling-point azeotrope
A solution of propanol in water having 71.69% $CH_3CH_2CH_2OH$ by mass—an azeotrope—has a lower boiling point than any other solution of these two components. In fractional distillation, solutions having less than 71.69% of the alcohol yield the azeotrope and water as ultimate products. Solutions with more than 71.69% of the alcohol yield the azeotrope and propanol. In each case, the azeotrope is drawn off through the condenser (Fig. 14-18), and the other component remains in the pot.

an predicted and boiling temperatures are correspondingly lower. In gure 14-5, we saw that the forces of attraction between unlike molecules are eater than those between like molecules in acetone–chloroform mixtures. It not unreasonable to expect the components in such solutions to show a duced tendency to vaporize and to have lower-than-predicted vapor pres- ıres. With acetone–carbon disulfide solutions, the situation is the reverse: ›rces of attraction between unlike molecules are weaker than between like olecules. This leads to greater tendencies for vaporization and higher vapor essures than predicted by Raoult's law.

If the departures from ideal solution behavior are sufficiently great, some lutions may have vapor pressures that pass through either a maximum or a inimum in vapor-pressure-composition graphs. Correspondingly, their ›iling points pass through either a minimum or maximum in boiling-point- mposition graphs. The solutions corresponding to these maxima or minima ›il at a constant temperature and produce a vapor having the *same* composi- ›n as the liquid. These solutions are called **azeotropes**. The boiling-point dia- am of a minimum boiling-point azeotrope is illustrated in Figure 14-19.

One of the most familiar azeotropes consists of 96.0% ethanol (C_2H_5OH) and 0% water, by mass, and has a boiling point of 78.174 °C. Pure ethanol has a boil- g point of 78.3 °C. Dilute ethanol–water solutions can be distilled to produce e azeotrope, but the remaining water cannot be removed by ordinary distilla- ›n. As a result, most ethanol used in the laboratory or in industry is only 96.0% $_2H_5OH$. To obtain absolute, or 100%, C_2H_5OH requires special measures.

4-7 Osmotic Pressure

the previous section our primary emphasis was on solutions containing a ›latile solvent and volatile solute. Another common type of solution is one ith a volatile solvent, such as water, but one or more *nonvolatile* solutes, such glucose, sucrose, or urea. Raoult's law still applies to the solvent in such ›lutions—the vapor pressure of the solvent is lowered.

Figure 14-20(a) pictures two aqueous solutions of a nonvolatile solute ithin the same enclosure. They are labeled A and B. The curved arrow indi- ıtes that water vaporizes from A and condenses into B. What is the driving ›rce behind this? It must be that the vapor pressure of H_2O above A is greater ıan that above B. Solution A is more dilute; it has a higher mole fraction of $_2O$. How long will this transfer of water continue? Solution A becomes more ›ncentrated as it loses water, and solution B becomes more dilute as it gains ater. When the mole fraction of H_2O is the same in both solutions, the *net* ansfer of H_2O stops.

Richard Megna/Fundamental Photographs

▲ FIGURE 14-20
Observing the direction of flow of water vapor
(a) Water passes, as vapor, from the more dilute solution (higher mole fraction of H_2O) to the more concentrated solution. **(b)** Water vapor in air condenses onto solid calcium chloride hexahydrate, $CaCl_2 \cdot 6\ H_2O$. The liquid water dissolves some of the solid. The eventual result could be an unsaturated solution.

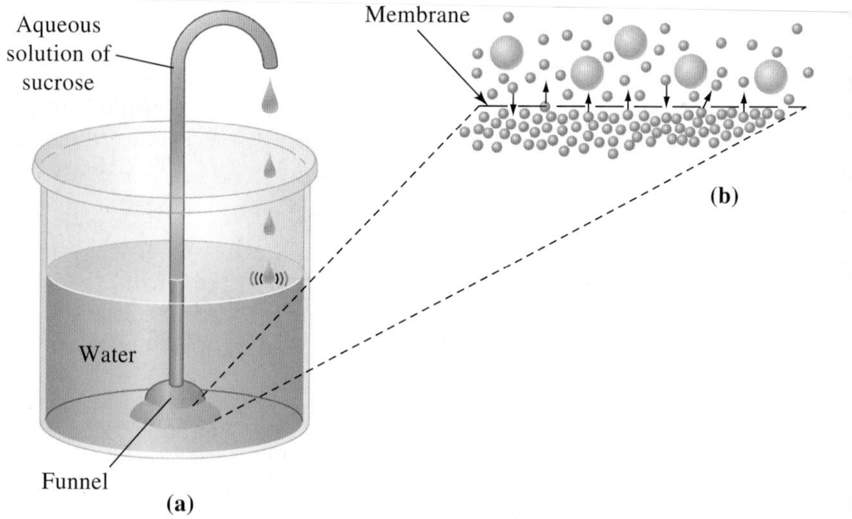

▲ FIGURE 14-21
Osmosis
(a) Water molecules pass through pores in the membrane and create a pressure within the funnel that causes the sucrose solution to rise, overflow, and fall into the pure water After a time, the solution inside the funnel becomes more dilute and the pure water in the beaker becomes a sucrose solution. Liquid flow stops when the compositions of the solutions separated by the membrane have become nearly equal. **(b)** Enlarged cross-section of the membrane demonstrating its semi-permeable properties: water molecule (represented as small blue spheres) freely cross the membrane, while the sucrose molecule (represented as large gray spheres) cannot cross the membrane.

A related phenomenon occurs when $CaCl_2 \cdot 6\,H_2O(s)$ is exposed to a (Fig. 14-20b). Water vapor from the air condenses on the solid, and the sol begins to dissolve, a phenomenon known as *deliquescence*. For a solid to del quesce, the partial pressure of water vapor in the air must be greater than th vapor pressure of water above a saturated aqueous solution of the solid. Th requirement is often met for certain solids under conditions of appropria relative humidity. The deliquescence of $CaCl_2 \cdot 6\,H_2O$ occurs when the relativ humidity exceeds 32%. (Relative humidity is described in Focus On 6- Earth's Atmosphere.)

▶ Semipermeable membranes are materials containing submicroscopic holes, such as a pig's bladder, parchment, or cellophane. The holes permit the passage of solvent molecules but not those of the solute.

Like the case just described, Figure 14-21 also pictures the flow of solve molecules. Here, however, the flow is not through the vapor phase. An aqu ous sucrose (sugar) solution in a long glass tube is separated from pure wat by a semipermeable membrane (permeable to water only). Water molecul can pass through the membrane in either direction, and they do. But becaus the concentration of water molecules is greater in the pure water than in th solution, there is a net flow from the pure water into the solution. This n flow, called **osmosis**, causes the solution to rise in the tube. The more conce trated the sucrose solution, the higher the solution level rises.

14-6 CONCEPT ASSESSMENT

Describe the similarities and differences between the phenomena depicted in Figures 14-20(a) and 14-21.

Applying pressure to the sucrose solution slows down the net flow of wat across the membrane into the solution. With a sufficiently high pressure, th net influx of water can be stopped altogether. The necessary pressure to sto osmotic flow is called the **osmotic pressure** of the solution. For a 20% sucros solution, this pressure is about 15 atm. The magnitude of osmotic pressu pends only on the number of solute particles per unit volume of solution. It es not depend on the identity of the solute. Properties of this sort, whose lues depend only on the concentration of solute particles in solution and *not* what the solute is, are called **colligative properties**. The following equation orks quite well for calculating osmotic pressures of *dilute* solutions of non- ctrolytes. The osmotic pressure is represented by the symbol π; R is the gas nstant (0.08206 L atm mol^{-1} K^{-1}); and T is the Kelvin temperature. The term epresents the amount of solute (in moles), and V is the volume (in liters) of ution. Notice that this equation is similar to the equation for the ideal gas law. this case, however, it is convenient to rearrange terms to yield equation (14.4). e ratio, n/V, then, is the *molarity* of the solution, represented by the symbol c.

◀ Vapor-pressure lowering, as expressed through Raoult's law for ideal solutions, is also a colligative property.

$$\pi V = nRT$$

$$\pi = \frac{n}{V}RT = c \times RT \qquad \textbf{(14.4)}$$

◀ The adjustment required to apply equation (14.4) to electrolyte solutions is discussed in Section 14-9.

The pressure difference of 18 mmHg that we calculate in Example 14-8 is sy to measure. (It corresponds to a solution height of about 25 cm.) This eans that we can easily use the measurement of osmotic pressure for deter- ining molar masses when we are dealing with very dilute solutions or lutes with high molar masses (or both). Example 14-9 shows how osmotic essure measurements can be used to determine molar mass.

EXAMPLE 14-8 Calculating Osmotic Pressure

What is the osmotic pressure at 25 °C of an aqueous solution that is 0.0010 M $C_{12}H_{22}O_{11}$ (sucrose)?

Analyze

We just need to substitute the data into equation (14.4).

Solve

$$\pi = \frac{0.0010 \text{ mol} \times 0.08206 \text{ atm L mol}^{-1} \text{ K}^{-1} \times 298 \text{ K}}{1 \text{ L}}$$

$$\pi = 0.024 \text{ atm (18 mmHg)}$$

Assess

At a very low concentration, there is an appreciable amount of osmotic pressure. This fact is used when measuring the molar mass of polymers and biopolymers.

PRACTICE EXAMPLE A: What is the osmotic pressure at 25 °C of an aqueous solution that contains 1.50 g $C_{12}H_{22}O_{11}$ in 125 mL of solution?

PRACTICE EXAMPLE B: What mass of urea, $CO(NH_2)_2$, would you dissolve in 225 mL of solution to obtain an osmotic pressure of 0.015 atm at 25 °C?

EXAMPLE 14-9 Establishing a Molar Mass from a Measurement of Osmotic Pressure

A 50.00 mL sample of an aqueous solution contains 1.08 g of human serum albumin, a blood-plasma protein. The solution has an osmotic pressure of 5.85 mmHg at 298 K. What is the molar mass of the albumin?

Analyze

We need to use osmotic pressure to determine the molar mass of a protein, human serum albumin, in a solution.

(continued)

Solve

Express osmotic pressure in atmospheres.

$$\pi = 5.85\ \text{mmHg} \times \frac{1\ \text{atm}}{760\ \text{mmHg}} = 7.70 \times 10^{-3}\ \text{atm}$$

Modify the basic equation for osmotic pressure, by showing moles of solute (n) as the mass of solute (m) divided by its molar mass (M), and solve for M.

$$\pi V = nRT \qquad \pi V = \frac{m}{M}RT \qquad M = \frac{mRT}{\pi V}$$

Obtain the value of M by substituting the given data into the last equation above, ensuring that units cancel to yield g/mol as the unit for M.

$$M = \frac{1.08\ \text{g} \times 0.08206\ \text{atm L mol}^{-1}\ \text{K}^{-1} \times 298\ \text{K}}{7.70 \times 10^{-3}\ \text{atm} \times 0.0500\ \text{L}} = 6.86 \times 10^{4}\ \text{g/mol}$$

Assess

Even though the solution is relatively dilute, knowing the osmotic pressure helped us determine the molar mass of human serum albumin.

PRACTICE EXAMPLE A: Creatinine is a by-product of nitrogen metabolism and can be used to provide an indication of renal function. A 4.04 g sample of creatinine is dissolved in enough water to make 100.0 mL of solution. The osmotic pressure of the solution is 8.73 mmHg at 298 K. What is the molar mass of creatinine?

PRACTICE EXAMPLE B: What would be the osmotic pressure of a solution containing 2.12 g of human serum albumin in 75.00 mL of water at 37.0 °C? Use the molar mass determined in Example 14-9.

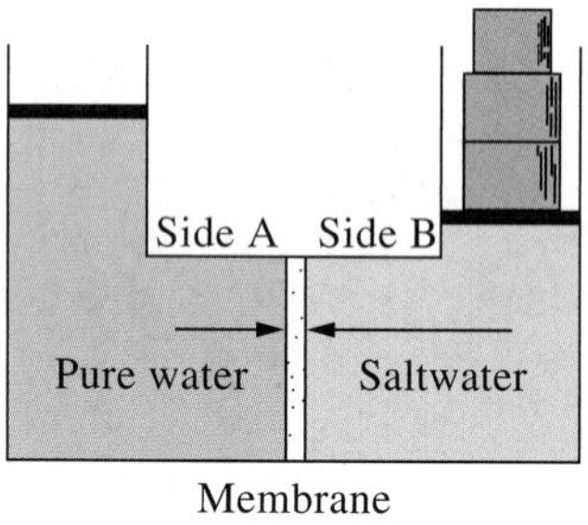

▲ FIGURE 14-22
Desalination of saltwater by reverse osmosis
The membrane is permeable to water but not to ions. The normal flow of water is from side A to side B. If a pressure is exerted on side B that exceeds the osmotic pressure of the saltwater, a net flow of water occurs in the *reverse* direction—from the saltwater to the pure water. The lengths of the arrows suggest the magnitudes of the flow of water molecules in each direction.

Practical Applications

Some of the best examples of osmosis are those associated with living orga isms. For instance, if red blood cells are placed in pure water, the cel expand and eventually burst as a result of water that enters through osm sis. The osmotic pressure associated with the fluid inside the cell is equiv lent to that of 0.92% (mass/volume) NaCl(aq). Thus, if cells are placed in sodium chloride (saline) solution of this concentration, there is no net flo of water through the cell membrane, and the cell remains stable. Such a sol tion is said to be *isotonic*. If cells are placed in a solution with a concentratic greater than 0.92% NaCl, water flows out of the cells, and the cells shrin Such a solution is said to be *hypertonic*. If the NaCl concentration is less tha 0.92%, the solution is *hypotonic*, and water flows into the cells. Fluids that a intravenously injected into patients to combat dehydration or to supp nutrients must be adjusted so that they are isotonic with blood. The osmot pressure of the fluids must be the same as that of 0.92% (mass/volum NaCl.

One recent application of osmosis goes to the very definition of osmot pressure. Suppose in the device shown in Figure 14-22, we apply a pressure the right side (side B) that is less than the osmotic pressure of the saltwate The net flow of water molecules through the membrane will be from side A side B, and ordinary osmosis occurs. If we apply a pressure greater than th

▶ A red blood cell in a hypertonic solution (left), an isotonic solution (center), and a hypotonic solution (right).

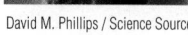

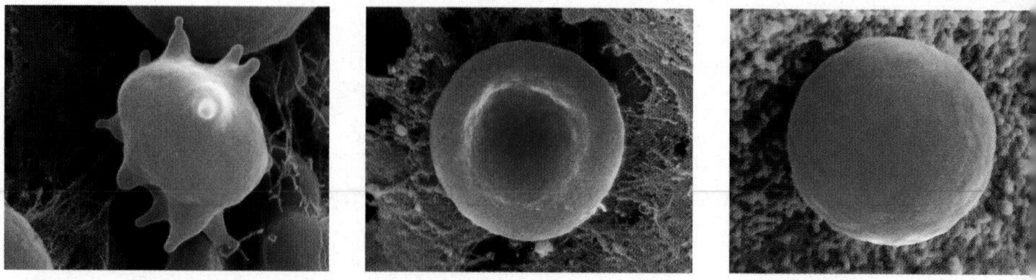

David M. Phillips / Science Source

smotic pressure to side B, we can cause a net flow of water in the *reverse* irection, from the saltwater into the pure water. This is the condition known ; **reverse osmosis**. Reverse osmosis can be used in the *desalination* of seawa-r to supply drinking water for emergency situations or as an actual source f municipal water. Another application of reverse osmosis is the removal of issolved materials from industrial or municipal wastewater before it is dis-harged into the environment.

▲ A small reverse-osmosis unit used to desalinize seawater.

4-8 Freezing-Point Depression and Boiling-Point Elevation of Nonelectrolyte Solutions

ı Section 14-6 we examined the lowering of the vapor pressure of a solvent pro-uced by a dissolved solute. Vapor pressure lowering is not measured as fre-uently as certain properties directly related to it. In Figure 14-23 the blue curves present the vapor pressure, fusion, and sublimation curves in the phase dia-ram for a pure solvent. The red curves represent the vapor pressure and fusion urves of the solvent in a solution. The sublimation curve for the solid solvent at freezes from the solution is shown in purple. Two assumptions are implicit Figure 14-23. One is that the solute is nonvolatile, the other is that the solid at freezes from a solution is pure solvent. For many mixtures, these require-ents are easily met.*

◀ In Figure 14-23, the solid and vapor chemical potentials do not change with the addition of a solute to the liquid phase. However, adding solute to the liquid does lower the chemical potential of the liquid solvent. Hence, to re-establish equilibrium, the S–L and the L–V curves shift to respectively lower and higher *T*. This can be thought of as an example of Le Châtelier's principle.

The vapor pressure curve of the solution (red) intersects the sublimation urve at a lower temperature than is the case for the pure solvent. The olid–liquid fusion curve, because it originates at the intersection of the subli-ation and vapor pressure curves, is also displaced to lower temperatures. ow recall how we establish normal melting points and boiling points in a hase diagram. They are the temperatures at which a line at $P = 1$ atm inter-cts the fusion and vapor pressure curves, respectively. Four points of inter-ction are highlighted in Figure 14-23—the freezing points and the boiling oints of the pure solvent and of the solvent in a solution. The freezing point of e solvent in solution is *depressed*, and the boiling point is *elevated*.

The extent to which the freezing point is lowered or the boiling point raised is roportional to the mole fraction of solute (just as is vapor pressure lowering).

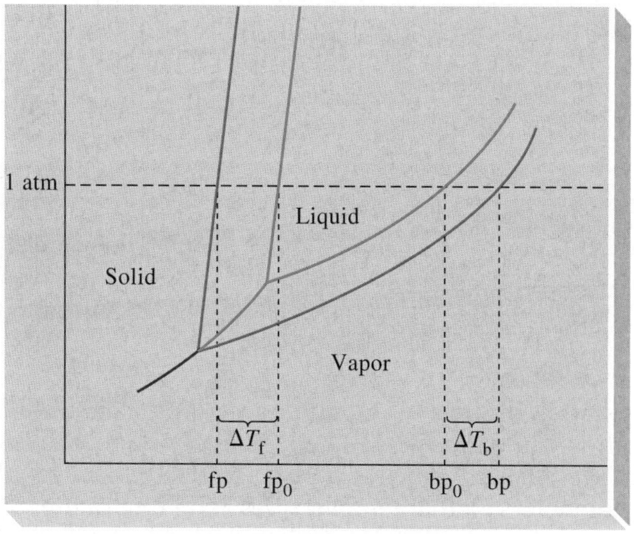

◀ FIGURE 14-23
Vapor-pressure lowering by a nonvolatlle solute
The normal freezing point and normal boiling point of the pure solvent are fp_0 and bp_0, respectively. The corresponding points for the solution are fp and bp. The freezing-point depression, ΔT_f, and the boiling-point elevation, ΔT_b, are indicated. Because the solute is assumed to be insoluble in the solid solvent, the sublimation curve of the solvent is unaffected by the presence of solute in the liquid solution phase. That is, the sublimation curve is the same for the two phase diagrams.

*Actually, the equation for freezing-point depression (14.5) applies even if the solute is volatile.

TABLE 14.3 Freezing-Point Depression and Boiling-Point Elevation Constants[a]

Solvent	Normal Freezing Point, °C	K_f, °C mol^{-1} kg	Normal Boiling Point, °C	K_b, °C mol^{-1} kg
Acetic acid	16.6	3.90	118	3.07
Benzene	5.53	5.12	80.10	2.53
Nitrobenzene	5.7	8.1	210.8	5.24
Phenol	41	7.27	182	3.56
Water	0.00	1.86	100.0	0.512

[a]Values correspond to freezing-point depressions and boiling-point elevations, in degrees Celsius, caused by 1 mol of solute particles dissolved in 1 kg of solvent in an ideal solution.

In *dilute* solutions, the solute mole fraction is proportional to its molality, and so we can write

$$\Delta T_f = -K_f \times m \qquad (14.5)$$

$$\Delta T_b = K_b \times m \qquad (14.6)$$

KEEP IN MIND

that freezing-point depression (ΔT_f) is defined as $T - T_f$, where T is the freezing point of the solution and T_f is the freezing point of the pure solvent, and similarly the boiling-point elevation (ΔT_b) is defined as $T - T_b$ where T_b is the boiling point of the pure solvent. So the need for the negative sign in equation (14.5) is evident.

In these equations, ΔT_f and ΔT_b are the freezing-point depression and boiling-point elevation, respectively; m is the solute molality; and K_f and K_b are proportionality constants. The value of K_f depends on the melting point, enthalpy of fusion, and molar mass of the solvent. The value of K_b depends on the boiling point, enthalpy of vaporization, and molar mass of the solvent. The units of K_f and K_b are °C mol^{-1} kg, and you can think of their values as representing the freezing-point depression and boiling-point elevation for a 1 molal (1 m) solution. In practice, though, equations (14.5) and (14.6) often fail for solutions as concentrated as 1 m. Table 14.3 lists some typical values of K_f and K_b.

Historically, chemists have used the group of colligative properties—vapor pressure lowering, freezing-point depression, boiling-point elevation, and osmotic pressure—for molecular mass determinations. In Example 14-9, we showed how this could be accomplished with osmotic pressure. Example 14-10 shows how freezing-point depression can be used to determine a molar mass and, with other information, a molecular formula. To help you understand how this is done, we present a three-step procedure in the form of answers to three separate questions. In other cases, you should be prepared to work out your own procedure.

14-7 CONCEPT ASSESSMENT

In what important way would Figure 14-23 change if it were based on the phase diagram of water rather than for the general case shown? Would a boiling-point elevation and a freezing-point depression still be expected?

▶ The adjustment required to apply these equations to electrolyte solutions is discussed in Section 14-9.

Molar mass determination by freezing-point depression or boiling-point elevation has its limitations. Equations (14.5) and (14.6) apply only to dilute solutions of nonelectrolytes, usually much less than 1 mol kg^{-1}. This requires the use of special thermometers so that temperatures can be measured very precisely, say to ±0.001 °C. Because boiling points depend on barometric pressure, precise measurements require that pressure be held constant. As a consequence, boiling-point elevation is not much used. The precision of the freezing-point depression method can be improved by using a solvent

EXAMPLE 14-10 Establishing a Molecular Formula with Freezing-Point Data

Nicotine, extracted from tobacco leaves, is a liquid completely miscible with water at temperatures below 60 °C. **(a)** What is the *molality* of nicotine in an aqueous solution that starts to freeze at −0.450 °C? **(b)** If this solution is obtained by dissolving 1.921 g of nicotine in 48.92 g H_2O, what must be the molar mass of nicotine? **(c)** Combustion analysis shows nicotine to consist of 74.03% C, 8.70% H, and 17.27% N, by mass. What is the molecular formula of nicotine?

Analyze

(a) We can establish the molality of the nicotine by using equation (14.5) with the value of K_f for water listed in Table 14.3. **(b)** Once we know the molality, we can use the definition of molality, but with a known molality (from part a) and an unknown molar mass of solute (M). **(c)** To establish the empirical formula of nicotine, we need to use the method of Example 3-5.

Solve

(a) Note that $T_f = -0.450\ °C$, and that $\Delta T_f = -0.450\ °C - 0.000\ °C = -0.450\ °C$.

$$\text{molality} = \frac{\Delta T_f}{-K_f} = \frac{-0.450\ °C}{-1.86\ °C\ \text{mol}^{-1}\ \text{kg}} = 0.242\ \frac{\text{mol}}{\text{kg}} = 0.242\ \text{mol kg}^{-1}$$

(b) Let's represent the molar mass as x g/mol. The amount, in moles, contained in a 1.921 g sample is $1.921\ \text{g} \times (1\ \text{mol}/x\ \text{g}) = (1.921/x)$ mol. Thus,

$$\text{molality} = \frac{1.921\ x^{-1}}{0.04892\ \text{kg water}} = 0.242\ \text{mol (kg water)}^{-1}$$

$$x = 162$$

The molar mass is 162 g/mol.

(c) This calculation is left as an exercise for you to do. The result you should obtain is C_5H_7N. The formula mass based on this empirical formula is 81 u. The molecular mass obtained from the molar mass in part (b) is exactly twice this value—162 u. The molecular formula is twice C_5H_7N, or $C_{10}H_{14}N_2$.

Assess

Using the freezing-point data is another experimental technique that can be used to obtain the chemical properties of a substance. In this example we were able to determine the molecular formula from the freezing-point depression and a known amount of substance dissolved in a solvent. Note that water was used as the solvent here; however, other solvents can be used in freezing-point experiments.

PRACTICE EXAMPLE A: Vitamin B_2, riboflavin, is soluble in water. If 0.833 g of riboflavin is dissolved in 18.1 g H_2O, the resulting solution has a freezing point of −0.227 °C. **(a)** What is the molality of the solution? **(b)** What is the molar mass of riboflavin? **(c)** What is the molecular formula of riboflavin if combustion analysis shows it to consist of 54.25% C, 5.36% H, 25.51% O, and 14.89% N?

PRACTICE EXAMPLE B: An aqueous solution that is 0.205 mol kg^{-1} urea, $CO(NH_2)_2$, is found to boil at 100.025 °C. Is the prevailing barometric pressure above or below 760.0 mmHg? [*Hint:* At what temperature should the solution begin to boil if atmospheric pressure is 760.0 mmHg?]

ith a larger K_f value than that of water. For example, for cyclohexane $_f = 20.0\ °C\ \text{mol}^{-1}$ kg and for camphor $K_f = 37.7\ °C\ \text{mol}^{-1}$ kg.

ractical Applications

ne typical automobile antifreeze is ethylene glycol, $HOCH_2CH_2OH$. It is a ood idea to leave the ethylene glycol–water mixture in the cooling system at l times to provide all-weather protection. In summer, the ethylene glycol elps by raising the boiling point of water and preventing the cooling system om boiling over.

Citrus growers faced with an impending freeze know they must take pre- entive measures only if the temperature drops below 0 °C by several degrees. he juice in the fruit has enough dissolved solutes to lower the freezing point y a degree or two. The growers also know they must protect lemons sooner

Wayne Eastep/The Image Bank/Getty Images

▲ Water sprayed on citrus fruit releases its heat of fusion as it freezes into a layer of ice that acts as a thermal insulator. For a time, the temperature remains at 0 °C. The juice of the fruit, having a freezing point below 0 °C, is protected from freezing.

Chalabala/Fotolia

▲ A typical aircraft deicer is propylene glycol, $CH_3CH(OH)CH_2OH$, diluted with water and applied as a hot, high-pressure spray.

Mark Joseph/Getty Images

▲ Lowering the freezing point of water on roads.

than oranges because lemons have a lower concentration of dissolved solut (sugars) than do oranges.

Salts, such as NaCl, can be used to prepare a *slush bath,* a mixture use to cool or freeze something. One example is the mixture of ice and NaCl(used to freeze ice cream in a home ice-cream maker. Because the slush bath at a temperature well below 0 °C, it is easy to freeze the sugar-and-milk mi ture that makes up the ice cream. NaCl is also useful for deicing roads. It effective in melting ice at temperatures as low as −21 °C (−6 °F). This is th lowest freezing point of a NaCl(aq) solution.

14-8 CONCEPT ASSESSMENT

Why do you suppose that the freezing point of NaCl(aq) is depressed no further than −21 °C, regardless of how much more NaCl(s) is added to water?

14-9 Solutions of Electrolytes

▶ Pure water does not conduct electricity. So why should we be careful with electricity when near water? It is not the water but the electrolytes that are dissolved in water that allow the water (solution) to conduct electricity.

The discussion of the electrical conductivities of solutions in Section 5-1 retrace some of the work done by Swedish chemist Svante Arrhenius for his doctor dissertation (1883). Prevailing opinion at the time was that ions form only wi the passage of electric current. Arrhenius, however, reached the conclusion th ions exist in a solid substance and become dissociated from each other when th solid dissolves in water. Such is the case with NaCl, for example. In other case as with HCl, ions do not exist in the substance but are formed when it dissolv in water. In any case, electricity is not required to produce ions.

Although Arrhenius developed his theory of electrolytic dissociation explain the electrical conductivities of solutions, he was able to apply it mo widely. One of his first successes came in explaining certain anomalous valu of colligative properties described by the Dutch chemist Jacobus van't Ho (1852–1911).

Anomalous Colligative Properties

▶ Later in this section, we explain why the experimentally determined i for 0.0100 m NaCl is 1.94 instead of 2.

Certain solutes produce a greater effect on colligative properties tha expected. For example, consider a 0.0100 m aqueous solution. The predicte freezing-point depression of this solution is

$$\Delta T_f = -K_f \times m = -1.86\ °C\ mol^{-1}\ kg \times 0.0100\ mol^{-1}\ kg = -0.0186\ °C$$

We expect the solution to have a freezing point of −0.0186 °C. If the 0.0100 solution is 0.0100 m urea, the measured freezing point is just about −0.0186 °

he solution is 0.0100 m NaCl, however, the measured freezing point is about .0361 °C.

Van't Hoff defined the factor i as the ratio of the measured value of a ligative property to the expected value if the solute is an electrolyte. For 100 m NaCl,

$$i = \frac{\text{measured } \Delta T_f}{\text{expected } \Delta T_f} = \frac{-0.0361\ °\text{C}}{-1.86\ °\text{C mol}^{-1}\text{ kg} \times 0.0100\ \text{mol}^{-1}\text{ kg}} = 1.94$$

Arrhenius's theory of electrolytic dissociation allows us to explain different lues of the van't Hoff factor i for different solutes. For such solutes as urea, cerol, and sucrose (all nonelectrolytes), $i = 1$. For a strong electrolyte such NaCl, which produces *two* moles of ions in solution per mole of solute dis-ved, we would expect the effect on freezing-point depression to be twice as eat as for a nonelectrolyte. We would expect that $i = 2$. Similarly, for $MgCl_2$, r expectation would be that $i = 3$. For the weak acid CH_3COOH (acetic d), which is only slightly ionized in aqueous solution, we expect i to be ghtly larger than one but not nearly equal to two.

This discussion suggests that equations (14.4), (14.5), and (14.6) should all rewritten in the form

$$\pi = i \times c \times RT$$

$$\Delta T_f = -i \times K_f \times m$$

$$\Delta T_b = i \times K_b \times m$$

these equations are used for nonelectrolytes, simply substitute $i = 1$. For ong electrolytes, predict a value of i as suggested in Example 14-11.

World History Archive / Alamy

▲ **Svante Arrhenius (1859–1927)**
At the time Arrhenius was awarded the Nobel Prize in chemistry (1903), his results were described thus: "Chemists would not recognize them as chemistry; nor physicists as physics. They have in fact built a bridge between the two." The field of physical chemistry had its origins in Arrhenius's work.

EXAMPLE 14-11 Predicting Colligative Properties for Electrolyte Solutions

Predict the freezing point of aqueous 0.00145 mol kg^{-1} $MgCl_2$.

Analyze

We will use a modified freezing-point depression equation in which the van't Hoff factor i is included. We first note that $MgCl_2$ is a salt that completely dissociates when it is dissolved in water. So we determine the value of i for $MgCl_2$. We can do this by writing an equation to represent the dissociation of $MgCl_2$(s). Then we use the appropriate freezing-point depression expression.

Solve

$$MgCl_2(s) \xrightarrow{H_2O} Mg^{2+}(aq) + 2\,Cl^-(aq)$$

Because *three* moles of ions are obtained per mole of formula units dissolved, we expect the value $i = 3$.

Now use the expression

$$\begin{aligned}\Delta T_f &= -i \times K_f \times m \\ &= -3 \times 1.86\ °\text{C mol}^{-1}\text{ kg} \times 0.00145\ \text{mol}^{-1}\text{ kg} \\ &= -0.0081\ °\text{C}\end{aligned}$$

The predicted freezing point is −0.0081 °C.

Assess

Because the value of i for $MgCl_2$ is not exactly 3, we are not justified in carrying more than one or two significant figures in our answer. If we had ignored the fact that $MgCl_2$ is a strong electrolyte, then our calculated freezing-point depression would have been three times as small as the experimental value. Always remember to include the van't Hoff factor when ionic compounds are given as part of the problem.

PRACTICE EXAMPLE A: What is the expected osmotic pressure of a 0.0530 M $MgCl_2$ solution at 25 °C?

PRACTICE EXAMPLE B: You want to prepare an aqueous solution that has a freezing point of −0.100 °C. How many milliliters of 12.0 M HCl would you use to prepare 250.0 mL of such a solution? [*Hint:* Note that in a dilute aqueous solution, molality and molarity are essentially numerically equal.]

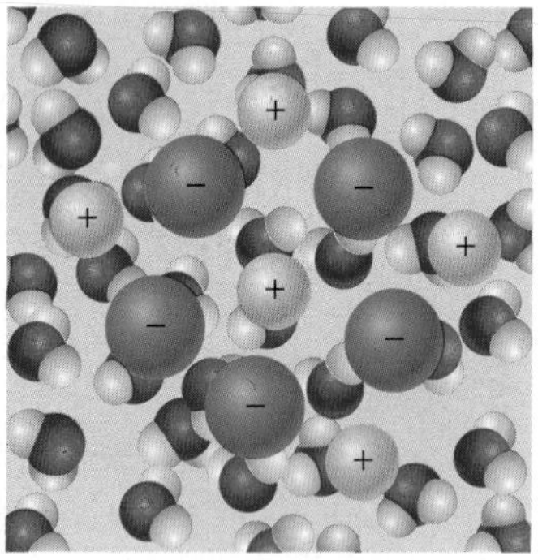

(a)

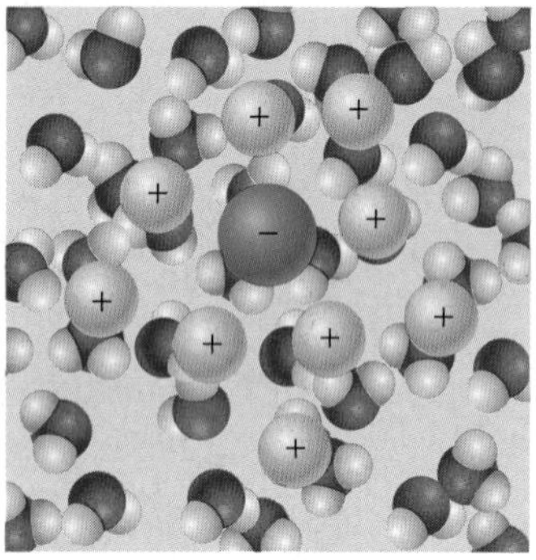

(b)

▲ FIGURE 14-24
Interionic attractions in aqueous solution
(a) A positive ion in aqueous solution is surrounded by a shell of negative ions. **(b)** A negative ion attracts positive ions to its immediate surroundings.

TABLE 14.4 Variation of the van't Hoff Factor, *i*, with Solution Molality

	Molality, mol kg^{-1}					
Solute	1.0	0.10	0.010	0.0010	· · ·	Inf. dil
NaCl	1.81	1.87	1.94	1.97	· · ·	2
$MgSO_4$	1.09	1.21	1.53	1.82	· · ·	2
$Pb(NO_3)_2$	1.31	2.13	2.63	2.89	· · ·	3

[a]The limiting values: $i = 2$, 2, and 3 are reached when the solution is infinitely dilute. Note that a solute whose ions are singly charged (for example, NaCl) approaches its limiting value more quickly than does a solute whose ions carry higher charges. Interionic attractio are greater in solutes with more highly charged ions.

Interionic Attractions

Despite its initial successes, there were apparent deficiencies in Arrheniu theory. The electrical conductivities of concentrated solutions of strong el trolytes are not as great as expected, and values of the van't Hoff facto depend on the solution concentrations, as shown in Table 14.4. For stro electrolytes that exist completely in ionic form in aqueous solutions, would expect $i = 2$ for NaCl, $i = 3$ for $MgCl_2$, and so on, regardless of t solution concentration.

These difficulties can be resolved with a theory of electrolyte solutions p posed by Peter Debye and Erich Hückel in 1923. This theory continues view strong electrolytes as existing only in ionic form in aqueous solutic but the ions in solution *do not behave independently of one another*. Instead, ea cation is surrounded by a cluster of ions in which anions predominate, a each anion is surrounded by a cluster in which cations predominate. In sho each ion is enveloped by an *ionic atmosphere* with a net charge opposite own (Fig. 14-24).

In an electric field, the mobility of each ion is reduced because of t attraction or drag exerted by its ionic atmosphere. Similarly, the magnitu of colligative properties are reduced. This explains why, for example, t value of i for 0.010 m NaCl is 1.94 rather than 2.00. What we can say is th each type of ion in an aqueous solution has two "concentrations." One called the *stoichiometric concentration* and is based on the amount of solu dissolved. The other is an "effective" concentration, called the *activity*, whi takes into account interionic attractions. Stoichiometric calculations of t type presented in Chapters 4 and 5 can be made with great accuracy usi stoichiometric concentrations. However, no calculations involving soluti properties are 100% accurate if stoichiometric concentrations are use Activities are needed instead. The activity of an ion in solution is related its stoichiometric concentration through a factor called an *activity coefficie* Activities were introduced in Chapter 13. In Chapter 15 their importance chemical equilibrium will be discussed in more detail.

14-10 Colloidal Mixtures

▶ Colloids are extremely important to the food industry. Gelatins are colloids, and many foods are colloidal mixtures for which the food industry exerts considerable research effort to prevent their separation.

In a mixture of sand and water, the sand (silica, SiO_2) quickly settles to t bottom of the container. Yet mixtures can be prepared containing up to 40 by mass of SiO_2, and the silica may remain dispersed in the aqueous mediu for many years. In these mixtures, the silica is *not* present as ions or mo cules. Rather, much larger particles of silica are present, though they are st submicroscopic in size. The mixture is said to be a **colloid**. To be classified *c loidal*, a material must have one or more of its dimensions (length, width,

◀ FIGURE 14-25
The Tyndall effect
The flashlight beam is not visible as it passes through a true solution (left), but it is readily seen as it passes through the colloidal dispersion of Fe_2O_3 (right).

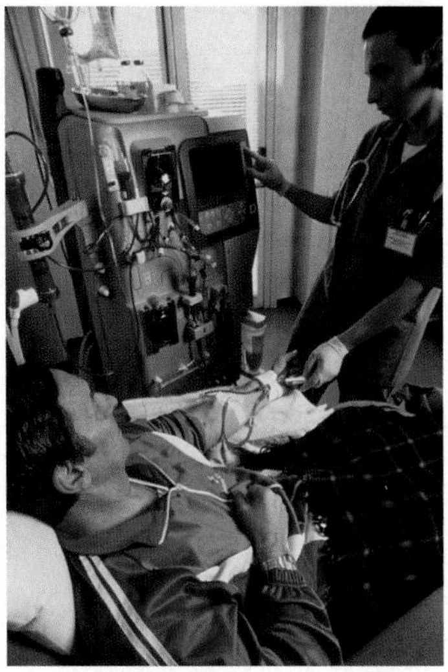

Mauro Fermariello/Science Photo Library

▲ **Hemodialysis**
An artificial kidney machine cleaning the blood of patients with impaired kidney function.

ickness) in the approximate range of 1–1000 nm. If all the dimensions are ıaller than 1 nm, the particles are of molecular size. If all the dimensions ceed 1000 nm, the particles are of ordinary, or macroscopic, size (even if ey are visible only under a microscope). One method of determining ıether a mixture is a true solution or a colloid is illustrated in Figure 14-25. hen light passes through a true solution, an observer viewing from a direcn perpendicular to the light beam sees no light. In a colloidal dispersion, ht is scattered in many directions and is readily seen. This effect, first stud-d by John Tyndall in 1869, is known as the *Tyndall effect*. A common example the scattering of light by dust particles in a flashlight beam.

The particles in colloidal silica have a spherical shape. Some colloidal parles are rod-shaped, and some, like gamma globulin in human blood plasma, ve a disc-like shape. Thin films, like an oil slick on water, are colloidal. And me colloids, such as cellulose fibers, are randomly coiled filaments.

What keeps the SiO_2 particles suspended in colloidal silica? The most portant factor is that the surfaces of the particles *adsorb*, or attach to themlves, ions from the solution, and they preferentially adsorb one type of ion er others. In the case of SiO_2, the preferred ions that are adsorbed are OH^- e Figure 14-26). As a result, the particles acquire a net negative charge. aving like charges, the particles repel one another. These mutual repulons overcome the force of gravity, and the particles remain suspended definitely.

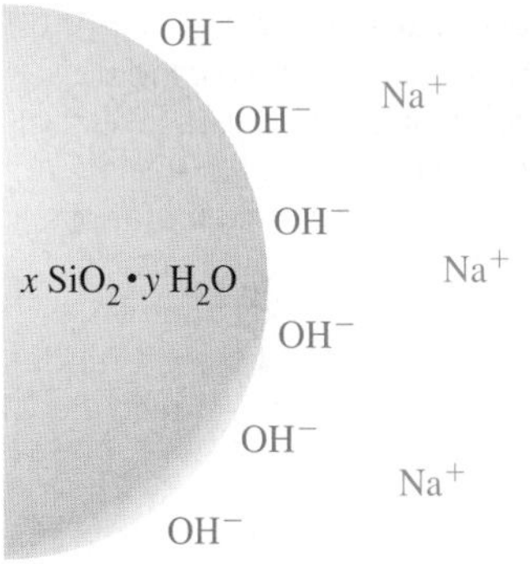

▲ FIGURE 14-26
Surface of SiO_2 particle in colloidal silica
The points made in this simplified drawing are: (1) The SiO_2 particles are hydrated; (2) OH^- ions are preferentially adsorbed on the surface; (3) In the immediate vicinity of the particle, negative ions outnumber positive ions and the particle carries a net negative charge. Not illustrated here are the facts that some of the negative charge comes from silicate anions (for example, SiO_3^{2-}), and as a whole, the solution in which these particles are found is electrically neutral.

Although electric charge can be important in stabilizing a colloid, a high ncentration of ions can also bring about the *coagulation*, or precipitation, a colloid (Fig. 14-27). The ions responsible for the coagulation are those rrying a charge opposite to that on the colloidal particles. *Dialysis*, a ocess similar to osmosis, can be used to remove excess ions from a colidal mixture. As suggested by Figure 14-28, molecules of solvent and molules or ions of solute pass through a semipermeable membrane, but the uch larger colloidal particles do not. In some cases, the process is more fective when carried out in an electric field. In *electrodialysis*, ions are tracted out of a colloidal mixture by an electrode carrying the opposite arge. A human kidney dialyzes blood, a colloidal mixture, to remove cess electrolytes produced by metabolic processes. Certain diseases cause e kidneys to lose this ability, but a dialysis machine, external to the body, n function for the kidneys.

Table 14.5 lists some common colloids, and as so aptly put by Wilder ıncroft, an American pioneer in the field of colloid chemistry, ". . . colloid emistry is essential to anyone who really wishes to understand . . . oils, eases, soaps, . . . glue, starch, adhesives, . . . paints, varnishes, lacquers, . . . eam, butter, cheese, . . . cooking, washing, dyeing, . . . colloid chemistry is e chemistry of life."

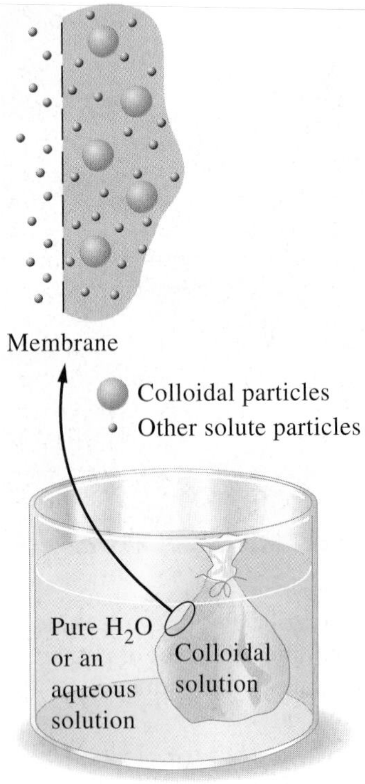

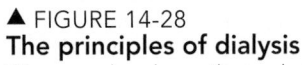

▲ FIGURE 14-28
The principles of dialysis
Water molecules, other solute molecules, and dissolved ions are all free to pass through the pores of the membrane (for example, a film of cellophane) in either direction. The direction of net flow of these species depends on their relative concentrations on either side of the membrane. Colloidal particles, however, cannot pass through the pores of the membrane.

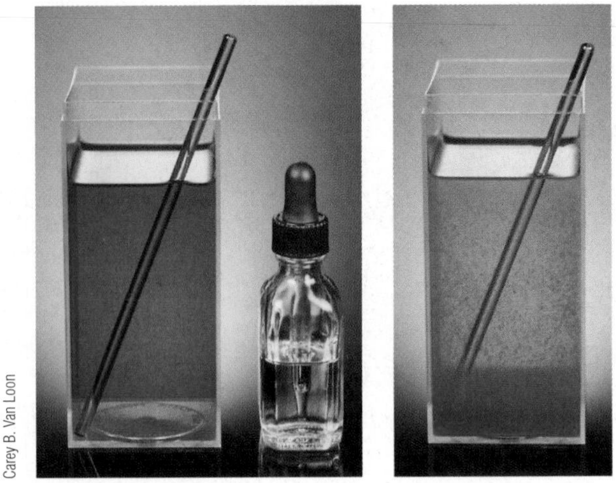

▲ FIGURE 14-27
Coagulation of colloidal iron oxide
On the left is red colloidal hydrous Fe_2O_3, obtained by adding $FeCl_3(aq)$ to boiling water. When a few drops of $Al_2(SO_4)_3(aq)$ are added, the suspended particles rapidly coagulate into a precipitate of $Fe_2O_3(s)$, as shown on the right.

TABLE 14.5 Some Common Types of Colloids

Dispersed Phase	Dispersion Medium	Type	Examples
Solid	Liquid	Sol	Clay sols,[a] colloidal silica, colloidal gold
Liquid	Liquid	Emulsion	Oil in water, milk, mayonnaise
Gas	Liquid	Foam	Soap and detergent suds, whipped cream, meringues
Solid	Gas	Aerosol[b]	Smoke, dust-laden air[c]
Liquid	Gas	Aerosol[b]	Fog, mist (as in aerosol products)
Solid	Solid	Solid sol	Ruby glass, certain natural and synthetic gems, blue rock salt, black diamond
Liquid	Solid	Solid emulsion	Opal, pearl
Gas	Solid	Solid foam	Pumice, lava, volcanic ash

[a]In water purification, it is sometimes necessary to precipitate clay particles or other suspended colloidal materials. This is often done by treating the water with an appropriate electrolyte. Clay sols are suspected of adsorbing organic substances, such as pesticides, and distributing them in the environment.
[b]Smogs are complex materials that are at least partly colloidal. The suspended particles are both solid and liquid. Other constituents of smog are molecular, for example, sulfur oxides, nitrogen oxides, and ozone.
[c]The bluish haze of tobacco smoke and the brilliant sunsets in desert regions are both attributable to the scattering of light by colloidal particles suspended in air.

MasteringCHEMISTRY — www.masteringchemistry.com

The separation of complex mixtures, such as gasoline, into their individual components can be accomplished using a technique known as chromatography. There are many variations of the technique, but they all take advantage of differences in the way molecules of different compounds interact with a common material. To find out more, go to the Focus On feature for Chapter 14, Chromatography, on the MasteringChemistry site.

Summary

-1 Types of Solutions: Some Terminology—In ɔlution, the **solvent**—usually the component present in atest amount—determines the state of matter in which solution exists (Table 14.1). A **solute** is a solution comnent dissolved in the solvent. Dilute solutions contain ıtively small amounts of solute and concentrated solu-ns, large amounts.

-2 Solution Concentration—Any description of composition of a solution must indicate the quantities solute and solvent (or solution) present. Solution contrations expressed as mass percent, volume percent, d mass/volume percent all have practical importance, do the units, parts per million **(ppm)**, parts per billion **b)**, and parts per trillion **(ppt)**. However, the more damental concentration units are mole fraction, larity, and molality. Molarity (moles of solute per liter solution) is temperature dependent, but mole fraction d **molality** (moles of solute per kilogram of solvent) not.

-3 Intermolecular Forces and the Solution ocess—Predictions about whether two substances will x to form a solution involve knowledge of intermolecu-forces between like and unlike molecules (Figs. 14-2 d 14-3). This approach makes it possible to identify an **al solution**, one whose properties can be predicted m properties of the individual solution components. st solutions are nonideal.

-4 Solution Formation and Equilibrium—nerally, a solvent has a limited ability to dissolve a ute. A solution containing the maximum amount of ute possible is a **saturated solution**. A solution with s than this maximum amount is an **unsaturated ution**. Under certain conditions a solution can be pre-red that contains more solute than a normal saturated ution; such solutions are called **supersaturated**. **ubility** refers to the concentration of solute in a satu-ed solution and depends on temperature. Graphs of ute solubility versus temperature (Fig. 14-10) can be d to devise conditions for recovering a pure solute m a solution of several solutes through **fractional crys-lization (recrystallization)** (Fig. 14-11).

-5 Solubilities of Gases—The solubilities of gases pend on pressure as well as temperature, and many niliar phenomena are related to gas solubilities. **nry's law** (equation 14.2) relates the concentration of a s in solution to its pressure above the solution.

-6 Vapor Pressures of Solutions—The vapor ssure of a solution depends on the vapor pressures of pure components. If the solution is ideal, **Raoult's law** uation 14.3) can be used to calculate the solution vapor ssure. Liquid–vapor equilibrium curves showing either ution vapor pressures (Fig. 14-16) or solution boiling ints (Fig. 14-17) as a function of solution composition help us to visualize **fractional distillation**, a common method of separating the volatile components of a solution. Such curves also illustrate the formation of azeotropes in some nonideal solutions. **Azeotropes** are solutions that boil at a constant temperature and produce vapor of the same composition as the liquid; they have boiling points that in some cases are greater than the boiling points of the pure components and in some cases, less (Fig. 14-19).

14-7 Osmotic Pressure—**Osmosis** is the spontaneous flow of solvent through a semipermeable membrane separating two solutions of different concentration. The net flow is from the less to the more concentrated solution (Fig. 14-21). Osmotic flow can be stopped by applying a pressure, called the **osmotic pressure**, to the more concentrated solution. In **reverse osmosis**, the direction of flow is reversed by applying a pressure that exceeds the osmotic pressure to the more concentrated solution. Both osmosis and reverse osmosis have important practical applications. Osmotic pressure can be calculated with a simple relationship (equation 14.4) resembling the ideal gas equation. **Colligative properties** are certain properties that depend only on the concentration of solute particles in a solution, and not on the identity of the solute. Vapor-pressure lowering (Section 14-6) is one such property, and osmotic pressure is another.

14-8 Freezing-Point Depression and Boiling-Point Elevation of Nonelectrolyte Solutions—Freezing-point depression and boiling-point elevation (Fig. 14-23) are colligative properties having many familiar practical applications. For reasonably dilute solutions, their values are proportional to the molality of the solution (equations 14.5 and 14.6). The proportionality constants are K_f and K_b, respectively (Table 14.3). Historically, freezing-point depression was a common method for determining molar masses.

14-9 Solutions of Electrolytes—Calculating colligative properties of electrolyte solutions is more difficult than for solutions of nonelectrolytes. The solute particles in electrolyte solutions are ions or ions and molecules. Calculations using equations (14.5) and (14.6) must be based on the total number of particles present, and the van't Hoff factor is introduced into these equations to reflect this number. In all but the most dilute solutions, composition must be in terms of activities—effective concentrations that take into account interionic forces.

14-10 Colloidal Mixtures—**Colloids** are an important intermediate state between a true solution and a heterogeneous mixture. Colloidal mixtures are responsible for some unusual phenomena (Fig. 14.25) and are encountered in a broad range of contexts, from fluids in living organisms to pollutants in large air masses (Table 14-5).

Integrative Example

A 50.00 g sample of a solution of naphthalene, $C_{10}H_8(s)$, in benzene, $C_6H_6(l)$, has a freezing point of 4.45 °C. Calculate mass percent $C_{10}H_8$ and the boiling point of this solution.

Analyze

Equation (14.5) relates freezing-point depression to the molality of a solution, but we will have to devise an algebr method that yields the mass of each solution component and consequently the mass percent composition. The boili point of the solution can be determined with a minimum of calculation through equation (14.6).

Solve

Substitute data for benzene from Table 14.3 ($K_f = 5.12$ °C kg mol^{-1}, fp = 5.53 °C) and the measured freezing point of the solution into equation (14.5), to obtain

$$\Delta T_f = -K_f \times m$$
$$(4.45 - 5.53)\ ^\circ\text{C} = -5.12\ \text{kg mol}^{-1}\ ^\circ\text{C} \times m$$
$$1.08 = 5.12\ \text{kg mol}^{-1} \times m$$

Express the molality of the solution in terms of the masses of naphthalene, x, and benzene, y, and the molar mass, 128.2 g $C_{10}H_8$/mol.

$$m = \frac{\dfrac{x\ \text{g}\ C_{10}H_8}{128.2\ \text{g}\ C_{10}H_8/\text{mol}}}{y\ \text{g}\ C_6H_6 \times 1\ \text{kg}/1000\ \text{g}\ C_6H_6}$$

Because the total mass is 50.00 g, $x + y = 50.00$ and the expression above reduces to

$$m = \frac{\text{moles}\ C_{10}H_8}{\text{kg}\ C_6H_6} = \frac{(x/128.2)}{(50.00 - x)/1000}\ \text{mol kg}^{-1}$$

Now, substituting this expression of m into the expression derived from equation (14.5), we obtain

$$1.08 = 5.12 \times \frac{(x/128.2)}{(50.00 - x)/1000}$$

which we solve for x, the mass of naphthalene in grams.

$$\frac{1.08 \times 128.2}{5.12 \times 1000} = \frac{x}{50.00 - x}$$
$$0.0270 = \frac{x}{50.00 - x}$$
$$1.0270x = 0.0270 \times 50.00$$
$$x = 1.31$$

The mass percent naphthalene in the solution is

$$\%\ C_{10}H_8 = \frac{1.31\ \text{g}\ C_{10}H_8}{50.00\ \text{g soln}} \times 100\% = 2.62\%$$

A simple way to find the boiling point of the solution is to first solve equation (14.5) for the molality of the solution.

$$\text{molality} = \frac{\Delta T_f}{-K_f} = \frac{-1.08\ ^\circ\text{C}}{-5.12\ ^\circ\text{C kg mol}^{-1}} = 0.211\ \text{mol kg}^{-}$$

Because the molality at the boiling point is the same as at the freezing point, substitute 0.211 m into equation (14.6) and solve for ΔT_b.

$$\Delta T_b = K_b \times m = 2.53\ ^\circ\text{C kg mol}^{-1} \times 0.211\ \text{mol kg}^{-1}$$
$$= 0.534\ ^\circ\text{C}$$

The boiling point of the solution is 0.534 °C higher than the normal boiling point of benzene (80.10 °C), that is,

$$80.10\ ^\circ\text{C} + 0.53\ ^\circ\text{C} = 80.63\ ^\circ\text{C}$$

Assess

The mass of solution can easily be determined to the nearest 0.01 g and expressed with four significant figures. T freezing point of the solution can also be established to the nearest 0.01 °C with good precision. However, the freezi point depression (−1.08 °C), the difference between two numbers of comparable magnitude, is valid only to about o part per hundred. Although significant-figure rules permit three significant figures in the remainder of the calculatic the actual precision of the calculated quantities is still only about one part per hundred. The final estimate of the bo ing point (80.63 °C) seems reasonably good since it required expressing ΔT_b only to two significant figures. V assume that the 0.211 m solution is dilute enough and close enough to ideal in behavior to make equations (14.5) a (14.6) applicable.

ACTICE EXAMPLE A: Water and phenol are completely miscible at temperatures above 66.8 °C but only partially scible at temperatures below 66.8 °C. In a mixture prepared at 29.6 °C from 50.0 g water and 50.0 g phenol, 32.8 g of a ase consisting of 92.50% water and 7.50% phenol by mass is obtained. This is a saturated solution of phenol in water. hat is the mass percent of water in the second phase—a saturated solution of water in phenol? What is the mole fraction phenol in the mixture at temperatures above 66.8 °C?

ACTICE EXAMPLE B: At a constant temperature of 25.00 °C, a current of dry air was passed through pure water d then through a drying tube, D_1, followed by passage through 1.00 m sucrose and another drying tube, D_2. After the periment, D_1 had gained 11.7458 g in mass and D_2 had gained 11.5057 g. Given that the vapor pressure of water is 76 mmHg at 25.00 °C, **(a)** what was the vapor pressure lowering in the 1.00 m sucrose solution, and **(b)** what was the pected lowering?

Exercises

omogeneous and Heterogeneous Mixtures

. Which of the following do you expect to be most water soluble, and why? $C_{10}H_8(s)$, $NH_2OH(s)$, $C_6H_6(l)$, $CaCO_3(s)$.

. Which of the following is moderately soluble both in water and in benzene, $C_6H_6(l)$, and why? **(a)** 1-butanol, $CH_3(CH_2)_2CH_2OH$; **(b)** naphthalene, $C_{10}H_8$; **(c)** hexane, C_6H_{14}; **(d)** $NaCl(s)$.

. Substances that dissolve in water generally do not dissolve in benzene. Some substances are moderately soluble in both solvents, however. One of the following is such a substance. Which do you think it is and why?

(a) *para*-Dichlorobenzene (a moth repellent)

(b) Salicyl alcohol (a local anesthetic)

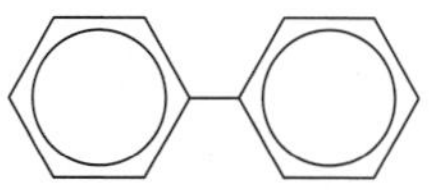

(c) Diphenyl (a heat transfer agent)

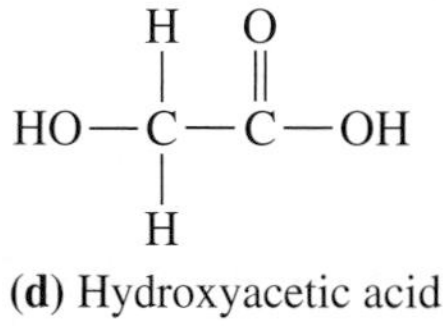

(d) Hydroxyacetic acid (used in textile dyeing)

Some vitamins are water soluble and some are fat soluble. (Fats are substances whose molecules have long hydrocarbon chains.) The structural formulas of two vitamins are shown here—one is water soluble and one is fat soluble. Identify which is which, and explain your reasoning.

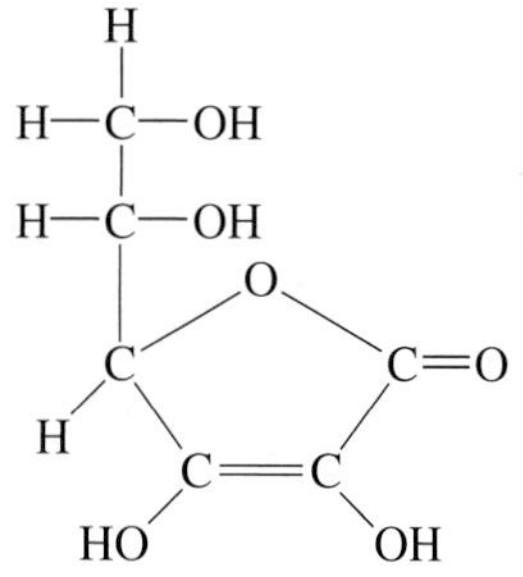

Vitamin C

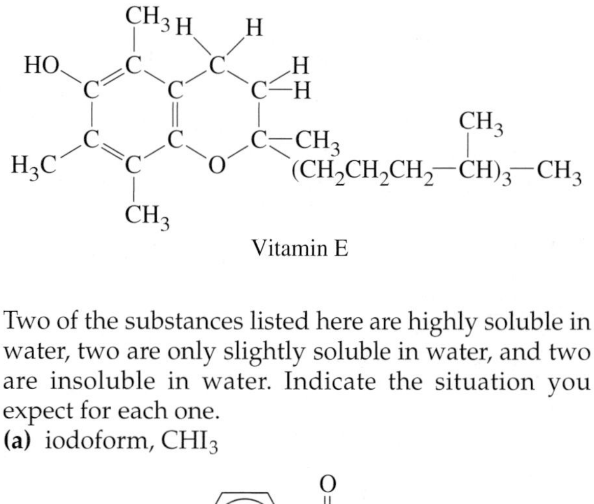

Vitamin E

5. Two of the substances listed here are highly soluble in water, two are only slightly soluble in water, and two are insoluble in water. Indicate the situation you expect for each one.
(a) iodoform, CHI_3
(b) benzoic acid, 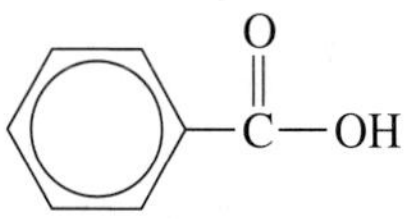
(c) formic acid, 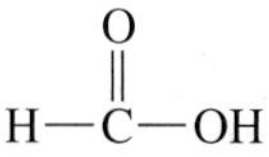
(d) 1-butanol, $CH_3CH_2CH_2CH_2OH$
(e) chlorobenzene, 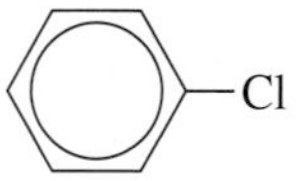
(f) propylene glycol, $CH_3CH(OH)CH_2OH$

6. Benzoic acid, C_6H_5COOH, is much more soluble in NaOH(aq) than it is in pure water. Can you suggest a reason for this? The structural formula for benzoic acid is given in Exercise 5(b).
7. In light of the factors outlined on pages 649–650, which of the following ionic fluorides would you expect to be most water soluble on a moles-per-li basis: MgF_2, NaF, KF, CaF_2? Explain your reasonin
8. Explain the observation that all metal nitrates water soluble but many metal sulfides are not. Amo metal sulfides, which would you expect to be m soluble?

Percent Concentration

9. A saturated aqueous solution of NaBr at 20 °C contains 116 g NaBr/100 g H_2O. Express this composition in the more conventional percent by mass, that is, as grams of NaBr per 100 grams of solution.
10. An aqueous solution with density 0.988 g/mL at 20 °C is prepared by dissolving 12.8 mL $CH_3CH_2CH_2OH$ (d = 0.803 g/mL) in enough water to produce 75.0 mL of solution. What is the percent $CH_3CH_2CH_2OH$ expressed as **(a)** percent by volume; **(b)** percent by mass; **(c)** percent (mass/volume)?
11. A certain brine has 3.87% NaCl by mass. A 75.0 mL sample weighs 76.9 g. How many liters of this solution should be evaporated to dryness to obtain 725 kg NaCl?
12. You are asked to prepare 125.0 mL of 0.0321 M $AgNO_3$. How many grams would you need of a sample known to be 99.81% $AgNO_3$ by mass?
13. According to Example 14-1, the mass percent ethanol in a particular aqueous solution is less than the volume percent in the same solution. Explain why this is also true for *all* aqueous solutions of ethanol. Woulc be true of all ethanol solutions, regardless of the otl component? Explain.
14. Blood cholesterol levels are generally expressed milligrams of cholesterol per deciliter of blood. Wh is the approximate mass percent cholesterol in a blo sample having a cholesterol level of 176? Why c you not give a more precise answer?
15. A certain vinegar is 6.02% acetic acid (CH_3COOH) mass. How many grams of CH_3COOH are contain in a 355 mL bottle of vinegar? Assume a density 1.01 g/mL.
16. 6.00 M sulfuric acid, H_2SO_4(aq), has a density 1.338 g/mL. What is the percent by mass of sulfu acid in this solution?
17. The sulfate ion level in a municipal water supply given as 46.1 ppm. What is [SO_4^{2-}] in this water?
18. A water sample is found to have 9.4 ppb of chlo form, $CHCl_3$. How many grams of $CHCl_3$ would found in a glassful (250 mL) of this water?

Molarity

19. An aqueous solution is 6.00% methanol (CH_3OH) by mass, with d = 0.988 g/mL. What is the molarity of CH_3OH in this solution?
20. A typical commercial grade aqueous phosphoric acid is 75% H_3PO_4 by mass and has a density of 1.57 g/mL. What is the molarity of H_3PO_4 in this solution?
21. How many milliliters of the ethanol–water solution described in Example 14-1 should be diluted with water to produce 825 mL of 0.235 M CH_3CH_2OH?
22. A 30.00%-by-mass solution of nitric acid, HNO_3, water has a density of 1.18 g/cm^3 at 20 °C. What is t molarity of HNO_3 in this solution?
23. What is the molarity of CO_2 in a liter of ocean water 25 °C that contains approximately 280 ppm of CC The density of ocean water is 1027 kg/m^3.
24. At 25 °C and 0% salinity the amount of oxygen in ocean is 5.77 mL/L. What is the molarity of oxygen the ocean at these conditions?

Molality

25. What is the molality of *para*-dichlorobenzene in a solution prepared by dissolving 2.65 g $C_6H_4Cl_2$ in 50.0 mL benzene (d = 0.879 g/mL)?
26. What is the molality of the sulfuric acid solution described in Exercise 16?
27. How many grams of iodine, I_2, must be dissolved in 725 mL of carbon disulfide, CS_2 (d = 1.261 g/mL), to produce a 0.236 m solution?
28. How many grams of water would you add to 1.00 kg of 1.38 mol kg^{-1} CH_3OH(aq) to reduce the molality to 1.00 mol kg^{-1} CH_3OH?
29. An aqueous solution is 34.0% H_3PO_4 by mass and h a density of 1.209 g/mL. What are the molarity a molality of this solution?
30. A 10.00%-by-mass solution of ethanol, CH_3CH_2O in water has a density of 0.9831 g/mL at 15 °C a 0.9804 g/mL at 25 °C. Calculate the molality of t ethanol–water solution at these two temperatur Does the molality differ at the two temperatures (th is, at 15 and 25 °C)? Would you expect the molarit to differ? Explain.

Mole Fraction, Mole Percent

31. A solution is prepared by mixing 1.28 mol C_7H_{16}, 2.92 mol C_8H_{18}, and 2.64 mol C_9H_{20}. What is the **(a)** mole fraction and **(b)** mole percent of each component of the solution?
32. Calculate the mole fraction of solute in the followi aqueous solutions: **(a)** 21.7% CH_3CH_2OH, by ma **(b)** 0.684 mol kg^{-1} $CO(NH_2)_2$ (urea).

Calculate the mole fraction of the solute in the following aqueous solutions: **(a)** 0.112 M $C_6H_{12}O_6$ (d = 1.006 g/mL); **(b)** 3.20% ethanol, by volume (d = 0.993 g/mL; pure CH_3CH_2OH, d = 0.789 g/mL).

Refer to Example 14-1. How many grams of CH_3CH_2OH must be added to 100.0 mL of the solution described in part (d) to increase the mole fraction of CH_3CH_2OH to 0.0525?

What volume of glycerol, $CH_3CH(OH)CH_2OH$ (d = 1.26 g/mL), must be added per kilogram of water to produce a solution with 4.85 mol % glycerol?

Two aqueous solutions of sucrose, $C_{12}H_{22}O_{11}$, are mixed. One solution is 0.1487 M $C_{12}H_{22}O_{11}$ and has d = 1.018 g/mL; the other is 10.00% $C_{12}H_{22}O_{11}$ by mass and has d = 1.038 g/mL. Calculate the mole percent $C_{12}H_{22}O_{11}$ in the mixed solution.

37. The Environmental Protection Agency has a limit of 15 ppm for the amount of lead in drinking water. If a 1.000 mL sample of water at 20 °C contains 15 ppm of lead, how many lead ions are there in this sample of water? What is the mole fraction of lead ion in solution?

38. The amount of CO_2 in the ocean is approximately 280 ppm. What is the mole fraction of CO_2 in a liter of ocean water?

lubility Equilibrium

Refer to Figure 14-10 and determine the molality of NH_4Cl in a saturated aqueous solution at 40 °C.

Refer to Figure 14-10 and estimate the temperature at which a saturated aqueous solution of $KClO_4$ is 0.200 m.

A solution of 20.0 g $KClO_4$ in 500.0 g of water is brought to a temperature of 40 °C.
(a) Refer to Figure 14-10 and determine whether the solution is unsaturated or supersaturated at 40 °C.
(b) Approximately what mass of $KClO_4$, in grams, must be added to saturate the solution (if originally unsaturated), or what mass of $KClO_4$ can be crystallized (if originally supersaturated)?

42. One way to recrystallize a solute from a solution is to change the temperature. Another way is to evaporate solvent from the solution. A 335 g sample of a saturated solution of $KNO_3(s)$ in water is prepared at 25.0 °C. If 55 g H_2O is evaporated from the solution at the same time as the temperature is reduced from 25.0 to 0.0 °C, what mass of $KNO_3(s)$ will recrystallize? (Refer to Figure 14-10.)

lubility of Gases

Under an $O_2(g)$ pressure of 1.00 atm, 28.31 mL of $O_2(g)$ dissolves in 1.00 L H_2O at 25 °C. What will be the molarity of O_2 in the saturated solution at 25 °C when the O_2 pressure is 3.86 atm? (Assume that the solution volume remains at 1.00 L.)

Using data from Exercise 43, determine the molarity of O_2 in an aqueous solution at equilibrium with air at normal atmospheric pressure. The volume percent of O_2 in air is 20.95%. [*Hint:* Recall equation (6.17).]

Natural gas consists of about 90% methane, CH_4. Assume that the solubility of natural gas at 20 °C and 1 atm gas pressure is about the same as that of CH_4, 0.02 g/kg water. If a sample of natural gas under a pressure of 20 atm is kept in contact with 1.00×10^3 kg of water, what mass of natural gas will dissolve?

At 1.00 atm, the solubility of O_2 in water is 2.18×10^{-3} M at 0 °C and 1.26×10^{-3} M at 25 °C. What volume of $O_2(g)$, measured at 25 °C and 1.00 atm, is expelled when 515 mL of water saturated with O_2 is heated from 0 to 25 °C?

The aqueous solubility at 20 °C of Ar at 1.00 atm is equivalent to 33.7 mL Ar(g), measured at 0 °C and 1.00 atm, per liter of water. What is the molarity of Ar in water that is saturated with air at 1.00 atm and 20 °C? Air contains 0.934% Ar by volume. Assume that the volume of water does not change when it becomes saturated with air.

48. The aqueous solubility of CO_2 at 20 °C and 1.00 atm is equivalent to 87.8 mL $CO_2(g)$, measured at 0 °C and 1.00 atm, per 100 mL of water. What is the molarity of CO_2 in water that is at 20 °C and saturated with air at 1.00 atm? The volume percent of CO_2 in air is 0.0360%. Assume that the volume of the water does not change when it becomes saturated with air.

49. Henry's law can be stated this way: The mass of a gas dissolved by a given quantity of solvent at a fixed temperature is directly proportional to the pressure of the gas. Show how this statement is related to equation (14.2).

50. Another statement of Henry's law is: At a fixed temperature, a given quantity of liquid dissolves the same volume of gas at all pressures. What is the connection between this statement and the one given in Exercise 49? Under what conditions is this second statement not valid?

oult's Law and Liquid–Vapor Equilibrium

What are the partial and total vapor pressures of a solution obtained by mixing 35.8 g benzene, C_6H_6, and 56.7 g toluene, $C_6H_5CH_3$, at 25 °C? At 25 °C the vapor pressure of C_6H_6 = 95.1 mmHg; the vapor pressure of $C_6H_5CH_3$ = 28.4 mmHg.

52. Determine the composition of the vapor above the benzene-toluene solution described in Exercise 51.

53. Calculate the vapor pressure at 25 °C of a solution containing 165 g of the *nonvolatile* solute, glucose, $C_6H_{12}O_6$, in 685 g H_2O. The vapor pressure of water at 25 °C is 23.8 mmHg.

54. Calculate the vapor pressure at 20 °C of a saturated solution of the *nonvolatile* solute, urea, $CO(NH_2)_2$, in methanol, CH_3OH. The solubility is 17 g urea/100 mL methanol. The density of methanol is 0.792 g/mL, and its vapor pressure at 20 °C is 95.7 mmHg.

55. Styrene, used in the manufacture of polystyrene plastics, is made by the extraction of hydrogen atoms from ethylbenzene. The product obtained contains about 38% styrene ($C_6H_5CH{=}CH_2$) and 62% ethylbenzene ($C_6H_5CH_2CH_3$), by mass. The mixture is separated by fractional distillation at 90 °C. Determine the composition of the vapor in equilibrium with this 38%–62% mixture at 90 °C. The vapor pressure of ethylbenzene is 182 mmHg and that of styrene is 134 mmHg.

56. Calculate $x_{C_6H_6}$ in a benzene–toluene liquid solution that is in equilibrium at 25 °C with a vapor phase that contains 62.0 mol % C_6H_6. (Use data from Exercise 51.)

57. A benzene–toluene solution with $x_{benz} = 0.300$ has a normal boiling point of 98.6 °C. The vapor pressure of pure toluene at 98.6 °C is 533 mmHg. What must be the vapor pressure of pure benzene at 98.6 °C? (Assu ideal solution behavior.)

58. The two NaCl(aq) solutions pictured are at the sa temperature.

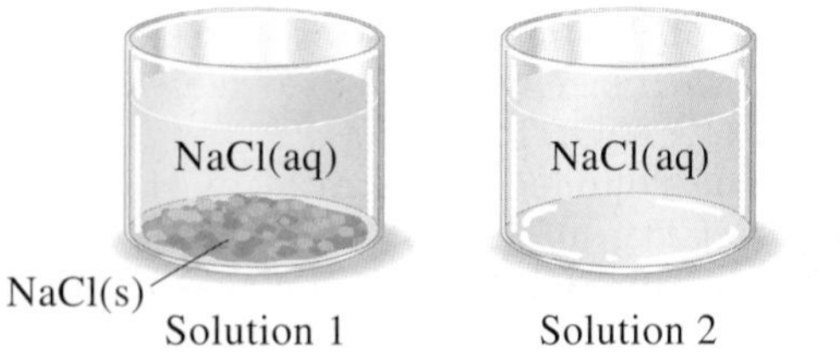

(a) Above which solution is the vapor pressure water, P_{H_2O}, greater? Explain.
(b) Above one of these solutions, the vapor pressu of water, P_{H_2O}, remains *constant*, even as water eva orates from solution. Which solution is this? Expla
(c) Which of these solutions has the higher boili point? Explain.

Osmotic Pressure

59. A 0.72 g sample of polyvinyl chloride (PVC) is dissolved in 250.0 mL of a suitable solvent at 25 °C. The solution has an osmotic pressure of 1.67 mmHg. What is the molar mass of the PVC?

60. Verify that a 20% aqueous solution by mass of sucrose ($C_{12}H_{22}O_{11}$) would rise to a height of about 150 m in an apparatus of the type pictured in Figure 14-21.

61. When the stems of cut flowers are held in concentrated NaCl(aq), the flowers wilt. In a similar solution a fresh cucumber shrivels up (becomes pickled). Explain the basis of these phenomena.

62. Some fish live in saltwater environments and some in freshwater, but in either environment they need water to survive. Saltwater fish drink water, but freshwater fish do not. Explain this difference between the two types of fish.

63. In what volume of water must 1 mol of a nonelectrolyte be dissolved if the solution is to have an osmotic pressure of 1 atm at 273 K? Which of the gas laws does this result resemble?

64. The molecular mass of hemoglobin is 6.86×10^4 u. What mass of hemoglobin must be present per 100.0 mL of a solution to exert an osmotic pressure of 7.25 mmHg at 25 °C?

65. At 25 °C a 0.50 g sample of polyisobutylene (a polymer used in synthetic rubber) in 100.0 mL of benzene solution has an osmotic pressure that supports a 5.1 mm column of solution ($d = 0.88$ g/mL). What is the molar mass of the polyisobutylene? (For H $d = 13.6$ g/mL.)

66. Use the concentration of an isotonic saline soluti 0.92% NaCl (mass/volume), to determine the osmo pressure of blood at body temperature, 37.0 °C. [*Hi* Assume that NaCl is completely dissociated in aq ous solutions.]

67. What approximate pressure is required in the reve osmosis depicted in Figure 14-22 if the saltwater co tains 2.5% NaCl, by mass? [*Hint:* Assume that NaC completely dissociated in aqueous solutions. Al assume a temperature of 25 °C.]

68. The two solutions pictured here are separated b semipermeable membrane that permits only the pa sage of water molecules. In what direction will a n flow of water occur, that is, from left to right or rig to left? Glycerol is $HOCH_2CH(OH)CH_2OH$; sucrose $C_{12}H_{22}O_{11}$.

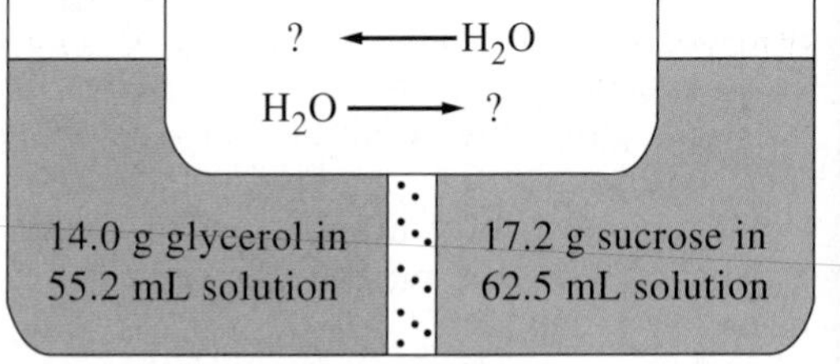

Freezing-Point Depression and Boiling-Point Elevation

69. 1.10 g of an unknown compound reduces the freezing point of 75.22 g benzene from 5.53 to 4.92 °C. What is the molar mass of the compound?

70. The freezing point of a 0.010 m aqueous solution of a nonvolatile solute is −0.072 °C. What would you expect the normal boiling point of this same solution to be?

71. Adding 1.00 g of benzene, C_6H_6, to 80.00 g cyclohexa C_6H_{12}, lowers the freezing point of the cyclohexa from 6.5 to 3.3 °C.
(a) What is the value of K_f for cyclohexane?
(b) Which is the better solvent for molar mass det minations by freezing-point depression, benzene cyclohexane? Explain.

The boiling point of water at 749.2 mmHg is 99.60 °C. What mass percent sucrose ($C_{12}H_{22}O_{11}$) should be present in an aqueous sucrose solution to raise the boiling point to 100.00 °C at this pressure?

A compound is 42.9% C, 2.4% H, 16.7% N, and 38.1% O, by mass. Addition of 6.45 g of this compound to 50.0 mL benzene, C_6H_6 ($d = 0.879$ g/mL), lowers the freezing point from 5.53 to 1.37 °C. What is the molecular formula of this compound?

Nicotinamide is a water-soluble vitamin important in metabolism. A deficiency in this vitamin results in the debilitating condition known as pellagra. Nicotinamide is 59.0% C, 5.0% H, 22.9% N, 13.1% O, by mass. Addition of 3.88 g of nicotinamide to 30.0 mL nitrobenzene, $C_6H_5NO_2$ ($d = 1.204$ g/mL), lowers the freezing point from 5.7 to −1.4 °C. What is the molecular formula of this compound?

Thiophene (fp = −38.3; bp = 84.4 °C) is a sulfur-containing hydrocarbon sometimes used as a solvent in place of benzene. Combustion of a 2.348 g sample of thiophene produces 4.913 g CO_2, 1.005 g H_2O, and 1.788 g SO_2. When a 0.867 g sample of thiophene is dissolved in 44.56 g of benzene (C_6H_6), the freezing point is lowered by 1.183 °C. What is the molecular formula of thiophene?

Coniferin is a glycoside (a derivative of a sugar) found in conifers, such as fir trees. When a 1.205 g sample of coniferin is subjected to combustion analysis (recall Section 3-3), the products are 0.698 g H_2O and 2.479 g CO_2. A 2.216 g sample is dissolved in 48.68 g H_2O, and the normal boiling point of this solution is found to be 100.068 °C. What is the molecular formula of coniferin?

77. Cooks often add some salt to water before boiling it. Some people say this helps the cooking process by raising the boiling point of the water. Others say not enough salt is usually added to make any noticeable difference. Approximately how many grams of NaCl must be added to a liter of water at 1 atm pressure to raise the boiling point by 2 °C? Is this a typical amount of salt that you might add to cooking water?

78. An important test for the purity of an organic compound is to measure its melting point. Usually, if the compound is not pure, it begins to melt at a *lower* temperature than the pure compound.
(a) Why is this the case, rather than the melting point being higher in some cases and lower in others?
(b) Are there any conditions under which the melting point of the impure compound is *higher* than that of the pure compound? Explain.

79. The freezing point of Arctic Ocean waters is about −1.94 °C. What is the molality of ions for a liter of ocean water?

80. If ocean water consisted of 3.5% salt, what would be the freezing point of an ocean?

rong Electrolytes, Weak Electrolytes, and Nonelectrolytes

Predict the approximate freezing points of 0.10 m solutions of the following solutes dissolved in water: **(a)** $CO(NH_2)_2$ (urea); **(b)** NH_4NO_3; **(c)** HCl; **(d)** $CaCl_2$; **(e)** $MgSO_4$; **(f)** CH_3CH_2OH (ethanol); **(g)** CH_3COOH (acetic acid).

Calculate the van't Hoff factors of the following weak electrolyte solutions:
(a) 0.050 m HCOOH, which begins to freeze at −0.0986 °C;
(b) 0.100 M HNO_2, which has a hydrogen ion (and nitrite ion) concentration of 6.91×10^{-3} M.

$NH_3(aq)$ conducts electric current only weakly. The same is true for acetic acid, CH_3COOH. When these solutions are mixed, however, the resulting solution conducts electric current very well. Propose an explanation.

An isotonic solution is described as 0.92% NaCl (mass/volume). Would this also be the required concentration for isotonic solutions of other salts, such as KCl, $MgCl_2$, or $MgSO_4$? Explain.

In the following diagrams, which representation demonstrates a weak electrolyte?

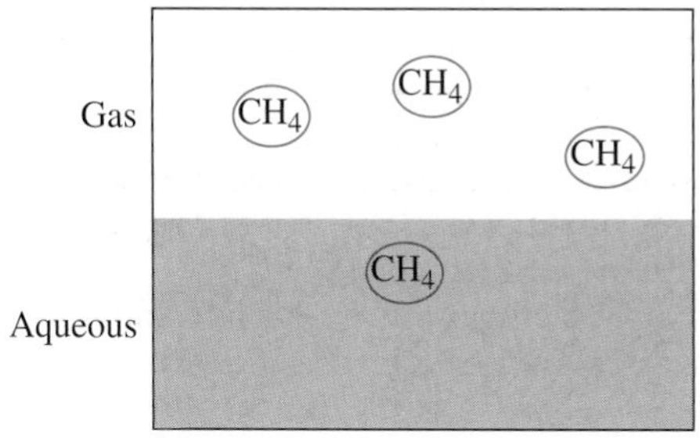

(a)

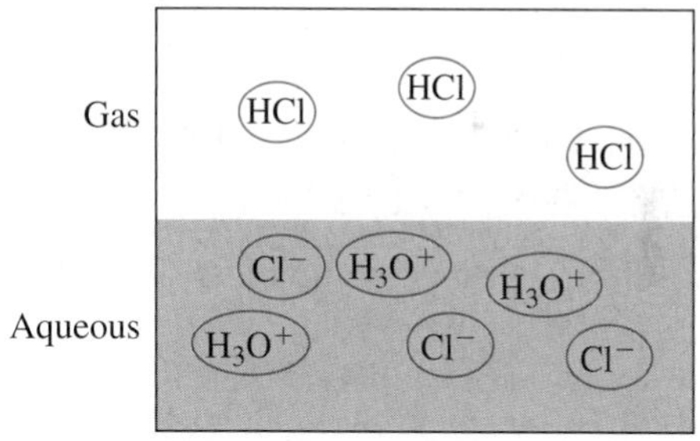

(b)

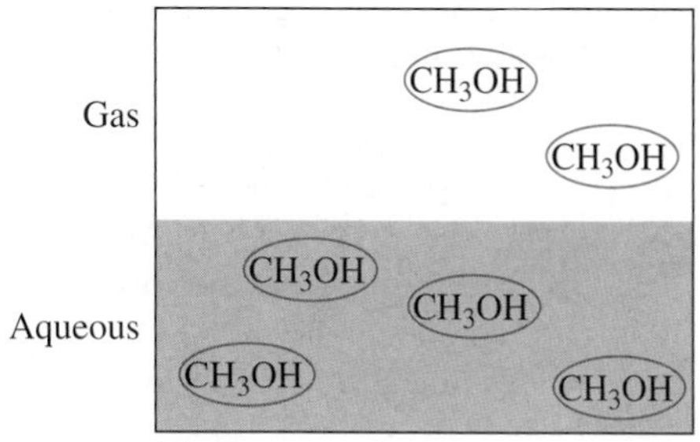

(c)

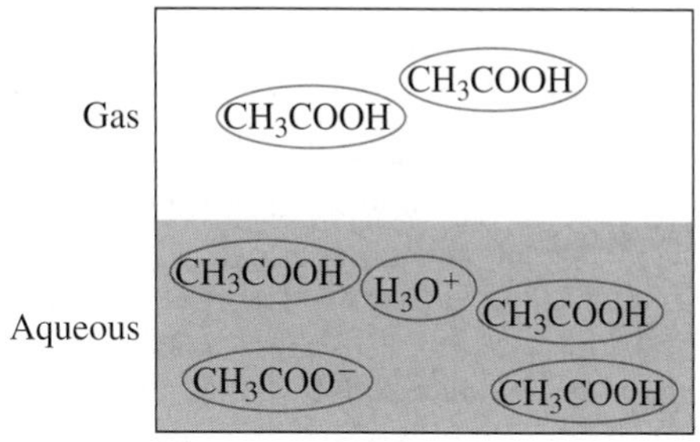

(d)

86. In the following diagrams, which representation demonstrates a strong electrolyte?

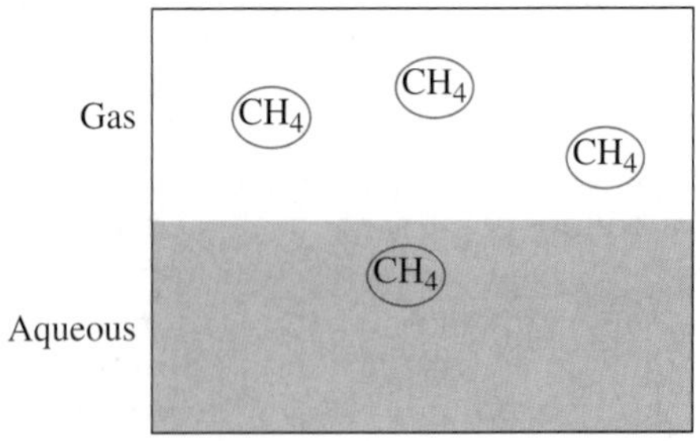

(a)

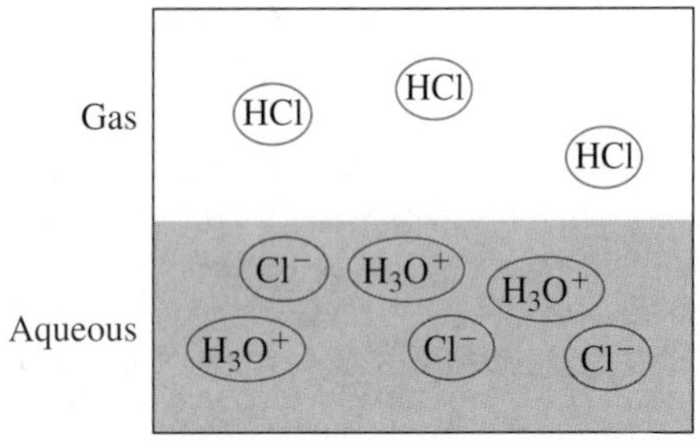

(b)

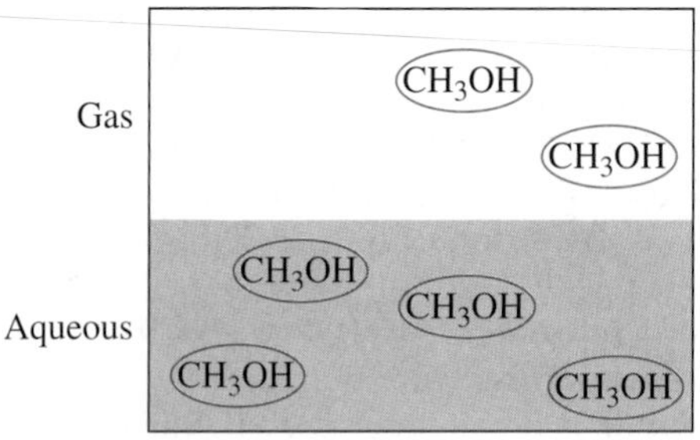

(c)

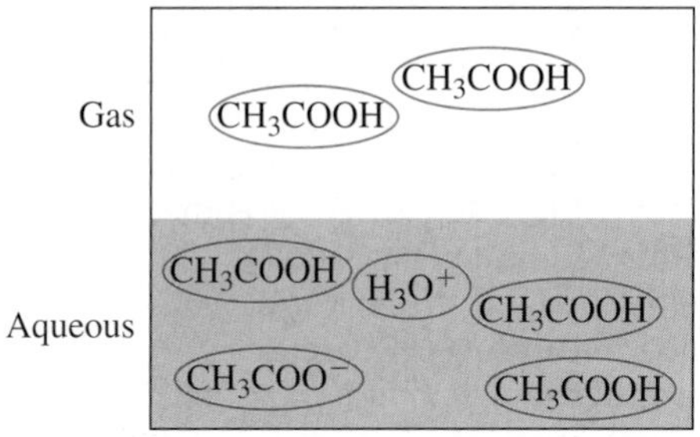

(d)

Integrative and Advanced Exercises

87. A typical root beer contains 0.13% of a 75% H_3PO_4 solution by mass. How many milligrams of phosphorus are contained in a 12 oz can of this root beer? Assume a solution density of 1.00 g/mL; also, 1 oz = 29.6 mL.

88. An aqueous solution has 109.2 g KOH/L solution. The solution density is 1.09 g/mL. Your task is to use 100.0 mL of this solution to prepare 0.250 m KOH. What mass of which component, KOH or H_2O, would you add to the 100.0 mL of solution?

89. The term "proof," still used to describe the ethanol content of alcoholic beverages, originated in seventeenth-century England. A sample of whiskey was poured on gunpowder and set afire. If the gunpowder ignited after the whiskey had burned off, this "proved" that the whiskey had not been watered down. The minimum ethanol content for a positive test was about 50%, by volume. The 50% ethanol solution became known as "100 proof." Thus, an 80-proof whiskey would be 40% CH_3CH_2OH by volume. Listed in the table below are some data for several aqueous solutions of ethanol. *With a minimum amount of calculation*, determine which of the solutions are more than 100 proof. Assume that the density of pure ethanol is 0.79 g/mL.

Molarity of Ethanol, M	Density of Solution, g/mL
4.00	0.970
5.00	0.963
6.00	0.955
7.00	0.947
8.00	0.936
9.00	0.926
10.00	0.913

90. Four aqueous solutions of acetone, CH_3COCH_3, prepared at different concentrations: **(a)** 0.10 CH_3COCH_3, by mass; **(b)** 0.100 M CH_3COC **(c)** 0.100 m CH_3COCH_3; and **(d)** $x_{acetone}$ = 0.1 Estimate the highest partial pressure of water at 25 to be found in the equilibrium vapor above th solutions. Also, estimate the lowest freezing poin be found among these solutions.

91. A solid mixture consists of 85.0% KNO_3 and 15. K_2SO_4, by mass. A 60.0 g sample of this solid is add to 130.0 g of water at 60 °C. Refer to Figure 14-10.
(a) Will all the solid dissolve at 60 °C?
(b) If the resulting solution is cooled to 0 °C, w mass of KNO_3 should crystallize?
(c) Will K_2SO_4 also crystallize at 0 °C?

92. Suppose you have available 2.50 L of a soluti (d = 0.9767 g/mL) that is 13.8% ethanol (CH_3CH_2O by mass. From this solution you would like to make *maximum* quantity of ethanol-water antifreeze solut that will offer protection to −2.0 °C. Would you a more ethanol or more water to the solution? What m of liquid would you add?

93. Hydrogen chloride is a colorless gas, yet when a b tle of concentrated hydrochloric acid [HCl(conc aq) opened, mist-like fumes are often seen to escape fr the bottle. How do you account for this?

94. Use the following information to confirm that triple point temperature of water is about 0.0098 °C
(a) The slope of the fusion curve of water in region of the normal melting point of ice (Fig. 12-30 −0.00750 °C/atm.
(b) The solubility of air in water at 0 °C and 1.00 at is 0.02918 mL of air per mL of water.

5. Stearic acid ($C_{18}H_{36}O_2$) and palmitic acid ($C_{16}H_{32}O_2$) are common fatty acids. Commercial grades of stearic acid usually contain palmitic acid as well. A 1.115 g sample of a commercial-grade stearic acid is dissolved in 50.00 mL benzene (d = 0.879 g/mL). The freezing point of the solution is found to be 5.072 °C. The freezing point of pure benzene is 5.533 °C, and K_f for benzene is 5.12 °C kg mol^{-1}. What is the mass percent of palmitic acid in the stearic acid sample?

6. Nitrobenzene, $C_6H_5NO_2$, and benzene, C_6H_6, are completely miscible in each other. Other properties of the two liquids are *nitrobenzene:* fp = 5.7 °C, K_f = 8.1 °C kg mol^{-1}; *benzene:* fp = 5.5 °C, K_f = 5.12 °C kg mol^{-1}. It is possible to prepare *two different* solutions with these two liquids having a freezing point of 0.0 °C. What are the compositions of these two solutions, expressed as mass percent nitrobenzene?

7. Refer to Figure 14-20(a). Initially, solution A contains 0.515 g urea, $CO(NH_2)_2$, dissolved in 92.5 g H_2O; solution B contains 2.50 g sucrose, $C_{12}H_{22}O_{11}$, dissolved in 85.0 g H_2O. What are the compositions of the two solutions when equilibrium is reached, that is, when the two have the same vapor pressure?

8. In Figure 14-21, why does the net transfer of water stop when the two solutions are of *nearly* equal concentrations rather than of *exactly* equal concentrations?

9. Shown below is a typical cooling curve for an aqueous solution. Why is there no horizontal straight-line portion comparable to that seen in the cooling curve for pure water in Figure 12-23?

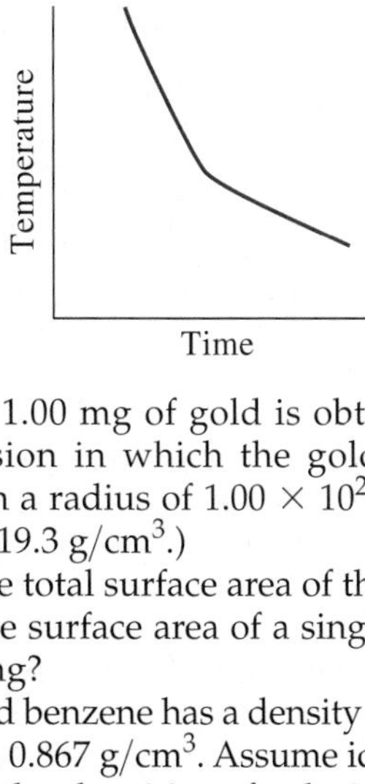

0. Suppose that 1.00 mg of gold is obtained in a colloidal dispersion in which the gold particles are spherical, with a radius of 1.00 × 10^2 nm. (The density of gold is 19.3 g/cm^3.)
(a) What is the total surface area of the particles?
(b) What is the surface area of a single cube of gold of mass 1.00 mg?

1. At 20 °C, liquid benzene has a density of 0.879 g/cm^3; liquid toluene, 0.867 g/cm^3. Assume ideal solutions.
(a) Calculate the densities of solutions containing 20, 40, 60, and 80 volume percent benzene.
(b) Plot a graph of density versus volume percent composition.
(c) Write an equation that relates the density (d) to the volume percent benzene (V) in benzene-toluene solutions at 20 °C.

2. The two compounds whose structures are depicted here are isomers. When derived from petroleum, they always occur mixed together. *meta*-Xylene is used in aviation fuels and in the manufacture of dyes and insecticides. The principal use of *para*-xylene is in the manufacture of polyester resins and fibers (for example, Dacron). Comment on the effectiveness of fractional distillation as a method of separating these two xylenes. What other method(s) might be used to separate them?

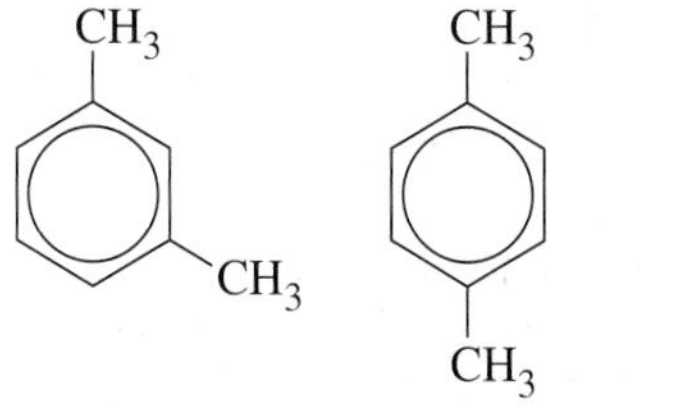

meta-xylene	*para*-xylene
fp, 47.9 °C	fp, 13.3 °C
bp, 139.1 °C	bp, 138.4 °C
d, 0.864 g/mL	*d*, 0.861 g/mL

103. Instructions on a container of antifreeze (ethylene glycol; fp, −12.6 °C, bp, 197.3 °C) give the following volumes of Prestone to be used in protecting a 12 qt cooling system against freeze-up at different temperatures (the remaining liquid is water): 10 °F, 3 qt; 0 °F, 4 qt; −15 °F, 5 qt; −34 °F, 6 qt. Since the freezing point of the coolant is successively lowered by using more antifreeze, why not use even more than 6 qt of antifreeze (and proportionately less water) to ensure the maximum protection against freezing?

104. Demonstrate that
(a) for a *dilute aqueous* solution, the numerical value of the molality is essentially equal to that of the molarity.
(b) in a *dilute* solution, the solute mole fraction is proportional to the molality.
(c) in a *dilute aqueous* solution, the solute mole fraction is proportional to the molarity.

105. At 25 °C and under an O_2(g) pressure of 1 atm, the solubility of O_2(g) in water is 28.31 mL/1.00 L H_2O. At 25 °C and under an N_2(g) pressure of 1 atm, the solubility of N_2(g) in water is 14.34 mL/1.00 L H_2O. The composition of the atmosphere is 78.08% N_2 and 20.95% O_2, by volume. What is the composition of air dissolved in water expressed as volume percents of N_2 and O_2?

106. We noted in Figure 14-17 that the liquid and vapor curves taken together outline a lens-shaped region when the normal boiling points of benzene-toluene solutions are plotted as a function of mole fraction of benzene. That is, unlike Figure 14-16, the liquid curve is not a straight line. Use data from Figure 14-17 and show by calculation that this should be the case.

107. A saturated solution prepared at 70 °C contains 32.0 g $CuSO_4$ per 100.0 g solution. A 335 g sample of this solution is then cooled to 0 °C and $CuSO_4 \cdot 5\,H_2O$ crystallizes out. If the concentration of a saturated solution at 0 °C is 12.5 g $CuSO_4$/100 g soln, what mass of $CuSO_4 \cdot 5\,H_2O$ would be obtained? [*Hint:* Note that the solution composition is stated in terms of $CuSO_4$ and that the solid that crystallizes is the hydrate $CuSO_4 \cdot 5\,H_2O$.]

108. The concentration of Ar in the ocean at 25 °C is 11.5 μM. The Henry's law constant for Ar is 1.5 × 10^{-3} mol L^{-1} atm^{-1}. Calculate the mass of Ar in a liter of ocean water. Calculate the partial pressure of Ar in the atmosphere.

109. The concentration of N_2 in the ocean at 25 °C is 445 μM. The Henry's law constant for N_2 is 0.61×10^{-3} mol L^{-1} atm^{-1}. Calculate the mass of N_2 in a liter of ocean water. Calculate the partial pressure of N_2 in the atmosphere.

110. A solution contains 750 g of ethanol and 85.0 g of sucrose (180 g mol^{-1}). The volume of the solution is 810.0 mL. Determine
(a) the density of the solution
(b) the percent of sucrose in the solution
(c) the mole fraction of sucrose
(d) the molality of the solution
(e) the molarity of the solution

111. What volume of ethylene glycol ($HOCH_2CH_2OH$, density = 1.12 g mL^{-1}) must be added to 20.0 L of water (K_f = 1.86 °C kg mol^{-1}) to produce a solution that freezes at −10 °C?

112. In the figure below, the open squares represent s vent molecules and the filled squares repres solute molecules.

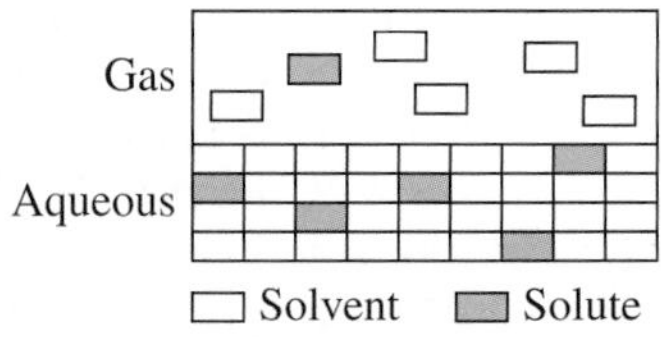

(a) What is the mole fraction of solute in the liq phase?
(b) Which component has the higher vapor pressu
(c) What is the percent solute in the vapor?

Feature Problems

113. Cinnamaldehyde is the chief constituent of cinnamon oil, which is obtained from the twigs and leaves of cinnamon trees grown in tropical regions. Cinnamon oil is used in the manufacture of food flavorings, perfumes, and cosmetics. The normal boiling point of cinnamaldehyde, $C_6H_5CH{=}CHCHO$, is 246.0 °C, but at this temperature it begins to decompose. As a result, cinnamaldehyde cannot be easily purified by ordinary distillation. A method that can be used instead is *steam distillation*. A heterogeneous mixture of cinnamaldehyde and water is heated until the sum of the vapor pressures of the two liquids is equal to barometric pressure. At this point, the temperature remains constant as the liquids vaporize. The mixed vapor condenses to produce two immiscible liquids; one liquid is essentially pure water and the other, pure cinnamaldehyde. The following vapor pressures of cinnamaldehyde are given: 1 mmHg at 76.1 °C; 5 mmHg at 105.8 °C; and 10 mmHg at 120.0 °C. Vapor pressures of water are given in Table 12.5.
(a) What is the approximate temperature at which the steam distillation occurs?
(b) The proportions of the two liquids condensed from the vapor is *independent* of the composition of the boiling mixture, as long as both liquids are present in the boiling mixture. Explain why this is so.
(c) Which of the two liquids, water or cinnamaldehyde, condenses in the greater quantity, by mass? Explain.

114. The phase diagram shown is for mixtures of HCl and H_2O at a pressure of 1 atm. The red curve represents the normal boiling points of solutions of HCl(aq) of various mole fractions. The blue curve represents the compositions of the vapors in equilibrium with boiling solutions.
(a) As a solution containing x_{HCl} = 0.50 begins to boil, will the vapor have a mole fraction of HCl equal to, less than, or greater than 0.50? Explain.
(b) In the boiling of a pure liquid, there is no change in composition. However, as a solution of HCl(aq)

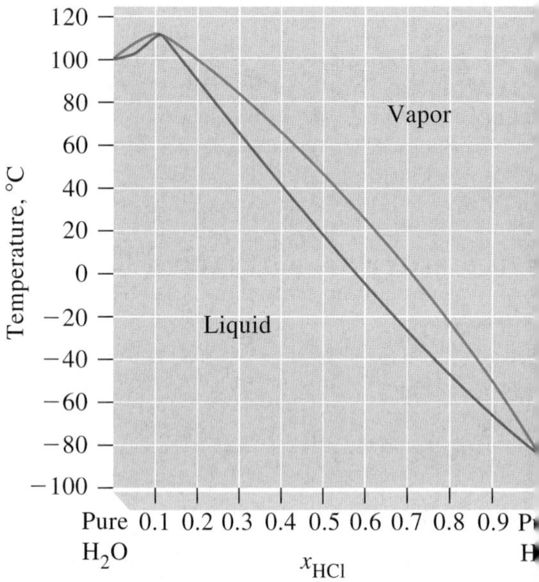

boils in an open container, the composition chang Explain why this is so.
(c) One particular solution (the azeotrope) is exception to the observation stated in part (b); t is, its composition remains *unchanged* during boili What are the approximate composition and boil point of this solution?
(d) A 5.00 mL sample of the azeotrope (1.099 g/mL) requires 30.32 mL of 1.006 M NaOH its titration in an acid-base reaction. Use these d to determine a more precise value of the compositi of the azeotrope, expressed as the mole fracti of HCl.

115. The laboratory device pictured on the following pa is called a *desiccator*. It can be used to maintain a co stant relative humidity within an enclosure. T material(s) used to control the relative humidity placed in the bottom compartment, and the su stance being subjected to a controlled relative hum ity is placed on the platform in the container.

(a) If the material in the bottom compartment is a saturated solution of NaCl(aq) in contact with NaCl(s), what will be the approximate relative humidity in the container at a temperature of 20 °C? Obtain the solubility of NaCl from Figure 14-10 and vapor pressure data for water from Table 12.5; use the definition of Raoult's law from page 660 and that of relative humidity from Focus On 6-1: Earth's Atmosphere; assume that the NaCl is completely dissociated into its ions.
(b) If the material placed on the platform is dry $CaCl_2 \cdot 6\,H_2O(s)$, will the solid deliquesce? Explain.
[*Hint:* Recall the discussion on pages 664–665.]
(c) To maintain $CaCl_2 \cdot 6\,H_2O(s)$ in the dry state in a desiccator, should the substance in the saturated solution in the bottom compartment be one with a high or a low water solubility? Explain.

116. Every year, oral rehydration therapy (ORT)—the feeding of an electrolyte solution—saves the lives of countless children worldwide who become severely dehydrated as a result of diarrhea. One requirement of the solution used is that it be *isotonic* with human blood.
(a) One definition of an isotonic solution given in the text is that it have the same osmotic pressure as 0.92% NaCl(aq) (mass/volume). Another definition is that the solution have a freezing point of −0.52 °C. Show that these two definitions are in reasonably close agreement given that we are using solution concentrations rather than activities.
(b) Use the freezing-point definition from part (a) to show that an ORT solution containing 3.5 g NaCl, 1.5 g KCl, 2.9 g $Na_3C_6H_5O_7$ (sodium citrate), and 20.0 g $C_6H_{12}O_6$ (glucose) per liter meets the requirement of being isotonic. [*Hint:* Which of the solutes are nonelectrolytes, and which are strong electrolytes?]

Self-Assessment Exercises

7. In your own words, define or explain the following terms or symbols: **(a)** x_B; **(b)** P^*_A; **(c)** K_f; **(d)** i; **(e)** activity.

8. Briefly describe each of the following ideas or phenomena: **(a)** Henry's law; **(b)** freezing-point depression; **(c)** recrystallization; **(d)** hydrated ion; **(e)** deliquescence.

9. Explain the important distinctions between each pair of terms: **(a)** molality and molarity; **(b)** ideal and nonideal solution; **(c)** unsaturated and supersaturated solution; **(d)** fractional crystallization and fractional distillation; **(e)** osmosis and reverse osmosis.

0. An aqueous solution is 0.010 M CH_3OH. The concentration of this solution is also very nearly **(a)** 0.010% CH_3OH (mass/volume); **(b)** 0.010 mol kg^{-1} CH_3OH; **(c)** $x_{CH_3OH} = 0.010$; **(d)** 0.990 M H_2O.

1. The most likely of the following mixtures to be an ideal solution is **(a)** $NaCl{-}H_2O$; **(b)** $CH_3CH_2OH{-}C_6H_6$; **(c)** $C_7H_{16}{-}H_2O$; **(d)** $C_7H_{16}{-}C_8H_{18}$.

2. The solubility of a nonreactive gas in water increases with **(a)** an increase in gas pressure; **(b)** an increase in temperature; **(c)** increases in both temperature and pressure; **(d)** an increase in the volume of gas in equilibrium with the available water.

3. Of the following aqueous solutions, the one with the lowest freezing point is **(a)** 0.010 mol kg^{-1} $MgSO_4$; **(b)** 0.011 mol kg^{-1} NaCl; **(c)** 0.018 mol kg^{-1} CH_3CH_2OH; **(d)** 0.0080 mol kg^{-1} $MgCl_2$.

4. An ideal liquid solution has two volatile components. In the vapor in equilibrium with the solution, the mole fractions of the components are **(a)** both 0.50; **(b)** equal, but not necessarily 0.50; **(c)** not very likely to be equal; **(d)** 1.00 for the solvent and 0.00 for the solute.

125. A solution prepared by dissolving 1.12 mol NH_4Cl in 150.0 g H_2O is brought to a temperature of 30 °C. Use Figure 14-10 to determine whether the solution is unsaturated or whether excess solute will crystallize.

126. NaCl(aq) isotonic with blood is 0.92% NaCl (mass/volume). For this solution, what is **(a)** $[Na^+]$; **(b)** the total molarity of ions; **(c)** the osmotic pressure at 37 °C; **(d)** the approximate freezing point? (Assume that the solution has a density of 1.005 g/mL.)

127. A solution (d = 1.159 g/mL) is 62.0% glycerol, $HOCH_2CH(OH)CH_2OH$, and 38.0% H_2O, by mass. Determine **(a)** the molarity of glycerol with H_2O as the solvent; **(b)** the molarity of H_2O with glycerol as the solvent; **(c)** the molality of H_2O in glycerol; **(d)** the mole fraction of glycerol; **(e)** the mole percent of H_2O.

128. Which aqueous solution from the column on the right has the property listed on the left? Explain your choices.

Property	Solution
1. lowest electrical conductivity	**a.** 0.10 mol kg^{-1} KCl(aq)
2. lowest boiling point	**b.** 0.15 mol kg^{-1} $C_{12}H_{22}O_{11}$(aq)
3. highest vapor pressure of water at 25 °C	**c.** 0.10 mol kg^{-1} CH_3COOH(aq)
4. lowest freezing point	**d.** 0.05 mol kg^{-1} NaCl

129. Which of the following represents $MgCl_2$ in solution?

(a)

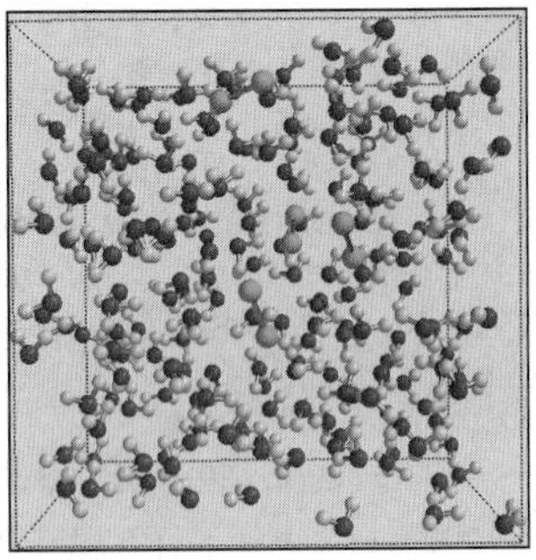

(b)

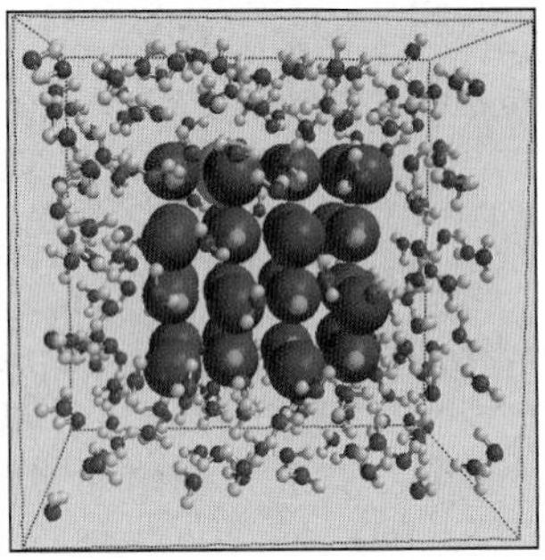

(c)

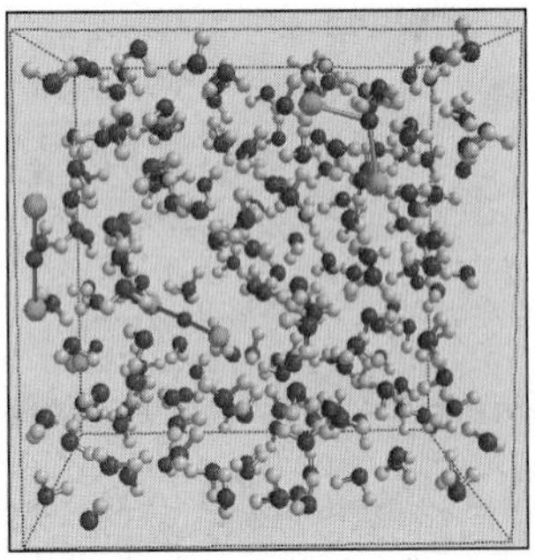

(d)

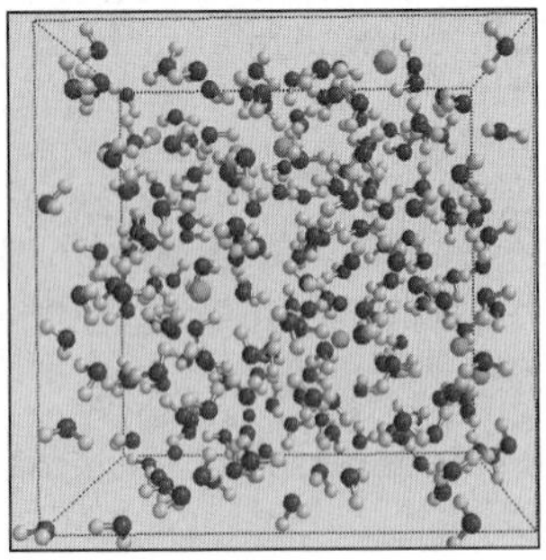

Legend: O ● H ● Cl ● Mg ●

130. Which of the following ions has the greater charge density? **(a)** Na^+; **(b)** F^-; **(c)** K^+; **(d)** Cl^-.

131. When NH_4Cl dissolves in a test tube of water, the test tube becomes colder. Is the magnitude of $\Delta H_{lattice}$ for NH_4Cl larger or smaller than the sum of $\Delta H_{hydration}$ of the ions?

132. In a saturated solution at 25 °C and 1 bar, for the following solutes, which condition will increase solubility? **(a)** Ar(g), decrease temperature; **(b)** NaCl(s), increase pressure; **(c)** N_2, decrease pressure; **(d)** CO_2, increase volume.

133. Which of the following represents a nonvolat solute?

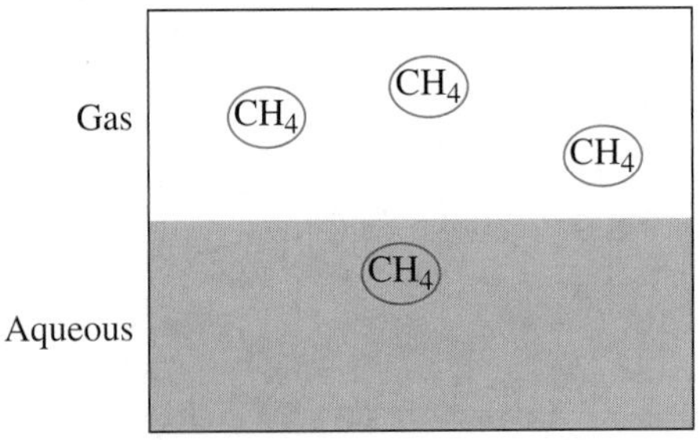

(a)

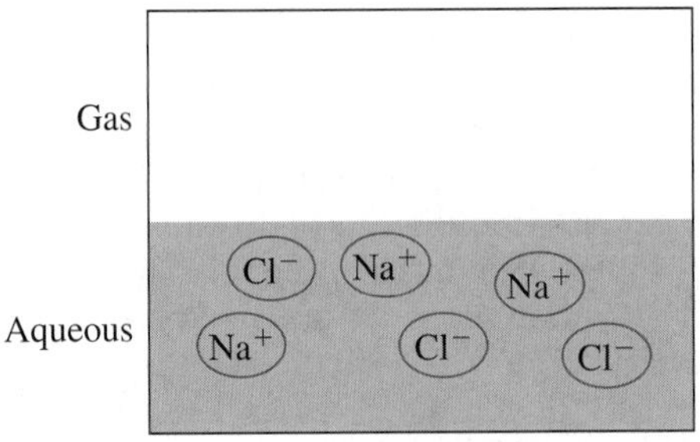

(b)

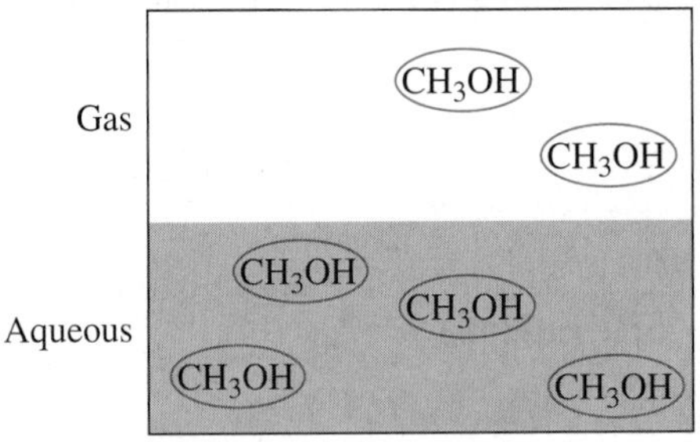

(c)

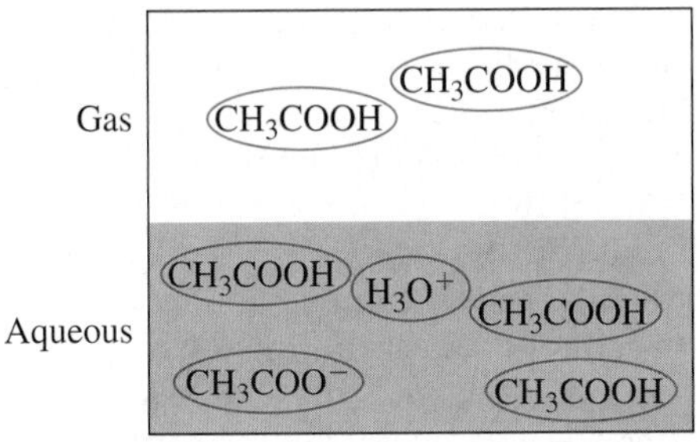

(d)

134. What is the mole fraction of a nonvolatile solute a hexane solution that has a vapor pressure 600 mmHg at 68.7 °C, hexane's normal boil point? **(a)** 0.21; **(b)** 0.11; **(c)** 0.27; **(d)** 0.79.

135. What is the osmotic pressure, in bar, of 15.2 L a 0.312 M starch solution at 75 °C? **(a)** 1372 **(b)** 194.5; **(c)** 355.7; **(d)** 0.0016; **(e)** 9.03.

136. What is the weight percent of 23.4 g of CaF_2 if solved in 10.5 mol of water? **(a)** 0.028; **(b)** 1.59; **(c)** 1 **(d)** 12.4; **(e)** none of these.

137. Using the method of concept mapping presentec Appendix E, construct a concept map showing relationships among the various concentration u presented in Section 14-2.

138. Construct a concept map for the relationships t exist among the colligative properties of Sections 1 14-7, 14-8, and 14-9.

ʔrinciples of Chemical :quilibrium

15

ƆNTENTS

LEARNING OBJECTIVES

15.1 Describe the meaning of the term *dynamic equilibrium.*

15.2 Discuss the relationships involving the reaction quotient (Q), the activities (a) of the reacting species, and the equilibrium constant (K).

15.3 Identify the changes to the equilibrium constant, K, when a chemical reaction is reversed or when the coefficients are multiplied or divided by a common factor.

15.4 Explain the significance of the magnitude of the equilibrium constant, K.

15.5 Compare the values of Q and K to determine the direction of net change.

15.6 Describe how the equilibrium position shifts in response to changes in volume, external pressure, or temperature.

15.7 Use an ICE table to organize information about a reaction system and determine either the equilibrium constant for the reaction or the equilibrium composition of the system.

ʼital natural reaction is in progress in the lightning bolt seen here: $(g) + O_2(g) \rightleftharpoons 2\,NO(g)$. Usually this reversible reaction does not occur to ʼ significant extent in the forward direction, but in the high-temperature lightning ʼt it does. At equilibrium at high temperatures, measurable conversion of $N_2(g)$ ɪ $O_2(g)$ to $NO(g)$ occurs. In this chapter we study the equilibrium condition in a ʼersible reaction and the factors affecting it.

Until now, we have emphasized reactions that go to completion and have used concepts of stoichiometry to calculate the outcomes of such reactions. Stoichiometric concepts, though useful, are some-ıat limited in their applicability because, as we established in ıapter 13, no reaction goes to completion. In this chapter, we will explore w to predict the outcome of almost any reaction, regardless of whether it curs to a very limited extent, goes almost to completion, or "stops" mewhere between, giving a mixture with appreciable amounts of both ıctants and products.

In Chapter 13, we examined chemical reactions—and the equilibrium ndition—from a thermodynamic perspective. We learned that the mposition of a system at equilibrium is governed by the equilibrium constant, K, for the reaction involved. We established that the equilibriu constant for a reaction provides a measure of the thermodynamic stability products relative to reactants: The larger the equilibrium constant, t greater the relative stability of products and the greater the tendency of t reaction to produce an equilibrium mixture containing relatively hi amounts of products.

In this chapter, we aim to deepen our understanding of chemical equili rium. We will begin with a brief discussion of the nature of the equilibriu state and then focus on some key relationships involving equilibrium co stants. Then we will make qualitative predictions about the condition of eq librium; finally, we will perform various equilibrium calculations. As we w discover throughout the remainder of the text, the equilibrium condition pla a role in numerous natural phenomena and affects the methods used to pr duce many important industrial chemicals.

15-1 The Nature of the Equilibrium State

Chemical reactions—like all processes—are governed by the laws of therm dynamics. These laws state that, for all processes, the combined energy o system and its surroundings must be conserved ($\Delta H_{sys} + \Delta H_{surr} = 0$, required by the first law of thermodynamics), and their combined entro cannot decrease ($\Delta S_{sys} + \Delta S_{surr} \geq 0$, as required by the second law of th modynamics). The consequences are profound and clear cut. There are or two options for a chemical reaction: It is either at equilibrium or spon neously approaching an equilibrium state. But what is the nature of th equilibrium state?

A system at **equilibrium** is in a state of balance. Macroscopically, the co position of the system is unchanging whereas, at the microscopic level, chan continues. The overall composition of the system does not change because, equilibrium, a balance has been achieved between two opposing process The conversion of small amounts of reactants into products is always perfec offset by the conversion of small amounts of products into reactants. This b ance maintains the lowest possible value for the Gibbs energy (G) of the sy tem and the maximum possible value for the combined entropy (S) of t system and its surroundings. Unless disturbed by an external influen the system remains indefinitely in this equilibrium state.

The following statements summarize two key ideas—from Chapter 13 concerning chemical reactions. The first emphasizes the spontaneity of t approach to an equilibrium state, and the second establishes the equilibriu condition, which can be represented symbolically as $Q = K$.

> In a closed reaction vessel at constant temperature, a reaction proceeds spontaneously toward equilibrium. At equilibrium, the reaction quotient Q attains the same constant value, K, irrespective of the starting amounts of reactants and products.

Let's explore the significance of this statement by focusing on a speci reaction.

▶ By writing an equation for a reaction with double arrows, each with a half arrowhead $\rightleftharpoons$, we are emphasizing that the reaction has reached equilibrium.

$$2\,Cu^{2+}(aq) + Sn^{2+}(aq) \rightleftharpoons 2\,Cu^{+}(aq) + Sn^{4+}(aq) \quad (15$$

Our first task is to write an expression for the reaction quotient Q for th reaction, and then to demonstrate that, for a fixed temperature, the value of attains the same value at equilibrium, irrespective of the initial amounts us

As discussed in Chapter 13, the reaction quotient is best defined in terms the *activities* of reactants and products. The activities are themselves express

TABLE 15.1 Activities of Some Substances

Substance	Activity[a]	Comment
Ideal gas, X(g)	$a = P_X/P°$	P_X is the partial pressure of the gas and $P° = 1$ bar $\approx$ 1 atm. When pressure is expressed in bar, the activity is equal to the numerical value of the pressure.
Pure solid or liquid, X(s) or X(l)	$a = 1$	For pressures normally encountered,[b] the activity of a pure solid (or a pure liquid) is equal to 1.
Solute in an ideal aqueous solution, X(aq)	$a = [X]/c°$	[X] is the concentration in mol/L and $c° = 1$ mol/L. When concentration is expressed in mol/L, the activity of a dissolved solute is equal to the numerical value of [X].

[a] To incorporate deviations from ideal behavior, we would write $a = \gamma P_X/P°$ for a gas and $a = \gamma[X]/c°$ for a solute in aqueous solution, where γ is an experimentally determined correction factor called the *activity coefficient*. More advanced treatments of this topic show how γ is related to the composition of the system.

[b] See Exercise 97.

terms of pressures or concentrations (Table 15.1). The expression for Q is obtained as follows.

1. Form a quotient in which the activities of products appear in the numerator and those of reactants appear in the denominator. In both the numerator and the denominator, the activities are combined by multiplying them. (The resulting expression is not yet the reaction quotient.)

$$\frac{a_{Cu^+} \times a_{Sn^{4+}}}{a_{Cu^{2+}} \times a_{Sn^{2+}}}$$

2. To obtain the reaction quotient expression, Q, raise each activity to a power equal to the corresponding coefficient from the balanced chemical equation.

$$Q = \frac{a^2_{Cu^+} \times a_{Sn^{4+}}}{a^2_{Cu^{2+}} \times a_{Sn^{2+}}} \quad (15.2)$$

3. If desired, rewrite the expression for Q by expressing the activities of reactants and products in terms of pressures or concentrations, as appropriate.

$$Q = \frac{a^2_{Cu^+} \times a_{Sn^{4+}}}{a^2_{Cu^{2+}} \times a_{Sn^{2+}}} = \frac{([Cu^+]/c°)^2 \times ([Sn^{4+}]/c°)}{([Cu^{2+}]/c°)^2 \times ([Sn^{2+}]/c°)} = \frac{[Cu^+]^2\,[Sn^{4+}]}{[Cu^{2+}]^2\,[Sn^{2+}]} \quad (15.3)$$

KEEP IN MIND

that the activity, a, provides a measure of the ability or potential of a substance to change the Gibbs energy, G, of the system. See page 624.

With an expression for Q, we can now demonstrate that, for a given temperature, Q has the same constant value at equilibrium. Table 15.2 summarizes the results of three different experiments, each carried out at 298 K and with a different set of starting concentrations. In the first experiment, only Cu^{2+}(aq) and Sn^{2+}(aq) are present initially; in the second experiment, only Cu^+(aq) and Sn^{4+}(aq); and in the third experiment, all species are present initially. The data in Table 15.2 can help us verify, for a given reaction system at a fixed (constant) temperature, the following:

1. **The reaction quotient Q has the same value at equilibrium no matter what the starting concentrations are.** Different starting concentrations were used in the three experiments summarized in Table 15.2, and different equilibrium mixtures were obtained. However, the value of Q at equilibrium was always close to 1.48.

TABLE 15.2 Three Approaches to Equilibrium in the Reaction $2\,Cu^{2+}(aq) + Sn^{2+}(aq) \rightleftharpoons 2\,Cu^{+}(aq) + Sn^{4+}(aq)$

	$[Cu^{2+}]$	$[Sn^{2+}]$	$[Cu^{+}]$	$[Sn^{4+}]$	Q
Experiment 1					
Initial:	0.100	0.100	0.000	0.000	$\frac{(0)^2(0)}{(0.100)^2(0.100)} = 0$
Equilibrium:	0.0360	0.0680	0.0640	0.0320	$\frac{(0.0640)^2(0.0320)}{(0.0360)^2(0.0680)} = 1.49$
Experiment 2					
Initial:	0.000	0.000	0.100	0.100	$\frac{(0.100)^2(0.100)}{(0)^2(0)} = \infty$[a]
Equilibrium:	0.0567	0.0283	0.0433	0.0717	$\frac{(0.0433)^2(0.0717)}{(0.0567)^2(0.0283)} = 1.48$
Experiment 3					
Initial:	0.100	0.100	0.100	0.100	$\frac{(0.100)^2(0.100)}{(0.100)^2(0.100)} = 1$
Equilibrium:	0.0922	0.0961	0.1078	0.1039	$\frac{(0.1078)^2(0.1039)}{(0.0922)^2(0.0961)} = 1.48$

[a] Strictly speaking, we cannot evaluate Q in this case. Any value divided by zero is undefined. By writing $Q = \infty$, we mean that as $[Cu^{2+}]$ and $[Sn^{2+}]$ approach zero, the value of Q approaches infinity.

2. **The equilibrium value of Q is represented by the symbol K and is cal the equilibrium constant.** Using the data in Table 15.2, we have $K = 1.48$ 298 K for reaction (15.1).

We will soon see how to use the equilibrium constant K for a reaction to c culate the outcome of a reaction for specified initial conditions. Thus, knowi the value of K for a reaction is highly desirable. Table 15.2 suggests an approa for determining the value of K for a reaction: (1) Prepare a mixture containi known amounts of reactants or products. (2) Wait for the system to reach eq librium at a specified temperature. (3) Evaluate the reaction quotient by usi the equilibrium amounts of reactants and products. To implement such approach, we must be able to decide if or when equilibrium has been reache

Figure 15-1 illustrates how the concentrations of Cu^{2+}, Sn^{2+}, Cu^{+}, and Sn change as the system approaches equilibrium for the three experiments list in Table 15.2. Notice that the concentrations of reactants and products rem constant with time once equilibrium has been reached, a necessary con quence of the equilibrium condition $Q = K$. However, we cannot say anythi about the time, t_e, it takes for the system to reach equilibrium, at least not ur we explore concepts of chemical kinetics—the study of chemical reacti rates—in Chapter 20.

15-1 CONCEPT ASSESSMENT

How would you write the reaction quotient expression for $Cu(s) + 2\,H^{+}(aq) \longrightarrow Cu^{2+}(aq) + H_2(g)$? Write this first in terms of activities and then convert to pressures and concentrations.

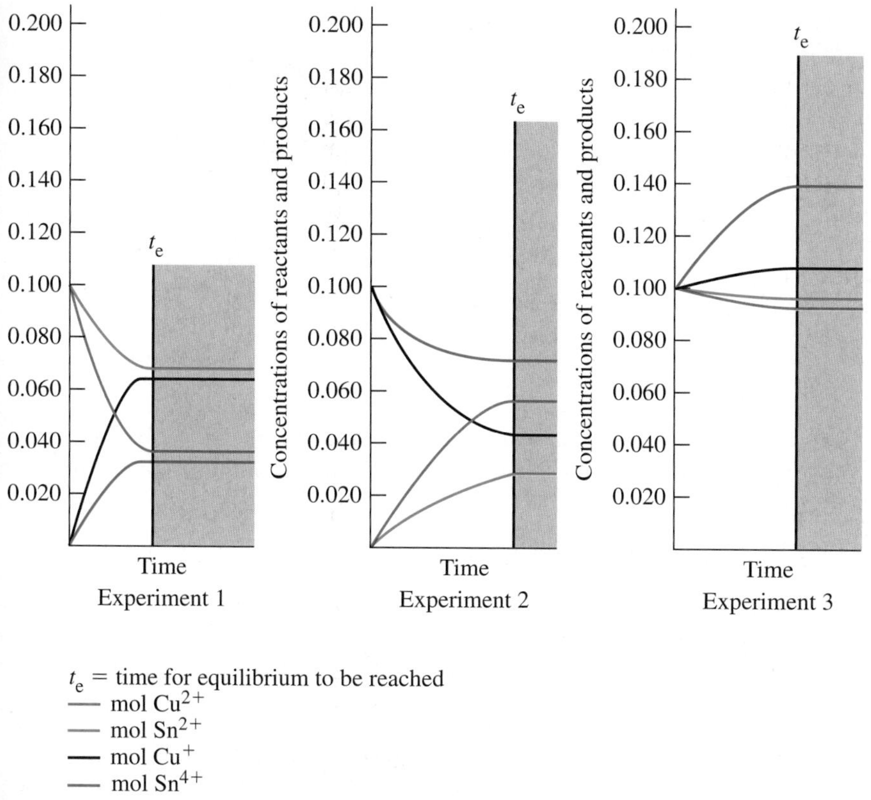

FIGURE 15-1
ree approaches to equilibrium in the reaction

$$2\ Cu^{2+}(aq) + Sn^{2+}(aq) \rightleftharpoons 2\ Cu^{+}(aq) + Sn^{4+}(aq)$$

e initial and equilibrium amounts for each of these three cases are listed in Table 15.2.

EXAMPLE 15-1 Relating Equilibrium Concentrations of Reactants and Products

These equilibrium concentrations are measured in reaction (15.1) at 298 K: $[Cu^+]_{eq} = 0.148$ M, $[Sn^{2+}]_{eq} = 0.124$ M, and $[Sn^{4+}]_{eq} = 0.176$ M. What is the equilibrium concentration of $Cu^{2+}(aq)$?

Analyze

First, we write the equilibrium constant expression for reaction (15.1) in terms of activities, along with the value of the equilibrium constant. Then, we convert the given concentrations to activities and substitute their values into the equilibrium constant expression. We solve the resulting expression for the activity of Cu^{2+}. Finally, we convert the activity of Cu^{2+} to concentration.

Solve

Write the equilibrium constant expression.

$$K = \frac{a^2_{Cu^+(aq)}\, a_{Sn^{4+}(aq)}}{a^2_{Cu^{2+}(aq)}\, a_{Sn^{2+}(aq)}} = 1.48$$

Convert the concentrations to activities by using $a_{X(aq)} = [X]/c° = [X]/1\,M$. As shown explicitly for $a_{Cu^+(aq)}$, this conversion simply involves removing the unit M.

$$a_{Cu^+(aq)} = \frac{0.148\ M}{1\ M} = 0.148 \qquad a_{Sn^{2+}(aq)} = 0.124$$

$$a_{Sn^{4+}(aq)} = 0.176$$

(continued)

Substitute the activities into the equilibrium constant expression.

$$K = \frac{0.148^2 \times 0.176}{a^2_{Cu^{2+}(aq)} \times 0.124} = 1.48$$

Solve for the unknown activity, $a_{Cu^{2+}(aq)}$.

$$a^2_{Cu^{2+}(aq)} = \frac{0.148^2 \times 0.176}{0.124 \times 1.48} = 0.0210$$

$$a_{Cu^{2+}(aq)} = \sqrt{0.0210} = 0.145$$

Convert from activity to concentration by using $[X] = a_{X(aq)} \times c° = a_{X(aq)} \times 1\ M$.

$$[Cu^{2+}] = 0.145 \times 1\ M = 0.145\ M$$

Assess

When solving equilibrium problems, we should examine the results to ensure they make sense. We can do this easily by placing the solution back into the equilibrium constant expression and calculating the equilibrium constant to see whether it agrees with the value stated in the problem, as shown below.

$$K = \frac{0.148^2 \times 0.176}{0.145^2 \times 0.124} = 1.48$$

PRACTICE EXAMPLE A: In another experiment also carried out at 298 K, equal concentrations of $[Cu^+]$, $[Sn^{4+}]$, and $[Sn^{2+}]$ are found to be in equilibrium in reaction (15.1). What must be the equilibrium concentration of $[Cu^{2+}]$?

PRACTICE EXAMPLE B: At 25 °C, $K = 9.14 \times 10^{-6}$ for the reaction $2\ Fe^{3+}(aq) + Hg_2{}^{2+}(aq) \rightleftharpoons 2\ Fe^{2+}(aq) + 2\ Hg^{2+}(aq)$. If the equilibrium concentrations of Fe^{3+}, Fe^{2+} and $Hg_2{}^{2+}$ are 0.015 M, 0.0025 M, and 0.0018 M, respectively, what is the equilibrium concentration of $Hg_2{}^{2+}$?

The Dynamic Nature of the Equilibrium Condition

The equilibrium condition $Q = K$ is a thermodynamic result, obtained Chapter 13 by applying the laws of thermodynamics (the second law, in p ticular). The thermodynamic equilibrium condition is indeed very useful— we will soon see—but it does not account for the following observation abo the equilibrium condition:

> The equilibrium condition is *dynamic*, with the forward and reverse reactions occurring not only indefinitely but also at exactly the same rate.

Let's illustrate the dynamic nature of the equilibrium condition by focusi on a specific example, namely an equilibrium mixture of AgI(s) and its sat rated solution.

$$AgI(s) \rightleftharpoons Ag^+(aq) + I^-(aq)$$

To prove that the equilibrium is dynamic, we could add to the equili rium mixture some radioactive iodine-131 as iodide ion, as illustrated Figure 15-2. If both the forward and reverse processes stopped at equili rium, radioactivity would be confined to the solution. What we fin though, is that radioactivity shows up in the solid as well. Over time, t radioactive iodide ions are distributed throughout the solution and und solved solid. The only way this can happen is if the dissolving and cryst lization processes continue indefinitely.

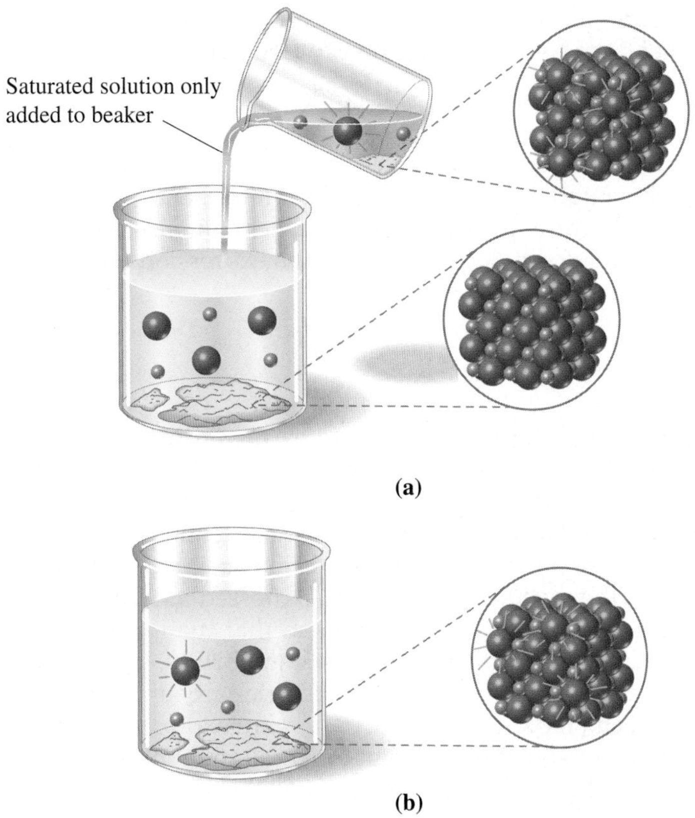

▲ FIGURE 15-2
Dynamic equilibrium illustrated
(a) A saturated solution of radioactive AgI is added to a saturated solution of AgI. **(b)** The radioactive iodide ions distribute themselves throughout the solution and the solid AgI, showing that the equilibrium is dynamic.

15-2 CONCEPT ASSESSMENT

Consider a hypothetical reaction in which a molecule, A, is converted to its somer, B, that is, the reversible reaction A $\rightleftharpoons$ B. Start with a flask containing ;4 molecules of A, represented by open circles. Convert the appropriate number of open circles to filled circles to represent the isomer B and portray the equilibrium condition if $K = 0.02$. Repeat the process for $K = 0.5$ and then for $K = 1$.

5-2 The Equilibrium Constant Expression

fore proceeding to other matters, we must emphasize that the reaction quo-nt expression for reaction (15.1) is just a specific example of a more general se. Let's focus on the following hypothetical, generalized reaction.

$$a\,\mathrm{A} + b\,\mathrm{B} + \cdots \rightleftharpoons c\,\mathrm{C} + d\,\mathrm{D} + \cdots \qquad \textbf{(15.4)}$$

this equation, the uppercase letters A, B, C, D, etc., refer to chemical sub-ances, and the lowercase letters a, b, c, d, etc., represent the coefficients quired to balance the equation. By applying the method described on pages 0 and 691, we obtain the following expression for the reaction quotient.

$$Q = \frac{a_{\mathrm{C}}^{c} \times a_{\mathrm{D}}^{d} \times \cdots}{a_{\mathrm{A}}^{a} \times a_{\mathrm{B}}^{b} \times \cdots} \qquad \textbf{(15.5)}$$

At equilibrium, we may substitute the equilibrium values of the activities i equation (15.5) and set the resulting expression equal to K.

$$K = \frac{a^c_{C,eq} \times a^d_{D,eq} \times \cdots}{a^a_{A,eq} \times a^b_{B,eq} \times \cdots} \qquad (15.6)$$

Although the notation used above is clear, few chemists use it. The numer subscripts make the expression appear rather cluttered. Many chemists wou eliminate the subscript "eq" and write the expression above in the followi abbreviated form.

$$K = \frac{a^c_{C} \times a^d_{D} \cdots}{a^a_{A} \times a^b_{B} \cdots} \qquad (15.7)$$

Unfortunately, equation (15.7), known as the **equilibrium constant** expr sion, "hides" the true meaning of what is intended and can be easily mis terpreted. It seems to imply that the value of K changes as the values of activities change. However, that is not the intended meaning. To the left the equal sign, we have **K, the equilibrium constant**. To the right of the eq sign, we have the reaction quotient, Q, expressed in terms of the activities. setting these two things equal, we mean that the reaction quotient has t value K, which is of course true only at equilibrium. Thus, when using equ tion (15.7), equilibrium values of the activities are implied.

We have seen (Table 15.1) that activities are expressed in different ways different types of substances. Therefore, equation (15.7) is also expressed different ways, depending on the types of substances involved.

Equilibria Involving Gases

Let's consider a reaction that involves only gases and assume ideal gas beh ior. For the activity of gas A, for example, we may write $a_{A(g)} = P_{A(g)}/P°$, wh $P° = 1$ bar. Similar expressions hold for the other gases. Thus, we c write equation (15.7) in the following form.

$$K = \frac{(P_C/P°)^c \times (P_D/P°)^d \cdots}{(P_A/P°)^a \times (P_B/P°)^b \cdots} = \frac{P^c_C \times P^d_D \cdots}{P^a_A \times P^b_B \cdots} \times \left(\frac{1}{P°}\right)^{\Delta\nu} \qquad (15.8)$$

where

$$\Delta\nu = \underbrace{(c + d + \cdots)}_{\text{The sum of coefficients for the products}} - \underbrace{(a + b + \cdots)}_{\text{The sum of coefficients for the reactants}}$$

Notice that $\Delta\nu$ is the sum of the product coefficients minus the sum of react coefficients. The expression above for K can be written in a simpler form if define

$$K_p = \frac{P^c_C \times P^d_D \cdots}{P^a_A \times P^b_B \cdots} \qquad (15.9)$$

K_p is an equilibrium constant expression written in terms of pressures inste of activities. It is important to remember that because equations (15.8) a (15.9) are equilibrium constant expressions, we must use equilibrium pr sures in these equations.

We can now express the relationship between K and K_p as

$$K = K_p \times (1/P°)^{\Delta\nu} \qquad (15.10)$$

▶ K and K_p will have the same numerical value for a reaction in which $\Delta\nu = 0$ because, for this value of $\Delta\nu$, the factor $(1/P°)^{\Delta\nu}$ has a numerical value of one. If $\Delta\nu \neq 0$, K and K_p have the same numerical value only if the pressures are expressed in bar.

In principle, when using equation (15.10) to evaluate K, we can express t pressures in any pressure unit, but using a unit other than bar requires extra calculation, that is, the evaluation of the factor $(1/P°)^{\Delta\nu}$. If we expre pressures in bar, then this factor has a numerical value of 1 and thus, K and ve the same numerical value. However, if we choose instead to express essures in atmospheres (atm), then $P° = 1\ \text{bar} = 1/1.01325\ \text{atm}$ and $P°)^{\Delta\nu}$ has a value of $(1.01325\ \text{atm})^{\Delta\nu}$. Therefore, when pressures are pressed in atmospheres, or any unit other than bar, K and K_p will not usually ve the same numerical value. This complication can be avoided entirely if choose judiciously to express all pressures in bar and substitute their vals *without units* into the expression for K_p. By doing so, we obtain the correct lue of K without having to worry about calculating the factor $(1/P°)^{\Delta\nu}$.

quilibria in Aqueous Solution

r a reaction that occurs in aqueous solution, the activities are expressed in rms of concentrations. For example, the activity of A(aq) is expressed as $= [A]/c°$, where $c° = 1\ \text{mol/L}$. We can write similar expressions for B(aq), aq), D(aq), etc. Thus, for a reaction in aqueous solution, we can write

$$K = \frac{([C]/c°)^c \times ([D]/c°)^d \cdots}{([A]/c°)^a \times ([B]/c°)^b \cdots} = \frac{[C]^c \times [D]^d \cdots}{[A]^a \times [B]^b \cdots} \times \left(\frac{1}{c°}\right)^{\Delta\nu} \tag{15.11}$$

Let's define

$$K_c = \frac{[C]^c\,[D]^d \cdots}{[A]^a\,[B]^b \cdots} \tag{15.12}$$

KEEP IN MIND

that equations (15.11) and (15.12) are equilibrium constant expressions, so we must use equilibrium concentrations in these equations.

Then, we can express the relationship between K and K_c as follows.

$$K = K_c \times (1/c°)^{\Delta\nu} \tag{15.13}$$

r reasons similar to those given on the previous page and above for reactions volving gases, we will obtain the correct value for K and avoid the complican of having to evaluate the factor $(1/c°)^{\Delta\nu}$ by always expressing concentrations mol/L and substituting their values *without units* into the expression for K_c.

quilibria Involving Pure Liquids and Solids

the previous cases, the substances involved in the reaction were either all ses or all dissolved in aqueous solution. Those systems are *homogeneous* cause all the substances are in the same phase and constitute a homogeous mixture. As established above, for a reaction in a homogeneous system, e equilibrium constant, K, may expressed in terms of either K_p or K_c. This is t necessarily the case for a heterogeneous system, where the substances exist different phases. To make this point clear, let us consider the following reacn, which involves substances in a variety of forms.

$$2\ \text{Al(s)} + 6\ \text{H}^+\text{(aq)} \rightleftharpoons 2\ \text{Al}^{3+}\text{(aq)} + 3\ \text{H}_2\text{(g)}$$

ecause the activities of H^+ and Al^{3+} are expressed in terms of concentrations, d that of H_2 in terms of pressure, the equilibrium constant K for this reaction nnot be related to K_c or K_p. (For a more detailed discussion of this point, see ge 615.) Also, because the activity of a pure solid or liquid is always equal to e (Table 15.1), neither solids nor liquids appear explicitly in equilibrium nstant expressions.

Pure solids and liquids are not included in equilibrium constant expressions.

nother way to think about this statement is that an equilibrium constant pression includes only those substances for which the activities, concentrans, or pressures can change during a chemical reaction. Since the activities pure solids and liquids are constant ($a = 1$), they are not included in the uilibrium constant expression.

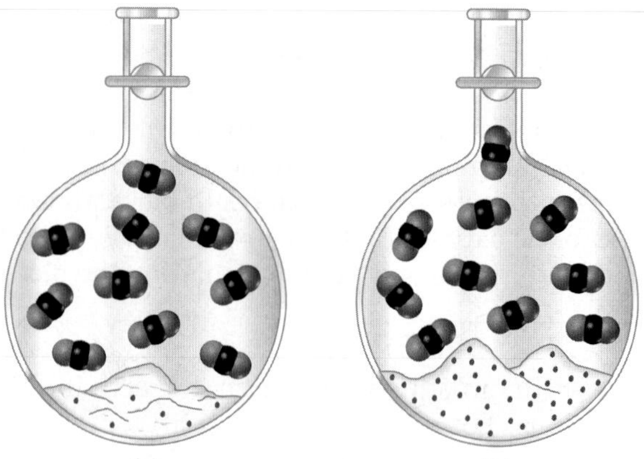

▶ FIGURE 15-3
Equilibrium in the reaction

$$CaCO_3(s) \rightleftharpoons CaO(s) + CO_2(g)$$

(a) Decomposition of $CaCO_3(s)$ upon heating in a closed vessel yields a few granules CaO(s), together with $CO_2(g)$, which soon exerts its equilibrium partial pressure. **(b)** Introduction of additional $CaCO_3(s)$ and/or more CaO(s) has no effect on the partial pressure of the $CO_2(g)$, which remains the same as in (a).

Another example of a heterogeneous reaction is the water–gas reactic which is used to make combustible gases from coal.

$$C(s) + H_2O(g) \rightleftharpoons CO(g) + H_2(g)$$

Although solid carbon must be present for the reaction to occur, the fir equilibrium constant expression for the reaction does not explicitly include C(

$$K = \frac{a_{CO(g)}a_{H_2(g)}}{a_{C(s)}a_{H_2O(g)}} = \frac{(P_{CO}/P°)(P_{H_2}/P°)}{(1)(P_{H_2O}/P°)} = \underbrace{\frac{P_{CO}P_{H_2}}{P_{H_2O}}}_{K_p} \times \frac{1}{P°} = K_p \times \left(\frac{1}{P°}\right)$$

The decomposition of calcium carbonate (limestone) is also a heterog neous reaction.

$$CaCO_3(s) \rightleftharpoons CaO(s) + CO_2(g)$$

For this reaction, the equilibrium constant expression depends explici only on P_{CO_2}.

$$K = \frac{a_{CO_2(g)}a_{CaO(s)}}{a_{CaCO_3(s)}} = \frac{(P_{CO_2}/P°)(1)}{(1)} = \underbrace{P_{CO_2}}_{K_p} \times \left(\frac{1}{P°}\right) = K_p \times \left(\frac{1}{P°}\right)$$

Because K is a constant, P_{CO_2} is also a constant equal to K_p. The value of P_{CO_2} independent of the quantities of $CaCO_3$ and CaO, as long as both solids a present. Figure 15-3 offers a conceptualization of this decomposition reactio.

EXAMPLE 15-2 Writing Equilibrium Constant Expressions for Reactions Involving Pure Solids or Liquids

At equilibrium in the following reaction at 60 °C, the partial pressures of the gases are found to be $P_{HI} = 3.70 \times 10^{-3}$ bar and $P_{H_2S} = 1.01$ bar. What is the value of K for the reaction?

$$H_2S(g) + I_2(s) \rightleftharpoons 2\,HI(g) + S(s) \qquad K = ?$$

Analyze

We need to first write the equilibrium constant expression in terms of activities, and then eliminate the activities of pure solids and pure liquids by setting their activities to 1.

Solve

Write the equilibrium constant expression in terms of activities. Note that activities for the iodine and sulfur are not included, since the activity of a pure solid is 1.

$$K = \frac{(a_{HI(g)})^2}{(a_{H_2S(g)})}$$

The partial pressures are given in bar. The activity of each gas is equal to the numerical value of its partial pressure.

$$a_{HI(g)} = 3.70 \times 10^{-3} \text{ and } a_{H_2S(g)} = 1.01$$

Substitute the given equilibrium data into the equilibrium constant expression.

$$K = \frac{(3.70 \times 10^{-3})^2}{1.01} = 1.36 \times 10^{-5}$$

Assess

Note that the equilibrium constant, K, has no units. You must remember that activities are dimensionless quantities, and that when the partial pressure of a gas is expressed in bar, the activity is equal to the numerical value of its pressure.

PRACTICE EXAMPLE A: Teeth are made principally from the mineral hydroxyapatite, $Ca_5(PO_4)_3OH$, which can be dissolved in acidic solution such as that produced by bacteria in the mouth. The reaction that occurs is $Ca_5(PO_4)_3OH(s) + 4\,H^+(aq) \rightleftharpoons 5\,Ca^{2+}(aq) + 3\,HPO_4^{2-}(aq) + H_2O(l)$. Write the equilibrium constant expression K_c for this reaction.

PRACTICE EXAMPLE B: The steam–iron process is used to generate $H_2(g)$, mostly for use in hydrogenating oils. Iron metal and steam $[H_2O(g)]$ react to produce $Fe_3O_4(s)$ and $H_2(g)$. Write an expression for K_p for this reaction.

5-3 Relationships Involving Equilibrium Constants

fore assessing an equilibrium situation, it may be necessary to make some eliminary calculations or decisions to get the appropriate equilibrium con- ınt expression. This section presents some useful ideas in working with uilibrium constants.

elationship of *K* to the Balanced Chemical Equation

e equilibrium constant expression and the value of K both depend on how write the equation for a reaction. Here are some general rules to keep in nd.

- When we *reverse* an equation, we *invert* the value of K.
- When we *multiply* the coefficients in a balanced equation by a common factor (2, 3, . . .), we raise the equilibrium constant to the *corresponding power* (2, 3, . . .).
- When we *divide* the coefficients in a balanced equation by a common factor (2, 3, . . .), we take the *corresponding root* of the equilibrium constant (square root, cube root, . . .).

illustrate these points, let us consider the synthesis of methanol (methyl ohol) from a carbon monoxide–hydrogen mixture called synthesis gas. ıis reaction is likely to become increasingly important as methanol and mixtures with gasoline find greater use as motor fuels. The balanced action is

$$CO(g) + 2\,H_2(g) \rightleftharpoons CH_3OH(g) \quad K = 9.23 \times 10^{-3}$$

Reuters/Corbis

▲ Methanol is actively being considered as an alternative fuel to gasoline.

Suppose that in discussing the synthesis of CH_3OH from CO and H_2, we d written the reverse reaction, that is,

$$CH_3OH(g) \rightleftharpoons CO(g) + 2\,H_2(g) \qquad K' = ?$$

ow, according to the generalized equilibrium constant expression (15.7), we ould write

$$K' = \frac{a_{CO(g)}a^2_{H_2(g)}}{a_{CH_3OH(g)}} = \frac{1}{\dfrac{a_{CH_3OH(g)}}{a_{CO(g)}\,a^2_{H_2(g)}}} = \frac{1}{K} = \frac{1}{9.23 \times 10^{-3}} = 1.08 \times 10^2$$

In the preceding expression, the equilibrium constant expression and K val for the reaction, as originally written, are printed in blue. We see that $K' = 1/$

Suppose that for a certain application we want an equation based on sy thesizing *two* moles of $CH_3OH(g)$.

$$2\,CO(g) + 4\,H_2(g) \rightleftharpoons 2\,CH_3OH(g) \qquad K'' = ?$$

Here, $K'' = K^2$. That is,

$$K'' = \frac{a^2_{CH_3OH(g)}}{a^2_{CO(g)}\,a^4_{H_2(g)}} = \left(\frac{a_{CH_3OH(g)}}{a_{CO(g)}\,a^2_{H_2(g)}}\right)^2 = (K)^2 = (9.23 \times 10^{-3})^2 = 8.52 \times 10^{-5}$$

15-3 CONCEPT ASSESSMENT

Can you conclude whether the numerical value of K for the reaction $2\,ICl(g) \rightleftharpoons I_2(g) + Cl_2(g)$ is greater or less than the numerical value of K for the reaction $ICl(g) \rightleftharpoons \frac{1}{2} I_2(g) + \frac{1}{2} Cl_2(g)$? Explain.

EXAMPLE 15-3 Relating K to the Balanced Chemical Equation

The following K value is given at 298 K for the synthesis of $NH_3(g)$ from its elements.

$$N_2(g) + 3\,H_2(g) \rightleftharpoons 2\,NH_3(g) \qquad K = 5.8 \times 10^5$$

What is the value of K at 298 K for the following reaction?

$$NH_3(g) \rightleftharpoons \frac{1}{2} N_2(g) + \frac{3}{2} H_2(g) \qquad K = ?$$

Analyze

The solution to this problem lies in recognizing that the reaction is the reverse and one-half of the given reaction. In this example we apply two of the rules given above that relate K to balanced chemical reactions.

Solve

First, reverse the given equation. This puts $NH_3(g)$ on the left side of the equation, where we need it.

$$2\,NH_3(g) \rightleftharpoons N_2(g) + 3\,H_2(g)$$

The equilibrium constant K' becomes

$$K' = 1/(5.8 \times 10^5) = 1.7 \times 10^{-6}$$

Then, to base the equation on 1 mol $NH_3(g)$, divide all coefficients by 2.

$$NH_3\,(g) \rightleftharpoons \frac{1}{2} N_2(g) + \frac{3}{2} H_2(g)$$

This requires the square root of K'.

$$K = \sqrt{1.7 \times 10^{-6}} = 1.3 \times 10^{-3}$$

Assess

Because the rules given on page 699 are used extensively throughout this book, memorizing them will be helpful.

PRACTICE EXAMPLE A: Use data from Example 15-2 to determine the value of K at 298 K for the reaction

$$\frac{1}{3} N_2(g) + H_2(g) \rightleftharpoons \frac{2}{3} NH_3(g)$$

PRACTICE EXAMPLE B: For the reaction $NO(g) + \frac{1}{2} O_2(g) \rightleftharpoons NO_2(g)$ at 184 °C, $K = 1.2 \times 10^2$. What is the value of K at 184 °C for the reaction $2\,NO_2(g) \rightleftharpoons 2\,NO(g) + O_2(g)$?

ombining Equilibrium Constant Expressions

Section 7-7, through Hess's law, we showed how to combine a series of uations into a single overall equation. The enthalpy change of the overall action was obtained by adding together the enthalpy changes of the individ-l reactions. A similar procedure can be used with equilibrium constants, but th this important difference:

When individual equations are combined (that is, added), their equilibrium constants are *multiplied* to obtain the equilibrium constant for the overall reaction.

Suppose we need the equilibrium constant for the reaction

$$N_2O(g) + \frac{1}{2}O_2(g) \rightleftharpoons 2\,NO(g) \qquad K = ? \qquad \textbf{(15.14)}$$

d know the K values of these two equilibria.

$$N_2(g) + \frac{1}{2}O_2(g) \rightleftharpoons N_2O(g) \qquad K = 5.4 \times 10^{-19} \qquad \textbf{(15.15)}$$

$$N_2(g) + O_2(g) \rightleftharpoons 2\,NO(g) \qquad K = 4.6 \times 10^{-31} \qquad \textbf{(15.16)}$$

uation (15.14) is obtained by reversing equation (15.15) and adding it to (15.16). is requires that we also take the *reciprocal* of the K value of equation (15.15).

$$N_2O(g) \rightleftharpoons N_2(g) + \frac{1}{2}O_2(g) \qquad K(a) = 1/(5.4 \times 10^{-19})$$

$$= 1.9 \times 10^{18}$$

$$N_2(g) + O_2(g) \rightleftharpoons 2\,NO(g) \qquad K(b) = 4.6 \times 10^{-31}$$

erall: $N_2O(g) + \frac{1}{2}O_2(g) \rightleftharpoons 2\,NO(g) \qquad K(\text{overall}) = ?$

e overall equation is expression (15.14), for which

$$K(\text{overall}) = \frac{a^2_{NO(g)}}{a_{N_2O(g)}\,a^{\frac{1}{2}}_{O_2(g)}} = \underbrace{\frac{a_{N_2(g)}a^{\frac{1}{2}}_{O_2(g)}}{a_{N_2O(g)}}}_{K(a)} \times \underbrace{\frac{a^2_{NO(g)}}{a_{N_2(g)}a_{O_2(g)}}}_{K(b)} = K(a) \times K(b)$$

$$= 1.9 \times 10^{18} \times 4.6 \times 10^{-31} = 8.5 \times 10^{-13}$$

15-4 CONCEPT ASSESSMENT

You want to calculate K for the reaction

$$CH_4(g) + 2\,H_2O(g) \rightleftharpoons CO_2(g) + 4\,H_2(g)$$

and you have available a K value for the reaction

$$CO_2(g) + H_2(g) \rightleftharpoons CO(g) + H_2O(g)$$

What additional K value do you need, assuming that all K values are at the same temperature?

elationship Between K_p and K_c for Reactions Involving Gases

o to this point, we have emphasized the use of partial pressures for specify-g the composition of a gas mixture. However, this is not the only possibility.

The composition of a gas mixture can also be specified by giving the concentrations, in moles per liter, of the component gases. Let's suppose we have n_A moles of gas A, n_B moles of gas B, n_C moles of gas C, and n_D moles of gas D in a container of volume V at temperature T. Assuming ideal gas behavior, the partial pressure of each gas is directly proportional to its concentration. For example, the partial pressure of gas A is

$$P_A = \frac{n_A RT}{V} = [A]RT$$

We can write similar expressions for P_B, P_C, and P_D. Thus, the equilibrium constant expression K_p for the following reaction

$$aA(g) + bB(g) + \cdots \rightleftharpoons cC(g) + dD(g) + \cdots$$

may also be written in terms of concentrations:

$$K_p = \frac{P_C^c \times P_D^d \cdots}{P_A^a \times P_B^b \cdots} = \frac{([C]RT)^c \times ([D]RT)^d \cdots}{([A]RT)^a \times ([B]RT)^b \cdots} = \frac{[C]^c \times [D]^d \cdots}{[A]^a \times [B]^b \cdots} \times (RT)^{\Delta\nu_{gas}}$$

$\Delta\nu_{gas}$ is the sum of coefficients for *gaseous* products minus the sum of coefficients for *gaseous* reactants. Because the quantity multiplying the factor $(RT)^{\Delta\nu_{gas}}$ is the equilibrium constant expression K_c, we may write

$$K_p = K_c \times (RT)^{\Delta\nu_{gas}} \quad \textbf{(15.17)}$$

To be consistent with the convention of expressing pressure in bar and concentrations in mol/L, the appropriate value of R to use in the equation above is $R = 0.083144598$ bar L K^{-1} mol^{-1}.

Relationships Among K, K_p, and K_c: A Summary

The thermodynamic equilibrium constant K is expressed in terms of activities (equation 15.7). It is a dimensionless quantity. Unlike K, the equilibrium constants K_p and K_c have units of $(\text{bar})^{\Delta\nu}$ and $(\text{mol/L})^{\Delta\nu}$, respectively. The factors $(1/P°)^{\Delta\nu}$ and $(1/c°)^{\Delta\nu}$ appearing in equations (15.10) and (15.13) ensure that K is a dimensionless quantity. Throughout this text, we will use K_p and K_c expressions when solving problems. To avoid the clutter of units in these calculations, and to simplify the conversion of a K_p or K_c value to a K value, we will always use only the numerical values of K_p, K_c, partial pressures, and concentrations in our calculations, that is, without explicitly including the units bar or mol/L.

EXAMPLE 15-4 Interrelating the Values of K_p and K_c

For the gas-phase reaction below, the value of K_c is 3.4 at 1000 K. What is the value of K_p at this temperature?

$$2\,SO_2(g) + O_2(g) \rightleftharpoons 2\,SO_3(g)$$

Analyze

K_c and K_p are related by equation (15.17), with $R = 0.08314$ bar L K^{-1} mol^{-1}. A key quantity appearing in equation (15.17) is $\Delta\nu_{gas}$, which is equal to the sum of coefficients for gaseous products minus the sum of coefficients for gaseous reactants. For the reasons given on page 696, units are omitted from our calculations.

Solve

Calculate $\Delta\nu_{gas}$ for the given chemical equation.

$$\Delta\nu_{gas} = 2 - (2 + 1) = -1$$

Substitute the values of $\Delta\nu_{gas}$ and K_c into equation (15.17), and solve for K_p.

$$K_p = K_c \times (RT)^{\Delta\nu_{gas}} = 2.8 \times 10^2 \times (0.08314 \times 1000)^{-1} = 3.4$$

Assess

As discussed above, when we express pressures in bar and concentrations in mol/L, the units of K_p and K_c are, respectively, $(\text{bar})^{\Delta\nu_{\text{gas}}}$ and $(\text{mol/L})^{\Delta\nu_{\text{gas}}}$. Because $\Delta\nu_{\text{gas}} = -1$ for this reaction, we have $K_p = 3.4\ \text{bar}^{-1}$ and $K_c = 2.8 \times 10^2\ (\text{mol/L})^{-1} = 2.8 \times 10^2\ \text{L mol}^{-1}$. For convenience, we often omit the units when specifying the value of K_p or K_c and when substituting pressures or concentrations into the corresponding equilibrium constant expressions, as we did in this example.

PRACTICE EXAMPLE A: For the reaction $2\ NH_3(g) \rightleftharpoons N_2(g) + 3\ H_2(g)$ at 298 K, $K_c = 2.8 \times 10^{-9}$. What is the value of K_p for this reaction?

PRACTICE EXAMPLE B: At 1065 °C, for the reaction $2\ H_2S(g) \rightleftharpoons 2\ H_2(g) + S_2(g)$, $K_p = 1.2 \times 10^{-2}$. What is the value of K_c for the reaction $H_2(g) + \frac{1}{2} S_2(g) \rightleftharpoons H_2S(g)$ at 1065 °C?

5-4 The Magnitude of an Equilibrium Constant

ble 15.3 lists equilibrium constants for several reactions mentioned in this apter or previously in the text. The first of these reactions is the synthesis of $_2O(l)$ from its elements. The equilibrium constant for this reaction is very rge: $K = 1.4 \times 10^{83}$ at 298 K. However, for the decomposition of $CaCO_3(s)$ at 8 K, the equilibrium constant is very small: $K = 1.9 \times 10^{-23}$. How do we terpret these values?

First, as we established in Chapter 13, the value of K provides a measure of e thermodynamic stability of products relative to that of reactants. The large lue of K for the water synthesis reaction signifies that, from a thermody- mic perspective, 2 mol of $H_2O(l)$ is much more stable than 2 mol $H_2(g)$ and mol $O_2(g)$. The small value of K for the decomposition of $CaCO_3(s)$ indicates at 1 mol $CaO(s)$ and 1 mol $CO_2(g)$ are much less stable than 1 mol $CaCO_3(s)$.

Second, because the equilibrium constant expression is a quotient with oducts appearing in the numerator and reactants in the denominator, a very rge value of K indicates that, at equilibrium, the products are present in uch greater amounts than the reactants are. To reach such an equilibrium ate, starting from an initial reaction mixture containing only reactants, the action goes almost to completion. By this we mean that one (or more) of the actants is almost completely consumed. If a reaction goes to completion, not ly is the limiting reactant completely used up but the maximum possible nount of product is also obtained.

◀ The Gibbs energy of formation, $\Delta_f G°$, of a compound provides a measure of the thermodynamic stability of a compound with respect to its elements.

A very large value of K signifies that the reaction, as written, exhibits a strong tendency to go to completion. An equilibrium mixture contains about as much product as can be formed from the given initial amounts of reactants.

ABLE 15.3 Equilibrium Constants of Some Common Reactions

Reaction	Equilibrium Constant, K
$2\ H_2(g) + O_2(g) \rightleftharpoons 2\ H_2O\ (l)$	1.4×10^{83} at 298 K
$CaCO_3(s) \rightleftharpoons CaO(s) + CO_2(g)$	1.9×10^{-23} at 298 K 1.0 at about 1200 K
$2\ SO_2(g) + O_2(g) \rightleftharpoons 2\ SO_3(g)$	3.4 at 1000 K
$C(s) + H_2O(g) \rightleftharpoons CO(g) + H_2(g)$	1.6×10^{-21} at 298 K 10.0 at about 1100 K

Let's explore this idea further by considering the water synthesis reacti and an initial mixture containing 1.0 mol each of $H_2(g)$ and $O_2(g)$. Because *K* very large, we may assume that, at equilibrium, the reaction mixture will co tain essentially as much H_2O as can be formed from 1.0 mol H_2 and 1.0 mol C For all the O_2 to react, we require 2.0 mol H_2. Because we do not have enou H_2 for all of the O_2 to react, we conclude that H_2 is the limiting reactant and is present in excess. Therefore, the maximum amount of product that can formed from this initial mixture is

$$1.0\ \text{mol H}_2 \times \frac{1\ \text{mol H}_2\text{O}}{2\ \text{mol H}_2} = 0.5\ \text{mol H}_2\text{O}$$

We predict that, at equilibrium, the mixture will contain about 0.5 mol H_2 0.5 mol O_2 (the excess amount of O_2), and very little H_2. In Section 15-7, v will be much more precise about the actual amount of H_2 present at equili rium.

What can we say about a reaction that has a very small value of *K*? Not on are the products much less stable than the reactants but the reaction al exhibits only a very small tendency to occur in the forward direction.

A very small value of *K* signifies that the reaction, as written, exhibits very little tendency to occur. An equilibrium mixture contains reactants, in essentially their initial amounts, and very small amounts of products.

Most reactions lie somewhere between the two extremes of very large ar very small values of *K*, with the consequences that (1) both the forward ar reverse reactions occur to an appreciable extent, and (2) an equilibrium mi ture contains appreciable amounts of both reactants and products.

In this text, we will use the following guideline to help us decide whether reaction goes to completion or to a very limited extent.

A reaction goes essentially to completion if $K > 10^{10}$ and not at all if $K < 10^{-10}$.

The reasons for choosing the values $K = 10^{10}$ and $K = 10^{-10}$ are explaine in Exercise 98.

The discussion above focuses only on the nature of the equilibrium state. makes no mention of the time required to reach equilibrium, t_e, which is a important factor, especially if the reaction occurs so slowly that the likelihoc of observing an equilibrium state is extremely low. The water synthesis rea tion is an excellent example of such a reaction. This reaction occurs so slow at 298 K that a mixture of hydrogen and oxygen gases is actually quite stab at room temperature. In other words, the water synthesis reaction does n occur to any great extent at room temperature, let alone to completion, eve though the equilibrium constant is very large. To reach a true equilibriu state at 298 K, this reaction would require our intervention (e.g., by heating th system to a higher temperature to increase the rate of reaction and then coc ing it back to 298 K or by adding a catalyst—a substance that increases the ra of reaction). We will discuss factors that affect the rates of chemical reactio in Chapter 20.

▶ In Chapter 20, we will discuss factors that influence the rate of a chemical reaction and learn that the rate of a reaction is strongly governed by the activation energy, E_a.

15-5 CONCEPT ASSESSMENT

Why is having a balanced equation a necessary condition for predicting the outcome of a chemical reaction, but often not a sufficient condition?

5-5 Predicting the Direction of Net Chemical Change

Chapter 13, we used the laws of thermodynamics, specifically the second v, to establish criteria (Tables 13.4 and 13.6) for predicting the direction net chemical change in a system from a given set of initial conditions. We nmarize some key results in Table 15.4.

ABLE 15.4 Using Q and K to Predict the Direction of Net Chemical Change

ondition	Direction of Net Chemical Change
$< K$	To the right ($\longrightarrow$)
$> K$	To the left ($\longleftarrow$)
$= K$	Equilibrium; no net change ($\rightleftharpoons$)

We can justify the relationships in Table 15.4 by considering the three exper-ents we discussed in Section 15-2. These three experiments involve the fol-ving reaction

$$2\,Cu^{2+}(aq) + Sn^{2+}(aq) \rightleftharpoons 2\,Cu^{+}(aq) + Sn^{4+}(aq)$$

which

$$Q = \frac{[Cu^{+}]^2\,[Sn^{4+}]}{[Cu^{2+}]^2\,[Sn^{2+}]}$$

Section 15-2, we found that Q must have the value $K = 1.48$ when this action reaches equilibrium at 298 K. To be able to predict the direction of net ange when $Q \neq K$, we must first establish how the value of Q changes hen there is net change in the *forward* direction (*to the right*) or in the *reverse* rection (to the *left*).

When there is net change to the right, $[Cu^{+}]$ and $[Sn^{4+}]$ increase whereas $u^{2+}]$ and $[Sn^{4+}]$ decrease. We deduce the combined effect of these changes examining the expression above for Q: The numerator increases, the nominator decreases, and the value of Q increases. Our conclusion is that *e value of Q increases when there is net change to the right.* Therefore, when $< K$, the equilibrium condition $Q = K$ can be achieved only if there is net ange in the forward direction or to the right.

Conversely, when there is net change to the left, $[Cu^{+}]$ and $[Sn^{4+}]$ decrease hereas $[Cu^{2+}]$ and $[Sn^{4+}]$ increase. The numerator in the expression for Q creases, the denominator increases, and thus, the value of Q decreases. *The lue of Q decreases when there is net change to the left.* We conclude that when $> K$, the equilibrium condition $Q = K$ can be achieved only if there is net ange in the reverse direction or to the left.

In Figure 15-4, different possibilities for the relationship involving the ini-al and equilibrium conditions are illustrated. Two possibilities for the initial ndition, pure reactants (Fig. 15-4a) and pure products (Fig. 15-4e), deserve me extra comment. Starting from pure reactants, we know that the direction net change must be in the forward direction (to the right) because, with ne of the products present, reaction to the left is impossible. Starting from ure products, the direction of net change is in the reverse direction (to the ft) because reaction to the right is impossible. However, the direction of net ange is not at all obvious when the initial reaction mixture contains both actants and products. In such situations, we must determine the direction of et change by first evaluating Q for the initial condition and then comparing e values of Q and K. We consider such a situation in Example 15-5.

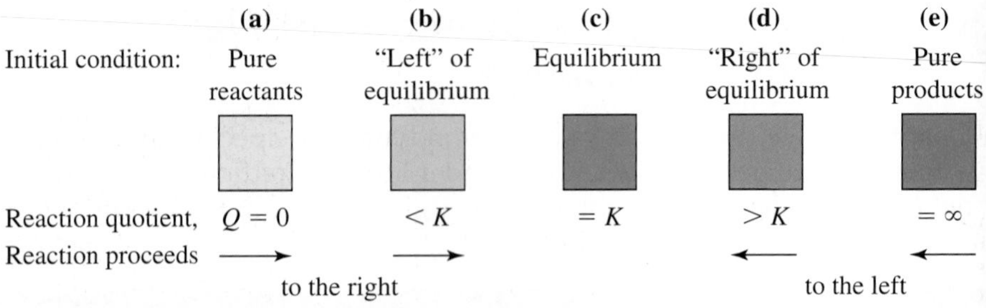

▲ FIGURE 15-4
Predicting the direction of net change
Five possibilities for the relationship of initial and equilibrium conditions are shown. From Table 15.2 and Figure 15-1, Experiment 1 corresponds to initial condition **(a)**; Experiment 2 to condition **(e)**; and Experiment 3 to **(b)**. The situation in Example 15-5 corresponds to condition **(d)**.

EXAMPLE 15-5 Predicting the Direction of a Net Chemical Change in Establishing Equilibrium

To increase the yield of $H_2(g)$ in the water–gas reaction—the reaction of $C(g)$ and $H_2O(g)$ to form $CO(g)$ and $H_2(g)$—a follow-up reaction called the "water–gas shift reaction" is generally used. In this reaction, some of the $CO(g)$ of the water gas is replaced by $H_2(g)$.

$$CO(g) + H_2O(g) \rightleftharpoons CO_2(g) + H_2(g)$$

$K_c = 1.00$ at about 1100 K. The following amounts of substances are brought together and allowed to react at this temperature: 1.00 mol CO, 1.00 mol H_2O, 2.00 mol CO_2, and 2.00 mol H_2. Compared with their initial amounts, which of the substances will be present in a greater amount and which in a lesser amount when equilibrium is established?

Analyze

Our task is to determine the direction of net change by evaluating Q_c and comparing it to K_c.

Solve

Write down the expression for Q_c.

$$Q_c = \frac{[CO_2][H_2]}{[CO][H_2O]}$$

Substitute concentrations into the expression for Q_c, by assuming an arbitrary volume V (which cancels out in the calculation).

$$Q_c = \frac{(2.00/V)(2.00/V)}{(1.00/V)(1.00/V)} = 4.00$$

Compare Q_c to K_c.

$$4.00 > 1.00$$

Because $Q_c > K_c$ (that is, $4.00 > 1.00$), a net change occurs to the *left*. When equilibrium is established, the amounts of CO and H_2O will be greater than the initial quantities and the amounts of CO_2 and H_2 will be less.

Assess

It is important to be able to determine the direction of reaction. As we will see in Section 15-7, this step must be completed before we attempt to determine what the equilibrium amounts will be.

PRACTICE EXAMPLE A: In Example 15-5, equal masses of CO, H_2O, CO_2, and H_2 are mixed at a temperature of about 1100 K. When equilibrium is established, which substance(s) will show an increase in quantity and which will show a decrease compared with the initial quantities?

PRACTICE EXAMPLE B: For the reaction $PCl_5(g) \rightleftharpoons PCl_3(g) + Cl_2(g)$, $K_c = 0.0454$ at 261 °C. If a vessel is filled with these gases such that the initial partial pressures are $P_{PCl_3} = 2.19$ bar, $P_{Cl_2} = 0.88$ bar, $P_{PCl_5} = 19.7$ bar, in which direction will a net change occur?

15-6 CONCEPT ASSESSMENT

mixture of 1.00 mol each of CO(g), $H_2O(g)$, and $CO_2(g)$ is placed in a 10.0 L ask at a temperature at which $K_p = 10.0$ in the reaction

$$CO(g) + H_2O(g) \rightleftharpoons CO_2(g) + H_2(g)$$

Vhen equilibrium is established, **(a)** the amount of $H_2(g)$ will be 1.00 mol; **(b)** the mounts of all reactants and products will be greater than 1.00 mol; **(c)** the amounts f all reactants and products will be less than 1.00 mol; **(d)** the amount of $CO_2(g)$ vill be greater than 1.00 mol and the amounts of CO(g), $H_2O(g)$, and $H_2(g)$ will be ess than 1.00 mol; **(e)** the amounts of reactants and products cannot be pre-dicted and can only be determined by analyzing the equilibrium mixture.

KEEP IN MIND

that volumes cancel in a reaction quotient or equilibrium constant expression whenever the sum of the exponents in the numerator equals that in the denominator. This can simplify problem solving in some instances.

5-6 Altering Equilibrium Conditions: Le Châtelier's Principle

times, we want only to make qualitative statements about the direction of a t change, whether the amount of a substance will have increased or creased when equilibrium is reached, and so on. Also, we may not have the ta needed for a quantitative calculation. In these cases, we can use a state-ent attributed to the French chemist Henri Le Châtelier (1884). **Le Châtelier's inciple** is hard to state unambiguously, but its essential meaning is stated re.

> When an equilibrium system is subjected to a change in temperature, pressure, or concentration of a reacting species, the system responds by attaining a new equilibrium that *partially* offsets the impact of the change.

As we will see in the examples that follow, it is generally not difficult to pre-ct the outcome of changing one or more variables in a system at equilibrium.

fect of Changing the Amounts of Reacting Species n Equilibrium

t's consider the following reaction

$$2\,SO_2(g) + O_2(g) \rightleftharpoons 2\,SO_3(g) \qquad K_c = 2.8 \times 10^2 \text{ at } 1000\text{ K}$$

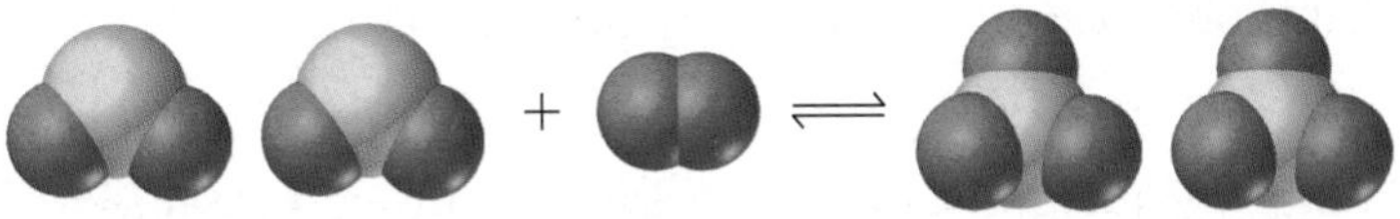

ppose we start with certain equilibrium amounts of SO_2, O_2, and SO_3, as ggested by Figure 15-5(a). Now let's create a disturbance in the equilibrium ixture by forcing an additional 1.00 mol SO_3 into the 10.0 L flask ig. 15-5b). How will the amounts of the reacting species change to re-establish uilibrium?

According to Le Châtelier's principle, if the system is to partially offset an tion that increases the equilibrium concentration of one of the reacting ecies, it must do so by favoring the reaction in which that species is con-med. In this case, this is the *reverse* reaction—conversion of some of the ded SO_3 to SO_2 and O_2. In the new equilibrium, there are greater amounts of l the substances than in the original equilibrium, but the additional amount SO_3 is less than the 1.00 mol that was added.

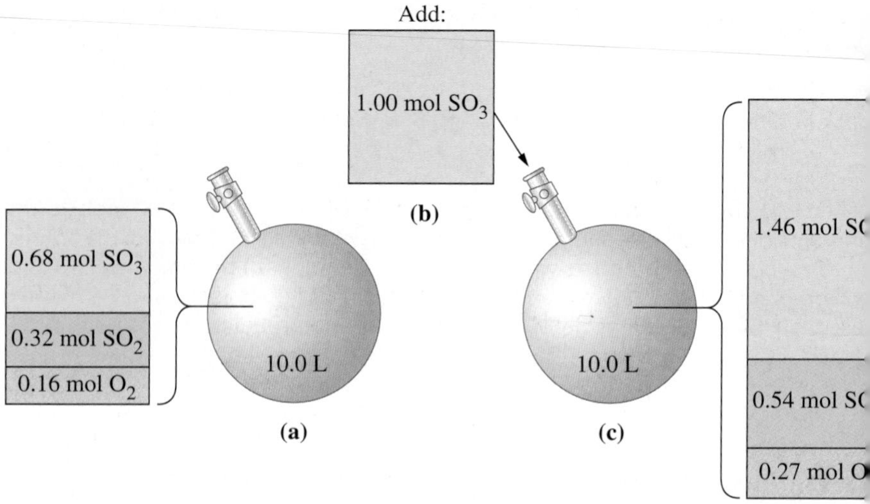

▲ FIGURE 15-5
Changing equilibrium conditions by changing the amount of a reacting specie

$$2\ SO_2(g) + O_2(g) \rightleftharpoons 2\ SO_3(g),\ K_c = 2.8 \times 10^2 \text{ at } 1000\text{ K}$$

(a) The original equilibrium condition. **(b)** Disturbance caused by adding 1.00 mol SC **(c)** The new equilibrium condition. The amount of SO_3 in the new equilibrium mixture 1.46 mol, is greater than the original 0.68 mol but it is not as great as immediately aft the addition of 1.00 mol SO_3. The effect of adding SO_3 to an equilibrium mixture is *partially* offset when equilibrium is restored.

Another way to look at the matter is to evaluate the reaction quotient imm diately after adding the SO_3.

Original equilibrium | *Immediately following disturbance*

$$Q_c = \frac{[SO_3]}{[SO_2]^2[O_2]} = K_c \qquad Q_c = \frac{[SO_3]}{[SO_2]^2[O_2]} > K_c$$

Adding any quantity of SO_3 to a *constant-volume* equilibrium mixture mak Q_c larger than K_c. A net change occurs in the direction that reduces $[SO_3]$, th is, to the left, or in the reverse direction. Notice that reaction in the rever direction increases $[SO_2]$ and $[O_2]$, further decreasing the value of Q_c.

We must point out that, for some reactions and certain initial conditior Le Châtelier's principle will give an incorrect prediction for the effect of addi more of a reactant to an equilibrium mixture (Exercise 96). For such cases, t correct result is obtained—as always—by comparing the values of Q and K.

EXAMPLE 15-6 Applying Le Châtelier's Principle: Effect of Adding More of a Reactant to an Equilibrium Mixture

Predict the effect of adding more $N_2(g)$ to a constant-volume equilibrium mixture of N_2, H_2, and NH_3.

$$N_2(g) + 3\ H_2(g) \rightleftharpoons 2\ NH_3(g)$$

Analyze

When a system at equilibrium is disturbed by adding more of one reactant at constant volume, the system responds by using up (consuming) some of the added reactant.

Solve

Increasing $[N_2]$ causes a net change in the direction that reduces $[N_2]$. Thus, the addition of N_2 causes the reaction to proceed in the forward direction. However, only a portion of the added N_2 is consumed in this reaction. When equilibrium is re-established, there will be more N_2 than was present originally, and also more NH_3, but the amount of H_2 will be *smaller*. Some of the original H_2 must be consumed in converting some of the added N_2 to NH_3.

Assess

Le Châtelier's principle can be applied here because the addition of N_2 is made at *constant volume*. Thus, the effect of adding more N_2 is always to decrease the value of Q_c and cause net change to the right. The situation is more complicated if the addition of N_2 is made at *constant pressure*. The addition of more N_2 at constant pressure causes an increase in the total volume of the system, and, therefore, the concentrations of all substances are affected. Somewhat surprisingly, for certain initial conditions, the addition of N_2 at constant pressure can cause net reaction to the left (see Exercise 96).

PRACTICE EXAMPLE A: Given the reaction $2\,CO(g) + O_2(g) \rightleftharpoons 2\,CO_2(g)$, what is the effect of adding $O_2(g)$ to a constant-volume equilibrium mixture?

PRACTICE EXAMPLE B: Calcination of limestone (decomposition by heating), $CaCO_3(s) \rightleftharpoons CaO(s) + CO_2(g)$, is the commercial source of quicklime, CaO(s). After this equilibrium has been established in a constant-temperature, constant-volume container, what is the effect on the equilibrium amounts of materials caused by *adding* some **(a)** CaO(s); **(b)** $CO_2(g)$; **(c)** $CaCO_3(s)$?

fect of Changes in Pressure or Volume on Equilibrium

ere are three ways to change the pressure of a constant-temperature equirium mixture.

. **Add or remove a gaseous reactant or product.** The effect of these actions on the equilibrium condition is simply that caused by adding or removing a reaction component, as described previously.

. **Add an inert gas to the constant-volume reaction mixture.** This has the effect of increasing the *total* pressure, but the partial pressures of the reacting species are all unchanged. An inert gas added to a constant-volume equilibrium mixture has no effect on the equilibrium condition.

. **Change the pressure by changing the volume of the system.** Decreasing the volume of the system increases the pressure, and increasing the system volume decreases the pressure. Thus, the effect of this type of pressure change is simply that of a volume change.

Let's explore the third situation first. Consider, again, the formation of)$_3$(g) from $SO_2(g)$ and $O_2(g)$.

$$2\,SO_2(g) + O_2(g) \rightleftharpoons 2\,SO_3(g) \qquad K_c = 2.8 \times 10^2 \text{ at } 1000\text{ K} \qquad \textbf{(15.18)}$$

e equilibrium mixture in Figure 15-6(a) has its volume reduced to one-tenth its original value by increasing the external pressure. To see how the equirium amounts of the gases change, let's first rearrange the equilibrium connt expression to the form

$$K_c = \frac{[SO_3]^2}{[SO_2]^2[O_2]} = \frac{(n_{SO_3}/V)^2}{(n_{SO_2}/V)^2(n_{O_2}/V)} = \frac{(n_{SO_3})^2}{(n_{SO_2})^2(n_{O_2})} \times V = 2.8 \times 10^2 \qquad \textbf{(15.19)}$$

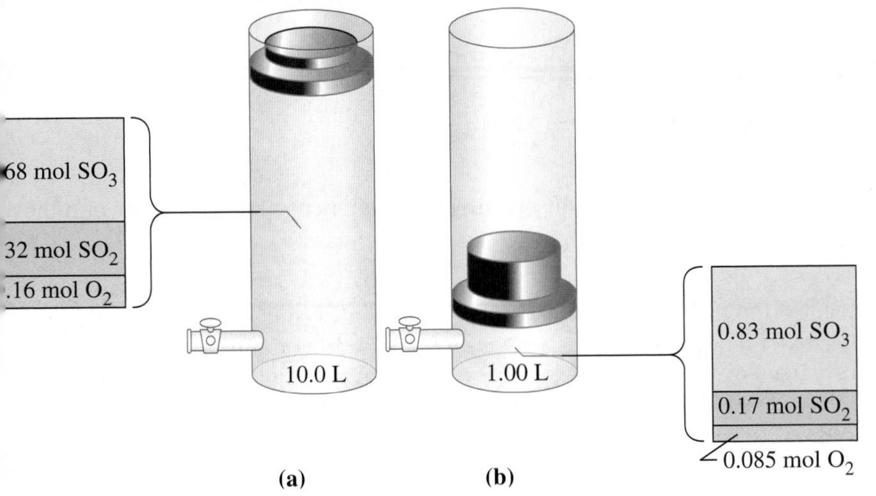

◀ FIGURE 15-6
Effect of pressure change on equilibrium in the reaction
$$2\,SO_2(g) + O_2(g) \rightleftharpoons 2\,SO_3(g)$$
An increase in external pressure causes a decrease in the reaction volume and a shift in equilibrium "to the right." (See Exercise 81 for a calculation of the new equilibrium amounts.)

From equation (15.19), we see that if *V* is *reduced* by a factor of 10, the ratio

$$\frac{(n_{SO_3})^2}{(n_{SO_2})^2(n_{O_2})}$$

must *increase* by a factor of 10. There is only one way in which the ratic moles of gases will increase in value: The number of moles of SO_3 m increase, and the numbers of moles of SO_2 and O_2 must decrease. The equi rium shifts in the direction producing more SO_3—to the right (Fig. 15-6b).

Notice that three moles of *gas* on the left produce two moles of *gas* on right in reaction (15.18). When compared at the same temperature and pr sure, two moles of $SO_3(g)$ occupy a smaller volume than does a mixture of t moles of $SO_2(g)$ and one mole of $O_2(g)$. Given this fact and the observat from equation (15.19) that a decrease in volume favors the production of ad tional SO_3, we can formulate a statement that is especially easy to apply.

> When the volume of an equilibrium mixture of gases is *reduced*, a net change occurs in the direction that produces *fewer moles of gas*. When the volume is *increased*, a net change occurs in the direction that produces *more moles of gas*.

Figure 15-6 suggests a way of decreasing the volume of gaseous mixture equilibrium—by increasing the external pressure. One way to increase the v ume is to lower the external pressure. Another way is to transfer the equil rium mixture from its original container to one of larger volume. A thi method is to add an inert gas at *constant pressure*; the volume of the mixtu must increase to make room for the added gas. The effect on the equilibriu however, is the same for all three methods: Equilibrium shifts in the directi of the reaction producing the greater number of moles of gas.

Equilibria between condensed phases are not affected much by changes external pressure because solids and liquids are not easily compressible. Al we cannot assess whether the forward or reverse reaction is favored by the changes by examining only the chemical equation.

KEEP IN MIND

that an inert gas has no effect on an equilibrium condition if the gas is added to a system maintained at constant volume, but it can have an effect if added at constant pressure.

15-7 CONCEPT ASSESSMENT

After the hypothetical reaction $A(g) + B(g) \rightleftharpoons C(g)$ reaches equilibrium in a closed container, 0.100 mol of the inert gas argon is added. In addition, the volume of the container is decreased. According to Le Châtelier's principle, will the reaction shift to the right or left? Explain.

EXAMPLE 15-7 Applying Le Châtelier's Principle: The Effect of Changing Volume

An equilibrium mixture of $N_2(g)$, $H_2(g)$, and $NH_3(g)$ is transferred from a 1.50 L flask to a 5.00 L flask. In which direction does a net change occur to restore equilibrium?

$$N_2(g) + 3\,H_2(g) \rightleftharpoons 2\,NH_3(g)$$

Analyze

Because the volume has increased, the reaction will move in the direction that increases the number of moles of gas.

Solve

When the gaseous mixture is transferred to the larger flask, the partial pressure of each gas and the total pressure drop. Whether we think in terms of a decrease in pressure or an increase in volume, we reach the same conclusion. Equilibrium shifts in such a way as to produce a larger number of moles of gas. Some of the NH_3 originally present decomposes back to N_2 and H_2. A net change occurs in the direction of the reverse reaction—to the left—in restoring equilibrium.

Assess

Whether we think in terms of a decrease in pressure or an increase in volume, the conclusion is the same.

PRACTICE EXAMPLE A: The reaction $N_2O_4(g) \rightleftharpoons 2\,NO_2(g)$ is at equilibrium in a 3.00 L cylinder. What would be the effect on the concentrations of $N_2O_4(g)$ and $NO_2(g)$ if the pressure were doubled (that is, cylinder volume decreased to 1.50 L)?

PRACTICE EXAMPLE B: How is the equilibrium amount of $H_2(g)$ produced in the water–gas shift reaction affected by changing the total gas pressure or the system volume? Explain.

$$CO(g) + H_2O(g) \rightleftharpoons CO_2(g) + H_2(g)$$

15-8 CONCEPT ASSESSMENT

The following reaction is brought to equilibrium at 700 °C.

$$2\,H_2S(g) + CH_4(g) \rightleftharpoons CS_2(g) + 4\,H_2(g)$$

Indicate whether each of the following statements is true, false, or not possible to evaluate from the information given.

(a) If the equilibrium mixture is allowed to expand into an evacuated larger container, the mole fraction of H_2 will increase.

(b) If several moles of Ar(g) are forced into the reaction container, the amounts of H_2S and CH_4 will increase.

(c) If the equilibrium mixture is cooled to 100 °C, the mole fractions of the four gases will likely change.

(d) If the equilibrium mixture is forced into a slightly smaller container, the partial pressures of the four gases will all increase.

Effect of Temperature on Equilibrium

We can think of changing the temperature of an equilibrium mixture in terms of adding heat (raising the temperature) or removing heat (lowering the temperature). According to Le Châtelier's principle, adding heat favors the reaction in which heat is absorbed (*endothermic* reaction). Removing heat favors the reaction in which heat is evolved (*exothermic* reaction). Stated in terms of changing temperature,

Raising the temperature of an equilibrium mixture shifts the equilibrium condition in the direction of the *endothermic* reaction. *Lowering the temperature* causes a shift in the direction of the *exothermic* reaction.

The principal effect of temperature on equilibrium is in changing the value of the equilibrium constant. In Chapter 13, we learned how to calculate equilibrium constants as a function of temperature. We also found the following:

For *endothermic* reactions, *K increases* as temperature increases. For *exothermic* reactions, *K decreases* as temperature increases.

EXAMPLE 15-8 Applying Le Châtelier's Principle: Effect of Temperature on Equilibrium

Consider the reaction

$$2\,SO_2(g) + O_2(g) \rightleftharpoons 2\,SO_3(g) \qquad \Delta_r H^\circ = -197.8 \text{ kJ mol}^{-1}$$

Will the amount of $SO_3(g)$ formed from given amounts of $SO_2(g)$ and $O_2(g)$ be greater at high or low temperatures?

Analyze

We must think of the impact made by changing the temperature. In general, an increase in temperature causes a shift in the direction of the endothermic reaction.

Solve

The sign of $\Delta_r H^\circ$ tells us that the forward reaction is exothermic. Thus, the reverse reaction is endothermic. In this case, increasing the temperature will favor the reverse reaction and lowering the temperature will favor the forward reaction. The conversion of SO_2 to SO_3 is favored at *low* temperatures.

Assess

Be sure not to confuse shifts in equilibrium with changes in reaction rates that result from temperature changes. That is, equilibria of exothermic and endothermic reactions will shift *differently* when temperatures are increased, but the rates of exothermic and endothermic reactions *both increase* with increasing temperature. Changing the temperature is somewhat different than other changes we have discussed in this section. Changing the temperature causes a shift in the equilibrium position and changes the value of the equilibrium constant.

PRACTICE EXAMPLE A: The reaction $N_2O_4(g) \rightleftharpoons 2\,NO_2(g)$ has $\Delta_r H^\circ = +57.2 \text{ kJ mol}^{-1}$. Will the amount of $NO_2(g)$ formed from $N_2O_4(g)$ be greater at high or low temperatures?

PRACTICE EXAMPLE B: The enthalpy of formation of NH_3 is $\Delta_f H^\circ[NH_3(g)] = -46.11$ kJ/mol. Will the concentration of NH_3 in an equilibrium mixture with its elements be greater at 100 or at 300 °C? Explain.

▲ **Sulfuric acid is produced from SO_3**

$$SO_3(g) + H_2O(l) \rightleftharpoons H_2SO_4(aq)$$

The catalyst used to speed up the conversion of SO_2 to SO_3 in the commercial production of sulfuric acid is $V_2O_5(s)$. What appears to be smoke coming from the cooling tower (in the rear) is in fact just water vapor.

Effect of a Catalyst on Equilibrium

A catalyst is a substance that, when added to a reaction mixture, speeds u both the forward and reverse reactions. Equilibrium is achieved more rapidl but the equilibrium amounts are unchanged by the catalyst. Consider agai reaction (15.18)

$$2\,SO_2(g) + O_2(g) \rightleftharpoons 2\,SO_3(g) \qquad K_c = 2.8 \times 10^2 \text{ at } 1000 \text{ K}$$

For a given set of reaction conditions, the equilibrium amounts of SO_2, O_2, an SO_3 have fixed values. This is true whether the reaction is carried out by a slo mogeneous reaction, catalyzed in the gas phase, or conducted as a heteroge-ous reaction on the surface of a catalyst. Stated another way, the presence of atalyst does not change the numerical value of the equilibrium constant.

A catalyst has no effect on the condition of equilibrium in a reversible reaction.

In Chapter 20, we will investigate in more detail the factors affecting reac-n rates, including the way in which a catalyst is able to speed up a reaction d reduce the time it takes for the reaction to reach equilibrium.

15-9 CONCEPT ASSESSMENT

wo students are performing the same experiment in which an endothermic eaction rapidly attains a condition of equilibrium. Student A does the reaction in beaker resting on the surface of the lab bench while student B holds the beaker n which the reaction occurs. Assuming that all other environmental variables are he same, which student should end up with more product? Explain.

5-7 Equilibrium Calculations: Some Illustrative Examples

e are now ready to tackle the problem of describing, in quantitative terms, e condition of equilibrium in a reversible reaction. Part of the approach we e may seem unfamiliar at first—it has an algebraic look to it. But as you just to this "new look," do not lose sight of the fact that we continue to use me familiar and important ideas—molar masses, molarities, and stoichio-etric factors from the balanced equation, for example.

The five numerical examples that follow apply the general equilibrium inciples described earlier in the chapter. The first four involve gases, while e fifth deals with equilibrium in an aqueous solution. (The study of equilib-in aqueous solutions is the principal topic of the next three chapters.) Each ample includes an assessment that summarizes the essential features of uilibrium calculations exemplified by that type of problem. You may find it lpful to return to these assessments from time to time as you encounter new uilibrium situations in later chapters.

Example 15-9 is relatively straightforward. It demonstrates how to deter-ine the equilibrium constant of a reaction when the equilibrium concentra-ons of the reactants and products are known.

Example 15-10 is somewhat more involved than Example 15-9. We are still terested in determining the equilibrium constant for a reaction, but we do not ve the same sort of information as in Example 15-9. We are given the initial ncentrations of all the reactants and products, but the equilibrium concentra-on of only one substance. This case requires a little algebra and some careful okkeeping. We will introduce a tabular system, sometimes called an **ICE table**, r keeping track of changing concentrations of reactants and products. The table ntains the *initial, change in*, and *equilibrium concentration* of each species. It is a lpful device that we will use throughout the next three chapters.

EXAMPLE 15-9 Determining a Value of K_c from the Equilibrium Quantities of Substances

Dinitrogen tetroxide, $N_2O_4(l)$, is an important component of rocket fuels—for example, as an oxidizer of liquid hydrazine in the Titan rocket. At 25 °C, N_2O_4 is a colorless gas that partially dissociates into NO_2, a red-brown gas. The color of an equilibrium mixture of these two gases depends on their relative proportions, which in turn depends on the temperature (Fig. 15-7).

(continued)

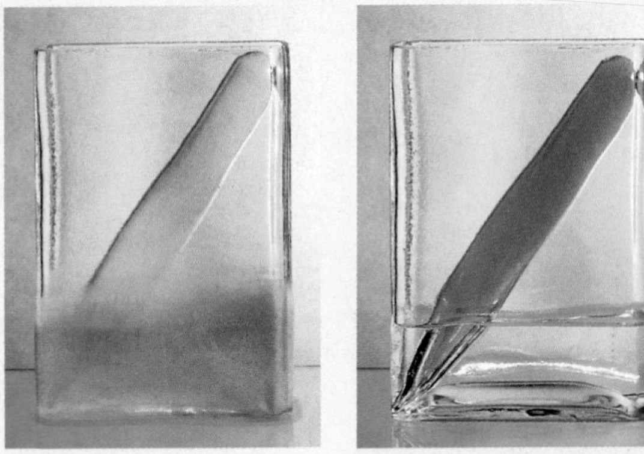

(a) (b)

▲ FIGURE 15-7
The equilibrium $N_2O_4(g) \rightleftharpoons 2\ NO_2(g)$
(a) At dry ice temperatures, N_2O_4 exists as a solid. The gas in equilibrium with the solid is mostly colorless N_2O_4, with only a trace of brown NO_2.
(b) When warmed to room temperature and above, the N_2O_4 melts and vaporizes. The proportion of $NO_2(g)$ at equilibrium increases over that at low temperatures, and the equilibrium mixture of $N_2O_4(g)$ and $NO_2(g)$ has a red-brown color.

Richard Megna/Fundamental Photographs, NYC

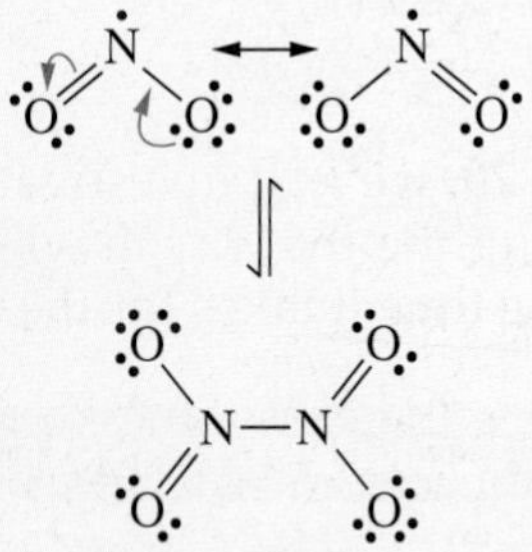

▲ **The Lewis structures of N_2O_4 and $NO_2(g)$**
Nitrogen dioxide is a free radical that combines exothermically to form dinitrogen tetroxide.

Equilibrium is established in the reaction $N_2O_4(g) \rightleftharpoons 2\ NO_2(g)$ at 25 °C. The quantities of the two gases present in a 3.00 L vessel are 7.64 g N_2O_4 and 1.56 g NO_2. What is the value of K_c for this reaction?

Analyze

We are given the equilibrium amounts (in terms of mass) of the reactants and products, along with the volume of the reaction vessel. We use these values to determine the equilibrium concentrations and plug them into the equilibrium constant expression.

Solve

Convert the mass of N_2O_4 to moles.

$$\text{mol } N_2O_4 = 7.64\text{ g } N_2O_4 \times \frac{1\text{ mol } N_2O_4}{92.01\text{ g } N_2O_4} = 8.303 \times 10^{-2}\text{ mol}$$

Convert moles of N_2O_4 to mol/L.

$$[N_2O_4] = \frac{8.303 \times 10^{2}\text{ mol}}{3.00\text{ L}} = 0.0277\text{ M}$$

Convert the mass of NO_2 to moles.

$$\text{mol } NO_2 = 1.56\text{ g } NO_2 \times \frac{1\text{ mol } NO_2}{46.01\text{ g } NO_2} = 3.391 \times 10^{-2}\text{ mol}$$

Convert moles of NO_2 to mol/L.

$$[NO_2] = \frac{3.391 \times 10^{-2}}{3.00\text{ L}} = 0.0113\text{ M}$$

Write the equilibrium constant expression, substitute the equilibrium concentrations, and solve for K_c.

$$K_c = \frac{[NO_2]^2}{[N_2O_4]} = \frac{(0.0113)^2}{(0.0277)} = 4.61 \times 10^{-3}$$

Assess

The quantities required in an equilibrium constant expression, K_c, are equilibrium *concentrations in moles per liter*, not simply equilibrium amounts in moles or masses in grams. It is helpful to organize all the equilibrium data and carefully label each item.

PRACTICE EXAMPLE A: Equilibrium is established in a 3.00 L flask at 1405 K for the reaction $2\ H_2S(g) \rightleftharpoons 2\ H_2(g) + S_2(g)$. At equilibrium, there is 0.11 mol $S_2(g)$, 0.22 mol $H_2(g)$, and 2.78 mol $H_2S(g)$. What is the value of K_c for this reaction?

PRACTICE EXAMPLE B: Equilibrium is established at 25 °C in the reaction $N_2O_4(g) \rightleftharpoons 2\ NO_2(g)$, $K_c = 4.61 \times 10^{-3}$. If $[NO_2] = 0.0236$ M in a 2.26 L flask, how many grams of N_2O_4 are also present?

EXAMPLE 15-10 **Determining a Value of K_p from Initial and Equilibrium Amounts of Substances: Relating K_c and K_p**

The equilibrium condition for $SO_2(g)$, $O_2(g)$, and $SO_3(g)$ is important in sulfuric acid production. When a 0.0200 mol sample of SO_3 is introduced into an evacuated 1.52 L vessel at 900 K, 0.0142 mol SO_3 is present at equilibrium. What is the value of K_p for the dissociation of $SO_3(g)$ at 900 K?

$$2\,SO_3(g) \rightleftharpoons 2\,SO_2(g) + O_2(g) \qquad K_p = ?$$

Analyze

Let's first determine K_c and then convert to K_p by using equation (15.17). In the ICE table below, the key term leading to the other data is the change in amount of SO_3: In progressing from 0.0200 mol SO_3 to 0.0142 mol SO_3, 0.0058 mol SO_3 is dissociated. The *negative sign* (−0.0058 mol) indicates that this amount of SO_3 is consumed in establishing equilibrium. In the row labeled "changes," the changes in amounts of SO_2 and O_2 must be related to the change in amount of SO_3. For this, we use the stoichiometric coefficients from the balanced equation: 2, 2, and 1. That is, *two* moles of SO_2 and *one* mole of O_2 are produced for every *two* moles of SO_3 that dissociate.

Solve

The reaction:	**$2\,SO_3(g)$**	$\rightleftharpoons$	**$2\,SO_2(g)$**	+	**$O_2(g)$**
initial amounts:	0.0200 mol		0.00 mol		0.00 mol
changes:	−0.0058 mol		+0.0058 mol		+0.0029 mol
equil amounts:	0.0142 mol		0.0058 mol		0.0029 mol
equil concns:	$[SO_3] = \dfrac{0.0142\text{ mol}}{1.52\text{ L}}$		$[SO_2] = \dfrac{0.0058\text{ mol}}{1.52\text{ L}}$		$[O_2] = \dfrac{0.0029\text{ mol}}{1.52\text{ L}}$
	$[SO_3] = 9.34 \times 10^{-3}$ M		$[SO_2] = 3.8 \times 10^{-3}$ M		$[O_2] = 1.9 \times 10^{-3}$ M

$$K_c = \frac{[SO_2]^2[O_2]}{[SO_3]^2} = \frac{(3.8 \times 10^{-3})^2(1.9 \times 10^{-3})}{(9.34 \times 10^{-3})^2} = 3.1 \times 10^{-4}$$

$$K_p = K_c(RT)^{\Delta\nu_{gas}} = 3.1 \times 10^{-4}\,(0.0831 \times 900)^{(2+1)-2}$$
$$= 3.1 \times 10^{-4}\,(0.0831 \times 900)^1 = 2.3 \times 10^{-2}$$

Assess

The chemical equation for a reversible reaction serves both to establish the form of the equilibrium constant expression and to provide the conversion factors (stoichiometric factors) to relate the equilibrium quantity of one species to equilibrium quantities of the others.

For equilibria involving gases, we can use either K_c or K_p. In general, if the data given involve amounts of substances and volumes, it is easier to work with K_c. If data are given as partial pressures, then work with K_p. Whether working with K_c or K_p or the relationship between them, we must always base these expressions on the given chemical equation, not on equations we may have used in other situations.

PRACTICE EXAMPLE A: A 5.00 L evacuated flask is filled with 1.86 mol NOBr. At equilibrium at 25 °C, there is 0.082 mol of Br_2 present. Determine K_c and K_p for the reaction $2\,NOBr(g) \rightleftharpoons 2\,NO(g) + Br_2(g)$.

PRACTICE EXAMPLE B: 0.100 mol SO_2 and 0.100 mol O_2 are introduced into an evacuated 1.52 L flask at 900 K. When equilibrium is reached, the amount of SO_3 found is 0.0916 mol. Use these data to determine K_p for the reaction $2\,SO_3(g) \rightleftharpoons 2\,SO_2(g) + O_2(g)$.

The methods used in Examples 15-9 and 15-10 are summarized in Figure 15-8. xample 15-11 demonstrates that we can often determine several pieces of useful formation about an equilibrium system from just the equilibrium constant and e reaction equation.

Example 15-12 brings back the ICE format, but with a twist. This time the nown values include the equilibrium constant and an initial amount of the reactant, but no information is given about the equilibrium amount of the reactant or e product. That means that we do not know how much the initial value will hange. We show this by using an "x" in that part of the table. The setup will be uite algebraic; in fact, we must use the quadratic formula to obtain a solution.

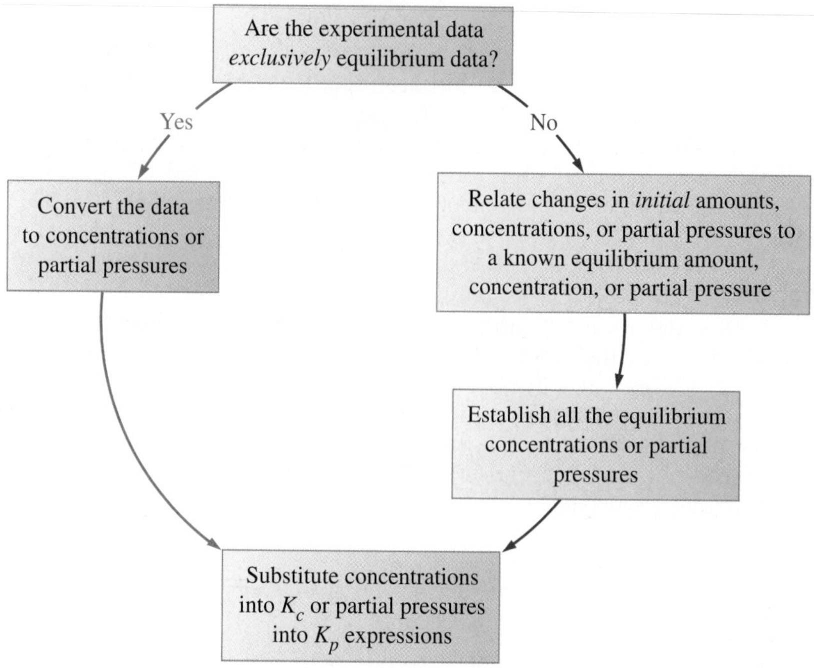

▶ FIGURE 15-8
Determining K_c or K_p from experimental data

EXAMPLE 15-11 Determining Equilibrium Partial and Total Pressures from a Value of K_p

Ammonium hydrogen sulfide, $NH_4HS(s)$, used as a photographic developer, is unstable and dissociates at room temperature.

$$NH_4HS(s) \rightleftharpoons NH_3(g) + H_2S(g) \qquad K_p = 0.108 \text{ at } 25\ °C$$

A sample of $NH_4HS(s)$ is introduced into an evacuated flask at 25 °C. What is the total gas pressure at equilibrium?

Analyze

We begin by writing the equilibrium constant expression in terms of pressure. The key step is to recognize that the pressure of ammonia is equal to the pressure of hydrogen sulfide. This will then allow us to determine the pressure of ammonia and hydrogen sulfide.

Solve

K_p for this reaction is just the product of the equilibrium partial pressures of $NH_3(g)$ and $H_2S(g)$ (NH_4HS is not included because it is a solid.) Because these gases are produced in equimolar amounts, $P_{NH_3} = P_{H_2S}$.

$$K_p = (P_{NH_3})(P_{H_2S}) = 0.108$$
$$K_p = (P_{NH_3})(P_{H_2S}) = (P_{NH_3})(P_{NH_3}) = (P_{NH_3})^2 = 0.108$$

Find P_{NH_3}. (Note that the unit bar appears because in the equilibrium expression the reference pressure $P°$ was implicitly included.)

$$P_{NH_3} = \sqrt{0.108} = 0.329 \text{ bar} \qquad P_{H_2S} = P_{NH_3} = 0.329 \text{ bar}$$

The total pressure is

$$P_{tot} = P_{NH_3} + P_{H_2S} = 0.329 \text{ bar} + 0.329 \text{ bar} = 0.658 \text{ bar}$$

Assess

When using K_p expressions, look for relationships among partial pressures of the reactants. If we need to relate the total pressure to the partial pressures of the reactants, we should be able to do this with some equations presented in Chapter 6 (for example, equations 6.15, 6.16, and 6.17).

PRACTICE EXAMPLE A: Sodium hydrogen carbonate (baking soda) decomposes at elevated temperatures and is one of the sources of $CO_2(g)$ when this compound is used in baking.

$$2\ NaHCO_3(s) \rightleftharpoons Na_2CO_3(s) + H_2O(g) + CO_2(g) \qquad K_p = 0.231 \text{ at } 100\ °C$$

What is the partial pressure of $CO_2(g)$ when this equilibrium is established starting with $NaHCO_3(s)$?

PRACTICE EXAMPLE B: If enough additional $NH_3(g)$ is added to the flask in Example 15-11 to raise its partial pressure to 0.500 bar at equilibrium, what will be the *total* gas pressure when equilibrium is re-established?

EXAMPLE 15-12 Calculating Equilibrium Concentrations from Initial Conditions

A 0.0240 mol sample of $N_2O_4(g)$ is allowed to come to equilibrium with $NO_2(g)$ in a 0.372 L flask at 25 °C. Calculate the amount of N_2O_4 present at equilibrium (Fig. 15-9).

$$N_2O_4(g) \rightleftharpoons 2\,NO_2(g) \qquad K_c = 4.61 \times 10^{-3} \text{ at } 25\ °C$$

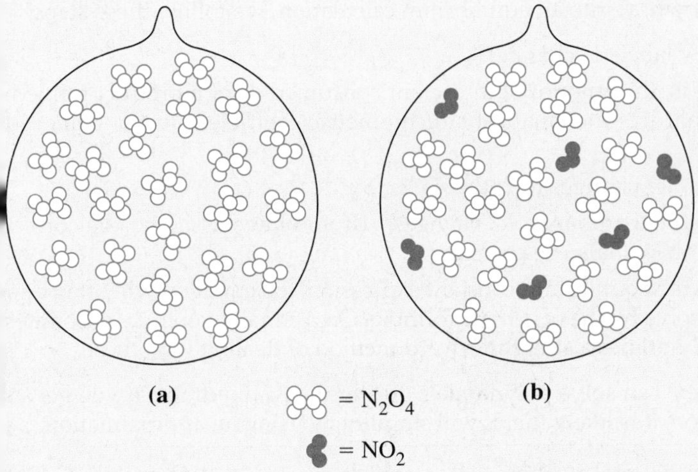

◀ FIGURE 15-9
Equilibrium in the reaction

$$N_2O_4(g) \rightleftharpoons 2\,NO_2(g)$$

at 25 °C—Example 15-12 illustrated
Each "molecule" illustrated represents 0.001 mol. **(a)** Initially, the bulb contains 0.024 mol N_2O_4, represented by 24 "molecules." **(b)** At equilibrium, some "molecules" of N_2O_4 have dissociated to NO_2. The 21 "molecules" of N_2O_4 and 6 of NO_2 correspond to 0.021 mol N_2O_4 and 0.006 mol NO_2 at equilibrium.

Analyze

We need to determine the amount of N_2O_4 that dissociates to establish equilibrium. For the first time, we introduce an algebraic unknown, x. Suppose we let $x =$ the number of moles of N_2O_4 that dissociate. In the following ICE table, we enter the value $-x$ into the row labeled "changes." The amount of NO_2 produced is $+2x$ because the stoichiometric coefficient of NO_2 is 2 and that of N_2O_4 is *1*.

Solve

The reaction:	$N_2O_4(g)$	$\rightleftharpoons$	$2\,NO_2(g)$
initial amounts:	0.0240 mol		0.00 mol
changes:	$-x$ mol		$+2x$ mol
equil amounts:	$(0.0240 - x)$ mol		$2x$ mol
equil concns:	$[N_2O_4] = (0.0240 - x)$ mol/0.372 L		$[NO_2] = 2x$ mol/0.372 L

$$K_c = \frac{[NO_2]^2}{[N_2O_4]} = \frac{\left(\dfrac{2x}{0.372}\right)^2}{\left(\dfrac{0.0240 - x}{0.372}\right)} = \frac{4x^2}{0.372(0.0240 - x)} = 4.61 \times 10^{-3}$$

$$4x^2 = 4.12 \times 10^{-5} - (1.71 \times 10^{-3})x$$

$$x^2 + (4.28 \times 10^{-4})x - 1.03 \times 10^{-5} = 0$$

$$x = \frac{-4.28 \times 10^{-4} \pm \sqrt{(4.28 \times 10^{-4})^2 + 4 \times 1.03 \times 10^{-5}}}{2}$$

$$= \frac{-4.28 \times 10^{-4} \pm \sqrt{(1.83 \times 10^{-7}) + 4.12 \times 10^{-5}}}{2}$$

$$x = \frac{-4.28 \times 10^{-4} \pm \sqrt{4.14 \times 10^{-5}}}{2}$$

$$= \frac{-4.28 \times 10^{-4} \pm 6.43 \times 10^{-3}}{2}$$

(continued)

$$= \frac{-4.28 \times 10^{-4} + 6.43 \times 10^{-3}}{2} = \frac{6.00 \times 10^{-3}}{2}$$

$$= 3.00 \times 10^{-3} \text{ mol } N_2O_4$$

The amount of N_2O_4 at equilibrium is $(0.0240 - x) = (0.0240 - 0.0030) = 0.0210$ mol N_2O_4.

Assess

When we need to introduce an algebraic unknown, x, into an equilibrium calculation, we follow these steps.

- Introduce x into the ICE setup in the row labeled "changes."
- Decide which change to label as x, that is, the amount of a reactant consumed or of a product formed. Usually, we base this on the species that has the smallest stoichiometric coefficient in the balanced chemical equation.
- Use stoichiometric factors to relate the other changes to x (that is, $2x, 3x, \ldots$).
- Consider that equilibrium amounts = initial amounts + "changes." (If you have assigned the correct signs to the changes, equilibrium amounts will also be correct.)
- After substitutions have been made into the equilibrium constant expression, the equation will often be a quadratic equation in x, which you can solve by the quadratic formula. Occasionally you may encounter a higher-degree equation. Appendix A-3 outlines a straightforward method of dealing with these.

Most of us can solve quadratic equations, but few can solve polynomials greater than a quadratic. If you get an equation that is a cubic or higher degree equation, it is likely that it will simplify by using an approximation.

PRACTICE EXAMPLE A: If 0.150 mol $H_2(g)$ and 0.200 mol $I_2(g)$ are introduced into a 15.0 L flask at 445 °C and allowed to come to equilibrium, how many moles of HI(g) will be present?

$$H_2(g) + I_2(g) \rightleftharpoons 2\,HI(g) \qquad K_c = 50.2 \text{ at } 445\ °C$$

PRACTICE EXAMPLE B: Suppose the equilibrium mixture of Example 15-12 is transferred to a 10.0 L flask. **(a)** Will the equilibrium amount of N_2O_4 increase or decrease? Explain. **(b)** Calculate the number of moles of N_2O_4 in the new equilibrium condition.

Our next example is similar to the previous one, but with this slight comp cation: Initially, we don't know whether a net change occurs to the right or the left to establish equilibrium. We can find out, though, by using the reactic quotient, Q_c, and proceeding in the manner suggested in Figure 15-10. Als because the reactants and products are in solution, we can work exclusive with concentrations in formulating the K_c expression.

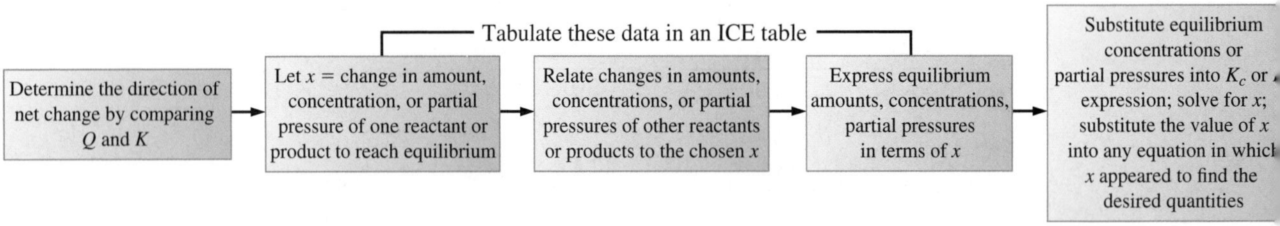

▲ FIGURE 15-10
Determining equilibrium concentrations and partial pressures

EXAMPLE 15-13 Using the Reaction Quotient, Q_c, in an Equilibrium Calculation

Solid silver is added to a solution with these initial concentrations: $[Ag^+] = 0.200$ M, $[Fe^{2+}] = 0.100$ M, and $[Fe^{3+}] = 0.300$ M. The following reversible reaction occurs.

$$Ag^+(aq) + Fe^{2+}(aq) \rightleftharpoons Ag(s) + Fe^{3+}(aq) \qquad K_c = 2.98$$

What are the ion concentrations when equilibrium is established?

Analyze

Because all reactants and products are present initially, we need to use the reaction quotient Q_c to determine the direction in which a net change occurs.

$$Q_c = \frac{[Fe^{3+}]}{[Ag^+][Fe^{2+}]} = \frac{0.300}{(0.200)(0.100)} = 15.0$$

Because Q_c (15.0) is larger than K_c (2.98), a net change must occur in the direction of the reverse reaction, *to the left*. Let's define x as the change in molarity of Fe^{3+}. Because the net change occurs *to the left*, we designate the changes for the species on the left side of the equation as positive and those on the right side as negative.

Solve

The reaction:	$Ag^+(aq)$ +	$Fe^{2+}(aq)$ $\rightleftharpoons$	$Ag(s) + Fe^{3+}(aq)$
initial concns:	0.200 M	0.100 M	0.300 M
changes:	$+x$ M	$+x$ M	$-x$ M
equil concns:	$(0.200 + x)$ M	$(0.100 + x)$ M	$(0.300 - x)$ M

$$K_c = \frac{[Fe^{3+}]}{[Ag^+][Fe^{2+}]} = \frac{(0.300 - x)}{(0.200 + x)(0.100 + x)} = 2.98$$

This equation, which is solved in Appendix A-3, is a quadratic equation for which the acceptable root is $x = 0.11$. To obtain the equilibrium concentrations, we substitute this value of x into the terms shown in the table of data.

$$[Ag^+]_{equil} = 0.200 + 0.11 = 0.31 \text{ M}$$
$$[Fe^{2+}]_{equil} = 0.100 + 0.11 = 0.21 \text{ M}$$
$$[Fe^{3+}]_{equil} = 0.300 - 0.11 = 0.19 \text{ M}$$

Assess

If we have done the calculation correctly, we should obtain a value very close to that given for K_c when we substitute the *calculated* equilibrium concentrations into the reaction quotient, Q_c. We do.

$$Q_c = \frac{[Fe^{3+}]}{[Ag^+][Fe^{2+}]} = \frac{(0.19)}{(0.31)(0.21)} = 2.9 \qquad (K_c = 2.98)$$

PRACTICE EXAMPLE A: Excess Ag(s) is added to 1.20 M $Fe^{3+}(aq)$. Given that

$$Ag^+(aq) + Fe^{2+}(aq) \rightleftharpoons Ag(s) + Fe^{3+}(aq) \qquad K_c = 2.98$$

what are the equilibrium concentrations of the species in solution?

PRACTICE EXAMPLE B: A solution is prepared with $[V^{3+}] = [Cr^{2+}] = 0.0100$ M and $[V^{2+}] = [Cr^{3+}] = 0.150$ M. The following reaction occurs.

$$V^{3+}(aq) + Cr^{2+}(aq) \rightleftharpoons V^{2+}(aq) + Cr^{3+}(aq) \qquad K_c = 7.2 \times 10^2$$

What are the ion concentrations when equilibrium is established? [*Hint*: The algebra can be greatly simplified by extracting the square root of both sides of an equation at the appropriate point.]

Solving Equilibrium Problems When *K* Is Very Small or Very Large

When the equilibrium constant for a reaction is very small ($K \ll 1$) or very large ($K \gg 1$), we can often use approximations to simplify our calculations because:

> At equilibrium, a reaction mixture will contain essentially only products if *K* is very large and only reactants if *K* is very small.

Between these two limiting cases, we should expect that the equilibrium mi ture will contain appreciable amounts of both reactants and products. He are some tips to guide you in the use of approximations to simplify equili rium calculations.

1. **$K \ll 1$:** Starting from an initial mixture containing only reactants, tl reaction will not go very far toward products. Thus, the equilibriu amounts of reactants will be close to their initial values (i.e., the concentr tions or pressures of reactants will typically decrease by very sma amounts). In this situation, we neglect changes in the reactant concentr tions. We use this approach to solve the problem in Example 15-14.
2. **$K \gg 1$:** In this case, we expect the reaction to go almost to completio Thus, if the initial reaction mixture contains only reactants, then the equ librium concentration of the limiting reactant will be very small. For suc cases, we can use the following strategy to simplify our calculations. V use this approach in Example 15-15.

   - Take the reaction to completion, that is, until the limiting reactant completely used up.
   - Take the reaction backward a little (from right to left) toward a true equ librium state.

The first step provides us with a reasonable approximation to the true equili rium state, and the second step adjusts for going a little too far towards product

To implement the approach for $K \gg 1$, we insert an additional row into ou ICE table. The additional row is placed immediately below the row containir the initial amounts. This additional row is used to show the amounts th would be obtained if the reaction went to completion. The next row is the used to show how the concentrations change when the reaction "backs up" little from completion, and the final row, as usual, is for the equilibrium co centrations.

EXAMPLE 15-14 Solving for Equilibrium Concentrations When the Equilibrium Constant Is Very Small

For the reaction $2\,H_2S(g) \rightleftharpoons 2\,H_2(g) + S_2(g)$, the equilibrium constant is $K_c = 4.20 \times 10^{-6}$ at 830 °C. What are the equilibrium concentrations when 0.500 mol H_2S is placed in an empty 1.0 L vessel at 830 °C? What fraction of the H_2S dissociated?

Analyze

The equilibrium constant for this reaction is quite small, so we anticipate that very little of the H_2S will react and the equilibrium concentration of H_2S will be close to the initial concentration: $[H_2S]_{eq} \approx [H_2S]_o$. This approximation may prove helpful for simplifying the equilibrium constant expression.

Solve

The following ICE table summarizes the situation.

	$2\,H_2S(g)$	$\rightleftharpoons$	$2\,H_2(g)$	+	$S_2(g)$
initial concns:	0.500 M		0 M		0 M
changes:	$-2x$ M		$+2x$ M		$+x$ M
equil concns:	$(0.500 - 2x)$ M		$2x$ M		x M

At equilibrium, we must have

$$K_c = \frac{[H_2]^2[S_2]}{[H_2S]^2} = \frac{(2x)^2(x)}{(0.500 - 2x)^2} = \frac{4x^3}{(0.500 - 2x)^2}$$

Solving the expression above is no easy task because it is a cubic equation. However, because $K_c \ll 1$, we expect that the reaction will not proceed very far toward products. Stated another way, we expect that the concentration of H_2S will decrease by only a very small amount. Therefore, it seems reasonable to assume that $2x$ will be very small compared with the 0.500 M. Let's assume that $0.500 - 2x \approx 0.500$. Then,

$$K_c \approx \frac{4x^3}{(0.500)^2} \quad \text{and} \quad x \approx \left[\frac{(0.500)^2 K_c}{4}\right]^{\frac{1}{3}} = \left[\frac{(0.250)(4.2 \times 10^{-6})}{4}\right]^{\frac{1}{3}} = 6.4 \times 10^{-3}$$

Does this result justify the assumption we made that very little of the H_2S reacts? In other words, are we justified in assuming that $2x$ is small compared with 0.500? To decide, we can use this value of x to estimate the fraction, or the percentage, of the H_2S that reacted.

$$\text{percentage of } H_2S \text{ that reacts} = \frac{\text{change in concn}}{\text{initial concn}} \times 100 = \frac{2x}{0.500} \times 100 = \frac{2(0.0064)}{0.500} \times 100 = 2.6$$

Because only a small percentage of the H_2S reacted, our assumption and the value of x obtained are considered acceptable. Therefore,

$[H_2S]_{eq} = (0.500 - 2x)\ M = 0.500\ M - 2(0.0064\ M) = 0.49\ M$
$[H_2]_{eq} = 2x\ M = 2 \times 0.0064\ M = 0.013\ M$
$[S_2]_{eq} = x\ M = 0.0064\ M$

Assess

When we substitute these values into the equilibrium constant expression, we obtain a value that is reasonably close to the actual value of K_c:

$$\frac{[H_2]^2[S_2]}{[H_2S]^2} = \frac{(0.013)^2(0.0064)}{(0.49)^2} = 4.5 \times 10^{-6}$$

This value is a little higher than $K_c = 4.2 \times 10^{-6}$, the difference between the two values arising from the use of the approximation that $2x$ is small compared with 0.500. We could obtain a more accurate result ($x = 6.3 \times 10^{-3}$) by solving the original cubic expression or by using the method of successive approximations, which is described in Appendix A-3.

EXAMPLE 15-15 Solving for Equilibrium Concentrations When the Equilibrium Constant Is Very Large

For the reaction $2\ NO(g) + Cl_2(g) \rightleftharpoons 2\ NOCl(g)$, the equilibrium constant is $K_c = 3.7 \times 10^8$ at 25 °C. What are the equilibrium amounts of all gases, at 25 °C, if 0.100 mol each of NO and Cl_2 are placed in a 1.00 L container?

Analyze

Because the equilibrium constant for this reaction is very large, we expect the reaction to go almost to completion. We will use the method described on page 720.

Solve

In the following equilibrium summary, we imagine the reaction reaches equilibrium by temporarily going to completion, and then reversing by a small extent. Starting from equal amounts of NO and Cl_2, and the assumption that the reaction goes temporarily to completion, NO is the limiting reactant and therefore, all of the NO reacts. The amount of Cl_2 that reacts is 0.100 mol NO × (1 mol Cl_2/2 mol NO) = 0.0500 mol Cl_2 and the amount of NOCl that is formed is 0.100 mol NO × (2 mol NOCl/2 mol NO) = 0.100 mol NOCl. The amount of Cl_2 that remains is (0.100 – 0.0500) mol Cl_2 = 0.0500 mol Cl_2.

	2 NO(g)	+	Cl_2(g)	$\rightleftharpoons$	2 NOCl(g)	
initial concns:	0.100 M		0.100 M		0 M	$Q_c = 0$
to completion:	0 M		0.0500 M		0.100 M	$Q_c = \infty$
changes:	$+2x$ M		$+x$ M		$-2x$ M	
equil concns:	$2x$ M		$(0.0500 + x)$ M		$(0.100 - 2x)$ M	$Q_c = K_c$

(continued)

At equilibrium, we must have

$$K_c = \frac{[NOCl]^2}{[NO]^2[Cl_2]} = \frac{(0.100 - 2x)^2}{(2x)^2(0.0500 + x)} = \frac{(0.100 - 2x)^2}{4x^2(0.500 + x)}$$

We can simplify the expression above considerably by assuming that, because the reaction needs to back up from completion by only a small extent to reach equilibrium, x will be small compared with 0.0500, and $2x$ will be small compared with 0.100. Thus, we may write

$$K_c \approx \frac{(0.100)^2}{4x^2(0.500)} \quad \text{and} \quad x \approx \sqrt{\frac{(0.100)^2}{4(0.0500)K_c}} = 1.16 \times 10^{-5}$$

Before using this value of x to calculate the equilibrium concentrations, we must first decide if the assumption we made is valid: Is x small compared with 0.0500? As the calculation below shows, x is only 0.023% of 0.0500. Clearly, we were justified in assuming that x is small.

$$\frac{x}{0.0500} \times 100\% = \frac{1.16 \times 10^{-5}}{0.0500} \times 100\% = 0.023\%$$

The equilibrium concentrations are

$$[NOCl]_{eq} = (0.100 - 2x)\ \text{M} = 0.100\ \text{M} - 2(1.16 \times 10^{-5})\ \text{M} = 0.100\ \text{M}$$

$$[NO]_{eq} = 2x\ \text{M} = 2(1.16 \times 10^{-5})\ \text{M} = 2.32 \times 10^{-5}\ \text{M}$$

$$[Cl_2]_{eq} = (0.0500 + x)\ \text{M} = (0.0500 + 1.16 \times 10^{-5})\ \text{M} = 0.0500\ \text{M}$$

Assess

Notice that, when the equilibrium constant for a reaction is very large, the equilibrium concentration of the limiting reactant is very small. In this case, the equilibrium concentration of NO is only 0.023% of the initial amount. In other words, 99.98% of the NO reacted. By following the method outlined on page 720, this equilibrium state was achieved by imagining the reaction goes temporarily to 100% completion and then backs up by a small extent (0.023%).

MasteringCHEMISTRY **www.masteringchemistry.com**

Reversible reactions play an important role in the conversion of elemental nitrogen, $N_2(g)$, into nitrogen compounds, both in Nature and in the chemical industry. For a discussion of both natural and industrial processes that convert elemental nitrogen to nitrogen compounds, go to the Focus On feature for Chapter 15, The Nitrogen Cycle and the Synthesis of Nitrogen Compounds, on the MasteringChemistry site.

Summary

15-1 The Nature of the Equilibrium State—When a reaction is carried out in a closed reaction vessel at constant temperature T, the system proceeds spontaneously toward **equilibrium**. At equilibrium, the reaction quotient Q attains the same constant value, K. Mathematically, the equilibrium condition is expressed as $Q = K$. An important property of the equilibrium state is that it is dynamic: The forward and reverse reactions continue to occur at equal rates after equilibrium is reached.

15-2 The Equilibrium Constant Expression—This condition of dynamic equilibrium is described through an **equilibrium constant expression**. The form of the equilibrium constant expression is established from the balanced chemical equation using activities (equation 15.7). The numerical value obtained from the equilibrium consta expression is referred to as the **equilibrium constant**, The thermodynamic equilibrium constant, K, is unitles The equilibrium constants K_c and K_p have units, althoug in most applications, we do not explicitly include the uni

15-3 Relationships Involving Equilibriu Constants—When the equation for a reversible reactio is written in the reverse order, the equilibrium consta expression and the value of K are both inverted from the original form. When two or more reactions are couple together, the equilibrium constant for the overall reactio is the product of the K values of the individual reaction The equilibrium constant of a reaction can have differe values depending on the reference state used. For K_c, ncentration reference state is used, while for K_p, a pressure reference state is used. The relationship between and K_p is given by equation (15.17).

5-4 The Magnitude of an Equilibrium Constant—The magnitude of the equilibrium constant n be used to determine the outcome of a reaction. For rge values of K the reaction goes to completion, with all actants converted to products. A very small equilibrium nstant, for example, a large negative power of ten, indicates that practically none of the reactants have been converted to products. Finally, equilibrium constants of an termediate value, for example, between 10^{-10} and 10^{10}, dicate that some of the reactants have been converted to oducts.

5-5 Predicting the Direction of Net Chemical Change—A comparison of the reaction quotient with e equilibrium constant makes it possible to predict the rection of net change leading to equilibrium (Fig. 15-4). $Q < K$, the forward reaction is favored, meaning that hen equilibrium is established the amounts of products ill have increased and the amounts of reactants will have decreased. If $Q > K$, the reverse reaction is favored until equilibrium is established. If $Q = K$, neither the forward nor reverse reaction is favored. The initial conditions are in fact equilibrium conditions.

15-6 Altering Equilibrium Conditions: Le Châtelier's Principle—**Le Châtelier's principle** is used to make qualitative predictions of the effects of different variables on an equilibrium condition. This principle describes how an equilibrium condition is modified, or "shifts," in response to the addition or removal of reactants or changes in reaction volume, external pressure, or temperature. Catalysts, by speeding up the forward and reverse reactions equally, have no effect on an equilibrium condition.

15-7 Equilibrium Calculations: Some Illustrative Examples—For quantitative equilibrium calculations, a few basic principles and algebraic techniques are required. A useful method employs a tabular system, called an **ICE table**, for keeping track of the *initial* concentrations of the reactants and products, *changes* in these concentrations, and the *equilibrium* concentrations.

Integrative Example

the manufacture of ammonia, the chief source of hydrogen gas is the following reaction for the reforming of methane high temperatures.

$$CH_4(g) + 2\,H_2O(g) \rightleftharpoons CO_2(g) + 4\,H_2(g) \qquad (15.20)$$

ne following data are also given.

(a) $CO(g) + H_2O(g) \rightleftharpoons CO_2(g) + H_2(g) \qquad \Delta_rH° = -40\ \text{kJ mol}^{-1};\ K_c = 1.4 \text{ at } 1000\ \text{K}$

(b) $CO(g) + 3\,H_2(g) \rightleftharpoons H_2O(g) + CH_4(g) \qquad \Delta_rH° = -230\ \text{kJ mol}^{-1};\ K_c = 190 \text{ at } 1000\ \text{K}$

t 1000 K, 1.00 mol each of CH_4 and H_2O are allowed to come to equilibrium in a 10.0 L vessel. Calculate the umber of moles of H_2 present at equilibrium. Would the yield of H_2 increase if the temperature were raised above 1000 K?

nalyze

rst, we should assemble the data needed to solve this problem. The amounts of substances and a reaction volume are ven, so we should be able to work with a K_c expression. However, because the K_c value for the reaction of interest is not ven, we will have to derive this value by combining the two equations for which data are given. This will yield values both K_c and Δ_rH for the reaction of interest.

To calculate the number of moles of H_2 at equilibrium we can use the ICE method, and to assess the effect of temperare on the equilibrium yield of H_2 we can apply Le Châtelier's principle.

olve

'e combine equations (a) and (b) to obtain the data needed in this problem.

(a)	$CO(g) + H_2O(g) \rightleftharpoons CO_2(g) + H_2(g)$	$\Delta_rH = -40\ \text{kJ mol}^{-1}$	$K_c = 1.4$
(b)	$CH_4(g) + H_2O(g) \rightleftharpoons CO(g) + 3\,H_2(g)$	$\Delta_rH = 230\ \text{kJ mol}^{-1}$	$K_c = 1/190$
verall:	$CH_4(g) + 2\,H_2O(g) \rightleftharpoons CO_2(g) + 4\,H_2(g)$	$\Delta_rH = 190\ \text{kJ mol}^{-1}$	$K_c = 1.4/190 = 7.4 \times 10^{-3}$

ext we set up an ICE table in which x represents the number of moles of CH_4 consumed in reaching equilibrium.

The reaction:	$CH_4(g)$	+	$2\,H_2O(g)$	$\rightleftharpoons$	$CO_2(g)$	+	$4\,H_2(g)$
initial amounts:	1.00 mol		1.00 mol		0.00 mol		0.00 mol
changes:	$-x$ mol		$-2x$ mol		x mol		$4x$ mol
equil amounts:	$(1.00 - x)$ mol		$(1.00 - 2x)$ mol		x mol		$4x$ mol
equil concns, M:	$(1.00 - x)/10.0$		$(1.00 - 2x)/10.0$		$x/10.0$		$4x/10.0$

Now we set up K_c and make substitutions into the expression.

$$K_c = \frac{[CO_2][H_2]^4}{[CH_4][H_2O]^2}$$

$$= \frac{(x/10.0)(4x/10.0)^4}{[(1.00 - x)/10.0][(1.00 - 2x)/10.0]^2}$$

$$= \frac{x(4x)^4}{100(1.00 - x)(1.00 - 2x)^2} = 7.4 \times 10^{-3}$$

The above equation reduces to

$$256x^5 = 0.74[(1.00 - x)(1.00 - 2x)^2]$$

and then to

$$256x^5 - 0.74[(1.00 - x)(1.00 - 2x)^2] = 0 \quad (15.$$

The solution to this equation is $x = 0.23$ mol. The number of moles of H_2 at equilibrium is $4x = 0.92$ mol.

Because the reaction is endothermic ($\Delta_r H$ = 190 kJ mol^{-1}), the forward reaction is favored at higher temperatures. The equilibrium yield of H_2 will increase if the temperature is raised above 1000 K.

Assess

Equation (15.21) looks impossibly difficult to solve, but it is not. It can be solved for x rather simply by the method of su cessive approximations. This is done in Appendix A-3, equation (A.2). An important clue as to the possible range of v ues for x can be found in the ICE table. Note that the equilibrium amount of $H_2O(g)$ is $1.00 - 2x$, meaning that $x < 0.$. or else all of the $H_2O(g)$ would be consumed. This marks a good place to start the approximations.

PRACTICE EXAMPLE A: Glycolysis involves ten biochemical reactions. The first two reactions of the glyc ysis cycle are

$$C_6H_{12}O_6(aq) + ATP(aq) \rightleftharpoons G6P(aq) + ADP(aq) \qquad \Delta_r H° = -19.74 \text{ kJ mol}^{-1}$$

$$G6P(aq) \rightleftharpoons F6P(aq) \qquad \Delta_r H° = 2.84 \text{ kJ mol}^{-1}$$

Calculate the equilibrium concentration of F6P(aq) generated in the glycolysis cycle at normal body temperature, 37 °C starting with $[C_6H_{12}O_6(aq)] = 1.20 \times 10^{-6}$ M; $[ATP(aq)] = 10^{-4}$ M; and $[ADP(aq)] = 10^{-2}$ M. The equilibrium co stant for the first reaction is 4.630×10^3; for the second reaction it is 2.76×10^{-1}. During a fever body temperatu increases. Will [G6P] increase or decrease with an increase with temperature?

PRACTICE EXAMPLE B: A procedure calls for adding 0.100 mol of $Br_2(g)$ at 25 °C to a reaction. The only source bromine in the laboratory is a bottle of liquid bromine. **(a)** Given the following data, what size container (in liters) mu be used to extract enough $Br_2(g)$ for this reaction? The equilibrium constant at 298 K for the reaction $Br_2(g) \rightleftharpoons 2\,Br($ is $K = 3.30 \times 10^{-29}$. The vapor pressure of liquid bromine is 0.289 atm. **(b)** At 1000 K the equilibrium constant for t reaction $Br_2(g) \rightleftharpoons 2\,Br(g)$ is 3.4×10^{-5}. What size vessel (in liters) will be needed if the temperature of the vapor raised to 1000 K?

Exercises

Writing Equilibrium Constant Expressions

1. Based on these descriptions, write a balanced equation and the corresponding K_c expression for each reversible reaction.
(a) Carbonyl fluoride, $COF_2(g)$, decomposes into gaseous carbon dioxide and gaseous carbon tetrafluoride.
(b) Copper metal displaces silver(I) ion from aqueous solution, producing silver metal and an aqueous solution of copper(II) ion.
(c) Peroxodisulfate ion, $S_2O_8{}^{2-}$, oxidizes iron(II) ion to iron(III) ion in aqueous solution and is itself reduced to sulfate ion.

2. Based on these descriptions, write a balanced equ tion and the corresponding K_p expression for ea reversible reaction.
(a) Oxygen gas oxidizes gaseous ammonia gaseous nitrogen and water vapor.
(b) Hydrogen gas reduces gaseous nitrogen dioxi to gaseous ammonia and water vapor.
(c) Nitrogen gas reacts with the solid sodium carbo ate and carbon to produce solid sodium cyanide a carbon monoxide gas.

3. Write equilibrium constant expressions, K_c, for the reactions
(a) $2\,NO(g) + O_2(g) \rightleftharpoons 2\,NO_2(g)$
(b) $Zn(s) + 2\,Ag^+(aq) \rightleftharpoons Zn^{2+}(aq) + 2\,Ag(s)$
(c) $Mg(OH)_2(s) + CO_3^{2-}(aq) \rightleftharpoons MgCO_3(s) + 2\,OH^-(aq)$

4. Write equilibrium constant expressions, K_p, for the reactions
(a) $CS_2(g) + 4\,H_2(g) \rightleftharpoons CH_4(g) + 2\,H_2S(g)$
(b) $Ag_2O(s) \rightleftharpoons 2\,Ag(s) + \frac{1}{2}O_2(g)$
(c) $2\,NaHCO_3(s) \rightleftharpoons Na_2CO_3(s) + CO_2(g) + H_2O(g)$

5. Write an equilibrium constant, K_c, for the formation from its gaseous elements of **(a)** 1 mol HF(g); **(b)** 2 mol $NH_3(g)$; **(c)** 2 mol $N_2O(g)$; **(d)** 1 mol $ClF_3(l)$.

6. Write an equilibrium constant, K_p, for the formation from its gaseous elements of **(a)** 1 mol NOCl(g); **(b)** 2 mol $ClNO_2(g)$; **(c)** 1 mol $N_2H_4(g)$; **(d)** 1 mol $NH_4Cl(s)$.

7. Determine values of K_c from the K_p values given.
(a) $SO_2Cl_2(g) \rightleftharpoons SO_2(g) + Cl_2(g)$
$K_p = 2.9 \times 10^{-2}$ at 303 K
(b) $2\,NO(g) + O_2(g) \rightleftharpoons 2\,NO_2(g)$
$K_p = 1.48 \times 10^4$ at 184 °C
(c) $Sb_2S_3(s) + 3\,H_2(g) \rightleftharpoons 2\,Sb(s) + 3\,H_2S(g)$
$K_p = 0.429$ at 713 K

8. Determine the values of K_p from the K_c values given.
(a) $N_2O_4(g) \rightleftharpoons 2\,NO_2(g)$
$K_c = 4.61 \times 10^{-3}$ at 25 °C
(b) $2\,CH_4(g) \rightleftharpoons C_2H_2(g) + 3\,H_2(g)$
$K_c = 0.154$ at 2000 K
(c) $2\,H_2S(g) + CH_4(g) \rightleftharpoons 4\,H_2(g) + CS_2(g)$
$K_c = 5.27 \times 10^{-8}$ at 973 K

9. The vapor pressure of water at 25 °C is 23.8 mmHg. Write K_p for the vaporization of water, with pressures in atmospheres. What is the value of K_c for the vaporization process?

10. If $K_c = 5.12 \times 10^{-3}$ for the equilibrium established between liquid benzene and its vapor at 25 °C, what is the vapor pressure of C_6H_6 at 25 °C, expressed in millimeters of mercury?

11. Determine K_c for the reaction

$$\frac{1}{2}N_2(g) + \frac{1}{2}O_2(g) + \frac{1}{2}Br_2(g) \rightleftharpoons NOBr(g)$$

from the following information (at 298 K).

$$2\,NO(g) \rightleftharpoons N_2(g) + O_2(g) \quad K_c = 2.1 \times 10^{30}$$

$$NO(g) + \frac{1}{2}Br_2(g) \rightleftharpoons NOBr(g) \quad K_c = 1.4$$

12. Given the equilibrium constant values

$$N_2(g) + \frac{1}{2}O_2(g) \rightleftharpoons N_2O(g) \quad K_c = 2.7 \times 10^{-18}$$

$$N_2O_4(g) \rightleftharpoons 2\,NO_2(g) \quad K_c = 4.6 \times 10^{-3}$$

$$\frac{1}{2}N_2(g) + O_2(g) \rightleftharpoons NO_2(g) \quad K_c = 4.1 \times 10^{-9}$$

Determine a value of K_c for the reaction

$$2\,N_2O(g) + 3\,O_2(g) \rightleftharpoons 2\,N_2O_4(g)$$

13. Use the following data to estimate a value of K_p at 1200 K for the reaction $2\,H_2(g) + O_2(g) \rightleftharpoons 2\,H_2O(g)$

$$C(graphite) + CO_2(g) \rightleftharpoons 2\,CO(g) \quad K_c = 0.64$$

$$CO_2(g) + H_2(g) \rightleftharpoons CO(g) + H_2O(g) \quad K_c = 1.4$$

$$C(graphite) + \frac{1}{2}O_2(g) \rightleftharpoons CO(g) \quad K_c = 1 \times 10^8$$

14. Determine K_c for the reaction $N_2(g) + O_2(g) + Cl_2(g) \rightleftharpoons 2\,NOCl(g)$, given the following data at 298 K.

$$\frac{1}{2}N_2(g) + O_2(g) \rightleftharpoons NO_2(g) \quad K_p = 1.0 \times 10^{-9}$$

$$NOCl(g) + \frac{1}{2}O_2(g) \rightleftharpoons NO_2Cl(g) \quad K_p = 1.1 \times 10^2$$

$$NO_2(g) + \frac{1}{2}Cl_2(g) \rightleftharpoons NO_2Cl(g) \quad K_p = 0.3$$

15. An important environmental and physiological reaction is the formation of carbonic acid, $H_2CO_3(aq)$, from carbon dioxide and water. Write the equilibrium constant expression for this reaction in terms of activities. Convert that expression into an equilibrium constant expression containing concentrations and pressures.

16. Rust, $Fe_2O_3(s)$, is caused by the oxidation of iron by oxygen. Write the equilibrium constant expression first in terms of activities, and then in terms of concentration and pressure.

Experimental Determination of Equilibrium Constants

17. 1.00×10^{-3} mol PCl_5 is introduced into a 250.0 mL flask, and equilibrium is established at 284 °C: $PCl_5(g) \rightleftharpoons PCl_3(g) + Cl_2(g)$. The quantity of $Cl_2(g)$ present at equilibrium is found to be 9.65×10^{-4} mol. What is the value of K_c for the dissociation reaction at 284 °C?

18. A mixture of 1.00 g H_2 and 1.06 g H_2S in a 0.500 L flask comes to equilibrium at 1670 K: $2\,H_2(g) + S_2(g) \rightleftharpoons 2\,H_2S(g)$. The equilibrium amount of $S_2(g)$ found is 8.00×10^{-6} mol. Determine the value of K_p at 1670 K for pressures expressed in atmospheres.

19. The two common chlorides of phosphorus, PCl_3 and PCl_5, both important in the production of other phosphorus compounds, coexist in equilibrium through the reaction

$$PCl_3(g) + Cl_2(g) \rightleftharpoons PCl_5(g)$$

At 250 °C, an equilibrium mixture in a 2.50 L flask contains 0.105 g PCl_5, 0.220 g PCl_3, and 2.12 g Cl_2. What are the values of **(a)** K_c and **(b)** K_p for this reaction at 250 °C? For (b), use partial pressures in atmospheres.

20. A 0.682 g sample of ICl(g) is placed in a 625 mL reaction vessel at 682 K. When equilibrium is reached between the ICl(g) and $I_2(g)$ and $Cl_2(g)$ formed by its dissociation, 0.0383 g I_2 is present. What is K_c for this reaction?

21. Write the equilibrium constant expression for the following reaction,

$$Fe(OH)_3 + 3H^+(aq) \rightleftharpoons Fe^{3+}(aq) + 3H_2O(l)$$

$$K = 9.1 \times 10^3$$

and compute the equilibrium concentration for $[Fe^{3+}]$ at pH = 7 (i.e., $[H^+] = 1.0 \times 10^{-7}$).

22. Write the equilibrium constant expression for the d solution of ammonia in water:

$$NH_3(g) \rightleftharpoons NH_3(aq) \qquad K = 57.5$$

Use this equilibrium constant expression to estim the partial pressure of $NH_3(g)$ over a solution conta ing 5×10^{-9} M $NH_3(aq)$. These are conditions simi to that found for acid rains with a high ammoni ion concentration.

Equilibrium Relationships

23. Equilibrium is established at 1000 K, where $K_c = 281$ for the reaction $2\,SO_2(g) + O_2(g) \rightleftharpoons 2\,SO_3(g)$. The equilibrium amount of $O_2(g)$ in a 0.185 L flask is 0.00247 mol. What is the ratio of $[SO_2]$ to $[SO_3]$ in this equilibrium mixture?

24. For the dissociation of $I_2(g)$ at about 1200 °C, $I_2(g) \rightleftharpoons 2\,I(g)$, $K_c = 1.1 \times 10^{-2}$. What volume flask should we use if we want 0.37 mol I to be present for every 1.00 mol I_2 at equilibrium?

25. In the Ostwald process for oxidizing ammonia, a variety of products is possible—N_2, N_2O, NO, and NO_2—depending on the conditions. One possibility is

$$NH_3(g) + \frac{5}{4}\,O_2(g) \rightleftharpoons NO(g) + \frac{3}{2}\,H_2O(g)$$

$$K_p = 2.11 \times 10^{19} \text{ at } 700 \text{ K}$$

For the decomposition of NO_2 at 700 K,

$$NO_2(g) \rightleftharpoons NO(g) + \frac{1}{2}\,O_2(g) \qquad K_p = 0.524$$

(a) Write a chemical equation for the oxidation of $NH_3(g)$ to $NO_2(g)$.
(b) Determine K_p for the chemical equation you have written.

26. At 2000 K, $K_c = 0.154$ for the reaction $2\,CH_4(g) \rightleftharpoons C_2H_2(g) + 3\,H_2(g)$. If a 1.00 L equilibrium mixture 2000 K contains 0.10 mol each of $CH_4(g)$ and $H_2(g)$,
(a) What is the mole fraction of $C_2H_2(g)$ present?
(b) Is the conversion of $CH_4(g)$ to $C_2H_2(g)$ favored high or low pressures?
(c) If the equilibrium mixture at 2000 K is transferr from a 1.00 L flask to a 2.00 L flask, will the numb of moles of $C_2H_2(g)$ increase, decrease, or rema unchanged?

27. An equilibrium mixture at 1000 K contai 0.276 mol H_2, 0.276 mol CO_2, 0.224 mol CO, a 0.224 mol H_2O.

$$CO_2(g) + H_2(g) \rightleftharpoons CO(g) + H_2O(g)$$

(a) Show that for this reaction, K_c is independent the reaction volume, V.
(b) Determine the value of K_c and K_p.

28. For the reaction $CO(g) + H_2O(g) \rightleftharpoons CO_2(g)$ $H_2(g)$, $K_p = 23.2$ at 600 K when pressures a expressed in atmospheres. Explain which of the fo lowing situations might be found at equilibriu **(a)** $P_{CO} = P_{H_2O} = P_{CO_2} = P_{H_2}$; **(b)** P_{H_2}/P_{H_2O} P_{CO_2}/P_{CO}; **(c)** $(P_{CO_2})(P_{H_2}) = (P_{CO})(P_{H_2}$ **(d)** $P_{CO_2}/P_{H_2O} = P_{H_2}/P_{CO}$.

Direction and Extent of Chemical Change

29. Can a mixture of 2.2 mol O_2, 3.6 mol SO_2, and 1.8 mol SO_3 be maintained indefinitely in a 7.2 L flask at a temperature at which $K_c = 100$ in this reaction? Explain.

$$2\,SO_2(g) + O_2(g) \rightleftharpoons 2\,SO_3(g)$$

30. Is a mixture of 0.0205 mol $NO_2(g)$ and 0.750 mol $N_2O_4(g)$ in a 5.25 L flask at 25 °C, at equilibrium? If not, in which direction will the reaction proceed—toward products or reactants?

$$N_2O_4(g) \rightleftharpoons 2\,NO_2(g) \qquad K_c = 4.61 \times 10^{-3} \text{ at } 25\text{ °C}$$

31. In the reaction $2\,SO_2(g) + O_2(g) \rightleftharpoons 2\,SO_3(g)$, 0.455 mol SO_2, 0.183 mol O_2, and 0.568 mol SO_3 are introduced simultaneously into a 1.90 L vessel at 1000 K.
(a) If $K_c = 2.8 \times 10^2$, is this mixture at equilibrium?
(b) If not, in which direction will a net change occur?

32. In the reaction $CO(g) + H_2O(g) \rightleftharpoons CO_2(g)$ $H_2(g)$, $K_c = 31.4$ at 588 K. Equal masses of each rea tant and product are brought together in a reacti vessel at 588 K.
(a) Can this mixture be at equilibrium?
(b) If not, in which direction will a net change occu

33. A mixture consisting of 0.150 mol H_2 and 0.150 mol is brought to equilibrium at 445 °C, in a 3.25 L flas What are the equilibrium amounts of H_2, I_2, and H

$$H_2(g) + I_2(g) \rightleftharpoons 2\,HI(g) \qquad K_c = 50.2 \text{ at } 445\text{ °C}$$

34. Starting with 0.280 mol $SbCl_3$ and 0.160 mol Cl_2, ho many moles of $SbCl_5$, $SbCl_3$, and Cl_2 are present wh equilibrium is established at 248 °C in a 2.50 L flask?

$$SbCl_5(g) \rightleftharpoons SbCl_3(g) + Cl_2(g)$$

$$K_c = 2.5 \times 10^{-2} \text{ at } 248$$

5. Starting with 0.3500 mol CO(g) and 0.05500 mol $COCl_2(g)$ in a 3.050 L flask at 668 K, how many moles of $Cl_2(g)$ will be present at equilibrium?

$$CO(g) + Cl_2(g) \rightleftharpoons COCl_2(g)$$

$$K_c = 1.2 \times 10^3 \text{ at } 668 \text{ K}$$

. 1.00 g *each* of CO, H_2O, and H_2 are sealed in a 1.41 L vessel and brought to equilibrium at 600 K. How many grams of CO_2 will be present in the equilibrium mixture?

$$CO(g) + H_2O(g) \rightleftharpoons CO_2(g) + H_2(g) \qquad K_c = 23.2$$

7. Equilibrium is established in a 2.50 L flask at 250 °C for the reaction

$$PCl_5(g) \rightleftharpoons PCl_3(g) + Cl_2(g) \qquad K_c = 3.8 \times 10^{-2}$$

How many moles of PCl_5, PCl_3, and Cl_2 are present at equilibrium, if

(a) 0.550 mol each of PCl_5 and PCl_3 are initially introduced into the flask?

(b) 0.610 mol PCl_5 alone is introduced into the flask?

. For the following reaction, $K_c = 2.00$ at 1000 °C.

$$2\,COF_2(g) \rightleftharpoons CO_2(g) + CF_4(g)$$

If a 5.00 L mixture contains 0.145 mol COF_2, 0.262 mol CO_2, and 0.074 mol CF_4 at a temperature of 1000 °C,

(a) Will the mixture be at equilibrium?

(b) If the gases are not at equilibrium, in what direction will a net change occur?

(c) How many moles of each gas will be present at equilibrium?

. In the following reaction, $K_c = 4.0$.

$$C_2H_5OH + CH_3COOH \rightleftharpoons CH_3COOC_2H_5 + H_2O$$

A reaction is allowed to occur in a mixture of 17.2 g C_2H_5OH, 23.8 g CH_3COOH, 48.6 g $CH_3COOC_2H_5$, and 71.2 g H_2O.

(a) In what direction will a net change occur?

(b) How many grams of each substance will be present at equilibrium?

. The N_2O_4–NO_2 equilibrium mixture in the flask on the left in the figure is allowed to expand into the evacuated flask on the right. What is the composition of the gaseous mixture when equilibrium is re-established in the system consisting of the two flasks?

$$N_2O_4(g) \rightleftharpoons 2\,NO_2(g) \qquad K_c = 4.61 \times 10^{-3} \text{ at } 25\text{ °C}$$

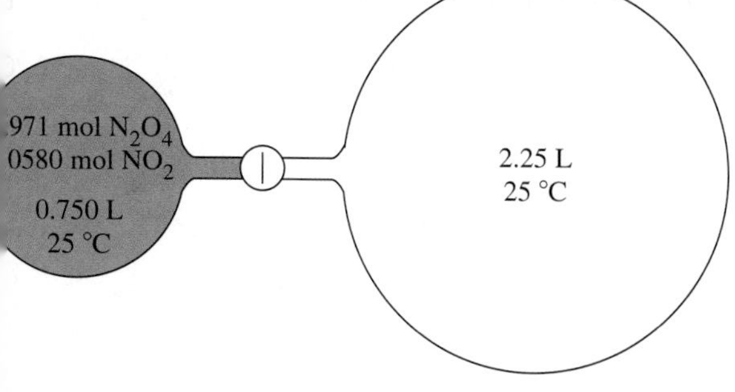

41. Formamide, used in the manufacture of pharmaceuticals, dyes, and agricultural chemicals, decomposes at high temperatures.

$$HCONH_2(g) \rightleftharpoons NH_3(g) + CO(g)$$

$$K_c = 4.84 \text{ at } 400 \text{ K}$$

If 0.186 mol $HCONH_2(g)$ dissociates in a 2.16 L flask at 400 K, what will be the *total* pressure at equilibrium?

42. A mixture of 1.00 mol $NaHCO_3(s)$ and 1.00 mol $Na_2CO_3(s)$ is introduced into a 2.50 L flask in which the partial pressure of CO_2 is 2.10 atm and that of $H_2O(g)$ is 715 mmHg. When equilibrium is established at 100 °C, will the partial pressures of $CO_2(g)$ and $H_2O(g)$ be greater or less than their initial partial pressures? Explain.

$$2\,NaHCO_3(s) \rightleftharpoons Na_2CO_3(s) + CO_2(g) + H_2O(g)$$

$$K_p = 0.23 \text{ at } 100\text{ °C (for pressures in atmospheres)}$$

43. Cadmium metal is added to 0.350 L of an aqueous solution in which $[Cr^{3+}] = 1.00$ M. What are the concentrations of the different ionic species at equilibrium? What is the minimum mass of cadmium metal required to establish this equilibrium?

$$2\,Cr^{3+}(aq) + Cd(s) \rightleftharpoons 2\,Cr^{2+}(aq) + Cd^{2+}(aq)$$

$$K_c = 0.288$$

44. Lead metal is added to 0.100 M $Cr^{3+}(aq)$. What are $[Pb^{2+}]$, $[Cr^{2+}]$, and $[Cr^{3+}]$ when equilibrium is established in the reaction?

$$Pb(s) + 2\,Cr^{3+}(aq) \rightleftharpoons Pb^{2+}(aq) + 2\,Cr^{2+}(aq)$$

$$K_c = 3.2 \times 10^{-10}$$

45. One sketch below represents an initial nonequilibrium mixture in the reversible reaction

$$SO_2(g) + Cl_2(g) \rightleftharpoons SO_2Cl_2(g) \qquad K_c = 4.0$$

Which of the other three sketches best represents an equilibrium mixture? Explain.

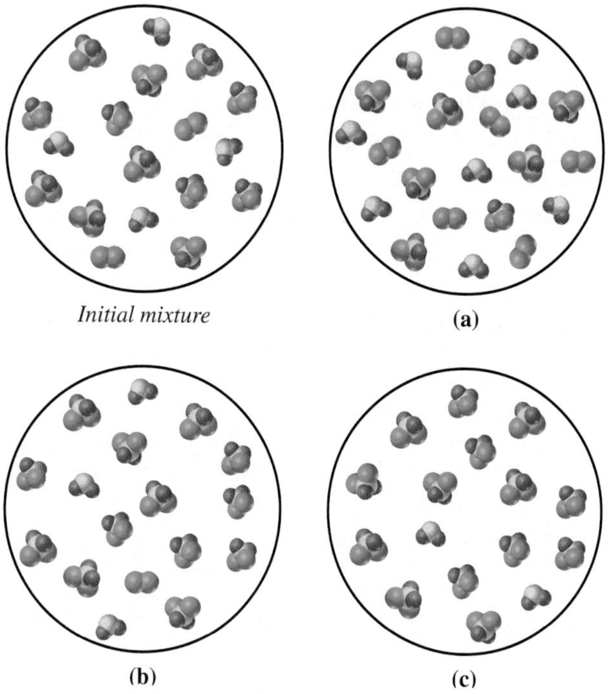

46. One sketch below represents an initial nonequilibrium mixture in the reversible reaction

$$2\,NO(g) + Br_2(g) \rightleftharpoons 2\,NOBr(g) \qquad K_c = 3.0$$

Which of the other three sketches best represents an equilibrium mixture? Explain.

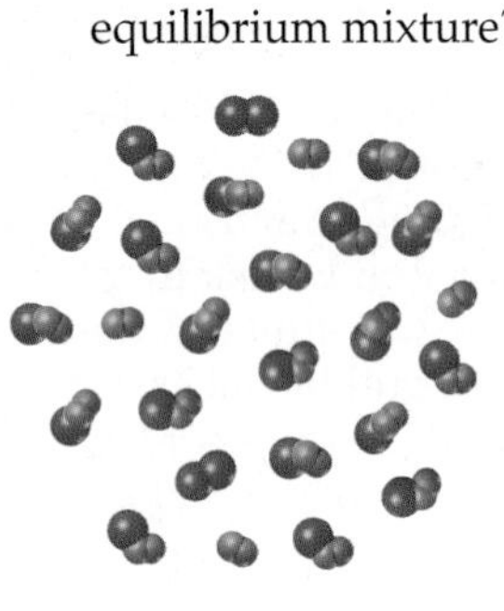

Initial mixture

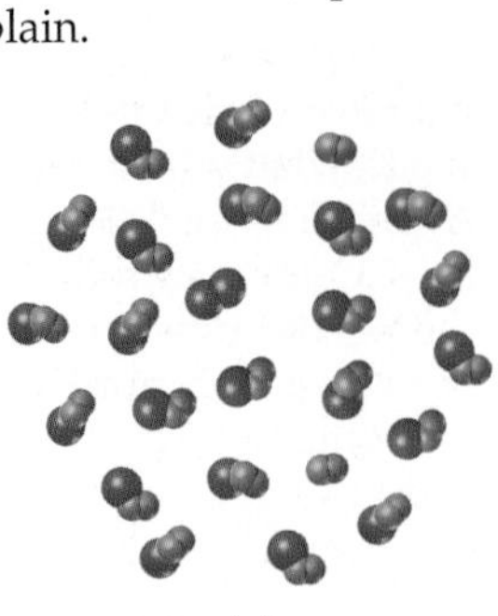

(a)

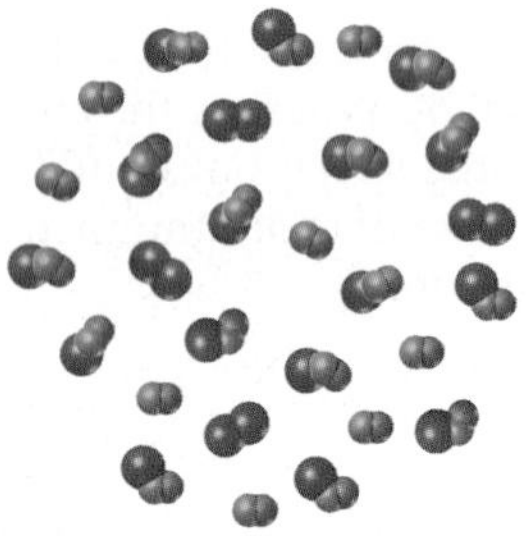

(b)

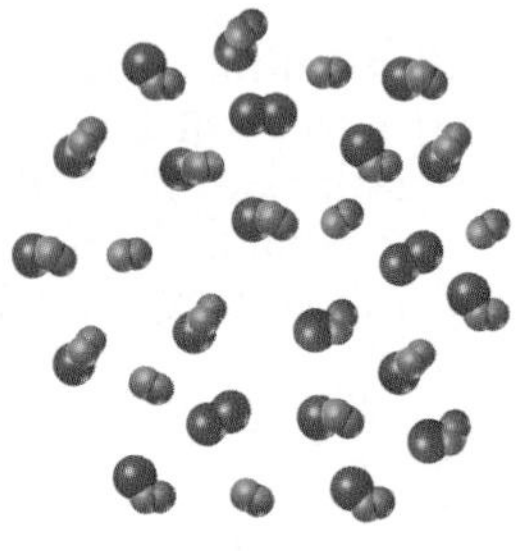

(c)

47. One important reaction in the citric acid cycle is

$$citrate(aq) \rightleftharpoons aconitate(aq) + H_2O(l) \qquad K = 0.0$$

Write the equilibrium constant expression f the above reaction. Given that the concentratio of [citrate(aq)] = 0.00128 M, [aconitate(aq)] = 4.0 10^{-5} M, and $[H_2O]$ = 55.5 M, calculate the reacti quotient. Is this reaction at equilibrium? If not, which direction will it proceed?

48. The following reaction is an important reaction in t citric acid cycle:

$$citrate(aq) + NAD_{ox}(aq) + H_2O(l) \rightleftharpoons$$
$$CO_2(aq) + NAD_{red} + oxoglutarate(aq) \quad K = 0.3$$

Write the equilibrium constant expression for the abo reaction. Given the following data for this reactic [citrate] = 0.00128 M, $[NAD_{ox}]$ = 0.00868 M, $[H_2O]$ 55.5 M, $[CO_2]$ = 0.00868 M, $[NAD_{red}]$ = 0.00132 and [oxoglutarate] = 0.00868 M, calculate the reacti quotient. Is this reaction at equilibrium? If not, in whi direction will it proceed?

Partial Pressure Equilibrium Constant, K_p

49. Refer to Example 15-2. $H_2S(g)$ at 747.6 mmHg pressure and a 1.85 g sample of $I_2(s)$ are introduced into a 725 mL flask at 60 °C. What will be the total pressure in the flask at equilibrium?

$$H_2S(g) + I_2(s) \rightleftharpoons 2\,HI(g) + S(s)$$

$K_p = 1.34 \times 10^{-5}$ at 60 °C (for pressures in atmospheres)

50. A sample of $NH_4HS(s)$ is placed in a 2.58 L flask containing 0.100 mol $NH_3(g)$. What will be the total gas pressure when equilibrium is established at 25 °C?

$$NH_4HS(s) \rightleftharpoons NH_3(g) + H_2S(g)$$

$K_p = 0.108$ at 25 °C (for pressures in atmospheres)

51. The following reaction is used in some self-contained breathing devices as a source of $O_2(g)$.

$$4\,KO_2(s) + 2\,CO_2(g) \rightleftharpoons 2\,K_2CO_3(s) + 3\,O_2(g)$$

$K_p = 28.5$ at 25 °C (for pressures in atmospheres)

Suppose that a sample of $CO_2(g)$ is added to an evacuated flask containing $KO_2(s)$ and equilibrium is established. If the equilibrium partial pressure of $CO_2(g)$ is found to be 0.0721 atm, what are the equilibrium pa tial pressure of $O_2(g)$ and the total gas pressure?

52. Concerning the reaction in Exercise 51, if $KO_2(s)$ a $K_2CO_3(s)$ are maintained in contact with air at 1.00 a and 25 °C, in which direction will a net change occ to establish equilibrium? Explain. [*Hint:* Recall equ tion (6.17). Air is 20.946% O_2 and 0.0379% CO_2 volume.]

53. Exactly 1.00 mol *each* of CO and Cl_2 are introduc into an evacuated 1.75 L flask, and the following eq librium is established at 668 K.

$$CO(g) + Cl_2(g) \longrightarrow COCl_2(g)$$

$K_p = 22.5$ (for pressures in atmospheres)

For this equilibrium, calculate **(a)** the partial pressu of $COCl_2(g)$; **(b)** the total gas pressure.

54. For the reaction $2\,NO_2(g) \rightleftharpoons 2\,NO(g) + O_2($ $K_c = 1.8 \times 10^{-6}$ at 184 °C. What is the value of K_p f this reaction at 184 °C, for pressures expressed atmospheres?

$$NO(g) + \frac{1}{2}O_2(g) \rightleftharpoons NO_2(g)$$

e Châtelier's Principle

5. Continuous removal of one of the products of a chemical reaction has the effect of causing the reaction to go to completion. Explain this fact in terms of Le Châtelier's principle.

5. We can represent the freezing of $H_2O(l)$ at 0 °C as $H_2O(l, d = 1.00\ g/cm^3) \rightleftharpoons H_2O(s, d = 0.92\ g/cm^3)$. Explain why increasing the pressure on ice causes it to melt. Is this the behavior you expect for solids in general? Explain.

7. Explain how each of the following affects the amount of H_2 present in an equilibrium mixture in the reaction

$$3\,Fe(s) + 4\,H_2O(g) \rightleftharpoons Fe_3O_4(s) + 4\,H_2(g) \qquad \Delta_rH° = -150\ kJ\ mol^{-1}$$

(a) Raising the temperature of the mixture; **(b)** introducing more $H_2O(g)$ at constant volume; **(c)** doubling the volume of the container holding the mixture; **(d)** adding an appropriate catalyst.

. In the gas phase, iodine reacts with cyclopentene (C_5H_8) by a free radical mechanism to produce cyclopentadiene (C_5H_6) and hydrogen iodide. Explain how each of the following affects the amount of HI(g) present in the equilibrium mixture in the reaction

$$I_2(g) + C_5H_8(g) \rightleftharpoons C_5H_6(g) + 2\,HI(g) \qquad \Delta_rH° = 92.5\ kJ\ mol^{-1}$$

(a) Raising the temperature of the mixture; **(b)** introducing more $C_5H_6(g)$ at constant volume; **(c)** doubling the volume of the container holding the mixture; **(d)** adding an appropriate catalyst; **(e)** adding an inert gas such as He to a constant-volume reaction mixture.

. The reaction $N_2(g) + O_2(g) \rightleftharpoons 2\,NO(g)$, $\Delta_rH° = +181\ kJ\ mol^{-1}$, occurs in high-temperature combustion processes carried out in air. Oxides of nitrogen produced from the nitrogen and oxygen in air are intimately involved in the production of photochemical smog. What effect does increasing the temperature have on **(a)** the equilibrium production of NO(g); **(b)** the rate of this reaction?

. Use data from Appendix D to determine whether the forward reaction is favored by high temperatures or low temperatures.

(a) $PCl_3(g) + Cl_2(g) \rightleftharpoons PCl_5(g)$

(b) $SO_2(g) + 2\,H_2S(g) \rightleftharpoons 2\,H_2O(g) + 3\,S(s)$

(c) $2\,N_2(g) + 3\,O_2(g) + 4\,HCl(g) \rightleftharpoons 4\,NOCl(g) + 2\,H_2O(g)$

. If the volume of an equilibrium mixture of $N_2(g)$, $H_2(g)$, and $NH_3(g)$ is reduced by doubling the pressure, will P_{N_2} have increased, decreased, or remained the same when equilibrium is re established? Explain.

$$N_2(g) + 3\,H_2(g) \rightleftharpoons 2\,NH_3(g)$$

. For the reaction

$$A(s) \rightleftharpoons B(s) + 2\,C(g) + \frac{1}{2}\,D(g) \qquad \Delta_rH° = 0$$

(a) Will K_p increase, decrease, or remain constant with temperature? Explain.
(b) If a *constant-volume* mixture at equilibrium at 298 K is heated to 400 K and equilibrium re-established, will the number of moles of D(g) increase, decrease, or remain constant? Explain.

63. What effect does increasing the volume of the system have on the equilibrium condition in each of the following reactions?

(a) $C(s) + H_2O(g) \rightleftharpoons CO(g) + H_2(g)$

(b) $Ca(OH)_2(s) + CO_2(g) \rightleftharpoons CaCO_3(s) + H_2O(g)$

(c) $4\,NH_3(g) + 5\,O_2(g) \rightleftharpoons 4\,NO(g) + 6\,H_2O(g)$

64. For which of the following reactions would you expect the extent of the forward reaction to increase with increasing temperatures? Explain.

(a) $NO(g) \rightleftharpoons \frac{1}{2}N_2(g) + \frac{1}{2}O_2(g) \quad \Delta_rH° = -90.2\ kJ\ mol^{-1}$

(b) $SO_3(g) \rightleftharpoons SO_2(g) + \frac{1}{2}O_2(g) \quad \Delta_rH° = +98.9\ kJ\ mol^{-1}$

(c) $N_2H_4(g) \rightleftharpoons N_2(g) + 2\,H_2(g) \quad \Delta_rH° = -95.4\ kJ\ mol^{-1}$

(d) $COCl_2(g) \rightleftharpoons CO(g) + Cl_2(g) \quad \Delta_rH° = +108.3\ kJ\ mol^{-1}$

65. The following reaction represents the binding of oxygen by the protein hemoglobin (Hb):

$$Hb(aq) + O_2(aq) \rightleftharpoons Hb{:}O_2(aq) \qquad \Delta_rH < 0$$

Explain how each of the following affects the amount of $Hb{:}O_2$: **(a)** increasing the temperature; **(b)** decreasing the pressure of O_2; **(c)** increasing the amount of hemoglobin.

66. In the human body, the enzyme carbonic anahydrase catalyzes the interconversion of CO_2 and HCO_3^- by either adding or removing the hydroxide anion. The overall reaction is endothermic. Explain how the following affect the amount of carbon dioxide: **(a)** increasing the amount of bicarbonate anion; **(b)** increasing the pressure of carbon dioxide; **(c)** increasing the amount of carbonic anhydrase; **(d)** decreasing the temperature.

67. A crystal of dinitrogen tetroxide (melting point, −9.3 °C; boiling point, 21.3 °C) is added to an equilibrium mixture of dintrogen tetroxide and nitrogen dioxide that is at 20.0 °C. Will the pressure of nitrogen dioxide increase, decrease, or remain the same? Explain.

68. When hydrogen iodide is heated, the degree of dissociation increases. Is the dissociation reaction exothermic or endothermic? Explain.

69. The standard enthalpy of reaction for the decomposition of calcium carbonate is $\Delta_rH° = 813.5\ kJ\ mol^{-1}$. As temperature increases, does the concentration of calcium carbonate increase, decrease, or remain the same? Explain.

70. Would you expect that the amount of N_2 to increase, decrease, or remain the same in a scuba diver's body as he or she descends below the water surface?

Reactions with Very Small or Very Large *K* Values

71. The equilibrium constant for the following reaction is $K = 6 \times 10^{-16}$ at 25 °C. What is the concentration of $Cu^{2+}(aq)$ when excess CuS(s) reaches equilibrium with a solution in which $[H_3O^+] = 0.30$ M and $[H_2S] = 0.10$ M?

$$CuS(s) + 2\,H_3O^+(aq) \rightleftharpoons Cu^{2+}(aq) + H_2S(aq) + 2\,H_2O(l)$$

72. For the reaction $C_2H_2(g) + 3\,H_2(g) \rightleftharpoons 2\,CH_4(g)$, the equilibrium constant is $K_c = 4.9 \times 10^{-11}$ at 1100 K. In an experiment, 0.100 mol C_2H_2 and 1.00 mol H_2 are added to an otherwise empty 5.00 L container, and the temperature is raised to 1100 K. What is the equilibrium concentration of CH_4?
73. The equilibrium constant for the following acid–base neutralization reaction is $K_c = 1.8 \times 10^9$ at 25 °C. What is the equilibrium concentration of CH_3COC if 0.100 moles each of CH_3COOH and NaOH a added to water to make 1.0 L of solution at 25 ° What fraction of the CH_3COOH reacts?

$$CH_3COOH(aq) + OH^-(aq) \rightleftharpoons CH_3COO^-(aq) + H_2O$$

74. The equilibrium constant for the following reacti has been estimated to be $K = 10^{20}$ at 25 °C. Estim the equilibrium concentration of NH_2^- in a soluti prepared by dissolving 0.0125 mol $NaNH_2$ in water make 1.00 L of solution at 25 °C. Is the result phy cally meaningful?

$$NH_2^-(aq) + H_2O(l) \rightleftharpoons NH_3(aq) + OH^-(l)$$

Integrative and Advanced Exercises

75. Explain why the percent of molecules that dissociate into atoms in reactions of the type $I_2(g) \rightleftharpoons 2\,I(g)$ *always* increases with an increase in temperature.
76. A 1.100 L flask at 25 °C and 1.00 atm pressure contains $CO_2(g)$ in contact with 100.0 mL of a saturated aqueous solution in which $[CO_2(aq)] = 3.29 \times 10^{-2}$ M.
(a) What is the value of K_c at 25 °C for the equilibrium $CO_2(g) \rightleftharpoons CO_2(aq)$?
(b) If 0.01000 mol of radioactive $^{14}CO_2$ is added to the flask, how many moles of the $^{14}CO_2$ will be found in the gas phase and in the aqueous solution when equilibrium is re-established? [*Hint:* The radioactive $^{14}CO_2$ distributes itself between the two phases in exactly the same manner as the nonradioactive $^{12}CO_2$.]
77. Refer to Example 15-13. Suppose that 0.100 L of the equilibrium mixture is diluted to 0.250 L with water. What will be the new concentrations when equilibrium is re-established?
78. In the equilibrium described in Example 15-12, the percent dissociation of N_2O_4 can be expressed as

$$\frac{3.00 \times 10^{-3}\ \text{mol}\ N_2O_4}{0.0240\ \text{mol}\ N_2O_4\ \text{initially}} \times 100\% = 12.5\%$$

What must be the total pressure of the gaseous mixture if $N_2O_4(g)$ is to be 10.0% dissociated at 298 K?

$$N_2O_4 \rightleftharpoons 2\,NO_2(g)$$

$$K_p = 0.113 \text{ at } 298\text{ K}$$

(for pressures in atmospheres)

79. Starting with $SO_3(g)$ at 1.00 atm, what will be the total pressure when equilibrium is reached in the following reaction at 700 K?

$$2\,SO_3(g) \rightleftharpoons 2\,SO_2(g) + O_2(g)$$

$$K_p = 1.6 \times 10^{-5}$$

(for pressures in atmospheres)

80. A sample of air with a mole ratio of N_2 to O_2 of 79:21 is heated to 2500 K. When equilibrium is established in a closed container with air initially at 1.00 atm, t mole percent of NO is found to be 1.8%. Calculate for the reaction below, assuming pressures a expressed in atmospheres.

$$N_2(g) + O_2(g) \rightleftharpoons 2\,NO(g)$$

81. Derive, by calculation, the equilibrium amounts SO_2, O_2, and SO_3 listed in **(a)** Figure 15-5(**(b)** Figure 15-6(b).
82. The decomposition of salicylic acid to phenol and c bon dioxide was carried out at 200.0 °C, a temperatu at which the reactant and products are all gaseous. 0.300 g sample of salicylic acid was introduced int 50.0 mL reaction vessel, and equilibrium was esta lished. The equilibrium mixture was rapidly cooled condense salicylic acid and phenol as solids; t $CO_2(g)$ was collected over mercury and its volu was measured at 20.0 °C and 730 mmHg. In two ide tical experiments, the volumes of $CO_2(g)$ obtain were 48.2 and 48.5 mL, respectively. Calculate K_p this reaction, for pressures in atmospheres.

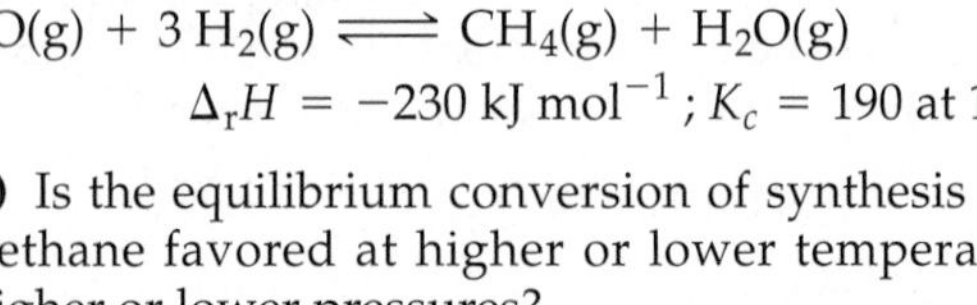

83. One of the key reactions in the gasification of coal the methanation reaction, in which methane is p duced from synthesis gas—a mixture of CO and H

$$CO(g) + 3\,H_2(g) \rightleftharpoons CH_4(g) + H_2O(g)$$

$$\Delta_r H = -230\ \text{kJ mol}^{-1};\ K_c = 190 \text{ at } 100($$

(a) Is the equilibrium conversion of synthesis gas methane favored at higher or lower temperature Higher or lower pressures?
(b) Assume you have 4.00 mol of synthesis gas wit 3:1 mole ratio of $H_2(g)$ to CO(g) in a 15.0 L fla What will be the mole fraction of $CH_4(g)$ at equil rium at 1000 K?

A sample of pure $PCl_5(g)$ is introduced into an evacuated flask and allowed to dissociate.

$$PCl_5(g) \rightleftharpoons PCl_3(g) + Cl_2(g)$$

If the fraction of PCl_5 molecules that dissociate is denoted by α, and if the total gas pressure is P, show that

$$K_p = \frac{\alpha^2 P}{1 - \alpha^2}$$

Nitrogen dioxide obtained as a cylinder gas is always a mixture of $NO_2(g)$ and $N_2O_4(g)$. A 5.00 g sample obtained from such a cylinder is sealed in a 0.500 L flask at 298 K. What is the mole fraction of NO_2 in this mixture?

$$N_2O_4(g) \rightleftharpoons 2\,NO_2(g) \qquad K_c = 4.61 \times 10^{-3}$$

What is the apparent molar mass of the gaseous mixture that results when $COCl_2(g)$ is allowed to dissociate at 395 °C and a total pressure of 3.00 atm?

$$COCl_2(g) \rightleftharpoons CO(g) + Cl_2(g)$$
$$K_p = 4.44 \times 10^{-2} \text{ at } 395\ °C \text{ (for pressures in atm)}$$

Think of the apparent molar mass as the molar mass of a hypothetical single gas that is equivalent to the gaseous mixture.

Show that in terms of mole fractions of gases and *total* gas pressure the equilibrium constant expression for

$$N_2(g) + 3\,H_2(g) \rightleftharpoons 2\,NH_3(g)$$

is

$$K_p = \frac{(x_{NH_3})^2}{(x_{N_2})(x_{H_2})^2} \times \frac{1}{(P_{tot})^2}$$

For the synthesis of ammonia at 500 K, $N_2(g) + 3\,H_2(g) \rightleftharpoons 2\,NH_3(g)$, $K_p = 9.06 \times 10^{-2}$ when the pressures are expressed in atmospheres. Assume that N_2 and H_2 are mixed in the mole ratio 1:3 and that the total pressure is maintained at 1.00 atm. What is the mole percent NH_3 at equilibrium? [*Hint:* Use the equation from Exercise 87.]

A mixture of $H_2S(g)$ and $CH_4(g)$ in the mole ratio 2:1 was brought to equilibrium at 700 °C and a total pressure of 1 atm. On analysis, the equilibrium mixture was found to contain 9.54×10^{-3} mol H_2S. The CS_2 present at equilibrium was converted successively to H_2SO_4 and then to $BaSO_4$; 1.42×10^{-3} mol $BaSO_4$ was obtained. Use these data to determine K_p at 700 °C for the reaction below. Assume pressures are expressed in atmospheres.

$$2\,H_2S(g) + CH_4(g) \rightleftharpoons CS_2(g) + 4\,H_2(g)$$
$$K_p \text{ at } 700\ °C = ?$$

A solution is prepared having these initial concentrations: $[Fe^{3+}] = [Hg_2^{2+}] = 0.5000$ M; $[Fe^{2+}] = [Hg^{2+}] = 0.03000$ M. The following reaction occurs among the ions at 25 °C.

$$2\,Fe^{3+}(aq) + Hg_2^{2+}(aq) \rightleftharpoons 2\,Fe^{2+}(aq) + 2\,Hg^{2+}(aq) \quad K_c = 9.14 \times 10^{-6}$$

What will be the ion concentrations at equilibrium?

91. Refer to the Integrative Example. A gaseous mixture is prepared containing 0.100 mol each of $CH_4(g)$, $H_2O(g)$, $CO_2(g)$, and $H_2(g)$ in a 5.00 L flask. Then the mixture is allowed to come to equilibrium at 1000 K in reaction (15.20). What will be the equilibrium amount, in moles, of each gas?

92. Concerning the reaction in Exercise 26 and the situation described in part (c) of that exercise, will the mole fraction of $C_2H_2(g)$ increase, decrease, or remain unchanged when equilibrium is re-established? Explain.

93. For the reaction $2\,NO(g) + Cl_2(g) \rightleftharpoons 2\,NOCl(g)$, $K_c = 3.7 \times 10^8$ at 298 K. In a 1.50 L flask, there are 4.125 mol of NOCl and 0.1125 mol of Cl_2 present at equilibrium (298 K).
(a) Determine the partial pressure of NO at equilibrium.
(b) What is the total pressure of the system at equilibrium?

94. At 500 K, a 10.0 L equilibrium mixture contains 0.424 mol N_2, 1.272 mol H_2, and 1.152 mol NH_3. The mixture is quickly chilled to a temperature at which the NH_3 liquefies, and the $NH_3(l)$ is completely removed. The 10.0 L gaseous mixture is then returned to 500 K, and equilibrium is re-established. How many moles of $NH_3(g)$ will be present in the new equilibrium mixture?

$$N_2(g) + 3\,H_2(g) \rightleftharpoons 2\,NH_3 \qquad K_c = 152 \text{ at } 500 \text{ K}$$

95. Recall the formation of methanol from synthesis gas, the reversible reaction at the heart of a process with great potential for the future production of automotive fuels (page 699).

$$CO(g) + 2\,H_2(g) \rightleftharpoons CH_3OH(g)$$
$$K_c = 14.5 \text{ at } 483 \text{ K}$$

A particular synthesis gas consisting of 35.0 mole percent CO(g) and 65.0 mole percent $H_2(g)$ at a total pressure of 100.0 atm at 483 K is allowed to come to equilibrium. Determine the partial pressure of $CH_3OH(g)$ in the equilibrium mixture.

96. For the reaction $N_2(g) + 3\,H_2(g) \rightleftharpoons 2\,NH_3(g)$, the equilibrium constant is $K_p = 36.5$ at 400 K. Two separate equilibrium mixtures have the following compositions at 400 K and a total pressure of 1.00 bar.

Equilibrium mixture A:

0.0424 mol N_2 0.136 mol H_2 0.176 mol NH_3

Equilibrium mixture B:

0.194 mol N_2 0.0403 mol H_2 0.0706 mol NH_3

(a) By expressing the partial pressures in the form $P_i = x_i P$, where x_i is the mole fraction of a particular gas and P is the total pressure, show that the reaction quotient for the reaction can be written as

$$Q_p = \frac{x_{NH_3}^2}{x_{N_2} x_{H_2}^3} \times \frac{1}{P^2}$$

[*Hint:* The mole fraction of N_2 is $x_{N_2} = n_{N_2}/n_{tot}$. Similar expressions hold for x_{H_2} and x_{NH_3}.]

(b) Calculate Q_p for each equilibrium mixture to verify that $Q_p = K_p$ in each case.

(c) Suppose that 0.100 mol N_2 is added *at constant pressure* to each mixture. By comparing the values of

Q_p and K_p, verify that the addition of N_2 causes a net reaction to the right for mixture A but a net reaction to the left for mixture B.

97. The activity of a pure solid or liquid is approximately $a = \exp\left[\frac{\overline{V}(P - P°)}{RT}\right]$, where $\overline{V}$ is the molar volume. For a pure solid or liquid under typical conditions, the quantity $\overline{V}(P - P°)/(RT)$ is quite small, primarily because $\overline{V}$ is small, and so $a \approx 1$. Use the following data, for liquid water at 300 K, to verify that $a \approx 1$ over a wide range of pressures. (The data are from *CRC Handbook of Chemistry and Physics*, 83rd ed.)

P, bar	1.0	10.0	100.0	1000.(
d, g mL^{-1}	0.99656	0.99696	1.0010	1.037:

98. For a reaction of the form $A + B \rightleftharpoons C + D$, a starting from an initial reaction mixture contain equal amounts of A and B, show that **(a)** 99.999% the reactants are *consumed* if $K = 10^{10}$ and **(b)** 99.99 of the reactants *remain* if $K = 10^{-10}$.

Feature Problems

99. A classic experiment in equilibrium studies dating from 1862 involved the reaction in solution of ethanol (C_2H_5OH) and acetic acid (CH_3COOH) to produce ethyl acetate and water.

$$C_2H_5OH + CH_3COOH \rightleftharpoons CH_3COOC_2H_5 + H_2O$$

The reaction can be followed by analyzing the equilibrium mixture for its acetic acid content.

$$2\,CH_3COOH(aq) + Ba(OH)_2(aq) \rightleftharpoons Ba(CH_3COO)_2(aq) + 2\,H_2O(l)$$

In one experiment, a mixture of 1.000 mol acetic acid and 0.5000 mol ethanol is brought to equilibrium. A sample containing exactly one-hundredth of the equilibrium mixture requires 28.85 mL 0.1000 M $Ba(OH)_2$ for its titration. Calculate the equilibrium constant, K_c, for the ethanol-acetic acid reaction based on this experiment.

100. The decomposition of HI(g) is represented by the equation

$$2\,HI(g) \rightleftharpoons H_2(g) + I_2(g)$$

HI(g) is introduced into five identical 400 cm^3 glass bulbs, and the five bulbs are maintained at 623 K. Each bulb is opened after a period of time and analyzed for I_2 by titration with 0.0150 M $Na_2S_2O_3$(aq).

$$I_2(aq) + 2\,Na_2S_2O_3(aq) \longrightarrow Na_2S_4O_6(aq) + 2\,NaI(aq)$$

Bulb Number	Initial Mass of HI(g), g	Time Bulb Opened, h	Volume 0.0150 M $Na_2S_2O_3$ Required for Titration, in mL
1	0.300	2	20.96
2	0.320	4	27.90
3	0.315	12	32.31
4	0.406	20	41.50
5	0.280	40	28.68

Data for this experiment are provided in the tal What is the value of K_c at 623 K?

101. In one of Fritz Haber's experiments to establish conditions required for the ammonia synthesis re tion, pure NH_3(g) was passed over an iron catalys 901 °C and 30.0 atm. The gas leaving the reactor v bubbled through 20.00 mL of a HCl(aq) solution this way, the NH_3(g) present was removed by re tion with HCl. The remaining gas occupied a volu of 1.82 L at 0 °C and 1.00 atm. The 20.00 mL of HCl(through which the gas had been bubbled requi 15.42 mL of 0.0523 M KOH for its titration. Anot 20.00 mL sample of the same HCl(aq) through wh no gas had been bubbled required 18.72 mL of 0.052: KOH for its titration. Use these data to obtain a valu K_p at 901 °C for the reaction $N_2(g) + 3\,H_2(g) \rightleftharpoons 2\,NH_3(g)$.

102. The following two equilibrium reactions can be w ten for aqueous carbonic acid, H_2CO_3(aq):

$$H_2CO_3(aq) \rightleftharpoons H^+(aq) + HCO_3^-(aq) \qquad K_1$$
$$HCO_3^-(aq) \rightleftharpoons H^+(aq) + CO_3^{2-}(aq) \qquad K_2$$

For each reaction write the equilibrium constant exp sion. By using Le Châtelier's principle we may naiv predict that by adding H_2CO_3 to the system, the c centration of CO_3^{2-} would increase. What we obse is that after adding H_2CO_3 to the equilibrium mixtu an increase in the concentration of CO_3^{2-} occurs wl $[CO_3^{2-}] \ll K_2$; however, the concentration of CO will decrease when $[CO_3^{2-}] \gg K_2$. Show that thi true by considering the ratio of $[H^+]/[HCO_3^-]$ bef and after adding a small amount of H_2CO_3 to the sc tion, and by using that ratio to calculate the $[CO_3^{2-}]$

103. In organic synthesis many reactions produce very li yield, that is $K \ll 1$. Consider the following hypoth cal reaction: $A(aq) + B(aq) \longrightarrow C(aq)$, $K = 1 \times 1($ We can extract product, C, from the aqueous layer adding an organic layer in which $C(aq) \longrightarrow C($ $K = 15$. Given initial concentrations of $[A] = 0.1$ $[B] = 0.1$, and $[C] = 0.1$, calculate how much C v be found in the organic layer. If the organic layer v not present, how much C would be produced?

Self-Assessment Exercises

104. In your own words, define or explain the following terms or symbols: **(a)** K_p; **(b)** Q_c; **(c)** $\Delta\nu_{gas}$.

105. Briefly describe each of the following ideas or phenomena: **(a)** dynamic equilibrium; **(b)** direction of a net chemical change; **(c)** Le Châtelier's principle; **(d)** effect of a catalyst on equilibrium.

106. Explain the important distinctions between each pair of terms: **(a)** reaction that goes to completion and reversible reaction; **(b)** K_c and K_p; **(c)** reaction quotient (Q) and equilibrium constant expression (K); **(d)** homogeneous and heterogeneous reaction.

107. In the reversible reaction $H_2(g) + I_2(g) \rightleftharpoons 2\,HI(g)$, an initial mixture contains 2 mol H_2 and 1 mol I_2. The amount of HI expected at equilibrium is **(a)** 1 mol; **(b)** 2 mol; **(c)** less than 2 mol; **(d)** more than 2 mol but less than 4 mol.

108. Equilibrium is established in the reaction $2\,SO_2(g) + O_2(g) \rightleftharpoons 2\,SO_3(g)$ at a temperature where $K_c = 100$. If the number of moles of $SO_3(g)$ in the equilibrium mixture is the same as the number of moles of $SO_2(g)$, **(a)** the number of moles of $O_2(g)$ is also equal to the number of moles of $SO_2(g)$; **(b)** the number of moles of $O_2(g)$ is half the number of moles of SO_2; **(c)** $[O_2]$ may have any of several values; **(d)** $[O_2] = 0.010$ M.

109. The volume of the reaction vessel containing an equilibrium mixture in the reaction $SO_2Cl_2(g) \rightleftharpoons SO_2(g) + Cl_2(g)$ is increased. When equilibrium is re-established, **(a)** the amount of Cl_2 will have increased; **(b)** the amount of SO_2 will have decreased; **(c)** the amounts of SO_2 and Cl_2 will have remained the same; **(d)** the amount of SO_2Cl_2 will have increased.

110. For the reaction $2\,NO_2(g) \rightleftharpoons 2\,NO(g) + O_2(g)$, $K_c = 1.8 \times 10^{-6}$ at 184 °C. At 184 °C, the value of K_c for the reaction $NO(g) + \frac{1}{2}O_2(g) \rightleftharpoons NO_2(g)$ is **(a)** 0.9×10^6; **(b)** 7.5×10^2; **(c)** 5.6×10^5; **(d)** 2.8×10^5.

111. For the dissociation reaction $2\,H_2S(g) \rightleftharpoons 2\,H_2(g) + S_2(g)$, $K_p = 1.2 \times 10^{-2}$ at 1065 °C. For this same reaction at 1000 K, **(a)** K_c is less than K_p; **(b)** K_c is greater than K_p; **(c)** $K_c = K_p$; **(d)** whether K_c is less than, equal to, or greater than K_p depends on the total gas pressure.

112. The following data are given at 1000 K: $CO(g) + H_2O(g) \rightleftharpoons CO_2(g) + H_2(g)$; $\Delta_rH° = -42$ kJ mol^{-1}; $K_c = 0.66$. After an initial equilibrium is established in a 1.00 L container, the equilibrium amount of H_2 can be increased by **(a)** adding a catalyst; **(b)** increasing the temperature; **(c)** transferring the mixture to a 10.0 L container; **(d)** in some way other than (a), (b), or (c).

113. Equilibrium is established in the reversible reaction $2\,A + B \rightleftharpoons 2\,C$. The equilibrium concentrations are $[A] = 0.55$ M, $[B] = 0.33$ M, $[C] = 0.43$ M. What is the value of K_c for this reaction?

114. The Deacon process for producing chlorine gas from hydrogen chloride is used in situations where HCl is available as a by-product from other chemical processes.

$$4\,HCl(g) + O_2(g) \rightleftharpoons 2\,H_2O(g) + 2\,Cl_2(g) \qquad \Delta_rH° = -114 \text{ kJ mol}^{-1}$$

A mixture of HCl, O_2, H_2O, and Cl_2 is brought to equilibrium at 400 °C. What is the effect on the equilibrium amount of $Cl_2(g)$ if

(a) additional $O_2(g)$ is added to the mixture at constant volume?

(b) $HCl(g)$ is removed from the mixture at constant volume?

(c) the mixture is transferred to a vessel of twice the volume?

(d) a catalyst is added to the reaction mixture?

(e) the temperature is raised to 500 °C?

115. For the reaction $SO_2(g) \rightleftharpoons SO_2(aq)$, $K = 1.25$ at 25 °C. Will the amount of $SO_2(g)$ be greater than or less than the amount of $SO_2(aq)$?

116. In the reaction $H_2O_2(g) \rightleftharpoons H_2O_2(aq)$, $K = 1.0 \times 10^5$ at 25 °C. Would you expect a greater amount of product or reactant?

117. An equilibrium mixture of SO_2, SO_3, and O_2 gases is maintained in a 2.05 L flask at a temperature at which $K_c = 35.5$ for the reaction

$$2\,SO_2(g) + O_2(g) \rightleftharpoons 2\,SO_3(g)$$

(a) If the numbers of moles of SO_2 and SO_3 in the flask are equal, how many moles of O_2 are present?
(b) If the number of moles of SO_3 in the flask is twice the number of moles of SO_2, how many moles of O_2 are present?

118. Using the method in Appendix E, construct a concept map of Section 15-6, illustrating the shift in equilibrium caused by the various types of disturbances discussed in that section.

16 Acids and Bases

LEARNING OBJECTIVES

16.1 Describe an acid–base reaction in the context of Brønsted–Lowry theory, and identify conjugate acid–base pairs.

16.2 Identify the relationship between the concentration of ions and pH.

16.3 Discuss the relationship between K_a (or pK_a) and the degree of ionization of an acid.

16.4 Describe the behavior of a strong acid or base in solution.

16.5 Describe the behavior of a weak acid or base in solution.

16.6 Describe the behavior of a polyprotic acid in solution.

16.7 Apply the concepts of material and charge balance in situations involving consecutive or simultaneous ionization reactions.

16.8 Predict and explain whether a solution of a salt is acidic, basic, or neutral.

16.9 Predict whether an acid–base neutralization reaction goes to completion or to a limited extent.

16.10 Identify the factors associated with the molecular structure that affect the strength of an acid or a base.

16.11 Identify the structure of the product of a Lewis acid–base reaction.

CONTENTS

Citrus fruit derives its acidic qualities from citric acid, $H_3C_6H_5O_7$, a type of acid (polyprotic) discussed in Section 16-6. Another important constituent of citrus fruit i ascorbic acid, or vitamin C (page 679), a dietary requirement to prevent scurvy.

The concepts of acids and bases are probably among the most fam iar chemistry concepts. The environmental problem of acid rain is popular topic in papers and magazines, and television commercia mention pH in relation to such products as deodorants, shampoos, ar antacids. For chemists, acid–base concepts are arguably among the mc important of all concepts in chemistry. The reason is that many chemic reactions can be characterized as some form of an acid–base reaction.

For many students, achieving mastery of acid–base concepts is a significant allenge, though well worth the effort. As we will see in this chapter and the xt, acid–base concepts can be used in many ways to solve a wide range of oblems.

This chapter and the next integrate concepts introduced in earlier chapters, cluding periodic trends (Chapter 9), molecular structure and shape hapter 10), chemical bonding (Chapters 10 and 11), thermodynamics hapters 7 and 13), and equilibrium (Chapter 15). Concepts from earlier chaprs will be combined with new concepts to help us rationalize observations out acids, bases, and their reactions. By the end of this chapter, we will have scovered the answers to some important questions, such as, Why are some ids or bases strong and others weak? Why do some acid–base reactions show a rong tendency to go to completion whereas others proceed to only a very limd extent? As always, our goal is to rationalize these observations in terms of e structures and properties of the substances involved. Finally, the ideas develed in this chapter will prove useful in subsequent chapters, including apters 26 and 27, which focus on the reactions of organic molecules.

6-1 Acids, Bases, and Conjugate Acid–Base Pairs

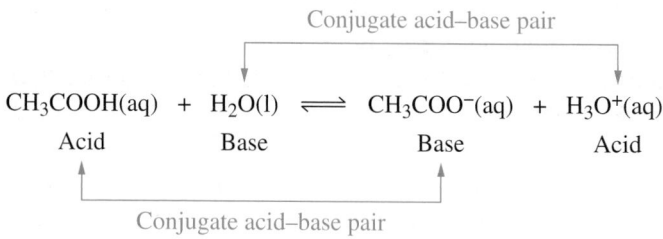

emists have been classifying substances as acids and bases for a long time. ntoine Lavoisier thought that the common element in all acids was oxygen, fact conveyed by its name. (*Oxygen* means "acid former" in Greek.) In 1810, umphry Davy showed that hydrogen instead is the element that acids have common. In 1884, Svante Arrhenius developed a theory of acids and bases part of his studies of electrolytic dissociation (Section 14-9). We discussed me aspects of the Arrhenius theory in Chapter 5. However, the Arrhenius eory focuses solely on the solute—the acid or base—and neglects entirely e key role played by the solvent. Consequently, the Arrhenius theory is not useful as more modern theories.

One of the most useful theories of acids and bases, particularly for scribing the reactions of acids and bases in aqueous solutions, is the **ønsted–Lowry theory**. In 1923, J. N. Brønsted and T. M. Lowry in Great itain independently proposed that an acid is a **proton donor** and a base is a **oton acceptor**. Let's use the Brønsted–Lowry theory to describe the ionizan of CH_3COOH in aqueous solution.

$$\underset{\text{Acid}}{CH_3COOH(aq)} + \underset{\text{Base}}{H_2O(l)} \rightleftharpoons \underset{\text{Base}}{CH_3COO^-(aq)} + \underset{\text{Acid}}{H_3O^+(aq)} \quad \textbf{(16.1)}$$

KEEP IN MIND

that a "proton donor" is a donor of H^+ ions. A hydrogen atom consists of one proton and one electron, and the hydrogen ion, H^+, is simply a proton.

In reaction (16.1), CH_3COOH acts as an *acid*. It gives up a proton, H^+, which taken up by H_2O. Thus, H_2O acts as a *base*. In the reverse reaction, the **dronium ion**, H_3O^+, acts as an acid and CH_3COO^- acts as a base.

When CH_3COOH loses a proton, it is converted into CH_3COO^-. Notice that e formulas of these two species differ by a single proton, H^+. Species that ffer by a single proton (H^+) constitute a **conjugate acid–base pair**. Within s pair, the species with the added H^+ is the acid, and the species without e H^+ is the base. Thus, for reaction (16.1), we can identify two conjugate id–base pairs.

Conjugate acid–base pair (H_2O / H_3O^+)

$$\underset{\text{Acid}}{CH_3COOH(aq)} + \underset{\text{Base}}{H_2O(l)} \rightleftharpoons \underset{\text{Base}}{CH_3COO^-(aq)} + \underset{\text{Acid}}{H_3O^+(aq)}$$

Conjugate acid–base pair (CH_3COOH / CH_3COO^-)

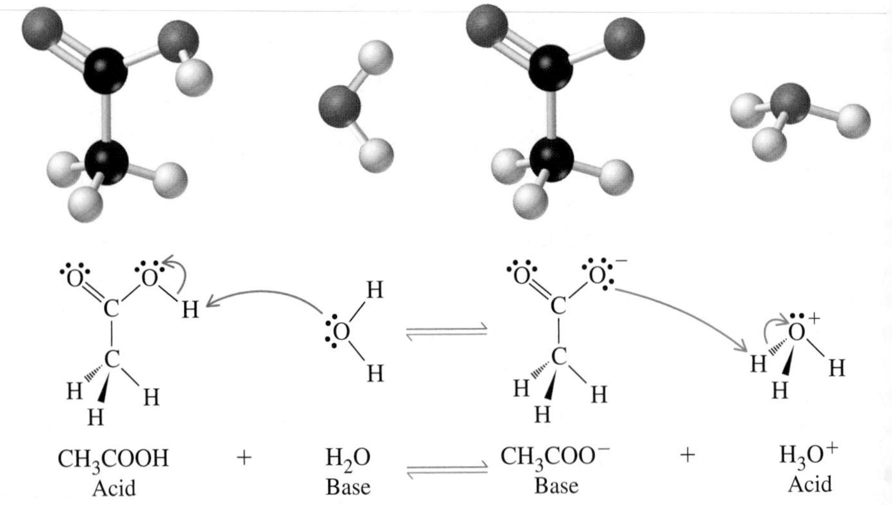

▲ FIGURE 16-1
The ionization of CH_3COOH in water
The curved arrows summarize our visualization of how electrons flow to form and brea
bonds in the ionization of acetic acid. The red arrows represent the forward reaction;
the blue arrows, the reverse reaction.

Figure 16-1 illustrates the proton transfer involved in reaction (16.1) an highlights another important aspect of the Brønsted–Lowry theory.

An acid contains at least one ionizable H atom, and a base contains an atom with a lone pair of electrons onto which a proton can bind.

Notice that, in the CH_3COOH molecule, the H atom in the —COOH grou is the ionizable (acidic) H atom. We will address the question of what makes hydrogen atom ionizable in Section 16-8. For now, we will simply accept t idea that a Brønsted–Lowry acid has at least one ionizable H atom.

Let's develop some more familiarity with these concepts by consideri another example: the ionization of NH_3 in water.

Conjugate acid–base pair

$$\underset{\text{Base}}{NH_3(aq)} + \underset{\text{Acid}}{H_2O(l)} \rightleftharpoons \underset{\text{Acid}}{NH_4^+(aq)} + \underset{\text{Base}}{OH^-(aq)} \qquad (16$$

Conjugate acid–base pair

This reaction is illustrated in Figure 16-2.

▶ Brønsted–Lowry theory is not restricted to the ionization of acids and bases in water. It is valid for any solvent.

As implied by the double arrows in reactions (16.1) and (16.2), the ioniz tion of an acid or a base in water is a reversible reaction that reaches a state dynamic equilibrium. In Section 16-3, we will consider some new ideas th will help us decide whether, for a given acid or base, equilibrium favors rea tants (the un-ionized acid or base) or products (the ionized form of the acid base). For now, it will be helpful to summarize some key aspects of t Brønsted–Lowry theory.

1. **An acid contains at least one ionizable H atom, and a base contains atom with a lone pair of electrons onto which a proton can bind.** For th

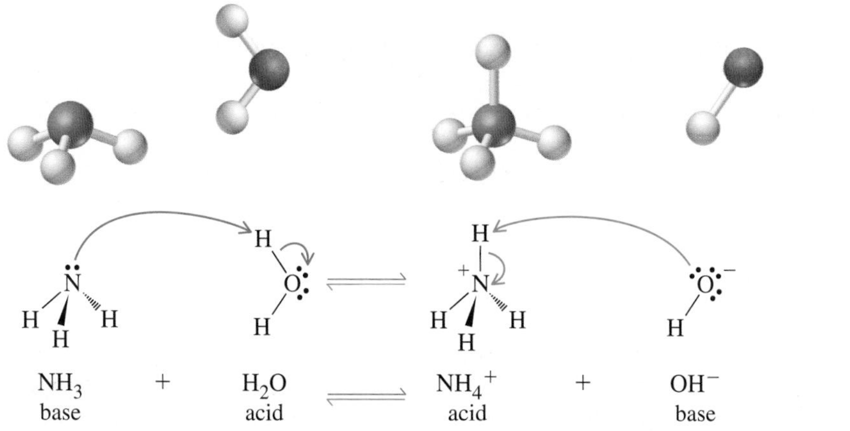

FIGURE 16-2
e ionization of NH_3 in water
e curved arrows summarize our visualization of how electrons flow to form and break nds. The red arrows represent the forward reaction; the blue arrows, the reverse ction.

reason, an acid may be represented in the Brønsted–Lowry theory by the general formula HA, H_2A, H_3A, etc., depending on the number of ionizable H atoms, and a base is represented by :B. There are substances that contain both an ionizable H atom and an atom with a lone pair of electrons. Such substances may behave as either an acid or a base, depending on the situation, and are said to be **amphiprotic**. For example, H_2O is amphiprotic. In reaction (16.1), H_2O acts as a base, whereas in reaction (16.2) it acts as an acid.

For a conjugate acid–base pair, the molecular formulas for the acid and base differ by a single proton (H^+). Therefore, to identify the species in a solution that constitute a conjugate acid–base pair, we need only identify those species that have molecular formulas that differ by one H^+ ion. Once such a pair has been identified, the species with the added H^+ is the acid, and the species without the H^+ is the base. For example, H_2O and OH^- are a conjugate acid–base pair because their formulas differ by one H^+. In this pair, H_2O is the acid and OH^- is the base. Similarly, because the formulas of NH_4^+ and NH_3 differ by one H^+, these two species constitute a conjugate acid–base pair, with NH_4^+ as the acid and NH_3 as the base.

When added to water, acids protonate water molecules to form hydronium (H_3O^+) ions and bases deprotonate water molecules to form hydroxide (OH^-) ions. The ability of the Brønsted–Lowry theory to account for the presence these ions in solution arises from its recognition of the role played by the solvent and makes it a more general and useful theory than the Arrhenius theory.

Before turning our attention to Example 16-1, in which we use the ɔnsted–Lowry theory to identify acids and bases in some typical acid–base ıctions, we will address one additional point: the nature of the hydrated ɔton. Because the H^+ ion is tiny, the positive charge of this ion is concented in a very small region; the ion has a very high positive charge density. ınsequently, the H^+ ion does not exist as a separate entity. It seeks out cen's of negative charge with which to bond (e.g., a lone pair of electrons). hen a H^+ ion attaches to a lone pair of electrons in an O atom in H_2O, a dronium (H_3O^+) ion is formed. The structure of the hydrated hydronium ı is actually not as simple as its formula suggests. This is because the H_3O^+ ı forms hydrogen bonds with several water molecules (Fig. 16-3).

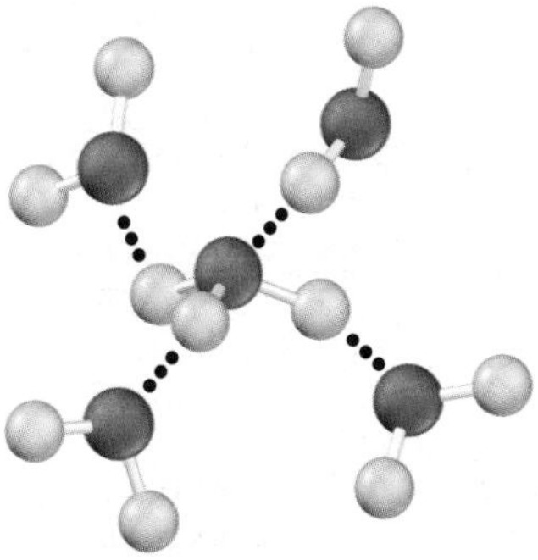

▲ FIGURE 16-3
A hydrated hydronium ion
This species, $H_{11}O_5^+$, consists of a central H_3O^+ ion hydrogen-bonded to four H_2O molecules.

EXAMPLE 16-1 Identifying Brønsted–Lowry Acids and Bases and Their Conjugates

For each of the following reactions, which occur in aqueous solution, identify the acids and bases in both the forward and reverse reactions.

(a) $HClO_2 + H_2O \rightleftharpoons ClO_2^- + H_3O^+$

(b) $OCl^- + H_2O \rightleftharpoons HOCl + OH^-$

(c) $NH_3 + H_2PO_4^- \rightleftharpoons NH_4^+ + HPO_4^{2-}$

(d) $HCl + H_2PO_4^- \rightleftharpoons Cl^- + H_3PO_4$

Analyze

In the Brønsted–Lowry theory, an acid is a proton (H^+) donor and a base is proton acceptor. The deprotonated form of an acid is the corresponding conjugate base. The protonated form of a base is the corresponding conjugate acid. A quick way to identify the members of a conjugate acid–base pair is to identify two species that have molecular formulas that differ by a single H^+ ion.

Solve

(a) In the reaction $HClO_2 + H_2O \rightleftharpoons ClO_2^- + H_3O^+$, we see that, in the forward reaction, a H^+ ion is transferred from $HClO_2$ to H_2O. Thus, $HClO_2$ acts as an acid and H_2O acts as a base. In the reverse reaction, a H^+ ion is transferred from H_3O^+ to ClO_2^-, so H_3O^+ acts as an acid and ClO_2^- acts as a base.

$$\underset{\text{Acid}}{HClO_2} + \underset{\text{Base}}{H_2O} \rightleftharpoons \underset{\text{Base}}{ClO_2^-} + \underset{\text{Acid}}{H_3O^+}$$

When the acid $HClO_2$ gives up a H^+ ion, it becomes the ClO_2^- ion. Their formulas differ by a single H^+, and, therefore, $HClO_2$ and ClO_2^- are a conjugate acid–base pair. When the base H_2O takes a proton, it becomes H_3O^+. Thus, H_3O^+ and H_2O constitute a conjugate acid–base pair. These associations are summarized below.

Conjugate acid–base pair (H_2O / H_3O^+)

$$\underset{\text{Acid}}{HClO_2} + \underset{\text{Base}}{H_2O} \rightleftharpoons \underset{\text{Base}}{ClO_2^-} + \underset{\text{Acid}}{H_3O^+}$$

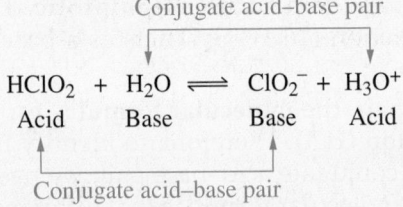

Conjugate acid–base pair ($HClO_2$ / ClO_2^-)

We use the same approach in parts (b)–(d) to identify acids and bases. However, in labeling the acids and bases below, we use the numbers 1 and 2 to identify the acid and base in the two conjugate acid–base pairs. That is, Acid(1) and Base(1) constitute one conjugate acid–base pair and Acid(2) and Base(2) the other.

(b) $$\underset{\text{Base(1)}}{OCl^-} + \underset{\text{Acid(2)}}{H_2O} \rightleftharpoons \underset{\text{Acid(1)}}{HOCl} + \underset{\text{Base(2)}}{OH^-}$$

(c) $$\underset{\text{Base(1)}}{NH_3} + \underset{\text{Acid(2)}}{H_2PO_4^-} \rightleftharpoons \underset{\text{Acid(1)}}{NH_4^+} + \underset{\text{Base(2)}}{HPO_4^{2-}}$$

(d) $$\underset{\text{Acid(1)}}{HCl} + \underset{\text{Base(2)}}{H_2PO_4^-} \rightleftharpoons \underset{\text{Base(1)}}{Cl^-} + \underset{\text{Acid(2)}}{H_3PO_4}$$

Assess

Notice that in (c), $H_2PO_4^-$ is acting as an acid but in (d), it is acting as a base. The conjugate base of $H_2PO_4^-$ is HPO_4^{2-} (the deprotonated form of $H_2PO_4^-$), and the conjugate acid of $H_2PO_4^-$ is H_3PO_4 (the protonated form of $H_2PO_4^-$). This is an example of the general rule that in a conjugate pair, the acid is the protonated form and the base is the deprotonated form.

PRACTICE EXAMPLE A: For each of the following reactions, identify the acids and bases in both the forward and reverse directions.

(a) $HF + H_2O \rightleftharpoons F^- + H_3O^+$

(b) $HSO_4^- + NH_3 \rightleftharpoons SO_4^{2-} + NH_4^+$

(c) $CH_3COO^- + HCl \rightleftharpoons CH_3COOH + Cl^-$

PRACTICE EXAMPLE B: Of the following species, one is acidic, one is basic, and one is amphiprotic in their reactions with water: HNO_2, PO_4^{3-}, HCO_3^-. Write the *four* equations needed to represent these facts.

16-1 CONCEPT ASSESSMENT

s it appropriate to describe each of the following as a conjugate acid–base pair? :xplain. **(a)** HCO_3^- and CO_3^{2-}; **(b)** HSO_3^- and SO_4^{2-}; **(c)** H_2CO_3 and $H_2C_2O_4$; **d)** HClO and ClO^-; **(e)** H_2S and S^{2-}.

6-2 Self-Ionization of Water and the pH Scale

Section 16-1, we learned that the H_2O molecule can act as either an acid action 16.2) or a base (reaction 16.1); it is amphiprotic. It should come as no rprise that amongst themselves water molecules can produce H_3O^+ and I$^-$ ions via the following **self-ionization** reaction or *autoionization* reaction:

$$2\,H_2O(l) \rightleftharpoons H_3O^+(aq) + OH^-(aq) \qquad \textbf{(16.3)}$$

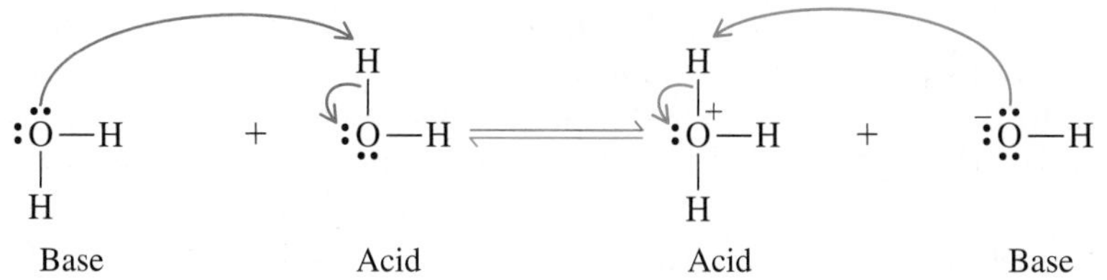

this reaction, one water molecule acts as an acid and the other acts as a base. Thus, even when it is pure, water contains H_3O^+ and OH^- ions, although concentrations are very low. The presence of these ions can be detected, wever, by using precise electrical conductivity measurements. The fact that self-ionization of water produces very low concentrations of H_3O^+ and I$^-$ ions is an indication that the equilibrium constant for reaction (16.3) is all, ranging from about 1.14×10^{-15} at 0 °C to about 5.45×10^{-13} at 100 °C. e thermodynamic equilibrium constant for reaction (16.3) is defined in ms of activities (see Section 15-1). It is given the symbol K_w and is called the **product of water**. Of course, equilibrium values of $[H_3O^+]$ and $[OH^-]$ ist be used in the expression for K_w.

$$K_w = \frac{a_{H_3O^+(aq)}\, a_{OH^-(aq)}}{a^2_{H_2O(l)}} \approx \frac{([H_3O^+]/c°)\,([OH^-]/c°)}{(1)^2} = \left(\frac{[H_3O^+]}{1\text{ M}}\right)\left(\frac{[OH^-]}{1\text{ M}}\right)$$

simplifying the expression above, we used the fact that $a = 1$ for $H_2O(l)$ and $= 1$ mol/L $= 1$ M. It is common practice to write the expression for K_w thout the units included, $K_w = [H_3O^+][OH^-]$, and to substitute only the merical values of $[H_3O^+]$ and $[OH^-]$ *without units* into this expression.

We are most interested in the value of K_w at 25 °C.

$$K_w = [H_3O^+]\,[OH^-] = 1.0 \times 10^{-14} \text{ (at 25 °C)} \qquad \textbf{(16.4)}$$

◀ A common misconception is that K_w is the acid ionization constant K_a for H_2O. Strictly speaking, $K_a = [H_3O^+]\,[OH^-]/[H_2O]$ has the value of $10^{-14}/55.55 = 1.8 \times 10^{-16}$ at 25 °C.

Reaction (16.3) indicates that $[H_3O^+]$ and $[OH^-]$ are equal in pure water. erefore,

$$\textit{pure water:}\quad [H_3O^+]/(1\text{ M}) = [OH^-]/(1\text{ M}) = \sqrt{K_w} = 1.0 \times 10^{-7} \text{ (at 25 °C)} \qquad \textbf{(16.5)}$$

Since reaction (16.3) reaches equilibrium not only in pure water but in all ueous solutions, we arrive at the following conclusion.

> In all aqueous solutions at 25 °C, the product of $[H_3O^+]$ and $[OH^-]$ always equals 1.0×10^{-14}.

If the concentration of H_3O^+ is increased by the addition of an acid, then the ncentration of OH^- must decrease to ensure the product of $[H_3O^+]$ and H$^-$] stays equal to 1.0×10^{-14}. Similarly, if the concentration of OH^- is increased by the addition of base, the concentration of H_3O^+ of must decrease ensure that $[H_3O^+] \times [OH^-]$ stays equal to 1.0×10^{-14}.

The self-ionization of water is an important reaction, from a conceptu point of view, because it reveals an important relationship between $[H_3O^+]$ and $[OH^-]$ that applies to all aqueous solutions. From a practical standpoi the reaction is not of much concern to us except when dealing with extreme dilute solutions. In fact, the self-ionization of water is partially suppressed the addition of acid or base to water.

> The self-ionization of water is partially suppressed by the addition of acid or base to water.

This statement is easily justified by applying Le Châtelier's principle reaction (16.3). When an acid is added to water, H_2O molecules are protonat and $[H_3O^+]$ increases. The increase in $[H_3O^+]$ causes net change to the left reaction (16.3), and, thus, the self-ionization of water is partially suppresse Similarly, the addition of a base to water increases $[OH^-]$, causes net change the left, and partially suppresses the self-ionization of water.

Given that the self-ionization of water is suppressed by the addition of acid or a base, we conclude that in any aqueous solution at 25 °C, the se ionization of water will contribute *less than* 1.0×10^{-7} M to $[H_3O^+]$ a $[OH^-]$. Clearly, such a small contribution will be important only extremely dilute solutions.

pH and pOH

KEEP IN MIND

that this definition of pH is one of the few scientific expressions that uses logarithms to the base 10 (log) rather than natural logarithms (ln).

In 1909, the Danish biochemist Søren Sørensen proposed the term **pH** to re to the "potential of hydrogen ion." He defined pH as the *negative of the lo rithm of* $[H^+]$. Restated in terms of $[H_3O^+]$,*

$$pH = -\log[H_3O^+] \quad (16.6)$$

Thus, in a solution that has $[H_3O^+] = 2.5 \times 10^{-3}$ M,

$$pH = -\log(2.5 \times 10^{-3}) = 2.60$$

To determine the $[H_3O^+]$ that corresponds to a particular pH value, we do inverse calculation. In a solution with pH = 4.50,

▶ The determination of logarithms and inverse logarithms (antilogarithms) is discussed in Appendix A. Significant figure rules for logarithms are also presented there.

$$\log[H_3O^+] = -4.50 \quad \text{and} \quad [H_3O^+] = 10^{-4.50} = 3.2 \times 10^{-5}\ \text{M}$$

The quantity **pOH** can be defined as

$$pOH = -\log[OH^-] \quad (16.7)$$

Another useful expression can be derived by taking the *negative logarithm* of t K_w expression (written for 25 °C) and introducing the symbol $pK_w = -\log K_w$

$$\begin{aligned} K_w &= [H_3O^+][OH^-] \\ -\log K_w &= -(\log[H_3O^+][OH^-]) \\ pK_w &= -(\log[H_3O^+] + \log[OH^-]) \\ &= -\log[H_3O^+] - \log[OH^-] \\ &= pH + pOH \end{aligned}$$

*Strictly speaking, we should use the *activity* of H_3O^+(aq), $a_{H_3O^+(aq)}$, a dimensionless quant But we will not use activities here. We will substitute the numerical value of the molarity H_3O^+ for its activity and recognize that some pH calculations may be only approximations.

ABLE 16.1 Acidic, Basic, and Neutral Solutions

	Neutral Solution	Acidic Solution	Basic Solution
elationship between			
H_3O^+] and [OH^-]	[H_3O^+] = [OH^-]	[H_3O^+] > [OH^-]	[H_3O^+] < [OH^-]
H_3O^+] at 25 °C	[H_3O^+] = 1.0×10^{-7} M	[H_3O^+] > 1.0×10^{-7} M	[H_3O^+] < 1.0×10^{-7} M
H at 25 °C	pH = 7	pH < 7	pH > 7

us, in general, the sum of the pH and pOH values must equal pK_w. At 25 °C, $_w = 1.0 \times 10^{-14}$ and $pK_w = 14.00$. Therefore,

$$\text{pH} + \text{pOH} = 14.00 \text{ (at 25 °C)} \qquad \textbf{(16.8)}$$

Thus, if we know the value of either pH or pOH, we can easily calculate the lue of the other.

cidic, Basic, and Neutral Solutions

pure water, the concentrations of H_3O^+ and OH^+ are equal. However, when acid or a base is added to water, the H_3O^+ and OH^- ions are no longer pre- nt in equal amounts. By comparing the values of [H_3O^+] and [OH^-], we can ssify a solution as *acidic, basic,* or *neutral* (Table 16.1). The classification can so be made, at 25 °C, by focusing on either [H_3O^+] or the pH.

The relationships between [H_3O^+], [OH^-], pH, and pOH, for acidic, basic, d neutral solutions, are illustrated in Figure 16-4. The pH values of a num- r of materials—some acidic and others basic—are depicted in Figure 16-5. ese values and the many examples in this chapter should help you become miliar with the pH concept. Later, we will consider two methods for mea- ring pH: by means of acid–base indicators (Section 17-3) and electrical mea- rements (Section 19-4).

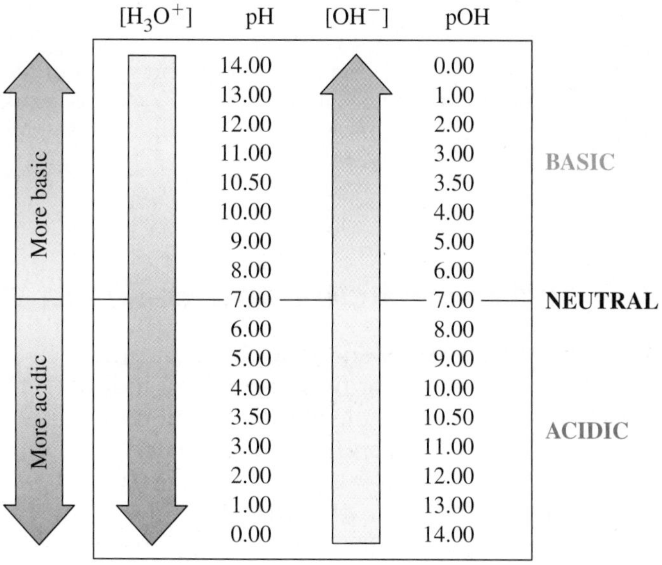

FIGURE 16-4
lating [H_3O^+], pH, [OH^-], and pOH
aqueous solutions at 25 °C, the sum of the pH and pOH values is always equal to 14.

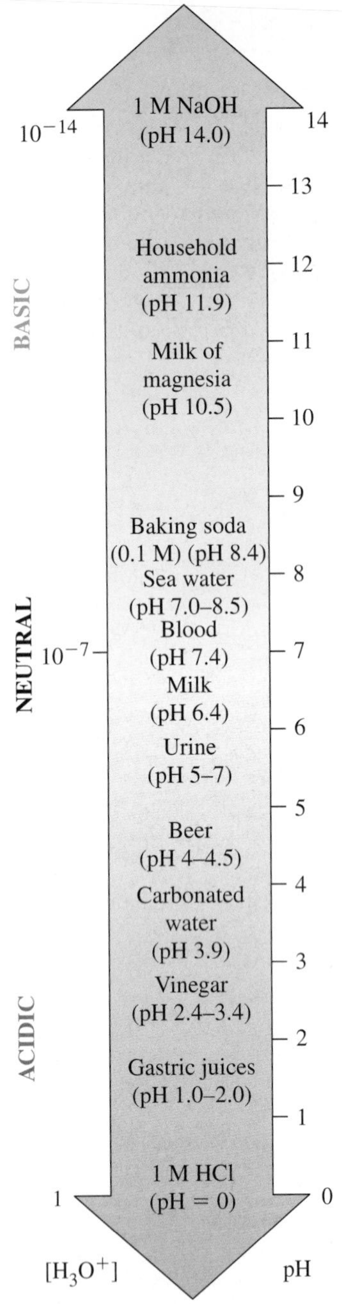

▲ FIGURE 16-5
The pH scale and pH values of some common materials
The scale shown here ranges from pH 0 to pH 14. Slightly negative pH values, perhaps to about −1 (corresponding to $[H_3O^+] \approx 10$ M), are possible. Also possible are pH values up to about 15 (corresponding to $[OH^-] \approx 10$ M). For practical purposes, however, the pH scale is useful only in the range $2 < \text{pH} < 12$, because the activities of H_3O^+ and OH^- in concentrated solutions may differ significantly from their molarities.

16-2 CONCEPT ASSESSMENT

At 40 °C, $K_w = 2.88 \times 10^{-14}$. If the pH of a solution is 7.00 at 40 °C, is the solution acidic, basic, or neutral?

EXAMPLE 16-2 Relating $[H_3O^+]$, $[OH^-]$, pH, and pOH

In a laboratory experiment at 25 °C, students measured the pH of samples of rainwater and household ammonia. Determine **(a)** $[H_3O^+]$ in the rainwater, with pH measured at 4.35; **(b)** $[OH^-]$ in the ammonia, with pH measured at 11.28.

Analyze

In this example we use the definition $\text{pH} = -\log[H_3O^+]$. For pOH we first determine pOH by using $\text{pOH} = 14 - \text{pH}$, and then by using $\text{pOH} = -\log[OH^-]$. To calculate concentration from a pH, we take the antilogarithm by raising 10 to minus the pH.

Solve

(a) $\log[H_3O^+] = -\text{pH} = -4.35$

$$[H_3O^+] = 10^{-4.35} = 4.5 \times 10^{-5}\text{ M}$$

(b) $\text{pOH} = 14.00 - \text{pH} = 14.00 - 11.28 = 2.72$

Now, use the definition $\text{pOH} = -\log[OH^-]$.

$$\log[OH^-] = -\text{pOH} = -2.72$$

$$[OH^-] = 10^{-2.72} = 1.9 \times 10^{-3}\text{ M}$$

Assess

The process of calculating hydronium ion concentration, $[H_3O^+]$, from pH is simply the antilogarithm of minus the pH value. To determine the concentration of $[OH^-]$ from a pH, we first calculated pOH, which is 14 − pH, and then calculated $[OH^-]$. Be careful when working these types of problems, and keep straight what you have to determine.

PRACTICE EXAMPLE A: Students found that a yogurt sample had a pH of 2.85. What are the $[H_3O^+]$ and $[OH^-]$ of the yogurt?

PRACTICE EXAMPLE B: The pH of a solution of acid is found to be 5.50 at 25 °C. What are $[H_3O^+]$ and $[OH^-]$ in this solution? What percentage of the H_3O^+ in this solution is produced by the self-ionization of water?

16-3 Ionization of Acids and Bases in Water

Figure 16-6 illustrates two ways of showing that ionization has occurred in solution of acid. One is by the color of an acid–base indicator; the other, t response of a pH meter. The pink color of the solution in Figure 16-7 tells the pH of the HCl solution is *less than 1.2*. (Unfortunately, the exact molarity this HCl solution is not known but it lies in the range 0.06 to 0.07 M.) The p meter registers a value of 1.20, indicating that $[H_3O^+] = 10^{-1.20}\text{ M} = 0.063$ in the HCl solution. The yellow color of the solution in Figure 16-7 indicat that the pH of 0.1 M CH_3COOH (acetic acid) is *2.8 or greater*. The pH me registers 2.80, and thus $[H_3O^+] = 10^{-2.80}\text{ M} = 0.0016$ M. Notice that the io ization of HCl generates a higher $[H_3O^+]$ than does the ionization CH_3COOH, even though the initial molarity of CH_3COOH (0.1 M) is grea than that of HCl (0.06 to 0.07 M). From this, we conclude that the ionization

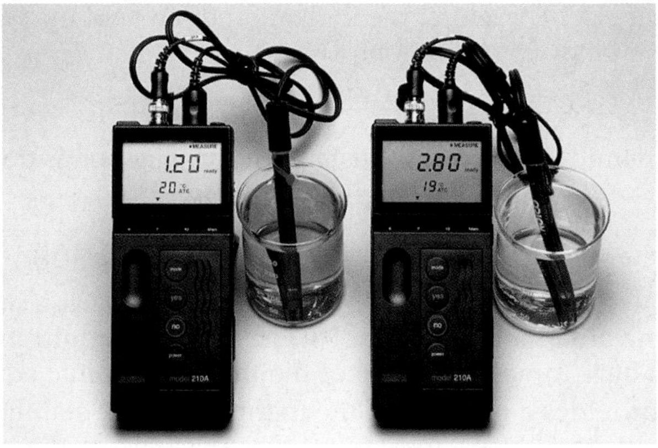

Tom Pantages

◀ Most laboratory pH meters can be read to the nearest 0.01 unit. Some pH meters for research work can be read to 0.001 unit, but unless unusual precautions are taken, the reading of the meter may not correspond to the true pH.

FIGURE 16-6
rong and weak acids compared
e color of thymol blue indicator, which is present in both solutions, depends on the
I of the solution.

$$\underset{\text{Red}}{pH} < 1.2 < \underset{\text{Orange}}{pH} < 2.8 < \underset{\text{Yellow}}{pH}$$

e principle of the pH meter is discussed in Section 19-4. (Left) The molarity of this
Cl solution is in the range 0.06 to 0.07 M, and the pH is 1.20. (Right) 0.1 M
H_3COOH has pH ≈ 2.8.

Cl occurs to a greater extent than does the ionization of CH_3COOH, an indi-tion that HCl is a much stronger acid than CH_3COOH. This conclusion is flected in the placements of HCl and CH_3COOH in Table 16.2 which ranks a ımber of acids and bases in order of increasing acid or base strengths. This dering is established by experiment.

The strength of an acid or a base is quantified by the value of the equilib-ım constant for the reaction describing its ionization in water. As discussed ı page 737, a monoprotic Brønsted–Lowry acid may be represented by the

ABLE 16.2 Relative Strengths of Some Common Brønsted–Lowry Acids and Bases

Acid		Conjugate Base	
Perchloric acid	$HClO_4$	Perchlorate ion	ClO_4^-
Hydroiodic acid	HI	Iodide ion	I^-
Hydrobromic acid	HBr	Bromide ion	Br^-
Hydrochloric acid	HCl	Chloride ion	Cl^-
Sulfuric acid	H_2SO_4	Hydrogen sulfate ion	HSO_4^-
Nitric acid	HNO_3	Nitrate ion	NO_3^-
Hydronium ion[a]	H_3O^+	Water[a]	H_2O
Hydrogen sulfate ion	HSO_4^-	Sulfate ion	SO_4^{2-}
Nitrous acid	HNO_2	Nitrite ion	NO_2^-
Acetic acid	CH_3COOH	Acetate ion	CH_3COO^-
Carbonic acid	H_2CO_3	Hydrogen carbonate ion	HCO_3^-
Ammonium ion	NH_4^+	Ammonia	NH_3
Hydrogen carbonate ion	HCO_3^-	Carbonate ion	CO_3^{2-}
Water	H_2O	Hydroxide ion	OH^-
Methanol	CH_3OH	Methoxide ion	CH_3O^-
Ammonia	NH_3	Amide ion	NH_2^-

Increasing acid strength (upward arrow, acid column); Increasing base strength (downward arrow, base column)

[a] The hydronium ion–water combination refers to the ease with which a proton is passed from one water molecule to another; that s, $H_3O^+ + H_2O \rightleftharpoons H_2O + H_3O^+$.

general formula HA. Therefore, the ionization of an acid may be represent generally by the following equation.

$$HA(aq) + H_2O(l) \rightleftharpoons A^-(aq) + H_3O^+(aq) \quad (16$$

The thermodynamic equilibrium constant for this reaction is defined as

$$K = \frac{a_{H_3O^+(aq)}\, a_{A^-(aq)}}{a_{HA(aq)}\, a_{H_2O(l)}} = \frac{([H_3O^+]/c^\circ)([A^-]/c^\circ)}{([HA]/c^\circ)(1)} = \frac{[H_3O^+][A^-]}{[HA]} \times \frac{1}{c^\circ}$$

where, as usual, the activities and concentrations appearing in this equati are equilibrium values and $c^\circ = 1$ mol/L. The quantity multiplying the fact ($1/c^\circ$) not only determines the value of the thermodynamic equilibrium co stant but also provides a measure of the thermodynamic stability of $A^-(aq)$ $H_3O^+(aq)$ relative to HA(aq) and $H_2O(l)$. For this reason, the quantity mul plying the factor ($1/c^\circ$) is given its own symbol, K_a, and is called the **acid io ization constant**.

KEEP IN MIND

that K_a is often called the acid dissociation constant. However, the terms *dissociation* and *ionization* have different meanings. Dissociation refers to the separation of an entity into two or more entities, or the separation of ion pairs into free ions. Ionization refers to the generation of one or more ions. Reaction (16.9) is clearly an ionization, and consequently we will refer to K_a as an acid ionization constant.

$$K_a = \frac{[H_3O^+][A^-]}{[HA]} \quad (16.1$$

K_a values span an enormous range. For example, K_a is about 10^9 for HI and less than 10^{-40} for CH_3CH_3. For this reason, we often use pK_a values instea The pK_a value of an acid is defined as follows.

$$pK_a = -\log K_a \text{ or } K_a = 10^{-pK_a} \quad (16.1$$

Thus, for example, pK_a is equal to -9 for HI and 40 for CH_3CH_3.

In a similar manner, we may write equations representing the ionization a base B, the **base ionization constant**, K_b, and pK_b.

$$B(aq) + H_2O(l) \rightleftharpoons BH^+(aq) + OH^-(aq) \quad (16.1$$

$$K_b = \frac{[BH^+][OH^-]}{[B]} \quad (16.1$$

$$pK_b = -\log K_b \text{ or } K_b = 10^{-pK_b} \quad (16.1$$

The value of K_a or K_b gives an indication of the strength of an acid or a bas The following points are worth remembering.

1. **A strong acid or base has a large ionization constant: K_a or K_b is mu greater than 1.** Therefore, we expect that the corresponding ionizatic reaction goes almost to completion. We will soon verify this point in tw ways (Figure 16-7 and Example 16-3). In most situations, we can safe assume that a strong acid or strong base is completely ionized in solutio Fortunately, there are relatively few common strong acids and stro bases (Table 16.3). Notice that the listing in Table 16.3 does not include or K_b values; these values are not needed. The main point is that the io ization constants are large enough to ensure that the acids and bases Table 16.3 are almost completely ionized in aqueous solution.

   The strong acids listed in Table 16.3 are molecular compounds where the strong bases are soluble ionic compounds called hydroxide Molecular compounds *ionize* in water: Neutral HA molecules produ H_3O^+ and A^- ions by reacting with water (equation 16.9). On the oth hand, soluble ionic hydroxides *dissociate* in water: Positive and negati ions (for example, Na^+ and OH^-), which are already present in the sol structure, enter the solution as free ions (Fig. 14-6).

   Memorizing the list in Table 16.3 can be extremely helpful. For exampl if the situation we are dealing with involves a strong acid or base, we ca safely assume the strong acid or base will react to completion. On th other hand, if the situation we are considering involves an acid and a ba that are *not* listed in Table 16.3, we can safely assume that the acid ar base are weak and react to a limited extent only.

TABLE 16.3 The Common Strong Acids and Strong Bases

Acids	Bases
HCl	LiOH
HBr	NaOH
HI	KOH
$HClO_4$	RbOH
HNO_3	CsOH
H_2SO_4[a]	$Mg(OH)_2$
	$Ca(OH)_2$
	$Sr(OH)_2$
	$Ba(OH)_2$

[a] H_2SO_4 ionizes in two distinct steps. It is a strong acid only in its first ionization (see page 760).

ABLE 16.4 Ionization Constants of Some Weak Acids and Weak Bases in Water at 25 °C

	Ionization Equilibrium	Ionization Constant K	pK
Acid		K_a =	pK_a =
Iodic acid	$HIO_3 + H_2O \rightleftharpoons H_3O^+ + IO_3^-$	1.6×10^{-1}	0.80
Chlorous acid	$HClO_2 + H_2O \rightleftharpoons H_3O^+ + ClO_2^-$	1.1×10^{-2}	1.96
Chloroacetic acid	$ClCH_2COOH + H_2O \rightleftharpoons H_3O^+ + ClCH_2COO^-$	1.4×10^{-3}	2.85
Nitrous acid	$HNO_2 + H_2O \rightleftharpoons H_3O^+ + NO_2^-$	7.2×10^{-4}	3.14
Hydrofluoric acid	$HF + H_2O \rightleftharpoons H_3O^+ + F^-$	6.6×10^{-4}	3.18
Formic acid	$HCOOH + H_2O \rightleftharpoons H_3O^+ + HCOO^-$	1.8×10^{-4}	3.74
Benzoic acid	$C_6H_5COOH + H_2O \rightleftharpoons H_3O^+ + C_6H_5COO^-$	6.3×10^{-5}	4.20
Hydrazoic acid	$HN_3 + H_2O \rightleftharpoons H_3O^+ + N_3^-$	1.9×10^{-5}	4.72
Acetic acid	$CH_3COOH + H_2O \rightleftharpoons H_3O^+ + CH_3COO^-$	1.8×10^{-5}	4.74
Hypochlorous acid	$HOCl + H_2O \rightleftharpoons H_3O^+ + OCl^-$	2.9×10^{-8}	7.54
Hydrocyanic acid	$HCN + H_2O \rightleftharpoons H_3O^+ + CN^-$	6.2×10^{-10}	9.21
Phenol	$C_6H_5OH + H_2O \rightleftharpoons H_3O^+ + C_6H_5O^-$	1.0×10^{-10}	10.00
Hydrogen peroxide	$H_2O_2 + H_2O \rightleftharpoons H_3O^+ + HO_2^-$	1.8×10^{-12}	11.74
Base		K_b =	pK_b =
Diethylamine	$(CH_3CH_2)_2NH + H_2O \rightleftharpoons (CH_3CH_2)_2NH_2^+ + OH^-$	6.9×10^{-4}	3.16
Ethylamine	$CH_3CH_2NH_2 + H_2O \rightleftharpoons CH_3CH_2NH_3^+ + OH^-$	4.3×10^{-4}	3.37
Ammonia	$NH_3 + H_2O \rightleftharpoons NH_4^+ + OH^-$	1.8×10^{-5}	4.74
Hydroxylamine	$HONH_2 + H_2O \rightleftharpoons HONH_3^+ + OH^-$	9.1×10^{-9}	8.04
Pyridine	$C_5H_5N + H_2O \rightleftharpoons C_5H_5NH^+ + OH^-$	1.5×10^{-9}	8.82
Aniline	$C_6H_5NH_2 + H_2O \rightleftharpoons C_6H_5NH_3^+ + OH^-$	7.4×10^{-10}	9.13

Acid strength (increasing upward)

Base strength (increasing upward)

. **A weak acid or base has a small ionization constant: K_a or K_b is much less than 1.** For a weak acid or base, the corresponding ionization reaction occurs to a limited extent, with a significant fraction of the acid or base *not* ionized. To determine the equilibrium composition of a solution of a weak acid or weak base, we need to solve an equilibrium problem, typically by using an ICE table and the value of the appropriate ionization constant, K_a or K_b. Ionization constants of some weak acids and weak bases are provided in Table 16.4. A more extensive list is given in Appendix D. Ionization constants of acids and bases are determined by experiment.

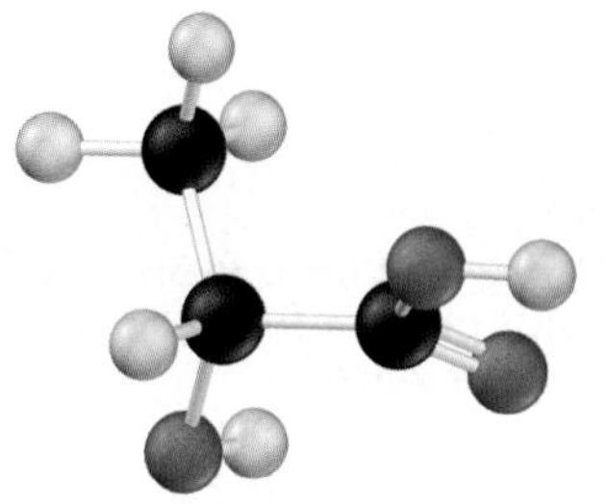

▲ Lactic acid, $CH_3CH(OH)COOH$.

Notice that several of the weak acids listed in Table 16.4 contain the group -COOH as part of the molecule. This grouping of atoms is called a *carboxyl* oup and is a common feature of many organic acids, including such biolog- ally important acids as lactic acid and all the amino acids, including glycine. rganic acids are also called carboxylic acids. The H atom of the carboxyl oup is removed as a proton when a carboxylic acid reacts with a base. We ill use a number of carboxylic acids as examples in this and later chapters.

The weak bases listed in Table 16.4 all contain an N atom. Not all weak ses contain an N atom, yet so many of them do that it worth noting the fol- wing points. For these weak bases, it is the N atom that is protonated when e base reacts with a Brønsted–Lowry acid. The protonation of the N atom is ustrated below for the ionization of CH_3NH_2 in water.

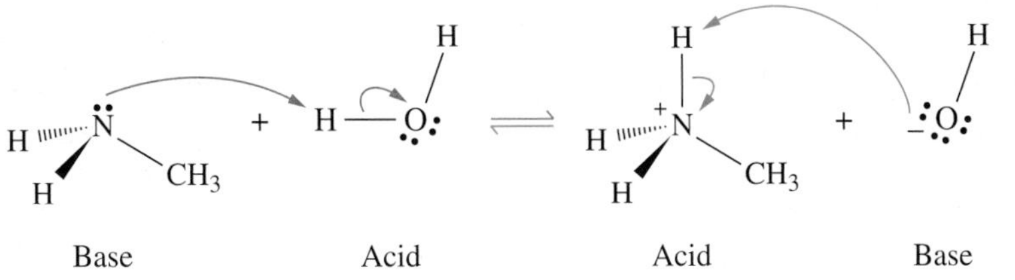

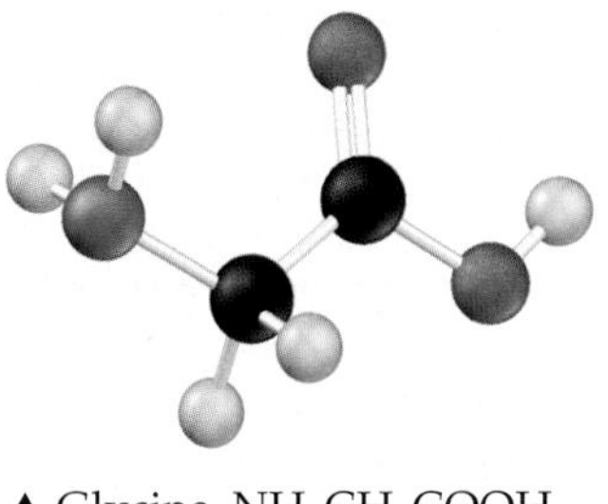

▲ Glycine, NH_2CH_2COOH.

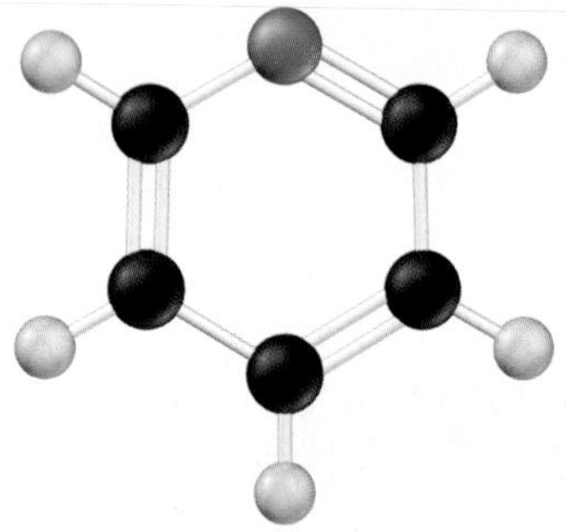

▲ Pyridine, C_5H_5N.

Moreover, most of the weak bases listed in Table 16.4 (all but pyridir C_5H_5N) are *amines*; they can be viewed as an ammonia molecule, NH_3, which some other group ($—C_6H_5$, $—CH_2CH_3$, $—OH$, or $—CH_3$) has been su stituted for one of the H atoms.

The extent to which acids or bases ionize in water is described in terms either the **degree of ionization (α)** or the **percent ionization**. For the ioniz tion of an acid, HA, we define the degree of ionization as follows.

$$\alpha = \frac{\text{the molarity of A}^- \text{ from the ionization of HA}}{\text{initial molarity of HA}} \quad \textbf{(16.15}$$

For the ionization of a base, B:

$$\alpha = \frac{\text{the molarity of BH}^+ \text{ from the ionization of B}}{\text{initial molarity of B}} \quad \textbf{(16.15}$$

The degree of ionization has a very simple interpretation: It is equal to t fraction of acid or base that exists at equilibrium in its ionized form. In prin ple, the degree of ionization can take on values from 0 (if none of the acid base ionizes) to 1 (if all of the acid or base ionizes). However, from a therm dynamic perspective, every reaction occurs to some extent and no reactic goes entirely to completion. Therefore, α is generally greater than 0 and le than 1. We can express the fraction of acid or base that ionizes as a percenta by multiplying α by 100.

$$\text{percent ionization} = 100 \times \alpha \quad \textbf{(16.1}$$

In general, the degree of ionization (or the percent ionization) of an acid a base in water depends on two factors: the value of the ionization consta and the initial molarity. In Figure 16-7, the degree of ionization is plotte against the ionization constant, assuming an initial molarity of 1.0 M. If, f example, K_a is greater than 20, the percent ionization of the acid will be great than 95%. If K_a is less than 10^{-2}, the percent ionization is less than 10%. T acids listed in Table 16.3 have K_a values ranging from about 20, for HNO_3, about 10^9, for HI. Consequently, the acids listed in Table 16.3 are essential completely ionized in water and are justifiably classified as strong acids.

Example 16-3 demonstrates the calculation of the percent ionization for tw different cases: a 1.00 M solution of a strong acid and a 1.00 M solution of weak acid. These kinds of calculations were done to construct Figure 16-7.

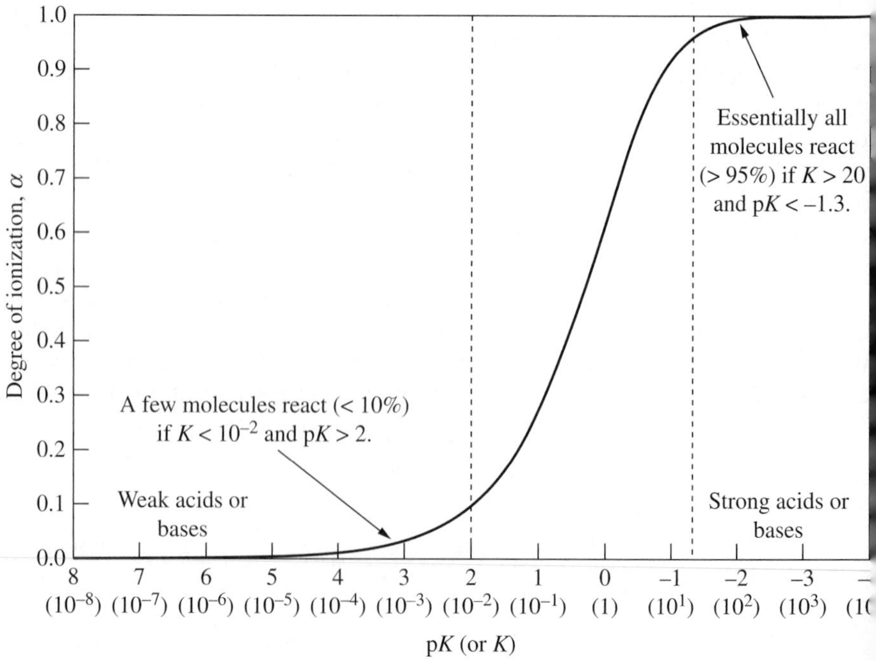

▶ FIGURE 16-7
Degree of ionization in a 1 M solution as a function of acid or base strength
The graph shows the fraction of molecules that react in a 1 M solution of acid or base at 298 K as a function of p*K*. The corresponding values of *K* are shown in parentheses below the p*K* values. For acids, the reaction is $HA(aq) + H_2O(l) \rightleftharpoons A^-(aq) + H_3O^+(aq)$ and $K = K_a$. For bases, the reaction is $B(aq) + H_2O(l) \rightleftharpoons BH^+(aq) + OH^-(aq)$ and $K = K_b$. In a 1 M solution of a strong acid or base at 298 K, more than 95% of the molecules react with water. In a 1 M solution of a weak acid or base, only a small fraction of molecules react with water.

EXAMPLE 16-3 Calculating Percent Ionization in an Aqueous Solution of an Acid

Calculate the equilibrium concentration and the percent ionization of HA in 1.00 M HA(aq) at 25 °C, assuming K_a is equal to **(a)** 1.00×10^5; **(b)** 1.00×10^{-5}.

Analyze

(a) A K_a value of 1.00×10^5 signifies that HA is a strong acid and that the ionization of HA will proceed almost nearly to completion. For such a situation, we can use the method described on page 720: (1) Take the reaction to completion. (2) Take the reaction backwards a little (from right to left) toward a true equilibrium state. **(b)** Here, K_a is rather small, an indication that the ionization of HA occurs to a rather limited extent. We expect the equilibrium concentration of HA to be nearly equal to the initial molarity.

Solve

(a) Because $K_a \gg 1$, we can use the approach described on page 720.

	HA(aq)	+	**$H_2O(l)$**	$\rightleftharpoons$	**$A^-(aq)$**	+	**$H_3O^+(aq)$**
initial concns:	1.00 M				0 M		0 M
to completion:	0 M				1.00 M		1.00 M
to the left:	$+x$ M				$-x$ M		$-x$ M
equil concns:	x M				$(1.00 - x)$ M		$(1.00 - x)$ M

Because K_a is large, we expect that most of the HA will react, and, thus, x represents a very small number. If we assume $x \ll 1$, we may write

$$K_a = \frac{[A^-][H_3O^+]}{[HA]} \approx \frac{(1.00)(1.00)}{x} \quad \text{and } x = \frac{(1.00)(1.00)}{K_a} = \frac{1.00}{1.00 \times 10^5} = 1.00 \times 10^{-5}$$

Therefore, $[HA] = x\text{ M} = 1.00 \times 10^{-5}\text{ M}$

The fraction of HA that remains at equilibrium is $(1.00 \times 10^{-5}\text{ M}/1.00\text{ M}) \times 100\% = 0.001\%$. Therefore, the percent ionization of HA is 99.999%.

(b) In this case, $K_a \ll 1$. The following equilibrium summary is appropriate.

	HA(aq)	+	**$H_2O(l)$**	$\rightleftharpoons$	**$A^-(aq)$**	+	**$H_3O^+(aq)$**
initial concns:	1.00 M						
changes:	$-x$ M				$+x$ M		$+x$ M
equil concns:	$(1.00 - x)$ M				x M		x M

Because K_a is small, we expect a very small amount of the HA to react. In others, we expect x to be much less than 1. By assuming $x \ll 1$, we obtain

$$K_a = \frac{[A^-][H_3O^+]}{[HA]} \approx \frac{(x)(x)}{1.00} = \frac{x^2}{1.00} \quad \text{and } x = \sqrt{1.00 \times K_a} = 3.16 \times 10^{-3}$$

Therefore: $[HA] = (1.00 - 0.00316)\text{ M} = 1.00\text{ M}$

$$\text{percent ionization} = \frac{3.16 \times 10^{-3}\text{ M}}{1.00\text{ M}} \times 100\% = 0.316\%$$

Assess

The results of these calculations verify the assertions we made earlier: In a solution of strong acid, essentially all of the acid ionizes. In a solution of a weak acid, only a small fraction of the acid ionizes. Finally, although we used the same symbol (x) for the unknown variable in both parts of this problem, two different meanings are associated with it. In part (a), x represents the amount of HA (on a mol/L basis) that does *not* react. In part (b), x represents the amount of HA that does react.

PRACTICE EXAMPLE A: For HNO_3, K_a has a value of about 20 at 25 °C. What is the degree of ionization of HNO_3 in 0.010 M HNO_3(aq) at 25 °C?

PRACTICE EXAMPLE B: For hypochlorous acid, HOCl, $K_a = 2.9 \times 10^{-8}$ at 25 °C. What is the degree of ionization of HOCl in 0.010 M HOCl(aq) at 25 °C? What is the pH of this solution?

In Example 16-3, our focus was on the relationship between the degree ionization and the strength of the acid. It is also instructive to consider hc the degree of ionization of a given acid varies with the initial concentrati (Figure 16-8). In a solution of a strong acid, the acid is essentially 100% ioniz for the concentrations typically encountered (1 M or less). The situation somewhat different for weak acids. Figure 16-8 implies that in a solution o weak acid, the acid is more highly ionized in dilute solutions than in conce trated solutions. For example, consider an acid HA with $K_a = 10^{-5}$. 1 M HA(aq), only a small fraction (less than 1%) of the HA molecules are ic ized, but in 1×10^{-7} M HA, more than 99% of the HA molecules are ionize The following statement generalizes this observation.

For a weak acid or a weak base, the degree of ionization increases with increasing dilution.

Example 16-4 shows by calculation that the percent ionization of weak acid or a weak base increases as the solution becomes more dilute fact that we can also demonstrate by a simple analysis of the ionizati reaction.

$$HA(aq) + H_2O(l) \rightleftharpoons A^-(aq) + H_3O^+(aq)$$

At equilibrium, n_{HA} moles of the acid HA, n_{A^-} moles of A^-, and n_{H_3} moles of H_3O^+ are present in a volume of V liters. The K_a expression is

$$K_a = \frac{[H_3O^+][A^-]}{[HA]} = \frac{(n_{H_3O^+}/V)(n_{A^-}/V)}{n_{HA}/V} = \frac{(n_{H_3O^+})(n_{A^-})}{n_{HA}} \times \frac{1}{V}$$

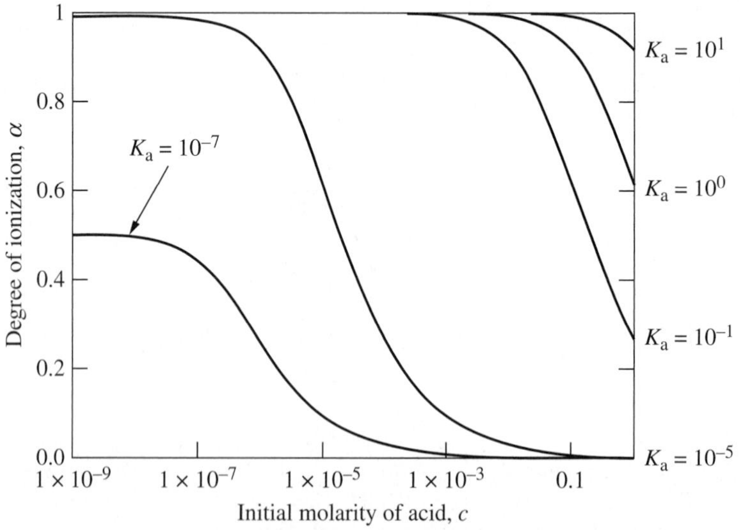

▲ FIGURE 16-8

Degree of ionization as a function of concentration for acids of varying strength

This graph shows that the degree of ionization increases as the initial molarity decreases. In other words, a given acid is ionized to a greater degree in a dilute solution than in a concentrated solution. We show, in Exercise 104, that in a very dilu solution, α approaches the value $K_a/(K_a + 10^{-7})$.

When we dilute the solution, V increases, $1/V$ decreases, and the ratio $(n_{H_3O^+})(n_{A^-})/n_{HA}$ must increase to maintain the constant value of K_a. In turn, $n_{H_3O^+}$ and n_{A^-} must increase and n_{HA} must decrease, signifying an increase in the degree of ionization.

EXAMPLE 16-4 Determining Percent Ionization as a Function of Initial Concentration

What is the percent ionization of acetic acid, CH_3COOH, in 1.0 M, 0.10 M, and 0.010 M CH_3COOH?

Analyze

It is important to recognize that we are dealing with a solution of a weak acid. In this particular case, we can arrive at this conclusion in two ways. (1) The formula CH_3COOH contains a —COOH group and signifies that CH_3COOH is an organic acid. Organic acids are typically weak acids. (2) The substance CH_3COOH does not appear on the list of common strong acids or strong bases, so it is not a strong acid or base. To ascertain whether or not it is a weak acid or a weak base, we should consult a table of acid or base ionization constants. In Table 16.4, CH_3COOH is identified as a weak acid with $K_a = 1.8 \times 10^{-5}$. This second approach is the approach we'd have to take if it is not obvious from the molecular formula whether the substance is an acid or a base (for example, to deduce that HCN, C_6H_5OH, and HN_3 are acids, not bases).

Having established that we are dealing with a solution of acid, the percent ionization is determined by dividing the amount of ionized acid by the initial acid concentration and multiplying by 100%.

Solve

Use the ICE format to describe 1.0 M CH_3COOH:

$$\mathbf{CH_3COOH + H_2O \rightleftharpoons H_3O^+ + CH_3COO^-}$$

	CH_3COOH	H_3O^+	CH_3COO^-
initial concns:	1.0 M	—	—
changes:	$-x$ M	$+x$ M	$+x$ M
equil concns:	$(1.0 - x)$ M	x M	x M

We need to calculate $[H_3O^+] = [CH_3COO^-] = x$ M. Because $K_a = 1.8 \times 10^{-5}$ is quite small, we anticipate that very little CH_3COOH ionizes. Therefore, let's assume that x is small and $1.0 - x \approx x$.

$$K_a = \frac{[H_3O^+][CH_3COO^-]}{[CH_3COOH]} = \frac{x \cdot x}{1.0 - x} \approx \frac{x^2}{1.0} = 1.8 \times 10^{-5}$$

$$x = \sqrt{1.8 \times 10^{-5}} = 4.2 \times 10^{-3}$$

We have $[H_3O^+] = [CH_3COO^-] = x\text{ M} = 4.2 \times 10^{-3}$ M. The percent ionization of CH_3COOH is

$$\%\ ionization = \frac{[H_3O^+]}{[CH_3COOH]} \times 100\% = \frac{4.2 \times 10^{-3}\text{ M}}{1.0\text{ M}} \times 100\% = 0.42\%$$

The assumption that x is small compared to 1.0 is clearly valid: x is only about 0.42% of 1.0. The calculations for 0.10 M CH_3COOH and 0.010 M CH_3COOH are very similar. In 0.10 M CH_3COOH, 1.3% of the acetic acid molecules are ionized and in 0.010 M CH_3COOH, 4.2% are ionized.

Assess

The purpose of calculating the percent ionization for three acetic acid solutions was to confirm the very important point made on page 748: For a weak acid, percent ionization increases with increasing dilution (Fig. 16-7). For very dilute solutions, the calculation of percent ionization is more complicated. (See Are You Wondering 16-2 on page 756.)

PRACTICE EXAMPLE A: What is the percent ionization of hydrofluoric acid in 0.20 M HF and in 0.020 M HF?

PRACTICE EXAMPLE B: In a 0.0284 M aqueous solution of lactic acid, a carboxylic acid that accumulates in the blood and muscles during physical activity, the acid is found to be 6.7% ionized. Determine K_a for lactic acid.

$$CH_3CH(OH)COOH + H_2O \rightleftharpoons H_3O^+ + CH_3CH(OH)COO^- \qquad K_a = ?$$

▶ Most people consider strong acids to be more dangerous than strong bases, but strong bases can cause serious burns and should be treated with as much care as acids.

▶ The dilution of strong acids and bases is usually exothermic. Never add water to concentrated strong acids and bases (particularly sulfuric acid) as the heat of dilution will boil the water and spatter concentrated acid.

16-4 Strong Acids and Strong Bases

A strong acid, such as HCl, is essentially completely ionized in aqueous sol tion.* Consequently, in the equation for the ionization of HCl, we use a rig arrow ($\longrightarrow$) instead of a double arrow ($\rightleftharpoons$):

$$HCl(aq) + H_2O(l) \longrightarrow Cl^-(aq) + H_3O^+(aq)$$

In a solution of HCl, there are two sources of H_3O^+: the ionization of HCl and t self-ionization of water. However, adding HCl to water suppresses the se ionization of water (see page 740), so the ionization of HCl is the only significa source of H_3O^+. The contribution from the self-ionization of water can general be ignored *unless the solution is extremely dilute*. Even for 1.0×10^{-6} M HCl(a the self-ionization of water contributes less than 1% to the total concentration H_3O^+ (see Exercise 102). For more concentrated solutions, the self-ionization water contributes much less than 1%, a fact we verify in Example 16-5.

EXAMPLE 16-5 Calculating Ion Concentrations in an Aqueous Solution of a Strong Acid

Calculate $[H_3O^+]$, $[Cl^-]$, and $[OH^-]$ in 0.015 M HCl(aq).

Analyze

Because HCl is a strong acid, all the HCl ionizes. The hydronium ion concentration is equal to the molarity of the solution. The hydroxide concentration is determined by using the ion product of water, K_w. The product of the hydronium ion concentration and hydroxide concentration must equal $K_w = 1.0 \times 10^{-14}$.

Solve

The ionization of HCl is represented by the following equation, the right arrow ($\longrightarrow$) emphasizing that the reaction goes essentially to completion.

$$HCl(aq) + H_2O(l) \longrightarrow Cl^-(aq) + H_3O^+(aq)$$

Therefore,

$$[H_3O^+] = 0.015\text{ M}$$

Because one Cl^- ion is produced for every H_3O^+ ion,

$$[Cl^-] = [H_3O^+] = 0.015\text{ M}$$

To calculate $[OH^-]$, we must use the following facts.

1. All the OH^- is derived from the self-ionization of water, by reaction (16.3).
2. $[OH^-]$ and $[H_3O^+]$ must have values consistent with K_w for water.

$$K_w = [H_3O^+][OH^-] = 1.0 \times 10^{-14}$$

So we have

$$[OH^-] = \frac{1.0 \times 10^{-14}}{1.5 \times 10^{-2}} = 6.7 \times 10^{-13}\text{ M}$$

Assess

The self-ionization of water contributes equal amounts of OH^- and H_3O^+ to the solution. The results of this example show that the self-ionization of water contributes only a small amount (6.7×10^{-13} M) of OH^- and H_3O^+. The self-ionization of water usually, but not always, plays a very minor role in determining the pH of a solution.

PRACTICE EXAMPLE A: A 0.0025 M solution of HI(aq) has $[H_3O^+] = 0.0025$ M. Calculate $[I^-]$, $[OH^-]$, and the pH of the solution.

PRACTICE EXAMPLE B: If 535 mL of *gaseous* HCl, at 26.5 °C and 747 mmHg, is dissolved in enough water to prepare 625 mL of solution, what is the pH of this solution?

*In very concentrated aqueous solutions, HCl does not exist exclusively as the separated ions H_3O^+ and Cl^-. One indication of this is that we can smell HCl in the vapor above such solutions.

Example 16-6 focuses on a solution of $Ca(OH)_2$, which we recognize as a rong base (Table 16.3). The dissolution of $Ca(OH)_2$ is represented by the folwing equation, the right arrow ($\longrightarrow$) signifying that the ions dissociate mpletely to produce free ions in solution.

$$Ca(OH)_2 \xrightarrow{H_2O} Ca^{2+}(aq) + 2\,OH^-(aq)$$

nis dissolution process is the only significant source of OH^- ions because, as e have already noted, the contribution from the self-ionization of water is gligible, except in extremely dilute solutions.

EXAMPLE 16-6 Calculating the pH of an Aqueous Solution of a Strong Base

Calcium hydroxide (slaked lime), $Ca(OH)_2$, is the cheapest strong base available. It is generally used for industrial operations in which a high concentration of OH^- is not required. $Ca(OH)_2(s)$ is soluble in water only to the extent of 0.16 g $Ca(OH)_2$/100.0 mL solution at 25 °C. What is the pH of saturated $Ca(OH)_2(aq)$ at 25 °C?

Analyze

Because the volume of solution is not specified, let's assume it is 100.0 mL = 0.1000 L. The resulting solution will be basic, so we should focus on the hydroxide ion. To solve this problem, we first calculate the molarity of the solution, and then determine the concentration of hydroxide ion in this solution. Finally, we calculate pOH and then pH.

Solve

Express the solubility of $Ca(OH)_2$ on a molar basis.

$$\text{molarity} = \frac{0.16\,\text{g Ca(OH)}_2 \times \dfrac{1\,\text{mol Ca(OH)}_2}{74.1\,\text{g Ca(OH)}_2}}{0.1000\,\text{L}} = 0.022\,\text{M Ca(OH)}_2$$

Relate the molarity of OH^- to the molarity of $Ca(OH)_2$.

$$[OH^-] = \frac{0.022\,\text{mol Ca(OH)}_2}{1\,\text{L}} \times \frac{2\,\text{mol OH}^-}{1\,\text{mol Ca(OH)}_2} = 0.044\,\text{M}$$

Calculate the pOH and, from it, the pH.

$$\text{pOH} = -\log[OH^-] = -\log 0.044 = 1.36$$

$$\text{pH} = 14.00 - \text{pOH} = 14.00 - 1.36 = 12.64$$

Assess

A common error is to neglect the factor 2 mol OH^-/1 mol $Ca(OH)_2$ in determining $[OH^-]$. When solving problems involving basic solutions, we often solve first for pOH. We must remember to finish the problem and convert from pOH to pH. Finally, although $Ca(OH)_2$ is a slightly soluble hydroxide salt, we observe that the pH of the solution is quite high.

PRACTICE EXAMPLE A: Milk of magnesia is a saturated solution of $Mg(OH)_2$. Its solubility is 9.63 mg $Mg(OH)_2$/100.0 mL solution at 20 °C. What is the pH of saturated $Mg(OH)_2$ at 20 °C?

PRACTICE EXAMPLE B: Calculate the pH of an aqueous solution that is 3.00% KOH, by mass, and has a density of 1.0242 g/mL.

16-1 ARE YOU WONDERING?

How do we calculate $[H_3O^+]$ in an extremely dilute solution of a strong acid?

The method of Example 16-5 won't work for calculating the pH of a solution as dilute as 1.0×10^{-8} M HCl. We would write $[H_3O^+] = 1.0 \times 10^{-8}$ M, and

(continued)

pH = 8.00. But how can a solution of a strong acid, no matter how dilute, have a pH greater than 7? The difficulty is that at this extreme dilution, we must consider two sources of H_3O^+: the ionization of the acid and the self-ionization of water. The ionization of HCl contributes 1.0×10^{-8} M to the total $[H_3O^+]$ and partially suppresses the ionization of water. To find the contribution from the self-ionization of water, we can use the following equilibrium summary. In the row for the initial concentrations, we enter 1.0×10^{-8} M, which is the contribution from the ionization of HCl.

	$2\,H_2O(l)$	$\rightleftharpoons$	$H_3O^+(aq)$	+	$OH^-(aq)$
initial concns:			1.0×10^{-8} M		
changes:			$+x$ M		$+x$ M
equil concns:			$(1.0 \times 10^{-8} + x)$ M		x M

At equilibrium, we must have $[H_3O^+][OH^-] = K_w$. Therefore,

$$[H_3O^+][OH^-] = (x + 1.0 \times 10^{-8})x = 1.0 \times 10^{-14}$$

This expression rearranges to the quadratic form

$$x^2 + (1.0 \times 10^{-8}x) - (1.0 \times 10^{-14}) = 0$$

The solution to this equation is $x = 9.5 \times 10^{-8}$. Therefore, we combine $[H_3O^+]$ from both sources to get $[H_3O^+] = (9.5 \times 10^{-8} + 1.0 \times 10^{-8})$ M $= 1.05 \times 10^{-7}$ M, and pH = 6.98.

From this result, we conclude that the pH is slightly less than 7, as expected for a very dilute solution of strong acid. Note that the self-ionization of water contributes nearly ten times as much hydronium ion to the solution as does the strong acid.

16-5 Weak Acids and Weak Bases

In this section, we will work through some additional examples, all of whic involve the ionization of a weak acid or a weak base in water. Typically, we a required to find the equilibrium concentrations or determine the pH. To do s we must solve an equilibrium problem.

For some students, equilibrium calculations involving aqueous solutio are among the most challenging in general chemistry. At times, the difficulty in sorting out what is relevant to a given problem. The number of types of ca culations seems very large, although in fact it is quite limited. The key to sol ing these equilibrium problems is to be able to imagine what is going on. He are some questions to ask yourself.

- Which are the principal species in solution?
- What are the chemical reactions that produce them?
- Can some reactions (for example, the self-ionization of water) b ignored?
- Can you make any assumptions that allow you to simplify the equili rium calculations?
- What is a reasonable answer to the problem? For instance, should th final solution be acidic (pH $<$ 7) or basic (pH $>$ 7)?

In short, first think through a problem *qualitatively*. At times, you may n even have to do a calculation. Next, organize the relevant data in a clear, log cal manner, as done in Example 16-7. In this way, many problems that at fir appear new to you will take on a familiar pattern. Look for other helpful hin as you proceed through this chapter and the following two chapters.

Examples 16-8 and 16-9 present a common problem involving weak cids and weak bases: calculating the pH of a solution of known molarity. he calculation invariably involves a quadratic equation, but very often we an make a simplifying assumption that leads to a shortcut that saves both ime and effort.

EXAMPLE 16-7 Determining a Value of K_a from the pH of a Solution of a Weak Acid

Butyric acid, $CH_3(CH_2)_2COOH$, is used to make compounds employed in artificial flavorings and syrups. A 0.250 M aqueous solution of butyric acid is found to have a pH of 2.72. Determine K_a for butyric acid.

$$CH_3(CH_2)_2COOH + H_2O \rightleftharpoons H_3O^+ + CH_3(CH_2)_2COO^- \qquad K_a = ?$$

▲ Butyric acid, $CH_3CH_2CH_2COOH$.

Analyze

For $CH_3(CH_2)_2COOH$, K_a is likely to be much larger than K_w. Therefore, we can assume that self-ionization of water is unimportant and that ionization of the butyric acid is the only source of H_3O^+. Let's treat the situation as if $CH_3(CH_2)_2COOH$ first dissolves in molecular form, and then the molecules ionize until equilibrium is reached. That is, we write the balanced chemical equation and use it as the basis for an ICE table, as discussed in Chapter 15 (page 713). We will represent the concentrations of H_3O^+ and $CH_3(CH_2)_2COO^-$ at equilibrium as x M.

Solve

	$CH_3(CH_2)_2COOH$ + H_2O ⇌	H_3O^+ +	$CH_3(CH_2)_2COO^-$
initial concns:	0.250 M	—	—
changes:	$-x$ M	$+x$ M	$+x$ M
equil concns:	$(0.250 - x)$ M	x M	x M

But x M is a known quantity. It is the $[H_3O^+]$ in solution, which we can determine from the pH.

$$\log[H_3O^+] = -pH = -2.72$$
$$[H_3O^+] = 10^{-2.72}\,M = 1.9 \times 10^{-3}\,M = x\,M$$

Now we can solve the following expression for K_a, by substituting in the value $x = 1.9 \times 10^{-3}$.

$$K_a = \frac{[H_3O^+][CH_3(CH_2)_2COO^-]}{[CH_3(CH_2)_2COOH]} = \frac{x \cdot x}{0.250 - x}$$
$$= \frac{(1.9 \times 10^{-3})(1.9 \times 10^{-3})}{0.250 - (1.9 \times 10^{-3})} = 1.5 \times 10^{-5}$$

Assess

Notice that our original assumption was correct: K_a is much larger than K_w. The $[OH^-]$ in this solution is $[OH^-] = 1.0 \times 10^{-14}/1.9 \times 10^{-3} = 5.3 \times 10^{-12}$ M. This calculation shows that the self-ionization of water contributes only 5.3×10^{-12} M to the total $[H_3O^+]$, a negligible contribution.

PRACTICE EXAMPLE A: Hypochlorous acid, HOCl, is used in water treatment and as a disinfectant in swimming pools. A 0.150 M solution of HOCl has a pH of 4.18. Determine K_a for hypochlorous acid.

PRACTICE EXAMPLE B: The much-abused drug cocaine is an alkaloid. Alkaloids are noted for their bitter taste, an indication of their basic properties. Cocaine, $C_{17}H_{21}O_4N$, is soluble in water to the extent of 0.17 g/100 mL solution, and a saturated solution has a pH = 10.08. What is the value of K_b for cocaine?

$$C_{17}H_{21}O_4N + H_2O \rightleftharpoons C_{17}H_{21}O_4NH^+ + OH^- \qquad K_b = ?$$

EXAMPLE 16-8 Calculating the pH of a Weak Acid Solution

Show by calculation that the pH of 0.100 M CH_3COOH should be about the value shown on the pH meter in Figure 16-6; that is, pH $\approx$ 2.8.

Analyze

Here, we know that K_a is much larger than K_w. Let's again treat the situation as if CH_3COOH first dissolves in molecular form and then ionizes until equilibrium is reached. In this case, the quantity x is an unknown that must be obtained by an algebraic solution.

Solve

	CH_3COOH	+ H_2O $\rightleftharpoons$	H_3O^+	+	CH_3COO^-
initial concns:	0.100 M		—		—
changes:	$-x$ M		$+x$ M		$+x$ M
equil concns:	$(0.100 - x)$ M		x M		x M

$$K_a = \frac{[H_3O^+][CH_3COO^-]}{[CH_3COOH]} = \frac{x \cdot x}{0.100 - x} = 1.8 \times 10^{-5}$$

We could solve this equation by using the quadratic formula, but let's instead make a simplifying assumption that is often valid. Assume that x is very small compared with 0.100. That is, assume that $(0.100 - x) \approx 0.100$.

$$x^2 = 0.100 \times 1.8 \times 10^{-5} = 1.8 \times 10^{-6}$$

$$x = \sqrt{1.8 \times 10^{-6}} = 1.3 \times 10^{-3}$$

Now, we must check our assumption: $0.100 - 0.0013 = 0.099 \approx 0.100$. Our assumption is good to about 1 part per 100 (1%) and is valid for a calculation involving two or three significant figures.

Finally, we have $[H_3O^+] = x$ M $= 1.3 \times 10^{-3}$ M and

$$pH = -\log[H_3O^+] = -\log(1.3 \times 10^{-3}) = -(-2.89) = 2.89$$

Assess

We observe that our answer is very close to the number on the pH meter. Therefore, our assumption to simplify the calculation was reasonable. This type of assumption may not always work and so we need to check the final answer until we are comfortable with knowing when and when not to apply the assumption to simplify the calculations.

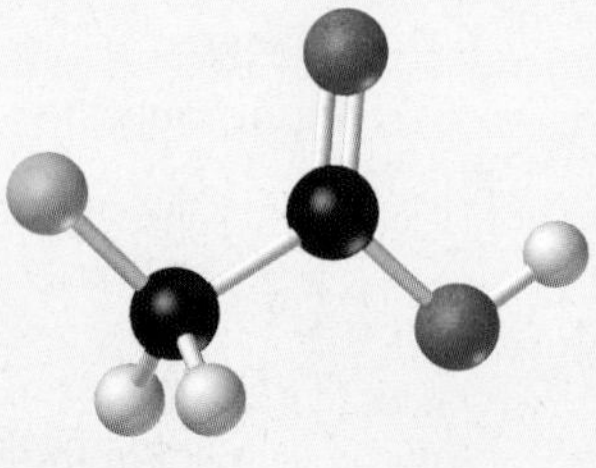

▲ Fluoroacetic acid, CH_2FCOOH.

PRACTICE EXAMPLE A: Substituting halogen atoms for hydrogen atoms bound to carbon increases the strength of carboxylic acids. Show that the pH of 0.100 M CH_2FCOOH, fluoroacetic acid, is lower than that calculated in Example 16-8 for 0.100 M CH_3COOH.

$$CH_2FCOOH + H_2O \rightleftharpoons H_3O^+ + CH_2FCOO^- \qquad K_a = 2.6 \times 10^{-3}$$

PRACTICE EXAMPLE B: Acetylsalicylic acid, $C_6H_4(OOCCH_3)COOH$, is an organic acid (general formula RCOOH) and the active component in aspirin. This acid is the cause of the stomach upset some people get when taking aspirin. Two extra-strength aspirin tablets, each containing 500 mg of acetylsalicylic acid, are dissolved in 325 mL of water. What is the pH of this solution?

$$RCOOH + H_2O \rightleftharpoons H_3O^+ + RCOO^- \qquad K_a = 3.3 \times 10^{-4}$$

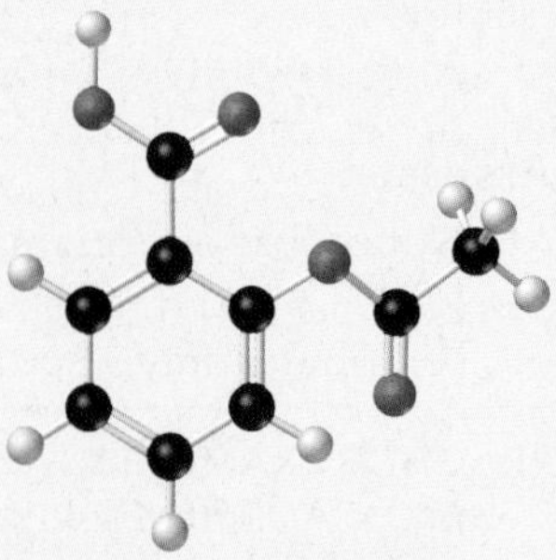

▲ Acetylsalicylic acid, $C_6H_4(OOCCH_3)COOH$.

EXAMPLE 16-9 Calculating the pH of a Solution of a Weak Base

What is the pH of a solution that is 0.00250 M $CH_3NH_2(aq)$? For methylamine, $K_b = 4.2 \times 10^{-4}$.

Analyze

In this example we will apply the same techniques as we did in Example 16-8. We will work the problem twice to see that the simplifying assumption breaks down for weak acids and weak bases at very low concentrations.

Solve

	CH_3NH_2	+	H_2O	$\rightleftharpoons$	$CH_3NH_3^+$	+	OH^-
initial concns:	0.00250 M				—		—
changes:	$-x$ M				$+x$ M		$+x$ M
equil concns:	$(0.00250 - x)$ M				x M		x M

$$K_b = \frac{[CH_3NH_3^+][OH^-]}{[CH_3NH_2]} = \frac{x \cdot x}{0.00250 - x} = 4.2 \times 10^{-4}$$

Now let's assume that x is very much less than 0.00250 and that $0.00250 - x \approx 0.00250$.

$$\frac{x^2}{0.00250} = 4.2 \times 10^{-4} \qquad x^2 = 1.1 \times 10^{-6} \qquad [OH^-] = x\,M = 1.0 \times 10^{-3}\,M$$

The value of x is nearly half as large as 0.00250—too large to ignore. This means using the *quadratic* formula.

$$\frac{x^2}{0.00250 - x} = 4.2 \times 10^{-4}$$

$$x^2 + (4.2 \times 10^{-4}x) - (1.1 \times 10^{-6}) = 0$$

$$x = \frac{(-4.2 \times 10^{-4}) \pm \sqrt{(4.2 \times 10^{-4})^2 + 4 \times 1.1 \times 10^{-6}}}{2}$$

The expression above provides two values of x, one positive and one negative. The negative value can be ignored because it yields a nonphysical result—a negative value for $[OH^-]$.

$$[OH^-] = x\,M = \frac{(-4.2 \times 10^{-4}) + (2.1 \times 10^{-3})}{2}\,M = 8.4 \times 10^{-4}\,M$$

$$pOH = -\log[OH^-] = -\log(8.4 \times 10^{-4}) = 3.08$$

$$pH = 14.00 - pOH = 14.00 - 3.08 = 10.92$$

Assess

After applying the simplifying assumption, if the value of x is a significant percentage of the initial concentration (for example, greater than 5%), then we should not use the simplifying assumption to obtain the hydronium concentration.

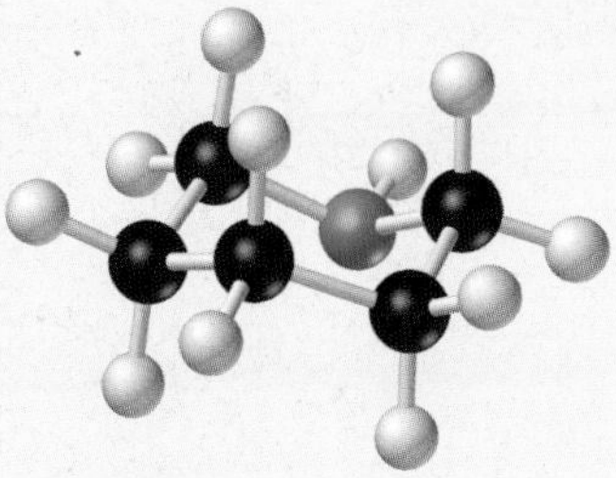

▲ Piperidine, $C_5H_{10}NH$.

PRACTICE EXAMPLE A: What is the pH of 0.015 M $CH_2FCOOH(aq)$?

$$CH_2FCOOH + H_2O \rightleftharpoons H_3O^+ + CH_2FCOO^- \qquad K_a = 2.6 \times 10^{-3}$$

PRACTICE EXAMPLE B: Piperidine is a base found in small amounts in black pepper. What is the pH of 315 mL of an aqueous solution containing 114 mg piperidine?

$$C_5H_{10}N + H_2O \rightleftharpoons C_5H_{10}NH^+ + OH^- \qquad K_b = 1.6 \times 10^{-3}$$

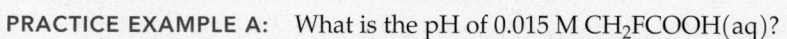

16-3 CONCEPT ASSESSMENT

Is it possible for a weak acid solution to have a lower pH than a strong acid solution? If not, why not? If it is possible, under what conditions?

16-2 ARE YOU WONDERING?

How do we calculate the pH of a very dilute solution of a weak acid?

Think of this as a companion question to the one posed in Are You Wondering 16-1, except here the acid in question is weak rather than strong. Because the ionization of a weak acid does not go to completion, we must consider two equilibria simultaneously: the ionization of the acid and the self-ionization of water. (In Are You Wondering 16-1, we considered only one equilibrium—the self-ionization of water—because we assumed that a strong acid ionizes completely.)

Let's consider a solution of HA, assuming that the initial concentration of HA is c M. There are two contributions to the H_3O^+ in solution: the ionization of HA, reaction (16.9), and the self-ionization of water, reaction (16.3). Let the contributions from the ionization of HA and the self-ionization of water be x M and y M, respectively. Our first task is to express the concentrations of all species in solution in terms of these two unknowns.

The ionization of HA causes the concentration of HA to *decrease* by x M and the concentrations of H_3O^+ and A^- to *increase* by x M. Similarly, the self-ionization of water causes the concentrations of H_3O^+ and OH^- to increase by y M. Therefore,

$$[HA] = (c - x)\text{ M} \quad [A^-] = x\text{ M} \quad [OH^-] = y\text{ M} \quad [H_3O^+] = (x + y)\text{ M}$$

The K_a and K_w expressions are

$$K_a = \frac{[H_3O^+][A^-]}{[HA]} = \frac{(x + y)(x)}{c - x} \qquad K_w = [H_3O^+][OH^-] = (x + y)(x)$$

Our next task is to solve this pair of equations simultaneously for x and y. We begin by rearranging the K_a expression to obtain expressions for $(x + y)$ and y:

$$(x + y) = \frac{K_a(c - x)}{x} \text{ and } y = \frac{K_a(c - x)}{x} - x$$

Now, we use these two equations to rewrite the expression for K_w in terms of x:

$$K_w = \frac{K_a(c - x)}{x} \times \left(\frac{K_a(c - x)}{x} - x\right)$$

This equation can then be rearranged as follows.

$$x^2 = \frac{K_a^2(c - x)^2}{K_w + K_a(c - x)}$$

For 1.0×10^{-5} M HCN, we would substitute $K_a = 6.2 \times 10^{-10}$ and $c = 1.0 \times 10^{-5}$ into the equation above and then solve for x by using the method of successive approximations (see Appendix A). We begin by assuming that $c - x \approx c$ and then solving the resulting expression for x. This assumption is reasonable because K_a is very small, which means that only a small fraction of the acid will ionize. We obtain $x = 4.87 \times 10^{-8}$. (For now, we will retain an extra digit and round off at the end.) We can improve this result by substituting this value of x into the right side of the equation above and solving again for x. We obtain $x = 4.85 \times 10^{-8}$. Repeating this process one more time, we obtain $x = 4.85 \times 10^{-8}$ again, which we take as our final result. By substituting this value for x into the equations for $x + y$ and y, we obtain $x + y = 1.27 \times 10^{-7}$ and $y = 7.86 \times 10^{-8}$. Therefore,

$$[H_3O^+] = (x + y)\text{ M} = 1.3 \times 10^{-7}\text{ M}$$
$$[CN^-] = x\text{ M} = 4.9 \times 10^{-8}\text{ M}$$
$$[OH^-] = y\text{ M} = 7.9 \times 10^{-8}\text{ M}$$

From this, we calculate $\text{pH} = -\log_{10}(1.3 \times 10^{-7}) = 6.89$, a reasonable result for a very dilute solution of a weak acid. Also, the degree of ionization of HCN is $\alpha = (4.9 \times 10^{-8}\text{ M})/(1.0 \times 10^{-5}\text{ M}) = 0.0049$, and the percent ionization is $\alpha \times 100\% = 0.49\%$. Finally, we note that the self-ionization of water makes a substantial contribution to $[H_3O^+]$: $(7.9 \times 10^{-8}\text{M})/(1.3 \times 10^{-7}\text{M}) \times 100\% = 61\%$. Clearly, the self-ionization of water cannot be ignored.

In Exercise 89, we solve this same problem by using a different approach.

ore on Simplifying Assumptions

e usual simplifying assumption is that of treating a weak acid or weak base though it remains essentially nonionized (so that $c - x \approx c$). In general, this sumption will work if the concentration (molarity) of the weak acid, c_A, or at of the weak base, c_B, exceeds the value of K_a or K_b by a factor of at least 0. That is,

$$\frac{c_A\,(\text{or } c_B)}{K_a\,(\text{or } K_b)} > 100$$

In any case, it is important to test the validity of any assumption that you ake. If the assumption is good to within a few percent (say, less than 5%), en it is generally valid. In Example 16-8, the simplifying assumption was od to about 1%, but in Example 16-9 it was off by 40%.

Carey B. Van Loon

▲ Phosphoric acid, H_3PO_4.

16-4 CONCEPT ASSESSMENT

You are given two bottles, each of which contains a 0.1 M solution of an unidentified acid. One bottle is labeled $K_a = 7.2 \times 10^{-4}$, and the other is abeled $K_a = 1.9 \times 10^{-5}$. Which bottle contains the more acidic solution? Which bottle has the acid with the larger pK_a?

6-6 Polyprotic Acids

l the acids listed in Table 16.4 are weak *monoprotic acids*, meaning that their olecules have only one ionizable H atom, even though several of these acids ntain more than one H atom. But some acids have more than one ionizable atom per molecule. These are **polyprotic acids**. Table 16.5 lists ionization nstants for several polyprotic acids. Additional listings can be found in ppendix D. We will focus on phosphoric acid, H_3PO_4.

The H_3PO_4 molecule has *three* ionizable H atoms; it is a *triprotic acid*. It iones in three steps. For each step, we can write an ionization equation and an id ionization constant with a distinctive value of K_a.

◀ Phosphoric acid ranks second only to sulfuric acid among the important commercial acids. It is used in the manufacture of phosphate fertilizers. Various sodium, potassium, and calcium phosphates are used in the food industry.

$$\text{) } H_3PO_4 + H_2O \rightleftharpoons H_3O^+ + H_2PO_4^- \qquad K_{a_1} = \frac{[H_3O^+][H_2PO_4^-]}{[H_3PO_4]} = 7.1 \times 10^{-3}$$

$$\text{) } H_2PO_4^- + H_2O \rightleftharpoons H_3O^+ + HPO_4^{2-} \qquad K_{a_2} = \frac{[H_3O^+][HPO_4^{2-}]}{[H_2PO_4^-]} = 6.3 \times 10^{-8}$$

$$\text{) } HPO_4^{2-} + H_2O \rightleftharpoons H_3O^+ + PO_4^{3-} \qquad K_{a_3} = \frac{[H_3O^+][PO_4^{3-}]}{[HPO_4^{2-}]} = 4.2 \times 10^{-13}$$

◀ The three ionization steps do not contribute equally to the total $[H_3O^+]$. In fact, the pH is dominated by only the first ionization as seen from the different K_a values here. This is only true, of course, for weak acids. The examples here illustrate the points well.

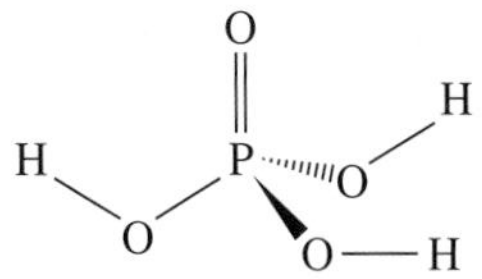

Phosphoric acid

There is a ready explanation for the relative magnitudes of the ionization nstants—that is, for the fact that $K_{a_1} > K_{a_2} > K_{a_3}$. When ionization occurs step (1), a proton (H^+) moves away from an ion with a -1 charge $I_2PO_4^-$). In step (2), the proton moves away from an ion with a -2 charge IPO_4^{2-}), a more difficult separation. As a result, the ionization constant in e second step is smaller than that in the first. Ionization is more difficult ill in step (3).

We can make three key statements about the ionization of phosphoric acid, illustrated in Example 16-10.

. K_{a_1} is so much larger than K_{a_2} and K_{a_3} that essentially all the H_3O^+ is produced in the first ionization step.

TABLE 16.5 Ionization Constants of Some Polyprotic Acids

Acid	Ionization Equilibria	Ionization Constants, *K*	p*K*
Hydrosulfuric[a]	$H_2S + H_2O \rightleftharpoons H_3O^+ + HS^-$	$K_{a_1} = 1.0 \times 10^{-7}$	$pK_{a_1} = 7.00$
Hydrosulfuric[a]	$HS^- + H_2O \rightleftharpoons H_3O^+ + S^{2-}$	$K_{a_2} = 1 \times 10^{-19}$	$pK_{a_2} = 19.0$
Carbonic[b]	$H_2CO_3 + H_2O \rightleftharpoons H_3O^+ + HCO_3^-$	$K_{a_1} = 4.4 \times 10^{-7}$	$pK_{a_1} = 6.36$
Carbonic[b]	$HCO_3^- + H_2O \rightleftharpoons H_3O^+ + CO_3^{2-}$	$K_{a_2} = 4.7 \times 10^{-11}$	$pK_{a_2} = 10.33$
Citric	$H_3C_6H_5O_7 + H_2O \rightleftharpoons H_3O^+ + H_2C_6H_5O_7^-$	$K_{a_1} = 7.5 \times 10^{-4}$	$pK_{a_1} = 3.12$
Citric	$H_2C_6H_5O_7^- + H_2O \rightleftharpoons H_3O^+ + HC_6H_5O_7^{2-}$	$K_{a_2} = 1.7 \times 10^{-5}$	$pK_{a_2} = 4.77$
Citric	$HC_6H_5O_7^{2-} + H_2O \rightleftharpoons H_3O^+ + C_6H_5O_7^{3-}$	$K_{a_3} = 4.0 \times 10^{-7}$	$pK_{a_3} = 6.40$
Phosphoric	$H_3PO_4 + H_2O \rightleftharpoons H_3O^+ + H_2PO_4^-$	$K_{a_1} = 7.1 \times 10^{-3}$	$pK_{a_1} = 2.15$
Phosphoric	$H_2PO_4^- + H_2O \rightleftharpoons H_3O^+ + HPO_4^{2-}$	$K_{a_2} = 6.3 \times 10^{-8}$	$pK_{a_2} = 7.20$
Phosphoric	$HPO_4^{2-} + H_2O \rightleftharpoons H_3O^+ + PO_4^{3-}$	$K_{a_3} = 4.2 \times 10^{-13}$	$pK_{a_3} = 12.38$
Oxalic	$H_2C_2O_4 + H_2O \rightleftharpoons H_3O^+ + HC_2O_4^-$	$K_{a_1} = 5.6 \times 10^{-2}$	$pK_{a_1} = 1.25$
Oxalic	$HC_2O_4^- + H_2O \rightleftharpoons H_3O^+ + C_2O_4^{2-}$	$K_{a_2} = 5.4 \times 10^{-5}$	$pK_{a_2} = 4.27$
Sulfurous[c]	$H_2SO_3 + H_2O \rightleftharpoons H_3O^+ + HSO_3^-$	$K_{a_1} = 1.3 \times 10^{-2}$	$pK_{a_1} = 1.89$
Sulfurous[c]	$HSO_3^- + H_2O \rightleftharpoons H_3O^+ + SO_3^{2-}$	$K_{a_2} = 6.2 \times 10^{-8}$	$pK_{a_2} = 7.21$
Sulfuric[d]	$H_2SO_4 + H_2O \rightleftharpoons H_3O^+ + HSO_4^-$	K_{a_1} = very large	$pK_{a_1} < 0$
Sulfuric[d]	$HSO_4^- + H_2O \rightleftharpoons H_3O^+ + SO_4^{2-}$	$K_{a_2} = 1.1 \times 10^{-2}$	$pK_{a_2} = 1.96$

Acid strength

[a]The value for K_{a_2} of H_2S most commonly found in older literature is about 1×10^{-14}, but current evidence suggests that the value is considerably smaller.

[b]H_2CO_3 cannot be isolated. It is in equilibrium with H_2O and dissolved CO_2. The value given for K_{a_1} is actually for the reaction

$$CO_2(aq) + 2\,H_2O \rightleftharpoons H_3O^+ + HCO_3^-$$

Generally, aqueous solutions of CO_2 are treated *as if* the $CO_2(aq)$ were first converted to H_2CO_3, followed by ionization of the H_2CO_3.

[c]H_2SO_3 is a hypothetical, nonisolatable species. The value listed for K_{a_1} is actually for the reaction

$$SO_2(aq) + 2\,H_2O \rightleftharpoons H_3O^+ + HSO_3^-$$

[d]H_2SO_4 is completely ionized in the first step.

2. So little of the $H_2PO_4^-$ formed in the first ionization step ionizes ar further that we can assume $[H_2PO_4^-] = [H_3O^+]$ in the solution.
3. $[HPO_4^{2-}] \approx K_{a_2}$, regardless of the molarity of the acid.*

Although statement (1) seems essential if statements (2) and (3) are to b valid, it is not as crucial as might first appear. Even for polyprotic acids whos K_a values do not differ greatly between successive ionizations, H_3O^+ often still determined almost exclusively by the K_{a_1} expression, and statemen (2) and (3) remain valid. As long as the polyprotic acid is weak in its fir ionization step, the concentration of the anion produced in this step will be s much less than the molarity of the acid that additional $[H_3O^+]$ produced in th second ionization remains negligible.

*If we assume that $[H_2PO_4^-] = [H_3O^+]$, the second ionization expression reduces to

$$\frac{\cancel{[H_3O^+]}[HPO_4^{2-}]}{\cancel{[H_2PO_4^-]}} = K_{a_2}$$

EXAMPLE 16-10 Calculating Ion Concentrations in a Polyprotic Acid Solution

For a 3.0 M H_3PO_4 solution, calculate **(a)** $[H_3O^+]$; **(b)** $[H_2PO_4^-]$; **(c)** $[HPO_4^{2-}]$; and **(d)** $[PO_4^{3-}]$.

Analyze

For a solution of a *weak* polyprotic acid, the first ionization step produces essentially all the H_3O^+ in solution, so we begin as we would for a weak monoprotic acid solution. The concentrations of the other species are obtained by using the expressions for K_{a_2} and K_{a_3}.

Solve

(a) For the reasons discussed above, let's assume that all the H_3O^+ forms in the *first* ionization step. This is equivalent to thinking of H_3PO_4 as though it were a monoprotic acid, ionizing only in the first step.

	$\mathbf{H_3PO_4 + H_2O \rightleftharpoons}$	$\mathbf{H_3O^+}$	$\mathbf{+\ H_2PO_4^-}$
initial concns:	3.0 M	—	—
changes:	$-x$ M	$+x$ M	$+x$ M
after first ionization:	$(3.0 - x)$ M	x M	x M

Following the usual assumption that x is much smaller than 3.0 and that $3.0 - x \approx 3.0$, we obtain

$$K_{a_1} = \frac{[H_3O^+][H_2PO_4^-]}{[H_3PO_4]} = \frac{x \cdot x}{(3.0 - x)} = \frac{x^2}{3.0} = 7.1 \times 10^{-3}$$

$$x^2 = 0.021 \qquad x = 0.14 \qquad [H_3O^+] = x\text{ M} = 0.14\text{ M}$$

In the assumption $3.0 - x \approx 3.0$, $x = 0.14$, which is 4.7% of 3.0. This is about the maximum error that can be tolerated for an acceptable assumption.

(b) From part (a), $x = [H_2PO_4^-] = [H_3O^+] = 0.14$ M.

(c) To determine $[H_3O^+]$ and $[H_2PO_4^-]$, we assumed that the second ionization is insignificant. Here we must consider the second ionization, no matter how slight; otherwise we would have no source of the ion HPO_4^{2-}. We can represent the second ionization, as shown in the following table. *Note especially how the results of the first ionization enter in.* We start with a solution in which $[H_2PO_4^-] = [H_3O^+] = 0.14$ M.

	$\mathbf{H_2PO_4^- + H_2O \rightleftharpoons}$	$\mathbf{H_3O^+}$	$\mathbf{+\ HPO_4^{2-}}$
from first ionization:	0.14 M	0.14 M	—
changes:	$-y$ M	$+y$ M	$+y$ M
after second ionization:	$(0.14 - y)$ M	$(0.14 + y)$ M	y M

If we assume that y is much smaller than 0.14, then $(0.14 + y) \approx (0.14 - y) \approx 0.14$.

We see from the calculation below that $y = K_{a_2} = 6.3 \times 10^{-8}$.

$$K_{a_2} = \frac{[H_3O^+][HPO_4^{2-}]}{[H_2PO_4^-]} = \frac{(0.14 + y)(y)}{(0.14 - y)} = \frac{\cancel{(0.14)}(y)}{\cancel{(0.14)}} = 6.3 \times 10^{-8}$$

$$[HPO_4^{2-}] = y\text{ M} = 6.3 \times 10^{-8}\text{ M}$$

Note that the assumption is valid.

(d) The PO_4^{3-} ion forms only in the third ionization step. When we write this acid ionization constant expression, we see that we have already calculated the ion concentrations other than $[PO_4^{3-}]$. We can simply solve the K_{a_3} expression for $[PO_4^{3-}]$.

$$K_{a_3} = \frac{[H_3O^+][PO_4^{3-}]}{[HPO_4^{2-}]} = \frac{0.14 \times [PO_4^{3-}]}{6.3 \times 10^{-8}} = 4.2 \times 10^{-13}$$

$$[PO_4^{3-}] = \frac{4.2 \times 10^{-13} \times 6.3 \times 10^{-8}}{0.14} = 1.9 \times 10^{-19}\text{ M}$$

Assess

In this example the major source of hydronium ions is from the first ionization step. In the second step the amount of hydronium ions is around 10^{-8} M, which is negligible compared with 0.14 M.

(continued)

PRACTICE EXAMPLE A: Malonic acid, HOOCCH$_2$COOH, is a diprotic acid used in the manufacture of barbiturates.

$$HOOCCH_2COOH + H_2O \rightleftharpoons H_3O^+ + HOOCCH_2COO^- \qquad K_{a_1} = 1.4 \times 10^{-3}$$
$$HOOCCH_2COO^- + H_2O \rightleftharpoons H_3O^+ + {}^-OOCCH_2COO^- \qquad K_{a_2} = 2.0 \times 10^{-6}$$

Calculate $[H_3O^+]$, $[HOOCCH_2COO^-]$, and $[{}^-OOCCH_2COO^-]$ in a 1.00 M solution of malonic acid.

PRACTICE EXAMPLE B: Oxalic acid, found in the leaves of rhubarb and other plants, is a diprotic acid.

$$H_2C_2O_4 + H_2O \rightleftharpoons H_3O^+ + HC_2O_4^- \qquad K_{a_1} = ?$$
$$HC_2O_4^- + H_2O \rightleftharpoons H_3O^+ + C_2O_4^{2-} \qquad K_{a_2} = ?$$

An aqueous solution that is 1.05 M $H_2C_2O_4$ has pH = 0.67. The free oxalate ion concentration in this solution is $[C_2O_4^{2-}] = 5.3 \times 10^{-5}$ M. Determine K_{a_1} and K_{a_2} for oxalic acid.

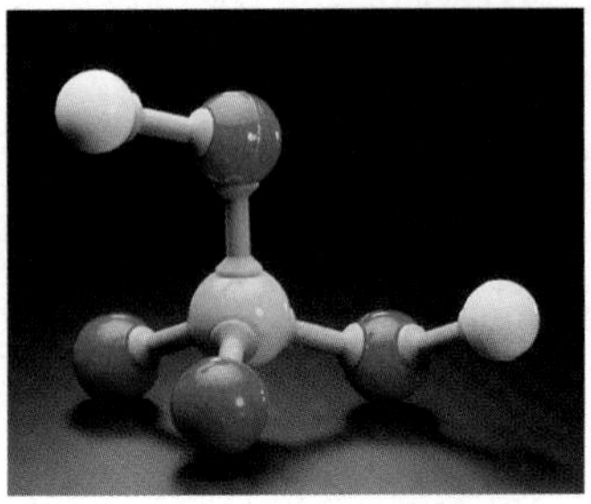

Carey B. Van Loon

▲ Sulfuric acid, H_2SO_4.

A Somewhat Different Case: H_2SO_4

Sulfuric acid differs from most polyprotic acids in this important respect:] is a *strong* acid in its first ionization and a *weak* acid in its second. Ionizatio is complete in the first step, which means that in most $H_2SO_4(aq)$ solution $[H_2SO_4] \approx 0$ M. Thus, if a solution is 0.50 M H_2SO_4, we can treat it a though it were 0.50 M H_3O^+ and 0.50 M HSO_4^- initially. Then we can dete mine the extent to which ionization of HSO_4^- produces additional H_3O and SO_4^{2-}, as illustrated in Example 16-11.

EXAMPLE 16-11 Calculating Ion Concentrations in Sulfuric Acid Solutions: Strong Acid Ionization Followed by Weak Acid Ionization

Calculate $[H_3O^+]$, $[HSO_4^-]$, and $[SO_4^{2-}]$ in 0.50 M H_2SO_4.

Analyze

We will modify the approach we used in Example 16-10 to incorporate the fact that for H_2SO_4 the first ionization step goes to completion.

Solve

	H_2SO_4	+ H_2O $\longrightarrow$	H_3O^+	+ HSO_4^-
initial concn:	0.50 M		—	—
changes:	−0.50 M		+0.50 M	+0.50 M
after first ionization:	≈ 0		0.50 M	0.50 M

	HSO_4^-	+ H_2O $\rightleftharpoons$	H_3O^+	+ SO_4^{2-}
from first ionization:	0.50 M		0.50 M	—
changes:	$-x$ M		$+x$ M	$+x$ M
after second ionization:	$(0.50 - x)$ M		$(0.50 + x)$ M	x M

We need to deal only with the ionization constant expression for K_{a_2}. If we assume that x is much smaller than 0.50, then $(0.50 + x) \approx (0.50 - x) \approx 0.50$ and

$$K_{a_2} = \frac{[H_3O^+][SO_4^{2-}]}{[HSO_4^-]} = \frac{(0.50 + x) \cdot x}{(0.50 - x)} = \frac{\cancel{0.50} \cdot x}{\cancel{0.50}} = 1.1 \times 10^{-2}$$

Our results, then, are

$$[H_3O^+] = 0.50 + x = 0.51\text{ M}; \qquad [HSO_4^-] = 0.50 - x = 0.49\text{ M}$$
$$[SO_4^{2-}] = x = K_{a_2} = 0.011\text{ M}$$

Assess

In obtaining these results, we assumed that x was much smaller than 0.50. This assumption is appropriate because $x = 0.011$ is only 2.2% of 0.50. Had x been greater than 5% of the initial molarity of the solution, then the assumption would not have been appropriate. Such a situation arises when dealing with more dilute solutions of H_2SO_4.

PRACTICE EXAMPLE A: Calculate $[H_3O^+]$, $[HSO_4^-]$, and $[SO_4^{2-}]$ in 0.20 M H_2SO_4.

PRACTICE EXAMPLE B: Calculate $[H_3O^+]$, $[HSO_4^-]$, and $[SO_4^{2-}]$ in 0.020 M H_2SO_4. [*Hint:* Is the assumption that $[HSO_4^-] = [H_3O^+]$ valid?]

16-7 Simultaneous or Consecutive Acid–Base Reactions: A General Approach

o far, we have used various simplifying assumptions when solving roblems. For example, when calculating the pH of 0.1 M CH_3COOH in xample 16-8, we assumed that the self-ionization of water makes a negligible ontribution and that only a small fraction of the CH_3COOH ionizes. When alculating the pH of 3.0 M H_3PO_4, we made similar assumptions and, in ddition, assumed that the first ionization step determines the total amount of H_3O^+ in solution. How should we proceed if we want to do these calculations *vithout* making these simplifying assumptions? The following approach can elp get us on the right track.

1. Identify the species present in solution (excluding H_2O). Write down balanced chemical equations for the reactions involving these species. Consider the concentrations of these species as unknowns.
2. Write equations that include these species. The number of equations involving these species should match the number of unknowns. The equations are of three types.
   (a) equilibrium constant expressions
   (b) material balance equations
   (c) an electroneutrality condition, also known as a charge balance equation
3. Solve the system of equations for the unknowns.

Let's apply this approach to set up the calculation of the pH of 0.1 M H_3PO_4. Ve know that H_3PO_4 is a weak triprotic acid that ionizes in three consecutive teps to produce $H_2PO_4^-$, HPO_4^{2-}, and PO_4^{3-} ions. The ionization of H_3PO_4 also roduces H_3O^+ ions. There will also be OH^- ions in solution because of the elf-ionization of water.

Species in solution

$$H_3PO_4,\ H_2PO_4^-,\ HPO_4^{2-},\ PO_4^{3-},\ H_3O^+,\ OH^-$$

Reactions and equilibrium constant expressions

$$H_3PO_4(aq) + H_2O(l) \rightleftharpoons H_3O^+(aq) + H_2PO_4^-(aq) \qquad K_{a_1} = \frac{[H_3O^+][H_2PO_4^-]}{[H_3PO_4]} = 7.1 \times 10^{-3}$$

$$H_2PO_4^-(aq) + H_2O(l) \rightleftharpoons H_3O^+(aq) + HPO_4^{2-}(aq) \qquad K_{a_2} = \frac{[H_3O^+][HPO_4^{2-}]}{[H_2PO_4^-]} = 6.3 \times 10^{-8}$$

$$HPO_4^{2-}(aq) + H_2O(l) \rightleftharpoons H_3O^+(aq) + PO_4^{3-}(aq) \qquad K_{a_3} = \frac{[H_3O^+][PO_4^{3-}]}{[HPO_4^{2-}]} = 4.2 \times 10^{-13}$$

$$2\,H_2O(l) \rightleftharpoons H_3O^+(aq) + OH^-(aq) \qquad K_w = [H_3O^+][OH^-]$$

The equilibrium constant expressions above give us *four* equations involvir *six* unknown concentrations: $[H_3PO_4]$, $[H_2PO_4^-]$, $[HPO_4^{2-}]$, $[PO_4^{3-}]$, $[H_3O^+$ and $[OH^-]$. We need two additional equations. We get them by writing a mat rial balance equation and a charge balance equation.

The following material balance equation accounts for the fact that the su of the concentrations of the phosphorus-containing species must equal the in tial or stoichiometric concentration of H_3PO_4.

Material balance equation (MBE)

$$0.10\ M = [H_3PO_4] + [H_2PO_4^-] + [HPO_4^{2-}] + [PO_4^{3-}]$$

The following charge balance equation (or electroneutrality condition) simpl verifies that the solution carries no net charge. The sum of the positive charg must equal the sum of the negative charges. We can sum these charges on mol/L basis.

Charge balance equation (CBE)

$$[H_3O^+] = [H_2PO_4^-] + 2 \times [HPO_4^{2-}] + 3 \times [PO_4^{3-}] + [OH^-]$$

KEEP IN MIND

that although this method gives you a quick way to set up a solution equilibrium calculation, more information may be needed for you to arrive at an answer without undue effort. For example, answers to the general questions posed on page 752 may point the way to simplifying the algebraic solution.

In the charge balance equation, the concentration of each species is multiplie by the magnitude of the charge on that species. For example, we multip $[PO_4^{3-}]$ by *three* because each PO_4^{3-} ion carries three units of negative charge

With the four equilibrium constant expressions, a material balance equatio and a charge balance equation, we have six equations involving si unknowns. In principle, this system of equations can be solved to find the si unknown concentrations, either by making appropriate simplifying approx mations or by computerized calculation.

The method outlined here is ideal for computerized calculation. Moreove because the additional manipulations required to convert concentrations t activities can be incorporated into the calculations, the results are generall both more accurate and more readily obtained than are those derived by tr ditional methods.

16-5 CONCEPT ASSESSMENT

Write material and charge balance equations for **(a)** 0.010 M H_2SO_4; **(b)** 0.025 M NH_3.

16-8 Ions as Acids and Bases

In our discussion to this point, we have emphasized the behavior of electri cally neutral molecules as acids (for example, HCl, CH_3COOH, H_3PO_4) or a bases (for example, NH_3, CH_3NH_2). We have also seen, however, that ion can act as acids or bases. For instance, in the second ionization step of H_3PO (part (c) of Example 16-10), the $H_2PO_4^-$ ion acts as an acid.

Let's think about how each of the following can be described as a acid–base reaction.

$$\underset{\text{Acid(1)}}{NH_4^+} + \underset{\text{Base(2)}}{H_2O} \rightleftharpoons \underset{\text{Base(1)}}{NH_3} + \underset{\text{Acid(2)}}{H_3O^+} \quad \textbf{(16.17)}$$

$$\underset{\text{Base(1)}}{CH_3COO^-} + \underset{\text{Acid(2)}}{H_2O} \rightleftharpoons \underset{\text{Acid(1)}}{CH_3COOH} + \underset{\text{Base(2)}}{OH^-} \quad \textbf{(16.18)}$$

In reaction (16.17), NH_4^+ is an *acid*, giving up a proton to water, a *base* Equilibrium in this reaction is described by means of the *acid ionization con stant* of the ammonium ion, NH_4^+.

$$K_a = \frac{[NH_3][H_3O^+]}{[NH_4^+]} = ? \quad \textbf{(16.19)}$$

vo of the concentrations in equation (16.19)—$[NH_3]$ and $[NH_4^+]$—are the me as in the K_b expression for NH_3, the conjugate base of NH_4^+. It seems at K_a for NH_4^+ and K_b for NH_3 should bear some relationship to each other, d they do. The easiest way to see this is to multiply both the numerator and e denominator of (16.19) by $[OH^-]$. The product $[H_3O^+] \times [OH^-]$ is the ion oduct of water, K_w, shown in red. The other concentrations, shown in blue, present the *inverse* of K_b for NH_3. The value obtained, 5.6×10^{-10}, is the issing value of K_a in expression (16.19).

$$K_a = \frac{[NH_3][H_3O^+][OH^-]}{[NH_4^+][OH^-]} = \frac{K_w}{K_b} = \frac{1.0 \times 10^{-14}}{1.8 \times 10^{-5}} = 5.6 \times 10^{-10}$$

nis result is an important consequence of the Brønsted–Lowry theory.

The product of the ionization constants of an acid and its conjugate base equals the ion product of water.

$$K_a\,(\text{acid}) \times K_b\,(\text{its conjugate base}) = K_w$$
$$K_b\,(\text{base}) \times K_a\,(\text{its conjugate acid}) = K_w \qquad (16.20)$$

◀ In many tabulations of ionization constants, only K_a values are listed, whether for neutral molecules or for ions. Equation (16.20) can be used to obtain the values of their conjugates.

In reaction (16.18), CH_3COO^- acts as a *base* by taking a proton from water, 1 *acid*. Here, equilibrium is described by means of the *base ionization constant* the acetate ion, CH_3COO^-. With expression (16.20) we can evaluate K_b.

$$K_b = \frac{[CH_3COOH][OH^-]}{[CH_3COO^-]} = \frac{K_w}{K_a(CH_3COOH)} = \frac{1.0 \times 10^{-14}}{1.8 \times 10^{-5}} = 5.6 \times 10^{-10}$$

From equation (16.20), we deduce that (1) the stronger the acid, the eaker its conjugate base; and (2) the weaker the acid, the stronger its conju- ate base. It is easy to misinterpret the second statement. It does *not* mean at the conjugate base of a weak acid is a strong base. When we compare the alues of the ionization constants for CH_3COOH and CH_3COO^-, it is clear at *the conjugate base of a weak acid is a weak base*. It is also true that the conju- ate acid of a weak base is a weak acid. The following statement summarizes ese relationships.

The conjugate of *weak* is *weak*.

Now, let's use equation (16.20) to calculate K_b for the conjugate base of a rong acid. The ionization constant of a strong acid is very large: $K_a \gg 1$ nd, therefore, K_b will be much smaller than K_w. For example, the ionization onstant of HI has been estimated to be about 10^9. Therefore,

$$K_b(I^-) = K_w/K_a = 10^{-14}/10^9 = 10^{-23}$$

Clearly, I^- is an extremely weak base. In fact, I^- is such a weak base that the H of a solution of NaI is no different from that of pure water (see Exercise 100). 1 other words, the conjugate base of a strong acid is an extremely weak base oo weak to affect the pH of a solution). Also, the conjugate acid of a strong base an extremely weak acid.

The conjugate of *strong* is *extremely weak*.

16-6 CONCEPT ASSESSMENT

A handbook that lists only pK_a values for weak electrolytes has the following entry for 1,2-ethanediamine, $NH_2CH_2CH_2NH_2$: pK_1 = 6.85(+2); pK_2 = 9.92(+1), and 2-aminopropanoic acid, $NH_2CH(CH_3)COOH$: pK_1 = 2.34(+1); pK_2 = 9.87(0). Interpret these handbook entries by writing equations for the ionization reactions to which these pK values apply. What are the corresponding values of the base ionization constants K_{b_1} and K_{b_2}?

Hydrolysis

In pure water at 25 °C, $[H_3O^+] = [OH^-] = 1.0 \times 10^{-7}$ M and pH = 7.00. *Pu water is pH neutral*. When NaCl dissolves in water at 25 °C, complete dissoci tion into Na^+ and Cl^- ions occurs, and the pH of the solution remains 7.0 This is because neither Na^+ nor Cl^- reacts with water.

$$Na^+ + H_2O \longrightarrow \text{no reaction} \qquad Cl^- + H_2O \longrightarrow \text{no reaction}$$

The fact that Cl^- does not react with water comes as no surprise. Becau the Cl^- ion is the conjugate base of a strong acid (HCl), the Cl^- ion is a extremely weak base, has little or no tendency to become protonated, and too weak to affect the pH of the solution. (In Section 16-11, we'll provide a explanation for why Na^+ does not affect the pH of a solution.)

As shown in Figure 16-9, when NH_4Cl is added to water, the pH fal below 7. This means that $[H_3O^+] > [OH^-]$ in the solution. A reaction pro ducing H_3O^+ must occur.

$$NH_4^+ + H_2O \rightleftharpoons NH_3 + H_3O^+$$

KEEP IN MIND

that many students find hydrolysis problems challenging. The equilibrium calculations are actually quite straightforward. The challenging aspect of these problems is recognizing *when* a hydrolysis reaction is the one on which to base the calculations.

The reaction between NH_4^+ and H_2O is fundamentally no different fro other acid–base reactions. A reaction between an ion and water, however, often called a **hydrolysis** reaction. We say that ammonium ion *hydrolyzes* (an chloride ion does not).

When sodium acetate is dissolved in water, the pH rises above 7 (se Figure 16-9). This means that $[OH^-] > [H_3O^+]$ in the solution. Here, aceta ion hydrolyzes.

$$CH_3COO^- + H_2O \rightleftharpoons CH_3COOH + OH^-$$

Carey B. Van Loon

▲ FIGURE 16-9
Ions as acids and bases
Each of these 1 M solutions contains bromthymol blue indicator, which has the following colors:

pH < 7	pH = 7	pH > 7
Yellow	Green	Blue

(Left) $NH_4Cl(aq)$ is acidic. (Center) NaCl(aq) is neutral. (Right) $NaCH_3COO(aq)$ is basic

e pH of Salt Solutions

e are now in a position to make both qualitative predictions and quantitative lculations concerning the pH values of aqueous solutions of salts. hichever of these tasks is called for, note that hydrolysis takes place only if ere is a chemical reaction producing a weak acid or weak base. The follow-g generalizations are useful.

- *Metal ions* with a +1 or +2 charge usually do not affect the pH of a solution. However, metal ions with higher charges (for example, Al^{3+}, Fe^{3+}, Cr^{3+}, etc.) may affect the pH of a solution. It may seem surprising that a metal ion, such as Al^{3+}, can affect the pH of a solution. We must remember that, in aqueous solution, ions are hydrated. If the charge on a metal ion is large (typically +3 or higher), then water molecules nearest the metal cation—those in the so-called primary hydration sphere—show an increased tendency to donate a proton to a water molecule in the bulk solution. The transfer of a proton from a water molecule in the primary hydration sphere to a water molecule in the bulk solution produces a H_3O^+ ion, thereby increasing $[H_3O^+]$ in the solution. (We will explore these ideas further in Section 16-10.)
- Some *polyatomic cations* act as acids in water. Important examples are NH_4^+ and the protonated forms of amines. Recall that amines are derived from NH_3, with one or more of the H atoms replaced by other groups. In a *primary* amine (general formula RNH_2), *one* of the H atoms of NH_3 has been replaced by another group, R, which is typically but not always a group of carbon and hydrogen atoms. Like NH_3, amines are weak bases. Therefore, the NH_4^+ and RNH_3^+ ions are weak acids.
- Many *anions* act as bases in water. When considering the nature of a negative ion, A^-, it is helpful to think about the acid strength of its conjugate acid, HA. This is because the tendency of A^- to act as a base is related to the acid strength of HA through the expression $K_b(A^-) = K_w/K_a(HA)$. If HA is weak acid, then A^- is a weak base: It will hydrolyze in water and affect the pH of a solution. On the other hand, if HA is a strong acid, then A^- is an extremely weak base and will not react with H_2O to any appreciable extent.

We must be extra careful when dealing with amphiprotic anions. As an xample, consider a solution of Na_2HPO_4. When this salt dissolves in water, a^+ and HPO_4^{2-} ions are produced. The Na^+ ion does not react with water, ut the HPO_4^{2-} ion can react with water in two ways: it can lose a proton to ater to become PO_4^{3-}, or it can gain a proton from H_2O to become $H_2PO_4^-$. ne HPO_4^{2-} ion is amphiprotic. To decide whether a solution of Na_2HPO_4 is idic or basic, we must determine whether the HPO_4^{2-} ion has a greater ten-ency to act as an acid or as a base. To make this determination, we can com-are the K_a and K_b values for the HPO_4^{2-} ion. The K_a and K_b values for the PO_4^{2-} ion are calculated from the ionization constants of H_3PO_4, as demon-rated below.

$$\text{) } HPO_4^{2-} + H_2O \rightleftharpoons PO_4^{3-} + H_3O^+ \quad K_a(HPO_4^{2-}) = ?$$

$$\text{) } HPO_4^{2-} + H_2O \rightleftharpoons H_2PO_4^- + OH^- \quad K_b(HPO_4^{2-}) = ?$$

eaction (1) is the third ionization of H_3PO_4. Therefore,

$$K_a(HPO_4^{2-}) = K_{a_3}(H_3PO_4) = 4.2 \times 10^{-13}$$

reaction (2), HPO_4^{2-} reacts as a base. Since $H_2PO_4^-$ and HPO_4^{2-} are a conju-ate acid–base pair, $K_b(HPO_4^{2-}) = K_W/K_a(H_2PO_4^-)$. By definition, $K_a(H_2PO_4^-)$ the equilibrium constant for the following reaction in which $H_2PO_4^-$ acts as acid in a reaction with water.

$$H_2PO_4^- + H_2O \rightleftharpoons HPO_4^{2-} + H_3O^+ \quad K_a(H_2PO_4^-) = ?$$

This reaction is also the second ionization of H_3PO_4. Therefore, $K_a(H_2PO_4^-)$ $K_{a_2}(H_3PO_4) = 6.3 \times 10^{-8}$. We have

$$K_b(HPO_4^{2-}) = \frac{K_W}{K_a(H_2PO_4^-)} = \frac{K_W}{K_{a_2}(H_3PO_4)} = \frac{1.0 \times 10^{-14}}{6.3 \times 10^{-8}} = 1.6 \times 10^{-7}$$

Since $K_b(HPO_4^{2-})$ is greater than $K_a(HPO_4^{2-})$, the HPO_4^{2-} ion exhibits greater tendency to react with water as a base than as an acid. Therefore, solution of Na_2HPO_4 is basic.

EXAMPLE 16-12 Making Qualitative Predictions About Hydrolysis Reactions

Predict whether each of the following solutions is acidic, basic, or pH neutral: **(a)** NaOCl(aq); **(b)** KCl(aq); **(c)** NH_4NO_3(aq).

Analyze

We need to recognize that all three salts are strong electrolytes and completely dissociate in water. Then, we can consider the ions separately and ask which will react (either as an acid or as a base) with water. Recall that the anions from strong acids (e.g., Cl^-) and metal cations with a +1 charge do not participate in hydrolysis.

Solve

(a) The ions present are Na^+, which does not hydrolyze, and OCl^-, which does. OCl^- is the conjugate base of HOCl and forms a basic solution.

$$OCl^- + H_2O \rightleftharpoons HOCl + OH^-$$

(b) Neither K^+ nor Cl^- hydrolyzes. KCl(aq) is neutral—that is, pH = 7 at 25 °C.

(c) NH_4^+ hydrolyzes, but NO_3^- does not (HNO_3 is a strong acid), so NO_3^- is an extremely weak base.

$$NH_4^+ + H_2O \rightleftharpoons NH_3 + H_3O^+$$

This reaction generates H_3O^+ and causes $[H_3O^+] > [OH^-]$. Thus, NH_4NO_3(aq) is acidic.

Assess

Recognizing that certain ions in solution can undergo hydrolysis in water will be an important concept in the next chapter. It is important to learn and understand here how this concept works.

PRACTICE EXAMPLE A: Predict whether each of the following 1.0 M solutions is acidic, basic, or pH neutral: **(a)** $CH_3NH_3^+NO_3^-$(aq); **(b)** NaI(aq); **(c)** $NaNO_2$(aq).

PRACTICE EXAMPLE B: Write equations for *two* reactions of $H_2PO_4^-$ with water, and explain which reaction occurs to the greater extent.

16-7 CONCEPT ASSESSMENT

Write a chemical equation showing how an HCO_3^- ion can act as both an acid and a base in aqueous solution. Without doing any pH calculations, determine whether 0.10 M $NaHCO_3$ is acidic, basic, or pH neutral. What about 0.10 M Na_2CO_3?

EXAMPLE 16-13 Evaluating Ionization Constants for Hydrolysis Reactions

Both sodium nitrite, $NaNO_2$, and sodium benzoate, NaC_6H_5COO, are used as food preservatives. If separate solutions of these two salts have the same molarity, which solution will have the *higher* pH?

Analyze

The anion in each salt is the conjugate base of a weak acid. Therefore, the anions will act as *weak bases*, making their solutions somewhat basic. The relevant reactions are

$$NO_2^- + H_2O \rightleftharpoons HNO_2 + OH^- \qquad K_b(NO_2^-) = ?$$
$$C_6H_5COO^- + H_2O \rightleftharpoons C_6H_5COOH + OH^- \qquad K_b(C_6H_5COO^-) = ?$$

To calculate the required K_b values, we will need to recall the relationship between K_a and K_b.

Solve

Our task is to determine the K_b values, neither of which is listed in a table in this chapter. Table 16.4 does list K_a for the conjugate acids, however. Equation (16.20) can be used to write

$$K_b \text{ of } NO_2^- = \frac{K_w}{K_a(HNO_2)} = \frac{1.0 \times 10^{-14}}{7.2 \times 10^{-4}} = 1.4 \times 10^{-11}$$

$$K_b \text{ of } C_6H_5COO^- = \frac{K_w}{K_a(C_6H_5COOH)} = \frac{1.0 \times 10^{-14}}{6.3 \times 10^{-5}} = 1.6 \times 10^{-10}$$

Because the K_b of $C_6H_5COO^-$ is larger than that of NO_2^-, the benzoate ion will hydrolyze to a greater extent than the nitrite ion and will give a solution with a higher $[OH^-]$. A sodium benzoate solution is more basic and has a higher pH than a sodium nitrite solution of the same concentration.

Assess

We could have reasoned out the answer without performing any calculations by focusing instead on the conjugate acids. Because HNO_2 is a stronger acid than C_6H_5COOH, the NO_2^- ion must be a weaker base than the $C_6H_5COO^-$ ion. This is all the information we need to decide which of the two solutions is more basic.

PRACTICE EXAMPLE A: The organic bases cocaine ($pK_b = 8.41$) and codeine ($pK_b = 7.95$) react with hydrochloric acid to form salts (similar to the formation of NH_4Cl by the reaction of NH_3 and HCl). If solutions of the following salts have the same molarity, which solution would have the higher pH: cocaine hydrochloride, $C_{17}H_{21}O_4NH^+Cl^-$, or codeine hydrochloride, $C_{18}H_{21}ClO_3NH^+Cl^-$?

PRACTICE EXAMPLE B: Predict whether the solution $NH_4CN(aq)$ is acidic, basic, or neutral; and explain the basis of your prediction.

EXAMPLE 16-14 Calculating the pH of a Solution in Which Hydrolysis Occurs

Sodium cyanide, NaCN, is extremely poisonous, but it has very useful applications in gold and silver metallurgy and in the electroplating of metals. Aqueous solutions of cyanides are especially hazardous if they become acidified, because toxic hydrogen cyanide gas, HCN(g), is released. Are NaCN(aq) solutions normally acidic, basic, or pH neutral? What is the pH of 0.50 M NaCN(aq)? Note that solutions containing cyanide ion must be handled with extreme caution. They should be handled only in a fume hood by an operator wearing protective clothing.

Analyze

Na^+ does not hydrolyze, but as represented below, CN^- does hydrolyze, producing a basic solution. The question now becomes a hydrolysis equilibrium problem.

Solve

In the tabulation of the concentrations of the species involved in the hydrolysis reaction, let $[OH^-] = x$ M.

	CN^-	+ $H_2O \rightleftharpoons$	HCN	+	OH^-
initial concns:	0.50 M		—		—
changes:	$-x$ M		$+x$ M		$+x$ M
equil concns:	$(0.50 - x)$ M		x M		x M

Use equation (16.20) to obtain a value of K_b.

$$K_b = \frac{K_w}{K_a(HCN)} = \frac{1.0 \times 10^{-14}}{6.2 \times 10^{-10}} = 1.6 \times 10^{-5}$$

(continued)

Now return to the tabulated data.

$$K_b = \frac{[HCN][OH^-]}{[CN^-]} = \frac{x \cdot x}{0.50 - x} = \frac{x^2}{0.50 - x} = 1.6 \times 10^{-5}$$

Assume: $x \ll 0.50$ and $0.50 - x \approx 0.50$.

$$x^2 = 0.50 \times 1.6 \times 10^{-5} = 0.80 \times 10^{-5} = 8.0 \times 10^{-6}$$

$$x = (8.0 \times 10^{-6})^{1/2} = 2.8 \times 10^{-3}$$

$$[OH^-] = x\,M = 2.8 \times 10^{-3}\,M$$

$$pOH = -\log[OH^-] = -\log(2.8 \times 10^{-3}) = 2.55$$

$$pH = 14.00 - pOH = 14.00 - 2.55 = 11.45$$

Assess

We see that in this example, the simplifying assumption works. We also note that the solution is fairly basic for a relatively dilute solution of a salt of a weak acid and a strong base.

PRACTICE EXAMPLE A: Sodium fluoride, NaF, is found in some toothpaste formulations as an anticavity agent. What is the pH of 0.10 M NaF(aq)?

PRACTICE EXAMPLE B: The pH of an aqueous solution of NaCN is 10.38. What is $[CN^-]$ in this solution?

16-9 Qualitative Aspects of Acid–Base Reactions

So far, we have considered only reactions in which an acid or a base reac with water. In this section, we will focus on the general case of a reacti involving an acid HA and a base B.

$$\underset{\text{Acid}}{HA} + \underset{\text{Base}}{B} \rightleftharpoons \underset{\text{Base}}{A^-} + \underset{\text{Acid}}{BH^+} \quad \text{(16.2}$$

Of course, the reaction between HA and B is reversible. In this section, we a interested in obtaining an answer to the question, Does the equilibrium fav the reactants, HA and B, or the products, A^- and BH^+?

For an acid–base reaction, equilibrium favors the formation of the weaker acid and the weaker base.

To justify this statement, we may proceed as follows. Reaction (16.21) is tl sum of the following reactions.

$$HA(aq) + H_2O(l) \rightleftharpoons A^-(aq) + H_3O^+(aq) \qquad K_1 = K_a(HA)$$

$$B(aq) + H_2O(l) \rightleftharpoons BH^+(aq) + OH^-(aq) \qquad K_2 = K_b(B)$$

$$H_3O^+(aq) + OH^-(aq) \rightleftharpoons 2\,H_2O(l) \qquad K_3 = 1/K_w$$

Therefore, the equilibrium constant for reaction (16.21) is K $K_a(HA)K_b(B)/K_w$. If we make the substitution $K_b(B) = K_w/K_a(BH^+)$, we obtai $K = K_a(HA)/K_a(BH^+)$. If instead we make the substitution $K_a(HA)$ $K_w/K_b(A^-)$, we obtain $K = K_b(B)/K_b(A^-)$. Consequently, the equilibrium cor stant for the reaction (16.21) is

$$K = \frac{K_a(HA)K_b(B)}{K_w} = \frac{K_a(HA)}{K_a(BH^+)} = \frac{K_b(B)}{K_b(A^-)} \quad \text{(16.2}$$

Suppose that HA is a stronger acid than BH^+. Then $K_a(HA) > K_b(BH^+)$ an $K > 1$. Because K is greater than 1, equilibrium favors the formation of BH weaker acid. If we assume instead that BH^+ is the stronger acid, then K ll be less than 1, and equilibrium favors the formation of HA, the weaker d. Using similar reasoning, but focusing instead on the bases B and A^-, arrive at the conclusion that *equilibrium always favors the formation of the aker base.*

Let's use the generalization above to predict the favored direction for the action of acetic acid, CH_3COOH, and pyridine, C_5H_5N. The $K_a(CH_3COOH)$ d $K_b(C_5H_5N)$ values are given in Table 16.4. The $K_b(CH_3COO^-)$ and $(C_5H_5NH^+)$ values are calculated by using equation (16.20).

$$\underset{\substack{\text{Acid(1)} \\ K_a = 1.8 \times 10^{-5}}}{CH_3COOH} + \underset{\substack{\text{Base(2)} \\ K_b = 1.5 \times 10^{-9}}}{C_5H_5N} \rightleftharpoons \underset{\substack{\text{Base(1)} \\ K_b = 5.6 \times 10^{-10}}}{CH_3COO^-} + \underset{\substack{\text{Acid(2)} \\ K_a = 6.7 \times 10^{-6}}}{C_5H_5NH^+}$$

Equilibrium favors the formation of CH_3COO^-, the weaker of the two ses, and $C_5H_5NH^+$, the weaker of the two acids.

We can use equation (16.22) to justify another important observation about id–base reactions, one that we will use repeatedly in the next chapter.

If the acid or base in an acid–base reaction is strong, they react essentially to completion.

r example, if $K_a(HA) = 10^2$, then K for the reaction of HA and B, reaction .21), is

$$K = \frac{K_a(HA)K_b(B)}{K_w} = \frac{10^2 K_b(B)}{10^{-14}} = 10^{16} \times K_b(B)$$

As long as K_b for the base is greater than 10^{-14}, the value of K will be large reater than 10^2) and the reaction between HA and B will go essentially to mpletion.

Similarly, if B is a strong base, the reaction between HA and B will go essenlly to completion, provided K_a for HA is greater than K_w.

16-8 CONCEPT ASSESSMENT

The equation representing the neutralization of acetic acid, CH_3COOH, by a base B is $CH_3COOH(aq) + B(aq) \rightleftharpoons CH_3COO^-(aq) + BH^+(aq)$. Of the bases listed in Table 16.4, which would be effective for neutralizing essentially all of the CH_3COOH in a sample, assuming that CH_3COOH and B are initially present in equal amounts?

6-10 Molecular Structure and Acid–Base Behavior

e have now dealt with a number of aspects of acid–base chemistry, both alitatively and quantitatively. Yet some very fundamental questions still main to be answered, such as these: Why is HCl a strong acid, whereas HF is weak acid? Why is acetic acid (CH_3COOH) a stronger acid than ethanol H_3CH_2OH) but a weaker acid than chloroacetic acid ($ClCH_2COOH$)?

These questions involve relative acid strengths. In this section, we will amine the relationship between molecular structure and the strengths of ids and bases.

Strengths of Binary Acids

Because the behavior of acids requires the loss of a proton through bo breakage, acid strength and bond strength appear to be related. In general, t stronger the H—X bond, the *weaker* the acid is. Stronger bonds are characte ized by short bond lengths and high bond dissociation energies. The appr priate bond dissociation energy to use is the ionization of the H—X bond the gas phase:

$$HX(g) \longrightarrow H^+(g) + X^-(g) \qquad (16.2$$

The bond dissociation energy for the gas phase ionization reacti (equation 16.23) can be obtained by using the following thermodynamic cyc

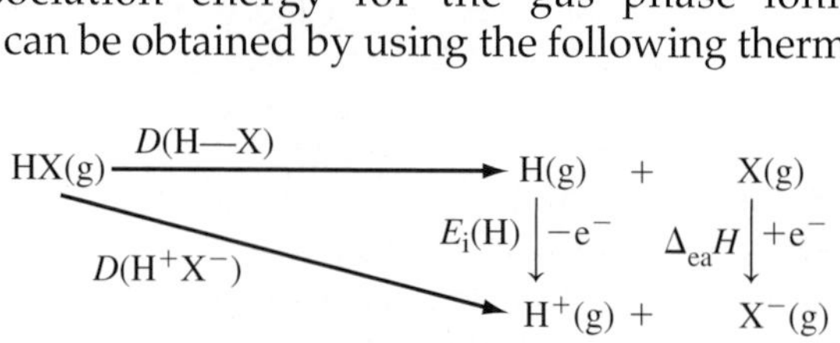

We can write $D(H^+X^-) = D(H{-}X) + E_i(H) + \Delta_{ea}H$, where $D(H{-}X)$ the bond dissociation energy for $HX(g) \rightarrow H(g) + X(g)$, $E_i(H)$ is the ionizati energy of the hydrogen atom, and $\Delta_{ea}H$ is the electron affinity of X, as defin on page 397. $D(H^+X^-)$ is called the *heterolytic bond dissociation energy.*

▶ The dissociation of a gas-phase molecule, AB, into A^+ and B^- is called *heterolysis* and the energy change for this process is called the *heterolytic bond dissociation energy*. The dissociation of a gas-phase molecule, AB, into A and B is called *homolysis*. Thus, the bond dissociation energy (*D*), introduced in Chapter 10, is more precisely called the *homolytic bond dissociation energy.*

Figure 16-10 shows $D(H^+X^-)$ values of binary acids formed by several el ments. For binary acids, acid strength increases as the heterolytic bond diss ciation energy decreases. Intuitively, this makes a lot of sense. The lower t energy requirement for converting an H—X molecule into H^+ and X^- ior the greater the acid strength. Can we explain the trend in acid strengths terms of (homolytic) bond dissociation energies, $D(H{-}X)$? Not really. F example, $D(H{-}X)$ values tend to increase from left to right in Figure 16-1 suggesting that the acid strength should decrease across the row. But the don't. The energy requirements for converting an H—X molecule into H ar X atoms are not reliable for predicting trends in acid strengths.

Trends in the strengths of binary acids are often explained by consideri variations in bond length and bond polarity. Such rationalizations are possib

increasing acid strength →

	H—CH$_3$	H—NH$_2$	H—OH	H—F
K_a	1×10^{-60}	1×10^{-34}	1.8×10^{-16}	6.6×10^{-4}
D(H—X)	414	389	464	565
D(H⁺X⁻)	**1717**	**1630**	**1598**	**1549**
			H—SH	**H—Cl**
K_a			1.0×10^{-7}	1×10^{6}
D(H—X)			368	431
D(H⁺X⁻)			**1458**	**1394**
			H—SeH	**H—Br**
K_a			1.3×10^{-4}	1×10^{8}
D(H—X)			335	364
D(H⁺X⁻)			**1434**	**1351**
			H—TeH	**H—I**
K_a			2.3×10^{-3}	1×10^{9}
D(H—X)			277	297
D(H⁺X⁻)			**1386**	**1314**

▶ FIGURE 16-10
Bond dissociation energies (kJ mol^{-1}) and K_a values for some binary acids
Homolytic bond dissociation energies, D(X—H), tend to increase from left to right and decrease from top to bottom in this table. Heterolytic bond dissociation energies, D(H⁺X⁻), decrease from left to right and from top to bottom in this table. The arrows indicate that acid strengths (K_a values) increase from left to right and from top to bottom. The K_a values for NH_3 and CH_4 are very small. These molecules do not behave as acids in water.

t a little tricky. Intuitively, we expect the acid strength of H—X to increase as e length and polarity of the bond increase. Longer bonds are weaker and eas- to break. Polar H—X bonds more readily produce H^+ and X^- ions because ch bonds already have partial ionic charges on the H and X atoms. As we ove from left to right across a row in Figure 16-10, the H—X bond length creases whereas the polarity of the bond increases. Because the acid strength $_a$ value) increases across the row, we arrive at the following conclusion.

When comparing binary acids of elements *in the same row* of the periodic table, acid strength increases as the polarity of the bond increases.

We arrive at a different conclusion if we compare binary acids from the same lumn in Figure 16-10. As we move from top to bottom in a column, both the nd length and acid strength of H—X increase whereas the polarity of the —X bond decreases. The following statement summarizes the situation.

When comparing binary acids of elements *in the same group* of the periodic table, acid strength increases as the length of the bond increases.

KEEP IN MIND

that electronegativity increases as we move from left to right across a period and decreases from top to bottom in a given group. Thus, the polarity of the H—X bond increases from left to right across a row in Figure 16-10 and decreases as we move from top to bottom down a column. Atomic radii show the opposite trend (decrease from left to right and increase from top to bottom) and so H—X bond lengths decrease from left to right and increase from top to bottom in Figure 16-10.

That HF is a weaker acid than the other hydrogen halides is expected, but at it should be so much weaker has always seemed an anomaly. Explanations this behavior center on the tendency for hydrogen bonding in HF (recall gure 12-5). For example, in HF(aq), ion pairs are held together by strong drogen bonds, which keeps the concentration of free H_3O^+ from being as rge as otherwise expected.

$$HF + H_2O \longrightarrow \underset{\text{Ion pair}}{(F^- \cdots H_3O^+)} \rightleftharpoons H_3O^+ + F^-$$

H_4 and NH_3 do not have acidic properties in water, but HF is an acid of oderate strength ($K_a = 6.6 \times 10^{-4}$).

trengths of Oxoacids

describe the relative strengths of oxoacids, we must focus on the attraction electrons from the O—H bond toward the central atom. The following fac- rs promote this electron withdrawal from O—H bonds: (1) a high elec- onegativity (EN) of the central atom and (2) a large number of terminal O oms in the acid molecule.

◀ The term *oxoacid* was defined in Chapter 3.

Neither HOCl nor HOBr has a terminal O atom. The major difference tween the two acids is one of electronegativity—Cl is slightly more elec- onegative than Br. As expected, HOCl is more acidic than HOBr.

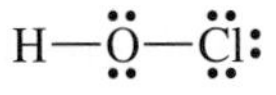

$EN_{Cl} = 3.0$

$K_a = 2.9 \times 10^{-8}$

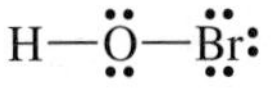

$EN_{Br} = 2.8$

$K_a = 2.1 \times 10^{-9}$

To compare the acid strengths of H_2SO_4 and H_2SO_3, we must look beyond e central atom, which is S in each acid.

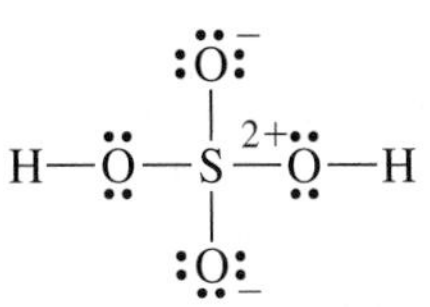

$K_{a_1} \approx 10^3$

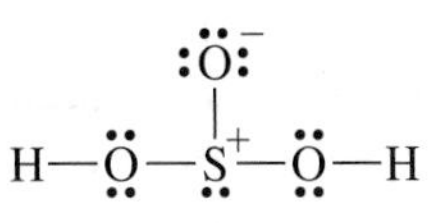

$K_{a_1} = 1.3 \times 10^{-2}$

A highly electronegative terminal O atom tends to withdraw electro from the O—H bonds, weakening the bonds and increasing the acidity of t molecule. Because H_2SO_4 has *two* terminal O atoms to only one in H_2SO_3, t electron-withdrawing effect is greater in H_2SO_4. As a result, H_2SO_4 is stronger acid than H_2SO_3.

Strengths of Organic Acids

This discussion of the relationship between molecular structure and ac strength concludes with a brief consideration of some organic compound Consider first the case of acetic acid and ethanol. Both have an O—H grou bonded to a carbon atom, but acetic acid is a much stronger acid tha ethanol.

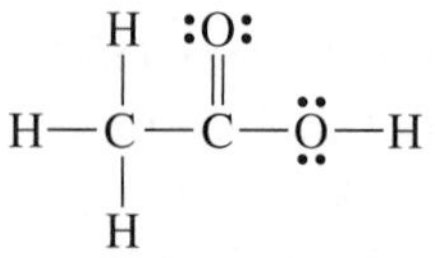

Acetic acid
$K_a = 1.8 \times 10^{-5}$

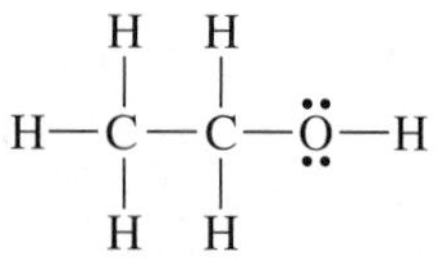

Ethanol
$K_a = 1.3 \times 10^{-16}$

One possible explanation for the large difference in acidity of these tw compounds is that the highly electronegative terminal O atom in acetic ac withdraws electrons from the O—H bond. The bond is weakened, and proton (H^+) is more readily taken from a molecule of the acid by a bas A more satisfactory explanation focuses on the anions formed in t ionization.

▶ Review the concept of resonance. Compounds or ions with more resonance structures are more stable.

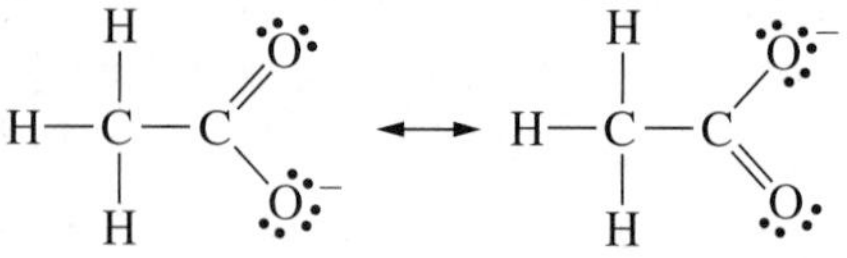

Acetate ion

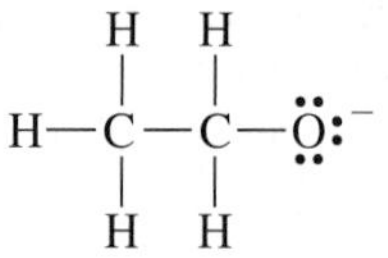

Ethoxide ion

There are two plausible structures for the acetate ion. These structur suggest that each carbon-to-oxygen bond is a "$\frac{3}{2}$" bond and that each O ato carries "$\frac{1}{2}$" unit of negative charge. In short, the excess unit of negati charge in CH_3COO^- is spread out. This arrangement reduces the ability either O atom to attach a proton and makes acetate ion only a wea Brønsted–Lowry base. In ethoxide ion, conversely, the unit of negati charge is localized on the single O atom. Ethoxide ion is a much strong base than is acetate ion. The stronger the conjugate base, the weaker the co responding acid.

The length of the carbon chain in a carboxylic acid has little effect on th acid strength, as in a comparison of acetic acid and octanoic acid.

CH_3COOH	$CH_3(CH_2)_6COOH$
Acetic acid	Octanoic acid
$K_a = 1.8 \times 10^{-5}$	$K_a = 1.3 \times 10^{-5}$

Yet, other atoms or groups of atoms substituted onto the carbon chain ma strongly affect acid strength. If a Cl atom is substituted for one of the H aton that is bonded to carbon in acetic acid, the result is chloroacetic acid.

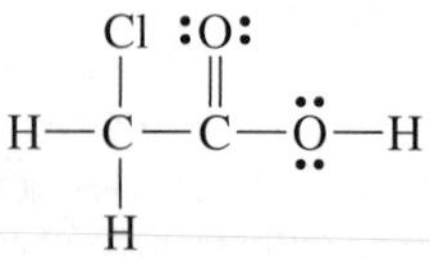

Chloroacetic acid
$K_a = 1.4 \times 10^{-3}$

e highly electronegative Cl atom helps draw electrons away from the —H bond. The O—H bond is weakened, the proton is lost more readily, d the acid is a stronger acid than acetic acid. This effect falls off rapidly as distance increases between the substituted atom or group and the O—H nd in an organic acid.

Example 16-15 illustrates some of the factors affecting acid strength that are cussed in this section.

EXAMPLE 16-15 Identifying Factors That Affect the Strengths of Acids

Explain which member of each of the following pairs is the stronger acid.

(a) (I) 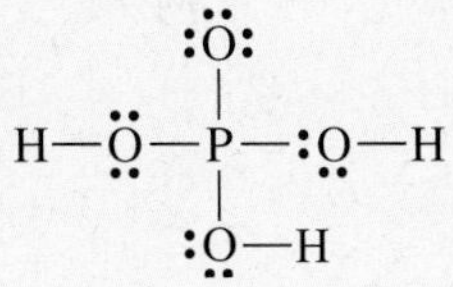 or (II) 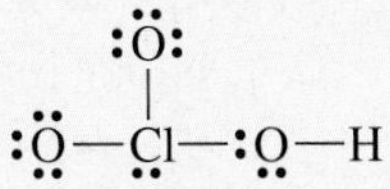

(b) (I) 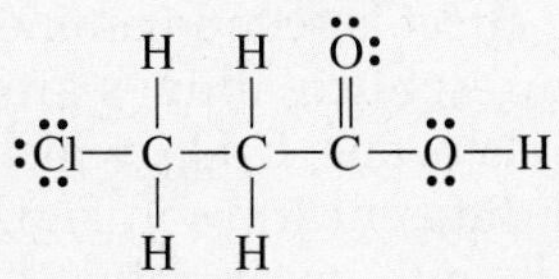 or (II) 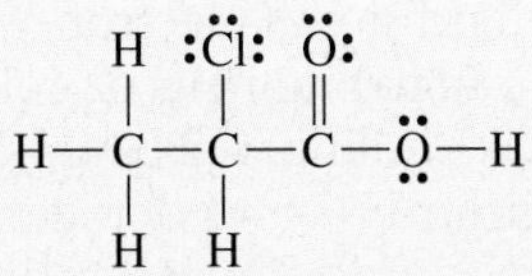

Analyze

In these types of questions we first identify the acidic proton(s), and then look for electronegative atoms or groups that pull electron density away from the acidic proton(s). The more electron density that is pulled away from the proton, the more acidic it is.

(a) Phosphoric acid, H_3PO_4, has four O atoms to the three in $HClO_3$, but it is the number of *terminal* O atoms that we must consider, not just the total number of O atoms in the molecule. $HClO_3$ has *two* terminal O atoms and H_3PO_4 has *one*. Also, the Cl atom (EN = 3.0) is considerably more electronegative than the P atom (EN = 2.1). These facts point to chloric acid (II) as being the stronger of the two acids. ($K_a \approx 5 \times 10^2$ for $HClO_3$ and $K_{a_1} = 7.1 \times 10^{-3}$ for H_3PO_4.)

(b) The Cl atom withdraws electrons more strongly when it is directly adjacent to the carboxyl group. Compound (II), 2-chloropropanoic acid ($K_a = 1.4 \times 10^{-3}$), is a stronger acid than compound (I), 3-chloropropanoic acid ($K_a = 1.0 \times 10^{-4}$).

Assess

This type of analysis is important in organic chemistry. To successfully solve these types of problems, we must know how to draw Lewis structures and we must understand the concept of electronegativity.

PRACTICE EXAMPLE A: Explain which is the stronger acid, HNO_3 or $HClO_4$; CH_2FCOOH or $CH_2BrCOOH$. [*Hint:* Draw plausible Lewis structures.]

PRACTICE EXAMPLE B: Explain which is the stronger acid, H_3PO_4 or H_2SO_3; CCl_3CH_2COOH or CCl_2FCH_2COOH. [*Hint:* Draw plausible Lewis structures.]

rengths of Amines as Bases

ne fundamental factor affecting the strength of an amine as a base concerns e ability of the lone pair of electrons on the N atom to bind a proton taken om an acid. When an atom or group of atoms more electronegative than H places one of the H atoms of NH_3, the electronegative group withdraws ectron density from the N atom. The lone-pair electrons cannot bind a pro- n as strongly, and the base is weaker. Thus, bromamine, in which the elec- onegative Br atom is attached to the amine group (NH_2), is a *weaker* base an ammonia.

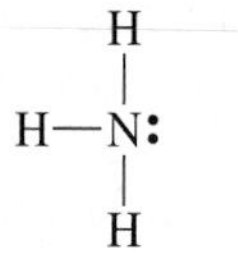

Ammonia
$pK_b = 4.74$

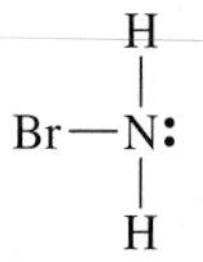

Bromamine
NH_2Br, $pK_b = 7.61$

Hydrocarbon chains have no electron-withdrawing ability. When they a attached to the amine group, the pK_b values are lower than for ammonia d to the electron-donating ability of CH_3 and CH_2CH_3.

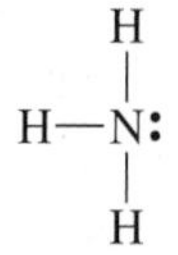

Ammonia
$pK_b = 4.74$

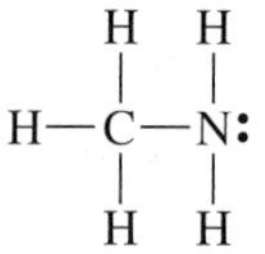

Methylamine
CH_3NH_2, $pK_b = 3.38$

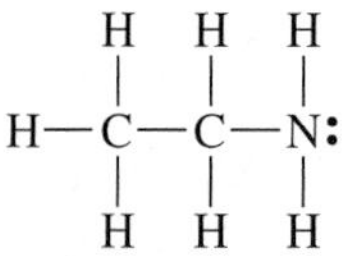

Ethylamine
$CH_3CH_2NH_2$, $pK_b = 3.37$

An additional electron-withdrawing effect is seen in amines that are bas on the benzene ring or related structures. Such amines are called *aroma* amines. Aniline, $C_6H_5NH_2$, is based on benzene, C_6H_6, which, as we learn in Section 11-7 and depicted in several different ways in Figures 11-29, 11- and 11-31, is a six-carbon ring molecule with unsaturation in the carbon- carbon bonds. The electrons associated with this unsaturation are said to *delocalized*. As suggested by the following structures, to some extent even t lone-pair electrons of the NH_2 group participate in the "spreading out" delocalized electrons. (The curved arrows suggest the progressive moveme of electrons around the ring.)

▶ Note that these are actually resonance Lewis structures.

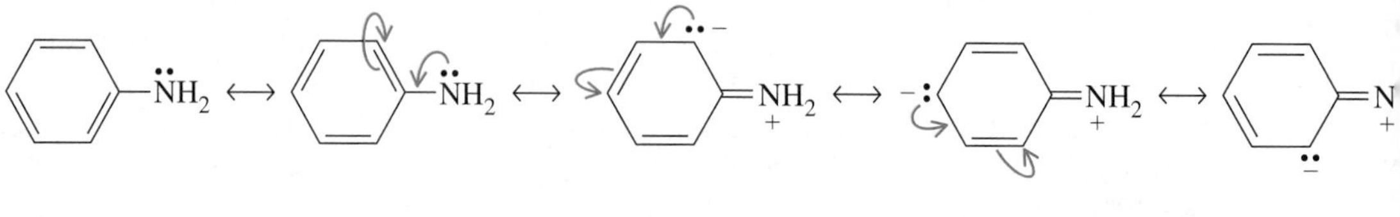

The withdrawal of electron charge density from the NH_2 group causes anili to be a much weaker base than is cyclohexylamine. (H atoms bonded to ri carbon atoms are not shown in the following structures.)

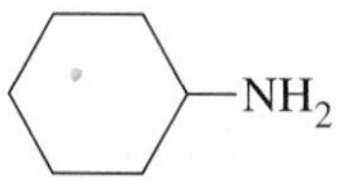

Cyclohexylamine, $pK_b = 3.36$

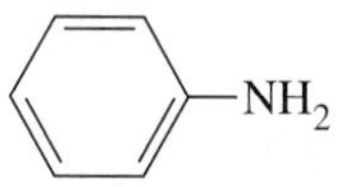

Aniline, $pK_b = 9.13$

▶ The pK_b for *meta*-chloroaniline is 10.66.

Replacement of a ring-bound H atom in aniline with an atom or grou that has a high electronegativity causes even more electron density to drawn away from the NH_2 group, further reducing the base strengt Also, the closer this ring substituent is to the NH_2 group, the greater is t effect.

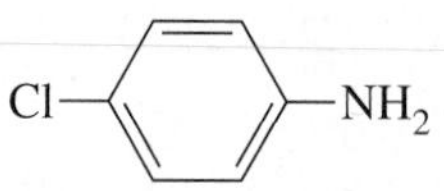

para-Chloroaniline, $pK_b = 10.01$

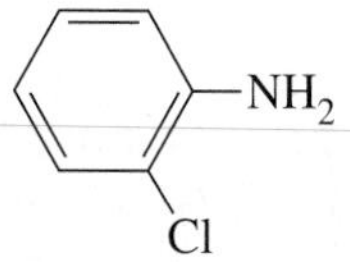

ortho-Chloroaniline, $pK_b = 11.36$

16-9 CONCEPT ASSESSMENT

Vould you expect pK_a of *ortho*-chlorophenol to be greater than, less than, or early the same as that of phenol? Explain.

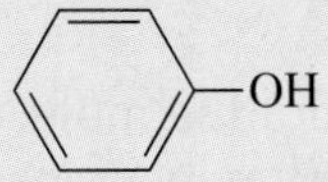

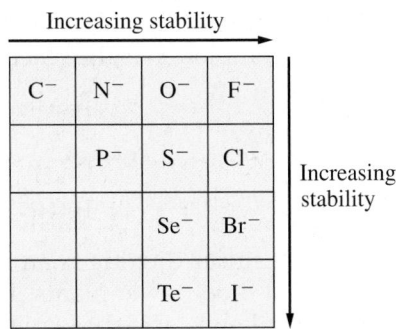

Phenol, pK_a = 10.00

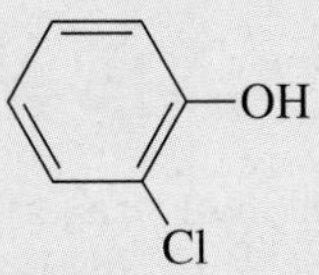

ortho-Chlorophenol, pK_a = ?

tionalization of Acid Strengths: An Alternative Approach

have already established that the acid strength of an acid, HA, and the base ength of its conjugate, A^-, are inversely related (page 763). For example, if HA s a strong tendency to lose a proton, then A^- exhibits a very weak tendency to ome protonated. That is, A^- is "stable", in the sense that it is not readily pro- ated. Because of the relationship between HA and A^-, we can take one of two proaches to rationalize the strength of an acid, HA. We can focus on

- factors that cause electron density to be drawn away from the H atom or
- factors that make A^- stable with respect to protonation.

pages 769–773, we focused primarily on the factors that cause electron nsity to be drawn away from the ionizable H atom of an acid. However, ganic chemists routinely focus on the factors that make A^- more stable and re difficult to protonate.

The stability of A^- depends on many factors. What is the atom bearing the gative charge? Is that atom highly electronegative? Is the atom small or large? at is the hybridization of the atom? Is the charge localized or delocalized? All se factors have a part in determining the stability of A^-. However, we will nsider only a few of these factors here. Here are a few generalizations.

Increasing stability →

C^-	N^-	O^-	F^-
	P^-	S^-	Cl^-
		Se^-	Br^-
		Te^-	I^-

Increasing stability ↓

- When comparing atoms *in the same period* (atoms of very similar sizes) in terms of their abilities to bear a negative charge, electronegativity is the important factor: **the more electronegative the atom is, the greater its ability to bear a negative charge.** For example, CH_3O^- is more stable than NH_2^-.
- When comparing atoms *from different periods* (atoms of very different sizes) in terms of their abilities to stabilize a negative charge, size is the important factor: **the larger the atom, the greater its ability to bear a negative charge.** For example, HS^- is more stable than HO^-.
- The stability of an anion increases as the number of electron-withdrawing groups increases (for example, $ClCH_2COO^-$ is more stable than CH_3COO^- but not as stable as Cl_3CCOO^-). Also, the closer the electron-withdrawing groups are to the atom bearing the negative charge, the more stable the anion. For example FCH_2COO^- is more stable than $FCH_2CH_2COO^-$.
- The stability of an anion increases as the number of atoms sharing the charge increases. However, in some cases, the number of contributing

structures is not the only consideration: The electronegativities and si: of the atoms sharing the negative charge may also play a role.

Exercise 103 focuses on the use of these generalizations.

16-11 Lewis Acids and Bases

▶ Bonding in the $H_3N—BF_3$ adduct can be described by the overlap of sp^3 orbitals on the N and B atoms, with the two electrons donated by the N atom.

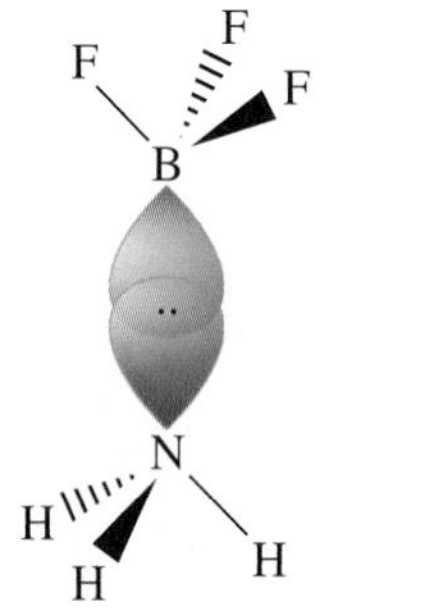

In the previous section, we presented ideas about the molecular structures acids and bases. In 1923, G. N. Lewis proposed an acid–base theory clos related to bonding and structure. The Lewis acid–base theory is not limited reactions involving H^+ and OH^-: It extends acid–base concepts to reactions gases and in solids. It is especially important in describing certain reactic between organic molecules.

A **Lewis acid** is a species (an atom, ion, or molecule) that is an *electron-p acceptor*, and a **Lewis base** is a species that is an *electron-pair donor*. A reacti between a Lewis acid (A) and a Lewis base (B:) results in the formation o covalent bond between them. The product of a Lewis acid–base reaction called an **adduct** (or *addition compound*). The reaction can be represented as

$$B\!: + A \longrightarrow B - A$$

where B:A is the adduct. The formation of a covalent chemical bond by o species donating a pair of electrons to another is called *coordination*, and t bond joining the Lewis acid and Lewis base is called a *coordinate covalent bo* (see page 415). *Lewis acids* are species with vacant orbitals that can accomm date electron pairs; *Lewis bases* are species that have lone-pair electrons ava able for sharing.

By these definitions, OH^-, a Brønsted–Lowry base, is also a Lewis ba because lone-pair electrons are present on the O atom. So too is NH_3 a Lew base. HCl, conversely, is not a Lewis acid: It is not an electron-pair accept We can think of HCl as producing H^+, however, and H^+ is a Lewis acid. I forms a coordinate covalent bond with an available electron pair.

Species with an incomplete valence shell are Lewis acids. When the Lev acid forms a coordinate covalent bond with a Lewis base, the octet is co pleted. A good example of octet completion is the reaction of BF_3 and NH_3.

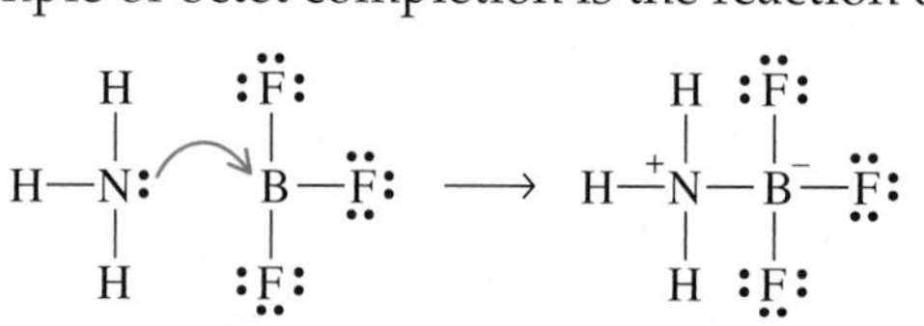

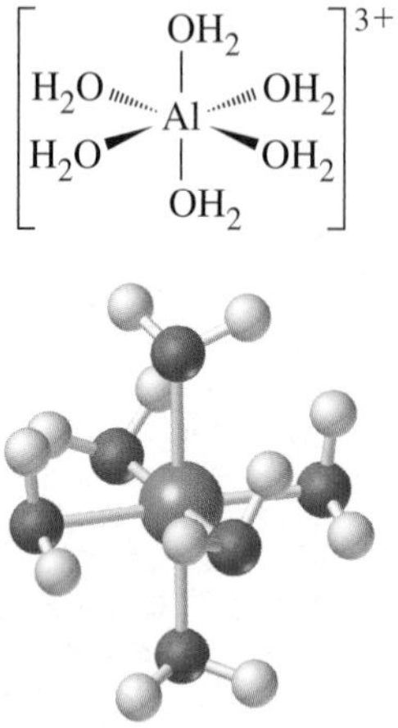

▲ FIGURE 16-11
The Lewis structure of $[Al(H_2O)_6]^{3+}$ and a ball-and-stick representation

The reaction of lime (CaO) with sulfur dioxide is an important reaction f reducing SO_2 emissions from coal-fired power plants. This reaction betwee solid and a gas underscores that Lewis acid–base reactions can occur in states of matter. The smaller curved red arrow in reaction (16.24) suggests th an electron pair in the Lewis structure is rearranged.

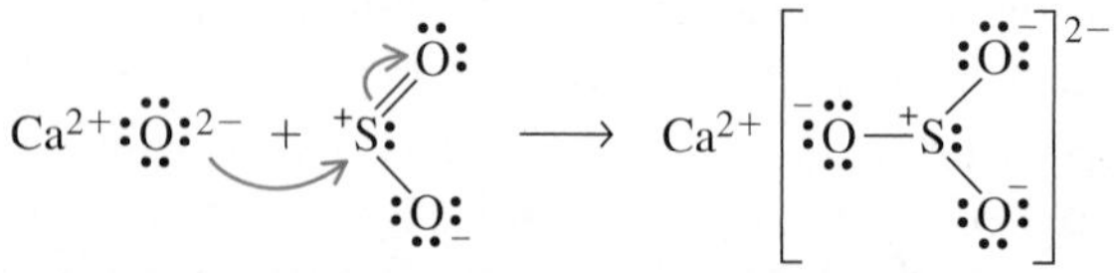

(16.2

An important application of the Lewis acid–base theory involves the form tion of *complex ions*. Complex ions are polyatomic ions that contain a centr metal ion to which other ions or small molecules are attached. *Hydrated metal io* form in aqueous solution because the water acts as a Lewis base and the me ion as a Lewis acid. The water molecules attach themselves to the metal ion means of coordinate covalent bonds. Thus, for example, when anhydrous AlC is added to water, the resultant solution becomes hot because of the heat evolv in the formation of the hydrated metal ion $[Al(H_2O)_6]^{3+}(aq)$ (Fig. 16-11).

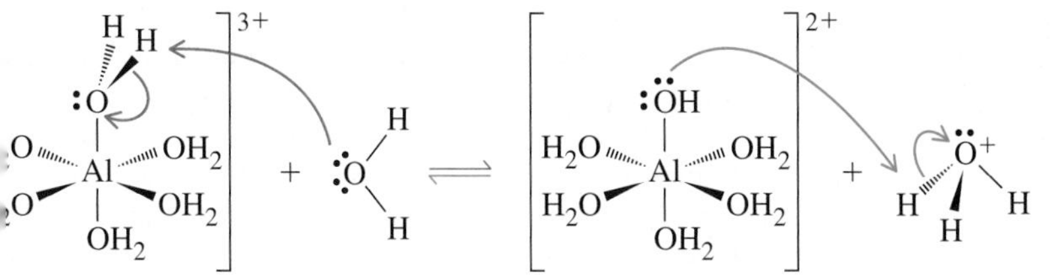

◀ FIGURE 16-12
Hydrolysis of $[Al(H_2O)_6]^{3+}$ to produce H_3O^+
An uncoordinated water molecule removes a proton from a coordinated water molecule.

The interaction between the metal ion and the water molecules is so strong ıt when the salt is crystallized from the solution, the water molecules crys- lize along with the metal ion, forming the hydrated metal salt $AlCl_3 \cdot 6\,H_2O$. aqueous solution, the hydrated metal ions can act as Brønsted–Lowry acids. r instance, the hydrolysis of hydrated Al^{3+} is given by

$$[Al(H_2O)_6]^{3+} + H_2O \rightleftharpoons [Al(OH)(H_2O)_5]^{2+} + H_3O^+$$

the hydrated metal ion, the OH bond in a water molecule becomes weak- ed. This happens because, in forming the coordinate covalent bond with the atom of the water, the metal ion causes electron density to be drawn toward it; nce, electron density is drawn away from the OH bond. As a consequence, coordinated H_2O molecule can donate a H^+ to a solvent H_2O molecule g. 16-12). The H_2O molecule that has ionized is converted to OH^-, which nains attached to the Al^{3+}; the charge on the complex ion is reduced from to 2+. The extent of ionization of $[Al(H_2O)_6]^{3+}$, measured by its K_a value d as pictured in Figure 16-13, is essentially the same as that of acetic acid $_a = 1.8 \times 10^{-5}$). Many other metal ions hydrolyze, especially the transition tal ions. These and other hydrated metal ions acting as acids are discussed later chapters.

Complex ions can also form between transition metal ions and other wis bases, such as NH_3. For instance, Zn^{2+} combines with NH_3 to form the mplex ion $[Zn(NH_3)_4]^{2+}$. The central Zn^{2+} ion accepts electrons from the wis base NH_3. to form coordinate covalent bonds; it is a Lewis acid. We ll discuss the application of Lewis acid–base theory to complex ions in apter 24.

16-3 ARE YOU WONDERING?

Why does $Na^+(aq)$ not act as an acid in aqueous solution?

Whether an aqueous solution of a metal ion is acidic depends on two principal factors. The first is the amount of charge on the cation; the second is the size of the ion. The greater the charge on the cation, the greater is the ability of the metal ion to draw electron density away from the O—H bond in a H_2O molecule in its hydration sphere, favoring the release of a H^+ ion. The smaller the cation, the more highly concentrated is the positive charge. Hence, for a given positive charge, the smaller the cation, the more acidic the solution.

The ratio of the charge on the cation to the volume of the cation is called the *charge density*.

$$\rho = \text{charge density} = \frac{\text{ionic charge}}{\text{ionic volume}}$$

The greater its charge density, the more effective a metal ion is at pulling electron density from the O—H bond and the more acidic is the hydrated cation (see the table and plot that follows). A highly concentrated positive charge on a small

(continued)

▲ FIGURE 16-13
Acidic properties of hydrated metal ions
The yellow color of bromthymol blue indicator in $Al_2(SO_4)_3(aq)$ denotes that the solution is acidic. The pH meter gives a more precise indication of the pH.

cation is better able to pull electron density from the O—H bond than is a less concentrated positive charge on a larger cation.

Thus the small, highly charged Al^{3+} ion produces acidic solutions, but the larger Na^+ cation, with a charge of just +1, does not increase the concentration of H_3O^+. In fact, none of the group 1 cations produces appreciably acidic solutions, and only Be^{2+} of the group 2 elements is small enough to do so ($pK_a = 5.4$).

Metal Cation	Ionic Radius, pm	$\rho \times 10^7$, Charge pm^{-3}	pK_a
Li^+	76	3.27	13.6
Na^+	102	1.53	14.2
K^+	138	0.680	14.5
Be^{2+}	45	23.2	5.4
Cu^{2+}	66	9.33	8.0
Ni^{2+}	69	8.35	9.9
Mg^{2+}	72	7.51	11.4
Zn^{2+}	74	7.00	9.0
Co^{2+}	74	7.00	9.7
Mn^{2+}	83	5.23	10.6
Ca^{2+}	100	3.22	12.8
Al^{3+}	53	23.8	5.0
Cr^{3+}	61	17.0	4.0
Ti^{3+}	67	13.5	2.2
Fe^{3+}	78	9.19	2.2

The pK_a of H_3O^+ is −1.7, and the pK_a of water is 15.7.

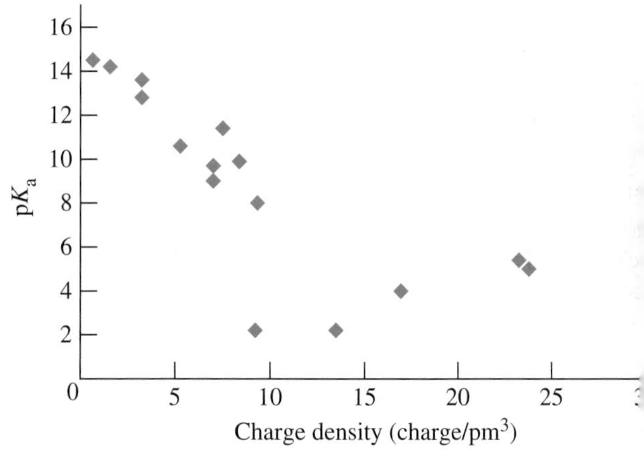

EXAMPLE 16-16 Identifying Lewis Acids and Bases

According to the Lewis theory, each of the following is an acid–base reaction. Which species is the acid and which is the base?

(a) $BF_3 + F^- \longrightarrow BF_4^-$

(b) $OH^-(aq) + CO_2(aq) \longrightarrow HCO_3^-(aq)$

Analyze

Recall that in Lewis theory an acid–base reaction involves the movement of electrons. The Lewis acid accepts electrons and the Lewis base donates electrons. In this example, we need to identify the species that is accepting the electrons and the one that is donating electrons.

Solve

(a) In BF_3, the B atom has a vacant orbital and an incomplete octet. The fluoride ion has an outer-shell octet of electrons. BF_3 is the electron-pair acceptor—the acid. F^- is the electron-pair donor—the base.

(b) We have already identified OH^- as a Lewis base, so we might suspect that it is the base and that $CO_2(aq)$ is the Lewis acid. The following Lewis structures show this to be so. As in reaction (16.24), a rearrangement of an electron pair at one of the double bonds is also required, as indicated by the smaller red arrow.

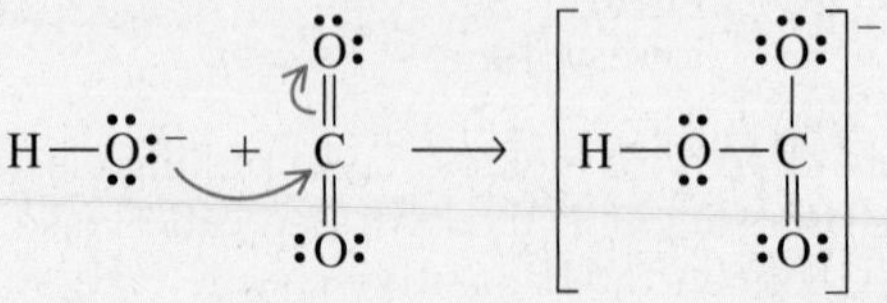

Assess

Typically, those species that have filled orbitals are Lewis bases, and those with vacant orbitals are Lewis acids. The transfer of electron density from a Lewis base to a vacant orbital on a Lewis acid is a recurring concept in chemistry. We will make use of this concept in the later chapters, as well as in organic chemistry. To describe the reaction in (b) in this way requires us to consider the electronic structure of CO_2 in terms of molecular orbital theory. Some of the $2p$ orbitals on the carbon and oxygen atoms in CO_2 combine to give bonding and antibonding π-type molecular orbitals. We describe a similar situation in Chapter 11 (see Figure 11-33). The vacant orbital in CO_2 that accepts the lone pair from OH^- is an antibonding π-type orbital.

PRACTICE EXAMPLE A: Identify the Lewis acids and bases in these reactions.

(a) $BF_3 + NH_3 \longrightarrow F_3BNH_3$

(b) $Cr^{3+} + 6\,H_2O \longrightarrow [Cr(H_2O)_6]^{3+}$

PRACTICE EXAMPLE B: Identify the Lewis acids and bases in these reactions.

(a) $Al(OH)_3 + OH^- \longrightarrow [Al(OH)_4]^-$

(b) $SnCl_4 + 2\,Cl^- \longrightarrow [SnCl_6]^{2-}$

16-10 CONCEPT ASSESSMENT

Liquid bromine in the presence of iron(III) tribromide forms a bromonium:iron(III) tribromide adduct. Propose a plausible mechanism for adduct formation and identify the Lewis acid and Lewis base. [*Hint*: What is the iron(III) electron configuration?]

MasteringCHEMISTRY www.masteringchemistry.com

Pure water is pH neutral, but rainwater is acidic. What causes rainwater to be acidic? One contributing factor is that carbon dioxide from the atmosphere reacts with water to form carbonic acid, H_2CO_3, a diprotic acid. For a discussion of the natural sources of acidity in rainwater, and how human activities also contribute, go to the Focus On feature for Chapter 16, Acid Rain, on the MasteringChemistry site.

Summary

16-1 Acids, Bases, and Conjugate Acid–Base Pairs—The **Brønsted–Lowry theory** describes an acid as a **proton donor** and a base as a **proton acceptor**. In an acid–base reaction, a base takes a proton (H^+) from an acid. In general, acid–base reactions are reversible. The conjugate base (A^-) is derived from the acid HA whereas the conjugate acid (HB^+) is derived from the base (B). The combinations of HA/A^- and B/HB^+ are known as **conjugate acid–base pairs**. The conjugate acid of H_2O is the **hydronium ion**, H_3O^+. Certain substances, for example water, are said to be **amphiprotic**. They can act as an acid or a base.

16-2 Self-Ionization of Water and the pH Scale—In pure water and in aqueous solutions, **self-ionization** of the water occurs to a very slight extent, producing H_3O^+ and OH^-, as described by the equilibrium constant K_w, known as the **ion product of water** (expression 16.4). The designations **pH** (expression 16.6) and **pOH** (expression 16.7) are often used to describe the concentrations of H_3O^+ and OH^- in aqueous solutions.

16-3 Ionization of Acids and Bases in Water—The equilibrium constant for the reaction of an acid with water is called the **acid ionization constant (K_a)**, equation (16.10), and that for the reaction of a base with water is called the **base ionization constant (K_b)**, equation (16.13). Strong acids and strong bases have large ionization constants (K_a or $K_b \gg 1$) and are essentially completely ionized in water. Weak acids and weak bases have small ionization constants (K_a or $K_b \ll 1$) and ionize to a limited extent in water. The extent to which acids or bases ionize in water is described in terms of either the **degree of ionization (α)** (equation 16.15) or the **percent ionization** (equation 16.16). For a weak acid or weak base, the degree of ionization increases with increasing dilution.

16-4 Strong Acids and Strong Bases—In aqueous solutions, strong acids ionize completely to produce H_3O^+, and strong bases dissociate completely to produce OH^-. Common strong acids and bases are given in Table 16.3 and can be easily memorized.

16-5 Weak Acids and Weak Bases—A weak acid or weak base ionizes to a limited extent in water. The extent of their ionization can be related to the ionization constants K_a and K_b or their logarithmic equivalents $pK_a = -\log K_a$ and $pK_b = -\log K_b$ (Table 16.4) by setting up and solving an equilibrium calculation. Calculations involving ionization equilibria are in many ways similar to those introduced in Chapter 15, although some additional considerations are necessary for polyprotic acids.

16-6 Polyprotic Acids—**Polyprotic acids** are acids with more than one ionizable H atom that undergo a stepwise ionization and have a different ionization constant, $K_{a_1}, K_{a_2}, \ldots,$ for each ionization step.

16-7 Simultaneous or Consecutive Acid–Base Reactions: A General Approach—In certain situations, it may be necessary to consider two or more ionization reactions. A general approach for handling such situations involves writing down all equilibrium constant expressions, one or more *material balance equations*, and a *charge balance equation* (also called an *electroneutrality condition*). A material balance equation indicates that the equilibrium concentrations of all forms of a given acid or base must be equal to the initial or stoichiometric concentration of the acid or base. The charge balance equation indicates that the solution carries no net charge.

16-8 Ions as Acids and Bases—In reactions betw ions and water—**hydrolysis** reactions—the ions react weak acids or weak bases. The pH of salt solutions deper on the anions and/or cations present. Anions from we acids act as bases whereas cations from weak bases as acids.

16-9 Qualitative Aspects of Acid–Ba Reactions—For an acid–base reaction, equilibri favors the formation of the weaker acid and the weal base. If the acid or base in an acid–base reaction is stro then the reaction goes essentially to completion.

16-10 Molecular Structure and Acid–Ba Behavior—Molecular composition and structure are keys to determining whether a substance is acidic, basic, amphiprotic. In addition, molecular structure affe whether an acid or a base is strong or weak. In assessi acid strength, for example, factors that affect the stren of the bond that must be broken to release H^+ must be co sidered. Alternatively, factors that affect the stability of anion formed by the acid can be considered. In assessi base strength, factors that affect the ability of lone-p electrons to bind a proton are of primary concern.

16-11 Lewis Acids and Bases—The Lewis acid–b theory views an electron-pair acceptor as a **Lewis acid** a an electron-pair donor as a **Lewis base**. The addition co pound of a Lewis acid–base reaction is referred to as **adduct**. The theory is most useful in situations that cann be described by means of proton transfers, for example, reactions involving gases and solids and in reactic between organic compounds (considered in Chapter 27

Integrative Example

Bromoacetic acid, $BrCH_2COOH$, has $pK_a = 2.902$. Calculate the expected values of **(a)** the freezing point 0.0500 M $BrCH_2COOH(aq)$ and **(b)** the osmotic pressure at 25 °C of 0.00500 M $BrCH_2COOH(aq)$.

Analyze

Freezing point and osmotic pressure are both colligative properties. As we saw in Chapter 14, the values of these prop ties depend on the total concentrations of particles (molecules and ions) in a solution, but not on the identity of the particles. We can use the ICE method for equilibrium calculations (Chapter 15) to determine the total concentrations particles (molecules and ions) in a weak electrolyte solution, as we learned to do in this chapter. Once we have the results we can turn to equations (14.4) and (14.5) to do the calculations required in parts (a) and (b).

Solve

A good place to begin is to convert the pK_a for bromoacetic acid to K_a.

$$pK_a = 2.902 = -\log K_a$$
$$K_a = 10^{-2.902} = 1.25 \times 10^{-3}$$

Next, write the equation for the reversible ionization reaction and the equilibrium constant expression.

$$BrCH_2COOH + H_2O \rightleftharpoons H_3O^+ + BrCH_2COO^-$$

$$K_a = \frac{[H_3O^+][BrCH_2COO^-]}{[BrCH_2COOH]} = 1.25 \times 10^{-3}$$

(a) Enter the relevant data into the ICE format under the equation for the ionization reaction.

$$\mathbf{BrCH_2COOH + H_2O \rightleftharpoons H_3O^+ + BrCH_2COO^-}$$

	$BrCH_2COOH$	H_3O^+	$BrCH_2COO^-$
initial concns:	0.0500 M	—	—
changes:	$-x$ M	$+x$ M	$+x$ M
equil concns:	$(0.0500 - x)$ M	x M	x M

The equilibrium constant expression based on these data is

$$K_a = \frac{[H_3O^+][BrCH_2COO^-]}{[BrCH_2COOH]} = \frac{x \cdot x}{0.0500 - x} = 1.25 \times 10^{-3}$$

Solve for x.

$$x^2 + 1.25 \times 10^{-3}x - 6.25 \times 10^{-5} = 0$$

$$x = \frac{-1.25 \times 10^{-3} + \sqrt{(1.25 \times 10^{-3})^2 + 4 \times 6.25 \times 10^{-5}}}{2}$$

$$x = \frac{-1.25 \times 10^{-3} + 1.59 \times 10^{-2}}{2} = 7.3 \times 10^{-3}$$

The total concentration of molecules and ions at equilibrium is

$$(0.0500 - x)\,\text{M} + x\,\text{M} + x\,\text{M} = (0.0500 + x)\,\text{M} = 0.0573\,\text{M}$$

Assuming that 0.0573 M = 0.0573 m, the freezing point depression of water caused by 0.0573 mol/L of particles is

$$\Delta T_f = -K_f \times m = -1.86\ °\text{C m}^{-1} \times 0.0573\ \text{m} = -0.107\ °\text{C}$$

The freezing point of 0.0500 M $BrCH_2COOH(aq)$ is 0.107 °C below the freezing point of water (0.000 °C), that is, −0.107 °C.

(b) Enter the relevant data into the ICE format under the equation for the ionization reaction.

	$BrCH_2COOH + H_2O \rightleftharpoons$	H_3O^+	$+ BrCH_2COO^-$
initial concns:	0.00500 M	—	—
changes:	$-x$ M	$+x$ M	$+x$ M
equil concns:	$(0.00500 - x)$ M	x M	x M

The equilibrium constant expression based on these data is

$$K_a = \frac{[H_3O^+][BrCH_2COO^-]}{[BrCH_2COOH]} = \frac{x \cdot x}{0.00500 - x} = 1.25 \times 10^{-3}$$

Solve for x.

$$x^2 + 1.25 \times 10^{-3}x - 6.25 \times 10^{-6} = 0$$

$$x = \frac{-1.25 \times 10^{-3} + \sqrt{(1.25 \times 10^{-3})^2 + 4 \times 6.25 \times 10^{-6}}}{2}$$

$$x = \frac{-1.25 \times 10^{-3} + 5.15 \times 10^{-3}}{2} = 1.95 \times 10^{-3}$$

The total concentration of molecules and ions at equilibrium is

$$c = (0.00500 - x)\,\text{M} + x\,\text{M} + x\,\text{M} = (0.00500 + x)\,\text{M} = 0.00695\,\text{M}$$

At 25.00 °C, the osmotic pressure of an aqueous solution with 0.00695 mol/L of particles (molecules and ions) is

$$\pi = c \times RT = 0.00695\ \text{mol L}^{-1} \times 0.08206\ \text{L atm mol}^{-1}\,\text{K}^{-1} \times 298.15\ \text{K} = 0.170\ \text{atm}$$

ssess

e pK_a is stated more precisely than in most previous equilibrium calculations, and this permitted us to carry three sig-ficant figures rather than the usual two in most of the calculations. The assumption in part (a) that 0.0573 M = 0.0573 m reasonable for a dilute aqueous solution with a density of essentially 1.00 g/mL. The mass of solvent (water) in one liter solution is very close to one kilogram, so that molarity (mol solute/L solution) and molality (mol solute/kg solvent) e essentially the same. The calculation in part (b) could have been done more easily by assuming that the concentration solute particles would be just 10% of that in part (a), that is, 0.00573 M compared to 0.0573 M. However, this would ve been a false assumption. Because the percent ionization of the acid is a function of its concentration, the total parti- concentration in part (b) was about 12% of that found in part (a), not 10%.

PRACTICE EXAMPLE A: The solubility of $CO_2(g)$ in H_2O at 25 °C and under a $CO_2(g)$ pressure of 1 atm is 1.4! CO_2/L. Air contains 0.037% CO_2 by volume. Use this information, together with data from Table 16.5, to show th rainwater saturated with CO_2 has a pH ≈ 5.6 (the normal pH for rainwater). [*Hint:* Recall Henry's law. What is t partial pressure of $CO_2(g)$ in air?]

PRACTICE EXAMPLE B: Often the following generalization applies to oxoacids with the formula $EO_m(OH)_n$ (wher is the central atom): If $m = 0$, $K_a \approx 10^{-7}$; if $m = 1$, $K_a \approx 10^{-2}$; if $m = 2$, K_a is large; and if $m = 3$, K_a is very large.

(a) Show that this generalization works well for the oxoacids of chlorine: HOCl, $pK_a = 7.52$; HOClO, $pK_a = 1.$ $HOClO_2$, $pK_a = -3$; $HOClO_3$, $pK_a = -8$.

(b) Estimate the value of K_{a_1} for H_3AsO_4.

(c) Write a Lewis structure for hypophosphorous acid, H_3PO_2, for which $pK_a = 1.1$.

Exercises

Brønsted–Lowry Theory of Acids and Bases

1. According to the Brønsted–Lowry theory, label each of the following as an acid or a base. **(a)** HNO_2; **(b)** OCl^-; **(c)** NH_2^-; **(d)** NH_4^+; **(e)** $CH_3NH_3^+$
2. Write the formula of the conjugate base in the reaction of each acid with water. **(a)** HIO_3; **(b)** C_6H_5COOH; **(c)** HPO_4^{2-}; **(d)** $C_2H_5NH_3^+$.
3. For each of the following, identify the acids and bases involved in both the forward and reverse directions.
   (a) $HOBr + H_2O \rightleftharpoons OBr^- + H_3O^+$
   (b) $HSO_4^- + H_2O \rightleftharpoons SO_4^{2-} + H_3O^+$
   (c) $HS^- + H_2O \rightleftharpoons H_2S + OH^-$
   (d) $C_6H_5NH_3^+ + OH^- \rightleftharpoons C_6H_5NH_2 + H_2O$
4. Which of the following species are *amphiprotic* in aqueous solution? For such a species, write one equation showing it acting as an acid, and another equation showing it acting as a base. OH^-, NH_4^+, H_2O, HS^-, NO_2^-, HCO_3^-, HBr.
5. With which of the following bases will the ionization of acetic acid, CH_3COOH, proceed furthest toward completion (to the right)? Explain your answ **(a)** H_2O; **(b)** NH_3; **(c)** Cl^-; **(d)** NO_3^-.
6. In a manner similar to equation (16.3), represent t self-ionization of the following liquid solven **(a)** NH_3; **(b)** HF; **(c)** CH_3OH; **(d)** CH_3COO **(e)** H_2SO_4.
7. With the aid of Table 16.2, predict the direction (fo ward or reverse) favored in each of the followi acid–base reactions.
   (a) $NH_4^+ + OH^- \rightleftharpoons H_2O + NH_3$
   (b) $HSO_4^- + NO_3^- \rightleftharpoons HNO_3 + SO_4^{2-}$
   (c) $CH_3OH + CH_3COO^- \rightleftharpoons CH_3COOH + CH_3($
8. With the aid of Table 16.2, predict the direction (fo ward or reverse) favored in each of the followi acid–base reactions.
   (a) $CH_3COOH + CO_3^{2-} \rightleftharpoons HCO_3^- + CH_3CO($
   (b) $HNO_2 + ClO_4^- \rightleftharpoons HClO_4 + NO_2^-$
   (c) $H_2CO_3 + CO_3^{2-} \rightleftharpoons HCO_3^- + HCO_3^-$

Strong Acids, Strong Bases, and pH

9. Calculate $[H_3O^+]$ and $[OH^-]$ for each solution: **(a)** 0.00165 M HNO_3; **(b)** 0.0087 M KOH; **(c)** 0.00213 M $Sr(OH)_2$; **(d)** 5.8×10^{-4} M HI.
10. What is the pH of each of the following solutions? **(a)** 0.0045 M HCl; **(b)** 6.14×10^{-4} M HNO_3; **(c)** 0.00683 M NaOH; **(d)** 4.8×10^{-3} M $Ba(OH)_2$.
11. Calculate $[H_3O^+]$ and pH in saturated $Ba(OH)_2(aq)$, which contains 3.9 g $Ba(OH)_2 \cdot 8\,H_2O$ per 100 mL of solution.
12. A saturated aqueous solution of $Ca(OH)_2$ has a pH of 12.35. What is the solubility of $Ca(OH)_2$, expressed in milligrams per 100 mL of solution?
13. What is $[H_3O^+]$ in a solution obtained by dissolving 205 mL HCl(g), measured at 23 °C and 751 mmHg, in 4.25 L of aqueous solution?
14. What is the pH of the solution obtained when 125 mL of 0.606 M NaOH is diluted to 15.0 L with water?
15. How many milliliters of concentrated HCl(aq) (36.0% HCl by mass, $d = 1.18$ g/mL) are required to produce 12.5 L of a solution with pH = 2.10?
16. How many milliliters of a 15.0%, by mass solution KOH(aq) ($d = 1.14$ g/mL) are required to produ 25.0 L of a solution with pH = 11.55?
17. What volume of 6.15 M HCl(aq) is required to exac neutralize 1.25 L of 0.265 M $NH_3(aq)$?
$$NH_3(aq) + H_3O^+(aq) \longrightarrow NH_4^+(aq) + H_2O(l)$$
18. A 28.2 L volume of HCl(g), measured at 742 mmI and 25.0 °C, is dissolved in water. What volume $NH_3(g)$, measured at 762 mmHg and 21.0 °C, mu be absorbed by the same solution to neutrali the HCl?
19. 50.00 mL of 0.0155 M HI(aq) is mixed with 75.00 mL 0.0106 M KOH(aq). What is the pH of the fir solution?
20. 25.00 mL of a $HNO_3(aq)$ solution with a pH of 2.12 mixed with 25.00 mL of a KOH(aq) solution with a p of 12.65. What is the pH of the final solution?

eak Acids, Weak Bases, and pH

se data from Table 16.4 as necessary.)

What are the $[H_3O^+]$ and pH of 0.143 M HNO_2?

What are the $[H_3O^+]$ and pH of 0.085 M $C_2H_5NH_2$?

For the ionization of phenylacetic acid,

$$C_6H_5CH_2CO_2H + H_2O \rightleftharpoons H_3O^+ + C_6H_5CH_2CO_2^-$$
$$K_a = 4.9 \times 10^{-5}$$

(a) What is $[C_6H_5CH_2CO_2^-]$ in 0.186 M $C_6H_5CH_2CO_2H$?
(b) What is the pH of 0.121 M $C_6H_5CH_2CO_2H$?

A 625 mL sample of an aqueous solution containing 0.275 mol propionic acid, $CH_3CH_2CO_2H$, has $[H_3O^+] = 0.00239$ M. What is the value of K_a for propionic acid?

$$CH_3CH_2CO_2H + H_2O \rightleftharpoons H_3O^+ + CH_3CH_2CO_2^- \quad K_a = ?$$

Fluoroacetic acid occurs in gifblaar, one of the most poisonous of all plants. A 0.318 M solution of the acid is found to have a pH = 1.56. Calculate K_a of fluoroacetic acid.

$$CH_2FCOOH(aq) + H_2O \rightleftharpoons H_3O^+(aq) + CH_2FCOO^-(aq) \quad K_a = ?$$

Caproic acid, $HC_6H_{11}O_2$, found in small amounts in coconut and palm oils, is used in making artificial flavors. A saturated aqueous solution of the acid contains 11 g/L and has pH = 2.94. Calculate K_a for the acid.

$$HC_6H_{11}O_2 + H_2O \rightleftharpoons H_3O^+ + C_6H_{11}O_2^- \quad K_a = ?$$

What mass of benzoic acid, C_6H_5COOH, would you dissolve in 350.0 mL of water to produce a solution with a pH = 2.85?

$$C_6H_5COOH + H_2O \rightleftharpoons H_3O^+ + C_6H_5COO^-$$
$$K_a = 6.3 \times 10^{-5}$$

What must be the molarity of an aqueous solution of trimethylamine, $(CH_3)_3N$, if it has a pH = 11.12?

$$(CH_3)_3N + H_2O \rightleftharpoons (CH_3)_3NH^+ + OH^-$$
$$K_b = 6.3 \times 10^{-5}$$

What are $[H_3O^+]$, $[OH^-]$, pH, and pOH of 0.55 M $HClO_2$?

What are $[H_3O^+]$, $[OH^-]$, pH, and pOH of 0.386 M CH_3NH_2?

The solubility of 1-naphthylamine, $C_{10}H_7NH_2$, a substance used in the manufacture of dyes, is given in a handbook as 1 g per 590 g H_2O. What is the approximate pH of a saturated aqueous solution of 1-naphthylamine?

$$C_{10}H_7NH_2 + H_2O \rightleftharpoons C_{10}H_7NH_3^+ + OH^-$$
$$pK_b = 3.92$$

32. A saturated aqueous solution of *o*-nitrophenol, $HOC_6H_4NO_2$, has pH = 4.53. What is the solubility of *o*-nitrophenol in water, in grams per liter?

$$HOC_6H_4NO_2 + H_2O \rightleftharpoons H_3O^+ + {}^-OC_6H_4NO_2$$
$$pK_a = 7.23$$

33. A particular vinegar is found to contain 5.7% acetic acid, CH_3COOH, by mass. What mass of this vinegar should be diluted with water to produce 0.750 L of a solution with pH = 4.52?

34. A particular household ammonia solution (d = 0.97 g/mL) is 6.8% NH_3 by mass. How many milliliters of this solution should be diluted with water to produce 625 mL of a solution with pH = 11.55?

35. A 275 mL sample of vapor in equilibrium with propan-1-amine at 25.0 °C is removed and dissolved in 0.500 L H_2O. For propan-1-amine, $pK_b = 3.43$ and v.p. = 316 Torr.
(a) What should be the pH of the aqueous solution?
(b) How many mg of NaOH dissolved in 0.500 L of water give the same pH?

36. One handbook lists a value of 9.5 for pK_b of quinoline, C_9H_7N, a weak base used as a preservative for anatomical specimens and to make dyes. Another handbook lists the solubility of quinoline in water at 25 °C as 0.6 g/100 mL. Use this information to calculate the pH of a saturated solution of quinoline in water.

37. In the diagram below, the sketch on the far left represents the $[H_3O^+]$ present in an acetic acid solution of molarity c. If the molarity of the solution is doubled, which of the sketches below best represents the resulting solution?

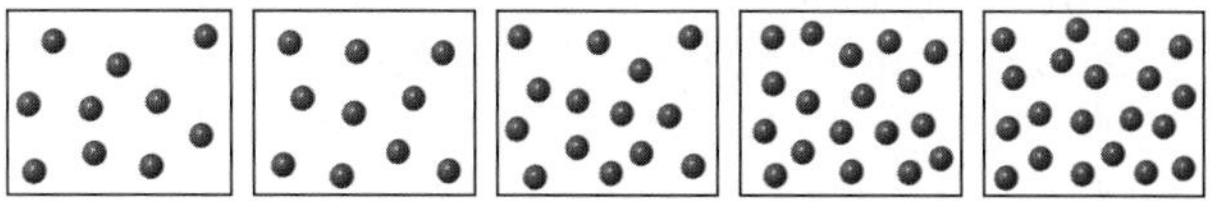

(a) (b) (c) (d)

38. In the diagram below, the sketch on the far left represents the $[OH^-]$ present in an ammonia solution of molarity c. If the solution is diluted to half its original molarity, which of the sketches below best represents the resulting solution?

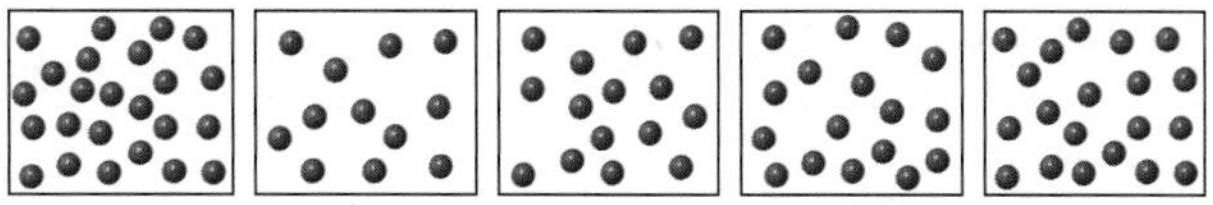

(a) (b) (c) (d)

Percent Ionization

39. What is the **(a)** degree of ionization and **(b)** percent ionization of propionic acid in a solution that is 0.45 M $CH_3CH_2CO_2H$?

$$CH_3CH_2CO_2H + H_2O \rightleftharpoons H_3O^+ + CH_3CH_2CO_2^- \qquad pK_a = 4.89$$

40. What is the **(a)** degree of ionization and **(b)** percent ionization of ethylamine, $C_2H_5NH_2$, in a 0.85 M aqueous solution?
41. What must be the molarity of an aqueous solution of NH_3 if it is 4.2% ionized?
42. What must be the molarity of an acetic acid solution if it has the same percent ionization as 0.100 M $CH_3CH_2CO_2H$ (propionic acid, $K_a = 1.3 \times 10^{-5}$)?
43. Continuing the dilutions described in Example 1 should we expect the percent ionization to be 1 in 0.0010 M CH_3COOH and 42% in 0.00010 CH_3COOH? Explain.
44. What is the **(a)** degree of ionization and **(b)** perc ionization of trichloroacetic acid in a 0.035 CCl_3COOH solution?

$$CCl_3COOH + H_2O \rightleftharpoons H_3O^+ + CCl_3COO^- \qquad pK_a = 0$$

Polyprotic Acids

(Use data from Table 16.5 as necessary.)

45. Explain why $[PO_4^{3-}]$ in 1.00 M H_3PO_4 is *not* simply $\frac{1}{3}[H_3O^+]$, but much, much less than $\frac{1}{3}[H_3O^+]$.
46. Cola drinks have a phosphoric acid content that is described as "from 0.057 to 0.084% of 75% phosphoric acid, by mass." Estimate the pH range of cola drinks corresponding to this range of H_3PO_4 content.
47. Determine $[H_3O^+]$, $[HS^-]$, and $[S^{2-}]$ for the following $H_2S(aq)$ solutions: **(a)** 0.075 M H_2S; **(b)** 0.0050 M H_2S; **(c)** 1.0×10^{-5} M H_2S.
48. For 0.045 M H_2CO_3, a weak diprotic acid, calculate **(a)** $[H_3O^+]$, **(b)** $[HCO_3^-]$, and **(c)** $[CO_3^{2-}]$. Use data from Table 16.5 as necessary.
49. Calculate $[H_3O^+]$, $[HSO_4^-]$, and $[SO_4^{2-}]$ in **(a)** 0.75 M H_2SO_4; **(b)** 0.075 M H_2SO_4; **(c)** 7.5×10^{-4} M H_2SO_4. [*Hint:* Check any assumptions that you make.]
50. Adipic acid, $HOOC(CH_2)_4COOH$, is among the top 50 manufactured chemicals in the United States (nearly 1 million metric tons annually). Its chief u is in the manufacture of nylon. It is a *diprotic a* having $K_{a_1} = 3.9 \times 10^{-5}$ and $K_{a_2} = 3.9 \times 10^{-6}$. A s urated solution of adipic acid is about 0.10 $HOOC(CH_2)_4COOH$. Calculate the concentration each ionic species in this solution.
51. The antimalarial drug quinine, $C_{20}H_{24}O_2N_2$, is *diprotic base* with a water solubility of 1.00 g/1900 of solution.
(a) Write equations for the ionization equilibria cor sponding to $pK_{b_1} = 6.0$ and $pK_{b_2} = 9.8$.
(b) What is the pH of saturated aqueous quinine?
52. For hydrazine, N_2H_4, $pK_{b_1} = 6.07$ and $pK_{b_2} = 15.$ Draw a structural formula for hydrazine, and wr equations to show the ionization of hydrazine in t distinctive steps. Calculate the pH of 0.245 $N_2H_4(aq)$.

Ions as Acids and Bases (Hydrolysis)

53. Codeine, $C_{18}H_{21}O_3N$, is an opiate, has analgesic and antidiarrheal properties, and is widely used. In water, codeine is a weak base. A handbook gives $pK_a = 6.05$ for protonated codeine, $C_{18}H_{21}O_3NH^+$. Write the reaction for $C_{18}H_{21}O_3NH^+$ and calculate pK_b for codeine.

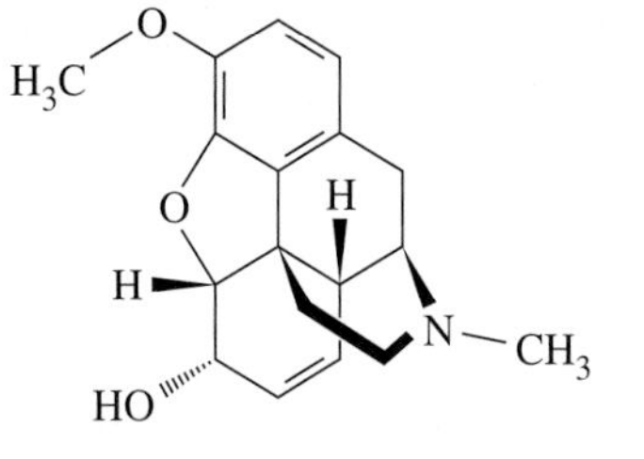

Codeine

54. Approximately 4 metric tons of quinoline, C_9H_7N, is produced annually. The principal source of quinoline is coal tar. Quinoline is a weak base in water. A har book gives $K_a = 6.3 \times 10^{-10}$ for protonated quir line, $C_9H_7NH^+$. Write the ionization reaction $C_9H_7NH^+$ and calculate pK_b for quinoline.

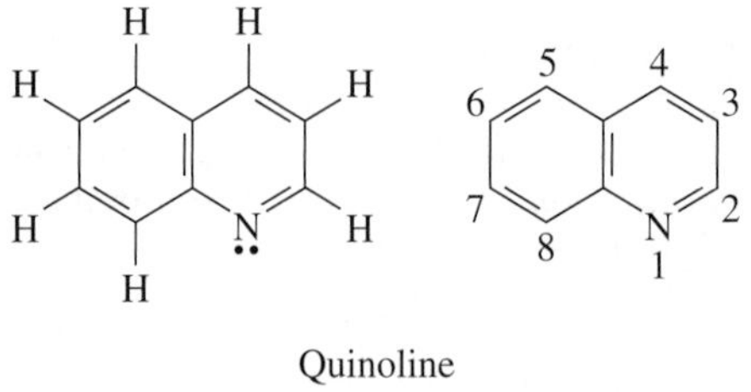

Quinoline

55. Complete the following equations in those instan in which a reaction (hydrolysis) will occur. If no re tion occurs, so state.
(a) $NH_4^+(aq) + NO_3^-(aq) + H_2O \longrightarrow$
(b) $Na^+(aq) + NO_2^-(aq) + H_2O \longrightarrow$
(c) $K^+(aq) + C_6H_5COO^-(aq) + H_2O \longrightarrow$

(d) $K^+(aq) + Cl^-(aq) + Na^+(aq) + I^-(aq) + H_2O \longrightarrow$

(e) $C_6H_5NH_3^+(aq) + Cl^-(aq) + H_2O \longrightarrow$

From data in Table 16.4, determine **(a)** K_a for $C_5H_5NH^+$; **(b)** K_b for $HCOO^-$; **(c)** K_b for $C_6H_5O^-$.

Predict whether a solution of each of the following salts is acidic, basic, or pH neutral: **(a)** KCl; **(b)** KF; **(c)** $NaNO_3$; **(d)** $Ca(OCl)_2$; **(e)** NH_4NO_2.

Arrange the following 0.010 M solutions in order of *increasing* pH: $NH_3(aq)$, $HNO_3(aq)$, $NaNO_2(aq)$, $CH_3COOH(aq)$, $NaOH(aq)$, $NH_4CH_3COO(aq)$, $NH_4ClO_4(aq)$.

What is the pH of an aqueous solution that is 0.089 M NaOCl?

What is the pH of an aqueous solution that is 0.123 M NH_4Cl?

Sorbic acid, $CH_2CH{=}CH{=}CHCH_2CO_2H$ ($pK_a = 4.77$), is widely used in the food industry as a preservative. For example, its potassium salt (potassium sorbate) is added to cheese to inhibit the formation of mold. What is the pH of 0.37 M potassium sorbate solution?

62. Pyridine, C_5H_5N ($pK_b = 8.82$), forms a salt, pyridinium chloride, as a result of a reaction with HCl. Write an ionic equation to represent the hydrolysis of the pyridinium ion, and calculate the pH of 0.0482 M $C_5H_5NH^+Cl^-(aq)$.

63. For each of the following ions, write two equations—one showing its ionization as an acid and the other as a base: **(a)** HSO_3^-; **(b)** HS^-; **(c)** HPO_4^-. Then use data from Table 16.5 to predict whether each ion makes the solution acidic or basic.

64. Suppose you wanted to produce an aqueous solution of pH = 8.65 by dissolving one of the following salts in water. Which salt would you use, and at what molarity? **(a)** NH_4Cl; **(b)** $KHSO_4$; **(c)** KNO_2; **(d)** $NaNO_3$.

olecular Structure and Acid–Base Behavior

Predict which is the stronger acid: **(a)** $HClO_2$ or $HClO_3$; **(b)** H_2CO_3 or HNO_2; **(c)** H_2SiO_3 or H_3PO_4. Explain.

Explain why trichloroacetic acid, CCl_3COOH, is a stronger acid than acetic acid, CH_3COOH.

Which is the stronger acid of each of the following pairs of acids? Explain your reasoning. **(a)** HBr or HI; **(b)** HOClO or HOBr; **(c)** $I_3CCH_2CH_2COOH$ or $CH_3CH_2CCl_2COOH$.

Indicate which of the following is the *weakest* acid, and give reasons for your choice: HBr; $CH_2ClCOOH$; CH_3CH_2COOH; CH_2FCH_2COOH; Cl_3COOH.

69. From the following bases, select the one with the *smallest* K_b and the one with the *largest* K_b, and give reasons for your choices.

(a) 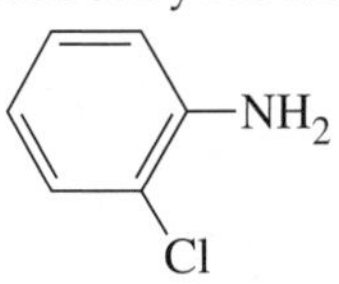

(b) 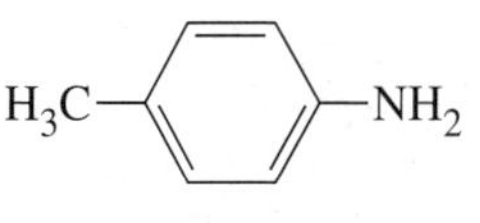

(c) $CH_3CH_2CH_2NH_2$ **(d)** $N{\equiv}CCH_2NH_2$

70. For the molecular models shown, write the formula of the species that is the most acidic and the one that is most basic, and give reasons for your choices.

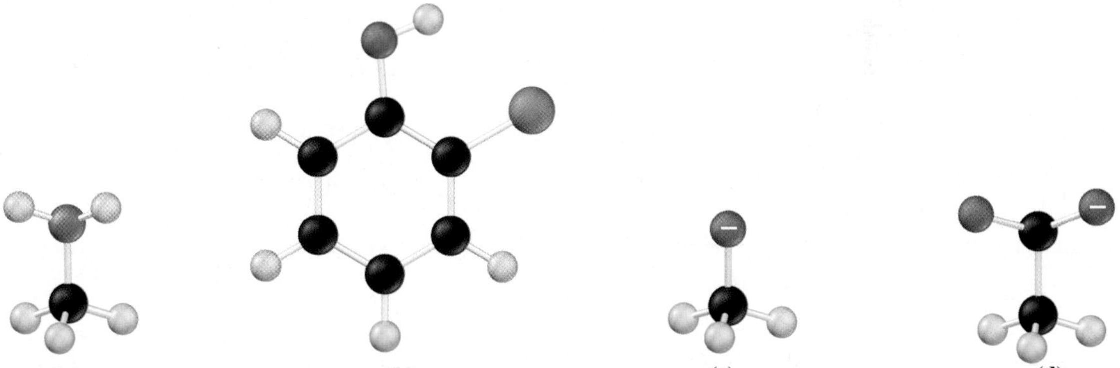

(a) (b) (c) (d)

wis Theory of Acids and Bases

For each reaction draw a Lewis structure for each species and indicate which is the acid and which is the base:
(a) $CO_2 + H_2O \longrightarrow H_2CO_3$
(b) $H_2O + BF_3 \longrightarrow H_2OBF_3$
(c) $O^{2-} + H_2O \longrightarrow 2\,OH^-$
(d) $S^{2-} + SO_3 \longrightarrow S_2O_3^{2-}$

In the following reactions indicate which is the Lewis acid and which is the Lewis base:
(a) $SOI_2 + BaSO_3 \longrightarrow Ba^{2+} + 2\,I^- + 2\,SO_2$
(b) $HgCl_3^- + Cl^- \longrightarrow HgCl_4^{2-}$

Indicate whether each of the following is a Lewis acid or base. **(a)** OH^-; **(b)** $(C_2H_5)_3B$; **(c)** CH_3NH_2.

74. Each of the following is a Lewis acid–base reaction. Which reactant is the acid, and which is the base? Explain.
(a) $SO_3 + H_2O \longrightarrow H_2SO_4$
(b) $Zn(OH)_2(s) + 2\,OH^-(aq) \longrightarrow [Zn(OH)_4]^{2-}(aq)$

75. The three following reactions are acid–base reactions according to the Lewis theory. Draw Lewis structures, and identify the Lewis acid and Lewis base in each reaction.
(a) $B(OH)_3 + OH^- \longrightarrow [B(OH)_4]^-$
(b) $N_2H_4 + H_3O^+ \longrightarrow N_2H_5^+ + H_2O$
(c) $(C_2H_5)_2O + BF_3 \longrightarrow (C_2H_5)_2OBF_3$

76. $CO_2(g)$ can be removed from confined quarters (such as a spacecraft) by allowing it to react with an alkali metal hydroxide. Show that this is a Lewis acid–base reaction. For example,

$$CO_2(g) + LiOH(s) \longrightarrow LiHCO_3(s)$$

77. The molecular solid $I_2(s)$ is only slightly soluble in water but will dissolve to a much greater extent in an aqueous solution of KI, because the I_3^- anion forms. Write an equation for the formation of the I_3^- anion, and indicate the Lewis acid and Lewis base.

78. The following very strong acids are formed by the reactions indicated:

$$HF + SbF_5 \longrightarrow HSbF_6$$

(called "super acid," hexafluoroantimonic acid)

$$HF + BF_3 \longrightarrow HBF_4$$

(tetrafluoroboric acid)

(a) Identify the Lewis acids and bases.
(b) To which atom is the H atom bonded in each ac

79. Use Lewis structures to diagram the following re tion in the manner of reaction (16.24).

$$H_2O + SO_2 \longrightarrow H_2SO_3$$

Identify the Lewis acid and Lewis base.

80. Use Lewis structures to diagram the following re tion in the manner of reaction (16.24).

$$2\,NH_3 + Ag^+ \longrightarrow [Ag(NH_3)_2]^+$$

Identify the Lewis acid and Lewis base.

Integrative and Advanced Exercises

81. The Brønsted–Lowry theory can be applied to acid–base reactions in nonaqueous solvents, where the relative strengths of acids and bases can differ from what they are in aqueous solutions. Indicate whether each of the following would be an acid, a base, or amphiprotic in pure liquid acetic acid, CH_3COOH, as a solvent. **(a)** CH_3COO^-; **(b)** H_2O; **(c)** CH_3COOH; **(d)** $HClO_4$. [*Hint:* Refer to Table 16.2.]

82. The pH of saturated $Sr(OH)_2(aq)$ is found to be 13.12. A 10.0 mL sample of saturated $Sr(OH)_2(aq)$ is diluted to 250.0 mL in a volumetric flask. A 10.0 mL sample of the diluted $Sr(OH)_2(aq)$ is transferred to a beaker, and some water is added. The resulting solution requires 25.1 mL of a HCl solution for its titration. What is the molarity of this HCl solution?

83. Several approximate pH values are marked on the following pH scale.

Some of the following solutions can be matched to one of the approximate pH values marked on the scale; others cannot. For solutions that can be matched to a pH value, identify each solution and its pH value. Identify the solutions that cannot be matched, and give reasons why matches are not possible. **(a)** 0.010 M H_2SO_4; **(b)** 1.0 M NH_4Cl; **(c)** 0.050 M KI; **(d)** 0.0020 M CH_3NH_2; **(e)** 1.0 M NaOCl; **(f)** 0.10 M C_6H_5OH; **(g)** 0.10 M HOCl; **(h)** 0.050 M $ClCH_2COOH$; **(i)** 0.050 M HCOOH.

84. Show that when $[H_3O^+]$ is reduced to half its original value, the pH of a solution increases by 0.30 unit, *regardless of the initial pH*. Is it also true that when any solution is diluted to half its original concentration, the pH increases by 0.30 unit? Explain.

85. Explain why $[H_3O^+]$ in a strong acid solution *doubles* as the total acid concentration doubles, whereas in a weak acid solution, $[H_3O^+]$ increases only by abou factor of $\sqrt{2}$.

86. Use data from Appendix D to determine whether ion product of water, K_w, increases, decreases, remains unchanged with increasing temperature.

87. From the observation that 0.0500 M vinylacetic acid a freezing point of -0.096 °C, determine K_a for this a

$$CH_2{=}CHCH_2CO_2H + H_2O \rightleftharpoons H_3O^+ + CH_2{=}CHCH_2CC$$

88. You are asked to prepare a 100.0 mL sample of a solut with a pH of 5.50 by dissolving the appropriate amo of a solute in water with pH = 7.00. Which of th solutes would you use, and in what quantity? Expl your choice. **(a)** 15 M $NH_3(aq)$; **(b)** 12 M HCl(a **(c)** $NH_4Cl(s)$; **(d)** glacial (pure) acetic acid, CH_3COC

89. Use material balance and an electroneutra condition (charge balance) to determine the pH **(a)** 1.0×10^{-5} M HCN and **(b)** 1.0×10^{-5} M C_6H_5N (aniline).

90. It is possible to write simple equations to relate p pK, and concentrations (*c*) of various solutio Three such equations are shown here.

Weak acid: $pH = \frac{1}{2}pK_a - \frac{1}{2}\log c$

Weak base: $pH = 14.00 - \frac{1}{2}pK_b + \frac{1}{2}\log c$

Salt of weak acid (pK_a) and strong base: $pH = 14.00 - \frac{1}{2}pK_w + \frac{1}{2}pK_a + \frac{1}{2}\log$

(a) Derive these three equations, and point out assumptions involved in the derivations.
(b) Use these equations to determine the pH 0.10 M $CH_3COOH(aq)$, 0.10 M $NH_3(aq)$, and 0.10 $NaCH_3COO(aq)$. Verify that the equations give corr

results by determining these pH values in the usual way.

91. A handbook lists the following formula for the percent ionization of a weak acid.

$$\% \text{ ionized} = \frac{100}{1 + 10^{(pK-pH)}}$$

(a) Derive this equation. What assumptions must you make in this derivation?
(b) Use the equation to determine the percent ionization of a formic acid solution, HCOOH(aq), with a pH of 2.50.
(c) A 0.150 M solution of propionic acid, CH_3CH_2COOH, has a pH of 2.85. What is K_a for propionic acid?

$$CH_3CH_2CO_2H + H_2O \rightleftharpoons H_3O^+ + CH_3CH_2CO_2^-$$

92. Oxalic acid, HOOCCOOH, a weak diprotic acid, has $pK_{a_1} = 1.25$ and $pK_{a_2} = 3.81$. A related diprotic acid, suberic acid, $HOOC(CH_2)_6COOH$ has $pK_{a_1} = 4.21$ and $pK_{a_2} = 5.40$. Offer a plausible reason as to why the *difference* between pK_{a_1} and pK_{a_2} is so much greater for oxalic acid than for suberic acid.

93. Here is a way to test the validity of the statement made on pages 761–762 in conjunction with the three key ideas governing the ionization of polyprotic acids. Determine the pH of 0.100 M succinic acid in two ways: first by assuming that H_3O^+ is produced only in the first ionization step, and then by allowing for the possibility that some H_3O^+ is also produced in the second ionization step. Compare the results, and discuss the significance of your finding.

$$H_2C_4H_4O_4 + H_2O \rightleftharpoons H_3O^+ + HC_4H_4O_4^- \qquad K_{a_1} = 6.2 \times 10^{-5}$$

$$HC_4H_4O_4^- + H_2O \rightleftharpoons H_3O^+ + C_4H_4O_4^{2-} \qquad K_{a_2} = 2.3 \times 10^{-6}$$

. What mass of acetic acid, CH_3COOH, must be dissolved per liter of aqueous solution if the solution is to have the same freezing point as 0.150 M $ClCH_2COOH$ (chloroacetic acid)?

. What is the pH of a solution that is 0.68 M H_2SO_4 and 1.5 M HCOOH (formic acid)?

. An aqueous solution of two weak acids has a stoichiometric concentration, *c*, in each acid. If one acid has a K_a value twice as large as the other, show that the pH of the solution is given by the equation $pH = -\frac{1}{2}\log(3c\,K_a)$, where K_a is the ionization constant of the weaker acid. Assume that the criteria for the simplifying assumption on page 757 are met.

97. Use the concept of hybrid orbitals to describe the bonding in the strong acids given in Exercise 78.

98. Phosphorous acid is listed in Appendix D as a *di*protic acid. Propose a Lewis structure for phosphorous acid that is consistent with this fact.

99. The following four equilibria lie to the right: $N_2H_5^+ + CH_3NH_2 \rightleftharpoons N_2H_4 + CH_3NH_3^+$; $H_2SO_3 + F^- \rightleftharpoons HSO_3^- + HF$; $CH_3NH_3^+ + OH^- \rightleftharpoons CH_3NH_2 + H_2O$; and $HF + N_2H_4 \rightleftharpoons F^- + N_2H_5^+$.
(a) Rank all the acids involved in order of decreasing acid strength.
(b) Rank all the bases involved in order of decreasing base strength.
(c) State whether each of the following two equilibria lies primarily to the right or to the left: **(i)** $HF + OH^- \rightleftharpoons F^- + H_2O$; **(ii)** $CH_3NH_3^+ + HSO_3^- \rightleftharpoons CH_3NH_2 + H_2SO_3$.

100. Given that K_a for HI is about 10^9, estimate the equilibrium concentrations of HI and I^- in **(a)** 1.0 M HI(aq); **(b)** 1.0 M NaI(aq).

101. In this problem, we will use material balance and charge balance concepts (pages 761–762) to calculate the equilibrium concentrations in an extremely dilute solution of a strong acid, 1.0×10^{-7} M HCl(aq).
(a) Write down the material balance equation for 1.0×10^{-7} M HCl(aq), assuming that HCl ionizes completely in aqueous solution.
(b) Write down the charge balance equation (an electroneutrality condition) for this solution.
(c) Use the material balance and charge balance equations, and the ion product for water, $K_w = [H_3O^+][OH^-]$, to show that $[H_3O^+]^2 - c_o[H_3O^+] - K_w = 0$, where $c_o = 1.0 \times 10^{-7}$ M.
(d) Use the quadratic formula to solve the equation in (c) for $[H_3O^+]$.
(e) Calculate $[OH^-]$ and compare this value with $[Cl^-]$. What conclusion can you draw from this comparison?

102. Follow the approach described in Exercise 101 for 1.0×10^{-6} M HCl(aq) to verify the claim made on page 750 that the self-ionization of water contributes less than 1% to the total amount of H_3O^+ in solution.

103. By focusing on the relative stabilities of the anions formed by ionization, rank the following compounds in each case in order of increasing acid strength (from weakest to strongest).
(a) CH_3COOH, CH_3OH, C_6H_5OH (phenol)
(b) HCN, HOCN, HCCH, HOClO

104. Show that in an extremely dilute solution of an acid, the degree of ionization, α, has the value $K_a/(K_a + 10^{-7})$ at 25 °C. [*Hint:* What is the pH of an extremely dilute solution at this temperature?]

Feature Problems

105. Maleic acid is a carbon–hydrogen–oxygen compound used in dyeing and finishing fabrics and as a preservative of oils and fats. In a combustion analysis, a 1.054 g sample of maleic acid yields 1.599 g CO_2 and 0.327 g H_2O. In a freezing-point depression experiment, a 0.615 g sample of maleic acid dissolved in 25.10 g of glacial acetic acid, $CH_3COOH(l)$ (which has the freezing-point depression constant $K_f = 3.90\ °C\ m^{-1}$ and in which maleic acid does not ionize), lowers the freezing point by 0.82 °C. In a titration experiment, a 0.4250 g sample of maleic acid is dissolved in water and requires 34.03 mL of 0.2152 M KOH for its complete neutralization. The pH of a 0.215 g sample of maleic acid dissolved in 50.00 mL of aqueous solution is found to be 1.80.

(a) Determine the empirical and molecular formulas of maleic acid. [*Hint:* Which experiment(s) provide the necessary data?]
(b) Use the results of part (a) and the titration data to rewrite the molecular formula to reflect the number of ionizable H atoms in the molecule.
(c) Given that the ionizable H atom(s) is(are) associated with the carboxyl group(s), write the plausible condensed structural formula of maleic acid.
(d) Determine the ionization constant(s) of maleic acid. If the data supplied are insufficient, indicate what additional data would be needed.
(e) Calculate the expected pH of a 0.0500 M aqueous solution of maleic acid. Indicate any assumptions required in this calculation.

106. In Example 16-9, rather than use the quadratic formula to solve the quadratic equation, we could have proceeded in the following way. Substitute the value yielded by our failed assumption—$x = 0.0010$—into the *denominator* of the quadratic equation; that is, use $(0.00250 - 0.0010)$ as the value of $[CH_3NH_2]$ and solve for a new value of x. Use this second value of x to re-evaluate $[CH_3NH$ $[CH_3NH_2] = (0.00250 -$ second value of x). So the simple quadratic equation for a third value x, and so on. After three or four trials, you w find that the value of x no longer changes. This the answer you are seeking. **(a)** Complete the cal lation of the pH of 0.00250 M CH_3NH_2 by t method, and show that the result is the same as t obtained by using the quadratic formula. **(b)** Use t method to determine the pH of 0.500 M $HClO_2$.

107. Apply the general method for solution equilibri calculations outlined on page 761–762 to determ the pH values of the following solutions. In apply the method, look for valid assumptions that m simplify the numerical calculations.
(a) a solution that is 0.315 M CH_3COOH and 0.25 $HCOOH$
(b) a solution that contains 1.55 g CH_3NH_2 and 12. NH_3 in 375 mL
(c) 1.0 M NH_4CN(aq)

Self-Assessment Exercises

108. In your own words, define or explain the following terms or symbols: **(a)** K_w; **(b)** pH; **(c)** pK_a; **(d)** hydrolysis; **(e)** Lewis acid.

109. Briefly describe each of the following ideas or phenomena: **(a)** conjugate base; **(b)** percent ionization of an acid or a base; **(c)** self-ionization; **(d)** amphiprotic behavior.

110. Explain the important distinctions between each pair of terms: **(a)** Brønsted–Lowry acid and base; **(b)** $[H_3O^+]$ and pH; **(c)** K_a for NH_4^+ and K_b for NH_3; **(d)** leveling effect and electron-withdrawing effect.

111. Of the following, the amphiprotic ion is **(a)** HCO_3^-; **(b)** CO_3^{2-}; **(c)** NH_4^+; **(d)** $CH_3NH_3^+$; **(e)** ClO_4^-.

112. The pH in 0.10 M CH_3CH_2COOH(aq) must be **(a)** equal to $[H_3O^+]$ in 0.10 M HNO_2(aq); **(b)** less than the pH in 0.10 M HI(aq); **(c)** greater than the pH in 0.10 M HBr(aq); **(d)** equal to 1.0.

113. In 0.10 M CH_3NH_2(aq), **(a)** $[H_3O^+] = 0.10$ M; **(b)** $[OH^-] = 0.10$ M; **(c)** pH < 7; **(d)** pH < 13.

114. The reaction of CH_3COOH(aq) proceeds furthest toward completion with a base when that base is **(a)** H_2O; **(b)** $CH_3NH_3^+$; **(c)** NH_4^+; **(d)** Cl^-; **(e)** CO_3^{2-}.

115. In 0.10 M H_2SO_4(aq), $[H_3O^+]$ is equal to **(a)** 0.050 M; **(b)** 0.10 M; **(c)** 0.11 M; **(d)** 0.20 M.

116. For H_2SO_3(aq), $K_{a_1} = 1.3 \times 10^{-2}$ and $K_{a_2} = 6.3 \times 10^{-8}$. In 0.10 M H_2SO_3(aq), **(a)** $[HSO_3^-] = 0.013$ M; **(b)** $[SO_3^{2-}] = 6.3 \times 10^{-8}$ M; **(c)** $[H_3O^+] = 0.10$ M; **(d)** $[H_3O^+] = 0.013$ M; **(e)** $[SO_3^{2-}] = 0.036$ M.

117. What is the pH of the solution obtained by mixing 24.80 mL of 0.248 M HNO_3 and 15.40 mL of 0.394 M KOH?

118. How many milliliters of a concentrated acetic acid solution (35.0% CH_3COOH by mass; $d = 1.044$ g/mL) must be diluted with water to produce 12.5 L of solution with pH 3.25?

119. Determine the pH of 2.05 M $NaCH_2ClCOO$. (U data from Table 16.4, as necessary.)

120. Several aqueous solutions are prepared. *Without c sulting any tables in the text,* arrange these ten so tions in order of *increasing* pH: 1.0 M NaBr, 0.05 CH_3COOH, 0.05 M NH_3, 0.02 M KCH_3COO, 0.05 $Ba(OH)_2$, 0.05 M H_2SO_4, 0.10 M HI, 0.06 M NaC 0.05 M NH_4Cl, and 0.05 M $CH_2ClCOOH$.

121. A solution is found to have pH = 5 × pOH. Is t solution acidic or basic? What is $[H_3O^+]$ in the so tion? Which of the following could be the solute this solution: NH_3, CH_3COOH, or NH_4CH_3CC and what would be its molarity?

122. Propionic acid, CH_3CH_2COOH, is 0.42% ionized 0.80 M solution. The K_a for this acid is **(a)** 1.42×10 **(b)** 1.42×10^{-7}; **(c)** 1.77×10^{-5}; **(d)** 6.15×1 **(e)** none of these.

123. The conjugate acid of HPO_4^{2-} is **(a)** PO_4 **(b)** $H_2PO_4^-$; **(c)** H_3PO_4; **(d)** H_3O^+; **(e)** none of the

124. The equilibria $OH^- + HClO \rightleftharpoons H_2O + Cl$ and $ClO^- + HNO_2 \rightleftharpoons HClO + NO_2^-$ both lie the right. Which of the following is a list of aci ranked in order of decreasing strength?
(a) $HClO > HNO_2 > H_2O$
(b) $ClO^- > NO_2^- > OH^-$
(c) $NO_2^- > ClO^- > OH^-$
(d) $HNO_2 > HClO > H_2O$
(e) none of these

125. 3.00 mol of calcium chlorite is dissolved in enou water to produce 2.50 L of solution. $K_a = 2.9 \times 1$ for HClO, and $K_a = 1.1 \times 10^{-2}$ for $HClO_2$. Comp the pH of the solution.

126. Appendix E describes a useful study aid known concept mapping. Using the method presented Appendix E, construct a concept map that summ rizes the material discussed in Section 16-8.

Additional Aspects of Acid–Base Equilibria

17

ONTENTS

LEARNING OBJECTIVES

17.1 Explain what is meant by the common-ion effect and how it relates to Le Châtelier's principle.

17.2 Explain how a buffer solution is able to resist attempts to change its pH.

17.3 Discuss the method by which a pH indicator helps determine the pH of a solution.

17.4 Describe the changes in composition that occur during an acid–base titration, and determine points on the titration curve.

17.5 Discuss the difficulties associated with calculating the determination of the pH of a solution containing a salt of a polyprotic acid.

17.6 Summarize the step-by-step process of performing acid–base equilibrium calculations.

OH(aq) is slowly added to an aqueous solution containing HCl(aq) and the icator phenolphthalein. The indicator color changes from colorless to red as the changes from 8.0 to 10.0. The equivalence point of the neutralization is reached en the solution turns a lasting pink (the pink seen here disappears when the flask wirled to mix the reactants). The selection of indicators for acid–base titrations is e of the topics considered in this chapter.

n our study of acid rain (see Focus On 16-1 on the MasteringChemistry website), we learned that a very small amount of atmospheric $CO_2(g)$ dissolves in rainwater. Yet this amount is sufficient to lower the pH of inwater by nearly 2 units. And when acid-forming air pollutants, such SO_2, SO_3, and NO_2, also dissolve in rainwater, it becomes even more idic. A chemist would say that water has no "buffer capacity"—that is, pH changes sharply when even small quantities of acids or bases are ssolved in it.

One of the main topics of this chapter is buffer solutions—solutions at can resist a change in pH when acids or bases are added to them. We ll consider how such solutions are prepared, how they maintain a nearly constant pH, and how they are used. At the end of the chapter, we w consider perhaps the most important buffer system to humans: the buff system that maintains the constant pH of blood.

Another topic that we will explore is acid–base titrations. Here, our aim w be to calculate how pH changes during a titration. We can use this informati to select an appropriate indicator for a titration and to determine, in gener which acid–base titrations work well and which do not. For the most part, w will find the calculations in this chapter to be extensions of those in Chapter 1

17-1 Common-Ion Effect in Acid–Base Equilibria

▶ The common-ion effect is not restricted to weak acids and weak bases. Buffers are the most important examples of the common-ion effect in weak acids and weak bases.

The questions answered in Chapter 16 were mostly of the type, "What is t pH of 0.10 M CH_3COOH, of 0.10 M NH_3, of 0.10 M H_3PO_4, of 0.10 M NH_4Cl In each of these cases, we think of dissolving a *single* substance in aqueo solution and determining the concentrations of the species present at equili rium. In most situations in this chapter, a solution of a weak acid or we base initially contains a second source of one of the ions produced in the io ization of the acid or base. The added ions are said to be *common* to the we acid or weak base. The presence of a common ion can have some importa consequences.

Carey B. Van Loon

▲ FIGURE 17-1
A weak acid–strong acid mixture
The solution pictured is 0.100 M CH_3COOH and 0.100 M HCl. The reading on the pH meter (1.0) indicates that essentially all the H_3O^+ comes from the strong acid HCl. The red color of the solution is that of thymol blue indicator. Compare this photo with Figure 16-6, in which the separate acids are shown.

Solutions of Weak Acids and Strong Acids

Consider a solution that is at the same time 0.100 M CH_3COOH and 0.100 HCl. We can write separate equations for the ionizations of the acids, o weak and the other strong. The ionization of CH_3COOH produces H_3O^+ a CH_3COO^- ions:

$$CH_3COOH + H_2O \rightleftharpoons CH_3COO^- + H_3O^+ \qquad (17.$$

The ionization of HCl produces H_3O^+ and Cl^- ions.

$$HCl + H_2O \longrightarrow Cl^- + H_3O^+ \qquad (17.2$$

Because H_3O^+ is formed in both ionization processes, we say that H_3O^+ is *common ion*. An important point concerning the common ion, H_3O^+, is th $[H_3O^+]$ appears in the equilibrium constant expressions for both reactior Therefore, the ionization of HCl affects the equilibrium position of reacti (17.1), and, in principle, the ionization of CH_3COOH affects the equilibriu position of reaction (17.2). We will now investigate the extent to which the io ization of each acid is affected by the other acid. To do this, we will write t ionization constant for a general acid, HA, in terms of the degree of ionizatio α, which we introduced in Section 16-3. Since α represents the fraction of H that exists as A^- at equilibrium, then $(1 - \alpha)$ is the fraction that exists as H Therefore, the equilibrium concentrations of A^- and HA may be represente as αc_o and $(1 - \alpha)c_o$, respectively, where c_o is the initial stoichiometric conce tration of HA. The K_a expression for HA is

$$K_a = \frac{[H_3O^+][A^-]}{[HA]} = \frac{[H_3O^+] \times \alpha c_o}{(1 - \alpha)c_o} = \frac{[H_3O^+] \times \alpha}{(1 - \alpha)}$$

By solving the expression above for α, we obtain

$$\alpha = \frac{K_a}{[H_3O^+] + K_a} = \frac{K_a}{10^{-pH} + K_a} \qquad (17.$$

With this expression, we can calculate the degree of ionization of a particul acid, provided we know the pH of the solution.

Figure 17-1 indicates that, for a solution that is simultaneously 0.100 CH_3COOH and 0.100 M HCl, the pH is equal to 1.0. Using $K_a \approx 10^6$ for HCl rough estimate), we find

$$\alpha = \frac{10^6}{10^{-1.0} + 10^6} = 0.9999999 \approx 1$$

/e see that HCl is completely ionized in this solution, irrespective of the presence of CH_3COOH. We use this fact in Example 17-1 to calculate the degree of ionization of CH_3COOH in a solution that is 0.100 M in both CH_3COOH and [Cl, and to illustrate that the ionization of a weak acid is significantly suppressed by the presence (or addition) of a strong acid. The approach used in xample 17-1 is recommended when the pH of the solution is not known in dvance.

EXAMPLE 17-1 **Demonstrating the Common-Ion Effect: A Solution of a Weak Acid and a Strong Acid**

(a) Determine $[H_3O^+]$ and $[CH_3COO^-]$ in 0.100 M CH_3COOH. **(b)** Then determine these same quantities in a solution that is 0.100 M in both CH_3COOH and HCl.

Analyze

In part (a) we must determine the species in a weak acid solution, and in part (b) we investigate the effects of the addition of a strong acid. The two acids have the common ion H_3O^+. The key to solving part (b) is recognizing that HCl, a strong acid, ionizes completely, irrespective of whether or not any other acids are present in the solution.

Solve

(a) This calculation was done in Example 16-8 (page 754). We found that in 0.100 M CH_3COOH, $[H_3O^+] = [CH_3COO^-] = 1.3 \times 10^{-3}$ M.

(b) Because HCl ionizes completely, the ionization of CH_3COOH is the only equilibrium we need to consider. However, the H_3O^+ produced by the ionization of HCl cannot be ignored. To account for the H_3O^+ produced by the ionization of HCl, we use 0.100 M for the initial concentration of H_3O^+ in the following equilibrium summary.

	CH_3COOH	+	H_2O	$\rightleftharpoons$	CH_3COO^-	+	H_3O^+
initial concns:	0.100 M				0 M		0.100 M
changes:	$-x$ M				$+x$ M		$+x$ M
equil concns:	$(0.100 - x)$ M				x M		$(0.100 + x)$ M

We begin by assuming that x is very small compared with 0.100. Thus, $0.100 - x \approx 0.100$ and $0.100 + x \approx 0.100$.

$$K_a = \frac{[H_3O^+][CH_3COO^-]}{[CH_3COOH]} = \frac{(0.100 + x) \cdot x}{0.100 - x} \approx \frac{\cancel{0.100} \cdot x}{\cancel{0.100}} = 1.8 \times 10^{-5}$$

We see that $x = 1.8 \times 10^{-5}$. Since $[CH_3COO^-] = x$ M and $[CH_3COOH] = (0.100 + x)$ M, then

$$[CH_3COO^-] = 1.8 \times 10^{-5}\ \text{M} \quad \text{and} \quad [CH_3COOH] = 0.100\ \text{M}$$

Notice that x is only 0.018% of 0.100, and so the assumption that x is small compared with 0.100 is valid.

Assess

It is instructive to compare $[CH_3COO^-]$ for a solution that is simultaneously 0.100 M CH_3COOH and 0.100 M HCl with that for 0.100 M CH_3COOH (see Example 16-3). For the former, we have $[CH_3COO^-] = 1.8 \times 10^{-5}$ M and for the latter, $[CH_3COO^-] = 1.3 \times 10^{-3}$ M. The much lower value of $[CH_3COO^-]$ for the solution containing both CH_3COOH and HCl indicates that the ionization of CH_3COOH is significantly suppressed by the addition of HCl. For the mixture of acids, we can use equation (17.3), with pH = 1.0 and $K_a = 1.8 \times 10^{-5}$, to calculate $\alpha = 1.8 \times 10^{-4}$. Consequently, the percent ionization is $100\alpha = 0.018\%$, which is much lower than the value of 1.3% we obtained in Example 16-3 for 0.100 M CH_3COOH. Clearly, the ionization of a weak acid is significantly suppressed by the presence (or addition) of strong acid.

(continued)

PRACTICE EXAMPLE A: Determine $[H_3O^+]$ and [HF] in 0.500 M HF. Then determine these concentrations in a solution that is 0.100 M HCl and 0.500 M HF.

PRACTICE EXAMPLE B: How many drops of 12 M HCl would you add to 1.00 L of 0.100 M CH_3COOH to make $[CH_3COO^-] = 1.0 \times 10^{-4}$ M? Assume that 1 drop = 0.050 mL and that the volume of solution remains 1.00 L after the 12 M HCl is diluted. [*Hint*: What must be the $[H_3O^+]$ in the solution?]

Example 17-1 illustrates that the ionization of a weak acid is significan suppressed by the presence (or addition) of a strong acid. It is also the ca that the ionization of a weak base is significantly suppressed by the preser or addition of a strong base. These statements can be justified by applyi Le Châtelier's principle. Let's first consider a solution of a weak acid, HA, th has reached equilibrium. The effect of adding strong acid is illustrated belo

When a strong acid supplies the common ion H_3O^+, the equilibrium shifts to form more HA

Added H_3O^+

$$HA + H_2O \rightleftharpoons H_3O^+ + A^-$$

The equilibrium shifts to the left.
The ionization of HA is suppressed.

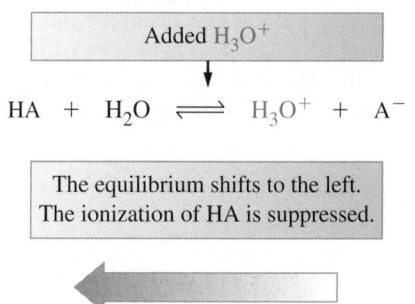

For a solution of a weak base, B, that has reached equilibrium, the effect adding a strong base can be described similarly.

When a strong base supplies the common ion OH^-, the equilibrium shifts to form more B.

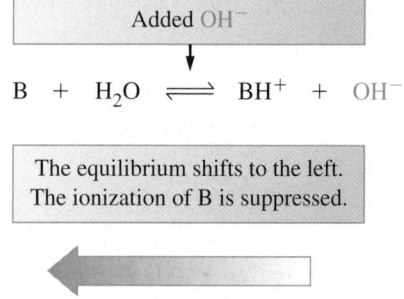

Added OH^-

$$B + H_2O \rightleftharpoons BH^+ + OH^-$$

The equilibrium shifts to the left.
The ionization of B is suppressed.

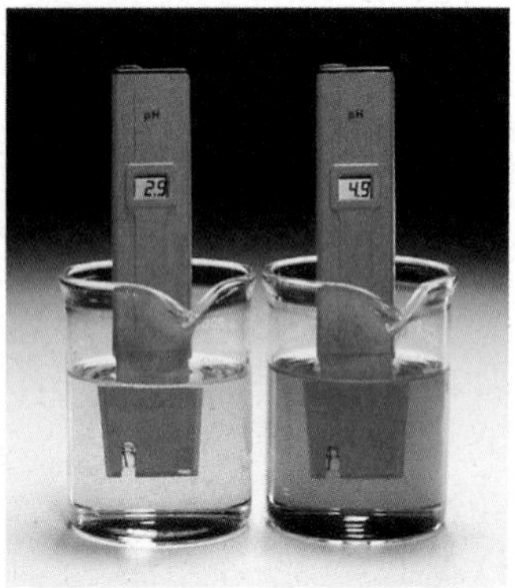

Carey B. Van Loon

(a) (b)

▲ FIGURE 17-2
A mixture of a weak acid and its salt
Bromphenol blue indicator is present in both solutions. Its color dependence on pH is

$$pH < 3.0 < pH < 4.6 < pH$$

Yellow Green Blue-Violet

(a) 0.100 M CH_3COOH has a calculated pH of 2.89, but **(b)** if the solution is also 0.100 M in $NaCH_3COO$, the calculated pH is 4.74. (The readability of the pH meters used here is 0.1 unit, and their accuracy is probably somewhat less than that. The discrepancy between 4.74 and the 4.9 value shown here is a result of their limited accuracy.)

Solutions of Weak Acids and Their Salts

Another important example of a situation involving the common ion eff arises when dealing with a solution containing a weak acid (HA) and a salt its conjugate base (for example, Na^+A^-). In such a situation, the anion of t salt is a common ion; it is produced by the ionization of the acid and b the dissociation of the salt. For example, in a solution that is simultaneous 0.100 M in both CH_3COOH and $NaCH_3COO$, there are two contributio to $[CH_3COO^-]$: the ionization of CH_3COOH and the dissociation $NaCH_3COO$. As indicated on the next page, the addition of $NaCH_3COO$ su presses the ionization of CH_3COOH.

When a salt supplies the common anion CH_3COO^-,
the equilibrium shifts to form more CH_3COOH.

$$NaCH_3COO(aq) \longrightarrow Na^+(aq) + CH_3COO^-(aq)$$

Added CH_3COO^-

$$CH_3COOH(aq) + H_2O(l) \rightleftharpoons H_3O^+(aq) + CH_3COO^-(aq)$$

The equilibrium shifts to the left.
The ionization of CH_3COOH is suppressed.

The common-ion effect of acetate ion on the ionization of acetic acid is depicted a Figure 17-2 and demonstrated in Example 17-2. In solving common-ion prob-ms, such as Example 17-2, assume that ionization of the weak acid (or base) does ot begin until both the weak acid (or base) and its salt have been placed in solu-on. Then consider that ionization occurs until equilibrium is reached.

EXAMPLE 17-2 Demonstrating the Common-Ion Effect: A Solution of a Weak Acid and a Salt of That Weak Acid

Calculate $[H_3O^+]$ and $[CH_3COO^-]$ in a solution that is 0.100 M in both CH_3COOH and $NaCH_3COO$.

Analyze

This example is very similar to Example 17-1; however, in this case we will be adding a salt of a weak acid and observing the shift in equilibrium. The setup shown here is very similar to that in Example 17-1(b), except that $NaCH_3COO$ is the source of the common ion.

Solve

The $NaCH_3COO$ dissociates completely. The CH_3COO^- produced by the dissociation of $NaCH_3COO$ is taken into account by using 0.100 M for the initial concentration of CH_3COO^- in the following equilibrium summary for the ionization of CH_3COOH.

	CH_3COOH	+	H_2O	$\rightleftharpoons$	CH_3COO^-	+	H_3O^+
initial concns:	0.100 M				0.100 M		—
changes:	$-x$ M				$+x$ M		$+x$ M
equil concns:	$(0.100 - x)$ M				$(0.100 + x)$ M		x M

Because the salt suppresses the ionization of CH_3COOH, we expect x to be very small and $0.100 - x \approx 0.100$ $+ x \approx 0.100$. This proves to be a valid assumption.

$$K_a = \frac{[H_3O^+][CH_3COO^-]}{[CH_3COOH]} = \frac{x \cdot (0.100 + x)}{0.100 - x} = \frac{x \cdot \cancel{0.100}}{\cancel{0.100}} = 1.8 \times 10^{-5}$$

$$[H_3O^+] = x\text{ M} = 1.8 \times 10^{-5}\text{ M} \qquad [CH_3COO^-] = (0.100 + x)\text{ M} = 0.100\text{ M}$$

Assess

The ionization of CH_3COOH is reduced about 100-fold because of the salt that was added. The calculations we performed in this example are very similar to those we did in Example 17-1(b). An important difference, however, is that here we solved for $[H_3O^+] = x$ M. In Example 17-1(b), we solved for $[CH_3COO^-] = x$ M. Note that when the concentrations of a weak acid and its conjugate base are the same, as is the case here, $K_a = [H_3O^+]$ and therefore, $pH = pK_a$.

PRACTICE EXAMPLE A: Calculate $[H_3O^+]$ and $[HCOO^-]$ in a solution that is 0.100 M HCOOH and 0.150 M NaHCOO.

PRACTICE EXAMPLE B: What mass of $NaCH_3COO$ should be added to 1.00 L of 0.100 M CH_3COOH to produce a solution with pH = 5.00? Assume that the volume remains 1.00 L.

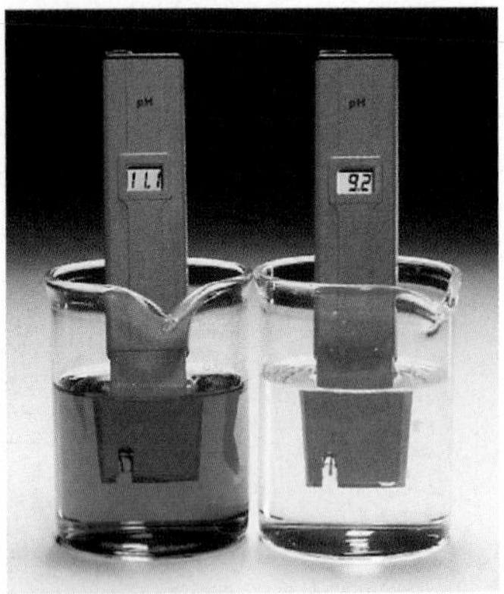

Carey B. Van Loon

(a) (b)

▲ FIGURE 17-3
A mixture of a weak base and its salt
Thymolphthalein indicator is blue if pH > 10 and colorless if pH < 10. **(a)** The pH of 0.100 M NH_3 is above 10 (calculated value: 11.11). **(b)** If the solution is also 0.100 M NH_4Cl, the pH drops below 10 (calculated value: 9.26). The ionization of NH_3 is suppressed in the presence of added NH_4^+. $[OH^-]$ decreases, $[H_3O^+]$ increases, and the pH is lowered.

Solutions of Weak Bases and Their Salts

The common-ion effect of a salt of a weak base is similar to the weak acid–anio situation just described. The suppression of the ionization of NH_3 by the com mon *cation*, NH_4^+, is pictured in Figure 17-3 and represented as follows.

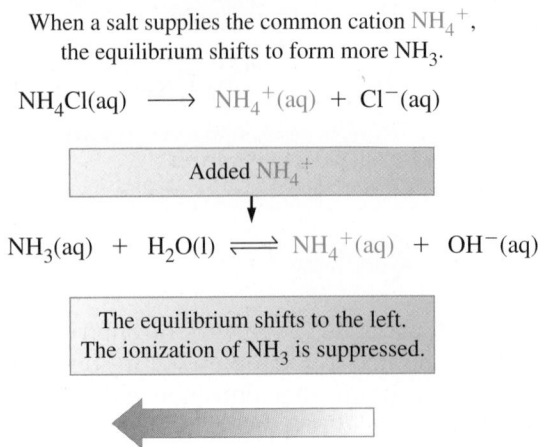

When a salt supplies the common cation NH_4^+, the equilibrium shifts to form more NH_3.

$$NH_4Cl(aq) \longrightarrow NH_4^+(aq) + Cl^-(aq)$$

Added NH_4^+

$$NH_3(aq) + H_2O(l) \rightleftharpoons NH_4^+(aq) + OH^-(aq)$$

The equilibrium shifts to the left.
The ionization of NH_3 is suppressed.

The key results of this section may be summarized as follows.

- The ionization of a weak acid, HA, is significantly suppressed by th addition of either a strong acid or a salt of its conjugate base (for exampl Na^+A^-).
- The ionization of a weak base, B, is significantly suppressed by the additic of either a strong base or a salt of its conjugate acid (for example, BH^+Cl^-

17-1 CONCEPT ASSESSMENT

Without doing detailed calculations, determine which of the following will *raise* the pH when added to 1.00 L of 0.100 M NH_3(aq): **(a)** 0.010 mol NH_4Cl(s); **(b)** 0.010 mol $(CH_3CH_2)_2NH$(l); **(c)** 0.010 mol HCl(g); **(d)** 1.00 L of 0.050 M NH_3(aq); **(e)** 1.00 g $Ca(OH)_2$(s). For NH_3, $pK_b = 4.74$; for $(CH_3CH_2)_2NH$, $pK_b = 3.16$.

17-2 Buffer Solutions

In Example 17-2, we considered a solution that is simultaneously 0.100 M i both CH_3COOH and $NaCH_3COO$ and verified by calculation that the ioniz tion of CH_3COOH is significantly suppressed in the presence of $NaCH_3CO$ Consequently, the solution contains appreciable equilibrium amounts of bo CH_3COOH and its conjugate base, CH_3COO^-. Such a solution is called **buffer solution** because of its ability to maintain a nearly constant pH whe for example, small amounts of a strong acid or a strong base are added to it when it is diluted with water. A buffer solution is able to resist changes in p because, as we will soon discover, it contains components capable of neutra izing other acids or bases but not each other. Buffer solutions are not just inte esting but also extremely useful.

We will begin our study of buffer solutions by first identifying the acti components of a buffer solution. Then, we will discuss how these componen give a buffer the ability to resist attempts to change its pH. Next, we wi develop an equation that highlights the relationship between the pH of buffer solution and the concentrations of the two active components. Finall we will consider the very practical matter of preparing a buffer solution wi a specified pH.

ecognizing a Buffer Solution

buffer solution contains either

- appreciable amounts of both a weak acid (HA) and its conjugate base (A^-), or
- appreciable amounts of both a weak base (B) and its conjugate acid (BH^+)

he term "appreciable" warrants some additional explanation. With both com-onents present in appreciable amounts, the solution will be able to neutralize ther an appreciable amount of an added acid or an appreciable amount of an dded base. A less obvious but equally important point is that to obtain appre-able amounts of the two active components, we must add two components to ie solution. The ionization of a weak acid HA never produces an appreciable mount of A^-. Similarly, the hydrolysis of A^- (a weak base) never produces an ppreciable amount of HA. Therefore, neither 0.100 M CH_3COOH nor 0.100 M aCH$_3$COO is a buffer solution, but a solution that is simultaneously 0.100 M H$_3$COOH and 0.100 M $NaCH_3COO$ is a buffer solution.

ow a Buffer Works

he solution described in Example 17-2 has $[CH_3COOH] = [CH_3COO^-]$ 0.100 M, $[H_3O^+] = 1.8 \times 10^{-5}$ M, and a pH of 4.74. Thus, the principal omponents of this solution are CH_3COOH, a weak acid, and CH_3COO^-, a eak base. This solution has the ability to resist attempts to change its pH ecause one component (CH_3COOH) is able to neutralize an added strong ase and the other component (CH_3COO^-) is able to neutralize a strong acid. he ability of a solution to neutralize either an added strong acid or an added rong base is another defining characteristic of a buffer solution.

CH_3COOH	+	OH^-	$\longrightarrow$	CH_3COO^-	+	H_2O	**(17.4)**
Weak acid		Strong base		Weak base			
rom the buffer)		(added to the buffer)		(released to the buffer)			

CH_3COO^-	+	H_3O^+	$\longrightarrow$	CH_3COOH	+	H_2O	**(17.5)**
Weak base		Strong acid		Weak acid			
rom the buffer)		(added to the buffer)		(released to the buffer)			

◀ A right arrow ($\longrightarrow$) is used in these neutralization reactions because, as established in Section 16-9, the reaction between an acid and a base goes essentially to completion if either the acid or base is strong.

et's focus first on reaction (17.5) to explain qualitatively what happens when *small* amount of a strong acid is added to a solution that is simultaneously 100 M in both CH_3COOH and $NaCH_3COO$. We begin by writing the K_a xpression for CH_3COOH and solving it for $[H_3O^+]$.

$$K_a = \frac{[H_3O^+][CH_3COO^-]}{[CH_3COOH]}$$

$$[H_3O^+] = \frac{[CH_3COOH]}{[CH_3COO^-]} \times K_a \qquad \textbf{(17.6)}$$

quation (17.6) indicates that $[H_3O^+]$ and therefore the pH depend on the atio $[CH_3COOH]/[CH_3COO^-]$. Before the addition of the acid, we have CH$_3$COOH] $\approx [CH_3COO^-]$ and the ratio $[CH_3COOH]/[CH_3COO^-]$ is pproximately equal to 1. After the addition of a small amount of strong acid, nd because of reaction (17.5), $[CH_3COOH]$ has increased slightly and CH$_3$COO$^-$] has decreased slightly. The ratio $[CH_3COOH]/[CH_3COO^-]$ is nly slightly greater than 1, and, therefore, $[H_3O^+]$ has barely changed, and ie pH remains very close to the original value of 4.74.

Now, let's imagine adding a *small* amount of a strong base to the original uffer solution. Since the added OH^- reacts with CH_3COOH (see equation 17.4), CH$_3$COOH] decreases slightly and $[CH_3COO^-]$ increases slightly. The ratio

◀ In a solution that is simultaneously 0.100 M in both CH_3COOH and $NaCH_3COO$, $[CH_3COOH]$ is approximately equal to $[CH_3COO^-]$. The two concentrations are not exactly equal because a very small amount of CH_3COOH ionizes. (See Example 17-2.)

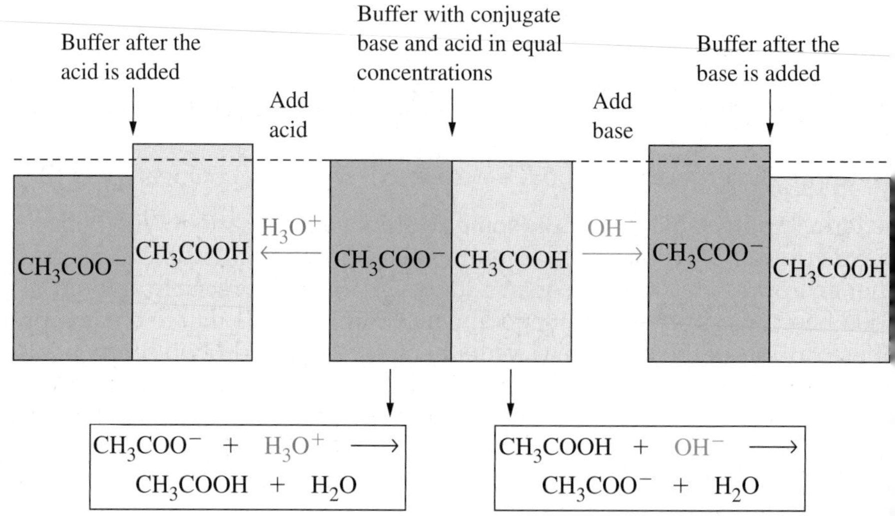

▲ FIGURE 17-4
How a buffer works
Acetate ion, the conjugate base of acetic acid, acts as a proton "sink" when strong aci is added. In this way, the ratio [conjugate base]/[acid] is kept approximately constant, there is a minimal change in pH. Similarly, acetic acid acts as a proton donor when strong base is added, keeping the ratio [conjugate base]/[acid] approximately constan and minimizing the change in pH.

$[CH_3COOH]/[CH_3COO^-]$ is only slightly smaller than 1, and again $[H_3O$ and pH have barely changed. The pH remains very close to the original valu of 4.74.

Figure 17-4 illustrates the variation of the concentration of the weak ac and its conjugate base that occurs when a strong acid or a strong base is adde to a solution initially containing equal concentrations of CH_3COOH ar $NaCH_3COO$.

Later in this section, we will be more specific about what constitutes *sma* additions of an acid or a base and *slight* changes in the concentrations of th buffer components and pH. Also, we will discover that a CH_3COOH $NaCH_3COO$ buffer is good only for maintaining a nearly constant pH in range of about 2 pH units centered on a $pH = pK_a = 4.74$. A buffer solution th maintains a nearly constant pH outside this range requires different buff components, as suggested in Example 17-3.

Finally, let us expand on a point made at the beginning of this section: buffer contains components that can neutralize an added acid or base but n each other. Perhaps it would have been better to say that the two componen coexist in the buffer with no net reaction between them. For example, in solution that is simultaneously 0.100 M in both CH_3COOH and $NaCH_3CO$ the "neutralization" of CH_3COOH by CH_3COO^- may actually occur. Wheth or not it occurs is of no concern because, as the equation below shows, th transfer of a proton from CH_3COOH to CH_3COO^- produces no net change.

$$CH_3COOH + CH_3COO^- \rightleftharpoons CH_3COO^- + CH_3COOH$$

EXAMPLE 17-3 Predicting Whether a Solution Is a Buffer Solution

Show that an NH_3–NH_4Cl solution is a buffer solution. Over what pH range would you expect it to function?

Analyze

To show that a solution has buffer properties, first identify a component in the solution that neutralizes acids and a component that neutralizes bases.

Solve

In this example, these components are NH_3 and NH_4^+, respectively. The NH_3 in solution neutralizes strong acid (H_3O^+) and NH_4^+ neutralizes strong base (OH^-), as shown in the equations below.

$$NH_3 + H_3O^+ \longrightarrow NH_4^+ + H_2O$$
$$NH_4^+ + OH^- \longrightarrow NH_3 + H_2O$$

For *all* aqueous solutions containing NH_3 and NH_4^+, we can write the following:

$$NH_3 + H_2O \rightleftharpoons NH_4^+ + OH^-$$
$$K_b = \frac{[NH_4^+][OH^-]}{[NH_3]} = 1.8 \times 10^{-5}$$

If a solution has approximately equal concentrations of NH_4^+ and NH_3, then $[OH^-] \approx 1.8 \times 10^{-5}$ M; pOH $\approx$ 4.74; and pH $\approx$ 9.26. Ammonia–ammonium chloride solutions are *basic* buffer solutions that function over the approximate pH range of 8 to 10.

Assess

As we will soon see, not all NH_3–NH_4Cl buffer solutions will be effective buffers. The best buffers have large values for $[NH_3]$ and $[NH_4^+]$, with $[NH_3] \approx [NH_4^+]$.

PRACTICE EXAMPLE A: Describe how a mixture of a strong acid (such as HCl) and the salt of a weak acid (such as $NaCH_3COO$) can result in a buffer solution. [*Hint:* What is the reaction that produces CH_3COOH?]

PRACTICE EXAMPLE B: Describe how a mixture of NH_3 and HCl can result in a buffer solution.

alculating the pH of a Buffer Solution

ne pH of a buffer solution is easily calculated by using the approach scribed in Example 17-4.

EXAMPLE 17-4 Calculating the pH of a Buffer Solution

What is the pH of a buffer solution prepared by dissolving 25.5 g $NaCH_3COO$ in a sufficient volume of 0.550 M CH_3COOH to make 500.0 mL of the buffer?

Analyze

This example is very similar to Example 17-2. The difference is that we need to calculate the molarity of the acetate ion before solving the equilibrium part.

Solve

The molarity of CH_3COO^- corresponding to 25.5 g $NaCH_3COO$ in 500.0 mL of solution is calculated as follows.

$$\text{amount of } CH_3COO^- = 25.5\text{ g } NaCH_3COO \times \frac{1\text{ mol } NaCH_3COO}{82.03\text{ g } NaCH_3COO} \times \frac{1\text{ mol } CH_3COO^-}{1\text{ mol } NaCH_3COO}$$
$$= 0.311\text{ mol } CH_3COO^-$$
$$[CH_3COO^-] = \frac{0.311\text{ mol } CH_3COO^-}{0.500\text{ L}} = 0.622\text{ M}$$

Equilibrium Calculation:

	CH_3COOH	+	H_2O	$\rightleftharpoons$	H_3O^+	+	CH_3COO^-
initial concns:							
weak acid:	0.550 M				—		—
salt:	—				—		0.622 M
changes:	$-x$ M				$+x$ M		$+x$ M
equil concns:	$(0.550 - x)$ M				$+x$ M		$(0.622 + x)$ M

(continued)

Let's assume that x is very small, so $0.550 - x \approx 0.550$ and $0.622 + x \approx 0.622$. We will find this assumption to be valid.

$$K_a = \frac{[H_3O^+][CH_3COO^-]}{[CH_3COOH]} = \frac{(x)(0.622)}{0.550} = 1.8 \times 10^{-5}$$

$$x = \frac{0.550}{0.622} \times 1.8 \times 10^{-5} = 1.6 \times 10^{-5}$$

$$[H_3O^+] = x\,M = 1.6 \times 10^{-5}\,M$$

$$pH = -\log[H_3O^+] = -\log(1.6 \times 10^{-5}) = 4.80$$

Notice that $x = 1.6 \times 10^{-5}$ is only 0.003% of 0.550, and so the assumption that x is small was justified.

Assess

We have seen that $pH = pK_a = 4.74$ when acetic acid and acetate ion are present in equal concentrations. Here the concentration of the conjugate *base* (acetate ion) is greater than that of the acetic acid. The solution should be somewhat more basic (less acidic) than $pH = 4.74$. A pH of 4.80 is a reasonable answer.

PRACTICE EXAMPLE A: What is the pH of a buffer solution prepared by dissolving 23.1 g NaHCOO in a sufficient volume of 0.432 M HCOOH to make 500.0 mL of the buffer?

PRACTICE EXAMPLE B: A handbook states that to prepare 100.0 mL of a particular buffer solution, mix 63.0 mL of 0.200 M CH_3COOH with 37.0 mL of 0.200 M $NaCH_3COO$. What is the pH of this buffer?

An important point worth noting in Example 17-4 is that if a solution is to b an effective buffer, the assumptions $(c - x) \approx c$ and $(c + x) \approx c$ will always b valid (c represents the numerical part of an expression of molarity). That i the *equilibrium* concentrations of the buffer components will be very nearly th same as their *stoichiometric* concentrations. As a result, in Example 17-4 we coul have gone directly from the stoichiometric concentrations of the buffer compo nents to the expression

KEEP IN MIND

that stoichiometric concentration is based on the amount of solute dissolved.

$$K_a = \frac{[H_3O^+][CH_3COO^-]}{[CH_3COOH]} = \frac{[H_3O^+](0.622)}{0.550} = 1.8 \times 10^{-5}$$

without setting up the ICE table.

An Equation for Buffer Solutions: The Henderson–Hasselbalch Equation

Although we can continue to use the format demonstrated in Example 17-4 fo buffer calculations, it is often useful to describe a buffer solution by means o an equation known as the **Henderson–Hasselbalch equation**. Biochemist and molecular biologists commonly use this equation. To derive this variatio of the ionization constant expression, let's consider a mixture of a hypothetica weak acid, HA (such as CH_3COOH), and its salt, NaA (such as $NaCH_3COO$ We start with the familiar expressions

$$HA + H_2O \rightleftharpoons H_3O^+ + A^-$$

$$K_a = \frac{[H_3O^+][A^-]}{[HA]}$$

and rearrange the right side of the K_a expression to obtain

$$K_a = [H_3O^+] \times \frac{[A^-]}{[HA]}$$

ext, we take the *negative logarithm* of each side of this equation.

$$-\log K_a = -\log[H_3O^+] - \log\frac{[A^-]}{[HA]}$$

ow, recall that $pH = -\log[H_3O^+]$ and that $pK_a = -\log K_a$, which gives

$$pK_a = pH - \log\frac{[A^-]}{[HA]}$$

lve for pH by rearranging the equation.

$$pH = pK_a + \log\frac{[A^-]}{[HA]}$$

is the conjugate base of the weak acid HA, so we can write the more gen-
al equation (17.7), the Henderson–Hasselbalch equation.

$$pH = pK_a + \log\frac{[\text{conjugate base}]}{[\text{acid}]} \qquad \textbf{(17.7)}$$

◀ When the [conjugate base] = [acid], then $pH = pK_a$. When we get to titrations, an indicator is chosen to change color at $pH = pK_a$ of the indicator.

To apply this equation to an acetic acid–sodium acetate buffer, we use pK_a for
H_3COOH and these concentrations: $[CH_3COOH]$ for [acid] and $[CH_3COO^-]$
r [conjugate base]. To apply it to an ammonia–ammonium chloride buffer, we
e pK_a for NH_4^+ and these concentrations: $[NH_4^+]$ for [acid] and $[NH_3]$ for
onjugate base].

Equation (17.7) is useful only when we can substitute *stoichiometric* or initial
ncentrations for equilibrium concentrations to give

$$pH = pK_a + \log\frac{[\text{conjugate base}]_{\text{initial}}}{[\text{acid}]_{\text{initial}}}$$

us avoiding the need to set up an ICE table. This constraint places important
nitations on the equation's validity, however. Later, we will see that there are
so conditions that must be met if a mixture is to be an effective buffer solution.
lthough the following rules may be overly restrictive in some cases, a reason-
le approach to the twin concerns of effective buffer action and the validity of
ing stoichiometric concentrations in equation (17.7) is to ensure that

KEEP IN MIND

that the Henderson–Hasselbalch equation is very useful but should probably not be committed to memory; it is easy to get the conjugate base and acid terms inverted. It is most important to understand the principles that lead to this equation, thereby avoiding the pitfalls of using the equation incorrectly or when it is not valid.

. the ratio of stoichiometric concentrations is within the following range

$$0.10 < \frac{[\text{conjugate base}]_{\text{initial}}}{[\text{acid}]_{\text{initial}}} < 10 \qquad \textbf{(17.8)}$$

. the stoichiometric concentration of each buffer component exceeds the value of K_a by a factor of at least 100

ewed another way, the use of stoichiometric concentrations in equation (17.7)
justified only if the calculated value of H_3O^+ is very small compared to
$\text{cid}]_{\text{initial}}$ and $[\text{conjugate base}]_{\text{initial}}$ (that is, only if the H_3O^+ from the ioniza-
on of the acid does not significantly change the stoichiometric concentrations
the buffer components).

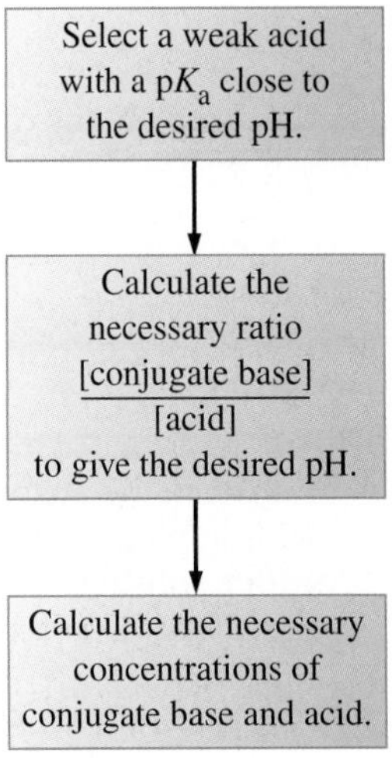

▲ One procedure to follow in making a buffer solution with a desired pH.

reparing Buffer Solutions

ppose we need a buffer solution with pH = 5.09. Equation (17.7) suggests
vo alternatives. One is to find a weak acid, HA, that has $pK_a = 5.09$ and pre-
are a solution with equal molarities of the acid and its salt.

$$pH = pK_a + \log\frac{\cancel{[A^-]}}{\cancel{[HA]}} = 5.09 + \log 1 = 5.09$$

Although this alternative is simple in concept, generally, it is not practic
We are not likely to find a readily available, water-soluble weak acid wi
exactly pK_a = 5.09. The second alternative, summarized in the margin of t
previous page, is to use a cheap, common weak acid such as acetic aci
CH_3COOH (pK_a = 4.74), and establish an appropriate ratio
$[CH_3COO^-]/[CH_3COOH]$ to obtain a pH of 5.09. Example 17-5 demonstrat
this second alternative.

EXAMPLE 17-5 Preparing a Buffer Solution of a Desired pH

What mass of $NaCH_3COO$ must be dissolved in 0.300 L of 0.25 M CH_3COOH to produce a solution with pH = 5.09? Assume that the solution volume remains constant at 0.300 L.

Analyze

Equilibrium among the buffer components is expressed by the equation

$$CH_3COOH + H_2O \rightleftharpoons H_3O^+ + CH_3COO^- \qquad K_a = 1.8 \times 10^{-5}$$

and by the ionization constant expression for acetic acid.

$$K_a = \frac{[H_3O^+][CH_3COO^-]}{[CH_3COOH]} = 1.8 \times 10^{-5}$$

Each of the three concentration terms appearing in a K_a expression should be an equilibrium concentration. The $[H_3O^+]$ corresponding to a pH of 5.09 is the equilibrium concentration. For $[CH_3COOH]$, we will assume that the equilibrium concentration is equal to the stoichiometric or initial concentration. The value of $[CH_3COO^-]$ that we calculate with the K_a expression is the equilibrium concentration, and we will assume that it is also the same as the stoichiometric concentration. Thus, we assume that neither the ionization of CH_3COOH to form CH_3COO^- nor the hydrolysis of CH_3COO^- to form CH_3COOH produces much of a difference between the stoichiometric (initial) and equilibrium concentrations of the buffer components. These assumptions work well if the conditions stated in expression (17.8) are met.

Solve

The relevant concentration terms, then, are

$$[H_3O^+] = 10^{-pH} = 10^{-5.09} = 8.1 \times 10^{-6}\text{ M}$$
$$[CH_3COOH] = 0.25\text{ M}$$
$$[CH_3COO^-] = ?$$

The required acetate ion concentration in the buffer solution is

$$[CH_3COO^-] = K_a \times \frac{[CH_3COOH]}{[H_3O^+]} = 1.8 \times 10^{-5} \times \frac{0.25}{8.1 \times 10^{-6}} = 0.56\text{ M}$$

We complete the calculation of the mass of sodium acetate with some familiar ideas of solution stoichiometry.

$$\text{mass} = 0.300\text{ L} \times \frac{0.56\text{ mol } CH_3COO^-}{1\text{ L}} \times \frac{1\text{ mol } NaCH_3COO}{1\text{ mol } CH_3COO^-} \times \frac{82.0\text{ g } NaCH_3COO}{1\text{ mol } NaCH_3COO} = 14\text{ g } NaCH_3COO$$

Assess

We check the answer by inserting the acetate ion and acetic acid concentrations, along with the pK_a of acetic acid, into equation (17.7) to obtain pH = 5.09. The method described in Example 17-5 is one way to obtain a buffer solution. Another approach involves adding an appropriate amount of strong base (e.g., 0.052 mol NaOH) to 0.300 L of 0.25 M $CH_3COOH(aq)$.

PRACTICE EXAMPLE A: How many grams of $(NH_4)_2SO_4$ must be dissolved in 0.500 L of 0.35 M NH_3 to produce a solution with pH = 9.00? (Assume that the solution volume remains at 0.500 L.)

PRACTICE EXAMPLE B: In Practice Example 17-3A, we established that an appropriate mixture of a strong acid and the salt of a weak acid is a buffer solution. Show that a solution made by adding 33.05 g $NaCH_3COO \cdot 3\,H_2O(s)$ to 300 mL of 0.250 M HCl should have pH ≈ 5.1.

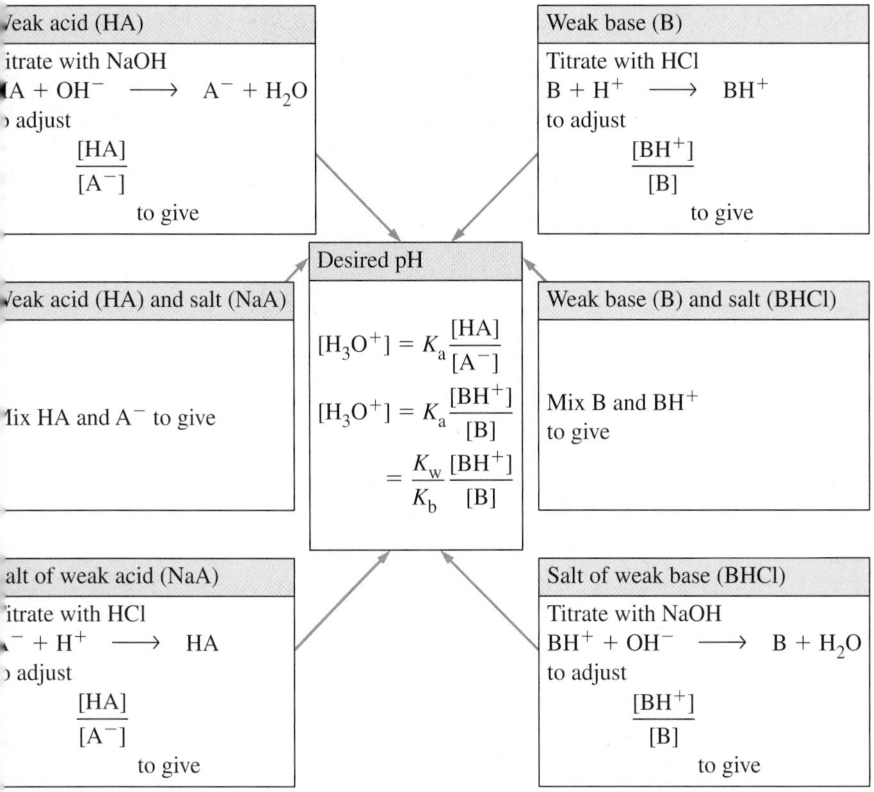

FIGURE 17-5
x methods for preparing buffer solutions
epending on the pH range required and the type of experiment the buffer is to be ,ed for, either a weak acid or a weak base can be used to prepare a buffer solution.

In Example 17-5, we achieved the desired ratio of $[CH_3COO^-]/[CH_3COOH]$ y adding 14 g of sodium acetate to the previously prepared 0.25 M CH_3COOH lution. This is a common method of obtaining a buffer solution. Other methods e sometimes useful as well. Sufficient NaOH(aq) could be added to $H_3COOH(aq)$ to neutralize some of the CH_3COOH, *producing* CH_3COO^- as a roduct. Or enough $NaCH_3COO(s)$ could be added to HCl(aq) to neutralize all e HCl, producing some CH_3COOH and leaving some CH_3COO^- in excess. As e saw in Chapter 16, amines are weak bases, so an aqueous mixture of an amine d its conjugate acid is a buffer solution. Buffer solutions based on amines can be repared in ways analogous to those based on weak acids. The methods available r making buffer solutions are summarized in Figure 17-5.

alculating pH Changes in Buffer Solutions

) calculate how the pH of a buffer solution changes when small amounts f a strong acid or base are added, we must first use *stoichiometric* principles) establish how much of one buffer component is consumed and how uch of the other component is produced. Then the new concentrations of eak acid (or weak base) and its salt can be used to calculate the pH of the uffer solution. Essentially, this problem is solved in two steps. First, we ssume that the neutralization reaction proceeds to *completion* and determine ew stoichiometric concentrations. Then these new stoichiometric concentra- ons are substituted into the equilibrium constant expression and the expres- on is solved for $[H_3O^+]$, which is converted to pH. This method is applied in xample 17-6 and illustrated in Figure 17-6.

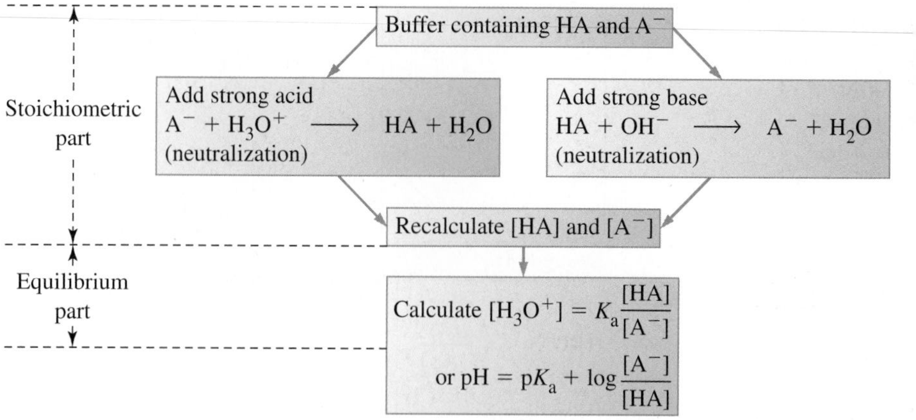

▲ FIGURE 17-6
Calculation of the new pH of a buffer after strong acid or base is added
The stoichiometric and equilibrium parts of the calculation are indicated. This scheme can also be applied to the conjugate acid–base pair BH^+/B, where B is a base.

EXAMPLE 17-6 Calculating pH Changes in a Buffer Solution

What are the effects on the pH of adding **(a)** 0.0060 mol HCl and **(b)** 0.0060 mol NaOH to 0.300 L of a buffer solution that is 0.250 M CH_3COOH and 0.560 M $NaCH_3COO$?

Analyze

In parts (a) and (b) we complete essentially the same calculations. We should recognize that we are adding a strong acid or a strong base to the buffer solution. To investigate this effect we make a stoichiometric calculation, followed by an equilibrium calculation. The stoichiometric calculation is necessary to account for the neutralization of the base or acid components of the buffer.

Solve

To judge the effect of adding either (a) acid or (b) base on the pH of the buffer, the value we must keep in mind is the pH of the original buffer. Because the initial (or stoichiometric) concentrations of CH_3COOH and CH_3COO^- are large and not too different, the initial pH of the buffer can be obtained by substituting the initial concentrations into equation (17.7). (See the discussion following equation (17.7) on page 799.)

$$pH = pK_a + \log \frac{[CH_3COO^-]}{[CH_3COOH]}$$

$$= 4.74 + \log \frac{0.560}{0.250} = 4.74 + 0.35 = 5.09$$

(a) Stoichiometric Calculation: Let's calculate amounts in moles, and assume that the neutralization goes to completion. Essentially, this is a limiting reactant calculation, but perhaps simpler than many of those in Chapter 4. In neutralizing the added H_3O^+, 0.0060 mol CH_3COO^- is converted to 0.0060 mol CH_3COOH.

	CH_3COO^-	+	H_3O^+	$\longrightarrow$	CH_3COOH	+	H_2O
original buffer:	$\underbrace{0.300\text{ L} \times 0.560\text{ M}}_{0.168\text{ mol}}$				$\underbrace{0.300\text{ L} \times 0.250\text{ M}}_{0.0750\text{ mol}}$		
add:			0.0060 mol				
changes:	−0.0060 mol		−0.0060 mol		+0.0060 mol		
final buffer:							
amounts:	0.162 mol		≈ 0		0.0810 mol		
concns:	$\underbrace{0.162\text{ mol}/0.300\text{ L}}_{0.540\text{ M}}$		≈ 0		$\underbrace{0.0810\text{ mol}/0.300\text{ L}}_{0.270\text{ M}}$		

Equilibrium Calculation: We can calculate the pH with equation (17.7), using the new equilibrium concentrations.

$$\mathrm{pH} = \mathrm{p}K_a + \log\frac{[CH_3COO^-]}{[CH_3COOH]}$$

$$= 4.74 + \log\frac{0.540}{0.270} = 4.74 + 0.30 = 5.04$$

(b) Stoichiometric Calculation: In neutralizing the added OH^-, 0.0060 mol CH_3COOH is converted to 0.0060 mol CH_3COO^-. The calculation of the new stoichiometric concentrations is shown on the last line of the following table.

	CH_3COOH	+	OH^-	$\longrightarrow$	CH_3COO^-	+	H_2O
original buffer:	0.300 L × 0.250 M = 0.0750 mol				0.300 L × 0.560 M = 0.168 mol		
add:			0.0060 mol				
changes:	−0.0060 mol		−0.0060 mol		+0.0060 mol		
final buffer:							
amounts:	0.0690 mol		≈0		0.174 mol		
concns:	0.0690 mol/0.300 L = 0.230 M		≈0		0.174 mol/0.300 L = 0.580 M		

Equilibrium Calculation: This is the same type of calculation as in part (a), but with slightly different concentrations.

$$\mathrm{pH} = 4.74 + \log\frac{0.580}{0.230} = 4.74 + 0.40 = 5.14$$

Assess

The addition of 0.0060 mol HCl *lowers* the pH from 5.09 to 5.04, which is only a small change in pH. The addition of 0.0060 mol OH^- *raises* the pH from 5.09 to 5.14—another small change. Had we instead added 0.0060 mol HCl or 0.0060 mol NaOH to 0.300 L of water, the pH would have changed by more than 5 pH units. The most important factors to confirm in a calculation of this type are that the magnitude of the pH change is small and that the change occurs in the correct direction: *lowering* of the pH by an acid and *raising* of the pH by a base. The results are indeed reasonable.

PRACTICE EXAMPLE A: A 1.00 L volume of buffer is made with concentrations of 0.350 M NaHCOO (sodium formate) and 0.550 M HCOOH (formic acid). **(a)** What is the initial pH? **(b)** What is the pH after the addition of 0.0050 mol HCl(aq)? (Assume that the volume remains 1.00 L.) **(c)** What would be the pH after the addition of 0.0050 mol NaOH to the original buffer?

PRACTICE EXAMPLE B: How many milliliters of 6.0 M HNO_3 would you add to 300.0 mL of the buffer solution of Example 17-6 to change the pH from 5.09 to 5.03?

Perhaps you have already noticed a way to simplify the calculation in xample 17-6. Because the buffer components are always present in the me solution of volume V, the numbers of moles can be substituted directly to equation (17.7) without regard for the particular value of V. Thus, in xample 17-6(b),

$$\mathrm{pH} = 4.74 + \log\frac{[CH_3COO^-]}{[CH_3COOH]} = 4.74 + \log\frac{0.174\ \cancel{\mathrm{mol}/V}}{0.0690\ \cancel{\mathrm{mol}/V}} = 4.74 + 0.40 = 5.14$$

This expression is also consistent with the observation that, on dilution, uffer solutions resist pH changes. Diluting a buffer solution means increasng its volume V by adding water. This action produces the same change in the umerator and the denominator of the ratio [conjugate base]/[acid]. The ratio self remains unchanged, as does the pH.

◀ Dilute and concentrated buffers will have the same pH, but as mentioned in the next section, a given volume of a dilute buffer will have a lower buffer capacity than the same volume of a more concentrated buffer.

▶ Our eyes can see an indicator color change over a range of about 2 pH units.

KEEP IN MIND

that, as a rule of thumb, the amounts of the buffer components should be at least ten times as great as the amount of acid or base to be neutralized.

▶ The buffer range for the ammonia–ammonium chloride solution is based on the pK_a of NH_4^+, 9.26.

▶ Two of the most important biological buffers are the phosphate and bicarbonate buffer systems. Proteins and nucleotides also function as buffers on the cellular level.

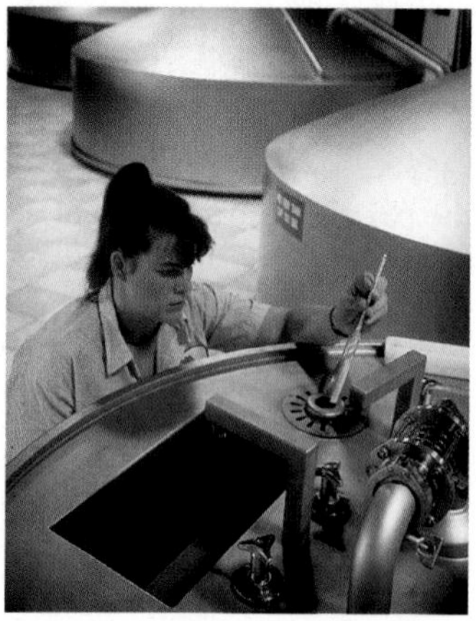

Tom Pantages

▲ A master brewer inspecting wort temperature and pH in the making of beer.

Buffer Capacity and Buffer Range

It is not difficult to see that if we add more than 0.0750 mol OH^- to the buf solution described in Example 17-6, the 0.0750 mol CH_3COOH will be co pletely converted to 0.0750 mol CH_3COO^-. An excess of OH^- will remain, a the solution will become strongly basic.

Buffer capacity refers to the amount of acid or base that a buffer can ne tralize before its pH changes appreciably. In general, the maximum buff capacity exists when the concentrations of a weak acid and its conjugate ba are kept *large* and *approximately equal to each other*. The **buffer range** is the p range over which a buffer effectively neutralizes added acids and bases a maintains a fairly constant pH. As equation (17.7) suggests,

$$pH = pK_a + \log\frac{[\text{conjugate base}]}{[\text{acid}]}$$

when the ratio [conjugate base]/[acid] = 1, pH = pK_a. When the ratio falls 0.10, the pH *decreases* by 1 pH unit from pK_a because log 0.10 = −1. If the rat increases to a value of 10, the pH *increases* by 1 unit because log 10 = 1. F practical purposes, this range of 2 pH units is the maximum range to which buffer solution should be exposed. For acetic acid–sodium acetate buffers, th effective range is about pH 3.7–5.7; for ammonia–ammonium chloride buffer it is about pH 8.3–10.3.

Applications of Buffer Solutions

An important example of a buffered system is that found in blood, whic must be maintained at a pH of 7.4 in humans. We consider the buffering blood in the Focus On feature for this chapter on the MasteringChemist website. But buffers have other important applications, too.

Protein studies often must be performed in buffered media because the stru tures of protein molecules, including the magnitude and kind of electric charg they carry, depend on the pH (see Section 28-4). The typical enzyme is a prote capable of catalyzing a biochemical reaction, so enzyme activity is closely linke to protein structure and hence to pH. Most enzymes in the body have their ma imum activity between pH 6 and pH 8. Studying enzyme activity in the labor tory usually means working with media buffered in this pH range.

The control of pH is often important in industrial processes. For example, i the mashing of barley malt, the first step of making beer, the pH of the sol tion must be maintained at 5.0 to 5.2, so that the protease and peptidas enzymes can hydrolyze the proteins from the barley. The inventor of the p scale, Søren Sørensen, was a research scientist in a brewery.

We will consider the importance of buffer solutions in solubility/precipit tion processes in Chapter 18.

17-2 CONCEPT ASSESSMENT

You are asked to make a buffer with a pH value close to 4 that would best resist an increase in pH. You can select one of the following acid–conjugate base pairs: acetic acid–acetate, $K_a = 1.8 \times 10^{-5}$; ammonium ion–ammonia, $K_a = 5.6 \times 10^{-10}$; or benzoic acid–benzoate, $K_a = 6.3 \times 10^{-5}$; and you can mix them in the following acid–conjugate base ratios: 1:1, 2:1, or 1:2. What combination would make the best buffer?

17-3 Acid–Base Indicators

An **acid–base indicator** is a substance whose color depends on the pH of th solution to which it is added. Several of the photographs in this and the pre ceding chapter have shown acid–base indicators in use. The indicator chose pended on just how acidic or basic the solution was. In this section, we will nsider how an acid–base indicator works and how an appropriate indicator selected for a pH measurement.

Acid–base indicators exist in two forms: (1) a weak acid, represented symlically as HIn and having one color, and (2) its conjugate base, represented In^- and having a different color. When just a small amount of indicator is ded to a solution, the indicator does not affect the pH of the solution. stead, the ionization equilibrium of the indicator is itself affected by the preiling $[H_3O^+]$ in solution.

$$\underset{\text{Acid color}}{HIn} + H_2O \rightleftharpoons H_3O^+ + \underset{\text{Base color}}{In^-}$$

om Le Châtelier's principle, we see that *increasing* $[H_3O^+]$ in a solution splaces the equilibrium to the left, increasing the proportion of HIn and nce the acid color. *Decreasing* $[H_3O^+]$ in a solution displaces the equilibım to the right, increasing the proportion of In^- and hence the base color. ne color of the solution depends on the relative proportions of the acid d base. The pH of the solution can be related to these relative proporons and to the pK_a of the indicator by means of an equation similar to uation (17.7).

$$pH = pK_{HIn} + \log\frac{[In^-]}{[HIn]} \qquad \textbf{(17.9)}$$

◀ The acid "color" of a few indicators is colorless.

In general, if 90% or more of an indicator is in the form HIn, the solution ill take on the acid color. If 90% or more is in the form In^-, the solution kes on the base (or anion) color. If the concentrations of HIn and In^- are out equal, the indicator is in the process of changing from one form to the her and has an intermediate color. The complete change in color occurs er a range of about *2 pH units*, with $pH = pK_{HIn}$ at about the middle of the nge. The colors and pH ranges of several acid–base indicators are shown Figure 17-7. A summary of these ideas is presented in Table 17.1, and an ample of their use is given below.

Bromthymol blue, $pK_{HIn} = 7.1$

$pH < 6.1$ (yellow) $pH \approx 7.1$ (green) $pH > 8.1$ (blue)

An acid–base indicator is usually prepared as a solution (in water, ethanol, some other solvent). In acid–base titrations, a few drops of the indicator olution are added to the solution being titrated. In other applications, porous aper is impregnated with an indicator solution and dried. When this paper is oistened with the solution being tested, it acquires a color determined by the H of the solution. This paper is usually called *pH test paper*.

TABLE 17.1 pH and the Colors of Acid–Base Indicators

Acid Color	Intermediate Color	Base Color
$[In^-]/[HIn] < 0.10$ $pH < pK_{HIn} + \log 0.10$ $pH < pK_{HIn} - 1$	$[In^-]/[HIn] \approx 1$ $pH \approx pK_{HIn} + \log 1$ $pH \approx pK_{HIn}$	$[In^-]/[HIn] > 10$ $pH > pK_{HIn} + \log 10$ $pH > pK_{HIn} + 1$

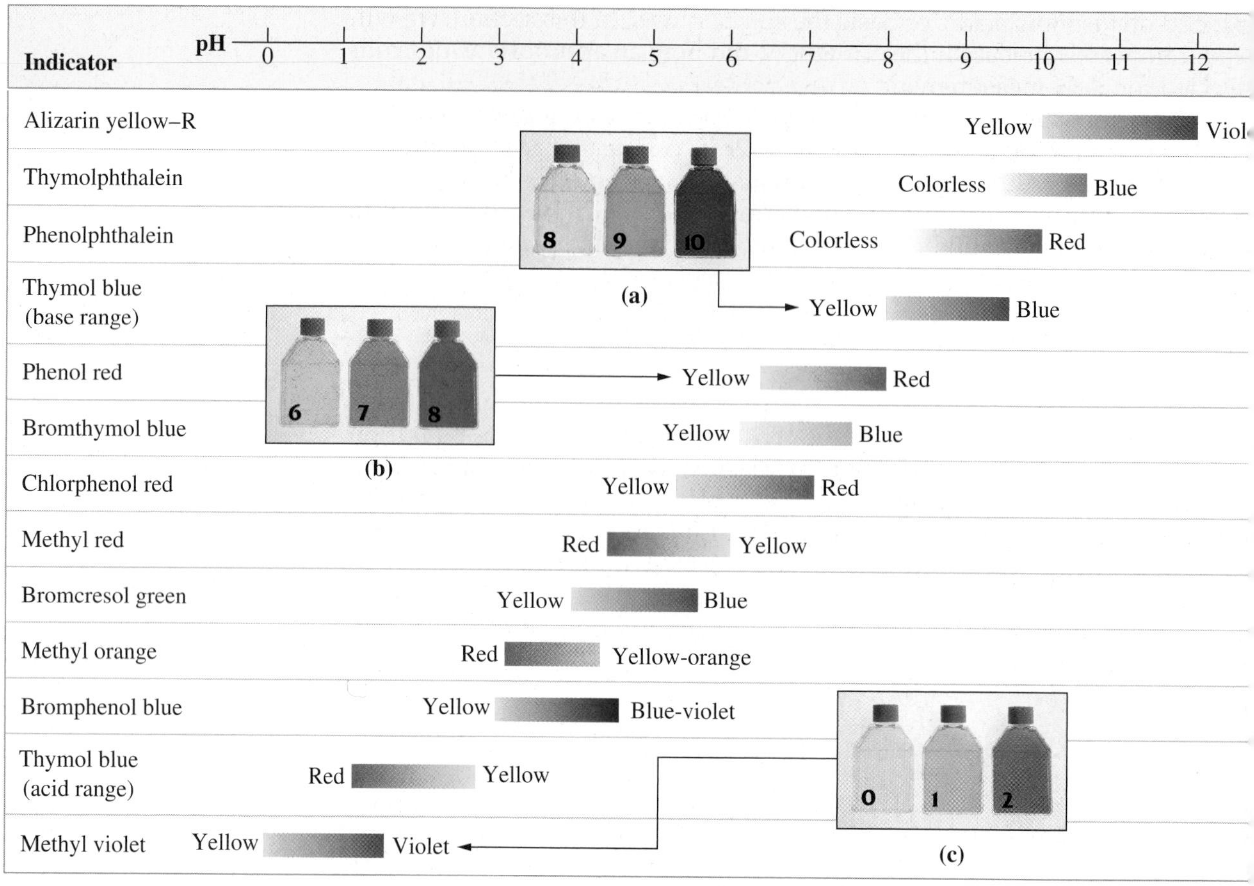

▲ FIGURE 17-7
pH and color changes for some common acid–base indicators
The indicators pictured and the pH values at which they change color are **(a)** thymol blue (pH 8–10); **(b)** phenol red (pH 6–8); and **(c)** methyl violet (pH 0–2).

17-3 CONCEPT ASSESSMENT

(1) Given that an indicator is itself a weak acid or base, why does adding it to a solution not change the nature of the equilibrium?

(2) Starting with about 10 mL of dilute NaCl(aq) containing a couple of drops of phenol red indicator, what color would the solution be for each of the following actions: **(a)** first 10 drops of 1.0 M HCl(aq) are added to the solution; **(b)** next 15 drops of 1.0 M $NaCH_3COO$(aq) are added to solution **(a)**; **(c)** then one drop of 1.0 M KOH(aq) is added to solution **(b)**; and **(d)** 10 more drops of 1.0 M KOH(aq) are added to solution **(c)**.

Cmspic/123 RF

▲ Testing swimming pool water for its chlorine content and pH.

Applications

Acid–base indicators are most useful when only an approximate pH determ nation is needed. For example, they are used in soil-testing kits to establish th approximate pH of soils. Soils are usually acidic in regions of high rainfall an heavy vegetation, and they are alkaline in more arid regions. The pH can va considerably with local conditions, however. If a soil is found to be too acid for a certain crop, its pH can be raised by adding slaked lime $[Ca(OH)_2]$. T reduce the pH of a soil, organic matter might be added.

In swimming pools, chlorinating agents are most effective at a pH of abo 7.4. At this pH, the growth of algae is avoided, and the corrosion of po plumbing is minimized. Phenol red (see Figure 17-7) is a common indicat ed in testing swimming pool water. If chlorination is carried out with $_2$(g), the pool water becomes acidic as a result of the reaction of Cl_2 with $_2$O: $Cl_2 + 2\,H_2O \longrightarrow H_3O^+ + Cl^- + HOCl$. In this case, a basic substance, ch as sodium carbonate, is used to raise the pH. Another widely used chlonating agent is sodium hypochlorite, NaOCl(aq), made by the reaction of $_2$(g) with excess NaOH(aq): $Cl_2 + 2\,OH^- \longrightarrow Cl^- + OCl^- + H_2O$. The cess NaOH raises the pH of the pool water. The pH is adjusted by adding an id, such as HCl or H_2SO_4.

7-4 Neutralization Reactions and Titration Curves

s we learned in the discussion of the stoichiometry of titration reactions ection 5-7), the **equivalence point** of a neutralization reaction is the point at hich both acid and base have been consumed and *neither* is in excess.

In a titration, one of the solutions to be neutralized—say, the acid—is laced in a flask or beaker, together with a few drops of an acid–base indicar. The other solution (the base) used in a titration is added from a buret and called the **titrant**. The titrant is added to the acid, first rapidly and then drop y drop, up to the equivalence point (recall Figure 5-18). The equivalence oint is located by noting the color change of the acid–base indicator. The oint in a titration at which the indicator changes color is called the **end point** f the indicator. The end point must match the equivalence point of the neualization. That is, if the indicator's end point is near the equivalence point of e neutralization, the color change marked by that end point will signal the tainment of the equivalence point. This match can be achieved by use of an dicator whose color change occurs over a pH range that includes the pH of e equivalence point.

A graph of pH versus volume of titrant (the solution in the buret) is called a **tration curve**. Titration curves are most easily constructed by measuring the H during a titration with a pH meter and plotting the data with a recorder. In is section we will emphasize calculating the pH at various points in a titraon. These calculations will serve as a review of aspects of acid–base equiliba considered earlier in this chapter and in the preceding chapter.

he Millimole

n a typical titration, the volume of solution delivered from a buret is less than 0 mL (usually about 20–25 mL). The molarity of the solution used for the tration is generally less than 1 M. The typical amount of OH^- (or H_3O^+) elivered from the buret during a titration is only a few thousandths of a ole—for example, 5.00×10^{-3} mol. In calculations it is often easier to work vith millimoles instead of moles. The symbol **mmol** stands for a **millimole**, vhich is one thousandth of a mole, or 10^{-3} mol.

Recall from Chapter 4 that molarity is defined as the number of moles per iter. We can use an alternative definition of molarity by converting from noles to millimoles and from liters to milliliters.

$$M = \frac{\text{mol}}{\text{L}} = \frac{\text{mol}/1000}{\text{L}/1000} = \frac{\text{mmol}}{\text{mL}}$$

hus, the expression from Chapter 4 that the amount of solute is the product of molarity and solution volume (page 123) can be based either on nol/L × L = mol or on mmol/mL × mL = mmol.

itration of a Strong Acid with a Strong Base

uppose that 25.00 mL of 0.100 M HCl (a strong acid) is placed in a small flask r beaker and that 0.100 M NaOH (a strong base) is added to it from a buret.

The pH of the accumulated solution can be calculated at different points in t titration, and these pH values can be plotted against the volume of NaC added. From this titration curve we can establish the pH at the equivalen point and identify an appropriate indicator for the titration. Some typical ca culations are outlined in Example 17-7.

EXAMPLE 17-7 Calculating Points on a Titration Curve: Strong Acid Titrated with a Strong Base

What is the pH at each of the following points in the titration of 25.00 mL of 0.100 M HCl with 0.100 M NaOH?

(a) before the addition of any NaOH (*initial pH*)

(b) after the addition of 24.00 mL 0.100 M NaOH (*before the equivalence point*)

(c) after the addition of 25.00 mL 0.100 M NaOH (*the equivalence point*)

(d) after the addition of 26.00 mL 0.100 M NaOH (*beyond the equivalence point*)

Analyze

Parts (a) to (d) correspond to four different stages of the titration. In part (a) we calculate the initial pH of the HCl solution before the titration begins. In part (b) most but not all the acid has been neutralized. In part (c) all the acid is neutralized, which corresponds to the equivalence point. In part (d) we are past the equivalence point and are dealing with a solution containing an unreacted strong base.

Solve

First, let's write the titration equation in the ionic and net ionic form.

Ionic form: $H_3O^+(aq) + Cl^-(aq) + Na^+(aq) + OH^-(aq) \longrightarrow Na^+(aq) + Cl^-(aq) + 2\,H_2O(l)$

Net ionic form: $H_3O^+(aq) + OH^-(aq) \longrightarrow 2\,H_2O(l)$

(a) Before any NaOH is added, we are dealing with 0.100 M HCl. This solution has $[H_3O^+] = 0.100$ M and pH = 1.00.

(b) The number of millimoles of H_3O^+ to be titrated is

$$25.00\text{ mL} \times \frac{0.100\text{ mmol }H_3O^+}{1\text{ mL}} = 2.50\text{ mmol }H_3O^+$$

The number of millimoles of OH^- present in 24.00 mL of 0.100 M NaOH is

$$24.00\text{ mL} \times \frac{0.100\text{ mmol}}{1\text{ mL}} = 2.40\text{ mmol }OH^-$$

Now we can represent the net ionic equation of the neutralization reaction in a familiar format.

	H_3O^+	+	OH^-	$\longrightarrow$	$2\,H_2O$
initially present:	2.50 mmol		—		
add:			2.40 mmol		
changes:	−2.40 mmol		−2.40 mmol		
after reaction:	0.10 mmol		≈0		

The remaining 0.10 mmol of H_3O^+ is present in 49.00 mL of solution (25.00 mL original + 24.00 mL added base).

$$[H_3O^+] = \frac{0.10\text{ mmol }H_3O^+}{49.00\text{ mL}} = 2.0 \times 10^{-3}\text{ M}$$

$$\text{pH} = -\log[H_3O^+] = -\log(2.0 \times 10^{-3}) = 2.70$$

(c) The equivalence point is the point at which the HCl is completely neutralized and no excess NaOH is present. As seen in the ionic form of the equation for the neutralization reaction, the solution at the equivalence point is simply NaCl(aq). And, as we learned in Section 16-7, because neither Na^+ nor Cl^- hydrolyzes in water, pH = 7.00.

(d) To determine the pH of the solution beyond the equivalence point, we can return to the format in (b), except that now OH^- is in excess. The amount of OH^- added is 26.00 mL × 0.100 mmol/L = 2.60 mmol.

	H_3O^+	+	OH^-	⟶	$2\,H_2O$
initially present:	2.50 mmol		—		
add:			2.60 mmol		
changes:	−2.50 mmol		−2.50 mmol		
after reaction:	≈0		0.10 mmol		

The excess 0.10 mmol of NaOH is present in 51.00 mL of solution (25.00 mL original acid + 26.00 mL added base). The concentration of OH^- in this solution is

$$[OH^-] = \frac{0.10 \text{ mmol } OH^-}{51.00 \text{ mL}} = 2.0 \times 10^{-3}\text{ M}$$

$$\text{pOH} = -\log(2.0 \times 10^{-3}) = 2.70 \qquad \text{pH} = 14.00 - 2.70 = 11.30$$

Assess

In strong acid–strong base titrations, there is an abrupt change in pH near the equivalence point (Fig. 17-8). For a strong acid–strong base titration, the pH at the equivalence point is equal to 7.

PRACTICE EXAMPLE A: For the titration of 25.00 mL of 0.150 M HCl with 0.250 M NaOH, calculate **(a)** the initial pH; **(b)** the pH when neutralization is 50.0% complete; **(c)** the pH when neutralization is 100.0% complete; and **(d)** the pH when 1.00 mL of NaOH is added beyond the equivalence point.

PRACTICE EXAMPLE B: For the titration of 50.00 mL of 0.00812 M $Ba(OH)_2$ with 0.0250 M HCl, calculate **(a)** the initial pH; **(b)** the pH when neutralization is 50.0% complete; **(c)** the pH when neutralization is 100.0% complete.

Figure 17-8 presents pH versus volume data and the titration curve for the HCl—NaOH titration. From this figure, we can establish these principal features of the titration curve for the titration of a *strong acid with a strong base*.

- The pH has a low value at the beginning of the titration.
- The pH changes slowly until just before the equivalence point.

Titration Data

mL NaOH(aq)	pH
0.00	1.00
10.00	1.37
20.00	1.95
22.00	2.19
24.00	2.70
25.00	7.00
26.00	11.30
28.00	11.75
30.00	11.96
40.00	12.36
50.00	12.52

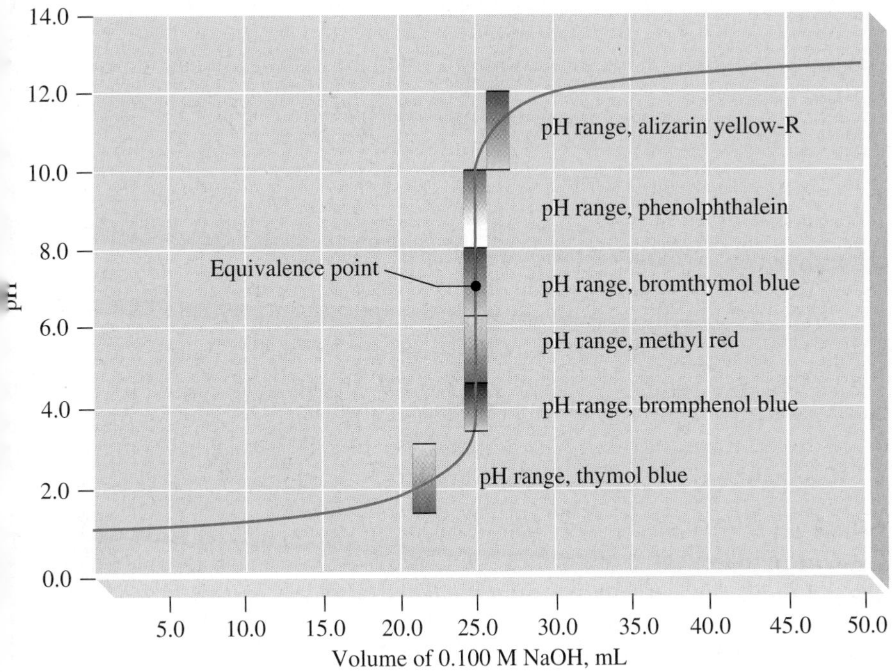

◀ FIGURE 17-8
Titration curve for the titration of a strong acid with a strong base—25.00 mL of 0.100 M HCl with 0.100 M NaOH
All indicators whose color ranges fall along the steep portion of the titration curve are suitable for this titration. Thymol blue changes color too soon; alizarin yellow-R, too late.

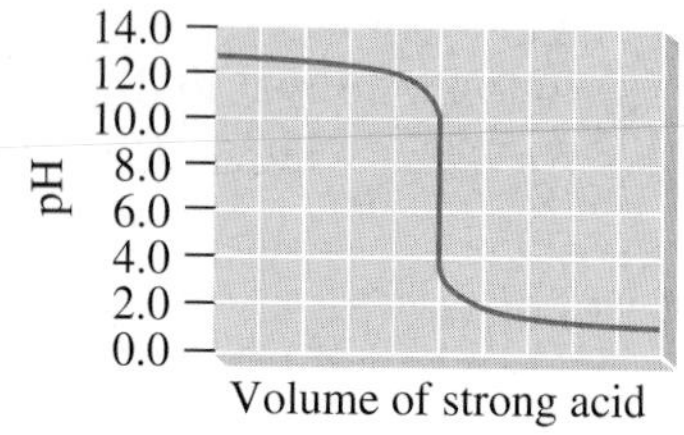

▲ FIGURE 17-9
Titration curve for the titration of a strong base with a strong acid

- At the equivalence point, the pH rises very sharply, perhaps by 6 uni for an addition of only 0.10 mL (2 drops) of base.
- Beyond the equivalence point, the pH again rises only slowly.
- Any acid–base indicator whose color changes in the pH range fro about 4 to 10 is suitable for this titration.

In the titration of a strong base with a strong acid, we can obtain a titratic curve essentially identical to Figure 17-8 by plotting pOH against volume titrant (the strong acid). Also, we can make a set of statements similar to tho listed above, except that pOH would be substituted for pH. Alternatively, if p is plotted against volume of titrant (the strong acid), the titration curve lool like Figure 17-8 flipped over from top to bottom, as shown in Figure 17-9.

Titration of a Weak Acid with a Strong Base

Several important differences exist between the titration of a weak acid with strong base and a strong acid with a strong base, but one feature is *unchang* when we compare the two titrations.

> For equal volumes of acid solutions of the same molarity, the volume of base required to titrate to the equivalence point is independent of the strength of the acid.

We can think of the neutralization of a weak acid, such as CH_3COOH, a involving the direct transfer of protons from CH_3COOH molecules t OH^- ions:

▶ The equation for the neutralization of CH_3COOH by OH^- can be represented as the sum of two equations, one for the ionization of CH_3COOH in water and the other for the reaction of H_3O^+ and OH^- to form water (the self-ionization of water written in reverse). Thus, the equilibrium constant for the neutralization reaction is the product of K_a and $1/K_w$.

$$CH_3COOH + OH^- \longrightarrow CH_3COO^- + H_2O \qquad K_{neutr.} = K_a/K_w = 1.8 \times 10^9 \gg 1$$

On the other hand, for the neutralization of a strong acid, the protons ar transferred from H_3O^+ ions to OH^- ions:

$$H_3O^+ + OH^- \longrightarrow 2\,H_2O \qquad K_{neutr.} = 1/K_w = 1.0 \times 10^{14} \gg 1$$

However, for both neutralization reactions, the acid and base react to comple tion (because $K_{neutr.} \gg 1$) and in a 1:1 mole ratio.

For a weak acid–strong base titration, the calculation of pH at differer stages of the titration requires a different approach than the one used i Example 17-7. Consider, for example, the titration of a solution of CH_3COOH with NaOH. Because the equilibrium constant for the neutralization c CH_3COOH by NaOH is extremely large, we can justifiably say that the neu tralization reaction

$$CH_3COOH + OH^- \longrightarrow CH_3COO^- + H_2O$$

goes essentially to completion. Therefore, starting from given initial amount of CH_3COOH and OH^-, the equilibrium amount of the limiting reactant (eithe CH_3COOH or OH^-) will be extremely small. However, if we were to base th calculation of the solution's pH on this reaction *and* on the assumption that i goes essentially to completion, we would quickly run into trouble.

To illustrate, let's consider adding 0.100 mol CH_3COOH and 0.030 mo NaOH to water to make 1.00 L of solution. In this case, NaOH is the limitin reactant. We expect most of the OH^- from NaOH to be consumed. Th equilibrium amounts of CH_3COOH and CH_3COO^- will be approximatel $(0.100 \times 0.030) = 0.070$ mol and 0.030 mol, respectively.

This summary assumes the reaction goes to completion.	CH_3COOH	+ OH^-	$\longrightarrow$	CH_3COO^-	+ H_2O
initial amounts:	0.100 mol	0.030 mol		0 mol	
changes:	−0.030 mol	−0.030 mol		+0.030 mol	
final amounts:	0.070 mol	0.030 mol		0.030 mol	

Titration Data

mL NaOH(aq)	pH
0.00	2.89
5.00	4.14
10.00	4.57
12.50	4.74
15.00	4.92
20.00	5.35
24.00	6.12
25.00	8.72
26.00	11.30
30.00	11.96
40.00	12.36
50.00	12.52

ne summary on the previous page gives accurate estimates of the equilib-um amounts of CH_3COOH and CH_3COO^- but an unsatisfactory result for e equilibrium concentration of OH^-. The equilibrium concentration of OH^- nnot be 0 mol/L. For example, we know that the product of $[H_3O^+]$ and $[OH^-]$ in any solution must always equal 1.0×10^{-14}, a fact that we have sed many times. This condition cannot be satisfied if $[OH^-] = 0$ mol/L. The ct that OH^- is not completely consumed is a reminder of an important point e've made before: No reaction goes all the way to completion.

We have several options for "adjusting" the estimates above. One option is imagine that the reaction above "backs up" a little bit to attain a true equi-brium state. (This is the approach we advocated in Example 15-15 to deal ith a reaction that goes nearly to completion.) Another option is follow up with a different equilibrium calculation. To identify the appro-riate equilibrium calculation to perform, we focus on the species produced nd left behind by assuming the neutralization reaction goes to completion. the present case, the neutralization reaction produces a solution that is, to very good approximation, 0.070 M in CH_3COOH and 0.030 M in CH_3COO^-. hus, the true equilibrium state may be determined by considering the ion-ation of CH_3COOH in the presence of an initial excess of $NaCH_3COO$. The ppropriate equilibrium summary is as follows.

	CH_3COOH	+	H_2O	$\rightleftharpoons$	CH_3COO^-	+	H_3O^+
itial concns:	0.070 M				0.030 M		
hanges:	$-x$ M				$+x$ M		$+x$ M
quil concns:	$(0.070 - x)$ M				$(0.030 + x)$ M		x M

Ve now employ a line of reasoning we've used many times before to simplify ubsequent calculations. That is, the presence of an excess of CH_3COO^- sup-resses the ionization of CH_3COOH, and consequently, we assume that x is mall (more specifically, that $x \ll 0.030$). Therefore,

$$K_a = \frac{[H_3O^+][CH_3COO^-]}{[CH_3COOH]} \approx \frac{[H_3O^+](0.030)}{(0.070)}$$

$$[H_3O^+] \approx \frac{0.070}{0.030} \times K_a = \frac{0.070}{0.030} \times 1.8 \times 10^{-5} = 4.2 \times 10^{-5}\ \text{M}$$

$$\text{pH} = -\log[H_3O^+] = -\log(4.2 \times 10^{-5}) = 4.44$$

)f course, we could have arrived at this result most directly (without setting p the equilibrium summary) by recognizing that the $CH_3COOH–CH_3COO^-$ olution is a buffer solution whose pH can be calculated with the Henderson–Hasselbalch equation.

◀ It is worth mentioning that, at this point, it doesn't matter that we actually prepared this solution by adding 0.100 mol CH_3COOH and 0.030 mol NaOH to water to make 1.00 L of solution. An identical solution can be prepared by adding 0.070 mol CH_3COOH and 0.030 mol $NaCH_3COO$ to water to make 1.00 L of solution.

In summary, the calculation of the pH at a given point in the titration of a veak acid by a strong base is typically divided into two parts.

1. **A stoichiometric calculation based on the neutralization reaction.** This calculation helps us to identify the principal components present in solution at equilibrium and provides accurate estimates of their concentrations. However, an additional (equilibrium) calculation is required to obtain satisfactory results for the concentrations of the other species present in solution.
2. **An equilibrium calculation involving the species produced by and left behind by the neutralization reaction.** This calculation is required to account for the fact that the neutralization reaction does not actually go all the way to completion.

'his approach is used in Example 17-8.

EXAMPLE 17-8 Calculating Points on a Titration Curve: Weak Acid Titrated with a Strong Base

What is the pH at each of the following points in the titration of 25.00 mL of 0.100 M CH_3COOH with 0.100 M NaOH?

(a) before the addition of any NaOH (*initial pH*)

(b) after the addition of 10.00 mL 0.100 M NaOH (*before equivalence point*)

(c) after the addition of 12.50 mL 0.100 M NaOH (*half-neutralization*)

(d) after the addition of 25.00 mL 0.100 M NaOH (*equivalence point*)

(e) after the addition of 26.00 mL 0.100 M NaOH (*beyond equivalence point*)

Analyze

Titrations between weak acids and strong bases or strong acids and weak bases have four regions of interest. The first is the initial pH, which we calculate in the same way we would calculate the pH for a solution of a weak acid or weak base. The second is the buffer region; the third is the hydrolysis region; and the fourth is beyond the equivalence point.

Solve

(a) The initial $[H_3O^+]$ is obtained by the calculation in Example 16-8 (page 754):

$$pH = -\log(1.3 \times 10^{-3}) = 2.89$$

(b) The number of millimoles of CH_3COOH initially present is

$$25.00\text{ mL} \times \frac{0.100\text{ mmol } CH_3COOH}{1\text{ mL}} = 2.50\text{ mmol } CH_3COOH$$

At this point in the titration, the number of millimoles of OH^- added is

$$10.00\text{ mL} \times \frac{0.100\text{ mmol } OH^-}{1\text{ mL}} = 1.00\text{ mmol } OH^-$$

The total solution volume = 25.00 mL original acid + 10.00 mL added base = 35.00 mL. We enter this information at appropriate points into the following setup.

Stoichiometric Calculation:

	CH_3COOH	+	OH^-	$\longrightarrow$	CH_3COO^-	+	H_2O
initially present:	2.50 mmol		—		—		
add:			1.00 mmol				
changes:	−1.00 mmol		−1.00 mmol		+1.00 mmol		
after reaction:							
mmol:	1.50 mmol				1.00 mmol		
concns:	1.50 mmol/35.00 mL		≈0		1.00 mmol/35.00 mL		
	0.0429 M				0.0286 M		

Equilibrium Calculation: The most direct approach is to recognize that the acetic acid–sodium acetate solution is a buffer solution whose pH can be calculated with the Henderson–Hasselbalch equation. We can use this equation for the reasons outlined on page 799: (1) The ratio $[CH_3COO^-]/[CH_3COOH]$ is 0.0286/0.0429 = 0.667 (satisfying the requirement that it be between 0.10 and 10), and (2) $[CH_3COO^-]$ and $[CH_3COOH]$ exceed $K_a = 1.8 \times 10^{-5}$ by the factors 1.6×10^3 and 2.4×10^3, respectively (satisfying the requirement that these factors exceed 100). So

$$pH = pK_a + \log\frac{[A^-]}{[HA]} = 4.74 + \log\frac{0.0286}{0.0429} = 4.74 - 0.18 = 4.56$$

Simpler still would be to substitute the numbers of millimoles of CH_3COO^- and CH_3COOH directly into the Henderson–Hasselbalch equation, without converting to molarities. That is,

$$pH = pK_a + \log\frac{[A^-]}{[HA]}$$

$$= 4.74 + \log\frac{1.00\text{ mmol}/\cancel{V}}{1.50\text{ mmol}/\cancel{V}} = 4.74 - 0.18 = 4.56$$

(c) When we have added 12.50 mL of 0.100 M NaOH, we have added $12.50 \times 0.100 = 1.25$ mmol OH^-. As the following setup shows, this is enough base to neutralize exactly *half* of the acid.

	CH_3COOH	+	OH^-	$\longrightarrow$	CH_3COO^-	+	H_2O
initially present:	2.50 mmol		—		—		
add:			1.25 mmol				
changes:	−1.25 mmol		−1.25 mmol		+1.25 mmol		
after reaction:	1.25 mmol		≈ 0		1.25 mmol		

Again, applying the Henderson–Hasselbalch equation, we get

$$\text{pH} = \text{p}K_a + \log\frac{[CH_3COO^-]}{[CH_3COOH]}$$

$$= 4.74 + \log\frac{1.25\ \cancel{\text{mmol}/V}}{1.25\ \cancel{\text{mmol}/V}} = 4.74 + \log 1 = 4.74$$

(d) At the equivalence point, neutralization is complete and 2.50 mmol $NaCH_3COO$ has been produced in 50.00 mL of solution, leading to 0.0500 M $NaCH_3COO$. The question becomes, "What is the pH of 0.0500 M $NaCH_3COO$?" To answer this question, we must recognize that CH_3COO^- hydrolyzes (and Na^+ does not). The hydrolysis reaction and value of K_b are

$$CH_3COO^- + H_2O \rightleftharpoons CH_3COOH + OH^-$$

$$K_b = \frac{K_w}{K_a} = \frac{1.0 \times 10^{-14}}{1.8 \times 10^{-5}} = 5.6 \times 10^{-10}$$

With a format similar to that used in the hydrolysis calculation of Example 16-14 (page 767), we obtain the following expression, where x M = $[OH^-]$ and $x \ll 0.0500$.

$$K_b = \frac{[CH_3COOH]}{[CH_3COO^-]} = \frac{x \cdot x}{0.0500 - x} = 5.6 \times 10^{-10}$$

$$x^2 = 2.8 \times 10^{-11} \qquad x = 5.3 \times 10^{-6}$$

$$[OH^-] = x\ \text{M} = 5.3 \times 10^{-6}\ \text{M}$$

$$\text{pOH} = -\log(5.3 \times 10^{-6}) = 5.28$$

$$\text{pH} = 14.00 - \text{pOH} = 14.00 - 5.28 = 8.72$$

(e) The amount of OH^- added is 26.00 mL × 0.100 mmol/mL = 2.60 mmol. The volume of solution is 25.00 mL acid + 26.00 mL base = 51.00 mL. The 2.60 mmol OH^- neutralizes the 2.50 mmol of available acid, and 0.10 mmol OH^- remains in *excess*. Beyond the equivalence point, the pH of the solution is determined by the excess strong base.

$$[OH^-] = \frac{0.10\ \text{mmol}}{51.00\ \text{mL}} = 2.0 \times 10^{-3}\ \text{M}$$

$$\text{pOH} = -\log(2.0 \times 10^{-3}) = 2.70 \qquad \text{pH} = 14.00 - 2.70 = 11.30$$

Assess

From part (c), we see that the pH at the equivalence point is *not* 7 as it was in the strong acid–strong base problem. We could have predicted this result before performing a single calculation. At the equivalence point, the principal species in solution is CH_3COO^-, the conjugate base of a weak acid (CH_3COOH). Recalling that the conjugate of weak is weak, we conclude that CH_3COO^- is a weak base and thus, the pH at the equivalence point must be greater than 7.

PRACTICE EXAMPLE A: A 20.00 mL sample of 0.150 M HF solution is titrated with 0.250 M NaOH. Calculate **(a)** the initial pH and the pH when neutralization is **(b)** 25.0%, **(c)** 50.0%, and **(d)** 100.0% complete. [*Hint*: What is the initial amount of HF, and what amount remains unneutralized at the points in question?]

PRACTICE EXAMPLE B: For the titration of 50.00 mL of 0.106 M NH_3 with 0.225 M HCl, calculate **(a)** the initial pH and the pH when neutralization is **(b)** 25.0% complete; **(c)** 50.0% complete; **(d)** 100.0% complete.

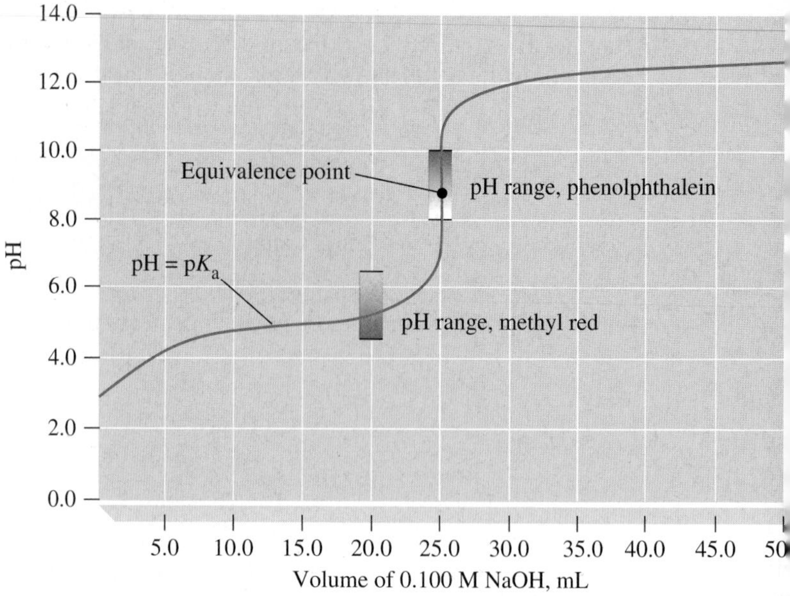

▶ FIGURE 17-10
Titration curve for the titration of a weak acid with a strong base—25.00 mL of 0.100 M CH_3COOH with 0.100 M NaOH
Phenolphthalein is a suitable indicator for this titration, but methyl red is not. When exactly half of the acid is neutralized, $[CH_3COOH] = [CH_3COO^-]$ and $pH = pK_a = 4.74$.

Here are the principal features of the titration curve for a weak acid titrate with a strong base (Fig. 17-10).

- The initial pH is higher (less acidic) than in the titration of a strong aci (The weak acid is only partially ionized.)
- There is an initial rather sharp increase in pH at the start of the titratio (The anion produced by the neutralization of the weak acid is a commc ion that reduces the extent of ionization of the acid.)
- Over a long section of the curve preceding the equivalence point, the pl changes only gradually. (Solutions corresponding to this portion of th curve are buffer solutions.)
- Because $[HA] = [A^-]$ at the point of half-neutralization, $pH = pK_a$.
- At the equivalence point, $pH > 7$. (The conjugate base of a weak aci hydrolyzes, producing OH^-.)
- Beyond the equivalence point, the titration curve is identical to that of strong acid with a strong base. (In this portion of the titration, the pH established entirely by the concentration of unreacted OH^-.)
- The steep portion of the titration curve at the equivalence point occu over a relatively short pH range (from about pH 7 to pH 10).
- The selection of indicators available for the titration is more limited tha in a strong acid–strong base titration. (Indicators that change color belo pH 7 cannot be used.)

As illustrated in Example 17-8 and suggested by Figure 17-11, the necessar calculations for a weak acid–strong base titration curve are of four distin

▶ FIGURE 17-11
Constructing the titration curve for a weak acid with a strong base
The calculations needed to plot this graph, illustrated in Example 17-8, can be divided into *four* types.

1. pH of a pure weak acid (initial pH)
2. pH of a buffer solution of a weak acid and its salt (over a broad range before the equivalence point)
3. pH of a salt solution undergoing hydrolysis (equivalence point)
4. pH of a solution of a strong base (over a broad range beyond the equivalence point)

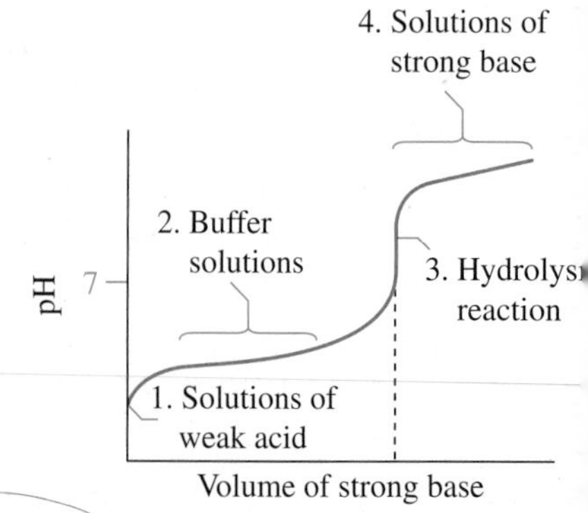

ypes, depending on the portion of the titration curve being described. One type f titration that generally cannot be performed successfully is that of a weak cid with a weak base (or vice versa). The equivalence point cannot be located recisely because the change in pH with volume of titrant is too gradual.

17-4 CONCEPT ASSESSMENT

To raise the pH of 1.00 L of 0.50 M HCl(aq) *significantly*, which of the following would you add to the solution and why? **(a)** 0.50 mol CH_3COOH; **(b)** 1.00 mol NaCl; **(c)** 0.60 mol $NaCH_3COO$; **(d)** 0.40 mol NaOH.

itration of a Weak Polyprotic Acid

he most striking evidence that a polyprotic acid ionizes in distinct steps comes y way of its titration curve. For a polyprotic acid, we expect to see a separate quivalence point for each acidic hydrogen. Thus, we expect to see three equivlence points when H_3PO_4 is titrated with NaOH(aq). In the neutralization of hosphoric acid by sodium hydroxide, essentially all the H_3PO_4 molecules are rst converted to the salt, NaH_2PO_4. Then all the NaH_2PO_4 is converted to Ja_2HPO_4; and finally the Na_2HPO_4 is converted to Na_3PO_4.

◀ This stepwise neutralization is observed only if successive ionization constants ($K_{a_1}, K_{a_2}, \ldots$) differ significantly in magnitude (for example, by a factor of 10^3 or more). Otherwise, the second neutralization step begins before the first step is completed, and so on.

The titration of 10.0 mL of 0.100 M H_3PO_4 with 0.100 M NaOH is pictured ı Figure 17-12. Notice that the first two equivalence points come at equal ıtervals on the volume axis, at 10.0 mL and at 20.0 mL. Although we expect a ıird equivalence point at 30.0 mL, it is not realized in this titration. The pH of ıe strongly hydrolyzed Na_3PO_4 solution at the third equivalence point— pproaching pH 13—is higher than can be reached by adding 0.100 M NaOH) water. Na_3PO_4(aq) is nearly as basic as the NaOH(aq) used in the titration ıs we will see in Section 17-5).

Let's focus on a few details of this titration. For each mole of H_3PO_4, 1 mol IaOH is required to reach the first equivalence point. At this first equivalence oint, the solution is essentially NaH_2PO_4(aq). This is an *acidic* solution ecause $K_{a_2} > K_b$ for $H_2PO_4^-$: the reaction that produces H_3O^+ predominates ver the one that produces OH^-.

$$H_2PO_4^- + H_2O \rightleftharpoons H_3O^+ + HPO_4^{2-} \qquad K_{a_2} = 6.3 \times 10^{-8}$$
$$H_2PO_4^- + H_2O \rightleftharpoons H_3PO_4 + OH^- \qquad K_b = 1.4 \times 10^{-12}$$

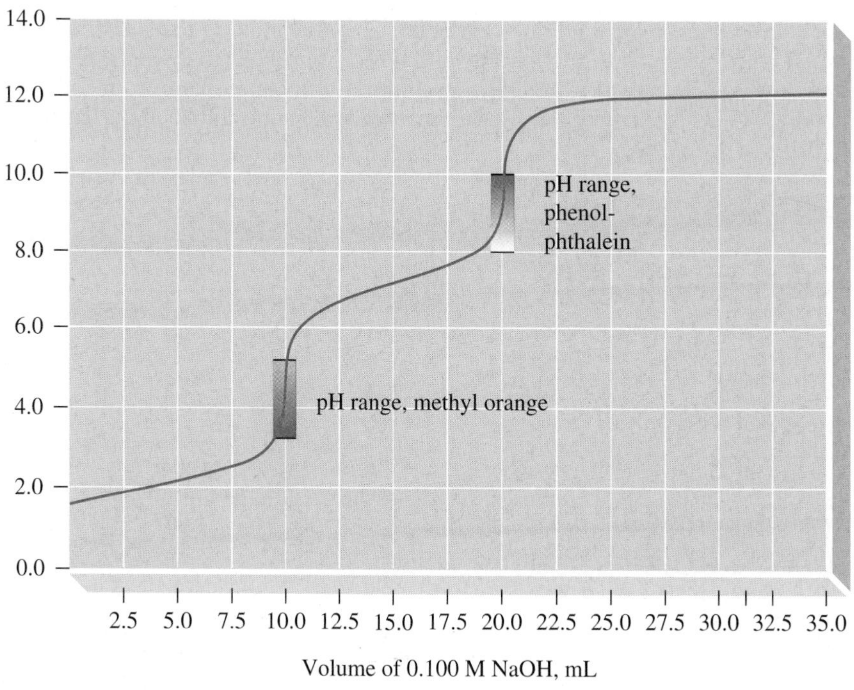

◀ FIGURE 17-12
Titration of a weak polyprotic acid—10.0 mL of 0.100 M H_3PO_4 with 0.100 M NaOH
A 10.0 mL volume of 0.100 M NaOH is required to reach the first equivalence point. The additional volume of 0.100 M NaOH required to reach the second equivalence point is also 10.0 mL. The pH does not increase sharply in the vicinity of the third equivalence point (30 mL).

The pH at the equivalence point falls within the pH range over which tl color of methyl orange indicator changes from red to orange.

An additional mole of NaOH is required to convert 1 mol $H_2PO_4^-$ to 1 m HPO_4^{2-}. At this second equivalence point in the titration of H_3PO_4, the sol tion is basic because $K_b > K_{a_3}$ for HPO_4^{2-}.

$$HPO_4^{2-} + H_2O \rightleftharpoons H_2PO_4^- + OH^- \qquad K_b = 1.6 \times 10^{-7}$$

$$HPO_4^{2-} + H_2O \rightleftharpoons H_3O^+ + PO_4^{3-} \qquad K_{a_3} = 4.2 \times 10^{-13}$$

Phenolphthalein is an appropriate indicator for this equivalence point; tl color of this indicator changes from colorless to light pink.

17-5 CONCEPT ASSESSMENT

Sketch the titration curve for ethane-1,2-diamine, $NH_2CH_2CH_2NH_2(aq)$, with HCl(aq) and label all important points on the titration curve. For ethane-1,2-diamine, $pK_{b_1} = 4.08$; $pK_{b_2} = 7.15$.

17-5 Solutions of Salts of Polyprotic Acids

In discussing the neutralization of phosphoric acid by a strong base, we foun that the first equivalence point should come in a somewhat acidic solutic and the second in a mildly basic solution. We reasoned that the third equiv lence point could be reached only in a strongly basic solution. The pH at th third equivalence point is not difficult to calculate. It corresponds to that $Na_3PO_4(aq)$, and PO_4^{3-} can ionize (hydrolyze) only as a base.

$$PO_4^{3-} + H_2O \rightleftharpoons HPO_4^{2-} + OH^- \qquad K_b = K_w/K_{a_3} = 2.4 \times 10^{-2}$$

EXAMPLE 17-9 Determining the pH of a Solution Containing the Anion (A^{n-}) of a Polyprotic Acid

Sodium phosphate, Na_3PO_4, is an ingredient of some preparations used to clean painted walls before they are repainted. What is the pH of 0.025 M $Na_3PO_4(aq)$?

Analyze

PO_4^{3-} is a weak base that will react with water to form HPO_4^{2-} and OH^-, thereby making the solution basic. We don't have to consider additional reactions, such as the reaction of HPO_4^{2-} and H_2O to give $H_2PO_4^-$ and OH^-, because, as discussed on pages 757–758, most of the OH^- in solution will come from the reaction of PO_4^{3-} and H_2O. The value of K_b for PO_4^{3-} is large enough, however, that we will *not* be able to make the usual simplifying approximation that $0.025 - x \approx 0.025$.

Solve

In the usual fashion, we can write

	PO_4^{3-} + H_2O ⇌	HPO_4^{2-} +	OH^-	$K_b = 2.4 \times 10^{-2}$
initial concns:	0.025 M	—	—	
changes:	$-x$ M	$+x$ M	$+x$ M	
equil concns:	$(0.025 - x)$ M	x M	x M	

$$K_b = \frac{[HPO_4^{2-}][OH^-]}{[PO_4^{3-}]} = \frac{x \cdot x}{0.025 - x} = 2.4 \times 10^{-2}$$

Solving the quadratic equation $x^2 + 0.024x - (0.025)(0.024) = 0$ yields $x = 0.15$ and thus, $[OH^-] = 0.015$ M.

$$pOH = -\log[OH^-] = -\log(0.015) = 1.82$$

$$pH = 14.00 - 1.82 = 12.18$$

Assess

Notice that more than half (about 61%) of the PO_4^{3-} reacts. In solving this problem, we assumed that most of the OH^- in solution comes from the reaction of PO_4^{3-} and H_2O. That is, the subsequent reaction of HPO_4^{2-} and H_2O does not contribute a significant amount of OH^-. Can we test the validity of this approximation? Of course we can. On page 816, we saw that

$$HPO_4^{2-} + H_2O \rightleftharpoons H_2PO_4^- + OH^- \qquad K_b = [H_2PO_4^-][OH^-]/[HPO_4^{2-}] = 1.6 \times 10^{-7}$$

Using $[HPO_4^{2-}] \approx [OH^-] \approx 0.015$ M (from above), we obtain $[H_2PO_4^-] \approx 1.6 \times 10^{-7}$. The amount of $H_2PO_4^-$ (and OH^-) generated from the reaction of HPO_4^{2-} and H_2O is, as predicted, very small. The approximation is valid.

PRACTICE EXAMPLE A: Using data from Table 16.5, calculate the pH of 1.0 M Na_2CO_3.

PRACTICE EXAMPLE B: Using data from Table 16.5, calculate the pH of 0.500 M Na_2SO_3.

It is more difficult to calculate the pH values of NaH_2PO_4(aq) and a_2HPO_4(aq) than of Na_3PO_4(aq). This is because with both $H_2PO_4^-$ and PO_4^{2-}, two equilibria must be considered *simultaneously*: ionization as an acid nd ionization as a base (hydrolysis). For solutions that are reasonably concenated (say, 0.10 M or greater), the pH values prove to be *independent* of the solution oncentration. Shown here (with pK_a values from Table 16.5) are general expresons, printed in blue, and their application to $H_2PO_4^-$(aq) and HPO_4^{2-}(aq):

for $H_2PO_4^-$: $$pH = \frac{1}{2}(pK_{a_1} + pK_{a_2}) = \frac{1}{2}(2.15 + 7.20) = 4.68 \qquad \textbf{(17.10)}$$

for HPO_4^{2-}: $$pH = \frac{1}{2}(pK_{a_2} + pK_{a_3}) = \frac{1}{2}(7.20 + 12.38) = 9.79 \qquad \textbf{(17.11)}$$

17-1 ARE YOU WONDERING?

How do we derive equations (17.10) and (17.11)?

Here is a good place to use the general problem-solving method introduced in Section 16–7 (page 761). Consider a solution of NaH_2PO_4, of molarity *c*. The principal concentrations that we must account for are $[Na^+]$, $[H_3O^+]$, $[H_3PO_4]$, $[H_2PO_4^-]$, $[HPO_4^{2-}]$, and $[OH^-]$. Of these, two have very simple values: $[Na^+] = c$, and $[OH^-] = K_w/[H_3O^+]$. Additionally, we can write the following equations.

1. *Acid Ionization*: $H_2PO_4^- + H_2O \rightleftharpoons H_3O^+ + HPO_4^{2-}$
$$K_{a_2} = \frac{[H_3O^+][HPO_4^{2-}]}{[H_2PO_4^-]}$$
2. *Hydrolysis*: $H_2PO_4^- + H_2O \rightleftharpoons H_3PO_4 + OH^-$
$$K_b = K_w/K_{a_1} = \frac{[H_3PO_4][OH^-]}{[H_2PO_4^-]}$$
3. *Material Balance*: The total concentration of the phosphorus-containing species is the stoichiometric molarity, *c*. Also, $[Na^+] = c$. We can write $[H_3PO_4] + [H_2PO_4^-] + [HPO_4^{2-}] + [PO_4^{3-}] = c = [Na^+]$
4. *Electroneutrality Condition*: $[H_3O^+] + [Na^+] = [H_2PO_4^-] + 2 \times [HPO_4^{2-}] + 3 \times [PO_4^{3-}] + [OH^-]$

The material balance equation and the electroneutrality condition can both be simplified by neglecting the terms involving $[PO_4^{3-}]$ and $[OH^-]$.

Solving the set of equations: Begin by substituting equation (3) into equation (4).

$$[H_3O^+] = [H_2PO_4^-] + 2 \times [HPO_4^{2-}] - [H_3PO_4] - [H_2PO_4^-] - [HPO_4^{2-}]$$
$$= [HPO_4^{2-}] - [H_3PO_4]$$

(continued)

◀ The K_a and K_b values for $H_2PO_4^-$ (see page 815) tell us that there is a greater tendency for $H_2PO_4^-$ to act as an acid. The solution will be acidic and $[OH^-]$ will be less than 10^{-7} M. Also, we expect $[PO_4^{3-}]$ to be very small. To obtain PO_4^{3-}, $H_2PO_4^-$ must ionize twice. However, because $H_2PO_4^-$ is a weak acid, only a small fraction will ionize to HPO_4^{2-} and an even smaller fraction will ionize further to PO_4^{3-}.

To continue from here, rearrange equation (1) to obtain $[HPO_4^{2-}]$ in terms of $[H_3O^+]$, $[H_2PO_4^-]$, and K_{a_2}; then rearrange equation (2) to obtain $[H_3PO_4]$ in terms of $[H_3O^+]$, $[H_2PO_4^-]$, and K_{a_1}. Next, substitute the results into the expression, $[H_3O^+] = [HPO_4^{2-}] - [H_3PO_4]$ to obtain an equation for $[H_3O^+]^2$ in terms of $[H_2PO_4^-]$, K_{a_1}, and K_{a_2}. Assume that $K_{a_1} + [H_2PO_4^-] \approx [H_2PO_4^-]$ and you will get an equation from which you can derive equation (17.10). The remainder of this derivation and the derivation of equation (17.11) are left for you to do (see Exercise 79).

17-6 Acid–Base Equilibrium Calculations: A Summary

In this and the preceding chapter, we have considered a variety of acid–ba equilibrium calculations. When you are faced with a new problem-solving s uation, you might find it helpful to relate the new problem to a type that yo have encountered before. It is best not to rely exclusively on "labeling" a pro lem, however. Some problems might not fit a recognizable category. Instea keep in mind some principles that apply regardless of the particular problem as suggested by these questions.

1. **Which species are potentially present in solution, and how large a their concentrations likely to be?**
   In a solution containing similar amounts of HCl and CH_3COOH, th only significant *ionic* species are H_3O^+ and Cl^-. HCl is a completely ioniz strong acid, and in the presence of a strong acid, the weak acid CH_3COO is only very slightly ionized because of the common-ion effect. In a mixtu containing similar amounts of two *weak* acids of similar strengths, such CH_3COOH and HNO_2, each acid partially ionizes. All of these concentr tions would be significant: $[CH_3COOH]$, $[CH_3COO^-]$, $[HNO_2]$, $[NO_2^-]$ and $[H_3O^+]$. In a solution containing phosphoric acid or a phosphate sa (or both), H_3PO_4, $H_2PO_4^-$, HPO_4^{2-}, PO_4^{3-}, OH^-, H_3O^+, and possibly oth cations might be present. If the solution is simply $H_3PO_4(aq)$, however, th only species present in significant concentrations are those associated wi the first ionization: H_3PO_4, H_3O^+, and $H_2PO_4^-$. If the solution is instea described as $Na_3PO_4(aq)$, the significant species are Na^+, PO_4^{3-}, and th ions associated with the hydrolysis of PO_4^{3-}, that is, HPO_4^{2-} and OH^-.

▶ Body temperature is 37 °C and at that temperature K_w does not equal 1.0×10^{-14}. Refer to Exercise 83 on page 827.

2. **Are reactions possible among any of the solution components; if s what is their stoichiometry?**
   Suppose that you are asked to calculate $[OH^-]$ in a solution that is pr pared to be 0.10 M NaOH and 0.20 M NH_4Cl. Before you answer th $[OH^-] = 0.10$ M, consider whether a solution can be *simultaneously* 0.10 M in OH^- and 0.20 M in NH_4^+. It cannot; any solution containing both NH_4^+ and OH^- must also contain NH_3. The OH^- and NH_4^+ react in a 1:1 mo ratio until OH^- is almost totally consumed:

$$NH_4^+ + OH^- \longrightarrow NH_3 + H_2O$$

   You are now dealing with the buffer solution 0.10 M NH_3–0.10 M NH_4^+.
3. **Which equilibrium equations apply to the particular situation? Whic are the most significant?**
   One equation that applies to all acids and bases in aqueous solutions $K_w = [H_3O^+][OH^-] = 1.0 \times 10^{-14}$. In many calculations, however, th equation is not significant compared with others. One situation in which is significant is in calculating $[OH^-]$ in an *acidic* solution or $[H_3O^+]$ in *basic* solution. After all, an acid does not produce OH^-, and a base doe not produce H_3O^+. Another situation in which K_w is likely to be signif cant is in a solution with a pH near 7.

Often you will find that the ionization equilibrium with the largest K value is e most significant, but this will not always be the case. The amounts of the var- us species in solution are also an important consideration. When one drop of)0 M H_3PO_4 ($K_{a_1} = 7.1 \times 10^{-3}$) is added to 1.00 L of 0.100 M CH_3COOH $_a = 1.8 \times 10^{-5}$), the acetic acid ionization is most important in establishing e pH of the solution. The solution contains far more acetic acid than it does osphoric acid.

17-6 CONCEPT ASSESSMENT

A solution is formed by mixing 200.0 mL of 0.100 M KOH with 100.0 mL of a solution that is both 0.200 M in CH_3COOH and 0.050 M in HI. *Without doing detailed calculations*, identify in the final solution **(a)** all the solute species present, **(b)** the two most abundant solute species, and **(c)** the two least abundant species.

MasteringCHEMISTRY www.masteringchemistry.com

Maintaining the proper pH in blood is important to one's health. To find out how the body maintains a normal pH in blood, go to the Focus On feature for Chapter 17, Buffers in Blood, on the MasteringChemistry site.

Summary

7-1 Common-Ion Effect in Acid–Base Equilibria—ne ionization of a weak electrolyte is suppressed by the dition of an ion that is the product of the ionization and is own as the *common-ion effect.* This effect is a manifesta- on of Le Châtelier's principle, introduced in Chapter 15.

7-2 Buffer Solutions—Solutions that resist changes pH upon the addition of small amounts of an acid or se are referred to as **buffer solutions**. The finite amount acid or base that a buffer solution can neutralize is own as the **buffer capacity**, and the pH range over hich the buffer solution neutralizes the added acid or se is referred to as **buffer range**. Key to the functioning a buffer solution is the presence of either a weak acid d its conjugate base or a weak base and its conjugate id (Fig. 17-4). Calculating the pH of a buffer solution can accomplished by using the ICE method developed Chapter 15 or by application of the **Henderson–asselbalch equation** (expression 17.7). Determination of e pH of a buffer solution after the addition of a strong id or base requires a stoichiometric calculation followed an equilibrium calculation (Fig. 17-6).

7-3 Acid–Base Indicators—Substances whose colors epend on the pH of a solution are known as **acid–base dicators**. Acid–base indicators exist in solution as a weak id (HIn) and its conjugate base (In^-). Each form has a fferent color and the proportions of the two forms deter- ine the color of the solution, which in turn depends on e pH of the solution. The pH range over which an id–base indicator changes color (Fig. 17-7) is determined the K_a of the specific indicator.

7-4 Neutralization Reactions and Titration urves—As described in Chapter 5, the concentration an acidic or basic solution of unknown concentration can be determined by titration with a base or acid of precisely known concentration. In this process a precisely measured volume of the solution of known concentration, the **titrant**, is added through a buret into a precisely measured quantity of the "unknown" contained in a beaker or flask. Typically, the amounts of reactants in a titration are of the order of 10^{-3} mol, that is, **millimoles (mmol)**. A **titration curve** is a graph of a measured property of the reaction mixture as a function of the volume of titrant added—pH for an acid–base titration (Fig. 17-8). The point at which neither reactant is in excess in a titration is known as the **equivalence point**. The **end point** of an acid–base titration can be located through the change in color of an indicator. The indicator must be chosen such that its color change occurs as close to the equivalence point as possible. Strong acid–strong base titration curves (Figs. 17-8 and 17-9) are different from weak acid–strong base titration curves (Fig. 17-10). The two main differences are seen in the latter type, a buffer region and a hydrolysis reaction at the equivalence point (Fig. 17-11).

17-5 Solutions of Salts of Polyprotic Acids—Calculating the pH of solutions containing the salts of polyprotic acids is made difficult by the fact that two or more equilibria occur simultaneously. Yet for certain solutions, the calculations can be reduced to a simple form (expressions 17.10 and 17.11).

17-6 Acid–Base Equilibrium Calculations: A Summary—As a general summary of acid–base equilibrium calculations, the essential factors are identifying all the species in solution, their concentrations, the possible reactions between them, and the stoichiometry and equilibrium constants of those reactions.

Integrative Example

The structural formula shown is *para*-hydroxybenzoic acid, a weak diprotic acid used as a food preservative. Titration 25.00 mL of a dilute aqueous solution of this acid requires 16.24 mL of 0.0200 M NaOH to reach the first equivalen point. The measured pH after the addition of 8.12 mL of the base is 4.57; after 16.24 mL, the pH is 7.02. Determine the va ues of pK_{a_1} and pK_{a_2} of *para*-hydroxybenzoic acid and the pH values for the two equivalence points in the titration.

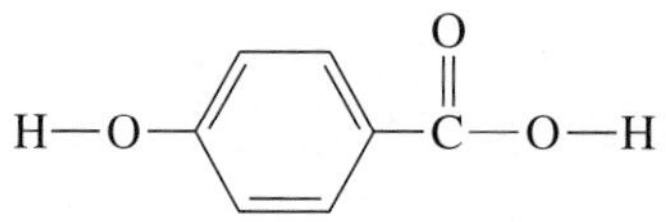

Analyze

This is a titration of a polyprotic weak acid with a strong base, and the titration curve for this problem should look ve similar to Figure 17-12. In the titration of a weak acid by a strong base we know that at the point of half-neutralizatio $pH = pK_a$ and therefore pK_{a_1} should be the pH at 8.12 mL. For pK_{a_2}, we will use expression (17.10) since at this poi in the titration we will have an aqueous solution of HOC_6H_4COONa, which is a salt of a polyprotic acid. The pH of th first equivalence point is given and to find the pH of the second equivalence point we must perform an ICE calculatio similar to the one in Example 16-14.

Solve

The volume of base needed to reach the first equivalence point is 16.24 mL; at this point, the pH = 7.02. A volume of 8.12 mL is needed to half-neutralize the acid in its first ionization step; at this point the $pH = pK_{a_1}$, that is, $pK_{a_1} = 4.57$.

At the first equivalence point, the solution is $HOC_6H_4COONa(aq)$ with pH = 7.02. Recognizing this as the salt produced in neutralizing a polyprotic acid in its first ionization, we use equation (17.10) to solve for pK_{a_2}. That is, the pH of an aqueous solution of the ion $HOC_6H_6COO^-$ is given by the expression

$$pH = \tfrac{1}{2}(pK_{a_1} + pK_{a_2}) = \tfrac{1}{2}(4.57 + pK_{a_2}) = 7.02$$

$$pK_{a_2} = (2 \times 7.02) - 4.57 = 9.47$$

Determining the pH at the second equivalence point involves additional calculations. We begin by noting that at the second equivalence point the solution is one of $NaOC_6H_4COONa$. The pH of the solution is established by the hydrolysis of $^-OC_6H_4COO^-$.

$$^-OC_6H_4COO^- + H_2O \rightleftharpoons HOC_6H_4COO^- + OH^-$$

$$K_b = K_w/K$$

To evaluate K_b, let's first obtain K_{a_2} from pK_{a_2}.

$$pK_{a_2} = -\log K_{a_2} = 9.47 \text{ and } K_{a_2} = 10^{-9.47} = 3.4 \times 10^{-10}$$

$$K_b = K_w/K_{a_2} = 1.0 \times 10^{-14}/3.4 \times 10^{-10} = 2.9 \times 10^{-5}$$

We can get the pH of this solution by first calculating $[OH^-]$ and pOH. However, to do this, we still need one more piece of data—the molarity of the $NaOC_6H_4COONa(aq)$ solution. We can get this from data for titration to the first equivalence point.

$$? \text{ mmol } OH^- = 16.24 \text{ mL} \times 0.0200 \text{ mmol } OH^-/\text{mL}$$

$$= 0.325 \text{ mmol } OH^-$$

$$? \text{ mmol } HOC_6H_4COOH = 0.325 \text{ mmol } OH^- \times 1 \text{ mmol}$$

$$HOC_6H_4COOH/\text{mmol } OH^- = 0.325 \text{ mmol } HOC_6H_4COOH$$

The amount of $^-OC_6H_4COO^-$ at the second equivalence point is the same as the amount of acid at the start of the titration.

$$0.325 \text{ mmol } ^-OC_6H_4COO^-$$

The volume of solution at the second equivalence point is

$$25.00 \text{ mL} + 16.24 \text{ mL} + 16.24 \text{ mL} = 57.48 \text{ mL}$$

hus,

$$[^{-}OC_6H_4COO^-] = 0.325 \text{ mmol } {}^{-}OC_6H_4COO^-/57.48 \text{ mL}$$
$$= 5.65 \times 10^{-3} \text{ M}$$

ow we can return to the hydrolysis equation and e expression for K_b, using the method of xample 16-14.

$$^{-}OC_6H_4COO^- + H_2O \rightleftharpoons HOC_6H_4COO^- + OH^-$$

	$^{-}OC_6H_4COO^-$	$HOC_6H_4COO^-$	OH^-
initial concns:	5.65×10^{-3} M	—	—
changes:	$-x$ M	$+x$ M	$+x$ M
equil concns:	$(5.65 \times 10^{-3} - x)$ M	x M	x M

$$K_b = \frac{x \cdot x}{(5.65 \times 10^{-3} - x)} = 2.9 \times 10^{-5}$$

ne solution to this equation is $x = 3.9 \times 10^{-4}$. hus, $[OH^-] = 3.9 \times 10^{-4}$ M, pOH = 3.41, and H = 10.59.

ssess

ne polyprotic acid, *para*-hydroxybenzoic acid, has two functional groups, a carboxylic acid and a phenolic group. Each oup has an ionizable proton. Given just two basic pieces of titration data, we used concepts from this and the preced- g chapter to determine the pK_a values for both ionizable groups as well as the pH at the two equivalence points. As a eck, note that the pK_a values of these groups are comparable to the values for their parent compounds, acetic acid and enol (see Table 16.4).

RACTICE EXAMPLE A: 7.500 g of a weak acid HA is added to sufficient distilled water to produce 500.0 mL of lution with pH = 2.716. This solution is titrated with NaOH(aq). Halfway to the equivalence point, pH = 4.602. What the freezing point of the solution?

RACTICE EXAMPLE B: The following titration curve was obtained as part of a general chemistry laboratory periment for an unknown that weighed 0.8 g. The titrant was either a 0.2 M strong base or 0.2 M strong acid. Estimate e molar mass of the unknown and its ionization constant.

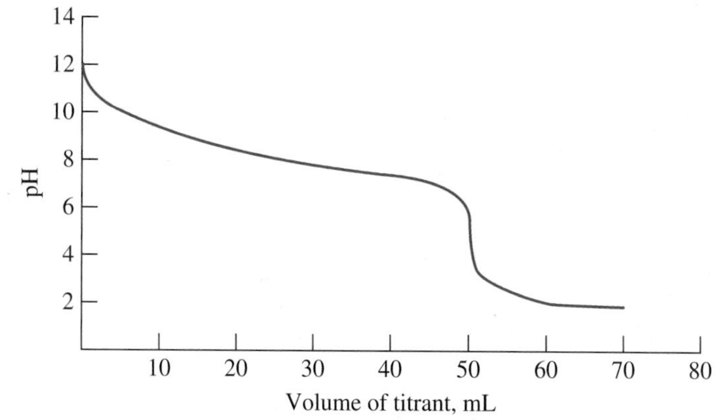

Exercises

he Common-Ion Effect

se data from Table 16.4 as necessary.)

. For a solution that is 0.275 M CH_3CH_2COOH (propionic acid, $K_a = 1.3 \times 10^{-5}$) and 0.0892 M HI, calculate **(a)** $[H_3O^+]$; **(b)** $[OH^-]$; **(c)** CH_3CH_2COO; **(d)** $[I^-]$.

. For a solution that is 0.164 M NH_3 and 0.102 M NH_4Cl, calculate **(a)** $[OH^-]$; **(b)** $[NH_4^+]$; **(c)** $[Cl^-]$; **(d)** $[H_3O^+]$.

. Calculate the *change* in pH that results from adding **(a)** 0.100 mol $NaNO_2$ to 1.00 L of 0.100 M HNO_2(aq); **(b)** 0.100 mol $NaNO_3$ to 1.00 L of 0.100 M HNO_3(aq). Why are the changes not the same?

4. In Example 16-4, we calculated the percent ionization of CH_3COOH in **(a)** 1.0 M; **(b)** 0.10 M; and **(c)** 0.010 M CH_3COOH solutions. Recalculate those percent ionizations if each solution also contains 0.10 M $NaCH_3COO$. Explain why the results are different from those of Example 16-4.

5. Calculate $[H_3O^+]$ in a solution that is **(a)** 0.035 M HCl and 0.075 M HOCl; **(b)** 0.100 M $NaNO_2$ and 0.0550 M HNO_2; **(c)** 0.0525 M HCl and 0.0768 M $NaCH_3COO$.

6. Calculate $[OH^-]$ in a solution that is **(a)** 0.0062 $Ba(OH)_2$ and 0.0105 M $BaCl_2$; **(b)** 0.315 M $(NH_4)_2S$ and 0.486 M NH_3; **(c)** 0.196 M NaOH and 0.264 NH_4Cl.

Buffer Solutions

(Use data from Tables 16.4 and 16.5 as necessary.)

7. What concentration of formate ion, $[HCOO^-]$, should be present in 0.366 M HCOOH to produce a buffer solution with pH = 4.06?

$$HCOOH + H_2O \rightleftharpoons H_3O^+ + HCOO^- \qquad K_a = 1.8 \times 10^{-4}$$

8. What concentration of ammonia, $[NH_3]$, should be present in a solution with $[NH_4^+] = 0.732$ M to produce a buffer solution with pH = 9.12? For NH_3, $K_b = 1.8 \times 10^{-5}$.

9. Calculate the pH of a buffer that is
(a) 0.012 M C_6H_5COOH ($K_a = 6.3 \times 10^{-5}$) and 0.033 M NaC_6H_5COO;
(b) 0.408 M NH_3 and 0.153 M NH_4Cl.

10. Lactic acid, $CH_3CH(OH)COOH$, is found in sour milk. A solution containing 1.00 g $NaCH_3CH(OH)COO$ in 100.0 mL of 0.0500 M $CH_3CH(OH)COOH$, has a pH = 4.11. What is K_a of lactic acid?

11. Indicate which of the following aqueous solutions are buffer solutions, and explain your reasoning. [*Hint:* Consider any reactions that might occur between solution components.]
(a) 0.100 M NaCl
(b) 0.100 M NaCl–0.100 M NH_4Cl
(c) 0.100 M CH_3NH_2–0.150 M $CH_3NH_3^+Cl^-$
(d) 0.100 M HCl–0.050 M $NaNO_2$
(e) 0.100 M HCl–0.200 M $NaCH_3COO$
(f) 0.100 M CH_3COOH–0.125 M $NaCH_3CH_2COO$

12. The $H_2PO_4^-$–HPO_4^{2-} combination plays a role in maintaining the pH of blood.
(a) Write equations to show how a solution containing these ions functions as a buffer.
(b) Verify that this buffer is most effective at pH 7.2.
(c) Calculate the pH of a buffer solution in which $[H_2PO_4^-] = 0.050$ M and $[HPO_4^{2-}] = 0.150$ M. [*Hint:* Focus on the second step of the phosphoric acid ionization.]

13. What is the pH of a solution obtained by adding 1.15 mg of aniline hydrochloride ($C_6H_5NH_3^+Cl^-$) to 3.18 L of 0.105 M aniline ($C_6H_5NH_2$)? [*Hint:* Check any assumptions that you make.]

14. What is the pH of a solution prepared by dissolving 8.50 g of aniline hydrochloride ($C_6H_5NH_3^+Cl^-$) in 750 mL of 0.215 M aniline ($C_6H_5NH_2$)? Would this solution be an effective buffer? Explain.

15. You wish to prepare a buffer solution with pH = 9.45.
(a) How many grams of $(NH_4)_2SO_4$ would you add to 425 mL of 0.258 M NH_3 to do this? Assume that the solution's volume remains constant.
(b) Which buffer component, and how much (in grams), would you add to 0.100 L of the buffer in part (a) to change its pH to 9.30? Assume that the solution's volume remains constant.

16. You prepare a buffer solution by dissolving 2.00 g ea of benzoic acid, C_6H_5COOH, and sodium benzoa NaC_6H_5COO, in 750.0 mL of water.
(a) What is the pH of this buffer? Assume that t solution's volume is 750.0 mL.
(b) Which buffer component, and how much grams), would you add to the 750.0 mL of buffer so tion to change its pH to 4.00?

17. If 0.55 mL of 12 M HCl is added to 0.100 L of the buf solution in Exercise 15(a), what will be the pH of t resulting solution?

18. If 0.35 mL of 15 M NH_3 is added to 0.750 L of t buffer solution in Exercise 16(a), what will be the p of the resulting solution?

19. You are asked to prepare a buffer solution with a pH 3.50. The following solutions, all 0.100 M, are availal to you: HCOOH, CH_3COOH, H_3PO_4, $NaHCO$ $NaCH_3COO$, and NaH_2PO_4. Describe how y would prepare this buffer solution. [*Hint:* What v umes of which solutions would you use?]

20. You are asked to reduce the pH of the 0.300 L of buf solution in Example 17-5 from 5.09 to 5.00. How ma milliliters of which of these solutions would y use: 0.100 M NaCl, 0.150 M HCl, 0.100 M $NaCH_3CO$ 0.125 M NaOH? Explain your reasoning.

21. Given 1.00 L of a solution that is 0.100 CH_3CH_2COOH and 0.100 M KCH_3CH_2COO,
(a) Over what pH range will this solution be an eff tive buffer?
(b) What is the buffer capacity of the solution? Th is, how many millimoles of strong acid or strong ba can be added to the solution before any significa change in pH occurs?

22. Given 125 mL of a solution that is 0.0500 M CH_3NH and 0.0500 M $CH_3NH_3^+Cl^-$,
(a) Over what pH range will this solution be an eff tive buffer?
(b) What is the buffer capacity of the solution? Th is, how many millimoles of strong acid or strong ba can be added to the solution before any significa change in pH occurs?

23. A solution of volume 75.0 mL contains 15.5 mm HCOOH and 8.50 mmol NaHCOO.
(a) What is the pH of this solution?
(b) If 0.25 mmol $Ba(OH)_2$ is added to the solutio what will be the pH?
(c) If 1.05 mL of 12 M HCl is added to the origin solution, what will be the pH?

24. A solution of volume 0.500 L contains 1.68 g NH_3 ar 4.05 g $(NH_4)_2SO_4$.
(a) What is the pH of this solution?
(b) If 0.88 g NaOH is added to the solution, what w be the pH?
(c) How many milliliters of 12 M HCl must be added 0.500 L of the original solution to change its pH to 9.0

. A handbook lists various procedures for preparing buffer solutions. To obtain a pH = 9.00, the handbook says to mix 36.00 mL of 0.200 M NH_3 with 64.00 mL of 0.200 M NH_4Cl.
(a) Show by calculation that the pH of this solution is 9.00.
(b) Would you expect the pH of this solution to remain at pH = 9.00 if the 100.00 mL of buffer solution were diluted to 1.00 L? To 1000 L? Explain.
(c) What will be the pH of the original 100.00 mL of buffer solution if 0.20 mL of 1.00 M HCl is added to it?
(d) What is the maximum volume of 1.00 M HCl that can be added to 100.00 mL of the original buffer solution so that the pH does not drop below 8.90?

26. An acetic acid–sodium acetate buffer can be prepared by the reaction

$$\underset{\text{(From NaCH}_3\text{COO)}}{CH_3COO^-} + \underset{\text{(From HCl)}}{H_3O^+} \longrightarrow CH_3COOH + H_2O$$

(a) If 12.0 g $NaCH_3COO$ is added to 0.300 L of 0.200 M HCl, what is the pH of the resulting solution?
(b) If 1.00 g $Ba(OH)_2$ is added to the solution in part (a), what is the new pH?
(c) What is the maximum mass of $Ba(OH)_2$ that can be neutralized by the buffer solution of part (a)?
(d) What is the pH of the solution in part (a) following the addition of 5.50 g $Ba(OH)_2$?

cid–Base Indicators

se data from Tables 16.4 and 16.5 as necessary.)

. A handbook lists the following data:

ndicator	K_{HIn}	Color Change Acid → Anion
romphenol blue	1.4×10^{-4}	yellow → blue
romcresol green	2.1×10^{-5}	yellow → blue
romthymol blue	7.9×10^{-8}	yellow → blue
,4-Dinitrophenol	1.3×10^{-4}	colorless → yellow
hlorphenol red	1.0×10^{-6}	yellow → red
hymolphthalein	1.0×10^{-10}	colorless → blue

(a) Which of these indicators change color in acidic solution, which in basic solution, and which near the neutral point?
(b) What is the approximate pH of a solution if bromcresol green indicator turns green? if chlorphenol red turns orange?

. With reference to the indicators listed in Exercise 27, what would be the color of each combination?
(a) 2,4-dinitrophenol in 0.100 M HCl(aq)
(b) chlorphenol red in 1.00 M NaCl(aq)
(c) thymolphthalein in 1.00 M NH_3(aq)
(d) bromcresol green in seawater (recall Figure 17-7)

. In the use of acid–base indicators,
(a) Why is it generally sufficient to use a *single* indicator in an acid–base titration, but often necessary to use *several* indicators to establish the approximate pH of a solution?
(b) Why must the quantity of indicator used in a titration be kept as small as possible?

. The indicator methyl red has a $pK_{HIn} = 4.95$. It changes from red to yellow over the pH range from 4.4 to 6.2.
(a) If the indicator is placed in a buffer solution of pH = 4.55, what percent of the indicator will be present in the acid form, HIn, and what percent will be present in the base or anion form, In^-?
(b) Which form of the indicator has the "stronger" (that is, more visible) color—the acid (red) form or base (yellow) form? Explain.

31. Phenol red indicator changes from yellow to red in the pH range from 6.6 to 8.0. *Without making detailed calculations*, state what color the indicator will assume in each of the following solutions: **(a)** 0.10 M KOH; **(b)** 0.10 M CH_3COOH; **(c)** 0.10 M NH_4NO_3; **(d)** 0.10 M HBr; **(e)** 0.10 M NaCN; **(f)** 0.10 M CH_3COOH–0.10 M $NaCH_3COO$.

32. Thymol blue indicator has *two* pH ranges. It changes color from red to yellow in the pH range from 1.2 to 2.8, and from yellow to blue in the pH range from 8.0 to 9.6. What is the color of the indicator in each of the following situations?
(a) The indicator is placed in 350.0 mL of 0.205 M HCl.
(b) To the solution in part (a) is added 250.0 mL of 0.500 M $NaNO_2$.
(c) To the solution in part (b) is added 150.0 mL of 0.100 M NaOH.
(d) To the solution in part (c) is added 5.00 g $Ba(OH)_2$.

33. In the titration of 10.00 mL of 0.04050 M HCl with 0.01120 M $Ba(OH)_2$ in the presence of the indicator 2,4-dinitrophenol, the solution changes from colorless to yellow when 17.90 mL of the base has been added. What is the approximate value of pK_{HIn} for 2,4-dinitrophenol? Is this a good indicator for the titration?

34. Solution **(a)** is 100.0 mL of 0.100 M HCl and solution **(b)** is 150.0 mL of 0.100 M $NaCH_3COO$. A few drops of thymol blue indicator are added to each solution. What is the color of each solution? What is the color of the solution obtained when these two solutions are mixed?

eutralization Reactions

. A 25.00 mL sample of H_3PO_4(aq) requires 31.15 mL of 0.2420 M KOH for titration to the second equivalence point. What is the molarity of the H_3PO_4(aq)?

36. A 20.00 mL sample of H_3PO_4(aq) requires 18.67 mL of 0.1885 M NaOH for titration from the first to the second equivalence point. What is the molarity of the H_3PO_4(aq)?

37. Two aqueous solutions are mixed: 50.0 mL of 0.0150 M H_2SO_4 and 50.0 mL of 0.0385 M NaOH. What is the pH of the resulting solution?

38. Two solutions are mixed: 100.0 mL of HCl(aq) wi pH 2.50 and 100.0 mL of NaOH(aq) with pH 11.(What is the pH of the resulting solution?

Titration Curves

39. Calculate the pH at the points in the titration of 25.00 mL of 0.160 M HCl when **(a)** 10.00 mL and **(b)** 15.00 mL of 0.242 M KOH have been added.
40. Calculate the pH at the points in the titration of 20.00 mL of 0.275 M KOH when **(a)** 15.00 mL and **(b)** 20.00 mL of 0.350 M HCl have been added.
41. Calculate the pH at the points in the titration of 25.00 mL of 0.132 M HNO_2 when **(a)** 10.00 mL and **(b)** 20.00 mL of 0.116 M NaOH have been added. For HNO_2, $K_a = 7.2 \times 10^{-4}$.

$$HNO_2 + OH^- \longrightarrow H_2O + NO_2^-$$

42. Calculate the pH at the points in the titration of 20.00 mL of 0.318 M NH_3 when **(a)** 10.00 mL and **(b)** 15.00 mL of 0.475 M HCl have been added. For NH_3, $K_b = 1.8 \times 10^{-5}$.

$$NH_3(aq) + HCl(aq) \longrightarrow NH_4^+(aq) + Cl^-(aq)$$

43. Explain why the volume of 0.100 M NaOH required to reach the equivalence point in the titration of 25.00 mL of 0.100 M HA is the same regardless of whether HA is a strong or a weak acid, yet the pH at the equivalence point is not the same.
44. Explain whether the equivalence point of each of the following titrations should be below, above, or at pH 7: **(a)** $NaHCO_3(aq)$ titrated with NaOH(aq); **(b)** HCl(aq) titrated with $NH_3(aq)$; **(c)** KOH(aq) titrated with HI(aq).
45. Sketch the titration curves of the following mixtures. Indicate the initial pH and the pH corresponding to the equivalence point. Indicate the volume of titrant required to reach the equivalence point, and select a suitable indicator from Figure 17-7.
    (a) 25.0 mL of 0.100 M KOH with 0.200 M HI
    (b) 10.0 mL of 1.00 M NH_3 with 0.250 M HCl
46. Determine the following characteristics of the titration curve for 20.0 mL of 0.275 M $NH_3(aq)$ titrated with 0.325 M HI(aq).
    (a) the initial pH
    (b) the volume of 0.325 M HI(aq) at the equivalence point
    (c) the pH at the half-neutralization point
    (d) the pH at the equivalence point
47. In the titration of 20.00 mL of 0.175 M NaOH, calculate the number of milliliters of 0.200 M HCl that must be added to reach a pH of **(a)** 12.55; **(b)** 10.80; **(c)** 4.: [*Hint:* Solve an algebraic equation in which the nu ber of milliliters is x. Which reactant is in excess each pH?]
48. In the titration of 25.00 mL of 0.100 M CH_3COO calculate the number of milliliters of 0.200 M NaC that must be added to reach a pH of **(a)** 3.85; **(b)** 5.: **(c)** 11.10. [*Hint:* Solve an algebraic equation in whi the number of milliliters is x. Which reactant is excess at each pH?]
49. Sketch a titration curve (pH versus mL of titrant) f each of the following three hypothetical weak aci when titrated with 0.100 M NaOH. Select suitat indicators for the titrations from Figure 17-7. [*Hi* Select a few key points at which to estimate the pH the solution.]
    (a) 10.00 mL of 0.100 M HX; $K_a = 7.0 \times 10^{-3}$
    (b) 10.00 mL of 0.100 M HY; $K_a = 3.0 \times 10^{-4}$
    (c) 10.00 mL of 0.100 M HZ; $K_a = 2.0 \times 10^{-8}$
50. Sketch a titration curve (pH versus mL of titrant) f each of the following hypothetical weak bases wh titrated with 0.100 M HCl. (Think of these bases involving the substitution of organic groups, R, f one of the H atoms of NH_3.) Select suitable indic tors for the titrations from Figure 17-7. [*Hint:* Selec few key points at which to estimate the pH of t solution.]
    (a) 10.00 mL of 0.100 M RNH_2; $K_b = 1 \times 10^{-3}$
    (b) 10.00 mL of 0.100 M $R'NH_2$; $K_b = 3 \times 10^{-6}$
    (c) 10.00 mL of 0.100 M $R''NH_2$; $K_b = 7 \times 10^{-8}$
51. For the titration of 25.00 mL of 0.100 M NaOH wi 0.100 M HCl, calculate the pOH at a few represen tive points in the titration, sketch the titration curve pOH versus volume of titrant, and show that it h exactly the same form as Figure 17-8. Then, using th curve and the simplest method possible, sketch t titration curve of pH versus volume of titrant.
52. For the titration of 25.00 mL 0.100 M NH_3 with 0.1 M HCl, calculate the pOH at a few representati points in the titration, sketch the titration curve pOH versus volume of titrant, and show that it h exactly the same form as Figure 17-10. Then, usi this curve and the simplest method possible, sket the titration curve of pH versus volume of titrant.

Salts of Polyprotic Acids

(Use data from Table 16.5 or Appendix D as necessary.)

53. Is a solution that is 0.10 M $Na_2S(aq)$ likely to be acidic, basic, or pH neutral? Explain.
54. Is a solution of sodium dihydrogen citrate, NaH_2Cit, likely to be acidic, basic, or neutral? Explain. Citric acid, H_3Cit, is $H_3C_6H_5O_7$.
55. Sodium phosphate, Na_3PO_4, is made commercially by first neutralizing phosphoric acid with sodium carbonate to obtain Na_2HPO_4. The Na_2HPO_4 is further neutralized to Na_3PO_4 with NaOH.
    (a) Write net ionic equations for these reactions.
    (b) Na_2CO_3 is a much cheaper base than is NaO Why do you suppose that NaOH must be used well as Na_2CO_3 to produce Na_3PO_4?
56. Both sodium hydrogen carbonate (sodium bicarbo ate) and sodium hydroxide can be used to neutrali acid spills. What is the pH of 1.00 M $NaHCO_3(aq)$ ar of 1.00 M NaOH(aq)? On a per-liter basis, do the two solutions have an equal capacity to neutrali acids? Explain. On a per-gram basis, do the tw

solids, $NaHCO_3(s)$ and $NaOH(s)$, have an equal capacity to neutralize acids? Explain. Why do you suppose that $NaHCO_3$ is often preferred to NaOH in neutralizing acid spills?

. The pH of a solution of 19.5 g of malonic acid in 0.250 L is 1.47. The pH of a 0.300 M solution of sodium hydrogen malonate is 4.26. What are the values of K_{a_1} and K_{a_2} for malonic acid?

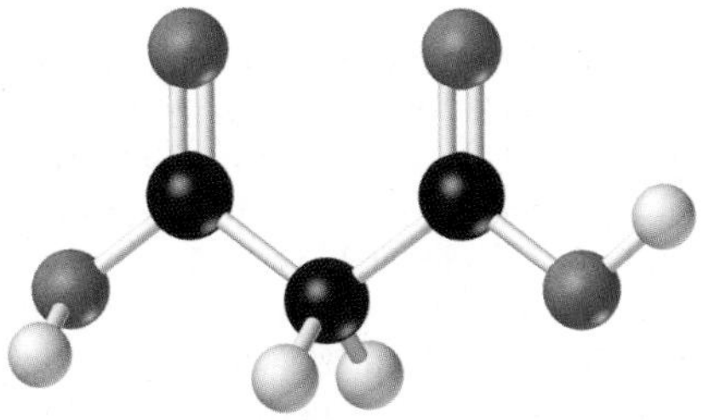

Malonic acid

. The ionization constants of *ortho*-phthalic acid are $K_{a_1} = 1.1 \times 10^{-3}$ and $K_{a_2} = 3.9 \times 10^{-6}$.

1. $C_6H_4(COOH)_2 + H_2O \rightleftharpoons H_3O^+ + C_6H_4(COOH)(COO^-)$

2. $C_6H_4(COOH)(COO^-) + H_2O \rightleftharpoons H_3O^+ + C_6H_4(COO^-)_2$

What are the pH values of the following aqueous solutions: **(a)** 0.350 M potassium hydrogen *ortho*-phthalate; **(b)** a solution containing 36.35 g potassium *ortho*-phthalate per liter?

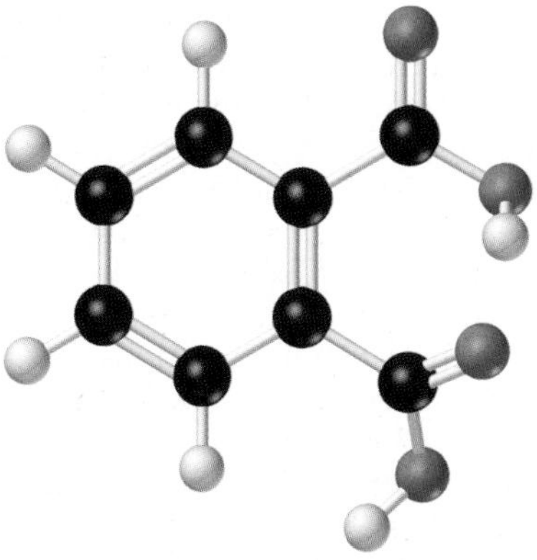

ortho-Phthalic acid

eneral Acid–Base Equilibria

. What stoichiometric concentration of the indicated substance is required to obtain an aqueous solution with the pH value shown: **(a)** $Ba(OH)_2$ for pH = 11.88; **(b)** CH_3COOH in 0.294 M $NaCH_3COO$ for pH = 4.52?

. What stoichiometric concentration of the indicated substance is required to obtain an aqueous solution with the pH value shown: **(a)** aniline, $C_6H_5NH_2$, for pH = 8.95; **(b)** NH_4Cl for pH = 5.12?

. Using appropriate equilibrium constants but *without doing detailed calculations*, determine whether a solution can be simultaneously:
(a) 0.10 M NH_3 and 0.10 M NH_4Cl, with pH = 6.07
(b) 0.10 M $NaCH_3COO$ and 0.058 M HI
(c) 0.10 M KNO_2 and 0.25 M KNO_3
(d) 0.050 M $Ba(OH)_2$ and 0.65 M NH_4Cl
(e) 0.018 M C_6H_5COOH and 0.018 M NaC_6H_5COO, with pH = 4.20
(f) 0.68 M KCl, 0.42 M KNO_3, 1.2 M NaCl, and 0.55 M $NaCH_3COO$, with pH = 6.4

62. This single equilibrium equation applies to different phenomena described in this or the preceding chapter.

$$CH_3COOH + H_2O \rightleftharpoons H_3O^+ + CH_3COO^-$$

Of these four phenomena, ionization of pure acid, common-ion effect, buffer solution, and hydrolysis, indicate which occurs if
(a) $[H_3O^+]$ and $[CH_3COOH]$ are high, but $[CH_3COO^-]$ is very low.
(b) $[CH_3COO^-]$ is high, but $[CH_3COOH]$ and $[H_3O^+]$ are very low.
(c) $[CH_3COOH]$ is high, but $[H_3O^+]$ and $[CH_3COO^-]$ are low.
(d) $[CH_3COOH]$ and $[CH_3COO^-]$ are high, but $[H_3O^+]$ is low.

Integrative and Advanced Exercises

. Sodium hydrogen sulfate, $NaHSO_4$, is an acidic salt with a number of uses, such as metal pickling (removal of surface deposits). $NaHSO_4$ is made by the reaction of H_2SO_4 with NaCl. To determine the percent NaCl impurity in $NaHSO_4$, a 1.016 g sample is titrated with NaOH(aq); 36.56 mL of 0.225 M NaOH is required.
(a) Write the net ionic equation for the neutralization reaction.
(b) Determine the percent NaCl in the sample titrated.
(c) Select a suitable indicator from Figure 17-7.

. You are given 250.0 mL of 0.100 M CH_3CH_2COOH (propionic acid, $K_a = 1.35 \times 10^{-5}$). You want to adjust its pH by adding an appropriate solution. What volume would you add of **(a)** 1.00 M HCl to lower the pH to 1.00; **(b)** 1.00 M $NaCH_3CH_2COO$ to raise the pH to 4.00; **(c)** water to raise the pH by 0.15 unit?

65. Even though the carbonic acid–hydrogen carbonate buffer system is crucial to the maintenance of the pH of blood, it has no practical use as a laboratory buffer solution. Can you think of a reason(s) for this? [*Hint:* Refer to data in Practice Example A of the Integrative Example in Chapter 16.]

66. Thymol blue in its acid range is not a suitable indicator for the titration of HCl by NaOH. Suppose that a

student uses thymol blue by mistake in the titration of Figure 17-8 and that the indicator end point is taken to be pH = 2.0.
(a) Would there be a sharp color change, produced by the addition of a single drop of NaOH(aq)?
(b) Approximately what percent of the HCl remains unneutralized at pH = 2.0?

67. Rather than calculate the pH for different volumes of titrant, a titration curve can be established by calculating the volume of titrant required to reach certain pH values. Determine the volumes of 0.100 M NaOH required to reach the following pH values in the titration of 20.00 mL of 0.150 M HCl: pH = **(a)** 2.00; **(b)** 3.50; **(c)** 5.00; **(d)** 10.50; **(e)** 12.00. Then plot the titration curve.

68. Use the method of Exercise 67 to determine the volume of titrant required to reach the indicated pH values in the following titrations.
(a) 25.00 mL of 0.250 M NaOH titrated with 0.300 M HCl; pH = 13.00, 12.00, 10.00, 4.00, 3.00
(b) 50.00 mL of 0.0100 M benzoic acid (C_6H_5COOH) titrated with 0.0500 M KOH: pH = 4.50, 5.50, 11.50 ($K_a = 6.3 \times 10^{-5}$)

69. A buffer solution can be prepared by starting with a weak acid, HA, and converting some of the weak acid to its salt (for example, NaA) by titration with a strong base. The *fraction* of the original acid that is converted to the salt is designated f.
(a) Derive an equation similar to equation (17.7) but expressed in terms of f rather than concentrations.
(b) What is the pH at the point in the titration of phenol, C_6H_5OH, at which $f = 0.27$ (pK_a of phenol = 10.00)?

70. You are asked to prepare a KH_2PO_4–Na_2HPO_4 solution that has the same pH as human blood, 7.40.
(a) What should be the ratio of concentrations $[HPO_4^{2-}]/[H_2PO_4^-]$ in this solution?
(b) Suppose you have to prepare 1.00 L of the solution described in part (a) and that this solution must be isotonic with blood (have the same osmotic pressure as blood). What masses of KH_2PO_4 and of $Na_2HPO_4 \cdot 12H_2O$ would you use? [*Hint:* Refer to the definition of isotonic on page 668. Recall that a solution of NaCl with 9.2 g NaCl/L solution is isotonic with blood, and assume that NaCl is completely ionized in aqueous solution.]

71. You are asked to bring the pH of 0.500 L of 0.500 M $NH_4Cl(aq)$ to 7.00. How many drops (1 drop = 0.05 mL) of which of the following solutions would you use: 10.0 M HCl or 10.0 M NH_3?

72. Because an acid–base indicator is a weak acid, it can be titrated with a strong base. Suppose you titrate 25.00 mL of a 0.0100 M solution of the indicator *p*-nitrophenol, $HOC_6H_4NO_2$, with 0.0200 M NaOH. The pK_a of *p*-nitrophenol is 7.15, and it changes from colorless to yellow in the pH range from 5.6 to 7.6.
(a) Sketch the titration curve for this titration.
(b) Show the pH range over which *p*-nitrophenol changes color.
(c) Explain why *p*-nitrophenol cannot serve as its own indicator in this titration.

73. The neutralization of NaOH by HCl is represented in equation (1), and the neutralization of NH_3 by HCl in equation (2).

1. $OH^- + H_3O^+ \rightleftharpoons 2\,H_2O \quad K = ?$
2. $NH_3 + H_3O^+ \rightleftharpoons NH_4^+ + H_2O \quad K = ?$

(a) Determine the equilibrium constant K for ea[ch] reaction.
(b) Explain why each neutralization reaction can [be] considered to go to completion.

74. The titration of a weak acid by a weak base is no[t a] satisfactory procedure because the pH does not increa[se] sharply at the equivalence point. Demonstrate this f[act] by sketching a titration curve for the neutralization [of] 10.00 mL of 0.100 M CH_3COOH with 0.100 M NH_3.

75. At times, a salt of a weak base can be titrated by [a] strong base. Use appropriate data from the text [to] sketch a titration curve for the titration of 10.00 mL [of] 0.0500 M $C_6H_5NH_3^+Cl^-$ with 0.100 M NaOH.

76. Sulfuric acid is a diprotic acid, strong in the first ioniz[a]tion step and weak in the second ($K_{a_2} = 1.1 \times 10^{-}$[…]). By using appropriate calculations, determine wheth[er] it is feasible to titrate 10.00 mL of 0.100 M H_2SO_4 to tw[o] distinct equivalence points with 0.100 M NaOH.

77. Carbonic acid is a weak diprotic acid (H_2CO_3) wi[th] $K_{a_1} = 4.43 \times 10^{-7}$ and $K_{a_2} = 4.73 \times 10^{-11}$. The equi[v]alence points for the titration come at approximate[ly] pH 4 and 9. Suitable indicators for use in titrating ca[r]bonic acid or carbonate solutions are methyl orange an[d] phenolphthalein.
(a) Sketch the titration curve that would be obtain[ed] in titrating a 10.0 mL sample of 1.00 M $NaHCO_3$(a[q]) with 1.00 M HCl.
(b) Sketch the titration curve for the titration of [a] 10.0 mL sample of 1.00 M Na_2CO_3(aq) with 1.00 M HC[l].
(c) What volume of 0.100 M HCl is required for t[he] complete neutralization of 1.00 g $NaHCO_3$(s)?
(d) What volume of 0.100 M HCl is required for t[he] complete neutralization of 1.00 g Na_2CO_3(s)?
(e) A sample of NaOH contains a small amount [of] Na_2CO_3. For titration to the phenolphthalein en[d] point, 0.1000 g of this sample requires 23.98 mL [of] 0.1000 M HCl. An additional 0.78 mL is required [to] reach the methyl orange end point. What is the pe[r]cent Na_2CO_3, by mass, in the sample?

78. Piperazine is a diprotic weak base used as a corrosi[on] inhibitor and an insecticide. Its ionization is describ[ed] by the following equations.

$$HN(C_4H_8)NH + H_2O \rightleftharpoons [HN(C_4H_8)NH_2]^+ + OH^- \quad pK_{b_1} = 4.…$$

$$[HN(C_4H_8)NH_2]^+ + H_2O \rightleftharpoons [H_2N(C_4H_8)NH_2]^{2+} + OH^- \quad pK_{b_2} = 8.…$$

The piperazine used commercially is a hexahydra[te], $C_4H_{10}N_2 \cdot 6H_2O$. A 1.00-g sample of this hexahydra[te] is dissolved in 100.0 mL of water and titrated wi[th] 0.500 M HCl. Sketch a titration curve for this titratio[n], indicating **(a)** the initial pH; **(b)** the pH at the ha[lf-]neutralization point of the first neutralization; **(c)** t[he] volume of HCl(aq) required to reach the first equiv[a]lence point; **(d)** the pH at the first equivalence poi[nt]; **(e)** the pH at the point at which the second step of t[he] neutralization is half-completed; **(f)** the volume [of] 0.500 M HCl(aq) required to reach the second equiv[a]lence point of the titration; **(g)** the pH at the seco[nd] equivalence point.

9. Complete the derivation of equation (17.10) outlined in Are You Wondering 17-1. Then derive equation (17.11).

0. Explain why equation (17.10) fails when applied to dilute solutions—for example, when you calculate the pH of 0.010 M NaH_2PO_4. [*Hint:* Refer also to Exercise 79.]

1. A solution is prepared that is 0.150 M CH_3COOH and 0.250 M NaHCOO.
(a) Show that this is a buffer solution.
(b) Calculate the pH of this buffer solution.
(c) What is the final pH if 1.00 L of 0.100 M HCl is added to 1.00 L of this buffer solution?

2. A series of titrations of lactic acid, $CH_3CH(OH)COOH$ ($pK_a = 3.86$) is planned. About 1.00 mmol of the acid will be titrated with NaOH(aq) to a final volume of about 100 mL at the equivalence point. **(a)** Which acid–base indicator from Figure 17-7 would you select for the titration? To assist in locating the equivalence point in the titration, a buffer solution is to be prepared having the same pH as that at the equivalence point. A few drops of the indicator in this buffer will produce the color to be matched in the titrations. **(b)** Which of the following combinations would be suitable for the buffer solutions: CH_3COOH–CH_3COO^-, $H_2PO_4^-$–HPO_4^{2-}, or NH_4^+–NH_3? **(c)** What ratio of conjugate base to acid is required in the buffer?

3. Hydrogen peroxide, H_2O_2, is a somewhat stronger acid than water. Its ionization is represented by the equation

$$H_2O_2 + H_2O \rightleftharpoons H_3O^+ + HO_2^-$$

In 1912, the following experiments were performed to obtain an approximate value of pK_a for this ionization at 0 °C. A sample of H_2O_2 was shaken together with a mixture of water and pentan-1-ol. The mixture settled into two layers. At equilibrium, the hydrogen peroxide had distributed itself between the two layers such that the water layer contained 6.78 times as much H_2O_2 as the pentan-1-ol layer. In a second experiment, a sample of H_2O_2 was shaken together with 0.250 M NaOH(aq) and pentan-1-ol. At equilibrium, the concentration of H_2O_2 was 0.00357 M in the pentan-1-ol layer and 0.259 M in the aqueous layer. In a third experiment, a sample of H_2O_2 was brought to equilibrium with a mixture of pentan-1-ol and 0.125 M NaOH(aq); the concentrations of the hydrogen peroxide were 0.00198 M in the pentan-1-ol and 0.123 M in the aqueous layer. For water at 0 °C, $pK_w = 14.94$. Find an approximate value of pK_a for H_2O_2 at 0 °C. [*Hint:* The hydrogen peroxide concentration in the aqueous layers is the total concentration of H_2O_2 and HO_2^-. Assume that the pentan-1-ol solutions contain no ionic species.]

. Sodium ammonium hydrogen phosphate, $NaNH_4HPO_4$, is a salt in which one of the ionizable H atoms of H_3PO_4 is replaced by Na^+, another is replaced by NH_4^+, and the third remains in the anion HPO_4^{2-}. Calculate the pH of 0.100 M $NaNH_4HPO_4$(aq).
[*Hint:* You can use the general method introduced on page 761. First, identify all the species that could be present and the equilibria involving these species. Then identify the two equilibrium expressions that will predominate and eliminate all the species whose concentrations are likely to be negligible. At that point, only a few algebraic manipulations are required.]

85. Consider a solution containing two weak monoprotic acids with dissociation constants K_{HA} and K_{HB}. Find the charge balance equation for this system, and use it to derive an expression that gives the concentration of H_3O^+ as a function of the concentrations of HA and HB and the various constants.

86. Calculate the pH of a solution that is 0.050 M acetic acid and 0.010 M phenylacetic acid.

87. A very common buffer agent used in the study of biochemical processes is the weak base TRIS, $(HOCH_2)_3CNH_2$, which has a pK_b of 5.91 at 25 °C. A student is given a sample of the hydrochloride of TRIS together with standard solutions of 10 M NaOH and HCl.
(a) Using TRIS, how might the student prepare 1 L of a buffer of pH = 7.79?
(b) In one experiment, 30 mmol of protons are released into 500 mL of the buffer prepared in part (a). Is the capacity of the buffer sufficient? What is the resulting pH?
(c) Another student accidentally adds 20 mL of 10 M HCl to 500 mL of the buffer solution prepared in part (a). Is the buffer ruined? If so, how could the buffer be regenerated?

88. The Henderson–Hasselbalch equation can be written as

$$pH = pK_a - \log\left(\frac{1}{\alpha} - 1\right) \quad \text{where } \alpha = \frac{[A^-]}{[A^-] + [HA]}.$$

Thus, the degree of ionization (α) of an acid can be determined if both the pH of the solution and the pK_a of the acid are known.
(a) Use this equation to plot the pH versus the degree of ionization for the second ionization constant of phosphoric acid ($K_a = 6.3 \times 10^{-8}$).
(b) If pH = pK_a what is the degree of ionization?
(c) If the solution had a pH of 6.0, what would the value of α be?

89. The pH of ocean water depends on the amount of atmospheric carbon dioxide. The dissolution of carbon dioxide in ocean water can be approximated by the following chemical reactions (Henry's Law constant for CO_2 is $K_H = [CO_2(aq)]/[CO_2(g)] = 0.8317$.) For reaction (2), $K = 2.8 \times 10^{-9}$:

$$CO_2(g) \rightleftharpoons CO_2(aq) \quad \textbf{(1)}$$
$$CaCO_3(s) \rightleftharpoons Ca^{2+}(aq) + CO_3^-(aq) \quad \textbf{(2)}$$
$$H_3O^+(aq) + CO_3^-(aq) \rightleftharpoons HCO_3^-(aq) + H_2O(l) \quad \textbf{(3)}$$
$$H_3O^+(aq) + HCO_3^-(aq) \rightleftharpoons CO_2(aq) + 2\,H_2O(l) \quad \textbf{(4)}$$

(a) Use the equations above to determine the hydronium ion concentration as a function of $[CO_2(g)]$ and $[Ca^{2+}]$.
(b) During preindustrial conditions, we will assume that the equilibrium concentration of $[CO_2(g)]$ = 280 ppm and $[Ca^{2+}]$ = 10.24 mM. Calculate the pH of a sample of ocean water.

90. A sample of water contains 23.0 g L^{-1} of Na^+(aq), 10.0 g L^{-1} of Ca^{2+}(aq), 40.2 g L^{-1} CO_3^{2-}(aq), and 9.6 g L^{-1} SO_4^{2-}(aq). What is the pH of the solution if the only other ions present are H_3O^+ and OH^-?

91. In 1922 Donald D. van Slyke (*J. Biol. Chem.*, **52**, 525) defined a quantity known as the buffer index: $\beta = dc_b/d(pH)$, where dc_b represents the increment of moles of strong base to one liter of the buffer. For the addition of a strong acid, he wrote $\beta = -dc_a/d(pH)$.

By applying this idea to a monoprotic acid and its conjugate base, we can derive the following expression:

$$\beta = 2.303\left(\frac{K_w}{[H_3O^+]} + [H_3O^+] + \frac{cK_a[H_3O^+]}{(K_a + [H_3O^+])^2}\right)$$

where c is the total concentration of monoprotic acid and conjugate base.

(a) Use the above expression to calculate the buffer index for the acetic acid buffer with a total acetic acid and acetate ion concentration of 2.0×10^{-2} and pH = 5.0.

(b) Use the buffer index from part (a) and calcula the pH of the buffer after the addition of of a stror acid. (*Hint:* Let $dc_a/d(\text{pH}) \approx \Delta c_a/\Delta\text{pH}$.)

(c) Make a plot of β versus pH for a 0.1 M acetic ac buffer system. Locate the maximum buffer index well as the minimum buffer indices.

Feature Problems

92. The graph below, which is related to a titration curve, shows the fraction (f) of the stoichiometric amount of acetic acid present as non-ionized CH_3COOH and as acetate ion, CH_3COO^-, as a function of the pH of the solution containing these species.
(a) Explain the significance of the point at which the two curves cross. What are the fractions and the pH at that point?
(b) Sketch a comparable set of curves for carbonic acid, H_2CO_3. [*Hint:* How many carbonate-containing species should appear in the graph? How many points of intersection should there be? at what pH values?]
(c) Sketch a comparable set of curves for phosphoric acid, H_3PO_4. [*Hint:* How many phosphate-containing species should appear in the graph? How many points of intersection should there be? at what pH values?]

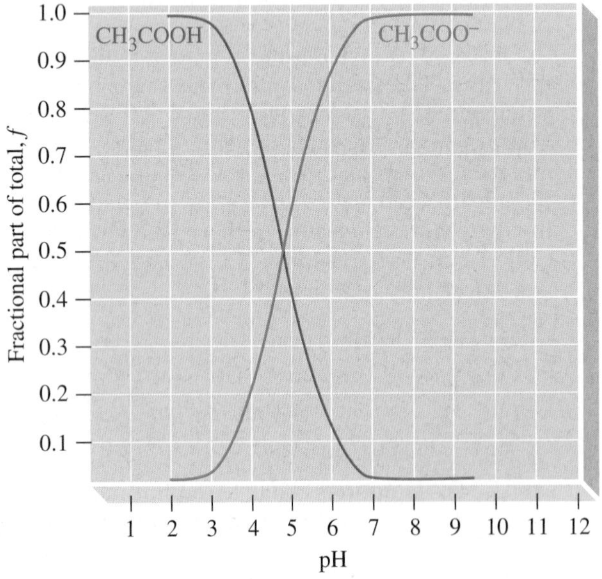

93. In some cases, the titration curve for a mixture of *two* acids has the same appearance as that for a single acid; in other cases it does not.
(a) Sketch the titration curve (pH versus volume of titrant) for the titration with 0.200 M NaOH of 25.00 mL of a solution that is 0.100 M in HCl and 0.100 M in HNO_3. Does this curve differ in any way from what would be obtained in the titration of 25.00 mL of 0.200 M HCl with 0.200 M NaOH? Explain.
(b) The titration curve shown was obtained when 10.00 mL of a solution containing both HCl and H_3PO_4 was titrated with 0.216 M NaOH. From th curve, determine the stoichiometric molarities of bo the HCl and the H_3PO_4.
(c) A 10.00 mL solution that is 0.0400 M H_3PO_4 ar 0.0150 M NaH_2PO_4 is titrated with 0.0200 M NaOI Sketch the titration curve.

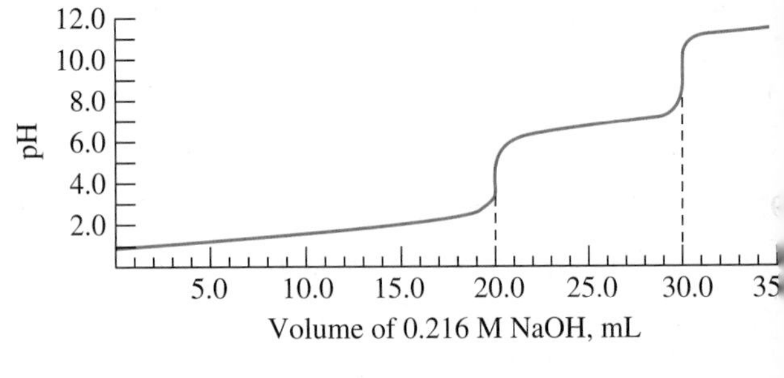

94. Amino acids contain both an acidic carboxylic ac group (—COOH) and a basic amino group (—NH The amino group can be *protonated* (that is, it has extra proton attached) in a strongly acidic solutio This produces a diprotic acid of the form H_2A^+, exemplified by the protonated amino acid alanine.

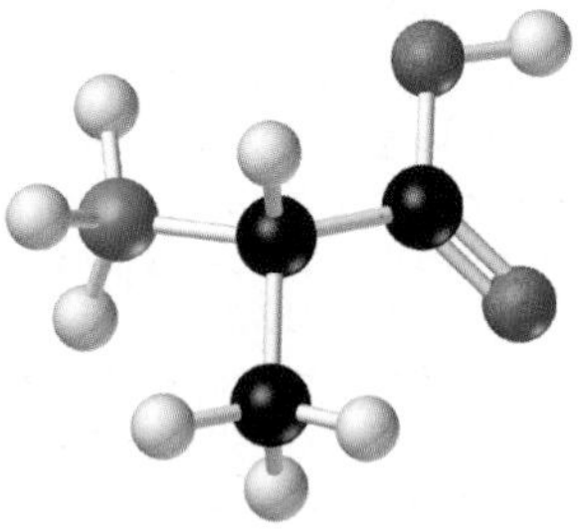

Protonated alanine

The protonated amino acid has two ionizable proto that can be titrated with OH^-.

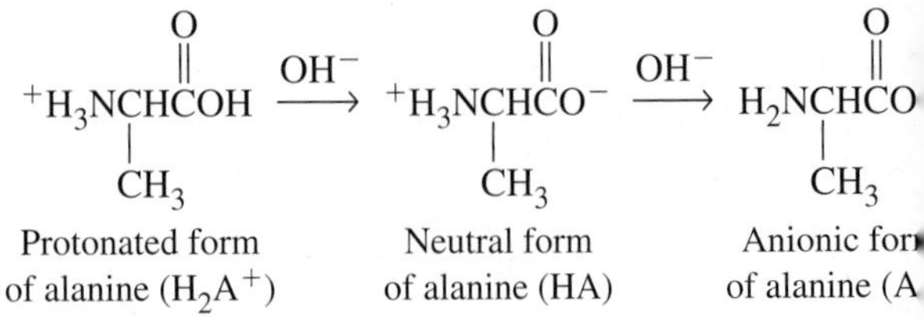

For the —COOH group, $pK_{a_1} = 2.34$; for the —NH group, $pK_{a_2} = 9.69$. Consider the titration of a 0.500 solution of alanine hydrochloride with 0.500 M NaC solution. What is the pH of **(a)** the 0.500 M alanine hydrochloride; **(b)** the solution at the first half-neutralization point; **(c)** the solution at the first equivalence point?

The dominant form of alanine present at the first equivalence point is electrically neutral despite the positive charge and negative charge it possesses. The point at which the neutral form is produced is called the *isoelectric point*. Confirm that the pH at the isoelectric point is

$$pH = \frac{1}{2}(pK_{a_1} + pK_{a_2})$$

What is the pH of the solution **(d)** halfway between the first and second equivalence points? **(e)** at the second equivalence point?

(f) Calculate the pH values of the solutions when the following volumes of the 0.500 M NaOH have been added to 50 mL of the 0.500 M alanine hydrochloride solution: 10.0 mL, 20.0 mL, 30.0 mL, 40.0 mL, 50.0 mL, 60.0 mL, 70.0 mL, 80.0 mL, 90.0 mL, 100.0 mL, and 110.0 mL.

(g) Sketch the titration curve for the 0.500 M solution of alanine hydrochloride, and label significant points on the curve.

Self-Assessment Exercises

5. In your own words, define or explain the following terms or symbols: **(a)** mmol; **(b)** HIn; **(c)** equivalence point of a titration; **(d)** titration curve.

6. Briefly describe each of the following ideas, phenomena, or methods: **(a)** the common-ion effect; **(b)** the use of a buffer solution to maintain a constant pH; **(c)** the determination of pK_a of a weak acid from a titration curve; **(d)** the measurement of pH with an acid–base indicator.

7. Explain the important distinctions between each pair of terms: **(a)** buffer capacity and buffer range; **(b)** hydrolysis and neutralization; **(c)** first and second equivalence points in the titration of a weak diprotic acid; **(d)** equivalence point of a titration and end point of an indicator.

8. Write equations to show how each of the following buffer solutions reacts with a small added amount of a strong acid or a strong base: **(a)** $HCOOH–KHCOO$; **(b)** $C_6H_5NH_2–C_6H_5NH_3^+Cl^-$; **(c)** $KH_2PO_4–Na_2HPO_4$.

9. Sketch the titration curves that you would expect to obtain in the following titrations. Select a suitable indicator for each titration from Figure 17-7.
(a) $NaOH(aq)$ titrated with $HNO_3(aq)$
(b) $NH_3(aq)$ titrated with $HCl(aq)$
(c) $CH_3COOH(aq)$ titrated with $KOH(aq)$
(d) $NaH_2PO_4(aq)$ titrated with $KOH(aq)$

0. A 25.00-mL sample of 0.0100 M C_6H_5COOH ($K_a = 6.3 \times 10^{-5}$) is titrated with 0.0100 M $Ba(OH)_2$. Calculate the pH **(a)** of the initial acid solution; **(b)** after the addition of 6.25 mL of 0.0100 M $Ba(OH)_2$; **(c)** at the equivalence point; **(d)** after the addition of a total of 15.00 mL of 0.0100 M $Ba(OH)_2$.

1. To *suppress* the ionization of formic acid, $HCOOH(aq)$, which of the following should be added to the solution? **(a)** NaCl; **(b)** NaOH; **(c)** NaHCOO; **(d)** $NaNO_3$.

2. To *increase* the ionization of formic acid, $HCOOH(aq)$, which of the following should be added to the solution? **(a)** NaCl; **(b)** NaHCOO; **(c)** H_2SO_4; **(d)** $NaHCO_3$.

103. To convert $NH_4^+(aq)$ to $NH_3(aq)$, **(a)** add H_3O^+; **(b)** raise the pH; **(c)** add $KNO_3(aq)$; **(d)** add NaCl.

104. During the titration of equal concentrations of a weak base and a strong acid, at what point would the $pH = pK_a$? **(a)** the initial pH; **(b)** halfway to the equivalence point; **(c)** at the equivalence point; **(d)** past the equivalence point.

105. Calculate the pH of the buffer formed by mixing equal volumes $[C_2H_5NH_2] = 1.49$ M with $[HClO_4] = 1.001$ M. $K_b = 4.3 \times 10^{-4}$.

106. Calculate the pH of a 0.5 M solution of $Ca(HSe)_2$, given that H_2Se has $K_{a_1} = 1.3 \times 10^{-4}$ and $K_{a_2} = 1 \times 10^{-11}$.

107. The effect of adding 0.001 mol KOH to 1.00 L of a solution that is 0.10 M NH_3–0.10 M NH_4Cl is to **(a)** raise the pH very slightly; **(b)** lower the pH very slightly; **(c)** raise the pH by several units; **(d)** lower the pH by several units.

108. The most acidic of the following 0.10 M salt solutions is **(a)** Na_2S; **(b)** $NaHSO_4$; **(c)** $NaHCO_3$; **(d)** Na_2HPO_4.

109. If an indicator is to be used in an acid–base titration having an equivalence point in the pH range 8 to 10, the indicator must **(a)** be a weak base; **(b)** have $K_a = 1 \times 10^{-9}$; **(c)** ionize in two steps; **(d)** be added to the solution only after the solution has become alkaline.

110. Indicate whether you would expect the equivalence point of each of the following titrations to be below, above, or at pH 7. Explain your reasoning. **(a)** $NaHCO_3(aq)$ is titrated with $NaOH(aq)$; **(b)** $HCl(aq)$ is titrated with $NH_3(aq)$; **(c)** $KOH(aq)$ is titrated with $HI(aq)$.

111. Using the method presented in Appendix E, construct a concept map relating the concepts in Sections 17-2, 17-3, and 17-4.

18 Solubility and Complex-Ion Equilibria

CONTENTS

LEARNING OBJECTIVES

18.1 Describe the meaning of the solubility product, K_{sp}, for a slightly soluble salt.

18.2 Determine the molar solubility of a salt given the value of the solubility product, K_{sp}.

18.3 Discuss how the addition or presence of a common ion affects the solubility of a slightly soluble salt.

18.4 Identify some limitations of the solubility product concept.

18.5 By comparing values of Q_{sp} and K_{sp}, predict whether or not precipitation will occur.

18.6 Describe how to use the technique of fractional precipitation to separate ions.

18.7 Discuss how the pH of a solution may affect the solubility of a salt.

18.8 Use the formation constant, K_f, of a complex ion to determine the concentrations of ions in solution.

18.9 Describe how to use precipitation, acid–base, redox, and complex-ion formation reactions in qualitative cation analysis.

Stalactite and stalagmite formations in a cavern in Liguria, Italy. Stalactites and stalagmites are formed from calcium salts deposited as underground water seeps int the cavern and evaporates.

The dissolution and precipitation of limestone ($CaCO_3$) underlie variety of natural phenomena, such as the formation of limesto caverns. Whether a solution containing Ca^{2+} and $CO_3{}^{2-}$ io undergoes precipitation depends on the concentrations of these ions. turn, the $CO_3{}^{2-}$ ion concentration depends on the pH of the solution. develop a better understanding of the conditions under which $CaCO_3$ d solves or precipitates, we must consider equilibrium relationshi between Ca^{2+} and $CO_3{}^{2-}$, and between $CO_3{}^{2-}$, H_3O^+, and $HCO_3{}^-$. Th requirement suggests a need to combine ideas about acid–base equilib from Chapters 16 and 17 with ideas about the new types of equilibria to introduced in this chapter.

Silver chloride is a familiar precipitate in the general chemistry labo tory, yet it does *not* precipitate from a solution that has a moderate to hi concentration of $NH_3(aq)$. The silver ion and ammonia combine inste form a species, called a *complex ion*, that remains in solution. Complex-ion rmation and equilibria involving complex ions are additional topics disssed in this chapter.

8-1 Solubility Product Constant, K_{sp}

ypsum ($CaSO_4 \cdot 2H_2O$) is a calcium mineral that is slightly soluble in ater. Groundwater that comes into contact with gypsum often contains me dissolved calcium sulfate. This water cannot be used for certain applitions, such as in evaporative cooling systems in power plants because the lcium sulfate might precipitate from the water and block pipes. The equirium between $Ca^{2+}(aq)$ and $SO_4^{2-}(aq)$ and undissolved $CaSO_4(s)$ can be presented as

$$CaSO_4(s) \rightleftharpoons Ca^{2+}(aq) + SO_4^{2-}(aq)$$

e thermodynamic equilibrium constant for this reaction is

$$K = \frac{a_{Ca^{2+}(aq)}a_{SO_4^{2-}(aq)}}{a_{CaSO_4(s)}} \approx \frac{([Ca^{2+}]/c°)([SO_4^{2-}]/c°)}{1} = [Ca^{2+}]\,[SO_4^{2-}] \times \left(\frac{1}{c°}\right)^2$$

here $c° = 1$ mol/L. The quantity multiplying $(1/c°)^2$ is called the **solubility oduct constant, K_{sp}**.

$$K_{sp} = [Ca^{2+}][SO_4^{2-}] = 9.1 \times 10^{-6} \text{ (at 25 °C)} \qquad \textbf{(18.1)}$$

ble 18.1 lists K_{sp} values and the equilibria to which they apply. The small K_{sp} lues reflect the fact that these salts produce very low concentrations of ions hen added to water (typically, much less than 0.01 mol/L).

KEEP IN MIND

that equilibrium concentrations must be used in the expressions for K and K_{sp}.

◀ Solubility product constants vary with temperature.

able 18.1 Several Solubility Product Constants at 25 °C[a]

olute	Solubility Equilibrium	K_{sp}
luminum hydroxide	$Al(OH)_3(s) \rightleftharpoons Al^{3+}(aq) + 3OH^-(aq)$	1.3×10^{-33}
arium carbonate	$BaCO_3(s) \rightleftharpoons Ba^{2+}(aq) + CO_3^{2-}(aq)$	5.1×10^{-9}
arium sulfate	$BaSO_4(s) \rightleftharpoons Ba^{2+}(aq) + SO_4^{2-}(aq)$	1.1×10^{-10}
alcium carbonate	$CaCO_3(s) \rightleftharpoons Ca^{2+}(aq) + CO_3^{2-}(aq)$	2.8×10^{-9}
alcium fluoride	$CaF_2(s) \rightleftharpoons Ca^{2+}(aq) + 2F^-(aq)$	5.3×10^{-9}
alcium sulfate	$CaSO_4(s) \rightleftharpoons Ca^{2+}(aq) + SO_4^{2-}(aq)$	9.1×10^{-6}
hromium(III) hydroxide	$Cr(OH)_3(s) \rightleftharpoons Cr^{3+}(aq) + 3OH^-(aq)$	6.3×10^{-31}
ron(III) hydroxide	$Fe(OH)_3(s) \rightleftharpoons Fe^{3+}(aq) + 3OH^-(aq)$	4×10^{-38}
ead(II) chloride	$PbCl_2(s) \rightleftharpoons Pb^{2+}(aq) + 2Cl^-(aq)$	1.6×10^{-5}
ead(II) chromate	$PbCrO_4(s) \rightleftharpoons Pb^{2+}(aq) + CrO_4^{2-}(aq)$	2.8×10^{-13}
ead(II) iodide	$PbI_2(s) \rightleftharpoons Pb^{2+}(aq) + 2I^-(aq)$	7.1×10^{-9}
agnesium carbonate	$MgCO_3(s) \rightleftharpoons Mg^{2+}(aq) + CO_3^{2-}(aq)$	3.5×10^{-8}
agnesium fluoride	$MgF_2(s) \rightleftharpoons Mg^{2+}(aq) + 2F^-(aq)$	3.7×10^{-8}
agnesium hydroxide	$Mg(OH)_2(s) \rightleftharpoons Mg^{2+}(aq) + 2OH^-(aq)$	1.8×10^{-11}
agnesium phosphate	$Mg_3(PO_4)_2(s) \rightleftharpoons 3Mg^{2+}(aq) + 2PO_4^{3-}(aq)$	1×10^{-25}
ercury(I) chloride	$Hg_2Cl_2(s) \rightleftharpoons Hg_2^{2+}(aq) + 2Cl^-(aq)$	1.3×10^{-18}
ilver bromide	$AgBr(s) \rightleftharpoons Ag^+(aq) + Br^-(aq)$	5.0×10^{-13}
ilver carbonate	$Ag_2CO_3(s) \rightleftharpoons 2Ag^+(aq) + CO_3^{2-}(aq)$	8.5×10^{-12}
ilver chloride	$AgCl(s) \rightleftharpoons Ag^+(aq) + Cl^-(aq)$	1.8×10^{-10}
ilver chromate	$Ag_2CrO_4(s) \rightleftharpoons 2Ag^+(aq) + CrO_4^{2-}(aq)$	1.1×10^{-12}
ilver iodide	$AgI(s) \rightleftharpoons Ag^+(aq) + I^-(aq)$	8.5×10^{-17}
trontium carbonate	$SrCO_3(s) \rightleftharpoons Sr^{2+}(aq) + CO_3^{2-}(aq)$	1.1×10^{-10}
trontium sulfate	$SrSO_4(s) \rightleftharpoons Sr^{2+}(aq) + SO_4^{2-}(aq)$	3.2×10^{-7}

A more extensive listing of K_{sp} values is given in Appendix D.

EXAMPLE 18-1 Writing Solubility Product Constant Expressions for Slightly Soluble Solutes

Write the solubility product constant expression for the solubility equilibrium of

(a) Calcium fluoride, CaF_2 (one of the substances used when a fluoride treatment is applied to teeth).

(b) Copper arsenate, $Cu_3(AsO_4)_2$ (used as an insecticide and fungicide).

Analyze

The equation for the solubility equilibrium is written for one mole of the slightly soluble solute. That is, the coefficient "1" is understood for the slightly soluble solute. The coefficients for the ions in solution are whatever is needed to balance the equation. The coefficients then establish the powers to which the ion concentrations are raised in the K_{sp} expression.

Solve

(a) $CaF_2(s) \rightleftharpoons Ca^{2+}(aq) + 2\,F^-(aq) \qquad K_{sp} = [Ca^{2+}][F^-]^2$

(b) $Cu_3(AsO_4)_2(s) \rightleftharpoons 3\,Cu^{2+}(aq) + 2\,AsO_4^{\;3-}(aq) \qquad K_{sp} = [Cu^{2+}]^3[AsO_4^{\;3-}]^2$

Assess

Note that the solid does not appear in the K_{sp} expression. The form of the K_{sp} expression is established by applying the rules for writing the thermodynamic equilibrium constant expression, K. According to these rules, the activity of a pure solid is equal to 1 and consequently, the activity of the solid effectively disappears from the expressions for K and K_{sp}.

PRACTICE EXAMPLE A: Write the solubility product constant expression for **(a)** $MgCO_3$ (one of the components of dolomite, a form of limestone) and **(b)** Ag_3PO_4 (used in photographic emulsions).

PRACTICE EXAMPLE B: A handbook lists $K_{sp} = 1 \times 10^{-7}$ for calcium hydrogen phosphate, a substance used in dentifrices and as an animal feed supplement. Write **(a)** the equation for the solubility equilibrium and **(b)** the solubility product constant expression for this slightly soluble solute.

Scimat Scimat/Getty Images

▲ Some calcium salts, such as calcium fluoride and calcium hydrogen phosphate, have beneficial uses, but another calcium salt, calcium oxalate (CaC_2O_4), can be harmful. The photo is a scanning electron microscope image of calcium oxalate crystals, a common type of kidney stone that can form in the human kidney.

18-1 CONCEPT ASSESSMENT

A large excess of $MgF_2(s)$ is maintained in contact with 1.00 L of pure water to produce a saturated solution of MgF_2. When an additional 1.00 L of pure water is added to the mixture and equilibrium re-established, compared with the original saturated solution, will $[Mg^{2+}]$ be **(a)** the same; **(b)** twice as large; **(c)** half as large; **(d)** some unknown fraction of the original $[Mg^{2+}]$? Explain.

18-2 Relationship Between Solubility and K_{sp}

Is there a relationship between the solubility product constant, K_{sp}, of a solu and the solute's *molar solubility*—its molarity in a saturated aqueous sol tion? As shown in Examples 18-2 and 18-3, there is a definite relationsh between them. As discussed in Section 18-4, calculations involving K_{sp} a generally more subject to error than are those involving other equilibriu constants, but the results are suitable for many purposes. In Example 18 we start with an experimentally determined solubility and obtain a val of K_{sp}.

The "inverse" of Example 18-2 is the calculation of the solubility of solute from its K_{sp} value. When this is done, as in Example 18-3, the result always a molar solubility—a molarity. Additional conversions are requir to obtain solubility in units other than moles per liter, as in Practi Example 18-3B.

EXAMPLE 18-2 Calculating K_{sp} of a Slightly Soluble Solute from Its Solubility

A handbook lists the aqueous solubility of $CaSO_4$ at 25 °C as 0.20 g $CaSO_4$/100 mL. What is the K_{sp} of $CaSO_4$ at 25 °C?

$$CaSO_4(s) \rightleftharpoons Ca^{2+}(aq) + SO_4^{2-}(aq) \qquad K_{sp} = ?$$

Analyze

We need to construct a conversion pathway that begins with finding $[Ca^{2+}]$ and $[SO_4^{2-}]$, which we can then substitute into the K_{sp} expression.

$$\text{g } CaSO_4/100 \text{ mL} \longrightarrow \text{mol } CaSO_4/\text{L} \longrightarrow [Ca^{2+}] \text{ and } [SO_4^{2-}] \longrightarrow K_{sp}$$

Solve

The first step is to convert the mass of $CaSO_4$ in a 100 mL volume to molar solubility. This is accomplished by using the inverse of the molar mass of $CaSO_4$ and replacing 100 mL with 0.100 L.

$$\frac{\text{mol } CaSO_4}{\text{L satd soln}} = \frac{0.20 \text{ g } CaSO_4}{0.100 \text{ L soln}} \times \frac{1 \text{ mol } CaSO_4}{136 \text{ g } CaSO_4}$$
$$= 0.015 \text{ M } CaSO_4$$

Using stoichiometric ratios (shown in blue), determine $[Ca^{2+}]$ and $[SO_4^{2-}]$.

$$[Ca^{2+}] = \frac{0.015 \text{ mol } CaSO_4}{1 \text{ L}} \times \frac{1 \text{ mol } Ca^{2+}}{1 \text{ mol } CaSO_4} = 0.015 \text{ M}$$

$$[SO_4^{2-}] = \frac{0.015 \text{ mol } CaSO_4}{1 \text{ L}} \times \frac{1 \text{ mol } SO_4^{2-}}{1 \text{ mol } CaSO_4} = 0.015 \text{ M}$$

Substitute these ion concentrations into the solubility product expression.

$$K_{sp} = [Ca^{2+}][SO_4^{2-}] = (0.015)(0.015) = 2.3 \times 10^{-4}$$

Assess

The K_{sp} result determined here is significantly different from the value in Table 18.1. The reason for this discrepancy is that ion activities need to be taken into account.

PRACTICE EXAMPLE A: A handbook lists the aqueous solubility of AgOCN as 7 mg/100 mL at 20 °C. What is the K_{sp} of AgOCN at 20 °C?

PRACTICE EXAMPLE B: A handbook lists the aqueous solubility of lithium phosphate at 18 °C as 0.034 g Li_3PO_4/100 mL soln. What is the K_{sp} of Li_3PO_4 at 18 °C?

EXAMPLE 18-3 Calculating the Solubility of a Slightly Soluble Solute from Its K_{sp} Value

Lead(II) iodide, PbI_2, is a dense, golden yellow, "insoluble" solid used in bronzing and in ornamental work requiring a golden color (such as mosaic gold). Calculate the molar solubility of lead(II) iodide in water at 25 °C, given that $K_{sp} = 7.1 \times 10^{-9}$.

Analyze

The solubility equilibrium equation

$$PbI_2(s) \rightleftharpoons Pb^{2+}(aq) + 2\,I^-(aq)$$

shows that for each mole of PbI_2 that dissolves, *one* mole of Pb^{2+} and *two* moles of I^- appear in solution. If we let *s* represent the number of moles of PbI_2 dissolved per liter of saturated solution, we have

$$[Pb^{2+}] = s \quad \text{and} \quad [I^-] = 2s$$

(continued)

Solve

These concentrations must also satisfy the K_{sp} expression.

$$K_{sp} = [Pb^{2+}][I^-]^2 = (s)(2s)^2 = 7.1 \times 10^{-9}$$
$$4s^3 = 7.1 \times 10^{-9}$$
$$s^3 = 1.8 \times 10^{-9}$$
$$s = (1.8 \times 10^{-9})^{1/3} = 1.2 \times 10^{-3}$$

The molar solubility of PbI_2 in water is 1.2×10^{-3} M.

Assess

One key part of this problem is making sure that we account for the correct number of moles of each species. In this case we had 2 moles of iodide ion for every 1 mole of lead.

PRACTICE EXAMPLE A: The K_{sp} of $Cu_3(AsO_4)_2$ at 25 °C is 7.6×10^{-36}. What is the molar solubility of $Cu_3(AsO_4)_2$ in H_2O at 25 °C?

PRACTICE EXAMPLE B: How many milligrams of $BaSO_4$ are dissolved in a 225 mL sample of saturated $BaSO_4$(aq)? $K_{sp} = 1.1 \times 10^{-10}$.

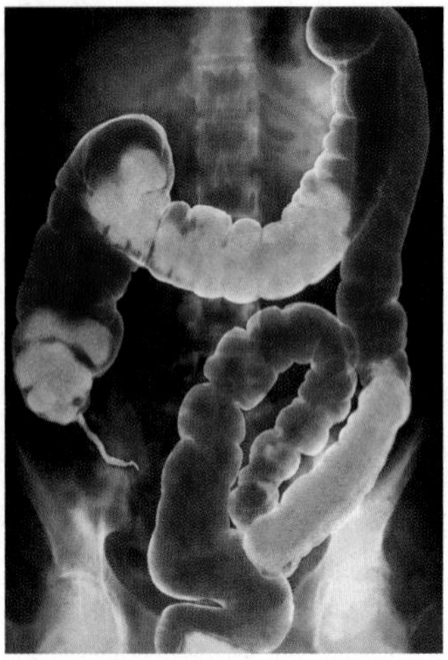

Cnri/Science Source

▲ Barium sulfate, $BaSO_4$, is a good absorber of X-rays. As a component of a "barium milk shake," $BaSO_4$ coats the intestinal tract so that this soft tissue will show up when X-rayed. Even though Ba^{2+}(aq) is poisonous, $BaSO_4$(s) is harmless because its aqueous solubility is very low.

18-1 ARE YOU WONDERING?

When comparing molar solubilities, is a solute with a larger value of K_{sp} always more soluble than one with a smaller value?

If the solutes being compared are of the same type (for example, all are of the type MX or all are of the type, MX_2), their molar solubilities will be related in the same way to their K_{sp} values. Then, the solute with the largest K_{sp} value will have the greatest molar solubility. Thus, AgCl ($K_{sp} = 1.8 \times 10^{-10}$) is more soluble than AgBr ($K_{sp} = 5.0 \times 10^{-13}$). For these particular solutes, the molar solubility is $s = \sqrt{K_{sp}}$.

If the solutes are *not* of the same type, you'll have to calculate, or at least estimate, each molar solubility and compare the results. Thus, even though its solubility product constant is smaller, Ag_2CrO_4 ($K_{sp} = 1.1 \times 10^{-12}$) is more soluble than AgCl ($K_{sp} = 1.8 \times 10^{-10}$). For Ag_2CrO_4, the molar solubility is $s = (K_{sp}/4)^{1/3} = 6.5 \times 10^{-5}$ M, whereas for AgCl, it is $s = \sqrt{K_{sp}} = 1.3 \times 10^{-5}$ M.

18-2 CONCEPT ASSESSMENT

Of the compounds CaF_2, $CaCl_2$, AgF, and AgCl, which would be considered insoluble? Explain.

18-3 Common-Ion Effect in Solubility Equilibria

In Examples 18-2 and 18-3, the ions in the saturated solutions came from a si gle source, the pure solid solute. Suppose that to the saturated solution of Pl in Example 18-3, we add some I^-—a *common ion*—from a source such as KI(a The situation is similar to our first encounter with the common-ion effect Chapter 17.

According to Le Châtelier's principle, an equilibrium mixture responds to forced increase in the concentration of one of its reactants by shifting in t direction in which that reactant is consumed. In the lead(II) iodide solubili uilibrium, addition of the common ion, I^-, causes the reverse reaction to be vored, leading to a new equilibrium.

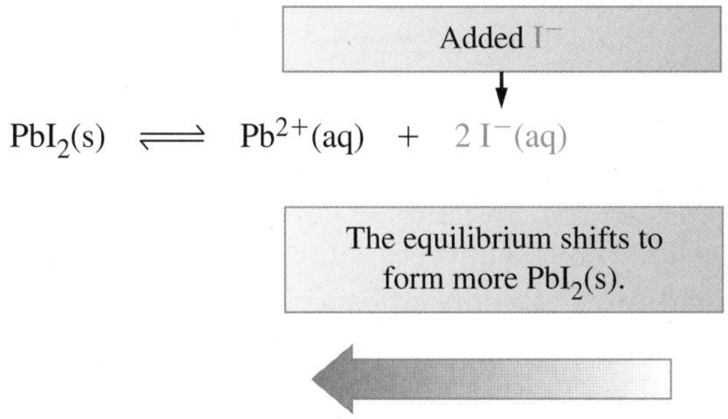

ıe addition of the common ion shifts the equilibrium of a slightly soluble nic compound toward the undissolved compound, causing more to precipi-te. Thus, the solubility of the compound is reduced.

> The solubility of a slightly soluble ionic compound is lowered in the presence of a second solute that furnishes a common ion.

The common-ion effect is illustrated in Figure 18-1, and it is applied quanti-tively in Example 18-4.

The solubility of PbI_2 in the presence of 0.10 M I^-, as calculated in ample 18-4, is about 2000 times less than its value in pure water xample 18-3). If you work out Practice Example 18-4A, you will see that the ect of added Pb^{2+} in reducing the solubility of PbI_2 is not as striking as that I^-, but it is significant nevertheless.

◀ Here K^+ does not take part in the process and is generally not included in the solubility equations. Such ions must always be present to ensure electrical neutrality of the solution. These ions are sometimes called "spectator ions."

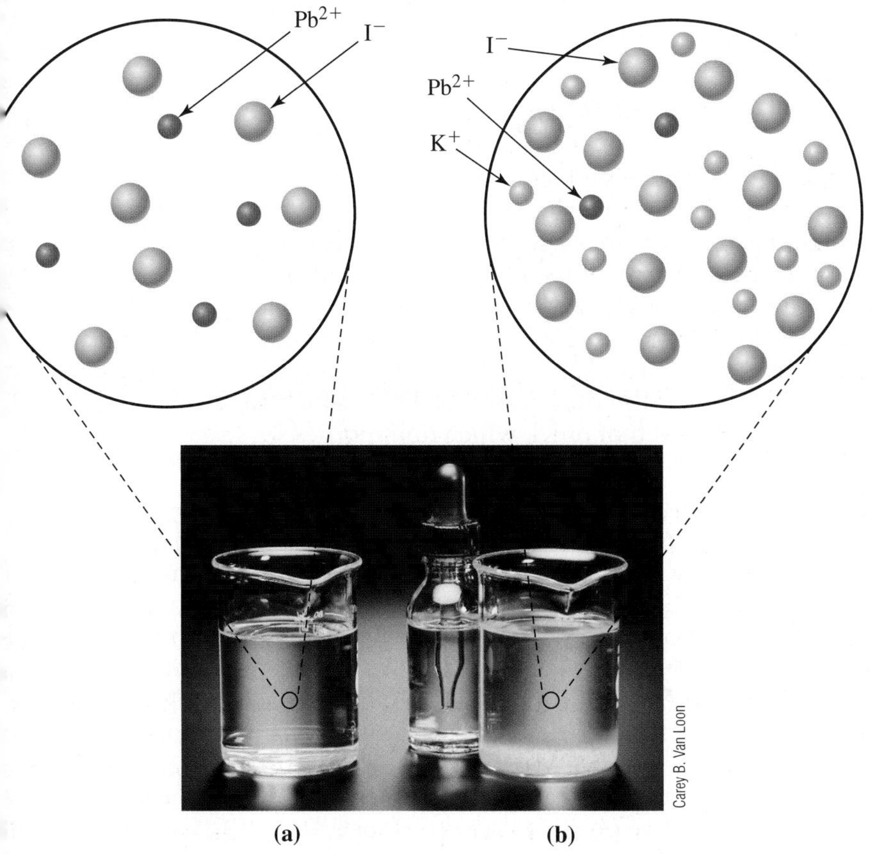

◀ FIGURE 18-1
The common-ion effect in solubility equilibrium
(a) A clear saturated solution of lead(II) iodide from which excess undissolved solute has been filtered off. **(b)** When a small volume of a concentrated solution of KI (containing the common ion, I^-) is added, a small quantity of $PbI_2(s)$ precipitates as a yellow solid. A common ion reduces the solubility of a sparingly soluble solute.

EXAMPLE 18-4 Calculating the Solubility of a Slightly Soluble Solute in the Presence of a Common Ion

What is the molar solubility of PbI_2 in 0.10 M KI(aq)?

Analyze

To solve this problem, let's set up an ICE table with *s* instead of *x* to represent changes in concentrations. Think of producing a saturated solution of PbI_2, but instead of using pure water as the solvent, we will use 0.10 M KI(aq). Thus, we begin with $[I^-] = 0.10$ M. Now let *s* represent the amount of PbI_2, in moles, that dissolves to produce 1 L of saturated solution. The additional concentrations appearing in this solution are *s* mol Pb^{2+}/L and 2*s* mol I^-/L.

Solve

The ICE table is

	$PbI_2(s) \rightleftharpoons$	$Pb^{2+}(aq)$ +	$2\,I^-(aq)$
initial concns, M:		0	0.10
from PbI_2, M:		s	$2s$
equil concns, M:		s	$(0.10 + 2s)$

The usual K_{sp} relationship must be satisfied.

$$K_{sp} = [Pb^{2+}][I^-]^2 = (s)(0.10 + 2s)^2 = 7.1 \times 10^{-9}$$

To simplify the solution to this equation, let's assume that 2*s* is much smaller than 0.10 M, so that $0.10 + 2s \approx 0.10$.

$$s(0.10)^2 = 7.1 \times 10^{-9}$$

$$s = \frac{7.1 \times 10^{-9}}{(0.10)^2} = 7.1 \times 10^{-7}$$

The molar solubility of PbI_2 in 0.10 M KI(aq) is 7.1×10^{-7} M.

Assess

First, our assumption is well justified: $2(7.1 \times 10^{-7})$ is much smaller than 0.10. Second, PbI_2 is much less soluble in 0.10 M KI(aq) than in water: 7.1×10^{-7} M versus 1.2×10^{-3} M (see Example 18-3).

PRACTICE EXAMPLE A: What is the molar solubility of PbI_2 in 0.10 M $Pb(NO_3)_2$(aq)? [*Hint:* To which ion concentration should the solubility be related?]

PRACTICE EXAMPLE B: What is the molar solubility of $Fe(OH)_3$ in a buffered solution with pH = 8.20?

A typical error in such problems as Example 18-4 is to double the common-i concentration—that is, to write $[I^-] = (2 \times 0.10)$ M instead of $[I^-] = 0.10$ Although it is true that in *any* aqueous solution of PbI_2, the $[I^-]$ derived fr PbI_2 is *twice* the molarity of the PbI_2, the $[I^-]$ that comes from a soluble stro electrolyte is determined only by the molarity of the strong electrolyte. Th $[I^-]$ in 0.10 M KI(aq) is 0.10 M. In 0.10 M KI(aq) that is also saturated wi PbI_2, the *total* $[I^-] = (0.10 + 2s)$ M. In short, no relationship exists betwe the stoichiometry of the dissolution of PbI_2, which requires a factor of 2 establishing $[I^-]$, and that of KI, which does not.

18-4 Limitations of the K_{sp} Concept

▶ The difference between stoichiometric concentration and activity increases as more solute is dissolved in the solvent. There are no interactions between solute particles in an ideal solution where stoichiometric and effective concentrations are equal. These interactions increase as more solute is dissolved.

We have repeatedly used the term *slightly soluble* in describing the solutes f which we have written K_{sp} expressions. You might wonder if we can write K expressions for moderately or highly soluble ionic compounds, such NaCl, KNO_3, and NaOH. The answer is yes, but the K_{sp} must be based ion *activities* rather than on concentrations. In ionic solutions of moderate high concentrations, activities and concentrations are generally far fro equal. For example, in 0.1 M KCl(aq) the activity is roughly 24% of t molarity. If we cannot use molarities in place of activities, much of the si plicity of the solubility product concept is lost. Thus, K_{sp} values are usua

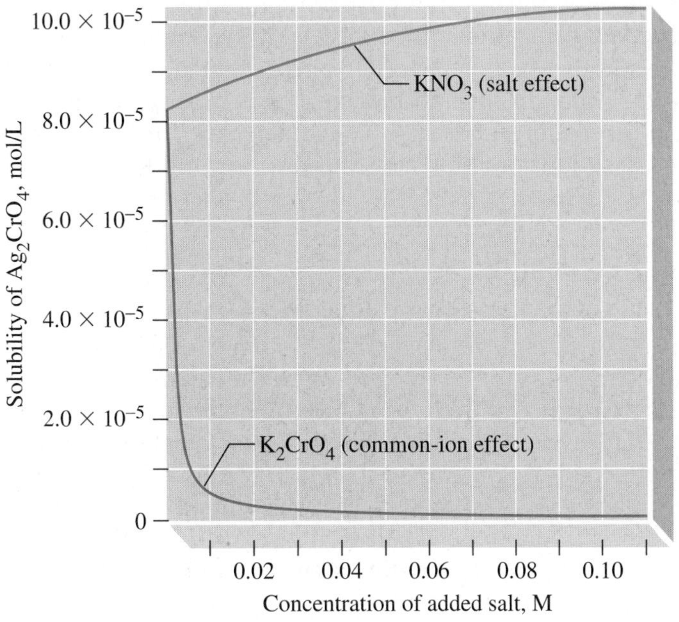

◀ FIGURE 18-2
Comparison of the common-ion effect and the salt effect on the molar solubility of Ag_2CrO_4
The presence of CrO_4^{2-} ions, derived from $K_2CrO_4(aq)$, reduces the solubility of Ag_2CrO_4 by a factor of about 35 over the concentration range shown (from 0 to 0.10 M added salt). Over the same concentration range, the solubility of Ag_2CrO_4 is increased by the presence of the diverse ions from KNO_3, but only by about 25%.

nited to slightly soluble (essentially insoluble) solutes, and ion molarities e used in place of activities. In addition, the K_{sp} concept has several other nitations, which are discussed in the following sections.

ne Diverse Noncommon Ion Effect: The Salt Effect

e have explored the effect of common ions on a solubility equilibrium, but hat effect do ions different from those involved in the equilibrium have on lute solubilities? The effect of "noncommon" or *diverse* ions is not as striking the common-ion effect. Moreover, diverse ions tend to increase rather than crease solubility. As the total ionic concentration of a solution increases, terionic attractions become more important. *Activities become smaller* than the oichiometric concentration. For the ions involved in the solution process, is means that higher concentrations must appear in solution before equilibm is established—the *solubility increases*. Figure 18-2 compares the effects of mmon ions and diverse ions.

The diverse ion effect is more commonly called the **salt effect**. As a result of e salt effect, the numerical value of a K_{sp} based on molarities will vary pending on the ionic atmosphere. Most tabulated values of K_{sp} are based on tivities rather than on molarities, thus avoiding the problem of the salt effect.

complete Dissociation of Solute into Ions

performing calculations involving K_{sp} values and solubilities, we have sumed that all the dissolved solute appears in solution as separated cations d anions. However, this assumption is often not valid. The solute might not 100% ionic, and some of the solute might enter solution in molecular form. ternatively, some ions in solution might join together into **ion pairs**. An ion ir is two oppositely charged ions that are held together by the electrostatic traction between the ions. For example, in a saturated solution of magneum fluoride, although most of the solute exists as Mg^{2+} and F^- ions, some ists as the ion pair MgF^+.

To the extent that a solution contains cations and anions of a solute as ion irs, the concentrations of the dissociated ions are reduced from the stoiiometric expectations. Thus, although the measured solubility of MgF_2 is out 4×10^{-3} M, we cannot safely assume that $[Mg^{2+}] = 4 \times 10^{-3}$ M, and at $[F^-] = 8 \times 10^{-3}$ M, because some of the Mg^{2+} and F^- ions are involved

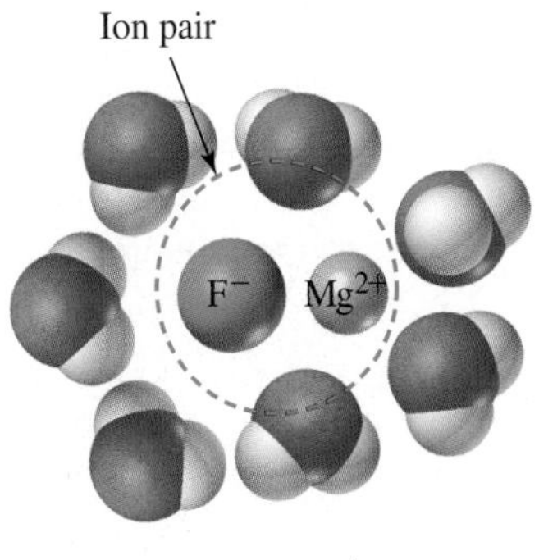

▲ An ion pair, MgF^+, in a magnesium fluoride aqueous solution. The water molecules surrounding the ion pair form what is referred to as a solvent cage.

in ion pairing. This means that additional solute must be present for t product of ion concentrations to equal K_{sp} of the solute, making the true s ubility of the solute greater than otherwise expected on the basis of K_{sp}.

The degree of ion-pair formation increases as mutual electrostatic attracti of the anion and cation increases. Hence, ion-pair formation is increasing likely when the cations or anions in solution carry multiple charges (for exa ple, Mg^{2+} or SO_4^{2-}).

Simultaneous Equilibria

The reversible reaction between a solid solute and its ions in aqueous soluti is never the sole process occurring. At the very least, the self-ionization water also occurs, although we can generally ignore it. Other equilibriu processes that may occur include reactions between solute ions and oth solution species. Two possibilities are acid–base reactions (Section 18-7) a complex-ion formation (Section 18-8). Calculations based on the K_{sp} expre sion may be in error if we fail to take into account other equilibrium process that occur simultaneously with solution equilibrium. We have encounter the dissolution of $PbI_2(s)$ several times already in this chapter. However, t dissolution process is, in fact, a lot more complex than we've shown. As su gested below, there are many competing processes:

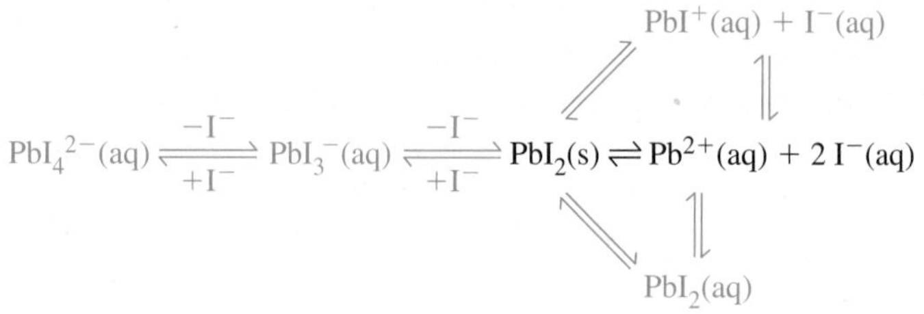

Assessing the Limitations of K_{sp}

Let's assess the importance of the effects discussed in this section, some which apply to $CaSO_4$. Recall that in Example 18-2 we calculated K_{sp} f $CaSO_4$ on the basis of its measured solubility. Our result was $K_{sp} = 2.3 \times 10^{-}$ This value is about 25 times larger than the value listed in Table 18.1, whi is $K_{sp} = 9.1 \times 10^{-6}$.

These conflicting results for $CaSO_4$ are understandable. The K_{sp} value list in Table 18.1 is based on ion activities, whereas the K_{sp} value calculated fro the experimentally determined solubility is based on ion concentratior assuming complete dissociation of the solute into ions and no ion-pair form tion. We will continue to substitute molarities for activities of ions, and the ca of $CaSO_4$ simply suggests that some of our results, although of the appropria general magnitude (that is, within a factor of 10 or 100), may not be high accurate. These order-of-magnitude results, however, still allow us to ma some correct predictions and to apply the K_{sp} concept in useful ways.

18-5 Criteria for Precipitation and Its Completeness

Silver iodide is a light-sensitive compound used in photographic film and al in cloud seeding to produce rain. Its solubility equilibrium and K_{sp} are repr sented as

$$AgI(s) \rightleftharpoons Ag^+(aq) + I^-(aq)$$
$$K_{sp} = [Ag^+][I^-] = 8.5 \times 10^{-17}$$

ppose we mix solutions of $AgNO_3(aq)$ and KI(aq) to obtain a mixed solu-n that has $[Ag^+] = 0.010$ M and $[I^-] = 0.015$ M. Is this solution unsatu-ed, saturated, or supersaturated?

Recall the reaction quotient, Q, that was described in detail in Chapter 15. It s the same form as an equilibrium constant expression but uses initial con-ntrations rather than equilibrium concentrations. Initially,

$$Q_{sp} = [Ag^+]_{init} \times [I^-]_{init} = (0.010)(0.015) = 1.5 \times 10^{-4} > K_{sp}$$

e fact that $Q_{sp} > K_{sp}$ indicates that the concentrations of Ag^+ and I^- are eady higher than they would be in a saturated solution and that a net ange should occur to the left. The solution is *supersaturated*. As is generally e case with supersaturated solutions, excess AgI should precipitate from ution. If we had found $Q_{sp} < K_{sp}$, the solution would have been *unsaturated*. precipitate would form from such a solution.

When applied to solubility equilibria, $\mathbf{Q_{sp}}$ is generally called the **ion product** cause its form is that of the product of ion concentrations raised to appropri-e powers. The criteria for determining whether ions in a solution will com-ne to form a precipitate require us to compare the ion product with K_{sp}.

- Precipitation *should occur* if $Q_{sp} > K_{sp}$.
- Precipitation *cannot occur* if $Q_{sp} < K_{sp}$.
- A solution is just saturated if $Q_{sp} = K_{sp}$.

These criteria are illustrated in Figure 18-3 and Example 18-5. The example phasizes the important point that *any possible dilutions must be considered ore the criteria for precipitation are applied.*

Precipitation of a solute is considered to be complete only if the amount of ute remaining in solution is very small. An arbitrary rule of thumb is that

(a) (b)

FIGURE 18-3
pplying the criteria for precipitation from solution—Example 18-5 ustrated
When three drops of 0.20 M KI are first added to 100.0 mL of 0.010 M $Pb(NO_3)_2$, precipitate forms because K_{sp} is exceeded in the immediate vicinity of the drops.) When the KI becomes uniformly mixed in the $Pb(NO_3)_2(aq)$, K_{sp} is no longer ceeded and the precipitate redissolves. The criteria for precipitation must be plied *after* dilution has occurred.

precipitation is complete if 99.9% or more of a particular ion has precipitate leaving less than 0.1% of the ion in solution. In Example 18-6, we will calcula the concentration of Mg^{2+} that remains in a solution from which $Mg(OH)_2($ has precipitated. We will compare this remaining $[Mg^{2+}]$ to the initial $[Mg^{2+}]$ determine the completeness of the precipitation.

EXAMPLE 18-5 Applying the Criteria for Precipitation of a Slightly Soluble Solute

Three drops of 0.20 M KI are added to 100.0 mL of 0.010 M $Pb(NO_3)_2$. Will a precipitate of lead(II) iodide form? (Assume 1 drop = 0.05 mL.)

$$PbI_2(s) \rightleftharpoons Pb^{2+}(aq) + 2\,I^-(aq) \qquad K_{sp} = 7.1 \times 10^{-9}$$

Analyze

We need to compare the product $[Pb^{2+}][I^-]^2$ formulated for the initial concentrations with the K_{sp} for PbI_2. For $[Pb^{2+}]$, the dilution caused by adding 3 drops (0.15 mL = 0.00015 L) to 0.100 L of solution is negligible and so we can simply use 0.010 M. For $[I^-]$, however, we must consider the great reduction in concentration that occurs when the three drops of 0.20 M KI are diluted to 100.0 mL.

Solve

Dilution Calculation:

$$\text{amount I}^- = 3\text{ drops} \times \frac{0.05\text{ mL}}{1\text{ drop}} \times \frac{1\text{ L}}{1000\text{ mL}} \times \frac{0.20\text{ mol KI}}{\text{L}} \times \frac{1\text{ mol I}^-}{1\text{ mol KI}} = 3 \times 10^{-5}\text{ mol I}^-$$

$$\text{Total volume} = 0.1000\text{ L} + \left(3\text{ drops} \times \frac{0.05\text{ mL}}{\text{drop}} \times \frac{1\text{ L}}{1000\text{ mL}}\right) = 0.1002\text{ L}$$

$$[I^-] = \frac{3 \times 10^{-5}\text{ mol I}^-}{0.1002\text{ L}} = 3 \times 10^{-4}\text{ M}$$

Applying Precipitation Criteria:

$$Q_{sp} = [Pb^{2+}][I^-]^2 = (0.010)(3 \times 10^{-4})^2 = 9 \times 10^{-10}$$

Because a Q_{sp} of 9×10^{-10} is *smaller than* a K_{sp} of 7.1×10^{-9}, we conclude that $PbI_2(s)$ should *not* precipitate.

Assess

In this calculation, as well as problems similar to this one, any possible dilutions must be considered *before* the criteria for precipitation are applied. Note that the addition of the three drops of KI(aq) does not change the value of $[Pb^{2+}]$: 0.010 M × (0.1000 L/0.1002 L) = 0.010 M (two significant figures).

PRACTICE EXAMPLE A: Three drops of 0.20 M KI are added to 100.0 mL of a 0.010 M solution of $AgNO_3$. Will a precipitate of silver iodide form?

PRACTICE EXAMPLE B: We saw in Example 18-5 that a 3-drop volume of 0.20 M KI is insufficient to cause precipitation in 100.0 mL of 0.010 M $Pb(NO_3)_2$. What minimum number of drops would be required to produce the first precipitate?

EXAMPLE 18-6 Assessing the Completeness of a Precipitation Reaction

The first step in a commercial process in which magnesium is obtained from seawater involves precipitating Mg^{2+} as $Mg(OH)_2(s)$. The magnesium ion concentration in seawater is about 0.059 M. If a seawater sample is treated so that its $[OH^-]$ is maintained at 2.0×10^{-3} M, **(a)** what will be $[Mg^{2+}]$ remaining in solution when precipitation stops ($K_{sp} = 1.8 \times 10^{-11}$)? **(b)** Is the precipitation of $Mg(OH)_2(s)$ complete under these conditions?

Analyze

In part (a) of this example we compare the Q_{sp} of the solution with the known K_{sp} for $Mg(OH)_2$ to determine whether precipitation will occur. If it will, then precipitation of $Mg(OH)_2(s)$ will continue as long as the ion product exceeds K_{sp} and will stop when that product is equal to K_{sp}. At the point at which the ion product equals K_{sp}, whatever Mg^{2+} is in solution remains in solution. In part (b) we need to compare the amount of Mg^{2+} remaining after precipitation with the original amount.

Solve

(a) There is no question that precipitation will occur, because the ion product $Q_{sp} = [Mg^{2+}][OH^-]^2 = (0.059)(2.0 \times 10^{-3})^2 = 2.4 \times 10^{-7}$ exceeds K_{sp}.

$$[Mg^{2+}][OH^-]^2 = [Mg^{2+}](2.0 \times 10^{-3})^2 = 1.8 \times 10^{-11} = K_{sp}$$

$$[Mg^{2+}]_{remaining} = \frac{1.8 \times 10^{-11}}{(2.0 \times 10^{-3})^2} = 4.5 \times 10^{-6}\ M$$

(b) $[Mg^{2+}]$ in seawater is reduced from 0.059 M to 4.5×10^{-6} M as a result of the precipitation reaction. Expressed as a percentage,

$$\%[Mg^{2+}]_{remaining} = \frac{4.5 \times 10^{-6}\ M}{0.059\ M} \times 100\% = 0.0076\%$$

Because less than 0.1% of the Mg^{2+} remains, we conclude that precipitation is essentially complete.

Assess

Because $[OH^-]$ is maintained at a constant value, the calculation of $[Mg^{2+}]_{remaining}$ is straightforward. (A method of maintaining a constant $[OH^-]$ during a precipitation is to carry out the precipitation from a buffer solution.) If $[OH^-]$ was initially 2.0×10^{-3} M but with no source of OH^- to replenish it, $[Mg^{2+}]_{remaining}$ would be 0.058 M. Verify this result yourself.

PRACTICE EXAMPLE A: A typical Ca^{2+} concentration in seawater is 0.010 M. Will the precipitation of $Ca(OH)_2$ be complete from a seawater sample in which $[OH^-]$ is maintained at 0.040 M?

PRACTICE EXAMPLE B: What $[OH^-]$ should be maintained in a solution if, after precipitation of Mg^{2+} as $Mg(OH)_2(s)$, the remaining Mg^{2+} is to be at a level of 1 μg Mg^{2+}/L?

18-2 ARE YOU WONDERING?

What conditions favor completeness of precipitation?

The key factors in determining whether the target ion is essentially completely removed from solution in a precipitation are (1) the value of K_{sp}, (2) the initial concentration of the target ion, and (3) the concentration of the common ion. In general, completeness of precipitation is favored by

- *a very small value of* K_{sp}. (The concentration of the target ion remaining in solution will be very small.)
- *a high initial concentration of the target ion*. (The concentration of the target ion remaining in solution will be only a very small fraction of the initial value.)
- *a concentration of common ion much larger than that of the target ion*. (The common-ion concentration will remain nearly constant during the precipitation.)

8-6 Fractional Precipitation

a large excess of $AgNO_3(s)$ is added to a solution containing the ions CrO_4^{2-} nd Br^-, a mixed precipitate of $Ag_2CrO_4(s)$ and AgBr(s) is obtained. There is a ethod of adding $AgNO_3$, however, that will cause AgBr(s) to precipitate but ave CrO_4^{2-} in solution. That method is fractional precipitation.

Fractional precipitation is a technique in which two or more ions in soluon, each capable of being precipitated by the same reagent, are separated by e proper use of that reagent: *One ion is precipitated, while the other(s) remains solution.* The primary condition for a successful fractional precipitation is

◀ Fractional precipitation is also called *selective precipitation.*

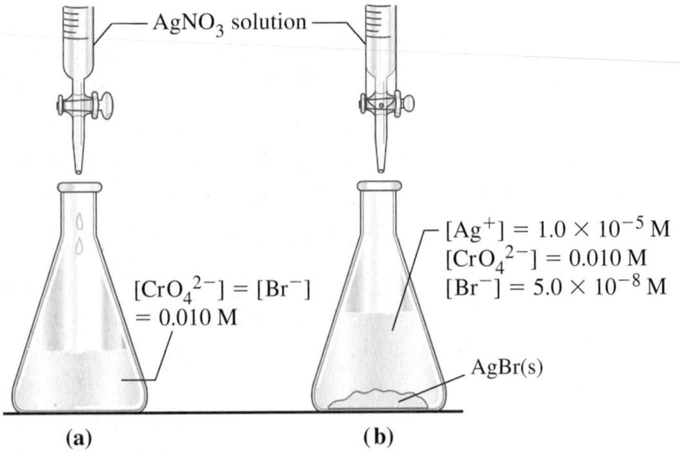

▶ FIGURE 18-4
Fractional precipitation—Example 18-7 illustrated
(a) $AgNO_3(aq)$ is slowly added to a solution that is 0.010 M in Br^- and 0.010 M in CrO_4^{2-}. **(b)** Essentially all the Br^- has precipitated as pale yellow AgBr(s), with $[Br^-] = 5.0 \times 10^{-8}$ M in solution. Red-brown $Ag_2CrO_4(s)$ is just about to precipitate.

that there be a significant difference in the solubilities of the substances bein separated. (Usually this means a significant difference in their K_{sp} values. The key to the technique is the slow addition of a concentrated solution of th precipitating reagent to the solution from which precipitation is to occur, a from a buret (Fig. 18-4).

Example 18-7 considers the separation of $CrO_4^{2-}(aq)$ and $Br^-(aq)$ throug the use of $Ag^+(aq)$.

EXAMPLE 18-7 Separating Ions by Fractional Precipitation

$AgNO_3(aq)$ is slowly added to a solution that has $[CrO_4^{2-}] = 0.010$ M and $[Br^-] = 0.010$ M.

(a) Show that AgBr(s) should precipitate before $Ag_2CrO_4(s)$ does.

(b) When $Ag_2CrO_4(s)$ begins to precipitate, what is $[Br^-]$ remaining in solution?

(c) Is complete separation of $Br^-(aq)$ and $CrO_4^{2-}(aq)$ by fractional precipitation feasible?

Analyze

First, write equations for the solubility equilibria and look up the corresponding K_{sp} values.

$$Ag_2CrO_4(s) \rightleftharpoons 2\,Ag^+(aq) + CrO_4^{2-}(aq) \qquad K_{sp} = 1.1 \times 10^{-12}$$
$$AgBr(s) \rightleftharpoons Ag^+(aq) + Br^-(aq) \qquad K_{sp} = 5.0 \times 10^{-13}$$

Next we determine the concentration of silver ion needed to precipitate either AgBr(s) or $Ag_2CrO_4(s)$. The one that requires the least amount of silver ion will precipitate first. The formation of the first precipitate helps keep $[Ag^+]$ below that required to form the second precipitate; $[Ag^+]$ will, however, slowly increase and, eventually, the second precipitate will form. We can use the value of $[Ag^+]$ at the point at which the second precipitate starts to form to calculate $[Br^-]$ and $[CrO_4^{2-}]$ and determine whether complete separation is possible (i.e., when 99.9% of the bromide ion has precipitated).

Solve

(a) The required values of $[Ag^+]$ for precipitation to start are

AgBr ppt: $Q_{sp} = [Ag^+][Br^-] = [Ag^+](0.010) = 5.0 \times 10^{-13} = K_{sp}$

$[Ag^+] = 5.0 \times 10^{-11}$ M

Ag_2CrO_4 ppt: $Q_{sp} = [Ag^+]^2[CrO_4^{2-}] = [Ag^+]^2(0.010)$

$= 1.1 \times 10^{-12} = K_{sp}$

$[Ag^+]^2 = 1.1 \times 10^{-10}$ and $[Ag^+] = 1.0 \times 10^{-5}$ M

Because the $[Ag^+]$ required to start the precipitation of AgBr(s) is much less than that for $Ag_2CrO_4(s)$, AgBr(s) precipitates first. As long as AgBr(s) is forming, the silver ion concentration can only slowly approach the value required for the precipitation of $Ag_2CrO_4(s)$.

(b) As AgBr(s) precipitates, $[Br^-]$ gradually decreases, and this permits $[Ag^+]$ to increase. When $[Ag^+]$ reaches 1.0×10^{-5} M, precipitation of $Ag_2CrO_4(s)$ begins. To determine $[Br^-]$ at the point at which $[Ag^+] = 1.0 \times 10^{-5}$ M, we use K_{sp} for AgBr and solve for $[Br^-]$.

$$K_{sp} = [Ag^+][Br^-] = (1.0 \times 10^{-5})[Br^-] = 5.0 \times 10^{-13}$$

$$[Br^-] = \frac{5.0 \times 10^{-13}}{1.0 \times 10^{-5}} = 5.0 \times 10^{-8}\,\text{M}$$

(c) Before $Ag_2CrO_4(s)$ begins to precipitate, $[Br^-]$ will have been reduced from 1.0×10^{-2} M to 5.0×10^{-8} M. Essentially, all the Br^- will have precipitated from solution as AgBr(s), whereas the CrO_4^{2-} remains in solution. Fractional precipitation is feasible for separating mixtures of Br^- and CrO_4^{2-}.

Assess

Even though the solubility products of these two compounds have similar values, we were able to separate the two ions. The concept of fractional precipitation is used to separate and identify unknown ions in solution.

PRACTICE EXAMPLE A: $AgNO_3(aq)$ is slowly added to a solution with $[Cl^-] = 0.115$ M and $[Br^-] = 0.264$ M. What percent of the Br^- remains unprecipitated at the point at which AgCl(s) begins to precipitate?

AgCl: $K_{sp} = 1.8 \times 10^{-10}$ AgBr: $K_{sp} = 5.0 \times 10^{-13}$

PRACTICE EXAMPLE B: A solution has $[Ba^{2+}] = [Sr^{2+}] = 0.10$ M. Use data from Appendix D to choose the best precipitating agent to separate these two ions. What is the concentration of the first ion to precipitate when the second ion begins to precipitate?

18-3 CONCEPT ASSESSMENT

In the fractional precipitation pictured in Figure 18-4 and described in Example 18-7, no mention was made of the concentration of the $AgNO_3(aq)$ used. Is this concentration immaterial? Explain.

8-7 Solubility and pH

he pH of a solution can affect the solubility of a salt to a large degree. That is specially true when the anion of the salt is the conjugate base of a weak acid r the base OH^- itself. An interesting example is the highly insoluble 1g(OH)$_2$(s), which, when suspended in water, is the popular antacid known s milk of magnesia. A **suspension** is a heterogeneous fluid containing solid articles that are sufficiently large for sedimentation and, unlike colloids, will ettle. Hydroxide ions from the dissolved magnesium hydroxide react with ydronium ions (in stomach acid) to form water.

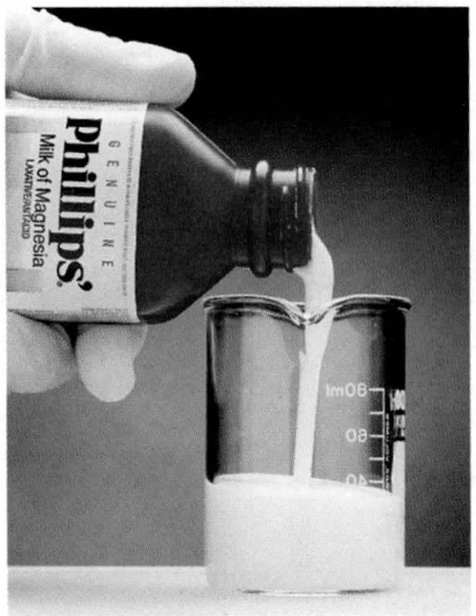

Richard Megna/Fundamental Photographs

▲ Milk of magnesia, an aqueous suspension of $Mg(OH)_2(s)$.

$$Mg(OH)_2(s) \rightleftharpoons Mg^{2+}(aq) + 2\,OH^-(aq) \qquad \textbf{(18.2)}$$

$$OH^-(aq) + H_3O^+(aq) \longrightarrow 2\,H_2O(l) \qquad \textbf{(18.3)}$$

According to Le Châtelier's principle, we expect reaction (18.2) to be dis- laced to the right—that is, additional $Mg(OH)_2$ would dissolve to replace)H$^-$ ions drawn off by the neutralization reaction (18.3). The overall net ionic quation can be obtained by doubling equation (18.3) and adding it to equa- on (18.2). At the same time, we can apply the method of combining equilib- um constants (page 701). The result is

$$Mg(OH)_2(s) \rightleftharpoons Mg^{2+}(aq) + \cancel{2\,OH^-(aq)} \qquad K_{sp} = 1.8 \times 10^{-11}$$

$$\cancel{OH^-(aq)} + 2\,H_3O^+(aq) \rightleftharpoons 4\,H_2O(l) \qquad K' = 1/K_w^2 = 1.0 \times 10^{28}$$

$$Mg(OH)_2(s) + 2\,H_3O^+(aq) \rightleftharpoons Mg^{2+}(aq) + 4\,H_2O(l) \qquad \textbf{(18.4)}$$

$$K = K_{sp}/K_w^2 = 1.8 \times 10^{-11} \times 1.0 \times 10^{28} = 1.8 \times 10^{17}$$

▶ In general, a water-insoluble salt of a weak acid can be dissolved by acid, but not the water-insoluble salt of a strong acid.

▶ Equilibrium constant expressions can be combined. See Section 15-3.

The large value of K for reaction (18.4) indicates that a greater amount c $Mg(OH)_2$ will react (dissolve) in acidic solution than in pure water.

Other slightly soluble solutes having basic anions (such as $ZnCO_3$, MgF and CaC_2O_4) also dissolve to a greater extent in acidic solutions. For thes solutes, we can write overall net ionic equations for solubility equilibria an corresponding K values based on K_{sp} for the solutes and K_a for the conjugat acids of the anions.

Although $Mg(OH)_2$ dissolves in acidic solution, in moderately or strongl basic solutions it does not. In Example 18-8, we calculate $[OH^-]$ in a solution c the weak base NH_3 and then use the criteria for precipitation to see $Mg(OH)_2(s)$ will precipitate. In Example 18-9, we determine how to adjus $[OH^-]$ to prevent precipitation of $Mg(OH)_2(s)$. This adjustment is made b adding NH_4^+ to the $NH_3(aq)$, thereby converting it to a buffer solution.

18-4 CONCEPT ASSESSMENT

Which will be affected more by the addition of a strong acid or a strong base: the solubility of CaF_2 or the solubility of $CaCl_2$? Explain.

EXAMPLE 18-8 Determining Whether a Precipitate Will Form in a Solution in Which There Is Also an Ionization Equilibrium

Should $Mg(OH)_2(s)$ precipitate from a solution that is 0.010 M $MgCl_2$ and also 0.10 M NH_3?

Analyze

The key here is in understanding that $[OH^-]$ is established by the ionization of $NH_3(aq)$.

$$NH_3(aq) + H_2O(l) \rightleftharpoons NH_4^+(aq) + OH^-(aq) \qquad K_b = 1.8 \times 10^{-5}$$

Solve

If we set this up in the usual way, the equilibrium values are $[NH_4^+] = [OH^-] = x$ M and $[NH_3] = (0.10 - x)$ M $\approx$ 0.10 M. Then we obtain

$$K_b = \frac{[NH_4^+][OH^-]}{[NH_3]} = \frac{x \cdot x}{0.10} = 1.8 \times 10^{-5}$$

$$x^2 = 1.8 \times 10^{-6} \qquad x = 1.3 \times 10^{-3} \qquad [OH^-] = x\,M = 1.3 \times 10^{-3}\,M$$

Now we can rephrase the original question: Should $Mg(OH)_2(s)$ precipitate from a solution in which $[Mg^{2+}] = 1.0 \times 10^{-2}$ M and $[OH^-] = 1.3 \times 10^{-3}$ M? We must compare the ion product, Q_{sp}, with K_{sp}.

$$Q_{sp} = [Mg^{2+}][OH^-]^2 = (1.0 \times 10^{-2})(1.3 \times 10^{-3})^2$$
$$= 1.7 \times 10^{-8} > K_{sp} = 1.8 \times 10^{-11}$$

Precipitation should occur.

Assess

In this example the ionization equilibrium was easy to identify. Always look for anions and cations of salts that can establish an ionization equilibrium (e.g., through hydrolysis).

PRACTICE EXAMPLE A: Should $Mg(OH)_2(s)$ precipitate from a solution that is 0.010 M $MgCl_2(aq)$ and also 0.10 M $NaCH_3COO$? $K_{sp}[Mg(OH)_2] = 1.8 \times 10^{-11}$; $K_a(CH_3COOH) = 1.8 \times 10^{-5}$. [*Hint:* What equilibrium expression establishes $[OH^-]$ in the solution?]

PRACTICE EXAMPLE B: Will a precipitate of $Fe(OH)_3$ form from a solution that is 0.013 M Fe^{3+} in a buffer solution that is 0.150 M CH_3COOH and 0.250 M $NaCH_3COO$?

EXAMPLE 18-9 Controlling an Ion Concentration, Either to Cause Precipitation or to Prevent It

What $[NH_4^+]$ must be maintained to prevent precipitation of $Mg(OH)_2(s)$ from a solution that is 0.010 M $MgCl_2$ and 0.10 M NH_3?

Analyze

The maximum value of the ion product, Q_{sp}, before precipitation occurs is 1.8×10^{-11}, the value of K_{sp} for $Mg(OH)_2$. This fact allows us to determine the maximum concentration of OH^- that can be tolerated.

Solve

$$[Mg^{2+}][OH^-]^2 = (1.0 \times 10^{-2})[OH^-]^2 = 1.8 \times 10^{-11}$$
$$[OH^-]^2 = 1.8 \times 10^{-9}$$
$$[OH^-] = 4.2 \times 10^{-5}\ M$$

Next let's determine what $[NH_4^+]$ must be present in 0.10 M NH_3 to maintain $[OH^-] = 4.2 \times 10^{-5}$ M.

$$NH_3(aq) + H_2O(l) \rightleftharpoons NH_4^+(aq) + OH^-(aq) \qquad K_b = 1.8 \times 10^{-5}$$

$$K_b = \frac{[NH_4^+][OH^-]}{[NH_3]} = \frac{[NH_4^+](4.2 \times 10^{-5})}{0.10} = 1.8 \times 10^{-5}$$

$$[NH_4^+] = \frac{0.10 \times 1.8 \times 10^{-5}}{4.2 \times 10^{-5}} = 0.043\ M$$

To keep the $[OH^-]$ at 4.2×10^{-5} M *or less*, and thus prevent the precipitation of $Mg(OH)_2(s)$, $[NH_4^+]$ should be maintained at 0.043 M or *greater*.

Assess

The reason for keeping $[NH_4^+]$ *greater* than 0.043 M is that at higher concentrations of NH_4^+, the equilibrium for the ionization of NH_3 shifts to the left, reducing the amount of OH^-. Practically, one would choose 0.10 M since the ratio $[NH_3]/[NH_4^+]$ would be 1.0 and the buffer capacity of the solution is at a maximum.

PRACTICE EXAMPLE A: What minimum $[NH_4^+]$ must be present to prevent precipitation of $Mn(OH)_2(s)$ from a solution that is 0.0050 M $MnCl_2$ and 0.025 M NH_3? For $Mn(OH)_2$, $K_{sp} = 1.9 \times 10^{-13}$.

PRACTICE EXAMPLE B: What is the molar solubility of $Mg(OH)_2(s)$ in a solution that is 0.250 M NH_3 and 0.100 M NH_4Cl? [*Hint:* Use equation (18.4).]

18-5 CONCEPT ASSESSMENT

Determine $[Mg^{2+}]$ in a saturated solution of $Mg(OH)_2$ at **(a)** pH = 10.00 and **(b)** pH = 5.00. Is each of these a plausible quantity? Explain.

8-8 Equilibria Involving Complex Ions

s shown in Figure 18-5, when moderately concentrated $NH_3(aq)$ is added to saturated solution of silver chloride in contact with undissolved AgCl(s), the olid dissolves. The key to this dissolving action is that Ag^+ ions from AgCl ombine with NH_3 molecules to form the ions $[Ag(NH_3)_2]^+$, which, together ith Cl^- ions, remain in solution as the soluble compound $Ag(NH_3)_2Cl$.

$$AgCl(s) + 2\,NH_3(aq) \longrightarrow [Ag(NH_3)_2]^+(aq) + Cl^-(aq) \qquad \textbf{(18.5)}$$

◀ In this reaction, the Ag^+ ion from AgCl acts as a Lewis acid and the NH_3 molecule act as a Lewis base.

ere $[Ag(NH_3)_2]^+$ is called a complex ion, and $Ag(NH_3)_2Cl$ is called a coordi- ation compound. A **complex ion** is a polyatomic cation or anion composed of a entral metal ion to which other groups (molecules or ions) called *ligands* are

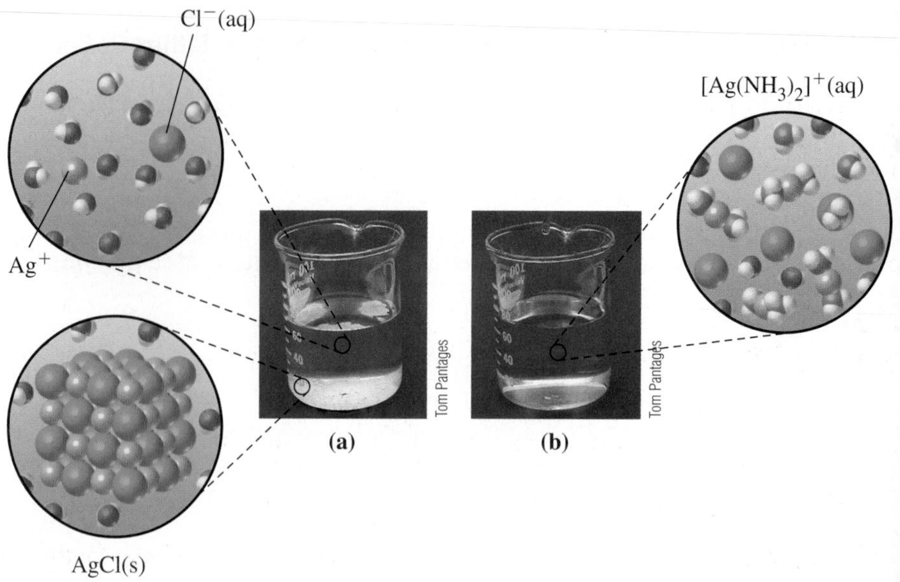

▲ FIGURE 18-5
Complex-ion formation: dissolution of AgCl(s) in NH_3(aq)
(a) A saturated solution of silver chloride in contact with excess AgCl(s). **(b)** When NH_3(aq) is added, the excess AgCl(s) dissolves through the formation of the complex ion $[Ag(NH_3)_2]^+$.

bonded. **Coordination compounds** are substances containing complex ions. helps to think of reaction (18.5) as involving two simultaneous equilibria.

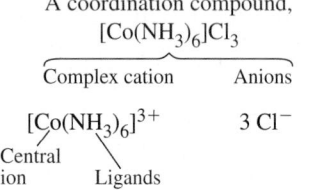

A coordination compound, $[Co(NH_3)_6]Cl_3$

Complex cation: $[Co(NH_3)_6]^{3+}$ (Central ion, Ligands) — Anions: 3 Cl^-

$$AgCl(s) \rightleftharpoons Ag^+(aq) + Cl^-(aq) \quad \text{(18.6}$$
$$Ag^+(aq) + 2\,NH_3(aq) \rightleftharpoons [Ag(NH_3)_2]^+(aq) \quad \text{(18.7}$$

The equilibrium in reaction (18.7) is shifted far to the right—$[Ag(NH_3)_2]^+$ is stable complex ion. The equilibrium concentration of Ag^+(aq) in (18.7) is kep so low that the ion product $[Ag^+][Cl^-]$ fails to exceed K_{sp} and AgCl remains i solution. Let's first apply additional qualitative reasoning of this sort i Example 18-10. Then we can turn to some of the quantitative calculations tha are also possible.

To describe the ionization of a weak acid, we use the ionization constant K For a solubility equilibrium, we use the solubility product constant K_{sp}. Th equilibrium constant that is used to deal with a complex-ion equilibrium i called the formation constant. The **formation constant, K_f,** of a complex ion i the equilibrium constant describing the formation of a complex ion from central ion and its attached groups. For reaction (18.7) this equilibrium con stant expression is

$$K_f = \frac{[[Ag(NH_3)_2]^+]}{[Ag^+][NH_3]^2} = 1.6 \times 10^7$$

Table 18.2 lists some representative formation constants, K_f.

One feature that distinguishes K_f from most other equilibrium constant previously considered is that K_f values are usually *large* numbers. This fact ca affect the way we approach certain calculations. With large K values, it i sometimes convenient to solve a problem in two steps. First, assume that th forward reaction goes to completion; second, assume that a small chang occurs in the reverse direction to establish the equilibrium. This approach i demonstrated in Example 18-11.

EXAMPLE 18-10 Predicting Reactions Involving Complex Ions

Predict what will happen if nitric acid is added to a solution of $[Ag(NH_3)_2]Cl$ in $NH_3(aq)$.

Analyze

We consider the equilibria represented by equations (18.6) and (18.7) and use Le Châtelier's principle to assess the effect of adding HNO_3 to the solution. An important consideration is that nitric acid will protonate the base, NH_3.

Solve

Since $HNO_3(aq)$ is a strong acid, we will represent the acid as $H_3O^+(aq)$ and write the protonation reaction as

$$H_3O^+(aq) + NH_3(aq) \longrightarrow NH_4^+(aq) + H_2O(l)$$

The formation reaction is

$$Ag^+(aq) + 2\,NH_3(aq) \rightleftharpoons [Ag(NH_3)_2]^+(aq)$$

To replace free NH_3 lost in this neutralization, equilibrium in the formation reaction shifts to the *left*. As a result, $[Ag^+]$ increases. When $[Ag^+]$ increases to the point at which the ion product $[Ag^+][Cl^-]$ exceeds K_{sp}, AgCl(s) precipitates (Fig. 18-6).

Assess

The addition of any acid to a solution of $[Ag(NH_3)_2]Cl$ will result in precipitation.

PRACTICE EXAMPLE A: Copper(II) ion forms both an insoluble hydroxide and the complex ion $[Cu(NH_3)_4]^{2+}$. Write equations to represent the expected reaction when **(a)** $CuSO_4(aq)$ and NaOH(aq) are mixed; **(b)** an excess of $NH_3(aq)$ is added to the product of part (a); and **(c)** an excess of $HNO_3(aq)$ is added to the product of part (b).

PRACTICE EXAMPLE B: Zinc(II) ion forms both an insoluble hydroxide and the complex ions $[Zn(OH)_4]^{2-}$ and $[Zn(NH_3)_4]^{2+}$. Write four equations to represent the reactions of **(a)** $NH_3(aq)$ with $ZnSO_4(aq)$, followed by **(b)** enough $HNO_3(aq)$ to make the product of part (a) acidic; **(c)** enough NaOH(aq) to make the product of part (b) slightly basic; **(d)** enough NaOH(aq) to make the product of part (c) strongly basic.

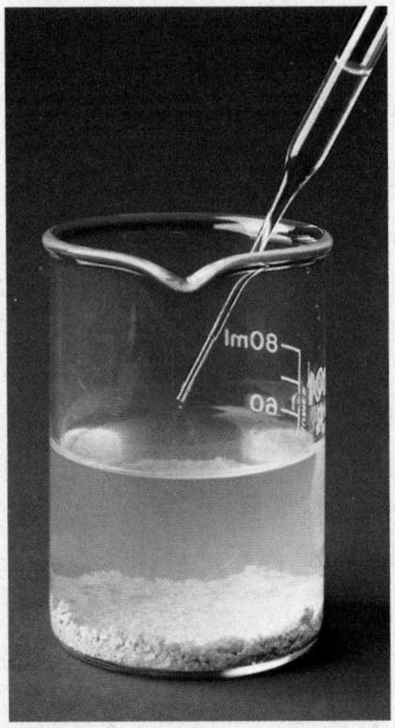

Richard Megna/Fundamental Photographs

▲ FIGURE 18-6
Reprecipitating AgCl(s)
The reagent being added to the solution containing $[Ag(NH_3)_2]^+$ and Cl^- is $HNO_3(aq)$. H_3O^+ from the acid reacts with $NH_3(aq)$ to form $NH_4^+(aq)$. As a result, equilibrium between $[Ag(NH_3)_2]^+$, Ag^+, and NH_3 is upset. The complex ion is destroyed, $[Ag^+]$ quickly rises to the point at which K_{sp} for AgCl is exceeded, and a precipitate forms.

Table 18.2 Formation Constants for Some Complex Ions[a]

Complex Ion	Equilibrium Reaction[b]	K_f
$[Co(NH_3)_6]^{3+}$	$Co^{3+} + 6\,NH_3 \rightleftharpoons [Co(NH_3)_6]^{3+}$	4.5×10^{33}
$[Cu(NH_3)_4]^{2+}$	$Cu^{2+} + 4\,NH_3 \rightleftharpoons [Cu(NH_3)_4]^{2+}$	1.1×10^{13}
$[Fe(CN)_6]^{4-}$	$Fe^{2+} + 6\,CN^- \rightleftharpoons [Fe(CN)_6]^{4-}$	1×10^{37}
$[Fe(CN)_6]^{3-}$	$Fe^{3+} + 6\,CN^- \rightleftharpoons [Fe(CN)_6]^{3-}$	1×10^{42}
$[Pb(OH)_3]^-$	$Pb^{2+} + 3\,OH^- \rightleftharpoons [Pb(OH)_3]^-$	3.8×10^{14}
$[PbCl_3]^-$	$Pb^{2+} + 3\,Cl^- \rightleftharpoons [PbCl_3]^-$	2.4×10^1
$[Ag(NH_3)_2]^+$	$Ag^+ + 2\,NH_3 \rightleftharpoons [Ag(NH_3)_2]^+$	1.6×10^7
$[Ag(CN)_2]^-$	$Ag^+ + 2\,CN^- \rightleftharpoons [Ag(CN)_2]^-$	5.6×10^{18}
$[Ag(S_2O_3)_2]^{3-}$	$Ag^+ + 2\,S_2O_3^{2-} \rightleftharpoons [Ag(S_2O_3)_2]^{3-}$	1.7×10^{13}
$[Zn(NH_3)_4]^{2+}$	$Zn^{2+} + 4\,NH_3 \rightleftharpoons [Zn(NH_3)_4]^{2+}$	4.1×10^8
$[Zn(CN)_4]^{2-}$	$Zn^{2+} + 4\,CN^- \rightleftharpoons [Zn(CN)_4]^{2-}$	1×10^{18}
$[Zn(OH)_4]^{2-}$	$Zn^{2+} + 4\,OH^- \rightleftharpoons [Zn(OH)_4]^{2-}$	4.6×10^{17}

[a]A more extensive tabulation is given in Appendix D.
[b]Tabulated here are *overall* formation reactions and the corresponding *overall* formation constants. In Section 24-8, we describe the formation of complex ions in a *stepwise* fashion and introduce formation constants for individual steps.

EXAMPLE 18-11 Determining Whether a Precipitate Will Form in a Solution Containing Complex Ions

A 0.10 mol sample of $AgNO_3$ is dissolved in 1.00 L of 1.00 M NH_3. If 0.010 mol NaCl is added to this solution, will AgCl(s) precipitate?

Analyze

We begin by first assuming that because the value of K_f for $[Ag(NH_3)_2]^+$ is very large, the formation reaction goes to completion. Then we use those results to perform an equilibrium calculation. Finally, from the equilibrium calculation we can determine a Q_{sp} for the reaction and determine whether precipitation will occur.

Solve

Assuming the formation reaction goes to completion, we obtain

	$Ag^+(aq)$	$+ 2\,NH_3(aq)$	$\longrightarrow [Ag(NH_3)_2]^+(aq)$
initial concns, M:	0.10	1.00	
changes, M:	−0.10	−0.20	+0.10
after reaction, M:	≈0	0.80	0.10

However, the concentration of uncomplexed silver ion, though very small, is not zero. To determine the value of $[Ag^+]$, let's start with $[[Ag(NH_3)_2]^+]$ and $[NH_3]$ in solution and establish $[Ag^+]$ at equilibrium.

	Ag^+	$+ 2\,NH_3$	$\rightleftharpoons [Ag(NH_3)_2]^+$
initial concns, M:	0	0.80	0.10
changes, M:	$+x$	$+2x$	$-x$
equil concns, M:	x	$0.80 + 2x$	$0.10 - x$

When substituting into the following expression, we make the assumption that $x \ll 0.10$, which we will find to be the case.

$$\frac{[[Ag(NH_3)_2]^+]}{[Ag^+][NH_3]^2} = \frac{0.10 - x}{x(0.80 + 2x)^2} \approx \frac{0.10}{x(0.80)^2} = 1.6 \times 10^7$$

$$x = [Ag^+] = \frac{0.10}{(1.6 \times 10^7)(0.80)^2} = 9.8 \times 10^{-9}\,M$$

Finally, we must compare $Q_{sp} = [Ag^+][Cl^-]$ with K_{sp} for AgCl (that is, 1.8×10^{-10}). $[Ag^+]$ is the value of x that we just calculated. Because the solution contains 0.010 mol NaCl/L, $[Cl^-] = 0.010\,M = 1.0 \times 10^{-2}\,M$, and

$$Q_{sp} = (9.8 \times 10^{-9})(1.0 \times 10^{-2}) = 9.8 \times 10^{-11} < 1.8 \times 10^{-10}$$

AgCl will not precipitate.

Assess

The formation reaction goes almost to completion, with only a small amount of silver ion remaining in solution. Our initial assumption that the formation reaction goes to completion was valid. The amount of silver ion remaining is not enough to cause precipitation.

PRACTICE EXAMPLE A: Will AgCl(s) precipitate from 1.50 L of a solution that is 0.100 M $AgNO_3$ and 0.225 M NH_3 if 1.00 mL of 3.50 M NaCl is added? [*Hint:* What are $[Ag^+]$ and $[Cl^-]$ immediately after the addition of the 1.00 mL of 3.50 M NaCl? Take into account the dilution of the NaCl(aq), but assume the total volume remains at 1.50 L.]

PRACTICE EXAMPLE B: A solution is prepared that is 0.100 M in $Pb(NO_3)_2$ and 0.250 M in the ethylenediaminetetraacetate anion, $EDTA^{4-}$. Together, Pb^{2+} and $EDTA^{4-}$ form the complex ion $[PbEDTA]^{2-}$. If the solution is also made 0.10 M in I^-, will PbI_2(s) precipitate? For PbI_2, $K_{sp} = 7.1 \times 10^{-9}$; for $[PbEDTA]^{2-}$, $K_f = 2 \times 10^{18}$.

Just as some precipitation reactions can be controlled by using a buffer solution (Example 18-9), precipitation from a solution of complex ions can be controlled by fixing the concentration of the complexing agent. Such a case i
demonstrated for AgCl in Example 18-12.

EXAMPLE 18-12 Controlling a Concentration to Cause or Prevent Precipitation from a Solution of Complex Ions

What is the *minimum* concentration of NH_3 needed to prevent AgCl(s) from precipitating from 1.00 L of a solution containing 0.10 mol $AgNO_3$ and 0.010 mol NaCl?

Analyze

To prevent AgCl(s) from precipitating, we must ensure that the solubility product for AgCl is not exceeded. We are given a fixed amount of chloride ion in solution, and so that means we need to determine the maximum concentration of silver ion that can exist without precipitation occurring. Finally, we can solve for the amount of NH_3 necessary to complex all the silver ion.

Solve

Solve for the amount of silver ion that can be in solution without precipitation in a solution containing 1.0×10^{-2} M Cl^-. That is, $[Ag^+][Cl^-] \le K_{sp}$.

$$[Ag^+](1.0 \times 10^{-2}) \le K_{sp} = 1.8 \times 10^{-10} \qquad [Ag^+] \le 1.8 \times 10^{-8}\ \text{M}$$

Thus, the maximum concentration of *uncomplexed* Ag^+ in solution is 1.8×10^{-8} M. This means that essentially all the Ag^+ (0.10 mol/L) must be tied up (complexed) in the complex ion $[Ag(NH_3)_2]^+$. We need to solve the expression at the right for $[NH_3]$.

$$K_f = \frac{[[Ag(NH_3)_2]^+]}{[Ag^+][NH_3]^2} = \frac{1.0 \times 10^{-1}}{1.8 \times 10^{-8}[NH_3]^2} = 1.6 \times 10^7$$

$$[NH_3]^2 = \frac{1.0 \times 10^{-1}}{1.8 \times 10^{-8} \times 1.6 \times 10^7} = 0.35 \qquad [NH_3] = 0.59\ \text{M}$$

The concentration just calculated is that of *free, uncomplexed* NH_3. Considering as well the 0.20 mol NH_3/L complexed in the 0.10 M $[Ag(NH_3)_2]^+$, we see that the total concentration of NH_3(aq) required is

$$[NH_3]_{tot} = 0.59\ \text{M} + 0.20\ \text{M} = 0.79\ \text{M}$$

Assess

In this example the precipitation from a solution is controlled by using a sufficiently large concentration of a complexing agent (i.e., NH_3).

PRACTICE EXAMPLE A: What $[NH_3]_{tot}$ is necessary to keep AgCl from precipitating from a solution that is 0.13 M $AgNO_3$ and 0.0075 M NaCl?

PRACTICE EXAMPLE B: What minimum concentration of thiosulfate ion, $S_2O_3^{2-}$, should be present in 0.10 M $AgNO_3$(aq) so that AgCl(s) does not precipitate when the solution is also made 0.010 M in Cl^-? For AgCl, $K_{sp} = 1.8 \times 10^{-10}$; for $[Ag(S_2O_3)_2]^{3-}$, $K_f = 1.7 \times 10^{13}$.

On page 845, there is a qualitative description of how the solubility of AgCl ıcreases in the presence of NH_3(aq). Example 18-13 shows how to calculate ıe actual solubility of AgCl in NH_3(aq).

18-6 CONCEPT ASSESSMENT

The measured molar solubility of AgCl in water is 1.3×10^{-5} M. In the presence of Cl^-(aq) as a common ion at different concentrations, the measured solubilities of AgCl are as listed below.

$[Cl^-]$, M:	0.0039	0.036	0.35	1.4	2.9	3.8
Solubility of AgCl, M:	7.2×10^{-7}	1.9×10^{-6}	1.7×10^{-5}	1.8×10^{-4}	1.0×10^{-2}	2.5×10^{-2}

Give a plausible explanation for the trends in the molar solubilities of the AgCl.

EXAMPLE 18-13 Determining the Solubility of a Solute When Complex Ions Form

What is the molar solubility of AgCl in 0.100 M $NH_3(aq)$?

Analyze

We need to have the equilibrium constant for the reaction that forms $[Ag(NH_3)_2]^+$ from AgCl(s) and $NH_3(aq)$. We can determine the equilibrium constant from the product of K_{sp} for AgCl(s) and K_f for the formation of $[Ag(NH_3)_2]^+$. By using that equilibrium constant, we can then determine the molar solubility.

Solve

As we have already seen, equation (18.5) describes the solubility equilibrium.

$$AgCl(s) + 2\,NH_3(aq) \rightleftharpoons [Ag(NH_3)_2]^+(aq) + Cl^-(aq) \qquad (18.5)$$

Let's base our calculation on the equilibrium constant K for reaction (18.5). There are two ways to obtain this value. By one method, equation (18.5) is written as the sum of equations (18.6) and (18.7) on page 846. Then its K value is obtained as the product of a K_{sp} and a K_f.

$$AgCl(s) \rightleftharpoons \cancel{Ag^+(aq)} + Cl^-(aq) \qquad K_{sp} = 1.8 \times 10^{-10}$$
$$\cancel{Ag^+(aq)} + 2\,NH_3(aq) \rightleftharpoons [Ag(NH_3)_2{}^+](aq) \qquad K_f = 1.6 \times 10^7$$
$$AgCl(s) + 2\,NH_3(aq) \rightleftharpoons [Ag(NH_3)_2{}^+](aq) + Cl^-(aq)$$
$$K = K_{sp} \times K_f = 1.8 \times 10^{-10} \times 1.6 \times 10^7 = 2.9 \times 10^{-3}$$

By a second method, the equilibrium constant expression for reaction (18.5) is written first, and then the numerator and denominator are multiplied by $[Ag^+]$.

$$K = \frac{[Ag(NH_3)_2{}^+][Cl^-]}{[NH_3]^2} = \frac{[Ag(NH_3)_2{}^+][Cl^-][Ag^+]}{[NH_3]^2[Ag^+]} = K_f \times K_{sp} = 2.9 \times 10^{-3}$$

The expression printed in red is K_f for $[Ag(NH_3)_2]^+$; the one in blue is K_{sp} for AgCl. The K value for reaction (18.5) is the product of the two.

According to equation (18.5), if s mol AgCl/L dissolves (the molar solubility), the expected concentrations of $[Ag(NH_3)_2]^+$ and Cl^- are also equal to s.

$$K = \frac{[[Ag(NH_3)_2]^+][Cl^-]}{[NH_3]^2} = \frac{s \cdot s}{(0.100 - 2s)^2} = \left(\frac{s}{0.100 - 2s}\right)^2 = 2.9 \times 10^{-3}$$

We can solve this equation by taking the square root of both sides.

$$\frac{s}{0.100 - 2s} = \sqrt{2.9 \times 10^{-3}} = 5.4 \times 10^{-2}$$

The molar solubility of AgCl(s) in 0.100 M $NH_3(aq)$ is 4.9×10^{-3} M.

$$s = 5.4 \times 10^{-3} - 0.11s$$
$$1.11s = 5.4 \times 10^{-3}$$
$$s = 4.9 \times 10^{-3}$$

Assess

The usual simplifying assumption, that is, $(0.100 - 2s) \approx 0.100$, would not have worked well in this calculation. If the simplifying assumption had been made, the value of s obtained would have been 5.4×10^{-3} and $2s$ would have been 10.8% of 0.100. That is, $0.100 - (2 \times 0.0054) \neq 0.100$. Also, the molar solubility is actually the *total* concentration of silver in solution: $[Ag^+] + [[Ag(NH_3)_2]^+]$. Only when K_f is large and the concentration of complexing agent sufficiently high can we ignore the concentration of uncomplexed metal ion, as was the case here.

PRACTICE EXAMPLE A: What is the molar solubility of $Fe(OH)_3$ in a solution containing 0.100 M $C_2O_4{}^{2-}$? For $[Fe(C_2O_4)_3]^{3-}$, $K_f = 2 \times 10^{20}$.

PRACTICE EXAMPLE B: *Without doing detailed calculations*, show that the order of *decreasing* solubility in 0.100 M $NH_3(aq)$ should be AgCl > AgBr > AgI.

18-9 Qualitative Cation Analysis

n *qualitative analysis*, we determine what substances are present in a mixture ut *not* their quantities. An analysis that aims at identifying the cations present n a mixture is called **qualitative cation analysis**. Qualitative cation analysis rovides us with many examples of precipitation (and dissolution) equilibria, cid–base equilibria, and oxidation–reduction reactions. Also, in the general hemistry laboratory it offers the challenge of unraveling a mystery—solving qualitative analysis "unknown."

In the scheme in Figure 18-7, about 25 common cations are divided into five roups, depending on differing solubilities of their compounds. The first cations eparated, Pb^{2+}, Hg_2^{2+}, and Ag^+, are those with insoluble *chlorides*. The reagent sed is HCl(aq). All other cations remain in solution because their chlorides are oluble. After the chloride group precipitate is removed, the remaining solution is reated with H_2S in an acidic medium. Under these conditions, a group of sulfides recipitates. They are known as the hydrogen sulfide group. Next, the solution ontaining the remaining cations is treated with H_2S in a buffer mixture of ammoia and ammonium ion, yielding a mixture of insoluble hydroxides and sulfides. This group is called the ammonium sulfide group. The sulfide of luminum(III) and chromium(III) are unstable, reacting with water to form he hydroxides.

◀ In the Hg_2^{2+} ion, a covalent bond links a Hg atom to a Hg^{2+} ion.

Treatment of the ammonium sulfide group filtrate with CO_3^{2-} yields he fourth group precipitate, which consists of the carbonates of Mg^{2+}, Ca^{2+}, Sr^{2+}, nd Ba^{2+}. It is called the carbonate group because the aqueous carbonate anion is he key reagent. At the end of this series of precipitations, the resulting solution ontains only Na^+, K^+, and NH_4^+, all of whose common salts are water soluble.

In this section, we will discuss the chemistry of the chloride and sulfide roups. The chemistry of metal carbonates is discussed in Chapter 21.

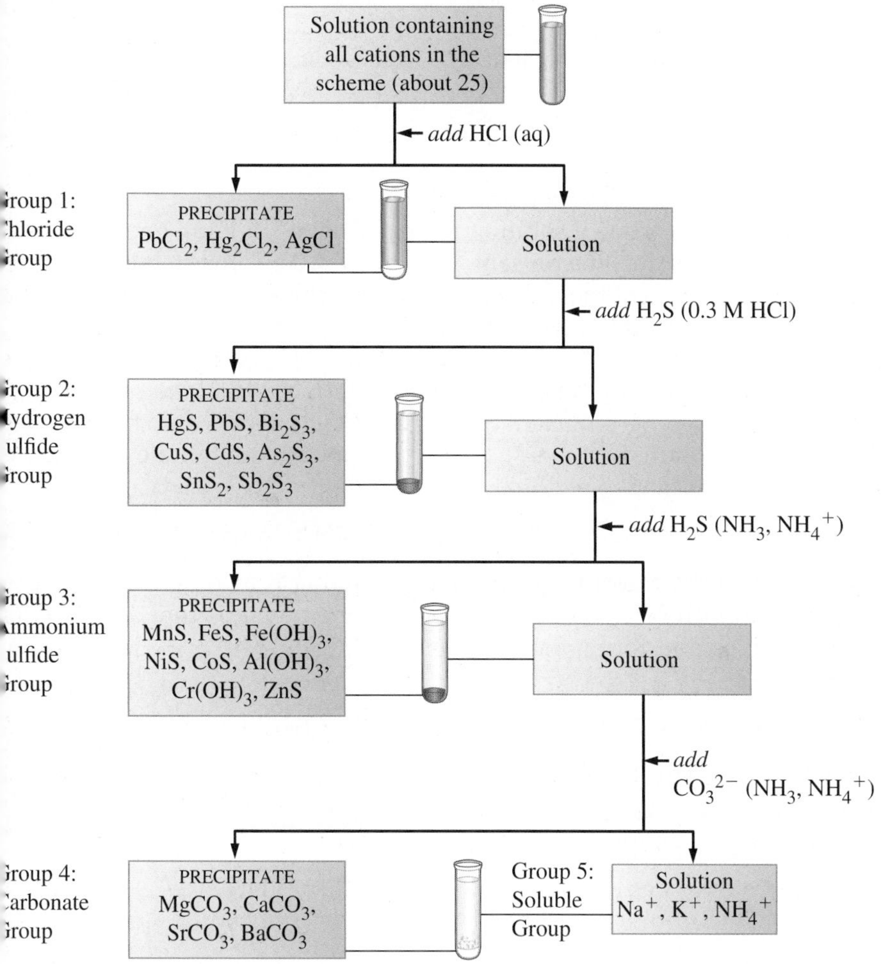

◀ FIGURE 18-7
Outline of a qualitative cation analysis
Various aspects of this scheme are described in the text. A sample containing all 25 cations can be separated into five groups by the indicated reagents.

18-3 ARE YOU WONDERING?

How can we test for the presence of Na^+, K^+, and NH_4^+ cations?

Tests by precipitation are difficult because the salts of these cations exhibit near-universal solubility. Both Na^+ and K^+ ions are most easily detected with a flame test. When a solution containing sodium ions is brought into contact with a flame, the characteristic orange-yellow color of the emission spectrum of sodium atoms is observed. For potassium atoms, a pale violet color results. To detect the presence of NH_4^+, we use the fact that the ammonium ion is the conjugate acid of the weak volatile base ammonia. Heating some of the original solution (not the final solution, which contains NH_4^+ ions added in the fractional precipitation scheme) with excess strong base will liberate ammonia.

$$NH_4^+(aq) + OH^- \longrightarrow NH_3(g) + H_2O(l)$$

The NH_3 is detected by its characteristic odor and its effect on the color of an acid–base indicator such as litmus.

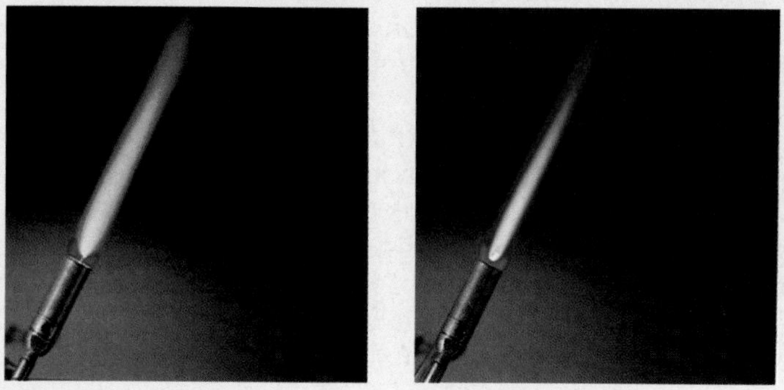

(a) (b)

▲ Flame colors of **(a)** sodium and **(b)** potassium.

Carey B. Van Loon

▶ The mechanism of light emission in flame tests is presented on page 981.

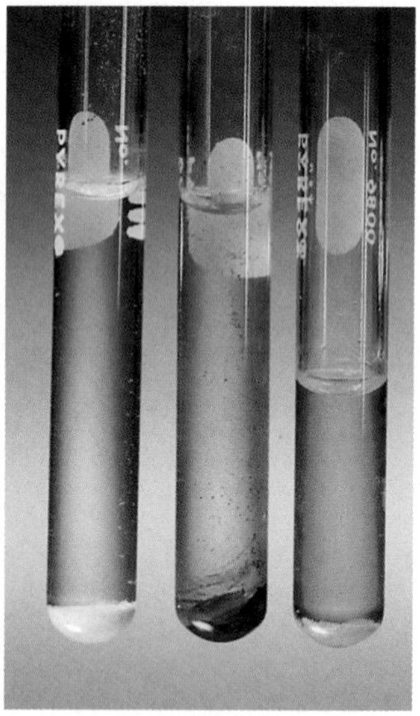

Carey B. Van Loon

(a) (b) (c)

▲ FIGURE 18-8
Chloride group precipitates
(a) Group 1 precipitate: a mixture of $PbCl_2$ (white), Hg_2Cl_2 (white), and AgCl (white). **(b)** Test for Hg_2^{2+}: a mixture of Hg (black) and $HgNH_2Cl$ (white). **(c)** Test for Pb^{2+}: a yellow precipitate of $PbCrO_4(s)$.

Cation Group 1: The Chloride Group

If a precipitate forms when a solution is treated with HCl(aq), one or more c the following cations must be present: Pb^{2+}, Hg_2^{2+}, Ag^+. To establish the pres ence or absence of each of these three cations, the chloride group precipitate i filtered off and subjected to further testing.

Of the three chlorides in the group precipitate, $PbCl_2(s)$ is the most solubl its K_{sp} is much larger than those of AgCl and Hg_2Cl_2. When the precipitate i washed with hot water, a sufficient quantity of $PbCl_2$ dissolves to permit a tes for Pb^{2+} in the solution. In this test, a lead compound less soluble than $PbCl_2$ such as lead(II) chromate, precipitates (Fig. 18-8).

$$Pb^{2+}(aq) + CrO_4^{2-}(aq) \longrightarrow PbCrO_4(s)$$

The portion of the chloride group precipitate that is insoluble in hot wate is then treated with $NH_3(aq)$. Two things happen. One is that any AgCl(s present dissolves and forms the complex ion $[Ag(NH_3)_2^+]$, as described b equation (18.5).

$$AgCl(s) + 2\,NH_3(aq) \longrightarrow [Ag(NH_3)_2]^+(aq) + Cl^-(aq)$$

At the same time, any $Hg_2Cl_2(s)$ present undergoes an oxidation–reductio reaction. One of the products of the reaction is finely divided black mercury. A black color is common for finely divided metals.

$$Hg_2Cl_2(s) + 2\,NH_3(aq) \longrightarrow \underbrace{Hg(l) + HgNH_2Cl(s)}_{\text{Dark gray}} + NH_4^+(aq) + Cl^-(aq)$$

ne appearance of a dark gray mixture of black mercury and white $HgNH_2Cl$ nercury(II) amidochloride] is the qualitative analysis confirmation of mer-ıry(I) ion (see Figure 18-8).

When the solution from reaction (18.5) is acidified with $HNO_3(aq)$, any sil-er ion present precipitates as AgCl(s). This is the reaction predicted in xample 18-10 and pictured in Figure 18-6.

18-7 CONCEPT ASSESSMENT

An unknown contains one or more of Pb^{2+}, $Hg_2{}^{2+}$, and Ag^+. It forms a white precipitate with HCl(aq). The precipitate is treated with hot water, yielding a solution that gives a yellow precipitate with $K_2CrO_4(aq)$. There is no change in color when the undissolved portion of the precipitate is treated with $NH_3(aq)$. Indicate for each chloride group cation whether it is present or absent, or about which there is some doubt.

ation Groups 2 and 3: Equilibria Involving Hydrogen Sulfide

igure 18-7 suggests that aqueous hydrogen sulfide (hydrosulfuric acid) is the ey reagent in the analysis of cation groups 2 and 3. In aqueous solution, H_2S a *weak diprotic acid*.

$$H_2S(aq) + H_2O(l) \rightleftharpoons H_3O^+(aq) + HS^-(aq) \qquad K_{a_1} = 1.0 \times 10^{-7}$$
$$HS^-(aq) + H_2O(l) \rightleftharpoons H_3O^+(aq) + S^{2-}(aq) \qquad K_{a_2} = 1 \times 10^{-19}$$

◀ $H_2S(g)$ has a familiar rotten egg odor, especially noticeable in volcanic areas and near sulfur hot springs. Because of the smell, any gas containing a significant amount of H_2S (4 ppm or more) is called "sour gas."

The extremely small value of K_{a_2} suggests that sulfide ion is a very strong ase, as seen in the following hydrolysis reaction and K_b value.

$$^{2-}(aq) + H_2O(l) \rightleftharpoons HS^-(aq) + OH^-(aq) \qquad K_b = K_w/K_{a_2}$$
$$= K_w/K_{a_2} = (1.0 \times 10^{-14})/(1 \times 10^{-19}) = 1 \times 10^5$$

he hydrolysis of S^{2-} should go nearly to completion, which means that very ttle S^{2-} can exist in an aqueous solution and that sulfide ion is probably not ne precipitating agent for sulfides.

One way to handle the precipitation and dissolution of sulfide precipitates to restrict the discussion to acidic solutions. Then we can write an equilib-ium constant expression in which the concentration terms for HS^- and S^{2-} re eliminated. This approach is reasonable because most sulfide separations re carried out in acidic solution.

Consider (1) the solubility equilibrium equation for PbS written to reflect ydrolysis of S^{2-}, (2) an equation written for the reverse of the first ionization f H_2S, and (3) an equation that is the reverse of the self-ionization of water. hese three equations can be combined into an overall equation that shows he dissolving of PbS(s) in an acidic solution. The equilibrium constant for this verall equation is generally referred to as K_{spa}.*

$$\text{1)} \quad PbS(s) + \cancel{H_2O(l)} \rightleftharpoons Pb^{2+}(aq) + \cancel{HS^-(aq)} + \cancel{OH^-(aq)} \qquad K_{sp} = 3 \times 10^{-28}$$
$$\text{2)} \quad H_3O^+(aq) + \cancel{HS^-(aq)} \rightleftharpoons H_2S(aq) + \cancel{H_2O(l)} \qquad 1/K_{a_1} = 1/1.0 \times 10^{-7}$$
$$\text{3)} \quad H_3O^+(aq) + \cancel{OH^-(aq)} \rightleftharpoons H_2O(l) + H_2O(l) \qquad 1/K_w = 1/1.0 \times 10^{-14}$$
$$\textit{Overall:}\ PbS(s) + 2\,H_3O^+(aq) \rightleftharpoons Pb^{2+}(aq) + H_2S(aq) + 2\,H_2O(l) \qquad K_{spa} = ?$$

$$K_{spa} = \frac{K_{sp}}{K_{a_1} \times K_w} = \frac{3 \times 10^{-28}}{(1.0 \times 10^{-7})(1.0 \times 10^{-14})} = 3 \times 10^{-7}$$

◀ $PbCl_2$ is sufficiently soluble that enough Pb^{2+} ions from cation group 1 remain in solution to precipitate again as PbS(s) in cation group 2.

*See R. J. Myers, *J. Chem. Educ.* **63,** 687 (1986).

Example 18-14 illustrates the use of K_{spa} in the type of calculation needed sort the sulfides into two qualitative analysis groups. Pb^{2+} is in qualitati analysis cation group 2, and Fe^{2+} is in group 3. The conditions cited in tl example are those generally used.

18-8 CONCEPT ASSESSMENT

The dissolution of PbS(s) in water is given by

$$PbS(s) + H_2O(l) \rightleftharpoons Pb^{2+}(aq) + HS^- + OH^-(aq)$$

In an apparent violation of Le Châtelier's principle, the addition of NaOH to the solution leads more PbS to dissolve. Explain.

EXAMPLE 18-14 Separating Metal Ions by Selective Precipitation of Metal Sulfides

Show that PbS(s) will precipitate but FeS(s) will not precipitate from a solution that is 0.010 M in Pb^{2+}, 0.010 M in Fe^{2+}, saturated in H_2S (0.10 M H_2S), and maintained with $[H_3O^+] = 0.30$ M. For PbS, $K_{spa} = 3 \times 10^{-7}$; for FeS, $K_{spa} = 6 \times 10^2$.

Analyze

We must determine whether, for the stated conditions, equilibrium is displaced in the forward or reverse direction in reactions of the type

$$MS(s) + 2\,H_3O^+(aq) \rightleftharpoons M^{2+}(aq) + H_2S(aq) + 2\,H_2O(l) \qquad \textbf{(18.8)}$$

where M represents either Pb or Fe. In each case, we can compare the Q_{spa} expression to the appropriate K_{spa} value for reaction (18.8). If $Q_{spa} > K_{spa}$, a net change will occur to the *left* and MS(s) will precipitate. If $Q_{spa} < K_{spa}$, a net change will occur to the *right*.

Solve

Calculating Q_{spa} we have

$$Q_{spa} = \frac{[M^{2+}][H_2S]}{[H_3O^+]^2} = \frac{0.010 \times 0.10}{(0.30)^2} = 1.1 \times 10^{-2}$$

For PbS: Q_{spa} of $1.1 \times 10^{-2} > K_{spa}$ of 3×10^{-7}. PbS(s) will precipitate.

For FeS: Q_{spa} of $1.1 \times 10^{-2} < K_{spa}$ of 6×10^2. FeS(s) will not precipitate.

Assess

The results here show that using hydrogen sulfide in an acidic medium is a good way to separate cations of groups 2 and 3. You may be wondering how the value for the K_{spa} for FeS was determined. The value of K_{spa} of FeS can be derived from K_{sp} of FeS (6×10^{-19}) by the method outlined for PbS.

PRACTICE EXAMPLE A: Show that $Ag_2S(s)$ ($K_{spa} = 6 \times 10^{-30}$) should precipitate and that FeS(s) ($K_{spa} = 6 \times 10^2$) should not precipitate from a solution that is 0.010 M Ag^+ and 0.020 M Fe^{2+}, but otherwise under the same conditions as in Example 18-14.

PRACTICE EXAMPLE B: What is the minimum pH of a solution that is 0.015 M Fe^{2+} and saturated in H_2S (0.10 M) from which FeS(s) ($K_{spa} = 6 \times 10^2$) can be precipitated?

18-9 CONCEPT ASSESSMENT

Both Cu^{2+} and Ag^+ are present in the same aqueous solution. Explain which of the following reagents would work best in separating these ions, precipitating one and leaving the other in solution: $(NH_4)_2CO_3(aq)$, $HNO_3(aq)$, $H_2S(aq)$, HCl(aq), $NH_3(aq)$, or NaOH(aq).

Dissolving Metal Sulfides

In the qualitative cation analysis, it is necessary to both precipitate and redissolve sulfides. Here, we look at several methods of dissolving metal sulfides. One way to increase the solubility of any sulfide is to allow it to react with an acid, as equation (18.8) suggests. According to Le Châtelier's principle, the solubility increases as the solution is made more acidic—equilibrium is shifted to the right. As a result, some water-insoluble sulfides, such as FeS, are readily soluble in strongly acidic solutions. Others, such as PbS and HgS, cannot be dissolved in acidic solutions because their K_{sp} values are too low. In these cases, $[H_3O^+]$ cannot be made high enough to force reaction (18.8) very far to the right.

Another way to promote the dissolving of metal sulfides is to use an *oxidizing* acid such as $HNO_3(aq)$. In this case, sulfide ion is oxidized to elemental sulfur and the free metal ion appears in solution, as in the dissolving of CuS(s).

$$3\,CuS(s) + 8\,H^+(aq) + 2\,NO_3^-(aq) \longrightarrow 3\,Cu^{2+}(aq) + 3\,S(s) + 2\,NO(g) + 4\,H_2O(l) \quad (18.9)$$

To render the $Cu^{2+}(aq)$ more visible, it is converted to the deeply colored complex ion $[Cu(NH_3)_4]^{2+}(aq)$ in a reaction in which NH_3 molecules replace H_2O molecules in a complex ion (Fig. 18-9).

$$\underset{\text{Pale blue}}{[Cu(H_2O)_4]^{2+}(aq)} + 4\,NH_3(aq) \longrightarrow \underset{\text{Deep violet}}{[Cu(NH_3)_4]^{2+}(aq)} + 4\,H_2O(l)$$

A few metal sulfides dissolve in a basic solution with a high concentration of HS^-, just as acidic oxides dissolve in solutions with a high concentration of OH^-. This property is used to advantage in separating the eight sulfides of cation group 2, the hydrogen sulfide group, into two subgroups. The subgroup consisting of HgS, PbS, CuS, CdS, and Bi_2S_3 remains unchanged after treatment with an alkaline solution with an excess of HS^-, but As_2S_3, SnS_2, and Sb_2S_3 dissolve.

Richard Megna/Fundamental Photographs

▲ FIGURE 18-9
Complex-ion formation: A test for $Cu^{2+}(aq)$
Dilute $CuSO_4(aq)$ (left) derives its pale blue color from the complex ions $[Cu(H_2O)_4]^{2+}$. When $NH_3(aq)$ is added (here labeled "conc ammonium hydroxide"), the color changes to a deep violet, signaling the presence of $[Cu(NH_3)_4]^{2+}$ (right). The deep violet color is detectable at much lower concentrations than the pale blue; the formation of $[Cu(NH_3)_4]^{2+}$ is a sensitive test for the presence of Cu^{2+}.

KEEP IN MIND

that $Cu^{2+}(aq)$ means that the copper(II) ion will be coordinated to several water molecules.

Mastering**CHEMISTRY** **www.masteringchemistry.com**

The formation of some of the Earth's minerals has been by chemical precipitation. Biological precipitation is responsible for the formation of certain marine shells. The Focus On feature for Chapter 18, Shells, Teeth, and Fossils, on the MasteringChemistry site describes the relationship between shells, teeth, and fossils.

Summary

18-1 Solubility Product Constant, K_{sp}—The equilibrium constant for the equilibrium between a solid ionic solute and its ions in a saturated aqueous solution is expressed through the **solubility product constant, K_{sp}** (expression 18.1, Table 18.1).

18-2 Relationship Between Solubility and K_{sp}—A solute's molarity in a saturated aqueous solution is known as its molar solubility. Molar solubility and K_{sp} are related to each other in a way that makes it possible to calculate one when the other is known.

18-3 Common-Ion Effect in Solubility Equilibria—The solubility of a slightly soluble ionic solute is greatly reduced in an aqueous solution containing an ion in common with the solubility equilibrium—a common ion.

18-4 Limitations of the K_{sp} Concept—The solubility product concept is most useful for slightly soluble ionic solutes, and especially in cases where activities can be replaced by molar concentrations. Moderately or highly soluble ionic compounds require the use of activities. The presence of common ions reduces the solubility of a slightly soluble ionic solute, generally to a significant degree.

Other factors also affect solute solubility. The presence of noncommon or diverse ions generally increases solute solubility, an effect called the **salt effect**. Solute activities, and hence solute solubilities, are strongly influenced by interionic attractions, which become especially significant at higher concentrations and for highly charged ions. Dissociation of ionic solutes may not be 100% complete, leading to the formation of **ion pairs** which act as single units, such as MgF^+ in $MgF_2(aq)$.

18-5 Criteria for Precipitation and Its Completeness—To determine whether a slightly soluble solute will precipitate from a solution, the **ion product, Q_{sp}**, is compared with the solubility product constant, K_{sp}. Q_{sp} is based on the initial ion concentrations in a solution. K_{sp}, on the other hand, is based on the equilibrium ion concentrations in a saturated solution. If $Q_{sp} > K_{sp}$, precipitation will occur; if $Q_{sp} < K_{sp}$, the solution will remain unsaturated. When $Q_{sp} = K_{sp}$, the solution is just saturated.

18-6 Fractional Precipitation—A comparison of K_{sp} values is a factor in determining the feasibility of **fractional precipitation**, a process in which one ionic species is removed by precipitation while others remain in solution.

18-7 Solubility and pH—The solubility of a slightly soluble solute is affected by pH if the anion is OH^- or derived from a weak acid. The solubility increases as the pH is lowered or decreases as the pH is raised. This can be illustrated through Le Châtelier's principle. Also, an equilibrium constant for the dissolution reaction can be obtained by combining the solubility equilibrium equation and the ionization equilibrium equation of the weak electrolyte. Some slightly soluble solutions form **suspensions**, which are heterogeneous fluids containing solid particles that will eventually settle.

18-8 Equilibria Involving Complex Ions—A **complex ion** is a polyatomic ion composed of a central metal ion bonded to two or more molecules or ions called ligands. The formation of a complex ion is an equilibrium process with an equilibrium constant called the **formation constant, K_f**. In general, if the formation constant is large, the concentration of uncomplexed metal ion in equilibrium with the complex ion is very small. Complex-ion formation can render certain insoluble materials quite soluble in appropriate aqueous solutions, such as AgCl in $NH_3(aq)$. A complex ion is either a cation or an anion depending upon the particular central metal ion and the particular ligands. When a complex ion and an oppositely charged ion combine they form a **coordination compound**.

18-9 Qualitative Cation Analysis—Precipitation, acid–base, oxidation–reduction, and complex-ion formation reactions are all used extensively in **qualitative cation analysis**. Such an analysis can provide a rapid means of determining the presence or absence of certain cations in an unknown material.

Integrative Example

Lime (quicklime), CaO, is obtained from the high-temperature decomposition of limestone ($CaCO_3$). Quicklime is the cheapest source of basic substances, but it is water insoluble. It does react with water, however, producing $Ca(OH)_2$ (slaked lime). Unfortunately, $Ca(OH)_2(s)$ has limited solubility in water.

$$Ca(OH)_2(s) \rightleftharpoons Ca^{2+}(aq) + 2\,OH^-(aq) \qquad K_{sp} = 5.5 \times 10^{-6}$$

When $Ca(OH)_2(s)$ reacts with a *soluble* carbonate, such as $Na_2CO_3(aq)$, however, a greater amount of $Ca(OH)_2(s)$ dissolves and the solution produced has a much higher pH. Equilibrium is displaced to the right in reaction (18.10) because $CaCO_3$ is much less soluble than $Ca(OH)_2$.

$$Ca(OH)_2(s) + CO_3^{2-}(aq) \rightleftharpoons CaCO_3(s) + 2\,OH^-(aq) \qquad \textbf{(18.10)}$$

Assume an initial $[CO_3^{2-}] = 1.0\,M$ in reaction (18.10), and show that the equilibrium pH should indeed be higher than that in saturated $Ca(OH)_2(aq)$.

Analyze

(1) Determine the pH of the saturated $Ca(OH)_2(aq)$. (2) Find the equilibrium constant, K, for reaction (18.10). (3) Calculate the equilibrium $[OH^-]$ in reaction (18.10). (4) Convert $[OH^-]$ to pOH and then to pH. Compare the pH in steps (1) and (4).

olve

Vrite the K_{sp} expression for a(OH)_2. Let s be the molar solu-ility, so the $[\text{OH}^-] = 2s$.

$$K_{sp} = [\text{Ca}^{2+}][\text{OH}^-]^2 = (s)(2s)^2 = 4s^3 = 5.5 \times 10^{-6}$$
$$s = (5.5 \times 10^{-6}/4)^{1/3} = 0.011\ \text{M}$$
$$[\text{OH}^-] = 2s = 0.022\ \text{M}$$

alculate the pOH.

$$\text{pOH} = -\log[\text{OH}^-] = -\log 0.022$$

)etermine the pH from the pOH y subtracting pOH from 14.00.

$$\text{pOH} = 1.66 \quad \text{pH} = 14.00 - \text{pOH} = 14.00 - 1.66 = 12.34$$

ombine solubility equilibrium quations for Ca(OH)_2 and aCO_3 to obtain the net ionic quation (18.10).

$$\text{Ca(OH)}_2(\text{s}) \rightleftharpoons \text{Ca}^{2+}(\text{aq}) + 2\,\text{OH}^-(\text{aq}) \qquad K_{sp} = 5.5 \times 10^{-6}$$
$$\text{Ca}^{2+}(\text{aq}) + \text{CO}_3{}^{2-}(\text{aq}) \rightleftharpoons \text{CaCO}_3(\text{s}) \qquad 1/K_{sp} = 1/(2.8 \times 10^{-9})$$
$$\text{Ca(OH)}_2(\text{s}) + \text{CO}_3{}^{2-}(\text{aq}) \rightleftharpoons \text{CaCO}_3(\text{s}) + 2\,\text{OH}^-(\text{aq}) \qquad \textbf{(18.10)}$$

Jse the K_{sp} values of Ca(OH)_2 nd CaCO_3 to obtain the overall K.

$$K = \frac{K_{sp}[\text{Ca(OH)}_2]}{K_{sp}[\text{CaCO}_3]}$$
$$K = 5.5 \times 10^{-6}/2.8 \times 10^{-9} = 2.0 \times 10^3$$

Vext, calculate the equilibrium $\text{OH}^-]$ in reaction (18.10), starting vith $[\text{CO}_3{}^{2-}] = 1.0\ \text{M}$, and pro-eeding in the familiar fashion o obtain a quadratic equation. he solution, to two significant igures, is

	$\text{Ca(OH)}_2(\text{s})$ +	$\text{CO}_3{}^{2-}(\text{aq})$	$\rightleftharpoons$ $\text{CaCO}_3(\text{s})$ +	$2\,\text{OH}^-(\text{aq})$
initial concns, M:		1.0		≈ 0
changes, M:		$-x$		$+2x$
equil concns, M:		$1.0 - x$		$2x$

$$K = \frac{[\text{OH}^-]^2}{[\text{CO}_3{}^{2-}]} = \frac{(2x)^2}{(1.0 - x)} = 2.0 \times 10^3$$
$$4x^2 + (2.0 \times 10^3)x - 2.0 \times 10^3 = 0$$
$$x = 1.0$$

rom the value of x, obtain $[\text{OH}^-]$, OH, and pH.

$$[\text{OH}^-] = 2x\ \text{M} = 2.0\ \text{M}$$
$$\text{pOH} = -\log[\text{OH}^-] = -\log 2.0 = -0.30$$
$$\text{pH} = 14.00 - \text{pOH} = 14.00 + 0.30 = 14.30$$

ssess

Ve have succeeded in showing that the solution produced by reaction (18.10) has a higher pH than that found in satu-ated $\text{Ca(OH)}_2(\text{aq})$, that is, a pH of 14.30 compared with 12.34. An interesting aspect of this calculation concerns solution f the quadratic equation $4x^2 + (2.0 \times 10^3)x - 2.0 \times 10^3 = 0$. By using the quadratic formula with no rounding of inter-nediate results, the value of $x = 0.998$, which rounds off to 1.0. Alternatively, note that by assuming that $x = 1.0$, the econd and third terms of the equation cancel, yielding the result $4(1.0^2) + (2.0 \times 10^3)(1.0) - 2.0 \times 10^3 = 4.0$. We con-lude that x must be very slightly less than 1.0. A value of $x = 0.998$ on the left side of the equation yields the result -0.016, very close to the 0 required for an exact solution.

RACTICE EXAMPLE A: Heavy fertilizer use can lead to phosphate pollution in lakes, causing an explosion of lant growth, particularly algae. Excess algae deplete the lake of the oxygen necessary for other plant growth and nimal life. A lake in the middle of a large farm was found to contain the phosphate ion in the concentration of $.13 \times 10^{-4}$ M. The lake measured 300 m × 150 m × 5 m. One method to neutralize $\text{PO}_4{}^{3-}$ is to add a calcium salt. What nass of $\text{Ca(NO}_3)_2$ must be added to lower the phosphate ion concentration to 1.00×10^{-12} M? $K_{sp} = 1.30 \times 10^{-32}$ for $\text{a}_3(\text{PO}_4)_2$.

RACTICE EXAMPLE B: In a laboratory procedure you are instructed to mix 350.0 mL of 0.200 M $\text{AgNO}_3(\text{aq})$ and 250.0 mL of 0.240 M $\text{Na}_2\text{SO}_4(\text{aq})$. What is the precipitate that you observe? To this mixture you add 400.0 mL of 0.500 M $\text{Na}_2\text{S}_2\text{O}_3$, causing the precipitate to dissolve. What is the mass of the remaining precipitate?

Exercises

(Use data from Chapters 16 and 18 and Appendix D, as needed.)

K_{sp} and Solubility

1. Write K_{sp} expressions for the following equilibria. For example, for the reaction $AgCl(s) \rightleftharpoons Ag^+(aq) + Cl^-(aq)$, $K_{sp} = [Ag^+][Cl^-]$.
   (a) $Ag_2SO_4(s) \rightleftharpoons 2\,Ag^+(aq) + SO_4^{2-}(aq)$
   (b) $Ra(IO_3)_2(s) \rightleftharpoons Ra^{2+}(aq) + 2\,IO_3^-(aq)$
   (c) $Ni_3(PO_4)_2(s) \rightleftharpoons 3\,Ni^{2+}(aq) + 2\,PO_4^{3-}(aq)$
   (d) $PuO_2CO_3(s) \rightleftharpoons PuO_2^{2+}(aq) + CO_3^{2-}(aq)$
2. Write solubility equilibrium equations that are described by the following K_{sp} expressions. For example, $K_{sp} = [Ag^+][Cl^-]$ represents $AgCl(s) \rightleftharpoons Ag^+(aq) + Cl^-(aq)$.
   (a) $K_{sp} = [Fe^{3+}][OH^-]^3$
   (b) $K_{sp} = [BiO^+][OH^-]$
   (c) $K_{sp} = [Hg_2^{2+}][I^-]^2$
   (d) $K_{sp} = [Pb^{2+}]^3[AsO_4^{3-}]^2$
3. The following K_{sp} values are found in a handbook. Write the solubility product expression to which each one applies. For example, $K_{sp}(AgCl) = [Ag^+][Cl^-] = 1.8 \times 10^{-10}$.
   (a) $K_{sp}(CrF_3) = 6.6 \times 10^{-11}$
   (b) $K_{sp}[Au_2(C_2O_4)_3] = 1 \times 10^{-10}$
   (c) $K_{sp}[Cd_3(PO_4)_2] = 2.1 \times 10^{-33}$
   (d) $K_{sp}(SrF_2) = 2.5 \times 10^{-9}$
4. Calculate the aqueous solubility, in moles per liter, of each of the following.
   (a) $BaCrO_4$, $K_{sp} = 1.2 \times 10^{-10}$
   (b) $PbBr_2$, $K_{sp} = 4.0 \times 10^{-5}$
   (c) CeF_3, $K_{sp} = 8 \times 10^{-16}$
   (d) $Mg_3(AsO_4)_2$, $K_{sp} = 2.1 \times 10^{-20}$
5. Arrange the following solutes in order of increasing molar solubility in water: $AgCN$, $AgIO_3$, AgI, $AgNO_2$, Ag_2SO_4. Explain your reasoning.
6. Which of the following saturated aqueous solutions would have the highest $[Mg^{2+}]$? Explain. **(a)** $MgCO_3$; **(b)** MgF_2; **(c)** $Mg_3(PO_4)_2$.
7. Fluoridated drinking water contains about 1 part per million (ppm) of F^-. Is CaF_2 sufficiently soluble in water to be used as the source of fluoride ion for th fluoridation of drinking water? Explain. [*Hint:* Thi of 1 ppm as signifying 1 g F^- per 10^6 g solution.]
8. In the qualitative cation analysis procedure, Bi^{3+} detected by the appearance of a white precipitate bismuthyl hydroxide, $BiOOH(s)$:

$$BiOOH(s) \rightleftharpoons BiO^+(aq) + OH^-(aq) \qquad K_{sp} = 4 \times 10^-$$

   Calculate the pH of a saturated aqueous solution BiOOH.
9. A solution is saturated with magnesium palmita $[Mg(C_{16}H_{31}O_2)_2$, a component of bathtub ring] 50 °C. How many milligrams of magnesium palm tate will precipitate from 965 mL of this solutic when it is cooled to 25 °C? For $Mg(C_{16}H_{31}O_2)$ $K_{sp} = 4.8 \times 10^{-12}$ at 50 °C and 3.3×10^{-12} at 25 °C.
10. A 725 mL sample of a saturated aqueous solution c calcium oxalate, CaC_2O_4, at 95 °C is cooled to 13 °C How many milligrams of calcium oxalate will pre cipitate? For CaC_2O_4, $K_{sp} = 1.2 \times 10^{-8}$ at 95 °C an 2.7×10^{-9} at 13 °C.
11. A 25.00 mL sample of a clear *saturated* solution of Pb requires 13.3 mL of a certain $AgNO_3(aq)$ for its titra tion. What is the molarity of this $AgNO_3(aq)$?

$$I^-(\text{satd } PbI_2) + Ag^+(\text{from } AgNO_3) \longrightarrow AgI(s)$$

12. A 250 mL sample of saturated $CaC_2O_4(aq)$ require 4.8 mL of 0.00134 M $KMnO_4(aq)$ for its titration in a acidic solution. What is the value of K_{sp} for CaC_2O obtained with these data? In the titration reactior $C_2O_4^{2-}$ is oxidized to CO_2 and MnO_4^- is reduce to Mn^{2+}.
13. To precipitate as $Ag_2S(s)$, all the Ag^+ present in 338 m of a saturated solution of $AgBrO_3$ requires 30.4 mL c $H_2S(g)$ measured at 23 °C and 748 mmHg. What is K_s for $AgBrO_3$?
14. Excess $Ca(OH)_2(s)$ is shaken with water to produce saturated solution. A 50.00 mL sample of the clea saturated solution is withdrawn and requires 10.7 m of 0.1032 M HCl for its titration. What is K_{sp} fo $Ca(OH)_2$?

The Common-Ion Effect

15. Calculate the molar solubility of $Mg(OH)_2$ ($K_{sp} = 1.8 \times 10^{-11}$) in **(a)** pure water; **(b)** 0.0862 M $MgCl_2$; **(c)** 0.0355 M KOH(aq).
16. How would you expect the presence of each of the following solutes to affect the molar solubility of $CaCO_3$ in water? Explain. **(a)** Na_2CO_3; **(b)** HCl; **(c)** $NaHSO_4$.
17. Describe the effects of the salts KI and $AgNO_3$ on the solubility of AgI in water.
18. Describe the effect of the salt KNO_3 on the solubility of AgI in water, and explain why it is different from the effects noted in Exercise 17.
19. A 0.150 M Na_2SO_4 solution that is saturated with Ag_2SO_4 has $[Ag^+] = 9.7 \times 10^{-3}$ M. What is the valu of K_{sp} for Ag_2SO_4 obtained with these data?
20. If 100.0 mL of 0.0025 M $Na_2SO_4(aq)$ is saturated with $CaSO_4$, how many grams of $CaSO_4$ would be presen in the solution? [*Hint:* Does the usual simplifying assumption hold?]
21. What $[Pb^{2+}]$ should be maintained in $Pb(NO_3)_2$(aq to produce a solubility of 1.5×10^{-4} mol PbI_2/L wher $PbI_2(s)$ is added?

22. What $[I^-]$ should be maintained in KI(aq) to produce a solubility of 1.5×10^{-5} mol PbI_2/L when $PbI_2(s)$ is added?

23. Can the solubility of Ag_2CrO_4 be lowered to 5.0×10^{-8} mol Ag_2CrO_4/L by using CrO_4^{2-} as the common ion? by using Ag^+? Explain.

24. A handbook lists the K_{sp} values 1.1×10^{-10} for $BaSO_4$ and 5.1×10^{-9} for $BaCO_3$. When saturated $BaSO_4(aq)$ is also made with 0.50 M $Na_2CO_3(aq)$, a precipitate of $BaCO_3(s)$ forms. How do you account for this fact, given that $BaCO_3$ has a larger K_{sp} than does $BaSO_4$?

25. A particular water sample that is saturated in CaF_2 has a Ca^{2+} content of 115 ppm (that is, 115 g Ca^{2+} per 10^6 g of water sample). What is the F^- ion content of the water in ppm?

26. Assume that, to be visible to the unaided eye, a precipitate must weigh more than 1 mg. If you add 1.0 mL of 1.0 M NaCl(aq) to 100.0 mL of a clear saturated aqueous AgCl solution, will you be able to see AgCl(s) precipitated as a result of the common-ion effect? Explain.

Criteria for Precipitation from Solution

27. Will precipitation of $MgF_2(s)$ occur if a 22.5 mg sample of $MgCl_2 \cdot 6\,H_2O$ is added to 325 mL of 0.035 M KF?

28. Will $PbCl_2(s)$ precipitate when 155 mL of 0.016 M KCl(aq) are added to 245 mL of 0.175 M $Pb(NO_3)_2(aq)$?

29. What is the minimum pH at which $Cd(OH)_2(s)$ will precipitate from a solution that is 0.0055 M in $Cd^{2+}(aq)$?

30. What is the minimum pH at which $Cr(OH)_3(s)$ will precipitate from a solution that is 0.086 M in $Cr^{3+}(aq)$?

31. Will precipitation occur in the following cases?
(a) 0.10 mg NaCl is added to 1.0 L of 0.10 M $AgNO_3(aq)$.
(b) One drop (0.05 mL) of 0.10 M KBr is added to 250 mL of a saturated solution of AgCl.
(c) One drop (0.05 mL) of 0.0150 M NaOH(aq) is added to 3.0 L of a solution with 2.0 mg Mg^{2+} per liter.

32. The electrolysis of $MgCl_2(aq)$ can be represented as

$$Mg^{2+}(aq) + 2\,Cl^-(aq) + 2\,H_2O(l) \longrightarrow Mg^{2+}(aq) + 2\,OH^-(aq) + H_2(g) + Cl_2(g)$$

The electrolysis of a 315 mL sample of 0.185 M $MgCl_2$ is continued until 0.652 L $H_2(g)$ at 22 °C and 752 mmHg has been collected. Will $Mg(OH)_2(s)$ precipitate when electrolysis is carried to this point? [*Hint:* Notice that $[Mg^{2+}]$ remains constant throughout the electrolysis, but $[OH^-]$ *increases*.]

33. Determine whether 1.50 g $H_2C_2O_4$ (oxalic acid: $K_{a_1} = 5.2 \times 10^{-2}$, $K_{a_2} = 5.4 \times 10^{-5}$) can be dissolved in 0.200 L of 0.150 M $CaCl_2$ without the formation of $CaC_2O_4(s)$ ($K_{sp} = 1.3 \times 10^{-9}$).

34. If 100.0 mL of a clear saturated solution of Ag_2SO_4 is added to 250.0 mL of a clear saturated solution of $PbCrO_4$, will any precipitate form? [*Hint:* Take into account the dilutions that occur. What are the possible precipitates?]

Completeness of Precipitation

35. When 200.0 mL of 0.350 M $K_2CrO_4(aq)$ are added to 200.0 mL of 0.0100 M $AgNO_3(aq)$, what percentage of the Ag^+ is left *unprecipitated*?

36. What percentage of the original Ag^+ remains in solution when 175 mL 0.0208 M $AgNO_3$ is added to 250 mL 0.0380 M K_2CrO_4?

37. If a constant $[Cl^-] = 0.100$ M is maintained in a solution in which the initial $[Pb^{2+}] = 0.065$ M, what percentage of the Pb^{2+} will remain in solution after $PbCl_2(s)$ precipitates? What $[Cl^-]$ should be maintained to ensure that only 1.0% of the Pb^{2+} remains unprecipitated?

38. The ancient Romans added calcium sulfate to wine to clarify it and to remove dissolved lead. What is the maximum $[Pb^{2+}]$ that might be present in wine to which calcium sulfate has been added?

Fractional Precipitation

39. Assume that the seawater sample described in Example 18-6 contains approximately 440 g Ca^{2+} per metric ton (1 metric ton $= 10^3$ kg; density of seawater $= 1.03$ g/mL).
(a) Should $Ca(OH)_2(s)$ precipitate from seawater under the stated conditions, that is, with $[OH^-] = 2.0 \times 10^{-3}$ M?
(b) Is the separation of Ca^{2+} from Mg^{2+} in seawater feasible?

40. Which one of the following solutions can be used to separate the cations in an aqueous solution in which $[Ba^{2+}] = [Ca^{2+}] = 0.050$ M: 0.10 M NaCl(aq), 0.05 M $Na_2SO_4(aq)$, 0.001 M NaOH(aq), or 0.50 M $Na_2CO_3(aq)$? Explain why.

41. KI(aq) is slowly added to a solution with $[Pb^{2+}] = [Ag^+] = 0.10$ M. For PbI_2, $K_{sp} = 7.1 \times 10^{-9}$; for AgI, $K_{sp} = 8.5 \times 10^{-17}$.
(a) Which precipitate should form first, PbI_2 or AgI?

(b) What $[I^-]$ is required for the *second* cation to begin to precipitate?
(c) What concentration of the first cation to precipitate remains in solution at the point at which the second cation begins to precipitate?
(d) Can $Pb^{2+}(aq)$ and $Ag^+(aq)$ be effectively separated by fractional precipitation of their iodides?

42. A solution is 0.010 M in both CrO_4^{2-} and SO_4^{2-}. To this solution, 0.50 M $Pb(NO_3)_2(aq)$ is slowly added.
(a) Which anion will precipitate first from solution?
(b) What is $[Pb^{2+}]$ at the point at which the second anion begins to precipitate?
(c) Are the two anions effectively separated by this fractional precipitation?

43. An aqueous solution that 2.00 M in $AgNO_3$ is slow added from a buret to an aqueous solution that 0.0100 M in Cl^- and 0.250 M in I^-.
(a) Which ion, Cl^- or I^-, is the first to precipitate?
(b) When the second ion begins to precipitate, what the remaining concentration of the first ion?
(c) Is the separation of Cl^- and I^- feasible by fra tional precipitation in this solution?

44. $AgNO_3(aq)$ is slowly added to a solution that is 0.250 M NaCl and also 0.0022 M KBr.
(a) Which anion will precipitate first, Cl^- or Br^-?
(b) What is $[Ag^+]$ at the point at which the secon anion begins to precipitate?
(c) Can the Cl^- and Br^- be separated effectively b this fractional precipitation?

Solubility and pH

45. Which of the following solids is (are) more soluble in an acidic solution than in pure water: KCl, $MgCO_3$, FeS, $Ca(OH)_2$, or C_6H_5COOH? Explain.

46. Which of the following solids is (are) more soluble in a basic solution than in pure water: $BaSO_4$, $H_2C_2O_4$, $Fe(OH)_3$, $NaNO_3$, or MnS? Explain.

47. The solubility of $Mg(OH)_2$ in a particular buffer solution is 0.65 g/L. What must be the pH of the buffer solution?

48. To 0.350 L of 0.150 M NH_3 is added 0.150 L of 0.100 M $MgCl_2$. How many grams of $(NH_4)_2SO_4$ should be present to prevent precipitation of $Mg(OH)_2(s)$?

49. For the equilibrium

$$Al(OH)_3(s) \rightleftharpoons Al^{3+}(aq) + 3\,OH^-(aq) \qquad K_{sp} = 1.3 \times 10^{-33}$$

(a) What is the *minimum pH* at which $Al(OH)_3(s)$ wi precipitate from a solution that is 0.075 M in Al^{3+}?
(b) A solution has $[Al^{3+}] = 0.075$ M and $[CH_3COOH] = 1.00$ M. What is the maximum quantity c $NaCH_3COO$ that can be added to 250.0 mL of thi solution before precipitation of $Al(OH)_3(s)$ begins?

50. Will the following precipitates form under the give conditions?
(a) $PbI_2(s)$, from a solution that is 1.05×10^{-3} M H 1.05×10^{-3} M NaI, and 1.1×10^{-3} M $Pb(NO_3)_2$.
(b) $Mg(OH)_2(s)$, from 2.50 L of 0.0150 M $Mg(NO_3)$ to which is added 1 drop (0.05 mL) of 6.00 M NH_3.
(c) $Al(OH)_3(s)$ from a solution that is 0.010 M i Al^{3+}, 0.010 M CH_3COOH, and 0.010 M $NaCH_3COO$.

Complex-Ion Equilibria

51. $PbCl_2(s)$ is considerably more soluble in HCl(aq) than in pure water, but its solubility in $HNO_3(aq)$ is not much different from what it is in water. Explain this difference in behavior.

52. Which of the following would be most effective, and which would be least effective, in reducing the concentration of the complex ion $[Zn(NH_3)_4]^{2+}$ in a solution: HCl, NH_3, or NH_4Cl? Explain your choices.

53. In a solution that is 0.0500 M in $[Cu(CN)_4]^{3-}$ and 0.80 M in free CN^-, the concentration of Cu^+ is 6.1×10^{-32} M. Calculate K_f of $[Cu(CN)_4]^{3-}$.

$$Cu^+(aq) + 4\,CN^-(aq) \rightleftharpoons [Cu(CN)_4]^{3-}(aq) \quad K_f = ?$$

54. Calculate $[Cu^{2+}]$ in a 0.10 M $CuSO_4(aq)$ solution that is also 6.0 M in free NH_3.

$$Cu^{2+}(aq) + 4\,NH_3(aq) \rightleftharpoons [Cu(NH_3)_4]^{2+}(aq) \qquad K_f = 1.1 \times 10^{13}$$

55. Can the following ion concentrations be maintaine in the same solution without a precipitate forming $[[Ag(S_2O_3)_2]^{3-}] = 0.048$ M, $[S_2O_3^{2-}] = 0.76$ M, an $[I^-] = 2.0$ M?

56. A solution is 0.10 M in *free* NH_3, 0.10 M in NH_4Cl, an 0.015 M in $[Cu(NH_3)_4]^{2+}$. Will $Cu(OH)_2(s)$ precipi tate from this solution? K_{sp} of $Cu(OH)_2$ is 2.2×10^{-20}

57. A 0.10 mol sample of $AgNO_3(s)$ is dissolved in 1.00 of 1.00 M NH_3. How many grams of KI can be dis solved in this solution without a precipitate of AgI(s forming?

58. A solution is prepared that has $[NH_3] = 1.00$ M an $[Cl^-] = 0.100$ M. How many grams of $AgNO_3$ can b dissolved in 1.00 L of this solution without a precipi tate of AgCl(s) forming?

Precipitation and Solubilities of Metal Sulfides

59. Can Fe^{2+} and Mn^{2+} be separated by precipitating FeS(s) and not MnS(s)? Assume $[Fe^{2+}] = [Mn^{2+}] = [H_2S] = 0.10$ M. Choose a $[H_3O^+]$ that ensures maximum precipitation of FeS(s) but not MnS(s). Will the separation be complete? For FeS, $K_{spa} = 6 \times 10^2$; fo MnS, $K_{spa} = 3 \times 10^7$.

60. A solution is 0.05 M in Cu^{2+}, in Hg^{2+}, and in Mn^{2+} Which sulfides will precipitate if the solution is made

to be 0.10 M $H_2S(aq)$ and 0.010 M HCl(aq)? For CuS, $K_{spa} = 6 \times 10^{-16}$; for HgS, $K_{spa} = 2 \times 10^{-32}$; for MnS, $K_{spa} = 3 \times 10^7$.

1. A buffer solution is 0.25 M CH_3COOH–0.15 M $NaCH_3COO$, saturated in H_2S (0.10 M), and with $[Mn^{2+}] = 0.15$ M.
(a) Show that MnS will *not* precipitate from this solution (for MnS, $K_{spa} = 3 \times 10^7$).
(b) Which buffer component would you increase in concentration, and to what minimum value, to ensure that precipitation of MnS(s) begins? Assume that the concentration of the other buffer component is held constant. [*Hint:* Recall equation (18.8).]

62. The following expressions pertain to the precipitation or dissolving of metal sulfides. Use information about the qualitative cation analysis scheme to predict whether a reaction proceeds to a significant extent in the forward direction and what the products are in each case.
(a) $Cu^{2+}(aq) + H_2S\,(\text{satd aq}) \longrightarrow$
(b) $Mg^{2+}(aq) + H_2S\,(\text{satd aq}) \xrightarrow{0.3\text{ M HCl}}$
(c) $PbS(s) + HCl\,(0.3\text{ M}) \longrightarrow$
(d) $ZnS(s) + HNO_3(aq) \longrightarrow$

Qualitative Cation Analysis

3. Suppose you did a group 1 qualitative cation analysis and treated the chloride precipitate with $NH_3(aq)$ without first treating it with hot water. What might you observe, and what valid conclusions could you reach about cations present, cations absent, and cations in doubt?

4. Show that in qualitative cation analysis group 1, if you obtain 1.00 mL of saturated $PbCl_2(aq)$ at 25 °C, sufficient Pb^{2+} should be present to produce a precipitate of $PbCrO_4(s)$. Assume that you use 1 drop (0.05 mL) of 1.0 M K_2CrO_4 for the test.

5. The addition of HCl(aq) to a solution containing several different cations produces a white precipitate. The filtrate is removed and treated with $H_2S(aq)$ in 0.3 M HCl. No precipitate forms. Which of the following conclusions is (are) valid? Explain.
(a) Ag^+ or Hg_2^{2+} (or both) is probably present.
(b) Mg^{2+} is probably not present.
(c) Pb^{2+} is probably not present.
(d) Fe^{2+} is probably not present.

66. Write net ionic equations for the following qualitative cation analysis procedures.
(a) precipitation of $PbCl_2(s)$ from a solution containing Pb^{2+}
(b) dissolution of $Zn(OH)_2(s)$ in a solution of NaOH(aq)
(c) dissolution of $Fe(OH)_3(s)$ in HCl(aq)
(d) precipitation of CuS(s) from an acidic solution of Cu^{2+} and H_2S

Integrative and Advanced Exercises

Use data from Chapters 16 and 18 and Appendix D as needed.)

7. A particular water sample has 131 ppm of $CaSO_4$ (131 g $CaSO_4$ per 10^6 g water). If this water is boiled in a teakettle, approximately what fraction of the water must be evaporated before $CaSO_4(s)$ begins to precipitate? Assume that the solubility of $CaSO_4(s)$ does not change much in the temperature range 0 to 100 °C.

8. A handbook lists the solubility of $CaHPO_4$ as 0.32 g $CaHPO_4 \cdot 2\,H_2O$/L and lists K_{sp} as 1×10^{-7}.

$$CaHPO_4(s) \rightleftharpoons Ca^{2+}(aq) + HPO_4^{2-}(aq)$$

(a) Are these data consistent? (That is, are the molar solubilities the same when derived in two different ways?)
(b) If there is a discrepancy, how do you account for it?

9. A 50.0 mL sample of 0.0152 M $Na_2SO_4(aq)$ is added to 50.0 mL of 0.0125 M $Ca(NO_3)_2(aq)$. What percentage of the Ca^{2+} remains unprecipitated?

70. What percentage of the Ba^{2+} in solution is precipitated as $BaCO_3(s)$ if equal volumes of 0.0020 M $Na_2CO_3(aq)$ and 0.0010 M $BaCl_2(aq)$ are mixed?

71. Determine the molar solubility of lead(II) azide, $Pb(N_3)_2$, in a buffer solution with pH = 3.00, given that

$$Pb(N_3)_2(s) \rightleftharpoons Pb^{2+}(aq) + 2\,N_3^-(aq) \qquad K_{sp} = 2.5 \times 10^{-9}$$

$$HN_3(aq) + H_2O(l) \rightleftharpoons H_3O^+(aq) + N_3^-(aq) \qquad K_a = 1.9 \times 10^{-5}$$

72. Calculate the molar solubility of $Mg(OH)_2$ in 1.00 M $NH_4Cl(aq)$.

73. The chief compound in marble is $CaCO_3$. Marble has been widely used for statues and ornamental work on buildings. However, marble is readily attacked by acids. Determine the solubility of marble (that is, $[Ca^{2+}]$ in a saturated solution) in **(a)** normal rainwater of pH = 5.6; **(b)** acid rainwater of pH = 4.20. Assume that the overall reaction that occurs is

$$CaCO_3(s) + H_3O^+(aq) \rightleftharpoons Ca^{2+}(aq) + HCO_3^-(aq) + H_2O(l)$$

74. What is the solubility of MnS, in grams per liter, in a buffer solution that is 0.100 M CH_3COOH – 0.500 M $NaCH_3COO$? For MnS, $K_{spa} = 3 \times 10^7$.

75. Write net ionic equations for each of the following observations.
(a) When concentrated $CaCl_2(aq)$ is added to $Na_2HPO_4(aq)$, a white precipitate forms that is 38.7% Ca by mass.
(b) When a piece of dry ice, $CO_2(s)$, is placed in a clear dilute solution of limewater [$Ca(OH)_2(aq)$], bubbles of gas evolve. At first, a white precipitate forms, but then it redissolves.

76. Concerning the reactions described in Exercise 75(b), **(a)** Will the same observations be made if $Ca(OH)_2(aq)$ is replaced by $CaCl_2(aq)$? Explain.
(b) Show that the white precipitate will redissolve if the $Ca(OH)_2(aq)$ is about 0.005 M, but not if the solution is saturated.

77. Reaction (18.10), described in the Integrative Example, is called a *carbonate transposition*. In such a reaction, anions of a slightly soluble compound (for example, hydroxides and sulfates) are obtained in a sufficient concentration in aqueous solution that they can be identified by qualitative analysis tests. Suppose that 3 M Na_2CO_3 is used and that an anion concentration of 0.050 M is sufficient for its detection. Predict whether carbonate transposition will be effective for detecting **(a)** SO_4^{2-} from $BaSO_4(s)$; **(b)** Cl^- from $AgCl(s)$; **(c)** F^- from $MgF_2(s)$.

78. For the titration in Example 18-7, verify the assertion that [Ag^+] increases very rapidly between the point at which AgBr has finished precipitating and Ag_2CrO_4 is about to begin.

79. Aluminum compounds are soluble in acidic solution, where aluminum(III) exists as the complex ion $[Al(H_2O)_6]^{3+}$, which is generally represented simply as $Al^{3+}(aq)$. They are also soluble in basic solutions, where the aluminum(III) is present as the complex ion $[Al(OH)_4]^-$. At certain intermediate pH values, th concentration of aluminum(III) that can exist solution is at a minimum. Thus, a plot of the total co centration of aluminum(III) in solution as a functio of pH yields a U-shaped curve. Demonstrate that th is the case with a few calculations.

80. The solubility of AgCN(s) in 0.200 M $NH_3(aq)$ 8.8×10^{-6} mol/L. Calculate K_{sp} for AgCN.

81. The solubility of $CdCO_3(s)$ in 1.00 M KI(aq) 1.2×10^{-3} mol/L. Given that K_{sp} of $CdCO_3$ 5.2×10^{-12}, what is K_f for $[CdI_4]^{2-}$?

82. Use K_{sp} for $PbCl_2$ and K_f for $[PbCl_3]^-$ to determine th molar solubility of $PbCl_2$ in 0.10 M HCl(aq). [*Hint:* Wh is the total concentration of lead species in solution?]

83. A mixture of $PbSO_4(s)$ and $PbS_2O_3(s)$ is shaken wi pure water until a saturated solution is formed. Bo solids remain in excess. What is [Pb^{2+}] in the sat rated solution? For $PbSO_4$, $K_{sp} = 1.6 \times 10^{-8}$; fo PbS_2O_3, $K_{sp} = 4.0 \times 10^{-7}$.

84. Use the method of Exercise 83 to determine [Pb^{2+}] i a saturated solution in contact with a mixture o $PbCl_2(s)$ and $PbBr_2(s)$.

85. A 2.50 g sample of $Ag_2SO_4(s)$ is added to a beake containing 0.150 L of 0.025 M $BaCl_2$.
(a) Write an equation for any reaction that occurs.
(b) Describe the final contents of the beaker—that i the masses of any precipitates present and the concen trations of the ions in solution.

86. How many moles of solid sodium fluoride should b added to 1.0 L of a saturated solution of barium fluo ride, BaF_2, at 25 °C to raise the fluoride concentratio to 0.030 mol/L? What mass of BaF_2 precipitates? Yo may ignore the hydrolysis of fluoride ion. [*Hint*: Th first part of this problem is most easily solved by firs writing down an electroneutrality condition. Se page 761.]

Feature Problems

87. In an experiment to measure K_{sp} of $CaSO_4$ [D. Masterman, *J. Chem. Educ.*, **64**, 409 (1987)], a saturated solution of $CaSO_4(aq)$ is poured into the ion-exchange column pictured (and described in Chapter 21). As the solution passes through the column, Ca^{2+} is retained by the ion-exchange medium and H_3O^+ is released; two H_3O^+ ions appear in the effluent solution for every Ca^{2+} ion. As the drawing suggests, a 25.00 mL sample is added to the column, and the effluent is collected and diluted to 100.0 mL in a volumetric flask. A 10.00 mL portion of the diluted solution requires 8.25 mL of 0.0105 M NaOH for its titration. Use these data to obtain a value of K_{sp} for $CaSO_4$.

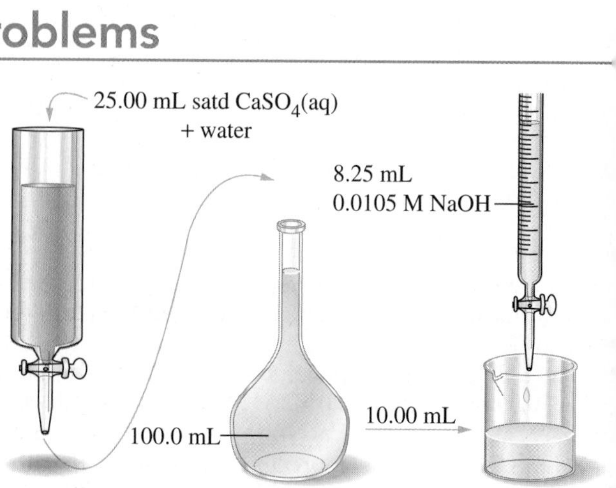

. In the *Mohr titration*, $Cl^-(aq)$ is titrated with $AgNO_3(aq)$ in solutions that are at about pH = 7. Thus, it is suitable for determining the chloride ion content of drinking water. The indicator used in the titration is $K_2CrO_4(aq)$. A red-brown precipitate of $Ag_2CrO_4(s)$ forms after all the Cl^- has precipitated. The titration reaction is $Ag^+(aq) + Cl^-(aq) \longrightarrow AgCl(s)$. At the equivalence point of the titration, the titration mixture consists of AgCl(s) and a solution having neither Ag^+ nor Cl^- in excess. Also, no $Ag_2CrO_4(s)$ is present, but it forms immediately after the equivalence point.
(a) How many milliliters of 0.01000 M $AgNO_3(aq)$ are required to titrate 100.0 mL of a municipal water sample having 29.5 mg Cl^-/L?
(b) What is $[Ag^+]$ at the equivalence point of the Mohr titration?
(c) What is $[CrO_4^{2-}]$ in the titration mixture to meet the requirement of no precipitation of $Ag_2CrO_4(s)$ until immediately after the equivalence point?
(d) Describe the effect on the results of the titration if $[CrO_4^{2-}]$ were (1) greater than that calculated in part (c) or (2) less than that calculated?
(e) Do you think the Mohr titration would work if the reactants were exchanged—that is, with $Cl^-(aq)$ as the titrant and $Ag^+(aq)$ in the sample being analyzed? Explain.

). The accompanying drawing suggests a series of manipulations starting with saturated $Mg(OH)_2(aq)$. Calculate $[Mg^{2+}(aq)]$ at each of the lettered stages.
(a) 0.500 L of saturated $Mg(OH)_2(aq)$ is in contact with $Mg(OH)_2(s)$.
(b) 0.500 L of H_2O is added to the 0.500 L of solution in part (a), and the solution is vigorously stirred. Undissolved $Mg(OH)_2(s)$ remains.
(c) 100.0 mL of the clear solution in part (b) is removed and added to 0.500 L of 0.100 M HCl(aq).
(d) 25.00 mL of the clear solution in part (b) is removed and added to 250.0 mL of 0.065 M $MgCl_2(aq)$.
(e) 50.00 mL of the clear solution in part (b) is removed and added to 150.0 mL of 0.150 M KOH(aq).

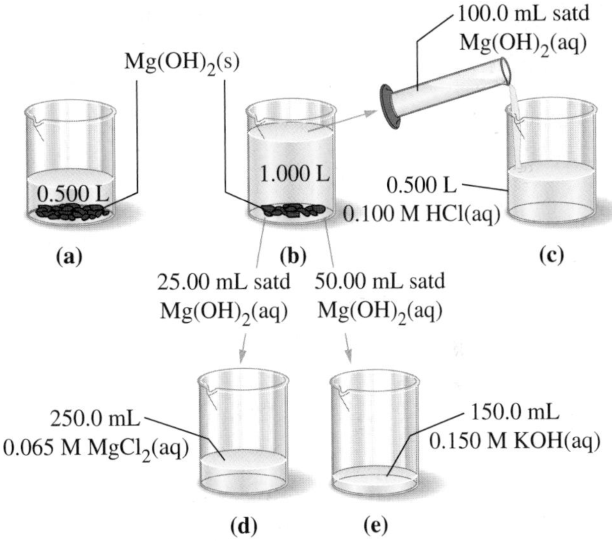

Self-Assessment Exercises

ɔ. In your own words, define the following terms or symbols: **(a)** K_{sp}; **(b)** K_f; **(c)** Q_{sp}; **(d)** complex ion.

1. Briefly describe each of the following ideas, methods, or phenomena: **(a)** common-ion effect in solubility equilibrium; **(b)** fractional precipitation; **(c)** ion-pair formation; **(d)** qualitative cation analysis.

2. Explain the important distinction between each pair of terms: **(a)** solubility and solubility product constant; **(b)** common-ion effect and salt effect; **(c)** ion pair and ion product.

3. Pure water is saturated with slightly soluble PbI_2. Which of the following is a correct statement concerning the lead ion concentration in the solution, and what is wrong with the others? **(a)** $[Pb^{2+}] = [I^-]$; **(b)** $[Pb^{2+}] = K_{sp}$ of PbI_2; **(c)** $[Pb^{2+}] = \sqrt{K_{sp}}$ of PbI_2; **(d)** $[Pb^{2+}] = 0.5[I^-]$.

4. Adding 1.85 g Na_2SO_4 to 500.0 mL of saturated aqueous $BaSO_4$: **(a)** reduces $[Ba^{2+}]$; **(b)** reduces $[SO_4^{2-}]$; **(c)** increases the solubility of $BaSO_4$; **(d)** has no effect.

5. The slightly soluble solute Ag_2CrO_4 is *most* soluble in **(a)** pure water; **(b)** 0.10 M K_2CrO_4; **(c)** 0.25 M KNO_3; **(d)** 0.40 M $AgNO_3$.

6. Cu^{2+} and Pb^{2+} are both present in an aqueous solution. To precipitate one of the ions and leave the other in solution, add **(a)** $H_2S(aq)$; **(b)** $H_2SO_4(aq)$; **(c)** $HNO_3(aq)$; **(d)** $NH_4NO_3(aq)$.

97. All but two of the following solutions yield a precipitate when the solution is also made 2.00 M in NH_3. Those two are **(a)** $MgCl_2(aq)$; **(b)** $FeCl_3(aq)$; **(c)** $(NH_4)_2SO_4(aq)$; **(d)** $Cu(NO_3)_2(aq)$; **(e)** $Al_2(SO_4)_3(aq)$.

98. To increase the molar solubility of $CaCO_3(s)$ in a saturated aqueous solution, add **(a)** ammonium chloride; **(b)** sodium carbonate; **(c)** ammonia; **(d)** more water.

99. The best way to ensure complete precipitation from saturated $H_2S(aq)$ of a metal ion, M^{2+}, as its sulfide, MS(s), is to **(a)** add an acid; **(b)** increase $[H_2S]$ in the solution; **(c)** raise the pH; **(d)** heat the solution.

100. Which of the following solids are likely to be more soluble in acidic solution and which in basic solution? Which are likely to have a solubility that is independent of pH? Explain. **(a)** $H_2C_2O_4$; **(b)** $MgCO_3$; **(c)** CdS; **(d)** KCl; **(e)** $NaNO_3$; **(f)** $Ca(OH)_2$.

101. Both Mg^{2+} and Cu^{2+} are present in the same aqueous solution. Which of the following reagents would work best in separating these ions, precipitating one and leaving the other in solution: NaOH(aq), HCl(aq), $NH_4Cl(aq)$, or $NH_3(aq)$? Explain your choice.

102. Will $Al(OH)_3(s)$ precipitate from a buffer solution that is 0.45 M CH_3COOH and 0.35 M $NaCH_3COO$ and also 0.275 M in $Al^{3+}(aq)$? For $Al(OH)_3$, $K_{sp} = 1.3 \times 10^{-33}$; for CH_3COOH, $K_a = 1.8 \times 10^{-5}$.

103. Saturated solutions of sodium phosphate, copper(II) chloride, and ammonium acetate are mixed together. The precipitate is **(a)** copper(II) acetate; **(b)** copper(II) phosphate; **(c)** sodium chloride; **(d)** ammonium phosphate; **(e)** nothing precipitates.

104. Which of the following has the highest molar solubility? **(a)** MgF_2, $K_{sp} = 3.7 \times 10^{-8}$; **(b)** $MgCO_3$, $K_{sp} = 3.5 \times 10^{-8}$; **(c)** $Mg_3(PO_4)_2$, $K_{sp} = 1 \times 10^{-25}$; **(d)** Li_3PO_4, $K_{sp} = 3.2 \times 10^{-9}$.

105. Lead(II) chloride is most soluble in **(a)** 0.100 M NaCl; **(b)** 0.100 $Na_2S_2O_3$; **(c)** 0.100 M $Pb(NO_3)_2$; **(d)** 0.100 M $NaNO_3$; **(e)** 0.100 $MnSO_4$.

106. Given the following ions in solution, Hg^{2+}, I^-, Ag^+, and NO_3^-, does the formation of a complex ion increase or decrease the amount of precipitate?

107. Will AgI(s) precipitate from a solution wi $[[Ag(CN)_2]^-] = 0.012$ M, $[CN^-] = 1.05$ M, ar $[I^-] = 2.0$ M? For AgI, $K_{sp} = 8.5 \times 10^{-17}$; f $[Ag(CN)_2]^-$, $K_f = 5.6 \times 10^{18}$.

108. Without performing detailed calculations, indica whether either of the following compounds is app ciably soluble in $NH_3(aq)$: **(a)** CuS, $K_{sp} = 6.3 \times 10^{-}$ **(b)** $CuCO_3$, $K_{sp} = 1.4 \times 10^{-10}$. Also use the fact th K_f for $[Cu(NH_3)_4]^{2+}$ is 1.1×10^{13}.

109. Appendix E describes a useful study aid known concept mapping. Using the methods presented Appendix E, construct a concept map that links t various factors affecting the solubility of slight soluble solutes.

lectrochemistry

19

ONTENTS

LEARNING OBJECTIVES

19.1 Describe the structure of an electrochemical cell, highlighting important features such as the salt bridge, cathode, and anode.

19.2 Describe the standard hydrogen electrode (SHE) and discuss the placement of half-cell reactions relative to the SHE.

19.3 Identify the relationships between standard Gibbs energy of reaction, standard cell potential, and the equilibrium constant *K*.

19.4 Using the Nernst equation, determine the spontaneous direction of reaction for given initial conditions.

19.5 Describe primary, secondary, reserve, and flow batteries and how they function as energy storage devices.

19.6 Describe the process of corrosion.

19.7 Within the context of electrolysis, describe what is meant by *overpotential*.

19.8 Identify a few industrial applications of electrolysis.

transit bus fitted with hydrogen–oxygen fuel cells. The use of fuel cells could amatically reduce urban air pollution. The conversion of chemical energy into ectrical energy is one of the main subjects of this chapter.

The mobile devices that many of us rely on—for example, our smartphones and laptop computers—and perhaps even the car or bus we use to get around town are powered by chemical reactions that prouce electricity. Such reactions are central to the area of chemistry known s electrochemistry.

Electrochemistry is concerned with *oxidation–reduction reactions*, a class f reactions we encountered in Chapter 5. In an oxidation–reduction reacon, the oxidation states of atoms in the reactants change. The changes in xidation states are imagined to occur as the result of electron transfer om one reactant to another. A classic example is the reaction of zinc metal, Zn(s), with aqueous copper sulfate, $CuSO_4$(aq), to form solid coppe Cu(s), and aqueous zinc sulfate, $ZnSO_4$(aq).

$$\underset{\text{silvery-gray}}{Zn(s)} + \underset{\text{blue}}{CuSO_4(aq)} \longrightarrow \underset{\text{reddish-brown}}{Cu(s)} + \underset{\text{colorless}}{ZnSO_4(aq)}$$

This reaction is often demonstrated by placing a zinc rod in a copper sulfa solution. The zinc rod becomes coated with a reddish-brown deposit of coppe and the blue of the solution fades (Figure 5-13). At the molecular level, Cu^2 ions in the solution are reduced to form Cu atoms that deposit onto the surfac of the zinc rod. At the same time, Zn atoms from the zinc rod are oxidized t produce Zn^{2+} ions that enter into the solution. The simultaneous oxidation (Zn to Zn^{2+} and reduction of Cu^{2+} to Cu involves the transfer of two electror from Zn to Cu^{2+}.

Another example of an oxidation–reduction reaction is the combination (oxygen and hydrogen to form water.

$$H_2(g) + 2\,O_2(g) \longrightarrow 2\,H_2O(l)$$

In this reaction, H atoms of molecular hydrogen are oxidized (from 0 to +1 and O atoms of molecular oxygen are reduced (from 0 to –2).

Under appropriate conditions, the electron transfer process in these chem cal reactions—and many others—can be used to our advantage, for exampl to run an electric motor, operate a phone, or initiate another chemical reactio

In this chapter, we will see how chemical reactions can be used to produc electricity and how electricity can be used to cause chemical reactions. Th practical applications of electrochemistry are countless, ranging from batte ies, fuel cells, and biological processes to the manufacture of key chemical the refining of metals, and methods for controlling corrosion. Before we ca understand such applications, we must first discuss how to carry out a oxidation–reduction reaction in an electrochemical cell and explore how th energy obtained from, or supplied to, an electrochemical cell is related to th conditions under which the cell operates.

19-1 Electrode Potentials and Their Measurement

The criteria for spontaneous change developed in Chapter 13 apply to rea tions of all types—precipitation, acid–base, and oxidation–reduction (redox We can devise an additional useful criterion for redox reactions, however.

Figure 19-1 shows that a redox reaction occurs between Cu(s) and Ag^+(aq but not between Cu(s) and Zn^{2+}(aq). Specifically, we see that silver ions a reduced to silver atoms on a copper surface, whereas zinc ions are *not* reduce

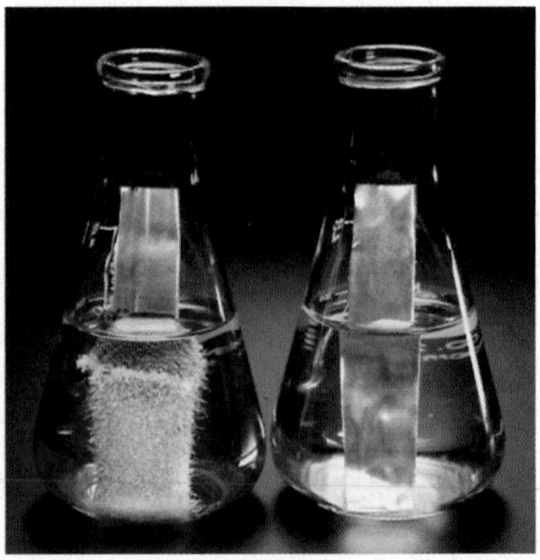

(a) (b)

▶ FIGURE 19-1
Behavior of Ag^+(aq) and Zn^{2+}(aq) in the presence of copper
(a) Copper metal displaces silver ions from colorless $AgNO_3$(aq) as a deposit of silver metal; the copper enters the solution as blue Cu^{2+}(aq).

$$Cu(s) + 2\,Ag^+(aq) \longrightarrow Cu^{2+}(aq) + 2\,Ag(s)$$

(b) Cu(s) *does not* displace colorless Zn^{2+} from $Zn(NO_3)_2$(aq).

$$Cu(s) + Zn^{2+}(aq) \longrightarrow \text{no reaction}$$

Carey B. Van Loon

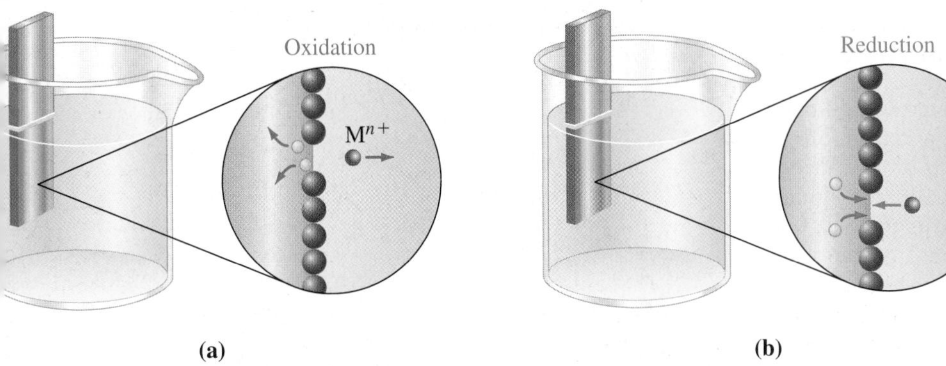

FIGURE 19-2
n electrochemical half-cell
e half-cell consists of a metal electrode, M(s), partially immersed in an aqueous lution of its ions, M^{n+}. (The anions required to maintain electrical neutrality in the lution are spectator ions and are not shown.) The situations illustrated here are limited metals that do not react with water. **(a)** In this oxidation process, a metal atom, M, on e electrode's surface loses electrons to the electrode and enters the solution as M^{n+} ns. **(b)** In this reduction process, a metal ion, M^{n+}, in the solution gains electrons from e electrode's surface and adds to the surface of the electrode as a metal atom, M.

zinc atoms on a copper surface. We can say that Ag^+ is more readily duced than is Zn^{2+}. In this section, we will introduce the *electrode potential,* a operty related to these reduction tendencies.

When used in electrochemical studies, a strip of metal, M, is called an **elec- ode**. An electrode immersed in a solution containing ions of the same metal, $^{n+}$, is called a **half-cell**. Two kinds of interactions are possible between metal oms on the electrode and metal ions in solution (Fig. 19-2):

◀ The term *electrode* is sometimes used for the entire half-cell assembly.

. A metal ion M^{n+} from solution may collide with the electrode, gain n electrons from it, and be converted to a metal atom M. *The ion is reduced.*

. A metal atom M on the surface may lose n electrons to the electrode and enter the solution as the ion M^{n+}. *The metal atom is oxidized.*

n equilibrium is quickly established between the metal and the solution, hich can be represented as

$$M(s) \underset{\text{reduction}}{\overset{\text{oxidation}}{\rightleftharpoons}} M^{n+}(aq) + ne^- \qquad \textbf{(19.1)}$$

owever, any changes produced at the electrode or in the solution as a conse- ence of this equilibrium are too slight to measure. Instead, measurements ust be based on a combination of *two different* half-cells. Specifically, we must easure the tendency for electrons to flow from the electrode of one half-cell the electrode of the other. Electrodes are classified according to whether oxi- ation or reduction takes place there. If oxidation takes place, the electrode is lled the **anode**. If reduction takes place, the electrode is called the **cathode**.

KEEP IN MIND

that although $M^{n+}(aq)$ and ne^- appear together on the right-hand side of this expression, only the ion M^{n+} enters the solution. The electrons remain on the electrode, M(s). Free electrons are never found in an aqueous solution.

Figure 19-3 depicts a combination of two half-cells, one with a Cu electrode contact with $Cu^{2+}(aq)$, and the other with an Ag electrode in contact with $g^+(aq)$. The two electrodes are joined by wires to an electric meter—here, a *ltmeter*. To complete the electric circuit, the two solutions must also be con- ected electrically. However, because charge is carried through solutions by e migration of *ions,* a wire cannot be used for this connection. The solutions ust either be in direct contact through a porous barrier or joined by a third lution in a U-tube called a **salt bridge**. The properly connected combination two half-cells is called an **electrochemical cell**.

Now, we will consider the changes that occur in the electrochemical cell in gure 19-3. As the arrows suggest, Cu atoms release electrons at the anode and nter the $Cu(NO_3)_2(aq)$ as Cu^{2+} ions. Electrons lost by the Cu atoms pass rough the wires and the voltmeter to the cathode, where they are gained by

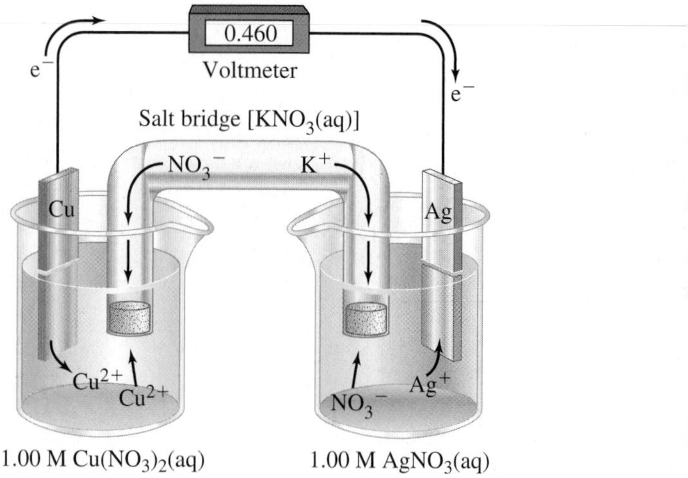

▶ FIGURE 19-3
Measurement of the electromotive force of an electrochemical cell
An electrochemical cell consists of two half-cells with electrodes joined by a wire and solutions joined by a salt bridge. (The ends of the salt bridge are plugged with a porous material that allows ions to migrate but prevents the bulk flow of liquid.) Electrons flow from the Cu electrode, the anode, where oxidation occurs to the Ag electrode, the cathode, where reduction occurs. For precise measurements, the amount of electric current drawn from the cell must be kept very small by means of either a specially designed voltmeter or a device called a potentiometer.

▶ Anions migrate toward the anode, and cations toward the cathode.

Ag^+ ions from the $AgNO_3(aq)$, producing a deposit of metallic silve Simultaneously, anions (NO_3^-) from the salt bridge migrate into the copp half-cell and neutralize the positive charge of the excess Cu^{2+} ions; cations (K^+ migrate into the silver half-cell and neutralize the negative charge of the exce NO_3^- ions. Each copper atom loses two electrons to produce Cu^{2+}; each Ag ion requires one electron to produce Ag(s); consequently, two silver atoms a produced for every Cu^{2+} ion formed. The overall reaction that occurs as th electrochemical cell spontaneously produces electric current is

KEEP IN MIND

that the overall reaction occurring in the electrochemical cell is identical to what happens in the direct addition of Cu(s) to $Ag^+(aq)$ pictured in Figure 19-1(a).

$$\begin{array}{lrcl} \textit{Oxidation:} & Cu(s) & \longrightarrow & Cu^{2+}(aq) + 2\,e^- \\ \textit{Reduction:} & 2\,\{Ag^+(aq) + e^- & \longrightarrow & Ag(s)\} \\ \hline \textit{Overall:} & Cu(s) + 2\,Ag^+(aq) & \longrightarrow & Cu^{2+}(aq) + 2\,Ag(s) \end{array} \quad \textbf{(19.}$$

The reading on the voltmeter (0.460 V) is significant. It is the **cell voltag** or the *potential difference* between the two half-cells. The unit of cell voltag **volt (V)**, is the energy per unit charge. Thus, a potential difference of one vo signifies an energy of one joule for every coulomb of charge passing throug an electric circuit: 1 V = 1 J/C. We can think of a voltage, or potential diffe ence, as the driving force for electrons; the greater the voltage, the greater th driving force. The flow of water from a higher to a lower level is analogous t this situation. The greater the difference in water levels, the greater the forc behind the flow of water. Cell voltage is also called **electromotive force (emf** or **cell potential**, and represented by the symbol E_{cell}.

▶ Such formulations as $Zn(s)/Zn^{2+}(aq)$ are called *couples* and are often used as abbreviations for half-cells.

Now let's return to the question raised by Figure 19-1: Why does copper n displace Zn^{2+} from solution? In an electrochemical cell consisting of $Zn(s)/Zn^{2+}(aq)$ half-cell and a $Cu^{2+}(aq)/Cu(s)$ half-cell, electrons flow *from th Zn to the Cu*. The spontaneous reaction in the electrochemical cell in Figure 19-4

$$\begin{array}{lrcl} \textit{Oxidation:} & Zn(s) & \longrightarrow & Zn^{2+}(aq) + 2\,e^- \\ \textit{Reduction:} & Cu^{2+}(aq) + 2\,e^- & \longrightarrow & Cu(s) \\ \hline \textit{Overall:} & Zn(s) + Cu^{2+}(aq) & \longrightarrow & Zn^{2+}(aq) + Cu(s) \end{array} \quad \textbf{(19.}$$

Because reaction (19.3) is a spontaneous reaction, the displacement of $Zn^{2+}(aq$ by Cu(s)—the *reverse* of reaction (19.3)—does *not* occur spontaneously. This the observation made in Figure 19-1. In Section 19-3, we will discuss how t predict the direction of spontaneous change for oxidation–reduction reaction

Cell Diagrams and Terminology

Drawing sketches of electrochemical cells, as in Figures 19-3 and 19-4, is help ful, but more often a simpler representation is used. A **cell diagram** shows th components of an electrochemical cell in a symbolic way. A cell diagram for a electrochemical cell has the following general form. (The black arrow, whic

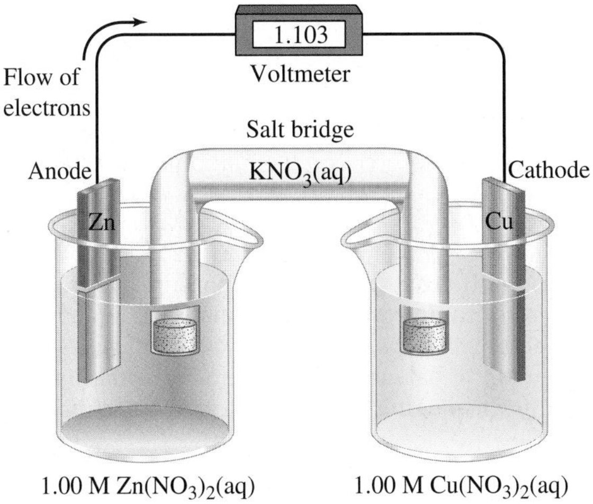

◀ FIGURE 19-4
The reaction
$Zn(s) + Cu^{2+}(aq) \longrightarrow Zn^{2+}(aq) + Cu(s)$ in an electrochemical cell

dicates the direction of electron flow in an external circuit, is implied. It is t usually included as part of the diagram.)

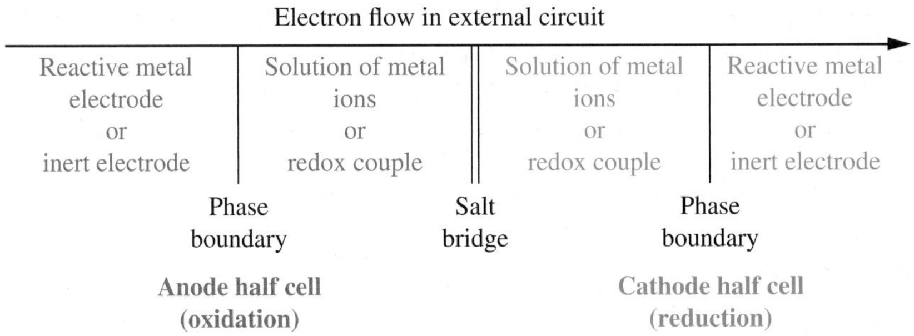

We will use the following generally accepted conventions in writing cell agrams.

- The anode, the electrode at which *oxidation* occurs, is placed at the *left* side of the diagram.
- The cathode, the electrode at which *reduction* occurs, is placed at the *right* side of the diagram.
- A boundary between different phases (for example, an electrode and a solution) is represented by a *single vertical line* (|).
- The boundary between half-cell compartments, commonly a salt bridge, is represented by a *double vertical line* (||). Species in aqueous solution are placed on either side of the double vertical line. Different species within the same solution are separated from each other by a comma. Although IUPAC recommends the use of double dashed vertical lines to represent the salt bridge, the recommendation has not yet been universally adopted by chemists. In this text, we will continue to use the double vertical line (||) for a salt bridge.
- When appropriate, concentrations, pressures, or activities for each species are given in parentheses.

◀ Several memory devices have been proposed for the oxidation/anode and reduction/cathode relationships. Perhaps the simplest is that in the oxidation/anode relationship, both terms begin with a vowel: *o*/*a*; in the reduction/cathode relationship, both begin with a consonant: *r*/*c*.

KEEP IN MIND

that the spectator ions are not shown in a cell diagram, but they are present. They pass through the salt bridge to maintain electrical neutrality.

ne cell diagram below (in black) is for the electrochemical cell shown in gure 19-4.

$$\text{ode} \longrightarrow \underset{\text{Half-cell (oxidation)}}{Zn(s)|Zn^{2+}(aq, 1.00\text{ M})} \underset{\text{Salt bridge}}{\|} \underset{\text{Half-cell (reduction)}}{Cu^{2+}(aq, 1.00\text{ M})|Cu(s)} \longleftarrow \text{cathode} \quad \underset{\text{Cell voltage}}{E_{cell} = 1.103\text{ V}} \tag{19.4}$$

The electrochemical cells of Figures 19-3 and 19-4 produce electricity as a result spontaneous chemical reactions; as such, they are called **voltaic**, or **galvanic**, **lls**. In Section 19-7 we will consider *electrolytic cells*—electrochemical cells in hich electricity is used to accomplish a nonspontaneous chemical change.

EXAMPLE 19-1 Representing a Redox Reaction by Means of a Cell Diagram

Aluminum metal displaces zinc(II) ion from aqueous solution.

(a) Write oxidation and reduction half-cell equations and an overall equation for this redox reaction.

(b) Write a cell diagram for a voltaic cell in which this reaction occurs.

Analyze

The term *displaces* means that aluminum goes into solution as $Al^{3+}(aq)$, forcing $Zn^{2+}(aq)$ out of solution as zinc metal. Al is oxidized to Al^{3+}, and Zn^{2+} is reduced to Zn. In combining the half-cell equations to produce the overall equation, we must take care to ensure that the *number of electrons involved in reduction equals the number involved in oxidation.* (This is the half-reaction method of balancing redox equations discussed in Section 5-5.) The cell diagram is written with the reduction half-cell equations as the right-hand electrode.

Solve

(a) The two half-cell equations are

$$\textit{Oxidation:} \quad Al(s) \longrightarrow Al^{3+}(aq) + 3\,e^-$$

$$\textit{Reduction:} \quad Zn^{2+}(aq) + 2\,e^- \longrightarrow Zn(s)$$

On inspecting these half-cell equations, we see that the number of electrons involved in oxidation and reduction are different. In writing the overall equation, the coefficients must be adjusted so that equal numbers of electrons are involved in oxidation and in reduction.

$$\textit{Oxidation:} \quad 2\,\{Al(s) \longrightarrow Al^{3+}(aq) + 3\,e^-\}$$

$$\textit{Reduction:} \quad 3\,\{Zn^{2+}(aq) + 2\,e^- \longrightarrow Zn(s)\}$$

$$\textit{Overall:} \quad 2\,Al(s) + 3\,Zn^{2+}(aq) \longrightarrow 2\,Al^{3+}(aq) + 3\,Zn(s)$$

(b) Al(s) is oxidized to $Al^{3+}(aq)$ in the anode half-cell (written on the left of the cell diagram), and $Zn^{2+}(aq)$ is reduced to Zn(s) in the cathode half-cell (written on the right of the cell diagram).

$$Al(s)|Al^{3+}(aq)||Zn^{2+}(aq)|Zn(s)$$

Assess

Whenever balancing redox equations, it is important to ensure that the number of electrons in the oxidation step equals the number of electrons in the reduction step. This is achieved by multiplying the entire half-cell equations(s) by the appropriate factor(s).

PRACTICE EXAMPLE A: Write the overall equation for the redox reaction that occurs in the voltaic cell $Sc(s)|Sc^{3+}(aq)||Ag^{+}(aq)|Ag(s)$.

PRACTICE EXAMPLE B: Draw a voltaic cell in which silver ion is displaced from solution by aluminum metal. Label the cathode, the anode, and other features of the cell. Show the direction of flow of electrons. Also, indicate the direction of flow of cations and anions from a $KNO_3(aq)$ salt bridge. Write an equation for the half-reaction occurring at each electrode, write a balanced equation for the overall cell reaction, and write a cell diagram.

EXAMPLE 19-2 Deducing the Balanced Redox Reaction from a Cell Diagram

The cell diagram for an electrochemical cell is written as

$$Ni(s)|NiCl_2(aq)||Ce(ClO_4)_4(aq), Ce(ClO_4)_3(aq)|Pt(s)$$

Write the equations for the half-cell reactions that occur at the electrodes. Balance the overall cell reaction.

Analyze

When inspecting a cell diagram, we first need to identify the species involved in oxidation and in reduction. Then we can write balanced half-cell equations. Finally, we can combine the half-cell equations to give the overall cell reaction. The new part to this example is the Ce^{4+}/Ce^{3+} couple in the presence of the inert platinum electrode at which the reduction takes place. Again, when balancing redox equations, it is important to ensure that the number of electrons in the oxidation step equals the number of electrons in the reduction step.

Solve

The cerium reduction reaction is

$$Ce^{4+}(aq) + e^- \longrightarrow Ce^{3+}(aq)$$

The nickel oxidation reaction is

$$Ni(s) \longrightarrow Ni^{2+}(aq) + 2\,e^-$$

In these half-cell equations the number of electrons involved in oxidation and reduction are different. In writing the overall equation, the coefficients must be adjusted so that equal numbers of electrons are involved in oxidation and in reduction.

$$\begin{array}{lrl} \textit{Oxidation:} & Ni(s) & \longrightarrow Ni^{2+}(aq) + 2\,e^- \\ \textit{Reduction:} & 2\,\{Ce^{4+}(aq) + e^- & \longrightarrow Ce^{3+}(aq)\} \\ \hline \textit{Overall:} & Ni(s) + 2\,Ce^{4+}(aq) & \longrightarrow Ni^{2+}(aq) + 2\,Ce^{3+}(aq) \end{array}$$

Assess

We can see the importance of balancing each half-cell equation with respect to charge and mass.

PRACTICE EXAMPLE A: The cell diagram for an electrochemical cell is written as

$$Sn(s)|SnCl_2(aq)||AgNO_3(aq)|Ag(s)$$

Write the equations for the half-cell reactions that occur at the electrodes. Balance the overall cell reaction.

PRACTICE EXAMPLE B: The cell diagram for an electrochemical cell is written as

$$In(s)|In(ClO_4)_3(aq)||CdCl_2(aq)|Cd(s)$$

Write the equations for the half-cell reactions that occur at the electrodes. Balance the overall cell reaction.

19-1 CONCEPT ASSESSMENT

Add appropriate arrows to Figure 19-4 to show the direction of migration of ions through the electrochemical cell.

9-2 Standard Electrode Potentials

ell voltages—*potential differences* between electrodes—are among the most recise scientific measurements possible. Potentials of individual electrodes, owever, cannot be precisely established. If we could make such measureents, cell voltages could be obtained just by subtracting one electrode potenal from another. The same result can be achieved by *arbitrarily* choosing a articular half-cell that is assigned an electrode potential of *zero*. Other halflls can then be compared with this reference. The commonly accepted refernce is the standard hydrogen electrode.

◀ This method is comparable to establishing standard enthalpies or Gibbs energies of formation on the basis of an arbitrary zero value.

The **standard hydrogen electrode (SHE)** is depicted in Figure 19-5. The HE involves equilibrium established on the surface of an inert metal (such as latinum) between H_3O^+ ions from a solution in which they are at unit activity hat is, $a_{H_3O^+} = 1$) and H_2 molecules from the gaseous state at a pressure of bar. The equilibrium reaction produces a particular potential on the metal urface, but this potential is arbitrarily taken to be *zero*.

$$2\,H^+(aq, a = 1) + 2\,e^- \overset{\text{on Pt}}{\rightleftharpoons} H_2(g, 1\text{ bar}) \qquad E° = 0 \text{ volt (V)} \qquad \textbf{(19.5)}$$

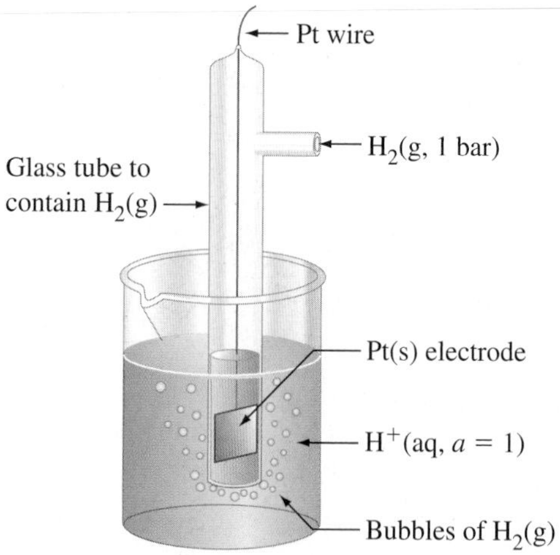

▲ FIGURE 19-5
The standard hydrogen electrode (SHE)
Because hydrogen is a gas at room temperature, electrodes cannot be constructed from it. The standard hydrogen electrode consists of a piece of platinum dipped into solution containing 1 M H^+(aq) with a stream of hydrogen passing over its surface. T platinum does not react but provides a surface for the reduction of H_3O^+(aq) to H_2(as well as the reverse oxidation half-cell reaction.

KEEP IN MIND

that we have adopted the use of 1 bar for standard pressure. Older textbooks and reference books have used 1 atm as standard pressure. The difference between $E°$ values defined with respect to the old and new standards of pressure are so small that the values can be used interchangeably. We justify this claim on page 873.

▶ In Chapter 5, we used the terms *half-equation* and *half-reaction* when referring to oxidation or reduction processes. In this chapter, we use the terms *half-cell equation* and *half-cell reaction* instead to emphasize that these processes occur within an electrochemical cell.

The diagram for this half-cell is

$$H^+(aq, a = 1)\,|\,H_2(g, 1\ bar)\,|\,Pt$$

The two vertical lines signify that three phases are present: solid platinu gaseous hydrogen, and aqueous hydrogen ion. For simplicity, we will usua write H^+ for H_3O^+, assume that unit activity (a = 1) exists at rough $[H^+]$ = 1 M.

By international agreement, a **standard electrode potential, $E°$**, measu the tendency for a *reduction* process to occur at an electrode. In all cases, t ionic species are present in aqueous solution at unit activity (approximate 1 M), and gases are at 1 bar pressure. Where no metallic substance is indicate the potential is established on an inert metallic electrode, such as platinum.

To emphasize that $E°$ refers to a reduction, we will write a reduction cou as a subscript to $E°$, as shown in half-cell reaction (19.6). The substance bei reduced is written on the left of the slash sign (/), and the chief reducti product on the right.

$$Cu^{2+}(1\ M) + 2\,e^- \longrightarrow Cu(s) \qquad E°_{Cu^{2+}/Cu} = ? \qquad (19$$

To determine the value of $E°$ for a standard electrode such as that to whi half-cell reaction (19.6) applies, we compare it with a standard hydrogen ele trode (SHE). In this comparison, the SHE is always taken as the electrode the *left* of the cell diagram—the anode—and the compared electrode is t electrode on the *right*—the cathode. In the following voltaic cell, the me sured potential difference is 0.340 V, with electrons flowing from the H_2 to t Cu electrode.

$$\underset{anode}{Pt|H_2(g, 1\ bar)|H^+(1\ M)}||\underset{cathode}{Cu^{2+}(1\ M)|Cu(s)} \qquad E°_{cell} = 0.340\ V \qquad (19$$

A **standard cell potential, $E°_{cell}$**, is the potential difference, or voltage, of a c formed from two *standard* electrodes. The *difference* is always taken in the fo lowing way:

$$E°_{cell} = \underset{(cathode)}{E°(right)} - \underset{(anode)}{E°(left)}$$

pplied to the cell diagram (19.7), we get

$$E^{\circ}_{cell} = E^{\circ}_{Cu^{2+}/Cu} - E^{\circ}_{H^+/H_2} = 0.340 \text{ V}$$
$$= E^{\circ}_{Cu^{2+}/Cu} - 0 \text{ V} = 0.340 \text{ V}$$
$$E^{\circ}_{Cu^{2+}/Cu} = 0.340 \text{ V}$$

us, the standard *reduction* half-cell reaction can be written as

$$Cu^{2+}(1 \text{ M}) + 2 e^- \longrightarrow Cu(s) \qquad E^{\circ}_{Cu^{2+}/Cu} = +0.340 \text{ V} \qquad \textbf{(19.8)}$$

e overall reaction occurring in the voltaic cell diagrammed in (19.7) can be presented as

$$H_2(g, 1 \text{ bar}) + Cu^{2+}(1 \text{ M}) \longrightarrow 2 H^+(1 \text{ M}) + Cu(s) \qquad E^{\circ}_{cell} = 0.340 \text{ V} \qquad \textbf{(19.9)}$$

ll reaction (19.9) indicates that $Cu^{2+}(1 \text{ M})$ is more easily reduced than is $^+(1 \text{ M})$.

Suppose the standard copper electrode in cell diagram (19.7) is replaced by standard zinc electrode, and the potential difference between the standard drogen and zinc electrodes is measured by using the same voltmeter conctions as in (19.7). In this case, the voltage is found to be −0.763 V. The gative sign indicates that electrons flow in the direction *opposite* that in 9.7)—that is, *from* the zinc electrode to the hydrogen electrode. Here $^+(1 \text{ M})$ is more easily reduced than is $Zn^{2+}(1 \text{ M})$. These findings are represented in the following cell diagram, in which the zinc electrode appears on e *right*.

$$Pt|H_2(g, 1 \text{ bar})|H^+(1 \text{ M})||Zn^{2+}(1 \text{ M})|Zn(s) \qquad E^{\circ}_{cell} = -0.763 \text{ V} \qquad \textbf{(19.10)}$$

e standard electrode potential for the Zn^{2+}/Zn couple can be written as

$$E^{\circ}_{cell} = E^{\circ}(\text{right}) - E^{\circ}(\text{left})$$
$$= E^{\circ}_{Zn^{2+}/Zn} - 0 \text{ V} = -0.763 \text{ V}$$
$$E^{\circ}_{Zn^{2+}/Zn} = -0.763 \text{ V}$$

us, the standard *reduction* half-cell reaction is

$$Zn^{2+}(1 \text{ M}) + 2 e^- \longrightarrow Zn(s) \qquad E^{\circ}_{Zn^{2+}/Zn} = -0.763 \text{ V} \qquad \textbf{(19.11)}$$

In summary, the potential of the standard hydrogen electrode is set at actly 0 V. Any electrode at which a reduction half-cell reaction shows a *eater* tendency to occur than does the reduction of H^+ (1 M) to H_2 (g, 1 bar) s a *positive* value for its standard electrode potential, E°. Any electrode at hich a reduction half-cell reaction shows a *lesser* tendency to occur than es the reduction of $H^+(1 \text{ M})$ to H_2 (g, 1 bar) has a *negative* value for its andard reduction potential, E°. Comparisons of the standard copper and nc electrodes to the standard hydrogen electrode are illustrated in Figure)-6. Table 19.1 on page 875 lists some common reduction half-cell reactions d their standard electrode potentials at 25 °C.

◀ A more extensive listing of reduction half-cell reactions and their potentials is given in Appendix D.

Notice that the standard reduction potentials listed in Table 19.1 refer to the d standard pressure of 1 atm instead of the new standard pressure of 1 bar, that Table 19.1 is consistent with other more extensive tabulations that still fer to the old standard of 1 atm. A reason for not changing the tables is that e difference between values defined with respect to the old and new standrds of pressure are so small that the values can be used interchangeably. The fference can be shown to be

$$E^{\circ} - E^{*} = 0.3382 \times \frac{\Delta\nu_{gas}}{z} \text{ mV}$$

here E° is the reduction potential defined with respect to 1 bar, E^{*} is the duction potential defined with respect to 1 atm, z is the number of electrons,

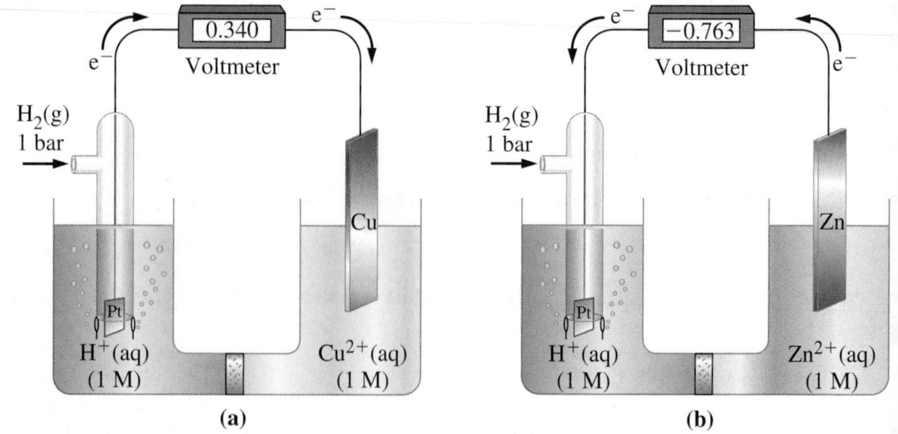

▲ FIGURE 19-6
Measuring standard electrode potentials
(a) A standard hydrogen electrode is the anode, and copper is the cathode. Contact between the half-cells occurs through a porous plate that prevents bulk flow of the solutions while allowing ions to pass. **(b)** This cell has the same connections as that in part (a), but with zinc substituting for copper. However, the electron flow is opposite that in (a), as noted by the *negative* voltage. (Zinc is the anode.)

and $\Delta\nu_{gas}$ is the sum of coefficients for gas-phase products minus the sum coefficients for gas-phase reactants:

$$\Delta\nu_{gas} = \sum|\nu_{gas}(\text{products})| - \sum|\nu_{gas}(\text{reactants})|$$

(Refer to Exercise 108 to see how the expression for $E° - E^*$ is derived.) Th expression shows that the difference $E° - E^*$ is typically only about 0.3 mil volts, which is usually smaller than the precision of the tabulated data.

19-2 CONCEPT ASSESSMENT

For Figure 19-6, describe any changes in mass that might be detected at the Pt, Cu, and Zn electrodes as electric current passes through the electrochemical cells.

KEEP IN MIND

that the $E°$ values in this formulation are for a reduction half-cell reaction, regardless of whether oxidation or reduction occurs in the half-cell.

▶ The placement of *oxidizing agents* in Table 19.1 is as follows: strongest oxidizing agents ($F_2, O_3, \ldots$), *left* sides, *top* of the list; weakest oxidizing agents ($Li^+, K^+, \ldots$), *left* sides, *bottom* of list. The placement of *reducing agents* is as follows: strongest reducing agents (Li, K, . . .), *right* sides, *bottom* of list; weakest reducing agents ($F^-, O_2, \ldots$), *right* sides, *top* of list.

Standard reduction potentials are used throughout this chapter for mar purposes. Our first objective will be to calculate standard cell potentials f redox reactions—$E°_{cell}$ values—from standard electrode potentials for half-ce reactions—$E°$ values. The procedure used is illustrated here for reaction (19. and cell diagram (19.4). Note that the first three equations are alternative wa of stating the same thing; we will generally not write all of them.

$$\begin{aligned} E°_{cell} &= E°(\text{right}) - E°(\text{left}) \\ &= E°(\text{cathode}) - E°(\text{anode}) \\ &= E°(\text{reduction half-cell}) - E°(\text{oxidation half-cell}) \\ &= E°_{Cu^{2+}/Cu} - E°_{Zn^{2+}/Zn} \\ &= 0.340\text{ V} - (-0.763\text{ V}) = 1.103\text{ V} \end{aligned}$$

Example 19-3 predicts $E°_{cell}$ for a new battery system. Example 19-4 us one known electrode potential and a measured $E°_{cell}$ value to determine a unknown $E°$.

TABLE 19.1 Standard Reduction Potentials at 298.15 K and 1 atm[a]

Reduction Half-Cell Reaction	Cell Notation	Standard Reduction Potential, V
Acidic solution		
$F_2(g) + 2e^- \rightleftharpoons 2F^-(aq)$	$F^- \mid F_2 \mid Pt$	+2.866
$O_3(g) + 2H^+(aq) + 2e^- \rightleftharpoons O_2(g) + H_2O(l)$	$H^+ \mid O_3, O_2 \mid Pt$	+2.075
$S_2O_8^{2-}(aq) + 2e^- \rightleftharpoons 2SO_4^{2-}(aq)$	$SO_4^{2-}, S_2O_8^{2-} \mid Pt$	+2.01
$H_2O_2(aq) + 2H^+(aq) + 2e^- \rightleftharpoons 2H_2O(l)$	$H_2O_2, H^+ \mid Pt$	+1.763
$Ce^{4+}(aq) + e^- \rightleftharpoons Ce^{3+}(aq)$	$Ce^{3+}, Ce^{4+} \mid Pt$	+1.72
$MnO_4^-(aq) + 4H^+(aq) + 3e^- \rightleftharpoons MnO_2(s) + 2H_2O(l)$	$MnO_2 \mid MnO_4^-, H^+ \mid Pt$	+1.70
$PbO_2(s) + SO_4^{2-}(aq) + 4H^+(aq) + 2e^- \rightleftharpoons PbSO_4(s) + 2H_2O(l)$	$PbSO_4, PbO_2 \mid SO_4^{2-}, H^+ \mid Pt$	+1.69
$Cl_2(g) + 2e^- \rightleftharpoons 2Cl^-(aq)$	$Cl^- \mid Cl_2 \mid Pt$	+1.36
$Cr_2O_7^{2-}(aq) + 14H^+(aq) + 6e^- \rightleftharpoons 2Cr^{3+}(aq) + 7H_2O(l)$	$Cr^{3+}, Cr_2O_7^{2-} \mid Pt$	+1.32
$MnO_2(s) + 4H^+(aq) + 2e^- \rightleftharpoons Mn^{2+}(aq) + 2H_2O(l)$	$Mn^{2+}, H^+ \mid MnO_2 \mid Pt$	+1.23
$O_2(g) + 4H^+(aq) + 4e^- \rightleftharpoons 2H_2O(l)$	$H^+ \mid O_2 \mid Pt$	+1.229
$IO_3^-(aq) + 12H^+(aq) + 10e^- \rightleftharpoons I_2(s) + 6H_2O(l)$	$IO_3^-, H^- \mid I_2 \mid Pt$	+1.20
$Br_2(l) + 2e^- \rightleftharpoons 2Br^-$	$Br^- \mid Br_2 \mid Pt$	+1.065
$NO_3^-(aq) + 4H^+(aq) + 3e^- \rightleftharpoons NO(g) + 2H_2O(l)$	$NO_3^-, H^+ \mid NO \mid Pt$	+0.956
$Ag^+(aq) + e^- \rightleftharpoons Ag(s)$	$Ag^+ \mid Ag$	+0.800
$Fe^{3+}(aq) + e^- \rightleftharpoons Fe^{2+}(aq)$	$Fe^{2+}, Fe^{3+} \mid Pt$	+0.771
$O_2(g) + 2H^+(aq) + 2e^- \rightleftharpoons H_2O_2(aq)$	$H_2O_2, H^+ \mid O_2 \mid Pt$	+0.695
$I_2(s) + 2e^- \rightleftharpoons 2I^-(aq)$	$I^- \mid I_2 \mid Pt$	+0.535
$Cu^+(aq) + e^- \rightleftharpoons Cu(s)$	$Cu^+ \mid Cu$	+0.520
$SO_4^{2-}(aq) + 4H^+(aq) + 2e^- \rightleftharpoons 2H_2O(l) + SO_2(g)$	$SO_4^{2-}, H^+ \mid SO_2 \mid Pt$	+0.17
$Sn^{4+}(aq) + 2e^- \rightleftharpoons Sn^{2+}(aq)$	$Sn^{2+}, Sn^{4+} \mid Pt$	+0.154
$S(s) + 2H^+(aq) + 2e^- \rightleftharpoons H_2S(g)$	$H^+ \mid H_2S \mid S$	+0.144
$2H^+(aq) + 2e^- \rightleftharpoons H_2(g)$	$H^+ \mid H_2 \mid Pt$	0.0
$Pb^{2+}(aq) + 2e^- \rightleftharpoons Pb(s)$	$Pb^{2+} \mid Pb$	−0.125
$Sn^{2+}(aq) + 2e^- \rightleftharpoons Sn(s)$	$Sn^{2+} \mid Sn$	−0.137
$Fe^{2+}(aq) + 2e^- \rightleftharpoons Fe(s)$	$Fe^{2+} \mid Fe$	−0.440
$Zn^{2+}(aq) + 2e^- \rightleftharpoons Zn(s)$	$Zn^{2+} \mid Zn$	−0.763
$Al^{3+}(aq) + 3e^- \rightleftharpoons Al(s)$	$Al^{3+} \mid Al$	−1.662
$Mg^{2+}(aq) + 2e^- \rightleftharpoons Mg(s)$	$Mg^{2+} \mid Mg$	−2.372
$Na^+(aq) + e^- \rightleftharpoons Na(s)$	$Na^+ \mid Na$	−2.714
$Ca^{2+}(aq) + 2e^- \rightleftharpoons Ca(s)$	$Ca^{2+} \mid Ca$	−2.868
$K^+(aq) + e^- \rightleftharpoons K(s)$	$K^+ \mid K$	−2.931
$Li^+(aq) + e^- \rightleftharpoons Li(s)$	$Li^+ \mid Li$	−3.05
Basic solution		
$O_3(g) + H_2O(l) + 2e^- \rightleftharpoons O_2(g) + 2OH^-(aq)$	$OH^- \mid O_2, O_3 \mid Pt$	+1.246
$OCl^-(aq) + H_2O(l) + 2e^- \rightleftharpoons Cl^-(aq) + 2OH^-(aq)$	$OH^-, OCl^-, Cl^- \mid Pt$	+0.890
$O_2(g) + 2H_2O(l) + 4e^- \rightleftharpoons 4OH^-(aq)$	$OH^- \mid O_2 \mid Pt$	+0.401
$2H_2O(l) + 2e^- \rightleftharpoons H_2(g) + 2OH^-(aq)$	$OH^- \mid H_2 \mid Pt$	−0.828

[a]The difference between the $E°$ values defined with respect to the old and new standards of pressure are so small that the values can ually be used interchangeably.

EXAMPLE 19-3 Combining $E°$ Values into $E°_{cell}$ for a Reaction

A new battery system currently under study for possible use in electric vehicles is the zinc–chlorine battery. The overall reaction producing electricity in this cell is $Zn(s) + Cl_2(g) \longrightarrow ZnCl_2(aq)$. What is $E°_{cell}$ of this voltaic cell?

Analyze

First we identify the species that are oxidized and reduced. Then we obtain the standard reduction potentials for the cathode and anode from Table 19.1 or Appendix D and calculate $E°_{cell}$.

Solve

The oxidation state of zinc changes from 0 to +2 and therefore is oxidized; consequently, the chlorine is reduced. The half-cell reactions are indicated below and are combined into the overall equation (19.12).

$$\begin{array}{lrcl} \textit{Oxidation:} & Zn(s) & \longrightarrow & Zn^{2+}(aq) + 2\,e^- \\ \textit{Reduction:} & Cl_2(g) + 2\,e^- & \longrightarrow & 2\,Cl^-(aq) \\ \hline \textit{Overall:} & Zn(s) + Cl_2(g) & \longrightarrow & Zn^{2+}(aq) + 2\,Cl^-(aq) \end{array} \quad \textbf{(19.12)}$$

$$\begin{aligned} E°_{cell} &= E°(\text{reduction half-cell}) - E°(\text{oxidation half-cell}) \\ &= 1.358\text{ V} - (-0.763\text{ V}) = 2.121\text{ V} \end{aligned}$$

Assess

Once the oxidized and reduced species are identified, we can establish $E°_{cell}$.

PRACTICE EXAMPLE A: What is $E°_{cell}$ for the reaction in which $Cl_2(g)$ oxidizes $Fe^{2+}(aq)$ to $Fe^{3+}(aq)$?

$$2\,Fe^{2+}(aq) + Cl_2(g) \longrightarrow 2\,Fe^{3+}(aq) + 2\,Cl^-(aq) \qquad E°_{cell} = ?$$

PRACTICE EXAMPLE B: Use data from Table 19.1 to determine $E°_{cell}$ for the redox reaction in which $Fe^{2+}(aq)$ is oxidized to $Fe^{3+}(aq)$ by $MnO_4^-(aq)$ in acidic solution.

EXAMPLE 19-4 Determining an Unknown $E°$ from an $E°_{cell}$ Measurement

Cadmium is found in small quantities wherever zinc is found. Unlike zinc, which in trace amounts is an essential element, cadmium is an environmental poison. To determine cadmium ion concentrations by electrical measurements, we need the standard electrode potential for the Cd^{2+}/Cd electrode. The voltage of the following voltaic cell is measured.

$$Cd(s)|Cd^{2+}(1\text{ M})||Cu^{2+}(1\text{ M})|Cu(s) \qquad E°_{cell} = 0.743\text{ V}$$

What is the standard electrode potential for the Cd^{2+}/Cd electrode?

Analyze

We know one half-cell potential and $E°_{cell}$ for the overall redox reaction. We can solve for the unknown standard electrode potential, $E°_{Cd^{2+}/Cd}$.

Solve

$$\begin{aligned} E°_{cell} &= E°(\text{right}) - E°(\text{left}) \\ 0.743\text{ V} &= E°_{Cu^{2+}/Cu} - E°_{Cd^{2+}/Cd} \\ &= 0.340\text{ V} - E°_{Cd^{2+}/Cd} \\ E°_{Cd^{2+}/Cd} &= 0.340\text{ V} - 0.743\text{ V} = -0.403\text{ V} \end{aligned}$$

Assess

Based on the entries in Table 19.1 we see that Cd(s) is a stronger reducing agent than Sn(s), but weaker than Fe(s). $Cd^{2+}(aq)$ is a weaker oxidizing agent than $Sn^{2+}(aq)$.

PRACTICE EXAMPLE A: In acidic solution, dichromate ion oxidizes oxalic acid, $H_2C_2O_4(aq)$, to $CO_2(g)$ in a reaction with $E°_{cell} = 1.81$ V.

$$Cr_2O_7^{2-}(aq) + 3\,H_2C_2O_4(aq) + 8\,H^+(aq) \longrightarrow 2\,Cr^{3+}(aq) + 7\,H_2O + 6\,CO_2(g)$$

Use the value of $E°_{cell}$ for this reaction, together with appropriate data from Table 19.1, to determine $E°$ for the $CO_2(g)/H_2C_2O_4(aq)$ electrode.

PRACTICE EXAMPLE B: In an acidic solution, $O_2(g)$ oxidizes $Cr^{2+}(aq)$ to $Cr^{3+}(aq)$. The $O_2(g)$ is reduced to $H_2O(l)$. E°_{cell} for the reaction is 1.653 V. What is the standard electrode potential for the couple Cr^{3+}/Cr^{2+}?

19-3 CONCEPT ASSESSMENT

For the half-cell reaction $ClO_4^-(aq) + 8\,H^+(aq) + 7\,e^- \longrightarrow \frac{1}{2}Cl_2(g) +$ 4 $H_2O(l)$, what are the standard-state conditions for the reactants and products?

9-3 E_{cell}, $\Delta_r G$, and K

hen a reaction occurs in a voltaic cell, the cell does work—electrical work. ink of this as the work of moving electric charges. The total work done is e product of three terms: (a) E_{cell}; (b) n, the number of moles of electrons ansferred between the electrodes; and (c) the electric charge per mole of elec- ons, called the **Faraday constant (*F*)**. The Faraday constant is equal to ,485 coulombs per mole of electrons (96,485 C/mol). Because the product lt × coulomb = joule, the unit of w_{elec} in equation (19.13) is joules (J).

$$w_{elec} = nFE_{cell} \qquad (19.13)$$

pression (19.13) applies only if the cell operates reversibly.* As discussed in ction 13-4, the work that can be derived from a process is equal to $-\Delta G$. For an ectrochemical reaction, we have $n = z\xi$, where z is the *electron number* (or *charge mber*) and ξ is the extent of reaction in moles (see Section 4-6). The electron mber is simply a number with no units. It is the number of electrons trans- red in the reaction as written. Substituting $n = z\xi$ into equation (19.13), we get $_{elec} = -\Delta G = z\xi FE_{cell}$, or $(\Delta G/\xi) = -zFE_{cell}$. The quantity $\Delta G/\xi$ is the Gibbs ergy change *per mole of reaction*, which is normally represented as $\Delta_r G$. Thus,

$$\Delta_r G = -zFE_{cell} \qquad (19.14)$$

the special case in which the reactants and products are in their standard states,

$$\Delta_r G^\circ = -zFE^\circ_{cell} \qquad (19.15)$$

e electron number, z, for a reaction depends on how the reaction is written. r the hydrogen electrode reactions below, we have $z = 2$ (for the reaction on e left) and $z = 1$ (for the reaction on the right).

$$2\,H^+(aq) + 2\,e^- \longrightarrow H_2(g) \quad \text{or} \quad H^+(aq) + e^- \longrightarrow \frac{1}{2}\,H_2(g)$$

owever, in considering an overall cell reaction, we must balance the elec- ons. Thus for the cell

$$Pt(s)|H_2(g)|H^+(aq, 1\,M)||Cu^{2+}(aq)|Cu(s)$$

e half-cell reactions can be written as

$$2\,H^+(aq) + 2\,e^- \longrightarrow H_2(g) \quad \text{and} \quad 2\,e^- + Cu^{2+}(aq) \longrightarrow Cu(s)$$

nus, the electron number is two and the overall electrochemical reaction is

$$H_2(g) + Cu^{2+}(aq) \longrightarrow 2\,H^+(aq) + Cu(s)$$

e standard reduction potential for this reaction is

$$E^\circ_{cell} = E^\circ(\text{right}) - E^\circ(\text{left})$$
$$= E^\circ_{Cu^{2+}/Cu} - 0\,V = 0.340\,V$$

Bettmann/Corbis

▲ **Michael Faraday (1791–1867)**
Faraday, an assistant to Humphry Davy and often called "Davy's greatest discovery," made many contributions to both physics and chemistry, including systematic studies of electrolysis.

The meaning of a reversible process was illustrated by Figure 7-12 on page 262. The reversible eration of a voltaic cell requires that electric current be drawn from the cell only very, very slowly.

The standard Gibbs energy of reaction is given by

$$\Delta_r G° = -zFE°_{cell} = 2 \times \frac{96{,}485\text{ C}}{\text{mol}} \times 0.340\text{ V}$$
$$= -6.5610 \times 10^4\text{ J mol}^{-1} = -65.6\text{ kJ mol}^{-1}$$

That is, 65.6 kJ of energy is generated when 1 mole of Cu^{2+} ions is reduced 2 moles of H^+ are produced. The process is accompanied by the passage two moles of electrons around the outer circuit. We could also have written t reactions as

$$\textit{Oxidation:}\quad \frac{1}{2}H_2(g) \longrightarrow H^+(aq) + e^-$$
$$\textit{Reduction:}\quad \frac{1}{2}Cu^{2+}(aq) + e^- \longrightarrow \frac{1}{2}Cu(s)$$
$$\textit{Overall:}\quad \frac{1}{2}H_2(g) + \frac{1}{2}Cu^{2+}(aq) \longrightarrow \frac{1}{2}Cu(s) + H^+(aq)$$

This reaction is represented by the same cell diagram given above, but t electron number is one; consequently, the Gibbs energy is one-half of that pr viously calculated, but the value of $E°_{cell}$ is the same. This result supports t fact that the standard reduction potential is an intensive property but t Gibbs energy is an extensive property. Finally, the reaction tells us that wh 0.5 mole of Cu^{2+} is reduced, 32.8 kJ of energy is released and one mole of ele trons passes from the anode to the cathode.

Our primary interest is not in calculating quantities of work but in usi expression (19.15) as a means of evaluating Gibbs energy changes from me sured cell potentials, as illustrated in Example 19-5.

EXAMPLE 19-5 Determining the Gibbs Energy of Reaction from a Cell Potential

Given that $E° = 2.121$ V, determine $\Delta_r G°$ for the reaction $Zn(s) + Cl_2(g, 1\text{ bar}) \longrightarrow ZnCl_2(aq, 1\text{ M})$.

Analyze

In this type of problem, the overall equation generally needs to be separated into two half-cell equations. Then the value of $E°_{cell}$ and the number of electrons (z) involved in the cell reaction can be determined. Refer to Example 19-3 to see that $E°_{cell} = 2.121$ V and $z = 2$.

Solve

We use $z = 2$ and $E° = 2.121$ V in equation (19.15).

$$\Delta_r G° = -zFE°_{cell} = -\left(2 \times \frac{96{,}485\text{ C}}{1\text{ mol}} \times 2.121\text{ V}\right) = -4.093 \times 10^5\text{ J mol}^{-1} = -409.3\text{ kJ mol}^{-1}$$

Assess

Since the overall cell reaction is the combination of elements to form a compound, $\Delta_r G°$ is equal to $\Delta_f G°$ for $ZnCl_2(aq)$.

PRACTICE EXAMPLE A: Use electrode potential data to determine $\Delta G°$ for the reaction

$$2\,Al(s) + 3\,Br_2(l) \longrightarrow 2\,Al^{3+}(aq, 1\text{ M}) + 6\,Br^-(aq, 1\text{ M}) \qquad \Delta_r G° = ?$$

PRACTICE EXAMPLE B: The hydrogen–oxygen fuel cell is a voltaic cell with a cell reaction of $2\,H_2(g) + O_2(g) \longrightarrow 2\,H_2O(l)$. Calculate $E°_{cell}$ for this reaction. [*Hint:* Use thermodynamic data from Appendix D (Table D-2).]

Combining Reduction Half-Cell Equations

Not only can equation (19.15) be used to determine $\Delta_r G°$ from $E°_{cell}$, as Example 19-5, but the calculation can be reversed and an $E°_{cell}$ value dete mined from $\Delta_r G°$. Moreover, equation (19.15) can be applied to half-cell rea tions and half-cell potentials—that is, to standard electrode potentials, $E°$. Th is what we must do, for example, to determine $E°$ for the half-cell reaction

$$Fe^{3+}(aq) + 3\,e^- \longrightarrow Fe(s)$$

Both in Table 19.1 and in Appendix D, the only entries that deal with Fe(s) d its ions are

$$^{2+}(aq) + 2\,e^- \longrightarrow Fe(s),\ E° = -0.440\ V \quad \text{and} \quad Fe^{3+}(aq) + e^- \longrightarrow Fe^{2+}(aq),\ E° = 0.771\ V$$

The half-cell equation we are seeking is simply the sum of these two half-uations, but the $E°$ value we are seeking is *not* the sum of -0.440 V and 0.771 V. hat we *can* add together, though, are the $\Delta_rG°$ values for the two known lf-cell reactions.

$$^{2+}(aq) + 2\,e^- \longrightarrow Fe(s); \qquad \Delta_rG° = -2 \times F \times (-0.440\ V)$$
$$Fe^{3+}(aq) + e^- \longrightarrow Fe^{2+}(aq); \qquad \Delta_rG° = -1 \times F \times (0.771\ V)$$
$$^{3+}(aq) + 3\,e^- \longrightarrow Fe(s); \qquad \Delta_rG° = (0.880F)\ V - (0.771F)\ V = (0.109F)\ V$$

Now, to get $E°_{Fe^{3+}/Fe}$, we can again use equation (19.15) and solve for $E°_{Fe^{3+}/Fe}$.

$$\Delta_rG° = -zFE°_{Fe^{3+}/Fe} = -3FE°_{Fe^{3+}/Fe} = (0.109F)\ V$$
$$E°_{Fe^{3+}/Fe} = (-0.109F/3F)\ V = -0.0363\ V$$

KEEP IN MIND

that Gibbs energy changes are functions of state and therefore $\Delta_rG°$ values can be combined to determine Gibbs energy changes for new reactions.

19-1 ARE YOU WONDERING?

How does the procedure for combining two $E°$ values to obtain an unknown $E°_{cell}$ relate to combining two $E°$ values to obtain an unknown $E°$?

We have just seen how to obtain an unknown $E°$ from two known values of $E°$ by working through the expression $\Delta_rG° = -zFE°$. As shown below for a hypothetical displacement reaction, we can similarly calculate an unknown $E°_{cell}$ through the expression $\Delta_rG° = -zFE°_{cell}$. (Note that for the oxidation half-cell reaction, $\Delta_rG°_{ox}$ is simply the negative of the value for the reverse half-cell reaction, $\Delta_rG°_{red}$.)

Reduction: $M^{z+}(aq) + z\,e^- \longrightarrow M(s) \qquad \Delta_rG°_{red} = -zFE°_{M^{z+}/M}$

Oxidation: $N(s) \longrightarrow N^{z+}(aq) + z\,e^-$

$$\Delta_rG°_{ox} = -(\Delta_rG°_{red}) = -(-zFE°_{N^{z+}/N}) = zFE°_{N^{z+}/N}$$

Overall: $M^{z+}(aq) + N(s) \longrightarrow M(s) + N^{z+}(aq)$

$$\Delta_rG° = \Delta_rG°_{red} + \Delta_rG°_{ox} = -zFE°_{cell} = -zFE°_{M^{z+}/M} + zFE°_{N^{z+}/N}$$

Dividing through the above equation by the term $-zF$, we obtain $E°_{cell}$ as the familiar difference in two electrode potentials.

$$E°_{cell} = E°_{M^{z+}/M} - E°_{N^{z+}/N}$$

We have been able to skip this calculation based on $\Delta_rG°$ values and proceed straight to the expression

$$E°_{cell} = E°(\text{reduction}) - E°(\text{oxidation})$$

because the term $-zF$ always cancels out. That is, z, the number of electrons, must have the same value for the oxidation and reduction half-cell reactions and the overall reaction. By contrast, when obtaining an unknown $E°$ from the known $E°$ values, the value for z will not be the same in all three places where it appears, and so we do have to work through the $\Delta_rG°$ expressions.

pontaneous Change in Oxidation–Reduction Reactions

ur main criterion for spontaneous change is that $\Delta_rG < 0$. According to uation (19.14), however, redox reactions have the property that, if $\Delta_rG < 0$, en $E_{cell} > 0$. That is, E_{cell} must be *positive* if Δ_rG is to be negative. Predicting e direction of spontaneous change in a redox reaction is a relatively simple atter by using the following ideas:

- If E_{cell} is *positive*, a reaction occurs spontaneously in the *forward* direction for the stated conditions. If E_{cell} is *negative*, the reaction occurs spontaneously in the *reverse* direction for the stated conditions. If $E_{cell} = 0$, the reaction is at equilibrium for the stated conditions.
- If a cell reaction is *reversed*, E_{cell} changes sign.

In the special case in which reactants and products are in their standard stat we work with $\Delta_r G°$ and $E°_{cell}$ values, as illustrated in Examples 19-6 and 19-

Even though we used electrode potentials and cell voltage to predict spontaneous reaction in Example 19-6, we do not have to carry out the rea tion in a voltaic cell. This is an important point to keep in mind. Thus, Cu^{2+} displaced from aqueous solution simply by adding aluminum metal, as show in Figure 19-7. Another point, illustrated by Example 19-7, is that qualitati answers to questions concerning redox reactions can be found without goi through a complete calculation of $E°_{cell}$.

Diane Hirsch/Fundamental Photographs

▲ FIGURE 19-7
Reaction of Al(s) and Cu^{2+}(aq)
Notice the holes in the foil where Al(s) has dissolved. Notice also the dark deposit of Cu(s) at the bottom of the beaker.

The Behavior of Metals Toward Acids

In the discussion of redox reactions in Chapter 5, it was noted that most me als react with an acid, such as HCl, but that a few do not. This observation c now be explained. When a metal, M, reacts with an acid, such as HCl, t metal is oxidized to the metal ion, such as M^{2+}. The reduction involves H being reduced to $H_2(g)$. These ideas can be expressed as

$$\begin{array}{lrcl} \textit{Oxidation:} & M(s) & \longrightarrow & M^{2+}(aq) + 2\,e^- \\ \textit{Reduction:} & 2\,H^+(aq) + 2\,e^- & \longrightarrow & H_2(g) \\ \hline \textit{Overall:} & M(s) + 2\,H^+(aq) & \longrightarrow & M^{2+}(aq) + H_2(g) \end{array}$$

$$E°_{cell} = E°_{H^+/H_2} - E°_{M^{2+}/M} = 0\text{ V} - E°_{M^{2+}/M} = -E°_{M^{2+}/M}$$

Metals with *negative* standard electrode potentials yield *positive* values of $E°_c$ in the above expression. These are the metals that should displace $H_2(g)$ fro acidic solutions. Thus, all the metals listed *below* hydrogen in Table 19 (Pb through Li) should react with acids.

In acids, such as HCl, HBr, and HI, the oxidizing agent is H^+ (that is, H_3O^+ Certain metals that will not react with HCl will react with an acid in the pre ence of an *anion* that is a better oxidizing agent than H^+. Nitrate ion is a goo oxidizing agent in acidic solution, and silver metal, which does not react wi HCl(aq), readily reacts with nitric acid, $HNO_3(aq)$.

$$3\,Ag(s) + NO_3^-(aq) + 4\,H^+(aq) \longrightarrow 3\,Ag^+(aq) + NO(g) + 2\,H_2O \quad E°_{cell} = 0.156$$

EXAMPLE 19-6 Applying the Criterion for Spontaneous Change in a Redox Reaction

Will aluminum metal displace Cu^{2+} ion from aqueous solution? That is, will a spontaneous reaction occur in the forward direction for the following reaction?

$$2\,Al(s) + 3\,Cu^{2+}(1\text{ M}) \longrightarrow 3\,Cu(s) + 2\,Al^{3+}(1\text{ M})$$

Analyze

We need to identify the species reduced in the reaction as it is written. We then calculate $E°_{cell}$. If $E°_{cell}$ is positive, then the reaction will occur spontaneously.

Solve

The cell diagram corresponding to the reaction is $Al(s)|Al^{3+}(aq)||Cu^{2+}(aq)|Cu(s)$, and $E°_{cell}$ is

$$\begin{aligned} E°_{cell} &= E°(\text{cathode}) - E°(\text{anode}) \\ &= E°_{Cu^{2+}/Cu} - E°_{Al^{3+}/Al} \\ &= 0.340\text{ V} - (-1.676\text{ V}) = 2.016\text{ V} \end{aligned}$$

Because $E°_{cell}$ is positive, the direction of spontaneous change is that of the forward reaction. Al(s) will displace Cu^{2+} from aqueous solution under standard-state conditions.

Assess

The positive value of $E°_{cell}$ means that the Gibbs energy change for the reaction is negative; hence, the reaction as written is spontaneous. Keep in mind that both $E°$ and $E°_{cell}$ are *intensive* properties. They do not depend on the quantities of materials involved, which means that their values are not affected by the choice of coefficients used to balance the equation for the cell reaction. We could just as well have written:

$$Al(s) + \frac{3}{2}Cu^{2+}(1\,M) \longrightarrow \frac{3}{2}Cu(s) + Al^{3+}(1\,M) \text{ or}$$

$$\frac{2}{3}Al(s) + Cu^{2+}(1\,M) \longrightarrow Cu(s) + \frac{2}{3}Al^{3+}(1\,M)$$

PRACTICE EXAMPLE A: Name one metal ion that Cu(s) will displace from aqueous solution, and determine E°_{cell} for the reaction.

PRACTICE EXAMPLE B: When sodium metal is added to seawater, which has $[Mg^{2+}] = 0.0512$ M, no magnesium metal is obtained. According to $E°$ values, should this displacement reaction occur? What reaction does occur?

EXAMPLE 19-7 Making Qualitative Predictions with Electrode Potential Data

Peroxodisulfate salts, such as $Na_2S_2O_8$, are oxidizing agents used in bleaching. Dichromates such as $K_2Cr_2O_7$ have been used as laboratory oxidizing agents. Which is the better oxidizing agent in acidic solution under standard conditions, $S_2O_8{}^{2-}$ or $Cr_2O_7{}^{2-}$?

Analyze

In a redox reaction, the oxidizing agent is reduced; the greater the tendency for this reduction to occur, the better the oxidizing agent. The reduction tendency, in turn, is measured by the $E°$ value.

Solve

Because the $E°$ value for the reduction of $S_2O_8{}^{2-}$(aq) to $S_2O_8{}^{2-}$(aq) (2.01 V) is larger than that for the reduction of $Cr_2O_7{}^{2-}$(aq) to Cr^{3+}(aq) (1.33 V), $S_2O_8{}^{2-}$(aq) should be the better oxidizing agent.

Assess

Inspection of standard reduction potentials enables us to qualitatively assess the spontaneity of a particular redox reaction.

PRACTICE EXAMPLE A: An inexpensive way to produce peroxodisulfates would be to pass O_2(g) through an acidic solution containing sulfate ion. Is this method feasible under standard conditions? [*Hint:* What would be the reduction half-cell reaction

PRACTICE EXAMPLE B: Consider the following observations: (1) Aqueous solutions of Sn^{2+} are difficult to maintain because atmospheric oxygen easily oxidizes Sn^{2+} to Sn^{4+}. (2) One way to preserve the Sn^{2+}(aq) solutions is to add some metallic tin. *Without doing detailed calculations,* explain these two statements by using $E°$ data.

he Relationship Between E°_{cell} and K

$_rG°$ and E°_{cell} were related through equation (19.15). In Chapter 13, $\Delta_r G°$ and were related through equation (13.17). The three quantities are thus related this way.

$$\Delta_r G° = -RT \ln K = -zFE^\circ_{cell}$$

nd therefore,

$$E^\circ_{cell} = \frac{RT}{zF}\ln K \qquad \textbf{(19.16)}$$

equation (19.16), R has a value of 8.3145 J mol^{-1} K^{-1} and z represents the umber of electrons involved in the reaction. If we then specify a temperature of °C = 298.15 K (the temperature at which electrode potentials are generally etermined), the combined terms "RT/F" in equation (19.16) can be replaced by single constant. This constant has the value 0.025693 J/C = 0.025693 V.

$$E^\circ_{cell} = \frac{RT}{zF}\ln K = \frac{8.3145\text{ J }\cancel{\text{mol}^{-1}}\ \cancel{\text{K}^{-1}} \times 298.15\ \cancel{\text{K}}}{z \times 96{,}485\text{ C }\cancel{\text{mol}^{-1}}}\ln K$$

◀ Note that any electrochemical cell, if left in a completed circuit, will eventually die as the redox reaction goes to completion. This means that the cell potential will eventually drop to zero. In Chapter 13, the relationship between Gibbs energy and equilibrium was established in a similar way.

◀ Equation 19.16 gives the expected result that reactions with equilibrium constants larger than one have a positive standard cell potential.

$$E^\circ_{cell} = \frac{0.025693\ \text{V}}{z} \ln K \quad (19.17)$$

The relationship between E°_{cell} and K is illustrated in Example 19-8. Als Figure 19-8 summarizes several important relationships from thermodynam ics, equilibrium, and electrochemistry.

EXAMPLE 19-8 Relating K to E°_{cell} for a Redox Reaction

What is the value of the equilibrium constant K for the reaction between copper metal and iron(III) ions in aqueous solution at 25 °C?

$$Cu(s) + 2\,Fe^{3+}(aq) \longrightarrow Cu^{2+}(aq) + 2\,Fe^{2+}(aq) \qquad K = ?$$

Analyze

We first identify the reactant that is reduced and the reactant that is oxidized. We then use the data in Table 19.1 and Appendix D to obtain the standard reduction potentials and hence the cell potential. Finally, we use equation (19.17) to obtain K from E°_{cell}.

Solve

First, we use data from Table 19.1 to determine E°_{cell}.

$$E^\circ_{cell} = E^\circ(\text{reduction half-cell}) - E^\circ(\text{oxidation half-cell})$$
$$= E^\circ_{Fe^{3+}/Fe^{2+}} - E^\circ_{Cu^{2+}/Cu}$$
$$= 0.771\ \text{V} - 0.340\ \text{V} = 0.431\ \text{V}$$

The charge number (z) for the cell reaction is 2.

$$E^\circ_{cell} = 0.431\ \text{V} = \frac{0.02569\ \text{V}}{2} \ln K$$

$$\ln K = \frac{2 \times 0.431\ \cancel{\text{V}}}{0.02569\ \cancel{\text{V}}} = 33.6$$

$$K = e^{33.6} = 4 \times 10^{14}$$

Assess

The positive value of the cell potential means that the equilibrium constant is greater than one. The value of the equilibrium constant is very large, and so we can expect this reaction to go to completion.

PRACTICE EXAMPLE A: Should the displacement of Cu^{2+} from aqueous solution by Al(s) go to completion? [*Hint:* Base your assessment on the value of K for the displacement reaction. We determined E°_{cell} for this reaction in Example 19-6.]

PRACTICE EXAMPLE B: Should the reaction of Sn(s) and Pb^{2+}(aq) go to completion? Explain.

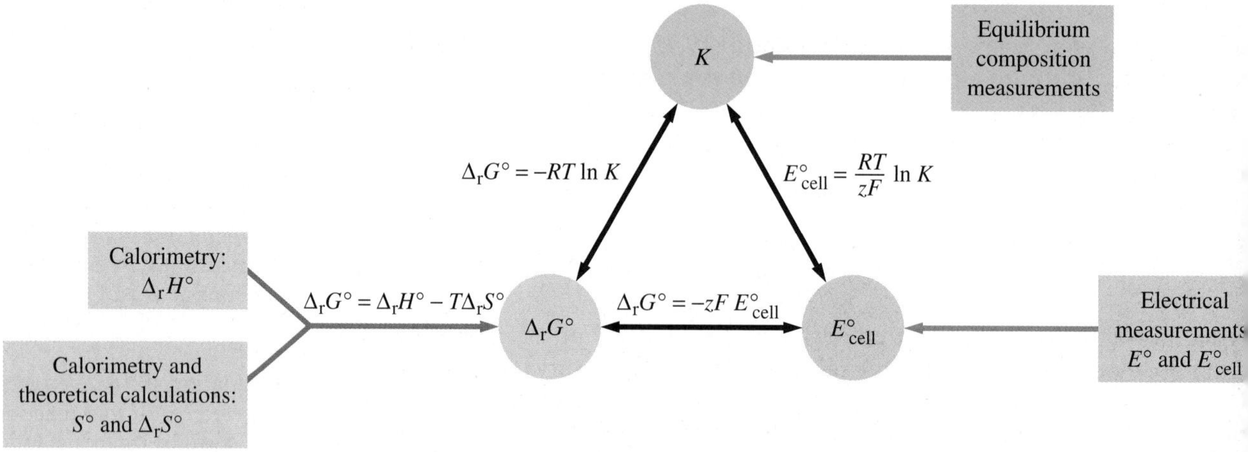

▲ FIGURE 19-8
A summary of important thermodynamic, equilibrium, and electrochemic relationships under standard conditions

19-4 CONCEPT ASSESSMENT

When two cell reactions with reactants and products in their standard states are compared, one is found to have a negative value of E°_{cell} and the other a positive value. Which cell reaction will proceed toward the formation of more products to establish equilibrium? Will a net cell reaction occur in the other case? If so, how will the equilibrium concentrations compare with the initial concentrations?

9-4 E_{cell} as a Function of Concentrations

hen we combine standard electrode potentials, we obtain a standard E°_{cell}, ch as $E^\circ_{cell} = 1.103$ V for the voltaic cell of Figure 19-4. For the following cell action at *nonstandard* conditions, however, the measured E_{cell} is not 1.103 V.

$$\text{Zn(s)} + \text{Cu}^{2+}(2.0\text{ M}) \longrightarrow \text{Zn}^{2+}(0.10\text{ M}) + \text{Cu(s)} \qquad E_{cell} = 1.142\text{ V}$$

xperimental measurements of cell potentials are often made for nonstandard nditions; these measurements have great significance, especially for perrming chemical analyses.

From Le Châtelier's principle, it would seem that *increasing* the concentran of a reactant (Cu^{2+}) while *decreasing* the concentration of a product (Zn^{2+}) ould favor the forward reaction. Zn(s) should displace Cu^{2+}(aq) even more adily than for standard-state conditions and $E_{cell} > 1.103$ V. E_{cell} is found to ry linearly with log ($[Zn^{2+}]/[Cu^{2+}]$), as illustrated in Figure 19-9.

It is not difficult to establish the relationship between the cell potential, E_{cell}, d the concentrations of reactants and products. We start from equation (13.15), hich involves the thermodynamic reaction quotient Q.

$$\Delta_r G = \Delta_r G^\circ + RT \ln Q$$

r $\Delta_r G$ and $\Delta_r G^\circ$, we can substitute $-zFE_{cell}$ and $-zFE^\circ_{cell}$, respectively.

$$-zFE_{cell} = -zFE^\circ_{cell} + RT \ln Q$$

ividing through by $-zF$ gives

$$E_{cell} = E^\circ_{cell} - \frac{RT}{zF} \ln Q$$

This equation was first proposed by Walther Nernst in 1889 and is known the **Nernst equation**.

By specifying a temperature of 298.15 K and replacing RT/F by 0.025693 V, in the development of equation (19.17), we find that the final form of the ernst equation is

$$E_{cell} = E^\circ_{cell} - \frac{0.0257\text{ V}}{z} \ln Q \qquad \textbf{(19.18)}$$

the Nernst equation, we make the usual substitutions into Q: $a = 1$ for the tivities of pure solids and liquids, partial pressures (bar) for the activities of ses, and molarities for the activities of solution components. Example 19-9 emonstrates that the Nernst equation makes it possible to calculate E_{cell} for y chosen concentrations, not just for standard conditions.

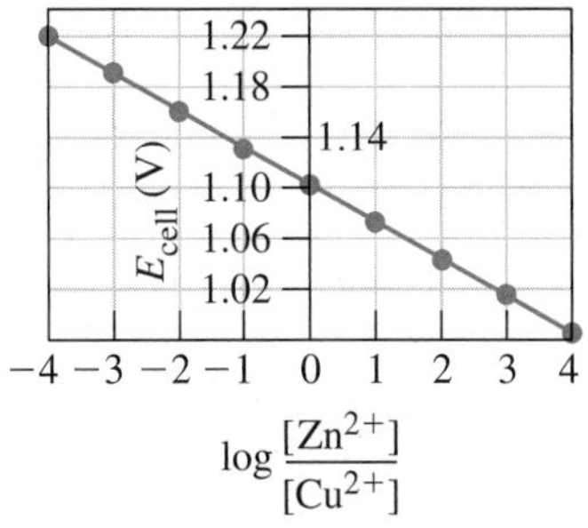

▲ FIGURE 19-9
Variation of E_{cell} with ion concentrations
The cell reaction is Zn(s) + Cu^{2+}(aq) ⟶ Zn^{2+}(aq) + Cu(s) and has $E^\circ_{cell} = 1.103$ V.

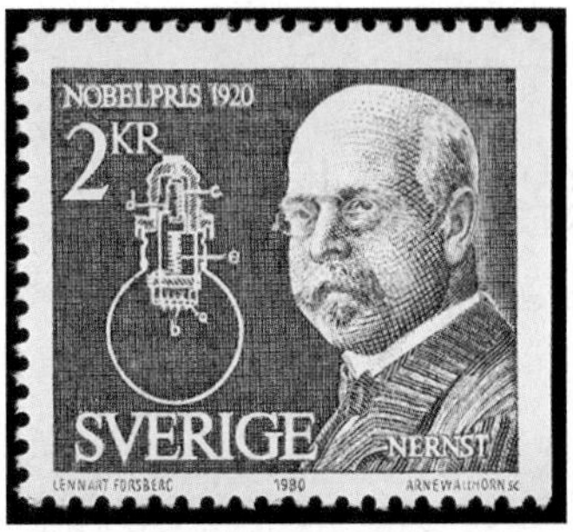

C. Marvin Lang/University of Wisconsin

▲ Walther Nernst (1864–1941)
Nernst was only 25 years old when he formulated his equation relating cell voltages and concentrations. He is also credited with proposing the solubility product concept in the same year. In 1906, he announced his "heat theorem," which we now know as the third law of thermodynamics.

◀ The Nernst equation can be expressed in terms of $\log_{10}$ by using the identity

$$\ln Q = (\log_{10} Q)/\log_{10} e = 2.303 \log_{10} Q$$

in equation (19.18). We obtain

$$E_{cell} = E^\circ_{cell} - \frac{0.0592\text{ V}}{z} \log_{10} Q$$

EXAMPLE 19-9 Applying the Nernst Equation for Determining E_{cell}

What is the value of E_{cell} for the voltaic cell pictured in Figure 19-10 and diagrammed as follows?

$$Pt|Fe^{2+}(0.10\ M), Fe^{3+}(0.20\ M)||Ag^{+}(1.0\ M)|Ag(s) \qquad E_{cell} = ?$$

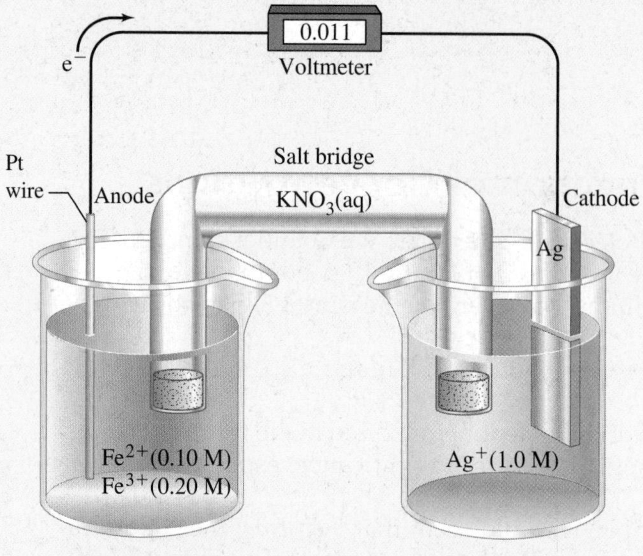

▲ FIGURE 19-10
A voltaic cell with nonstandard conditions—Example 19-9 illustrated

Analyze

To use the Nernst equation we need to establish $E°_{cell}$ and the reaction to which the cell diagram corresponds so that the form of the reaction quotient (Q) can be revealed (see Example 19-2). Once we have determined the form of the Nernst equation, we can insert the concentration of the species.

Solve

Two steps are required when using the Nernst equation. First, to determine $E°_{cell}$, use data from Table 19.1 to write

$$E°_{cell} = E°(\text{cathode}) - E°(\text{anode})$$
$$= E°_{Ag^+/Ag} - E°_{Fe^{3+}/Fe^{2+}} \qquad (19.19)$$
$$= 0.800\ V - 0.771\ V = 0.029\ V$$

Now, to determine E_{cell} for the reaction

$$Fe^{2+}(0.10\ M) + Ag^{+}(1.0\ M) \longrightarrow Fe^{3+}(0.20\ M) + Ag(s) \quad E_{cell} = ? \qquad (19.20)$$

substitute appropriate values into the Nernst equation (19.18), starting with $E°_{cell} = 0.029$ V and $z = 1$,

$$E_{cell} = 0.029\ V - \frac{0.0257\ V}{1}\ln\frac{[Fe^{3+}]}{[Fe^{2+}][Ag^{+}]}$$

and for concentrations $[Fe^{2+}] = 0.10$ M; $[Fe^{3+}] = 0.20$ M; $[Ag^{+}] = 1.0$ M.

$$E_{cell} = 0.029\ V - 0.0257\ V \times \ln\frac{0.20}{0.10 \times 1.0}$$
$$= 0.029\ V - 0.0257\ V \times \ln 2 = 0.029\ V - 0.018\ V$$
$$= 0.011\ V$$

Assess

The E_{cell} is positive so that the reaction is spontaneous in the direction of the reduction of silver.

PRACTICE EXAMPLE A: Calculate E_{cell} for the following voltaic cell.

$$Al(s)|Al^{3+}(0.36\ M)||Sn^{4+}(0.086\ M), Sn^{2+}(0.54\ M)|Pt$$

PRACTICE EXAMPLE B: Calculate E_{cell} for the following voltaic cell.

$$Pt(s)|Cl_2(1\ bar)|Cl^{-}(1.0\ M)||Pb^{2+}(0.050\ M), H^{+}(0.10\ M)|PbO_2(s)$$

19-5 CONCEPT ASSESSMENT

Describe *two* sets of conditions under which the measured E_{cell} for a reaction is equal to E°_{cell}.

In Section 19-3, we developed a criterion for spontaneous change $E_{cell} > 0$), but we used the criterion only with $E°$ data from Table 19.1 ualitative conclusions reached with E°_{cell} values often hold over a broad range f nonstandard conditions as well. However, when E°_{cell} is within a few hundredths of a volt of zero, it is sometimes necessary to determine E_{cell} for non- andard conditions in order to apply the criterion for spontaneity of redox eactions, as illustrated in Example 19-10.

EXAMPLE 19-10 Predicting Spontaneous Reactions for Nonstandard Conditions

Will the cell reaction proceed spontaneously as written for the following cell?

$$Ag(s)|Ag^+(0.075\text{ M})||Hg^{2+}(0.85\text{ M})|Hg(l)$$

Analyze

To decide whether a reaction is spontaneous, we need to calculate E_{cell} by using equation (19.18) with the concentrations given. We then identify the oxidized and reduced species and look up the appropriate standard half-cell potentials. Then we construct the chemical equation that corresponds to the cell diagram, choosing an appropriate electron number (z).

Solve

To determine E°_{cell} from $E°$ data we write

$$\begin{array}{lrcl} \textit{Oxidation:} & 2\,Ag(s) & \longrightarrow & 2\,Ag^+(aq) + 2\,e^- \\ \textit{Reduction:} & Hg^{2+}(aq) + 2\,e^- & \longrightarrow & Hg(l) \\ \hline \textit{Overall:} & 2\,Ag(s) + Hg^{2+}(aq) & \longrightarrow & 2\,Ag^+(aq) + Hg(l) \end{array}$$

$$\begin{aligned} E^{\circ}_{cell} &= E°(\text{reduction half-cell}) - E°(\text{oxidation half-cell}) \\ &= 0.854\text{ V} - (0.800\text{ V}) = 0.054\text{ V} \end{aligned}$$

The overall reaction that we have written has an electron number $z = 2$, so that the Nernst equation is

$$E_{cell} = 0.054\text{ V} - \frac{0.0257\text{ V}}{2}\ln\frac{[Ag^+]^2}{[Hg^{2+}]}$$

By using the concentrations $[Ag^+] = 0.075$ M and $[Hg^{2+}] = 0.85$ M provided, we obtain

$$\begin{aligned} E_{cell} &= 0.054\text{ V} - 0.0129\text{ V}\ln\frac{[0.075]^2}{[0.85]} \\ &= 0.054\text{ V} - 0.0129\text{ V}\ln(0.0066) = 0.054\text{ V} - 0.0129\text{ V} \times (-5.021) \\ &= 0.054\text{ V} + 0.065\text{ V} = 0.119\text{ V} \end{aligned}$$

Because $E_{cell} > 0$, we conclude that the reaction as written is spontaneous.

Assess

If we had used an electron number of $z = 1$, the overall reaction would have been

$$\textit{Overall } (z = 1)\text{: } Ag(s) + \frac{1}{2}Hg^{2+}(aq) \longrightarrow Ag^+(aq) + \frac{1}{2}Hg(l)$$

The corresponding Nernst equation is

$$E_{cell} = 0.054\text{ V} - \frac{0.0257\text{ V}}{1}\ln\frac{[Ag^+]}{[Hg^{2+}]^{1/2}}$$

(continued)

By rearranging slightly and using a property of logarithms,

$$E_{cell} = 0.054\text{ V} - \frac{0.0257\text{ V}}{1}\ln\left(\frac{[Ag^+]^2}{[Hg^{2+}]}\right)^{1/2} = 0.054\text{ V} - \frac{0.0257\text{ V}}{2}\ln\frac{[Ag^+]^2}{[Hg^{2+}]}$$

we have recovered the Nernst equation for the cell reaction by using an electron number (z) of 2. We conclude that as long as we balance charge and the electron number correctly, we will always get the correct result.

PRACTICE EXAMPLE A: Will the cell reaction proceed spontaneously as written for the following cell?

$$Cu(s)|Cu^{2+}(0.15\text{ M})||Fe^{3+}(0.35\text{ M}), Fe^{2+}(0.25\text{ M})|Pt(s)$$

PRACTICE EXAMPLE B: For what ratio of $[Ag^+]^2/[Hg^{2+}]$ will the cell reaction in Example 19-10 not be spontaneous in either direction?

19-6 CONCEPT ASSESSMENT

The following cell is set up under standard-state conditions.

$$Pb(s)|Pb^{2+}||Cu^{2+}|Cu(s) \qquad E^\circ_{cell} = 0.47\text{ V}$$

When sodium sulfate is added to the anode half-cell, formation of a white precipitate is observed, accompanied by a change in the value of E_{cell}. Explain these observations, and predict whether the new E_{cell} is greater or less than E°_{cell}.

Concentration Cells

The voltaic cell in Figure 19-11 consists of two hydrogen electrodes. One is standard hydrogen electrode (SHE), and the other is a hydrogen electro[de] immersed in a solution of unknown $[H^+]$, less than 1 M. The cell diagram is

$$Pt|H_2(g, 1\text{ bar})|H^+(x\text{ M})||H^+(1\text{ M})|H_2(g, 1\text{ bar})|Pt$$

The reaction occurring in this cell is

$$\begin{aligned} &\textit{Reduction:} && 2\,H^+(1\text{ M}) + \cancel{2e^-} \longrightarrow \cancel{H_2(g, 1\text{ bar})} \\ &\textit{Oxidation:} && \cancel{H_2(g, 1\text{ bar})} \longrightarrow 2\,H^+(x\text{ M}) + \cancel{2e^-} \\ &\textit{Overall:} && 2\,H^+(1\text{ M}) \longrightarrow 2\,H^+(x\text{ M}) \end{aligned} \qquad \textbf{(19.}$$

$$E^\circ_{cell} = E^\circ_{H^+/H_2} - E^\circ_{H^+/H_2} = 0\text{ V}$$

The voltaic cell in Figure 19-11 is called a concentration cell. A **concentrati[on] cell** consists of two half-cells with *identical electrodes* but different ion conce[n]trations. Because the electrodes are identical, the standard electrode potentia[ls] are numerically equal and subtracting one from the other leads to the val[ue] $E^\circ_{cell} = 0$. However, because the ion concentrations differ, there is a potenti[al]

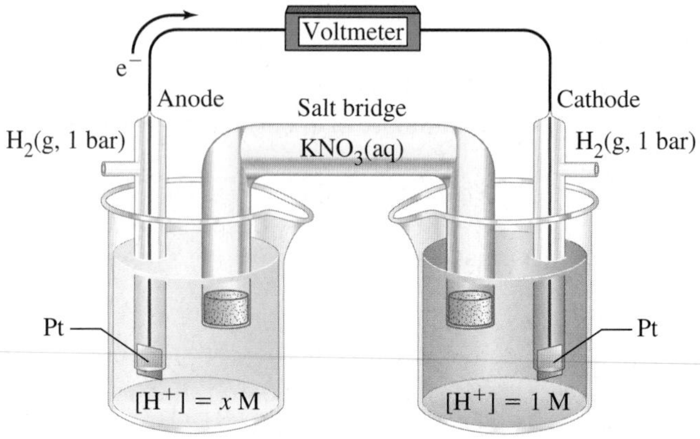

▶ FIGURE 19-11
A concentration cell
The cell consists of two hydrogen electrodes. The electrode on the right is a SHE. Oxidation occurs at the anode on the left, where $[H^+]$ is less than 1 M. The reading on the voltmeter is directly proportional to the pH of the solution in the anode compartment.

fference between the two half-cells. The spontaneous change in a concentra-on cell always occurs such that the concentrated solution becomes more lute, and the dilute solution becomes more concentrated. The final result is if the solutions were simply mixed. In a concentration cell, however, the tural tendency for entropy to increase in a mixing process is used as a eans of producing electricity.

The Nernst equation for reaction (19.21) takes the form

$$E_{cell} = E^\circ_{cell} - \frac{0.0257\ \text{V}}{2}\ln\frac{x^2}{1^2}$$

hich simplifies to

$$E_{cell} = 0 - \frac{0.0257\ \text{V}}{2} \times 2\ln\frac{x}{1} = -0.0257\ \text{V}\ln x$$

Since x is $[H^+]$, we can rewrite the expression above in terms of $H = -\log[H^+]$ by using $\ln[H^+] = -2.303\log[H^+]$ to switch from natural common logarithms. The final result is

$$E_{cell} = (0.0592\ \text{pH})\ \text{V} \qquad \textbf{(19.22)}$$

here the pH is that of the unknown solution. If an unknown solution has a H of 3.50, for example, the measured cell voltage in Figure 19-11 will be $_{ell} = (0.0592 \times 3.50)\ \text{V} = 0.207\ \text{V}$.

Constructing and using a hydrogen electrode is difficult. The Pt metal surface ust be specially prepared and maintained, gas pressure must be controlled, d the electrode cannot be used in the presence of strong oxidizing or reducing gents. The solution to these problems is discussed later in this chapter.

19-7 CONCEPT ASSESSMENT

Write a cell diagram for a possible voltaic cell in which the cell reaction is $Cl^-(0.50\ M) \longrightarrow Cl^-(0.10\ M)$. What would be E_{cell} for this reaction?

leasurement of K_{sp}

he difference in concentration of ions in the two half-cells of a concentration ll accounts for the observed E_{cell}. It also provides a basis for determining K_{sp} alues for sparingly soluble ionic compounds. Consider the following concen-ation cell.

$$Ag(s)|Ag^+(\text{satd AgI})||Ag^+(0.100\ M)|Ag \qquad E_{cell} = 0.417\ \text{V}$$

t the anode, a silver electrode is placed in a saturated aqueous solution of sil-er iodide. At the cathode, a second silver electrode is placed in a solution ith $[Ag^+] = 0.100$ M. The two half-cells are connected by a salt bridge, and e measured cell voltage is 0.417 V (Fig. 19-12). The cell reaction occurring in is *concentration cell* is

$$\begin{array}{lrl} \textit{Reduction:} & Ag^+(0.100\ M) + e^- & \longrightarrow Ag(s) \\ \textit{Oxidation:} & Ag(s) & \longrightarrow Ag^+(\text{satd AgI}) \\ \hline \textit{Overall:} & Ag^+(0.100\ M) & \longrightarrow Ag^+(\text{satd AgI}) \end{array} \qquad \textbf{(19.23)}$$

he calculation of K_{sp} of silver iodide is completed in Example 19-11.

lternative Standard Electrodes

he standard hydrogen electrode is not the most convenient to use because it quires highly flammable hydrogen gas to be bubbled over the platinum elec-ode. Other electrodes can be used as secondary standard electrodes, such as the

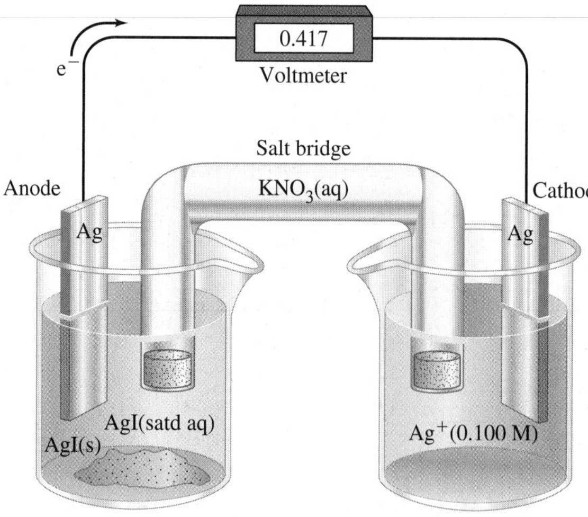

▶ FIGURE 19-12
A concentration cell for determining K_{sp} of AgI
The silver electrode in the anode compartment is in contact with a saturated solution of AgI. In the cathode compartment, $[Ag^+] = 0.100$ M.

EXAMPLE 19-11 Using a Voltaic Cell to Determine K_{sp} of a Slightly Soluble Solute

With the data given for reaction (19.23), calculate K_{sp} for AgI.

$$AgI(s) \rightleftharpoons Ag^+(aq) + I^-(aq) \qquad K_{sp} = ?$$

Analyze

Once we have determined the concentration of Ag^+ ions from the Nernst equation for the cell, we can calculate the equilibrium constant by using the expression for the solubility product.

Solve

$[Ag^+]$ in saturated silver iodide solution can be represented as x. The Nernst equation is then applied to reaction (19.23). (To simplify the equations that follow, we have dropped the unit V, which would otherwise appear in several places.)

$$E_{cell} = E^\circ_{cell} - \frac{0.0257}{z}\ln\frac{[Ag^+]_{satd\ AgI}}{[Ag^+]_{0.100\ M\ soln}}$$

$$= E^\circ_{cell} - \frac{0.0257}{1}\ln\frac{x}{0.100}$$

$$0.417 = 0 - 0.0257(\ln x - \ln 0.100)$$

Divide both sides of the equation by 0.0257.

$$\frac{0.417}{0.0257} = -\ln x + \ln 0.100$$

$$\ln x = \ln 0.100 - \frac{0.417}{0.0257} = -2.30 - 16.2 = -18.5$$

$$x = [Ag^+] = e^{-18.5} = 9.2 \times 10^{-9}\ M$$

Because in saturated AgI the concentrations of Ag^+ and I^- are equal,

$$K_{sp} = [Ag^+][I^-] = (9.2 \times 10^{-9})(9.2 \times 10^{-9}) = 8.5 \times 10^{-17}$$

Assess

Apart from the use of the Nernst equation, the other essential aspect is the realization that the only source of Ag^+ and I^- is from the AgI present; the saturated AgI(s) electrode has $[Ag^+] = [I^-]$.

PRACTICE EXAMPLE A: K_{sp} for AgCl $= 1.8 \times 10^{-10}$. What would be the measured E_{cell} for the voltaic cell in Example 19-11 if the contents of the anode half-cell were saturated AgCl(aq) and AgCl(s)?

PRACTICE EXAMPLE B: Calculate the K_{sp} for PbI_2 given the following concentration cell information.

$$Pb(s)|Pb^{2+}(\text{satd } PbI_2)||Pb^{2+}(0.100\ M)|Pb(s) \qquad E_{cell} = 0.0567\ V$$

silver–silver chloride electrode, in which a silver wire is covered with a layer insoluble solid silver chloride. The silver-chloride-coated silver wire is immerse in a 1 M potassium chloride solution (see Figure 19-13a), giving the electrode

$$Ag(s)|AgCl(s)|Cl^-(1.0\ M)$$

ith a half-cell reaction of

$$AgCl(s) + e^- \longrightarrow Ag(s) + Cl^-(aq)$$

nis electrode has been measured against the standard hydrogen elec-ode, and the electrode potential has been found to be 0.22233 V at 25 °C. nce all components of this electrode are in their standard states, the stan-ard electrode potential of the silver–silver chloride electrode is 0.22233 V 25 °C.

An alternative electrode is the calomel electrode, illustrated in Figure 19-13(b). this electrode, mercurous chloride (calomel, Hg_2Cl_2) is mixed with mercury form a paste, which is in contact with liquid mercury, Hg(l), and the whole tup is immersed in either a 1.0 M solution of potassium chloride or a satu-ted solution of potassium chloride. The electrode is

$$Hg(l)|Hg_2Cl_2(s)|Cl^-(1.0\,M)$$

d the half-cell reaction is

$$\frac{1}{2}Hg_2Cl_2(s) + e^- \longrightarrow Hg(l) + Cl^-(aq)$$

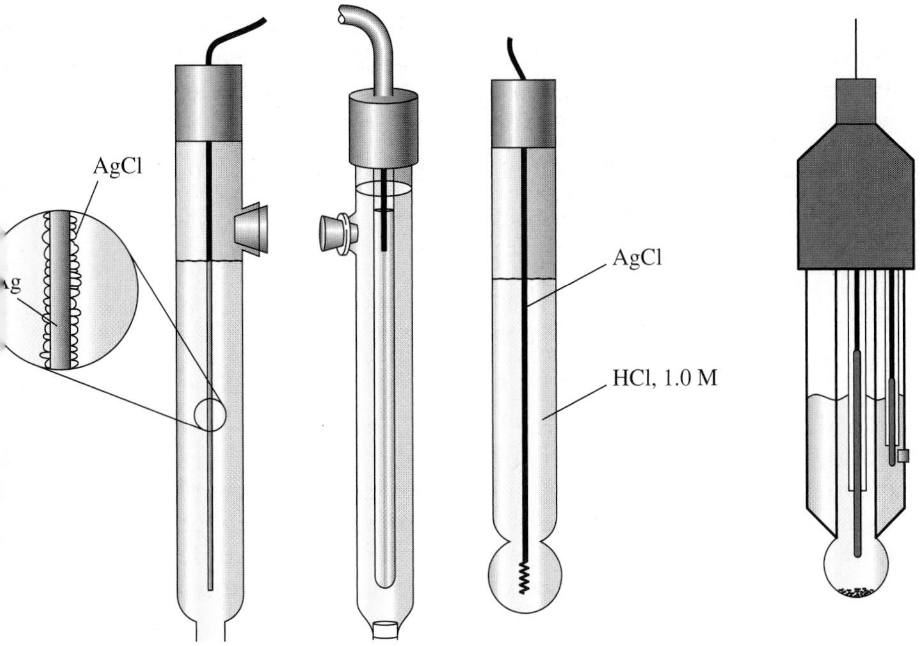

FIGURE 19-13
chematic diagrams of some common electrodes
) The silver–silver chloride electrode. Silver wire is coated with silver chloride and mersed in a 1 M aqueous solution of KCl. At the bottom of the tube is a fritted orous) disc to allow contact with a solution of interest. **(b)** The standard calomel ectrode is a tube containing a paste of calomel and mercury, immersed in a 1.0 M olution of KCl. Contact with an external circuit is made with a Pt wire inserted in the ner tube; the inner tube makes contact with the outer 1.0 M solution of KCl through a nall hole in the bottom of the inner tube. **(c)** A glass electrode consists of tube with a ery thin glass bulb at the end and a Ag–AgCl electrode immersed in a 1.0 M HCl olution. When the glass electrode is dipped into a solution, ions interact with the embrane. The potential established on the silver wire depends on the solution being sted. **(d)** A modern pH electrode consists of a glass electrode and an internal g–AgCl reference electrode. There is a small sintered disc in the side of the outer be that acts as a salt bridge between the electrode and the unknown solution.

The standard electrode potential at 25 °C is 0.2680 V. If, however, a saturate solution of KCl is used, as opposed to one of 1 M, the reduction potential 0.2412 V. This electrode is known as the saturated calomel electrode (SCE) ar is often used as a reference.

In practice a variety of reference electrodes are used; therefore, it is nece sary to quote reduction potentials with respect to a specific reference.

19-8 CONCEPT ASSESSMENT

Why do the calomel standard reduction potential and the saturated calomel electrode have different potentials?

The Glass Electrode and the Electrochemical Measurement of pH

To measure the pH of a solution electrochemically, we need an electrode th responds to changes in $[H^+(aq)]$. We noted that the standard hydrogen ele trode is difficult to use for this purpose, and so, for routine use, a simpler an safer electrode is needed. Such an electrode is the *glass electrode*, which consis of a very thin walled glass bulb (see Figure 19-13c) at the end of a tube that co tains a silver–silver chloride electrode and a HCl solution of known compositic (e.g., 1 M). When the bulb is placed in a solution of unknown pH, a potenti develops because of the concentration difference across the membrane, anal gous to a concentration cell. To measure this potential difference, a referenc electrode is used, which can be either a saturated calomel electrode or a secon silver–silver chloride electrode, as in the combination electrode shown i Figure 19-13(d). The overall cell can be represented as

$$Ag(s)|AgCl(s)|Cl^-(1.0\ M), H^+(1.0\ M)|\text{glass membrane}|H^+(\text{unknown})||Cl^-(1.0\ M)|AgCl(s)|Ag(s$$

where the two electrodes are connected by a salt bridge. The half-cell rea tions are

$$\begin{aligned} Ag(s) + Cl^-(aq) &\longrightarrow AgCl(s) + e^- \\ H^+(1.0\ M) &\longrightarrow H^+(\text{unknown}) \\ AgCl(s) + e^- &\longrightarrow Ag(s) + Cl^-(aq) \end{aligned}$$

The half-cell potentials of the two half-cell reactions for the silver–silv chloride electrodes cancel each other out and make no contribution to the ce potential. The difference in the molar Gibbs energy between the two half-cel corresponding to the dilution of protons from a known concentration of 1.0 M to the unknown solution is the source of the potential difference across th glass membrane. The difference in the molar Gibbs energy across the men brane, using $G_m = G^\circ_m + RT \ln[H^+]$, is

▶ Recall that G°_m and G_m represent the standard molar Gibbs energy and the molar Gibbs energy, respectively. $G_m = G/n$ was discussed in Chapter 13 on page 623.

$$\begin{aligned} \Delta G_m &= G_m(\text{unknown}) - G_m(1.0\ M) \\ &= G^\circ_m + RT \ln[\text{unknown}] - G^\circ_m - RT \ln 1.0 \\ &= RT \ln[\text{unknown}] \end{aligned}$$

Converting this to a potential by dividing by $-zF$, $z = 1$, and assumin $T = 298.15$ K, we obtain

$$E_{cell} = 0.0592\ V \times pH$$

after converting the logarithm to base 10 and using the definition o $pH = -\log[\text{unknown}]$. The cell potential is measured with a pH meter, voltage measuring device that electronically converts E_{cell} to pH and display the result in pH units.

The glass electrode was devised in 1906 by German biologist Max Creme and it was the prototype for a large number of membrane electrodes that ar selective for a particular ion, such as the ions K^+, NH_4^+, Cl^-, and many other ıch electrodes are known collectively as *ion-selective electrodes*, and they have any applications in environmental chemistry and biochemistry.

9-5 Batteries: Producing Electricity Through Chemical Reactions

battery is a device that stores chemical energy for later release as electric-. Some batteries consist of a single voltaic cell with two electrodes and the ›propriate electrolyte(s); an example is a flashlight cell. Other batteries nsist of two or more voltaic cells joined in series fashion—plus to inus—to increase the total voltage; an example is an automobile battery. this section, we will consider three types of voltaic cells and the batteries ısed on them.

◀ Batteries are vitally important to modern society. Annual production in developed nations has been estimated at more than 10 batteries per person per year.

- **Primary cells.** The cell reaction in a primary cell is not reversible. When the reactants have been mostly converted to products, no more electricity is produced and a battery employing a primary cell(s) is dead.
- **Secondary cells.** The cell reaction in a secondary cell *can* be reversed by passing electricity through the cell (charging). A battery employing secondary cells can be used through several hundred or more cycles of discharging followed by charging.
- **Reserve batteries.** The electrolyte in a reserve battery is isolated from the rest of the battery to reduce self-discharge or the chemical degradation typically found in other batteries. The battery is designed as a long-term storage battery, which can be used to deliver high power over a relatively short time.
- **Flow batteries** and **fuel cells.** Materials (reactants, products, and electrolytes) pass through the battery, which is simply a converter of chemical energy to electric energy. These types of batteries can be run indefinitely as long as they are supplied by electrolytes.

◀ Cell phones, laptop computers, and many other devices rely heavily on rechargeable batteries. Advances in electrochemistry and engineering are leading to the development of batteries that weigh less, last longer, and provide more power for portable electronic devices.

he Leclanché (Dry) Cell

he most common form of voltaic cell is the *Leclanché cell*, invented by the ench chemist Georges Leclanché (1839–1882) in the 1860s. Popularly called a *y cell*, because no free liquid is present, or *flashlight battery*, the Leclanché cell diagrammed in Figure 19-14. In this cell, oxidation occurs at a zinc anode ıd reduction at an inert carbon (graphite) cathode. The electrolyte is a moist aste of MnO_2, $ZnCl_2$, NH_4Cl, and carbon black (soot). The maximum cell oltage is 1.55 V. The anode (oxidation) half-cell reaction is simple.

$$\textit{Oxidation:}\quad Zn(s) \longrightarrow Zn^{2+}(aq) + 2\,e^-$$

he reduction is more complex. Essentially, it involves the reduction of MnO_2 compounds having Mn in a +3 oxidation state, for example,

$$\textit{Reduction:}\quad 2\,MnO_2(s) + H_2O(l) + 2\,e^- \longrightarrow Mn_2O_3(s) + 2\,OH^-(aq)$$

n acid–base reaction occurs between NH_4^+ (from NH_4Cl) and OH^-.

$$NH_4^+(aq) + OH^-(aq) \longrightarrow NH_3(g) + H_2O(l)$$

buildup of $NH_3(g)$ around the cathode would disrupt the current because ıe $NH_3(g)$ adheres to the cathode. That buildup is prevented by a reaction etween Zn^{2+} and $NH_3(g)$ to form the complex ion $[Zn(NH_3)_2]^{2+}$, which crys-llizes as the chloride salt.

$$Zn^{2+}(aq) + 2\,NH_3(g) + 2\,Cl^-(aq) \longrightarrow [Zn(NH_3)_2]Cl_2(s)$$

The Leclanché cell is a *primary cell*; it cannot be recharged. This cell is heap to make, but it has some drawbacks. When current is drawn rapidly

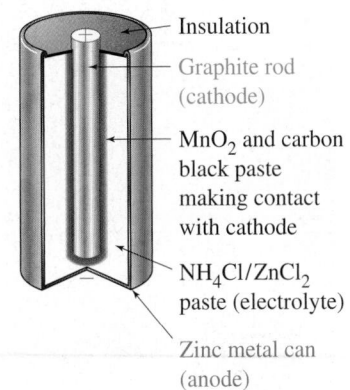

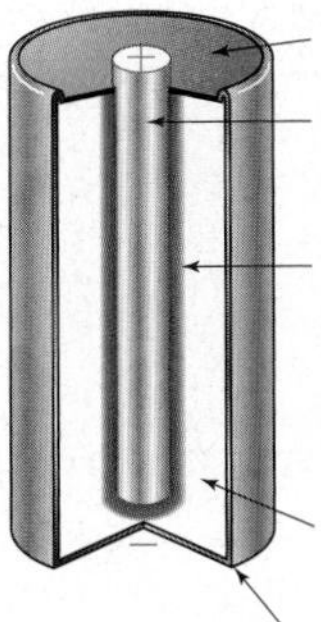

▲ FIGURE 19-14
The Leclanché (dry) cell
The chief components of the cell are a graphite (carbon) rod serving as the cathode, a zinc container serving as the anode, and an electrolyte.

▲ **E°_{cell} is an intensive property** The voltage of a dry cell battery does not depend on the size of the battery—all of those pictured here are 1.5 V batteries. Although these batteries deliver the same voltage, the total energy output of each battery is different.

from the cell, products, such as NH_3, build up on the electrodes, causing th voltage to drop. Also, because the electrolyte medium is acidic, zinc met slowly dissolves.

A superior form of the Leclanché cell is the *alkaline cell*, which uses NaO or KOH in place of NH_4Cl as the electrolyte. The reduction half-cell reactio is the same as that shown above, but the oxidation half-cell reactio involves the formation of $Zn(OH)_2(s)$, which can be thought of as occurrin in two steps.

$$\begin{aligned} Zn(s) &\longrightarrow Zn^{2+}(aq) + 2\,e^- \\ Zn^{2+}(aq) + 2\,OH^-(aq) &\longrightarrow Zn(OH)_2(s) \\ \hline Zn(s) + 2\,OH^-(aq) &\longrightarrow Zn(OH)_2(s) + 2\,e^- \end{aligned}$$

The advantages of the alkaline cell are that zinc does not dissolve as readily a basic (alkaline) medium as in an acidic medium, and the cell does a bett job of maintaining its voltage as current is drawn from it.

The Lead–Acid (Storage) Battery

Secondary cells are commonly encountered joined together in series in th *lead–acid battery*, or *storage battery*, which has been used in automobile since about 1915 (Fig. 19-15). A storage battery is capable of repeated us because its chemical reactions are reversible. That is, the discharged energ can be restored by supplying electric current to recharge the cells in th battery.

The reactants in a lead–acid cell are spongy lead packed into a lead grid the anode, red-brown lead(IV) oxide packed into a lead grid at the cathod and an electrolyte solution consisting of dilute sulfuric acid (about 35 H_2SO_4, by mass). In this strongly acidic medium, the ionization of H_2SO

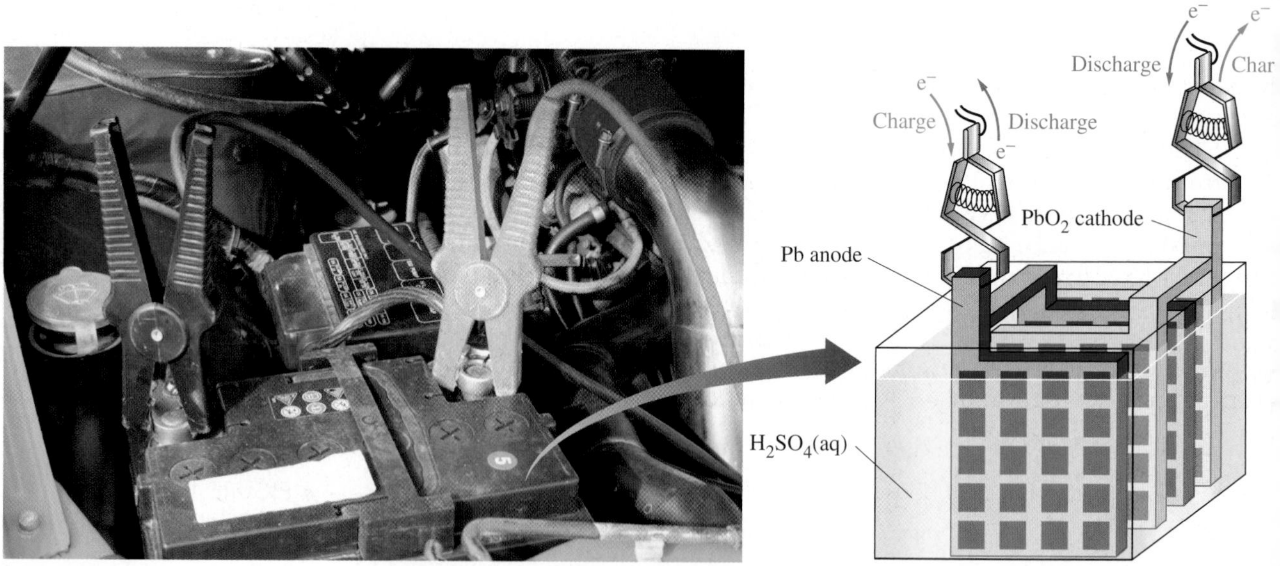

▲ FIGURE 19-15
A lead–acid (storage) battery
The composition of the electrodes is described in the text. The reaction that occurs as the battery is discharged is given in equation (19.24). For the battery depicted in the diagram on the right, two anode plates are connected in parallel, as are two cathode plates. This arrangement produces 4.04 V. The photo on the left shows a typical car battery which is composed of six cells in parallel to produce 12 V.

oes not go to completion. Both $HSO_4^-(aq)$ and $SO_4^{2-}(aq)$ are present, but $ISO_4^-(aq)$ predominates. The half-cell reactions and overall reaction are

$$\begin{array}{ll} \textit{Reduction:} & PbO_2(s) + 3\,H^+(aq) + HSO_4^-(aq) + 2\,e^- \longrightarrow PbSO_4(s) + 2\,H_2O(l) \\ \textit{Oxidation:} & Pb(s) + HSO_4^-(aq) \longrightarrow PbSO_4(s) + H^+(aq) + 2\,e^- \\ \hline \textit{Overall:} & PbO_2(s) + Pb(s) + 2\,H^+(aq) + 2\,HSO_4^-(aq) \longrightarrow 2\,PbSO_4(s) + 2\,H_2O(l) \end{array}$$

$$E_{cell} = E_{PbO_2/PbSO_4} - E_{PbSO_4/Pb} = 1.74\text{ V} - (-0.28\text{ V}) = 2.02\text{ V} \qquad \textbf{(19.24)}$$

◀ You can think of the half-cell reactions as occurring in two steps: (1) oxidation of Pb(s) to $Pb^{2+}(aq)$ and reduction of $PbO_2(s)$ to $Pb^{2+}(aq)$, followed by (2) precipitation of $PbSO_4(s)$ at each electrode.

When an automobile engine is started, the battery is at first discharged.)nce the car is in motion, an alternator powered by the engine constantly echarges the battery. At times, the plates of the battery become coated with $bSO_4(s)$ and the electrolyte becomes sufficiently diluted with water that the attery must be recharged by connecting it to an external electric source. This orces the reverse of reaction (19.24), a nonspontaneous reaction.

$$2\,PbSO_4(s) + 2\,H_2O(l) \longrightarrow Pb(s) + PbO_2(s) + 2\,H^+(aq) + 2\,HSO_4^-(aq) \qquad E_{cell} = -2.02\text{ V}$$

◀ In spite of its usefulness and ability to deliver a strong current, the lead storage battery is also a pollution hazard. All batteries should be disposed of properly and should not be dumped in land fills or garbage disposal sites.

To prevent the anode and cathode from coming into contact with each other, ausing a *short circuit*, sheets of an insulating material are used to separate lternating anode and cathode plates. A group of anodes is connected together lectrically, as is a group of cathodes. This parallel connection increases the lectrode area in contact with the electrolyte solution and increases the current-elivering capacity of the cell. Cells are then joined in a series fashion, positive o negative, to produce a battery. The typical 12 V battery consists of six cells, ach cell with a potential of about 2 V.

◀ Lead-acid storage batteries are also used to power golf carts, wheelchairs, and passenger carts in airport terminals.

he Silver–Zinc Cell: A Button Battery

he cell diagram of a *silver–zinc cell* (Fig. 19-16) is

$$Zn(s), ZnO(s)|KOH(satd)|Ag_2O(s), Ag(s)$$

◀ Rechargeable silver oxide batteries have been developed and provide alternatives to lithium-ion batteries.

he half-cell reactions on discharging are

$$\begin{array}{ll} \textit{Reduction:} & Ag_2O(s) + H_2O(l) + 2\,e^- \longrightarrow 2\,Ag(s) + 2\,OH^-(aq) \\ \textit{Oxidation:} & Zn(s) + 2\,OH^-(aq) \longrightarrow ZnO(s) + H_2O(l) + 2\,e^- \\ \hline \textit{Overall:} & Zn(s) + Ag_2O(s) \longrightarrow ZnO(s) + 2\,Ag(s) \end{array} \qquad \textbf{(19.25)}$$

Because no solution species is involved in the cell reaction, the quantity of lectrolyte is very small and the electrodes can be maintained very close ogether. The cell voltage is 1.8 V, and its storage capacity is six times greater han that of a lead–acid battery of the same size. These characteristics nake batteries, such as the silver–zinc cell, useful in button batteries. These

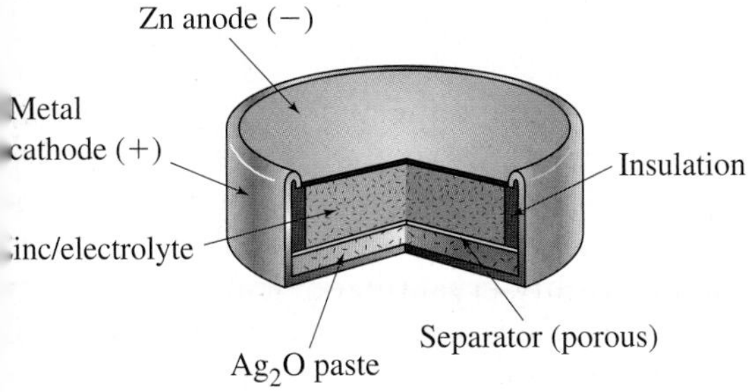

◀ FIGURE 19-16
A silver–zinc button (miniature) cell

miniature batteries are used in watches, hearing aids, and cameras. In adc tion, silver–zinc batteries fulfill the requirements of spacecraft, satellit missiles, rockets, space launch vehicles, torpedoes, underwater vehicles, aı life-support systems. On the Mars *Pathfinder* mission, the rover and t cruise system were powered by solar cells. The energy storage requiremen of the lander were met by modified silver–zinc batteries with about thr times the storage capacity of the standard nickel–cadmium rechargeab battery.

Tatiana Popova / Shutterstock

▲ A rechargeable nickel–cadmium cell, or nicad battery.

The Nickel–Cadmium Cell: A Rechargeable Battery

The *nickel–cadmium cell* (or *nicad battery*) is commonly used in cordless electı devices, such as electric shavers and handheld calculators. The anode in th cell is cadmium metal, and the cathode is the Ni(III) compound NiO(OI supported on nickel metal. The half-cell reactions for a nickel–cadmium ba tery during discharge are

$$\begin{array}{ll} \textit{Reduction:} & 2\,NiO(OH)(s) + 2\,H_2O(l) + 2\,e^- \longrightarrow 2\,Ni(OH)_2(s) + 2\,OH^-(aq) \\ \textit{Oxidation:} & Cd(s) + 2\,OH^-(aq) \longrightarrow Cd(OH)_2(s) + 2\,e^- \\ \hline \textit{Overall:} & Cd(s) + 2\,NiO(OH)(s) + 2\,H_2O(l) \longrightarrow 2\,Ni(OH)_2(s) + Cd(OH)_2(s) \end{array}$$

This cell gives a fairly constant voltage of 1.4 V. When the cell is recharged l connection to an external voltage source, the reactions above are reverse Nickel–cadmium batteries can be recharged many times because the soli products adhere to the surface of the electrodes.

In primary cells the positive and negative electrodes are known as th cathode, where reduction takes place, and the anode, where oxidation tak place. In rechargeable systems, however, we have either a charging mode (a discharging mode, and so depending whether electrons are flowing out (the cell or flowing into the cell, the notion of the anode and th cathode changes. On the discharge of a nicad battery, the NiO(OH) ele trode is the cathode because reduction is taking place, but on the charge, is the anode because oxidation is taking place (the reverse reaction). In di charge mode the NiO(OH) electrode electrons are removed from the ele trode because of the reduction process, and so this electrode is positivel charged. In the charging mode electrons are being removed from this ele trode by the oxidation process; this is the anode and it is positively charge Therefore, regardless of charging or discharging, the NiO(OH) electrod is positive.

The negative electrode, the cadmium electrode in a nicad battery, is th anode on discharging (oxidation) and the cathode (reduction) on charging. I both charging and discharging, the anode is the electrode from which ele trons exit the battery, and the cathode is the electrode at which electrons ent the battery.

In summary, when dealing with rechargeable batteries, it is better to spea of the positive and negative electrodes and avoid the terms *cathode* and *anode*

The Lithium-Ion Battery

Lithium-ion batteries are a type of rechargeable battery now commonly use in consumer electronics, such as cell phones, laptop computers, and MP3 play ers. In a lithium-ion battery, the lithium ion moves between the positive an negative electrodes. The positive electrode consists of lithium cobalt(III) oxid $LiCoO_2$, and the negative electrode is highly crystallized graphite. To com plete the battery an electrolyte is needed, which can consist of an organic so vent and ions, such as $LiPF_6$. The structure of $LiCoO_2$ and graphite electrode is illustrated in Figure 19-17. In the charging cycle at the positive electrode lithium ions are released into the electrolyte solution as electrons are remove

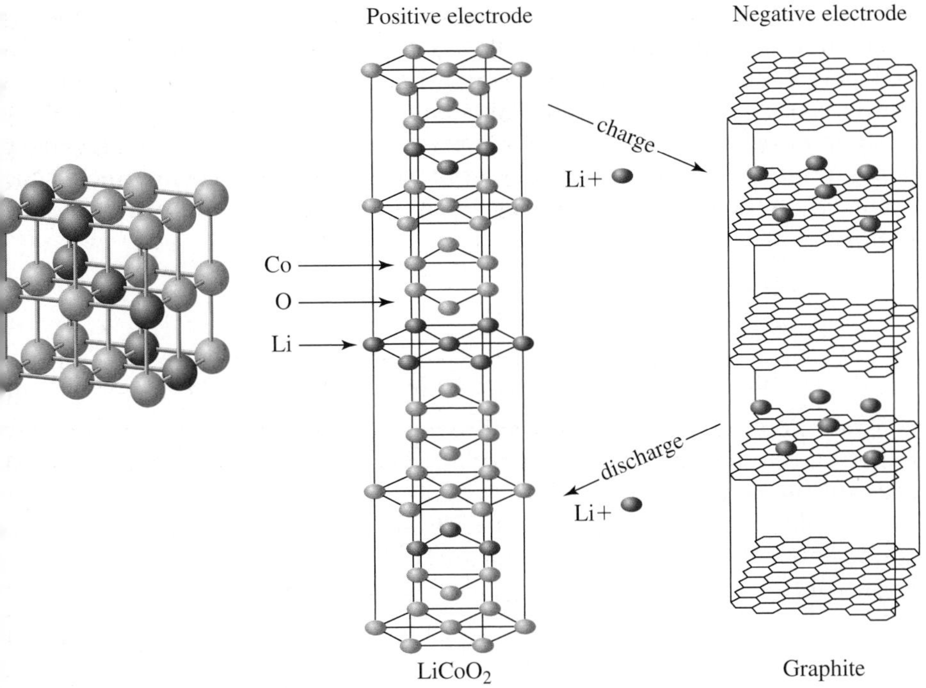

▲ FIGURE 19-17
he electrodes of a lithium-ion battery
he layered graphite electrode is shown with lithium ions (violet) intercalated. The iCoO$_2$ is shown as a face-centered cubic lattice, with the oxygen atoms (red) ccupying the corners and the faces, the cobalt atoms (pink) occupying half of the dges, and the lithium atoms occupying half of the edges and the central octahedral ole. This arrangement leads to planes of oxygen, cobalt, oxygen, lithium, oxygen, obalt, and oxygen atoms, as indicated in the figure.

rom the electrode. To maintain a charge balance, one cobalt(III) ion is oxi-lized to cobalt(IV) for each lithium ion released:

$$LiCoO_2(s) \longrightarrow Li_{(1-x)}CoO_2(s) + xLi^+(\text{solvent}) + x\,e^-$$

At the negative electrode, lithium ions enter between the graphite layers and re reduced to lithium metal. This insertion of a guest atom into a host solid s called *intercalation*, and the resulting product is called an *intercalation ompound*:

$$C(s) + x\,Li^+(\text{solvent}) + x\,e^- \longrightarrow Li_xC(s)$$

In the operation of a lithium-ion battery the source of the electrons is the xidation of the Co(III) to Co(IV). The lithium ion takes these electrons to the raphite electrode during charging and returns them to the positive electrode luring discharge.

Many other lithium batteries exist that use many different materials for he positive electrode, while graphite is the most common negative elec-rode. A major development is in the use of conducting polymers as the elec-rolyte, which has led to a whole range of *lithium-ion polymer batteries*. The levelopment of new batteries based on lithium ions is currently an area of reat interest.

Reserve Battery

Reserve batteries are designed to become active when a particular activation vent occurs. Some of the earliest reserve batteries were water-activated. They were constructed dry, stored under dry conditions, and activated by water. B Telephone Laboratories developed the magnesium/silver chloride seawat activated battery for military purposes. Their work lead to the development power sources for sonobuoys, weather balloons, air-sea rescue equipment, a emergency lights. For example, aviation and marine life jackets are powered a magnesium/copper(I) chloride reserve battery. As shown below, the reacti occurring in the magnesium/copper(I) chloride battery involves the oxidati of Mg by CuCl.

Anode:	$Mg(s) \longrightarrow Mg^{2+}(aq) + 2\,e^-$	
Cathode:	$2\,CuCl(s) + 2\,e^- \longrightarrow 2\,Cu(s) + 2\,Cl^-(aq)$	
Overall	$Mg(s) + 2\,CuCl(s) \longrightarrow Mg^{2+}(aq) + 2\,Cl^-(aq) + 2\,Cu(s)$	

The following simple diagram illustrates the basic structure of a reser battery. The magnesium metal of the anode and the copper(I) chloride of t cathode are held apart by a series of plastic separators. For the battery to ope ate (for electrons to flow), an electrolyte solution must fill gap between t anode and cathode. In marine applications, the battery is activated when se water fills the gap.

▶ The voltage is 0.00 V until the reserve battery comes in contact with a solution containing ions.

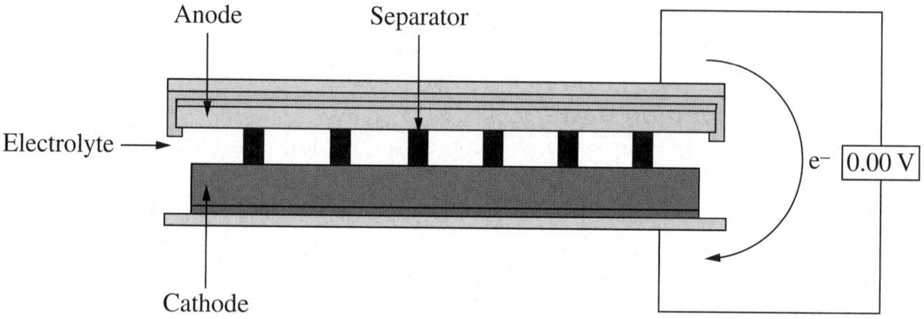

Source: Adapted from *Understanding Batteries*, by R. M. Dell and David Anthony James Rand, (Royal Society of Chemistry, 2001, pp. 87–94).

The lifetime of a magnesium/copper(I) chloride reserve battery is any where from 30 minutes to 15 hours.

Fuel Cells

The three types of cells considered in the remainder of this section fall into th fourth category mentioned on page 891; they are found in flow batteries.

For most of the twentieth century, scientists explored the possibility of co verting the chemical energy of fuels directly to electricity. The essenti process in a fuel cell is *fuel + oxygen* $\longrightarrow$ *oxidation products*. The first fuel cel were based on the reaction of hydrogen and oxygen. Figure 19-18 represent such a fuel cell. The overall change is that $H_2(g)$ and $O_2(g)$ in an alkalin medium produce $H_2O(l)$.

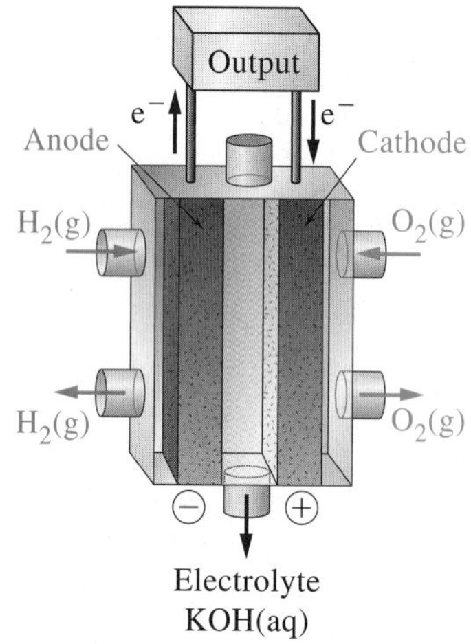

▲ FIGURE 19-18
A hydrogen–oxygen fuel cell
A key requirement in fuel cells is porous electrodes that allow for easy access of the gaseous reactants to the electrolyte. The electrodes chosen should also catalyze the electrode reactions.

Reduction:	$O_2(g) + 2\,H_2O(l) + 4\,e^- \longrightarrow 4\,OH^-(aq)$	
Oxidation:	$2\,\{H_2(g) + 2\,OH^-(aq) \longrightarrow 2\,H_2O(l) + 2\,e^-\}$	
Overall:	$2\,H_2(g) + O_2(g) \longrightarrow 2\,H_2O(l)$	(19.26

$$E^\circ_{cell} = E^\circ_{O_2/OH^-} - E^\circ_{H_2O/H_2} = 0.401\ V - (-0.828\ V) = 1.229\ V$$

The theoretical maximum energy available as electric energy in an electrochemical cell is equal to $\Delta_r G^\circ$ for the reaction. The maximum energ release when a fuel is burned is $\Delta_r H^\circ$. One of the measures used to evaluate fuel cell is the *efficiency value*, $\varepsilon = \Delta_r G^\circ/\Delta_r H^\circ$. For the hydrogen–oxygen fue cell, $\varepsilon = -474.4\ kJ\ mol^{-1}/-571.6\ kJ\ mol^{-1} = 0.83$.

The day is fast approaching when fuel cells based on the direct oxidation of ›mmon fuels will become a reality. For example, the half-cell reaction and cell ›action for a fuel cell using methane (natural gas) are

$$\begin{array}{lrl} \textit{Reduction:} & 2\,\{O_2(g) + 4\,H^+ + 4\,e^- & \longrightarrow 2\,H_2O(l)\} \\ \textit{Oxidation:} & CH_4(g) + 2\,H_2O(l) & \longrightarrow CO_2(g) + 8\,H^+ + 8\,e^- \\ \hline \textit{Overall:} & CH_4(g) + 2\,O_2(g) & \longrightarrow CO_2(g) + 2\,H_2O(l) \end{array}$$

$$\Delta_r H° = -890\ \text{kJ mol}^{-1} \qquad \Delta_r G° = -818\ \text{kJ mol}^{-1} \qquad \varepsilon = 0.92 \qquad \textbf{(19.27)}$$

Jeffrey Blackler / Alamy

▲ A methane fuel cell car that runs on methane generated from biowaste.

ı an automobile powered by a methane fuel cell, the engine operates by .) vaporizing a liquid hydrocarbon, (2) partially oxidizing the fuel vapor to O(g), (3) converting CO(g) to CO_2(g) and H_2(g) in the presence of steam and catalyst, and (4) producing electricity by feeding H_2(g) and air through the ıel cell.

A fuel cell should actually be called an *energy converter* rather than a battery. ›s long as fuel and O_2(g) are available, the cell will produce electricity. It does ›ot have the limited capacity of a primary battery or the fixed storage capacity f a secondary battery. Fuel cells based on reaction (19.26) have had their most ›otable successes as energy sources in space vehicles. (Water produced in the ell reaction is also a valuable product of the fuel cell.)

◀ Fuel cells are environmentally friendly. Oxygen and hydrogen are readily available. Hydrogen, although dangerous, can now be transported safely by the use of special materials that can adsorb large volumes.

›ir Batteries

ı a fuel cell, O_2(g) is the oxidizing agent that oxidizes a fuel such as H_2(g) or ᶜH_4(g). Another kind of flow battery, because it uses O_2(g) from air, is .nown as an *air battery*. The substance that is oxidized in an air battery is ›pically a metal.

One heavily studied battery system is the aluminum–air battery in which ›xidation occurs at an aluminum anode and reduction at a carbon–air cathode. 'he electrolyte circulated through the battery is NaOH(aq). Because it is in the ›resence of a high concentration of OH^-, Al^{3+} produced at the anode forms he complex ion $[Al(OH)_4]^-$. The operation of the battery is suggested by 'igure 19-19. The half-cell reactions and the overall cell reaction are

$$\begin{array}{lrl} \textit{:eduction:} & 3\,\{O_2(g) + 2\,H_2O(l) + 4\,e^- & \longrightarrow 4\,OH^-(aq)\} \\ \textit{)xidation:} & 4\,\{Al(s) + 4\,OH^-(aq) & \longrightarrow [Al(OH)_4]^-(aq) + 3\,e^-\} \\ \hline \textit{)verall:} & 4\,Al(s) + 3\,O_2(g) + 6\,H_2O(l) + 4\,OH^-(aq) & \longrightarrow 4[Al(OH)_4]^-(aq) \end{array} \qquad \textbf{(19.28)}$$

The battery is kept charged by feeding chunks of Al and water into it. A ypical air battery can power an automobile several hundred kilometers ›efore refueling is necessary. The electrolyte is circulated outside the battery, vhere $Al(OH)_3$(s) is precipitated from the $[Al(OH)_4]^-$(aq). This $Al(OH)_3$(s) s collected and can then be converted back to aluminum metal at an aluninum manufacturing facility.

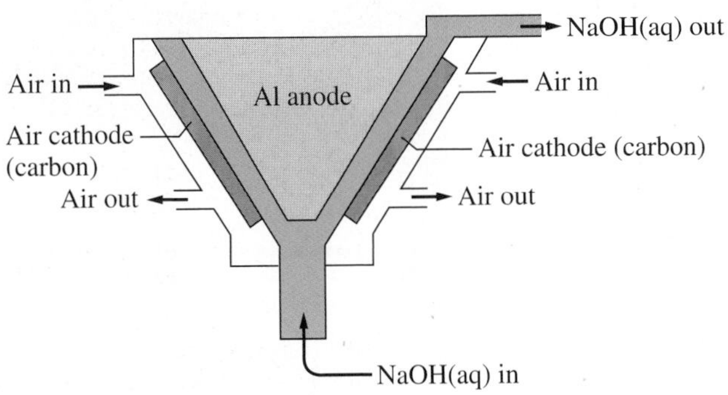

◀ FIGURE 19-19
A simplified aluminum–air battery

19-9 CONCEPT ASSESSMENT

Why do dry cells and lead-acid cells "run down" during use? Do you think a fuel cell runs down? Explain.

19-6 Corrosion: Unwanted Voltaic Cells

The reactions occurring in voltaic cells (batteries) are important sources electricity, but similar reactions also underlie corrosion processes. First, w will consider the electrochemical basis of corrosion, and then we will see ho electrochemical principles can be applied to control corrosion.

Figure 19-20(a) demonstrates the basic processes in the corrosion of an ir nail. The nail is embedded in an agar gel in water. The gel contains the com mon acid–base indicator phenolphthalein and potassium ferricyanid $K_3[Fe(CN)_6]$. Within hours of starting the experiment, a deep blue precipita forms at the head and tip of the nail. Along the body of the nail, the agar g turns pink. The blue precipitate, Turnbull's blue, establishes the presence iron(II). The pink color is that of phenolphthalein in basic solution. From the observations, we write two simple half-cell equations.

$$\textit{Reduction:} \quad O_2(g) + 2\,H_2O(l) + 4\,e^- \longrightarrow 4\,OH^-(aq)$$
$$\textit{Oxidation:} \quad 2\,Fe(s) \longrightarrow 2\,Fe^{2+}(aq) + 4\,e^-$$

The potential difference for these two half-cell reactions is

$$E^\circ_{cell} = E^\circ_{O_2/OH^-} - E^\circ_{Fe^{2+}/Fe} = 0.401\text{ V} - (-0.440\text{ V}) = 0.841\text{ V}$$

indicating that the corrosion process should be spontaneous when reactan and products are in their standard states. Typically, the corrosion medium ha $[OH^-] \ll 1$ M, the reduction half-cell reaction is even more favorable, an E_{cell} is even greater than 0.841 V. Corrosion is especially significant in acid solutions, in which the reduction half-cell reaction is

$$O_2(g) + 4\,H^+(aq) + 4\,e^- \longrightarrow 2\,H_2O(l) \qquad E^\circ_{O_2/H_2O} = 1.229\text{ V}$$

In the corroding nail of Figure 19-20(a), oxidation occurs at the head an tip. Electrons given up in the oxidation move along the nail and are used t reduce dissolved O_2. The reduction product, OH^-, is detected by the phe nolphthalein. In the bent nail in Figure 19-20(b), oxidation occurs at *thre* points: the head and tip and also the bend. The nail is preferentially oxidize

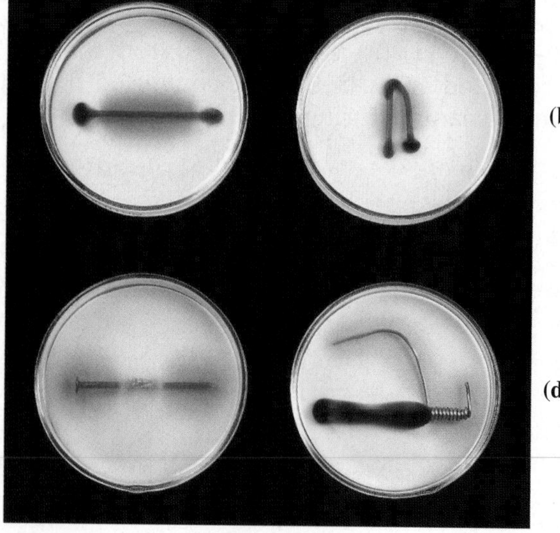

▶ FIGURE 19-20
Demonstration of corrosion and methods of corrosion protection
The pink color results from the indicator phenolphthalein in the presence of base; the dark blue color results from the formation of Turnbull's blue $KFe[Fe(CN)_6]$. Corrosion (oxidation) of the nail occurs at strained regions: **(a)** the head and tip and **(b)** a bend in the nail. **(c)** Contact with zinc protects the nail from corrosion. Zinc is oxidized instead of the iron (forming the faint white precipitate of zinc ferricyanide). **(d)** Copper does not protect the nail from corrosion. Electrons lost in the oxidation half-cell reaction distribute themselves along the copper wire, as seen by the pink color that extends the full length of the wire.

Carey B. Van Loon

these points because the strained metal is more active (more anodic) than e unstrained metal. This situation is similar to the preferential rusting of a ented automobile fender.

Some metals, such as aluminum, form corrosion products that adhere ghtly to the underlying metal and protect it from further corrosion. Iron xide (rust), however, flakes off and constantly exposes fresh surface. This dif- rence in corrosion behavior explains why cans made of iron deteriorate pidly in the environment, whereas aluminum cans have an almost unlim- ed lifetime. The simplest method of protecting a metal from corrosion is to over it with paint or some other protective coating impervious to water, an nportant reactant and solvent in corrosion processes.

Another method of protecting an iron surface is to plate it with a thin layer f a second metal. Iron can be plated with copper by electroplating or with tin y dipping the iron into molten tin. In either case, the underlying metal is pro- cted as long as the coating remains intact. If the coating is cracked, as when "tin" can is dented, the underlying iron is exposed and begins to corrode. on, being more active than copper and tin, undergoes oxidation; the reduc- on half-cell reaction occurs on the plating (Figs. 19-20d and 19-21).

When iron is coated with zinc (galvanized iron), the situation is different. inc is more active than iron. If a break occurs in the zinc plating, the iron is ill protected because the zinc is oxidized instead of the iron, and corrosion roducts protect the zinc from further corrosion (Figs. 19-20c and 19-21).

Still another method is used to protect large iron and steel objects in contact ith water or moist soils—ships, storage tanks, pipelines, plumbing systems. his method involves connecting a chunk of magnesium or some other active etal to the object, either directly or through a wire. Oxidation occurs at the ctive metal, which slowly dissolves. The iron surface acquires electrons from e oxidation of the active metal; the iron acts as a cathode and supports a *duction* half-cell reaction. As long as some of the active metal remains, the on is protected. This type of protection is called **cathodic protection**, and the ctive metal is called, appropriately, a *sacrificial anode*. Millions of pounds of agnesium are used annually in the United States in sacrificial anodes.

Gary Woodard/Getty Images

▲ **Galvanized nails**

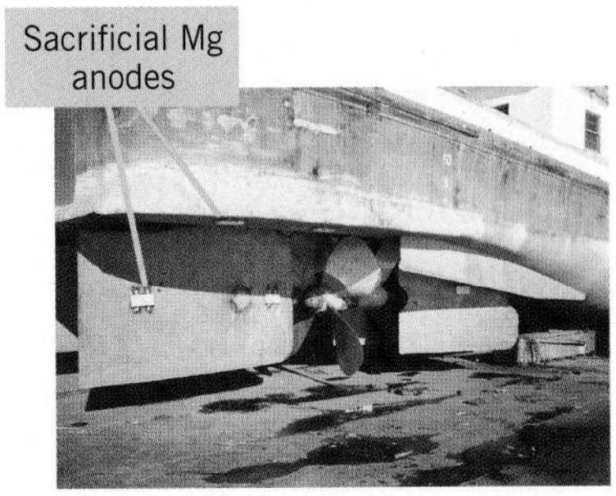

▲ **Magnesium sacrificial anodes** The small cylindrical bars of magnesium attached to the steel ship provide cathodic protection against corrosion.

Missouri Dry Dock

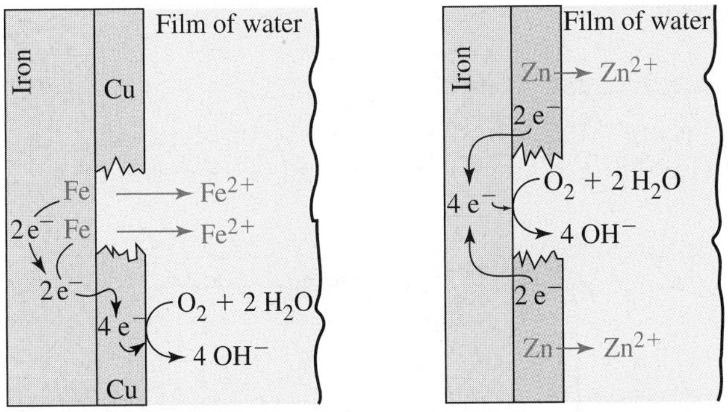

(a) Copper-plated iron **(b)** Galvanized iron

▲ FIGURE 19-21
Protection of iron against electrolytic corrosion
n the *anodic* reaction, the metal that is more easily oxidized loses electrons to produce netal ions. In **(a)**, this is iron; in **(b)**, it is zinc. In the *cathodic* reaction, oxygen gas, which is dissolved in a thin film of water on the metal, is reduced to OH^-. Rusting of on occurs in (a), but it does not in (b). When iron corrodes, Fe^{2+} and OH^- ions from he half-cell reactions initiate these further reactions.

$$Fe^{2+}(aq) + 2\,OH^-(aq) \longrightarrow Fe(OH)_2(s)$$

$$4\,Fe(OH)_2(s) + O_2(g) + 2\,H_2O(l) \longrightarrow 4\,Fe(OH)_3(s)$$

$$2\,Fe(OH)_3(s) \longrightarrow \underset{\text{rust}}{Fe_2O_3 \cdot H_2O(s)} + H_2O(l)$$

19-10 CONCEPT ASSESSMENT

Of the metals Al, Cu, Ni, and Zn, which could act as a sacrificial anode for iron?

19-7 Electrolysis: Causing Nonspontaneous Reactions to Occur

KEEP IN MIND

that voltaic (galvanic) and electrolytic cells are the two categories subsumed under the more general term *electrochemical cell.*

Until now, the emphasis has been on voltaic (galvanic) cells, electrochemic cells in which chemical change is used to produce electricity. Another type c electrochemical cell—the **electrolytic cell**—uses electricity to produce a *nor spontaneous* reaction. The process in which a nonspontaneous reaction is dr ven by the application of electric energy is called **electrolysis**.

Let's explore the relationship between voltaic and electrolytic cells b returning briefly to the cell shown in Figure 19-4. When the cell functior spontaneously, electrons flow from the zinc to the copper and the overa chemical change in the voltaic cell is

$$Zn(s) + Cu^{2+}(aq) \longrightarrow Zn^{2+}(aq) + Cu(s) \qquad E^{\circ}_{cell} = 1.103\ V$$

Now suppose the same cell is connected to an external electric source of vol age greater than 1.103 V (Fig. 19-22). That is, the connection is made so th electrons are forced into the zinc electrode (now the cathode) and remove from the copper electrode (now the anode). The overall reaction in this case i the *reverse* of the voltaic cell reaction, and E°_{cell} is *negative*.

$$\begin{array}{lrcl} \textit{Reduction:} & Zn^{2+}(aq) + 2\,e^{-} & \longrightarrow & Zn(s) \\ \textit{Oxidation:} & Cu(s) & \longrightarrow & Cu^{2+}(aq) + 2\,e^{-} \\ \hline \textit{Overall:} & Cu(s) + Zn^{2+}(aq) & \longrightarrow & Cu^{2+}(aq) + Zn(s) \end{array}$$

$$E^{\circ}_{cell} = E^{\circ}_{Zn^{2+}/Zn} - E^{\circ}_{Cu^{2+}/Cu} = -0.763\ V - 0.340\ V = -1.103\ V$$

Thus, reversing the direction of the electron flow changes the voltaic cell int an electrolytic cell.

Predicting Electrolysis Reactions

For the cell in Figure 19-22 to function as an electrolytic cell with reactants an products in their standard states, the external voltage has to exceed 1.103 V.

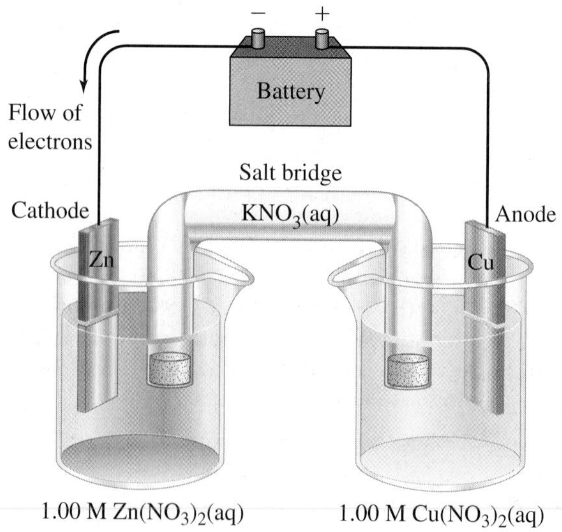

▶ FIGURE 19-22
An electrolytic cell
The direction of electron flow is the reverse of that in the voltaic cell of Figure 19-4, and so is the cell reaction. Now the zinc electrode is the *cathode* and the copper electrode, the *anode*. The battery must have a voltage in excess of 1.103 V in order to force electrons to flow in the reverse (nonspontaneous) direction.

Ve can make similar calculations for other electrolyses. What actually happens, owever, does not always correspond to these calculations. Four complicating ctors must be considered:

1. A voltage significantly in excess of the calculated value, an **overpotential**, may be necessary to cause a particular electrode reaction to occur. Overpotentials are needed to overcome interactions at the electrode surface and are particularly common when gases are involved. For example, the overpotential for the discharge of $H_2(g)$ at a mercury cathode is approximately 1.5 V; the overpotential on a platinum cathode is practically zero.
2. Competing electrode reactions may occur. In the electrolysis of *molten* sodium chloride with inert electrodes, only one oxidation and one reduction are possible.

$$\textit{Reduction:}\quad 2\,Na^+ + 2\,e^- \longrightarrow 2\,Na(l)$$

$$\textit{Oxidation:}\quad 2\,Cl^- \longrightarrow Cl_2(g) + 2\,e^-$$

In the electrolysis of *aqueous* sodium chloride with inert electrodes, there are *two* possible reduction half-cell reactions and *two* possible oxidation half-cell reactions.

$$\textit{eduction:}\quad 2\,Na^+(aq) + 2\,e^- \longrightarrow Na(s) \qquad E^\circ_{Na^+/Na} = -2.71\ V \qquad \textbf{(19.29)}$$

$$2\,H_2O(l) + 2\,e^- \longrightarrow H_2(g) + 2\,OH^-(aq) \qquad E^\circ_{H_2O/H_2} = (-0.83\ V) \qquad \textbf{(19.30)}$$

$$\textit{xidation:}\quad 2\,Cl^-(aq) \longrightarrow Cl_2(g) + 2\,e^- \qquad -E^\circ_{Cl_2/Cl^-} = -(1.36\ V) \qquad \textbf{(19.31)}$$

$$2\,H_2O(l) \longrightarrow O_2(g) + 4\,H^+(aq) + 4\,e^- \qquad -E^\circ_{O_2/H_2O} = -(1.23\ V) \qquad \textbf{(19.32)}$$

Reaction (19.29) can be eliminated as a possible reduction half-cell reactions: Unless the overpotential for $H_2(g)$ is unusually high, the reduction of Na^+ is far more difficult to accomplish than that of H_2O. This leaves two possibilities for the cell reaction.

◀ We have written a *minus* sign in front of the electrode potentials in (19.31) and (19.32) as a way of emphasizing the *oxidation* rather than the reduction tendency.

Half-cell reaction (19.30) + half-cell reaction (19.31):

$$\textit{Reduction:}\quad 2\,H_2O(l) + 2\,e^- \longrightarrow H_2(g) + 2\,OH^-(aq)$$

$$\textit{Oxidation:}\quad 2\,Cl^-(aq) \longrightarrow Cl_2(g) + 2\,e^-$$

$$\textit{Overall:}\quad 2\,Cl^-(aq) + 2\,H_2O(l) \longrightarrow Cl_2(g) + H_2(g) + 2\,OH^-(aq) \qquad \textbf{(19.33)}$$

$$E^\circ_{cell} = E^\circ_{H_2O/H_2} - E^\circ_{Cl_2/Cl^-} = -0.83\ V - (1.36\ V) = -2.19\ V$$

Half-cell reaction (19.30) + half-cell reaction (19.32):

$$\textit{Reduction:}\quad 2\,\{2\,H_2O(l) + 2\,e^- \longrightarrow H_2(g) + 2\,OH^-(aq)\}$$

$$\textit{Oxidation:}\quad 2\,H_2O(l) \longrightarrow O_2(g) + 4\,H^+(aq) + 4\,e^-$$

$$\textit{Overall:}\quad 2\,H_2O(l) \longrightarrow 2\,H_2(g) + O_2(g) \qquad \textbf{(19.34)}$$

$$E^\circ_{cell} = E^\circ_{H_2O/H_2} - E^\circ_{O_2/H_2O} = -0.83\ V - (1.23\ V) = -2.06\ V$$

Charles D. Winters / Science Source

▲ The electrolysis of water into $H_2(g)$ and $O_2(g)$, shown by bubbles at the electrodes and the gases collecting in the test tubes—twice the volume of $H_2(g)$ as $O_2(g)$.

In the electrolysis of NaCl(aq), $H_2(g)$ is the product expected at the *cathode*. Because cell reactions (19.33) and (19.34) have E°_{cell} values that are so similar, a mixture of $Cl_2(g)$ and $O_2(g)$ would be the expected product at the *anode*. Actually, because of the high overpotential of $O_2(g)$ compared to $Cl_2(g)$, cell reaction (19.33) predominates; $Cl_2(g)$ is essentially the only product at the anode.

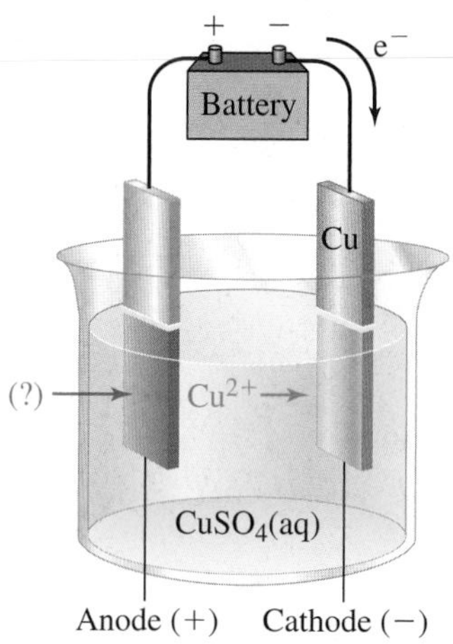

▲ FIGURE 19-23
Predicting electrode reactions in electrolysis—Example 19-11 illustrated
Electrons are forced onto the copper cathode by the external source (battery). Cu^{2+} ions are attracted to the cathode and are reduced to Cu(s). The oxidation half-reaction depends on the metal used for the anode.

3. The reactants very often are in nonstandard states. In the industrial ele trolysis of NaCl(aq), $[Cl^-] \approx 5.5$ M, not the unit activity ($[Cl^-] \approx 1$ M implied in half-cell reaction (19.31); therefore $E_{Cl_2/Cl^-} = 1.31$ V (*not* 1.3 V). Also, the pH in the anode half-cell is adjusted to 4, not the unit acti ity ($[H_3O^+] \approx 1$ M) implied in half-cell reaction (19.32); hen $E_{O_2/H_2O} = 0.99$ V (*not* 1.23 V). The net effect of these nonstandard cond tions is to favor the production of O_2 at the anode. In practice, howeve the Cl_2(g) obtained contains less than 1% O_2(g), indicating the overpov ering effect of the high overpotential of O_2(g). Not surprisingly, the pr portion of O_2(g) increases significantly in the electrolysis of very dilu NaCl(aq).

4. The nature of the electrodes matters. An *inert electrode*, such as platinum provides a surface on which an electrolysis half-cell reaction occurs, bu the reactants themselves must come from the electrolyte solution. A *active electrode* is one that can itself participate in the oxidation or reductio half-reaction. The distinction between inert and active electrodes explored in Figure 19-23 and Example 19-12.

Quantitative Aspects of Electrolysis

We have seen how to calculate the theoretical voltage required for electrolysi Equally important are calculations of the quantities of reactants consume and products formed in an electrolysis. For these calculations, we will cor tinue to use stoichiometric factors from the chemical equation, but anothe

19-2 ARE YOU WONDERING?

Why is the anode (+) in an electrolytic cell but (−) in a voltaic cell?

Assigning the terms *anode* and *cathode* is not based on the electrode charges; it is based on the half-cell reactions at the electrode surfaces. Specifically,

- *Oxidation* always occurs at the *anode* of an electrochemical cell. Because of the buildup of electrons freed in the oxidation half-cell reaction, the anode of a *voltaic* cell is (−). Because electrons are withdrawn from it, the anode in an *electrolytic* cell is (+). For both cell types, the anode is the electrode from which electrons *exit* the cell.
- *Reduction* always occurs at the cathode of an electrochemical cell. Because of the *removal* of electrons by the reduction half-cell reaction, the cathode of a *voltaic* cell is (+). Because of the electrons forced onto it, the cathode of an *electrolytic* cell is (−). For both cell types, the cathode is the electrode at which electrons *enter* the cell.

The following table summarizes the relationship between a voltaic cell and an electrolytic cell.

	Voltaic Cell			Electrolytic Cell	
Oxidation:	$A \longrightarrow A^+ + e^-$	Anode (negative)	*Oxidation:*	$B \longrightarrow B^+ + e^-$	Anode (positive)
Reduction:	$B^+ + e^- \longrightarrow B$	Cathode (positive)	*Reduction:*	$A^+ + e^- \longrightarrow A$	Cathode (negative)
Overall:	$A + B^+ \longrightarrow A^+ + B$ $\Delta_r G < 0$ Spontaneous redox reaction releases energy		*Overall:*	$A^+ + B \longrightarrow A + B^+$ $\Delta_r G > 0$ Nonspontaneous redox reaction absorbs energy to drive it	
	The system (the cell) does work on the surroundings			The surroundings (the source of energy) do work on the system	

Note that the sign of each electrode in an electrolytic cell is the same as the sign of the battery electrode to which it is attached.

EXAMPLE 19-12 Predicting Electrode Half-Reactions and Overall Reactions in Electrolysis

Refer to Figure 19-23. Predict the electrode reactions and the overall reaction when the anode is made of **(a)** copper and **(b)** platinum.

Analyze

In both cases we have to decide on the likely oxidation and reduction processes. The low reduction potential of $Cu^{2+}(aq)$ makes this the likely reduction process in both cases. What about oxidation processes? The possibilities are in **(a)** oxidation of the copper electrode (anode) ($E° = 0.340$ V), oxidation of sulfate anion (2.01 V), and oxidation of water (1.23 V). Thus, the most easily oxidized is the copper at the anode. In **(b)**, the platinum electrode is inert and is not easily oxidized. Of the other two candidates, sulfate anion and water, water has the lower oxidation potential.

Solve

Reduction: $$Cu^{2+}(aq) + 2\,e^- \longrightarrow Cu(s) \qquad E°_{Cu^{2+}/Cu} = 0.340\text{ V}$$

(a) At the cathode we have the reduction of $Cu^{2+}(aq)$. At the anode, Cu(s) can be oxidized to $Cu^{2+}(aq)$, as represented by

Oxidation: $$Cu(s) \longrightarrow Cu^{2+}(aq) + 2\,e^-$$

If the oxidation and reduction half-cell equations are added, $Cu^{2+}(aq)$ cancels out. The electrolysis reaction is simply

$$Cu(s)[\text{anode}] \longrightarrow Cu(s)[\text{cathode}] \qquad (19.35)$$

$$E°_{cell} = E°_{Cu^{2+}/Cu} - E°_{Cu^{2+}/Cu} = 0.340\text{ V} - 0.340\text{ V} = 0$$

(b) The oxidation that occurs most readily is that of H_2O, shown in reaction (19.32).

Oxidation: $$2\,H_2O(l) \longrightarrow O_2(g) + 4\,H^+(aq) + 4\,e^-$$
$$-E°_{O_2/H_2O} = -1.23\text{ V}$$

The electrolysis reaction and its $E°_{cell}$ are

$$2\,Cu^{2+}(aq) + 2\,H_2O(l) \longrightarrow 2\,Cu(s) + 4\,H^+(aq) + O_2(g) \qquad (19.36)$$

$$\begin{aligned} E°_{cell} &= E°(\text{reduction half-cell}) - E°(\text{oxidation half-cell}) \\ &= E°_{Cu^{2+}/Cu} - E°_{O_2/H_2O} \\ &= 0.340\text{ V} - 1.23\text{ V} = -0.89\text{ V} \end{aligned}$$

Assess

(a) Only a very small voltage is needed to overcome the resistance in the electric circuit for this electrolysis. For every Cu atom that enters the solution at the anode, an active electrode, one Cu^{2+} ion, deposits as a Cu atom at the cathode. Copper is transferred from the anode to the cathode through the solution as Cu^{2+} and the concentration of $CuSO_4(aq)$ remains unchanged.

(b) A potential greater than 0.89 V is required to electrolyze water and deposit copper. Keep in mind that when calculating $E°_{cell}$ as a *difference* between two $E°$ values, the $E°$ values are reduction potentials. Because $-E°$ corresponds to the half-cell potential for the oxidation process, the *difference* between two reduction potentials is equivalent to the *sum* of a reduction potential and an oxidation potential.

PRACTICE EXAMPLE A: Use data from Table 19.1 to predict the probable products when Pt electrodes are used in the electrolysis of KI(aq).

PRACTICE EXAMPLE B: In the electrolysis of $AgNO_3(aq)$, what are the expected electrolysis products if the anode is silver metal and the cathode is platinum?

actor enters in as well: the quantity of electric charge associated with one nole of electrons. This factor is provided by the Faraday constant, which we an write as

$$1\text{ mol }e^- = 96{,}485\text{ C}$$

Generally, electric charge is not measured directly; instead, it is the electr current that is measured. One *ampere* (A) of electric current represents the pa sage of 1 coulomb of charge per second (C/s). The product of current and tim yields the total quantity of charge transferred.

$$\text{charge (C)} = \text{current (C/s)} \times \text{time (s)}$$

To determine the number of moles of electrons involved in an electrolys reaction, we can write

$$\text{number of mol e}^- = \text{current}\left(\frac{\text{C}}{\text{s}}\right) \times \text{time (s)} \times \frac{1\ \text{mol e}^-}{96{,}485\ \text{C}}$$

As illustrated in Example 19-13, to determine the mass of a product in a electrolysis reaction, follow this conversion pathway.

$$\text{C/s} \longrightarrow \text{C} \longrightarrow \text{mol e}^- \longrightarrow \text{mol product} \longrightarrow \text{g product}$$

EXAMPLE 19-13 Calculating Quantities Associated with Electrolysis Reactions

The electrodeposition of copper can be used to determine the copper content of a sample. The sample is dissolved to produce $Cu^{2+}(aq)$, which is electrolyzed. At the cathode, the reduction half-cell reaction is $Cu^{2+}(aq) + 2\ e^- \longrightarrow Cu(s)$. What mass of copper can be deposited in 1.00 hour by a current of 1.62 A?

Analyze

To find the mass of copper deposited, we first need to determine the number of moles of electrons generated in the given time. Because we know that for each copper(II) ion we need two electrons, we can calculate the mass by using the number of moles of electrons.

Solve

First, we determine the number of moles of electrons involved in the electrolysis in the manner outlined above:

$$1.00\ \text{h} \times \frac{60\ \text{min}}{1\ \text{h}} \times \frac{60\ \text{s}}{1\ \text{min}} \times \frac{1.62\ \text{C}}{1\ \text{s}} \times \frac{1\ \text{mol e}^-}{96{,}485\ \text{C}} = 0.0604\ \text{mol e}^-$$

The mass of Cu(s) produced at the cathode by this number of moles of electrons is calculated as follows:

$$\text{mass of Cu} = 0.0604\ \text{mol e}^- \times \frac{1\ \text{mol Cu}}{2\ \text{mol e}^-} \times \frac{63.5\ \text{g Cu}}{1\ \text{mol Cu}} = 1.92\ \text{g Cu}$$

Assess

The key factor in this calculation, relating moles of copper to moles of electrons, is printed in blue. This type of conversion is very similar to the one you learned when doing stoichiometric problems.

PRACTICE EXAMPLE A: If 12.3 g of Cu is deposited at the cathode of an electrolytic cell after 5.50 h, what was the current used?

PRACTICE EXAMPLE B: For how long would the electrolysis in Example 19-13 have to be carried out, using Pt electrodes and a current of 2.13 A, to produce 2.62 L $O_2(g)$ at 26.2 °C and 738 mmHg pressure at the anode?

19-8 Industrial Electrolysis Processes

Modern industry could not function in its present form without electro lysis reactions. A number of elements are produced almost exclusively by electrolysis—for example, aluminum, magnesium, chlorine, and fluorine Among chemical compounds produced industrially by electrolysis are NaOH $K_2Cr_2O_7$, $KMnO_4$, $Na_2S_2O_8$, and a number of organic compounds.

◀ The refining of copper by electrolysis.

lectrorefining

he *electrorefining* of metals involves the deposition of pure metal at a cathode, om a solution containing the metal ion. Copper produced by the smelting of opper ores is of sufficient purity for some uses, such as plumbing, but it is not ure enough for applications in which high electrical conductivity is required. or these applications, the copper must be more than 99.5% pure. The electroly-s reaction (19.35) on page 903 is used to obtain such high-purity copper. A hunk of impure copper is the anode and a thin sheet of pure copper is the cath-de. During the electrolysis, Cu^{2+} produced at the anode migrates through an queous sulfuric acid–copper(II) sulfate solution to the cathode, where it is educed to Cu(s). The pure copper cathode increases in size as the impure chunk f copper is consumed. As noted in Example 19-12(a), the electrolysis is carried ut at a low voltage—from 0.15 to 0.30 V. Under these conditions, Ag, Au, and Pt mpurities are not oxidized at the anode, and they drop to the bottom of the tank s a sludge called *anode mud.* Sn, Bi, and Sb are oxidized, but they precipitate as xides or hydroxides; Pb is oxidized but precipitates as $PbSO_4(s)$. As, Fe, Ni, o, and Zn are oxidized but form water-soluble species. Recovery of Ag, Au, nd Pt from the anode mud helps offset the cost of the electrolysis.

lectroplating

n *electroplating,* one metal is plated onto another, often less expensive, metal y electrolysis. This procedure is done for decorative purposes or to protect he underlying metal from corrosion. Silver-plated flatware, for example, onsists of a thin coating of metallic silver on an underlying base of iron. In lectroplating, the item to be plated is the cathode in an electrolytic cell. The lectrolyte contains ions of the metal to be plated, which are attracted to the athode, where they are reduced to metal atoms.

In copper plating, the electrolyte is usually copper sulfate. In silver plating, t is commonly $K[Ag(CN)_2](aq)$. The concentration of free silver ion in a solu-ion of the complex ion $[Ag(CN)_2]^-(aq)$ is very low, and electroplating under hese conditions promotes a strongly adherent microcrystalline deposit of the netal. Chromium plating is useful for its resistance to corrosion as well as its ppearance. Steel can be chromium-plated from an aqueous solution of CrO_3 and H_2SO_4. The plating obtained, however, is thin and porous and ends to develop cracks. In practice, the steel is first plated with a thin coat of opper or nickel, and then the chromium plating is applied. Chromium plating r cadmium plating is used to weatherproof machine parts. Metal plating can ven be applied to some plastics. The plastic must first be made electrically onductive—for example, by coating it with graphite powder. Copper plating f plastics has been used to improve the quality of some microelectronic cir-uit boards. Electroplating is even used, quite literally, to make money. The J.S. penny is no longer copper throughout. A zinc plug is electroplated with a hin coat of copper, and the copper-plated plug is stamped to create a penny.

Sam Ogden/Science Photo Library

▲ A rack of metal parts being lifted from the electrolyte solution after electroplating.

Electrosynthesis

Electrosynthesis is a method of producing substances through electrolysis rea tions. It is useful for certain syntheses in which reaction conditions must be car fully controlled. Manganese dioxide occurs naturally as the mineral *pyrolusit* but small crystal size and lattice imperfections make this material inadequa for certain modern applications, such as alkaline batteries. The electrosynthes of MnO_2 is carried out in a solution of $MnSO_4$ in $H_2SO_4(aq)$. Pure $MnO_2(s)$ formed by the oxidation of Mn^{2+} at an inert anode, such as graphite.

$$\textit{Oxidation:}\quad Mn^{2+}(aq) + 2\,H_2O(l) \longrightarrow MnO_2(s) + 4\,H^+(aq) + 2\,e^-$$

The reaction at the cathode is the reduction of H^+ to $H_2(g)$, and the overa electrolysis reaction is

$$Mn^{2+}(aq) + 2\,H_2O(l) \longrightarrow MnO_2(s) + 2\,H^+(aq) + H_2(g)$$

An example of electrosynthesis in organic chemistry is the reduction c acrylonitrile, $CH_2{=}CH{-}C{\equiv}N$, to adiponitrile, $N{\equiv}C(CH_2)_4C{\equiv}N$, at lead cathode (chosen because of the high overpotential of H_2 on lead). Oxyge is released at the anode.

$$\textit{Reduction:}\quad 2\,CH_2{=}CH{-}C{\equiv}N + 2\,H_2O + 2\,e^- \longrightarrow N{\equiv}C(CH_2)_4C{\equiv}N + 2\,OH$$

The commercial importance of this electrolysis is that adiponitrile can b readily converted to two other compounds: hexamethylenediamin $H_2NCH_2(CH_2)_4CH_2NH_2$, and adipic acid, $HOOCCH_2(CH_2)_2CH_2COOH$ These two compounds are the monomers used to make the polymer *Nylon 6* (page 1313).

The Chlor–Alkali Process

On page 901 we described the electrolysis of NaCl(aq) through the reductio half-cell reaction (19.30) and the oxidation half-cell reaction (19.31).

$$2\,Cl^-(aq) + 2\,H_2O(l) \longrightarrow 2\,OH^-(aq) + H_2(g) + Cl_2(g) \qquad E^\circ_{cell} = -2.19\ V$$

When conducted on an industrial scale, this electrolysis is called the *chlor–alka process*, named after the two principal products: *chlorine* and the *alka* NaOH(aq). The chlor–alkali process is one of the most important of all elec trolytic processes because of the high value of these products.

In the *diaphragm cell* depicted in Figure 19-24, $Cl_2(g)$ is produced in th anode compartment, and $H_2(g)$ and NaOH(aq) in the cathode compartmen If $Cl_2(g)$ comes in contact with NaOH(aq), the Cl_2 disproportionates int $ClO^-(aq)$, $ClO_3^-(aq)$, and $Cl^-(aq)$. The purpose of the diaphragm is to pre vent this contact. The NaCl(aq) in the anode compartment is kept at a slightl higher level than that in the cathode compartment. This disparity creates a gradual flow of NaCl(aq) between the compartments and reduces the back flow of NaOH(aq) into the anode compartment. The solution in the cathod compartment, about 10%–12% NaOH(aq) and 14%–16% NaCl(aq), is concen trated and purified by evaporating some water and crystallizing the NaCl(s) The final product is 50% NaOH(aq), with up to 1% NaCl(aq).

The theoretical voltage required for this electrolysis is 2.19 V. However, a a result of the internal resistance of the cell and overpotentials at the elec trodes, a voltage of about 3.5 V is used. The current is kept very high, typicall about 1×10^5 A.

The NaOH(aq) produced in a diaphragm cell is not pure enough for certai uses, such as rayon manufacture. A higher purity is achieved if electrolysis i carried out in a mercury cell, illustrated in Figure 19-25. This cell takes advantag of the high overpotential for the reduction of $H_2O(l)$ to $H_2(g)$ and $OH^-(aq)$ at a

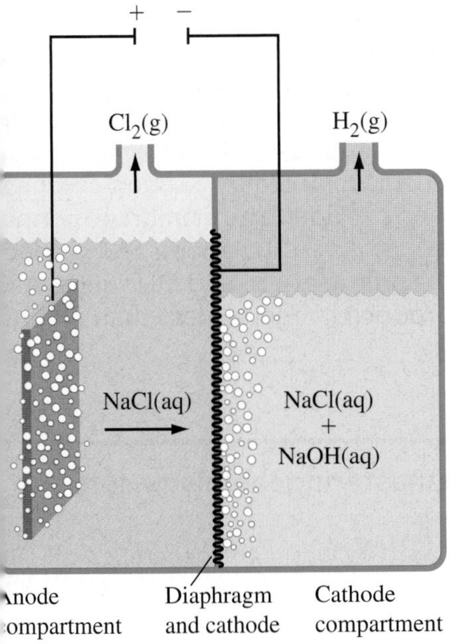

◀ FIGURE 19-24
A diaphragm chlor–alkali cell
The anode may be made of graphite or, in more modern technology, specially treated titanium metal. The diaphragm and cathode are generally fabricated as a composite unit consisting of asbestos or an asbestos–polymer mixture deposited on a steel wire mesh. To avoid the use of asbestos, a more modern development substitutes a fluorocarbon mesh for the asbestos.

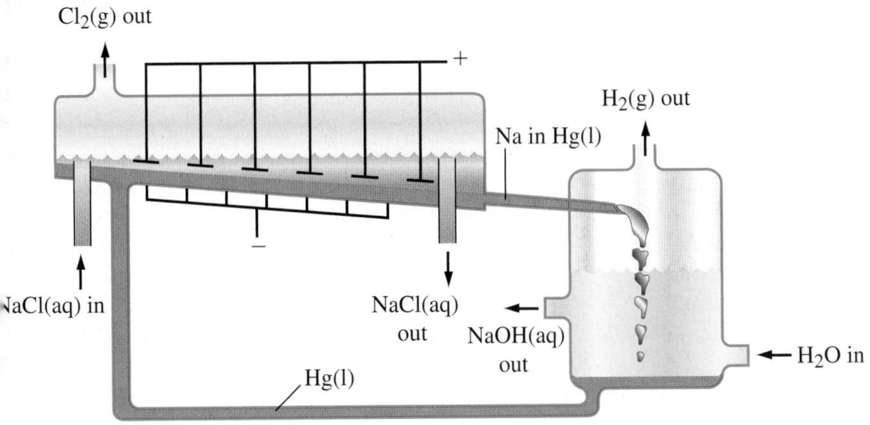

▲ FIGURE 19-25
he mercury-cell chlor–alkali process
he cathode is a layer of Hg(l) that flows along the bottom of the tank. Anodes, at hich $Cl_2(g)$ forms, are situated in the NaCl(aq) just above the Hg(l). Sodium formed t the cathode dissolves in the Hg(l), and the sodium amalgam is decomposed with ater to produce NaOH(aq) and $H_2(g)$. The regenerated Hg(l) is recycled.

nercury cathode. The reduction that occurs instead is that of $Na^+(aq)$ to Na, vhich dissolves in Hg(l) to form an amalgam (Na–Hg alloy) with about 0.5% Na by mass.

$$2\,Na^+(aq) + 2\,Cl^-(aq) \longrightarrow 2\,Na(in\ Hg) + Cl_2(g) \qquad E^\circ_{cell} = -3.20\ V$$

Vhen the Na amalgam is removed from the cell and treated with water, NaOH(aq) forms,

$$2\,Na(in\ Hg) + 2\,H_2O(l) \longrightarrow 2\,Na^+(aq) + 2\,OH^-(aq) + H_2(g) + Hg(l)$$

nd the liquid mercury is recycled back to the electrolytic cell.

Although the mercury cell has the advantage of producing concentrated igh-purity NaOH(aq), it has some disadvantages. The mercury cell requires a igher voltage (about 4.5 V) than does the diaphragm cell (3.5 V) and consumes nore electrical energy, about 3400 kWh/ton Cl_2 in a mercury cell, compared vith 2500 kWh/ton Cl_2 in a diaphragm cell. Another serious drawback is the need to control mercury effluents to the environment. Mercury losses, which one time were as high as 200 g Hg per ton Cl_2, have been reduced to abo 0.25 g Hg per ton of Cl_2 in older plants and half this amount in new plants.

The ideal chlor–alkali process is one that is energy-efficient and does n use mercury. A type of cell offering these advantages is the *membrane cell,* i which the porous diaphragm of Figure 19-24 is replaced by a cation-exchang membrane, normally made of a fluorocarbon polymer. The membrane permi hydrated cations (Na^+ and H_3O^+) to pass between the anode and cathod compartments but severely restricts the backflow of Cl^- and OH^- ions. As result, the sodium hydroxide solution produced contains less than 50 pp chloride ion contaminant.

MasteringCHEMISTRY www.masteringchemistry.com

The concepts presented in this chapter can be used to explain the roles played by ions in the generation of biological electric currents. Electric currents in biological systems are generated in muscle contraction and neuron activity, for example. For a discussion of the source of biological electric currents, go to the Focus On feature for Chapter 19, Membrane Potentials, on the MasteringChemistry site.

Summary

19-1 Electrode Potentials and Their Measurement—In an **electrochemical cell,** electrons in an oxidation–reduction reaction are transferred at metal strips called **electrodes** and conducted through an external circuit. The oxidation and reduction half-cell reactions occur in separate regions called **half-cells.** In a half-cell, an electrode is immersed in a solution. The electrodes of the two half-cells are joined by a wire, and an electrical connection between the solutions is also made, as through a **salt bridge** (Fig. 19-3). The cell reaction involves oxidation at one electrode called the **anode** and reduction at the other electrode called the **cathode.** A **voltaic (galvanic) cell** produces electricity from a spontaneous oxidation–reduction reaction. The difference in electric potential between the two electrodes is the **cell voltage**; the unit of cell voltage is the **volt (V).** The cell voltage is also called the **cell potential** or **electromotive force (emf)** and designated as E_{cell}. A **cell diagram** displays the components of a cell in a symbolic way (expression 19-4).

19-2 Standard Electrode Potentials—The reduction occurring at a **standard hydrogen electrode (SHE),** $2\,H^+(aq, a = 1) + 2\,e^- \xrightarrow{\text{on Pt}} H_2(g, 1\text{ bar})$, is arbitrarily assigned a potential of zero. A half-cell has a **standard electrode potential,** $E°$, in which all reactants and products are at unit activity. A half-cell reaction with a *positive* standard electrode potential ($E°$) occurs more readily than does reduction of H^+ ions at the SHE. A *negative* standard electrode potential signifies a lesser tendency to undergo reduction. The **standard cell potential** ($E°_{cell}$) of a voltaic cell is the *difference* between $E°$, of the cathode and $E°$, of the anode; that is, $E°_{cell} = E°(\text{cathode}) - E°(\text{anode})$.

19-3 E_{cell}, $\Delta_r G$, and K—Cell voltages based on standard electrode potentials are $E°_{cell}$ values. The electrical work that can be obtained from a cell depends on the number of electrons involved in the cell reaction, the cell potential, and the **Faraday constant (F),** which is the number of coulombs of charge per mole of electrons—96,485 C/mol e⁻. An important relationship exists between $E°_{cell}$ and $\Delta_r G°$, namely, $\Delta_r G° = -zFE°_{cell}$. The equilibrium constant of the cell reaction K is related to $E°_{cell}$ through the expression $\Delta_r G° = -RT \ln K = -zFE°_{cell}$.

19-4 E_{cell} as a Function of Concentrations—In the **Nernst equation,** E_{cell} for nonstandard conditions is related to $E°_{cell}$ and the reaction quotient Q (equation 19.18). If $E_{cell} > 0$, the cell reaction is spontaneous in the forward direction for the stated conditions; if $E_{cell} < 0$, the forward reaction is *nonspontaneous* for those conditions. A **concentration cell** consists of two half-cells with identical electrodes but different solution concentrations.

19-5 Batteries: Producing Electricity Through Chemical Reactions—An important application of voltaic cells is found in various battery systems. A **battery** stores chemical energy so that it can be released as energy. Batteries consist of one or more voltaic cells and are divided into four major classes: **primary** (Leclanché), **secondary** (lead–acid, silver–zinc, nicad, and lithium-ion), **reserve batteries** (magnesium/copper (I) chloride), and **flow batteries** or **fuel cells** in which reactants, such as hydrogen and oxygen, are continuously fed into the battery and chemical energy is converted to electric energy.

19-6 Corrosion: Unwanted Voltaic Cells—Electrochemistry plays a key role in corrosion and its control. Oxidation half-cell reactions produce anodic regions and reduction half-cell reactions cathodic regions. **Cathodic protection** is achieved when a more active metal is attached to the metal being protected from corrosion. The more tive metal, a "sacrificial" anode, is preferentially oxidized nile the protected metal is a cathode at which a harmless duction half-cell reactions occurs.

9-7 Electrolysis: Causing Nonspontaneous eactions to Occur—In **electrolysis**, a nonsponta-ous chemical reaction occurs as electrons from an ternal source are forced to flow in a direction opposite at in which they would flow spontaneously. The elec-ochemical cell in which electrolysis is conducted is lled an **electrolytic cell**. $E°$, values are used to establish e theoretical voltage requirements for an electrolysis. ometimes, particularly when a gas is liberated at an electrode, the voltage requirement for the electrode reaction exceeds the value of $E°$. The additional voltage requirement is called the **overpotential**. The amounts of reactants and products involved in an electrolysis can be calculated from the amount of electric charge passing through the electrolytic cell. The Faraday constant is featured in these calculations.

19-8 Industrial Electrolysis Processes—Electrolysis has many industrial applications, including electroplating, refining of metals, and production of substances such as $NaOH(aq)$, $H_2(g)$, and $Cl_2(g)$.

Integrative Example

vo electrochemical cells are connected as shown.

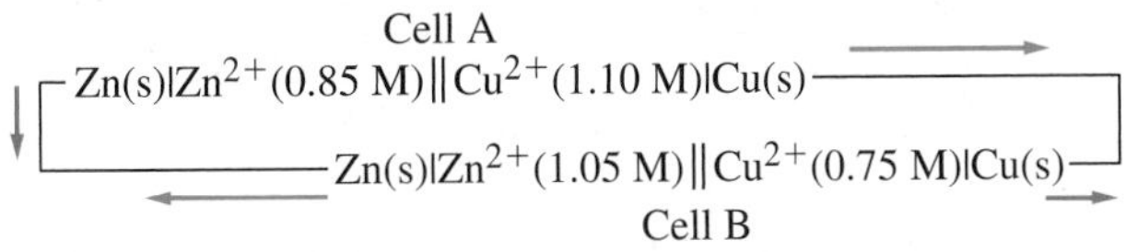

) Do electrons flow in the direction of the red arrows or the blue arrows?

) What are the ion concentrations in the half-cells at the point at which current ceases to flow?

nalyze

ne two cells differ only in their ion concentrations, which means that they have the same $E°_{cell}$ value but different E_{cell} alues. In part (a), use the Nernst equation (19.18) to determine which cell has the greater E_{cell} value when functioning as voltaic cell. This will establish the direction that electrons flow. In part (b), write and solve an equation relating ion con-ntrations to the condition where the two cells have the same voltage but, being connected in opposition to each other, roduce no net electric current.

olve

) In each voltaic cell zinc is the anode, copper is the cathode, and the cell reaction is

$$Zn(s) + Cu^{2+}(aq) \longrightarrow Zn^{2+}(aq) + Cu(s)$$

The E_{cell} values are given by the Nernst equation.

$$E_{cell} = E°_{cell} - (0.0257/2)\ \ln\ [Zn^{2+}]/[Cu^{2}] \qquad \textbf{(19.37)}$$

Note that for Cell A,

$$[Zn^{2+}]/[Cu^{2+}] = 0.85\ M/1.10\ M < 1$$
$$\ln\ [Zn^{2+}]/[Cu^{2+}] < 0, \quad \text{and} \quad E_{cell} > E°_{cell}$$

For Cell B,

$$[Zn^{2+}]/[Cu^{2+}] = 1.05\ M/0.75\ M > 1$$
$$\ln\ [Zn^{2+}]/[Cu^{2+}] > 0, \text{ and } E_{cell} < E°_{cell}$$

The voltage of Cell A is greater than that of Cell B.

In the connection of the two cells shown in the diagram, Cell A is a *voltaic cell* and Cell B is an *electrolytic cell*. There is an emf from Cell B that resists that of Cell A, but on balance, the electron flow is in the direction of the red arrows.

) As electrons flow between the two cells, the overall reaction in Cell A causes $[Zn^{2+}]$ to increase, $[Cu^{2+}]$ to decrease, and E_{cell} to decrease. The overall reaction in Cell B causes $[Zn^{2+}]$ to decrease, $[Cu^{2+}]$ to increase, and the back emf to increase. When the back emf from Cell B equals E_{cell} of Cell A, electrons cease to flow.

The cell diagrams when this condition is reached, with x representing *changes* in concentrations, are

Cell A: $Zn(s)|Zn^{2+}(0.85 + x)\ M||Cu^{2+}(1.10 - x)\ M|Cu(s)$
Cell B: $Zn(s)|Zn^{2+}(1.05 - x)\ M||Cu^{2+}(0.75 + x)\ M|Cu(s)$

Use equation (19.37) to obtain E_{cell} for each cell.

$$Cell\ A:\quad E_{cell} = E^\circ_{cell} - \frac{0.0257\ V}{2}\ln\frac{(0.85 + x)}{(1.10 - x)}$$

$$Cell\ B:\quad E_{cell} = E^\circ_{cell} - \frac{0.0257\ V}{2}\ln\frac{(1.05 - x)}{(0.75 + x)}$$

Set the two expressions equal to one another, cancel the terms E°_{cell} and (0.0257 V)/2, to obtain

$$\ln\frac{(0.85 + x)}{(1.10 - x)} = \ln\frac{(1.05 - x)}{(0.75 + x)}$$

Because the logarithms of the quantities on the two sides are equal, so too are the quantities themselves.

$$\frac{(0.85 + x)}{(1.10 - x)} = \frac{(1.05 - x)}{(0.75 + x)}$$

The expression to be solved is a quadratic equation

$$(0.85 + x)(0.75 + x) = (1.10 - x)(1.05 - x)$$

Cancel x^2 on each side

$$0.64 + 1.60x + \cancel{x^2} = 1.16 - 2.15x + \cancel{x^2}$$

$$0.64 + 1.60x = 1.16 - 2.15x$$

Solve to obtain

$$3.75x = 0.52 \quad \text{and} \quad x = 0.14$$

When electrons no longer flow, the ion concentrations are as follows:

Cell A: $[Zn^{2+}] = 0.99\ M$; $[Cu^{2+}] = 0.96\ M$.
Cell B: $[Zn^{2+}] = 0.91\ M$; $[Cu^{2+}] = 0.89\ M$.

Assess

Once the direction of electron flow had been established, it was possible to decide in which cell $[Zn^{2+}]$ would increa and in which cell it would decrease. At equilibrium the two cell potentials became equal. That the calculated equilibriu concentrations are correct can be seen in $0.99/0.96 \approx 0.91/0.89$.

PRACTICE EXAMPLE A: Current fuel cells use the reaction of $H_2(g)$ and $O_2(g)$ to form $H_2O(l)$. Often, the $H_2(g)$ obtained by the steam reforming of a hydrocarbon, such as $C_3H_8(g) + 3\,H_2O(g) \longrightarrow 3\,CO(g) + 7\,H_2(g)$. A futu possibility is a fuel cell that converts a hydrocarbon, such as propane, directly to $CO_2(g)$ and $H_2O(l)$:

$$C_3H_8(g) + 5\,O_2(g) \longrightarrow 3\,CO_2(g) + 4\,H_2O(l)$$

Based on this reaction, use data from Table 19.1 and Appendix D to determine E° for the reduction of $CO_2(g)$ to $C_3H_8($g in an acidic solution.

PRACTICE EXAMPLE B: A battery system that may be used to power automobiles in the future is the aluminum–a battery. This is a *flow battery* in which oxidation occurs at an aluminum anode and reduction at a carbon–air cathod The electrolyte circulated through the battery is NaOH(aq); the ultimate reaction product is $Al(OH)_3(s)$, which removed from the battery as it is formed. In operation the battery can be kept charged by feeding Al anode slugs an water into it; oxygen is drawn from the air (see Figure 19-18). The battery can power an automobile several hundre kilometers between charges. The $Al(OH)_3(s)$ removed from the battery can be converted back to aluminum in a aluminum manufacturing facility.

(a) In actual practice Al^{3+} produced at the anode does not precipitate as $Al(OH)_3(s)$ but is obtained as the complex io $[Al(OH)_4]^-$ in the presence of NaOH(aq). $Al(OH)_3$ is precipitated from the circulating NaOH(aq) electrolyte *outsic* the battery. Write plausible equations for oxidation and reduction half-cell reactions and for the net reaction th occurs in the battery.

(b) The theoretical voltage of the aluminum–air cell is +2.73 V. Use this information and data from Table 19.1 to obtai E° for the reduction

$$[Al(OH)_4]^-(aq) + 3\,e^- \longrightarrow Al(s) + 4\,OH^-(aq) \quad E^\circ = ?$$

(c) Given that E°_{cell} for the reaction is +2.73 V, that $\Delta_f G^\circ[OH^-(aq)] = -157\ kJ\ mol^{-1}$, and that $\Delta_f G^\circ[H_2O(l)] =$ $-237.2\ kJ\ mol^{-1}$, determine the Gibbs energy of formation, $\Delta_f G^\circ$, of the aqueous aluminate ion, $[Al(OH)_4]^-$.

(d) What mass of aluminum is consumed if 10.0 A of electric current is drawn from the battery for 4.00 h?

Exercises

(Use data from Table 19.1 and Appendix D as necessary.)

Standard Electrode Potentials

1. From the observations listed, estimate the value of $E°$ for the half-cell reaction $M^{2+}(aq) + 2\,e^- \longrightarrow M(s)$.
(a) The metal M reacts with $HNO_3(aq)$, but not with HCl(aq); M displaces $Ag^+(aq)$, but not $Cu^{2+}(aq)$.
(b) The metal M reacts with HCl(aq), producing $H_2(g)$, but displaces neither $Zn^{2+}(aq)$ nor $Fe^{2+}(aq)$.

2. You must estimate $E°$ for the half-cell reaction $In^{3+}(aq) + 3\,e^- \longrightarrow In(s)$. You have no electrical equipment, but you do have all of the metals listed in Table 19.1 and aqueous solutions of their ions, as well as In(s) and $In^{3+}(aq)$. Describe the experiments you would perform and the accuracy you would expect in your result.

3. $E°_{cell} = 0.201$ V for the reaction

$$3\,Pt(s) + 12\,Cl^-(aq) + 2\,NO_3^-(aq) + 8\,H^+(aq) \longrightarrow 3[PtCl_4]^{2-}(aq) + 2\,NO(g) + 4\,H_2O(l)$$

What is $E°$ for the reduction of $[PtCl_4]^{2-}$ to Pt in acidic solution?

4. Ascorbic acid ($C_6H_8O_6$, also commonly known as vitamin C, can be used to reduce a wide variety of transition metal ions. Given that $E°_{cell} = 0.71$ V for the reaction $C_6H_8O_6(aq) + 2\,Fe^{3+}(aq) \rightarrow C_6H_6O_6(aq) + 2\,Fe^{2+}(aq) + 2\,H^+(aq)$, what is $E°$ for the half-cell reaction $C_6H_6O_6(aq) + 2\,H^+(aq) + 2\,e^- \rightarrow C_6H_8O_6(aq)$?

5. Given that $E°_{cell}$ for the aluminum-air battery is 2.71 V, what is $E°$ for the reduction half-cell reaction $[Al(OH)_4]^-(aq) + 3\,e^- \longrightarrow Al(s) + 4\,OH^-(aq)$? [*Hint:* Refer to cell reaction (19.28).]

6. The theoretical $E°_{cell}$ for the methane–oxygen fuel cell is 1.06 V. What is $E°$ for the reduction half-cell reaction $CO_2(g) + 8\,H^+(aq) + 8\,e^- \longrightarrow CH_4(g) + 2\,H_2O(l)$? [*Hint:* Refer to cell reaction (19.27).]

7. The following sketch is of a voltaic cell consisting of two standard electrodes for two metals, M and N:

$$M^{z+}(aq) + z\,e^- \longrightarrow M(s) \qquad E°_{M^{z+}/M}$$
$$N^{z+}(aq) + z\,e^- \longrightarrow N(s) \qquad E°_{N^{z+}/N}$$

Use the standard reduction potentials of these half-reactions to answer the questions that follow:

$$Ag^+(aq) + e^- \longrightarrow Ag(s)$$
$$Zn^{2+}(aq) + 2\,e^- \longrightarrow Zn(s)$$
$$Cu^{2+}(aq) + 2\,e^- \longrightarrow Cu(s)$$
$$Al^{3+}(aq) + 3\,e^- \longrightarrow Al(s)$$

(a) Determine which pair of these half-cell reactions leads to a cell reaction with the largest positive cell potential, and calculate its value. Which couple is at the anode and which at the cathode?
(b) Determine which pair of these half-cell reactions leads to the cell with the smallest positive cell potential, and calculate its value. Which couple is at the anode and which is at the cathode?

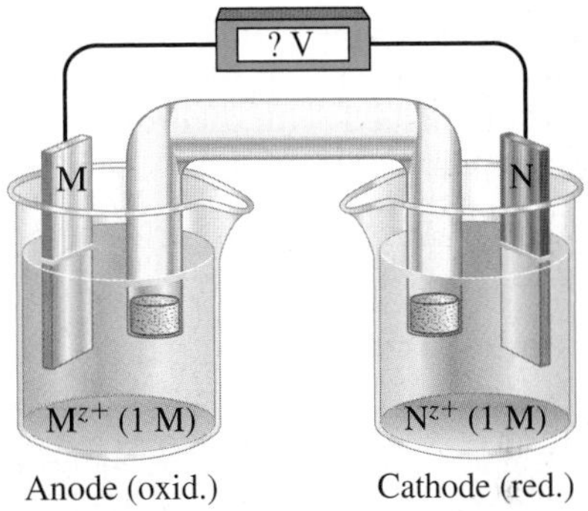

8. Given these half-cell reactions and associated standard reduction potentials, answer the questions that follow:

$$[Zn(NH_3)_4]^{2+}(aq) + 2\,e^- \longrightarrow Zn(s) + 4\,NH_3(aq) \qquad E° = -1.015\text{ V}$$
$$Ti^{3+}(aq) + e^- \longrightarrow Ti^{2+}(aq) \qquad E° = -0.37\text{ V}$$
$$VO^{2+}(aq) + 2\,H^+(aq) + e^- \longrightarrow V^{3+}(aq) + H_2O(l) \qquad E° = 0.340\text{ V}$$
$$Sn^{2+}(aq) + 2\,e^- \longrightarrow Sn(aq) \qquad E° = -0.14\text{ V}$$

(a) Determine which pair of half-cell reactions leads to a cell reaction with the largest positive cell potential, and calculate its value. Which couple is at the anode and which is at the cathode?
(b) Determine which pair of these half-cell reactions leads to the cell with the smallest positive cell potential, and calculate its value. Which couple is at the anode and which is at the cathode?

Predicting Oxidation–Reduction Reactions

9. Ni^{2+} has a more positive reduction potential than Cd^{2+}.
(a) Which ion is more easily reduced to the metal?
(b) Which metal, Ni or Cd, is more easily oxidized?

10. Use standard reduction potentials to predict which metal in each of the following pairs is the stronger reducing agent under standard conditions: **(a)** zinc or magnesium; **(b)** sodium or tin.

11. Assume that all reactants and products are in their standard states, and use data from Table 19.1 to predict whether a spontaneous reaction will occur in the forward direction in each case.
(a) $Sn(s) + Pb^{2+}(aq) \longrightarrow Sn^{2+}(aq) + Pb(s)$
(b) $Cu^{2+}(aq) + 2\,I^-(aq) \longrightarrow Cu(s) + I_2(s)$
(c) $4\,NO_3^-(aq) + 4\,H^+(aq) \longrightarrow 3\,O_2(g) + 4\,NO(g) + 2\,H_2O(l)$
(d) $O_3(g) + Cl^-(aq) \longrightarrow OCl^-(aq) + O_2(g)$ (basic solution)

12. For the reduction half-cell reactions $Hg_2^{2+}(aq) + 2\,e^- \longrightarrow 2\,Hg(l)$, $E° = 0.797$ V. Will Hg(l) react with and dissolve in HCl(aq)? in $HNO_3(aq)$? Explain.

13. Use data from Table 19.1 to predict whether, to any significant extent,
(a) Mg(s) will displace Pb^{2+} from aqueous solution;
(b) Sn(s) will react with and dissolve in 1 M HCl;
(c) SO_4^{2-} will oxidize Sn^{2+} to Sn^{4+} in acidic solution;
(d) $MnO_4^-(aq)$ will oxidize $H_2O_2(aq)$ to $O_2(g)$ in acidic solution;
(e) $I_2(s)$ will displace $Br^-(aq)$ to produce $Br_2(l)$.

14. Consider the reaction $Co(s) + Ni^{2+}(aq) \longrightarrow Co^{2+}(aq) + Ni(s)$, with $E°_{cell} = 0.02$ V. If Co(s) is added to a solution with $[Ni^{2+}] = 1$ M, should the reaction go to completion? Explain.

15. Dichromate ion ($Cr_2O_7^{2-}$) in acidic solution is a good oxidizing agent. Which of the following oxidations can be accomplished with dichromate ion in acidic solution? Explain.
(a) $Sn^{2+}(aq)$ to $Sn^{4+}(aq)$
(b) $I_2(s)$ to $IO_3^-(aq)$
(c) $Mn^{2+}(aq)$ to $MnO_4^-(aq)$

16. The standard electrode potential for the reduction $Eu^{3+}(aq)$ to $Eu^{2+}(aq)$ is -0.43 V. Use the data Appendix D to determine which of the followi is capable of reducing $Eu^{3+}(aq)$ to $Eu^{2+}(aq)$ und standard-state conditions: Al(s), Co(s), H_2O_2(a Ag(s), $H_2C_2O_4(aq)$.

17. Predict whether the following metals will react wi the acid indicated. If a reaction does occur, write t net ionic equation for the reaction. Assume that rea tants and products are in their standard states. **(a)** A in $HNO_3(aq)$; **(b)** Zn in HI(aq); **(c)** Au in HNO_3 (f the couple Au^{3+}/Au, $E° = 1.52$ V).

18. Predict whether, to any significant extent,
(a) Fe(s) will displace $Zn^{2+}(aq)$;
(b) $MnO_4^-(aq)$ will oxidize $Cl^-(aq)$ to $Cl_2(g)$ acidic solution;
(c) Ag(s) will react with 1 M HCl(aq);
(d) $O_2(g)$ will oxidize $Cl^-(aq)$ to $Cl_2(g)$ in acid solution.

Galvanic Cells

19. Write cell reactions for the electrochemical cells diagrammed here, and use data from Table 19.1 to calculate $E°_{cell}$ for each reaction.
(a) $Al(s)|Al^{3+}(aq)||Sn^{2+}(aq)|Sn(s)$
(b) $Pt(s)|Fe^{2+}(aq), Fe^{3+}(aq)||Ag^+(aq)|Ag(s)$
(c) $Cr(s)|Cr^{2+}(aq)||Au^{3+}(aq)|Au(s)$
(d) $Pt(s)|O_2(g)|H^+(aq)||OH^-(aq)|O_2(g)|Pt(s)$

20. Write the half-cell reactions and the balanced chemical equation for the electrochemical cells diagrammed here. Use data from Table 19.1 and Appendix D to calculate $E°_{cell}$ for each reaction.
(a) $Cu(s)|Cu^{2+}(aq)||Cu^+(aq)|Cu(s)$
(b) $Ag(s)|AgI(s)|I^-(aq)||Cl^-(aq)|AgCl(s)|Ag(s)$
(c) $Pt|Ce^{4+}(aq), Ce^{3+}(aq)||I^-(aq)|\,I_2(s)|C(s)$
(d) $U(s)|U^{3+}(aq)||V^{2+}(aq)\,|V(s)$

21. Use the data in Appendix D to calculate the standard cell potential for each of the following reactions. Which reactions will occur spontaneously?
(a) $H_2(g) + F_2(g) \longrightarrow 2\,H^+(aq) + 2\,F^-(aq)$
(b) $Cu(s) + Ba^{2+}(aq) \longrightarrow Cu^{2+}(aq) + Ba(s)$
(c) $3\,Fe^{2+}(aq) \longrightarrow Fe(s) + 2\,Fe^{3+}(aq)$
(d) $Hg(l) + HgCl_2(aq) \longrightarrow Hg_2Cl_2(s)$

22. In each of the following examples, sketch a voltaic cell that uses the given reaction. Label the anode and cathode; indicate the direction of electron flow; wri a balanced equation for the cell reaction; and calcula $E°_{cell}$.
(a) $Cu(s) + Fe^{3+}(aq) \longrightarrow Cu^{2+}(aq) + Fe^{2+}(aq)$
(b) $Pb^{2+}(aq)$ is displaced from solution by Al(s)
(c) $Cl_2(g) + H_2O(l) \longrightarrow Cl^-(aq) + O_2(g) + H^+$(a
(d) $Zn(s) + H^+ + NO_3^- \longrightarrow Zn^{2+} + H_2O(l) + NO$(

23. Use the data in Appendix D to calculate the standa cell potential for each of the following reaction Which reactions will occur spontaneously?
(a) $Fe^{3+}(aq) + Ag(s) \longrightarrow Fe^{2+}(aq) + Ag^+(aq)$
(b) $Sn(s) + Sn^{4+}(aq) \longrightarrow 2\,Sn^{2+}(aq)$
(c) $2\,Hg^{2+}(aq) + 2\,Br^-(aq) \longrightarrow Hg_2^{2+}(aq) + Br_2$(l
(d) $2\,NO_3^-(aq) + 4\,H^+(aq) + Zn(s) \longrightarrow Zn^{2+}(aq) + 2\,NO_2(g) + 2\,H_2O$(

24. Write a cell diagram and calculate the value of $E°_{cell}$ f a voltaic cell in which
(a) $Cl_2(g)$ is reduced to $Cl^-(aq)$ and Fe(s) is oxidize to $Fe^{2+}(aq)$;
(b) $Ag^+(aq)$ is displaced from solution by Zn(s);
(c) The cell reaction is $2\,Cu^+(aq) \longrightarrow Cu^{2+}(aq)$ Cu(s);
(d) $MgBr_2(aq)$ is produced from Mg(s) and $Br_2(l)$.

$\Delta_rG°$, $E°_{cell}$, and K

25. Determine the values of $\Delta_rG°$ for the following reactions carried out in voltaic cells.
(a) $2\,Al(s) + 3\,Cu^{2+}(aq) \longrightarrow 2\,Al^{3+}(aq) + 3\,Cu(s)$
(b) $O_2(g) + 4\,I^-(aq) + 4\,H^+(aq) \longrightarrow 2\,H_2O(l) + 2\,I_2(s)$
(c) $Cr_2O_7^{2-}(aq) + 14\,H^+(aq) + 6\,Ag(s) \longrightarrow 2\,Cr^{3+}(aq) + 6\,Ag^+(aq) + 7\,H_2O(l)$

26. Write the equilibrium constant expression for each the following reactions, and determine the value of at 25 °C. Use data from Table 19.1.
(a) $2\,V^{3+}(aq) + Ni(s) \longrightarrow 2\,V^{2+}(aq) + Ni^{2+}(aq)$
(b) $MnO_2(s) + 4\,H^+(aq) + 2\,Cl^-(aq) \longrightarrow Mn^{2+}(aq) + 2\,H_2O(l) + Cl_2$(g
(c) $2\,OCl^-(aq) \longrightarrow 2\,Cl^-(aq) + O_2(g)$ (basic solutio

7. For the reaction

$$MnO_4^-(aq) + 8\,H^+(aq) + 5\,Ce^{3+}(aq) \longrightarrow 5\,Ce^{4+}(aq) + Mn^{2+}(aq) + 4\,H_2O(l)$$

use data from Table 19.1 to determine **(a)** E°_{cell}; **(b)** $\Delta_r G^\circ$; **(c)** K; **(d)** whether the reaction goes substantially to completion when the reactants and products are initially in their standard states.

8. Consider the voltaic cell below.

$$Pt(s)|Cr^{3+}(aq), Cr^{2+}(aq)||V^{3+}(aq), V^{2+}(aq)|C(s)$$

Use data from Appendix D to determine **(a)** the equation for the cell reaction; **(b)** E°_{cell}; **(c)** $\Delta_r G^\circ$; **(d)** K; **(e)** whether the reaction goes essentially to completion, or to a limited extent only, when the reactants and products are initially in their standard states.

9. For the reaction $2\,Cu^+(aq) + Sn^{4+}(aq) \longrightarrow 2\,Cu^{2+}(aq) + Sn^{2+}(aq)$, $E^\circ_{cell} = -0.0050$ V,
(a) can a solution be prepared at 298 K that is 0.500 M in each of the four ions?
(b) If not, in which direction will a reaction occur?

30. For the reaction $2\,H^+(aq) + BrO_4^-(aq) + 2\,Ce^{3+}(aq) \longrightarrow BrO_3^-(aq) + 2\,Ce^{4+}(aq) + H_2O(l)$, $E^\circ_{cell} = -0.017$ V, answer the following questions:
(a) Can a solution be prepared at 298 K that has $[BrO_4^-] = [Ce^{4+}] = 0.675$ M, $[BrO_3^-] = [Ce^{3+}] = 0.600$ M and pH = 1?
(b) If not, in which direction will a reaction occur?

31. Use thermodynamic data from Appendix D to calculate a theoretical voltage of the silver–zinc button cell described on page 893.

32. The theoretical voltage of the aluminum–air battery is $E^\circ_{cell} = 2.71$ V. Use data from Appendix D and equation (19.28) to determine $\Delta_f G^\circ$ for $Al[(OH)_4]^-$.

33. By the method of combining reduction half-cell reactions illustrated on page 878, determine $E^\circ_{IrO_2/Ir}$, given that $E^\circ_{Ir^{3+}/Ir} = 1.156$ V and $E^\circ_{IrO_2/Ir^{3+}} = 0.223$ V.

34. Determine $E^\circ_{MoO_2/Mo^{3+}}$, given that $E^\circ_{H_2MoO_4/MoO_2} = 0.646$ V and $E^\circ_{H_2MoO_4/Mo^{3+}} = 0.428$ V. (See page 878).

oncentration Dependence of E_{cell}—The Nernst Equation

5. A voltaic cell represented by the following cell diagram has $E_{cell} = 1.250$ V. What must be $[Ag^+]$ in the cell?

$$Zn(s)|Zn^{2+}(1.00\text{ M})||Ag^+(x\text{ M})|Ag(s)$$

6. For the cell pictured in Figure 19-11, what is E_{cell} if the unknown solution in the half-cell on the left **(a)** has pH = 5.25; **(b)** is 0.0103 M HCl; **(c)** is 0.158 M CH_3COOH ($K_a = 1.8 \times 10^{-5}$)?

7. Use the Nernst equation and Table 19.1 to calculate E_{cell} for each of the following cells.
(a) $Al(s)|Al^{3+}(0.18\text{ M})||Fe^{2+}(0.85\text{ M})|Fe(s)$
(b) $Ag(s)|Ag^+(0.34\text{ M})||Cl^-(0.098\text{ M}), Cl_2(g, 0.55\text{ bar})|Pt(s)$

8. Use the Nernst equation and data from Appendix D to calculate E_{cell} for each of the following cells.
(a) $Mn(s)|Mn^{2+}(0.40\text{ M})||Cr^{3+}(0.35\text{ M}), Cr^{2+}(0.25\text{ M})|Pt(s)$
(b) $Mg(s)|Mg^{2+}(0.016\text{ M})||[Al(OH)_4]^-(0.25\text{ M}), OH^-(0.042\text{ M})|Al(s)$

9. Consider the reduction half-cell reactions listed in Appendix D, and give plausible explanations for the following observations:
(a) For some half-cell reactions E depends on pH; for others, it does not.
(b) Whenever H^+ appears in a half-cell equation it is on the *left* side.
(c) Whenever OH^- appears in a half-cell equation it is on the *right* side.

0. Write an equation to represent the oxidation of $Cl^-(aq)$ to $Cl_2(g)$ by $PbO_2(s)$ in an acidic solution. Will this reaction occur spontaneously in the forward direction if all other reactants and products are in their standard states and **(a)** $[H^+] = 6.0$ M; **(b)** $[H^+] = 1.2$ M; **(c)** pH = 4.25? Explain.

1. If $[Zn^{2+}]$ is maintained at 1.0 M,
(a) what is the minimum $[Cu^{2+}]$ for which reaction (19.3) is spontaneous in the forward direction?
(b) Should the displacement of $Cu^{2+}(aq)$ by Zn(s) go to completion? Explain.

42. Can the displacement of Pb(s) from 1.0 M $Pb(NO_3)_2$ be carried to completion by tin metal? Explain.

43. A concentration cell is constructed of two hydrogen electrodes: one immersed in a solution with $[H^+] = 1.0$ M and the other in 0.65 M KOH.
(a) Determine E_{cell} for the reaction that occurs.
(b) Compare this value of E_{cell} with E° for the reduction of H_2O to $H_2(g)$ in basic solution, and explain the relationship between them.

44. If the 0.65 M KOH of Exercise 43 is replaced by 0.65 M NH_3,
(a) will E_{cell} be higher or lower than in the cell with 0.65 M KOH?
(b) What will be the value of E_{cell}?

45. Consider the voltaic cell $Mg(s)\,|\,Mg^{2+}(\text{satd } Mg_3(PO_4)_2)||Mg^{2+}(0.125\text{ M})|Mg(s)$. What is the value of E_{cell}? For $Mg_3(PO_4)_2$, $K_{sp} = 1.0 \times 10^{-25}$.

46. A voltaic cell, with $E_{cell} = 0.180$ V, is constructed as follows:

$$Ag(s)|Ag^+(\text{satd } Ag_3PO_4)||Ag^+(0.140\text{ M})|Ag(s)$$

What is the K_{sp} of Ag_3PO_4?

47. For the voltaic cell,

$$Sn(s)|Sn^{2+}(0.075\text{ M})||Pb^{2+}(0.600\text{ M})|Pb(s)$$

(a) what is E_{cell} initially?
(b) If the cell is allowed to operate spontaneously, will E_{cell} increase, decrease, or remain constant with time? Explain.
(c) What will be E_{cell} when $[Pb^{2+}]$ has fallen to 0.500 M?
(d) What will be $[Sn^{2+}]$ at the point at which $E_{cell} = 0.020$ V?
(e) What are the ion concentrations when $E_{cell} = 0$?

48. For the voltaic cell,

$$Ag(s)|Ag^+(0.015\ M)||Fe^{3+}(0.055\ M), Fe^{2+}(0.045\ M)|Pt(s)$$

(a) what is E_{cell} initially?
(b) As the cell operates, will E_{cell} increase, decrease, or remain constant with time? Explain.
(c) What will be E_{cell} when $[Ag^+]$ has increased to 0.020 M?
(d) What will be $[Ag^+]$ when $E_{cell} = 0.010$ V?
(e) What are the ion concentrations when $E_{cell} = 0$?

49. Show that the oxidation of $Cl^-(aq)$ to $Cl_2(g)$ by $Cr_2O_7^{2-}(aq)$ in acidic solution, with reactants and products in their standard states, does not occur spontaneously. Explain why it is still possible to use this method to produce $Cl_2(g)$ in the laboratory. What experimental conditions would you use?

50. Derive a balanced equation for the reaction occurri in the cell:

$$Fe(s)|Fe^{2+}(aq)||Fe^{3+}(aq), Fe^{2+}(aq)|Pt(s)$$

(a) If $E^\circ_{cell} = 1.21$ V, calculate $\Delta_r G^\circ$ and the equili rium constant for the reaction.
(b) Use the Nernst equation to determine the pote tial for the cell:

$$Fe(s)|Fe^{2+}(aq, 1.0 \times 10^{-3}\ M)||Fe^{3+}(aq, 1.0 \times 10^{-3}\ M, Fe^{2+}(aq, 0.10\ M)|Pt($$

(c) In light of (a) and (b), what is the likelihood being able to observe the disproportionation Fe^{2+} into Fe^{3+} and Fe under standard conditions?

Batteries and Fuel Cells

51. The iron–chromium redox battery makes use of the reaction

$$Cr^{2+}(aq) + Fe^{3+}(aq) \longrightarrow Cr^{3+}(aq) + Fe^{2+}(aq)$$

occurring at a chromium anode and an iron cathode.
(a) Write a cell diagram for this battery.
(b) Calculate the theoretical voltage of the battery.

52. Refer to the discussion of the Leclanché cell (page 891).
(a) Combine the several equations written for the operation of the Leclanché cell into a single overall equation.
(b) Given that the voltage of the Leclanché cell is 1.55 V, estimate the electrode potentials, E, for each of the half-cell reactions. Why are your values only estimates?

53. What is the theoretical standard cell voltage, E°_{cell}, of each of the following voltaic cells? **(a)** the hydrogen–oxygen fuel cell described by equation (19.26); **(b)** the zinc–air battery; **(c)** a magnesium–iodine battery.

54. For the alkaline Leclanché cell (page 891)
(a) write the overall cell reaction.
(b) Determine E°_{cell} for that cell reaction.

55. One of the advantages of the aluminum-air battery over the iron–air and zinc–air batteries is the greater quantity of charge transferred per unit mass of metal consumed. Show that this is indeed the case. Assume that zinc and iron are oxidized to oxidation state +2 in air batteries.

56. Describe how you might construct batteries with ea of the following voltages: **(a)** 0.10 V; **(b)** 2.5 V; **(c)** 10.0 Be as specific as you can about the electrodes ar solution concentrations you would use, and indica whether the battery would consist of a single cell two or more cells connected in series.

57. A lithium battery, which is different from a lithiur ion battery, uses lithium metal as one electrode ar carbon in contact with MnO_2 in a paste of KOH as th other electrode. The electrolyte is lithium perchlora in a nonaqueous solvent, and the construction is sim lar to the silver battery. The half-cell reactions involv the oxidation of lithium and the reaction

$$MnO_2(s) + 2\,H_2O(l) + e^- \longrightarrow Mn(OH)_3(s) + OH^-(aq) \qquad E^\circ = -0.20$$

Draw a cell diagram for the lithium battery, identi the negative and positive electrodes, and estimate th cell potential under standard conditions.

58. For each of the following potential battery system describe the electrode reactions and the net cell rea tion you would expect. Determine the theoretic voltage of the battery.
(a) $Zn–Br_2$
(b) $Li–F_2$

Electrochemical Mechanism of Corrosion

59. Refer to Figure 19-20, and describe in words or with a sketch what you would expect to happen in each of the following cases.
(a) Several turns of copper wire are wrapped around the head and tip of an iron nail.
(b) A deep scratch is filed at the center of an iron nail.
(c) A galvanized nail is substituted for the iron nail.

60. When an iron pipe is partly submerged in water, the iron dissolves more readily below the waterline than at the waterline. Explain this observation by relating to the description of corrosion given in Figure 19-21.

61. Natural gas transmission pipes are sometimes pro tected against corrosion by the maintenance of a sma potential difference between the pipe and an ine electrode buried in the ground. Describe how th method works.

62. In the construction of the Statue of Liberty, a fram work of iron ribs was covered with thin sheets c copper less than 2.5 mm thick. A layer of asbestos separated the copper skin and iron framework. Over time, the asbestos wore away and the iron ribs corroded. Some of the ribs lost more than half their mass in the 100 years before the statue was restored. At the same time, the copper skin lost only about 4% of its thickness. Use electrochemical principles to explain these observations.

lectrolysis Reactions

3. How many grams of metal are deposited at the cathode by the passage of 2.15 A of current for 75 min in the electrolysis of an aqueous solution containing **(a)** Zn^{2+}; **(b)** Al^{3+}; **(c)** Ag^{+}; **(d)** Ni^{2+}?

4. A quantity of electric charge brings about the deposition of 3.28 g Cu at a cathode during the electrolysis of a solution containing $Cu^{2+}(aq)$. What volume of $H_2(g)$, measured at 28.2 °C and 763 mmHg, would be produced by this same quantity of electric charge in the reduction of $H^+(aq)$ at a cathode?

5. Which of the following reactions occur spontaneously, and which can be brought about only through electrolysis, assuming that all reactants and products are in their standard states? For those requiring electrolysis, what is the *minimum* voltage required?
(a) $2\,H_2O(l) \longrightarrow 2\,H_2(g) + O_2(g)$ [in 1 M $H^+(aq)$]
(b) $Zn(s) + Fe^{2+}(aq) \longrightarrow Zn^{2+}(aq) + Fe(s)$
(c) $2\,Fe^{2+}(aq) + I_2(s) \longrightarrow 2\,Fe^{3+}(aq) + 2\,I^-(aq)$
(d) $Cu(s) + Sn^{4+}(aq) \longrightarrow Cu^{2+}(aq) + Sn^{2+}(aq)$

6. An aqueous solution of K_2SO_4 is electrolyzed by means of Pt electrodes.
(a) Which of the following gases should form at the *anode:* O_2, H_2, SO_2, SO_3? Explain.
(b) What product should form at the *cathode*? Explain.
(c) What is the *minimum* voltage required? Why is the actual voltage needed likely to be higher than this value?

7. If a lead storage battery is charged at too high a voltage, gases are produced at each electrode. (It is possible to recharge a lead-storage battery only because of the high overpotential for gas formation on the electrodes.)
(a) What are these gases?
(b) Write a cell reaction to describe their formation.

8. A dilute aqueous solution of Na_2SO_4 is electrolyzed between Pt electrodes for 3.75 h with a current of 2.83 A. What volume of gas, saturated with water vapor at 25 °C and at a total pressure of 742 mmHg, would be collected at the *anode*? Use data from Table 12.5, as required.

9. Calculate the quantity indicated for each of the following electrolyses.
(a) the mass of Zn deposited at the cathode in 42.5 min when 1.87 A of current is passed through an aqueous solution of Zn^{2+}
(b) the time required to produce 2.79 g I_2 at the anode if a current of 1.75 A is passed through KI(aq)

0. Calculate the quantity indicated for each of the following electrolyses.
(a) $[Cu^{2+}]$ *remaining* in 425 mL of a solution that was originally 0.366 M $CuSO_4$, after passage of 2.68 A for 282 s and the deposition of Cu at the cathode
(b) the time required to reduce $[Ag^+]$ in 255 mL of $AgNO_3(aq)$ from 0.196 to 0.175 M by electrolyzing the solution between Pt electrodes with a current of 1.84 A

71. A *coulometer* is a device for measuring a quantity of electric charge. In a silver coulometer, $Ag^+(aq)$ is reduced to Ag(s) at a Pt cathode. If 1.206 g Ag is deposited in 1412 s by a certain quantity of electricity, **(a)** how much electric charge (in C) must have passed, and **(b)** what was the magnitude (in A) of the electric current?

72. Electrolysis is carried out for 2.00 h in the following cell. The platinum *cathode*, which has a mass of 25.0782 g, weighs 25.8639 g after the electrolysis. The platinum *anode* weighs the same before and after the electrolysis.
(a) Write plausible equations for the half-cell reactions that occur at the two electrodes.
(b) What must have been the magnitude of the current used in the electrolysis (assuming a constant current throughout)?
(c) A gas is collected at the anode. What is this gas, and what volume should it occupy if (when dry) it is measured at 23 °C and 755 mmHg pressure?

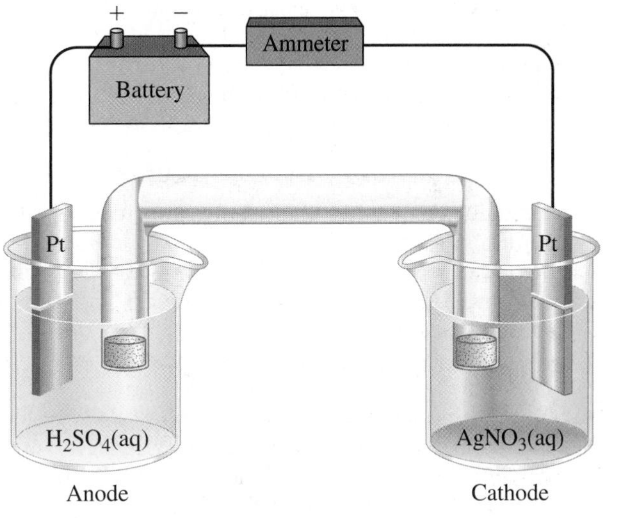

73. A solution containing both Ag^+ and Cu^{2+} ions is subjected to electrolysis. **(a)** Which metal should plate out first? **(b)** Plating out is finished after a current of 0.75 A is passed through the solution for 2.50 hours. If the total mass of metal is 3.50 g, what is the mass percent of silver in the product?

74. A solution containing a mixture of a platinum(II) salt contaminated by approximately 10 mole % of another oxidation state is electrolyzed at 1.20 A for 32.0 minutes, at which point no more platinum is deposited. **(a)** What is the oxidation state of the contaminant? **(b)** What is the composition of the mixture?

Integrative and Advanced Exercises

75. Two voltaic cells are assembled in which the following reactions occur.

$$V^{2+}(aq) + VO^{2+}(aq) + 2\,H^+(aq) \longrightarrow 2\,V^{3+}(aq) + H_2O(l) \quad E^\circ_{cell} = 0.616\ V$$

$$V^{3+}(aq) + Ag^+(aq) + H_2O(l) \longrightarrow VO^{2+}(aq) + 2\,H^+(aq) + Ag(s) \quad E^\circ_{cell} = 0.439\ V$$

Use these data and other values from Table 19.1 to calculate E° for the half-cell reaction $V^{3+}(aq) + e^- \longrightarrow V^{2+}(aq)$.

76. Suppose that a fully charged lead–acid battery contains 1.50 L of 5.00 M H_2SO_4. What will be the concentration of H_2SO_4 in the battery after 2.50 A of current is drawn from the battery for 6.0 h?

77. The energy consumption in electrolysis depends on the product of the charge and the voltage [volt × coulomb = V·C = J(joules)]. Determine the theoretical energy consumption per 1000 kg Cl_2 produced in a diaphragm chlor–alkali cell (page 906) that operates at 3.45 V. Express this energy in **(a)** kJ; **(b)** kilowatt-hours, kWh.

78. For the half-cell reaction $V^{3+}(aq) + e^- \rightarrow V^{2+}(aq)$, $E^\circ = -0.255$ V. If excess Ni(s) is added to a solution in which $[V^{3+}(aq)] = 0.500$ M, what will be the concentration of $Ni^{2+}(aq)$ when the equilibrium is reached at 298 K?

$$Ni(s) + 2\,V^{3+}(aq) \rightleftharpoons Ni^{2+}(aq) + 2\,V^{2+}(aq)$$

79. A voltaic cell is constructed based on the following reaction and initial concentrations:

$$Fe^{2+}(0.0050\ M) + Ag^+(2.0\ M) \rightleftharpoons Fe^{3+}(0.0050\ M) + Ag(s)$$

Calculate $[Fe^{2+}]$ when the cell reaction reaches equilibrium.

80. To construct a voltaic cell with $E_{cell} = 0.0860$ V, what $[Cl^-]$ must be present in the cathode half-cell to achieve this result?

$$Ag(s)|Ag^+(satd\ AgI)||Ag^+(satd\ AgCl,\ x\ M\ Cl^-)|Ag(s)$$

81. Describe a laboratory experiment that you could perform to evaluate the Faraday constant, *F*, and then show how you could use this value to determine the Avogadro constant.

82. The hydrazine fuel cell is based on the reaction

$$N_2H_4(aq) + O_2(g) \longrightarrow N_2(g) + 2\,H_2O(l)$$

The theoretical E°_{cell} of this fuel cell is 1.559 V. Use this information and data from Appendix D to calculate a value of $\Delta_f G^\circ$ for $[N_2H_4(aq)]$.

83. It is sometimes possible to separate two metal ions through electrolysis. One ion is reduced to the free metal at the cathode, and the other remains in solution. In which of these cases would you expect complete or nearly complete separation? Explain. **(a)** Cu^{2+} and K^+; **(b)** Cu^{2+} and Ag^+; **(c)** Pb^{2+} and Sn^{2+}.

84. Show that for some fuel cells the efficiency value, $\varepsilon = \Delta_r G^\circ/\Delta_r H^\circ$, can have a value greater than 1.00. Can you identify one such reaction? [*Hint:* Use data from Appendix D.]

85. In one type of Breathalyzer (alcohol meter), the qua tity of ethanol in a sample is related to the amount electric current produced by an ethanol–oxygen fu cell. Use data from Table 19.1 and Appendix D determine **(a)** E°_{cell} and **(b)** E° for the reduction $CO_2(g)$ to $CH_3CH_2OH(g)$.

86. You prepare 1.00 L of a buffer solution that is 1.00 NaH_2PO_4 and 1.00 M Na_2HPO_4. The solution divided in half between the two compartments of electrolytic cell. Both electrodes used are Pt. Assun that the only electrolysis is that of water. If 1.25 A current is passed for 212 min, what will be the pH each cell compartment at the end of the electrolysis

87. Assume that the volume of each solution in Figure 19- is 100.0 mL. The cell is operated as an electrolytic ce using a current of 0.500 A. Electrolysis is stopped aft 10.00 h, and the cell is allowed to function as a volta cell. What is E_{cell} at this point?

88. A common reference electrode consists of a silver wi coated with AgCl(s) and immersed in 1 M KCl.

$$AgCl(s) + e^- \longrightarrow Ag(s) + Cl^-(1\ M)\ E^\circ = 0.2223$$

(a) What is E°_{cell} when this electrode is a cathode in cor bination with a standard zinc electrode as an anode? **(b)** Cite several reasons why this electrode should easier to use than a standard hydrogen electrode. **(c)** By comparing the potential of this silver–silv chloride electrode with that of the silver–silver ic electrode, determine K_{sp} for AgCl.

89. The electrodes in the following electrochemical ce are connected to a voltmeter as shown. The half-ce on the right contains a standard silver–silver chloric electrode (see Exercise 88). The half-cell on the le contains a silver electrode immersed in 100.0 mL 1.00 × 10^{-3} M $AgNO_3(aq)$. A porous plug throug which ions can migrate separates the half-cells. **(a)** What is the initial reading on the voltmeter? **(b)** What is the voltmeter reading after 10.00 mL 0.0100 M K_2CrO_4 has been added to the half-cell c the left and the mixture has been thoroughly stirred **(c)** What is the voltmeter reading after 10.00 mL 10.0 M NH_3 has been added to the half-cell described part (b) and the mixture has been thoroughly stirred?

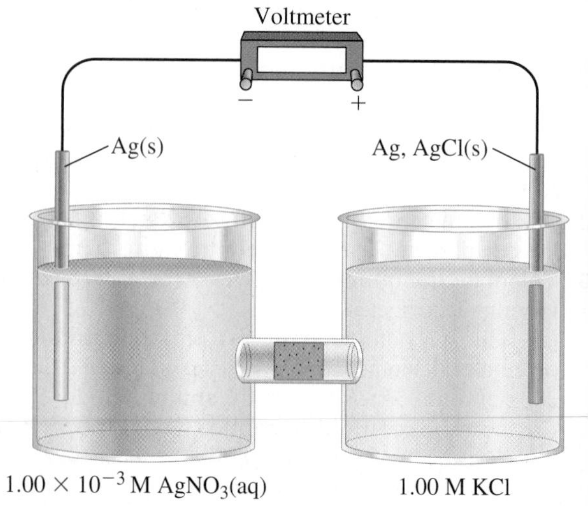

An important source of Ag is recovery as a by-product in the metallurgy of lead. The percentage of Ag in lead was determined as follows. A 1.050-g sample was dissolved in nitric acid to produce $Pb^{2+}(aq)$ and $Ag^{+}(aq)$. The solution was then diluted to 500.0 mL with water, a Ag electrode was immersed in the solution, and the potential difference between this electrode and a SHE was found to be 0.503 V. What was the percent Ag by mass in the lead metal?

A test for completeness of electrodeposition of Cu from a solution of $Cu^{2+}(aq)$ is to add $NH_3(aq)$. A blue color signifies the formation of the complex ion $[Cu(NH_3)_4]^{2+}$ ($K_f = 1.1 \times 10^{13}$). Let 250.0 mL of 0.1000 M $CuSO_4(aq)$ be electrolyzed with a 3.512 A current for 1368 s. At this time, add a sufficient quantity of $NH_3(aq)$ to complex any remaining Cu^{2+} and to maintain a free $[NH_3] = 0.10$ M. If $[Cu(NH_3)_4]^{2+}$ is detectable at concentrations as low as 1×10^{-5} M, should the blue color appear?

2. A solution is prepared by saturating 100.0 mL of 1.00 M $NH_3(aq)$ with AgBr. A silver electrode is immersed in this solution, which is joined by a salt bridge to a standard hydrogen electrode. What will be the measured E_{cell}? Is the standard hydrogen electrode the anode or the cathode?

3. The electrolysis of $Na_2SO_4(aq)$ is conducted in two separate half-cells joined by a salt bridge, as suggested by the cell diagram $Pt|Na_2SO_4(aq)||Na_2SO_4(aq)|Pt$.
(a) In one experiment, the solution in the anode compartment becomes more acidic, and that in the cathode compartment more basic, during the electrolysis. When the electrolysis is discontinued and the two solutions are mixed, the resulting solution has pH = 7. Write half-cell equations and the overall electrolysis equation.
(b) In a second experiment, a 10.00-mL sample of an unknown concentration of $H_2SO_4(aq)$ and a few drops of phenolphthalein indicator are added to the $Na_2SO_4(aq)$ in the cathode compartment. Electrolysis is carried out with a current of 21.5 mA (milliamperes) for 683 s, at which point, the solution in the cathode compartment acquires a lasting pink color. What is the molarity of the unknown $H_2SO_4(aq)$?

4. A Ni anode and an Fe cathode are placed in a solution with $[Ni^{2+}] = 1.0$ M and then connected to a battery. The Fe cathode has the shape shown. How long must electrolysis be continued with a current of 1.50 A to build a 0.050-mm-thick deposit of nickel on the iron? (Density of nickel = 8.90 g/cm^3.)

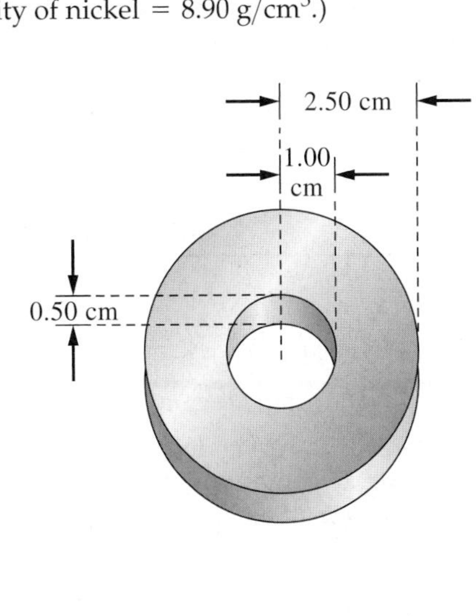

95. Initially, each of the half-cells in Figure 19-21 contained a 100.0-mL sample of solution with an ion concentration of 1.000 M. The cell was operated as an electrolytic cell, with copper as the anode and zinc as the cathode. A current of 0.500 A was used. Assume that the only electrode reactions occurring were those involving Cu/Cu^{2+} and Zn/Zn^{2+}. Electrolysis was stopped after 10.00 h, and the cell was allowed to function as a voltaic cell. What was E_{cell} at that point?

96. Silver tarnish is mainly Ag_2S:
$$Ag_2S(s) + 2\,e^- \longrightarrow 2\,Ag(s) + S^{2-}(aq) \qquad E° = -0.691\text{ V}$$
A tarnished silver spoon is placed in contact with a commercially available metallic product in a glass baking dish. Boiling water, to which some $NaHCO_3$ has been added, is poured into the dish, and the product and spoon are completely covered. Within a short time, the removal of tarnish from the spoon begins.
(a) What metal or metals are in the product?
(b) What is the probable reaction that occurs?
(c) What do you suppose is the function of the $NaHCO_3$?
(d) An advertisement for the product appears to make two claims: (1) No chemicals are involved, and (2) the product will never need to be replaced. How valid are these claims? Explain.

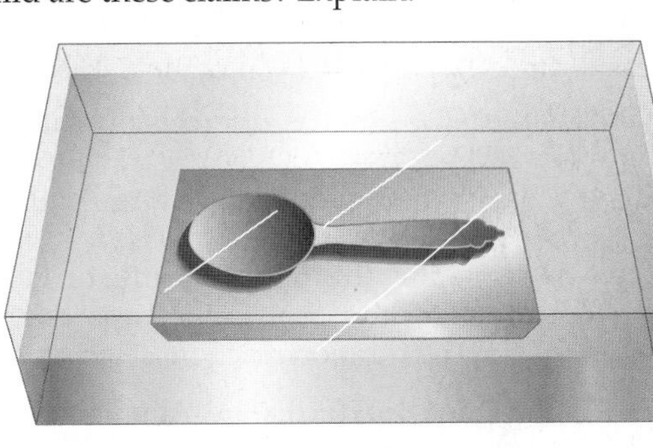

97. Your task is to determine $E°$ for the reduction of $CO_2(g)$ to $C_3H_8(g)$ in two different ways and to explain why each gives the same result. **(a)** Consider a fuel cell in which the cell reaction corresponds to the complete combustion of propane gas. Write the half-cell reactions and the overall reaction. Determine $\Delta_r G°$ and $E°_{cell}$ for the reaction, then obtain $E°_{CO_2/C_3H_8}$. **(b)** Without considering the oxidation that occurs simultaneously, obtain $E°_{CO_2/C_3H_8}$ directly from tabulated thermodynamic data for the reduction half-cell reaction.

98. Equation (19.15) gives the relationship between the standard Gibbs energy of a reaction and the standard cell potential. We know how the Gibbs energy varies with temperature.
(a) Making the assumption that $\Delta_r H°$ and $\Delta_r S°$ do not vary significantly over a small temperature range, derive an equation for the temperature variation of $E°_{cell}$.
(b) Calculate the cell potential of a Daniell cell at 50 °C under standard conditions. The overall cell reaction for a Daniell cell is $Zn(s) + Cu^{2+}(aq) \rightarrow Zn^{2+}(aq) + Cu(s)$.

99. Show that for nonstandard conditions the temperature variation of a cell potential is
$$E(T_1) - E(T_2) = (T_1 - T_2)\frac{(\Delta_r S° - R\ln Q)}{zF}$$

where $E(T_1)$ and $E(T_2)$ are the cell potentials at T_1 and T_2, respectively. We have assumed that the value of Q is maintained at a constant value. For the nonstandard cell below, the potential drops from 0.394 V at 50.0 °C to 0.370 V at 25.0 °C. Calculate Q, $\Delta_r H°$, and $\Delta_r S°$ for the reaction, and calculate K for the two temperatures.

$$Cu(s)|Cu^{2+}(aq)||Fe^{3+}(aq), Fe^{2+}(aq)|Pt(s)$$

Choose concentrations of the species involved in the cell reaction that give the value of Q that you have calculated, and then determine the equilibrium concentrations of the species at 50.0 °C.

100. Show that for a combination of half-cell reactions that produce a standard reduction potential for a half-cell that is not directly observable, the standard reduction potential is

$$E° = \frac{\sum n_i E_i^\circ}{\sum n_i}$$

where n_i is the number of electrons in each half-cell rea tion of potential E_i°. Use the following half-reactions

$$H_5IO_6(aq) + H^+(aq) + 2\,e^- \longrightarrow IO_3^-(aq) + 3\,H_2O(l) \qquad E° = 1.60$$

$$IO_3^-(aq) + 6\,H^+(aq) + 5\,e^- \longrightarrow \frac{1}{2} I_2(s) + 3\,H_2O(l \qquad E° = 1.19$$

$$2\,HIO(aq) + 2\,H^+(aq) + 2\,e^- \longrightarrow I_2(s) + 2\,H_2O(\qquad E° = 1.45$$

$$I_2(s) + 2\,e^- \longrightarrow 2\,I^-(aq) \qquad E° = 0.535$$

Calculate the standard reduction potential for

$$H_6IO_6(aq) + 5\,H^+(aq) + 2\,I^-(aq) + 3\,e^- \longrightarrow \frac{1}{2} I_2(s) + 4\,H_2O(l) + 2\,HIO(a$$

Feature Problems

101. Consider the following electrochemical cell:

$$Pt(s)|H_2(g, 1\ bar)|H^+(1\ M)||Ag^+(x\ M)|Ag(s)$$

(a) What is $E°_{cell}$—that is, the cell potential when $[Ag^+] = 1$ M?
(b) Use the Nernst equation to write an equation for E_{cell} when $[Ag^+] = x$.
(c) Now imagine titrating 50.0 mL of 0.0100 M $AgNO_3$ in the cathode half-cell compartment with 0.0100 M KI. The titration reaction is

$$Ag^+(aq) + I^-(aq) \longrightarrow AgI(s)$$

Calculate $[Ag^+]$ and then E_{cell} after addition of the following volumes of 0.0100 M KI: **(i)** 0.0 mL; **(ii)** 20.0 mL; **(iii)** 49.0 mL; **(iv)** 50.0 mL; **(v)** 51.0 mL; **(vi)** 60.0 mL.
(d) Use the results of part (c) to sketch the titration curve of E_{cell} versus volume of titrant.

102. Ultimately, $\Delta_f G°$ values must be based on experimental results; in many cases, these experimental results are themselves obtained from $E°$ values. Early in the twentieth century, G. N. Lewis conceived of an experimental approach for obtaining standard potentials of the alkali metals. This approach involved using a solvent with which the alkali metals do not react. Ethylamine was the solvent chosen. In the following cell diagram, Na(amalg, 0.206%) represents a solution of 0.206% Na in liquid mercury.

1. $Na(s)|Na^+$(in ethylamine)$|$Na(amalg, 0.206%) $\qquad E_{cell} = 0.8453$ V

Although Na(s) reacts violently with water to produce $H_2(g)$, at least for a short time, a sodium amalgam electrode does not react with water. This makes it possible to determine E_{cell} for the following voltaic cell.

2. Na(amalg, 0.206%)$|Na^+(1\ M)||H^+(1\ M)|H_2(g, 1\ bar)$ $\qquad E_{cell} = 1.8673$ V

(a) Write equations for the cell reactions that occu in the voltaic cells (1) and (2).
(b) Use equation (19.14) to establish $\Delta_r G$ for the ce reactions written in part (a).
(c) Write the overall equation obtained by combin ing the equations of part (a), and establish $\Delta_r G°$ fo this overall reaction.
(d) Use the $\Delta_r G°$ value from part (c) to obtain $E°_{ce}$ for the overall reaction. From this result, obtai $E°_{Na^+/Na}$. Compare your result with the value liste in Appendix D.

103. The following sketch is called an electrode potenti diagram. Such diagrams summarize electrod potential data more efficiently than do listings suc as that in Appendix D. In this diagram for bromir and its ions in basic solution,

$$BrO_4^-(aq) \xrightarrow{1.025\ V} BrO_3^-(aq)$$

signifies

$$BrO_4^-(aq) + H_2O(l) + 2\,e^- \longrightarrow BrO_3^-(aq) + 2\,OH^-(aq), E°_{BrO_4^-/BrO_3^-} = 1.025$$

Similarly,

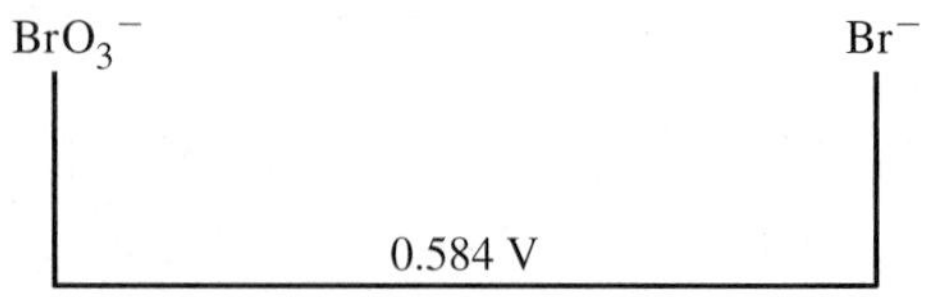

signifies

$$BrO_3^-(aq) + 3\,H_2O(l) + 6\,e^- \longrightarrow Br^-(aq) + 6\,OH^-(aq), E°_{BrO_3^-/Br^-} = 0.584$$

With reference to Appendix D and to the method determining $E°$ values outlined on page 878, supp the missing data in the following diagram.

Basic solution ($[OH^-] = 1$ M):

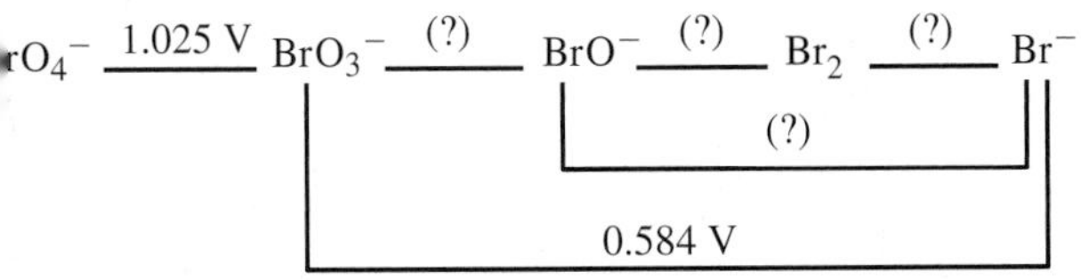

4. Only a tiny fraction of the diffusible ions move across a cell membrane in establishing a Nernst potential (see Focus On 19–1: Membrane Potentials), so there is no detectable concentration change. Consider a typical cell with a volume of 10^{-8} cm^3, a surface area (A) of 10^{-6} cm^2, and a membrane thickness (l) of 10^{-6} cm. Suppose that $[K^+] = 155$ mM inside the cell and $[K^+] = 4$ mM outside the cell and that the observed Nernst potential across the cell wall is 0.085 V. The membrane acts as a charge-storing device called a *capacitor*, with a *capacitance*, C, given by

$$C = \frac{\varepsilon_0 \varepsilon A}{l}$$

where ε_0 is the *dielectric constant* of a vacuum and the product $\varepsilon_0\varepsilon$ is the dielectric constant of the membrane, having a typical value of $3 \times 8.854 \times 10^{-12}$ $C^2\,N^{-1}\,m^{-2}$ for a biological membrane. The SI unit of capacitance is the *farad*, $1\text{ F} = 1$ coulomb per volt $= 1\text{ C V}^{-1} = 1 \times C^2\,N^{-1}\,m^{-1}$.
(a) Determine the capacitance of the membrane for the typical cell described.
(b) What is the net charge required to maintain the observed membrane potential?
(c) How many K^+ ions must flow through the cell membrane to produce the membrane potential?
(d) How many K^+ ions are in the typical cell?
(e) Show that the fraction of the intracellular K^+ ions transferred through the cell membrane to produce the membrane potential is so small that it does not change $[K^+]$ within the cell.

05. When deciding whether a particular reaction corresponds to a cell with a positive standard cell potential, which of the following thermodynamic properties would you use to get your answer without performing any calculations? Which would you not use? Explain. **(a)** $\Delta_r G°$; **(b)** $\Delta_r S°$; **(c)** $\Delta_r H°$; **(d)** $\Delta_r U°$; **(e)** K.

06. Consider two cells involving two metals X and Y

$$X(s)|X^+(aq)||H^+(aq), H_2(g, 1\text{ bar})|Pt(s)$$

$$X(s)|X^+(aq)||Y^{2+}(aq)|Y(s)$$

In the first cell electrons flow from the metal X to the standard hydrogen electrode. In the second cell electrons flow from metal X to metal Y. Is $E°_{X^+/X}$ greater or less than zero? Is $E°_{X^+/X} > E_{Y^{2+}/Y}$? Explain.

07. Describe in words how you would calculate the standard potential of the $Fe^{2+}/Fe(s)$ couple from those of Fe^{3+}/Fe^{2+} and $Fe^{3+}/Fe(s)$.

108. In 1982, the International Union of Pure and Applied Chemistry (IUPAC) redefined the standard state pressure to be 1 bar. As a result, standard reduction potentials for half-cell reactions must be adjusted accordingly. In this problem, we illustrate the approach to adjusting these values. Since the hydrogen electrode reaction is assigned a value of 0 V, for both the old (1 atm) and new (1 bar) standards of pressure, the *change* in reduction potential for a half-reaction should be calculated from the balanced equation that includes the following hydrogen electrode half-cell reaction.

$$\frac{1}{2}H_2(g) \rightarrow H^+(aq) + e^-$$

For example, to calculate the change in reduction potential for the half-cell reaction $NO_3^-(aq) + 4\,H^+(aq) + 3\,e^- \rightarrow NO(g) + 2\,H_2O(l)$, we focus on the overall equation obtained by combining these two half-reactions.

$$\text{Oxidation:} \quad 3\left\{\frac{1}{2}H_2(g) \rightarrow H^+(aq) + e^-\right\}$$

$$\text{Reduction:} \quad NO_3^-(aq) + 4\,H^+(aq) + 3\,e^- \rightarrow NO(g) + 2\,H_2O(l)$$

$$\text{Overall:} \quad \frac{3}{2}H_2(g) + NO_3^-(aq) + H^+(aq) \rightarrow NO(g) + 2\,H_2O(l)$$

Since the cell potential is directly proportional to the Gibbs energy of reaction, $\Delta_r G = -zFE_{cell}$, the change in cell potential that results from a change in pressure (from 1 atm to 1 bar) is obtained by first calculating the corresponding change in the Gibbs energy of reaction.
(a) Starting from equation (13.34), show that $\Delta_r G° = \Delta_r G^* - \Delta\nu_{gas} \ln(P°/P^*)$ where $\Delta_r G°$ and $\Delta_r G^*$ are the Gibbs energy of reaction for $P° = 1$ bar and $P^* = 1$ atm, respectively, and $\Delta\nu_{gas}$ is the sum of coefficients for gas-phase products minus the sum of coefficients for gas-phase reactants. [*Hint*: For gases, $\mu° = \mu^* + RT\ln(P°/P^*)$ whereas $\mu° \approx \mu^*$ for liquids, solids, and dissolved solutes.]
(b) Use the result from **(a)** and equations (19.14) and (19.15) to show that $E°_{cell} = E^*_{cell} + (3.382 \times 10^{-4}\text{ V}) \times \Delta\nu_{gas}/z$.
(c) Calculate $E°_{cell} - E^*_{cell}$ for the overall reaction above.
(d) Given that $E^*_{cell} = 0.956$ V for the overall reaction above (see Table 19.1), what is the corresponding value of $E°_{cell}$? [Note: Since the hydrogen electrode half-cell reaction is assigned a value of 0 V, this value of $E°_{cell}$ is equal to the standard reduction potential for $NO_3^-(aq) + 4\,H^+(aq) + 3\,e^- \rightarrow NO(g) + 2\,H_2O(l)$ at 1 bar.]

109. Some electrochemical cells employ large biological molecules known as enzymes. An enzyme increases the rate of a biochemical reaction. Some enzymes perform oxidation reactions, and some others perform reduction reactions. An electrochemical cell based on the use of enzymes is given in the following illustration.

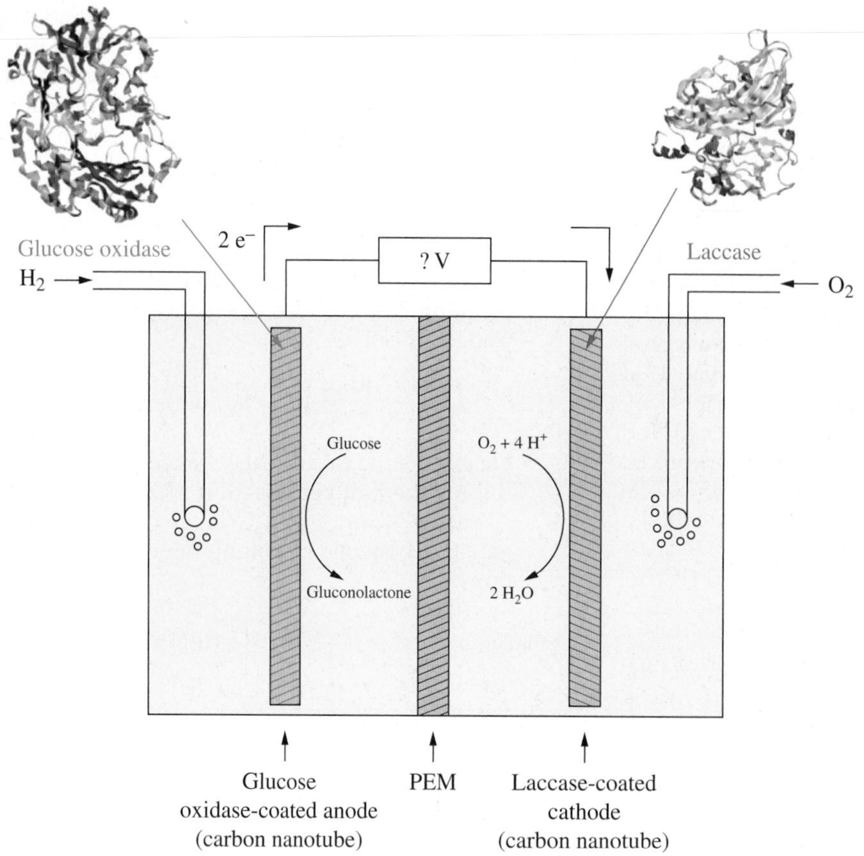

The half-cell reactions for electrochemical cell and the corresponding reduction potentials are given below. The symbol $E°'$ is used here because these values refer to the biological standard state: 37 °C, $a_{H^+} = 10^{-7}$, all other activities equal to 1.

Anode: $C_6H_{10}O_6(aq) + 2\,H^+(aq) + 2\,e^- \rightleftharpoons C_6H_{12}O_6(aq)$ $E°' = -0.3583$ V

Cathode: $O_2(g) + 4\,H^+(aq) + 4\,e^- \rightleftharpoons 2\,H_2O(l)$ $E°' = 0.82$ V

(a) Write an equation for the overall reaction occurring in this electrochemical cell.
(b) Write a cell diagram for this battery.
(c) Calculate the standard cell potential for the electrochemical reaction.

110. Electricity can be produced by the action of microbes on organic matter in soil or waste water. For example, in the fuel cell shown on the right, the anode produces electrons by the oxidation of organic compounds in soil.
The anode is coated with a film containing *geobacter metallireducens*, a bacterium that can oxidize organic compounds, such as acetic acid, to carbon dioxide. The electrons produced at the anode are then used at the cathode to reduce dissolved oxygen to water. The redox couple CO_2/CH_3COO^- has a biological standard reduction potential (see Exercise 109) of – 0.29 V. The redox couple O_2/H_2O has a biological standard reduction potential of 0.82 V.
(a) Write the anode half-cell reaction.
(b) Write the cathode half-cell reaction.
(c) Calculate the overall standard cell potential.
(d) What role does the microbe play in this electrochemical process?

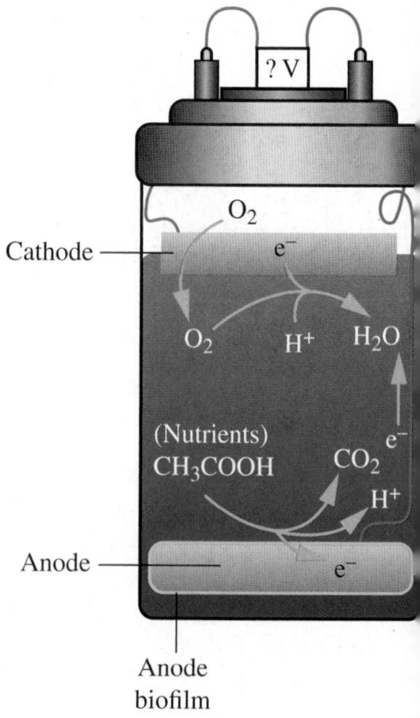

Source: Based on http://en.wikipedia.org/wiki/Microbial_fuel_cell#mediaviewer/File:SoilMFC.png.

Self-Assessment Exercises

1. In your own words, define the following symbols or terms: **(a)** $E°$; **(b)** F; **(c)** anode; **(d)** cathode.

2. Briefly describe each of the following ideas, methods, or devices: **(a)** salt bridge; **(b)** standard hydrogen electrode (SHE); **(c)** cathodic protection; **(d)** fuel cell.

3. Explain the important distinctions between each pair of terms: **(a)** half-cell reaction and overall cell reaction; **(b)** voltaic cell and electrolytic cell; **(c)** primary battery and secondary battery; **(d)** E_{cell} and $E°_{cell}$.

4. Of the following statements concerning electrochemical cells, the correct ones are: **(a)** The cathode is the negative electrode in both voltaic and electrolytic cells. **(b)** The function of a salt bridge is to permit the migration of electrons between the half-cell compartments of an electrochemical cell. **(c)** The anode is the negative electrode in a voltaic cell. **(d)** Electrons leave the cell from either the cathode or the anode, depending on what electrodes are used. **(e)** Reduction occurs at the cathode in both voltaic and electrolytic cells. **(f)** If electric current is drawn from a voltaic cell long enough, the cell becomes an electrolytic cell. **(g)** The cell reaction is an oxidation–reduction reaction.

15. For the half-cell reaction $Hg^{2+}(aq) + 2\,e^- \longrightarrow Hg(l)$, $E° = 0.854$ V. This means that **(a)** Hg(l) is more readily oxidized than $H_2(g)$; **(b)** $Hg^{2+}(aq)$ is more readily reduced than $H^+(aq)$; **(c)** Hg(l) will dissolve in 1 M HCl; **(d)** Hg(l) will displace Zn(s) from an aqueous solution of Zn^{2+} ion.

16. The value of $E°_{cell}$ for the reaction $Zn(s) + Pb^{2+}(aq) \longrightarrow Zn^{2+}(aq) + Pb(s)$ is 0.66 V. This means that for the reaction $Zn(s) + Pb^{2+}(0.01\ M) \longrightarrow Zn^{2+}(0.10\ M) + Pb(s)$, E_{cell} equals **(a)** 0.72 V; **(b)** 0.69 V; **(c)** 0.66 V; **(d)** 0.63 V.

17. For the reaction $Co(s) + Ni^{2+}(aq) \longrightarrow Co^{2+}(aq) + Ni(s)$, $E°_{cell} = 0.03$ V. If cobalt metal is added to an aqueous solution in which $[Ni^{2+}] = 1.0$ M, **(a)** the reaction will not proceed in the forward direction at all; **(b)** the displacement of Ni(s) from the $Ni^{2+}(aq)$ will go to completion; **(c)** the displacement of Ni(s) from the solution will proceed to a considerable extent, but the reaction will not go to completion; **(d)** there is no way to predict how far the reaction will proceed.

18. The gas evolved at the *anode* when $K_2SO_4(aq)$ is electrolyzed between Pt electrodes is most likely to be **(a)** O_2; **(b)** H_2; **(c)** SO_2; **(d)** SO_3; **(e)** a mixture of sulfur oxides.

19. The quantity of electric charge that will deposit 4.5 g Al at a cathode will also produce the following volume at STP of $H_2(g)$ from $H^+(aq)$ at a cathode: **(a)** 44.8 L; **(b)** 22.4 L; **(c)** 11.2 L; **(d)** 5.6 L.

20. If a chemical reaction is carried out in a fuel cell, the maximum amount of useful work that can be obtained is **(a)** $\Delta_r G$; **(b)** $\Delta_r H$; **(c)** $\Delta_r G/\Delta_r H$; **(d)** $T\Delta_r S$.

21. For the reaction $Zn(s) + H^+(aq) + NO_3^-(aq) \longrightarrow Zn^{2+}(aq) + H_2O(l) + NO(g)$, describe the voltaic cell in which it occurs, label the anode and cathode, use a table of standard electrode potentials to evaluate $E°_{cell}$, and balance the equation for the cell reaction.

122. The following voltaic cell registers an $E_{cell} = 0.108$ V. What is the pH of the unknown solution?

$$Pt|H_2(g, 1\ bar)|H^+(x\ M)||H^+(1.00\ M)|H_2(g, 1\ bar)|Pt$$

123. $E°_{cell} = -0.0050$ V for the reaction, $2\,Cu^+(aq) + Sn^{4+}(aq) \longrightarrow 2\,Cu^{2+}(aq) + Sn^{2+}(aq)$.
(a) Can a solution be prepared that is 0.500 M in each of the four ions at 298 K?
(b) If not, in what direction must a net reaction occur?

124. For each of the following combinations of electrodes (A and B) and solutions, indicate
- the overall cell reaction
- the direction in which electrons flow spontaneously (from A to B, or from B to A)
- the magnitude of the voltage read on the voltmeter, V

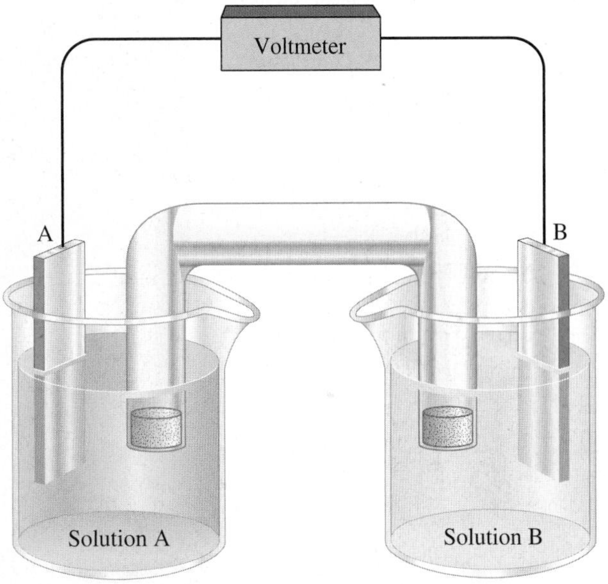

	A	Solution A	B	Solution B
(a)	Cu	1.0 M Cu^{2+}	Fe	1.0 M Fe^{2+}
(b)	Pt	1.0 M Sn^{2+}/1.0 M Sn^{4+}	Ag	1.0 M Ag^+
(c)	Zn	0.10 M Zn^{2+}	Fe	1.0×10^{-3} M Fe^{2+}

125. Use data from Table 19.1, as necessary, to predict the probable products when Pt electrodes are used in the electrolysis of **(a)** $CuCl_2(aq)$; **(b)** $Na_2SO_4(aq)$; **(c)** $BaCl_2(l)$; **(d)** KOH(aq).

126. Using the method presented in Appendix E, construct a concept map showing the relationship between electrochemical cells and thermodynamic properties.

127. Construct a concept map illustrating the relationship between batteries and electrochemical ideas.

128. Construct a concept map illustrating the principles of electrolysis and its industrial applications.

20 Chemical Kinetics

CONTENTS

LEARNING OBJECTIVES

20.1 Express the rate of reaction in terms of the rate of appearance or the rate of disappearance of a substance.

20.2 Explain how a graph of concentration versus time can be used to determine the rate of reaction at any instant.

20.3 Describe how the method of initial rates can be used to determine the rate law for a reaction.

20.4 Use the rate law for a zero-order reaction, together with experimental data, to obtain the rate constant for a zero-order reaction.

20.5 Use the rate law for a first-order reaction to determine the rate constant or half-life of a reaction.

20.6 Identify the integrated rate law for a second-order reaction, and derive the relationship between half-life and rate constant.

20.7 Summarize the differences between zero-order, first-order, and second-order reactions.

20.8 Discuss the significance of the activation energy for a reaction, and describe the reaction profile for both endothermic and exothermic reactions.

20.9 Discuss the Arrhenius equation and how it can be used to estimate the activation energy for a reaction.

20.10 Describe the general structure of a reaction profile for a two-step mechanism, and describe what is meant by the rate-determining step.

20.11 Explain the role played by a catalyst, and describe how enzymes perform this task.

Although stable at room temperature, ammonium dichromate decomposes very rapidly once ignited:

$$(NH_4)_2Cr_2O_7(s) \xrightarrow{\Delta} N_2(g) + 4\,H_2O(g) + Cr_2O_3(s)$$

The rates of chemical reactions and the effect of temperature on those rates are among several key concepts explored in this chapter.

Rocket fuel is designed to give a rapid release of gaseous produc and energy to provide a rocket maximum thrust. Milk is stored in refrigerator to slow down the chemical reactions that cause it spoil. Current strategies to reduce the rate of deterioration of the ozo layer try to deprive the ozone-consuming reaction cycle of key intermed ates that come from chlorofluorocarbons (CFCs). Catalysts are used reduce the harmful emissions from internal combustion engines that co tribute to smog. These examples illustrate the importance of the rates chemical reactions. Moreover, how fast a reaction occurs depends on th

action mechanism—the step-by-step molecular pathway leading from reac-
nts to products. Thus, *chemical kinetics* concerns how rates of chemical
actions are measured, how they can be predicted, and how reaction-rate
ata are used to deduce probable reaction mechanisms.

The chapter begins with a discussion of the meaning of a rate of reaction and
me ideas about measuring rates of reaction. This is followed by the introduc-
on of mathematical equations, called rate laws, that relate the rates of reac-
ons to the concentrations of the reactants. Finally, with this information as
ckground, we will turn to one of our central purposes of this chapter: relat-
g rate laws to plausible reaction mechanisms.

0-1 Rate of a Chemical Reaction

ate, or speed, refers to something that happens in a unit of time. A car travel-
g at 60 km/h, for example, covers a distance of 60 kilometers in one hour.
or chemical reactions, the rate of reaction describes how fast the concentra-
on of a reactant or product changes with time.

To illustrate, let's consider the reaction that begins immediately after the
ns Fe^{3+} and Sn^{2+} are simultaneously introduced into water.

$$2\,Fe^{3+}(aq) + Sn^{2+}(aq) \longrightarrow 2\,Fe^{2+}(aq) + Sn^{4+}(aq) \qquad (20.1)$$

ippose that 38.5 s after the reaction starts, $[Fe^{2+}]$ is found to be 0.0010 M.
uring the period of time, $\Delta t = 38.5$ s, the *change* in concentration of Fe^{2+},
hich we designate as $\Delta[Fe^{2+}]$, is $\Delta[Fe^{2+}] = 0.0010\,M - 0\,M = 0.0010\,M$.
he *average* rate at which Fe^{2+} is
rmed in this interval is the change in concentration of Fe^{2+} divided by the
ange in time.

◀ Recall that the symbol [ ] means "concentration." Also, Δ means "the change in," that is, the final value minus the initial value.

$$\text{rate of formation of } Fe^{2+} = \frac{\Delta[Fe^{2+}]}{\Delta t} = \frac{0.0010\,M}{38.5\,s} = 2.6 \times 10^{-5}\,M\,s^{-1}$$

ow has the concentration of Sn^{4+} changed during the 38.5 s we were moni-
ring the Fe^{2+}? Can you see that in 38.5 s, $\Delta[Sn^{4+}]$ will be 0.00050 M − 0 M=
00050 M? Because only *one* Sn^{4+} ion is produced for every *two* Fe^{2+} ions, the
ildup of $[Sn^{4+}]$ will be only one-half that of $[Fe^{2+}]$. Consequently the rate of
rmation of Sn^{4+} is

$$\text{rate of formation of } Sn^{4+} = \frac{0.00050\,M}{38.5\,s} = 1.3 \times 10^{-5}\,M\,s^{-1}$$

We can also follow the course of the reaction by monitoring the concentra-
ons of the reactants. Thus, the amount of Fe^{3+} consumed is the same as the
nount of Fe^{2+} produced. The *change* in concentration of Fe^{3+} is
$[Fe^{3+}] = -0.0010$ M. The concentration change is negative because Fe^{3+} is
nsumed by the reaction. Thus,

$$\frac{\Delta[Fe^{3+}]}{\Delta t} = \frac{-0.0010\,M}{38.5\,s} = -2.6 \times 10^{-5}\,M\,s^{-1}$$

ne quantity above is the average rate of change of change of $[Fe^{3+}]$ in this inter-
al. It is a negative quantity because $[Fe^{3+}]$ decreases with time. The average rate
disappearance of Fe^{3+} is defined as follows.

$$\text{rate of disappearance of } Fe^{3+} = -\frac{\Delta[Fe^{3+}]}{\Delta t} = 2.6 \times 10^{-5}\,M\,s^{-1}$$

'hy is a negative sign incorporated into the definition of rate in this case? It is
ecause the term "rate of disappearance" implies that $[Fe^{3+}]$ decreases with
me. When told the rate of disappearance of Fe^{3+} is $2.6 \times 10^{-5}\,M\,s^{-1}$, we

know the rate of change of *concentration* must be $-2.6 \times 10^{-5}\,M\,s^{-1}$.

In the same way that we related the rate of formation of Sn^{4+} to that of Fe^{2} we can relate the rate of disappearance of Sn^{2+} to that of Fe^{3+}. That is, the rate disappearance of Sn^{2+} is half that of Fe^{3+}, giving

$$\text{rate of disappearance of } Sn^{2+} = 1.3 \times 10^{-5}\,M\,s^{-1}$$

When we refer to the rate of reaction (20.1), which of the four quantiti described should we use? To avoid confusion in this matter, the Internatior Union of Pure and Applied Chemistry (IUPAC) recommends that we use a *ge eral* rate of reaction, which, for the hypothetical reaction represented by the b anced equation,

$$a\,A + b\,B + \cdots \longrightarrow g\,G + h\,H + \cdots$$

is

$$\text{rate of reaction} = -\frac{1}{a}\frac{\Delta[A]}{\Delta t} = -\frac{1}{b}\frac{\Delta[B]}{\Delta t} = \frac{1}{g}\frac{\Delta[G]}{\Delta t} = \frac{1}{h}\frac{\Delta[H]}{\Delta t} \qquad (20$$

In this expression, we take the negative value of $\Delta[X]/\Delta t$, when X refers to reactant to ensure that the **rate of reaction** is a positive quantity. To obtain a si gle, positive quantity it is necessary to divide all rates by the appropriate st chiometric coefficients. If we apply this expression to reaction (20.1), we obtai

$$\text{rate of reaction} = -\frac{1}{2}\frac{\Delta[Fe^{3+}]}{\Delta t} = -\frac{\Delta[Sn^{2+}]}{\Delta t}$$

$$= \frac{1}{2}\frac{\Delta[Fe^{2+}]}{\Delta t} = \frac{\Delta[Sn^{4+}]}{\Delta t} = 1.3 \times 10^{-5}\,M\,s^{-1}$$

EXAMPLE 20-1 Expressing the Rate of a Reaction

Suppose that at some point in the reaction

$$A + 3\,B \longrightarrow 2\,C + 2\,D$$

$[B] = 0.9986\,M$, and that 13.20 min later $[B] = 0.9746\,M$. What is the average rate of reaction during this time period, expressed in $M\,s^{-1}$?

Analyze

This is a straightforward application of the definition for rate of reaction, expression (20.2). To formulate the rate, we use $\Delta[B] = 0.9746\,M - 0.9986\,M = -0.0240\,M$ and $\Delta t = 13.20$ min.

Solve

The solution is

$$\text{average rate of reaction} = -\frac{1}{3}\frac{\Delta[B]}{\Delta t} = -\frac{1}{3} \times \frac{-0.0240\,M}{13.20\,\text{min}} = 6.06 \times 10^{-4}\,M\,\text{min}^{-1}$$

To express the rate of reaction in moles per liter per second, we must convert from min^{-1} to s^{-1}. We can do this with the conversion factor 1 min/60 s.

$$\text{rate of reaction} = 6.06 \times 10^{-4}\,M\,\text{min}^{-1} \times \frac{1\,\text{min}}{60\,s} = 1.01 \times 10^{-5}\,M\,s^{-1}$$

Alternatively, we could have converted 13.20 min to 792 s and used $\Delta t = 792$ s in evaluating the rate of reaction.

Assess

By monitoring changes in concentration over a time period, we can obtain the average rate of reaction. Remember that the rate of reaction can be defined in terms of any reactant or product.

PRACTICE EXAMPLE A: At some point in the reaction $2\,A + B \longrightarrow C + D$, $[A] = 0.3629\,M$. At a time 8.25 min later $[A] = 0.3187\,M$. What is the average rate of reaction during this time interval, expressed in $M\,s^{-1}$?

PRACTICE EXAMPLE B: In the reaction $2\,A \longrightarrow 3\,B$, $[A]$ drops from 0.5684 M to 0.5522 M in 2.50 min. What is the average rate of formation of B during this time interval, expressed in $M\,s^{-1}$?

20-1 CONCEPT ASSESSMENT

In the reaction of gaseous nitrogen and hydrogen to form gaseous ammonia, what are the relative rates of disappearance of the two reactants? How is the rate of formation of the product related to the rates of disappearance of the reactants?

0-2 Measuring Reaction Rates

› determine a rate of reaction, we need to measure changes in concentration 'er time. A change in time can be measured with a stopwatch or other timing :vice, but how do we measure concentration changes during a chemical action? Also, why is the term *average* used in referring to a rate of reaction? ıese are two of the questions that are answered in this section.

ollowing a Chemical Reaction

ıe decomposition of hydrogen peroxide, a common antiseptic, results in the rmation of water and $O_2(g)$*

$$H_2O_2(aq) \longrightarrow H_2O(l) + \frac{1}{2}O_2(g) \quad \textbf{(20.3)}$$

'e can follow the progress of the reaction by focusing either on the formation $O_2(g)$ or on the disappearance of H_2O_2. For example, we can

- Measure the volumes of $O_2(g)$ produced at different times and relate these volumes to decreases in concentration of H_2O_2 (Fig. 20-1).
- Remove small samples of the reaction mixture from time to time, and analyze these samples for their H_2O_2 content. One way to do this is by titration with $KMnO_4$ in acidic solution. The net ionic equation for this oxidation–reduction reaction is

$$MnO_4^-(aq) + 5\,H_2O_2(aq) + 6\,H^+(aq) \longrightarrow 2\,Mn^{2+}(aq) + 8\,H_2O(l) + 5\,O_2(g) \quad \textbf{(20.4)}$$

ıble 20.1 lists typical data for the decomposition of H_2O_2, and Figure 20-2 splays comparable data graphically.

KEEP IN MIND

that according to expression (20.2), the rate of this reaction can be expressed either as rate of disappearance of H_2O_2 or 2 × rate of formation of $O_2(g)$.

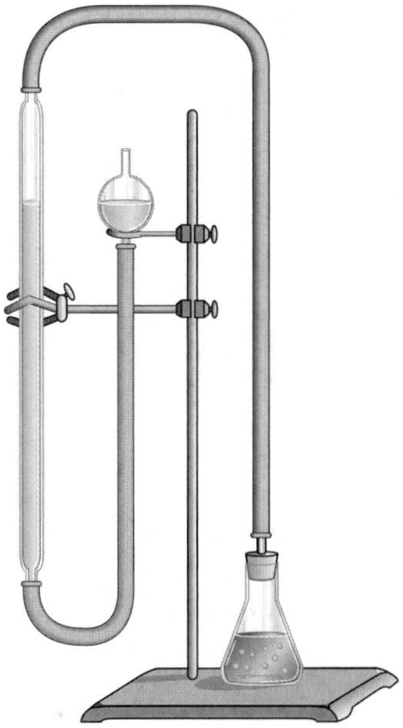

▲ FIGURE 20-1
Experimental setup for determining the rate of decomposition of H_2O_2
Oxygen gas given off by the reaction mixture is trapped, and its volume is measured in the gas buret. The amount of H_2O_2 consumed and the remaining concentration of H_2O_2 can be calculated from the measured volume of $O_2(g)$.

ate of Reaction Expressed as Concentration hange over Time

›me data extracted from Figure 20-2 are listed in Table 20.2 (in blue). Column III ;ts the concentration of H_2O_2 at the times shown in column I. Column II lists ıe *arbitrary* time interval between data points—400 s. Column IV reports the ›ncentration changes that occur for each 400 s interval. The rates of reaction, :pressed as the rate of disappearance of H_2O_2, are shown in column V. The ıta show that the reaction rate is not constant—the lower the remaining con- :ntration of H_2O_2, the more slowly the reaction proceeds.

TABLE 20.1 Decomposition of H_2O_2

Time, s	$[H_2O_2]$, M
0	2.32
200	2.01
400	1.72
600	1.49
1200	0.98
1800	0.62
3000	0.25

ate of Reaction Expressed as the Slope of a Tangent Line

'hen the rate of reaction is expressed as $-\Delta[H_2O_2]/\Delta t$, the result is an *average* ılue for the time interval Δt. For example, in the interval from 1200 to 1600 s, ıe rate averages $6.3 \times 10^{-4}\ M\ s^{-1}$ (fourth entry in column V of Table 20.2). This

*Although reaction (20.3) goes to completion, it does so very slowly. Generally, a catalyst is used speed up the reaction. We describe the catalysis in this reaction in Section 20-11. When $_2O_2(aq)$ is applied to an open wound, the enzyme catalase in blood catalyzes its decomposition.

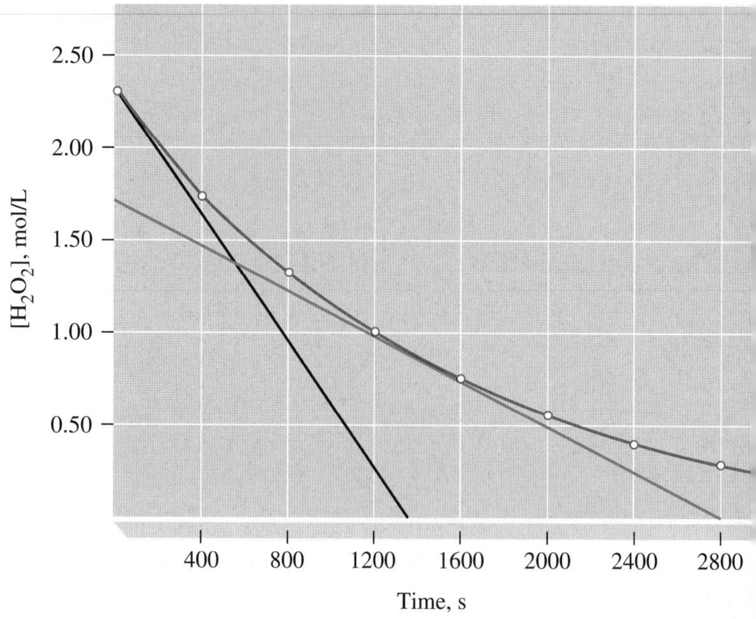

▶ FIGURE 20-2
Graphical representation of kinetic data for the reaction:

$$H_2O_2(aq) \longrightarrow H_2O(l) + \frac{1}{2}O_2(g)$$

This is the usual form in which concentration–time data are plotted. Reaction rates are determined from the slopes of the tangent lines. The blue line has a slope of $-1.70\ M/2800\ s = -6.1 \times 10^{-4}\ M\ s^{-1}$. The slope of the black line and its relation to the initial rate of reaction are described in Example 20-2.

TABLE 20.2 Decomposition of H_2O_2—Derived Rate Data

I	II	III	IV	V
Time, s	Δt, s	$[H_2O_2]$, M	$\Delta[H_2O_2]$, M	Reaction Rate $-\Delta[H_2O_2]/\Delta t$, $M\ s^{-1}$
0		2.32		
	400		−0.60	15.0×10^{-4}
400		1.72		
	400		−0.42	10.5×10^{-4}
800		1.30		
	400		−0.32	8.0×10^{-4}
1200		0.98		
	400		−0.25	6.3×10^{-4}
1600		0.73		
	400		−0.19	4.8×10^{-4}
2000		0.54		
	400		−0.15	3.8×10^{-4}
2400		0.39		
	400		−0.11	2.8×10^{-4}
2800		0.28		

can be thought of as the reaction rate at about the middle of the interval—1400 We could just as well have chosen $\Delta t = 200$ s in Table 20.2. In this case, to obta the rate of reaction at $t = 1400$ s, we would use concentration data at $t = 130($ and $t = 1500$ s. The rate of reaction would be slightly less than $6.3 \times 10^{-4}\ M\ s^{-}$ The smaller the time interval used, the closer the result is to the actual rate $t = 1400$ s. As Δt approaches 0 s, the rate of reaction approaches the *negati of the slope of the tangent line* to the curve of Figure 20-2. The rate of reactic determined from the slope of a tangent line to a concentration–time curve the **instantaneous rate of reaction** at the point where the tangent line touch the curve. The best estimate of the rate of reaction of H_2O_2 at 1400 s, the is obtained as follows from the slope of the blue tangent line in Figure 20- $-(-6.1 \times 10^{-4}\ M\ s^{-1}) = 6.1 \times 10^{-4}\ M\ s^{-1}$.

To better understand the difference between average and instantaneous reac-
n rates, think of taking a 190 km highway trip in 2.00 h. The *average* speed is
km/h. The *instantaneous* speed is the speedometer reading at any instant.

20-1 ARE YOU WONDERING?

How do we differentiate between the average rate of reaction and the instantaneous rate?

If you are familiar with differential calculus, you probably know the answer. If the rate of reaction (20.3) is written as

$$\lim_{\Delta t \longrightarrow 0} \frac{-\Delta[H_2O_2]}{\Delta t}$$

the delta quantities can be replaced by the differentials $-d[H_2O_2]$ and dt, leading to the expression

$$\frac{-d[H_2O_2]}{dt}$$

Thus, the delta notation (not taken to the limit of $\Delta t \longrightarrow 0$) signifies an *average* rate and the differential notation, an *instantaneous* rate.

itial Rate of Reaction

metimes we simply want to find the rate of reaction when the reactants are
st brought together—the **initial rate of reaction**. This rate can be obtained from
e tangent line to the concentration–time curve at $t = 0$. An alternative way is to
easure the concentration of the chosen reactant as soon as possible after mix-
g, in this way obtaining Δ[reactant] for a very short time interval (Δt) at
sentially $t = 0$. These two approaches give the same result if the time interval
ed is limited to that in which the tangent line and the concentration-time curve
actically coincide. In Figure 20-2 this condition occurs for about the first 200 s.

EXAMPLE 20-2 Determining and Using an Initial Rate of Reaction

From the data in Table 20.2 and Figure 20-2 for the decomposition of H_2O_2, determine **(a)** the initial rate of reaction, and **(b)** $[H_2O_2]_t$ at $t = 100$ s, assuming that the initial rate is constant for at least 100 s.

Analyze

To determine the initial rate of reaction from the slope of the tangent line in part (a), we use the intersection of the black tangent line with the axis. In part (b) we will assume that the rate determined in (a) remains essentially constant for at least 100 s.

Solve

(a) $t = 0$, $[H_2O_2] = 2.32$ M; $t = 1360$ s, $[H_2O_2] = 0$.

$$\text{initial rate of reaction} = -(\text{slope of tangent line}) = \frac{-(0 - 2.32)\ \text{M}}{(1360 - 0)\ \text{s}}$$

$$= 1.71 \times 10^{-3}\ \text{M s}^{-1}$$

An alternative method is to use data from Table 20.1: $[H_2O_2] = 2.32$ M at $t = 0$ and $[H_2O_2] = 2.01$ M at $t = 200$ s.

$$\text{initial rate} = \frac{-\Delta[H_2O_2]}{\Delta t} = \frac{-(2.01 - 2.32)\ \text{M}}{200\ \text{s}}$$

$$= 1.55 \times 10^{-3}\ \text{M s}^{-1}$$

(continued)

(b) Since

$$\text{rate of reaction} = \frac{-\Delta[H_2O_2]}{\Delta t}$$

then

$$1.71 \times 10^{-3}\,\text{M s}^{-1} = \frac{-\Delta[H_2O_2]}{100\,\text{s}}$$

$$-(1.71 \times 10^{-3}\,\text{M s}^{-1})(100\,\text{s}) = \Delta[H_2O_2] = [H_2O_2]_t - [H_2O_2]_0$$

$$-1.71 \times 10^{-1}\,\text{M} = [H_2O_2]_t - 2.32\,\text{M}$$

$$[H_2O_2]_t = 2.32\,\text{M} - 0.17\,\text{M} = 2.15\,\text{M}$$

Assess

In part (a) the agreement between the two methods is fairly good, although it might be better if the time interval were shorter than 200 s. Of the two results, the one based on the tangent line is presumably more precise because it is expressed with more significant figures. However, the reliability of the tangent line depends on how carefully the tangent line is constructed. Another reason for favoring a graphical method is that it tends to minimize the effect of errors that may be found in individual data points. In part (b) if we had used initial rate $= 1.6 \times 10^{-3}\,\text{M s}^{-1}$, we would have calculated $[H_2O_2]_t = 2.16\,\text{M}$.

PRACTICE EXAMPLE A: For reaction (20.3), determine **(a)** the *instantaneous* rate of reaction at 2400 s and **(b)** $[H_2O_2]$ at 2450 s. [*Hint:* Assume that the instantaneous rate of reaction at 2400 s holds constant for the next 50 s.]

PRACTICE EXAMPLE B: Use data *only* from Table 20.2 to determine $[H_2O_2]$ at $t = 100$ s. Compare this value with the one calculated in Example 20-2(b). Explain the reason for the difference.

20-2 CONCEPT ASSESSMENT

As shown later in the chapter, for certain reactions the initial and instantaneous rates of reaction are equal throughout the course of the reaction. What must be the shape of the concentration–time graph for such a reaction?

20-3 Effect of Concentration on Reaction Rates: The Rate Law

▶ Rate laws are obtained experimentally and thus, are empirical.

One of the goals in a chemical kinetics study is to derive an equation that ca be used to predict the relationship between the rate of reaction and the co centrations of reactants. Such an experimentally determined equation is call a **rate law**, or **rate equation**.

Consider the hypothetical reaction

$$a\,A + b\,B + \cdots \longrightarrow g\,G + h\,H + \cdots \qquad \text{(20}$$

where $a, b, \ldots$ stand for coefficients in the balanced equation. The rate law f reaction (20.5) can often be expressed in the following general form.*

$$\text{rate of reaction} = k[A]^m[B]^n \cdots \qquad \textbf{(20.6)}$$

The terms $[A], [B], \ldots$ represent reactant molarities. The required exponen $m, n, \ldots$ are generally small, positive whole numbers, although in some cas

*We assume that reaction (20.5) goes to completion. If it is reversible, the form of the rate equ tion is more complex than that given in equation (20.6). Even for reversible reactions, thoug equation (20.6) applies to the *initial* rate of reaction because in the early stages of the reacti there are not enough products for a significant reverse reaction to occur.

ey may be zero, fractional, or negative. They must be determined by experi-ent and are generally *not* related to stoichiometric coefficients $a, b, \ldots$. That often $m \neq a$, $n \neq b$, and so on.

The term *order* is related to the exponents in the rate law and is used in two ays: (1) If $m = 1$, we say that the reaction is *first order in A*. If $n = 2$, the reac-n is *second order in B*, and so on. (2) The *overall* **order of reaction** is the sum all the exponents: $m + n + \cdots$. The proportionality constant k relates the te of reaction to reactant concentrations and is called the **rate constant** of the action. Its value depends on the specific reaction, the presence of a catalyst any), and the temperature. *The larger the value of k, the faster a reaction goes.* e order of the reaction establishes the general form of the rate law and the propriate units of k (that is, depending on the values of the exponents).

◀ A common misconception is that rate and k are the same. As equation (20.6) clearly shows, rate $\neq k$ in general. The rate constant, k, is one of the factors affecting the rate of reaction but not the only one.

With the rate law for a reaction, we can

- calculate rates of reaction for known concentrations of reactants
- derive an equation that expresses a reactant concentration as a function of time

t how do we establish the rate law? We need to use *experimental* data of the type scribed in Section 20-2. The method we describe next works especially well.

ethod of Initial Rates

s its name implies, this method requires us to work with *initial* rates of reac-n. As an example, let's look at a specific reaction: that between mercury(II) loride and oxalate ion.

$$2\,HgCl_2(aq) + C_2O_4^{2-}(aq) \longrightarrow 2\,Cl^-(aq) + 2\,CO_2(g) + Hg_2Cl_2(s) \quad \textbf{(20.7)}$$

e tentative rate law that we can write for this reaction is

$$\text{rate of reaction} = k[HgCl_2]^m[C_2O_4^{2-}]^n \quad \textbf{(20.8)}$$

e can follow the reaction by measuring the quantity of $Hg_2Cl_2(s)$ formed as unction of time. Some representative data are given in Table 20.3, which we n assume are based on either the rate of formation of Hg_2Cl_2 or the rate of sappearance of $C_2O_4^{2-}$. In Example 20-3, we will use some of these data to ustrate the method of initial rates.

ABLE 20.3 Kinetic Data for the Reaction:
$2\,HgCl_2 + C_2O_4^{2-} \longrightarrow 2\,Cl^- + 2\,CO_2 + Hg_2Cl_2$

xperiment	$[HgCl_2]$, M	$[C_2O_4^{2-}]$, M	Initial Rate, M min^{-1}
	$[HgCl_2]_1 = 0.105$	$[C_2O_4^{2-}]_1 = 0.15$	$R_1 = 1.8 \times 10^{-5}$
	$[HgCl_2]_2 = 0.105$	$[C_2O_4^{2-}]_2 = 0.30$	$R_2 = 7.1 \times 10^{-5}$
	$[HgCl_2]_3 = 0.052$	$[C_2O_4^{2-}]_3 = 0.30$	$R_3 = 3.5 \times 10^{-5}$

EXAMPLE 20-3 Establishing the Order of a Reaction by the Method of Initial Rates

Use data from Table 20.3 to establish the order of reaction (20.7) with respect to $HgCl_2$ and $C_2O_4^{2-}$ and also the overall order of the reaction.

Analyze

We need to determine the values of m and n in equation (20.8). In comparing Experiment 2 with Experiment 3, note that $[HgCl_2]$ is essentially doubled ($0.105\text{ M} \approx 2 \times 0.052\text{ M}$) while $[C_2O_4^{2-}]$ is held constant (at 0.30 M). Note also that $R_2 = 2 \times R_3$ ($7.1 \times 10^{-5} \approx 2 \times 3.5 \times 10^{-5}$). Rather than use the actual concentrations and rates in the following rate equation, let's work initially with their symbolic equivalents.

(continued)

Solve

We begin by writing

$$R_2 = k \times [HgCl_2]_2^m \times [C_2O_4^{2-}]_2^n = k \times (2 \times [HgCl_2]_3)^m \times [C_2O_4^{2-}]_3^n$$

$$R_3 = k \times [HgCl_2]_3^m \times [C_2O_4^{2-}]_3^n$$

$$\frac{R_2}{R_3} = \frac{2 \times \cancel{R_3}}{\cancel{R_3}} = 2 = \frac{\cancel{k} \times 2^m \times \cancel{[HgCl_2]_3^m} \times \cancel{[C_2O_4^{2-}]_3^n}}{\cancel{k} \times \cancel{[HgCl_2]_3^m} \times \cancel{[C_2O_4^{2-}]_3^n}} = 2^m$$

In order that $2^m = 2, m = 1$.

To determine the value of n, we can form the ratio R_2/R_1. Now, $[C_2O_4^{2-}]$ is doubled and $[HgCl_2]$ is held constant. This time, let's use actual concentrations instead of symbolic equivalents. Also, we now have the value $m = 1$.

$$R_2 = k \times [HgCl_2]_2^1 \times [C_2O_4^{2-}]_2^n = k \times (0.105)^1 \times (2 \times 0.15)^n$$

$$R_1 = k \times [HgCl_2]_1^1 \times [C_2O_4^{2-}]_1^n = k \times (0.105)^1 \times (0.15)^n$$

$$\frac{R_2}{R_1} = \frac{7.1 \times 10^{-5}}{1.8 \times 10^{-5}} \approx 4 = \frac{\cancel{k} \times \cancel{(0.105)^1} \times 2^n \times \cancel{(0.15)^n}}{\cancel{k} \times \cancel{(0.105)^1} \times \cancel{(0.15)^n}} = 2^n$$

In order that $2^n = 4, n = 2$.

Assess

In summary, the reaction is *first* order in $HgCl_2$ ($m = 1$), *second* order in $C_2O_4^{2-}$ ($n = 2$), and *third* order overall ($m + n = 1 + 2 = 3$). In this example we found n by solving the equation $2^n = 4$. In some cases we may need to solve equations containing a noninteger number, such as in $2^n = 1.4143$. To solve this kind of equation, we take the logarithm of both sides, for example, $\log(2^n) = \log(1.4143)$, and rearrange to get $n = \log(1.4143)/\log(2)$. The answer is $n = 0.5$. Consult Appendix A-2 if you are unfamiliar with logarithms.

PRACTICE EXAMPLE A: The decomposition of N_2O_5 is given by the following equation:

$$2\,N_2O_5 \longrightarrow 4\,NO_2 + O_2$$

At an initial $[N_2O_5] = 3.15$ M, the initial rate of reaction $= 5.45 \times 10^{-5}$ M s^{-1}, and when $[N_2O_5] = 0.78$ M, the initial rate of reaction $= 1.35 \times 10^{-5}$ M s^{-1}. Determine the order of this decomposition reaction.

PRACTICE EXAMPLE B: Consider a hypothetical Experiment 4 in Table 20.3, in which the initial conditions are $[HgCl_2]_4 = 0.025$ M and $[C_2O_4^{2-}]_4 = 0.045$ M. Predict the initial rate of reaction.

We made an important observation in Example 20-3: If a reaction is *fir* *order* in one of the reactants, doubling the initial concentration of that reacta causes the initial rate of reaction to double. Following is the general effect *doubling* the initial concentration of a particular reactant (with other reacta concentrations held constant).

▶ Reactions that are too fast for traditional methods of analysis are followed by spectroscopic methods.

- *Zero order* in the reactant—there is *no effect* on the initial rate of reactio
- *First order* in the reactant—the initial rate of reaction *doubles*.
- *Second order* in the reactant—the initial rate of reaction *quadruples*.
- *Third order* in the reactant—the initial rate of reaction *increases eightfold*

As previously mentioned, the order of a reaction, as indicated through th rate law, establishes the units of the rate constant, k. That is, if on the left sic of the rate law the rate of reaction has the units M (time)$^{-1}$, on the right sid the units of k must provide for the cancellations that also lead to M (time)$^{-}$ Thus, for the rate law established in Example 20-3,

rate law:	rate of reaction =	$k \times$	$[HgCl_2] \times$	$[C_2O_4^{2-}]^2$
units:	M min^{-1}	M^{-2} min^{-1}	M	M^2

Once we have the exponents in a rate equation, we can determine the valı of the rate constant, k. To do this, all we need is the rate of reaction correspon ing to known initial concentrations of reactants, as illustrated in Example 20-4

EXAMPLE 20-4 Using the Rate Law

Use the results of Example 20-3 and data from Table 20.3 to establish the value of k in the rate law (20.8).

Analyze

We can use data from any one of the three experiments of Table 20.3, together with the values $m = 1$ and $n = 2$.

Solve

First, we solve equation (20.8) for k.

$$k = \frac{R_1}{[HgCl_2][C_2O_4^{2-}]^2} = \frac{1.8 \times 10^{-5}\,M\,min^{-1}}{0.105\,M \times (0.15)^2\,M^2}$$
$$= 7.6 \times 10^{-3}\,M^{-2}\,min^{-1}$$

Assess

If the rate data in Table 20.3 were based on the disappearance of $HgCl_2$ instead of $C_2O_4^{2-}$, R_1 in this setup would be twice as great and k for the general rate of reaction would be based on $-\frac{1}{2} \times \Delta[HgCl_2]/\Delta t$. Note the units on the rate constant are $M^{-2}\,min^{-1}$, which are appropriate units for a third-order rate constant. Checking the units in an answer is one way to ensure that we have not made any mistakes.

PRACTICE EXAMPLE A: A reaction has the rate law: rate $= k[A]^2[B]$. When $[A] = 1.12$ M and $[B] = 0.87$ M, the rate of reaction $= 4.78 \times 10^{-2}\,M\,s^{-1}$. What is the value of the rate constant, k?

PRACTICE EXAMPLE B: What is the rate of reaction (20.7) at the point where $[HgCl_2] = 0.050$ M and $[C_2O_4^{2-}] = 0.025$ M?

20-3 CONCEPT ASSESSMENT

The rate of decomposition of gaseous acetaldehyde, CH_3CHO, to gaseous methane and carbon monoxide is found to increase by a factor of 2.83 when the initial concentration of acetaldehyde is doubled. What is the order of this reaction?

0-4 Zero-Order Reactions

n overall **zero-order reaction** has a rate law in which the sum of the expo- nts, $m + n \cdots$, is equal to 0. As an example, let's take a reaction in which a ngle reactant A decomposes to products.

$$A \longrightarrow \text{products}$$

the reaction is zero order, the rate law is

$$\text{rate of reaction} = k[A]^0 = k = \text{constant} \qquad \textbf{(20.9)}$$

Other features of this zero-order reaction are:

- The concentration–time graph is a straight line with a *negative* slope (Fig. 20-3).
- The rate of reaction, which is equal to k and remains constant throughout the reaction, is the *negative* of the slope of this line.
- The units of k are the same as the units of the rate of a reaction: mol L^{-1} (time)$^{-1}$, for example, mol $L^{-1}\,s^{-1}$ or $M\,s^{-1}$.

Equation (20.9) is the rate law for a zero-order reaction. Another useful uation, called an **integrated rate law**, expresses the concentration of a actant as a function of time. This equation can be established rather easily

▶ The dependence of k on temperature is discussed in Section 20-9.

20-2 ARE YOU WONDERING?

What is the difference between the rate and the rate *constant*?

Many students have difficulty with this distinction. Remember, a *rate of reaction* tells how the concentration of a reactant or product changes with time and is often expressed as moles per liter per second. A rate of reaction can be established through an expression of the type $-\Delta[A]/\Delta t$, from a tangent to a concentration–time curve, and by calculation from a rate law. In most cases, the rate of a reaction strongly depends on reactant concentrations and, therefore, changes continuously during a reaction.

The *rate constant* of a reaction (k) relates the rate of a reaction to reactant concentrations. Generally, it is not itself a rate of reaction, but it can be used to calculate rates of reaction. Once the value of k at a given temperature has been established, *this value stays fixed* for the given reaction, regardless of the reactant concentrations. The units of reaction rate do not depend on the order of a reaction, but those of k do, as we shall see in this and the following two sections.

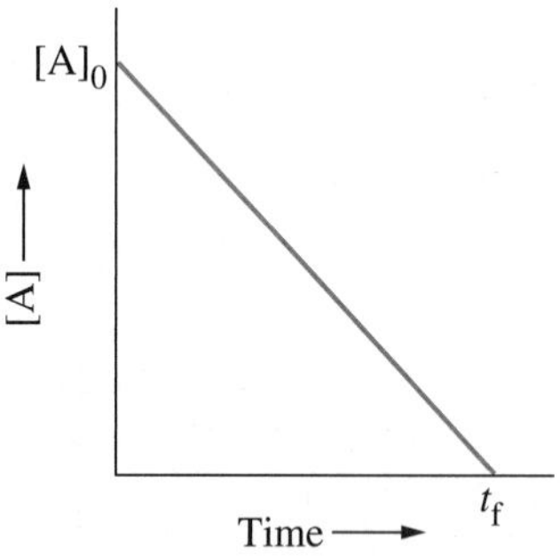

▲ FIGURE 20-3
A zero-order reaction: A ⟶ products
The initial concentration of the reactant A is $[A]_0$, that is, $[A] = [A]_0$ at $t = 0$. $[A]$ decreases until the reaction stops. This occurs at the time, t_f, where $[A] = 0$. The slope of the line is

$$\frac{(0 - [A]_0)}{(t_f - 0)} = -\frac{[A]_0}{t_f}.$$

The rate constant is the *negative* of the slope:

$$k = -\text{slope} = \frac{[A]_0}{t_f}.$$

from the graph in Figure 20-3. Let's start with the general equation for straight line

$$y = mx + b$$

and substitute $y = [A]_t$ (the concentration of A at some time t); $x = t$ (time); $b = [A]_0$ (the initial concentration of A at time $t = 0$); and $m = -k$ (m, the slope of the straight line, is obtained as indicated in the caption to Figure 20-3).

$$[A]_t = -kt + [A]_0 \tag{20.10}$$

▶ Perhaps the most important examples of zero-order reactions are found in the action of enzymes. Enzyme-catalyzed reactions are discussed in Section 20-11.

20-3 ARE YOU WONDERING?

Does the term "integrated" rate law have anything to do with integral calculus?

Not surprisingly, it does. The rate of reaction in a rate law, such as equation (20.9), is an *instantaneous* rate, which we learned in Are You Wondering 20-1 can be represented through differentials. When $-d[A]/dt$ is substituted for the rate of reaction in the rate law for a zero-order reaction, we get the equation $-d[A]/dt = k$. We can separate the differentials to obtain $d[A] = -k\,dt$. At this point we can apply the calculus procedure of integration to obtain, successively, the following expressions. The final one is equation (20.10), the integrated rate law for a zero-order reaction.

$$\int_{[A]_0}^{[A]_t} d[A] = -k\int_0^t dt, \qquad [A]_t - [A]_0 = -kt, \qquad [A]_t = -kt + [A]_0$$

20-5 First-Order Reactions

An overall **first-order reaction** has a rate law in which the sum of the exponents, $m + n + \cdots$, is equal to 1. A particularly common type of first-order reaction, and the only type we will consider, is one in which a single reacta

ecomposes into products. The decomposition of H_2O_2, a reaction we encoun-
red in Section 20-2, is a first-order reaction.

$$H_2O_2(aq) \longrightarrow H_2O(l) + \frac{1}{2}O_2(g)$$

ne rate of reaction depends on the concentration of H_2O_2 raised to the *first* ɔwer, that is,

$$\text{rate of reaction} = k[H_2O_2] \qquad \textbf{(20.11)}$$

It is easy to establish that the reaction is first order by the method of initial ıtes, but there are also other ways of recognizing a first-order reaction.

n Integrated Rate Law for a First-Order Reaction

et us begin our discussion of first-order reactions as we did zero-order reac-ons, by examining a hypothetical reaction

$$A \longrightarrow \text{products}$$

ɔr a first-order reaction, the rate law is

$$\text{rate of reaction} = -\frac{\Delta[A]}{\Delta t} = k[A] \qquad \textbf{(20.12)}$$

◀ Most processes in nature follow first-order chemical kinetics. If you had a bank that compounded interest continuously, then your funds would grow exponentially.

'e can obtain the *integrated* rate law for this first-order reaction by applying ıe calculus technique of *integration* to equation (20.12). The result of this erivation (shown in Are You Wondering 20-4) is

$$\ln\frac{[A]_t}{[A]_0} = -kt \quad \text{or} \quad \ln[A]_t = -kt + \ln[A]_0 \qquad \textbf{(20.13)}$$

A]$_t$ is the concentration of A at time t, $[A]_0$ is its concentration at $t = 0$, and k the rate constant. Because the logarithms of numbers are dimensionless ıave no units), the product $-k \times t$ must also be dimensionless. This means ıat the unit of k in a first-order reaction is $(\text{time})^{-1}$, such as s^{-1} or min^{-1}. quation (20.13) is that of a straight line.

KEEP IN MIND

that scatter of experimental data on a straight-line graph often results in a value of k calculated from equation (20.13) less reliable than a value obtained from the slope of the straight line.

Equation of straight line

$$\underbrace{\ln[A]_t}_{y} = \underbrace{(-k)t}_{m \cdot x} + \underbrace{\ln[A]_0}_{b}$$

20-4 ARE YOU WONDERING?

How do we obtain the integrated rate law for a first-order reaction?

If the reaction A $\longrightarrow$ products is first order, then the rate law is, in differential form, $d[A]/dt = -k[A]$.

Separation of the differentials leads to the expression $d[A]/[A] = -k\,dt$.

Integration of this expression between the limits $[A]_0$ at time $t = 0$ and $[A]_t$ at time t is indicated through the expression

$$\int_{[A]_0}^{[A]_t} \frac{d[A]}{[A]} = -k\int_0^t dt$$

The result of the integration is equation (20.13), the integrated rate law.

$$\ln\frac{[A]_t}{[A]_0} = -kt$$

EXAMPLE 20-5 Using the Integrated Rate Law for a First-Order Reaction

$H_2O_2(aq)$, initially at a concentration of 2.32 M, is allowed to decompose. What will $[H_2O_2]$ be at $t = 1200$ s? Use $k = 7.30 \times 10^{-4}\,s^{-1}$ for this first-order decomposition.

Analyze

We have values for three of the four quantities in equation (20.13):

$$k = 7.30 \times 10^{-4}\,s^{-1} \qquad t = 1200\,s$$
$$[H_2O_2]_0 = 2.32\,M \qquad [H_2O_2]_t = ?$$

We need to solve for the fourth quantity.

Solve

We substitute into the expression

$$\begin{aligned}\ln[H_2O_2]_t &= -kt + \ln[H_2O_2]_0 \\ &= -(7.30 \times 10^{-4}\,s^{-1} \times 1200\,s) + \ln(2.32) \\ &= -0.8760 + 0.8415 = -0.0345 \\ [H_2O_2]_t &= e^{-0.0345} = 0.966\,M\end{aligned}$$

Assess

This calculated value agrees well with the experimentally determined value of 0.98 M, shown below.

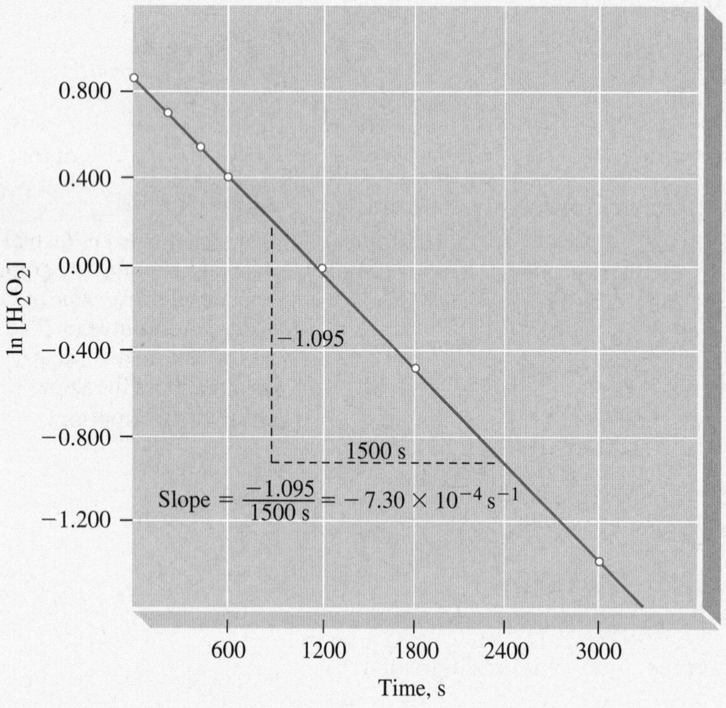

◀ FIGURE 20-4
Test for a first-order reaction: Decomposition of $H_2O_2(aq)$
Plot of ln H_2O_2 versus t. The data are based on Table 20.1 and are listed below. The slope of the line is used in the text.

t, s	$[H_2O_2]$, M	ln $[H_2O_2]$
0	2.32	0.842
200	2.01	0.698
400	1.72	0.542
600	1.49	0.399
1200	0.98	−0.020
1800	0.62	−0.48
3000	0.25	−1.39

PRACTICE EXAMPLE A: The reaction $A \longrightarrow 2B + C$ is first order. If the initial $[A] = 2.80\,M$ and $k = 3.02 \times 10^{-3}\,s^{-1}$, what is the value of $[A]$ after 325 s?

PRACTICE EXAMPLE B: Use data tabulated in Figure 20-4, together with equation (20.13), to show that the decomposition of H_2O_2 is a first-order reaction. [*Hint:* Use a pair of data points for $[H_2O_2]_0$ and $[H_2O_2]_t$ and their corresponding times to solve for k. Repeat this calculation using other sets of data. How should the results compare?]

An easy test for a first-order reaction is to plot the natural logarithm of a actant concentration versus time and see if the graph is linear. The data from ble 20.1 are plotted in Figure 20-4, and the rate constant k is derived from the ope of the line: $k = -\text{slope} = -(-7.30 \times 10^{-4}\,\text{s}^{-1}) = 7.30 \times 10^{-4}\,\text{s}^{-1}$. An ernative, nongraphical approach, illustrated in Practice Example 20-5B, is to bstitute data points into equation (20.13) and solve for k.

Although until now we have used only molar concentrations in kinetics uations, we can sometimes work directly with the masses of reactants. nother possibility is to work with a fraction of reactant consumed, as is done the concept of half-life.

The **half-life** of a reaction is the time required for one-half of a reactant to be nsumed. It is the time during which the amount of reactant or its concentra-n decreases to one-half of its initial value. That is, at $t = t_{1/2}$, $[\text{A}]_t = \frac{1}{2}[\text{A}]_0$. this time, equation (20.13) takes the form

$$\ln\frac{[\text{A}]_t}{[\text{A}]_0} = \ln\frac{\frac{1}{2}\cancel{[\text{A}]_0}}{\cancel{[\text{A}]_0}} = \ln\frac{1}{2} = -\ln 2 = -k \times t_{1/2}$$

$$t_{1/2} = \frac{\ln 2}{k} = \frac{0.693}{k} \qquad \textbf{(20.14)}$$

uation (20.14) is valid only for first-order reactions. We will derive half-life pressions for other types of reactions as we encounter them.

For the decomposition of $H_2O_2(aq)$, the reaction described in Section 20-2, e conclude that the half-life is

$$t_{1/2} = \frac{0.693}{7.30 \times 10^{-4}\,\text{s}^{-1}} = 9.49 \times 10^2\,\text{s} = 949\,\text{s}$$

uation (20.14) indicates that *the half-life is constant for a first-order reaction.* us, regardless of the value of $[\text{A}]_0$ at the time we begin to follow a reaction, $t = t_{1/2}$, $[\text{A}] = \frac{1}{2}[\text{A}]_0$. After *two* half-lives, that is, at $t = 2 \times t_{1/2}$, $[\text{A}] = \frac{1}{2} \times$ $\text{A}]_0 = \frac{1}{4}[\text{A}]_0$. At $t = 3 \times t_{1/2}$, $[\text{A}] = \frac{1}{8}[\text{A}]_0$, and so on.

◀ Half-life is constant for first-order reactions but not constant for zero-order or second-order reactions. For a zero-order reaction, half-life decreases with decreasing reactant concentration. For a second-order reaction, half-life increases with decreasing reactant concentration.

The constancy of the half-life and its independence of the initial concentra-on can be used as a test for a first-order reaction. Try it with the simple ncentration–time graph of Figure 20-2. That is, starting with $[H_2O_2] = 2.32$ M $t = 0$, at what time is $[H_2O_2]$ equal to 1.16 M? 0.58 M? 0.29 M? Starting with $_2O_2] = 1.50$ M at $t = 600$ s, at what time is $[H_2O_2] = 0.75$ M?

In the discussion above we made the assumption that the stoichiometric efficient $a = 1$. What would the rate equations look like if $a \neq 1$? In this case e would write

$$a\,\text{A} \longrightarrow \text{products}$$

r which the rate law is

$$\text{rate of reaction} = -\frac{1}{a}\frac{\Delta[\text{A}]}{\Delta t} = k[\text{A}]$$

hich can be rearranged to give

$$-\frac{\Delta[\text{A}]}{\Delta t} = ak[\text{A}]$$

e expression above is very similar to equation (20.12) except that ak has ken the place of k in that equation. As a result, the integrated rate law d the half-life for the reaction $a\,\text{A} \longrightarrow$ products can be obtained from uations (20.13) and (20.14) by replacing k with ak. We obtain

$$\ln[\text{A}]_t = -akt + \ln[\text{A}]_0 \quad \text{and} \quad t_{1/2} = \frac{0.693}{ak}$$

As illustrated in Example 20-6, a first-order reaction can also be described terms of the percent of a reactant consumed or remaining.

EXAMPLE 20-6 Expressing Fraction (or Percent) of Reactant Consumed in a First-Order Reaction

Use a value of $k = 7.30 \times 10^{-4}\,s^{-1}$ for the first-order decomposition of $H_2O_2(aq)$ to determine the percent H_2O_2 that has decomposed in the first 500.0 s after the reaction begins.

Analyze

The ratio $[H_2O_2]_t/[H_2O_2]_0$ represents the fractional part of the initial amount of H_2O_2 that remains *unreacted* at time t. Our problem is to evaluate this ratio at $t = 500.0$ s. This is done by making use of equation (20.13), which relates the concentration at $t = 0$ to the concentration at some other time.

Solve

Substituting into equation (20.13),

$$\ln\frac{[H_2O_2]_t}{[H_2O_2]_0} = -kt = -7.30 \times 10^{-4}\,s^{-1} \times 500.0\,s = -0.365$$

$$\frac{[H_2O_2]_t}{[H_2O_2]_0} = e^{-0.365} = 0.694 \qquad \text{and} \qquad [H_2O_2]_t = 0.694[H_2O_2]_0$$

The fractional part of the H_2O_2 remaining is 0.694, or 69.4%. The percent of H_2O_2 that has decomposed is 100.0% − 69.4% = 30.6%.

Assess

The integrated form of the rate law is used for determining concentrations as a function of time. The integrated form of the rate law for first-order reactions is useful for many types of reactions and events. We will see this equation later when we talk about radioactivity (Section 25-5).

PRACTICE EXAMPLE A: Consider the first-order reaction A $\longrightarrow$ products, with $k = 2.95 \times 10^{-3}\,s^{-1}$. What percent of A remains after 150 s?

PRACTICE EXAMPLE B: At what time after the start of the reaction is a sample of $H_2O_2(aq)$ two-thirds decomposed?

Reactions Involving Gases

For gaseous reactions, rates are often measured in terms of gas pressures. Fo the hypothetical reaction, $A(g) \longrightarrow$ products, the initial partial pressur $(P_A)_0$, and the partial pressure at some time t, $(P_A)_t$, are related through th expression

$$\ln\frac{(P_A)_t}{(P_A)_0} = -kt \qquad \textbf{(20.1}$$

To see how this equation is derived, start with the ideal gas equation writte for reactant A: $P_AV = n_ART$. Note that the ratio n_A/V is the same as [A So, $[A]_0 = (P_A)_0/RT$ and $[A]_t = (P_A)_t/RT$. Substitute these terms into equa tion (20.13), and note that the RT terms cancel in the numerator and denom nator and leave the simple ratio $(P_A)_t/(P_A)_0$.

Di-*t*-butyl peroxide (DTBP) is used as a catalyst in the manufacture o polymers. In the gaseous state, DTBP decomposes into acetone and ethane b a first-order reaction.

$$\underset{\text{DTBP}}{C_8H_{18}O_2(g)} \longrightarrow \underset{\text{acetone}}{2\,CH_3COCH_3(g)} + \underset{\text{ethane}}{CH_3CH_3(g)} \qquad \textbf{(20.1}$$

Partial pressures of DTBP are plotted as a function of time in Figure 20-5, an the half-life of the reaction is indicated.

As shown in Are You Wondering 20-5, the partial pressure of the reactan DTBP can be obtained from two experimentally determined quantities: th initial pressure P_0, and the total pressure P_{total}.

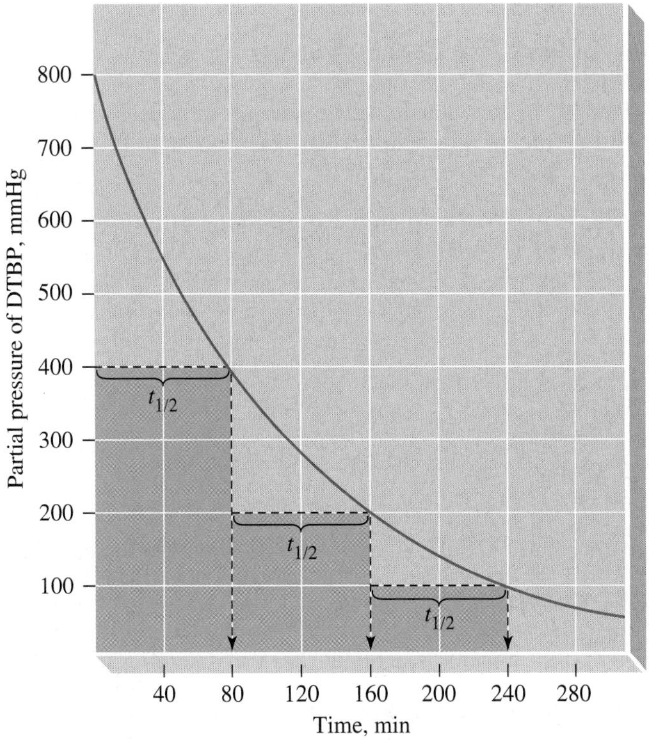

FIGURE 20-5
ecomposition of di-*t*-butyl peroxide (DTBP) at 147 °C
ne decomposition reaction is described through equation (20.16). In this graph the partial pressure of DTBP as a function of time, three successive half-life eriods of 80 min each are indicated. This constancy of the half-life is proof at the reaction is first order.

20-5 ARE YOU WONDERING?

How can the partial pressure be obtained experimentally in a reaction involving gases?

In the study of a reaction like the decomposition of DTBP, the *total* pressure is typically measured as a function of time. At any time t, the total pressure is

$$(P_{\text{total}})_t = (P_{\text{DTBP}})_t + (P_{\text{acetone}})_t + (P_{\text{ethane}})_t$$

If the initial presure of DTBP is P_0, then using the stoichiometry of the balanced chemical equation, the partial pressure of DTBP is $P_{\text{DTBP}} = (P_0 - P_{\text{ethane}})$ since for every mole of DTBP that decomposes, a mole of ethane is produced. The partial pressure of acetone is $P_{\text{acetone}} = 2\,P_{\text{ethane}}$, again using the stoichiometry of the reaction. The total pressure is then given by

$$P_{\text{total}} = (P_0 - P_{\text{ethane}}) + 2\,P_{\text{ethane}} + P_{\text{ethane}} = P_0 + 2\,P_{\text{ethane}}$$

and

$$P_{\text{ethane}} = \frac{P_{\text{total}} - P_0}{2}$$

So that

$$P_{\text{DTBP}} = (P_0 - P_{\text{ethane}}) = \frac{2\,P_0 - (P_{\text{total}} - P_0)}{2} = \frac{3\,P_0 - P_{\text{total}}}{2}$$

◀ Note that the total pressure in reaction (20.16) increases from P_0 at the start of the reaction to $3P_0$ when P_{DTBP} has fallen to zero.

EXAMPLE 20-7 Applying First-Order Kinetics to a Reaction Involving Gases

Reaction (20.16) is started with pure DTBP at 147 °C and 800.0 mmHg pressure in a flask of constant volume. **(a)** What is the value of the rate constant *k*? **(b)** At what time will the partial pressure of DTBP be 50.0 mmHg?

Analyze

In part (a) we observe from Figure 20-5 that $t_{1/2} = 8.0 \times 10^1$ min. Recall that for a first-order reaction the relationship between $t_{1/2}$ and the rate constant is $t_{1/2} = 0.693/k$. In part (b) the final DTBP partial pressure of 50.0 mmHg is $\frac{1}{16}$ of the starting pressure of 800.0 mmHg; that is, $P_{DTBP} = (\frac{1}{2})^4 \times 800.0 = 50.0$ mmHg. The reaction must go through *four* half-lives.

Solve

(a) Substituting in the appropriate values we have

$k = 0.693/t_{1/2} = 0.693/8.0 \times 10^1 \text{ min} = 8.7 \times 10^{-3} \text{ min}^{-1}$

(b) Therefore, $t = 4 \times t_{1/2} = 4 \times 8.0 \times 10^1 \text{ min} = 3.2 \times 10^2 \text{ min}$.

Assess

This type of analysis is useful only if we have the half-life data. If we do not, we will need to use equation (20.13).

PRACTICE EXAMPLE A: Start with DTBP at a pressure of 800.0 mmHg at 147 °C. What will be the pressure of DTBP at $t = 125$ min, if $t_{1/2} = 8.0 \times 10^1$ min? [*Hint:* Because 125 min is not an exact multiple of the half-life, you must use equation (20.15). Can you see that the answer is between 200 and 400 mmHg?]

PRACTICE EXAMPLE B: Use data from Table 20.4 to determine **(a)** the partial pressure of ethylene oxide, and **(b)** the total gas pressure after 30.0 h in a reaction vessel at 415 °C if the initial partial pressure of $(CH_2)_2O(g)$ is 782 mmHg.

TABLE 20.4 Some First-Order Processes

Process	Half-Life, $t_{1/2}$	Rate Constant k, s^{-1}
Radioactive decay of $^{238}_{92}U$	4.51×10^9 yr	4.87×10^{-18}
Radioactive decay of $^{14}_{6}C$	5.73×10^3 yr	3.83×10^{-12}
Radioactive decay of $^{32}_{15}P$	14.3 d	5.61×10^{-7}
Radioactive decay of $^{131}_{53}I$	8.04 d	9.98×10^{-7}
$C_{12}H_{22}O_{11}(aq) + H_2O(l) \xrightarrow{15\,°C} C_6H_{12}O_6(aq) + C_6H_{12}O_6(aq)$ (sucrose → glucose + fructose)	8.4 h	2.3×10^{-5}
$(CH_2)_2O(g) \xrightarrow{415\,°C} CH_4(g) + CO(g)$ (ethylene oxide)	56.3 min	2.05×10^{-4}
$N_2O_5(g) \xrightarrow[45\,°C]{\text{in } CCl_4} N_2O_4(g) + \frac{1}{2}O_2(g)$	18.6 min	6.21×10^{-4}
$CH_3COOH(aq) \longrightarrow H^+(aq) + CH_3COO^-(aq)$	8.9×10^{-7}s	7.8×10^5

▶ We will explore radioactive decay in some detail in Chapter 25.

Examples of First-Order Reactions

One of the most familiar examples of a first-order process is radioactive decay. For example, the radioactive isotope iodine-131, used in treating thyroid disorders, has a half-life of 8.04 days. Whatever number of iodine-131 atoms are in a sample at a given moment, there will be half that number in 8.04 days; one-quarter of that number in 8.04 + 8.04 = 16.08 days; and so on. The rate constant for the decay is $k = 0.693/t_{1/2}$, and in equation (20.13) we can use numbers of atoms, that is, N_t for $[A]_t$ and N_0 for $[A]_0$. Table 20.4 lists several examples of first-order processes. Note the great range of values of $t_{1/2}$ and k. The processes range from very slow to ultrafast.

20-4 CONCEPT ASSESSMENT

Without doing calculations, sketch **(a)** concentration versus time plots for two first-order reactions, one having a large k and the other a small k, and **(b)** ln k versus time plots for the same two situations. Describe the similarities and differences between the two plots in each case.

20-6 Second-Order Reactions

An overall **second-order reaction** has a rate law with the sum of the exponents, $m + n \cdots$, equal to 2. As with zero- and first-order reactions, our discussion will be limited to reactions involving the decomposition of a single reactant

$$\text{A} \longrightarrow \text{products}$$

If the reaction is second order, then we can write

$$\text{rate of reaction} = k[\text{A}]^2 \qquad \textbf{(20.17)}$$

Again, our primary interest will be in the integrated rate law that is derived from the rate law. For the reaction we are considering, the integrated rate law is

$$\frac{1}{[\text{A}]_t} = kt + \frac{1}{[\text{A}]_0} \qquad \textbf{(20.18)}$$

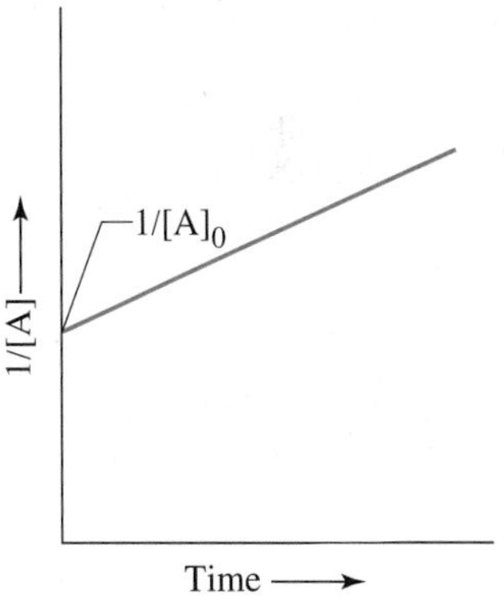

▲ FIGURE 20-6
A straight-line plot for the second-order reaction A ⟶ products
The reciprocal of the concentration, 1/[A], is plotted against time. As the reaction proceeds, [A] decreases and 1/[A] increases in a linear fashion. The slope of the line is the rate constant k.

Figure 20-6 is a plot of $1/[\text{A}]_t$ against time. The slope of the line is k, and the intercept is $1/[\text{A}]_0$. From the graph, we can see that the units of k must be the *reciprocal* of concentration divided by time: $\text{M}^{-1}/(\text{time})$ or $\text{M}^{-1}(\text{time})^{-1}$—for example, $\text{M}^{-1}\,\text{s}^{-1}$ or $\text{M}^{-1}\,\text{min}^{-1}$. We can reach this same conclusion by determining the units of k that produce the required units for the rate of a reaction, that is, moles per liter per unit time. From equation (20.17):

$$\textit{rate law:} \quad \text{rate of reaction} = k \times [\text{A}]^2$$
$$\textit{units:} \quad \text{M time}^{-1} \quad \text{M}^{-1}\,\text{time}^{-1}\,\text{M}^2$$

For the half-life of the second-order reaction A ⟶ products, we can substitute $t = t_{1/2}$ and $[\text{A}] = \frac{1}{2}[\text{A}]_0$ into equation (20.18).

$$\frac{1}{[\text{A}]_0/2} = kt_{1/2} + \frac{1}{[\text{A}]_0}; \quad \frac{2}{[\text{A}]_0} = kt_{1/2} + \frac{1}{[\text{A}]_0} \qquad \textbf{(20.19a)}$$

and

$$t_{1/2} = \frac{1}{k[\text{A}]_0} \qquad \textbf{(20.19b)}$$

From equation (20.19a, b), we see that the half-life depends on both the rate constant *and* the initial concentration $[\text{A}]_0$. The *half-life is not constant*. Its value depends on the concentration of reactant at the start of each half-life interval. Because the starting concentration is always one-half that of the previous half-life, each successive half-life is twice as long as the one before it.

Pseudo-First-Order Reactions

At times, it is possible to simplify the kinetic study of complex reactions b getting them to behave like reactions of a lower order. Then their rate law become easier to work with. Consider the hydrolysis of ethyl acetate, which i second order overall.

$$\underset{\text{Ethyl acetate}}{CH_3COOCH_2CH_3} + H_2O \longrightarrow \underset{\text{Acetic acid}}{CH_3COOH} + \underset{\text{Ethanol}}{CH_3CH_2OH}$$

Suppose we follow the hydrolysis of 1 L of aqueous 0.01 M ethyl acetate t completion. $[CH_3COOCH_2CH_3]$ decreases from 0.01 M to essentially zer This means that 0.01 mol $CH_3COOCH_2CH_3$ is consumed, and along with i 0.01 mol H_2O. Now consider what happens to the molarity of the H_2C Initially, the solution contains about 1000 g H_2O, or about 55.5 mol H_2C When the reaction is completed, there is still 55.5 mol H_2O (that i $55.5 - 0.01 \approx 55.5$). The molarity of the water remains essentially constar throughout the reaction—55.5 M. The rate of reaction does not appear t depend on $[H_2O]$. So, the reaction appears to be *zero* order in H_2O, *first* orde in $CH_3COOCH_2CH_3$, and *first* order overall. A second-order reaction that i made to behave like a first-order reaction by holding one reactant concentra tion constant is called a *pseudo*-first-order reaction. We can treat the reactio with the methods of first-order reaction kinetics. Other reactions of highe order can be made to behave like reactions of lower order under certain con ditions. Thus, a third-order reaction might be converted to pseudo-second order or even to pseudo-first-order.

▶ The rate constant obtained from a study that used an excess of one reactant, sometimes called Ostwald's isolation method after the kineticist who invented it, is a pseudo-rate constant and is concentration-dependent.

20-6 ARE YOU WONDERING?

How do we obtain the integrated rate law for the second-order reaction A ⟶ products?

In differential form, the rate law for a second-order reaction, A ⟶ products, is $d[A]/dt = -k[A]^2$.

Separation of the differentials leads to the expression $d[A]/[A]^2 = -k\,dt$.

Integration of this expression between the limits $[A]_0$ at time $t = 0$ and $[A]_t$ at time t is indicated through the expression

$$\int_{[A]_0}^{[A]_t} \frac{d[A]}{[A]^2} = -\int_0^t k\,dt$$

The result of the integration is equation (20.18), the integrated rate law.

$$-\frac{1}{[A]_t} + \frac{1}{[A]_0} = -kt \quad \text{or} \quad \frac{1}{[A]_t} = kt + \frac{1}{[A]_0}$$

20-7 Reaction Kinetics: A Summary

Let's pause briefly to review what we have learned about rates of reactior rate constants, and reaction orders. Although a problem often can be solved i several different ways, these approaches are generally most direct.

1. To calculate a rate of reaction when the rate law is known, use this expres sion: rate of reaction $= k[A]^m[B]^n \cdots$
2. To determine a rate of reaction when the rate law is not given, use
   - the slope of an appropriate tangent line to the graph of $[A]$ versus t
   - the expression $-\Delta[A]/\Delta t$, with a short time interval Δt

TABLE 20.5 Reaction Kinetics: A Summary for the Hypothetical Reaction a A ⟶ Products

Order	Rate Law[a]	Integrated Rate Equation[b]	Straight Line	$k =$	Units of k	Half-Life[b]
0	rate = k	$[A]_t = -akt + [A]_0$	$[A]$ v. time	$-\frac{1}{a} \times$ slope	$\text{mol L}^{-1}\text{ s}^{-1}$	$\frac{[A]_0}{2ak}$
1	rate = $k[A]$	$\ln[A]_t = -akt + \ln[A]_0$	$\ln[A]$ v. time	$-\frac{1}{a} \times$ slope	s^{-1}	$\frac{0.693}{ak}$
2	rate = $k[A]^2$	$\frac{1}{[A]_t} = akt + \frac{1}{[A]_0}$	$\frac{1}{[A]}$ v. time	$\frac{1}{a} \times$ slope	$\text{L mol}^{-1}\text{ s}^{-1}$	$\frac{1}{ak[A]_0}$

[a]rate $= -\left(\frac{1}{a}\right)\frac{\Delta[A]}{\Delta t}$

[b]To obtain the expressions for a A ⟶ products, we replace k with ak in the expressions given in the text for A ⟶ products.

3. To determine the order of a reaction, use one of the following methods.
   - Use the method of initial rates if the experimental data are given in the form of reaction rates at different initial concentrations.
   - Find the graph of rate data that yields a straight line (Table 20.5).
   - Test for the constancy of the half-life (good only for a first-order reaction).
   - Substitute rate data into integrated rate laws to find the one that gives a constant value of k.
4. To find the rate constant k for a reaction, use one of the following methods.
   - Obtain k from the slope of a straight-line graph.
   - Substitute concentration–time data into the appropriate integrated rate law.
   - Obtain k from the half-life of the reaction (good only for a first-order reaction).
5. To relate reactant concentrations and times, use the appropriate integrated rate law after first determining k.

EXAMPLE 20-8 Graphing Data to Determine the Order of a Reaction

The data listed in Table 20.6 were obtained for the decomposition reaction A ⟶ products. **(a)** Establish the order of the reaction. **(b)** What is the rate constant, k? **(c)** What is the half-life, $t_{1/2}$, if $[A]_0 = 1.00$ M?

TABLE 20.6 Kinetic Data for Example 20-8

Time, min	[A], M	ln [A]	1/[A]
0	1.00	0.00	1.00
5	0.63	−0.46	1.6
10	0.46	−0.78	2.2
15	0.36	−1.02	2.8
25	0.25	−1.39	4.0

Analyze

To determine the order of the reaction, we need to plot the data according to the integrated rate law for the different reaction orders. The plot yielding a straight line shows the overall reaction order. Recall that the slope of the straight line is related to the rate constant. The half-life of the reaction is also dependent on the overall reaction order.

Solve

(a) Plot the following three graphs.

1. $[A]$ versus time. (If a straight line, reaction is zero order.)

(continued)

2. ln[A] versus time. (If a straight line, reaction is first order.)
3. 1/[A] versus time. (If a straight line, reaction is second order.)

These graphs are plotted in Figure 20-7. The reaction is second order.

(b) The slope of graph 3 in Figure 20-7 is

$$k = \frac{(4.00 - 1.00)\,\text{L/mol}}{25\ \text{min}} = 0.12\ \text{M}^{-1}\,\text{min}^{-1}$$

(c) According to equation (20.19),

$$t_{1/2} = \frac{1}{k[\text{A}]_0} = \frac{1}{0.12\ \text{M}^{-1}\text{min}^{-1} \times 1.00\ \text{M}} = 8.3\ \text{min}$$

Assess

We can check the result in this example by examining the half-life results on the plot of [A] versus time. The plot shows that [A] decreases to 0.5 M between $t = 5$ min and $t = 10$ min, and so $t_{1/2} = 8.3$ min is a reasonable answer.

PRACTICE EXAMPLE A: In the decomposition reaction B $\longrightarrow$ products, the following data are obtained: $t = 0$ s, [B] = 0.88 M; 25 s, 0.74 M; 50 s, 0.62 M; 75 s, 0.52 M; 100 s, 0.44 M; 150 s, 0.31 M; 200 s, 0.22 M; 250 s, 0.16 M. What are the order of this reaction and its rate constant k?

PRACTICE EXAMPLE B: The following data are obtained for the reaction A $\longrightarrow$ products: $t = 0$ min, [A] = 0.250 M; 4.22 min, 0.210 M; 6.60 min, 0.188 M; 10.61 min, 0.150 M; 14.48 min, 0.114 M; 18.00 min, 0.083 M. What are the order of this reaction and its rate constant, k?

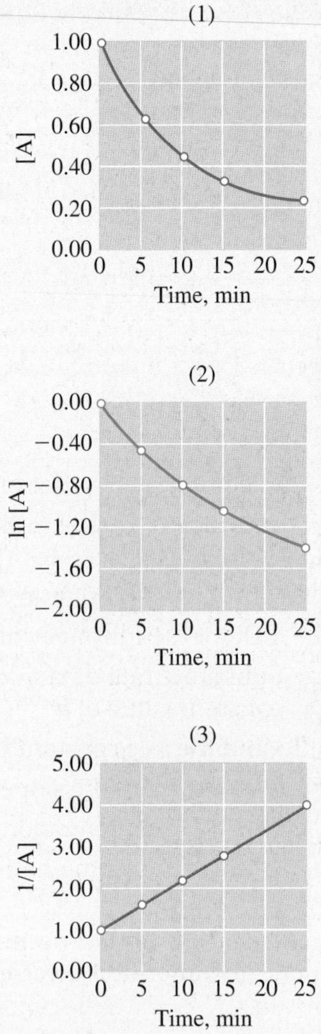

▶ FIGURE 20-7
Testing for the order of a reaction—Example 20-8 illustrated
The straight-line plot is obtained for 1/[A] versus t, graph (3). The reaction is second order.

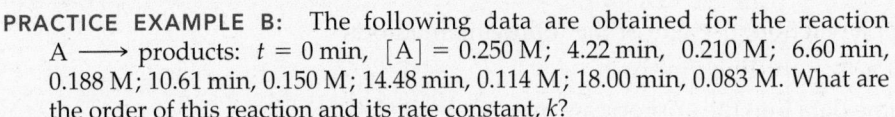

20-5 CONCEPT ASSESSMENT

How could you ascertain, by examining only a plot of [A] versus time, whether a reaction is **(a)** zero order; **(b)** first order; **(c)** second order?

20-8 Theoretical Models for Chemical Kinetics

Practical aspects of reaction kinetics—rate laws, rate constants—can b described without considering the behavior of individual molecules However, insight into the processes involved requires examination at th molecular level. For example, experiments show that the decomposition o H_2O_2 is first order, but *why* is this so? The remainder of the chapter consider the theoretical aspects of chemical kinetics that help answer such questions.

▶ Not all collisions are effective in causing a reaction. The average time between collisions in a gas at STP is about 10^{-10} s and the time that the collision is actually taking place is about 10^{-13} s.

Collision Theory

In our discussion of kinetic–molecular theory in Chapter 6 our emphasis was o molecular speeds. A further aspect of the theory with relevance to chemical kinet ics is collision density, the number of collisions per unit volume per unit time ı a typical gas-phase reaction, the calculated collision density is of the order of 0^{32} collisions per liter per second. If each collision yielded product molecules, ıe rate of reaction would be about 10^6 M s^{-1}, an extremely rapid rate. The ypical gas-phase reaction would go essentially to completion in a fraction of a econd. Gas-phase reactions generally proceed at a much slower rate, perhaps n the order of 10^{-4} M s^{-1}. This must mean that, generally, *only a fraction of the ollisions among gaseous molecules lead to chemical reaction.* This is a reasonable onclusion; we should not expect every collision to result in a reaction.

For a reaction to occur following a collision between molecules, there must e a redistribution of energy that puts enough energy into certain key bonds) break them. We would not expect two slow-moving molecules to bring nough kinetic energy into their collision to permit bond breakage. We would xpect two fast-moving molecules to do so, however, or perhaps one extremely ast molecule colliding with a slow-moving one. The **activation energy** of a eaction is the minimum kinetic energy that molecules must bring to their colsions for a chemical reaction to occur.

The kinetic–molecular theory can be used to establish the fraction of all the ıolecules in a mixture that possess certain kinetic energies. The results of this alculation are depicted in Figure 20-8. On this graph, a hypothetical energy is oted and the fraction of all molecules having energies in excess of this value s identified. Let's assume that these are the molecules whose molecular colliions are most likely to lead to chemical reaction. The rate of a reaction, then, epends on the product of the collision frequency *and* the fraction of these activated" molecules—in other words, on how often molecules with suffiient kinetic energy to react are likely to collide with each other. Because the raction of high-energy molecules is generally so small, the rate of reaction is sually much smaller than the collision frequency. Moreover, the *higher* the ctivation energy of a reaction, the *smaller* is the fraction of energetic collisions nd the slower the reaction.

Another factor that can strongly affect the rate of a reaction is the *orientation* f molecules at the time of their collision. In a reaction in which two hydrogen toms combine to form a hydrogen molecule (see margin) no bonds are broen and a H—H bond forms

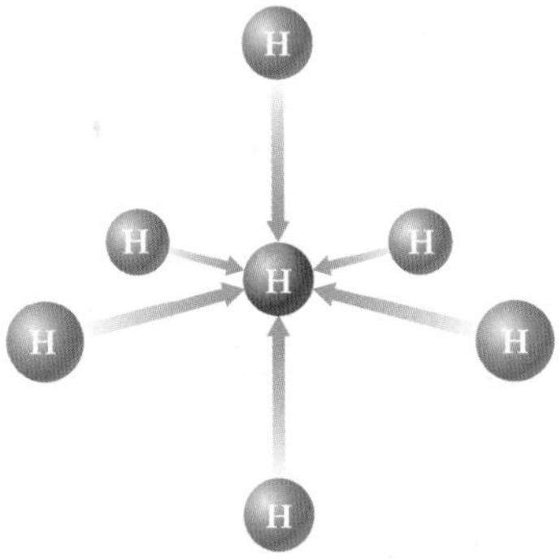

$$\text{H}\cdot + \cdot\text{H} \longrightarrow \text{H}_2$$

The H atoms are spherically symmetrical, and all approaches of one H atom o another prior to collision are equivalent. Orientation is *not* a factor, and the eaction occurs about as rapidly as the atoms collide. Orientation of the collidng molecules, however, is a crucial matter in the reaction of N_2O and NO, repesented here in an equation highlighting chemical bonds.

$$\text{N}{\equiv}\overset{+}{\text{N}}{-}\text{O}^- + \text{N}{=}\text{O} \longrightarrow \text{N}{\equiv}\text{N} + {}^-\text{O}{-}\overset{+}{\text{N}}{=}\text{O} \qquad \textbf{(20.20)}$$

The fundamental changes that occur during a successful collision are that the N—O bond in N_2O breaks and a new O—N bond is established to the NO

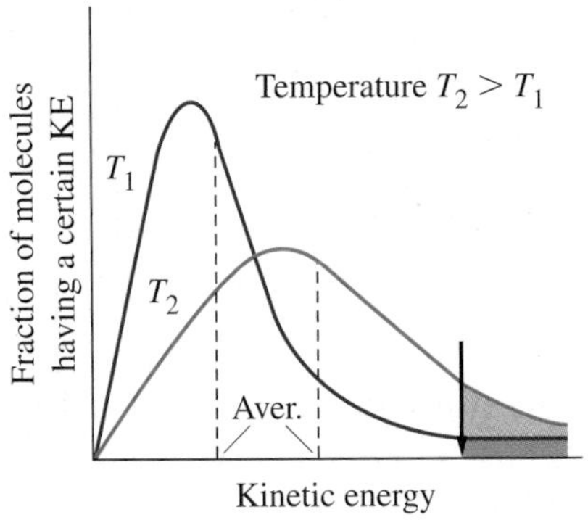

◀ FIGURE 20-8
Distribution of molecular kinetic energies
At both temperatures, the fraction of all molecules having kinetic energies in excess of the value marked by the heavy black arrow is small. (Note the shaded areas on the right.) At the higher temperature T_2 (red), however, this fraction is considerably larger than at the lower temperature T_1 (blue).

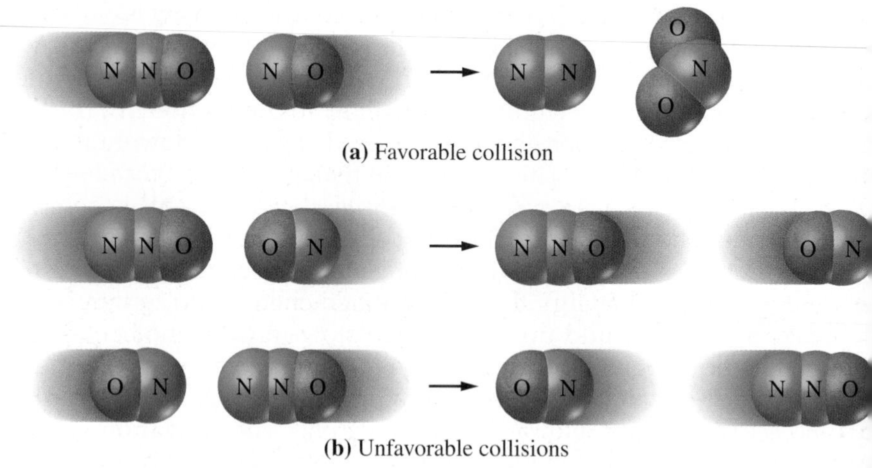

▶ FIGURE 20-9
Molecular collisions and chemical reactions
(a) A favorable collision between N_2O and NO molecules, resulting in the products N_2 and NO_2. **(b)** Two unfavorable collisions between N_2O and NO molecules; no reaction follows the collisions.

molecule. As a result of the collision, the molecules N_2 and NO_2 are formed. As suggested by Figure 20-9, a favorable collision requires the N atom of the NO molecule to strike the O atom of N_2O during a collision. Other orientations, such as the N atom of NO striking the terminal N atom of N_2O, do not produce a reaction. The number of unfavorable collisions in the reaction mixture exceeds the number of favorable ones.

Transition State Theory

In a theory proposed by Henry Eyring (1901–1981) and others, special emphasis is placed on a hypothetical species believed to exist in a transitory state that lies between the reactants and the products. We call this state the **transition state**, and the hypothetical species, the **activated complex**. The activated complex, formed through collisions, either dissociates back into the original reactants or forms product molecules. We can represent an activated complex for reaction (20.20) in this way.

▶ Modern Raman laser spectroscopic methods can be used to study reactions that occur on the femtosecond (10^{-15} s) scale. On this very short time scale, even rapid molecule motions (e.g., bond vibrations and rotations about a bond) can be resolved. Such techniques are used to study activated complexes.

$$\underset{\text{Reactants}}{N{\equiv}\overset{+}{N}{-}O^- + N{=}O} \rightleftharpoons \underset{\text{Activated complex}}{N{\equiv}N\cdots O\cdots N{=}O} \longrightarrow \underset{\text{Products}}{N{\equiv}N + {}^-O{-}\overset{+}{N}{=}O}$$

In the reactants, there is no bond between the O atom of N_2O and the N atom of NO. In the activated complex, the O atom has been partially removed from the N_2O molecule and partially joined to the NO molecule, as indicated by the *partial bonds* ($\cdots$). The formation of the activated complex is a reversible process. Once formed, some molecules of the activated complex may dissociate back into the reactants, but others may dissociate into the product molecules, where the partial bond of the O atom in N_2O has been completely severed and the partial bond between the O atom and NO has become a complete bond.

Figure 20-10, called a reaction profile, is a graphical way of looking at activation energy. In a **reaction profile**, energies are plotted on the vertical axis against a quantity called "progress of reaction," or simply reaction progress on the horizontal axis. Think of the progress of reaction as representing the extent of the reaction. That is, the reaction starts with reactants on the left, progresses through a transition state, and ends with products on the right.

The difference in energies between the reactants and products is Δ_rH for the reaction. Reaction (20.20) is an exothermic reaction with $\Delta_rH = -139$ kJ. The difference in energy between the activated complex and the reactants, 209 kJ mol^{-1}, is the *activation energy* of the reaction. Thus, a large energy barrier separates the reactants from the products, and only very energetic molecules can pass over this barrier. Figure 20-11 suggests an analogy to activation energy and the reaction profile.

KEEP IN MIND

that the difference in potential energy between reactants and product is Δ_rU of a reaction. For this reaction $\Delta_rU = \Delta_rH$ because the number of product gas molecules is equal to the number of reactant gas molecules. As we learned in Chapter 7, even when this is not the case, the differences between Δ_rU and Δ_rH are usually quite small.

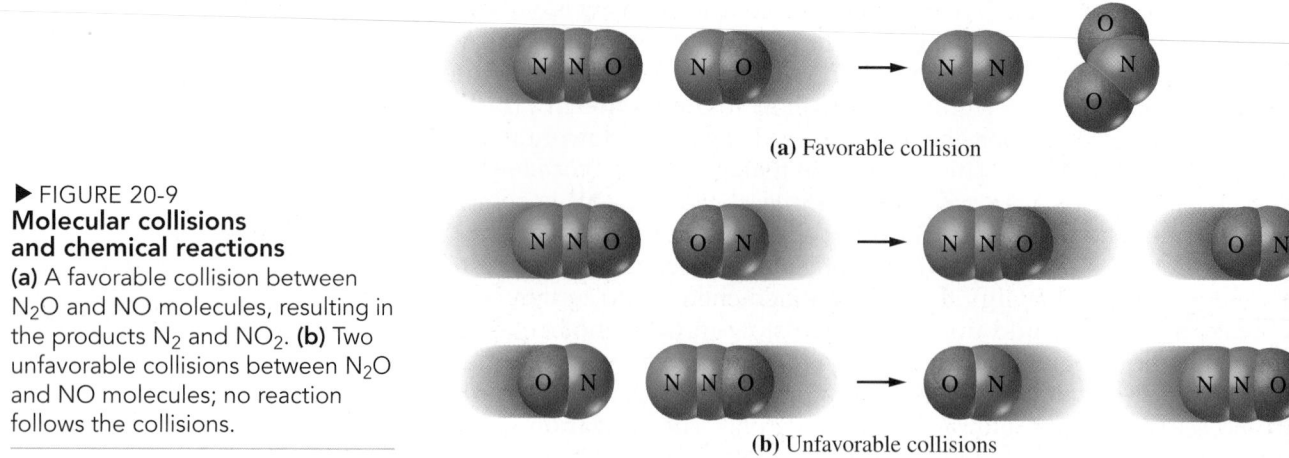

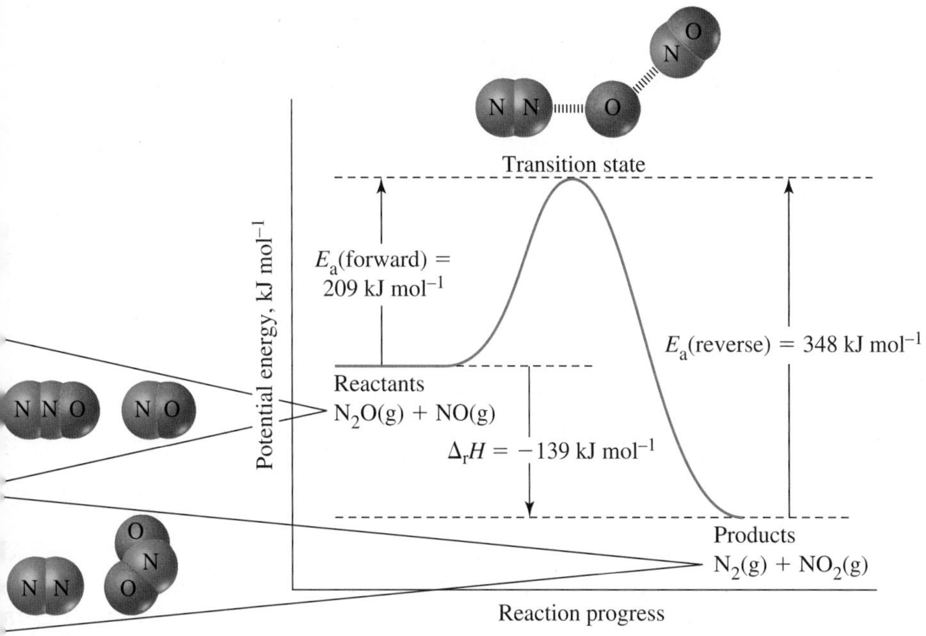

▲ FIGURE 20-10
A reaction profile for the reaction

$$N_2O(g) + NO(g) \longrightarrow N_2(g) + NO_2(g)$$

The simplified reaction profile traces energy changes during the course of reaction (20.20). The reactant and product molecules are depicted by the molecular models, as is the activated complex.

▲ FIGURE 20-11
An analogy for a reaction profile and activation energy
A hike (red path) is taken from the valley on the left (reactants) over the ridge to the valley on the right (products). The ridge above the starting point corresponds to the transition state. It is probably the height of this ridge (activation energy) more than anything else that determines how many people are willing to take the hike, regardless of the fact that it is all downhill on the other side.

Figure 20-10 describes both the forward reaction and its reverse—the reaction of N_2 and N_2O to form N_2O and NO. The activation energy for the reverse reaction is 348 kJ mol^{-1}; this reverse reaction is highly endothermic. Figure 20-10 also illustrates two useful ideas. (1) The enthalpy change of a reaction is equal to the difference in activation energies of the forward and reverse reactions. (2) For an

20-7 ARE YOU WONDERING?

Is there a molecular interpretation of reaction progress?

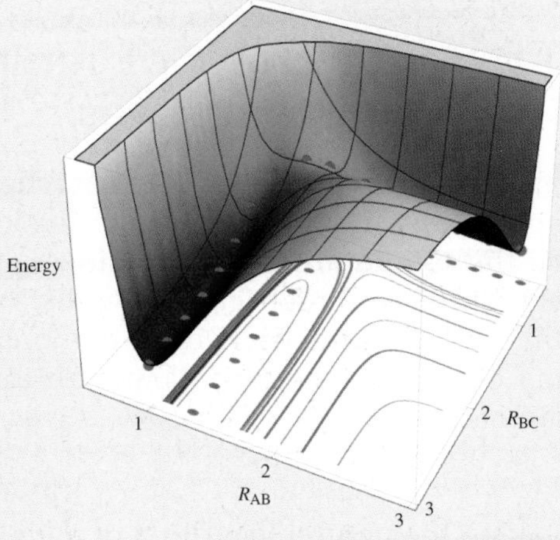

Simply stated, reaction progress refers to the minimum energy path (MEP) that leads from reactants to products. To understand MEP, we must know how the potential energy (PE) of the system depends on the relative positions of all the particles in the reaction. Let us consider a very simple endothermic gas phase reaction: $AB(g) + C(g) \longrightarrow A(g) + BC(g)$. The PE of the system is illustrated in the graph. Such a representation of the PE is commonly referred to as a potential energy surface (PES).

For this reaction, the PE can be expressed in terms of two distances: R_{AB}, the distance between atoms A and B, and R_{BC}, the distance between atoms B and C.

(continued)

The reaction progress on the PES is highlighted by the red dots, and the path defined by the red dots represents the MEP. The highest point along the MEP is called the *transition state*. A simplified one-dimensional plot of the MEP, which we call the *reaction profile*, is given below. Note that the highest point on the reaction profile is not necessarily the highest point on the PES for the reaction.

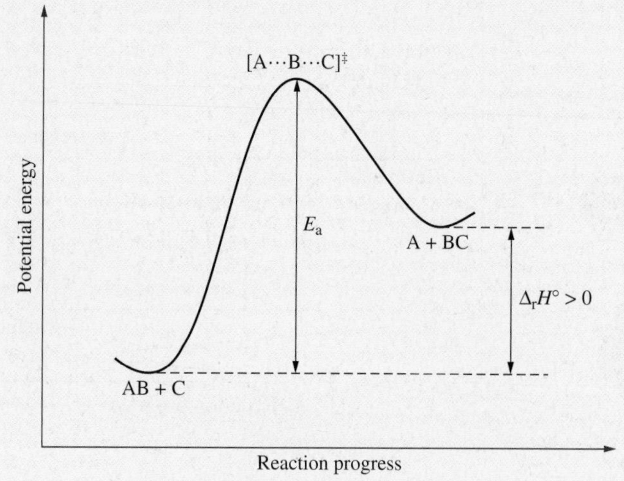

endothermic reaction, the activation energy must be equal to or greater than the enthalpy of reaction (and usually it is greater).

Attempts at purely theoretical predictions of rate constants have not been very successful. The principal value of reaction-rate theories is to help us explain experimentally observed reaction-rate data. For example, in the next section we will see how the concept of activation energy enters into a discussion of the effect of temperature on reaction rates.

20-6 CONCEPT ASSESSMENT

Indicate whether the following conditions can exist for a chemical reaction: **(a)** $\Delta_r H < 0 < E_a$; **(b)** $0 < \Delta_r H < E_a$; **(c)** $0 < E_a < \Delta_r H$; **(d)** $0 = \Delta_r H < E_a$; **(e)** $E_a < \Delta_r H < 0$. Where the conditions can exist, describe the reaction profile.

20-9 The Effect of Temperature on Reaction Rates

From practical experience, we expect chemical reactions to go faster at higher temperatures. To speed up the biochemical reactions involved in cooking, we raise the temperature, and to slow down other reactions, we lower the temperature—as in refrigerating milk to prevent it from souring.

In 1889, Svante Arrhenius demonstrated that the rate constants of many chemical reactions vary with temperature in accordance with the expression

$$k = Ae^{-E_a/RT} \tag{20.21}$$

▶ This equation indicates that a rate constant *increases* as the temperature *increases* and as the activation energy *decreases*.

By taking the natural logarithm of both sides of this equation, we obtain the following expression.

$$\ln k = -\frac{E_a}{RT} + \ln A$$

A graph of ln k versus $1/T$ is a straight line, thus giving us a graphical ethod for determining the activation energy of a reaction, as shown in gure 20-12. We can also derive a useful variation of the equation by writ- g it twice—each time for a different value of k and the corresponding mperature—and then eliminating the constant ln A. The result, also lled the Arrhenius equation, is

◀ This is the same technique used for the Clausius–Clapeyron equation on page 536 and illustrated in Appendix A-4.

$$\ln \frac{k_2}{k_1} = -\frac{E_a}{R}\left(\frac{1}{T_2} - \frac{1}{T_1}\right) \qquad \textbf{(20.22)}$$

equation (20.22)* T_2 and T_1 are two Kelvin temperatures; k_2 and k_1 are the te constants at these temperatures; and E_a is the activation energy in joules er mole. R is the gas constant expressed as 8.3145 J mol^{-1} K^{-1}.

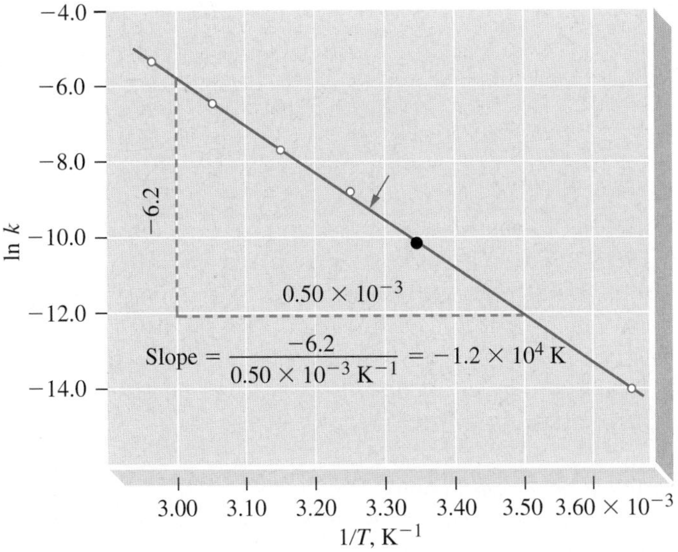

FIGURE 20-12
emperature dependence of the rate constant k for the reaction

$$N_2O_5\ (\text{in } CCl_4) \longrightarrow N_2O_4\ (\text{in } CCl_4) + \frac{1}{2} O_2(g)$$

ata are plotted as follows, for the representative point in black.

$$T = 298\ \text{K}$$
$$1/T = 1/298 = 0.00336 = 3.36 \times 10^{-3}\ \text{K}^{-1}$$
$$k = 3.46 \times 10^{-5}\ \text{s}^{-1};\ \ln k = \ln 3.46 \times 10^{-5} = -10.272$$

o evaluate E_a,

$$\text{slope of line} = -E_a/R = -1.2 \times 10^4\ \text{K}$$
$$E_a = 8.3145\ \text{J mol}^{-1}\ \text{K}^{-1} \times 1.2 \times 10^4\ \text{K}$$
$$= 1.0 \times 10^5\ \text{J/mol} = 1.0 \times 10^2\ \text{kJ/mol}$$

A more precise plot yields a value of E_a = 106 kJ/mol. The arrow points to data eferred to in Example 20-9.)

*To see that this expression is dimensionally correct, note that on the right side the units the numerator are J mol^{-1} (for E_a) and K^{-1} (for the quantity in parentheses) and that in he denominator the units are J mol^{-1} K^{-1} (for R). Cancellation yields a dimensionless quantity on he right, to match the dimensionless logarithmic term on the left.

Equation (20.22) can be written in exponential form.

$$\frac{k_2}{k_1} = e^{-\frac{E_a}{R}\left(\frac{1}{T_2} - \frac{1}{T_1}\right)}$$

Which equation is used is a matter of calculational convenience.

Arrhenius established equation (20.22) by fitting experimental data into hi equation. This was before the collision theory of chemical reactions had bee developed, but his equation is consistent with the collision theory. In the pre ceding section, we discussed the importance of (1) the frequency of molecula collisions, (2) the fraction of collisions energetic enough to produce a reactior

EXAMPLE 20-9 Applying the Arrhenius Equation

Use data from Figure 20-12 to determine the temperature at which $t_{1/2}$ for the first-order decomposition of N_2O_5 in CCl_4 is 2.00 h.

Analyze

First, find the rate constant k corresponding to a 2.00 h half-life. This can be done by using the half-life for a *first-order reaction,*

$$k = \frac{\ln 2}{t_{1/2}} = \frac{0.693}{2.00\ \text{h}} = \frac{0.693}{7200\ \text{s}} = 9.63 \times 10^{-5}\ \text{s}^{-1}$$

Now, the temperature at which $k = 9.63 \times 10^{-5}\ \text{s}^{-1}$ can be determined in two ways: graphically and analytically.

Solve

Graphical Method. The temperature at which $\ln k = \ln 9.63 \times 10^{-5} = -9.248$ is marked by the red arrow in Figure 20-12. The value of $1/T$ corresponding to $\ln k = -9.248$ is $1/T = 3.28 \times 10^{-3}\ \text{K}^{-1}$, which means that

$$T = \left(\frac{1}{3.28 \times 10^{-3}\ \text{K}^{-1}}\right) = 305\ \text{K}$$

Analytical Method. Take T_2 to be the temperature at which $k = k_2 = 9.63 \times 10^{-5}\ \text{s}^{-1}$. T_1 is some other temperature at which a value of k is known. Suppose we take $T_1 = 298$ K and $k_1 = 3.46 \times 10^{-5}\ \text{s}^{-1}$, a point referred to in the caption of Figure 20-12. The activation energy is 106 kJ/mol $= 1.06 \times 10^5$ J/mol (the more precise value given in Figure 20-12). Now we can solve equation (20.22) for T_2. (For simplicity, we have omitted units below, but the temperature is obtained in kelvin.)

$$\ln\frac{k_2}{k_1} = -\frac{E_a}{R}\left(\frac{1}{T_2} - \frac{1}{T_1}\right)$$

$$\ln\frac{9.63 \times 10^{-5}}{3.46 \times 10^{-5}} = -\frac{1.06 \times 10^5}{8.3145}\left(\frac{1}{T_2} - \frac{1}{298}\right)$$

$$1.024 = -1.27 \times 10^4\left(\frac{1}{T_2} - 0.00336\right) = 42.7 - \frac{1.27 \times 10^4}{T_2}$$

$$\frac{1.27 \times 10^4}{T_2} = 42.7 - 1.024 = 41.7$$

$$T_2 = \frac{(1.27 \times 10^4)}{41.7} = 305\ \text{K}$$

Assess

Both methods agree extremely well in this case. Depending on the circumstance one method may be preferred over the other.

PRACTICE EXAMPLE A: What is the half-life of the first-order decomposition of N_2O_5 at 75.0 °C? Use data from Example 20-9.

PRACTICE EXAMPLE B: At what temperature will it take 1.50 h for two-thirds of a sample of N_2O_5 in CCl_4 to decompose in Example 20-9?

ıd (3) the need for favorable orientations during collisions. Let's represent e collision frequency by the symbol Z_0. From kinetic–molecular theory, the action of sufficiently energetic collisions proves to be $e^{-E_a/RT}$. The probability f favorable orientations of colliding molecules is p and commonly referred to s the steric factor. Typical steric factors are given in Table 20.7. Notice that the eric factor decreases as the complexity of the reactant molecules increases. his reflects the fact that fewer collisions will occur with the proper orientaon to produce chemical reactions.

In collision theory, the rate constant of a reaction can be expressed as the product of Z_0, $e^{-E_a/RT}$, and p. If the product $Z_0 \times p$ is replaced by A, collision theory ields a result identical to Arrhenius's experimentally determined equation.

$$k = Z_0 \cdot p \cdot e^{-E_a/RT} = Ae^{-E_a/RT}$$

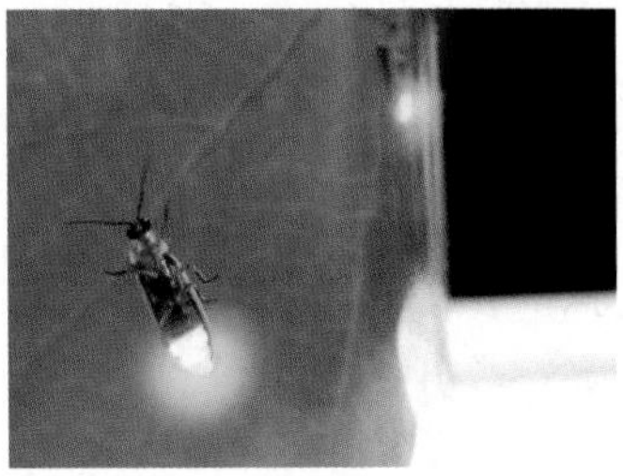

Anita Patterson Peppers/Shutterstock

▲ Both the rate of chirping of tree crickets and the flashing of fireflies roughly double for a 10 °C temperature rise. This corresponds to an activation energy of about 50 kJ/mol and suggests that the physiological processes governing these phenomena involve chemical reactions.

TABLE 20.7 Steric Factors (p) for Selected Reactions

Reaction	Steric Factor, p
$H(g) + H(g) \longrightarrow H_2(g)$	1
$O(g) + N_2(g) \longrightarrow N_2O(g)$	0.8
$2\,CH_3(g) \longrightarrow C_2H_6(g)$	0.073
$SO(g) + O_2(g) \longrightarrow SO_2(g) + O(g)$	2.4×10^{-3}
$CH_3(g) + C_2H_6(g) \longrightarrow CH_4(g) + C_2H_5(g)$	7.3×10^{-4}
$H_2(g) + C_2H_4(g) \longrightarrow C_2H_6(g)$	2.5×10^{-6}

20-7 CONCEPT ASSESSMENT

Without doing calculations, use a ln k versus $1/T$ graph to explain whether the rate of change of the rate constant with temperature is affected more strongly by a low or high energy of activation.

20-10 Reaction Mechanisms

$NO_2(g)$ is known to play a key role in the formation of photochemical smog (see page 957), but it is unlikely that very much of this gas is formed in the tmosphere by the direct reaction

$$2\,NO(g) + O_2(g) \longrightarrow 2\,NO_2(g) \quad \textbf{(20.23)}$$

◀ Although much progress has been made in the theoretical understanding of reaction mechanisms, by far most of the data and rate constants are obtained experimentally.

For this reaction to occur in a single step in the manner suggested by quation (20.23), *three* molecules would have to collide simultaneously, or very early so. A three-molecule collision is an unlikely event. The reaction appears o follow a different mechanism or pathway. One of the main purposes in letermining rate laws of chemical reactions is to relate them to probable reacion mechanisms.

A **reaction mechanism** is a step-by-step detailed description of a chemical eaction. Each step in a mechanism is called an **elementary process**, which lescribes any molecular event that significantly alters a molecule's energy or eometry or produces a new molecule. Two requirements of a plausible reacion mechanism are that it must

- be consistent with the stoichiometry of the overall reaction
- account for the *experimentally determined* rate law

In this section, we will first explore the nature of elementary processes and hen apply these processes to two simple types of reaction mechanisms.

Elementary Processes

The characteristics of elementary processes are as follows:

1. Elementary processes are either **unimolecular**—a process in which a single molecule dissociates—or **bimolecular**—a process involving the collision of two molecules. A *termolecular* process, which would involve the simultaneous collision of three molecules, is relatively rare as an elementary process.
2. The exponents of the concentrations in the rate law for an elementary process are the *same* as the stoichiometric coefficients in the balanced equation for the process (see Table 20.8). (Note that this is unlike the case of the overall rate law, for which the exponents are *not* necessarily related to the stoichiometric coefficients in the overall equation.)
3. Elementary processes are reversible, and some may reach a condition of equilibrium in which the rates of the forward and reverse processes are equal.
4. Certain species are produced in one elementary process and consumed in another. In a proposed reaction mechanism, such intermediates must not appear in either the overall chemical equation or the overall rate law.
5. One elementary process may occur much more slowly than all the others and in some cases may determine the rate of the overall reaction. Such a process is called the **rate-determining step** (Fig. 20-13).

Keep these characteristics in mind as we will apply them in our analysis of different mechanisms below.

TABLE 20.8 Rate Laws for Elementary Processes

Elementary Process[a]	Rate Law	Molecularity[b]
$A \longrightarrow$ products	$Rate = k[A]$	1 (unimolecular)
$A + A \longrightarrow$ products	$Rate = k[A]^2$	2 (bimolecular)
$A + B \longrightarrow$ products	$Rate = k[A][B]$	2 (bimolecular)
$A + A + A \longrightarrow$ products	$Rate = k[A]^3$	3 (termolecular)
$2\,A + B \longrightarrow$ products	$Rate = k[A]^2[B]$	3 (termolecular)
$A + B + C \longrightarrow$ products	$Rate = k[A][B][C]$	3 (termolecular)

[a]An elementary process is a specific collisional event or molecular process.
[b]Molecularity refers to the number of molecular entities involved in an elementary process.

Multistep Reactions and the Rate-Determining Step

A multistep reaction involves two or more elementary processes. Thus, the reaction profile for a multistep reaction will include two or more transition states, one transition state for each step as suggested by Figure 20-13.

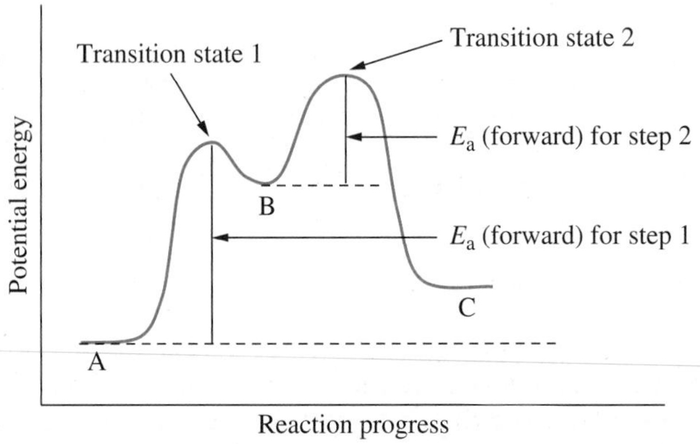

▶ FIGURE 20-13
Reaction profile for a hypothetical two-step reaction
The first step in this mechanism is $A \longrightarrow B$, and the second step is $B \longrightarrow C$. The overall reaction is $A \longrightarrow C$. The species B is a reaction intermediate; it is formed during the reaction but does not appear in the overall chemical equation for the reaction.

For a multistep reaction, the rate-determining step corresponds to the elementary step having the transition state of *highest energy* (i.e., highest point along the reaction profile). The step with the highest activation energy is not necessarily the rate-determining step.

or the reaction depicted in Figure 20-13, there are two reversible steps, the rst of which produces a reaction intermediate, B. A **reaction intermediate** is a pecies that is formed during a multistep reaction but does not appear in the verall chemical equation.

Step 1:	$A \underset{k_{-1}}{\overset{k_1}{\rightleftarrows}} B$
Step 2:	$B \underset{k_{-2}}{\overset{k_2}{\rightleftarrows}} C$
Overall:	$A \rightleftarrows C$

◀ The IUPAC convention for labeling rate constants is as follows: The rate constants for the steps in a mechanism are numbered sequentially, k_1, k_2, k_3, and so on. If step n is a reversible step, then the rate constant for the reverse reaction is denoted k_{-n}. This is just a labeling convention and no mathematical relationship is implied. Also, according to IUPAC, the symbol $\rightleftarrows$ indicates that a reaction is reversible, but not necessarily at equilibrium. The symbol $\rightleftharpoons$ indicates that a reaction is at equilibrium.

tep 1 has the highest activation energy and, thus, it has the smallest rate contant. The conversion of A into B is slow, but step 1 is not the slowest or rateetermining step. Let's consider why this is the case.

Once B is formed, it will be converted back into A more rapidly than it will e converted into C. (Compare the activation energies for B $\longrightarrow$ C and $\longrightarrow$ A.) As a result, the concentration of B is kept relatively small, and the ate of conversion of B into C will be relatively slow. Therefore, the rate of eaction is ultimately determined by the rate of conversion of B into C.

Interestingly, for the reaction profile given in Figure 20-13, the rateetermining step is actually step 2, even though step 1 has the smallest rate onstant.

Mechanism with a Slow Step Followed by a Fast Step

he reaction between gaseous iodine monochloride and gaseous hydrogen roduces iodine and hydrogen chloride as gaseous products.

$$H_2(g) + 2\,ICl(g) \longrightarrow I_2(g) + 2\,HCl(g)$$

he experimentally determined rate law for this reaction is

$$\text{rate of reaction} = k[H_2][ICl]$$

Let's begin with a mechanism that seems plausible, such as the following wo-step mechanism.

(1) Slow:	$H_2 + ICl \longrightarrow HI + HCl$
(2) Fast:	$HI + ICl \longrightarrow I_2 + HCl$
Overall:	$H_2 + 2\,ICl \longrightarrow I_2 + 2\,HCl$

This scheme seems plausible for two reasons: (1) The sum of the two steps yields the experimentally observed *overall* reaction. (2) As we have noted, uninolecular and bimolecular elementary processes are most plausible, and each step in the above mechanism is bimolecular. Because each step is an elemenary process, we can write

$$\text{rate (1)} = k_1[H_2][ICl] \quad \text{and} \quad \text{rate (2)} = k_2[HI][ICl]$$

Now, note that our mechanism proposes that step (1) occurs slowly but step (2) occurs rapidly. This suggests that HI is consumed in the second step ust as fast as it is formed in the first. The first step is the rate-determining step,

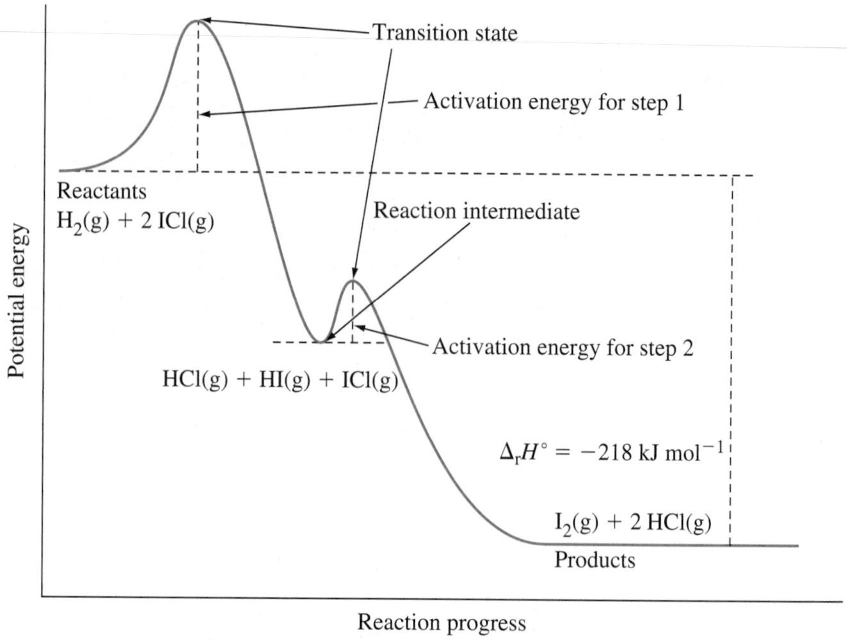

▶ FIGURE 20-14
Reaction profile for the reaction $H_2(g) + 2\,ICl(g) \longrightarrow I_2(g) + 2\,HCl(g)$

and the rate of the overall reaction is governed just by the rate at which HI is formed in this first step, that is, by rate (1). This explains why the observed rate law for the net reaction is rate of reaction $= k[H_2][ICl]$. The proposed mechanism gives a rate law that is in agreement with experiment, as it should if we have made a reasonable proposal.

The species HI is a reaction intermediate; it does not appear in the experimental rate law. In this case, the intermediate species is a well-known stable molecule. Often, when postulating mechanisms, we have to invoke less well-known and less stable species; and in these instances, we have to rely on the chemical reasonableness of the basic assumptions. The presence of a reaction intermediate leads to a slightly more complicated reaction profile. The reaction profile for the two steps in the proposed mechanism is shown in Figure 20-14. We see that there are two transition states and one reaction intermediate. Since the transition state for the first step is the highest point on the reaction profile, the first step is the rate-determining step. It is important to understand the difference between a transition state (activated complex) and a reaction intermediate. The transition state represents the highest energy structure involved in a reaction (or step in a mechanism). Transition states exist only momentarily and can never be isolated, whereas reaction intermediates can sometimes be isolated. Transition states have *partially formed bonds*, whereas reaction intermediates have *fully formed bonds*.

▶ A transition state is not a "real" species. It is only hypothetical.

A Mechanism with a Fast Reversible First Step Followed by a Slow Step

The rate law for the reaction of NO(g) and $O_2(g)$

$$2\,NO(g) + O_2(g) \longrightarrow 2\,NO_2(g) \tag{20.23}$$

is found to be

$$\text{rate of reaction} = k[NO]^2[O_2] \tag{20.24}$$

Even though it is consistent with this rate law, we have already noted that the one-step *termolecular* mechanism suggested by equation (20.23) is highly improbable. Let's explore instead the following mechanism.

$$\text{Fast:} \quad 2\,NO(g) \underset{k_{-1}}{\overset{k_1}{\rightleftharpoons}} N_2O_2(g) \tag{20.25}$$

$$\text{Slow:} \quad N_2O_2(g) + O_2(g) \xrightarrow{k_2} 2\,NO_2(g) \tag{20.26}$$

$$\text{Overall:} \quad 2\,NO(g) + O_2(g) \longrightarrow 2\,NO_2(g) \tag{20.23}$$

n this mechanism, there is a rapid equilibrium as the first step, but some of he N_2O_2 is slowly drawn off and consumed in the second, slow step. The rate w for the slow, or *rate-determining*, step (20.26) is

$$\text{rate of reaction} = k_2[N_2O_2][O_2] \tag{20.27}$$

ecause N_2O_2 is an *intermediate*, however, we must eliminate it from the rate w. We are told that the first step of the mechanism consists of a fast eversible reaction, so we can assume that this step progresses rapidly to equi-brium. If this is the case, the forward and reverse rates of reaction in the first ep become equal and we write

$$\text{rate of forward reaction} = \text{rate of reverse reaction}$$
$$k_1[NO]^2 = k_{-1}[N_2O_2]$$

Jow, let us arrange this equation into an expression having a ratio of rate onstants on one side and a ratio of concentration terms on the other. Also, we an replace the ratio of rate constants by a single constant, which we will epresent as K_1.

$$K_1 = \frac{k_1}{k_{-1}} = \frac{[N_2O_2]}{[NO]^2}$$

he above expression is known as an equilibrium constant expression; the umerical constant, K_1, is an equilibrium constant. Next, we rearrange the xpression to solve for the term $[N_2O_2]$.

$$[N_2O_2] = K_1[NO]^2$$

hen, substituting this into equation (20.27) we obtain the experimentally bserved rate law.

$$\text{rate of reaction} = k_2[N_2O_2][O_2] = k_2K_1[NO]^2[O_2]$$

he experimentally observed rate constant, k, is related to the other constants n the proposed mechanism, as follows:

$$k = k_2K_1 = k_2 \times \frac{k_1}{k_{-1}}$$

◀ Equilibrium constant expressions are of fundamental importance throughout chemistry. Here, we see their significance in chemical kinetics. In Chapter 13, we explored their thermodynamic basis. In Chapter 15, we described the experimental basis of equilibrium constant expressions and their application to the stoichiometry of reversible reactions.

The type of mechanism described here with a *rapid pre-equilibrium* is a very ommon mechanism and is to be expected when the overall stoichiometry uggests an unlikely termolecular collision.

We have just shown that the proposed mechanism is consistent with 1) the reaction stoichiometry and (2) the experimentally determined rate aw. Whether this mechanism is the actual reaction path, we cannot say, owever. All that we can say is that it is *plausible*; it has not been ruled out y kinetics. We reinforce this point in Example 20-10, where we demon-strate that there is another plausible mechanism for reaction (20.23).

The Steady-State Approximation

The reaction mechanisms considered so far have had one particular rate-determining step, and the rate law of the reaction could be deduced from the ate of this step after the relationships for the concentrations of any intermedi-ates were established. In complex multistep reaction mechanisms, however, more than one step may control the rate of a reaction.

EXAMPLE 20-10 Testing a Reaction Mechanism

An alternative mechanism of the reaction $2\,NO(g) + O_2(g) \longrightarrow 2\,NO_2(g)$ follows. Show that this mechanism is consistent with the rate law, equation (20.24).

$$\text{Fast:} \quad NO(g) + O_2(g) \underset{k_{-1}}{\overset{k_1}{\rightleftharpoons}} NO_3(g)$$

$$\text{Slow:} \quad NO_3(g) + NO(g) \xrightarrow{k_2} 2\,NO_2(g)$$

$$\text{Overall:} \quad 2\,NO(g) + O_2(g) \longrightarrow 2\,NO_2(g)$$

Analyze

In this type of problem we begin by identifying the slow step (which is typically given) and using it to write the rate of reaction. Because the fast step shown above is given as an equilibrium, we can assume that the equilibrium is rapidly established. The common species between the two reactions, NO_3, is an intermediate, is not found in the rate law, and so it can be eliminated by using the reaction equilibrium constant expression for the fast step.

Solve

The rate equation for the rate-determining step is

$$\text{rate of reaction} = k_2[NO_3][NO]$$

Eliminate $[NO_3]$, by assuming that the pre-equilibrium is rapidly established.

$$\text{rate of forward reaction} = \text{rate of reverse reaction}$$
$$k_1[NO][O_2] = k_{-1}[NO_3]$$

Rearrange the previous equation to develop an expression for an equilibrium constant (K) in terms of the rate constants k_1 and k_{-1}.

$$K = \frac{k_1}{k_{-1}} = \frac{[NO_3]}{[NO][O_2]}$$

Rearrange this expression for K to solve for $[NO_3]$.

$$[NO_3] = K[NO][O_2]$$

Finally, substitute the value of $[NO_3]$ into the rate equation to obtain the observed rate law (equation 20.24).

$$\text{rate} = k_2[NO_3][NO]$$
$$\text{rate of reaction} = k_2K[NO]^2[O_2] = k_2\frac{k_1}{k_{-1}}[NO]^2[O_2] = k[NO]^2[O_2]$$

Assess

The final rate law obtained, rate $= k[NO]^2[O_2]$, is consistent with the experimental rate law. This alternative mechanism is plausible based on this analysis. However, it does not mean that this is the reaction mechanism.

PRACTICE EXAMPLE A: In a proposed two-step mechanism for the reaction $CO(g) + NO_2(g) \longrightarrow CO_2(g) + NO(g)$, the second, fast step is $NO_3(g) + CO(g) \longrightarrow NO_2(g) + CO_2(g)$. What must be the *slow* step? What would you expect the rate law of the reaction to be? Explain.

PRACTICE EXAMPLE B: Show that the proposed mechanism for the reaction $2\,NO_2(g) + F_2(g) \longrightarrow 2\,NO_2F(g)$ is plausible. The rate law is rate $= k[NO_2][F_2]$.

$$\text{Fast:} \quad NO_2(g) + F_2(g) \rightleftharpoons NO_2F_2(g)$$

$$\text{Slow:} \quad NO_2F_2(g) \longrightarrow NO_2F(g) + F(g)$$

$$\text{Fast:} \quad F(g) + NO_2(g) \longrightarrow NO_2F(g)$$

To illustrate, let's reconsider the first mechanism presented for the reaction of nitric monoxide with oxygen, but this time we will make no assumptions about the relative rates of the steps in the mechanism. The proposed mechanism is

$$NO + NO \xrightarrow{k_1} N_2O_2$$

$$N_2O_2 \xrightarrow{k_{-1}} NO + NO$$

$$N_2O_2 + O_2 \xrightarrow{k_2} 2\,NO_2$$

or clarity, the first, reversible reaction is written as two separate steps.

We choose one of the steps of the mechanism that provides a convenient elationship to the observed rate of reaction. In this case, we will use the last tep, since it involves the disappearance of O_2. Therefore, the rate of the reacon from this mechanism is

$$\text{rate of reaction} = k_2[N_2O_2][O_2] \qquad \textbf{(20.27)}$$

As before, the intermediate N_2O_2 must be eliminated from this rate law. We an do this by assuming that the $[N_2O_2]$ reaches a *steady-state condition* in hich N_2O_2 is produced and consumed at equal rates. That is, $[N_2O_2]$ emains constant throughout most of the reaction. We can use the steady-state ssumption to express $[N_2O_2]$ in terms of $[NO]$.

KEEP IN MIND

that if the rate of change of the concentration of a substance is zero, then the concentration of that substance is constant.

$$\Delta[N_2O_2]/\Delta t = \text{rate of formation of } N_2O_2 - \text{rate of disappearance of } N_2O_2 = 0$$
$$\text{rate of formation of } N_2O_2 = \text{rate of disappearance of } N_2O_2$$

he rate of disappearance of N_2O_2 is made up of two parts—the reverse step f equation (20.25) and the forward step of (20.26)—so we write

$$\text{rate of disappearance of } N_2O_2 = k_{-1}[N_2O_2] + k_2[N_2O_2][O_2]$$

he two rates for the steps depleting the concentration of N_2O_2 have been dded. Now, as dictated by the steady-state assumption, the rate of disappearnce of N_2O_2 is equated with the rate of appearance of N_2O_2, which is $_1[NO]^2$.

$$k_1[NO]^2 = k_{-1}[N_2O_2] + k_2[N_2O_2][O_2] = [N_2O_2](k_{-1} + k_2[O_2])$$

earranging to solve for $[N_2O_2]$, we have

$$[N_2O_2] = \frac{k_1[NO]^2}{k_{-1} + k_2[O_2]}$$

Ve now substitute this into equation (20.27) to obtain

$$\text{rate} = k_2[O_2][N_2O_2] = k_2[O_2]\left(\frac{k_1[NO]^2}{k_{-1} + k_2[O_2]}\right)$$

r

$$\text{rate} = \frac{k_1k_2[O_2][NO]^2}{k_{-1} + k_2[O_2]}$$

◀ This rate law does not have the general form of equation (20.6), or the usual nth order rate law seen in Table 20.5.

This is the rate law for the proposed mechanism based on our steady-state nalysis. This rate law is more complicated than the observed rate law. What happened? In carrying out the steady-state calculation, we did not make any assumptions about the relative rates of the three steps in the mechanism. If we now make the assumption that the rate of disappearance of N_2O_2 in the second step of the proposed mechanism is greater than the rate of disappearance of N_2O_2 in the third step of the proposed mechanism, then

$$k_{-1}[N_2O_2] > k_2[N_2O_2][O_2]$$

which means

$$k_{-1} > k_2[O_2]$$

and

$$k_{-1} + k_2[O_2] \approx k_{-1}$$

so that

$$[N_2O_2] = \frac{k_1[NO]^2}{k_{-1}}$$

If we substitute this value of $[N_2O_2]$ into equation (20.27) and replace k_1k_2/k_{-1} by k, we obtain for the overall reaction

$$\text{rate of reaction} = \frac{k_1k_2}{k_{-1}}[NO]^2[O_2] = k[NO]^2[O_2]$$

The result of a steady-state analysis of any mechanism in which no rate-determining step can be identified will often be a complicated rate law. The use of this type of rate law is illustrated in the section on enzyme catalysis later in this chapter.

Relationship Between the Equilibrium Constant and Rate Constants

Given the requirement that the rates of the forward and reverse reactions become equal at equilibrium, it seems that a relationship should exist between the equilibrium constant and the rate constants for the forward and reverse reactions. That such a relationship does exist can be demonstrated easily for elementary reactions. Consider again the hypothetical generalized reaction

$$a\,A + b\,B + \cdots \underset{k_{-1}}{\overset{k_1}{\rightleftharpoons}} g\,G + h\,H + \cdots$$

k_1 and k_{-1} are the rate constants for the forward and reverse reactions. *With the assumption that both the forward and reverse reactions are elementary reactions,* we can write

$$\text{rate of forward reaction} = k_1[A]^a[B]^b\cdots$$
$$\text{rate of reverse reaction} = k_{-1}[G]^g[H]^h\cdots$$

At equilibrium, these two rates become equal. Thus, we can write

$$k_1[A]^a[B]^b\cdots = k_{-1}[G]^g[H]^h\cdots$$

which can be rearranged into an expression having rate constants on one side and concentrations on the other:

$$\frac{k_1}{k_{-1}} = \frac{[G]^g[H]^h\cdots}{[A]^a[B]^b\cdots}$$

Because the right side of the equation above is the equilibrium constant expression for the reaction, we arrive at the following result:

$$\frac{k_1}{k_{-1}} = K$$

Keep in mind that this result is based on the assumption that the forward and reverse reactions are elementary reactions. For reactions that involve a multistep mechanism, the relationship between k and the rate constants is more complicated. For a mechanism involving n steps, it can be demonstrated that the relationship between k and the rate constants is

$$\frac{k_1}{k_{-1}} \times \frac{k_2}{k_{-2}} \times \cdots \times \frac{k_n}{k_{-n}} = K$$

Although the expression above shows that there is indeed a relationship between the equilibrium constant and rate constants, it is generally easier to obtain K directly from measurements at equilibrium than to attempt a calculation based on rate constants.

mog—An Environmental Problem with Roots
Chemical Kinetics

bout 100 years ago, a new word entered the English language—*smog*. It ferred to a condition, common in London, in which a combination of *smoke* ıd *fog* obscured visibility and produced health hazards (including death). ıese conditions are often associated with heavy industry, and this type of nog is now called *industrial* smog.

◀ Air quality in London has been greatly improved by control measures, such as elimination of coal as a household fuel, introduced after a severe smog episode in 1952.

A more familiar form of air pollution commonly thought of as smog results om the action of sunlight on the products of combustion. Chemical reactions ought about by light are called *photochemical* reactions, and the smog formed imarily as a result of photochemical reactions is **photochemical smog**. This pe of smog is associated with high-temperature combustion processes, such those that occur in internal combustion engines. Because the combustion of otor fuels takes place in air rather than in pure oxygen, oxides of nitrogen, incipally NO(g), are inevitably found in the exhaust from motor vehicles. ther products found in the exhaust are hydrocarbons (unburned gasoline) ıd partially oxidized hydrocarbons. These, then, are the starting materials— ıe precursors—of photochemical smog.

Many substances have been identified in smoggy air, including NO, O_2, O_3 (ozone, an allotrope of oxygen described on page 432 and disıssed further in Section 22-4), and a variety of organic compounds derived om gasoline hydrocarbons. Ozone is very reactive and is largely responsile for the breathing difficulties that some people experience during smog pisodes. Another noxious substance found in smog is an organic ompound known as peroxyacetyl nitrate (PAN). PAN is a powerful crimator—that is, it causes tear formation in the eyes. Photochemical nog components cause heavy crop damages (to oranges, for example) and ıe deterioration of rubber goods. And, of course, the best-known symbol of hotochemical smog is the hazy brown air that results in reduced visibility ig. 20-15).

Frontpage/Shutterstock

▲ FIGURE 20-15
Smog in Mexico City
At times, the topographical features, climatic conditions, traffic congestion, and heavy industrial pollution combine to create severe smog conditions in Mexico City.

Chemists who have been studying photochemical smog formation over the ast several decades have determined that certain precursors are converted to ıe observable smog components through the action of sunlight. Because the hemical reactions involved are very complex and still not totally understood, e will give only a very brief, simplified reaction mechanism showing how hotochemical smog is formed.

Smog formation begins with NO(g), produced by reaction (20.28).

$$N_2(g) + O_2(g) \xrightarrow{\Delta} 2\,NO(g) \tag{20.28}$$

O(g) is subsequently converted to $NO_2(g)$, which then absorbs ultraviolet adiation from sunlight and decomposes.

$$NO_2 + \text{sunlight} \longrightarrow NO + O \tag{20.29}$$

his is followed by a reaction forming ozone, O_3.

$$O + O_2 \longrightarrow O_3 \tag{20.30}$$

hus, a large buildup of ozone in photochemical smog requires a plentiful ource of NO_2. At one time, this source was believed to be reaction (20.31).

$$2\,NO + O_2 \longrightarrow 2\,NO_2 \tag{20.31}$$

t is now well established, however, that reaction (20.31) occurs much too lowly to yield the required levels of NO_2 in photochemical smog. Instead, O is rapidly converted to NO_2 when it reacts with O_3,

$$O_3 + NO \longrightarrow O_2 + NO_2 \tag{20.32}$$

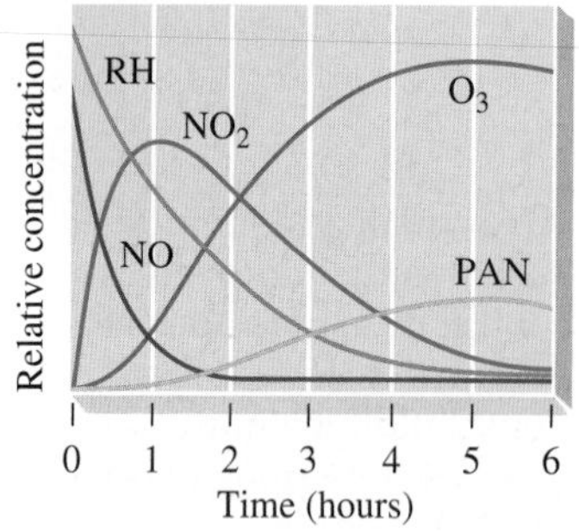

▲ FIGURE 20-16
Smog component profile
Data from a smog chamber show how the concentrations of smog components change with time. For example, the concentrations of hydrocarbon (RH) and nitrogen monoxide (NO) fall continuously, whereas that of nitrogen dioxide (NO_2) rises to a maximum and then drops off. The concentrations of ozone (O_3) and peroxyacetyl nitrate (PAN) build up more slowly. Any reaction scheme proposed to explain smog formation must be consistent with observations such as these. Under actual smog conditions, the pattern of concentration changes shown here repeats itself on a daily basis.

Even though reaction (20.32) accounts for the formation of NO_2, it leads to tl destruction of ozone. Thus, photochemical smog formation cannot occur ju through the reaction sequence: (20.28), (20.29), (20.30), and (20.31). The ozor would be consumed as quickly as it was formed, and there would be no ozor buildup at all.

It is now known that organic compounds, particularly unburned hydroca bons in automotive exhaust, provide a pathway for the conversion of NO t NO_2. The reaction sequence that follows involves some highly reactiv molecular fragments. Recall (page 434) that these fragments are known as *fr radicals* and are represented by formulas written with a bold dot. R represents a hydrocarbon molecule, and R· is a free-radical fragment of hydrocarbon molecule. Oxygen atoms, fragments of the O_2 molecule, are als represented as free radicals, as are hydroxyl groups, fragments of the H_2 molecule.

$$RH + O\cdot \longrightarrow R\cdot + \cdot OH$$
$$RH + \cdot OH \longrightarrow R\cdot + H_2O$$
$$R\cdot + O_2 \longrightarrow RO_2\cdot$$
$$RO_2\cdot + NO \longrightarrow RO\cdot + NO_2$$

The final step in this reaction mechanism accounts for the rapid conversion c NO to NO_2 that seems essential to smog formation.

The role of NO_2 in the formation of the smog component PAN is suggeste by the equation

$$CH_3C(=O)-O-O\cdot + NO_2 \longrightarrow \underset{\text{PAN}}{CH_3C(=O)-O-ONO_2}$$

The details of smog formation have been worked out in part through th use of smog chambers. By varying experimental conditions in these chambers scientists have been able to create polluted atmospheres very similar to smog For example, they have found that if hydrocarbons are omitted from the start ing materials in the smog chamber, no ozone is formed. The reaction schem just proposed is consistent with this observation. Figure 20-16 gives a typica result from a smog chamber.

To control smog, automobiles are now provided with *catalytic converters* CO and hydrocarbons are oxidized to CO_2 and H_2O in the presence of an oxi dation catalyst such as platinum or palladium metal. NO must be reduced t N_2, and this requires a reduction catalyst. A dual-catalyst system uses both types of catalysts. Alternatively, the air-fuel ratio of the engine is set to pro duce some CO and unburned hydrocarbons; these then act as reducing agent to reduce NO to N_2.

$$2\,CO(g) + 2\,NO(g) \longrightarrow 2\,CO_2(g) + N_2(g)$$

Next, the exhaust gases are passed through an oxidation catalyst to oxidize the remaining hydrocarbons and CO to CO_2 and H_2O. Future smog-control mea sures may include the use of alternative fuels, such as methanol or hydrogen and the development of electric-powered automobiles.

20-11 Catalysis

A reaction can generally be made to go faster by increasing the tempera ture. Another way to speed up a reaction is to use a catalyst. A **catalys** provides an alternative reaction pathway of lower activation energy. The catalyst participates in a chemical reaction but does not itself undergo a ermanent change. As a result, the formula of a catalyst does not appear in e overall chemical equation (its formula is generally placed over the reaction arrow).

The success of a chemical process often hinges on finding the right catalyst, s in the manufacture of nitric acid. By conducting the oxidation of $NH_3(g)$ ery quickly (less than 1 ms) in the presence of a Pt–Rh catalyst, $NO(g)$ can be btained as a product instead of $N_2(g)$. The formation of $HNO_3(aq)$ from $O(g)$ then follows easily.

In this section, the two basic types of catalysis—homogeneous and eterogeneous—are described first. This is followed by discussions of the atalyzed decomposition of $H_2O_2(aq)$ and the biological catalysts called nzymes.

omogeneous Catalysis

igure 20-17 shows reaction profiles for the decomposition of formic acid HCOOH). In the uncatalyzed reaction, a H atom must be transferred from ne part of the formic acid molecule to another, shown by the blue, dotted rrow. Then a C—O bond breaks. Because the energy requirement for this om transfer is high, the activation energy is high and the reaction is slow. In ne acid-catalyzed decomposition of formic acid, a hydrogen ion from solution ttaches itself to the O atom that is singly bonded to the C atom to form $[HCOOH_2]^+$. The C—O bond breaks, and a H atom attached to a carbon atom the intermediate species $[HCO]^+$ is released to the solution as H^+.

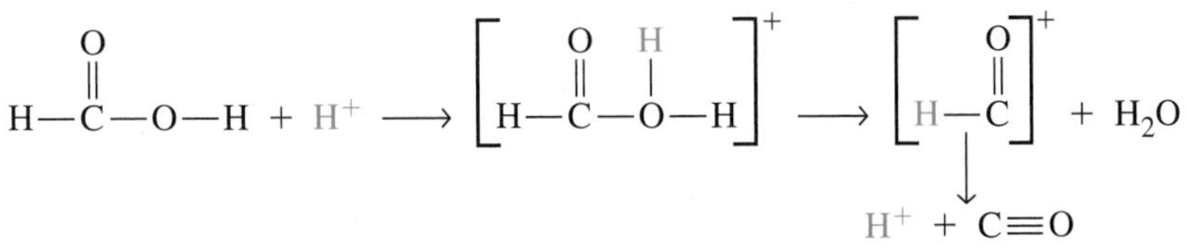

his catalyzed reaction pathway does not require a H atom to be transferred ithin the formic acid molecule. It has a lower activation energy than does ne uncatalyzed reaction and proceeds at a faster rate. Because the reactants nd products of this reaction are all present throughout the solution, r homogeneous mixture, this type of catalysis is called **homogeneous atalysis**.

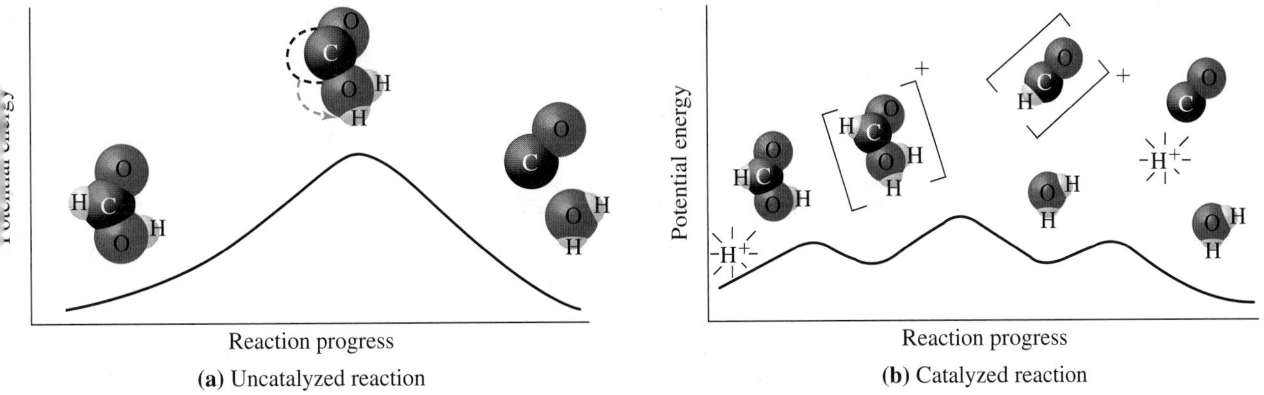

FIGURE 20-17
n example of homogeneous catalysis
he activation energy is lowered in the presence of H^+, a catalyst for the ecomposition of HCOOH.

Heterogeneous Catalysis

Many reactions can be catalyzed by allowing them to occur on an appropria solid surface. Essential reaction intermediates are found on the surface. Th type of catalysis is called **heterogeneous catalysis** because the catalyst is pr sent in a different phase of matter than are the reactants and product Catalytic activity is associated with many transition elements and their con pounds. The precise mechanism of heterogeneous catalysis is not totall understood, but in many cases the availability of electrons in *d* orbitals in su face atoms may play a role.

A key feature of heterogeneous catalysis is that reactants from a gaseous c solution phase are adsorbed, or attached, to the surface of the catalyst. Not a surface atoms are equally effective for catalysis; those that are effective a called **active sites**. Basically, heterogeneous catalysis involves (1) *adsorption* c reactants; (2) *diffusion* of reactants along the surface; (3) *reaction* at an active sit to form adsorbed product; and (4) *desorption* of the product.

An interesting reaction is the oxidation of CO to CO_2 and the reduction c NO to N_2 in automotive exhaust gases as a smog-control measure. Figure 20-1 shows how this reaction is thought to occur on the surface of rhodium metal i a catalytic converter. In general, the reaction profile for a surface-catalyze reaction resembles that shown in Figure 20-19.

The Catalyzed Decomposition of Hydrogen Peroxide

As previously noted (see the footnote on page 925), the decomposition c $H_2O_2(aq)$ is a slow reaction and generally must be catalyzed. Iodide ion is good catalyst that seems to function by means of the following two-ste mechanism.

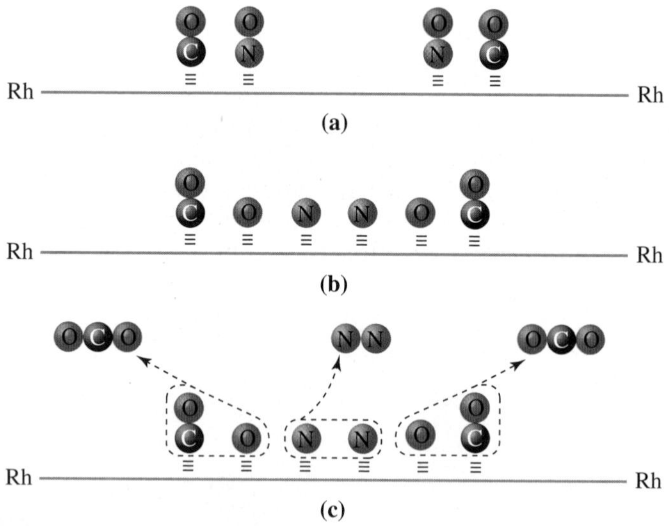

▲ FIGURE 20-18
Heterogeneous catalysis in the reaction

$$2\,CO + 2\,NO \xrightarrow{Rh} 2\,CO_2 + N_2$$

(a) Molecules of CO and NO are adsorbed on the rhodium surface. **(b)** The adsorbed NO molecules dissociate into adsorbed N and O atoms. **(c)** Adsorbed CO molecules and O atoms combine to a form CO_2 molecules, which desorb into the gaseous state. Two N atoms combine and are desorbed as a N_2 molecule.

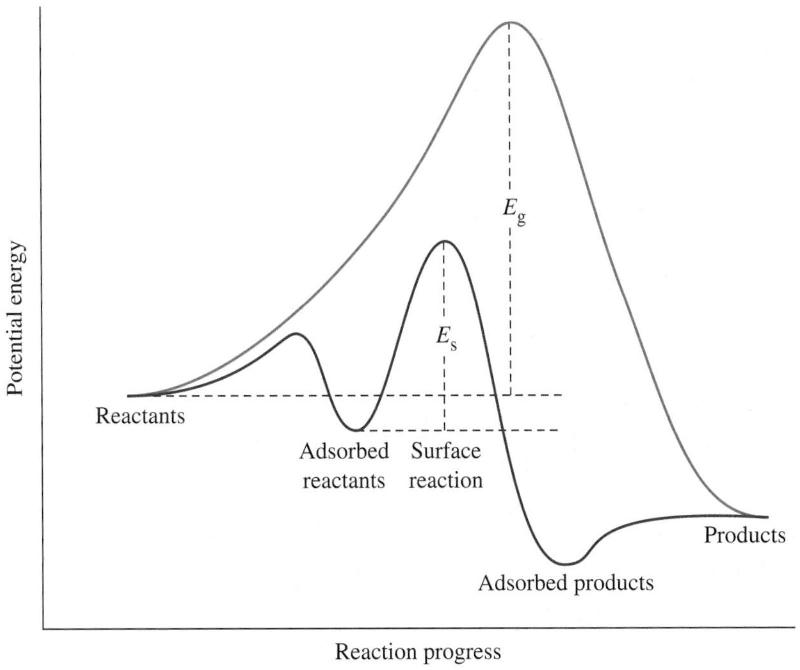

FIGURE 20-19
eaction profile for a surface-catalyzed reaction
the reaction profile (blue) for the surface-catalyzed reaction, the activation energy or the reaction step, E_s, is considerably less than in the reaction profile (red) for the ncatalyzed gas-phase reaction, E_g.

$$\begin{array}{lrcl} \textit{Slow:} & H_2O_2 + I^- & \longrightarrow & OI^- + H_2O \\ \textit{Fast:} & H_2O_2 + OI^- & \longrightarrow & H_2O + I^- + O_2(g) \\ \hline \textit{Overall:} & 2\,H_2O_2 & \longrightarrow & 2\,H_2O + O_2(g) \end{array}$$

As required for a catalyzed reaction, the formula of the catalyst does not ppear in the overall equation. Neither does the intermediate species OI^-. The ate of reaction of H_2O_2 is determined by the rate of the slow first step.

$$\text{rate of reaction } H_2O_2 = k[H_2O_2][I^-] \tag{20.33}$$

◀ Although catalysts are not in the overall chemical equation, they do participate in the chemical reaction and do appear in the mechanism.

◀ The decomposition of hydrogen peroxide, H_2O_2, to H_2O and O_2 is a highly exothermic reaction that is catalyzed by platinum metal.

Because I^- is constantly regenerated, its concentration is constant throughout a given reaction. If the product of the constant terms $k[I^-]$ is replaced by a new constant, k', the rate law can be rewritten as

$$\text{rate of reaction of } H_2O_2 = k'[H_2O_2] \quad \textbf{(20.34)}$$

Equation (20.33) indicates that the rate of decomposition of $H_2O_2(aq)$ is affected by the initial concentration of I^-. For each initial concentration of I^- we obtain a different rate constant, k', in equation (20.34).

We have just described the homogeneous catalysis of the decomposition of hydrogen peroxide. The decomposition can also be catalyzed by heterogeneous catalysis, as seen at the bottom of page 961.

20-8 CONCEPT ASSESSMENT

In the production of ammonia from nitrogen and hydrogen, the rate of reaction can be increased by using a catalyst or by increasing the temperature. Is the means by which these two different methods increase the rate of reaction the same? Explain.

Enzymes as Catalysts

Unlike platinum, which catalyzes a wide variety of reactions, the catalytic action of high-molar-mass proteins known as **enzymes** is very specific. For example, in the digestion of milk, lactose, a more complex sugar, breaks down into two simpler ones, glucose and galactose. This occurs in the presence of the enzyme *lactase*.

▶ Many people lose the ability to produce lactase when they become adults. In such cases, lactose passes through the small intestine into the colon, where it ferments and may cause severe gastric disturbances.

$$\underset{\text{"Milk sugar"}}{\text{lactose}} \xrightarrow{\text{lactase}} \text{glucose} + \text{galactose}$$

Biochemists describe enzyme activity with the "lock-and-key" model (Fig. 20-20). The reacting substance, the **substrate** (S), attaches itself to the enzyme (E) at a particular point called an *active site* to form the enzyme–substrate complex (ES). The complex decomposes to form products (P) and regenerates the enzyme.

$$E + S \underset{k_{-1}}{\overset{k_1}{\rightleftharpoons}} ES$$

$$ES \xrightarrow{k_2} E + P$$

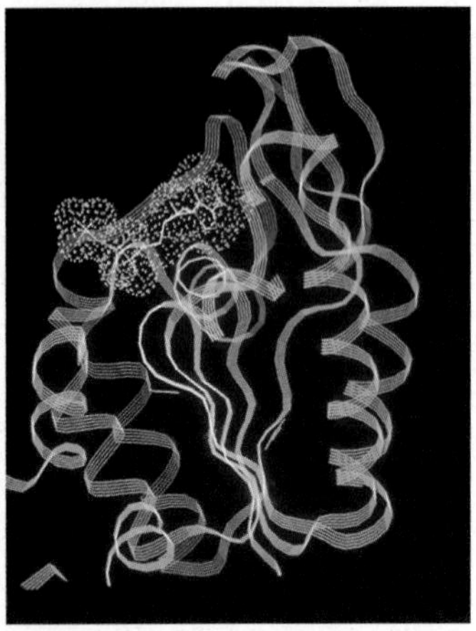

Oxford Molecular Biophysics Labarotary/Science photo Library

▲ A computer graphics representation of the enzyme phosphoglycerate kinase (carbon backbone shown as blue ribbon). A molecule of ATP, the substrate, is shown in green.

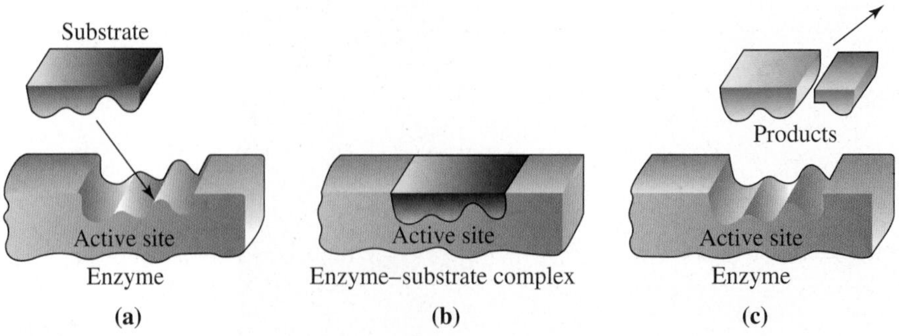

▲ FIGURE 20-20
Lock-and-key model of enzyme action
(a) The substrate attaches itself to an active site on an enzyme molecules. **(b)** Reaction occurs. **(c)** Product molecules detach themselves from the site, freeing the enzyme molecule to attach another molecule of substrate. The substrate and enzyme must have complementary structures to produce a complex, hence the term *lock and key.*

Most human enzyme-catalyzed reactions proceed fastest at about 37 °C •ody temperature). If the temperature is raised much higher than that, the ructure of the enzyme changes, the active sites become distorted, and the ıtalytic activity is lost.

Determining the rates of enzyme-catalyzed reactions is an important part of ızyme studies. Figure 20-21, a plot of reaction rate against substrate concenation, illustrates what we generally observe. Along the rising portion of the raph, the rate of reaction is proportional to the substrate concentration, [S]. he reaction is first order: rate of reaction $= k[\mathrm{S}]$. At high substrate concentraons, the rate is independent of [S]. The reaction follows a zero-order rate quation: rate of reaction $= k$.

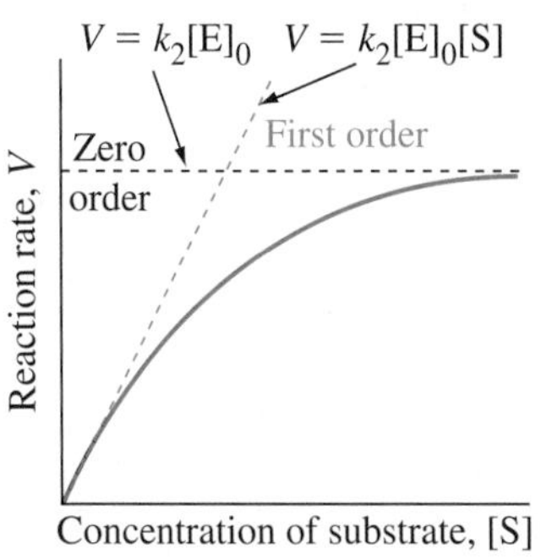

▲ FIGURE 20-21
Effect of substrate concentration on the rate of an enzyme reaction

This behavior can be understood in terms of the three-step mechanism iven above. The rate at which the product appears, which is often called the elocity (V) of the reaction by biochemists, is given by

$$\text{rate of production of P} = V = k_2[\mathrm{ES}]$$

o proceed further, we need an expression for the enzyme–substrate complex, nd we can get this by applying the steady-state approximation to the concenation of the enzyme–substrate complex.

$$\text{rate of formation of ES} = \text{rate of destruction of ES}$$

$$k_1[\mathrm{E}][\mathrm{S}] = (k_{-1} + k_2)[\mathrm{ES}] \tag{20.35}$$

Ve can solve this equation for [ES], but the solution contains the concentraon of free enzyme E, which is unknown. However, we do know the *total* conentration of enzyme in an experiment, $[\mathrm{E}]_0$. Then, by the condition known as *aterial balance*, we have

$$[\mathrm{E}]_0 = [\mathrm{E}] + [\mathrm{ES}]$$

olving this equation for [E] and substituting the result into equation (20.35), ve get

$$k_1[\mathrm{S}]([\mathrm{E}]_0 - [\mathrm{ES}]) = (k_{-1} + k_2)[\mathrm{ES}]$$

$$[\mathrm{ES}] = \frac{k_1[\mathrm{E}]_0[\mathrm{S}]}{(k_{-1} + k_2) + k_1[\mathrm{S}]}$$

or the concentration of the enzyme–substrate complex. Substituting this alue into the rate of reaction, we get

$$V = \frac{k_2 k_1[\mathrm{E}]_0[\mathrm{S}]}{(k_{-1} + k_2) + k_1[\mathrm{S}]}$$

his equation can be put into a more convenient form by dividing the numertor and denominator by k_1 and replacing a ratio of rate constants by the single onstant K_M.

$$K_\mathrm{M} = \frac{k_{-1} + k_2}{k_1}$$

)ur final result is

$$V = \frac{k_2[\mathrm{E}]_0[\mathrm{S}]}{K_\mathrm{M} + [\mathrm{S}]} \tag{20.36}$$

◀ The reaction mechanism outlined here was proposed by Michaelis and Menten in 1913, accounting for the subscript M in the constant K_M.

Now, let's test whether the reaction velocity V, given by equation (20.36), epends on the substrate concentration in the manner suggested by Figure 20-21. t sufficiently low concentrations of S, we have the inequality

$$K_\mathrm{M} \gg [\mathrm{S}]$$

We can ignore [S] with respect to K_M in the denominator and obtain the fo lowing for the reaction velocity.

$$V = \frac{k_2}{K_M}[E]_0[S]$$

Because the total enzyme concentration is constant, the rate law is first ord with respect to substrate, as observed experimentally.

The other limiting case is that in which the velocity of the reaction become independent of substrate concentration. At sufficiently high concentratior of substrate,

$$[S] \gg K_M$$

and

$$V = k_2[E]_0$$

Here, for a chosen concentration of enzyme, the reaction velocity is constar and is the maximum reaction velocity attainable for the particular enzyme. Th velocity corresponds to the experimentally observed plateau at high substrat concentrations in a plot such as Figure 20-21. Thus, there is a satisfactory agre ment between the predictions of the postulated mechanism and the experimer tal results. As is typical of the scientific method, postulated mechanisms ar continually tested by subsequent experiment and modified when necessary.

MasteringCHEMISTRY **www.masteringchemistry.com**

A typical explosion is a combustion reaction that proceeds at an ever-increasing rate. For a discussion of how combustion reactions can become explosive and ways to prevent such occurrences, go to the Focus On feature for Chapter 20, Combustion and Explosions, on the MasteringChemistry site.

Summary

20-1 Rate of a Chemical Reaction—The **rate of reaction** reflects the rate of change in the concentrations of the reactants and products of a reaction. A *general rate of reaction* (equation 20.2) is defined so that the same value is obtained no matter which reactant or product serves as the basis of kinetic measurements.

20-2 Measuring Reaction Rates—An **initial rate of reaction** is the rate measured over a short time interval at the start of a reaction. An **instantaneous rate of reaction** is assessed over an infinitesimal time interval at any point in the reaction, often through the slope of a tangent line to a graph of concentration versus time (Fig. 20-2). Reaction rates measured over longer time intervals are simply *average* rates, since in almost all cases reaction rates continuously decrease as a reaction proceeds.

20-3 Effect of Concentration on Reaction Rates: The Rate Law—The relationship between the rate of a reaction and the concentrations of the reactants is called the **rate law (rate equation)** (equation 20.6); it has the form: rate of reaction = $k[A]^m[B]^n \cdots$. The **order of a reaction** refers to the exponents $m, n, \ldots$ in the rate law. If $m = 2$, the reaction is second order in A; if $n = 1$, the reac tion is first order in B; and so on. The *overall order* of a rea tion is given by the sum $m + n + \cdots$. One method c establishing the rate law of a reaction is by the *method* *initial* rates. The **rate constant** relates the rate of reaction t reactant concentrations.

20-4 Zero-Order Reactions—A reaction havin $m + n + \cdots = 0$ is a **zero-order reaction** (equation 20.9 A useful equation expressing concentration of a reactar as a function of time is called an **integrated rate la** (equation 20.10). A plot of concentration as a function c time for the zero-order reaction A $\longrightarrow$ products is straight line with a slope of $-k$ (Fig. 20-3).

20-5 First-Order Reactions—A reaction havin $m + n + \cdots = 1$ is a **first-order reaction** (equations 20.1 and 20.13). A graph of ln[A] versus time for the first-orde reaction A $\longrightarrow$ products is a straight line with a slope c $-k$ (Fig. 20-4). The **half-life** of a reaction is the tim required for the amount of a reactant to be reduced to on half its initial value. For a first-order reaction, the half-li is a constant (equation 20.14). Many important reaction e first-order processes, including the decay of radioac- re nuclides (Table 20.4).

)-6 Second-Order Reactions—A reaction having $+ n + \cdots = 2$ is a **second-order reaction** (equations).17 and 20.18). A graph of $1/[A]$ versus time for the cond-order reaction, $A \longrightarrow$ products, is a straight line ith a slope of k (Fig. 20-6). For a second-order reaction, e half-life is *not* constant; each successive half-life period twice as long as the one preceding it (equations 20.19a ıd 20.19b).

Some second-order reactions, called pseudo-first-order actions, can be treated as first-order if one of the reactants present at such a high concentration that its concentration mains essentially constant during the reaction.

0-7 Reaction Kinetics: A Summary—A helpful ımmary of some basic ideas of reaction kinetics can be und on pages 940–942 and in Table 20.5.

0-8 Theoretical Models for Chemical Kinetics— he rate of reaction depends on the number of molecular)llisions per unit volume per unit time, the proportion of olecules having energies in excess of the **activation energy** ig. 20-8) and proper orientation of molecules for effective)llisions (Fig. 20-9). **A reaction profile** (Fig. 20-10) traces e progress of a reaction, highlighting the energy states of e reactants, the products, and the **activated complex**, a ypothetical transitory species that exists in a high-energy **ansition state** between reactants and products. For a **ultistep reaction**, the reaction proceeds through multi- le transition states (one for each step), and the rate- etermining step will be the one that proceeds through ıe transition state of highest energy (Fig. 20-13).

0-9 The Effect of Temperature on Reaction ates—The principal basis for describing the effect of mperature on the rate of a chemical reaction is the rrhenius equation (20.21) or a variant of it (20.22). A raph of ln k versus $1/T$ is linear, with the slope of the line qual to $-E_a/R$ (Fig. 20-12).

20-10 Reaction Mechanisms—A **reaction mechanism** is a step-by-step description of a chemical reaction consisting of a series of **elementary processes**. Rate laws are written for the elementary processes and combined into a rate law for the overall reaction. To be plausible, the reaction mechanism must be consistent with the stoichiometry of the overall reaction and its experimentally determined rate law.

The most common elementary processes are **unimolecular** (one molecule dissociates) and **bimolecular** (two molecules collide). Some of the species in elementary processes may be **reaction intermediates**, species produced in one elementary process and consumed in another. One of the elementary processes may be the **rate-determining step**. When a single rate-determining step cannot be identified, the mechanism can often be established by the *steady-state approximation*. Reaction mechanisms can also be depicted through reaction profiles (Fig. 20-14).

Photochemical smog forms through the action of sunlight on the combustion products of internal combustion engines. The mechanism of its formation has been extensively studied by the methods of chemical kinetics.

20-11 Catalysis—A **catalyst** provides an alternative reaction mechanism with a lower activation energy, thereby speeding up the overall reaction, but the catalyst is not itself changed by the reaction. In **homogeneous catalysis**, the catalytic reaction occurs within a single phase (Fig. 20-17). In **heterogeneous catalysis**, the catalytic action occurs on a surface separating two phases (Figs. 20-18 and 20-19). In biochemical reactions, the catalysts are high-molecular-mass proteins called **enzymes**. The reactant, called the **substrate**, attaches to the **active site** on the enzyme, where reaction occurs (Fig. 20-20). The rate of an enzyme-catalyzed reaction can be calculated with an equation (20.36) based on a generally accepted mechanism of enzyme catalysis.

Integrative Example

eroxyacetyl nitrate (PAN) is an air pollutant produced in photochemical smog by the reaction of hydrocarbons, oxides nitrogen, and sunlight. PAN is unstable and dissociates into peroxyacetyl radicals and $NO_2(g)$. Its presence in polluted ir is like a reservoir for NO_2 storage.

$$\underset{\text{PAN}}{CH_3\overset{\overset{O}{\|}}{C}OONO_2} \longrightarrow \underset{\text{Peroxyacetyl radical}}{CH_3\overset{\overset{O}{\|}}{C}OO\cdot} + NO_2$$

The first-order decomposition of PAN has a half-life of 35 hours at 0 °C and 30.0 min at 25 °C. At what temperature ill a sample of air containing 5.0×10^{14} PAN molecules per liter decompose at the rate of 1.0×10^{12} PAN molecules per ter per minute?

nalyze

his problem, which requires four principal tasks, is centered on the relationship between rate constants and temperature quation 20.22) and between a rate constant and the rate of a reaction (equation 20.6). Specifically, we will need to) convert the two half-lives to values of k; (2) use those values of k and their related temperatures to determine the acti- ation energy of the reaction; (3) find the value of k corresponding to the decomposition rate specified; and (4) calculate ıe temperature at which k has the value determined in (3).

Solve

Determine the value of k at 0 °C for the first-order reaction.

$$k = 0.693/t_{1/2}$$

$$k = \frac{0.693}{35\text{ h}} \times \frac{1\text{ h}}{60\text{ min}} = 3.3 \times 10^{-4}\text{ min}^{-1}$$

At 25 °C,

$$k = \frac{0.693}{30.0\text{ min}} = 2.31 \times 10^{-2}\text{ min}^{-1}$$

To determine the activation energy of the reaction, substitute these data

$$k_2 = 2.31 \times 10^{-2}\text{ min}^{-1};\ T_2 = 25\text{ °C} = 298\text{ K}$$

$$k_1 = 3.3 \times 10^{-4}\text{ min}^{-1};\ T_1 = 0\text{ °C} = 273\text{ K}$$

into equation (20.22)

$$\ln\frac{k_2}{k_1} = -\frac{E_a}{R}\left(\frac{1}{T_2} - \frac{1}{T_1}\right)$$

$$\ln\frac{2.31 \times 10^{-2}\text{ min}^{-1}}{3.3 \times 10^{-4}\text{ min}^{-1}} = \frac{-E_a}{8.3145\text{ J mol}^{-1}\text{ K}^{-1}} \times \left(\frac{1}{298\text{ K}} - \frac{1}{273\text{ K}}\right)$$

$$= \frac{-E_a}{8.3145\text{ J mol}^{-1}}(0.00336 - 0.00366) = 4.2$$

$$E_a = \frac{8.3145\text{ J mol}^{-1} \times 4.25}{0.00030} = 1.2 \times 10^5\text{ J mol}^{-}$$

Because the reaction is first order, the rate law is rate $= k[\text{PAN}]$, which can be rearranged to give

$$k = \text{rate}/[\text{PAN}]$$

Because mol/L and molecules/L are related through the Avogadro constant, [PAN] can be expressed as molecules/L.

$$k = \frac{\text{rate of reaction}}{[\text{PAN}]} = \frac{1.0 \times 10^{12}\text{ molecules L}^{-1}\text{ min}^{-1}}{5.0 \times 10^{14}\text{ molecules L}^{-1}}$$

$$= 2.0 \times 10^{-3}\text{ min}^{-1}$$

To determine the unknown temperature, choose one of the known combinations of k and T as k_2 and T_2. For example,

$$k_2 = 2.31 \times 10^{-2}\text{ min}^{-1};\ T_2 = 298\text{ K}$$

The unknown combination is

$$k_1 = 2.0 \times 10^{-3}\text{ min}^{-1};\ T_1 = ?$$

Substitute these values into equation (20.22), together with the known value, $E_a = 1.2 \times 10^5$ J mol^{-1} to obtain

$$\ln\frac{2.31 \times 10^{-2}\text{ min}^{-1}}{2.0 \times 10^{-3}\text{ min}^{-1}} = -\frac{1.2 \times 10^5\text{ J mol}^{-1}}{8.3145\text{ J mol}^{-1}\text{ K}^{-1}} \times \left(\frac{1}{298\text{ K}} - \frac{1}{T_1}\right)$$

$$\ln 12 = -1.4 \times 10^4\text{ K} \times \left(3.36 \times 10^{-1}\text{ K} - \frac{1}{T_1}\right)$$

$$2.5 = 47 - \frac{1.4 \times 10^4\text{ K}}{T_1}$$

$$50T_1 = 1.4 \times 10^4\text{ K}$$

$$T_1 = 2.8 \times 10^2\text{ K}$$

ssess

o check that the final answer is reasonable, we note that based on the values $k = 3.3 \times 10^{-4}\ \text{min}^{-1}$ at 273 K and $= 2.31 \times 10^{-2}\ \text{min}^{-1}$ at 298 K, the temperature at which $k = 2.0 \times 10^{-3}\ \text{min}^{-1}$ should be somewhere between 273 K and 98 K. Our result is just that: 2.8×10^2 K.

RACTICE EXAMPLE A: At room temperature (20 °C), milk turns sour in about 64 hours. In a refrigerator at 3 °C, ilk can be stored three times as long before it sours. **(a)** Estimate the activation energy of the reaction that causes the ouring of milk. **(b)** How long should it take milk to sour at 40 °C?

RACTICE EXAMPLE B: The following mechanism can be used to account for the change in apparent order of nimolecular reactions, such as the conversion of cyclopropane (A) into propene (P), where A* is an energetic form of yclopropane that can either react or return to unreacted cyclopropane.

$$\text{A} + \text{A} \underset{k_{-1}}{\overset{k_1}{\rightleftharpoons}} \text{A}^* + \text{A}$$

$$\text{A}^* \xrightarrow{k_2} \text{P}$$

how that at low pressures of cyclopropane, the rate law is second order in A and at high pressures, it is first order in A.

Exercises

ates of Reactions

1. In the reaction $2\,\text{A} + \text{B} \longrightarrow \text{C} + 3\,\text{D}$, reactant A is found to disappear at the rate of $6.2 \times 10^{-4}\ \text{M s}^{-1}$. **(a)** What is the rate of reaction at this point? **(b)** What is the rate of disappearance of B? **(c)** What is the rate of formation of D?
2. From Figure 20-2 estimate the rate of reaction at **(a)** $t = 800$ s; **(b)** the time at which $[H_2O_2] = 0.50$ M.
3. In the reaction $\text{A} \longrightarrow$ products, [A] is found to be 0.485 M at $t = 71.5$ s and 0.474 M at $t = 82.4$ s. What is the average rate of the reaction during this time interval?
4. In the reaction $\text{A} \longrightarrow$ products, at $t = 0$, [A] = 0.1565 M. After 1.00 min, [A] = 0.1498 M, and after 2.00 min, [A] = 0.1433 M.
(a) Calculate the average rate of the reaction during the first minute and during the second minute.
(b) Why are these two rates not equal?
5. In the reaction $\text{A} \longrightarrow$ products, 4.40 min after the reaction is started, [A] = 0.588 M. The rate of reaction at this point is rate $= -\Delta[\text{A}]/\Delta t = 2.2 \times 10^{-2}\ \text{M min}^{-1}$. Assume that this rate remains constant for a short period of time.
(a) What is [A] 5.00 min after the reaction is started?
(b) At what time after the reaction is started will [A] = 0.565 M?
6. Refer to Experiment 2 of Table 20.3 and to reaction (20.7) and rate law (20.8). Exactly 1.00 h after the reaction is started, what are **(a)** $[HgCl_2]$ and **(b)** $[C_2O_4^{2-}]$ in the mixture?
7. For the reaction $\text{A} + 2\,\text{B} \longrightarrow 2\,\text{C}$, the rate of reaction is $1.76 \times 10^{-5}\ \text{M s}^{-1}$ at the time when [A] = 0.3580 M.
(a) What is the rate of formation of C?
(b) What will [A] be 1.00 min later?
(c) Assume the rate remains at $1.76 \times 10^{-5}\ \text{M s}^{-1}$. How long would it take for [A] to change from 0.3580 to 0.3500 M?
8. If the rate of reaction (20.3) is $5.7 \times 10^{-4}\ \text{M s}^{-1}$, what is the rate of production of O_2(g) from 1.00 L of the H_2O_2(aq), expressed as **(a)** mol $O_2\ \text{s}^{-1}$; **(b)** mol $O_2\ \text{min}^{-1}$; **(c)** mL O_2(STP) min^{-1}?
9. In the reaction $\text{A(g)} \longrightarrow 2\,\text{B(g)} + \text{C(g)}$, the *total* pressure increases while the *partial* pressure of A(g) decreases. If the initial pressure of A(g) in a vessel of constant volume is 1.000×10^3 mmHg,
(a) What will be the total pressure when the reaction has gone to completion?
(b) What will be the total gas pressure when the partial pressure of A(g) has fallen to 8.00×10^2 mmHg?
10. At 65 °C, the half-life for the first-order decomposition of N_2O_5(g) is 2.38 min.

$$N_2O_5(g) \longrightarrow 2\,NO_2(g) + \frac{1}{2}O_2(g)$$

If 1.00 g of N_2O_5 is introduced into an evacuated 15 L flask at 65 °C,
(a) What is the initial partial pressure, in mmHg, of N_2O_5(g)?
(b) What is the partial pressure, in mmHg, of N_2O_5(g) after 2.38 min?
(c) What is the total gas pressure, in mmHg, after 2.38 min?

Method of Initial Rates

11. The initial rate of the reaction $A + B \longrightarrow C + D$ is determined for different initial conditions, with the results listed in the table.
(a) What is the order of reaction with respect to A and to B?
(b) What is the overall reaction order?
(c) What is the value of the rate constant, k?

Expt	[A], M	[B], M	Initial Rate, M s^{-1}
1	0.185	0.133	3.35×10^{-4}
2	0.185	0.266	1.35×10^{-3}
3	0.370	0.133	6.75×10^{-4}
4	0.370	0.266	2.70×10^{-3}

12. For the reaction $A + B \longrightarrow C + D$, the following initial rates of reaction were found. What is the rate law for this reaction?

Expt	[A], M	[B], M	Initial Rate, M min^{-1}
1	0.50	1.50	4.2×10^{-3}
2	1.50	1.50	1.3×10^{-2}
3	3.00	3.00	5.2×10^{-2}

13. The following rates of reaction were obtained in thr experiments with the reaction $2\,NO(g) + Cl_2(g)$ —— $2\,NOCl(g)$.

Expt	Initial [NO], M	Initial [Cl_2], M	Initial Rate of Reaction, M s^{-1}
1	0.0125	0.0255	2.27×10^{-5}
2	0.0125	0.0510	4.55×10^{-5}
3	0.0250	0.0255	9.08×10^{-5}

What is the rate law for this reaction?

14. The following data are obtained for the initial rates reaction in the reaction $A + 2\,B + C \longrightarrow 2\,D + E$.

Expt	Initial [A], M	Initial [B], M	[C], M	Initial Rate
1	1.40	1.40	1.00	R_1
2	0.70	1.40	1.00	$R_2 = \frac{1}{2} \times R_1$
3	0.70	0.70	1.00	$R_3 = \frac{1}{4} \times R_2$
4	1.40	1.40	0.50	$R_4 = 16 \times R_3$
5	0.70	0.70	0.50	$R_5 = ?$

(a) What are the reaction orders with respect to A, and C?
(b) What is the value of R_5 in terms of R_1?

First-Order Reactions

15. One of the following statements is true and the other is false regarding the first-order reaction $2\,A \longrightarrow B + C$. Identify the true statement and the false one, and explain your reasoning.
(a) The rate of the reaction decreases as more and more of B and C form.
(b) The time required for one-half of substance A to react is directly proportional to the quantity of A present initially.

16. One of the following statements is true and the other is false regarding the first-order reaction $2\,A \longrightarrow B + C$. Identify the true statement and the false one, and explain your reasoning.
(a) A graph of [A] versus time is a straight line.
(b) The rate of the reaction is one-half the rate of disappearance of A.

17. The first-order reaction $A \longrightarrow$ products has $t_{1/2} = 180$ s.
(a) What percent of a sample of A remains *unreacted* 900 s after a reaction has been started?
(b) What is the rate of reaction when [A] = 0.50 M?

18. The reaction $A \longrightarrow$ products is first order in A. Initially, [A] = 0.800 M; and after 54 min, [A] = 0.100 M.
(a) At what time is [A] = 0.025 M?
(b) What is the rate of reaction when [A] = 0.025 M?

19. The reaction $A \longrightarrow$ products is first order in A.
(a) If 1.60 g A is allowed to decompose for 38 min, th mass of A remaining undecomposed is found to b 0.40 g. What is the half-life, $t_{1/2}$, of this reaction?
(b) Starting with 1.60 g A, what is the mass of remaining undecomposed after 1.00 h?

20. In the first-order reaction $A \longrightarrow$ product [A] = 0.816 M initially and 0.632 M after 16.0 min.
(a) What is the value of the rate constant, k?
(b) What is the half-life of this reaction?
(c) At what time will [A] = 0.235 M?
(d) What will [A] be after 2.5 h?

21. In the first-order reaction $A \longrightarrow$ products, it found that 99% of the original amount of reactant decomposes in 137 min. What is the half-life, $t_{1/2}$, this decomposition reaction?

22. The half-life of the radioactive isotope phosphorus-3 is 14.3 days. How long does it take for a sample phosphorus-32 to lose 99% of its radioactivity?

23. Acetoacetic acid, CH_3COCH_2COOH, a reagent use in organic synthesis, decomposes in acidic solutio producing acetone and $CO_2(g)$.

$$CH_3COCH_2COOH(aq) \longrightarrow CH_3COCH_3(aq) + CO_2(g$$

This first-order decomposition has a half-life of 144 min

(a) How long will it take for a sample of acetoacetic acid to be 65% decomposed?
(b) How many liters of $CO_2(g)$, measured at 24.5 °C and 748 Torr, are produced as a 10.0 g sample of CH_3COCH_2COOH decomposes for 575 min? [Ignore the aqueous solubility of $CO_2(g)$.]

. The following first-order reaction occurs in $CCl_4(l)$ at 45 °C: $N_2O_5 \longrightarrow N_2O_4 + \frac{1}{2}O_2(g)$. The rate constant is $k = 6.2 \times 10^{-4}\ s^{-1}$. An 80.0 g sample of N_2O_5 in $CCl_4(l)$ is allowed to decompose at 45 °C.
(a) How long does it take for the quantity of N_2O_5 to be reduced to 2.5 g?
(b) How many liters of O_2, measured at 745 mmHg and 45 °C, are produced up to this point?

. For the reaction A $\longrightarrow$ products, the following data give [A] as a function of time: $t = 0$ s, [A] = 0.600 M; 100 s, 0.497 M; 200 s, 0.413 M; 300 s, 0.344 M; 400 s, 0.285 M; 600 s, 0.198 M; 1000 s, 0.094 M.
(a) Show that the reaction is first order.
(b) What is the value of the rate constant, k?
(c) What is [A] at $t = 750$ s?

26. The decomposition of dimethyl ether at 504 °C is

$$(CH_3)_2O(g) \longrightarrow CH_4(g) + H_2(g) + CO(g)$$

The following data are partial pressures of dimethyl ether (DME) as a function of time: $t = 0$ s, $P_{DME} =$ 312 mmHg; 390 s, 264 mmHg; 777 s, 224 mmHg; 1195 s, 187 mmHg; 3155 s, 78.5 mmHg.
(a) Show that the reaction is first order.
(b) What is the value of the rate constant, k?
(c) What is the total gas pressure at 390 s?
(d) What is the total gas pressure when the reaction has gone to completion?
(e) What is the total gas pressure at $t = 1000$ s?

eactions of Various Orders

ree different sets of data of [A] versus time are given in e following table for the reaction A $\longrightarrow$ products. 'int: There are several ways of arriving at answers for ch of the following six questions.]

Data for Exercises 27–32

I		II		III	
Time, s	[A], M	Time, s	[A], M	Time, s	[A], M
0	1.00	0	1.00	0	1.00
25	0.78	25	0.75	25	0.80
50	0.61	50	0.50	50	0.67
75	0.47	75	0.25	75	0.57
100	0.37	100	0.00	100	0.50
150	0.22			150	0.40
200	0.14			200	0.33
250	0.08			250	0.29

'. Which of these sets of data corresponds to a **(a)** zero-order, **(b)** first-order, **(c)** second-order reaction?

3. What is the value of the rate constant k of the zero-order reaction?

). What is the approximate half-life of the first-order reaction?

). What is the approximate initial rate of the second-order reaction?

L. What is the approximate rate of reaction at $t = 75$ s for the **(a)** zero-order, **(b)** first-order, **(c)** second-order reaction?

2. What is the approximate concentration of A remaining after 110 s in the **(a)** zero-order, **(b)** first-order, **(c)** second-order reaction?

3. The reaction A + B $\longrightarrow$ C + D is second order in A and zero order in B. The value of k is 0.0103 $M^{-1}\ min^{-1}$. What is the rate of this reaction when [A] = 0.116 M and [B] = 3.83 M?

34. A reaction is 50% complete in 30.0 min. How long after its start will the reaction be 75% complete if it is **(a)** first order; **(b)** zero order?

35. The decomposition of HI(g) at 700 K is followed for 400 s, yielding the following data: at $t = 0$, [HI] = 1.00 M; 100 s, 0.90 M; 200 s, 0.81 M; 300 s, 0.74 M; 400 s, 0.68 M. What are the reaction order and the rate constant for the reaction:

$$HI(g) \longrightarrow \frac{1}{2}H_2(g) + \frac{1}{2}I_2(g)?$$

Write the rate law for the reaction at 700 K.

36. For the disproportionation of p-toluenesulfonic acid,

$$3\ ArSO_2H \longrightarrow ArSO_2SAr + ArSO_3H + H_2O$$

(where Ar = p-$CH_3C_6H_4$—), the following data were obtained: $t = 0$ min, $[ArSO_2H] = 0.100$ M; 15 min, 0.0863 M; 30 min, 0.0752 M; 45 min, 0.0640 M; 60 min, 0.0568 M; 120 min, 0.0387 M; 180 min, 0.0297 M; 300 min, 0.0196 M.
(a) Show that this reaction is second order.
(b) What is the value of the rate constant, k?
(c) At what time would $[ArSO_2H] = 0.0500$ M?
(d) At what time would $[ArSO_2H] = 0.0250$ M?
(e) At what time would $[ArSO_2H] = 0.0350$ M?

37. For the reaction A $\longrightarrow$ products, the following data were obtained: $t = 0$ s, [A] = 0.715 M; 22 s, 0.605 M; 74 s, 0.345 M; 132 s, 0.055 M. **(a)** What is the order of this reaction? **(b)** What is the half-life of the reaction?

38. The following data were obtained for the dimerization of 1,3-butadiene, $2\,C_4H_6(g) \longrightarrow C_8H_{12}(g)$, at 600 K: $t = 0$ min, $[C_4H_6] = 0.0169$ M; 12.18 min, 0.0144 M; 24.55 min, 0.0124 M; 42.50 min, 0.0103 M; 68.05 min, 0.00845 M.
(a) What is the order of this reaction?
(b) What is the value of the rate constant, k?
(c) At what time would $[C_4H_6] = 0.00423$ M?
(d) At what time would $[C_4H_6] = 0.0050$ M?

39. For the reaction A $\longrightarrow$ products, the data tabulated below are obtained.
(a) Determine the initial rate of reaction (that is, $-\Delta[A]/\Delta t$) in each of the two experiments.
(b) Determine the order of the reaction.

First Experiment	
$[A] = 1.512$ M	$t = 0$ min
$[A] = 1.490$ M	$t = 1.0$ min
$[A] = 1.469$ M	$t = 2.0$ min

Second Experiment	
$[A] = 3.024$ M	$t = 0$ min
$[A] = 2.935$ M	$t = 1.0$ min
$[A] = 2.852$ M	$t = 2.0$ min

40. For the reaction A $\longrightarrow$ 2 B + C, the following data are obtained for [A] as a function of time: $t = 0$ min, $[A] = 0.80$ M; 8 min, 0.60 M; 24 min, 0.35 M; 40 min, 0.20 M.
(a) By suitable means, establish the order of the reaction.
(b) What is the value of the rate constant, k?
(c) Calculate the rate of formation of B at $t = 30$ min.

41. In three different experiments, the following resu were obtained for the reaction A $\longrightarrow$ produc $[A]_0 = 1.00$ M, $t_{1/2} = 50$ min; $[A]_0 = 2.00$ M, $t_{1/}$ 25 min; $[A]_0 = 0.50$ M, $t_{1/2} = 100$ min. Write the r equation for this reaction, and indicate the value of
42. Ammonia decomposes on the surface of a hot tur sten wire. Following are the half-lives that w obtained at 1100 °C for different initial concentratic of NH_3: $[NH_3]_0 = 0.0031$ M, $t_{1/2} = 7.6$ min; 0.0015 3.7 min; 0.00068 M, 1.7 min. For this decompositi reaction, what is **(a)** the order of the reaction; **(b)** rate constant, k?
43. The half-lives of both zero-order and second-orc reactions depend on the initial concentration, as w as on the rate constant. In one case, the half-life g longer as the initial concentration increases, and in t other it gets shorter. Which is which, and why is the situation the same for both?
44. Consider three hypothetical reactions, A — products, all having the same numerical value of t rate constant k. One of the reactions is zero order, c is first order, and one is second order. What must the initial concentration $[A]_0$ if **(a)** the zero- and fir order; **(b)** the zero- and second-order; **(c)** the first- a second-order reactions are to have the same half-lif

Collision Theory; Activation Energy

45. Explain why
(a) A reaction rate cannot be calculated from the collision frequency alone.
(b) The rate of a chemical reaction may increase dramatically with temperature, whereas the collision frequency increases much more slowly.
(c) The addition of a catalyst to a reaction mixture can have such a pronounced effect on the rate of a reaction, even if the temperature is held constant.
46. If even a tiny spark is introduced into a mixture of $H_2(g)$ and $O_2(g)$, a highly exothermic explosive reaction occurs. Without the spark, the mixture remains unreacted indefinitely.
(a) Explain this difference in behavior.
(b) Why is the nature of the reaction independent of the size of the spark?
47. For the reversible reaction A + B $\rightleftharpoons$ C + D, the enthalpy change of the forward reaction is +21 kJ/mol. The activation energy of the forward reaction is 84 kJ/mol.
(a) What is the activation energy of the reverse reaction?
(b) In the manner of Figure 20-10, sketch the reaction profile of this reaction.
48. By an appropriate sketch, indicate why there is some relationship between the enthalpy change and the activation energy for an endothermic reaction but not for an exothermic reaction.
49. By inspection of the reaction profile for the reaction to D given below, answer the following questions.

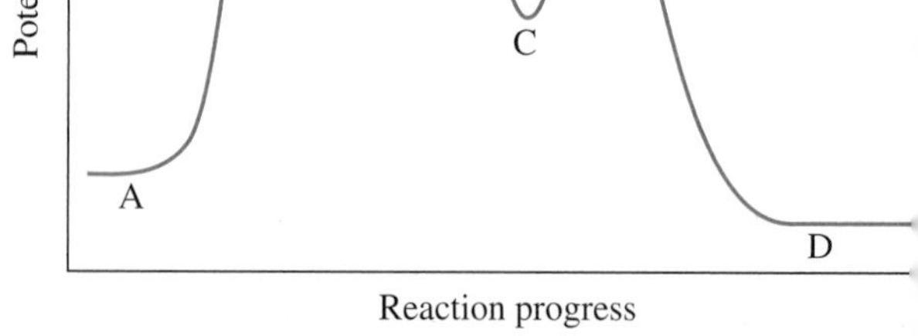

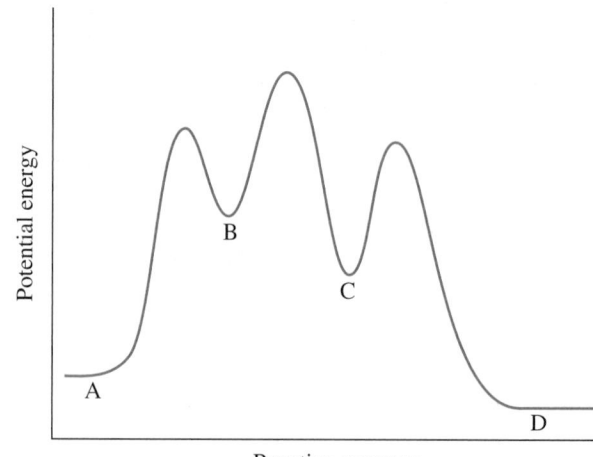

(a) How many intermediates are there in the reactio
(b) How many transition states are there?
(c) Which step has the largest rate constant?
(d) Which step has the smallest rate constant?
(e) Is the first step of the reaction exothermic endothermic?
(f) Is the overall reaction exothermic or endothermi

. By inspection of the reaction profile for the reaction A to D given, answer the following questions.
(a) How many intermediates are there in the reaction?
(b) How many transition states are there?
(c) Which step has the largest rate constant?
(d) Which step has the smallest rate constant?
(e) Is the first step of the reaction exothermic or endothermic?
(f) Is the overall reaction exothermic or endothermic?

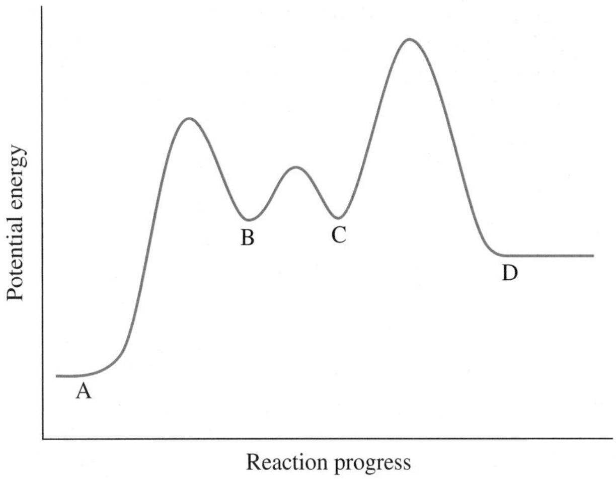

ffect of Temperature on Rates of Reaction

. The rate constant for the reaction $H_2(g) + I_2(g) \longrightarrow 2\,HI(g)$ has been determined at the following temperatures: 599 K, $k = 5.4 \times 10^{-4}\,M^{-1}\,s^{-1}$; 683 K, $k = 2.8 \times 10^{-2}\,M^{-1}\,s^{-1}$. Calculate the activation energy for the reaction.

. At what temperature will the rate constant for the reaction in Exercise 51 have the value $k = 5.0 \times 10^{-3}\,M^{-1}\,s^{-1}$?

. A treatise on atmospheric chemistry lists the following rate constants for the decomposition of PAN described in the Integrative Example on page 965: 0 °C, $k = 5.6 \times 10^{-6}\,s^{-1}$; 10 °C, $3.2 \times 10^{-5}\,s^{-1}$; 20 °C, $1.6 \times 10^{-4}\,s^{-1}$; 30 °C, $7.6 \times 10^{-4}\,s^{-1}$.
(a) Construct a graph of ln k versus $1/T$.
(b) What is the activation energy, E_a, of the reaction?
(c) Calculate the half–life of the decomposition reaction at 40 °C.

. The reaction $C_2H_5I + OH^- \longrightarrow C_2H_5OH + I^-$ was studied in an ethanol (C_2H_5OH) solution, and the following rate constants were obtained: 15.83 °C, $k = 5.03 \times 10^{-5}$; 32.02 °C, 3.68×10^{-4}; 59.75 °C, 6.71×10^{-3}; 90.61 °C, $0.119\,M^{-1}\,s^{-1}$.
(a) Determine E_a for this reaction by a graphical method.
(b) Determine E_a by the use of equation (20.22).
(c) Calculate the value of the rate constant k at 100.0 °C.

. The first-order reaction A $\longrightarrow$ products has a half-life, $t_{1/2}$, of 46.2 min at 25 °C and 2.6 min at 102 °C.
(a) Calculate the activation energy of this reaction.
(b) At what temperature would the half-life be 10.0 min?

56. For the first-order reaction

$$N_2O_5(g) \longrightarrow 2\,NO_2(g) + \frac{1}{2}O_2(g)$$

$t_{1/2}$ = 22.5 h at 20 °C and 1.5 h at 40 °C.
(a) Calculate the activation energy of this reaction.
(b) If the Arrhenius constant $A = 2.05 \times 10^{13}\,s^{-1}$, determine the value of k at 30 °C.

57. A commonly stated rule of thumb is that reaction rates double for a temperature increase of about 10 °C. (This rule is very often wrong.)
(a) What must be the approximate activation energy for this statement to be true for reactions at about room temperature?
(b) Would you expect this rule of thumb to apply at room temperature for the reaction profiled in Figure 20-10? Explain.

58. Concerning the rule of thumb stated in Exercise 57, estimate how much faster cooking occurs in a pressure cooker with the vapor pressure of water at 2.00 atm instead of in water under normal boiling conditions. [*Hint:* Refer to Table 12.5.]

atalysis

. The following statements about catalysis are not stated as carefully as they might be. What slight modifications would you make in them?
(a) A catalyst is a substance that speeds up a chemical reaction but does not take part in the reaction.
(b) The function of a catalyst is to lower the activation energy for a chemical reaction.

. The following substrate concentration [S] versus time data were obtained during an enzyme-catalyzed reaction: t = 0 min, [S] = 1.00 M; 20 min, 0.90 M; 60 min, 0.70 M; 100 min, 0.50 M; 160 min, 0.20 M. What is the order of this reaction with respect to S in the concentration range studied?

61. What are the similarities and differences between the catalytic activity of platinum metal and of an enzyme?

62. Certain gas-phase reactions on a heterogeneous catalyst are first order at low gas pressures and zero order at high pressures. Can you suggest a reason for this?

63. The graph shows the effect of enzyme concentration on the rate of an enzyme reaction. What reaction conditions are necessary to account for this graph?

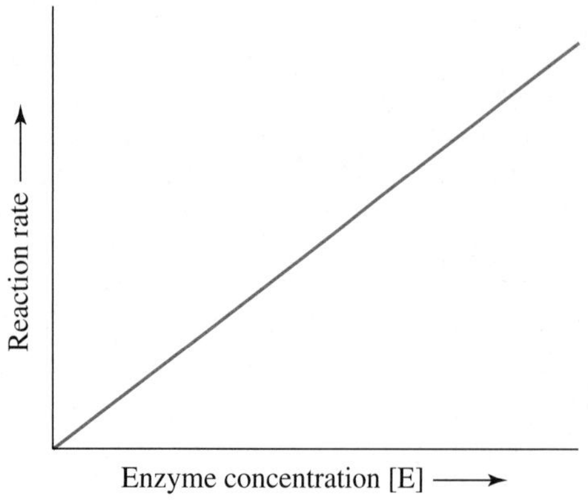

64. The graph shows the effect of temperature on enzyme activity. Explain why the graph has the general shape shown. For human enzymes, at what temperature would you expect the maximum in the curve to appear?

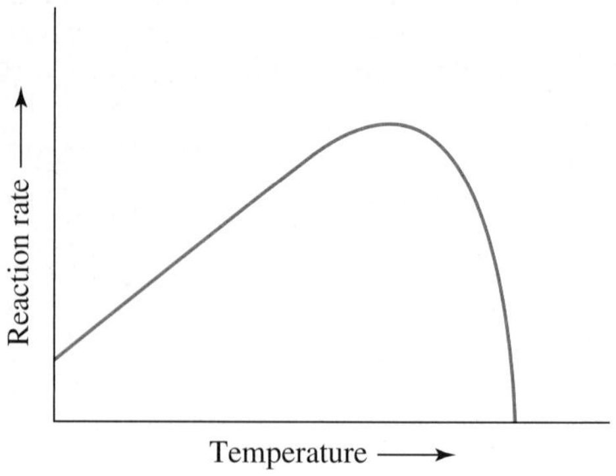

Reaction Mechanisms

65. We have used the terms *order of a reaction* and *molecularity of an elementary process* (that is, unimolecular, bimolecular). What is the relationship, if any, between these two terms?

66. According to collision theory, chemical reactions occur through molecular collisions. A unimolecular elementary process in a reaction mechanism involves dissociation of a *single* molecule. How can these two ideas be compatible? Explain.

67. The reaction $2\,NO + 2\,H_2 \longrightarrow N_2 + 2\,H_2O$ is second order in $[NO]$ and first order in $[H_2]$. A *three-step* mechanism has been proposed. The first, fast step is the elementary process given as equation (20.25) on page 953. The third step, also fast, is $N_2O + H_2 \longrightarrow N_2 + H_2O$. Propose an entire three-step mechanism, and show that it conforms to the experimentally determined reaction order.

68. The mechanism proposed for the reaction of $H_2(g)$ and $I_2(g)$ to form $HI(g)$ consists of a fast reversible first step involving $I_2(g)$ and $I(g)$, followed by a slow step. Propose a two-step mechanism for the reaction $H_2(g) + I_2(g) \longrightarrow 2\,HI(g)$, which is known to be first order in H_2 and first order in I_2.

69. The reaction $2\,NO + Cl_2 \longrightarrow 2\,NOCl$ has the rate law: rate of reaction $= k[NO]^2[Cl_2]$. Propose a two-step mechanism for this reaction consisting of a fast reversible first step, followed by a slow step.

70. A simplified rate law for the reaction $2\,O_3(g) \longrightarrow 3\,O_2(g)$ is

$$\text{rate} = k\frac{[O_3]^2}{[O_2]}$$

For this reaction, propose a two-step mechanism that consists of a fast, reversible first step, followed by a slow second step.

71. One proposed mechanism for the formation of a double helix in DNA is given by

$$(S_1 + S_2) \rightleftharpoons (S_1\!:\!S_2)^* \quad \text{(fast)}$$
$$(S_1\!:\!S_2)^* \longrightarrow S_1\!:\!S_2 \quad \text{(slow)}$$

where S_1 and S_2 represent strand 1 and 2, and $(S_1\!:\!S_2)^*$ represents an unstable helix. Write the rate of reaction expression for the formation of the double helix.

72. One proposed mechanism for the condensation of propanone, $(CH_3)_2CO$, is as follows:

$$(CH_3)_2CO(aq) + OH^-(aq) \rightleftharpoons CH_3C(O)CH_2^-(aq) + H_2O$$
$$CH_3C(O)CH_2^-(aq) + (CH_3)_2CO(aq) \longrightarrow \text{products}$$

Use the steady-state approximation to determine the rate of formation for the product.

Integrative and Advanced Exercises

73. Suppose that the reaction in Example 20-8 is first order with a rate constant of $0.12\ \text{min}^{-1}$. Starting with $[A]_0 = 1.00\ \text{M}$, will the curve for $[A]$ versus t for the first-order reaction cross the curve for the second-order reaction at some time after $t = 0$? Will the two curves cross if $[A]_0 = 2.00\ \text{M}$? In each case, if the curves are found to cross, at what time will this happen?

74. $[A]_t$ as a function of time for the reaction $A \longrightarrow$ products is plotted in the following graph. Use data from this graph to determine **(a)** the order of the reaction; **(b)** the rate constant, k; **(c)** the rate of the reaction at $t = 3.5$ min, using the results of parts (a) and (b); **(d)** the rate of the reaction at $t = 5.0$ min, from the slope of the tangent line; **(e)** the initial rate of the reaction.

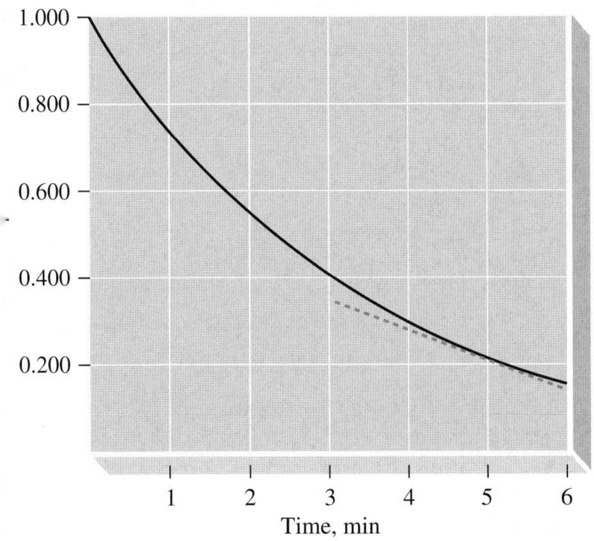

75. Exactly 300 s after decomposition of $H_2O_2(aq)$ begins (reaction 20.3), a 5.00 mL sample is removed and immediately titrated with 37.1 mL of 0.1000 M $KMnO_4$. What is $[H_2O_2]$ at this 300 s point in the reaction?

76. Use the method of Exercise 75 to determine the volume of 0.1000 M $KMnO_4$ required to titrate 5.00 mL samples of $H_2O_2(aq)$ for each of the entries in Table 20.1. Plot these volumes of $KMnO_4(aq)$ as a function of time, and show that from this graph you can get the same rate of reaction at 1400 s as that obtained in Figure 20-2.

77. The initial rate of reaction (20.3) is found to be $1.7 \times 10^{-3}\ M\ s^{-1}$. Assume that this rate holds for 2 minutes. Start with 175 mL of 1.55 M $H_2O_2(aq)$ at $t = 0$. How many milliliters of $O_2(g)$, measured at 24 °C and 757 mmHg, are released from solution in the first minute of the reaction?

78. We have seen that the units of k depends on the overall order of a reaction. Derive a general expression for the units of k for a reaction of any overall order, based on the order of the reaction (o) and the units of concentration (M) and time (s).

79. Hydroxide ion is involved in the mechanism of the following reaction but is not consumed in the overall reaction.

$$OCl^- + I^- \xrightarrow{OH^-} OI^- + Cl^-$$

(a) From the data given, determine the order of the reaction with respect to OCl^-, I^-, and OH^-.
(b) What is the overall reaction order?
(c) Write the rate equation, and determine the value of the rate constant, k.

$[OCl^-]$, M	$[I^-]$, M	$[OH^-]$, M	Rate Formation OI^-, $M\ s^{-1}$
0.0040	0.0020	1.00	4.8×10^{-4}
0.0020	0.0040	1.00	5.0×10^{-4}
0.0020	0.0020	1.00	2.4×10^{-4}
0.0020	0.0020	0.50	4.6×10^{-4}
0.0020	0.0020	0.25	9.4×10^{-4}

80. The half-life for the first-order decomposition of nitramide, $NH_2NO_2(aq) \longrightarrow N_2O(g) + H_2O(l)$, is 123 min at 15 °C. If 165 mL of a 0.105 M NH_2NO_2 solution is allowed to decompose, how long must the reaction proceed to yield 50.0 mL of $N_2O(g)$ collected over water at 15 °C and a barometric pressure of 756 mmHg? (The vapor pressure of water at 15 °C is 12.8 mmHg.)

81. The decomposition of ethylene oxide at 690 K is monitored by measuring the *total* gas pressure as a function of time. The data obtained are $t = 10$ min, $P_{tot} =$ 139.14 mmHg; 20 min, 151.67 mmHg; 40 min, 172.65 mmHg; 60 min, 189.15 mmHg; 100 min, 212.34 mmHg; 200 min, 238.66 mmHg; ∞, 249.88 mmHg. What is the order of the reaction $(CH_2)_2O(g) \longrightarrow CH_4(g) + CO(g)$?

82. Refer to Example 20-7. For the decomposition of di-*t*-butyl peroxide (DTBP), determine the time at which the *total* gas pressure is 2100 mmHg.

83. The following data are for the reaction $2\,A + B \longrightarrow$ products. Establish the order of this reaction with respect to A and to B.

Expt 1, [B] = 1.00 M		Expt 2, [B] = 0.50 M	
Time, min	[A], M	Time, Min	[A], M
0	1.000×10^{-3}	0	1.000×10^{-3}
1	0.951×10^{-3}	1	0.975×10^{-3}
5	0.779×10^{-3}	5	0.883×10^{-3}
10	0.607×10^{-3}	10	0.779×10^{-3}
20	0.368×10^{-3}	20	0.607×10^{-3}

84. Show that the following mechanism is consistent with the rate law established for the iodide–hypochlorite reaction in Exercise 79.

Fast: $$OCl^- + H_2O \underset{k_{-1}}{\overset{k_1}{\rightleftharpoons}} HOCl + OH^-$$

Slow: $$I^- + HOCl \xrightarrow{k_2} HOI + Cl^-$$

Fast: $$HOI + OH^- \underset{k_{-3}}{\overset{k_3}{\rightleftharpoons}} H_2O + OI^-$$

85. In the hydrogenation of a compound containing a carbon-to-carbon triple bond, two products are possible, as in the reaction

$$CH_3{-}C{\equiv}C{-}CH_3 + H_2 \longrightarrow$$

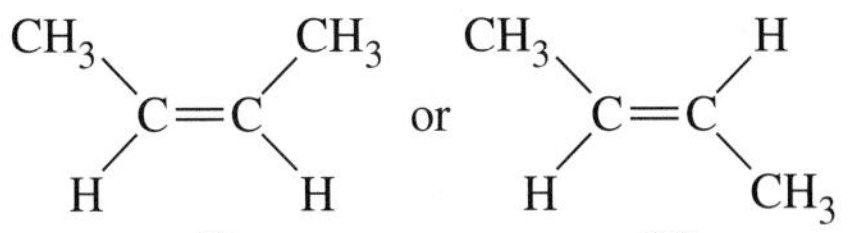

(I) (II)

The amount of each product can be controlled by using an appropriate catalyst. The Lindlar catalyst is a *heterogeneous* catalyst that produces only one of these products. Which product, I or II, do you think is produced and why? Draw a sketch of how the reaction might occur.

86. Derive a plausible mechanism for the following reaction in aqueous solution, $Hg_2^{2+} + Tl^{3+} \longrightarrow 2\,Hg^{2+} + Tl^{+}$, for which the observed rate law is: rate $= k[Hg_2^{2+}][Tl^{3+}]/[Hg^{2+}]$.

87. The following three-step mechanism has been proposed for the reaction of chlorine and chloroform.

$$\text{(1)}\ Cl_2(g) \underset{k_{-1}}{\overset{k_1}{\rightleftharpoons}} 2\,Cl(g)$$

$$\text{(2)}\ Cl(g) + CHCl_3(g) \xrightarrow{k_2} HCl(g) + CCl_3(g)$$

$$\text{(3)}\ CCl_3(g) + Cl(g) \xrightarrow{k_3} CCl_4(g)$$

The numerical values of the rate constants for these steps are $k_1 = 4.8 \times 10^3$; $k_{-1} = 3.6 \times 10^3$; $k_2 = 1.3 \times 10^{-2}$; $k_3 = 2.7 \times 10^2$. Derive the rate law and the magnitude of k for the overall reaction.

88. For the reaction $A \longrightarrow$ products, derive the integrated rate law and an expression for the half-life if the reaction is third order.

89. The reaction $A + B \longrightarrow$ products is first order in A, first order in B, and second order overall. Consider that the starting concentrations of the reactants are $[A]_0$ and $[B]_0$, and that x represents the decrease in these concentrations at the time t. That is, $[A]_t = [A]_0 - x$ and $[B]_t = [B]_0 - x$. Show that the integrated rate law for this reaction can be expressed as shown below.

$$\ln\frac{[A]_0 \times [B]_t}{[B]_0 \times [A]_t} = ([B]_0 - [A]_0) \times kt$$

90. The rate of the reaction

$$2\,CO(g) \longrightarrow CO_2(g) + C(s)$$

was studied by injecting CO(g) into a reaction vessel and measuring the total pressure at constant volume.

P_{total}, Torr	Time, s
250	0
238	398
224	1002
210	1801

What is the rate constant of this reaction?

91. The kinetics of the decomposition of phosphine at 950 K

$$4\,PH_3(g) \longrightarrow P_4(g) + 6\,H_2(g)$$

was studied by injecting $PH_3(g)$ into a reaction vessel and measuring the total pressure at constant volume.

P_{total}, Torr	Time, s
100	0
150	40
167	80
172	120

What is the rate constant of this reaction?

92. The rate of an enzyme-catalyzed reaction can b slowed down by the presence of an inhibitor (I) th reacts with the enzyme in a rapid equilibrium proces

$$E + I \rightleftharpoons EI$$

By adding this step to the mechanism for enzym catalysis on page 962, determine the effect of adding th concentration $[I]_0$ on the rate of an enzyme-catalyze reaction.

93. By taking the reciprocal of both sides of equa tion 20.36, obtain an expression for $1/V$. Using th resulting equation, suggest a strategy for determinin the Michaelis–Menten constant, K_M, and the value of k

94. You want to test the following proposed mechanis for the oxidation of HBr.

$$HBr + O_2 \xrightarrow{k_1} HOOBr$$

$$HOOBr + HBr \xrightarrow{k_2} 2\,HOBr$$

$$HOBr + HBr \xrightarrow{k_3} H_2O + Br_2$$

You find that the rate is first order with respect to HB and to O_2. You cannot detect HOBr among the product
(a) If the proposed mechanism is correct, which mu be the rate-determining step?
(b) Can you prove the mechanism from thes observations?
(c) Can you disprove the mechanism from thes observations?

95. The decomposition of nitric oxide occurs through tw parallel reactions:

$$NO(g) \longrightarrow \frac{1}{2}N_2(g) + \frac{1}{2}O_2(g) \qquad k_1 = 25.7\,s^{-}$$

$$NO(g) \longrightarrow \frac{1}{2}N_2O(g) + \frac{1}{4}O_2(g) \qquad k_2 = 18.2\,s^{-}$$

(a) What is the reaction order for these reactions?
(b) Which reaction is the slow reaction?
(c) If the initial concentration of NO(g) is 2.0 M, wh is the concentration of $N_2(g)$ after 0.1 seconds?
(d) If the initial concentration of NO(g) is 4.0 M, wh is the concentration of $N_2O(g)$ after 0.025 seconds?

Feature Problems

96. Benzenediazonium chloride decomposes by a first-order reaction in water, yielding $N_2(g)$ as one product.

$$C_6H_5N_2Cl(aq) \longrightarrow C_6H_5Cl(aq) + N_2(g)$$

The reaction can be followed by measuring the vo ume of $N_2(g)$ as a function of time. The data in th table on the next page were obtained for the decomp sition of a 0.071 M solution at 50 °C, where $t = \infty$ co responds to the completed reaction.

Time, min	$N_2(g)$, mL	Time, min	$N_2(g)$, mL
0	0	18	41.3
3	10.8	21	44.3
6	19.3	24	46.5
9	26.3	27	48.4
12	32.4	30	50.4
15	37.3	∞	58.3

(a) Convert the information given here into a table with one column for time and the other for $[C_6H_5N_2Cl]$.
(b) Construct a table similar to Table 20.2, in which the time interval is $\Delta t = 3$ min.
(c) Plot graphs similar to Figure 20-2, showing both the formation of $N_2(g)$ and the disappearance of $C_6H_5N_2Cl$ as a function of time.
(d) From the graph of part (c), determine the rate of reaction at $t = 21$ min, and compare your result with the reported value of $1.1 \times 10^{-3}\ M\ min^{-1}$.
(e) Determine the initial rate of reaction.
(f) Write the rate law for the first-order decomposition of $C_6H_5N_2Cl$, and estimate a value of k based on the rate determined in parts (d) and (e).
(g) Determine $t_{1/2}$ for the reaction by estimation from the graph of the rate data and by calculation.
(h) At what time would the decomposition of the sample be three-fourths complete?
(i) Plot $\ln[C_6H_5N_2Cl]$ versus time, and show that the reaction is indeed first order.
(j) Determine k from the slope of the graph of part (i).

7. The object is to study the kinetics of the reaction between peroxodisulfate and iodide ions.

(a) $S_2O_8^{2-}(aq) + 3\,I^-(aq) \longrightarrow 2\,SO_4^{2-}(aq) + I_3^-(aq)$

The I_3^- formed in reaction (a) is actually a complex of iodine, I_2, and iodide ion, I^-. Thiosulfate ion, $S_2O_3^{2-}$, also present in the reaction mixture, reacts with I_3^- just as fast as it is formed.

(b) $2\,S_2O_3^{2-}(aq) + I_3^-(aq) \longrightarrow S_4O_6^{2-} + 3\,I^-(aq)$

When all of the thiosulfate ion present initially has been consumed by reaction (b), a third reaction occurs between $I_3^-(aq)$ and starch, which is also present in the reaction mixture.

(c) $I_3^-(aq) + \text{starch} \longrightarrow \text{blue complex}$

The rate of reaction (a) is inversely related to the time required for the blue color of the starch–iodine complex to appear. That is, the faster reaction (a) proceeds, the more quickly the thiosulfate ion is consumed in reaction (b), and the sooner the blue color appears in reaction (c). One of the photographs shows the initial colorless solution and an electronic timer set at $t = 0$; the other photograph shows the very first appearance of the blue complex (after 49.89 s). Tables I and II list some actual student data obtained in this study.

Richard Megna/Fundamental Photographs

Richard Megna/Fundamental Photographs

TABLE I

Reaction conditions at 24 °C: 25.0 mL of the $(NH_4)_2S_2O_8(aq)$ listed, 25.0 mL of the KI(aq) listed, 10.0 mL of 0.010 M $Na_2S_2O_3(aq)$, and 5.0 mL starch solution are mixed. The time is that of the first appearance of the starch–iodine complex.

	Initial Concentrations, M		
Experiment	$(NH_4)_2S_2O_8$	KI	Time, s
1	0.20	0.20	21
2	0.10	0.20	42
3	0.050	0.20	81
4	0.20	0.10	42
5	0.20	0.050	79

TABLE II

Reaction conditions: those listed in Table I for Experiment 4, but at the temperatures listed.

Experiment	Temperature, °C	Time, s
6	3	189
7	13	88
8	24	42
9	33	21

(a) Use the data in Table I to establish the order of reaction (a) with respect to $S_2O_8^{2-}$ and to I^-. What is the overall reaction order? [*Hint:* How are the times required for the blue complex to appear related to the actual rates of reaction?]
(b) Calculate the initial rate of reaction in Experiment 1, expressed in $M\ s^{-1}$. [*Hint:* You must take into account the dilution that occurs when the various solutions are mixed, as well as the reaction stoichiometry indicated by equations (a), (b), and (c).]
(c) Calculate the value of the rate constant, k, based on experiments 1 and 2.
(d) Calculate the rate constant, k, for the four different temperatures in Table II.
(e) Determine the activation energy, E_a, of the peroxodisulfate–iodide ion reaction.
(f) The following mechanism has been proposed for reaction (a). The first step is slow, and the others are fast.

$$I^- + S_2O_8^{2-} \longrightarrow IS_2O_8^{3-}$$
$$IS_2O_8^{3-} \longrightarrow 2\,SO_4^{2-} + I^+$$
$$I^+ + I^- \longrightarrow I_2$$
$$I_2 + I^- \longrightarrow I_3^-$$

Show that this mechanism is consistent with both the stoichiometry and the rate law of reaction (a). Explain why it is reasonable to expect the first step in the mechanism to be slower than the others.

Self-Assessment Exercises

98. In your own words, define or explain the following terms or symbols: **(a)** $[A]_0$; **(b)** k; **(c)** $t_{1/2}$; **(d)** zero-order reaction; **(e)** catalyst.

99. Briefly describe each of the following ideas, phenomena, or methods: **(a)** the method of initial rates; **(b)** activated complex; **(c)** reaction mechanism; **(d)** heterogeneous catalysis; **(e)** rate-determining step.

100. Explain the important distinctions between each pair of terms: **(a)** first-order and second-order reactions; **(b)** rate law and integrated rate law; **(c)** activation energy and enthalpy of reaction; **(d)** elementary process and overall reaction; **(e)** enzyme and substrate.

101. The rate equation for the reaction $2\,A + B \longrightarrow C$ is found to be rate $= k[A][B]$. For this reaction, we can conclude that **(a)** the units of $k = s^{-1}$; **(b)** $t_{1/2}$ is constant; **(c)** the value of k is independent of the values of [A] and [B]; **(d)** the rate of formation of C is twice the rate of disappearance of A.

102. A first-order reaction, $A \longrightarrow$ products, has a half-life of 75 s, from which we can draw two conclusions. Which of the following are those two? **(a)** The reaction goes to completion in 150 s; **(b)** the quantity of A remaining after 150 s is half of what remains after 75 s; **(c)** the same quantity of A is consumed for every 75 s of the reaction; **(d)** one-quarter of the original quantity of A is consumed in the first 37.5 s of the reaction; **(e)** twice as much A is consumed in 75 s when the initial amount of A is doubled; **(f)** the amount of A consumed in 150 s is twice as much as is consumed in 75 s.

103. A first order reaction $A \longrightarrow$ products has a half-life of 13.9 min. The rate at which this reaction proceeds when $[A] = 0.40\ M$ is **(a)** $0.020\ mol\ L^{-1}\ min^{-1}$; **(b)** $5.0 \times 10^{-2}\ mol\ L^{-1}\ min^{-1}$; **(c)** $8.0\ mol\ L^{-1}\ min^{-1}$; **(d)** $0.125\ mol\ L^{-1}\ min^{-1}$.

104. The reaction $A \longrightarrow$ products is second order. The initial rate of decomposition of A when $[A]_0 = 0.50\ M$ is **(a)** the same as the initial rate for any other value of $[A]_0$; **(b)** half as great as when $[A]_0 = 1.00\ M$; **(c)** five times as great as when $[A]_0 = 0.10\ M$; **(d)** four times as great as when $[A]_0 = 0.25\ M$.

105. The rate of a chemical reaction generally increases rapidly, even for small increases in temperature, because of a rapid increase in **(a)** collision frequency; **(b)** fraction of reactant molecules with very high kinetic energies; **(c)** activation energy; **(d)** average kinetic energy of the reactant molecules.

106. For the reaction $A + B \longrightarrow 2\,C$, which proceeds by a single-step bimolecular elementary process, **(a)** $t_{1/2} = 0.693/k$; **(b)** rate of appearance of C = −rate of disappearance of A; **(c)** rate of reaction = $k[A][B]$; **(d)** $\ln[A]_t = -kt + \ln[A]_0$.

107. In the first-order decomposition of substance A the following concentrations are found at the indicated times: $t = 0$ s, $[A] = 0.88$ M; 50 s, 0.62 M; 100 s 0.44 M; 150 s, 0.31 M. Calculate the instantaneo rate of decomposition at $t = 100$ s.

108. A reaction is 50% complete in 30.0 min. How lo after its start will the reaction be 75% complete if it **(a)** first order; **(b)** zero order?

109. A kinetic study of the reaction $A \longrightarrow$ produc yields the data: $t = 0$ s, $[A] = 2.00$ M; 500 s, 1.00 M 1500 s, 0.50 M; 3500 s, 0.25 M. *Without performi detailed calculations*, determine the order of this rea tion and indicate your method of reasoning.

110. For the reaction $A \longrightarrow$ products the following da are obtained.

Experiment 1		Experiment 2	
$[A] = 1.204$ M	$t = 0$ min	$[A] = 2.408$ M	$t = 0$ min
$[A] = 1.180$ M	$t = 1.0$ min	$[A] = ?$	$t = 1.0$ mi
$[A] = 0.602$ M	$t = 35$ min	$[A] = ?$	$t = 30$ mi

(a) Determine the initial rate of reaction Experiment 1.
(b) If the reaction is second order, what will be [A at $t = 1.0$ min in Experiment 2?
(c) If the reaction is first order, what will be [A] 30 min in Experiment 2?

111. For the reaction $A + 2\,B \longrightarrow C + D$, the rate law rate of reaction $= k[A][B]$.
(a) Show that the following mechanism is consiste with the stoichiometry of the overall reaction ar with the rate law.

$$A + B \longrightarrow I \quad \text{(slow)}$$
$$I + B \longrightarrow C + D \quad \text{(fast)}$$

(b) Show that the following mechanism is consiste with the stoichiometry of the overall reaction, but n with the rate law.

$$2\,B \underset{k_{-1}}{\overset{k_1}{\rightleftharpoons}} B_2 \quad \text{(fast)}$$
$$A + B_2 \xrightarrow{k_2} C + D \quad \text{(slow)}$$

112. If the plot of the reactant concentration versus time nonlinear, but the concentration drops by 50% eve 10 seconds, then the order of the reaction is **(a)** ze order; **(b)** first order; **(c)** second order; **(d)** third ord

113. If the plot of the reactant concentration versus time linear, then the order of the reaction is **(a)** zero ord **(b)** first order; **(c)** second order; **(d)** third order.

114. One example of a zero-order reaction is the decor position of ammonia on a hot platinum wir $2\,NH_3(g) \longrightarrow N_2(g) + 3\,H_2(g)$. If the concentr tion of ammonia is doubled, the rate of the reacti will **(a)** be zero; **(b)** double; **(c)** remain the sam **(d)** exponentially increase.

115. Using the method presented in Appendix E, co struct a concept map illustrating the concepts Sections 20-8 and 20-9.

Chemistry of the Main-Group Elements I: Groups 1, 2, 13, and 14

21

ONTENTS

me of the dramatic colors seen in fireworks displays are the flame colors of some the groups 1 and 2 metals. These colors, as we will see, are related to the ectronic structures of those metal atoms.

LEARNING OBJECTIVES

21.1 Describe how the charge density of ions varies with ionic radius and ionic charge.

21.2 Identify compounds that can be prepared from NaCl.

21.3 Describe the reactivity of the group 2 alkaline earth metals and its relationship to charge density.

21.4 Describe the bonding and structure of the boron family elements and the way in which they commonly react.

21.5 Describe the two most common allotropes of carbon and the one structure of silicon.

The chemistry of the elements is best described by using the periodic table as a basis. The trends within the groups and across the periods allow us to organize our thinking about the chemistry of the ele-ents by using patterns. The chemistry of each group has both similarities d differences that can be understood in terms of the underlying princi-es that organize the periodic table. As we explore the chemistry of these oups, we will discover that the first member of a group is often arkedly different from the others in its physical and chemical properties. pically, the second member of the group exhibits properties that are ost representative of the group.

In this chapter, our focus is on groups 1, 2, 13, and 14—the first four oups of the main group of elements. We will start with the chemistry the group 1 metals: the alkali metals. They are the first members of the block elements. Atoms of the group 1 elements have ground state

KEEP IN MIND
that hydrogen is often placed in group 1 of the periodic table, but this element is not an alkali metal. Francium is an alkali metal, but this highly radioactive metal is so rare that few of its properties have been measured.

valence configurations that consist of a single electron in an *s* orbital. As di cussed in Chapter 9, the group 1 atom in a given period is always the large (having the largest atomic radius) and is the most easily ionized (having t lowest first ionization energy). As a result, the group 1 elements have low den ties (some have densities of less than 1 g cm^{-3}), are easily oxidized, and a highly reactive. The reactivity of the alkali metals is evident in their viole reactions with water.

Next, we will discuss the group 2 metals: the alkaline earth metals. The grou metals are considered *s*-block elements because atoms of these metals ha valence configurations that consist of two electrons in an *s* orbital. Like th group 1 neighbors, the metals of group 2 are highly reactive and less dense th a typical metal, though they are less reactive and denser than the alkali meta Unlike the group 1 metals, the group 2 metals react only slowly, or not at a with water, and all of them have densities greater than that of water.

Atoms of the *p*-block elements are characterized by valence configuratio involving electrons in *p*-orbitals. In this chapter we will discuss groups and 14. Atoms of these elements have the configurations ns^2np^1 and ns^2n respectively. These are the first groups in which both metals and nonmet are encountered. Boron is the only nonmetal in group 13 and has interesti chemistry because it tends to form molecules with incomplete octets arou the central boron atoms. Aluminum is the most abundant of the metals a one of the most widely used. Aluminum metal is obtained from its co pounds by using electrolysis. Because aluminum production requires pro gious quantities of electricity, aluminum-production plants are often locat near plentiful sources of hydroelectricity.

The remaining elements of group 13—gallium, indium, and thallium—a all metals. The chemistry of group 13 is dominated by boron and aluminu and we will mention the heavier elements only briefly in this chapter. Group contains a nonmetal (carbon), two metalloids (silicon and germanium), a two metals (tin and lead). Carbon has the most important chemistry of t group since it occurs in all living systems and we devote three chapters (26, 2 and 28) to the chemistry of carbon. Silicon is found in numerous minera forming many different and interesting oxoanions. In contrast to aluminum, t and lead can be obtained by chemical reduction using methods known sin ancient times.

This chapter and the remaining chapters offer many opportunities to rel new information to principles presented earlier in the text. Ideas of atom structure, periodic trends in atomic and ionic radii, chemical bondin and thermodynamics will help us to understand the chemical behavior of t elements.

21-1 Periodic Trends and Charge Density

The chemistry of the elements can, to some extent, be rationalized in terms the periodic trends that we have covered previously in this text. The atoms each group of the periodic table have similar electronic configurations a consequently, the elements in a given group have similar—but not exactly t same—chemical properties. The first member of a group is the lightest a often has features that are different from the remaining members of the grou In this section we will briefly review trends in atomic properties and introdu a new ionic property. With these ideas as a basis we will begin to understa the trends in the chemistry of the elements.

The atomic properties that are responsible for the chemistry of an eleme are atomic radius, ionization energy, electron affinity, and polarizability. T electronegativity of an atom is also an important consideration. Before w attempt to explain the chemistry of the elements in terms of atomic properti it will be helpful to examine Figure 21-1, which summarizes the period

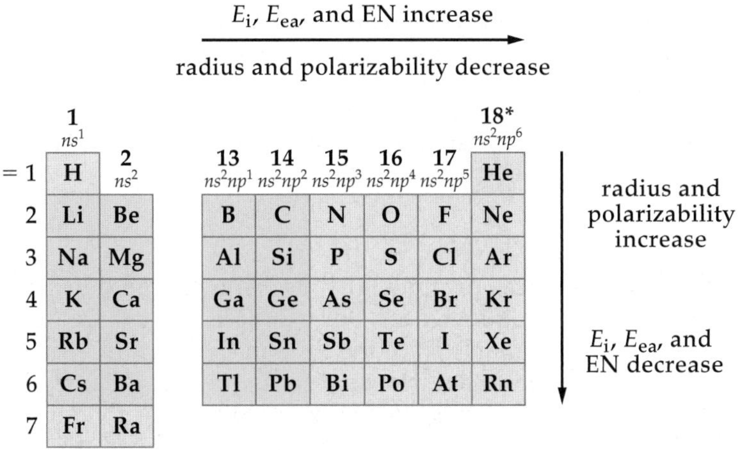

*The ground state electronic configuration of He is $1s^2$.

FIGURE 21-1
summary of trends in atomic radius, first ionization energy, electron finity, electronegativity, and atomic polarizability
omic radius and polarizability decrease from left to right in a given period and crease from top to bottom in a given group. Ionization energy (*Ei*), electron affinity $_{ea}$), and electronegativity (EN) increase from left to right in a given period and decrease om top to bottom in a given group. The shaded elements are the focus of this chapter. omic radii, ionization energies, and electron affinities were discussed in Chapter 9. ectronegativities were discussed in Chapter 10 and polarizabilities in Chapter 12.

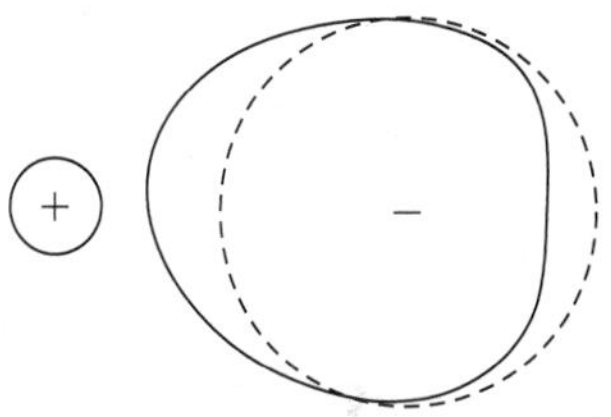

▲ FIGURE 21-2
Polarization of the electron cloud of an anion by a cation
A cation with a high charge density distorts the electron cloud around the anion. The undistorted electron cloud is shown with a dashed line and the distorted electron cloud is shown with a solid line. The greater the distortion of the electron cloud, the greater the degree of covalency of the bond between the anion and cation.

ends we have discussed previously in this text. The figure shows that atomic dii and polarizabilities decrease from left to right in a period and increase om top to bottom in a group. First ionization energies, electron affinities, and ectronegativities show the opposite trend: these quantities increase across a eriod and decrease down a group. It is also important to remember that ionic dii follow the same trend as atomic radii and that cations are smaller than e parent atoms while anions are larger. Explanations for these trends have een given elsewhere in the text (see Chapters 9 and 10).

Most of the elements we consider in this chapter are metals, and when a etal combines with a nonmetallic element to form a compound, a metal om is converted into a cation. When a cation interacts with an anion, the ectron cloud of the anion is distorted into the internuclear region toward e cation, as suggested by Figure 21-2. As a result of this distortion, the ond between the cation and the anion has some covalent character in addion to its ionic character; the greater the distortion, the greater the covalent aracter.

The polarizability of the anion and the polarizing power of the cation determine the extent to which the electron cloud of the anion is distorted. It is genally true that anions derived from atoms that are lower down in a group are rger and more polarizable. For example, the I^- ion is much more polarizable an F^-. The polarizing power of a cation is related to its **charge density**. The arge density, ρ, can be defined as charge per unit volume. For a metal cation, $^{z+}$, with ionic radius r, the charge density is calculated by using the formula elow

$$\rho = \frac{ze}{V} = \frac{(1.60 \times 10^{-19}\,\text{C})(z)}{\frac{4}{3}\pi r^3} \qquad \textbf{(21.1)}$$

In equation (21.1), the numerator represents the total charge, in coulombs, the cation; the denominator represents the volume of the cation, which is ssumed to be spherical. If the ionic radius is expressed in millimeters, then

KEEP IN MIND

that a variety of different definitions have been used for charge density. For example, some authors have defined charge density as the amount of charge distributed over the *surface* of the cation and thus, they calculate the charge density as $ze/(4\pi r^2)$, with $e = 1.60 \times 10^{-19}$ C. Both definitions provide a measure of the charge-to-size ratio for ions. The use of charge-to-size ratios can be traced back to at least 1928, when G. H. Cartledge introduced the *ionic potential, z/r,* in an attempt to explain variations in the properties of compounds.

the charge density is typically between 1 and 1000 C mml^{-3}. As an exampl consider the lithium cation, which has a charge of +1 and an ionic radi of 73 pm:

$$\rho_{Li^+} = \frac{(1.60 \times 10^{-19}\,C)(1)}{\frac{4}{3}\pi\,(73 \times 10^{-9}\,mm)^3} = 98\,C\,mm^{-3}$$

The ionic radius decreases dramatically, and the charge density increase as the charge on the cation increases. For example, the charge density of tl Al^{3+} ion is much greater than that of the Li^+ ion (770 C mm^{-3} for Al^{3+} vers 98 C mm^{-3} for Li^+) because the charge is much greater (+3 for Al versus +1 for Li^+) and the ionic radius is much smaller (53 for Al^{3+} vers 73 pm for Li^+).

We anticipate that the higher the charge density, the greater the polarizi power of the cation and the greater the ability of a cation to distort the electrc cloud of an anion toward itself. For example, because the charge density of tl Al^{3+} ion is much greater than that of the Li^+ ion, we expect the interactic between Al^{3+} and I^- to have a greater degree of covalency than does the inte action between Li^+ and I^-.

Throughout this chapter, we will use the charge density concept to rationa ize certain observations. For example, we will use it to help us understar why sometimes there are dramatic differences in the properties of elements the same group and interesting similarities in the properties of elements in di ferent groups. However, a single quantity—such as charge density—cannot l used to rationalize all things and it should never be used as a substitute f careful consideration of all contributing factors.

TABLE 21.1 Abundances of Group 1 Elements

Elements	Abundances, ppm[a]	Rank
Li	20	32
Na	23,600	6
K	20,900	8
Rb	90	22
Cs	3	46
Fr	Trace	—

[a]Grams per 1000 kg of solid crust.

Source: Data from the *CRC Handbook of Chemistry and Physics*, 95th ed.

21-1 CONCEPT ASSESSMENT

One way of distinguishing ionic behavior from covalent behavior is by comparing melting points. Ionic compounds tend to have higher melting points than covalent compounds. Which of the two halides of aluminum, AlF_3 and AlI_3, is expected to have the lowest melting point?

21-2 Group 1: The Alkali Metals

As Table 21.1 indicates, the group 1 elements, the **alkali metals**, are relative abundant. Some of their compounds have been known and used since prehi toric times. Yet these elements were not isolated in pure form until abo 200 years ago. The compounds of the alkali metals are difficult to decompo by ordinary chemical means, so discovery of the elements had to await ne scientific developments. Sodium (1807) and potassium (1807) were discovere through electrolysis. Lithium was discovered in 1817. Cesium (1860) an rubidium (1861) were identified as new elements through their emissic spectra. Francium (1939) was isolated in the radioactive decay products actinium.

▲ The mineral spodumene, $LiAl(SiO_3)_2$.

Because most alkali metal compounds are water soluble, a number of L Na, and K compounds, including chlorides, carbonates, and sulfates, can l obtained from natural brines. A few alkali metal compounds, such as NaC KCl, and Na_2CO_3, can be mined as solid deposits. Sodium chloride is als obtained from seawater. An important source of lithium is the miner *spodumene*, $LiAl(SiO_3)_2$, shown in the margin. Rubidium and cesium a obtained as by-products in the processing of lithium ores.

hysical Properties of the Alkali Metals

y any measure, the group 1 elements are the most active metals. Table 21.2 lists veral of their properties, and a few of these properties are discussed next.

ame Colors The energy differences between the valence-shell *s* and *p* bitals of the group 1 atoms match those of certain wavelengths of visible ght. As a result, when heated in a flame, group 1 compounds produce char- teristic flame colors, as was shown in Figure 8.10. For example, when NaCl vaporized in a flame, ion pairs are converted to gaseous atoms. Sodium oms, Na(g), are excited to higher energies, and light with a wavelength of 39 nm (yellow) is emitted as the excited atoms (Na*) revert to their ground- ate electron configurations.

$$Na^+Cl^-(g) \longrightarrow Na(g) + Cl(g)$$

$$\underset{[Ne]3s^1}{Na(g)} \longrightarrow \underset{[Ne]3p^1}{Na^*(g)}$$

$$Na^*(g) \longrightarrow Na(g) + h\nu \text{ (589 nm; yellow)}$$

lkali metal compounds are used in pyrotechnic displays—fireworks.

ensities and Melting Points Atoms of the group 1 elements are the largest in eir respective periods, and the atomic radii increase from the top to the bot- m within this group, as was described in Chapter 9. These large atoms make r a relatively low mass per unit volume—that is, low density. The lighter of e alkali metals (Li, Na, and K) will float on water. The large atomic sizes, gether with the fact that each of these atoms has only one valence electron, ads to rather weak metallic bonding. This property, in turn, leads to soft met- s with low melting points. A bar of sodium has the consistency of a stick of itter and is easily cut with a knife, as shown in the photo in the margin.

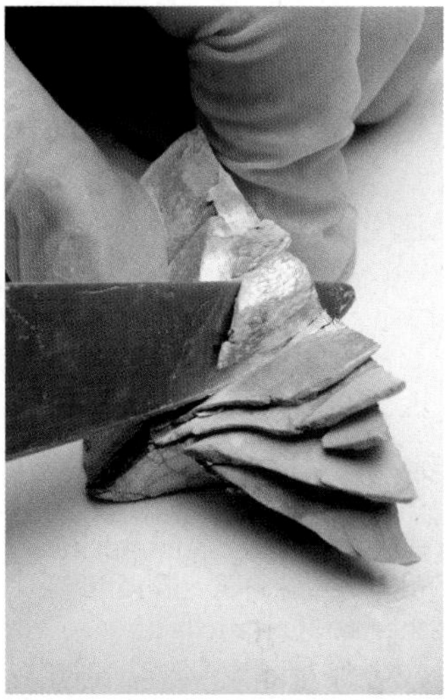

Chip Clark/Fundamental Photographs

▲ The cutting of metallic sodium
The sodium, an active metal, is covered with a thick oxide coating.

TABLE 21.2 Some Properties of the Group 1 (Alkali) Metals

	Li	Na	K	Rb	Cs
Atomic number	3	11	19	37	55
Valence-shell electron configuration	$2s^1$	$3s^1$	$4s^1$	$5s^1$	$6s^1$
Atomic (metallic) radius, pm	152	186	227	248	265
Ionic (M^+) radius, pm[a]	73	116	152	166	181
Electronegativity	1.0	0.9	0.8	0.8	0.8
Charge density of M^+, C mm^{-3}	98	24	11	8	6
First ionization energy, kJ mol^{-1}	520.2	495.8	418.8	403.0	375.7
Electrode potential $E°$, V[b]	−3.040	−2.713	−2.924	−2.924	−2.923
Melting point, °C	180.54	97.81	63.65	39.05	28.4
Boiling point, °C	1347	883.0	773.9	687.9	678.5
Density, g cm^{-3} at 20 °C	0.534	0.971	0.862	1.532	1.873
Hardness[c]	0.6	0.4	0.5	0.3	0.2
Electrical conductivity[d]	17.1	33.2	22.0	12.4	7.76
Flame color	Carmine	Yellow	Violet	Bluish red	Blue
Principal visible emission lines, nm	610, 671	589	405, 767	780, 795	456, 459

[a] The values given here assume a coordination number of 4 for Li^+ and 6 for the others.
[b] For the reduction $M^+(aq) + e^- \longrightarrow M(s)$.
[c] Hardness measures the ability of substances to scratch, abrade, or indent one another. On the Mohs scale, ten minerals are ranked by hardness, ranging from that of talc (0) to diamond (10). Other values: wax (0 °C), 0.2; asphalt, 1–2; fingernail, 2.5; copper, 2.5–3; iron, 4–5; chromium, 9. Each substance can scratch only other substances with hardness values lower than its own.
[d] On a scale relative to silver as 100.

Electrode Potentials A good indicator of the extreme metallic character of th group 1 elements is their standard reduction potentials, which are large, neg tive quantities. The ions $M^+(aq)$ are very difficult to reduce to the metals M(s and in turn, the metals are very easily oxidized to $M^+(aq)$. All the alkali me als easily reduce water to $H_2(g)$.

$$2\,M(s) + 2\,H_2O(l) \longrightarrow 2\,M^+(aq) + 2\,OH^-(aq) + H_2(g) \qquad (21.$$

$$E^\circ_{cell} = E^\circ_{H_2O/H_2} - E^\circ_{M^+/M}$$
$$= -0.828\text{ V} - E^\circ_{M^+/M}$$

Using the electrode potentials given in Table 21.2, we can calculate the follow ing E°_{cell} values.

$$E^\circ_{cell} = 2.212\text{ V (for Li)} \quad 1.885\text{ V (for Na)} \quad 2.096\text{ V (for K)}$$
$$2.096\text{ V (for Rb)} \quad 2.095\text{ V (for Cs)}$$

The E°_{cell} values indicate that of the alkali metals, lithium is the stronge reducing agent in aqueous solution. However, all the alkali metals are stron reducing agents in aqueous solution. If we convert the E°_{cell} values given abov to equilibrium constants (K) by using the equation $\ln K = zFE^\circ/RT$, we fin that the values of K range from 7×10^{31} for the reaction of Na(s) and water t 2×10^{37} for the reaction of Li(s) and water. The equilibrium position for rea tion (21.2) is always very far to the right, and so reaction (21.2) goes essentiall to completion regardless of which alkali metal is involved.

KEEP IN MIND

that the equilibrium position for a reaction is controlled by thermodynamic factors—such as E°_{cell} and K. The rate of a reaction, and the time it takes to reach equilibrium, is controlled by kinetic factors.

Experiments show that lithium reacts more slowly and less vigorousl with water than do any of the other alkali metals. To explain this observatio it is necessary to consider what happens to the energy that is released by th reaction as the metal is oxidized by water. As the metal reacts, energ released by the reaction is used to heat the system, including unreacte metal. For all the alkali metals except lithium, energy released by the reactic is sufficient to melt the unreacted metal. The melting of the metal increase the rate of reaction because it causes more metal atoms to come into conta with water molecules. Because lithium metal does not melt as the reactio proceeds, the reaction of lithium and water is neither as fast nor as vigorou as it is for the other alkali metals.

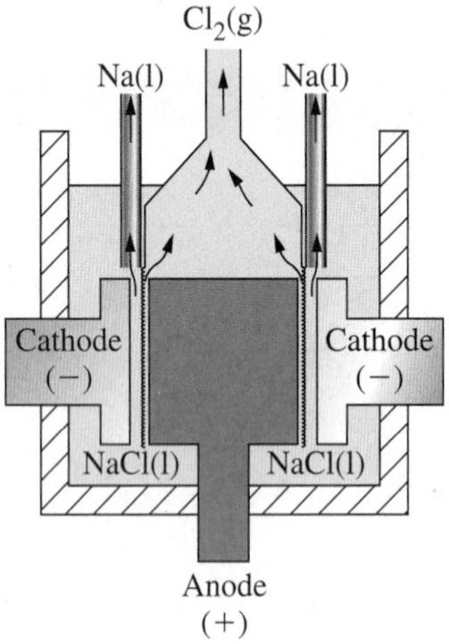

▲ FIGURE 21-3
The Downs cell used for the production of sodium
The electrolysis cell shown here is the Downs cell. The electrolyte is molten NaCl(l) to which $CaCl_2$ has been added to lower the melting point of NaCl(s). Liquid sodium metal forms at the steel cathode and $Cl_2(g)$ forms at the graphite anode. The chlorine and sodium are kept apart by a steel gauze diaphragm.

Production and Uses of the Alkali Metals

Lithium and sodium are produced from their molten chlorides by electroly sis, Figure 21-3. The electrolysis of NaCl(l), for example, is carried out a about 600 °C.

$$2\,NaCl(l) \xrightarrow{\text{electrolysis}} 2\,Na(l) + Cl_2(g) \qquad (21.$$

The melting point of NaCl is 801 °C, which is too high a temperature to carr this electrolysis economically. Adding $CaCl_2$ to the mixture reduces the meltin point. (Calcium metal, also produced in the electrolysis, precipitates out fro the Na(l) as the liquid metal is cooled. The final product is 99.95% Na.)

Potassium metal is produced by the reduction of molten KCl by liqui sodium.

$$KCl(l) + Na(l) \xrightarrow{850\,°C} NaCl(l) + K(g) \qquad (21.$$

Reaction (21.4) is reversible; at low temperatures, most of the KCl(l) remai unreacted. At 850 °C, however, the equilibrium is displaced far to the right a K(g) escapes from the molten mixture (an application of Le Châtelier's princ ple). The K(g) is freed of any Na(g) present by condensation of the vapor, fo lowed by fractional distillation of the liquid metals. Rb and Cs can be produce in much the same way, with Ca metal as the reducing agent.

Because sodium metal is so easily oxidized, its most important use is as reducing agent—for example, in obtaining such metals as titanium, zirco nium, and hafnium.

For example, titanium metal can be obtained from the reduction of $TiCl_4$ by a, as shown below.

$$TiCl_4 + 4\,Na \xrightarrow{\Delta} Ti + 4\,NaCl$$

◀ We introduced the symbol $\xrightarrow{\Delta}$ in Chapter 4 (page 115) to indicate that a reaction is carried out at an elevated temperature.

odium is also used as a heat-transfer medium in nuclear reactors. Liquid dium is especially good for this purpose because it has a low melting point, a gh boiling point, and a low vapor pressure. Also, it has better thermal con-uctivity and a higher specific heat than do most liquid metals. Finally, its low ensity and low viscosity make it easy to pump. Sodium is also used in sodium apor lamps, which are very popular for outdoor lighting. Because each lamp ses only a few milligrams of Na, however, the total quantity consumed in this pplication is rather small.

Lithium metal is used as an alloying agent to make high-strength, low-ensity alloys with aluminum and with magnesium. These alloys are used in e aerospace and aircraft industries. The use of lithium as an anode material batteries is on the increase, in part because of its ease of oxidation and in art because a small mass of lithium produces a large number of electrons. nly 6.94 g Li (1 mol) needs to be consumed to produce one mole of elec-ons. Lithium batteries are particularly useful where the installed battery ust have high reliability and a long lifetime, as in cardiac pacemakers. An -ray photograph of a pacemaker is shown in the margin.

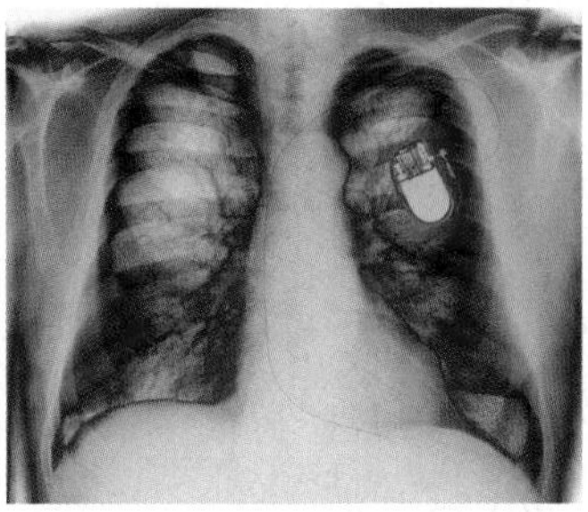

Spl/ Science Source

▲ An X-ray photograph showing a heart pacemaker powered by a lithium battery.

21-1 ARE YOU WONDERING?

Why is lithium the most easily oxidized of the group 1 elements on the basis of *E*° values but not on the basis of ionization energies?

Being the smallest of the alkali metal atoms, Li has the highest first ionization energy. It is the most difficult to oxidize in the reaction $M(g) \longrightarrow M^+(g) + e^-$. However, oxidation to produce $M^+(aq)$ is a different matter. We can think of it as the overall result of a hypothetical three-step process.

Sublimation:	$M(s)$	$\longrightarrow$	$\cancel{M(g)}$
Ionization:	$\cancel{M(g)}$	$\longrightarrow$	$\cancel{M^+(g)} + e^-$
Hydration:	$\cancel{M^+(g)}$	$\longrightarrow$	$M^+(aq)$
Overall:	$M(s)$	$\longrightarrow$	$M^+(aq) + e^-$

Thus, to compare tendencies to form $M^+(aq)$ by oxidation of the metals, we must compare tendencies in each of these three steps. The data given in Exercise 60 reveal that Li^+ has an unusually large hydration energy—enough to make $Li^+(aq)$ especially hard to reduce and Li(s) especially easy to oxidize. The large hydration energy is a consequence of the small size of the Li^+ ion, which allows a close approach to surrounding water molecules and strong attractive ion–dipole forces between them, as shown in Figure 21-4.

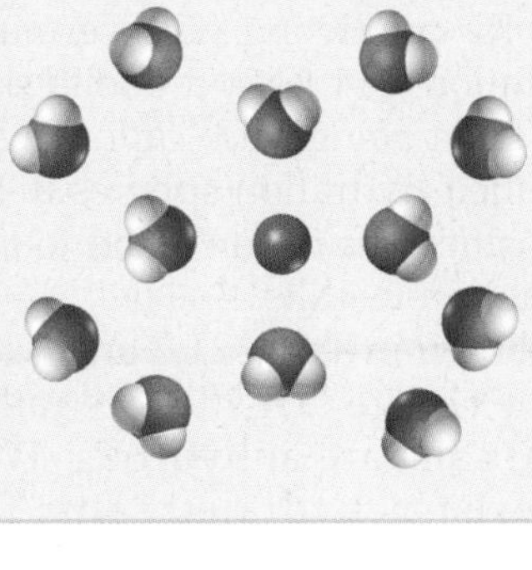

◀ FIGURE 21-4
Hydration of a Li^+ ion
Electrostatic forces hold a small number of H_2O molecules around a Li^+ ion in a primary hydration sphere. These molecules, in turn, hold other molecules, but more weakly, in a secondary hydration sphere.

KEEP IN MIND

that the more negative the value of $E°$, the more difficult is the reduction but the more easily does the reverse process—the oxidation half-reaction—occur. The value of $E°_{Li^+/Li}$, −3.040 V, comes at the very bottom of a listing of electrode potentials (recall Table 19.1), making Li(s) the easiest substance to oxidize.

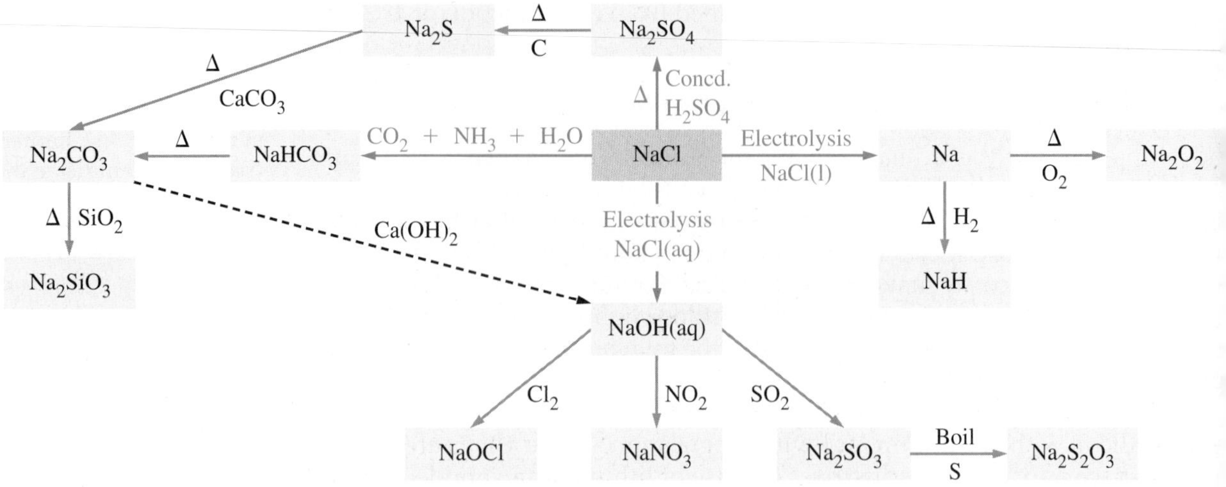

▲ FIGURE 21-5
Preparation and reactions of sodium compounds
This is a simple and common method of summarizing important reactions, starting from a compound of central importance. Some of these reactions are described in this section. A number of these compounds may be prepared by alternative methods. The conversion of Na_2CO_3 to NaOH (dashed arrow) is no longer of commercial importance

Group 1 Compounds

Group 1 metals, Li–Fr, have a valence configuration of ns^1 and occur exclusively in the +1 oxidation state. Most of the compounds of group 1 metals ar stable, ionic solids. When studying the chemistry of the elements, we are faced with a large amount of information that is challenging to organize. Figure 21- introduces a useful diagrammatic format for summarizing reaction chemistr Although this diagram deals with sodium compounds, some of which are discussed in this section, similar diagrams can be constructed for other compounds. Such diagrams note a compound of central importance (NaCl i Figure 21-5) and show how several other compounds can be obtained from i Some of these conversions occur in one step, such as the reaction of NaCl wit H_2SO_4 to form Na_2SO_4 (discussed on page 990). Other conversions involv two or more consecutive reactions, such as the preparation of sodium carbonate, Na_2CO_3. Example 21-1 illustrates the multistep synthesis of sodium carbonate. The principal reactants required for the conversions are written nea the connecting arrows. If a reaction mixture must be heated, a Δ symbol is use to indicate this requirement. The reactions may also produce by-products tha are not noted in the diagram.

Hydration of Salts When salts are dissolved in water, the cations are hydrated as suggested by Figure 21-4. The anions are similarly hydrated but with th slightly positive hydrogen atoms of the water molecules directed toward th anion. When a salt crystallizes from an aqueous solution, the salt that is obtaine may or may not contain water molecules—called water of crystallization—a part of the solid structure. No simple rule exists for predicting with certaint whether the ions will retain all or part of their hydration spheres in the solid stat because a number of factors must be considered. That being said, cations wit high charge densities tend to retain all or part of their hydration spheres in th solid state. When the cations have low charge densities, the cations tend to los their hydration spheres; thus, they tend to form anhydrous salts.

The charge densities of the alkali metals are shown in Table 21.2 and, wit the exception of lithium and perhaps sodium, the charge densities are rathe low; consequently, the majority of alkali metal salts are anhydrous. Wit lithium and sodium, salts are most likely to exist as hydrated salts. Th

EXAMPLE 21-1 Writing Chemical Equations from a Summary Diagram of Reaction Chemistry

Use the information provided in Figure 21-5 to write balanced chemical equations for the reactions involved in synthesizing sodium carbonate from sodium chloride.

Analyze

Consult Figure 21-5 to find a route from NaCl to Na_2CO_3. One possible route involves the conversions $NaCl \longrightarrow Na_2SO_4 \longrightarrow Na_2S \longrightarrow Na_2CO_3$. For each conversion, the other necessary reactants (H_2SO_4, C, $CaCO_3$) are noted. Use this information to write balanced chemical equations for each conversion, keeping in mind that other reaction products (such as HCl, CO and CaS) must be included.

Solve

First, $Na_2SO_4(s)$ is produced from NaCl(s) and concentrated sulfuric acid.

$$2\,NaCl(s) + H_2SO_4(concd\ aq) \xrightarrow{\Delta} Na_2SO_4(s) + 2\,HCl(g)$$

Next, the Na_2SO_4 is reduced to Na_2S with carbon.

$$Na_2SO_4(s) + 4\,C(s) \xrightarrow{\Delta} Na_2S(s) + 4\,CO(g)$$

The final step is a reaction between Na_2S and $CaCO_3$.

$$Na_2S(s) + CaCO_3(s) \xrightarrow{\Delta} CaS(s) + Na_2CO_3(s)$$

Assess

There are three conversions and thus three separate chemical equations. Each chemical equation is balanced and produces one of the sodium compounds on the route chosen.

PRACTICE EXAMPLE A: Write chemical equations for the reactions involved in synthesizing sodium nitrate from sodium chloride.

PRACTICE EXAMPLE B: Write chemical equations for the reactions involved in synthesizing sodium thiosulfate from sodium chloride.

mpirical formulas of the alkali metal perchlorates illustrate nicely the reater tendency of Li and Na to form hydrated salts: $LiClO_4 \cdot 3\,H_2O$, $aClO_4 \cdot H_2O$, $KClO_4$, $RbClO_4$, $CsClO_4$.

alides All alkali metals react vigorously, sometimes explosively, with halo-ens to produce ionic halides, the most important of which are NaCl and KCl. odium chloride—salt—is the most used of all minerals for the production of nemicals. It is not listed among the top chemicals, however, because it is con-dered a raw material, not a manufactured chemical. Large quantities of NaCl an be obtained by evaporation of seawater, as shown in the photograph. nnual use of NaCl in the United States amounts to about 50 million metric ons. Salt is used to preserve meat and fish, control ice on roads, and regener-te water softeners. In the chemical industry, NaCl is a source of many chemi-als, including sodium metal, chlorine gas, hydrochloric acid, and sodium ydroxide.

Potassium chloride, KCl, is obtained from naturally occurring brines (con-entrated solutions of salts). It is most extensively used in plant fertilizers ecause potassium is a major essential element for plant growth. Potassium

◀ Sea salt (sodium chloride) stacks that have been harvested by evaporation of seawater.

chloride is also used as a raw material in the manufacture of KOH, KNO_3, an other industrially important potassium compounds.

Alkali Metal Hydrides When an alkali metal (M) is heated in the presence c hydrogen gas, an ionic hydride is formed:

$$2\,M(s) + H_2(g) \xrightarrow{\Delta} 2\,MH(s)$$

Alkali metal hydrides contain the hydride ion, H^-, and have the sodium chlc ride structure. All the alkali metal hydrides are very reactive. For exampl they react readily with water to give a hydroxide salt and hydrogen gas:

$$MH(s) + H_2O(l) \longrightarrow MOH(aq) + H_2(g)$$

Alkali metal hydrides also react with metal halides. An important example is th reaction of lithium hydride and aluminum chloride, $AlCl_3$, which produce lithium aluminum hydride, $LiAlH_4$, and lithium chloride. Lithium aluminur hydride (LAH) is a powerful reducing agent used in organic chemistr $LiAlH_4$ can be made using the following reaction:

$$4\,LiH + AlCl_3 \xrightarrow{(C_2H_5)_2O} LiAlH_4 + 3\,LiCl$$

The reaction is carried out by adding finely divided LiH(s) to a solution c $AlCl_3$ in a nonaqueous solvent, such as diethyl ether, $(C_2H_5)_2O$. A nonaque ous solvent is used because both $LiAlH_4$ and LiH react vigorously with wate $LiAlH_4$ is obtained as a white solid by careful and controlled evaporation c the solvent.

Oxides and Hydroxides The alkali metals react rapidly with oxygen to produc several different ionic oxides. Under appropriate conditions—generally by car fully controlling the supply of oxygen—the oxide M_2O can be prepared for eac of the alkali metals. Lithium reacts with excess oxygen to give Li_2O and a sma amount of lithium *peroxide*, Li_2O_2. Sodium reacts with excess oxygen to giv mostly the peroxide Na_2O_2 and a small amount of Na_2O. Potassium, rubidiun and cesium react to form the *superoxides*, MO_2. The oxides, peroxides, an superoxides are ionic compounds. The Lewis symbol for the O^{2-} ion and th Lewis structures for the O_2^{2-} and O_2^- ions are shown below:

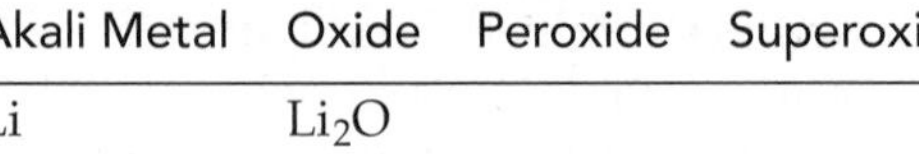

Oxide ion   Peroxide ion   Superoxide ion

The principal products of the reactions of the alkali metals with excess oxyge are summarized in the following table.

Principal Combustion Product			
Akali Metal	Oxide	Peroxide	Superoxide
Li	Li_2O		
Na		Na_2O_2	
K			KO_2
Rb			RbO_2
Cs			CsO_2

▶ These are not the only oxides formed by the alkali metals. About nine different cesium oxides are known.

Notice that the principal combustion product shifts from the oxide (M_2O) t the superoxide (MO_2) as we move down the group from Li to Cs. If we assum that the principal combustion product is the compound that is *most stable wit respect to the starting materials,* then we conclude that $Li_2O(s)$ is more stabl than $Li_2O_2(s)$ or $LiO_2(s)$, but for the larger alkali metal ions, $MO_2(s)$ is mor stable than $M_2O(s)$ or $M_2O_2(s)$. In Figure 21-6, we use $\Delta_f G°$ values to compar the stabilities of the various oxides, relative to the elements M(s) and O_2(g

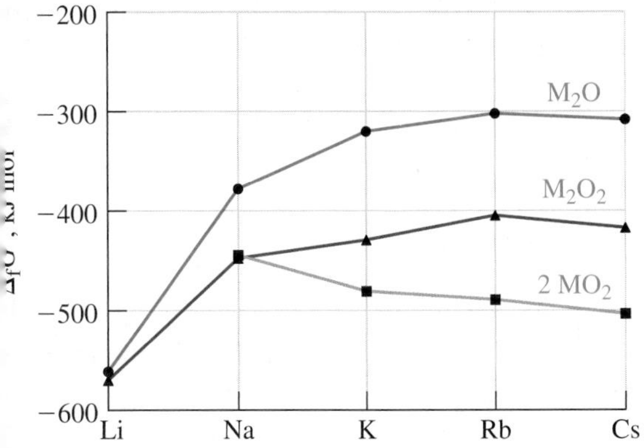

◀ FIGURE 21-6
Relative stabilities of M_2O, M_2O_2, and $2MO_2$ for the alkali metals
The values of $\Delta_f G°$ at 298 K for forming $M_2O(s)$,$M_2O_2(s)$, and 2 $MO_2(s)$ from M(s) and $O_2(g)$ are plotted for the alkali metals. For K, Rb, and Cs, the superoxides are more stable relative to M(s) and $O_2(g)$ than are the peroxides and monoxides. The $\Delta_f G°$ values for $Li_2O(s)$ and $Li_2O_2(s)$ are very similar at 298 K. At higher temperatures, $\Delta_f G°$ for $Li_2O(s)$ is more negative than that of $Li_2O_2(s)$.

or K, Rb, and Cs, the superoxide is most stable (lowest in energy) and for Li nd Na, the oxides (M_2O) and peroxides (M_2O_2) are most stable.

Perhaps it comes as a surprise that the principal combustion product in iese reactions is MO_2 (or M_2O_2) and not M_2O. Because the O^{2-} ion has a omplete octet, it is tempting to jump to the conclusion that O^{2-} is a very stale ion. However, stability can be measured only relative to some set of refernce conditions. Compared with oxygen molecules, the O^{2-} ion is not very able. The formation of the O^{2-} ion from O_2 is a very endothermic process ecause the O=O bond must be broken and two electrons must be added to n O atom. The formation of one mole of O^{2-} ions from O_2 molecules requires 52 kJ of energy, as shown below:

▲ The data in Figure 21-6 show that the $\Delta_f G°$ values for $Li_2O(s)$ and $Li_2O_2(s)$ are almost the same at 298 K. In a combustion reaction, heat is not dissipated immediately, and the system will be heated temporarily to a very high temperature. At high temperatures, $\Delta_f G°$ for $Li_2O(s)$ is more negative than it is for $Li_2O_2(s)$—see Exercise 59—and the formation of $Li_2O(s)$ is thermodynamically favored.

$$\frac{1}{2}O_2(g) \longrightarrow O(g) \qquad \Delta_r H = \frac{1}{2}(498\ \text{kJ mol}^{-1})$$
$$O(g) + e^- \longrightarrow O^-(g) \qquad \Delta_r H = -141\ \text{kJ mol}^{-1}$$
$$O^-(g) + e^- \longrightarrow O^{2-}(g) \qquad \Delta_r H = +744\ \text{kJ mol}^{-1}$$
$$\frac{1}{2}O_2(g) + 2e^- \longrightarrow O^{2-}(g) \qquad \Delta_r H = 852\ \text{kJ mol}^{-1}$$

he formation of O_2^- or O_2^{2-} ions requires much less energy because the =O bond does not have to be broken. Because the energy requirement for orming O^{2-} ions is so large, it is somewhat surprising that $M_2O(s)$ is formed t all.

The energy consumed in the formation of M^+ and O^{2-} ions is offset by the nergy released when M^+ and O^{2-} ions combine to give $M_2O(s)$, as shown elow.

$$2\,M^+(g) + O^{2-}(g) \longrightarrow M_2O(s) \qquad \Delta_r H < 0$$

he enthalpy change for the process above is the lattice energy (see Chapter 12). he formation of $Li_2O(s)$ starting from Li(s) and $O_2(s)$ is very favorable ecause the lattice energy of Li_2O is very large (very negative). The lattice nergy of Li_2O is large because the Li^+ ion is small and can pack around the $)^{2-}$ ions very efficiently. The lattice energies of Na_2O, K_2O, Rb_2O, and Cs_2O re much smaller (less negative) than that of Li_2O, and so the formation of the A_2O lattice is much less favorable. The larger alkali metal ions pack more efficiently around the larger anions (O_2^{2-} or O_2^-), and thus the heavier alkali netals react with excess oxygen to give either M_2O_2 or MO_2.

The combustion of Li(s) in excess oxygen produces some $Li_2O_2(s)$ as a ninor product, and this suggests that $Li_2O_2(s)$ is a reasonably stable compound. Solid Li_2O_2 decomposes to Li_2O, as shown below, when heated to bout 300 °C:

$$Li_2O_2(s) \xrightarrow{\Delta} Li_2O(s) + \frac{1}{2}O_2(g)$$

The other alkali metal peroxides must be heated to higher temperature (>500 °C) before decomposition occurs. The decomposition of $Li_2O_2(s)$ t $Li_2O(s)$ is thermodynamically favorable because of large differences in the la tice energies of the two solids, but the decomposition may also be assiste kinetically, as suggested below:

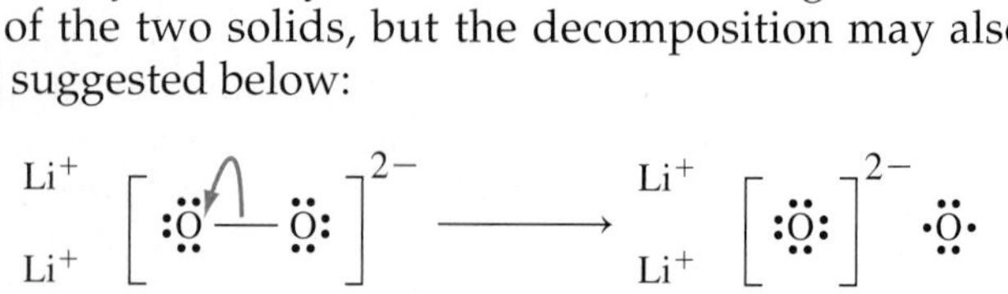

Because of the high polarizing power of the Li^+ ion, electrons in the O_2^{2-} bon are dragged from the bond to one of the oxygen atoms, yielding an O^{2-} ion an an oxygen atom. The oxygen atom combines with another oxygen atom to forı an O_2 molecule.

The peroxides have many important uses. For example, sodium peroxide i used as a bleaching agent and a powerful oxidant. Lithium peroxide c sodium peroxide are sometimes used in emergency breathing devices in sub marines and spacecraft because these compounds react with carbon dioxide t produce oxygen, as shown below:

$$2\,M_2O_2(s) + 2\,CO_2(s) \longrightarrow 2\,M_2CO_3(s) + O_2(g) \qquad (M = Li, Na) \qquad \textbf{(21.5}$$

Potassium superoxide, KO_2, can also be used for this purpose. The oxide peroxides, and superoxides of the alkali metals react with water to form basi solutions. The reaction of an alkali metal oxide with water is an acid–bas reaction that produces the alkali metal hydroxide.

The chemical equation and net ionic equation for the reaction are give below:

$$M_2O(s) + H_2O(l) \longrightarrow 2\,MOH(aq)$$

$$O^{2-}(aq) + H_2O(l) \longrightarrow 2\,OH^-(aq)$$

The resulting solution is quite basic because one mole of the oxide produce two moles of hydroxide ions.

The peroxide ion reacts with water in a similar manner to produce hydrox ide ion and hydrogen peroxide.

$$O_2^{2-}(aq) + 2\,H_2O(l) \longrightarrow 2\,OH^-(aq) + H_2O_2(aq)$$

Hydrogen peroxide then slowly disproportionates into water and oxyge (page 175).

The superoxide ion reacts with water to give hydroxide ions, hydrogen pe oxide, and oxygen.

$$2\,O_2^-(aq) + 2\,H_2O(l) \longrightarrow 2\,OH^-(aq) + H_2O_2(aq) + O_2(g)$$

The hydroxides of the group 1 metals are strong bases because they dissociat to release hydroxide ions in aqueous solution. As we learned in Section 19- sodium hydroxide is produced commercially by the electrolysis of NaCl(aq $Na^+(aq)$ goes through the electrolysis unchanged; $Cl^-(aq)$ is oxidized t $Cl_2(g)$; and H_2O is reduced to $H_2(g)$. Potassium hydroxide and lithiun hydroxide are made in a similar fashion. Alkali metal hydroxides can also b prepared by the reaction of the group 1 metals with water (equation 21.2 Alkali hydroxides are important in the manufacture of soaps and detergent described later in this section.

Carbonates and Sulfates Except for Li_2CO_3, all the alkali metal carbonate (M_2CO_3) are soluble in water and can be heated to very high temperature (>800 °C) before decomposing to $M_2O(s)$ and $CO_2(g)$. The lower aqueou solubility of Li_2CO_3 and its lower stability with respect to the oxide are specifi xamples of a more general observation: *compounds derived from the first member a group are often markedly different from compounds derived from the other members the group.* Typically, the second member of the group exhibits chemistry ost representative of the group. No single explanation accounts for the fact at the first member is not representative of the group. However, in many ses, the distinctive chemistry of the first member of the group can be attrib-ted to the small atomic size, inability to expand the valence shell and, espe-ally for the *p*-block elements, extensive π-bonding.

Lithium carbonate is used in the treatment of individuals who have bipolar isorder. Daily dosages of 1–2 g Li_2CO_3 maintain a level of Li^+ in the blood of out one millimole per liter. This treatment apparently influences the balance Na^+ and K^+, that of Mg^{2+} and Ca^{2+}, or both, across cell membranes.

Sodium carbonate (soda ash) is used primarily in the manufacture of glass. he Na_2CO_3 produced in the United States now comes mostly from natural ources, such as the mineral *trona*, $Na_2CO_3 \cdot NaHCO_3 \cdot n\,H_2O$, found in dry kes in California and in immense deposits in western Wyoming. In the past, odium carbonate was manufactured mostly from NaCl, $CaCO_3$, and NH_3, sing a process introduced by the Belgian chemist Ernest Solvay in 1863.

◀ The term *ash* signifies a product obtained by heating or burning. The earliest alkaline materials (notably, potassium carbonate, *potash*) were extracted from the ashes of burned plants. The first commercial source of *soda ash* was the Leblanc process, featured in Example 21-1.

The great success of the *Solvay process* over the synthetic method outlined in xample 21-1 lies in the efficient use of certain raw materials through recycling. n outline of the process is shown in Figure 21-7. The key step involves the reac-on of $NH_3(g)$ and $CO_2(g)$ in saturated NaCl(aq). Of the possible ionic com-ounds that could precipitate from such a mixture (NaCl, NH_4Cl, $NaHCO_3$, d NH_4HCO_3), the least soluble is *sodium hydrogen carbonate* (sodium bicarbon-te). The following equation gives a simplified description of the process.

▲ The mineral *trona*, from Green River, Wyoming, is currently the principal source of Na_2CO_3 in the United States.

John Cancalosi/National Geographic Society/Corbis

$$Na^+(aq) + Cl^-(aq) + NH_3(g) + CO_2(g) + H_2O(l) \longrightarrow NaHCO_3(s) + NH_4^+(aq) + Cl^-(aq) \quad (21.6)$$

n industry, the reaction above is actually carried out in two stages. In the first age, ammonia is bubbled into a concentrated brine (NaCl) solution, and in

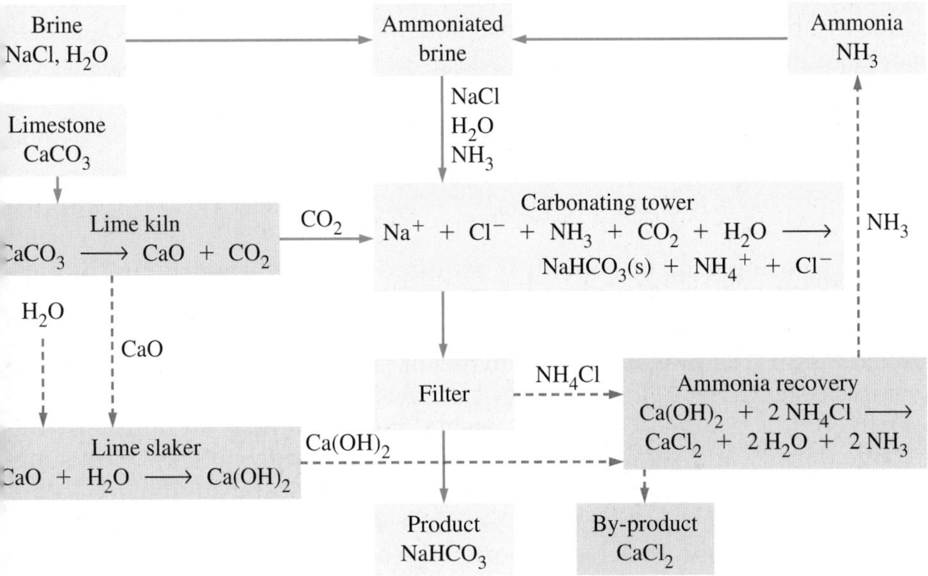

▲ FIGURE 21-7

he Solvay process for the manufacture of $NaHCO_3$

he main reaction sequence is traced by solid arrows. Recycling reactions are shown by ashed arrows.

the second stage, CO_2 is bubbled through the ammoniated brine. The sodiu
bicarbonate is isolated and sold, or it is converted to sodium carbonate b
heating, as shown below in reaction (21.7).

$$2\,NaHCO_3(s) \xrightarrow{\Delta} Na_2CO_3(s) + H_2O(g) + CO_2(g) \qquad \textbf{(21.}$$

One noteworthy feature of the Solvay process is that it involves only simp precipitation and acid–base reactions. Another feature is that materials pr duced in one step are efficiently recycled into a subsequent step. Recyclin makes good economic sense: A process that recycles materials minimizes th use of raw materials (an expense in purchasing) and cuts down on the pr duction of by-products (an expense in disposal). Thus, when limeston ($CaCO_3$) is heated to produce the reactant CO_2, the other reaction produc CaO, is also used. It is converted to $Ca(OH)_2$, which is then used to conve NH_4Cl (another reaction by-product) to $NH_3(g)$. The $NH_3(g)$ is recycled int the production of ammoniated brine.

The Solvay process has ultimately only one by-product—$CaCl_2$—fc which the demand is very limited. In the past, some $CaCl_2$ was used for dei ing roads in the winter (page 672) and for dust control on dirt roads in th summer (by means of the deliquescence of $CaCl_2 \cdot 6\,H_2O$, page 666). The bul of the $CaCl_2$, however, was dumped into local lakes and streams (notabl Onondaga Lake, near Solvay, New York), resulting in extensive environmer tal contamination. Environmental regulations no longer permit such dump ing. Partly because of these regulations, but mainly for economic reason natural sources of sodium carbonate have supplanted the Solvay process i the United States. The process is still widely used elsewhere in the worl however.

Sodium sulfate, Na_2SO_4, is obtained partly from natural sources, partly fro neutralization reactions, and partly through a process discovered by Johan Rudolf Glauber in 1625. Glauber's process is based on the following reaction

▶ Equations (21.8) are often replaced by the overall equation, $H_2SO_4(\text{concd aq}) + 2\,NaCl(s) \longrightarrow Na_2SO_4(s) + 2\,HCl(g)$

$$\begin{aligned} H_2SO_4(\text{concd aq}) + NaCl(s) &\xrightarrow{\Delta} NaHSO_4(s) + HCl(g) \\ NaHSO_4(s) + NaCl(s) &\xrightarrow{\Delta} Na_2SO_4(s) + HCl(g) \end{aligned} \qquad \textbf{(21.8}$$

The strategy behind reactions (21.8) is the production of a *volatile* acid (HCl) b heating one of its salts (NaCl) with a *nonvolatile* acid (H_2SO_4). Several othe acids can be produced by similar reactions. The major use of Na_2SO_4 is in th paper industry. For instance, in the kraft process for papermaking, undesi able lignin is removed from wood by digesting the wood in an alkaline solu tion of Na_2S. The Na_2S required for this step is produced by the reduction c Na_2SO_4 with carbon.

$$Na_2SO_4(s) + 4\,C(s) \xrightarrow{\Delta} Na_2S(s) + 4\,CO(g)$$

About 45 kilograms of Na_2SO_4 is required for every metric ton of pape produced.

Alkali Metal Complexes Most alkali metals show little or no tendency to forn complexes with neutral Lewis bases, such as NH_3, because most of them hav relatively low charge densities. However, the Li^+ ion has an unusually hig charge density and shows a greater tendency to form complexes with simpl Lewis bases. For example, in the presence of ammonia, the lithium ion forms tetrahedral complex, $[Li(NH_3)_4]^+$, which is shown in the margin.

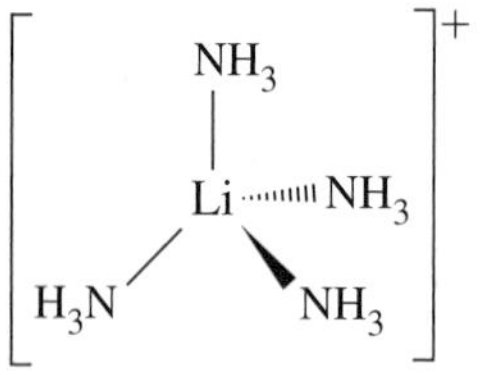

▲ The $[Li(NH_3)_4]^+$ complex

In 1967, Charles J. Pedersen reported the discovery of a type of Lewis bas that could coordinate the "recalcitrant alkali metal ions," a quote from hi 1987 Nobel lecture. The structure of the Lewis base is that of a ring of oxyge atoms connected by $-CH_2CH_2-$ units, as shown in Figure 21-8(a). Th

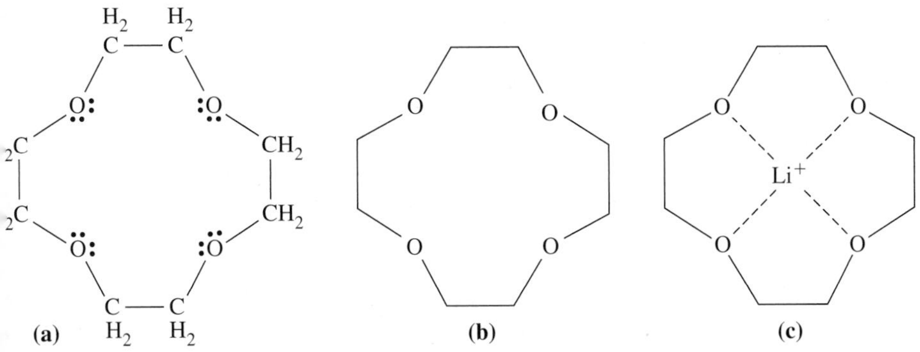

FIGURE 21-8
2-crown-4 ether and the $[Li(12\text{-crown-}4)]^+$ complex
) The 12-crown-4 ether consists of a ring of 12 atoms, of which 4 are oxygen atoms. **)** A simplified illustration of 12-crown-4. **(c)** In the $[Li(12\text{-crown-}4)]^+$ complex, lectron density is donated from the oxygen atoms to the Li^+ ion.

ng structures were called *crown ethers* by Pedersen because they resemble a onarch's crown. Crown ethers are given a name that reflects its structure in very intuitive way. For example, the crown ether shown in Figure 21-8 is alled 12-crown-4 because the ring consists of 12 carbon and oxygen atoms, nd of those 12 ring atoms, 4 are oxygen. A line-angle formula of 12-crown-4 shown in Figure 21-8(b). When a crown ether forms a complex with a etal ion, it encapsulates the metal ion, as shown in Figure 21-8(c). he metal ion is held in place by the donation of electron density from the xygen atoms.

The cavity size of a crown ether depends on the number of $-OCH_2CH_2-$ units in the ring. This is illustrated in Figure 21-9. Cavity size one of the factors that determine which cations will bind to a particular rown ether. The diameter of a K^+ ion is about 304 pm, and thus a K^+ ion fits icely in the cavity of the 18-crown-6 ether. Thus, the K^+ ion binds preferenally to 18-crown-6, whereas Li^+ and Na^+ bind less effectively. A chemical quation representing the binding of a metal cation and 18-crown-6 is given elow:

$$M^+ + 18\text{-crown-}6 \rightleftharpoons [M(18\text{-crown-}6)]^+$$

Equilibrium constants for the reaction of the alkali metal ions with 8-crown-6 are plotted in Figure 21-10. The figure shows that the equilibrium onstant is greatest when the cation is K^+, partly because K^+ fits best into the avity of 18-crown-6. However, the values of the equilibrium constants are not ifferent enough to make 18-crown-6 selective for just K^+; the other alkali etal ions also form complexes with 18-crown-6 to varying degrees.

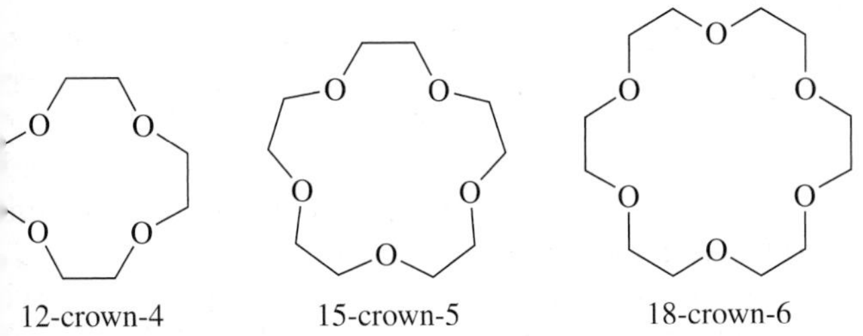

FIGURE 21-9
hree different crown ethers
he cavity size of a crown ether increases with the number of atoms in the ring.

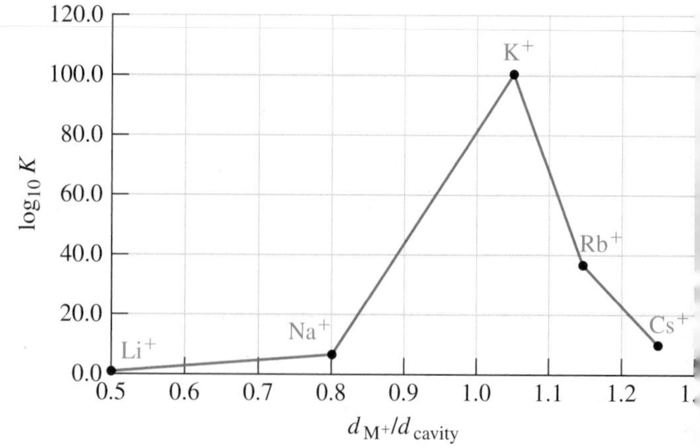

▶ FIGURE 21-10
Selectivity of 18-crown-6
The $\log_{10} K$ values for the reaction of alkali metal ions (M^+) with 18-crown-6 in water at 25 °C are plotted against the ratio of diameters, d_{M^+}/d_{cavity}. The equilibrium constant is largest for K^+ because the radius of this cation most closely matches that of the cavity.

The formation of crown ether complexes has been exploited in syntheti organic chemistry to dissolve ionic reagents in nonpolar solvents, such as ben zene. For example, $KMnO_4$ is not soluble in pure benzene but [K(18-crown-6) [MnO_4] is soluble. The $[K(18\text{-crown-}6)]^+$ complex is soluble in nonpolar sol vents because most of the complex is made up of a hydrocarbon skeletor which is compatible with other nonpolar molecules. The MnO_4^- ion remain in the vicinity of the complexed cation because of electrostatic attraction to th positive charge on the complex. However, the MnO_4^- ion does not interac strongly with the nonpolar benzene molecules, and thus it is hardly solvate by benzene molecules. Consequently, MnO_4^- is much more reactive in ben zene than it is in water.

▶ An emulsion is a suspension of one liquid in another (see Table 14.5). Soap is an emulsifier because it facilitates the formation of an emulsion of oil droplets in water.

Alkali Metal Detergents and Soaps A **detergent** is a cleansing agent use primarily because it can emulsify oils. Although the term *detergent* include common soaps, it is used primarily to describe certain synthetic products such as sodium lauryl sulfate, whose manufacture involves the followin conversions.

$$\underset{\text{Lauryl alcohol}}{CH_3(CH_2)_{10}CH_2OH} \longrightarrow \underset{\text{Lauryl hydrogen sulfate}}{CH_3(CH_2)_{10}CH_2OSO_3H} \longrightarrow \underset{\text{Sodium lauryl sulfate}}{CH_3(CH_2)_{10}CH_2OSO_3^-\ Na^+}$$

Notice that the lauryl sulfate anion has a long, nonpolar tail and a highl polar (negatively charged) head. These structural features are common to a detergents and soaps, as described in more detail below.

A **soap** is a specific kind of detergent that is the salt of a metal hydroxid and a fatty acid. An example is sodium palmitate, which is the product of th reaction of palmitic acid and NaOH.

$$\underset{\text{Palmitic acid}}{CH_3(CH_2)_{14}-\overset{\overset{\displaystyle O}{\|}}{C}-O-H} + Na^+ + OH^- \longrightarrow \underset{\substack{\text{Sodium palmitate}\\ \text{(a soap)}}}{CH_3(CH_2)_{14}-\overset{\overset{\displaystyle O}{\|}}{C}-O^-\ Na^+} + H_2O \qquad \textbf{(21.9}$$

Again, notice that the anion of the soap has a long, nonpolar tail and a pola head. Figure 21-11 illustrates the detergent action of sodium palmitate.

Sodium soaps are the familiar hard (bar) soaps. Potassium soaps have low melting points and are soft soaps. Lithium stearate ($C_{17}H_{35}CO_2Li$), a soap not seen in ordinary household use, is used to thicken some oils into greases. These greases have excellent water-repellent and lubricating prop erties at both high and low temperatures. The greases remain in contac with moving metal parts under conditions in which oil by itself woulc run off.

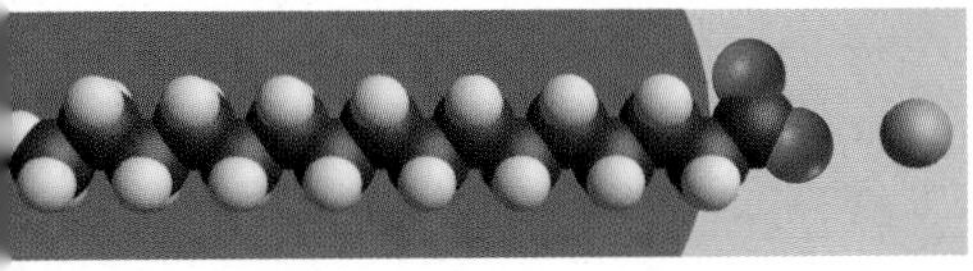

(a)

(b)

FIGURE 21-11
llustration of soap molecules and their leaning action
a) The sodium palmitate molecule has a long onpolar portion buried in a droplet of oil nd a polar head projecting into an aqueous edium. **(b)** Electrostatic attractions between hese polar heads and water molecules cause he droplet to be emulsified, or solubilized.

21-2 CONCEPT ASSESSMENT

The nitrates MNO_3 (M = Na, K, Rb, Cs) decompose to the nitrites (MNO_2) on heating; while in contrast, $LiNO_3$ decomposes to Li_2O. Suggest a reason for this difference in behavior. Write balanced equations for both reactions.

21-3 Group 2: The Alkaline Earth Metals

As a group, the alkaline earth metals of group 2 are as common as the elements of group 1. Table 21.3 shows that calcium and magnesium are particularly abundant. Even beryllium, the least abundant member of group 2, is accessible because it occurs in deposits of the mineral *beryl*, $Be_3Al_2Si_6O_{18}$, which is shown in the margin. The other group 2 elements are found primarily as carbonates, sulfates, and silicates. Radium, like its group 1 neighbor, francium, is a radioactive element found only in trace amounts. Radium is more interesting for its radioactive properties than for its chemical similarities to the other group 2 elements.

TABLE 21.3 Abundances of Group 2 Elements

Elements	Abundances, ppm[a]	Rank
Be	2.8	48
Mg	23,300	7
Ca	41,500	5
Sr	370	15
Ba	425	14
Ra	Trace	—

[a]Grams per 1000 kg of solid crust.

Even though the group 2 metal oxides and hydroxides are only very slightly soluble in water, they are basic, or *alkaline*. At one time, insoluble substances that do not decompose on heating were called "earths." This term is the basis of the group 2 name: **alkaline earth metals**.

From a chemical standpoint (for example, in their abilities to react with water and acids and to form ionic compounds), the heavier group 2 metals—Ca, Sr, Ba, and Ra—are nearly as active as the group 1 metals. In terms of certain physical properties (for example, density, hardness, and melting point), all the group 2 elements are more typically metallic than the group 1 elements, as we can see by comparing Tables 21.2 and 21.4.

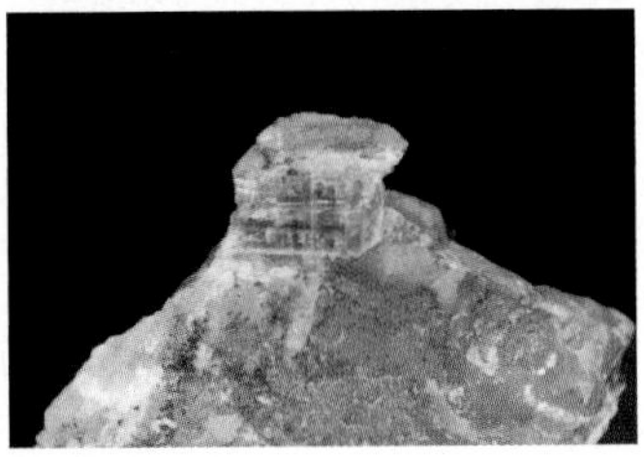

▲ An emerald crystal, which is based on the mineral *beryl*, embedded in a calcite matrix.

Table 21.4 shows that beryllium is out of step with the other group 2 elements in some of its physical properties. For instance, it has a higher melting point and is much harder than the others. Its chemical properties also differ significantly. For example,

- Be is quite unreactive with air and water;
- BeO does not react with water, whereas the other MO oxides form $M(OH)_2$;

TABLE 21.4 Some Properties of Group 2 (Alkaline Earth) Metals

	Be	Mg	Ca	Sr	Ba
Atomic number	4	12	20	38	56
Atomic (metallic) radius, pm	111	160	197	215	222
Ionic (M^{2+}) radius, pm[a]	41	86	114	132	149
Electronegativity	1.5	1.2	1.0	1.0	0.9
Charge density of M^{2+} ($C\ mm^{-3}$)	1108	120	52	33	23
First ionization energy, $kJ\ mol^{-1}$	899.4	737.7	589.7	549.5	502.8
Electrode potential $E°$, V[b]	−1.85	−2.356	−2.84	−2.89	−2.92
Melting point, °C	1278	648.8	839	769	729
Boiling point, °C	2970[c]	1090	1483.6	1383.9	1637
Density, $g\ cm^{-3}$ at 20 °C	1.85	1.74	1.55	2.54	3.60
Hardness[d]	~5	2.0	1.5	1.8	~2
Electrical conductivity[d]	39.7	35.6	40.6	6.90	3.20
Flame color	None	None	Orange-red	Scarlet	Green

[a]Ionic radii are for a coordination number of 6, except for Be^{2+}, for which the coordination number is 4.
[b]For the reduction $M^{2+}(aq) + 2\,e^- \longrightarrow M(s)$.
[c]Boiling point at 5 mmHg pressure.
[d]See footnotes c and d of Table 21.2.

- Be and BeO dissolve in strongly basic solutions to form the io $[Be(OH)_4]^{2-}$;
- $BeCl_2$ and BeF_2 in the molten state are poor conductors of electricity; the are covalent substances.

The unusual chemical behavior of beryllium is related to the high charg density of the beryllium cation. Beryllium shows only a limited tendency t form ionic compounds because the small Be^{2+} ion polarizes any nearby anior drawing electron density toward itself, creating a bond with significant cova lent character. Thus, compounds of beryllium show some properties that ar more typical of covalent solids. For example, while the other oxides of group are basic, BeO is an *amphoteric* oxide, reacting with both strong acids and base to yield complex ions:

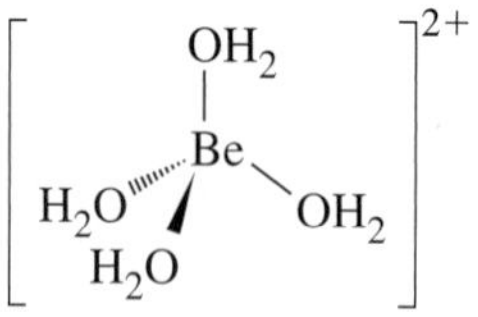

▲ Tetrahedral shape of the $[Be(OH_2)_4]^{2+}$ ion.

$$BeO(s) + H_2O(l) + 2\,H_3O^+(aq) \longrightarrow [Be(H_2O)_4]^{2+}(aq)$$

$$BeO(s) + H_2O(l) + 2\,OH^-(aq) \longrightarrow [Be(OH)_4]^{2-}(aq)$$

In these complex ions, electron density is donated from H_2O or OH^- to Be^{2} because of the high polarizing power of the Be^{2+} ion and the resulting com plex has a well-defined structure. The figure in the margin shows that th $[Be(H_2O)_4]^{2+}$ ion, for example, has a tetrahedral structure.

The reaction of beryllium metal with aqueous acids yields hydrogen an ionic compounds, such as $BeCl_2 \cdot 4\,H_2O$. In $BeCl_2 \cdot 4\,H_2O$, water molecules ar covalently bonded to Be^{2+} ions, producing the complex cations $[Be(OH_2)_4]^{2}$ that, together with the anions Cl^-, form the crystal lattice. In covalent com pounds, Be atoms appear to use hybrid orbitals—*sp* orbitals in $BeCl_2(g)$ an sp^3 orbitals in $BeCl_2(s)$ (Fig. 21-12).

21-3 CONCEPT ASSESSMENT

Why, on going from monomeric $BeCl_2$ to dimeric $(BeCl_2)_2$ to polymeric $(BeCl_2)_n$, does the atomic arrangement around the beryllium change from linear to trigonal planar to tetrahedral?

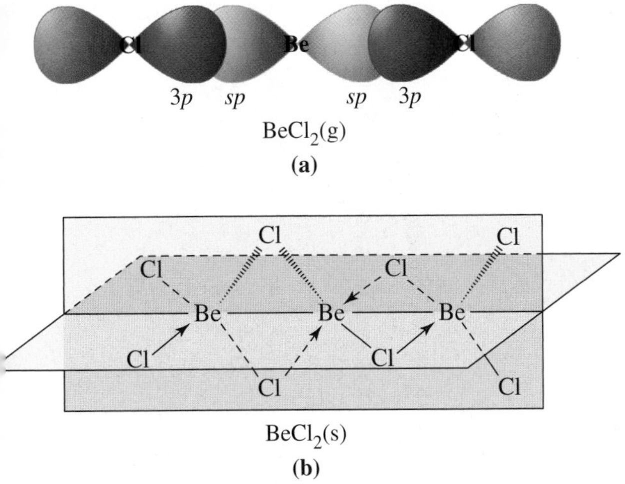

◀ FIGURE 21-12
Covalent bonds in $BeCl_2$
(a) In gaseous $BeCl_2$, discrete molecules exist with the bonding scheme shown. **(b)** In solid $BeCl_2$, two Cl atoms are bonded to one Be atom by normal covalent bonds. Two other Cl atoms are bonded by coordinate covalent bonds, using lone-pair electrons on the Cl atoms. Once formed, these two types of bonds are indistinguishable. $BeCl_2$ units are linked into long, chain-like polymeric molecules as $(BeCl_2)_n$.

roduction and Uses of the Alkaline Earth Metals

he preferred method of producing the group 2 metals (except Mg) is by educing their salts with other active metals. Beryl, $Be_3Al_2Si_6O_{18}$, is the natural source of beryllium compounds. This mineral is processed to produce eF_2, which is then reduced with Mg to give Be(s). Beryllium metal is used as n alloying agent when low density is a primary requirement. Because Be can vithstand metal fatigue, an alloy of copper with about 2% Be is used in prings, clips, and electrical contacts. The Be atom does not readily absorb -rays or neutrons, so beryllium is also used to make windows for X-ray tubes nd for various components in nuclear reactors. Beryllium and its compounds re limited in their use, however, because they are toxic. In addition, they are uspected of being carcinogens even at a level as low as 0.002 ppm in air.

Calcium, strontium, and barium are obtained by the reduction of their xides with aluminum; Ca and Sr also can be obtained by electrolysis of their nolten chlorides. Calcium metal is used primarily as a reducing agent to prepare, from their oxides or fluorides, other metals such as U, Pu, and most of he lanthanides. Strontium and barium have limited use in alloys, but some of heir compounds (discussed later) are quite important. Some salts of Sr and Ba rovide vivid colors for pyrotechnic displays.

Magnesium metal is obtained by electrolysis of the molten chloride in the *ow process*. The Dow process is outlined in Figure 21-13, and the electrolysis of

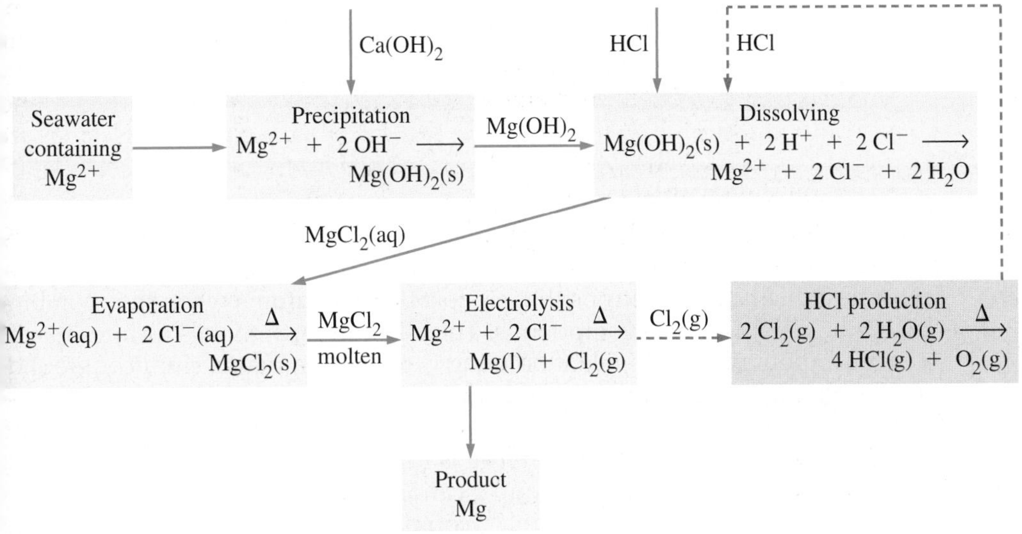

◀ FIGURE 21-13
The Dow process for the production of Mg
The main reaction sequence is traced by solid arrows. The recycling of $Cl_2(g)$ is shown by dashed arrows.

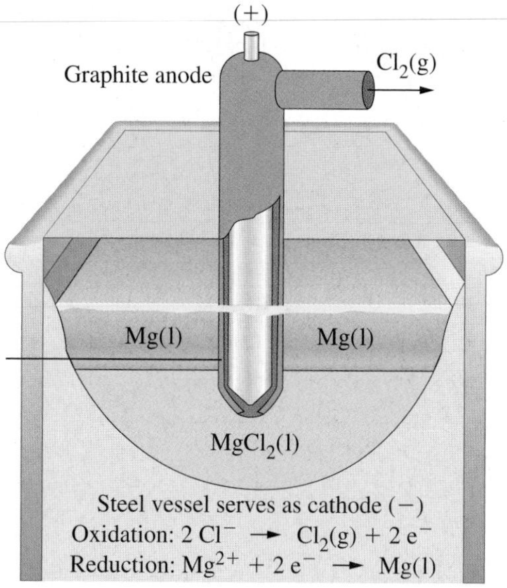

▶ FIGURE 21-14
The electrolysis of molten $MgCl_2$
The electrolyte is a mixture of molten NaCl, $CaCl_2$, and $MgCl_2$. This mixture has a lower melting point and higher electrical conductivity than $MgCl_2$ alone, but if the voltage is carefully controlled, only Mg^{2+} is reduced in the electrolysis.

▶ In Example 18-6, we demonstrated that this precipitation goes to completion.

$MgCl_2(l)$ is pictured in Figure 21-14. Like the Solvay process for makin $NaHCO_3$, the Dow process takes advantage of simple chemistry and recyclin

The source of magnesium is seawater or natural brines. The abundance c Mg^{2+} in seawater is about 1350 mg/L. The first step in the Dow process is th precipitation of $Mg(OH)_2(s)$ with slaked lime, $Ca(OH)_2$, as the source c OH^-. Slaked lime is formed by the reaction of quicklime (CaO) with wate The precipitated $Mg(OH)_2(s)$ is washed, filtered, and dissolved in HCl(aq The resulting concentrated $MgCl_2(aq)$ is dried by evaporation, melted, an electrolyzed, yielding pure Mg metal and $Cl_2(g)$. The $Cl_2(g)$ is converted t HCl, which is recycled.

Magnesium has a lower density than that of any other metal used for struc tural purposes. Lightweight objects, such as aircraft parts, are manufacture from magnesium alloyed with aluminum and other metals. Magnesium i a good reducing agent and is used in a number of metallurgical processe such as the production of beryllium, as mentioned earlier. The ease wit which magnesium is oxidized also underlies its use in sacrificial anodes fo corrosion protection (page 899). Magnesium's most spectacular use may be i firework, as it burns in air with a brilliant white light.

21-4 CONCEPT ASSESSMENT

The good reducing properties of magnesium are illustrated by the fact that the metal burns in an atmosphere of pure carbon dioxide. Write a plausible equation(s) for this reaction.

Group 2 Compounds

The alkaline earth metals always exist in the +2 oxidation state in their com pounds. Recall that atoms of the group 2 metals have an ns^2 valence configura tion and it is the ns^2 electrons that are lost by these atoms when they combin with nonmetals to form compounds. The alkaline earth metals form primaril ionic compounds, but covalent bonding is evident in magnesium compound and especially in beryllium compounds.

The properties of group 2 compounds differ from those of group 1 com pounds. In some cases, this difference is attributable to the smaller ionic siz and the larger ionic charge of group 2 cations. For example, the lattice energ of $Mg(OH)_2$ is about $-3000\ kJ\ mol^{-1}$, compared with about $-900\ kJ\ mol^{-1}$ fo

[N]aOH. This difference in lattice energy helps explain why NaOH is very soluble [i]n water—up to about 20 M NaOH(aq)—while $Mg(OH)_2$ is only sparingly [s]oluble, with $K_{sp} = 1.8 \times 10^{-11}$. The heavier group 2 hydroxides are some[w]hat more soluble than $Mg(OH)_2$. Other alkaline earth compounds that are [o]nly slightly soluble include the carbonates, fluorides, and oxides.

[H]alides The group 2 metals react directly with the halogens to form halides:

$$M(s) + X_2 \longrightarrow MX_2(s) \qquad (21.10)$$

(M = a group 2 metal, and X = F, Cl, Br, or I)

[H]owever, the standard method for preparing $MX_2(s)$ in anhydrous form is to [d]ehydrate the hydrates obtained from the reaction of the metal and aqueous [h]ydrohalic acid, HX(aq). This method of preparation cannot be used to pre[p]are beryllium halides because the hydrates of beryllium halides decompose [t]o $Be(OH)_2(s)$ on heating. For example, when $[Be(H_2O)_4]Cl_2(s)$ is heated, the [f]ollowing reaction occurs:

$$[Be(H_2O)_4]Cl_2(s) \xrightarrow{\Delta} Be(OH)_2(s) + 2\,H_2O(g) + 2\,HCl(g)$$

[I]nstead anhydrous $BeCl_2$ is prepared from BeO and CCl_4, as shown below.

$$2\,BeO + CCl_4 \xrightarrow{1070\text{ K}} 2\,BeCl_2 + CO_2$$

[T]he halides have varied uses. For example, $MgCl_2$ is used in the preparation [o]f magnesium metal, in fireproofing wood, in special cements, in ceramics, in [t]reating fabrics, and as a refrigeration brine.

[O]xides and Hydroxides All the group 2 metals burn in air to produce oxides, [M]O(s):

$$2\,M(s) + O_2(g) \longrightarrow 2\,MO(s)$$

[I]n the presence of excess oxygen, the heavier group 2 metals—such as barium—[f]orm peroxides:

$$Ba(s) + O_2(g) \longrightarrow BaO_2(s)$$

[T]he peroxides of Mg, Ca, and Sr are also known but they are less stable, presum[a]bly because the charge densities of Mg^{2+}, Ca^{2+}, and Sr^{2+} are high enough to [a]ssist the decomposition of $O_2^{\,2-}$, as suggested on page 988. On heating, the [p]eroxides decompose to give MO(s) and $O_2(g)$:

$$MO_2(s) \xrightarrow{\Delta} MO(s) + \frac{1}{2}O_2(g)$$

Although oxides of the group 2 metals can be obtained from burning the [m]etals in air or oxygen (or by thermal decomposition of the peroxides), the [c]hief method of preparing them, except for BeO(s), is the thermal decomposi[t]ion of the respective carbonate:

$$MCO_3(s) \xrightarrow{\Delta} MO(s) + CO_2(g) \quad (M = Mg, Ca, Sr, and Ba) \qquad (21.11)$$

[A] particularly useful oxide is calcium oxide, CaO(s), also known as **quicklime**. [I]t is used in water treatment, in the removal of $SO_2(g)$ from the smokestack [g]ases in electric power plants, and in the making of $Ca(OH)_2$—an important [a]nd inexpensive strong base. The conversion of CaO to $Ca(OH)_2$ is a specific [e]xample of the reaction that converts a group 2 oxide into a hydroxide:

$$MO(s) + H_2O(l) \longrightarrow M(OH)_2(aq) \qquad (21.12)$$

All the hydroxides of the group 2 metals are strong bases. Calcium hydrox[i]de, $Ca(OH)_2$, also known as **slaked lime**, is the cheapest commercial strong [b]ase. It is not very soluble in water, but it is used in a variety of applications, [s]uch as the Solvay and Dow processes.

A mixture of slaked lime, sand, and water composes the mortar used i bricklaying. Excess water in the mortar is absorbed by the bricks and then lo by evaporation. In the final setting of the mortar, $CO_2(g)$ from the air reac with $Ca(OH)_2(s)$ and converts it to $CaCO_3(s)$, as shown below:

$$Ca(OH)_2(s) + CO_2(g) \longrightarrow CaCO_3(s) + H_2O(g) \qquad \textbf{(21.1}$$

The final form of the mortar is a complex mixture of hydrated calciu carbonate and silicate (from the sand).

Reaction (21.13) is general for all group 2 hydroxides. Art conservators hav used this reaction to preserve art objects. For example, an aqueous solution o $Ba(NO_3)_2$ is sprayed onto cracking frescos (paintings embedded in plaster When the solution has had time to fill small cracks and spaces, an aqueou ammonia solution is applied to the surface of the fresco. The ammonia raise the pH of the solution, resulting in formation of $Ba(OH)_2$. As excess wate evaporates, carbon dioxide from the air reacts with the barium hydroxid Insoluble barium carbonate is produced, binding the cracking fresco togethe and strengthening it without affecting the delicate colors.

21-5 CONCEPT ASSESSMENT

Comment on the statement "The best way to way to prepare BeO is to heat the compound $BeCO_3(s)$." If you disagree with this statement suggest an alternative.

Hydration of Salts Another common characteristic of alkaline earth com pounds is the formation of hydrates. Typical hydrates are $MX_2 \cdot 6\,H_2O$, wher M = Mg, Ca, or Sr, and X = Cl or Br. The Ba^{2+} ion has a low charge density an shows little or no tendency to retain its hydration sphere in the solid state. Th formulas for the hydrates of the alkaline earth nitrates illustrate that the degre of hydration typically decreases as the charge density of the metal ion decrease $Mg(NO_3)_2 \cdot 6\,H_2O$, $Ca(NO_3)_2 \cdot 4\,H_2O$, $Sr(NO_3)_2 \cdot 4\,H_2O$, $Ba(NO_3)_2$.

Carbonates and Sulfates The group 2 carbonates are insoluble in water, a are the sulfates of Ca, Sr, and Ba. Because of this insolubility, these compound are the most important minerals of the group 2 metals. The most familia is $CaCO_3$, the principal component of the rock *limestone*. If limestone contain more than 5% $MgCO_3$, it is usually called dolomite limestone, or *dolomite* Some clay, sand, or quartz may also be present in limestone. The primary us of limestone (about 70%) is as a building stone. Among other applications limestone is used in the manufacture of quicklime and slaked lime, as a ingredient in glass, and as a flux in metallurgical processes. A *flux* is a materia that combines with impurities and removes them as a free-flowing liquid— *slag*—during production of a metal (page 1104).

Vladimir Kant/Shutterstock

▲ The decomposition (calcination) of limestone is carried out in a long rotary kiln, whether for the production of quicklime, CaO, or for the manufacture of Portland cement.

Portland cement, another important product of limestone, is a complex mix ture of calcium silicates and aluminates. It is produced in long rotary kilns lik the one shown in the photograph in the margin. In the kiln, mixtures of lime stone, clay, and sand are heated to progressively higher temperatures as the slowly move down the inclined kiln. First, moisture and then chemicall bound water are driven off. This process is followed by decomposition (calci nation) of the limestone to $CaO(s)$ and $CO_2(g)$. Finally, CaO combines wit silica (SiO_2) and alumina (Al_2O_3) from the sand and clay to form silicates an aluminates. Pure cement does not have much strength. When mixed wit sand, gravel, and water, however, it sets into the familiar rock-like mass calle *concrete*. Portland cement is an especially valuable material for building bridg piers and other underwater structures because it hardens even when under water—it is a *hydraulic* cement.

Pure, white $CaCO_3$ is used in a wide variety of products. For example, it is sed in papermaking to impart brightness, opacity, smoothness, and good ık-absorbing qualities to paper. It is particularly suited to newer papermak- ıg processes that produce acid-free (alkaline) paper with an expected shelf fe of 300 years or more. $CaCO_3$ is used as a filler in plastics, rubber, floor tiles, utties, and adhesives, as well as in foods and cosmetics. It is also used as an ntacid and as a dietary supplement for the prevention of osteoporosis, a con- ition in which the bones become porous and brittle and break easily.

Three steps are required to obtain pure $CaCO_3$ from limestone: (1) thermal ecomposition of limestone (called **calcination**), (2) reaction of CaO with vater (slaking), and (3) conversion of an aqueous suspension of $Ca(OH)_2(s)$ o precipitated $CaCO_3$ (carbonation).

Calcination: $$CaCO_3(s) \xrightarrow{\Delta} CaO(s) + CO_2(g) \quad \textbf{(21.14)}$$

Slaking: $$CaO(s) + H_2O(l) \longrightarrow Ca(OH)_2(s) \quad \textbf{(21.15)}$$

Carbonation: $$Ca(OH)_2(s) + CO_2(aq) \xrightarrow{aq} CaCO_3(s) + H_2O(l) \quad \textbf{(21.16)}$$

he calcination of limestone, reaction (21.14), is highly reversible at room tem- erature. Thus, a high temperature must be used and $CO_2(g)$ must be contin- ously removed from the kiln to prevent the reverse reaction.

Limestone ($CaCO_3$) is also responsible for the beautiful natural formations ound in limestone caves. Natural groundwater is slightly acidic because of lissolved $CO_2(g)$ and is essentially a solution of carbonic acid, H_2CO_3.

$$CO_2(aq) + 2\,H_2O(l) \rightleftharpoons H_3O^+(aq) + HCO_3^-(aq) \quad K_{a_1} = 4.4 \times 10^{-7} \quad \textbf{(21.17)}$$

$$HCO_3^-(aq) + H_2O(l) \rightleftharpoons H_3O^+(aq) + CO_3^{2-}(aq) \quad K_{a_2} = 4.7 \times 10^{-11} \quad \textbf{(21.18)}$$

lthough the carbonates are not very soluble in water, they are bases; thus, hey dissolve readily in acidic solutions. As mildly acidic groundwater seeps hrough limestone beds, insoluble $CaCO_3$ is converted to soluble $Ca(HCO_3)_2$.

$$CaCO_3(s) + H_2O(l) + CO_2(aq) \rightleftharpoons Ca(HCO_3)_2(aq) \quad K = 2.6 \times 10^{-5} \quad \textbf{(21.19)}$$

ver time, this dissolving action can produce a large cavity in the limestone ed—a limestone cave. Reaction (21.19) is reversible, however, and evapora- ion of the solution causes a loss of both water and CO_2 and conversion of $Ca(HCO_3)_2(aq)$ back to $CaCO_3(s)$. This process occurs very slowly. Yet over a eriod of many years, as $Ca(HCO_3)_2(aq)$ drips from the ceiling of a cave, $CaCO_3(s)$ remains as icicle-like deposits called **stalactites**. Some of the drip- ing solution may hit the floor of the cave before decomposition occurs; in this vay, limestone deposits build up from the floor in formations called **stalag- mites**. Eventually, some stalactites and stalagmites grow together into lime- stone columns, as shown in Figure 21-15.

Douglas Knight/Shutterstock

▲ FIGURE 21-15
Stalactites and stalagmites
Stalactites hang from the top; stalagmites rise from the ground.

Another important calcium-containing mineral is *gypsum*, $CaSO_4 \cdot 2\,H_2O$. n the United States, about 50 million metric tons of gypsum are consumed annually. About half of this is converted to **plaster of Paris**, $CaSO_4 \cdot \frac{1}{2}H_2O$, a hemihydrate.

$$CaSO_4 \cdot 2\,H_2O(s) \xrightarrow{\Delta} CaSO_4 \cdot \tfrac{1}{2}H_2O(s) + \tfrac{3}{2}H_2O(g) \quad \textbf{(21.20)}$$

When mixed with water, plaster of Paris reverts to gypsum. Because it expands as it sets, a mixture of plaster of Paris and water is useful in making castings where sharp details of an object must be retained. Plaster of Paris is used extensively in jewelry making and in dental work. The most important application, though, is in producing gypsum wallboard (drywall), which has all but supplanted other interior wall coverings in the construction industry.

Barium sulfate has had important applications in medical imaging because barium is opaque to X-rays. Although barium ion is toxic, the compound $BaSO_4$ is so insoluble as to be safe to use as a "barium milkshake" to coat the stomach or upper gastrointestinal tract or in a "barium enema" for the lower tract.

SuperStock

▲ Plaster of Paris castings and the molds used to form them.

21-6 CONCEPT ASSESSMENT

Suggest a reason why the hydrated beryllium ion is $[Be(H_2O)_4]^{2+}$, whereas the hydrated magnesium ion is $[Mg(H_2O)_6]^{2+}$.

Diagonal Relationship of Lithium and Magnesium

We have described trends in the periodic table either vertically (down group) or horizontally (across periods), but an investigation of the chemistr of lithium and magnesium reveals a **diagonal relationship**. Let us look a these similarities.

The following list contains some of the chemical similarities betwee lithium and magnesium that lead to the diagonal relationship between them

- Lithium and magnesium combine with O_2 to give the oxide rather tha the peroxide.
- The carbonates of lithium and magnesium can be decomposed thermall to give the oxide and carbon dioxide. The carbonates of the remainin group 1 metals are stable to thermal decomposition.
- As mentioned earlier, the salts of lithium and magnesium are strongl hydrated.
- The fluorides of lithium and magnesium are sparingly soluble in wate whereas the later group 1 fluorides are soluble.
- LiOH is the least soluble of the group 1 hydroxides and $Mg(OH)_2$ is onl sparingly soluble in water.

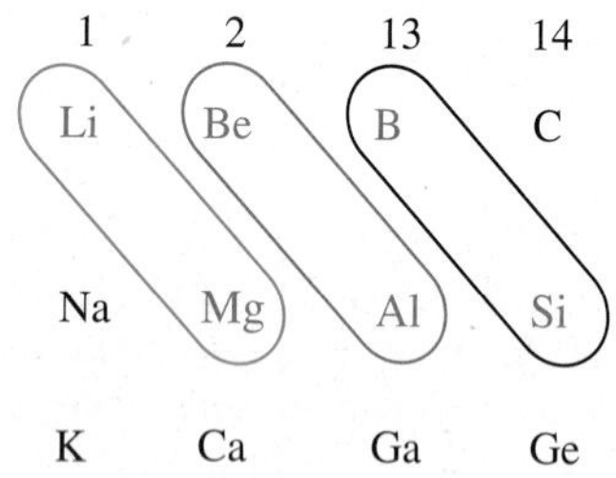

▲ FIGURE 21-16
Diagonal relationships
The two elements in each encircled pair exhibit many similar properties.

The diagonal relationship between lithium and magnesium can be under stood in terms of charge densities. The similarities in charge density arise from the increase in size of the Mg^{2+} relative to Li^+, which is counterbalanced b the increase in charge. Both Li^+ and Mg^{2+} have high polarizing power and thei compounds exhibit a high degree of covalency. Diagonal relationships als exist between Be and Al and between B and Si also. These relationships ar emphasized in Figure 21-16.

21-7 CONCEPT ASSESSMENT

Lithium and magnesium combine with N_2 to give the nitrides Li_3N and Mg_3N_2, respectively. Suggest a reason for this and write balanced chemical equations for the reactions.

EXAMPLE 21-2 Identifying Elements and Compounds from a Description of Reaction Chemistry

In an experiment, 0.1 moles of M, a group 1 metal, react with sufficient oxygen to give 0.05 moles of compound X. Compound X is then allowed to react with water, and a hydroxide is the only product. In a separate experiment, 0.1 moles of the metal reacts with water to give 0.1 moles of a hydroxide and 0.05 moles of a gas, Y. Identify the metal and compounds X and Y. Write balanced chemical equations for the reactions described.

Analyze

Three reactions are described and some information is provided for each reaction. One way to tackle this problem is to write partial chemical equations for these reactions, and then use the information provided to complete the equations.

Solve

We are told that M is a group 1 metal, so the hydroxide has the formula MOH. The partial chemical equations are as follows:

$$M + O_2 \longrightarrow X$$
$$X + H_2O \longrightarrow MOH$$
$$M + H_2O \longrightarrow MOH + \frac{1}{2}Y$$

All the alkali metals react with water to give MOH and H_2. Thus, Y must be H_2. In the first reaction, X could be M_2O, M_2O_2, or MO_2. However, the second reaction has only one product. Because M_2O_2 and MO_2 react with water to give MOH and other products, we know that X cannot be M_2O_2 or MO_2. Thus, X must be M_2O. The only alkali metal that reacts with oxygen to give M_2O as the principal product is lithium, so M must be Li. The complete chemical equations are

$$2\,Li(s) + \frac{1}{2}O_2(g) \longrightarrow Li_2O(s)$$
$$Li_2O(s) + H_2O(l) \longrightarrow 2\,LiOH(s)$$
$$Li(s) + H_2O(l) \longrightarrow LiOH(aq) + \frac{1}{2}H_2(g)$$

Assess

The key to solving this problem was in recognizing that only the normal oxide, M_2O, reacts with water to give MOH as the only product and that lithium is the only alkali metal that gives M_2O.

PRACTICE EXAMPLE A: Sodium nitrite (0.1 mol) reacts with sodium metal (0.3 mol) to give compound X (0.2 mol) and nitrogen gas (0.05 mol). Compound X (0.1 mol) reacts with oxygen (0.05 mol) to give compound Y (0.1 mol) only. Identify X and Y, and write balanced chemical equations for the reactions described.

PRACTICE EXAMPLE B: A group 2 metal (0.1 mol) is heated with carbon (0.2 mol) at 1100 °C to produce a single compound X (0.1 mol). When compound X (0.1 mol) is heated in excess $N_2(g)$ at 1100 °C, solid carbon (0.1 mol) is produced along with compound Y (0.1 mol). The group 2 metal sulfate is used in plaster of Paris and the anion in compound Y is isoelectronic with CO_2. Identify compounds X and Y, write balanced chemical equations for the reactions described, and describe the shape of the anion in Y.

21-4 Group 13: The Boron Family

The only group 13 element that is almost exclusively nonmetallic in its physical and chemical properties is boron. The remaining members of group 13—Al, Ga, In, and Tl—are metals and will be discussed later in this section. In group 13 we find for the first time elements possessing more than one oxidation state. All the elements of this group exhibit both the +1 and +3 oxidation states. Boron is a nonmetal and forms primarily covalent bonds. The other members of the group, despite being metals, commonly form covalent bonds. The tendency for forming covalent bonds can be attributed to the high charge densities (Table 21.5) of the group 13 ions. The high charge density means that the group 13 ions have a high polarizing power that leads to covalent bonding between cation and an anion.

TABLE 21.5 Charge Densities of Group 13 Elements in the +3 Oxidation State

Element	Charge Density, C mm^{-3}
B	1663
Al	770
Ga	261
In	138
Tl	105

Boron and Its Compounds

Many boron compounds lack an octet of electrons about the central boron atom, which makes the compounds *electron deficient*. This deficiency also makes them strong Lewis acids. The electron deficiency of some boron compounds leads to bonding of a type that we have not previously encountered. This type of bonding occurs in the boron hydrides.

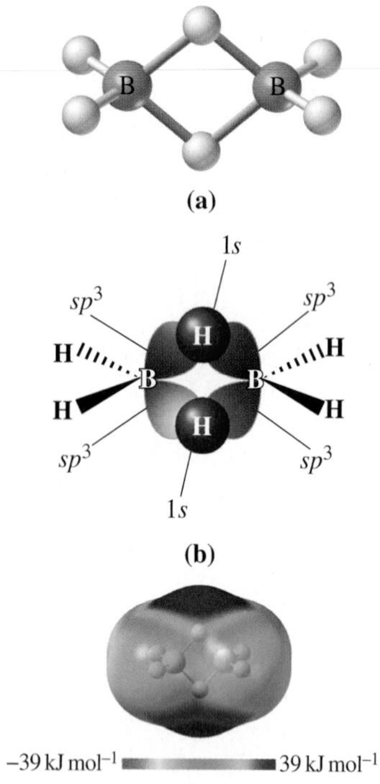

▲ FIGURE 21-17
Structure of diborane, B_2H_6
(a) The molecular structure.
(b) Bonding.
(c) Electrostatic potential map.

Boron Hydrides The molecule BH_3 (borane) may exist as a reaction intermed ate, but it has not been isolated as a stable compound. The B atom in BH_3 lack a complete octet—it has only six electrons in its valence shell. The simple boron hydride that has been isolated is diborane, B_2H_6, but this molecu defies simple description. In the following structural formula, what holds th two borane units together?

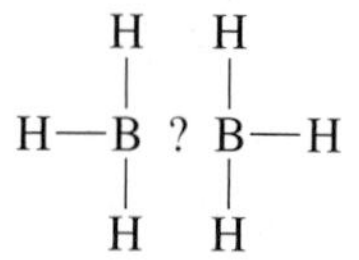

To explain the structure and bonding in B_2H_6, we need to use molecula orbital theory because simpler bonding theories fail for this molecule. Th problem is this: The B_2H_6 molecule has only *12* valence electrons (three eac from the two B atoms, and one each from the six H atoms). The minimu number of valence-shell atomic orbitals required to make a Lewis structur for B_2H_6 resembling that of C_2H_6, however, is *14* (four each from the two atoms, and one each from the six H atoms).

The structure of diborane is illustrated in Figure 21-17. The two B atom and four of the H atoms lie in the same plane (a plane perpendicular to th plane of the page). The orbitals used by the B atoms to bond these particula four H atoms can be viewed as sp^3. Eight electrons are involved in these fou bonds. Four electrons are left to bond the two remaining H atoms to the two atoms and also to bond the B atoms together. This is accomplished if each o the two H atoms simultaneously bonds to *both* B atoms.

Atom bridges are actually fairly common, although we have not had muc occasion to deal with them before. (See the discussion of Al_2Cl_6, page 1009. The B—H—B bridges are unusual, however, in having only *two* electron shared among *three* atoms. For this reason, these bonds are referred to as *three center* two-electron bonds.

We can rationalize the bonding in these three-center bonds with molecu lar orbital theory. The *six* atomic orbitals shown in Figure 21-17(b)—two *sp* orbitals from each B atom and an *s* orbital from each bridging H atom—ar combined into *six* molecular orbitals in these two bridge bonds. Of these si molecular orbitals, *two* are bonding orbitals, and these are the orbitals int which the four electrons are placed. The concept of bridge bonds can b extended to B—B—B bonds to describe the structure of higher boranes such as B_5H_9 (Fig. 21-18).

Boron hydrides are widely used in reactions for synthesizing organic com pounds. They continue to provide new and exciting developments in chemistry

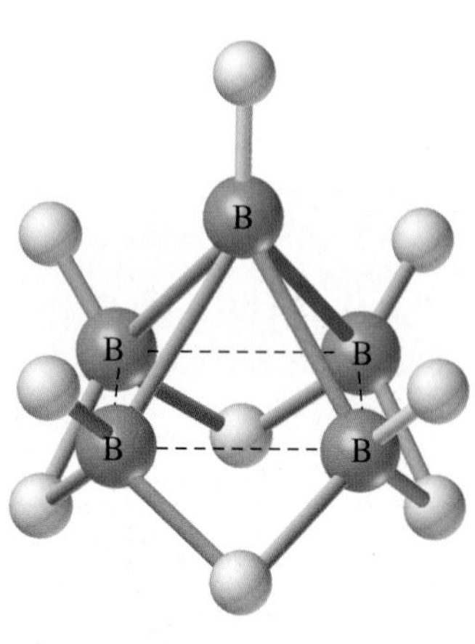

▲ FIGURE 21-18
Structure of pentaborane, B_5H_9
The boron atoms are joined by multicenter B—B—B bonds. Five of the H atoms are bonded to one B atom each. The other four H atoms bridge pairs of B atoms.

Other Boron Compounds Boron compounds are widely distributed i Earth's crust, but concentrated ores are found in only a few locations—in Italy Russia, Tibet, Turkey, and the desert regions of California. Typical of these ore is the hydrated borate *borax*, $Na_2B_4O_7 \cdot 10\,H_2O$. Figure 21-19 illustrates how borax can be converted to a variety of boron compounds.

One useful compound that can be obtained by crystallization from a solutio of borax and hydrogen peroxide is *sodium perborate*, $NaBO_3 \cdot 4\,H_2O$. This formula is deceptively simple; a more precise formula is $Na_2[B_2(O_2)_2(OH)_4] \cdot 6\,H_2O$ Sodium perborate contains the perborate ion, $[B_2(O_2)_2(OH)_4]^{2-}$, the struc ture of which is shown in Figure 21-20. Sodium perborate is a bleach alterna tive used in many color-safe bleaches. The key to the bleaching action is th presence of the two peroxo groups (—O—O—) bridging the boron atoms i the $[B_2(O_2)_2(OH)_4]^{2-}$ ion.

One of the key compounds from which other boron compounds can be syn thesized is *boric acid*, $B(OH)_3$. The weakly acidic nature of boric acid come about in a rather unusual way. The electron-deficient $B(OH)_3$ molecule accepts a

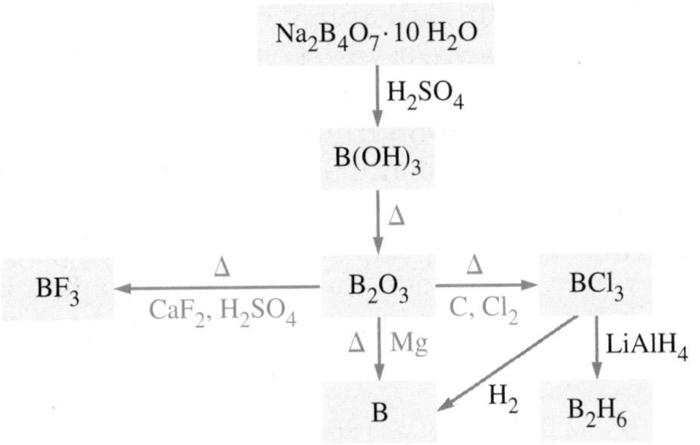

▲ FIGURE 21-19
Preparation of some boron compounds
$Na_2B_4O_7 \cdot 10\ H_2O$ (borax) is converted to $B(OH)_3$ by reaction with H_2SO_4. When heated strongly, $B(OH)_3$ is converted to B_2O_3. A variety of boron-containing compounds and boron itself can be prepared from B_2O_3.

OH^- ion from the self-ionization of water, forming the complex ion $[B(OH)_4]^-$. Thus, the source of the H_3O^+ in $B(OH)_3(aq)$ is the water itself. This ionization scheme, together with the fact that $B(OH)_3$ is a *mono*protic, not *tri*protic, acid suggests that the best formula of boric acid is $B(OH)_3$, not H_3BO_3.

$$B(OH)_3(aq) + 2\,H_2O(l) \longrightarrow H_3O^+(aq) + B(OH)_4^-(aq) \qquad K_a = 5.6 \times 10^{-10}$$

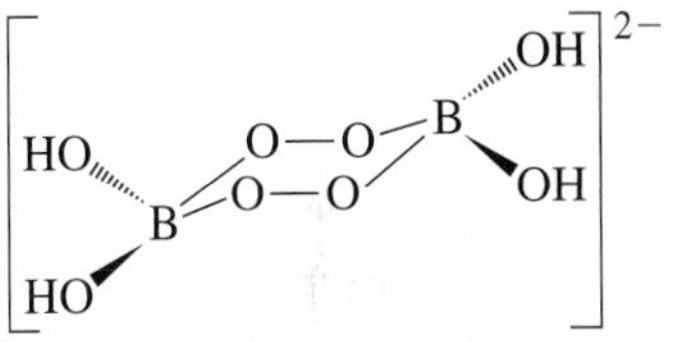

▲ FIGURE 21-20
Perborate ion

Borate salts, as expected of the salts of a weak acid, produce basic solutions by hydrolysis, which accounts for their use in cleaning agents. Boric acid also acts as an insecticide, particularly to kill roaches, and as an antiseptic in eyewash solutions. Boron compounds are used in products as varied as adhesives, cement, disinfectants, fertilizers, fire retardants, glass, herbicides, metallurgical fluxes, and textile bleaches and dyes.

The halides of boron, particularly BF_3 and BCl_3, provide examples of the Lewis acid behavior of boron compounds. For example, BF_3 can react with diethyl ether, $(C_2H_5)_2O$, to form an **adduct**:

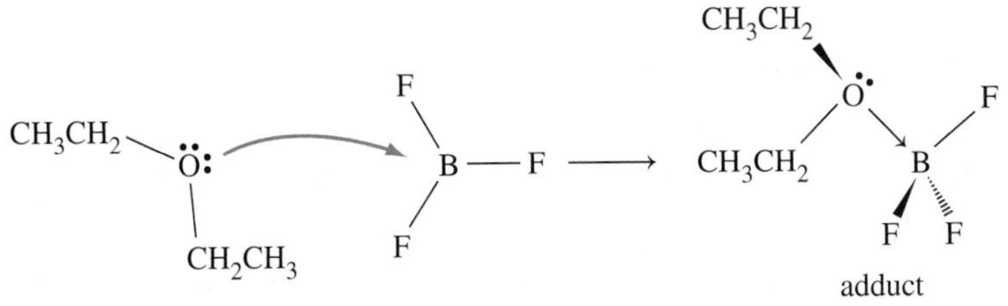

In the formation of an adduct, a coordinate covalent bond is formed between a Lewis acid and a Lewis base, with the pair of electrons for the bond coming from the Lewis base. In the reaction above, the transfer of electron density from the Lewis base to the Lewis acid is shown by the red arrow. Although a coordinate covalent bond, once formed, is indistinguishable from a regular covalent bond, we sometimes use an arrow, rather than a straight line, to identify a coordinate covalent bond. Notice that the hybridization of the boron atom changes from sp^2 to sp^3 when BF_3 and $(C_2H_5)_2O$ form an adduct.

Unlike BH_3, BF_3 does not dimerize, and the boron–fluorine bond energy in BF_3 is extremely high (646 kJ mol^{-1}) and is comparable to the bond energies of many double bonds. A possible explanation for these observations is that in a BF_3 molecule, some π bonding occurs in addition to σ bonding. A delocalized π system involving the empty $2p$ orbital on the boron atom and one full $2p$ orbital on each of the fluorine atoms can be formed (Figure 21-21).

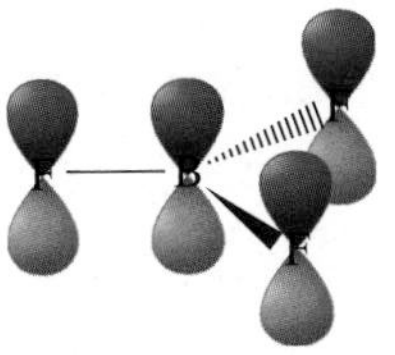

▲ FIGURE 21-21
Delocalized π system in BF_3
The $2p$ orbitals on the B and F atoms overlap to form a delocalized π system.

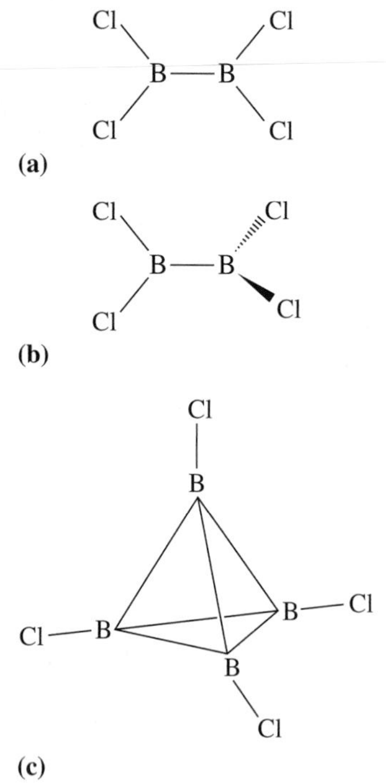

▲ FIGURE 21-22
Structures of (a) planar B_2Cl_4, (b) nonplanar B_2Cl_4, (c) tetrahedral B_4Cl_4.

Evidence for π bonding in BF_3 is that the B—F bond length increases (from 130 pm to 145 pm) when BF_3 reacts with F^- to form the BF_4^-, ion. In BF_4^-, the 2*s* and 2*p* orbitals on the B atom are used in σ bonding and are not available for π bonding. In the other boron trihalides, such as BCl_3, π bonding also occurs but to a lesser extent. Evidence that supports less π bonding in BCl_3 is that BCl_3 is a stronger Lewis acid than BF_3 and shows a greater tendency than BF_3 to form adducts.

Boron forms a variety of halides, many of which contain B—B bonds. A few such compounds are shown in Figure 21-22. One of these compounds, B_2Cl_4, is obtained from the reaction of BCl_3 and Cu:

$$2\,BCl_3 + 2\,Cu \longrightarrow B_2Cl_4 + 2\,CuCl$$

B_2Cl_4 is an interesting compound because, in the solid phase (mp −92.6 °C) the B_2Cl_4 molecule adopts a planar geometry (Fig. 21-22a) but in the gas phase (bp 65.5 °C), it is nonplanar (Fig. 21-22b).

Another interesting chloride of boron is B_4Cl_4, which consists of a tetrahedral arrangement of boron atoms with a chlorine atom bonded to each B atom (Fig. 21-22c). The B_4Cl_4 molecule is highly electron deficient and the bonding is usually described in terms of three-center two-electron bonds.

21-8 CONCEPT ASSESSMENT

Boron forms a compound with the formula $B_2H_2(CH_3)_4$. Suggest a probable structure.

EXAMPLE 21-3 Writing Chemical Equations from a Summary Diagram of Reaction Chemistry

Using Figure 21-19, write chemical equations for the successive conversions of borax to **(a)** boric acid, **(b)** B_2O_3, and **(c)** impure boron metal.

Analyze

Figure 21-19 lists the key substances involved in each reaction. We can write an incomplete chemical equation for each reaction and then identify other plausible reactants and products.

Solve

(a) The conversion of a borax, a salt, to boric acid, $B(OH)_3$, requires H_2SO_4. An incomplete chemical equation for the reaction is as follows:

$$Na_2B_4O_7 \cdot 10\,H_2O(aq) + H_2SO_4(aq) \longrightarrow B(OH)_3(aq)$$

This is an acid–base reaction. The other products of the reaction are Na_2SO_4 and H_2O. The reaction does not involve changes in oxidation states, and thus the equation can be balanced by inspection. The balanced chemical equation is given below:

$$Na_2B_4O_7 \cdot 10\,H_2O(aq) + H_2SO_4(aq) \longrightarrow 4\,B(OH)_3(aq) + Na_2SO_4(aq) + 5\,H_2O(l)$$

(b) $B(OH)_3$ can be converted to B_2O_3 by heating. The conversion of a hydroxide to an oxide requires that H_2O be driven off. The balanced chemical equation for the conversion of $B(OH)_3$ to B_2O_3 is given below:

$$2\,B(OH)_3(s) \xrightarrow{\Delta} B_2O_3(s) + 3\,H_2O(g)$$

(c) B is obtained when B_2O_3 is heated with Mg. An incomplete chemical equation for the reaction is as follows:

$$B_2O_3(s) + Mg(s) \xrightarrow{\Delta} B(s)$$

The other product must be MgO. The balanced chemical equation is given below.

$$B_2O_3(s) + 3\,Mg(s) \xrightarrow{\Delta} 2\,B(s) + 3\,MgO(s)$$

Assess

The reaction summary diagram identifies only some of the substances involved in the various reaction pathways. Writing down an incomplete chemical equation for a reaction can often make it easier to identify other reactants and products that are involved.

PRACTICE EXAMPLE A: Using Figure 21-19, write chemical equations for the sequence of reactions by which borax is converted to diborane.

PRACTICE EXAMPLE B: Using Figure 21-19, write chemical equations for the sequence of reactions by which borax is converted to BF_3.

Properties and Uses of Group 13 Metals

In their appearance and physical properties and in most of their chemical behaviors, aluminum, gallium, indium, and thallium are metallic. Some properties of the group 13 metals are listed in Table 21.6.

The most important of the group 13 metals is aluminum, which is used mainly in lightweight alloys. Aluminum, like most of the other main-group metals, is an active metal. Because it is easily oxidized to the +3 ion, aluminum is an excellent reducing agent—for example, reacting with acids to reduce $H^+(aq)$ to $H_2(g)$.

$$2\,Al(s) + 6\,H^+(aq) \longrightarrow 2\,Al^{3+}(aq) + 3\,H_2(g) \qquad \textbf{(21.21)}$$

Aluminum also reacts in basic solutions as shown below:

$$2\,Al(s) + 2\,OH^-(aq) + 6\,H_2O(l) \longrightarrow 2\,[Al(OH)_4]^-(aq) + 3\,H_2(g) \qquad \textbf{(21.22)}$$

◀ Certain drain cleaners are a mixture of NaOH and Al(s). When they are added to water, reaction (21.22) occurs. The evolved $H_2(g)$ helps unplug a stopped-up drain. The heat of reaction melts fats and grease, and the NaOH(aq) solubilizes them.

Air or other oxidants easily oxidize powdered aluminum in highly exothermic reactions. The reaction of Al and O_2 yields Al_2O_3:

$$2\,Al(s) + \tfrac{3}{2}O_2(g) \longrightarrow Al_2O_3(s) \qquad \Delta_r H = -1676\ \text{kJ mol}^{-1} \qquad \textbf{(21.23)}$$

Aluminum is such a good reducing agent that it will extract oxygen from other metal oxides, producing aluminum oxide while liberating the other metal in its free state. The following reaction, known as the **thermite reaction**, produces liquid iron and is used in the on-site welding of large metal objects.

$$Fe_2O_3(s) + 2\,Al(s) \longrightarrow Al_2O_3(s) + 2\,Fe(l) \qquad \textbf{(21.24)}$$

TABLE 21.6 Some Properties of the Group 13 Metals

	Al	Ga	In	Tl
Atomic number	13	31	49	81
Atomic (metallic) radius, pm	143	122	163	170
Ionic (M^{3+}) radius, pm	53	62	79	88
Electronegativity	1.5	1.6	1.7	1.8
First ionization energy, kJ mol^{-1}	577.6	578.8	558.3	589.3
Electrode potential $E°$, V[a]	−1.676	−0.56	−0.34	+0.72
Melting point, °C	660.37	29.78	156.17	303.55
Boiling point, °C	2467	2403	2080	1457
Density, g cm^{-3} at 20 °C	2.698	5.907	7.310	11.85
Hardness[b]	2.75	1.5	1.2	1.25
Electrical conductivity[b]	59.7	9.1	19.0	8.82

[a] For the reduction $M^{3+}(aq) + 3\,e^- \longrightarrow M(s)$.
[b] See footnotes c and d of Table 21.2.

Richard Megna/Fundamental Photographs

▲ Thermite reaction.

The thermite reaction is highly exothermic and visually spectacular. It i shown in the photograph in the margin.

Gallium metal is of great importance in the electronics industry. It is used t make gallium arsenide (GaAs), a compound that can convert light directl into electricity (photoconduction). This semiconducting material is also use in light-emitting diodes (LEDs) (see Focus On 21: Gallium Arsenide) and i solid-state devices such as transistors.

Indium is a soft silvery metal used to make low-melting alloys. Like GaA InAs also finds use in low-temperature transistors and as a photoconductor i optical devices.

Thallium and its compounds are extremely toxic; as a result, they have fe industrial uses. One possible use, however, is in high-temperature supe conductors. For example, a thallium-based ceramic with the approximat formula $Tl_2Ba_2Ca_2Cu_3O_{8+x}$ exhibits superconductivity at temperatures a high as 125 K.

▶ As discussed on page 1119 a superconducting material loses its electrical resistance below a certain temperature. Metals typically become superconducting only a few degrees above 0 K.

Oxidation States of Group 13 Metals

Aluminum, at the top of the group of four metals in group 13, occurs almos exclusively in the +3 oxidation state in its compounds. Gallium also favors th +3 oxidation state. Indium compounds can be found with +3 and +1 oxida tion states, though the +3 is more common. In thallium, this preference i reversed. For example, thallium forms the oxide Tl_2O, the hydroxide TlOH and the carbonate Tl_2CO_3. These compounds are ionic and in some respect resemble group 1 compounds. Thus, TlOH is both very soluble and a stron base in aqueous solution. The higher stability of the +1 over the +3 oxidatio state of thallium is often described as the **inert pair effect**. Thallium has th electron configuration $[Xe]4f^{14}5d^{10}6s^26p^1$. In forming the Tl^+ ion, a Tl aton loses the $6p$ electron and retains two electrons in its $6s$ subshell. It is this pair o electrons—$6s^2$—that is called the inert pair. The electron configuratio $(n-1)s^2(n-1)p^6(n-1)d^{10}ns^2$ is commonly encountered in ions of th post-transition elements. One explanation of the inert pair effect is that th small bond energies and lattice energies associated with the large atoms an ions at the bottom of a group are not sufficiently great to offset the ionizatio energies of the ns^2 electrons.

Aluminum

Aluminum is the third most abundant element and composes 8.3% by mass o Earth's solid crust. On average, more than 5 million metric tons of aluminun are produced per year in the United States.

Production of Aluminum When an aluminum cap was placed atop th Washington Monument in 1884, aluminum was still a semiprecious metal. A that time it cost $1 per ounce to produce, equivalent to the daily wage of a skilled laborer working a 10-hour day. As a result, aluminum was used mainl in jewelry and artwork. But just two years later, all this changed. In 1886 Charles Martin Hall in the United States and Paul Héroult in France indepen dently discovered an economically feasible method of producing aluminur from Al_2O_3 by electrolysis.

▶ Stimulated by a remark by one of his professors, Charles Martin Hall invented the electrolytic process for the production of aluminum at the age of 23, eight months after his graduation from Oberlin College. Paul Héroult, a student of Le Châtelier, and also at the age of 23, invented the identical process in the same year.

The manufacture of Al involves several interesting principles. The chief ore *bauxite*, contains Fe_2O_3 as an impurity that must be removed. The principl used in the separation is that Al_2O_3 is an *amphoteric* oxide and dissolves ir NaOH(aq), whereas the iron oxide is a *basic* oxide and does not. Wher $Al_2O_3(s)$ is added to NaOH(aq), $Al_2O_3(s)$ dissolves because the following reaction converts the solid to soluble $[Al(OH)_4]^-$.

$$Al_2O_3(s) + 2\,OH^-(aq) + 3\,H_2O(l) \longrightarrow 2\,[Al(OH)_4]^-(aq)$$

Vhen the solution containing $[Al(OH)_4]^-$ is slightly acidified, $Al(OH)_3(s)$ recipitates. Pure Al_2O_3 is obtained by heating the $Al(OH)_3$.

$$[Al(OH)_4]^-(aq) + H_3O^+(aq) \longrightarrow Al(OH)_3(s) + 2\,H_2O(l)$$
$$2\,Al(OH)_3(s) \xrightarrow{\Delta} Al_2O_3(s) + 3\,H_2O(g)$$

he separation of a mixture of $Fe^{3+}(aq)$ and $Al^{3+}(aq)$, which makes use of nese reactions, is illustrated in Figure 21-23.

Al_2O_3 has a very high melting point (2020 °C) and produces a liquid that is a oor electrical conductor. Thus, its electrolysis is not feasible without a better onducting solvent. That was the crux of the discovery by Hall and Héroult. hey found, independently, that up to 15% Al_2O_3, by mass, can be dissolved in ne molten mineral *cryolite*, Na_3AlF_6, at about 1000 °C. Cryolite consists of Ja$^+$ and $AlF_6{}^{3-}$ ions. In the molten state, it is a good electrical conductor. In the Iall–Héroult process, aluminum metal is produced by electrolysis of Al_2O_3 in nolten cryolite. A typical electrolysis cell, pictured in Figure 21-24, produces luminum of 99.6%–99.8% purity. The electrode reactions are not known with ertainty, but the overall electrolysis reaction is

Oxidation: $3\,\{C(s) + 2\,O^{2-} \longrightarrow CO_2(g) + 4\,e^-\}$

Reduction: $4\,\{Al^{3+} + 3\,e^- \longrightarrow Al(l)\}$

Overall: $3\,C(s) + 4\,Al^{3+} + 6\,O^{2-} \longrightarrow 4\,Al(l) + 3\,CO_2(g)$ **(21.25)**

The energy consumed to produce aluminum by electrolysis is very high, bout 15 kWh per kg of Al. This is more than three times the amount of energy onsumed per kg of Na in the electrolysis of NaCl(aq). Because of the high nergy requirements for producing Al(s), aluminum production facilities are enerally located near low-cost hydroelectric power sources. The energy equired to recycle Al is only about 5% of that to produce the metal from bauxte, and currently about 45% of the Al produced in the United States is btained from the recycling of scrap aluminum.

Why is so much energy consumed in the electrolytic production of aluninum? Any process that must be carried out at a high temperature requires

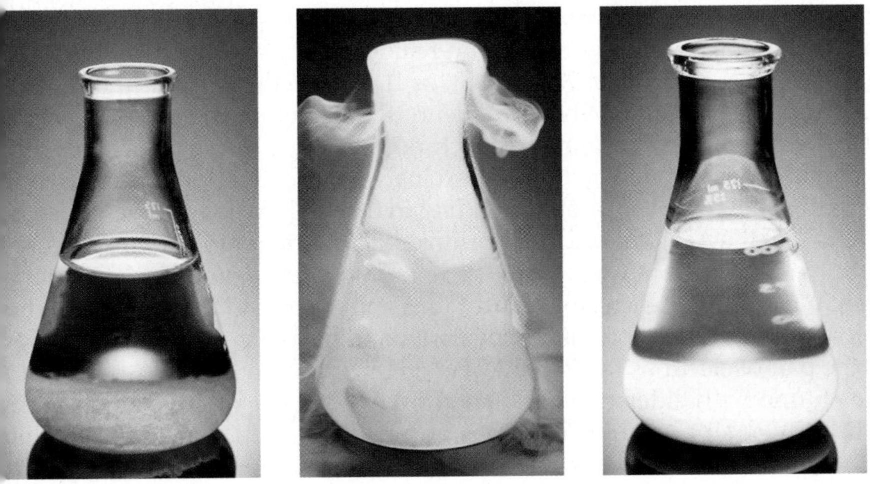

(a) (b) (c)

▲ FIGURE 21-23
Purifying bauxite
(a) When an excess of $OH^-(aq)$ is added to a solution containing $Al^{3+}(aq)$ and $Fe^{3+}(aq)$, the Fe^{3+} precipitates as $Fe(OH)_3(s)$ and the $Al(OH)_3(s)$ first formed edissolves to produce $[Al(OH)_4]^-(aq)$. **(b)** The $Fe(OH)_3(s)$ is filtered off, and the $[Al(OH)_4]^-(aq)$ is made slightly acidic through the action of CO_2, here added as dry ce. **(c)** The precipitated $Al(OH)_3(s)$ collects at the bottom of a clear, colorless solution.

Carey B. Van Loon

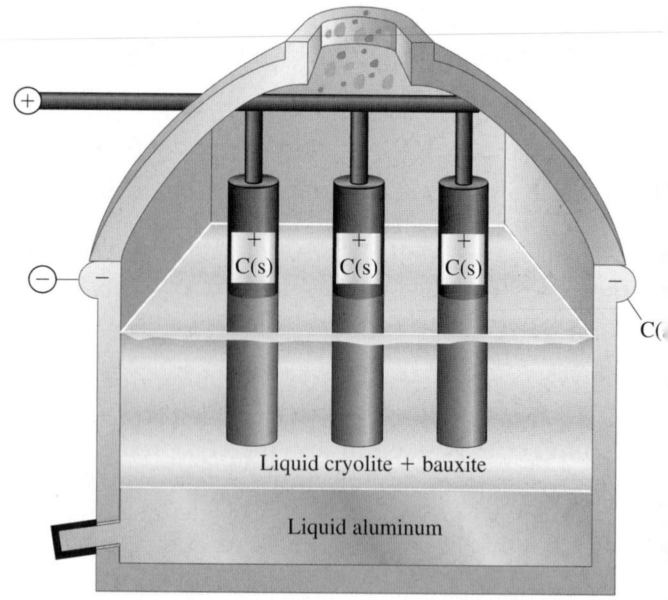

▶ FIGURE 21-24
Electrolysis cell for aluminum production
The cathode is a carbon lining in a steel tank. The anodes are also made of carbon. Liquid aluminum is denser than the electrolyte medium and collects at the bottom of the tank.

large amounts of energy for heating. In the electrolytic production of Al, th electrolysis bath must be kept at about 1000 °C, which is done by means o electric heating. Two other factors, however, are also involved in the larg energy consumption. First, to produce one mole of Al, *three* moles of electron must be transferred: $Al^{3+} + 3\,e^- \longrightarrow Al(l)$. Additionally, the molar mass o Al is relatively low, 27 g mol^{-1}. The electric current equivalent to the passag of one mole of electrons produces only 9 g Al. In contrast, one mole of elec trons produces 12 g Mg, 20 g Ca, or 108 g Ag. Yet the same factors that make A production a significant energy consumer make Al an outstanding energ producer when it is used in a battery. (Recall the aluminum–air batter described on page 897.)

Aluminum Halides Aluminum fluoride, AlF_3, has considerable ionic character. I has a high melting point (1290 °C) and when molten it is a conductor of electri current. In contrast, the other aluminum halides exist as *molecular* species wit the formula Al_2X_6 (for which X = Cl, Br, or I). We can think of this molecule a comprising two AlX_3 units. When two identical units combine, the resulting mol ecule is called a **dimer**. The dimeric structure of Al_2Cl_6 is shown in Figure 21-25 Notice that two Cl atoms are bonded exclusively to each Al atom and two C atoms bridge the two metal atoms. Bonding in this molecule can be described b assuming that the Al atoms are sp^3 hybridized. Each bridging Cl atom appears t bond to two Al atoms in two ways. The bond to one Al atom is a conventiona covalent bond because each atom contributes one electron to the bond. The bon to a second Al atom is a coordinate covalent bond, where the chlorine atom pro vides the pair of electrons for the bond, noted by arrows in Figure 21-25.

The aluminum halides, just like the boron halides, are reactive Lewis acids They readily accept a pair of electrons and form adducts. For example, they form adducts with ethers, as does BF_3. In another example, addition of Cl^- t $AlCl_3$ produces the tetrahedral $[AlCl_4]^-$. The formation of $[AlCl_4]^-$ is impor tant in the use of $AlCl_3$ as a Lewis acid catalyst in the Friedel–Crafts reaction The most common reaction of this type involves the addition of an alky group, such as the ethyl group, CH_3CH_2—, to a benzene ring, as shown below

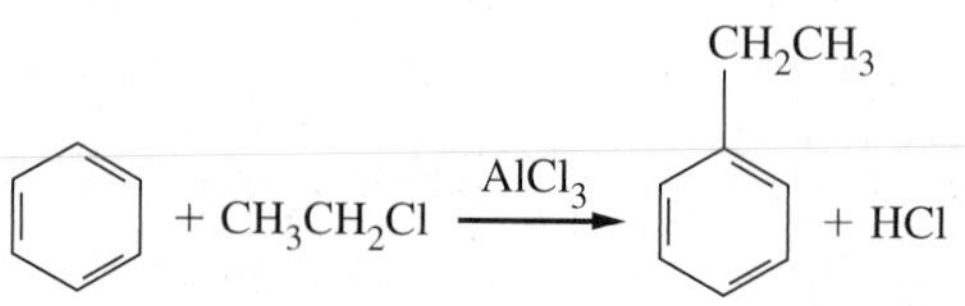

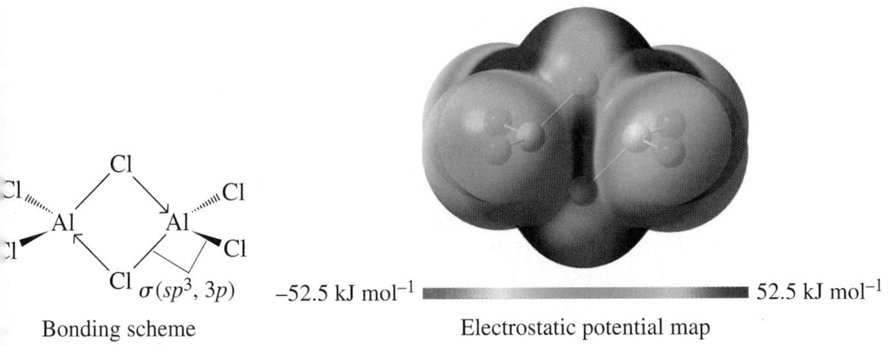

◀ FIGURE 21-25
Bonding in Al_2Cl_6
Two Cl atoms bridge the $AlCl_3$ units to form Al_2Cl_6. Electrons donated by these Cl atoms to Al atoms are indicated by arrows.

n this reaction, $AlCl_3$ acts as a Lewis acid and assists in generating the posi-ive cation $[CH_3CH_2]^+$:

$$CH_3CH_2Cl + AlCl_3 \longrightarrow [CH_3CH_2]^+ + [AlCl_4]^-$$

The $[CH_3CH_2]^+$ cation attacks the benzene ring, liberating a proton that eacts with $[AlCl_4]^-$ to regenerate $AlCl_3$ and HCl.

As pointed out on page 1007, a very important halide complex of aluminum s cryolite, Na_3AlF_6. Natural deposits of cryolite were discovered in Greenland n 1794 and occur almost nowhere else. For aluminum production, natural ryolite has been largely displaced by cryolite synthesized in a lead-clad ves-el by using the following reaction:

$$6\,HF + Al(OH)_3 + 3\,NaOH \longrightarrow Na_3AlF_6 + 6\,H_2O \qquad \textbf{(21.26)}$$

◀ Corundum, when pure, is also known as the gemstone *white sapphire*. Certain other gemstones consist of corundum with small amounts of transition metal ions as impurities: Cr^{3+} in ruby and Fe^{2+} and Ti^{4+} in blue sapphire, for example. Artificial gemstones are made by fusing corundum with carefully controlled amounts of other oxides.

luminum Oxide and Hydroxide Aluminum oxide is often referred to as *alu-nina*; when in crystalline form, it is called *corundum*.

The bonding and crystal structure of alumina account for its physical prop-rties. The small Al^{3+} ions and small O^{2-} ions form a very stable ionic lattice. he crystal has a cubic closest packed structure of O^{2-} ions, with Al^{3+} ions ccupying octahedral holes. Alumina is a very hard material and is often used s an abrasive. It is also resistant to heat (mp 2020 °C) and is used in linings for igh-temperature furnaces and as a catalyst support in industrial chemical rocesses. Aluminum oxide is relatively unreactive except at very high tem-eratures. Its stability at high temperatures classifies it as a *refractory* material.

As mentioned previously, aluminum is protected against reaction with vater in the pH range 4.5–8.5 by a thin, impervious coating of Al_2O_3. This coat-ng can be purposely thickened to enhance the corrosion resistance of the metal y a process known as *anodizing*. An aluminum object is used for the anode and graphite electrode is used for the cathode in an electrolyte bath of $H_2SO_4(aq)$. he half-reaction occurring at the anode during electrolysis is shown below:

$$2\,Al(s) + 3\,H_2O(l) \longrightarrow Al_2O_3(s) + 6\,H^+(aq) + 6\,e^-$$

l_2O_3 coatings of varying porosity and thickness can be obtained. Also, the xide can be made to absorb pigments or other additives. Anodized aluminum s used to make everyday items, such as the drinking cups shown in the photo-raph in the margin, and is used in architectural components of buildings, uch as bronze or black window frames.

Aluminum hydroxide is *amphoteric*. It reacts with acids to form $Al(H_2O)_6]^{3+}$, as shown below:

$$Al(OH)_3(s) + 3\,H_3O^+(aq) \longrightarrow [Al(H_2O)_6]^{3+}(aq) \qquad \textbf{(21.27)}$$

lso, it reacts with bases to form $[Al(OH)_4]^-$:

$$Al(OH)_3(s) + OH^-(aq) \longrightarrow [Al(OH)_4]^-(aq) \qquad \textbf{(21.28)}$$

Nicola Stratford/Getty Images

▲ Drinking cups made of anodized aluminum.

▶ A disadvantage of the use of aluminum sulfate in sizing paper is that its acidic character contributes to deterioration of the paper. In contrast, calcium carbonate maintains an alkaline medium in paper (page 999).

Aluminum Sulfate and Alums Aluminum sulfate, $Al_2(SO_4)_3$, is the mos important commercial aluminum compound. It is prepared by the reaction c hot concentrated $H_2SO_4(aq)$ on $Al_2O_3(s)$. The product that crystallizes fror solution is $Al_2(SO_4)_3 \cdot 18\, H_2O$. More than 1 million metric tons of aluminur sulfate are produced annually in the United States, about half of which is use in water purification. In this application, the pH of the water is adjusted s that $Al(OH)_3(s)$ precipitates when aluminum sulfate is added. As th $Al(OH)_3(s)$ settles, it removes suspended solids in the water. Another impor tant use is in the sizing of paper. *Sizing* refers to incorporating materials, such a waxes, glues, or synthetic resins, into paper to make the paper more wate resistant. $Al(OH)_3$ precipitated from $Al_2(SO_4)_3(aq)$ helps deposit the sizin agent in the paper.

▶ In the industrial world, *alum* usually refers to simple aluminum sulfate; terms such as *potash* (potassium) *alum* and *ammonium alum* designate the double salts.

When an aqueous solution of equimolar amounts of $Al_2(SO_4)_3$ and K_2SO is allowed to crystallize, the crystals obtained are of potassium aluminum su fate, $KAl(SO_4)_2 \cdot 12\, H_2O$. This is just one of a large class of double salts calle alums. **Alums** have the formula $M(I)M(III)(SO_4)_2 \cdot 12\, H_2O$, where M(I) is a uni positive cation (other than Li^+) and M(III) is a tripositive cation—Al^{3+}, Ga^{3+} In^{3+}, Ti^{3+}, V^{3+}, Cr^{3+}, Mn^{3+}, Fe^{3+}, Co^{3+}, Re^{3+}, or Ir^{3+}. The actual ions presen in the alums are $[M(H_2O)_6]^+$, $[M(H_2O)_6]^{3+}$, and SO_4^{2-}. The most common alum have M(I) = K^+, Na^+, or NH_4^+ and M(III) = Al^{3+}. Li^+ does not form alum because the ion is too small to accommodate six water molecules. Sodium alu minum sulfate is the leavening acid in baking powders, and potassiun aluminum sulfate is used in dyeing. The fabric to be dyed is dipped into solution of the alum and heated in steam. Hydrolysis of $[Al(H_2O)_6]^{3}$ deposits $Al(OH)_3$ into the fibers of the material, and the dye is adsorbed o the $Al(OH)_3$ that deposits.

Diagonal Relationship of Beryllium and Aluminum

The charge densities of Be^{2+} and Al^{3+} are very similar and this similarity ha been used to rationalize the following observations:

- The Be^{2+} ion is hydrated in aqueous solutions, forming $[Be(H_2O)_4]^{2+}$ Similarly, Al^{3+} is also hydrated as $[Al(H_2O)_6]^{3+}$. Because the Al^{3+} io is larger than the Be^{2+} ion, it can accommodate a greater number of wate molecules in its primary hydration sphere. In both cases, the cation are sufficiently polarizing that the following reactions occur in aqueou solutions:

$$[Be(H_2O)_4]^{2+}(aq) + H_2O(l) \rightleftharpoons [Be(H_2O)_3(OH)]^{+}(aq) + H_3O^{+}(aq)$$
$$[Al(H_2O)_6]^{3+}(aq) + H_2O(l) \rightleftharpoons [Al(H_2O)_5(OH)]^{2+}(aq) + H_3O^{+}(aq)$$

  As a result, aqueous solutions of beryllium and aluminum salts, such a $Be(NO_3)_2$ or $Al(NO_3)_3$, are slightly acidic.

- The hydroxides of beryllium and aluminum are both amphoteric an form tetrahydroxido complexes, $[Be(OH)_4]^{2-}$ or $[Al(OH)_4]^{-}$, in highl basic solutions.

- In air, both metals form a strong oxide coating, which protects the metal from reaction.

- Both metals form carbides (Be_2C and Al_4C_3) containing the C^{4-} ion. Th carbides react with water to form methane:

$$Be_2C(s) + 4\, H_2O(l) \longrightarrow 2\, Be(OH)_2(s) + CH_4(g)$$
$$Al_4C_3(s) + 12\, H_2O(l) \longrightarrow 4\, Al(OH)_3(s) + 3\, CH_4(g)$$

- Both Be and Al form halides, such as $BeCl_2$ or $AlCl_3$, that can act as Lewi acids and Friedel–Crafts catalysts.

21-9 CONCEPT ASSESSMENT

AlF_3 is almost insoluble in anhydrous HF, but it dissolves if KF is present. Passage of BF_3 through the resulting solution causes AlF_3 to precipitate. How can you explain these observations?

21-5 Group 14: The Carbon Family

The properties of the group 14 elements vary dramatically within the group. Tin and lead, at the bottom of the group, have mainly metallic properties. Germanium is a *metalloid* (or *semi-metal*), and it exhibits semiconductor behavior. Silicon is mostly nonmetallic in its chemical behavior but is sometimes classified as a metalloid. Silicon also exhibits semiconductor behavior. Carbon, the first member of group 14, is a nonmetal. We will first discuss carbon and silicon and then tin and lead; germanium is mentioned only briefly.

The essential differences between carbon and silicon, as outlined in Table 21.7, are perhaps the most striking between any second- and third-period elements within a group in the periodic table. As suggested by the approximate bond energies, strong C—C and C—H bonds account for the central role of carbon–atom chains and rings in establishing the chemical behavior of carbon. A study of these chains and rings and their attached atoms is the focus of organic chemistry (Chapters 26 and 27) and biochemistry (Chapter 28). The Si—Si and Si—H bonds are much weaker than the Si—O bond. The strength of the Si—O bond accounts for the predominance of the silicates and related compounds among silicon compounds.

Carbon

Carbon is so much the central element to the study of organic and biochemistry that the rich and important inorganic chemistry of carbon is sometimes

TABLE 21.7 Comparison of Carbon and Silicon

Carbon	Silicon
Two principal allotropes: graphite and diamond	One stable, diamond-type crystalline structure
Forms two stable *gaseous* oxides, CO and CO_2, and several less stable ones, such as C_3O_2	Forms only one *solid* oxide (SiO_2) that is stable at room temperature; a second oxide (SiO) is stable only in the temperature range of 1180–2480 °C
Insoluble in alkaline media	Reacts in alkaline media, forming $H_2(g)$and $SiO_4^{4-}(aq)$
Principal oxoanion is CO_3^{2-}, which has a trigonal-planar shape	Principal oxoanion is SiO_4^{2-}, which has a tetrahedral shape
Strong tendency for catenation,[a] with straight and branched chains and rings containing up to hundreds of C atoms	Less tendency for catenation,[a] with silicon atom chains limited to about six Si atoms
Readily forms multiple bonds through use of the orbital sets $sp^2 + p$ and $sp + 2p$	Multiple bond formation much less common than with carbon
Approximate single-bond energies, kJ mol^{-1}: C—C, 347 C—H, 414 C—O, 360	Approximate single-bond energies, kJ mol^{-1}: Si—Si, 226 Si—H, 318 Si—O, 464

[a]Catenation is the joining together of like atoms into chains.

overlooked. It is the inorganic aspects of the chemistry of carbon that we emphasize in this section.

Production and Uses of Carbon Graphite is a form of carbon that is widely distributed in Earth's crust, some of it in deposits rich enough for commercial exploitation. The bulk of industrial graphite, however, is synthesized from carbon-containing materials. The key requirement is to heat the high-carbon content material to a temperature of about 3000 °C in an electric furnace. In this process, a contiguous array of sp^2 hybridized carbon atoms form symmetrically conjoined 6-member rings in a planar aromatic structure ultimately leading to the graphite as shown in Figure 12-33.

Graphite has excellent lubricating properties, even when dry. The planes of carbon atoms in graphite are held together by relatively weak forces and can easily slip past one another. This property is handy in pencil "lead," which is actually a thin rod made from a mixture of graphite and clay that glides easily on paper. Graphite's principal use is based on its ability to conduct electric current; it is used for electrodes in batteries and industrial electrolysis. Graphite's use in foundry molds, furnaces, and other high-temperature environments is based on its ability to withstand high temperatures.

▲ Fabric for use in composite materials woven from carbon fibers.

A newly developed use of graphite is in the manufacture of strong, lightweight composites consisting of graphite fibers, shown in the margin, and various plastics. These composites are used in products ranging from tennis rackets to lightweight aircraft. When carbon-based fibers, such as rayon, are carefully heated to a very high temperature, all volatile matter is driven off, leaving a carbon residue with the graphite structure.

As indicated in the phase diagram for carbon in Figure 21-26, graphite is the more stable form of carbon, not only at room temperature and normal atmospheric pressure but at temperatures up to 3000 °C and pressures of 10^4 atm and higher. Diamond is the more stable form of carbon at very high pressures. Diamonds can be synthesized from graphite by heating the graphite to temperatures of 1000–2000 °C and subjecting it to pressures of 10^5 atm or more. Usually the graphite is mixed with a metal, such as iron. The metal melts, and the graphite is converted to diamond within the liquid metal. Diamonds can then be picked out of the solidified metal.

According to Figure 21-26, we might expect diamond to revert to graphite at room temperature and pressure. Fortunately for the jewelry industry and for those who treasure diamonds as gems, many phase changes that require a rearrangement in bond type and crystal structure occur extremely slowly. That is very much the case with the diamond–graphite transition.

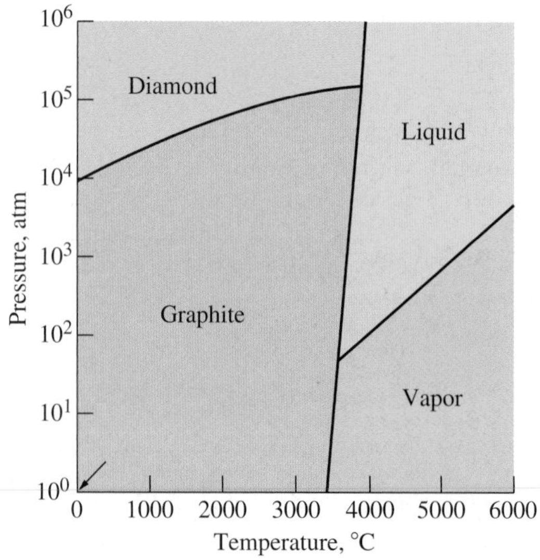

▶ FIGURE 21-26
Simplified phase diagram for carbon
Notice that pressure is plotted on a logarithmic scale. The arrow marks the point 1 atm, 25 °C.

Most diamonds that are used as gemstones are natural diamonds. For ıdustrial purposes, inferior specimens of natural diamonds or, increasingly, ynthetic diamonds, shown in the margin, are used. The industrial use epends on two key properties. Diamonds are used as abrasives because they re extremely hard (10 on the Mohs hardness scale). No harder substance is nown. Diamonds also have a high thermal conductivity (they dissipate heat uickly), so they are used in drill bits for cutting steel and other hard materi-ls. The rapid dissipation of heat makes the drilling process faster and ıcreases the lifetime of the bit. A recent development has been the creation of iamond films that can be deposited directly onto metals, which imparts some f the properties of diamond to the metal. For example, when a metal is coated ith a thin diamond film, the resulting material has a high thermal conductiv-y. Such materials have been used in heat sinks for computer chips to help issipate the heat that computer chips generate.

◀ Because of their expected future importance, diamond films were named "Molecule of the Year" by the journal *Science* in 1990. In 1991, the fullerenes earned that honor.

Carbon can be obtained in several other forms, consisting of mixed crys-ılline or amorphous structures. Incomplete combustion of natural gas, such as ı an improperly adjusted Bunsen burner in the laboratory, produces a smoky ame. The smoke can be deposited as finely divided soot called **carbon black**. arbon black is used as filler in rubber tires (several kilograms per tire), as a pig-ıent in printing inks, and as the transfer material in carbon paper, typewriter ibbons, laser printers, and photocopying machines. Recently, new allotropes of arbon have been isolated from the decomposition of graphite. These allotropes, nown as fullerenes and nanotubes, were presented in Chapter 12 (page 549). ecall that the molecule C_{60} is remarkably stable and has a shape resembling a occer ball (12 pentagonal faces, 20 hexagonal faces, and 60 vertices). Other ullerenes include C_{70}, C_{74}, and C_{82}. Fullerenes are typically produced by laser ecomposition of graphite under a helium atmosphere. Because nitrogen and xygen interfere with the process of forming fullerenes, soot, which is formed y the combustion of hydrocarbons in air, does not contain fullerenes.

▲ **Synthetic diamonds**

In Chapter 12, we discussed various forms of carbon: graphite, diamond, ullerenes, and nanotubes. In this chapter we focus on graphene. **Graphene** is sheet of carbon atoms only one atom thick. The layers of graphite are raphene. A carbon nanotube is a sheet of graphene rolled into a cylinder. A ullerene is obtained when an appropriate number of hexagonal rings in raphene are replaced by pentagonal rings; the presence of the pentagonal ings causes the flat sheet to pucker and form a spherical ball. The relationship etween graphene and other forms of carbon is illustrated in Figure 21-27. raphene has very interesting electronic properties because electrons in these heets are moving very quickly—at approximately 1/300 of the speed of light. t is expected that graphene will play an important role in the development of lectronic devices, possibly replacing silicon.

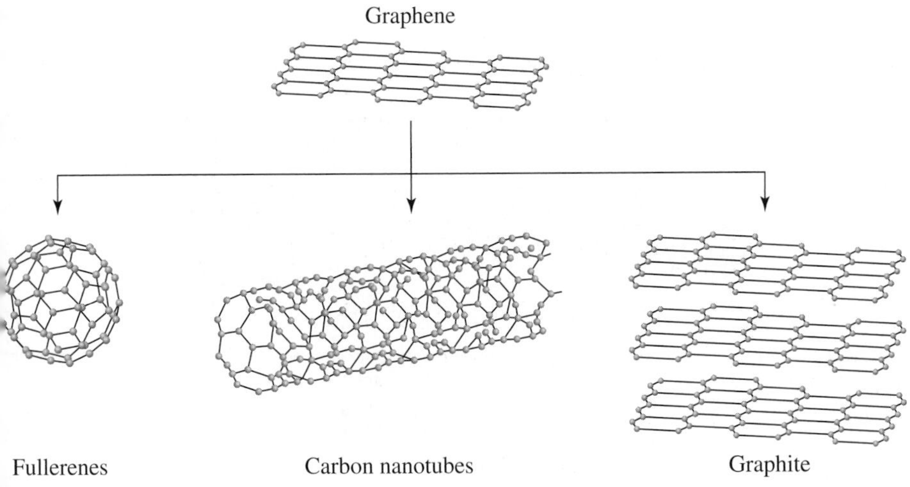

◀ FIGURE 21-27
The relationship between graphene and the other forms of carbon

It was once thought impossible that a sheet of carbon, only one atom thic could be made or isolated. However, scientists in the United Kingdom (And Geim and Kostya Novosalov at Manchester University) were able to isola graphene in 2004 by using a technique called micromechanical cleavag Essentially, the process is just like drawing with a pencil, the "lead" of whic contains graphite, and looking among the traces for graphene pieces in th trail left by the pencil. Another way to isolate graphene is to use adhesive tap repeatedly to peel away layers of carbon atoms from a graphite surface in process called exfoliation. The exfoliation of graphite is illustrated schemat cally in Figure 21-28. These methods produce flakes that contain 1 to 10 laye of graphene and are up to 100 μm thick. The single-layer flakes have to b found among the thousands of thicker flakes.

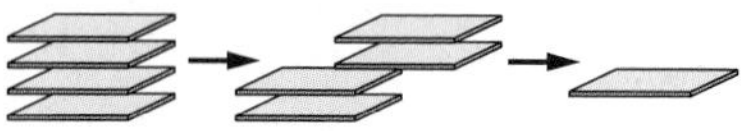

▲ FIGURE 21-28
Exfoliation of graphite

Two other noteworthy carbon-containing materials are coke and charcoa When coal is heated in the absence of air, volatile substances are driven off, lea ing a high-carbon residue called *coke*. This same type of destructive distillation wood produces *charcoal*. Currently, coke is the principal metallurgical reducin agent. It is used in blast furnaces, for instance, to reduce iron oxide to iron meta

Carbon Dioxide The chief oxides of carbon are carbon monoxide, CO, and ca bon dioxide, CO_2. There are about 380 ppm of CO_2 in air (0.038% by volume CO occurs to a much lesser extent. Although they are only minor constituents air, these two oxides are important in many ways.

Carbon dioxide is the only oxide of carbon formed when carbon or carbon containing compounds are burned in an *excess* of air (providing an abundanc of O_2). This condition exists when a fuel-lean mixture is burned in an autom bile engine. Thus, for the combustion of the gasoline component octane,

$$C_8H_{18}(l) + \frac{25}{2}O_2(g) \longrightarrow 8\,CO_2(g) + 9\,H_2O(l) \qquad \textbf{(21.2}$$

If the combustion occurs in a *limited* quantity of air, carbon monoxide is als produced. This condition prevails when a fuel-rich mixture is burned in a automobile engine. One possibility for the incomplete combustion of octane i

$$C_8H_{18}(l) + 12\,O_2(g) \longrightarrow 7\,CO_2(g) + CO(g) + 9\,H_2O(l) \qquad \textbf{(21.3}$$

CO as an air pollutant comes chiefly from the incomplete combustion of foss fuels in automobile engines. CO is an inhalation poison because CO molecule bond irreversibly to Fe atoms in hemoglobin in blood and displace the O molecules that the hemoglobin normally carries, as illustrated in Figure 21-2

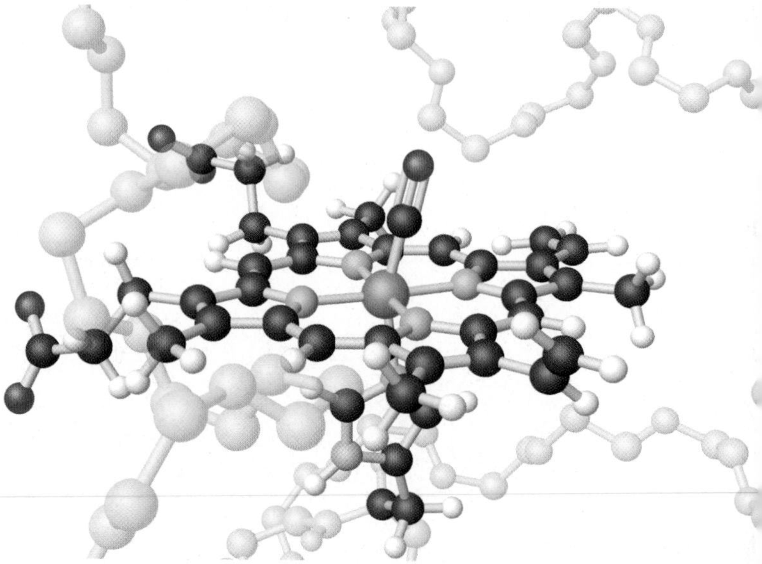

▶ FIGURE 21-29
CO bound to hemoglobin
Carbon monoxide binds to the iron atoms in hemoglobin more strongly than does oxygen. Thus, carbon monoxide's toxicity arises because it prevents hemoglobin from binding with oxygen. The portion of the hemoglobin molecule shown here is called a *heme* group. An iron atom (brown) is at the center of the group and is surrounded by four nitrogen atoms. In hemoglobin, an O_2 molecule projects above the plane of the iron and nitrogen atoms, but here it has been replaced by a CO molecule (black and red).

TABLE 21.8 Some Industrial Methods of Preparing CO_2

Method	Chemical Reaction
Recovery from exhaust stack gases in the combustion of carbonaceous fuels, such as the combustion of coke	$C(s) + O_2(g) \longrightarrow CO_2(g)$
Recovery in ammonia plants from steam-reforming reactions used to produce hydrogen	$CH_4(g) + 2\,H_2O(g) \longrightarrow CO_2(g) + 4\,H_2(g)$
Decomposition (calcination) of limestone at about 900 °C	$CaCO_3(s) \longrightarrow CaO(s) + CO_2(g)$
Fermentation by-product in the production of ethanol	$\underset{\text{a sugar}}{C_6H_{12}O_6(aq)} \longrightarrow 2\,C_2H_5OH(aq) + 2\,CO_2(g)$

Not only does incomplete combustion of gasoline contribute to air pollution, ut also it represents a loss of efficiency. A given quantity of gasoline evolves ss heat if $CO(g)$ is formed as a combustion product rather than $CO_2(g)$.

Although carbon dioxide can be obtained directly from the atmosphere as a y-product of the liquefaction of air, this is not an important source. Some of ne principal commercial sources of CO_2 are summarized in Table 21.8.

The major use of carbon dioxide (about 50%) is as a refrigerant in the form f dry ice for freezing, preserving, and transporting food. Carbonated beverges account for about 20% of CO_2 consumption. Other important uses are in il recovery in oil fields and in fire-extinguishing systems. Of course, the najor use is not by humans but by algae and plants.

Atmospheric CO_2 is the source of all the carbon-containing compounds organic compounds) synthesized by green plants. Here we briefly consider the *arbon cycle*—the major exchanges that occur between the atmosphere and the urface of Earth. A portion of the carbon cycle is represented in Figure 21-30. tmospheric CO_2 is the only source of carbon available to plants for making

21-10 CONCEPT ASSESSMENT

With a minimum of calculation, determine how much less heat is produced per mol of $C_8H_{18}(l)$ burned in reaction (21.30) than in reaction (21.29).

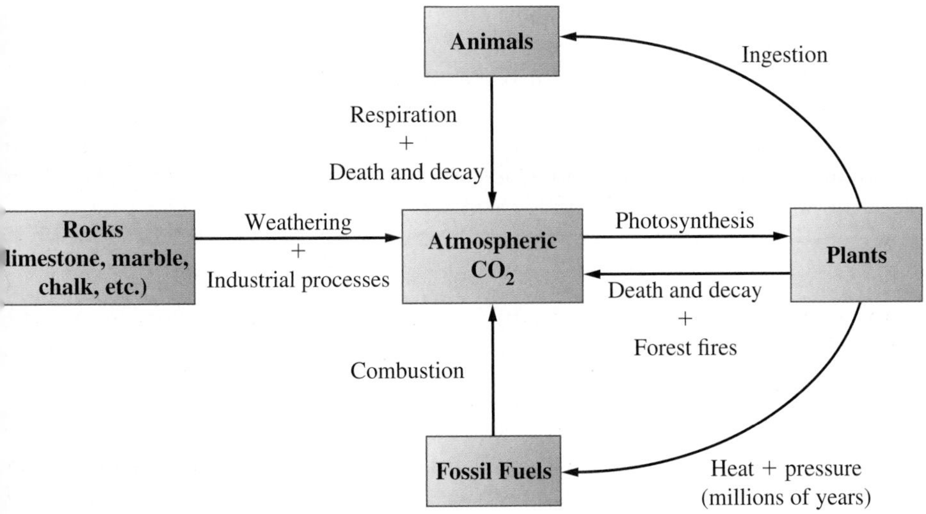

FIGURE 21-30
he carbon cycle

▶ Melvin Calvin won the Nobel Prize in Chemistry in 1961 for his research on the assimilation of carbon dioxide in plants.

organic compounds by the process of *photosynthesis*. The process is extreme complicated, and its details have been known for only a few decades. involves up to 100 sequential steps for the conversion of 6 mol CO_2 to 1 m $C_6H_{12}O_6$ (glucose). The overall change is represented by the chemical equ tion below:

$$6\,CO_2(g) + 6\,H_2O(l) \xrightarrow[\text{sunlight}]{\text{chlorophyll}} C_6H_{12}O_6(s) + 6\,O_2(g) \quad \Delta_rH = +2.8 \times 10^3\ \text{kJ mol}^{-}$$

The overall reaction is highly endothermic. The required energy comes fro sunlight. Chlorophyll, a green pigment in plants, is crucial to the proces Atmospheric oxygen is a by-product of the reaction.

Here are some of the ideas illustrated in Figure 21-30. When animals co sume plants, carbon atoms pass to the animals. Some carbon is returned to th atmosphere as CO_2 when the animals breathe and when they expel g (methane). Additional CO_2 returns to the atmosphere as plants and anima die and their remains are broken down by bacteria. Some carbon in decayir organic matter is converted to coal, petroleum, and natural gas. This carbon unavailable for photosynthesis.

Not represented in the drawing is the cycle of CO_2 through the oceans the world. Phytoplankton (small, floating green organisms) also carry on ph tosynthesis, converting CO_2 to organic compounds. Phytoplankton are at th bottom of the ocean food chain, directly and indirectly supporting all the an mals in the oceans.

Huge quantities of carbon have accumulated in the form of carbona rocks (mostly $CaCO_3$). These come from the shells of decayed mollusks i ancient seas.

Human activities now play a far more significant role in the carbon cyc than in preindustrial times. The combustion of fossil fuels is replacin stored carbon by carbon dioxide to an even greater extent. We have alread seen the possible consequences of this distortion of the carbon cycle: a increased level of atmospheric CO_2 and a future global warming (page 280 Human disruption of the natural carbon cycle has become a widely debate issue.

Carbon Monoxide A modern method of making carbon monoxide is the stea reforming of natural gas, which is based on the following chemical reaction:

$$CH_4(g) + H_2O(g) \longrightarrow CO(g) + 3\,H_2(g) \qquad \textbf{(21.3}$$

Steam refers to gaseous water. *Reforming* refers to the restructuring of a carbo compound, such as CH_4 to CO. The reforming of natural gas (mostly CH_4) i an important source of $H_2(g)$ for use in the synthesis of NH_3 (page 1068).

There are three main uses of carbon monoxide. One is in synthesizing othe compounds. For example, a mixture of CO and H_2 produced by reformin methane or some other hydrocarbon and known as *synthesis gas* can be co verted to a new organic chemical product, such as methanol, CH_3OH:

$$CO(g) + 2\,H_2(g) \longrightarrow CH_3OH(l)$$

Another use of CO is as a reducing agent. For instance, CO can be used t reduce iron oxide to iron, as shown below:

$$Fe_2O_3(s) + 3\,CO(g) \longrightarrow 2\,Fe(l) + 3\,CO_2(g)$$

The reaction can be carried out by heating Fe_2O_3 and coke—a form of pur carbon—in a blast furnace. Carbon is first converted to CO and then C reduces Fe_2O_3 to Fe.

A third use of CO is as a fuel, usually mixed with CH_4, H_2, and other com bustible gases. This was discussed in Section 7-9.

ther Inorganic Carbon Compounds Carbon combines with metals to form *rbides*. In many cases, the carbon atoms occupy the holes or voids, also called terstitial sites, in metal structures, forming *interstitial carbides*. With active etals, the carbides are ionic. *Calcium carbide* forms in the high-temperature action of lime and coke:

$$CaO(s) + 3\,C(s) \xrightarrow{2000\,^{\circ}C} CaC_2(s) + CO(g) \qquad (21.32)$$

alcium carbide is important because acetylene ($HC\equiv CH$) can easily be made om it, and acetylene can be used to synthesize many chemical compounds:

$$CaC_2(s) + 2\,H_2O(l) \longrightarrow Ca(OH)_2(s) + C_2H_2(g) \qquad (21.33)$$

olid CaC_2 can be regarded as a face-centered cubic array of Ca^{2+} ions with e C_2^{2-} ions in the octahedral holes, as shown in the figure in the margin.

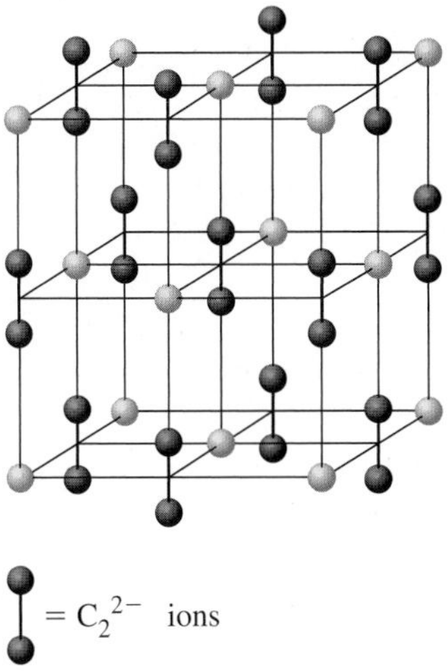

▲ **Structure of calcium carbide**

Carbon disulfide, CS_2, can be synthesized by the reaction of methane and sulır vapor in the presence of a catalyst:

$$CH_4(g) + 4\,S(g) \longrightarrow CS_2(l) + 2\,H_2S(g) \qquad (21.34)$$

arbon disulfide is a highly flammable, volatile liquid that acts as a solvent for ılfur, phosphorus, bromine, iodine, fats, and oils. Its uses as a solvent are ecreasing, however, because CS_2 is poisonous. Other important uses are in e manufacture of rayon and cellophane.

Carbon tetrachloride, CCl_4, can be prepared by the direct chlorination of ethane, as shown below:

$$CH_4(g) + 4\,Cl_2(g) \longrightarrow CCl_4(l) + 4\,HCl(g) \qquad (21.35)$$

lthough CCl_4 has been extensively used as a solvent, dry-cleaning agent, nd fire extinguisher, these uses have been steadily declining because CCl_4 uses liver and kidney damage and is a known carcinogen.

Certain groupings of atoms, several containing C atoms, have some of the aracteristics of a halogen atom. They are called *pseudohalogens* and include e following groupings of atoms:

—CN (cyanide) —OCN (cyanate) —SCN (thiocyanate)

he *cyanide* ion, CN^-, is similar to the halide ions, X^-, in that it forms an insolble silver salt, AgCN, and an acid, HCN. Hydrocyanic acid, HCN, is a liquid at boils at about room temperature. It is a very weak acid, unlike HCl. espite its extreme toxicity, HCN has important uses in the manufacture of lastics. The combination of two cyanide groups produces *cyanogen*, $(CN)_2$. his gas resembles chlorine gas in undergoing a disproportionation reaction basic solution:

$$(CN)_2 + 2\,OH^-(aq) \longrightarrow CN^-(aq) + OCN^-(aq) + H_2O(l) \qquad (21.36)$$

yanogen is used in organic synthesis, as a fumigant, and as a rocket propellant.

ilicon

mong the elements, silicon is second only to oxygen in its abundance in arth's crust. Silicon is to the mineral world what carbon is to the living orld—the backbone element.

roduction and Uses of Silicon Elemental silicon is produced when quartz or and (SiO_2) is reduced by reaction with coke in an electric arc furnace. The alanced chemical equation for the reaction is given below:

$$SiO_2 + 2\,C \xrightarrow{\Delta} Si + 2\,CO(g) \qquad (21.37)$$

ery high purity Si for solar cells can be made by reducing Na_2SiF_6 with etallic Na. The Na_2SiF_6 required for this process is obtained as a by-product

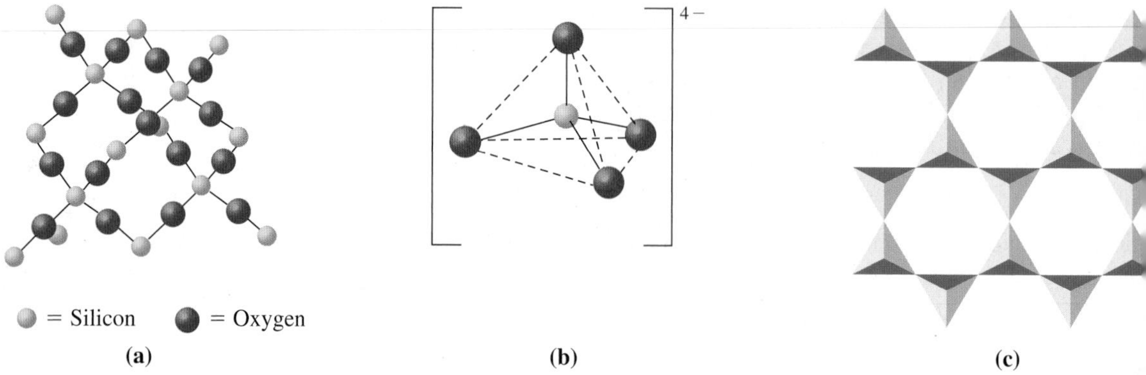

▲ FIGURE 21-31
Structures of silica and silicates
(a) A three-dimensional network of bonds in silica, SiO_2. **(b)** The silicate anion, SiO_4^{4-} commonly found in silicate materials. The Si atom is in the center of the tetrahedron and is surrounded by four O atoms. **(c)** A depiction of the structure of a mica, using tetrahedra to represent the SiO_4 units. The cations K^+ and Al^{3+} are also present but are not shown. This view, looking down on top of a tetrahedron, is the usual way that materials scientists represent silicates and similar materials.

of the formation of phosphate fertilizers (page 1077). High-purity silicon also required in the manufacture of transistors and other semiconducto devices.

Oxides of Silicon; Silicates Silica, SiO_2, is the only stable oxide of silico Silica is a network covalent solid (not a molecular solid, like CO_2). In silic each Si atom is bonded to *four* O atoms and each O atom to *two* Si atoms. Th structure is that of a network covalent solid, as suggested by Figure 21-31(a This structure is reminiscent of the diamond structure, and silica has certai properties that resemble those of diamond. For example, quartz, a form of si ica, is fairly hard (with a Mohs hardness of 7) but not as hard as diamon (with a Mohs hardness of 10), has a high melting point (about 1700 °C), and a nonconductor of electricity. Silica is the basic raw material of the glass an ceramics industries.

The central feature of all *silicates* is the SiO_4^{4-} tetrahedron depicted i Figure 21-31(b). These tetrahedra may be arranged in a wide variety of way Here are just a few examples:

- *Simple SiO_4 tetrahedra.* Typical minerals in which the anions are simp SiO_4^{4-} tetrahedra are *thorite* ($ThSiO_4$) and *zircon* ($ZrSiO_4$).
- *Two SiO_4 tetrahedra joined end-to-end.* The silicon atoms in the two tetrahe dra share an O atom between them in the anion $Si_2O_7^{6-}$, found in th mineral *thortveitite* ($Sc_2Si_2O_7$). See Figure 21-32(a).
- *SiO_4 tetrahedra joined into long chains.* Each Si atom shares an O atom wit the Si atom in an adjacent tetrahedron on either side. An example is *spo dumene*, the principal source of lithium and lithium compounds. It empirical formula is $LiAl(SiO_3)_2$. See Figure 21-32(b).
- *SiO_4 tetrahedra joined into a double chain.* Half the Si atoms share three o their four O atoms with Si atoms in adjacent tetrahedra, and half shar only two. In *chrysotile asbestos*, double chains are held together by cation (chiefly Mg^{2+}); this mineral has a fibrous appearance. The empirical fo mula is $Mg_3(Si_2O_5)(OH)_4$. See Figure 21-32(c).
- *SiO_4 tetrahedra are bonded together in two-dimensional sheets.* Each Si ator shares O atoms with the Si atoms in three adjacent tetrahedra. Thi structure is depicted in Figure 21-31(c). In *muscovite mica,* with the empiric

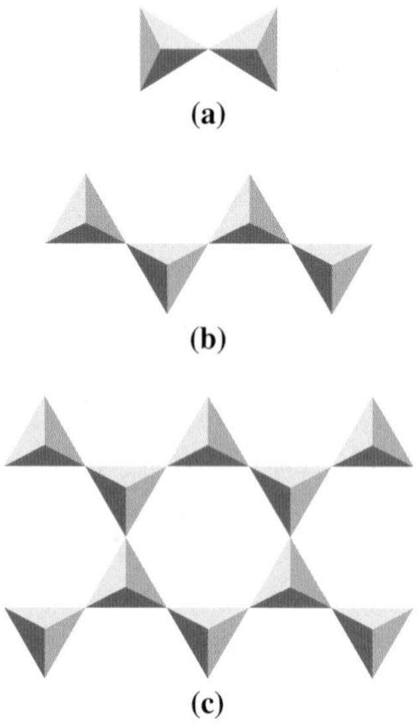

▲ FIGURE 21-32
Linking of SiO_4 tetrahedra
In **(a)**, two tetrahedra are linked at one corner to form the $Si_2O_7^{6-}$ ion. In **(b)**, tetrahedra are linked at two corners to form a long chain. The empirical formula of the chain is SiO_3^{2-}. In **(c)**, SiO_4 tetrahedra are joined into a double chain.

formula, $KAl_2(AlSi_3O_{10})(OH)_2$, the counterions that bond sheets together in layers are mostly K^+ and Al^{3+}. Bonding within the sheets is stronger than that between the sheets, so muscovite mica flakes easily.

- *SiO_4 tetrahedra are bonded together in three-dimensional structures.* Three-dimensional arrays of tetrahedra result when each Si atom shares all four O atoms with the Si atoms in four adjoining tetrahedra (as shown in Figure 21-31a). This is the most common arrangement, occurring in silica (quartz) and in the majority of silicate minerals.

21-11 CONCEPT ASSESSMENT

The empirical formula of the mineral *beryl* is $Be_3Al_2Si_6O_{18}$. By using the several descriptions of silicate minerals just given as a guide, describe the structure of the silicate anion in beryl.

SiO_2 is a weakly acidic oxide and slowly dissolves in strong bases. It forms series of silicates, such as Na_4SiO_4 (sodium orthosilicate) and Na_2SiO_3 sodium metasilicate). These compounds are somewhat soluble in water and, s a result, are sometimes referred to as "water glass."

Silicate anions are bases; when acidified, they produce silicic acids, which re unstable and decompose to silica. The silica obtained, however, is not a crystalline solid or powder. Depending on the acidity of the solution, the silica obtained as a colloidal dispersion, a gelatinous precipitate, or a solid-like gel n which all the water is entrapped. These hydrated silicates are polymers of ilica formed by the elimination of water molecules between neighboring molcules of silicic acid. The process begins with the following reaction:

$$SiO_4{}^{4-}(aq) + 4\,H^+(aq) \longrightarrow Si(OH)_4$$

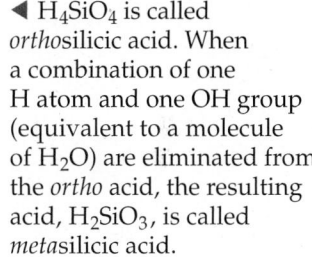

◀ H_4SiO_4 is called *ortho*silicic acid. When a combination of one H atom and one OH group (equivalent to a molecule of H_2O) are eliminated from the *ortho* acid, the resulting acid, H_2SiO_3, is called *meta*silicic acid.

21-2 ARE YOU WONDERING?

Why is the structure of SiO_2 different from that of CO_2?

Because both silicon and carbon are in group 14 of the periodic table and have four valence electrons, we might expect them to form oxides with similar properties. In CO_2, the side-to-side overlap of $2p$ orbitals of the C and O atoms is extensive and consequently, the carbon-to-oxygen double bond in CO_2 is stronger (799 kJ mol^{-1}) than two single bonds (2×360 kJ mol^{-1}). This results in the familiar Lewis structure of CO_2.

$$:\ddot{O}=C=\ddot{O}:$$

Silicon, being in the third period, would have to use $3p$ orbitals to form double bonds with oxygen. The side-to-side overlap of these orbitals with the $2p$ orbitals of oxygen is quite limited. In terms of energy, a stronger bonding arrangement results if the Si atoms form *four* single bonds with O atoms (bond energy: 4×464 kJ mol^{-1} = 1856 kJ mol^{-1}) rather than *two* double bonds (bond energy: 2×640 kJ mol^{-1} = 1280 kJ mol^{-1}). Because each O atom must simultaneously bond to two Si atoms, the result is a network of —Si—O—Si— bonds (see Figure 21-31a).

On page 994 we contrasted Be and Mg, the second- and third-period members of group 2. Here is another example of how the second-period member of a group (that is, carbon) differs from the higher period members.

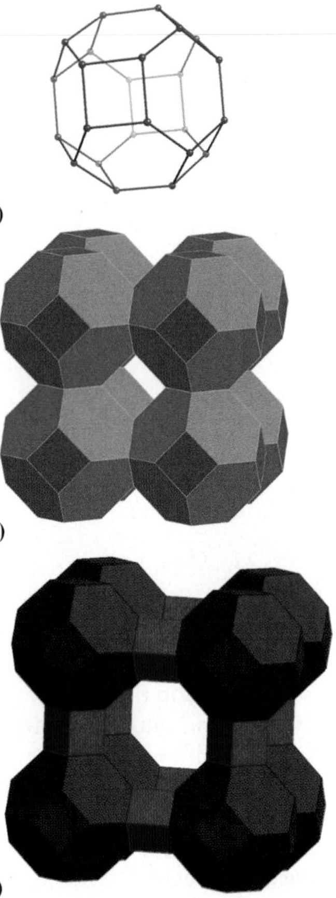

(a) (b) (c)

▲ FIGURE 21-33
Structures of zeolites
(a) The basic structural unit in many zeolites is a β-cage, which consists of six-membered and four-membered rings.
(b) This is the structure obtained when eight β-cages are joined together by sharing faces of the four-membered rings. This structural unit is found in sodalite, a naturally occurring zeolite. **(c)** This is the structure obtained when the four-membered rings of the β-cages are joined together by bridges. The resulting structure is more open than the structure shown in (b). Such structures are found in zeolite-A, a synthetic zeolite.

▶ The dehydrated form of a zeolite is obtained by heating the zeolite under vacuum. The heating process drives off the waters of hydration from the zeolite structure, leaving behind a structure that has a high affinity for water.

It is followed by the reaction below:

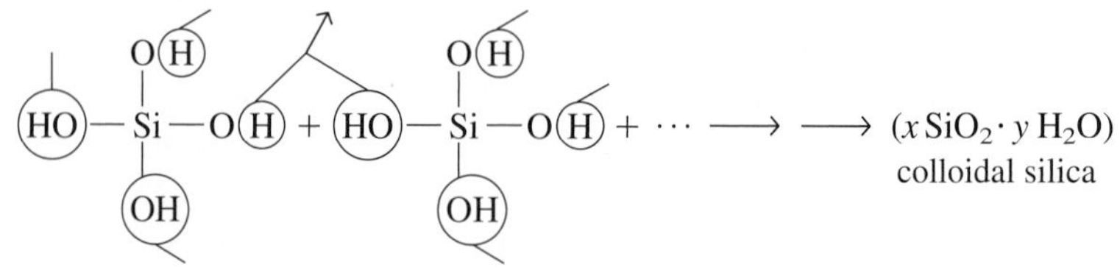

Notice that —H and —OH combine to form water (HOH) and that Si—O—S bridges are produced.

Zeolites: An Important Class of Aluminosilicates

A **zeolite** is a three-dimensional network of SiO_4 and AlO_4 tetrahedra. Many c the zeolite structures contain a ring based on $Si_6O_{18}{}^{12-}$. The ring of Si atom some of which can be replaced by Al atoms, is represented by a hexagon. Th hexagons can be joined together to give the structure in Figure 21-33(a). Bear i mind that the hexagons shown in Figure 21-33 emphasize the positions of th Si and Al atoms. The straight lines do not represent bonds; the atoms at the ve tices of the hexagons are, in fact, connected by bent Si—O—Si or Al—O—A bridges.

The structure shown in Figure 21-33(a) is called a β-cage and it is present i a naturally occurring zeolite known as *sodalite*. In the β-cage, there are fou and six-membered rings of nonoxygen atoms. In sodalite, eight of the β-cages a joined together by sharing the faces of the four-membered rings to give the cub structure shown in Figure 21-33(b). An alternative is to bridge oxygen at eac corner of the four ring faces. This produces the structure shown in Figure 21-33(c which is found in a synthetic zeolite, $Na_{12}(AlO_2)_{12}(SiO_2)_{12} \cdot 27\,H_2O$, know as zeolite-A. The Na^+ ions in the formula of zeolite-A are required to offset th decrease in positive charge that occurs when Al^{3+} ions replace Si^{4+} ions i the lattice.

The important consequence of the arrangements described above, and illus trated in Figure 21-33, is the presence of channels and cavities in zeolite stru tures. These channels and cavities give zeolites their important properties. Fo example, because small molecules are able to diffuse into cavities and large molecules are excluded, zeolites have been used as molecular sieves to remov molecules of certain sizes from a mixture. Zeolites, in their dehydrated form have also been used for removing water from gases or organic solvents. Fo example, in the drying of benzene, water molecules diffuse into the zeolite la tice while benzene molecules, which are too large, are excluded. The zeolite ca be separated from the benzene and can be regenerated by heating.

Another important application of zeolites is as an ion exchange material i the treatment of hard water, as illustrated in Figure 21-34. Hard water contain significant concentrations of ions, especially Ca^{2+}, Mg^{2+}, or Fe^{2+}, and a zeolit can be used to exchange these ions with Na^+ ions. It is desirable to remov Ca^{2+}, Mg^{2+}, and Fe^{2+} ions from water because these ions react with $CO_3{}^{2}$ ions or anions of soaps to form insoluble precipitates. For exampl Ca^{2+}, Mg^{2+}, and Fe^{2+} ions react with $CO_3{}^{2-}$ ions to form a mixed precipitate c $CaCO_3$, $MgCO_3$, and rust called *boiler scale*. The formation of boiler scale low ers the efficiency of water heaters and builds up in pipes or on the inside of containers used for boiling water. Ca^{2+} and Mg^{2+} ions in hard water com bine with anions of soaps—such as the palmitate ion, equation (21.9)—to form insoluble precipitates that accumulate on the surfaces of bathtubs or shower and contribute to the formation of *soap scum* and *bathtub rings*.

The equation below represents the exchange that occurs when, for exampl Ca^{2+} ions (in hard water) are exchanged with Na^+ ions (in a zeolite):

$$Na_2[\text{zeolite}](s) + Ca^{2+}(aq) \longrightarrow Ca[\text{zeolite}](s) + 2\,Na^+(aq)$$

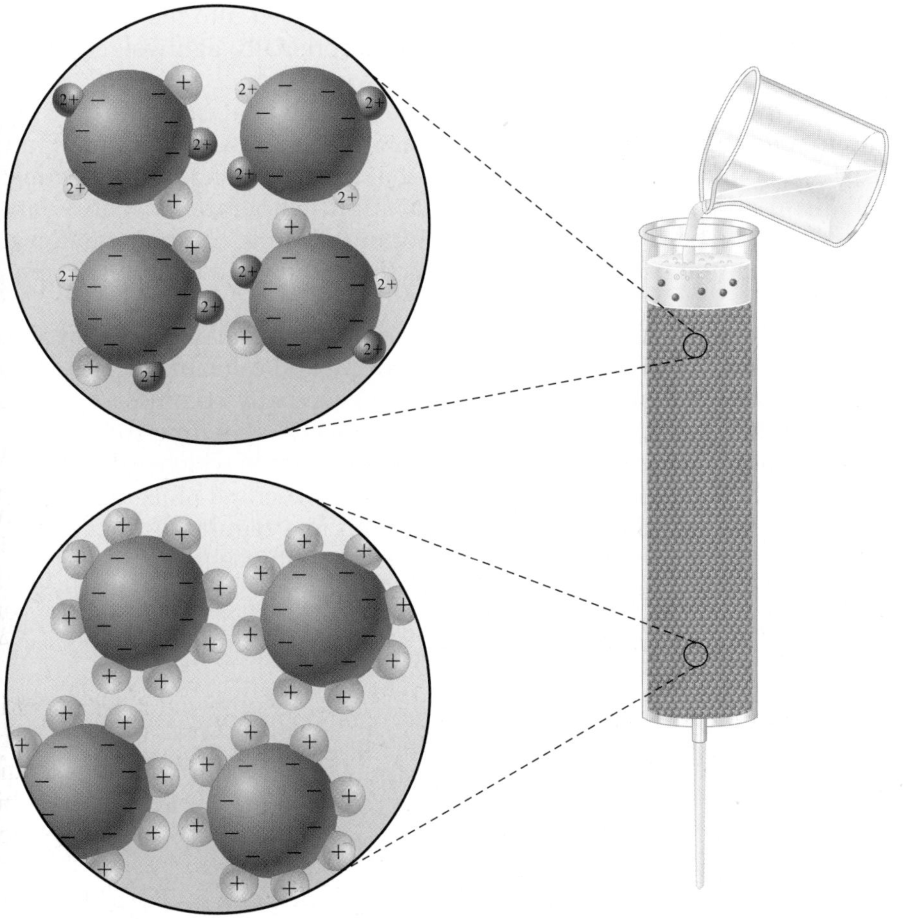

FIGURE 21-34
on exchange
ne resin shown here is a cation-exchange resin (a zeolite, for example). Multivalent ations (green, Fe^{2+}, yellow, Ca^{2+}) in the solution replace Na^+ (orange) at the top of e resin column. By the time the water has reached the bottom of the column, all the ultivalent ions have been removed and only Na^+ ions remain as counterions. The xchange can be represented as $2\ NaR + M^{2+} \rightleftharpoons MR_2 + 2\ Na^+$. The reaction ccurs in the forward direction during water softening. As expected from Le Châtelier's rinciple, in the presence of concentrated NaCl(aq), the reverse reaction is favored and ne resin is recharged.

ecause sodium compounds are generally soluble, the replacement of Ca^{2+} ons by Na^+ ions prevents the formation of insoluble precipitates.

Zeolites are used not only in ion exchange resins but also in detergents to elp remove any Ca^{2+} and Mg^{2+} ions that might be present in water used for ashing clothes. The removal of these ions helps the detergents foam better nd also helps to prevent the formation of insoluble calcium and magnesium ompounds.

Modern automobile engines require fuels containing short-chain hydrocarons with low boiling points, and some require high octane fuels containing ranched-chain hydrocarbons. Zeolite catalysts are used to speed up the conersion of long-chain hydrocarbons—found in crude oil—into short-chain or ranched-chain hydrocarbons. Consequently, zeolites play an important role n the petroleum industry.

ilicates in Ceramics and Glass Hydrated silicate polymers are important in ne ceramics industry. A colloidal dispersion of particles in a liquid is called a *ol*. The sol can be poured into a mold and, following removal of some of the

▲ Ceramic components of an automobile engine. (Photo courtesy of Kyocera Industrial Ceramics Corp./Vancouver, WA)

Photo courtesy of Kyocera Industrial Ceramics Corp./Vancouver

liquid, is converted to a *gel*. The gel is then processed into the final ceram product. This sol-gel process can produce exceptionally lightweight ceram materials.

Uses of these advanced ceramics fall into two general categories: (1) ele trical, magnetic, or optical applications (as in the manufacture of integrate circuit components) and (2) applications that take advantage of the ceramic mechanical and structural properties at high temperatures. These latt properties have been explored in developing ceramic components for g turbines and automotive engines, such as those shown in the photograph i the margin.

If sodium and calcium carbonates are mixed with sand and fused at abo 1500 °C, the result is a liquid mixture of sodium and calcium silicates. Whe cooled, the liquid becomes more viscous and eventually becomes a solid th is transparent to light; this solid is called a **glass**. Crystalline solids have long-range order, whereas glasses are *amorphous solids* in which order is foun over relatively short distances only. Think of the structural units in glass (sil cate anions) as being in a jumbled rather than in a regular arrangement. glass and a crystalline solid also differ in their melting behavior. A glass sof ens and melts over a broad temperature range, whereas a crystalline solid h a definite, sharp melting point. Different types of glass and methods of mal ing them are described later in this section.

Silanes and Silicones Several silicon–hydrogen compounds are known, b because Si—Si single bonds are not particularly strong, the chain length i these compounds, called *silanes*, is limited to six.

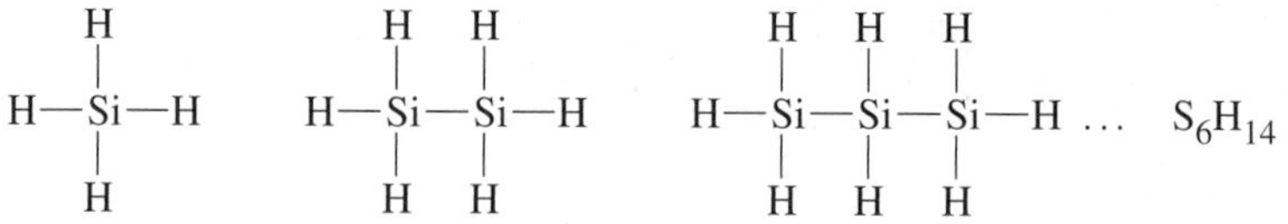

Monosilane Disilane Trisilane Hexasilane

Other atoms or groups of atoms can be substituted for H atoms in silanes t produce compounds called *organosilanes*. Typical is the direct reaction of S and methyl chloride, CH_3Cl. The equation for the reaction is given below:

$$2\,CH_3Cl + Si \longrightarrow (CH_3)_2SiCl_2$$

The reaction of $(CH_3)_2SiCl_2$, dichlorodimethylsilane, with water produces a interesting compound, *dimethylsilanol*, $(CH_3)_2Si(OH)_2$. Dimethylsilanol unde goes a polymerization reaction in which H_2O molecules are eliminated fro among large numbers of silanol molecules. The result of this polymerization is material consisting of molecules with long silicon-oxygen chains: **silicones**.

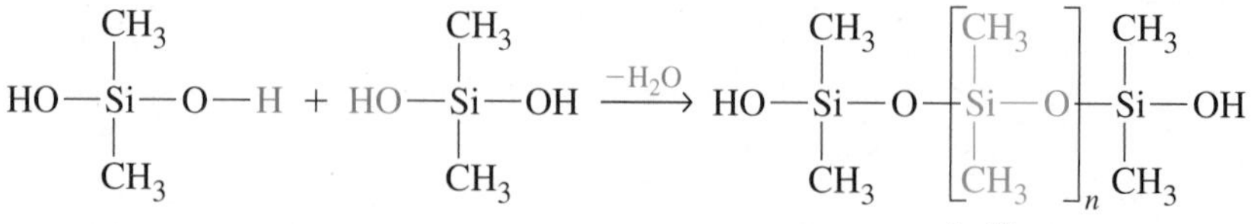

A silicone

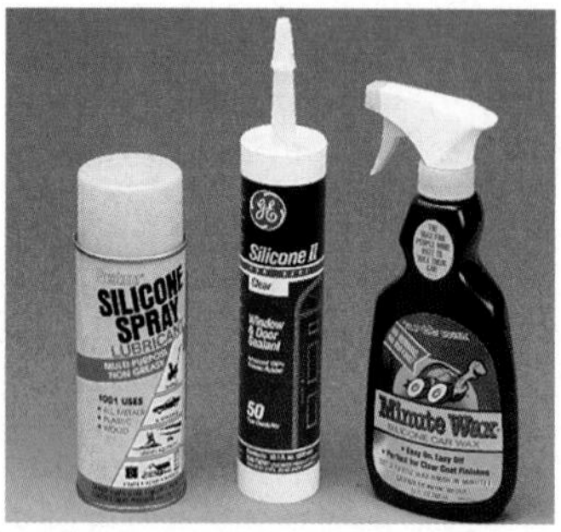

Frank Labua

▲ Some common applications of silicones.

Silicones are important polymers because they are versatile. They are use to make a variety of useful products, such as those shown in the photograp in the margin. Silicones can be obtained either as oils or as rubber-like mater als. Silicone oils are not volatile and do not decompose when heated. Als they can be cooled to low temperatures without solidifying or becomin viscous. Silicone oils are excellent high-temperature lubricants. In contras ydrocarbon oils break down at high temperatures, become very viscous, and ıen solidify at low temperatures. Silicone rubbers retain their elasticity at low mperatures and are chemically resistant and thermally stable. This makes ıem useful in caulking around windows, for instance.

licon Halides Silicon reacts readily with the halogens (X_2) to form the prod-:ts SiX_4. As we might expect from their molecular structures and size (polar-ability), at room temperature SiF_4 is a gas, $SiCl_4$ and $SiBr_4$ are liquids, and I_4 is a solid. Both SiF_4 and $SiCl_4$ are readily hydrolyzed with water, but the ırmer is only partially hydrolyzed. The hydrolysis reactions for F_4 and $SiCl_4$ are as follows:

$$SiF_4(g) + 4\,H_2O(l) \rightleftharpoons SiO_2(s) + 2\,H_3O^+(aq) + [SiF_6]^{2-}(aq) + 2\,HF(aq) \quad (21.38)$$

$$SiCl_4(l) + 2\,H_2O(l) \longrightarrow SiO_2(s) + 4\,HCl(aq) \quad (21.39)$$

The ability of the Si atom to expand its valence shell is illustrated by the for-ıation of the ion $[SiF_6]^{2-}$. The corresponding $[SiCl_6]^{2-}$ has not been prepared, robably because the chloride ion is too big for six of them to fit around the atom.

$SiCl_4$ is manufactured on a large scale to produce finely divided silica (reac-on 21.39), used as a reinforcing filler in silicone rubber, and very pure silicon ır transistors used in computer chips.

lass Making Glass and the art of glassmaking have been known for millen-ia. Beautiful stained-glass windows can be seen in medieval and modern ıurches; ancient glass containers for perfume and oil are displayed in many ıuseums. Today, glass is indispensable in almost every facet of life.

Soda–lime glass is the oldest form of glass. The starting material in its manu-cture is a mixture of sodium carbonate (*soda* ash), calcium carbonate (which ecomposes to form quick*lime* when heated), and silicon dioxide. The mixture ın be fused at a relatively low temperature (1300 °C) compared with the ıelting point of pure silica (1710 °C), and it is easy to form into the shapes eeded. The effect of the sodium ions is to break up the crystalline lattice of ıe SiO_2; the calcium ions render the glass insoluble in water, so it can be used ır such items as drinking glasses and windows. At the high temperatures nployed, chemical reactions occur that produce a mixture of sodium and cal-um silicates as the ultimate glass product.

Robin Holden Sr/Shutterstock

Stained-glass windows form the 20-meter-high ceiling, called the Glory Window, side the Chapel of Thanksgiving in Dallas, Texas.

Glass containing even small amounts of FeO has a distinctive green col (bottle glass). Glass can be made colorless by incorporating MnO_2 into th glassmaking process. The MnO_2 oxidizes green $FeSiO_3$ to yellow $Fe_2(SiO_3$ and is itself reduced to Mn_2O_3, which imparts a violet color. The yellow ar violet are complementary colors, so the glass appears colorless. Where desire color can be imparted by means of appropriate additives, such as CoO f cobalt blue glass. To produce an opaque glass, such additives as calcium pho phate are used. In Bohemian crystal, most of the Na^+ is replaced by K^+; a gla with exceptional transparency can be made by incorporating lead oxides.

A problem with soda–lime glass is its high coefficient of thermal expa sion—its dimensions change significantly with temperature. The glass cann withstand thermal shock. This limitation posed a particular problem for th lanterns used in the early days of railroads. In the rain, the hot glass in the lanterns would easily shatter. Adding B_2O_3 to the glass solved the problem. *borosilicate* glass has a low coefficient of thermal expansion and is thus resi tant to thermal shock. This is the glass more commonly known by its trac name, Pyrex. It is widely used in chemical laboratories and for cookware the home.

Most glass has small bubbles or impurities in it that decrease its ability transmit light without scattering—a phenomenon observed in the distorte images produced by the thick bottoms of drinking glasses. In modern *fibe optic* cables, sound waves are converted to electrical impulses, which a transmitted as laser light beams. The light must be transmitted over long di tances without distortion or loss of signal. For this purpose, a special gla made of pure silica is required. The key to making this glass is in purifying si ica, which can be done by a series of chemical reactions. First, impure quar or sand is reduced to silicon by using coke as a reducing agent. The silicon then allowed to react with $Cl_2(g)$ to form $SiCl_4(g)$. The reactions involved the conversion of SiO_2 to $SiCl_4$ are as follows:

$$SiO_2(s) + 2\,C(s) \longrightarrow Si(s) + 2\,CO(g)$$
$$Si(s) + 2\,Cl_2(g) \longrightarrow SiCl_4(g)$$

Finally, the $SiCl_4$ is burned in a methane–oxygen flame. SiO_2 deposits as a fir ash, and chlorocarbon compounds escape as gaseous products. The SiO_2, wit impurity levels reduced to parts per billion, can then be melted and drawn int the fine filaments required in fiber-optic cable. Tens of millions of kilometers fiber-optic cable are currently produced annually in the United States.

Diagonal Relationship of Boron and Silicon

The following similarities between B and Si are listed here:

- Boron forms a solid acidic oxide, B_2O_3, like that of silicon, SiO_2. In co trast Al_2O_3 is amphoteric and CO_2 is acidic.
- Boric acid, H_3BO_3, is a weak acid similar to silicic acid H_4SiO_4.
- There is a wide range of polymeric borates and silicates, based on share oxygen atoms.
- Both boron and silicon form gaseous hydrides.

This diagonal relationship is not readily understood and cannot be inte preted in terms of charge density since the bonding in boron compounds an in silicon compounds is exclusively covalent. The elements are, howeve both metalloids, have similar electronegativities, and have similar sizes leac ing to similar chemical behavior.

Properties and Uses of Tin and Lead

The data in Table 21.9 suggest that tin and lead are rather similar to each othe Both are soft and malleable and melt at low temperatures. The ionizatio

TABLE 21.9 Some Properties of Tin and Lead

	Sn	Pb
Atomic number	50	82
Atomic (metallic) radius, pm	141	175
Ionic (M^{2+}) radius, pm[a]	93	118
First ionization energy, kJ mol^{-1}	709	716
Electrode potential $E°$, V		
$[M^{2+}(aq) + 2e^- \longrightarrow M(s)]$	−0.137	−0.125
$[M^{4+}(aq) + 2e^- \longrightarrow M^{2+}(aq)]$	+0.154	+1.5
Melting point, °C	232	327
Boiling point, °C	2623	1751
Density, g cm^{-3} at 20 °C	5.77 (α, gray) 7.29 (β, white)	11.34
Hardness[b]	1.6	1.5
Electrical conductivity[b]	14.4	7.68

[a]For coordination number four.
[b]See footnotes c and d of Table 21.2.

nergies and standard electrode potentials of the two metals are also about the ame. This means that their tendencies to be oxidized to the +2 oxidation state re comparable.

The fact that both tin and lead can exist in two oxidation states, +2 and +4, is n example of the inert pair effect (page 1006). In the +2 oxidation state, the nert pair ns^2 is not involved in bond formation, whereas in the +4 oxidation tate, the pair does participate. Tin displays a stronger tendency to exist in the 4 oxidation state than does lead. That tendency is consistent with the trend bserved in group 13, in which the lower oxidation state is favored farther own a group.

Another difference between tin and lead is that tin exists in two common rystalline forms (α and β), whereas lead has but a single solid form. The α gray), or nonmetallic, form of tin is stable below 13 °C; the β (white), or netallic, form of tin is stable above 13 °C. Ordinarily, when a sample of β tin is ooled, it must be kept below 13 °C for a long time before the transition to α tin ccurs. Once it does begin, however, the transformation takes place rather apidly and with dramatic results. Because α tin is less dense than the β variety, the tin expands and crumbles to a powder. This transformation leads to he disintegration of objects made of tin. It has been a particular problem in hurches in colder climates because some organ pipes are made of tin or tin lloys. The transformation is known in northern Europe as *tin disease, tin pest,* r *tin plague.*

The chief tin ore is tin(IV) oxide, SnO_2, known as *cassiterite*. After initial urification, the tin(IV) oxide is reduced with carbon (coke) to produce tin netal, as shown below:

$$SnO_2(s) + C(s) \xrightarrow{\Delta} Sn(l) + CO_2(g) \qquad \textbf{(21.40)}$$

Nearly 50% of the tin metal produced is used in tinplate, especially in plating iron for use in cans for storing foods. The next most important use (about 5% of the total produced) is in the manufacture of **solders**—low-melting lloys used to join wires or pieces of metal. Other important alloys of tin are *ronze* (90% Cu, 10% Sn) and *pewter* (85% Sn, 7% Cu, 6% Bi, 2% Sb). Alloys of n and Pb are used to make organ pipes.

Lead is found chiefly as lead(II) sulfide, PbS, an ore known as *galena*. ead(II) sulfide is first converted to lead(II) oxide by heating it strongly in air, a process called **roasting**. The oxide is then reduced with coke to produce th metal. The reactions are as follows:

$$2\,PbS(s) + 3\,O_2(g) \xrightarrow{\Delta} 2\,PbO(s) + 2\,SO_2(g) \quad (21.4$$

$$2\,PbO(s) + C(s) \xrightarrow{\Delta} 2\,Pb(l) + CO_2(g) \quad (21.4$$

More than half the lead produced is used in lead–acid (storage) batterie Other uses include the manufacture of solder and other alloys, ammunitio and radiation shields (to protect against X-rays).

Compounds of Group 14 Metals

TABLE 21.10 Charge Densities of Tin and Lead Ions

Ion	Charge Density, C mm^{-3}
Sn^{2+}	54
Pb^{2+}	32
Sn^{4+}	267
Pb^{4+}	196

As mentioned earlier, both tin and lead exhibit the +2 and +4 oxidation state The charge densities of these ions are shown in Table 21.10. The charge densit of Sn^{2+} is such that many of the compounds containing tin in the +2 oxidatio state are covalent; however, a few ionic solids containing the Sn^{2+} ion ar known. Lead(II) exists in many ionic solids. When the metals are in the + oxidation state, they form covalent bonds, not ionic bonds, because thes Sn^{4+} and Pb^{4+} have very large charge densities and draw electron densit from surrounding anions toward themselves.

▶ Another mixed-valence oxide that we have encountered on several occasions in this text is Fe_3O_4 (see, for example, page 86).

Oxides Tin forms two primary oxides: SnO and SnO_2. By heating SnO in ai it can be converted to SnO_2. One use of SnO_2 is as a jewelry abrasive.

Lead forms a number of oxides, and the chemistry of some of these oxide is not completely understood. The best known oxides of lead are yellow PbC *litharge*; red-brown lead dioxide, PbO_2; and a mixed-valence oxide known a *red lead*, Pb_3O_4. Lead oxides are used in the manufacture of lead–acid (storage batteries, glass, ceramic glazes, cements (PbO), metal-protecting pain (Pb_3O_4), and matches (PbO_2). Other lead compounds are generally mad from the oxides.

Because lead tends to be in the +2 oxidation state, lead(IV) compound tend to undergo reduction to compounds of lead(II) and are therefore goo oxidizing agents. A case in point is PbO_2. In Chapter 19, we noted its use a the cathode in lead–acid storage cells. The reduction of $PbO_2(s)$ can be repre sented by the half-equation

$$PbO_2(s) + 4\,H^+(aq) + 2\,e^- \longrightarrow Pb^{2+}(aq) + 2\,H_2O(l) \qquad E° = +1.455\text{ V}$$

$PbO_2(s)$ is a better oxidizing agent than $Cl_2(g)$ and nearly as good a $MnO_4^-(aq)$. For example, $PbO_2(s)$ can oxidize HCl(aq) to $Cl_2(g)$, as show below:

$$PbO_2(s) + 4\,HCl(aq) \longrightarrow PbCl_2(aq) + 2\,H_2O(l) + Cl_2(g) \qquad E°_{cell} = 0.097\text{ V}$$

Halides Tin(IV) chloride is a covalent compound. It is an oily liquid that react with moisture in the air as follows:

$$SnCl_4(l) + 4\,H_2O(l) \longrightarrow Sn(OH)_4(s) + 4\,HCl(g)$$

Lead(IV) chloride is also a covalent compound and a yellow oil that react with moisture in the air in a manner similar to $SnCl_4$.

Tin(II) and lead(II) chlorides are quite different. Lead(II) chloride is a whit insoluble ionic solid, whereas tin(II) chloride is a covalent solid that is solubl in organic solvents. In the gas phase, $SnCl_2$ is a V-shaped molecule, a expected from VSEPR theory. We might expect $SnCl_2$ to act as a Lewis bas because of the presence of the lone pair. However, it acts as a Lewis acid. Fo example, $SnCl_2$ reacts with Cl^- to form $SnCl_3^-$.

Both chlorides of tin—$SnCl_2$ and $SnCl_4$—have important uses. Tin(II) chlo ride, $SnCl_2$, is a good reducing agent and is used in the quantitative analysi of iron ores to reduce iron(III) to iron(II) in aqueous solution. Tin(IV) chloride $SnCl_4$, is formed by the direct reaction of tin and $Cl_2(g)$; it is the form in whic tin is recovered from scrap tinplate. Tin(II) fluoride, SnF_2 (stannous fluoride as used as an anticavity additive to toothpaste but has largely been replaced y NaF in gel toothpastes.

ther Compounds As we should expect from the solubility guidelines able 5.1), one of the few soluble lead compounds is lead(II) nitrate, $Pb(NO_3)_2$. is formed in the reaction of PbO_2 with nitric acid, as shown below:

$$2\,PbO_2(s) + 4\,HNO_3(aq) \longrightarrow 2\,Pb(NO_3)_2(aq) + 2\,H_2O(l) + O_2(g)$$

he addition of a soluble chromate salt to $Pb(NO_3)_2(aq)$ produces lead(II) hromate $PbCrO_4$, a yellow pigment known as chrome yellow. Another lead-ased pigment used in ceramic glazes and once extensively used in the manu-cture of paint is *white lead*, $2\,PbCO_3 \cdot Pb(OH)_2$.

ead Poisoning

eginning with the ancient Romans and continuing to fairly recent times, lead as been used in plumbing systems, including those designed to transport rinking water. Exposure to lead has also occurred through cooking and eat-ng utensils and pottery glazes made with lead. In colonial times, lead poison-ng was clearly diagnosed as the cause of "dry bellyache" suffered by some orth Carolinians who consumed rum made in New England. The distilling quipment used in the manufacture of the rum had components made of lead.

Mild forms of lead poisoning produce nervousness and depression. More evere cases can lead to permanent nerve, brain, and kidney damage. Lead terferes with the biochemical reactions that produce the iron-containing eme group in hemoglobin. As little as 10–15 μg Pb/dL in blood seems to pro-uce physiological effects, especially in small children. The phaseout of aded gasoline has resulted in a dramatic drop in average lead levels in lood. The drop in blood lead levels is evident in the graph shown in igure 21-35 and parallels the decline in the use of lead in gasoline. The princi-al sources of lead contamination now seem to be lead-based painted surfaces und in old buildings and soldered joints in plumbing systems. Lead has een eliminated from modern plumbing solder, which is now a mixture of 5% Sn and 5% Sb. Because lead is toxic, its disposal is closely monitored. ecycling provides about three-quarters of the current lead metal production.

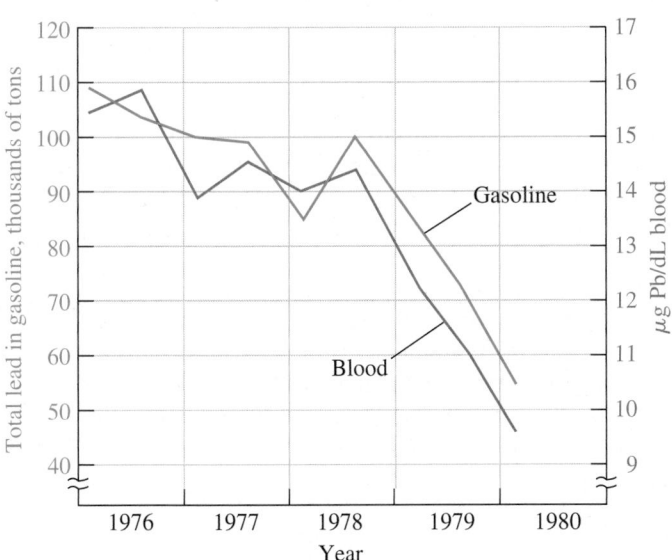

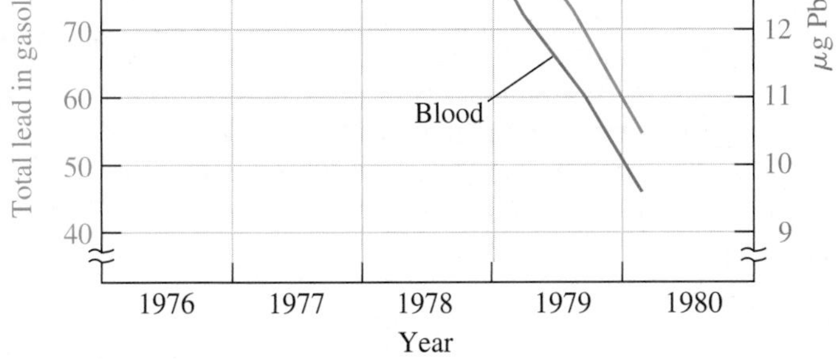

FIGURE 21-35
ead in gasoline and in blood
he level of lead in the blood of a representative human population showed a dramatic ecline that paralleled the decline in the use of lead additives in gasoline in the 1970s. ata source: Environmental Protection Agency, Office of Policy Analysis, 1984.)

21-12 CONCEPT ASSESSMENT

To prevent the air oxidation of aqueous solutions of Sn^{2+} to Sn^{4+}, metallic tin is sometimes kept in contact with the $Sn^{2+}(aq)$. Suggest how this contact helps prevent the oxidation.

MasteringCHEMISTRY www.masteringchemistry.com

Gallium arsenide, GaAs, may be one of the most versatile high-tech materials of our time. For a discussion of some of its properties and uses, go to the Focus On feature for Chapter 21, Gallium Arsenide, on the MasteringChemistry site.

Summary

21-1 Periodic Trends and Charge Density—Trends in atomic or ionic radii, ionization energy, electron affinity, electronegativity, and polarizability (Fig. 21-1), and the **charge density** (equation 21.1) of an ion can be used to rationalize the chemistry of the elements. The greater the charge density, the greater is the tendency for that cation to form bonds with covalent character.

21-2 Group 1: The Alkali Metals—The **alkali metals** (group 1) are the most active of the metals, as indicated by their low ionization energies and large negative electrode potentials. Lithium exhibits some properties similar to magnesium in group 2. Most of the alkali metals are prepared by the electrolysis of their molten salts. Electrolysis of NaCl(aq) produces NaOH(aq), from which many other sodium compounds can be prepared (Fig. 21-5). Na_2CO_3 can be produced from NaCl, NH_3, and $CaCO_3$ by the Solvay process (Fig. 21-7). A **detergent** is a cleansing agent, often sodium salts with long-chain hydrocarbon terminated with an anionic sulfate group. A **soap** is also a detergent, but soaps are the sodium salts of long-chain fatty acids terminating in the carboxylate group.

21-3 Group 2: The Alkaline Earth Metals—The **alkaline earth metals** (group 2), like group 1 metals, are also very active. Some group 2 metals are prepared by the electrolysis of a molten salt (Fig. 21-14) and some by chemical reduction. Among the most important of the alkaline earth compounds are the carbonates, especially $CaCO_3$. The oxide of calcium, CaO, called **quicklime**, is formed by the high-temperature decomposition (**calcination**) of limestone. The hydroxide of calcium, $Ca(OH)_2$, called **slaked lime**, is formed in the reaction of quicklime and water. Reversible reactions involving CO_3^{2-}, HCO_3^-, $CO_2(g)$, and H_2O account for the formation of limestone caves and interior features, such as **stalactites** and **stalagmites** (Fig. 21-15). **Plaster of Paris** is the hemihydrate of calcium sulfate, $CaSO_4 \cdot \frac{1}{2}H_2O$, formed by heating the mineral gypsum, $CaSO_4 \cdot 2\,H_2O$. The similar properties exhibited by the pairs of elements (Li, Mg), (Be, Al), and (B, Si) are known as **diagonal relationships**.

21-4 Group 13: The Boron Family—Group 13 contains one nonmetal, B, and the metals Al, Ga, In, and Tl. Boron compounds are often electron deficient (fewer than an octet of electrons around the B atoms). Three-cent two-electron bonds are used in describing the bonding i diborane (Fig. 21-17). Aluminum occurs almost exclusive in the +3 oxidation state, whereas thallium is mostly in th +1 oxidation state. This is a manifestation of the presen of a pair of *s* electrons in the valence shells of certain pos transition elements. These electrons do not participate i chemical bonding, a consequence referred to as the **ine pair effect**. The principal metal of group 13 is aluminur whose large-scale use is made possible by an effectiv method of production (Fig. 21-24). The amphoterism o Al_2O_3 is the basis for separating Al_2O_3 from impuritie mostly Fe_2O_3. Electrolysis is carried out in molten Na_3Al with Al_2O_3 as a solute. Aluminum is a powerful reducin agent in the **thermite reaction**, in which Fe_2O_3 is reduce to the metal. Aluminum chloride forms a **dimer**, in whic two bridging chlorine atoms join together two AlC units (Fig. 21-25). $AlCl_3$ is very reactive. For example, molecule of $AlCl_3$, a Lewis acid, attaches itself to a Lew base to form an addition compound called an **adduc Alums** are a class of double salts with the formu $M(I)M(III)(SO_4)_2 \cdot 12\,H_2O$. Gallium has gained impo tance in the electronics industry because of the desirab semiconductor properties of gallium arsenide (GaAs).

21-5 Group 14: The Carbon Family—Group 1 contains one nonmetal (C), two metalloids (Si and Ge), an the metals Sn and Pb. Group 14 is notable for the significa differences between the first two members (Table 21.7 Carbon is found in several different physical forms i addition to its allotropes; one of these is **carbon black**. A interesting form of carbon is **graphene**, a sheet of carbo atoms only one atom thick. Carbon monoxide and carbo dioxide are both formed in the combustion of fossil fuel Carbon monoxide is a poison (Fig. 21-29) and carbon dio ide plays a key role in the carbon cycle (Fig. 21-30). All th elements of group 14 form halides of the type MX_4. All bu carbon can employ expanded valence shells to form com pounds of the form species, such as the anion $[MX_6]^{2-}$ Progressing down group 14, the halide MX_2 become more stable than MX_4 because of the inert pair effect. Silic (SiO_2) and various silicate anions are ubiquitous compo nents of the mineral world. They are also constituents o ceramic materials and **glass**. Organosilanes, compound which the hydrogen atoms of silanes are replaced by rganic groups, can be polymerized to form **silicones**. **eolites** are aluminosilicates that are used as molecular eves and in treating hard water. Tin and lead in group 14 ave some similarities (Table 21.9). They are both soft mets with low melting points. They also have some differnces, including the fact that tin acquires the oxidation state +4 rather easily, whereas the +2 oxidation state is favored by lead. One of the most important uses of tin is in low-melting-point alloys known as **solders**. Both elements are obtained by the reduction of their oxides. The metallurgical method of converting a sulfide ore to an oxide, typified by the conversion of PbS to PbO on strong heating, is called **roasting**.

Integrative Example

Vithout performing detailed calculations, demonstrate that reaction (21.19) correctly describes the dissolving action of ainwater on limestone (for $CaCO_3$, $K_{sp} = 2.8 \times 10^{-9}$).

nalyze

Vrite the equations that, when combined, yield equation (21.19) for the overall reaction occurring when rainwater acts on mestone. Evaluate the sources and relative importance of the different species involved in the dissolution of the limeone. Describe the overall result.

olve

he relevant equations to describe the dissolution of calium carbonate in rainwater are the solubility product xpression for $CaCO_3$ and two equations describing the nization of CO_2 in water (that is, carbonic acid). The ree equations are as follows:

$$CaCO_3(s) \rightleftharpoons Ca^{2+}(aq) + CO_3^{2-}(aq) \qquad K_{sp} = 2.8 \times 10^{-9}$$

$$CO_2 + 2\,H_2O \rightleftharpoons H_3O^+ + HCO_3^- \qquad K_{a_1} = 4.4 \times 10^{-7} \quad \textbf{(21.17)}$$

$$HCO_3^- + H_2O \rightleftharpoons H_3O^+ + CO_3^{2-} \qquad K_{a_2} = 4.7 \times 10^{-11} \quad \textbf{(21.18)}$$

quation (21.19) is related to these three equations in this ay:

K_{sp} expression + equation (21.17) − equation (21.18)

Vhich leads to

$$CaCO_3(s) + H_2O + CO_2 \rightleftharpoons Ca^{2+}(aq) + 2\,HCO_3^-(aq)$$
$$K = (K_{sp} \times K_{a_1})/K_{a_2} = 2.6 \times 10^{-5} \quad \textbf{(21.19)}$$

Next, consider which is the greater source of CO_3^{2-} ns in solution: a saturated aqueous solution of $CaCO_3$ or an aqueous carbonic acid solution. In saturated $CaCO_3(aq)$,

$$[Ca^{2+}] = [CO_3^{2-}] \quad K_{sp} = [Ca^{2+}][CO_3^{2-}] = [CO_3^{2-}]^2 = 2.8 \times 10^{-9}$$

$$[CO_3^{2-}] = (2.8 \times 10^{-9})^{\frac{1}{2}} = 5.3 \times 10^{-5}\,M$$

In a carbonic acid solution, H_2CO_3 is a weak *diprotic* acid with $K_{a_2} \ll K_{a_1}$. Thus,

$$[H_3O^+] = [HCO_3^-], \text{ and } [CO_3^{2-}] = K_{a_2} = 4.7 \times 10^{-11}\,M$$

Note that the carbonate ion concentration in a saturated $CaCO_3$ solution is much greater than the carbonate ion concentration observed in a carbonic acid solution. This means that when the two processes occur simultaneously, carbonate ion from the dissolution of $CaCO_3$ acts as a common ion in the carbonic acid equilibrium. This displaces reaction (21.18) *to the left*, converting CO_3^{2-} to HCO_3^- while consuming H_3O^+. Removal of H_3O^+ in reaction (21.18) stimulates reaction (21.17) to shift *to the right*, producing more H_3O^+ and, simultaneously, more HCO_3^-. The overall effect is that H_2O, CO_2, and CO_3^{2-} (from $CaCO_3$) are consumed and HCO_3^- is produced, just as suggested by equation (21.19).

ssess

Ve have assessed the qualitative correctness of reaction (21.19). To calculate the quantitative extent of the dissolution of $CaCO_3(s)$ in rainwater is somewhat more difficult. The calculation centers on the combined equilibrium constant expreson for reaction (21.19), $K = 2.6 \times 10^{-5}$, and is affected by the partial pressure of atmospheric CO_2 in equilibrium with ainwater. Typical data are given in Chapter 16, Practice Example A, page 782.

RACTICE EXAMPLE A: Write chemical equations for the reactions that occur when NaCN is dissolved in water and vhen $Al(NO_3)_3$ is dissolved in water. Then, use data from Appendix D to explain why a precipitate of $Al(OH)_3$ forms vhen equal volumes of 1.0 M aqueous solutions of NaCN and $Al(NO_3)_3$ are mixed.

RACTICE EXAMPLE B: The compound $BeCl_2 \cdot 4\,H_2O$ cannot be dehydrated by heating and it dissolves in water to ive an acidic solution. Conversely, $CaCl_2 \cdot 6\,H_2O$ can be dehydrated by heating and it dissolves in water to give a olution with neutral pH. Explain these observations and write chemical equations for the reactions that occur, if any, vhen the salts are heated and when they are dissolved in water.

Exercises

Group 1: The Alkali Metals

1. Use information from the chapter to write chemical equations to represent each of the following:
(a) reaction of cesium metal with chlorine gas
(b) formation of sodium peroxide (Na_2O_2)
(c) thermal decomposition of lithium carbonate
(d) reduction of sodium sulfate to sodium sulfide
(e) combustion of potassium to form potassium superoxide

2. Use information from the chapter to write chemical equations to represent each of the following:
(a) reaction of rubidium metal with water
(b) thermal decomposition of aqueous $KHCO_3$
(c) combustion of lithium metal in oxygen gas
(d) action of concentrated aqueous H_2SO_4 on KCl(s)
(e) reaction of lithium hydride with water

3. Describe a simple test for determining whether a pure white solid is LiCl or KCl.

4. Describe two methods for determining the identity of an unknown compound that is either Li_2CO_3 or K_2CO_3.

5. Arrange the following compounds in the expected order of increasing solubility in water, and give the basis for your arrangement: Li_2CO_3, Na_2CO_3, $MgCO_3$.

6. The first electrolytic process to produce sodium metal used molten NaOH as the electrolyte. Write probable half-equations and an overall equation for this electrolysis.

7. A 1.26 L sample of KCl(aq) is electrolyzed for 3.50 min with a current of 0.910 A.
(a) Calculate the pH of the solution after electrolysis.
(b) Why doesn't the result depend on the initial concentration of the KCl(aq)?

8. A lithium battery used in a cardiac pacemaker has a voltage of 3.0 V and a capacity of 0.50 A h (ampere hour). Assume that 5.0 μW of power is needed to regulate the heartbeat. [*Hint:* See Appendix B.]
(a) How long will the implanted battery last?
(b) How many grams of lithium must be present i the battery for the lifetime calculated in part (a)?

9. An analysis of a Solvay-process plant shows that fc every 1.00 kg of NaCl consumed, 1.03 kg of NaHCC are obtained. The quantity of NH_3 consumed in th overall process is 1.5 kg.
(a) What is the percent efficiency of this process fc converting NaCl to $NaHCO_3$?
(b) Why is so little NH_3 required?

10. Consider the reaction $Ca(OH)_2(s) + Na_2SO_4(aq$ $\rightleftharpoons CaSO_4(s) + 2\,NaOH(aq)$.
(a) Write a net ionic equation for this reaction.
(b) Will the reaction essentially go to completion?
(c) What will be $[SO_4^{2-}]$ and $[OH^-]$ *at equilibriu* if a slurry of $Ca(OH)_2(s)$ is mixed with 1.00 M $Na_2SO_4(aq)$?

11. The standard Gibbs energies of formation, $\Delta_f G$ for $Na_2O(s)$ and $Na_2O_2(s)$ are -379.09 kJ mol^{-1} an -449.63 kJ mol^{-1}, respectively, at 298 K. Calculate th equilibrium constant for the reaction below at 298 K Is $Na_2O_2(s)$ thermodynamically stable with respect t $Na_2O(s)$ and $O_2(g)$ at 298 K?

$$Na_2O_2(s) \longrightarrow Na_2O(s) + \frac{1}{2}O_2(g)$$

12. The standard Gibbs energies of formation, $\Delta_f G°$, fc $KO_2(s)$ and $K_2O(s)$ are -240.59 kJ mol^{-1} and -322.09 mol^{-1}, respectively, at 298 K. Calculate the equilib rium constant for the reaction below at 298 K. I $KO_2(s)$ thermodynamically stable with respect t $K_2O(s)$ and$O_2(g)$ at 298 K?

$$2\,KO_2(s) \longrightarrow K_2O(s) + \frac{3}{2}O_2(g)$$

Group 2: The Alkaline Earth Metals

13. In the manner used to construct Figure 21-5, complete the diagram outlined. Specifically, indicate the reactants (and conditions) you would use to produce the indicated substances from $Ca(OH)_2$.

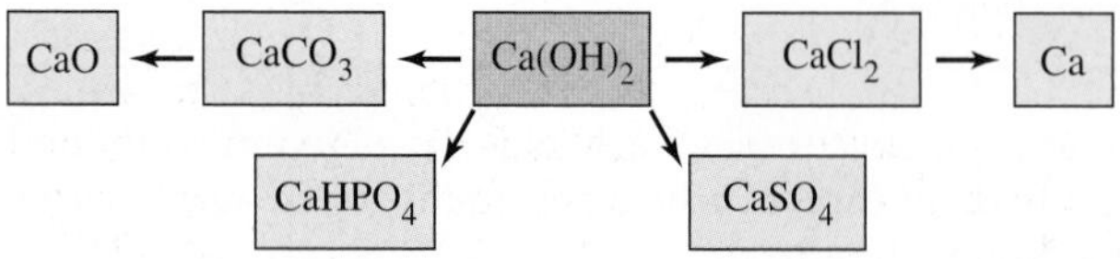

14. Replace the calcium-containing substances shown in the diagram accompanying Exercise 13 by their magnesium-containing equivalents. Then describe the reactants (and conditions) you would use to produce the indicated substances from $MgSO_4$.

15. In the Dow process (Fig. 21-13), the starting materia is Mg^{2+} in seawater and the final product is Mg meta This process seems to violate the principle of conser vation of charge. Does it? Explain.

16. Which has the **(a)** higher melting point, MgO or BaC **(b)** greater solubility in water, MgF_2 or $MgCl_2$? Explair

17. Write chemical equations to represent the following
(a) reduction of BeF_2 to Be metal with Mg as a reduc ing agent
(b) reaction of barium metal with $Br_2(l)$
(c) reduction of uranium(IV) oxide to uranium meta with calcium as the reducing agent
(d) calcination of dolomite, a mixed calcium magne sium carbonate ($MgCO_3 \cdot CaCO_3$)
(e) complete neutralization of phosphoric acid wit quicklime

8. Write chemical equations for the reactions you would expect to occur when
(a) $Mg(HCO_3)_2(s)$ is heated to a high temperature
(b) $BaCl_2(l)$ is electrolyzed
(c) Sr(s) is added to cold dilute HBr(aq)
(d) $Ca(OH)_2(aq)$ is added to $H_2SO_4(aq)$
(e) $CaSO_4 \cdot 2\,H_2O(s)$ is heated

9. *Without performing detailed calculations,* indicate whether equilibrium is displaced either far to the left or far to the right for each of the following reactions. Use data from Appendix D as necessary.
(a) $BaSO_4(s) + CO_3^{2-}(aq) \rightleftharpoons BaCO_3(s) + SO_4^{2-}(aq)$
(b) $Mg_3(PO_4)_2(s) + 3\,CO_3^{2-}(aq) \rightleftharpoons 3\,MgCO_3(s) + 2\,PO_4^{3-}(aq)$
(c) $Ca(OH)_2(s) + 2\,F^-(aq) \rightleftharpoons CaF_2(s) + 2\,OH^-(aq)$

20. *Without performing detailed calculations,* indicate why you would expect each of the following reactions to occur to a significant extent as written. Use data from Appendix D as necessary.
(a) $BaCO_3(s) + 2\,CH_3CO_2H(aq) \longrightarrow Ba^{2+}(aq) + 2\,CH_3CO_2^-(aq) + H_2O(l) + CO_2(g)$
(b) $Ca(OH)_2(s) + 2\,NH_4^+(aq) \longrightarrow Ca^{2+}(aq) + 2\,NH_3(aq) + 2\,H_2O(l)$
(c) $BaF_2(s) + 2\,H_3O^+(aq) \longrightarrow Ba^{2+}(aq) + 2\,HF(aq) + 2\,H_2O(l)$

21. With respect to decomposition to MO(s) and $SO_3(g)$, which of the group 2 sulfates, $MSO_4(s)$, do you expect to be least stable? Explain your answer.

22. With respect to the decomposition to MO(s) and $CO_2(g)$, which of the group 2 carbonates, $MCO_3(s)$, do you expect to be most stable? Explain your answer.

Group 13: The Boron Family

23. The molecule tetraborane has the formula B_4H_{10}.
(a) Show that this is an electron-deficient molecule.
(b) How many bridge bonds must occur in the molecule?
(c) Show that butane, C_4H_{10}, is not electron deficient.

24. Write Lewis structures for the following species, both of which involve coordinate covalent bonding:
(a) tetrafluoroborate ion, BF_4^-, used in metal cleaning and in electroplating baths
(b) boron trifluoride ethylamine, used in curing epoxy resins (ethylamine is $C_2H_5NH_2$)

25. Write chemical equations to represent the following:
(a) the preparation of boron from BBr_3
(b) the formation of BF_3 from B_2O_3
(c) the combustion of boron in hot $N_2O(g)$

26. Assign oxidation states to all the atoms in a perborate ion based on the structure on page 1003.

27. Write chemical equations to represent the
(a) reaction of Al(s) with HCl(aq);
(b) reaction of Al(s) with NaOH(aq);
(c) oxidation of Al(s) to $Al^{3+}(aq)$ by an aqueous solution of sulfuric acid; the reduction product is $SO_2(g)$.

28. Write plausible equations for the
(a) reaction of Al(s) with $Br_2(l)$;
(b) production of Cr from $Cr_2O_3(s)$ by the thermite reaction, with Al as the reducing agent;
(c) separation of Fe_2O_3 impurity from bauxite ore.

29. In some foam-type fire extinguishers, the reactants are $Al_2(SO_4)_3(aq)$ and $NaHCO_3(aq)$. When the extinguisher is activated, these reactants mix, producing $Al(OH)_3(s)$ and $CO_2(g)$. The $Al(OH)_3$–CO_2 foam extinguishes the fire. Write a net ionic equation to represent this reaction.

30. Some baking powders contain the solids $NaHCO_3$ and $NaAl(SO_4)_2$. When water is added to this mixture of compounds, $CO_2(g)$ and $Al(OH)_3(s)$ are two of the products. Write plausible net ionic equations for the formation of these two products.

31. The maximum resistance to corrosion of aluminum is between pH 4.5 and 8.5. Explain how this observation is consistent with other facts about the behavior of aluminum presented in this text.

32. Describe a series of *simple* chemical reactions that you could use to determine whether a particular metal sample is "aluminum 2S" (99.2% Al) or "magnalium" (70% Al, 30% Mg). You are permitted to destroy the metal sample in the testing.

33. In the purification of bauxite ore, a preliminary step in the production of aluminum, $[Al(OH)_4]^-(aq)$ can be converted to $Al(OH)_3(s)$ by passing $CO_2(g)$ through the solution. Write an equation for the reaction that occurs. Could HCl(aq) be used instead of $CO_2(g)$? Explain.

34. In 1825, Hans Oersted produced aluminum chloride by passing chlorine over a heated mixture of carbon and aluminum oxide. In 1827, Friedrich Wöhler obtained aluminum by heating aluminum chloride with potassium. Write plausible equations for these reactions.

35. A description for preparing potassium aluminum alum calls for dissolving aluminum foil in KOH(aq). The solution obtained is treated with $H_2SO_4(aq)$, and the alum is crystallized from the resulting solution. Write plausible equations for these reactions.

36. Handbooks and lists of chemicals do not contain entries under the formulas $Al(HCO_3)_3$ and $Al_2(CO_3)_3$. Explain why these compounds do not exist.

37. Compound $FB(BF_2)_2$ disproportionates at –30 °C to give BF_3 and very unstable B_8F_{12}. The B_8F_{12} molecule is unstable because it contains a strained four-membered ring of boron atoms. Write a balanced chemical equation for this process, and draw a plausible structure for B_8F_{12}. [*Hint:* The B_8F_{12} contains six BF_2 units, only two of which are part of the four-membered ring.]

38. Gallium trichloride ($GaCl_3$) is a very active catalyst for a number of organic transformations. In the solid state, $GaCl_3$ exists as a dimer with the formula Ga_2Cl_6. Draw a plausible structure for the Ga_2Cl_6 molecule, and describe the bonding. [*Hint:* Two chlorine atoms are simultaneously bonded to two gallium atoms.]

Group 14: The Carbon Family

39. Comment on the accuracy of a jeweler's advertising that "diamonds last forever." In what sense is the statement true, and in what ways is it false?

40. A temporary fix for a "sticky" lock is to scrape a pencil point across the notches on the key and to work the key in and out of the lock a few times. What is the basis of this fix?

41. Write a chemical equation to represent
(a) the reduction of silica to elemental silicon by aluminum;
(b) the preparation of potassium metasilicate by the high-temperature fusion of silica and potassium carbonate;
(c) the reaction of Al_4C_3 with water to produce methane.

42. Write a chemical equation to represent
(a) the reaction of potassium cyanide solution with silver nitrate solution;
(b) the combustion of Si_3H_8 in an excess of oxygen;
(c) the reaction of dinitrogen with calcium carbide to give calcium cyanamide (CaNCN).

43. Describe what is meant by the terms *silane* and *silanol*. What is their role in the preparation of silicones?

44. Describe and explain the similarities and differences between the reaction of a silicate with an acid and that of a carbonate with an acid.

45. Methane and sulfur vapor react to form carbon disulfide and hydrogen sulfide. Carbon disulfide reacts with $Cl_2(g)$ to form carbon tetrachloride and S_2Cl_2. Further reaction of carbon disulfide and S_2Cl_2 produces additional carbon tetrachloride and sulfur. Write a series of equations for the reactions described here.

46. In a manner similar to that outlined on page 1022,
(a) write equations to represent the reaction of $(CH_3)_3SiCl$ with water, followed by the elimination of H_2O from the resulting silanol molecules.
(b) Does a silicone polymer form from part (a)?
(c) What would be the corresponding product obtained from CH_3SiCl_3?

47. Show that the empirical formula given for muscovi mica is consistent with the expected oxidation stat of the elements present.

48. Show that the empirical formula given for crysoti asbestos is consistent with the expected oxidatic states of the elements present.

49. Write plausible chemical equations for the **(a)** dissol ing of lead(II) oxide in nitric acid; **(b)** heating $SnCO_3(s)$; **(c)** reduction of lead(II) oxide by carbo **(d)** reduction of $Fe^{3+}(aq)$ to $Fe^{2+}(aq)$ by Sn^{2+}(ac **(e)** formation of lead(II) sulfate during high-temperatu roasting of lead(II) sulfide.

50. Write plausible chemical equations for prepari each compound from the indicated starting materia **(a)** $SnCl_2$ from SnO; **(b)** $SnCl_4$ from Sn; **(c)** PbCr(from PbO_2. What reagents (acids, bases, salts) ar equipment commonly available in the laboratory a needed for each reaction?

51. Lead(IV) oxide, PbO_2, is a good oxidizing agent. U appropriate data from Appendix D to determi whether $PbO_2(s)$ in a solution with $[H_3O^+] = 1$ M a sufficiently good oxidizing agent to carry the fc lowing oxidations to the point at which the conce tration of the species being oxidized decreases one-thousandth of its initial value.
(a) $Fe^{2+}(1\ M)$ to Fe^{3+}
(b) $SO_4^{2-}(1\ M)$ to $S_2O_8^{2-}$
(c) $Mn^{2+}(1 \times 10^{-4}\ M)$ to MnO_4^-

52. Aqueous tin(II) ion, $Sn^{2+}(aq)$, is a good reducing ager Use data from Appendix D to determine wheth $Sn^{2+}(aq)$ is a sufficiently good reducing agent to redu **(a)** $I_2(s)$ to $I^-(aq)$; **(b)** $Fe^{3+}(aq)$ to $Fe^{2+}(aq)$; **(c)** Zn^{2+}(a to Zn(s); and **(d)** $Pb^{2+}(aq)$ to Pb(s). Assume that a reactants and products are in their standard states.

53. Would you expect the reaction of Sn(s) and $Cl_2(g)$ yield $SnCl_2$ or $SnCl_4$?

54. Would you expect the reaction of Ge(s) and $F_2(g)$ yield GeF_2, with germanium in the +2 oxidation stat or GeF_4, with germanium in the +4 oxidation state?

Integrative and Advanced Exercises

55. A chemical that should exist as a crystalline solid is seen to be a mixture of a solid and liquid in a container on a storeroom shelf. Give a plausible reason for that observation. Should the chemical be discarded or is it still useful for some purposes?

56. The following series of observations is made: (1) a small piece of dry ice $[CO_2(s)]$ is added to 0.005 M $Ca(OH)_2(aq)$. (2) Initially, a white precipitate forms. (3) After a short time the precipitate dissolves.
(a) Write chemical equations to explain these observations.
(b) If the 0.005 M $Ca(OH)_2(aq)$ is replaced by 0.005 M $CaCl_2(aq)$, would a precipitate form? Explain.
(c) If the 0.005 M $Ca(OH)_2(aq)$ is replaced by 0.010 M $Ca(OH)_2(aq)$, a precipitate forms but does not re-dissolve. Explain why.

57. The melting point of NaCl(s) is 801 °C, much higher than that of NaOH (322 °C). More energy is consumed to melt and maintain molten NaCl than NaOH. Yet th preferred commercial process for the production (sodium is electrolysis of NaCl(l) rather than NaOH(Give a reason or reasons for this discrepancy.

58. Although the triiodide ion, I_3^-, is known to exist i aqueous solutions, the ion is stable in only certai ionic solids. For example, CsI_3 is stable with respect decomposition to CsI and I_2, but LiI_3 is not stable wit respect to LiI and I_2. Draw a Lewis structure for the I_3 ion and suggest a reason why CsI_3 is stable wit respect to decomposition to the iodide but LiI_3 is not.

59. At 298 K, the $\Delta_f G°$ values for $Li_2O(s)$ and Li_2O_2(s suggest that $Li_2O_2(s)$ is thermodynamically mor stable than $Li_2O(s)$. At 1000 K, however, the situatio is reversed. The standard Gibbs energies of formatio $\Delta_f G°$, for $Li_2O(s)$ and $Li_2O_2(s)$ are -466.40 kJ mol^- and -419.02 kJ mol^{-1}, respectively, at 1000 K. Calcula the equilibrium constant for the following reaction

1000 K and the equilibrium partial pressure of $O_2(g)$ above a sample of $Li_2O_2(s)$ at 1000 K.

$$Li_2O_2(s) \longrightarrow Li_2O(s) + \frac{1}{2}O_2(g)$$

60. The chemical equation for the hydration of an alkali metal ion is $M^+(g) \rightarrow M^+(aq)$. The standard Gibbs energy changes and the enthalpy changes for the process are denoted by $\Delta_{hydr}G°$ and $\Delta_{hydr}H°$ respectively. $\Delta_{hydr}G°$ and $\Delta_{hydr}H°$ values are given below for the alkali metal ions.

M^+	Li^+	Na^+	K^+	Rb^+	Cs^+
$\Delta_{hydr}H°$	−522	−407	−324	−299	−274 kJ mol^{-1}
$\Delta_{hydr}G°$	−481	−375	−304	−281	−258 kJ mol^{-1}

Use the data above to calculate $\Delta_{hydr}S°$ values for the hydration process. Explain the trend in the $\Delta_{hydr}S°$ values.

61. Lithium superoxide, $LiO_2(s)$, has never been isolated. Use ideas from Chapter 12, together with data from this chapter and Appendix D, to estimate $\Delta_fH°$ for $LiO_2(s)$ and assess whether $LiO_2(s)$ is thermodynamically stable with respect to $Li_2O(s)$ and $O_2(g)$.
(a) Use the Kapustinskii equation, along with appropriate data below, to estimate the lattice energy, *U*, for $LiO_2(s)$. (See Exercise 127 in Chapter 12.) The ionic radii for Li^+ and O_2^- are 73 pm and 144 pm, respectively.
(b) Use your result from part (a) in the Born–Fajans–Haber cycle to estimate $\Delta_fH°$ for $LiO_2(s)$. [*Hint:* For the process $O_2(g) + e^- \rightarrow O_2^-(g)$, $\Delta_rH° = -43$ kJ mol^{-1}. See Table 21.2 and Appendix D for the other data that are required.]
(c) Use your result from part (b) to calculate the enthalpy of reaction for the decomposition of $LiO_2(s)$ to $Li_2O(s)$ and $O_2(g)$. For $Li_2O(s)$, $\Delta_fH° = -598.73$ kJ mol^{-1}.
(d) Use your result from part (c) to decide whether $LiO_2(s)$ is thermodynamically stable with respect to $Li_2O(s)$ and $O_2(g)$. Assume that entropy effects can be neglected.

62. When a 0.250 g sample of Ca is heated in air, 0.325 g of product is obtained. Assume that all the Ca appears in the product.
(a) If the product were pure CaO, what mass should have been obtained?
(b) Show that the 0.325 g product could be a mixture of CaO and Ca_3N_2.
(c) What is the mass percent of CaO in the CaO–Ca_3N_2 mixed product?

63. Comment on the feasibility of using a reaction similar to (21.4) to produce **(a)** lithium metal from LiCl; **(b)** cesium metal from CsCl, with Na(l) as the reducing agent in each case. [*Hint:* Consider data from Table 21.2.]

64. Concerning the thermite reaction,
(a) use data from Appendix D to calculate $\Delta_rH°$ at 298 K for the reaction below.

$$2\ Al(s) + Fe_2O_3(s) \longrightarrow 2\ Fe(s) + Al_2O_3(s)$$

(b) Write an equation for the reaction when $MnO_2(s)$ is substituted for $Fe_2O_3(s)$, and calculate $\Delta_rH°$ for this reaction.
(c) Show that if MgO were substituted for Fe_2O_3, the reaction would be *endothermic*.

65. Use data from Appendix D (Table D-2) to calculate a value of $E°$ for the reduction of $Li^+(aq)$ to Li(s), and compare your result with the value listed in Table 21.2.

66. The electrolysis of 0.250 L of 0.220 M $MgCl_2$ is conducted until 104 mL of gas (a mixture of H_2 and water vapor) is collected at 23 °C and 748 mmHg. Will $Mg(OH)_2(s)$ precipitate if electrolysis is carried to this point? (Use 21 mmHg as the vapor pressure of the solution.)

67. A particular water sample contains 56.9 ppm SO_4^{2-} and 176 ppm HCO_3^-, with Ca^{2+} as the only cation.
(a) How many parts per million of Ca^{2+} does the water contain?
(b) How many grams of CaO are consumed in removing HCO_3^-, from 602 kg of the water?
(c) Show that the Ca^{2+} remaining in the water after the treatment described in part (b) can be removed by adding Na_2CO_3.
(d) How many grams of Na_2CO_3 are required for the precipitation referred to in part (c)?

68. An aluminum production cell of the type pictured in Figure 21-24 operates at a current of 1.00×10^5 A and a voltage of 4.5 V. The cell is 38% efficient in using electrical energy to produce chemical change. (The rest of the electrical energy is dissipated as thermal energy in the cell.)
(a) What mass of Al can be produced by this cell in 8.00 h?
(b) If the electrical energy required to power this cell is produced by burning coal (85% C; heat of combustion of C = 32.8 kJ/g) in a power plant with 35% efficiency, what mass of coal must be burned to produce the mass of Al determined in part (a)?

69. Use data from Appendix D (Table D-2) to estimate the minimum voltage required to electrolyze Al_2O_3 in the Hall-Héroult process, reaction (21.25). Use $\Delta_fG°[Al_2O_3(l)] = -1520$ kJ mol^{-1}. Show that the oxidation of the graphite anode to $CO_2(g)$ permits the electrolysis to occur at a lower voltage than if the electrolysis reaction were $Al_2O_3(l) \longrightarrow 2\ Al(l) + \frac{3}{2}O_2(g)$.

70. At 20 °C, a saturated aqueous solution of $Pb(NO_3)_2$ maintains a relative humidity of 97%. What must be the composition of this solution, expressed as g $Pb(NO_3)_2$/100.0 g H_2O?

71. Use information from this chapter and elsewhere in this text to explain why the compounds $PbBr_4$ and PbI_4 do *not* exist.

72. The reaction of borax, calcium fluoride, and concentrated sulfuric acid yields sodium hydrogen sulfate, calcium sulfate, water, and boron trifluoride as products. Write a balanced equation for this reaction.

73. The dissolution of $MgCO_3(s)$ in $NH_4^+(aq)$ can be represented as

$$MgCO_3(s) + NH_4^+(aq) \rightleftharpoons Mg^{2+}(aq) + HCO_3^-(aq) + NH_3(aq)$$

Calculate the molar solubility of $MgCO_3$ in each of the following solutions: **(a)** 1.00 M $NH_4Cl(aq)$; **(b)** a buffer that is 1.00 M NH_3 and 1.00 M NH_4Cl; **(c)** a buffer that is 0.100 M NH_3 and 1.00 M NH_4Cl.

74. Show that, in principle, $Na_2CO_3(aq)$ can be converted almost completely to NaOH(aq) by the reaction

$$Ca(OH)_2(s) + Na_2CO_3(aq) \longrightarrow CaCO_3(s) + 2\,NaOH(aq)$$

75. Assume that the packing of spherical atoms in crystalline metals is the same for Li, Na, and K, and explain why Na has a higher density than *both* Li and K. [*Hint:* Use data from Table 21.2.]

76. Would you expect the lattice energy of MgS(s) to be less than, greater than, or about the same as that of MgO(s)? Use appropriate data from various locations in this text to obtain the values of the two lattic energies. Use a value of 456 kJ mol^{-1} for the proce: $S^-(g) + e^- \longrightarrow S^{2-}(g)$.

77. There has been some interest in the alkali metal fu lerides, $M_nC_{60}(s)$, because at low temperatures, som of these compounds become superconducting. Th alkali metal fullerides are ionic crystals comprisin M^+ ions and $C_{60}{}^{n-}$ ions. The value of n can b deduced from the crystal structure. If M_nC_{60} consis of a cubic closest packed array of fulleride ions, wit M^+ ions occupying all the octahedral and tetrahedr holes in the fulleride lattice, then what is the value o n and what is the empirical formula of the fulleride?

Feature Problems

78. In Chapter 19, we examined the relationship of electrode potentials to thermodynamic data. In fact, electrode potentials can be calculated from tabulated thermodynamic data (many of which, in turn, were established from electrochemical measurements). To demonstrate this fact and to pursue further the discussion in Are You Wondering 21-1, combine the three steps for the oxidation of Li(s) with a corresponding set of three steps for the reduction of H^+ (1 M) to $H_2(g)$. Obtain $\Delta_rH°$ for the overall reaction.
(a) Neglect entropy changes that occur (that is, assume that $\Delta_rG° \approx \Delta_rH°$), and estimate the value of $E°_{Li^+/Li}$.
(b) Combine the calculated $\Delta_rH°$ value with a value of $\Delta_rS°$ to obtain another estimate of $E°_{Li^+/Li}$. [*Hint:* Hydration energies for $Li^+(g)$ and $H^+(g)$ to form 1 M solutions are –522 and –1094 kJ mol^{-1}, respectively. Also, use data from various locations in this text.]

79. Mono Lake in eastern California is a rather unusual salt lake. The lake has no outlets; water leaves only by evaporation. The rate of evaporation is great enough that the lake level would be lowered by three meters per year if not for fresh water entering through underwater springs and streams originating in the nearby Sierra Nevada mountains. The principal salts in the lake are the chlorides, bicarbonates, and sulfates of sodium. An approximate "recipe" for simulating the lake water is to dissolve 18 tablespoons of sodium bicarbonate, 10 tablespoons of sodium chloride, and 8 teaspoons of Epsom salt (magnesium sulfate heptahydrate) in 4.5 liters of water (although the lake water actually contains only trace amounts of magne sium ion). Assume that 1 tablespoon of any of the sal weighs about 10 g. (1 tablespoon = 3 teaspoons.)
(a) Expressed as grams of salt per liter, what is th approximate salinity of Mono Lake? How does thi salinity compare with seawater, which is approx mately 0.438 M NaCl and 0.0512 M $MgCl_2$?
(b) Estimate an approximate pH for Mono Lak water. How does your estimate compare with th observed pH of about 9.8? Actually, the recipe for th lake water also calls for a pinch of borax. How woul its presence affect the pH? [Borax is a sodium sal $Na_2B_4O_7 \cdot 10\,H_2O$, related to the weak monoprot boric acid ($pK_a = 9.25$).]
(c) Mono Lake has some unusual limestone forma tions called *tufa*. They form at the site of underwate springs and grow only underwater, although som project above water, having formed at a time whe the lake level was higher. Explain how the tufa form. [*Hint*: What chemical reaction(s) is(are) involved?]

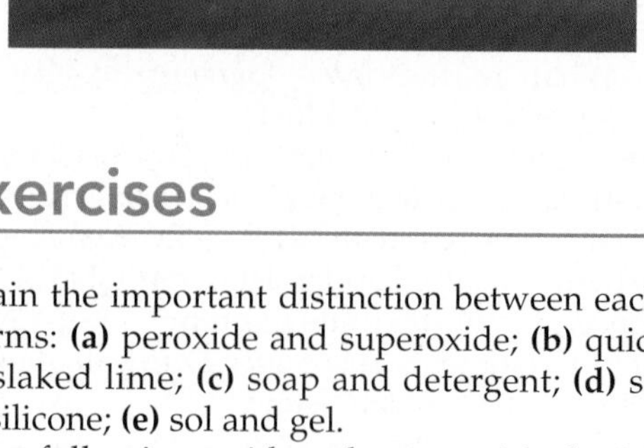

Beppesensation/Fotolia

Self-Assessment Exercises

80. In your own words, define the following terms: **(a)** dimer; **(b)** adduct; **(c)** calcination; **(d)** amphoteric oxide; **(e)** three-center two-electron bond.

81. Briefly describe each of the following ideas, methods, or phenomena: **(a)** diagonal relationship; **(b)** preparation of deionized water by ion exchange; **(c)** thermite reaction; **(d)** inert pair effect.

82. Explain the important distinction between each pai of terms: **(a)** peroxide and superoxide; **(b)** quicklim and slaked lime; **(c)** soap and detergent; **(d)** silicat and silicone; **(e)** sol and gel.

83. Of the following oxides, the one with the highes melting point is **(a)** Li_2O; **(b)** BaO; **(c)** MgO; **(d)** SiO_2.

84. The best oxidizing agent of the following oxides is **(a)** Li_2O; **(b)** MgO; **(c)** Al_2O_3; **(d)** CO_2; **(e)** SnO_2; **(f)** PbO_2.

85. Predict the products of the following reactions:
(a) $BCl_3(g) + NH_3(g)$
(b) $K(s) + O_2(g)$
(c) $Li(s) + O_2(g)$
(d) $BaO_2(s) + H_2O(l)$

86. A chemist knows that aluminum is more reactive toward oxygen than is iron, but many people believe the opposite. What evidence was provided in this chapter that demonstrates that aluminum is a much stronger reducing agent (and is much more reactive) than iron? If aluminum is so reactive, then why is it that aluminum can be used for making products—pots, pans, aluminum foil, siding for houses—that last so long?

87. Listed are several pairs of substances. For some pairs, one or both members of the pair react individually with water to produce a gas. For others, neither member of the pair reacts with water. The pair for which *each* member reacts with water and yields the *same* gaseous product is **(a)** Al(s) and Ba(s); **(b)** Ca(s) and $CaH_2(s)$; **(c)** Na(s) and $Na_2O_2(s)$; **(d)** K(s) and $KO_2(s)$; **(e)** $NaHCO_3(s)$ and HCl(aq).

88. Complete and balance the following. Write the simplest equation possible. If no reaction occurs, so state.
(a) $Li_2CO_3(s) \xrightarrow{\Delta}$
(b) $CaCO_3(s) + HCl(aq) \longrightarrow$
(c) $Al(s) + NaOH(aq) \longrightarrow$
(d) $BaO(s) + H_2O(l) \longrightarrow$
(e) $Na_2O_2(s) + CO_2(g) \longrightarrow$

89. Assuming that water, common reagents (acids, bases, salts), and simple laboratory equipment are available, give a practical method to prepare **(a)** $MgCl_2$ from $MgCO_3(s)$; **(b)** $NaAl(OH)_4$ from Na(s) and Al(s); and **(c)** Na_2SO_4 from NaCl(s).

90. Write the simplest chemical equation to represent the reaction of **(a)** $K_2CO_3(aq)$ and $Ba(OH)_2(aq)$; **(b)** $Mg(HCO_3)_2(aq)$ on heating; **(c)** tin(II) oxide when heated with carbon; **(d)** $CaF_2(s)$ and H_2SO_4(concd aq); **(e)** $NaHCO_3(s)$ and HCl(aq); **(f)** $PbO_2(s)$ and HBr(aq); and **(g)** the reduction of SiF_4 to pure Si, by using Na as the reducing agent.

91. Write an equation to represent the reaction of gypsum, $CaSO_4 \cdot 2\,H_2O$, with ammonium carbonate to produce ammonium sulfate (a fertilizer), calcium carbonate, and water.

92. Write chemical equations to represent the most probable outcome in each of the following. If no reaction is likely to occur, so state.
(a) $B(OH)_3 \xrightarrow{\Delta}$
(b) $Al_2O_3(s) \xrightarrow{\Delta}$
(c) $CaSO_4 \cdot 2\,H_2O(s) \xrightarrow{\Delta}$

93. A chemical dictionary gives the following descriptions of the production of some compounds. Write plausible chemical equations based on these descriptions.
(a) lead(II) carbonate: adding a solution of sodium bicarbonate to a solution of lead nitrate
(b) lithium carbonate: reaction of lithium oxide with ammonium carbonate solution
(c) hydrogen peroxide: by the action of dilute sulfuric acid on barium peroxide
(d) lead(IV) oxide: action of an alkaline solution of calcium hypochlorite on lead(II) oxide

94. Name the chemical compound(s) you would expect to be the *primary* constituent(s) of **(a)** stalactites; **(b)** gypsum; **(c)** "barium milkshake"; **(d)** blue sapphires.

95. How many cubic meters of $CO_2(g)$ at 102 kPa and 288 K are produced in the calcination of 5.00×10^3 kg of the mineral dolomite, $CaMg(CO_3)_2$?

22 Chemistry of the Main-Group Elements II: Groups 18, 17, 16, 15, and Hydrogen

CONTENTS

LEARNING OBJECTIVES

22.1 Discuss general trends in the bonding and acid–base properties of the binary oxides formed by the second-period and third-period elements.

22.2 Describe the synthesis and properties of the most common xenon fluorides, and explain why their melting points do not follow the trend expected of nonpolar molecules.

22.3 Discuss the properties of the halogens and the general trends in their reactivity.

22.4 Discuss the main differences between oxygen and sulfur, and the molecular compounds that they tend to form.

22.5 Describe the synthesis of ammonia from elemental nitrogen, discuss the industrial importance of this synthesis, and describe the allotropes of phosphorus.

22.6 Identify the types of hydrides formed by hydrogen and some important uses of hydrogen.

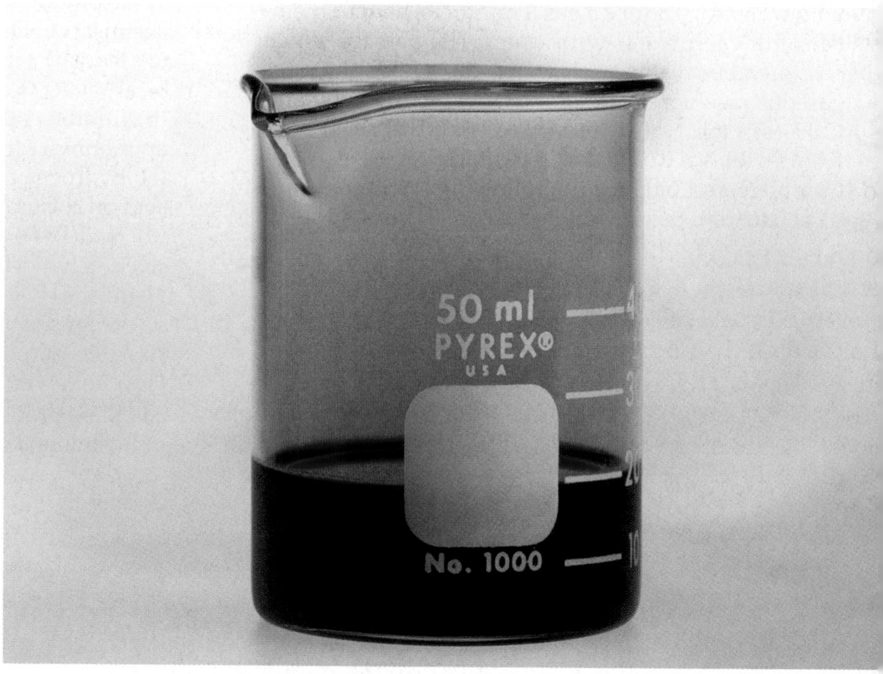

Bromine, a nonmetal, is a reactant in the synthesis of flame retardants for use in plastics

In this chapter we continue our discussion of the p-block elements. We wil progress from right to left through the periodic table and begin with a sur vey of the noble gases (group 18), the atoms of which have filled valenc shells. The filled valence shells of the group 18 atoms makes these gase special—or noble—in the sense that they are particularly unreactive, thoug not totally so, as we will soon see. We next examine the halogens (group 17 and then proceed to groups 16 and 15. In each successive group, proceedin from right to left, the number of electrons in the p-subshell of an aton decreases by one and as a consequence, the tendency to form more than on covalent bond increases. Finally, in this chapter we will consider hydrogen Hydrogen has unique chemistry and is not easily placed in any group of th periodic table. To emphasize the uniqueness of hydrogen, some chemist prefer to use a version of the periodic table, such as that shown at the to of the next page, in which hydrogen is separated from the rest of the table.

																	18
1	2						H					13	14	15	16	17	He
Li	Be											B	C	N	O	F	Ne
Na	Mg	3	4	5	6	7	8	9	10	11	12	Al	Si	P	S	Cl	Ar
K	Ca	Sc	Ti	V	Cr	Mn	Fe	Co	Ni	Cu	Zn	Ga	Ge	As	Se	Br	Kr
Rb	Sr	Y	Zr	Nb	Mo	Tc	Ru	Rh	Pd	Ag	Cd	In	Sn	Sb	Te	I	Xe
Cs	Ba	La–Lu	Hf	Ta	W	Re	Os	Ir	Pt	Au	Hg	Tl	Pb	Bi	Po	At	Rn
Fr	Ra	Ac–Lr	Rf	Db	Sg	Bh	Hs	Mt	Ds	Rg	Cn		Fl		Lv		

▲ **An alternative version of the periodic table**
This version differs slightly from the one given on the inside front cover in that hydrogen is separated from the rest of the elements. Also, it does not explicitly show the lanthanides (La–Lr) or the actinides (Ac–Lr). Some chemists prefer to put H on its own because of the uniqueness of hydrogen and the difficulty in placing it in a specific group.

Among the fundamental concepts emphasized in this chapter are atomic, physical, and thermodynamic properties; bonding and structure; acid–base chemistry; and oxidation states, redox reactions, and electrochemistry.

22-1 Periodic Trends in Bonding

To uncover trends in bonding it is instructive to consider a series of binary compounds with a common element, such as fluorine or oxygen. First, we will consider the fluorides of the second- and third-row elements, with the general formula AF_n.

As we progress from left to right through the periodic table, the metallic character of the elements decreases (see Figure 9-19) and the bonding in the fluorides, AF_n, changes from ionic bonding to covalent bonding. In Table 22.1 we show the formulas, bonding types, and phases at room temperature of the binary fluorides of the second- and third-period elements. In the fluorides, we observe a transition from ionic bonding to covalent bonding as we move left to right across a period. Notice that BeF_2 and AlF_3 are network covalent solids. In a network covalent solid, each atom is bonded to one or more atoms by covalent bonds to form a giant molecule. Most of the fluorides are molecular compounds, with covalent bonds holding together the atoms of the molecule and with intermolecular forces, such as dipole–dipole or London dispersion forces, contributing to the attraction among individual molecules. In Table 22.1, such compounds are called *molecular* covalent compounds to distinguish them from the *network* covalent compounds.

The diagonal relationship between beryllium and aluminum, discussed previously in Chapter 21, is evident in Table 22.1. Both beryllium (in group 2) and aluminum (in group 13) form network covalent solids with fluorine.

TABLE 22.1 Binary Fluorides, AF_n, of the Second- and Third-Row Elements[a]

Group	1	2	13	14	15	16	17
Formula	LiF(s)	BeF_2(s)	BF_3(g)	CF_4(g)	NF_3(g)	OF_2(g)	F_2(g)
Bonding	Ionic	Network Covalent	Molecular Covalent	Molecular Covalent	Molecular Covalent	Molecular Covalent	Molecular Covalent
Formula	NaF(s)	MgF_2(s)	AlF_3(s)	SiF_4(g)	PF_5(g)	SF_6(g)	ClF_5(g)
Bonding	Ionic	Ionic	Network Covalent	Molecular Covalent	Molecular Covalent	Molecular Covalent	Molecular Covalent

[a]For each element, the substance shown is the one that has the greatest number of fluorine atoms per atom of A. The phase indicated is the phase at 25 °C.

The phase of the compound at room temperature is a reflection of the type of bonding in the compound. Ionic solids melt at very high temperatures because the melting process requires the breaking of strong ionic bonds in the crystal lattice. Similarly, network covalent solids have very high melting temperatures because covalent bonds have to be broken in the melting process. In the molecular covalent substances, only weak intermolecular forces contribute to the attraction between the molecules; thus, these species are gases at room temperature.

For the fluorides of the second-row elements, the formulas are relatively easy to predict and explain. The number of fluorine atoms per formula unit is equal to the number of valence electrons an atom must gain or lose to attain a noble gas configuration. For the fluorides of groups 14 to 16, the number of fluorine atoms per formula unit is equal to the number of valence electrons required to fill the valence shell of carbon, nitrogen, or oxygen. For example, a carbon atom with configuration $[He]2s^22p^2$ requires four electrons to complete its valence shell, and because each fluorine atom has only one unpaired valence electron, carbon and fluorine combine in the ratio 1:4 to give CF_4. For the fluorides of groups 1, 2, and 13, however, the number of fluorine atoms per formula unit is equal to the number of valence electrons a lithium, beryllium, or boron atom would have to lose to attain a ns^2 configuration. For example, lithium must lose only one electron to attain a ns^2 configuration, and because a fluorine atom needs only one electron to complete its valence shell, lithium and fluorine combine 1:1 to give LiF. The formulas of BeF_2 and BF_3 are related to the number of valence electrons a Be or B atom must lose to attain a noble gas (He) configuration, even though neither Be nor B actually loses any of its valence electrons—both Be and B form covalent bonds with fluorine.

For the fluorides of the third-row elements, the number of fluorine atoms per formula unit is more difficult to predict. For the fluorides shown in Table 22.1, the oxidation state of the element bonded to fluorine increases in steps of 1 as we move left to right across the third period, except in going from sulfur to chlorine. Chlorine has an oxidation state of +7 in some of its compounds, as is the case in $HClO_4$ and Cl_2O_7, but it does not form the heptafluoride (ClF_7) with Cl in the +7 oxidation state. Presumably, this occurs because the chlorine atom is not large enough to interact simultaneously with seven fluorine atoms. This argument is supported by the observation that iodine, also in group 17 but a much larger atom, forms the heptafluoride IF_7.

Similar trends in bonding are observed for the oxides of the second- and third-row elements, as can be seen in Table 22.2. Also shown in Table 22.2 are the acid–base properties of the oxides. Half of these oxides are acidic, only three

TABLE 22.2 Binary Oxides of the Second- and Third-Row Elements[a]

Group	1	2	13	14	15	16	17
Formula	$Li_2O(s)$	$BeO(s)$	$B_2O_3(s)$	$CO_2(g)$	$N_2O_5(g)$	$O_2(g)$	$OF_2(g)$[b]
Bonding	Ionic	Ionic	Network Covalent	Molecular Covalent	Molecular Covalent	Molecular Covalent	Molecular Covalent
Acid–Base Properties	Basic	Amphoteric	Acidic	Acidic	Acidic	Neutral	Neutral
Formula	$Na_2O(s)$	$MgO(s)$	$Al_2O_3(s)$	$SiO_2(s)$	$P_4O_{10}(s)$	$SO_3(l)$[c]	$Cl_2O_7(l)$
Bonding	Ionic	Ionic	Ionic	Network Covalent	Molecular Covalent	Molecular Covalent	Molecular Covalent
Acid–Base Properties	Basic	Basic	Amphoteric	Acidic	Acidic	Acidic	Acidic

[a]Some elements form more than one compound with oxygen. The substance shown is the one having the element in its most highly oxidized form. The phase indicated is the phase at 25 °C.
[b]OF_2 is included in the table even though OF_2 is not an oxide. It is a fluoride with oxygen having an oxidation state of +2.
[c]In the liquid phase, SO_3 consists of a trimer of SO_3 molecules, as shown in Figure 22-14.

re basic (Li_2O, Na_2O, and MgO), and two are amphoteric (BeO and Al_2O_3). The diagonal relationships between Be and Al and between B and Si are evident. Both Be and Al form amphoteric oxides and both B and Si form network covalent oxides that are acidic. A transition occurs from basic oxides to acidic oxides as we proceed from metallic to nonmetallic elements, or, stated another way, as we proceed from the least electronegative elements to the most electronegative elements. For example, sodium oxide reacts with water to give sodium hydroxide, while sulfur trioxide reacts with water to give sulfuric acid, as shown below:

$$Na_2O(s) + H_2O(l) \longrightarrow 2\,NaOH(aq)$$
$$SO_3(l) + H_2O(l) \longrightarrow H_2SO_4(l)$$

With the exception of OF_2, the oxides of the elements in groups 14 to 17 react with water to give oxoacids, as shown above for SO_3. OF_2 reacts with water, but the reaction gives a binary acid (HF) and not the oxoacid (HOF). That the acid–base behavior of OF_2 is different from the oxides of other *p*-block elements comes as no surprise; OF_2 is not an oxide. It is a fluoride with oxygen having an oxidation state of +2.

The oxides of Be and Al, as we saw in Chapter 21, are amphoteric because these oxides react with both acids and bases. Beryllium oxide, BeO, reacts as follows:

$$BeO(s) + 2\,H_3O^+(aq) + H_2O(l) \longrightarrow [Be(OH_2)_4]^{2+}(aq)$$
$$BeO(s) + 2\,OH^-(aq) + H_2O(l) \longrightarrow [Be(OH)_4]^{2-}(aq)$$

We should not forget the point made in Chapter 21: *as we move down a group, the metallic character of an element increases.* This has an important consequence for the oxides of the group 14 elements. The oxides of C and Si are acidic, the oxide of Sn is amphoteric, and the oxide of Pb is basic. The amphoterism of SnO is evident in the reactions of SnO with HCl(aq) and NaOH(aq), which are given below:

$$SnO(s) + 2\,HCl(aq) \longrightarrow SnCl_2(aq) + H_2O(l)$$
$$SnO(s) + NaOH(aq) + H_2O(l) \longrightarrow Na^+(aq) + [Sn(OH)_3]^-(aq)$$

As we discuss the chemistry of the elements of groups 15 through 18, we will see these trends repeated many times. Explanations of these trends are based on differences in electronegativities and differences in the sizes of the atoms involved.

22-2 Group 18: The Noble Gases

In 1785, Henry Cavendish, the discoverer of hydrogen, passed electric discharges through air to form oxides of nitrogen. (A similar process occurs during lightning storms.) He then dissolved these oxides in water to form nitric acid. Even by using excess oxygen, Cavendish was unable to get all the air to react. He suggested that air contained an unreactive gas making up "not more than 1/120 of the whole." John Rayleigh and William Ramsay isolated this gas one century later (1894) and named it argon. The name is derived from the Greek *argos*, "lazy one," meaning "inert." Its *inability* to form chemical compounds with any of the other elements—its chemical inertness—was found to be argon's most notable feature. Because argon resembled no other known element, Ramsay placed it in a separate group of the periodic table and reasoned that there should be other members of this group.

Ramsay then began a systematic search for other inert gases. In 1895, he extracted helium from a uranium mineral. A few years later, by very carefully distilling liquid argon, he was able to extract three additional inert gases: neon, krypton, and xenon. The final member of the group of inert gases, a radioactive element called radon, was discovered in 1900. In 1962, compounds of Xe were first prepared, and so the inert gases proved to be not completely

▲ **William Ramsay (1852–1916)** This distinguished Scottish chemist received the 1904 Nobel Prize in Chemistry for his work on the noble gases.

Graeme Dawes/ Shutterstock

▲ Gaseous atoms in the region of an electric discharge emit light. The light emitted by neon atoms is red-orange. Neon lights of other colors use other noble gases or mixtures of gases.

inert after all. Since that time, this group of gases has been called the **nob gases**. They are found in group 18 of the periodic table shown on the insid front cover.

Occurrence

Air contains 0.000524% He, 0.001818% Ne, and 0.934% Ar, by volume. Th proportion of Kr is about 1 ppm by volume, and that of Xe, 0.05 ppn The atmosphere is the only source of all these gases except helium. The mai source of helium is certain natural gas wells in the western United State that produce natural gas containing up to 8% He by volume. It is cost-effectiv to extract helium from natural gas even down to levels of about 0.3% Underground helium accumulates as a result of α-particle emission b radioactive elements in Earth's crust. Whereas the abundance of He on Eart is very limited, it is second only to hydrogen in the universe as a whole.

Most of the noble gases have escaped from the atmosphere since Earth wa formed, but Ar is an exception. The concentration of Ar remains quite hig because it is constantly being formed by the radioactive decay of potassium-4(a reasonably abundant, naturally occurring radioactive isotope. Helium is als constantly produced through α-particle emissions by radioactive deca processes, but because the molar mass of He is 10 times less than that of Ar, escapes from the atmosphere into outer space at a higher rate.

Properties and Uses

The lighter noble gases are commercially important, in part because they ar chemically inert. The efficiency and life of electric lightbulbs are increase when they are filled with an argon–nitrogen mixture. Electric discharg through neon-filled glass or plastic tubes produces a distinctive red ligh (neon light). Krypton and xenon are used in lasers and in flashlamps in pho tography. Helium has several unique physical properties. Best known of thes is that it exists as a liquid at temperatures approaching 0 K. All other sub stances freeze to solids at temperatures well above 0 K. (The melting poin of solid H_2, for example, is 14 K.) Because of their inertness, both He and A are used to protect materials from nitrogen and oxygen in the air; conse quently, He and Ar are used for this purpose in certain types of welding, i metallurgical processes, and in the preparation of ultrapure Si and Ge an other semiconductor materials. Helium mixed with oxygen is used as breathing mixture for deep-sea diving and in certain medical applications Large quantities of liquid helium are used to maintain low temperature (cryogenics). Some metals essentially lose their electrical resistivity at liqui He temperatures and become *superconductors*. Powerful magnets can be mad by immersing the coils of electromagnets in liquid helium. Such magnets ar used in particle accelerators and in nuclear fusion research. More familiar use of large electromagnets cooled by liquid helium are nuclear magnetic reso nance (NMR) instruments in research laboratories and magnetic resonanc imaging (MRI) devices in hospitals. Helium is also used to fill lighter-than-ai airships (blimps). Some of these applications are highlighted in the margin.

▶ As we learned in Chapter 14 (page 659), using helium–oxygen mixtures instead of air for deep-sea diving prevents the condition known as the *bends* because helium is more rapidly and smoothly expelled from the blood than is nitrogen.

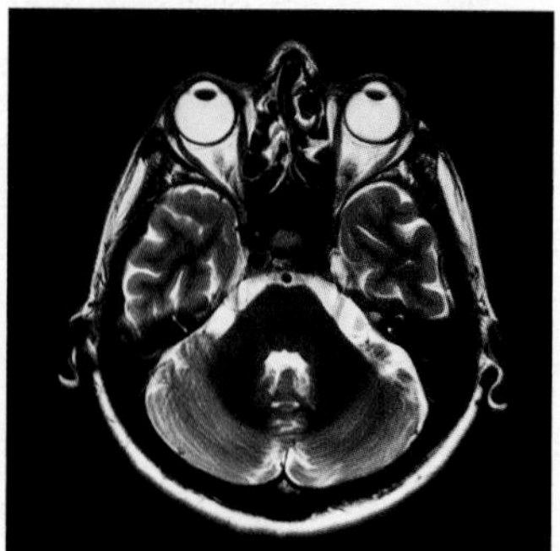

Mandritoiu/Shutterstock

▲ An MRI image of a head.

Xenon Compounds

In this section, we focus exclusively on compounds of xenon because most o the known noble gas compounds contain xenon. A few compounds of kryp ton, such as KrF_2, have been synthesized and well characterized. Rador is expected to form compounds even more readily than xenon, because o its lower ionization energy, but the chemistry of radon is complicated by it radioactivity.

Initially, the noble gases were thought to be chemically inert. This apparen inertness helped provide a theoretical framework for the Lewis theory o bonding. Subsequently it was found that compounds of xenon can be mad ıirly easily, and these compounds have added significantly to our knowledge f chemical bonding.

In the 1930s, Linus Pauling did theoretical calculations that suggested xenon hould form oxide and fluoride compounds, but attempts to make them failed t that time. In 1962, Neil Bartlett and Derek H. Lohmann discovered that $)_2$ and PtF_6 would combine in a 1 : 1 mole ratio to form the compound O_2PtF_6. 'roperties of this compound suggest it to be ionic: $[O_2]^+[PtF_6]^-$. The energy equired to extract an electron from O_2 is 1177 kJ mol^{-1}, almost identical to the irst ionization energy of Xe, 1170 kJ mol^{-1}. The size of the Xe atom is also oughly the same as that of the dioxygen molecule. Thus, it was reasoned that he compound $XePtF_6$ might also exist. Bartlett and Lohmann were able to prepare a yellow crystalline solid with that composition.*

Soon thereafter, chemists around the world synthesized several additional oble gas compounds. In general, the conditions necessary to form noble gas ompounds are as Pauling predicted. The formation of a noble gas compound equires:

- a readily ionizable (therefore, high atomic number) noble gas atom
- highly electronegative atoms (such as F or O) to bond to it

Xenon compounds have been synthesized that have Xe in four possible oxiation states, as summarized below:

	+2	+4	+6	+8
Examples:	XeF_2	XeF_4, $XeOF_2$	XeF_6, XeO_3	XeO_4, H_4XeO_6

Because it is difficult to oxidize Xe to these positive oxidation states, we would xpect Xe compounds to be easily reduced, making them very strong oxidizing gents. For example, in aqueous acidic solution, XeF_2 is an excellent oxidizing gent, as indicated by the large $E°$ value for the following half-reaction:

$$XeF_2(aq) + 2\,H^+(aq) + 2\,e^- \longrightarrow Xe(g) + 2\,HF(aq) \qquad E° = +2.64\text{ V}$$

The significance of this large $E°$ value is that XeF_2 is not very stable in aqueous olution. In aqueous solution, XeF_2 oxidizes the water, producing $O_2(g)$, as hown below:

$$\begin{array}{lrl} \textit{Reduction:} & 2\{XeF_2(aq) + 2\,H^+(aq) + 2\,e^- & \longrightarrow Xe(g) + 2\,HF(aq)\} \\ \textit{Oxidation:} & 2\,H_2O(l) & \longrightarrow 4\,H^+(aq) + O_2(g) + 4\,e^- \\ \hline \textit{Overall:} & 2\,XeF_2(aq) + 2\,H_2O(l) & \longrightarrow 2\,Xe(g) + 4\,HF(aq) + O_2(g) \end{array}$$

$$\begin{aligned} E°_{cell} &= E°(\text{reduction}) - E°(\text{oxidation}) \\ &= E°_{XeF_2/Xe} - E°_{O_2/H_2O} \\ &= 2.64\text{ V} - (1.229\text{ V}) = 1.41\text{ V} \end{aligned}$$

Xenon reacts directly only with F_2. Three different xenon fluorides can be produced by heating xenon and fluorine in nickel reaction vessels, but the product obtained depends on the experimental conditions, as summarized below:

$$Xe(g) + F_2(g) \xrightarrow[400\,°C,\ 1\text{ atm}]{Xe:F_2\ =\ 2:1} XeF_2(s)$$

$$Xe(g) + 2\,F_2(g) \xrightarrow[400\,°C,\ 6\text{ atm}]{Xe:F_2\ =\ 1:5} XeF_4(s)$$

$$Xe(g) + 3\,F_2(g) \xrightarrow[300\,°C,\ 50\text{ atm}]{Xe:F_2\ =\ 1:20} XeF_6(s)$$

All three of these xenon fluorides are colorless, volatile solids. Notice that, for each synthesis, the Xe : F_2 mole ratio is significantly different from the

*It has since been established that this solid is more complicated than first thought. It has the formula $Xe(PtF_6)_n$, where n is between 1 and 2.

Layna Bowers / Alamy

▲ Because of the explosive nature of hydrogen, helium is now used in airships.

◀ In 2000, chemists at the University of Helsinki reported† they had made HArF molecules that were stable at very low temperatures (below 27 K). The report has generated much discussion—and debate—among chemists about the existence of argon compounds and the possibilities for synthesizing argon compounds that are stable at higher temperatures.

†L. Khriachtchev, M. Pettersson, N. Runenburg, J. Lundell, and M. Rasanen, *Nature*, **406**, 874 (2000).

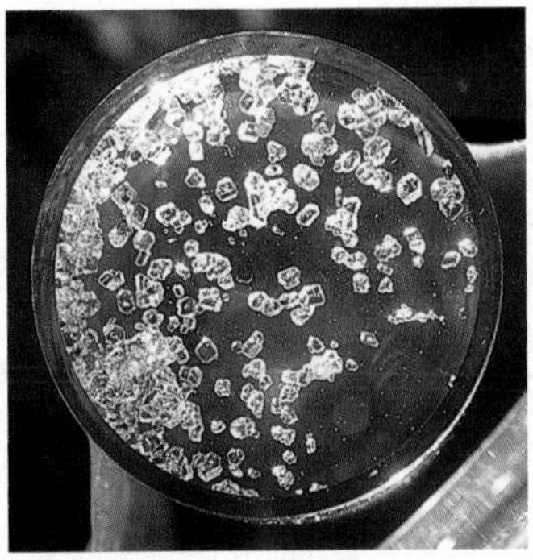

Photograph Courtesy of Argonne National Laboratory

▲ Crystals of xenon tetrafluoride under magnification.

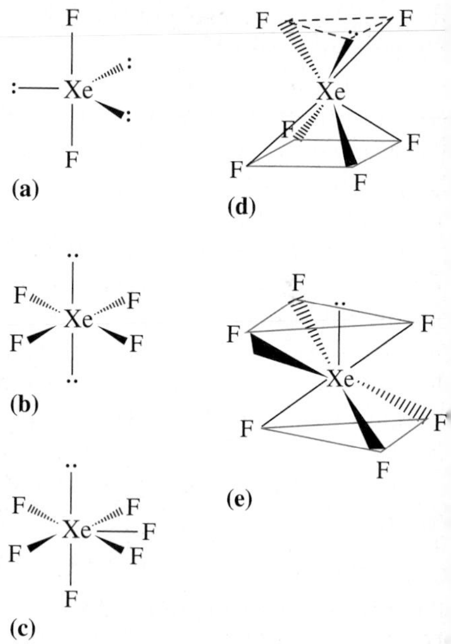

▶ FIGURE 22-1
Electron group geometries for XeF_2, XeF_4, and XeF_6
(a) XeF_2 trigonal bipyramidal; **(b)** XeF_4 octahedral; **(c)** XeF_6 pentagonal bipyramid; **(d)** XeF_6 capped trigonal prism; **(e)** XeF_6 capped octahedron.

stoichiometry of the reaction. For example, the synthesis of XeF_2 requires a excess of Xe, and the synthesis of XeF_6 requires large excess of F_2. A majo difficulty in synthesizing the xenon fluorides is that all three products tend to form simultaneously. The most difficult to prepare in pure form, starting from Xe and F_2, is XeF_4. Xenon tetrafluoride can be prepared in high purity by the reaction below, which uses $O_2F_2(g)$ instead of $F_2(g)$:

$$Xe(g) + 2\,O_2F_2(g) \longrightarrow XeF_4(g) + 2\,O_2(g)$$

The shapes of the XeF_2, XeF_4, and XeF_6 molecules are shown in Figure 22-1. The XeF_2 and XeF_4 molecules have shapes that are easily understood in terms of VSEPR theory, but the shape of the XeF_6 molecule is a little more difficult to interpret. In the XeF_6 molecule, the xenon atom is surrounded by seven electron pairs, and three possible arrangements of these pairs are shown in Figure 22-1. These are pentagonal bipyramid, capped trigonal prism and capped octahedron. These structures are expected to have nearly the same energy and the preferred arrangement depends strongly on the exact conditions. In the gas phase, the XeF_6 molecule has a capped octahedral structure (Figure 22-1e). In this structure, the six fluorine atoms form a distorted octahedron, and the lone pair on xenon is directed toward the center of one of the triangular faces. In the solid phase, XeF_6 exists as XeF_5^+, which is square pyramidal and F^-, with the F^- ions forming bridges between XeF_5^+ ions.

To conform to the shapes indicated by VSEPR theory (or experiment, where the structure is known), bonding theory based on hybridization of orbitals requires the hybrid orbitals sp^3d for XeF_2, sp^3d^2 for XeF_4, and sp^3d^3 for XeF_6. However, in light of the observed bond lengths and bond energies in these fluorides and the high energy (estimated to be 1000 kJ mol^{-1}) required to promote an electron from a $5p$ to a $5d$ orbital, there is doubt as to whether d orbitals are involved in the bond formation. A molecular orbital description can be constructed that does not involve the participation of the xenon $5d$ orbitals, but in its simplest form this description cannot explain the nonoctahedral shape of XF_6. The situation described here reinforces the caveat that approximate theories of chemical bonding must be viewed critically.

▶ Actually, the XeF_6 molecule does possess a small dipole moment (about 0.03 debye). Compare this value with that for H_2O (1.84 debye) and HCl (1.03 debye).

Some properties of XeF_2, XeF_4, and XeF_6 are listed in Table 22.3. Because of their molecular structures, the XeF_2, XeF_4, and XeF_6 molecules are all nonpolar, and thus we expect that in each of these fluorides, the molecules interact with each other primarily through London dispersion forces. However, the melting points of these fluorides decrease as the molecular size increases,

TABLE 22.3 Some Properties[a] of XeF_2, XeF_4, and XeF_6

	XeF_2	XeF_4	XeF_6
Melting point, °C	129	117	49
$\Delta_f H^\circ_{solid}$, kJ mol^{-1}	−163	−267	−338
$\Delta_f H^\circ_{gas}$, kJ mol^{-1}	−107	−206	−279
$\Delta_f G^\circ_{gas}$, kJ mol^{-1}	−96	−138	—
Xe—F bond energy, kJ mol^{-1}	133	131	126
Xe—F bond length, pm	200	195	189

[a] The $\Delta_f H^\circ$ and $\Delta_f G^\circ$ values are for 25 °C. The bond energies and bond lengths are average values.

contrary to what we would expect if only London dispersion forces contributed to the attraction among molecules. The trend in melting points indicates that, in the xenon fluorides, there are additional interactions to consider. It has been suggested that, while the XeF_2 and XeF_4 molecules are nonpolar, the polar Xe—F bonds interact with each other in the manner suggested in Figure 22-2. The electrostatic potential maps shown in Figure 22-2 indicate that the Xe—F bonds in XeF_2 are more polar than they are in XeF_4; thus, the interactions of the bond dipoles are more important in $XeF_2(s)$ than in $XeF_4(s)$. The interactions of bond dipoles help to explain not only why the xenon fluorides have high melting points but also why the melting point of $XeF_2(s)$ is higher than that of $XeF_4(s)$.

Xenon forms other compounds, in which xenon is bonded to chlorine, oxygen, or nitrogen. Some of the simpler compounds include $XeCl_2$, XeO_3, XeO_4, $XeOF_2$, XeO_2F_2, and $XeOF_4$. However, many of these compounds must be prepared by using an indirect route. For example, XeO_3 can be made from XeF_6 by using the following reaction:

$$2\,XeF_6(s) + 3\,SiO_2(s) \longrightarrow 2\,XeO_3(s) + 3\,SiF_4(g)$$

Why is it that Xe(g) reacts directly with only $F_2(g)$? We can formulate an answer to this question by focusing on the following synthesis:

$$Xe(g) + F_2(g) \longrightarrow XeF_2(s)$$

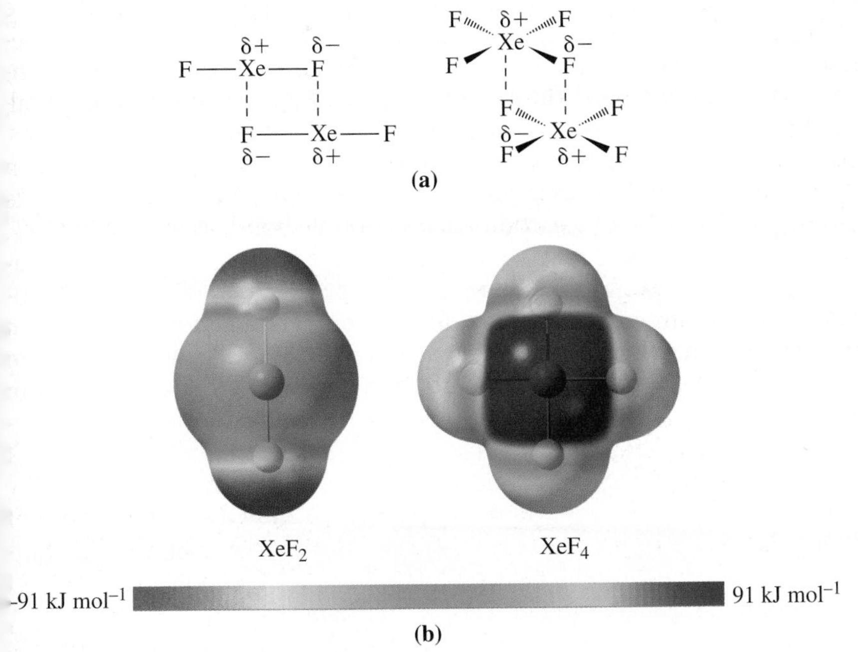

◀ FIGURE 22-2
Interactions and electrostatic potential maps of XeF_2 and XeF_4
(a) Possible interactions between bond dipoles in XeF_2 and XeF_4 lead to relatively high melting points. **(b)** The electrostatic potential maps of XeF_2 and XeF_4 show that the Xe—F bond dipole is smaller in XeF_4 than it is in XeF_2.

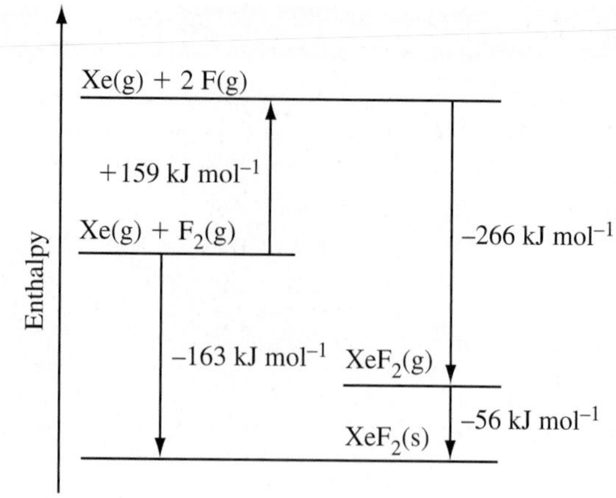

▶ FIGURE 22-3
Enthalpy diagram for the formation of $XeF_2(s)$

▶ The low bond energy of the F—F bond is probably the result of repulsions involving the lone-pair electrons on the two F atoms. The repulsions are rather strong because the atomic radius of fluorine is small, and thus the six lone pairs are in close proximity. The weakness of the F—F bond contributes to the reactivity of F_2.

The enthalpy diagram shown in Figure 22-3 shows that the reaction of Xe(g and $F_2(g)$ is exothermic (and energetically favorable) because the reactio involves the breaking of relatively weak F—F bonds (159 kJ mol^{-1}) and th formation of twice as many Xe—F bonds, each of which is almost as strong a an F—F bond. (For the synthesis of all the xenon fluorides, starting fro Xe(g) and $F_2(g)$, the number of Xe—F bonds formed is always two times th number of F—F bonds broken.) None of the other noble gases react directl with fluorine because, for the other noble gases, the bond energy of the bon formed with F is not large enough to offset the energy requirements of break ing the F—F bond. For example, the bond energy of the Kr—F bond, in KrF_2 is only about 50 kJ mol^{-1}.

A similar analysis can also be used to help us to understand why xeno does not react directly with Cl_2 or O_2 to form chlorides or oxides. Such a analysis (see Exercise 109) reveals that the Xe—Cl or Xe—O bond energy i much too small to offset the energy requirements for breaking the Cl—C or O=O bonds. (The bond energies of Cl_2 and O_2 are 243 kJ mol^{-1} an 498 kJ mol^{-1}, respectively.)

All the xenon fluorides react with water to form various products. Fo example, in aqueous solution, xenon hexafluoride is first hydrolyzed to xeno oxide tetrafluoride, $XeOF_4$, which is further hydrolyzed to xenon trioxide. Th reactions are as follows:

$$XeF_6(s) + H_2O(l) \longrightarrow XeOF_4(aq) + 2\,HF(aq)$$
$$XeOF_4(l) + 2\,H_2O(l) \longrightarrow XeO_3(aq) + 4\,HF(aq)$$

The xenon fluorides are good fluorinating agents. Xenon difluoride is some times used in organic chemistry to add fluorine atoms to carbon compounds An advantage of using XeF_2 for this purpose is that the by-product, Xe(g), i easily separated from the desired product. For example, XeF_2 can be used t add F atoms to either side of a carbon–carbon double bond, as shown below:

$$XeF_2(s) + CH_2{=}CH_2(g) \longrightarrow CH_2FCH_2F(g) + Xe(g)$$

The xenon fluorides are also strong oxidizing agents and can be used to oxi dize fluorides of other elements. In the reaction below, XeF_4 is used to oxidiz SF_4 to SF_6. In the reaction below, the oxidation state of sulfur changes fro +4 to +6.

$$XeF_4(s) + 2\,SF_4(g) \longrightarrow 2\,SF_6(g) + Xe(g)$$

22-1 CONCEPT ASSESSMENT

Which is likely to be the more stable, XeF_2 or $XeCl_2$? Explain.

22-3 Group 17: The Halogens

he word *halogen* was introduced in 1811 to describe the ability of chlorine to orm ionic compounds with metals. The name is based on the Greek words *alos* and *gen* meaning "salt former." The name was later extended to include luorine, bromine, and iodine as well. Mendeleev placed the halogens in Group VII of his periodic table (page 378), which is now group 17 in the UPAC table (page 52). Current interest in the halogens extends far beyond heir ability to form metallic salts.

roperties

he **halogens** exist as diatomic molecules, symbolized by X_2, where X is a generic symbol for a halogen atom. That these elements occur as nonpolar liatomic molecules accounts for their relatively low melting and boiling oints (Table 22.4). As expected, melting and boiling points increase as we nove down the group from the smallest and lightest member of the group, luorine, to the largest and heaviest, iodine. Conversely, chemical reactivity oward other elements and compounds increases in the *opposite* order, with luorine being the most reactive and iodine the least reactive.

All the halogen atoms have large electron affinities (see Table 22.4 and igure 9-16), and show a strong tendency to gain electrons. Consequently, the alogens are rather good oxidizing agents.

We have mentioned previously, and we will see again, that the elements of eriod 2 have distinctly different chemistry from the rest of the group ecause of their small sizes and inability to expand their valence shells. However, for the halogens, the differences between the second-row element fluorine) and the members of the group are much less dramatic. Still, fluoine differs from the other halogens in a few ways. For example, a fluorine tom almost always forms just one covalent bond, whereas chlorine, romine, and iodine atoms typically form more than one bond and as many s seven in some of their compounds. Although all the halogens are quite eactive and are found in nature only as compounds, fluorine is considerably nore reactive than the other members of the group. It reacts directly with lmost all the elements, except for oxygen, nitrogen, and the lighter noble ases, and forms compounds with even the most unreactive metals. It reacts with almost all materials, especially organic compounds, to produce fluorides. The reactivity of fluorine can be attributed to the weakness of the fluorine–fluorine bond in F_2, which arises, as mentioned earlier, because of the small size of the fluorine atom and the repulsion between the lone pairs on the fluorine atoms.

◀ In the perhalic acids, HXO_4, where X = Cl, Br, or I, the halogen atom forms seven bonds, four of which are sigma bonds and three of which are π bonds. The Lewis structure for perchloric acid is shown below.

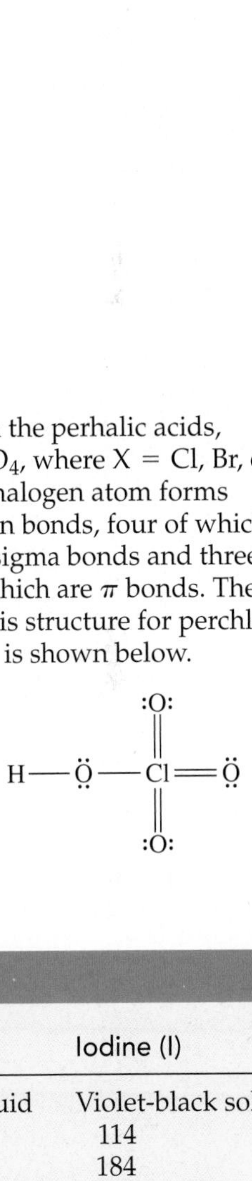

TABLE 22.4 Group 17 Elements: The Halogens

	Fluorine (F)	Chlorine (Cl)	Bromine (Br)	Iodine (I)
Physical form at room temperature	Pale yellow gas	Yellow-green gas	Dark red liquid	Violet-black solid
Melting point, °C	−220	−101	−7.2	114
Boiling point, °C	−188	−35	58.8	184
Electron configuration	$[He]2s^22p^5$	$[Ne]3s^23p^5$	$[Ar]3d^{10}4s^24p^5$	$[Kr]4d^{10}5s^25p^5$
Covalent radius, pm	71	99	114	133
Ionic (X^-) radius, pm	133	181	196	220
First ionization energy, kJ mol^{-1}	1681	1251	1140	1008
Electron affinity, kJ mol^{-1}	−328.0	−349.0	−324.6	−295.2
Electronegativity	4.0	3.0	2.8	2.5
Standard electrode potential, V ($X_2 + 2e^- \longrightarrow 2X^-$)	2.866	1.358	1.065	0.535

Fluorine differs from the other halogens in that it shows a much greater tendency to form ionic bonds with metals. Perhaps this is most evident when we look at the binary compounds formed by the halogens and the group 13 metals. The trifluorides of Al, Ga, and In are all ionic compounds, with very high lattice energies and very high melting points (>1000 °C) whereas the trichlorides are volatile compounds with much lower melting points (<600 °C). For $AlCl_3$, $GaCl_3$, and $InCl_3$, the bonding is largely covalent because chloride ions are much larger, and much more polarizable, than fluoride ions. Also, in the solid state, the chlorides of the group 13 metals contain dimers, M_2Cl_6, whereas the fluorides of the group 13 metals are all ionic lattices containing M^{3+} and F^- ions.

Another important difference between fluorine and the other halogens is that fluorine shows the ability to stabilize other elements in very high oxidation states. For example, fluorine reacts with sulfur to give SF_6, with sulfur in the +6 oxidation state, whereas chlorine reacts directly with molten sulfur to give S_2Cl_2, with sulfur in the +1 oxidation state.

Much of the reaction chemistry of the halogens involves oxidation–reduction reactions in aqueous solutions. For these reactions, standard electrode potentials are helpful for understanding the reactivity of the halogens. Among the properties of the halogens listed in Table 22.4 are potentials for the following half-reaction:

KEEP IN MIND

that the halogens are extremely toxic.

$$X_2 + 2\,e^- \longrightarrow 2\,X^-(aq)$$

By this measure, fluorine is clearly the most reactive element of the group ($E° = 2.866$ V). Of all the elements, it shows the greatest tendency to gain electrons and is therefore the most easily reduced. Given this fact, it is not surprising that fluorine occurs naturally only in combination with other elements, and only as the fluoride ion, F^-. Although both chlorine and bromine can exist in a variety of positive oxidation states, they are mostly found in naturally occurring compounds as chloride and bromide ions. Natural deposits of chlorate and perchlorate, sources of positive oxidation states, are found worldwide. There are, however, naturally occurring compounds in which iodine is in a positive oxidation state (such as the iodate ion IO_3^-, in $NaIO_3$). In the case of iodine, the tendency for I_2 to be reduced to I^- is not particularly great ($E° = 0.535$ V).

Electrode Potential Diagrams

When we summarize the reduction tendencies of main-group metals and their ions, generally one or, at most, a few $E°$ values tell the story, and these values are easily incorporated into tables, such as that in Appendix D. However, the oxidation–reduction chemistry of some of the nonmetals is much richer and involves a larger number of relevant $E°$ values. In these cases, *electrode potential diagrams* are particularly useful for summarizing $E°$ data. Partial diagrams for chlorine are shown in Figure 22-4. In these diagrams, a number written above a line segment is the $E°$ value for reduction of the species on the left (higher oxidation state) to the one on the right (lower oxidation state). For a reduction involving species not joined by a line segment, we generally can calculate the appropriate value of $E°$ by the method illustrated in Example 22-1.

Production and Uses

Although the existence of fluorine had been known since early in the nineteenth century, no one was able to devise a chemical reaction to extract the free element from its compounds. Finally, in 1886, Henri Moissan succeeded in preparing $F_2(g)$ by an electrolysis reaction. Moissan's method, which is still the only important commercial method for fluorine extraction, involves the

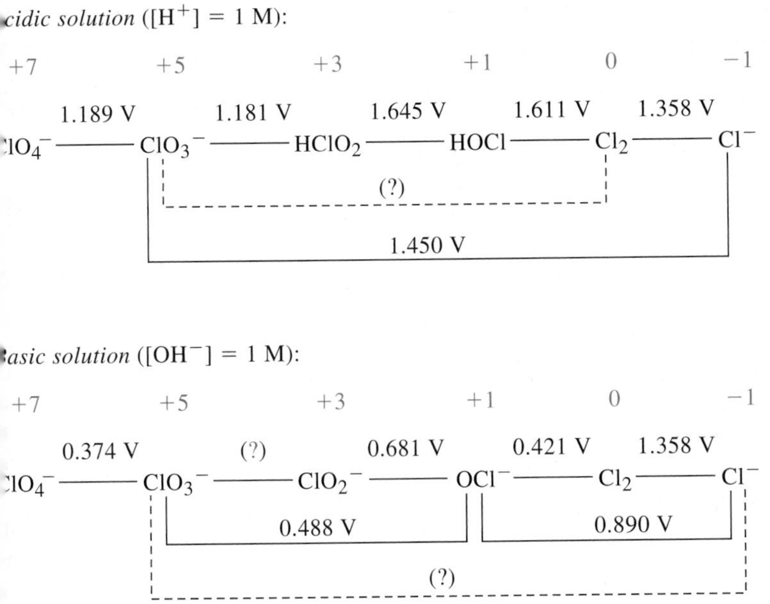

◀ FIGURE 22-4
Standard electrode potential diagrams for chlorine
The numbers in red are the oxidation states of the Cl atom. $E°$ values are written above the line segments for the reduction of the species on the left to the species on the right. All reactants and products are at unit activity. Because of the basic properties of the ClO_2^- and OCl^- ions, the weak acids $HClO_2$ and HOCl form in acidic solution.

EXAMPLE 22-1 Using an Electrode Potential Diagram to Determine $E°$ for a Half-Reaction

Determine $E°$ for the reduction of ClO_3^- to ClO_2^- in a basic solution, which is marked (?) in Figure 22-4.

Analyze

One way we can approach this problem is to identify a route that (1) converts ClO_3^- to ClO_2^- and (2) involves $E°$ values that are known. One such route involves converting ClO_3^- to OCl^-, for which $E° = 0.488$ V, and then converting OCl^- to ClO_2^-, for which $E° = -0.681$ V. (The $E°$ value for converting ClO_2^- to OCl^- is given as 0.681 V in Figure 22-4. For the reverse process, the $E°$ value has the opposite sign.) First, we should write balanced equations for $ClO_3^- \longrightarrow ClO^-$ and $ClO^- \longrightarrow ClO_2^-$, and then focus on how we combine the equations and corresponding $E°$ values to obtain the equation and $E°$ value for the desired conversion, $ClO^- \longrightarrow ClO_2^-$. Because the desired process is a reduction process (and not an oxidation–reduction process), we cannot simply add or subtract the $E°$ values to obtain $E°$ for the desired process (see page 879). We must first convert the given $E°$ values to $\Delta_r G°$ values, then combine those $\Delta_r G°$ values to obtain a $\Delta_r G°$ value for the desired process, and finally convert this $\Delta_r G°$ value back to a value of $E°$.

Solve

Figure 22-4 identifies only the oxidized and reduced forms of the chlorine species. By using the method described on page 174, we obtain the following balanced equations for the conversions of ClO_3^- to ClO^- and ClO^- to ClO_2^-:

$$ClO_3^-(aq) + 2\,H_2O(l) + 4\,e^- \longrightarrow OCl^-(aq) + 4\,OH^-(aq) \quad E° = 0.488\text{ V} \quad \Delta_r G° = -4F \times 0.488\text{ V}$$
$$OCl^-(aq) + 2\,OH^-(aq) \longrightarrow ClO_2^-(aq) + H_2O(l) + 2\,e^- \quad E° = -0.681\text{ V} \quad \Delta_r G° = -2F \times (-0.681\text{ V})$$

The overall equation for the desired process is the sum of the two equations above, and the corresponding $\Delta_r G°$ value is the sum of the $\Delta_r G°$ values.

$$ClO_3^-(aq) + H_2O(l) + 2\,e^- \longrightarrow ClO_2^-(aq) + 2\,OH^-(aq) \quad \Delta_r G° = -F\,(4 \times 0.488 - 2 \times 0.681)\text{ V}$$

The $\Delta_r G°$ value for the process above is related to the $E°$ value by the expression $\Delta_r G° = -2FE°$, and so, the $E°$ value is calculated as follows.

$$E° = \frac{F[(2 \times 0.681) - (4 \times 0.488)]\text{ V}}{-2F} = 0.295\text{ V}$$

Assess

When adding half-equations to give another half-equation, we must add the $\Delta_r G°$ values, not the $E°$ values.

PRACTICE EXAMPLE A: Determine the missing $E°$ value for the dashed line that joins ClO_3^- and Cl^- in basic solutions in Figure 22-4.

PRACTICE EXAMPLE B: Determine the missing $E°$ value for the dashed line that joins ClO_3^- and Cl_2 in acidic solutions in Figure 22-4.

▶ The hydrogen difluoride ion in KHF_2 features a strong hydrogen bond, with a H^+ midway between two F^- ions: $[F—H—F]^-$.

electrolysis of HF dissolved in molten KHF_2. The chemical equation for th reaction is given below:

$$2\,HF \xrightarrow[KF \cdot 2\,HF(l)]{\text{electrolysis}} H_2(g) + F_2(g) \qquad (22.1)$$

The corresponding half-reactions are as follows:

$$\textit{Anode:} \quad 2\,F^- \longrightarrow F_2(g) + 2\,e^-$$
$$\textit{Cathode:} \quad 2\,H^+ + 2\,e^- \longrightarrow H_2(g)$$

Moissan also developed the electric furnace and was honored for both thes achievements with the Nobel Prize in chemistry in 1906. Nevertheless, th challenge of producing fluorine by means of a chemical reaction remained. I 1986, one century after Moissan isolated fluorine, the chemical synthesis c fluorine was announced (see Exercise 14).

Although chlorine can be prepared by several chemical reactions, electroly sis of NaCl(aq) is the usual industrial method, as we have mentioned previ ously (see page 906). The electrolysis reaction is

$$2\,Cl^-(aq) + 2\,H_2O(l) \xrightarrow{\text{electrolysis}} 2\,OH^-(aq) + H_2(g) + Cl_2(g) \qquad (22.2)$$

Victor B/Fotolia

▲ **Salt formation in the Dead Sea**
The high concentrations of salts in the Dead Sea make it a good source of bromine and a number of other chemicals.

Bromine can be extracted from seawater, where it occurs in concentration of about 70 ppm as Br^-, or from inland brine sources. Seawater from the Dead Sea, highlighted in the margin, is a good source of bromine. The seawater o brine solution is adjusted to pH 3.5 and treated with $Cl_2(g)$, which oxidize Br^- to Br_2 in the following displacement reaction:

$$Cl_2(g) + 2\,Br^-(aq) \longrightarrow Br_2(l) + 2\,Cl^-(aq) \qquad E^\circ_{cell} = 0.293\ V \qquad (22.3)$$

The liberated Br_2 is swept from seawater with a current of air or from brine with steam. A dilute bromine vapor forms and can be concentrated by variou methods. The reaction above also forms the basis of a test for the presence o Br^-, as described in the margin.

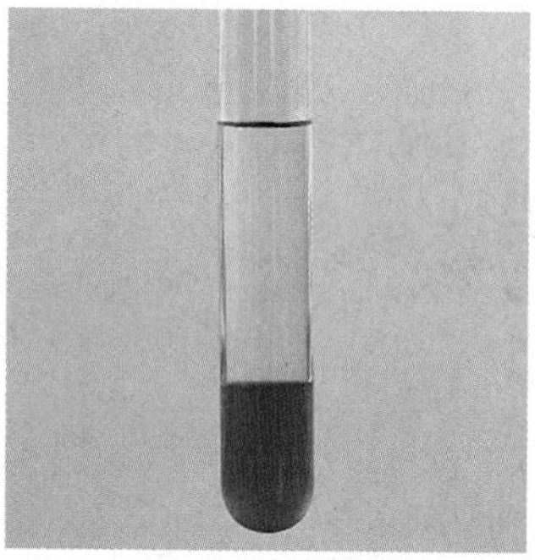

Tom Pantages

▲ **A test for Br^-(aq)**
Reaction (22.3) is used as a qualitative test in the laboratory. The liberated Br_2 is extracted from the top aqueous layer into the bottom layer, here the organic solvent chloroform, $CHCl_3$.

Certain marine plants, such as seaweed, absorb and concentrate I^- selec tively in the presence of Cl^- and Br^-. Iodine is obtained in small quantitie from such plants. In the United States, I_2 is obtained from inland brines by a process similar to that for the production of Br_2. Another abundant natura source of iodine is $NaIO_3$, found in large deposits in Chile. Because the oxi dation state of iodine must be reduced from +5 in IO_3^- to 0 in I_2, the conver sion of IO_3^- to I_2 requires the use of a reducing agent. Aqueous sodium hydrogen sulfite (sodium bisulfite) is used as the reducing agent in the firs part of a two-step procedure, followed by the reaction of I^- with additiona IO_3^- to produce I_2. The net ionic equations for the reactions are giver below:

$$IO_3^-(aq) + 3\,HSO_3^-(aq) \longrightarrow I^-(aq) + 3\,SO_4^{2-}(aq) + 3\,H^+(aq) \qquad (22.4)$$

$$5\,I^-(aq) + IO_3^-(aq) + 6\,H^+(aq) \longrightarrow 3\,I_2(s) + 3\,H_2O(l) \qquad (22.5)$$

The halogen elements form a variety of useful compounds, and the ele- ments themselves are largely used to produce these compounds. All the halogens are used to make halogenated organic compounds. In the past, flu- orine was used to make chlorofluorocarbons (CFCs), which were used as refrigerants, but international treaties have banned the production of CFCs in most countries because they damage the stratospheric ozone layer (see Focus On 22–1, www.masteringchemistry.com). Now fluorine is used to make hydrochlorofluorocarbons (HCFCs), which are more environmentally benign alternatives to CFCs. Fluorinated organic compounds tend to be chemically inert, and it is this inertness that makes them useful as components in harsh chemical environments. Fluorine is a key element in a variety of useful inor- ganic compounds. Some of these compounds and their uses are listed in Table 22.5.

TABLE 22.5 Important Inorganic Compounds of Fluorine

Compound	Uses
Na_3AlF_6	Manufacture of aluminum
BF_3	Catalyst
CaF_2	Optical components, manufacture of HF, metallurgical flux
ClF_3	Fluorinating agent, reprocessing nuclear fuels
HF	Manufacture of F_2, AlF_3, Na_3AlF_6, and fluorocarbons
LiF	Ceramics manufacture, welding, and soldering
NaF	Fluoridating water, dental prophylaxis, insecticide
SF_6	Insulating gas for high-voltage electrical equipment
PO_3F^{2-}	Manufacture of toothpaste
UF_6	Manufacture of uranium fuel for nuclear reactors

With an annual production of more than 13 million metric tons, elemental hlorine ranks about eighth in quantity among manufactured chemicals in he United States. It has three main commercial uses: (1) production of chloinated organic compounds (about 70%), chiefly ethylene dichloride, CH_2ClCH_2Cl, and vinyl chloride, $CH_2{=}CHCl$ (the monomer of polyvinyl hloride, PVC); (2) as a bleach in the paper and textile industries and for the reatment of swimming pools, municipal water, and sewage (about 20%); nd (3) production of dozens of chlorine-containing inorganic chemicals about 10%).

◀ The major industrial source of HCl is the chlorination of organic compounds. For example, HCl is obtained as a by-product in the chlorination of methane, as shown below:

$$CH_4 + Cl_2 \longrightarrow CH_3Cl + HCl$$

Bromine is used to make brominated organic compounds. Some of these are ised as fire retardants and pesticides. Others are used extensively as dyes and harmaceuticals. An important inorganic bromine compound is AgBr, the prinary light-sensitive agent used in photographic film.

Iodine is of much less commercial importance than chlorine. Iodine and its compounds, however, do have applications as catalysts, antiseptics, and germicides and in the preparation of pharmaceuticals and photographic emulsions (as AgI).

Hydrogen Halides

We have encountered the hydrogen halides from time to time throughout this text. In aqueous solution, they are called the *hydrohalic acids*. Except for HF, hydrohalic acids are strong acids in water. An explanation for why HF is a weak acid was given on page 771.

One well-known property of HF is its ability to etch (and ultimately to dissolve) glass, an application that is highlighted in the margin. The reaction is similar to one between HF and silica, SiO_2.

$$SiO_2(s) + 4\,HF(aq) \longrightarrow 2\,H_2O(l) + SiF_4(g) \qquad (22.6)$$

Because HF reacts with glass, it must be stored in special containers coated with a lining of Teflon or polyethylene.

Hydrogen fluoride is commonly produced by a method discussed in Section 21-2. When a halide salt (such as fluorite, CaF_2) is heated with a *nonvolatile* acid, such as concentrated $H_2SO_4(aq)$, a sulfate salt and the volatile hydrogen halide are produced.

$$CaF_2(s) + H_2SO_4(\text{concd aq}) \xrightarrow{\Delta} CaSO_4(s) + 2\,HF(g) \qquad (22.7)$$

Richard Megna/Fundamental Photographs

▲ Glass etched with hydrofluoric acid.

This method also works for preparing HCl(g) but not for HBr(g) or HI(g). Concentrated $H_2SO_4(aq)$ is a sufficiently strong oxidizing agent to oxidize Br^- to Br_2 and I^- to I_2. For example, the reaction of NaBr(s) and concentrated $H_2SO_4(aq)$ yields $Br_2(g)$ and not HBr(g).

$$2\,NaBr(s) + 2\,H_2SO_4(\text{concd aq}) \xrightarrow{\Delta} Na_2SO_4(s) + 2\,H_2O(l) + Br_2(g) + SO_2(g) \qquad (22.8)$$

TABLE 22.6 Standard Gibbs Energy of Formation of Hydrogen Halides at 298 K

	$\Delta_f G°$, kJ mol^{-1}
HF(g)	−273.2
HCl(g)	−95.30
HBr(g)	−53.45
HI(g)	+1.70

We can get around this difficulty by using a *nonoxidizing* nonvolatile aci such as phosphoric acid, H_3PO_4. Also, all the hydrogen halides can be forme by the direct combination of the elements, as shown below:

$$H_2(g) + X_2(g) \longrightarrow 2\,HX(g) \qquad (22.$$

The reaction of $H_2(g)$ and $F_2(g)$ is very fast, however, occurring with explo sive violence under some conditions. With $H_2(g)$ and $Cl_2(g)$, the reaction als proceeds rapidly (explosively) in the presence of light (photochemically init ated), although some HCl is made this way commercially. With Br_2 and I_2, th reaction occurs more slowly and a catalyst is required.

The data in Table 22.6 show that the standard free energies of formation o HF(g), HCl(g), and HBr(g) are large and negative, suggesting that for ther reaction (22.9) goes to completion. For HI(g), $\Delta_f G°$ is small and positive. Thi suggests that even at room temperature HI(g) should dissociate to some exter into its elements. Because the dissociation of HI(g) has a high activatio energy, however, the dissociation occurs only very slowly in the absence of catalyst. As a result, HI(g) is stable at room temperature.

EXAMPLE 22-2 Determining K_p for a Dissociation Reaction from Gibbs Energies of Formation

What is the value of *K* for the dissociation of HI(g) into its elements at 298 K?

Analyze

Dissociation of HI(g) into its elements is the reverse of the formation reaction, and $\Delta_r G°$ for the dissociation reaction is the negative of $\Delta_f G°$ listed in Table 22.6. The value of *K* for the dissociation reaction can be calculated by using the expression $\Delta_r G° = -RT \ln K$.

Solve

The dissociation reaction and the value of $\Delta_r G°$ are as follows:

$$HI(g) \longrightarrow \frac{1}{2}H_2(g) + \frac{1}{2}I_2(s) \qquad \Delta_r G° = -1.70 \text{ kJ mol}^{-1} \qquad (22.10)$$

Now we can use the relationship $\Delta_r G° = -RT \ln K$.

$$\ln K = \frac{-\Delta_r G°}{RT} = \frac{-(-1.70 \times 10^3 \text{ J mol}^{-1})}{8.3145 \text{ J mol}^{-1} \text{ K}^{-1} \times 298 \text{ K}} = 0.686$$

$$K = e^{0.686} = 1.99$$

Assess

The value of *K* is reasonable at just slightly larger than 1 because $\Delta_r G°$ for the dissociation reaction is just slightly less than zero. When $\Delta_r G° = 0$, the value of *K* is equal to 1. When $\Delta_r G°$ is a small negative number, the forward reaction is slightly favored and *K* is just slightly larger than one.

PRACTICE EXAMPLE A: What is the value of *K* for the dissociation of HF(g) into its elements at 298 K? Use data from Table 22.6.

PRACTICE EXAMPLE B: Use data from Table 22.6 to determine *K* and the percent dissociation of HCl(g) into its elements at 298 K.

Oxoacids and Oxoanions

Fluorine, the most electronegative element, adopts the −1 oxidation state in its compounds. The other halogens, when bonded to a more electronegative element such as oxygen, can have any one of several positive oxidation states: +1, +3, +5, or +7. This variability of oxidation states, which was illustrated through electrode potential diagrams for chlorine (Fig. 22-4), is emphasized again by the oxoacids listed in Table 22.7. The structures of some of these

TABLE 22.7 Oxoacids of the Halogens[a]

Oxidation State of Halogen	Chlorine	Bromine	Iodine
+1	HOCl	HOBr	HOI
+3	$HClO_2$	—	—
+5	$HClO_3$	$HBrO_3$	HIO_3
+7	$HClO_4$	$HBrO_4$	HIO_4; H_5IO_6

[a]In all these acids, H atoms are bonded to O atoms—as shown for HOCl, HOBr, and HOI—not to the central halogen atom. More accurate representations of the other acids would be HOClO (instead of $HClO_2$), $HOClO_2$ (instead of $HClO_3$), and so on.

xoacids are shown in Figure 22-5. Chlorine forms a complete set of oxoacids n all these oxidation states, but bromine and iodine do not. Only a few of the xoacids can be isolated in pure form ($HClO_4$, HIO_3, HIO_4, H_5IO_6); the rest re stable only in aqueous solution.

The shapes of the oxoacids (and the corresponding oxoanions) are based on tetrahedral arrangement of the electron pairs around the chlorine atom. The tructures of the acids and the electrostatic potential maps for the anions are lisplayed in Figure 22-5. Two things are noteworthy in Figure 22-5. First, in he structures for $HClO_3$ and $HClO_4$, the bonds shown as dashed wedges are louble bonds. The chlorine-oxygen double bonds, with a bond length of bout 141 pm, are much shorter than a Cl—O single bond, which has a bond ength of about 164 pm. Second, the electrostatic potential maps indicate that, s the number of oxygen atoms increases, the negative charge on the anion ets progressively delocalized. In OCl^-, the oxygen atom is shown in red, vhich indicates that the negative charge is highly localized on the oxygen tom. In the other anions, the oxygens are shown in yellow or green, which ndicates that they are much less negative. (See Figure 10-5 for a description of he color scheme used in the electrostatic potential maps.) Generally speaking,

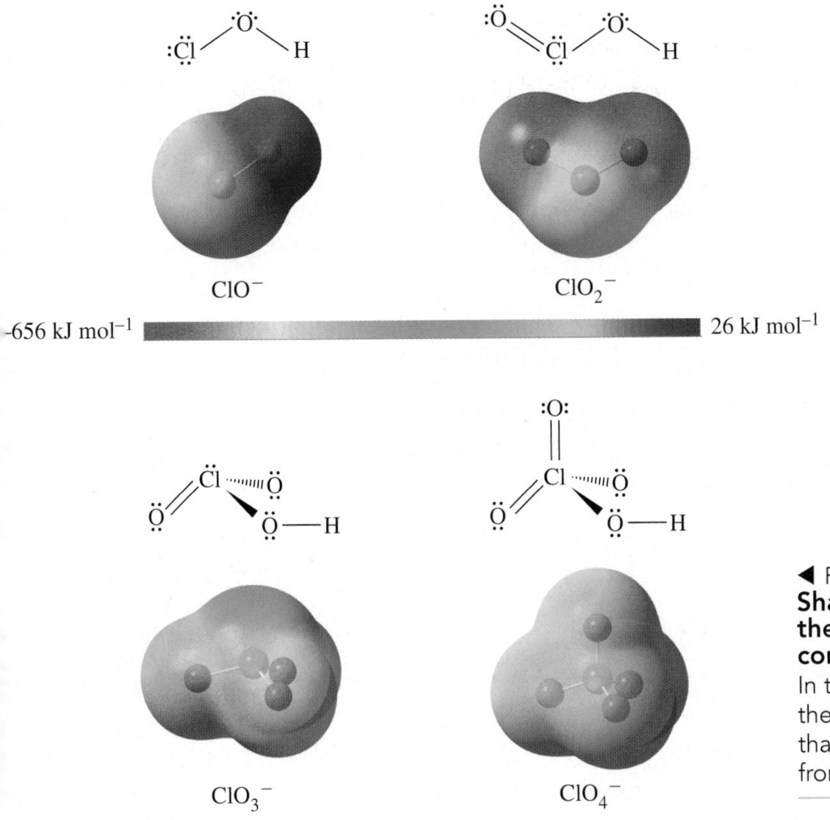

◀ FIGURE 22-5
Shapes of the oxoacids of chlorine and the electrostatic potential maps of the corresponding anions
In the structures shown for $HClO_3$ and $HClO_4$, the dashed wedges represent double bonds that point behind the plane of the page, away from the viewer.

the more delocalized the charge, the more stable the anion is in aqueous solution and the weaker it is as a Brønsted–Lowry base. Because the charge of the anion is delocalized to the greatest extent in ClO_4^-, we conclude that ClO_4^- is the weakest base, and $HClO_4$, its conjugate acid, is the strongest acid.

▶ Never mix bleach with ammonia or household cleaners because the formation of toxic gases, such as chloramine (NH_2Cl) or chlorine (Cl_2), can result.

An easily prepared oxidizing agent for laboratory use is an aqueous solution of chlorine ("chlorine water"). The solution is not just one of Cl_2, however, because $Cl_2(aq)$ disproportionates into $HOCl(aq)$ and $HCl(aq)$.

$$\begin{array}{ll} \textit{Reduction:} & Cl_2(g) + 2\,e^- \longrightarrow 2\,Cl^-(aq) \\ \textit{Oxidation:} & Cl_2(g) + 2\,H_2O(l) \longrightarrow 2\,HOCl(aq) + 2\,H^+(aq) + 2\,e^- \\ \hline \textit{Overall:} & Cl_2(g) + H_2O(l) \longrightarrow HOCl(aq) + H^+(aq) + Cl^-(aq) \end{array} \quad (22.11)$$

$$E^\circ = E^\circ_{Cl_2/Cl^-} - E^\circ_{HOCl/Cl_2} = 1.358\ V - 1.611\ V = -0.253\ V$$
$$\Delta_r G^\circ = -zFE^\circ = 48.8\ kJ\ mol^{-1}$$

KEEP IN MIND

that E_{Cl_2/Cl^-} is independent of pH, but both E_{HOCl/Cl_2} and E_{OCl^-/Cl_2} are pH dependent. They become smaller as the pH increases, increasing E_{cell} and making $\Delta_r G$ less positive and eventually negative. In terms of Le Châtelier's principle, the smaller $[H^+]$ (greater $[OH^-]$), the more the disproportionation reaction is favored.

Although the disproportionation of Cl_2 in water is nonspontaneous when all reactants and products are in their standard states, the reaction does occur to a limited extent in solutions that are not strongly acidic, as the E° and $\Delta_r G$ values indicate.

In contrast, the disproportionation is spontaneous for standard-state conditions in basic solution, because $\Delta_r G^\circ$ for the reaction is negative, as demonstrated below:

$$\begin{array}{ll} \textit{Reduction:} & Cl_2(g) + 2\,e^- \longrightarrow 2\,Cl^-(aq) \\ \textit{Oxidation:} & Cl_2(g) + 4\,OH^-(aq) \longrightarrow 2\,OCl^-(aq) + 2\,H_2O(l) + 2\,e^- \\ \hline \textit{Overall:} & Cl_2(g) + 2\,OH^-(aq) \longrightarrow OCl^-(aq) + Cl^-(aq) + H_2O(l) \end{array} \quad (22.12)$$

$$E^\circ = E^\circ_{Cl_2/Cl^-} - E^\circ_{OCl/Cl_2} = 1.358\ V - 0.421\ V = 0.937\ V$$
$$\Delta_r G^\circ = -zFE^\circ = -181\ kJ\ mol^{-1}$$

$HOCl(aq)$ is an effective germicide used, for example, in water purification and the treatment of swimming pools. Aqueous solutions of *hypochlorite* salts, notably $NaOCl(aq)$, are used as common household bleaches. The bleaching action of $NaOCl(aq)$ is highlighted in the margin. Reaction (22.12) is used to make solid household bleaches, such as $Ca(OCl)Cl$, which is a mixed salt containing both OCl^- and Cl^- ions. An aqueous solution of $Ca(OCl)Cl$ is obtained from the reaction of $Ca(OH)_2(aq)$ and Cl_2:

$$Ca(OH)_2(aq) + Cl_2(g) \longrightarrow Ca(OCl)Cl(aq) + H_2O(l)$$

Solid $Ca(OCl)Cl$ is obtained when the water evaporates.

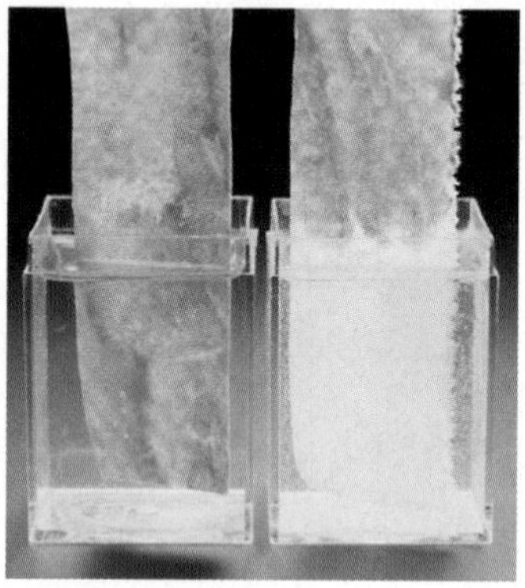

Carey B. Van Loon

▲ **Bleaching with hypochlorite ion**
Both strips of cloth are heavily stained with tomato sauce. Pure water (left) has little ability to remove the stain. NaOCl (right) rapidly bleaches (oxidizes) the colored components of the sauce to colorless products.

22-1 ARE YOU WONDERING?

Are there oxoacids of fluorine?

The hypothetical oxoacid of fluorine, HFO_2(HOFO), would have a *positive* formal charge on the central F atom.

$$\overset{-1}{:\ddot{\underset{..}{O}}}-\overset{+1}{\ddot{\underset{..}{F}}}-\ddot{\underset{..}{O}}-H$$

This is not something that we would expect of fluorine, the most electronegative of all the elements. Hypothetical oxoacids with more O atoms would have the F atom with even higher positive formal charges. Fluorine does form HOF, an oxoacid with no formal charges $H-\ddot{\underset{..}{O}}-\ddot{\underset{..}{F}}:$, but HOF exists only in the solid and liquid states. In water, HOF decomposes to HF, H_2O_2, and O_2.

Chlorine dioxide, ClO_2, is an important bleach for paper and fibers. Its eduction with peroxide ion in aqueous solution produces *chlorite* salts:

$$2\,ClO_2(g) + O_2^{2-}(aq) \longrightarrow 2\,ClO_2^-(aq) + O_2(g) \qquad (22.13)$$

odium chlorite is used as a bleaching agent for textiles.

Chlorate salts, which contain the ClO_3^- ion, form when $Cl_2(g)$ disproportionates in hot alkaline solutions, as shown below. (Hypochlorites form in cold lkaline solutions; recall reaction 22.12.)

$$3\,Cl_2(g) + 6\,OH^-(aq) \longrightarrow 5\,Cl^-(aq) + ClO_3^-(aq) + 3\,H_2O(l) \qquad (22.14)$$

hlorates are good oxidizing agents. Also, solid chlorates produce oxygen gas hen they decompose, which makes them useful in matches and fireworks. A imple laboratory method of producing $O_2(g)$ involves heating $KClO_3(s)$ in e presence of $MnO_2(s)$, a catalyst.

$$2\,KClO_3(s) \xrightarrow[MnO_2]{\Delta} 2\,KCl(s) + 3\,O_2(g) \qquad (22.15)$$

similar reaction is used as a source of emergency oxygen in aircraft and ubmarines.

Perchlorate salts are prepared mainly by electrolyzing chlorate solutions.)xidation of ClO_3^- occurs at a Pt anode through the half-reaction

$$ClO_3^-(aq) + H_2O(l) \longrightarrow ClO_4^-(aq) + 2\,H^+(aq) + 2\,e^- \qquad E° = -1.189\text{ V} \quad (22.16)$$

ompared with the other oxoacid salts, perchlorates are relatively stable. For xample, they do not disproportionate, because no oxidation state higher than -7 is available to chlorine. At elevated temperatures or in the presence of a eadily oxidizable compound, however, perchlorate salts may react explosively, o caution is advised when using them. Mixtures of ammonium perchlorate and owdered aluminum are used as the propellant in some solid-fuel rockets, such s those used on the Space Shuttle. Ammonium perchlorate is especially dangerous to handle, because an explosive reaction may occur when the oxidizing gent ClO_4^- acts on the reducing agent NH_4^+.

◀ A particularly destructive explosion of ammonium perchlorate occurred at a rocket fuel plant in Henderson, Nevada, in 1988.

nterhalogen Compounds

nterhalogen compounds contain two different halogens. The general formuas for interhalogens are XY, XY_3, XY_5, or XY_7, where X and Y represent two diferent halogens. Some of the more common ones are listed in Table 22.8. The nolecular structures of the interhalogen compounds feature the large, less lectronegative halogen as the central atom and the smaller halogen atoms as erminal atoms. Molecular shapes of the interhalogen compounds agree quite vell with VSEPR theory predictions. Structures for IF_x ($x = 1, 3, 5, 7$) are hown in Figure 22-6, including one type that we have not seen previously. In F_7, *seven* electron pairs are distributed around the central I atom in the form f a pentagonal-bipyramid.

Most interhalogen compounds are very reactive. For example, ClF_3 and BrF_3 eact explosively with water, organic materials, and some inorganic materials.

TABLE 22.8 Some Interhalogen Compounds

XY	XY_3	XY_5	XY_7
ClF(g)[a]	$ClF_3(g)$	$ClF_5(g)$	
BrF(g)	$BrF_3(l)$	$BrF_5(l)$	
BrCl(g)	$IF_3(s)$	$IF_5(l)$	$IF_7(g)$
ICl(s)	$ICl_3(s)$		
IBr(s)			

[a]The states of matter are given for 25 °C and 1 atm, except IF_3 which decomposes above −28 °C.

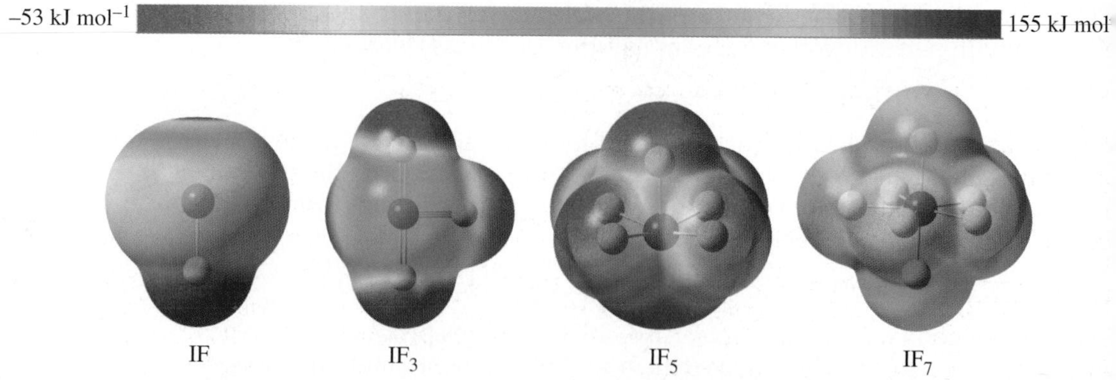

▲ FIGURE 22-6
Structures and electrostatic potential maps of the interhalogen compounds of iodine and fluorine

These two fluorides are used to make fluorinated compounds. For exampl ClF_3 can be used to convert U(s) to UF_6(g), so that the isotopes of uranium ca be separated from each other by gaseous diffusion (see page 228).

$$U(s) + 3\,ClF_3(g) \longrightarrow UF_6(l) + 3\,ClF(g)$$

ICl is used as an iodination reagent in organic chemistry.

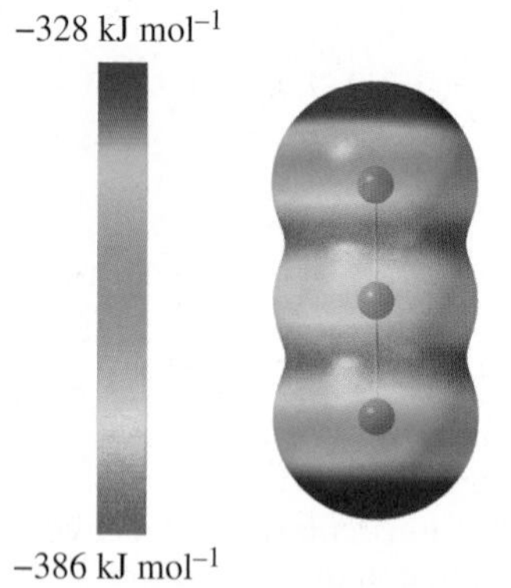

▲ FIGURE 22-7
Structure of the I_3^- ion
Notice that in the electrostatic potential map for I_3^-, the three lone pairs around the central iodine atom lead to a neutral charge distribution at this atom, not the negative charge suggested by formal charge arguments.

Polyhalide Ions

The triiodide ion, I_3^-, is one of a group of species called **polyhalide ions** tha are produced by the reaction of a halide ion with a halogen molecule. In th reaction below, the I^- ion acts as a *Lewis base* (an electron-pair donor) and the I molecule as a *Lewis acid* (an electron-pair acceptor).

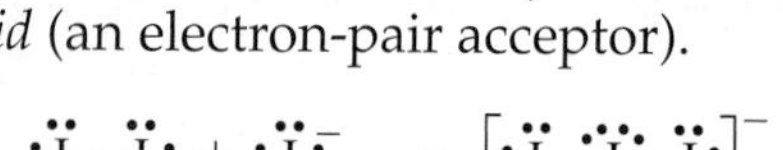

(22.17

The structure of the I_3^- ion is shown in Figure 22-7. The familiar iodin solutions used as antiseptics typically contain triiodide ion, and triiodide io solutions are widely used in analytical chemistry.

22-2 CONCEPT ASSESSMENT

Do you expect the ions ICl_2^+ and ICl_2^- to have the same shape? Explain.

22-4 Group 16: The Oxygen Family

Of the group 16 elements, oxygen and sulfur are clearly nonmetallic in thei behavior, but the heavier elements have some metallic properties.

Properties

On the basis of electron configurations alone, we expect oxygen and sulfur t be similar. Both elements form ionic compounds with active metals, and bot form similar covalent compounds, such as H_2S and H_2O, CS_2 and CO_2, SCl_2 and Cl_2O.

Even so, there are important differences between oxygen and sulfur com pounds. For example, H_2O has a very high boiling point (100 °C) for a com pound of such low molecular mass (18 u), whereas the boiling point of H_2S (molecular mass, 34 u) is much lower (−61 °C). This difference in behavior ca be explained in terms of the extensive hydrogen bonding that occurs in H_2O but not in H_2S (see page 522). Table 22.9 is a comparison of some propertie

TABLE 22.9 Comparisons of Oxygen and Sulfur

Oxygen	Sulfur
$O_2(g)$ at 298 K and 1 atm	$S_8(s)$ at 298 K and 1 atm
Two allotropes: $O_2(g)$ and $O_3(g)$	Two solid crystalline forms and many different molecular species in liquid and gaseous states
Principal oxidation states: $-2, -1, 0\ (-\frac{1}{2} \text{ in } O_2^-)$	Possible oxidation states: all values from -2 to $+6$
$O_2(g)$ and $O_3(g)$ are very good oxidizing agents	$S_8(s)$ is a poor oxidizing agent
Forms, with metals, oxides that are mostly ionic in character	Forms ionic sulfides with the most active metals, but many metal sulfides have partial covalent character
O^{2-} completely hydrolyzes in water, producing OH^-	S^{2-} strongly hydrolyzes in water to HS^- (and OH^-)
O is not often the central atom in a structure and can never have more than four atoms bonded to it; more commonly it has two (as in H_2O) or occasionally three (as in H_3O^+)	S is the central atom in many structures; can easily accommodate up to six electron pairs around itself (e.g., SO_3, SO_4^{2-}, SF_6)
Can form only two-atom and three-atom chains, as in H_2O_2 and O_3; compounds with O—O bonds decompose readily	Can form molecules with up to six S atoms per chain in compounds such as H_2S_n, Na_2S_n, $H_2S_nO_6$
Forms the oxide CO_2, which reacts with NaOH(aq) to produce $Na_2CO_3(aq)$	Forms the sulfide CS_2, which reacts with NaOH(aq), producing $Na_2CS_3(aq)$ and $Na_2CO_3(aq)$
Forms, with hydrogen, the compound H_2O, which is a liquid at 298 K and 1 atm; is extensively hydrogen bonded; has a large dipole moment; is an excellent solvent for ionic solids; forms hydrates and aqua complexes; is oxidized with difficulty	Forms, with hydrogen, the compound H_2S, which is a (poisonous) gas at 298 K and 1 atm; is not hydrogen bonded; has a small dipole moment; is a poor solvent; forms no complexes; is easily oxidized

of sulfur and oxygen. In general, the differences can be attributed to the following characteristics of the oxygen atom: (1) small size, (2) high electronegativity, and (3) its inability to employ an expanded valence shell in Lewis structures.

As indicated in Table 22.9, the principal oxidation states of oxygen are -2, -1, and 0. Sulfur, conversely, can exhibit all oxidation states from -2 to $+6$, including several "mixed" oxidation states, such as $+2.5$ in the tetrathionate ion, $S_4O_6^{2-}$.

Occurrence, Production, and Uses

Oxygen Oxygen is the most abundant element in Earth's crust, making up 45.5% by mass. It is also the most abundant element in seawater, accounting for nearly 90% of the mass. In the atmosphere, it is second only to nitrogen in abundance, accounting for 23.15% by mass and 21.04% by volume. Although it is obtained to a limited extent by the decomposition of oxygen-containing compounds and the electrolysis of water, the principal commercial source of oxygen is the fractional distillation of liquid air, which also produces nitrogen, argon, and other noble gases. This process involves only physical changes and is described in Figure 22-8.

Annually, oxygen is one of the principal chemicals manufactured in the United States. The uses of oxygen gas are summarized in Table 22.10. With its commercial availability, O_2 is not commonly prepared in the laboratory. In a submarine, in spacecraft, and in an emergency breathing apparatus, however, it is necessary to generate small quantities of oxygen from solids. The reaction of potassium *superoxide* with CO_2 works well for this purpose because, as the reaction below shows, KO_2 removes CO_2 while O_2 is being formed:

$$4\,KO_2(s) + 2\,CO_2(g) \longrightarrow 2\,K_2CO_3(s) + 3\,O_2(g)$$

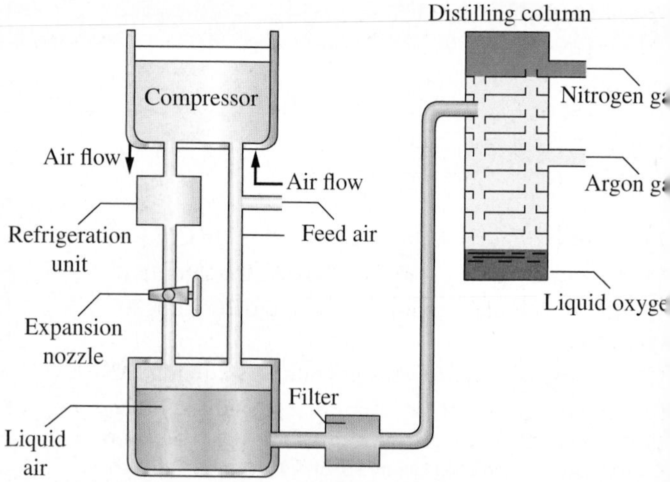

▶ FIGURE 22-8
The fractional distillation of liquid air—a simplified representation
Clean air is fed into a compressor and cooled by refrigeration. The cold air then expands through a nozzle and is cooled still further—enough to cause it to liquefy. The liquid air is filtered to remove solid CO_2 and hydrocarbons and then distilled. Liquid air enters the top of the column where nitrogen, the most volatile component (lowest boiling point), passes off as a gas. In the middle of the column, gaseous argon is removed. Liquid oxygen, the least volatile component, collects at the bottom. The normal boiling points of nitrogen, argon, and oxygen are 77.4, 87.5, and 90.2 K, respectively.

TABLE 22.10 Uses of Oxygen Gas

Manufacture of iron and steel
Manufacture and fabrication of other metals (cutting and welding)
Chemicals manufacture and other oxidation processes
Water treatment
Oxidizer of rocket fuels
Medicinal uses
Petroleum refining

22-3 CONCEPT ASSESSMENT

Which of the following dilute aqueous solutions can be used to obtain oxygen and hydrogen gases simultaneously through electrolysis using platinum electrodes: $H_2SO_4(aq)$, $CuSO_4(aq)$, $NaOH(aq)$, $KNO_3(aq)$, $NaI(aq)$? Explain.

Reza/Webistan/Getty Image

▲ These vast formations of solid sulfur were formed by the solidification of liquid sulfur obtained by the Frasch process.

Sulfur Sulfur is the sixteenth most abundant element in Earth's crust, accounting for 0.0384% by mass. Sulfur occurs as elemental sulfur, as mineral sulfides and sulfates, as $H_2S(g)$ in natural gas, and as organosulfur compounds in oil and coal. Extensive deposits of elemental sulfur are found in Texas and Louisiana, some of them in offshore sites. This sulfur is mined using the **Frasch process,** which is illustrated in Figure 22-9. Superheated water (at about 160 °C and 16 atm) is forced down the outermost of three concentric pipes into an underground bed of sulfur-containing rock. The sulfur melts and forms a liquid pool. Compressed air (at 20–25 atm) is pumped down the innermost pipe and forces the liquid sulfur–water mixture up the remaining pipe. The evaporation of water from the mixture yields formations of solid sulfur such as those shown in the margin.

Although the Frasch process was once the principal source of elemental sulfur, that is no longer the case. This change has been brought about by the need to control sulfur emissions from industrial operations. Today, most elemental sulfur is obtained from H_2S, which is a common impurity in oil and natural gas. After being removed from the fuel, H_2S is reduced to elemental sulfur in a two-step process. A stream of H_2S gas is split into two parts. One part (about one-third of the stream) is burned to convert H_2S to SO_2. The streams are rejoined in a catalytic converter at 200–300 °C, where the following reaction occurs:

$$2\,H_2S(g) + SO_2(g) \longrightarrow 3\,S(g) + 2\,H_2O(g) \quad (22.18)$$

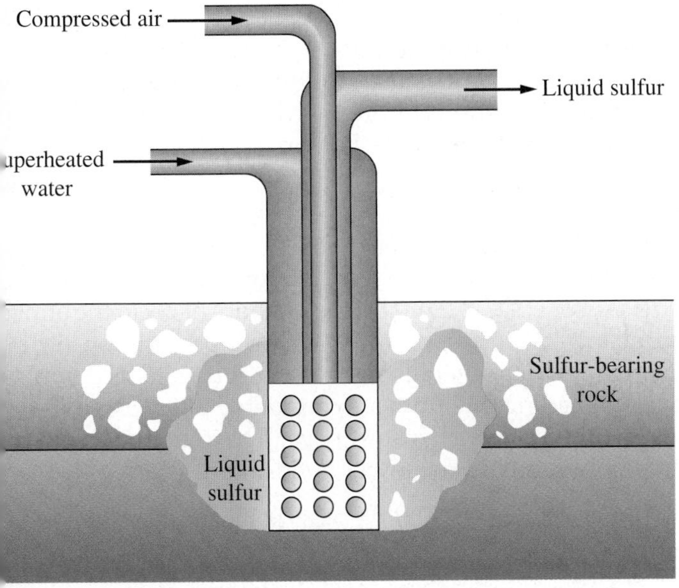

◀ FIGURE 22-9
The Frasch process
Sulfur is melted by using superheated water, and liquid sulfur is forced to the surface.

About 90% of all the sulfur produced is burned to form $SO_2(g)$, and in turn, most $SO_2(g)$ is converted to sulfuric acid, H_2SO_4. The conversion of S to H_2SO_4 is only one of several possibilities identified in Figure 22-10. Elemental sulfur does have a few uses of its own, however. One of these is in vulcanizing rubber (page 1311); another is as a fungicide used for dusting grapevines.

Selenium and Tellurium Selenium and tellurium have properties similar to those of sulfur, but they are more metallic. For example, sulfur is an electrical insulator, whereas selenium and tellurium are semiconductors. Selenium and tellurium are obtained mostly as by-products of metallurgical processes, such as in the anode mud deposited in the electrolytic refining of copper (page 905). Although there is not much use for tellurium compounds, selenium is used in the manufacture of rectifiers (devices that convert alternating to direct electric current). Both Se and Te are employed in the preparation of alloys, and their compounds are used as additives to control the color of glass.

Selenium also displays the property of *photoconductivity*: The electrical conductivity of selenium increases in the presence of light. This property is used, for example, in photocells in cameras. In some modem photocopying machines, the light-sensitive element is a thin film of Se deposited on aluminum. The light and dark areas of the image being copied are converted into a distribution of charge on the light-sensitive element. A dry black powder (toner) coats the charged portions of the light-sensitive element, and this image is transferred to a sheet of paper. Next, the dry powder is fused to the paper. In the final step, the electrostatic charge on the light-sensitive element is neutralized to prepare it for the next cycle.

Clive Streeter/Dorling Kindersley

▲ A selenium-coated light-sensitive element from a photocopier.

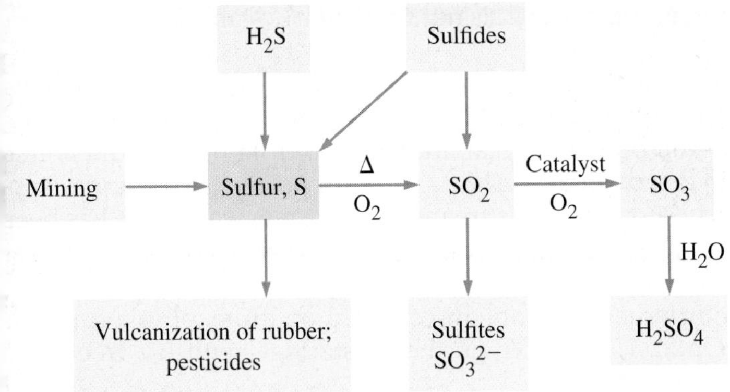

◀ FIGURE 22-10
Sources and uses of sulfur and its oxides

Polonium Polonium is a very rare, radioactive metal, and the only elemer that crystallizes in a simple cubic lattice. Because it is extremely low in abur dance, it has not found much practical use. Polonium was the first nev radioactive element isolated from uranium ore by Marie and Pierre Curie i 1898. Madame Curie named it after her native Poland.

Allotropy of Oxygen: Ozone

As we first learned in Chapter 3, allotropy refers to the existence of an element i two or more different molecular forms, and we were introduced to the allotrope c ordinary dioxygen, O_2; it is ozone, O_3. Normally, the quantity of O_3 in the atmos phere is quite limited at low altitudes, about 0.04 parts per million (ppm However, its level increases (perhaps severalfold), in smog situations as describe on page 957. Ozone levels exceeding 0.12 ppm are considered unhealthful.

The reaction below produces $O_3(g)$ from $O_2(g)$. The reaction is highl endothermic and occurs only rarely in the lower atmosphere.

$$3\,O_2(g) \longrightarrow 2\,O_3(g) \qquad \Delta_r H° = +285 \text{ kJ mol}^{-1}$$

This reaction does occur in high-energy environments, such as electrica storms. The pungent odor you may at times smell around heavy-dut electrical equipment or xerographic office copiers is probably O_3. Th chief method of producing ozone in the laboratory, in fact, is to pass an electri discharge (high-energy electrons) through $O_2(g)$. Because ozone is unstabl and decomposes back to $O_2(g)$, it is always generated at the point of use.

Ozone is an excellent oxidizing agent. Its oxidizing ability is surpassed b few other substances (two are F_2 and OF_2). The following standard reductio potentials show that, in acid solution, $O_3(g)$ is a much stronger oxidizin agent than $O_2(g)$, but not as strong as $F_2(g)$ or $OF_2(g)$.

Half-Equation	$E°$, V
$F_2(g) + 2\,e^- \longrightarrow 2\,F^-(aq)$	+2.87
$OF_2(g) + 2\,H^+(aq) + 4\,e^- \longrightarrow H_2O(l) + 2\,F^-(aq)$	+2.0
$O_3(g) + 2\,H^+(aq) + 2\,e^- \longrightarrow O_2(g) + H_2O(l)$	+2.07
$O_2(g) + 4\,H^+(aq) + 4\,e^- \longrightarrow 2\,H_2O(l)$	+1.23

Christian F. Schönbein, the discoverer of ozone, was a German-borr chemist working at the University of Basel in Switzerland. In 1866, he notec that when ozone is passed through a concentrated aqueous KOH solution, a red color forms. Subsequently, it was established that an ozonide, containin O_3^- ions, is formed by the following redox reaction:

$$2\,KOH(aq) + 5\,O_3(g) \longrightarrow 2\,KO_3(aq) + 5\,O_2(g) + H_2O(l)$$

The ozonide subsequently reacts with water to give KOH(aq) and $O_2(g)$. By using other methods of preparation, it has been possible to prepare othe alkali metal ozonides. However, the stability of these ozonides decreases as the size of the metal cation decreases because, as discussed in Chapter 21 a large polarizable anion, such as O_3^-, is not particularly stable in the presence of small, highly polarizing cations.

The most important use of ozone is as a substitute for chlorine in purifying drinking water. Its advantages are that it does not impart a taste to the water and it does not form the potentially carcinogenic chlorination products tha chlorine can. Its main disadvantage is that O_3 decomposes quickly and disap pears from water fairly soon after it is treated. Over time, water that has beer treated with ozone is not as well protected against bacterial contamination as is water treated with chlorine.

An important environmental problem centered on atmospheric ozone is discussed in Focus On 22-1, available on the MasteringChemistry website at www.masteringchemistry.com.

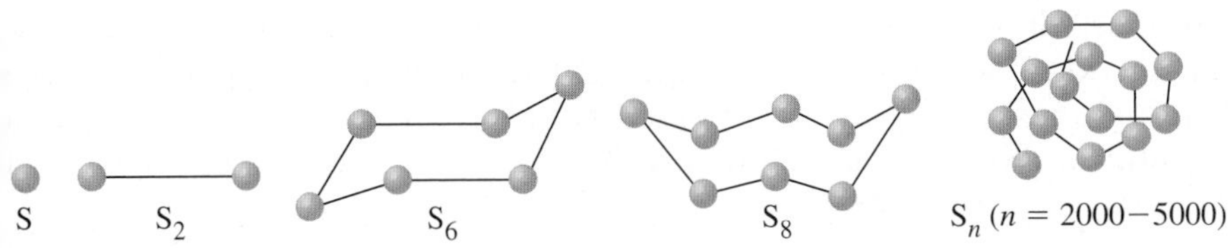

◀ FIGURE 22-11
Some molecular forms of sulfur

22-4 CONCEPT ASSESSMENT

Would you expect the shape of the ozonide ion to be the same as that of ozone? Explain.

llotropy and Polymorphism of Sulfur

ulfur has more allotropes than most elements (Figure 22-11). The most common ructural unit in the solid state is the S_8 ring, although an additional half-dozen yclic structures are known with up to 20 S atoms per ring. In sulfur vapor, S, 2, S_4, S_6, and S_8 can all exist under the appropriate conditions. Long-chain mol-cules of sulfur atoms are found in liquid sulfur. *Rhombic sulfur* (S_α), the stable olid at room temperature, is made up of cyclic S_8 molecules. At 95.5 °C, it con-erts to *monoclinic sulfur* (S_β), which is also made up of S_8 molecules but has a dif-erent crystal structure than S_α. At 119 °C, S_β melts, yielding *liquid sulfur* (S_λ), a traw-colored liquid comprising mostly S_8 molecules but with other cyclic mole-ules containing from 6 to 20 atoms. At 160 °C, the cyclic molecules open up and ecombine into long spiral-chain molecules, producing another form of *liquid sul-ur* (S_μ), a dark, viscous liquid. The chain length and viscosity reach a maximum t about 180 °C. At higher temperatures, the chains break up and the viscosity ecreases. At 445 °C, the liquid boils, producing *sulfur vapor*. At the boiling point, 8 molecules predominate in the vapor but they break down into smaller mole-ules at higher temperatures. *Plastic sulfur* forms if liquid S_μ is poured into cold ater. Plastic sulfur consists of long, spiral-chain molecules and has rubberlike roperties. On standing, it reverts to rhombic sulfur, a brittle solid.

The following sequence summarizes the phase transitions that occur in ulfur as the temperature increases:

$$S_\alpha \xrightarrow{95.5\ °C} S_\beta \xrightarrow{119} S_\lambda \xrightarrow{160} S_\mu \xrightarrow{445} S_8(g) \longrightarrow S_6 \longrightarrow S_4 \xrightarrow{1000} S_2 \xrightarrow{2000} S$$

ome of these forms of sulfur are shown in Figure 22-12.

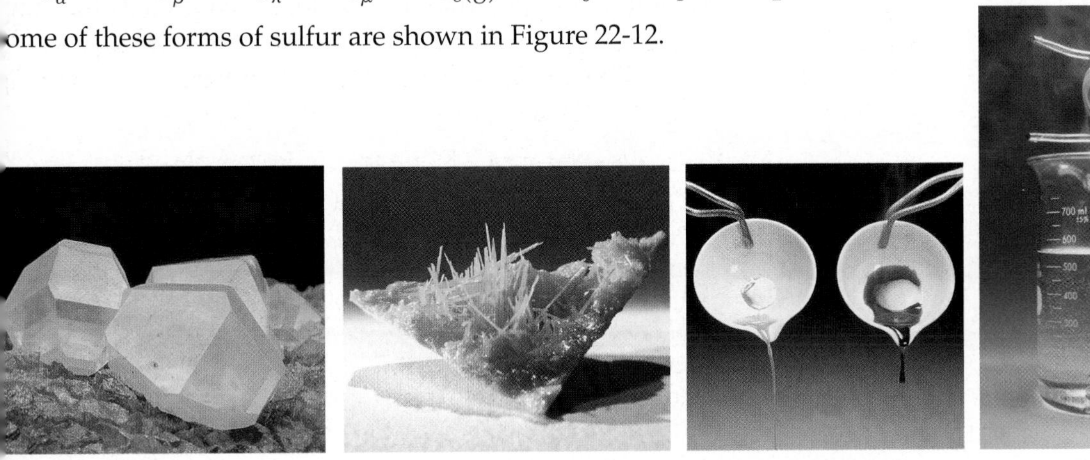

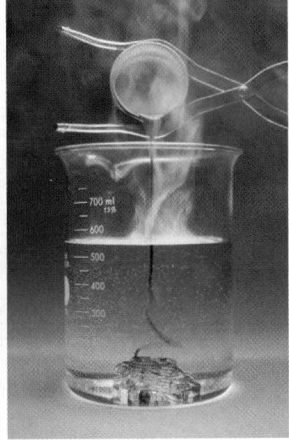

(a) (b) (c) (d)

▲ FIGURE 22-12
everal macroscopic forms of sulfur
a) Rhombic sulfur. **(b)** Monoclinic sulfur. **(c)** On the left, monoclinic sulfur has just melted o form an orange liquid. On the right, after continued heating, the liquid becomes red nd more viscous. **(d)** Liquid sulfur is poured into water to produce plastic sulfur.

) Jeffrey A. Scovil Photography; **(b)** Tom Bochsler; **(c)** Carey B. Van Loon; **(d)** Charles D. Winters/Science Source

Because some of these transitions, especially those in the solid state, a sluggish, additional phenomena are occasionally seen. For example, if rhom bic sulfur is heated rapidly, it may melt at 113 °C and fail to convert to mono clinic sulfur. However, monoclinic sulfur may then freeze from this liqui only to melt again at 119 °C.

Oxygen Compounds

Oxygen is so central to the study of chemistry that we constantly refer to i physical and chemical properties in developing a framework of chemica principles. For instance, our discussion of stoichiometry began with com bustion reactions—reactions of substances with $O_2(g)$ to form produc such as $CO_2(g)$, $H_2O(l)$, and $SO_2(g)$. Combustion reactions also figure prominently in the presentation of thermochemistry. Many of the molecule and polyatomic anions described in the chapters on chemical bonding wer oxygen-containing species. Water was a primary subject in the discussion o liquids, solids, and intermolecular forces as well as in the study of acid–bas and other solution equilibria. The dual roles of hydrogen peroxide as an ox dizing agent and a reducing agent were described in Section 5- and the kinetics of the decomposition of H_2O_2 was examined in detail i Chapter 20.

A systematic study of oxygen compounds is generally done in conjunctio with the study of the other elements. Thus, the oxides of boron were consid ered in the discussion of boron chemistry in Chapter 21; the oxides of carbo were also considered there. The survey of the chemistry of the alkali and alka line earth metals in Chapter 21 provided an opportunity to describe norma oxides, peroxides, and superoxides; and the important oxides of sulfur, nitro gen, and phosphorus are discussed in this chapter.

Sulfur Compounds

As in the corresponding discussion for the halogens (Section 22-3), oxidation reduction chemistry is a primary concern here. To assist in this discussion, w provide electrode potential diagrams for some important sulfur-containin species in Figure 22-13.

Sulfur Dioxide and Sulfur Trioxide More than a dozen oxides of sulfur hav been reported, but only sulfur dioxide, SO_2, and sulfur trioxide, SO_3 are commonly encountered. Typical structures are shown in Figure 22-1

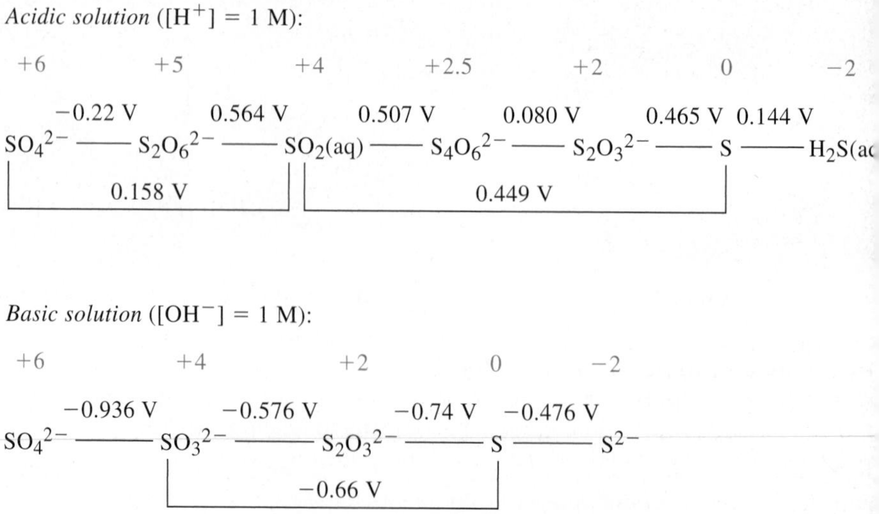

▶ FIGURE 22-13
Electrode potential diagrams for sulfur

he main commercial methods of producing $SO_2(g)$ are the direct combustion f sulfur, reaction (22.19), and the roasting of metal sulfides, reaction (22.20):

$$S(s) + O_2(g) \xrightarrow{\Delta} SO_2(g) \qquad (22.19)$$

$$2\,ZnS(s) + 3\,O_2(g) \xrightarrow{\Delta} 2\,ZnO(s) + 2\,SO_2(g) \qquad (22.20)$$

The main use of SO_2 is in the synthesis of SO_3 to make sulfuric acid, H_2SO_4. n the *contact process*, $SO_2(g)$ is formed by reaction (22.19) or (22.20). Then ulfur trioxide is produced by oxidizing $SO_2(g)$ in an exothermic, reversible eaction:

$$2\,SO_2(g) + O_2(g) \rightleftharpoons 2\,SO_3(g) \qquad (22.21)$$

eaction (22.21) is the key step in the process, but it occurs very slowly unless atalyzed. The principal catalyst is V_2O_5 mixed with alkali metal sulfates. The atalysis involves adsorption of the $SO_2(g)$ and $O_2(g)$ on the catalyst, followed y reaction at active sites and desorption of SO_3.

ulfuric Acid SO_3 reacts with water to form H_2SO_4, but the direct reaction f $SO_3(g)$ and water produces a fine mist of $H_2SO_4(aq)$ droplets with unre-cted $SO_3(g)$ trapped inside the droplets. This misting would result in a reat loss of product and a tremendous pollution problem. To avoid these utcomes, $SO_3(g)$ is instead bubbled through 98% H_2SO_4 in towers packed vith a ceramic material. The $SO_3(g)$ readily dissolves in the sulfuric acid nd reacts with the small amount of water present to increase the concentra-ion of the sulfuric acid. The result is a form of sulfuric acid sometimes called *leum* but more commonly called *fuming sulfuric acid*. In a sense, the product s greater than 100% H_2SO_4. Sufficient water is added to the circulating acid n the tower to maintain the required concentration. Later, sulfuric acid of he strength desired is produced by dilution with water. If we use the for-nula $H_2S_2O_7$ (disulfuric acid) as an example of a particular oleum, the reac-ions are

$$SO_3(g) + H_2SO_4(l) \longrightarrow H_2S_2O_7(l) \qquad (22.22)$$

$$H_2S_2O_7(l) + H_2O(l) \longrightarrow 2\,H_2SO_4(l) \qquad (22.23)$$

$$H_2SO_4(l) \xrightarrow{H_2O} H_2SO_4(aq) \qquad (22.24)$$

Dilute sulfuric acid, $H_2SO_4(aq)$, enters into all the common reactions of a trong acid, such as neutralizing bases. It reacts with metals to produce $H_2(g)$ nd dissolves carbonates to liberate $CO_2(g)$.

Concentrated sulfuric acid has some distinctive properties. It has a very trong affinity for water, strong enough that it will even remove H and O toms (in the proportion H_2O) from some compounds. In the reaction of con-entrated sulfuric acid with a carbohydrate such as sucrose, all the H and O toms are removed and a residue of pure carbon is left, as shown in the photo n the next page. The chemical equation for the reaction that occurs is given elow:

$$C_{12}H_{22}O_{11}(s) \xrightarrow{H_2SO_4(concd)} 12\,C(s) + 11\,H_2O(l) \qquad (22.25)$$

oncentrated sulfuric acid is a moderately good oxidizing agent and is able, or example, to react with copper.

$$Cu(s) + 2\,H_2SO_4(concd) \longrightarrow Cu^{2+}(aq) + SO_4^{2-}(aq) + 2\,H_2O(l) + SO_2(g) \qquad (22.26)$$

For a very long time, sulfuric acid has ranked among the top manufactured hemicals, with annual production in the United States of about 45 million netric tons. Sulfuric acid continues to have many uses, but the bulk of H_2SO_4 s used in the manufacture of fertilizers. Other uses are found in various met-llurgical processes, oil refining, and the manufacture of the white pigment itanium dioxide. Also familiar is its use as the electrolyte in storage batteries

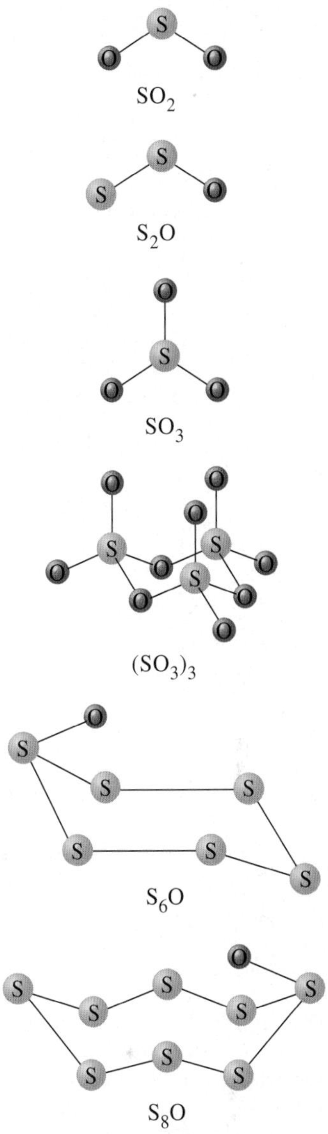

▲ FIGURE 22-14
Structures of some sulfur oxides
To conform to the observed structures of SO_2, S_2O, and SO_3, the hybridization scheme proposed for the central S atom is sp^2. S_2O has a similar structure to SO_2 but with a S atom substituted for one O atom. The monomeric SO_3 exists in equilibrium with the trimer $(SO_3)_3$, in which the O—S—O angle is approximately tetrahedral and the hybridization of the S atom should be sp^3. The oxides S_6O and S_8O illustrate sulfur's ability to form ring compounds.

(a) (b)

▶ **(a)** Concentrated sulfuric acid is added to cane sugar. **(b)** Carbon is produced in the reaction.

Tom Bochsler

for automobiles and emergency power supplies. We might say that sulfuri acid was the workhorse of the "old economy" but that it has a lesser role in th "new economy."

When $SO_2(g)$ reacts with water, it produces $H_2SO_3(aq)$, but this acid, *su furous acid*, has never been isolated in pure form. Salts of sulfurous acid, *sulfite* are good reducing agents and are easily oxidized by $O_2(g)$. In aqueous solutior sulfite ions, SO_3^{2-}, are oxidized by O_2 to SO_4^{2-} ions, as shown below:

$$O_2(g) + 2\,SO_3^{2-}(aq) \longrightarrow 2\,SO_4^{2-}(aq) \qquad (22.27$$

Interestingly, sulfite ion can also act as an oxidizing agent, as in this reactio with H_2S.

$$2\,H_2S(g) + 2\,H^+(aq) + SO_3^{2-}(aq) \longrightarrow 3\,H_2O(l) + 3\,S(s) \qquad (22.28$$

Both H_2SO_3 and H_2SO_4 are diprotic acids. They ionize in two steps anc produce two types of salts, one in each ionization step. The term **acid salt** i sometimes used for salts such as $NaHSO_3$ and $NaHSO_4$ because their anion undergo a further *acid* ionization. H_2SO_3 is a weak acid in both ionizatio steps, whereas H_2SO_4 is strong in the first step and somewhat weak ir the second. If a solution of H_2SO_4 is sufficiently dilute, however (les than about 0.001 M), we can treat the acid as if both ionization steps go t completion.

▶ Actually, $NaHSO_3$ cannot be isolated as a solid. When we attempt to crystallize this salt from an aqueous solution containing HSO_3^-, the reaction $2\,HSO_3^- \longrightarrow S_2O_5^{2-} + H_2O$ occurs. The product obtained is sodium *metabisulfite*, $Na_2S_2O_5$.

Sulfates and Sulfites Sulfate and sulfite salts have a number of importan uses. Calcium sulfate dihydrate (gypsum) is used to make the hemihydrat (plaster of Paris) for the building industry (page 999). Aluminum sulfat is used in water treatment and in sizing paper (page 1010). Copper(II) sulfat is employed as a fungicide and an algicide and in electroplating. The chie application of sulfites is in the pulp and paper industry. Sulfites solubiliz *lignin*, a polymeric substance that coats the cellulose fibers in wood. This treat ment frees the fibers for processing into wood pulp and then paper. Sulfite are also used as reducing agents in photography and as scavengers of O_2(aq in treating boiler water (reaction 22.27). Compounds of sulfur(IV) have lon been used as food preservatives and antioxidants. For example, exposure t $SO_2(g)$ prevents the discoloration of dried fruits, and soluble sulfites act a anti-microbial agents in winemaking.

Thiosulfates In addition to sulfite and sulfate ions, another important sulfur-oxygen ion is thiosulfate ion, $S_2O_3^{2-}$. The prefix *thio* signifies that a S aton replaces an O atom in a compound. Thus, the thiosulfate ion can be viewed a a sulfate ion, SO_4^{2-}, in which a S atom replaces one of the O atoms. The forma xidation state of S in $S_2O_3^{2-}$ is +2, but as Figure 22-15 indicates, the two atoms are not equivalent: The central S atom is in the oxidation state +6, and ıe terminal S atom, −2. The structures of several other thio anions are also ıown in Figure 22-15.

Thiosulfates can be prepared by boiling elemental sulfur in an alkaline ›lution of sodium sulfite. The sulfur is oxidized and the sulfite ion is reduced, oth to thiosulfate ion.

$$SO_3^{2-}(aq) + S(s) \longrightarrow S_2O_3^{2-}(aq) \qquad \textbf{(22.29)}$$

Thiosulfate solutions are important in photographic film processing (see age 1159). They are also common analytical reagents, often used in conjunction ith iodine. For example, in one method of analysis for copper, an excess of ›dide ion is added to $Cu^{2+}(aq)$, producing CuI(s) and triiodide ion, I_3^-.

$$2\,Cu^{2+}(aq) + 5\,I^-(aq) \longrightarrow 2\,CuI(s) + I_3^-(aq) \qquad \textbf{(22.30)}$$

he excess triiodide ion is then titrated with a standard solution of $a_2S_2O_3(aq)$, forming I^- and $S_4O_6^{2-}$, the *tetrathionate* ion.

$$I_3^-(aq) + 2\,S_2O_3^{2-}(aq) \longrightarrow 3\,I^-(aq) + S_4O_6^{2-}(aq) \qquad \textbf{(22.31)}$$

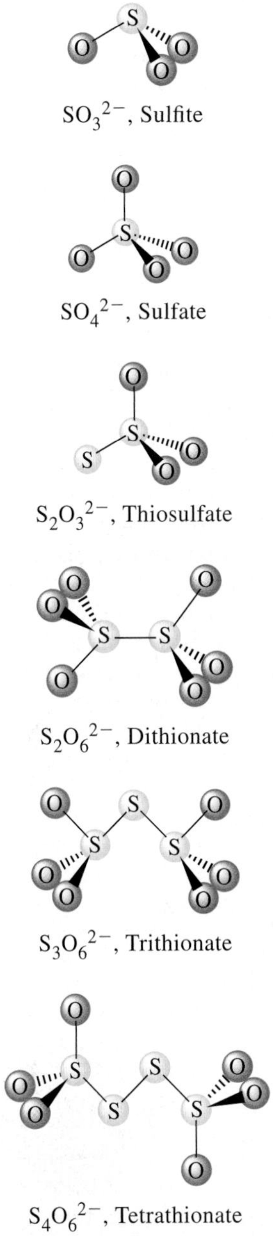

▲ FIGURE 22-15
Structures of some oxoanions of sulfur

›xygen and Sulfur Halides

oth oxygen and sulfur form a number of interesting compounds with the alogens. Oxygen, for example, forms the fluorides OF_2 and O_2F_2, which have tructures similar to those of water and hydrogen peroxide but which are much ıore reactive. Sulfur also forms compounds with the halogens; the analogous ompounds SF_2 and S_2F_2 are known, as are the compounds SF_4 and SF_6. The eactivities of SF_4 and SF_6 are quite different. SF_6 is a colorless, odorless, and nreactive gas. It is so unreactive that it can be safely inhaled, in small quanti- es, resulting in a very deep voice. (As you may already know, helium gas has ıe opposite effect when inhaled in small quantities.) Conversely, SF_4 is a very eactive gas and a powerful fluorinating agent. In the following reaction, F_4 converts BCl_3 to BF_3.

$$3\,SF_4 + 4\,BCl_3 \longrightarrow 4\,BF_3 + 3\,SCl_2 + 3\,Cl_2$$

Sulfur and chlorine also form the compounds S_2Cl_2 and SCl_4, but the best- nown halide is SCl_2. It is a foul-smelling, red liquid (melting point, −122 °C; oiling point, 59 °C) that has been used in the production of the notorious, and ighly poisonous, mustard gas, $S(CH_2CH_2Cl)_2$. The production of mustard as is based on the following reactions:

$$SCl_2 + 2\,CH_2CH_2 \longrightarrow S(CH_2CH_2Cl)_2$$

/lustard gas is not actually a gas, but a volatile liquid (melting point, 13 °C; oiling point, 235 °C). During World War I it was sprayed as a mist that stayed lose to the ground and was blown by wind onto the enemy. Exposure to mus- ard gas causes blistering of the skin, internal and external bleeding, blind- ess, and, after four or five weeks, death.

O_2 Emissions and the Environment

ndustrial smog consists primarily of particles (ash and smoke), $SO_2(g)$, nd H_2SO_4 mist. A variety of industrial operations produce significant quantities f $SO_2(g)$, but the main contributors to atmospheric releases of $SO_2(g)$ are ›ower plants burning coal or high-sulfur fuel oils. SO_2 can oxidize to SO_3, specially when the reaction is catalyzed on the surfaces of airborne particles r through reaction with NO_2:

$$SO_2(g) + NO_2(g) \longrightarrow SO_3(g) + NO(g) \qquad \textbf{(22.32)}$$

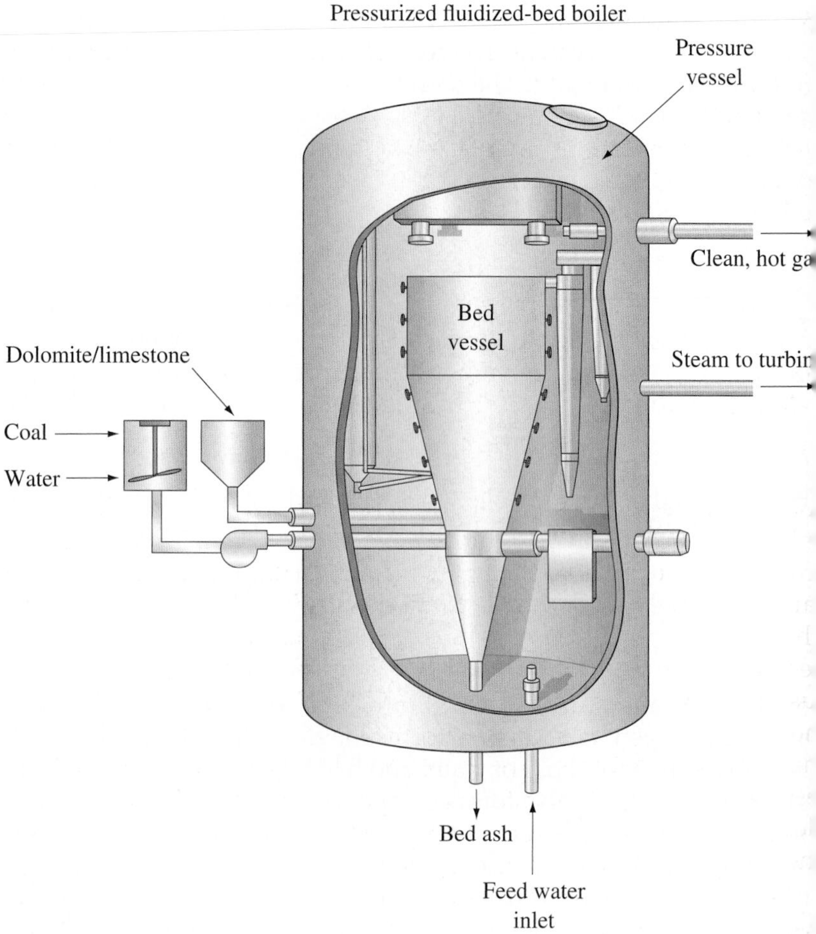

▶ FIGURE 22-16
Fluidized-bed combustion
Powdered coal, limestone, and air are introduced into a combustion chamber where water circulating through coils is converted to steam. Combustion is carried out at a relatively low temperature (760–860 °C), which minimizes the production of NO(g) from $N_2(g)$ and $O_2(g)$. At the same time, $SO_2(g)$ from sulfur in the coal reacts with CaO(s) from decomposition of the limestone, forming $CaSO_3(s)$ in a Lewis acid–base reaction.

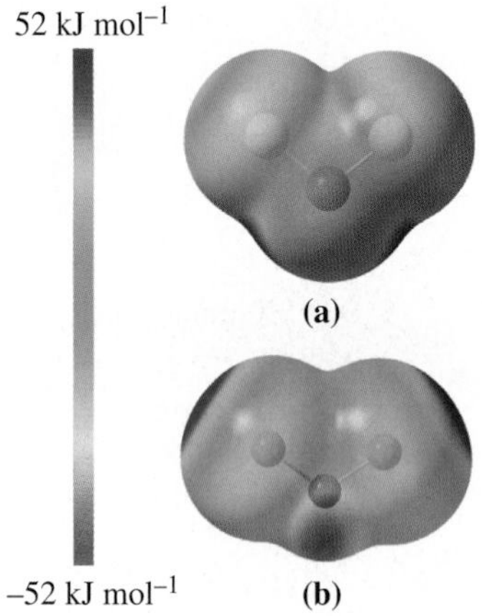

▲ Electrostatic potential maps of **(a)** OF_2 and **(b)** OCl_2.

In turn, SO_3 can react with water vapor in the atmosphere to produc H_2SO_4 mist, a component of acid rain. Also, the reaction of H_2SO_4 with ai borne NH_3 produces particles of $(NH_4)_2SO_4$. The details of the effect of lo concentrations of SO_2 and H_2SO_4 on the body are not well understood, but is clear that these substances are respiratory irritants. Levels above 0.10 ppr are considered potentially harmful.

The control of industrial smog and acid rain hinges on the removal of sulfu from fuels and the control of $SO_2(g)$ emissions. Dozens of processes have bee proposed for removing SO_2 from smokestack gases, one of which is illustrate in Figure 22-16. In this process, $SO_2(g)$ from the coal reacts with CaO(s) t form deposits of $CaSO_3(s)$.

22-5 CONCEPT ASSESSMENT

Electrostatic potential maps for OF_2 and OCl_2 are shown in the margin. Explain the differences.

22-5 Group 15: The Nitrogen Family

The chemistry of the group 15 elements, especially that of nitrogen and phos phorus, is an extensive subject. We will discuss the special significance c these two elements to living matter later in the text, but even here you shoul get a sense of the richness of their chemistry. For example, nitrogen atoms ca exist in many oxidation states, which is evident from the variety of nitroger containing species identified in Figure 22-17.

cidic solution ($[H^+] = 1$ M):

5 +4 +3 +2 +1 0 −1 −2 −3

0.803 V 1.065 V 0.996 V 1.591 V 1.766 V −1.87 V 1.42 V 1.275 V

O_3^- — N_2O_4 — HNO_2 — NO — N_2O — N_2 — NH_3OH^+ — $N_2H_5^+$ — NH_4^+

asic solution ($[OH^-] = 1$ M):

5 +4 +3 +2 +1 0 −1 −2 −3

−0.86 V 0.867 V −0.46 V 0.76 V 0.94 V −3.04 V 0.73 V 0.10 V

O_3^- — N_2O_4 — NO_2^- — NO — N_2O — N_2 — NH_2OH — N_2H_4 — NH_3

◀ FIGURE 22-17
Electrode potential diagrams for nitrogen

letallic–Nonmetallic Character in Group 15

ll the elements in group 15 have the valence-shell electron configuration s^2np^3. This configuration suggests nonmetallic behavior and doesn't give ny clues as to metallic character that might exist. Table 22.11, however, indi-ates the usual decrease of ionization energy with increasing atomic number. hese values, taken together with physical properties from the table, do sug-est the order of metallic character within the group. Nitrogen and phospho-us are nonmetallic, arsenic and antimony are metalloids, and bismuth is etallic. The first ionization energy of bismuth is actually somewhat less han that of magnesium, and its third ionization energy (2466 kJ mol^{-1}) is ss than the third ionization energy of aluminum (2745 kJ mol^{-1}). The ectronegativities indicate a high degree of nonmetallic character for nitrogen nd less so for the remaining members of the group.

Three group 15 elements—phosphorus, arsenic, and antimony—exhibit lotropy. The common forms of phosphorus at room temperature, both non-etallic, are white and red phosphorus. For arsenic and antimony, the more able allotropic forms are the metallic ones. These forms have high densities, oderate thermal conductivities, and limited abilities to conduct electricity. ismuth is a metal despite its low electrical conductivity, which, nevertheless, better than that of manganese and almost as good as that of mercury. The onmetals and metals in group 15 are also distinguishable by their oxides. The xides of nitrogen and phosphorus (for example, N_2O_3 and P_4O_6) are acidic

TABLE 22.11 Selected Properties of Group 15 Elements

Element	Covalent Radius, pm	Electronegativity	First Ionization Energy kJ mol^{-1}	Common Physical Form(s)	Density of Solid, g cm^{-3}	Comparative Electrical Conductivity[a]
N	75	3.0	1402	Gas	1.03 (−252 °C)	—
P	110	2.1	1012	Wax-like white solid;	1.82	—
				Red solid	2.20	10^{-17}
As	121	2.0	947	Yellow solid;	2.03	—
				Gray solid with metallic luster	5.78	6.1
Sb	140	1.9	834	Yellow solid;	5.3	—
				Silvery white metallic solid	6.69	4.0
Bi	155	1.9	703	Pinkish white metallic solid	9.75	1.5

[a]These values are relative to an assigned value of 100 for silver.

when they react with water. As with the nonmetals in groups 16 and 17, this i typical for nonmetal oxides. Arsenic(III) oxide and antimony(III) oxide ar amphoteric, whereas bismuth(III) oxide acts only as a base, a property typica of metal oxides.

Occurrence, Production, and Uses

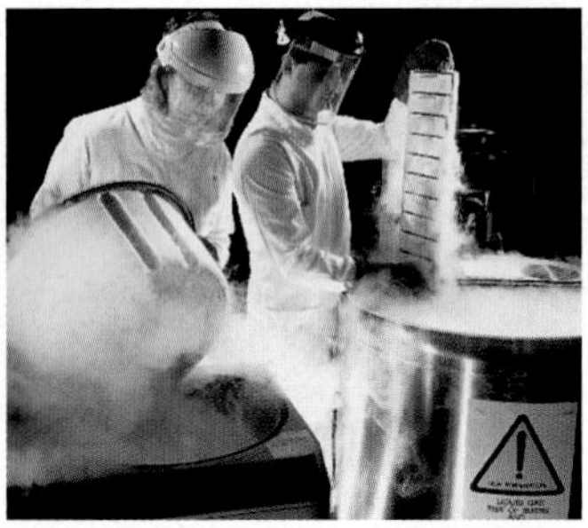

Geoff Tompkinson/Science Photo Library

▲ Viruses for use in medical research are frozen in liquid nitrogen.

Nitrogen Nitrogen occurs mainly in the atmosphere. Its abundance in Earth solid crust is only 0.002% by mass. The only important nitrogen-containin minerals are KNO_3 (niter, or saltpeter) and $NaNO_3$ (soda niter, or Chile sal peter), found in a few desert regions. Other natural sources of nitroger containing compounds are plant and animal protein and the fossilize remains of ancient plant life, such as coal.

Until about 100 years ago, sources of pure nitrogen and its compound were quite limited. This all changed with the invention of a process for th liquefaction of air in 1895 (see Figure 22-8) and the development of the Haber Bosch process for converting nitrogen to ammonia in 1908 (page 1068). A hos of nitrogen compounds can be made from ammonia. Nitrogen has man important uses of its own in addition to being a precursor of manufacture nitrogen compounds. Some of these uses are listed in Table 22.12.

Phosphorus Although phosphorus is the eleventh most abundant elemer and makes up about 0.11% of Earth's crust by mass, it was not discovere until 1669. It was originally isolated from putrefied urine—an effective if nc particularly pleasant source. Today the principal source of phosphorus con pounds is phosphate rock, a class of minerals known as *apatites*, such as fluc rapatite $Ca_5(PO_4)_3F$ or $3\,Ca_3(PO_4)_2 \cdot CaF_2$. Elemental phosphorus is prepare by heating phosphate rock, silica (SiO_2), and coke (C) in an electric furnac The overall change that occurs is

$$2\,Ca_3(PO_4)_2(s) + 10\,C(s) + 6\,SiO_2(s) \xrightarrow{\Delta} 6\,CaSiO_3(l) + 10\,CO(g) + P_4(g) \quad \textbf{(22.3}$$

The $P_4(g)$ is condensed, collected, and stored under water as white phosphoru

Although compounds of phosphorus are vitally important to living orgar isms (DNA and phosphates in bones and teeth, for example), the element itse is not widely used. Almost all the elemental phosphorus produced is reox dized to give P_4O_{10} for the manufacture of high-purity phosphoric acid. Th rest is used to make organophosphorus compounds and phosphorus sulfide (P_4S_3) in match heads.

Arsenic, Antimony, and Bismuth Arsenic is obtained by heating arsenic containing metal sulfides. For example, FeAsS yields FeS and As(g). Th As(g) deposits as As(s), which can be used to make other compounds. Som arsenic is also obtained by the reduction of arsenic(III) oxide with CO(g Antimony is obtained mainly from its sulfide ores. Bismuth is obtained as by-product of the refining of other metals.

Both As and Sb are used in making alloys of other metals. For example, th addition of As and Sb to Pb produces an alloy that has desirable properties fc use as electrodes in lead–acid batteries. Arsenic and antimony are used to prc duce semiconductor materials, such as GaAs, GaSb, and InSb, in electroni devices.

TABLE 22.12 Uses of Nitrogen Gas

Provide a blanketing (inert) atmosphere for the production of chemicals and electronic components
Pressurized gas for enhanced oil recovery
Metals treatment
Refrigerant (e.g., fast freezing of foods)

litrogen Compounds

he substance from which all nitrogen compounds are ultimately derived, $_2$(g), is unusually stable. One explanation of the limited reactivity of ıe N_2 molecule is based on its electronic structure. As discussed in hapter 10, the bond between the two N atoms in N_2 is a *triple* covalent ond, which is unusually strong and difficult to break. In thermochemical erms, the enthalpy change associated with breaking the bonds in one ıole of N_2 molecules is very high—the dissociation reaction is highly ndothermic.

$$N{\equiv}N(g) \longrightarrow 2\,N(g) \quad \Delta_r H^\circ = +945.4 \text{ kJ mol}^{-1}$$

lso, the standard Gibbs energies of formation of many nitrogen compounds re positive, which means that their formation reactions are not spontaneous. or NO(g),

$$\frac{1}{2}N_2(g) + \frac{1}{2}O_2 \longrightarrow NO(g) \qquad \Delta_f G^\circ = 86.55 \text{ kJ mol}^{-1}$$

eactions with a positive $\Delta_r G^\circ$, such as the formation of NO(g) from its ele-ıents, do not occur to any significant extent at normal temperatures and tmospheric conditions. Just imagine the situation if $\Delta_f G^\circ[NO(g)] =$ ·86.55 kJ mol^{-1} instead of +86.55 kJ mol^{-1}. The reaction of N_2(g) and O_2(g)) form NO(g) would proceed to a far greater extent. With an atmosphere epleted in O_2(g) and rich in noxious NO(g), life as we know it would not be ossible.

litrides Nitrogen forms binary compounds with most other elements, and ıese compounds can be grouped into four categories. In ionic (salt-like) itrides, the nitrogen is present as the N^{3-} ion. These compounds form with thium and the group 2 metals. Thus, when magnesium is burned in air (see igure 2-1), a small quantity of magnesium nitride forms, together with the rincipal product, magnesium oxide.

$$3\,Mg(s) + N_2(g) \xrightarrow{\Delta} Mg_3N_2(s) \qquad \textbf{(22.34)}$$

he nitride ion is a very strong base. In aqueous solution, it accepts protons ·om water molecules to form ammonia molecules and hydroxide ions.

$$N^{3-}(aq) + 3\,H_2O(l) \longrightarrow NH_3(aq) + 3\,OH^-(aq)$$

ı the reaction of magnesium nitride with water, magnesium and hydroxide ons combine to form insoluble $Mg(OH)_2$, and the ammonia is released as a as, which is easily detectable by its odor.

$$Mg_3N_2(s) + 6\,H_2O(l) \longrightarrow 3\,Mg(OH)_2(s) + 2\,NH_3(g)$$

When nitrogen combines with other typical nonmetals, it does so by form-ıg covalent bonds, yielding covalent nitrides. Bonding in these nitrides can e described in terms of the general principles presented in Chapters 10 and 1. Some binary covalent nitrides are $(CN)_2$, P_3N_5, As_4N_4, S_2N_2, and S_4N_4. Jitrogen combines with elements of group 13, producing compounds of ıe form MN (where M = B, Al, Ga, In, or Tl). These compounds have olid structures that resemble those of graphite or diamond, with M and N toms bonded to form planes of hexagonal rings (the graphite-like form) or diamond-like lattice. A fourth type of binary nitrides are the metallic itrides with formulas such as MN, M_3N, and M_4N. These are *interstitial ompounds*, in which N atoms occupy some or all of the interstices (voids) in ıe structure of the metal. They are hard, chemically inert, high-melting-point olids often used to harden and protect surfaces. Typical examples are 'iN, VN, and UN, with melting points of 2950 °C, 2050 °C, and 2800 °C, espectively.

Gary J. Shulfer University of Wisconsin

▲ **Fritz Haber (1868–1934)** Haber's perfection of the ammonia synthesis reaction, which made the manufacture of inexpensive explosives possible, was of critical importance to Germany during World War I. After the war, Haber again applied his chemical knowledge for his country's benefit by attempting, unsuccessfully, to extract gold from seawater for use in paying war reparations. Despite his past services, this Jewish scientist was driven from his academic post by the Nazi regime in 1933.

▲ Anhydrous liquid ammonia being applied directly to the soil.

Photo by Lynn Betts, USDA Natural Resources Conservation Service

Ammonia and Related Compounds The Haber–Bosch process for the synthe sis of ammonia involves the reaction below:

$$N_2(g) + 3\,H_2(g) \rightleftharpoons 2\,NH_3(g) \qquad \textbf{(22.3}$$

As we noted in an earlier encounter with this reaction (see Focus On 15- www.masteringchemistry.com), a high yield of ammonia requires (1) a hig temperature (400 °C), (2) a catalyst to speed up the reaction, and (3) a high pre sure (about 200 atm). The key to achieving essentially 100% yield is continuou removal of NH_3 and recycling of the unreacted $N_2(g)$ and $H_2(g)$. The NH_3 i removed by liquefaction. The Haber–Bosch process is outlined in Figure 22-1 A critical aspect of the process is having a source of $H_2(g)$. Mostly, this is mad by the re-forming of natural gas (see page 1016).

Ammonia is the starting material in the manufacture of most other nitroge compounds, but it has some direct uses of its own. Its most important use is as fertilizer. The highest concentration in which nitrogen fertilizer can be applie to fields is as pure liquid NH_3, known as "anhydrous ammonia." $NH_3(aq)$ i also applied in a variety of household cleaning products, such as commerci glass cleaners. In these products, the ammonia acts as an inexpensive base t produce $OH^-(aq)$. The $OH^-(aq)$ reacts with grease and oil molecules to co vert them into compounds that are more soluble in water. In addition, the aqu ous ammonia solution dries quickly, leaving few streaks on glass.

Because ammonia is a base, a simple approach to producing certain nitroge compounds is to neutralize ammonia with an appropriate acid. The acid–bas reaction that forms ammonium sulfate, an important solid fertilizer, is

$$2\,NH_3(aq) + H_2SO_4(aq) \longrightarrow (NH_4)_2SO_4(aq) \qquad \textbf{(22.3}$$

Ammonium chloride, made by the reaction of $NH_3(aq)$ and $HCl(aq)$, is use in the manufacture of dry-cell batteries, in cleaning metals, and as an age to help solder flow smoothly when soldering metals. Ammonium nitrat made by the reaction of $NH_3(aq)$ and $HNO_3(aq)$, is used both as a fertiliz and as an explosive. The explosive power of ammonium nitrate was n fully appreciated until a shipload of this material exploded in Texas Cit Texas, in 1947, killing many people. More recently, mixtures of ammoniu

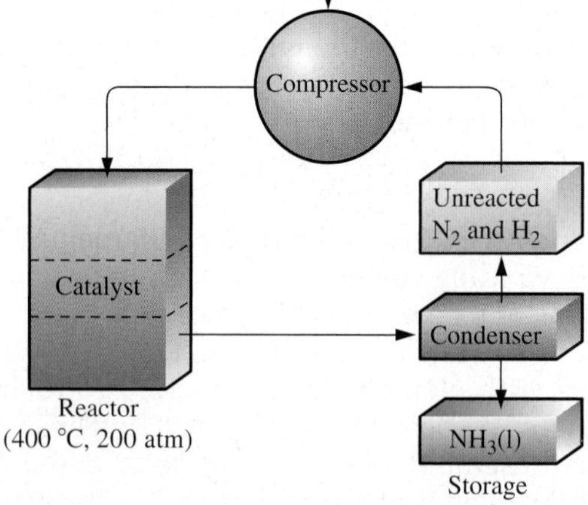

▲ FIGURE 22-18
Ammonia synthesis reaction—the Haber–Bosch process
The gaseous N_2–H_2 mixture is introduced into a reactor at high temperature and pressure in the presence of a catalyst. The gaseous N_2–H_2–NH_3 mixture leaves the reactor and is cooled as it passes through a condenser. Liquefied NH_3 is removed, and the remaining N_2–H_2 mixture is compressed and returned to the reactor. The yield is essentially 100%.

itrate and fuel oil were used as explosives in the terrorist attacks on the Vorld Trade Center in New York City in 1993 and the Alfred P. Murrah ederal Building in Oklahoma City in 1995. The reaction of $NH_3(aq)$ and $H_3PO_4(aq)$ yields ammonium phosphates, such as $NH_4H_2PO_4$ and $(NH_4)_2HPO_4$. These compounds are good fertilizers because they supply two ital plant nutrients, N and P; they are also used as fire retardants.

Urea, which contains 46% nitrogen by mass, is often manufactured at mmonia synthesis plants by using the following reaction:

$$2\,NH_3(g) + CO_2(g) \longrightarrow CO(NH_2)_2(s) + H_2O(l) \qquad \textbf{(22.37)}$$

he structure of the urea molecule is shown in the margin. Urea is an excellent ertilizer, either as a pure solid, as a solid mixed with ammonium salts, or in a ery concentrated aqueous solution mixed with NH_4NO_3 or NH_3 (or both). rea is also used as a feed supplement for cattle and in the production of polyners and pesticides.

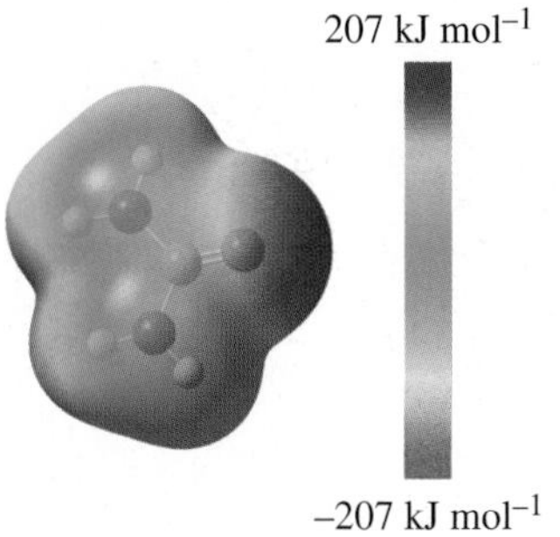

▲ Urea.

ther Hydrides of Nitrogen A great deal has been said in this text about the rincipal hydride of nitrogen: ammonia, NH_3. Here we describe some lessernown hydrides. If a H atom in NH_3 is replaced by the group $—NH_2$, ne resulting molecule is $H_2N—NH_2$ or N_2H_4, *hydrazine* ($pK_{b_1} = 6.07$; $K_{b_2} = 15.05$). Replacement of a H atom in NH_3 by —OH produces NH_2OH, *ydroxylamine* ($pK_b = 8.04$). Hydrazine and hydroxylamine are weak bases. ecause it has two N atoms, N_2H_4 ionizes in two steps. Hydrazine and ydroxylamine form salts analogous to ammonium salts, such as $N_2H_5^+NO_3^-$, $N_2H_6^{2+}SO_4^{2-}$, and $NH_3OH^+Cl^-$. As expected, these salts hydrolyze in water) yield acidic solutions.

Hydrazine and some of its derivatives burn in air with the evolution of rge quantities of heat; they are used as rocket fuels (see margin). For the ombustion of hydrazine,

$$N_2H_4(l) + O_2(g) \longrightarrow N_2(g) + 2\,H_2O(l) \qquad \Delta_rH° = -622.2\ \text{kJ mol}^{-1}\ \textbf{(22.38)}$$

eaction (22.38) can also be used to remove dissolved $O_2(g)$ from boiler water. Iydrazine is particularly valued for this purpose because no salts (ionic comounds) form that would be objectionable in the water. The industrial preparaon of hydrazine is described in Focus On 4-1, www.masteringchemistry.com.

Both hydrazine and hydroxylamine can act as either oxidizing or reducing gents (usually the latter), depending on the pH and the substances with hich they react. The oxidation of hydrazine in acidic solution by nitrite ion roduces *hydrazoic acid*, $HN_3(aq)$.

▲ The lifting thrusters of a space shuttle, shown here in an in-flight test, use methylhydrazine, CH_3NHNH_2, as a fuel.

Msfc/NASA

$$N_2H_5^+(aq) + NO_2^-(aq) \longrightarrow HN_3(aq) + 2\,H_2O(l) \qquad \textbf{(22.39)}$$

Pure HN_3 is a colorless liquid that boils at 37 °C. It is very unstable and will etonate when subjected to shock. Resonance structures for the HN_3 molecule re shown below:

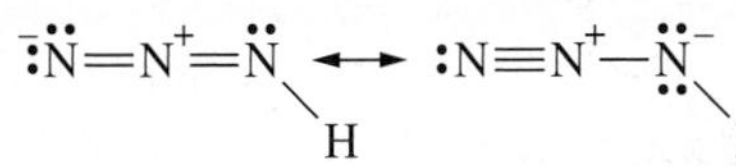

n aqueous solution, HN_3 is a weak acid; its salts are called *azides*. Azides conin the azide ion, N_3^-, and resemble chlorides in some properties (for nstance, AgN_3 is insoluble in water), but they are unstable. Some azides (such s lead azide) are used to make detonators. The release of $N_2(g)$ by the decomosition of sodium azide, NaN_3, is the basis of the air-bag safety system in utomobiles (page 212).

Oxides of Nitrogen

Nitrogen forms a series of oxides in which the oxidation state of N can have very value ranging from +1 to +5 (Table 22.13). All these oxides are gases at

TABLE 22.13 Oxides of Nitrogen

O.S. of N	Formula
+1	N_2O
+2	NO
+3	N_2O_3
+4	NO_2
+4	N_2O_4
+5	N_2O_5

25 °C, except N_2O_5, which is a solid with a sublimation pressure of 1 atm a 32.5 °C. It is impossible to obtain either brown $NO_2(g)$ or its colorless dime $N_2O_4(g)$, as a pure gas at temperatures between about −10 °C and 140 ° because of the equilibrium between them (review Example 15-9). At lowe temperatures, N_2O_4 can be obtained as a pure solid, and above 140 °C, th gas-phase equilibrium strongly favors $NO_2(g)$. In the solid state, N_2O_3 is pa blue; in the liquid state, it is bright blue.

All nitrates decompose on heating, but only NH_4NO_3 yields $N_2O(g$ Nitrates of active metals, such as $NaNO_3$, yield the corresponding nitrite, suc as $NaNO_2$, and $O_2(g)$. Nitrates of less active metals, such as $Pb(NO_3)_2$, yiel the metal oxide, $NO_2(g)$ and $O_2(g)$. These methods of preparing oxides c nitrogen are outlined in Table 22.14.

As mentioned in our assessment of the metallic–nonmetallic character c the group 15 elements, the oxides of nitrogen are acidic, and they react wit water to give acidic solutions. For example, the reaction of N_2O_5 and H_2 yields HNO_3, as shown below, and thus N_2O_5 is the acid anhydride of HNO

$$N_2O_5(s) + H_2O(l) \longrightarrow 2\,HNO_3(aq)$$

The acid anhydride of nitrous acid, HNO_2, is N_2O_3. Nitrogen dioxid NO_2, produces both HNO_3 and NO when it reacts with water. But N_2O is nc an acid anhydride in the usual sense; it is related to *hyponitrous* acid, H_2N_2C (HON=NOH), which yields N_2O and H_2O on decomposition.

$$H_2N_2O_2 \longrightarrow N_2O + H_2O$$

Among the oxides of nitrogen, N_2O (laughing gas) has anesthetic properti and finds some use in dentistry and in providing pain relief during childbirt N_2O has also been used as a propellant in pressured cans of fats (such a whipped cream) and as a power booster in combustion engines. The powe boosting capability of N_2O arises because at high temperatures—such as those i a combustion chamber—two moles of N_2O decompose to give a total of thre moles of N_2 and O_2. The increase in the number of moles of gas increases th force on the pistons in the engine, leading ultimately to increased acceleration.

Other oxides of nitrogen include NO_2, which is employed in the manufa ture of nitric acid; N_2O_4, which is used extensively as an oxidizer in rock fuels; and NO, which is arguably the most important oxide of nitrogen, at lea from a biological standpoint. Among its many biological functions, NO help to protect the heart, stimulate the brain, and kill bacteria. It has been called th "miracle molecule" because it helps to regulate the health of almost every ce in the body.

The discovery of the role of NO in biological systems followed a somewh tortuous route. In 1829, it was discovered that the compound trinitroglycerin $C_3H_5(NO_3)_3$, a highly explosive compound, helped dilate (enlarge) blood ve sels and relieved symptoms of a heart attack. Later acetylcholine, $C_7H_{16}NO_2$

TABLE 22.14 Preparation of Oxides of Nitrogen

Oxide	A Method of Preparation
N_2O	$NH_4NO_3(s) \xrightarrow{\Delta} N_2O(g) + 2\,H_2O(g)$
NO	$3\,Cu(s) + 8\,H^+(aq) + 2\,NO_3^-(aq) \longrightarrow 3\,Cu^{2+}(aq) + 2\,NO(g) + 4\,H_2O(l)$
N_2O_3	$2\,NO(g) + N_2O_4(g) \xrightarrow{-20\,°C} 2\,N_2O_3(l)$
NO_2	$2\,Pb(NO_3)_2(s) \xrightarrow{\Delta} 2\,PbO(s) + 4\,NO_2(g) + O_2(g)$
	$2\,NO(g) + O_2(g) \rightleftharpoons 2\,NO_2(g) \quad K_p = 1.6 \times 10^{12}$ (at 298 K)
N_2O_4	$2\,NO_2(g) \rightleftharpoons N_2O_4(g) \quad K_p = 8.84$ (at 298 K)
N_2O_5	$4\,HNO_3(l) + P_4O_{10}(s) \xrightarrow{-10\,°C} 4\,HPO_3(s) + 2\,N_2O_5(s)$

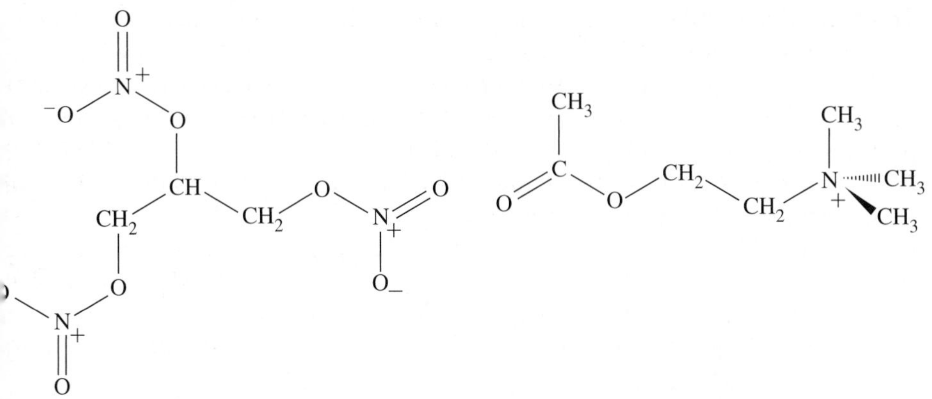

(a) **(b)**

Structures of **(a)** trinitroglycerine, and **(b)** acetylcholine.

'as discovered to have a similar effect on blood vessels. Lewis structures of iese two dissimilar molecular species are shown above. Surprisingly, the link etween them and their action on blood vessels was not explained until 1987. ı that year, two reports were published by two research groups, one lead by ouis Ignarro in the United States and the other lead by Salvador Moncada in ıe United Kingdom, which demonstrated that the link is the nitric oxide mol-:ule, NO. It is now known that acetylcholine causes enzymes in the blood essel to release NO, which in turn causes other enzymes to relax the muscle f the vessel. Trinitroglycerine is converted to NO(g) by metabolic processes ıd NO(g) acts as a signaling agent in blood vessels and causes the muscle in ıe vessels to relax.

The NO molecule has also been implicated in the human response to infec-on. When infection occurs, the immune system produces both NO and O_2^-. hese two radicals react as follows to produce the peroxynitrite anion, ONO_2^-:

$$NO(g) + O_2^-(g) \longrightarrow O{=}N{-}O{-}O^-(g)$$

The peroxynitrite anion is a strong and versatile oxidant that can break own the structures of, and thus kill, cells of invading species. Over time, ‹cess peroxynitrite anions isomerize to harmless nitrate anions.

One of the interesting features of the oxides of nitrogen is that their stan-ard Gibbs energies of formation are all *positive* quantities. This feature sug-ests that the oxides, such as $N_2O(g)$, are thermodynamically unstable, and ıat decomposition to the elements is spontaneous under standard conditions.

$$2\,N_2O(g) \longrightarrow 2\,N_2(g) + O_2(g) \qquad \Delta_rG° = -208\ \text{kJ mol}^{-1}\ (\text{at } 298\ \text{K}) \qquad \textbf{(22.40)}$$

Actually, $N_2O(g)$ is quite stable at room temperature because the activa-on energy for its decomposition is very high—about 250 kJ mol^{-1}. At higher ːmperatures (about 600 °C), its rate of decomposition becomes appreciable. eaction (22.40) accounts for the ability of N_2O to support combustion ecause the $O_2(g)$ necessary for the combustion is produced by the decom-osition of N_2O. Overall reactions of the following sort occur.

$$H_2(g) + N_2O(g) \longrightarrow H_2O(l) + N_2(g)$$
$$Cu(s) + N_2O(g) \longrightarrow CuO(s) + N_2(g)$$

Iotice that one of the combustion products is the normal combustion product ıd the other is N_2.

Nitrogen monoxide (nitric oxide), NO(g), is produced commercially by the 'stwald process, in which $NH_3(g)$ is oxidized in the presence of a catalyst:

$$4\,NH_3(g) + 5\,O_2(g) \xrightarrow[850\,°C]{Pt} 4\,NO(g) + 6\,H_2O(g) \qquad \textbf{(22.41)}$$

Tom Pantages

▲ The combustion of copper gauze in $N_2O(g)$.

The oxidation of NH_3 to NO is the first step in converting NH_3 to a number c other nitrogen compounds.

Another source of NO, usually unwanted, is in high-temperature combus tion processes, such as those that occur in automobile engines and in electri power plants. At the same time that the fuel combines with oxygen from air t produce a high temperature, $N_2(g)$ and $O_2(g)$ in the hot air combine to a lim ited extent to form NO(g).

$$N_2(g) + O_2(g) \xrightarrow{\Delta} 2\,NO(g) \qquad (22.42$$

Brown nitrogen dioxide, $NO_2(g)$, is often seen in reactions involving nitri acid. An example is the reaction of Cu(s) with warm concentrated $HNO_3(aq$

$$Cu(s) + 4\,H^+(aq) + 2\,NO_3^-(aq) \longrightarrow Cu^{2+}(aq) + 2\,H_2O(l) + 2\,NO_2(g) \qquad (22.43$$

Of particular interest to atmospheric chemists is the key role of $NO_2(g)$ in th formation of photochemical smog (page 957).

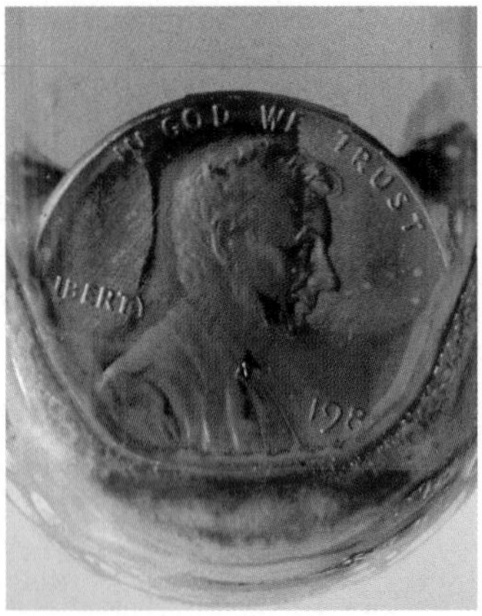

Richard Megna/Fundamental Photographs

▲ A copper penny reacting with nitric acid. The reaction is that given by equation (22.43). The blue-green color of the solution is due to $Cu^{2+}(aq)$, and the reddish brown color is that of nitrogen dioxide, $NO_2(g)$.

Nitric Acid and Nitrates The commercial synthesis of nitric acid does not us N_2O_5, as might be expected. It involves the following three reactions, the firs of which—the Ostwald process—was described previously. NO(g) from reac tion (22.45) is recycled into reaction (22.44).

$$4\,NH_3(g) + 5\,O_2(g) \xrightarrow[850\,°C]{Pt} 4\,NO(g) + 6\,H_2O(g) \qquad (22.41$$

$$2\,NO(g) + O_2(g) \longrightarrow 2\,NO_2(g) \qquad (22.44$$

$$3\,NO_2(g) + H_2O(l) \longrightarrow 2\,HNO_3(aq) + NO(g) \qquad (22.45$$

Nitric acid is used in the preparation of various dyes; drugs; fertilizer (ammonium nitrate); and explosives, such as nitroglycerin, nitrocellulose, an trinitrotoluene (TNT). It is also used in metallurgy and in reprocessing spen nuclear fuels. Nitric acid is about twelfth, by mass, among the top chemica produced in the United States.

Nitric acid is also a good oxidizing agent. For example, copper reacts wit dilute $HNO_3(aq)$, producing primarily NO or, with concentrated $HNO_3(aq$ NO_2 (reaction 22.43). With a more active metal, such as Zn, the reductio product has N in one of its lower oxidation states, for example, NH_4^+. Nitrate can be made by neutralizing nitric acid with appropriate bases.

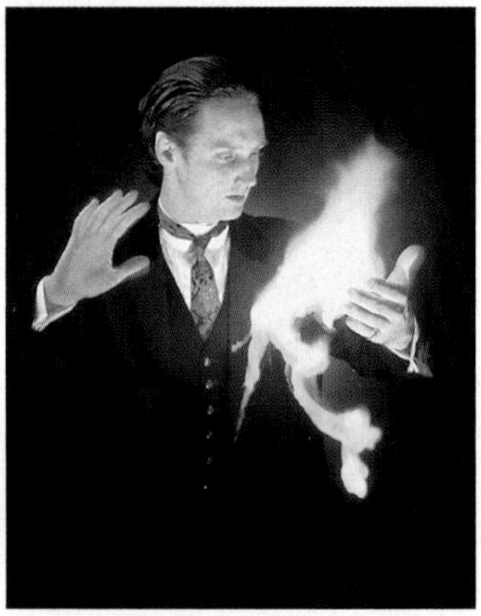

Richard Megna/ Fundamental Photographs

▲ Flash paper is used by magicians for dramatic effect. It can be made by treating paper with nitric and sulfuric acids. This process converts the cellulose fibers into nitrocellulose, which burns cleanly and rapidly.

Nitrogen Halides

Nitrogen forms halides with the elements of group 17. Nitrogen trifluoride, NF_3 can be made by the fluorination of ammonia in the presence of a Cu catalyst.

$$4\,NH_3(g) + 3\,F_2(g) \xrightarrow{Cu} NF_3(g) + 3\,NH_4F(s)$$

Nitrogen trifluoride is a colorless, odorless gas and is one of the few nitro gen compounds that is thermodynamically stable with respect to the ele ments ($\Delta_f G° = -83.3$ kJ mol^{-1}). Although the nitrogen atom in NF_3 has lone pair of electrons, NF_3 has very little tendency to act as a Lewis base unlike ammonia, but it can be made to react, in the gas phase, with oxygen t give NF_3O, a stable but somewhat unusual molecule. In the NF_3O molecule the nitrogen atom is the central atom, and the nitrogen–oxygen bond i much shorter than the nitrogen–fluorine bonds. The following resonanc structures have been proposed, and the structure on the right appears to b an important contributor.

In contrast to nitrogen trifluoride, nitrogen trichloride, NCl_3, is neither a as nor stable. It is an oily, yellow, highly explosive liquid and care must be ken in handling or making this compound. Its explosive nature can be attrib-ted to its endothermic enthalpy of formation ($\Delta_f H° = 230.0 \text{ kJ mol}^{-1}$), hich indicates that the decomposition to $N_2(g)$ and $Cl_2(g)$ is highly exother-ic and thermodynamically favorable. Nitrogen trichloride is not prepared om the direct reaction of $N_2(g)$ and $Cl_2(g)$ but from the reaction of ammo-ium chloride with chlorine, as shown below:

$$NH_4Cl(s) + 3\,Cl_2(g) \rightleftharpoons NCl_3(l) + 4\,HCl(g)$$

he equilibrium is shifted to the right-hand side by dissolving the NCl_3 in an rganic solvent. In contrast to NF_3, nitrogen trichloride reacts with water to ive ammonia.

$$NCl_3(aq) + 3\,H_2O(l) \longrightarrow NH_3(aq) + 3\,HOCl(aq)$$

his reaction produces HOCl, a bleaching agent, and thus nitrogen trichloride, iluted in air, has been used to bleach flour. The compounds NBr_3 and NI_3 are lso known, and are even more reactive and explosive (and thus, more dan-erous to handle) than NCl_3. NI_3 is so unstable that it detonates with the ightest contact, even if touched with a feather or breathed on.

Two other fluorides of nitrogen are known: N_2F_4 and N_2F_2. The structures f the N_2F_4 and N_2F_2 molecules are shown in Figure 22-19. Dinitrogen tetra-uoride, N_2F_4, interconverts between two conformations—staggered and auche—because the two NF_2 units can rotate independently about the N—N bond. (We will encounter these conformations again in organic chem-try in Chapters 26 and 27.) Dinitrogen difluoride, N_2F_2, exists in two geo-etrical forms, called geometrical isomers, that are not easily interconverted. ecause the double bond prevents twisting of the molecule about the nitro-en– nitrogen bond axis, the N_2F_2, molecule exists as either the *cis* isomer or e *trans* isomer. The *cis* isomer has both fluorine atoms on the same side of the ouble bond, whereas the *trans* isomer has the fluorine atoms on opposite ides of the double bond. (We will meet this type of isomerism again in hapter 26.)

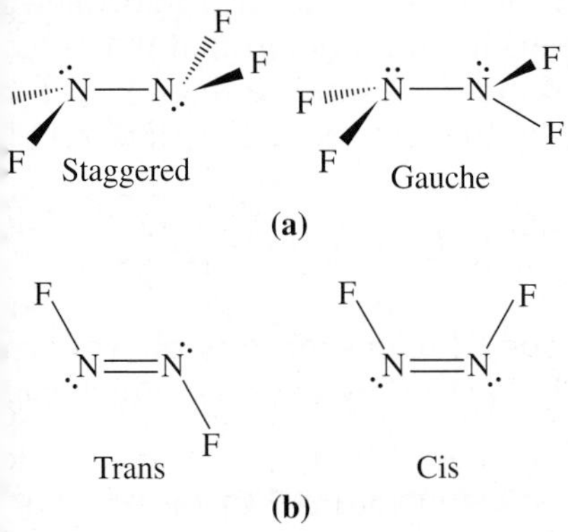

◀ FIGURE 22-19
Structures of N_2F_4 and N_2F_2
(a) In the staggered conformation of N_2F_4, the lone pairs on the nitrogen atoms are diametrically opposed to each other. The gauche conformation is obtained from the staggered conformation by rotating one NF_2 group by 60° with respect to the other. **(b)** Because the double bond prevents the molecule from twisting about the nitrogen–nitrogen bond axis, there are two geometrical isomers for the F—N=N—F molecule.

llotropes of Phosphorus

hite phosphorus is a white, waxy, phosphorescent solid that can be cut with a nife. (A phosphorescent material glows in the dark.) It is a nonconductor of lectricity, can ignite spontaneously in air (hence it is stored under water), and insoluble in water but soluble in some nonpolar solvents, such as CS_2. The olid has P_4 molecules as its basic structural units (Fig. 22-20a). The P_4 molecule tetrahedral, with a P atom at each corner. The phosphorus-to-phosphorus onds in P_4 appear to involve the overlap of 3*p* orbitals almost exclusively. uch overlap normally produces 90° bond angles, but in P_4 the P—P—P ond angles are 60°. The bonds are strained, and as might be expected, species ith strained bonds are reactive.

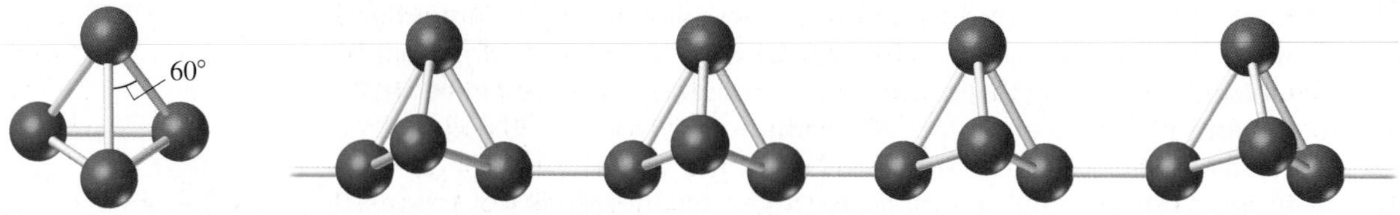

(a) White phosphorus

(b) Red phosphorus

▲ FIGURE 22-20
Two forms of phosphorus
(a) Structure of white phosphorus: the P_4 molecule. **(b)** Structure of red phosphorus.

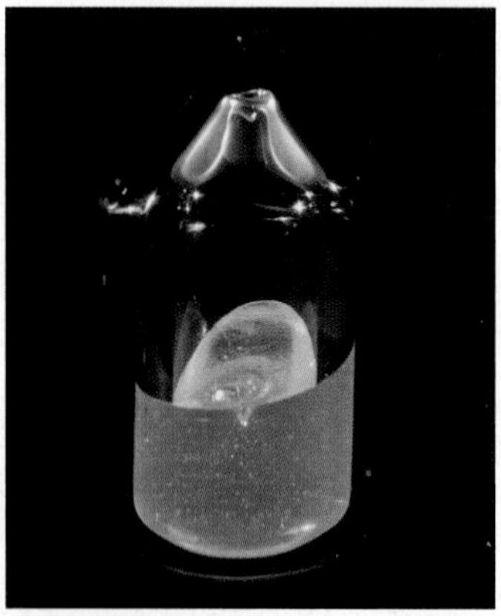

Theodore W. Gray

▲ The glow of white phosphorus gave the element its name—*phos*, light, and *phorus*, bringing. The solid has a relatively high vapor pressure, and the glow results from the slow reaction between phosphorus vapor and oxygen in air.

When white phosphorus is heated to about 300 °C out of contact with air, transforms to *red phosphorus*. What appears to happen is that one P—P bon per P_4 molecule breaks, and the resulting fragments join together into lor chains, as suggested in Figure 22-20(b). Red P is less reactive than white Because they have a different atomic arrangement in their basic structur units, red and white phosphorus are allotropic forms of phosphorus rath than just different solid phases. These two allotropes of phosphorus are show in the photograph below. The triple point of red phosphorus is 590 °C an 43 atm. Thus, red phosphorus sublimes without melting (at about 420 °C).

Despite the fact that white P is the form obtained by condensing $P_4(g)$ an that the conversion of white P to red P is a very slow process at ambient ten peratures, red P is actually the more thermodynamically stable of the tw forms at 298.15 K. Nevertheless, white P is assigned values of 0 for $\Delta_f H°$ an $\Delta_f G°$, and for red P, these values are negative.

Phosphorus Compounds

In Chapter 28 (found on MasteringChemistry) we will discover that con pounds of phosphorus are essential to all living organisms. However, the is also a significant "inorganic" chemistry of phosphorus, as we will see i this section.

▶ Phosphine is extremely poisonous and has been used as a fumigant against rodents and insects.

Phosphine The most important compound of phosphorus and hydrogen phosphine, PH_3. This compound is analogous to ammonia in that it acts as base and forms phosphonium (PH_4^+) compounds. Unlike ammonia, PH_3 thermally unstable. Phosphine is produced by the disproportionation of P_4 i aqueous base.

$$P_4(s) + 3\,OH^-(aq) + 3\,H_2O(l) \longrightarrow 3\,H_2PO_2^-(aq) + PH_3(g)$$

Phosphorus Trichloride Another phosphorus(III) compound is phosphoru trichloride, PCl_3. One typical reaction of PCl_3 is its hydrolysis, which pro duces hydrochloric and phosphorous acids.

$$PCl_3(l) + 6\,H_2O(l) \longrightarrow H_3PO_3(aq) + 3\,H_3O^+(aq) + 3\,Cl^-(aq)$$

▶ The white and red allotropes of phosphorus.

Cl_3 is the most important phosphorus halide, and a variety of phosphorus(III) mpounds are made from it. It is produced by the direct action of $Cl_2(g)$ on emental phosphorus. Although you may never see PCl_3, chemicals made om it are everywhere—soaps and detergents, plastics and synthetic rubber, ylon, motor oils, and insecticides and herbicides. A variety of organic groups n replace one or more chlorine atoms of PCl_3 to give a family of phosphine-ke compounds. These are good Lewis bases and can act as ligands in mplex-ion formation.

The halide PCl_5 is obtained by the reaction of Cl_2 with PCl_3 in tetra-loromethane (CCl_4). In the gas phase, PCl_5 exists as discrete trigonal pyramidal molecules. In the solid state, it exists as $[PCl_4]^+[PCl_6]^-$, in which e ions are tetrahedral and octahedral, respectively.

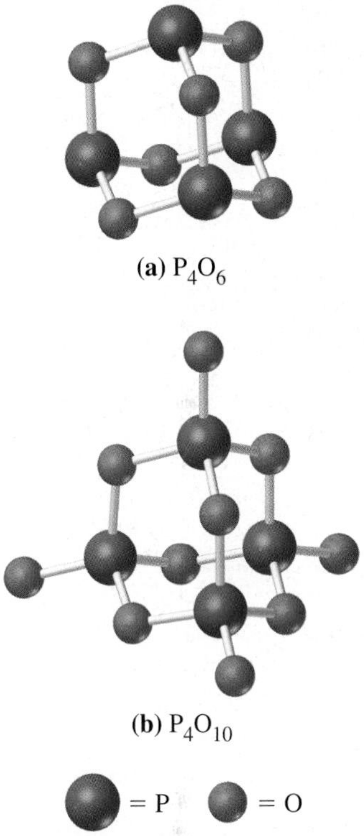

▲ FIGURE 22-21
Molecular structures of P_4O_6 and P_4O_{10}

22-6 CONCEPT ASSESSMENT

In the solid phase, PCl_5 forms PCl_4^+ and PCl_6^+. However, PBr_5 forms $PBr_4^+Br^-$. Suggest a reason for this difference in structure.

xides of Phosphorus The simplest formulas we can write for the oxides at have phosphorus in the oxidation states +3 and +5 are P_2O_3 and P_2O_5, spectively. The corresponding names are "phosphorus trioxide" and hosphorus pentoxide." P_2O_3 and P_2O_5 are only empirical formulas, how-ver. The true molecular formulas of the oxides are double those—that is, $_4O_6$ and P_4O_{10}.

The structure of each oxide molecule is based on the P_4 tetrahedron and so ust have *four* P atoms, not two. As shown in Figure 22-21(a), in P_4O_6 one O om bridges each pair of P atoms in the P_4 tetrahedron, which means that ere are *six* O atoms per P_4 tetrahedron. The structure of P_4O_{10} is shown in igure 22-21(b). In addition to the six bridging O atoms, one O atom is bonded each corner P atom. This means that there are a total of *ten* O atoms per P_4 trahedron.

The reaction of P_4 with a limited quantity of $O_2(g)$ produces P_4O_6. If an cess of $O_2(g)$ is used, P_4O_{10} is obtained. Both oxides react with water to rm oxoacids—both are *acid anhydrides.*

$$P_4O_6(l) + 6\,H_2O(l) \longrightarrow \underset{\text{phosphorous acid}}{4\,H_3PO_3(aq)}$$

$$P_4O_{10}(s) + 6\,H_2O(l) \longrightarrow \underset{\text{phosphoric acid}}{4\,H_3PO_4(aq)} \qquad (22.46)$$

ecause the formulas of phosphoric acid (H_3PO_4) and phosphorous acid H_3PO_3) are very similar, it is tempting to think that the acids are somewhat milar. It turns out that these two acids have structures that are different in a ery significant way. The structures and electrostatic potential maps of $_3PO_4$ and H_3PO_3 are shown on page 1076. In H_3PO_4, the P atom is bonded four oxygen atoms and each of the H atoms is bonded to an oxygen atom, ut in H_3PO_3, the P atom is bonded to three oxygen atoms and one H atom. ll three H atoms in phosphoric acid are ionizable, and so H_3PO_4 is a tripro-c acid. Only two of the H atoms in H_3PO_3 are ionizable and so H_3PO_3 is a iprotic acid.

◀ Ionizable H atoms in oxoacids are associated with the linkage E—O—H, where E represents an electronegative central atom with an attached terminal O. See pages 771–772.

22-7 CONCEPT ASSESSMENT

Write condensed structural formulas for phosphoric acid and phosphorous acid. Condensed structural formulas are discussed on page 70.

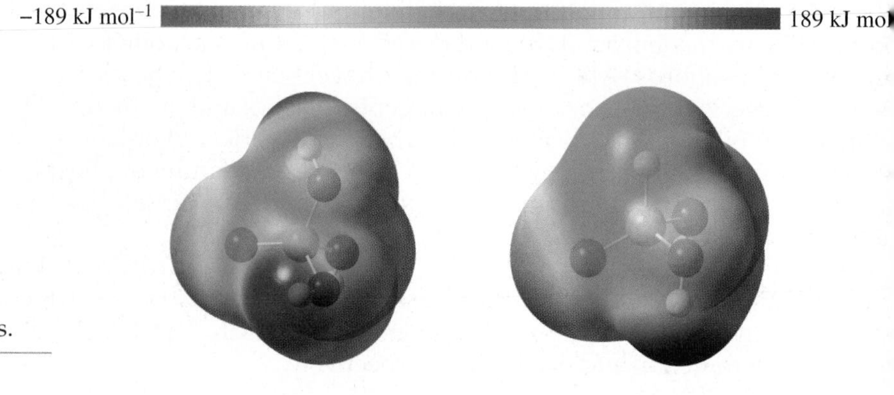

▶ Phosphoric and phosphorous acids.

Phosphoric Acid Phosphoric acid, H_3PO_4, ranks about seventh among th chemicals manufactured in the United States, with an annual production o more than 13 million tons. It is used mainly to make fertilizers, but it is als used to treat metals to make them more corrosion-resistant. Phosphoric aci has many uses in the food industry: It is used to make baking powders an instant cereals, in cheese making, in curing hams, and in soft drinks to impa tartness.

If P_4O_{10} and H_2O are combined in a 1 : 6 mole ratio in reaction (22.46), th liquid product should be pure H_3PO_4, that is, 100% H_3PO_4, a compoun called *orthophosphoric* acid. An analysis of the liquid, however, shows to be only about 87.3% H_3PO_4. The "missing" phosphorus is still present in th liquid, but as $H_4P_2O_7$, a compound called either *diphosphoric acid* or *pyropho phoric acid*. A molecule of diphosphoric acid forms when a molecule of H_2O eliminated from between two molecules of orthophosphoric acid, as shown i Figure 22-22. If a third molecule of orthophosphoric acid joins in by the elimina tion of another H_2O molecule, the product is $H_5P_3O_{10}$, triphosphoric acid, an so on. As a class, the chain-like phosphoric acid structures are called *polypho phoric* acids, and their salts are called *polyphosphates*. Two especially importan derivatives of the polyphosphoric acids present in living organisms are the sub stances known as ADP and ATP. The "A" portion of the acronym stands fo adenosine, a combination of an organic base called adenine and a five-carbo

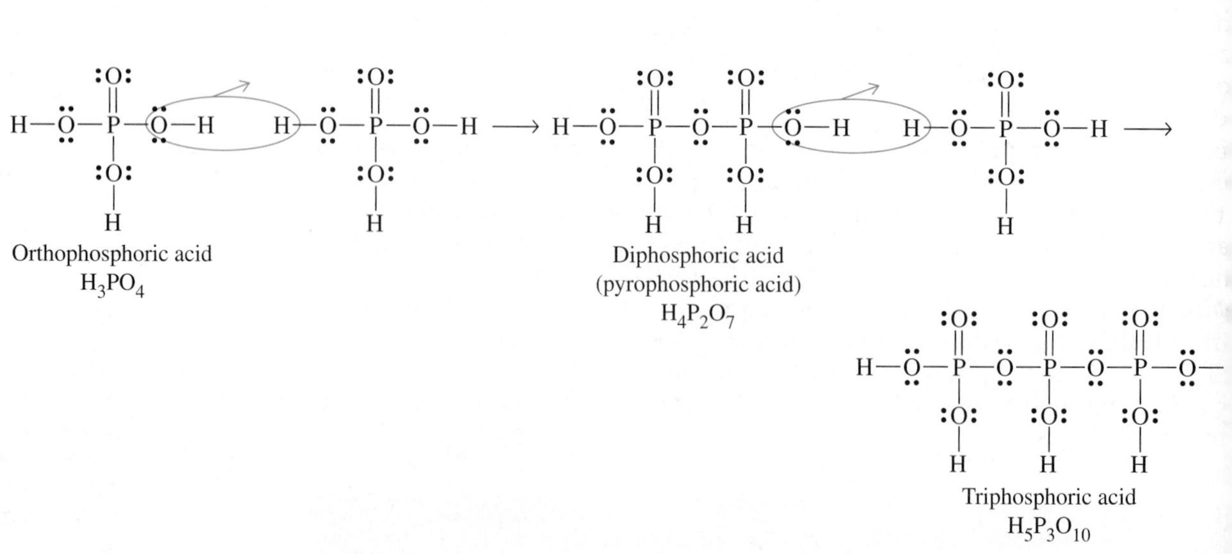

▲ FIGURE 22-22
Formation of polyphosphoric acids
Removal of H_2O molecules results in P—O—P bridges.

ugar called ribose. If this adenosine combination is linked to a diphosphate ion, he product is ADP, adenosine diphosphate. Addition of a phosphate ion to ADP yields ATP; adenosine triphosphate. These polyphosphates are described n more detail in Chapter 28.

Most phosphoric acid is made by the action of sulfuric acid on phosphate ock. The chemical equation for the reaction is given below:

$$\underset{\text{Fluorapatite}}{Ca_3(PO_4)_2 \cdot CaF_2(s)} + 10\,H_2SO_4(\text{concd aq}) + 20\,H_2O(l) \longrightarrow$$
$$6\,H_3PO_4(aq) + 10\,\underset{\text{Gypsum}}{CaSO_4 \cdot 2\,H_2O(s)} + 2\,HF(aq) \qquad \textbf{(22.47)}$$

he HF is converted to insoluble Na_2SiF_6, and the gypsum is filtered off along vith other insoluble impurities. The phosphoric acid is concentrated by evap- ration. Phosphoric acid obtained by this "wet process" contains a variety of netal ions as impurities and is dark green or brown. Nevertheless, it is satis- actory for the manufacture of fertilizers and for metallurgical operations.

If H_3PO_4 from reaction (22.47) is used in place of H_2SO_4 to treat phosphate ock, the principal product is calcium dihydrogen phosphate. This compound, a ertilizer containing 20–21% P, is marketed under the name *triple superphosphate*:

$$Ca_3(PO_4)_2 \cdot CaF_2(s) + 14\,H_3PO_4(\text{concd aq}) + 10\,H_2O(l) \longrightarrow$$
$$\underset{\text{Triple superphosphate}}{10\,Ca(H_2PO_4)_2 \cdot H_2O(s)} + 2\,HF(aq) \qquad \textbf{(22.48)}$$

hosphorus and the Environment

hosphates are widely used as fertilizers because phosphorus is an essential utrient for plant growth. However, heavy fertilizer use may lead to phos- hate pollution of lakes, ponds, and streams, causing an explosion of plant rowth, particularly algae. The algae deplete the oxygen content of the ater, eventually killing fish. This type of change, occurring in freshwater odies as a result of their enrichment by nutrients, is called **eutrophication**. utrophication is a natural process that occurs over geological time periods, ut it can be greatly accelerated by human activities, as shown in the photo- raph in the margin.

Lance Rider/Shutterstock

▲ The natural eutrophication of a lake is greatly accelerated by phosphates in wastewater and the agricultural runoff of fertilizers.

Natural sources of plant nutrients include animal wastes, decomposition of ead organic matter, and natural nitrogen fixation. Human sources include dustrial wastes and municipal sewage plant effluents, in addition to fertil- er runoff. One way to reduce phosphate discharges into the environment is remove them from the wastewater in sewage treatment plants. In the pro- ssing of sewage, polyphosphates are degraded to orthophosphates by bacte- al action. The orthophosphates can then be precipitated, either as iron(III) osphates, aluminum phosphates, or as calcium phosphate or hydroxyap- ite [$Ca_5(OH)(PO_4)_3$]. The precipitating agents are generally aluminum sul- te, iron(III) chloride, or calcium hydroxide (slaked lime). In a fully equipped odern sewage treatment plant, up to 98% of the phosphates in sewage can removed.

2-6 Hydrogen: A Unique Element

e first period has only two elements: hydrogen and helium. Hydrogen is ite reactive, but helium is inert. In the case of helium there is no difficulty ating its electronic structure and chemical properties to those of the other ble gases in group 18. Conversely, the chemical and physical properties of drogen cannot be correlated with any of the main groups in the periodic le. Hydrogen is truly unique and is best considered on its own.

The ground state electron configuration of a hydrogen atom ($1s^1$) is consister with the ground state electron configuration of the alkali metal atoms (ns^1), so seems logical to place hydrogen in group 1. However, such a placement suggest that hydrogen will have properties similar to those of the alkali metals. This i not the case. Alkali metal atoms show a tendency to form M^+ ions. Although H is known in acid–base chemistry, a hydrogen atom has a much greater tendenc to form a covalent bond through the sharing of a pair of electrons. Hydrogen' electron configuration also resembles that of the halogens in being just one ele tron short of that of a noble gas. But unlike the halogens, hydrogen rarely form H^- ions, except with the most active metals.

In still other ways hydrogen is like the group 14 elements—both have hal filled valence shells and similar electronegativity values. Thus, the group H— and CH_3— have one unpaired electron and can form compounds suc as LiH and $LiCH_3$. In spite of all of this, hydrogen is best treated as a group o its own.

Think how important hydrogen has been in the study of chemistry. Joh Dalton based atomic masses on a value of 1 for the H atom. Humphry Dav (1810) proposed that hydrogen is the key element in acids. We saw in Chapter that theoretical studies of the H atom provided us with our modern vie of atomic structure. In Chapter 11 we found that the H_2 molecule was the star ing point for modern theories of chemical bonding. For all of its theoretic significance, though, hydrogen is also of great practical importance, as we wi emphasize in this section.

Occurrence and Preparation

Hydrogen is a very minor component of the atmosphere, about 0.5 ppm a Earth's surface. At altitudes above 2500 km, the atmosphere is mostly atomi hydrogen at extremely low pressures. In the universe as a whole, hydroge accounts for about 90% of the atoms and 75% of the mass. On Earth, *hydroge occurs in more compounds than does any other element.*

The free element can easily be produced, but from only a few of its com pounds. Our first choice might be H_2O—the most abundant hydrogen com pound. To extract hydrogen from water means reducing the oxidation state of H from +1 in H_2O to 0 in H_2. This requires an appropriate *reducing* agent, such a carbon (coal or coke), carbon monoxide, or a hydrocarbon—particularl methane (natural gas). The first pair of reactions that follow are called th **water gas** reactions; they represent a way of making combustible gases—C and H_2—from steam.

Water gas reactions: $$C(s) + H_2O(g) \longrightarrow CO(g) + H_2(g) \quad (22.4$$

$$CO(g) + H_2O(g) \longrightarrow CO_2(g) + H_2(g) \quad (22.5$$

Re-forming of methane: $$CH_4(g) + H_2O(g) \longrightarrow CO(g) + 3\,H_2(g)$$

Another source of $H_2(g)$ is as a by-product in petroleum refining.

Often we use methods in the chemical laboratory that are not commerciall feasible. Electrolysis of water is one useful laboratory method for producin small quantities of $H_2(g)$. Another involves the reaction of active metals i acidic solutions, an example of which is given below.

$$Zn(s) + 2\,H^+(aq) \longrightarrow Zn^{2+}(aq) + H_2(g)$$

Hydrogen Compounds

Hydrogen forms binary compounds, called **hydrides**, with most of the othe elements. Binary hydrides are usually grouped into three broad categorie covalent, ionic, and metallic. *Covalent hydrides* are those formed betwee ydrogen and nonmetals. Some of these hydrides are simple molecules that ın be formed by the direct union of hydrogen and the second element. Two ‹amples are given below:

$$H_2(g) + Cl_2(g) \longrightarrow 2\,HCl(g)$$
$$3\,H_2(g) + N_2(g) \longrightarrow 2\,NH_3(g)$$

Ionic hydrides form between hydrogen and the most active metals, particu- .rly those of groups 1 and 2. In these compounds hydrogen exists as the ydride *ion*, H^-.

$$2\,M(s) + H_2(g) \longrightarrow 2\,MH(s) \qquad M(s) + H_2(g) \longrightarrow MH_2(s)$$

(M is any group 1 metal) (M is Ca, Sr, or Ba)

Richard Megna/Fundamental Photographs

▲ **Reaction of CaH_2 with water** The pink color of phenolphthalein indicator added to the water signals the production of $Ca(OH)_2$ in reaction (22.51).

mic hydrides react vigorously with water to produce $H_2(g)$. CaH_2, a gray olid, has been used as a portable source of $H_2(g)$ for filling weather observa- on balloons.

$$CaH_2(s) + 2\,H_2O(l) \longrightarrow Ca(OH)_2(s) + 2\,H_2(g) \qquad \textbf{(22.51)}$$

he reaction between CaH_2 and water is highlighted in the margin.

Metal hydrides feature prominently in organic chemistry. For example, ecause CaH_2 reacts readily with water, it is often used to remove water from rganic solvents. Sodium hydride (NaH) is used as a strong base in the syn- ıesis of many organic compounds. Lithium hydride (LiH) is used to make thium aluminum hydride ($LiAlH_4$), a powerful reducing agent used in rganic chemistry (see page 986).

Metallic hydrides are commonly formed with the transition elements— roups 3 to 12. A distinctive feature of these hydrides is that in many cases ıey are *nonstoichiometric*—the ratio of H atoms to metal atoms is variable, not xed. This is because H atoms can enter the voids or holes among the metal toms in a crystalline lattice and fill some but not others.

ses of Hydrogen

Iydrogen is not listed among the top chemicals produced, because only a mall percentage is ever sold to customers. Most hydrogen is produced and sed on the spot. In these terms, its most important use (about 42%) is in the ıanufacture of NH_3 (reaction 22.35). The next most important use of H_2 ıbout 38%) is in petroleum refining, where it is produced in some operations nd consumed in others, such as in the production of the high-octane gasoline omponent, isooctane, from diisobutylene.

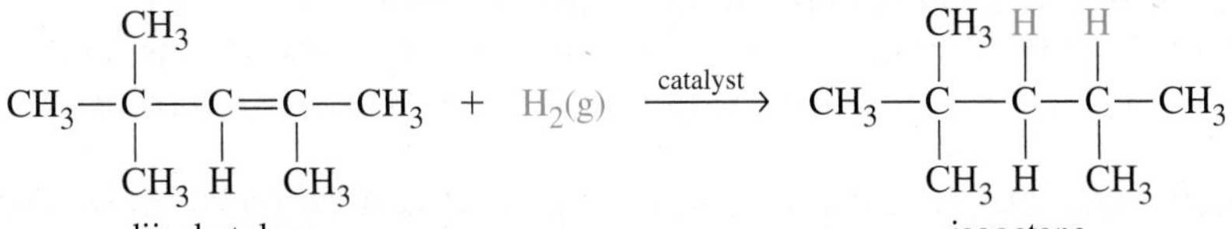

Richard Megna/Fundamental Photographs

▲ Liquid vegetable oils contain long molecules with some carbon-to-carbon double bonds. When some of these double bonds in the molecules are hydrogenated to give carbon-to-carbon single bonds, the result is conversion of the liquid to a solid "partially hydrogenated vegetable oil."

In similar reactions, called **hydrogenation reactions**, hydrogen atoms, in the resence of a catalyst, can be added to double or triple bonds in other molecules. 'his type of reaction, for example, converts liquid oleic acid, $C_{17}H_{33}COOH$, to olid stearic acid, $C_{17}H_{35}COOH$.

$$\text{:}H_3(CH_2)_7CH{=}CH(CH_2)_7COOH + H_2(g) \xrightarrow{Ni} CH_3(CH_2)_{16}COOH \qquad \textbf{(22.52)}$$

Oleic acid Stearic acid

imilar reactions serve as the basis for converting oils that contain carbon- arbon double bonds, such as vegetable oils, into solid or semisolid fats, such s shortening, as described in the margin.

Another important chemical manufacturing process that uses hydrogen the synthesis of methyl alcohol (methanol), an alternative fuel.

$$CO(g) + 2\,H_2(g) \xrightarrow{\text{catalyst}} CH_3OH(g)$$

Hydrogen gas is an excellent reducing agent and in some cases is used produce metals from their oxide ores. For example, the following reaction used, at 850 °C, to produce tungsten metal from its oxide:

$$WO_3(s) + 3\,H_2(g) \longrightarrow W(s) + 3\,H_2O(g)$$

The uses of hydrogen described here, together with several others, are listed Table 22.15.

TABLE 22.15 Some Uses of Hydrogen

Synthesis of ammonia, NH_3 hydrogen chloride, HCl methanol, CH_3OH
Hydrogenation reactions in petroleum refining converting oils to fats
Reduction of metal oxides, such as those of iron, cobalt, nickel, copper, tungsten, molybdenum
Metal cutting and welding with atomic and oxyhydrogen torches
Rocket fuel, usually $H_2(l)$ in combination with $O_2(l)$
Fuel cells for generating electricity, in combination with $O_2(g)$

Hydrogen and the Environment—A Hydrogen Economy

As we contemplate the eventual decline of the world's supplies of fossil fuel hydrogen emerges as an attractive means of storing, transporting, and usin energy. For example, when an automobile engine burns hydrogen rather tha gasoline, its exhaust is essentially pollution free. The range of supersoni aircraft could be increased if they used liquid hydrogen as a fuel. A hypersoni airplane (the space plane) might also become possible. As we learned i Chapter 19, one method of using hydrogen that is already available combine H_2 and O_2 to form H_2O in an electrochemical fuel cell, which converts chemi cal energy directly to electricity. The subsequent conversion of electrica energy to mechanical energy (work) can be carried out much more efficientl than can the conversion of heat to mechanical energy.

The basic problems are in finding a cheap source of hydrogen and a effective means of storing it. One possibility is to use hydrogen made by th electrolysis of seawater. This possibility requires an abundant energy sourc however—perhaps nuclear fusion energy if it can be developed. Anothe alternative is the thermal decomposition of water. The problem here is tha even at 2000 °C, water is only about 1% decomposed. What is needed is a ther mochemical cycle, a series of reactions that have as their overall reactior $2\,H_2O(l) \longrightarrow 2\,H_2(g) + O_2(g)$. Ideally, no single reaction in the cycle woul require a very high temperature. Still another alternative being studie involves the use of solar energy to decompose water—photodecomposition.

Storage of gaseous hydrogen is difficult because of the bulk of the gas When liquefied, hydrogen occupies a much smaller volume, but because of it very low boiling point (−253 °C), the $H_2(l)$ must be stored at very low tem peratures. Also, hydrogen must be maintained out of contact with oxygen o air, with which it forms explosive mixtures. One approach may be to dissolv $H_2(g)$ in a metal or metal alloy, such as an iron–titanium alloy. The gas ca be released by mild heating. In an automobile, this storage system woul eplace the gasoline tank. The heat required to release hydrogen from the netal would come from the engine exhaust.

If the problems described here can be solved, not only could hydrogen be sed to supplant gasoline as a fuel for transportation but it could also replace atural gas for space heating. Because H_2 is a good reducing agent, it could eplace carbon (as coal or coke) in metallurgical processes and, of course, it vould be abundantly available for reaction with N_2 to produce NH_3 for the nanufacture of fertilizers. The combination of all these potential uses of ydrogen could lead to a fundamental change in our way of life and give rise o what is called a **hydrogen economy**.

MasteringCHEMISTRY www.masteringchemistry.com

Ozone plays a vital role in protecting life on Earth because it absorbs potentially harmful ultraviolet radiation and also helps to maintain a heat balance in the atmosphere. For a discussion of some ozone-producing and ozone-destroying reactions occurring in the atmosphere, and the impact made by human activities, go to the Focus On feature for Chapter 22, The Ozone Layer and Its Environmental Role, on the MasteringChemistry site.

Summary

2-1 Periodic Trends in Bonding—The bonding in he fluorides changes from ionic bonding to covalent bondng as we move from left to right across the periodic table. n the transition from ionic bonding to covalent bonding, ve pass through a group of elements that react with fluoine to give network covalent compounds. The bonding in he oxides exhibits a similar transition. The acid–base charcter of the oxides also changes as we move left to right cross the periodic table. Oxides of the metallic elements on the left of the periodic table) are generally basic, and xides of nonmetallic elements are generally acidic. Between these two extremes are the amphoteric oxides erived from some of the elements in groups 2 and 13.

2-2 Group 18: The Noble Gases—The **noble gases** group 18) form very few compounds. The chemistry of this roup is concerned mainly with compounds of xenon and he two most electronegative elements, F and O.

2-3 Group 17: The Halogens—The **halogens** group 17) are among the most reactive elements, forming ompounds with all elements in the periodic table excludng some of the noble gases. Electrode potential diagrams re a way to summarize the oxidation–reduction chemistry f an element, and are introduced for the oxidation states f chlorine. Often electrode potentials for reduction rocesses not specifically represented in a diagram can be btained by means of calculations based on the relationhip $\Delta_r G° = -zFE°$. The oxoacids and oxoanions of chloine are described in terms of the methods required to repare them, their acid–base properties, their strengths as xidizing or reducing agents, and their structures. In this roup we observe several differences between the first and igher members of a group of the periodic table. Thus, for example, we note the failure of fluorine to form stable oxoacids. The later members of the group form **interhalogens**, such as ICl, and **polyhalide ions**, such as I_3^-.

22-4 Group 16: The Oxygen Family—The oxygen family (group 16) includes two reactive nonmetals (O and S), two metalloids (Se and Te), and a metal (Po). Again note the difference in properties between the first and second member of this group. The ability to form strong π bonds with $2p$ orbitals by oxygen contrasts with the single bond chain-like structures favored by sulfur. Sulfur is extracted from underground sources by the **Frasch process**. Sulfur forms the important acids H_2SO_4 and H_2SO_3. The salt $NaHSO_4$ is an example of an **acid salt** since the anion HSO_4^- possesses an ionizable proton.

22-5 Group 15: The Nitrogen Family—The nitrogen family (group 15) shows the progression of properties from nonmetallic to metallic within a group. The chemistry of nitrogen and phosphorus is wide and diverse. The ammonium salts of nitric acid are used in fertilizers. Phosphate ions are also important in fertilizers and in living systems as ADP or ATP. Also, phosphates are implicated in the pollution process of **eutrophication**.

22-6 Hydrogen: A Unique Element—Hydrogen is a unique element that does not fit into any group. Hydrogen can occur as $H^+(aq)$ in aqueous solutions and as H^- in **hydrides,** for example, NaH(s). Hydrogen is a good reducing agent and can be used in **hydrogenation reactions** (equation 22.52). Hydrogen can be obtained from water by electrolysis and also by the **water gas** reaction (equation 22.49). Hydrogen gas is potentially a very promising nonpolluting means of storing, transporting, and using energy and could lead to the so-called **hydrogen economy**.

Integrative Example

For use in analytical chemistry, sodium thiosulfate solutions must be carefully prepared. In particular, the solutions must be kept from becoming acidic. In strongly acidic solutions, thiosulfate ion disproportionates into $SO_2(g)$ and $S_8(s)$.

Carey B. Van Loon

▲ **Decomposition of thiosulfate ion**
When an aqueous solution of $Na_2S_2O_3$ is acidified, the sulfur is in the colloidal state when first formed (right).

Show that the disproportionation of $S_2O_3^{2-}(aq)$ is spontaneous for standard-state conditions in acidic solution, but not in basic solution.

Analyze

Begin by writing the half-equations and an overall equation for the disproportionation reaction. Determine E°_{cell} for the reaction and thus whether the reaction is spontaneous for standard-state conditions in acidic solution. Then make a qualitative assessment of whether the reaction is likely to be more spontaneous or less spontaneous in basic solution.

Solve

Base the overall equation on the verbal description of the reaction.

Reduction:

$$4\,S_2O_3^{2-}(aq) + 24\,H^+(aq) + 16\,e^- \longrightarrow S_8(s) + 12\,H_2O(l)$$

Oxidation:

$$4\{S_2O_3^{2-}(aq) + H_2O(l) \longrightarrow 2\,SO_2(g) + 2\,H^+(aq) + 4\,e^-\}$$

Overall:

$$8\,S_2O_3^{2-}(aq) + 16\,H^+(aq) \longrightarrow S_8(s) + 8\,SO_2(g) + 8\,H_2O(l) \quad \textbf{(22.53)}$$

To determine E°_{cell} for the reaction (22.53), use data fror Figure 22-13. That figure gives an E° value for the reductio half-reaction (0.465 V) but no value for the oxidation. To obtain this missing E°, use additional data from Figure 22-1 together with the method of Example 22-1. That is, the sum of the half-equation

$$4\,SO_2(g) + 4\,H^+(aq) + 6\,e^- \longrightarrow S_4O_6^{2-}(aq) + 2\,H_2O($$

$$\Delta_r G^\circ = -6FE^\circ = -6F \times 0.507$$

and the half-equation

$$S_4O_6^{2-}(aq) + 2\,e^- \longrightarrow 2\,S_2O_3^{2-}(aq)$$

$$\Delta_r G^\circ = -2FE^\circ = -2F \times 0.080\ \text{V}$$

yields the desired new half-equation and its E° value.

$$4\,SO_2(g) + 4\,H^+(aq) + 8\,e^- \longrightarrow 2\,S_2O_3^{2-}(aq) + 2\,H_2O(l$$

$$\Delta_r G^\circ = -F[(6 \times 0.507) + (2 \times 0.080)]\ \text{V}$$

$$\Delta_r G^\circ = -8FE^\circ = -F(3.202)\ \text{V}$$

$$E^\circ = (3.202/8)\ \text{V} = 0.400\ \text{V}$$

Now we can calculate E°_{cell} for reaction (22.53).

$$E^\circ_{cell} = E^\circ(\text{reduction}) - E^\circ(\text{oxidation})$$

$$= 0.465\ \text{V} - 0.400\ \text{V} = 0.065\ \text{V}$$

The disproportionation is spontaneous for standard-state conditions in acidic solution.

Increasing $[OH^-]$, as would be the case in making the solution basic, means decreasing $[H^+]$. In fact $OH^- = 1$ M corresponds to $[H^+] = 1 \times 10^{-14}$ M. Because equation (22.53) has $H^+(aq)$ on the *left* side of the equation, a decrease in $[H^+]$ favors the *reverse* reaction (by Le Châtelier's principle). At some point before the solution becomes basic, the forward reaction is no longer spontaneous.

Assess

This calculation demonstrated in a qualitative way that $S_2O_3^{2-}(aq)$ is stable in basic solutions and spontaneously disproportionates in acidic solutions. To determine the pH at which the disproportionation become spontaneous, one can use the Nernst equation, as seen in Exercise 100.

PRACTICE EXAMPLE A: Use information from Figure 22-17 to decide whether the nitrite anion, NO_2^- disproportionates spontaneously in basic solution to NO_3^- and NO. Assume standard-state conditions.

PRACTICE EXAMPLE B: Does HNO_2 spontaneously disproportionate to NO_3^- and NO in acidic solution Assume standard-state conditions. [*Hint*: Use data from Figure 22-17.]

Exercises

Periodic Trends in Bonding and Acid–Base Character of Oxides

1. Give the formula of the stable fluoride formed by Li, Be, B, C, N, and O. For these fluorides, describe the variation in the bonding that occurs as we move from left to right across the period.
2. Fluorine is able to stabilize elements in very high oxidation states. For each of the elements Na, Mg, Al, Si, P, S, and Cl, give the formula of the highest oxidation-state of fluoride that is known to exist. Then, describe the variation in bonding that occurs as we move from left to right across the period.
3. The oxides of the phosphorus(III), antimony(III), and bismuth(III) are P_4O_6, Sb_4O_6, and Bi_2O_3. Only one of these oxides is amphoteric. Which one? Which of these oxides is most acidic? Which is most basic?
4. The oxides of the selenium(IV) and tellurium(IV) are SeO_2 and TeO_2. One of these oxides is amphoteric and one is acidic. Which is which?

The Noble Gases

5. A 55 L cylinder contains Ar at 145 atm and 26 °C. What minimum volume of air at STP must have been liquefied and distilled to produce this Ar? Air contains 0.934% Ar, by volume.
6. Some sources of natural gas contain 8% He by volume. How many liters of such a natural gas must be processed at STP to produce 5.00 g of He?
7. Use VSEPR theory to predict the probable geometric structures of **(a)** XeO_3; **(b)** XeO_4; **(c)** XeF_5^+.
8. Use VSEPR theory to predict the probable geometric structures of the molecules **(a)** O_2XeF_2; **(b)** O_3XeF_2; **(c)** $OXeF_4$.
9. Write a chemical equation for the hydrolysis of XeF_4 that yields XeO_3, Xe, O_2, and HF as products.
10. Write a chemical equation for the hydrolysis in alkaline solution of XeF_6 that yields XeO_6^{4-}, Xe, O_2, F^-, and H_2O as products.
11. Provide an explanation for the observation that helium, neon, and argon do not react directly with fluorine.
12. Provide an explanation for the inability of O_2 to react directly with xenon.

The Halogens

13. Freshly prepared solutions containing iodide ion are colorless, but over time they usually turn yellow. Describe a plausible chemical reaction (or reactions) to account for this observation.
14. Fluorine can be prepared by the reaction of hexafluoromanganate(IV) ion, MnF_6^{2-}, with antimony pentafluoride to produce manganese(IV) fluoride and SbF_6^-, followed by the disproportionation of manganese(IV) fluoride to manganese(III) fluoride and $F_2(g)$. Write chemical equations for these two reactions.
15. Make a general prediction about which of the halogen elements, F_2, Cl_2, Br_2, or I_2, displaces other halogens from a solution of halide ions. Which of the halogens is able to displace $O_2(g)$ from water? Which is able to displace $H_2(g)$ from water?
16. The following properties of astatine have been measured or estimated: **(a)** covalent radius; **(b)** ionic radius (At^-); **(c)** first ionization energy; **(d)** electron affinity; **(e)** electronegativity; **(f)** standard reduction potential. Based on periodic relationships and data in Table 22.4, what values would you expect for these properties?
17. The abundance of F^- in seawater is 1 g F^- per ton of seawater. Suppose that a commercially feasible method could be found to extract fluorine from seawater.
(a) What mass of F_2 could be obtained from 1 km^3 of seawater (d = 1.03 g cm^{-3})?
(b) Would the process resemble that for extracting bromine from seawater? Explain.
18. Fluorine is produced chiefly from fluorite, CaF_2. Fluorine can also be obtained as a by-product of the production of phosphate fertilizers, derived from phosphate rock $[3\,Ca_3(PO_4)_2 \cdot CaF_2]$. What is the maximum mass of fluorine that could be extracted as a by-product from 1.00×10^3 kg of phosphate rock?
19. Show by calculation whether the disproportionation of chlorine gas to chlorate and chloride ions will occur under standard-state conditions in an acidic solution.
20. Show by calculation whether the reaction
$2\,HOCl(aq) \longrightarrow HClO_2(aq) + H^+(aq) + Cl^-(aq)$
will go essentially to completion as written for standard-state conditions.
21. Predict the geometric structures of **(a)** $\underline{Br}F_3$; **(b)** $\underline{I}F_5$; **(c)** $Cl_3\underline{I}F^-$. (The central atom is underlined.)
22. Which of the following species has a linear structure: $\underline{Cl}F_2^+$, $\underline{I}BrF^-$, $\underline{O}Cl_2$, $\underline{Cl}F_3$, or $\underline{S}F_4$? (The central atom is underlined.) Do any two of these species have the same structure?
23. When iodine is added to an aqueous solution of iodide ion, the I_3^- ion is formed, according to the reaction below:

$$I_2(aq) + I^-(aq) \rightleftharpoons I_3^-(aq)$$

The equilibrium constant for the reaction above is $K = 7.7 \times 10^2$ at 25 °C.
(a) What is $E°$ for the reaction above?
(b) If a 0.0010 mol sample of I_2 is added to 1.0 L of 0.0050 M NaI(aq) at 25 °C, then what fraction of the I_2 remains unreacted at equilibrium?

24. The trichloride ion, Cl_3^-, is not very stable in aqueous solution. The equilibrium constant for the following dissociation reaction is 5.5 at 25 °C:

$$Cl_3^-(aq) \rightleftharpoons Cl^-(aq) + Cl_2(aq)$$

(a) Draw a Lewis structure for the Cl_3^- ion and pre dict the geometry.
(b) Calculate the equilibrium concentration of Cl_3^- i 0.0010 moles each of KCl and Cl_2 are dissolved i water at 25 °C to make 1.0 L of solution.

Oxygen

25. Each of the following compounds decomposes to produce $O_2(g)$ when heated: **(a)** $HgO(s)$; **(b)** $KClO_4(s)$. Write plausible equations for these reactions.

26. Ozone is a powerful oxidizing agent. Using ozone as the oxidant, write equations to represent oxidation of **(a)** $Br^-(aq)$ to $BrO^-(aq)$ (hypobromite); **(b)** $NO(g)$ to $NO_2(g)$; **(c)** $Fe^{2+}(aq)$ to $Fe^{3+}(aq)$ in acidic solution; **(d)** $Ag(s)$ to $AgO(s)$; **(e)** $[Fe(CN)_6]^{4-}$ to $[Fe(CN)_6]^{3-}$ in basic solution. [*Hint*: The half-equations for the reduction of O_3 in acidic and basic solutions are given in Table D.4 of Appendix D.]

27. *Without performing detailed calculations*, determine which of the following compounds has the greatest percent oxygen by mass: dinitrogen tetroxide, aluminum oxide, tetraphosphorus hexoxide, or carbon dioxide.

28. *Without performing detailed calculations*, determine which decomposition yields the most $O_2(g)$ **(a)** per mole and **(b)** per gram of substance.
(1) ammonium nitrate $\longrightarrow$ nitrogen + oxygen + water
(2) hydrogen peroxide $\longrightarrow$ oxygen + water
(3) potassium chlorate $\longrightarrow$ potassium chloride + oxygen

29. The natural abundance of O_3 in unpolluted air at ground level is about 0.04 parts per million (ppm) by volume. What is the approximate partial pressure of O_3 under these conditions, expressed in millimeters of mercury?

30. A typical concentration of O_3 in the ozone layer is 5×10^{12} O_3 molecules cm^{-3}. What is the partial pressure of O_3, expressed in millimeters of mercury, in that layer? Assume a temperature of 220 K.

31. Explain why the volumes of $H_2(g)$ and $O_2(g)$ obtained in the electrolysis of water are not the same.

32. In the electrolysis of a sample of water 22.83 mL of $O_2(g)$ was collected at 25.0 °C at an oxygen partial pressure of 736.7 mmHg. Determine the mass of water that was decomposed.

33. Hydrogen peroxide is a somewhat stronger acid than water. For the ionization

$$H_2O_2(aq) + H_2O(l) \longrightarrow H_3O^+(aq) + HO_2^-(aq)$$

$pK_a = 11.75$. Calculate the expected pH of a typical antiseptic solution that is 3.0% H_2O_2 by mass.

34. In water, $O^{2-}(aq)$ is a strong base. If 100.0 mg of $K_2O(s)$ is dissolved in 1.25 L of aqueous solution, what will be the pH of the solution?

35. The conversion of $O_2(g)$ to $O_3(g)$ can be accomplished in an electric discharge, $3\,O_2(g) \longrightarrow 2\,O_3(g)$. Use a bond dissociation energy of 498 kJ mol^{-1} for $O_2(g$ and data from Appendix D to calculate an averag oxygen-to-oxygen bond energy in $O_3(g)$.

36. Estimate the average bond energy in $O_3(g)$ from th structure on page 432 and data in Table 10.3. Compar this result with that of Exercise 35.

37. Use Lewis structures and other information to explai the observation that
(a) H_2S is a gas at room temperature, whereas H_2O i a liquid.
(b) O_3 is diamagnetic.

38. Use Lewis structures and other information to explai the observation that
(a) the oxygen-to-oxygen bond lengths in O_2, O_3 and H_2O_2 are 121, 128, and 148 pm, respectively.
(b) the oxygen-to-oxygen bond length of O_2 is 121 pm and for O_2^+ is 112 pm. Why is the bond length for O_2 so much shorter than for O_2?

39. Which of the following reactions are likely to go to completion or very nearly so?
(a) $H_2O_2(aq) + 2\,I^-(aq) + 2\,H^+(aq) \longrightarrow I_2(s) + 2\,H_2O(l$
(b) $O_2(g) + 2\,H_2O(l) + 4\,Cl^-(aq) \longrightarrow 2\,Cl_2(g) + 4\,OH^-(aq$
(c) $O_3(g) + Pb^{2+}(aq) + H_2O(l) \longrightarrow PbO_2(s) + 2\,H^+(aq) + O_2(g$
(d) $HO_2^-(aq) + 2\,Br^-(aq) + H_2O(l) \longrightarrow 3\,OH^-(aq) + Br_2(l$

40. Each of the following compounds produces $O_2(g$ when strongly heated: **(a)** $Hg(NO_3)_2(s)$; **(b)** $H_2O_2(aq)$ Write a plausible equation for the reaction that occur in each instance.

41. In the laboratory, small quantities of oxygen gas ca be prepared by heating potassium chlorate, $KClO_3(s)$ in the presence of $MnO_2(s)$, a catalyst. What volum of oxygen, measured at 25 °C and 101 kPa, is obtaine from the decomposition of 1.0 g $KClO_3(s)$? [*Hint:* Th other product of the reaction is KCl(s).]

42. Joseph Priestley, a British chemist, was credite with the discovering oxygen in 1774. In his experi ments, he generated oxygen gas by heating HgO(s) The other product of the decomposition reaction i Hg(l). What volume of wet $O_2(g)$ is obtained fron the decomposition of 1.0 g HgO(s), if the gas i collected over water at 25 °C and a barometric pres sure of 756 mmHg? The vapor pressure of water i 23.76 mmHg at 25 °C.

ulfur

3. Give an appropriate name to each of the following compounds: **(a)** MgS; **(b)** H_2S; **(c)** $Ca(HSO_3)_2$; **(d)** $Na_2S_2O_3$; **(e)** S_4N_4.

4. Give an appropriate formula for each of the following compounds: **(a)** calcium sulfate dihydrate; **(b)** hydrosulfuric acid; **(c)** sodium hydrogen sulfate; **(d)** disulfuric acid.

5. Give a specific example of a chemical equation that illustrates the
(a) reaction of a metal sulfide with HCl(aq);
(b) action of a *nonoxidizing* acid on a metal sulfite;
(c) oxidation of $SO_2(aq)$ to $SO_4^{2-}(aq)$ by $MnO_2(s)$ in acidic solution;
(d) disproportionation of $S_2O_3^{2-}$ in acidic solution.

6. Show how you would use elemental sulfur, chlorine gas, metallic sodium, water, and air to produce aqueous solutions containing **(a)** Na_2SO_3; **(b)** Na_2SO_4; **(c)** $Na_2S_2O_3$. [*Hint:* You will have to use information from other chapters as well as this one.]

7. Describe a chemical test you could use to determine whether a white solid is Na_2SO_4 or $Na_2S_2O_3$. Explain the basis of this test using a chemical equation or equations.

8. Explain why sulfur can occur naturally as sulfates, but not as sulfites.

49. $Mg(HSO_4)_2$ is a very efficient acid catalyst that is used for a variety of organic transformations. What is the pH of a 250 mL aqueous solution containing 16.5 g of $Mg(HSO_4)_2$? For H_2SO_4, $K_{a_2} = 1.1 \times 10^{-2}$.

50. What mass of Na_2SO_3 was present in a sample that required 26.50 mL of 0.0510 M $KMnO_4$ for its oxidation to Na_2SO_4 in an acidic solution? MnO_4^- is reduced to Mn^{2+}.

51. A 1.500 g sample of iron ore is dissolved, and the $Fe^{3+}(aq)$ is treated with excess KI. The liberated $I_2(aq)$ requires 15.25 mL of 0.100 M $Na_2S_2O_3$ for its titration. What is the mass percent of iron in the ore?

52. A 25.0 L sample of a natural gas, measured at 25 °C. and 740.0 Torr, is bubbled through $Pb^{2+}(aq)$, yielding 0.535 g of PbS(s). What mass of sulfur can be recovered per cubic meter of this natural gas?

53. What is the oxidation state of sulfur in the following compounds? **(a)** $MgSO_3$; **(b)** SCl_4; **(c)** $MgSO_4$; **(d)** S_2I_2; **(e)** S_4N_4.

54. What is the oxidation state of sulfur in the following compounds? **(a)** S_2Br_2; **(b)** SCl_2; **(c)** $Na_2S_2O_3$; **(d)** $(NH_4)_2S_4O_6$.

Nitrogen Family

5. Write balanced equations for the following important commercial reactions involving nitrogen and its compounds.
(a) the principal artificial method of fixing atmospheric N_2
(b) oxidation of ammonia to NO
(c) preparation of nitric acid from NO

6. When heated, each of the following substances decomposes to the products indicated. Write balanced equations for these reactions.
(a) $NH_4NO_3(s)$ to $N_2(g)$, $O_2(g)$, and $H_2O(g)$
(b) $NaNO_3(s)$ to sodium nitrite and oxygen gas
(c) $Pb(NO_3)_2(s)$ to lead(II) oxide, nitrogen dioxide, and oxygen

7. Sodium nitrite can be made by passing oxygen and nitrogen monoxide gases into an aqueous solution of sodium carbonate. Write a balanced equation for this reaction.

8. Concentrated $HNO_3(aq)$ used in laboratories is usually 15 M HNO_3 and has a density of 1.41 g mL^{-1}. What is the percent by mass of HNO_3 in this concentrated acid?

9. In 1968, before pollution controls were introduced, over 75 billion gallons of gasoline were used in the United States as a motor fuel. Assume an emission of oxides of nitrogen of 5 grams per vehicle mile and an average mileage of 15 miles per gallon of gasoline. How many kilograms of nitrogen oxides were released into the atmosphere in the United States in 1968?

0. One reaction that competes with reaction (22.41), the Ostwald process, is the reaction of gaseous ammonia and nitrogen monoxide to produce gaseous nitrogen and gaseous water. Use data from Appendix D to determine $\Delta_r H°$ for this reaction, per mole of ammonia consumed.

61. Use information from this chapter and previous chapters to write chemical equations to represent the following:
(a) equilibrium between nitrogen dioxide and dinitrogen tetroxide in the gaseous state
(b) the reduction of nitrous acid by $N_2H_5^+$ forming hydrazoic acid, followed by the reduction of additional nitrous acid by the hydrazoic acid, yielding nitrogen and dinitrogen monoxide
(c) the neutralization of $H_3PO_4(aq)$ to the second equivalence point by $NH_3(aq)$

62. Use information from this chapter and previous chapters to write plausible chemical equations to represent the following:
(a) the reaction of silver metal with $HNO_3(aq)$
(b) the complete combustion of the rocket fuel, unsymmetrical dimethylhydrazine, $(CH_3)_2NNH_2$
(c) the preparation of sodium triphosphate by heating a mixture of sodium dihydrogen phosphate and sodium hydrogen phosphate.

63. Draw plausible Lewis structures for
(a) monomethylhydrazine, $CH_3(NH)NH_2$;
(b) dimethylhydrazine, $(CH_3)_2NNH_2$;
(c) dinitrogen tetroxide, N_2O_4;
(d) phosphoric acid, H_3PO_4;
(e) nitryl chloride, $ClNO_2$ (N is the central atom).

64. Both nitramide and hyponitrous acid have the formula $H_2N_2O_2$. Hyponitrous acid is a weak diprotic acid; nitramide contains the amide group ($-NH_2$). Draw plausible Lewis structures for these two substances.

65. Supply an appropriate name for each of the following: **(a)** HPO_4^{2-}; **(b)** $Ca_2P_2O_7$; **(c)** $H_6P_4O_{13}$.

66. Write an appropriate formula for each of the following: **(a)** hydroxylamine; **(b)** calcium hydrogen phosphate; **(c)** lithium nitride.
67. Use Figure 22-17 to establish $E°$ for the reduction of N_2O_4 to NO in an acidic solution.
68. Use Figure 22-17 to establish $E°$ for the reduction of NO_3^- to NO_2^- in a basic solution.
69. All the group 15 elements form trifluorides, but nitrogen is the only group 15 element that does not form a pentafluoride.
(a) Suggest a reason why nitrogen does not form pentafluoride.
(b) The observed bond angle in NF_3 is approximatel 102.5 °C. Use VSEPR theory to rationalize the struc ture of the NF_3 molecule.
70. The structures of the NH_3 and NF_3 molecules are sim ilar, yet the dipole moment for the NH_3 molecule i rather large (1.47 debye) and that of the NF_3 molecul is rather small (0.24 debye). Provide an explanatio for this difference in the dipole moments.

Hydrogen

71. Use data from Table 7.2 (page 273) to calculate the standard enthalpies of combustion of the four alkane hydrocarbons listed there.
72. Based on the results of Exercise 71, which alkane evolves the greatest amount of heat upon combustion on **(a)** a *per mole* basis and **(b)** a *per gram* basis? Which is the most desirable alkane from the standpoint of reducing the emission of carbon dioxide to the atmosphere? Explain.
73. Write chemical equations for the following reactions:
(a) the displacement of $H_2(g)$ from HCl(aq) by Al(s)
(b) the re-forming of propane gas (C_3H_8) with steam
(c) the reduction of $MnO_2(s)$ to Mn(s) with $H_2(g)$
74. Write equations to show how to prepare $H_2(g)$ from each of the following substances: **(a)** H_2O; **(b)** HI(aq); **(c)** Mg(s); **(d)** CO(g). Use other common laboratory reactants as necessary, that is, water, acids or bases, metals, and so on.
75. $CaH_2(s)$ reacts with water to produce $Ca(OH)_2$ and $H_2(g)$. Ca(s) reacts with water to produce the same products. Na(s) reacts with water to form NaOH and $H_2(g)$. *Without doing detailed calculations,* determine **(a)** which of these reactions produces the most H_2 pe liter of water used, and **(b)** which solid—CaH_2, Ca, o Na—produces the most $H_2(g)$ *per gram* of the solid.
76. What volume of $H_2(g)$ at 25 °C and 752 mmHg i required to hydrogenate oleic acid, $C_{17}H_{33}COOH(l)$ to produce one mole of stearic acid, $C_{17}H_{35}COOH(s)$ Assume reaction (22.52) proceeds with a 95% yield.
77. *Without doing detailed calculations,* explain in which c the following materials you would expect to find th greatest mass percent of hydrogen: seawater, th atmosphere, natural gas (CH_4), ammonia.
78. How many grams of $CaH_2(s)$ are required to generat sufficient $H_2(g)$ to fill a 235 L weather observatio balloon at 722 mmHg and 19.7 °C?

$$CaH_2(s) + 2\,H_2O(l) \longrightarrow Ca(OH)_2(aq) + 2\,H_2(g)$$

79. The amide anion NH_2^- is a very strong base. On th basis of molecular orbital theory, would you expec NH_2^- to be linear or bent?
80. On the basis of molecular orbital theory, would yo expect NH_2^+ to be linear or bent?

Integrative and Advanced Exercises

81. The boiling points of oxygen and argon are −183 °C and −189 °C, respectively. Because the boiling points are so similar, argon obtained from the fractional distillation of liquid air is contaminated with oxygen. The following three-step procedure can be used to obtain pure argon from the oxygen-contaminated sample:
(1) Excess hydrogen is added to the mixture and then the mixture is ignited.
(2) The mixture from step (1) is then passed over hot copper(II) oxide.
(3) The mixture from step (2) is passed over a dehydrated zeolite material (see Chapter 21).
Explain the purpose of each step, writing chemical equations for any reactions that occur.
82. In 1988, G. J. Schrobilgen, professor of chemistry at McMaster University in Canada, reported the synthesis of an ionic compound, $[HCNKrF][AsF_6]$, which consists of $HCNKrF^+$ and AsF_6^- ions. In the $HCNKrF^+$ ion, the krypton is covalently bonded to both fluorine and nitrogen. Draw Lewis structures for these ions, and estimate the bond angles.
83. Suppose that no attempt is made to separate the $H_2(g)$ and $O_2(g)$ produced by the electrolysis of water. What volume of a H_2—O_2 mixture, saturated with $H_2O(g)$ and collected at 23 °C and 755 mmHg would be produced by electrolyzing 17.3 g water Assume that the water vapor pressure of the dilut electrolyte solution is 20.5 mmHg.
84. The photograph was taken after a few drops of a deep-purple acidic solution of $KMnO_4(aq)$ wer added to $NaNO_3(aq)$ (left) and to $NaNO_2(aq)$ (right) Explain the difference in the results shown.

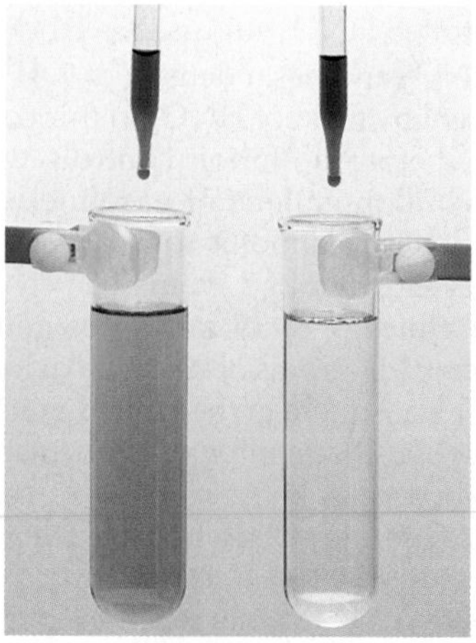

Tom Pantages

85. Zn can reduce NO_3^- to $NH_3(g)$ in basic solution. (The following equation is not balanced.)

$$NO_3^-(aq) + Zn(s) + OH^-(aq) + H_2O(l) \longrightarrow [Zn(OH)_4]^{2-}(aq) + NH_3(g)$$

The NH_3 can be neutralized with an excess of HCl(aq). Then, the unreacted HCl can be titrated with NaOH. In this way a quantitative determination of NO_3^- can be achieved. A 25.00 mL sample of nitrate solution was treated with zinc in basic solution. The $NH_3(g)$ was passed into 50.00 mL of 0.1500 M HCl. The excess HCl required 32.10 mL of 0.1000 M NaOH for its titration. What was the $[NO_3^-]$ in the original sample?

86. Oxygen atoms are an important constituent of the thermosphere, a layer of the atmosphere with temperatures up to 1500 K. Calculate the average translational kinetic energy of O atoms at 1500 K.

87. One reaction for the production of adipic acid, $HOOC(CH_2)_4COOH$, used in the manufacture of nylon, involves the oxidation of cyclohexanone, $C_6H_{10}O$, in a nitric acid solution. Assume that dinitrogen monoxide is also formed, and write a balanced equation for this reaction.

88. Despite the fact that it has the higher molecular mass, XeO_4 exists as a gas at 298 K, whereas XeO_3 is a solid. Give a plausible explanation for this observation.

89. The text mentions that ammonium perchlorate is an explosion hazard. Assuming that NH_4ClO_4 is the sole reactant in the explosion, write a plausible equation(s) to represent the reaction that occurs.

90. The following bond energies are given for 298 K: O_2, 498; N_2, 946; F_2, 159; Cl_2, 243; ClF, 251; OF (in OF_2), 213; ClO (in Cl_2O), 205; and NF (in NF_3), 280 kJ mol^{-1}. Calculate $\Delta_f H°$ at 298 K for **(a)** ClF(g); **(b)** $OF_2(g)$; **(c)** $Cl_2O(g)$; **(d)** $NF_3(g)$.

91. The standard electrode potential of fluorine cannot be measured directly because F_2 reacts with water, displacing O_2. Using thermodynamic data from Appendix D, obtain a value of $E°_{F_2/F^-}$.

92. Polonium is the only element known to crystallize in the simple cubic form. In this structure, the interatomic distance between a Po atom and each of its six nearest neighbors is 335 pm. Use this description of the crystal structure to estimate the density of polonium.

93. Refer to Figure 11-25 to arrange the following species in the expected order of increasing **(a)** bond length and **(b)** bond strength (energy): $O_2, O_2^+, O_2^-, O_2^{2-}$. State the basis of your expectation.

94. One reaction of a chlorofluorocarbon implicated in the destruction of stratospheric ozone is $CFCl_3 + h\nu \longrightarrow \cdot CFCl_2 + Cl\cdot$
(a) What is the energy of the photons $(h\nu)$ required to bring about this reaction, expressed in kilojoules per mole?
(b) What is the frequency and wavelength of the light necessary to produce the reaction? In what portion of the electromagnetic spectrum is this light found?

95. The composition of a phosphate mineral can be expressed as % P, % P_4O_{10}, or % BPL [bone phosphate of lime, $Ca_3(PO_4)_2$].
(a) Show that % P = 0.436 × (% P_4O_{10}) and % BPL = 2.185 × (% P_4O_{10}).
(b) What is the significance of a % BPL greater than 100?
(c) What is the % BPL of a typical phosphate rock?

96. Estimate the percent dissociation of $Cl_2(g)$ into Cl(g) at 1 atm total pressure and 1000 K. Use data from Appendix D and equations found elsewhere in this text, as necessary.

97. *Peroxonitrous acid* is an unstable intermediate formed in the oxidation of HNO_2 by H_2O_2. It has the same formula as *nitric acid,* HNO_3. Show how you would expect peroxonitrous and nitric acids to differ in structure.

98. The structure of $N(SiH_3)_3$ involves a planar arrangement of N and Si atoms, whereas that of the related compound $N(CH_3)_3$ has a pyramidal arrangement of N and C atoms. Propose bonding schemes for these molecules that are consistent with this observation.

99. In the extraction of bromine from seawater (reaction 22.3), seawater is first brought to a pH of 3.5 and then treated with $Cl_2(g)$. In practice, the pH of the seawater is adjusted with H_2SO_4, and the mass of chlorine used is 15% in excess of the theoretical. Assuming a seawater sample with an initial pH of 7.0, a density of 1.03 g cm^{-3}, and a bromine content of 70 ppm by mass, what masses of H_2SO_4 and Cl_2 would be used in the extraction of bromine from 1.00×10^3 L of seawater?

100. Refer to the Integrative Example on page 1082. Assume that the disproportionation of $S_2O_3^{2-}$ is no longer spontaneous when the partial pressure of $SO_2(g)$ above a solution with $[S_2O_3^{2-}] = 1$ M has dropped to 1×10^{-6} atm. Show that this condition is reached while the solution is still acidic.

101. The bond energies of Cl_2 and F_2 are 243 and 159 kJ mol^{-1}, respectively. Use these data to explain why XeF_2 is a much more stable compound than $XeCl_2$. [*Hint:* Recall that Xe exists as a monatomic gas.]

102. Write plausible half-equations and a balanced oxidation–reduction equation for the disproportionation of XeF_4 to Xe and XeO_3 in aqueous acidic solution. Xe and XeO_3 are produced in a 2:1 mole ratio, and $O_2(g)$ is also produced.

103. A handbook gives the value $E° = 0.174$ V for the reduction half-reaction $S + 2H^+ + 2e^- \longrightarrow H_2S(g)$. In Figure 22-13, the value given for the segment S—$H_2S(aq)$ is 0.144 V. Why are these $E°$ values different? Can both be correct?

104. The solubility of $Cl_2(g)$ in water is 6.4 g L^{-1} at 25 °C. Some of this chlorine is present as Cl_2, and some is found as HOCl or Cl^-. For the hydrolysis reaction

$$Cl_2(aq) + 2H_2O(l) \longrightarrow HOCl(aq) + H_3O^+(aq) + Cl^-(aq) \qquad K_c = 4.4 \times 10^{-4}$$

For a saturated solution of Cl_2 in water, calculate $[Cl_2]$, $[HOCl]$, $[H_3O^+]$, and $[Cl^-]$.

105. Not shown in Figure 22-17 are electrode potential data involving hydrazoic acid. Given that $E° = -3.09$ V for the reduction of HN_3 to N_2 in acidic solution, what is $E°$ for the reduction of HN_3 to NH_4^+ in acidic solution?

106. The heavier halogens (Cl, Br, and I) form compounds in which the central halogen atom, X, is bonded directly to oxygen and to fluorine. Several examples are known, including those with formulas of the type FXO_2, FXO_3, and F_3XO. The structures of these molecules are all consistent with the VSEPR model. Draw Lewis structures and predict the geometries of **(a)** chloryl fluoride, $FClO_2$; **(b)** perchloryl fluoride, $FClO_3$; **(c)** F_3ClO.

107. Draw Lewis structures for O_3 and SO_2. In what ways are the structures similar? In what ways do they differ?

108. Chemists have successfully synthesized the ionic compound $[N_5][SbF_6]$, which consists of N_5^+ and SbF_6^- ions. Draw Lewis structures for these ions an assign formal charges to the atoms in your structure Describe the structures of these ions. [*Hint:* The skele ton structure for N_5^+ is N—N—N—N—N an several resonance structures can be drawn.]

109. Refer to Figure 22-3 and then construct an enthalp diagram for forming $XeO_3(g)$ from Xe(g) and O_2(g For XeO_3, the average bond enthalpy is abou 36 kJ mol^{-1}, and for O_2, the bond enthalpy is 498 k mol^{-1}. What is $\Delta_f H°$ for $XeO_3(g)$? Does your resu support the observation that Xe(g) does not reac directly with $O_2(g)$ to form $XeO_3(g)$?

Feature Problems

110. Various thermochemical cycles are being explored as possible sources of $H_2(g)$. The object is to find a series of reactions that can be conducted at moderate temperatures (about 500 °C) and that results in the decomposition of water into H_2 and O_2. Show that the following series of reactions meets these requirements.

$$FeCl_2 + H_2O \longrightarrow Fe_3O_4 + HCl + H_2$$
$$Fe_3O_4 + HCl + Cl_2 \longrightarrow FeCl_3 + H_2O + O_2$$
$$FeCl_3 \longrightarrow FeCl_2 + Cl_2$$

111. The decomposition of aqueous hydrogen peroxide is catalyzed by Fe^{3+}(aq). A proposed mechanism for this catalysis involves two reactions. In the first reaction, Fe^{3+} is reduced by H_2O_2. In the second, the iron is oxidized back to its original form, while hydrogen peroxide is reduced. Write an equation for the overall reaction and show that the overall reaction is indeed spontaneous. What are the minimum and maximum values of $E°$ for a catalyst to function in this way? Which of the following should be able to catalyze the decomposition of hydrogen peroxide by the mechanism outlined here: **(a)** Cu^{2+}; **(b)** Br_2; **(c)** Al^{3+}; **(d)** Au^{3+}? In the reaction between iodic acid and hydrogen peroxide in the presence of starch indicator, the color of the reaction mixture oscillates between deep blue and colorless. What is the basis of these changes in color? Will this oscillation of color continue indefinitely? Explain. [*Hint:* Refer to Exercise 97 in Chapter 20, particularly to equation (c).]

112. Both in this chapter and in Chapter 19, we have stressed the relationship between $E°$ values and thermodynamic properties. We can use this relationship to add some missing features to an electrode potential diagram. For example, note that $ClO_2(g)$, which has Cl in the oxidation state +4, is not included in Figure 22-4. Using data from Figure 22-4 and Appendix D, add $ClO_2(g)$ to the electrode potential diagram for acidic solutions, and indicate the $E°$ values that link $ClO_2(g)$ to $ClO_3^-(aq)$ and to $HClO_2(aq)$.

113. A yellow-brown saturated solution of I_2 in water (top layer in (a)) is initially brought into contact with colorless CCl_4(l) (bottom layer in (a)). Once the system reaches equilibrium (b) we observe that I_2 i considerably more soluble in CCl_4(l) (violet, bottom layer) than it is in H_2O(l) (colorless, top layer). The concentration of I_2 in its saturated aqueous solution is 1.33×10^{-3} M, and the equilibrium achieved when I_2 distributes itself between H_2O and CCl_4 is

Carey B. Van Loon

(a) (b)

$$I_2(aq) \rightleftharpoons I_2(CCl_4) \qquad K_c = 85.5$$

(a) A 10.0 mL sample of saturated I_2(aq) is shaken with 10.0 mL CCl_4. After equilibrium is established the two liquid layers are separated. How many milligrams of I_2 will be in the aqueous layer?
(b) If the 10.0 mL of aqueous layer from part (a) is extracted with a second 10.0 mL portion of CCl_4 how many milligrams of I_2 will remain in the aqueous layer when equilibrium is reestablished?
(c) If the 10.0 mL sample of saturated I_2(aq) in part (a had originally been extracted with 20.0 mL CCl_4 would the mass of I_2 remaining in the aqueous layer have been less than, equal to, or greater than that in part (b)? Explain.

114. The so-called pyroanions, $X_2O_7^{n-}$, form a series of structurally similar polyatomic anions for the elements Si, P, and S.
(a) Draw the Lewis structures of these anions, and predict the geometry of the anions. What is the maximum number of atoms that can lie in a plane?
(b) Each pyroanion in part (a) corresponds to a pyroacid, $X_2O_7H_n$. Compare each pyroacid to the acid

containing only one atom of the element in its maximum oxidation state. From this comparison, suggest a strategy for the preparation of these pyroacids.
(c) What is the chlorine analogue of the pyroanions? For which acid is this species the anhydride?

115. A description of bonding in XeF_2 based on the valence bond model requires the 5*d* orbitals of Xe. A more satisfactory description uses a molecular orbital approach involving three-center bonds. Assume that bonding involves the $5p_z$ orbital of Xe and the $2p_z$ orbitals of the two F atoms. These three atomic orbitals combine to give three molecular orbitals: one bonding, one nonbonding, and one antibonding. Recall that for bonding to occur, atomic orbitals with the same phase must overlap to form bonding molecular orbitals (see Chapter 11).
(a) Construct diagrams similar to Figure 11-33 to indicate the overlap of the three atomic orbitals in forming the three molecular orbitals. Assume that the order of energy of the molecular orbitals is bonding MO < nonbonding MO < antibonding MO.
(b) Sketch a molecular orbital energy-level diagram, and assign the appropriate number of electrons from fluorine and xenon to the molecular orbitals. What is the bond order?
(c) With the aid of VSEPR theory, show that this molecular orbital description based on three-center bonds works well for XeF_4 but not for XeF_6.

116. The sketch is a portion of the phase diagram for the element sulfur. The transition between solid orthorhombic (S_α) and solid monoclinic (S_β) sulfur, in the presence of sulfur vapor, is at 95.3 °C. The triple point involving monoclinic sulfur, liquid sulfur, and sulfur vapor is at 119 °C.
(a) How would you modify the phase diagram to represent the melting of orthorhombic sulfur that is sometimes observed at 113 °C? [*Hint:* What would the phase diagram look like if the monoclinic sulfur did not form?]
(b) Account for the observation that if a sample of rhombic sulfur is melted at 113 °C and then heated, the liquid sulfur freezes at 119 °C upon cooling.

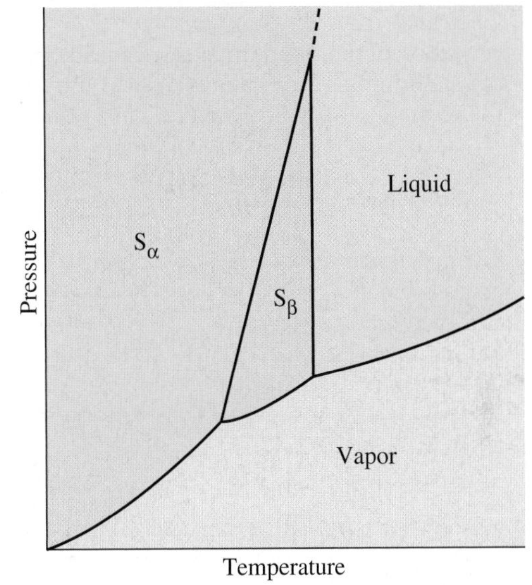

Self-Assessment Exercises

117. In your own words, define the following terms: **(a)** polyhalide ion; **(b)** polyphosphate; **(c)** interhalogen; **(d)** disproportionation.

118. Briefly describe each of the following terms: **(a)** Frasch process; **(b)** water gas reactions; **(c)** eutrophication; **(d)** electrode potential diagram.

119. Explain the important distinctions between each pair of terms: **(a)** acid salt and acid anhydride; **(b)** azide and nitride; **(c)** white phosphorus and red phosphorus; **(d)** ionic hydride and metallic hydride.

120. Which of the following can oxidize Br^- to Br_2 in solution? **(a)** $I_2(aq)$; **(b)** $Cl_2(aq)$; **(c)** $H_2(g)$; **(d)** $Cl^-(aq)$; **(e)** $I_3^-(aq)$.

121. All of the following compounds yield $O_2(g)$ when heated to about 1000 K except **(a)** $KClO_3$; **(b)** $KClO_4$; **(c)** N_2O; **(d)** $CaCO_3$; **(e)** $Pb(NO_3)_2$.

122. All of the following substances are bases except for **(a)** H_2NNH_2; **(b)** NH_3; **(c)** HN_3; **(d)** NH_2OH; **(e)** CH_3NH_2.

123. The best reducing agent of the following substances is **(a)** H_2S; **(b)** O_3; **(c)** H_2SO_4; **(d)** NaF; **(e)** H_2O.

124. Of the following substances, the one that is unimportant is the production of fertilizers is **(a)** NH_3; **(b)** phosphate rock; **(c)** HNO_3; **(d)** Na_2CO_3; **(e)** H_2SO_4.

125. All of the following have a tetrahedral shape except **(a)** SO_4^{2-}; **(b)** XeF_4; **(c)** CCl_4; **(d)** XeO_4; **(e)** NH_4^+.

126. Two of the following, through a reaction occurring in a weakly acidic solution, produce the same gaseous product. They are **(a)** $CaH_2(s)$; **(b)** $Na_2O_2(s)$; **(c)** NaOH(s); **(d)** Al(s); **(e)** $NaHCO_3(s)$; **(f)** $N_2H_4(l)$.

127. Write a plausible chemical equation to represent the reaction of **(a)** $Cl_2(g)$ with cold NaOH(aq); **(b)** NaI(s) with hot H_2SO_4(concd aq); **(c)** $Cl_2(g)$ with $KI_3(aq)$; **(d)** NaBr(s) with hot H_3PO_4 (concd aq); **(e)** $NaHSO_3(aq)$ with $MnO_4^-(aq)$ in dilute $H_2SO_4(aq)$.

128. Give a practical laboratory method that you might use to produce small quantities of the following gases and comment on any difficulties that might arise: **(a)** O_2; **(b)** NO; **(c)** H_2; **(d)** NH_3; **(e)** CO_2.

129. Complete and balance equations for these reactions.
(a) $LiH(s) + H_2O(l) \longrightarrow$
(b) $C(s) + H_2O(g) \xrightarrow{\Delta}$
(c) $NO_2(g) + H_2O(l) \longrightarrow$

130. If Br^- and I^- occur together in an aqueous solution, I^- can be oxidized to IO_3^- with an excess of $Cl_2(aq)$. Simultaneously, Br^- is oxidized to Br_2, which is extracted with $CS_2(l)$. Write chemical equations for the reactions that occur.

131. Suppose that the sulfur present in seawater as SO_4^{2-} (2650 mg L^{-1}) could be recovered as elemental sulfur. If this sulfur were then converted to H_2SO_4, how many cubic kilometers of seawater would have to be processed to yield the average U.S. annual consumption of about 45 million tons of H_2SO_4?

132. Although relatively rare, all of the following compounds exist. Based on what you know about related compounds (for example, from the periodic table), propose a plausible name or formula for each compound: **(a)** silver(I) astatide; **(b)** Na_4XeO_6; **(c)** magnesium polonide; **(d)** H_2TeO_3; **(e)** potassium thioselenate; **(f)** $KAtO_4$.

133. A portion of the standard electrode potential diagram of selenium is given below. What is the $E°$ value for the reduction of H_2SeO_3 to H_2Se in 1 M acid?

$$SeO_4^{2-} \xrightarrow{1.15\ V} H_2SeO_3 \xrightarrow{0.74\ V} Se \xrightarrow{-0.35\ V} H_2Se$$

$$H_2SeO_3 \xrightarrow{(?)} H_2Se$$

134. What is the acid anhydride of **(a)** H_2SO_4; **(b)** H_2SO **(c)** $HClO_4$; **(d)** HIO_3?

135. Use the following electrode potential diagram fc basic solutions to classify each of the statemen below as true or false. Assume standard conditions

$$SO_4^{2-} \xrightarrow{-0.936\ V} SO_3^{2-} \xrightarrow{-0.576\ V}$$

$$S_2O_3^{2-} \xrightarrow{-0.74\ V} S \xrightarrow{-0.476\ V} S^{2}$$

(a) Sulfate (SO_4^{2-}) is a stronger oxidant than thic sulfate ($S_2O_3^{2-}$) in basic solution.
(b) S^{2-} can be used as a reducing agent in basi solutions.
(c) $S_2O_3^{2-}$ is stable with respect to disproportiona tion to SO_3^{2-} and S in basic solution.

The Transition Elements

23

CONTENTS

LEARNING OBJECTIVES

23.1 Describe in general terms the properties of the first-row transition metals.

23.2 Explain the two main processes of extractive metallurgy: pyrometallurgy and hydrometallurgy.

23.3 Explain what *pig iron* is and how it is transformed into steel.

23.4 Discuss the unique properties of the first-row transition metals from scandium to manganese in the context of their oxidation state.

23.5 Describe the electron configurations for the iron triad elements in their most common oxidation states.

23.6 Discuss the reason that Cu, Ag, and Au are termed the *coinage metals*.

23.7 Describe some uses of the group 12 metals and some important compounds they form.

23.8 Discuss why lanthanides are so difficult to separate from each other, based on their reactivity and active orbitals.

23.9 Identify the formula and describe the structure of the yttrium, barium, copper, and oxygen high-temperature superconductor.

Whiskers of rutile, TiO_2, in quartz (left) and titanium ore (right), a source of rutile. Titanium metal, obtained from rutile, is used in industry because it has low density and high strength. Pure TiO_2 is a bright white pigment used in paints and specialty papers.

There are more transition elements—members of the *d* and *f* blocks—than main-group elements. Although some of the transition elements are rare and of limited use, others play crucial roles in many aspects of modern life. All the transition elements are metals; among them are both the chief structural metal, iron (Fe), and important alloying metals in the manufacture of steel (V, Cr, Mn, Co, Ni, Mo, W). The best electrical conductors (Ag, Cu) are transition metals. The compounds of several transition metals (Ti, Fe, Cr) are the primary constituents of paint pigments. Compounds of silver (Ag) provide the essential material for photographic film. Specialized materials for modern applications, such as color television screens, use compounds of the *f*-block elements (lanthanide oxides). Nine of the transition metals are essential elements for living organisms.

The chemistry of the *d*-block and *f*-block elements has both theoretical an practical significance. These elements and their compounds provide insigh into fundamental aspects of bonding, magnetism, and reaction chemistry.

23-1 General Properties

The high melting points, good electrical conductivity, and moderate-to-extrem hardness of the transition elements result from the ready availability of elec trons and orbitals for metallic bonding (see the Chapter 11 appendix a www.masteringchemistry.com). Some similarities are found among the transi tion elements, but each element also has some unique properties that make i and its compounds useful in particular ways. Table 23.1 lists properties of th fourth-period transition elements—the first transition series.

Atomic (Metallic) Radii

In Table 23.1, with the exception of Sc and Ti, we find little variation among th atomic radii across the first transition series. The chief difference in atomi structure between successive elements involves one unit of positive charge o the nucleus and one electron in an orbital of an *inner* electron shell. This is not a major difference and does not cause much of a change in atomic radius, espe cially in the middle of the series.

When an element of the first transition series is compared with elements o the second and third series within the same group, important difference appear. Consider the members of group 6—Cr, Mo, and W. As we migh expect, the atomic radius of Mo is larger than that of Cr; but contrary to ou expectation, the atomic radius of W is the same as that of Mo, not larger. In th aufbau process, 18 electrons are added in progressing from Cr to Mo, and al of them enter *s*, *p*, and *d* subshells. Between Mo and W, however, 32 electron must be added, and 14 of them enter the 4*f* subshell. Electrons in an *f* subshel are not very effective in screening outer-shell electrons from the nucleus. As a result, the outer-shell electrons are held more tightly by the nucleus than we

TABLE 23.1 Selected Properties of Elements of the First Transition Series

	Sc	Ti	V	Cr	Mn	Fe	Co	Ni	Cu	Zn
Atomic number	21	22	23	24	25	26	27	28	29	30
Electron config.[a]	$3d^14s^2$	$3d^24s^2$	$3d^34s^2$	$3d^54s^1$	$3d^54s^2$	$3d^64s^2$	$3d^74s^2$	$3d^84s^2$	$3d^{10}4s^1$	$3d^{10}4s^2$
Metallic radius, pm	161	145	132	125	124	124	125	125	128	133
Ioniz. energy, kJ mol^{-1}										
First	631	658	650	653	717	759	758	737	745	906
Second	1235	1310	1414	1592	1509	1561	1646	1753	1958	1733
Third	2389	2653	2828	2987	3248	2957	3232	3393	3554	3833
$E°$, V[b]	−2.03	−1.63	−1.13	−0.90	−1.18	−0.440	−0.277	−0.257	+0.340	−0.763
Common positive oxidation states[c]	3	2, 3, 4	2, 3, 4, 5	2, 3, 6	2, 3, 4, 7	2, 3, 6	2, 3	2, 3	1, 2	2
mp, °C	1397	1672	1710	1900	1244	1530	1495	1455	1083	420
Density, g cm^{-3}	3.00	4.50	6.11	7.14	7.43	7.87	8.90	8.91	8.95	7.14
Hardness[d]	—	—	—	9.0	5.0	4.5	—	—	2.8	2.5
Electrical conductivity[e]	3	4	6	12	1	16	25	23	93	27

[a]Each atom has an argon inner-core configuration.
[b]For the reduction process, $M^{2+}(aq) + 2e^- \longrightarrow M(s)$ [except for scandium, where the ion is $Sc^{3+}(aq)$].
[c]The most important oxidation states are printed in red.
[d]Hardness values are on the Mohs scale (see Table 21.2).
[e]Electrical conductivity compared with an arbitrarily assigned value of 100 for silver.

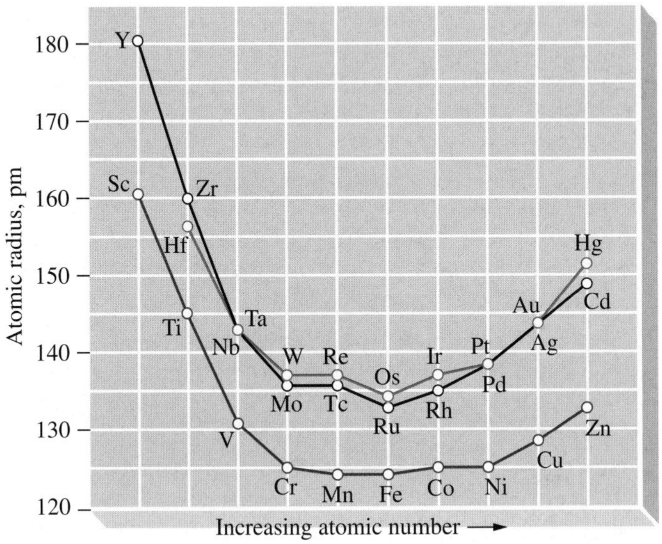

◀ FIGURE 23-1
Atomic radii of the *d*-block elements
The radii of the fourth-period transition elements (blue) are smaller than those of the corresponding group members in the succeeding periods. This trend is not seen between the fifth-period (black) and sixth-period (red) members, illustrating the contraction in atomic radii associated with the lanthanide series.

would otherwise expect. Atomic radii do not increase. In fact, in the series of elements in which the $4f$ subshell is filled, atomic radii decrease somewhat. This phenomenon occurs in the lanthanide series ($Z = 58$ to 71) and is called the **lanthanide contraction**. The lanthanide contraction is made more apparent in the graphs in Figure 23-1.

Electron Configurations and Oxidation States

The elements of the first transition series have electron configurations with the following characteristics:

- an inner core of electrons in the argon configuration
- two electrons in the $4s$ orbital for eight members and one $4s$ electron for the remaining two (Cr and Cu)
- a number of $3d$ electrons, ranging from one in Sc to ten in Cu and Zn

As we have seen for some of the main-group elements in Chapters 21 and 22, an element may display multiple oxidation states. Often, however, one particular oxidation state is the most common for an element. Ti atoms, with the electron configuration $[Ar]3d^24s^2$, tend to use all four electrons beyond the argon core in compound formation and display the oxidation state +4. It is also possible, however, for Ti atoms to use fewer electrons, as through the loss of the $4s^2$ electrons to form the ion Ti^{2+}. With Ti, then, we note two features: (1) several possible oxidation states, as shown in Figure 23-2, and (2) a maximum oxidation state corresponding to the group number, 4. These two features continue with V, Cr, and Mn, for which the maximum oxidation states are +5, +6 and +7, respectively. A shift in behavior occurs in groups 8–12, however. Thus, although Fe, Co, and Ni can all exist in more than one oxidation state, they do not display the wide variety found in the earlier members of the first transition series. Nor do they exhibit a maximum oxidation state corresponding to their group number. In crossing the first transition series, the nuclear charge, number of d electrons, and energy requirement for the successive ionization of d electrons increase. Involvement of a large number of d electrons in bond formation becomes increasingly unfavorable energetically, and only the lower oxidation states are commonly encountered for these later elements of the first transition series.

Although the transition elements display a variety of oxidation states, they differ in the ease with which these oxidation states can be attained and in their stabilities. The stability of an oxidation state for a given transition metal depends on a number of factors—other atoms to which the transition metal atom is bonded; whether the compound is in solid form or in solution; and the pH of the

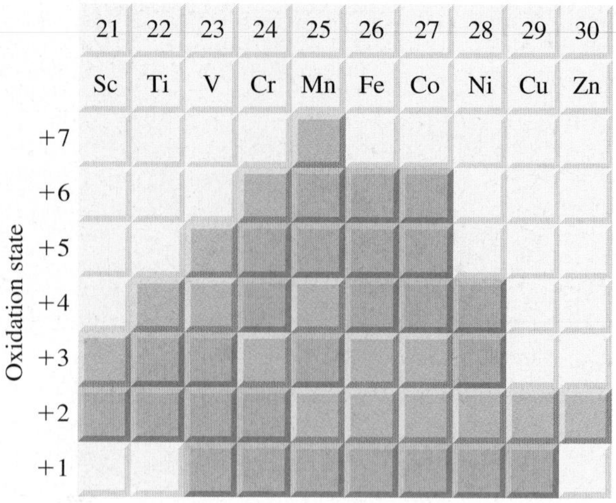

▶ FIGURE 23-2
Positive oxidation states of the elements of the first transition series
Common oxidation states are shown in red and less common ones in gray. Some oxidation states are rather rare, and a zero or negative oxidation state is occasionally found, chiefly in transition metal complexes. For example, the oxidation state of Cr is −2 in $Na_2[Cr(CO)_5]$, −1 in $Na_2[Cr_2(CO)_{10}]$, and 0 in $Cr(CO)_6$.

solution. For example, $TiCl_2$ is a well-characterized compound as a solid, but the Ti^{2+} ion is oxidized to Ti^{3+} by dissolved oxygen in aqueous solutions and even by the water itself. Conversely, Co^{3+}(aq) readily oxidizes water to O_2(g) and is itself reduced to Co^{2+}(aq). Co^{3+}(aq) can be stabilized, however, in certain complex ions. Generally speaking, higher oxidation states in transition metals are stabilized when oxide or fluoride ions are bound to the metal. In Figure 23-2 and elsewhere, the term *common* oxidation state refers to an oxidation state often found in aqueous solution.

Another feature of the transition metals is the progressively increasing stability of higher oxidation states in descending a group of the periodic table, the reverse of the trend often seen for main-group elements. Consider Cr, Mo and W in group 6, all of which can exhibit oxidation states ranging from +6 to −2. The number of compounds with Cr in the +6 oxidation state is rather limited, whereas those of Mo and W abound. Chromium is encountered in the oxidation states +5 and +4 mostly in unstable intermediates, whereas Mo and W exhibit a rich chemistry in these states. The most stable oxidation state of chromium is +3. Although it is a strong reducing agent, Cr^{2+}(aq) nevertheless is readily obtainable, whereas Mo and W are not obtainable as the simple +2 cation. This trend favoring lower oxidation states for the first group member and higher oxidation states for the later members is also found in other groups of transition metals. For instance, although Fe does not exhibit an oxidation state corresponding to the group number, Os does form the stable oxide OsO_4 with Os in the +8 oxidation state.

Ionization Energies and Electrode Potentials

Ionization energies are fairly constant across the first transition series. Values of the first ionization energies are about the same as for the group 2 metals. Standard electrode potentials gradually increase in value across the series. With the exception of the oxidation of Cu to Cu^{2+}, however, all these elements are more readily oxidized than hydrogen. This means these metals reduce H^+(aq) to H_2(g). Additional comments on electrode potentials, some supported by electrode potential diagrams, are found throughout the chapter.

Ionic and Covalent Compounds

We tend to think of metals as forming ionic compounds with nonmetals. This is certainly the case with group 1 and most group 2 metal compounds. However, some metal compounds have significant covalent character; $BeCl_2$ and $AlCl_3$ (Al_2Cl_6), for example, are molecular compounds. Transition metal compounds display both ionic and covalent character. In general, compounds with the transition metal in lower oxidation states are essentially ionic, while those in igher oxidation states have covalent character. As an example, MnO is a green onic solid with a melting point of 1785 °C, whereas Mn_2O_7 is a dark red, oily, nolecular liquid that boils at room temperature and is highly explosive. nother feature of ionic compounds of the transition metals is that the metal toms often occur in polyatomic cations or anions rather than as the simple nonatomic ion. Some common examples are VO_2^+, MnO_4^-, and $Cr_2O_7^{2-}$.

◀ The oxidation state of manganese is +7 in Mn_2O_7. The bonding in this compound is not ionic, because an ion with a charge of +7 would have a very high charge density and would be strongly polarizing. Mn_2O_7 is a molecular compound with covalent bonds between Mn and O atoms.

Catalytic Activity

An unusual ability to adsorb gaseous species makes some transition metals, uch as Ni and Pt, good heterogeneous catalysts. The possibility of multiple xidation states seems to account for the ability of some transition metal ions o serve as catalysts in certain oxidation–reduction reactions. In still other ypes of catalysis, complex-ion formation may play an important role. As we aw in a limited way in Chapter 18 and will explore more fully in Chapter 24, omplex-ion formation is a particularly distinctive feature of transition metal hemistry.

Catalysis is an essential aspect of about 90% of all chemical manufacturing rocesses, and the transition metals are often the key elements in the catalysts ised. For example, Ni is used in the hydrogenation of oils (page 1079); Pt, Pd, nd Rh are used in catalytic converters in automobiles (page 958); Fe_3O_4 is the nain component of the catalyst used in the synthesis of ammonia (Focus On 5-1, www.masteringchemistry.com) and V_2O_5 is used in the conversion of $SO_2(g)$ to $SO_3(g)$ in the manufacture of sulfuric acid (page 1061).

Transition metal catalysts are used in both homogeneous and heterogeneous atalysis. In homogeneous catalysis (page 959), the reactants, products, and catalyst are all in the same phase (often liquid or gas) and the transition metal is art of a compound or a complex. In homogeneous catalysis, the transition netal atoms or ions serve as electron banks that lend out electrons at the appropriate time or store them for later use. In heterogeneous catalysis page 960), the catalyst is in a different phase from the reactants or the products, nd, typically, the catalyst provides a surface on which the reaction occurs. The ydrogenation of oils (page 1079) makes use of heterogeneous catalysis. The letails concerning how a catalyst functions in a hydrogenation reaction depend ot only on the metal used but also on the experimental conditions (for example, the amounts of reactants used and the temperature). In Figure 23-3, the vents that are believed to occur in the hydrogenation of C_2H_4 are illustrated. A C_2H_4 molecule is adsorbed onto the metal surface, with electron density being ransferred from the π bond in C_2H_4 to metal atoms in the surface (a Lewis cid–base reaction). Hydrogen molecules may also be adsorbed onto the surface

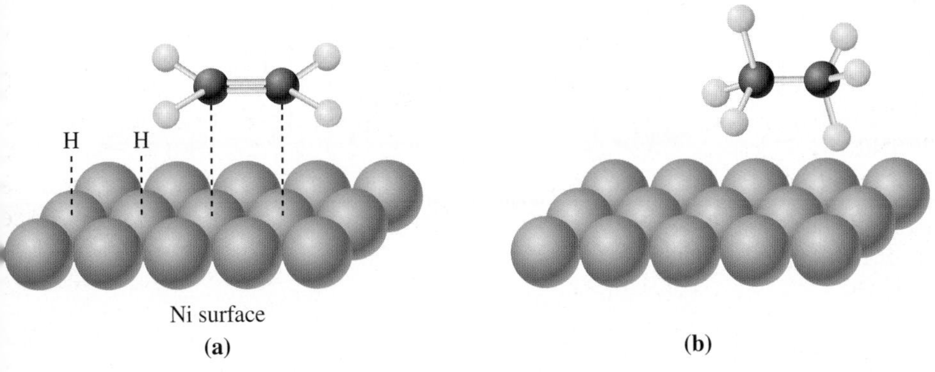

▲ FIGURE 23-3
Schematic representation of the metal-catalyzed hydrogenation of C_2H_4
(a) H_2 and C_2H_4 molecules adsorb onto the metal surface, causing weakening of the H—H bond and the π bond in C_2H_4. **(b)** H atoms become bonded to carbon atoms, converting C_2H_4 to C_2H_6, which desorbs from the metal surface. The exact details of the mechanism are a matter of debate.

near the C_2H_4 molecule, causing a weakening of the H—H bond. Hydrogen atoms are then transferred to the carbon atoms in the C_2H_4 molecule, producing C_2H_6 molecules, which detach from the metal surface.

▲ Color-enhanced image of magnetic domains in a ferromagnetic garnet film.

Color and Magnetism

As we will explain in Section 24-6, the five *d* orbitals in a transition metal atom (or ion) do not all have the same energy when the atom is part of a compound. Electronic transitions that occur within the *d* orbitals impart color to transition compounds and their solutions. Consequently, transition metal compounds and their solutions exhibit a wide variety of colors.

Because most transition elements have partially filled *d* subshells, many transition metals and their compounds are paramagnetic—that is, they have unpaired electrons. This description certainly fits Fe, Co, and Ni, but these three metals are unique among the elements in displaying a special magnetic property: the ability to be made into permanent magnets, a property known as **ferromagnetism**. A key feature of ferromagnetism is that in the solid state the metal atoms are thought to be grouped together into small regions—called *domains*—containing rather large numbers of atoms. Instead of the individual magnetic moments of the atoms within a domain being randomly oriented, all the magnetic moments are directed in the same way. In an unmagnetized piece of iron, the domains are oriented in several directions and their magnetic effects cancel. When the metal is placed in a magnetic field, however, the domains line up and a strong resultant magnetic effect is produced. This alignment of domains may actually involve the growth of domains with favorable orientations at the expense of those with unfavorable orientations (rather like a recrystallization of the material). The ordering of domains can persist when the object is removed from the magnetic field, and thus permanent magnetism results. Paramagnetism and ferromagnetism are compared in Figure 23-4.

The key factors in ferromagnetism are that (1) the atoms involved have unpaired electrons (a property possessed by many atoms), and (2) interatomic distances are of just the right magnitude to make possible the ordering of atoms into domains. If atoms are too large, interactions among them are too weak to produce this ordering. With small atoms, the tendency is for atoms to pair and their magnetic moments to cancel. This critical factor of atomic size is just met in Fe, Co, and Ni. It is possible, however, to prepare *alloys* of other metals in which this condition is also met. Some examples are Al–Cu–Mn, Ag–Al–Mn, and Bi–Mn.

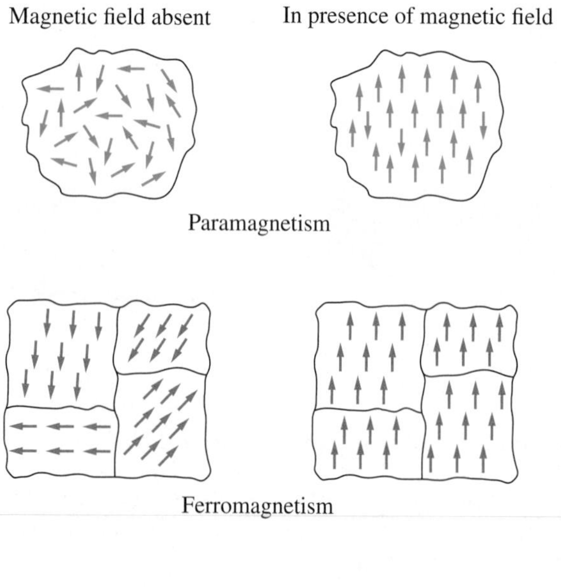

▶ FIGURE 23-4
Ferromagnetism and paramagnetism compared
In a paramagnetic material, the effect of a magnetic field is to align the magnetic moments of the individual atoms. In a ferromagnetic material, the magnetic moments are aligned within domains even in the absence of a magnetic field, but the direction of the alignment varies from one domain to another. The effect of the magnetic field is to change the orientation of these varied alignments into a single direction—the direction of the magnetic field.

:omparison of Transition and Main-Group Elements

Vith the main-group elements, the s and p orbitals of the outermost electron hell are the most important in determining the nature of the chemical bonding hat occurs. Participation in bonding by d orbitals is nonexistent for second-eriod elements and for group 1 and 2 metals. With the transition elements, d rbitals are as important as s and p orbitals. Most of the observed behavioral dif-erences between the transition and main-group elements—multiple versus ingle oxidation states, complex-ion formation, color, magnetic properties, and atalytic activity—can be traced to the orbitals that are most involved in bond ormation.

23-2 Principles of Extractive Metallurgy

Many of the transition elements have important uses related to their metallic properties—iron for its structural strength and copper for its excellent electri-al conductivity, for example. Unlike the more chemically reactive metals of groups 1 and 2 and aluminum in group 3, which are produced mainly by nodern methods of electrolysis, the transition metals are obtained by proce-lures developed over many centuries.

The term *metallurgy* describes the general study of metals. **Extractive metal-urgy** describes the winning of metals from their ores. There is no single nethod of extractive metallurgy, but a few basic operations generally apply. Let us illustrate them with the extractive metallurgy of zinc.

:oncentration In mining operations, the desired mineral from which a metal s to be extracted often constitutes only a small percentage (or occasionally just a fraction of a percent) of the material mined. It is necessary to separate the lesired ore from waste rock before proceeding with other metallurgical oper-ations. One useful method, *flotation*, is described in Figure 23-5.

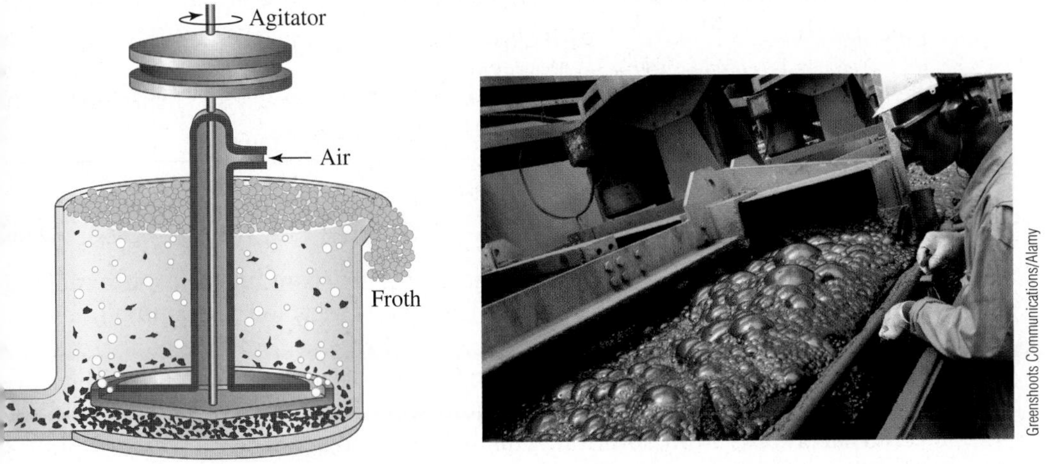

(a) (b)

▲ FIGURE 23-5
Concentration of an ore by flotation
(a) Powdered ore is suspended in water in a large vat, together with suitable additives, and the mixture is agitated with air. Particles of ore become attached to air bubbles, rise to the top of the vat, and are collected in the overflow froth. Particles of undesired waste rock (gangue) fall to the bottom. **(b)** The froth formed in the flotation process.

▶ Smelting is the combination of roasting and the addition of a chemical reducing agent to decompose the ore, driving off other elements as gases or slag and leaving just the metal behind.

Roasting An ore is roasted (heated to a high temperature) to convert a metal compound to its oxide, which can then be reduced. For zinc, the commercially important ores are $ZnCO_3$ (smithsonite) and ZnS (sphalerite). $ZnCO_3(s)$, like the carbonates of the group 2 metals, decomposes to ZnO(s) and $CO_2(g)$ when it is strongly heated. When strongly heated in air, ZnS(s) reacts with $O_2(g)$ producing ZnO(s) and $SO_2(g)$. In modern smelting operations, $SO_2(g)$ is converted to sulfuric acid rather than being vented to the atmosphere.

$$ZnCO_3(s) \xrightarrow{\Delta} ZnO(s) + CO_2(g) \quad (23.1)$$

$$2\,ZnS(s) + 3\,O_2(g) \xrightarrow{\Delta} 2\,ZnO(s) + 2\,SO_2(g) \quad (23.2)$$

Reduction Because it is inexpensive and easy to handle, carbon, in the form of coke or powdered coal, is used as the reducing agent whenever possible. Several reactions occur simultaneously in which both C(s) and CO(g) act as reducing agents. The reduction of ZnO is carried out at about 1100 °C, a temperature above the boiling point of zinc. The zinc is obtained as a vapor and condensed to the liquid.

$$ZnO(s) + C(s) \xrightarrow{\Delta} Zn(g) + CO(g) \quad (23.3)$$

$$ZnO(s) + CO(g) \xrightarrow{\Delta} Zn(g) + CO_2(g) \quad (23.4)$$

Refining The metal produced by chemical reduction is usually not pure enough for its intended uses. Impurities must be removed; that is, the metal must be refined. The refining process chosen depends on the nature of the impurities. The impurities in zinc are mostly Cd and Pb, which can be removed by the fractional distillation of liquid zinc.

Most of the zinc produced worldwide, however, is refined electrolytically usually in a process that combines reduction and refining. ZnO from the roasting step is dissolved in $H_2SO_4(aq)$. This is represented by the ionic equation

$$ZnO(s) + 2\,H^+(aq) + SO_4^{2-}(aq) \longrightarrow Zn^{2+}(aq) + SO_4^{2-}(aq) + H_2O(l) \quad (23.5)$$

Powdered Zn is added to the solution to displace less active metals, such as Cd. Then the solution is electrolyzed. The electrode reactions are

Cathode: $$Zn^{2+}(aq) + 2\,e^- \longrightarrow Zn(s)$$

Anode: $$H_2O(l) \longrightarrow \tfrac{1}{2}O_2(g) + 2\,H^+(aq) + 2\,e^-$$

Unchanged: $$SO_4^{2-}(aq) \longrightarrow SO_4^{2-}(aq)$$

Overall: $$Zn^{2+}(aq) + SO_4^{2-}(aq) + H_2O(l) \longrightarrow Zn(s) + 2\,H^+(aq) + SO_4^{2-}(aq) + \tfrac{1}{2}O_2(g) \quad (23.6)$$

▶ The sulfate anion is a spectator ion in this reaction and could be canceled.

Note that in the overall electrolysis reaction, Zn^{2+} is reduced to pure metallic zinc and sulfuric acid is regenerated. The acid is recycled in reaction (23.5).

Zone Refining In discussing freezing-point depression (Section 14-8), we assumed that a solute is *soluble* in a liquid solvent and *insoluble* in the solid solvent that freezes from solution. This behavior suggests a particularly simple way to purify a solid: Melt the solid and then refreeze a portion of it. Impurities remain in the liquid phase, and the solid that freezes is pure. In practice, the method is not quite so simple because the solid that freezes is wet with unfrozen liquid and thereby retains some impurities. Also, one or more solutes (impurities) might be slightly soluble in the solid solvent. In any case, the impurities do distribute themselves between the solid and liquid, concentrating in the liquid phase. If the solid that freezes from a liquid is remelted and the molten material refrozen, the solid obtained in the second freezing is purer than that in the first. Repeating the melting and refreezing procedure hundreds of times produces a very pure solid product.

EXAMPLE 23-1 Writing Chemical Equations for Metallurgical Processes

Write chemical equations to represent the **(a)** roasting of galena, PbS; **(b)** reduction of $Cu_2O(s)$ with charcoal as a reducing agent; **(c)** deposition of pure silver from an aqueous solution of Ag^+.

Analyze

When answering these questions we have to make sure we understand the meaning of the terms *roasting*, *reduction*, and *deposition*; these terms have all been discussed in the text.

Solve

(a) We expect this process to be essentially the same as reaction (23.2).

$$2\,PbS(s) + 3\,O_2(g) \xrightarrow{\Delta} 2\,PbO(s) + 2\,SO_2(g)$$

(b) The simplest possible equation is

$$Cu_2O(s) + C(s) \xrightarrow{\Delta} 2\,Cu(l) + CO(g)$$

(c) The reduction half-reaction is

$$Ag^+(aq) + e^- \longrightarrow Ag(s)$$

Assess

Roasting converts a metal sulfide to a metal oxide. When carbon is used as a reducing agent, CO(g) is produced, not $CO_2(g)$. The deposition of Ag(s) involves a reduction half-reaction. The accompanying oxidation half-reaction is not specified and neither is it specified whether this is an electrolysis process or whether silver is displaced by a more active metal, but these distinctions are not important in answering the question.

PRACTICE EXAMPLE A: Write plausible chemical equations to represent the **(a)** roasting of Cu_2S; **(b)** reduction of WO_3 with $H_2(g)$; **(c)** thermal decomposition of HgO to its elements.

PRACTICE EXAMPLE B: Write chemical equations to represent the **(a)** reduction of Cr_2O_3 to chromium with silicon as the reducing agent; **(b)** conversion of $Co(OH)_3(s)$ to $Co_2O_3(s)$ by roasting; **(c)** production of pure $MnO_2(s)$ from $MnSO_4(aq)$ at the anode in an electrolysis cell. [*Hint:* A few simple products are not specifically mentioned; propose plausible ones.]

The purification procedure just described implies that the melting and refreezing is done in batches, but in practice it is done *continuously*. In the method known as **zone refining**, a cylindrical rod of material is alternately melted and refrozen as a series of heating coils passes along the rod (Fig. 23-6). Impurities concentrate in the molten zones, and the portions of the rod behind

Photography by Sol and Seymour Mednick

◀ FIGURE 23-6
Zone refining
As a heating coil moves up the rod of material, melting occurs. Impurities concentrate in the molten zone. The portion of the rod below the molten zone is purer than the portion in or above the molten zone. With each successive passage of the heating coil, the rod becomes purer.

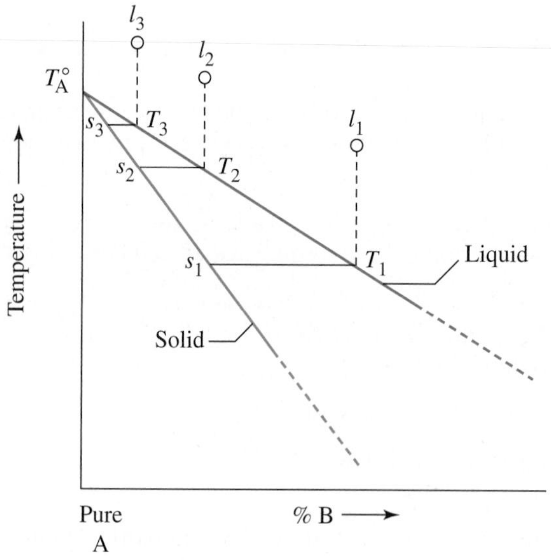

▲ FIGURE 23-7
The principle of zone refining
The red line shows the freezing points of solutions of impurity B in substance A. The blue line gives the composition of the solid that freezes from these solutions. In some cases, the blue line is nearly coincident with the temperature axis. When a solution of composition l_1 is cooled to temperature T_1, it freezes to produce a solid of composition s_1. If a small quantity of this solid is removed from the solution and melted, it produces a new liquid, l_2. The freezing point of l_2 is T_2, and the composition of the solid freezing from the solution is s_2. When removed from the solution, a small portion of this solid produces liquid l_3 and so on. With each melting/freezing cycle, the melting point increases and the point representing the composition of the solid moves closer to pure A. In zone refining the melting/freezing cycles are conducted continuously, not in batches as described here.

these zones are somewhat purer than the portions in front of the zones. Eventually, impurities are swept to the end of the rod, which is cut off. The principle of zone refining is shown graphically in Figure 23-7. This process is capable of producing materials in which the impurity levels are as low as 10 parts per billion (ppb), a common requirement of substances used in semiconductors.

Thermodynamics of Extractive Metallurgy

It is interesting to think of the reduction of zinc oxide by carbon, as shown in reaction (23.3), as a competition between zinc and carbon for O atoms. Zinc has them initially in ZnO, and carbon acquires them in forming CO. To establish the conditions under which carbon will reduce zinc oxide to zinc, we start by comparing the relative tendencies for zinc and carbon to undergo oxidation. We can assess these tendencies through Gibbs energy changes.

$$\text{(a)}\quad 2\,C(s) + O_2(g) \longrightarrow 2\,CO(g) \qquad \Delta_r G^\circ_{(a)}$$
$$\text{(b)}\quad 2\,Zn(s) + O_2(g) \longrightarrow 2\,ZnO(s) \qquad \Delta_r G^\circ_{(b)}$$

To determine if the reduction of zinc oxide to zinc by carbon is a spontaneous reaction, we need the value of $\Delta_r G^\circ$ for the overall reaction, represented below by reversing equation (b) and adding it to equation (a).

$$\text{(a)}\qquad 2\,C(s) + O_2(g) \longrightarrow 2\,CO(g) \qquad \Delta_r G^\circ_{(a)}$$
$$-\text{(b)}\qquad 2\,ZnO(s) \longrightarrow 2\,Zn(s) + O_2(g) \qquad -\Delta_r G^\circ_{(b)}$$
$$\textit{Overall:}\quad 2\,ZnO(s) + 2\,C(s) \longrightarrow 2\,Zn(s) + 2\,CO(g) \qquad \Delta_r G^\circ = \Delta_r G^\circ_{(a)} - \Delta_r G^\circ_{(b)}$$

KEEP IN MIND
that a nonspontaneous process can sometimes be achieved by *coupling* with a spontaneous process. In the reduction of ZnO with C, the nonspontaneous $2\,ZnO \longrightarrow 2\,Zn + O_2$ is coupled with the spontaneous $2\,C + O_2 \longrightarrow 2\,CO$.

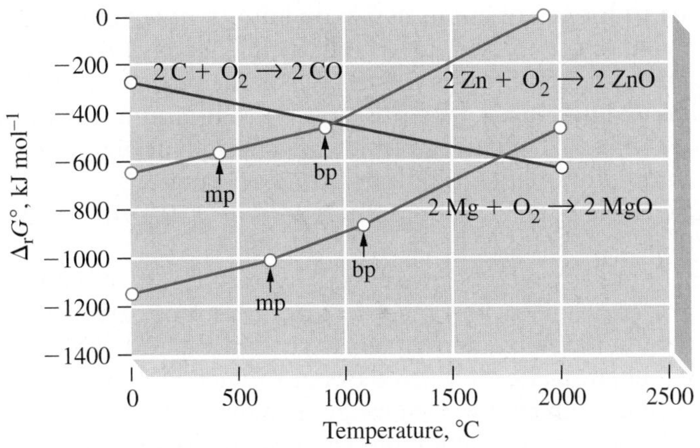

◀ FIGURE 23-8
$\Delta_r G°$ as a function of temperature for some reactions of extractive metallurgy
The points on the lines marked by arrows indicate the melting points and boiling points of Zn and Mg. At these points, the states of matter in which the metal exists change from (s) to (l) and from (l) to (g), respectively.

Ve still need some numerical data to complete our assessment. In Figure 23-8, ie blue line gives $\Delta_r G°_{(a)}$ as a function of temperature, and the top red line ives $\Delta_r G°_{(b)}$.

Figure 23-8 shows that at low temperatures, $\Delta_r G°_{(b)}$ is much more negative ıan $\Delta_r G°_{(a)}$. This makes $\Delta_r G°$ for the overall reaction positive and the reaction *onspontaneous*. At high temperatures, the situation is reversed: $\Delta_r G°_{(a)}$ is more egative than $\Delta_r G°_{(b)}$, and the overall reaction is *spontaneous*. The switchover om nonspontaneous to spontaneous occurs at the point of intersection of the lue line and the red line—about 950 °C. There, $\Delta_r G°$ for the overall reaction *zero*.

When we make a similar assessment for the reduction $2\,MgO(s) + C(s) \longrightarrow 2\,Mg(g) + 2\,CO(g)$, we conclude that the reaction does not ecome spontaneous until a temperature in excess of 1700 °C is reached. This an exceedingly high temperature at which to carry out a chemical reaction, nd it is not used in the metallurgy of magnesium.

lternative Methods in Extractive Metallurgy Some common variations of he methods previously discussed are worth mentioning. First, many ores ontain several metals, and it is not always necessary to separate them. For xample, a major use of vanadium, chromium, and manganese is in making lloys with iron. Obtaining each metal by itself is not commercially important. 'hus, the principal chromium ore *chromite*, $Fe(CrO_2)_2$, can be reduced to give n alloy of Fe and Cr called *ferrochrome*. Ferrochrome may be added directly to ron, together with other metals, to produce one type of steel. Vanadium and nanganese can be isolated as the oxides V_2O_5 and MnO_2, respectively. When ron-containing compounds are added to these oxides and the mixtures are educed, ferrovanadium and ferromanganese alloys form.

23-1 ARE YOU WONDERING?

Why is the slope of the blue line in Figure 23-8 negative, whereas the other slopes are positive?

Recall from Chapter 13 that $\Delta_r G° = \Delta_r H° - T\Delta_r S°$. If $\Delta_r H°$ does not change appreciably with temperature, then the temperature variation of $\Delta_r G°$ is determined primarily by the $-T\Delta_r S°$ term. We expect $\Delta_r S°$ to be negative for the reaction $2\,Zn(s) + O_2(g) \longrightarrow 2\,ZnO(s)$, because a mole of gas is lost. The term $-T\Delta_r S°$ is *positive*, and $\Delta_r G°$ *increases* with temperature. Conversely, in the reaction $2\,C(s) + O_2(g) \longrightarrow 2\,CO(g)$, an additional mole of gas forms, resulting in a positive value of $\Delta_r S°$. As a consequence, $-T\Delta_r S°$ is *negative* and $\Delta_r G°$ *decreases* with temperature.

Wah Chang

▲ Vacuum-distilled metallic titanium sponge produced by the Kroll process.

Extensive production of titanium was an important development of the latt half of the twentieth century, spurred first by the needs of the military and the by the aircraft industry. Steel is unsuitable as the structural metal for aircra because it has a high density ($d = 7.8\ \mathrm{g\ cm^{-3}}$). Aluminum has the advantage o a low density ($d = 2.70\ \mathrm{g\ cm^{-3}}$), but it loses strength at high temperatures. Fo certain aircraft components, titanium is a good alternative to aluminum an steel because its density is moderately low ($d = 4.50\ \mathrm{g\ cm^{-3}}$) and it does no lose strength at high temperatures.

Titanium metal cannot be produced by reduction of TiO_2 with carbo because the metal and carbon react to form titanium carbides. Also, at hig temperatures the metal reacts with air to form TiO_2 and TiN. The metallurg of titanium, then, must be conducted out of contact with air and with an activ metal rather than with carbon as a reducing agent.

The first step in the production of Ti is the conversion of *rutile* ore (TiO_2) t $TiCl_4$ by reaction with carbon and $Cl_2(g)$.

$$TiO_2(s) + 2\,C(s) + 2\,Cl_2(g) \xrightarrow{800\ ^\circ C} TiCl_4(g) + 2\,CO(g) \qquad (23.7)$$

The purified $TiCl_4$ is next reduced to Ti with a good reducing agent. The *Kro process* uses Mg.

$$TiCl_4(g) + 2\,Mg(l) \xrightarrow{1000\ ^\circ C} Ti(s) + 2\,MgCl_2(l) \qquad (23.8)$$

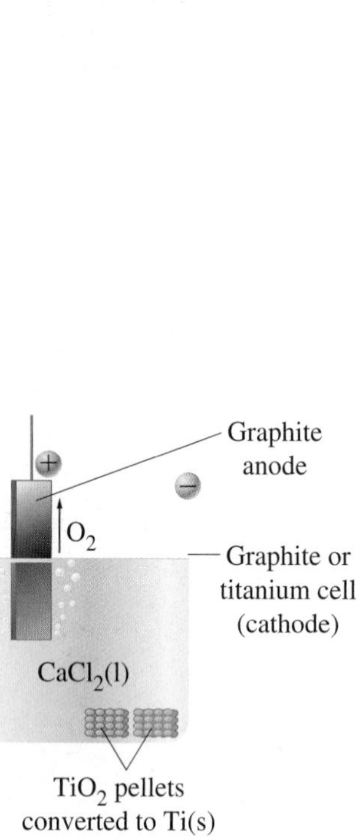

▲ Electrolytic production of Ti(s) from $TiO_2(s)$.

The $MgCl_2(l)$ is removed and electrolyzed to produce Cl_2 and Mg, which ar recycled in reactions (23.7) and (23.8), respectively. The Ti is obtained as a sin tered (fused) mass called *titanium sponge*. This sponge must be subjected t further treatment and alloying with other metals before it can be used.

The Kroll process is slow; it takes a week to produce a few tons of Ti. It is als demanding in health and safety terms because it requires high-temperatur vacuum distillation to remove the Mg and $MgCl_2$ from the titanium. Recentl an electrolytic process has been suggested for the production of Ti from rutile Porous pellets of TiO_2 are placed at the cathode of an electrolytic cell contain ing molten calcium chloride. The pellets dissolve in the electrolyte and oxid ions (O^{2-}) are discharged as oxygen at a graphite anode. The Ti(IV) is reduce at the cathode, which is the vessel containing the electrolytic cell and is made o either graphite or titanium. The titanium metal is obtained as sponge. Thi method, devised by Derek Fray, George Chen, and Tom Farthing in the Unite Kingdom, is being developed as a commercial process for the production o Ti(s) at a substantially lower cost than the Kroll process.

Metallurgy of Copper The extraction of copper from its ores (generally sul fides) is rather complicated, chiefly because the copper ores usually contai iron sulfides. The scheme of extractive metallurgy previously discussed pro duces copper contaminated with iron. For some metals, such as V, Cr, and Mn contamination with iron is not a problem because the metals are mostly use in the manufacture of steel. Copper, however, is prized commercially for th properties of the pure metal. To avoid contamination with iron, severa changes to the usual metallurgical methods are necessary.

Concentration of copper is done by flotation, and roasting converts iron sul fides to iron oxides. The copper remains as the sulfide if the temperature is kep below 800 °C. Reduction of the roasted ore in a furnace at 1400 °C causes th material to melt and separate into two layers. The bottom layer, called *coppe matte*, consists chiefly of the molten sulfides of copper and iron. The top layer i a silicate slag formed by the reaction of oxides of Fe, Ca, and Al with SiO (which typically is present in the ore or can be added). For example,

Nordroden/Shutterstock

▲ Slag formed during the smelting of copper ore.

$$FeO(s) + SiO_2(s) \xrightarrow{\Delta} FeSiO_3(l) \qquad (23.9)$$

A process called *conversion* is carried out in another furnace, where air i blown through the molten copper matte. First, the remaining iron sulfide is con verted to the oxide, followed by formation of slag [$FeSiO_3(l)$]. The slag i poured off, and air is again blown through the furnace. The following reactions occur and yield a product that is about 98%–99% Cu:

$$2\,Cu_2S(l) + 3\,O_2(g) \xrightarrow{\Delta} 2\,Cu_2O(l) + 2\,SO_2(g) \qquad \textbf{(23.10)}$$

$$2\,Cu_2O(l) + Cu_2S(l) \xrightarrow{\Delta} 6\,Cu(l) + SO_2(g) \qquad \textbf{(23.11)}$$

The product of reaction (23.11) is called *blister copper* because frozen bubbles of $SO_2(g)$ are present. Blister copper can be used where high purity is not required (as in plumbing).

Refining blister copper to obtain high-purity copper is done electrolytically by the method outlined on page 905. High-purity copper is essential in electrical applications.

Pyrometallurgical Processes The metallurgical method based on roasting an ore, followed by reduction of the oxide to the metal, is called **pyrometallurgy**, the prefix *pyro-* suggesting that high temperatures are involved. Some of the characteristics of pyrometallurgy are as follows:

- large quantities of waste materials produced in concentrating low-grade ores
- high energy consumption to maintain high temperatures necessary for roasting and reduction of ores
- gaseous emissions that must be controlled, such as $SO_2(g)$ in roasting

Many of the metallurgical processes described earlier fall into this category.

Hydrometallurgical Processes In **hydrometallurgy**, the materials handled are water and aqueous solutions at moderate temperatures rather than dry solids at high temperatures. Generally, three steps are involved in hydrometallurgy:

1. *Leaching:* Metal ions are extracted (leached) from the ore by a liquid. Leaching agents include water, acids, bases, and salt solutions. Oxidation–reduction reactions may also be involved.
2. *Purification and concentration:* Impurities are separated, and the solution produced by leaching may be made more concentrated. Methods include the adsorption of impurities on the surface of activated charcoal, ion exchange, and the evaporation of water.
3. *Precipitation:* The desired metal ions are either precipitated in an ionic solid or reduced to the free metal, often electrolytically.

Hydrometallurgy has long been used in obtaining silver and gold from natural sources. A typical gold ore currently being processed in the United States has only about 10 g Au per metric ton of ore. The leaching step in gold processing is known as *cyanidation*. The process is based on the following reaction:

$$4\,Au(s) + 8\,CN^-(aq) + O_2(g) + 2\,H_2O(l) \longrightarrow 4[Au(CN)_2]^-(aq) + 4\,OH^-(aq) \qquad \textbf{(23.12)}$$

The $[Au(CN)_2]^-(aq)$ is then filtered and concentrated. This is followed by displacement of Au(s) from solution by an active metal, such as zinc.

$$2[Au(CN)_2]^-(aq) + Zn(s) \longrightarrow 2\,Au(s) + [Zn(CN)_4]^{2-}(aq) \qquad \textbf{(23.13)}$$

In one hydrometallurgical process for zinc, a zinc sulfide ore is leached with a sulfuric acid solution at 150 °C and an oxygen pressure of about 7 atm. The overall reaction is

$$ZnS(s) + H_2SO_4(aq) + \tfrac{1}{2}O_2(g) \longrightarrow ZnSO_4(aq) + S(s) + H_2O(l) \qquad \textbf{(23.14)}$$

In this process, there is no $SO_2(g)$ emission. Also, mercury impurities in the ZnS ore are retained in the leaching solution rather than being emitted with $SO_2(g)$ as in the traditional roasting process. Following the leaching process,

$ZnSO_4(aq)$ is electrolyzed to produce pure Zn at the cathode, and $H_2SO_4(aq)$ is regenerated; see reaction (23.6). The $H_2SO_4(aq)$ is recycled into the leaching operation.

23-1 CONCEPT ASSESSMENT

The text describes the production of pure zinc from zinc sulfide ore both by a pyrometallurgical and by a hydrometallurgical process. Which aspects of the two processes are different and which are similar?

23-3 Metallurgy of Iron and Steel

▶ In the United States, slightly more than half of all iron and steel production comes from recycled iron and steel.

Iron is the most widely used metal from Earth's crust, and for this reason, we use this section to explore the metallurgy of iron and its principal alloy—steel—somewhat more fully. A type of steel called *wootz steel* was first produced in India about 3000 years ago; that same steel became famous in ancient times as Damascus steel, prized for making swords because of its suppleness and ability to hold a cutting edge. Many technological advances have been made since ancient times. These include introduction of the blast furnace around A.D. 1300, the Bessemer converter in 1856, the open-hearth furnace in the 1860s, and the basic oxygen furnace in the 1950s. A true understanding of the iron- and steelmaking processes has developed only within the past few decades, however. This understanding is based on concepts of thermodynamics, equilibrium, and kinetics.

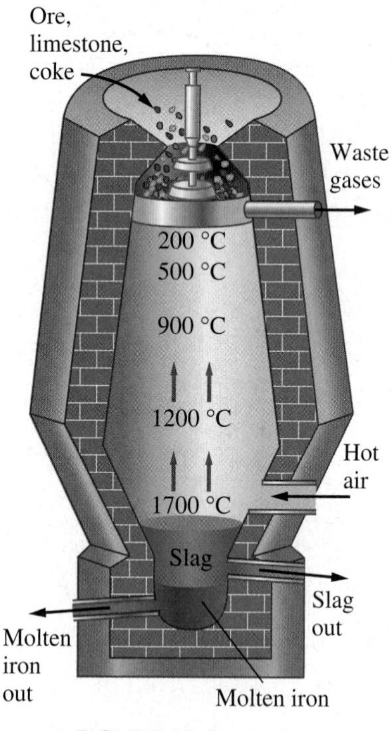

▲ FIGURE 23-9
Typical blast furnace
Iron ore, coke, and limestone are added at the top of the furnace, and hot air is introduced through the bottom. Maximum temperatures are attained near the bottom of the furnace where molten iron and slag are drained off. The principal reactions occurring in the blast furnace are outlined in Table 23.2.

Pig Iron

The reactions that occur in a blast furnace are complex. A highly simplified representation of the reduction of iron ore to impure iron is

$$Fe_2O_3(s) + 3\,CO(g) \longrightarrow 2\,Fe(l) + 3\,CO_2(g) \qquad \textbf{(23.15)}$$

A more complete description of the blast furnace reactions, including the removal of impurities as slag, is given in Table 23.2. Approximate temperatures are given for these reactions so that you can key them to regions of the blast furnace pictured in Figure 23-9.

TABLE 23.2 Some Blast Furnace Reactions

Formation of gaseous reducing agents CO(g) and $H_2(g)$:
$C + H_2O \longrightarrow CO + H_2$ (>600 °C)
$C + CO_2 \longrightarrow 2\,CO$ (1700 °C)
$2\,C + O_2 \longrightarrow 2\,CO$ (1700 °C)
Reduction of iron oxide:
$3\,CO + Fe_2O_3 \longrightarrow 2\,Fe + 3\,CO_2$ (900 °C)
$3\,H_2 + Fe_2O_3 \longrightarrow 2\,Fe + 3\,H_2O$ (900 °C)
Slag formation to remove impurities from ore:
$CaCO_3 \longrightarrow CaO + CO_2$ (800–900 °C)
$CaO + SiO_2 \longrightarrow CaSiO_3(l)$ (1200 °C)
$6\,CaO + P_4O_{10} \longrightarrow 2\,Ca_3(PO_4)_2(l)$ (1200 °C)
Impurity formation in the iron:
$MnO + C \longrightarrow Mn + CO$ (1400 °C)
$SiO_2 + 2\,C \longrightarrow Si + 2\,CO$ (1400 °C)
$P_4O_{10} + 10\,C \longrightarrow 4\,P + 10\,CO$ (1400 °C)

The blast furnace charge—that is, the solid reactants—consists of iron ore, oke, a slag-forming flux, and perhaps some scrap iron. The exact proportions epend on the composition of the iron ore and its impurities. The common ores f iron are the oxides and carbonate: hematite (Fe_2O_3), magnetite (Fe_3O_4), monite ($2\,Fe_2O_3 \cdot 3\,H_2O$), and siderite ($FeCO_3$). The purpose of the flux is to naintain the proper ratio of acidic oxides (SiO_2, Al_2O_3, and P_4O_{10}) to basic xides (CaO, MgO, and MnO) to obtain an easily liquefied silicate, aluminate, r phosphate slag. Because acidic oxides predominate in most ores, the flux enerally employed is limestone, $CaCO_3$, or dolomite, $CaCO_3 \cdot MgCO_3$.

The iron obtained from a blast furnace is called **pig iron**. It contains about 5% Fe, 3%–4% C, and varying quantities of other impurities. *Cast iron* can be btained by pouring pig iron directly into molds of the desired shape. Cast ron is very hard and brittle and is used only where it is not subjected to nechanical or thermal shock, such as in engine blocks, brake drums, and ransmission housings in automobiles.

▲ Pouring pig iron.

◀ Thermal shock occurs when an object undergoes a rapid change of temperature. Engine blocks get quite hot, but because they cool slowly, they are not usually subject to thermal shock.

teel

hree fundamental changes must be made to convert pig iron to **steel**:

1. reduction of the carbon content from 3%–4% in pig iron to 0%–1.5% in steel
2. removal, through slag formation, of Si, Mn, and P (each present in pig iron to the extent of 1% or so), together with other minor impurities
3. addition of alloying elements (such as Cr, Ni, Mn, V, Mo, and W) to give the steel its desired end properties

◀ Steel made with 18% Cr and 8% Ni resists corrosion and is commonly known as *stainless steel*.

The most important method of steelmaking today is the **basic oxygen process**. xygen gas at about 10 atm pressure and a stream of powdered limestone are fed hrough a water-cooled tube (called a *lance*) and discharged above the molten pig ron (Fig. 23-10). The reactions that occur (Table 23.3) accomplish the first two bjectives. A typical reaction time is 22 minutes. The reaction vessel is tilted to our off the liquid slag floating on top of the iron, and then the desired alloying lements are added.

Steelmaking has been undergoing rapid technological changes. It is now ossible to make iron and steel directly from iron ore in a single-step, contin-ous process at temperatures below the melting point of any of the materials sed in the process. In the *direct reduction of iron* (DRI), CO(g) and H_2(g), btained in the reaction of steam with natural gas, are used as reducing gents. The economic viability of the DRI process depends on an abundant upply of natural gas. Currently, only a small percentage of the world's iron roduction is by direct reduction, but this is a fast-growing component of the ron and steel industry, particularly in the Middle East and South America.

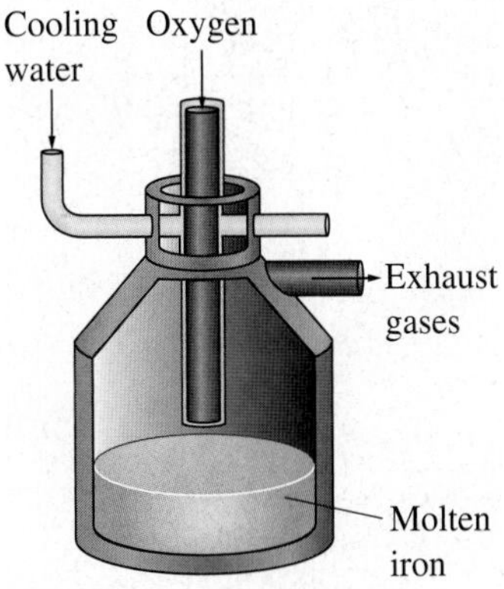

▲ FIGURE 23-10
A basic oxygen furnace

TABLE 23.3 Some Reactions Occurring in Steelmaking Processes

$$2\,C + O_2 \longrightarrow 2\,CO$$
$$2\,FeO + Si \longrightarrow 2\,Fe + SiO_2$$
$$FeO + Mn \longrightarrow Fe + MnO$$
$$FeO + SiO_2 \longrightarrow \underset{\text{slag}}{FeSiO_3}$$
$$MnO + SiO_2 \longrightarrow \underset{\text{slag}}{MnSiO_3}$$
$$4\,P + 5\,O_2 \longrightarrow P_4O_{10}$$
$$6\,CaO + P_4O_{10} \longrightarrow \underset{\text{slag}}{2\,Ca_3(PO_4)_2}$$

23-2 CONCEPT ASSESSMENT

Write a balanced chemical equation for the direct reduction of iron(III) oxide by hydrogen gas.

23-4 First-Row Transition Metal Elements: Scandium to Manganese

The properties and uses of the first-row transition metals span a wide rang strikingly illustrating periodic behavior despite the small variation in some the atomic properties listed in Table 23.1. The preparation, uses, and reactior of the compounds of these metals illustrate concepts we have previously dis cussed, including the variability of oxidation states.

Scandium

Scandium is a rather obscure metal, though not especially rare. It constitute about 0.0025% of Earth's crust, which makes it more abundant than many bette known metals, including lead, uranium, molybdenum, tungsten, antimon silver, mercury, and gold. Its principal mineral form is *thortveitite*, $Sc_2Si_2O_7$. Mo scandium is obtained from uranium ores, however, in which it occurs only to th extent of about 0.01% Sc by mass. The commercial uses of scandium are limite and its production is measured in gram or kilogram quantities, not in tons. On application is in high-intensity lamps. The pure metal is usually prepared by th electrolysis of a fused mixture of $ScCl_3$ with other chlorides.

Because of its noble-gas electron configuration, the Sc^{3+} ion lacks some of th characteristic properties of transition metal ions. For instance, the ion is colorles and diamagnetic, as are most of its salts. In its chemical behavior, Sc^{3+} mos closely resembles Al^{3+}, as in the hydrolysis of $[Sc(H_2O)_6]^{3+}(aq)$ to yield acidi solutions and in the formation of an amphoteric gelatinous hydroxide, $Sc(OH)_3$

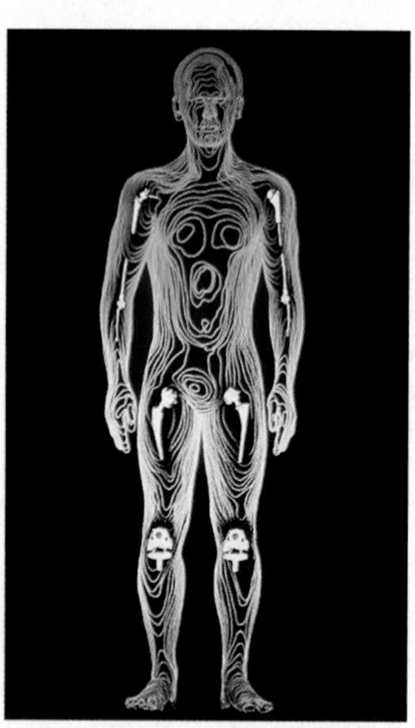

PhotoLink/Photodisc/Getty Images

▲ A computer-generated representation of titanium joint implants at the shoulders, elbows, hips, and knees.

lucato/Getty Images

▲ White $TiO_2(s)$, mixed with other components to produce the desired color, is the leading pigment used in paints.

Titanium

Titanium is the ninth most abundant element, constituting 0.6% of Earth's soli crust. The metal is greatly valued for its low density, high structural strengt and corrosion resistance. The first two properties account for its extensive us in the aircraft industry and the third for its uses in the chemical industry: i pipes, component parts of pumps, and reaction vessels. Titanium is also use in dental and other bone implants. The metal provides a strong support an bone bonds directly to a titanium implant, making it a part of the body.

Several compounds of titanium are of particular commercial importanc *Titanium tetrachloride*, $TiCl_4$, is the starting material for preparing other T compounds and plays a central role in the metallurgy of titanium. $TiCl_4$ is als used to formulate catalysts for the production of plastics. The usual method o preparing $TiCl_4$ involves the reaction of naturally occurring rutile (TiO_2) wit carbon and $Cl_2(g)$ [reaction (23.7)].

$TiCl_4$ is a colorless liquid (mp −24 °C; bp 136 °C). In the +4 oxidation state all the valence-shell electrons of Ti atoms are employed in bond formation. I this oxidation state, Ti bears a strong resemblance to the group 14 elements with some properties and a molecular shape (tetrahedral) similar to those o CCl_4 and $SiCl_4$. The hydrolysis of $TiCl_4$, when carried out in moist air, is th basis for a type of smoke grenade in which $TiO_2(s)$ is the smoke.

$$TiCl_4(l) + 2\,H_2O(l) \longrightarrow TiO_2(s) + 4\,HCl(g)$$

$SiCl_4$ also fumes in moist air in a similar reaction.

Titanium dioxide, TiO_2, is bright white, opaque, inert, and nontoxic. Becaus of these properties and its relative low cost, it is now the most widely use hite pigment for paints. In this application, TiO_2 has displaced toxic basic ad carbonate—so-called white lead. TiO_2 is also used as a paper whitener nd in glass, ceramics, floor coverings, and cosmetics.

To produce pure TiO_2 for these and other uses, a gaseous mixture of $TiCl_4$ nd O_2 is passed through a silica tube at about 700 °C:

$$TiCl_4(g) + O_2(g) \xrightarrow{\Delta} TiO_2(s) + 2\,Cl_2(g)$$

anadium

anadium is a fairly abundant element (0.02% of Earth's crust) found in several ozen ores. Its principal ores are rather complex, such as *vanadinite*, $b_5(VO_4)_3Cl$. The metallurgy of vanadium is not simple, but vanadium of igh purity (99.99%) is obtainable. For most of its applications, though, vana- ium is prepared as an iron–vanadium alloy, *ferrovanadium*, containing from 5% to 95% V. About 80% of the vanadium produced is for the manufacture of teel. Vanadium-containing steels are used in applications requiring strength nd toughness, such as in springs and high-speed machine tools.

The most important compound of vanadium is the pentoxide, V_2O_5, used ainly as a catalyst, as in the conversion of $SO_2(g)$ to $SO_3(g)$ in the contact ethod for the manufacture of sulfuric acid. The activity of V_2O_5 as an oxida- ion catalyst may be linked to its reversible loss of oxygen occurring from 00 to 1100 °C.

In its compounds, vanadium can exist in a variety of oxidation states. In ach of these oxidation states, vanadium forms an oxide or ion. The ions dis- lay distinctive colors in aqueous solution (Fig. 23-11). The acid–base proper- ies of vanadium oxides are in accord with factors established earlier in the ext: If the central metal atom is in a low oxidation state, the oxide acts as a ase; in higher oxidation states for the central atom, acidic properties become mportant. Vanadium oxides with V in the +2 and +3 oxidation states are asic, whereas those in the +4 and +5 oxidation states are amphoteric.

Most compounds with vanadium in its highest oxidation state (+5) are ood oxidizing agents. In its +2 oxidation state, vanadium (as V^{2+}) is a good educing agent. The oxidation–reduction relationships between the ionic pecies pictured in Figure 23-11 are summarized in Table 23.4.

Richard Megna/Fundamental Photographs

▲ FIGURE 23-11
Some vanadium species in solution
The yellow solution has vanadium in the +5 oxidation state, as VO_2^+. In the blue solution, the oxidation state is +4, in VO^{2+}. The green solution contains V^{3+}, and the violet solution contains V^{2+}.

hromium

lthough it is found only to the extent of 122 parts per million (0.0122%) in arth's crust, chromium is one of the most important industrial metals. The pro- uction of ferrochrome from *chromite*, $Fe(CrO_2)_2$, was discussed in Section 23-2. hromium metal is hard and maintains a bright surface through the protective ction of an invisible oxide coating. Because it is resistant to corrosion, hromium is extensively used in plating other metals.

◀ The word *chromium* is derived from the Greek *chroma*, meaning "color"—an apt name, given the range of colors found in chromium compounds.

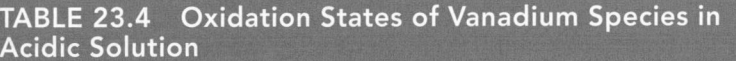

TABLE 23.4 Oxidation States of Vanadium Species in Acidic Solution

O.S. Change	Reduction Half-Reaction	$E°$
$+5 \longrightarrow +4$:	$VO_2^+(aq) + 2\,H^+(aq) + e^- \longrightarrow VO^{2+}(aq) + H_2O(l)$ (yellow) (blue)	1.000 V
$+4 \longrightarrow +3$:	$VO^{2+}(aq) + 2\,H^+(aq) + e^- \longrightarrow V^{3+}(aq) + H_2O(l)$ (green)	0.337 V
$+3 \longrightarrow +2$:	$V^{3+}(aq) + e^- \longrightarrow V^{2+}(aq)$ (violet)	−0.255 V
$+2 \longrightarrow 0$:	$V^{2+}(aq) + 2\,e^- \longrightarrow V(s)$	−1.13 V

EXAMPLE 23-2 Using Electrode Potential Data to Predict an Oxidation–Reduction Reaction

Can $MnO_4^-(aq)$ be used to oxidize $VO^{2+}(aq)$ to $VO_2^+(aq)$ for standard-state conditions in an acidic solution? If so, write a balanced equation for the redox reaction.

Analyze

We need to start by writing two half-equations, one for the reduction of $MnO_4^-(aq)$ to $Mn^{2+}(aq)$ the other for the oxidation of $VO^{2+}(aq)$ to $VO_2^+(aq)$. Both occur in acidic solution. We find one $E°$ value in Table 23.4 and the other in Table 19.1. We then combine the $E°$ values to obtain a value for $E°_{cell}$.

Reduction: $MnO_4^-(aq) + 8\,H^+(aq) + 5\,e^- \longrightarrow Mn^{2+}(aq) + 4\,H_2O(l)$

Oxidation: $5\{VO^{2+}(aq) + H_2O(l) \longrightarrow VO_2^+(aq) + 2\,H^+(aq) + e^-\}$

Overall: $5\,VO^{2+}(aq) + MnO_4^-(aq) + H_2O(l) \longrightarrow 5\,VO_2^+(aq) + Mn^{2+}(aq) + 2\,H^+(aq)$

$$E°_{cell} = E°_{MnO_4^-/Mn^{2+}} - E°_{VO_2^+/VO^{2+}} = 1.51\text{ V} - 1.000\text{ V} = 0.51\text{ V}$$

Because $E°_{cell}$ is positive, we predict that $MnO_4^-(aq)$ should oxidize $VO^{2+}(aq)$ to $VO_2^+(aq)$ for standard-state conditions in acidic solution.

Assess

By using equation (19.17), we can easily verify that K for the overall reaction is very large (greater than 1×10^{43}), which is another indication that $MnO_4^-(aq)$ could be effectively used for oxidizing $VO^{2+}(aq)$ to $VO_2^+(aq)$.

PRACTICE EXAMPLE A: Use data from Tables 19.1 and 23.4 to determine whether nitric acid can be used to oxidize $V^{3+}(aq)$ to $VO^{2+}(aq)$ for standard-state conditions. If so, write a balanced equation for the reaction.

PRACTICE EXAMPLE B: Select a *reducing agent* from Table 19.1 that can be used to reduce $VO^{2+}(aq)$ to $V^{2+}(aq)$ for standard-state conditions in acidic solution. Consider that the reduction occurs in two stages: $VO^{2+}(aq) \longrightarrow V^{3+}(aq) \longrightarrow V^{2+}(aq)$, but note that the $V^{2+}(aq)$ must not be reduced to V(s).

Steel is chrome-plated from an aqueous solution containing CrO_3 anc H_2SO_4. The plating obtained is thin and porous. It tends to develop crack unless the steel is first plated with copper or nickel, which provides the tru protective coating. Then chromium is plated over this layer for extra protec tion and decorative purposes. The efficiency of chrome-plating is limite by the fact that reduction of Cr(VI) to Cr(0) produces only $\frac{1}{6}$ mol Cr per mole o electrons. In other words, large quantities of electric energy are required fo chrome-plating relative to other types of metal plating.

Chromium, like vanadium, has a variety of oxidation states in aqueou solution, each having a different color.

▶ The colors may also depend on other species present in solution. For example, if $[Cl^-]$ is high, $[Cr(H_2O)_6]^{3+}$ is converted to $[CrCl_2(H_2O)_4]^+$ and the violet color changes to green.

O.S. + 2: $[Cr(H_2O)_6]^{2+}$, blue

O.S. + 3: (acidic) $[Cr(H_2O)_6]^{3+}$, violet (basic) $[Cr(OH)_4]^-$, green

O.S. + 6: (acidic) $Cr_2O_7^{2-}$, orange (basic) CrO_4^{2-}, yellow

The oxides and hydroxides of chromium conform to the general principles o acid–base behavior: CrO is basic, Cr_2O_3 is amphoteric, and CrO_3 is acidic.

Pure chromium reacts with dilute HCl(aq) or $H_2SO_4(aq)$ to produce $Cr^{2+}(aq)$. Nitric acid and other oxidizing agents alter the surface of the meta (perhaps by formation of an oxide coating). They render the metal resistant to further attack—it becomes *passive*. A better source of chromium compound than the pure metal is the alkali metal chromates, which contain Cr(VI) anc can be obtained directly from chromite ore by reactions such as

$$4\,Fe(CrO_2)_2(s) + 8\,Na_2CO_3(s) + 7\,O_2(g) \xrightarrow{\Delta} 2\,Fe_2O_3(s) + 8\,Na_2CrO_4(s) + 8\,CO_2(g) \quad \textbf{(23.16}$$

The *sodium chromate*, $Na_2CrO_4(s)$, produced by this reaction is the source o many industrially important chromium compounds.

The Cr(VI) oxidation state is also observed in the red oxide, CrO_3. As expected, this oxide dissolves in water to produce a strongly acidic solution

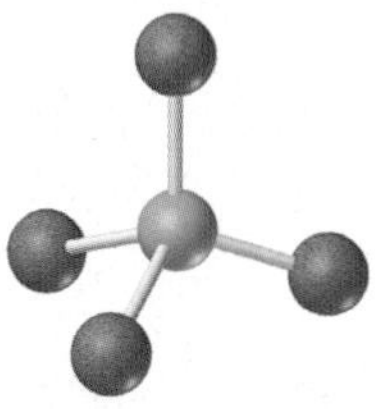

Chromate ion (CrO_4^{2-})

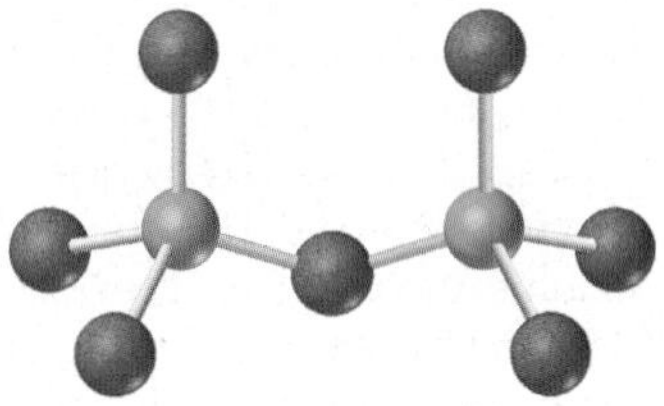

Dichromate ion ($Cr_2O_7^{2-}$)

he product of the reaction is not the expected chromic acid, H_2CrO_4, however, /hich has never been isolated in the pure state. Instead, the observed reaction is

$$2\,CrO_3(s) + H_2O(l) \longrightarrow 2\,H^+(aq) + Cr_2O_7^{2-}(aq) \qquad \textbf{(23.17)}$$

is possible to crystallize a *dichromate* salt from an aqueous solution of CrO_3. the solution is made basic, the color turns from orange to yellow. From basic olutions, only *chromate* salts can be crystallized. Thus, whether a solution ontains Cr(VI) as $Cr_2O_7^{2-}$ or CrO_4^{2-} or a mixture of the two depends on the H. The relevant equations follow.

$$2\,CrO_4^{2-}(aq) + 2\,H^+(aq) \rightleftharpoons Cr_2O_7^{2-}(aq) + H_2O(l) \qquad \textbf{(23.18)}$$

$$K_c = \frac{[Cr_2O_7^{2-}]}{[CrO_4^{2-}]^2[H^+]^2} = 3.2 \times 10^{14} \qquad \textbf{(23.19)}$$

Le Châtelier's principle predicts that the forward reaction of (23.18) is avored in acidic solutions and that the predominant Cr(VI) species is $Cr_2O_7^{2-}$. n basic solution, H^+ ions are removed and the reverse reaction is favored, orming CrO_4^{2-} as the principal species. Careful control of the pH is necessary /hen $Cr_2O_7^{2-}$ is used as an oxidizing agent or CrO_4^{2-} as a precipitating agent. n addition, equation (23.19) can be used to calculate the relative amounts of e two ions as a function of $[H^+]$.

Chromate ion in basic solution can be used to precipitate metal chromates uch as $BaCrO_4(s)$ and $PbCrO_4(s)$. It is not a good oxidizing agent, however; is not readily reduced.

$$CrO_4^{2-}(aq) + 4\,H_2O(l) + 3\,e^- \longrightarrow [Cr(OH)_4]^-(aq) + 4\,OH^-(aq) \quad E^\circ = -0.13\text{ V}$$

Dichromates are poor precipitating agents but excellent oxidizing agents, /hich are used in a variety of industrial processes. In the chrome leather taning process, for example, animal hides are immersed in $Na_2Cr_2O_7(aq)$, which s then reduced by $SO_2(g)$ to soluble basic chromic sulfate, $Cr(OH)SO_4$. ollagen, a protein in hides, reacts to form an insoluble chromium complex. he hides become *leather*, a tough, pliable material resistant to biological attack.

Dichromates are easily reduced to Cr_2O_3. In the case of ammonium dichromate, simply heating the compound produces Cr_2O_3 in a dramatic reaction Fig. 23-12).

Carey B. Van Loon

▲ FIGURE 23-12
Decomposition of $(NH_4)_2Cr_2O_7$
mmonium dichromate (left) contains both an oxidizing agent, $Cr_2O_7^{2-}$, and a reducing gent, NH_4^+. The products of the reaction between these two ions are $Cr_2O_3(s)$ (right), $N_2(g)$, and $H_2O(g)$. Considerable heat and light are also evolved (center).

▶ FIGURE 23-13
Relationship between Cr^{2+} and Cr^{3+}
The solution on the left, containing blue $Cr^{2+}(aq)$, is prepared by dissolving chromium metal in HCl(aq). Within minutes, the $Cr^{2+}(aq)$ is oxidized to green $Cr^{3+}(aq)$ by atmospheric oxygen (right). The green color is that of the complex ion $[CrCl_2(H_2O)_4]^+(aq)$.

Carey B. Van Loon

Chromium(II) compounds can be prepared by the reduction of Cr(III) compounds with zinc in acidic solution or electrolytically at a lead cathode. The most distinctive feature of Cr(II) compounds is their reducing power.

$$Cr^{3+}(aq) + e^- \longrightarrow Cr^{2+}(aq) \qquad E° = -0.424\ V$$

That is, the oxidation of $Cr^{2+}(aq)$ occurs readily. In fact, Cr(II) solutions can be used to purge gases of trace amounts of $O_2(g)$, through the following reaction illustrated in Figure 23-13.

$$4\,Cr^{2+}(aq) + O_2(g) + 4\,H^+(aq) \longrightarrow 4\,Cr^{3+}(aq) + 2\,H_2O(l) \qquad E°_{cell} = +1.653\ V$$

Pure Cr can be obtained in small amounts by reducing Cr_2O_3 with Al in a reaction similar to the thermite reaction.

$$Cr_2O_3(s) + 2\,Al(s) \longrightarrow Al_2O_3(s) + 2\,Cr(l) \qquad (23.20)$$

Manganese

Manganese is a fairly abundant element, constituting about 1% of Earth's crust. Its principal ore is *pyrolusite*, MnO_2. Like V and Cr, Mn is most important in steel production, generally as the iron–manganese alloy, *ferromanganese*. Ferromanganese can be obtained by the reduction of a mixture of pyrolusite and hematite iron ores with carbon.

$$MnO_2(s) + Fe_2O_3(s) + 5\,C(s) \xrightarrow{\Delta} \underbrace{Mn(s) + 2\,Fe(s)}_{\text{Ferromanganese}} + 5\,CO(g)$$

Mn participates in the purification of iron by reacting with sulfur and oxygen and removing them through slag formation. In addition, Mn increases the hardness of steel. Steel containing high proportions of Mn is extremely tough and wear-resistant in such applications as railroad rails, bulldozers, and road scrapers.

The electron configuration of Mn is $[Ar]3d^54s^2$. By employing first the two 4s electrons and then, consecutively, up to all five of its unpaired 3d electrons, manganese exhibits all oxidation states from +2 to +7. The most important reactions of manganese compounds are oxidation–reduction reactions. Standard electrode potential diagrams are given in Figure 23-14. These diagrams help explain the following observations:

- $Mn^{3+}(aq)$ is unstable; that is, its disproportionation is spontaneous.

$$2\,Mn^{3+}(aq) + 2\,H_2O(l) \longrightarrow Mn^{2+}(aq) + MnO_2(s) + 4\,H^+(aq)$$
$$E°_{cell} = 0.54\ V \qquad (23.21)$$

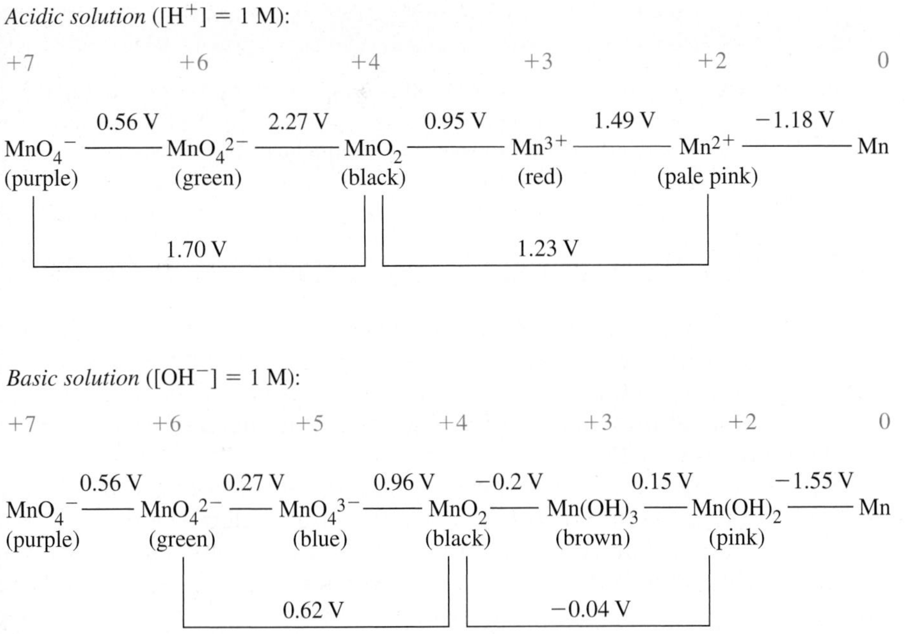

◀ FIGURE 23-14
Electrode potential diagrams for manganese

- Manganate ion, MnO_4^{2-}, is also unstable in acidic solution; its disproportionation is spontaneous.

$$3\,MnO_4^{2-}(aq) + 4\,H^+(aq) \longrightarrow MnO_2(s) + 2\,MnO_4^-(aq) + 2\,H_2O(l) \qquad E^\circ_{cell} = 1.70\ V \quad \textbf{(23.22)}$$

- If $[OH^-]$ is kept sufficiently high, the following reaction can be reversed; thus, manganate ion can be maintained as a stable species in a strongly basic medium.

$$3\,MnO_4^{2-}(aq) + 2\,H_2O(l) \longrightarrow MnO_2(s) + 2\,MnO_4^-(aq) + 4\,OH^-(aq) \qquad E^\circ_{cell} = 0.04\ V \quad \textbf{(23.23)}$$

Manganese dioxide is used in dry-cell batteries, in glass and ceramic glazes, nd as a catalyst; it is also the principal source of manganese compounds. Vhen MnO_2 is heated in the presence of an alkali and an oxidizing agent, a nanganate salt is produced.

$$3\,MnO_2 + 6\,KOH + KClO_3 \xrightarrow{\Delta} 3\,K_2MnO_4 + KCl + 3\,H_2O$$

$_2MnO_4$ is extracted from the fused mass with water and can then be oxidized o $KMnO_4$, *potassium permanganate* (with Cl_2 as an oxidizing agent, for nstance). Potassium permanganate, $KMnO_4$, is an important laboratory oxiizing agent. For chemical analyses, it is generally used in acidic solutions in vhich it is reduced to Mn^{2+}(aq). In the analysis of iron by MnO_4^-, a sample of e^{2+} is prepared by dissolving iron in an acid and reducing any Fe^{3+} back to e^{2+}. Then the sample is titrated with MnO_4^-(aq).

$$5\,Fe^{2+}(aq) + MnO_4^-(aq) + 8\,H^+(aq) \longrightarrow 5\,Fe^{3+}(aq) + Mn^{2+}(aq) + 4\,H_2O(l) \quad \textbf{(23.24)}$$

Mn^{2+}(aq) has a barely discernible pale pink color. MnO_4^-(aq) is an intense urple color. At the end point of the titration reaction (23.24), the solution cquires a lasting light purple color with just one drop of excess MnO_4^-(aq) recall Figure 5-19). MnO_4^-(aq) is less satisfactory for titrations in alkaline olutions because the insoluble reduction product, brown MnO_2(s), obscures he end point.

23-3 CONCEPT ASSESSMENT

The dichromate ion is formed by the condensation of chromate ion in an acidic solution. Further condensation occurs at low pH to form $[Cr_3O_{10}]^{2-}$. Suggest the structures for these anions. What other elements form similar anions?

23-5 The Iron Triad: Iron, Cobalt, and Nickel

Tom Pantages

▲ Cobalt–samarium magnets are used in high-efficiency motors.

The transition elements iron, cobalt, and nickel comprise the *iron triad*. Iron with an annual worldwide production of more than 1.1 billion metric tons*, i the most important metal in modern civilization. It is widely distributed i Earth's crust at an abundance of 4.7%. The major commercial use of iron is t make steel (see Section 23-3).

Cobalt is among the rarer elements. It comprises only about 0.0020% c Earth's crust, but it occurs in sufficiently concentrated deposits (ores) so th its annual production runs into the millions of pounds. Cobalt is used primar ily in alloys with other metals. Like iron, cobalt is ferromagnetic. One alloy c cobalt, Co_5Sm, makes a particularly strong and lightweight permanent mag net. Because of the strength of its magnetic field, magnets of this alloy are use in the manufacture of miniature electronic devices.

Nickel ranks twenty-fourth in abundance among the elements in Earth' crust. Its ores are mainly the sulfides, oxides, silicates, and arsenides Particularly large deposits are found in Canada. Of the 136 million kilogram of nickel consumed annually in the United States, about 80% goes to the pro duction of alloys. Another 15% is used in electroplating, and the remainder fo miscellaneous purposes (for example, as catalysts).

Oxidation States

Variability of oxidation state is seen in the iron triad, even if not to the sam degree as with vanadium, chromium, and manganese. The +2 oxidation stat is commonly encountered in all three metals.

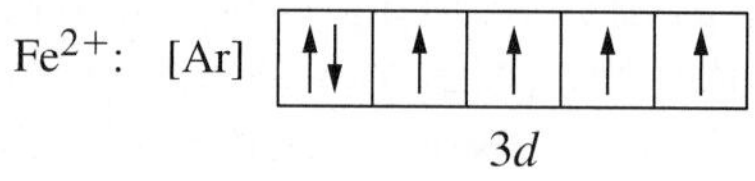

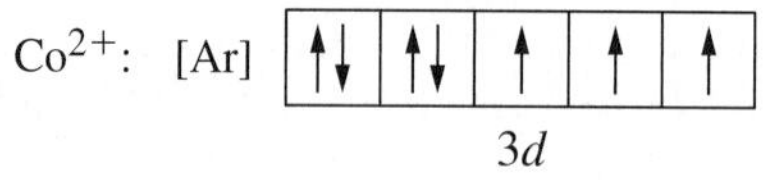

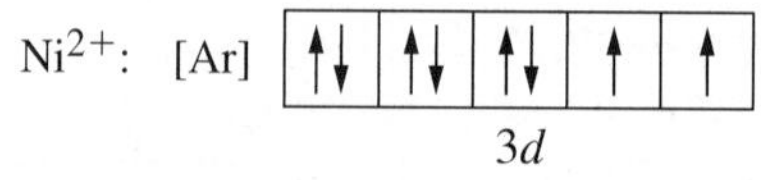

For cobalt and nickel, the +2 oxidation state is the most stable, but for iron, the most stable is +3.

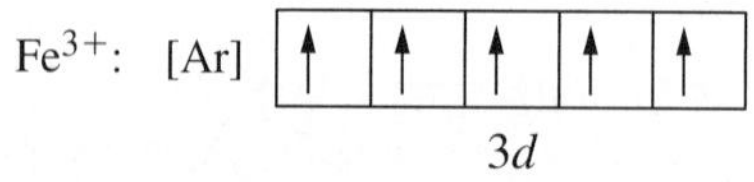

The following observations confirm that, for iron, the +3 oxidation state i more stable than the +2 oxidation state but, for cobalt and nickel, the situatio is reversed. Consequently, neither Co^{3+} nor Ni^{3+} is readily formed from the + oxidation state.

* Source: www.worldsteel.org

- $Fe^{2+}(aq)$ is spontaneously oxidized to $Fe^{3+}(aq)$ by $O_2(g)$ at 1 atm in a solution with $[H^+] = 1$ M.

$$4\,Fe^{2+}(aq) + O_2(g) + 4\,H^+(aq) \longrightarrow 4\,Fe^{3+}(aq) + 2\,H_2O(l)$$
$$E^\circ_{cell} = 0.44\text{ V} \quad \textbf{(23.25)}$$

  This reaction is still spontaneous with lower O_2 partial pressures and less acidic solutions.

- For the half-reaction $Co^{3+}(aq) + e^- \longrightarrow Co^{2+}(aq)$, $E^\circ = 1.82$ V. The reduction of $Co^{3+}(aq)$ to $Co^{2+}(aq)$ occurs readily; conversely, the oxidation of $Co^{2+}(aq)$ to Co^{3+} occurs only with difficulty. As will be noted in Chapter 24, however, the oxidation state +3 can be attained when Co^{3+} is the central metal ion in very stable complex ions.
- Nickel(III) compounds are used in batteries. For example, in the nickel–cadmium (Nicad) cell, the cathode half-reaction is $NiO(OH) + H^+ + e^- \longrightarrow Ni(OH)_2$. It is the ease of this reduction, when combined with the oxidation half-reaction, $Cd(s) + 2\,OH^-(aq) \longrightarrow Cd(OH)_2(s) + 2\,e^-$, that produces a cell with a voltage of about 1.5 V.

Some Reactions of the Iron Triad Elements

The reactions of the iron triad elements are many and varied. The metals are more active than hydrogen and liberate $H_2(g)$ from an acidic solution. Hydrated Co^{2+} and Ni^{2+} are red and green, respectively. In aqueous solution, Fe^{2+} is pale green and fully hydrated Fe^{3+} is purple. Generally, however, solutions of $Fe^{2+}(aq)$ are yellow to brown, but this color probably results from the presence of species formed in the hydrolysis of $Fe^{3+}(aq)$. Like the hydrolysis of $Al^{3+}(aq)$, described on page 777, that of $Fe^{3+}(aq)$ produces an acidic solution.

$$[Fe(H_2O)_6]^{3+}(aq) + H_2O(l) \rightleftharpoons [FeOH(H_2O)_5]^{2+}(aq) + H_3O^+(aq)$$
$$K_a = 8.9 \times 10^{-4} \quad \textbf{(23.26)}$$

Some reactions that can be used to identify and distinguish between $Fe^{2+}(aq)$ and $Fe^{3+}(aq)$ are summarized in Table 23.5.

An interesting set of reactions involves the complex ions $[Fe(CN)_6]^{4-}$ and $[Fe(CN)_6]^{3-}$. These ions are commonly called *ferrocyanide* and *ferricyanide*, respectively. $Fe^{3+}(aq)$ yields a dark blue precipitate called *Prussian blue* when treated with potassium ferrocyanide, $K_4[Fe(CN)_6](aq)$, whereas $Fe^{2+}(aq)$ yields a similar blue precipitate, called *Turnbull's blue*, when treated with potassium ferricyanide, $K_3[Fe(CN)_6](aq)$. Together, these two and similar related precipitates are known commercially as *iron blue*. Iron blue is used as a pigment for paints, printing inks, laundry bluing, art colors, cosmetics (eye shadow), and blueprinting. An additional sensitive test for $Fe^{3+}(aq)$ is the formation of a blood-red complex ion with thiocyanate ion, $SCN^-(aq)$.

◀ In Chapter 24, we will learn to write the systematic names for ferrocyanide and ferricyanide ions; they are hexacyanidoferrate(II) and hexacyanidoferrate(III), respectively.

◀ The blue precipitate that establishes the presence of $Fe^{2+}(aq)$ from the corroding nails in Figure 19-20 is Turnbull's blue.

$$[Fe(H_2O)_6]^{3+} + SCN^-(aq) \longrightarrow [FeSCN(H_2O)_5]^{2+} + H_2O(l)$$

TABLE 23.5 Some Qualitative Tests for $Fe^{2+}(aq)$ and $Fe^{3+}(aq)$

Reagent	$Fe^{2+}(aq)$	$Fe^{3+}(aq)$
NaOH(aq)	Green precipitate	Red-brown precipitate
$K_4[Fe(CN)_6]$	White precipitate, turning blue rapidly	Prussian blue precipitate
$K_3[Fe(CN)_6]$	Turnbull's blue precipitate	Red-brown (no precipitate)
KSCN(aq)	No color	Deep red

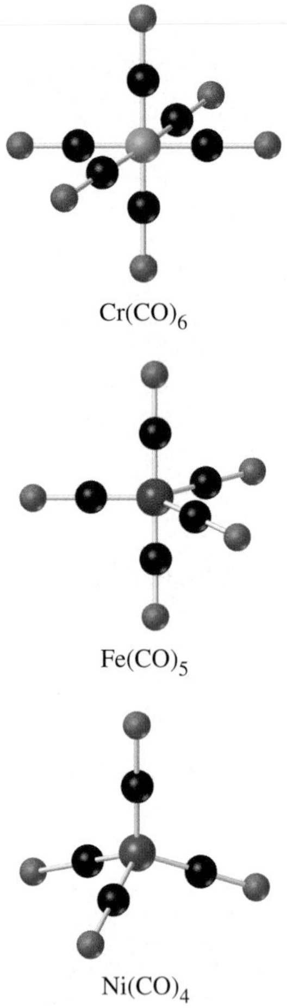

▲ FIGURE 23-15
Structure of some simple carbonyls

TABLE 23.6 Three Metal Carbonyls

	Number of e^-		
	From Metal	From CO	Total
$Cr(CO)_6$	24	12	36
$Fe(CO)_5$	26	10	36
$Ni(CO)_4$	28	8	36

With only a few exceptions, the transition metals form compounds witl carbon monoxide (CO), called **metal carbonyls**. In the simple metal carbonyl listed in Table 23.6,

- each CO molecule contributes an electron pair to an empty orbital of th metal atom
- all electrons are paired (most metal carbonyls are diamagnetic)
- the metal atom acquires the electron configuration of the noble gas Kr

The structures of the simple carbonyls in Figure 23-15 are those that we woulc predict from VSEPR theory (based on a number of electron pairs around th central atom equal to the number of CO molecules).

Metal carbonyls are produced in several ways. Nickel combines with CO(g at ordinary temperatures and pressures in a reversible reaction.

$$Ni(s) + 4\,CO(g) \rightleftharpoons Ni(CO)_4(l)$$

The reaction above forms the basis of an important industrial process callec the Mond process for obtaining nickel metal from its oxides. In the Monc process, CO(g) is passed over a mixture of metal oxides. Nickel is carried off a $Ni(CO)_4(g)$, while the other oxides are reduced to the metals. Whe $Ni(CO)_4(g)$ is subsequently heated to about 250 °C, the carbonyl comple decomposes, yielding Ni(s). With iron, it is necessary to use higher tempera tures (200 °C) and CO(g) pressures (100 atm).

$$Fe(s) + 5\,CO(g) \rightleftharpoons Fe(CO)_5(g)$$

In other cases, the carbonyl is obtained by reducing a metal compound in th presence of CO(g).

Carbon monoxide poisoning results from a reaction similar to carbonyl for mation. CO molecules coordinate with Fe atoms in hemoglobin in the blood displacing the O_2 molecules normally carried by hemoglobin. The metal car bonyls themselves are also very poisonous.

23-4 CONCEPT ASSESSMENT

What is the oxidation state of iron in the compound $Fe(CO)_5$?

23-6 Group 11: Copper, Silver, and Gold

Throughout the ages, Cu, Ag, and Au have been the preferred metals for coin because they are so durable and resistant to corrosion. The data in Table 23. help us understand why this is so. The metal ions are easy to reduce to the fre metals, which means that the metals are difficult to oxidize.

In Mendeleev's periodic table, the alkali metals (group 1) and the coinag metals (group 11) appear together as group I. The only similarity betwee the two subgroups, however, is that both have a single *s* electron in the valenc

TABLE 23.7 Some Properties of Copper, Silver, and Gold

	Cu	Ag	Au
Electron configuration	$[Ar]3d^{10}4s^1$	$[Kr]4d^{10}5s^1$	$[Xe]4f^{14}5d^{10}6s^1$
Metallic radius, pm	128	144	144
First ioniz. energy, kJ mol^{-1}	745	731	890
Electrode potential, V			
$M^+(aq) + e^- \longrightarrow M(s)$	+0.520	+0.800	+1.83
$M^{2+}(aq) + 2\,e^- \longrightarrow M(s)$	+0.340	+1.39	—
$M^{3+}(aq) + 3\,e^- \longrightarrow M(s)$	—	—	+1.52
Oxidation states[a]	+1, +2	+1, +2	+1, +3

[a]The most common oxidation states are shown in red.

hells of their atoms. More significant are the differences between the group 1 nd group 11 metals. For example, the first ionization energies for the group 11 ıetals are much larger than for the group 1 metals, and the standard electrode otentials, $E°$, are positive for the group 11 metals and negative for the group 1 ıetals.

Like the other transition elements that precede them in the periodic table, ıe group 11 metals can use *d* electrons in chemical bonding. Thus they can xist in different oxidation states, exhibit paramagnetism and color in some of ıeir compounds, and form complex ions. They also possess to a high degree ome of the distinctive physical properties of metals—malleability, ductility, nd excellent electrical and thermal conductivity.

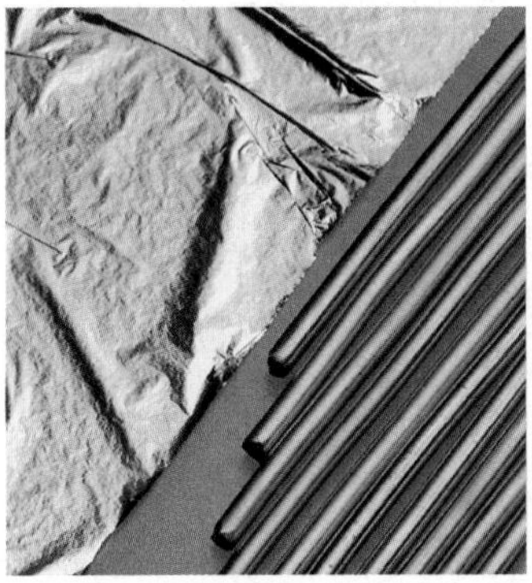

▲ **Gold leaf and copper wire**

Copper, silver, and gold—the coinage metals—are used in jewelry making nd the decorative arts. Gold, for instance, is extraordinarily malleable and an be pounded into thin translucent sheets known as gold leaf. The coinage ıetals are valued by the electronics industry for their ability to conduct elecricity. Silver has the highest electrical conductivity of any pure element, but oth copper and gold are more often used as electrical conductors because opper is inexpensive and gold does not readily corrode. The most important se of gold is as the monetary reserve of nations throughout the world.

Generally, the coinage metals are resistant to air oxidation, although silver vill tarnish through reactions with sulfur compounds in air to produce black Ag_2S. In moist air, copper corrodes to produce green basic copper carbonate. 'his is the green color associated with copper roofing and gutters and bronze tatues. (Bronze is an alloy of Cu and Sn.) Fortunately, this corrosion product orms a tough adherent coating that protects the underlying metal. The corroion reaction is complex but may be summarized as

$$2\,Cu(s) + H_2O(g) + CO_2(g) + O_2(g) \longrightarrow \underset{\text{Basic copper carbonate}}{Cu_2(OH)_2CO_3(s)} \qquad \textbf{(23.27)}$$

The group 11 metals do not react with HCl(aq), but both Cu and Ag react vith concentrated $H_2SO_4(aq)$ or $HNO_3(aq)$. The metals are oxidized to Cu^{2+} nd Ag^+, respectively, and the reduction products are $SO_2(g)$ in $H_2SO_4(aq)$ nd either NO(g) or $NO_2(g)$ in $HNO_3(aq)$.

Au does not react with either acid, but it will react with "royal water"—*aqua egia* (1 part HNO_3 and 3 parts HCl). The $HNO_3(aq)$ oxidizes the metal and Cl^- from the HCl(aq) promotes the formation of the stable complex ion $[AuCl_4]^-$.

$$Au(s) + 4\,H^+(aq) + NO_3^-(aq) + 4\,Cl^-(aq) \longrightarrow [AuCl_4]^-(aq) + 2\,H_2O(l) + NO(g) \qquad \textbf{(23.28)}$$

Trace amounts of Cu are essential to life, but larger quantities are toxic, espeially to bacteria, algae, and fungi. Among the many copper compounds used

as pesticides are the basic acetate, carbonate, chloride, hydroxide, and sulfat
Commercially, the most important copper compound is $CuSO_4 \cdot 5\,H_2O$. I
addition to its agricultural applications, $CuSO_4$ is employed in batteries an
electroplating, in the preparation of other copper salts, and in a variety
industrial processes.

Silver(I) nitrate is the principal silver compound of commerce and is also a
important laboratory reagent for the precipitation of anions, most of whic
form insoluble silver salts. These precipitation reactions can be used for th
quantitative determination of anions, either gravimetrically (by weighin
precipitates) or volumetrically (by titration). Most other Ag compounds ar
derived from $AgNO_3$. Ag compounds are used in electroplating, in the manu
facture of batteries, as catalysts, and in cloud seeding (AgI). Silver halide
such as AgBr, are used in photography (see Section 24-11), although this
diminishing now because of the advent of digital cameras.

Gold compounds are used in electroplating, photography, medicinal chem
istry (as anti-inflammatory agents for severe rheumatoid arthritis, for example
and the manufacture of special glasses and ceramics (such as ruby glass).

Tom Pantages

▲ Gold plating on a space antenna.

23-5 CONCEPT ASSESSMENT

What is the ground state electron configuration of gold(III) ion?

23-7 Group 12: Zinc, Cadmium, and Mercury

The properties of the group 12 elements are consistent with elements havin
full subshells, $(n-1)d^{10}ns^2$; some of those properties are summarized i
Table 23.8. The low melting and boiling points of the group 12 metals can prob
ably be attributed to the fact that, with only the ns^2 electrons participatin
metallic bonding is weak. Mercury is the only metal that exists as a liquid a
room temperature and below (although liquid gallium can easily be supe
cooled to room temperature). Mercury differs from Zn and Cd in a number o
ways in addition to physical appearance.

- Mercury has little tendency to combine with oxygen. Its oxide, HgO,
thermally unstable.
- Very few mercury compounds are water soluble, and most are no
hydrated.

TABLE 23.8 Some Properties of the Group 12 Metals

	Zn	Cd	Hg
Density, g cm^{-3}	7.14	8.64	13.59 (liquid)
Melting point, °C	419.6	320.9	−38.87
Boiling point, °C	907	765	357
Electron configuration	$[Ar]3d^{10}4s^2$	$[Kr]4d^{10}5s^2$	$[Xe]4f^{14}5d^{10}6s^2$
Atomic radius, pm	133	149	160
Ionization energy, kJ mol^{-1}			
First	906	867	1006
Second	1733	1631	1809
Principal oxidation state(s)	+2	+2	+1, +2
Electrode potential $E°$, V			
$[M^{2+}(aq) + 2\,e^- \longrightarrow M]$	−0.763	−0.403	+0.854
$[M_2{}^{2+}(aq) + 2\,e^- \longrightarrow 2\,M]$	—	—	+0.796

- Many mercury compounds are covalent. Except for HgF_2, mercury halides are only slightly ionized in aqueous solution.
- Mercury(I) forms a common diatomic ion with a metal–metal covalent bond, $Hg_2{}^{2+}$.
- Mercury will not displace $H_2(g)$ from $H^+(aq)$.

Some of these differences exhibited by mercury can probably be attributed) the *relativistic effect* discussed in Chapter 9 (Focus On 9-1, www nasteringchemistry.com). Because the speed of the 6*s* electrons reaches a sigificant fraction of the speed of light, as they approach the high positive iarge of the mercury nucleus, their masses increase (as predicted by instein's theory of relativity) and the 6*s* orbital shrinks in size. The closer pproach of the 6*s* electrons to the nucleus subjects them to a greater attractive rce than that experienced by the ns^2 electrons in Zn and Cd. As a result, for xample, the first ionization energy of Hg is greater than that of Zn or Cd.

Mary Evans / Heinz Zinram / The Image Works

▲ A brass seagoing chronometer made by John Harrison in the eighteenth century.

ses of the Group 12 Metals and Their Compounds

bout one-third of the zinc produced is used in coating iron to give it corrosion rotection (Section 19-6). The product is called *galvanized iron*. Large quantities f Zn are consumed in the manufacture of alloys. For example, about 20% of ie production of Zn is used in *brass*, a copper alloy having 20%–45% Zn and nall quantities of Sn, Pb, and Fe. Brass is a good electrical conductor and is rrosion resistant. Zinc is also employed in the manufacture of dry-cell batters, in printing (lithography), in the construction industry (roofing materials), nd as sacrificial anodes in corrosion protection (Section 19-6).

Although poisonous, cadmium is substituted for zinc as a shiny and protecve plating on iron in special applications. It is used in bearing alloys, in lowelting solders, in aluminum solders, and as an additive to impart strength to pper. Another application, based on its neutron-absorbing capacity, is in ntrol rods and shielding for nuclear reactors.

The principal uses of mercury take advantage of its metallic and liquid roperties and its high density. It is used in thermometers, barometers, gasressure regulators, and electrical relays and switches, and as electrodes, as in ie chlor–alkali process (Section 19-8). Mercury vapor is used in fluorescent ibes and street lamps. Mercury alloys, called **amalgams**, can be made with iost metals, and some of these amalgams are of commercial importance. A lver dental filling is an amalgam of mercury with a silver alloy containing bout 70% Ag, 26% Sn, 3% Cu, and 1% Zn.

◀ Iron is one of the few metals that does not form an amalgam. Mercury is generally stored and shipped in iron containers.

Table 23.9 lists a few important compounds of the group 12 metals nd some of their uses. Some of the most interesting compounds are the

TABLE 23.9 Some Important Compounds of the Group 12 Metals

Compound	Uses
ZnO	Reinforcing agent in rubber; pigment; cosmetics; dietary supplement; photoconductors in copying machines
ZnS	Phosphors in X-ray and television screens; pigment; luminous paints
$ZnSO_4$	Rayon manufacture; animal feeds; wood preservative
CdO	Electroplating; batteries; catalyst
CdS	Solar cells; photoconductor in photocopying; phosphors; pigment
$CdSO_4$	Electroplating; standard voltaic cells (Weston cell)
HgO	Polishing compounds; dry cells; antifouling paints; fungicide; pigment
$HgCl_2$	Manufacture of Hg compounds; disinfectant; fungicide; insecticide; wood preservative
Hg_2Cl_2	Electrodes; pharmaceuticals; fungicide

semiconductors ZnO, CdS, and HgS, which are also the artist's pigments zinc white, cadmium yellow, and vermilion (red), respectively. Like all semiconductor materials, these compounds have an electronic structure consisting of a valence band and a conduction band (see the Chapter 11 appendix at www.masteringchemistry.com). When light interacts with these compounds, electrons from the valence band may absorb photons and be excited into the conduction band. The energy of the light absorbed must equal or exceed the energy difference between the bands, called the *band gap*. The characteristic colors of these materials depend on the widths of the band gaps, as described in Focus On 21-1, www.masteringchemistry.com.

Mercury and Cadmium Poisoning

Accumulations of mercury in the body affect the nervous system and cause brain damage. One form of chronic mercury poisoning, "hatter's disease," was fairly common in the nineteenth century. Mercury compounds were used to convert fur to felt for making hats. Many hat makers of the time worked in hot, cramped spaces and used these compounds without special precautions. The hatters inadvertently ingested or inhaled the toxic mercury compounds while they worked.

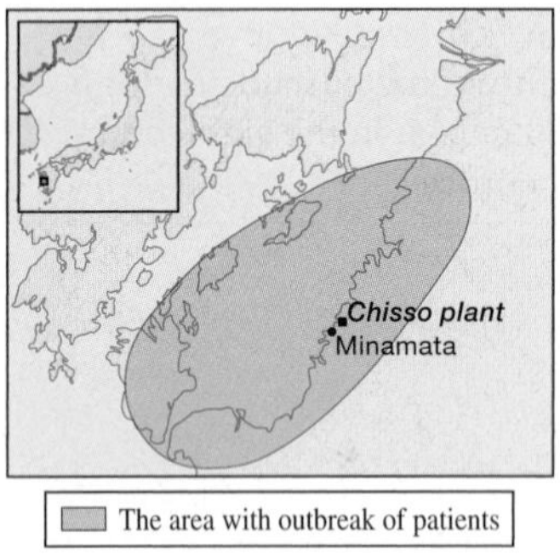

▲ The area of contamination in Japan where mercury poisoning was observed.

One proposed mechanism of mercury poisoning, based on the fact that Hg has a high affinity for sulfur, involves interference with the functioning of sulfur-containing enzymes. Organic mercury compounds are generally more poisonous than inorganic ones and much more toxic than the element itself. An insidious aspect of mercury poisoning is that certain microorganisms have the ability to convert inorganic mercury(II) compounds to methylmercury (CH_3Hg^+) compounds, which then concentrate in the food chains of fish and other aquatic life. An early discovery of the environmental hazard of mercury was in Japan in the 1950s. Dozens of cases of mercury poisoning, including more than 40 deaths, occurred among residents of the shores of Minamata Bay. Local seafood with up to 20 ppm of mercury was a major component of the victims' diet. The source of contamination was traced to a chemical plant discharging mercury waste into the bay.

In the free state, mercury is most poisonous as a vapor. Levels of mercury that exceed 0.05 mg Hg/m^3 air are considered unsafe. Although we think of mercury as having a low vapor pressure, the concentration of Hg in its saturated vapor far exceeds this limit, and mercury vapor levels sometimes exceed safe limits where mercury is used—as in chlor–alkali plants, thermometer factories, and smelters.

Although zinc is an essential element in trace amounts, cadmium, which so closely resembles zinc, is a poison. One effect of cadmium poisoning is an extremely painful skeletal disorder known as *itai-itai kyo* (Japanese for "ouch-ouch" disease). This disorder was discovered in an area of Japan where effluents from a zinc mine became mixed with irrigation water used in rice fields, and cadmium poisoning was discovered in people who ate the rice. Cadmium poisoning can also cause liver damage, kidney failure, and pulmonary disease. The mechanism of cadmium poisoning may involve substitution in certain enzymes of Cd (a poison) for Zn (an essential element). Concern over cadmium poisoning has increased with an awareness that some cadmium is almost always found in zinc and zinc compounds, materials that have many commercial applications.

23-6 CONCEPT ASSESSMENT

Suggest the geometry of the cadmium anion $[CdCl_5]^{3-}$.

23-8 Lanthanides

he elements from lanthanum ($Z = 57$) through lutetium ($Z = 71$) are vari-usly called the *lanthanide, lanthanoid,* or *rare earth elements*. The rare earth lements are "rare" only relative to the alkaline earth metals (group 2).)therwise, they are not particularly rare. Ce, Nd, and La, for example, are more bundant than lead, and Tm is about as abundant as iodine. The lanthanides ccur primarily as oxides, and mineral deposits containing them are found in arious locations. Large deposits near the California–Nevada border are being eveloped to provide oxides of the lanthanides for use as phosphors in color ıonitors and television sets.

Because the differences in electron configuration among the lanthanides are ıainly in 4*f* orbitals, and because 4*f* electrons play a minor role in chemical onding, strong similarities are found among these elements. For example, $E°$ alues for the reduction process $M^{3+}(aq) + 3\,e^- \longrightarrow M(s)$ do not show ıuch variation. All fall between −2.38 V (La) and −1.99 V (Eu). The differ-nces in properties that do exist among the lanthanides arise mostly from the ınthanide contraction discussed in Section 23-1. This contraction is best illus-rated in the radii of the ions M^{3+}. These radii decrease regularly by about -2 pm for each unit increase in atomic number, from a radius of 106 pm for a^{3+} to 85 pm for Lu^{3+}.

The lanthanides, which we can represent by the general symbol Ln, are eactive metals that liberate $H_2(g)$ from hot water and from dilute acids by ndergoing oxidation to $Ln^{3+}(aq)$. The lanthanides combine with $O_2(g)$, sul-ır, the halogens, $N_2(g)$, $H_2(g)$, and carbon in much the same way as expected ɔr metals about as active as the alkaline earths. The pure metals can be pre-ared by electrolytic reduction of Ln^{3+} in a molten salt.

The most common oxidation state for the lanthanides is +3. About half ıe lanthanides can also be obtained in the oxidation state +2; the other half, -4. The reasons for the range of oxidation states, and the predominance of ıe +3 oxidation state, are not totally clear. Most of the lanthanide ions are aramagnetic and colored in aqueous solution.

The lanthanide elements are extremely difficult to extract from their nat-ral sources and to separate from one another. All the methods for doing so re based on the following principle: Species that are strongly dissimilar can ften be completely separated in a one-step process, such as separating $g^+(aq)$ and $Cu^{2+}(aq)$ by adding $Cl^-(aq)$; AgCl is insoluble. At best, species ıat are very similar can only be fractionated in a one-step process. That is, he ratio of the concentration of one species to that of another is altered ightly. To achieve a complete separation means repeating the same basic tep hundreds or even thousands of times. To achieve the separation of the ınthanides, the methods of fractional crystallization, fractional precipitation, olvent extraction, and ion exchange were brought to their highest level of erformance.

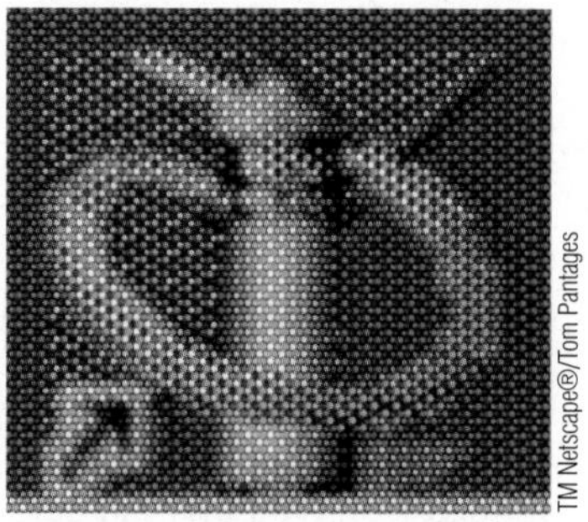

▲ The individual elements of a CRT computer screen are phosphors that contain the lanthanide elements. The phosphors glow red, blue, or green depending on their composition. (TM Netscape ®)

23-9 High-Temperature Superconductors

Лagnetically levitated trains, magnetic resonance imaging (MRI) for medical iagnoses, and particle accelerators used in high-energy physics all require high ıagnetic fields generated by superconducting electromagnets. Superconductors ffer no resistance to an electric current, so electricity is conducted with no loss of nergy (Fig. 23-16).

If cooled to near absolute zero, all metals become superconducting. Several ıetals and alloys superconduct even at marginally higher temperatures of 0–15 K. To maintain a superconductor at these extremely low temperatures equires liquid helium (bp, 4 K) as a coolant.

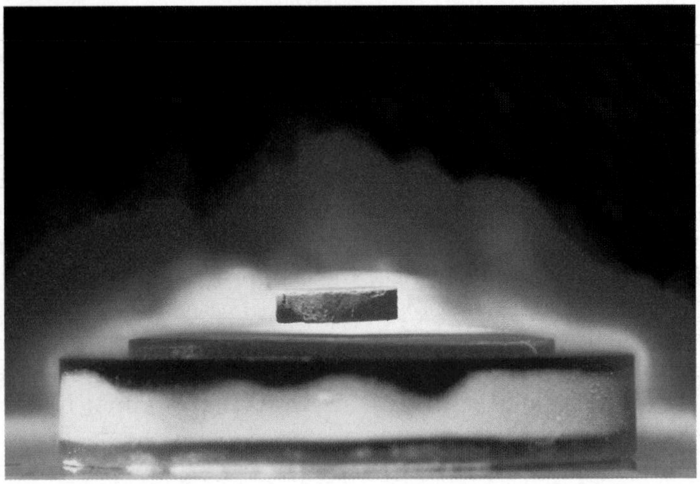

David Parker/IMI/UNIV. OF BIRMINGHAM HIGH TC CONSORTIUM/Science Photo Library

▶ FIGURE 23-16
Magnet levitation using a superconductor
The small magnet induces an electric current in the superconductor below it. Associated with this current is another magnetic field that opposes the field of the small magnet, causing it to be repelled. The magnet remains suspended above the superconductor as long as the superconducting current is present, and the current persists as long as the temperature of the superconductor is maintained at 10–15 K.

In the mid-1980s, materials made of lanthanum, strontium, copper, anc oxygen were found to become superconducting at 30 K. This was a mucl higher temperature for superconductivity than had been previously achieved More surprising, the new materials were not metals but *ceramics*! In shor order, other types of ceramic superconductors were discovered.

One of these new types was particularly easy to make. When a stoichio metric mixture of yttrium oxide (Y_2O_3), barium carbonate ($BaCO_3$), anc copper(II) oxide (CuO) is heated in a stream of $O_2(g)$, a ceramic is producec with the approximate formula $YBa_2Cu_3O_x$ (where x is slightly less than 7) which is an oxygen-deficient version of the ceramic $YBa_2Cu_3O_7$ shown ir Figure 23-17). This so-called YBCO ceramic becomes superconducting at th remarkably high temperature of 92 K. Although a temperature of 92 K is stil quite low, it is far above the boiling point of helium. In fact, it is above the boil ing point of nitrogen (77 K). Thus, inexpensive liquid nitrogen can be used a the coolant.

Many variations of the basic YBCO formula are possible. Almost any lan thanide element can be substituted for yttrium, and combinations of group elements can be substituted for barium. All these variations yield material that are superconducting at relatively high temperatures, but of the group, th yttrium compound is superconducting at the highest temperature.

The record high temperature for superconductivity set by the YBCO ceram ics was soon eclipsed by another group of ceramics containing bismuth anc

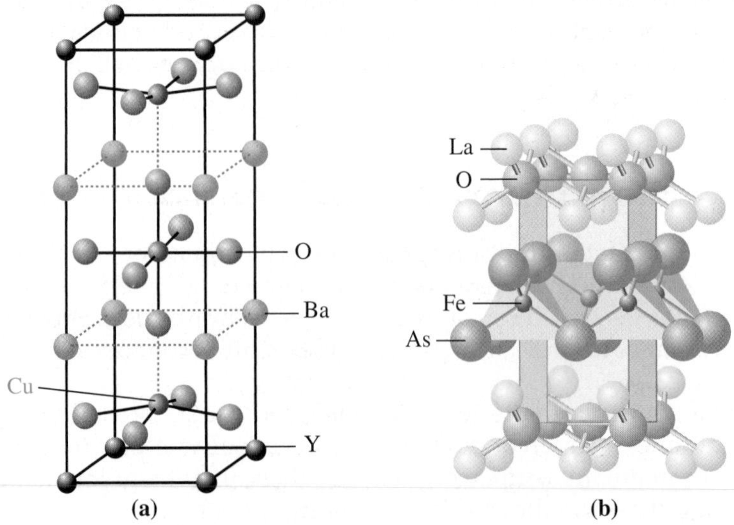

▶ FIGURE 23-17
Structures of $YBa_2Cu_3O_7$ and LaOFeAs
Two important classes of superconductors are based on the structures of **(a)** $YBa_2Cu_3O_7$ and **(b)** LaOFeAs.

opper, such as $Bi_2Sr_2CaCu_2O_8$. One of these is superconducting at 110 K, but his record was also short-lived. A ceramic containing thallium and copper, vith the approximate formula $TlBa_2Ca_3Cu_4O_y$ (where *y* is slightly larger than 0), was found to become superconducting at 125 K. Now, the search contin-es for materials that might become superconducting at room temperature about 293 K).

To date, many of the ceramic superconductors contain copper and share a ommon structural feature: copper and oxygen atoms bonded together in pla-ar sheets. In YBCO superconductors, the Cu—O planes are widely separated. n bismuth superconductors, the Cu—O planes occur in "sandwiches" consist-ng of two closely spaced sheets separated by a layer of group 2 ions. These andwiches are separated from one another by several layers of bismuth oxide. n the thallium superconductors, the Cu—O planes are stacked in groups of hree, like triple-decker sandwiches.

Recently,* the first copper-free ceramic superconductors were discovered. his new class of superconducting ceramics, with approximate formula $aO_{1-x}F_xFeAs$, is based on iron, a metal that is much more abundant than opper. These superconducting materials are based on LaOFeAs, shown in igure 23-17(b), but some of the O^{2-} ions are replaced by F^- ions. LaOFeAs has stack of alternating layers of LaO and FeAs, as illustrated in Figure 23-17(b). he LaO layer is positively charged, consisting of La^{3+} and O^{2-} ions, and the eAs layer is negatively charged, with the bonding between Fe and As being redominantly covalent. Replacing some of the O^{2-} ions with F^- ions gives a naterial that superconducts up to about 26 K. Although this temperature is not articularly high, the discovery has generated much excitement because it has pened up new possibilities for developing high-temperature superconductors.

The current theory of superconductivity, developed in the 1950s, explains he superconducting behavior of metals at very low temperatures but not the igher-temperature superconductivity of ceramics. It seems that the electrons n all known superconductors move through the material in pairs—a sort of uddy system that allows the electrons to move without resistance. The mech-nism by which electron pairs form in high-temperature superconductors, owever, is clearly different from that in low-temperature superconductors. ack of a suitable theory complicates the search for higher-temperature super-onductors. When the mechanism for high-temperature superconductors is etter understood, new breakthroughs might be easier to accomplish. Perhaps room-temperature superconducting material will be possible.

Despite this less-than-complete understanding of high-temperature super-onductors, engineers are already building devices that use the new materials. Vires have been made that are superconducting at liquid nitrogen tempera-ures, and new devices for precise magnetic field measurements using eramic superconductors are now being produced. Ultimately, ceramic uperconductors may find application in low-cost, energy-efficient electric ower transmission.

MasteringCHEMISTRY www.masteringchemistry.com

Nanoparticles are chemical structures whose dimensions are in the range 1–100 nm. One example of a nanoparticle is a quantum dot, a cluster of atoms with a diameter of tens of nanometers. For a discussion of the interesting and unusual properties of quantum dots, go to the Focus On feature for Chapter 23, Nanotechnology and Quantum Dots, on the MasteringChemistry site.

*Y. Kamihara, T. Watanabe, M. Hirano, and H. Hosono, *J. Am. Chem. Soc.*, **130**, 11 (2008).

Summary

23-1 General Properties—More than half the elements are the metals known as transition elements. Most are more reactive than hydrogen. Transition metals tend to exist in several different oxidation states in their compounds (see Table 23.1), and they readily form complex ions (discussed in Chapter 24). Many transition metals and their compounds are paramagnetic, and certain of the metals (Fe, Co, and Ni) and their alloys exhibit **ferromagnetism**. Within a group of *d*-block elements, the members of the second and third transition series resemble one another more than they do the group member in the first transition series. This is a consequence of the phenomenon known as the **lanthanide contraction** occurring in the sixth period.

23-2 Principles of Extractive Metallurgy—The extraction of metals from their ores is called **extractive metallurgy**. **Pyrometallurgy** relies on roasting an ore with subsequent reduction of the oxide to the metal. In **hydrometallurgy** metal ions are leached from ores using aqueous solutions of acid or bases. Various methods can be used to refine impure metals. One technique for producing an ultrapure metal is **zone refining**, which involves the continuous melting and refreezing of the metal to concentrate the impurities in one region of the sample, which is then discarded (Fig. 23-6). The thermodynamics of metallurgy can be viewed as the appropriate coupling of reactions into a spontaneous process (Fig. 23-8).

23-3 Metallurgy of Iron and Steel—The conversion of iron(III) oxide into iron by reduction with carbon monoxide in a blast furnace (Fig. 23-9) leads to the formation of **pig iron**. Pig iron is converted to **steel** through the **basic oxygen process** (Fig. 23-10).

23-4 First-Row Transition Elements: Scandium to Manganese—Oxidation–reduction reactions are commonly encountered with transition metal compounds of Sc, Ti, V, Cr, and Mn. Two common types of oxidizing agents are the dichromates and permanganates. In aqueous solution, dichromate ion is in equilibrium with chromate ion, which is a good precipitating agent for a number of metal ions. Most oxides and hydroxides of the transition metals are basic if the metal is in one of its lower oxidation states. In higher oxidation states, some transition metal oxides and hydroxy compounds are amphoteric, and in the highest oxidation states, a few are acidic (as is CrO_3, for example).

23-5 The Iron Triad: Iron, Cobalt, and Nickel—The iron triad elements (Fe, Co, and Ni) exhibit a variability in oxidation state with +2 the most common. Like other transition elements these metals form compounds with carbon monoxide called **metal carbonyls**.

23-6 Group 11: Copper, Silver, and Gold—The metals Cu, Ag, and Au are resistant to corrosion and are widely used in coins and jewelry. These so-called coinage metals are excellent conductors of heat and electricity and are highly malleable. They also find many uses in the electronics industry.

23-7 Group 12: Zinc, Cadmium, and Mercury—These group 12 elements have chemical properties consistent with a filled *d* subshell. Zn is used in an alloy with copper to produce brass, which is a good electrical conductor and is resistant to corrosion. Most metals form alloys with mercury called **amalgams**.

23-8 Lanthanides—This first series of *f*-block elements have long been known as the rare earth elements not because of their scarcity but because they are difficult to separate from one another. Their chemical behavior is strongly influenced by the lanthanide contraction, which produces similarities in their atomic and ionic sizes.

23-9 High-Temperature Superconductors—High-temperature superconducting materials have been made using some of the transition elements. In particular a ceramic material that is superconducting at 92 K is made from yttrium, barium, copper, and oxygen. These ceramic materials have the potential for many applications, for example, more efficient electric power transmission.

Integrative Example

Although a number of slightly soluble copper(I) compounds (such as CuCN) can exist in contact with water, it is not possible to prepare a solution with a high concentration of Cu^+ ion.

Show that $Cu^+(aq)$ disproportionates to $Cu^{2+}(aq)$ and Cu(s), and explain why a high $[Cu^+]$ cannot be maintained in aqueous solution.

Analyze

Write a plausible equation describing the disproportionation reaction. Then determine E°_{cell} for the reaction, followed by the equilibrium constant *K*, and see what conclusions can be drawn from the numerical value of *K*.

Solve

The half-equations and overall equation for the disproportionation are

$$\begin{aligned}
&\textit{Reduction:} && Cu^+(aq) + e^- \longrightarrow Cu(s)\\
&\textit{Oxidation:} && Cu^+(aq) \longrightarrow Cu^{2+}(aq) + e^-\\
&\textit{Overall:} && 2\,Cu^+(aq) \longrightarrow Cu^{2+}(aq) + Cu(s)
\end{aligned}$$

Find E° values for the couples $Cu^+/Cu(s)$ and Cu^{2+}/Cu^+ in Appendix D and combine them to obtain E°_{cell}.

$$\begin{aligned}
E^\circ_{cell} &= E^\circ(\text{reduction}) - E^\circ(\text{oxidation})\\
&= E^\circ_{Cu^+/Cu(s)} - E^\circ_{Cu^{2+}/Cu^+} = 0.520\text{ V} - 0.159\text{ V} = 0.361\text{ V}
\end{aligned}$$

se equation (19.17) to obtain the value of K. In the quation $n = 1$,

$$E^\circ_{cell} = \frac{0.025693\ V}{n} \ln K$$

$$\ln K = \frac{n \times E^\circ_{cell}}{0.025693\ V} = \frac{1 \times 0.361\ V}{0.025693\ V} = 14.1$$

$$K = e^{14.1} = 1.3 \times 10^6$$

hus, for the disproportionation reaction,

$$K = \frac{[Cu^{2+}]}{[Cu^+]^2} = 1.3 \times 10^6$$

nd

$$[Cu^{2+}] = 1.3 \times 10^6 \times [Cu^+]^2$$

maintain $[Cu^+] = 1$ M in solution, $[Cu^{2+}]$ would ave to be more than 1×10^6 M—a clear impossibility.

ssess

s a practical matter, we could not maintain $[Cu^+]$ at much more than 0.002 M, for even this would require that $[Cu^{2+}] \approx 5$ M.

RACTICE EXAMPLE A: Consider a galvanic cell based on the following half-reactions. Assuming the cell operates nder standard conditions at 25 °C, what is the spontaneous cell reaction? Under what nonstandard conditions is the pontaneous formation of $[PtCl_6]^{2-}$ favored? Do you think that, in practical terms, a significant amount of $PtCl_6{}^{2-}$ can be btained by altering the concentrations in a galvanic cell?

$$PtCl_6{}^{2-} + 2\,e^- \longrightarrow PtCl_4{}^{2-} + 2\,Cl^- \qquad E^\circ = 0.68\ V$$
$$V^{3+} + e^- \longrightarrow V^{2+} \qquad E^\circ = -0.255\ V$$

RACTICE EXAMPLE B: Because metallic titanium exhibits excellent corrosion resistance, it is often desirable to coat on objects with a thin coating of titanium metal. One approach involves production of Ti(s) from electrolysis of molten ixtures of NaCl and $TiCl_2$. The production of Ti(s) involves the disproportionation of Ti^{2+} to Ti^{3+} and Ti. Write a balanced hemical equation for the disproportionation reaction and use the following half-reactions to decide whether e disproportionation reaction is spontaneous under standard conditions. $Ti^{2+} + 2e^- \longrightarrow Ti$, $E^\circ = -1.630$ V; $Ti^{3+} + e^- \longrightarrow Ti^{2+}$, $E^\circ = -0.369$ V.

Exercises

roperties of the Transition Elements

1. By means of orbital diagrams, write electron configurations for the following transition element atom and ions: **(a)** V; **(b)** Cr^{3+}; **(c)** Mn^{2+}; **(d)** Fe^{2+}; **(e)** Cu^{2+}; **(f)** Ni^{2+}.
2. Arrange the following species according to the number of unpaired electrons they contain, starting with the one that has the greatest number: Fe, Sc^{3+}, Ti^{2+}, Mn^{4+}, Cr, Cu^{2+}.
3. Describe how the transition elements compare with main-group metals (such as group 2) with respect to oxidation states, formation of complexes, colors of compounds, and magnetic properties.
4. With only minor irregularities, the melting points of the first series of transition metals rise from that of Sc to that of Cr and then fall to that of Zn. Give a plausible explanation for this phenomenon based on atomic structure.
5. Why do the atomic radii vary so much more for two main-group elements that differ by one unit in atomic number than they do for two transition elements that differ by one unit?
6. The metallic radii of Ni, Pd, and Pt are 125, 138, and 139 pm, respectively. Why is the difference in radius between Pt and Pd so much less than between Pd and Ni?
7. Which of the first transition series elements exhibits the greatest number of different oxidation states in its compounds? Explain.
8. Why is the number of common oxidation states for the elements at the beginning and those at the end of the first transition series less than for elements in the middle of the series?
9. As a group, the lanthanides are more reactive metals than are those in the first transition series. How do you account for this difference?
10. The maximum difference in standard reduction potential, $E^\circ_{M^{2+}/M(s)}$, among members of the first transition series is about 2.4 V. For the lanthanides, the maximum difference in $E^\circ_{M^{3+}/M(s)}$ is only about 0.4 V. How do you account for this fact?

Reactions of Transition Metals and Their Compounds

11. Complete and balance the following equations. If no reaction occurs, so state.
    (a) $TiCl_4(g) + Na(l) \xrightarrow{\Delta}$
    (b) $Cr_2O_3(s) + Al(s) \xrightarrow{\Delta}$
    (c) $Ag(s) + HCl(aq) \longrightarrow$
    (d) $K_2Cr_2O_7(aq) + KOH(aq) \longrightarrow$
    (e) $MnO_2(s) + C(s) \xrightarrow{\Delta}$
12. By means of a chemical equation, give an example to represent the reaction of **(a)** a transition metal with a nonoxidizing acid; **(b)** a transition metal oxide with NaOH(aq); **(c)** an inner transition metal with HCl(aq).
13. Write balanced chemical equations for the following reactions described in the chapter.
    (a) the reaction of $Sc(OH)_3(s)$ with HCl(aq)
    (b) oxidation of $Fe^{2+}(aq)$ by $MnO_4^-(aq)$ in basic solution to give $Fe^{3+}(aq)$ and $MnO_2(s)$
    (c) the reaction of $TiO_2(s)$ with molten KOH to form K_2TiO_3
    (d) oxidation of Cu(s) to $Cu^{2+}(aq)$ with H_2SC (concd aq) to form $SO_2(g)$
14. Write balanced equations for the following reactio described in the chapter.
    (a) Sc(l) is produced by the electrolysis of Sc_2O_3 di solved in $Na_3ScF_6(l)$.
    (b) Cr(s) reacts with HCl(aq) to produce a blue sol tion containing $Cr^{2+}(aq)$.
    (c) $Cr^{2+}(aq)$ is readily oxidized by $O_2(g)$ to Cr^{3+}(aq
    (d) Ag(s) reacts with concentrated $HNO_3(aq)$, ar $NO_2(g)$ is evolved.
15. Suggest a series of reactions, using common chem cals, by which each of the following syntheses can b performed.
    (a) $Fe(OH)_3(s)$ from FeS(s)
    (b) $BaCrO_4(s)$ from $BaCO_3(s)$ and $K_2Cr_2O_7(aq)$
16. Suggest a series of reactions, using common chem cals, by which each of the following syntheses can b performed.
    (a) $Cu(OH)_2(s)$ from CuO(s)
    (b) $CrCl_3(aq)$ from $(NH_4)_2Cr_2O_7(s)$

Extractive Metallurgy

17. One of the simplest metals to extract from its ores is mercury. Mercury vapor is produced by roasting cinnabar ore (HgS) in air. Alternatives to this simple roasting, designed to reduce or eliminate SO_2 emissions, is to roast the ore in the presence of a second substance. For example, when cinnabar is roasted with quicklime, the products are mercury vapor and calcium sulfide and calcium sulfate. Write equations for the two reactions described here.
18. According to Figure 23-8, $\Delta_r G°$ decreases with temperature for the reaction $2\,C(s) + O_2(g) \longrightarrow 2\,CO(g)$. How would you expect $\Delta_r G°$ to vary with temperature for the following reactions?
    (a) $C(s) + O_2(g) \longrightarrow CO_2(g)$
    (b) $2\,CO(g) + O_2(g) \longrightarrow 2\,CO_2(g)$
19. Calcium will reduce MgO(s) to Mg(s) at all temper tures from 0 to 2000 °C. Use this fact, together with th melting point (839 °C) and boiling point (1484 °C) calcium, to sketch a plausible graph of $\Delta_r G°$ as a fun tion of temperature for the reaction $2\,Ca(s)$ $O_2(g) \longrightarrow 2\,CaO(s)$.
20. One method of obtaining chromium metal fro chromite ore is as follows. After reaction (23.16 sodium chromate is reduced to chromium(III) oxic by carbon. Then the chromium(III) oxide is reduced chromium metal by silicon. Write plausible equatio to describe these two reactions.

Oxidation–Reduction

21. Write plausible half-equations to represent each of the following in acidic solution.
    (a) CuO as an oxidizing agent
    (b) FeO as a reducing agent
22. Write plausible half-equations to represent each of the following in basic solution.
    (a) oxidation of $Fe(OH)_3(s)$ to FeO_4^{2-}
    (b) reduction of $[Ag(CN)_2]^-$ to silver metal
23. Use electrode potential data from this chapter or Appendix D to predict whether each of the following reactions will occur to any significant extent under standard-state conditions.
    (a) $2\,VO_2^+ + 6\,Br^- + 8\,H^+ \longrightarrow 2\,V^{2+} + 3\,Br_2(l) + 4\,H_2O$
    (b) $VO_2^+ + Fe^{2+} + 2\,H^+ \longrightarrow VO^{2+} + Fe^{3+} + H_2O$
    (c) $MnO_2(s) + H_2O_2 + 2\,H^+ \longrightarrow Mn^{2+} + 2\,H_2O + O_2(g)$
24. You are given these three reducing agents: Zn(s $Sn^{2+}(aq)$, and $I^-(aq)$. Use data from Appendix D t determine which of them can, under standard-sta conditions in acidic solution, reduce
    (a) $Cr_2O_7^{2-}(aq)$ to $Cr^{3+}(aq)$
    (b) $Cr^{3+}(aq)$ to $Cr^{2+}(aq)$
    (c) $SO_4^{2-}(aq)$ to $SO_2(g)$
25. Refer to Example 23-2. Select a reducing agent (fro Table 23.1 or Appendix D) that will reduce VO^{2+} t V^{3+} and no further in acidic solution.

26. The electrode potential diagram for manganese in acidic solutions in Figure 23-14 does not include a value of $E°$ for the reduction of MnO_4^- to Mn^{2+}. Use other data in the figure to establish this $E°$, and compare your result with the value found in Table 19.1.

27. Use data from the text to construct a standard electrode potential diagram relating the following chromium species in acidic solution.

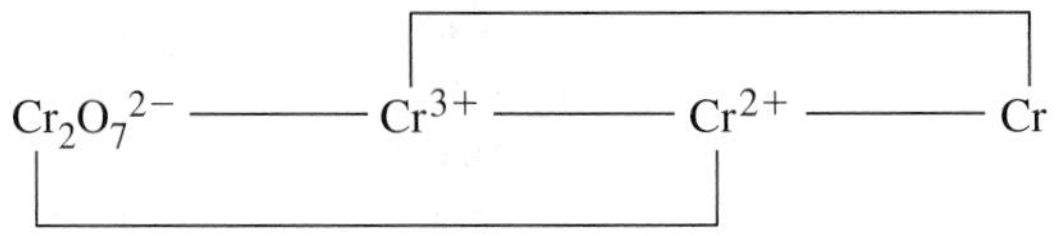

28. Use data from the text to construct a standard electrode potential diagram relating the following vanadium species in acidic solution.

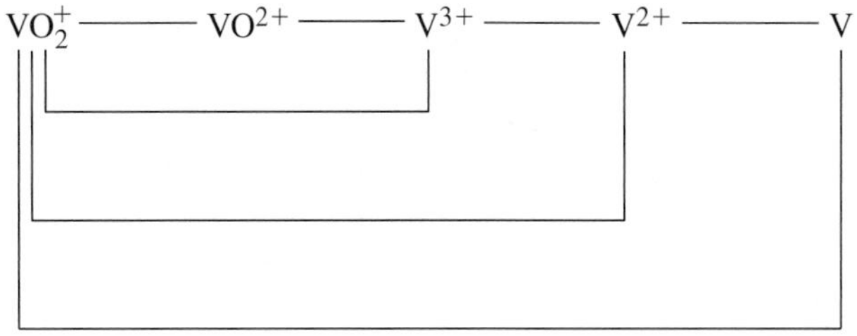

Chromium and Chromium Compounds

29. When a soluble lead compound is added to a solution containing primarily *orange* dichromate ion, *yellow* lead chromate precipitates. Describe the equilibria involved.

30. When *yellow* $BaCrO_4$ is dissolved in HCl(aq), a *green* solution is obtained. Write a chemical equation to account for the color change.

31. When Zn(s) is added to $K_2Cr_2O_7$ dissolved in HCl(aq), the color of the solution changes from orange to green, then to blue, and, over a period of time, back to green. Write equations for this series of reactions.

32. If $CO_2(g)$ under pressure is passed into $Na_2CrO_4(aq)$, $Na_2Cr_2O_7(aq)$ is formed. What is the function of the $CO_2(g)$? Write a plausible equation for the net reaction.

33. Use equation (23.19) to determine $[Cr_2O_7^{2-}]$ in a solution that has $[CrO_4^{2-}] = 0.20$ M and pH of **(a)** 7.12 and **(b)** 9.15.

34. If a solution is prepared by dissolving 1.505 g Na_2CrO_4 in 345 mL of a buffer solution with pH = 7.55, what will be $[CrO_4^{2-}]$ and $[Cr_2O_7^{2-}]$?

35. How many grams of chromium would be deposited on an object in a chrome-plating bath (see page 1108) after 1.00 h at a current of 3.4 A?

36. How long would an electric current of 3.5 A have to pass through a chrome-plating bath (see page 1108) to produce a chromium deposit 0.0010 mm thick on an object with a surface area of 0.375 m^2? (The density of Cr is 7.14 g cm^{-3}.)

37. Why is it reasonable to expect the chemistry of dichromate ion to involve mainly oxidation–reduction reactions and that of chromate ion to involve mainly precipitation reactions?

38. What products are obtained when $Mg^{2+}(aq)$ and $Cr^{3+}(aq)$ are each treated with a limited amount of NaOH(aq)? With an excess of NaOH(aq)? Why are the results different in these two cases?

The Iron Triad

39. Will reaction (23.25) still be spontaneous in the forward direction in a solution containing equal concentrations of Fe^{2+} and Fe^{3+}, a pH of 3.25, and under an $O_2(g)$ partial pressure of 0.20 atm?

40. Based on the description of the nickel–cadmium cell on page 1113, and with appropriate data from Appendix D, estimate $E°$ for the reduction of NiO(OH) to $Ni(OH)_2$.

41. Write a net ionic equation to represent the precipitation of Prussian blue, described on page 1113.

42. The reaction to form Turnbull's blue (page 1113) appears to occur in two stages. First, $Fe^{2+}(aq)$ is oxidized to $Fe^{3+}(aq)$ and ferricyanide ion is reduced to ferrocyanide ion. Then, the $Fe^{3+}(aq)$ and ferrocyanide ion combine. Write equations for these reactions.

Group 11 Metals

43. Write plausible equations for the following reactions occurring in the hydrometallurgy of the coinage metals.
(a) Copper is precipitated from a solution of copper(II) sulfate by treatment with $H_2(g)$.
(b) Gold is precipitated from a solution of Au^+ by adding iron(II) sulfate.
(c) Copper(II) chloride solution is reduced to copper(I) chloride when treated with $SO_2(g)$ in acidic solution.

44. In the metallurgical extraction of silver and gold, an alloy of the two metals is often obtained. The alloy can be separated into Ag and Au either with concentrated HNO_3 or boiling concentrated H_2SO_4, in a process called *parting*. Write chemical equations to show how these separations work.

45. Use the result of the Integrative Example to determine whether a solution can be prepared with $[Cu^+]$ equal to **(a)** 0.20 M; **(b)** 1.0×10^{-10} M.

46. Show that the corrosion reaction in which Cu is converted to its basic carbonate (reaction 23.27) can be thought of in terms of a combination of oxidation–reduction, acid–base, and precipitation reactions.

Group 12 Metals

47. Use data from Table 23.8 to determine $E°$ for the reduction of Hg^{2+} to Hg_2^{2+} in aqueous solution.
48. At 400 °C, $\Delta_r G° = -25$ kJ mol^{-1} for the reaction $2\ Hg(l) + O_2(g) \longrightarrow 2\ HgO(s)$. If a sample of HgO(s) is heated to 400 °C, what will be the equilibrium partial pressure of $O_2(g)$?
49. Use Figure 23-8 to estimate for the reaction $ZnO(s) + C(s) \rightleftharpoons Zn(l) + CO(g)$, at 800 °C, **(a)** a value of K_p and **(b)** the equilibrium pressure of CO(g).
50. The vapor pressure of Hg(l) as a function of temperature is $\log P(\text{mmHg}) = (-0.05223\,a/T) + b$, where $a = 61{,}960$ and $b = 8.118$; T is the Kelvin temperature. Show that at 25 °C, the concentration of Hg(g) i equilibrium with Hg(l) greatly exceeds the maximu permissible level of 0.05 mg Hg/m^3 air.
51. In ZnO, the band gap between the valence an conduction bands is 325 kJ mol^{-1}, and in CdS it 250 kJ mol^{-1}. Show that CdS absorbs some visib light but ZnO does not. Explain the observed color ZnO is white and CdS is yellow.
52. CdS is yellow, HgS is red, and CdSe is black. Whic of these materials has the largest band gap? th smallest? How does the band gap relate to th observed color?

Integrative and Advanced Exercises

53. Although Au reacts with and dissolves in aqua regia (3 parts HCl + 1 part HNO_3), Ag does not dissolve. What is (are) the likely reason(s) for this difference?
54. The text mentions that scandium metal is obtained from its molten chloride by electrolysis, and that titanium is obtained from its chloride by reduction with magnesium. Why are these metals not obtained by the reduction of their oxides with carbon (coke), as are metals such as zinc and iron?
55. The text notes that in small quantities, zinc is an essential element (though it is toxic in higher concentrations). Tin is considered to be a toxic metal. Can you think of reasons why, for food storage, tinplate instead of galvanized iron is used in cans?
56. In an atmosphere polluted with industrial smog, Cu corrodes to a basic sulfate, $Cu_2(OH)_2SO_4$. Propose a series of chemical reactions to describe this corrosion.
57. What formulas would you expect for the metal carbonyls of **(a)** molybdenum, **(b)** osmium; **(c)** rhenium? Note that the simple carbonyls shown in Figure 23-15 have one metal atom per molecule. Some metal carbonyls are *binuclear*; that is, they have two metal atoms bonded together in the carbonyl structure. Also, **(d)** explain why iron and nickel carbonyls are liquids at room temperature, whereas that of cobalt is a solid, and **(e)** describe the probable nature of the bonding in the compound $Na[V(CO)_6]$.
58. For the straight-line graphs in Figure 23-8, explain why **(a)** breaks occur at the melting points and boiling points of the metals; **(b)** the slopes of the lines become more positive at these breaks; **(c)** the break at the boiling point is sharper than at the melting point.
59. Attempts to make CuI_2 by the reaction of $Cu^{2+}(aq)$ and $I^-(aq)$ produce CuI(s) and $I_3^-(aq)$ instead. Without performing detailed calculations, show why this reaction should occur.

$$2\ Cu^{2+}(aq) + 5\ I^-(aq) \longrightarrow 2\ CuI(s) + I_3^-(aq)$$

60. Without performing detailed calculations, show that significant disproportionation of AuCl occurs if you attempt to make a saturated aqueous solution. Use data from Table 23.7 and $K_{sp}(AuCl) = 2.0 \times 10^{-13}$.
61. In acidic solution, silver(II) oxide first dissolves to pro duce $Ag^{2+}(aq)$. This is followed by the oxidation c $H_2O(l)$ to $O_2(g)$ and the reduction of Ag^{2+} to Ag^+.
(a) Write equations for the dissolution and oxidation reduction reactions.
(b) Show that the oxidation–reduction reaction i indeed spontaneous.
62. Equation (23.18), which represents the chromate dichromate equilibrium, is actually the sum of tw equilibrium expressions. The first is an acid–bas reaction, $H^+ + CrO_4^{2-} \rightleftharpoons HCrO_4^-$. The secon reaction involves elimination of a water molecul between two $HCrO_4^-$ ions (a dehydration reaction $2\ HCrO_4^- \rightleftharpoons Cr_2O_7^{2-} + H_2O$. If the ionizatio constant, K_a, for $HCrO_4^-$ is 3.2×10^{-7}, what is th value of K for the dehydration reaction?
63. Show that under the following conditions, Ba^{2+}(aq can be separated from $Sr^{2+}(aq)$ and $Ca^{2+}(aq)$ by pre cipitating $BaCrO_4(s)$ with the other ions remainin in solution:

$$[Ba^{2+}] = [Sr^{2+}] = [Ca^{2+}] = 0.10\ M$$
$$[CH_3COOH] = [CH_3COO^-] = 1.0\ M$$
$$[Cr_2O_7^{2-}] = 0.0010\ M$$
$$K_{sp}(BaCrO_4) = 1.2 \times 10^{-10}$$
$$K_{sp}(SrCrO_4) = 2.2 \times 10^{-5}$$

Use data from this and previous chapters, as necessary

64. A 0.960 g sample of impure hematite (Fe_2O_3) is treate with 1.752 g of oxalic acid ($H_2C_2O_4 \cdot 2\ H_2O$) in a acidic medium (reaction 1). Following this, the exces oxalic acid is titrated with 35.16 mL of 0.100 M $KMnO_4$ (reaction 2). What is the mass percent o Fe_2O_3 in the impure sample of hematite? The follow ing equations are neither complete nor balanced.

(1) $H_2C_2O_4(aq) + Fe_2O_3 \longrightarrow Fe^{2+}(aq) + CO_2(g)$
(2) $H_2C_2O_4(aq) + MnO_4(aq) \longrightarrow Mn^{2+}(aq) + CO_2(g$

65. Both $Cr_2O_7^{2-}(aq)$ and $MnO_4^-(aq)$ can be used t titrate $Fe^{2+}(aq)$ to $Fe^{3+}(aq)$. Suppose you have avail able as titrants two solutions: 0.1000 M $Cr_2O_7^{2-}$(aq and 0.1000 M $MnO_4^-(aq)$.

(a) For which solution would the greater volume of titrant be required for the titration of a particular sample of Fe^{2+}(aq)? Explain.
(b) How many mL of 0.1000 M MnO_4^-(aq) would be required for a titration if the same titration requires 24.50 mL of 0.1000 M $Cr_2O_7^{2-}$(aq)?

66. The only important compounds of Ag(II) are AgF_2 and AgO. Why would you expect these two compounds to be stable, but not other silver(II) compounds such as $AgCl_2$, $AgBr_2$, and AgS?

67. A certain steel is to be analyzed for Cr and Mn. By suitable treatment, the Cr in the steel is oxidized to $Cr_2O_7^{2-}$(aq) and the Mn to MnO_4^-(aq). A 10.000 g sample of steel is used to produce 250.0 mL of a solution containing $Cr_2O_7^{2-}$(aq) and MnO_4^-(aq). A 10.00 mL portion of this solution is added to $BaCl_2$(aq), and by proper adjustment of the pH, the chromium is completely precipitated as $BaCrO_4$(s); 0.549 g is obtained. A second 10.00 mL portion of the solution requires exactly 15.95 mL of 0.0750 M Fe^{2+}(aq) for its titration in acidic solution. Calculate the % Cr and % Mn in the steel sample. [*Hint:* In the titration MnO_4^-(aq) is reduced to Mn^{2+}(aq) and $Cr_2O_7^{2-}$(aq) is reduced to Cr^{3+}(aq); the Fe^{2+}(aq) is oxidized to Fe^{3+}(aq).]

68. The palladium content of a steel sample was determined as follows. A 16.312 g steel sample was dissolved in concentrated HCl(aq). The solution obtained was treated to remove interfering ions, to establish the proper pH, and to obtain a final solution volume of 250.0 mL. A 10.00 mL sample of this solution was then treated with dimethylglyoxime to convert all of the palladium to palladium dimethylglyoximate, a chemical compound that is 31.61% Pd (by mass), 28.54% C, 4.19% H, 19.01% O, and 16.64% N. The mass of purified, dry palladium dimethylglyoximate obtained was 0.0784 g.
(a) What is the empirical formula of palladium dimethylglyoximate?
(b) What is the mass percent of palladium in the steel sample?

69. A solution is believed to contain one or more of the following ions: Cr^{3+}, Zn^{2+}, Fe^{3+}, Ni^{2+}. When the solution is treated with excess NaOH(aq), a precipitate forms. The solution in contact with the precipitate is colorless. The precipitate is dissolved in HCl(aq), and the resulting solution is treated with NH_3(aq). No precipitation occurs. Based solely on these observations, what conclusions can you draw about the ions present in the original solution? That is, which ion(s) are likely present, which are most likely not present, and about which can we not be certain? [*Hint:* Refer to Appendix D for solubility product and complex-ion formation data.]

70. Nearly all mercury(II) compounds exhibit covalent bonding. Mercury(II) chloride is a covalent molecule that dissolves in warm water. The stability of this compound is exploited in the determination of the levels of chloride ion in blood serum. Typical human blood serum levels range from 90 to 115 mmol L^{-1}. The chloride concentration is determined by titration with $Hg(NO_3)_2$. The indicator used in the titration is diphenylcarbazone, $C_6H_5N{=}NCONHNHC_6H_5$, which complexes with the mercury(II) ion after all the chloride has reacted with the mercury(II). Free diphenylcarbazone is pink in solution, and when it is complexed with mercury(II), it is blue. Thus, the diphenylcarbazone acts as an indicator, changing from pink to blue when the first excess of mercury(II) appears. In an experiment, $Hg(NO_3)_2$(aq) solution is standardized by titrating 2.00 mL of 0.0108 M NaCl solution. It takes 1.12 mL of $Hg(NO_3)_2$(aq) to reach the diphenylcarbazone end point. A 0.500 mL serum sample is treated with 3.50 mL water, 0.50 mL of 10% sodium tungstate solution, and 0.50 mL of 0.33 M H_2SO_4(aq) to precipitate proteins. After the proteins are precipitated, the sample is filtered and a 2.00 mL aliquot of the filtrate is titrated with $Hg(NO_3)_2$ solution, requiring 1.23 mL. Calculate the concentration of Cl^-. Express your answer in mmol L^{-1}. Does this concentration fall in the normal range?

71. Covalent bonding is involved in many transition metal compounds. Draw Lewis structures, showing any nonzero formal charges, for the following molecules or ions: **(a)** Hg_2^{2+}; **(b)** Mn_2O_7; **(c)** OsO_4. [*Hint:* In (b), there is one Mn—O—Mn linkage in the molecule.]

72. For a coordination number of four, the radius of Mn^{7+} has been estimated to be 39 pm. Estimate the charge density for the Mn^{7+} ion. Express your answer in C mm^{-3}. How does this compare with the charge density of Be^{2+} given in Table 21.4? Would you expect the bonding in Mn_2O_7 to be primarily ionic or primarily covalent? Explain.

73. Nitinol is a nickel–titanium alloy known as *memory metal*. The name nitinol is derived from the symbols for nickel (Ni), titanium (Ti), and the acronym for the Naval Ordinance Laboratory (NOL), where it was discovered. If an object made out of nitinol is heated to about 500 °C for about an hour and then allowed to cool, the original shape of the object is "remembered," even if the object is deformed into a different shape. The original shape can be restored by heating the metal. Because of this property, nitinol has found many uses, especially in medicine and orthodontics (for braces). Nitinol exists in a number of different solid phases. In the so-called austerite phase, the metal is relatively soft and elastic. The crystal structure for the austerite phase can be described as a simple cubic lattice of Ti atoms with Ni atoms occupying cubic holes in the lattice of Ti atoms. What is the empirical formula of nitinol and what is the percent by mass of titanium in the alloy?

Feature Problems

74. As a continuation of Exercise 101 of Chapter 13 and the discussion on page 1100, consider the three graphs of $\Delta_r G°$ as a function of temperature on the next page.
(a) Explain the slopes of the three functions. Specifically, why is one line essentially parallel to the temperature axis, why does one have a positive slope, and why does one have a negative slope?

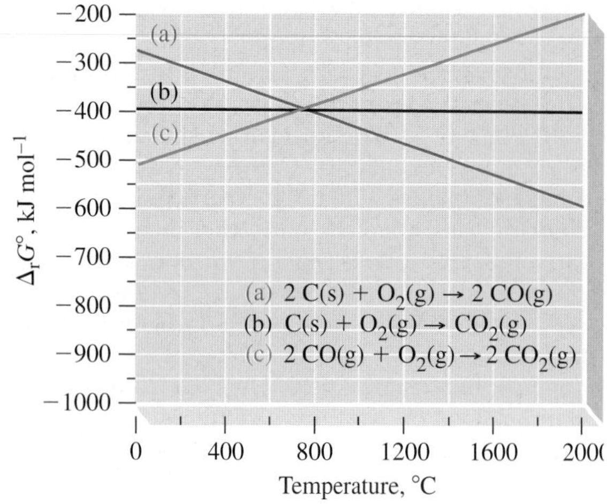

(b) Table 23.2 lists as an additional blast furnace reaction, $C(s) + CO_2(g) \longrightarrow 2\,CO(g)$. Determine how $\Delta_r G°$ for this reaction is related to the three reactions shown in the figure, and plot $\Delta_r G°$ for this reaction as a function of temperature. If an equilibrium is established in this reaction at 1000 °C and the partial pressure of $CO_2(g)$ is 0.25 atm, what should be the equilibrium partial pressure of CO(g)?

75. Several transition metal ions are found in cation group 3 of the qualitative analysis scheme outlined in Figure 18-7. At one point in the separation and testing of this group, a solution containing Fe^{3+}, Co^{2+}, Ni^{2+}, Al^{3+}, Cr^{3+}, and Zn^{2+} is treated with an excess c NaOH(aq), together with $H_2O_2(aq)$.

(1) The excess NaOH(aq) causes *three* of the cations t precipitate as hydroxides and *three* to form hydroxid complex ions.

(2) In the presence of $H_2O_2(aq)$, the cation in one c the insoluble hydroxides is oxidized from the +2 t the +3 oxidation state, and one of the hydroxido com plex ions is also oxidized.

(3) The three insoluble hydroxides are found as dark precipitate.

(4) The solution above the precipitate has a yellow colo

(5) The dark precipitate from (3) reacts with HCl(aq and all the cations return to solution; one of the cation is reduced from the +3 to the +2 oxidation state.

(6) The solution from (5) is treated with 6 M NH_3(aq and a precipitate containing one of the cations forms.

(a) Write equations for the reactions referred to i item (1).

(b) Write an equation for the most likely reaction i which a hydroxide precipitate is oxidized in item (2)

(c) What is the ion responsible for the yellow color of th solution in item (4)? Write an equation for its formation.

(d) Write equations for the dissolution of the precipi tate and the reduction of the cation in item (5).

(e) Write an equation for the precipitate formatio in item (6). [*Hint:* You may need solubility produc and complex-ion formation data from Appendix D together with descriptive information from thi chapter and from elsewhere in the text.]

Self-Assessment Exercises

76. In your own words, define the following terms: **(a)** domain; **(b)** flotation; **(c)** leaching; **(d)** amalgam.

77. Briefly describe each of the following ideas, phenomena, or methods: **(a)** lanthanide contraction; **(b)** zone refining; **(c)** basic oxygen process; **(d)** slag formation.

78. Explain the important distinctions between each pair of terms: **(a)** ferromagnetism and paramagnetism; **(b)** roasting and reduction; **(c)** hydrometallurgy and pyrometallurgy; **(d)** chromate and dichromate.

79. Describe the chemical composition of the material called **(a)** pig iron; **(b)** ferromanganese alloy; **(c)** chromite ore; **(d)** brass; **(e)** aqua regia; **(f)** blister copper; **(g)** stainless steel.

80. Three properties expected for transition elements are **(a)** low melting points; **(b)** high ionization energies; **(c)** colored ions in solution; **(d)** positive standard electrode (reduction) potentials; **(e)** diamagnetism; **(f)** complex ion formation; **(g)** catalytic activity.

81. The only diamagnetic ion of the following group is **(a)** Cr^{2+}; **(b)** Zn^{2+}; **(c)** Fe^{3+}; **(d)** Ag^{2+}; **(e)** Ti^{3+}.

82. All of the following elements have an ion displaying the +6 oxidation state except **(a)** Mo; **(b)** Cr; **(c)** Mn; **(d)** V; **(e)** S.

83. The best oxidizing agent of the following group of ions is **(a)** $Ag^+(aq)$; **(b)** $Cl^-(aq)$; **(c)** $H^+(aq)$; **(d)** $Na^+(aq)$; **(e)** $OH^-(aq)$.

84. To separate Fe^{3+} and Ni^{2+} from an aqueous solutio containing both ions, with one cation forming a precipitate and the other remaining in solution add to the solution **(a)** NaOH(aq); **(b)** H_2S(g **(c)** HCl(aq); **(d)** NH_3(aq).

85. Of the following, the two solids that will liberat $Cl_2(g)$ when heated with HCl(aq) are **(a)** NaCl(s **(b)** $ZnCl_2$(s); **(c)** MnO_2(s); **(d)** CuO(s); **(e)** $K_2Cr_2O_7$(s **(f)** NaOH(s).

86. Provide the missing name or formula for the following

(a) chromium(VI) oxide	_____
(b) _____	K_2MnO_4
(c) _____	$Cr(CO)_6$
(d) barium dichromate	_____
(e) _____	$La_2(SO_4)_3 \cdot 9\,H_2C$
(f) gold(III) cyanide trihydrate	_____

87. Balance the following oxidation–reduction equations

(a) $Fe_2S_3(s) + H_2O(l) + O_2(g) \longrightarrow Fe(OH)_3(s) + S$

(b) $Mn^{2+}(aq) + S_2O_8^{2-}(aq) + H_2O(l) \longrightarrow MnO_4^-(aq) + SO_4^{2-}(aq) + H^+(aq$

(c) $Ag(s) + CN^-(aq) + O_2(g) + H_2O(l) \longrightarrow [Ag(CN)_2]^-(aq) + OH^-(aq$

88. Explain why Zn, Cd, and Hg resemble the group 2 metals in some of their properties.

89. Explain why gold dissolves in aqua regia but not in $HNO_3(aq)$.

90. Explain why 1.0 M $Fe(NO_3)_3(aq)$ is acidic.

Complex Ions and Coordination Compounds

24

CONTENTS

Turquoise is a mineral of copper, $CuAl_6(PO_4)_4(OH)_8 \cdot 4\ H_2O$. The distinctive color of this gemstone and many others is a consequence of the nature of metal–ligand bonding in complex ions, a central topic of this chapter.

LEARNING OBJECTIVES

24.1 Describe how ligands and metals form complexes, and explain the meaning of coordination number.

24.2 Distinguish between a monodentate and a bidentate ligand, and describe chelation.

24.3 Use nomenclature rules to assign the names of common transition metal complexes.

24.4 Give examples of structural isomers (ionization, coordination, and linkage) and stereoisomers (geometric and optical).

24.5 Describe what happens to the metal *d*-orbitals in an octahedral ligand field, and explain what is meant by spectrochemical series.

24.6 Correlate the structure and *d*-electron counts of complexes to their magnetic properties (paramagnetic versus diamagnetic).

24.7 Discuss how colors are displayed by coordination complexes and the relationship of primary, secondary, and complementary colors.

24.8 Describe how to quantify the rate of complex-ion formation and the driving force behind the chelate effect.

24.9 Describe how complex ions may behave as acids and bases.

24.10 Describe the meaning and physical effects of labile ligands.

24.11 Discuss some important applications of coordination complexes.

Chapter 23 included discussions of several situations involving a succession of color changes that were attributed to changes in oxidation state. The color changes discussed in this chapter are not generally caused by oxidation–reduction reactions. Instead, they are observed with changes in the groups (ligands) bound to a metal center, even though the oxidation state of the metal remains unchanged. Explanation of this observation requires a fuller exploration of the nature of complex ions and coordination compounds, a subject that was briefly introduced in Chapter 18. One topic we will consider is the geometric structure of complex ions, which reveals new possibilities for isomerism: the existence of compounds having identical compositions but different structures and properties. We will also examine the nature of the bondin between ligands and the metal centers to which they are attached. It i through an understanding of bonding in complex ions that we can gain som insight into the origin of their colors.

24-1 Werner's Theory of Coordination Compounds: An Overview

Interfoto/ Alamy

▲ **Alfred Werner (1866–1919)** Werner's success in explaining coordination compounds came in large part through his application of new ideas: the theory of electrolytic dissociation and principles of structural chemistry.

Prussian blue (page 1113), accidentally discovered early in the eighteenth cen tury, was perhaps the first known coordination compound—the type explore in this chapter. However, nearly a century passed before the uniqueness o these compounds came to be appreciated. In 1798, B. M. Tassaert obtained ye low crystals of a compound having the formula $CoCl_3 \cdot 6\,NH_3$ from a mixtur of $CoCl_3$ and $NH_3(aq)$. What seemed unusual was that both $CoCl_3$ and NH are stable compounds capable of independent existence, yet they combine t form still another stable compound. Such compounds made up of two simple compounds came to be called **coordination compounds**.

In 1851, another coordination compound of $CoCl_3$ and NH_3 was discov ered. This one had the formula $CoCl_3 \cdot 5\,NH_3$ and formed purple crystals. Th two compounds are shown in Figure 24-1. Their formulas follow.

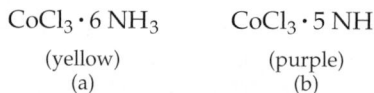

$CoCl_3 \cdot 6\,NH_3$	$CoCl_3 \cdot 5\,NH_3$
(yellow)	(purple)
(a)	(b)

The mystery of coordination compounds deepened as more were discovere and studied. For example, when treated with $AgNO_3(aq)$, compound (a formed *three* moles of AgCl(s), as expected, but compound (b) formed only *tw* moles of AgCl(s).

Inorganic coordination chemistry was a hot field of research in the secon half of the nineteenth century, and all the pieces started to fall into place wit the work of the Swiss chemist Alfred Werner. Werner's theory of coordinatio compounds explained the reactions of compounds (a) and (b) wit $AgNO_3(aq)$ by considering that, in aqueous solutions, these two compound ionize in the following way:

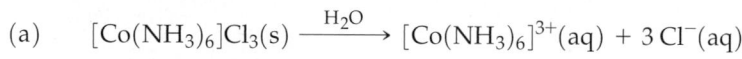

$$\text{(a)}\quad [Co(NH_3)_6]Cl_3(s) \xrightarrow{H_2O} [Co(NH_3)_6]^{3+}(aq) + 3\,Cl^-(aq)$$

$$\text{(b)}\quad [CoCl(NH_3)_5]Cl_2(s) \xrightarrow{H_2O} [CoCl(NH_3)_5]^{2+}(aq) + 2\,Cl^-(aq)$$

Thus, compound (a) produces the three moles of Cl^- per mole of compoun necessary to precipitate three moles of AgCl(s), while compound (b) produce only two moles of Cl^-. Werner's proposal of this ionization scheme was base on extensive studies of the electrical conductivity of coordination compounds

Carey B. Van Loon

▲ FIGURE 24-1
Two coordination compounds
The compound on the left is $[Co(NH_3)_6]Cl_3$. The compound on the right is $[CoCl(NH_3)_5]Cl_2$.

ompound (a) is a better conductor than compound (b), consistent with roducing four ions per formula unit compared to three for compound (b). $oCl_3 \cdot 4\,NH_3$ is a still poorer conductor, corresponding to the formula $CoCl_2(NH_3)_4]Cl$. $CoCl_3 \cdot 3\,NH_3$ is a *nonelectrolyte*, corresponding to the for- ıula $[CoCl_3(NH_3)_3]$.

The heart of Werner's theory, proposed in 1893, was that certain metal toms, primarily those of transition metals, have two types of valence or bond- ıg capacity. One, the *primary* valence, is based on the number of electrons the tom loses in forming the metal ion. A *secondary* valence is responsible for the onding of other groups, called **ligands**, to the central metal ion.

In modern usage, a **complex** is any species involving coordination of gands to a metal center. The metal center can be an atom or an ion, and the omplex can be a cation, an anion, or a neutral molecule. In a chemical for- ıula, a complex—a metal center and attached ligands—is set off by square rackets, [ ]. Compounds that are complexes or contain complex ions are nown as coordination compounds.

$[Co(NH_3)_6]^{3+}$	$[CoCl_4(NH_3)_2]^-$	$[CoCl_3(NH_3)_3]$	$K_4[Fe(CN)_6]$
Complex cation	Complex anion	Neutral complex	Coordination compound

The **coordination number** of a complex is the number of points around the etal center at which bonds to ligands can form. Coordination numbers rang- ng from 2 to 12 have been observed, although 6 is by far the most common umber, followed by 4. Coordination number 2 is limited mostly to complexes f Cu(I), Ag(I), and Au(I). Coordination numbers greater than 6 are not often ound in members of the first transition series, but are more common in those f the second and third series. Stable complexes with coordination numbers 3 nd 5 are rare. The coordination number observed in a complex depends on a umber of factors, such as the ratio of the radius of the central metal atom or on to the radii of the attached ligands.

Coordination numbers of some common ions are listed in Table 24.1. The our most commonly observed geometric shapes of complex ions are shown in igure 24-2. One practical use of the coordination number is to assist in writ- ng and interpreting formulas of complexes, as illustrated in Example 24-1.

TABLE 24.1 Some Common Coordination Numbers of Metal Ions

Cu^+	2, 4
Ag^+	2
Au^+	2, 4
Fe^{2+}	6
Co^{2+}	4, 6
Ni^{2+}	4, 6
Cu^{2+}	4, 6
Zn^{2+}	4
Pt^{2+}	4
Al^{3+}	4, 6
Sc^{3+}	6
Cr^{3+}	6
Fe^{3+}	6
Co^{3+}	6
Au^{3+}	4
Pt^{4+}	6

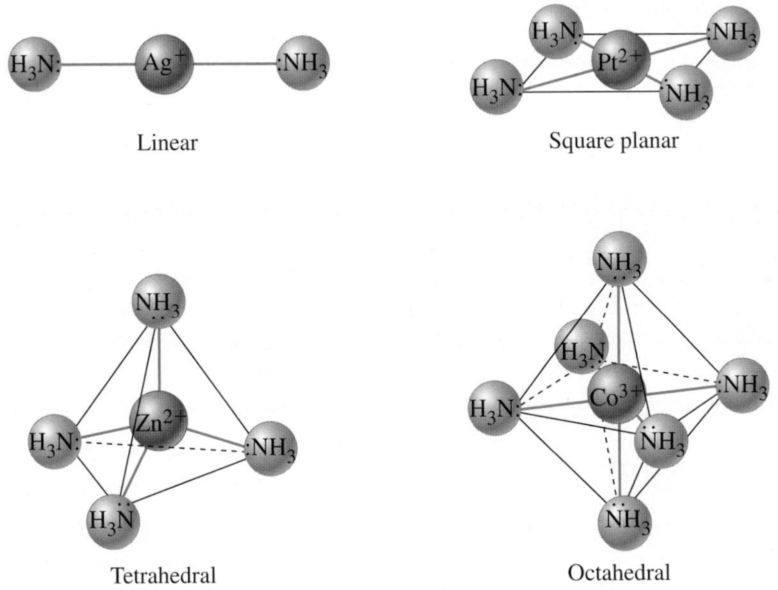

▲ FIGURE 24-2
Structures of some complex ions
Attachment of the NH_3 molecules occurs through the lone-pair electrons on the N atoms. n each complex, all the ligands are the same and hence no distortions of these shapes are observed.

EXAMPLE 24-1 Relating the Formula of a Complex to the Coordination Number and Oxidation State of the Central Metal

What are the coordination number and oxidation state of Co in the complex ion $[CoCl(NO_2)(NH_3)_4]^+$?

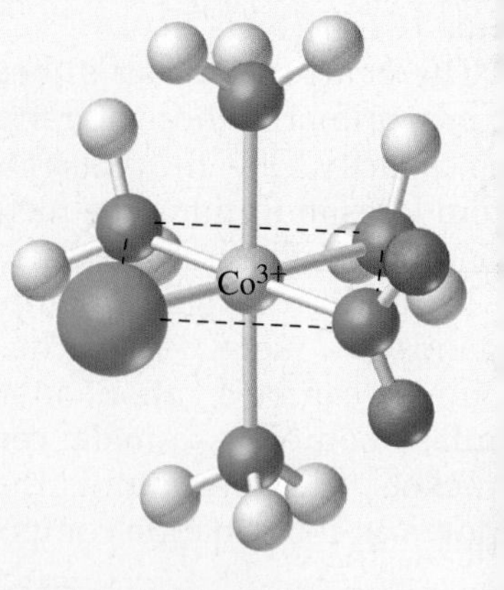

▲ The complex ion $[CoCl(NO_2)(NH_3)_4]^+$

Analyze

In determining the oxidation state of the metal ion in a complex, it is important to recognize which ligands are charged (invariably negatively charged) and which are neutral.

Solve

The complex ion has as ligands *one* Cl^- ion, *one* NO_2^- ion, and *four* NH_3 molecules. The coordination number is 6. Of these six ligands, *two* carry a charge of −1 each (the Cl^- and NO_2^- ions) and four are *neutral* (the NH_3 molecules). The total contribution of the anions to the net charge on the complex ion is −2. Because the net charge on the complex ion is +1, the oxidation state of the central cobalt ion is +3. Diagrammatically, we can write

Oxidation state = x

Charge of −1 on Cl^- } Total negative charge: −2
Charge of −1 on NO_2^-

$[CoCl(NO_2)(NH_3)_4]^+$

Coordination number = 6

Net charge on complex ion

$$x - 2 = +1$$
$$x = +3$$

Assess

The oxidation state of the metal can be determined by using the following relationship:

charge on the complex ion = oxidation state of the metal + sum of charges on the ligands

In this case,

$$(+1) = \text{oxidation state of the metal} + (-1) + (-1)$$
$$\text{oxidation state of the metal} = +3$$

PRACTICE EXAMPLE A: What are the coordination number and oxidation state of nickel in the ion $[Ni(CN)_4I]^{3-}$?

PRACTICE EXAMPLE B: Write the formula of a complex with cyanide ion ligands, an iron ion with an oxidation state of +3, and a coordination number of 6.

24-1 CONCEPT ASSESSMENT

A complex of Al(III) can be formulated as $AlCl_3 \cdot 3\,H_2O$. The coordination number is not known but is expected to be 4 or 6. Describe how Werner's methods, that is, reaction with $AgNO_3(aq)$ or conductivity measurements, could help elucidate the actual coordination number.

KEEP IN MIND

that the covalent bond formed in the Lewis acid–base reaction is a *coordinate* covalent bond. Thus we can think of the coordination number of a transition metal ion in a complex as the number of coordinate covalent bonds in the complex.

24-2 Ligands

A common feature shared by the ligands in coordination complexes is the abilit to donate electron pairs to central metal atoms or ions. Ligands are *Lewis bases.* I accepting electron pairs, central metal atoms or ions act as *Lewis acids.* A liganc that uses one pair of electrons to form one point of attachment to the central meta atom or ion is called a **monodentate** ligand. Some examples of monodentat ligands are monatomic anions such as the halide ions, polyatomic anions such a hydroxide ion, simple molecules, such as ammonia (called *ammine* when it is ligand), and more complex molecules, such as methanamine, CH_3NH_2 (some times called *methylamine*). Other examples are given in Table 24.2.

TABLE 24.2 Some Common Monodentate Ligands

Formula	Name as Ligand	Formula	Name as Ligand[a]	Formula	Name as Ligand
Neutral molecules		Anions		Anions	
H_2O	Aqua	F^-	Fluorido	SO_4^{2-}	Sulfato
NH_3	Ammine	Cl^-	Chlorido	$S_2O_3^{2-}$	Thiosulfato
CO	Carbonyl	Br^-	Bromido	NO_2^-	Nitrito-*N*-[b]
NO	Oxidonitrogen	I^-	Iodido	ONO^-	Nitrito-*O*-[b]
CH_3NH_2	Methanamine	O^{2-}	Oxido	SCN^-	Thiocyanato-*S*-[c]
C_5H_5N	Pyridine	OH^-	Hydroxido	NCS^-	Thiocyanato-*N*-[c]
		CN^-	Cyanido		

[a]Before 2005, these ligands were named as follows: F^-, fluoro; Cl^-, chloro; Br^-, bromo; I^-, iodo; O^{2-}, oxo; OH^-, hydroxo; CN^-, cyano.
[b]If the nitrite ion is attached through the N atom ($-NO_2$), the designation *nitrito-N-* is used; if attached through an O atom (—ONO), *nitrito-O-*.
[c]If the thiocyanate ion is attached through the S atom (—SCN), the name *thiocyanato-S-* is used; if attachment is through the N atom (—NCS), *thiocyanato-N-*.

KEEP IN MIND

that the lone pairs of electrons of a polydentate ligand must be far enough apart to attach to the metal center at two or more points; the donated pairs of electrons must be on different atoms.

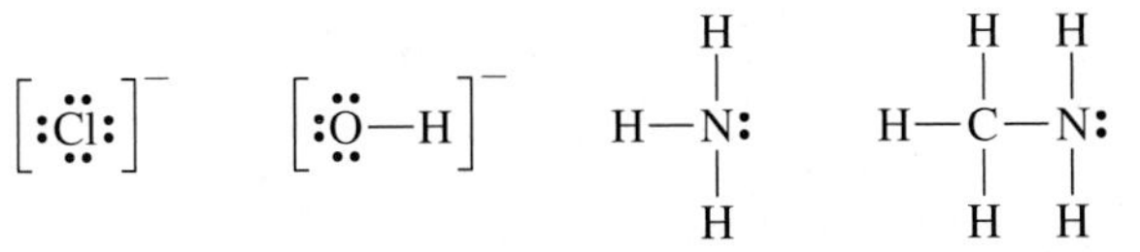

Ligand name: Chlorido Hydroxido Ammine Methanamine

Some ligands are capable of donating more than a single electron pair from *different* atoms in the ligand and to *different* sites in the geometric structure of a complex. These are called **polydentate** ligands. The molecule *ethylenediamine* (en) can donate two electron pairs, one from each N atom. Since en attaches to the metal center at two points, it is called a **bidentate** ligand.

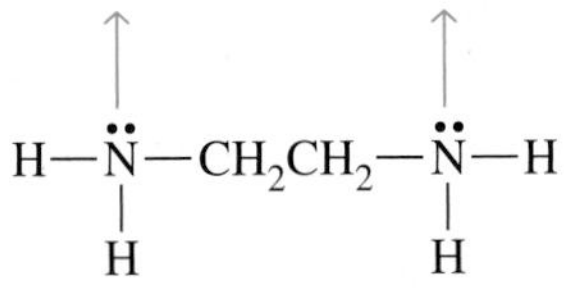

KEEP IN MIND

that the $EDTA^{4-}$ ligand is not planar as suggested in Table 24.3. Use VSEPR theory to gain an insight into the shape of this polyatomic anion. Also, refer to Figure 24-23.

Three common polydentate ligands are shown in Table 24.3.

TABLE 24.3 Some Common Polydentate Ligands (Chelating Agents)

Abbreviation	Name	Formula
en	Ethylenediamine	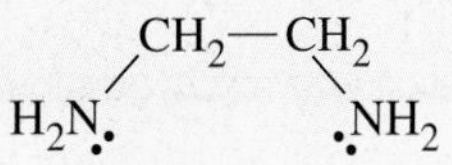
ox^{2-}[a]	Oxalato	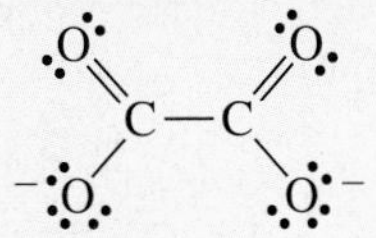
$EDTA^{4-}$[b]	Ethylenediaminetetraacetato	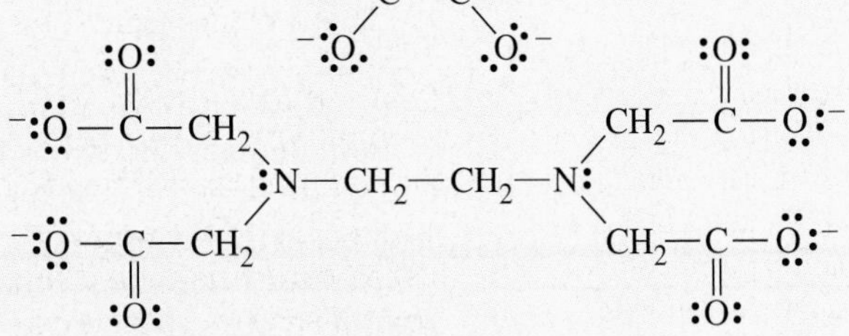

[a]Oxalic acid is a diprotic acid denoted H_2ox. It is the ox^{2-} anion that binds as a bidentate ligand.
[b]Ethylenediaminetetraacetic acid, a tetraprotic acid, is denoted H_4EDTA.

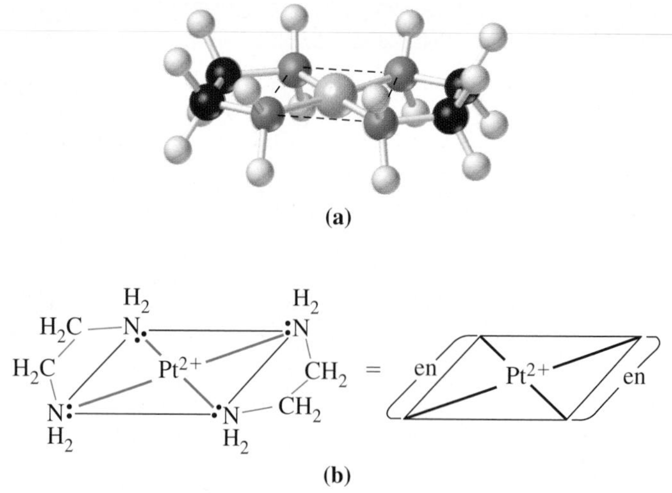

▶ FIGURE 24-3
Three representations of the chelate $[Pt(en)_2]^{2+}$
(a) Overall structure. **(b)** The ligands attach at adjacent corners along an edge of the square. They do *not* bridge the square by attaching to opposite corners. Bonds are shown in red, and the square-planar shape is indicated by the black parallelogram.

Figure 24-3 represents the attachment of two ethylenediamine (en) ligands to a Pt^{2+} ion. Here is how we can establish that each ligand is attached to two positions in the coordination sphere around the Pt^{2+} ion.

- Because Pt^{2+} ion exhibits a coordination number of 4 with monodentate ligands, and because $[Pt(en)_2]^{2+}$ is unable to attach additional ligands such as NH_3, H_2O, or Cl^-, we conclude that each en group must be attached at two points.
- The en ligands in the complex ion exhibit no further basic properties. They cannot accept protons from water to produce OH^-, as they would if they had an available lone pair of electrons. Both $—NH_2$ groups of each en molecule must be tied up in the complex ion.

Note the two five-member rings (pentagons) outlined in Figure 24-3(b). They consist of Pt, N, and C atoms. When the bonding of a polydentate ligand to a metal ion produces a ring (normally with five or six members), we refer to the complex as a **chelate** (pronounced KEY-late). The polydentate ligand is called a **chelating agent**, and the process of chelate formation is called *chelation*.

24-2 CONCEPT ASSESSMENT

The ligand diethylenetriamine (abbreviated det), $H_2NCH_2CH_2NHCH_2CH_2NH_2$, can form complexes. Classify this ligand as mono-, bi-, tri-, or quadridentate.

24-1 ARE YOU WONDERING?

How did such terms as *ligand, monodentate,* and *chelate* get into the vocabulary of chemistry?

Ligand comes from the Latin word *ligare*, which means to bind. It is quite appropriate to describe groups that are bound to a metal center as ligands. Dentate is also derived from a Latin word, *dens*, meaning tooth. Figuratively speaking, a monodentate ligand has one tooth; a bidentate ligand has two teeth; and a polydentate ligand has several. A ligand attaches itself to the metal center in accordance with the number of "teeth" it possesses. This is an easily remembered and colorful metaphor. Chelate is derived from the Greek word *chela*, which means a crab's claw. The way in which a chelating agent attaches itself to a metal ion resembles a crab's claw—another colorful metaphor.

24-3 Nomenclature

The system of naming complexes originated with Werner, but it has been modified several times over the years. Even today, usage varies somewhat. Our approach will be to consider a few general rules that permit us to relate names and formulas of simple complexes. We will not consider any of the complicated cases for which writing names and formulas is more challenging.

1. *Anions as ligands are named by using the ending* -o. As implied by Table 24.2, normally *-ide* endings change to *-ido*, *-ite* to *-ito*, and *-ate* to *-ato*.
2. *Neutral molecules as ligands generally carry the unmodified name.* For example, the name ethylenediamine is used both for the free molecule and for the molecule as a ligand. Aqua, ammine, carbonyl, and oxidonitrogen are important exceptions (see Table 24.2).
3. *The number of ligands of a given type is denoted by a prefix.* The usual prefixes are *mono* = 1, *di* = 2, *tri* = 3, *tetra* = 4, *penta* = 5, and *hexa* = 6. As in many other cases, the prefix *mono-* is often not used. If the ligand name is a composite name that itself contains a numerical prefix, such as ethylene*di*amine, place parentheses around the name and precede it with *bis* = 2, *tris* = 3, *tetrakis* = 4, and so on. Thus, dichlorido signifies *two* Cl^- ions as ligands, and *penta*aqua signifies *five* H_2O molecules. To indicate the presence of *two* ethylenediamine (en) ligands, we write *bis*(ethylenediamine).
4. *When we name a complex, ligands are named first, in alphabetical order, followed by the name of the metal center. The oxidation state of the metal center is denoted by a Roman numeral. If the complex is an anion, the ending* -ate *is attached to the name of the metal.* Prefixes (*di, tri, bis, tris,...*) are ignored in establishing the alphabetical order. Thus, $[CrCl_2(H_2O)_4]^+$ is called tetraaquadichloridochromium(III) ion; $[CoCl_2(en)_2]^+$ is dichloridobis(ethylenediamine)cobalt(III) ion; and $[Cr(OH)_4]^-$ is tetrahydroxidochromate(III) ion. For complex anions of a few of the metals, the English name is replaced by the Latin name given in Table 24.4. Thus, $[CuCl_4]^{2-}$ is the tetrachloridocuprate(II) ion.
5. *When we write the formula of a complex, the chemical symbol of the metal center is written first, followed by the ligand symbols (formulas, abbreviations, or acronyms) in alphabetical order.* Let's illustrate the application of this rule with a few examples. The formula of the tetraamminechloridonitrito-*N*-cobalt(III) ion should be written as $[CoCl(NH_3)_4(NO_2)]^+$, not as $[Co(NH_3)_4Cl(NO_2)]^+$ as the name itself suggests. For the $[Fe(CN)_2Cl_4]^{3-}$ ion, the order of ligands is determined by the atomic symbols of the donor atoms of the ligands (C of CN^- and Cl in Cl^-). The cyanido ligand is listed first because, alphabetically, C precedes Cl.
6. *In names and formulas of coordination compounds, cations come first followed by anions.* This is the same order as in simple ionic compounds like NaCl for sodium chloride. For example, the formula $[Pt(NH_3)_4][PtCl_4]$ represents the coordination compound tetraammineplatinum(II) tetrachloridoplatinate(II).

TABLE 24.4 Names for Some Metals in Complex Anions

Iron	⟶	Ferrate
Copper	⟶	Cuprate
Tin	⟶	Stannate
Silver	⟶	Argentate
Lead	⟶	Plumbate
Gold	⟶	Aurate

◀ Occasionally, the metal center will be in the oxidation state 0, as in $[W(CO)_6]$ hexacarbonyltungsten(0).

◀ Before 2005, the rule was to list anionic ligands first, in alphabetical order, followed by neutral ligands, also in alphabetical order. The new rule is simpler: Ligands are listed in alphabetical order irrespective of charge.

Although most complexes are named in the manner just outlined, some common, or trivial, names are still in use. Two such trivial names are ferrocyanide for $[Fe(CN)_6]^{4-}$ and ferricyanide for $[Fe(CN)_6]^{3-}$. These common names suggest the oxidation state of the central metal ions through the *o* and *i* designations (*o* for the ferrous ion, Fe^{2+}, in $[Fe(CN)_6]^{4-}$ and *i* for the ferric ion, Fe^{3+}, in $[Fe(CN)_6]^{3-}$). These trivial names do not indicate that the metal ions have a coordination number of 6, however. The systematic names—hexacyanidoferrate(II) and hexacyanidoferrate(III)—are more informative.

EXAMPLE 24-2 Relating Names and Formulas of Complexes

(a) What is the name of the complex $[CoCl_3(NH_3)_3]$? **(b)** What is the formula of the compound pentaaquachloridochromium(III) chloride? **(c)** What is the name of the compound $K_3[Fe(CN)_6]$?

Analyze

To translate a chemical formula into the appropriate IUPAC name, we need to determine the oxidation state of the metal center (as done in Example 24-1), and then focus on naming the ligands. Translating an IUPAC name to a chemical formula is a simpler exercise, provided we know the names of common ligands and understand the meanings of the various prefixes.

Solve

(a) $[CoCl_3(NH_3)_3]$ consists of *three* ammonia molecules and *three* chloride ions attached to a central Co^{3+} ion; it is electrically neutral. The name of this neutral complex is triamminetrichloridocobalt(III).

(b) The central metal ion is Cr^{3+}. There are five H_2O molecules and one Cl^- ion as ligands. The complex ion carries a net charge of 2+. Two Cl^- ions are required to neutralize the charge on this complex cation. The formula of the coordination compound is $[CrCl(H_2O)_5]Cl_2$.

(c) This compound consists of K^+ cations and complex anions having the formula $[Fe(CN)_6]^{3-}$. Each cyanide ion carries a charge of 1−, so the oxidation state of the iron must be +3. The Latin-based name "ferrate" is used because the complex ion is an anion. The name of the anion is hexacyanidoferrate(III) ion. The coordination compound is potassium hexacyanidoferrate(III).

Assess

Part (b) of this example gives us an opportunity to make a couple of additional, general remarks. First, in the name pentaaquachloridochromium(III) chloride, the number of chloride ions (the counter ions) is not explicitly stated. This is because, in general, the number of counter ions can be derived from the name. Second, by writing the formula of the compound in (b) as $[CrCl(H_2O)_5]Cl_2$, we have inadvertently broken one of IUPAC's rules, albeit one we have not yet mentioned: *In the formula of a complex, the formula for a given ligand should be written such that the atomic symbol of the donor atom is nearest the central atom.* By IUPAC's rules, the formula of the compound in (b) should be written as $[CrCl(OH_2)_5]Cl_2$ because the O atom of a water molecule is the one that coordinates with the central chromium ion. However, because chemists almost always write the formula for water as H_2O, we have adopted the method of representing water as H_2O, not as OH_2, in coordination complexes.

PRACTICE EXAMPLE A: What is the formula of the compound potassium hexachloridoplatinate(IV)?

PRACTICE EXAMPLE B: What is the name of the compound $[Co(NH_3)_5(SCN)]$?

24-3 CONCEPT ASSESSMENT

A student named a coordination complex tripotassium dichloridodibromidodihydroxidoiron. The student's instructor pointed out that although the correct formula for the compound could be deduced, this name violates the IUPAC convention. How does this name violate IUPAC rules, and what is the correct name?

24-4 Isomerism

As previously noted (page 95), *isomers* are substances that have the same formulas but differ in their structures and properties. Several kinds of isomerism are found among complex ions and coordination compounds. These can be lumped into two broad categories: **Structural isomers** differ in basic structure or bond type—what ligands are bonded to the metal center and through which atoms. **Stereoisomers** have the same number and types of ligands and the same mode of attachment, but they differ in the way in which the ligands occupy the space around the metal center. Of the following five examples, the first three are types of structural isomerism (ionization, coordination, and linkage isomerism), and the remaining two are types of stereoisomerism (geometric and optical isomerism).

▶ Structural isomers are also known as constitutional isomers.

Ionization Isomerism

The two coordination compounds whose formulas are shown here have the same central ion (Cr^{3+}), and five of the six ligands (NH_3 molecules) are the same. The compounds differ in that one has SO_4^{2-} ion as the sixth ligand, with a Cl^- ion to neutralize the charge of the complex ion, whereas the other has Cl^- as the sixth ligand and SO_4^{2-} to neutralize the charge of the complex ion.

(a)	(b)
$[Cr(NH_3)_5SO_4]Cl$	$[CrCl(NH_3)_5]SO_4$
Pentaamminesulfatochromium(III) chloride	Pentaamminechloridochromium(III) sulfate

Coordination Isomerism

A situation somewhat similar to that just described can arise when a coordination compound is composed of both complex cations and complex anions. The ligands can be distributed differently between the two complex ions, as are NH_3 and CN^- in these two compounds.

(a)	(b)
$[Co(NH_3)_6][Cr(CN)_6]$	$[Cr(NH_3)_6][Co(CN)_6]$
Hexaamminecobalt(III) hexacyanidochromate(III)	Hexaamminechromium(III) hexacyanidocobaltate(III)

Linkage Isomerism

Some ligands may attach to the central metal ion of a complex ion in different ways. For example, the nitrite ion, a monodentate ligand, has electron pairs available for coordination both on the N and O atoms.

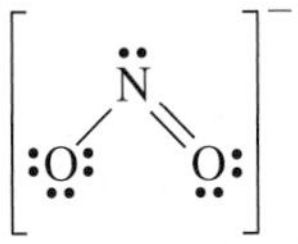

Whether attachment of this ligand is through the N or an O atom, the formula of the complex ion is unaffected. The properties of the complex ion, however, may be affected. When attachment occurs through the N atom, the ligand can be referred to as nitro or, more properly, nitrito-*N*- when naming the complex. Coordination through an O atom can be referred to as nitrito or, more properly, nitrito-*O*- when naming the complex.

(a)	(b)
$[Co(NO_2)(NH_3)_5]^{2+}$	$[Co(NH_3)_3(ONO)]^{2+}$
Pentaamminenitrito-*N*-cobalt(III) ion	Pentaamminenitrito-*O*-cobalt(III) ion

The structures of these compounds are illustrated in Figure 24-4.

Geometric Isomerism

If a single Cl^- ion is substituted for an NH_3 molecule in the square-planar complex ion $[Pt(NH_3)_4]^{2+}$ in Figure 24-2, it does not matter at which corner of the square this substitution is made. All four possibilities are alike. If a *second* Cl^- is substituted for an NH_3, there are two distinct possibilities (Fig. 24-5b). The two Cl^- ions can either be along the same edge of the square (***cis***) or on opposite corners, across from each other (***trans***). To distinguish clearly between these two possibilities, we must either draw a structure or refer to the appropriate name. The formula alone will not distinguish between them. (Note that this complex is a neutral species, not an ion.)

$[PtCl_2(NH_3)_2]$

cis-diamminedichloridoplatinum(II)

or

trans-diamminedichloridoplatinum(II)

◀ The term *cis* means "on this side" in Latin, and *trans* means "across." Domestic airline flights in the United States are *cis*atlantic, whereas flights to Europe are *trans*atlantic.

The type of isomerism described here in which the positions of the ligands produce distinct isomers is called **geometric isomerism**. Interestingly, if a *third*

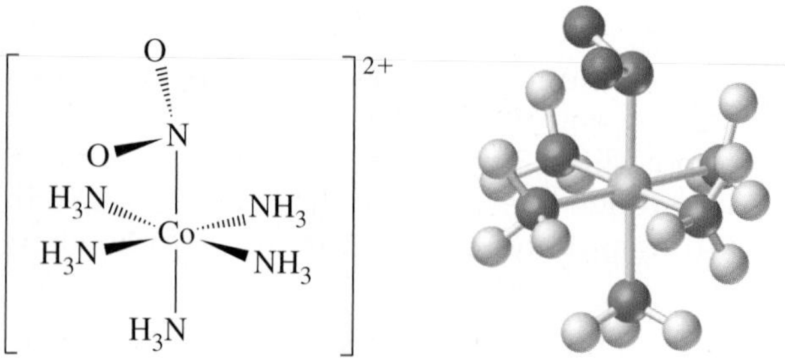

(a) $[Co(NH_3)_5NO_2]^{2+}$

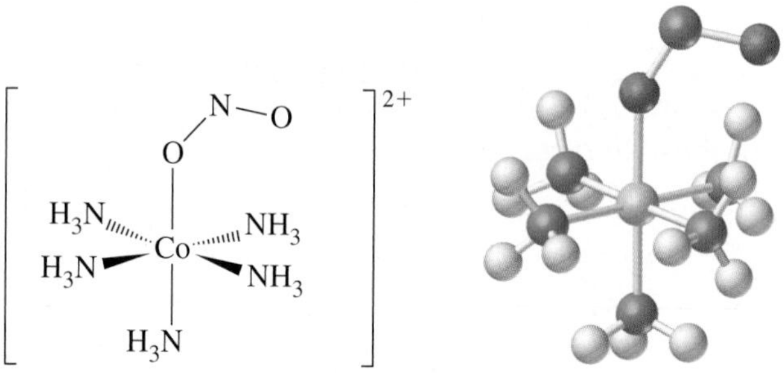

(b) $[Co(NH_3)_5(ONO)]^{2+}$

▶ FIGURE 24-4
Linkage isomerism—illustrated
(a) Pentaamminenitrito-*N*-cobalt(III) cation.
(b) Pentaamminenitrito-*O*-cobalt(III) cation.

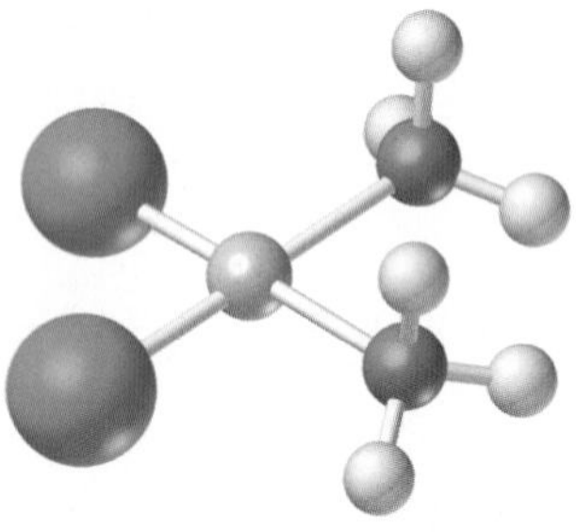

cis-$[PtCl_2(NH_3)_2]$

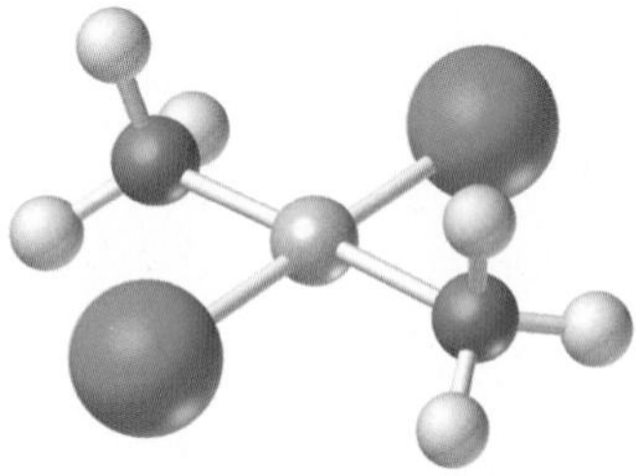

trans-$[PtCl_2(NH_3)_2]$

▲ The geometric isomers of $[PtCl_2(NH_3)_2]$

Cl^- ion is substituted, isomerism disappears (Fig. 24-5c). There is only one complex ion with the formula $[PtCl_3(NH_3)]^-$.

With an octahedral complex, the situation is a bit more complicated. Take the complex ion $[Co(NH_3)_6]^{3+}$ of Figure 24-2 as an example. If one Cl^- is substituted for an NH_3, a single structure results. With the substitution of *two* Cl^- ions for NH_3 molecules, *cis* and *trans* isomers result. The *cis* isomer has two Cl^- ions along the same edge of the octahedron (Fig. 24-6a). The *trans* isomer has two Cl^- ions on opposite corners, that is, at opposite ends of a line drawn through the central metal ion (Fig. 24-6b). One difference between the two is that the *cis* isomer has a purple color and the *trans* has a bright green color.

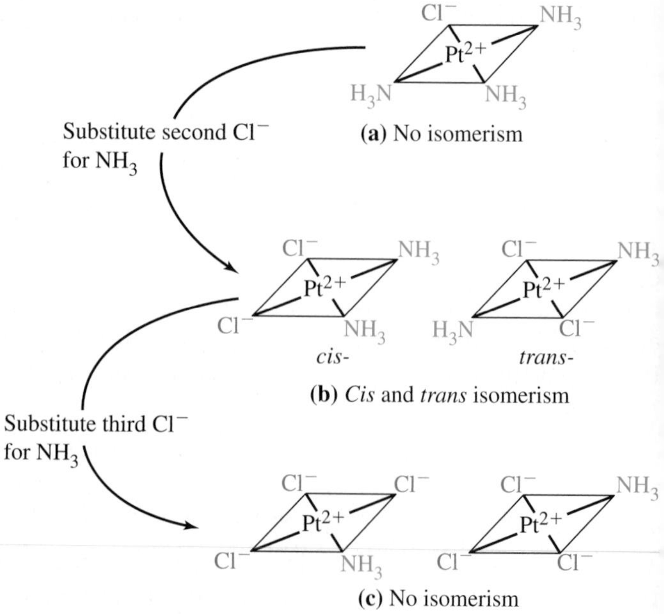

▶ FIGURE 24-5
Geometric isomerism—illustrated
For the square-planar complexes shown here, isomerism exists only when two Cl^- ions have replaced NH_3 molecules.

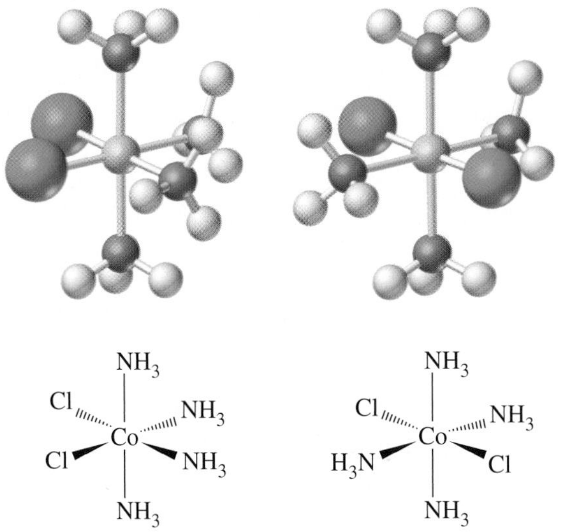

cis-$[CoCl_2(NH_3)_4]^+$ (purple) **(a)**

trans-$[CoCl_2(NH_3)_4]^+$ (green) **(b)**

▲ FIGURE 24-6
***Cis* and *trans* isomers of an octahedral complex**
The Co^{3+} ion is at the center of the octahedron, and NH_3 and Cl^- ligands are at the vertices.

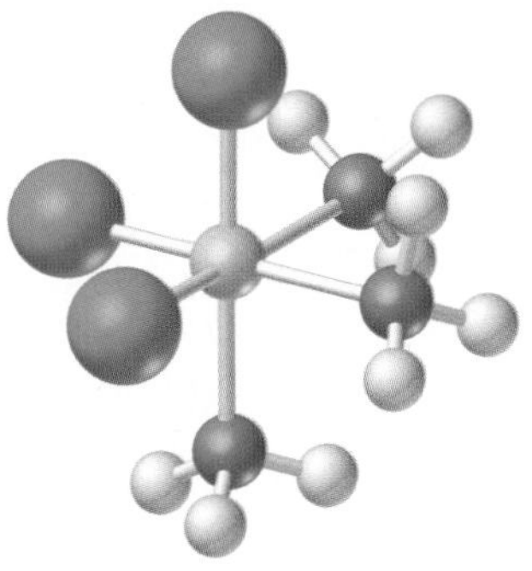

fac-$[CoCl_3(NH_3)_3]$

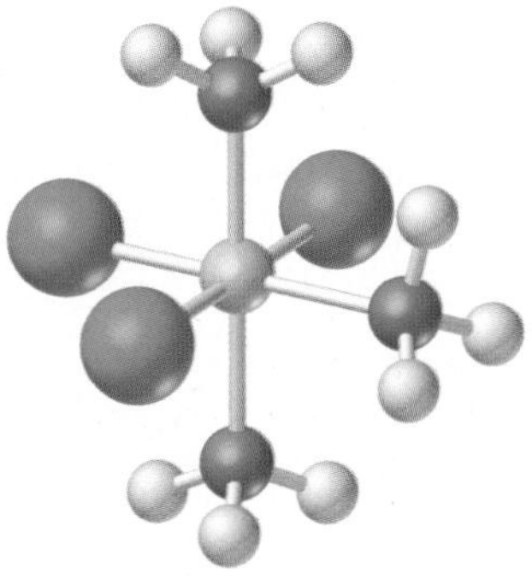

mer-$[CoCl_3(NH_3)_3]$

▲ The geometric isomers of $[CoCl_3(NH_3)_3]$

If a *third* Cl^- is substituted for an NH_3 in Figure 24-6(a), two possibilities exist. If this third substitution is at either the top or bottom of the structure, the result is that three Cl^- ions appear on the same face of the octahedron. This is called a *fac* (facial) isomer. If the third substitution is at either of the other two positions, the result is three Cl^- ions around a perimeter or meridian of the octahedron. This is a *mer* (meridional) isomer.

24-4 CONCEPT ASSESSMENT

Explain why the substitution of a *fourth* Cl^- for an NH_3 in *mer*-$[CoCl_3(NH_3)_3]$ may produce some of the same product as the substitution of a fourth Cl^- for an NH_3 in *fac*-$[CoCl_3(NH_3)_3]$, or it may also produce a different product.

EXAMPLE 24-3 Identifying Geometric Isomers

Sketch structures of all the possible isomers of $[CoCl(NH_3)_3(ox)]$.

Analyze

The Co^{3+} ion exhibits a coordination number of 6. The structure is octahedral. Recall that ox (oxalate ion) is a bidentate ligand carrying a charge of 2– (see Table 24.3). Also, as shown in Figure 24-3, a bidentate ligand must be attached in *cis* positions, not *trans*.

Solve

Once the ox ligand is placed, any position is available to the Cl^-. This leaves two possibilities for the three NH_3 molecules. They can be situated (1) on the same face of the octahedron (*fac* isomer) or (2) around a perimeter of the octahedron (*mer* isomer).

(continued)

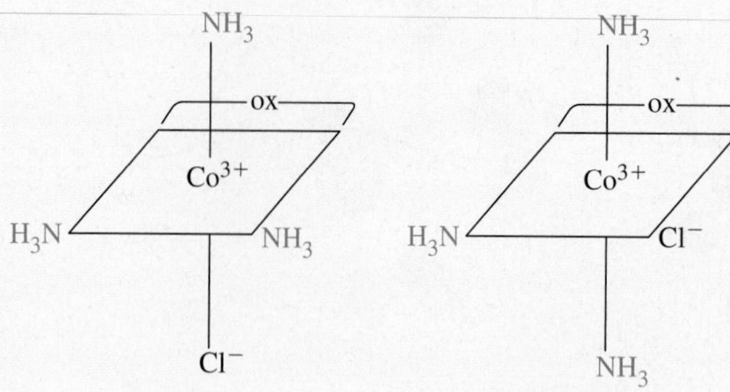

fac isomer *mer* isomer

Assess

We can also draw these stereoisomers by using the dashed and solid wedge symbols, together with a line diagram for the oxalate anion.

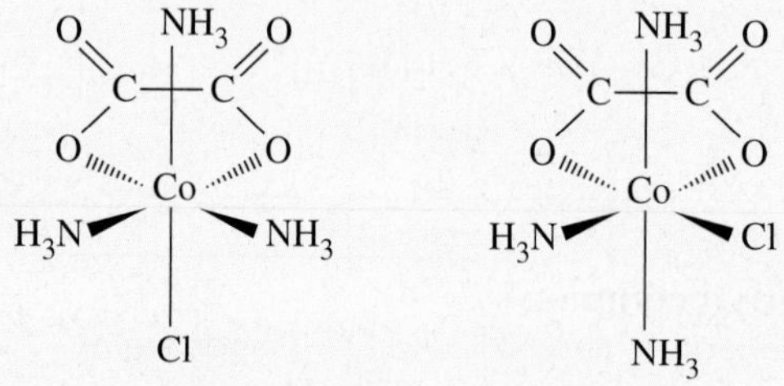

fac isomer *mer* isomer

PRACTICE EXAMPLE A: Sketch the geometric isomers of $[CoCl_2(NH_3)_2(ox)]^-$.

PRACTICE EXAMPLE B: Sketch the geometric isomers of $[Mo(C_5H_5N)_2(CO)_2Cl_2]^+$.

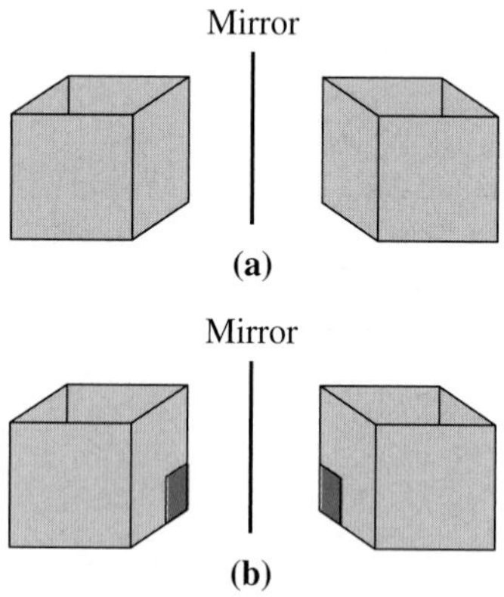

▲ FIGURE 24-7
Superimposable and nonsuperimposable objects—an open-top box
(a) You can place the box into its mirror image (hypothetically) in several ways. The box and its mirror image are superimposable. **(b)** No matter how you place the box in its mirror image, the stickers will not appear in the same position. This box and its mirror image are nonsuperimposable.

Optical Isomerism

To understand optical isomerism, we need to understand the relationship between an object and its mirror image. Features on the right side of the object appear on the left side of its image in a mirror, and vice versa. Certain objects can be rotated in such a way as to be *superimposable* on their mirror images, but other objects are *nonsuperimposable* on their mirror images. An unmarked tennis ball is superimposable on its mirror image, but a left hand is nonsuperimposable on its mirror image (a right hand).

Consider the open-top cardboard box pictured in Figure 24-7(a). There are a number of hypothetical ways in which the box can be superimposed on its mirror image. Now imagine that a distinctive sticker is placed at a corner of one side of the box (Fig. 24-7b). In this case, there is no way that the box and its mirror image can be superimposed; they are clearly different. This is equivalent to saying that there is no way that a formfitting left glove can be worn on a right hand (turning it inside out is not allowed).

The two structures of $[Co(en)_3]^{3+}$ depicted in Figure 24-8 are related to each other as are an object and its image in a mirror. Furthermore, the two structures are *nonsuperimposable*, like a left and a right hand. The two structures represent two different complex ions; they are isomers.

Structures that are nonsuperimposable mirror images of each other are called **enantiomers** and are said to be **chiral** (pronounced KYE-rull). (Structures that are superimposable are *achiral*.) Whereas other types of isomers may differ significantly in their physical and chemical properties,

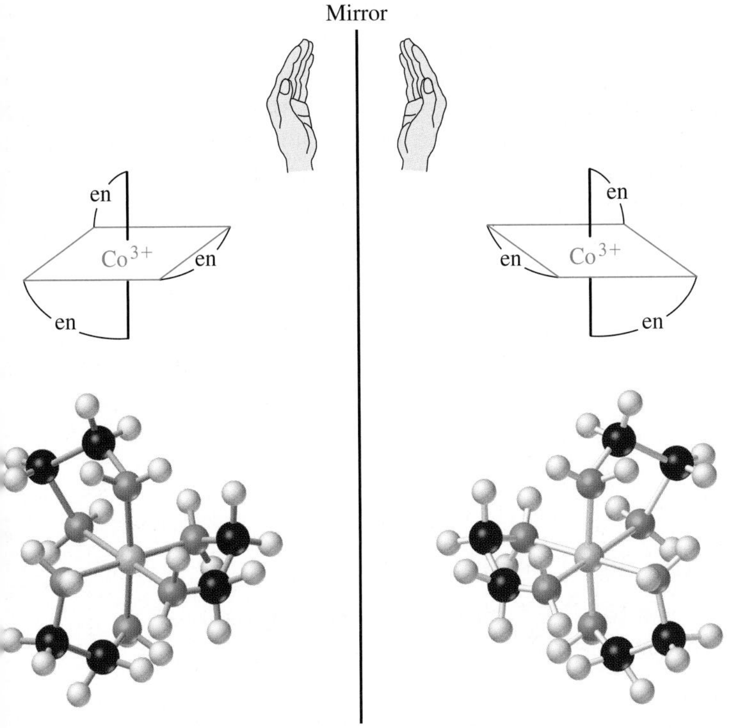

◀ FIGURE 24-8
Optical isomers
The two structures are nonsuperimposable mirror images. Like a right hand and a left hand, one structure cannot be superimposed onto the other.

enantiomers have identical properties except in a few specialized situations. These exceptions involve phenomena that are directly linked to chirality, or handedness, at the molecular level. An example is *optical activity*, pictured in Figure 24-9.

Interactions between a beam of polarized light and the electrons in an enantiomer cause the plane of the polarized light to rotate. One enantiomer rotates the plane of polarized light to the right (clockwise) and is said to be *dextrorotatory* (designated + or *d*). The other enantiomer rotates the plane of polarized light to the same extent, but to the left (counterclockwise). It is said to be *levorotatory* (− or *l*). Because they can rotate the plane of polarized

◀ The prefixes *dextro* and *levo* are derived from the Latin words *dexter*, "right," and *levo*, "left."

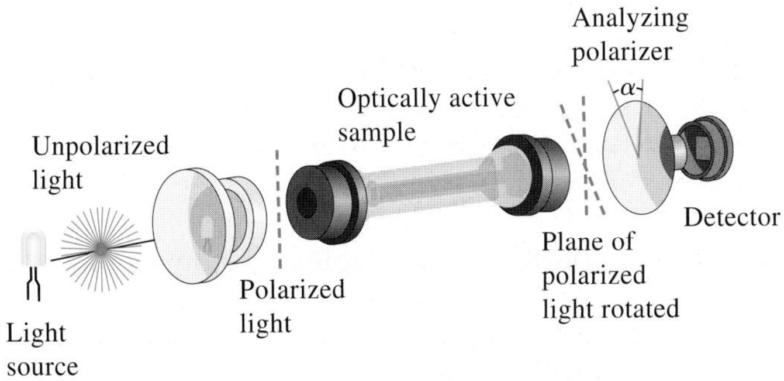

▲ FIGURE 24-9
Optical activity
Light from an ordinary source consists of electromagnetic waves vibrating in all planes; it is unpolarized. This light is passed through a polarizer, a material that screens out all waves except those vibrating in a particular plane. The plane of polarization of transmitted polarized light is then changed by passage through an optically active substance. The angle through which the plane of polarization has been rotated is determined by rotating an analyzer (a second polarizer) to the extent that all the polarized light is absorbed.

◀ Rotation of plane polarized light is a physical property. Chemical activity, or lack of it, is not a physical property. The physical and chemical properties, however, both depend on the existence of a center of asymmetry.

▶ A process based on chirality would be creating the maximum number of matched pairs of gloves from a bin of gloves of identical sizes of which 90 are for the right hand, but only 10 are for the left hand.

light, the enantiomers are said to be *optically active* and are referred to a **optical isomers**.

When an optically active complex is synthesized, a mixture of the tw optical isomers (enantiomers) is obtained, such as the two $[Co(en)_3]^{3+}$ iso mers in Figure 24-8. The optical rotation of one isomer just cancels that o the other. The mixture, called a *racemic mixture*, produces no net rotation o the plane of polarized light. Separating the *d* and *l* isomers of a racemi mixture is called *resolution*. This separation can sometimes be achieve through chemical reactions controlled by chirality, with the two enantiomer behaving differently. Many phenomena of the living state, such as the effec tiveness of a drug, the activity of an enzyme, and the ability of a microor ganism to promote a reaction, involve chirality. We will discuss chirality again in Chapters 26, 27, and 28.

24-2 ARE YOU WONDERING?

How can we tell if a molecule is superimposable on its mirror image?

The complex ion $[CrBr_2(H_2O)_2(NH_3)_2]^+$ has five isomers; one is shown below with its mirror image. To test if the mirror image is superimposable on the original, imagine rotating the mirror image about the vertical axis ($H_2O—Cr—NH_3$) by 180° so that the two Br^- ligands are in the same position as in the original molecule. We see that the NH_3 and H_2O ligands that are in the same plane as the two Br^- ligands are in reversed positions when compared with the original molecule. Thus the molecule and its mirror image are not superimposable. The molecule is potentially optically active.

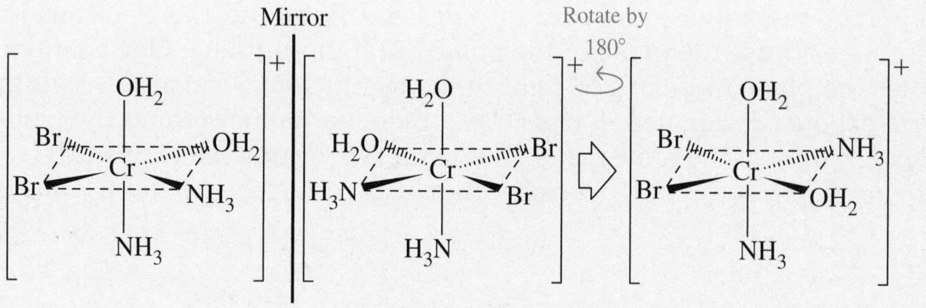

▲ FIGURE 24-10
Hypothetical structures for $[CoCl_2(NH_3)_4]^+$
If the complex ion had this hexagonal structure, there should be *three* distinct isomers, but only two isomers exist—*cis* and *trans.*

Isomerism and Werner's Theory

The study of isomerism played a crucial role in the development of Werner's theory of coordination chemistry. Werner proposed that complexes with coordination number 6 have an octahedral structure, but other possibilities were also proposed. For example, Figure 24-10 shows a hypothetical hexagonal structure for $[CoCl_2(NH_3)_4]^+$. However, this hexagonal structure would require the existence of *three* isomers, but a third isomer was never found; the only two isomers are those pictured in Figure 24-6. Additional direct evidence came with the discovery of optical isomerism in the tris(ethylenediamine) cobalt(III) ion (Fig. 24-8). Neither the hexagonal structure nor alternative structures can account for this isomerism—but, as we have seen, the octahedral structure does. Werner even succeeded in preparing an optically active octahedral complex with only inorganic ligands, to overcome objections that the optical activity of tris(ethylenediamine)cobalt(III) ion owed its optical activity to its carbon atoms and not to its geometric structure.

24-5 CONCEPT ASSESSMENT

Of the following complex cations, which are identical, which are geometrical isomers, which are enantiomers?

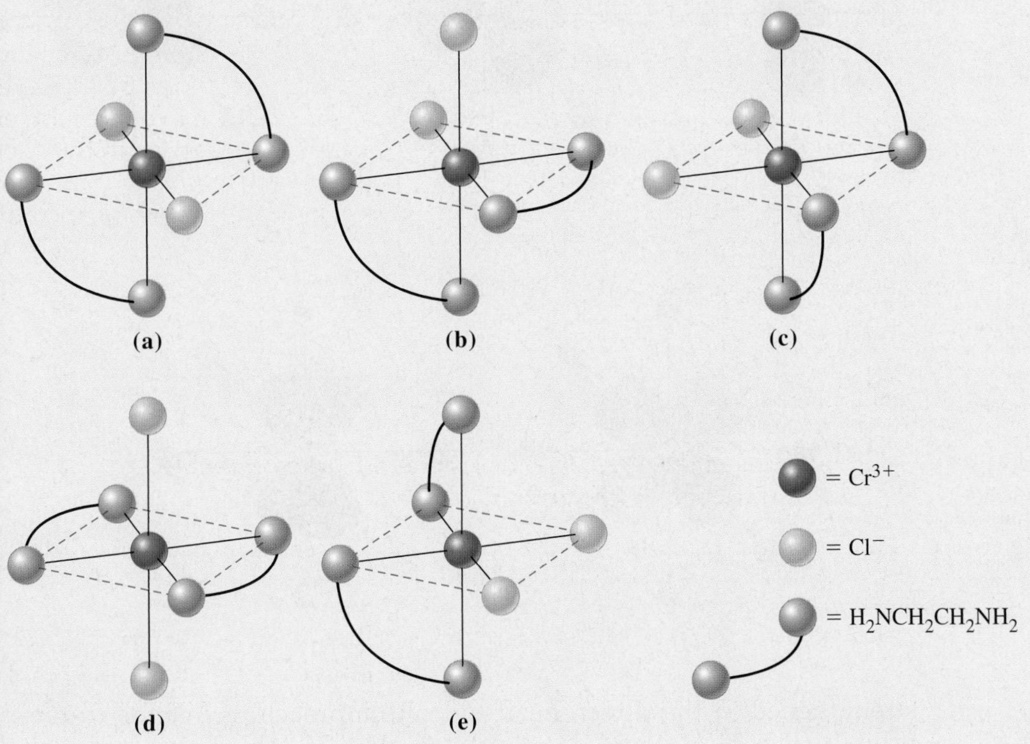

24-5 Bonding in Complex Ions: Crystal Field Theory

The theories of chemical bonding that were so useful in earlier chapters do not help much in explaining the characteristic colors and magnetic properties of complex ions. In transition metal ions, we need to focus on how the electrons in the *d* orbitals of a metal ion are affected when they are in a complex. A theory that provides that focus and an explanation of these properties is crystal field theory.

In the **crystal field theory**, bonding in a complex ion is considered to be an electrostatic attraction between the positively charged nucleus of the central metal ion and electrons in the ligands. Repulsions also occur between the ligand electrons and electrons in the central ion. In particular, the crystal field theory focuses on the repulsions between ligand electrons and *d* electrons of the central ion.

◀ Modifications of the simple crystal field theory that take into account such factors as the partial covalency of the metal–ligand bond are called *ligand field theory*. This term is often used to signify both the purely electrostatic crystal field theory and its modifications.

First, a reminder about the *d* orbitals introduced in Figure 8-30: All five of the orbitals are alike in energy when in an isolated atom or ion, but they are *unlike* in their spatial orientations. One of them, d_{z^2}, is directed along the *z* axis, and another, $d_{x^2-y^2}$, has lobes along the *x* and *y* axes. The remaining three have lobes extending into regions between the perpendicular *x*, *y*, and *z* axes. In the presence of ligands, because repulsions exist between ligand electrons and *d* electrons, the *d*-orbital energy levels of the central metal ion are raised. As we will soon see, however, they are not all raised to the same extent.

Figure 24-11 depicts six anions (ligands) approaching a central metal ion along the *x*, *y*, and *z* axes. This direction of approach leads to an octahedral complex. Repulsions between ligand electrons and *d*-orbital electrons are

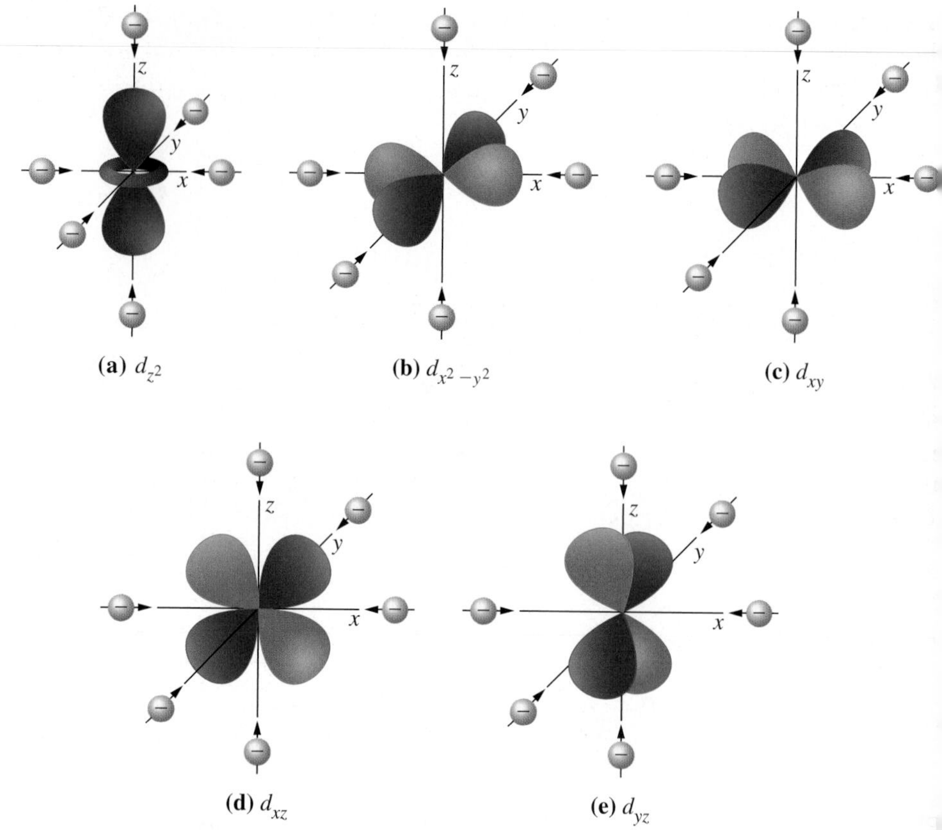

▶ FIGURE 24-11
Approach of six anions to a metal ion to form a complex ion with octahedral structure
The ligands (anions, in this case) approach the central metal ion along the *x*, *y*, and *z* axes. Maximum repulsion occurs with the d_{z^2} and $d_{x^2-y^2}$ orbitals, and their energies are raised. Repulsions with the other *d* orbitals are not as great. A difference in energy results between the two sets of *d* orbitals.

strengthened in the direct, head-to-head approach of ligands to the d_{z^2} orbitals (Fig. 24-11a) and $d_{x^2-y^2}$ orbitals (Fig. 24-11b). These two orbitals have their energy raised with respect to an average *d*-orbital energy for a central metal ion in the field of the ligands. For the other three orbitals (d_{xy}, d_{xz}, and d_{yz}, Figure 24-11c–e), ligands approach between the lobes of the orbitals and there is a gain in stability over the head-to-head approach; these orbital energies are lowered with respect to the average *d*-orbital energy. The difference in energy between the two groups of *d* orbitals is called *crystal field splitting* and is represented by the symbol Δ_o, with the subscript o emphasizing that the crystal field splitting shown in Figure 24-12 is for an octahedral complex.

The removal of the degeneracy of the *d* orbitals by the crystal field has important consequences for the electron configurations of transition metal ions having between 4 and 7 *d* electrons. Consider the transition metal ion Cr^{2+} with a d^4 configuration. If the four *d* electrons are assigned to the orbitals of lowest energy first, the first three electrons go into the d_{xy}, d_{xz}, and d_{yz} orbitals according to Hund's rule (page 354)—but what about the fourth

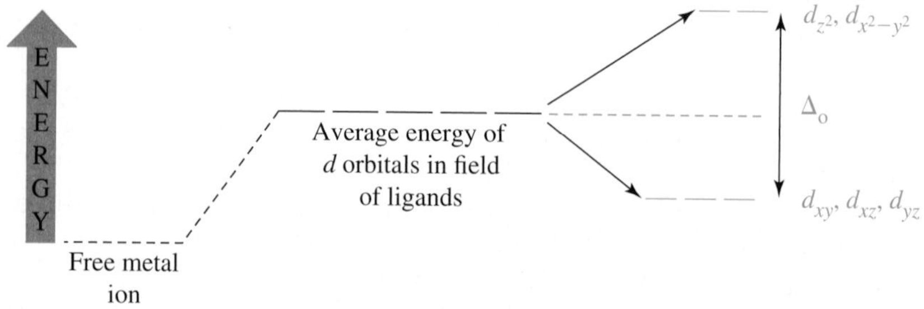

▲ FIGURE 24-12
Splitting of *d* energy levels in the formation of an octahedral complex ion
The *d*-orbital energy levels of the free central ion are raised in the presence of ligands to the average level shown, but the five levels are split into two groups.

lectron? The aufbau process (page 355) suggests that the fourth electron hould pair up with any one of the three electrons already in the d_{xy}, d_{xz}, and $_{yz}$ orbitals.

> **KEEP IN MIND**
>
> that the configurations $d_{xy}^2 d_{xz}^1 d_{yz}^1$, $d_{xy}^1 d_{xz}^2 d_{yz}^1$, and $d_{xy}^1 d_{xz}^1 d_{yz}^2$ all represent the same energy state and are indistinguishable. Thus, the paired electrons in the diagram at the left can be placed in any one of the three *d* orbitals in the lower-energy subset.

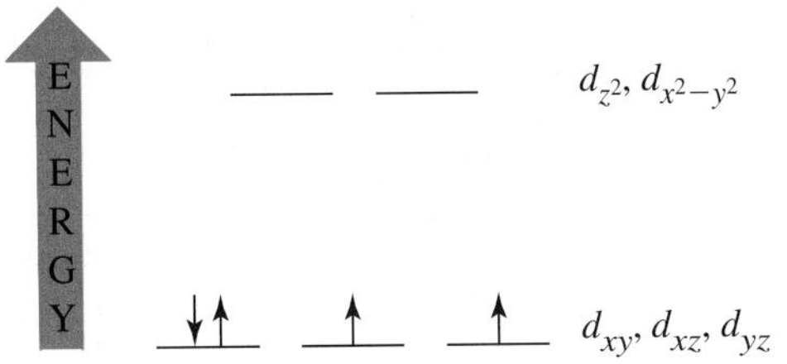

'lacing the fourth electron in the lower level confers extra stability (lower nergy) on the complex, but some of this stability is offset because it requires nergy, called the **pairing energy** (*P*), to force an electron into an orbital that is lready occupied by an electron. Alternatively, the electron could be assigned o either the $d_{x^2-y^2}$ or d_{z^2} orbital, avoiding the pairing energy.

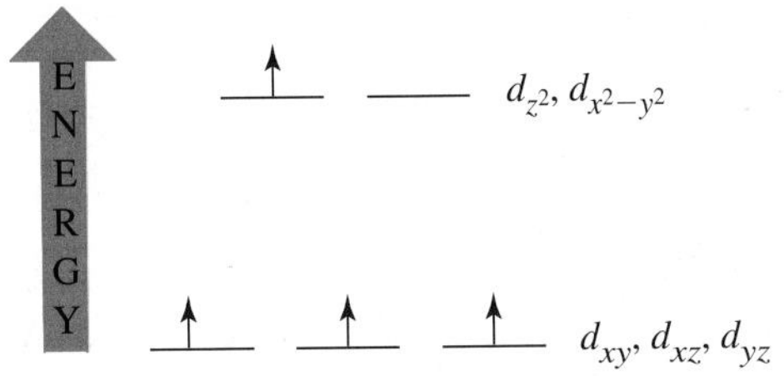

Placing the fourth electron in the upper level requires energy and offsets the xtra stability acquired by placing the first three electrons in the lower level. To pair or not to pair, that is the question.

Whether the fourth electron enters the lowest level and becomes paired or, instead, enters the upper level with the same spin as the first three electrons depends on the magnitude of Δ_o. If Δ_o is greater than the pairing energy, *P*, greater stability is obtained if the fourth electron is paired with one in the lower level. If Δ_o is less than the pairing energy, greater stability is obtained by keeping the electrons unpaired. Thus, for octahedral chromium(II) complexes, there are two possibilities for the number of unpaired electrons. In one case, there are four unpaired electrons when $\Delta_o < P$; this situation corresponds to the maximum number of unpaired electrons and is referred to as **high spin**. Ligands such as H_2O and F^- produce only a small crystal field splitting, leading to high-spin complexes; such ligands are said to be *weak-field ligands*. As an example, $[Cr(H_2O)_6]^{2+}$ is a weak-field complex. In the other case, when $\Delta_o > P$, there are two unpaired electrons; this corresponds to the minimum number of unpaired electrons and is referred to as **low spin**. Ligands, such as NH_3 and CN^-, produce large crystal field splitting, leading to low-spin complexes; such ligands are said to be *strong-field ligands*. $[Cr(CN)_6]^{4-}$ is a strong-field complex.

> ◀ These two possibilities exist for complex ions because the crystal field splitting and pairing energies are small and of comparable value. In considering the electron configurations in atoms, the spacing between energy levels is much greater than the pairing energy.

Different ligands can be arranged in order of their abilities to produce a splitting of the *d* energy levels. This arrangement is known as the **spectrochemical series**.

Strong field
(large Δ_o)

$CN^- > NO_2^- >$ en $>$ py $\approx NH_3 > EDTA^{4-} > SCN^- > H_2O >$
$ONO^- > ox^{2-} > OH^- > F^- > SCN^- > Cl^- > Br^- > I^-$

(small Δ_o)
Weak field

The red color indicates the donor atom.

To summarize, consider these two complexes of Co(III): $[CoF_6]^{3-}$ and $[Co(NH_3)_6]^{3+}$. The F^- ion is a weak-field ligand, whereas NH_3 is a strong-field

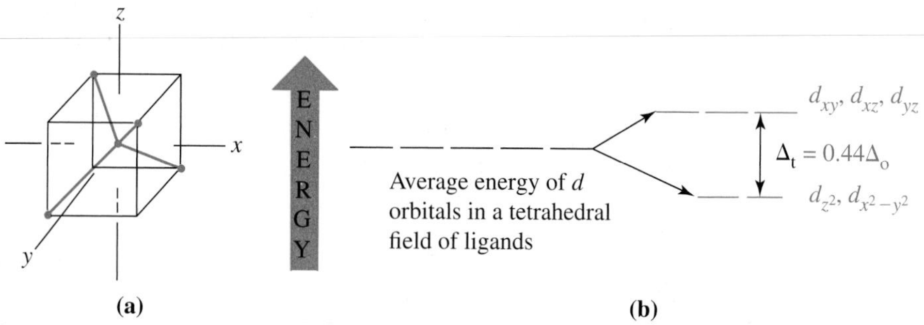

▲ FIGURE 24-13
Crystal field splitting in a tetrahedral complex ion
(a) The positions of attachment of ligands to a metal ion leading to the formation of a tetrahedral complex ion. **(b)** Interference with the *d* orbitals directed along the *x*, *y*, and *z* axes is not as great as with those that lie between the axes (see Figure 24-11). As a result, the pattern of crystal field splitting is reversed from that of an octahedral complex. Δ_t denotes the energy separation for a tetrahedral complex.

ligand. Because the crystal field splitting for NH_3 is greater than the pairing energy for Co^{3+}, we have the following situations.

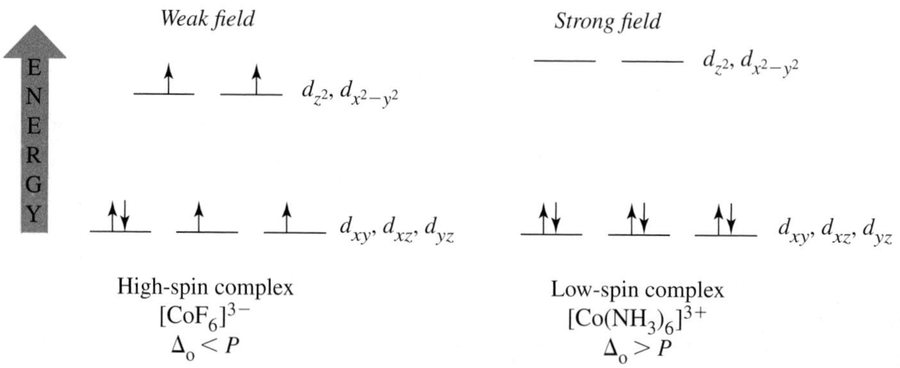

So far, we have considered just octahedral complexes. In the formation of complex ions of other geometric structures, ligands approach from different directions and produce different patterns of splitting of the *d* energy level. Figure 24-13 shows the pattern for tetrahedral complexes, and Figure 24-14 shows the pattern for square-planar complexes. In comparisons of hypothetical complexes of different structures having the same combinations of ligands, metal ions, and metal–ligand distances, we find the greatest energy separation of the *d* levels for the square-planar complex and the smallest energy separation for the tetrahedral complex. Because the tetrahedral splitting is small, almost all tetrahedral complexes are high spin.

24-3 ARE YOU WONDERING?

Are the two groups of orbitals split equally with respect to the average energy of the *d* orbitals?

The answer is no. The reason is that the total energy must be constant. If we consider a d^{10} ion such as Zn^{2+} in an octahedral field, the destabilization caused by the four electrons in the $d_{x^2-y^2}$ and d_{z^2} orbitals must be offset by the stabilization gained by the six electrons in the d_{xy}, d_{xz}, and d_{yz}, orbitals. This requires that

$$(6 \times \text{the energy of } d_{xy}, d_{xy}, d_{yz} \text{ orbitals}) + (4 \times \text{the energy of } d_{x^2-y^2}, d_{z^2}) = 0$$

That is, the energy gained equals the energy lost. Also, the energy difference between the orbitals is

$$(\text{the energy of } d_{x^2-y^2}, d_{z^2}) - (\text{the energy of } d_{xy}, d_{xy}, d_{yz} \text{ orbitals}) = \Delta_o$$

To satisfy these two relationships, it is necessary that

$$\text{the energy of } d_{xy}, d_{xz}, d_{yz} \text{ orbitals} = -0.4\Delta_o$$
$$\text{the energy of } d_{x^2-y^2}, d_{z^2} = 0.6\Delta_o$$

The splitting is as shown.

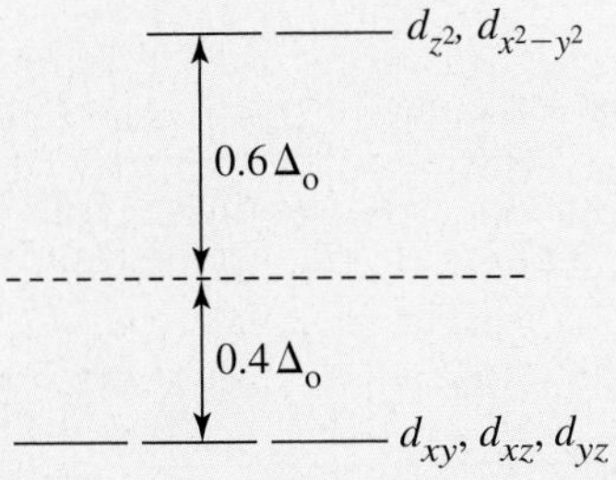

The splitting of the two groups of orbitals is not equal with respect to the average energy of the *d* orbitals.

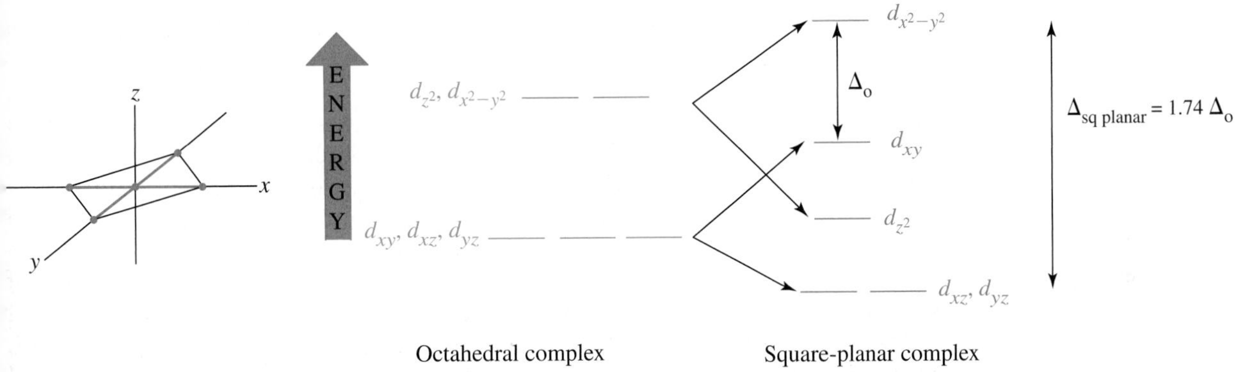

(a) **(b)**

▲ FIGURE 24-14
Comparison of crystal field splitting in a square-planar and an octahedral complex
(a) The positions of attachment of ligands to a metal ion leading to the formation of a square-planar complex. **(b)** Splitting of the *d* energy level in a square-planar complex can be related to that of the octahedral complex. There are no ligands along the *z* axis in a square-planar complex, so we expect the repulsion between ligands and d_{z^2} electrons to be much less than in an octahedral complex. The d_{z^2} energy level is lowered considerably from that in an octahedral complex. Similarly, the energy levels of the d_{xz} and d_{yz} orbitals are lowered slightly because the electrons in these orbitals are concentrated in planes perpendicular to that of the square-planar complex. The energy of the $d_{x^2-y^2}$ orbital is raised because the *x* and *y* axes represent the direction of approach of four ligands to the central ion. The energy of the d_{xy} orbital is also raised because this orbital lies in the plane of the ligands in the square-planar complex. The energy difference between the d_{xy} and $d_{x^2-y^2}$ orbitals in a square-planar complex is the same as in an octahedral complex because these orbitals are equally affected by ligand repulsions in both complexes. The maximum energy difference between *d* orbitals in a square-planar complex is $\Delta_{sq\ planar} = 1.74\Delta_o$.

24-6 CONCEPT ASSESSMENT

From the following crystal field splitting diagrams identify: **(i)** a tetrahedral Mn^{2+} complex; **(ii)** a strong-field octahedral complex of Co^{3+}; **(iii)** a weak-field octahedral complex of Fe^{2+}; **(iv)** a tetrahedral Ni^{2+} complex; **(v)** a high-spin octahedral complex of Fe^{3+}.

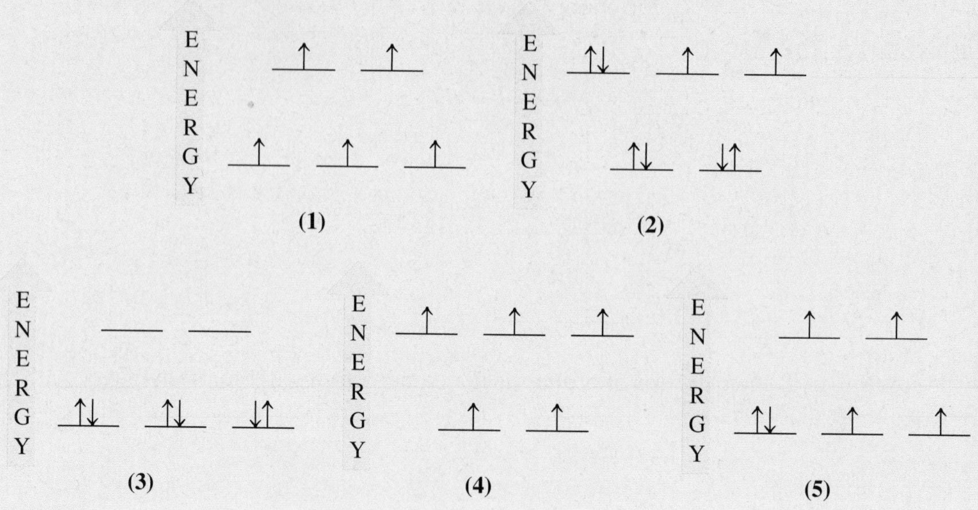

24-6 Magnetic Properties of Coordination Compounds and Crystal Field Theory

The paramagnetism of the dioxygen molecule was dramatically illustrated in Chapter 10 (Figure 10-3, page 417) by the interaction of liquid oxygen with the poles of a strong magnet. The origin of the paramagnetism is the existence of unpaired electrons in the molecule. Transition metal coordination compounds exhibit varying degrees of paramagnetism. Some are diamagnetic. A paramagnetic substance is pulled into, and a diamagnetic substance is pushed out of, a magnetic field. A straightforward way to measure magnetic properties is to weigh a substance "in" and "out" of a magnetic field, as illustrated in Figure 24-15. The mass of the substance is the

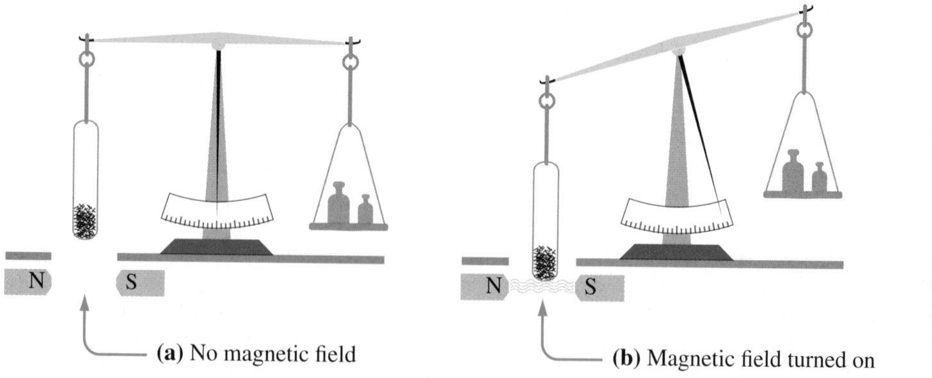

▲ FIGURE 24-15
Paramagnetism—illustrated
(a) A sample is weighed in the absence of a magnetic field. **(b)** When the field is turned on, the balanced condition is upset. The sample gains weight because it is now subjected to two attractive forces: the force of gravity *and* the force of interaction of the external magnetic field and the unpaired electrons.

same whatever magnetic property the substance possesses. However, if the substance is diamagnetic, it is slightly repelled by a magnetic field and *weighs* less within the field. If the substance is paramagnetic, it *weighs* more within the field.

The degree to which a substance weighs more in the magnetic field depends on the number of unpaired electrons. In the previous section, we saw that a high-spin d^n complex has more unpaired electrons than a low-spin d^n complex. Thus, measuring the change in weight of the complex in a magnetic field allows us to determine whether a complex is high or low spin. The magnetic properties of a complex depend on the magnitude of the crystal field splitting. Strong-field ligands tend to form low-spin, weakly paramagnetic, or even diamagnetic, complexes. Weak-field ligands tend to form high-spin, strongly paramagnetic complexes. The results of measuring the magnetic properties of coordination compounds can therefore be interpreted from crystal field theory, as demonstrated in Examples 24-4 and 24-5.

EXAMPLE 24-4 Using the Spectrochemical Series to Predict Magnetic Properties

How many unpaired electrons would you expect to find in the octahedral complex $[Fe(CN)_6]^{3-}$?

Analyze

We expect complexes with strong field ligands (that is, ligands that are high in the spectrochemical series) to be low spin.

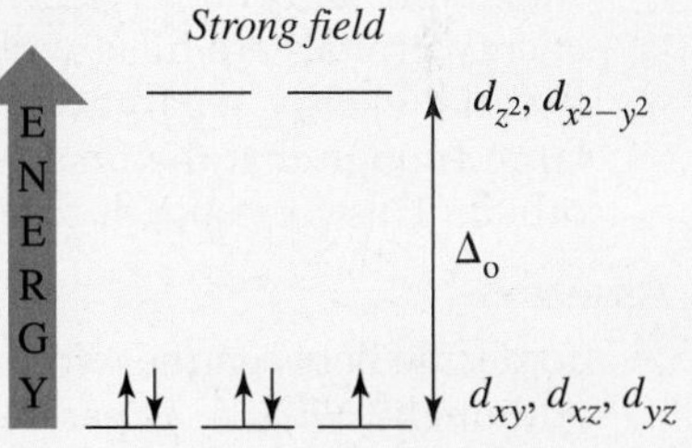

Low-spin complex
$[Fe(CN)_6]^{3-}$
$\Delta_o > P$

Solve

The Fe atom has the electron configuration $[Ar]3d^6 4s^2$. The Fe^{3+} ion has the configuration $[Ar]3d^5$. CN^- is a strong-field ligand. Because of the large energy separation in the *d* levels of the metal ion produced by this ligand, we expect all the electrons to be in the lowest energy level. There should be only one unpaired electron.

Assess

When dealing with ligands at the extremes of the spectrochemical series, it is easier to determine whether the complex is high or low spin. Occasionally, the crystal field splitting is comparable to the pairing energy; in such cases, changes in pressure or temperature can cause a crossover from low spin to high spin. Such a phenomenon is called *spin crossover*. A complex that displays spin crossover is tris(2-picolylamine)iron(II).

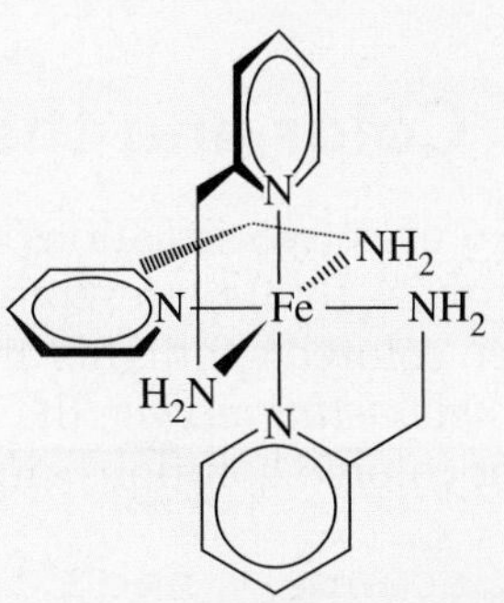

Tris(2-picolylamine)iron(II)

PRACTICE EXAMPLE A: How many unpaired electrons would you expect to find in the octahedral complex $[MnF_6]^{2-}$?

PRACTICE EXAMPLE B: How many unpaired electrons would you expect to find in the tetrahedral complex $[CoCl_4]^{2-}$? Would you expect more, fewer, or the same number of unpaired electrons as in the octahedral complex $[Co(H_2O)_6]^{2+}$?

EXAMPLE 24-5 Using the Crystal Field Theory to Predict the Structure of a Complex from Its Magnetic Properties

The complex ion $[Ni(CN)_4]^{2-}$ is diamagnetic. Use ideas from the crystal field theory to speculate on its probable structure.

Analyze

We can eliminate an octahedral structure because the coordination number is 4 (not 6). Our choice is between tetrahedral and square-planar geometries. We compare the crystal field splitting diagrams for each.

Solve

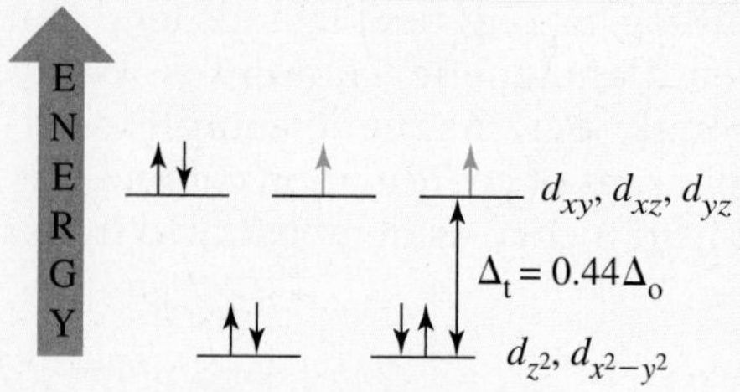

The electron configuration of Ni is $[Ar]3d^84s^2$, and that of Ni(II) is $[Ar]3d^8$. Because the complex ion is diamagnetic, all $3d$ electrons must be paired. Let us see how we would distribute these $3d$ electrons if the structure were tetrahedral (recall Figure 24-13). We would place four electrons (all paired) into the two lowest d levels. We would then distribute the remaining four electrons among the three higher-level d orbitals. Two of the electrons would be unpaired, and the complex ion would be paramagnetic.

Because the tetrahedral structure would be paramagnetic, we can conclude that the structure of the diamagnetic $[Ni(CN)_4]^{2-}$ ion must be square-planar. Let's also demonstrate that this is a reasonable conclusion based on the d-orbital energy-level diagram for a square-planar complex (recall Figure 24-14). First, the three lowest-energy orbitals are filled with electrons (six), and then we assume that the energy separation between the d_{xy} and $d_{x^2-y^2}$ orbitals is large enough that the final two electrons remain paired in the d_{xy} orbital. This corresponds to a diamagnetic complex ion.

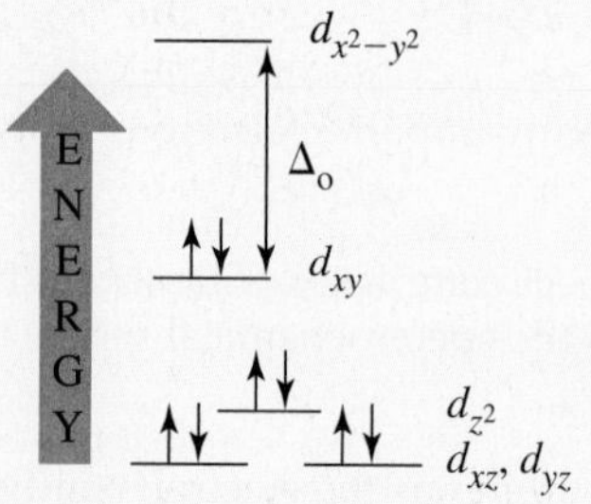

Assess

To decide between the two possible geometries, we needed to know the magnetic properties of the complex. Although $[NiCl_4]^{2-}$ is paramagnetic and tetrahedral, the stronger field ligands in $[Ni(CN)_4]^{2-}$ increase the energy separation of the d orbitals, making the square-planar geometry more stable. (VSEPR theory predicts that the tetrahedral geometry is more stable.)

PRACTICE EXAMPLE A: The complex ion $[Co(CN)_4]^{2-}$ is paramagnetic with three unpaired electrons. Use ideas from the crystal field theory to speculate on its probable structure.

PRACTICE EXAMPLE B: Would you expect $[Cu(NH_3)_4]^{2+}$ to be diamagnetic or paramagnetic? Can you use this information about the magnetic properties of $[Cu(NH_3)_4]^{2+}$ to help you determine whether the structure of $[Cu(NH_3)_4]^{2+}$ is tetrahedral or square-planar? Explain.

24-7 Color and the Colors of Complexes

Figure 24-16 uses two situations for mixing color to help clarify the nature of colors. Figure 24-16(a) represents *additive mixing*, the type of color mixing that occurs when colored spotlights are superimposed. Figure 24-16(b), conversely, represents *subtractive mixing*, the type of color mixing that occurs when paint pigments or colored solutions are mixed.

Primary, Secondary, and Complementary Colors

▶ Additive color mixing is used in color monitors and color television screens. Subtractive color mixing is used for all the photographs and art in this printed text.

For the additive mixing of light beams in Figure 24-16(a), the **primary colors** are defined as any three colors that, when combined, yield white light (W). The colors used in the figure are *red* (R), *green* (G), and *blue* (B); their sum can be represented as R + G + B = W. The **secondary colors** are those produced by combining two primary colors. Figure 24-16(a) indicates that the secondary colors are *yellow* (Y = R + G), *cyan* (C = G + B), and *magenta* (M = B + R).

Each secondary color is a **complementary color** of one of the primary colors. Again, from Figure 24-16(a), *cyan* (C) is the complementary color of red (R);

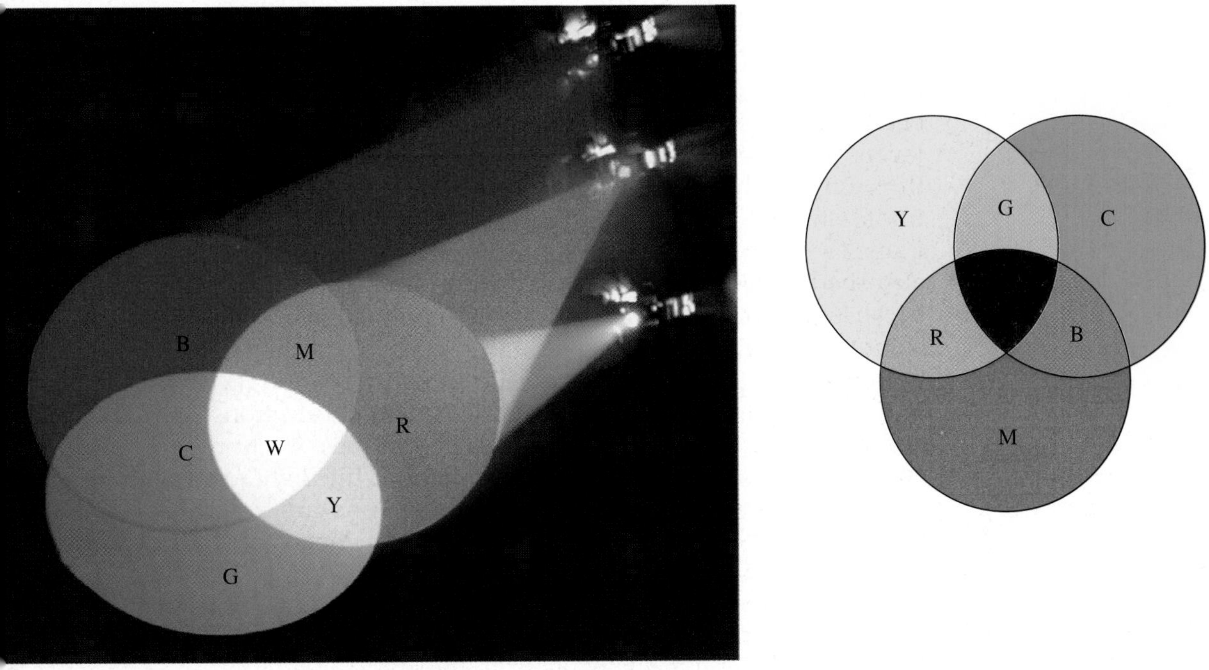

(**a**) Additive color mixing

(**b**) Subtractive color mixing

▲ FIGURE 24-16
The mixing of colors
(a) The additive mixing of three beams of colored light: red (R), green (G), and blue (B). The secondary colors—yellow (Y), cyan (C), and magenta (M)—are produced in regions where two of the beams overlap. The overlap of all three beams produces white light (W). **(b)** The subtractive mixing of three pigments with the primary colors magenta (M), yellow (Y), and cyan (C). Here, the secondary colors—red (R), green (G), and blue (B)—form when two of the primary colors are mixed. A mixture of all three primary colors produces a very dark brown to black color.

In four-color printing, such as in this book and in color printers in computer systems, the basic inks used are magenta, yellow, cyan, and black. In photographs and drawings in this book, very small dots of these four basic colors, printed singly and in various combinations, produce the colored images you see. (With a magnifying glass, you can see individual dots of color.)

magenta (M), of green (G); and yellow (Y), of blue (B). Figure 24-16(a) shows that when a primary color and its complementary color are mixed, the result is white light. This must be the case: Because cyan, for example, is itself the combination of two of the primary colors—green and blue (C = G + B)—the combination of cyan and red (C + R) is the same as the combination of the three primary colors: G + B + R = W.

In subtractive color mixing, some of the wavelength components of white light are removed by absorption, and the reflected light (for example, from a painted surface or colored fabric) or transmitted light (as seen through glass or a solution) is deficient in some wavelength components. The reflected or transmitted light is colored. In subtractive color mixing (Fig. 24-16b), there are also primary, secondary, and complementary colors. In this case, *magenta* (M), *yellow* (Y), and *cyan* (C) are the primary colors and *green* (G), *blue* (B), and *red* (R) are the secondary colors. If a material absorbs all three primary colors, there is essentially no light left to be reflected or transmitted; the material appears black, or nearly so. If a material absorbs one color, primary or secondary, the reflected or transmitted light is the complementary color. Thus, a magenta sweater has that color because the dye it contains strongly absorbs green light and reflects magenta, the complement of green. A solution of red food dye has that color because the dye absorbs cyan light and transmits red.

Colored Solutions

Now let's direct our attention specifically to colored solutions. Colored solutions contain species that can absorb photons of visible light and use the energy of those photons to promote electrons in the species to higher energy levels. The energies of the photons must just match the energy differences through which the electrons are to be promoted. Because the energies of photons are related to the frequencies (and wavelengths) of light (recall Planck's equation, $E = h\nu$), only certain wavelength components are absorbed as white light passes through the solution. The emerging light, because it is lacking some wavelength components, is no longer white; it is colored.

Ions having (1) a noble-gas electron configuration, (2) an outer shell of 18 electrons, or (3) the "18 + 2" configuration (18 electrons in the $n - 1$ shell and two in the n, or outermost, shell) do not have electron transitions in the energy range corresponding to visible light. White light passes through these solutions without being absorbed; these ions are colorless in solution. Examples are the alkali and alkaline earth metal ions, the halide ions, Zn^{2+}, Al^{3+}, and Bi^{3+}.

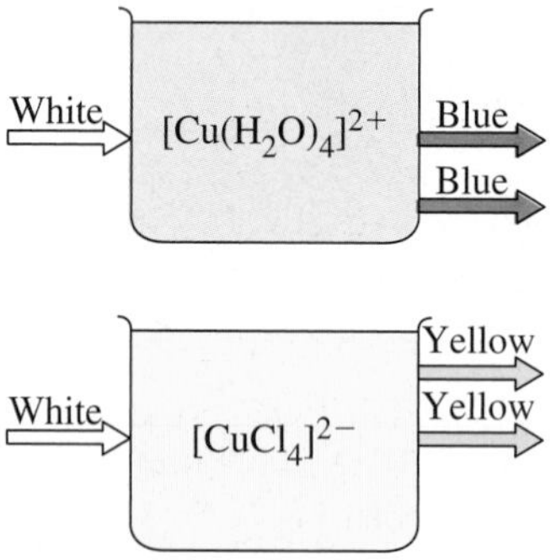

▲ FIGURE 24-17
Light absorption and transmission
$[Cu(H_2O)_4]^{2+}$ absorbs in the yellow region of the spectrum and transmits blue light. $[CuCl_4]^{2-}$ absorbs in the blue region of the spectrum and transmits yellow light.

Crystal field splitting of the *d* energy levels produces the energy difference, Δ, that accounts for the colors of complex ions. Promotion of an electron from a lower to a higher *d* level results from the absorption of the appropriate components of white light; the transmitted light is colored. Subtracting one color from white light leaves the complementary color. A solution containing $[Cu(H_2O)_4]^{2+}$ absorbs most strongly in the yellow region of the spectrum (about 580 nm). The wavelength components of the light transmitted combine to produce the color *blue*. Thus, aqueous solutions of copper(II) compounds usually have a characteristic blue color. In the presence of high concentrations of Cl^-, copper(II) forms the complex ion $[CuCl_4]^{2-}$. This species absorbs strongly in the blue region of the spectrum. The transmitted light is *yellow*, as is the color of the solution. Figure 24-17 suggests light absorption by these solutions. The colors of some complex ions of chromium are given in Table 24.5. The colors of six related coordination compounds of cobalt(III) are shown in Figure 24-18.

TABLE 24.5 Some Coordination Compounds of Cr^{3+} and Their Colors

Isomer	Color
$[Cr(H_2O)_6]Cl_3$	Violet
$[CrCl(H_2O)_5]Cl_2$	Blue-green
$[Cr(NH_3)_6]Cl_3$	Yellow
$[CrCl(NH_3)_5]Cl_2$	Purple

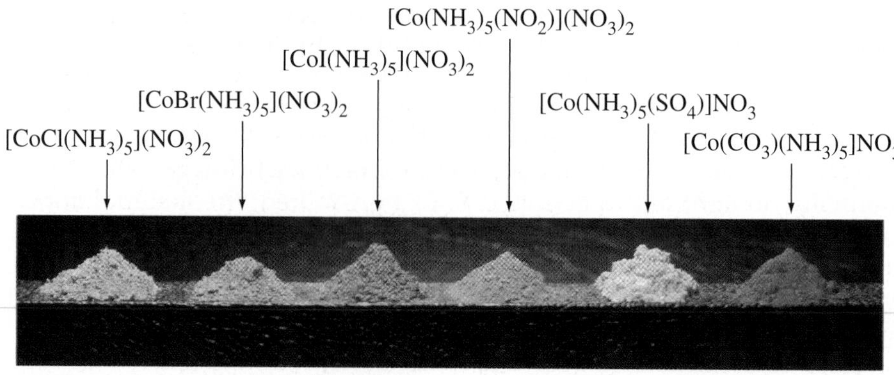

Carey B. Van Loon

▶ FIGURE 24-18
Effect of ligands on the colors of coordination compounds
These compounds all consist of a six-coordinate cobalt complex ion in combination with nitrate ions. In each case, the complex ion has five NH_3 molecules and one other group as ligands.

EXAMPLE 24-6 Relating the Colors of Complexes to the Spectrochemical Series

Table 24.5 lists the color of $[Cr(H_2O)_6]Cl_3$ as violet, whereas that of $[Cr(NH_3)_6]Cl_3$ is yellow. Explain this difference in color.

Analyze

Both chromium(III) complexes are octahedral, and the electron configuration of Cr^{3+} is $[Ar]3d^3$. From these facts, we can construct the energy-level diagram shown here. The three unpaired electrons go into the three lower-energy *d* orbitals. When a photon of light is absorbed, an electron is promoted from a *d* orbital of lower energy to a *d* orbital of higher energy. The quantity of energy required for this promotion depends on the energy level separation, Δ_o.

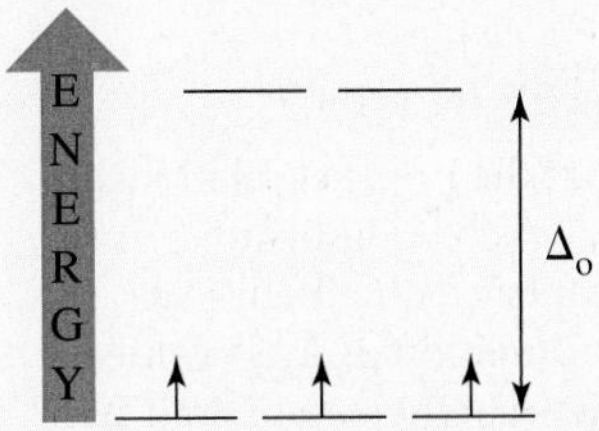

Solve

According to the spectrochemical series, NH_3 produces a greater splitting of the *d* energy level than does H_2O. We should expect $[Cr(NH_3)_6]^{3+}$ to absorb light of a *shorter* wavelength (higher energy) than does $[Cr(H_2O)_6]^{3+}$. Thus, $[Cr(NH_3)_6]^{3+}$ absorbs in the violet region of the spectrum, and the transmitted light is *yellow*. $[Cr(H_2O)_6]^{3+}$ absorbs in the yellow region of the spectrum, and the transmitted light is *violet*.

Assess

If we measure the wavelength, λ, of the photon absorbed, we can then calculate the splitting energy, $\Delta_o = hc/\lambda$. Stronger field ligands increase the magnitude of Δ_o; therefore, shorter wavelengths of light are absorbed and longer wavelengths are transmitted.

PRACTICE EXAMPLE A: The color of $[Co(H_2O)_6]^{2+}$ is pink, whereas that of tetrahedral $[CoCl_4]^{2-}$ is blue. Explain this difference in color.

PRACTICE EXAMPLE B: One of the following solids is yellow, and the other is green: $Fe(NO_3)_2 \cdot 6\,H_2O$; $K_4[Fe(CN)_6] \cdot 3\,H_2O$. Indicate which is which, and explain your reasoning.

24-8 Aspects of Complex-Ion Equilibria

In Chapter 18, we learned that complex-ion formation can have a great effect on the solubilities of substances, as in the ability of $NH_3(aq)$ to dissolve rather large quantities of AgCl(s). To calculate the solubility of AgCl(s) in $NH_3(aq)$, we had to use the formation constant, K_f, of $[Ag(NH_3)_2]^+$. Equations (24.1) and (24.2) illustrate how we dealt with formation constants in Chapter 18, with $[Zn(NH_3)_4]^{2+}$ as an example.

$$Zn^{2+}(aq) + 4\,NH_3(aq) \rightleftharpoons [Zn(NH_3)_4]^{2+}(aq) \qquad \textbf{(24.1)}$$

$$K_f = \frac{[[Zn(NH_3)_4]^{2+}]}{[Zn^{2+}][NH_3]^4} = 4.1 \times 10^8 \qquad \textbf{(24.2)}$$

In fact, cations in aqueous solution exist mostly in *hydrated* form. That is, $Zn^{2+}(aq)$ is actually $[Zn(H_2O)_4]^{2+}$. As a result, when NH_3 molecules bond to Zn^{2+} and form an ammine complex ion, they do not enter an empty coordination sphere. They must displace H_2O molecules. This displacement occurs in a stepwise fashion. The reaction

$$[Zn(H_2O)_4]^{2+} + NH_3 \rightleftharpoons [Zn(H_2O)_3NH_3]^{2+} + H_2O \qquad \textbf{(24.3)}$$

for which

$$K_1 = \frac{[[Zn(H_2O)_3NH_3]^{2+}]}{[[Zn(H_2O)_4]^{2+}][NH_3]} = 3.9 \times 10^2 \qquad \textbf{(24.4)}$$

is followed by

$$[Zn(H_2O)_3NH_3]^{2+} + NH_3 \rightleftharpoons [Zn(H_2O)_2(NH_3)_2]^{2+} + H_2O \qquad \textbf{(24.5)}$$

for which

$$K_2 = \frac{[[Zn(H_2O)_2(NH_3)_2]^{2+}]}{[[Zn(H_2O)_3NH_3]^{2+}][NH_3]} = 2.1 \times 10^2 \quad (24.6)$$

and so on.

KEEP IN MIND

that the product relationship between equilibrium constants is equivalent to a summation of $\Delta_r G°$ values. Hence, when an overall $\Delta_r G°$ is used to evaluate K_n

$\Delta_r G°$ (overall) $=$
$\Delta_r G_1° + \Delta_r G_2° + \Delta_r G_3° + \cdots$
$= -RT \ln K_1 - RT \ln K_2 - RT \ln K_3 - \cdots$
$= -RT \ln(K_1 \times K_2 \times K_3 \cdots)$
$= -RT \ln \beta_n = -RT \ln K_f$

The value of K_1 in equation (24.4) is often designated as β_1 and called the formation constant for the complex ion $[Zn(H_2O)_3NH_3]^{2+}$. The formation of $[Zn(H_2O)_2(NH_3)_2]^{2+}$ is represented by the *sum* of equations (24.3) and (24.5),

$$[Zn(H_2O)_4]^{2+} + 2\,NH_3 \rightleftharpoons [Zn(H_2O)_2(NH_3)_2]^{2+} + 2\,H_2O \quad (24.7)$$

The formation constant β_2, in turn, is given by the *product* of equations (24.4) and (24.6).

$$\beta_2 = \frac{[[Zn(H_2O)_2(NH_3)_2]^{2+}]}{[[Zn(H_2O)_4]^{2+}][NH_3]^2} = K_1 \times K_2 = 8.2 \times 10^4 \quad (24.8)$$

For the next ion in the series, $[Zn(H_2O)(NH_3)_3]^{2+}$, $\beta_3 = K_1 \times K_2 \times K_3$. For the final member, $[Zn(NH_3)_4]^{2+}$, $\beta_4 = K_1 \times K_2 \times K_3 \times K_4$, and it is this product of constants that we called the formation constant in Section 18-8 and listed as K_f in Table 18.2. Additional stepwise formation constants are presented in Table 24.6.

The large numerical value of K_1 for reaction (24.3) indicates that Zn^{2+} has a greater affinity for NH_3 (a stronger Lewis base) than it does for H_2O. Displacement of ligand H_2O molecules by NH_3 occurs even if the number of NH_3 molecules present in aqueous solution is much smaller than the number of H_2O molecules, as in dilute $NH_3(aq)$. The fact that the successive K values decrease regularly in the displacement process, at least for displacements involving neutral molecules as ligands, is due in part to statistical factors. An NH_3 molecule has a better chance of replacing a H_2O molecule in $[Zn(H_2O)_4]^{2+}$, in which each coordination position is occupied by H_2O, than in $[Zn(H_2O)_3NH_3]^{2+}$, in which one of the positions is already occupied by NH_3. Also, once the degree of substitution of NH_3 for H_2O has become large, the chances improve for H_2O molecules to replace NH_3 molecules in a reverse reaction. Again, this tends to reduce the value of K. If irregularities arise in the succession of K values, it is often because of a change in structure of the complex ion at some point in the series of displacement reactions.

If the ligand in a substitution process is polydentate, it displaces as many H_2O molecules as there are points of attachment. Thus, ethylenediamine (en)

TABLE 24.6 Stepwise and Overall Formation (Stability) Constants for Several Complex Ions

Metal[a] Ion	Ligand	K_1	K_2	K_3	K_4	K_5	K_6	β_n (or K_f)[b]
Ag^+	NH_3	2.0×10^3	7.9×10^3					1.6×10^7
Zn^{2+}	NH_3	3.9×10^2	2.1×10^2	1.0×10^2	5.0×10^1			4.1×10^8
Cu^{2+}	NH_3	1.9×10^4	3.9×10^3	1.0×10^3	1.5×10^2			1.1×10^{13}
Ni^{2+}	NH_3	6.3×10^2	1.7×10^2	5.4×10^1	1.5×10^1	5.6	1.1	5.3×10^8
Cu^{2+}	en	5.2×10^{10}	2.0×10^9					1.0×10^{20}
Ni^{2+}	en	3.3×10^7	1.9×10^6	1.8×10^4				1.1×10^{18}
Ni^{2+}	EDTA	4.2×10^{18}						4.2×10^{18}

[a]In many tabulations in the chemical literature, formation-constant data are presented as logarithms: that is, $\log K_1$, $\log K_2, \ldots$, and $\log \beta_n$.
[b]The β_n listed is for the number of steps shown: e.g., for $[Ag(NH_3)_2]^+$, $\beta_2 = K_f = K_1 \times K_2$; for $[Ni(en)_3]^{2+}$, $\beta_3 = K_f = K_1 \times K_2 \times K_3$; and for $[Ni(EDTA)]^{2-}$, $\beta_1 = K_f = K_1$.

isplaces H_2O molecules in $[Ni(H_2O)_6]^{2+}$ two at a time, in three steps. The irst step is

$$Ni(H_2O)_6]^{2+} + en \rightleftharpoons [Ni(en)(H_2O)_4]^{2+} + 2\,H_2O \qquad K_1(\beta_1) = 3.3 \times 10^7 \qquad \textbf{(24.9)}$$

Jote from Table 24.6 that the complex ions with polydentate ligands have nuch larger formation constants than do those with monodentate ligands. For xample, K_f (that is, β_3) for $[Ni(en)_3]^{2+}$ is 1.1×10^{18}, whereas K_f (that is, β_6) or $[Ni(NH_3)_6]^{2+}$ is 5.3×10^8. The additional stability of chelates over com- lexes with monodentate ligands is known as the **chelation (or chelate) effect.** 'his effect can be partly attributed to the increase in entropy associated with helation. In the displacement of H_2O by NH_3, the entropy change is small two particles on each side of an equation such as (24.3)]. An ethylenediamine nolecule, conversely, displaces *two* H_2O molecules [two particles on the left nd three on the right of equation (24.9)]. The larger, positive value of $\Delta_r S°$ for he displacement by ethylenediamine means a more negative $\Delta_r G°$ and a larger K.

24-7 CONCEPT ASSESSMENT

The Cr(III) cation in aqueous solution can form 6-coordinate complexes with ethylenediamine and $EDTA^{4-}$. Which complex do you expect to be the most stable if all water molecules in the six-coordinate $Cr^{3+}(aq)$ ion are replaced?

24-9 Acid–Base Reactions of Complex Ions

Ve have described complex-ion formation in terms of Lewis acids and bases. Complex ions may also exhibit acid–base properties in the Brønsted–Lowry ense; that is, they may act as proton donors or acceptors. Figure 24-19 repre- ents the ionization of $[Fe(H_2O)_6]^{3+}$ as an acid. A proton from a *ligand* water nolecule in hexaaquairon(III) ion is transferred to a *solvent* water molecule. The H_2O ligand is converted to OH^-.

$$[Fe(H_2O)_6]^{3+} + H_2O \rightleftharpoons [FeOH(H_2O)_5]^{2+} + H_3O^+ \quad K_{a_1} = 9 \times 10^{-4} \qquad \textbf{(24.10)}$$

The second ionization step is

$$[FeOH(H_2O)_5]^{2+} + H_2O \rightleftharpoons [Fe(OH)_2(H_2O)_4]^+ + H_3O^+ \quad K_{a_2} = 5 \times 10^{-4} \qquad \textbf{(24.11)}$$

From these K_a values, we see that $Fe^{3+}(aq)$ is fairly acidic (compared, for example, with acetic acid, with $K_a = 1.8 \times 10^{-5}$). To suppress ionization (hydrolysis) of $[Fe(H_2O)_6]^{3+}$, we need to maintain a low pH by the addition of acids such as HNO_3 or $HClO_4$. The ion $[Fe(H_2O)_6]^{3+}$ is violet in color, but aqueous solutions of $Fe^{3+}(aq)$ are generally yellow because of the presence of hydroxido complex ions.

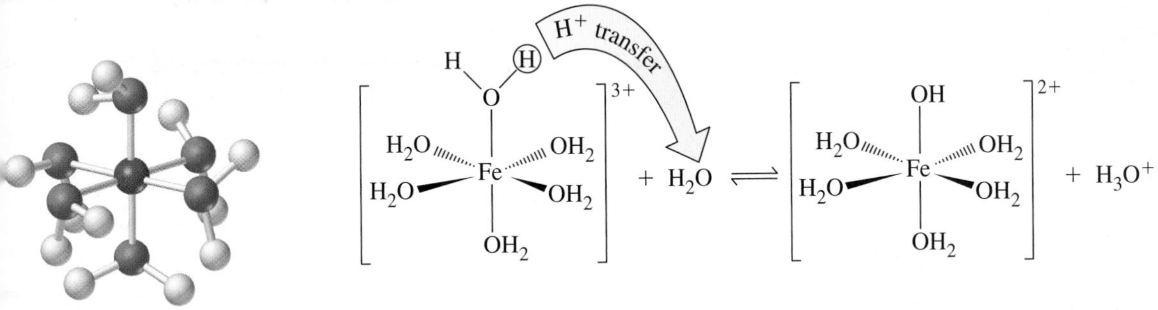

$[Fe(H_2O)_6]^{3+}$

▲ FIGURE 24-19
Ionization of $[Fe(H_2O)_6]^{3+}$

The ions Cr^{3+} and Al^{3+} behave in a manner similar to Fe^{3+} except that, wit them, hydroxido complex-ion formation can continue until complex anion are produced. $Cr(OH)_3$ and $Al(OH)_3$, as we previously noted, are soluble i alkaline as well as acidic solutions; they are amphoteric.

Regarding the acid strengths of aqua complex ions, a critical factor is th charge-to-radius ratio of the central metal ion. Thus, the small, highly charge Fe^{3+} attracts electrons away from an O—H bond in a ligand water molecul more strongly than does Fe^{2+}. Hence, $[Fe(H_2O)_6]^{3+}$ is a stronger aci ($K_{a_1} = 9 \times 10^{-4}$) than is $[Fe(H_2O)_6]^{2+}$ ($K_{a_1} = 1 \times 10^{-7}$).

24-10 Some Kinetic Considerations

When $NH_3(aq)$ is added to an aqueous solution containing Cu^{2+}, there is change in color from pale blue to very deep blue. The reaction involves NH molecules displacing H_2O molecules as ligands.

$$\underset{\text{(pale blue)}}{[Cu(H_2O)_4]^{2+}} + 4\,NH_3 \longrightarrow \underset{\text{(very deep blue)}}{[Cu(NH_3)_4]^{2+}} + 4\,H_2O$$

This reaction occurs very rapidly—as rapidly as the two reactants can b brought together. The addition of HCl(aq) to an aqueous solution of Cu^{2+} pro duces an immediate color change from pale blue to green, or even yellow i the HCl(aq) is sufficiently concentrated.

$$\underset{\text{(pale blue)}}{[Cu(H_2O)_4]^{2+}} + 4\,Cl^- \longrightarrow \underset{\text{(yellow)}}{[CuCl_4]^{2-}} + 4\,H_2O$$

▶ The terms *labile* and *inert* are not related to the thermodynamic stabilities of complex ions or to the equilibrium constants for ligand-substitution reactions. The terms are *kinetic* terms, referring to the rates at which ligands are exchanged.

Complex ions in which ligands can be interchanged rapidly are said to b **labile**. $[Cu(H_2O)_4]^{2+}$, $[Cu(NH_3)_4]^{2+}$, and $[CuCl_4]^{2-}$ are all labile (Fig. 24-20).

In freshly prepared $CrCl_3(aq)$, the ion *trans*-$[CrCl_2(H_2O)_4]^+$ produces green color, but the color gradually turns to violet (Fig. 24-21). This colo change results from the very slow exchange of H_2O for Cl^- ligands. A comple ion that exchanges ligands slowly is said to be *nonlabile*, or **inert**. In general complex ions of the first transition series, except for those of Cr(III) and Co(III), are kinetically labile. Those of the second and third transition serie are generally kinetically inert. Whether a complex ion is labile or inert affect the ease with which it can be studied. The inert ones are easiest to isolate and characterize, which may explain why so many of the early studies of comple ions were based on Cr(III) and Co(III).

Carey B. Van Loon

▲ FIGURE 24-21
Inert complex ions
The green solid $CrCl_3 \cdot 6\,H_2O$ produces the green aqueous solution on the left. The color is due to *trans*-$[CrCl_2(H_2O)_4]^+$. A slow exchange of H_2O for Cl^- ligands leads to a violet solution of $[Cr(H_2O)_6]^{3+}$ in one or two days (right).

Carey B. Van Loon

▲ FIGURE 24-20
Labile complex ions
The exchange of ligands in the coordination sphere of Cu^{2+} occurs very rapidly. The solution at the extreme left is formed by dissolving $CuSO_4$ in concentrated HCl(aq). Its yellow color is due to $[CuCl_4]^{2-}$. When a small amount of water is added, the mixture of $[Cu(H_2O)_4]^{2+}$ and $[CuCl_4]^{2-}$ ions produces a yellow-green color. When $CuSO_4$ is dissolved in water, a light blue solution of $[Cu(H_2O)_4]^{2+}$ forms. NH_3 molecules readily displace H_2O molecules as ligands and produce deep blue $[Cu(NH_3)_4]^{2+}$ (extreme right).

24-11 Applications of Coordination Chemistry

he applications of coordination chemistry are numerous and varied. They ange from analytical chemistry to biochemistry. The several brief examples in his section give some idea of this diversity.

Cisplatin: A Cancer-Fighting Drug

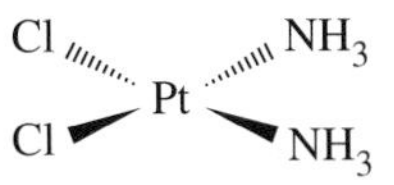

cis-$[PtCl_2(NH_3)_2]$
(cisplatin)

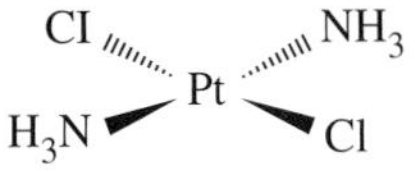

trans-$[PtCl_2(NH_3)_2]$
(transplatin)

hemotherapy is a treatment used for some types of cancer. The treatment mploys anticancer drugs to destroy cancer cells. An important cancer-fighting rug is cisplatin, which is commonly used to treat testicular, bladder, lung, sophagus, stomach, and ovarian cancers.

Cisplatin was first synthesized in 1845 by Michel Peyrone, an Italian doctor, ut its structure was not elucidated until almost fifty years later, in 1893, by Verner. The anticancer activity of cisplatin was discovered in the 1960s by Barrett Rosenberg, a professor of chemistry at Michigan State University. Not only did Rosenberg discover the anticancer activity of cisplatin, but he was also the first to eport that the *trans* isomer (transplatin) was ineffective in killing cancer cells.

Before examining the anticancer activity of cisplatin, let's first consider its ynthesis. One method for making cisplatin starts from $K_2[PtCl_4]$, which is onverted to $K_2[PtI_4]$, by treatment with an aqueous solution of KI:

$$K_2[PtCl_4] + 4\,KI \longrightarrow K_2[PtI_4] + 4\,KCl$$

n the next step, NH_3 is added, forming a yellow compound, *cis*-$[PtI_2(NH_3)_2]$. The formation of *cis*-$[PtI_2(NH_3)_2]$ is a key step, which occurs in two stages.

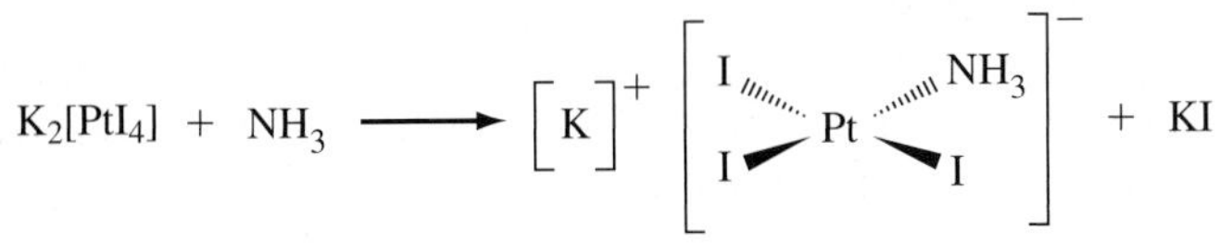

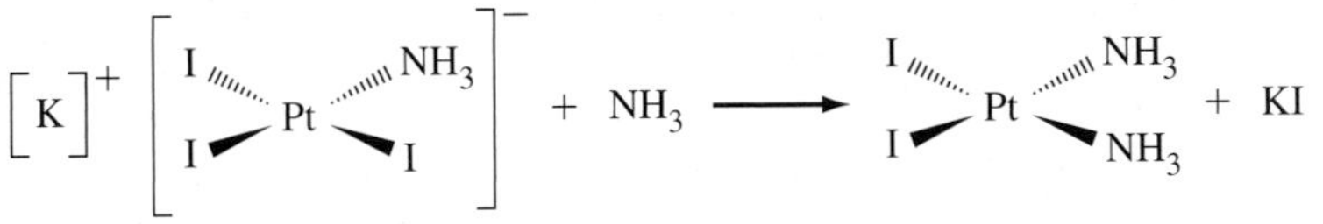

◀ Empirically, certain ligands are found to have a greater tendency than others to direct an incoming ligand toward a *trans* position. For the monodentate ligands we have encountered, the approximate ordering from strongest *trans* directors to weakest *trans* directors is $CN^-, CO > NO_2^-, I^- > Br^- > Cl^- > NH_3 > OH^- > H_2O$.

It might seem strange that only one isomer (the *cis* isomer) is obtained. One way to rationalize why this happens is to consider the ligands in $[PtI_3(NH_3)]^-$ n terms of their tendencies to direct an incoming ligand toward the *trans* position. Empirical studies indicate that NH_3 is a weaker *trans* director than is I^-, nd so the second NH_3 molecule is preferentially directed to a position that is *trans* to I^-, rather than being directed to a position that is *trans* to NH_3. As pointed out in the margin note, Cl^- is a weaker *trans* director than is I^-, so reatment of $K_2[PtCl_4]$ with NH_3 will give a lower yield of the *cis* isomer of $PtCl_2(NH_3)_2$. Consequently, the conversion of $K_2[PtCl_4]$ to $K_2[PtI_4]$ is an mportant step in the synthesis.

The remaining steps in preparing cisplatin are as follows. When *cis*-$[PtI_2(NH_3)_2]$ is treated with $AgNO_3$, insoluble AgI precipitates, leaving behind a solution of *cis*-$[Pt(NH_3)_2(OH_2)_2]^{2+}$. Finally, treatment with KCl gives *cis*-$[PtCl_2(NH_3)_2]$ as a yellow precipitate.

Let's now consider the essential features of the anticancer activity of cisplatin. Cisplatin enters cancer cells mainly by diffusion. Once inside the cell, one of the chloride ions in cisplatin is replaced by a water molecule.

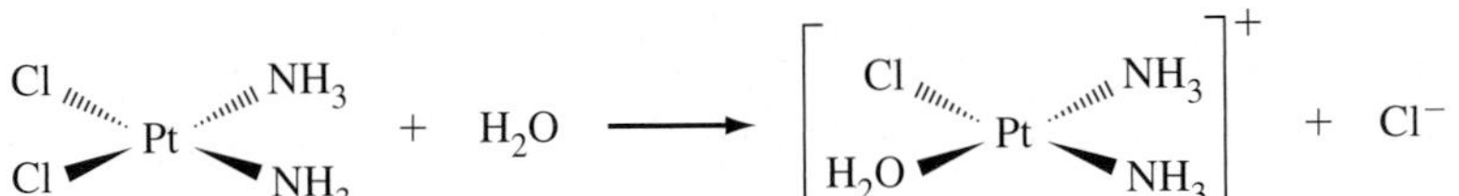

Cisplatin

The anticancer activity of cisplatin is associated with the binding c $[Pt(Cl)(H_2O)(NH_3)_2]^+$ to a cellular DNA molecule. (We will discuss DNA i more detail in Chapter 28.) When $[Pt(Cl)(H_2O)(NH_3)_2]^+$ binds to a DNA mol ecule, structural deformations occur in the DNA molecule and these deforma tions, if not repaired by proteins in the cell, lead ultimately to cell death.

It may seem surprising that the *cis* and *trans* isomers of $PtCl_2(NH_3)_2$ exhibi a dramatic difference in anticancer activity, given their structural similaritie Overall, the *trans* isomer (transplatin) is more reactive and potentially mor potent than cisplatin but the higher reactivity leads ultimately to lower anti cancer activity. Because of its increased reactivity, transplatin can underg many side reactions before it reaches its target; thus, transplatin is less effec tive in killing cancer cells.

The discovery of the anticancer activity of cisplatin was nothing less tha monumental. To date, thousands of platinum-containing compounds hav been investigated as potential chemotherapy drugs. Worldwide annual sales o platinum-based anticancer drugs are currently in excess of $2 billion.

Hydrates

When a compound is crystallized from an aqueous solution of its ions, the crys tals obtained are often hydrated. As originally described in Chapter 3, a hydrat is a substance that has a fixed number of water molecules associated with eac formula unit. In some cases, the water molecules are ligands bonded directly t a metal ion. The coordination compound $[Co(H_2O)_6](ClO_4)_2$ may be repre sented as the hexahydrate, $Co(ClO_4)_2 \cdot 6\,H_2O$. In the hydrate $CuSO_4 \cdot 5\,H_2O$ four H_2O molecules are associated with copper in the complex io $[Cu(H_2O)_4]^{2+}$, and the fifth with the $SO_4{}^{2-}$ anion by hydrogen bonding Another possibility for hydrate formation is that the water molecules may b incorporated into definite positions in the solid crystal but not associated wit any particular cations or anions, as in $BaCl_2 \cdot 2\,H_2O$. This is called *lattice water* Finally, part of the water may be coordinated to an ion and part of it may be lat tice water, as appears to be the case with alums, such as $KAl(SO_4)_2 \cdot 12\,H_2O$.

Stabilization of Oxidation States

The standard electrode potential for the reduction of Co(III) to Co(II) is

$$Co^{3+}(aq) + e^- \longrightarrow Co^{2+}(aq) \quad E° = +1.82\text{ V}$$

This large positive value suggests that $Co^{3+}(aq)$ is a strong oxidizing agent strong enough to oxidize water to $O_2(g)$.

$$4\,Co^{3+}(aq) + 2\,H_2O(l) \longrightarrow 4\,Co^{2+}(aq) + 4\,H^+(aq) + O_2(g) \quad E°_{cell} = +0.59\text{ V} \tag{24.12}$$

Yet one of the complex ions featured in this chapter has been $[Co(NH_3)_6]^{3+}$ This ion is stable in water solution, even though it contains cobalt in the +3 oxidation state. Reaction (24.12) will not occur if the concentration of Co^{3+} is sufficiently low and $[Co^{3+}]$ is kept very low because of the great stability of the complex ion.

$$Co^{3+}(aq) + 6\,NH_3(aq) \rightleftharpoons [Co(NH_3)_6]^{3+}(aq) \qquad \beta_6 = K_f = 4.5 \times 10^{33}$$

In fact, the concentration of free Co^{3+} is so low that for the half-reaction

$$[Co(NH_3)_6]^{3+} + e^- \longrightarrow [Co(NH_3)_6]^{2+}$$

$E°$ is only +0.10 V. As a consequence, not only is $[Co(NH_3)_6]^{3+}$ stable but also $[Co(NH_3)_6]^{2+}$ is rather easily oxidized to the Co(III) complex.

The ability of strong electron-pair donors (strong Lewis bases) to stabilize high oxidation states in the way that NH_3 does in Co(III) complexes and O^{2-} in Mn(VII) complexes (such as $MnO_4{}^-$) affords a means of attaining certain oxidation states that might otherwise be difficult or impossible to attain.

hotography: Fixing a Photographic Film

black-and-white photographic film is an emulsion of a finely divided silver alide (typically AgBr) coated on a strip of polymer, such as a modified cellu-se. In the *exposure* step, the film is exposed to light and some of the tiny gran-les of AgBr(s) absorb photons. The photons promote the oxidation of Br^- to r and the reduction of Ag^+ to Ag. The Ag and Br atoms remain in the crys-alline lattice of AgBr(s) as "defects," in numbers that depend on the intensity f the light absorbed: The brighter the light, the more Ag atoms. Because the ctual number of Ag atoms produced in the exposure is not large, the silver is nvisible to the eye. The pattern of distribution of the Ag atoms, however, cre-tes a *latent* image of the object photographed. To obtain a visible image, the ilm is developed.

In the *developing* step, the exposed film is placed in a solution of a mild educing agent such as hydroquinone, $C_6H_4(OH)_2$. An oxidation–reduction eaction occurs in which Ag^+ ions are reduced to Ag and the hydroquinone is xidized. The action of the developer is such that reduction of Ag^+ to Ag ccurs just in those granules of AgBr(s) that contain Ag atoms of the latent mage. As a result, the number of Ag atoms in the film is greatly increased, and he latent image becomes visible. Bright regions of the photographed object ppear as dark regions in the photographic image. At this stage, the film is a hotographic *negative*. This negative cannot be exposed to light, however, ecause reduction of Ag^+ to Ag could still occur in the previously unexposed ranules of AgBr(s). The negative must be fixed.

The *fixing* step requires that the black metallic silver of the negative remain n the film and the unexposed AgBr(s) be removed. A common fixer is an queous solution of sodium thiosulfate (also known as sodium hyposulfite or ypo). Because the complex ion $[Ag(S_2O_3)_2]^{3-}$ has a large formation constant, he following reaction is driven to completion—the AgBr(s) dissolves.

$$AgBr(s) + 2\,S_2O_3^{2-}(aq) \longrightarrow [Ag(S_2O_3)_2]^{3-}(aq) + Br^-(aq) \qquad \textbf{(24.13)}$$

Once the negative has been fixed, it is used to produce a *positive* image, he final photograph. This is done by projecting light through the negative nto a piece of photographic paper. Regions of the negative that are dark ransmit little light to the photographic paper and will appear light when he photographic paper is subsequently developed and fixed. Conversely, ight areas of the negative will appear dark in the final print. In this way, the reas of light and dark in the final print are the same as in the photographed bject.

ualitative Analysis

n discussing qualitative cation analysis in Section 18-9 (page 851), we showed ow the group 1 precipitate—AgCl(s), $PbCl_2(s)$, or $Hg_2Cl_2(s)$—is separated y taking advantage of the stable complex ion formed by $Ag^+(aq)$ and $NH_3(aq)$.

$$AgCl(s) + 2\,NH_3(aq) \longrightarrow [Ag(NH_3)_2]^+(aq) + Cl^-(aq)$$

The qualitative analysis scheme abounds in other examples of complexion for-nation. For example, at a point in the procedure for cation group 3, a test is eeded for Co^{2+}. In the presence of SCN^- ion, Co^{2+} forms a blue thiocyanato omplex ion, $[Co(SCN)_4]^{2-}$ (Fig. 24-22a). A problem develops, however, if even trace amount of Fe^{3+} is present in the solution. Fe^{3+} reacts with SCN^- to pro-duce $[Fe(H_2O)_5SCN]^{2+}$, a strongly colored, blood-red complex ion (Fig. 24-22b.) Fortunately, this complication can be resolved by treating a solution contain-ng both Co^{2+} and Fe^{3+} with an excess of F^-. The Fe^{3+} is converted to the xtremely stable, pale yellow $[FeF_6]^{3-}$. The complex ion $[CoF_4]^{2-}$, being much ess stable than $[Co(SCN)_4]^{2-}$, does not form. As a result, the $[Co(SCN)_4]^{2-}(aq)$ an be detected via the blue-green solution color (Fig. 24-22c).

(a) (b) (c)

▶ FIGURE 24-22
Qualitative tests for Co^{2+} and Fe^{3+}
(a) $[Co(SCN)_4]^{2-}$ complex ion.
(b) $[Fe(H_2O)_5SCN]^{2+}$ complex ion.
(c) Mixture of $[FeF_6]^{3-}$ and $[Co(SCN)_4]^{2-}$.

Sequestering Metal Ions

Metal ions can act as unintended catalysts in promoting undesirable chemical reactions in a manufacturing process, or they may alter the properties of the material being manufactured. Thus, for many industrial purposes, it is imperative to remove mineral impurities from water. Often these impurities, such as Cu^{2+}, are present only in trace amounts, and precipitation of metal ions is feasible only if K_{sp} for the precipitate is very small. An alternative is to treat the water with a chelating agent. This reduces the free cation concentrations to the point at which the cations can no longer enter into objectionable reactions. The cations are said to be *sequestered*. Among the chelating agents widely employed are the salts of *ethylenediaminetetraacetic acid* (H_4EDTA), usually as the sodium salt.

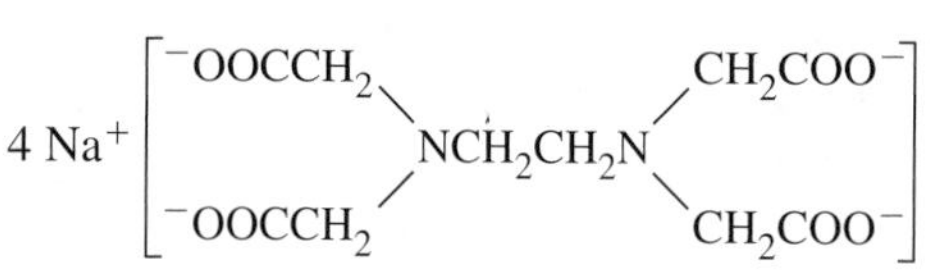

A representative complex ion formed by a metal ion, M^{n+}, with the hexadentate $EDTA^{4-}$ anion is depicted in Figure 24-23. The high stability of such complexes can be attributed to the presence of five, five-member chelate rings. The stability of the complexes can also be attributed to the chelate effect.

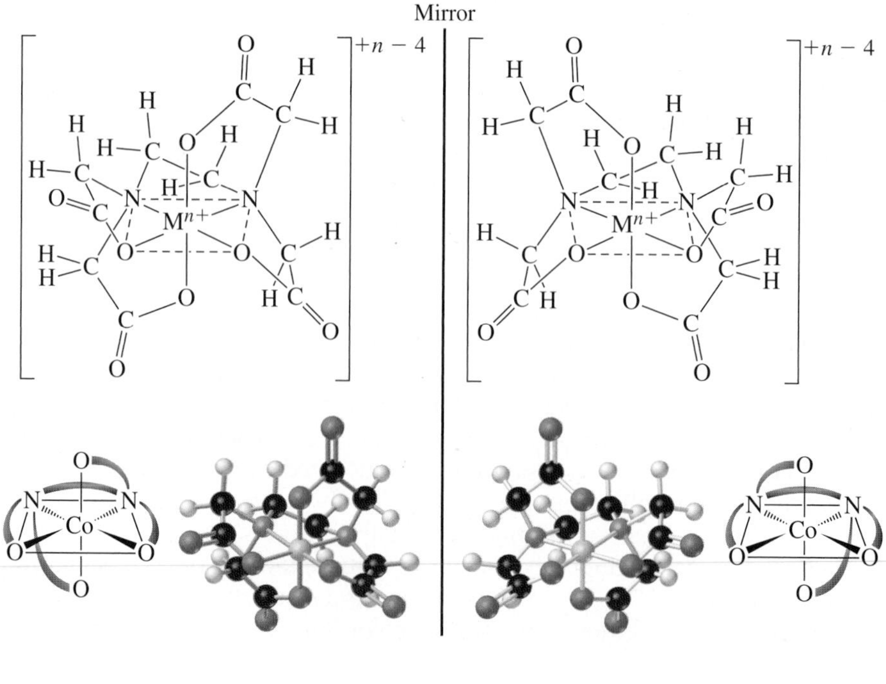

▶ FIGURE 24-23
Structure of a metal–EDTA complex
The central metal ion M^{n+} (pale green) can be Ca^{2+}, Mg^{2+}, Fe^{2+}, Fe^{3+}, and so on. The ligand is $EDTA^{4-}$, and the net charge on the complex is $+n - 4$. Structural diagrams and ball-and-stick models are shown for the two optical isomers of an $[MEDTA]^{n-4}$ complex.

In the presence of $EDTA^{4-}(aq)$, Ca^{2+}, Mg^{2+}, and Fe^{3+} in hard water are nable to form boiler scale or to precipitate as insoluble soaps. The cations re sequestered in the complex ions: $[Ca(EDTA)]^{2-}$, $[Mg(EDTA)]^{2-}$, and $Fe(EDTA)]^{-}$, having K_f values of 4.0×10^{10}, 4.0×10^{8}, and 1.7×10^{24}, espectively.

Chelation with EDTA can be used in treating some cases of metal poisonng. If a person with lead poisoning is fed $[Ca(EDTA)]^{2-}$, the following xchange occurs because $[Pb(EDTA)]^{2-}$ ($K_f = 2 \times 10^{18}$) is even more stable han $[Ca(EDTA)]^{2-}$ ($K_f = 4 \times 10^{10}$).

$$Pb^{2+} + [Ca(EDTA)]^{2-} \longrightarrow [Pb(EDTA)]^{2-} + Ca^{2+}$$

he body excretes the lead complex, and the Ca^{2+} remains as a nutrient. A imilar method can be used to rid the body of radioactive isotopes, as in the reatment of plutonium poisoning.

Some plant fertilizers contain EDTA chelates of metals such as Cu^{2+} as a solble form of the metal ion for the plant to use. Metal ions can catalyze reacions that cause mayonnaise and salad dressings to spoil, and the addition of DTA reduces the concentration of metal ions by chelation.

Tom Pantages

▲ Some products containing EDTA.

iological Applications: Porphyrins

he structure in Figure 24-24 is commonly found in both plant and animal natter. If the eight R groups are all H atoms, the molecule is called *porphin*. The entral N atoms can give up their H atoms, and a metal atom can coordinate imultaneously with all four N atoms. The porphin is a tetradentate ligand for he central metal, and the metal–porphin complex is called a *porphyrin*. Specific orphyrins differ in their central metals and in the R groups on the porphin rings.

In *photosynthesis*, carbon dioxide and water, in the presence of inorganic salts, catalytic agent called *chlorophyll*, and sunlight, combine to form carbohydrates.

$$n\,CO_2 + n\,H_2O \xrightarrow[\text{chlorophyll}]{\text{sunlight}} \underbrace{(CH_2O)_n}_{\text{carbohydrates}} + n\,O_2 \qquad \textbf{(24.14)}$$

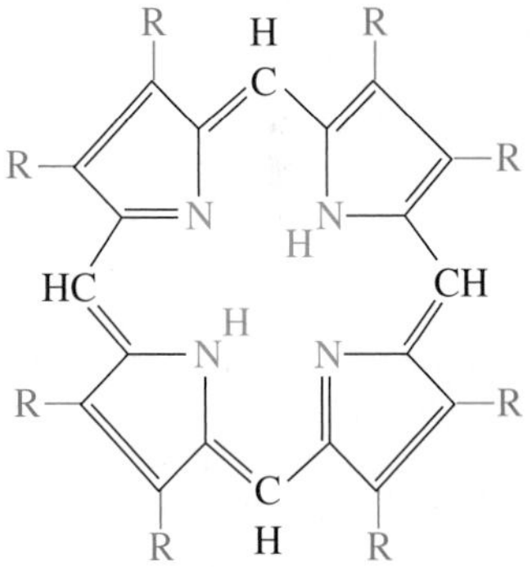

▲ FIGURE 24-24
The porphyrin structure

arbohydrates are the main structural materials of plants. Chlorophyll is a reen pigment that absorbs sunlight and directs the storage of this energy into he chemical bonds of the carbohydrates. The structure of one type of chlorohyll is shown in Figure 24-25; it is a porphyrin. The central metal ion is Mg^{2+}.

Green is the complementary color of magenta—a purplish red—so we hould expect chlorophyll to absorb light in the red region of the spectrum about 670–680 nm). This suggests that green plants should grow more readily n red light than in light of other colors, and some experimental evidence indiates that this is the case. For example, the maximum rate of formation of $O_2(g)$ by reaction (24.14) occurs with red light.

In Chapter 28, we will consider another porphyrin structure that is essential o life—hemoglobin.

KEEP IN MIND

magnesium is not a transition metal.

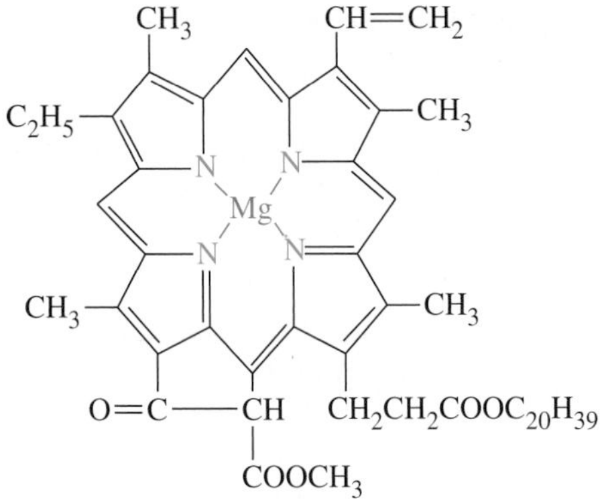

◀ FIGURE 24-25
Structure of chlorophyll *a*

MasteringCHEMISTRY www.masteringchemistry.com

Transition metal ions are responsible for the brilliant colors in many gemstones. For a discussion of the ions responsible for the colors of rubies and emeralds, and the mechanisms that allow these gemstones to maintain their colors indefinitely, go to the Focus On feature for Chapter 24, Colors in Gemstones, on the MasteringChemistry site.

Summary

24-1 Werner's Theory of Coordination Compounds: An Overview—Many metal atoms or ions, particularly among the transition elements, have the ability to bond with **ligands** (electron-pair donors) to form a **complex**, a species in which there are coordinate covalent bonds between ligands and metal centers. The number of electron pairs donated to the central metal atom or ion by the ligands is the **coordination number**. A compound having one or more complexes as a constituent is called a **coordination compound**.

24-2 Ligands—A ligand that attaches to a central metal atom or ion by donating a single electron pair is said to be **monodentate**. A **bidentate** ligand binds to the central metal atom or ion with two pairs of electrons (Fig. 24-3). **Polydentate** ligands are able to attach simultaneously to two or more positions at the metal center. This multiple attachment produces complexes with five- or six-member rings of atoms—**chelates**. A polydentate ligand is also referred to as a **chelating agent**.

24-3 Nomenclature—The name of a complex conveys information about the number and kinds of ligands; the oxidation state of the metal center; and whether the complex is a neutral species, a cation, or an anion.

24-4 Isomerism—The positions at which ligands are attached to the metal center are not always equivalent, and as a consequence isomerism occurs in coordination complexes. **Structural isomers** differ in what ligands are attached to the metal ion and through which atoms, one example being linkage isomerism (Fig. 24-4). In **geometric isomerism**, different structures with different properties result, depending on where the attachment of ligands occurs. In coordination complexes in which there are two ligands of one type and the remainder of another type, the *cis* isomer has the two identical ligands on the same side of a square in a planar structure (Fig. 24-5) or the same edge of an octahedral structure (Fig. 24-6). The ***trans*** isomer has the two ligands on opposite corners of a square (Fig. 24-5) or diametrically opposed in an octahedral structure (Fig. 24-6). **Optical isomers** differ by being nonsuperimposable mirror images of each other (Fig. 24-8). The two isomers so related are called **enantiomers** and they are said to be **chiral**. Optical isomerism and geometrical isomerism are forms of **stereoisomerism**, that is, isomerism based on the way ligands occupy the three-dimensional space around the metal center.

24-5 Bonding in Complex Ions: Crystal Field Theory—**Crystal field theory** is a bonding theory useful in explaining the magnetic properties and characteristic colors of complex ions (Fig. 24-11). This theory emphasizes the splitting of the *d* energy level of the central metal ion as a result of repulsions between *d*-orbital electrons of the central ion and electrons of the ligand (Fig. 24-12). Whether or not electrons are accommodated in the upper *d* orbitals depends on whether the crystal field splitting, Δ, is greater than the **pairing energy** (P). In d^n complexes, where $4 \le n \le 7$, **low spin** complexes occur when $P > \Delta$, and **high spin** complexes occur when $P < \Delta$. A prediction of the magnitude of *d*-level splitting produced by a ligand can be made via the ranking known as the **spectrochemical series.**

24-6 Magnetic Properties of Coordination Compounds and Crystal Field Theory—The magnetic properties of coordination compounds—whether diamagnetic or paramagnetic, and to what degree—can be assessed by measuring the change in weight of the substance when placed in a magnetic field (Fig. 24-15).

24-7 Color and the Colors of Complexes—The colors displayed by coordination complexes (Fig. 24-18) arise from the absorption of certain wavelength components of white light as an electron is excited from a lower energy *d* orbital to a higher energy *d* orbital in a crystal field. The transmitted light is deficient in the absorbed wavelengths and hence of a different color. The relationship of **primary, secondary**, and **complementary colors** is described and presented visually in Figure 24-16.

24-8 Aspects of Complex-Ion Equilibria—Formation of a coordination complex can be viewed as a stepwise equilibrium process in which other ligands displace H_2O molecules from aqua complex ions. The stepwise constants can be combined into an overall formation constant for the complex ion, K_f. When a chelating agent binds to a metal ion, an increased stability is observed because of the entropy changes produced when the chelating agent displaces water molecules from binding sites in the coordination sphere. This greater stability achieved is called the **chelation (chelate) effect**.

24-9 Acid–Base Reactions of Complex Ions—The ability of ligand water molecules to ionize causes some aqua complexes to exhibit acidic properties and helps explain amphoterism.

24-10 Some Kinetic Considerations—Also important in determining properties of a complex ion is the rate at which the ion exchanges ligands between its coordination sphere and the solution. Exchange is rapid in a **labile** complex and slow in an **inert** complex.

24-11 Applications of Coordination Chemistry— Complex-ion formation can be used to stabilize certain oxidation states, such as Co(III). Other applications include dissolving precipitates, such as AgCl by $NH_3(aq)$ in the qualitative analysis scheme and AgBr by $Na_2S_2O_3(aq)$ in the photographic process, and sequestering ions by chelation, as with EDTA.

Integrative Example

Absorbance is a measure of the proportion of monochromatic (single-color) light that is absorbed as the light passes through a solution. An *absorption spectrum* is a graph of absorbance as a function of wavelength. High absorbances correspond to large proportions of the light entering a solution being absorbed. Low absorbances signify that large proportions of the light are transmitted. The absorption spectrum of $[Ti(H_2O)_6]^{3+}(aq)$ is shown in Figure 24-26(a).

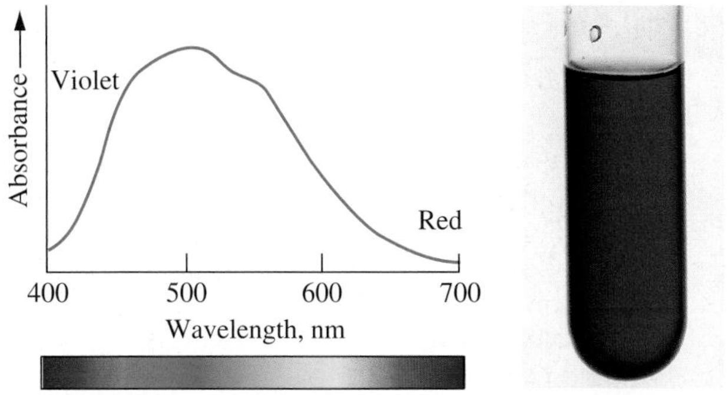

(a) (b)

Richard Megna/Fundamental Photographs

◀ FIGURE 24-26
The color of $[Ti(H_2O)_6]^{3+}(aq)$
(a) The absorption spectrum of $[Ti(H_2O)_6]^{3+}(aq)$.
(b) A solution containing $[Ti(H_2O)_6]^{3+}(aq)$.

(a) Describe the color of light that $[Ti(H_2O)_6]^{3+}(aq)$ absorbs most strongly, and the color of the solution. **(b)** Describe the electron transition responsible for the absorption peak, and determine the energy associated with this absorption.

Analyze

The highest absorbances in the spectrum shown in Figure 24-26(a) come at about 500 nm. We use the electromagnetic spectrum in Figure 8-3 to determine the color of the absorbed light. We determine that the electron configuration of Ti^{3+}, the central ion in the complex ion, is $[Ar]3d^1$. In the ground state, the 3*d* electron is in one of the three degenerate lower levels in the *d*-orbital splitting diagram for an octahedral complex (Fig. 24-12).

Solve

(a) The electromagnetic spectrum in Figure 8-3 indicates that the absorbed light should be dark green. Figure 24-16(b) indicates that, in subtractive color mixing, the complementary color of green is magenta. Thus the color of the transmitted light (and hence the observed color of the solution) is a blend of red and blue.

(b) The quantity of energy we seek is that corresponding to electromagnetic radiation at the peak of the absorption spectrum: 500 nm. First we can establish the frequency of this light from $c = \nu \times \lambda$.

$$\nu = \frac{c}{\lambda} = \frac{2.998 \times 10^8 \text{ m s}^{-1}}{500 \times 10^{-9} \text{ m}} = 6.00 \times 10^{14} \text{ s}^{-1}$$

Then, we use Planck's equation to determine *E*:

$$E = h\nu = (6.626 \times 10^{-34} \text{ J s}) \times (6.00 \times 10^{14} \text{ s}^{-1}) = 3.98 \times 10^{-19} \text{ J}$$

This is the energy per photon. If we want the energy on a per-mole basis, we can write

$$E = (3.98 \times 10^{-19} \text{ J}) \times (6.022 \times 10^{23} \text{ mol}^{-1}) \times \frac{1 \text{ kJ}}{1000 \text{ J}}$$
$$= 240 \text{ kJ mol}^{-1}$$

Assess

Using ideas of subtractive color mixing, we ascertained the wavelength components of white light that the complex cation absorbs. The broadness of the absorption band is due to vibrations in the metal to oxygen bonds. By using the maximum of the absorption curve, we calculated the most probable energy of the absorbed photons.

PRACTICE EXAMPLE A: The compound $[CoCl(en)_2(NO_2)_2]$ has been prepared in a number of isomeric forms. One form undergoes no reaction with either $AgNO_3$ or en and is optically inactive. A second form reacts with $AgNO_3$ to form a white precipitate, does not react readily with en, and is optically inactive. A third form is optically active and reacts both with en and $AgNO_3$. Assuming a coordination number of 6 for the cobalt ion, identify each of the three isomeric forms by name, and sketch each of the structures.

PRACTICE EXAMPLE B: A compound is analyzed and found to contain 46.2% Pt, 33.6% Cl, 16.6% N, and 3.6% H. The freezing point of a 0.1 M aqueous solution of the compound is −0.74 °C. What is the structural formula of the compound? What possible isomeric forms are there for this compound?

Exercises

Nomenclature

1. Write the formula and name of
(a) a complex ion having Cr^{3+} as the central ion and two NH_3 molecules and four Cl^- ions as ligands
(b) a complex ion of iron(III) having a coordination number of 6 and CN^- as ligands
(c) a coordination compound comprising two types of complex ions: one a complex of Cr(III) with ethylenediamine (en), having a coordination number of 6; the other, a complex of Ni(II) with CN^-, having a coordination number of 4
2. What are the coordination number and the oxidation state of the central metal ion in each of the following complexes? Name each complex.
(a) $[Co(NH_3)_6]^{2+}$
(b) $[AlF_6]^{3-}$
(c) $[Cu(CN)_4]^{2-}$
(d) $[CrBr_2(NH_3)_4]^+$
(e) $[Co(ox)_3]^{4-}$
(f) $[Ag(S_2O_3)_2]^{3-}$
3. Supply acceptable names for the following:
(a) $[Ag(NH_3)_2]Cl$
(b) $[Cu(H_2O)_2(NH_3)_4]SO_4$
(c) $PtCl_2(en)$
(d) $[CrBr(H_2O)_5]^{2+}$
(e) $Rb[AgF_4]$
(f) $Na_2[Fe(CN)_5NO]$
4. Write appropriate formulas for the following.
(a) potassium hexacyanidoferrate(III)
(b) bis(ethylenediamine)copper(II) ion
(c) pentaaquahydroxidoaluminum(III) chloride
(d) amminechloridobis(ethylenediamine) chromium(III) sulfate
(e) tris(ethylenediamine)iron(III) hexacyanidoferrate(II)

Bonding and Structure in Complex Ions

5. Draw Lewis structures for the following ligands:
(a) H_2O
(b) CH_3NH_2
(c) ONO^-
(d) SCN^-
6. Draw Lewis structures for the following ligands:
(a) hydroxido
(b) sulfato
(c) oxalato
(d) thiocyanato-*N*-
7. Draw a plausible structure to represent:
(a) $[FeBr(ox)_2]^{2-}$
(b) $[CoBr_4(NH_3)_2]^-$
(c) $[Cu(EDTA)]^{2-}$
(d) $[CrBr_2(H_2O)_4]^+$
(e) $[PtBr_6]^{2-}$
8. Draw plausible structures of the following chelate complexes.
(a) $[Pt(ox)_2]^{2-}$
(b) $[Cr(ox)_3]^{3-}$
(c) $[Fe(EDTA)]^{2-}$
9. Draw plausible structures corresponding to each of the following names.
(a) diamminediaquabromidochloridocobalt(III) ion
(b) hexacarbonylmanganese(I) ion
(c) diamminetetrachloridoplatinum(IV)
(d) diammineaquatrichloridocobalt(III)
10. Draw plausible structures corresponding to each of the following names.
(a) pentamminenitrito-*N*-cobalt(III) ion
(b) ethylenediaminedithiocyanato-*S*-copper(II)
(c) hexaaquanickel(II) ion

Isomerism

11. Which of these general structures for a complex ion would you expect to exhibit *cis* and *trans* isomerism? Explain.
    (a) tetrahedral
    (b) square-planar
    (c) linear
12. Which of these octahedral complexes would you expect to exhibit *geometric* isomerism? Explain.
    (a) $[Cr(NH_3)_5OH]^{2+}$
    (b) $[CrCl_2(H_2O)(NH_3)_3]^+$
    (c) $[CrCl_2(en)_2]^+$
    (d) $[CrCl_4(en)]^-$
    (e) $[Cr(en)_3]^{3+}$
13. If A, B, C, and D are four different ligands,
    (a) how many geometric isomers will be found for square-planar $[PtABCD]^{2+}$?
    (b) Will tetrahedral $[ZnABCD]^{2+}$ display optical isomerism?
14. Write the names and formulas of three coordination isomers of $[Co(en)_3][Cr(ox)_3]$.
15. Draw a structure for *cis*-dichloridobis(ethylenediamine)cobalt(III) ion. Is this ion chiral? Is the *trans* isomer chiral? Explain.
16. The structures of four complex ions are given. Each has Co^{3+} as the central ion. The ligands are H_2O, NH_3, and oxalate ion, $C_2O_4{}^{2-}$. Determine which, *if any*, of these complex ions are isomers (geometric or optical); which, *if any*, are identical (that is, have identical structures); and which, *if any*, are distinctly different.

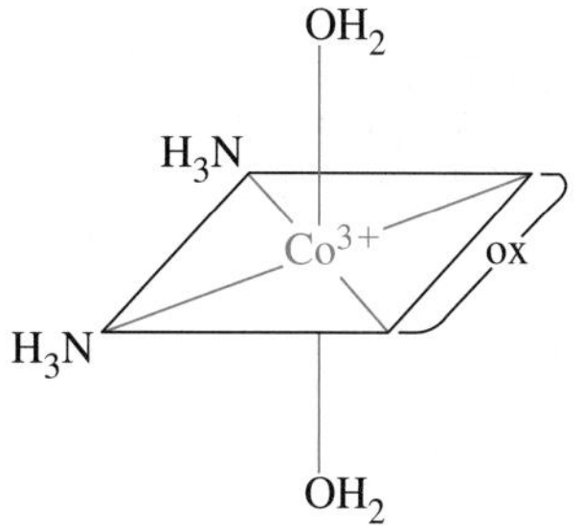

(a)

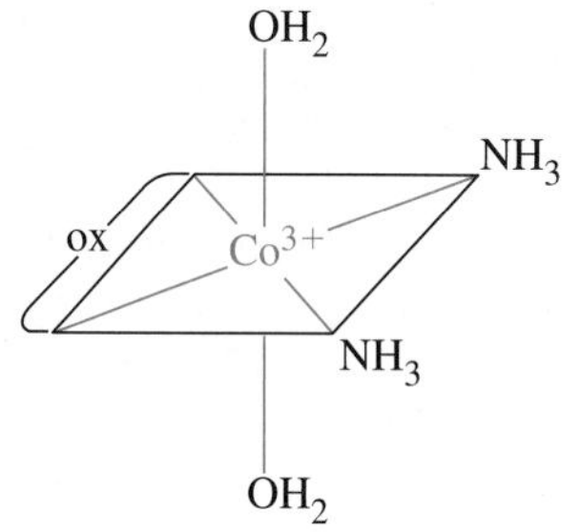

(b)

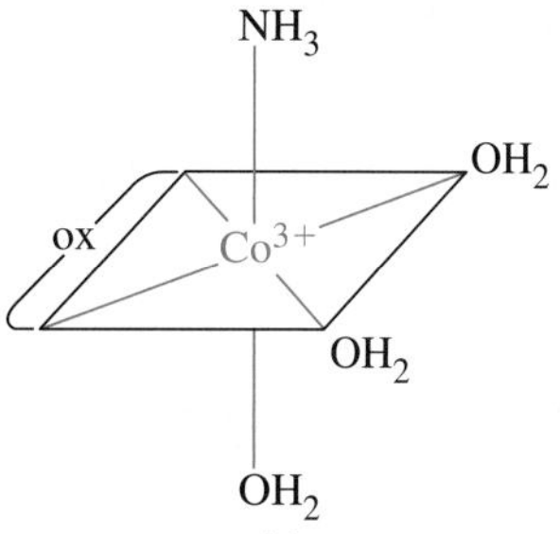

(c)

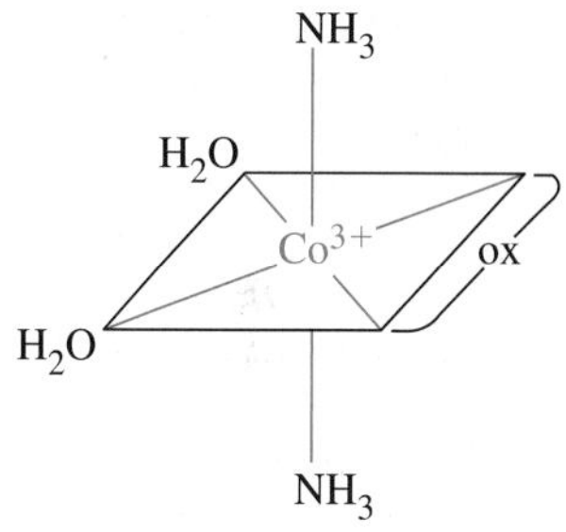

(d)

Crystal Field Theory

17. Describe how the crystal field theory explains the fact that so many transition metal compounds are colored.
18. Cyanido complexes of transition metal ions (such as Fe^{2+} and Cu^{2+}) are often yellow, whereas aqua complexes are often green or blue. Explain the basis for this difference in color.
19. If the ion Co^{2+} is linked with strong-field ligands to produce an octahedral complex, the complex has *one* unpaired electron. If Co^{2+} is linked with weak-field ligands, the complex has *three* unpaired electrons. How do you account for this difference?
20. In contrast to the case of Co^{2+} considered in Exercise 19, no matter what ligand is linked to Ni^{2+} to form an octahedral complex, the complex always has *two* unpaired electrons. Explain this fact.
21. Predict:
    (a) which of the complex ions, $[MoCl_6]^{3-}$ and $[Co(en)_3]^{3+}$, is diamagnetic and which is paramagnetic;
    (b) the number of unpaired electrons expected for the tetrahedral complex ion $[CoCl_4]^{2-}$.
22. Predict:
    (a) whether the square-planar complex ion $[Cu(py)_4]^{2+}$ is diamagnetic or paramagnetic
    (b) whether octahedral $[Mn(CN)_6]^{3-}$ or tetrahedral $[FeCl_4]^-$ has the greater number of unpaired electrons.
23. In Example 24-5, we chose between a tetrahedral and a square-planar structure for $[Ni(CN)_4]^{2-}$ based on magnetic properties. Could we similarly use magnetic properties to establish whether the ammine complex of Ni(II) is octahedral $[Ni(NH_3)_6]^{2+}$ or tetrahedral $[Ni(NH_3)_4]^{2+}$? Explain.
24. In both $[Fe(H_2O)_6]^{2+}$ and $[Fe(CN)_6]^{4-}$ ions, the iron is present as Fe(II); however, $[Fe(H_2O)_6]^{2+}$ is paramagnetic, whereas $[Fe(CN)_6]^{4-}$ is diamagnetic. Explain this difference.

Complex-Ion Equilibria

25. Write equations to represent the following observations.
    (a) A mixture of $Mg(OH)_2(s)$ and $Zn(OH)_2(s)$ is treated with $NH_3(aq)$. The $Zn(OH)_2$ dissolves, but the $Mg(OH)_2(s)$ is left behind.
    (b) When NaOH(aq) is added to $CuSO_4(aq)$, a pale blue precipitate forms. If $NH_3(aq)$ is added, the precipitate redissolves, producing a solution with an intense deep blue color. If this deep blue solution is made acidic with $HNO_3(aq)$, the color is converted back to pale blue.
26. Write equations to represent the following observations.
    (a) A quantity of $CuCl_2(s)$ is dissolved in concentrated HCl(aq) and produces a yellow solution. The solution is diluted to twice its volume with water and assumes a green color. On dilution to ten times its original volume, the solution becomes pale blue.

(b) When chromium metal is dissolved in HCl(aq), a blue solution is produced that quickly turns green. Later the green solution becomes blue-green and then violet.

27. Which of the following complex ions would you expect to have the largest overall K_f, and why? $[Cu(H_2O)_6]^{2+}$; $[Cu(H_2O)_3(NH_3)_3]^{2+}$; $[Cu(en)_3]^{2+}$; $[Cu(en)_2(H_2O)_2]^{2+}$; $[Cu(en)(NH_3)_4]^{2+}$.

28. Use data from Table 24.6 to determine values of **(a)** β_4 for the formation of $[Zn(NH_3)_4]^{2+}$; **(b)** β_4 for the formation of $[Ni(H_2O)_2(NH_3)_4]^{2+}$.

29. Write a series of equations to show the stepwise displacement of H_2O ligands in $[Fe(H_2O)_6]^{3+}$ by ethylenediamine, for which $\log_{10} K_1 = 4.34$, $\log_{10} K_2 = 3.31$, and $\log_{10} K_3 = 2.05$. What is the overall formation constant, $\beta_3 = K_f$, for $[Fe(en)_3]^{3+}$?

30. For the overall reaction $[Cu(H_2O)_6]^{2+} + 4NH_3 \rightleftharpoons [Cu(H_2O)_2(NH_3)_4]^{2+} + 4H_2O$, a tabulation of formation constants lists the following $\log_{10} K$ values: $\log_{10} K_1 = 4.25$, $\log_{10} K_2 = 3.61$, $\log_{10} K_3 = 2.98$, and $\log_{10} K_4 = 2.24$. What is the overall formation constan $\beta_4 = K_f$ for $[Cu(H_2O)_2(NH_3)_4]^{2+}$?

31. Explain the following observations in terms o complex-ion formation.
(a) $Al(OH)_3(s)$ is soluble in NaOH(aq) but insolubl in $NH_3(aq)$.
(b) $ZnCO_3(s)$ is soluble in $NH_3(aq)$, but ZnS(s) is no
(c) The molar solubility of AgCl in pure water is abou 1×10^{-5} M; in 0.04 M NaCl(aq), it is about 2×10^{-6} M but in 1 M NaCl(aq), it is about 8×10^{-5} M.

32. Explain the following observations in terms c complex-ion formation.
(a) $CoCl_3$ is unstable in aqueous solution, bein reduced to $CoCl_2$ and liberating $O_2(g)$. Ye $[Co(NH_3)_6]Cl_3$ can be easily maintained in aqueou solution.
(b) AgI is insoluble in water and in dilute $NH_3(aq)$ but AgI will dissolve in an aqueous solution o sodium thiosulfate.

Acid–Base Properties

33. Which of the following would you expect to react as a Brønsted–Lowry acid: $[Cu(NH_3)_4]^{2+}$, $[FeCl_4]^-$, $[Al(H_2O)_6]^{3+}$, or $[Zn(OH)_4]^{2-}$? Why?

34. Write simple chemical equations to show how th complex ion $[Cr(H_2O)_5(OH)]^{2+}$ acts as **(a)** an acid **(b)** a base.

Applications

35. From data in Chapter 18,
(a) Derive an equilibrium constant for reaction (24.13), and explain why this reaction (the fixing of photographic film) is expected to go essentially to completion.
(b) Explain why $NH_3(aq)$ cannot be used in the fixing of photographic film.

36. Show that the oxidation of $[Co(NH_3)_6]^{2+}$ to $[Co(NH_3)_6]^{3+}$ referred to on page 1158 should occur spontaneously in alkaline solution with H_2O_2 as an oxidizing agent.

37. Explain why $K_2[PtCl_4]$ is first converted to $K_2[PtI_4]$ i the synthesis of the anticancer drug cisplatin, *cis*-$PtCl_2(NH_3)_2$.

38. Draw dashed and solid wedge diagrams of transplatin *trans*-$PtCl_2(NH_3)_2$, and cisplatin, *cis*-$PtCl_2(NH_3)_2$ Then, explain how transplatin can be more reactive ye less effective at killing cancer cells than is cisplatin.

Integrative and Advanced Exercises

39. From each of the following names, you should be able to deduce the formula of the complex ion or coordination compound intended. Yet, these are not the best systematic names that can be written. Replace each name with one that is more acceptable: **(a)** cupric tetraammine ion; **(b)** tetraamminedichlorido cobaltic chloride; **(c)** platinic(IV) hexachloride ion; **(d)** disodium copper tetrachloride; **(e)** dipotassium antimony(III) pentachloride.

40. Magnus's green salt has the empirical formula $PtCl_2 \cdot 2\,NH_3$. It is a coordination compound consisting of both complex cations and complex anions. Write the probable formula of this coordination compound according to Werner's theory, and assign it a systematic name.

41. How many isomers are there of the complex io $[CoCl_2(en)(NH_3)_2]^+$? Sketch their structures.

42. Explain the following observations through a series o equations. The green solid $CrCl_3 \cdot 6\,H_2O$ dissolves i water to form a green solution. The solution slowly turns blue-green; after a day or two, the solution i violet. When the violet solution evaporates to dry ness, a green solid remains.

43. The *cis* and *trans* isomers of $[CoCl_2(en)_2]^+$ can be dis tinguished via a displacement reaction with oxalate ion. What difference in reactivity toward oxalate ior would you expect between the *cis* and *trans* isomers Explain.

44. Write half-equations and an overall equation to represent the oxidation of tetraammineplatinum(II) ion to *trans*-tetraamminedichloridoplatinum(IV) ion by Cl_2. Then make sketches of the two complex ions.

45. We learned in Chapter 16 that for polyprotic acids, ionization constants for successive ionization steps decrease rapidly. That is, $K_{a_1} \gg K_{a_2} \gg K_{a_3}$. The ionization constants for the first two steps in the ionization of $[Fe(H_2O)_6]^{3+}$ (reactions 24.10 and 24.11) are more nearly equal in magnitude. Why does this multistep ionization seem not to follow the pattern for polyprotic acids?

46. Following are the names of five coordination compounds containing complexes with platinum(II) as the central metal ion and ammonia molecules and/or chloride ions as ligands: **(a)** potassium amminetrichloridoplatinate(II); **(b)** diamminedichloridoplatinum(II); **(c)** triamminechloridoplatinum(II) chloride; **(d)** tetraammineplatinum(II) chloride; **(e)** potassium tetrachloridoplatinate(II). Make a rough sketch of the expected graph when electric conductivity is plotted as a function of the number of chlorido ligands. [*Hint:* Your graph should be based on five points, but no quantitative data are given.]

47. For a solution that is 0.100 M in $[Fe(H_2O)_6]^{3+}$,
(a) assuming that ionization of the aqua complex ion proceeds only through the first step, equation (24.10), calculate the pH of the solution.
(b) Calculate $[[Fe(H_2O)_5(OH)]^{2+}]$ if the solution is also 0.100 M $HClO_4$. (ClO_4^- does not complex with Fe^{3+}.)
(c) Can the pH of the solution be maintained so that $[[Fe(H_2O)_5(OH)]^{2+}]$ does not exceed 1×10^{-6} M? Explain.

48. A solution that is 0.010 M in Pb^{2+} is also made to be 0.20 M in a salt of EDTA (that is, having a concentration of the $EDTA^{4-}$ ion of 0.20 M). If this solution is now made 0.10 M in H_2S and 0.10 M in H_3O^+, will PbS(s) precipitate?

49. Without performing detailed calculations, show why you would expect the concentrations of the various ammine–aqua complex ions to be negligible compared with that of $[Cu(NH_3)_4]^{2+}$ in a solution having a total Cu(II) concentration of 0.10 M and a total concentration of NH_3 of 1.0 M. Under what conditions would the concentrations of these ammine–aqua complex ions (such as $[Cu(H_2O)_3NH_3]^{2+}$) become more significant relative to the concentration of $[Cu(NH_3)_4]^{2+}$? Explain.

50. Verify the statement on page 1161 that neither Ca^{2+} nor Mg^{2+} found in natural waters is likely to precipitate from the water on the addition of other reagents if the ions are complexed with EDTA. Assume reasonable values for the total metal ion concentration and that of *free* EDTA, such as 0.10 M each.

51. Estimate the total $[Cl^-]$ required in a solution that is initially 0.10 M $CuSO_4$ to produce a visible yellow color.

$$\underset{\text{(blue)}}{[Cu(H_2O)_4]^{2+}} + 4\,Cl^- \rightleftharpoons \underset{\text{(yellow)}}{[CuCl_4]^{2-}} + 4\,H_2O$$

$$K_f = 4.2 \times 10^5$$

Assume that 99% conversion of $[Cu(H_2O)_4]^{2+}$ to $[CuCl_4]^{2-}$ is sufficient for this to happen, and ignore the presence of any mixed aqua–chlorido complex ions.

52. Refer to the stability of $[Co(NH_3)_6]^{3+}(aq)$ on page 1158, and
(a) verify that E°_{cell} reaction (24.12) is +0.59 V.
(b) Calculate $[Co^{3+}]$ in a solution at equilibrium that has a total concentration of cobalt of 0.1 M and $[NH_3] = 0.1$ M.
(c) Show that for the value of $[Co^{3+}]$ calculated in part (b), reaction (24.12) will not occur. [*Hint:* Assume a low, but reasonable, concentration of Co^{2+} (say, 1×10^{-4} M) and a partial pressure of $O_2(g)$ of 0.2 atm.]

53. A Cu electrode is immersed in a solution that is 1.00 M NH_3 and 1.00 M in $[Cu(NH_3)_4]^{2+}$. If a standard hydrogen electrode is the cathode, E_{cell} is +0.08 V. What is the value obtained by this method for the formation constant, K_f, of $[Cu(NH_3)_4]^{2+}$?

54. The following concentration cell is constructed.

$$Ag|Ag^+(0.10\ M[Ag(CN)_2]^-, 0.10\ M\ CN^-)\ ||Ag^+(0.10\ M)|Ag$$

If K_f for $[Ag(CN)_2]^-$ is 5.6×10^{18}, what value would you expect for E_{cell}? [*Hint:* Recall that the anode is on the left.]

55. The compound $CoCl_2 \cdot 2\,H_2O \cdot 4\,NH_3$ may be one of the hydrate isomers $[CoCl(H_2O)(NH_3)_4]Cl \cdot H_2O$ or $[Co(H_2O)_2(NH_3)_4]Cl_2$. A 0.10 M aqueous solution of the compound is found to have a freezing point of −0.56 °C. Determine the correct formula of the compound. The freezing-point depression constant for water is 1.86 mol kg^{-1} °C, and for aqueous solutions, molarity and molality can be taken as approximately equal.

56. Explain why aqueous solutions of $[Sc(H_2O)_6]Cl_3$ and $[Zn(H_2O)_4]Cl_2$ are colorless, but an aqueous solution of $[Fe(H_2O)_6]Cl_3$ is not.

57. Provide a valence bond description of the bonding in the $Cr(NH_3)_6^{3+}$ ion. According to the valence bond description, how many unpaired electrons are there in the $Cr(NH_3)_6^{3+}$ complex? How does this prediction compare with that of crystal field theory?

58. A tabulation of formation constants lists the following $\log_{10} K$ values for the formation of $[Cu(NH_3)_4]^{2+}$: $\log_{10} K_1 = 4.28$, $\log_{10} K_2 = 3.59$, $\log_{10} K_3 = 3.00$, and $\log_{10} K_4 = 2.18$. Calculate the mole fractions of Cu^{2+}, $[Cu(NH_3)]^{2+}$, $[Cu(NH_3)_2]^{2+}$, $[Cu(NH_3)_3]^{2+}$, and $[Cu(NH_3)_4]^{2+}$ in solution when $[NH_3]_{eq} = 2.0$ M.

59. Acetyl acetone undergoes an isomerization to form a type of alcohol called an *enol*.

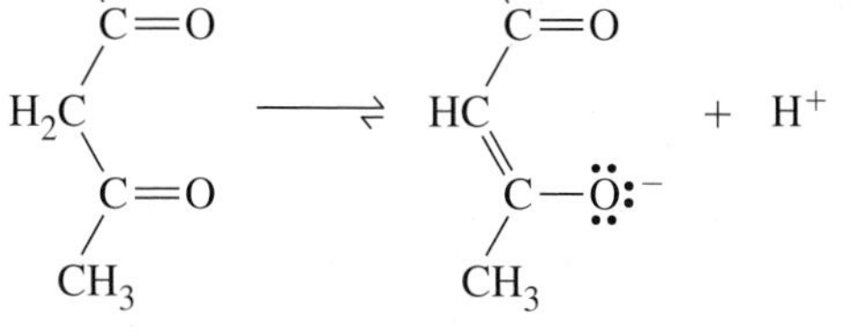

The enol, abbreviated acacH, can act as a bidentate ligand as the anion $acac^-$. Which of the following compounds are optically active: $Co(acac)_3$; *trans*-$[Co(acac)_2(H_2O)_2]Cl_2$; *cis*-$[Co(acac)_2(H_2O)_2]Cl_2$?

60. We have seen that complex formation can stabilize oxidation states. An important illustration of this fact is the oxidation of water in acidic solutions by $Co^{3+}(aq)$ but not by $[Co(en)_3]^{3+}$. Use the following data.

$$[Co(H_2O)_6]^{3+} + e^- \longrightarrow [Co(H_2O)_6]^{2+} \qquad E° = 1.82\ V$$

$$[Co(H_2O)_6]^{2+} + 3\ en \longrightarrow [Co(en)_3]^{2+} + 6\ H_2O(l) \qquad \log \beta_3 = 12.18$$

$$[Co(H_2O)_6]^{3+} + 3\ en \longrightarrow [Co(en)_3]^{3+} + 6\ H_2O(l) \qquad \log \beta_3 = 47.30$$

Calculate $E°$ for the reaction

$$[Co(en)_3]^{3+} + e^- \longrightarrow [Co(en)_3]^{2+}$$

Show that $[Co(en)_3]^{3+}$ is stable in water but $Co^{3+}(aq)$ is not.

61. The amino acid glycine ($NH_2CH_2CO_2H$, denoted Hgly) binds as an anion and is a bidentate ligand. Draw and name all possible isomers of $[Co(gly)_3]$. How many isomers are possible for the compound $[CoCl(gly)_2(NH_3)]$? [*Hint:* $NH_2CH_2CO_2^-$ is the glycinate anion.]

62. The structure of $K_2[PtCl_6]$ in the solid state is shown in the next column. Identify the type of cubic unit cell, and describe the structure in terms of the holes occupied by the various ions.

63. The graph that follows represents the molar conductivity of some Pt(IV) complexes. The ligands in these complexes are NH_3 molecules or Cl^- ions, the coordination number of Pt(IV) is 6, and the counter ions (for balancing charge) are K^+ or Cl^-. Write formulas for the coordination compounds corresponding to each poir in the graph. (Molar conductivity is the electrica conductivity, under precisely defined conditions, of a aqueous solution containing one mole of a compound

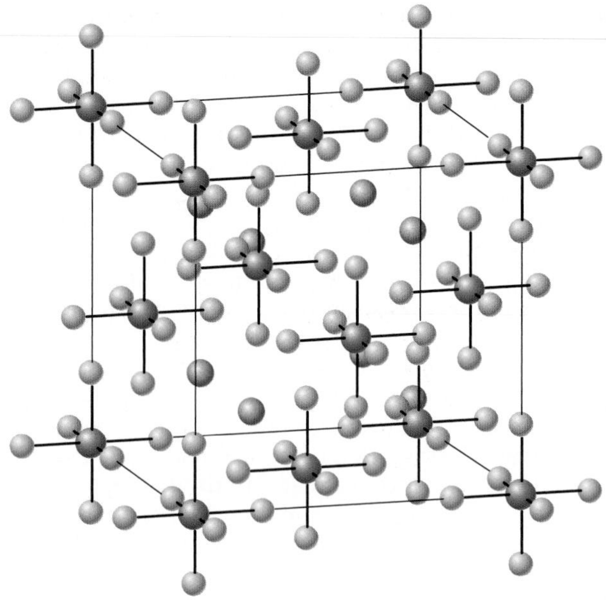

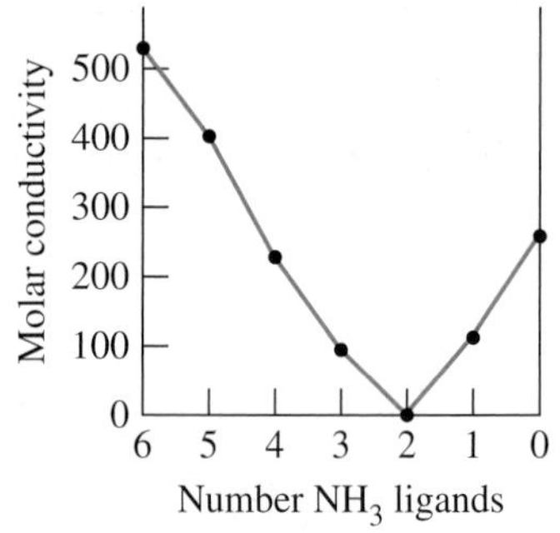

Feature Problems

64. A structure that Werner examined as a possible alternative to the octahedron is the trigonal prism.

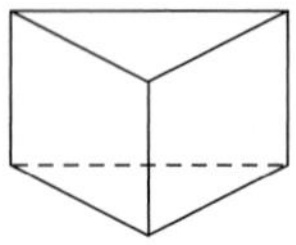

(a) Does this structure predict the correct number of isomers for the complex ion $[CoCl_2(NH_3)_4]^+$? If not, why not?

(b) Does this structure account for optical isomerism in $[Co(en)_3]^{3+}$? Explain.

65. Werner demonstrated that octahedral complexes can exhibit optical isomerism, and to Werner's satisfaction, this confirmed the octahedral arrangement of ligands. However, skeptics of his theory said that because the ligands contained carbon atoms, he could not rule out carbon as the source of the optical activity. Werner devised and prepared the following compound in which the OH^- groups act as bridging groups.

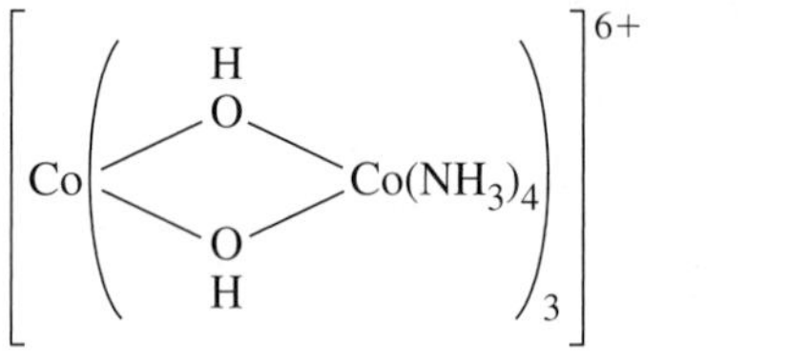

Werner resolved this compound into its optical isc mers, confirming his theory and confounding his crit ics. What are the oxidation states of the Co ions? If th complex is low spin, what is the number of unpaire electrons in the molecule? Draw the structures of th two optical isomers.

66. The crystal field model discussed in the text describe how the degeneracy of the *d* orbitals is removed by a octahedral field of ligands. We have seen that th d_{xy}, d_{xz}, and d_{yz} orbitals are stabilized (lower energy with respect to the average energy of the *d* orbita and that the $d_{x^2-y^2}$ and d_{z^2} orbitals are destabilize As described in Are You Wondering 24-3, the stab lization is $-0.4\Delta_o$ and destabilization is $0.6\Delta_o$. Th crystal field stabilization energy (CFSE) can be defined as CFSE = [(number of electrons in the d_{xy}, d_{xz}, and d_{yz} orbitals) × $(-0.4\Delta_o)$] + [(number of electrons in the $d_{x^2-y^2}$ and d_{z^2} orbitals) × $(0.6\Delta_o)$]. The following table contains the enthalpy of hydration for the reaction

$$M^{2+}(g) + 6\,H_2O(l) \longrightarrow [M(H_2O)_6]^{2+}(aq)$$

(a) Plot the hydration energies as a function of the atomic number of the metals shown.
(b) Assuming that all the hexaaqua complexes are high spin, which ions have zero CFSE?
(c) If lines are drawn between those ions with CFSE = 0, a line of negative slope is obtained. Can you explain this finding?
(d) The ions that do not have CFSE = 0 have heats of hydration that are more negative than the lines drawn in part (c). What is the explanation for this?

Dipositive Metal Ion	Hydration Energy, kJ mol^{-1}
Ca	−2468
Sc	−2673
Ti	−2750
V	−2814
Cr	−2799
Mn	−2743
Fe	−2843
Co	−2904
Ni	−2986
Cu	−2989
Zn	−2939

(e) Estimate the value of Δ_o for the Fe(II) ion in an octahedral field of water molecules.
(f) What wavelength of light would the $[Fe(H_2O)_6]^{2+}$ ion absorb?

Self-Assessment Exercises

7. In your own words, describe the following terms or symbols: **(a)** coordination number; **(b)** Δ_o; **(c)** ammine complex; **(d)** enantiomer.

8. Briefly describe each of the following ideas, phenomena, or methods: **(a)** spectrochemical series; **(b)** crystal field theory; **(c)** optical isomer; **(d)** structural isomerism.

9. Explain the important distinction between each of the following pairs: **(a)** coordination number and oxidation number; **(b)** monodentate and polydentate ligands; **(c)** *cis* and *trans* isomers; **(d)** dextrorotatory and levorotatory compounds; **(e)** low-spin and high-spin complexes.

0. The oxidation state of Ni in the complex ion $[Ni(CN)_4I]^{3-}$ is **(a)** −3; **(b)** −2; **(c)** 0; **(d)** +2; **(e)** +3.

1. The coordination number of Pt in the complex ion $[PtCl_2(en)_2]^{2+}$ is **(a)** 2; **(b)** 3; **(c)** 4; **(d)** 5; **(e)** 6.

2. Of the following complex ions, the one that exhibits isomerism is **(a)** $[Ag(NH_3)_2]^+$; **(b)** $[CoNO_2(NH_3)_5]^{2+}$; **(c)** $[Pt(en)(NH_3)_2]^{2+}$; **(d)** $[CoCl(NH_3)_5]^{2+}$; **(e)** $[PtCl_6]^{2-}$.

3. Of the following complex ions, the one that is optically active is **(a)** *cis*-$[CoCl_2(en)_2]^+$; **(b)** $[CoCl_2(NH_3)_4]^+$; **(c)** $[CoCl_4(NH_3)_2]^-$; **(d)** $[CuCl_4]^-$.

4. The number of unpaired electrons in the complex ion $[Cr(NH_3)_6]^{2+}$ is **(a)** 5; **(b)** 4; **(c)** 3; **(d)** 2; **(e)** 1.

5. Of the following, the one that is a Brønsted–Lowry acid is **(a)** $[Cu(NH_3)_4]^{2+}$; **(b)** $[FeCl_4]^-$; **(c)** $[Fe(H_2O)_6]^{3+}$; **(d)** $[Zn(OH)_4]^-$.

6. The most soluble of the following solids in $NH_3(aq)$ is **(a)** $Ca(OH)_2$; **(b)** $Cu(OH)_2$; **(c)** $BaSO_4$; **(d)** $MgCO_3$; **(e)** Fe_2O_3.

7. Name the following coordination compounds. Which one(s) exhibit(s) stereoisomerism? Explain.
(a) $[CoBr(NH_3)_5]SO_4$
(b) $[Cr(NH_3)_6][Co(CN)_6]$
(c) $Na_3[Co(NO_2)_6]$
(d) $[Co(en)_3]Cl_3$

78. Write appropriate formulas for the following species.
(a) dicyanidoargentate(I) ion
(b) triamminenitrito-*N*-platinum(II) ion
(c) aquachloridobis(ethylenediamine)cobalt(III) ion
(d) potassium hexacyanidochromate(II)

79. Draw structures to represent these four complex ions: **(a)** $[PtCl_4]^{2-}$; **(b)** $[FeCl_4(en)]^-$; **(c)** *cis*-$[FeCl_2(en)(ox)]^-$; **(d)** *trans*-$[CrCl(NH_3)_4(OH)]^+$.

80. How many different structures are possible for each of the following complex ions?
(a) $[Co(H_2O)(NH_3)_5]^{3+}$
(b) $[Co(H_2O)_2(NH_3)_4]^{3+}$
(c) $[Co(H_2O)_3(NH_3)_3]^{3+}$
(d) $[Co(H_2O)_4(NH_3)_2]^{3+}$

81. Indicate what type of isomerism may be found in each of the following cases. If no isomerism is possible, so indicate.
(a) $[Zn(NH_3)_4][CuCl_4]$
(b) $[Fe(CN)_5SCN]^{4-}$
(c) $[NiCl(NH_3)_5]^+$
(d) $[PtBrCl_2(py)]^-$
(e) $[Cr(NH_3)_3(OH)_3]^-$

82. Indicate what type of isomerism may be found in each of the following cases. If no isomerism is possible, so indicate.
(a) $[CrBr_2(en)_2]^+$
(b) $[CoBr(ox)_2(SCN)]^{3-}$
(c) $[NiCl_4(en)]^{2-}$
(d) $[PtBrCl(ox)]^-$
(e) $[Cr(Cl)_3(det)]$, det is $H_2N(CH_2)_2NH(CH_2)_2NH_2$

83. Of the complex ions $[Co(H_2O)_6]^{3+}$ and $[Co(en)_3]^{3+}$, one has a yellow color in aqueous solution; the other, blue. Match each ion with its expected color, and state your reason for doing so.

84. Using the method presented in Appendix E, construct a concept map depicting the essential ideas of crystal field theory, and how the theory explains the colors and magnetic properties of transition metal complexes.

25 Nuclear Chemistry

CONTENTS

LEARNING OBJECTIVES

25.1 Distinguish between alpha and beta particles and positrons.

25.2 Describe the radioactive decay series and some of its applications.

25.3 Determine the chemical equation of a nuclear reaction involving the bombardment of a given element to produce a radioactive nuclide.

25.4 Discuss how the transuranium elements were generated.

25.5 Use the mathematical expression for the rate of decay of radioactive particles to determine their half-life.

25.6 Describe the meaning of nuclear binding energy and how it relates to nuclear fusion and fission.

25.7 Identify the factors that affect nuclear stability.

25.8 Describe a nuclear reactor within the context of nuclear fission.

25.9 Discuss the potential applications of nuclear fusion and its limitations and dangers.

25.10 Discuss the types of interactions radiation may have with matter.

25.11 Describe a few applications of radioisotopes.

The thin strands of nebulosity are the remains of a star that ended its life in an enormous supernova explosion around 11,000 years ago. Near the center of this large shell of gas is a rapidly spinning neutron star believed to be the surviving core of the original star. Elementary particles, such as neutrons, and nuclear reactions, such as those occurring in stars, are discussed in this chapter.

The origin of the elements is the stars, including our sun. Nuclear fusion in stars creates heavier elements from lighter ones. The heaviest elements, those with atomic numbers greater than 83, have unstable nuclei—that is, they are *radioactive*.

Certain isotopes of the lighter elements are also radioactive. The isotope carbon-14, for example, has chemical and physical properties that are essentially identical to those of the much more abundant isotopes carbon-12 and carbon-13. Carbon-14, however, is radioactive, and it is this property that is used in the technique known as radiocarbon dating.

▶ When radioactivity was first discovered, its dangers were not appreciated. The first workers in the area did not take the precautions that are routine today.

Until now our discussion of chemistry and chemical reactions has revolved round the exchange of electrons between atoms. In this chapter, we will conider a variety of phenomena that originate within the nuclei of atoms. ollectively, we refer to these phenomena as *nuclear chemistry*. Although stars re the natural source of all the elements, we will discuss how new heavy elements and radioactive isotopes of existing lighter elements can be made artifiially. We will also discuss the effects of ionizing radiation on matter. These ffects can have both positive and negative outcomes and are a subject of socity's continuing nuclear debate.

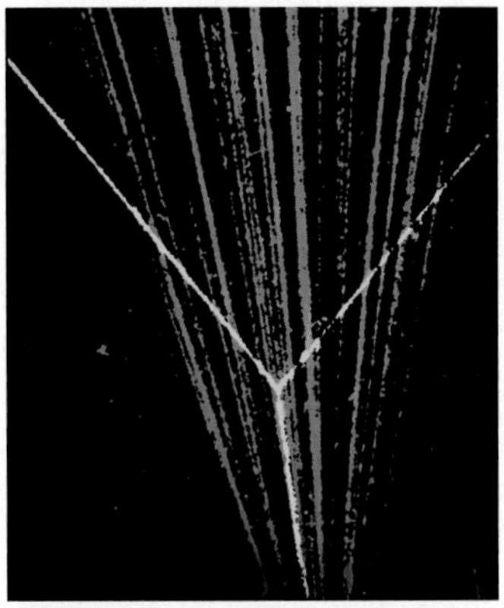

Science Photo Library

▲ Alpha particles leave a trail of liquid droplets, artificially colored green in this photograph, as they pass through a supersaturated vapor in a detector known as a cloud chamber. The chamber also contains He(g), and the trail of one α particle (yellow) is striking the nucleus of a He atom. Following the collision, the α particle and the He atom move apart along lines at about a 90° angle.

25-1 Radioactivity

he term *radioactivity* was proposed by Marie Curie to describe the emission f ionizing radiation by some of the heavier elements. Ionizing radiation, as he name implies, interacts with matter to produce ions. This means that the adiation is sufficiently energetic to break chemical bonds. Some ionizing raditon is particulate (consisting of particles), and some is nonparticulate. We ntroduced α, β, and γ radiation in Section 2-2. Let's describe them again in nore detail, together with two other nuclear processes.

Alpha Particles

Alpha (α) particles are the nuclei of helium-4 atoms, ${}^{4}_{2}He^{2+}$, ejected spontaeously from the nuclei of certain radioactive atoms. We can think of α-particle mission as a process in which a bundle of two protons and two neutrons is emited by an unstable nucleus, resulting in a lighter nucleus. Alpha particles prouce large numbers of ions via their collisions and near collisions with atoms as hey travel through matter, but their penetrating power is low. (Generally, a few heets of paper can stop them.) Because they have a positive charge, α particles re deflected by electric and magnetic fields (recall Figure 2-10).

We can represent the production of α particles by means of a nuclear equaion. A **nuclear equation is** written to conform to two rules:

1. The sum of mass numbers must be the same on both sides.
2. The sum of atomic numbers must be the same on both sides.

n equation (25.1), the alpha particle is represented as ${}^{4}_{2}He$.

$$ {}^{238}_{92}U \longrightarrow {}^{234}_{90}Th + {}^{4}_{2}He \qquad \textbf{(25.1)} $$

Mass numbers total 238, and atomic numbers total 92. The loss of an α particle esults in a *decrease* of 2 in the atomic number and 4 in the mass number of the ucleus.

KEEP IN MIND

that in a *nuclear* equation we represent only the nuclei of atoms, not the atoms as a whole. Although we do not keep track of electrons, electric charge is conserved by the requirement that the sums of the atomic numbers be the same on the two sides of the equation.

Beta Particles

Beta (β^-) particles are deflected by electric and magnetic fields in the *opposite* lirection from α particles. They are less massive than α particles, so they are leflected more strongly than α particles (recall Figure 2-10). Also, they have a greater penetrating power through matter than do α particles (a book, rather han just a few sheets of paper, may be required to stop them). Beta (β^-) paricles are electrons, but they are electrons that originate from the nuclei of toms in nuclear decay processes and are therefore extremely energetic. Electrons that surround the nucleus are given the familiar symbol, e^-.

The simplest decay process producing a β^- particle is the decay of a free neutron, which is unstable outside the nucleus of an atom.

$$ {}^{1}_{0}n \longrightarrow {}^{1}_{1}p + {}^{0}_{-1}\beta + \nu \qquad \textbf{(25.2)} $$

◀ A neutrino has no charge and its detection was not easy. Theoretically, to balance nuclear equations, it was known that such a particle must exist. Many elementary particles have been found through experiments based on symmetry arguments.

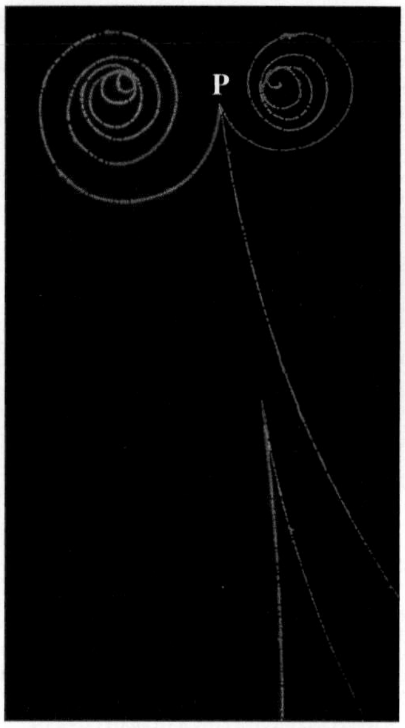

Lawrence Berkeley Laboratory/Science Photo Library

▲ **A colorized cloud chamber photograph**
Point P marks an atomic nucleus that interacts with a γ-ray photon (not visible), producing a β^- particle and a positron (spiral green and red tracks, respectively). The photon also dislodges an orbital electron (vertical green track).

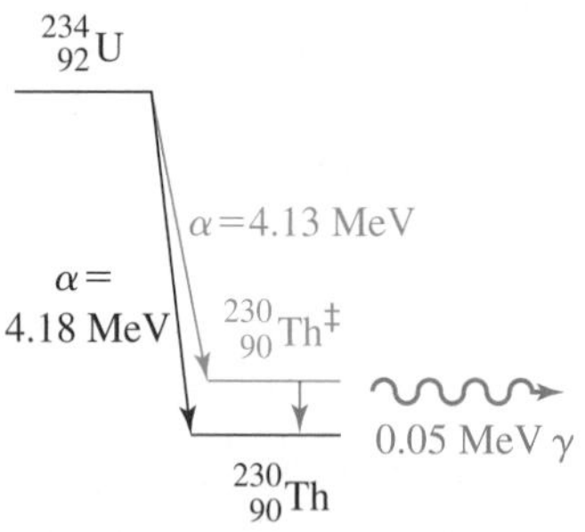

▲ FIGURE 25-1
Production of γ rays
The transition of a $^{230}_{90}Th$ nucleus between the two energy states shown results in the emission of a 0.05 MeV gamma ray.

An electronvolt (eV) is the energy acquired by an electron when it falls through an electric potential difference of 1 volt:

$$1 \text{ eV} = 1.6022 \times 10^{-19} \text{ J}$$
$$1 \text{ MeV} = 1 \times 10^{6} \text{ eV}$$

A β^- particle does not have an atomic number, but its -1 charge is equivaler to an atomic number of -1. In nuclear equations, the β^- particle is represente as $^{0}_{-1}\beta$. Also, a β^- particle is small enough, compared to protons and neutron that its mass can be ignored in most calculations. Equation (25.2) introduc the symbol ν to represent an entity called a *neutrino*. This particle was fir postulated in the 1930s as necessary for the conservation of certain properti during the β^- decay process. Because they interact so weakly with matte neutrinos were not detected until the 1950s. Even today, little is known of the properties, including their rest mass. (Rest mass is discussed on page 310).

For a typical β^- decay process, as represented by equation (25.3), we ca think of a neutron *within* the nucleus of an atom spontaneously converting t a proton and an electron. This proton remains in the nucleus, whereas the ele tron is emitted as a β^- particle. Because of the extra proton, the atomic numb *increases* by one unit, while the mass number is unchanged. The elusive neu trino is generally not included in the nuclear equation.

$$^{234}_{90}Th \longrightarrow {}^{234}_{91}Pa + {}^{0}_{-1}\beta \qquad (25.$$

In a similar manner, in some decay processes a *proton* within the nucleus converted to a neutron, and a β^+ particle and a neutrino* are emitted.

$$^{1}_{1}p \longrightarrow {}^{1}_{0}n + {}^{0}_{+1}\beta + \nu \qquad (25.4$$

The β^+ particle, also called a **positron**, has properties similar to the β^- particl except that it carries a *positive* charge. (See the photograph in the margin.) Th particle is also known as a *positive electron* and is designated $^{0}_{+1}\beta$ in nuclea equations. Positron emission is commonly encountered with artificially pro duced radioactive nuclei of the lighter elements. For example,

$$^{30}_{15}P \longrightarrow {}^{30}_{14}Si + {}^{0}_{+1}\beta \qquad (25.5$$

Electron Capture

Another process that achieves the same effect as positron emission is **electro capture (EC)**. In this case, an electron from an inner electron shell (usually th shell $n = 1$) is absorbed by the nucleus, where it converts a proton to a neu tron. When an electron from a higher quantum level drops to the energy leve vacated by the captured electron, an X-ray is emitted. For example,

$$^{202}_{81}Tl + {}^{0}_{-1}e \longrightarrow {}^{202}_{80}Hg \text{ (followed by an X-ray)} \qquad (25.6$$

Gamma Rays

Some radioactive decay processes that yield α or β^- particles leave th nucleus in an excited state. The nucleus then loses energy in the form of elec tromagnetic radiation called gamma rays. **Gamma (γ) rays** are a highly pene trating form of radiation that are *undeflected* by electric and magnetic field (recall Figure 2-10). (Lead bricks more than several centimeters thick may b required to stop them.) In the radioactive decay of $^{234}_{92}U$, 77% of the nucle emit α particles having an energy of 4.18 MeV. The remaining 23% of the $^{234}_{92}U$ nuclei produce α particles with energies of 4.13 MeV. In the latter case, th $^{230}_{90}Th$ nuclei are left with an excess energy of 0.05 MeV. This energy i released as γ rays. If the unstable excited Th nucleus is denoted as $^{230}_{90}Th^{\ddagger}$, w can write

$$^{234}_{92}U \longrightarrow {}^{230}_{90}Th^{\ddagger} + {}^{4}_{2}He \qquad (25.7$$

$$^{230}_{90}Th^{\ddagger} \longrightarrow {}^{230}_{90}Th + \gamma \qquad (25.8$$

This γ-emission process is represented diagrammatically in Figure 25-1.

*There appear to be two related entities: the neutrino and antineutrino. Neutrinos accompan positron emission and electron capture; antineutrinos are associated with β^- emission.

25-1 ARE YOU WONDERING?

How does an α particle get out of a nucleus?

The answer has to do with quantum theory and the nature of the forces involved. Consider the potential energy diagram shown below. The blue line represents the potential energy, where we imagine an α particle as a separate particle within a nucleus such as $^{238}_{92}U$. Region A represents the potential energy of the α particle when it is held within the nucleus by the forces inside the uranium nucleus. Region C represents the potential energy of the α particle when it is free of the nucleus. The potential energy along the downward curving portion of the blue line represents the Coulomb (electrostatic) repulsion between the positively charged α particle and the nucleus remaining after the α particle has escaped ($^{234}_{90}Th$).

To get to region C, the α particle must get past the barrier in region B. The potential energy just beyond A, the radius of the nucleus, is greater than the energy of the α particle. (We know this from the measured energy of the α particle.) The α particle could not escape the nucleus if it were governed by classical physics because this would require an input of energy equal to the height of the barrier. Radioactive nuclei decay spontaneously, however, without an input of energy. How can the α particle get from region A to region C?

It passes through the barrier in a process known as *tunneling*. Classically, to go from A to C, the α particle would violate the principle of the conservation of energy. The α particle, however, possesses wave-like properties, as seen through the wave function at the bottom of the figure. Quantum mechanics predicts a finite probability of finding the α particle in a classically forbidden region. The wave function for the α particle trails off in the barrier region (B) and then reaches the outside, where it appears as a wave with much smaller amplitude. There is a finite probability (ψ^2) of finding the α particle outside the nucleus: The α particle has tunneled through the barrier.

Moreover, an alternative form of the uncertainty principle tells us that energy conservation can be violated by an amount ΔE for the length of time Δt given by

$$\Delta E \times \Delta t = \frac{h}{4\pi}$$

That is, the wave–particle duality of quantum theory allows the conservation of energy to be violated for brief periods—long enough for an α particle to tunnel through the barrier. ΔE corresponds to the energy difference between the barrier height and the α particle's energy, and Δt corresponds to the time required to pass through the barrier. The higher and wider the potential energy barrier, the less time the particle has to escape and the less likely it will do so. Thus, the height and width of the barrier control the rate of decay.

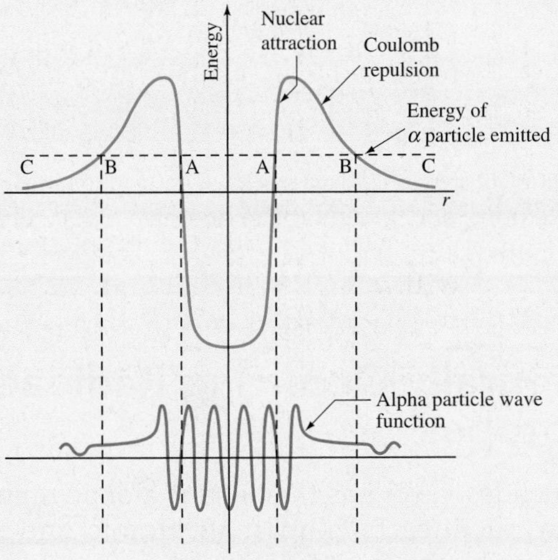

▲ Tunneling of an α particle out of the nucleus.

EXAMPLE 25-1 Writing Nuclear Equations for Radioactive Decay Processes

Write nuclear equations to represent (a) α-particle emission by ^{222}Rn and (b) radioactive decay of bismuth-215 to polonium-215.

Analyze

In part (a) we can identify two of the species involved in this process from the information given and we can write an incomplete equation. The unknown element in the incomplete equation is identified by determining the atomic number (Z) and mass number (A) that will balance the incomplete equation. Part (b) is done in a way very similar to that in part (a).

Solve

(a) Since the $^{222}_{86}$Rn nucleus ejects an α particle, $^{4}_{2}$He, as shown in the following incomplete nuclear equation.

$$^{222}_{86}\text{Rn} \longrightarrow ? + {}^{4}_{2}\text{He}$$

Because the ejected α particle contains two protons, the unknown product must contain two fewer protons than $^{222}_{86}$Rn: $Z = 86 - 2 = 84$. This atomic number identifies the element as polonium, $_{84}$Po. The mass number (A) of the product can be obtained by subtracting the mass number of the α particle from that of the radon isotope: $A = 222 - 4 = 218$. The completed nuclear equation is

$$^{222}_{86}\text{Rn} \longrightarrow {}^{218}_{84}\text{Po} + {}^{4}_{2}\text{He}$$

(b) The atomic number of bismuth is 83 and that of polonium is 84. We can approach this problem as we did part (a).

$$^{215}_{83}\text{Bi} \longrightarrow {}^{215}_{84}\text{Po} + ?$$

There is no change in mass number, so the particle has a zero mass number. Its atomic number is $Z = 83 - 84 = -1$. Only a $^{0}_{-1}\beta$ particle fits these parameters: Beta (β^-) decay is the only type of emission leading to an increase of one unit in atomic number without a change in the mass number.

$$^{215}_{83}\text{Bi} \longrightarrow {}^{215}_{84}\text{Po} + {}^{0}_{-1}\beta$$

Assess

It is interesting that when starting from different elements, we can arrive at the same element through different types of particle emission. Note that even though we came to the same decayed element, we ended with different isotopes of that element.

PRACTICE EXAMPLE A: Write a nuclear equation to represent β^- particle emission by $^{241}_{94}$Pu.

PRACTICE EXAMPLE B: Write a nuclear equation to represent the decay of a radioactive nucleus to produce ^{58}Ni and a positron.

25-1 CONCEPT ASSESSMENT

Which type(s) of radioactive decay transform(s) the nucleus of an atom to that of a different element, and which type(s) do not?

KEEP IN MIND

that the term *nuclide*, first introduced in Chapter 2 (see page 45), is used for "a species of atom characterized by the constitution of its nucleus, in particular by the numbers of protons and neutrons in its nucleus" [T. P. Kohman, *Am. J. Phys.*, **15**, 356 (1947)]. The term is derived from the word nuclear and the Greek word for "species," *eidos*.

25-2 Naturally Occurring Radioactive Isotopes

Of the stable nuclides, $^{209}_{83}$Bi has the highest atomic number and mass number. All known nuclides beyond it in atomic and mass numbers are radioactive. Naturally occurring $^{238}_{92}$U is radioactive and disintegrates by the loss of α particles.

$$^{238}_{92}\text{U} \longrightarrow {}^{234}_{90}\text{Th} + {}^{4}_{2}\text{He}$$

$^{34}_{90}$Th is also radioactive; it decays by β^- emission.

$$^{234}_{90}\text{Th} \longrightarrow {}^{234}_{91}\text{Pa} + {}^{0}_{-1}\beta$$

$^{34}_{91}$Pa also decays by β^- emission to produce $^{234}_{92}$U, which is also radioactive.

$$^{234}_{91}\text{Pa} \longrightarrow {}^{234}_{92}\text{U} + {}^{0}_{-1}\beta$$

'he term *daughter* is commonly used to describe the new nuclide produced in radioactive decay. Thus, ^{234}Th is the daughter of ^{238}U, ^{234}Pa is the daughter f ^{234}Th, and so on.

The chain of radioactive decay that begins with $^{238}_{92}$U continues through a umber of steps of α and β^- emission until it eventually terminates with a table isotope of lead—$^{206}_{82}$Pb. The entire scheme is outlined in Figure 25-2. All aturally occurring radioactive nuclides of high atomic number belong to ne of three **radioactive decay series**: the *uranium* series just described, the *horium* series, or the *actinium* series. (The actinium series actually begins vith uranium-235, which was once called actino-uranium.)

Even though some of the daughters in natural radioactive decay schemes ave very short half-lives, all are present because they are constantly forming s well as decaying. It is likely that only about one gram of radium-226 was resent in several tons of uranium ore processed by Marie Curie in her dis-overy of radium in 1898. Nevertheless, she was successful in isolating it. The re also contained only a fraction of a milligram of polonium, which she was ble to detect but not isolate.

Radioactive decay schemes can be used to determine the ages of rocks and hereby the age of Earth (see Section 25-5). The appearance of certain radioactive ubstances in the environment can also be explained through radioactive decay eries. The nuclides ^{210}Po and ^{210}Pb have been detected in cigarette smoke. These radioactive isotopes are derived from ^{238}U, found in trace amounts in the hosphate fertilizers used in tobacco fields. These α-emitting isotopes have been mplicated in the link between cigarette smoking and cancer and heart disease.

Bettmann/Corbis

▲ **Marie Sklodowska Curie (1867–1934)**
Marie Curie shared the 1903 Nobel Prize in physics for studies on radiation phenomena. In 1911, she won the Nobel Prize in chemistry for her discovery of polonium and radium.

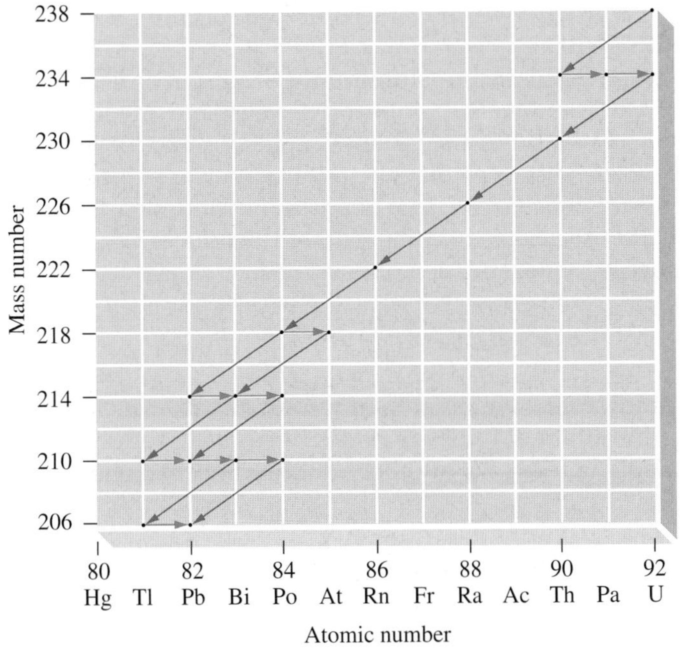

▲ FIGURE 25-2
The natural radioactive decay series for $^{238}_{92}$U (uranium series)
The long arrows pointing down and to the left correspond to α-particle emissions. The short horizontal arrows represent β^- emissions. Other natural decay series originate with $^{232}_{90}$Th (thorium series) and $^{235}_{92}$U (actinium series).

Radioactivity, which is so common among isotopes of high atomic numbe is a relatively rare phenomenon among the naturally occurring lighte isotopes. Even so, ^{40}K is a radioactive isotope, as are ^{50}V and ^{138}La. ^{40}K decay by β^- emission and by electron capture.

$$^{40}_{19}K \longrightarrow {}^{40}_{20}Ca + {}^{0}_{-1}\beta \quad \text{and} \quad {}^{40}_{19}K + {}^{0}_{-1}e \longrightarrow {}^{40}_{18}Ar$$

At the time Earth was formed ^{40}K was much more abundant than it is now. I is believed that the high argon content of the atmosphere (0.934% by volum and almost all of it as ^{40}Ar) is derived from the radioactive decay of ^{40}K Aside from ^{40}K and ^{14}C (produced by cosmic radiation), the most importan radioactive isotopes of the lighter elements are produced *artificially*.

25-2 CONCEPT ASSESSMENT

Explain why francium, the heaviest of the alkali metals (group 1), is not found in minerals containing the other alkali metals and is also one of the rarest elements.

Bettmann/Corbis

▲ **Iréne Joliot-Curie (1897–1956)**
Iréne Joliot-Curie and her husband, Frédéric Joliot, shared the 1935 Nobel Prize in chemistry for the artificial production of radioactive nuclides.

25-3 Nuclear Reactions and Artificially Induced Radioactivity

Ernest Rutherford discovered that atoms of one element can be transforme into atoms of another element. He did this in 1919 by bombarding $^{14}_{7}N$ nucle with α particles, producing $^{17}_{8}O$ and protons. In this way, he was able to obtai protons outside atomic nuclei. The process can be represented as

$$^{14}_{7}N + {}^{4}_{2}He \longrightarrow {}^{17}_{8}O + {}^{1}_{1}H \qquad (25.9$$

In reaction (25.9), instead of a nucleus disintegrating spontaneously, it mus be struck by another small particle to induce a nuclear reaction. $^{17}_{8}O$ is a natu rally occurring *nonradioactive* isotope of oxygen (0.037% natural abundance) The situation with $^{30}_{15}P$, which can also be produced by a nuclear reaction, is somewhat different.

In 1934, when bombarding aluminum with α particles, Iréne Joliot-Curie (daughter of Marie and Pierre Curie) and her husband, Frédéric Joliot, observec the emission of two types of particles: neutrons and positrons. The Joliots observed that when bombardment by α particles was stopped, the emission o neutrons also stopped; the emission of positrons continued, however. Their con clusion was that the nuclear bombardment produces $^{30}_{15}P$, which undergoes radioactive decay by the emission of positrons.

$$^{27}_{13}Al + {}^{4}_{2}He \longrightarrow {}^{30}_{15}P + {}^{1}_{0}n$$
$$^{30}_{15}P \longrightarrow {}^{30}_{14}Si + {}^{0}_{+1}\beta$$

The first radioactive nuclide obtained by artificial means was $^{30}_{15}P$. Now over 1000 artificially radioactive nuclides have been produced, and their num ber considerably exceeds the number of nonradioactive ones (about 280).

EXAMPLE 25-2 Writing Equations for Nuclear Bombardment Reactions

Write a nuclear equation for the production of ^{56}Mn by bombardment of ^{59}Co with neutrons.

Analyze

In this example we proceed in a manner similar to that in Example 25-1: We first write an incomplete nuclear equation from the given information. We must also realize that a particle is produced along with ^{56}Mn. To find the mass number (A) of the unknown particle, we must subtract the mass number of the Mn atom from that of the Co atom plus the neutron that initiates the reaction. Thus, for the unknown particle, $A = 59 + 1 - 56 = 4$. Subtracting the atomic number of Mn from that of Co provides the atomic number of the unknown particle: $Z = 27 - 25 = 2$.

Solve

The unknown particle must have $A = 4$ and $Z = 2$; it is an α particle.

$$^{59}_{27}\text{Co} + ^{1}_{0}\text{n} \longrightarrow ^{56}_{25}\text{Mn} + ^{4}_{2}\text{He}$$

Assess

By using the concept that a balanced nuclear equation has the same overall atomic number and mass number on both sides of the equation, we can identify the unknown particle that has been ejected during bombardment.

PRACTICE EXAMPLE A: Write a nuclear equation for the production of ^{147}Eu by bombardment of ^{139}La with ^{12}C.

PRACTICE EXAMPLE B: Write a nuclear equation for the production of ^{124}I by bombardment of ^{121}Sb with α particles. Also, write an equation for the subsequent decay of ^{124}I by positron emission.

25-4 Transuranium Elements

Until 1940, the only known elements were those that occur naturally. In 1940, bombardment of $^{238}_{92}\text{U}$ atoms with neutrons produced the first synthetic element. First, the unstable nucleus $^{239}_{92}\text{U}$ forms. This nucleus then undergoes β^- decay, yielding the element neptunium, with $Z = 93$.

$$^{238}_{92}\text{U} + ^{1}_{0}\text{n} \longrightarrow ^{239}_{92}\text{U} + \gamma$$

$$^{239}_{92}\text{U} \longrightarrow ^{239}_{93}\text{Np} + ^{0}_{-1}\beta$$

Bombardment by neutrons is an effective way to produce nuclear reactions because these heavy uncharged particles are not repelled as they approach a nucleus.

Since 1940, all the elements from $Z = 93$ to 118 have been synthesized. Many of the new elements of high atomic number have been formed by bombarding transuranium atoms with the nuclei of lighter elements. For example, an isotope of the element $Z = 105$ can be produced by bombarding atoms of $^{249}_{98}\text{Cf}$ with $^{15}_{7}\text{N}$ nuclei.

$$^{249}_{98}\text{Cf} + ^{15}_{7}\text{N} \longrightarrow ^{260}_{105}\text{Db} + 4\,^{1}_{0}\text{n} \quad \textbf{(25.10)}$$

To bring about nuclear reactions such as (25.10) requires bombarding atomic nuclei with energetic particles. Such energetic particles can be obtained in an accelerator. A type of accelerator known as a cyclotron is illustrated in Figure 25-3.

A *charged-particle accelerator*, as the name implies, can produce only beams of charged particles (such as $^{1}_{1}\text{H}^+$) as projectiles. In many cases, neutrons are most effective as projectiles for nuclear bombardment. The neutrons required can be generated through a nuclear reaction produced by a charged-particle beam. In the following reaction, $^{2}_{1}\text{H}$ represents a beam of deuterons, the nucleus of a deuterium atom (actually, $^{2}_{1}\text{H}^+$), from an accelerator.

$$^{9}_{4}\text{Be} + ^{2}_{1}\text{H} \longrightarrow ^{10}_{5}\text{B} + ^{1}_{0}\text{n}$$

Another important source of neutrons for nuclear reactions is a nuclear reactor (as we will see in Section 25-8).

David Parker/Science Photo Library

▲ Stanford Linear Accelerator Center (SLAC0 PEP-II collider). Electrons (blue) and positrons (pink) circulate in opposite directions along the two rings before being forced to collide. These collisions produce large quantities of subatomic particles (e.g., "B mesons and anti-B mesons").

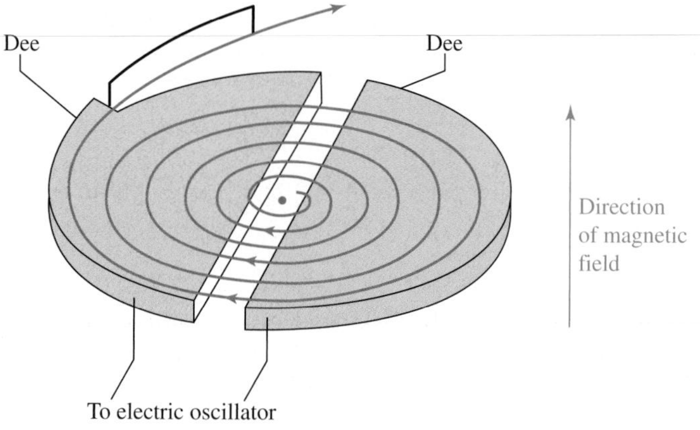

▲ FIGURE 25-3
The cyclotron
This type of accelerator consists of two hollow, flat, semicircular boxes, called *dees*, that are kept electrically charged. The entire assembly is maintained within a magnetic field. The particles to be accelerated, in the form of positive ions, are produced at the center of the opening between the dees. They are then attracted into the negatively charged dee and forced into a circular path by the magnetic field. When the particles leave the dee and enter the gap, the electric charges on the dees are reversed, so that the particles are attracted into the opposite dee. The particles are accelerated as they pass the gap and travel a wider circular path in the new dee. This process is repeated many times until the particles are brought to the required energy.

25-5 Rate of Radioactive Decay

In time, we can expect every atomic nucleus of a radioactive nuclide to disintegrate, but it is impossible to predict when any one nucleus will do so. Although we cannot make predictions for a particular atom, we can use statistical methods to make predictions for a collection of atoms. Based on experimental observations, a **radioactive decay law** has been established.

The rate of disintegration of a radioactive material—called the activity, *A*, or the decay rate—is directly proportional to the number of atoms present.

In mathematical terms,

$$\text{rate of decay} \propto N \quad \text{and} \quad \text{rate of decay} = A = \lambda N \qquad \textbf{(25.11)}$$

The activity is expressed in atoms per unit time, such as atoms per second. N is the number of atoms in the sample being observed; $\boldsymbol{\lambda}$ is the **decay constant**, which has units of time^{-1}. Consider the case of a 1,000,000-atom sample disintegrating at the rate of 100 atoms per second. In such a case, $N = 1.0 \times 10^6$ and

$$\lambda = A/N = 100 \text{ atom s}^{-1}/1.0 \times 10^6 \text{ atom} = 1.0 \times 10^{-4} \text{ s}^{-1}$$

Radioactive decay is a *first-order* process. To relate it to the first-order kinetics that we studied in Chapter 20, think of the activity as corresponding to a rate of reaction; the number of atoms as corresponding to the concentration of a reactant; and the decay constant, λ, as corresponding to a rate constant, k. This correspondence can be carried further by writing an integrated radioactive decay law and a relationship between the decay constant and the **half-life** of he process—the length of time required for half of a radioactive sample to lisintegrate.

$$\ln\left(\frac{N_t}{N_0}\right) = -\lambda t \quad (25.12)$$

$$t_{1/2} = \frac{\ln(2)}{\lambda} \quad (25.13)$$

n these equations, N_0 represents the number of atoms at some initial time $t = 0$); N_t is the number of atoms at some later time, t; λ is the decay con- tant; and $t_{1/2}$ is the half-life.

Recall from Chapter 20 (page 935) that the half-life of a first-order process is *constant*. Thus, if half the atoms of a radioactive sample disintegrate in .5 min, the number of atoms remaining will be reduced to one-fourth the riginal number in 5.0 min, one-eighth in 7.5 min, and so on. The shorter the alf-life, the larger the value of λ and the faster the decay process. Half-lives of adioactive nuclides range from extremely short to very long, as suggested by he representative data in Table 25.1.

You may be wondering whether, like first-order chemical reactions, adioactive decay is temperature dependent. We learned in Chapter 20 that an mportant determinant of the rate of a chemical reaction is the height of the nergy barrier between the reactants and products (the activation energy). The igher the temperature, the greater the number of molecules that can sur- nount the barrier as a result of collisions and the faster the reaction proceeds. Although there is also an energy barrier that confines nuclear particles to the ucleus, molecular collisions do not invest any energy in nuclear particles. Moreover, in radioactive decay, nuclear particles do not escape the nucleus by urmounting an energy barrier—they tunnel through it. Thus, the rates of adioactive decay processes are independent of temperature.

TABLE 25.1 Some Representative Half-Lives

Nuclide	Half-Life[a]	Nuclide	Half-Life[a]	Nuclide	Half-Life[a]
$^{3}_{1}H$	12.33 a	$^{40}_{19}K$	1.26×10^9 a	$^{214}_{84}Po$	1.64×10^{-4} s
$^{14}_{6}C$	5715 a	$^{80}_{35}Br$	17.6 min	$^{222}_{86}Rn$	3.823 d
$^{13}_{8}O$	8.7×10^{-3} s	$^{90}_{38}Sr$	29.1 a	$^{226}_{88}Ra$	1.599×10^3 a
$^{28}_{12}Mg$	21 h	$^{131}_{53}I$	8.021 d	$^{234}_{90}Th$	24.10 d
$^{32}_{15}P$	14.3 d	$^{137}_{55}Cs$	30.2 a	$^{238}_{92}U$	4.47×10^9 a
$^{35}_{16}S$	87.9 d				

[a]s, second; min, minute; h, hour; d, day; a, year.
Source: CRC Handbook of Chemistry and Physics, 92 ed., by W. M. Haynes - Editor in Chief.

EXAMPLE 25-3 Using the Half-Life Concept and the Radioactive Decay Law to Describe the Rate of Radioactive Decay

The phosphorus isotope ^{32}P is used in biochemical studies to determine the pathways of phosphorus atoms in living organisms. Its presence is detected through its emission of β^- particles. **(a)** What is the decay constant for ^{32}P, expressed in the unit s^{-1}? **(b)** What is the activity of a 1.00 mg sample of ^{32}P (that is, how many atoms disintegrate per second)? **(c)** Approximately what mass of ^{32}P will remain in the original 1.00 mg sample after 57 days? (See Table 25.1.) **(d)** What will be the rate of radioactive decay after 57 days?

Analyze

To solve these types of problems, we need to determine the decay constant, λ, of the radioactive species, which is related to the concept of half-life through equation (25.13). After determining the decay constant from the half-life of the sample, we can then use it to determine the activity for part (b).

(continued)

Solve

(a) We can determine λ from $t_{1/2}$ with equation (25.13). The first result we get has the unit d^{-1}. We must convert this unit to h^{-1}, min^{-1}, and s^{-1}.

$$\lambda = \frac{0.693}{14.3\ d} \times \frac{1\ d}{24\ h} \times \frac{1\ h}{60\ min} \times \frac{1\ min}{60\ s} = 5.61 \times 10^{-7}\ s^{-1}$$

(b) First, let us find the number of atoms, N, in 1.00 mg of ^{32}P.

$$N\ (^{32}P\ \text{atoms}) = 0.00100\ g \times \frac{1\ mol\ ^{32}P}{32.0\ g} \times \frac{6.022 \times 10^{23}\ ^{32}P\ \text{atoms}}{1\ mol\ ^{32}P}$$
$$= 1.88 \times 10^{19}\ ^{32}P\ \text{atoms}$$

Then, we can multiply this number by the decay constant to get the activity or decay rate.

$$\text{activity} = \lambda N = 5.61 \times 10^{-7}\ s^{-1} \times 1.88 \times 10^{19}\ \text{atoms}$$
$$= 1.05 \times 10^{13}\ \text{atoms}\ s^{-1}$$

(c) A period of 57 days is 57/14.3 = 4.0 half-lives. As shown in Figure 25-4, the quantity of radioactive material decreases by one-half for every half-life. The quantity remaining is $(\frac{1}{2})^4$ of the original quantity.

$$?\ mg\ ^{32}P = 1.00\ mg \times \left(\frac{1}{2}\right)^4 = 1.00\ mg \times \frac{1}{16} = 0.063\ mg\ ^{32}P$$

(d) The activity is directly proportional to the number of radioactive atoms remaining (activity $= \lambda N$), and the number of atoms is directly proportional to the mass of ^{32}P. When the mass of ^{32}P has dropped to one-sixteenth its original mass, the number of ^{32}P atoms also falls to one-sixteenth the original number, and the rate of decay is one-sixteenth the original activity.

$$\text{rate of decay} = \frac{1}{16} \times 1.05 \times 10^{13}\ \text{atoms}\ s^{-1} = 6.56 \times 10^{11}\ \text{atoms}\ s^{-1}$$

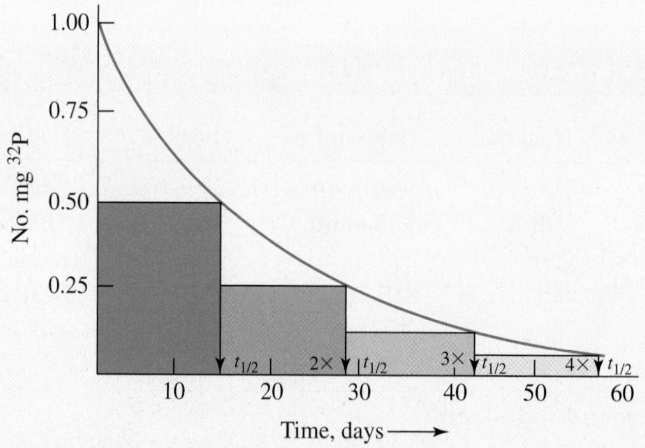

▲ FIGURE 25-4
Radioactive decay of a hypothetical ^{32}P sample—Example 25-3 illustrated

Assess

Phosphorus-32 isotope is an ideal radionuclide to use because of its short half-life. This time is long enough to carry out experiments, yet short enough that disposal is not a major problem. We see in this example that after 57 days, approximately 6% of the original mass of phosphorus remains.

PRACTICE EXAMPLE A: ^{131}I is a β^- emitter used as a tracer for radioimmunoassays in biological systems. Use information in Table 25.1 to determine **(a)** the decay constant in s^{-1}; **(b)** the activity of a 2.05 mg sample of ^{131}I; **(c)** the percentage of ^{131}I remaining after 16 days; and **(d)** the rate of β^- emission after 16 days.

PRACTICE EXAMPLE B: ^{223}Ra has a half-life of 11.43 days. How long would it take for the activity associated with a sample of ^{223}Ra to decrease to 1.0% of its current value?

25-3 CONCEPT ASSESSMENT

Why are radioactive nuclides with intermediate half-lives generally more hazardous than those with either extremely short or extremely long half-lives?

Radiocarbon Dating

In the upper atmosphere, ${}^{14}_{6}C$ is formed at a constant rate by the bombardment of ${}^{14}_{7}N$ with neutrons.

$$ {}^{14}_{7}N + {}^{1}_{0}n \longrightarrow {}^{14}_{6}C + {}^{1}_{1}H $$

The neutrons are produced by cosmic rays. ${}^{14}_{6}C$ disintegrates by β^- emission.

Carbon-containing compounds in living organisms are in equilibrium with ^{14}C in the atmosphere—that is, these organisms replace ^{14}C atoms that have undergone radioactive decay with "fresh" ^{14}C atoms through interactions with their environment. The ^{14}C isotope is radioactive and has a half-life of 5730 years. The activity associated with ^{14}C that is in equilibrium with its

EXAMPLE 25-4 Applying the Integrated Rate Law for Radioactive Decay: Radiocarbon Dating

A wooden object found in an ancient burial mound is subjected to radiocarbon dating. The activity associated with its ^{14}C content is 10 dis min^{-1} g^{-1}. What is the age of the object? In other words, how much time has elapsed since the tree from which the wood came was cut down?

Analyze

The solution requires three equations: (25.11), (25.12), and (25.13).

Solve

Equation (25.13) is used to determine the decay constant.

$$ \lambda = \frac{0.693}{5730\ \text{a}} = 1.21 \times 10^{-4}\ \text{a}^{-1} $$

Next, equation (25.11) relates to the actual number of atoms: N at $t = 0$ (the time when the ^{14}C equilibrium was destroyed) and N_t at time t (the present time). As discussed on page 1182, the activity just before the ^{14}C equilibrium was destroyed was 15 dis min^{-1} g^{-1}; at the time of the measurement, it is 10 dis min^{-1} g^{-1}. The corresponding numbers of atoms are equal to these activities divided by λ.

$$ N_0 = A_0/\lambda = 15/\lambda \quad \text{and} \quad N_t = A_t/\lambda = 10/\lambda $$

Finally, we substitute into equation (25.12).

$$ \ln\frac{N_t}{N_0} = \ln\frac{10/\lambda}{15/\lambda} = \ln\frac{10}{15} = -(1.21 \times 10^{-4}\ \text{a}^{-1})t $$

$$ -0.41 = -(1.21 \times 10^{-4}\ \text{a}^{-1})t $$

$$ t = \frac{0.41}{1.21 \times 10^{-4}\ \text{a}^{-1}} = 3.4 \times 10^{3}\ \text{a} $$

Assess

In the previous example, we observed that the activity depends on the amount of material. Therefore, the results of radiocarbon dating depend on knowing the activity at the time the equilibrium between ^{14}C and the other nonradioactive carbon isotopes ceases. If at the time equilibrium was destroyed the activity was 14 dis min^{-1} g^{-1}, we would have determined the object to be 2.8×10^3 years old, which is approximately a 17% error.

PRACTICE EXAMPLE A: What is the age of a mummy, given a ^{14}C activity of 8.5 dis min^{-1} g^{-1}?

PRACTICE EXAMPLE B: What should be the current activity, in dis min^{-1} g^{-1}, of a wooden object believed to be 1100 years old?

▶ The remains of a man, referred to as Ötzi, frozen in a glacier in the Austrian Alps have been dated by the radiocarbon method as 5300 years old.

environment is about 0.25 Bq (Bq is the unit becquerel which represents disin tegrations per second) per gram of carbon. When an organism dies (fo instance, when a tree is cut down), this equilibrium is destroyed and the disin tegration rate falls off because the dead organism no longer absorbs new ^{14}C From the measured disintegration rate at some later time, the age can be esti mated (that is, the elapsed time since the ^{14}C equilibrium was disrupted).

The Age of Earth

NASA

▲ A lunar rock that has been radiometrically dated to be about 4.6 billion years old.

The natural radioactive decay scheme of Figure 25-2 suggests the eventua fate of all $^{238}_{92}U$ in nature—conversion to lead. Naturally occurring uraniun minerals always have associated with them some nonradioactive leac formed by radioactive decay. From the mass ratio of $^{206}_{82}Pb$ to $^{238}_{92}U$ in such a mineral, it is possible to estimate the age of the igneous rock containing the mineral. The age of the rock refers to the time elapsed since molter magma solidified to form the rock. One assumption of this method is that the initial radioactive nuclide, the final stable nuclides, and all the products of a decay series remain in the rock. Another assumption is that any lead present in the rock initially consisted of the several isotopes of lead in their present naturally occurring abundances.

The half-life of $^{238}_{92}U$ is 4.5×10^9 years. According to the natural decay scheme of Figure 25-2, the basic change that occurs as atoms of $^{238}_{92}U$ and its daughters pass through the entire sequence of steps is

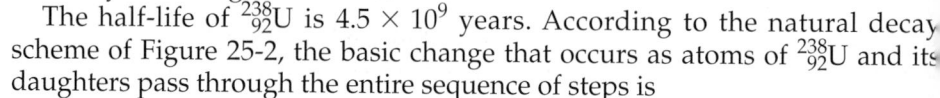

$$^{238}_{92}U \longrightarrow {}^{206}_{82}Pb + 8\,^{4}_{2}He + 6\,^{0}_{-1}\beta$$

The decay sequence for $^{238}U \longrightarrow {}^{206}Pb$ has 14 steps. The first step, however has a much longer half-life than any of the other steps in the series and can thus be thought of as the rate-determining step, with the subsequent steps being "fast." As discussed in Chapter 20, we can ignore the effect on the overall rate of fast steps that occur after the slow, rate-determining step. Thus, the half-life for ^{238}U is essentially equal to the time it takes to convert half the initial ^{238}U to the ^{206}Pb isotope. Discounting the mass associated with the β^- particles, for every 238 g of uranium that undergoes complete decay, 206 g of lead and 32 g of helium are produced.

Suppose that in a hypothetical rock containing no lead initially, 1.000 g of $^{238}_{92}U$ had disintegrated through one half-life, 4.51×10^9 years. At the end of that time, 0.500 g $^{238}_{92}U$ would have disintegrated and another 0.500 g would remain. The quantity of $^{206}_{82}Pb$ now present in the rock would be

$$0.500\text{ g }^{238}_{92}U \times \frac{206\text{ g }^{206}_{82}Pb}{238\text{ g }^{238}_{92}U} = 0.433\text{ g }^{206}_{82}Pb$$

The ratio of lead-206 to uranium-238 in the rock would be

$$^{206}_{82}Pb/^{238}_{92}U = 0.433/0.500 = 0.866$$

If the $^{206}_{82}Pb/^{238}_{92}U$ mass ratio is less than 0.866, the age of the rock is less than one half-life of $^{238}_{92}U$. A higher ratio indicates a greater age for the rock. The est estimates of the age of the oldest rocks, and presumably of Earth itself, re about 4.5×10^9 years. These estimates are based on the $^{206}_{82}Pb/^{238}_{92}U$ ratio nd on ratios for other pairs of isotopes from natural radioactive decay series.

Modern Radioactive Dating (Geochronology)

Aodern radioactive dating techniques use mass spectrometry to analyze parent r daughter nuclides, or both. A mass spectrometer (described in Chapter 2) an be used to determine the amount of the various isotopes found in geolog- cal material (e.g., rocks), and that data can be used to calculate the age of he material. One particular pair of isotopes currently used in dating geolog- cal material is potassium-40 and argon-40. Potassium-40 is radioactive and nainly undergoes β-decay to calcium-40; however, some potassium-40 ındergoes electron capture and decays to argon-40. Attempts have been nade to perform radiodating for the potassium-40–calcium-40 pair, but this s not currently possible because of the large natural abundance of calcium.

Let us look at how isotope pairs can be used to determine the age of geo- ogical material. We begin by rewriting equation (25.12) as

$$N = N_0 e^{-\lambda t} \tag{25.14}$$

vhere N is the number of atoms currently present, N_0 is the initial number of ıtoms at $t = 0$, and λ is the decay constant for ^{40}K. Since we don't know how nany atoms were present at $t = 0$, we can determine that by recognizing that

$$N_0 = N + D \tag{25.15}$$

vhere D is the number of daughter atoms, that is, the atoms resulting from the decay of an isotope. An example of a daughter atom is ^{40}Ar, which is a daugh- er atom of ^{40}K. Combining equations (25.14) and (25.15) and solving for t we have

$$t = \frac{1}{\lambda} \ln\left(\frac{D}{N} + 1\right) \tag{25.16}$$

Equation (25.16) neglects the presence of daughter atoms, D_0, that may have been present at $t = 0$. To account for this we rewrite equation (25.16) as

$$t = \frac{1}{\lambda} \ln\left(\frac{D_t - D_0}{N} + 1\right) \tag{25.17}$$

where D_t are the total daughter atoms at the current time. Equation (25.17) cannot be used in its current form to accurately determine the age of geologi- cal material. To accurately determine the geological age, we must consider the branching decay. Branching decay occurs when an isotope can decay through different pathways to other nuclides. The figure below illustrates branching for potassium-40.

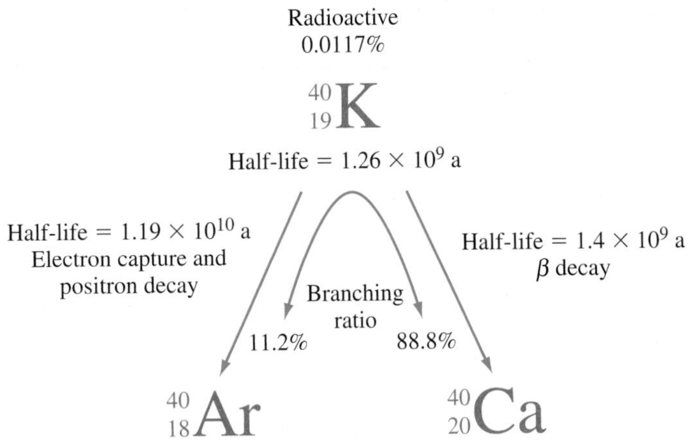

When branching is taken into account, we obtain

$$t = \frac{1}{\lambda_\varepsilon + \lambda_\beta}\ln\left[\frac{^{40}Ar_{rad}}{^{40}K}\left(\frac{\lambda_\varepsilon + \lambda_\beta}{\lambda_\varepsilon} + 1\right)\right] \quad (25.18)$$

where λ_ε is the decay constant for the electron capture branch of ^{40}K decay and λ_β is the decay constant for the β-decay branch. $^{40}Ar_{rad}$ is the argon-40 produced by the decay of potassium-40 and corrected for the presence of trapped atmospheric argon-40. Several assumptions are made in deriving equation (25.18). The assumption most likely to cause problems is what the source of the argon is in the sample. The argon may be from the potassium-40 decay, it may be atmospheric argon, or it may be a mixture in which it is possible to identify and subtract any atmospheric argon-40 that contaminates the sample.

25-6 Energetics of Nuclear Reactions

The energy change accompanying a nuclear reaction can be described by using the mass–energy equivalence derived by Albert Einstein:

KEEP IN MIND

that the mass in equation (25.19) is the rest mass of the particle, as discussed on page 310.

$$E = mc^2 \quad (25.19)$$

An energy change in a process is always accompanied by a mass change, and the constant that relates them is the square of the speed of light. In chemical reactions energy changes are so small that the equivalent mass changes are undetectable (though real nevertheless). In fact, we base the balancing of equations and stoichiometric calculations on the principle that mass is conserved (unchanged) in a chemical reaction. In nuclear reactions, energies are orders of magnitude greater than in chemical reactions. Perceptible changes in mass do occur.

If the exact masses of atoms are known, the energy of a nuclear reaction can be calculated with equation (25.19). The term m is the net change in mass, in kilograms, and c is expressed in meters per second. The resulting energy is in joules. Another common unit for expressing nuclear energy is the megaelectronvolt (MeV). (Recall from page 1172 that 1 eV = 1.6022×10^{-19} J.)

$$1\ \text{MeV} = 1.6022 \times 10^{-13}\ \text{J} \quad (25.20)$$

Equation (25.20) is a conversion factor between megaelectronvolts and joules. A conversion factor between atomic mass units (u) and joules (J) is also helpful. This relationship can be established by determining the energy associated with a mass of 1 u. This calculation is based on carbon-12, and note that 1 u is exactly $\frac{1}{12}$ of the mass of a carbon-12 atom. The mass, in grams, corresponding to 1 u can be calculated as follows:

$$1\ \text{u} \times \frac{1\ ^{12}\text{C atom}}{12\ \text{u}} \times \frac{1\ \text{mol}\ ^{12}\text{C}}{6.0221 \times 10^{23}\ \text{atoms}\ ^{12}\text{C}} \times \frac{12\ \text{g}}{1\ \text{mol}\ ^{12}\text{C}} = 1.6606 \times 10^{-24}\ \text{g}$$

Converting this value of m to kilograms and using it in equation (25.19) gives

$$E = 1\ \text{u} \times \frac{1.6606 \times 10^{-24}\ \text{g}}{\text{u}} \times \frac{1\ \text{kg}}{1000\ \text{g}} \times (2.9979 \times 10^8)^2 \frac{\text{m}^2}{\text{s}^2}$$
$$= 1.4924 \times 10^{-10}\ \text{J}$$

Thus, the energy equivalent of 1 u is

$$1\ \text{atomic mass unit (u)} = 1.4924 \times 10^{-10}\ \text{J} \quad (25.21)$$

Finally, to express this energy in MeV,

$$1\ \text{atomic mass unit (u)} = 1.4924 \times 10^{-10}\ \text{J} \times \frac{1\ \text{MeV}}{1.6022 \times 10^{-13}\ \text{J}} = 931.5\ \text{MeV} \quad (25.22)$$

hese conversion factors are used in Example 25-5, together with the principle hat the total mass–energy of the products of a nuclear reaction is equal to the otal mass–energy of the reactants. The masses required in calculations based n nuclear reactions are *nuclear* masses. The relationship of a nuclear mass to a uclidic (atomic) mass is

$$\text{nuclear mass} = \text{nuclidic (atomic) mass} - \text{mass of extranuclear electrons}$$

EXAMPLE 25-5 Calculating the Energy of a Nuclear Reaction with the Mass–Energy Relationship

What is the energy, in joules and in megaelectronvolts, associated with the α decay of ^{238}U?

$$^{238}_{92}\text{U} \longrightarrow {}^{234}_{90}\text{Th} + {}^{4}_{2}\text{He}$$

The nuclidic (atomic) masses in atomic mass units (u) are from Table D.5 in Appendix D:

$$^{238}_{92}\text{U} = 238.0508\ \text{u} \qquad ^{234}_{90}\text{Th} = 234.0437\ \text{u} \qquad ^{4}_{2}\text{He} = 4.0026\ \text{u}$$

Analyze

The key concept here is the fact that, during a nuclear reaction, a loss or gain of mass is balanced by a gain or loss of energy. We need to determine the loss or gain of mass and then use the conversion factors (25.21) and (25.22) to convert the loss or gain of mass to the corresponding amount of energy.

Solve

The net change in mass that accompanies the decay of a single nucleus of ^{238}U is shown below. Note that the masses of the extranuclear electrons do not enter into the calculation of the net change in mass.

$$\begin{aligned}\text{change in mass}&\\ &= \text{nuclear mass of } {}^{234}_{90}\text{Th} + \text{nuclear mass of } {}^{4}_{2}\text{He} - \text{nuclear mass of } {}^{238}_{92}\text{U}\\ &= [234.0437\ \text{u} - (90 \times \text{mass e}^-)] + [4.0026\ \text{u} - (2 \times \text{mass e}^-)] - [238.0508\ \text{u} - (92 \times \text{mass e}^-)]\\ &= 234.0437\ \text{u} + 4.0026\ \text{u} - 238.0508\ \text{u} - 92 \times \text{mass e}^- + 92 \times \text{mass e}^-\\ &= -0.0045\ \text{u}\end{aligned}$$

We can use this loss of mass and conversion factors (25.21) and (25.22) to write

$$E = -0.0045\ \text{u} \times \frac{1.49 \times 10^{-10}\ \text{J}}{\text{u}} = -6.7 \times 10^{-13}\ \text{J}$$

or

$$E = -0.0045\ \text{u} \times \left(\frac{931.5\ \text{MeV}}{\text{u}}\right) = -4.2\ \text{MeV}$$

Assess

The negative sign denotes that energy is lost in the nuclear reaction. This is the kinetic energy of the departing α particle. Note that you can use either nuclear or nuclidic (atomic) masses in calculations based on a nuclear equation. The change in mass will be the same in either case because the masses of the electrons will cancel out.

PRACTICE EXAMPLE A: What is the energy associated with the α decay of ^{146}Sm (145.913053 u) to ^{142}Nd (141.907719 u)? Use 4.002603 u as the mass of ^4He.

PRACTICE EXAMPLE B: The decay of ^{222}Rn by α-particle emission is accompanied by a loss of 5.590 MeV of energy. What quantity of mass, in atomic mass units (u), is converted to energy in this process?

Figure 25-5 suggests formation of the nucleus of a ^4_2He atom from two protons and two neutrons. In this process, there is a *mass defect* of 0.0305 u. That is, the experimentally determined mass of a ^4_2He nucleus is 0.0305 u *less* than the combined mass of two protons and two neutrons. This "lost" mass is liberated as energy. With expression (25.22), we can show that 0.0305 u of mass is equivalent to an energy of 28.4 MeV. Because this is the energy released in forming a ^4_2He nucleus, it is called the **nuclear binding energy**. Viewed another way, a ^4_2He nucleus would have to absorb 28.4 MeV to cause its protons and neutrons

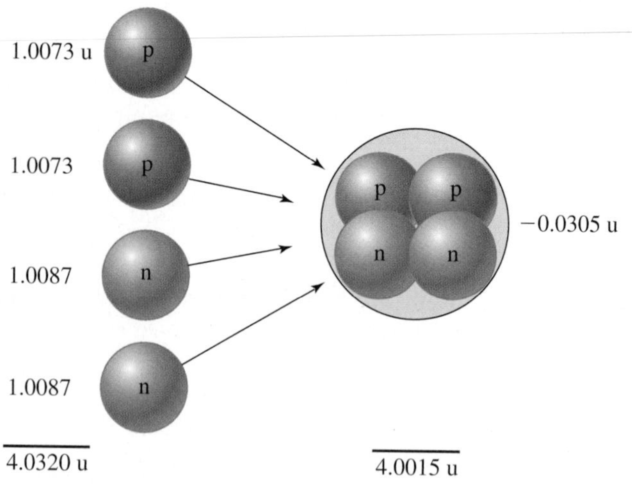

▲ FIGURE 25-5
Nuclear binding energy in ^4_2He
The mass of a helium nucleus (^4_2He) is 0.0305 u (atomic mass unit) less than the combined masses of two protons and two neutrons. The energy equivalent to this loss of mass (called the mass defect) is the nuclear energy that binds the nuclear particles together.

▶ Only four forces are known: gravity, electromagnetic forces, weak nuclear force, and strong nuclear force. The nuclear forces act only within a nucleus, whereas gravity and EM can operate over essentially infinite distances.

▶ Nucleons are nuclear particles: protons and neutrons.

to become separated. If the binding energy is considered to be apportioned equally among the two protons and two neutrons in ^4_2He, the binding energy is 7.10 MeV per nucleon.

Let us calculate the binding energy per nucleon on the following elements carbon-12, iron-56, and uranium-235. For carbon-12 we first calculate the change in mass:

$$\begin{aligned}\text{change in mass} &= [(6 \times \text{mass } ^1\text{H atom}) + (6 \times \text{mass neutron})] - \text{mass } ^{12}\text{C atom} \\ &= [6 \times 1.007825 \text{ u} + 6 \times 1.008665 \text{ u}] - 12.0000 \text{ u} \\ &= 0.09894 \text{ u}\end{aligned}$$

The mass for the ^1H atom and neutron were obtained from Appendix D. The next step is to convert this to the binding energy per nucleon:

$$\text{binding energy per nucleon} = \frac{0.09894 \text{ u} \times \dfrac{931.5 \text{ MeV}}{\text{u}}}{12 \text{ nucleons}} = 7.680 \text{ MeV/nucleon}$$

For iron-56 we find the binding energy per nucleon is

$$\text{binding energy per nucleon} = \frac{0.52846 \text{ u} \times \dfrac{931.5 \text{ MeV}}{\text{u}}}{56 \text{ nucleons}} = 8.790 \text{ MeV/nucleon}$$

and for uranium-235

$$\text{binding energy per nucleon} = \frac{1.915072 \text{ u} \times \dfrac{931.5 \text{ MeV}}{\text{u}}}{235 \text{ nucleons}} = 7.591 \text{ MeV/nucleon}$$

At this point we notice that iron-56 has the larger binding energy per nucleon, meaning that more energy is required to separate the iron-56 nucleons than those of carbon-12 and uranium-235. In other words, iron-56 is more stable than the other two elements.

Figure 25-6 indicates that the maximum binding energy per nucleon is found in a nucleus with a mass number of approximately 60. This finding leads to two interesting conclusions: (1) If small nuclei are combined into a heavier one (up to about $A = 60$), the binding energy per nucleon increases and a

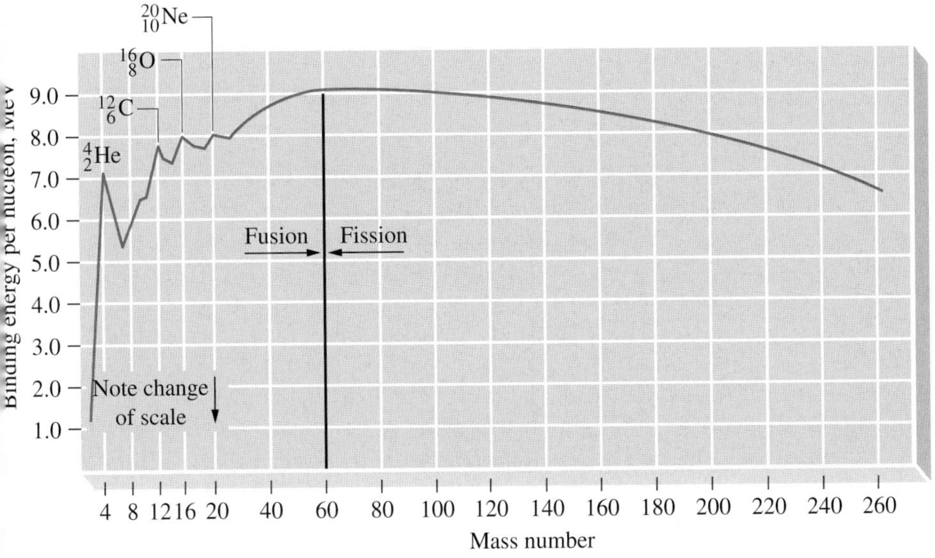

▲ FIGURE 25-6
Average binding energy per nucleon as a function of mass number
The average binding energy per nucleon is a measure of nuclear stability. An increase in the average binding energy per nucleon is achieved by combining lighter nuclides (fusion) to obtain a nuclide with a mass number less than 60 or by splitting heavier nuclides (fission) to obtain a nuclide with a mass number greater than 60.

certain quantity of mass must be converted to energy. The nuclear reaction is highly exothermic. This *fusion* process serves as the basis of the hydrogen bomb. (2) For nuclei with mass numbers above 60, the addition of extra nucleons to the nucleus would require the expenditure of energy (since the binding energy per nucleon decreases). However, the *disintegration* of heavier nuclei into lighter ones is accompanied by the release of energy. This nuclear *fission* process serves as the basis of the atomic bomb and conventional nuclear power reactors. Nuclear fission and fusion will be discussed after we examine Figure 25-6 and consider the question of nuclear stability.

25-7 Nuclear Stability

A number of basic questions have probably occurred to you as nuclear decay processes have been described: Why do some radioactive nuclei decay by α emission, some by β^- emission, and so on? Why do the lighter elements have so few naturally occurring radioactive nuclides, whereas all those of the heavier elements seem to be radioactive?

Our first clue to answers for such questions comes from Figure 25-6, in which several nuclides are specifically noted. These nuclides have higher binding energies per nucleon than those of their neighbors. Their nuclei are especially stable. This observation is consistent with a theory of nuclear structure known as the *shell theory*. In the formation of a nucleus, protons and neutrons are believed to occupy a series of nuclear shells. This process is analogous to building up the electronic structure of an atom by the successive addition of electrons to electronic shells. Just as the aufbau process periodically produces electron configurations of exceptional stability, so do certain nuclei acquire a special stability as nuclear shells are closed. This condition of special stability of an atomic nucleus occurs for certain numbers of protons or neutrons known as **magic numbers** (Table 25.2).

Another observation concerning nuclei is that among stable nuclei, the number of protons and the number of neutrons is most commonly *even*. There are fewer stable nuclei with odd numbers of protons and of neutrons.

TABLE 25.2 Magic Numbers for Nuclear Stability

Number of Protons	Number of Neutrons
2	2
8	8
20	20
28	28
50	50
82	82
114	126
	184

TABLE 25.3 Distribution of Naturally Occurring Stable Nuclides

Combination	Number of Nuclides
Z even–N even	163
Z even–N odd	55
Z odd–N even	50
Z odd–N odd	4

The relationship between numbers of protons (Z), numbers of neutrons (N), and the stability of isotopes is summarized in Table 25.3. Note particularly that stable atoms with the combination Z odd–N odd are very rare. This combination is found only in the nuclides $^{2}_{1}H$, $^{6}_{3}Li$, $^{10}_{5}B$, and $^{14}_{7}N$. Still another observation is that elements of *odd* atomic number generally have only one or two stable isotopes, whereas those of *even* atomic number have several. Thus, F ($Z = 9$) and I ($Z = 53$) each have only one stable nuclide, and Cl ($Z = 17$) and Cu ($Z = 29$) each have two. On the other hand, O ($Z = 8$) has three and Ca ($Z = 20$) has six.

Neutrons are thought to provide a nuclear force to bind protons and neutrons together into a stable unit. Without neutrons, the electrostatic forces of repulsion between positively charged protons would cause the nucleus to fly apart. For the elements of lower atomic numbers (up to about $Z = 20$), the required number of neutrons for a stable nucleus is about equal to the number of protons, for example, $^{4}_{2}He$, $^{12}_{6}C$, $^{16}_{8}O$, $^{28}_{14}Si$, $^{40}_{20}Ca$. For higher atomic numbers, because of increasing repulsive forces between protons, larger numbers of neutrons are required and the neutron-to-proton (N/Z) ratio increases. For bismuth, the ratio is about 1.5 : 1. Above atomic number 83, no matter how many neutrons are present, the nucleus is unstable. Thus, all isotopes of the known elements with $Z > 83$ are radioactive. Figure 25-7 indicates roughly the range of N/Z ratios as a function of atomic number for stable atoms.

Using the ideas outlined here, nuclear scientists have predicted the possible existence of atoms of high atomic number that should have very long half-lives, a belt of stability in Figure 25-7. After a search of many years, such atoms have been created. In 1999, the bombardment of a plutonium-242 target with calcium-48 ions produced the isotopes ^{287}Fl and ^{289}Fl, with half-lives of 0.5 s and 3.0 s, respectively. Although these half-lives may appear short, they are practically an eternity compared with the half-lives of other superheavy atoms, which are in the microsecond range (Table 25.4).

25-4 CONCEPT ASSESSMENT

What proton number is found in a greater number of isotopes than any other proton number? What is the corresponding situation for neutron numbers?

TABLE 25.4 Longest Lived Isotopes of Elements 100–118

Number	Name	Longest-Lived Isotope	Half-Life
100	Fermium	^{257}Fm	100.5 days
101	Mendelevium	^{258}Md	51.5 days
102	Nobelium	^{259}No	58 minutes
103	Lawrencium	^{262}Lr	3.6 hours
104	Rutherfordium	^{266}Rf	~10 hours
105	Dubnium	^{268}Db	26 hours
106	Seaborgium	^{269}Sg	~2.1 minutes
107	Bohrium	^{270}Bh	61 seconds
108	Hassium	^{277m}Hs	~12 minutes
109	Meitnerium	^{278}Mt	~5 seconds
110	Darmstadtium	^{281m}Ds	~3.7 minutes
111	Roentgenium	^{281}Rg	22 seconds
112	Copernium	^{285m}Cn	~8.9 minutes
113	Ununtrium	^{286}Uut	~0.9 seconds
114	Flerovium	^{288}Fl	~0.8 seconds
115	Ununpentium	^{288}Uup	173 milliseconds
116	Livermorium	^{293}Lv	80 milliseconds
117	Ununseptium	^{294}Uus	~50 milliseconds
118	Ununoctium	^{294}Uuo	0.7 milliseconds

Note: The superscript m indicates that the half-life is for an excited state of the isotope. This excited state is sometimes referred to as a metastable state or isomer of the nuclide.

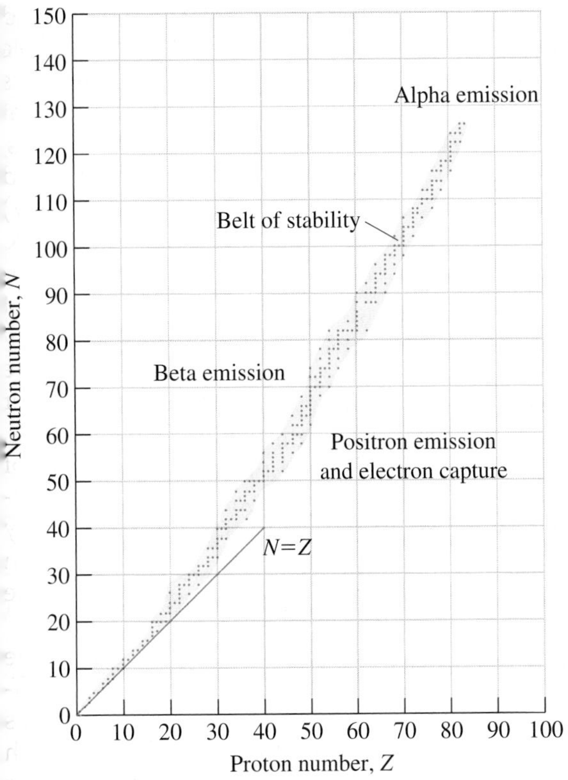

◀ FIGURE 25-7
Neutron-to-proton ratio and the stable nuclides up to $Z = 83$
The points within the belt of stability identify the stable nuclides. Some radioactive nuclides are also in this belt, but most lie outside, and the mode of their radioactive decay is indicated. The stable nuclides of low atomic numbers lie on or near the line $N = Z$; they have a neutron-to-proton ratio of one, or nearly so. At higher atomic numbers, the neutron-to-proton ratios increase to about 1.5.

EXAMPLE 25-6 Predicting Which Nuclei Are Radioactive

Which of the following nuclides would you expect to be stable, and which radioactive? **(a)** ^{82}As; **(b)** ^{118}Sn; **(c)** ^{214}Po.

Analyze

This question requires us to apply the rules of nuclear stability. We may also use Figure 25-7 to determine whether the nuclide lies within the belt of stability.

Solve

(a) Arsenic-82 has $Z = 33$ and $N = 49$. This is an odd–odd combination that is found in only four of the lighter elements. ^{82}As is radioactive. (Note also that this nuclide is outside the belt of stability in Figure 25-7.)

(b) Tin has an atomic number of 50—a magic number. The neutron number is 68 in the nuclide ^{118}Sn. This is an even–even combination, and we should expect the nucleus to be stable. Moreover, Figure 25-7 shows that this nuclide is within the belt of stability. ^{118}Sn is a stable nuclide.

(c) ^{214}Po has an atomic number of 84. All known atoms with $Z > 83$ are radioactive. ^{214}Po is radioactive.

Assess

When considering whether a nuclide is radioactive, we must remember that all elements with $Z > 83$ are radioactive, and those with $Z < 84$ are radioactive if their Z and N form odd–odd combinations. A few exceptions exist to these rules, specifically when nuclides have atomic numbers that are one of the magic numbers.

PRACTICE EXAMPLE A: Which of the following nuclides would you expect to be stable, and which radioactive? **(a)** ^{88}Sr; **(b)** ^{118}Cs; **(c)** ^{30}S.

PRACTICE EXAMPLE B: Write plausible nuclear equations to represent the radioactive decay of the fluorine isotopes ^{17}F and ^{22}F.

25-5 CONCEPT ASSESSMENT

Of the radioactive nuclides among the following, which one is most likely to decay by β^- emission, and which one by β^+ emission: ^{44}Ca, ^{57}Cu, ^{100}Zr, ^{235}U? Explain.

▶ Ida Tacke Nodack, codiscoverer of the element rhenium, was the first person to suggest that Fermi's proposed experiments had produced fission. Her explanation was not generally accepted, however, until several years later.

25-8 Nuclear Fission

In 1934, Enrico Fermi proposed that transuranium elements might be produced by bombarding uranium with neutrons. He reasoned that the successive loss of β^- particles would cause the atomic number to increase, perhaps to as high as 96. When such experiments were carried out, it was found that, in fact, the product did emit β^- particles. But in 1938, Otto Hahn, Lise Meitner, and Fritz Strassman found by chemical analysis that the products did not correspond to elements with $Z > 92$. Neither were they the neighboring elements of uranium—Ra, Ac, Th, and Pa. Instead, the products were radioisotopes of much lighter elements, such as Sr and Ba. Neutron bombardment of uranium nuclei causes certain of them to undergo **fission** into smaller fragments, as suggested by Figure 25-8.

The energy equivalent of the mass destroyed in a fission event is somewhat variable, but the average energy is approximately 3.20×10^{-11} J (200 MeV).

$$^{235}_{92}\text{U} + ^{1}_{0}\text{n} \longrightarrow ^{236}_{92}\text{U} \longrightarrow \text{fission fragments} + \text{neutrons} + 3.20 \times 10^{-11}\ \text{J}$$

An energy of 3.20×10^{-11} J may seem small, but this energy is for the fission of a *single* $^{235}_{92}U$ nucleus. What if 1.00 g $^{235}_{92}U$ were to undergo fission?

$$? \text{kJ} = 1.00\ \text{g}\ ^{235}\text{U} \times \frac{1\ \text{mol}\ ^{235}\text{U}}{235\ \text{g}\ ^{235}\text{U}} \times \frac{6.022 \times 10^{23}\ \text{atoms}\ ^{235}\text{U}}{1\ \text{mol}\ ^{235}\text{U}} \times \frac{3.20 \times 10^{-11}\ \text{J}}{1\ \text{atoms}\ ^{235}\text{U}}$$
$$= 8.20 \times 10^{10}\ \text{J} = 8.20 \times 10^{7}\ \text{kJ}$$

This is an enormous quantity of energy! To release that quantity of energy would require the complete combustion of nearly three tons of coal.

Nuclear Reactors

In the fission of $^{235}_{92}U$, on average, 2.5 neutrons are released per fission event. These neutrons, on average, produce two or more fission events. The neutrons produced by the second round of fission produce another four or five events, and so on. The result is a *chain reaction*. If the reaction is uncontrolled, the released energy causes an explosion; this is the basis of the atomic bomb. Fission leading to an uncontrolled explosion occurs only if the quantity of

Dobresum/Shutterstock

(a)

Victor Korotayev/Reuters/Corbis

(b)

Kyodo/Reuters/Corbis

(c)

▲ **(a)** The Three Mile Island nuclear reactor, site of a small nuclear accident in 1979. **(b)** The Chernobyl nuclear reactor, site of a major nuclear accident in 1986. **(c)** Fukushima nuclear power plant in Japan before the nuclear accident in 2011.

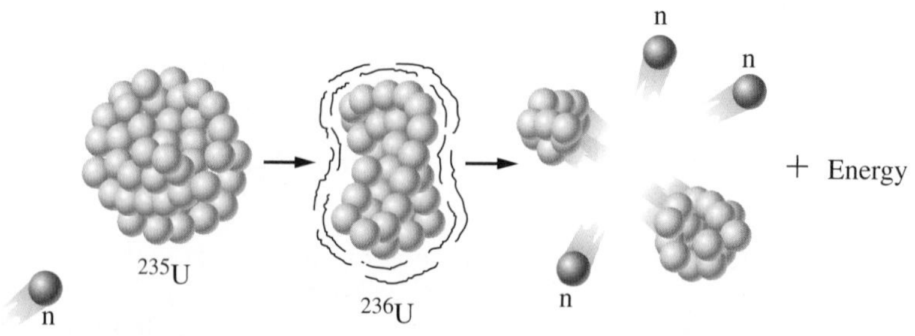

▲ FIGURE 25-8
Nuclear fission of $^{235}_{92}U$ with thermal neutrons
A neutron possessing ordinary thermal energy strikes a $^{235}_{92}U$ nucleus. First, the unstable nucleus $^{236}_{92}U$ is produced; this then breaks up into a light fragment, a heavy fragment, and several neutrons. Various nuclear fragments are possible, but the most probable mass numbers are 97 for the light fragment and 137 for the heavy one.

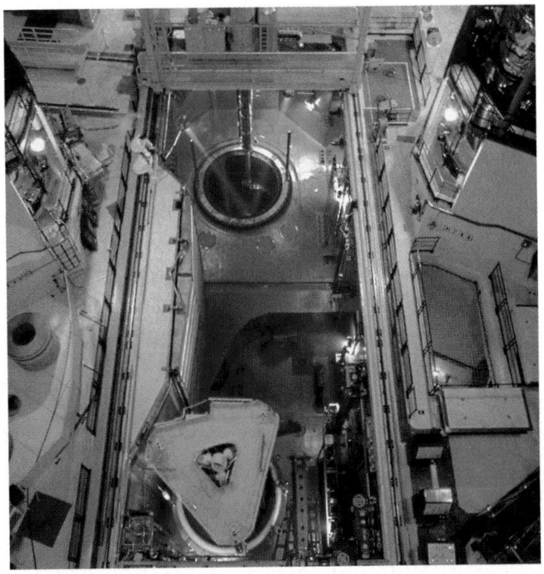

Yann Arthus-Bertrand/Eureka/Corbis

▶ **The core of a nuclear reactor** The characteristic blue glow in the water surrounding the core is called *Cerenkov radiation*. It results when charged particles pass through a transparent medium faster than does light in the same medium. The charged particles are produced by nuclear fission. The radiation is analogous to the shock wave produced in a sonic boom.

^{235}U exceeds the critical mass. The *critical mass* is the quantity of ^{235}U sufficiently large to retain enough neutrons to sustain a chain reaction. Quantities smaller than this are subcritical; neutrons escape at too great a rate to produce a chain reaction.

In a nuclear reactor, the release of fission energy is controlled. One common design, called the *pressurized water reactor* (PWR), is pictured in Figure 25-9. In the core of the reactor, rods of uranium-rich fuel are suspended in water maintained under a pressure of 70 to 150 atm. The water serves a dual purpose. First, it slows down the neutrons from the fission process so that they possess only normal thermal energy. These thermal neutrons are better able to induce fission than highly energetic ones. In this capacity, the water acts as a **moderator**. Water also functions as a heat transfer medium. Fission energy maintains the water at a high temperature (about 300 °C). The high-temperature water is brought in contact with colder water in a heat exchanger. The colder water is converted to steam, which drives a turbine, which in turn drives an electric generator. A final

◀ The water used in a pressurized water reactor is heavy water. The chemical formula for heavy water is $^{2}H_2O$ or D_2O, where D represents the deuterium atom, ^{2}H.

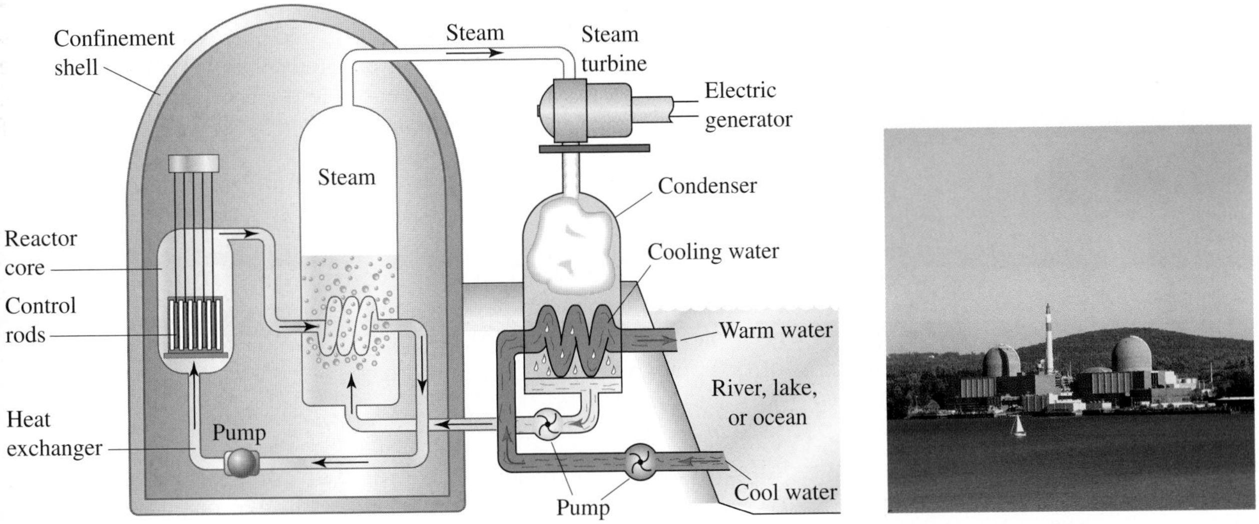

(a) (b)

Alex Bartel/Science Photo Library

▲ FIGURE 25-9
Pressurized water nuclear reactor
(a) Schematic of a nuclear power plant. **(b)** A nuclear power plant in New York State.

25-2 ARE YOU WONDERING?

Are nuclear reactors safe?

There have been three notorious accidents at nuclear power stations. The first occurred at the Three Mile Island (TMI) generating plant near Middleton, Pennsylvania, in 1979. The TMI reactor is of the light-water type, in which water is used as the moderator and coolant. In this accident, some coolant (also the moderator) was lost; the chain reaction in the reactor stopped because there were too few slow neutrons. However, radioactive decay of the fission fragments continued, causing the fuel rods to get very hot. A partial meltdown resulted and, in turn, caused a fracture in one of the reactors. The fracture permitted the venting of a small amount of radioactive steam into the atmosphere. The reactor is now sealed, but electronic robots have discovered substantial damage to the fuel rods.

The second incident occurred at Chernobyl, in Ukraine, in 1986. Graphite was used as the moderator there. When the coolant was turned off because of human error, the chain reaction went out of control. A tremendous rise in temperature followed, leading to a meltdown. During the meltdown, the graphite moderator surrounding the rods burned and radioactive smoke spewed out of the reactor. Radioactive materials were dispersed over much of Europe, Canada, and the United States. Although only a few dozen people were killed in the Chernobyl accident, many more will eventually die of cancer because of the related radiation. The type of accident observed at Chernobyl cannot happen in a light-water reactor, in which the coolant is the moderator.

A third incident occurred on 11 March 2011 in the Japanese Fukushima prefecture when a 9.0-magnitude earthquake generated a 15-meter tsunami. The boiling water reactors survived the earthquake; however, the tsunami caused significant damage to backup generators, which prevented proper cooling and water recirculation functions. Without these functions, four reactors overheated, the end result of which was several explosions that leaked radioactive material. The main radionuclides released from this incident were I-131, Cs-134, and Cs-137. It has been estimated that 970 PBq (I-131 equivalent) was released into the atmosphere and water. The Japanese government shut down the four reactors and either capped or, as of 2015, is in the process of encapsulating them. To prevent further spread of the radionuclides, the Japanese government has sprayed a dust-suppressing polymer resin over the ground near the reactors, purified seawater with zeolites, and coated an 18 hectare area of seabed in the plant's port with cement and bentonite. There were no deaths resulting directly from this incident. Decontamination of the region was continuing in 2015, with as many as 81,000 people still displaced from their homes. Despite the severity of these incidents, the safety record of nuclear reactors still remains particularly high, but the need for the storage of the nuclear waste produced is a vexing problem. Ways of dealing with these problems are discussed in the Focus On feature for Chapter 25 on the Mastering Chemistry website.

component of the nuclear reactor is a set of **control rods**, usually cadmium metal, whose function is to absorb neutrons. When the rods are lowered into the reactor, the fission process is slowed down. When the rods are raised, the density of neutrons and the rate of fission increase.

Breeder Reactors

All that is required to initiate the fission of ${}^{235}_{92}\mathrm{U}$ are neutrons of ordinary thermal energies. Conversely, nuclei of ${}^{238}_{92}\mathrm{U}$, the abundant nuclide of uranium (99.28%), undergo the following reactions only when struck by energetic neutrons.

$$ {}^{238}_{92}\mathrm{U} + {}^{1}_{0}\mathrm{n} \longrightarrow {}^{239}_{92}\mathrm{U} $$
$$ {}^{239}_{92}\mathrm{U} \longrightarrow {}^{239}_{93}\mathrm{Np} + {}^{0}_{-1}\beta $$
$$ {}^{239}_{93}\mathrm{Np} \longrightarrow {}^{239}_{94}\mathrm{Pu} + {}^{0}_{-1}\beta $$

A fissionable nuclide such as $^{235}_{92}U$ is called *fissile*; $^{239}_{94}Pu$ is also fissile. A nuclide such as $^{238}_{92}U$, which can be converted into a fissile nuclide, is said to be *fertile*. In a *breeder nuclear reactor*, a small quantity of fissile nuclide provides the neutrons that convert a large quantity of a fertile nuclide into a fissile one. (The newly formed fissile nuclide then participates in a self-sustaining chain reaction.)

An obvious advantage of the breeder reactor is that the amount of uranium fuel available immediately jumps by a factor of about 100. This is the ratio of naturally occurring $^{238}_{92}U$ to $^{235}_{92}U$. But the potential advantage is even greater than this. Breeder reactors might use as nuclear fuels materials that have even very low uranium contents, such as shale deposits with about 0.006% U by mass.

There are, however, important disadvantages to breeder reactors. This is especially true of the type known as the *liquid-metal-cooled fast breeder reactor* (LMFBR). Systems must be designed to handle a liquid metal, such as sodium, which becomes highly radioactive in the reactor. Also, both the rates of heat and neutron production are greater in the LMFBR than in the PWR, so materials deteriorate more rapidly. Perhaps the greatest unsolved problems are those of handling radioactive wastes and reprocessing plutonium fuel. Plutonium is one of the most toxic substances known. It can cause lung cancer when inhaled in even microgram (10^{-6} g) amounts. Furthermore, because plutonium has a long half-life (24,000 a), any accident involving it could leave an affected area almost permanently contaminated.

25-9 Nuclear Fusion

The **fusion** of atomic nuclei is the process that produces energy in the sun. An uncontrolled fusion reaction is the basis of the hydrogen bomb. A controlled fusion reaction could provide an almost unlimited source of energy. The nuclear reaction that holds the most immediate promise is the deuterium–tritium reaction.

$$^{2}_{1}H + ^{3}_{1}H \longrightarrow ^{4}_{2}He + ^{1}_{0}n + \text{energy}$$

The difficulties in developing a fusion energy source are probably without parallel in the history of technology. In fact, the feasibility of a controlled fusion reaction has yet to be fully demonstrated. There are a number of problems. To permit their fusion, the nuclei of deuterium and tritium must be forced into close proximity. Because atomic nuclei repel one another, this close approach requires the nuclei to have very high thermal energies. At the temperatures necessary to initiate a fusion reaction, gases are completely ionized into a mixture of atomic nuclei and electrons known as a *plasma*. Still higher plasma temperatures—more than 40,000,000 K—are required to initiate a self-sustaining reaction (one that releases more energy than is required to get it started). A method must be devised to confine the plasma out of contact with other materials. The plasma loses thermal energy to any material it strikes. Also, a plasma must be at a sufficiently high density for a sufficient time to permit the fusion reaction to occur. The two methods receiving greatest attention are confinement in a magnetic field and heating of a frozen deuterium-tritium pellet with laser beams. Another series of technical problems involves the handling of liquid lithium, which is the anticipated heat-transfer medium and tritium ($^{3}_{1}H$) source.

◀ In the hydrogen bomb, these high temperatures are attained by exploding an atomic (fission) bomb, which triggers the fusion reaction.

KEEP IN MIND

that *both* nuclear fusion and nuclear fission reactions release energy.

$$^{7}_{3}Li + \underset{\text{(fast)}}{^{1}_{0}n} \longrightarrow ^{4}_{2}He + ^{3}_{1}H + \underset{\text{(slow)}}{^{1}_{0}n}$$

Finally, for the magnetic containment method, the magnetic field must be produced by superconducting magnets, which currently are very expensive to operate.

EFDA-JET/Science Photo Library

▲ The plasma chamber of a fusion reactor of the magnetic confinement type (called a *tokamak*). The chamber walls are lined with carbon-fiber composite tiles to protect against the high-temperature plasma.

A fusion reactor is currently being developed in France by an internationa consortium consisting of the European Union (EU), India, Japan, People' Republic of China, Russia, South Korea, the United States, and Portugal. Th project, titled ITER (International Thermonuclear Experimental Reactor), i based on the tokamak design shown in the photo in the margin. The goal of th project is to generate and sustain 500 MW of fusion power for 1000 second through the fusion of 0.5 g of a deuterium–tritium mixture. ITER is expected t generate 5 to 10 times as much energy as is needed to initiate the reaction at it fusion temperature. The first operation of the fusion reactor should take plac in 2018.

The advantages of fusion over fission could be enormous. Since deuteriun constitutes about one in every 6500 H atoms, the oceans of the world can sup ply an almost limitless amount of nuclear fuel. It is estimated that there is suf ficient lithium on Earth to provide a source of tritium for about 1 millior years. Also, nuclear fusion would not pose the vexing problems of radioactiv waste storage and disposal associated with nuclear fission.

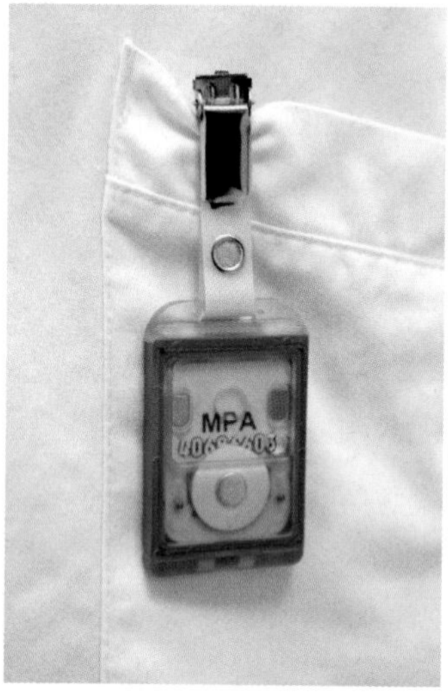

Jürgen Schulzki / Alamy

▲ A film badge used for detecting radiation.

25-10 Effect of Radiation on Matter

Although there are substantial differences in the way in which α particles β^- and β^+ particles, and γ rays interact with matter, they share an importan feature: They dislodge electrons from atoms and molecules to produce ions The ionizing power of radiation may be described in terms of the number o ion pairs formed per centimeter of path through a material. An *ion pair* consists of an ionized electron and the resulting positive ion. Alpha particles have the greatest ionizing power, followed by β particles and then γ rays. The ionized electrons produced directly by the collisions of particles of radiation with atoms are called *primary* electrons. These electrons may themselves possess sufficient energies to cause *secondary* ionizations.

Not all interactions between radiation and matter produce ion pairs. In some cases, electrons are simply raised to higher atomic or molecular energy levels. The return of these electrons to their normal states is then accompanied by radiation—X-rays, ultraviolet light, or visible light, depending on the energies involved.

Some of the possibilities described here are depicted in Figure 25-10.

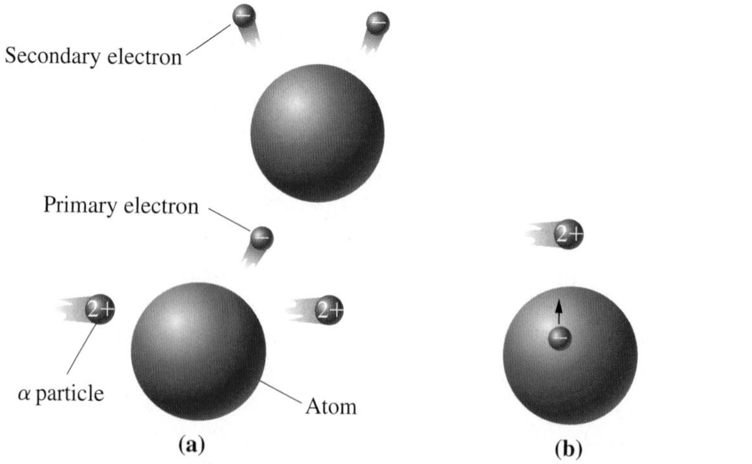

▲ FIGURE 25-10
Some interactions of radiation with matter
(a) The production of primary and secondary electrons by collisions. **(b)** The excitation of an atom by the passage of an α particle. An electron is raised to a higher energy level within the atom. The excited atom reverts to its normal state by emitting radiation.

Radiation Detectors

The interactions of radiation with matter can serve as bases for the detection of radiation and the measurement of its intensity. One of the simplest methods is that used by Henri Becquerel in his discovery of radioactivity—the exposure of a photographic film, as in film badge radiation detectors. The effect of α and β particles, and γ rays on a photographic emulsion is similar to that of X-rays.

One type of detector used to study high-energy radiation such as γ rays is the *bubble chamber*. In this device, a liquid, usually hydrogen, is kept just at its boiling point. As ion pairs are produced by the transit of an ionizing ray, bubbles of vapor form around the ions. The tracks of bubbles can be photographed and analyzed, and different types of radiation produce different tracks. Charged particles, for example, can be detected by their deflection in a magnetic field.

The most common device for detecting and measuring ionizing radiation is the *Geiger–Müller* (G–M) *counter*, depicted in Figure 25-11. The G–M counter consists of a cylindrical cathode with a wire anode running along its axis. The anode and cathode are sealed in a gas-filled glass tube. Ionizing radiation passing through the tube produces primary ions, and this is followed by secondary ionization. The positive ions are attracted to the cathode and the electrons to the anode, leading to a pulse of electric current. The tube is quickly recharged in preparation of the next ionizing event. The pulses of electric current are counted.

A detector widely used in biological studies is the *scintillation counter*. It is especially useful in detecting radiation that is not energetic enough to cause ionization. The radiation excites certain atoms in the detecting medium; when these atoms revert to their ground state, they emit pulses of light that can be counted. The light emission is similar to that produced when phosphors in a television screen are struck by cathode rays.

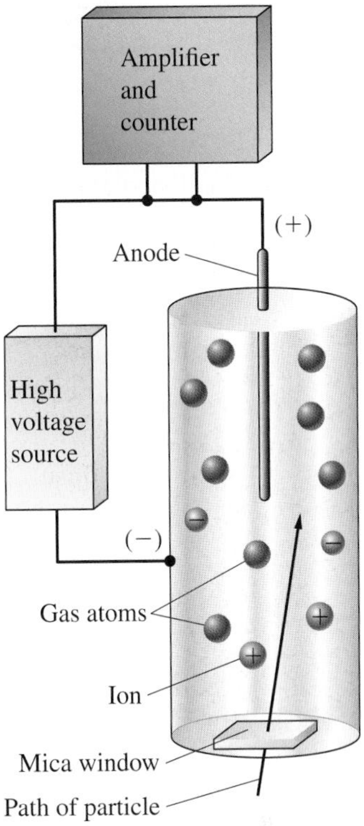

▲ FIGURE 25-11
A Geiger–Müller counter
Radiation enters the G–M tube through the mica window. The radiation ionizes some of the gas (usually argon) in the tube. A pulse of electric current passes through the electric circuit and is counted.

Effect of Ionizing Radiation on Living Matter

All life exists against a background of naturally occurring ionizing radiation—cosmic rays, ultraviolet light, and emanations from radioactive elements, such as uranium in rocks. The level of this radiation varies from point to point on Earth, being greater, for instance, at higher elevations. Only in recent times have humans been able to create situations in which organisms might be exposed to radiation at levels significantly higher than natural background radiation.

The interactions of radiation with living matter are the same as with other forms of matter—ionization, excitation, and dissociation of molecules. There is no question of the effect of large doses of ionizing radiation on organisms—the organisms are killed. But even slight exposures to ionizing radiation can cause changes in cell chromosomes. Thus it is believed that, even at low dosage rates, ionizing radiation can result in birth defects, leukemia, bone cancer, and other forms of cancer. The nagging question that has eluded any definitive answers is how great an increase in the incidence of birth defects and cancers might be caused by certain levels of radiation.

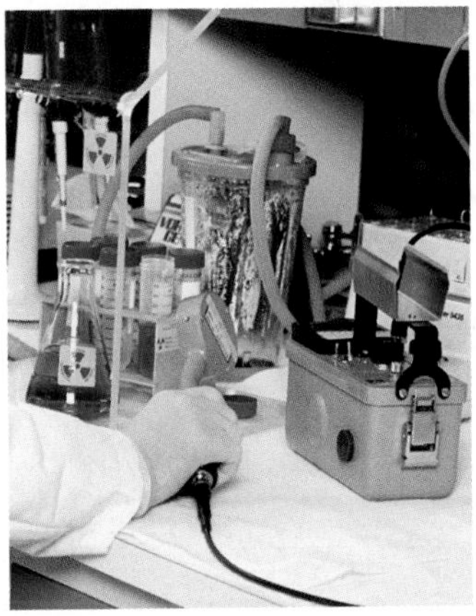

Stephen Agricola / The Image Works

▲ Laboratory technician checking for radioactive isotopes.

Radiation Dosage

One unit long used to describe exposure to radiation is the rad. One **rad** (*r*adiation *a*bsorbed *d*ose) corresponds to the absorption of 1×10^{-2} J of energy per kilogram of matter. The effect of a dose of one rad on living matter is variable, however, and a better unit is one that takes this variability into account. The **rem** (*r*adiation *e*quivalent for *m*an) is the rad multiplied by the *relative biological effectiveness* (Q). The factor Q takes into account that equal doses of radiation of different types may have differing effects. Table 25.5 summarizes several radiation units.

It is thought that a dose of 1000 rem absorbed in a short time interval will kill 100% of the population. A short-term dose of 450 rem would probably

TABLE 25.5 Radiation Units[a]

	Unit	Definition
Radioactive decay	Becquerel, Bq	s^{-1} (disintegrations per second)
	Curie, Ci	An amount of radioactive material decaying at the same rate as 1 g of radium (3.70×10^{10} dis s^{-1}) 1 Ci = 3.70×10^{10} Bq
Absorbed dose	Gray, Gy	One gray of radiation deposits one joule of energy per kilogram of matter
	Rad	1 rad = 0.01 Gy
Equivalent dose	Sievert, Sv	1 Sv = 100 rem
	Rem	1 rem = 1 rad × *Q* The quality factor, *Q*, is about 1 for X-rays, γ rays, and β^- particles; 3 for slow neutrons; 10 for protons and fast neutrons; and 20 for α particles

[a]SI units are shown in blue. Sources of α radiation are relatively harmless when external to the body and extremely hazardous when taken internally, as in the lungs or stomach. Other forms of radiation (X-rays, γ rays), because they are highly penetrating, are hazardous even when external to the body.

result in death within 30 days of about 50% of the population. A single dose of 1 rem delivered to 1 million people would probably produce about 100 cases of cancer within 20–30 years. The total body radiation received by most of the world's population from normal background sources is about 0.13 rem [130 millirem (mrem)] per year. The dose delivered in a chest X-ray examination is about 20 mrem.

Some of the foregoing statements about radiation dosages and their anticipated effects are based on (1) medical histories of the survivors of the Hiroshima and Nagasaki atomic blasts, (2) the incidence of leukemia and other cancers in children whose mothers received diagnostic radiation during pregnancy, and (3) the occurrence of lung cancers among uranium miners in the United States. What does all of this tell us about a safe level of radiation exposure? One approach has been to extrapolate from these high doses to the lower doses affecting the general population. This has led the U.S. National Council on Radiation Protection and Measurements to recommend that the dosage for the general population be limited to 0.17 rem (170 mrem) per year from all sources above background level. Experts disagree, however, on how the data observed for high dosages should be extrapolated to low doses. Some experts believe that the 0.17 rem per year figure is too high. If they are right, an additional dosage of 0.17 rem per year above normal background levels might cause statistically significant increases in the incidence of birth defects and cancers.

An Environmental Issue Involving Radon

All atoms with an atomic number greater than 83 are radioactive. The nuclei of these atoms are unstable and emit α, β, and γ radiation, eventually breaking down to more stable elements with lower atomic numbers. Radon-222, a colorless, odorless gas, is produced by the loss of α particles from radium-226, which in turn results from the radioactive decay, through several steps, of uranium-238.

In December 1984, a worker at a nuclear power plant in New Jersey registered high readings on a radiation detector during a routine safety check. But the radiation he had been exposed to didn't come from within the plant—it came from his own home. This incident led to the recognition that people can be exposed to high levels of radioactivity from radon. An estimated 21,000 lung cancer deaths per year are attributable to residential radon exposure. It is also estimated that 7 million homes have radon limits in excess of the U.S. Environmental Protection Agency's reference level of 150 Bq m^{-3}. The possible harmful effects of this exposure, primarily an increased risk of lung cancer, are fairly well documented, but the topic remains one of continuing research and debate.

In some instances, the source of radon is in wastes from uranium mining or phosphate production. In most cases, it is emitted by the radioactive decay of ^{238}U present in small amounts in rocks and soils. Because radon is a gas, it readily passes through air passages in the body and is breathed in and out. The product formed when a ^{222}Rn atom gives up an α particle is the isotope polonium-218, which also emits α particles. Unlike radon, polonium is a solid. Health hazards posed by radon seem to be from ^{218}Po and other radioactive decay products becoming attached to dust particles in the air and then being breathed into the lungs.

Fortunately, indoor radon can be rather easily detected because of its radioactivity. The chief method of reducing radon levels is through improved ventilation and by venting subsoil radon to keep it from concentrating within a building. Minimizing indoor radon should be a part of building construction.

25-11 Applications of Radioisotopes

We have described both the destructive capacity of nuclear reactions and the potential of these reactions to provide new sources of energy. Less heralded but equally important are a variety of practical applications of radioactivity. We close this chapter with a brief survey of some uses of radioisotopes.

Cancer Therapy

Ionizing radiation in low doses can induce cancers, but this same radiation, particularly γ rays, is also used in the treatment of cancer. Although ionizing radiation tends to destroy all cells, cancerous cells are more easily destroyed than normal ones. Thus, a carefully directed beam of γ rays or high-energy X-rays of the appropriate dosage may be used to arrest the growth of cancerous cells. Also coming into use for some forms of cancer is radiation therapy that employs beams of protons or neutrons.

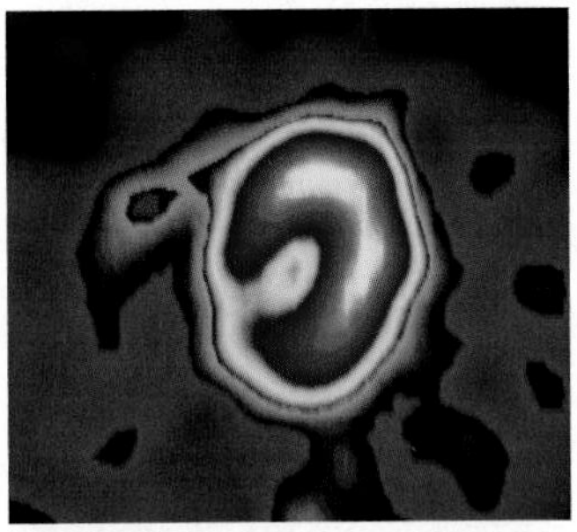

Cnri/Science Source

▲ Nuclear medicine provides methods for diagnosing potential life-threatening conditions this
In this thallium-201 scan of a normal heart, γ rays released in the radioactive decay of ^{201}Tl are detected and used to provide an image of blood flow in the heart muscle. Normal ^{201}Tl uptake by healthy heart muscle appears as a pink and red donut-shaped feature in this image. Comparing scans with those made during exercise can reveal blood supply issues.

Radioactive Tracers

The small mass differences between the isotopes of an element may produce small differences in physical properties. The differing rates of diffusion of $^{238}UF_6(g)$ and $^{235}UF_6(g)$ result from such a mass difference (page 226). Generally speaking, though, the physical and chemical properties of the various isotopes of an element are practically identical. However, if one of the isotopes is radioactive, its actions can be easily followed with radiation detectors. This is the principle behind the use of *radioactive tracers*, or tagged atoms. For example, if a small quantity of radioactive ^{32}P (as a phosphate) is added to a nutrient solution fed to plants, the uptake of all the phosphorus atoms—radioactive and nonradioactive alike—can be followed by charting the regions of the plant that become radioactive. Similarly, the fate of iodine (as iodide ion) in the body can be determined by having a person drink a solution of dissolved iodides containing a small quantity of a radioactive iodide as a tracer. Abnormalities in the thyroid gland can be detected in this way. Because I^- concentrates in the thyroid, people can protect themselves by ingesting nonradioactive iodides before being exposed to a radioactive iodide. The thyroid becomes saturated with the nonradioactive iodide and rejects the radioactive iodide. Iodide tablets were distributed to the general population in some regions of Europe before the arrival of fallout from the Chernobyl nuclear accident in 1986.

Industrial applications of tracers are also numerous. The fate of a catalyst in a chemical plant can be followed by incorporating a radioactive tracer in the catalyst, such as ^{192}Ir in a Pt-Ir catalyst. Monitoring the activity of the ^{192}Ir makes it possible to determine the rate at which the catalyst is being carried away and to which parts of the plant it is being carried.

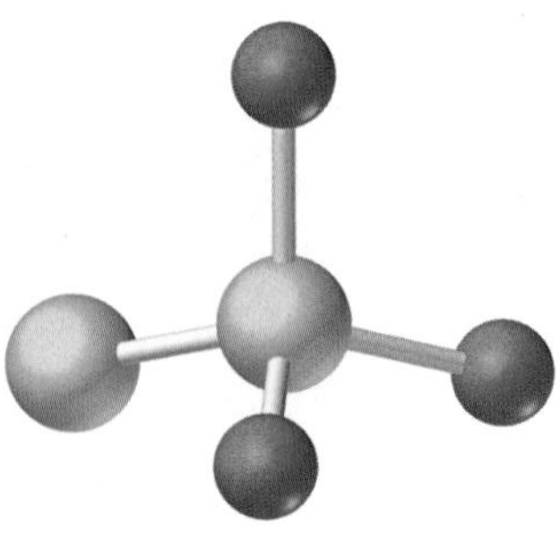

▲ FIGURE 25-12
Structure of thiosulfate ion, $S_2O_3^{2-}$
The central S atom is in the +6 oxidation state. The terminal S atom is in the −2 oxidation state.

Structures and Mechanisms

Often the mechanism of a chemical reaction or the structure of a species can be inferred from experiments using radioisotopes as tracers. Consider the following experimental proof that the two S atoms in the thiosulfate ion, $S_2O_3^{2-}$, are not equivalent.

$S_2O_3^{2-}$ is prepared from radioactive sulfur (^{35}S) and sulfite ion containing the nonradioactive isotope ^{32}S.

$$^{35}S + {}^{32}SO_3^{2-} \longrightarrow {}^{35}S^{32}SO_3^{2-} \qquad (25.23)$$

When the thiosulfate ion is decomposed by acidification, all the radioactivity appears in the precipitated sulfur and none in the $SO_2(g)$. The bonds of the ^{35}S atoms must be different from those of the ^{32}S atoms (Fig. 25-12).

$$^{35}S^{32}SO_3^{2-} + 2\,H^+ \longrightarrow H_2O + {}^{32}SO_2(g) + {}^{35}S(s) \qquad (25.24)$$

In reaction (25.25), nonradioactive KIO_4 is added to a solution containing iodide ion labeled with the radioisotope ^{128}I. All the radioactivity appears in the I_2 and none in the IO_3^-. This proves that all the IO_3^- is produced by reduction of IO_4^- and none by oxidation of I^-.

$$IO_4^- + 2\,{}^{128}I^- + H_2O \longrightarrow {}^{128}I_2 + IO_3^- + 2\,OH^- \qquad (25.25)$$

Analytical Chemistry

The usual procedure of analyzing a substance by precipitation involves filtering, washing, drying, and weighing a pure precipitate. An alternative is to incorporate a radioactive isotope in the precipitating reagent. By measuring the activity of the precipitate and comparing it with that of the original solution, the amount of precipitate can be calculated without having to purify, dry, and weigh it.

Another method of importance in analytical chemistry is *neutron activation analysis*. In this procedure, the sample to be analyzed, normally nonradioactive, is bombarded with neutrons; the element of interest is converted to a radioisotope. The conversion of a stable isotope (X) to an unstable isotope (X*) by neutron capture can be represented as

$$^{A}X + {}^{1}_{0}n \longrightarrow {}^{(A+1)}X^* \longrightarrow {}^{(A+1)}X + \gamma$$

In this example, the excited nucleus that is formed decays with the emission of a γ ray with a distinctive energy. Neutron-activated nuclei may decay by other modes, however, such as β decay. The activity of the radioisotope formed is measured. This measurement, together with such factors as the rate of neutron bombardment, the half-life of the radioisotope, and the efficiency of the radiation detector, can be used to calculate the quantity of the element in the sample. The method is especially attractive because (1) trace quantities of elements can be determined (sometimes in parts per billion or less); (2) a sample can be tested without destroying it; and (3) the sample can be in any state of matter, including biological materials. Neutron activation analysis has been used, for instance, to determine the authenticity of old paintings. Old masters formulated their own paints. Differences between formulations are easily detected through the trace elements they contain.

SuperStock

▲ Neutron activation was used to establish that this painting was not painted by Rembrandt, but by an artist in the school of Rembrandt.

Radiation Processing

Radiation processing describes industrial applications of ionizing radiation—γ rays from ^{60}Co or electron beams from electron accelerators. The ionizing radiation is used in the production of certain materials or to modify their properties. Its most extensive current use is in breaking, re-forming, and cross-linking polymer chains to affect the physical and mechanical properties of plastics used in foamed products, electrical insulation, and packaging materials. Ionizing radiation is used to sterilize medical supplies, such as sutures, syringes, and hospital garb. In sewage treatment plants, radiation processing has been used to decrease the settling time of sewage sludge and to kill pathogens. Radiation is now being used in some instances in the preservation of foods as an alternative to canning, freeze-drying, or refrigeration. In radiation processing, the irradiated material is not rendered radioactive, although the ionizing radiation may produce some chemical changes.

Cordelia Molloy / Science Source

▲ The strawberries on the left have been preserved by irradiation.

MasteringCHEMISTRY www.masteringchemistry.com

The use of nuclear reactions is an attractive "clean" approach to meeting today's energy consumption. However, the waste products created by nuclear power plants are not easily disposed. For a discussion of different methods of handling nuclear waste, go to the Focus On feature for Chapter 25, Radioactive Waste Disposal, on the Mastering Chemistry site.

Summary

25-1 Radioactivity—All the heavier elements ($Z > 83$) and a few of the lighter ones have naturally occurring nuclides that produce ionizing radiation. **Alpha (α) particles** emanating from radioactive nuclei are identical to the nuclei of helium-4 atoms, $^{4}_{2}He^{2+}$. **Beta (β^-) particles** are electrons that originate from the conversion of a neutron to a proton in the radioactive nucleus; **positrons (β^+)** are positively charged particles otherwise identical to electrons that originate in the nucleus in the conversion of a proton to a neutron. Beta (β^-) particles and positrons have a greater penetrating power through matter than do α particles. **Electron capture (EC)** occurs when an electron from a low-lying energy level is absorbed by the nucleus. The electron and a proton combine to form a neutron and liberate X radiation. **Gamma (γ)** rays are a nonparticulate form of electromagnetic radiation released when a nucleus in an excited state returns to the ground state. **Nuclear equations** to represent nuclear processes must be balanced, which requires that (1) the sum of the mass numbers and (2) the sum of the atomic numbers must be the same on both sides of the nuclear equation (equation 25.1).

25-2 Naturally Occurring Radioactive Isotopes—All naturally occurring radioactive nuclides of high atomic number are members of a **radioactive decay series** that originates with a long-lived isotope of high atomic number and terminates with a stable isotope, such as ^{206}Pb (Fig. 25-2).

25-3 Nuclear Reactions and Artificially Induced Radioactivity—Radioactive nuclides can be synthesized by nuclear reactions in which target nuclei are bombarded with energetic particles such as α particles or neutrons. More than 1000 radioactive nuclides have been produced in this way. Many of them have important practical applications.

25-4 Transuranium Elements—Elements beyond $Z = 92$, commonly referred to as the transuranium elements, have been synthesized by bombarding naturally occurring isotopes with charged particles produced in a charged-particle accelerator (Fig. 25-3).

25-5 Rate of Radioactive Decay—The **radioactive decay law**, based on experimental observations, states that the decay rate is directly proportional to the number of atoms present; hence it is a first-order process (equation 25.11). The proportionality constant between the decay rate and the number of atoms is the **decay constant, λ**. The **half-life,** $t_{1/2}$, is the time it takes for half of the radioactive atoms in a sample to decay. Important applications of the radioactive decay law are radiocarbon dating to determine the age of materials derived from once-living matter and determining the age of rocks and of Earth itself.

25-6 Energetics of Nuclear Reactions—The energy changes in nuclear reactions are a consequence of Einstein's discovery of a mass–energy equivalence (equation 25.19).

Nuclear binding energy is the energy released when nucleons fuse together into a nucleus; the "lost" mass is known as the mass defect. A plot of binding energy per nucleon versus mass number passes through a maximum at about $A = 60$. Thus, the *fusion* of nucleons is favored at lower mass numbers, and the *fission* of nuclei into lighter fragments is favored at higher mass numbers (Fig. 25-6).

25-7 Nuclear Stability—The stability of a nuclide depends on the ratio N/Z, on whether the numbers of neutrons (N) and protons (Z) are odd or even, and whether either is a **magic number** arising from nuclear shell theory (Table 25.2). A plot of N versus Z shows that all stable nuclides lie within a belt of stability originating along the line $N = Z$, and expanding above the line at higher values of Z (Fig. 25-7). Most radioactive nuclides lie outside the belt and undergo radioactive decay of a type that moves daughter nuclides into the belt.

25-8 Nuclear Fission—In the process of **fission**, the nucleus of a heavy nuclide, such as ^{235}U, splits into two smaller fragments after being struck by a thermal neutron (Fig. 25-8). Also released are two or three neutrons that can trigger the fission of other nuclei in a chain reaction. Essential components of a nuclear power reactor (Fig. 25-9) are the fissionable (*fissile*) nuclide, a **moderator** (such as water) to slow down the neutrons released during fission, **control rods** (such as cadmium) to control the fission by absorbing neutrons, and a heat-transfer medium (such a water).

25-9 Nuclear Fusion—The **fusion** of lighter nucle into heavier ones converts small quantities of mass int enormous amounts of energy. This process occurs contin uously in stars and in the hydrogen bomb. A controlle fusion reaction has yet to be achieved, but fusion researc is being actively pursued because of the enormous poten tial of fusion as an energy source.

25-10 Effect of Radiation on Matter—The ioniz ing power of radiation (Fig. 25-10) is the basis bot of radiation's effects on matter and methods used t detect radiation. Radiation detectors include simple film badges, Geiger–Müller (G–M) counters for routin measurements (Fig. 25-11), and scintillation counters i biomedical studies. Two units of measure are used t quantify exposure to radioactivity. One—the **rad** (radia tion absorbed dose)—is related to the amount of radiatio energy absorbed, while the other—the **rem** (radiatio equivalent for man)—takes into account the differin effects of the various types of radiation (Table 25.5).

25-11 Applications of Radioisotopes—Radio active nuclides have important uses in diagnostic medi cine, in cancer therapy, in studying the mechanisms o chemical and biochemical reactions, and as tracers in vari ous scientific and industrial settings. Additionally, radia tion processing is used in processing and preserving food

Integrative Example

On April 26, 1986, an explosion at the nuclear power plant at Chernobyl, Ukraine, released the greatest quantity of radioactive material ever associated with an industrial accident. (See the photo and discussion on pages 1190–1192.) One of the radioisotopes in this emission was ^{131}I, a β^- emitter with a half-life of 8.021 days.

Assume that the total quantity of ^{131}I released was 250 g, and determine the number of *curies* associated with this ^{131}I 30.0 days after the accident.

Analyze

From the initial mass of ^{131}I and the known half-life of ^{131}I determine how many grams of ^{131}I remain after 30 days. Convert this mass to a number of atoms using the molar mass of ^{131}I and Avogadro's number. Finally, the product of the number of atoms and the decay constant yields the activity in disintegrations per second, which can be converted to curies.

Solve

In equation (25.12), use $\lambda = 0.693/8.021$ d and $t = 30.0$ d. Because the mass of ^{131}I is directly proportional to the number of ^{131}I atoms, we can replace N_0 by $mass_0 = 250$ g, and N_t by $mass_t$, the mass we seek.

$$\ln\frac{mass_t}{mass_0} = \ln\frac{mass_t}{250\text{ g}} = \frac{-0.693}{8.021\text{ d}} \times 30.0\text{ d} = -2.59$$

$$mass_t = 250\text{ g} \times e^{-2.59} = 250\text{ g} \times 0.0750 = 18.8\text{ g}$$

To determine the number of iodine-131 atoms, use the molar mass, 131 g ^{131}I/mol ^{131}I, and the Avogadro constant.

$$N_t = 18.8\text{ g }^{131}I \times \frac{1\text{ mol }^{131}I}{131\text{ g }^{131}I} \times \frac{6.022 \times 10^{23}\ ^{131}I\text{ atoms}}{1\text{ mol }^{131}I}$$

$$= 8.64 \times 10^{22}\ ^{131}I\text{ atoms}$$

To establish the decay constant in s^{-1} and to obtain λ from $t_{1/2}$, use the method outlined in Example 25-3a.

$$\lambda = 0.693/t_{1/2} = \frac{0.693}{8.04\text{ d}} \times \frac{1\text{ d}}{24\text{ h}} \times \frac{1\text{ h}}{60\text{ min}} \times \frac{1\text{ min}}{60\text{ s}}$$

$$= 9.98 \times 10^{-7}\text{ s}^{-1}$$

To determine the decay rate (activity) after 30.0 days, use the two previous results in equation (25.11).

$$A = \lambda N = 9.98 \times 10^{-7}\text{ s}^{-1} \times 8.64 \times 10^{22}\text{ atoms}$$

$$= 8.62 \times 10^{16}\text{ dis s}^{-1}$$

Finally, to express the activity in curies use the definition of a curie given in Table 25.5.

$$A = 8.62 \times 10^{16} \text{ dis s}^{-1} \times \frac{1 \text{ Ci}}{3.70 \times 10^{10} \text{ dis s}^{-1}}$$
$$= 2.33 \times 10^{6} \text{ Ci}$$

Assess

In this calculation we could also have shown that in 30 days (about $4 \times t_{1/2}$) approximately 93% of the ^{131}I had disintegrated. In the process, approximately 2×10^6 Ci of radiation was emitted. If we also knew the energy of the β^- particles and something about the exposed population, we could also have estimated radiation doses in rads.

PRACTICE EXAMPLE A: ^{40}K undergoes radioactive decay by electron capture to ^{40}Ar and by β^- emission to ^{40}Ca. The fraction of the decay that occurs by electron capture is 0.110. The half-life of ^{40}K is 1.26×10^9 years. Assuming that a rock in which ^{40}K has undergone decay retains all the ^{40}Ar produced, what would be the ^{40}Ar/^{40}K mass ratio in a rock that is 1.5×10^9 years old?

PRACTICE EXAMPLE B: Most nuclear power plants use zirconium in the fuel rods because zirconium maintains its structural integrity under exposure to the radiation in the nuclear reactor. The nuclear accidents at Three Mile Island, Chernobyl, and Fukushima involved the evolution of hydrogen gas from the reduction of water. The half-cell reduction potential for Zr is

$$ZrO_2(s) + 4\,H_3O^+(aq) + 4\,e^- \longrightarrow Zr(s) + 6\,H_2O(l) \qquad E° = -1.43 \text{ V}$$

(a) Can Zr reduce water under standard-state conditions?
(b) Calculate the equilibrium constant for the reduction of water by zirconium.
(c) Is the reaction spontaneous if the pH = 7 and Zr, ZrO_2, and water are in their standard states?
(d) Was the reduction of water by Zr the culprit in the reactor accidents mentioned above?

Exercises

Radioactive Processes

1. What nucleus is obtained in each process?
(a) $^{55}_{26}$Fe decays by β^- emission.
(b) $^{238}_{92}$U decays by α emission.
(c) $^{222}_{86}$Rn decays by two successive α emissions.
(d) $^{64}_{29}$Cu decays by two successive β^- emissions.
2. What nucleus is obtained in each process?
(a) $^{214}_{82}$Pb decays through two successive β^- emissions.
(b) $^{226}_{88}$Ra decays through three successive α emissions.
(c) $^{69}_{33}$As decays by β^+ emission.
3. Based on a favorable N:Z ratio for the product nucleus, write the most plausible equation for the decay of $^{14}_{6}$C.
4. Write a plausible equation for the decay of tritium, $^{3}_{1}$H, the radioactive isotope of hydrogen.

Radioactive Decay Series

5. The natural decay series starting with the radionuclide $^{232}_{90}$Th follows the sequence represented here. Construct a graph of this series, similar to Figure 25-2.

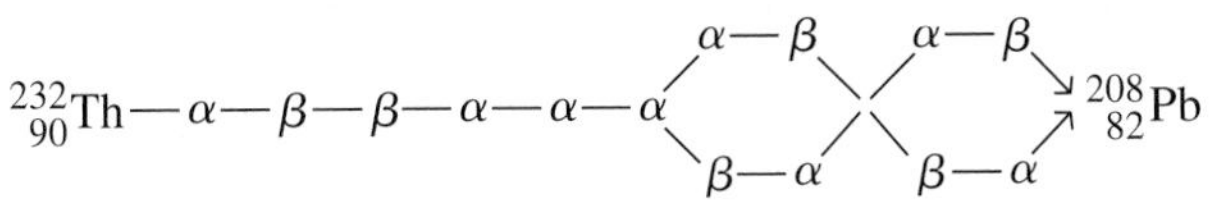

6. The natural decay series starting with the radionuclide $^{235}_{92}$U follows the sequence represented here. Construct a graph of this series, similar to Figure 25-2.

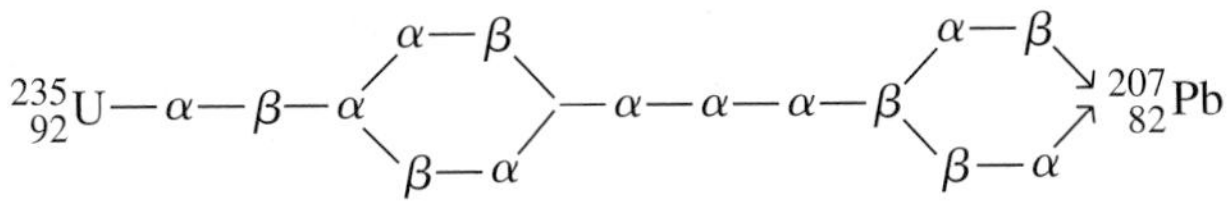

7. The uranium series described in Figure 25-2 is also known as the "4n + 2" series because the mass number of each nuclide in the series can be expressed by the equation $A = 4n + 2$, where n is an integer. Show that this equation is indeed applicable to the uranium series.
8. Just as the uranium series is called the "4n + 2" series, the thorium series can be called the "4n" series and the actinium series the "4n + 3" series. A "4n + 1" series has also been established, with $^{241}_{94}$Pu as the parent nuclide. To which series does each of the following belong? **(a)** $^{214}_{83}$Bi; **(b)** $^{216}_{84}$Po; **(c)** $^{215}_{85}$At; **(d)** $^{235}_{92}$U.

Nuclear Reactions

9. Supply the missing information in each of the following nuclear equations representing a radioactive decay process.
   (a) ${}^{160}_{?}\text{W} \longrightarrow {}^{?}_{?}\text{Hf} + ?$
   (b) ${}^{38}_{?}\text{Cl} \longrightarrow {}^{?}_{?}\text{Ar} + ?$
   (c) ${}^{214}_{?}? \longrightarrow {}^{?}_{?}\text{Po} + {}^{0}_{-1}\beta$
   (d) ${}^{32}_{17}\text{Cl} \longrightarrow {}^{?}_{16}? + ?$
10. Complete the following nuclear equations.
   (a) ${}^{23}_{13}\text{Al} + {}^{4}_{2}\text{He} \longrightarrow ? + {}^{1}_{0}\text{n}$
   (b) ${}^{63}_{29}\text{Cu} + ? \longrightarrow {}^{63}_{30}\text{Zn} + 2{}^{1}_{0}\text{n}$
   (c) $? + {}^{1}_{0}\text{n} \longrightarrow {}^{95}_{42}\text{Mo} + {}^{140}_{50}\text{Sn} + 2{}^{1}_{0}\text{n}$
   (d) ${}^{31}_{15}\text{P} + {}^{1}_{1}\text{H} \longrightarrow {}^{28}_{14}\text{Si} + ?$
   (e) ${}^{11}_{5}\text{B} + {}^{4}_{2}\text{He} \longrightarrow ? + {}^{1}_{0}\text{n}$
11. Write equations for the following nuclear reactions.
   (a) bombardment of ${}^{7}\text{Li}$ with protons to produce ${}^{8}\text{Be}$ and γ rays
   (b) bombardment of ${}^{9}\text{Be}$ with ${}^{2}_{1}\text{H}$ to produce ${}^{10}\text{B}$
   (c) bombardment of ${}^{14}\text{N}$ with neutrons to produce ${}^{14}\text{C}$
12. Write equations for the following nuclear reactions.
   (a) bombardment of ${}^{238}\text{U}$ with α particles to produce ${}^{239}\text{Pu}$
   (b) bombardment of tritium (${}^{3}_{1}\text{H}$) with ${}^{2}_{1}\text{H}$ to produce ${}^{4}\text{He}$
   (c) bombardment of ${}^{33}\text{S}$ with neutrons to produce ${}^{33}\text{P}$
13. Write nuclear equations to represent the formation of an isotope of element 83 with a mass number of 218 by the bombardment of uranium-238 by hydrogen-2 nuclei followed by a succession of five α-particle emissions.
14. Write nuclear equations to represent the formation of a hypothetical isotope of element 118 with a mass number of 293 by the bombardment of lead-208 by krypton-86 nuclei, followed by a chain of α-particle emissions to the element seaborgium.
15. Scientists from Dubna, Russia, observed the existence of elements 118 and 116 at the Joint Institute for Nuclear Research U400 cyclotron in 2005. This was the result of bombarding calcium-48 ions on a californium-249 target. Write a complete nuclear equation for this reaction.
16. The immediate decay product of element 118 is thought to be element 116. Write a complete nuclear equation for this reaction.
17. Element 120 is located in a region of the neutron versus proton map known as the island of stability. Write a nuclear equation for the generation of element 120 by bombarding iron isotopes on a plutonium target.
18. Another possible nuclear reaction leading to the formation of element 120 is between uranium-238 and nickel-64. Write a nuclear equation for this nuclear reaction.

Rate of Radioactive Decay

19. Of the radioactive nuclides in Table 25.1,
   (a) which one has the largest value for the decay constant, λ?
   (b) Which one loses 75% of its radioactivity in approximately one month?
   (c) Which ones lose more than 99% of their radioactivity in one month?
20. In a comparison of two radioisotopes, isotope A requires 18.0 hours for its decay rate to fall to $\frac{1}{16}$ its initial value, while isotope B has a half-life that is 2.5 times that of A. How long does it take for the decay rate of isotope B to decrease to $\frac{1}{32}$ of its initial value?
21. The disintegration rate for a sample containing ${}^{101}_{45}\text{Rh}$ as the only radioactive nuclide is 4650 dis h^{-1}. The half-life of ${}^{101}_{45}\text{Rh}$ is 3.30 years. Estimate the number of atoms of ${}^{101}_{45}\text{Rh}$ in the sample.
22. How many years must the radioactive sample of Exercise 21 be maintained before the disintegration rate falls to 101 dis min^{-1}?
23. A sample containing ${}^{224}_{88}\text{Ra}$, which decays by α-particle emission, disintegrates at the following rate, expressed as disintegrations per minute or counts per minute (cpm): $t = 0$, 1000 cpm; $t = 1$ h, 992 cpm; $t = 10$ h, 924 cpm; $t = 100$ h, 452 cpm; $t = 250$ h, 138 cpm. What is the half-life of this nuclide?
24. Iodine-129 is a product of nuclear fission, whether from an atomic bomb or a nuclear power plant. It is a β^{-} emitter with a half-life of 1.7×10^{7} years. How many disintegrations per second would occur in a sample containing 1.00 mg ${}^{129}\text{I}$?
25. Suppose that a sample containing ${}^{32}\text{P}$ has an activity 1000 times the detectable limit. How long would an experiment have to be run with this sample before the radioactivity could no longer be detected?
26. What mass of carbon-14 must be present in a sample to have an activity of 1.00 mCi?

Age Determinations with Radioisotopes

27. A wooden object is claimed to have been found in an Egyptian pyramid and is offered for sale to an art museum. Radiocarbon dating of the object reveals a disintegration rate of 10.0 dis $\text{min}^{-1}\ \text{g}^{-1}$. Do you think the object is authentic? Explain.
28. The lowest level of ${}^{14}\text{C}$ activity that seems possible for experimental detection is 0.03 dis $\text{min}^{-1}\ \text{g}^{-1}$. What is the maximum age of an object that can be determined by the carbon-14 method?

29. What should be the mass ratio $^{208}Pb/^{232}Th$ in a meteorite that is approximately 2.7×10^9 years old? The half-life of ^{232}Th is 1.40×10^{10} years. [*Hint:* One ^{208}Pb atom is the final decay product of one ^{232}Th atom.]
30. Concerning the decay of ^{232}Th described in Exercise 29, a certain rock has a $^{208}Pb/^{232}Th$ mass ratio of 0.25/1.00. Estimate the age of the rock.
31. A lunar rock was analyzed for argon by mass spectrometry and for potassium by atomic absorption. The results of these analyses showed that the sample contained 3.02×10^{-5} mL g^{-1} at STP of argon and 0.083% of potassium. The half-life of potassium-40 is 1.26×10^9 years. Calculate the age of the lunar rock.
32. What is the age of a piece of volcanic rock that has a mass ratio of argon-40 to potassium-40 of 1.9? The half-life of potassium-40 by β decay is 1.26×10^9 years and by electron capture, 1.4×10^9 years.

Energetics of Nuclear Reactions

33. Using appropriate equations in the text, determine **(a)** the energy in joules corresponding to the destruction of 6.02×10^{-23} g of matter; **(b)** the energy in megaelectronvolts that would be released if one α particle were completely destroyed.
34. The measured mass of the nucleus of an atom of silver-107 is 106.879289 u. For this atom, determine the binding energy per nucleon in megaelectronvolts.
35. Use the electron mass from Table 2.1 and the measured mass of the nuclide $^{19}_{9}F$, 18.998403 u, to determine the binding energy per nucleon (in megaelectronvolts) of this atom.
36. Use the electron mass from Table 2.1 and the measured mass of the nuclide $^{56}_{26}Fe$, 55.934939 u, to determine the binding energy per nucleon (in megaelectronvolts) of this atom.
37. Calculate the energy, in megaelectronvolts, released in the nuclear reaction

$$^{10}_{5}B + ^{4}_{2}He \longrightarrow ^{13}_{6}C + ^{1}_{1}H$$

The nuclidic masses are $^{10}_{5}B$ = 10.01294 u; $^{4}_{2}He$ = 4.00260 u; $^{13}_{6}C$ = 13.00335 u; $^{1}_{1}H$ = 1.00783 u.
38. You are given the following nuclidic masses: $^{6}_{3}Li$ = 6.01513 u; $^{4}_{2}He$ = 4.00260 u; $^{3}_{1}H$ = 3.01604 u; $^{1}_{0}n$ = 1.008665 u. How much energy, in megaelectronvolts, is released in the following nuclear reaction?

$$^{6}_{3}Li + ^{1}_{0}n \longrightarrow ^{4}_{2}He + ^{3}_{1}H$$

39. Calculate the number of neutrons that could be created with 6.75×10^6 MeV of energy.
40. When β^+ and β^- particles collide, they annihilate each other, producing two γ rays that move away from each other along a straight line. What is the approximate energies of these two γ rays, in MeV?

Nuclear Stability

41. Which member of the following pairs of nuclides would you expect to be most abundant in natural sources? Explain your reasoning. **(a)** $^{20}_{10}Ne$ or $^{22}_{10}Ne$; **(b)** $^{17}_{8}O$ or $^{18}_{8}O$; **(c)** $^{6}_{3}Li$ or $^{7}_{3}Li$.
42. Which member of the following pairs of nuclides would you expect to be most abundant in natural sources? Explain your reasoning. **(a)** $^{40}_{20}Ca$ or $^{42}_{20}Ca$; **(b)** $^{31}_{15}P$ or $^{32}_{15}P$; **(c)** $^{63}_{30}Zn$ or $^{64}_{30}Zn$.
43. One member each of the following pairs of radioisotopes decays by β^- emission, and the other by positron (β^+) emission: **(a)** $^{29}_{15}P$ and $^{33}_{15}P$; **(b)** $^{120}_{53}I$ and $^{134}_{53}I$. Which is which? Explain your reasoning.
44. Each of the following isotopes is radioactive: **(a)** $^{28}_{15}P$; **(b)** $^{45}_{19}K$; **(c)** $^{73}_{30}Zn$. Which would you expect to decay by β^+ emission?
45. Some nuclides are said to be doubly magic. What do you suppose this term means? Postulate some nuclides that might be doubly magic, and locate them in Figure 25-7.
46. Both β^- and β^+ emissions are observed for artificially produced radioisotopes of low atomic numbers, but only β^- emission is observed with naturally occurring radioisotopes of high atomic number. Why do you suppose this is so?

Fission and Fusion

47. Refer to the Integrative Example. In contrast to the Chernobyl accident, the 1979 nuclear accident at Three Mile Island released only 170 curies of ^{131}I. How many milligrams of ^{131}I does this represent?
48. Explain why more energy is released in a fusion process than in a fission process.

Effect of Radiation on Matter

49. Explain why the rem is more satisfactory than the rad as a unit for measuring radiation dosage.
50. Discuss briefly the basic difficulties in establishing the physiological effects of low-level radiation.

51. ^{90}Sr is both a product of radioactive fallout and a radioactive waste in a nuclear reactor. This radioisotope is a β^- emitter with a half-life of 29.1 years. Suggest reasons why ^{90}Sr is such a potentially hazardous substance.

52. ^{222}Rn is an α-particle emitter with a half-life of 3.823 days. Is it hazardous to be near a flask containing this isotope? Under what conditions might ^{222}Rn be hazardous?

Applications of Radioisotopes

53. Describe how you might use radioactive materials to find a leak in the $H_2(g)$ supply line in an ammonia synthesis plant.

54. Explain why neutron activation analysis is so useful in identifying trace elements in a sample, in contrast to ordinary methods of quantitative analysis, such as precipitation or titration.

55. A small quantity of NaCl containing radioactive $^{24}_{11}Na$ is added to an aqueous solution of $NaNO_3$. The solution is cooled, and $NaNO_3$ is crystallized from the solution. Would you expect the $NaNO_3(s)$ to be radioactive? Explain.

56. The following reactions are carried out with HCl(aq) containing some tritium (^3_1H) as a tracer. Would you expect any of the tritium radioactivity to appear in the $NH_3(g)$? In the H_2O? Explain.

$$NH_3(aq) + HCl(aq) \longrightarrow NH_4Cl(aq)$$

$$NH_4Cl(aq) + NaOH(aq) \longrightarrow NaCl(aq) + H_2O(l) + NH_3(g)$$

Integrative and Advanced Exercises

57. In some cases, the most abundant isotope of an element can be established by rounding off the atomic mass to the nearest whole number, as in ^{39}K, ^{85}Rb, and ^{88}Sr. But in other cases, the isotope corresponding to the rounded-off atomic mass does not even occur naturally, as in ^{64}Cu. Explain the basis of this observation.

58. The overall change in the radioactive decay of $^{238}_{92}U$ to $^{206}_{82}Pb$ is the emission of eight α particles. Show that if this loss of eight α particles were not also accompanied by six β^- emissions, the product nucleus would still be radioactive.

59. Use data from the text to determine how many metric tons (1 metric ton = 1000 kg) of bituminous coal (85% C) would have to be burned to release as much energy as is produced by the fission of 1.00 kg $^{235}_{92}U$.

60. One method of dating rocks is based on their $^{87}Sr/^{87}Rb$ ratio. ^{87}Rb is a β^- emitter with a half-life of 4.88×10^{10} years. A certain rock has a mass ratio $^{87}Sr/^{87}Rb$ of 0.004/1.00. What is the age of the rock?

61. How many millicuries of radioactivity are associated with a sample containing 5.10 mg ^{229}Th, which has a half-life of 7.9×10^3 years?

62. What mass of ^{90}Sr, with a half-life of 29.1 years, is required to produce 1.00 millicurie of radioactivity?

63. Refer to the Integrative Example. Another radioisotope produced in the Chernobyl accident was ^{137}Cs. If a 1.00 mg sample of ^{137}Cs is equivalent to 89.8 millicuries, what must be the half-life (in years) of ^{137}Cs?

64. The percent natural abundance of ^{40}K is 0.0117%. The radioactive decay of ^{40}K atoms occurs 89% by β^- emission; the rest is by electron capture and β^+ emission. The half-life of ^{40}K is 1.26×10^9 years. Calculate the number of β^- particles produced per second by the ^{40}K present in a 1.00 g sample of the mineral *microcline*, $KAlSi_3O_8$.

65. The carbon-14 dating method is based on the assumption that the rate of production of ^{14}C by cosmic ray bombardment has remained constant for thousands of years and that the ratio of ^{14}C to ^{12}C has also remained constant. Can you think of any effects of human activities that could invalidate this assumption in the future?

66. Calculate the minimum kinetic energy (in megaelectronvolts) that α particles must possess to produce the nuclear reaction

$$^4_2He + ^{14}_7N \longrightarrow ^{17}_8O + ^1_1H$$

The nuclidic masses are ^4_2He = 4.00260 u; $^{14}_7N$ = 14.00307 u; ^1_1H = 1.00783 u; $^{17}_8O$ = 16.99913 u.

67. Hydrogen gas is spiked with tritium to the extent of 5.00% by mass. What is the activity in curies of a 4.65 L sample of this gas at 25.0 °C and 1.05 atm pressure? [*Hint:* Use 3.02 u as the atomic mass of tritium and data from elsewhere in the text, as necessary.]

68. A certain shale deposit containing 0.006% U by mass is being considered for use as a potential fuel in a breeder reactor. Assuming a density of 2.5 g/cm^3, how much energy could be released from 1.00×10^3 cm^3 of this material? Assume a fission energy of 3.20×10^{-11} J per fission event (that is, per U atom).

69. An ester forms from a carboxylic acid and an alcohol.

$$RCO_2H + HOR' \longrightarrow RCO_2R' + H_2O$$

This reaction is superficially similar to the reaction of an acid with a base such as sodium hydroxide. The mechanism of the reaction can be followed by using the tracer ^{18}O. This isotope is not radioactive, but other physical measurements can be used to detect its presence. When the *esterification* reaction is carried out with the alcohol containing oxygen-18 atoms, no oxygen-18 beyond its naturally occurring abundance is found in the water produced. How does this result affect the perception that this reaction is like an acid–base reaction?

70. The conversion of CO_2 into carbohydrates by plants via photosynthesis can be represented by the reaction

$$6\,CO_2(g) + 6\,H_2O \xrightarrow{\text{light}} C_6H_{12}O_6 + 6\,O_2(g)$$

To study the mechanism of photosynthesis, algae were grown in water containing ^{18}O, that is, $H_2{}^{18}O$. The oxygen evolved contained oxygen-18 in the same ratio to the other oxygen isotopes as the water in which the reaction was carried out. In another experiment, algae were grown in water containing only ^{16}O, but with oxygen-18 present in the CO_2. The oxygen evolved in this experiment contained no oxygen-18. What conclusion can you draw about the mechanism of photosynthesis from these experiments?

71. Assume that when Earth formed, uranium-238 and uranium-235 were equally abundant. Their current percent natural abundances are 99.28% uranium-238 and 0.72% uranium-235. Given half-lives of 4.5×10^9 years for uranium-238 and 7.1×10^8 years for uranium-235, determine the age of Earth corresponding to this assumption.

Feature Problems

72. The *packing fraction* of a nuclide is related to the fraction of the total mass of a nuclide that is converted to nuclear binding energy. It is defined as the fraction $(M - A)/A$, where M is the actual nuclidic mass and A is the mass number. Use data from a handbook (such as the *Handbook of Chemistry and Physics*, published by the CRC Press) to determine the packing fractions of some representative nuclides. Plot a graph of packing fraction versus mass number, and compare it with Figure 25-6. Explain the relationship between the two.

73. For medical uses, radon-222 formed in the radioactive decay of radium-226 is allowed to collect over the radium metal. Then, the gas is withdrawn and sealed into a glass vial. Following this, the radium is allowed to disintegrate for another period, when a new sample of radon-222 can be withdrawn. The procedure can be continued indefinitely. The process is somewhat complicated by the fact that radon-222 itself undergoes radioactive decay to polonium-218, and so on. The half-lives of radium-226 and radon-222 are 1.60×10^3 years and 3.82 days, respectively.

(a) Beginning with pure radium-226, the number of radon-222 atoms present starts at zero, increases for a time, and then falls off again. Explain this behavior. That is, because the half-life of radon-222 is so much shorter than that of radium-226, why doesn't the radon-222 simply decay as fast as it is produced, without ever building up to a maximum concentration?

(b) Write an expression for the rate of change (dD/dt) in the number of atoms (D) of the radon-222 daughter in terms of the number of radium-226 atoms present initially (N_0) and the decay constants of the parent (λ_p) and daughter (λ_d).

(c) Integration of the expression obtained in part (b) yields the following expression for the number of atoms of the radon-222 daughter (D) present at a time t.

$$D = \frac{N_0 \lambda_p (e^{-\lambda_p \times t} - e^{-\lambda_d \times t})}{\lambda_d - \lambda_p}$$

Starting with 1.00 g of pure radium-226, approximately how long will it take for the amount of radon-222 to reach its maximum value: one day, one week, one year, one century, or one millennium?

74. Radioactive decay and mass spectrometry are often used to date rocks after they have cooled from a magma. ^{87}Rb has a half-life of 4.88×10^{10} years and follows the radioactive decay

$$^{87}Rb \longrightarrow {}^{87}Sr + \beta^-$$

A rock was dated by assaying the product of this decay. The mass spectrum of a homogenized sample of rock showed the $^{87}Sr/^{86}Sr$ ratio to be 2.25. Assume that the original $^{87}Sr/^{86}Sr$ ratio was 0.700 when the rock cooled. Chemical analysis of the rock gave 15.5 ppm Sr and 265.4 ppm Rb, using the average atomic masses from a periodic table. The other isotope ratios were $^{86}Sr/^{88}Sr = 0.119$ and $^{84}Sr/^{88}Sr = 0.007$. The isotopic ratio for $^{87}Rb/^{85}Rb$ is 0.330. The isotopic masses are as follows:

Isotope	Atomic Mass, u
^{87}Rb	86.909
^{85}Rb	84.912
^{88}Sr	87.906
^{86}Sr	85.909
^{84}Sr	83.913
^{87}Sr	86.909

Calculate the following:

(a) the average atomic mass of Sr in the rock
(b) the original concentration of Rb in the rock in ppm
(c) the percentage of rubidium-87 decayed in the rock
(d) the time since the rock cooled

Self-Assessment Exercises

75. In your own words, define the following symbols: **(a)** α; **(b)** β^-; **(c)** β^+; **(d)** γ; **(e)** $t_{1/2}$.

76. Briefly describe each of the following ideas, phenomena, or methods: **(a)** radioactive decay series; **(b)** charged-particle accelerator; **(c)** neutron-to-proton ratio; **(d)** mass–energy relationship; **(e)** background radiation.

77. Explain the important distinctions between each pair of terms: **(a)** electron and positron; **(b)** half-life and decay constant; **(c)** mass defect and nuclear binding energy; **(d)** nuclear fission and nuclear fusion; **(e)** primary and secondary ionization.

78. Which of the following types of radiation is deflected in a magnetic field? **(a)** X-ray; **(b)** γ ray; **(c)** β ray; **(d)** neutrons.

79. A process that produces a one-unit increase in atomic number is **(a)** electron capture; **(b)** β^- emission; **(c)** α emission; **(d)** γ-ray emission.

80. Of the following nuclides, the one most likely to be radioactive is **(a)** ^{31}P; **(b)** ^{66}Zn; **(c)** ^{35}Cl; **(d)** ^{108}Ag.

81. One of the following elements has eight naturally occurring *stable* isotopes. We should expect that one to be **(a)** Ra; **(b)** Au; **(c)** Cd; **(d)** Br.

82. Of the following nuclides, the highest nuclear binding energy per nucleon is found in **(a)** $^{3}_{1}H$; **(b)** $^{16}_{8}O$; **(c)** $^{56}_{26}Fe$; **(d)** $^{235}_{92}U$.

83. The most radioactive of the isotopes of an element is the one with the largest value of its **(a)** half-life, $t_{1/2}$; **(b)** neutron number, N; **(c)** mass number, Z; **(d)** radioactive decay constant, λ.

84. Given a radioactive nuclide with $t_{1/2} = 1.00$ h and a current disintegration rate of 1000 atoms s^{-1}, three hours from now the disintegration rate will be **(a)** 1000 atoms s^{-1}; **(b)** 333 atoms s^{-1}; **(c)** 250 atoms s^{-1}; **(d)** 125 atoms s^{-1}.

85. Write nuclear equations to represent
(a) the decay of ^{214}Ra by α-particle emission
(b) the decay of ^{205}At by positron emission
(c) the decay of ^{212}Fr by electron capture
(d) the reaction of two deuterium nuclei (deuterons) to produce a nucleus of $^{3}_{2}He$
(e) the production of $^{243}_{97}Bk$ by the α-particle bombardment of $^{241}_{95}Am$
(f) a nuclear reaction in which thorium-232 is bombarded with α particles, producing a new nuclide and four neutrons.

86. ^{223}Ra has a half-life of 11.43 d. How long would it take for the radioactivity associated with a sample of ^{223}Ra to decrease to 1% of its current value?

87. A sample of radioactive $^{35}_{16}S$ disintegrates at a rate of 1.00×10^3 atoms min^{-1}. The half-life of $^{35}_{16}S$ is 87.9 d. How long will it take for the activity of this sample to decrease to the point of producing **(a)** 253, **(b)** 104, and **(c)** 52 dis min^{-1}?

88. Neutron bombardment of ^{23}Na produces an isotope that is a β emitter. After β emission, the final product is **(a)** ^{24}Na; **(b)** ^{24}Mg; **(c)** ^{23}Ar; **(d)** ^{24}Ar; **(e)** none of these.

89. A nuclide has a decay rate of $2.00 \times 10^{10}\ s^{-1}$. After 25.0 days, its decay rate is $6.25 \times 10^{8}\ s^{-1}$. What is the nuclide's half-life? **(a)** 25.0 d; **(b)** 12.5 d; **(c)** 50.0 d; **(d)** 5.00 d; **(e)** none of these.

90. A nuclide has a half-life of 1.91 years. Its decay constant has a numerical value of **(a)** 1.32; **(b)** 2.76; **(c)** 0.363; **(d)** 0.524; **(e)** none of these.

91. A nuclide has a decay constant of $4.28 \times 10^{-4}\ h^{-1}$. If the activity of a sample is $3.14 \times 10^{5}\ s^{-1}$, how many atoms of the nuclide are present in the sample? **(a)** 2.64×10^{12}; **(b)** 7.34×10^{8}; **(c)** 2.04×10^{5}; **(d)** 4.40×10^{10}; **(e)** none of these.

92. Based on magic numbers, which nuclide is the least stable? **(a)** ^{59}Ni; **(b)** ^{51}V; **(c)** ^{122}Sb; **(d)** ^{16}O; **(e)** ^{12}C.

93. Is radioactivity temperature dependent? Explain.

Structures of Organic Compounds

26

CONTENTS

LEARNING OBJECTIVES

26.1 Name organic hydrocarbons containing functional groups.

26.2 Use Newman projections to represent the possible conformations of an alkane.

26.3 Describe and draw the chair conformation of cyclohexane, and identify the axial and equatorial hydrogens.

26.4 Use the *R, S* system of nomenclature to name chiral organic compounds.

26.5 Name and identify stereoisomers of alkenes by using the *E, Z* system of nomenclature.

26.6 Identify some key characteristics of aromatic hydrocarbons, and name them according to IUPAC rules.

26.7 Discuss the structure, function, and synthesis of organic compounds containing key functional groups.

26.8 Given a molecular formula, determine the degree of unsaturation and suggest a plausible molecular structure.

Coffee beans contain an alkaloid compound commonly known as caffeine. Caffeine is a stimulant of the central nervous system producing alertness and heightened concentration.

To early nineteenth-century chemists, organic chemistry meant the study of compounds obtainable only from living matter, which was thought to have the "vital force" needed to make these compounds. In 1828, Friedrich Wöhler set out to synthesize ammonium cyanate, NH_4OCN, using the reaction below:

$$AgOCN(s) + NH_4Cl(aq) \xrightarrow{\Delta} AgCl(s) + NH_4OCN(aq)$$

The white crystalline solid he obtained from the solution had none of the properties of ammonium cyanate, even though it had the same composition. The compound was not NH_4OCN but $(NH_2)_2CO$—*urea*, an organic compound. As Wöhler excitedly reported to J. J. Berzelius, "I must tell you that I can make urea without the use of kidneys, either man or dog. Ammonium cyanate is urea."

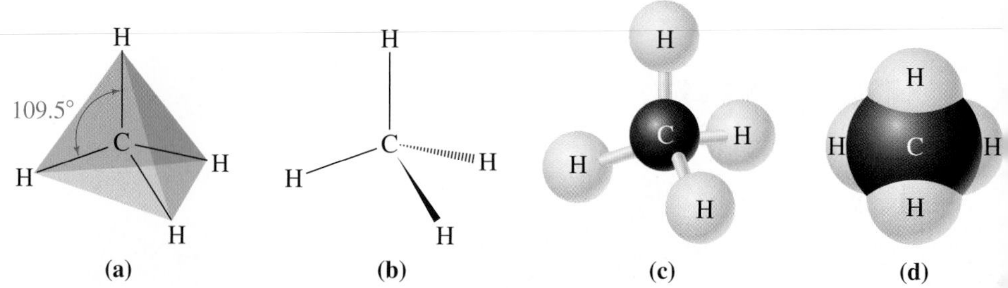

▲ FIGURE 26-1
Representation of the methane molecule
(a) Tetrahedral structure showing bond angle. **(b)** Dashed-wedged line structure convention used to suggest a three-dimensional structure through a structural formula. The solid lines represent bonds in the plane of the page. The dashed wedge projects *away* from the viewer (behind the plane of the page), and the heavy wedge projects *toward* the viewer (out of the page). **(c)** Ball-and-stick model. **(d)** Space-filling model.

KEEP IN MIND

that the hybridization of the C atoms in ethane and propane is sp^3, as in methane. The tetrahedral geometry at the C atoms in alkanes means that the propane chain is not linear.

Since that time, chemists have synthesized millions of organic compounds, and today organic compounds represent about 98% of all known chemical substances. In this chapter, we build on the introduction to organic compounds in Chapter 3 by exploring some of the principal types of organic compounds. In this chapter, we will focus on the structures and properties of organic compounds, and we will consider the preparation and some uses of these compounds. In the next chapter, we will turn our attention to reactions that interconvert these compounds.

26-1 Organic Compounds and Structures: An Overview

As we learned in Chapter 3, organic compounds contain carbon and hydrogen atoms or carbon and hydrogen in combination with a few other types of atoms, such as oxygen, nitrogen, and sulfur. Carbon is singled out for special study because the ability of C atoms to form strong covalent bonds with one another allows them to join together into straight chains, branched chains, and rings. The nearly infinite number of possible bonding arrangements of C atoms accounts for the vast number and variety of organic compounds.

The simplest organic compounds are those of carbon and hydrogen—**hydrocarbons**—and the simplest hydrocarbon is methane, CH_4, the chief constituent of natural gas.

From VSEPR theory we expect the electron group geometry around the central C atom in CH_4 to be tetrahedral, as illustrated in Figure 26-1(a). The four H atoms are equivalent: They are equidistant from the C atom and attached to it by covalent bonds of equal strength. The angle between any two C—H bonds is close to 109.5°. In Figure 26-1(b), the three-dimensional structure of CH_4 is represented by using the **dashed and solid wedge line notation** we introduced in Chapter 10. Other commonly employed depictions of methane are also shown in Figure 26-1.

The bonding in CH_4 is most easily described in terms of valence bond theory. As shown in Figure 11-8 (page 473), each of the carbon–hydrogen bonds is a σ bond formed by the overlap of a 1*s* orbital on hydrogen with an sp^3 orbital on carbon.

The removal of one H atom from a CH_4 molecule leaves the $—CH_3$ group. Now imagine forming a covalent bond between two $—CH_3$ groups. The resulting molecule is *ethane*, C_2H_6 (Fig. 26-2). By increasing the number of C atoms in the chain, we can obtain still more hydrocarbons. The three-carbon molecule propane, C_3H_8, is pictured in Figure 26-3.

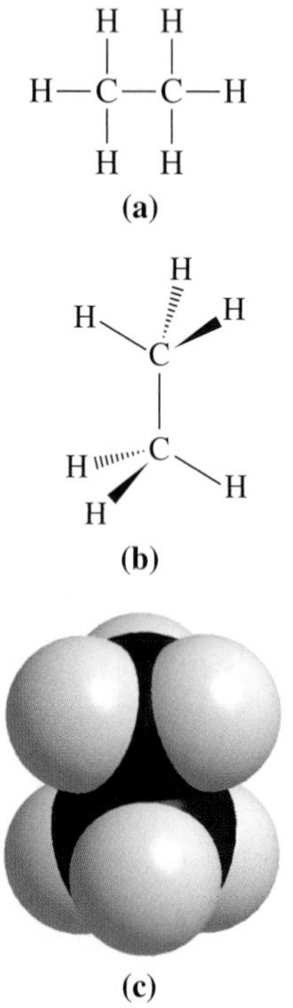

▲ FIGURE 26-2
The ethane molecule, C_2H_6
(a) Structural formula. **(b)** Dashed-wedged line structure. **(c)** Space-filling model.

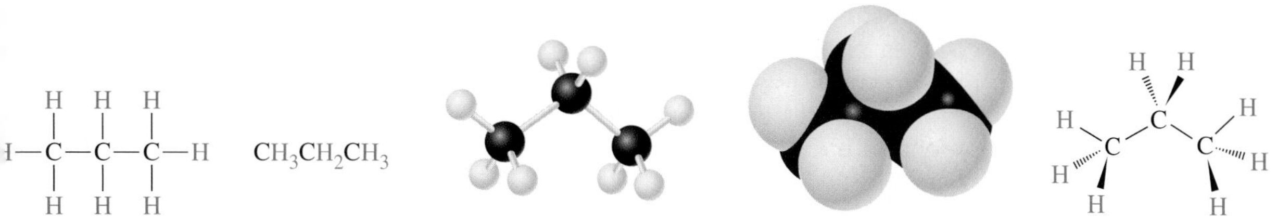

▲ FIGURE 26-3
The propane molecule, C_3H_8
(a) Structural formula. **(b)** Condensed structural formula. **(c)** Ball-and-stick model. **(d)** Space-filling model. **(e)** Dashed-wedged line notation.

Constitutional Isomerism in Organic Compounds

As we have previously learned, compounds that have the same molecular formula but different structural formulas are called *isomers*. Forms of isomerism abound in organic chemistry. We will encounter two types of isomers in this chapter—constitutional isomers and stereoisomers—but our focus in this section is on constitutional isomers. **Constitutional isomers** have different bond connectivities and thus different skeletal structures. For example, C_4H_{10} has two constitutional isomers, as shown below:

◀ Because constitutional isomers have different structural formulas, they are sometimes called structural isomers. However, the term is not recommended by IUPAC.

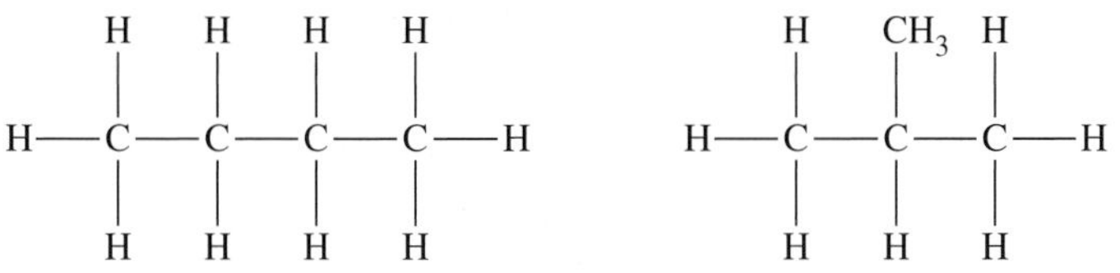

Butane Methylpropane

Butane has a single chain of four carbon atoms. Methylpropane has a three-carbon chain with a $-CH_3$ group bonded to the second carbon. Butane is called a straight-chain hydrocarbon—although the molecule does not have a straight shape—and methylpropane is an example of a branched-chain hydrocarbon. Because butane and methylpropane have different structural formulas, they are different compounds and have different physical properties. For example, the boiling point of butane is −0.5 °C and that of methylpropane is −11.7 °C.

The number of constitutional isomers increases rapidly with the number of carbon atoms. For example, C_5H_{12} has only three isomers (see Example 26-1) whereas $C_{10}H_{22}$ has 75 and $C_{20}H_{42}$ has more than 300,000.

In Chapter 3, we mentioned a way of greatly simplifying the writing of organic structures. We draw lines to represent chemical bonds, and wherever a line ends or meets another line, there is a C atom. We then assume that enough H atoms are bonded to the C atoms to satisfy the need for each C atom to form *four* bonds. Such structural formulas are called *line-angle formulas*, or *line structures*. Line-angle formulas for the three isomers in Example 26-1 are given in the margin. In these formulas, the lines representing the bonds between carbon atoms in the longest continuous chain are arranged in a zigzag fashion, which is consistent with the three-dimensional structures of these molecules. When present, shorter side chains are attached to the appropriate carbon atom of the longest chain, as is the case for structures (2) and (3).

(1)

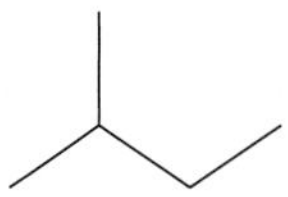

(2)

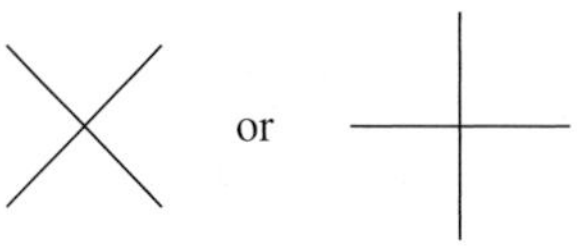

(3)

Nomenclature

Early organic chemists often assigned names related to the origin or properties of new compounds. Some of these names are still in common use. Citric acid is found in citrus fruit; uric acid is present in urine; formic acid is found in ants (from the Latin word for ant, *formica*); and morphine induces sleep (from *Morpheus*, the ancient Greek god of sleep). As thousands upon thousands of new compounds were synthesized, it became apparent that a system of common names was unworkable. Following several interim systems, one recommended by the International Union of Pure and Applied Chemistry (IUPAC) was adopted.

26-1 CONCEPT ASSESSMENT

In 1874, van't Hoff and Le Bel published separate papers advancing the hypothesis that the four bonds from a central carbon extend tetrahedrally. This marked the beginning of the field of stereochemistry. Part of their justification was that there was only one known compound with the formula CH_2X_2. Using the compound CH_2F_2 as an example, determine the expected number of isomers if the orientation of the bonds to a carbon atom is tetrahedral and the expected number if the orientation is square-planar.

EXAMPLE 26-1 Identifying Isomers

Write structural formulas for all the constitutional isomers with the molecular formula C_5H_{12}.

Analyze

First we write the longest chain of C atoms and add an appropriate number of H atoms (12, in this case) to give each C atom four bonds. Next, we look for isomers that have fewer carbon atoms in the longest chain.

Solve

The longest chain of carbon atoms that we can draw has five carbon atoms in it. When we add hydrogen atoms to give each carbon atom four bonds, we obtain structure (1).

(1)

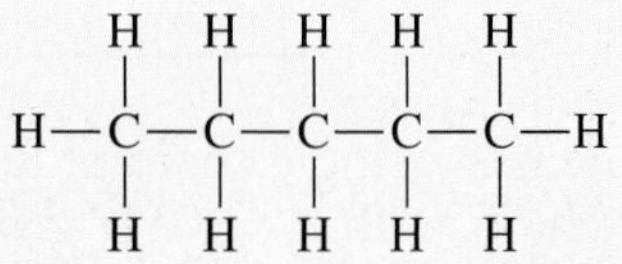

Now, let's look for isomers with four C atoms in the longest chain and one C atom as a branch (five C atoms in all). There is only *one* possibility. Notice that if structure (2) is flipped from left to right, the identical structure (2′) is obtained.

(2)

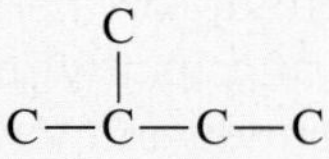

(2′)

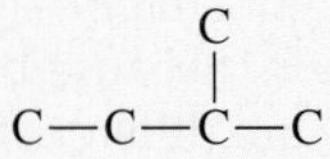

Again, we complete the structure of this isomer by adding H atoms.

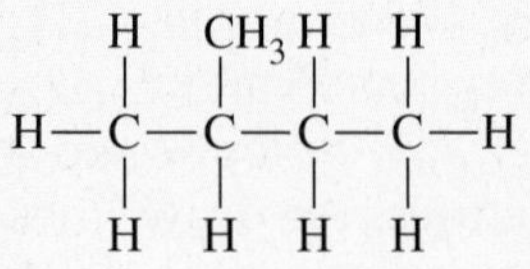

Finally, let's consider a three-carbon chain with two one-carbon branches. Again, there is only *one* possibility.

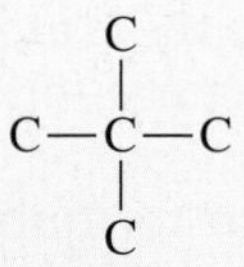

that is,

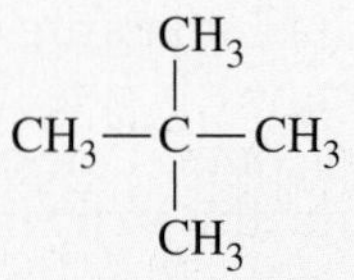

The number of isomers with the formula C_5H_{12} is three.

Assess

We must be careful to recognize when two possible structures are actually the same structure, as was the case for structures (2) and (2′).

PRACTICE EXAMPLE A: Write condensed structural formulas for the five constitutional isomers with the formula C_6H_{14}.

PRACTICE EXAMPLE B: Write condensed structural formulas for the nine constitutional isomers with the formula C_7H_{16}.

In this introduction to nomenclature, we will consider only hydrocarbons with all carbon-to-carbon bonds as single bonds. These are known as **saturated hydrocarbons**, or alkanes. We have no trouble naming the first few: CH_4, methane; C_2H_6, ethane; C_3H_8, propane. We encounter our first difficulty with C_4H_{10}, which has two constitutional isomers. This problem is resolved by assigning the name *butane* to the straight-chain isomer, $CH_3CH_2CH_2CH_3$, and *isobutane* to the branched-chain isomer, $CH_3CH(CH_3)_2$. This method is inadequate for the C_5H_{12} alkanes, for which there are three structural isomers (Example 26-1), and it is even less satisfactory for alkanes with greater numbers of C atoms.

To be able to name molecules that have even greater complexity, we must first consider the nature of some of the possible side chains. A side chain is an alkane with one hydrogen atom removed. The resulting group of atoms is called an **alkyl group**, which is named by replacing the ending *-ane* in the corresponding alkane with *-yl*. For example, $—CH_3$ is the methyl group and $—CH_2CH_2CH_3$ is the propyl group. An alkyl side chain is also called a substituent alkyl group because it replaces (substitutes for) a hydrogen atom in the main chain. Table 26.1 shows some common alkyl groups. We see that some of the names incorporate prefixes, such as *sec-*, an abbreviation for secondary, or *tert-*, an abbreviation for tertiary. We use the terms *primary, secondary*, and *tertiary* to provide information about the nature of carbon atoms in organic molecules. A **primary carbon** is attached to one other carbon atom. Carbon atoms at the ends of an alkane chain are always primary carbons. The hydrogen atoms attached to a primary carbon atom are called **primary hydrogen atoms**, and an alkyl group formed by removing a primary hydrogen atom is a primary group. A **secondary carbon** is attached to two other carbon atoms, and a **tertiary carbon**, to three others. Their hydrogen atoms are labeled similarly. As we can see from Table 26.1, the removal of a secondary hydrogen results in the formation of a secondary alkyl group, and the removal of a tertiary hydrogen results in a tertiary alkyl group. Finally, a carbon attached to four carbon atoms is called **quaternary**. The classification of carbon and hydrogen atoms is illustrated in Figure 26-4.

The following rules enable us to name branched-chain hydrocarbons unambiguously as long as we apply the rules in sequence.

1. Select the *longest* continuous carbon chain in the molecule and use the hydrocarbon name of this chain as the base name. Except for the common names methane, ethane, propane, and butane, standard Greek prefixes relate the name to the number of C atoms in the chain, as in *pent*ane (C_5), *hex*ane (C_6), *hep*tane (C_7), *oct*ane (C_8), ...
2. Consider every branch of the main branch to be a substituent alkyl group. Table 26.1 gives the names of common alkyl substituents. When the substituent is more complex, we use the rules given previously to name the side group, bearing in mind that we would change the *-ane* ending to *-yl*.
3. Number the C atoms of the continuous base chain so that the substituents appear *at the lowest numbers* possible.
4. Name each substituent according to its chemical identity and the number of the C atom to which it is attached. For identical substituents use *di, tri, tetra*, and so on, and write the appropriate carbon number for each substituent.
5. Separate numbers from one another with commas but no spaces, and separate numbers from letters with hyphens.
6. List the substituents *alphabetically* by name. When determining alphabetical order, the prefixes *di-*, *tri-*, *sec-*, and *tert-* are ignored. Thus, *tert*-butyl, precedes methyl in the name 4-*tert*-butyl-2-methylheptane. However, the prefix *iso-* is not ignored when deciding the alphabetical order.

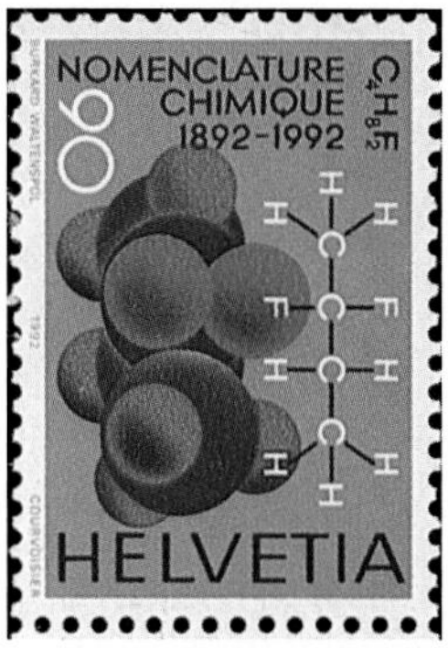

Stamp from the private collection of C.M. Lang. Photography by Gary J. Shulfer, University of Wisconsin, Stevens Point. "1992, Switzerland (Scott #913)"; Scott Standard Postage Stamp Catalogue, Scott Pub. Co., Sidney, Ohio

▲ A Swiss stamp commemorating the 100th anniversary of an international congress in Geneva, at which a systematic nomenclature of organic compounds was adopted.

◀ When distinguishing the different types of H and C atoms, the symbols 1°, 2°, 3°, and 4° are often used in place of the words primary, secondary, tertiary, and quaternary.

◀ Do not try to memorize these rules at the outset. Refer to them as you proceed through the examples, and they will become part of your vocabulary of organic chemistry.

26-2 CONCEPT ASSESSMENT

Is it possible that an organic molecule contains a quaternary hydrogen atom?

TABLE 26.1 Some Common Alkyl Groups

Common Name	IUPAC Name	Structural Formula
Methyl	Methyl	$-CH_3$
Ethyl	Ethyl	$-CH_2CH_3$
Propyl[a]	Propyl	$-CH_2CH_2CH_3$
Isopropyl	1-Methylethyl	CH_3CHCH_3 (bond from the central C)
Butyl[a]	Butyl	$-CH_2CH_2CH_2CH_3$
Isobutyl	2-Methylpropyl	$-CH_2CH(CH_3)CH_3$ (CH_3 drawn above the CH)
sec-Butyl[b]	1-Methylpropyl	$CH_3CHCH_2CH_3$ (bond from the second C)
tert-Butyl[c]	1,1-Dimethylethyl	$CH_3C(CH_3)CH_3$ (CH_3 drawn above the central C; bond from it)

[a]In the past, the prefix *normal* or *n*- was used for a straight-chain alkyl group, such as *n*-propyl or *n*-butyl.
[b]*sec* = secondary
[c]*tert* = tertiary

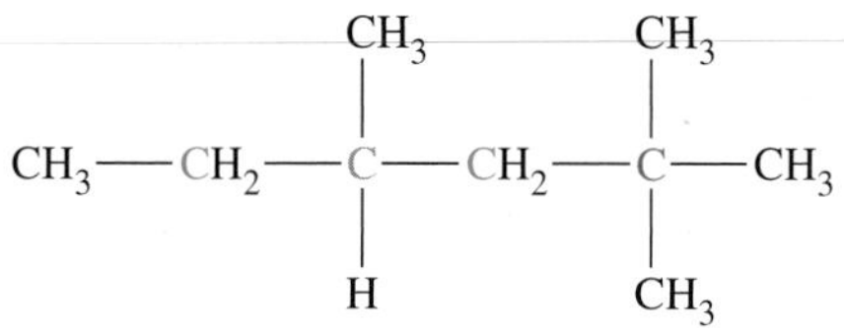

2,2,4-trimethylhexane

C = primary carbon
C = secondary carbon
C = tertiary carbon
C = quaternary carbon

▲ FIGURE 26-4
Classification of carbon and hydrogen atoms
In 2,2,4-trimethylhexane, there are five primary carbons (shown in black), two secondary carbon atoms (shown in blue), one tertiary carbon atom (shown in gray), and one quaternary carbon atom (shown in red). The hydrogen atoms bonded to a primary carbon atom are called primary hydrogen atoms. Similarly, secondary or tertiary hydrogens are bonded, respectively, to secondary or tertiary carbon atoms.

Functional Groups

Organic compounds typically contain elements in addition to carbon and hydrogen. These elements occur as distinctive groupings of one or several atoms called **functional groups**. We have already encountered (in Chapter 3) a

EXAMPLE 26-2 Naming an Alkane Hydrocarbon

Give an appropriate IUPAC name for the following compound, an important constituent of gasoline.

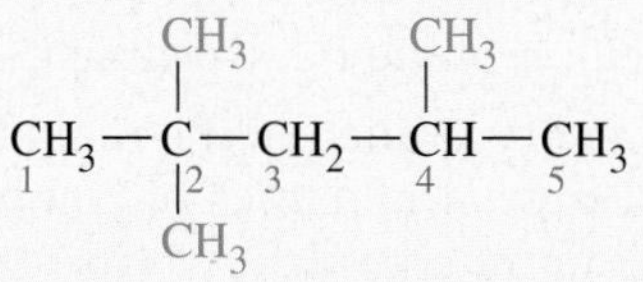

Analyze

We apply the rules listed above. With practice, you will be able to apply the rules without referring back to the list.

Solve

The C atoms are numbered in red, and the side-chain substituents to be named are shown in blue. The longest chain of C atoms is five, and the carbons are numbered so that the one with two substituent groups is number 2 instead of number 4. Each substituent is a methyl group, $-CH_3$. Two methyl groups are on the second C atom, and one methyl group is on the fourth C atom. The correct name is 2,2,4-trimethylpentane.

Assess

If we had numbered the C atoms from right to left, we would have obtained the name 2,4,4-trimethylpentane. This is *not* an acceptable name because it does not use the *smallest* numbers possible.

PRACTICE EXAMPLE A: Give an IUPAC name for $CH_3CH_2CH(CH_3)CH_2CH_2C(CH_3)_2CH_2CH_3$. (*Hint:* Any CH_3 group enclosed in parentheses is bonded only to the C atom preceding it.)

PRACTICE EXAMPLE B: Give an IUPAC name for $CH_3CH_2CH(CH_3)CH_2CH_2CH(CH_3)CH_2CH_3$. (*Hint:* See the hint given in Practice Example A.)

EXAMPLE 26-3 Writing the Formula to Correspond to the Name of an Alkane Hydrocarbon

Write a condensed structural formula for 4-*tert*-butyl-2-methylheptane.

Analyze

The name tells us that the compound is a heptane with two substituent groups, *tert*-butyl and methyl, positioned at carbons 4 and 2, respectively.

Solve

Because the compound is a heptane, the longest chain of C atoms is seven.

C—C—C—C—C—C—C

Starting on the left, we attach a methyl group to the second C atom.

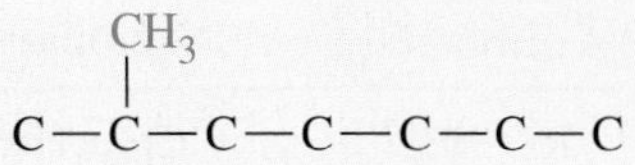

Next, we attach a *tert*-butyl group to the fourth C atom.

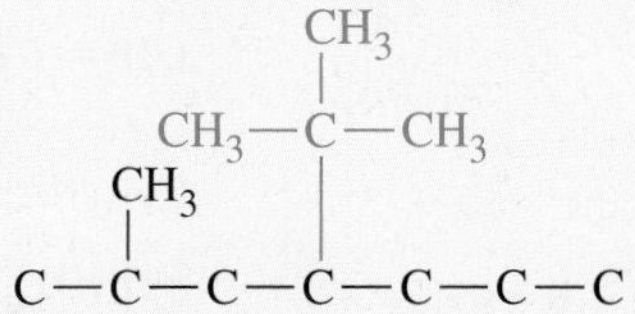

Finally, we add the remaining hydrogen atoms to give each C atom four bonds.

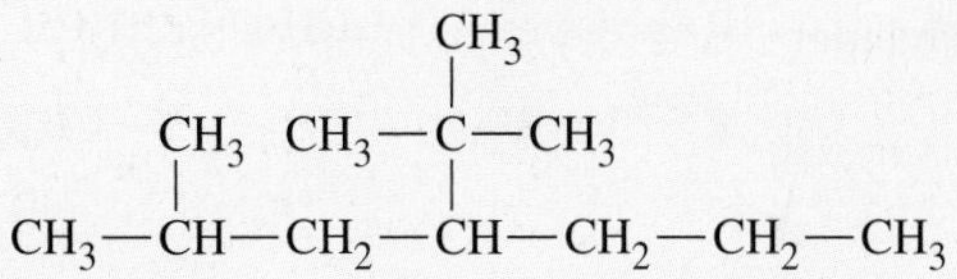

Assess

To check the answer, we use the nomenclature rules given on page 1211 to name the structure we've drawn. We should obtain the name that was given.

PRACTICE EXAMPLE A: Write a condensed structural formula for 3-ethyl-2,6-dimethylheptane.

PRACTICE EXAMPLE B: Write a condensed structural formula for 3-ethyl-2,4-dimethylpentane.

few functional groups, such as the —OH group in alcohols and the —COOH group in carboxylic acids. Table 26.2 lists the major types of organic compounds with their distinctive functional groups shown in red. The physical and chemical properties of organic molecules generally depend on the particular functional groups present. Compounds with the same functional group generally have similar chemical properties. Thus, a convenient way to study organic chemistry is to consider the properties associated with specific functional groups.

In some cases, a functional group simply takes the place of an H atom in a hydrocarbon chain or ring. Such is the case with alcohols and alkyl halides (also called haloalkanes). When naming alcohols and alkyl halides, we must identify the position of the —OH group or halogen atom in the molecule. For example, consider the following possibilities that arise when a Br atom takes the place of a hydrogen atom in pentane:

$CH_3CH_2CH_2CH_2CH_2Br$

1-Bromopentane

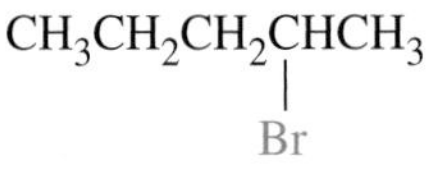

2-Bromopentane

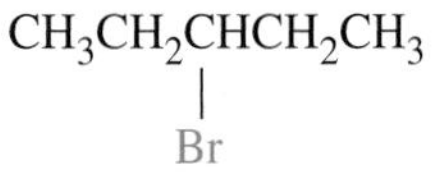

3-Bromopentane

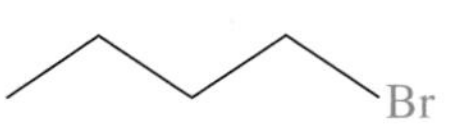

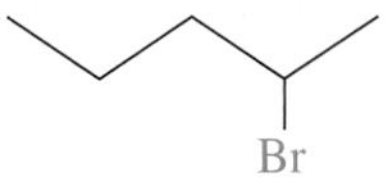

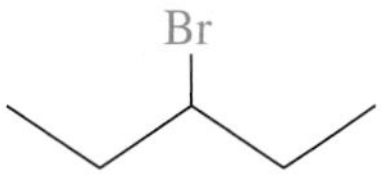

These three monobromopentanes possess the same carbon skeleton and differ only in the position of the bromine atom on the carbon chain. Notice that we include the appropriate carbon number for the Br substituent when naming the compound. In general, the carbon number is placed immediately before the part of the name to which it relates. Here, the carbon number refers to the position of the bromine atom, and so the number is placed immediately before the prefix *bromo-*.

TABLE 26.2 Some Classes of Organic Compounds and Their Functional Groups

Class	General Structural Formula[a]	Example	Preferred IUPAC Name[b]	Other Name(s)
Alkane	R—H	$CH_3CH_2CH_2CH_2CH_2CH_3$	Hexane	—
Alkene	>C=C<	$CH_2{=}CHCH_2CH_2CH_3$	Pent-1-ene	1-Pentene
Alkyne	—C≡C—	$CH_3C{\equiv}CCH_2CH_2CH_2CH_2CH_3$	Oct-2-yne	2-Octyne
Alcohol	R—OH	$CH_3CH_2CH_2CH_2OH$	Butan-1-ol	1-Butanol
Alkyl halide	R—X[c]	$CH_3CH_2CH_2CH_2CH_2CH_2Br$	1-Bromohexane	—
Ether	R—O—R′	$CH_3{-}O{-}CH_2CH_2CH_3$	1-Methoxypropane	Methyl propyl ether[d]
Amine	$R{-}NH_2$	$CH_3CH_2CH_2{-}NH_2$	Propan-1-amine	1-Propanamine, 1-aminopropane, Propylamine[d]
Aldehyde	R—C(=O)—H	$CH_3CH_2CH_2C({=}O){-}H$	Butanal	Butyraldehyde[d]
Ketone	R—C(=O)—R′	$CH_3CH_2C({=}O)CH_2CH_2CH_3$	Hexan-3-one	3-Hexanone, ethyl propyl ketone[d]
Carboxylic acid	R—C(=O)—OH	$CH_3CH_2CH_2C({=}O){-}OH$	Butanoic acid	Butyric acid[d]
Ester	R—C(=O)—OR′	$CH_3CH_2CH_2C({=}O){-}OCH_3$	Methyl butanoate	Methyl butyrate[d]
Amide	$R{-}C({=}O){-}NH_2$	$CH_3CH_2CH_2C({=}O){-}NH_2$	Butanamide	Butyramide
Arene	Ar—H[e]	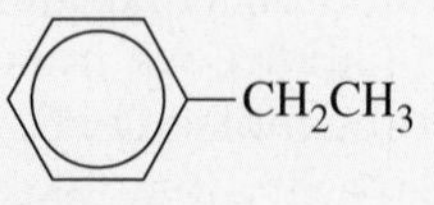	Ethylbenzene	—
Aryl halide	Ar—X[c]	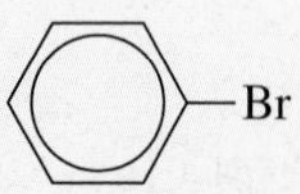	Bromobenzene	—
Phenol	Ar—OH	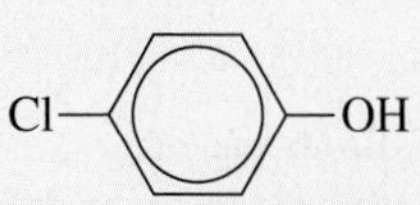	4-Chlorophenol	*p*-Chlorophenol

[a]The functional group is shown in red. R and R′ represent alkyl groups.
[b]In the preferred IUPAC name, the carbon number is placed immediately before the part of the name to which it relates. For example, it appears before *ene* in the name of an alkene; before *-yne* in an alkyne; before *-ol* in the name of an alcohol; etc.
[c]X stands for a halogen atom: F, Cl, Br, or I.
[d]Common name.
[e]Ar stands for an aromatic (*aryl*) group, such as the benzene ring.

We will discuss several classes of organic compounds in this chapter, focusing primarily on their structures and properties. In the next chapter, we will focus on some of their characteristic reactions.

26-3 CONCEPT ASSESSMENT

What is the IUPAC name for $CH_3CH_2CH_2CH(OH)CH_3$? You may find it useful to review Section 3-7.

26-2 Alkanes

In this section we will explore some properties of the alkanes. The essential characteristic of **alkane** hydrocarbon molecules is that they have only single covalent bonds. The bonds in these compounds are said to be *saturated*.

The alkanes range in complexity from methane, CH_4, to molecules containing fifty C atoms or more. Most have the formula C_nH_{2n+2}, and each alkane differs from the preceding one in a sequence by a $-CH_2-$, or *methylene* group. Substances whose molecules differ only by a constant unit such as $-CH_2-$ are said to form a **homologous series**. Members of such a series usually have closely related chemical and physical properties. The data in Table 26.3 indicate that boiling points of alkanes are related to polarizabilities and shapes in the ways discussed in Section 12-1. Intermolecular attractions between the straight-chain molecules are strongest, and these molecules have the highest boiling points. Isomers with more compact structures have lower boiling points.

◀ Alkanes are nonpolar, water-insoluble compounds with relatively low melting points and boiling points.

26-4 CONCEPT ASSESSMENT

Illustrate the meaning of homologous series by writing a homologous series for alkyl halides.

Conformations

With ball-and-stick models, we can visualize an important type of motion in alkane molecules—rotation of groups with respect to one another about the σ bond connecting them. **Conformations** are different spatial arrangements that are possible in a molecule. One conformation can be converted into another by rotations about σ bonds. In Figure 26-5, we focus on one of the many possible conformations of the CH_3CH_3 molecule. The spatial arrangement of the H atoms can be seen more clearly if we view the molecule along the C—C axis, as suggested in Figure 26-5(a). When the molecule is viewed in this way, the carbon atom toward the rear is obscured by the one in front, as shown in Figure 26-5(b), but the bonds to the hydrogen atoms are clearly seen. The view along the C—C bond is shown in a slightly different way in Figure 26-5(c).

TABLE 26.3 Boiling Points of Some Isomeric Alkanes

Formula	Isomer	Boiling Point, °C	Formula	Isomer	Boiling Point, °C
C_4H_{10}	Butane	−0.5	C_6H_{14}	Hexane	68.7
	Methylpropane	−11.7		3-Methylpentane	63.3
C_5H_{12}	Pentane	36.1		2-Methylpentane	60.3
	2-Methylbutane	27.9		2,3-Dimethylbutane	58.0
	2,2-Dimethylpropane	9.5		2,2-Dimethylbutane	49.7

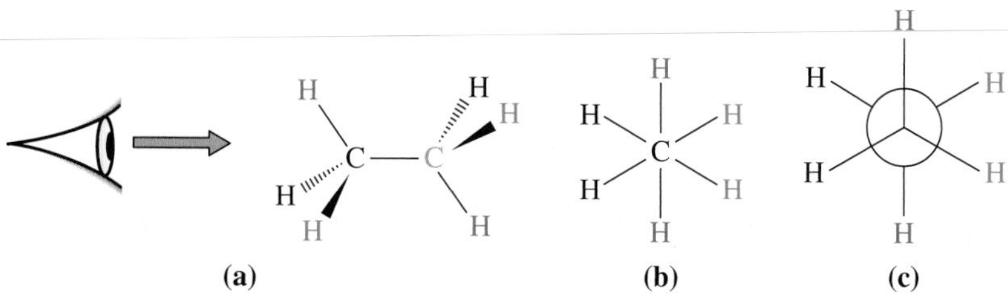

▲ FIGURE 26-5
Staggered conformation of ethane
When the ethane molecule is viewed along the C—C axis, as suggested in **(a)**, the rear carbon atom is obscured by the carbon atom in front, as shown in **(b)**. In the Newman projection, shown in **(c)**, the front carbon is located at the intersection of the three arms of the inverted Y and the rear carbon is represented by a circle.

▶ Newman projections are named after an organic chemistry professor, Melvin S. Newman, from Ohio State University. He introduced these representations in 1952 to help students understand conformations, stereochemistry, and symmetry of organic molecules.

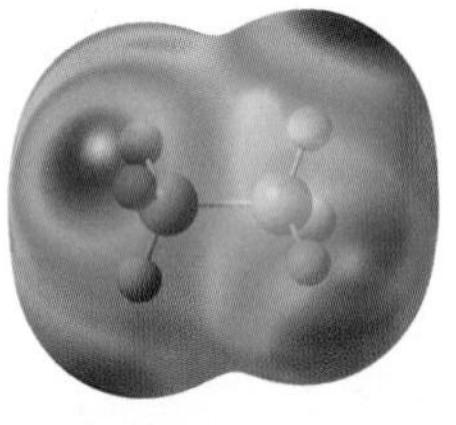

Staggered conformation

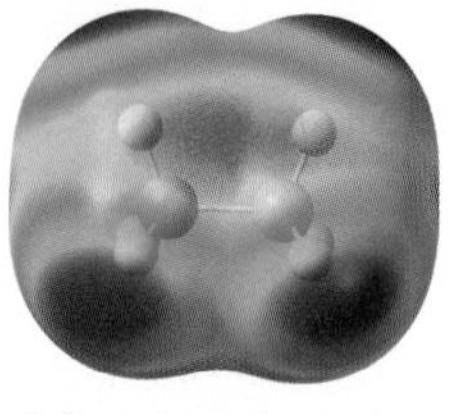

Eclipsed conformation

KEEP IN MIND

that the barrier to rotation in ethane (12 kJ mol^{-1}) is comparable to the strengths of various intermolecular forces we discussed in Chapter 12.

In this representation, called a *Newman projection,* the carbon atom toward the front is located at the point where the lines representing the three carbon–hydrogen bonds intersect with each other. The carbon atom toward the rear is depicted by a circle and its bonds project from the outer edge of the circle. Newman projections are used to represent the many different spatial arrangements of atoms that result from rotations about a σ bond. The arrangement (or conformation) shown in Figure 26-5 is called the *staggered conformation*. In this conformation, the carbon–hydrogen bonds in one $-CH_3$ group are positioned exactly halfway between those of the other $-CH_3$ group and the H atoms are located a maximum distance apart.

A second conformation can be obtained by rotating one of the methyl groups in the staggered conformation by 60° about the C—C axis, as suggested in Figure 26-6. When the resulting conformation is viewed along the C—C axis, all the hydrogen atoms on the first carbon atom are directly in front of those on the second carbon atom. This is called the *eclipsed* conformation. To make the three rear hydrogen atoms more visible in the Newman projection, they are drawn slightly out of the perfectly eclipsed position. The eclipsed and staggered conformations represent two extremes and all the possible conformations in between are called collectively *skew conformations*.

Electrostatic potential maps for the staggered and eclipsed conformations of ethane are shown in the margin. In the staggered conformation, the hydrogen atoms (in the blue regions) are as far apart from one another as possible and in the eclipsed conformation, they are as close as possible.

Energy is required to convert from the staggered conformation to the eclipsed conformation, and for ethane, the energy required is about 12.0 kJ mol^{-1}. Because of this energy requirement, there is a *barrier to internal rotation*, and thus the $-CH_3$ groups in ethane do not rotate entirely freely about the C—C bond. However, the barrier to internal rotation is small enough that, at room

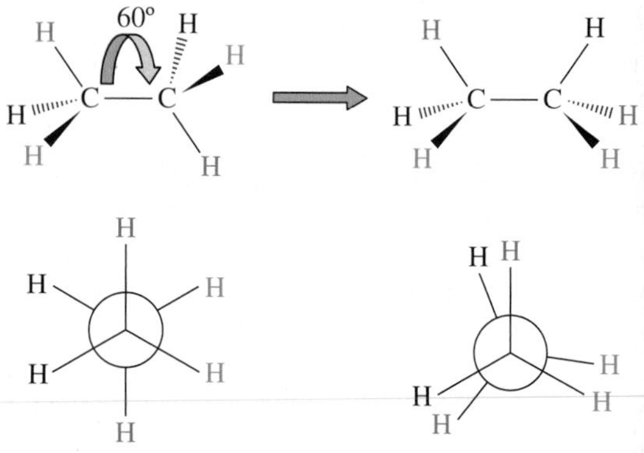

Staggered Eclipsed

▶ FIGURE 26-6
Staggered and eclipsed conformations of ethane
When the methyl group on the right is rotated by 60° about the C—C bond, the staggered conformation of ethane is converted into the eclipsed conformation.

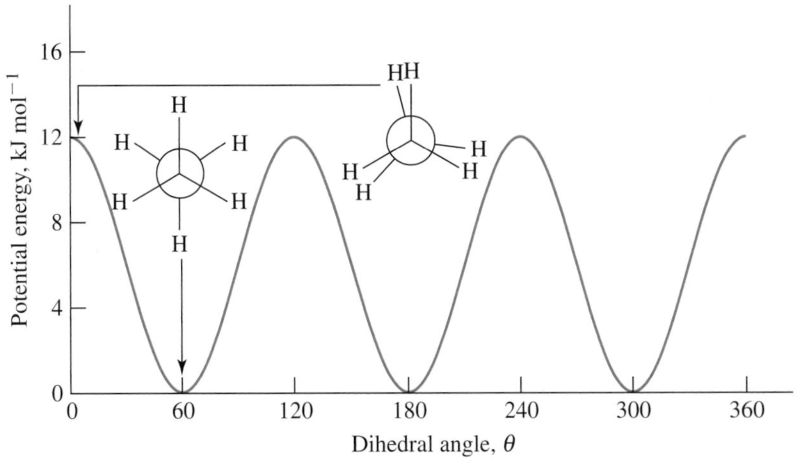

▲ FIGURE 26-7
Potential energy diagram for the internal rotation of the methyl groups in ethane

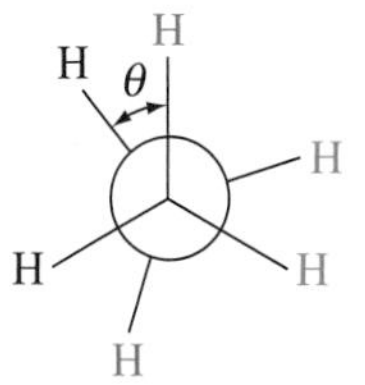

The dihedral angle, θ, refers to the angle of rotation about a carbon–carbon bond.

temperature, molecules interconvert rapidly from one conformation to another. For this reason, rotation about the C—C bond is sometimes called *free rotation.* The barrier to internal rotation arises because as the molecule converts from the staggered conformation through all the skew conformations to the eclipsed conformation, the C—H bonds of one —CH_3 group draw closer to the C—H bonds of the other —CH_3 group and the electrons in these bonds experience increased repulsion. The eclipsed conformation is highest in energy because in this conformation, the electrons in the C—H bonds of one methyl group are close to the electrons in the C—H bonds of the other methyl group.

The conversion from one conformation to another can be followed by using the dihedral angle, θ, which refers to the angle of rotation about the carbon–carbon bond, as shown in the margin. When $\theta = 0°$, the molecule is in the eclipsed conformation, and when $\theta = 60°$, it is in the staggered conformation. The energy changes that occur as the ethane molecule converts from one conformation to another are shown graphically in Figure 26-7. The difference in energy between the eclipsed and staggered conformations is called the rotational or **torsional energy**. In the eclipsed conformation, there are three C—H bond interactions, and thus each C—H bond interaction contributes 4.0 kJ mol^{-1}.

Let's consider rotation about a C—C bond in propane, $CH_3CH_2CH_3$, which is the next member of the homologous series. The potential energy diagram of propane is similar to that of ethane. However, an important difference is that the energy difference between the eclipsed and staggered conformations, and thus the barrier to rotation about a C—C bond, is slightly greater in propane (13.6 kJ mol^{-1}) than it is in ethane (12.0 kJ mol^{-1}). Newman projections for the conformations of propane can help us understand this difference. The Newman projections for the staggered and eclipsed conformations of propane, shown in the margin, are similar to those of ethane except that one of the H atoms has been replaced by a methyl group. The eclipsed conformation of propane has two C—H bond interactions and an interaction involving a C—H bond and a carbon–methyl bond. If we assume that each C—H bond interaction contributes 4.0 kJ mol^{-1} to the barrier to rotation, as was the case in ethane, then the interaction between the carbon– hydrogen bond and the carbon-methyl bond contributes $(13.6 - 2 \times 4.0) = 5.6$ kJ mol^{-1}. Thus, the interaction between a carbon–hydrogen bond and a carbon-methyl bond is slightly more repulsive than the interaction of two C—H bonds.

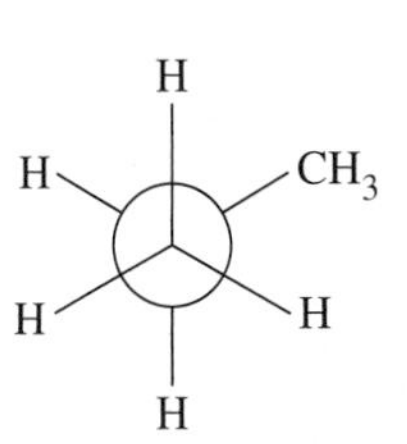

Staggered

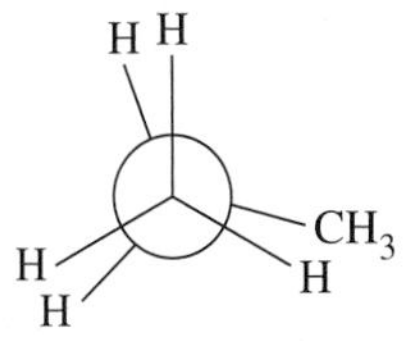

Eclipsed

The next member in the homologous series is butane, $CH_3CH_2CH_2CH_3$. If we number the carbon atoms 1 through 4, then we have two distinct C—C bonds, C1—C2 and C2—C3, about which conformers can be formed. (The C3—C4 bond axis is equivalent to the C1—C2 bond axis because it does not matter whether we number the carbon atoms from left to right or right t left.) When the butane molecule is viewed along the C1—C2 bond, w obtain the following Newman projections for the staggered and eclipse conformations:

▶ When drawing Newman projections for molecules, it is customary to put the lower-numbered carbon atom at the front of the Newman projection.

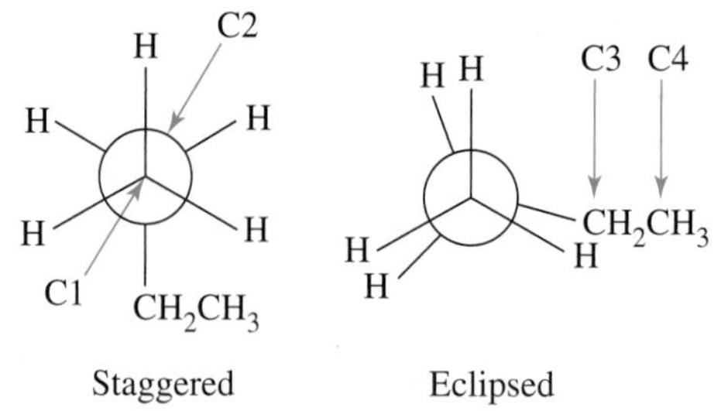

26-5 CONCEPT ASSESSMENT

Will all the eclipsed conformers for rotation about the C1—C2 axis in butane have the same energy?

When the butane molecule is viewed along the C2—C3 bond, severa eclipsed and staggered conformations can be identified (Figure 26-8). Ther

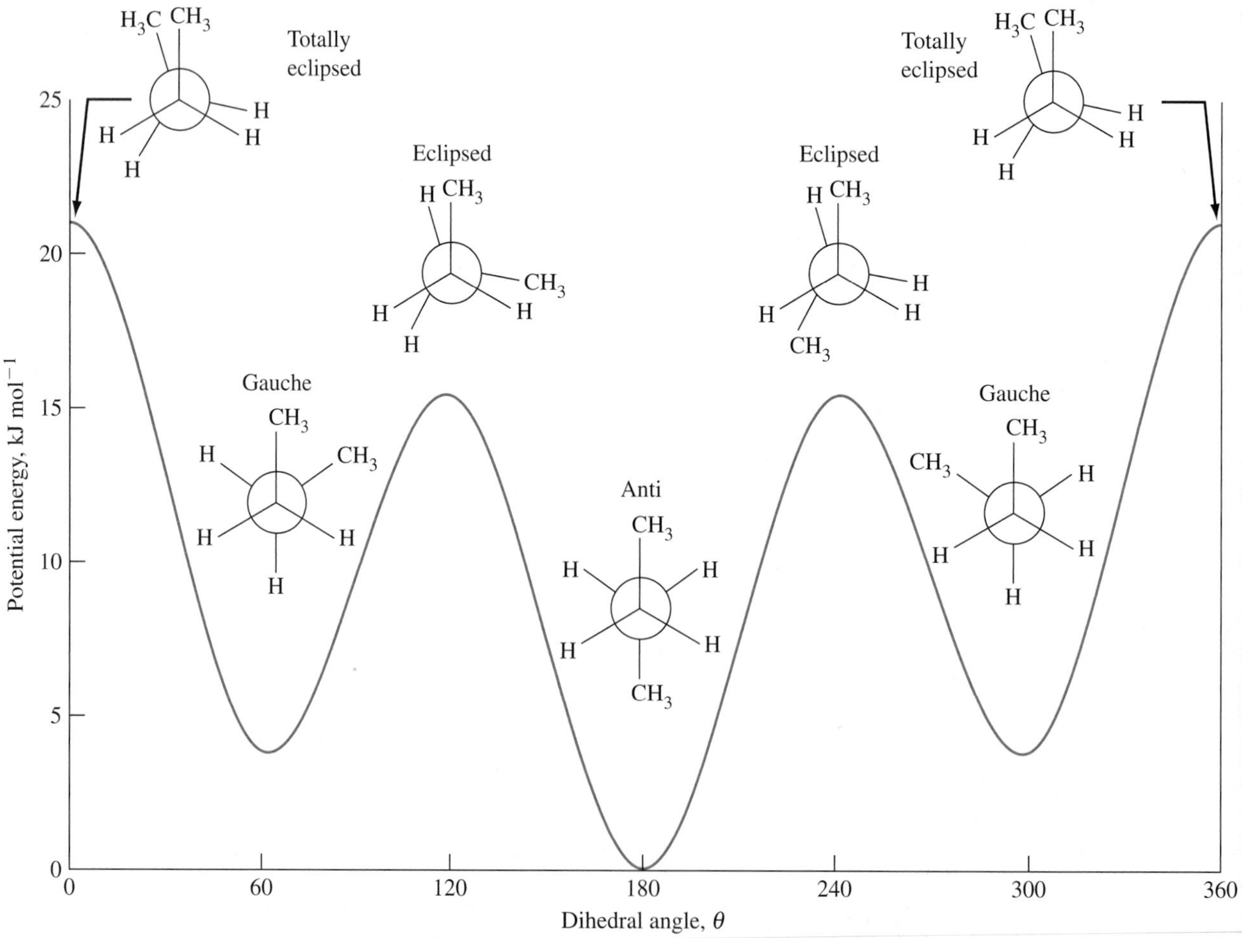

▲ FIGURE 26-8
Conformations and their potential energies of butane for rotation about the C2—C3 bond

are two distinct eclipsed conformations, eclipsed and totally eclipsed, and two distinct staggered conformations, anti and gauche. The totally eclipsed is highest in energy because in this conformation, the two largest constituents are eclipsed and interfere with each other, as suggested in Figure 26-9. The interference or crowding of large substituents is called *steric hindrance*.

Of the conformations shown in Figure 26-8, the anti conformation is the lowest in energy because the methyl groups are as far apart as possible. This conformation is called the **anti** conformation because the methyl groups are diagonally opposite each other. The other staggered conformation has the methyl groups to the left and right of each other and is called the **gauche** conformation. (*Gauche* is a French word meaning "left" or "awkward.")

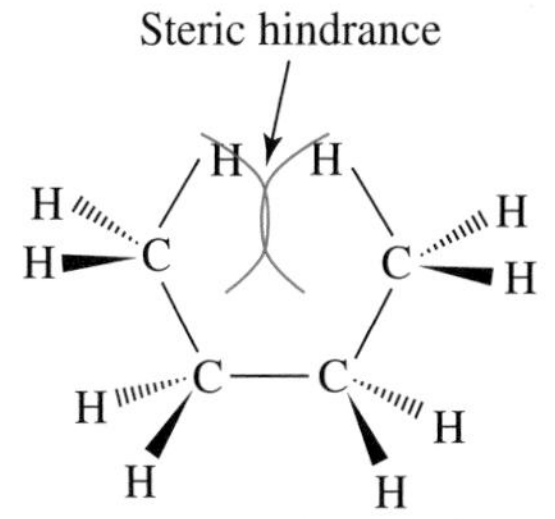

▲ FIGURE 26-9
Steric hindrance in a totally eclipsed conformation of butane
In the totally eclipsed conformation, the two methyl groups interfere with each other.

EXAMPLE 26-4 Choosing the Most Stable Conformer of an Alkane

Draw the conformation of 2,3-dimethylpentane that is lowest in energy when the molecule is viewed along C2—C3 axis.

Analyze

We must first draw a structural diagram for the molecule and identify the C2—C3 axis and then draw a Newman projection corresponding to a view along the C2—C3 bond. The lowest energy structure is the one that minimizes the repulsions among the substituents.

Solve

A structural diagram for 2,3-dimethylpentane is given here.

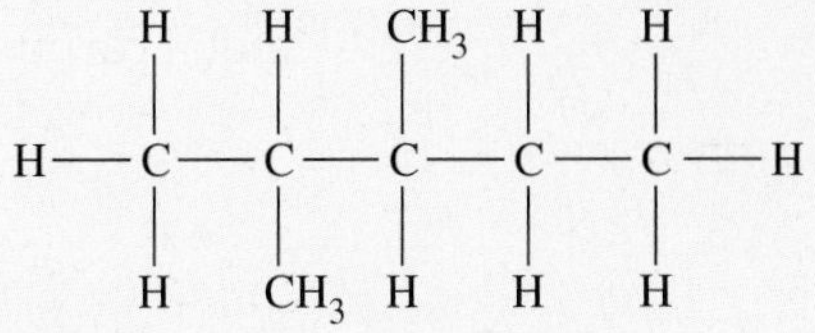

A hydrogen atom and two methyl groups are bonded to C2. Three different groups are bonded to C3: a hydrogen atom, a methyl group, and an ethyl group. The conformation of lowest energy is the one in which the alkyl groups are staggered. In addition to this, we should minimize the number of gauche interactions between the larger groups.

We must now construct a Newman projection corresponding to the view along the C2—C3 bond. First, we draw a circle to represent the rear carbon, and then we add lines for the bonds formed by the front carbon atom. Finally, we add lines for the bonds formed by the rear carbon atom.

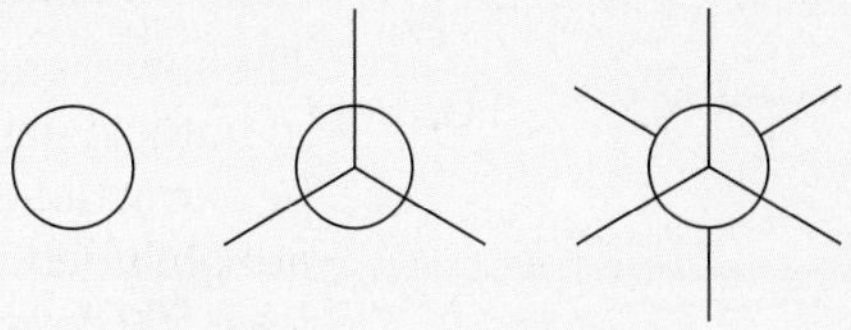

We are now ready to attach the groups to construct 2,3-dimethylpentane. Let us add two methyl groups and a hydrogen atom to the second carbon atom. Note it does not matter where we place the groups.

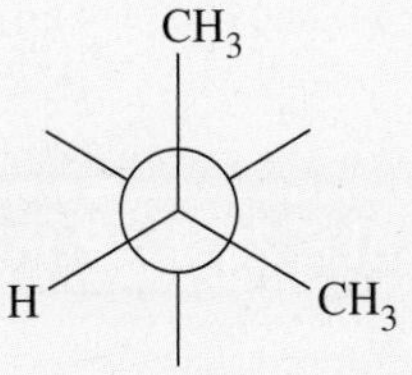

We must now add a hydrogen atom, as well as ethyl and methyl groups to the rear carbon. There are three possible staggered conformations:

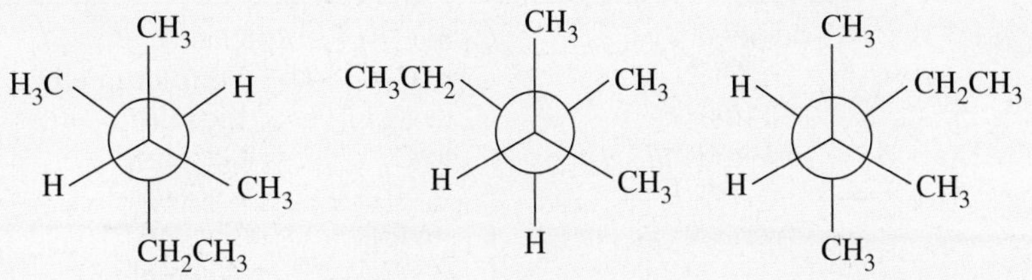

(a) (b) (c)

(continued)

Which one of these conformations is lowest in energy? Conformation (a) has two gauche interactions, but conformations (b) and (c) have three such interactions. Therefore, conformation (a) has the lowest energy.

Assess

In solving this problem, we considered rotation about a particular carbon–carbon bond. Keep in mind that rotations about other carbon–carbon bonds also occur simultaneously, a complication that we will not consider.

PRACTICE EXAMPLE A: Draw Newman projections for the staggered conformations of 2-methylpentane when the molecule is viewed along the C1—C2 bond. Rank the conformations in order of increasing energy (from lowest to highest).

PRACTICE EXAMPLE B: Sketch a potential energy diagram for rotation about the C1—C2 axis in 1-chloropropane.

Preparation of Alkanes

The chief source of alkanes is petroleum, as we describe at the end of this section, but several laboratory methods are also available for their preparation. In the presence of a metal catalyst such as Pt, Pd, or Ni, unsaturated hydrocarbons, whether containing double or triple bonds, may be converted to alkanes by the addition of H atoms to the multiple bond systems.

▶ Figure 23-3 gives a schematic representation of the role played by the catalyst in equation (26.1).

$$CH_2{=}CH_2 + H_2 \xrightarrow{\text{Pt, Pd, or Ni}} CH_3{-}CH_3 \quad (26.1)$$

In another type of reaction, halogenated hydrocarbons react with alkali metals to produce alkanes of double the carbon content (equation 26.2). Finally, alkali metal salts of carboxylic acids can be fused with alkali metal hydroxides. Sodium carbonate and an alkane with one carbon fewer than the metal carboxylate are formed (equation 26.3).

$$2\,CH_3CH_2Br + 2\,Na \xrightarrow{\text{heat/pressure}} 2\,NaBr + CH_3CH_2CH_2CH_3 \quad (26.2)$$

$$CH_3{-}\overset{\overset{\displaystyle O}{\|}}{C}{-}ONa + NaOH \xrightarrow{\Delta} Na_2CO_3 + CH_4 \quad (26.3)$$

Alkanes from Petroleum

The lower molecular mass alkanes, methane and ethane, are found principally in natural gas. Propane and butane are found dissolved in petroleum; they can be extracted as gases and sold as liquefied petroleum gas (LPG). Higher alkanes are obtained by the fractional distillation of petroleum, a complex mixture of at least 500 compounds. The main petroleum fractions are listed in Table 26.4. (The liquid–vapor equilibrium principles that underlie fractional distillation were discussed on page 663.)

TABLE 26.4 Principal Petroleum Fractions

Boiling Range, °C	Composition	Fraction	Uses
Below 0	C_1 to C_4	Gas	Gaseous fuel
0–50	C_5 to C_7	Petroleum ether	Solvents
50–100	C_6 to C_8	Ligroin	Solvents
70–150	C_6 to C_9	Gasoline	Motor fuel
150–300	C_{10} to C_{16}	Kerosene	Jet fuel, diesel oil
Over 300	C_{16} to C_{18}	Gas–oil	Diesel oil, cracking stock
—	C_{18} to C_{20}	Wax–oil	Lubricating oil, mineral oil, cracking stock
—	C_{21} to C_{40}	Paraffin wax	Candles, wax paper
—	above C_{40}	Residuum	Roofing tar, road materials, waterproofing

Not all the hydrocarbons found in gasoline are equally desirable because some hydrocarbons burn more smoothly than others. (Explosive burning results in engine knocking.) The hydrocarbon *2,2,4-trimethylpentane*, a structural isomer of octane, has excellent engine performance, and it is given an octane rating of 100. *Heptane* has poor engine performance; its octane rating is 0. These two hydrocarbons serve as a reference system for rating gasoline. By identifying the composition of the *n*-heptane and 2,2,4-trimethylpentane (isooctane) mixture that matches the performance characteristics of the gasoline being tested, an octane rating can be assigned. For example, a gasoline that gives the same performance as a mixture of 87% 2,2,4-trimethylpentane and 13% heptane is assigned an octane number of 87. In general, branched-chain hydrocarbons have higher octane ratings (burn more smoothly) than their straight-chain counterparts.

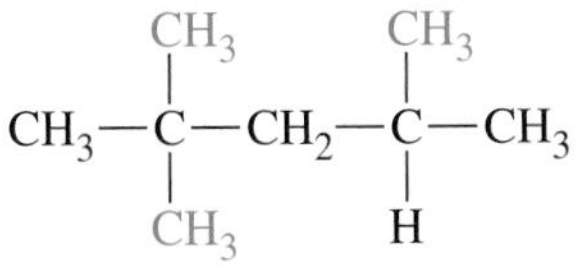

2,2,4-Trimethylpentane
Octane rating: 100

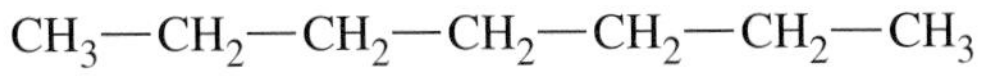

n-Heptane
Octane rating: 0

▲ A catalytic cracking unit (cat cracker) at a petroleum refinery.

Gasoline obtained by fractional distillation of petroleum has an octane number of 50–55 and is not acceptable for use in automobiles, which require fuels with octane numbers near 90. Extensive modifications of the gasoline fraction are required. In *thermal cracking*, large hydrocarbon molecules (called *cracking stock*) are broken down into molecules in the gasoline range, and the presence of special catalysts promotes the production of branched-chain hydrocarbons. For example, the molecule $C_{15}H_{32}$ might be broken down into C_8H_{18} and C_7H_{14}. The process of *re-forming*, or isomerization, converts straight-chain to branched-chain hydrocarbons and other types of hydrocarbons having higher octane numbers. In thermal and catalytic cracking, some low-molecular mass hydrocarbons are rejoined into higher molecular mass hydrocarbons by a process known as *alkylation*.

◀ Not only do the processes of cracking, re-forming, and alkylation produce a higher grade of gasoline but they also increase the yield of gasoline obtained from crude oil.

The octane rating of gasoline can be further improved by adding antiknock compounds to prevent premature combustion. At one time, the preferred additive was tetraethyllead, $(C_2H_5)_4Pb$. Lead additives have been phased out of gasoline in most countries because lead is toxic, and substitutes such as the oxygenated hydrocarbons methanol and ethanol are used instead.

26-3 Cycloalkanes

Alkanes in chain structures have the formula C_nH_{2n+2}. However, alkanes can also exist in ring, or cyclic, structures; such alkanes are called *cycloalkanes* and are said to be *alicyclic*. For simple cycloalkanes, we can think of the rings as having formed by the joining of the two ends of a straight-chain alkane after the elimination of an H atom from each end. Simple cycloalkanes have the formula C_nH_{2n}. Here are a few examples:

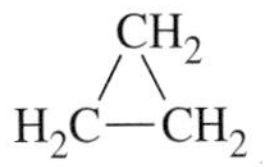

Cyclopropane
C_3H_6

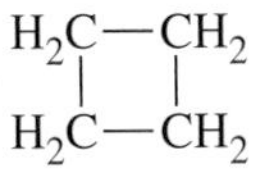

Cyclobutane
C_4H_8

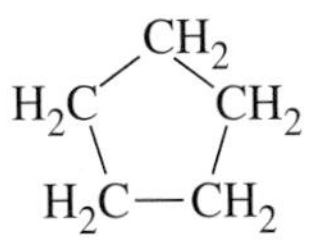

Cyclopentane
C_5H_{10}

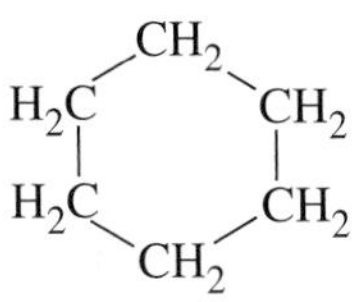

Cyclohexane
C_6H_{12}

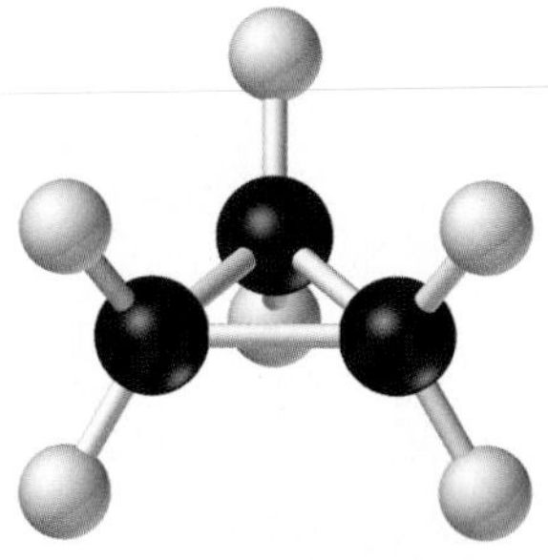

(a)

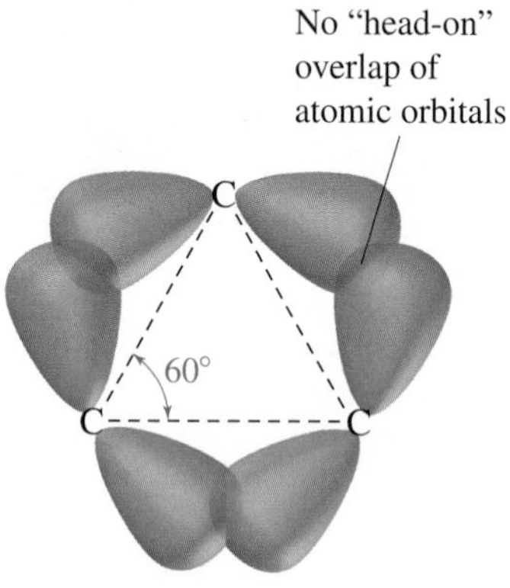

(b)

▲ FIGURE 26-10
Ring strain in cyclopropane
(a) Ball-and-stick model.
(b) Valence-bond picture of cyclopropane using sp^3 hybrid orbitals, which leads to poor overlap of the orbitals and hence weak bonds.

▶ The C—C bond energy in cyclopropane is about 289 kJ mol^{-1}. This is substantially smaller than the C—C bond energy of 347 kJ mol^{-1} given in Table 10.3.

The line-angle representations shown at the bottom of page 1221 might lea you to believe that the carbon atoms in cycloalkanes all lie in the same plane However, this is not generally the case. As we will soon see, cyclopropane is th only cycloalkane in which the carbon atoms form a planar ring.

We can use the nomenclature rules on page 1211 to name a cycloalkan having substituent groups. For example, we would name the following com pound 1,1,3-trimethylcyclopentane not 1,3,3-trimethylcyclopentane:

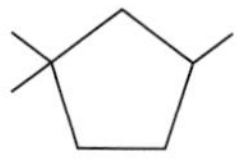

Ring Strain in Cycloalkanes

Some discussion of the bonding in cycloalkanes is warranted because, in these molecules, the sp^3-hybridized carbons do not necessarily form bonds that ar 109.5° apart. For example, the C—C—C bond angles in cyclopropane ar only 60° (Fig. 26-10). Because the bond angle in cyclopropane is much smalle than the ideal bond angle of 109.5°, the cyclopropane molecule is "strained." A illustrated in Figure 26-10(b), the carbon sp^3 orbitals in propane do not overlap as extensively as they do in straight-chain alkanes. As a result of the poo orbital overlap, the C—C bonds in cyclopropane are substantially weake than they are in propane and other straight-chain alkanes, and cyclopropane is considerably more reactive than a straight-chain alkane.

We can use heats of combustion to estimate the amount of ring strain in cycloalkanes. The heats of combustion of propane, butane, pentane, and hexane are given in Table 26.5 along with those of the corresponding cycloalkanes. For the straight-chain alkanes, the heat of combustion change by about 658 kJ mol^{-1} with each CH_2 group added. This suggests that each CH_2 group contributes, on average, 658 kJ mol^{-1} to the heat of combustion The formula of a cycloalkane may be written as $(CH_2)_n$ and, if we assume tha the C—C bonds that link together CH_2 groups in cycloalkanes are the same as they are in straight-chain alkanes, then the heat of combustion of a cycloalkane should be about $-n \times 658$ kJ mol^{-1}. The energy associated with ring strain is released as heat when the compound is burned and thus, the heat of combustion is more negative than expected. The estimated and experimen tal heats of combustion for a few cycloalkanes are given in Table 26.5. The dif ference between the estimated and experimental values provides a measure of the ring strain in the cycloalkanes, and these values are also shown in the table. The data in Table 26.5 show that, as we proceed from cyclopropane to cyclohexane, the amount of ring strain decreases. Interestingly, the cyclo hexane ring is essentially free of ring strain. We will soon see why.

The data in Table 26.5 show that cyclopropane and cyclobutane experience a significant amount of ring strain. However, the *ring strain per CH_2 group* is

TABLE 26.5 Heats of Combustion and Ring Strain in Some Cycloalkanes (All Values in kJ mol^{-1})

Alkane	Experimental $\Delta_{comb}H°$	Contribution to $\Delta_{comb}H°$ from Each Additional CH_2 Group	Cycloalkane	Experimental $\Delta_{comb}H°$	Estimated $\Delta_{comb}H°$[b]	Ring Strain[c]
propane	−2220		cyclopropane	−2092	−1974	118
butane	−2877	−657[a]	cyclobutane	−2744	−2632	112
pentane	−3536	−659	cyclopentane	−3320	−3290	30
hexane	−4194	−658	cyclohexane	−3948	−3948	0

[a]The difference between successive values of $\Delta_{comb}H°$ is the contribution to $\Delta_{comb}H°$ made by adding another CH_2 group. For example, $\Delta_{comb}H°$(butane) − $\Delta_{comb}H°$(propane) = −657 kJ mol^{-1}.
[b]If there is no ring strain, then the heat of combustion of a cycloalkane should be $-n \times 658$ kJ mol^{-1}.
[c]The ring strain is equal to the difference between the estimated and experimental values of $\Delta_{comb}H°$.

ubstantially higher in cyclopropane (39 kJ mol^{-1}) than it is in cyclobutane 28 kJ mol^{-1}). The cyclopropane ring is strained not only because of poor rbital overlap but also because the carbon–hydrogen bonds in this molecule re eclipsed. This is most evident from the Newman projection of cyclo-ropane, which is shown in Figure 26-11.

In cyclopropane, the carbon atoms lie in the same plane, but in other ycloalkanes, they do not. For example, both the cyclobutane and the cyclopen-ane molecules buckle slightly to give rings that are puckered rather than planar, s shown below.

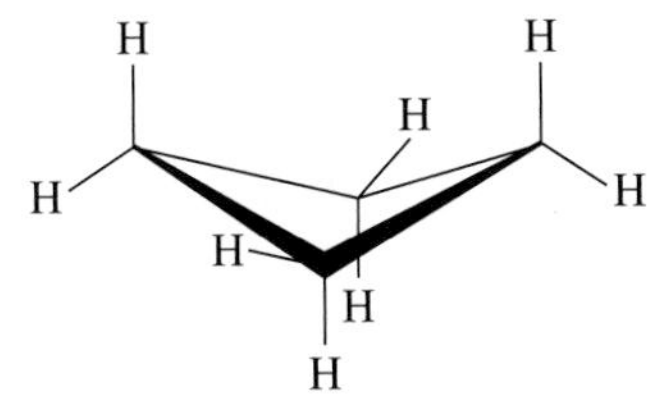

Cyclobutane

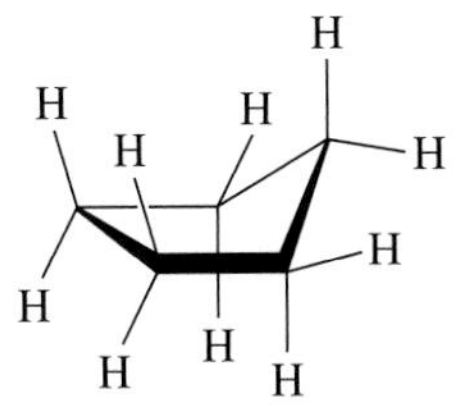

Cyclopentane

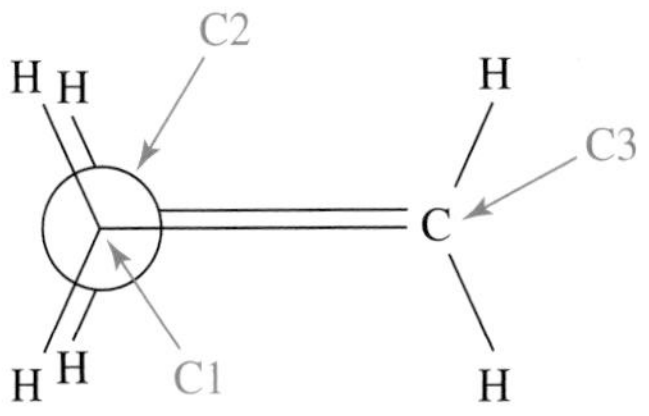

▲ FIGURE 26-11
Newman projection for cyclopropane
The carbon atom in front (C1) and the carbon atom in the rear (C2) are both bonded to the other carbon atom (C3). The H—C1 and H—C2 bonds are eclipsed.

When these molecules buckle, some or all of the hydrogen atoms move lightly out of the totally eclipsed conformation, relieving repulsion associated vith the eclipsing of carbon–hydrogen bonds but at the expense of decreasing he already strained C—C—C bond angles even further.

For cyclohexane, two conformations of the molecules are important. They are hown as ball-and-stick models in Figure 26-12. These are called the *boat* and the *hair conformations*. The boat conformation is less stable (29 kJ mol^{-1} higher in nergy) than the chair conformation. There are other conformations of cyclo-exane, but we will not discuss them. You may encounter them in more dvanced organic chemistry courses.

Cis–Trans Isomerism in Disubstituted Cycloalkanes

The ring in a cycloalkane has two distinct faces, and the substituents onded to a particular carbon atom are adjacent to opposite faces of the ing. Let's consider, for example, the ball-and-stick model of cyclopropane, vhich was shown in Figure 26-10. If we focus on the hydrogen atoms onded to the rearmost carbon, for example, then we see that one of the ydrogen atoms is above the plane of the three carbon atoms and the other s below the plane. Stated another way, one hydrogen atom is adjacent to he upper face and the other is adjacent to the lower face. If hydrogen atoms

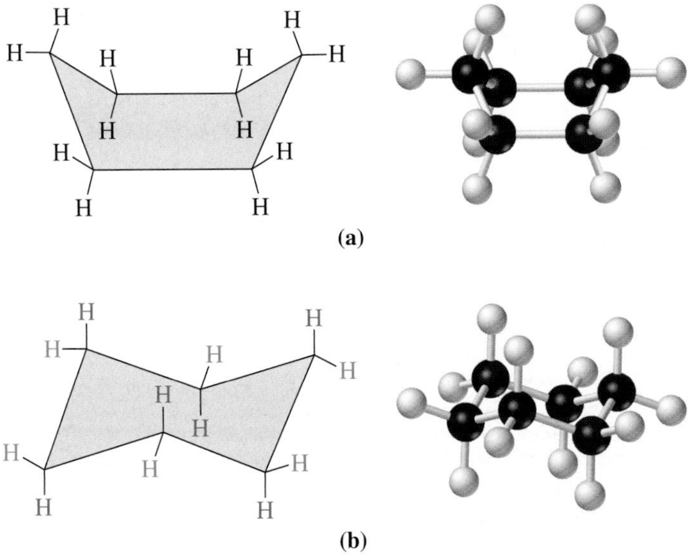

◀ FIGURE 26-12
Two important conformations of cyclohexane
(a) Boat form. **(b)** Chair form. The H atoms extending laterally from the ring—equatorial H atoms—are shown in blue. The H atoms projecting above and below the ring—axial H atoms—are in red.

on different carbon atoms are replaced by other substituents, as is the case in 1,2-dimethylcyclopropane or 1,2-dimethylhexane, then various isomers are possible. The *cis* and *trans* isomers of 1,2-dimethylcyclopropane are shown below. (*cis* is a Latin word meaning "on this side"; *trans* is Latin for "across.")

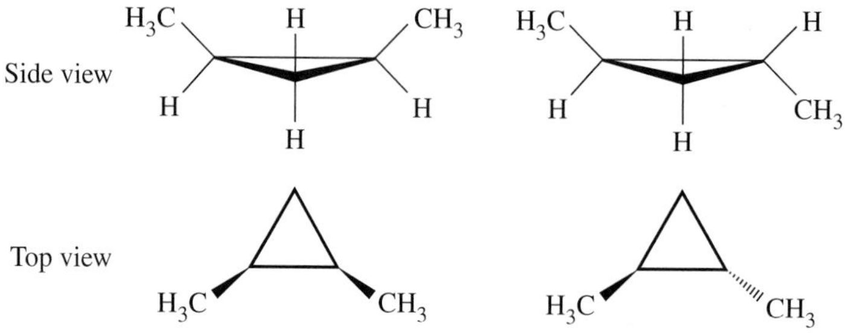

cis-1,2-dimethylcyclopropane *trans*-1,2-dimethylcyclopropane

The *cis* and *trans* isomers of 1,2-dimethylcyclohexane are

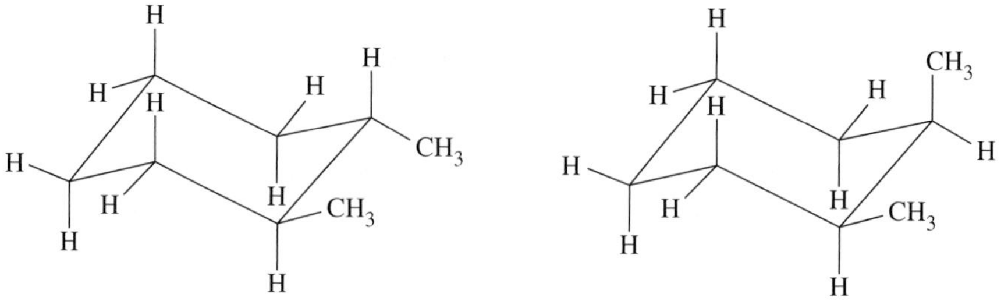

trans-1,2-dimethylcyclohexane *cis*-1,2-dimethylcyclohexane

▶ Conformers are interconverted by rotation about bonds. However, isomers can be interconverted only by breaking and reforming bonds.

In the *cis* isomers, the two substituents are adjacent to the same face of the ring and in the *trans* isomers they are adjacent to opposite faces. To convert the *cis* isomer into the *trans* isomer (or vice versa), bonds must be broken and reformed. ***cis–trans* isomerism** is only one type of a more general kind of isomerism known as **stereoisomerism**. (This term is derived from the Greek word *stereos*, meaning "solid, or three-dimensional, in nature"). For stereoisomers, the number and types of atoms and bonds are the same, but certain atoms are oriented differently in space.

For disubstituted cycloalkanes, many isomers are possible. Consider chloromethylcyclohexane, for example. It has eight isomers in total, three of which are shown below:

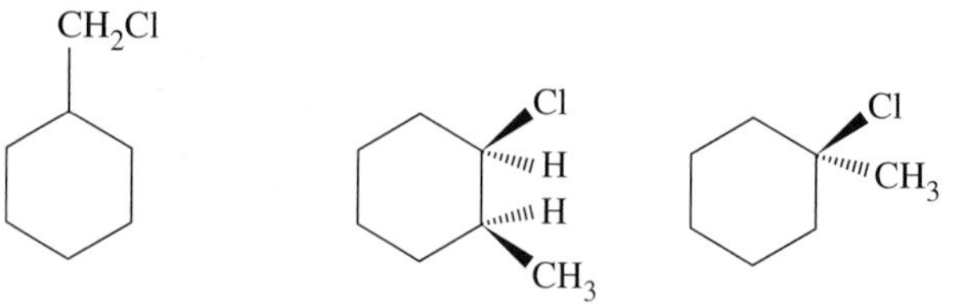

Chloromethylcyclohexane *cis*-1-chloro-2-methylcyclohexane 1-chloro-1-methylcyclohexane

26-6 CONCEPT ASSESSMENT

Draw dashed and solid wedge line structures for the *trans* isomers of chloromethylcyclohexane.

A Closer Look at Cyclohexane

The bond angles are approximately 109.5° in both the chair and the boat forms of cyclohexane. However, the chair form of cyclohexane is slightly lower in energy. Why is this? To answer this question, we will first describe how to draw the chair form of cyclohexane and then use a Newman projection to provide the insight needed.

1. Draw two parallel lines that are slightly tilted.

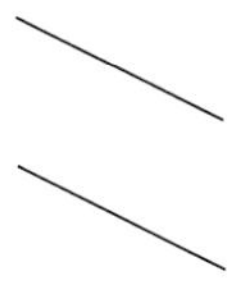

2. Connect the lower ends with a cap that points upward.

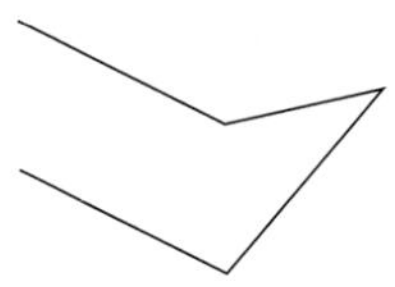

3. Connect the upper ends with a cap that points downward.

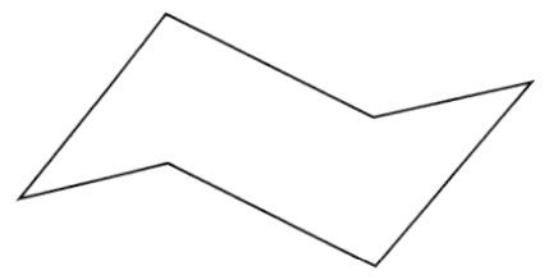

4. Add an axial hydrogen atom to each carbon atom. The bonds to the axial hydrogen atoms point up when the carbon atom points upward, and down when the carbon atom points downward. They are parallel to an imaginary line that passes through the center of the cyclohexane ring.

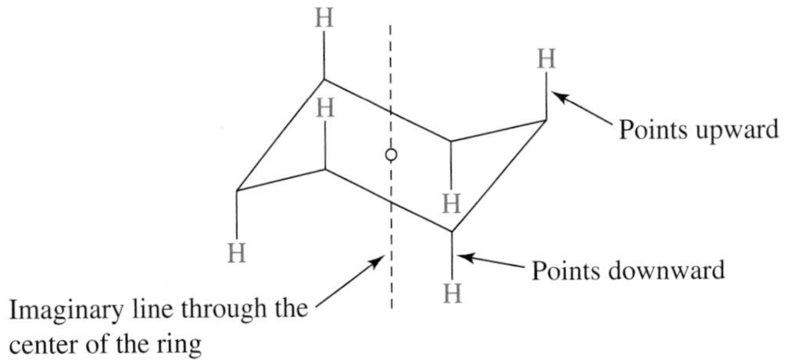

5. Add an equatorial hydrogen atom to each carbon atom. The bonds to the equatorial hydrogen atoms point sideways and complete the tetrahedral arrangement of bonds around each carbon atom. In this diagram, the axial hydrogens are shown in red and the equatorial hydrogens are shown in blue.

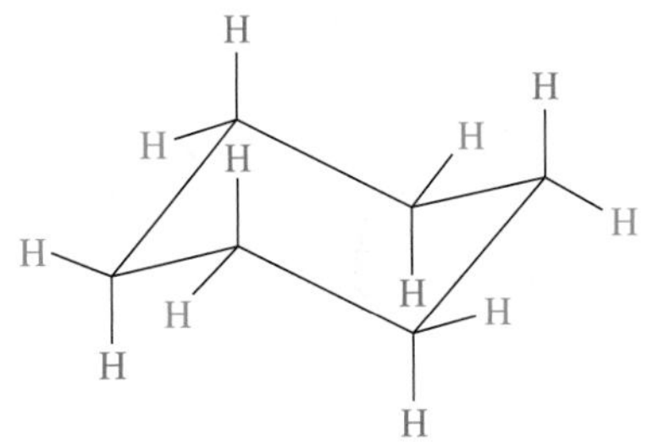

As we can see from these drawings, there are two types of hydrogen atoms: *axial* and *equatorial*. The bonds to the axial hydrogen atoms are directed vertically up or down and are parallel to an imaginary axis that passes through the center of the ring. The bonds to the equatorial hydrogen atoms are directed sideways from the ring. As we move around the ring, the axial hydrogen atoms alternate such that one points upward and the next points downward. There is a similar alternation for the equatorial hydrogen atoms. For the equatorial hydrogen atoms, it is instructive to note which pairs of C—H and C—C bonds are parallel.

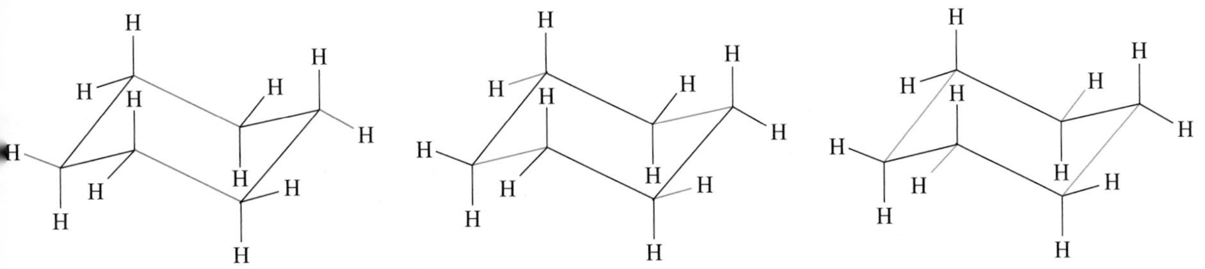

It is possible to construct a Newman projection for the chair form of cyclo-hexane if we view the molecule along a pair of C—C bonds that are parallel to each other, as suggested below:

▶ Building a molecular model will help you visualize many of the concepts being discussed in this section. Build a model of cyclohexane to convince yourself that the same Newman projection is obtained when the molecule is viewed along the other two pairs of parallel C—C bonds.

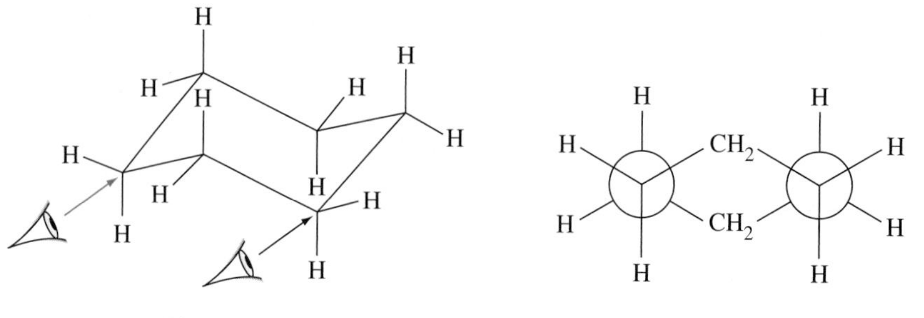

Line structure Newman projection

The Newman projection above shows that the carbon–hydrogen bonds are staggered. Some of the carbon–hydrogen bonds in the boat form are eclipsed and consequently the boat form is higher in energy (and is less stable) than the chair form.

The cyclohexane molecule interconverts rapidly between two stable chair conformations, as suggested in Figure 26-13. This interconversion is called a ring flip. When the ring flips from one chair conformation to another, every axial hydrogen in one conformation becomes an equatorial hydrogen in the other conformation, and vice versa. At room temperature, the cyclohexane ring undergoes approximately 100,000 ring flips per second.

The two chair conformations of cyclohexane shown in Figure 26-13 are equiv-alent and have exactly the same energy. However, when H atoms in cyclo-hexane are replaced by substituents, the chair conformations no longer have the same energy. Consider, for example, the following chair conformations for methylcyclohexane:

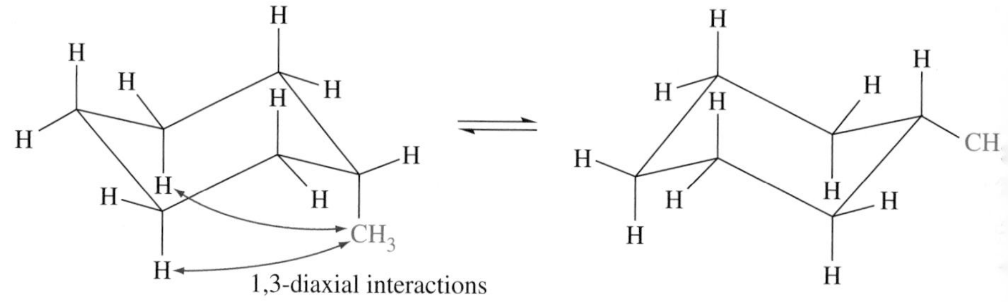

Less stable More stable

The conformer with the methyl group in the equatorial position is of lower energy than the one with the methyl group in the axial position because, when

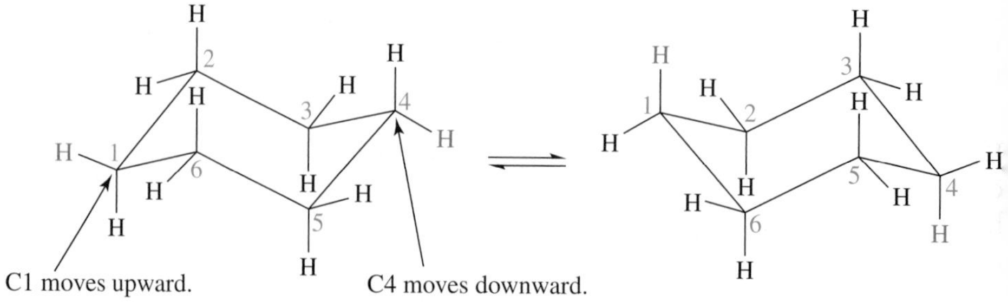

▲ FIGURE 26-13
The interconversion of two chair conformations in cyclohexane
In the figure, two of the H atoms are shown in red to emphasize that when the cyclohexane ring converts from one chair conformation to another, the equatorial hydrogen atoms are converted into axial hydrogen atoms and vice versa. The interconversion of the two chair forms proceeds through other conformations, including the boat form.

the methyl group is in an axial position, it interacts simultaneously (blue arrows) with two axial H atoms. Each interaction is called a *1,3-diaxial interaction* because it involves a substituent on the carbon atom that is numbered C3, assuming the carbon atom bonded to the methyl group is C1.

26-7 CONCEPT ASSESSMENT

If you could measure the individual heats of combustion of axial and equatorial methylcyclohexane, which conformer would burn more exothermically?

The energy differences between the axial and the equatorial forms of several mono-substituted cyclohexanes have been measured. Some of these energy differences are given in Table 26.6. Table 26.6 shows that for alkyl groups the energy difference between the axial and equatorial forms increases with the size of the group, a direct consequence of increasingly unfavorable steric interactions. This effect is particularly pronounced in *tert*-butylcyclohexane. Only about 0.01% of these molecules exist as the axial conformer at room temperature. The molecule is effectively locked with the *tert*-butyl group in the equatorial position!

TABLE 26.6 Gibbs Energy Differences Between Axial and Equatorial Conformers of Mono-Substituted Cyclohexanes

Substituent	$\Delta G°$, kJ mol^{-1}
$-H$	0
$-CH_3$	7.1
$-CH_2CH_3$	7.3
$-CH(CH_3)_2$	9.2
$-C(CH_3)_3$	21

If there are two substituents, they compete for the equatorial position. In general, the conformation of lowest energy has the bulkier group in the equatorial position.

Let us compare some isomers of dimethylcyclohexane. In 1,1-dimethylcyclohexane, one methyl group is axial and the other equatorial. A ring flip produces a conformation of equal energy, as shown below:

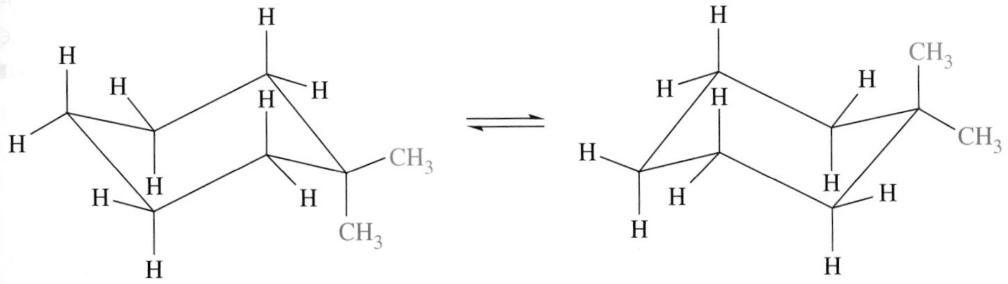

Now consider *cis*-1,4-dimethylcyclohexane. Both of the chair conformations shown below have one axial methyl group and one equatorial group. The two conformations have the same energy:

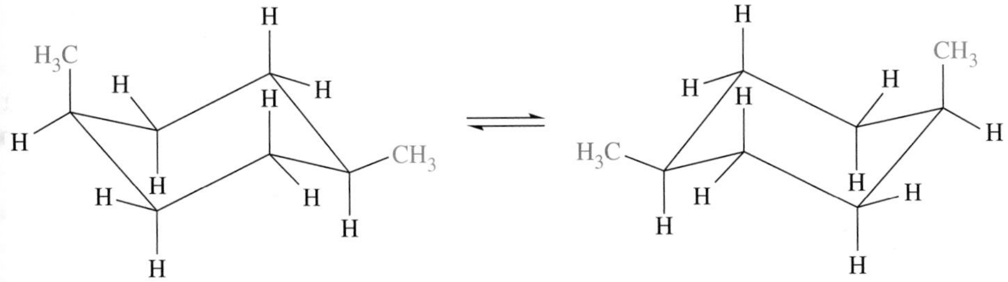

Unlike the *cis* isomer, the *trans* isomer can exist in one of two different chair conformations. One conformation has two axial methyl groups and the other has two equatorial groups, as shown below:

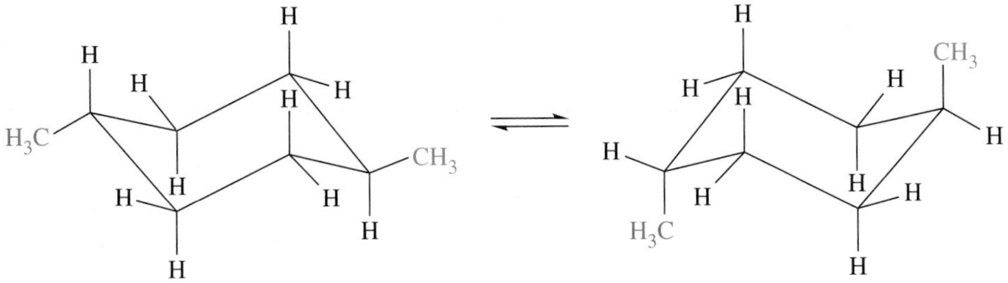

The conformation that has both methyl groups in equatorial positions is 14.2 kJ mol^{-1} lower in energy than the conformation that has the methyl groups in axial positions. This energy difference is twice that given in Table 26.6 for methylcyclohexane.

26-8 CONCEPT ASSESSMENT

Which chair conformation of *cis*-1-fluoro-4-methylcyclohexane is lower in energy?

EXAMPLE 26-5 Predicting Which Conformer of a Disubstituted Cycloalkane Is Lowest in Energy

Draw the lowest energy conformation of *cis*-1,3-dimethylcyclohexane.

Analyze

First, we draw a cyclohexane ring showing both the axial and the equatorial bonds but without the substituents added to the ring. Then, we consider the placement of the substituents. A useful tip is to realize that when the carbon atoms in a cyclohexane ring are numbered, there is a relationship between the bonds on the odd- and even-numbered carbon atoms. When the odd-numbered carbons have their *up* bonds axial, the even-numbered carbon atoms have their *down* bonds axial. When the odd-numbered carbons have their *up* bonds equatorial, the even-numbered carbons have their *down* bonds equatorial. The relationship is illustrated in the diagram below:

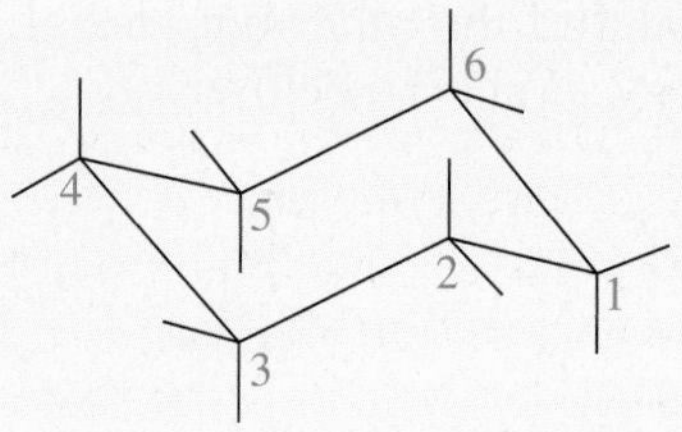

Solve

We are dealing with a *cis* isomer, and so both methyl groups are adjacent to the same face of the ring. This is possible only if the two methyl groups are bonded to the two carbon atoms with both bonds up or both bonds down. The conformation of lowest energy will be the one that has the methyl groups in the equatorial positions.

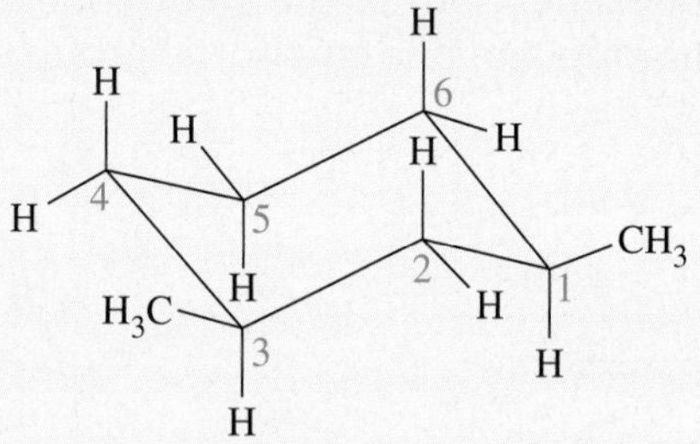

Assess

If the molecule above undergoes a ring flip, the methyl groups will be in the axial positions and the resulting conformation will be higher in energy.

PRACTICE EXAMPLE A: Draw the lower energy conformation of *trans*-1,4-dimethylcyclohexane.

PRACTICE EXAMPLE B: Draw the lower energy conformation of *cis*-1-*tert*-butyl-2-methylcyclohexane.

26-4 Stereoisomerism in Organic Compounds

In Figure 26-14, we have summarized different forms of isomerism that we encounter in organic chemistry. We have already discussed constitutional isomerism and learned about one form of stereoisomerism, namely *cis–trans*

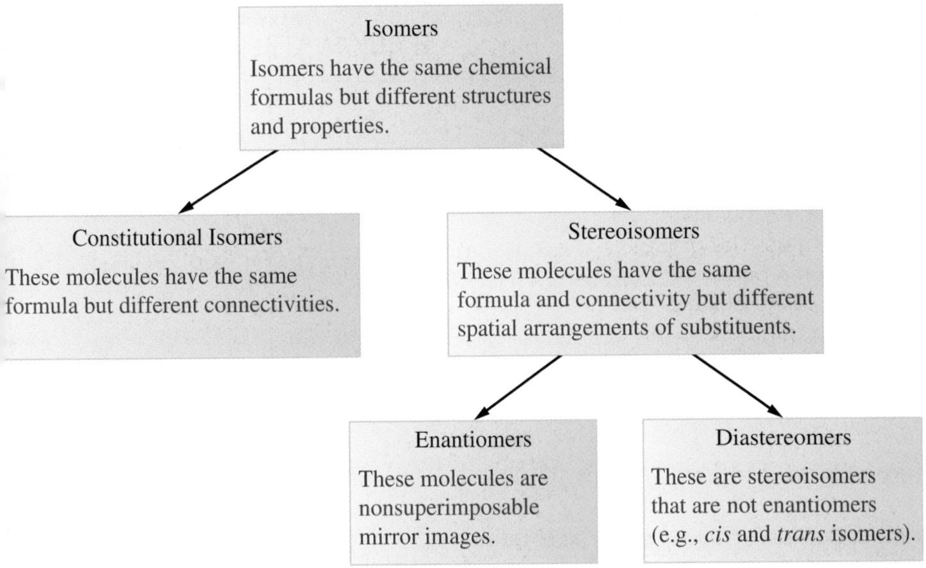

◀ FIGURE 26-14
Isomerism summarized

isomerism in cycloalkanes. As illustrated in Figure 26-14, *cis–trans* isomers are also known as **diastereomers**. In this section we will focus on another form of stereoisomerism—**enantiomerism**—which often arises when an organic compound has an asymmetric carbon. An **asymmetric** carbon (also called a **chiral** carbon) is one that is bonded to four different groups. Molecules with an asymmetric carbon exist as stereoisomers that cannot be interconverted without breaking and reforming bonds. We will see (in Chapter 28) that molecules with asymmetric carbon atoms figure prominently in biochemistry.

Chirality

We saw in Chapter 24 that a solution of an optically active compound can rotate the plane of polarized light. The requirement for optical activity is that the molecule be asymmetric—that is, its mirror image cannot be superimposed on the original molecule. This situation can arise at a tetrahedral C atom when all four groups attached to the C atom are different. Consider the molecule 3-methylhexane shown in Figure 26-15. The illustration shows that there are two nonsuperimposable isomers of 3-methylhexane related as mirror images. The two isomers are said to be *enantiomers*. A molecule that is not superimposable on its mirror image is said to be *chiral*. Compounds whose structures are superimposable on their mirror images are **achiral**. Examples of chiral and achiral molecules are shown here, using the dashed and solid wedge line notation:

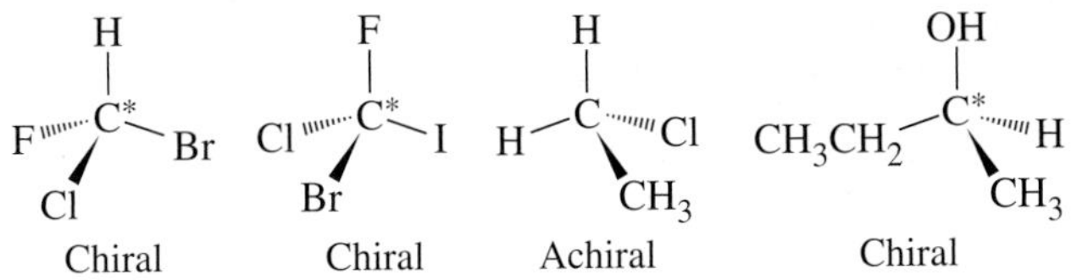

All the chiral molecules shown contain an atom that is connected to four different substituent groups. The C atom to which the four different groups are attached is said to be **asymmetric**, or a **stereocenter**. Centers of this type are sometimes denoted by an asterisk. Molecules with one stereocenter are always chiral. As we will see in Chapter 28, molecules incorporating more than one stereocenter need not be chiral, and many chiral molecules occur in nature. As described in Chapter 27, the existence of chirality can play an important role in establishing some reaction mechanisms.

KEEP IN MIND

that a solution of one enantiomer rotates the plane of polarized light in one direction whereas a solution of the other enantiomer rotates the light in the opposite direction. With a 50:50 mixture of the two enantiomers—a racemic mixture—no rotation of the plane of polarized light is observed.

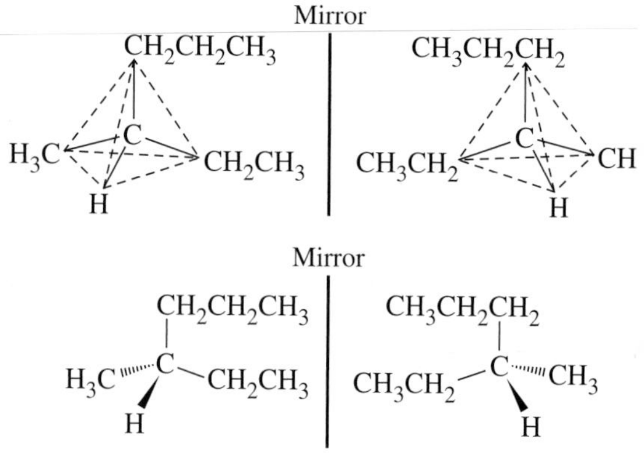

▶ FIGURE 26-15
Nonsuperimposable mirror images of 3-methylhexane
Notice in the diagram that we have adopted the convention used by organic chemists in drawing chiral centers: The groups attached to the central carbon in the plane of the paper are connected with solid lines, the group in front of the plane of the paper is attached by a solid wedge, and the group behind the plane of the paper is indicated by a dashed wedge.

EXAMPLE 26-6 Identifying a Chiral Molecule

Predict whether either 2-chloropentane or 3-chloropentane is chiral.

Analyze

To decide whether a molecule is chiral, we look for a C atom that has four different groups attached.

Solve

The two compounds are shown below.

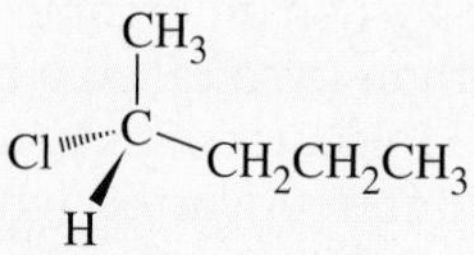
2-chloropentane

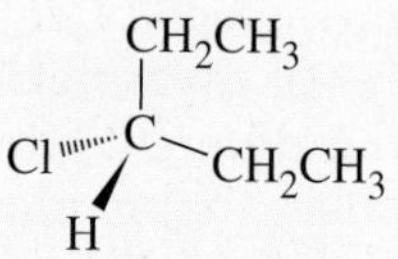
3-chloropentane

We see that 2-chloropentane contains a C atom that has four different groups attached; hence, 2-chloropentane is chiral. However, 3-chloropentane does not have such a C atom; its structure is identical to its mirror image and thus, 3-chloropentane is achiral.

Assess

In drawing the structures, we focused only on the carbons to which the Cl atoms were attached. All the other carbons are bonded to at least two H atoms (two groups are the same) and cannot possibly be chiral.

PRACTICE EXAMPLE A: Which of the following chlorofluorohydrocarbons is chiral: **(a)** $CF_3CH_2CCl_3$; **(b)** $CF_2HCHFCCl_3$; **(c)** $CClFHCHHCCl_2F$?

PRACTICE EXAMPLE B: Which of the following chloroalcohols is chiral: **(a)** $CH_2ClCH_2CH_2OH$; **(b)** $CH_2ClCH(OH)CH_3$; **(c)** $CH(OH)ClCH_2CH_3$?

26-9 CONCEPT ASSESSMENT

How many chiral centers are there in 2,3-dibromopentane? How many different stereoisomers are there?

EXAMPLE 26-7 Identifying Chiral Carbon Atoms in Cycloalkanes

Identify any chiral carbon atoms in the molecules to the right.

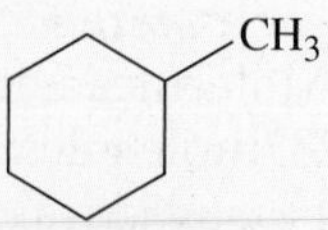
Methylcyclohexane

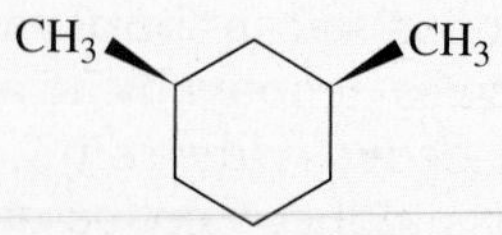
cis-1,3-dimethylcyclohexane

Analyze

A carbon atom is chiral if it is bonded to four different groups. To determine whether a carbon atom in a ring structure is chiral, we must go around the ring in each direction (clockwise and counterclockwise) to see if there is a point of difference. If there is a point of difference, then the carbon atom is chiral.

Solve

When considering carbon atoms in the rings of these molecules, we need only focus on carbon atoms where hydrogen atoms have been replaced by other substituents. If there is no substituent, then the carbon atom has two identical groups (two hydrogen atoms) and thus, cannot be chiral. Thus, for methylcyclohexane, we focus only on C1, the carbon bonded to the methyl substituent. The clockwise and counterclockwise paths starting from C1 have identical constitutions ($—CH_2CH_2CH_2CH_2CH_2—$) and so, we conclude that C1 in methylcyclohexane is not chiral.

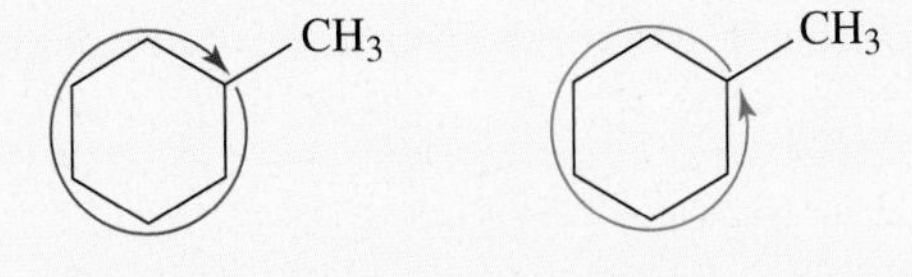

Clockwise path   Counterclockwise path

When we compare the clockwise and counterclockwise paths, starting from C1, in 1,3-dimethylcyclohexane, we notice that there is a point of difference. The constitution of the clockwise path (blue) is $—CH_2CH_2CH_2CH(CH_3)CH_2—$ and that of the counterclockwise path (red) is $—CH_2CH(CH_3)CH_2CH_2CH_2CH_2—$. The two paths are different; thus, C1 is chiral. Following similar reasoning, we find that C3 is also chiral. Thus, in cis-1,3-dimethylcyclohexane, C1 and C3 are chiral.

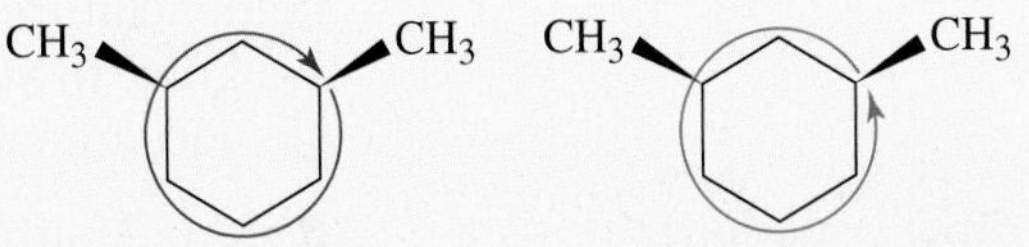

Clockwise path   Counterclockwise path

Assess

In most of the examples we have considered so far, there has been only one chiral atom per molecule. Such molecules are optically active. In this example, we see that 1,3-dimethylcyclohexane has two chiral carbon atoms. A molecule with two or more chiral carbon atoms may or may not be optically active. Optical activity of molecules containing two or more chiral atoms is discussed in advanced organic chemistry courses.

PRACTICE EXAMPLE A: Identify the chiral carbon atoms in the molecule shown in the diagram on the right.

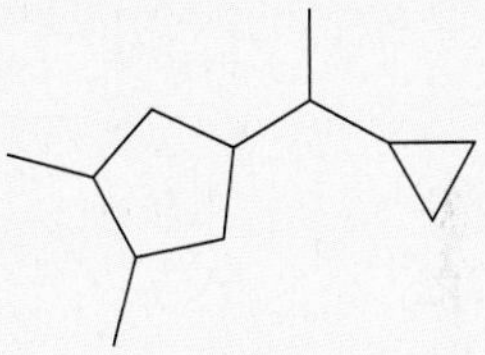

PRACTICE EXAMPLE B: How many chiral atoms are there in 1,1,3-trimethylcyclohexane?

Naming Enantiomers: The *R*, *S* System of Nomenclature

To differentiate two enantiomers, we need a system of nomenclature that indicates the arrangement or configuration of the four groups about the chiral center. Such a system was developed by three chemists, R. S. Cahn, C. Ingold, and V. Prelog. The first step is to rank the four substituents in order of decreasing priority. The rules for assigning priorities will be described shortly.

Suppose that in the hypothetical compound *Cabcd*, substituent *a* has the highest priority, *b* the second highest, *c* the third highest, and *d* the lowest. Now position the molecule (mentally, on paper, or with a model set) so that the lowest priority substituent is as far away from the viewer as possible (Fig. 26-16). This procedure results in two possible arrangements for the remaining substituents (one for each enantiomer). In the **R, S system**, if the sequence from *a* to *b* to *c*, as viewed toward the substituent of lowest priority, is clockwise, the configuration of the stereocenter is named *R* (*rectus*, Latin, meaning right). Conversely, if the sequence is counterclockwise, the stereocenter is named *S* (*sinister*, Latin, meaning left). The symbol (*R*) or (*S*) is added as a prefix to the name of the chiral compound, as in (*R*)-2-chlorobutane

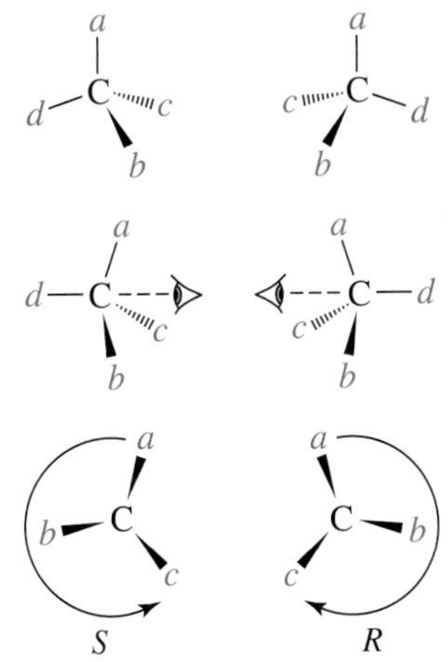

▲ FIGURE 26-16
Assignment of *R* and *S* configuration at a tetrahedral stereocenter
The group of lowest priority is placed as far away from the viewer as possible.

and (*S*)-2-chlorobutane, shown below. A racemic mixture of the enantiomer is designated (*R*, *S*) as in (*R*, *S*)-2-chlorobutane.

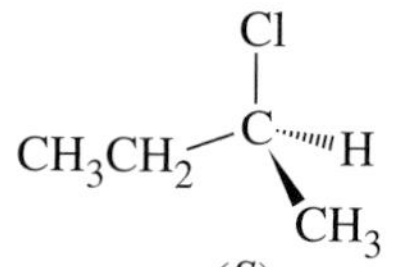

(*S*)

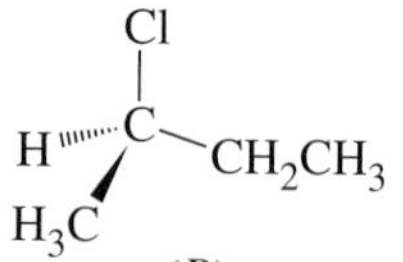

(*R*)

Rules for Assigning Priorities to Substituents In order to apply the *R*, nomenclature to a stereocenter, we must first describe how the priorities are assigned to substituents. When looking at the atoms attached directly to the stereocenter, the rules devised by Cahn, Ingold, and Prelog are as follows:

▶ Strictly speaking, the parameter that decides priority is atomic mass, not atomic number. However, atomic mass increases as the atomic number increases, so unless we are comparing isotopes of the same element (same atomic number, different atomic masses), we can assign priorities by focusing on atomic number.

Rule 1. A substituent atom of higher atomic number takes precedence over one of lower atomic number. Consider the enantiomer of 1-chloro-1-iodoethane shown below.

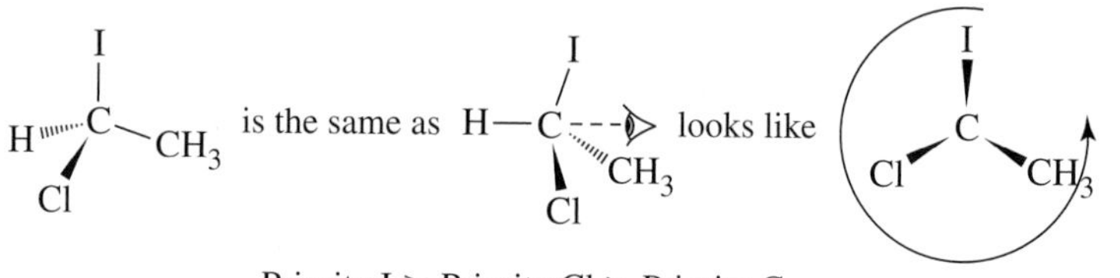

Priority I > Priority Cl > Priority C
(*S*)-1-chloro-1-iodoethane

The order of priority, as viewed toward the H atom (which has the lowest priority), is counterclockwise. Therefore the enantiomer is (*S*)-1-chloro-1-iodoethane.

Rule 2. If two substituent atoms attached to the stereocenter have the same priority, proceed along the two substituent chains until a point of difference is reached. The atom of higher atomic number at this point establishes the priority. Thus, an ethyl group takes priority over a methyl group for the following reason. Although at the point of attachment to the stereocenter each substituent has a C atom, equal in priority, beyond these C atoms the methyl group has an H atom and the ethyl group has a higher-priority C atom.

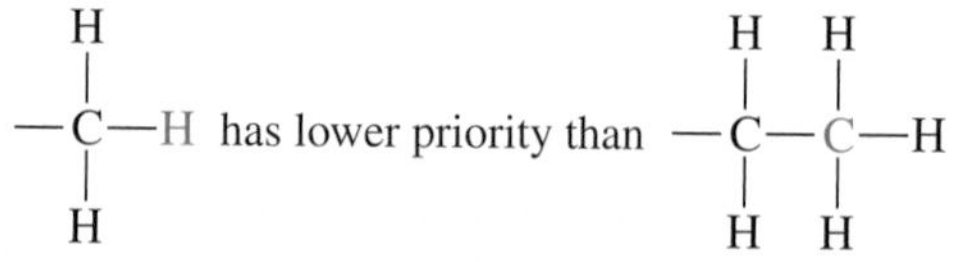

It is important to understand that the decision on priority is made at the *first* point of difference along otherwise similar substituent chains. When that point has been reached, the constitution of the remainder of the chain is immaterial.

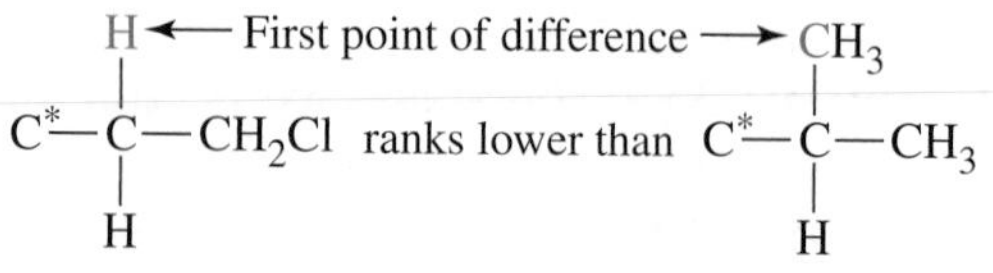

Rule 3. Double bonds and triple bonds are treated as if they were single, and the atoms in them are duplicated or triplicated at each end by the particular atoms at the *other end of the multiple bond*. For example,

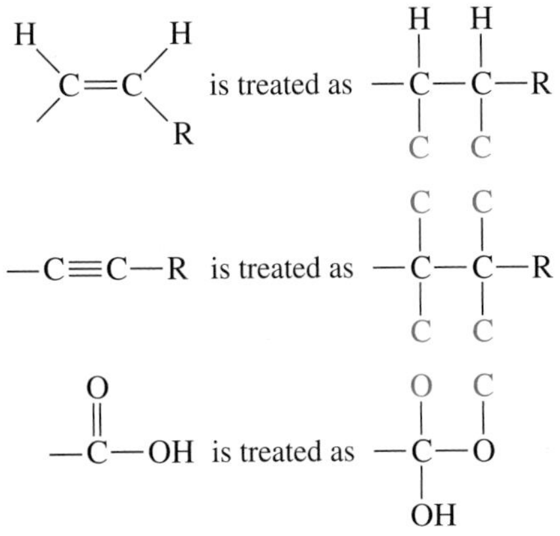

Note that the atoms shown in red are added simply for the purpose of assigning priority to the groups containing a multiple bond; they are *not* really there! To illustrate, the —CH_2OH group has lower priority than —CHO.

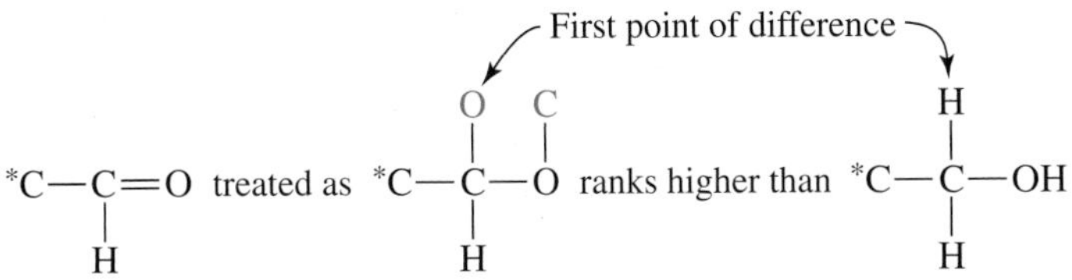

The assignment of priorities and configuration of a chiral center is illustrated in Example 26-8.

EXAMPLE 26-8 Assignment of Priorities and Configuration of a Chiral Center

Name each of the following compounds, including the assignment of configuration.

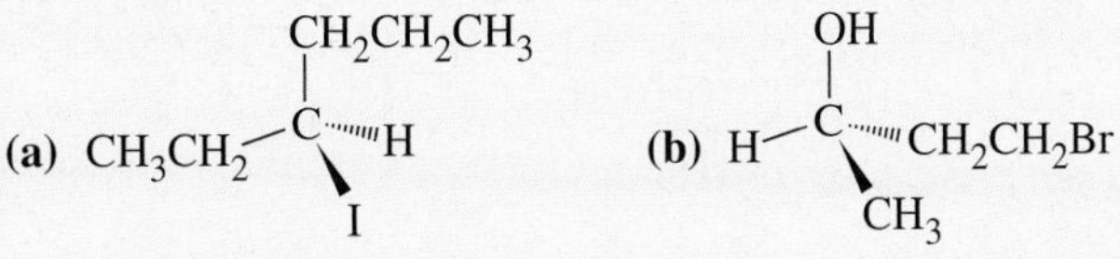

Analyze

To assign the configuration at the stereocenter, we must first assign the priorities of the substituents. Then we determine the *R* or *S* configuration by viewing the molecule toward the atom of lowest priority.

Solve

(a) This molecule is 3-iodohexane with C3 as the chiral center. The order of priorities of atoms attached to C3 is

$$I > C\ (ethyl) = C\ (propyl) > H$$

(continued)

In order to decide the ranking of the ethyl group relative to the propyl group, we go to the first point of difference in the chains of these substituents.

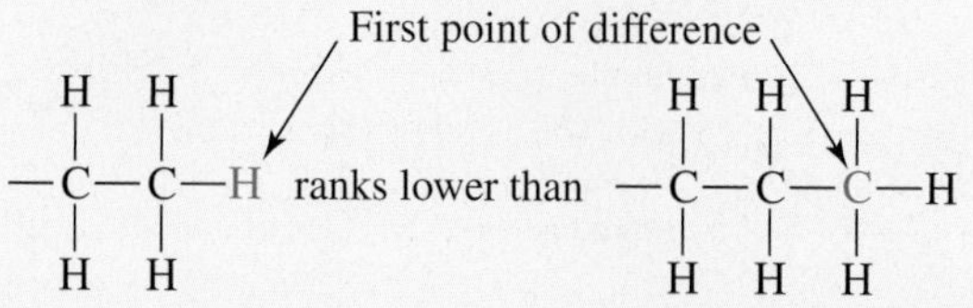

We see that the propyl group ranks higher than the ethyl group, so the order of priorities is

$$\text{I} > \text{C (propyl)} > \text{C (ethyl)} > \text{H}$$

Viewing the molecule toward the lowest priority H atom, we see that the priorities decrease in a counterclockwise manner, so the configuration at the stereo center is *S*. The complete name of the molecule is (*S*)-3-iodohexane.

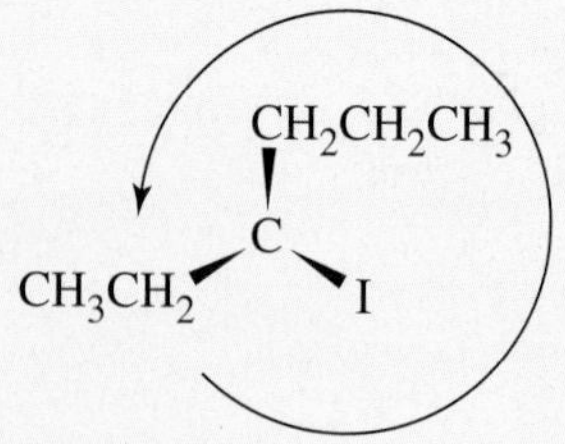

(b) This molecule is 4-bromobutan-2-ol (see Are You Wondering 26-1), with C2 as the chiral center. The order of priorities of atoms attached to C2 is

$$\text{O} > \text{C (bromoethyl)} = \text{C (methyl)} > \text{H}$$

In order to decide the ranking of the bromoethyl group relative to the methyl group, we go to the first point of difference in the chains of these substituents.

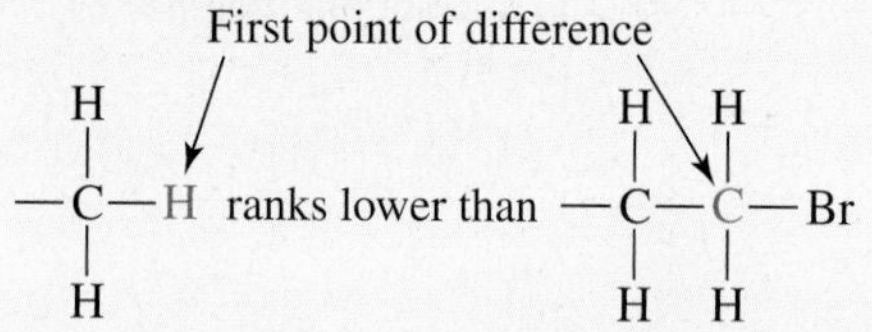

We see that the bromoethyl group ranks higher than the methyl group, so the order of priorities is

$$\text{I} > \text{C (bromoethyl)} > \text{C (methyl)} > \text{H}$$

Viewing the molecule toward the lowest priority H atom is not quite as straightforward as in part (a). This is because the H atom is in the plane of the paper and thus not as far away from the viewers as possible. We can tackle this problem in one of two ways. The first requires some three-dimensional "vision," or "stereoperception," in that we imagine picking up the molecule by the methyl group and bromoethyl group and reorienting the molecule so that the H atom points away from us, to give

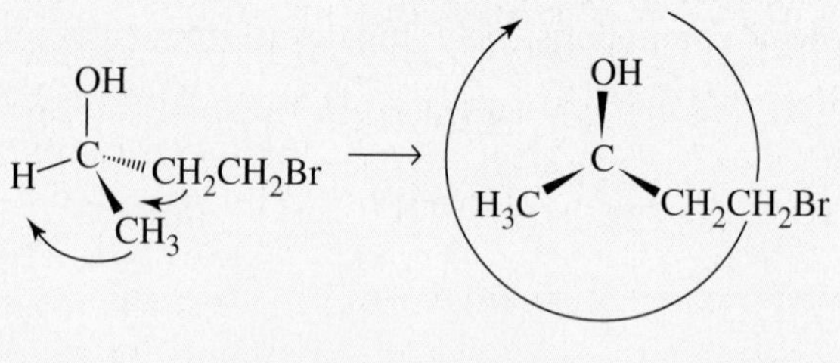

where we can now discern the sequence of priorities as clockwise, so the compound is (*R*)-4-bromobutan-2-ol.

Alternatively, we can switch a pair of groups so that the group of lowest priority is bonded by a dashed wedge.

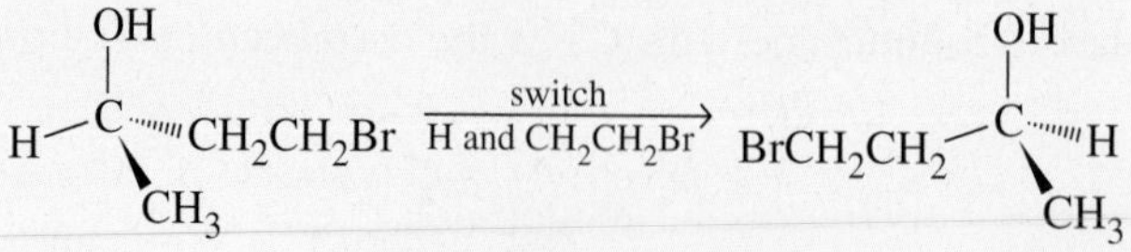

Now draw the view of the molecule toward the group of lowest priority. Here the sequence of priorities is counterclockwise. Because we switched two groups, we created the enantiomer of the molecule whose configuration we are actually seeking. Thus, although in the switched molecule the order of priorities corresponds to an *S* configuration, the *original* molecule corresponds to the *R* configuration. Therefore, as determined previously, the molecule is (*R*)-4-bromobutan-2-ol.

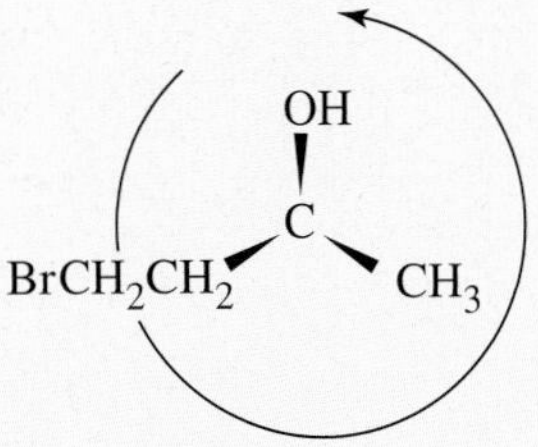

Assess

When assigning priorities to groups, it may be necessary, as it was in (b), to work backward along the chain to find the first point of difference and then compare the atomic numbers of the atoms at that point.

PRACTICE EXAMPLE A: Indicate whether each of the following structures has the *R* configuration or the *S* configuration.

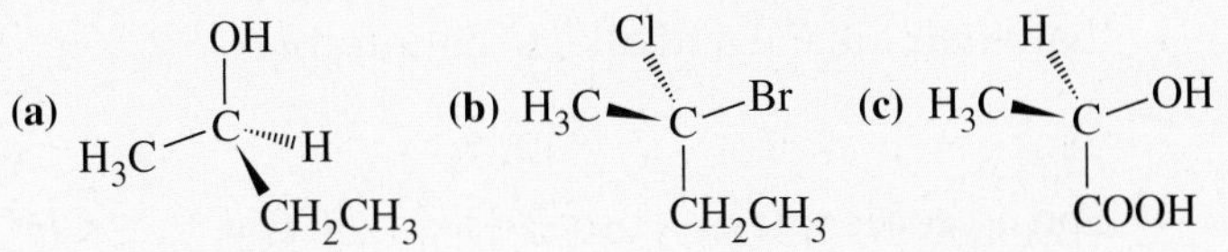

PRACTICE EXAMPLE B: Do the structures in each of the following pairs represent identical molecules or pairs of enantiomers?

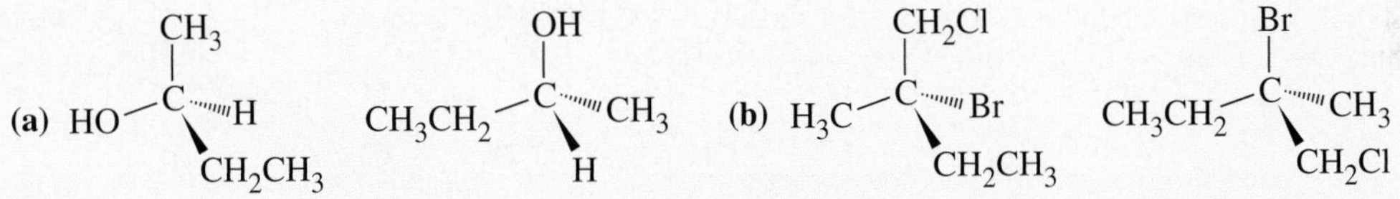

26-5 Alkenes and Alkynes

A straight- or branched-chain alkane, with formula C_nH_{2n+2}, has the maximum number of H atoms possible for its number of C atoms. In other classes of hydrocarbons, compounds with the same number of C atoms but fewer H atoms, the C atoms must join into rings, form carbon-to-carbon multiple bonds, or do both to ensure that each C atom forms a total of four bonds. We have already discussed some aspects of ring structures (in Section 26-3). In this section, we focus on hydrocarbons whose molecules contain some double or triple bonds between C atoms. Such molecules are said to be **unsaturated**. If the molecule has *one double bond*, the hydrocarbons are the simple **alkenes**, or **olefins**; they have the general formula C_nH_{2n}. Simple alkynes have *one triple bond* in their molecules and have the general formula C_nH_{2n-2}.

Systematic names are given below for a few alkenes and alkynes. The names given in parentheses are also commonly used.

$CH_2{=}CH_2$ Ethene (ethylene)	$CH_3CH_2CH{=}CH_2$ But-1-ene	$CH_3CH_2CH_2C(CH_2CH_3){=}CH_2$ 2-Ethylpent-1-ene
$HC{\equiv}CH$ Ethyne (acetylene)	$CH_3CH_2C{\equiv}CH$ But-1-yne (ethylacetylene)	$CH_3CH(CH_3)C{\equiv}CCH_3$ 4-Methylpent-2-yne (isopropylmethylacetylene)

The following are the modifications of the rules on page 1211 required to name alkenes and alkynes:

1. Select as the base chain the longest chain containing the multiple bond.
2. Number the C atoms of the chain to place the multiple bond at the lowest possible number.
3. Use the ending *-ene* for alkenes and *-yne* for alkynes. Place the number for the position of the multiple bond immediately before the *-ene* or *-yne* ending.

Thus, in the name 2-ethylpent-1-ene, the longest chain *containing the multiple bond* is a five-carbon chain (*pent-*). This is *not* a substituted hexane, and it is not a hexene. The double bond makes the molecule an alkene, and its location between the first and second carbon makes it a pent-1-ene. The ethyl group is attached to the second C atom, and so the alkene is named 2-ethylpent-1-ene.

The alkenes are similar to the alkanes in physical properties. At room temperature, those containing 2 to 4 C atoms are gases; those with 5 to 18 are liquids; those with more than 18 are solids. In general, alkynes have higher boiling points than their alkane and alkene counterparts.

Stereoisomerism in Alkenes

The molecules but-2-ene, $CH_3CH{=}CHCH_3$, and but-1-ene $CH_2{=}CHCH_2CH_3$, differ in the position of the double bond and are constitutional isomers. However, another type of isomerism is possible in 2-butene, as shown in the two structures below:

KEEP IN MIND

that *cis* means "on the same side" and *trans* means "across."

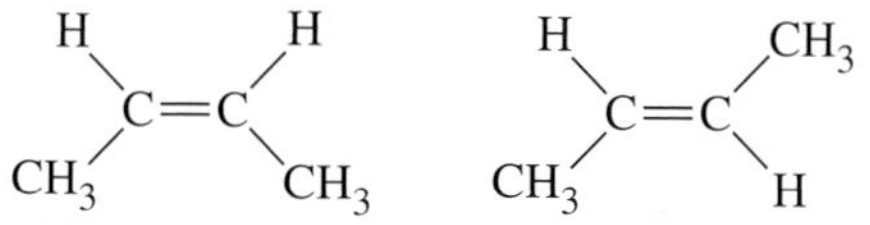

(a) *cis*-but-2-ene **(b)** *trans*-but-2-ene

The two isomers of but-2-ene are stereoisomers because they have exactly the same *constitution* but a different spatial arrangement of their atoms in space. As we learned in Section 11-4, a double bond between C atoms consists of the overlap of hybrid orbitals to form a σ bond *and* the sideways overlap of *p* orbitals to form a π bond. Because of the π bond, rotation about a double bond is severely restricted. Molecule (a) cannot be converted into molecule (b) simply by twisting one end of the molecule through 180°, so the two molecules are distinctly different above. To differentiate these two molecules, we call molecule (a) *cis*-but-2-ene and we call molecule (b) *trans*-but-2-ene. Because of differences in their molecular structures, the compounds have different physical properties. For example, the melting points are −139 °C for *cis*-but-2-ene and −106 °C for *trans*-but-2-ene; the boiling points are 3.7 °C for *cis*-but-2-ene and 0.9 °C for *trans*-but-2-ene.

Preparation and Uses of Alkenes and Alkynes

The general laboratory preparation of alkenes uses an *elimination reaction*, a reaction in which atoms are removed from adjacent positions on a carbon chain. Elimination reactions will be examined in more detail in Chapter 27. For now, it is enough to know that, in an elimination reaction, a small molecule is produced, and an additional bond is formed between the C atoms. For example, H_2O is eliminated in the following reaction:

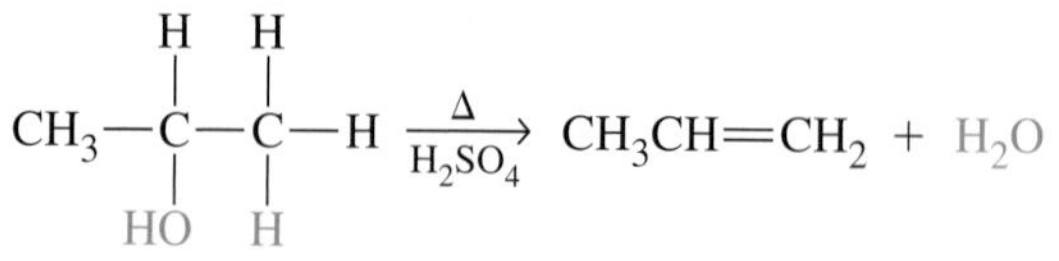

(26.4)

The principal alkene of the chemical industry is ethene (ethylene). Its chief use is in the manufacture of polymers (Chapter 27), although it is also used to manufacture other organic chemicals. Reaction (26.4) is relatively unimportant in the commercial production of ethylene, which is obtained mainly by thermal cracking of other hydrocarbons.

The simplest alkyne is ethyne (acetylene), which can be prepared from coal, water, and limestone in a three-step process:

$$CaCO_3 \xrightarrow{\Delta} CaO + CO_2$$

$$CaO + 3\,C \xrightarrow[2000\ °C]{\text{electric furnace}} \underset{\substack{\text{calcium acetylide}\\ \text{(calcium carbide)}}}{CaC_2} + CO$$

$$CaC_2 + 2\,H_2O \longrightarrow \underset{\text{acetylene}}{HC{\equiv}CH} + Ca(OH)_2 \qquad (26.5)$$

Most other alkynes are prepared from acetylene by taking advantage of the acidity of the C—H bond. In the presence of a very strong base, such as sodium amide ($NaNH_2$), the amide anion removes the proton from acetylene to form ammonia and the salt sodium acetylide. The acetylide can then react with an alkyl halide, such as CH_3Br:

$$H{-}C{\equiv}C{-}H + Na^{+}\,{}^{-}NH_2 \longrightarrow H{-}C{\equiv}C^{-}Na^{+} + NH_3$$

$$H{-}C{\equiv}C^{-}Na^{+} + Br{-}CH_3 \longrightarrow H{-}C{\equiv}C{-}CH_3 + Na^{+}Br^{-} \qquad (26.6)$$

By continuing this reaction, the triple bond can be positioned as desired in the chain, as in the synthesis of pent-2-yne:

$$CH_3C{\equiv}C{-}H + Na^{+}\,{}^{-}NH_2 \longrightarrow CH_3C{\equiv}C^{-}Na^{+} + NH_3$$

$$CH_3C{\equiv}C^{-}Na^{+} + Br{-}CH_2CH_3 \longrightarrow CH_3C{\equiv}CCH_2CH_3 + Na^{+}Br^{-} \qquad (26.7)$$

At one time, acetylene was one of the most important organic raw materials in the chemical industry. At present, the chief use of acetylene is in the manufacture of other chemicals for polymer production, such as vinyl chloride, $H_2C{=}CHCl$, which is polymerized to polyvinyl chloride (PVC). We will discuss polymers and polymerization reactions in the next chapter.

Acetylene is also used to produce high-temperature flames that are used in a variety of applications. For example, the combustion of acetylene in excess oxygen is the basis of oxyacetylene torches used for cutting and welding metals.

$$HC{\equiv}CH(g) + \frac{5}{2}\,O_2(g) \longrightarrow 2\,CO_2(g) + H_2O(l) \qquad \Delta_r H = -1300\ \text{kJ mol}^{-1}$$

The large negative enthalpy of combustion of acetylene results from acetylene's large *positive* enthalpy of formation: $\Delta_f H°[C_2H_2(g)] = +226.7\ \text{kJ mol}^{-1}$.

Alkenes and alkynes are used by chemists to make other compounds. The reactive part of these compounds are the π bonds and the characteristic reaction is an addition reaction, in which atoms or grouping of atoms add to the carbon atoms on either side of a double or triple bond. We have already seen one example of an addition reaction, reaction (26.1), in which hydrogen atoms add across a carbon–carbon bond of an alkene to give an alkane. Although we will examine reactions of alkenes in more detail in Chapter 27, it is worth mentioning now that certain addition reactions form the basis of simple qualitative tests that can be used to determine whether a compound is an alkene or an alkyne. For example, $H_2C{=}CH_2$ will absorb H_2 in the presence of a metal catalyst, reaction (26.1), and it will decolorize a solution of bromine water, $Br_2(aq)$, because of the following reaction:

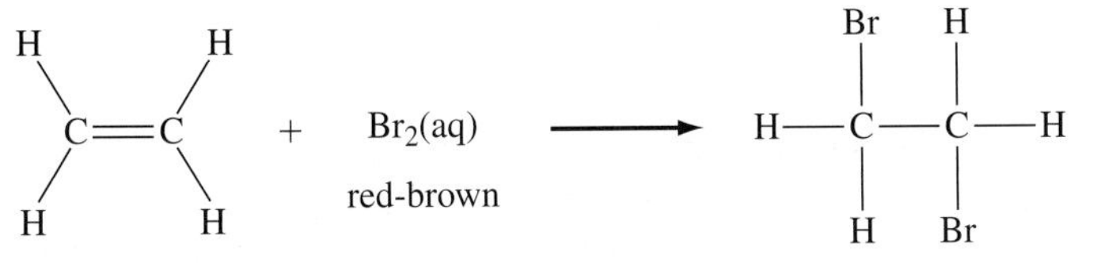

(26.8)

The decolorization of bromine by cyclohexene can be seen in the photo in the margin on the next page.

Bogdan Vasilescu/Fotolia

▲ Cutting steel with an oxyacetylene torch.

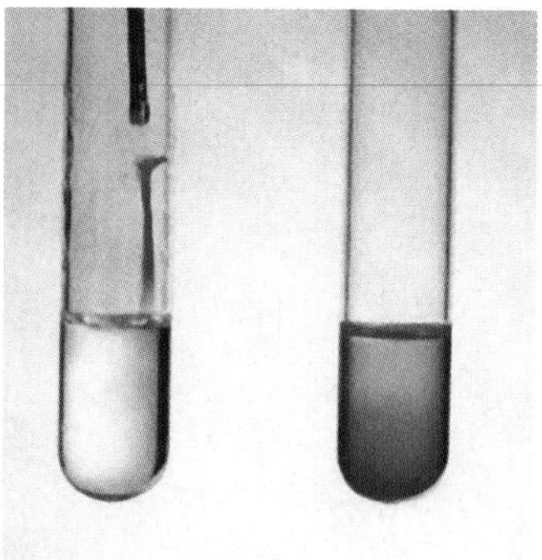

Tom Bochsler

▲ The test tube on the left contains cyclohexene and the one on the right contains cyclohexane. When bromine Br_2, is added to the tube containing cyclohexene, the red-brown color disappears because Br_2 adds across the double bond. The red-brown color persists in the tube filled with cyclohexane.

Naming the Stereoisomers of Highly Substituted Alkenes: The *E*, *Z* System of Nomenclature

Unfortunately, the *cis–trans* nomenclature is not useful for naming highly substituted alkenes. Consider for example the following stereoisomers of 1-bromo-1-chloro-2-fluoroethene:

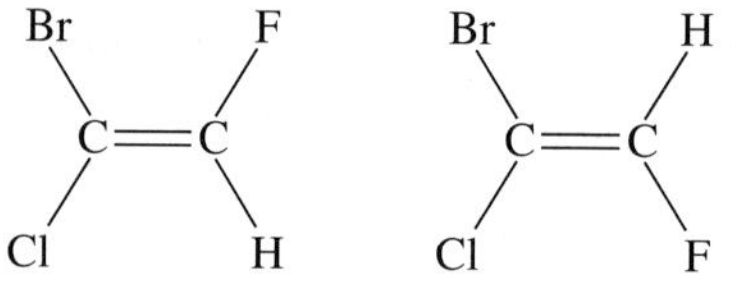

Which is *cis*? Which is *trans*?

In the structure on the left, the F atom is *cis* to Br and *trans* to Cl; in the structure on the right, F is *cis* to Cl and *trans* to Br. Clearly, the *cis–trans* descriptors are not very helpful for naming highly substituted alkenes. An alternative system for naming such alkenes has been adopted by IUPAC: the ***E*, *Z* system**. The Cahn–Ingold–Prelog rules, discussed previously, are used systematically to assign priorities to the substituents on the carbon atoms of the double bond. The stereochemistry about the double bond is assigned *Z* (from the German word *zusammen*, meaning "together") if the two groups of higher priority *at each end of the double bond* are on the same side of the molecule. If the two groups of higher priority are on opposite sides of the double bond, the configuration is denoted by an *E* (from the German word *entgegen*, meaning "opposite"). These ideas are summarized in the diagram below:

▶ In the *Z* isomer, the two groups of higher priority are on the zame zide (same side) of the double bond but, in the *E* isomer, they are on either side.

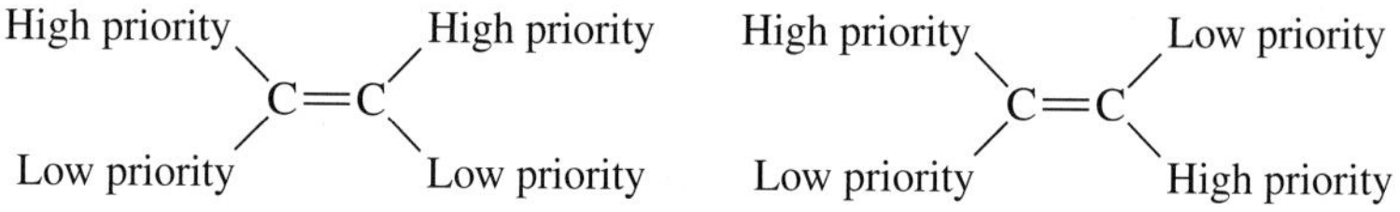

The *Z* isomer The *E* isomer

We can use the *E*, *Z* system to name the two stereoisomers of 1-bromo-1-chloro-2-fluoroethene:

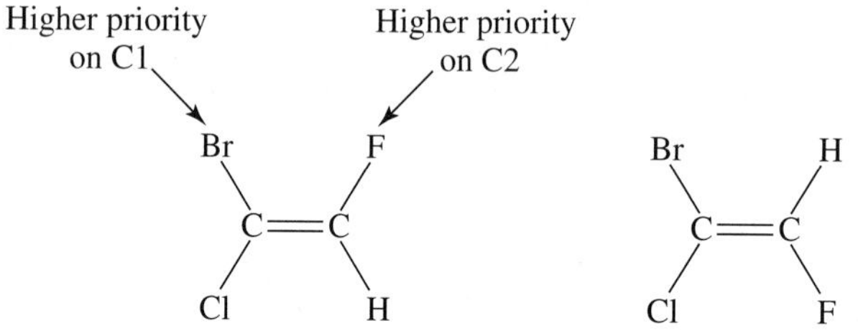

(*Z*)-1-bromo-1-chloro-2-fluoroethene (*E*)-1-bromo-1-chloro-2-fluoroethene

Note that, as in *R*, *S* nomenclature, the *E* or *Z* is placed in parentheses. The *E*, *Z* system of nomenclature can be used for all alkene stereoisomers; consequently, the IUPAC recommends that this system be used exclusively. However, many chemists continue to use the *cis* and *trans* designations for simple alkenes.

EXAMPLE 26-9 Assignment of Configurations of Alkenes

Name each of the following compounds, including the assignment of configuration.

(a)

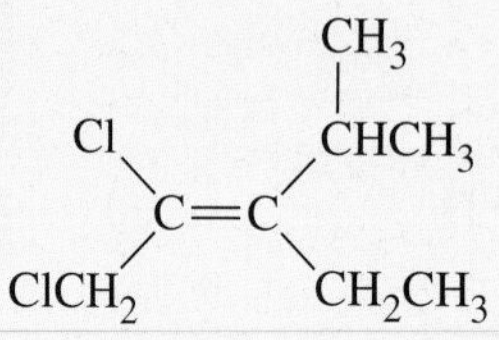

(b)

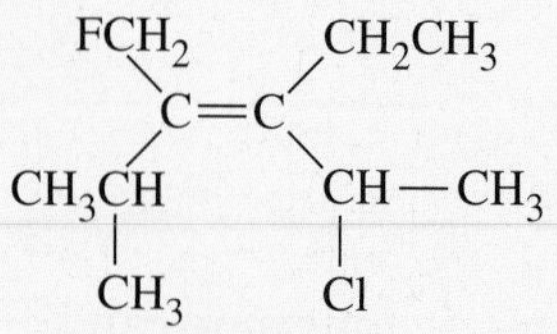

Analyze

To assign the configuration of the alkene, we must first assign the priorities to the substituents attached to each carbon in the double bond. Once this is done, we can assign the *E* or *Z* designation.

Solve

(a) One of the sp^2 carbon atoms is bonded to a Cl atom and a C atom; thus, the chlorine atom has highest priority. The other sp^2 carbon is bonded to an ethyl group and an isopropyl group. The carbon of the isopropyl group is bonded to C, C, and H, and the carbon of the ethyl group to C, H, and H. A carbon can be canceled in each group. Of the remaining atoms, the carbon has the highest priority. Thus, the isopropyl group takes precedence. The groups of highest priority are on the same side of the double bond. The complete name of the molecule is (*Z*)-1,2-dichloro-3-ethyl-4-methylpent-2-ene.

(b) One of the sp^2 carbon atoms is bonded to a fluoromethyl group and an isopropyl group. The carbon of the fluoromethyl group is bonded to F, H, and H; the carbon of the isopropyl group to C, C, and H. Of these six atoms, the fluorine has the highest priority. Thus, the fluoromethyl group takes precedence. The other sp^2 carbon atom is bonded to an ethyl group and a chloroethyl group. Because the first point of difference on these substituent chains is a Cl atom, the chloroethyl group takes precedence. The groups of highest priority are on opposite sides of the double bond. The complete name of the molecule is (*E*)-2-chloro-3-ethyl-4-fluoromethyl-5-methylhex-3-ene.

Assess

For (b), the carbon atom bonded to chlorine is chiral and so, the molecule exists as either the *R* or *S* enantiomer. It takes quite a bit of practice to master organic nomenclature and, admittedly, it is not the most exciting of topics. However, it is an important part of organic chemistry. There could very well be someone else who shares your name, but there is only one (*Z*)-1,2-dichloro-3-ethyl-4-methylpent-2-ene.

PRACTICE EXAMPLE A: Assign a configuration to each of the following alkenes:

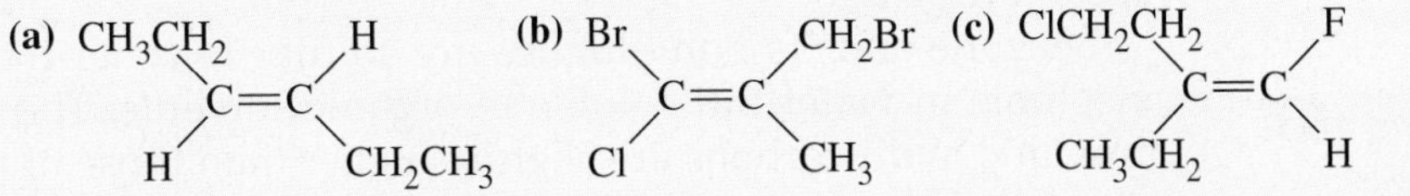

PRACTICE EXAMPLE B: Draw and label the *E* and *Z* isomers of the following compounds:

(a) $CH_3CH_2CH{=}CHCH_3$ (b) $CH_3CH_2\underset{|\atop Cl}{C}{=}CHCH_2CH_3$ (c) $CH_3CH_2\underset{|\atop CH_3}{C}{=}CHCH_3$

26-6 Aromatic Hydrocarbons

Aromatic hydrocarbons have ring structures with unsaturation (multiple-bond character) in the carbon-to-carbon bonds in the rings. Most aromatic hydrocarbons are based on the molecule benzene, C_6H_6. In Section 11-6, bonding in the benzene molecule was discussed in some detail. Several ways of representing the molecule were shown, including those in the margin.

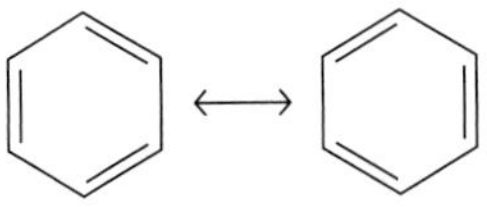

Kekulé structures

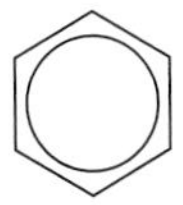

Simplified molecular orbital representation

▲ Neither the C atoms at the vertices nor the H atoms bonded to them are shown in these structures.

Other examples of aromatic hydrocarbons include

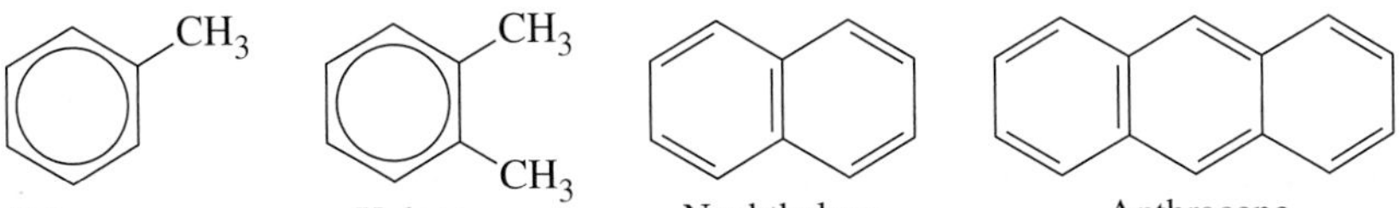

Toluene *o*-Xylene Naphthalene Anthracene

Toluene and *o*-xylene are *substituted benzenes*, and naphthalene and anthracene feature *fused benzene rings*. When rings are fused together, the resultant structure has two C and four H atoms fewer than the starting structures. Thus, the formula for naphthalene is $C_6H_6 + C_6H_6 - 2\,C - 4\,H = C_{10}H_8$; for anthracene: $C_{10}H_8 + C_6H_6 - 2\,C - 4\,H = C_{14}H_{10}$. In this text, we use the inscribed circle for the simple benzene ring and alternating single and double bonds for fused rings. For the fused-ring systems, the bond arrangement represents one of the possible resonance structures for the molecule.

▲ August Kekulé (1829–1896) proposed the hexagonal ring structure for benzene in 1865. His representation of this molecule is still widely used.

Stamp from the private collection of C.M. Lang. Photography by Gary J. Shulfer, University of Wisconsin, Stevens Point. "1996, Belgium (Scott #624)"; Scott Standard Postage Stamp Catalogue, Scott Pub. Co., Sidney, Ohio

Characteristics of Aromatic Hydrocarbons

Aromatic hydrocarbons are highly flammable and should always be handled with care. Prolonged inhalation of benzene vapor results in a decreased production of both red blood cells and white blood cells, which can be fatal. Also benzene is a carcinogen. Benzene and some other toxic aromatic compounds have been isolated in the tar formed by burning cigarettes, in polluted air, and as a decomposition product of grease in the charcoal grilling of meat.

A close examination of the structures of aromatic molecules shows that they all share two common features.

- They contain carbon atoms arranged in a *planar* (flat) ring with a *conjugated* bonding system—a bonding scheme among the ring atoms that in valence bond theory, consists of alternating single and double bonds. The system must extend throughout the ring and cause a lowering of the electronic energy.
- The π electron clouds associated with the double bonds must involve $(4n + 2)$ electrons, where $n = 0, 1, 2, \ldots$.

The benzene molecule has six electrons in the π electron clouds: $(4 \times 1) + 2 = 6$. The naphthalene molecule has 10: $(4 \times 2) + 2 = 10$. And the anthracene molecule has 14: $(4 \times 3) + 2 = 14$.

$CH_2{=}CH{-}CH{=}CH{-}CH{=}CH_2$

Hexa-1,3-5-triene (preferred) or 1,3-5 hexatriene

Cyclopenta-1,3-diene (preferred) or 1,3-cyclopentadiene

Neither of the two molecules depicted in the margin is aromatic. The 1,3,5-hexatriene molecule has six π electrons in its conjugated bonding system, but it is not cyclic. The 1,3-cyclopentadiene molecule is cyclic, but has only four π electrons in a conjugated bonding system that does not extend completely around the ring.

Benzene and its homologues are similar to other hydrocarbons in being insoluble in water but soluble in organic solvents. The boiling points of the aromatic hydrocarbons are slightly higher than those of the alkanes of similar carbon content. For example, hexane, C_6H_{14}, boils at 69 °C, whereas benzene boils at 80 °C. This can be explained by the planar structure and delocalized electron charge density of benzene, which increases the attractive forces between molecules. The symmetrical structure of benzene permits closer packing of molecules in the crystalline state and results in a higher melting point than for hexane. Benzene melts at 5.5 °C and hexane melts at −95 °C.

Two important aromatic functional groups are the phenyl and benzyl groups. A **phenyl group** is obtained when one of the six equivalent H atoms of a benzene molecule is removed. A **benzyl group** is obtained by replacing one of the H atoms in a methyl group with a phenyl group.

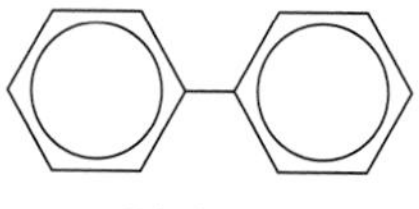

Biphenyl

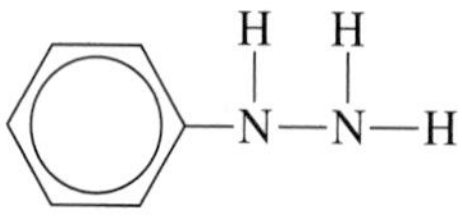

Phenylhydrazine

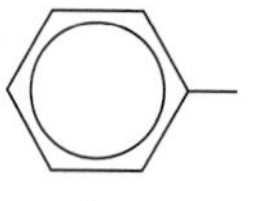

Phenyl group

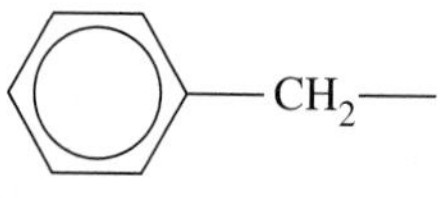

Benzyl group

Two phenyl groups may bond together, as in biphenyl, or phenyl groups may be substituents in other molecules, as in phenylhydrazine, used in the detection of sugars. The structures of biphenyl and phenylhydrazine are shown in the margin.

26-10 CONCEPT ASSESSMENT

What is the structure of the molecule with the name (*E*)-3-benzyl-2,5-dichloro-4-methylhex-3-ene?

Naming Aromatic Hydrocarbons

Other atoms or groups may be substituted for H atoms on the benzene molecule, and to name these compounds, we use a numbering system for the C atoms in the ring. If the name of an aromatic compound is based on a common name other than benzene (such as toluene), the characteristic substituent group (for example, he $-CH_3$ in toluene) is assigned position "1" on the benzene ring. Otherwise, he substituents are listed alphabetically and the carbon atoms in the ring are umbered so that the substituents appear at the lowest numbers possible, as hown below for 1-bromo-2-chlorobenzene.

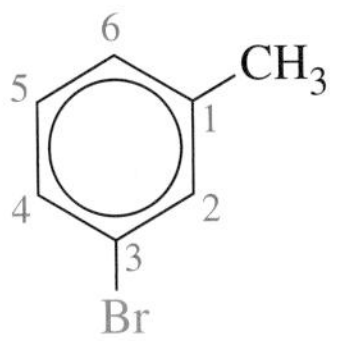

3-Bromotoluene
(*m*-bromotoluene)

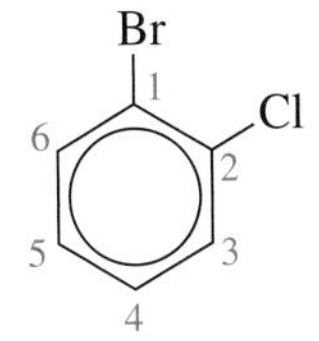

1-Bromo-2-chlorobenzene
(*o*-bromochlorobenzene)

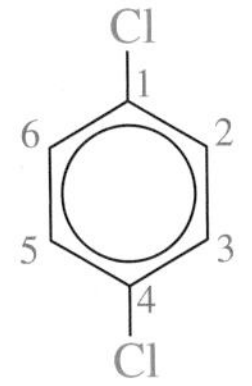

1,4-Dichlorobenzene
(*p*-dichlorobenzene)

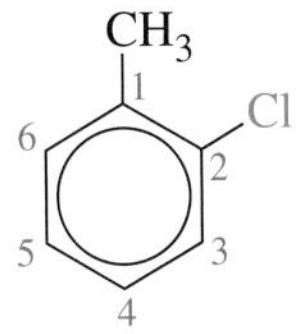

2-Chlorotoluene
(*o*-chlorotoluene)

◀ *p*-dichlorobenzene is used in mothballs. Naphthalene has also been used in mothballs but is being replaced by *p*-dichlorobenzene because *p*-dichlorobenzene has a less intense smell.

The terms *ortho, meta,* and *para* (*o*-, *m*-, *p*-) can be used when there are two substituents on the benzene ring. **Ortho** refers to substituents on adjacent carbon atoms, **meta** to substituents with one carbon atom between them, and **para** to substituents opposite one another on the ring.

Uses of Aromatic Hydrocarbons

Over 90% of the billions of pounds of benzene produced annually in the United States is derived from petroleum. The process involves dehydrogenation and cyclization of hexane to the aromatic hydrocarbon. The most important use of the petroleum-produced benzene is in manufacturing vinylbenzene for the production of styrene plastics. Other applications include the manufacture of phenol, the synthesis of dodecylbenzene (for detergents), and as an octane enhancer in gasoline. The production of aromatic compounds by dehydrogenation (removal of hydrogen) of alkanes yields large amounts of hydrogen gas, which is an important reactant in the synthesis of ammonia (page 1068).

26-7 Organic Compounds Containing Functional Groups

In this section we will describe the structure and nomenclature of organic compounds containing functional groups. We will defer a discussion of the chemical transformations between some of them until the next chapter. As we proceed you will see how the stereochemistry that we have discussed is also applied to these molecules.

◀ In the nineteenth century, chemists used the terms *aliphatic* (meaning fat-like) or *aromatic* (having a pleasant odor) to describe organic compounds. These days, we use the term *aromatic* to describe compounds made up of molecules with the features described on page 1240, regardless of their odors. The term *aliphatic* now refers to compounds that are not aromatic.

Alcohols and Phenols

Alcohols and **phenols** are characterized by the **hydroxyl** group, $-OH$. In alcohols, the hydroxyl group is bound to an sp^3 hybridized carbon atom. If this C atom also has only *one* other C atom (and two H atoms) bonded to it, the alcohol is a *primary* alcohol. If it has two other C atoms (and one H atom), the alcohol is a *secondary* alcohol. Finally, if it bonded to *three* other C atoms (and no H atoms), it is a *tertiary* alcohol. The systematic naming of alcohols uses the suffix *-ol* and was discussed in Chapter 3. In phenols, the hydroxyl group is attached to a benzene ring. Phenols are more acidic than alcohols because the anion formed by a phenol is stabilized by resonance but the anion formed by an alcohol is not (see page 772).

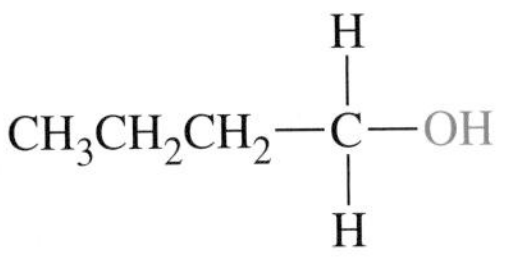

Butan-1-ol
(butyl alcohol)
(a *primary* alcohol)

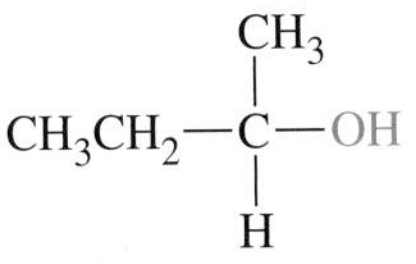

Butan-2-ol
(*sec*-butyl alcohol)
(a *secondary* alcohol)

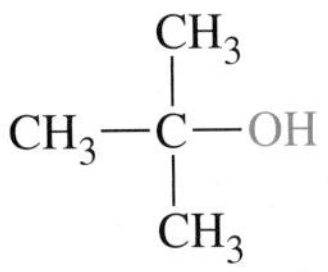

2-Methylpropan-2-ol
(*tert*-butyl alcohol)
(a *tertiary* alcohol)

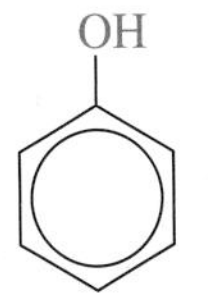

Phenol

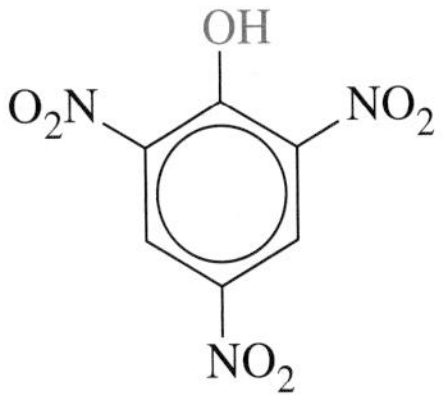

2,4,6-Trinitrophenol

A molecule may have more than one —OH group. Those with two —OH groups are known as *diols* (or glycols), and those with more than two —OH groups are called *polyols*. Ethylene glycol, a diol, is used in automobil antifreeze solutions; glycerol, a polyol, is an important biological molecul used as part of the body's mechanism for fat storage.

Ethane-1,2-diol (ethylene glycol) | Propane-1,2,3-triol (glycerol)

Physical properties of aliphatic alcohols are strongly influenced by hydrogen bonding. As the chain length increases, however, the influence o the polar hydroxyl group on the properties of the molecule diminishes. Th molecule becomes less like water and more like a hydrocarbon. As a consequence, low-molecular mass alcohols tend to be water soluble, whereas high molecular mass alcohols are not. The boiling points and solubilities of th phenols vary widely, depending on the nature of the other substituents on th benzene ring.

26-11 CONCEPT ASSESSMENT

Explain why the name *sec*-pentyl alcohol does not unambiguously identify a compound whereas the name *sec*-butyl alcohol does.

Preparation and Uses of Alcohols Two methods by which alcohols can be synthesized are by the hydration of alkenes and the hydrolysis of alkyl halides.

$$\underset{\text{propene (propylene)}}{CH_3CH{=}CH_2} + H_2O \xrightarrow{H_2SO_4} \underset{\text{propan-2-ol (isopropyl alcohol)}}{CH_3\overset{\overset{OH}{|}}{C}HCH_3} \qquad (26.9)$$

$$\underset{\text{1-bromopropane}}{CH_3CH_2CH_2Br} + OH^- \longrightarrow \underset{\text{propan-1-ol}}{CH_3CH_2CH_2OH} + Br^- \qquad (26.10)$$

Reaction (26.9) is an example of an *addition reaction* and reaction (26.10) is an example of a *substitution reaction*. In an addition reaction, one or more atoms add to a molecule. In a substitution reaction, an atom or a grouping of atoms is replaced by another atom or group of atoms. We will have a closer look at these types of reactions in Chapter 27.

Methanol (wood alcohol) is the simplest alcohol. It is a highly toxic substance that can lead to blindness or death if ingested. Most methanol is manufactured from carbon monoxide and hydrogen.

$$CO(g) + 2\,H_2(g) \xrightarrow[200\text{ atm}]{350\text{ °C}} CH_3OH(g)$$
$$ZnO, Cr_2O_3$$

Methanol is the most extensively produced alcohol. It is used in the synthesis of other organic chemicals and as a solvent, but potentially its most important use may be as a motor fuel (recall Section 7-9).

Ethanol, CH_3CH_2OH, is grain alcohol, which is found in alcoholic beverages. It is easily produced by the fermentation of the juices of sugarcane or other materials that contain natural sugars. The industrial method involves the hydration of ethylene with sulfuric acid catalyst (similar to reaction 26.9).

Ethylene glycol, $HOCH_2CH_2OH$, is water soluble and has a higher boiling point (197 °C) than water. These properties make it an excellent permanent, nonvolatile antifreeze for automobile radiators. It is also used in the manufacture of solvents, paint removers, and plasticizers (softeners).

Glycerol (glycerin), $HOCH_2CH(OH)CH_2OH$, is obtained commercially as a by-product in the manufacture of soap. It is a sweet, syrupy liquid that is miscible with water in all proportions. Because it takes up moisture from the air, glycerol can be used to keep skin moist and soft and is found in lotions and cosmetics.

An interesting and useful derivative of an alcohol is the alkoxide ion, RO^-, a deprotonated form of the alcohol. The alkoxide ion is formed by the reaction of sodium or potassium metal with the alcohol. For example, the methoxide ion, CH_3O^-, is produced in the following reaction:

$$CH_3OH + Na \longrightarrow CH_3O^- + Na^+ + \frac{1}{2}H_2(g) \qquad \textbf{(26.11)}$$

Reaction (26.11) provides the basis of a simple qualitative test that is sometimes used to confirm whether a compound is an alcohol. For alcohols, the addition of sodium metal results in the formation of $H_2(g)$ bubbles. Reaction (26.11) is more commonly used to produce alkoxide ions that can be used in other reactions. For example, an alkoxide ion can react with a haloalkane, as suggested below:

$$CH_3O^- + CH_3CH_2Cl \longrightarrow CH_3OCH_2CH_3 + Cl^-$$

This reaction is yet another example of a substitution reaction (Chapter 27). The compound formed is an ether, discussed next.

Peter Dean/Shutterstock

▲ Giant soap bubble
A soap bubble consists of an air pocket enclosed in a thin film of soap in water. As the water evaporates, the film breaks and the bubble bursts. Glycerol added to the soap–water mixture forms hydrogen bonds to both the soap and water molecules. This slows the rate of evaporation of the water and increases the strength of the film, allowing the production of very large bubbles.

Ethers

An **ether** is a compound with the general formula R—O—R′. Structurally, ethers can be pure aliphatic, pure aromatic, or mixed.

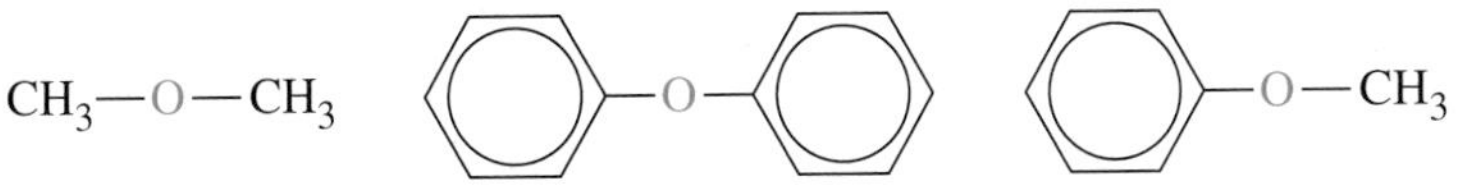

Dimethyl ether   Diphenyl ether   Methyl phenyl ether (anisole)

Dimethyl ether has the same formula as ethanol, but the two substances have quite different physical and chemical properties. They have different properties because each has a different functional group (see Table 26.2). Dimethyl ether and ethanol are constitutional isomers.

Ethers can be viewed as alkane or aromatic compounds containing the group RO—; this group is known as the *alkoxy group*. The IUPAC system for naming ethers treats them as alkanes that bear an alkoxy substituent, that is, as alkoxyalkanes. The smaller substituent is considered part of the alkoxy group and the larger substituent defines the stem. Thus, for example, the IUPAC name for ethylmethyl ether is methoxyethane, and for anisole, it is methoxybenzene.

Ethers can also be cyclic. For example, if an oxygen atom takes the place of a carbon atom in a cyclohexane molecule, we obtain the cyclic ether shown in the margin. The oxygen atom in this structure is called a *heteroatom* (because it is not a carbon atom) and the compound is called a **heterocyclic compound**. The simplest system for naming cyclic ethers is based on using the prefix *oxa-* in front of the name of the cycloalkane. The prefix *oxa-* indicates that a carbon atom has been replaced by an oxygen atom. Thus, the name of the compound shown in the margin is oxacyclohexane.

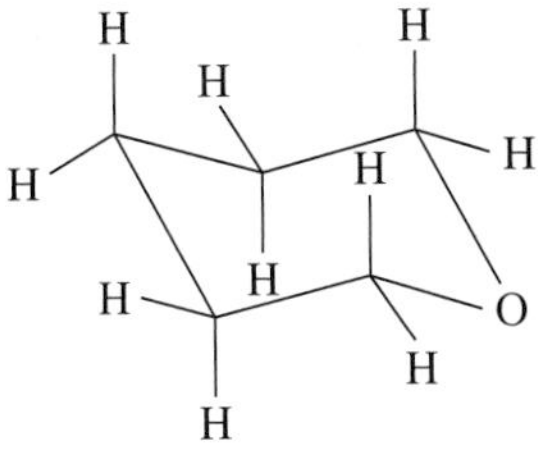

Preparation and Uses of Ethers Symmetrical ethers, such as diethyl ether, can be prepared by the elimination of a water molecule from between two alcohol molecules with a strong dehydrating agent, such as concentrated H_2SO_4.

$$CH_3CH_2OH + HOCH_2CH_3 \xrightarrow[140\ ^\circ C]{\text{conc. } H_2SO_4} CH_3CH_2OCH_2CH_3 + H_2O \qquad \textbf{(26.12)}$$

▶ The molecule below should be named *tert*-butyl methyl ether (TBME), but has been referred to by industrial chemists as MTBE for so long that methyl *tert*-butyl ether has become the commonly used name for this compound.

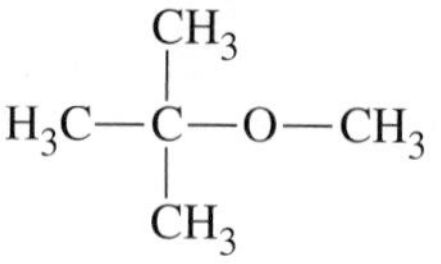

Methyl *tert*-butyl ether (MTBE)

Chemically, ethers are comparatively unreactive. The ether linkage is stable in the presence of most oxidizing and reducing agents, as well as dilute acids and alkalis.

Diethyl ether has been used extensively as a general anesthetic. It is easy to administer and produces excellent relaxation of the muscles. Also, it affects the pulse rate, rate of respiration, and blood pressure only slightly. However, it is somewhat irritating to the respiratory passages and produces nausea. Methyl propyl ether (neothyl) is also used as an anesthetic and is less irritating to the respiratory passages. Dimethyl ether, a gas at room temperatures, is used as a propellant for aerosol sprays. Higher molecular mass ethers are used as solvents for varnishes and lacquers. Methyl *tert*-butyl ether, an unsymmetrical ether, marketed under the name MTBE, has been used as an octane enhancer in gasoline. However, because of its relatively high solubility in water (≈ 5 g/100 g H_2O), in some localities it has caused rather extensive groundwater pollution through leakage from underground storage tanks; it is currently being phased out of use.

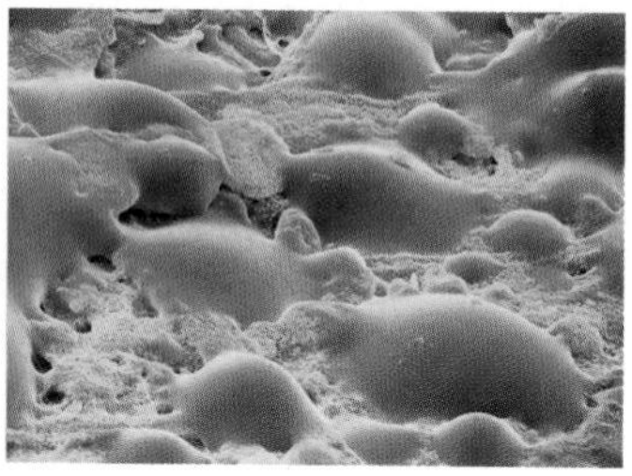

Science Photo Library

▲ False-color scanning electron micrograph of the sticky surface of a 3M Post-it note. The bubbles are 15–40 μm in diameter and consist of a urea–formaldehyde adhesive. Each time the note is pressed to a surface, fresh adhesive is released. The note can be reattached to a surface as long as some of the bubbles remain.

Aldehydes and Ketones

Aldehydes and ketones contain the **carbonyl group**.

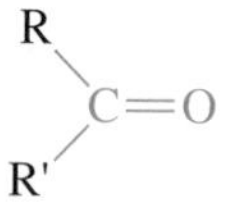

If either the R or R′ group is an H atom, the compound is an **aldehyde**. If the R and R′ groups are alkyl or aromatic (aryl) groups, the compound is a **ketone**.

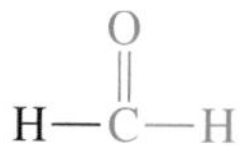

Methanal (formaldehyde)

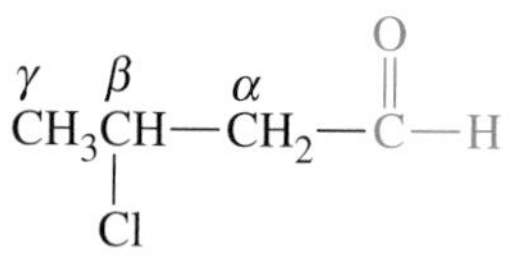

3-Chlorobutanal (β-chlorobutyraldehyde)

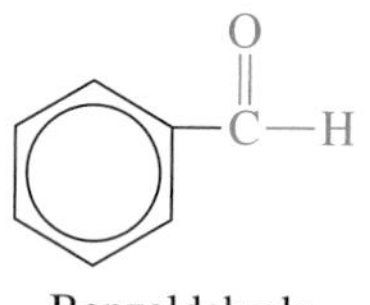

Benzaldehyde

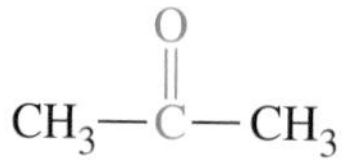

Propanone (acetone)

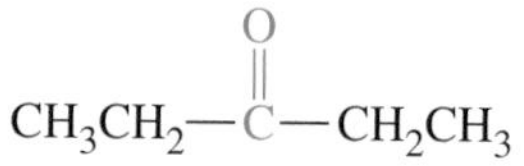

Pentan-3-one (diethyl ketone)

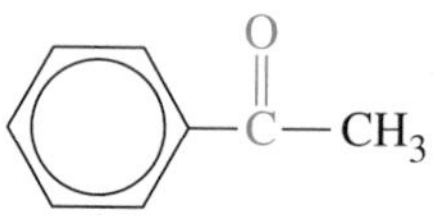

1-Phenylethanone (acetophenone)

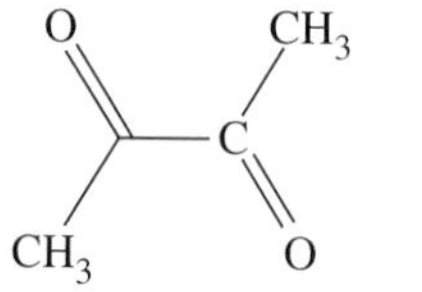

Butanedione, a diketone

Aldehydes and ketones often have characteristic and recognizable odors. For example, 2-heptanone is a liquid with a clove-like odor that accounts for the odors of many fruits and dairy products. Some aldehydes and ketones find use as flavoring agents. For example, vanillin, the compound responsible for vanilla flavor, is an aldehyde. Alpha-demascone and 2-octanone are ketones responsible for berry and mushroom flavors, respectively. Butanedione, shown in the margin, is a yellow liquid with a cheese-like smell that gives butter its flavor.

The IUPAC naming system for aldehydes uses the suffix *-al*. The parent chain is the longest chain containing the aldehyde group. Accordingly, the four-carbon aldehyde is called butanal because the name is derived from the alkane (butane) with the *-e* replaced by *-al*. The numbering of the chain starts at the carbon of the aldehyde group; it is always at position one and need not be specified. Thus 3-chloro-2-methylbutanal is

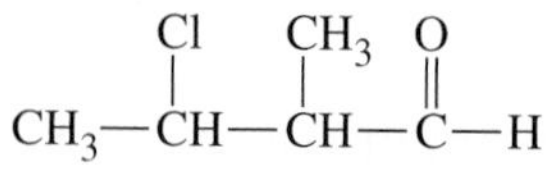

KEEP IN MIND

that the CHO functional group can be only at the terminus of a carbon chain, whereas the CO functional group cannot be at the end of a carbon chain.

The IUPAC naming system for ketones uses the suffix *-one*. The parent chain must include the carbonyl group and be numbered from the end of the chain that reaches the carbonyl group first; this ensures, as required by IUPAC rules, that the number for the carbonyl group is as low as possible. For example, the molecule

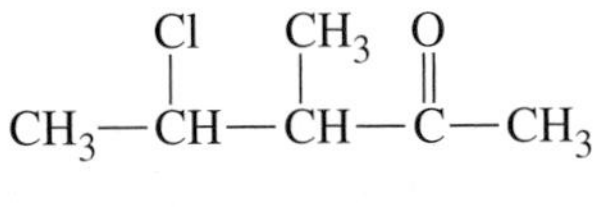

is 4-chloro-3-methylpentan-2-one. The number for the position of the carbonyl group is placed immediately before the *-one* ending.

Preparation and Uses of Aldehydes and Ketones Aldehydes can be produced by the oxidation of a *primary* alcohol with an oxidizing agent such as dichromate ion in acidic solution. However, the aldehyde so formed is readily oxidized further to a carboxylic acid.

$$\underset{\substack{\text{ethanol}\\\text{(a primary alcohol)}}}{CH_3CH_2OH} \xrightarrow[H^+]{Cr_2O_7^{2-}} \underset{\substack{\text{acetaldehyde}\\\text{(an aldehyde)}}}{CH_3CHO} \xrightarrow[H^+]{Cr_2O_7^{2-}} \underset{\substack{\text{acetic acid}\\\text{(an acid)}}}{CH_3CO_2H} \qquad (26.13)$$

A milder oxidizing agent in a nonaqueous medium is required to stop the oxidation at the aldehyde. This partial oxidation can be done with the reagent pyridinium chlorochromate (PCC) in an organic solvent such as dichloromethane. For example,

$$CH_3(CH_2)_8CH_2OH \xrightarrow{PCC,\ CH_2Cl_2} CH_3(CH_2)_8CHO$$

◀ PCC is formed as yellow-orange crystals by the reaction between CrO_3 and HCl in pyridine.

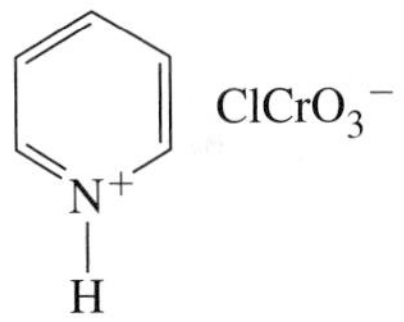

Oxidation of a *secondary* alcohol produces a ketone.

$$\underset{\substack{\text{propan-2-ol}\\\text{(a secondary alcohol)}}}{CH_3CH(OH)CH_3} \xrightarrow[H^+]{Cr_2O_7^{2-}} \underset{\substack{\text{acetone}\\\text{(a ketone)}}}{CH_3\overset{\overset{\displaystyle O}{\|}}{C}CH_3} \qquad (26.14)$$

KEEP IN MIND

that tertiary alcohols cannot be oxidized to aldehydes or ketones because a tertiary carbon is bonded to three alkyl groups and cannot form a double bond with an oxygen atom unless one of the carbon–carbon bonds is broken. Oxidation of a tertiary alcohol is typically difficult and requires a very strong oxidizing agent.

Ketones are much more resistant to oxidation than are alcohols and aldehydes.

Aldehydes and ketones occur widely in nature. Typical natural sources are

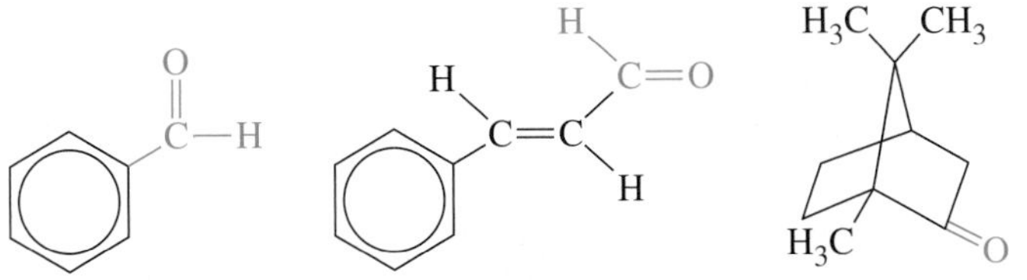

Benzaldehyde (almonds) | Cinnamaldehyde (cinnamon) | Camphor (obtained from camphor tree)

Aldehydes and ketones can be reduced to primary and secondary alcohols, respectively, by sodium borohydride, $NaBH_4$. For example, hexan-2-one is reduced to hexan-2-ol by $NaBH_4$:

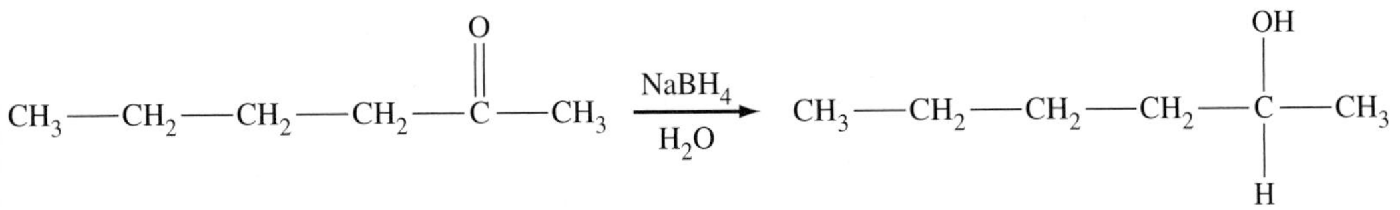

The reaction effectively adds two H atoms across the carbon–oxygen double bond. However, H atoms are not directly involved. The reaction involves two distinct steps (1) attack of the carbonyl carbon atom by a hydride (H^-) ion; (2) protonation of the carbonyl oxygen atom.

Aldehydes and ketones find use as starting materials and reagents for the synthesis of other organic compounds, and, generally speaking, aldehydes are more reactive than ketones. The carbon atom in the carbonyl group is slightly positive and is prone to attack by species that are attracted to centers of positive charge.

The simplest aldehyde is formaldehyde ($H_2C{=}O$), a colorless gas that dissolves readily in water. Billions of kilograms of formaldehyde are used each year in the manufacture of synthetic resins. A polymer of formaldehyde called paraformaldehyde is used as an antiseptic and an insecticide.

Acetone is the most important of the ketones. It is a volatile liquid (boiling point, 56 °C) and highly flammable. Acetone is a good solvent for a variety of organic compounds and is widely used in solvents for varnishes, lacquers, and plastics. Unlike many common organic solvents, acetone is miscible with water in all proportions.

Carboxylic Acids

▶ The carboxyl group is represented either as —COOH or $—CO_2H$.

As we learned in Chapter 3, **carboxylic acids** contain the **carboxyl group** (*carb*onyl and hydr*oxyl*).

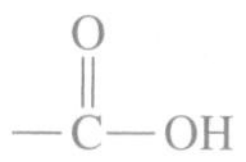

Carboxylic acids have the general formula RCOOH. If two carboxyl groups are found on the same molecule, the acid is called a dicarboxylic acid. In the *fatty acids* found in naturally occurring fats and oil, R is a high-molecular mass alkyl group. The carboxyl group can also be found attached to the benzene ring.

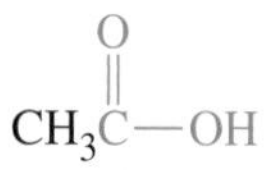

Acetic acid
(an aliphatic acid)

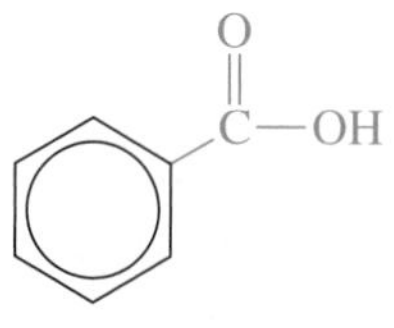

Benzoic acid
(an aromatic acid)

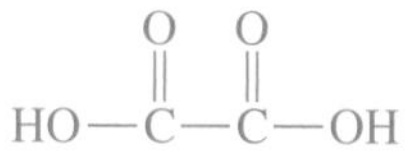

Oxalic acid
(an aliphatic dicarboxylic acid)

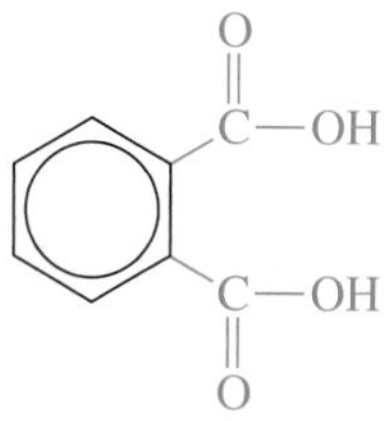

Phthalic acid
(an aromatic dicarboxylic acid)

Straight- or branched-chain acids can be named either by their IUPAC names or by using Greek letters in conjunction with common names. Cycloalkanes with a —COOH substituent are named as *cycloalkanecarboxylic acids*; that is, the ending "carboxylic acid" is combined with the name of the cycloalkane. Aromatic acids are named as derivatives of benzoic acid. Here are a few examples:

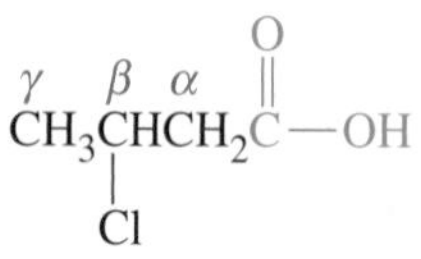

3-Chlorobutanoic acid
β-chlorobutyric acid

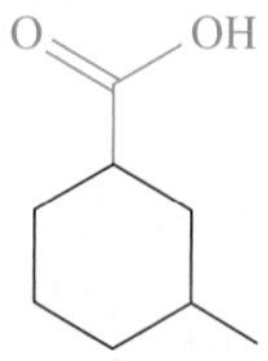

3-methylcyclohexanecarboxylic acid

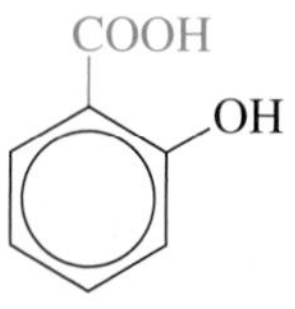

2-Hydroxybenzoic acid
o-hydroxybenzoic acid
(also salicylic acid)

Notice that in straight- or branched-chain carboxylic acids, the C atom in the —COOH group is C1, but in cyclic and aromatic acids, the C atom bonded to the —COOH group is C1.

Carboxylic acids are found widely in nature. Spinach, rhubarb, and other green leafy vegetables are rich in *oxalic acid* (see margin). Sour milk and sore muscles have elevated levels of *lactic acid*. Citrus fruits are rich in *citric acid*. Line-angle formulas for these acids are shown in the margin.

Carboxylic acids, especially those of low molecular weight, exhibit characteristic odors. Ethanoic acid (acetic acid) is the acid that gives vinegar its characteristic smell. The presence of butanoic acid contributes to the strong flavor and aroma of many cheeses, and (*E*)-3-methylhex-2-enoic acid has been identified as the principal compound responsible for the smell of human sweat.

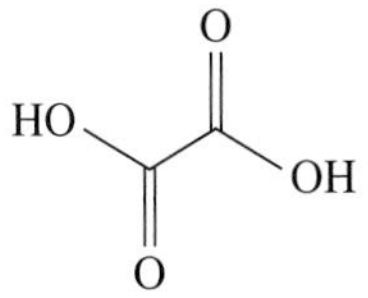

Oxalic acid

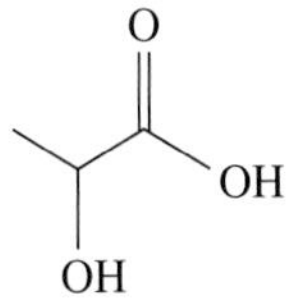

Lactic acid

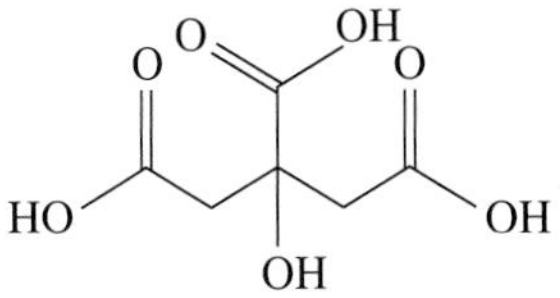

Citric acid

26-12 CONCEPT ASSESSMENT

Draw the structure of (*E*)-3-methylhex-2-enoic acid.

Some properties of carboxylic acids can be rationalized in terms of the ability of carboxylic acid molecules to form hydrogen bonds among themselves or with other molecules (such as water). For example, low-molecular-weight carboxylic acids are soluble in water because of their ability to hydrogen bond with water. Also, carboxylic acids also have relatively high melting and boiling points because of hydrogen bonding.

As discussed in Chapter 16, soluble carboxylic acids behave as weak acids when they dissolve in water. A simple way to test whether a compound is a carboxylic acid is to add it to an aqueous solution of sodium hydrogen carbonate, $NaHCO_3(aq)$, or sodium carbonate, $Na_2CO_3(aq)$. If the compound is an acid, bubbles of $CO_2(g)$ will be visible because of the following reaction.

$$2\,RCOOH(aq) + Na_2CO_3(aq) \longrightarrow 2\,RCOONa(aq) + CO_2(g) + H_2O(l) \quad \textbf{(26.15)}$$

Preparation and Uses of Carboxylic Acids Carboxylic acids can be obtained in the laboratory by the oxidation of a primary alcohol or an aldehyde. For this purpose, the oxidizing agent is generally $KMnO_4(aq)$ in an alkaline medium. Because the medium is alkaline, the product is the potassium salt, but the free carboxylic acid can be regenerated by making the medium acidic.

$$CH_3CH_2OH \xrightarrow[OH^-,\ heat]{KMnO_4} CH_3COO^-K^+ \xrightarrow{H^+} CH_3COOH + K^+ \quad \textbf{(26.16)}$$

$$CH_3CH_2CH_2OH \xrightarrow[OH^-,\ heat]{KMnO_4} CH_3CH_2COO^-K^+ \xrightarrow{H^+} CH_3CH_2COOH + K^+ \quad \textbf{(26.17)}$$

We have seen (earlier in this section) that primary alcohols and aldehydes can be oxidized to carboxylic acids. Carboxylic acids can also be prepared by the hydrolysis of nitriles. The hydrolysis of a nitrile can be carried out in either acidic or basic solution, as suggested below:

$$R{-}C{\equiv}N + 2\,H_2O \xrightarrow[\Delta]{H^+ \text{ or } OH^-} RCOOH + NH_3$$

If the reaction is carried out in basic solution, the carboxylate anion, $RCOO^-$, is produced. To obtain RCOOH, the solution must be acidified.

Carboxylic acids are used extensively in organic chemistry to make other compounds and we will examine some of these reactions next (and in Chapter 27). Because many derivatives of carboxylic acids involve simple replacement of the hydroxyl groups, the following grouping of atoms is encountered frequently:

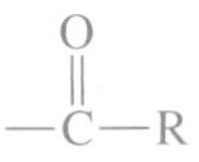

The grouping above is called an *acyl group*. The names of acyl groups are derived from the acid names by replacing *-ic acid* with the suffix *-yl*, as shown below:

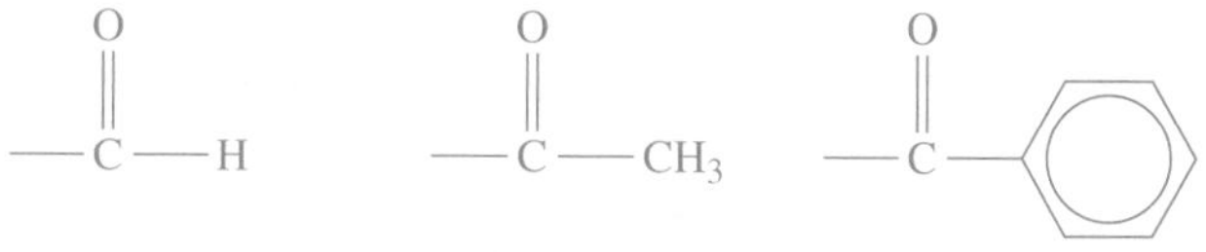

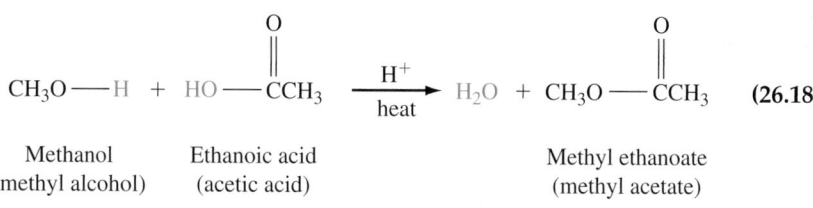

IUPAC name:	Methanoyl (from methanoic acid)	Ethanoyl (from ethanoic acid)	Benzoyl (from benzoic acid)
Common name:	Formyl (from formic acid)	Acetyl (from acetic acid)	Benzoyl

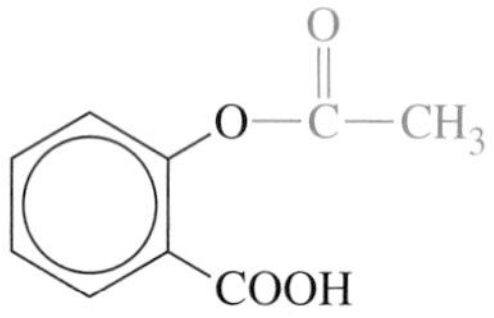

Acetylsalicylic acid (aspirin)

The IUPAC names for these acyl groups are almost never used. The acetyl derivative of *o*-hydroxybenzoic acid (salicylic acid) is what we know as ordinary aspirin, acetylsalicylic acid (ASA), the structure of which is shown in the margin.

Esters

As shown in Table 26.2, the general formula of an ester is RCOOR′. Comparing the general formula of an ester with that of a carboxylic acid, we see that an **ester** is a molecule in which the hydroxyl group (—OH) of a carboxylic acid functional group is replaced by an alkoxyl group (—OR′). In the lab, esters can be prepared by the reaction of a carboxylic acid and an alcohol. The products of the reaction are an ester and a water molecule.

Robin W/ Shutterstock

▲ The distinctive aroma and flavor of oranges are due in part to the ester octyl acetate, $CH_3(CH_2)_6CH_2OOCCH_3$.

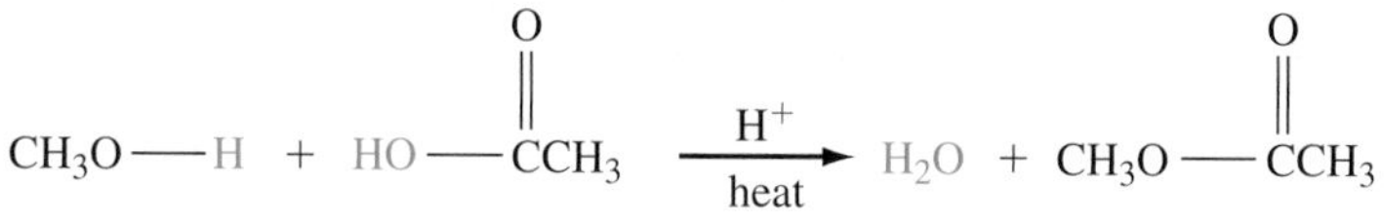

(26.18)

Methanol (methyl alcohol) Ethanoic acid (acetic acid) Methyl ethanoate (methyl acetate)

Because the reaction above is reversible, an excess of alcohol is often used to ensure a high yield of the ester.

Esters have two-part names. The first part is the alkyl designation of the —OR′ group. The second part is the carboxylic acid name with *-ic acid* changed to *-ate*. For example, the combination based on ethanoic acid and the methoxy group is *methyl ethanoate*. A few more examples are shown here.

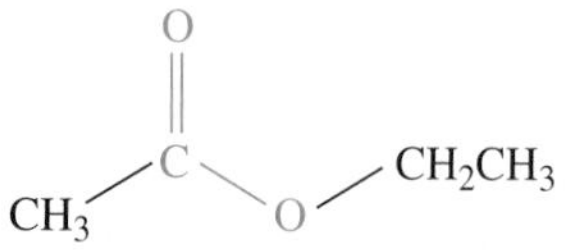

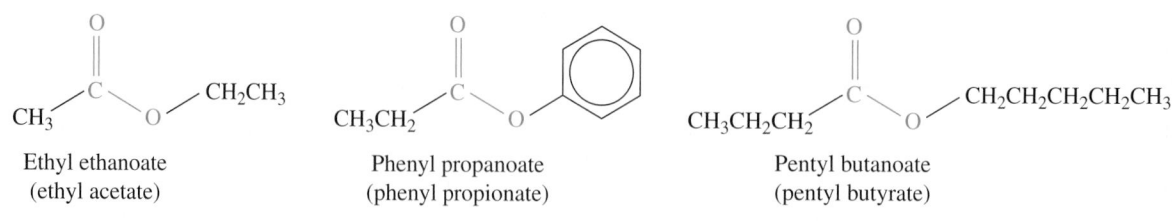

Ethyl ethanoate (ethyl acetate)

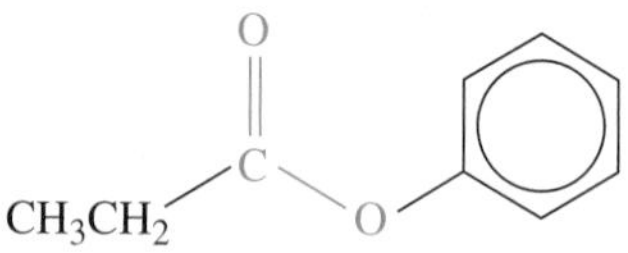

Phenyl propanoate (phenyl propionate)

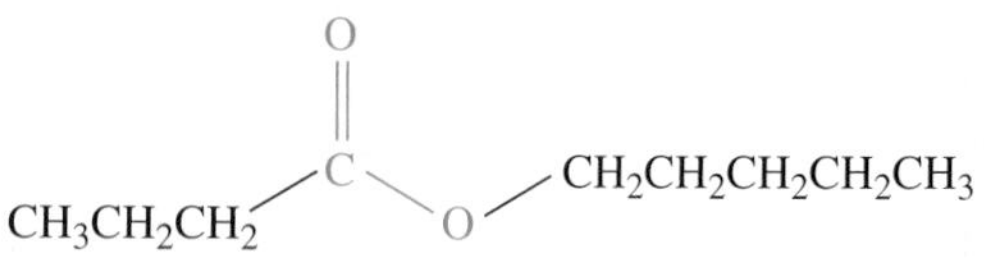

Pentyl butanoate (pentyl butyrate)

Unlike the pungent odors of the carboxylic acids from which they are derived, esters have very pleasant aromas. The characteristic fragrances of many flowers and fruits can be traced to the esters they contain. Esters are used in perfumes and in the manufacture of flavoring agents for the confectionery and soft drink industries. Most esters are colorless liquids that are insoluble in water. Their melting points and boiling points are generally lower than those of alcohols and acids of comparable carbon content. This is because of the absence of hydrogen bonding in esters.

Amides

The replacement of the hydroxyl group in the carboxylic acid functional group with an $-NH_2$ group produces an **amide**. For example, the molecule butanamide (shown in margin) is formed by replacing the hydroxyl group in ethanoic acid by the $-NH_2$ group.

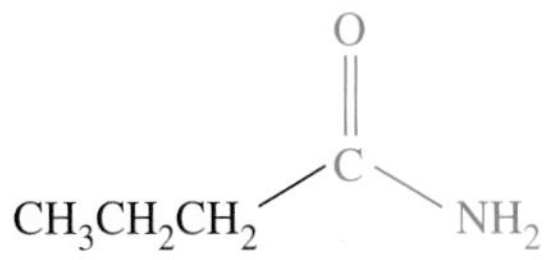

Butanamide

The name of an amide is constructed from the alkane part of the acid name and the suffix *-amide* is added. If hydrogen atoms on the nitrogen atom are replaced with other groups, then a substituted amide is obtained. For example, in the molecules below, one or both of the H atoms in the $-NH_2$ group of ethanamide have been replaced by other groups. Here are a few examples.

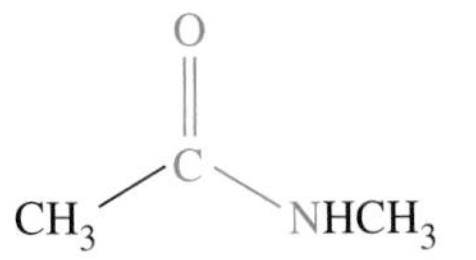

N-methylethanamide

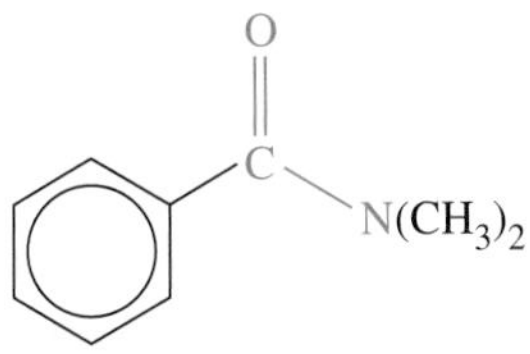

N,N-dimethylbenzamide

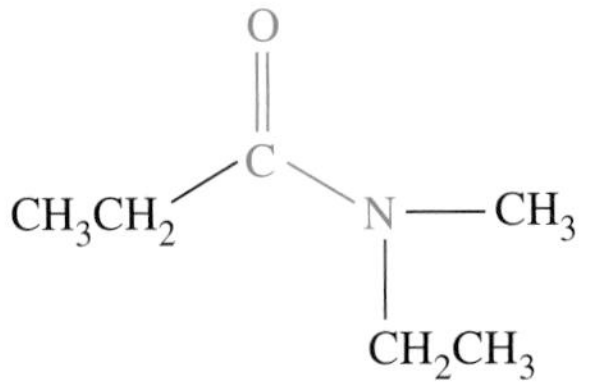

N-ethyl-*N*-methylpropanamide

As the examples above show, we name a substituted amide by using the prefix *N*- in front of each nitrogen substituent. The replacement of one H atom gives a *N*-substituted amide. The active ingredient in Tylenol® is an *N*-substituted amide (see margin). The $-CO-NH-$ linkage present in *N*-substituted amides is called a *peptide bond*, and it is a key bond type in proteins, as we will discover in Chapter 28.

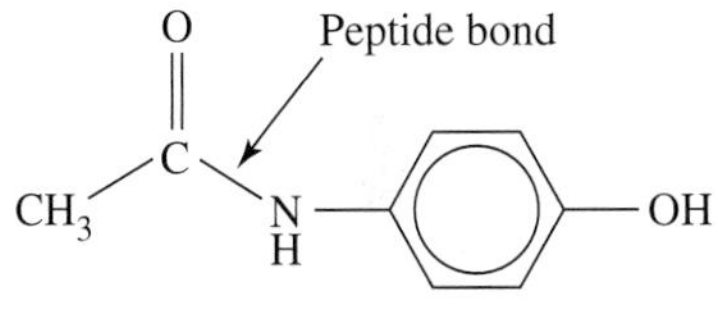

N-(4-hydroxyphenyl)ethanamide
(Tylenol®)

Despite the presence of the $-NH_2$ group in simple amides, they are not Brønsted–Lowry bases like amines or ammonia because the carbonyl group promotes the resonance structures shown below.

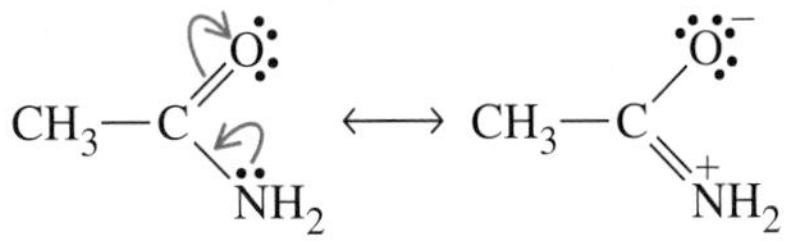

◀ Resonance in an amide.

The lone pair of electrons on the nitrogen atom is delocalized over the carbonyl group toward the more electronegative oxygen atom. In fact, as seen from the resonance structures, it is the carbonyl group that is most likely to be protonated in acidic conditions. The protonation of the oxygen in a carbonyl group is often an important first step in reactions of esters and amides under acidic conditions.

Preparation and Uses of Amides One way of making an amide is to treat a carboxylic acid with ammonia, forming an ammonium salt, followed by heating. The following sequence summarizes a pathway for making ethanamide.

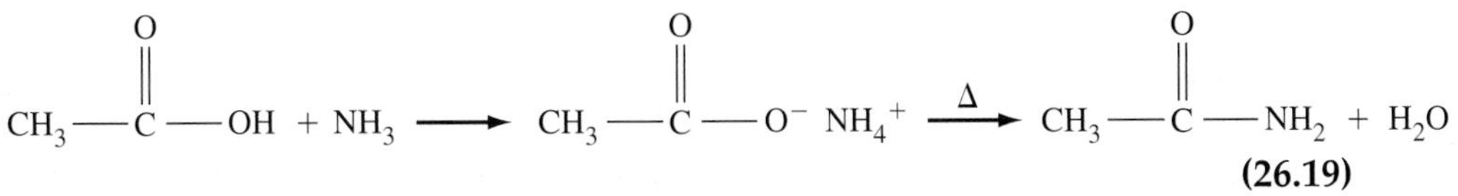

$$CH_3COOH + NH_3 \longrightarrow CH_3COO^- NH_4^+ \xrightarrow{\Delta} CH_3CONH_2 + H_2O \quad (26.19)$$

However, amides are seldom prepared as described above. As we will see in Chapter 27, a more effective way of converting CH_3COOH to CH_3CONH_2 is to treat the acid with $SOCl_2$ first, which converts the acid to a more reactive *acid chloride* (CH_3COCl), and then treat the acid chloride with ammonia.

▶ A polyamide is a polymer of amides joined by peptide bonds. Polymers are discussed in Chapter 27.

Amides can be converted into other compounds, but compared with other carboxylic acid derivatives, amides are relatively unreactive. The stability of biological structures and the properties of *polyamides*, such as silk and nylon, depend, in part, on the presence of amide linkages. Amides can, however, be converted into nitriles, R—C≡N, using a strong dehydrating agent such as P_4O_{10}.

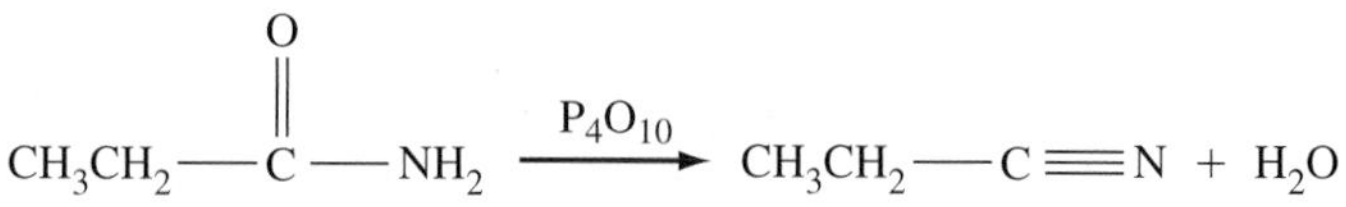

Amides are also routinely used to make amines, a class of compounds we discuss next.

$CH_3CH_3NH_2$
Ethanamine
(a primary amine)

$CH_3CH_2CH_2NCH_3$ (with H on N)
N-methylpropan-1-amine
(a secondary amine)

$CH_3CH_2CH_2NCH_3$ (with CH_2CH_3 on N)
N-ethyl-*N*-methylpropan-1-amine
(a tertiary amine)

Amines

Amines are organic derivatives of ammonia, NH_3, in which one or more organic groups (R) are substituted for H atoms. Their classification is based on the number of R groups bonded to the nitrogen atom—one for primary amines, two for secondary, and three for tertiary (see margin). Primary amines are named by identifying the parent chain (the longest carbon chain) containing the NH_2 group and replacing the *e* of the parent chain with the suffix *-amine*. If required, a number is placed before the suffix to indicate the position of the NH_2 group. Secondary and tertiary amines are named in a similar manner. However, the symbol *N* may be used to indicate the replacement of H atoms in the —NH_2 group by other substituents, as illustrated by the examples in the margin.

One of the chief methods of preparing an amine is by the reduction of a nitro compound.

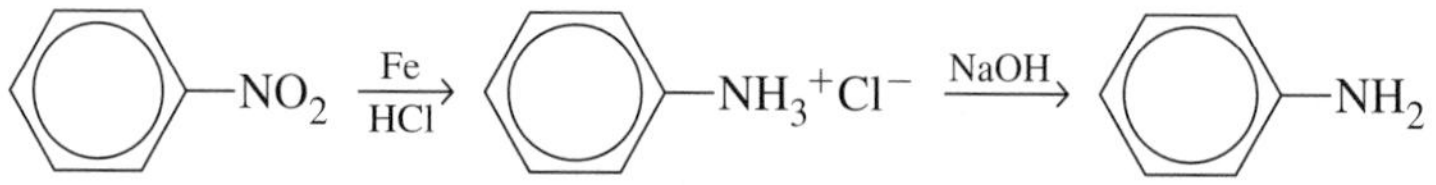
(26.20)

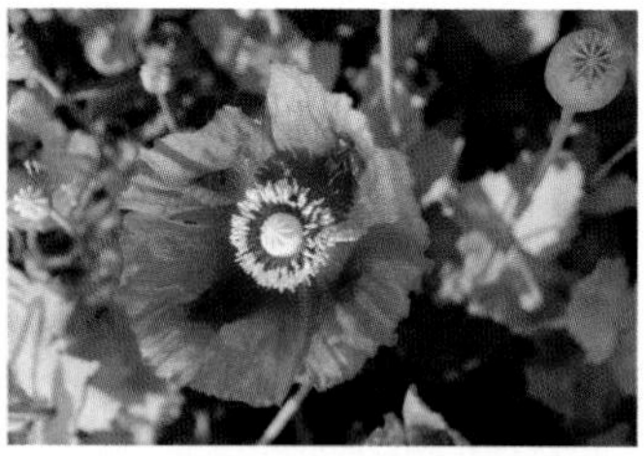
Karel Gallas/Shutterstock

An alternative method of preparing an amine involves the reaction of ammonia with an alkyl halide. The methylammonium ion, $CH_3NH_3^+$, reacts with additional ammonia (an acid-base reaction) to give the amine.

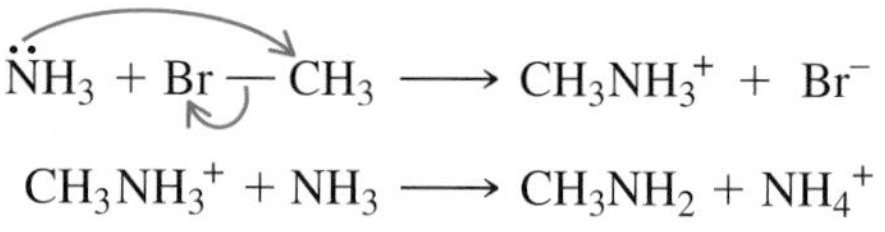

The formation of pure primary amines is difficult by this method because when the primary amine is formed, it can take part in a further substitution reaction with the alkyl halide to form a secondary amine, for example,

$$CH_3NH_2 + CH_3Br \longrightarrow CH_3NHCH_3 + HBr \quad (26.21)$$

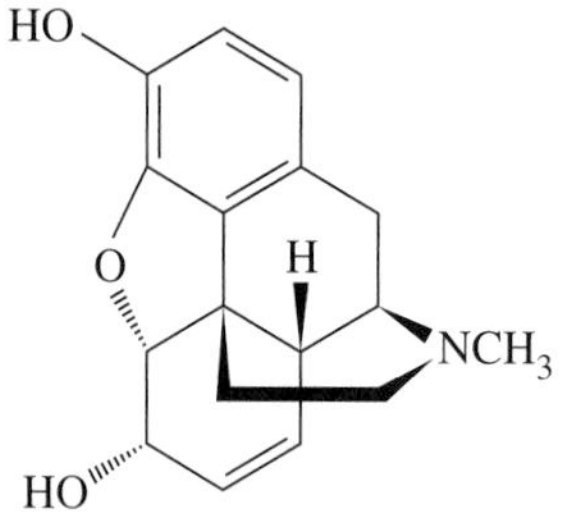
Morphine

▲ Morphine, a very powerful and addictive painkiller, can be isolated from the opium poppy (*Papaver somniferum*).

The compound formed here is dimethylamine. Further reaction with bromomethane produces the tertiary amine, trimethylamine.

Amines of low molecular mass are gases that are readily soluble in water, yielding basic solutions. The volatile members have odors similar to ammonia, but more fish-like. Primary and secondary amines form hydrogen bonds, but these bonds are weaker than those in water because nitrogen is less electronegative than oxygen. Like ammonia, amines have trigonal pyramidal structures with a lone pair of electrons on the N atom. Also like ammonia, amines owe their basicity to these lone-pair electrons. In aromatic amines, because of unsaturation in the benzene ring, electrons are drawn into the ring, which reduces the electron charge density on the nitrogen atom. As a result, aromatic amines are *weaker* bases than ammonia. Aliphatic amines are somewhat stronger bases than ammonia.

$$NH_3 + H_2O \rightleftharpoons NH_4^+ + OH^- \qquad K_b = 1.8 \times 10^{-5}$$

$$\underset{\text{methanamine}}{CH_3NH_2} + H_2O \rightleftharpoons CH_3NH_3^+ + OH^- \qquad K_b = 4.2 \times 10^{-5}$$

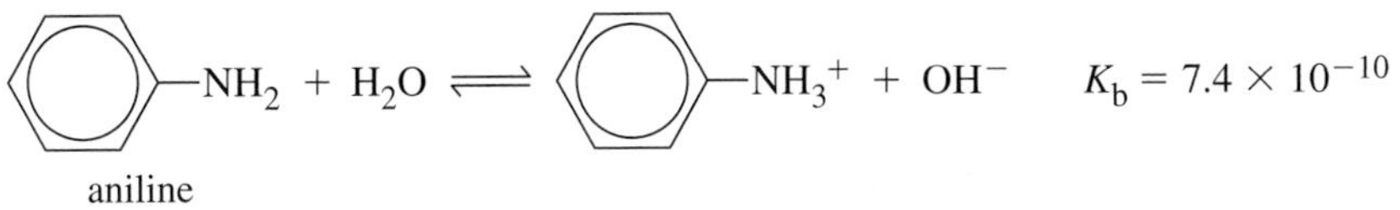

aniline

Dimethylamine is an accelerator for the removal of hair from hides in the processing of leather. Butanamines and pentanamines are used as antioxidants, corrosion inhibitors, and in the manufacture of oil-soluble soaps. Dimethylamine and trimethylamine are used in the manufacture of ion-exchange resins. Additional applications are found in the manufacture of disinfectants, insecticides, herbicides, drugs, dyes, fungicides, soaps, cosmetics, and photographic developers. Many amines demonstrate biological activity, including cocaine, nicotine, morphine (see margin, previous page), quinine, and vitamin B6, to name just a few.

Important derivatives of ammonia are the *tetraalkyl ammonium salts*. The cation in these salts has four organic groups attached to the nitrogen atom and is called a *quaternary ammonium ion*. In a quaternary ammonium ion, the nitrogen atom has a formal charge of +1. Two examples of such cations are shown in the margin.

The acetylcholine cation is similar to the much simpler tetramethylammonium cation, except that one of the methyl groups has been replaced by a $CH_3COOCH_2CH_2—$ group. Quaternary ammonium ions, such as acetylcholine, are involved in the systems that transmit nerve impulses in the human body. Also, many poisons that affect the central nervous system contain a quaternary ammonium functional group.

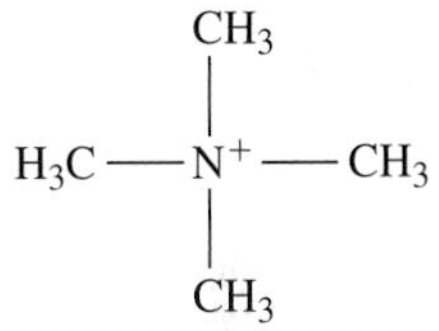

Tetramethylammonium ion

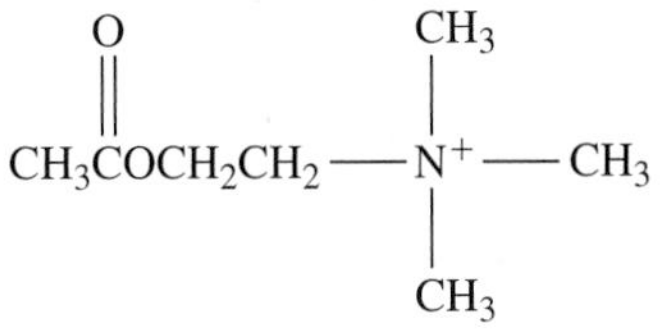

Acetylcholine

Heterocyclic Compounds

In most of the ring structures considered to this point, all the ring atoms have been carbon; these structures are said to be carbocyclic. However, in many compounds, both natural and synthetic, one or more of the atoms in a ring structure is not carbon. These ring structures are said to be **heterocyclic**. The heterocyclic systems most commonly encountered contain N, O, and S atoms, and the rings are of various sizes.

Pyridine is a nitrogen analogue of benzene (Fig. 26-17). Unlike benzene, it is water soluble and has basic properties (the unshared pair of electrons on the N atom is not part of the π electron cloud of the ring system). Pyridine, a liquid with a disagreeable odor, was once obtained exclusively from coal tar, but it is now used so extensively that several synthetic methods have been developed for its production. It is used in the production of pharmaceuticals such as sulfa drugs and antihistamines, as a denaturant for ethyl alcohol, as a solvent for organic chemicals, and in the preparation of waterproofing agents for textiles. A number of other examples of heterocyclic compounds are discussed in Chapter 28.

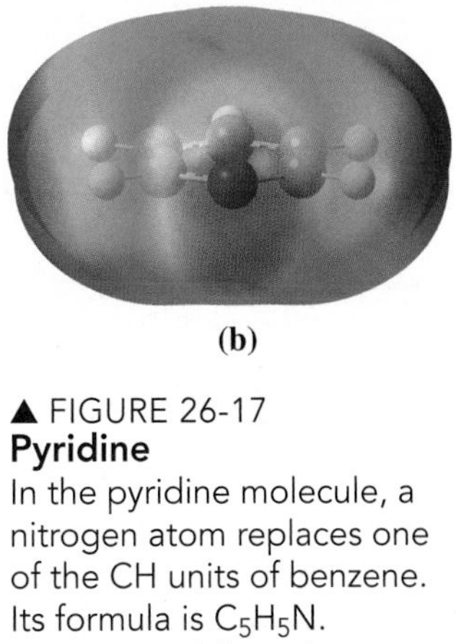

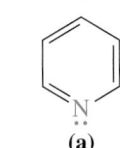

▲ FIGURE 26-17
Pyridine
In the pyridine molecule, a nitrogen atom replaces one of the CH units of benzene. Its formula is C_5H_5N.
(a) Structural formula
(b) Electrostatic potential map

IUPAC Functional Group Priority List

Increasing priority ↑

- carboxylic acid
- ester
- amide
- aldehyde
- ketone
- alcohol
- amine
- alkene
- alkyne
- ether
- alkyl halide

26-1 ARE YOU WONDERING?

How do we name compounds with more than one functional group?

In our discussion of organic compounds we restricted our discussion to molecules with only one functional group. How do we name the following compounds, each of which has more than one functional group?

(a) 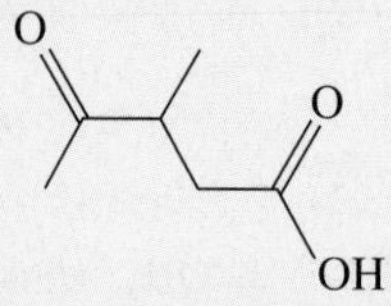

(b) 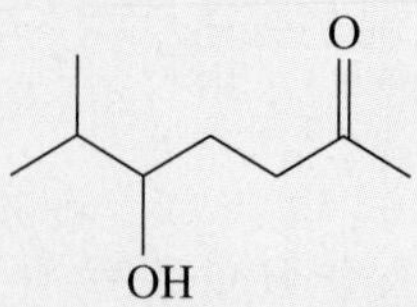

The IUPAC rules provide a solution to this problem by assigning priorities to functional groups. The priorities of some of the common functional groups are provided in the marginal table.

Notice that functional groups with two heteroatoms (carboxylic acids, esters, and amides) have higher priorities than do those with only one heteroatom.

The priority order enables us to decide that molecule (a) is an acid, in fact a pentanoic acid. But what prefix do we use for the carbonyl group? The prefix *oxo-* is used to indicate either an aldehyde or a ketone as a substituent. We name the molecule 3-methyl-4-oxopentanoic acid.

The prefixes for the —OH and $—NH_2$ substituents are *hydroxy* and *amino* respectively. Thus, the name of molecule (b) is 5-hydroxy-6-methylheptan-2-one.

A more complete discussion of nomenclature is normally covered in more advanced organic chemistry courses.

26-8 From Molecular Formula to Molecular Structure

Given only a molecular formula (perhaps derived from an experimentally determined percentage composition and a molecular weight determination) and some chemical evidence that suggests what functional groups might be present, how might we proceed to propose a possible structure?

First, it is helpful to determine how many elements of unsaturation are in a molecule. An element of unsaturation is a structural feature, such as a carbon–carbon double or triple bond or a ring structure, that causes the number of hydrogen atoms in a molecule to be less than the maximum number possible. We have learned that a saturated hydrocarbon with n carbon atoms has $2n + 2$ hydrogen atoms, the maximum number possible. We have also learned that the general formula is C_nH_{2n} for a simple alkene (a hydrocarbon having only one double bond) or a cycloalkane. Thus, a simple alkene or a cycloalkane has two fewer H atoms than the maximum number possible. We conclude that the number of H atoms decreases by two for each carbon–carbon π bond or ring structure that is present. Consider, for example, a molecule with the formula C_5H_{10}. The molecule has two fewer hydrogen atoms than the maximum possible and it could be one of the following molecules. (There are other isomers too. Try to draw structures for them.)

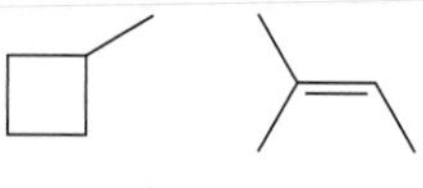

Each of these molecules contains one structural element (a π bond or a ring structure) that contributes to it being unsaturated. For each of these structures, the **degree of unsaturation** is equal to one. When the degree of unsaturation is equal to one, there will be two fewer H atoms than the maximum possible. When there are four hydrogen atoms less than the maximum, the degree of unsaturation is two.

> If an organic molecule with *n* carbon atoms has fewer than $2n + 2$ hydrogen atoms, then it contains elements of unsaturation, such as π bonds or ring structures.

KEEP IN MIND

that benzene, C_6H_6, has eight fewer H atoms than the maximum possible. Thus, the degree of unsaturation associated with a benzene ring is four, a result which is consistent with one ring plus the equivalent of three delocalized π bonds. A high degree of unsaturation is often accounted for by a benzene ring.

Up to this point, we have focused on molecules containing only carbon and hydrogen. The determination of the degree of unsaturation in molecules containing O, N, or Cl atoms requires some explanation. In organic compounds, halogen atoms, such as chlorine, are typically terminal atoms, not central atoms. In this regard, halogen atoms are similar to hydrogen atoms. When determining the degree of unsaturation, halogen atoms are counted as if they were hydrogen atoms. Stated another way, each halogen atom effectively adds one hydrogen to the hydrogen count. Thus, C_5H_9Cl has one degree of unsaturation because $9 + 1$ is two less than the maximum number of hydrogen atoms possible. Some possible structures for C_5H_9Cl are shown below. (Again, there are other possible structures. Draw structures for the other isomers.)

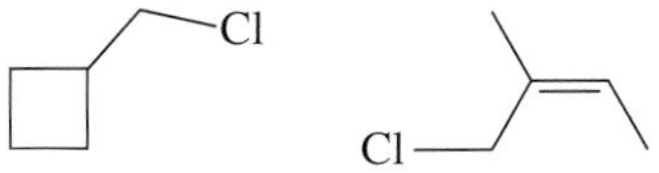

Now let us consider the situations that can arise if oxygen atoms are present in a molecule. It turns out that the degree of unsaturation in a carbon–hydrogen–oxygen compound can be established by considering only the number of carbon and hydrogen atoms in the molecule; that is, we can ignore the oxygen atoms. To understand why, let's consider a molecule that contains only one oxygen atom. The following linkages are possible:

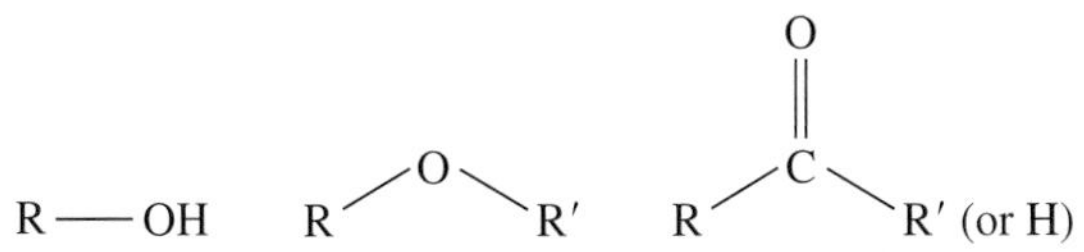

Alcohol Ether Ketone (or aldehyde)

For an alcohol or ether, we see that the oxygen atom is inserted either between a carbon atom and a hydrogen atom or between two carbon atoms. Neither of these linkages causes a decrease in the number of hydrogen atoms and thus, neither is considered an element of unsaturation. Turning this idea around, we can say that if the formula of a molecule is $C_nH_{2n+2}O$, then it must be an alcohol or an ether with no π bonds or ring structures. In a carbonyl compound, however, the oxygen atom takes the place of two hydrogen atoms. Thus, a carbonyl compound has one element of unsaturation and a general formula of $C_nH_{2n}O$.

To illustrate these ideas, consider a molecule having the formula C_3H_6O. For three carbon atoms, the maximum number of hydrogen atoms is eight. Ignoring the oxygen atom, we see that the number of hydrogen atoms is two less than the maximum, and so there is one degree of unsaturation. Three possible isomers are shown below. (Try to identify and draw structures for the other isomers.)

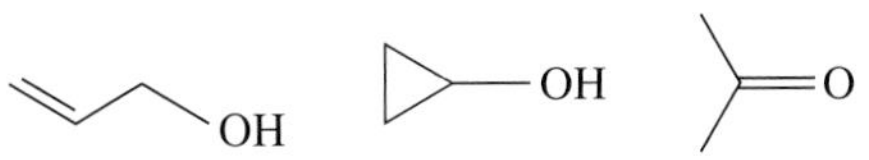

Now consider the molecule C_3H_8O. Ignoring the oxygen atoms, we determine that there are no elements of unsaturation. Consequently, there are no π bonds or rings in the molecule. Two possible isomers are shown below:

The ideas discussed above are easily extended to molecules containing more than one oxygen atom. For example, let us consider a molecule with the formula $C_3H_6O_2$. Again, in determining the degree of unsaturation, we ignore the oxygen atoms. Because there are only six hydrogen atoms and not eight, we conclude that the degree of unsaturation is one. Two possible isomers are a carboxylic acid or an ester, which are shown below:

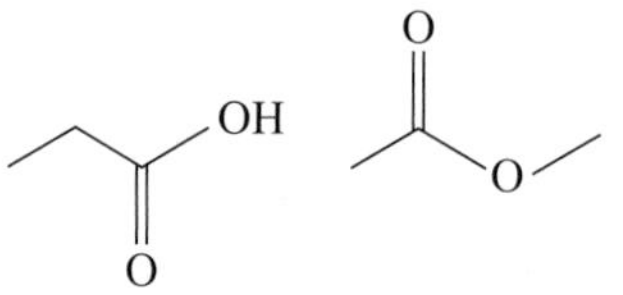

26-13 CONCEPT ASSESSMENT

Is it possible one of the isomers of $C_3H_6O_2$ is a dialdehyde?

Nitrogen atoms also appear in organic compounds. How do we deal with a nitrogen atom in determining the degree of unsaturation? Let's consider the molecule C_3H_9N. Two isomers are shown below:

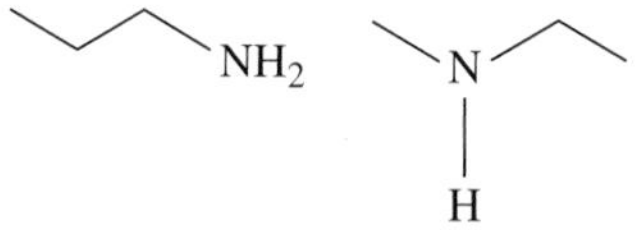

Both molecules are saturated and contain the maximum number of hydrogen atoms possible. How can we deduce the degree of unsaturation from the formula of this molecule? The key idea is that for each nitrogen atom present in the formula, the hydrogen atom count is diminished by one. Thus, the number of H atoms in C_3H_9N is considered to be 8 (not 9), which is equal to the number of H atoms expected in a saturated molecule containing three carbon atoms. For a compound with the formula C_4H_9N, the number of H atoms is effectively 8. A saturated molecule containing 4 carbon atoms should contain $2 \times 4 + 2 = 10$ hydrogen atoms. Because the molecule effectively contains only 8 hydrogen atoms (two less than the maximum), we know the degree of unsaturation is one. Two possible isomers for C_4H_9N are shown below:

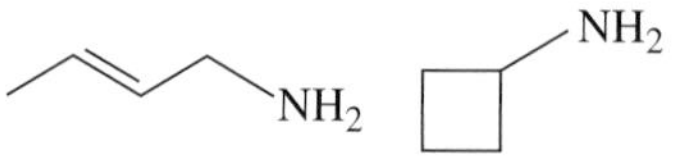

EXAMPLE 26-10 Determining the Degree of Unsaturation and Suggesting Possible Structures for a Molecule

What different types of compounds are possible for a compound with the molecular formula $C_6H_{10}O$?

Analyze

We need to establish the degree of unsaturation in the molecule and then construct one example of each type of molecule that can be formed.

Solve

As we described above, we can ignore the oxygen atom in determining the degree of unsaturation. The maximum number of hydrogen atoms is $2 \times 6 + 2 = 14$. The molecular formula has only 10 hydrogen atoms, so we know the degree of unsaturation is two. What types of molecules can be formed? We might start by considering dienes (alkenes with two double bonds) with either an alcohol or an ether functional group.

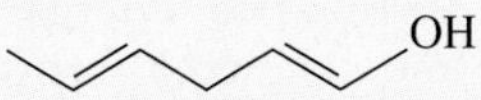

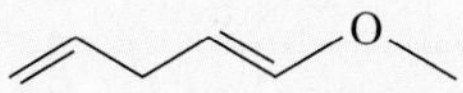

Alternatively, we can have a cyclic alkane with a carbonyl group or a cyclic alkene with an alcohol functional group. Two possible structures are shown here.

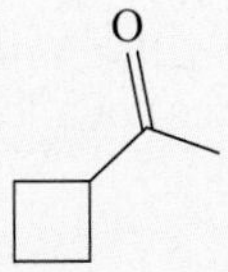

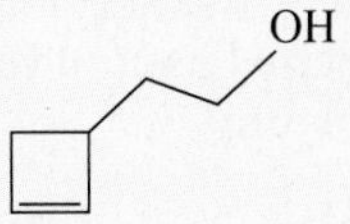

Two other possibilities are a ketone with an alkene side chain or an aldehyde with an alkene side chain. Two possible structures are shown here.

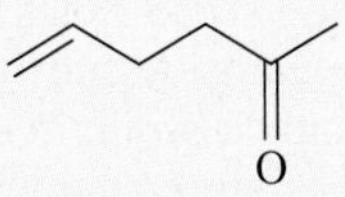

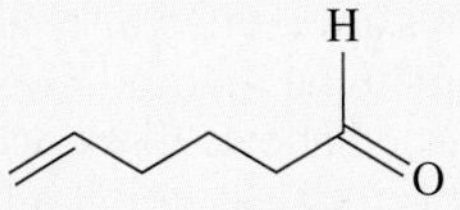

Assess

We have provided an example of an ether (with two degrees of unsaturation in the alkane chain), an aldehyde (with one degree of unsaturation in a side chain), a ketone, and some alcohols. This list is not exhaustive and you should be able to find other isomers.

PRACTICE EXAMPLE A: What different types of compounds are obtainable for a compound with the molecular formula $C_5H_{11}N$?

PRACTICE EXAMPLE B: What different types of compounds are obtainable for a compound with the molecular formula $C_5H_{10}O_2$?

MasteringCHEMISTRY **www.masteringchemistry.com**

Obtaining enantiomerically pure samples of chiral molecules is extremely important when synthesizing pharmaceuticals. When a synthesis yields a mixture of enantiomers, the enantiomers may be separated from each other using a process called chemical resolution. For a discussion of this method, go to the Focus On feature for Chapter 26, Chemical Resolution of Enantiomers, on the MasteringChemistry site.

Summary

26-1 Organic Compounds and Structures: An Overview—Organic chemistry deals with compounds of carbon, the simplest of which, called **hydrocarbons**, contain only carbon and hydrogen atoms. The carbon atoms are bonded to one another in straight- or branched-chains or in rings. In **saturated hydrocarbons** all chemical bonds are single bonds. Isomers that differ in their structural skeletons are **constitutional isomers**. Alkane substituents that branch off a main hydrocarbon chain are called **alkyl groups** (Table 26.1). Carbon atoms in alkane chains are classified as **primary**, **secondary**, **tertiary**, and **quaternary** depending on the number of carbon atoms attached to them. Hydrogen atoms are similarly classified depending on the type of carbon atom to which they are attached; for example, hydrogen atoms attached to a primary carbon atom are called **primary hydrogen atoms**. A systematic nomenclature based on IUPAC rules is used to name alkanes. Distinctive groupings of atoms are referred to as **functional groups** (Table 26.2). Organic compounds are typically classified according to their functional groups. A three-dimensional representation of hydrocarbons can be accomplished using the **dashed and solid wedge line notation**. In this notation, solid lines represent bonds in the plane of the paper, solid wedges show bonds sticking out toward the viewer, and dashed wedges show bonds directed away from the viewer.

26-2 Alkanes—Among saturated hydrocarbons, also known as **alkanes**, those based on a straight- or branched-chains of carbon atoms have the formula C_nH_{2n+2}. Alkanes that differ in sequence by a constant unit form a **homologous series**. Hydrocarbons can adopt different **conformations**, or arrangements, of their atoms. Alkanes have eclipsed and staggered conformations. Newman projections are used to depict these conformations (Figure 26-6). The energy difference between the eclipsed and staggered forms of an alkane is called the **torsional energy**. The variation in potential energy as one end of the molecule is rotated with respect to the other can be displayed in the form of a potential energy diagram (Fig. 26-8). The preference for certain conformations is driven in part by steric interactions involving bulky groups. In butane, for example, the **anti** conformation is lower in energy than the **gauche** conformation. The principal natural sources of alkanes are natural gas and petroleum.

26-3 Cycloalkanes—Hydrocarbons that form ring structures are termed alicyclic. Ring strain exists in cycloalkanes containing 3, 4, and 5 carbon atoms. A study of heats of combustion shows that cyclohexane is strain free. Disubstituted cycloalkanes exhibit a form of **stereoisomerism** known as ***cis–trans*** **isomerism**. Cyclohexanes undergo rapid ring flips (Fig. 26-13). In substituted cyclohexanes, substituents compete for the equatorial positions.

26-4 Stereoisomerism in Organic Compounds—A carbon atom that is bonded to four different groups is called an **asymmetric** carbon atom. One asymmetric carbon atom is also called a **stereocenter**. A molecule containing an asymmetric carbon atom is not superimposable on its mirror image and is **chiral**. Chiral molecules are optically active—they rotate the plane of polarized light. Molecules with superimposable mirror images are **achiral**. Molecules with sp^3 stereocenters are named by using the ***R, S*** **system**. The rules for an *R, S* designation require assigning priorities to the substituents on a chiral C atom and "reading" the descending priority order of the substituents when the molecule is viewed in a particular orientation. If the descending priority order is clockwise, the molecule is *R*; if it is counterclockwise, the molecule is *S*. The two molecules with *R* and *S* configurations are known as **enantiomers**. Stereoisomers that are not enantiomers are called **diastereomers**. *Cis* and *trans* isomers are examples of diastereomers.

26-5 Alkenes and Alkynes—Hydrocarbons whose molecules contain fewer hydrogen atoms than found in the corresponding alkane are **unsaturated** hydrocarbons. The unsaturation arises because of the presence of one or more carbon-to-carbon double or triple bonds. Hydrocarbons containing one or more double bonds are **alkenes**, or **olefins**, whereas those containing one or more triple bonds are **alkynes**. The placement of double or triple bonds in an unsaturated hydrocarbon can lead to different constitutional isomers, as in but-2-ene versus but-1-ene, for example. In but-2-ene the orientation of the two end methyl groups leads to *cis* and *trans* isomers, a type of isomerism called *cis–trans* isomerism. Highly substituted alkenes are named by the ***E, Z*** **system**. Molecules are given the Z designation when higher priority substituents at each end of a double bond are on the same side of the double bond and the *E* designation when they are on opposite sides of the double bond.

26-6 Aromatic Hydrocarbons—Many unsaturated hydrocarbons with ring structures are classified as **aromatic hydrocarbons**; most are based on the benzene molecule, C_6H_6. Two systems are commonly used to name substituents on a benzene ring. One uses a numbering system, such as 1,2-dichlorobenzene. The other uses the relative positions of substituents. Substituents on adjacent carbon atoms are referred to as **ortho (*o*)**; substituents separated by one carbon are **meta (*m*)**; and substituents opposite each other on the ring are **para (*p*)**. A **phenyl** group is a benzene ring with one hydrogen atom removed. A **benzyl** group is a methyl group with one hydrogen atom replaced by a phenyl group.

26-7 Organic Compounds Containing Functional Groups—*Alcohols and phenols:* Hydrocarbons that contain an —OH (**hydroxyl**) group are classified as **alcohols** in aliphatic compounds and **phenols** in aromatic compounds. Aliphatic alcohols are also classified as primary, secondary, or tertiary, depending on how many other substituents (excluding H atoms) are attached to the same carbon atom as the hydroxyl group: one for primary, two for secondary, and three for tertiary. Diols and polyols are molecules with two or more hydroxyl groups, respectively.

Ethers: The general formula of an **ether** is R—O—R′. The ether linkage is very stable and ethers are generally unreactive, resistant to both oxidation and reduction. One of the major uses of ethers is as solvents.

Aldehydes and ketones: The **carbonyl group** has a carbon atom attached by a double bond to an oxygen atom. If one of the two substituents attached to the carbonyl group is an H atom, the compound is an **aldehyde**, otherwise it is a **ketone**. In IUPAC nomenclature, the suffix *-al* is used for aldehydes and *-one* for ketones.

Carboxylic acids: The **carboxyl group** is a combination of the carbonyl and hydroxyl groups. **Carboxylic acids** have the general formula RCOOH and can be prepared by the oxidation of aldehydes. An acetyl group results when the —OH of a carboxyl group is replaced by a methyl group.

Esters: An **ester** is formed when an alkoxy group replaces the —OH of a carboxyl group. The general formula of an ester is RCOOR′. Most esters are colorless liquids that are insoluble in water and have pleasant fragrances.

Amides: The general formula of a primary **amide** is $RCONH_2$. A primary amide has two H atoms bonded to N and results when the hydroxyl group of a carboxylic acid is replaced by $—NH_2$. In secondary or tertiary amides, one or two H atoms in the $—NH_2$ are replaced by alkyl or aryl groups. Because of resonance, amides are much weaker bases than are ammonia or amines.

Amines: **Amines** are organic derivatives of ammonia. They are classified according to the number of organic groups (R) substituted for hydrogen atoms in NH_3. A primary amine has one substituent; a secondary amine, two substituents; and a tertiary amine, three substituents. Amines are weak bases, with the aliphatic amines being stronger bases than ammonia and the aromatic amines being weaker. Tetraalkyl ammonium salts are organic derivatives of the ammonium cation with a suitable anion.

Heterocyclic compounds: In a **heterocyclic compound**, at least one of the ring atoms is a heteroatom (not carbon). The most common heteroatoms are N, O, and S. Heterocyclic compounds tend to be more soluble in water than the parent compound because of the polarity of the heterocycle.

26-8 From Molecular Formula to Molecular Structure—A useful idea in determining a suitable structure is the **degree of unsaturation**, which is equal to the total number of π bonds and ring structures in a molecule. The degree of unsaturation is determined by comparing the number of H atoms in the molecule to the maximum number possible for the number of carbon atoms present. In counting the number of H atoms in a molecule, each halogen atom effectively increases the hydrogen count by one; oxygen atoms can be ignored; and each nitrogen atom, in effect, decreases the hydrogen atom count by one.

Integrative Example

An acyclic organic compound with the formula $C_6H_{12}O$ is found to be optically active, does not decolorize Br_2 (in CCl_4), and does not undergo reaction when treated with a mixture of $Na_2Cr_2O_7$ and H_2SO_4. However, a reaction does occur when the compound is treated with $NaBH_4$. Identify a compound having these physical and chemical properties. Then draw structures for the two stereoisomers, using dashed and solid wedge symbols. Finally, give acceptable names for the two stereoisomers.

Analyze

There is only one oxygen atom per molecule, so the compound must be an alcohol, ether, aldehyde, or ketone. Also, because there are only 12 hydrogen atoms per molecule, rather than $2 \times 6 + 2 = 14$, we know that the molecule must contain either a π bond or a ring. We are told the compound is acyclic, and thus the molecule must contain a π bond and not a ring. Alcohols (except tertiary alcohols) and aldehydes are rather easily oxidized, so we suspect the molecule is neither an alcohol (though possibly a tertiary alcohol) nor an aldehyde. The ether linkage is stable in the presence of most oxidizing and reducing agents, and because the compound we need can be reduced, we suspect that that the compound is not an ether. At this point, we must decide whether the molecule is a ketone or a tertiary alcohol containing a carbon–carbon double bond. We can rule out a carbon–carbon double bond because the compound does not react with Br_2, a reaction characteristic of alkenes. The compound we seek is a ketone.

Solve

We have reasoned that the compound is a ketone. The linkage associated with a ketone is

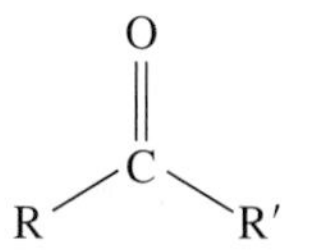

The five remaining carbon atoms are distributed among the two alkyl groups. The carbon of a carbonyl group is bonded to only three groups, and thus it cannot be chiral. The chiral carbon is part of one of the alkyl groups. Both R and R′ must contain at least one carbon atom and so, the maximum number of carbon atoms in R or R′ is four.

Let's choose R to be the alkyl group that contains the chiral carbon. A chiral carbon atom must be bonded to four different groups and yet R contains no more than four carbon atoms. The only possibility for R is

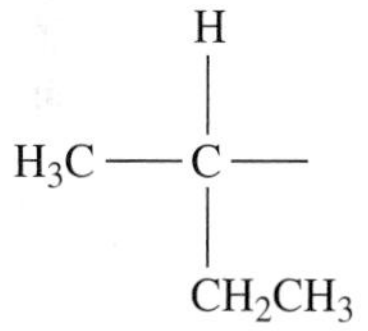

We have accounted for five of the six carbon atoms (R contains four carbon atoms and the carbonyl group contains one). Thus, R′ represents a methyl group. A skeletal structure for the molecule is

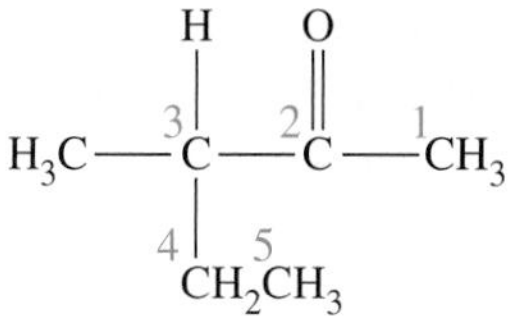

3-methylpentan-2-one

The longest carbon chain has five carbon atoms with a carbonyl group bonded to C2 and a methyl group bonded to C3. The molecule is 3-methylpentan-2-one. Because C3 is chiral, two enantiomers are possible.

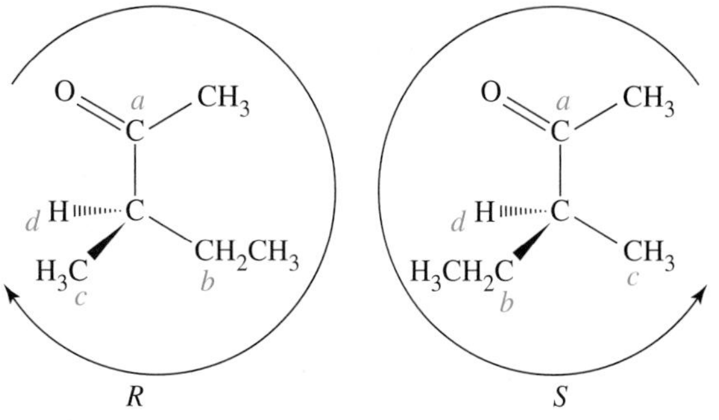

(*R*)-3-methylpentan-2-one (*S*)-3-methylpentan-2-one

Assess

It may not be obvious that there is only one possibility for R. Try drawing structures of other ketones having the formula $C_6H_{12}O$ to convince yourself that there is only one possibility. Had we not been told the compound was acyclic, a unique identification would not have been possible.

PRACTICE EXAMPLE A: Compound A with the formula C_3H_8O is soluble in water and reacts with sodium metal, producing bubbles of gas. When compound A is treated with chromic acid (a mixture of $Na_2Cr_2O_7$ and H_2SO_4), compound B is formed. Compound B dissolves readily in $Na_2CO_3(aq)$ and reacts with ethanol, yielding compound C which has a fruity fragrance. Identify compounds A, B, and C.

PRACTICE EXAMPLE B: The following molecules all have the molecular formula $C_5H_{10}O$. You suspect that you have a sample of one of these compounds. What tests could you perform to ascertain which of these compounds you have?

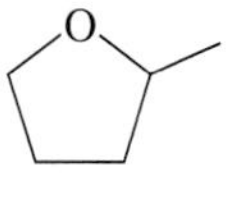

(a)

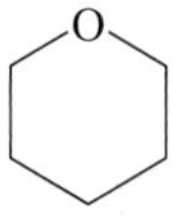

(b)

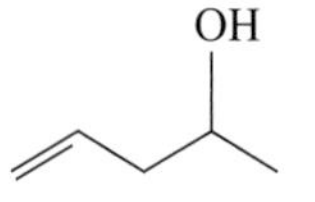

(c)

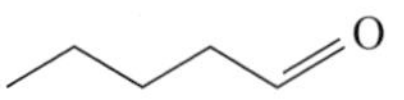

(d)

Exercises

Organic Structures

1. Write structural formulas corresponding to the following condensed formulas.
   (a) $CH_3CH(CH_3)CH_2CBr_2CH_2CHClCH_3$
   (b) $CH_2{=}CHCH_2CHBrCH(CH_3)CH_3$
   (c) $CH_3CH_2CH_2C(CH_3)_2CH_3$
   (d) $CH_3C(CH_3)_2CH{=}C(CH_3)CH_2CH_3$
2. Draw a structural formula for each of the following compounds and provide the preferred IUPAC name.
   (a) 3-isopropyloctane; **(b)** 2-chloro-3-methylpentane; **(c)** 2-pentene; **(d)** dipropyl ether.
3. Supply a structural formula for each of the following compounds.
   (a) 1,3,5-trimethylbenzene; **(b)** *p*-nitrophenol; **(c)** 3-amino-2,5-dichlorobenzoic acid (a plant-growth regulator).
4. Write structural formulas corresponding to these condensed formulas.
   (a) $(CH_3)_3CCH_2CH(CH_3)CH_2CH_2CH_3$
   (b) $(CH_3)_2CHCH_2C(CH_3)_2CH_2Br$
   (c) $Cl_3CCH_2CH(CH_3)CH_2Cl$
5. Draw Lewis structures for the following simple organic molecules: **(a)** $CH_3CBr_2CH_2CH_3$; **(b)** $HOCH(CH_3)CH_2CH_3$; **(c)** CF_3COOH; **(d)** $CH_2BrC(O)CH_2CH_3$.
6. Draw Lewis structures of the following simple organic molecules: **(a)** CH_3CH_2COOH; **(b)** H_3CCN; **(c)** $CH_3CH_2NH_2$.
7. With appropriate sketches, represent chemical bonding in terms of the overlap of hybridized and unhybridized atomic orbitals in the following molecules.
   (a) C_4H_{10}; **(b)** $H_2C{=}CHCl$; **(c)** $CH_3C{\equiv}CH$
8. With appropriate sketches, represent chemical bonding in terms of the overlap of hybridized and unhybridized atomic orbitals in the following molecules.
   (a) $CH_3\overset{O}{\overset{\|}{C}}CH_3$ **(b)** $CH_3\overset{O}{\overset{\|}{C}}{-}OH$; **(c)** $H_2C{=}C{=}CH_2$

Isomers

9. What is the relationship, if any, between the molecules in each of the following pairs? The relationship may be any of identical structures, constitutional isomers, stereoisomers, or no relationship.

(a)

$CH_3CH_2CH_2Cl$ and $ClCH_2CH_2CH_3$

(b)

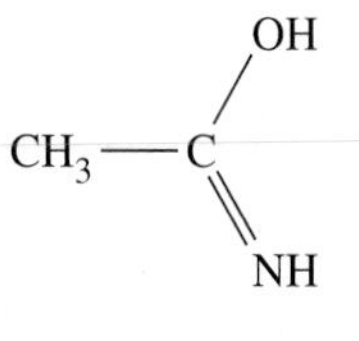

and

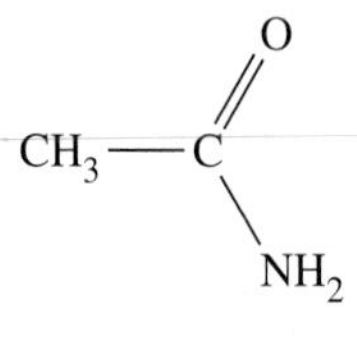

(c)

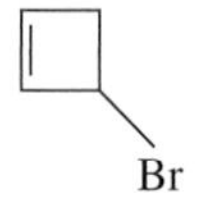

and

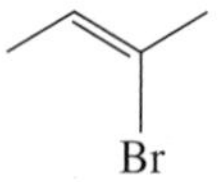

(d)

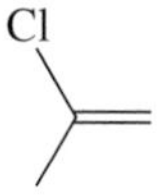

and

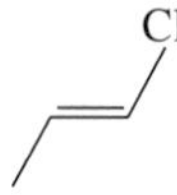

(e)

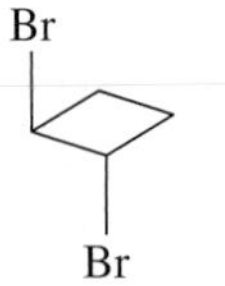

and

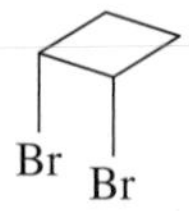

10. What is the relationship, if any, between the molecules in each of the following pairs? The relationship may be any of identical structures, constitutional isomers, stereoisomers, or no relationship.

(a) 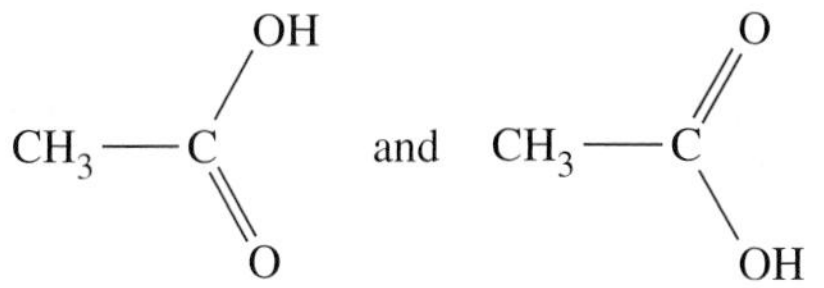

(b) 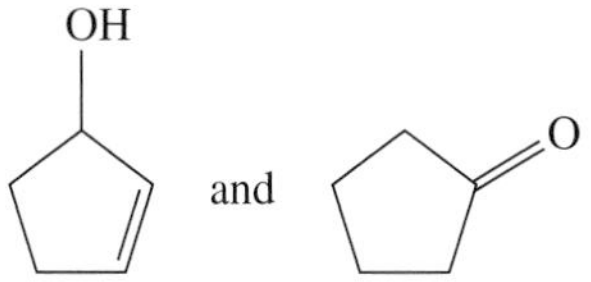

(c) 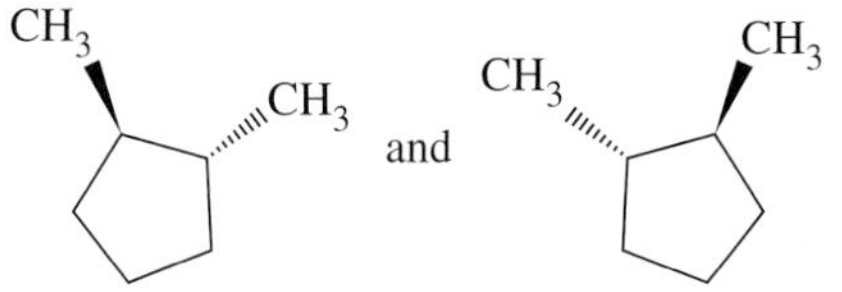

(d) 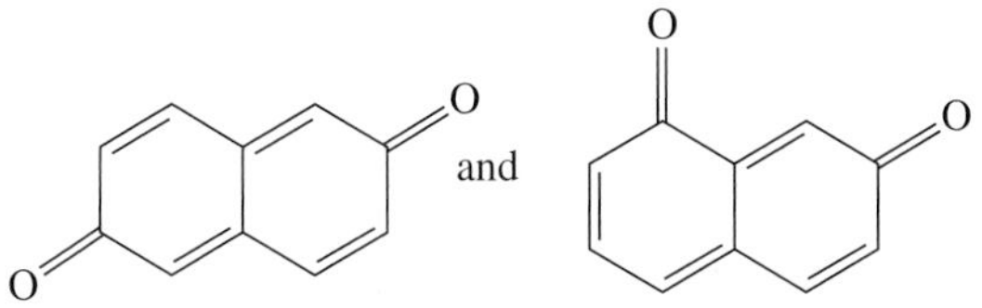

(e) 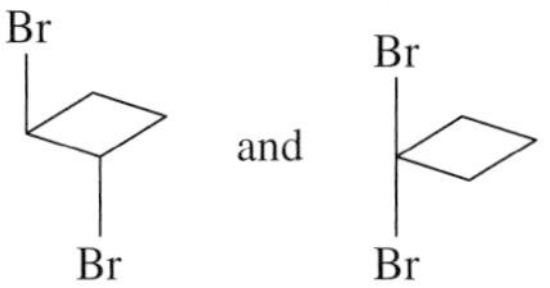

11. Draw structural formulas for all isomers of pentanol, $C_5H_{11}OH$.
12. Draw and name all the isomers of **(a)** C_6H_{14}; **(b)** C_4H_8; **(c)** C_4H_6. [*Hint:* Do not forget double bonds, rings, and combinations of these.]

13. Identify the chiral carbon atoms, if any, in the following compounds.

(a) 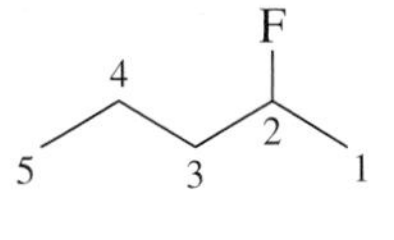 **(b)** 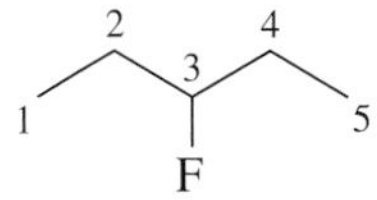

(c) 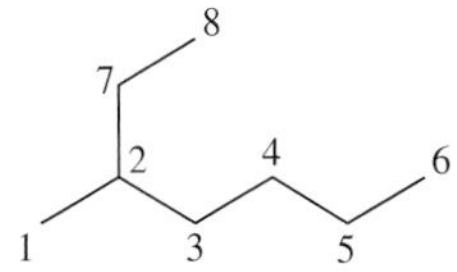

14. Identify the chiral carbon atoms, if any, in the following compounds.

(a) 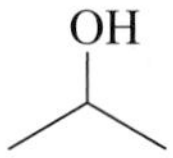 **(b)** 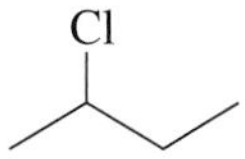 **(c)** 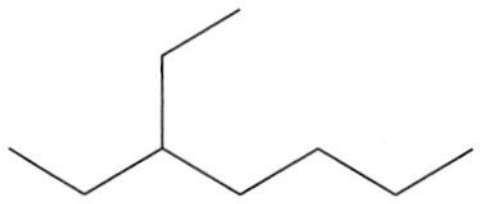

15. Identify the chiral carbon atoms, if any, in the following compounds:

(a) 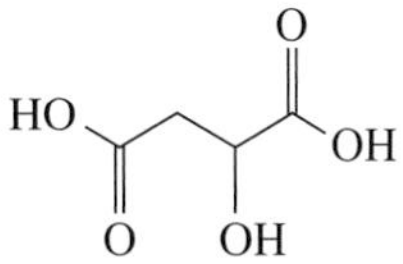 **(b)** 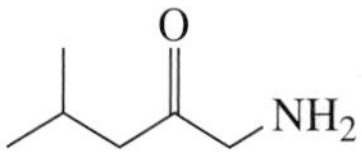 **(c)** 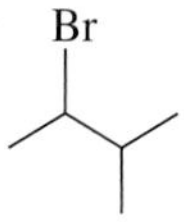

16. Identify the chiral carbon atoms, if any, in the following compounds.

(a) 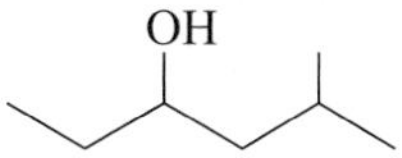 **(b)** 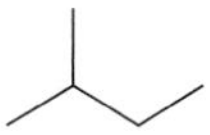 **(c)** 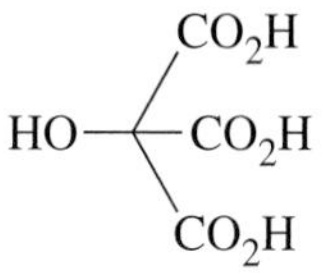

Functional Groups

17. Classify each compound by its functional group (i.e., alcohol, amine, etc.).
    (a) $CH_3CHBrCH_2CH_3$
    (b) $C_6H_5CH_2CHO$
    (c) $CH_3COCH_2CH_3$
    (d) $C_6H_4(OH)_2$
18. Classify each compound by its functional group (i.e., alcohol, amine, etc.).
    (a) $CH_3CH_2CH_2CH(OH)CH_3$
    (b) $CH_3C(CH_3)_2CH_2COOH$
    (c) $CH_3CH(CH_3)COOCH_2CH_3$
    (d) $CH_3C(CH_3)_2NH_2$
    (e) $CH_3CH_2CH_2CHO$
19. The functional groups in each of the following pairs have certain features in common, but what is the essential difference between them?
    (a) carbonyl and carboxyl
    (b) aldehyde and ketone
    (c) acetic acid and acetyl group
20. By name or formula, give one example of each of the following types of compounds: **(a)** aromatic nitro compound; **(b)** aliphatic amine; **(c)** chlorophenol; **(d)** aliphatic diol; **(e)** unsaturated aliphatic alcohol; **(f)** alicyclic ketone; **(g)** halogenated alkane; **(h)** aromatic dicarboxylic acid.

21. Identify and name the functional groups in each of the following.

(a) 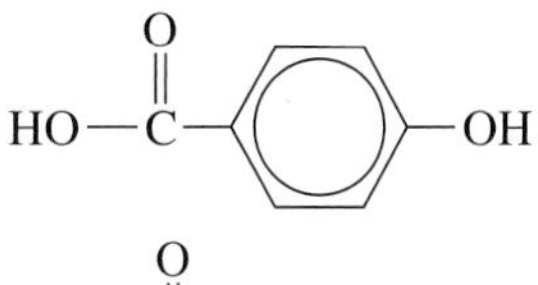

(b) 

(c) 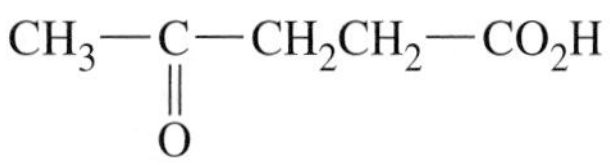

(d) 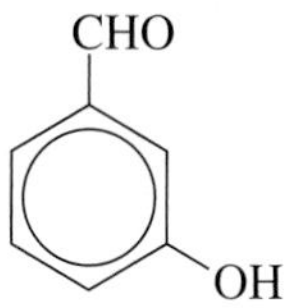

22. Identify and name the functional groups in each of the following.

(a) 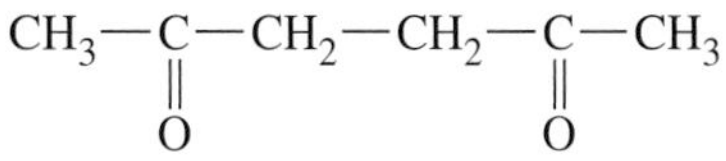

(b) $CH_3—CH_2—CONHCH_3$

(c) 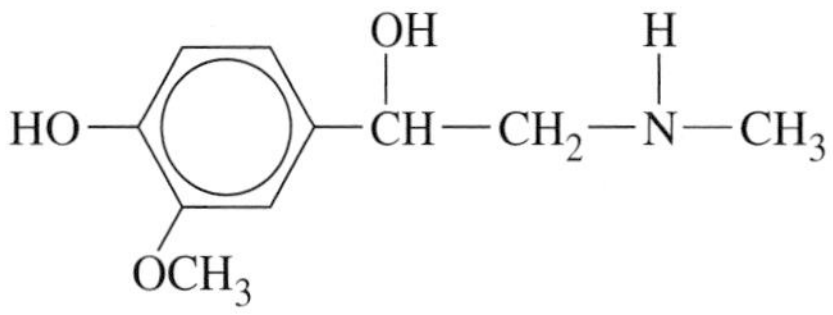

(d) 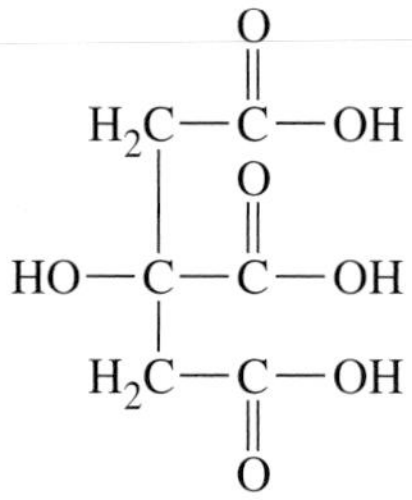

23. Give the isomers of $C_4H_{10}O$ that are ethers.
24. Give the isomers of $C_5H_{12}O$ that are ethers.
25. Give the isomers of the carboxylic acid with molecular formula $C_5H_{10}O_2$.
26. Give the isomers of the carboxylic acid with the molecular formula $C_4H_8O_2$.
27. Give the isomers of the esters having the molecular formula $C_5H_{10}O_2$.
28. Give the isomers of the esters having the molecular formula $C_4H_8O_2$.
29. Give the noncyclic isomers with molecular formula $C_4H_8O_2$ that contain more than one functional group.
30. Give the isomers with molecular formula $C_5H_{10}O_2$ that contain more than one functional group.

Nomenclature and Formulas

31. Give an acceptable name for each of the following.

(a) 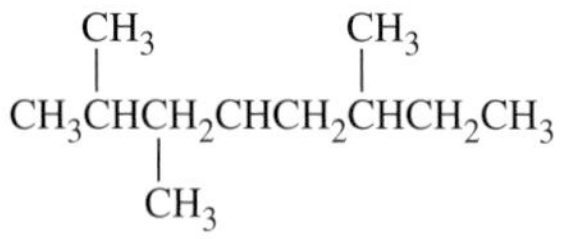 (b) 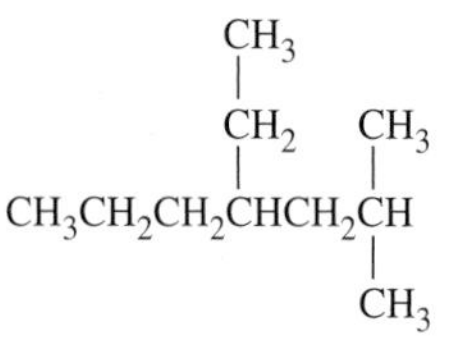

(c) 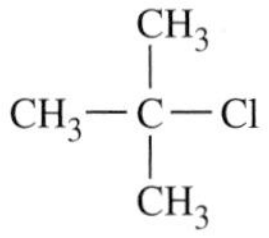 (d) 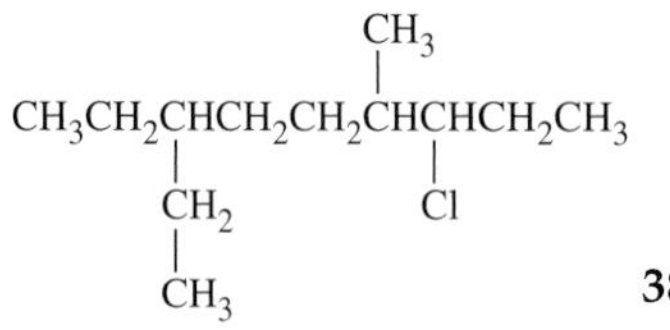

32. Give an acceptable name for each of the following.

(a) 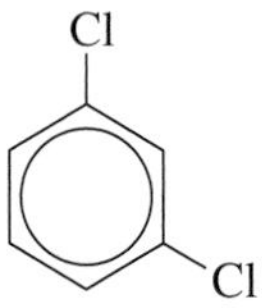 (b) 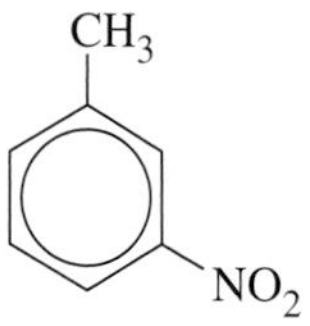

(c) 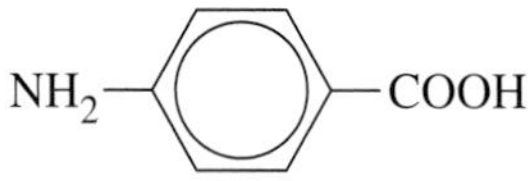

33. Give an acceptable name for each of the following structures.
    (a) $CH_3CH_2C(CH_3)_3$
    (b) $(CH_3)_2C{=}CH_2$
    (c) 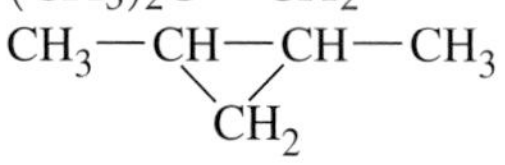
    (d) $CH_3C{\equiv}CCH(CH_3)_2$
    (e) $CH_3CH(C_2H_5)CH(CH_3)CH_2CH_3$
    (f) $CH_3CH(CH_3)CH(CH_3)CCH_2CH_2CH{=}CH_2$
34. Draw a condensed structure to correspond to each of the following names.
    (a) methylbutane; **(b)** cyclohexene;
    (c) 2-methyl-3-hexyne; **(d)** 2-butanol;
    (e) ethyl isopropyl ether; **(f)** propanal
35. Does each of the following names convey sufficient information to suggest a specific structure? Explain.
    (a) pentene; **(b)** butanone; **(c)** butyl alcohol;
    (d) methylaniline; **(e)** methylcyclopentane;
    (f) dibromobenzene
36. Indicate why each of these names is incorrect, and give a correct name.
    (a) 3-pentene; **(b)** pentadiene; **(c)** 1-propanone;
    (d) bromopropane; **(e)** 2,6-dichlorobenzene;
    (f) 2-methyl-3-pentyne
37. Supply condensed structural formulas for the following substances.
    (a) 2,4,6-trinitrotoluene (TNT—an explosive)
    (b) methyl salicylate (oil of wintergreen) [*Hint:* Salicylic acid is *o*-hydroxybenzoic acid.]
    (c) 2-hydroxy-1,2,3-propanetricarboxylic acid (citric acid, $C_6H_8O_7$)
38. Supply condensed structural formulas for the following substances.
    (a) *o*-*tert*-butylphenol (an antioxidant in aviation gasoline)
    (b) 1-phenylpropan-2-amine (benzedrine—an amphetamine, ingredient in "pep pills")
    (c) 2-methylheptadecane (a sex pheromone of tiger moths—a chemical used for communication among members of the species) [*Hint:* Heptadeca means 17.]
39. Name the following amines.
    (a) 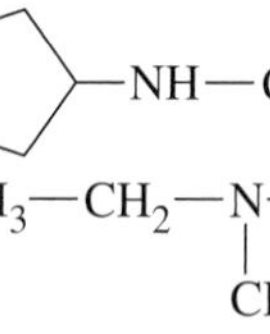
    (b) 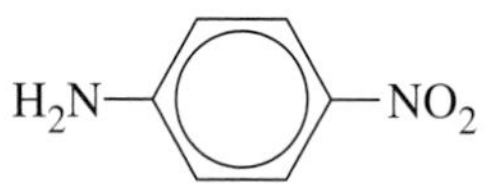
    (c) 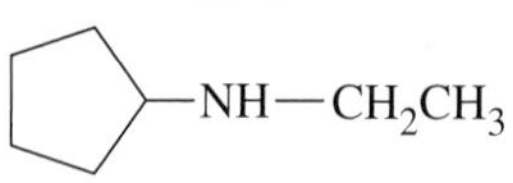
    (d) 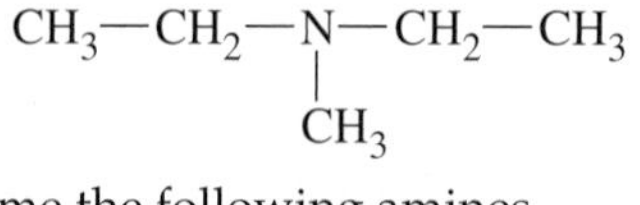
40. Name the following amines.
    (a) $CH_3{-}CH_2{-}NH_2$
    (b) 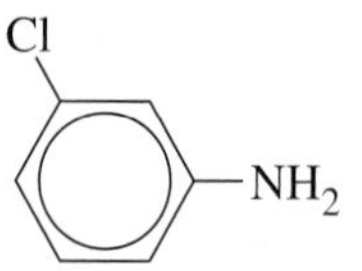
    (c) 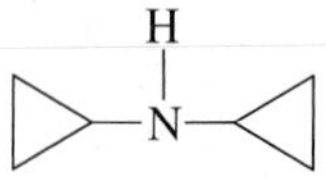
    (d) $Cl{-}CH_2{-}CH_2{-}NH_2$

Alkanes and Cycloalkanes

41. Classify the carbon atoms in **(a)** methylbutane, and **(b)** 2,2-dimethylpropane as methyl, primary (1°), secondary (2°), tertiary (3°), or quaternary (4°).
42. Classify the carbon atoms in **(a)** 2,4-dimethylpentane, and **(b)** ethylcyclobutane as methyl, primary (1°), secondary (2°), tertiary (3°), or quaternary (4°).
43. Draw Newman projections for the staggered and eclipsed conformations of pentane for rotation about the C2—C3 bond. Which conformation is lowest in energy?
44. Draw Newman projections for the staggered and eclipsed conformations of 2-methylpentane for rotation about the C2—C3 bond. Which conformation is lowest in energy?
45. Draw the most stable conformation for the molecule below:

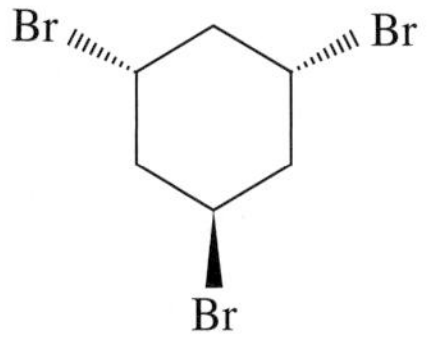

46. Draw the most stable conformation for the molecule below:

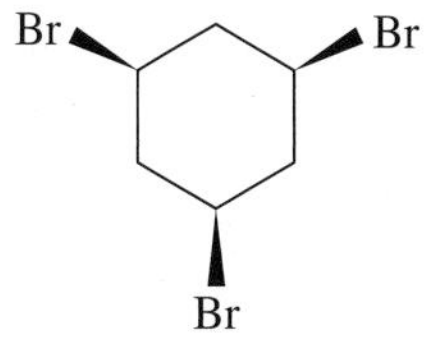

47. For each of the following substituted cyclohexanes, draw the two possible chair conformations, label each substituent as axial or equatorial, and identify the more stable conformer.
    (a) cyclohexanol
    (b) *trans*-3-methylcyclohexanol
48. For each of the following substituted cyclohexanes, draw the two possible chair conformations, label each substituent as axial or equatorial, and identify the more stable conformer.
    (a) *cis*-1-isopropyl-3-methylcyclohexane
    (b) *cis*-4-*tert*-butylcyclohexanol

Alkenes

49. Why is it not necessary to refer to ethene and propene as eth-1-ene and prop-1-ene? Can the same be said for butene?
50. Alkenes (olefins) and cyclic alkanes (alicyclics) each have the generic formula C_nH_{2n}. In what important ways do these types of compounds differ structurally?
51. Assign a configuration (*E* or *Z*) to each of the following molecules.

(a) 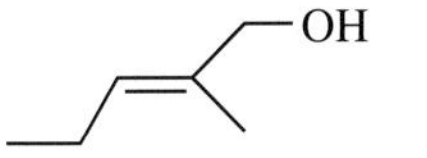

(d) 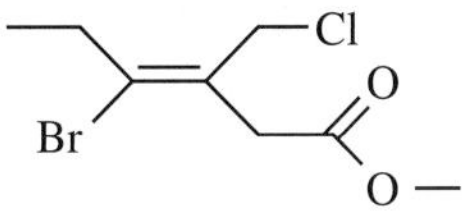

(b) 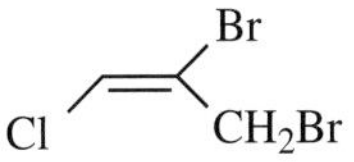

(e) 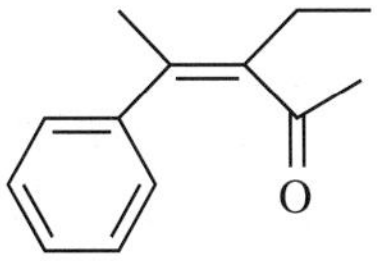

(c) 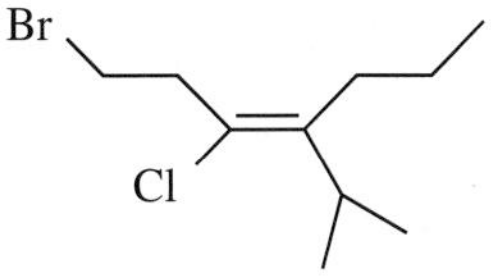

52. Assign a configuration (*E* or *Z*) to each of the following molecules.

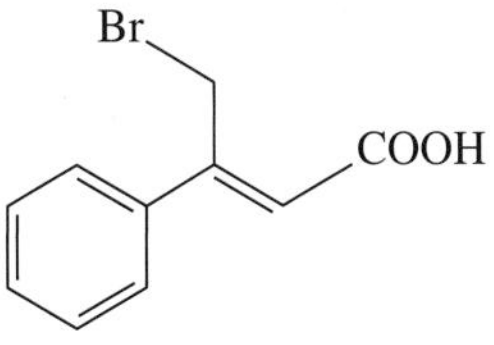
(a)

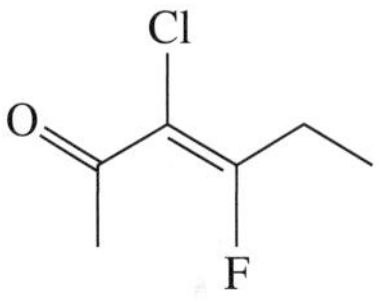
(b)

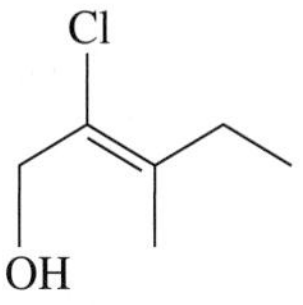
(c)

53. Draw the *E* and *Z* isomers of **(a)** 2-chlororbut-2-ene; **(b)** 3-methylpent-2-ene
54. Draw the *E* and *Z* isomers of **(a)** 3-methylhex-3-ene; **(b)** 3-fluoro-2-methylhex-3-ene

Aromatic Compounds

55. Supply a name or structural formula for each of the following.
    (a) phenylacetylene
    (b) *m*-dichlorobenzene
    (c) 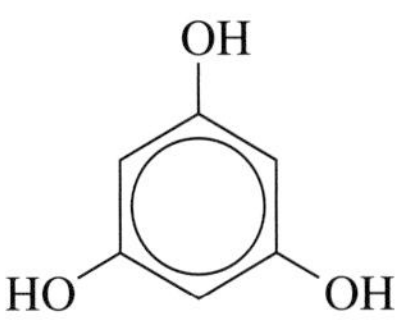
56. Supply a name or structural formula for each of the following.
    (a) *p*-phenylphenol
    (b) 3-hydroxy-4-isopropyltoluene (thymol—flavor constituent of the herb thyme)
    (c) 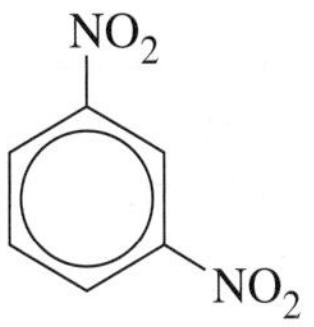

Organic Stereochemistry

57. Draw suitable structural formulas to show that there are *four* constitutional isomers of $C_3H_6Cl_2$.
58. Which of the following pairs of molecules are constitutional isomers and which are not? Explain.
    (a) $CH_3CH_2CH_2CH_3$ and $CH_3CH{=}CHCH_3$
    (b) $CH_3(CH_2)_5CH(CH_3)_2$ and $CH_3(CH_2)_4CH(CH_3)CH_2CH_3$
    (c) $CH_3CHClCH_2CH_3$ and $CH_3CH_2CH_2CH_2CH_2Cl$

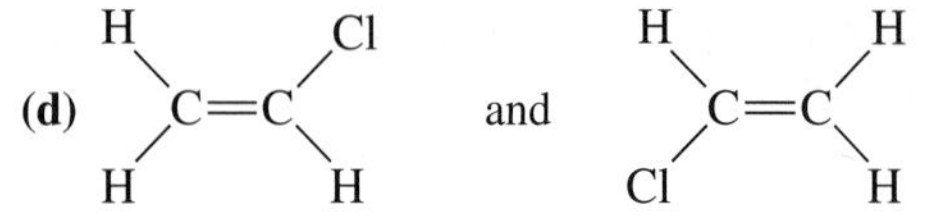

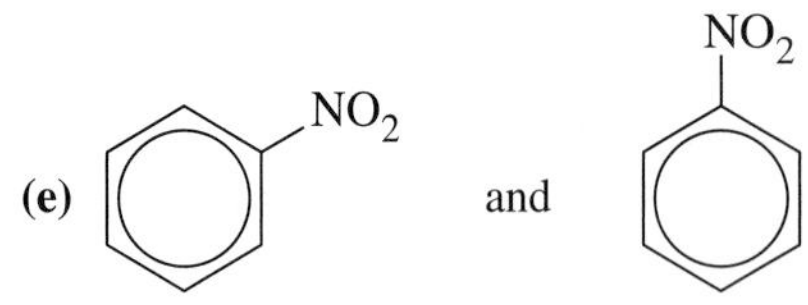

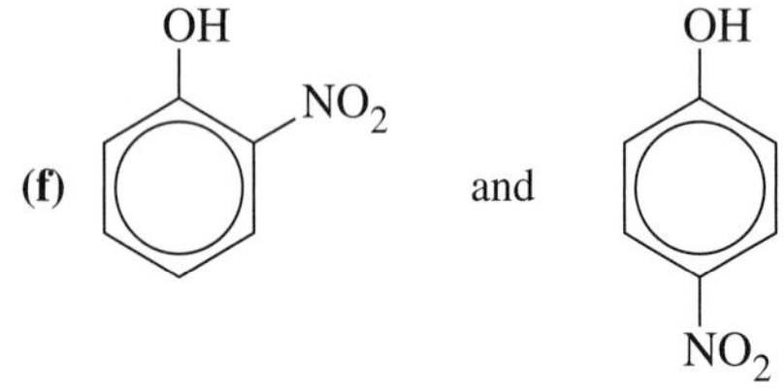

59. For each pair of structures shown below, indicate whether the two species are identical molecules, enantiomers, or isomers of some other sort.

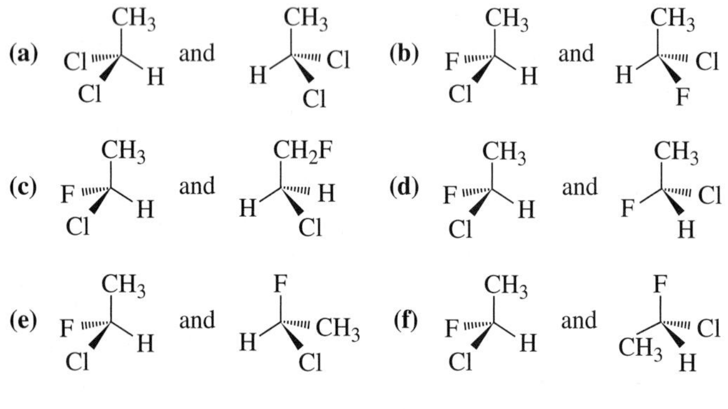

60. For each pair of structures shown below, indicate whether the two species are identical molecules, enantiomers, or isomers of some other sort.

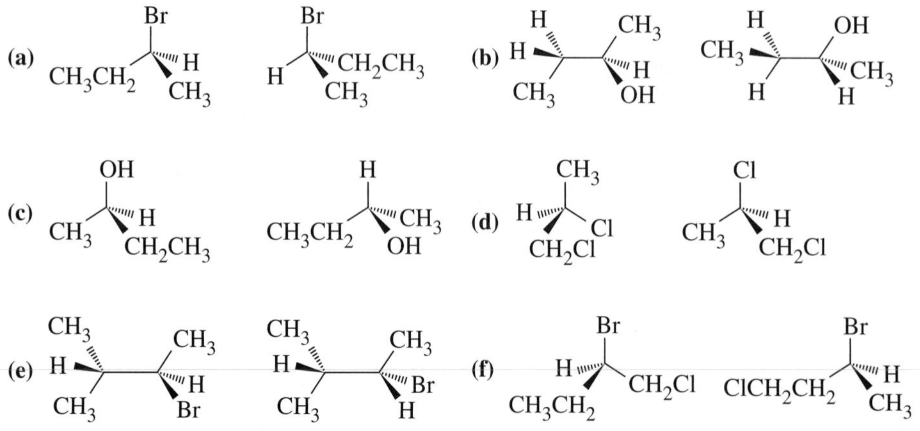

61. Name the following molecules with the appropriate stereochemical designation.

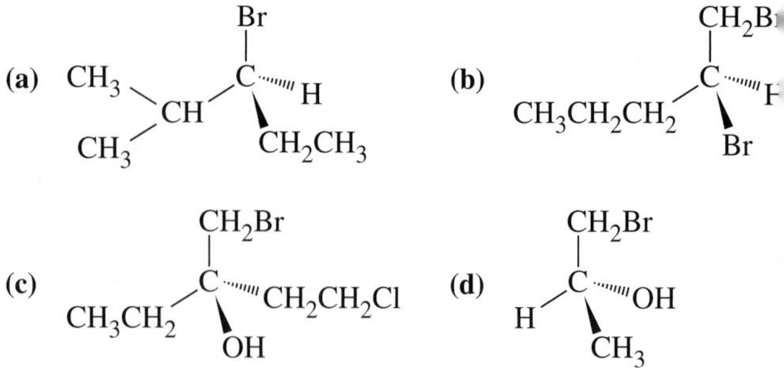

62. Name the following molecules with the appropriate stereochemical designation.

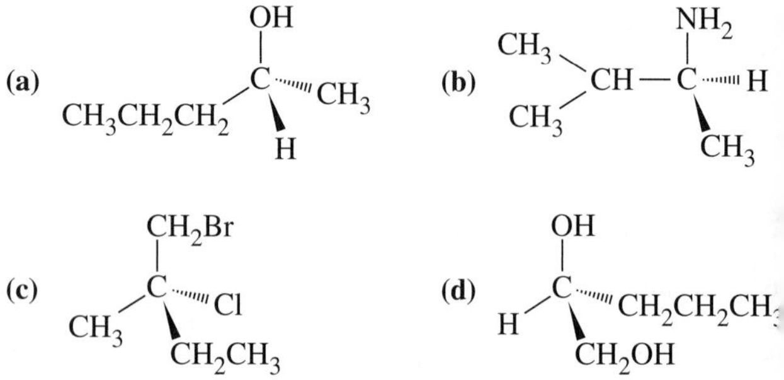

63. Name the following molecules with the appropriate stereochemical designation.

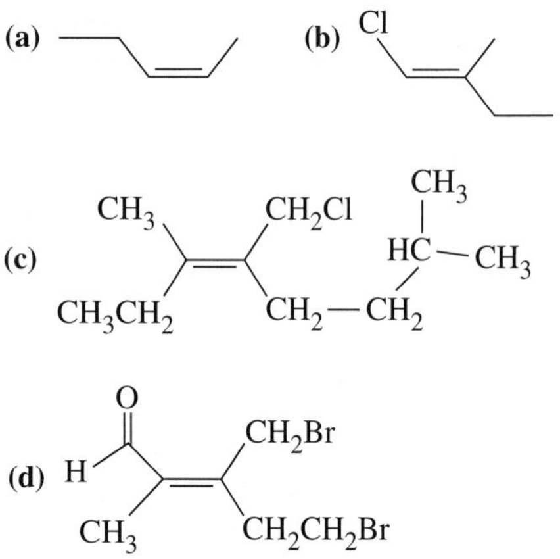

64. Name the following molecules with the appropriate stereochemical designation.

(a) 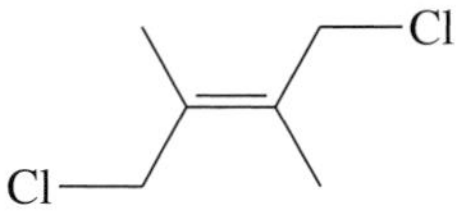 **(b)** 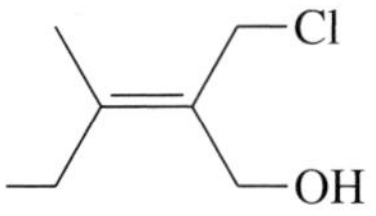

(c) 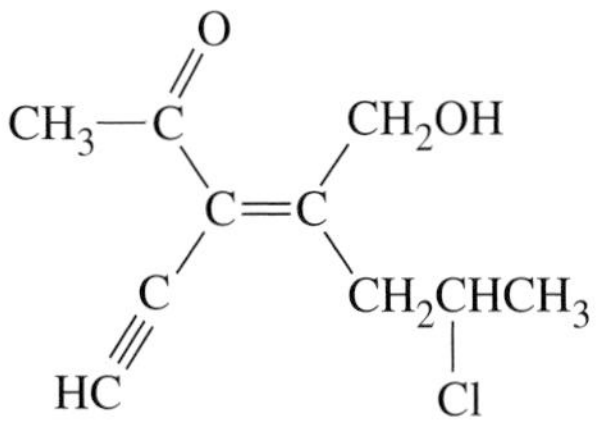

(d) 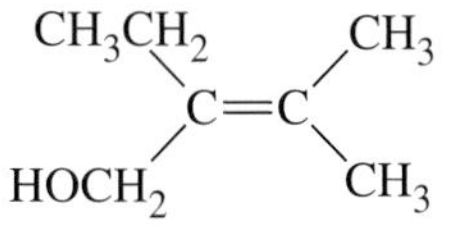

65. Draw the structure for each of the following.
(a) (*Z*)-1,3,5-tribromopent-2-ene
(b) (*E*)-1,2-dibromo-3-methylhex-2-ene
(c) (*S*)-1-bromo-1-chlorobutane
(d) (*R*)-1,3-dibromohexane
(e) (*S*)-1-chloropropan-2-ol

66. Draw the structure for each of the following.
(a) (*R*)-1-bromo-1-chloroethane
(b) (*E*)-2-bromopent-2-ene
(c) (*Z*))-1-chloro-3-ethylhept-3-ene
(d) (*R*)-2-hydroxypropanoic acid
(e) (*S*)-2-aminopropanoate anion

Structures and Properties of Organic Compounds

67. Consider the following molecular formulas. How many elements of unsaturation are there in each case? **(a)** $C_4H_{11}N$; **(b)** C_4H_6O; **(c)** $C_9H_{15}ClO$.

68. Consider the following molecular formulas. How many elements of unsaturation are there in each case? **(a)** C_5H_9NO **(b)** $C_5H_8O_3$; **(c)** C_5H_9ClO.

69. How many elements of unsaturation are there in the molecule below? What is the molecular formula?

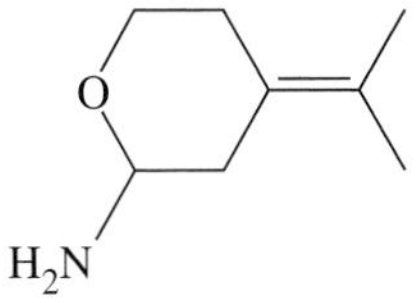

70. How many elements of unsaturation are there in the molecule below? What is the molecular formula?

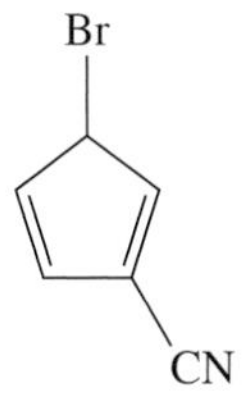

71. Match the following compounds with the chemical properties in the next column. Write a chemical equation for the reactions described in (a)–(d).

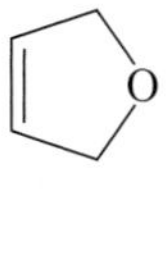 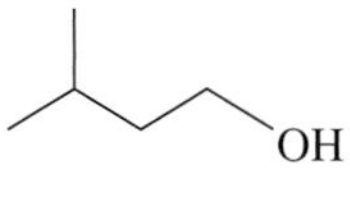 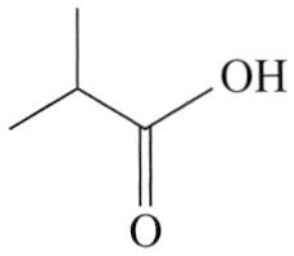

(1) **(2)** **(3)**

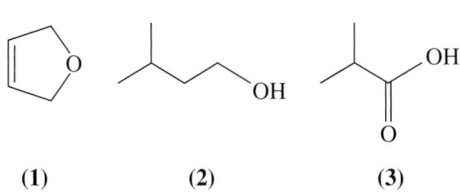

(a) is easily oxidized
(b) decolorizes bromine water
(c) generates bubbles of gas when treated with $Na_2CO_3(aq)$
(d) generates bubbles of gas when sodium metal is added

72. Match the following compounds with the chemical properties given below. Write a chemical equation for the reactions described in (a)–(d).

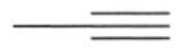 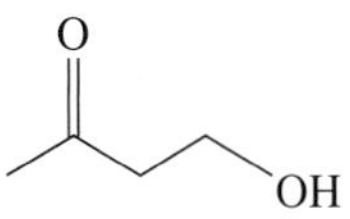 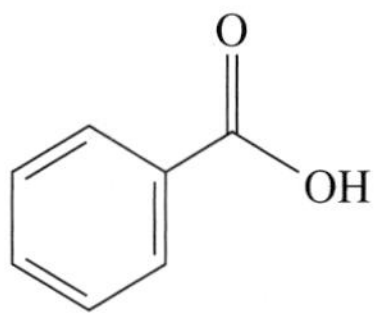

(1) **(2)** **(3)**

(a) forms an ester with ethanol
(b) absorbs H_2 in the presence of a metal catalyst
(c) neutralizes NaOH
(d) forms an ether when heated strongly with H_2SO_4

73. Draw as many structural isomers as you can for cyclic ethers (no —OH groups) having the formula C_4H_6O. Try to draw at least six. (There are more than six.)

74. Draw as many structural isomers as you can for cyclic alcohols having the formula C_4H_6O. Try to draw at least five. (There are more than five.)

Integrative and Advanced Exercises

75. Supply condensed or structural formulas for the following substances.
(a) cycloocta-1,5-diene (an intermediate in the manufacture of resins)
(b) 3,7,11-trimethyl-2,6,10-dodecatriene-1-ol (farnesol—odor of lily of the valley) [*Hint:* Dodeca means 12.]
(c) 2,6-dimethyl-5-hepten-1-al (used in the manufacture of perfume)
76. Draw structural formulas for all the isomers listed in Table 26.3, and show that, indeed, the substances with more compact structures have lower boiling points.
77. By drawing suitable structural formulas, establish that there are *17* isomers of $C_6H_{13}Cl$. [*Hint:* Refer to Example 26-1.]
78. The symbol:

is often used to represent benzene. It is also the structural formula of cyclohexatriene. Are benzene and cyclohexatriene the same substance? Explain.
79. Use the half-reaction method to balance the following redox equations.
(a) $C_6H_5NO_2 + Fe + H^+ \longrightarrow C_6H_5NH_3^+ + Fe^{3+} + H_2O$
(b) $C_6H_5CH_2OH + Cr_2O_7^{2-} + H^+ \longrightarrow C_6H_5CO_2H + Cr^{3+} + H_2O$
(c) $CH_3CH{=}CH_2 + MnO_4^- + H_2O \longrightarrow CH_3CHOHCH_2OH + MnO_2 + OH^-$
80. A 10.6 g sample of benzaldehyde was allowed to react with 5.9 g $KMnO_4$ in an excess of KOH(aq). After filtration of the $MnO_2(s)$ and acidification of the solution, 6.1 g of benzoic acid was isolated. What was the percent yield of the reaction? [*Hint:* Write half-equations for the oxidation and reduction half-reactions.]
81. Combustion of a 0.1908 g sample of a compound gave 0.2895 g CO_2 and 0.1192 g H_2O. Combustion of a second sample weighing 0.1825 g yielded 40.2 mL of $N_2(g)$, collected over 50% KOH(aq) (vapor pressure = 9 mmHg) at 25 °C and 735 mmHg barometric pressure. When 1.082 g of compound was dissolved in 26.00 g benzene (mp 5.50 °C, $K_f = 5.12$ °C mol kg^{-1}), the solution had a freezing point of 3.66 °C. What is the molecular formula of this compound?
82. Draw and name all derivatives of benzene having the formula **(a)** C_8H_{10}; **(b)** C_9H_{12}.
83. In the monochlorination of hydrocarbons, a hydrogen atom is replaced by a chlorine atom. How many different monochloro derivatives of 2-methylbutane are possible?
84. A particular colorless organic liquid is known to be one of the following compounds: butan-1-ol diethyl ether, methyl propyl ether, butyraldehyde, or propionic acid. Can you identify which it is, based on the following tests? If not, what additional test would you perform? (1) A 2.50 g sample dissolved in 100.0 g water has a freezing point of −0.7 °C. (2) An aqueous solution of the liquid does not change the color of blue litmus paper. (3) When alkaline $KMnO_4(aq)$ is added to the liquid and the mixture is heated, the purple color of the MnO_4^- disappears.
85. Write structural formulas for the following.
(a) 2,4-dimethylpenta-1,4-diene
(b) 2,3-dimethylpentane
(c) 1,2,4-tribromobenzene
(d) methyl ethanoate
(e) butanone
86. Give the systematic names, including any stereochemical designations, for each of the following.

(a) 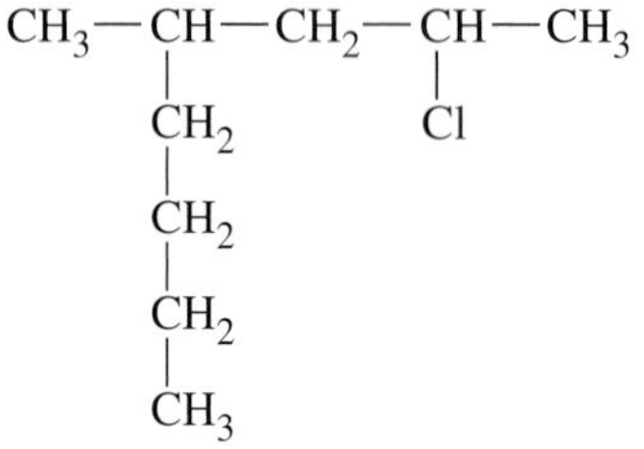

(b) 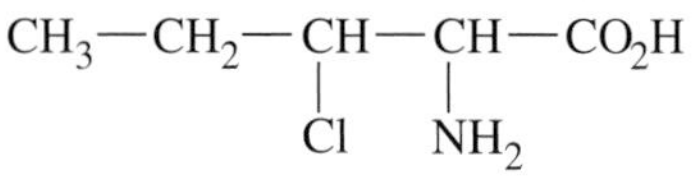

(c) $CH_3OCH_2CH_3$

(d) 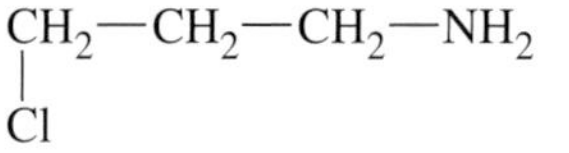

(e) 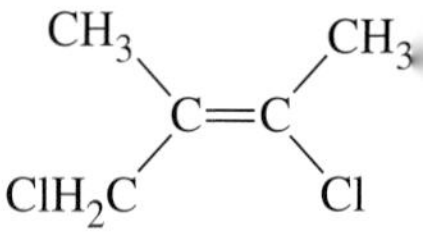

87. Write structural formulas for all the isomers of C_4H_7Cl. Indicate any enantiomers or diastereomers that occur.
88. Compound A is an alcohol of formula $C_5H_{12}O$ that can be resolved into enantiomers.
(a) Draw *three* possible structures of compound A.
(b) Treatment of A with CrO_3/pyridine gives compound B, which also exhibits optical activity. What are the structural formulas of A and of B? Name and draw the enantiomers of A and B.
89. Levomethadyl acetate (shown below) is used in the treatment of narcotic addiction.

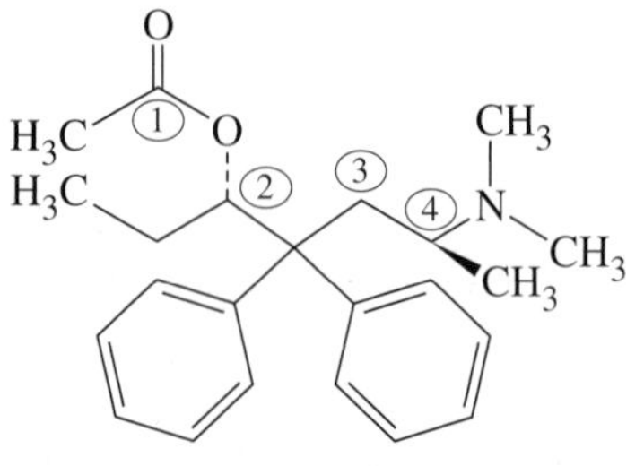

(a) Name the functional groups in levomethadyl acetate. **(b)** What is the hybridization of the numbered carbon atoms and the nitrogen atom? **(c)** Which, if any, of the numbered carbon atoms are chiral?

90. Thiamphenicol (shown below) is an antibacterial agent.

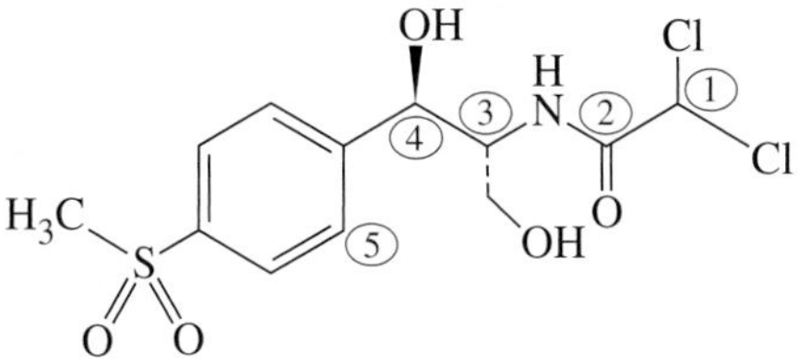

(a) Name the functional groups of thiamphenicol.
(b) What is the hybridization of the numbered carbon atoms and the nitrogen atom?
(c) Which, if any, of the numbered carbon atoms are chiral?

91. Ephedrine (shown below) is used as a decongestant in cold remedies.

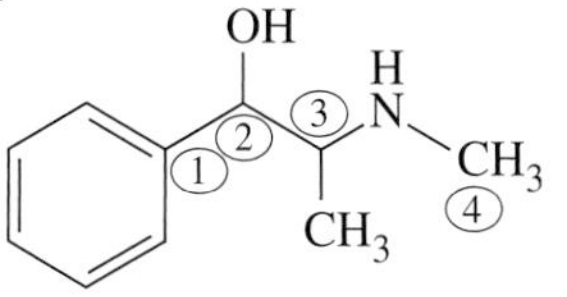

(a) Name the functional groups of ephedrine. **(b)** What is the hybridization of the numbered carbon atoms and the nitrogen atom? **(c)** Which, if any, of the numbered carbon atoms are chiral? **(d)** The pH of a solution of 1 g of ephedrine in 200 g of water is 10.8. What is the pK_b of ephedrine?

92. For each of the following molecules, determine the hybridization of each carbon atom, the total number of carbon–carbon σ bonds and the total number of carbon–carbon π bonds. Also, among the sp^3-hybridized carbon atoms, which ones are primary (1°), secondary (2°), tertiary (3°), or quaternary (4°)?

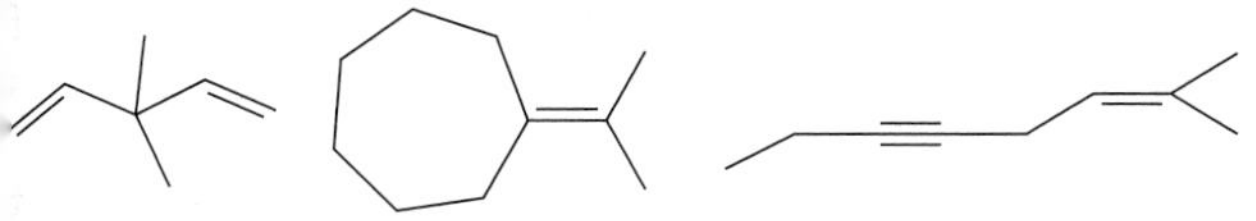

(a) **(b)** **(c)**

93. Determine the configuration, *R* or *S*, of each chiral carbon atom in the molecules that follow. (Ph represents a phenyl group.)

(a)

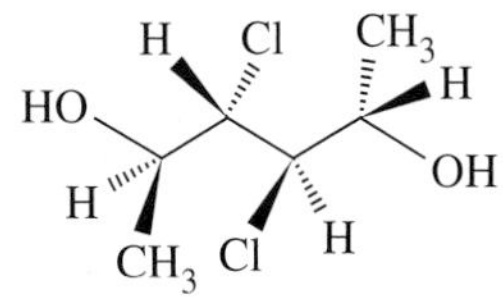

(b)

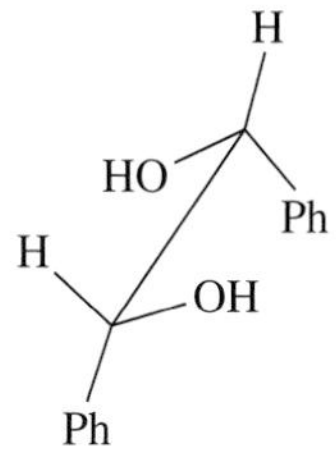

94. Among all ethers with the formula C_4H_6O, draw structures for
(a) two ethers with two sp^2 and two sp^3 carbon atoms
(b) an ether with four sp^2 carbon atoms
(c) an ether with two *sp* and two sp^3 carbon atoms

95. A structural formula for cholesterol is shown below. How many chiral carbon atoms are there in the cholesterol molecule? What is the configuration, *R* or *S*, of the carbon atom bonded to the —OH group? What is the configuration, *E* or *Z*, of the double bond?

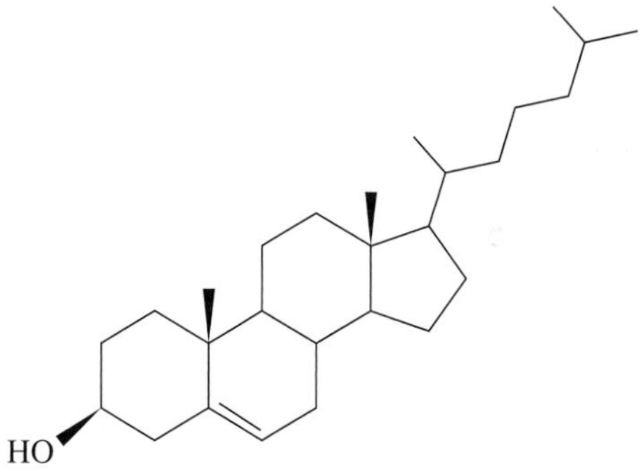

Cholesterol

96. For each of the following molecules **(a)** draw the two chair conformations and specify which conformation is more stable; **(b)** determine the number of chiral carbon atoms and for each chiral carbon, determine its configuration (*R* or *S*).

(1)

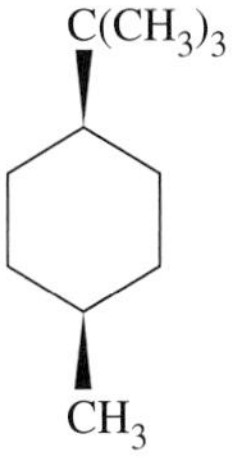

(2)

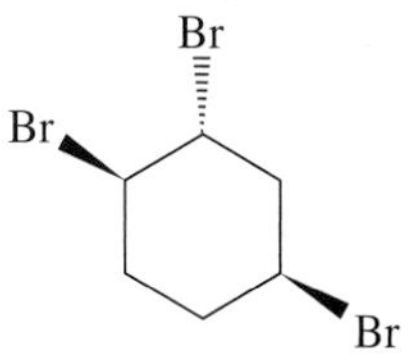

Feature Problem

97. Organic chemists use a variety of methods to help them identify the functional groups in a molecule. In this chapter, we mentioned a few simple chemical ways to test for alkenes, alcohols, carboxylic acids, and so on. Such tests are quick and easy, but today organic chemists rely heavily on instrumental techniques. *Infrared (IR) spectroscopy* is an instrumental technique for identifying functional groups in a molecule. When infrared radiation is absorbed by a molecule, it causes atoms in bonds to vibrate back and forth with increased amplitude. We saw (in Chapter 8) that the energies of electrons in atoms are quantized and so too are the vibrations of atoms in molecules. Because each functional group has a particular grouping of atoms, there is a characteristic infrared absorption associated with each type of functional group. Some characteristic infrared absorptions are summarized in the following table. Infrared absorptions of molecules are identified by specifying the *wavenumber* of the light that is absorbed. The wavenumber is simply the reciprocal of wavelength: wavenumber $= 1/\lambda = \nu/c$. (The definitions of λ, ν, and c were given in Chapter 8.) The SI unit

for wavenumber is m^{-1} but it is often given in units of cm^{-1}. The wavenumber represents the number of cycles of the wave in each meter or centimeter along the light beam.

The data in the table indicate that molecules containing the carbonyl group (C=O) absorb light with wavenumbers between 1680 and 1750 cm^{-1}. Molecules containing a carbon–carbon triple bond (C≡C) absorb light with wavenumbers between 2100 and 2200 cm^{-1}. To identify the functional groups present in a molecule, the *infrared spectrum* of the molecule is obtained by using an instrument called an *infrared spectrometer*. A schematic diagram of an infrared spectrometer is shown below.

The light from the infrared light source is directed through a monochromator, which can be set to select a specific wavelength of light. The light then passes through a beam splitter, which splits the light into two separate beams, a reference beam and an incident beam. If the wavenumber of the incident beam matches one of the characteristic absorptions of the molecule, the sample absorbs the light, producing molecules that vibrate with greater energy. Because light has been absorbed by the sample, the intensity of the transmitted beam is less than that of the reference beam. The decrease in intensity is detected by the detector. By varying the wavenumber of light that reaches the sample and monitoring the percentage of light that is transmitted, an infrared spectrum is obtained (see graphs). An infrared spectrum is a plot of percent transmittance versus wavenumber. One hundred percent transmittance means none of the incident light was absorbed and 0% transmittance means all of the incident light was absorbed.

Type of bond	Wavenumber, cm^{-1}
Single bonds	
—C—H	2850–3300
=C—H	3000–3100
≡C—H	≈3300
N—H	3300–3500
O—H	3200–3600
Double bonds	
C=C	1620–1680
C=N	1500–1650
C=O	1680–1750
Triple bonds	
C≡C	2100–2200
C≡N	2200–2300

(a) The infrared absorptions given in the table above range from 1500 cm^{-1} to 3600 cm^{-1}. Calculate the corresponding ranges of wavelength and frequency to verify that these absorptions correspond to the infrared region of the electromagnetic spectrum. (See Figure 8-3.)
(b) Identify the bonds responsible for the absorptions labeled A, B, C, and D in the two infrared spectra shown here. One spectrum is for acetone and the other is for 1-propanol, both of which are colorless liquids. Which of the two spectra is that of acetone?
(c) An isomer of acetone exhibits a strong IR absorption at 1645 cm^{-1} and also absorbs strongly from 2860 through 3600 cm^{-1}. The compound decolorizes bromine water, $Br_2(aq)$, and produces bubbles of gas when sodium metal is added to it. What is the structure of this compound?

Infrared spectrometer

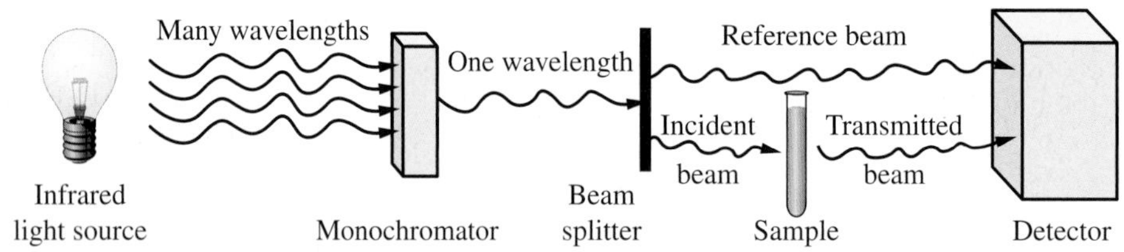

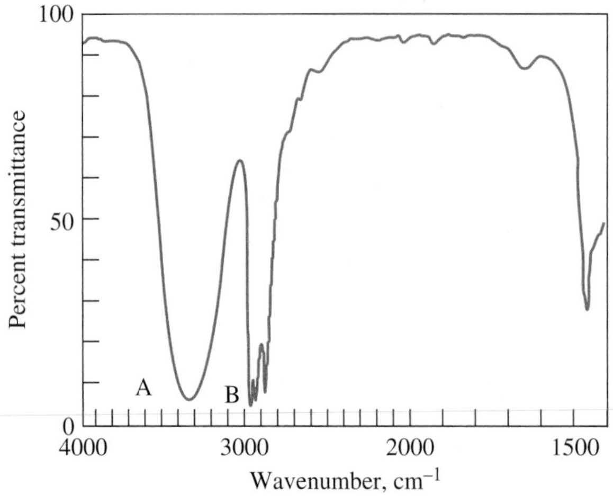

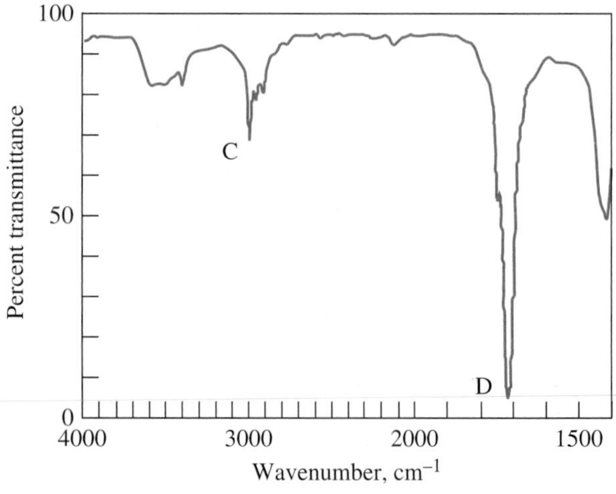

Self-Assessment Exercises

98. In your own words, define the following terms or symbols:
(a) *tert-* **(b)** R— **(c)** 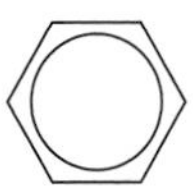 **(d)** carbonyl group **(e)** primary amine

99. Explain the important distinctions between each pair of terms: **(a)** alkane and alkene; **(b)** aliphatic and aromatic compound; **(c)** alcohol and phenol; **(d)** ether and ester; **(e)** amine and ammonia.

100. Describe the characteristics of each of the following types of isomers: **(a)** constitutional; **(b)** stereoisomer; **(c)** cis; **(d)** ortho.

101. The compound isoheptane is best represented by the formula **(a)** C_7H_{14}; **(b)** $CH_3(CH_2)_5CH_3$; **(c)** $(CH_3)_2CH(CH_2)_3CH_3$; **(d)** $C_6H_{11}CH_3$.

102. A compound with the same hydrogen-to-carbon ratio as cyclobutane is **(a)** C_4H_{10}; **(b)** $CH_3CH{=}CHCH_3$; **(c)** $CH_3C{\equiv}CCH_3$; **(d)** C_6H_6.

103. Three isomers exist of the hydrocarbon **(a)** C_3H_8; **(b)** C_4H_8; **(c)** C_4H_{10}; **(d)** C_6H_6; **(e)** C_5H_{12}.

104. Give names for the following molecules.

(a) $(CH_3)_2CBrCH_2CHClCH_2CH(CH_3)_2$

(b) 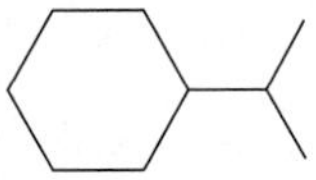 **(c)** 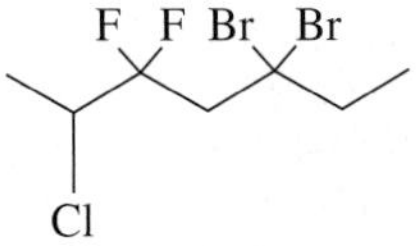

105. Assign configurations, *R* or *S*, to the chiral carbons in the molecules below. Then identify **(a)** any two identical structures; **(b)** any two constitutional isomers; **(c)** any two diastereomers; **(d)** a pair of enantiomers.

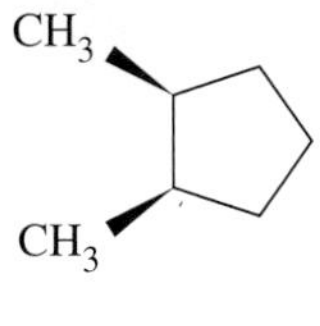

(1)

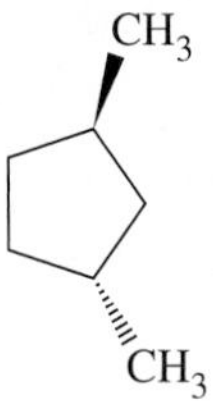

(2)

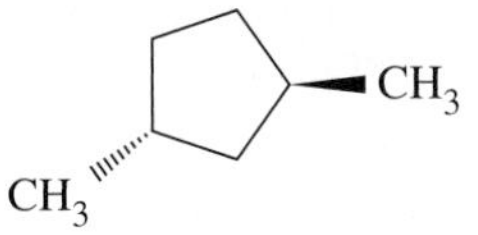

(3)

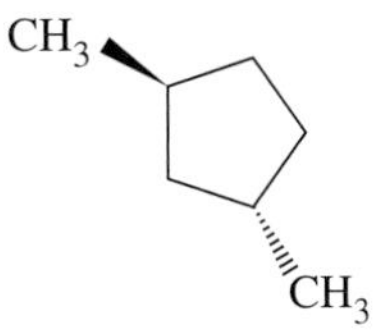

(4)

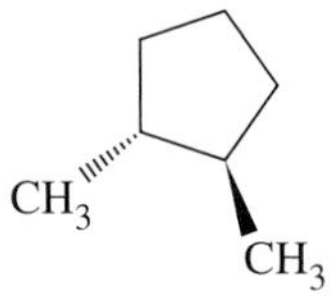

(5)

106. Consider the following pairs of structures. In each case, are the structures different conformers or are they isomers? If they are isomers, then state whether they are constitutional isomers, diastereomers, or enantiomers.

(a) 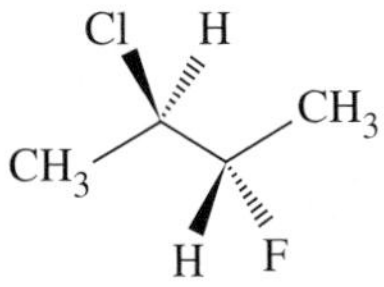

and 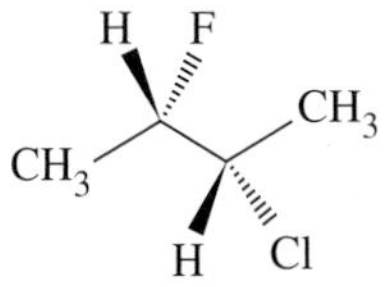

(b) 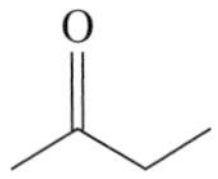

and 

(c) 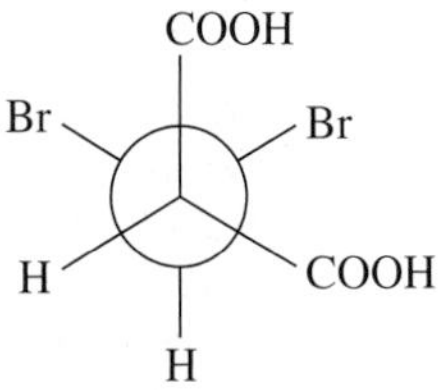

and 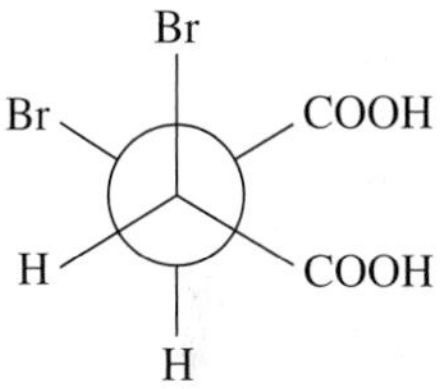

(d) 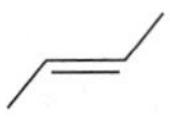 and 

107. Draw a Newman projection for the conformation of lowest energy for viewing 2-methylhexane along the C2—C3 bond.

108. To prepare methyl ethyl ketone, one should oxidize **(a)** propan-2-ol; **(a)** butan-1-ol; **(c)** butan-2-ol; **(d)** *tert*-butyl alcohol.

109. Which hydrocarbon has the greater number of isomers, C_4H_8 or C_4H_{10}? Explain your choice.

110. For each of the following pairs, indicate which substance has
(a) the higher boiling point, C_6H_{12} or C_2H_4
(b) the greater solubility in water, C_3H_7OH or $C_7H_{15}OH$
(c) the greater acidity in aqueous solution, C_6H_5CHO or C_6H_5COOH

111. Draw the structures of
(a) (*E*)-3-benzyl-2,5-dichloro-4-methylhex-3-ene
(b) 1-nitro-4-vinylbenzene [*Hint:* The vinyl group is $H_2C{=}CH^-$.]
(c) *trans*-1-(4-bromophenyl)-2-methylcyclohexane

27 Reactions of Organic Compounds

CONTENTS

LEARNING OBJECTIVES

27.1 Distinguish between substitution, addition, and elimination reactions.

27.2 Distinguish between S_N1 and S_N2 mechanisms for substitution reactions.

27.3 Distinguish between E1 and E2 mechanisms for elimination reactions.

27.4 Discuss the mechanisms for substitution and elimination reactions of alcohols.

27.5 Describe the mechanism for the addition of water across the double bond of an alkene.

27.6 Describe the mechanism of electrophilic aromatic substitution and the role of substituents on the aromatic ring.

27.7 Describe the trend in reactivities of alkanes toward bromine.

27.8 Distinguish between chain-reaction polymerization and condensation polymerization.

27.9 Discuss the idea of retrosynthesis in determining the proper reaction sequence to obtain a desired organic product.

A branch of the Pacific Yew tree (*Taxus brevifolia*), which grows in forests along the Pacific coast of British Columbia and Washington. The needles and seeds of the Pacific Yew are poisonous but the bark contains a compound called paclitaxel (formerly known as taxol), an effective chemotherapy drug. Unfortunately, the bark of a 100-year-old Pacific yew tree yields about 3 kilograms of bark and only 300 mg of paclitaxel—barely enough for a single dose. Paclitaxel, $C_{47}H_{51}NO_{14}$, is a complicated organic molecule, the structure of which was determined in 1971 (see margin). In 1994, two different groups of organic chemists, one lead by Professor Robert A. Holton of Florida State University and another lead by Professor Kyriacos C. Nicolau of the University of California (San Diego), announced independently that they were able to synthesize paclitaxel from simple starting materials.

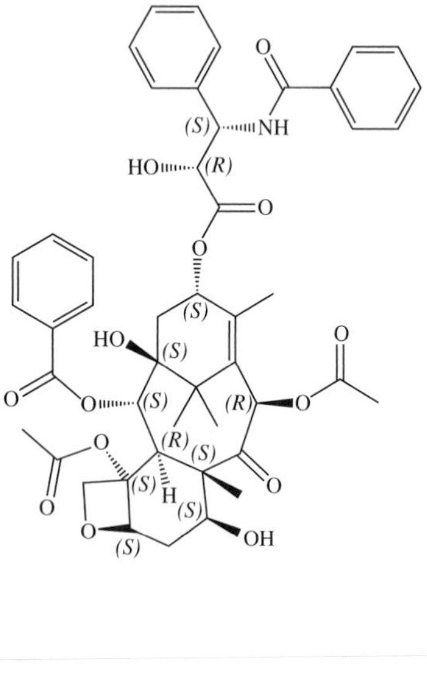

In Chapter 26, we focused primarily on the structures of organic compounds and learned that organic compounds can be categorized into a relatively small number of families: alkanes, haloalkanes, alcohols, alkenes, and so on. In this chapter, we are interested in the characteristic reactions of some of these compounds and the mechanisms that describe how these reactions occur. Given that millions of organic compounds are known, and countless reactions involving them are possible, it may seem surprising that the reactions of organic compounds can be rationalized in terms of a relatively small number of reaction mechanisms. In this chapter, we will explore a few of these mechanisms.

We cannot possibly provide a complete perspective of organic reactions and their mechanisms in one chapter. We will focus instead on core concepts that will not only help us understand the reactions but also serve us well in understanding the rationale of organic reactions and that they occur in very predictable ways. Throughout this chapter, we will emphasize the molecular aspects of chemical reactions—molecules coming together, with electron-rich regions being drawn naturally toward electron-poor regions; old bonds breaking and new bonds forming. We will use ideas about the structures and properties of molecules—gained in earlier chapters—to help us understand some important ideas about reactions of organic compounds.

We begin with a brief overview of the different types of organic reactions and then discuss these reaction types in terms of the structures of the molecules involved, the acidic or basic character of the reacting species, and the relative stabilities of reactants, products, and reaction intermediates. We will rely heavily on concepts from earlier chapters but will apply them in new ways. Also, we will make much more use of line-angle structures in this chapter to emphasize the functional groups and the prominent role they play in organic reactions.

27-1 Organic Reactions: An Introduction

Organic compounds undergo a variety of different reactions, including substitution, addition, elimination, and rearrangement reactions. In a **substitution reaction**, an atom, an ion, or a group in one molecule is replaced by (substituted with) another. Here are a few examples of substitution reactions:

$$CH_3Cl + NaOH \xrightarrow{H_2O} CH_3OH + NaCl$$

$$(CH_3)_3CBr + H_2O \xrightarrow[\Delta]{(CH_3)_2CO} (CH_3)_3COH + HBr$$

$$C_6H_6 + HNO_3 \xrightarrow{H_2SO_4} C_6H_5NO_2 + H_2O$$

In each equation, the solvent used in the reaction is written above the arrow. In the first reaction, the Cl atom in a CH_3Cl molecule is substituted with an $-OH$ group. In the second reaction, the Br atom in a $(CH_3)_3CBr$ molecule is substituted with an $-OH$ group, and in the third reaction, an H atom in a C_6H_6 molecule is substituted with an $-NO_2$ group. Although each reaction is a substitution reaction, we will soon see that they occur via different mechanisms.

In an **addition reaction**, a molecule adds across a double or triple bond in another molecule. Here are two examples:

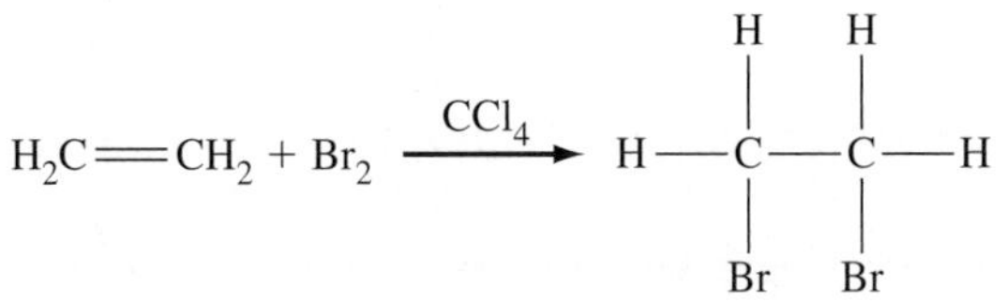

$$HC{\equiv}CH + HBr \longrightarrow H_2C{=}CHBr$$

In an **elimination reaction**, atoms or groups that are bonded to adjacent atoms are eliminated as a small molecule. Typically, the order of the bond between the two adjacent atoms increases as a result of the elimination, as shown in the example below:

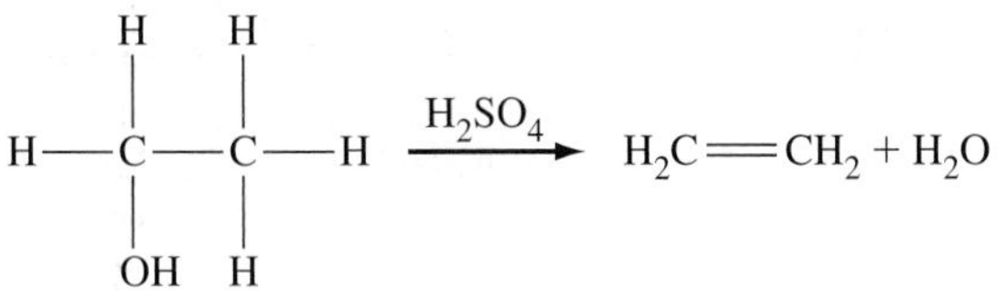

The following reaction, in which two molecules of HBr are eliminated to yield an alkyne, is an example of a double elimination reaction.

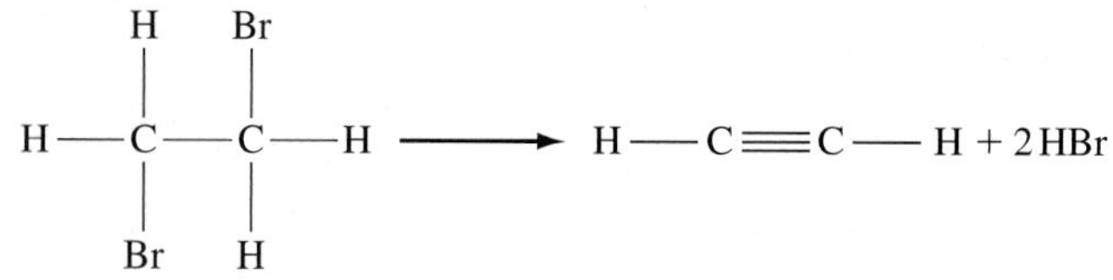

When an organic compound undergoes a **rearrangement reaction** (or an isomerization reaction), the carbon skeleton of a molecule is rearranged.

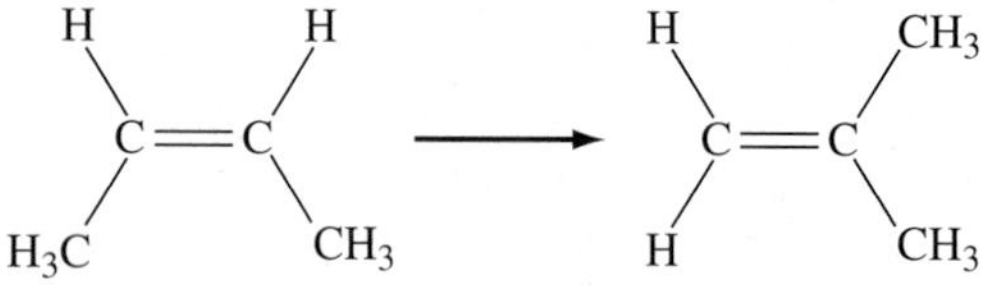

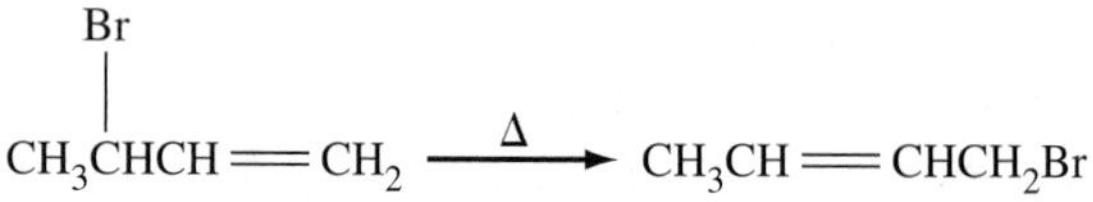

EXAMPLE 27-1 Identifying Types of Organic Reactions

State whether each of the following reactions is a substitution, an addition, an elimination, or a rearrangement reaction. (In the equations below, H^+ and Pd are catalysts.)

(a)

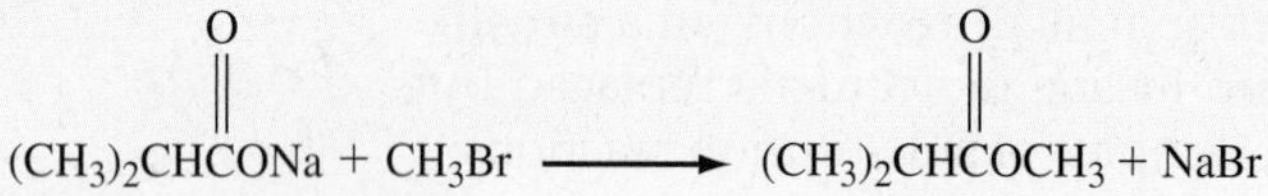

(b)

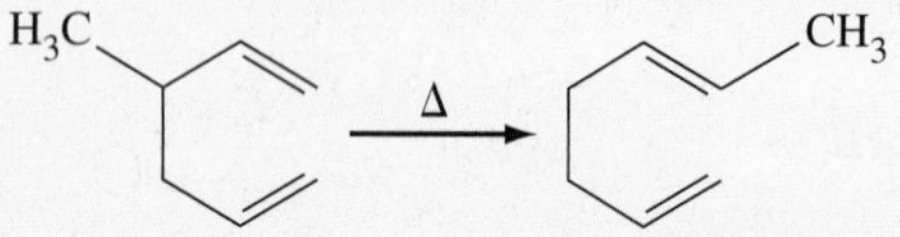

(c) $2\,CH_3OH \xrightarrow[\Delta]{H^+} CH_3OCH_3 + H_2O$

(d) cyclohexanol $\xrightarrow[\Delta]{H^+}$ cyclohexene + water

(e) but-2-yne + $2\,H_2 \xrightarrow{Pd}$ butane

Analyze

In a substitution reaction, an atom or a group of atoms is replaced with another atom or group of atoms. In an addition reaction, atoms add across a double or triple bond, causing a reduction in bond order between two atoms. In an elimination reaction, atoms or groups of atoms that are bonded to adjacent atoms are eliminated, causing an increase in bond order between two atoms. Typically, a rearrangement reaction involves a change in constitution or stereochemical configuration.

Solve

(a) Because a Br atom in CH_3Br is replaced by a $(CH_3)_2CHCOO$ group, the reaction is a substitution reaction.

(b) The reaction involves only a change in the skeletal structure (constitution), and so the reaction is a rearrangement reaction.

(c) When we write the reaction out as $CH_3OH + CH_3OH \longrightarrow CH_3OCH_3 + H_2O$, we see that an —OH group in one molecule is replaced by a —CH_3O group. The reaction is a substitution reaction.

(d) Here, we need to convert the names of the compounds into structural formulas.

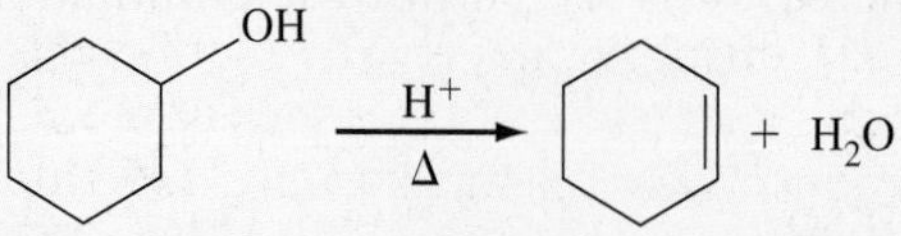

The reaction involves elimination of OH and H from adjacent carbon atoms in cyclohexanol. Thus, the reaction is an elimination reaction.

(e) In this reaction, an alkyne is converted into an alkane. Because there is a reduction in bond order, we know that the reaction is an addition reaction involving the addition of H atoms across the triple bond. Our prediction is confirmed by writing the reaction in terms of structural formulas:

$$CH_3C{\equiv}CCH_3 + 2H_2 \longrightarrow CH_3CH_2CH_2CH_3$$

Assess

To identify the reaction type correctly, it is often helpful to write out structural formulas for all molecules involved.

PRACTICE EXAMPLE A: Classify the following reactions as substitution, addition, elimination, or rearrangement reactions.

(a)

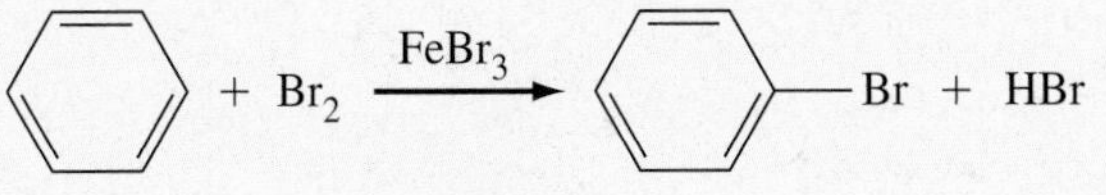

(b)

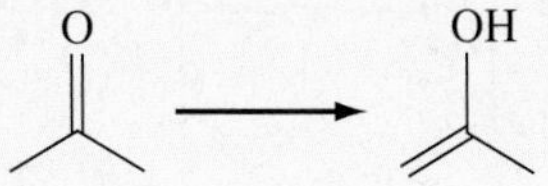

(c)

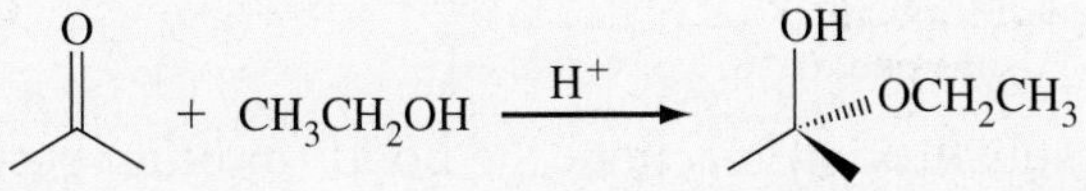

PRACTICE EXAMPLE B: Classify each of the following reactions as a substitution, an addition, an elimination, or a rearrangement reaction.

(a) $CH_3CH_2CH_2CH_2Br + (CH_3)_3CO^-K^+ \longrightarrow CH_3CH_2CH{=}CH_2 + KBr + (CH_3)_3COH$

(b)

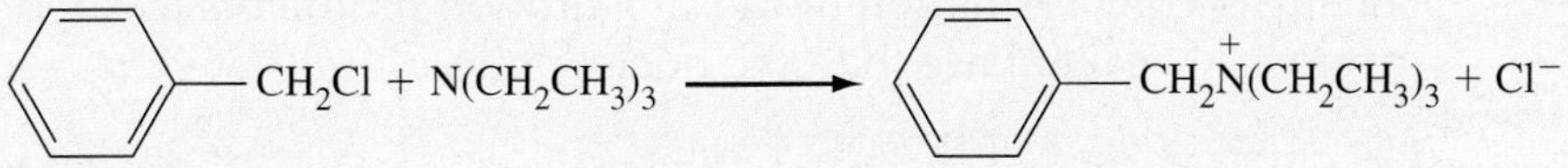

(c) $CH_3CH_2C{\equiv}C^-Na^+ + CH_3CH_2Br \longrightarrow CH_3CH_2C{\equiv}CCH_2CH_3 + NaBr$

27-2 Introduction to Nucleophilic Substitution Reactions

In this section, we consider the essential features of substitution reactions involving compounds in which the functional group is bonded to an sp^3 hybridized carbon atom. We will focus primarily on substitution reactions of haloalkanes. In a haloalkane, the carbon atom bonded to the halogen is sp^3 hybridized. With appropriate modifications, the concepts learned in this section can be used to understand substitution reactions of other compounds.

An example of a substitution reaction involving an sp^3 hybridized carbon is the replacement of a Cl atom in chloromethane by a hydroxide group to give methanol:

$$CH_3Cl + OH^- \longrightarrow CH_3OH + Cl^- \qquad (27.1)$$

▲ Volcanic eruptions produce significant quantities of chloromethane, CH_3Cl.

D. Harlow/U.S. Geological Survey, Denver

This type of chemical reaction is called a **nucleophilic substitution reaction** because the hydroxide ion is a **nucleophile**—a reactant that seeks sites of low electron density in a molecule. The nucleophile is electron rich, and it is the species that donates a pair of electrons to another molecule. The C atom in CH_3Cl is considered electron poor because the more electronegative Cl atom draws electron density away from the C atom; the C atom in CH_3Cl is said to be electrophilic (electron seeking). The **electrophile**, also known as the **substrate**, is the species that is being attacked by the nucleophile and accepts a pair of electrons in forming a new bond with the nucleophile. Electrophiles contain atoms that have relatively low electron densities. In reaction (27.1), the hydroxide ion attacks at the electron poor C atom, displacing the chloride ion, which is called the **leaving group**.

Nucleophiles, often denoted by the abbreviation Nu, can be negatively charged or neutral, but every nucleophile contains a pair of electrons (often an unshared pair) that are used for forming a bond with an electrophilic atom. Nucleophiles are, in fact, Lewis bases. The nucleophilic substitution of a haloalkane can be described by either of two general equations:

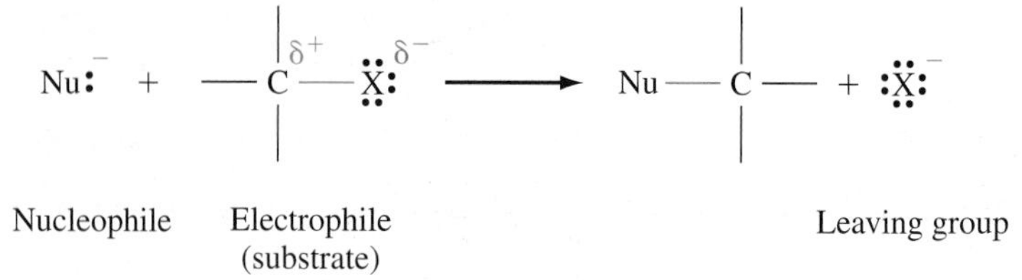

(27.2a)

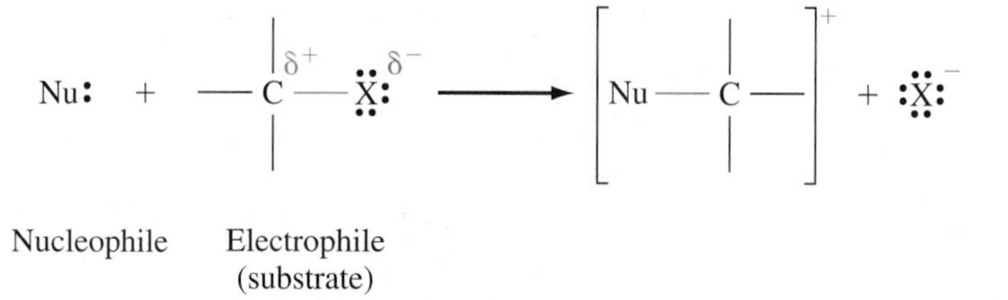

(27.2b)

In a nucleophilic substitution reaction, the bond (in blue) between carbon and the leaving group is broken and a new bond (in red) is formed by using a lone pair from the nucleophile. The electron pair from the bond that is broken ends up as a lone pair on the leaving group. In light of the discussion of acid–base chemistry in Section 16-9 and Section 27A on www.masteringchemistry.com, formation of the substitution product is favored if the leaving group is a weaker base than the nucleophile.

> Mastery of acid-base concepts is extremely important because all chemical reactions—including those involving organic molecules—can be characterized as some form of acid–base reaction. For a discussion of acid–base concepts, with a focus on organic reactions, see Section 27A, *Organic Acids and Bases*, on the MasteringChemistry site (www.masteringchemistry.com).
>
> MasteringCHEMISTRY

As we will soon see, nucleophilic substitution of haloalkanes occurs via one of two mechanisms. Before examining the mechanisms, let's consider a few specific examples of nucleophilic substitution reactions.

In the reaction below, the $CH_3C{\equiv}C^-$ ion is the nucleophile and CH_3CH_2Br is the electrophile:

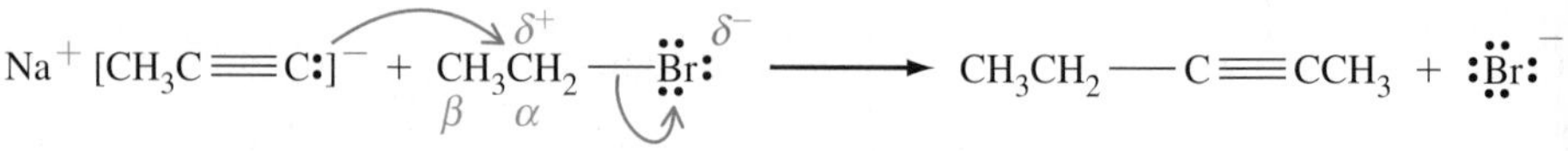

▶ In a haloalkane, the carbon bonded to the halogen atom is called the alpha (α) carbon. A carbon bonded to the α carbon is called a beta (β) carbon. A carbon bonded to a β carbon is called a gamma (γ) carbon, and so on. This system of labeling carbon atoms is important when discussing reactions.

The α carbon in CH_3CH_2Br is electron poor (δ^+); it is the atom of the electrophile that is attacked by the nucleophile. Because both the nucleophile and the leaving group are Lewis bases, we can use acid–base concepts to establish that the reaction above will occur as written. The $CH_3C{\equiv}C^-$ ion has the negative charge localized on C, a very small atom compared with Br, and so $CH_3C{\equiv}C^-$ is a much stronger base than Br^-. Because the reaction involves the formation of a weaker base, the reaction will occur as written.

In contrast to the example above, the following substitution reactions will not occur as written because they both involve the formation of a base that is stronger than the nucleophile:

$$HO^- + H_3C{-}H \rightleftharpoons H_3C{-}OH + H^- \ (H^- \text{ is a stronger base than } OH^-)$$

$$Br^- + CH_3CH_2{-}OH \rightleftharpoons CH_3CH_2{-}Br + OH^- \ (OH^- \text{ is a stronger base than } Br^-)$$

27-1 ARE YOU WONDERING?

What is the difference between a base and a nucleophile?

We have seen that a molecule or an ion having a lone pair of electrons can act as an electron pair donor in a reaction with either a proton or an electrophilic carbon atom. In both cases, a new bond is formed. When the bond is to a proton, the electron pair donor has reacted as a base. When the bond is to a carbon atom, the electron pair donor has reacted as a nucleophile. Chemists use the terms *basicity* and *nucleophilicity* when referring to the tendency of an electron pair donor to act as a base or a nucleophile in a reaction. **Basicity** is a measure of the tendency of an electron pair donor to react with a proton. **Nucleophilicity** is a measure of how readily (how fast) a nucleophile attacks an electrophilic carbon atom. We use equilibrium constants (K_b) when comparing molecules or ions in terms of their basicity, and thus basicity is an equilibrium (thermodynamic) property. When comparing molecules or ions in terms of their nucleophilicity, we are comparing them in terms of the rates at which they attack an electrophilic carbon, and thus nucleophilicity is a kinetic property.

What can make things a little confusing is that a molecule or an ion with a lone pair can act as either a base or a nucleophile. For example, the methoxide ion, CH_3O^-, is both a strong base and a good nucleophile.

Nucleophile Electrophile

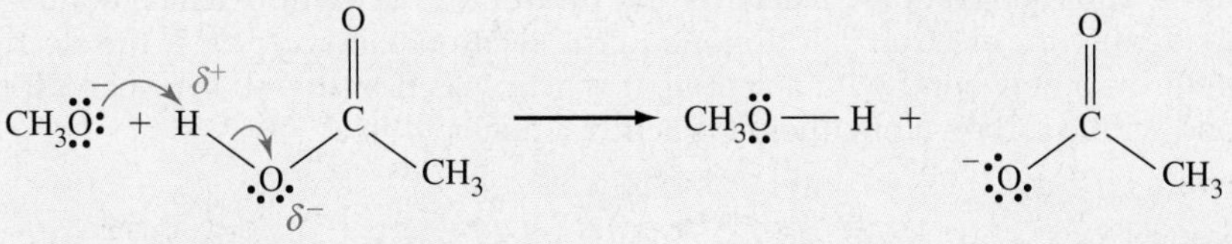

Base Acid

Thus, it is important to understand factors affecting not only basicity but also nucleophilicity. Factors affecting basicity are discussed in Section 16-10 in the textbook and in Section 27A on www.masteringchemistry.com. Factors affecting nucleophilicity are discussed on page 1281.

▲ Sir Robert Robinson (1886–1975) was awarded the 1947 Nobel Prize in Chemistry for his contributions to the synthesis of natural products. He is also credited with introducing the use of curved arrows to show the movement of electron pairs in chemical reactions.

The S_N1 and S_N2 Mechanisms of Nucleophilic Substitution Reactions

Chemical research, starting in the 1890s, has shown that nucleophilic substitution reactions can involve two types of mechanisms. The reaction between chloromethane and the hydroxide ion (equation 27.1) has a rate law that is first order in both the nucleophile and the electrophile. That is,

$$\text{rate} = k[OH^-][CH_3Cl]$$

The mechanism for this reaction involves a bimolecular rate-determining step in which the nucleophilic group approaches the electrophilic carbon atom and, simultaneously, the leaving group departs. The entering and leaving groups act at the same time in what is known as a *concerted step*, as depicted in this schematic representation:

KEEP IN MIND

that the electrophile accepts the pair of electrons. Thus, the nucleophile acts as a Lewis base, and the electrophile as a Lewis acid.

EXAMPLE 27-2 Identifying Electrophiles, Nucleophiles, Leaving Groups, Acids, and Bases

It is important to be able to distinguish when an electron pair donor in a reaction is acting as a base or as a nucleophile. The following reactions are elementary reactions. In each case, determine whether the electron pair donor is acting as a Brønsted–Lowry base or as a nucleophile. Then, identify the acid, electrophile, and leaving group, as appropriate. Finally, use arrows to indicate the movement of electrons.

(a)

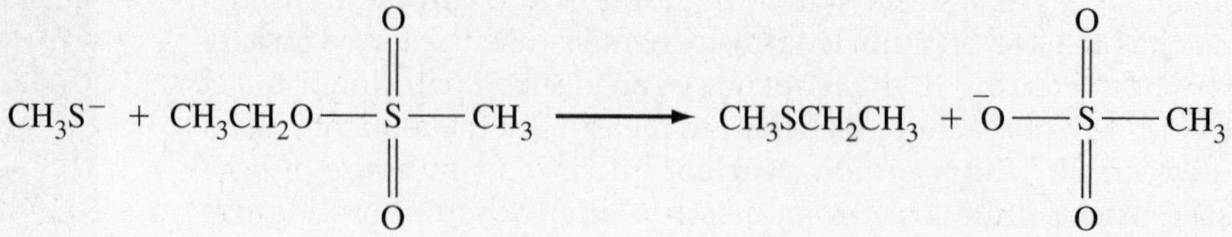

(b)

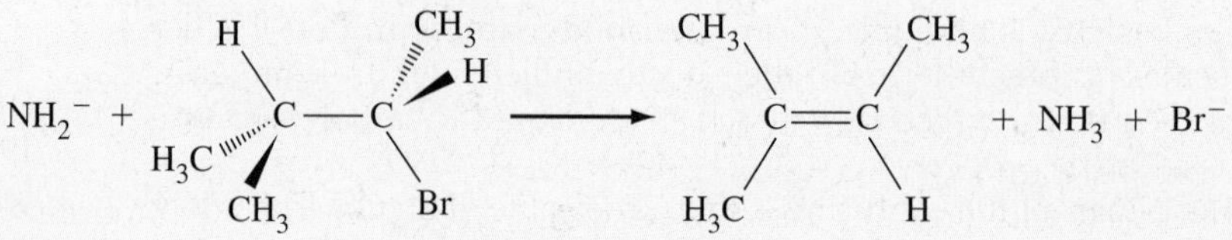

(c)

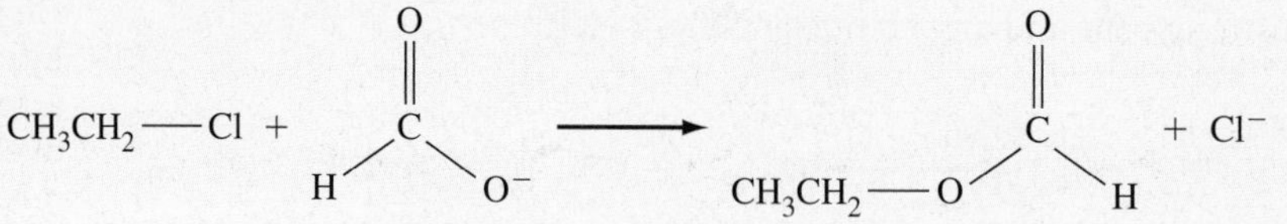

Analyze

First, we put lone pairs on the atoms, and then compare reactants and products to determine which bonds are formed and which are broken. We identify the electron pair donor and the electron pair acceptor. If the electron pair donor (the attacking species) forms a new bond with a proton, it is acting as Brønsted–Lowry base. If the attacking species forms a new bond with a carbon atom, then it is acting a nucleophile.

Solve

(a) In the equation below, reactants and products are shown with all lone pairs. We see that a new bond is formed between sulfur and carbon, and a bond between carbon and oxygen is broken. CH_3S^- is acting as a nucleophile, $CH_3CH_2OSO_2CH_3$ is the electrophile (or substrate), and $^-OSO_2CH_3$ is the leaving group. The red arrows indicate the movement of electrons.

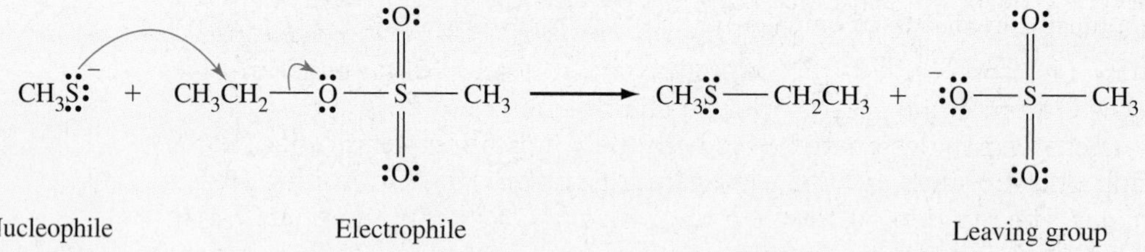

(b) In this reaction, NH_2^- reacts to form NH_3. Because the chemical formulas of these species differ by a single proton, we know that NH_3 and NH_2^- are a conjugate acid–base pair. NH_2^- is acting a Brønsted–Lowry base and $(CH_3)_2CHCHBrCH_3$ is acting as an acid.

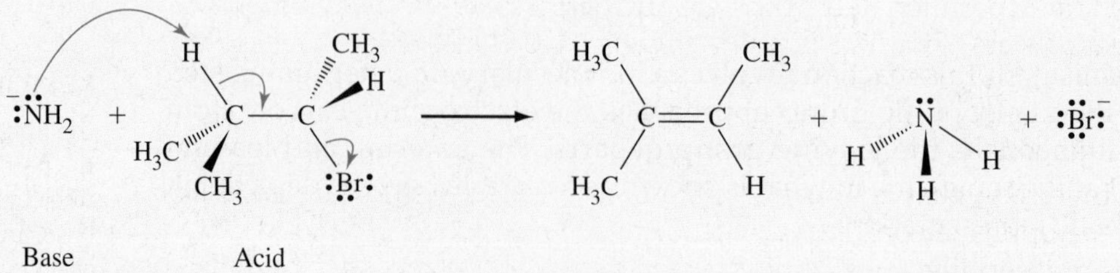

(c) We see from the equation below that a new carbon–oxygen bond is formed and a carbon–chlorine bond is broken. The $HCOO^-$ ion is acting as a nucleophile, CH_3CH_2Cl is the electrophile, and Cl^- is the leaving group.

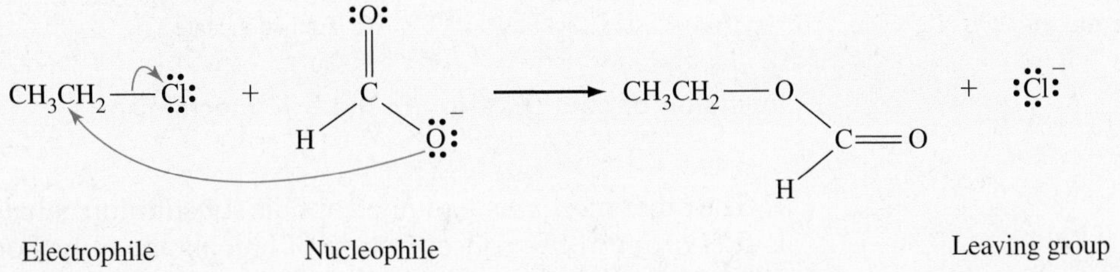

Assess

We can assess whether or not the substitution reactions in this example are feasible. For example, in (a), we can say that $^-OSO_2CH_3$ is a weaker (more stable) base than CH_3S^- because the negative charge in $^-OSO_2CH_3$ is highly delocalized; it is shared equally by three electronegative oxygen atoms. Thus, equilibrium favors products and we predict that the substitution will occur to a significant extent. For (c), we must compare the stabilities of the bases $HCOO^-$ and Cl^-. The conjugate acid of $HCOO^-$ is HCOOH, a weak acid, and so $HCOO^-$ is a weak base. Cl^- is an extremely weak base ($pK_b \approx 21$). Because the weaker base appears on the right side of the equation, we expect reaction (c) will occur as written.

PRACTICE EXAMPLE A: In each of the following reactions, determine whether the electron pair donor is acting as a Brønsted–Lowry base or a nucleophile. Identify the acid, electrophile, and leaving group, as appropriate, and use arrows to indicate the movement of electrons.

(a) $CH_3I + NaOH \longrightarrow CH_3OH + NaI$

(b) $NaOH + CH_3CH_2CH_2CO_2H \longrightarrow CH_3CH_2CH_2CO_2Na + H_2O$

PRACTICE EXAMPLE B: In each of the following reactions, determine whether the electron pair donor is acting as a Brønsted–Lowry base or a nucleophile. Identify the acid, electrophile, and leaving group, as appropriate, and use arrows to indicate the movement of electrons.

(a) $NH_2Na + CH_3CH_2OH \longrightarrow CH_3CH_2ONa + NH_3$

(b) $(CH_3)_2CHCH_2CH_2Br + CH_3SNa \longrightarrow (CH_3)_2CHCH_2CH_2SCH_3 + NaBr$

The curved red arrows in the examples we've considered illustrate how electrons flow to form and break bonds. Thus, the arrow starts from an electron-rich center (the electrons on the hydroxyl group in this case) to an electron-poor region (the C atom in the polar carbon-to-chlorine bond).

This mechanism is designated **S_N2**, with the S indicating substitution, the N indicating nucleophilic, and the 2 indicating that the rate-determining step is bimolecular. The reaction profile for an S_N2 reaction is shown in Figure 27-1, in which the transition state formed by the HO^- ion and the chloromethane is shown. The hydroxide attacks on the side opposite the Cl atom, and the C—Cl bond starts to break as the C—O bond simultaneously starts to form, producing the transition state shown. In the S_N2 mechanism, the nucleophile donates a pair of electrons to the electrophile to form a covalent bond.

Is there any experimental evidence to support this mechanism and postulated transition state? First, the rate law is suggestive of a bimolecular step; second, an observation by Paul Walden in 1893 confirmed the formation of the transition state as shown in Figure 27-2. Walden discovered that if the C atom bonded to the halogen is stereogenic, or chiral, the configuration at the chiral carbon is inverted—that is, the molecular structure is inverted and an enantiomer of the opposite configuration is formed. Thus, when (*S*)-2-iodobutane undergoes nucleophilic substitution by the hydroxide group, the compound (*R*)-butan-2-ol is formed (Fig. 27-2). This *inversion of configuration* is taken as confirmation of the proposed bimolecular mechanism and the five-coordinate transition state.

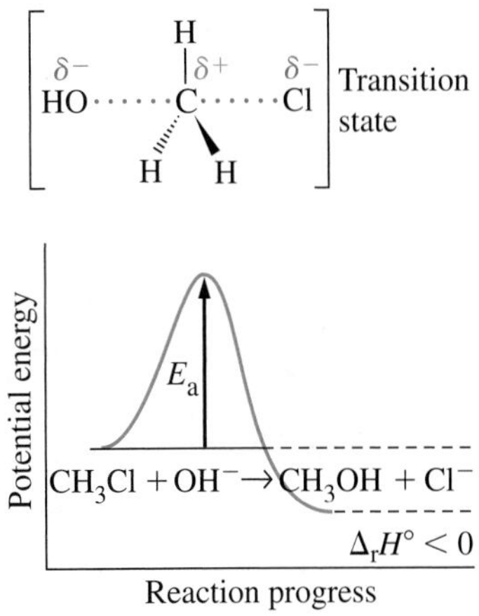

▲ FIGURE 27-1
Reaction profile for an S_N2 reaction
The bonds that are forming and breaking in the transition state are shown by dotted lines.

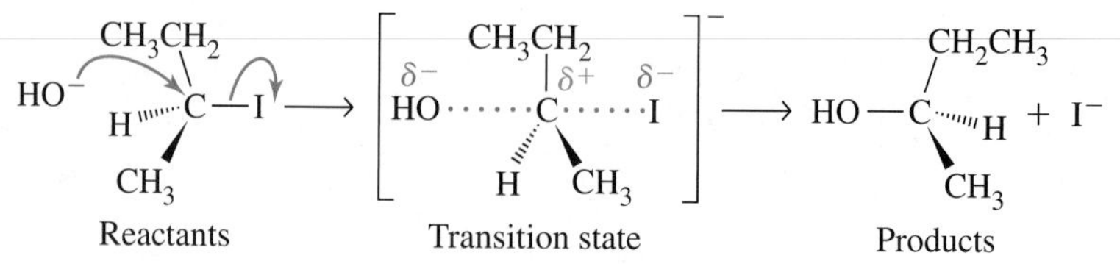

▲ FIGURE 27-2
Inversion of configuration in the S_N2 mechanism

The other mechanism of nucleophilic substitution at haloalkanes has a rate that is first order in the concentration of haloalkane only. For the reaction between 2-bromo-2-methylpropane and water, in which H_2O acts as the nucleophile,

$$(CH_3)_3CBr + H_2O \longrightarrow (CH_3)_3COH + HBr$$

the rate law is

$$\text{rate} = k[(CH_3)_3CBr]$$

This rate law suggests that the rate-determining step is unimolecular. The mechanism for this reaction is shown in Figure 27-3. The first step is a slow unimolecular step in which the haloalkane ionizes to form a bromide ion and a *carbocation*. The carbocation has a planar geometry and the positively charged carbon atom is sp^2 hybridized. In the second step, the carbocation reacts rapidly with the nucleophile, a water molecule in this case, producing the conjugate acid of an alcohol (a protonated alcohol). The protonated alcohol dissociates rapidly in the presence of excess water, forming the neutral alcohol and a hydronium ion.

This mechanism is designated **S_N1**. Again, the S indicates substitution and the N nucleophilic; in this case, the 1 indicates that the rate-determining step is unimolecular. The reaction profile for an S_N1 reaction is shown in Figure 27-4, which includes three transition states and two intermediates.

Apart from the rate law, what other evidence is there for the planar carbocation? Again, we can use a chiral haloalkane and investigate the chirality of the product. When one of the enantiomers of 3-bromo-3-methylhexane reacts

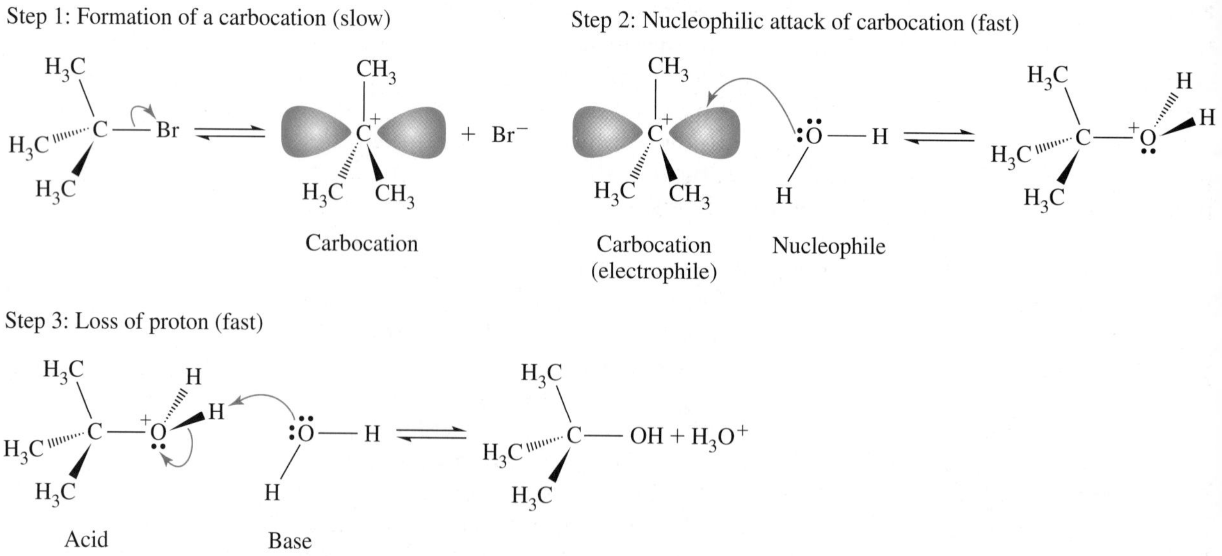

▲ FIGURE 27-3
The S_N1 mechanism
The first step is the formation of a carbocation. It is slow and rate determining. In the second step, the electron pair donor acts as a nucleophile and attacks the electrophilic carbon atom of the carbocation, forming a σ bond.

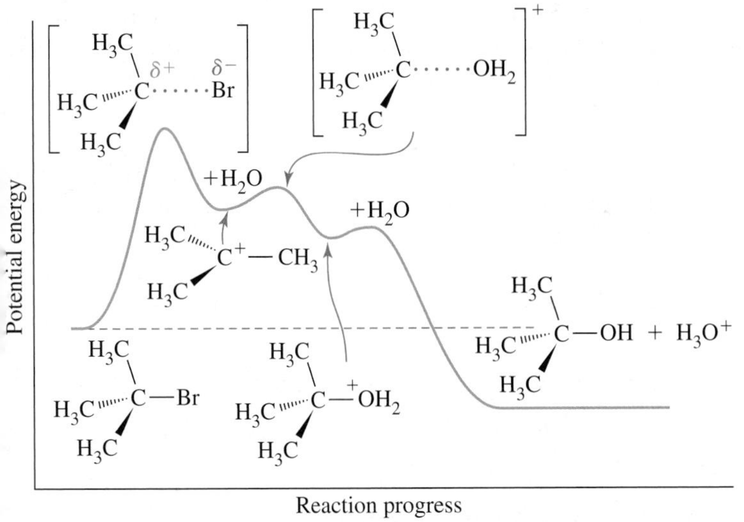

◀ FIGURE 27-4
Reaction profile for the reaction between $(CH_3)_3CBr$ and water
The carbocation formed in the rate-determining step is planar. The central C atom in the carbocation intermediate is sp^2 hybridized. The remaining *p* orbital is used in the formation of the new chemical bond. The central C atom of the 2,2-dimethylethyloxonium ion (the second intermediate) becomes hybridized subsequently.

with water, the product is a racemic mixture of the enantiomers of 3-methylhexan-3-ol. The carbocation intermediate formed in the rate-determining step is planar. The water molecule (nucleophile) can form a new bond on either face of the carbocation intermediate. This results in a mixture of the product enantiomers (Fig. 27-5). The formation of a racemic mixture provides confirmation of the unimolecular rate-determining step in the S_N1 mechanism.

When is a mechanism S_N1 and when is it S_N2? To answer this question, we need to briefly discuss some kinetic studies carried out by Christopher Ingold and Edward Hughes in 1937. Measurement of rates of reaction under different sets of experimental conditions can provide insight into reaction mechanisms. For example, consider the rate of the following second-order reaction between a bromoalkane and the nucleophile Cl^-:

$$R{-}Br + Cl^- \xrightarrow{S_N2} R{-}Cl + Br^-$$

From a thermodynamic point of view, the substitution of Br^- by Cl is favorable because Br^- is a weaker base than the Cl^- ion. The *rate* of reaction shows a dramatic dependence on the degree of branching of the α carbon.

◀ In the same manner that alcohols are classified as primary, secondary, and tertiary, so are haloalkanes. Thus, 2-bromopropane is a secondary haloalkane.

	CH_3Br >	CH_3CH_2Br >	$CH_3CH_2CH_2Br$ >	$CH_3CH(CH_3)Br$ >	$(CH_3)_3CBr$
Relative rate (for S_N2)	1200	40	16	1	Too slow to measure

In the S_N2 mechanism, the nucleophile attacks the electrophilic center on the side *opposite* the leaving group, often called *backside attack*. Because the

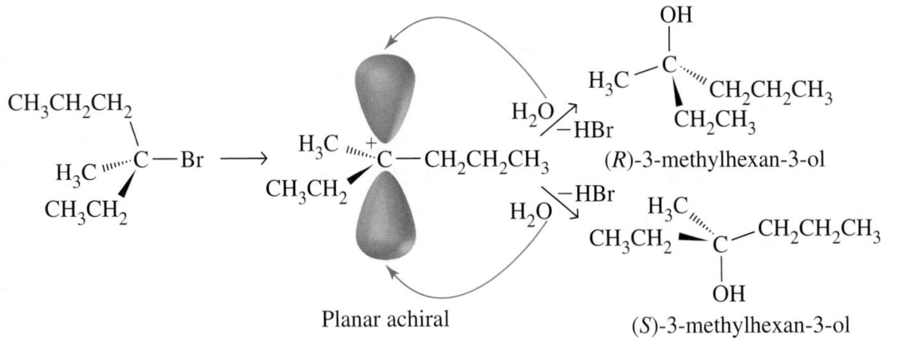

◀ FIGURE 27-5
Formation of a racemic mixture in an S_N1 reaction

nucleophile attacks the backside of the α carbon, bulky substituents bonded to the α carbon will make it harder for the nucleophile to get to the backside. The obstruction of the nucleophile from interacting with an electrophilic carbon atom is an example of steric hindrance. Backside attack of CH_3Br is fastest because there is very little steric hindrance (Fig. 27-6). However, the backside of $(CH_3)_3CBr$ is almost completely blocked from nucleophilic attack, and thus this compound does not undergo an S_N2 reaction; nucleophilic substitution occurs by an S_N1 mechanism instead.

What effect does alkyl substitution have on the rate of an S_N1 reaction? When we look at a series of bromoalkanes reacting with water, a very different order of reactivity is observed:

$$R{-}Br + H_2O \longrightarrow R{-}OH + HBr$$

$$S_N1 \text{ reactivity: } CH_3Br < CH_3CH_2Br < (CH_3)_2CHBr < (CH_3)_3CBr$$

α carbon: methyl, primary, secondary, tertiary

The order of reactivity, which is the opposite of that observed for the S_N2 reaction, follows the order of carbocation stability. The reactivity is related to the stability of the carbocation because the rate-determining step in the S_N1 reaction is the formation of the carbocation (Fig. 27-4). The order of carbocation stability is as follows:

$$\text{Relative stability: } CH_3^+ < CH_3CH_2^+ < (CH_3)_2CH^+ < (CH_3)_3C^+$$

Least stable — Most stable

Thus, reagents or reaction conditions that favor the formation of a carbocation will increase the rate of the S_N1 reaction.

It is important to emphasize, however, that carbocations are not particularly stable, at least not in the usual sense. Carbocations are reactive intermediates with rather short lifetimes. A tertiary carbocation, such as $(CH_3)_3C^+$, has a lifetime of about 10^{-10} s in water whereas a secondary carbocation, such as $(CH_3)_2CH^+$, has a lifetime of about 10^{-12} s in water. What is the explanation for the relative stabilities of methyl, primary, secondary, and tertiary carbocations? Alkyl groups stabilize carbocations and, in that role, they appear to be electron donating. However, the explanation of how alkyl groups stabilize a carbocation is a matter of some debate. One explanation is based on the concept of *hyperconjugation*, which is illustrated in Figure 27-7 for $CH_3CH_2^+$, a primary carbocation. The positive carbon is sp^2 hybridized. Around the positively charged carbon atom is a trigonal planar arrangement of atoms and an empty 2*p* orbital perpendicular to the plane of hybridization. The adjacent carbon atom is sp^3 hybridized and forms a σ bond with a hydrogen atom. The C—H bond can donate electron density to the empty 2*p* orbital, as suggested by the dashed arrow in Figure 27-7. If more alkyl groups are present in a carbocation, then more interactions of this type occur, and a greater degree of stabilization is the result. Thus, tertiary carbocations are more stable than secondary carbocations, which are, in turn, more stable than primary carbocations. Because no alkyl groups are present in a methyl carbocation, methyl carbocations are least stable; in fact, it is assumed they are never formed.

Other explanations for the stabilization of carbocations by alkyl groups are also used. However, all explanations have a common theme: the donation of electron density from filled orbitals that are aligned (or partially aligned) with the empty 2*p* orbital on the positively charged carbon atom of the carbocation. Irrespective of the explanation, alkyl groups help to stabilize a carbocation; thus, as the number of alkyl groups bonded to the α carbon increases, the stability of the carbocation also increases, as does the speed of the S_N1 reaction.

▲ FIGURE 27-6
Steric hindrance in a series of bromoalkanes
The models are viewed from the vantage point of a nucleophile attacking the α carbon from the backside. As the number of methyl groups bonded to the α carbon increases, access to the partial positive charge on the α carbon decreases.

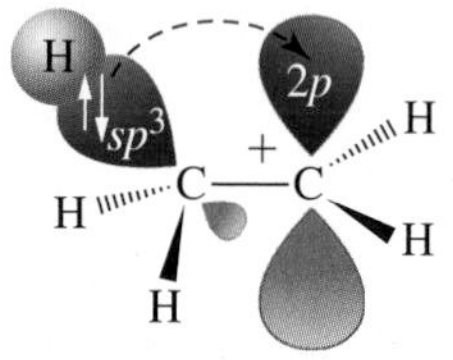

▲ FIGURE 27-7
Stabilization of a carbocation through hyperconjugation
The carbocation is stabilized through the donation of electron density from a C—H bond to an empty 2*p* orbital on the positively charged carbon atom.

EXAMPLE 27-3 Recognizing S_N1 and S_N2 Reactions and Predicting the Products

Consider the following combinations of reactants. In each case, predict whether a substitution reaction will occur. If so, identify the products and suggest the likely mechanism.

(a) $CN^- + CH_3CH_2CH_2Cl \longrightarrow$
(b) $Br^- + CH_3CH_2OH \longrightarrow$
(c) $CH_3OH + (CH_3)_3CCl$ (in methanol) $\longrightarrow$

Analyze

To decide whether a reaction of this type will take place, we first identify the electrophile, nucleophile, and leaving group. Keep in mind that the order of reactivity for S_N2 is methyl > primary > secondary > tertiary and that the order is the opposite for S_N1.

Solve

(a) The nucleophile is the CN^- ion, the electrophile is the chloropropane, and the leaving group is the Cl^- ion. The CN^- ion is a much stronger base than the Cl^- ion, and so the equilibrium constant for the reaction should be large. The chloropropane is a primary haloalkane, so the likely mechanism is S_N2 and the product will be $CH_3CH_2CH_2CN$.

(b) The nucleophile is the Br^- ion, the electrophile is the ethanol, and the potential leaving group is the OH^- ion. The OH^- ion is a much stronger base than the Br^- ion, so the equilibrium constant for the reaction is much *less* than one. No reaction is expected.

(c) The nucleophile is the CH_3OH molecule, the electrophile is the *t*-butyl chloride, and the leaving group is the Cl^- ion. In this case we have to know the relative basicities of CH_3OH and Cl^- to decide in which direction the reaction will proceed. If we assume that the basicities of methanol and chloride ion are about the same (both are very weak bases), we expect an equilibrium to be established. However, the fact that we are using a large excess of methanol (methanol is the solvent) shifts the equilibrium toward the product (both are very weak bases). The product is $(CH_3)_3COCH_3$, and the likely mechanism is S_N1 since the electrophile is a tertiary haloalkane.

Assess

In Section 27-3 we will see that other products, not just substitution products, are possible in some of these reactions.

PRACTICE EXAMPLE A: Predict whether the following reactions will take place, and suggest the likely mechanism:

(a) $CH_3CC^- + CH_3Br \longrightarrow$
(b) $Cl^- + CH_3CH_2CN \longrightarrow$
(c) $CH_3NH_2 + (CH_3)_3CCl \longrightarrow$

PRACTICE EXAMPLE B: From the following information, state whether the reaction proceeds by an S_N1 or S_N2 mechanism. Then, write a detailed mechanism for the reaction.

(a) The following reaction is carried out in acetone (propanone):

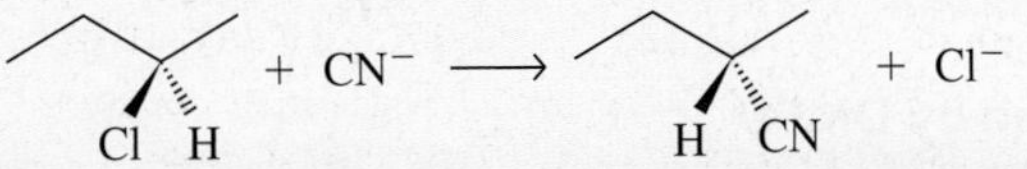

(b) A sample that contains only one enantiomer (either *R* or *S*) of an optically active compound is said to be *optically pure*. An optically pure sample of 2-iodobutane is dissolved in methanol. The resultant solution of 2-methoxybutane is optically inactive.

Solvent Effects in S_N1 and S_N2 Reactions

The nucleophile in a substitution reaction is typically a negatively charged ion or a polar molecule. Thus, to dissolve the starting materials, we must generally use polar solvents. (Recall from Chapter 14 that like dissolves like.) However, the nature of the solvent—in particular, the way in which solvent

Some polar aprotic solvents

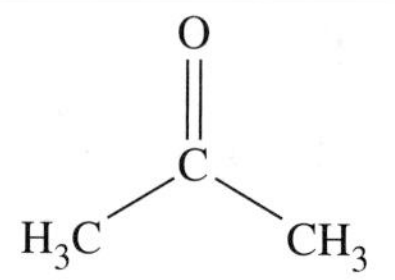

Propanone (acetone)

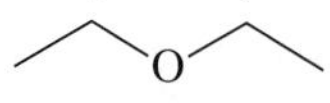

Diethylether

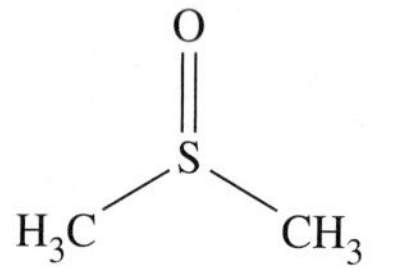

Dimethylsulfoxide (DMSO)

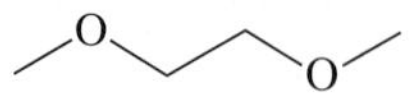

Dimethoxyethane (DME)

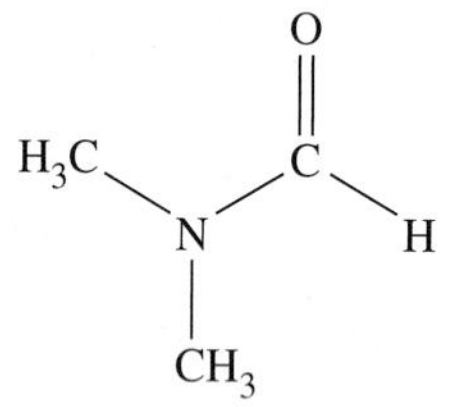

Dimethylformamide (DMF)

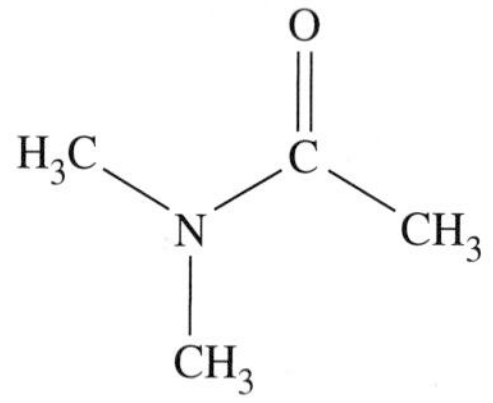

Dimethylacetamide (DMA)

molecules interact with the nucleophile—plays a critical role in determining whether a reaction will follow an S_N1 or an S_N2 mechanism. To understand solvent effects, it is helpful to differentiate between protic and aprotic solvents. **Protic solvents** are solvents whose molecules have hydrogen atoms bonded to electronegative atoms, such as oxygen and nitrogen. Water, methanol, ethanol, ethanoic acid (acetic acid), and methanamine (methylamine) are examples of *polar* protic solvents. (Not only do these molecules have H atoms bonded to O or N, but they are also polar.) **Aprotic solvents** are solvents whose molecules do not have a hydrogen atom bonded to a highly electronegative atom, such as O or N. Aprotic solvents can be polar or nonpolar, depending on whether the solvent molecules are polar or nonpolar. A few examples of polar aprotic solvents are shown in the margin. Another example (not shown) is hexamethylphosphoric triamide (HMPT), $[(CH_3)_2N]_3P{=}O$. Examples of nonpolar aprotic solvents are hexane and benzene.

What type of solvent favors the S_N1 pathway? The rate-determining step in an S_N1 reaction is the formation of a carbocation. If the solvent does not stabilize the ions formed, an S_N1 reaction will not occur. Polar protic solvents, such as water and methanol, promote S_N1 reactions because the molecules of such solvents stabilize the carbocation. Typically, molecules of a polar protic solvent stabilize carbocations through the donation of lone pairs from an oxygen or a nitrogen atom. They stabilize anions through the formation of hydrogen bonds. Aprotic solvents are able to stabilize the cation through the donation of lone pairs. However, aprotic solvents cannot stabilize the anions that may be formed because they lack the hydrogen atom needed to form a hydrogen bond with the anion.

What type of solvent promotes an S_N2 reaction? Polar aprotic solvents work best for S_N2 reactions. The pathway is a concerted reaction in which the nucleophile attacks an electrophile. The reactivity of the nucleophile is significantly reduced if the solvent molecules interact strongly with the nucleophile (for example, through the formation of hydrogen bonds). The stronger the interactions between the solvent molecules and the nucleophile, the greater the amount of energy needed to remove the shell of solvent molecules around the nucleophile and the more difficult it is for the nucleophile to shed the solvent molecules in its solvent shell. Figure 27-8 illustrates the idea that a given nucleophile (F^-, for example) will be more strongly solvated, and less reactive as a nucleophile, in a polar protic solvent than in a polar aprotic solvent. Strong intermolecular interactions, such as hydrogen bonding, between the nucleophile and the solvent molecules are avoided by using aprotic solvents, and thus aprotic solvents are used to promote S_N2 reactions.

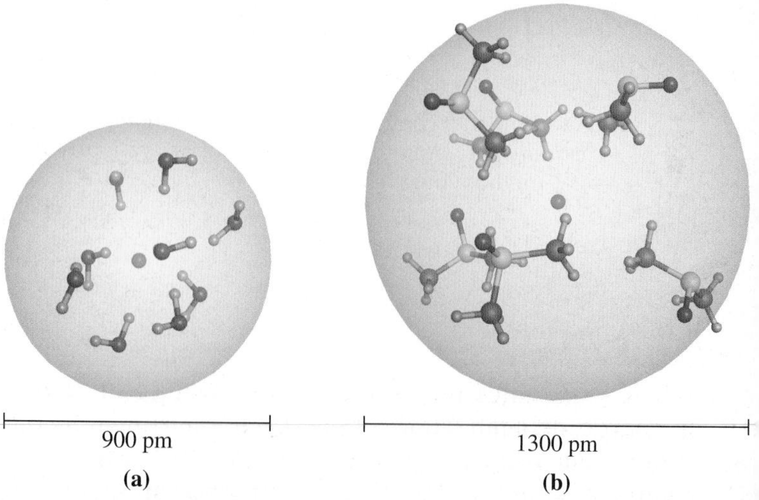

▶ FIGURE 27-8
Solvation of F^- by water and by DMSO
Computer simulations indicate that the F^- ion, shown in green, is more strongly solvated by water **(a)** than by DMSO **(b)**. The spheres enclosing the F^- ion and the surrounding solvent molecules emphasize that solvent molecules are, on average, closer to the F^- ion when the solvent is water. For the hypothetical process of transferring a fluoride ion from water to DMSO, $F^-(aq) \rightarrow F^-(DMSO)$, $\Delta_r G°$ has been estimated to be +73 kJ/mol [G. T. Hefter, *Pure Appl. Chem.*, **63**, 1749 (1991)]. The positive value of $\Delta_r G°$ indicates that F^- is thermodynamically more stable, and thus less reactive as a nucleophile, when solvated by water.

27-1 CONCEPT ASSESSMENT

Are the following solvents protic or aprotic? acetonitrile, CH_3CN; ammonia, NH_3; trimethylamine, $(CH_3)_3N$; formamide, $HCONH_2$; acetone, CH_3COCH_3.

Factors Affecting Nucleophilicity

Nucleophilicity is a measure of how readily (how fast) a nucleophile attacks an electrophilic carbon atom bearing a leaving group. But what makes a good nucleophile? It is tempting to think that there is a simple relationship between nucleophilicity and basicity because both involve the donation of an electron pair to an electrophile. No simple relationship exists because, as pointed out in Are You Wondering 27-1, nucleophilicity and basicity are fundamentally different properties. Nucleophilicity is a kinetic property (related to a rate of reaction) and basicity is a thermodynamic property (related to the thermodynamic stability of a base). Many texts correlate trends in nucleophilicity with trends in basicity. It is true that such correlations exist, but the correlations are of little use without an understanding of the underlying reasons for these correlations. The correlations exist because the explanation of trends in nucleophilicity is based, in part, on factors we use for explaining trends in basicities. For example, both nucleophilicity and basicity depend on the electronegativity, size, and hybridization of the atom bearing the lone pair; charge delocalization; and the effect of electron-withdrawing or electron-donating groups. However, trends in nucleophilicity depend on other factors too, such as the interactions between the nucleophile and solvent molecules, steric effects, and the nature of the electrophile. Therefore, trends in nucleophilicity do not always follow trends in basicity.

Let us consider a few principles that we can use for understanding trends in nucleophilicity. When applying these principles, our point of reference is the nucleophilic atom, which possesses the lone pair that is used to form a bond with the electrophile. As we work our way through these principles, we will see similarities to explanations given earlier (Section 27A on www.masteringchemistry.com), but it is important that we take careful note of the differences.

1. All else being equal, a negatively charged nucleophile will react faster than an uncharged nucleophile. For example, HO^- is a stronger nucleophile than H_2O, and CH_3O^- is a stronger nucleophile than CH_3OH. A negatively charged atom is more strongly attracted to an electrophilic center than is an atom having no charge or a partial negative charge.
2. When comparing molecules or ions in which the nucleophilic atoms are from the *same row of the periodic table*, the electronegativity of the nucleophilic atom is an important factor because it affects how available the lone pair will be for bonding to the electrophilic atom. In the following anions, the nucleophilic atoms are from the second period.

Increasing electronegativity of the nucleophilic atom →

$$H_3C^- > H_2N^- > HO^- > F^-$$

← Increasing nucleophilicity

F^- is not as good a nucleophile as H_3C^- because F is more electronegative than C, and thus a lone pair on F^- is less available for bonding than is the lone pair on H_3C^-.

It is sometimes possible to predict the effect of other factors, such as charge delocalization or the presence of electron-withdrawing or electron-donating groups. Consider, for example, the following series of nucleophiles, arranged in order of decreasing nucleophilicity (from strongest to weakest). All the nucleophilic atoms in these ions are from the second period.

$$H_2N^- > CH_3O^- > HO^- > HCOO^-$$

← Increasing nucleophilicity

H_2N^- is the strongest nucleophile because N is less electronegative than O, and thus the lone pair on N is more available for bonding than is a lone pair on O. CH_3O^- is a stronger nucleophile than HO^- because, in solution, the CH_3 group is electron donating and this helps to make the lone pairs on O less stable and more reactive. $HCOO^-$ is a weaker electrophile than HO^- because the negative charge is delocalized (shared equally by both oxygens), and thus $HCOO^-$ will be less reactive than HO^-.

3. Negatively charged nucleophiles are more reactive in polar aprotic solvents than they are in polar protic solvents. In a polar protic solvent, anions are more highly solvated and thus less reactive (Fig. 27-8).

   The importance of considering how the solvent affects nucleophilicity cannot be overstated. The trend in reactivities of a series of nucleophiles can, in some instances, be completely reversed by changing the solvent.

Protic solvent: $I^- > Br^- > Cl^- > F^-$

Aprotic solvent: $F^- > Cl^- > Br^- > I^-$

← Increasing nucleophilicity

4. When comparing nucleophiles in which the nucleophilic atoms are from *the same group*, nucleophilicity *usually* increases as the size (and thus, polarizability) of the nucleophilic atom increases. The larger and more polarizable the nucleophilic atom, the easier it is for the charge cloud of the nucleophile to be distorted toward the electrophilic carbon atom. The distortion of the charge cloud toward the electrophilic carbon atom helps to lower the energy of the transition state and increases the rate of reaction. (See Figure 27-9.)

5. Bulky groups adjacent to the nucleophilic atom reduce the reactivity of a nucleophile because these groups hinder the approach of the nucleophile toward the electrophilic atom. For example, consider the relative reactivities of the methoxide and *tert*-butoxide ions:

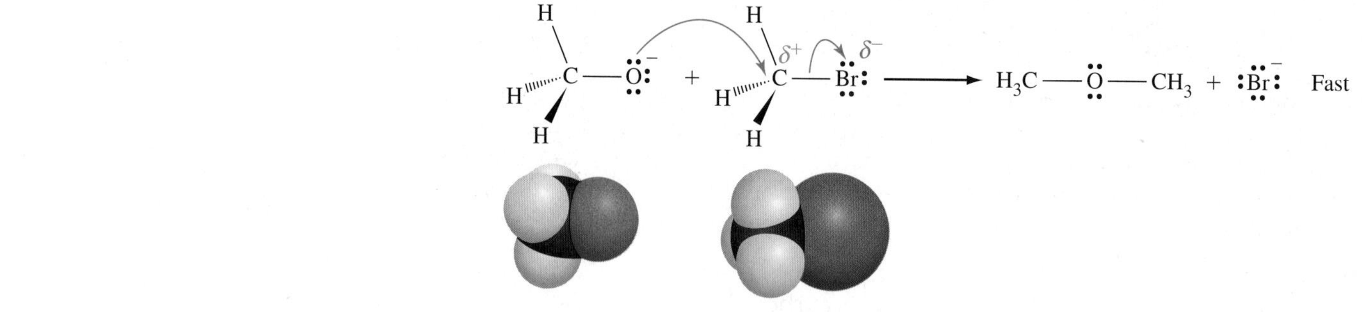

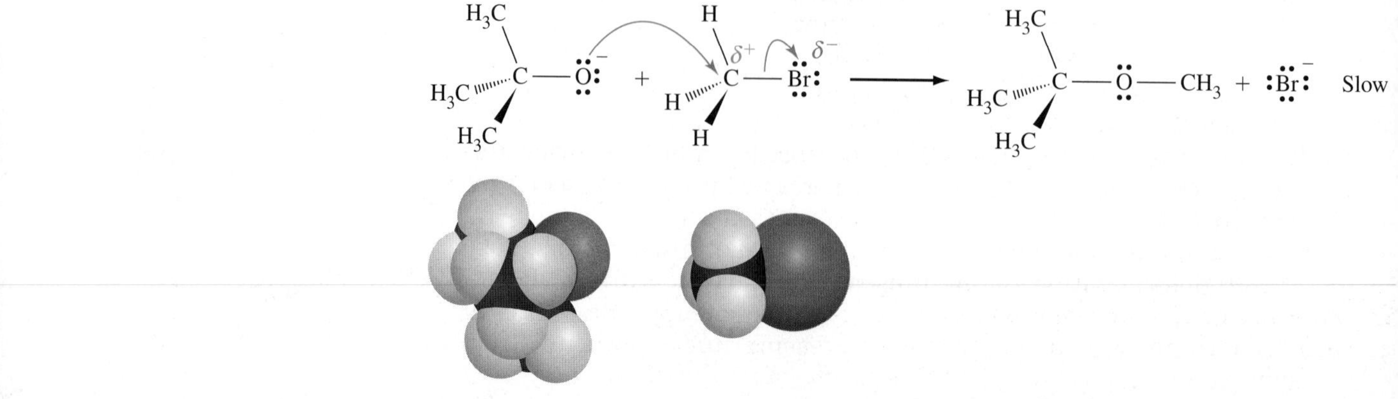

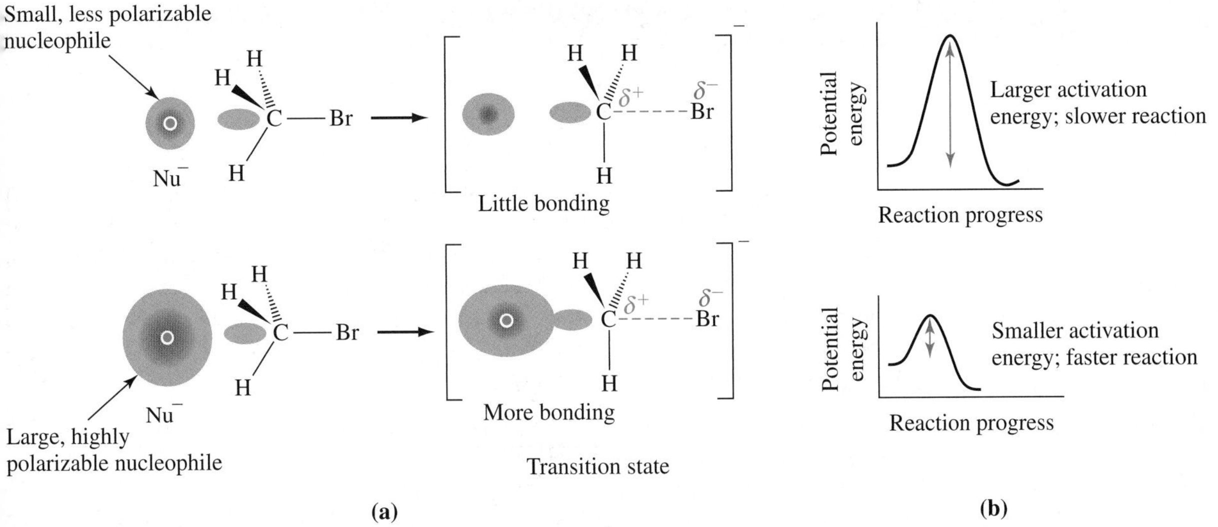

▲ FIGURE 27-9
The effect of the polarizability of the nucleophile in an S_N2 reaction
For simplicity, the nucleophile is represented by a spherical charge distribution. **(a)** The formation of the transition state in the S_N2 mechanism involves transfer of electron density from the nucleophile to a vacant orbital localized on C. For the larger and more polarizable nucleophile, the transfer of electron density occurs sooner (at larger distances) and the resulting transition state is of lower energy, as shown in **(b)**. Because the rate of reaction increases as the activation energy decreases, the more polarizable nucleophile reacts faster. Keep in mind that solvent-nucleophile interactions (not shown) also affect the ability of the nucleophile to bond with the electrophilic carbon atom.

As indicated by the space-filling models of the $(CH_3)_3CO^-$ and CH_3O^- ions, the nucleophilic oxygen atom of $(CH_3)_3CO^-$ is significantly hindered from closely approaching the electrophilic carbon atom of CH_3Br but the oxygen atom of CH_3O^- is not. Consequently, $(CH_3)_3CO^-$ is a weaker nucleophile than CH_3O^-.

In Table 27.1 we classify some commonly encountered nucleophiles as excellent (strong), good, and fair. However, it is important that we use the information in this table cautiously. The preceding discussion makes it clear that the nucleophilicity of an electron pair donor is influenced by several factors, only some of which are related to the structure of the nucleophile itself. Nucleophilicity depends also on the solvent used and the nature of the electrophile. A given electron pair donor may behave as a strong nucleophile in a particular situation, but it may not be a strong nucleophile in another.

A Summary of S_N1 and S_N2 Reactions

In this section, we considered two different mechanisms by which a substitution reaction at an sp^3 hybridized carbon atom occurs. It is important to remember that (1) S_N1 and S_N2 reactions compete with each other, and (2) the exact details of any particular reaction might place it somewhere between these two mechanistic extremes. With an appropriate selection of reactants and reaction conditions, we might be able to push a particular reaction toward one mechanism or the other. Ultimately, the molecules in a system will always react by the lowest energy pathway. Depending on the reactants and reaction conditions, the lowest energy pathway for a substitution reaction may be the S_N1 reaction or the S_N2 reaction or something in between these two extremes. Table 27.2 summarizes some key ideas from this section and identifies the combinations of reactants and reaction conditions that push a substitution reaction toward either the S_N1 or S_N2 mechanism.

TABLE 27.1 Classification of Common Nucleophiles

Excellent Nucleophiles		
		Rate[a]
NC^-	Cyanide	126,000
HS^-	Thiolate	126,000
I^-	Iodide	80,000
Good Nucleophiles		
		Rate
HO^-	Hydroxide	16,000
Br^-	Bromide	10,000
N_3^-	Azide	8,000
NH_3	Ammonia	8,000
NO_2^-	Nitrite	5,000
Fair Nucleophiles		
		Rate
Cl^-	Chloride	1,000
CH_3COO^-	Acetate	630
F^-	Fluoride	80
CH_3OH	Methanol	1
H_2O	Water	1

[a]The rate is a relative rate. A rate of 100 means that the nucleophile reacts 100 times as fast as water does.

TABLE 27.2 Relative Reactivities of Haloalkanes

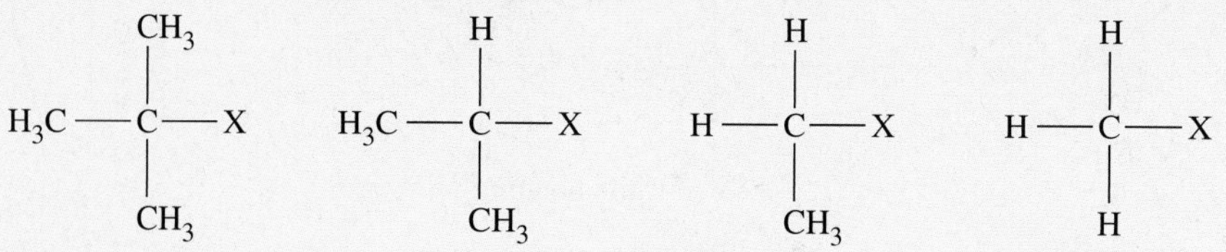

Electrophile	3° 2°	1° Methyl
Stability of Carbocation	Forms a relatively stable carbocation	Form relatively unstable carbocations
S_N1 Reactivity	← increasing S_N1 reactivity	No S_N1
S_N2 Reactivity	No S_N2	increasing S_N2 reactivity →
α Carbon	Sterically hindered	Not sterically hindered
Solvent	Use a polar protic solvent to promote the S_N1 reaction	Use a polar aprotic solvent to promote the S_N2 reaction

▲ The symbols 1°, 2°, and 3° stand for primary, secondary, and tertiary, respectively.

EXAMPLE 27-4 Predicting Whether a Substitution Reaction Proceeds by an S_N1 or S_N2 Mechanism

Does the substitution reaction below follow an S_N1 or an S_N2 mechanism? What are the products? Write the steps of the mechanism and use arrows to show the movement of electrons.

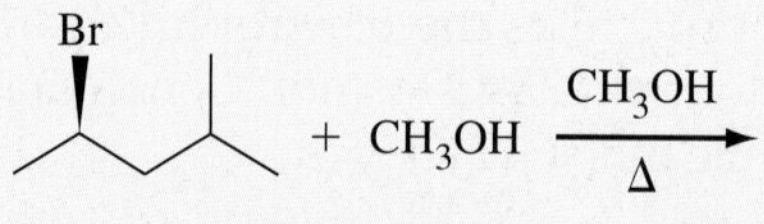

(*R*)-2-bromo-4-methylpentane

Analyze

The haloalkane is a 2° haloalkane. Secondary haloalkanes undergo substitution reactions by either the S_N1 or S_N2 depending on the nucleophile and solvent. (See Table 27.2.)

Solve

Methanol is the nucleophile and the haloalkane is the electrophile. Because the nucleophile is uncharged, its nucleophilicity is determined primarily by the polarizability of the nucleophilic atom (O). The O atom is relatively small and not very polarizable; thus, CH_3OH is a weak nucleophile. A weak nucleophile disfavors an S_N2 reaction. Also, the solvent is polar protic and will help to stabilize a carbocation. With a weak nucleophile and a polar protic solvent, we expect the substitution reaction to occur by an S_N1 mechanism. The carbocation that is formed reacts with a solvent molecule CH_3OH (a *solvolysis reaction*) to form a protonated ether. The final product is an ether, which is obtained when a proton is transferred from the protonated ether to a CH_3OH molecule from the solvent. We will obtain two products, the *R* and *S* stereoisomers, because CH_3OH can attack the carbocation from either side. The steps are as follows:

Step 1: Formation of a carbocation

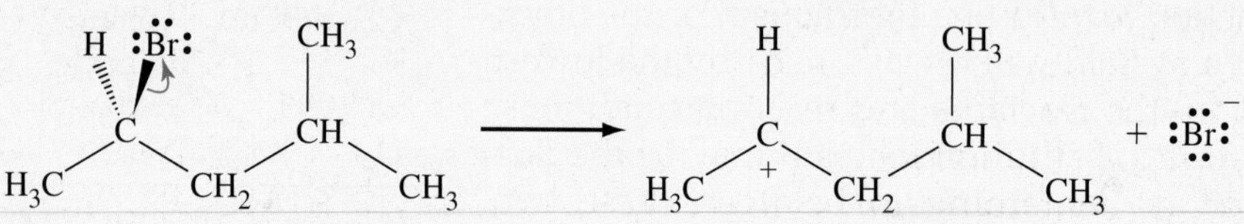

Step 2: Nucleophilic attack by CH_3OH

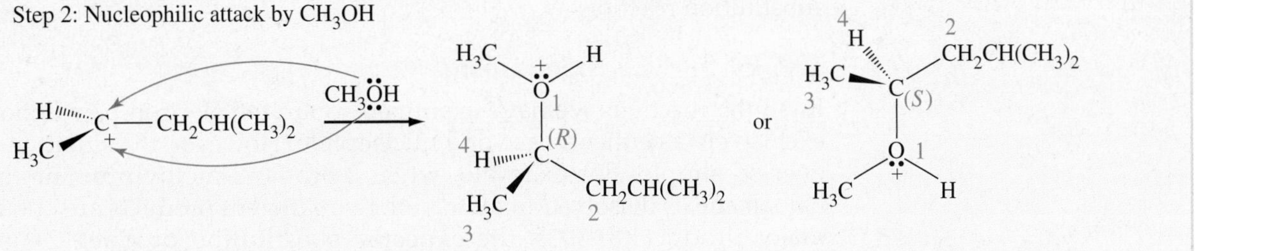

Step 3: Loss of proton to solvent (ignoring stereochemistry)

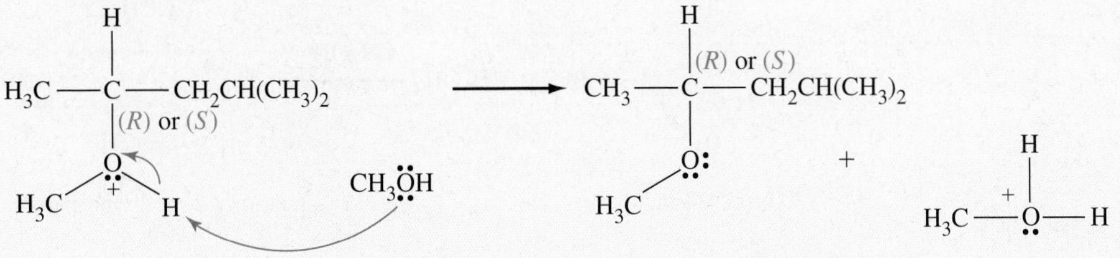

Thus, the reaction will produce a racemic mixture consisting of the (*R*) and (*S*) stereoisomers of 2-methoxy-4-methylpentane.

Assess

To name the products, you may find it helpful to review the nomenclature rules given in Chapter 26. The reaction considered in this example is also called a solvolysis reaction, because the solvent acts as the nucleophile.

PRACTICE EXAMPLE A: Predict whether the substitution product obtained in the reaction below follows an S_N1 or an S_N2 mechanism. Show all steps in the mechanism, using arrows to indicate the movement of electrons. Name the product.

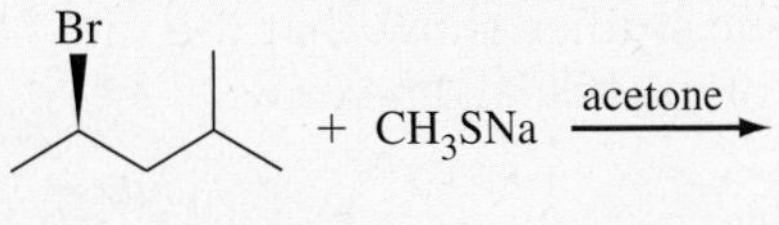

(*R*)-2-bromo-4-methylpentane

PRACTICE EXAMPLE B: Does the substitution reaction below occur by an S_N1 or an S_N2 mechanism?

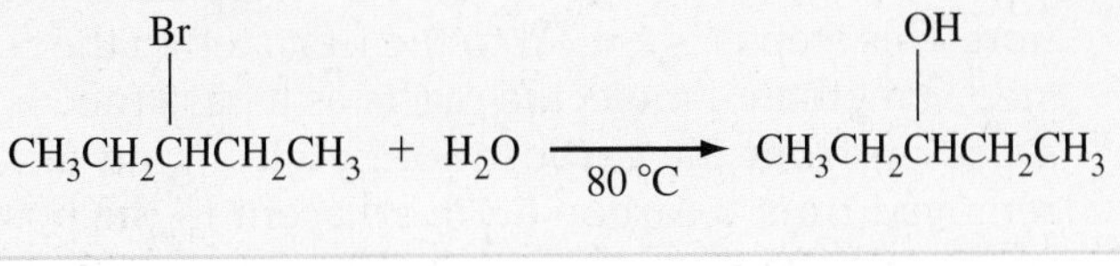

27-3 Introduction to Elimination Reactions

In the previous section, we saw that a haloalkane can undergo a substitution reaction in which the halogen atom is replaced by another group. Haloalkanes can also undergo elimination reactions, in which the halogen atom and a hydrogen atom bonded to the β carbon are removed from the molecule.

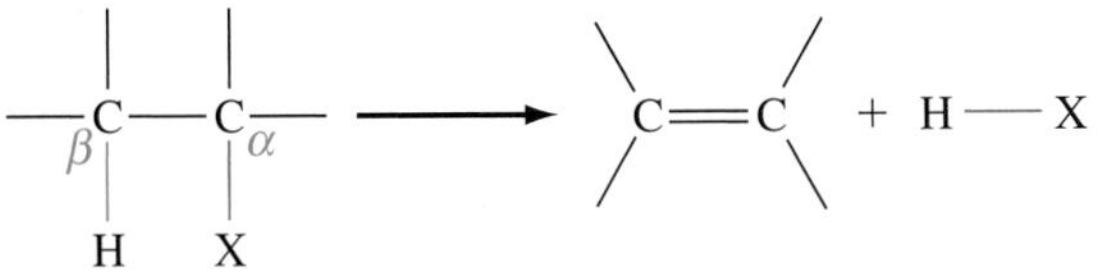

(27.3)

As is true for substitution reactions, elimination reactions of haloalkanes can occur by different mechanisms. If the rate-determining step is unimolecular, then the mechanism is called **E1**. If the rate-determining step is bimolecular, the mechanism is called **E2**. In this section, we will explore these mechanisms in some detail and discuss the competition that occurs between elimination and substitution reactions.

The E1 and E2 Mechanisms

In all the reactions we have examined so far, the electron pair donor reacted exclusively as either a base or a nucleophile. However, the situation is hardly ever so simple. For example, when 2-bromo-2-methylpropane, a tertiary haloalkane, is dissolved in methanol, two different products are obtained. The major product (81%) is the expected substitution product, 2-methoxy-2-methylpropane. The reaction proceeds via the S_N1 mechanism.

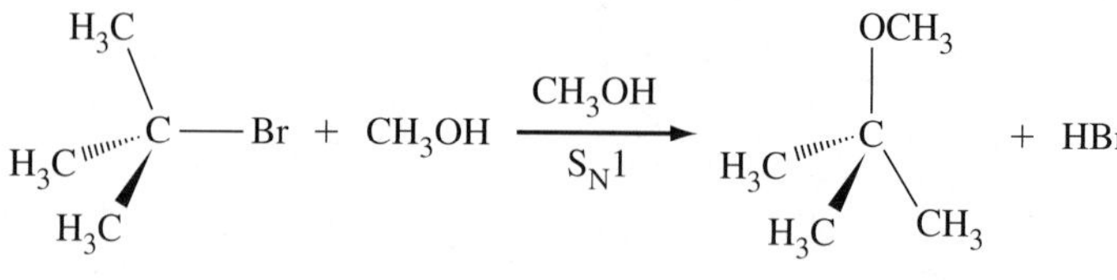

The minor product (19%) is 2-methylprop-1-ene. A balanced chemical equation for the formation of the minor product is given below:

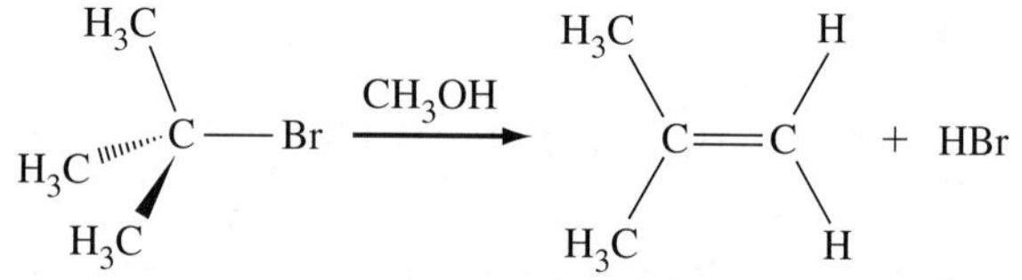

Kinetic studies show that the rate of this reaction depends on the concentration of the haloalkane only:

$$rate = k[(CH_3)_3CBr]$$

Because the reaction is an elimination reaction following first-order kinetics, it is called an **E1 reaction**. The mechanism for this E1 reaction is shown in Figure 27-10. The first step is a slow step involving the formation of a carbocation. This step is identical to the first step in the S_N1 mechanism. In the second step, a methanol molecule reacts as a base (not as a nucleophile) and removes a proton from the carbocation, yielding an alkene. Notice that a proton is removed from a carbon atom adjacent to the positively charged carbon atom of the carbocation, a β carbon. The reaction profile is shown in Figure 27-11.

When we compare the E1 mechanism to the S_N1 mechanism (Fig. 27-3), we see why the reaction of $(CH_3)_3CBr$ and CH_3OH yields a mixture of two

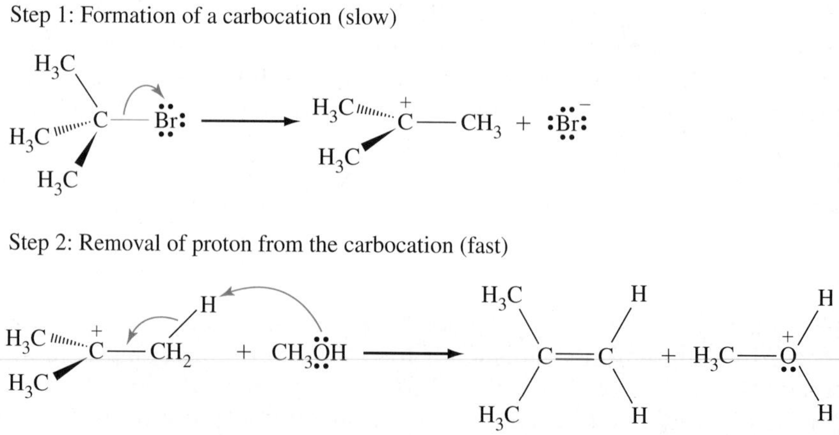

Step 1: Formation of a carbocation (slow)
Step 2: Removal of proton from the carbocation (fast)

▶ FIGURE 27-10
The E1 mechanism
The first step is the formation of a carbocation. It is slow and rate determining. The second step is an acid–base reaction in which a proton is removed from a carbon atom adjacent to the positively charged carbon atom of the carbocation.

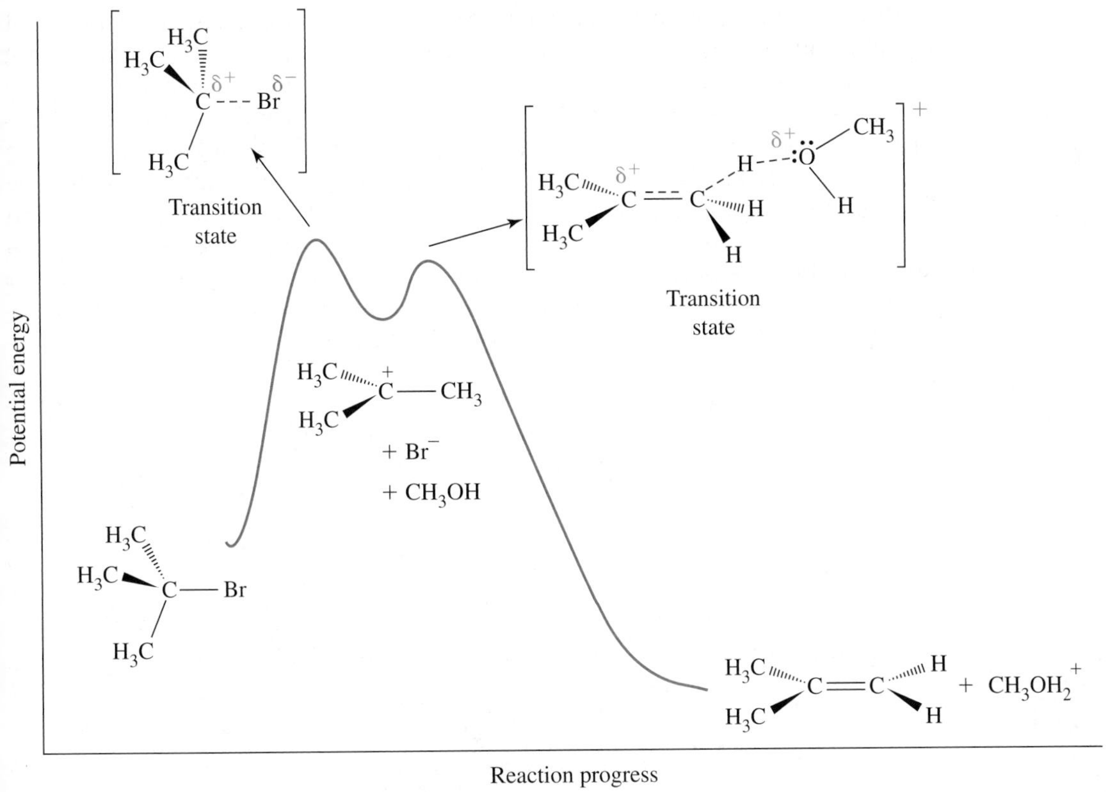

▲ FIGURE 27-11
Reaction profile for the E1 reaction between $(CH_3)_3CBr$ and methanol
The first step is a highly endothermic process involving the formation of a carbocation. In the transition state for the first step, the C—Br bond is partially broken. In the second step, CH_3OH acts as a base and removes a proton from the β carbon. In the transition state for the second step, a new π bond is forming between two C atoms. Simultaneously, a C—H bond is breaking and a new H—O bond is forming.

products. After the formation of the carbocation, methanol can act like a nucleophile and attack the electrophilic carbon atom (producing the S_N1 product) or it can act as a base and remove a proton to produce an alkene. The essential difference between the S_N1 and E1 mechanism is the role played by the electron pair donor in the second step. In the S_N1 mechanism, the electron pair donor acts as a nucleophile and forms a σ bond with the electrophilic carbon atom of the carbocation. In the E1 mechanism, the electron pair donor acts as a base and forms a σ bond with a proton.

In the example we just considered, the nucleophile was CH_3OH. The CH_3OH molecule is a weak nucleophile (it is a neutral molecule and the nucleophilic atom is relatively small and not very polarizable) and also a weak base. Let us consider what would happen in the reaction of $(CH_3)_3CBr$ and CH_3CH_2ONa in ethanol, bearing in mind that the $CH_3CH_2O^-$ ion is a stronger base than CH_3OH. We anticipate that the amount of alkene in the product mixture will increase, and as the equation below shows, this is in fact what is observed:

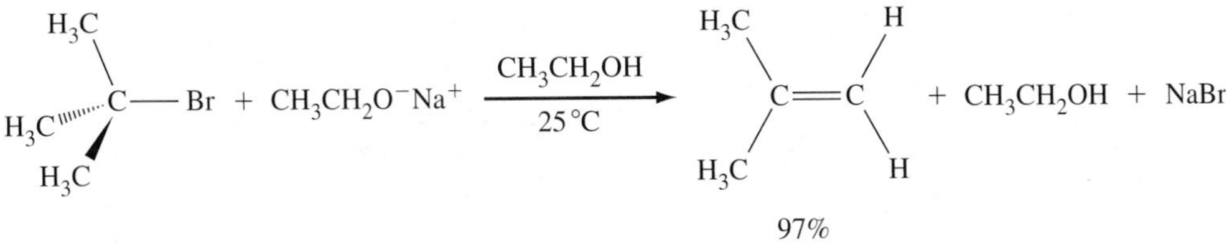

A study of the kinetics of this reaction reveals, however, that the rate of elimination depends on the concentrations of the substrate and the base:

$$rate = k[(CH_3)_3CBr][CH_3CH_2O^-]$$

Whereas the use of a stronger base promotes elimination at the expense of substitution, we can see that the mechanism of the elimination reaction cannot be the E1 mechanism. (Recall that the rate of an E1 reaction depends only on the concentration of the substrate.) Because the kinetics of this elimination reaction are second order, the rate-determining step must be bimolecular. Thus, the reaction is called an **E2 reaction**. Detailed studies of E2 reactions reveal that the E2 mechanism consists of a single step that proceeds through a single transition state where three changes are occurring simultaneously: (1) removal of a proton from the β carbon; (2) departure of the leaving group; (3) formation of a π bond between the α and β carbon atoms. The reaction profile for the E2 reaction between $(CH_3)_3CBr$ and $CH_3CH_2O^-$ is illustrated in Figure 27-12.

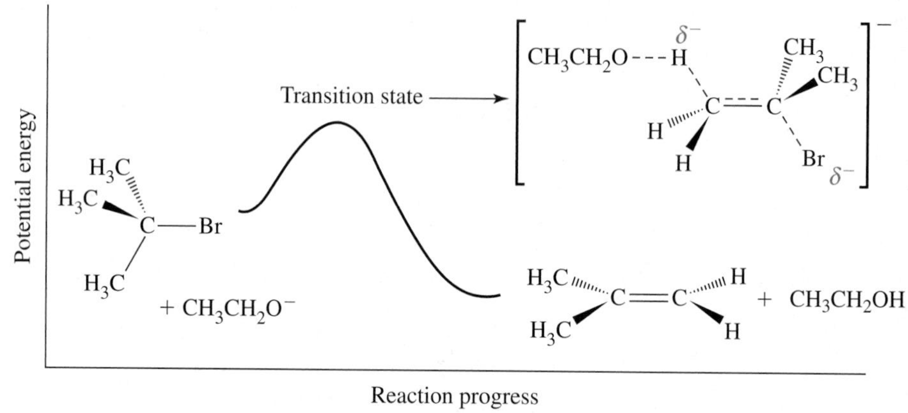

▶ FIGURE 27-12
Reaction profile for the E2 reaction of $(CH_3)_3CBr$ and $CH_3CH_2O^-$ in ethanol

27-2 CONCEPT ASSESSMENT

One product in the reaction of $(CH_3)_3CBr$ and $NaCH_3CH_2O$ in ethanol is an alkene. What is the other product and by what mechanism is it formed?

Predicting the Major Elimination Product

In the elimination reactions that we have studied so far, only one elimination product was possible. However, it is often the case that more than one elimination product can be produced. For example, in the reaction of 2-bromobutane with sodium methoxide in methanol, three E2 products are possible:

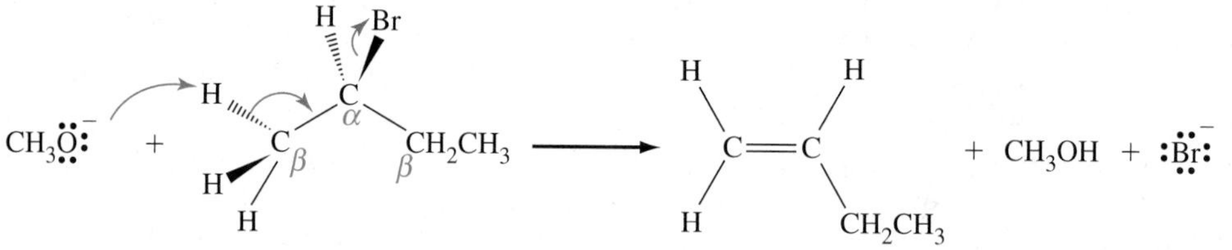

but-1-ene (18%) **(27.4a)**

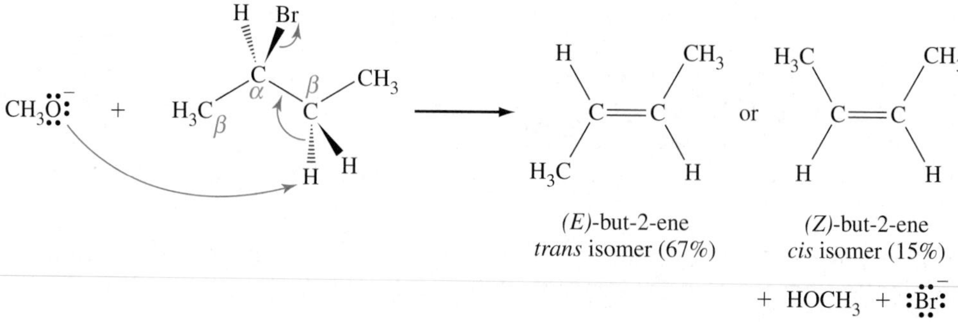

(27.4b)

KEEP IN MIND

that the product mixture for an elimination reaction will be richer in the product that is formed the fastest. The relative rates of the elimination reactions are determined by the energies of the corresponding transition states. For an elimination reaction, the transition state has some double bond character, and so the factors known to stabilize an alkene will also stabilize the transition state. As a consequence, the pathway that leads to the more stable alkene usually also passes through the transition state of lowest energy; thus, the reaction leading to the most stable alkene is not only the most exothermic but also the fastest.

In but-1-ene, the double bond is at the end of the carbon chain (a terminal C=C bond) and in the *cis* and *trans* isomers of but-2-ene, the C=C bond is an internal double bond. Notice that there is a strong preference or selectivity for forming the internal double bond (82% versus 18%).

When more than one elimination product is possible, can we reliably predict which will be the major product? Usually, the major product of an elimination reaction is the one that is most stable. But what can we say about the relative stabilities of alkenes? In the diagram below, alkenes are ranked in order of increasing stability:

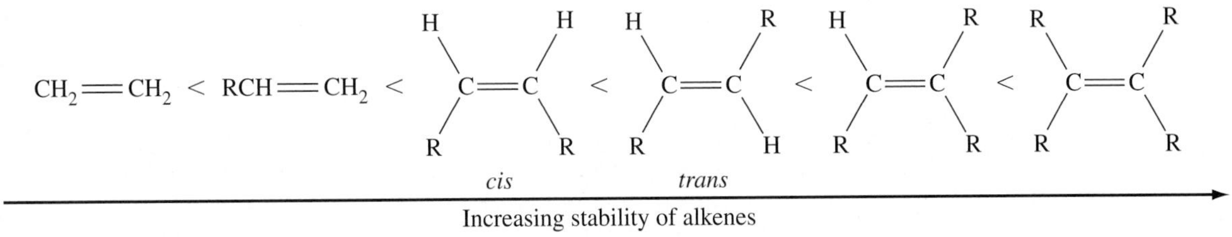

This order of stability is based on experimental data (heats of hydrogenation). The following statement summarizes the trend in stabilities:

> The greater the number of alkyl substituents bonded to the *sp*^{2} carbons of an alkene, the greater its stability.

◀ The explanation of why the stability of an alkene increases with the degree of alkyl substitution is a matter of some conjecture. A common explanation is based on the concept of *hyperconjugation*. According to IUPAC, in the formalism that separates bonds into σ and π types, hyperconjugation is the interaction of σ bonds (e.g., C—H, C—C, etc.) with a π network. Other explanations are based on the interactions of filled and empty molecular orbitals and call into question the appropriateness of using hyperconjugation as an explanation. Even in the absence of an explanation, the experimental facts are unambiguous: The stability of an alkene increases with the number of alkyl groups on the carbon atoms of the double bond.

In general, more highly substituted alkenes are more stable than less highly substituted alkenes. Also, bulky alkyl groups like to be as far as apart as possible, and so, for example, the *trans* isomer is more stable than the *cis* isomer.

Let's return to reaction (27.4) and rationalize the product distribution that is obtained. The alkenes having an internal double bond are the major products because they are more highly substituted and the transition states leading to these alkenes are lower in energy than the transition state leading to the less highly substituted alkene. Thus, the alkenes having the internal double bonds are formed the fastest and are the major components of the product mixture. Of the two alkenes having an internal double bond, the *trans* stereoisomer is the more stable (and the transition state leading to this stereoisomer is lower in energy) and thus, the *trans* isomer is the major product.

A word of caution is in order. Often, the major product in an elimination reaction is the most highly substituted alkene but important exceptions exist to this rule. The formation of the most highly substituted alkene requires that the base remove a proton from a secondary or tertiary carbon atom. Access to the hydrogen atoms on a secondary or tertiary carbon atom is hindered by the adjacent groups. Small bases, such as OH^-, CH_3O^-, or $CH_3CH_2O^-$, can get past these groups fairly easily but larger bases, such as $(CH_3)_3CO^-$, cannot. A large base will preferentially attack the more exposed hydrogen atoms, yielding the least highly substituted alkene. In the following reactions, the same substrate reacts with different bases. In the first reaction, a smaller, less hindered base is used and the major elimination product is the more highly substituted alkene. In the second reaction, a larger, more hindered base is used and the major elimination product is the less highly substituted alkene.

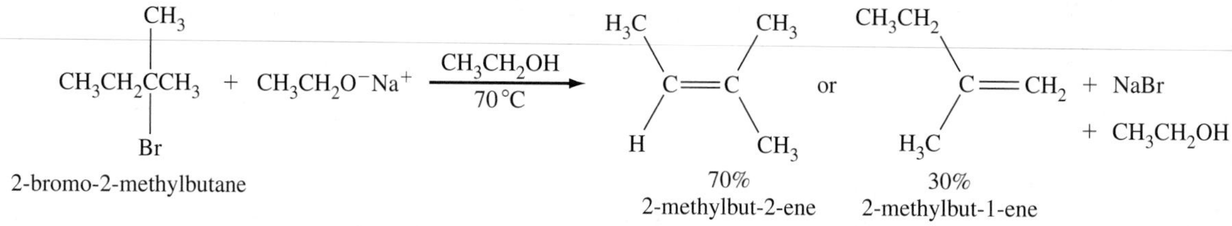

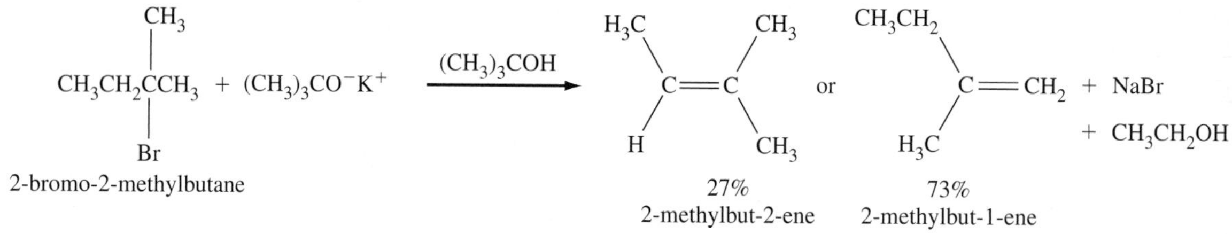

For an E2 reaction, the H atom on the β carbon must be "anti" to the leaving group. This requirement places a restriction on the elimination products that can be formed from a 2° haloalkane. For a more detailed discussion and a worked example involving an elimination reaction, see Section 27B, *A Closer Look at the E2 Mechanism*, on the MasteringChemistry site (www.masteringchemistry.com).

MasteringCHEMISTRY

Substitution and Elimination Reactions: A Summary

We have seen that haloalkanes can undergo a variety of reactions: S_N2, S_N1, E2, or E1. In principle, all these reactions compete with one another, and a variety of products are possible. When determining whether a reaction will proceed through an S_N2, S_N1, E2, or E1 mechanism, we must consider many factors. What is the nature of the electron pair donor? Is it a good nucleophile? Is it a strong base? Is it sterically hindered? What is the nature of the electrophile? Is it sterically hindered? Does it have a good leaving group? And what about the solvent? Does it promote carbocation formation (S_N1/E1 reactions)? Does it increase or decrease the nucleophilicity of the electron pair donor?

With all these factors to consider, it is natural to ask: Is it possible to predict reliably the outcome of a reaction and the mechanisms that are involved? The answer to this question is a tentative yes. There are some guiding principles, which we present in Figure 27-13, but remember that exceptions to these guidelines exist. To make the best use of these guidelines, the following stepwise approach is recommended:

1. Show all lone pairs, formal charges, and partial charges on atoms that have them.
2. Identify the electron pair donor and the electron pair acceptor (the electrophile).
3. Consider the electrophile. Is it primary, secondary, or tertiary?
4. Consider the electron pair donor. Is it a strong or weak nucleophile? Is it a strong or weak base? Is it sterically hindered?
5. Is the solvent protic or aprotic?
6. Determine whether the dominant reaction will be S_N2, E2, or S_N1 and E1. (Remember that S_N1 and E1 always occur at the same time.)

Before we attempt to apply this approach to specific cases, let's have a look at the guidelines summarized in Figure 27-13. The guidelines are presented in the form of a decision tree, with the first consideration being the base strength of the electron pair donor, :B. Additional considerations include steric hindrance, nucleophilicity, and solvent effects.

For primary alkanes (Figure 27-13a), the possible reactions are S_N2 and E2. An S_N1 reaction is not a likely possibility because primary carbocations are not

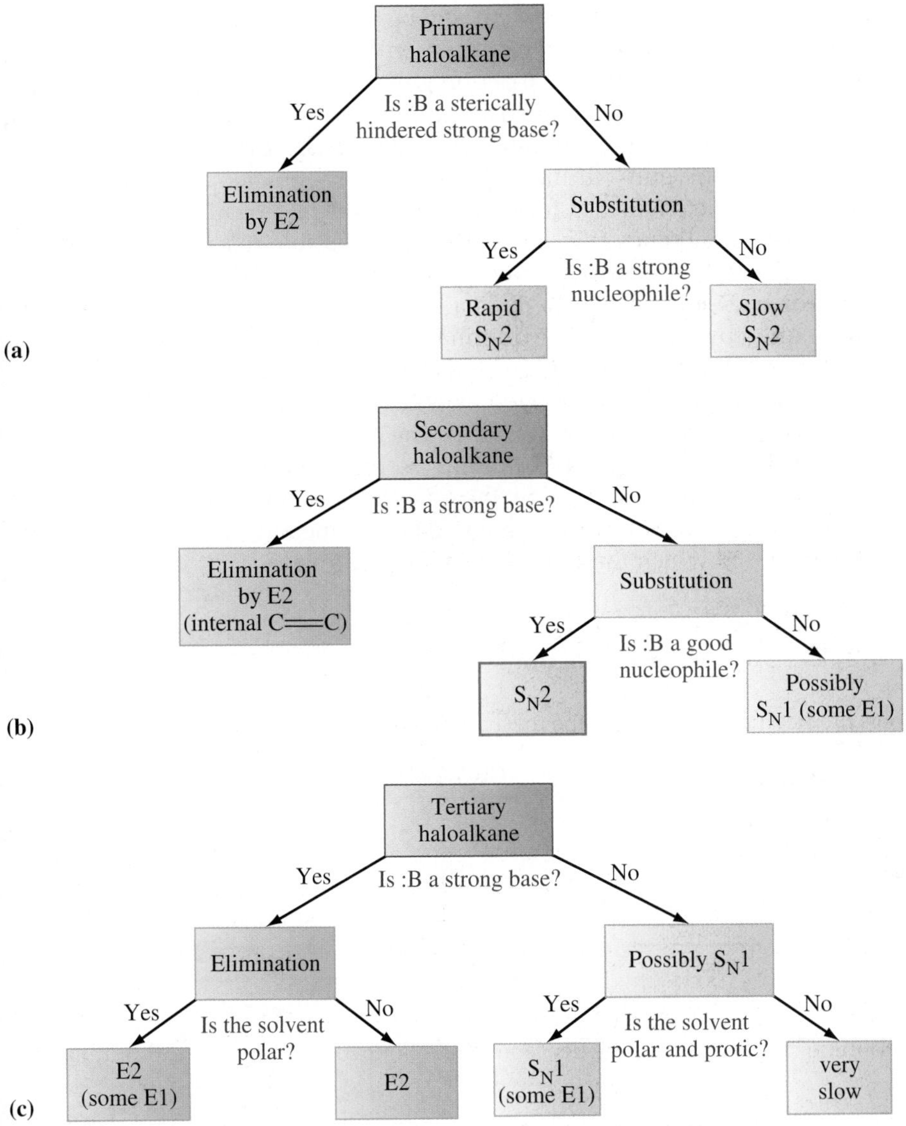

▲ FIGURE 27-13
Summary of S_N2, E2, S_N1, and E1 reactions for primary, secondary, and tertiary haloalkanes
(a) Primary haloalkanes undergo S_N2 or E2 reactions. **(b)** Secondary haloalkanes undergo S_N2, E2, S_N1, and E1 reactions. The red box indicates there are additional considerations. For an alkene with internal C=C bond, the possibility of (*E*) and (*Z*) stereoisomers exists. When a secondary haloalkane undergoes an S_N2 reaction, watch for an inversion of configuration, (R) ↔ (S). In polar protic solvents, a secondary haloalkane can react with a weak nucleophile to give S_N1 and E1 products. **(c)** Tertiary haloalkanes undergo S_N1, E1, and E2 reactions. Methyl halides (CH_3X) are not included because the methyl halides undergo only S_N2 reactions.

stable. If the electron donor, :B, is a sterically hindered strong base, such as R_3CO^-, then the E2 reaction will dominate because the electron donor will not be able to attack the electrophilic carbon atom from the backside. Instead, a proton will be removed from a β carbon. If :B is not sterically hindered, then it will act as a nucleophile and attack the electrophilic carbon from the backside in an S_N2 reaction. If :B is a strong nucleophile, such as I^-, ^-CN, RS^-, or RO^-, then the S_N2 reaction will occur quite rapidly. If :B is a weak nucleophile, then the S_N2 reaction will occur rather slowly.

The greatest range of possibilities occurs for secondary haloalkanes (Fig. 27-13b). If the electron donor is a strong base, then elimination by E2 is the main reaction and the major product is an alkene with an internal C=C bond. If :B is a weak base but a strong or good nucleophile, such as I^-, ^-CN,

RS^-, or RCO_2^-, then the main reaction will be S_N2, particularly if a polar aprotic solvent is used. It is important to remember that if the α carbon in the haloalkane is chiral, then the S_N2 reaction will invert the configuration of the α carbon. If :B is a weak base and also a poor nucleophile, such as H_2O or ROH, then the S_N1 reaction will dominate provided a polar protic solvent is used. Molecules of the solvent will most likely act as the nucleophile in this case. Typically, E1 products will also form along with the S_N1 products.

For tertiary haloalkanes (Figure 27-13c), the possible reactions are S_N1, E1, or E2. An S_N2 reaction is not very likely because the α carbon is too sterically hindered for backside attack. If :B is a strong base, such as R_3CO^- or RO^-, then elimination by E2 will be the dominant reaction. If :B is a weak base (and a poor nucleophile), such as H_2O or ROH, then the main reaction will be S_N1 with some E1 products being formed as well, provided a polar protic solvent is used. It is important to remember that if the α carbon is chiral, the S_N1 products will consist of a mixture of (*R*) and (*S*) stereoisomers.

In Example 27-5, we illustrate how the guidelines summarized in Figure 27-13 can be used to predict the products that will be obtained in a given situation and the mechanisms by which these products will be formed.

EXAMPLE 27-5 Predicting S_N2, E2, S_N1, and E1 Reactions

For each of the following reactions, predict the products and the mechanisms by which the products are formed.

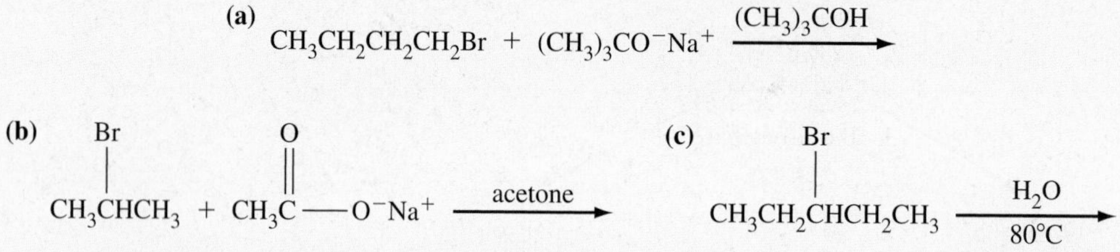

Analyze

We follow the six-step approach outlined above and use the decision tree in Figure 27-13 to guide our thinking.

Solve

(a) We identify the electrophile as $CH_3CH_2CH_2CH_2Br$, which is a primary haloalkane. The nucleophile is $(CH_3)_3CO^-$, a sterically hindered strong base.

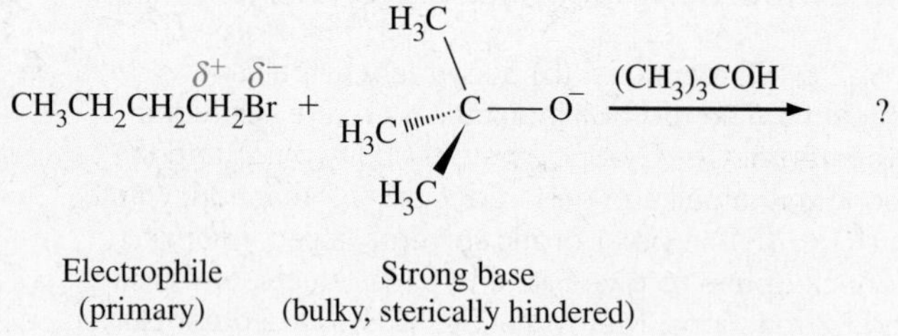

Although a strong base is usually a strong nucleophile, the bulkiness of this base disfavors substitution, and we expect that elimination by E2 will provide the major product.

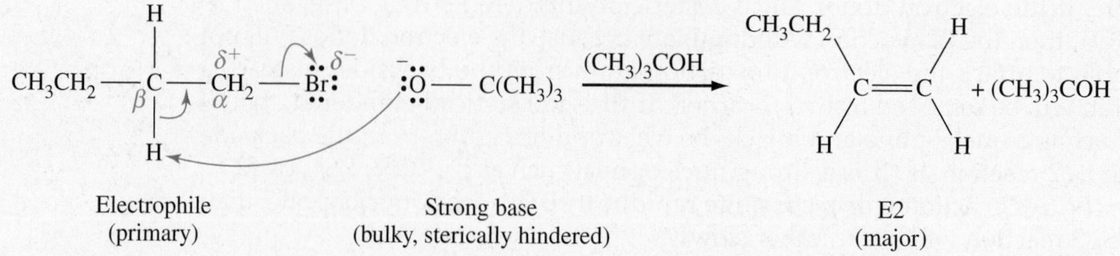

Substitution by S_N2 will yield $CH_3CH_2CH_2CH_2OC(CH_3)_3$, but the substitution product will be the minor product.

(b) The electrophile is a secondary haloalkane. The electron pair donor is a weak base (the conjugate base of a weak acid) and, because it is negatively charged, it is a reasonably good nucleophile, especially in a polar aprotic solvent. We expect that the main reaction will be S_N2.

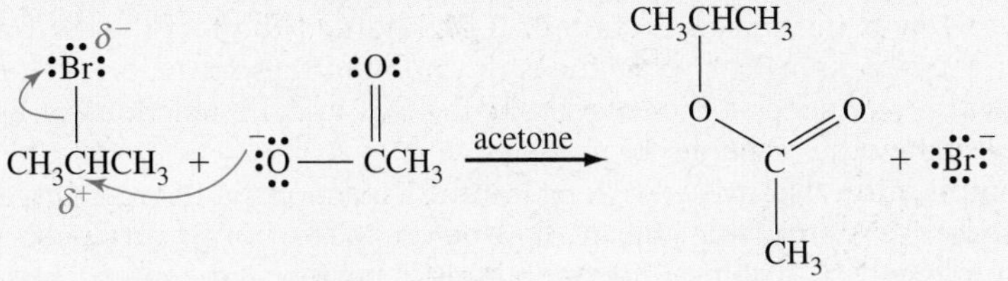

Electrophile (secondary) | Weak base (good nucleophile) | S_N2 product (major)

(c) Molecules of the solvent also serve as the electron pair donor. The reaction involves a secondary haloalkane with a weak base/weak nucleophile in a polar protic solvent. These conditions do not favor S_N2 or E2 but rather S_N1 and E1. S_N1 and E1 always occur together. Secondary carbocations are relatively stable, especially in a polar protic solvent, such as water. We expect both S_N1 and E1 products. Because H_2O is a very weak base and a fair nucleophile, substitution will dominate over elimination.

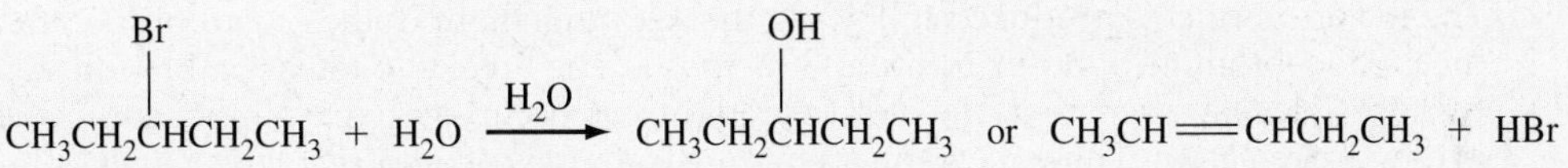

Electrophile (secondary) | Weak base (weak nucleophile) | S_N1 (major) | E1 (minor)

Assess

The mechanism for the reaction in (c) was not explicitly shown. We must make sure we are able to show the steps involved in forming the S_N1 and E1 products, including using arrows to show the movement of electrons. Also in (c), (*E*) and (*Z*) isomers of pent-2-ene are possible.

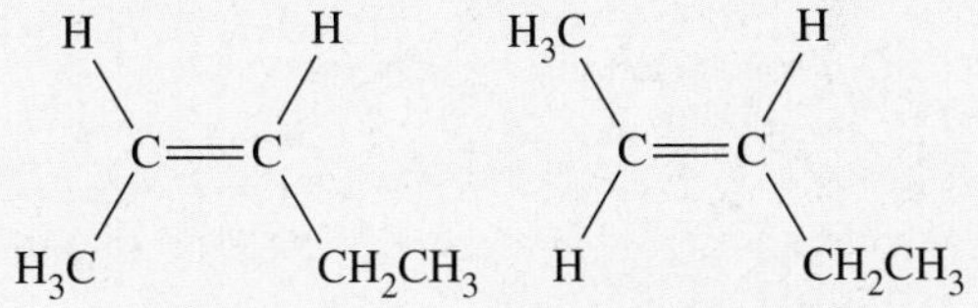

(*Z*)-pent-2-ene | (*E*)-pent-2-ene

PRACTICE EXAMPLE A: For the following reaction, predict the major product and the mechanism by which it is formed:

$$CH_3CH_2CH_2Br + CH_3S^-Na^+ \xrightarrow{\text{acetone}}$$

PRACTICE EXAMPLE B: Predict the substitution and elimination products that are possible in the following reaction. Which is the major product?

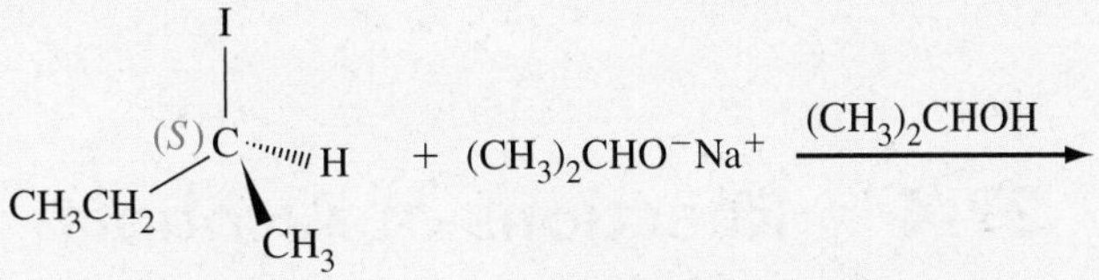

27-2 ARE YOU WONDERING?

Are S_N1/E1 reactions of haloalkanes used in organic synthesis?

The summary given at the end of Section 27-3 highlighted the fact that S_N1 and E1 reactions always occur together. We also know that once the carbocation in these reactions is formed, it is susceptible to attack from a variety of species. Consequently, the S_N1 and E1 reactions of haloalkanes are not often used for synthesizing other organic compounds.

Another complication that arises with S_N1 and E1 reactions is the rearrangement of the carbocation intermediate to form a more stable intermediate before the S_N1 or E1 product is formed. Such a situation arises, for example, when 2-bromo-3-methylbutane is heated with water. The major S_N1 product is not 3-methylbutan-2-ol but rather 2-methylbutan-2-ol.

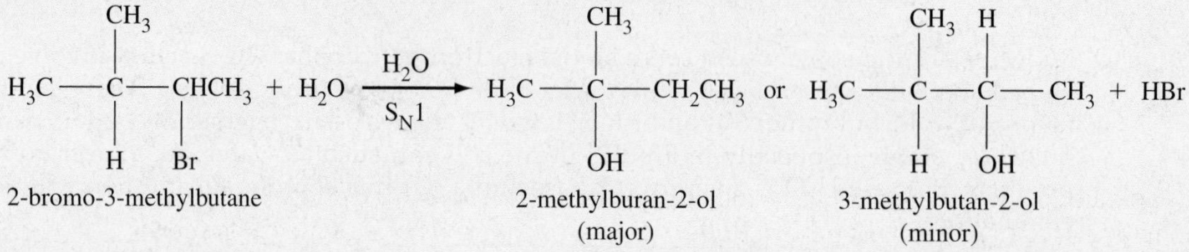

The major product is produced following the rearrangement of the 2° carbocation that is formed when the carbon–bromine bond in 2-bromo-3-methylbutane breaks to release a Br^- ion. As suggested below, two possible rearrangements, labeled (a) and (b), can occur. Both rearrangements involve the movement of a hydrogen atom. These rearrangements are examples of a *hydride shift*.

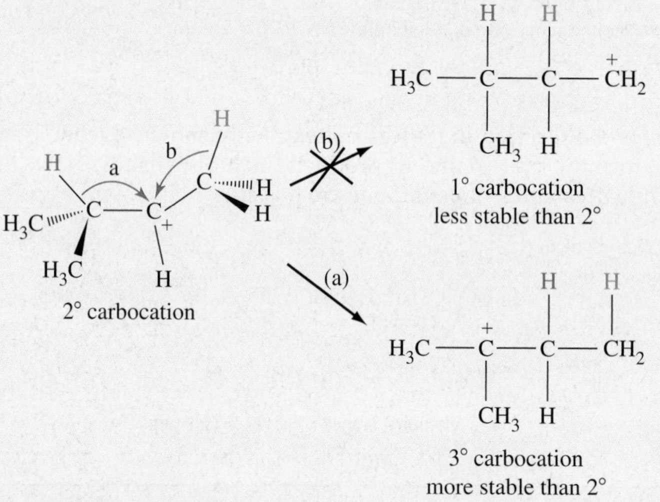

Rearrangement (b) does not occur because it results in a 1° carbocation, which is highly unstable. Rearrangement (a) converts a 2° carbocation into a more stable 3° carbocation. The 3° carbocation is the one that leads to the major substitution product.

Anytime a carbocation ion forms, rearrangement of the carbocation is always a possibility. To make matters worse, the rearrangement can also involve the movement of methyl groups or alkyl chains.

Given the myriad possibilities that can arise in S_N1 and E1 reactions, and the difficulty in predicting or controlling what will happen, it comes as no surprise that S_N1 and E1 reactions are not used as often as other reactions for organic synthesis.

27-4 Reactions of Alcohols

Alcohols figure prominently in organic synthesis because they can be readily converted into other compounds. In Chapter 26, we saw that primary alcohols could be oxidized to aldehydes or carboxylic acids and secondary alcohols could be oxidized to ketones. We also mentioned that alcohols react with carboxylic acids to form esters. In Section 27-2, we saw examples of reactions in

TABLE 27.3 Some Reactions of Alcohols

Type of Reaction	Equation
Deprotonation	$ROH + Na \longrightarrow RO^-Na^+ + \frac{1}{2}H_2$
Oxidation[a]	$ROH \xrightarrow{[O]}$ aldehyde, ketone or carboxylic acid
Esterification	$ROH + R'COOH \underset{\Delta}{\xrightarrow{H^+}} R'COOR + H_2O$
Substitution	$ROH^b + R'X \longrightarrow ROR' + HX$ $ROH^c + HX \longrightarrow RX + H_2O$
Elimination	$ROH \underset{\Delta}{\xrightarrow{H^+}}$ alkenes

[a]In the equation for oxidation, [O] represents the oxidizing agent, such as $Na_2Cr_2O_7/H_2SO_4$ or PCC/CH_2Cl_2. PCC is pyridinium chlorochromate. See page 1245.
[b]ROH is acting as a nucleophile in this reaction. See Section 27-2.
[c]ROH is acting as an electrophile in this reaction.

which an alcohol (ROH) reacted as a nucleophile in a substitution reaction with a haloalkane (R′X) to form an ether (ROR′). Some of the reactions of alcohols are summarized in Table 27.3.

In this section, we focus on reactions in which the —OH group is replaced by a halogen atom (in a substitution reaction) or eliminated as H_2O (in an elimination reaction).

Substitution and Elimination Reactions of Alcohols

The —OH group of an alcohol can be replaced by a halogen atom or eliminated as H_2O under appropriate conditions. Let's consider the feasibility of the following substitution reactions by using concepts we've discussed previously in this chapter.

$$CH_3CH_2CH_2OH + NaI \longrightarrow CH_3CH_2CH_2I + NaOH \quad \text{(does not occur)}$$
$$CH_3CH_2CH_2OH + HI \xrightarrow{\Delta} CH_3CH_2CH_2I + H_2O \quad \text{(occurs slowly)} \qquad (27.5)$$

In both of these reactions, I^- is the nucleophile and OH^- is the leaving group. Although the I^- ion is a good nucleophile, the OH^- ion is strongly basic and a poor leaving group. Consequently, the first reaction does not occur. Why does the second reaction occur? In the presence of strong acid, the oxygen atom of the —OH group is protonated, forming $R{-}^+OH_2$. For the protonated alcohol, the leaving group is H_2O instead of OH^-. Because H_2O is a much weaker base than OH^-, it is a much better leaving group. In reaction (27.5), the conditions are just right for an S_N2 reaction: the α carbon atom in $CH_3CH_2CH_2OH$ is primary (not sterically hindered) and I^- is a very good nucleophile. The mechanism of the reaction is shown in Figure 27-14. It involves a reversible protonation step followed by an S_N2 reaction. Recall that an S_N2 reaction is a concerted reaction involving backside attack at the α carbon.

KEEP IN MIND

that the weaker the base, the better it is as a leaving group. I^- is a very weak base and a good leaving group. OH^- is a strong base and a poor leaving group.

27-3 CONCEPT ASSESSMENT

Draw the structure of the transition state for the second step shown in Figure 27-14.

If a tertiary alcohol, such as $(CH_3)_3COH$, is used in place of the primary alcohol in reaction (27.5), the substitution occurring in the second step occurs by an S_N1 reaction, not by S_N2. Substitution occurs by S_N1 because the backside of

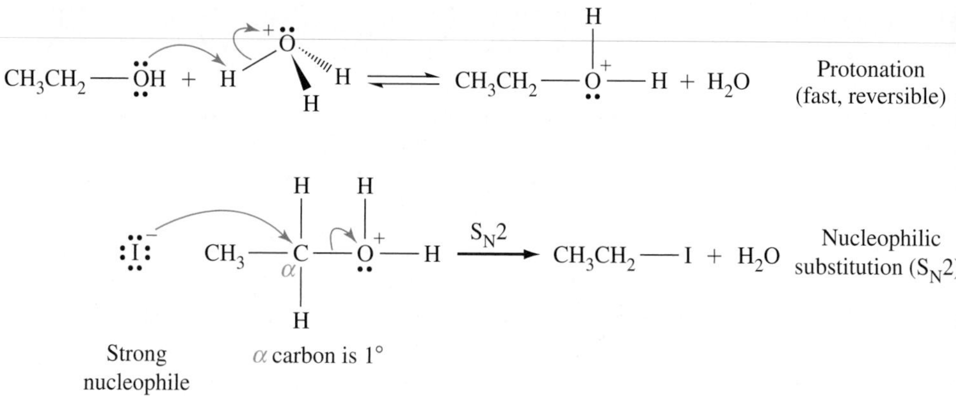

▲ FIGURE 27-14
Mechanism for the reaction of CH_3CH_2OH and HI
The first step is a reversible step in which the oxygen atom of the alcohol functional group is protonated. The second step is a nucleophilic attack of the α carbon. Because the α carbon is primary and I^- is a strong nucleophile, the substitution occurs by the S_N2 mechanism.

the α carbon in a tertiary alcohol is sterically hindered and inaccessible to a nucleophile. As shown in Figure 27-15, the substitution involves the formation of a carbocation followed by nucleophilic attack by I^-. The carbocation is stabilized by the electron-donating alkyl groups.

As we saw in the previous sections, elimination reactions compete with substitution reactions. So, it should come as no surprise that alcohols can also undergo elimination reactions. The elimination of water (H—OH) from an alcohol is an important method of synthesizing alkenes. The elimination of water from an alcohol is also called a **dehydration reaction**. The general form of a dehydration reaction is

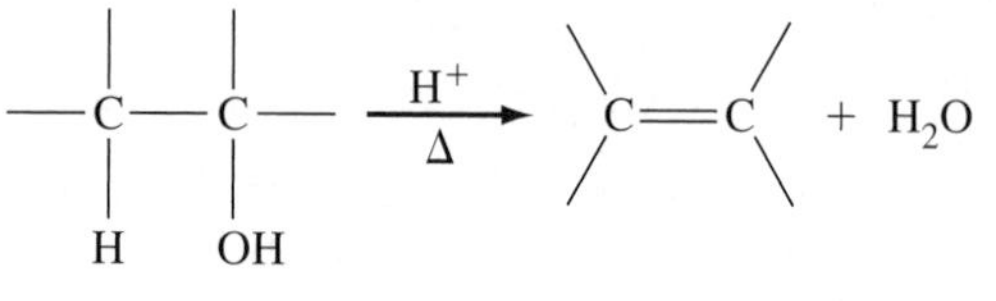

(27.6)

The dehydration of an alcohol requires an acid catalyst. The acid catalyst protonates the alcohol so that the leaving group will be H_2O (a good leaving group) rather than OH^- (a poor leaving group). To promote elimination (dehydration)

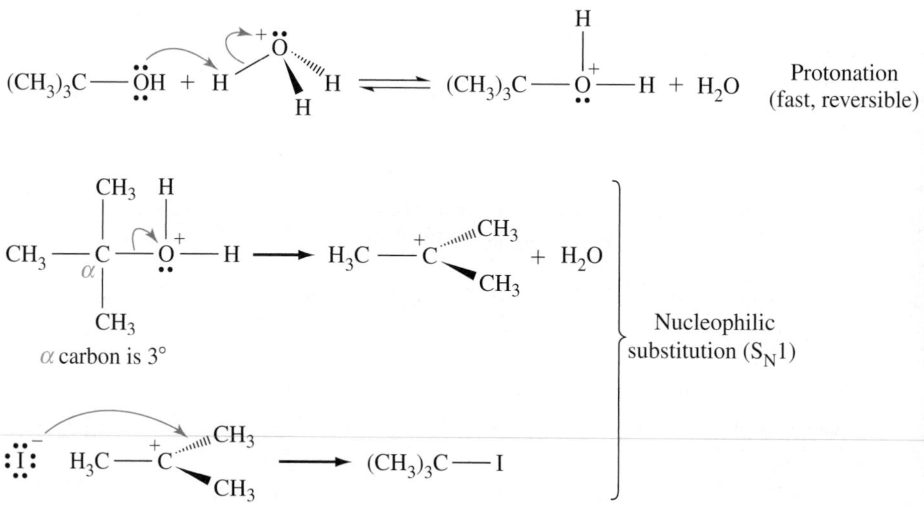

▶ FIGURE 27-15
Mechanism for the reaction of $(CH_3)_3COH$ and HI
The first step is a reversible step in which the oxygen atom of the alcohol functional group is protonated. Because the α carbon is tertiary, the backside of the α carbon is not susceptible to backside attack and so substitution by S_N2 does not occur. Instead, substitution occurs by S_N1.

over substitution, concentrated H_2SO_4 or H_3PO_4 is used rather than HI or HBr. (Remember, I^- and Br^- are strong nucleophiles, and if HI or HBr is used in place of H_2SO_4 or H_3PO_4, substitution will dominate over elimination.) Dehydration of tertiary alcohols occurs by an E1 reaction. Consider, for example, the dehydration of 2-methylbutan-2-ol, a tertiary alcohol:

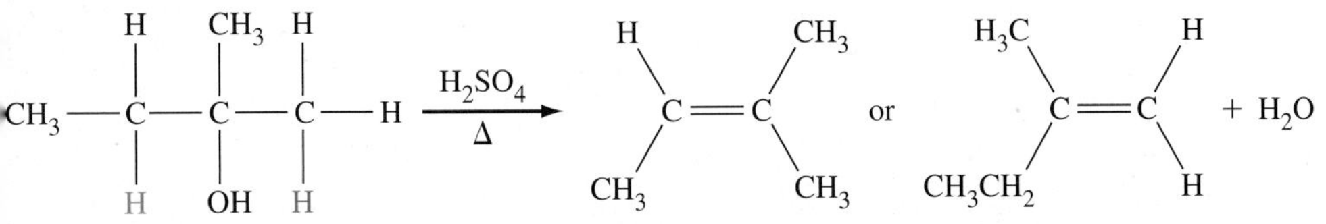

2-methylbutan-2-ol

2-methylbut-2-ene (major product)

2-methylbut-1-ene (minor product)

The mechanism for this reaction is shown in Figure 27-16. The first step (Fig. 27-16a) involves protonation of the alcohol. After protonation, dehydration occurs by E1 (Fig. 27-16b). Elimination of H—OH yields 2-methylbut-2-ene and elimination of HO—H yields 2-methylbut-1-ene. As discussed in Section 27-3, the major product in an elimination reaction is usually the more highly substituted alkene.

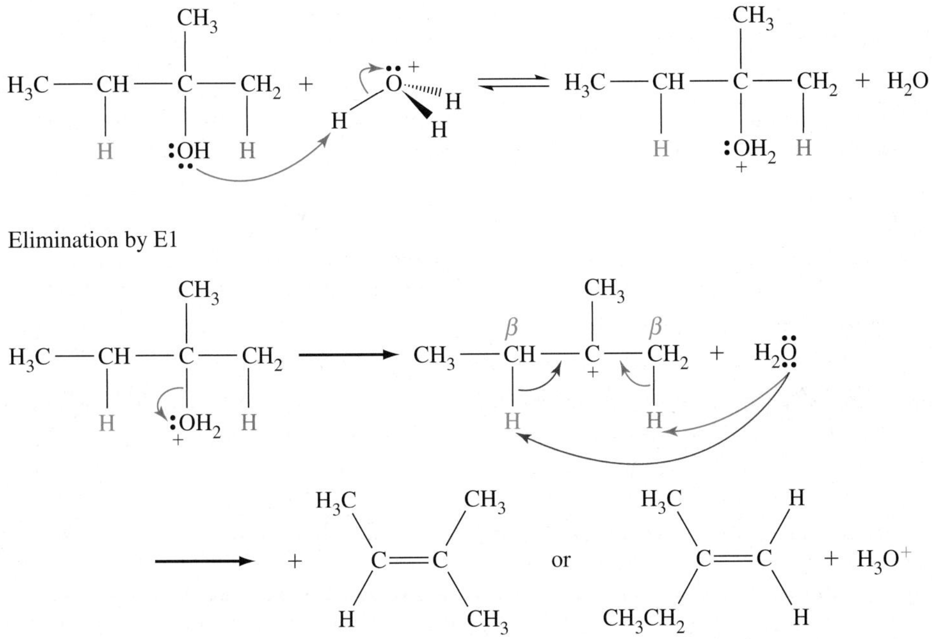

▲ FIGURE 27-16
Mechanism for the acid-catalyzed dehydration of 2-methylbutan-2-ol
(a) In the presence of strong acid, the alcohol is protonated. **(b)** Elimination involves the formation of a carbocation, followed by the removal of a proton from one of the β carbons. The blue arrows show the movement of electrons when H is attacked. The red arrows show the movement of electrons when H is attacked. The major product is the more highly substituted alkene. Notice that the H_3O^+ ion consumed in the protonation step is regenerated in the last step.

Dehydration of secondary alcohols usually occurs by E1, but it can occur by E2. Ethanol, a primary alcohol, undergoes dehydration by a concerted E2 reaction, as shown below:

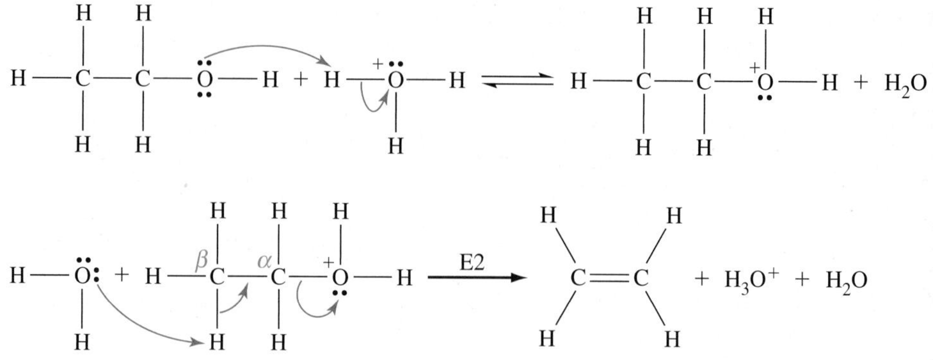

EXAMPLE 27-6 Predicting the Products of a Reaction Involving an Alcohol

Predict the products of the following reactions. If appropriate, suggest a mechanism by which the reaction occurs by using arrows to show the movement of electrons.

(a) propan-1-ol + sodium hydroxide $\xrightarrow{H_2O}$

(b) (*R*)-2-bromo-3-methylbutane + ethanol $\xrightarrow{\text{ethanol}}$

Analyze

First, we write condensed structural formulas for the reactants, and then we consider the role of the alcohol in each case. Alcohols can be oxidized or deprotonated. They can act as either a nucleophile or an electrophile in a substitution reaction or they can be dehydrated. To make a decision, we must consider the α carbon atoms in the reactants and classify them as primary, secondary, or tertiary. We must also consider the other reactants, the solvent, and the reaction conditions.

Solve

(a) The condensed structural formula for propan-1-ol is $CH_3CH_2CH_2OH$. It is a primary alcohol. Neither substitution nor elimination (dehydration) is possible under the conditions specified. We know that sodium hydroxide is a relatively strong base, and so we should consider the feasibility of the following acid–base reaction:

$$CH_3CH_2CH_2OH + NaOH \longrightarrow CH_3CH_2CH_2O^-Na^+ + H_2O$$

We must compare the strengths of the acids (or the bases) on either side of the equation. We know that ^-OH and $CH_3CH_2CH_2O^-$ are both considered strong bases, but which is stronger? Because both bases have the negative charge localized on oxygen, we expect them to be fairly similar in strength. Thus, we predict that the reaction will not go to completion. At equilibrium, significant amounts of all species will be present.

(b) The condensed structural formulas for 2-bromo-3-methylbutane and ethanol are $CH_3CHBrCH(CH_3)_2$ and CH_3CH_2OH. The carbon bonded to bromine in the haloalkane is a 2° carbon atom and Br^- is a very weak base and a good leaving group. CH_3CH_2OH is a weak base and also a weak nucleophile. The solvent is polar protic. On the basis of this information, we conclude that S_N1 and E1 reactions will occur (see Figure 27-13). The first step in both the S_N1 and E1 mechanisms is the formation of the carbocation:

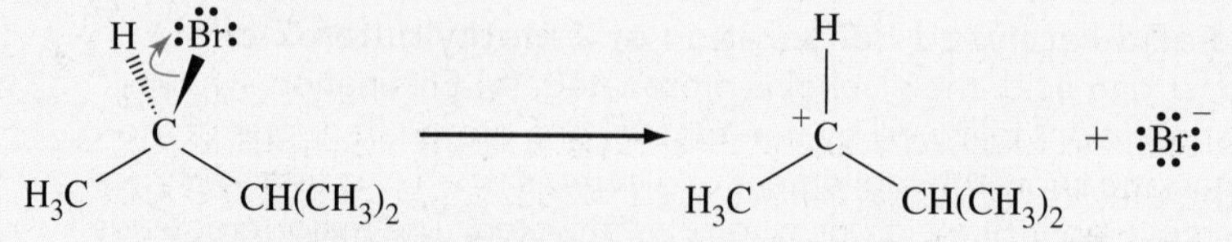

(*R*)-2-bromo-3-methylbutane

Once the carbocation is formed, either substitution or elimination can occur. The steps involved in the substitution are as follows:

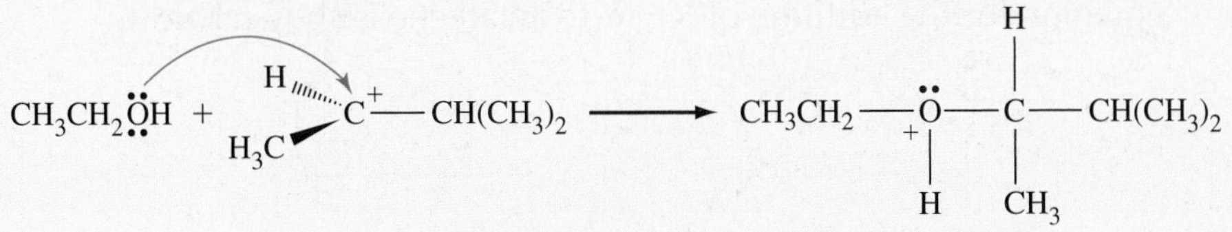

(*R*) and (*S*)

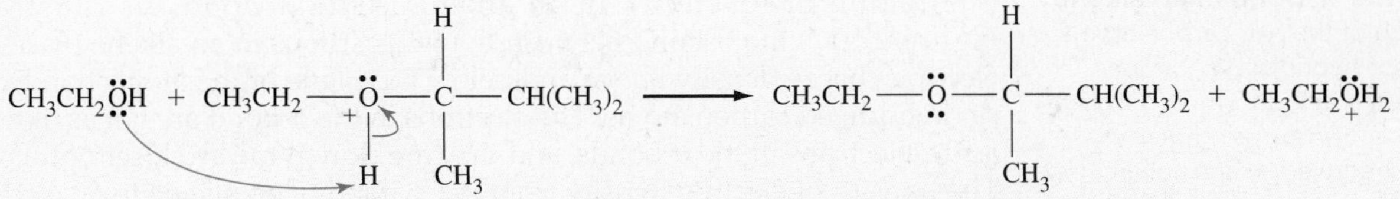

(*R*) and (*S*)

In the diagram above, the nucleophile is shown attacking the carbocation from above, but it can attack from either above or below. Consequently, both (*R*) and (*S*) stereoisomers are obtained.

The steps involved in the elimination are shown below. The elimination reaction produces two alkenes, depending on which H atom (red or blue) is abstracted by the CH_3CH_2OH molecule. The major product is the most highly substituted alkene.

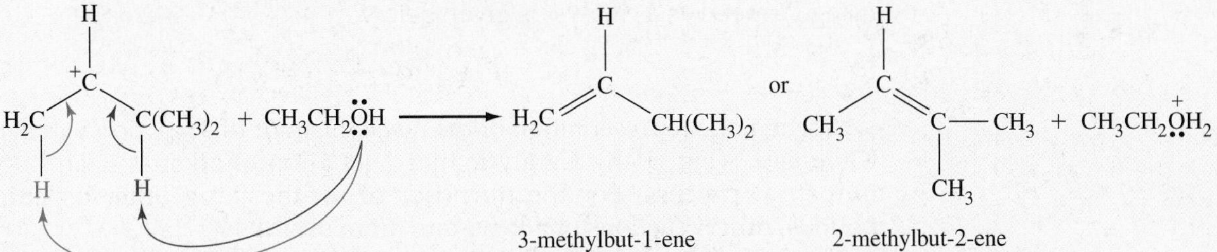

3-methylbut-1-ene (minor elimination product) 2-methylbut-2-ene (major elimination product)

Assess

As shown in (a), the deprotonation of CH_3CH_2OH cannot be accomplished effectively by using NaOH. However, the deprotonation can be accomplished by using $NaNH_2$. The pK_a for NH_3 is about 34, and so pK_b for NH_2^- is about $14 - 34 = -20$. For part (b), keep in mind that a carbocation can undergo a rearrangement (see Are You Wondering 27-2). Consequently, a variety of products are obtained in substitution and elimination reactions involving secondary and tertiary alcohols. We will not explore this complication.

PRACTICE EXAMPLE A: Predict the products of the following reactions. If appropriate, suggest a mechanism by which the reaction occurs, using arrows to show the movement of electrons.

(a) (*R*)-butan-2-ol $\xrightarrow[H_2SO_4]{Na_2Cr_2O_7}$ **(b)** propan-1-ol $\xrightarrow[\Delta]{\text{conc. } H_2SO_4}$

PRACTICE EXAMPLE B: Predict the products of the following reactions. If appropriate, suggest a mechanism by which the reaction occurs, using arrows to show the movement of electrons.

(a) 3-methylbutan-1-ol + hydrogen iodide $\longrightarrow$ **(b)** 2-methylpropan-2-ol + sodium $\longrightarrow$

27-5 Introduction to Addition Reactions: Reactions of Alkenes

In the previous sections, we focused on reactions involving Lewis bases or nucleophiles in which the electron pair to be donated was a lone pair of electrons. In this section, we focus on a few reactions of alkenes, such as $(CH_3)_2C{=}CH_2$, and will see that the π bond in these molecules acts as the Lewis base or nucleophile in those reactions.

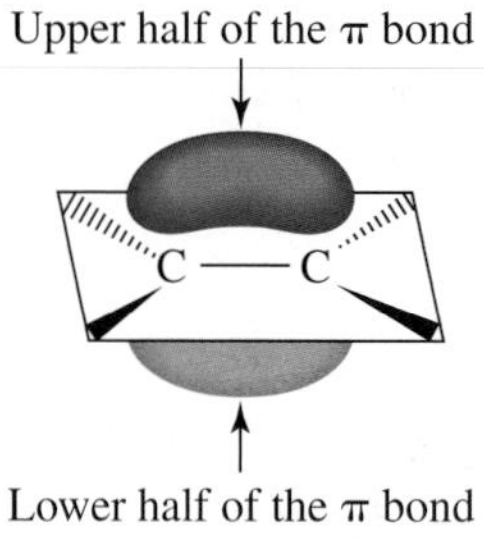

▲ FIGURE 27-17
Schematic representation of the π bond in an alkene
Each of the two carbon atoms is sp^2 hybridized and forms three σ bonds. The σ bonds are shown by using the dashed-wedge and solid-wedge notation. The π bond in an alkene places electron density above and below the plane of the atoms bonded to the carbon atoms by σ bonds.

The characteristic reaction of alkenes (and alkynes) is addition of two substituents, X and Y, to each of the carbons in the double or triple bond. In this section, we will focus primarily on addition reactions of alkenes. The general equation for the addition of X—Y to an alkene is shown below.

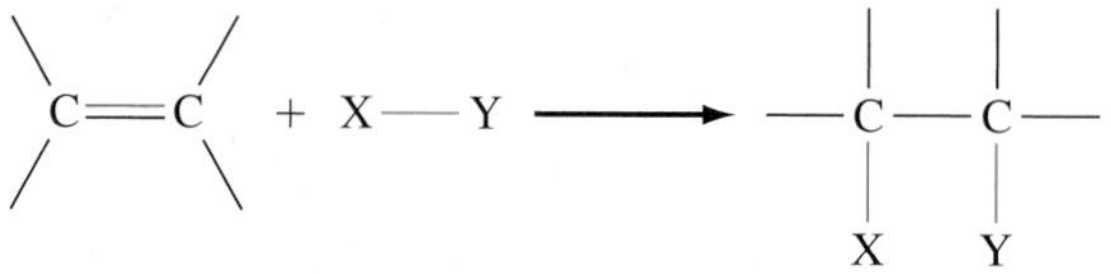

(27.7)

The region of reactivity in an alkene is the π bond. As illustrated in Figure 27-17, the π bond between the two carbons in an alkene places a high electron charge density above and below the plane of the atoms bonded to the sp^2 hybridized carbon atoms. The electrons in the π bond are not as tightly held as the electrons in the σ bonds, and they are drawn toward electrophilic atoms. The transfer of electron density from the π bond of an alkene to an electrophile is a common feature in the reactions we discuss in this section.

Addition of Hydrogen: Hydrogenation

The addition of H_2 across the double bond of an alkene yields an alkane. The reaction is very slow unless a finely divided metal catalyst, such as Ni, Pd, Pt, or Rh, is used. The equation for the hydrogenation of ethene, $H_2C{=}CH_2$, by using Pt metal as a catalyst is given below:

$$H_2C{=}CH_2 + H_2 \xrightarrow{Pt} CH_3CH_3$$

A schematic representation of the mechanism of this reaction was shown in Chapter 23 (Fig. 23-3). Catalytic hydrogenation of alkenes is an important industrial process. For the purposes of synthesizing other organic compounds, other reactions of alkenes are more useful.

The hydrogenation of alkynes is similar to that for alkenes. The metal-catalyzed hydrogenation of an alkyne also yields an alkane, unless a special catalyst called Lindlar's catalyst is used. Lindlar's catalyst consists of barium sulfate coated with palladium metal and "poisoned" with quinoline, a heterocyclic aromatic compound shown in the margin. In Lindlar's catalyst, the catalytic activity of palladium is decreased and the hydrogenation of an alkyne stops at the alkene. The chemical equations are given below for the hydrogenation of but-2-yne by using Pt as a catalyst and Lindlar's catalyst. Notice that when a "pure" metal catalyst is used, the hydrogenation of an alkyne consumes two moles of hydrogen per mole of alkyne. When Lindlar's catalyst is used, only the *cis* (*Z*) stereoisomer is obtained. The two H atoms add from the same side. In contrast, using sodium in liquid ammonia as a hydrogenation agent leads to the formation of the *trans* (*E*) alkene.

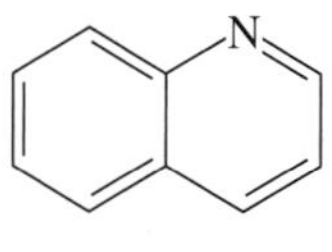

▲ Quinoline is used in Lindlar's catalyst to lower the catalytic activity of palladium metal.

$$H_3C{-}C{\equiv}C{-}CH_3 + 2\,H_2 \xrightarrow{Pt} CH_3CH_2CH_2CH_3$$

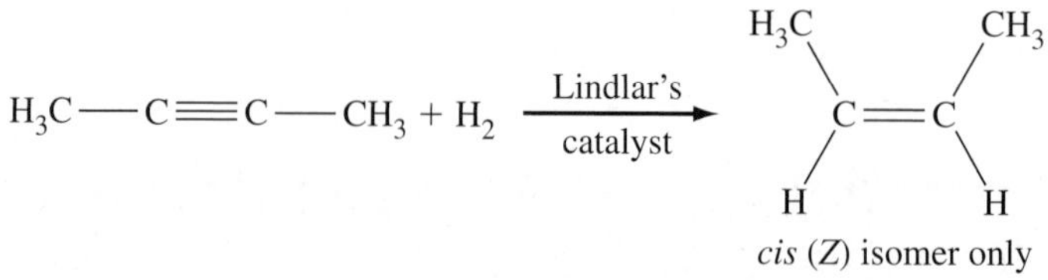

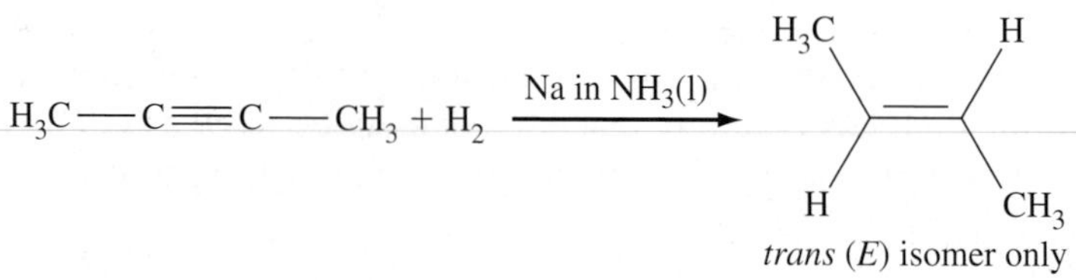

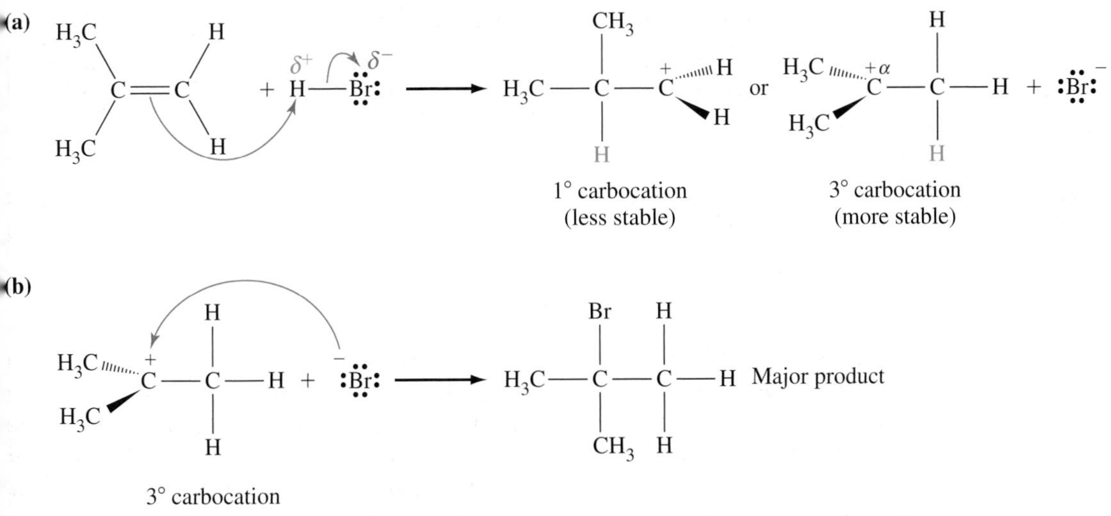

▲ FIGURE 27-18
Mechanism for electrophilic addition of HBr to 2-methylprop-1-ene
(a) The alkene acts a nucleophile and attacks the partially positive hydrogen atom of HBr. In principle, the H atom can add to either carbon atom of the double bond, but there is a strong preference for adding to the least-substituted carbon atom. When H adds to the least substituted carbon atom, the resulting carbocation is more stable. **(b)** The Br^- ion attacks the positive carbon atom of the carbocation.

Addition of Hydrogen Halides (HX)

The addition of HX to an alkene can be represented by the following general equation:

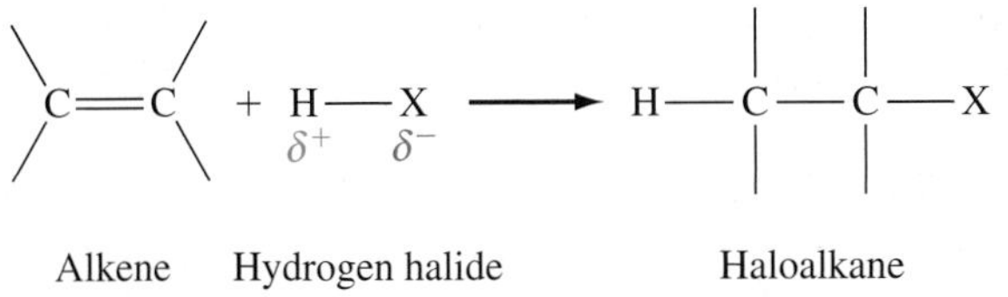

(27.8)

The mechanism for the reaction consists of two steps, as illustrated in Figure 27-18 for the reaction of $(CH_3)_2C{=}CH_2$ and HBr.

In the first step (Fig. 27-18a), the partially positive H atom adds to one of the carbon atoms of the double bond. This step produces a carbocation. In principle, the H atom can add to either one of the carbon atoms but, in practice, it adds preferentially to the least-substituted carbon atom. When H adds to the least substituted carbon, the positive charge of the carbocation ends up on the carbon atom most able to stabilize the positive charge, that is, on the most highly substituted carbon. (Recall that alkyl groups help to stabilize a carbocation.) In the second step (Fig. 27-18b), the Br^- ion acts as a nucleophile and attacks the positively charged carbon atom of the carbocation.

When H—X adds to the double bond of an alkene, the H atom adds to the carbon atom having the smallest number of alkyl groups.

If the reaction is carried out in an aqueous solution, a mixture of products is obtained. To avoid this complication, the reaction can be carried out by bubbling HCl, HBr, or HI through the pure alkene or by carrying out the reaction in a solvent whose molecules are not nucleophilic.

27-4 CONCEPT ASSESSMENT

What are the major and minor products obtained in the reaction of (*E*)-3-methylpent-2-ene and HBr?

Addition of Water: Hydration

The addition of H_2O to an alkene is represented by the following general equation. An alcohol is formed.

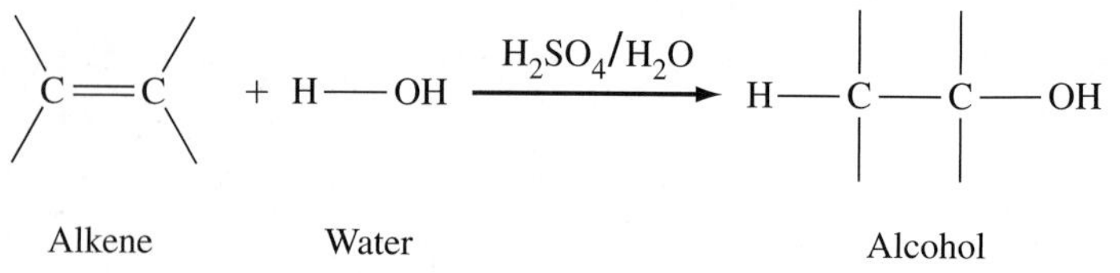

(27.9)

The reaction occurs only in acidic solution and can be carried out in a mixture of H_2SO_4 and H_2O (typically 50% H_2SO_4 by volume). The mechanism is similar to the one given previously for the addition of HX to an alkene. It involves the formation of a carbocation followed by nucleophilic attack. The mechanism for the reaction of $(CH_3)_2C{=}CH_2$ and H_2O is shown in Figure 27-19. In an acidic solution, there is an excess of H_3O^+ ions, and these ions are the source of the electrophile, H^+. In the first step, an H atom from H_3O^+ adds to the sp^2 hybridized carbon that has the greatest number of H atoms, yielding a carbocation (Fig. 27-19a). In the second step, a water molecule acts as a nucleophile and attacks the positively charged carbon of the carbocation (Fig. 27-19b). In the third step, a proton is removed, yielding an alcohol (Fig. 27-19c).

27-5 CONCEPT ASSESSMENT

Would hydration of (*E*)-but-2-ene yield exclusively (*R*)-butan-2-ol, (*S*)-butan-2-ol, or a mixture of the (*R*) and (*S*) enantiomers? Explain.

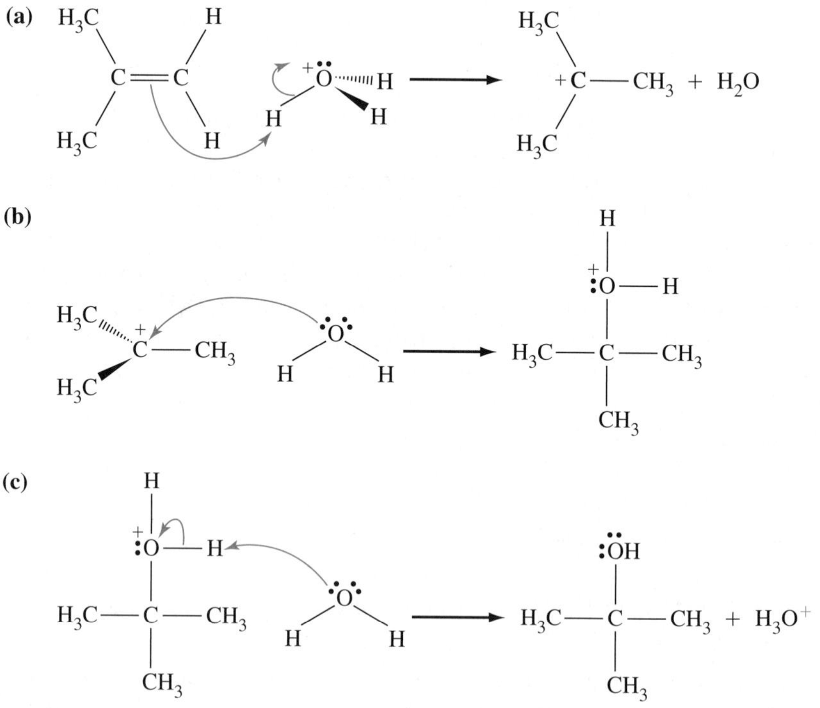

▶ FIGURE 27-19
Mechanism for the acid-catalyzed addition of H_2O to 2-methylprop-1-ene
(a) In an acidic solution, an H atom from an H_3O^+ ion can add to a carbon in the double bond of an alkene. As discussed in the text, there is a preference for the H atom to add to the least-substituted carbon atom. The other carbocation that can be formed in this step is less stable and is not shown.
(b) A water molecule attacks the carbocation, forming a protonated alcohol.
(c) The protonated alcohol transfers a proton to a water molecule.

Addition of Halogens: Halogenation

When a halogen, represented by X_2, adds across the double bond of an alkene, the product is a dihalide. The halogen atoms are bonded to adjacent carbons. Such a dihalide is called a **vicinal dihalide**. (When two halogen atoms are bonded to the same carbon, the compound is a **geminal**

dihalide.) The general equation for the addition of X_2 to an alkene is shown in equation (27.10):

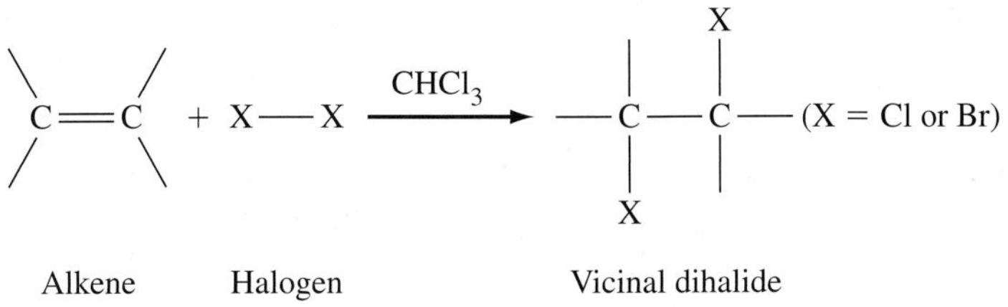

(27.10)

The reaction above occurs readily at room temperature when the halogen is Cl_2 or Br_2. A nonaqueous solvent, such as $CHCl_3$ or CCl_4, must be used to prevent the formation of other products, including alcohols. The mechanism for the addition of X_2 to an alkene is significantly different from the mechanisms considered previously for the addition of HX and H_2O. The first step of the mechanism is the formation of a **bridged halonium ion**, rather than a carbocation (Fig. 27-20a). In the halonium ion, the halogen atom is bonded to two carbon atoms and it is the atom that carries the positive charge. Notice that the carbon atoms and the halogen atom in the halonium ion all have complete octets. Take careful note of the movement of electrons that occurs in the formation of the halonium ion. The electron pair in the π bond of the alkene is directed toward a chlorine atom. As this occurs, an electron pair from a chlorine atom is directed toward one of the carbon atoms and the Cl—Cl bond is breaking. In the next step (Fig. 27-20b), a Cl^- ion attacks a carbon atom in the bridge from the backside because the bridge blocks the carbon atoms from a frontside attack.

27-6 CONCEPT ASSESSMENT

Apply the ideas discussed above to the bromination of cyclopentene. What is the product obtained in this reaction? Specify stereochemistry, if relevant.

Carboxylic acids and their derivatives are important organic compounds. They are found widely in nature, and biological systems, and are used extensively in organic synthesis. The reactions of these compounds, and the mechanisms by which they occur, are discussed in Section 27C, *Carboxylic Acids and Their Derivatives: The Addition-Elimination Mechanism*, on the MasteringChemistry site (www.masteringchemistry.com).

MasteringCHEMISTRY

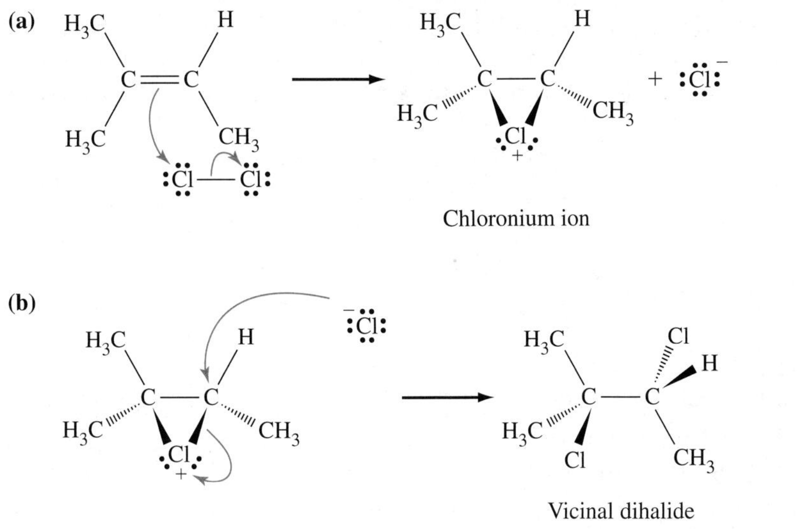

◀ FIGURE 27-20
Mechanism for chlorination of 2-methylbut-2-ene
(a) A chlorine atom adds to the alkene by bonding simultaneously to both carbon atoms of the double bond. The structure formed in this step is called a chloronium ion. Notice that the chlorine atom bridges two carbon atoms and is the atom carrying the positive charge. **(b)** In the second step, a Cl^- attacks one of the carbon atoms in the bridge from the backside. A vicinal dihalide is formed.

27-6 Electrophilic Aromatic Substitution

We saw in the previous section that alkenes typically undergo addition reactions, in which substituents add across a carbon–carbon double bond. Because we often represent the structure of benzene as a six-membered ring with alternating single and double bonds, it is tempting to think that benzene would also undergo addition reactions in which substituents add across one of the carbon–carbon double bonds. As illustrated below, benzene does not undergo addition reactions:

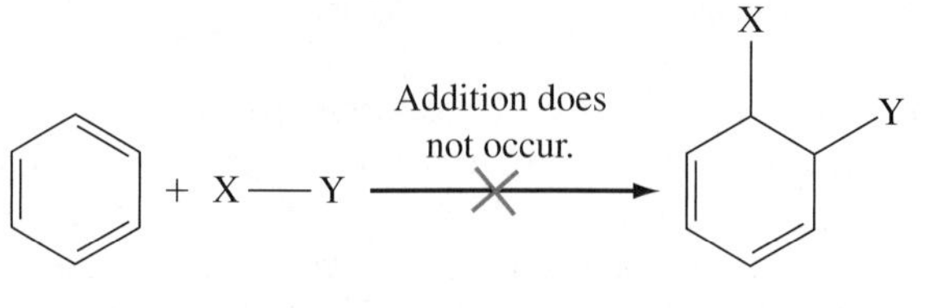

Benzene (aromatic) Hypothetical addition product (not aromatic)

Benzene and its derivatives typically react with electrophiles in substitution reactions. The general equation for the reaction of benzene and an electrophilic species, E—Y, is shown below:

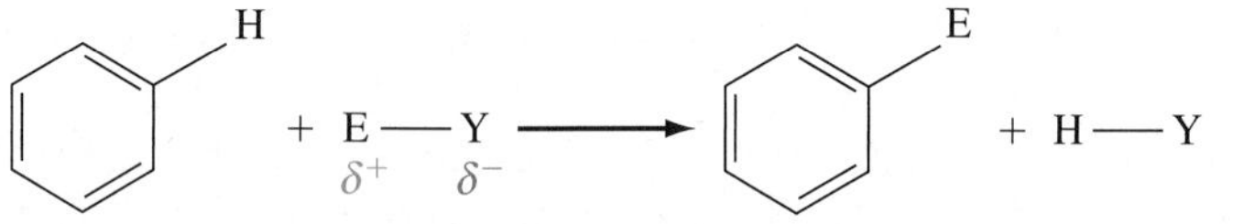

(27.11)

The reaction is an **electrophilic substitution reaction** because one of the hydrogen atoms in benzene is replaced by, or substituted with, an electrophile, E. The mechanism is shown Figure 27-21 and it involves two steps. In the first step, the electrophile accepts an electron pair from the π system of the benzene ring to form a carbocation, $C_6H_6E^+$, called an **arenium ion**. In the second step, the arenium ion loses a proton. In essence, electrophilic aromatic substitution follows an addition-elimination sequence. Notice that in the second step, Y^- acts as a base and attacks the H atom on the carbon that is bonded to E. It is tempting (but wrong) to think that Y^- would act as a nucleophile and attack the positively charged carbon atom of the arenium ion. Y^- does not attack the positively charged carbon for two reasons. First, the positive charge of the arenium ion is not localized on a particular carbon atom. It is shared equally by three carbons, as suggested in Figure 27-22, and thus these carbon atoms are only partially positive. Second, and more important, if Y^- attacks and forms a bond with one of the partially positive carbon atoms, the resulting product is not aromatic and thus is much less stable.

Various electrophilic aromatic substitutions are possible, each of which involves a different electrophilic species. We will examine a few cases, placing special consideration on the methods used for generating the electrophile.

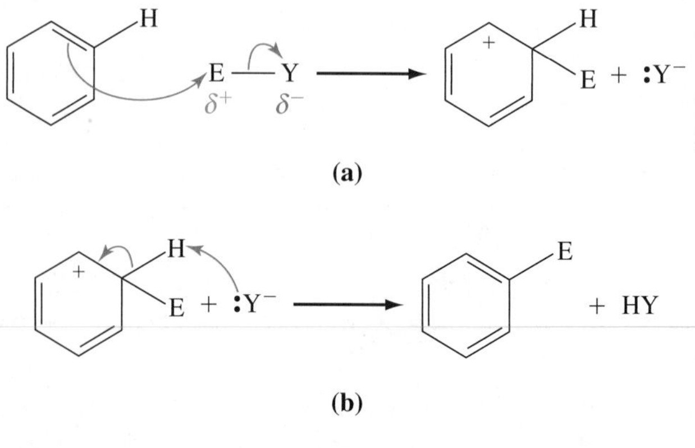

▶ FIGURE 27-21
General mechanism for electrophilic aromatic substitution of benzene
(a) The electrophile accepts an electron pair and forms a bond with a carbon atom in the benzene ring. The aromaticity of the ring is lost. **(b)** A proton is removed from the carbocation, restoring the aromaticity of the ring.

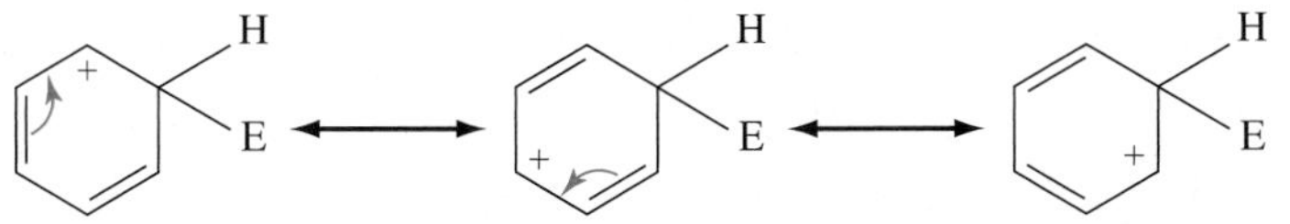

▲ FIGURE 27-22
Three equivalent contributing resonance structures for $C_6H_6E^+$
There are three equivalent contributing resonance structures for the $C_6H_6E^+$ carbocation. The positive charge is shared equally by three carbon atoms in the ring.

Nitration: Substitution of —H with —NO_2

To replace an H atom in benzene with a nitro group, —NO_2, benzene is treated with a mixture of sulfuric acid (H_2SO_4) and nitric acids (HNO_3). The reaction of H_2SO_4 and HNO_3 produces a nitronium ion, NO_2^+:

$$HNO_3 + 2\,H_2SO_4 \longrightarrow O{=}N^+{=}O + H_3O^+ + 2\,HSO_4^-$$

The mechanism for the nitration of benzene involves two steps. In the first step, the nitrogen atom of the nitronium ion accepts a pair of electrons from the π system of the benzene ring. A carbon–nitrogen bond is formed:

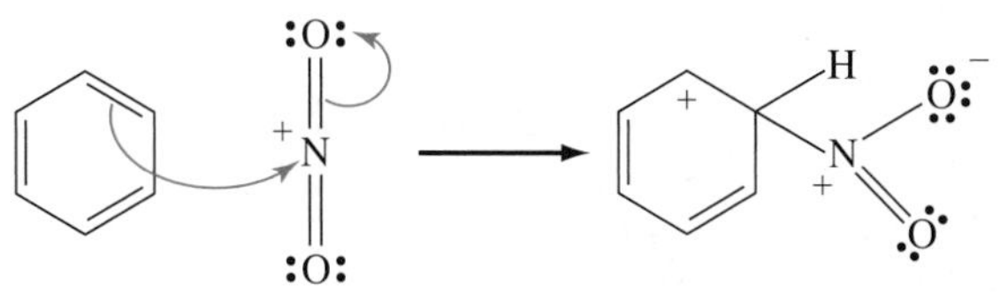

In the second step, the arenium ion is deprotonated by HSO_4^-, yielding nitrobenzene:

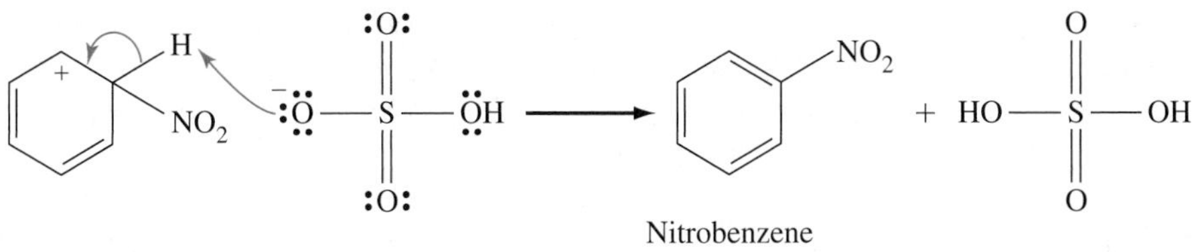

Halogenation

Benzene can be converted into chlorobenzene or bromobenzene by treating benzene with either Cl_2 or Br_2 in the presence of an appropriate catalyst. The chemical equations for these conversions are shown below:

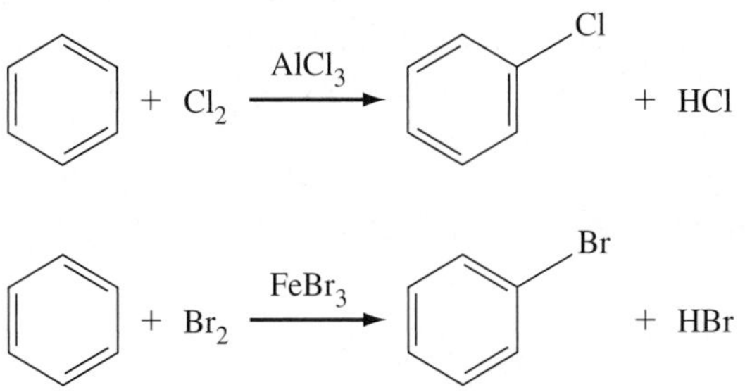

In these reactions, the catalyst reacts with Cl_2 or Br_2 to form an intermediate species that reacts with benzene. For example, Cl_2 and $AlCl_3$ react as follows:

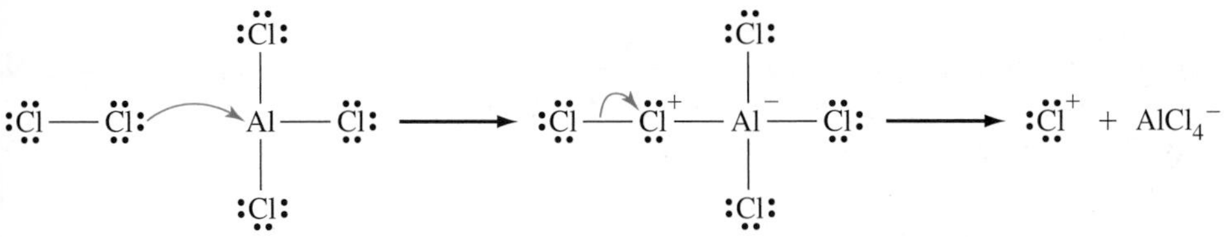

Following the formation of $Cl^+AlCl_4^-$, the chlorination of benzene proceeds as follows.

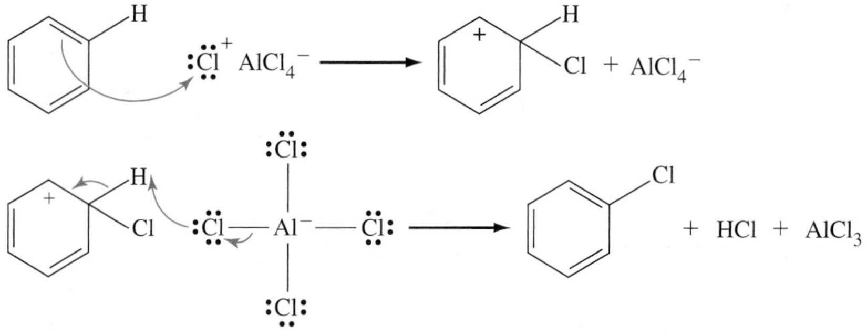

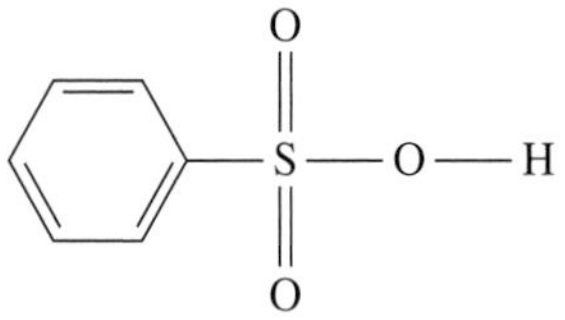

Benzenesulfonic acid

Many other substitutions are possible. For example, treating benzene with either concentrated sulfuric acid or *fuming sulfuric acid* (SO_3 in concentrated H_2SO_4) replaces a hydrogen atom with the $-SO_3H$ group to make benzenesulfonic acid (see Exercise 38). Methods and reagents are available for replacing an H atom with an alkyl group, R (see Exercise 37), or an acyl group, RC=O.

Ortho, Para-directing Substituents and Meta-directing Substituents

The substitution of a single atom or group, X, for an H atom in benzene can occur at any one of the six positions on the ring. We say that the six positions are equivalent. If a group Y is substituted for an H atom in C_6H_5X, this question arises: To which of the remaining five positions does the Y group go? If all the sites on the benzene ring were equally preferred, the distribution of the products would be a purely statistical one. That is, Y can be substituted in five possible positions, and we should get 20% of each one. Since two possibilities lead to an *ortho* isomer and two lead to a *meta* isomer, however, we should expect the distribution of products to be 40% ortho, 40% meta, and 20% para:

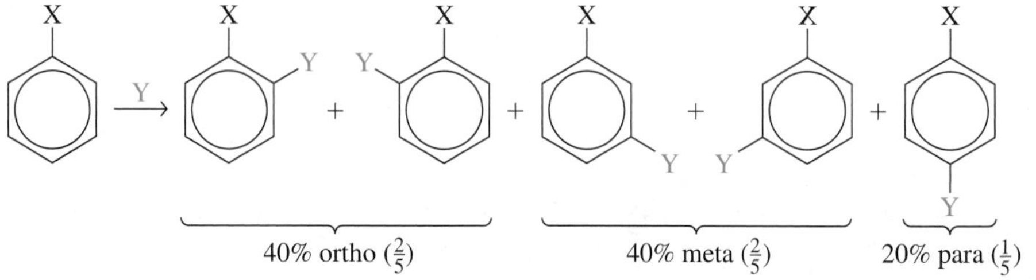

Momo_leif/Fotolia

▲ A Chinese rose has the aroma of black tea, primarily because of 1,3-dimethoxy-5-methylbenzene, a derivative of benzene.

The following scheme describes the products resulting from nitration followed by chlorination (reaction 27.12) and chlorination followed by nitration (reaction 27.13). It shows that the substitution is *not random*. The $-NO_2$ group directs Cl to a meta position. Almost no *ortho* or *para* isomer is formed in reaction (27.12). The Cl group, on the other hand, is an *ortho, para* director. Essentially no *meta* isomer is produced in reaction (27.13).

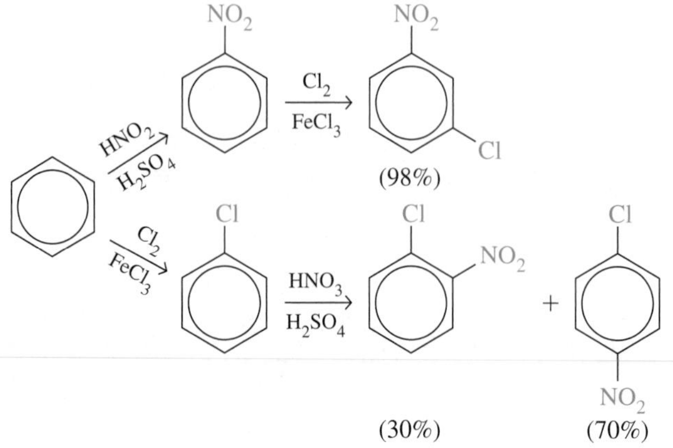

(27.12)

(27.13)

Whether a group is an *ortho, para,* or *meta* director depends on how the presence of one substituent alters the electron distribution in the benzene ring. As a result, attack by a second group is more likely at one type of position than another. Examination of many reactions leads to the following order:

Ortho, para directors: —NH_2, —OR, —OH, —OCOR, —R, —X (X = halogen) (from strongest to weakest)

Meta directors: —NO_2, —CN, —SO_3H, —CHO, —COR, —COOH, —COOR (from strongest to weakest)

When two groups of the same type (both *o*-, *p*-directing or both *m*-directing) are present, the stronger director wins out. When two groups of different type (one *o*-, *p*-directing and one *m*-directing) are present, then the *o*-, *p*-directing group guides the reaction.

EXAMPLE 27-7 Predicting the Products of an Aromatic Substitution Reaction

Predict the products of the mononitration of

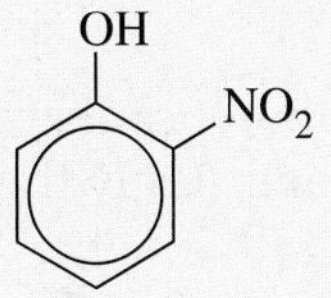

Analyze

The groups are different, and so the *o*-, *p*-directing group (—OH) guides the reaction.

Solve

Because the —OH group is an *o*-, *p*-directing group and guides the reaction, we expect the following products:

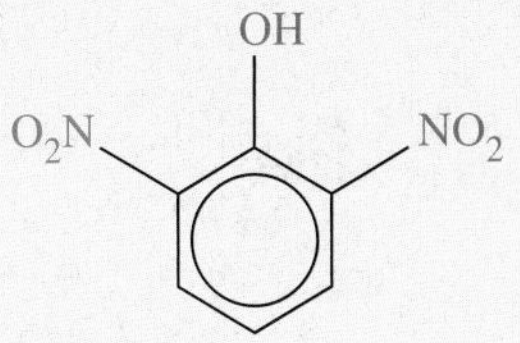

2,6-dinitrophenol

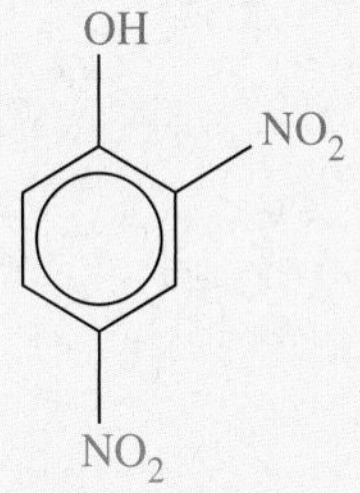

2,4-dinitrophenol

Assess

In the diagrams below, we show the flow of electron density that occurs during the formation of 2,4-dinitrophenol:

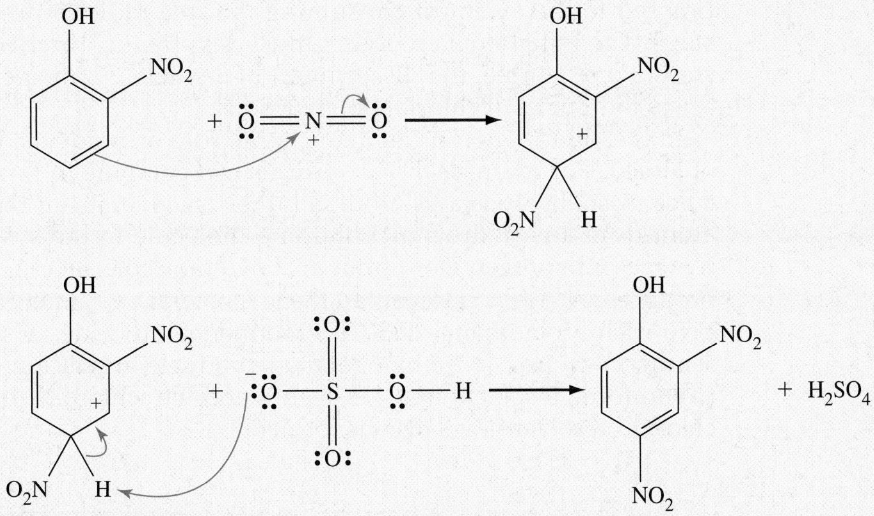

PRACTICE EXAMPLE A: Predict the major product(s) of the mononitration of benzaldehyde, C_6H_5CHO.

PRACTICE EXAMPLE B: Predict the major product(s) of the mononitration of 1,3-dichlorobenzene.

27-7 Reactions of Alkanes

Saturated hydrocarbons have little affinity for most chemical reactants. They are nonpolar substances that are insoluble in water and unreactive toward acids, bases, or oxidizing agents. The most commonly encountered reaction of alkanes is with oxygen. Alkanes burn; the *oxidation* of hydrocarbons underlies their important use as fuels. For example, octane reacts with oxygen as follows:

$$C_8H_{18}(l) + \frac{25}{2}O_2(g) \longrightarrow 8\,CO_2(g) + 9\,H_2O(l) \quad \Delta_r H^\circ = -5.48 \times 10^3 \text{ kJ mol}^{-1} \quad \textbf{(27.14)}$$

Alkanes, however, will also react with halogens under the right conditions. For example, alkanes react only slowly with halogens at room temperature, but at higher temperatures, particularly in the presence of light, halogenation occurs. The reaction between an alkane and a halogen is a substitution reaction, with the halogen atom replacing a hydrogen atom. For example, in the reaction below, a chlorine atom replaces a hydrogen atom in methane:

$$CH_4 + Cl_2 \xrightarrow[\text{light}]{\text{heat or}} CH_3Cl + HCl \quad \textbf{(27.15)}$$

The substitution occurs by a *chain reaction*, written as follows for the chlorination of methane. (Only the electrons involved in bond breakage or formation are shown.)

Initiation:

$$Cl{:}Cl \xrightarrow[\text{light}]{\text{heat or}} 2\,Cl\cdot$$

Propagation:

$$H_3C{:}H + Cl\cdot \longrightarrow H_3C\cdot + H{:}Cl$$

$$H_3C\cdot + Cl{:}Cl \longrightarrow H_3C{:}Cl + Cl\cdot$$

Termination:

$$Cl\cdot + Cl\cdot \longrightarrow Cl{:}Cl$$

$$H_3C\cdot + Cl\cdot \longrightarrow H_3C{:}Cl$$

$$H_3C\cdot + H_3C\cdot \longrightarrow H_3C{:}CH_3$$

The reaction is initiated when some Cl_2 molecules absorb sufficient energy to dissociate into Cl atoms (represented above as $Cl\cdot$). Cl atoms collide with CH_4 molecules to produce methyl *free radicals* ($H_3C\cdot$), which combine with Cl_2 molecules to form CH_3Cl molecules. When any or all of the last three reactions proceed to the extent of consuming the free radicals present, the reaction stops. The initiation step occurs much less frequently than the propagation steps. For example, the dissociation of a single Cl_2 molecule probably produces thousands of chlorination reactions.

In the chlorination of methane, a mixture of products, not just CH_3Cl, is obtained. For example, $H_3C\cdot$ radicals can combine to form CH_3CH_3 molecules. Also, because a Cl atom is rather reactive, it can remove a hydrogen atom from any hydrogen-containing molecule in the system. If a Cl atom removes a hydrogen atom from a CH_3Cl molecule, an $H_2ClC\cdot$ radical will be produced. A $H_2ClC\cdot$ radical can then react with a Cl_2 molecule or a Cl atom to give dichloromethane, CH_2Cl_2 (methylene chloride, a solvent and paint remover). More highly halogenated products, including trichloromethane (chloroform, a solvent and fumigant), and tetrachloromethane (carbon tetrachloride, a solvent) will also be formed.

27-7 CONCEPT ASSESSMENT

In the chlorination of methane, small amounts of chloroethane will also be produced. Suggest how the formation of chloroethane might occur.

Not all the halogens show the same reactivity with methane. Iodine is not very reactive and fluorine reacts explosively with methane unless special precautions are taken.

$$I_2 \ll Br_2 < Cl_2 \ll F_2 \xrightarrow[\text{Increasing reactivity toward } CH_4]{}$$

How can we rationalize this trend in reactivities? Let's compare the energy changes involved in the propagation steps (Table 27.4).

The activation energy for hydrogen removal is rather small for fluorine (only 5 kJ/mol) and quite large for iodine (140 kJ mol^{-1}). Thus, the rate of hydrogen removal is highest for fluorine and lowest for iodine. However, activation energies don't tell the whole story. The enthalpy change for the halogenation process also plays a role. For example, heat released by the propagation steps is not immediately dissipated to the surroundings and is absorbed by the reacting system, causing the temperature to rise. The greater the amount of heat released, the greater the temperature rise and the greater the increase in the rate of halogenation. For the fluorination of methane, the propagation steps release a large quantity of heat ($\Delta H° = -432$ kJ mol^{-1}), leading ultimately to a large increase in the rate of halogenation, whereas in the iodination of methane, the propagation steps absorb heat.

Just as the halogens show different reactivities, so too do the hydrogen atoms in a molecule. Consider, for example, the bromination of methylpropane:

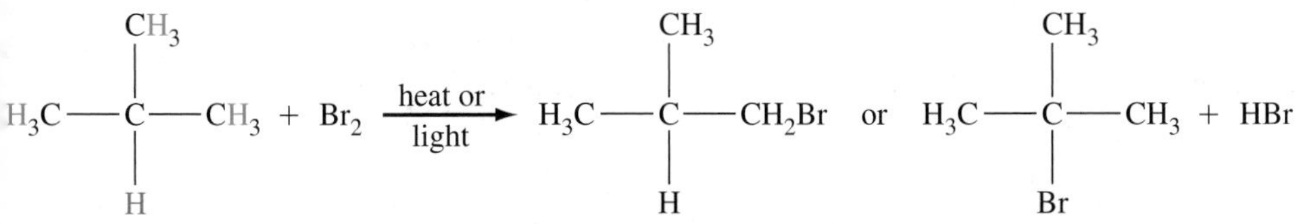

Methylpropane — 1-bromo-2-methylpropane (<1%) — 2-bromo-2-methylpropane (>99%)

Methylpropane has two types of H atoms. The H atoms shown in blue are primary H atoms (see Figure 26-4) and the H atom shown in red is a tertiary H atom. From a statistical standpoint, we should expect 1-bromo-2-methylpro-prane to be the major product. (There are nine primary H atoms but only one tertiary H atom.) However, the major product is the one formed when the tertiary H atom is replaced. The selectivity of bromine for different H atoms is summarized on the next page. (Chlorine is more reactive and not as selective as bromine.)

TABLE 27.4 Activation Energy and Enthalpy Change in kJ mol^{-1} for Steps Involved in the Halogenation of Methane

	Fluorination		Chlorination		Bromination		Iodination	
	E_a	$\Delta_r H°$	E_a	$\Delta_r H°$	E_a	$\Delta_r H°$	E_a	$\Delta_r H°$
$CH_4 + X\cdot \longrightarrow CH_3\cdot + HX$	5.0	−130	16	8	78	74	140	142
$CH_3\cdot + X_2 \longrightarrow CH_3X + X\cdot$	≈0	−302	≈0	−109	≈0	−100	≈0	−89
$CH_4 + X_2 \longrightarrow CH_3X + HX$		−432		−101		−26		53

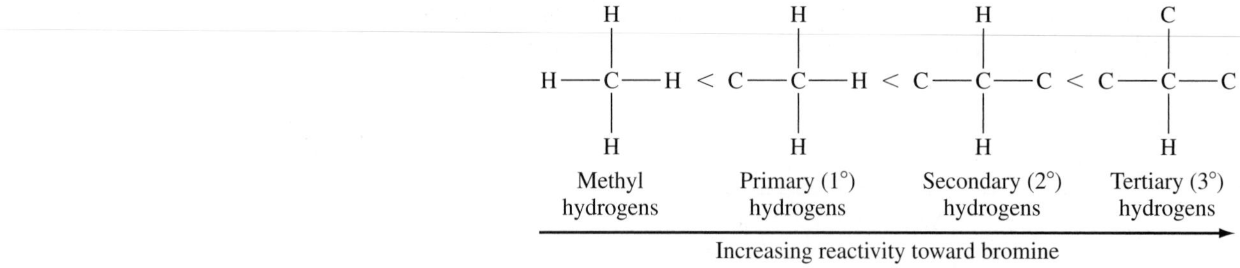

The trend in reactivities can be rationalized in terms of the stabilities of the radicals that are formed when a particular H atom is removed. For example, the removal of a tertiary H atom yields a 3° radical, whereas the removal of a primary H atom yields a 1° radical. Like carbocations, radicals are stabilized by alkyl groups, and so the relative stabilities of radicals follow those of carbocations: $3° > 2° > 1° \gg$ methyl.

27-8 CONCEPT ASSESSMENT

What is the major product obtained in the bromination of $(CH_3)_2CHCH_2CH_3$?

27-8 Polymers and Polymerization Reactions

An Overview

Polymers are made up of simple molecules with low molecular masses joined together into extremely large molecules. Polymers with molecular masses below about 20,000 u are called low polymers and those above 20,000 u are called high polymers.

One familiar polymer is *polyethylene*. As its name implies, its basic unit, or *monomer*, is the ethylene molecule, which has the Lewis structure

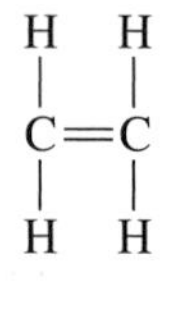

We can imagine that the polymerization of ethylene begins with the "opening up" of the double bonds in ethylene molecules.

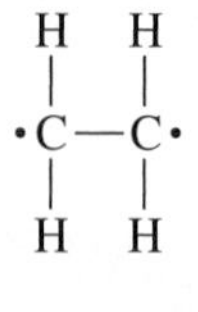

Then each C atom in the resulting molecular fragment (radical) forms an additional single covalent bond with a C atom in another molecular fragment, and so on, producing the structure shown below:

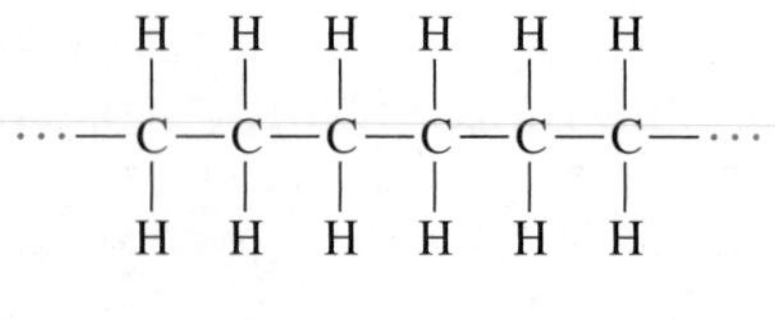

In the following notation, the monomer unit is enclosed in square brackets and the subscript n signifies the number of monomers present in the final *macromolecule*. Typically, n might range from several hundred to several thousand.

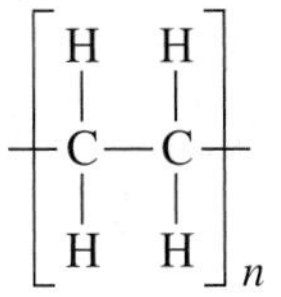

Another polymer in which monomer units join end to end is *latex*—natural rubber:

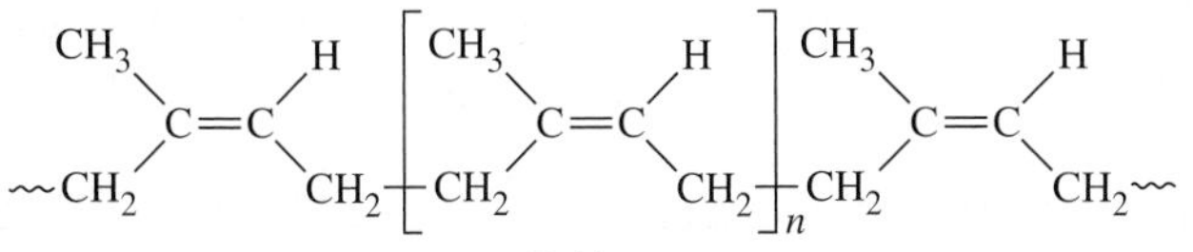

Rubber

Early rubber products were of limited use because they were sticky in hot weather and stiff in cold weather. In 1839, Charles Goodyear accidentally discovered that by heating a sulfur–rubber mixture, a product could be made that was stronger, more elastic, and more resistant to heat and cold than natural rubber. This process is now called vulcanization (after *Vulcan*, the Roman god of fire). The purpose of vulcanization is to form *cross-links* between long polymer chains. An example of a cross-link through two sulfur atoms is shown here:

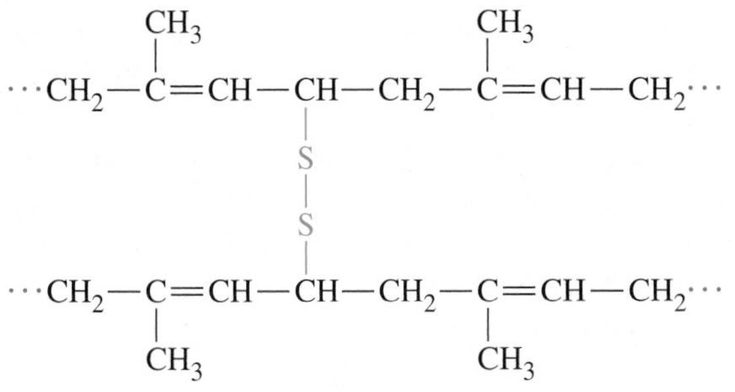

Polymers are familiar products in the modern world. Nylon, one of the first polymers developed, is like an artificial silk and is used in making clothing, ropes, and sails. The fluorine-containing polymer Teflon (polytetrafluoroethylene) is used in nonstick frying and baking pans. Polyvinyl chloride (PVC) is used in food wrap, hoses, pipes, and floor tile. In all, the polymer industry is huge. It has been estimated that about half of all chemists work with polymers. Of particular interest to them are the reactions that can be used to make polymers. We will briefly survey the main types of polymerization reactions.

Chain-Reaction Polymerization

Monomers with carbon-to-carbon double bonds typically undergo **chain-reaction polymerization**. The net result is that the double bonds open up and monomer units add to growing chains. As with other chain reactions, the mechanism involves three characteristic steps: initiation, propagation, and termination. Let us illustrate this mechanism for the formation of the polymer polyethylene from the monomer ethylene (ethene). The key to the polymerization reaction is the free-radical initiator. In reaction (27.16), an organic peroxide dissociates into two radicals. The radicals add to the double bonds of ethylene molecules to form radical intermediates that attack more ethylene molecules and form new

◀ Molecules of a *poly*mer are formed by the joining together of many simple molecules called *mono*mers.

James Holmes / Zedcor /Science Source

▲ Extruding polyethylene film.

intermediates of longer and longer length, as in reaction (27.17). The chains terminate as a result of reactions such as (27.18) and (27.19).

Initiation:

$$\mathrm{R{-}O{:}O{-}R \longrightarrow 2\,R{-}O\cdot} \quad \text{(an organic peroxide)} \tag{27.16}$$

followed by

Propagation:

$$\begin{aligned} \mathrm{CH_2{=}CH_2 + RO\cdot} &\longrightarrow \mathrm{R{-}O{-}CH_2{-}CH_2\cdot} \\ \mathrm{ROCH_2CH_2\cdot + CH_2{=}CH_2} &\longrightarrow \mathrm{ROCH_2CH_2CH_2CH_2\cdot} \\ \mathrm{RO(CH_2)_3CH_2\cdot + CH_2{=}CH_2} &\longrightarrow \mathrm{RO(CH_2)_5CH_2\cdot} \end{aligned} \tag{27.17}$$

Termination:

$$\mathrm{RO(CH_2)_{\mathit{x}}CH_2\cdot + RO\cdot \longrightarrow RO(CH_2)_{\mathit{x}}CH_2OR} \tag{27.18}$$

or

$$\mathrm{RO(CH_2)_{\mathit{x}}CH_2\cdot + RO(CH_2)_{\mathit{y}}CH_2\cdot \longrightarrow RO(CH_2)_{\mathit{x}}CH_2CH_2(CH_2)_{\mathit{y}}OR} \tag{27.19}$$

Several polymers formed by chain-reaction polymerization are listed in Table 27.5.

Notice that the termination steps, reactions (27.18) and (27.19), produce a polymer with —OR groups at either end of the chain. The —OR groups have no effect on the properties of the polymer because the polymer has hundreds, possibly thousands, of monomer units, and the —OR groups are only at the ends of each polymer strand. The long chain of monomer units gives the polymer its particular properties, and the group serves only to initiate the reaction and to terminate the polymer chains.

Step-Reaction Polymerization

In **step-reaction polymerization**, also called *condensation polymerization*, the monomers typically have two or more functional groups that react to join the two molecules together. Usually, this involves the elimination of a small molecule, such as H_2O. In chain-reaction polymerization, the reaction of a monomer can occur only at the end of a growing polymer chain, but in step-reaction polymerization, any pair of monomers is free to join into a *dimer*; the dimer can join with a monomer to form a *trimer*; two dimers can join to form a *tetramer*; and so on. Step-reaction polymerization tends to occur slowly and produces polymers of only moderately high molecular masses (less than 10^5 u). The formation of polyethylene glycol terephthalate, Dacron, is illustrated in

TABLE 27.5 Some Polymers Produced by Chain-Reaction Polymerization

Name	Monomer	Polymer	Uses
Polyethylene	$CH_2{=}CH_2$	$\text{+}CH_2{-}CH_2\text{+}_n$	Bags, bottles, tubing, packaging film
Polypropylene	$CH_2{=}CHCH_3$	$\left(\text{+}CH_2{-}\underset{\mid\atop CH_3}{CH}\text{+}\right)_n$	Laboratory and household ware, artificial turf, surgical casts, toys
Poly(vinyl chloride) PVC	$CH_2{=}CHCl$	$\left(\text{+}CH_2{-}\underset{\mid\atop Cl}{CH}\text{+}\right)_n$	Bottles, floor tile, food wrap, piping, hoses
Poly(tetrafluoroethylene), Teflon	$CF_2{=}CF_2$	$\text{+}CF_2{-}CF_2\text{+}_n$	Bearings, insulation, nonstick surfaces, gaskets, industrial ware
Polystyrene	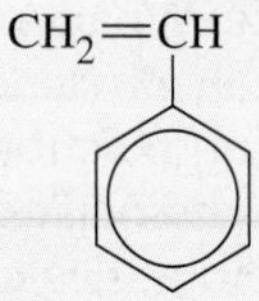	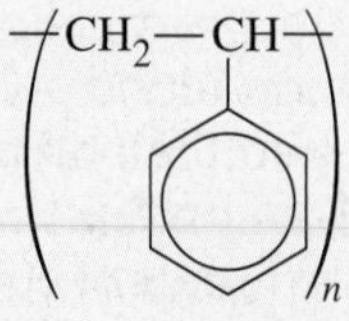	Packaging, refrigerator doors, cups, ice buckets, and coolers (as foam)

TABLE 27.6 Some Polymers Produced by Step-Reaction Polymerization

Name	Monomer	Polymer	Uses
Poly(ethylene glycol terephthalate) (Dacron)	$HOCH_2CH_2OH$ and 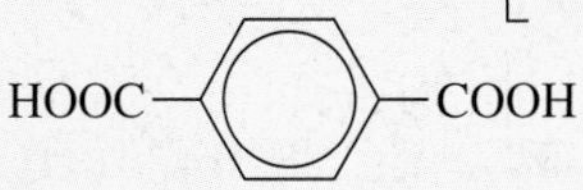	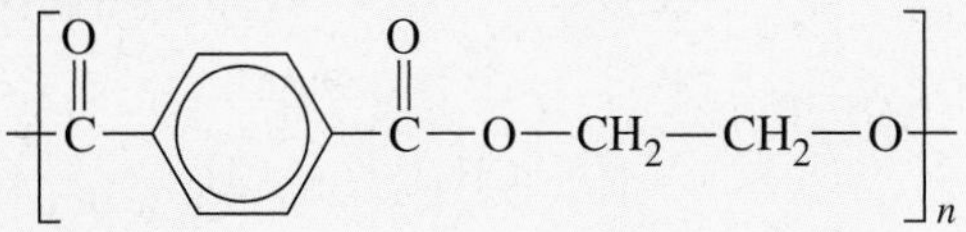	Textile fabrics, twine and rope, fire hoses, plastic containers
Poly(hexamethyleneadipamide) (Nylon 66)	$H_2N(CH_2)_6NH_2$ and $HOOC(CH_2)_4COOH$	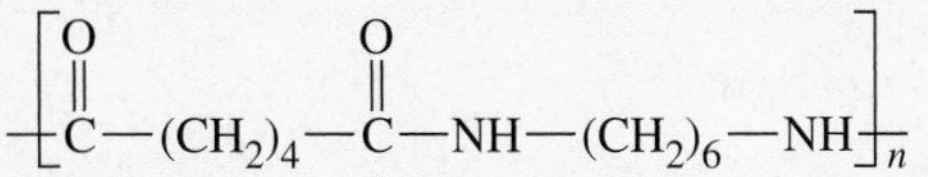	Hosiery, rope, tire cord, parachutes, artificial blood vessels
Polyurethane	$HO(CH_2)_4OH$ and $OCN(CH_2)_6NCO$	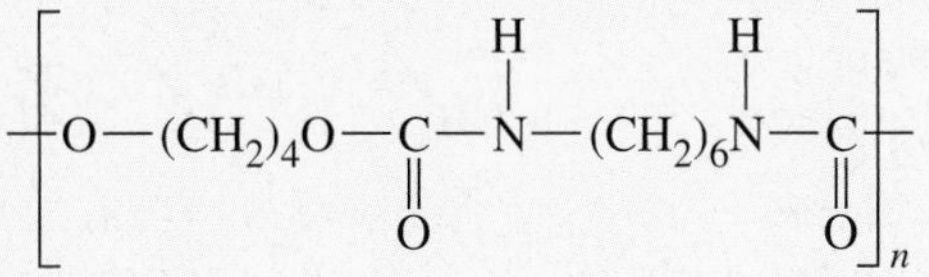	Spandex fibers, bristles for brushes, cushions and mattresses (as foam)

the following reaction, and several other polymers formed by this method are listed in Table 27.6.

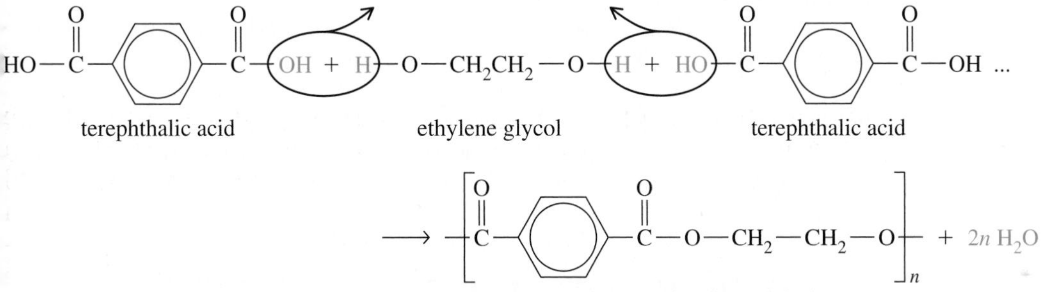

poly(ethylene glycol terephthalate)
(Dacron)

Stereospecific Polymers

The physical properties of a polymer are determined by a number of factors, such as the average length (average molecular mass) of the polymer chains and the strength of intermolecular forces between chains. Another important factor is whether the polymer chains display any crystallinity—that is, an ordered geometry and spacing of the atoms between polymer chains. In general, amorphous polymers are glass-like or rubbery. A high-strength fiber, on the other hand, must possess some crystallinity. Many polymers have both crystalline and amorphous regions. The relative amount of each type of region affects the physical properties of the polymer.

The usual designation of a polymer, as applied to polypropylene, for example, is not very revealing about the structure of the polymer:

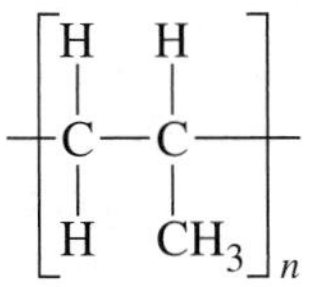

It does not indicate the orientation of the groups along the polymer chain. If propylene is polymerized by the method shown for ethylene on page 1312, the orientation of the groups is random (Fig. 27-23). A polymer of this type is called *atactic*. Because there is no regularity to the structure, atactic polymers are amorphous. In an *isotactic* polymer, all $-CH_3$ groups have the same orientation, and

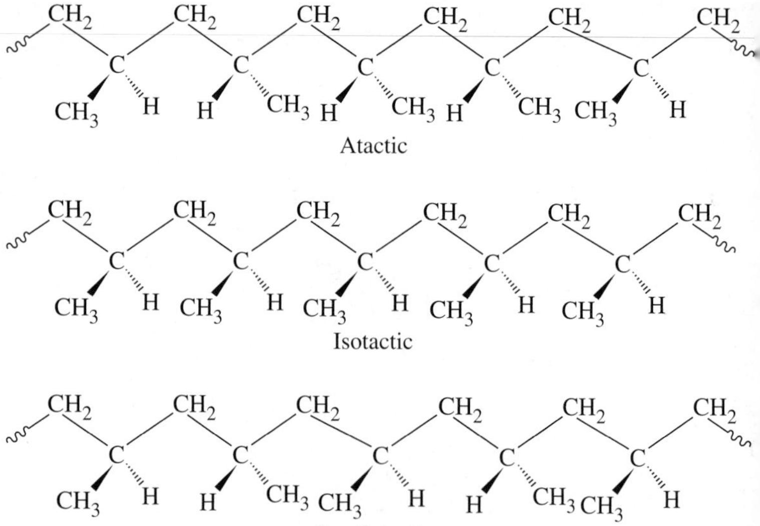

▶ FIGURE 27-23
Three representations of a polypropylene chain
In the *atactic* polymer, the $-CH_3$ groups are randomly distributed on the chain. In the *isotactic* polymer, all $-CH_3$ groups are shown coming out of the plane of the page. In the *syndiotactic* polymer, $-CH_3$ groups alternate along the chain—one in front of the plane of the page, the next one behind the plane, and so on.

in a *syndiotactic* polymer, the $-CH_3$ groups alternate back and forth along the chain. Because of their structural regularity, isotactic and syndiotactic polymers possess crystallinity, which makes them stronger and more resistant to chemical attack than an atactic polymer.

In the 1950s, Karl Ziegler and Giulio Natta developed procedures for controlling the spatial orientation of substituent groups on a polymer chain by using special catalysts, such as $(CH_3CH_2)_3Al + TiCl_4$. This discovery, recognized through the award of a Nobel Prize in 1963, revolutionized polymer chemistry. Through stereospecific polymerization, it is possible to literally tailor-make large molecules.

27-9 Synthesis of Organic Compounds

Originally, all organic compounds were isolated from natural sources. However, as chemists developed an understanding of the chemical behavior of organic compounds, they began to devise methods of synthesizing compounds from simple starting materials. Moreover, some of the compounds synthesized by the newly developed methods had never been observed to occur naturally.

In organic synthesis, chemists attempt to transform simple, readily available compounds into more complex molecules with desirable physical and chemical properties. Some syntheses are designed to make biologically active compounds that are otherwise available only from natural sources and in only small quantities at high cost. Other synthetic approaches are designed to make new compounds, similar to naturally occurring ones but even more powerful in biological activity, such as medications to fight disease.

The approach that a synthetic organic chemist takes is to apply a knowledge of a wide variety of reaction types and reaction mechanisms to devise a synthetic scheme for assembling simple molecules into more complex structures. Figure 27-24 summarizes some of the chemical transformations between functional groups considered in this chapter (and Chapter 26). This summary stresses that each type of reaction has three components: the starting materials, the products, and the required reagents. For any reaction, we can complete a chemical equation if we have information about two of the three components of the reaction and if we know the type of reaction that converts the starting materials to the products. For example, in the following nucleophilic substitution reaction,

$$CH_3Br + ? \longrightarrow CH_3CN + ?$$

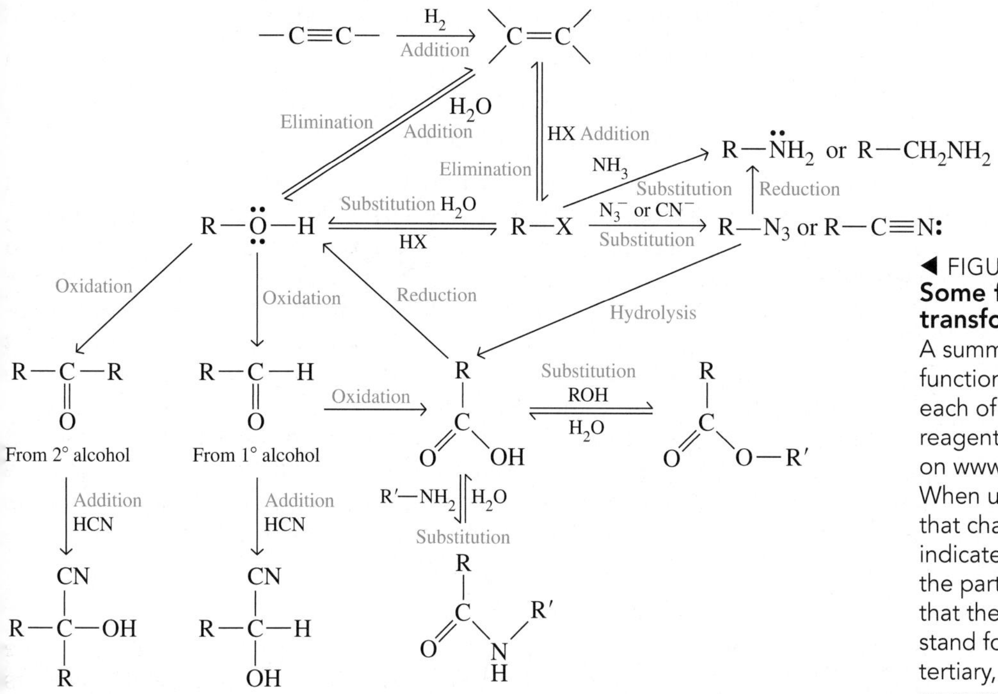

◀ FIGURE 27-24
Some functional group transformations
A summary of some important functional group transformations, each of which requires a specific reagent as described in the text or on www.masteringchemistry.com. When using this diagram, remember that changes in the R group are not indicated because this depends on the particular group involved. Recall that the symbols 1°, 2°, and 3° stand for primary, secondary, and tertiary, respectively.

we see that CN has been substituted for Br as a functional group. To achieve this transformation, the nucleophile CN^- is required, and we complete the equation:

$$CH_3Br + CN^- \longrightarrow CH_3CN + Br^-$$

A somewhat different, and more common, situation is that represented by

$$? + Br^- \longrightarrow CH_3Br + ?$$

where we need to work backward to decide on the electrophile in a nucleophilic substitution reaction to get the desired product, CH_3Br. The electrophile will need to be connected to the leaving group (for example, another halogen atom) in the same manner and place as the Br atom will be in the product. An appropriate complete equation is

$$CH_3I + Br^- \longrightarrow CH_3Br + I^-$$

Let's apply the approach outlined above to the synthesis of ethyl ethanoate, starting only from inorganic substances and carbon (coke). First, we break the desired compound down into its constituent molecules, namely ethanol and ethanoic acid, which can be combined to give the ester:

$$CH_3COOH + HOCH_2CH_3 \longrightarrow H_2O + CH_3COOCH_2CH_3$$

We therefore need to be able to produce ethanol, which can, in turn, be converted into ethanoic acid:

$$HOCH_2CH_3 \xrightarrow{?} CH_3COOH$$

The reagent required to oxidize ethanol is $K_2Cr_2O_7$ in acidic solution. We now need a route to the synthesis of ethanol from carbon. One inorganic reagent that we have at our disposal that contains a —OH group is water. We therefore write the equation

$$? + H_2O \longrightarrow HOCH_2CH_3$$

A consideration of the reactions in Figure 27-24 indicates that one possibility is the addition of water to the double bond in ethene in the presence of sulfuric acid as a catalyst. Thus, we write

$$H_2C{=}CH_2 + H_2O \xrightarrow{H_2SO_4} HOCH_2CH_3$$

To continue, the ethene molecule has to be generated. Ethene is related to ethyne by a hydrogenation reaction:

$$H{-}C{\equiv}C{-}H + H_2 \xrightarrow{Pt} H_2C{=}CH_2$$
heat and pressure

Our source of carbon is the element carbon and we need a way of producing a molecule with a carbon-chain length of two. The production of calcium carbide and its subsequent hydrolysis to give ethyne is described on page 1237. We have now completed the description of the desired synthesis. The underlying strategy is to break the desired molecule down into its constituent molecules and work backward, using the chemical transformations available. This approach is known as *retrosynthesis* and is the method of choice in modern synthetic organic chemistry.

Often, in the design of a synthetic pathway, one or more routes are possible. For example, in the synthesis described above, one of the steps required the synthesis of ethanol. We chose to make it by adding H_2O across the double bond of ethene. However, we could have taken another approach. For example, we could have used the following substitution reaction:

$$CH_3CH_2Cl + OH^- \longrightarrow CH_3CH_2OH + Cl^-$$

To obtain the chloroethane required for this reaction, we could substitute a Cl atom for an H atom in CH_3CH_3 which, in turn, can be produced by the hydrogenation of ethyne.

How do we decide which synthetic pathway is best? Frequently, the chosen pathway is the one that produces the greatest yield of the desired product. However, other factors may need to be considered. What are the yields of the various reactions involved? Are the required reagents readily available and what is their cost? Are the required reagents or intermediate compounds toxic or dangerous, and, if so, are there methods or equipment in place to be able to work with these compounds? Clearly, knowledge of the physical and chemical properties of organic compounds figure prominently in the design of synthetic pathways.

MasteringCHEMISTRY www.masteringchemistry.com

Ionic liquids are organic salts with room temperature melting points. They are environmentally friendly, are able to dissolve most organic molecules, have high thermal stabilities and are nonflammable—properties that make these liquids attractive as solvents for organic synthesis. For a discussion of ionic liquids, go to the Focus On feature for Chapter 27, Green Chemistry and Ionic Liquids, on the MasteringChemistry site.

Summary

27-1 Organic Reactions: An Introduction—Organic compounds undergo a variety of reactions. In a **substitution reaction**, an atom, an ion, or a group in one molecule is replaced by another. In an **elimination reaction**, atoms or groups that are bonded to adjacent atoms are removed as a small molecule. In an **addition reaction**, a molecule adds across a double or triple bond in another molecule. In a **rearrangement reaction**, the carbon skeleton (or constitution) of a molecule is rearranged.

27-2 Introduction to Nucleophilic Substitution Reactions—A common reaction involving sp^3 hybridized carbon atoms is a **nucleophilic substitution reaction**, in which a **nucleophile** attacks the **electrophilic** carbon atom of the **substrate** yielding the product and a **leaving group**. The leaving group is the substituent that is displaced from the electrophilic carbon atom during the reaction. The nucleophile contains a pair of electrons that seeks the positive carbon atom of the other reactant. **Basicity** is a measure of the tendency of an electron pair donor to react with a proton. The **nucleophilicity** of an electron pair donor depends on a number of factors, including whether the nucleophile is neutral or negatively charged; the electronegativity and size of the nucleophilic atom; whether the nucleophilic atom is sterically hindered; and the solvent. Two mechanisms for nucleophilic substitution reactions are $\mathbf{S_N1}$ and $\mathbf{S_N2}$. The S_N1 mechanism involves two steps: (1) a unimolecular rate-determining step involving the formation of a carbocation, followed by (2) nucleophilic attack of the carbocation (Figs. 27-3 and 27-4). S_N1 reactions occur most rapidly for reactions involving tertiary substrates and weak nucleophiles reacting in a polar **protic solvent**. Methyl and primary substrates do not undergo S_N1 reactions (Table 27.2). The S_N2 mechanism is a concerted bimolecular reaction in which a nucleophile attacks the electrophilic carbon atom from the backside (Figs. 27-1 and 27-2). S_N2 reactions occur most rapidly for reactions involving methyl and primary substrates and good or strong nucleophiles reacting in a polar **aprotic solvent**. If the electrophilic carbon atom is chiral, and if both the nucleophile and leaving group are high in the *R/S* priority order, then backside attack (S_N2) leads to an inversion of configuration ($R \leftrightarrow S$) at the chiral carbon.

27-3 Introduction to Elimination Reactions—Under appropriate conditions, haloalkanes undergo elimination reactions to form alkenes. Usually, the major elimination product is the one that is the most highly substituted. An **E1 reaction** is an elimination reaction with a unimolecular rate-determining step, involving carbocation formation, followed by removal of a proton from a β carbon atom (Figs. 27-10 and 27-11). An **E2 reaction** is a concerted, bimolecular reaction (Fig. 27-12). Elimination reactions compete with nucleophilic substitution reactions. S_N1 and E1 reactions always occur together. The use of a strong base favors elimination by E2 and the use of a weak base favors substitution by either S_N1 or S_N2 (Fig. 27-13).

27-4 Reactions of Alcohols—Alcohols undergo a variety of reactions (Table 27.3). Like haloalkanes, alcohols can undergo substitution or elimination reactions. These reactions require the use of an acid catalyst (Figs. 27-14 and 27-15). When an alcohol acts as the substrate in a substitution reaction, the —OH group is replaced by another group. When an alcohol acts as the substrate in an elimination reaction, an alkene and a water molecule are produced (reaction 27.6). The elimination of water from an alcohol is also called a **dehydration reaction**.

27-5 Introduction to Addition Reactions: Reactions of Alkenes—The characteristic reaction of alkenes is the addition of one substituent to each of the carbon atoms in the double bond (reaction 27.7). A variety of addition reactions are possible: hydrogenation; addition of hydrogen halides (reaction 27.8); hydration (reaction 27.9), producing a **vicinal dihalide** or a **geminal dihalide**; and halogenation (reaction 27.10). In these reactions, a transfer of electron density occurs from the π bond of the alkene to an electrophilic atom. In all these reactions—except halogenation—the electrophilic atom tends to add to the least-substituted carbon atom of the double bond, yielding a carbocation that has the positive charge localized on the more highly substituted carbon atom. In a halogenation reaction, a **bridged halonium ion** is formed in which a halogen atom is bonded to both carbon atoms (Fig. 27-20). The halonium ion is then attacked from the backside by a halide ion.

27-6 Electrophilic Aromatic Substitution—Benzene reacts with electrophiles in substitution reactions in which any of the six H atoms is replaced by another atom or group. In a nitration reaction, $-H$ is replaced by $-NO_2$; in a halogenation reaction, $-H$ is replaced by a halogen atom ($-X$). Nitration and halogenation of benzene are examples of an **electrophilic substitution reaction** (reaction 27.11). An electrophilic substitution reaction involves two steps: (1) an electrophilic attack of a carbon atom in the benzene ring, forming a carbocation called an **arenium ion**; and (2) the loss of a proton to a base (Fig. 27-21). Derivatives of benzene can also undergo substitution reactions. A substituent that is bonded to the carbon atom in a benzene ring can be *ortho, para* directing, or *meta* directing.

27-7 Reactions of Alkanes—Alkanes and other hydrocarbons react with oxygen during combustion to form carbon dioxide and water (reaction 27.14). Alkanes are relatively unreactive except with the halogens, which can substitute for hydrogen atoms in a substitution reaction (reaction 27.15). The halogenation of alkanes occurs by a chain reaction that involves initiation, propagation, and termination steps. Among the halogens, bromine shows the greatest selectivity for which hydrogen atom in an alkane is replaced: $3° > 2° > 1° >$ methyl.

27-8 Polymers and Polymerization Reactions—Polymers are a very important class of substances in modern life. One method of polymerization, the joining of monomers to form polymers, is called **step-reaction polymerization**, or condensation polymerization. It produces polymers of moderate molecular mass. Another type of polymerization mechanism is **chain-reaction polymerization**, a three-step process involving initiation (reaction 27.16), propagation (reaction 27.17), and termination (reaction 27.18), producing polymers of high molecular mass.

27-9 Synthesis of Organic Compounds—The synthesis of larger and more complex organic compounds from simpler precursors can be achieved through a sequence of reactions of different types and varied mechanisms. Figure 27-24 summarizes a number of functional group transformations that can be used in designing synthetic strategies. One powerful strategy is to work backward from the desired product to the starting materials, a method known as retrosynthesis.

Integrative Example

The female of a species of worm produces the sex attractant *spodoptol*, which has the following structure.

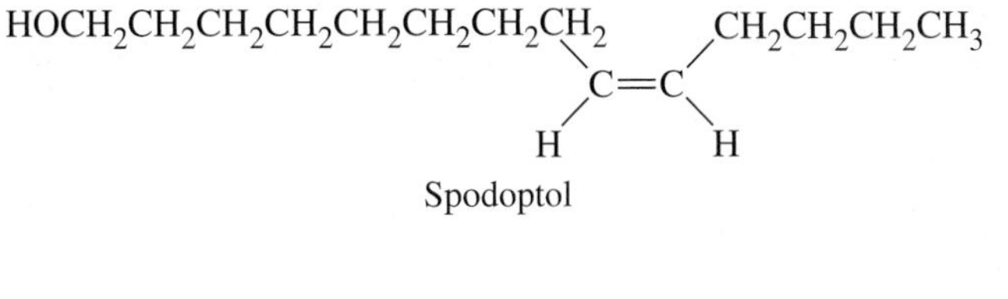

Spodoptol

Spodoptol is a pheromone that attracts the male worms of the species. Synthetic spodoptol in traps can be used to control the population of worms. Devise a synthesis of spodoptol starting from the alcohol shown below.

$$HOCH_2CH_2CH_2CH_2CH_2CH_2CH_2CH_2C{\equiv}CH$$

Analyze

In designing the synthesis, we break down the spodoptol molecule into the starting material and the necessary alkyl group that has to be attached to a double bond, namely a *n*-butyl group. In order to produce the alkene skeleton structure, we will have to convert the triple bond in the starting material to a double bond with *cis* stereochemistry.

Solve

Working backward, this *cis* isomer is obtained by using Lindlar's catalyst (recall page 1300).

$$HOCH_2(CH_2)_6CH_2C{\equiv}C(CH_2)_3CH_3 \xrightarrow[\text{Lindlar's catalyst}]{H_2}$$

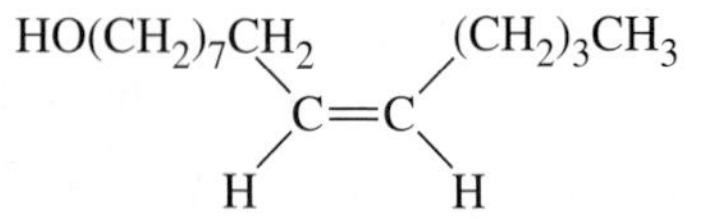

Now we need to make the alkyne needed in the above hydrogenation. The starting material provided lacks the butyl group at the triple bond. To attach the alkyl group, we can perform a nucleophilic substitution at 1-bromobutane with the acetylide ion generated from the starting material, by using sodium amide:

$$^{-}O(CH_2)_8C{\equiv}C{:}^{-} + CH_3(CH_2)_3Br \longrightarrow {^{-}O(CH_2)_8C{\equiv}C(CH_2)_3CH_3} + Br^{-}$$

Because the alcohol proton of the acetylide is also acidic, we need to add an *excess* of sodium amide to give the combined alkoxide and acetylide. (Ammonia is also produced in the reaction.)

$$HO(CH_2)_8C{\equiv}CH \xrightarrow[NaNH_2]{\text{excess}} {^{-}O(CH_2)_8C{\equiv}C{:}^{-}}$$

We can now combine these steps into a complete synthesis where we have converted the alkoxide to the alcohol in the next-to-last step by adding ethanol. The ethanol gives up a proton to the alkoxide we have synthesized (the ethoxide ion is the weaker base).

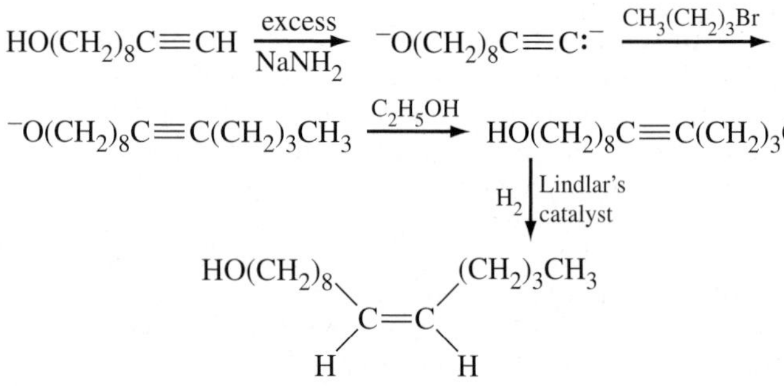

Assess

By using the backward synthetic approach—retrosynthesis—we have established a possible synthetic pathway for the synthesis of spodoptol from the starting material. This is a common approach in the development of a synthetic route to a target compound. A proposed synthesis should always be carefully examined to assess whether side reactions are possible because they can decrease the yield of the desired product. In the present case, a side reaction is possible in the second step above. The $—O(CH_2)_8CC—$ ion could also react with $CH_3(CH_2)_3Br$ to give an ether.

PRACTICE EXAMPLE A: How could cyclohexa-1,3-diene be synthesized from cyclohexane?

PRACTICE EXAMPLE B: How could the compound at right be prepared from benzene by using electrophilic aromatic substitution reactions discussed in this chapter?

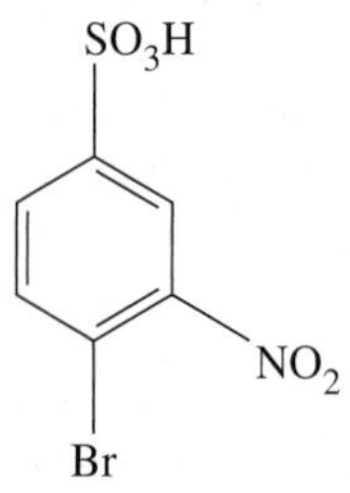

Exercises

Types of Organic Reactions

1. Describe what is meant by each of the following reaction types, and illustrate with an example: **(a)** nucleophilic substitution reaction; **(b)** electrophilic substitution reaction; **(c)** addition reaction; **(d)** elimination reaction; **(e)** rearrangement reaction.
2. Describe what is meant by each of the following reaction types, and illustrate with an example from the text: **(a)** dehydration; **(b)** hydrolysis; **(c)** solvolysis; **(d)** hydration of an alkene.
3. Identify the following types of reactions.

(a) 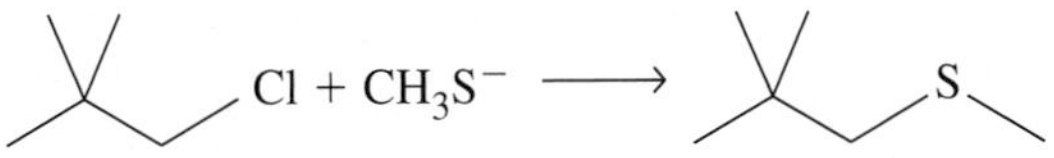

(b) 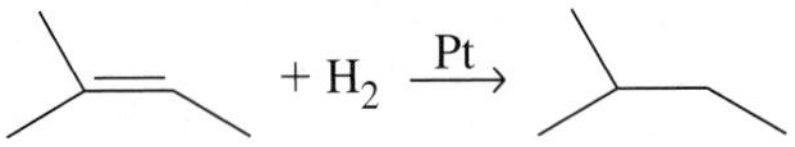

(c) 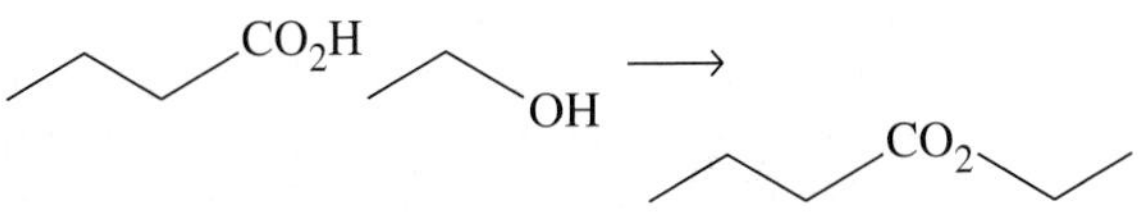

4. Identify the following types of reactions.

(a) 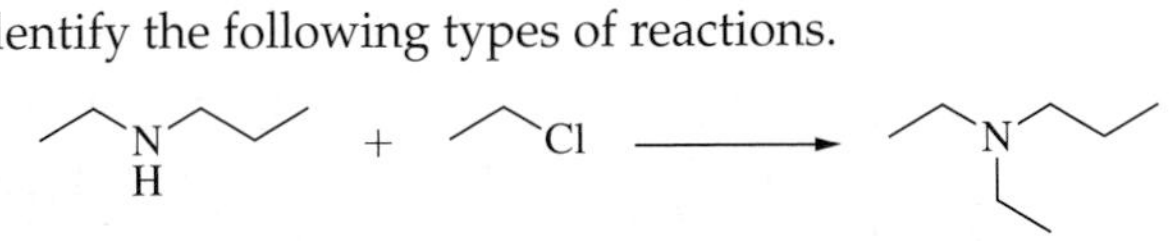

(b) 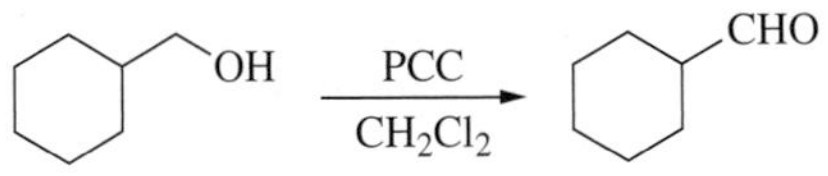

(c) 

(d) 

Note: PCC is pyridinium chlorochromate (see page 1245).

5. Write a balanced chemical equation for the reaction that is described and then classify the reaction as a substitution, an elimination, an addition, or a rearrangement reaction:
(a) Ethene and Br_2 react in carbon tetrachloride to give 1,2-dibromoethane.
(b) Iodoethane reacts with KOH(aq) yielding ethene, water, and potassium bromide.
(c) Chloromethane reacts with NaOH(aq) to give methanol and sodium chloride.
6. Write a balanced chemical equation for the reaction that is described and then classify the reaction as a substitution, an elimination, an addition, or a rearrangement reaction.
(a) 3,3-dimethylbut-1-ene reacts in acid solution to yield 2,3-dimethylbut-2-ene.
(b) 1-iodo-2,2-dimethylpropane reacts with water to produce, 2,2-dimethylpropan-1-ol, and HI(aq).
(c) 2-chloro-2-methylpropane reacts with NaOH(aq) to give 2-methylprop-1-ene, sodium chloride, and water.

Substitution and Elimination Reactions

7. Write equations for the substitution reaction of 1-bromobutane, a typical primary haloalkane, with the following reagents: **(a)** NaOH; **(b)** NH_3; **(c)** NaCN; **(d)** CH_3CH_2ONa.
8. Write equations for the substitution reaction of 1-bromo-3-methylbutane, a primary haloalkane, with the following reagents: **(a)** NaN_3; **(b)** CH_3CH_2ONa; **(c)** $CH_3C{\equiv}CNa$; **(d)** CH_3CH_2SNa; **(e)** CH_3COONa.
9. Answer the following questions for this S_N2 reaction:

$$CH_3CH_2CH_2CH_2Br + NaOH \longrightarrow CH_3CH_2CH_2CH_2OH + NaBr$$

(a) What is the rate expression for the reaction?
(b) Draw the reaction profile for the reaction. Label all parts. Assume that the products are lower in energy than the reactants.
(c) What is the effect on the rate of the reaction of doubling the concentration of 1-bromobutane?
(d) What is the effect on the rate of the reaction of halving the concentration of sodium hydroxide?

10. Answer the following questions for this S_N1 reaction:

$$CH_3CH_2CH_2C(CH_3)_2Br + CH_3CH_2OH \longrightarrow CH_3CH_2CH_2C(CH_3)_2OCH_2CH_3 + HBr$$

(a) What is the rate expression for the reaction?
(b) Draw the reaction profile for the reaction. Label all parts. Assume that the products are lower in energy than the reactants.
(c) What is the effect on the rate of the reaction of doubling the concentration of 2-bromo-2-methylpentane?
(d) The solvent for the reaction is ethanol. What is the effect on the rate of the reaction of adding more ethanol?

11. Answer the following questions for this E2 reaction:

$$CH_3CH_2CH_2Br + NaOH \longrightarrow CH_3CH{=}CH_2 + NaBr + H_2O$$

(a) What is the rate expression for the reaction?
(b) Draw the reaction profile for the reaction. Label all parts. Assume that the products are lower in energy than the reactants.
(c) What is the effect on the rate of the reaction of doubling the concentration of $CH_3CH_2CH_2Br$?
(d) What is the effect on the rate of the reaction of halving the concentration of NaOH?

12. Answer the following questions for this E1 reaction:

$$CH_3CH_2C(CH_3)_2Br + CH_3OH \longrightarrow CH_3CH{=}C(CH_3)_2 + Br^- + CH_3OH_2^+$$

(a) What is the rate expression for the reaction?
(b) Draw the reaction profile for the reaction. Label all parts. Assume that the products are lower in energy than the reactants.
(c) What is the effect on the rate of the reaction of doubling the concentration of $CH_3CH_2C(CH_3)_2Br$?
(d) What is the effect on the rate of the reaction of doubling the concentration of CH_3OH?

13. Identify the nucleophile, electrophile, and leaving group in each of the following substitution reactions. Predict whether equilibrium favors the reactants or products:
(a) $CH_3CH_2ONa + CH_3CH_2CH_2I \rightleftharpoons CH_3CH_2CH_2OCH_2CH_3 + NaI$
(b) $CH_3CH_2NH_3^+ + KI \rightleftharpoons NH_3 + CH_3CH_2I + K^+$

14. Identify the nucleophile, electrophile, and leaving group in each of the following substitution reactions. Predict whether equilibrium favors the reactants or products:
(a) $CH_3CH_2CH_2CH_2I + NaN_3 \longrightarrow CH_3CH_2CH_2CH_2N_3 + NaI$
(b) $CH_3CH_2NH_2 + NaBr \longrightarrow CH_3CH_2Br + NaNH_2$

15. Molecule (a) below reacts faster in the S_N2 reaction than does molecule (b). Explain this observation. [*Hint*: Draw chair structures for the most stable conformations of the two molecules.]

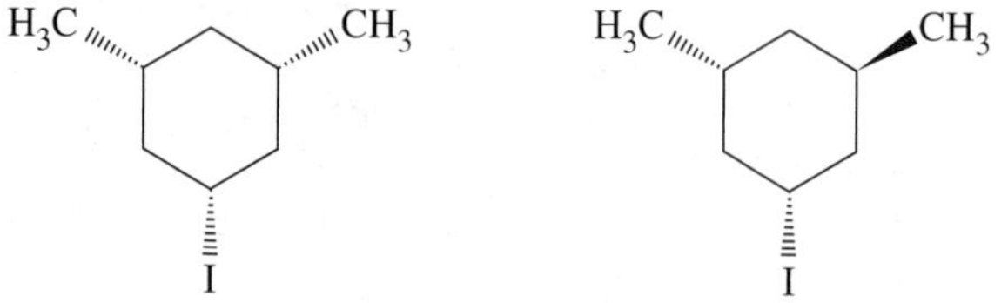

(a) (b)

16. Molecule (a) below reacts faster in the S_N1 reaction than does molecule (b). Explain this observation. [*Hint*: Draw chair structures for the two molecules.]

(a) (b)

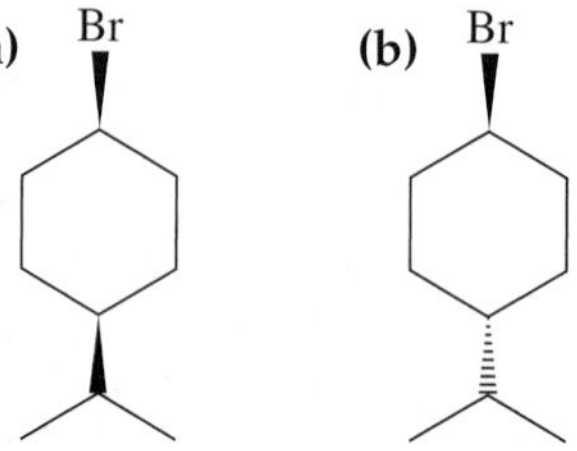

17. A sample of (*S*)-$CH_3CH_2CH(CH_3)Cl$ is hydrolyzed by water, and the resulting solution is optically inactive. **(a)** Write the formula of the product. **(b)** By which nucleophilic substitution reaction mechanism does this reaction occur?

18. A sample of (*R*)-$CH_3CH_2CH(CH_3)Cl$ reacts with CH_3O^- in dimethyl sulfoxide, $(CH_3)_2SO$, a convenient solvent for organic reactions. The resulting solution is optically active. **(a)** Write the formula of the product. **(b)** By which mechanism does this nucleophilic substitution reaction occur?

19. A sample of (*S*)-$CH_3CH(Cl)CH_2CH_3$ is dissolved in ethanol, and the resulting solution is optically inactive. **(a)** Write the formula of the product. **(b)** By which mechanism does this nucleophilic substitution reaction occur?

20. A sample of (*R*)-2-iodobutane, (*R*)-$CH_3CH_2CH(I)CH_3$, reacts with KCN in aqueous medium, and the resulting solution is optically active. **(a)** Write the formula of the product. **(b)** By which mechanism does this nucleophilic substitution reaction occur?

21. Give the major products obtained in each of the following reactions and indicate which mechanisms are involved:

(a)

(b)

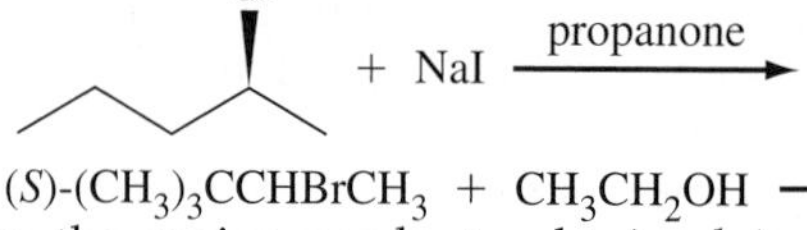

(c) (*S*)-$(CH_3)_3CCHBrCH_3 + CH_3CH_2OH \longrightarrow$

22. Give the major products obtained in each of the following reactions and indicate which mechanisms are involved:

(a)

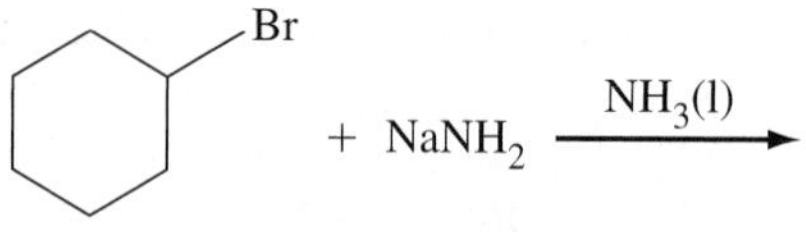

(b) $(CH_3)_2CHCH_2CH_2CH_2Br + NaOCH_2CH_3 \xrightarrow{CH_3CH_2OH}$

(c) (*R*)-2-iodobutane + NaHS $\xrightarrow{DMSO}$

Alcohols and Alkenes

23. Predict the major organic product obtained in each of the following reactions. Assume that [O] represents $Na_2Cr_2O_7$ in H_2SO_4:
(a) $(CH_3)_2CHCH_2OH + HBr \longrightarrow$
(b) $(CH_3)_3COH + K \longrightarrow$
(c) $(CH_3)_2CHOH \xrightarrow{[O]}$
(d) $CH_3CH_2CH_2OH + CH_3CH_2I \longrightarrow$

24. Predict the products of the following reactions. Assume that [O] represents $Na_2Cr_2O_7$ in H_2SO_4:
(a) $(CH_3)_2CHCH_2OH \xrightarrow{[O]}$
(b) $CH_3CH_2OH + (CH_3)_2CHCOOH \xrightarrow[\Delta]{H^+}$
(c) 3-methylbut-2-ol + NaCl $\xrightarrow{H_2SO_4}$
(d) $CH_3CH_2CH_2OH \xrightarrow[\Delta]{H_2SO_4}$

25. Explain how you could carry out the following conversion. Write a mechanism for the reaction:

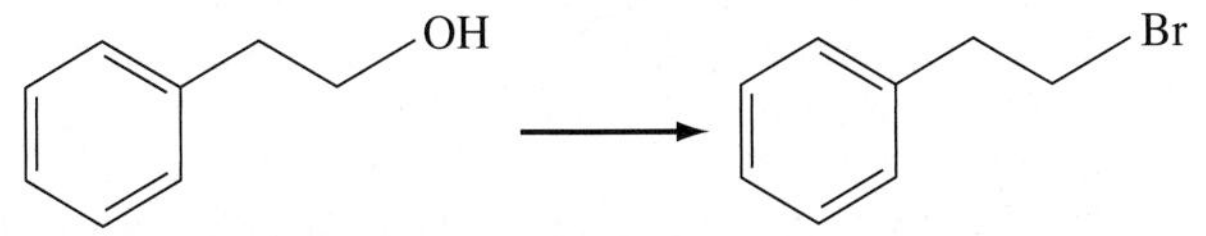

26. Explain how you could carry out the following conversion. Write a mechanism for the reaction:

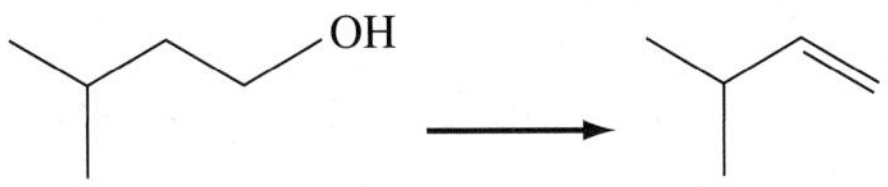

27. As discussed in Are You Wondering 27-2, carbocation rearrangements occur in some substitution and elimination reactions. What elimination product would you expect in the following reaction *if no rearrangement of the carbocation occurs*? Experiment reveals that the main product is 2,3-dimethylpent-2-ene. Suggest a mechanism that shows how 2,3-dimethylpent-2-ene is formed. [*Hint:* Consider the formation and rearrangement of a carbocation.]

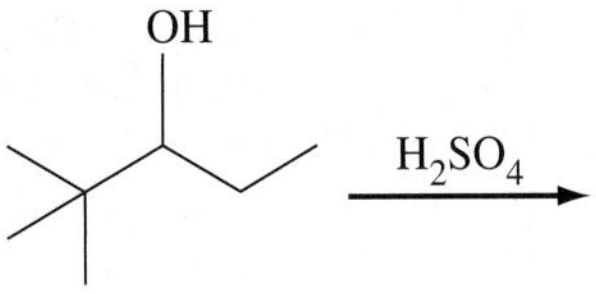

28. The molecule in the next column is 2,2-dimethylpropan-1-ol, a primary alcohol. Because no H atoms are bonded to the β carbon atom in this molecule, dehydration seems unlikely. However, when 2,2-dimethylpropan-1-ol is heated with an acid catalyst, 2-methylbut-2-ene is obtained. Suggest a mechanism that shows how 2-methylbut-2-ene is formed. [*Hint:* Consider the formation and rearrangement of a carbocation. See Are You Wondering 27-2.]

29. Draw the structures of the products of each of the following reactions:
(a) propene + hydrogen (Pt, heat)
(b) butan-2-ol + heat (in the presence of sulfuric acid)
30. Predict the product(s) of the reaction of:
(a) HCl with 2-chloroprop-1-ene
(b) HCN with $CH_3CH{=}CH_2$
(c) HCl with $CH_3CH{=}C(CH_3)_2$
31. Give the structures of the main organic product(s) in each of the following reactions.
(a)

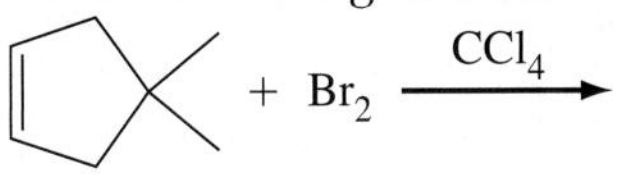

(b) (Z)-but-2-ene + HI $\longrightarrow$
(c) 2-methylbut-1-ene + $H_2O \xrightarrow{H_2SO_4}$
32. Give the structure of the main organic product(s) in each of the following reactions:
(a) (E)-but-2-ene + HBr $\longrightarrow$
(b) 2-methylpent-2-ene + $H_2O \xrightarrow{H_2SO_4}$
(c)

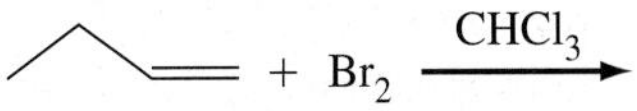

33. Give the major product that forms when (Z)-3-methylpent-2-ene reacts with each of the following reagents: **(a)** HI; **(b)** H_2 in the presence of a platinum catalyst; **(c)** H_2O in H_2SO_4; **(d)** Br_2 in CCl_4.
34. Give the major product that forms when 1-ethylcyclohex-1-ene reacts with each of the following reagents: **(a)** HI; **(b)** H_2 in the presence of a platinum catalyst; **(c)** H_2O in H_2SO_4; **(d)** Br_2 in CCl_4.

Electrophilic Aromatic Substitution

35. Sketch the reaction profile for the mechanism depicted in Figure 27-21, by assuming the overall reaction is exothermic. Also, experimental evidence suggests that the activation energy for the first step is greater than the activation energy for the second step. Draw structures for the intermediate and transition states.
36. The chlorination of benzene can be carried out by allowing benzene to react with Cl_2 in the presence of $AlCl_3$. As discussed on page 1305, Cl_2 and $AlCl_3$ react to form $Cl^+AlCl_4^-$. Sketch the reaction profile for the reaction of benzene and $Cl^+AlCl_4^-$. Use the same assumptions as in Exercise 35.
37. Alkylation of benzene can be accomplished by treating benzene with haloalkane (RX) in the presence of $AlCl_3$. The reaction is known as a Friedel–Crafts alkylation reaction. (The reaction is named after Charles Friedel, a French chemist, and James M. Crafts, an American chemist, who discovered this method of making alkylbenzenes in 1877.) An example of a Friedel–Crafts alkylation reaction is shown below:

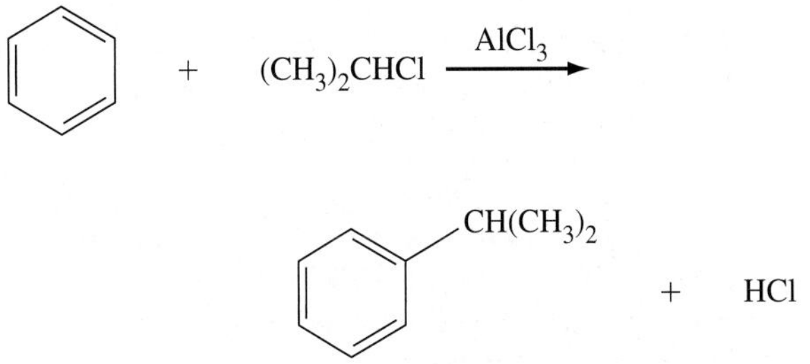

Cumene

The mechanism for this reaction involves the following steps. First, $(CH_3)_2CHCl$ and $AlCl_3$ react in a Lewis acid–base reaction to form an adduct, $(CH_3)_2CH{-}Cl{-}AlCl_3$, in which a chlorine atom is bonded to both carbon and aluminum. The adduct then dissociates to $(CH_3)_2CH^+$, a carbocation, and $AlCl_4^-$. The carbocation acts as an electrophile in a reaction with benzene, forming an arenium ion. Finally, a proton is removed from the arenium ion by $AlCl_4^-$, yielding an alkylbenzene, HCl, and $AlCl_3$. Write chemical equations for the elementary processes involved in forming cumene and HCl from $(CH_3)_2CHCl$ and benzene. Use curved arrows to show the movement of electrons.

38. Treating benzene with fuming sulfuric acid yields benzenesulfonic acid, which is formed by the following reaction:

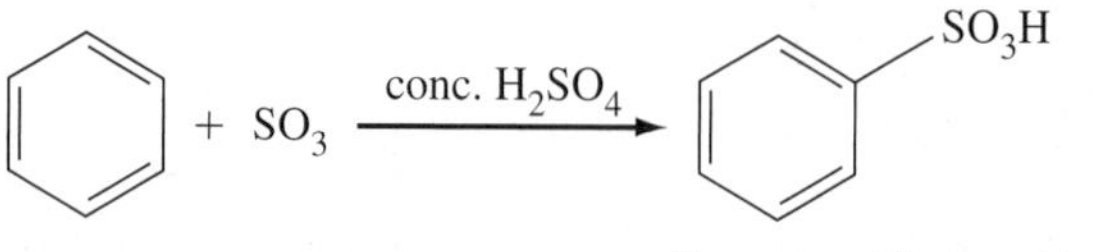

Benzenesulfonic acid

The SO_3 that participates in the reaction above is formed by the reaction of sulfuric acid molecules:

$$2\,H_2SO_4 \longrightarrow SO_3 + HSO_4^- + H_3O^+$$

The reaction of benzene and SO_3 to give benzenesulfonic acid involves the following elementary processes. SO_3 acts as an electrophile in a reaction with benzene, forming an arenium ion. A proton is then transferred from the arenium ion to HSO_4^-, forming the benzenesulfonate ion, $C_6H_5SO_3^-$, and H_2SO_4. Finally, $C_6H_5SO_3^-$ is protonated by H_3O^+ to give benzenesulfonic acid and a water molecule. Write balanced chemical equations for these elementary processes, using curved arrows to show the movement of electrons.

39. Predict the main product(s) of **(a)** the mononitration of chlorobenzene; **(b)** the monosulfonation of nitrobenzene; **(c)** the monochlorination of 1-methyl-2-nitrobenzene.

40. Predict the main product(s) of **(a)** the mononitration of benzoic acid; **(b)** the monosulfonation of phenol; **(c)** the monobromination of 2-nitrobenzaldehyde.

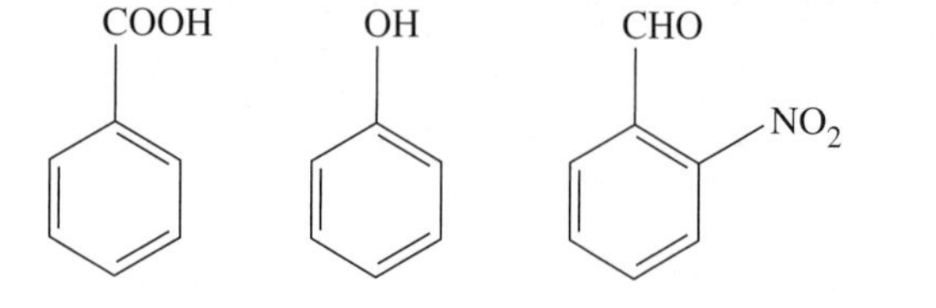

Benzoic acid Phenol 2-nitrobenzaldehyde

Reactions of Alkanes

41. What is major product expected from the monobromination of 2,2,3-trimethylpentane?

42. What are the major products expected from the monobromination of methylcyclohexane? What about monochlorination?

43. **(a)** Write the initiation, propagation, and termination steps involved in the monofluorination of 2,3-dimethylbutane to give 1-fluoro-2,3-dimethylbutane.
(b) Explain why in the monofluorination of 2,3-dimethylbutane the major product is 1-fluoro-2,3-dimethylbutane, not 2-fluoro-2,3-dimethylbutane.

44. Write the initiation, propagation, and termination steps involved in the monobromination of 2,3-dimethylbutane to give 2-bromo-2,3-dimethylbutane.

Polymerization Reactions

45. In referring to the molecular mass of a polymer, we can speak only of the average molecular mass. Explain why the molecular mass of a polymer is not a unique quantity, as it is for a substance like benzene.

46. Explain why Dacron is called a polyester. What is the percent oxygen, by mass, in Dacron?

47. Nylon 66 is produced by the reaction of hexane-1,6-diamine with adipic acid, $HOOC(CH_2)_4COOH$. A different nylon polymer is obtained if sebacoyl chloride is substituted for the adipic acid. What is the basic repeating unit of this nylon structure?

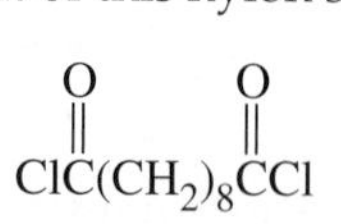

48. Would you expect a polymer to be formed by the reaction of terephthalic acid with ethyl alcohol in place of ethylene glycol? With glycerol in place of ethylene glycol? Explain.

Synthesis of Organic Compounds

49. Starting with acetylene as the only source of carbon, together with any inorganic reagents required, devise a method to synthesize acetaldehyde.

50. Starting with acetylene as the only source of carbon, together with any inorganic reagents required, devise a method to synthesize 1,1,2,2-tetrabromoethane.

51. How would you synthesize (*E*)- and (*Z*)-hept-3-ene from acetylene and any other chemicals?

52. How would you synthesize (*R*)-butan-2-amine from (*S*)-butan-2-ol?

53. The azide anion is a nucleophile and when attached to a carbon atom, undergoes reduction to the amino group and free nitrogen. Suggest a method of preparation of the primary amine propanamine.

54. The cyanide anion is a nucleophile and when attached to a carbon atom, is reduced to a primary amine. Suggest a method of preparing propanamine from chloroethane.

Integrative and Advanced Exercises

55. Draw a structure to represent the principal product of each of the following reactions:
(a) pentan-1-ol + dichromate ion (acid solution)
(b) butyric acid + ethanol (acid solution)
(c) 2-methylbut-1-ene + HBr

56. Predict the products of the monobromination of **(a)** *m*-dinitrobenzene; **(b)** aniline; **(c)** *p*-bromoanisole.

57. Write the formulas of the products formed from the reaction of propene with each of the following substances: **(a)** H_2 with Pt catalyst; **(b)** Cl_2; **(c)** HCl; **(d)** H_2O (in acid).

58. Write the formulas of the products formed from the reaction of but-2-ene with each of the following substances: **(a)** H_2 with Pt catalyst; **(b)** Cl_2; **(c)** HCl; **(d)** H_2O (in acid).

59. Which of the following species gives the reaction indicated? Write the structures of the reaction products.

(1) $CH_3CH_2\overset{\overset{\displaystyle O}{\|}}{C}{-}OCH_3$ (2) $CH_3CH_2\overset{\overset{\displaystyle O}{\|}}{C}{-}OH$ (3) $CH_3\overset{\overset{\displaystyle O}{\|}}{C}{-}O^-$

(a) reacts with dilute HCl
(b) hydrolyzes
(c) neutralizes dilute NaOH

60. Which of the following species gives the reaction indicated? Write the structures of the reaction products.

(1) $CH_3CH_2\overset{\overset{\displaystyle O}{\|}}{C}{-}NH_2$

(2) $CH_3CH_2NH_2$ (3) $CH_3CH_2CH_2NH_3^+Cl^-$

(a) neutralizes dilute HCl
(b) hydrolyzes
(c) neutralizes dilute NaOH

61. Write the formulas of the products expected to form in the following situations. If no reaction occurs, write N.R.
(a) $CH_3CH_2NH_2(aq) + HCl(aq) \longrightarrow$
(b) $(CH_3)_3N(aq) + HBr(aq) \longrightarrow$
(c) $CH_3CH_2NH_3^+(aq) + H_3O^+(aq) \longrightarrow$
(d) $CH_3CH_2NH_3^+(aq) + OH^-(aq) \longrightarrow$

62. Write the formulas of the products expected to form in the following situations. If no reaction occurs, write N.R.

(a) 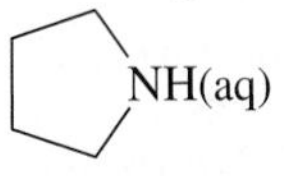 $+ HCl(aq) \longrightarrow$

(b) $(CH_3)_4N^+(aq) + HCl(aq) \longrightarrow$
(c) $CH_3CH_2NH_2(aq) + OH^-(aq) \longrightarrow$
(d) $(CH_3)_3NH^+ + OH^- \longrightarrow$

63. To prepare butan-2-one, which of these compounds would you oxidize: propan-2-ol, butan-1-ol, butan-2-ol, or 2-methylpropan-2-ol? Explain.

64. Indicate the principal product(s) you would expect in **(a)** treating $CH_3CH_2CH{=}CH_2$ with *dilute* $H_2SO_4(aq)$; **(b)** exposing a mixture of chlorine and propane gases to ultraviolet light; **(c)** heating a mixture of isopropyl alcohol and benzoic acid; **(d)** oxidizing butan-2-ol with $Cr_2O_7^{2-}$ in acidic solution.

65. Match the following compounds with the chemical properties given below. Write the structure of the products of the reactions described in (a) to (e).

(1)  (2) 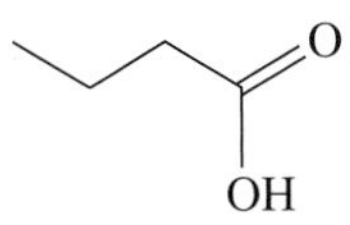

(3) 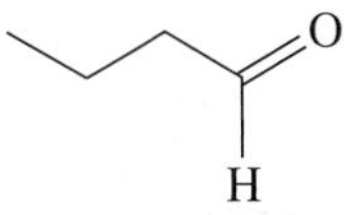

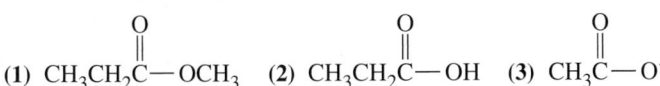

(a) is easily oxidized
(b) neutralizes NaOH(aq)
(c) forms an ester with ethanol
(d) can be oxidized to a carboxylic acid
(e) can be reduced to an alcohol

66. Match the following compounds with the chemical properties given below. Write the structure of the products of the reactions described in (a) to (e).

(1) 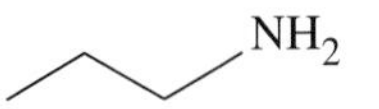 (2) 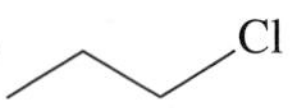 (3) 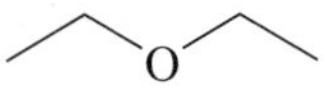

(a) neutralizes HCl(aq)
(b) neutralizes NaOH(aq)
(c) forms an amide with ethanoic acid
(d) reacts with ammonia
(e) reacts with $CN^-(aq)$

67. Write the structures of the isomers you would expect to obtain in the mononitration of *m*-methoxybenzaldehyde:

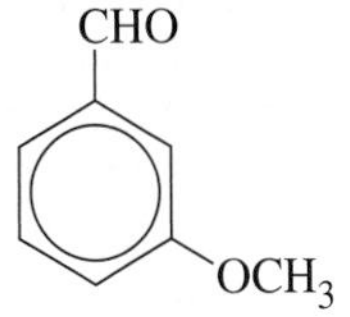

68. In the chlorination of CH_4, some CH_3CH_2Cl is obtained as a product. Explain why this should be so.

69. The three isomeric tribromobenzenes, I, II, and III, when nitrated, form three, two, and one mononitrotribromobenzenes, respectively. Assign correct structures to I, II, and III.

70. Write the name and structure of the benzene derivatives described below.
(a) Formula: C_8H_{10}; forms three monochlorination products when treated with Cl_2 and $FeCl_3$
(b) Formula: C_9H_{12}; forms one monochlorination product when treated with Cl_2 and $FeCl_3$
(c) Formula: C_9H_{12}; forms four monochlorination products when treated with Cl_2 and $FeCl_3$

71. For the monochlorination of hydrocarbons, the following ratio of reactivities has been found: 3° > 2° > 1°, 4.3:3:1. How many different monochloro derivatives of 2-methylbutane are possible, and what percentage of each would you expect to find?

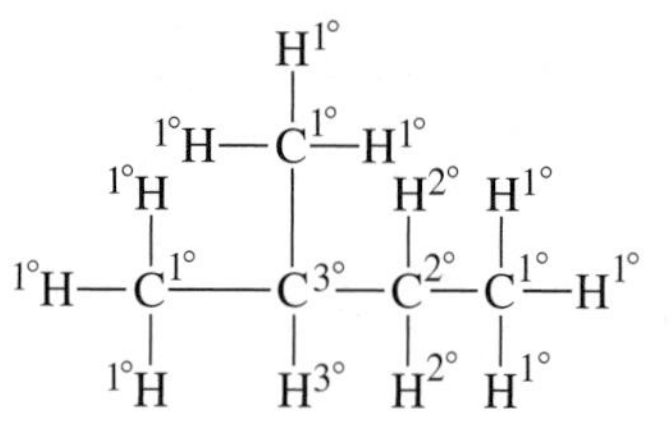

72. The cyanide anion is a nucleophile and, when attached to a carbon atom, undergoes hydrolysis under basic conditions to the carboxylate anion. Suggest a method of preparing sodium butanoate from chloropropane. How can sodium butanoate be converted into butanoic acid?

73. The iodide ion cannot displace the —OH group in ethanol, but excess HI will react to produce ethyl iodide. Explain.

74. Starting with the compounds chloromethane, chloroethane, sodium azide, sodium cyanide, and a reducing agent, suggest how the following compounds could be synthesized.
(a) *N*-methylpropanamide
(b) ethyl ethanoate
(c) *N*-methylethanamine
(d) tetramethylammonium chloride

75. Predict and name the product(s) obtained from the following reaction. Write out the mechanism for the reaction and use curved arrows to show the movement of electrons.

$$(CH_3)_2CHCH{=}CH_2 \xrightarrow{H_2SO_4,\ H_2O}$$

76. Predict and name the product(s) obtained from the following reaction. Write out the mechanism for the reaction and use curved arrows to show the movement of electrons.

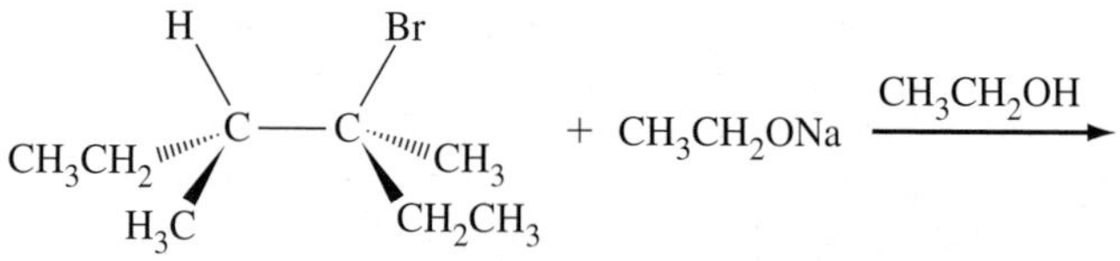

77. Starting with benzene and methane, and suitable inorganic reagents, suggest how the following compound might be synthesized. [*Hint:* See Exercise 37 for a description of how an alkyl group can be added to the benzene ring.]

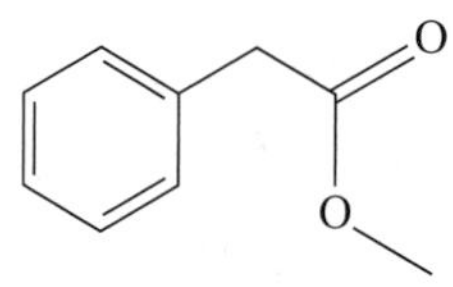

Feature Problem

78. The reduction of aldehydes and ketones with a suitable hydride-containing reducing agent is a good way of synthesizing alcohols. This approach would be even more effective if, instead of a hydride, we could use a source of nucleophilic carbon. Attack by a carbon atom on a carbonyl group would give an alcohol and simultaneously form a carbon-to-carbon bond. How can we make a C atom in an alkane nucleophilic? This was achieved by Victor Grignard, who created the organometallic reagent R—MgBr, with the following reaction in diethyl ether:

$$R{-}Br + Mg \longrightarrow R{-}MgBr$$

The Grignard reagent is rarely isolated. It is formed in solution and used immediately in the desired reaction. The alkylmetal bond is highly polar, with the partial negative charge on the C atom, which makes the C atom highly nucleophilic. The Grignard reagent (R—MgBr) can attack a carbonyl group in an aldehyde or ketone as follows:

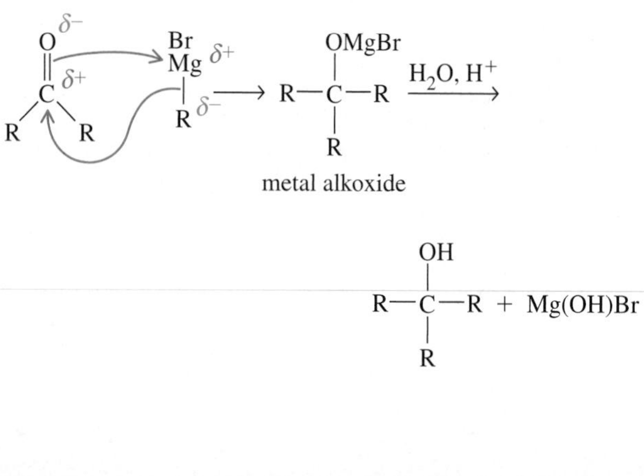

Addition of dilute aqueous acid solution to the metal alkoxide furnishes the alcohol. The important synthetic consequence of this procedure is that we have prepared a product with more carbon atoms than present in the starting material. A simple starting material can be transformed into a more complex molecule.

(a) What is the product of the reaction between methanal and the Grignard reagent formed from 1-bromobutane after the addition of dilute acid?
(b) By using a Grignard reagent, devise a synthesis for hexan-2-ol.
(c) By using a Grignard reagent, devise a synthesis for 2-methylhexan-2-ol.
(d) Grignard reagents can also be formed with aryl halides, such as chlorobenzene. What would be the product of the reaction between the Grignard reagent of chlorobenzene and propanone? Can you think of an alternative synthesis of this product, again using a Grignard reagent?
(e) The basicity of the C atom bound to the magnesium in the Grignard reagent can be used to make Grignard reagents of terminal alkynes. Write the equation of the reaction between ethylmagnesium bromide and hex-1-yne. [*Hint:* Ethane is evolved.]
(f) By using a Grignard reagent, suggest a synthesis for hept-2-yn-1-ol.

Self-Assessment Exercises

79. Explain the important distinctions between each pair of terms: **(a)** nucleophilic substitution and electrophilic aromatic substitution; **(b)** addition and elimination; **(c)** S_N1 and S_N2; **(d)** E1 and E2.
80. Explain the important distinctions between each pair of terms: **(a)** base and nucleophile; **(b)** α carbon and β carbon; **(c)** polar protic solvent and polar aprotic solvent; **(d)** carbocation and radical.
81. **(a)** Which of the nucleophiles CN^- or Cl^- reacts faster with CH_3CH_2I in an S_N2 reaction?
    (b) Which of the substrates, $CH_3CH_2CH(CH_3)CH_2I$ or CH_3I, reacts faster with OH^- in an S_N2 reaction?
82. Which of the following is the strongest nucleophile for an S_N2 reaction? **(a)** H_2O; **(b)** CH_3CH_2OH; **(c)** $CH_3CH_2O^-$; **(d)** $CH_3CO_2^-$; **(e)** CH_3S^-.
83. Which of the following reactions would give a better yield of $CH_3OCH(CH_3)_2$? Explain.

$$CH_3ONa + (CH_3)_2CHI \longrightarrow CH_3OCH(CH_3)_2 + NaI$$
$$(CH_3)_2CHONa + CH_3I \longrightarrow CH_3OCH(CH_3)_2 + NaI$$

84. What is the major elimination product obtained in the following reactions?

(a)

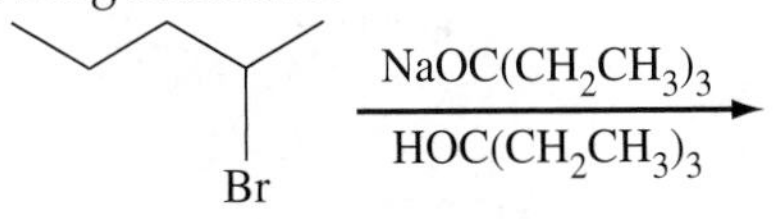

(b)

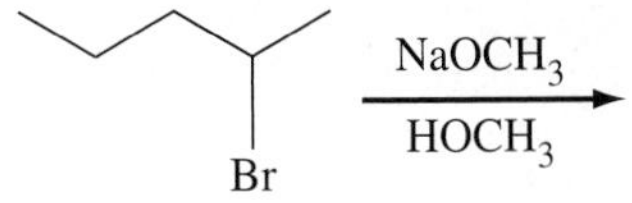

85. What is the major organic product obtained in the following reactions?

(a)

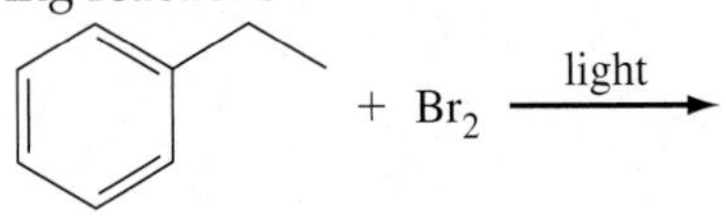

(b)

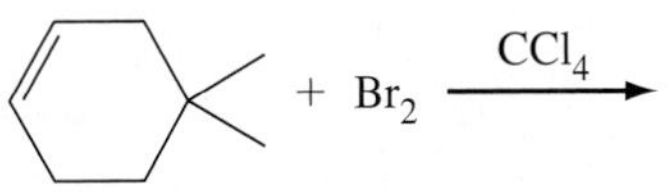

(c)

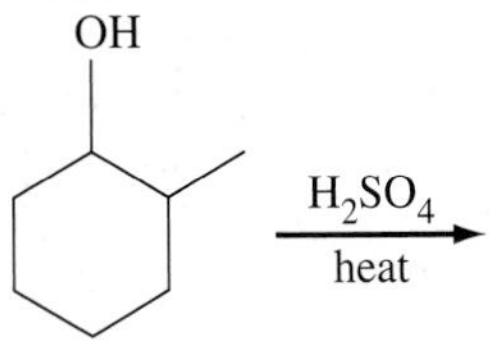

(d)

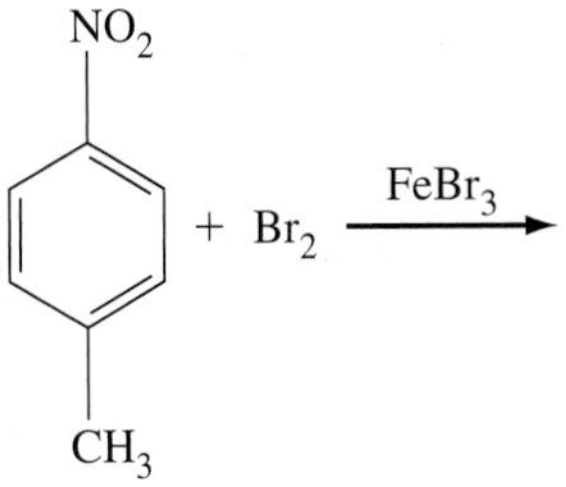

(e)

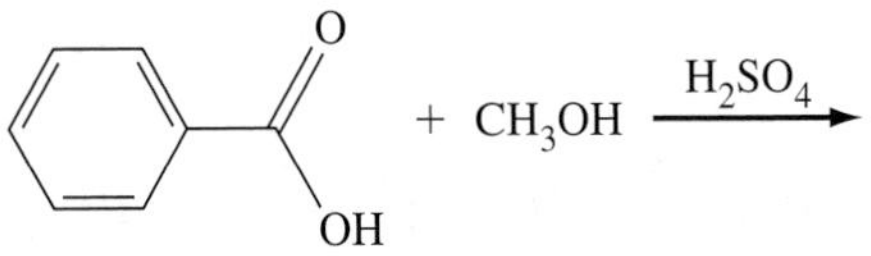

86. When $(CH_3CH_2)_3CBr$ is added to CH_3OH at room temperature, the major product is $CH_3O(CH_2CH_3)_3$ and a minor product is $CH_3CH{=}C(CH_2CH_3)_2$. Write out the mechanisms for the reactions leading to these products and use curved arrows to show the movement of electrons.

28 Chemistry of the Living State

on MasteringChemistry (www.masteringchemistry.com)

CONTENTS

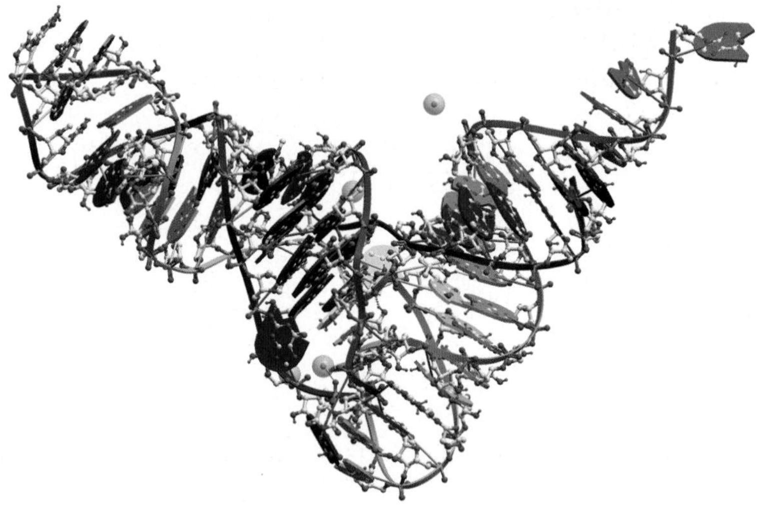

Transfer ribonucleic acid, tRNA, shown as a ribbon model. A ball and stick representation of tRNA is superimposed on the ribbon model. The transparent blue spheres represent magnesium ions, and the green, blue, yellow, orange, and red objects represent the nucleic acid bases. As we learn in this chapter, nucleic acids are important components in the chemistry of life.

The planet Earth teems with life—from the tiny bacterium to the enormous whale. In spite of extraordinary differences in outward appearance, all living organisms share similar needs and chemical structures. Raw materials (food) are required for building cells and providing the energy of metabolism. The cells of almost all forms of living things on Earth use the same types of complicated structures to perform specialized functions. In this chapter, we will stress the structures of the macromolecules common to all organisms. Our discussion will emphasize how fundamental principles of chemistry as presented throughout this textbook can contribute to a knowledge of the living state.

An overview of the molecules that make up living matter—their composition, chemical and physical properties, and their structure—is presented in this chapter. For a complete discussion of these topics, go to Chapter 28, Chemistry of the Living State, on the MasteringChemistry site (www.masteringchemistry.com).

MasteringCHEMISTRY

A

Appendix

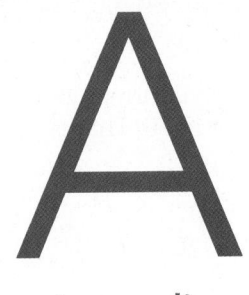

Mathematical Operations

A-1 Exponential Arithmetic

Measured quantities in this text range from very small to very large. For example, the mass of an individual hydrogen atom is 0.00000000000000000000000167 g, and the number of molecules in 18.0153 g of pure water is 602,214,000,000,000,000,000,000. These numbers are difficult to write in conventional form and are even more cumbersome to handle in numerical calculations. We can greatly simplify them by expressing them in exponential form. The *exponential form* of a number consists of a coefficient (a number with value between 1 and 10) multiplied by a power of 10.

The number 10^n is the *nth power* of 10. If n is a *positive* quantity, 10^n is *greater than 1*. If n is a *negative* quantity, 10^n is *between 0 and 1*. The value of 10^0 is 1.

Positive powers	*Negative powers*
$10^0 = 1$	$10^0 = 1$
$10^1 = 10$	$10^{-1} = \dfrac{1}{10} = 0.1$
$10^2 = 10 \times 10 = 100$	$10^{-2} = \dfrac{1}{10 \times 10} = \dfrac{1}{10^2} = 0.01$
$10^3 = 10 \times 10 \times 10 = 1000$	$10^{-3} = \dfrac{1}{10 \times 10 \times 10} = \dfrac{1}{10^3} = 0.001$

To express the number 3170 in exponential form, we write

$$3170 = 3.17 \times 1000 = 3.17 \times 10^3$$

For the number 0.00046 we write

$$0.00046 = 4.6 \times 0.0001 = 4.6 \times 10^{-4}$$

A simpler method of converting a number to exponential form that avoids intermediate steps is illustrated below.

$$3\underbrace{170}_{3\,2\,1} = 3.17 \times 10^3$$

$$0.\underbrace{0004}_{1\,2\,3\,4}6 = 4.6 \times 10^{-4}$$

That is, to convert a number to exponential form,

- Move the decimal point to obtain a coefficient with value between 1 and 10.
- The exponent (power) of 10 is equal to the number of places the decimal point is moved.
- If the decimal point is moved *to the left*, the exponent of 10 is *positive*.
- If the decimal point is moved *to the right*, the exponent of 10 is *negative*.

To convert a number from exponential form to conventional form, move the decimal point the number of places indicated by the power of 10. That is,

$$6.1 \times 10^6 = 6.\underbrace{100000}_{123456} = 6{,}100{,}000$$

$$8.2 \times 10^{-5} = \underbrace{00000}_{54321}8.2 = 0.000082$$

▶ The instructions given here are for a typical electronic calculator. The keystrokes required with your calculator may be somewhat different. Look for specific instructions in the instruction manual supplied with the calculator.

Electronic calculators designed for scientific and engineering work easily accommodate exponential numbers. A typical procedure is to key in the number, followed by the key "EXP" or "EE." Thus, the keystrokes required for the number 6.57×10^3 are

[6] [.] [5] [7] [EXP] [3]

and the result displayed is [6.57^{03}] or [6.57E03] or [$6.57^{\times 10^3}$] or something similar. For the number 6.25×10^{-4}, the keystrokes are

[6] [.] [2] [5] [EXP] [4] [±]

and the result displayed is [6.25^{-04}] or [6.25E − 04] or [$6.25^{\times 10^{-4}}$], or something similar.

Some calculators have a mode setting that automatically converts all numbers and calculated results to the exponential form, regardless of the form in which numbers are entered. In this mode setting you can generally also set the number of significant figures to be carried in displayed results.

Addition and Subtraction

To add or subtract numbers written in exponential form, first express each quantity as *the same power of 10*. Then add and/or subtract the coefficients as indicated. That is, treat the power of 10 as you would a unit common to the terms being added and/or subtracted. In the example that follows, convert 3.8×10^{-3} to 0.38×10^{-2} and use 10^{-2} as the common power of 10.

$$(5.60 \times 10^{-2}) + (3.8 \times 10^{-3}) - (1.52 \times 10^{-2}) = (5.60 + 0.38 - 1.52) \times 10^{-2}$$
$$= 4.46 \times 10^{-2}$$

Multiplication

Consider the numbers $a \times 10^y$ and $b \times 10^z$. Their product is $a \times b \times 10^{(y+z)}$. *Coefficients are multiplied, and exponents are added.*

$$0.0220 \times 0.0040 \times 750 = (2.20 \times 10^{-2})(4.0 \times 10^{-3})(7.5 \times 10^2)$$
$$= (2.20 \times 4.0 \times 7.5) \times 10^{(-2-3+2)} = 66 \times 10^{-3}$$
$$= 6.6 \times 10^1 \times 10^{-3} = 6.6 \times 10^{-2}$$

Division

Consider the numbers $a \times 10^y$ and $b \times 10^z$. Their quotient is

$$\frac{a \times 10^y}{b \times 10^z} = (a/b) \times 10^{(y-z)}$$

Coefficients are divided, and the exponent of the denominator is subtracted from the exponent of the numerator.

$$\frac{20.0 \times 636 \times 0.150}{0.0400 \times 1.80} = \frac{(2.00 \times 10^1)(6.36 \times 10^2)(1.50 \times 10^{-1})}{(4.00 \times 10^{-2}) \times 1.80}$$
$$= \frac{2.00 \times 6.36 \times 1.50 \times 10^{(1+2-1)}}{(4.00 \times 1.80) \times 10^{-2}} = \frac{19.1 \times 10^2}{7.20 \times 10^{-2}}$$
$$= 2.65 \times 10^{(2-(-2))} = 2.65 \times 10^4$$

Raising an Exponential Number to a Power

To "square" the number $a \times 10^y$ means to determine the value $(a \times 10^y)^2$, or the product $(a \times 10^y)(a \times 10^y)$. According to the rule for multiplication, this product is $(a \times a) \times 10^{(y+y)} = a^2 \times 10^{2y}$. When an exponential number is raised to a power, *the coefficient is raised to that power and the exponent is multiplied by the power*. For example,

$$(0.0034)^3 = (3.4 \times 10^{-3})^3 = (3.4)^3 \times 10^{(3)(-3)} = 39 \times 10^{-9} = 3.9 \times 10^{-8}$$

Extracting the Root of an Exponential Number

To extract the root of a number is the same as raising the number to a fractional power. This means that the square root of a number is the number to the one-half power; the cube root is the number to the one-third power; and so on. Thus,

$$\sqrt{a \times 10^y} = (a \times 10^y)^{1/2} = a^{1/2} \times 10^{y/2}$$

$$\sqrt{156} = \sqrt{1.56 \times 10^2} = (1.56)^{1/2} \times 10^{2/2} = 1.25 \times 10^1 = 12.5$$

In the following example, where the cube root is sought, the exponent (-5) is not divisible by 3; the number is rewritten with an exponent (-6) that is divisible by 3.

$$(3.52 \times 10^{-5})^{1/3} = (35.2 \times 10^{-6})^{1/3} = (35.2)^{1/3} \times 10^{-6/3} = 3.28 \times 10^{-2}$$

A-2 Logarithms

The *common* logarithm (log) of a number (N) is the exponent (x) to which the base 10 must be raised to yield the number N. That is, $\log N = x$ means that $N = 10^x = 10^{\log N}$. For simple powers of ten, for example,

$$\log 1 = \log 10^0 = 0$$

$$\log 10 = \log 10^1 = 1 \qquad \log 0.10 = \log 10^{-1} = -1$$

$$\log 100 = \log 10^2 = 2 \qquad \log 0.01 = \log 10^{-2} = -2$$

Most of the numbers that result from measurements and appear in calculations are not simple powers of 10, but it is not difficult to obtain logarithms of these numbers with an electronic calculator. To find the logarithm of a number, enter the number, followed by the "LOG" key.

$$\log 734 = 2.866$$

$$\log 0.0150 = -1.824$$

Another common example requires us to find the number having a certain logarithm. This number is often called the *antilogarithm* or the *inverse* logarithm. For example, if $\log N = 4.350$, what is N? N, the antilogarithm, is simply $10^{4.350}$, and to find its value we enter 4.350, followed by the key "10^x". Depending on the calculator used, it is usually necessary to press the key "INV" or "2nd F" before the log key.

$$\log N = 4.350$$

$$N = 10^{4.350}$$

$$N = 2.24 \times 10^4$$

If the task is to find the antilogarithm of -4.350, we again note that $N = 10^{-4.350}$, and $N = 4.47 \times 10^{-5}$. The required keystrokes on a typical electronic calculator are

[4] [.] [3] [5] [0] [±] [INV] [log]

and the display, to three significant figures, is or [4.47E – 05] or [$4.47^{\times 10^{-05}}$], etc.

Some Useful Relationships

From the definition of a logarithm we can write $M = 10^{\log M}$, $N = 10^{\log N}$, and $M \times N = 10^{\log (M \times N)}$. This means that

$$\log (M \times N) = \log M + \log N$$

Similarly, it is not difficult to show that

$$\log \frac{M}{N} = \log M - \log N$$

Finally, because $N^2 = N \times N$, $10^{\log N^2} = 10^{\log N} \times 10^{\log N}$, and

$$\log N^2 = \log N + \log N = 2 \log N$$

Or, in more general terms,

$$\log N^a = a \log N$$

This expression is especially useful for extracting the roots of numbers. Thus, to determine $(2.5 \times 10^{-8})^{1/5}$, we write

$$\log (2.5 \times 10^{-8})^{1/5} = \tfrac{1}{5}\log (2.5 \times 10^{-8}) = \tfrac{1}{5}(-7.60) = -1.52$$
$$(2.5 \times 10^{-8})^{1/5} = 10^{-1.52} = 0.030$$

Significant Figures in Logarithms

To establish the number of significant figures to use in a logarithm or antilogarithm, use this fundamental rule: All digits to the *right* of the decimal point in a logarithm are significant. Digits to the *left* are used to establish the power of 10. Thus, the logarithm −2.08 is expressed to *two* significant figures. The antilogarithm of −2.08 should also be expressed to *two* significant figures; it is 8.3×10^{-3}. To help settle this point, take the antilogarithms of −2.07, −2.08, and −2.09. You will find these antilogs to be 8.5×10^{-3}, 8.3×10^{-3}, and 8.1×10^{-3}, respectively. Only *two* significant figures are justified.

Natural Logarithms

Logarithms can be expressed to a base other than 10. For instance, because $2^3 = 8$, $\log_2 8 = 3$ (read as, "the logarithm of 8 to the base 2 is equal to 3"). Similarly, $\log_2 10 = 3.322$. Several equations in this text are derived by the methods of calculus and involve logarithms. These equations require that the logarithm be a "natural" one. A *natural* logarithm has the base $e = 2.71828\ldots$. A logarithm to the base "*e*" is usually denoted as *ln*.

The relationship between a "natural" and "common" logarithm simply involves the factor $\log_e 10 = 2.303$. That is, for the number N, $\ln N = 2.303 \log N$. The methods and relationships described for logarithms and antilogarithms to the base 10 all apply to the base e as well, except that the relevant keys on an electronic calculator are "ln" and "e^x" rather than "LOG" and "10^x."

A-3 Algebraic Operations

An algebraic equation is solved when one of the quantities, the unknown, is expressed in terms of all the other quantities in the equation. This effect is achieved when the unknown is present, *alone*, on one side of the equation, and the rest of the terms are on the other side. To solve an equation, a rearrangement of terms may be necessary. The basic principle governing these rearrangements is quite simple. *Whatever is done to one side of the equation must be done to the other as well.*

$3x^2 + 6 = 33$	***Solve for x.***
$3x^2 + \cancel{6} - \cancel{6} = 33 - 6$	(1) Subtract 6 from each side.
$3x^2 = 27$	
$\frac{3x^2}{3} = \frac{27}{3}$	(2) Divide each side by 3.
$x^2 = 9$	
$\sqrt{x^2} = \sqrt{9}$	(3) Extract the square root of each side.
$x = 3$	(4) Simplify. The square root of 9 is 3.

Quadratic Equations

A quadratic equation has the form $ax^2 + bx + c = 0$, where a, b, and c are constants (a cannot be equal to 0). A number of calculations in the text require us to solve a quadratic equation. At times, quadratic equations are of the form

$$(x + n)^2 = m^2$$

Such equations can be solved by extracting the square root of each side.

$$x + n = \pm m \quad \text{and} \quad x = m - n \quad \text{or} \quad x = -m - n$$

More likely, however, the *quadratic formula* will be needed.

$$x = \frac{-b \pm \sqrt{b^2 - 4ac}}{2a}$$

In Example 15-13 on page 718, the following equation must be solved.

$$\frac{(0.300 - x)}{(0.200 + x)(0.100 + x)} = 2.98$$

This is a quadratic equation, but before the quadratic formula can be applied, the equation must be rearranged to the standard form: $ax^2 + bx + c = 0$. This is accomplished in the steps that follow.

$$(0.300 - x) = 2.98(0.200 + x)(0.100 + x)$$
$$0.300 - x = 2.98(0.0200 + 0.300x + x^2)$$
$$0.300 - x = 0.0596 + 0.894x + 2.98x^2$$
$$2.98x^2 + 1.894x - 0.240 = 0 \qquad \textbf{(A.1)}$$

Now we can apply the quadratic formula.

$$x = \frac{-1.894 \pm \sqrt{(1.894)^2 + (4 \times 2.98 \times 0.240)}}{2 \times 2.98}$$
$$= \frac{-1.894 \pm \sqrt{3.587 + 2.86}}{2 \times 2.98}$$
$$= \frac{-1.894 \pm \sqrt{6.45}}{2 \times 2.98} = \frac{-1.894 \pm 2.54}{2 \times 2.98}$$
$$= \frac{-1.894 + 2.54}{2 \times 2.98} = \frac{0.65}{5.96} = 0.11$$

Note that only the (+) value of the (±) sign was used in solving for x. If the (−) value had been used, a negative value of x would have resulted. However, for the given situation a negative value of x is meaningless.

The Method of Successive Approximations

The quadratic equation that was just solved using the quadratic formula can be solved by an alternative method that can be extended to equations of higher order, such as the cubic, quartic, and quintic equations often encountered in solving equilibrium problems. To illustrate the method, suppose we wish to solve expression (A.1) without recourse to the quadratic formula. We can rewrite the equation as follows

$$x = \frac{2.98x^2 - 0.240}{-1.894}$$

and make a guess at the value of x, which we substitute into the right-hand side of the equation to calculate a new value of x. If we guess 0.15, which is reasonable given the starting concentrations involved in Example 15-13, we calculate

$$x = \frac{2.98 \times (0.15)^2 - 0.240}{-1.894} = 0.091$$

We can now use this value of x to calculate a new one.

$$x = \frac{2.98 \times (0.091)^2 - 0.240}{-1.894} = 0.114$$

Repeating this procedure one more time, we get

$$x = \frac{2.98 \times (0.114)^2 - 0.240}{-1.894} = 0.106$$

One more attempt gives a value of 0.11, which is in agreement with the answer previously obtained. The method that we have just used is called the method of successive approximations.

Let us now apply the method of successive approximations to the equation obtained in the Integrative Example of Chapter 15, namely

$$256x^5 - 0.74[(1.00 - x)(1.00 - 2x)^2] = 0 \qquad \textbf{(A.2)}$$

The approach we can take here is to guess a value of x; evaluate the expression to see how close to zero it comes; and then adjust the value of x accordingly. Let us start with a guess of 0.40. The result is

$$256(0.40)^5 - 0.74[(1.00 - 0.40)(1.00 - 2 \times 0.40)^2] = 2.60$$

Clearly the value of 0.40 is too large. If we now try 0.10 we obtain a value of −0.42. We have overshot the value of x. We can now try a value of 0.25 (halfway between our two previous guesses) and obtain 0.11. We realize now that we have to reduce the guessed value of x slightly to get closer to zero. If we try 0.20 we obtain −0.13, and if we next try the value $x = 0.225$ we obtain −0.03. We are now very close to our goal of finding the value of x that satisfies the expression. One final guess of 0.23 gives a value of −0.001, a very satisfactory result.

An alternative approach is to rewrite expression (A.2) as

$$x^5 = \frac{0.74[(1.00 - x)(1.00 - 2x)^2]}{256}$$

and evaluate x as the fifth root of this new expression. If we substitute a value of $x = 0.40$ on the right side, we obtain $x = 0.15$ on the left side. Now by using this value on the right we calculate a new value of $x = 0.26$ on the left. By using this value we obtain $x = 0.22$ on the left, and finally with this last value we obtain $x = 0.23$, in agreement with our previous procedure. Which method we use is a matter of convenience, but the second method provides a new value of x whereas the first method may require more trial and error. When using the method of successive approximations, it is often a useful strategy to take the average of two results in order to speed up the convergence. The method of successive approximations can be very useful, but sometimes, depending how the equation is set up, the convergence to the correct answer may be slow or the process may even diverge. In such circumstances, the expression can be graphed as a function of x to ascertain where the solutions occur. In any event, we must always make sure that any answer obtained is reasonable from a chemical or physical point of view.

A-4 Graphs

Suppose the following sets of numbers are obtained for two quantities x and y by laboratory measurement.

$$x = 0, 1, 2, 3, 4, \ldots$$

$$y = 2, 4, 6, 8, 10, \ldots$$

The relationship between these sets of numbers is not difficult to establish.

$$y = 2x + 2$$

Ideally, the results of experimental measurements are best expressed through a mathematical equation. Sometimes, however, an exact equation cannot be written, or its form is not clear from the experimental data. The graphing of data is very useful in such cases. In Figure A-1 the points listed above are located on a coordinate grid in which x values are placed along the horizontal axis (abscissa) and y values along the vertical axis (ordinate). For each point the x and y values are indicated in parentheses.

The data points are seen to define a straight line. A mathematical equation of a straight line has the form

$$y = mx + b$$

Values of m, the *slope* of the line, and b, the *intercept*, can be obtained from the straight-line graph.

When $x = 0$, $y = b$. The intercept is the point where the straight line intersects the y-axis. The slope can be obtained from two points on the graph.

$$y_2 = mx_2 + b \quad \text{and} \quad y_1 = mx_1 + b$$

$$y_2 - y_1 = m(x_2 - x_1) + \cancel{b} - \cancel{b}$$

$$m = \frac{y_2 - y_1}{x_2 - x_1}$$

From the straight line in Figure A-1, can you establish that $m = b = 2$?

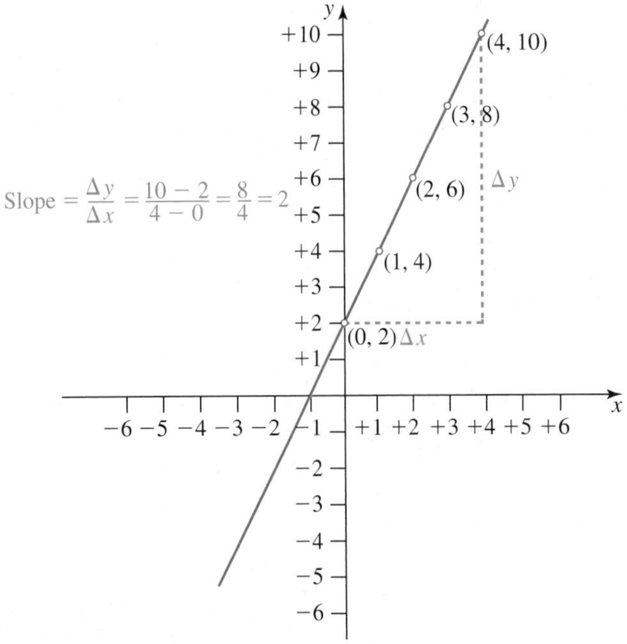

▲ FIGURE A-1
A straight-line graph: $y = mx + b$

The technique used above to eliminate the constant b is applied to logarithmic functions in several places in the text. For example, the expression written below is from page 536. In this expression P is a pressure, T is a Kelvin temperature, and A and B are constants. The equation is that of a straight line.

$$\ln P = -A\left(\frac{1}{T}\right) + B$$

equation of straight line: $$y = m \times x + b$$

We can write this equation twice, for the point (P_1,T_1) and the point (P_2,T_2).

$$\ln P_1 = -A\left(\frac{1}{T_1}\right) + B \quad \text{and} \quad \ln P_2 = -A\left(\frac{1}{T_2}\right) + B$$

The difference between these equations is

$$\ln P_2 - \ln P_1 = -A\left(\frac{1}{T_2}\right) + \cancel{B} + A\left(\frac{1}{T_1}\right) - \cancel{B}$$

$$\ln\frac{P_2}{P_1} = -A\left(\frac{1}{T_2} - \frac{1}{T_1}\right) = A\left(\frac{1}{T_1} - \frac{1}{T_2}\right)$$

A-5 Using Conversion Factors (Dimensional Analysis)

Some calculations in general chemistry require that a quantity measured in one set of units be converted to another set of units. Consider this fact.

$$1\text{ m} = 100\text{ cm}$$

Divide each side of the equation by 1 m.

$$\frac{1\,\cancel{\text{m}}}{1\,\cancel{\text{m}}} = \frac{100\text{ cm}}{1\text{ m}}$$

On the left side of the equation, the numerator and denominator are identical; they cancel.

$$1 = \frac{100\text{ cm}}{1\text{ m}} \tag{A.3}$$

On the right side they are not identical, but they are equal because they do represent the *same length*. The ratio 100 cm/1 m, when multiplied by a length in meters, converts that length to centimeters. The ratio is called a *conversion factor*.

Consider the question, "how many centimeters are there in 6.22 m?" The measured quantity is 6.22 m, and multiplying this quantity by 1 does not change its value.

$$6.22\text{ m} \times 1 = 6.22\text{ m}$$

Now replace the factor "1" by its equivalent—the conversion factor (A.3). Cancel the unit, m, and carry out the multiplication.

$$6.22\,\cancel{\text{m}} \times \underbrace{\frac{100\text{ cm}}{1\,\cancel{\text{m}}}}_{\substack{\text{this factor}\\ \text{converts}\\ \text{m to cm}}} = 622\text{ cm}$$

Next consider the question, "how many meters are there in 576 cm?" If we use the same factor (A.3) as before, the result is nonsensical.

$$576\text{ cm} \times \frac{100\text{ cm}}{1\text{ m}} = 5.76 \times 10^4\text{ cm}^2/\text{m}$$

Factor (A.3) must be rearranged to 1 m/100 cm.

$$576\ \cancel{\text{cm}} \times \underbrace{\frac{1\ \text{m}}{100\ \cancel{\text{cm}}}}_{\substack{\text{this factor}\\ \text{converts}\\ \text{cm to m}}} = 5.76\ \text{m}$$

This second example emphasizes two points.

1. There are two ways to write a conversion factor—in one form or its reciprocal (inverse). Because a conversion factor is equal to 1, its value is not changed by the inversion, but:
2. A conversion factor must be used in such a way as to produce the necessary cancellation of units.

◀ Because of the importance of the cancellation of units, this problem-solving method is often called *unit analysis* or *dimensional analysis*.

Calculations based on conversion factors are always of the form

$$\textit{information sought} = \textit{information given} \times \textit{conversion factor(s)} \qquad \textbf{(A.4)}$$

Often several conversions must be made in sequence in order to get to the desired result. For example, if we want to know how many yards (yd) there are in 576 cm, we find that there is no direct cm → yd conversion factor available. From the inside back cover of the text, however, we do find a conversion factor for cm → in. Thus, we can develop a *conversion pathway*, that is, a series of conversion factors that will take us from centimeters to yards:

$$\text{cm} \rightarrow \text{in.} \rightarrow \text{ft} \rightarrow \text{yd}$$

$$?\ \text{yards} = 576\ \cancel{\text{cm}} \times \frac{1\ \cancel{\text{in.}}}{2.54\ \cancel{\text{cm}}} \times \frac{1\ \cancel{\text{ft}}}{12\ \cancel{\text{in.}}} \times \frac{1\ \text{yd}}{3\ \cancel{\text{ft}}}$$

$$= 6.30\ \text{yd}$$

We can use the same idea of a conversion pathway to deal with the somewhat more challenging situation faced when the units are squared (or cubed). Consider the question, "how many square feet (ft^2) correspond to an area of 1.00 square meter (m^2), given that 1 m = 39.37 in and 12 in = 1 ft?" Here, it may be helpful to begin by drawing a sketch or outline of the situation. Figure A-2 represents an area of 1.00 m^2. Think of it as a square with sides 1 m long. Figure A-2 also represents the length 1 ft and an area of 1.00 ft^2. Do you see that there is somewhat more than 9 ft^2 in 1 m^2?

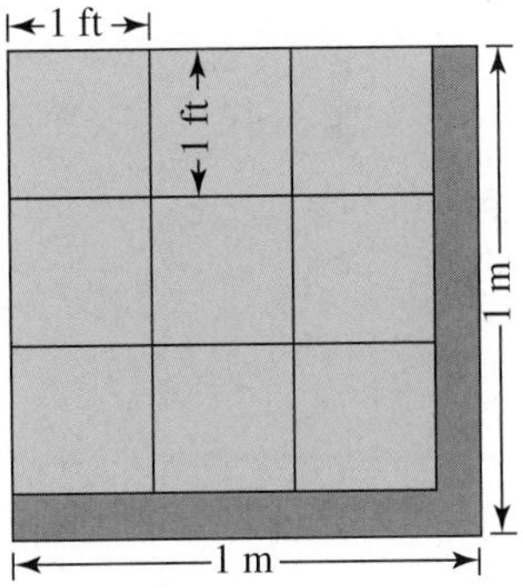

▲ FIGURE A-2
Comparison of one square foot and one square meter
One meter is slightly longer than 3 ft; 1 m^2 is somewhat larger than 9 ft^2.

We can write expression (A.4) as follows:

$$?\ \text{ft}^2 = 1.00\ \text{m}^2 \times \underbrace{\left(\frac{39.37\ \text{in.}}{1\ \text{m}}\right)\left(\frac{39.37\ \text{in.}}{1\ \text{m}}\right)}_{\substack{\text{to convert}\\ \text{m}^2\ \text{to in}^2}} \times \underbrace{\left(\frac{1\ \text{ft}}{12\ \text{in.}}\right)\left(\frac{1\ \text{ft}}{12\ \text{in.}}\right)}_{\substack{\text{to convert}\\ \text{in}^2\ \text{to ft}^2}}$$

This is the same as writing

$$?\ \text{ft}^2 = 1.00\ \cancel{\text{m}^2} \times \frac{(39.37)^2\ \cancel{\text{in.}^2}}{1\ \cancel{\text{m}^2}} \times \frac{1\ \text{ft}^2}{(12)^2\ \cancel{\text{in.}^2}} = 10.8\ \text{ft}^2$$

Another way to look at the problem is to convert the length 1.00 m to feet,

$$?\ \text{ft} = 1.00\ \cancel{\text{m}} \times \frac{39.37\ \cancel{\text{in.}}}{1\ \cancel{\text{m}}} \times \frac{1\ \text{ft}}{12\ \cancel{\text{in.}}} = 3.28\ \text{ft}$$

and square the result

$$?\ \text{ft}^2 = 3.28\ \text{ft} \times 3.28\ \text{ft} = 10.8\ \text{ft}^2$$

Our last example incorporates several of the ideas we have discussed. Here we will examine the situation in which the units in both the numerator and denominator must be converted. Consider the question, "how many meters per second (m/s) correspond to a speed of 63 mph, given that 1 mi = 5280 ft?"

We need to convert from miles to meters in the numerator and from hours to seconds in the denominator. We will need to use conversion factors from elsewhere in Section A-5 in addition to the given value. Also, we must be careful that our conversion factors produce the correct cancellation of units.

$$? \frac{\text{m}}{\text{s}} = \frac{63\ \cancel{\text{mi}}}{1\ \cancel{\text{h}}} \times \frac{1\ \cancel{\text{h}}}{60\ \cancel{\text{min}}} \times \frac{1\ \cancel{\text{min}}}{60\ \text{s}} \times \frac{5280\ \cancel{\text{ft}}}{1\ \cancel{\text{mi}}} \times \frac{12\ \cancel{\text{in.}}}{1\ \cancel{\text{ft}}} \times \frac{1\ \text{m}}{39.37\ \cancel{\text{in.}}}$$

$$= 28\ \frac{\text{m}}{\text{s}}$$

In an alternative approach we break down the problem into three steps: (1) Convert 63 miles to a distance in meters; (2) convert 1 hour to a time in seconds; and (3) express the speed as a ratio of distance over time.

Step 1.

$$\text{distance} = 63\ \cancel{\text{mi}} \times \frac{5280\ \cancel{\text{ft}}}{1\ \cancel{\text{mi}}} \times \frac{12\ \cancel{\text{in.}}}{1\ \cancel{\text{ft}}} \times \frac{1\ \text{m}}{39.37\ \cancel{\text{in.}}} = 1.0 \times 10^5\ \text{m}$$

Step 2.

$$\text{time} = 1\ \cancel{\text{h}} \times \frac{60\ \cancel{\text{min}}}{1\ \cancel{\text{h}}} \times \frac{60\ \text{s}}{1\ \cancel{\text{min}}} = 3.6 \times 10^3\ \text{s}$$

Step 3.

$$\text{speed} = \frac{\text{distance}}{\text{time}} = \frac{1.0 \times 10^5\ \text{m}}{3.6 \times 10^3\ \text{s}} = 28\ \frac{\text{m}}{\text{s}}$$

In conclusion, we have shown (1) how to make a conversion factor; (2) that a conversion factor may be inverted; (3) that a series of conversion factors may be used to make a conversion pathway; (4) that conversion factors may be raised to powers, if necessary; and (5) that conversions of values with units in both the numerator and the denominator (such as miles per hour or pounds per square inch) can be performed in one step or in several steps.

Appendix B

Some Basic Physical Concepts

B-1 Velocity and Acceleration

Time elapses as an object moves from one point to another. The *velocity* of the object is defined as the distance traveled per unit of time. An automobile that travels a distance of 60.0 km in exactly one hour has a velocity of 60.0 km/h (or 16.7 m/s).

Table B.1 contains data on the velocity of a free-falling object. For this motion, velocity is not constant—it increases with time. The falling object "speeds up" continuously. The rate of change of velocity with time is called *acceleration*. Acceleration has the units of distance per unit time per unit time. With the methods of calculus, mathematical equations can be derived for the velocity (u) and distance (d) traveled in a time (t) by an object that has a constant acceleration (a).

$$u = at \qquad \textbf{(B.1)}$$

$$d = \tfrac{1}{2}at^2 \qquad \textbf{(B.2)}$$

For a free-falling object, the constant acceleration, called the *acceleration due to gravity*, is $a = g = 9.8\ \text{m/s}^2$. Equations (B.1) and (B.2) can be used to calculate the velocity and distance traveled by a free-falling object.

TABLE B.1 Velocity and Acceleration of a Free-Falling Body

Time Elapsed, s	Total Distance, m	Velocity, m/s	Acceleration, m/s^2
0	0		
		4.9	
1	4.9		9.8
		14.7	
2	19.6		9.8
		24.5	
3	44.1		9.8
		34.3	
4	78.4		

B-2 Force and Work

Newton's *first law* of motion states that an object at rest remains at rest, and that an object in motion remains in uniform motion unless acted upon by an external force. The tendency for an object to remain at rest or in uniform motion is called *inertia; a force* is what is required to overcome inertia. Since the application of a force either gives an object motion or changes its motion, the actual effect of a force is to change the velocity of an object. Change in velocity is an *acceleration*, so force is what provides an object with acceleration.

Newton's *second law* of motion describes the force F required to produce an acceleration a in an object of mass m.

$$F = ma \qquad \textbf{(B.3)}$$

The basic unit of force in the SI system is the *newton* (N). It is the force required to provide a one-kilogram mass with an acceleration of one meter per second per second.

$$1\ \mathrm{N} = 1\ \mathrm{kg} \times 1\ \mathrm{m\ s^{-2}} \qquad \textbf{(B.4)}$$

The force of gravity on an object (its weight) is the product of the mass of the object and the acceleration due to gravity, g.

$$F = mg \qquad \textbf{(B.5)}$$

Work (w) is performed when a force acts through a distance.

$$\text{work } (w) = \text{force } (F) \times \text{distance } (d) \qquad \textbf{(B.6)}$$

The *joule* (J) is the amount of work associated with a force of one newton (N) acting through a distance of one meter.

$$1\ \mathrm{J} = 1\ \mathrm{N} \times 1\ \mathrm{m} \qquad \textbf{(B.7)}$$

From the definition of the newton in expression (B.4), we can also write

$$1\ \mathrm{J} = 1\ \mathrm{kg} \times 1\ \mathrm{m\ s^{-2}} \times 1\ \mathrm{m} = 1\ \mathrm{kg\ m^2\ s^{-2}} \qquad \textbf{(B.8)}$$

B-3 Energy

Energy is defined as the capacity to do work, but there are other useful descriptions of energy as well. For example, a moving object possesses a kind of energy known as *kinetic energy*. We can obtain a useful equation for kinetic energy by combining some of the other simple equations in this appendix. Thus, because work is the product of a force and distance (equation B.6), and force is the product of a mass and acceleration (equation B.3), we can write

$$w \text{ (work)} = m \times a \times d \qquad \textbf{(B.9)}$$

Now, if we substitute equation (B.2) relating acceleration (a), distance (d), and time (t), into equation (B.9), we obtain

$$w \text{ (work)} = m \times a \times \frac{1}{2}at^2 \qquad \textbf{(B.10)}$$

Finally, let's substitute expression (B.1) relating acceleration (a) and velocity (u) into (B.10). That is, because $a = u/t$,

$$w \text{ (work)} = \frac{1}{2}m\left(\frac{u}{t}\right)^2 t^2 \qquad \textbf{(B.11)}$$

Think of the work in (B.11) as the amount of work necessary to produce a velocity of u in an object of mass m. This amount of work is the energy that appears in the object as kinetic energy (E_k).

$$E_k \text{ (kinetic energy)} = \frac{1}{2}mu^2 \qquad \textbf{(B.12)}$$

An object at rest may also have the capacity to do work by changing its position. The energy it possesses, which can be transformed into actual work, is called *potential energy*. Think of potential energy as energy "stored" within an object. Equations can be written for potential energy, but the exact forms of these equations depend on the manner in which the energy is "stored."

B-4 Magnetism

Attractive and repulsive forces associated with a magnet are centered at regions called *poles*. A magnet has a north and a south pole. If two magnets are aligned such that the north pole of one is directed toward the south pole of the second, an attractive force develops. If the alignment brings like poles into proximity, either both north or both south, a repulsive force develops. *Unlike poles attract; like poles repel.*

A *magnetic field* exists in that region surrounding a magnet in which the influence of the magnet can be felt. Internal changes produced within an iron object by a magnetic field, not produced in a field-free region, are responsible for the attractive force that the object experiences.

B-5 Static Electricity

Another property with which certain objects may be endowed is *electric charge*. Analogous to the case of magnetism, *unlike charges attract, and like charges repel* (recall Figure 2-4). The unit of charge is called a *coulomb*, C. In Coulomb's law, stated below, a *positive* force between electrically charged objects is *repulsive; a negative* force is *attractive*.

$$F = \frac{Q_1 Q_2}{4\pi \varepsilon r^2} \qquad \textbf{(B.13)}$$

where Q_1 is the charge on object 1,

Q_2 is the charge on object 2,

r is the distance between the objects, and

ε is the *dielectric constant*, whose numerical value reflects the effect that the medium separating two charged objects has on the force existing between them. For a vacuum, $\varepsilon = \varepsilon_0 = 8.85419 \times 10^{-12}\ \mathrm{C^2\,N^{-1}\,m^{-2}} = 8.85419 \times 10^{-12}\ \mathrm{C^2\,J^{-1}\,m^{-1}}$; for other media, ε_0 is greater than 1 (for example, for water $\varepsilon = 78.5\varepsilon_0$).

An *electric field* exists in that region surrounding an electrically charged object in which the influence of the electric charge is felt. If an uncharged object is brought into the field of a charged object, the uncharged object may undergo internal changes that it would not experience in a field-free region. These changes may lead to the production of electric charges in the formerly uncharged object, a phenomenon called *induction* (illustrated in Figure B-1).

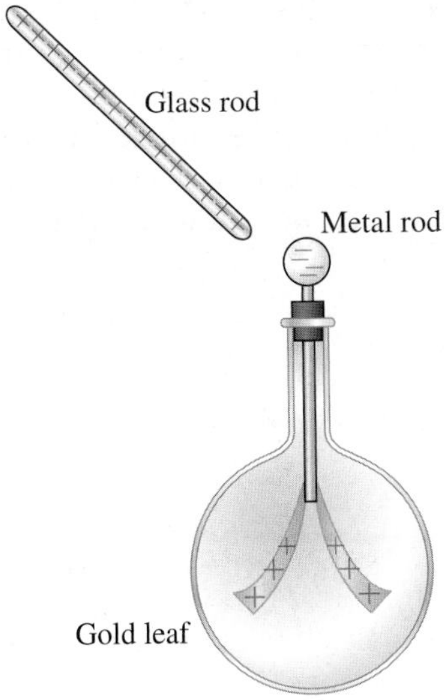

◀ FIGURE B-1
Production of electric charges by induction in a gold-leaf electroscope
The glass rod acquires a positive electric charge by being rubbed with a silk cloth. As the rod is brought near the electroscope, a separation of charge occurs in the electroscope. The leaves become positively charged and repel one another. Negative charge is attracted to the spherical terminal at the end of the metal rod. If the glass rod is removed, the charges on the electroscope redistribute themselves, and the leaves collapse. If the spherical ball is touched by an electric conductor before the glass rod is removed, negative charge is removed from the ball. The electroscope retains a net positive charge, and the leaves remain outstretched.

The potential energy (*PE*) associated with the interaction of two charged objects is given by

$$PE = \frac{Q_1Q_2}{4\pi\varepsilon r} \tag{B.14}$$

PE is equal to the work done when the distance between the two objects is decreased from infinity to r.

B-6 Current Electricity

Current electricity is a flow of electrically charged particles. In electric currents in metallic conductors, the charged particles are electrons; in molten salts or in aqueous solutions, the particles are both negatively and positively charged ions.

As pointed out in Section B-5, the unit of electric charge is called a *coulomb* (C). The unit of electric current known as the *ampere* (A) is defined as a flow of one coulomb per second through an electrical conductor. Two variables determine the magnitude of the electric current I flowing through a conductor. These are the potential difference, or voltage drop, E, along the conductor, and the electrical resistance of the conductor, R. The units of voltage and resistance are the *volt* (V) and *ohm*, respectively. The relationship of electric current, voltage, and resistance is given by Ohm's law.

$$I = \frac{E}{R} \tag{B.15}$$

One joule of energy is associated with the passage of one coulomb of electric charge through a potential difference (voltage) of one volt. That is, one joule = one volt-coulomb. Electric *power* refers to the rate of production (or consumption) of electric energy. It has the unit *watt* (W).

$$1\ \text{W} = 1\ \text{J s}^{-1} = 1\ \text{V C s}^{-1}$$

Since one coulomb per second is a current of one ampere,

$$1\ \text{W} = 1\ \text{V} \times 1\ \text{A} \tag{B.16}$$

Thus, a 100-watt light bulb operating at 110 V draws a current of 100 W/110 V = 0.91 A.

B-7 Electromagnetism

The relationship between electricity and magnetism is an intimate one. Interactions of electric and magnetic fields result in (1) magnetic fields associated with the flow of electric current (as in electromagnets), (2) forces experienced by current-carrying conductors when placed in a magnetic field (as in electric motors), and (3) electric current being induced when an electric conductor is moved through a magnetic field (as in electric generators). Several observations described in this text can be understood in terms of electromagnetic phenomena.

C

SI Units

Appendix

The system of units that will in time be used universally for expressing all measured quantities is Le Système International d'Unités (The International System of Units), adopted in 1960 by the Conference Générale des Poids et Measures (General Conference of Weights and Measures). A summary of some of the provisions of the SI convention is provided here.

C-1 SI Base Units

A single unit has been established for each of the basic quantities involved in measurement. These are as follows:

Physical Quantity	Unit	Symbol
Length	Meter	m
Mass	Kilogram	kg
Time	Second	s
Electric current	Ampere	A
Temperature	Kelvin	K
Luminous intensity	Candela	cd
Amount of substance	Mole	mol
Plane angle	Radian	rad
Solid angle	Steradian	sr

C-2 SI Prefixes

Distinctive prefixes are attached to the base unit to express quantities that are *multiples* (greater than) or *submultiples* (less than) of the base unit. The multiples and submultiples are obtained by multiplying the base unit by powers of ten.

Multiple	Prefix	Symbol	Submultiple	Prefix	Symbol
10^{12}	tera	T	10^{-1}	deci	d
10^{9}	giga	G	10^{-2}	centi	c
10^{6}	mega	M	10^{-3}	milli	m
10^{3}	kilo	k	10^{-6}	micro	μ
10^{2}	hecto	h	10^{-9}	nano	n
10^{1}	deka	da	10^{-12}	pico	p
			10^{-15}	femto	f
			10^{-18}	atto	a

C-3 Derived SI Units

A number of quantities must be derived from measured values of the SI base quantities [for example, volume has the unit $(\text{length})^3$]. Two sets of derived units are given, those whose names follow directly from the base units and those that are given special names. Notice that the units used in the text differ in some respects from those in the table. For example, for the most part, the text expresses density as g cm^{-3}, molar mass as g mol^{-1}, molar volume as mL mol^{-1} or L mol^{-1}, and molar concentration (molarity) as mol L^{-1}, or M.

▶ Two other SI conventions are illustrated through this table: (1) Units are written in singular form—meter or m, *not* meters or ms; (2) negative exponents are preferred to the shilling bar or solidus (/), that is, m s^{-1} and m s^{-2}, *not* m/s and m/s/s. In some chapters both ways of expressing units have been used as you need to be comfortable with both systems.

Physical Quantity	Unit	Symbol
Area	Square meter	m^2
Volume	Cubic meter	m^3
Velocity	Meter per second	m s^{-1}
Acceleration	Meter per second squared	m s^{-2}
Density	Kilogram per cubic meter	kg m^{-3}
Molar mass	Kilogram per mole	kg mol^{-1}
Molar volume	Cubic meter per mole	m^3 mol^{-1}
Molar concentration	Mole per cubic meter	mol m^{-3}

Physical Quantity	Unit	Symbol	In SI Units
Frequency	hertz	Hz	s^{-1}
Force	newton	N	kg m s^{-2}
Pressure	pascal	Pa	N m^{-2}
Energy	joule	J	kg m^2 s^{-2}
Power	watt	W	J s^{-1}
Electric charge	coulomb	C	A s
Electric potential difference	volt	V	J A^{-1} s^{-1}
Electric resistance	ohm	Ω	V A^{-1}

C-4 Units to Be Discouraged or Abandoned

There are several commonly used units whose use is to be discouraged and ultimately abandoned. Their gradual disappearance is to be expected, though each is used in this text. A few such units are listed.

▶ Another SI convention is implied here. No commas are used in expressing large numbers. Instead, spaces are left between groupings of three digits, that is, 101 325 rather than 101,325. Decimal points are written either as periods or commas. Numbers in this text retained the comma separators in numbers of at least five digits.

Physical Quantity	Unit	Symbol	Definition of SI Units
Length	ångstrom	Å	1×10^{-10} m
Force	dyne	dyn	1×10^{-5} N
Energy	erg	erg	1×10^{-7} J
Energy	calorie	cal	4.184 J
Pressure	atmosphere	atm	101 325 Pa
Pressure	millimeter of mercury	mmHg	133.322 Pa
Pressure	torr	Torr	133.322 Pa

D

Appendix

Data Tables

TABLE D.1 Ground-State Electron Configurations

Z	Element	Configuration	Z	Element	Configuration	Z	Element	Configuration
1	H	$1s^1$	37	Rb	[Kr] $5s^1$	72	Hf	[Xe] $4f^{14}5d^2 6s^2$
2	He	$1s^2$	38	Sr	[Kr] $5s^2$	73	Ta	[Xe] $4f^{14}5d^3 6s^2$
3	Li	[He] $2s^1$	39	Y	[Kr] $4d^1 5s^2$	74	W	[Xe] $4f^{14}5d^4 6s^2$
4	Be	[He] $2s^2$	40	Zr	[Kr] $4d^2 5s^2$	75	Re	[Xe] $4f^{14}5d^5 6s^2$
5	B	[He] $2s^2 2p^1$	41	Nb	[Kr] $4d^4 5s^1$	76	Os	[Xe] $4f^{14}5d^6 6s^2$
6	C	[He] $2s^2 2p^2$	42	Mo	[Kr] $4d^5 5s^1$	77	Ir	[Xe] $4f^{14}5d^7 6s^2$
7	N	[He] $2s^2 2p^3$	43	Tc	[Kr] $4d^5 5s^2$	78	Pt	[Xe] $4f^{14}5d^9 6s^1$
8	O	[He] $2s^2 2p^4$	44	Ru	[Kr] $4d^7 5s^1$	79	Au	[Xe] $4f^{14}5d^{10} 6s^1$
9	F	[He] $2s^2 2p^5$	45	Rh	[Kr] $4d^8 5s^1$	80	Hg	[Xe] $4f^{14}5d^{10} 6s^2$
10	Ne	[He] $2s^2 2p^6$	46	Pd	[Kr] $4d^{10}$	81	Tl	[Xe] $4f^{14}5d^{10} 6s^2 6p^1$
11	Na	[Ne] $3s^1$	47	Ag	[Kr] $4d^{10} 5s^1$	82	Pb	[Xe] $4f^{14}5d^{10} 6s^2 6p^2$
12	Mg	[Ne] $3s^2$	48	Cd	[Kr] $4d^{10} 5s^2$	83	Bi	[Xe] $4f^{14}5d^{10} 6s^2 6p^3$
13	Al	[Ne] $3s^2 3p^1$	49	In	[Kr] $4d^{10} 5s^2 5p^1$	84	Po	[Xe] $4f^{14}5d^{10} 6s^2 6p^4$
14	Si	[Ne] $3s^2 3p^2$	50	Sn	[Kr] $4d^{10} 5s^2 5p^2$	85	At	[Xe] $4f^{14}5d^{10} 6s^2 6p^5$
15	P	[Ne] $3s^2 3p^3$	51	Sb	[Kr] $4d^{10} 5s^2 5p^3$	86	Rn	[Xe] $4f^{14}5d^{10} 6s^2 6p^6$
16	S	[Ne] $3s^2 3p^4$	52	Te	[Kr] $4d^{10} 5s^2 5p^4$	87	Fr	[Rn] $7s^1$
17	Cl	[Ne] $3s^2 3p^5$	53	I	[Kr] $4d^{10} 5s^2 5p^5$	88	Ra	[Rn] $7s^2$
18	Ar	[Ne] $3s^2 3p^6$	54	Xe	[Kr] $4d^{10} 5s^2 5p^6$	89	Ac	[Rn] $6d^1 7s^2$
19	K	[Ar] $4s^1$	55	Cs	[Xe] $6s^1$	90	Th	[Rn] $6d^2 7s^2$
20	Ca	[Ar] $4s^2$	56	Ba	[Xe] $6s^2$	91	Pa	[Rn] $5f^2 6d^1 7s^2$
21	Sc	[Ar] $3d^1 4s^2$	57	La	[Xe] $5d^1 6s^2$	92	U	[Rn] $5f^3 6d^1 7s^2$
22	Ti	[Ar] $3d^2 4s^2$	58	Ce	[Xe] $4f^2 6s^2$	93	Np	[Rn] $5f^4 6d^1 7s^2$
23	V	[Ar] $3d^3 4s^2$	59	Pr	[Xe] $4f^3 6s^2$	94	Pu	[Rn] $5f^6 7s^2$
24	Cr	[Ar] $3d^5 4s^1$	60	Nd	[Xe] $4f^4 6s^2$	95	Am	[Rn] $5f^7 7s^2$
25	Mn	[Ar] $3d^5 4s^2$	61	Pm	[Xe] $4f^5 6s^2$	96	Cm	[Rn] $5f^7 6d^1 7s^2$
26	Fe	[Ar] $3d^6 4s^2$	62	Sm	[Xe] $4f^6 6s^2$	97	Bk	[Rn] $5f^9 7s^2$
27	Co	[Ar] $3d^7 4s^2$	63	Eu	[Xe] $4f^7 6s^2$	98	Cf	[Rn] $5f^{10} 7s^2$
28	Ni	[Ar] $3d^8 4s^2$	64	Gd	[Xe] $4f^7 5d^1 6s^2$	99	Es	[Rn] $5f^{11} 7s^2$
29	Cu	[Ar] $3d^{10} 4s^1$	65	Tb	[Xe] $4f^9 6s^2$	100	Fm	[Rn] $5f^{12} 7s^2$
30	Zn	[Ar] $3d^{10} 4s^2$	66	Dy	[Xe] $4f^{10} 6s^2$	101	Md	[Rn] $5f^{13} 7s^2$
31	Ga	[Ar] $3d^{10} 4s^2 4p^1$	67	Ho	[Xe] $4f^{11} 6s^2$	102	No	[Rn] $5f^{14} 7s^2$
32	Ge	[Ar] $3d^{10} 4s^2 4p^2$	68	Er	[Xe] $4f^{12} 6s^2$	103	Lr	[Rn] $5f^{14} 6d^1 7s^2$
33	As	[Ar] $3d^{10} 4s^2 4p^3$	69	Tm	[Xe] $4f^{13} 6s^2$	104	Rf	[Rn] $5f^{14} 6d^2 7s^2$
34	Se	[Ar] $3d^{10} 4s^2 4p^4$	70	Yb	[Xe] $4f^{14} 6s^2$	105	Db	[Rn] $5f^{14} 6d^3 7s^2$
35	Br	[Ar] $3d^{10} 4s^2 4p^5$	71	Lu	[Xe] $4f^{14} 5d^1 6s^2$	106	Sg	[Rn] $5f^{14} 6d^4 7s^2$
36	Kr	[Ar] $3d^{10} 4s^2 4p^6$						

The electron configurations printed in red are those of the noble gases. Each noble gas configuration serves as the core of the electron configurations of the elements that follow it, until the next noble gas is reached. Thus, [He] represents the core configuration of the second period elements; [Ne], the third period; [Ar], the fourth period; [Kr], the fifth period; [Xe], the sixth period; and [Rn], the seventh period.

TABLE D.2 Thermodynamic Properties of Substances at 298.15 K.* Substances are at 1 bar pressure. For aqueous solutions, solutes are at unit activity (roughly 1 M). Data for ions in aqueous solution are relative to values of *zero* for $\Delta_f H°$, $\Delta_f G°$, and $S°$ for H^+

Inorganic Substances

	$\Delta_f H°$, kJ mol^{-1}	$\Delta_f G°$, kJ mol^{-1}	$S°$, J mol^{-1} K^{-1}	C_p, J mol^{-1} K^{-1}
Aluminum				
$Al(s)$	0	0	28.33	24.2
$Al^{3+}(aq)$	−531	−485	−321.7	—
$AlCl_3(s)$	−704.2	−628.8	110.7	91.1
$Al_2Cl_6(g)$	−1291	−1220.	490.	157.72
$AlF_3(s)$	−1504	−1425	66.44	75.1
Al_2O_3(α solid)	−1676	−1582	50.92	79.0
$Al(OH)_3(s)$	−1276	—	—	93.1
$Al_2(SO_4)_3(s)$	−3441	−3100.	239	259.4
Barium				
$Ba(s)$	0	0	62.8	28.1
$Ba^{2+}(aq)$	−537.6	−560.8	9.6	—
$BaCO_3(s)$	−1216	−1138	112.1	85.35
$BaCl_2(s)$	−858.6	−810.4	123.7	75.1
$BaF_2(s)$	−1207	−1157	96.36	71.2
$BaO(s)$	−553.5	−525.1	70.42	47.3
$Ba(OH)_2(s)$	−944.7	—	—	101.6
$Ba(OH)_2 \cdot 8\,H_2O(s)$	−3342	−2793	427	—
$BaSO_4(s)$	−1473	−1362	132.2	101.8
Beryllium				
$Be(s)$	0	0	9.50	16.4
$BeCl_2$(α solid)	−490.4	−445.6	82.68	62.4
BeF_2(α solid)	−1027	−979.4	53.35	51.8
$BeO(s)$	−609.6	−580.3	14.14	25.6
Bismuth				
$Bi(s)$	0	0	56.74	25.5
$BiCl_3(s)$	−379.1	−315.0	177.0	105.0
$Bi_2O_3(s)$	−573.9	−493.7	151.5	113.5
Boron				
$B(s)$	0	0	5.86	11.1
$BCl_3(l)$	−427.2	−387.4	206.3	106.7
$BF_3(g)$	−1137	−1120.	254.1	50.45
$B_2H_6(g)$	35.6	86.7	232.1	56.7
$B_2O_3(s)$	−1273	−1194	53.97	62.8
Bromine				
$Br(g)$	111.9	82.40	175.0	20.8
$Br^-(aq)$	−121.6	−104.0	82.4	−141.8
$Br_2(g)$	30.91	3.11	245.5	36.0
$Br_2(l)$	0	0	152.2	75.7
$BrCl(g)$	14.64	−0.98	240.1	35.0
$BrF_3(g)$	−255.6	−229.4	292.5	66.6
$BrF_3(l)$	−300.8	−240.5	178.2	124.6

[a]Data for inorganic substances and for organic compounds with up to two carbon atoms per molecule are adapted from D. D. Wagman, et al., "The NBS Tables of Chemical Thermodynamic Properties: Selected Values for Inorganic and C_1 and C_2 Organic Substances in SI Units" *Journal of Physical and Chemical Reference Data* 11 (1982) Supplement 2. Data for other organic compounds are from J. A. Dean, *Lange's Handbook of Chemistry*. 15th ed., McGraw-Hill, 1999, and other sources.

Inorganic Substances

	$\Delta_f H°$, kJ mol^{-1}	$\Delta_f G°$, kJ mol^{-1}	$S°$, J mol^{-1} K^{-1}	C_p, J mol^{-1} K^{-1}
Cadmium				
$Cd(s)$	0	0	51.76	26.0
$Cd^{2+}(aq)$	−75.90	−77.61	−73.2	—
$CdCl_2(s)$	−391.5	−343.9	115.3	74.7
$CdO(s)$	−258.2	−228.4	54.8	43.4
Calcium				
$Ca(s)$	0	0	41.42	25.9
$Ca^{2+}(aq)$	−542.8	−553.6	−53.1	—
$CaCO_3(s)$	−1207	−1129	92.9	80.6
$CaCl_2(s)$	−795.8	−748.1	104.6	72.9
$CaF_2(s)$	−1220.	−1167	68.87	67.0
$CaH_2(s)$	−186.2	−147.2	42	41.0
$Ca(NO_3)_2(s)$	−938.4	−743.1	193.3	149.4
$CaO(s)$	−635.1	−604.0	39.75	42.0
$Ca(OH)_2(s)$	−986.1	−898.5	83.39	87.5
$Ca_3(PO_4)_2(s)$	−4121	−3885	236.0	227.8
$CaSO_4(s)$	−1434	−1322	106.7	99.7
Carbon (See also the table of organic substances.)				
$C(g)$	716.7	671.3	158.0	20.8
$C(diamond)$	1.90	2.90	2.38	6.1
$C(graphite)$	0	0	5.74	8.5
$CCl_4(g)$	−102.9	−60.59	309.9	83.3
$CCl_4(l)$	−135.4	−65.21	216.4	130.7
$C_2N_2(g)$	309.0	297.4	241.9	56.8
$CO(g)$	−110.5	−137.2	197.7	29.1
$CO_2(g)$	−393.5	−394.4	213.7	37.1
$CO_3^{2-}(aq)$	−677.1	−527.8	−56.9	—
$C_3O_2(g)$	−93.72	−109.8	276.5	67.0
$C_3O_2(l)$	−117.3	−105.0	181.1	—
$COCl_2(g)$	−218.8	−204.6	283.5	57.7
$COS(g)$	−142.1	−169.3	231.6	41.5
$CS_2(l)$	89.70	65.27	151.3	76.4
Chlorine				
$Cl(g)$	121.7	105.7	165.2	21.8
$Cl^-(aq)$	−167.2	−131.2	56.5	−136.4
$Cl_2(g)$	0	0	223.1	33.9
$ClF_3(g)$	−163.2	−123.0	281.6	63.9
$ClO_2(g)$	102.5	120.5	256.8	42.0
$Cl_2O(g)$	80.3	97.9	266.2	45.4
Chromium				
$Cr(s)$	0	0	23.77	23.4
$[Cr(H_2O)_6]^{3+}(aq)$	−1999	—	—	—
$Cr_2O_3(s)$	−1140.	−1058	81.2	118.7
$CrO_4^{2-}(aq)$	−881.2	−727.8	50.21	—
$Cr_2O_7^{2-}(aq)$	−1490.	−1301	261.9	—
Cobalt				
$Co(s)$	0	0	30.04	24.8
$CoO(s)$	−237.9	−214.2	52.97	55.2
$Co(OH)_2$(pink solid)	−539.7	−454.3	79	68.8

(continued)

Inorganic Substances

	$\Delta_f H°$, kJ mol^{-1}	$\Delta_f G°$, kJ mol^{-1}	$S°$, J mol^{-1} K^{-1}	C_p, J mol^{-1} K^{-1}
Copper				
$Cu(s)$	0	0	33.15	24.4
$Cu^{2+}(aq)$	64.77	65.49	−99.6	—
$CuCO_3 \cdot Cu(OH)_2(s)$	−1051	−893.6	186.2	—
$CuO(s)$	−157.3	−129.7	42.63	42.3
$Cu(OH)_2(s)$	−449.8	—	—	95.19
$CuSO_4 \cdot 5\,H_2O(s)$	−2280.	−1880.	300.4	—
Fluorine				
$F(g)$	78.99	61.91	158.8	22.7
$F^-(aq)$	−332.6	−278.8	−13.8	−106.7
$F_2(g)$	0	0	202.8	31.3
Helium				
$He(g)$	0	0	126.2	20.8
Hydrogen				
$H(g)$	218.0	203.2	114.7	20.8
$H^+(aq)$	0	0	0	0
$H_2(g)$	0	0	130.7	28.8
$HBr(g)$	−36.40	−53.45	198.7	29.1
$HCl(g)$	−92.31	−95.30	186.9	29.1
$HCl(aq)$	−167.2	−131.2	56.5	−136.4
$HClO_2(aq)$	−51.9	5.9	188.3	—
$HCN(g)$	135.1	124.7	201.8	35.9
$HF(g)$	−271.1	−273.2	173.8	—
$HI(g)$	26.48	1.70	206.6	29.2
$HNO_3(l)$	−174.1	−80.71	155.6	109.9
$HNO_3(aq)$	−207.4	−111.3	146.4	−86.6
$H_2O(g)$	−241.8	−228.6	188.8	33.6
$H_2O(l)$	−285.8	−237.1	69.91	75.3
$H_2O_2(g)$	−136.3	−105.6	232.7	43.1
$H_2O_2(l)$	−187.8	−120.4	109.6	89.1
$H_2S(g)$	−20.63	−33.56	205.8	34.2
$H_2SO_4(l)$	−814.0	−690.0	156.9	138.9
$H_2SO_4(aq)$	−909.3	−744.5	20.1	−293.0
Iodine				
$I(g)$	106.8	70.25	180.8	20.8
$I^-(aq)$	−55.19	−51.57	111.3	−142.3
$I_2(g)$	62.44	19.33	260.7	36.9
$I_2(s)$	0	0	116.1	54.4
$IBr(g)$	40.84	3.69	258.8	36.4
$ICl(g)$	17.78	−5.46	247.6	35.6
$ICl(l)$	−23.89	−13.58	135.1	135.1
Iron				
$Fe(s)$	0	0	27.28	25.1
$Fe^{2+}(aq)$	−89.1	−78.90	−137.7	—
$Fe^{3+}(aq)$	−48.5	−4.7	−315.9	—
$FeCO_3(s)$	−740.6	−666.7	92.9	82.1
$FeCl_3(s)$	−399.5	−334.0	−142.3	96.7
$FeO(s)$	−272.0	—	—	49.91
$Fe_2O_3(s)$	−824.2	−742.2	87.40	103.9
$Fe_3O_4(s)$	−1118	−1015	146.4	143.4
$Fe(OH)_3(s)$	−823.0	−696.5	106.7	101.7

Inorganic Substances

	$\Delta_f H°$, kJ mol^{-1}	$\Delta_f G°$, kJ mol^{-1}	$S°$, J mol^{-1} K^{-1}	C_p, J mol^{-1} K^{-1}
Lead				
Pb(s)	0	0	64.81	26.4
Pb^{2+}(aq)	−1.7	−24.43	10.5	—
PbI_2(s)	−175.5	−173.6	174.9	77.4
PbO_2(s)	−277.4	−217.3	68.6	64.6
$PbSO_4$(s)	−919.9	−813.1	148.6	103.2
Lithium				
Li(g)	159.4	126.7	138.8	20.8
Li(s)	0	0	29.12	24.8
Li^+(aq)	−278.5	−293.3	13.4	68.6
LiCl(s)	−408.6	−384.4	59.33	48.0
LiOH(s)	−484.9	−439.0	42.80	49.6
$LiNO_3$(s)	−483.1	−381.1	90.0	—
Magnesium				
Mg(s)	0	0	32.68	24.9
Mg^{2+}(aq)	−466.9	−454.8	−138.1	—
$MgCl_2$(s)	−641.3	−591.8	89.62	71.4
$MgCO_3$(s)	−1096	−1012	65.7	75.5
MgF_2(s)	−1123	−1070	57.24	61.6
MgO(s)	−601.7	−569.4	26.94	37.2
$Mg(OH)_2$(s)	−924.5	−833.5	63.18	77.0
MgS(s)	−346.0	−341.8	50.33	45.6
$MgSO_4$(s)	−1285	−1171	91.6	96.5
Manganese				
Mn(s)	0	0	32.01	26.3
Mn^{2+}(aq)	−220.8	−228.1	−73.6	50.0
MnO_2(s)	−520.0	−465.1	53.05	54.1
MnO_4^-(aq)	−541.4	−447.2	191.2	−82.0
Mercury				
Hg(g)	61.32	31.82	175.0	20.8
Hg(l)	0	0	76.02	28.0
HgO(s)	−90.83	−58.54	70.29	44.1
Nitrogen				
N(g)	472.7	455.6	153.3	20.8
N_2(g)	0	0	191.6	29.1
NF_3(g)	−124.7	−83.2	260.7	53.4
NH_3(g)	−46.11	−16.45	192.5	35.1
NH_3(aq)	−80.29	−26.50	111.3	—
NH_4^+(aq)	−132.5	−79.31	113.4	79.9
NH_4Br(s)	−270.8	−175.2	113	96.0
NH_4Cl(s)	−314.4	−202.9	94.6	84.1
NH_4F(s)	−464.0	−348.7	71.96	65.3
NH_4HCO_3(s)	−849.4	−665.9	120.9	—
NH_4I(s)	−201.4	−112.5	117	—
NH_4NO_3(s)	−365.6	−183.9	151.1	139.3
NH_4NO_3(aq)	−339.9	−190.6	259.8	−6.7
$(NH_4)_2SO_4$(s)	−1181	−901.7	220.1	187.5
N_2H_4(g)	95.40	159.4	238.5	48.4
N_2H_4(l)	50.63	149.3	121.2	98.9
NO(g)	90.25	86.55	210.8	29.9
N_2O(g)	82.05	104.2	219.9	38.6

(continued)

Inorganic Substances	$\Delta_f H°$, kJ mol^{-1}	$\Delta_f G°$, kJ mol^{-1}	$S°$, J mol^{-1} K^{-1}	C_p, J mol^{-1} K^{-1}
NO_2(g)	33.18	51.31	240.1	37.2
N_2O_4(g)	9.16	97.89	304.3	79.2
N_2O_4(l)	−19.50	97.54	209.2	142.7
N_2O_5(g)	11.3	115.1	355.7	95.3
NO_3^-(aq)	−205.0	−108.7	146.4	−86.6
NOBr(g)	82.17	82.42	273.7	45.5
NOCl(g)	51.71	66.08	261.7	44.7
Oxygen				
O(g)	249.2	231.7	161.1	21.9
O_2(g)	0	0	205.1	29.4
O_3(g)	142.7	163.2	238.9	39.2
OH^-(aq)	−230.0	−157.2	−10.75	−148.5
OF_2(g)	24.7	41.9	247.4	43.3
Phosphorus				
P(α white)	0	0	41.09	23.8
P(red)	−17.6	−12.1	22.80	21.2
P_4(g)	58.91	24.44	280.0	67.2
PCl_3(g)	−287.0	−267.8	311.8	71.8
PCl_5(g)	−374.9	−305.0	364.6	112.8
PH_3(g)	5.4	13.4	210.2	37.1
P_4O_{10}(s)	−2984	−2698	228.9	211.71
PO_4^{3-}(aq)	−1277	−1019	−222	—
Potassium				
K(g)	89.24	60.59	160.3	20.8
K(s)	0	0	64.18	29.6
K^+(aq)	−252.4	−283.3	102.5	21.8
KBr(s)	−393.8	−380.7	95.90	52.4
KCN(s)	−113.0	−101.9	128.5	66.3
KCl(s)	−436.7	−409.1	82.59	51.3
$KClO_3$(s)	−397.7	−296.3	143.1	100.3
$KClO_4$(s)	−432.8	−303.1	151.0	112.4
KF(s)	−567.3	−537.8	66.57	49.0
KI(s)	−327.9	−324.9	106.3	52.9
KNO_3(s)	−494.6	−394.9	133.1	96.4
KOH(s)	−424.8	−379.1	78.9	68.9
KOH(aq)	−482.4	−440.5	91.6	−126.8
K_2SO_4(s)	−1438	−1321	175.6	131.5
Silicon				
Si(s)	0	0	18.83	20.0
SiH_4(g)	34.3	56.9	204.6	42.8
Si_2H_6(g)	80.3	127.3	272.7	80.8
SiO_2(quartz)	−910.9	−856.6	41.84	44.4
Silver				
Ag(s)	0	0	42.55	25.4
Ag^+(aq)	105.6	77.11	72.68	21.8
AgBr(s)	−100.4	−96.90	107.1	52.4
AgCl(s)	−127.1	−109.8	96.2	50.8
AgI(s)	−61.84	−66.19	115.5	56.8
$AgNO_3$(s)	−124.4	−33.41	140.9	93.1
Ag_2O(s)	−31.05	−11.20	121.3	65.9
Ag_2SO_4(s)	−715.9	−618.4	200.4	131.4

Inorganic Substances

	$\Delta_f H°$, kJ mol^{-1}	$\Delta_f G°$, kJ mol^{-1}	$S°$, J mol^{-1} K^{-1}	C_p, J mol^{-1} K^{-1}
Sodium				
Na(g)	107.3	76.76	153.7	20.8
Na(s)	0	0	51.21	28.2
Na^+(aq)	−240.1	−261.9	59.0	46.4
Na_2(g)	142.1	103.9	230.2	37.6
NaBr(s)	−361.1	−349.0	86.82	51.4
Na_2CO_3(s)	−1131	−1044	135.0	112.3
$NaHCO_3$(s)	−950.8	−851.0	101.7	87.6
NaCl(s)	−411.2	−384.1	72.13	50.5
NaCl(aq)	−407.3	−393.1	115.5	−90.0
$NaClO_3$(s)	−365.8	−262.3	123.4	—
$NaClO_4$(s)	−383.3	−254.9	142.3	111.3
NaF(s)	−573.6	−543.5	51.46	46.9
NaH(s)	−56.28	−33.46	40.02	36.4
NaI(s)	−287.8	−286.1	98.53	52.1
$NaNO_3$(s)	−467.9	−367.0	116.5	92.9
$NaNO_3$(aq)	−447.5	−373.2	205.4	−40.2
Na_2O_2(s)	−510.9	−447.7	95.0	89.2
NaOH(s)	−425.6	−379.5	64.46	59.5
NaOH(aq)	−470.1	−419.2	48.1	−102.1
NaH_2PO_4(s)	−1537	−1386	127.5	−116.86
Na_2HPO_4(s)	−1748	−1608	150.5	135.3
Na_3PO_4(s)	−1917	−1789	173.8	153.47
$NaHSO_4$(s)	−1126	−992.8	113.0	—
Na_2SO_4(s)	−1387	−1270	149.6	128.2
Na_2SO_4(aq)	−1390.	−1268	138.1	−201.0
$Na_2SO_4 \cdot 10\,H_2O$(s)	−4327	−3647	592.0	—
$Na_2S_2O_3$(s)	−1123	−1028	155	—
Sulfur				
S(g)	278.8	238.3	167.8	23.7
S(rhombic)	0	0	31.80	22.6
S_8(g)	102.3	49.63	431.0	156.06
S_2Cl_2(g)	−18.4	−31.8	331.5	124.3
SF_6(g)	−1209	−1105	291.8	97.0
SO_2(g)	−296.8	−300.2	248.2	39.9
SO_3(g)	−395.7	−371.1	256.8	50.7
SO_4^{2-}(aq)	−909.3	−744.5	20.1	−293.0
$S_2O_3^{2-}$(aq)	−648.5	−522.5	67	—
SO_2Cl_2(g)	−364.0	−320.0	311.9	77.0
SO_2Cl_2(l)	−394.1	—	—	−134.0
Tin				
Sn(white)	0	0	51.55	27.0
Sn(gray)	−2.09	0.13	44.14	25.8
$SnCl_4$(l)	−511.3	−440.1	258.6	165.3
SnO(s)	−285.8	−256.9	56.5	44.3
SnO_2(s)	−580.7	−519.6	52.3	52.6
Titanium				
Ti(s)	0	0	30.63	25.0
$TiCl_4$(g)	−763.2	−726.7	354.9	95.4
$TiCl_4$(l)	−804.2	−737.2	252.3	145.2
TiO_2(s)	−944.7	−889.5	50.33	55.0

(continued)

Inorganic Substances

	$\Delta_f H°$, kJ mol^{-1}	$\Delta_f G°$, kJ mol^{-1}	$S°$, J mol^{-1} K^{-1}	C_p, J mol^{-1} K^{-1}
Uranium				
$U(s)$	0	0	50.21	27.7
$UF_6(g)$	−2147	−2064	377.9	129.6
$UF_6(s)$	−2197	−2069	227.6	166.8
$UO_2(s)$	−1085	−1032	77.03	63.6
Zinc				
$Zn(s)$	0	0	41.63	25.4
$Zn^{2+}(aq)$	−153.9	−147.1	112.1	46.0
$ZnO(s)$	−348.3	−318.3	43.64	40.3

Organic Substances

	Name	$\Delta_f H°$, kJ mol^{-1}	$\Delta_f G°$, kJ mol^{-1}	$S°$, J mol^{-1} K^{-1}	C_p, J mol^{-1} K^{-1}
$CH_4(g)$	Methane(g)	−74.81	−50.72	186.3	35.7
$C_2H_2(g)$	Acetylene(g)	226.7	209.2	200.9	44.0
$C_2H_4(g)$	Ethylene(g)	52.26	68.15	219.6	42.9
$C_2H_6(g)$	Ethane(g)	−84.68	−32.82	229.6	52.5
$C_3H_8(g)$	Propane(g)	−103.8	−23.3	270.3	73.6
$C_4H_{10}(g)$	Butane(g)	−125.6	−17.1	310.2	97.5
$C_6H_6(g)$	Benzene(g)	82.6	129.8	269.3	82.4
$C_6H_6(l)$	Benzene(l)	49.0	124.5	173.4	136.0
$C_6H_{12}(g)$	Cyclohexane(g)	−123.4	32.0	298.4	106.3
$C_6H_{12}(l)$	Cyclohexane(l)	−156.4	26.9	204.4	154.9
$C_{10}H_8(g)$	Naphthalene(g)	150.6	224.2	333.2	131.9
$C_{10}H_8(s)$	Naphthalene(s)	77.9	201.7	167.5	165.7
$CH_2O(g)$	Formaldehyde(g)	−108.6	−102.5	218.8	35.4
$CH_3CHO(g)$	Acetaldehyde(g)	−166.2	−128.9	250.3	55.3
$CH_3CHO(l)$	Acetaldehyde(l)	−192.3	−128.1	160.2	89.0
$CH_3OH(g)$	Methanol(g)	−200.7	−162.0	239.8	44.1
$CH_3OH(l)$	Methanol(l)	−238.7	−166.3	126.8	81.1
$CH_3CH_2OH(g)$	Ethanol(g)	−235.1	−168.5	282.7	65.6
$CH_3CH_2OH(l)$	Ethanol(l)	−277.7	−174.8	160.7	112.3
$C_6H_5OH(s)$	Phenol(s)	−165.1	−50.4	144.0	127.4
$(CH_3)_2CO(g)$	Acetone(g)	−216.6	−153.0	295.0	74.5
$(CH_3)_2CO(l)$	Acetone(l)	−247.6	−155.6	200.5	126.3
$CH_3COOH(g)$	Acetic acid(g)	−432.3	−374.0	282.5	63.4
$CH_3COOH(l)$	Acetic acid(l)	−484.5	−389.9	159.8	123.3
$CH_3COOH(aq)$	Acetic acid(aq)	−485.8	−396.5	178.7	−6.3
$C_6H_5COOH(s)$	Benzoic acid(s)	−385.2	−245.3	167.6	146.8
$CH_3NH_2(g)$	Methylamine(g)	−22.97	32.16	243.4	50.1
$C_6H_5NH_2(g)$	Aniline(g)	86.86	166.8	319.3	107.9
$C_6H_5NH_2(l)$	Aniline(l)	31.6	149.2	191.3	191.9

TABLE D.3 Equilibrium Constants

A. Ionization Constants of Weak Acids at 25 °C

Name of Acid	Formula	K_a	Name of Acid	Formula	K_a
Acetic	CH_3COOH	1.8×10^{-5}	Hyponitrous	$HON{=}NOH$	8.9×10^{-8}
Acrylic	$HC_3H_3O_2$	5.5×10^{-5}		$HON{=}NO^-$	4×10^{-12}
Arsenic	H_3AsO_4	6.0×10^{-3}	Iodic	HIO_3	1.6×10^{-1}
	$H_2AsO_4^-$	1.0×10^{-7}	Iodoacetic	CH_2ICOOH	6.7×10^{-4}
	$HAsO_4^{2-}$	3.2×10^{-12}	Malonic	$H_2C_3H_2O_4$	1.5×10^{-3}
Arsenous	H_3AsO_3	6.6×10^{-10}		$HC_3H_2O_4^-$	2.0×10^{-6}
Benzoic	C_6H_5COOH	6.3×10^{-5}	Nitrous	HNO_2	7.2×10^{-4}
Bromoacetic	$CH_2BrCOOH$	1.3×10^{-3}	Oxalic	$H_2C_2O_4$	5.4×10^{-2}
Butyric	$HC_4H_7O_2$	1.5×10^{-5}		$HC_2O_4^-$	5.3×10^{-5}
Carbonic	H_2CO_3	4.4×10^{-7}	Phenol	C_6H_5OH	1.0×10^{-10}
	HCO_3^-	4.7×10^{-11}	Phenylacetic	$HC_8H_7O_2$	4.9×10^{-5}
Chloroacetic	$CH_2ClCOOH$	1.4×10^{-3}	Phosphoric	H_3PO_4	7.1×10^{-3}
Chlorous	$HClO_2$	1.1×10^{-2}		$H_2PO_4^-$	6.3×10^{-8}
Citric	$H_3C_6H_5O_7$	7.4×10^{-4}		HPO_4^{2-}	4.2×10^{-13}
	$H_2C_6H_5O_7^-$	1.7×10^{-5}	Phosphorous	H_3PO_3	3.7×10^{-2}
	$HC_6H_5O_7^{2-}$	4.0×10^{-7}		$H_2PO_3^-$	2.1×10^{-7}
Cyanic	$HOCN$	3.5×10^{-4}	Propionic	CH_3CH_2COOH	1.3×10^{-5}
Dichloroacetic	$CHCl_2COOH$	5.5×10^{-2}	Pyrophosphoric	$H_4P_2O_7$	3.0×10^{-2}
Fluoroacetic	CH_2FCOOH	2.6×10^{-3}		$H_3P_2O_7^-$	4.4×10^{-3}
Formic	$HCOOH$	1.8×10^{-4}		$H_2P_2O_7^{2-}$	2.5×10^{-7}
Hydrazoic	HN_3	1.9×10^{-5}		$HP_2O_7^{3-}$	5.6×10^{-10}
Hydrocyanic	HCN	6.2×10^{-10}	Selenic	H_2SeO_4	strong acid
Hydrofluoric	HF	6.6×10^{-4}		$HSeO_4^-$	2.2×10^{-2}
Hydrogen peroxide	H_2O_2	2.2×10^{-12}	Selenous	H_2SeO_3	2.3×10^{-3}
Hydroselenic	H_2Se	1.3×10^{-4}		$HSeO_3^-$	5.4×10^{-9}
	HSe^-	1×10^{-11}	Succinic	$H_2C_4H_4O_4$	6.2×10^{-5}
Hydrosulfuric	H_2S	1.0×10^{-7}		$HC_4H_4O_4^-$	2.3×10^{-6}
	HS^-	1×10^{-19}	Sulfuric	H_2SO_4	strong acid
Hydrotelluric	H_2Te	2.3×10^{-3}		HSO_4^-	1.1×10^{-2}
	HTe^-	1.6×10^{-11}	Sulfurous	H_2SO_3	1.3×10^{-2}
Hypobromous	$HOBr$	2.5×10^{-9}		HSO_3^-	6.2×10^{-8}
Hypochlorous	$HOCl$	2.9×10^{-8}	Thiophenol	C_6H_5SH	3.2×10^{-7}
Hypoiodous	HOI	2.3×10^{-11}	Trichloroacetic	CCl_3COOH	3.0×10^{-1}

B. Ionization Constants of Weak Bases at 25 °C

Name of Base	Formula	K_b	Name of Base	Formula	K_b
Ammonia	NH_3	1.8×10^{-5}	Isoquinoline	C_9H_7N	2.5×10^{-9}
Aniline	$C_6H_5NH_2$	7.4×10^{-10}	Methylamine	CH_3NH_2	4.2×10^{-4}
Codeine	$C_{18}H_{21}O_3N$	8.9×10^{-7}	Morphine	$C_{17}H_{19}O_3N$	7.4×10^{-7}
Diethylamine	$(C_2H_5)_2NH$	6.9×10^{-4}	Piperdine	$C_5H_{11}N$	1.3×10^{-3}
Dimethylamine	$(CH_3)_2NH$	5.9×10^{-4}	Pyridine	C_5H_5N	1.5×10^{-9}
Ethylamine	$C_2H_5NH_2$	4.3×10^{-4}	Quinoline	C_9H_7N	6.3×10^{-10}
Hydrazine	NH_2NH_2	8.5×10^{-7}	Triethanolamine	$C_6H_{15}O_3N$	5.8×10^{-7}
	$NH_2NH_3^+$	8.9×10^{-16}	Triethylamine	$(C_2H_5)_3N$	5.2×10^{-4}
Hydroxylamine	NH_2OH	9.1×10^{-9}	Trimethylamine	$(CH_3)_3N$	6.3×10^{-5}

(continued)

C. Solubility Product Constants[a]

Name of Solute	Formula	K_{sp}
Aluminum hydroxide	$Al(OH)_3$	1.3×10^{-33}
Aluminum phosphate	$AlPO_4$	6.3×10^{-19}
Barium carbonate	$BaCO_3$	5.1×10^{-9}
Barium chromate	$BaCrO_4$	1.2×10^{-10}
Barium fluoride	BaF_2	1.0×10^{-6}
Barium hydroxide	$Ba(OH)_2$	5×10^{-3}
Barium sulfate	$BaSO_4$	1.1×10^{-10}
Barium sulfite	$BaSO_3$	8×10^{-7}
Barium thiosulfate	BaS_2O_3	1.6×10^{-5}
Bismuthyl chloride	$BiOCl$	1.8×10^{-31}
Bismuthyl hydroxide	$BiOOH$	4×10^{-10}
Cadmium carbonate	$CdCO_3$	5.2×10^{-12}
Cadmium hydroxide	$Cd(OH)_2$	2.5×10^{-14}
Cadmium sulfide[b]	CdS	8×10^{-28}
Calcium carbonate	$CaCO_3$	2.8×10^{-9}
Calcium chromate	$CaCrO_4$	7.1×10^{-4}
Calcium fluoride	CaF_2	5.3×10^{-9}
Calcium hydroxide	$Ca(OH)_2$	5.5×10^{-6}
Calcium hydrogen phosphate	$CaHPO_4$	1×10^{-7}
Calcium oxalate	CaC_2O_4	4×10^{-9}
Calcium phosphate	$Ca_3(PO_4)_2$	2.0×10^{-29}
Calcium sulfate	$CaSO_4$	9.1×10^{-6}
Calcium sulfite	$CaSO_3$	6.8×10^{-8}
Chromium(II) hydroxide	$Cr(OH)_2$	2×10^{-16}
Chromium(III) hydroxide	$Cr(OH)_3$	6.3×10^{-31}
Cobalt(II) carbonate	$CoCO_3$	1.4×10^{-13}
Cobalt(II) hydroxide	$Co(OH)_2$	1.6×10^{-15}
Cobalt(III) hydroxide	$Co(OH)_3$	1.6×10^{-44}
Copper(I) chloride	$CuCl$	1.2×10^{-6}
Copper(I) cyanide	$CuCN$	3.2×10^{-20}
Copper(I) iodide	CuI	1.1×10^{-12}
Copper(II) arsenate	$Cu_3(AsO_4)_2$	7.6×10^{-36}
Copper(II) carbonate	$CuCO_3$	1.4×10^{-10}
Copper(II) chromate	$CuCrO_4$	3.6×10^{-6}
Copper(II) ferrocyanide	$Cu_2[Fe(CN)_6]$	1.3×10^{-16}
Copper(II) hydroxide	$Cu(OH)_2$	2.2×10^{-20}
Copper(II) sulfide[b]	CuS	6×10^{-37}
Iron(II) carbonate	$FeCO_3$	3.2×10^{-11}
Iron(II) hydroxide	$Fe(OH)_2$	8.0×10^{-16}
Iron(II) sulfide[b]	FeS	6×10^{-19}
Iron(III) arsenate	$FeAsO_4$	5.7×10^{-21}
Iron(III) ferrocyanide	$Fe_4[Fe(CN)_6]_3$	3.3×10^{-41}
Iron(III) hydroxide	$Fe(OH)_3$	4×10^{-38}
Iron(III) phosphate	$FePO_4$	1.3×10^{-22}
Lead(II) arsenate	$Pb_3(AsO_4)_2$	4.0×10^{-36}
Lead(II) azide	$Pb(N_3)_2$	2.5×10^{-9}
Lead(II) bromide	$PbBr_2$	4.0×10^{-5}
Lead(II) carbonate	$PbCO_3$	7.4×10^{-14}
Lead(II) chloride	$PbCl_2$	1.6×10^{-5}
Lead(II) chromate	$PbCrO_4$	2.8×10^{-13}
Lead(II) fluoride	PbF_2	2.7×10^{-8}
Lead(II) hydroxide	$Pb(OH)_2$	1.2×10^{-15}
Lead(II) iodide	PbI_2	7.1×10^{-9}
Lead(II) sulfate	$PbSO_4$	1.6×10^{-8}
Lead(II) sulfide[b]	PbS	3×10^{-28}
Lithium carbonate	Li_2CO_3	2.5×10^{-2}
Lithium fluoride	LiF	3.8×10^{-3}
Lithium phosphate	Li_3PO_4	3.2×10^{-9}
Magnesium ammonium phosphate	$MgNH_4PO_4$	2.5×10^{-13}
Magnesium carbonate	$MgCO_3$	3.5×10^{-8}
Magnesium fluoride	MgF_2	3.7×10^{-8}
Magnesium hydroxide	$Mg(OH)_2$	1.8×10^{-11}
Magnesium phosphate	$Mg_3(PO_4)_2$	1×10^{-25}
Manganese(II) carbonate	$MnCO_3$	1.8×10^{-11}
Manganese(II) hydroxide	$Mn(OH)_2$	1.9×10^{-13}
Manganese(II) sulfide[b]	MnS	3×10^{-14}
Mercury(I) bromide	Hg_2Br_2	5.6×10^{-23}
Mercury(I) chloride	Hg_2Cl_2	1.3×10^{-18}
Mercury(I) iodide	Hg_2I_2	4.5×10^{-29}
Mercury(II) sulfide[b]	HgS	2×10^{-53}
Nickel(II) carbonate	$NiCO_3$	6.6×10^{-9}
Nickel(II) hydroxide	$Ni(OH)_2$	2.0×10^{-15}
Scandium fluoride	ScF_3	4.2×10^{-18}
Scandium hydroxide	$Sc(OH)_3$	8.0×10^{-31}
Silver arsenate	Ag_3AsO_4	1.0×10^{-22}
Silver azide	AgN_3	2.8×10^{-9}
Silver bromide	$AgBr$	5.0×10^{-13}
Silver carbonate	Ag_2CO_3	8.5×10^{-12}
Silver chloride	$AgCl$	1.8×10^{-10}
Silver chromate	Ag_2CrO_4	1.1×10^{-12}
Silver cyanide	$AgCN$	1.2×10^{-16}
Silver iodate	$AgIO_3$	3.0×10^{-8}
Silver iodide	AgI	8.5×10^{-17}
Silver nitrite	$AgNO_2$	6.0×10^{-4}
Silver sulfate	Ag_2SO_4	1.4×10^{-5}
Silver sulfide[b]	Ag_2S	6×10^{-51}
Silver sulfite	Ag_2SO_3	1.5×10^{-14}
Silver thiocyanate	$AgSCN$	1.0×10^{-12}
Strontium carbonate	$SrCO_3$	1.1×10^{-10}
Strontium chromate	$SrCrO_4$	2.2×10^{-5}
Strontium fluoride	SrF_2	2.5×10^{-9}
Strontium sulfate	$SrSO_4$	3.2×10^{-7}
Thallium(I) bromide	$TlBr$	3.4×10^{-6}
Thallium(I) chloride	$TlCl$	1.7×10^{-4}
Thallium(I) iodide	TlI	6.5×10^{-8}
Thallium(III) hydroxide	$Tl(OH)_3$	6.3×10^{-46}
Tin(II) hydroxide	$Sn(OH)_2$	1.4×10^{-28}
Tin(II) sulfide[b]	SnS	1×10^{-26}
Zinc carbonate	$ZnCO_3$	1.4×10^{-11}
Zinc hydroxide	$Zn(OH)_2$	1.2×10^{-17}
Zinc oxalate	ZnC_2O_4	2.7×10^{-8}
Zinc phosphate	$Zn_3(PO_4)_2$	9.0×10^{-33}
Zinc sulfide[b]	ZnS	2×10^{-25}

D. Complex-Ion Formation Constants[c, d]

Formula	K_f	Formula	K_f	Formula	K_f
$[Ag(CN)_2]^-$	5.6×10^{18}	$[Co(ox)_3]^{3-}$	10^{20}	$[HgI_4]^{2-}$	6.8×10^{29}
$[Ag(EDTA)]^{3-}$	2.1×10^{7}	$[Cr(EDTA)]^-$	10^{23}	$[Hg(ox)_2]^{2-}$	9.5×10^{6}
$[Ag(en)_2]^+$	5.0×10^{7}	$[Cr(OH)_4]^-$	8×10^{29}	$[Ni(CN)_4]^{2-}$	2×10^{31}
$[Ag(NH_3)_2]^+$	1.6×10^{7}	$[CuCl_3]^{2-}$	5×10^{5}	$[Ni(EDTA)]^{2-}$	3.6×10^{18}
$[Ag(SCN)_4]^{3-}$	1.2×10^{10}	$[Cu(CN)_4]^{3-}$	2.0×10^{30}	$[Ni(en)_3]^{2+}$	2.1×10^{18}
$[Ag(S_2O_3)_2]^{3-}$	1.7×10^{13}	$[Cu(EDTA)]^{2-}$	5×10^{18}	$[Ni(NH_3)_6]^{2+}$	5.5×10^{8}
$[Al(EDTA)]^-$	1.3×10^{16}	$[Cu(en)_2]^{2+}$	1×10^{20}	$[Ni(ox)_3]^{4-}$	3×10^{8}
$[Al(OH)_4]^-$	1.1×10^{33}	$[Cu(NH_3)_4]^{2+}$	1.1×10^{13}	$[PbCl_3]^-$	2.4×10^{1}
$[Al(ox)_3]^{3-}$	2×10^{16}	$[Cu(ox)_2]^{2-}$	3×10^{8}	$[Pb(EDTA)]^{2-}$	2×10^{18}
$[CdCl_4]^{2-}$	6.3×10^{2}	$[Fe(CN)_6]^{4-}$	10^{37}	$[PbI_4]^{2-}$	3.0×10^{4}
$[Cd(CN)_4]^{2-}$	6.0×10^{18}	$[Fe(EDTA)]^{2-}$	2.1×10^{14}	$[Pb(OH)_3]^-$	3.8×10^{14}
$[Cd(en)_3]^{2+}$	1.2×10^{12}	$[Fe(en)_3]^{2+}$	5.0×10^{9}	$[Pb(ox)_2]^{2-}$	3.5×10^{6}
$[Cd(NH_3)_4]^{2+}$	1.3×10^{7}	$[Fe(ox)_3]^{4-}$	1.7×10^{5}	$[Pb(S_2O_3)_3]^{4-}$	2.2×10^{6}
$[Co(EDTA)]^{2-}$	2.0×10^{16}	$[Fe(CN)_6]^{3-}$	10^{42}	$[PtCl_4]^{2-}$	1×10^{16}
$[Co(en)_3]^{2+}$	8.7×10^{13}	$[Fe(EDTA)]^-$	1.7×10^{24}	$[Pt(NH_3)_6]^{2+}$	2×10^{35}
$[Co(NH_3)_6]^{2+}$	1.3×10^{5}	$[Fe(ox)_3]^{3-}$	2×10^{20}	$[Zn(CN)_4]^{2-}$	1×10^{18}
$[Co(ox)_3]^{4-}$	5×10^{9}	$[Fe(SCN)]^{2+}$	8.9×10^{2}	$[Zn(EDTA)]^{2-}$	3×10^{16}
$[Co(SCN)_4]^{2-}$	1.0×10^{3}	$[HgCl_4]^{2-}$	1.2×10^{15}	$[Zn(en)_3]^{2+}$	1.3×10^{14}
$[Co(EDTA)]^-$	10^{36}	$[Hg(CN)_4]^{2-}$	3×10^{41}	$[Zn(NH_3)_4]^{2+}$	4.1×10^{8}
$[Co(en)_3]^{3+}$	4.9×10^{48}	$[Hg(EDTA)]^{2-}$	6.3×10^{21}	$[Zn(OH)_4]^{2-}$	4.6×10^{17}
$[Co(NH_3)_6]^{3+}$	4.5×10^{33}	$[Hg(en)_2]^{2+}$	2×10^{23}	$[Zn(ox)_3]^{4-}$	1.4×10^{8}

[a]Data are at various temperatures around "room" temperature, from 18 to 25 °C.
[b]For a solubility equilibrium of the type $MS(s) + H_2O \rightleftharpoons M^{2+}(aq) + HS^-(aq) + OH^-(aq)$.
[c]The ligands referred to in this table are monodentate: Cl^-, CN^-, I^-, NH_3, OH^-, SCN^-, $S_2O_3^{2-}$; bidentate: ethylenediamine (en), oxalate ion (ox); tetradentate: ethylenediaminetetraacetato ion, $EDTA^{4-}$.
[d]The K_f values are cumulative or overall formation constants (see page 1154).

TABLE D.4 Standard Electrode (Reduction) Potentials at 25 °C[a]

Reduction Half-Reaction	Cell Notation	Standard Reduction Potential, V
$F_2(g) + 2e^- \rightleftharpoons 2F^-(aq)$	$F^- \mid F_2 \mid Pt$	+2.866
$OF_2(g) + 2H^+(aq) + 4e^- \rightleftharpoons H_2O(l) + 2F^-(aq)$	$F^-, H^+ \mid OF_2 \mid Pt$	+2.1
$O_3(g) + 2H^+(aq) + 2e^- \rightleftharpoons O_2(g) + H_2O(l)$	$H^+, H_2O \mid O_3, O_2 \mid Pt$	+2.075
$S_2O_8^{2-}(aq) + 2e^- \rightleftharpoons 2SO_4^{2-}(aq)$	$SO_4^{2-}, S_2O_8^{2-} \mid Pt$	+2.01
$Ag^{2+}(aq) + 2e^- \rightleftharpoons Ag^+(aq)$	$Ag^+, Ag^{2+} \mid Pt$	+1.98
$Co^{3+}(aq) + e^- \rightleftharpoons Co^{2+}(aq)$	$Co^{2+}, Co^{3+} \mid Pt$	+1.92
$H_2O_2(aq) + 2H^+(aq) + 2e^- \rightleftharpoons 2H_2O(l)$	$H_2O_2, H^+ \mid Pt$	+1.763
$Ce^{4+}(aq) + e^- \rightleftharpoons Ce^{3+}(aq)$	$Ce^{3+}, Ce^{4+} \mid Pt$	+1.72
$MnO_4^-(aq) + 4H^+(aq) + 3e^- \rightleftharpoons MnO_2(s) + 2H_2O(l)$	$MnO_2 \mid MnO_4^-, H^+ \mid Pt$	+1.70
$PbO_2(s) + SO_4^{2-}(aq) + 4H^+(aq) + 2e^- \rightleftharpoons PbSO_4(s) + 2H_2O(l)$	$PbSO_4, PbO_2 \mid SO_4^{2-}, H^+ \mid Pt$	+1.69
$Mn^{3+}(aq) + e^- \rightleftharpoons Mn^{2+}(aq)$	$Mn^{2+}, Mn^{3+} \mid Pt$	+1.54
$Au^{3+}(aq) + 3e^- \rightleftharpoons Au(s)$	$Au^{3+} \mid Au$	+1.52
$MnO_4^-(aq) + 8H^+(aq) + 5e^- \rightleftharpoons Mn^{2+}(aq) + 4H_2O(l)$	$Mn^{2+}, MnO_4^- \mid Pt$	+1.51
$2BrO_3^-(aq) + 12H^+(aq) + 10e^- \rightleftharpoons Br_2(l) + 6H_2O(l)$	$BrO_3^-, H^+ \mid Br_2 \mid Pt$	+1.478
$PbO_2(s) + 4H^+(aq) + 2e^- \rightleftharpoons Pb^{2+}(aq) + 2H_2O(l)$	$Pb^{2+}, H^+ \mid PbO_2 \mid Pt$	+1.455
$ClO_3^-(aq) + 6H^+(aq) + 6e^- \rightleftharpoons Cl^-(aq) + 2H_2O(l)$	$Cl^-, ClO_3^-, H^+ \mid Pt$	+1.450
$Au^{3+}(aq) + 2e^- \rightleftharpoons Au^+(aq)$	$Au^+, Au^{3+} \mid Pt$	+1.36

(continued)

Reduction Half-Reaction	Cell Notation	Standard Reduction Potential, V
$Cl_2(g) + 2e^- \rightleftharpoons 2\,Cl^-(aq)$	$Cl^- \mid Cl_2 \mid Pt$	+1.36
$Cr_2O_7^{2-}(aq) + 14\,H^+(aq) + 16\,e^- \rightleftharpoons 2\,Cr^{3+}(aq) + 7\,H_2O(l)$	$Cr^{3+}, Cr_2O_7^{2-} \mid Pt$	+1.32
$Tl^{3+}(aq) + 2e^- \rightleftharpoons Tl^+(aq)$	$Tl^+, Tl^{3+} \mid Pt$	+1.252
$MnO_2(s) + 4\,H^+(aq) + 2\,e^- \rightleftharpoons Mn^{2+}(aq) + 2\,H_2O(l)$	$Mn^{2+}, H^+ \mid MnO_2 \mid Pt$	+1.23
$O_2(g) + 4\,H^+(aq) + 4\,e^- \rightleftharpoons 2\,H_2O(l)$	$H_2O \mid O_2 \mid Pt$	+1.229
$IO_3^-(aq) + 12\,H^+(aq) + 10\,e^- \rightleftharpoons I_2(s) + 6\,H_2O(l)$	$IO_3^-, H^+ \mid I_2 \mid Pt$	+1.20
$ClO_4^-(aq) + 2\,H^+(aq) + 2\,e^- \rightleftharpoons ClO_3^-(aq) + H_2O(l)$	$ClO_4^-, H^+, ClO_3^- \mid Pt$	+1.189
$ClO_3^-(aq) + 2\,H^+(aq) + e^- \rightleftharpoons ClO_2(g) + H_2O(l)$	$ClO_3, H^+ \mid ClO_2^- \mid Pt$	+1.175
$NO_2(g) + H^+(aq) + e^- \rightleftharpoons HNO_2(aq)$	$HNO_2, H^+ \mid NO \mid Pt$	+1.07
$Br_2(l) + 2\,e^- \rightleftharpoons 2\,Br^-$	$Br^-, Br_2 \mid Pt$	+1.065
$NO_2(g) + 2\,H^+(aq) + 2\,e^- \rightleftharpoons NO(g) + H_2O(l)$	$NO_2, NO \mid Pt$	+1.03
$AuCl_4^-(aq) + 3\,e^- \rightleftharpoons Au(s) + 4\,Cl^-(aq)$	$AuCl_4^-, Cl^- \mid Au$	+1.002
$VO_2^+(aq) + 2\,H^+(aq) + e^- \rightleftharpoons VO^{2+}(aq) + H_2O(l)$	$VO^{2+}, VO_2^+, H^+ \mid Pt$	+1.000
$NO_3^-(aq) + 4\,H^+(aq) + 3\,e^- \rightleftharpoons NO(g) + 2\,H_2O(l)$	$NO_3^-, H^+ \mid NO \mid Pt$	+0.956
$Cu^{2+}(aq) + I^-(aq) + e^- \rightleftharpoons CuI(s)$	$Cu^{2+}, I^- \mid CuI \mid Pt$	+0.86
$2\,Hg^{2+}(aq) + 2\,e^- \rightleftharpoons Hg_2^{2+}(aq)$	$Hg_2^{2+}, Hg^{2+} \mid Pt$	+0.92
$2\,Hg^{2+}(aq) + 2\,e^- \rightleftharpoons Hg(l)$	$Hg^{2+} \mid Hg$	+0.854
$Ag^+(aq) + e^- \rightleftharpoons Ag(s)$	$Ag^+ \mid Ag$	+0.800
$Fe^{3+}(aq) + e^- \rightleftharpoons Fe^{2+}(aq)$	$Fe^{2+}, Fe^{3+} \mid Pt$	+0.771
$O_2(g) + 2\,H^+(aq) + 2\,e^- \rightleftharpoons H_2O_2(aq)$	$H_2O_2, H^+ \mid O_2 \mid Pt$	+0.695
$2\,HgCl_2(aq) + 2\,e^- \rightleftharpoons Hg_2Cl_2(s) + Cl^-(aq)$	$Cl^-, HgCl_2 \mid Hg_2Cl_2 \mid Pt$	+0.63
$MnO_4^-(aq) + e^- \rightleftharpoons MnO_4^{2-}(aq)$	$MnO_4^{2-}, MnO_4^- \mid Pt$	+0.56
$I_2(s) + 2\,e^- \rightleftharpoons 2\,I^-(aq)$	$I^- \mid I_2 \mid Pt$	+0.535
$Cu^+(aq) + e^- \rightleftharpoons Cu(s)$	$Cu^+ \mid Cu$	+0.520
$H_2SO_3(aq) + 4\,H^+(aq) + 4\,e^- \rightleftharpoons S(s) + 2\,H_2O(l)$	$H_2SO_3, H^+ \mid S \mid Pt$	+0.449
$O_2(g) + 2\,H_2O(1) + 4e^- \rightleftharpoons 4\,OH^-$	$OH^- \mid O_2 \mid Pt$	+0.401
$C_2N_2(g) + 2\,H^+(aq) + 2\,e^- \rightleftharpoons 2\,HCN(aq)$	$HCN, H^+ \mid C_2N_2 \mid Pt$	+0.37
$Fe(CN)_6^{3-}(aq) + e^- \rightleftharpoons Fe(CN)_6^{4-}(aq)$	$Fe(CN)_6^{4-}, Fe(CN)_6^{3-} \mid Pt$	+0.361
$Cu^{2+}(aq) + 2\,e^- \rightleftharpoons Cu(s)$	$Cu^{2+} \mid Cu$	+0.340
$VO^{2+}(aq) + 2\,H^+(aq) + e^- \rightleftharpoons V^{3+}(aq) + H_2O(l)$	$V^{3+}, VO^{2+}, H^+ \mid Pt$	+0.337
$PbO_2(s) + 2\,H^+(aq) + 2\,e^- \rightleftharpoons PbO(s) + H_2O(l)$	$H^+ \mid PbO_2, PbO \mid Pt$	+0.28
$Hg_2Cl_2(s) + 2\,e^- \rightleftharpoons 2\,Hg(l) + 2\,Cl^-(aq)$	$Cl^- \mid Hg_2Cl_2 \mid Hg \mid Pt$	+0.2676
$HAsO_2(aq) + 3\,H^+(aq) + 3\,e^- \rightleftharpoons As(s) + 2\,H_2O(l)$	$HAsO_2, H^+ \mid As$	+0.240
$AgCl(s) + e^- \rightleftharpoons Ag(s) + Cl^-$	$Cl^- \mid AgCl \mid Ag$	+0.2223
$SO_4^{2-}(aq) + 4\,H^+(aq) + 2\,e^- \rightleftharpoons 2\,H_2O(l) + SO_2(g)$	$SO_4^{2-}, H^+ \mid SO_2 \mid Pt$	+0.17
$Cu^{2+}(aq) + e^- \rightleftharpoons Cu^+(aq)$	$Cu^+, Cu^{2+} \mid Pt$	+0.159
$Sn^{4+}(aq) + 2\,e^- \rightleftharpoons Sn^{2+}(aq)$	$Sn^{2+}, Sn^{4+} \mid Pt$	+0.154
$S(s) + 2\,H^+(aq) + 2\,e^- \rightleftharpoons H_2S(g)$	$H^+ \mid H_2S \mid S$	+0.144
$AgBr(s) + e^- \rightleftharpoons Ag(s) + Br^-(aq)$	$Br^- \mid AgBr \mid Ag$	+0.071
$2\,H^+(aq) + 2\,e^- \rightleftharpoons H_2(g)$	$H^+ \mid H_2 \mid Pt$	0.0
$Pb^{2+}(aq) + 2\,e^- \rightleftharpoons Pb(s)$	$Pb^{2+} \mid Pb$	−0.125
$Sn^{2+}(aq) + 2\,e^- \rightleftharpoons Sn(s)$	$Sn^{2-} \mid Sn$	−0.137
$AgI(s) + e^- \rightleftharpoons Ag(s) + I^-(aq)$	$I^+ \mid AgI \mid Ag$	−0.152
$V^{3+}(aq) + e^- \rightleftharpoons V^{2+}(aq)$	$V^{3+}, V^{2+} \mid Pt$	−0.255
$Ni^{2+}(aq) + 2\,e^- \rightleftharpoons Ni(s)$	$Ni^{2+} \mid Ni$	−0.257
$H_3PO_4(aq) + 2\,H^+(aq) + 2\,e^- \rightleftharpoons H_3PO_3(aq) + H_2O(l)$	$H_3PO_3, H_3PO_4, H^+ \mid Pt$	−0.276
$Co^{2+}(aq) + 2\,e^- \rightleftharpoons Co(s)$	$Co^{2+} \mid Co$	−0.277
$Tl^+ + e^- \rightleftharpoons Tl$	$Tl^+ \mid Tl$	−0.336
$In^{3+}(aq) + 3\,e^- \rightleftharpoons In(s)$	$In^{3+} \mid In$	−0.338

Reduction Half-Reaction	Cell Notation	Standard Reduction Potential, V
$PbSO_4(s) + 2\,e^- \rightleftharpoons Pb(s) + SO_4^{2-}$	$SO_4^{2-}\,|\,PbSO_4\,|\,Pb$	−0.356
$Cd^{2+}(aq) + 2\,e^- \rightleftharpoons Cd$	$Cd^{2+}\,|\,Cd$	−0.403
$Cr^{3+}(aq) + e^- \rightleftharpoons Cr^{2+}(aq)$	$Cr^{2+}, Cr^{3+}\,|\,Pt$	−0.424
$Fe^{2+}(aq) + 2\,e^- \rightleftharpoons Fe(s)$	$Fe^{2+}\,|\,Fe$	−0.440
$2\,CO_2(aq) + 2\,H^+(aq) + 2\,e^- \rightleftharpoons H_2C_2O_4(aq)$	$H_2C_2O_4, CO_2, H^+\,|\,Pt$	−0.49
$Zn^{2+}(aq) + 2\,e^- \rightleftharpoons Zn(s)$	$Zn^{2+}\,|\,Zn$	−0.763
$2\,H_2O(l) + 2\,e^- \rightleftharpoons H_2(g) + 2\,OH^-(aq)$	$OH^-\,|\,H_2\,|\,Pt$	−0.828
$Cr^{2+}(aq) + 2\,e^- \rightleftharpoons Cr(s)$	$Cr^{2+}\,|\,Cr$	−0.90
$Mn^{2+}(aq) + 2\,e^- \rightleftharpoons Mn(s)$	$Mn^{2+}\,|\,Mn$	−1.180
$Al^{3+}(aq) + 3\,e^- \rightleftharpoons Al(s)$	$Al^{3+}\,|\,Al$	−1.662
$Mg^{2+}(aq) + 2\,e^- \rightleftharpoons Mg(s)$	$Mg^{2+}\,|\,Mg$	−2.372
$Na^+(aq) + e^- \rightleftharpoons Na(s)$	$Na^+\,|\,Na$	−2.714
$Ca^{2+}(aq) + 2\,e^- \rightleftharpoons Ca(s)$	$Ca^{2+}\,|\,Ca$	−2.868
$Sr^{2+}(aq) + 2\,e^- \rightleftharpoons Sr(s)$	$Sr^{2+}\,|\,Sr$	−2.899
$Ba^{2+}(aq) + 2\,e^- \rightleftharpoons Ba(s)$	$Ba^{2+}\,|\,Ba$	−2.905
$K^+(aq) + e^- \rightleftharpoons K(s)$	$K^+\,|\,K$	−2.931
$Li^+(aq) + e^- \rightleftharpoons Li(s)$	$Li^+\,|\,Li$	−3.05
Basic Solution		
$O_3(g) + H_2O(l) + 2\,e^- \rightleftharpoons O_2(g) + 2\,OH^-(aq)$	$OH^-\,|\,O_3, O_2\,|\,Pt$	+1.246
$ClO^-(aq) + H_2O(l) + 2\,e^- \rightleftharpoons Cl^-(aq) + 2\,OH^-(aq)$	$OH^-, ClO_3^-, Cl^-\,|\,Pt$	+0.890
$H_2O_2(aq) + 2\,e^- \rightleftharpoons 2\,OH^-(aq)$	$OH^-, H_2O_2\,|\,Pt$	+0.88
$BrO^-(aq) + H_2O(l) + 2\,e^- \rightleftharpoons Br^-(aq) + 2\,OH^-(aq)$	$OH^-, BrO^-, Br^-\,|\,Pt$	+0.766
$ClO_3^-(aq) + 3\,H_2O(l) + 6\,e^- \rightleftharpoons Cl^-(aq) + 6\,OH^-(aq)$	$OH^-, ClO_3^-, Cl^-\,|\,Pt$	+0.622
$2\,AgO(s) + H_2O(l) + 2\,e^- \rightleftharpoons Ag_2O(s) + 2\,OH^-(aq)$	$OH^-\,|\,Ag_2O\,|\,AgO\,|\,Pt$	+0.604
$MnO_4^-(aq) + 2\,H_2O(l) + 3\,e^- \rightleftharpoons MnO_2(s) + 4\,OH^-(aq)$	$OH^-, MnO_4^-\,|\,MnO_2\,|\,Pt$	+0.60
$BrO_3^-(aq) + 3\,H_2O(l) + 6\,e^- \rightleftharpoons Br^-(aq) + 6\,OH^-(aq)$	$OH^-, BrO_3^-, Br^-\,|\,Pt$	+0.584
$2\,BrO^-(aq) + 2\,H_2O(l) + 2\,e^- \rightleftharpoons Br_2(l) + 4\,OH^-(aq)$	$OH^-, BrO^-\,|\,Br_2\,|\,Pt$	+0.455
$2\,IO^-(aq) + 2\,H_2O(l) + 2\,e^- \rightleftharpoons I_2(s) + 4\,OH^-(aq)$	$OH^-, IO^-\,|\,I_2\,|\,Pt$	+0.42
$O_2(g) + 2\,H_2O(l) + 4\,e^- \rightleftharpoons 4\,OH^-(aq)$	$OH^-\,|\,O_2\,|\,Pt$	+0.401
$Ag_2O(s) + H_2O(l) + 2\,e^- \rightleftharpoons 2\,Ag(s) + 2\,OH^-(aq)$	$OH^-, Ag_2O\,|\,Ag$	+0.342
$Co(OH)_3(s) + e^- \rightleftharpoons Co(OH)_2(s) + OH^-(aq)$	$OH^-\,|\,Co(OH)_3\,|\,Co(OH)_2\,|\,Pt$	+0.17
$2\,MnO_2(s) + H_2O(l) + 2\,e^- \rightleftharpoons Mn_2O_3(s) + 2\,OH^-(aq)$	$OH^-, MnO_2\,|\,Mn_2O_3\,|\,Pt$	+0.118
$NO_3^-(aq) + H_2O(l) + 2\,e^- \rightleftharpoons NO_2^-(aq) + 2\,OH^-(aq)$	$OH^-, NO_2^-, NO_3^-\,|\,Pt$	+0.01
$CrO_4^{2-}(aq) + 4\,H_2O(l) + 3\,e^- \rightleftharpoons Cr(OH)_3(s) + 5\,OH^-(aq)$	$OH^-, CrO_4^{2-}\,|\,Cr(OH)_3\,|\,Pt$	−0.11
$S(s) + 2\,e^- \rightleftharpoons S^{2-}(aq)$	$S^{2-}\,|\,S\,|\,Pt$	−0.48
$HPbO_2^-(aq) + H_2O(l) + 2\,e^- \rightleftharpoons Pb(s) + 3\,OH^-(aq)$	$OH^-, HPbO_2^-\,|\,Pb$	−0.54
$HCHO(aq) + 2\,H_2O(l) + 2\,e^- \rightleftharpoons CH_3OH(aq) + 2\,OH^-(aq)$	$OH^-, HCHO, CH_3\,OH\,|\,Pt$	−0.59
$SO_3^{2-}(aq) + 3\,H_2O(l) + 4\,e^- \rightleftharpoons S(s) + 6\,OH^-(aq)$	$OH^-, SO_3^{2-}\,|\,S\,|\,Pt$	−0.66
$AsO_4^{3-}(aq) + 2\,H_2O(l) + 2\,e^- \rightleftharpoons AsO_2^-(aq) + 4\,OH^-(aq)$	$OH^-, AsO_4^{3-}, AsO_2^-\,|\,Pt$	−0.67
$AsO_2^-(aq) + 2\,H_2O(l) + 3\,e^- \rightleftharpoons As(s) + 4\,OH^-(aq)$	$OH^-, SO_2^-\,|\,As\,|\,Pt$	−0.68
$Cd(OH)_2(s) + 2\,e^- \rightleftharpoons Cd(s) + 2\,OH^-(aq)$	$OH^-\,|\,Cd(OH)_2\,|\,Cd\,|\,Pt$	−0.824
$2\,H_2O(l) + 2\,e^- \rightleftharpoons H_2(g) + 2\,OH^-(aq)$	$OH^-\,|\,H_2\,|\,Pt$	−0.828
$OCN^-(aq) + H_2O(l) + 2\,e^- \rightleftharpoons CN^-(aq) + 2\,OH^-(aq)$	$OH^-, CN^-, OCN^-\,|\,Pt$	−0.97
$As(s) + 3\,H_2O(l) + 3\,e^- \rightleftharpoons AsH_3(g) + 3\,OH^-(aq)$	$OH^-\,|\,As\,|\,AsH_3\,|\,Pt$	−1.21
$Zn(OH)_2(s) + 2\,e^- \rightleftharpoons Zn(s) + 2\,OH^-(aq)$	$OH^-\,|\,Zn(OH)_2\,|\,Zn$	−1.246
$Sb(s) + 3\,H_2O(l) + 3\,e^- \rightleftharpoons SbH_3(g) + 3\,OH^-(aq)$	$OH^-\,|\,Sb\,|\,SbH_3\,|\,Pt$	−1.338
$Al(OH)_4^-(aq) + 3\,e^- \rightleftharpoons Al(s) + 4\,OH^-(aq)$	$OH^-, Al(OH)_4^-\,|\,Al$	−2.310
$Mg(OH)_2(s) + 2\,e^- \rightleftharpoons Mg(s) + 2\,OH^-(aq)$	$OH^-\,|\,Mg(OH)_2\,|\,Mg$	−2.687

[a] The difference between the $E°$ values defined with respect to 1 atm and 1 bar are so small that the values can usually be used interchangeably.

TABLE D.5 Isotopic Masses and Their Abundance[a]

Z	Name	Symbol	Mass of Atom, u	% Abundance
1	Hydrogen	^{1}H	1.007825	99.9885
	Deuterium	^{2}H	2.014102	0.0115
	Tritium	^{3}H	3.016049	—
2	Helium	^{3}He	3.016029	0.000137
		^{4}He	4.002603	99.999863
3	Lithium	^{6}Li	6.015122	7.59
		^{7}Li	7.016004	92.41
4	Beryllium	^{9}Be	9.012182	100
5	Boron	^{10}B	10.012937	19.9
		^{11}B	11.009305	80.1
6	Carbon	^{12}C	12.000000	98.93
		^{13}C	13.003355	1.07
		^{14}C	14.003242	—
7	Nitrogen	^{14}N	14.003074	99.632
		^{15}N	15.000109	0.368
8	Oxygen	^{16}O	15.994915	99.757
		^{17}O	16.999132	0.038
		^{18}O	17.999160	0.205
9	Fluorine	^{19}F	18.998403	100
10	Neon	^{20}Ne	19.992440	90.48
		^{21}Ne	20.993847	0.27
		^{22}Ne	21.991386	9.25
11	Sodium	^{23}Na	22.989770	100
12	Magnesium	^{24}Mg	23.985042	78.99
		^{25}Mg	24.985837	10.00
		^{26}Mg	25.982593	11.01
13	Aluminum	^{27}Al	26.981538	100
14	Silicon	^{28}Si	27.976927	92.2297
		^{29}Si	28.976495	4.6832
		^{30}Si	29.973770	3.0872
15	Phosphorus	^{31}P	30.973762	100
16	Sulfur	^{32}S	31.972071	94.93
		^{33}S	32.971458	0.76
		^{34}S	33.967867	4.29
		^{36}S	35.967081	0.02
17	Chlorine	^{35}Cl	34.968853	75.78
		^{37}Cl	36.965903	24.22
18	Argon	^{36}Ar	35.967546	0.3365
		^{38}Ar	37.962732	0.0632
		^{40}Ar	39.962383	99.6003
19	Potassium	^{39}K	38.963707	93.2581
		^{40}K	39.963999	0.0117
		^{41}K	40.961826	6.7302
20	Calcium	^{40}Ca	39.962591	96.941
		^{42}Ca	41.958618	0.647
		^{43}Ca	42.958767	0.135
		^{44}Ca	43.955481	2.086
		^{46}Ca	45.953693	0.004
		^{48}Ca	47.952534	0.187

[a]The isotopic mass data are from G. Audi and A. H. Wapstra, and M. Dedieu, *Nuclear Physics A*, volume 565, pages 1-65 (1993) and G. Audi and A. H. Wapstra, *Nuclear Physics A*, volume 595, pages 409-480 (1995). The percent natural abundance data are from K.J.R. Rosman and P.D.P. Taylor, *Pure and Applied Chemistry*, volume 70, pages 217-235 (1998).

Z	Name	Symbol	Mass of Atom, u	% Abundance
21	Scandium	^{45}Sc	44.955910	100
22	Titanium	^{46}Ti	45.952629	8.25
		^{47}Ti	46.951764	7.44
		^{48}Ti	47.947947	73.72
		^{49}Ti	48.947871	5.41
		^{50}Ti	49.944792	5.18
23	Vanadium	^{50}V	49.947163	0.250
		^{51}V	50.943964	99.750
24	Chromium	^{50}Cr	49.946050	4.345
		^{52}Cr	51.940512	83.789
		^{53}Cr	52.940654	9.501
		^{54}Cr	53.938885	2.365
25	Manganese	^{55}Mn	54.938050	100
26	Iron	^{54}Fe	53.939615	5.845
		^{56}Fe	55.934942	91.754
		^{57}Fe	56.935399	2.119
		^{58}Fe	57.933280	0.282
27	Cobalt	^{59}Co	58.933200	100
28	Nickel	^{58}Ni	57.935348	68.0769
		^{60}Ni	59.930791	26.2231
		^{61}Ni	60.931060	1.1399
		^{62}Ni	61.928349	3.6345
		^{64}Ni	63.927970	0.9256
29	Copper	^{63}Cu	62.929601	69.17
		^{65}Cu	64.927794	30.83
30	Zinc	^{64}Zn	63.929147	48.63
		^{66}Zn	65.926037	27.90
		^{67}Zn	66.927131	4.10
		^{68}Zn	67.924848	18.75
		^{70}Zn	69.925325	0.62
31	Gallium	^{69}Ga	68.925581	60.108
		^{71}Ga	70.924705	39.892
32	Germanium	^{70}Ge	69.924250	20.84
		^{72}Ge	71.922076	27.54
		^{73}Ge	72.923459	7.73
		^{74}Ge	73.921178	36.28
		^{76}Ge	75.921403	7.61
33	Arsenic	^{75}As	74.921596	100
34	Selenium	^{74}Se	73.922477	0.89
		^{76}Se	75.919214	9.37
		^{77}Se	76.919915	7.63
		^{78}Se	77.917310	23.77
		^{80}Se	79.916522	49.61
		^{82}Se	81.916700	8.73
35	Bromine	^{79}Br	78.918338	50.69
		^{81}Br	80.916291	49.31
36	Krypton	^{78}Kr	77.920386	0.35
		^{80}Kr	79.916378	2.28
		^{82}Kr	81.913485	11.58
		^{83}Kr	82.914136	11.49
		^{84}Kr	83.911507	57.00
		^{86}Kr	85.910610	17.30

(continued)

Z	Name	Symbol	Mass of Atom, u	% Abundance
37	Rubidium	^{85}Rb	84.911789	72.17
		^{87}Rb	86.909183	27.83
38	Strontium	^{84}Sr	83.913425	0.56
		^{86}Sr	85.909262	9.86
		^{87}Sr	86.908879	7.00
		^{88}Sr	87.905614	82.58
39	Yttrium	^{89}Y	88.905848	100
40	Zirconium	^{90}Zr	89.904704	51.45
		^{91}Zr	90.905645	11.22
		^{92}Zr	91.905040	17.15
		^{94}Zr	93.906316	17.38
		^{96}Zr	95.908276	2.80
41	Niobium	^{93}Nb	92.906378	100
42	Molybdenum	^{92}Mo	91.906810	14.84
		^{94}Mo	93.905088	9.25
		^{95}Mo	94.905841	15.92
		^{96}Mo	95.904679	16.68
		^{97}Mo	96.906021	9.55
		^{98}Mo	97.905408	24.13
		^{100}Mo	99.907477	9.63
43	Technetium	^{98}Tc	97.907216	—
44	Ruthenium	^{96}Ru	95.907598	5.54
		^{98}Ru	97.905287	1.87
		^{99}Ru	98.905939	12.76
		^{100}Ru	99.904220	12.60
		^{101}Ru	100.905582	17.06
		^{102}Ru	101.904350	31.55
		^{104}Ru	103.905430	18.62
45	Rhodium	^{103}Rh	102.905504	100
46	Palladium	^{102}Pd	101.905608	1.02
		^{104}Pd	103.904035	11.14
		^{105}Pd	104.905084	22.33
		^{106}Pd	105.903483	27.33
		^{108}Pd	107.903894	26.46
		^{110}Pd	109.905152	11.72
47	Silver	^{107}Ag	106.905093	51.839
		^{109}Ag	108.904756	48.161
48	Cadmium	^{106}Cd	105.906458	1.25
		^{108}Cd	107.904183	0.89
		^{110}Cd	109.903006	12.49
		^{111}Cd	110.904182	12.80
		^{112}Cd	111.902757	24.13
		^{113}Cd	112.904401	12.22
		^{114}Cd	113.903358	28.73
		^{116}Cd	115.904755	7.49
49	Indium	^{113}In	112.904061	4.29
		^{115}In	114.903878	95.71
50	Tin	^{112}Sn	111.904821	0.97
		^{114}Sn	113.902782	0.66
		^{115}Sn	114.903346	0.34

Z	Name	Symbol	Mass of Atom, u	% Abundance
50	Tin (continued)	^{116}Sn	115.901744	14.54
		^{117}Sn	116.902954	7.68
		^{118}Sn	117.901606	24.22
		^{119}Sn	118.903309	8.59
		^{120}Sn	119.902197	32.58
		^{122}Sn	121.903440	4.63
		^{124}Sn	123.905275	5.79
51	Antimony	^{121}Sb	120.903818	57.21
		^{123}Sb	122.904216	42.79
52	Tellurium	^{120}Te	119.904020	0.09
		^{122}Te	121.903047	2.55
		^{123}Te	122.904273	0.89
		^{124}Te	123.902819	4.74
		^{125}Te	124.904425	7.07
		^{126}Te	125.903306	18.84
		^{128}Te	127.904461	31.74
		^{130}Te	129.906223	34.08
53	Iodine	^{127}I	126.904468	100
54	Xenon	^{124}Xe	123.905896	0.09
		^{126}Xe	125.904269	0.09
		^{128}Xe	127.903530	1.92
		^{129}Xe	128.904779	26.44
		^{130}Xe	129.903508	4.08
		^{131}Xe	130.905082	21.18
		^{132}Xe	131.904154	26.89
		^{134}Xe	133.905395	10.44
		^{136}Xe	135.907220	8.87
55	Cesium	^{133}Cs	132.905447	100
56	Barium	^{130}Ba	129.906310	0.106
		^{132}Ba	131.905056	0.101
		^{134}Ba	133.904503	2.417
		^{135}Ba	134.905683	6.592
		^{136}Ba	135.904570	7.854
		^{137}Ba	136.905821	11.232
		^{138}Ba	137.905241	71.698
57	Lanthanum	^{138}La	137.907107	0.090
		^{139}La	138.906348	99.910
58	Cerium	^{136}Ce	135.907144	0.185
		^{138}Ce	137.905986	0.251
		^{140}Ce	139.905434	88.450
		^{142}Ce	141.909240	11.114
59	Praseodymium	^{141}Pr	140.907648	100
60	Neodymium	^{142}Nd	141.907719	27.2
		^{143}Nd	142.909810	12.2
		^{144}Nd	143.910083	23.8
		^{145}Nd	144.912569	8.3
		^{146}Nd	145.913112	17.2
		^{148}Nd	147.916889	5.7
		^{150}Nd	149.920887	5.6
61	Promethium	^{145}Pm	144.912744	—

(continued)

Z	Name	Symbol	Mass of Atom, u	% Abundance
62	Samarium	^{144}Sm	143.911995	3.07
		^{146}Sm	145.913041	
		^{147}Sm	146.914893	14.99
		^{148}Sm	147.914818	11.24
		^{149}Sm	148.917180	13.82
		^{150}Sm	149.917271	7.38
		^{152}Sm	151.919728	26.75
		^{154}Sm	153.922205	22.75
63	Europium	^{151}Eu	150.919846	47.81
		^{153}Eu	152.921226	52.19
64	Gadolinium	^{152}Gd	151.919788	0.20
		^{154}Gd	153.920862	2.18
		^{155}Gd	154.922619	14.80
		^{156}Gd	155.922120	20.47
		^{157}Gd	156.923957	15.65
		^{158}Gd	157.924101	24.84
		^{160}Gd	159.927051	21.86
65	Terbium	^{159}Tb	158.925343	100
66	Dysprosium	^{156}Dy	155.924278	0.06
		^{158}Dy	157.924405	0.10
		^{160}Dy	159.925194	2.34
		^{161}Dy	160.926930	18.91
		^{162}Dy	161.926795	25.51
		^{163}Dy	162.928728	24.90
		^{164}Dy	163.929171	28.18
67	Holmium	^{165}Ho	164.930319	100
68	Erbium	^{162}Er	161.928775	0.14
		^{164}Er	163.929197	1.61
		^{166}Er	165.930290	33.61
		^{167}Er	166.932045	22.93
		^{168}Er	167.932368	26.78
		^{170}Er	169.935460	14.93
69	Thulium	^{169}Tm	168.934211	100
70	Ytterbium	^{168}Yb	167.933894	0.13
		^{170}Yb	169.934759	3.04
		^{171}Yb	170.936322	14.28
		^{172}Yb	171.936378	21.83
		^{173}Yb	172.938207	16.13
		^{174}Yb	173.938858	31.83
		^{176}Yb	175.942568	12.76
71	Lutetium	^{175}Lu	174.940768	97.41
		^{176}Lu	175.942682	2.59
72	Hafnium	^{174}Hf	173.940040	0.16
		^{176}Hf	175.941402	5.26
		^{177}Hf	176.943220	18.60
		^{178}Hf	177.943698	27.28
		^{179}Hf	178.945815	13.62
		^{180}Hf	179.946549	35.08
73	Tantalum	^{180}Ta	179.947466	0.012
		^{181}Ta	180.947996	99.988

Z	Name	Symbol	Mass of Atom, u	% Abundance
74	Tungsten	^{180}W	179.946706	0.12
		^{182}W	181.948206	26.50
		^{183}W	182.950224	14.31
		^{184}W	183.950933	30.64
		^{186}W	185.954362	28.43
75	Rhenium	^{185}Re	184.952956	37.40
		^{187}Re	186.955751	62.60
76	Osmium	^{184}Os	183.952491	0.02
		^{186}Os	185.953838	1.59
		^{187}Os	186.955748	1.96
		^{188}Os	187.955836	13.24
		^{189}Os	188.958145	16.15
		^{190}Os	189.958445	26.26
		^{192}Os	191.961479	40.78
77	Iridium	^{191}Ir	190.960591	37.3
		^{193}Ir	192.962924	62.7
78	Platinum	^{190}Pt	189.959930	0.014
		^{192}Pt	191.961035	0.782
		^{194}Pt	193.962664	32.967
		^{195}Pt	194.964774	33.832
		^{196}Pt	195.964935	25.242
		^{198}Pt	197.967876	7.163
79	Gold	^{197}Au	196.966552	100
80	Mercury	^{196}Hg	195.965815	0.15
		^{198}Hg	197.966752	9.97
		^{199}Hg	198.968262	16.87
		^{200}Hg	199.968309	23.10
		^{201}Hg	200.970285	13.18
		^{202}Hg	201.970626	29.86
		^{204}Hg	203.973476	6.87
81	Thallium	^{203}Tl	202.972329	29.524
		^{205}Tl	204.974412	70.476
82	Lead	^{204}Pb	203.973029	1.4
		^{206}Pb	205.974449	24.1
		^{207}Pb	206.975881	22.1
		^{208}Pb	207.976636	52.4
83	Bismuth	^{209}Bi	208.980383	100
84	Polonium	^{209}Po	208.982416	—
85	Astatine	^{210}At	209.987131	—
86	Radon	^{222}Rn	222.017570	—
87	Francium	^{223}Fr	223.019731	—
88	Radium	^{226}Ra	226.025403	—
89	Actinium	^{227}Ac	227.027747	—
90	Thorium	^{232}Th	232.038050	100
		^{234}Th	234.043601	
91	Protactinium	^{231}Pa	231.035879	100
92	Uranium	^{234}U	234.040946	0.0055
		^{235}U	235.043923	0.7200
		^{238}U	238.050783	99.2745
93	Neptunium	^{237}Np	237.048167	—
94	Plutonium	^{244}Pu	244.064198	—
95	Americium	^{243}Am	243.061373	—

(continued)

Z	Name	Symbol	Mass of Atom, u	% Abundance
96	Curium	^{247}Cm	247.070347	—
97	Berkelium	^{247}Bk	247.070299	—
98	Californium	^{251}Cf	251.079580	—
99	Einsteinium	^{252}Es	252.082972	—
100	Fermium	^{257}Fm	257.095099	—
101	Mendelevium	^{258}Md	258.098425	—
102	Nobelium	^{259}No	259.101024	—
103	Lawrencium	^{262}Lr	262.109692	—
104	Rutherfordium	^{263}Rf	263.118313	—
105	Dubnium	^{262}Db	262.011437	—
106	Seaborgium	^{266}Sg	266.012238	—
107	Bohrium	^{264}Bh	264.012496	—
108	Hassium	^{269}Hs	269.001341	—
109	Meitnerium	^{268}Mt	268.001388	—
110	Darmstadtium	^{272}Ds	272.001463	—
111	Roentgenium	^{272}Rg	272.001535	—
112	Copernicium	^{277}Cn	(277)	—
114	Flerovium	^{289}Fl	(289)	—
116	Livermorium	^{289}Lv	(289)	—
118	Ununoctium	^{293}Uuo	(293)	—

E Concept Maps

Appendix

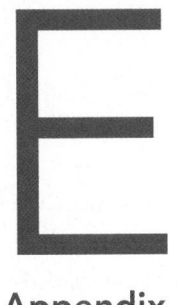

As you study chemistry by reading this book and attending class, you will encounter many ideas and concepts. The task of linking them together can be quite daunting. An effective way to accomplish this task is through "concept mapping." A concept map is a visual map presenting the relationships among a set of connected concepts and ideas. It is a tangible way to display how your mind perceives a particular topic. By constructing a concept map, you reflect on what you understand and what you do not understand. In a concept map, each concept, usually represented by a word or two in a box, is connected to other concept boxes by lines or arrows. A word or brief phrase adjacent to the line or arrow defines the relationship between the connected concepts. Each major concept box has lines to and from several other concept boxes, thereby generating a network, or map.

E-1 How to Construct a Concept Map

1. To create a concept map, construct a list of facts, terms, and ideas that you think are in any way associated with the topic, based on your reading and class attendance. Start by asking yourself, what was the class or reading assignment about? The answer to this question will provide the initial (most general) concepts. The list of concepts will grow as you think further about the answer to this question. You can review the chapter summaries, which emphasize the important points of the chapters, as well as the key terms of that chapter.
2. Review the concepts in your list, and categorize them from most general to most specific. Keep in mind that several of the concepts may have the same level of generality. At other times, it may be difficult to determine the relative importance of two related concepts; to get around this dilemma, try posing the following question: Which concept can be understood without reference to the other? The answer is likely the more general concept.
3. Once the categories have been decided, center the most general concept at the top of the page, and draw a box around it.
4. Arrange the next-most-general rank of concepts below the most general concept. Draw boxes around these concepts, and draw lines linking them to the most general concept. The links should have arrowheads to show the directions in which they should be read.
5. The next step is to label the linkages with short phrases, or even single words, which properly relate the linked concepts. When you place concept 1, a linkage phrase, and concept 2 in sequence, a sensible phrase should result. For example, measurements (concept 1) generate (linkage phrase) numbers (concept 2) that have (linkage phrase) uncertainty (concept 3). The inclusion of linkage labels is important. The appropriate linkage phrase shows that you understand the relationship between the concepts.
6. Proceed down the page, adding rows of ever-more-specific concepts. The most specific concepts should end up at the bottom of your map.

7. Throughout the map, search for cross-links between closely related concepts appearing on the same line. Use dashed lines with double arrowheads to indicate the cross-links.
8. As a last step, assess the map and redraw it if necessary to produce a more logical and neat map.

Once you have constructed the map, check that a concept appears only once and that you have labeled all linkages. Finally, remember that there is no one *correct* concept map for a collection of concepts. However, some concept maps are much more effective than others at displaying the relationships among a given set of concepts.

The diagram in this appendix represents a concept map for the scientific method and measurements. Notice that the concept of SI units could be further connected to such concepts as fundamental units and derived units.

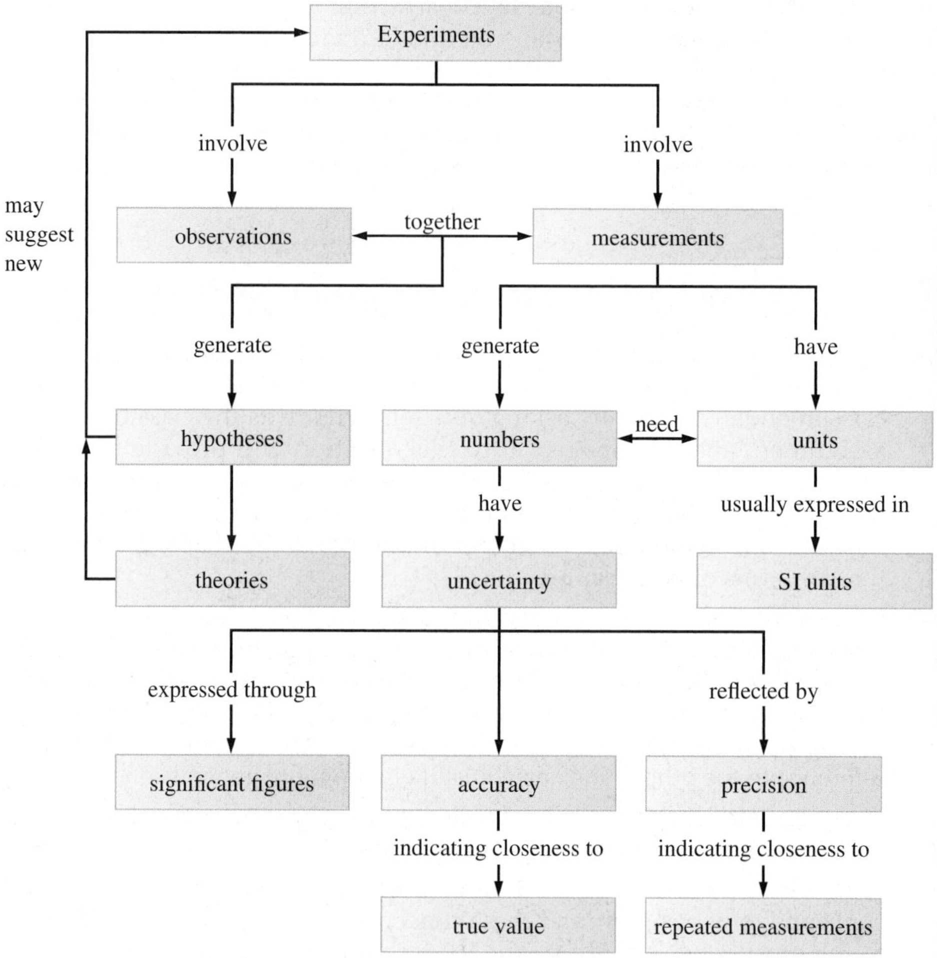

Appendix F

Glossary

Absolute configuration refers to the spatial arrangement of the groups attached to a chiral carbon atom. The two possibilities are D and L.

Absorption spectroscopy is concerned with the absorption of electromagnetic radiation by atoms or molecules.

Accuracy is the "closeness" of a measured value to the true or accepted value of a quantity.

Acetyl group (See **acyl group**.)

An **achiral** molecule has a structure that is superimposable on its mirror image. (See also **chiral**.)

An **acid** is (1) a hydrogen-containing compound that can produce hydrogen ions, H^+ (Arrhenius theory); (2) a proton donor (Brønsted–Lowry theory); (3) an atom, ion, or molecule that can accept a pair of electrons to form a covalent bond (Lewis theory).

An **acid–base indicator** is a substance used to measure the pH of a solution or to signal the equivalence point in an acid–base titration. The nonionized weak acid form has one color and the anionic form, a different color.

An **acid ionization constant, K_a,** is the equilibrium constant for the ionization reaction of a weak acid.

An **acid salt** contains an anion that can act as an acid (proton donor); examples are $NaHSO_4$ and NaH_2PO_4.

The **actinides** are a series of radioactive elements $(Z = 90 - 103)$ characterized by partially filled $5f$ orbitals in their atoms.

An **activated complex** is an intermediate in a chemical reaction formed through collisions between energetic molecules. Once formed, it dissociates either into the products or back to the reactants.

Activation energy is the minimum total kinetic energy that molecules must bring to their collisions for a chemical reaction to occur.

Active sites are the locations at which catalysis occurs, whether on the surface of a heterogeneous catalyst or an enzyme.

The **activity *(a)*** of a substance provides a measure of the ability or potential of a substance to change the Gibbs energy (G) of a system. By definition, it is related to the chemical potential (μ) and the standard chemical potential ($\mu°$) of a substance through the equation $\mu = \mu° + RT \ln a$. In practical applications, the activity of a substance is often expressed as the product of an activity coefficient and the ratio of the stoichiometric concentration or pressure to that of a reference state.

The **actual yield** is the *measured* quantity of a product obtained in a chemical reaction. (See also **theoretical yield** and **percent yield**.)

The **acyl group** is $-\overset{\overset{\displaystyle O}{\|}}{C}-R$. If $R = H$, this is called the **formyl** group; $R = CH_3$, **acetyl**; and $R = C_6H_5$, **benzoyl**.

An **addition–elimination reaction** is the overall reaction that occurs when compounds are interconverted. It involves (1) a nucleophilic addition to the carbonyl carbon to form a tetrahedral intermediate, followed by (2) an elimination reaction that regenerates the carbonyl group.

In an **addition reaction**, a molecule adds across a double or triple bond in another molecule.

An **adduct** is a compound formed by joining together two simpler molecules through a coordinate covalent bond, such as the adduct of BF_3 and $(C_2H_5)_2O$ pictured on page 1003.

Adenosine diphosphate (ADP) and **adenosine triphosphate (ATP)** are agents involved in energy transfers during metabolism. The hydrolysis of ATP produces ADP, the ion HPO_4^{2-}, and a release of energy.

Adhesive forces are intermolecular forces between unlike molecules, such as molecules of a liquid and of a surface with which it is in contact.

ADP (See **adenosine diphosphate**.)

Alcohols contain the functional group $-OH$ and have the general formula ROH.

Aldehydes have the general formula

$$R-\overset{\overset{\displaystyle O}{\|}}{C}-H\ .$$

Alicyclic hydrocarbon molecules have their carbon atom skeletons arranged in rings and resemble aliphatic (rather than aromatic) hydrocarbons.

Aliphatic hydrocarbon molecules have their carbon atom skeletons arranged in straight or branched chains.

Alkali metals is the family name for the group 1 elements of the periodic table.

Alkaline earth metals is the family name for the group 2 elements of the periodic table.

Alkane hydrocarbon molecules have only single covalent bonds between carbon atoms. In their chain structures alkanes have the general formula C_nH_{2n+2}.

Alkene hydrocarbons have one or more carbon-to-carbon double bonds in their molecules. The simple alkenes have the general formula C_nH_{2n}.

Alkyl groups are alkane hydrocarbon molecules from which one hydrogen atom has been extracted. For example, the group $-CH_3$ is the **methyl** group; $-CH_2CH_3$ is the **ethyl** group.

Alkyne hydrocarbons have one or more carbon-to-carbon triple bonds in their molecules. The simple alkynes have the general formula C_nH_{2n-2}.

An **alloy** is a mixture of two or more metals. Some alloys are solid solutions, some are heterogeneous mixtures, and some are intermetallic compounds.

An **alpha (α) particle** is a combination of two protons and two neutrons identical to the helium ion, that is, $^4He^{2+}$. Alpha particles are emitted in some radioactive decay processes.

Alums are sulfates of the general formula, $M(I)M(III)(SO_4)_2 \cdot 12\ H_2O$. M(I) is most commonly an alkali metal or ammonium ion, and M(III) is most commonly Al^{3+}, Fe^{3+}, or Cr^{3+}.

Amalgams are metal alloys containing mercury. Depending on their compositions, some are liquid and some are solid.

An **amide** is derived from the ammonium salt of a carboxylic acid and has the general formula

$$R-\overset{\displaystyle O}{\overset{\|}{C}}-NH_2 .$$

An **amine** is an organic base having the formula RNH_2 (primary), R_2NH (secondary), or R_3N (tertiary), depending on the number of hydrogen atoms of an NH_3 molecule that are replaced by R groups.

An **α-amino acid** is a carboxylic acid that has an amino group ($-NH_2$) attached to the carbon atom adjacent to the **carboxyl group** ($-COOH$).

An **amorphous solid** is one for which there is no long-range order in the arrangement of atoms, molecules, or ions.

Amplitude is the height of the crest of a wave above the center line of the wave.

Amphiprotic substances can act either as an acid or as a base.

An **angular wave function,** $Y(\theta, \phi)$, is the part of a wave function that depends on the angles θ and ϕ when the Schrüdinger wave equation is expressed in spherical polar coordinates. (See also **radial wave function.**)

An **anion** is a negatively charged ion. An anion migrates toward the anode in an electrochemical cell.

The **anode** is the electrode in an electrochemical cell at which an oxidation half-cell reaction occurs.

Groups bonded to adjacent carbon atoms (in a carbon–carbon single bond) are said to be in an **anti conformation** if the **dihedral angle** between the groups is 180°.

An **antibonding molecular orbital** describes regions in a molecule in which there is a zero electron density somewhere between two bonded atoms.

An **aprotic solvent** is a solvent whose molecules do not have a hydrogen atom bonded to an electronegative element.

An **arenium ion** is a cationic species with structural formula

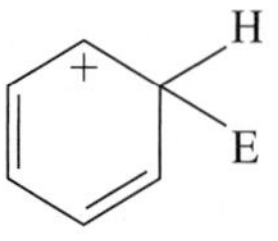

. An arenium ion is formed when an electrophile (E^+) accepts an electron pair from the π system of the benzene ring.

Aromatic hydrocarbons are organic substances whose carbon atom skeletons are arranged in hexagonal rings, based on benzene, C_6H_6.

Asymmetric is the term used to describe a C atom with four different substituent groups. A molecule with such a C atom is chiral.

One **atmosphere** (atm) is the pressure exerted by a column of mercury exactly 760 mm high when the density of mercury is 13.5951 g/cm^3 and the acceleration due to gravity is $g = 9.80665\ m/s^2$.

The **atom** is the basic building block of matter. The number of different atoms currently known is 114. A chemical element consists of a single type of atom, and a chemical compound consists of two or more different kinds of atoms.

The **atomic mass (weight)** of an element is the average of the isotopic masses weighted according to the isotopic abundances of the isotopes of the element and relative to the value of exactly 12 u for a carbon-12 atom.

The **atomic mass interval** is used for certain elements to indicate the range of values expected for the atomic mass because of observed variations in the isotopic abundances of these elements. The interval is expressed in the form $[a, b]$, which indicates that the atomic mass is expected to be between a and b atomic mass units.

An **atomic mass unit, u,** is used to express the masses of individual atoms. One u is 1/12 the mass of a carbon-12 atom.

The **atomic number, Z,** is the number of protons in the nucleus of an atom. It is also the number of electrons outside the nucleus of an electrically neutral atom.

Atomic (line) spectra are produced by dispersing light emitted by excited gaseous atoms. Only a discrete set of wavelength components (seen as colored lines) is present in a line spectrum.

ATP (See **adenosine triphosphate.**)

The **aufbau process** is a method of writing electron configurations. Each element is described as differing from the preceding one in terms of the orbital to which the one additional electron is assigned.

An **average bond energy** is the average of bond-dissociation energies for a number of different species containing a particular covalent bond. (See also **bond dissociation energy.**)

The **Avogadro constant,** N_A, has a value of $6.02214 \times 10^{23}\ mol^{-1}$. It is the number of elementary units in one mole.

Avogadro's law (hypothesis) states that at a fixed temperature and pressure, the volume of a gas is directly proportional to the amount of gas and that equal volumes of different gases, compared under identical conditions of temperature and pressure, contain equal numbers of molecules.

An **azeotrope** is a solution that boils at a constant temperature, producing vapor of the same composition as the liquid. In some cases, the azeotrope boils at a lower temperature than the solution components, in other cases, at a higher temperature.

A **balanced equation** has the same number of atoms of each type on both sides. (See also **chemical equation.**)

One **bar** is equal to 100 kilopascals (1 bar = 100 kPa).

A **barometer** is a device used to measure the pressure of the atmosphere.

Barometric pressure is the prevailing pressure of the atmosphere as indicated by a barometer.

A **base** is (1) a compound that produces hydroxide ions, OH^-, in water solution (Arrhenius theory); (2) a proton acceptor (Brønsted–Lowry theory); (3) an atom, ion, or molecule that can donate a pair of electrons to form a covalent bond (Lewis theory).

A **base ionization constant,** K_b, is the equilibrium constant for the ionization reaction of a weak base.

Basicity is a measure of the tendency of an electron pair donor to react with a proton.

The **basic oxygen process** is the principal process used to convert impure iron (pig iron) into steel.

A **battery** is a voltaic cell [or a group of voltaic cells connected in series (+ to −)] used to produce electricity from chemical change.

bcc (See **body-centered cubic.**)

A **benzyl** group is a methyl group with one hydrogen atom replaced by a phenyl group.

A **beta** (β^-) **particle** is an electron emitted as a result of the conversion of a neutron to a proton in certain atomic nuclei undergoing radioactive decay.

A **bidentate** ligand attaches itself to the central atom of a complex at two points in the coordination sphere.

A **bimolecular process** is an elementary process involving the collision of two molecules.

Binary compounds are compounds composed of *two* elements.

A **body-centered cubic (bcc)** crystal structure is one in which the unit cell has structural units at each corner and one in the center of the cube.

Boiling is a process in which vaporization occurs throughout a liquid. It occurs when the vapor pressure of a liquid is equal to barometric pressure.

A **bomb calorimeter** is a device used to measure the heat of a combustion reaction. The quantity measured is the heat of reaction at constant volume, $q_V = \Delta U$.

A **bond angle** is the angle between two covalent bonds. It is the angle between hypothetical lines joining the nuclei of two atoms to the nucleus of a third atom to which they are covalently bonded.

Bond-dissociation energy, D, is the quantity of energy required to break one mole of covalent bonds in a gaseous species, usually expressed in kJ mol^{-1}. (See also **average bond energy**.)

Bond length (bond distance) is the distance between the centers of two atoms joined by a covalent bond.

Bond order is one-half the difference between the numbers of electrons in bonding and in antibonding molecular orbitals in a covalent bond. A single bond has a bond order of 1; a double bond, 2; and a triple bond, 3.

A **bond pair** is a pair of electrons involved in covalent bond formation.

A **bonding molecular orbital** describes regions of high electron density in the internuclear region between two bonded atoms.

Boyle's law states that the volume of a fixed amount of gas at a constant temperature is inversely proportional to the gas pressure.

In a **bridged halonium ion**, a halogen atom (X) is bonded to (bridges) two carbon atoms that are bonded to each other:

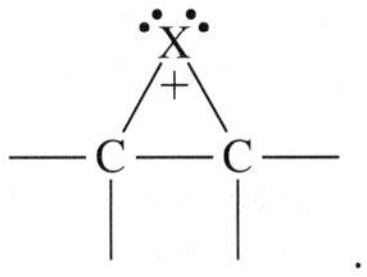

. The halogen atom has a complete octet, comprising two bonding pairs and two lone pairs, and it bears a formal charge of +1. If the halogen atom is chlorine or bromine, the ion is called a chloronium or bromonium ion.

Buffer capacity refers to the amount of acid and/or base that a buffer solution can neutralize while maintaining an essentially constant pH.

Buffer range is the range of pH values over which a buffer solution can maintain a fairly constant pH.

A **buffer solution** resists a change in its pH. It contains components capable of neutralizing small added amounts of acids and base.

By-products are substances produced along with the principal product in a chemical process, either through the main reaction or a side reaction.

Calcination refers to the decomposition of a solid by heating at temperatures below its melting point, such as the decomposition of calcium carbonate to calcium oxide and $CO_2(g)$.

The thermochemical **calorie (cal)** is defined to be 4.184 J. Historically, the calorie was any of several approximately equal units of heat, each measured as the quantity of heat required to raise the temperature of one gram of water, at a specified temperature, by one degree Celsius.

A **calorimeter** is a device (of which there are numerous types) used to measure a quantity of heat.

A **carbohydrate** is a polyhydroxy aldehyde, a polyhydroxy ketone, a derivative of these, or a substance that yields them upon hydrolysis. Carbohydrates can be viewed as "hydrates" of carbon, in the sense that their general formulas are $C_x(H_2O)_y$.

Carbon black is a finely divided amorphous form of carbon prepared by the incomplete combustion of hydrocarbons.

The carbonyl group is found in aldehydes, ketones, and carboxylic acids

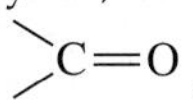

.

The **carboxyl group** is

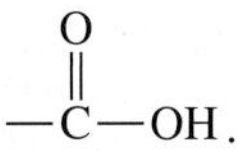

.

A **carboxylic acid** has one or more carboxyl groups attached to a hydrocarbon chain or ring structure.

A **catalyst** provides an alternative mechanism of lower activation energy for a chemical reaction. The rate of reaction is increased, and the catalyst is regenerated.

The **cathode** is the electrode of an electrochemical cell where a reduction half-reaction occurs.

Cathode rays are negatively charged particles (electrons) emitted at the negative electrode (cathode) in the passage of electricity through gases at very low pressures.

Cathodic protection is a method of corrosion control in which the metal to be protected is joined to a more active metal that corrodes instead. The protected metal acts as the cathode of a voltaic cell.

A **cation** is a positively charged ion. A cation migrates toward the cathode in an electrochemical cell.

The **cell** is the fundamental unit of living organisms.

A **cell diagram** is a symbolic representation of an electrochemical cell that indicates the substances entering into the cell reaction, electrode materials, solution concentrations, etc.

The **cell voltage (cell potential), E_{cell}**, is the potential difference (voltage) between the two electrodes of an electrochemical cell.

The **Celsius** temperature scale is based on a value of 0 °C for the normal melting point of ice and 100 °C for the normal boiling point of water.

A **central atom** in a structure is an atom that is bonded to two or more other atoms.

In **chain-reaction polymerization**, a reaction is initiated by "opening up" a carbon-to-carbon double bond. Monomer units add to free-radical intermediates to produce a long-chain polymer.

The **charge density**, ρ, is the charge per unit volume in an ion.

Charles's law states that the volume of a fixed amount of gas at a constant pressure is directly proportional to the Kelvin (absolute) temperature.

A **chelate** results from the attachment of polydentate ligands to the central atom of a complex ion.

A **chelating agent** is a polydentate ligand. It simultaneously attaches to two or more positions in the coordination sphere of the central atom of a complex ion.

Chelation is the process of chelate formation.

The **chelation effect** refers to an exceptional stability conferred to a complex ion when polydentate ligands are present.

Chemical change (See **chemical reaction**.)

Chemical energy is the energy associated with chemical bonds and intermolecular forces.

A **chemical equation** is a symbolic representation of a chemical reaction. Symbols and formulas are used to represent reactants and products, and stoichiometric coefficients are used to balance the equation. (See also **balanced equation**.)

A **chemical formula** represents the relative numbers of atoms of each kind in a substance through symbols and numerical subscripts.

The **chemical potential (μ)** of a substance provides a measure of how much the Gibbs energy (G) of a system changes if the amount of that substance changes while all other system variables are kept constant. By definition, it is the rate of change of G with respect to the amount n of the substance.

A **chemical property** is the ability (or inability) of a sample of matter to undergo a particular chemical reaction.

A **chemical reaction** is a process in which one set of substances (reactants) is transformed into a new set of substances (products).

Chemical symbols are abbreviations of the names of the elements consisting of one or two letters (e.g., N = nitrogen and Ne = neon).

Chiral refers to a molecule with a structure that is not superimposable on its mirror image. (See also **enantiomers**.)

The term ***cis*** describes geometric isomers in which two groups are attached on the same side of a double bond in an organic molecule, or along the same edge of a square in a square-planar complex, or at two adjacent vertices of an octahedral complex. (See also **geometric isomerism**.)

cis–trans **isomerism** is a type of stereoisomerism.

A **closed system** is one that can exchange energy but not matter with its surroundings.

Cohesive forces are intermolecular forces between like molecules, such as within a drop of liquid.

Coke is a relatively pure form of carbon produced by heating coal out of contact with air (destructive distillation).

Colligative properties —vapor pressure lowering, freezing point depression, boiling point elevation, osmotic pressure—have values that depend on the number of solute particles in a solution but not on their identity.

A **colloid** is a mixture that contains particles that are larger than ions or molecules but are still submicroscopic.

The **common-ion effect** describes the effect on an equilibrium by a substance that furnishes ions that can participate in the equilibrium.

A **complementary color** is a secondary color that mixes with the opposite primary color on the color wheel to produce white light in additive color mixing or black in subtractive color mixing.

A **complex** is a polyatomic cation, anion, or neutral molecule in which groups (molecules or ions) called ligands are bonded to a central metal atom or ion.

A **complex ion** is a complex having a net electrical charge.

Composition refers to the components and their relative proportions in a sample of matter.

A **compound** is a substance made up of two or more elements. It does not change its identity in physical changes, but it can be broken down into its constituent elements by chemical changes.

Concentration (1) refers to the composition of a solution. (2) See **extractive metallurgy**.

In a **concentration cell** identical electrodes are immersed in solutions of different concentrations. The voltage (emf) of the cell is a function simply of the concentrations of the two solutions.

Condensation is the passage of molecules from the gaseous state to the liquid state.

A **condensed structural formula** is a simplified representation of a structural formula.

Conformations refer to the different spatial arrangements possible in a molecule. Examples are the "boat" and "chair" forms of cyclohexane.

A **conjugate acid** is formed when a Brønsted–Lowry base gains a proton. Every base has a conjugate acid.

A **conjugate acid–base pair** is pair of molecules or ions for which the chemical formulas differ by a single proton: H^+ (e.g., H_3O^+ and H_2O; H_2O and OH^-; NH_4^+ and NH_3; H_3PO_4 and $H_2PO_4^-$).

A **conjugate base** remains after a Brønsted–Lowry acid has lost a proton. Every acid has a conjugate base.

Consecutive reactions are two or more reactions carried out in sequence. A product of each reaction becomes a reactant in a following reaction until a final product is formed.

Constitutional isomers have different bond connectivities, and thus different skeletal structures.

The **contact process** is a process for the manufacture of sulfuric acid having as its key reaction the oxidation of $SO_2(g)$ to $SO_3(g)$ in contact with a catalyst.

Control rods are neutron-absorbing metal rods (e.g., Cd) that are used to control the neutron flux in a nuclear reactor and thereby control the rate of the fission reaction.

The **conventional atomic mass** is provided for elements that have their atomic masses defined in terms of an atomic mass interval and may be used in situations when a representative value of the atomic mass is required. (See also **atomic mass interval**.)

In a **coordinate covalent bond**, electrons shared between two atoms are contributed by just one of the atoms. As a result, the bonded atoms exhibit formal charges.

Coordination compounds are neutral complexes or compounds containing complex ions.

Coordination number is the number of positions around a central atom where ligands can be attached in the formation of a complex. Applied to a crystalline solid, coordination number signifies the number of nearest neighboring atoms (or ions of opposite charge) to any given atom (or ion) in a crystal.

Coupled reactions are sets of chemical reactions that occur together. One (or more) of the reactions taken alone is (are) *nonspontaneous* and other(s), *spontaneous*. The overall reaction is *spontaneous*.

A **covalent bond** is formed when electrons are shared between a pair of atoms. In valence bond theory, the sharing of the electrons is said to occur in the region in which atomic orbitals overlap.

Covalent radius is one-half the distance between the centers of two atoms that are bonded covalently. It is the atomic radius associated with an element in its covalent compounds.

The **critical point** refers to the temperature and pressure at which a liquid and its vapor become identical. It is the highest temperature point on the vapor pressure curve.

A **crystal** is a solid for which the atoms, ions, or molecules are in an ordered arrangement extending in all three spatial dimensions.

Crystal field theory describes bonding in complexes in terms of electrostatic attractions between ligands and the nucleus of the central metal. Particular attention is focused on the splitting of the d energy level of the central metal.

Cubic closest packed is one of the two ways in which spheres can be packed to minimize the amount of free space or voids among them.

Dalton's law of partial pressures states that in a mixture of gases, the total pressure is the sum of the partial pressures of the gases present. (See also **partial pressure**.)

The **dashed and solid wedge line notation** is a method of conveying a three-dimensional perspective to a structure plotted in a plane.

The ***d*** **block** refers to that section of the periodic table in which the process of orbital filling (aufbau process) involves a *d* subshell.

A **decay constant** is a first-order rate constant describing radioactive decay.

Degree of ionization refers to the extent to which molecules of a weak acid or weak base ionize. The degree of ionization increases as the weak electrolyte solution is diluted. (See also **percent ionization**.)

The **degree of unsaturation** is equal to the total number of π bonds and ring structures in a molecule.

Degenerate orbitals are orbitals that are at the same energy level.

A **delocalized molecular orbital** describes a region of high electron density that extends over three or more atoms.

Denaturation refers to the loss of a protein's secondary and tertiary structures.

Density is a physical property obtained by dividing the mass of a material or object by its volume (i.e., mass per unit volume).

Deoxyribonucleic acid (DNA) is the substance that makes up the genes of the chromosomes in the nuclei of cells.

Deposition is the passage of molecules from the gaseous to the solid state.

Detergents are cleansing agents that act by emulsifying oils. Most common among synthetic detergents are the salts of organic sulfonic acids, $RSO_3^-Na^+$.

Dextrorotatory means the ability to rotate the plane of polarized light to the *right*, designated (+).

Diagonal relationships refer to similarities that exist between certain pairs of elements in different groups and periods of the periodic table, such as Li and Mg, Be and Al, and B and Si.

A **diamagnetic** substance has all its electrons paired and is slightly repelled by a magnetic field.

Diastereomers are optically active isomers of a compound, but their structures are *not* mirror images (as are enantiomers).

Diffraction is the process by which waves (e.g., light) are spread out as a result of passing through a narrow opening or across an edge. The diffracted waves subsequently interfere with each other to produce a diffraction pattern.

Diffusion refers to the spreading of a substance (usually a gas or liquid) into a region where it is not originally present as a result of random molecular motion.

A **dihedral angle** is used to describe the relative orientation of four atoms linearly connected as in A–B–C–D. The dihedral angle about the B–C bond is defined as the angle between a plane formed by the atoms A–B–C and a second plane defined by the atoms B–C–D. The dihedral angle may be used for describing the relative orientation of groups bonded to adjacent carbon atoms in a carbon–carbon single bond. In this context, it refers to the angle of rotation about the carbon–carbon bond.

A **dimer** is a molecule comprised of two simpler formula units, such as Al_2Cl_6, which is a dimer of $AlCl_3$.

Dipole moment, μ, is a measure of the extent to which a separation exists between the centers of positive and negative charge within a molecule. The unit used to measure dipole moment is the **debye**, 3.34×10^{-30} C m.

Dissociation refers to the separation of an entity into two or more entities.

Distillation: In simple distillation, a volatile liquid is separated from nonvolatile solutes. In **fractional distillation**, volatile components of a solution are separated based on their different volatilities.

Dispersion (London) forces are intermolecular forces associated with instantaneous and induced dipoles.

In a **disproportionation reaction,** the same substance is both oxidized and reduced.

In a **double covalent bond,** *two pairs* of electrons are shared between bonded atoms. The bond is represented by a double-dash sign (=).

An **E1 reaction** is an elimination reaction in which the rate-determining step is unimolecular.

An **E2 reaction** is an elimination reaction in which the rate-determining step is bimolecular.

Effective nuclear charge, Z_{eff}, is the positive charge acting on a particular electron in an atom. Its value is the charge on the nucleus, reduced to the extent that other electrons screen the particular electron from the nucleus.

Effusion is the escape of a gas through a tiny hole in its container.

An **electrochemical cell** is a device in which the electrons transferred in an oxidation–reduction reaction are made to pass through an electrical circuit. (See also **electrolytic cell** and **voltaic cell**.)

An **electrode** is a surface on which an oxidation or a reduction half-reaction occurs.

Electrolysis is the decomposition of a substance, either in the molten state or in an electrolyte solution, by means of electric current.

An **electrolyte** is a substance that provides ions when dissolved in water.

An **electrolytic cell** is an electrochemical cell in which a nonspontaneous reaction is carried out by electrolysis.

Electromagnetic radiation is a form of energy propagated as mutually perpendicular electric and magnetic fields. It includes visible light, infrared, ultraviolet, X ray, and radio waves.

Electromotive force (emf) is the potential difference between two electrodes in a voltaic cell, expressed in volts.

Electron affinity (E_{ea}) is the energy change associated with the gain of an electron by a neutral gaseous atom.

Electron capture (EC) is a form of radioactive decay in which an electron from an inner electronic shell is absorbed by a nucleus. In the nucleus the electron is used to convert a proton to a neutron.

An **electron configuration** is a designation of how electrons are distributed among various orbitals in an atom.

Electronegativity (EN) is a measure of the electron-attracting power of a bonded atom; metals have low electronegativities, and nonmetals have high electronegativities.

The **electronegativity difference** between two atoms that are bonded together is used to assess the degree of polarity in the bond.

Electron-group geometry refers to the geometrical distribution about a central atom of the electron pairs in its valence shell.

Electron spin is a characteristic of electrons giving rise to the magnetic properties of atoms. According to quantum theory, electron spin is characterized by two quantum numbers, s and m_s, with $s = \frac{1}{2}$ and $m_s = +\frac{1}{2}$ or $-\frac{1}{2}$. The value of s indicates that an electron is a "spin $\frac{1}{2}$ particle." The values of m_s are associated with the two possible orientations for the magnetic field arising from an electron's spin.

Electrons are particles carrying the fundamental unit of negative electric charge. They are found outside the nuclei of all atoms.

Electron-withdrawing substituents are atoms or groups of atoms that draw electron density toward themselves. Highly electronegative atoms, such as F, O, N, and Cl, are examples.

An **electrophile** contains an electron-deficient, and thus an electron-attracting, region (an electrophilic center) and is a reagent that forms a bond to its reaction partner (the nucleophile) by accepting both bonding electrons from that reaction partner.

An **electrophilic** center in a molecule is an electron-deficient, electron-attracting region.

In an **electrophilic substitution reaction**, an electrophile replaces another atom or group in a molecule. An example of an electrophilic substitution reaction is the replacement of an H atom in benzene with a nitro ($—NO_2$) group.

An **electrostatic potential map** uses colors to show the variation of electrostatic potential that occurs as a positive point charge is moved from point to point on a surface of constant electron charge density.

An **element** is a substance composed of a single type of atom. It cannot be broken down into simpler substances by chemical reactions.

An **elementary process** is an event that significantly alters a molecule's energy or geometry or produces a new molecule(s). It represents a single step in a reaction mechanism.

In an **elimination reaction**, atoms or groups that are bonded to adjacent atoms are eliminated as a small molecule (e.g., H_2O) and an additional bond is formed between carbon atoms.

Emission spectroscopy is concerned with the emission of electromagnetic radiation by atoms or molecules.

An **empirical formula** is the simplest chemical formula that can be written for a compound, that is, having the smallest integral subscripts possible.

Enantiomers (optical isomers) are molecules whose structures are nonsuperimposable mirror images. The molecules are optically active, that is, able to rotate the plane of polarized light.

An **endothermic reaction** results in a lowering of the temperature of an isolated system or the absorption of heat by a system that interacts with its surroundings.

The **end point** is the point in a titration where the indicator used changes color. A properly chosen indicator has its end point coming as closely as possible to the equivalence point of the titration.

Energy is the capacity to do work. (See also **work**.)

An **energy-level diagram** is a representation of the allowed energy levels for a given system (for example, the allowed energy levels of the hydrogen atom). Typically, the levels are shown as horizontal lines, along a vertical (energy) axis, to illustrate the relative energy-ordering of, and spacing between, the various energy levels.

The **English system** of measurement has the yard as its unit of length, the pound as its unit of mass, and the second as its unit of time.

Enthalpy, *H*, is a thermodynamic function used to describe constant-pressure processes: $H = U + PV$, and at constant pressure, $\Delta H = \Delta U + P\,\Delta V = q_p$.

Enthalpy change, ΔH, is the difference in enthalpy between two states of a system. For a chemical reaction carried out at constant temperature and pressure and with work limited to pressure–volume work, the enthalpy change is called the *heat of reaction at constant pressure*.

An **enthalpy diagram** is a diagrammatic representation of the enthalpy changes in a process.

Enthalpy (heat) of formation (See **standard enthalpy of formation**.)

Enthalpy of reaction (See **standard enthalpy of reaction.**)

Entropy, *S*, is a thermodynamic property related to the number of energy levels among which the energy of a system is spread. The greater the number of energy levels for a given total energy, the greater the entropy.

Entropy change, ΔS, is the difference in entropy between two states of a system.

An **enzyme** is a high molar mass protein that catalyzes biological reactions.

An **equation of state** is a mathematical expression relating the amount, volume, temperature, and pressure of a substance (usually applied to gases).

Equilibrium refers to a condition where (1) the forward and reverse processes proceed at equal rates; (2) for reactions at constant T and constant P, the Gibbs energy of reaction, $\Delta_r G$, is equal to zero. For a system that has reached equilibrium, no further net change occurs. For example, amounts of reactants and products in a reversible reaction remain constant over time.

The **equilibrium constant** is the numerical value of the equilibrium constant expression.

An **equilibrium constant expression** describes the relationship among the activities, concentrations, or partial pressures of the substances present in a system at equilibrium.

The **equivalence point** of a titration is the condition in which the reactants are in stoichiometric proportions. They consume each other, and neither reactant is in excess.

An **ester** is the product of the elimination of H_2O from between an acid and an alcohol molecule. Esters have the general formula $R{-}\overset{\overset{\displaystyle O}{\|}}{C}{-}O{-}R'$.

An **ether** has the general formula $R{-}O{-}R'$.

Eutrophication is the deterioration of a freshwater body caused by nutrients such as nitrates and phosphates, which stimulate algae growth, oxygen depletion, and fish kills.

Evaporation is the physical process of a liquid changing to a vapor. (See also **vaporization**.)

In an **excited (electronic) state** of an atom, one or more electrons are promoted to a higher energy level than in the ground state. (See also **ground state**.)

An **exothermic reaction** produces an increase in temperature in an isolated system or, for a system that interacts with its surroundings, the evolution of heat.

Expanded valence shell is a term used to describe Lewis structures in which certain atoms in the third or higher period of the periodic table appear to require 10 or 12 electrons in their valence shells.

An **extensive property** is one, like mass or volume, whose value depends on the quantity of matter observed.

The **extent of reaction** (ξ) is a quantity used to describe the progress of a chemical reaction. It is defined as $\xi = \Delta n_k/\nu_k$, where Δn_k is the change in the amount of a given reactant or product and ν_k is the corresponding stoichiometric number. (See also **stoichiometric number**.)

Extractive metallurgy refers to the process of extracting a metal from its ores. Generally this occurs in four steps. **Concentration** separates the ore from waste rock (gangue). **Roasting** converts the ore to the metal oxide. **Reduction** (usually with carbon) converts the oxide to the metal. **Refining** removes impurities from the metal.

The **E, Z system** is a system of nomenclature used to describe the arrangement of substituent groups bonded to the C atoms of a carbon–carbon double bond.

A **face-centered cubic (fcc)** crystal structure is one in which the unit cell has structural units at the eight corners and in the center of each face of the unit cell. It is derived from the cubic closest packed arrangement of spheres.

The **Fahrenheit** temperature scale is based on a value of 32 °F as the melting point of ice and 212 °F as the boiling point of water.

A **family** of elements is a numbered group from the periodic table, sometimes carrying a distinctive name. For example, group 17 is the halogen family.

The **Faraday constant, F,** is the magnitude (absolute value) of the charge associated with one mole of electrons, 96,485 C/mol.

Fats are triglycerides in which saturated fatty acid components predominate.

The ***f* block** is that portion of the periodic table where the process of filling of electron orbitals (aufbau process) involves f subshells. These are the lanthanide and actinide elements.

fcc (See **face-centered cubic**.)

Ferromagnetism is a property that permits certain materials (notably Fe, Co, and Ni) to be made into permanent magnets. The magnetic moments of individual atoms are aligned into domains. In the presence of a magnetic field, these domains orient themselves to produce a permanent magnetic moment.

A **Fischer projection formula** is a two-dimensional representation of a three-dimensional structural formula. It shows how the stereochemistry at a chiral carbon atom is represented in two dimensions, and how the carbon-chain backbone is arranged on the page.

The **first law of thermodynamics**, expressed as $\Delta U = q + w$, is an alternative statement of the law of conservation of energy. (See also **law of conservation of energy**.)

A **first-order reaction** is one for which the sum of the concentration-term exponents in the rate equation is 1.

Fission (See **nuclear fission**.)

A **flow battery** is a battery in which materials (reactants, products, electrolytes) pass continuously through the battery. The battery is simply a converter of chemical to electrical energy.

Formal charge is the charge an atom would acquire if the bonding electrons in each bond were divided equally between the two atoms involved. It is the number of outer-shell (valence) electrons minus the number of electrons assigned to that atom in a Lewis structure.

The **formation constant, K_f**, describes equilibrium among a complex ion, the free metal ion, and ligands.

Formula mass is the mass of a formula unit of a compound, relative to a mass of exactly 12 u for carbon-12.

A **formula unit** is the smallest collection of atoms or ions from which the empirical formula of a compound can be established.

Fractional crystallization (recrystallization) is a method of purifying a substance by crystallizing the pure solid from a saturated solution while impurities remain in solution.

Fractional distillation (See **distillation**.)

Fractional precipitation is a technique in which two or more ions in solution, each capable of being precipitated by the same reagent, are separated by the use of that reagent.

The **Frasch process** is a method of extracting sulfur from underwater deposits. It is based on the use of superheated water to melt the sulfur.

A **free radical** is an atom, a molecule, or a molecular fragment with one or more unpaired electrons.

Freezing is the conversion of a liquid to a solid that occurs at a fixed temperature known as the **freezing point**.

The **frequency** of a wave motion is the number of wave crests or troughs that pass through a given point in a unit of time. It is expressed by the unit time^{-1} (e.g., s^{-1}, also called a hertz, Hz).

A **fuel cell** is a voltaic cell in which the cell reaction is the equivalent of the combustion of a fuel. Chemical energy of the fuel is converted to electricity.

A **function of state (state function)** is a property that assumes a unique value when the state or present condition of a system is defined. This value is *independent* of how the state is attained.

A **functional group** is an atom or grouping of atoms attached to a hydrocarbon residue, R. The functional group often confers specific chemical and physical properties to an organic molecule.

Fusion (See **nuclear fusion**.)

Galvanic cell (See **voltaic cell**.)

Gamma (γ) rays are a form of electromagnetic radiation of high penetrating power emitted by certain radioactive nuclei.

In a **gas**, atoms or molecules are generally much more widely separated than in liquids and solids. A gas assumes the shape of its container and expands to fill the container, thus having neither definite shape nor volume.

The **gas constant, R**, is the numerical constant appearing in the ideal gas equation ($PV = nRT$) and in several other equations as well.

Groups bonded to adjacent carbon atoms (in a carbon–carbon single bond) are said to be in the **gauche conformation** if the dihedral angle between the groups is 60°.

In a **geminal dihalide**, two halogen atoms are bonded to the same carbon.

The **general gas equation** is an expression based on the ideal gas equation and written in the form $P_1V_1/n_1T_1 = P_2V_2/n_2T_2$.

Geometric isomerism in organic compounds refers to the existence of nonequivalent structures (*cis* and *trans*) that differ in the positioning of substituent groups relative to a double bond. In complexes, the nonequivalent structures are based on the positions at which ligands are attached to the metal center.

Gibbs energy, G, is a thermodynamic function designed to produce a criterion for spontaneous change when considering processes that occur at constant temperature and constant pressure. It is defined through the equation $G = H - TS$.

Gibbs energy change, ΔG, is the change in Gibbs energy that accompanies a process and can be used to indicate the direction of spontaneous change. For a spontaneous process at constant temperature and pressure, $\Delta G < 0$. (See also **standard Gibbs energy change**.)

Gibbs energy of reaction is the rate of change of Gibbs energy with respect to the extent of reaction.

Glass is a transparent, amorphous solid consisting of Na^+ and Ca^{2+} ions in a network of SiO_4^{4-} anions. It is made by fusing together a mixture of sodium and calcium carbonates with sand.

Global warming refers to the warming of Earth that results from an accumulation in the atmosphere of gases such as CO_2 that absorb infrared radiation radiated from Earth's surface.

Graham's law states that the rates of effusion or diffusion of two different gases are inversely proportional to the square roots of their molar masses.

The **ground (electronic) state** is the lowest energy state for the electrons in an atom or molecule.

A **group** is a vertical column of elements in the periodic table. Members of a group have similar properties.

A **half-cell** is a combination of an electrode and a solution. An electro chemical cell is a combination of two half-cells.

The **half-life** ($t_{1/2}$) of a reaction is the time required for one-half of a reactant to be consumed. In a nuclear decay process, it is the time required for one-half of the atoms present in a sample to undergo radioactive decay.

A **half-reaction** describes one portion of an overall oxidation–reduction reaction, either the oxidation or the reduction.

Halogens (group 17) are the most reactive nonmetals, having the electron configuration ns^2np^5 in the electronic shell of highest principal quantum number.

Hard water contains dissolved minerals in significant concentrations. If the hardness is primarily due to HCO_3^- and associated cations, the water has **temporary hardness**. If the hardness is due to anions other than HCO_3^- (e.g., SO_4^{2-}), the water has **permanent hardness**.

hcp (See **hexagonal closest packed**.)

Heat is a transfer of thermal energy as a result of a temperature difference.

Heat capacity is the quantity of heat required to change the temperature of an object or substance by one degree, usually expressed as J °C^{-1} or cal °C^{-1}. **Specific heat capacity** is the heat capacity per gram of substance, i.e., J °C^{-1} g^{-1}, and **molar heat capacity** is the heat capacity per mole, i.e., J °C^{-1} mol^{-1}.

A **heat of reaction** is energy converted from chemical to thermal (or vice versa) in a reaction. In an isolated system, this energy conversion causes a temperature change, and in a system that interacts with its surroundings, heat (q) is either evolved to or absorbed from the surroundings.

The **Heisenberg uncertainty principle** states that it is impossible to know with absolute certainty both the position and the momentum of a particle.

The **Henderson–Hasselbalch equation** has the form, $pH = pK_a + \log$ [conjugate base]/[acid], in which stoichiometric concentrations of the weak acid and its conjugate base are used in place of the equilibrium concentrations. There are limitations on its validity.

Henry's law relates the solubility of a gas to the gas pressure maintained above a solution of the gaseous solute. The solubility is directly proportional to the pressure of the gas above the solution.

The **hertz (Hz)** is the SI unit of frequency, equal to s^{-1}.

Hess's law states that the enthalpy change for an overall or net process is the sum of enthalpy changes for individual steps in the process.

Heterocyclic compounds are based on hydrocarbon ring structures in which one or more C atoms are replaced by atoms such as N, O, or S.

Heterogeneous catalysis is catalytic action that takes place on a surface separating two phases.

In a **heterogeneous mixture**, components separate into physically distinct regions of differing properties and often differing composition.

Hexagonal closest packed is one of the two ways in which spheres can be packed to minimize the amount of free space or voids among them. The crystal structure based on this type of packing is referred to as **hcp**.

In a **high-spin complex,** weak crystal field splitting leads to a maximum number of unpaired electrons in the *d* subshell of the central metal atom or ion.

Homogeneous catalysis refers to a catalytic reaction taking place in a single phase.

A **homogeneous mixture (solution)** is a mixture of elements and/or compounds that has a uniform composition and properties within a given sample. However, the composition and properties may vary from one sample to another.

A **homologous series** is a group of compounds that differ in composition by some constant unit ($—CH_2$ in the case of alkanes).

Hund's rule (rule of maximum multiplicity) states that whenever orbitals of equal energy are available, electrons occupy these orbitals singly before any pairing of electrons occurs.

A **hybrid orbital** is one of a set of identical orbitals reformulated from pure atomic orbitals and used to describe certain covalent bonds.

Hybridization refers to combining pure atomic orbitals to generate hybrid orbitals in the valence bond approach to covalent bonding.

A **hydrate** is a compound in which a fixed number of water molecules is associated with each formula unit, such as $CuSO_4 \cdot 5\,H_2O$.

Hydrides are compounds of hydrogen, usually divided into the categories of covalent (e.g., H_2O and HCl), ionic (e.g., LiH and CaH_2), and metallic (mostly nonstoichiometric compounds with the transition metals).

A **hydrocarbon** is a compound containing the two elements carbon and hydrogen. The C atoms are arranged in straight or branched chains or ring structures.

A **hydrogen bond** is an intermolecular force of attraction in which an H atom covalently bonded to one atom is attracted simultaneously to another highly nonmetallic atom of the same or a nearby molecule.

In a **hydrogenation reaction**, H atoms are added to multiple bonds between carbon atoms, converting carbon-to-carbon double bonds to single bonds and carbon-to-carbon triple bonds to double or single bonds. It is a reaction, for example, that converts an unsaturated to a saturated fatty acid.

Hydrolysis is a special name given to acid–base reactions in which ions act as acids or bases. As a result of hydrolysis, many salt solutions are not pH neutral.

Hydrometallurgy refers to metallurgical procedures where water and aqueous solutions are used to extract metals from their ores. In the first step, **leaching**, the target metal is obtained in soluble form in aqueous solution. Other steps include purifying the leached solution and depositing the metal from solution.

Hydronium ion, H_3O^+, is the form in which protons are found in aqueous solution. The terms "hydrogen ion" and "hydronium ion" are often used synonymously.

The **hydroxyl** group is $—OH$ and is usually found attached to a straight or branched hydrocarbon chain (an alcohol) or a ring structure (a phenol).

A **hypothesis** is a tentative explanation of a series of observations or of a natural law.

An **ICE table** is a format for organizing the data in an equilibrium calculation. It is based on the initial concentrations of reactants and products, changes in concentrations to attain equilibrium, and equilibrium concentrations.

An **ideal (perfect) gas** is one whose behavior can be predicted by the ideal gas equation.

Ideal gas constant (See **gas constant**.)

The **ideal gas equation** relates the pressure, volume, temperature, and number of moles of ideal gas (n) through the expression $PV = nRT$.

An **ideal solution** has $\Delta_{soln}H = 0$ and certain properties (notably vapor pressure) that are predictable from the properties of the solution components.

An **indicator** is an added substance that changes color at the equivalence point in a titration.

The **inductive effect** refers to the shifting of electron density from one atom toward another through the chain of σ bonds that connects them.

Industrial smog is air pollution in which the chief pollutants are $SO_2(g)$, $SO_3(g)$, H_2SO_4 mist, and smoke.

Inert complex is the term used to describe a complex ion in which the exchange of ligands occurs very slowly.

The **inert pair effect** refers to the tendency of the electron pair in the outermost *s* orbital of certain post-transition elements to remain un-ionized or unshared.

The **initial rate of a reaction** is the rate of a reaction immediately after the reactants are brought together.

An **inorganic compound** is any combination of elements that does not fit the category of organic compound. (See also **organic compound**.)

An **instantaneous rate of reaction** is the rate of a reaction at some precise point in the reaction. It is obtained from the slope of a tangent line to a concentration–time graph.

An **integrated rate law (equation)** is derived from a rate law (equation) by the calculus technique of integration. It relates the concentration of a reactant (or product) to elapsed time from the start of a reaction. The equation has different forms depending on the order of the reaction.

An **intensive property** is *independent* of the quantity of matter involved in the observation. Density and temperature are examples of intensive properties.

An **interhalogen** compound is a covalent compound between two or more halogen elements, such as ICl and BrF_3.

An **intermediate** is the product of one reaction that is consumed in a following reaction in a process that proceeds through several steps.

The **internal energy, *U*,** of a system is the total energy attributed to the particles of matter and their interactions within a system.

An **ion** is a charged species consisting of a single atom or a group of atoms. It is formed when a neutral atom or a covalently bonded group of atoms either gains or loses electrons.

Ion exchange is a process in which ions held to the surface of an ion exchange material are exchanged for other ions in solution. For example, Na^+ may be exchanged for Ca^{2+} and Mg^{2+}, or OH^- may be exchanged for SO_4^{2-}.

An **ionic solid** is composed of ions held in place through interionic forces of attraction.

An **ion pair** is an association of a cation and an anion in solution. Such combinations, when they occur, can have a significant effect on solution equilibria.

An **ion product, Q_{sp},** is formulated in the same manner as a solubility product constant, K_{sp}, but with *nonequilibrium* values of activities or concentrations. A comparison of Q_{sp} and K_{sp} provides a criterion for precipitation from solution.

The **ion product of water, K_w,** is the product of $[H_3O^+]$ and $[OH^-]$ in pure water or in an aqueous solution. This product has a unique value that depends only on temperature. At 25 °C, $K_w = 1.0 \times 10^{-14}$.

An **ionic bond** results from the transfer of electrons between metal and nonmetal atoms. Positive and negative ions are formed and held together by electrostatic attractions.

An **ionic compound** is a compound consisting of positive and negative ions that are held together by electrostatic forces of attraction.

Ionic radius is the radius of a spherical ion. It is the atomic radius associated with an element in its ionic compounds.

Ionization refers to the generation of one or more ions.

The first **ionization energy** is the energy required to remove the most loosely held electron from a *gaseous* atom. The second ionization energy is the energy required to remove an electron from a gaseous unipositive ion, and so on.

An **irreversible** process takes place in one or several finite steps such that the system is not in equilibrium with its surroundings.

The **isoelectric point, *pI*,** of an amino acid is the pH at which the dipolar structure or **zwitterion** predominates.

Isoelectronic species have the same number of electrons (usually in the same configuration). Na^+ and Ne are isoelectronic, as are CO and N_2.

An **isolated system** is one that exchanges neither energy nor matter with its surroundings.

Isomers are two or more compounds having the same formula but different structures and therefore different chemical and/or physical properties.

Isotopes of an element are atoms with different numbers of neutrons in their nuclei. That is, isotopes of an element have the same atomic numbers but different mass numbers.

Isotopic mass (See **nuclidic mass.**)

IUPAC (or **IUC**) refers to the International Union of Pure and Applied Chemistry.

K_c is the relationship among the *concentrations* of the reactants and products in a reversible reaction at equilibrium. Concentrations are expressed as molarities.

K_p, the partial pressure equilibrium constant, is the relationship that exists among the partial pressures of gaseous reactants and products in a reversible reaction at equilibrium. Partial pressures are expressed in bar (after 1982) or atm (before 1982).

The **Kelvin** temperature is an **absolute** temperature. That is, the lowest attainable temperature is 0 K = −273.15 °C (the temperature at which molecular motion ceases). Kelvin and Celsius temperatures are related through the expression $T\,(\text{K}) = t(°\text{C}) + 273.15$.

A **ketone** has the general formula

$$R-\overset{\overset{\displaystyle O}{\|}}{C}-R'.$$

A **kilopascal (kPa)** is a unit of pressure equal to 1000 pascals (Pa) or 1000 N m^{-2}. One atmosphere of pressure is 101.325 kPa.

Kinetic energy is energy of motion. The kinetic energy of an object with mass *m* and velocity *u* is $\text{K.E.} = \frac{1}{2}mu^2$.

The **kinetic-molecular theory of gases** is a model for describing gas behavior. It is based on a set of assumptions and yields equations from which various properties of gases can be deduced.

Labile complex is the term used to describe a complex ion in which a rapid exchange of ligands occurs.

The **lanthanide contraction** refers to the decrease in atomic size in a series of elements in which an *f* subshell fills with electrons (an inner transition series). It results from the ineffectiveness of *f* electrons in shielding outer-shell electrons from the nuclear charge of an atom.

The **lanthanides** are the elements ($Z = 58 - 71$) characterized by a partially filled $4f$ subshell in their atoms. Because lanthanum resembles them, La ($Z = 57$) is generally considered together with them.

Lattice energy is the quantity of energy released in the formation of one mole of a crystalline ionic solid from its separated gaseous ions.

Gay-Lussac's law of combining volumes states that, when compared at the same temperature and pressure, the volumes of *gases* involved in a reaction are in the ratio of small whole numbers.

The **law of conservation of energy** states that energy can neither be created nor destroyed in ordinary processes.

The **law of conservation of mass** states that the total mass of the products of a chemical reaction is the same as the total mass of the reactants entering into the reaction.

The **law of constant composition (definite proportions)** states that all samples of a compound have the same composition, that is, the same proportions by mass of the constituent elements.

The **law of multiple proportions** states that if two elements form two or more compounds, the masses of one element combined with a fixed mass of the second are in the ratio of small whole numbers when the different compounds are compared.

The **leaving group** is the species expelled from an electrophilic molecule following attack by a nucleophile.

Le Châtelier's principle states that an action that tends to change the temperature, pressure, or concentrations of reactants in a system at equilibrium stimulates a response that partially offsets the change while a new equilibrium condition is established.

Levorotatory means the ability to rotate the plane of polarized light to the *left*, designated (−).

Lewis acid (See **acid.**)

Lewis base (See **base.**)

A **Lewis structure** is a combination of Lewis symbols that depicts the transfer or sharing of electrons in a chemical bond.

In the **Lewis symbol** of an element, valence electrons are represented by dots placed around the chemical symbol of the element.

The **Lewis theory** refers to a description of chemical bonding through Lewis symbols and Lewis structures in accordance with a particular set of rules.

Ligands are the groups that are coordinated (bonded) to the central atom in a complex by using a pair of electrons that is donated to the central atom.

The **limiting reactant (reagent)** in a reaction is the reactant that is consumed completely. The quantity of product(s) formed depends on the quantity of the limiting reactant.

Line-angle formulas are shorthand representations of organic molecules in which bond lines are drawn, but chemical symbols are written only for elements other than carbon and hydrogen.

A **line spectrum** is produced from the emission of light produced from excited atom or ions. The spectrum contains lines at discrete wavelengths which arise from transitions from one energy level to another.

Lipids include a variety of naturally occurring substances (e.g., fats and oils) sharing the property of solubility in solvents of low polarity [such as in $CHCl_3$, CCl_4, C_6H_6, and $(C_2H_5)_2O$].

In a **liquid**, atoms or molecules are in close proximity (although generally not as close as in a solid). A liquid occupies a definite volume, but has the ability to flow and assume the shape of its container.

London forces (See **dispersion forces**.)

A **lone pair** is a pair of electrons found in the valence shell of an atom and *not* involved in bond formation.

In a **low-spin complex**, strong crystal field splitting leads to a minimum number of unpaired electrons in the *d* subshell of the central metal atom or ion.

Magic numbers is a term used to describe numbers of protons and neutrons that confer a special stability to an atomic nucleus.

The **main-group elements** are those in which *s* or *p* subshells are being filled in the aufbau process. They are also referred to as the *s*-block and *p*-block elements. They are found in groups 1, 2, and 13–18 in the periodic table (the A groups).

A **manometer** is a device used to measure the pressure of a gas, usually by comparing the gas pressure with barometric pressure.

Mass describes the quantity of matter in an object.

The **mass number, *A*,** is the total of the number of protons and neutrons in the nucleus of an atom.

A **mass spectrometer (mass spectrograph)** is a device used to separate and to measure the quantities and masses of different ions in a beam of positively charged gaseous ions.

Matter is anything that occupies space, has the property known as mass, and displays inertia.

Melting is the transition of a solid to a liquid and occurs at the **melting point**. The melting point and freezing point of a substance are identical.

A **meta (*m*-) isomer** has two substituents on a benzene ring separated by one C atom.

Metabolism refers to the totality of the chemical reactions occurring in living organisms.

A **metal** is an element whose atoms have small numbers of electrons in the outermost electronic shell. Removal of an electron(s) from a metal atom occurs without great difficulty, producing a positive ion (cation). Metals generally have a lustrous appearance, are malleable and ductile, and are able to conduct heat and electricity.

Metal carbonyls are complexes with *d*-block metals as central atoms and CO molecules as ligands, e.g., $Ni(CO)_4$.

Metallic radius is one-half the distance between the centers of adjacent atoms in a solid metal.

Metal atoms form a **metallic solid** through the delocalization of electrons. The delocalized electrons freely move throughout the solid, lending to it various properties, such as conductivity.

A **metalloid** is an element that may display both metallic and nonmetallic properties under the appropriate conditions.

A **microstate** is a specific microscopic configuration describing how the particles of a system are distributed among the available energy levels.

A **millimeter of mercury (mmHg)** is a unit of pressure, usually applied to gases. For example, standard atmospheric pressure is equal to the pressure exerted by a 760-mm column of mercury.

A **millimole (mmol)** is one-thousandth of a mole (0.001 mol). It is especially useful in titration calculations.

A **mixture** is any sample of matter that is not pure, that is, not an element or compound. The composition of a mixture, unlike that of a substance, can be varied. Mixtures are either *homogeneous* or *heterogeneous*.

Moderator control slows down energetic neutrons from a fission process so that they are able to induce additional fission.

Molality, *m*, is a solution concentration expressed as the amount of solute, in moles, divided by the mass of solvent, in kg.

Molar mass, *M*, is the mass of one mole of atoms, formula units, or molecules of a substance.

A **mole** is an amount of substance containing Avogadro's number (6.02214×10^{23}) of atoms, formula units, or molecules.

Mole fraction describes a mixture in terms of the fraction of all the molecules that are of a particular type. It is the amount of one component, in moles, divided by the total amount of all the substances in the mixture.

A **mole percent** is a mole fraction expressed on a percentage basis, that is, mole fraction $\times$ 100%.

A **molecular compound** is a compound comprised of discrete molecules.

A **molecular formula** denotes the numbers of the different atoms present in a molecule. In some cases the molecular formula is the same as the empirical formula; in others it is an integral multiple of that formula.

Molecular geometry refers to the geometric shape of a molecule or polyatomic ion. In a species in which all electron pairs are bond pairs, the molecular geometry is the same as the electron-group geometry. In other cases, the two properties are related but not the same.

Molecular mass is the mass of a molecule relative to a mass of exactly 12 u for carbon-12.

Molecular orbital theory describes the covalent bonds in a molecule by replacing atomic orbitals of the component atoms by molecular orbitals belonging to the molecule as a whole. A set of rules is used to assign electrons to these molecular orbitals, thereby yielding the electronic structure of the molecule.

A **molecular solid** is a solid that is composed of discrete molecules.

A **molecule** is a group of bonded atoms held together by covalent bonds and existing as a separate entity. A molecule is the smallest entity having the characteristic proportions of the constituent atoms present in a substance.

A **monodentate ligand** is a ligand that is able to attach to a metal center in a complex at only one position and using just one lone pair of electrons.

A **monosaccharide** is a single, simple molecule having the structural features of a carbohydrate. It can also be called a simple sugar.

A **multiple covalent bond** is a bond in which more than two electrons are shared between the bonded atoms.

A **natural law** is a concise statement, often in mathematical terms, that summarizes observations of certain natural phenomena.

The **Nernst equation** is used to relate E_{cell}, E°_{cell} and the activites of the reactants and products in a cell reaction.

A **net ionic equation** represents a reaction between ions in solution in such a way that all nonparticipant (spectator) ions are eliminated from the equation. The equation must be balanced both atomically and for net electric charge.

A **network-covalent solid** is a substance in which covalent bonds extend throughout the crystal, making the covalent bond both an *intra*molecular and an *inter*molecular force.

In a **neutralization reaction**, an acid and a base react to form a salt. Often, water is also produced.

Neutrons are electrically neutral fundamental particles of matter found in all atomic nuclei except that of the simple hydrogen atom, protium, ^{1}H.

The **neutron number** is the number of neutrons in the nucleus of an atom. It is equal to the mass number (A) minus the atomic number (Z).

Noble gases are the elements of group 18. Their atoms have the electron configuration ns^2np^6 in the electronic shell of highest principal quantum number. (The noble gas helium has the configuration $1s^2$.)

A **nonelectrolyte** is a substance that is essentially non-ionized, both in the pure state and in solution.

A **nonmetal** is an element whose atoms tend to gain small numbers of electrons to form negative ions (anions) with the electron configuration of a noble gas. Nonmetal atoms may also alter their electron configurations by sharing electrons. Nonmetals are mostly gases, liquid (bromine), or low melting point solids and are very poor conductors of heat and electricity.

A **nonspontaneous process** is one that will not occur naturally. A nonspontaneous process can be brought about only by intervention from outside the system, as in the use of electricity to decompose a chemical compound (electrolysis).

The **normal boiling point** is the temperature at which the vapor pressure of a liquid is 1 atm. It is the temperature at which the liquid boils in a container open to the atmosphere at a pressure of 1 atm.

A **nuclear equation** represents the changes that occur during a nuclear process. The target nucleus and bombarding particle are represented on the left side of the equation, and the product nucleus and ejected particle on the right side.

Nuclear fission is a radioactive decay process in which a heavy nucleus breaks up into two lighter nuclei and several neutrons, accompanied by the release of energy.

In **nuclear fusion** small atomic nuclei are fused into larger ones, with some of their mass being converted to energy.

Nucleic acids are cell components comprised of purine and pyrimidine bases, pentose sugars, and phosphoric acid.

A **nucleophile** is a reactant that seeks out, and bonds to, an electrophile (a center of low electron density). The nucleophile donates both bonding electrons to the electrophile.

A **nucleophilic substitution reaction** is a reaction between a nucleophile and an electrophile. The nucleophile attacks at a center of low electron density on the electrophile, and the leaving group is ejected from another point.

Nucleophilicity is a measure of how readily (how fast) a nucleophile attacks an electrophilic carbon atom bearing a leaving group.

Nuclide is a term used to designate an atom with a specific atomic number and mass number. It is represented by the symbolism $^{A}_{Z}E$.

Nuclidic mass is the mass, in atomic mass units, u, of an individual atom relative to an arbitrarily assigned value of exactly 12 u for the nuclidic mass of carbon-12.

An **octet** refers to *eight* electrons in the outermost (valence) electronic shell of an atom in a Lewis structure.

The **octet rule** states that the number of electrons associated with bond pairs and lone pairs of electrons for each of the Lewis symbols (except H) in a Lewis structure will be eight (an **octet**).

Oils are triglycerides in which unsaturated fatty acid components predominate.

Olefins are organic compounds that contain one or more carbon-to-carbon double bonds.

Oligosaccharides are carbohydrates consisting of two to ten monosaccharide units. (See also **sugar**.)

An **open system** is one that can exchange both matter and energy with its surroundings.

Optical isomerism results from the presence of a chiral atom in a structure, leading to a pair of optical isomers that differ only in the direction that they rotate the plane of polarized light. (See also **enantiomers**.)

Optical isomers, also called enantiomers (nonsuperimposable mirror images), are isomers that differ only in the direction they rotate the plane of polarized light.

An atomic **orbital** is a mathematical function, the square of which gives the probability density in the vicinity of a particular point. It can be used to describe regions in an atom where the electron charge density or the probability of finding an electron is high. The several kinds of atomic orbitals ($s, p, d, f, \ldots$) differ from one another in the shapes of the regions of high electron density they describe.

An **orbital diagram** is a representation of an electron configuration in which the most probable orbital designation and spin of each electron in the atom are indicated.

The **order of a reaction** relates to the exponents of the concentration factors in the rate law for a chemical reaction. The order can be stated with respect to a particular reactant (first order in A, second order in B, . . .) or, more commonly, as the overall order. The overall order is the sum of the exponents.

An **organic compound** is made up of carbon and hydrogen or carbon, hydrogen, and a small number of other elements, such as oxygen, nitrogen, and sulfur.

An **ortho (*o*-) isomer** has two substituents attached to adjacent C atoms in a benzene ring.

Osmosis is the net flow of solvent molecules through a semipermeable membrane, from a more dilute solution (or from the pure solvent) into a more concentrated solution.

Osmotic pressure is the pressure that would have to be applied to a solution to stop the passage through a semipermeable membrane of solvent molecules from the pure solvent.

An **overall reaction** or **overall equation** is the overall or net change that occurs when a process is carried out in more than one step.

An **overpotential** is the voltage in excess of the theoretical value required to produce a particular electrode reaction in electrolysis.

Oxidation is a process in which electrons are "lost" and the oxidation state of some atom increases. (Oxidation can occur only in combination with reduction.)

In an **oxidation–reduction (redox)** reaction certain atoms undergo changes in oxidation state. The substance containing atoms whose oxidation states *increase* is **oxidized.** The substance containing atoms whose oxidation states *decrease* is **reduced.**

An **oxidation state** relates to the number of electrons an atom loses, gains, or shares in combining with other atoms to form molecules or polyatomic ions.

An **oxidizing agent (oxidant)** makes possible an oxidation process by itself being *reduced.*

An **oxoacid** is an acid in which an ionizable hydrogen atom is bonded through an oxygen atom to a central atom, that is, E—O—H. Other groups bonded to the central atom are either additional —OH groups or O atoms (or in a few cases H atoms).

An **oxoanion** is a polyatomic anion containing a nonmetal, such as Cl, N, P, or S, in combination with some number of oxygen atoms.

Pairing energy is the energy requirement to force an electron into an orbital that is already occupied by one electron.

A **para (*p*-) isomer** has two substituents bonded to C atoms that are diagonally opposite to one another on a benzene ring.

A **paramagnetic** substance has one or more unpaired electrons in its atoms or molecules. It is attracted into a magnetic field.

A **partial pressure** is the pressure exerted by an individual gas in a mixture, independently of other gases. Each gas in the mixture expands to fill the container and exerts its own partial pressure.

A **pascal (Pa)** is a pressure of one N/m^2.

The **Pauli exclusion principle** states that no two electrons may have an identical set of quantum numbers. This limits occupancy of an orbital to two electrons with opposing spins.

The ***p* block** is that portion of the periodic table in which the filling of electron orbitals (aufbau process) involves *p* subshells.

A **peptide bond** is formed by the elimination of a water molecule from between two amino acid molecules. The H atom comes from the $-NH_2$ group of one amino acid and the —OH group, from the —COOH group of the other acid.

The **percent ionization** of a weak acid or a weak base is the percent of its molecules that ionize in an aqueous solution.

Percent isotopic abundances refer to the relative proportions, expressed as percentages by number, in which the isotopes of an element are found in various sources.

Percent yield is the percent of the theoretical yield of product that is actually obtained in a chemical reaction. (See also **actual yield** and **theoretical yield.**)

A **perfect gas** is one whose behavior can be predicted by the ideal gas equation. It is also used to describe a gas whose molecules are "point masses" that do not interact with one another. (See also **ideal gas.**)

A **period** is a horizontal row of the periodic table.

The **periodic law** refers to the periodic recurrence of certain physical and chemical properties when the elements are considered in terms of increasing atomic number.

The **periodic table** is an arrangement of the elements, by atomic number, in which elements with similar physical and chemical properties are grouped together in vertical columns.

Permanent hard water (See **hard water.**)

The **peroxide** ion has the structure

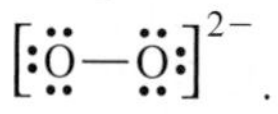

The **pH** of a solution is defined in terms of the activity, *a*, of the H_3O^+ ion through the expression $pH = -\log a_{H_3O^+}$. For sufficiently dilute solutions, $a_{H_3O^+} = [H_3O^+]/(1\,M)$ and $pH = -\log\{[H_3O^+]/(1\,M)\}$. The pH provides a measure of the $[H_3O^+]$ in a solution because, for example, $[H_3O^+] = 1\,M \times 10^{-pH}$.

A **phase diagram** is a graphical representation of the conditions of temperature and pressure at which solids, liquids, and gases (vapors) exist, either as single phases or states of matter or as two or more phases in equilibrium.

A **phenol** has the functional group —OH as part of an aromatic hydrocarbon structure.

A **phenyl group** is a benzene ring from which one H atom has been removed: $-C_6H_5$.

Photochemical smog is air pollution resulting from reactions involving sunlight, ozone, hydrocarbons, and oxides of nitrogen.

The **photoelectric effect** is the emission of electrons by certain materials when their surfaces are struck by electromagnetic radiation of the appropriate frequency.

A **photon** is a "particle" of light. The energy of a beam of light is concentrated into these photons.

In a **physical change,** one or more physical properties of a sample of matter change, but the composition remains unchanged.

A **physical property** is a characteristic that a substance can display without undergoing a change in its composition.

A **pi (π) bond** results from the side-to-side overlap of *p* orbitals, producing a high electron density above and below the line joining the bonded atoms.

Pig iron is an impure form of iron (about 95% Fe and 3–4% C, together with small quantities of Mn, Si, and P) produced in a blast furnace.

p*K* is a shorthand designation for an ionization constant: $pK = -\log K$. p*K* values are useful when comparing the relative strengths of acids or bases.

Planck's constant, *h*, is the proportionality constant that relates the energy of a photon of light to its frequency. Its value is 6.626×10^{-34} J s.

Plaster of Paris, $CaSO_4 \cdot \frac{1}{2}H_2O$, is a hemihydrate of calcium sulfate obtained by heating gypsum, $CaSO_4 \cdot 2\,H_2O$. It is a widely used material in the construction industry.

The **pOH** of a solution is defined in terms of the activity, *a*, of the OH^- ion through the expression $pOH = -\log a_{OH^-}$. For sufficiently dilute solutions, $a_{OH^-} = [OH^-]/(1\,M)$ and $pOH = -\log\{[OH^-]/(1\,M)\}$. The pOH provides a measure of the $[OH^-]$ in a solution because, for example, $[OH^-] = 1\,M \times 10^{-pOH}$.

A **polar covalent bond** is the result of a difference in electronegativity between the two atoms forming a covalent bond. The atom with the higher electronegativity pulls electron density away from the atom with a lower electronegativity. This shift in electron density leads to a polarized bond, with one end of the bond labeled by $\delta-$ and the other by $\delta+$. These symbols represent an excess ($\delta-$) or a deficiency ($\delta+$) of electron density on the atom relative to an isolated atom.

In a **polar molecule,** the presence of one or more polar covalent bonds leads to a separation of the positive and negative charge centers for the molecule as a whole. A polar molecule has a resultant dipole moment.

Polarizability describes the ease with which the electron cloud of an atom or molecule can be distorted in an electric field, that is, the ease with which a dipole can be induced.

A **polyatomic ion** is a combination of two or more covalently bonded atoms that exists as an ion.

A **polydentate ligand** is capable of donating more than a single electron pair to the metal center of a complex, from different atoms in the ligand and to different sites in the geometric structure.

In a **polyhalide ion** two or more halogen atoms are covalently bonded into a polyatomic anion, e.g., I_3^-.

Polymorphism refers to the existence of a solid substance in more than one crystalline form.

In a **polypeptide**, a large number of amino acid units join together through peptide bonds.

A **polyprotic acid** is capable of losing more than a single proton per molecule in acid–base reactions. Protons are lost in a stepwise fashion, with the first proton being the most readily lost.

A **polysaccharide** is a carbohydrate (such as starch or cellulose) consisting of more than ten monosaccharide units.

A **positron (β^+)** is a *positive* electron emitted as a result of the conversion of a proton to a neutron in a radioactive nucleus.

Potential energy is energy due to position or arrangement. It is the energy associated with forces of attraction and repulsion between objects.

The term **ppb** (parts per billion) refers to the number of parts of a component to one billion parts of the medium in which it is found.

The term **ppm** (parts per million) refers to the number of parts of a component to one million parts of the medium in which it is found.

The term **ppt** (parts per trillion) refers to the number of parts of a component to one trillion parts of the medium in which it is found.

A **precipitate** is an insoluble solid that deposits from a solution as a result of a chemical reaction.

Precision is the degree of reproducibility of a measured quantity—the closeness of agreement among repeated measurements.

Pressure is a force per unit area. Applied to gases, pressure is most easily understood in terms of the height of a liquid column that can be maintained by the gas.

Pressure–volume work is work associated with the expansion or compression of gases.

A **primary carbon** is attached to one other carbon atom.

A **primary battery** produces electricity from a chemical reaction that cannot be reversed. As a result the battery cannot be recharged.

A **primary color** is one of a set of colors that when added together as light produce white light. Subtractive mixing leads to an absence of color (black). Red, yellow, and blue are a set of primary colors.

Hydrogen atoms attached to a primary carbon atom are called **primary hydrogen atoms**.

Primary structure refers to the sequence of amino acids in the polypeptide chains that make up a protein.

A **principal electronic shell (level)** refers to the collection of all orbitals having the same value of the principal quantum number, n. For example, the $3s$, $3p$, and $3d$ orbitals comprise the third principal shell ($n = 3$).

The **products** are the substances formed in a chemical reaction.

Properties are qualities or attributes that can be used to distinguish one sample of matter from others.

A **protein** is a large polypeptide, that is, having a molecular mass of 10,000 u or more.

In a **protic solvent** the molecules have hydrogen atoms bonded to electronegative atoms, such as oxygen or nitrogen.

A **proton acceptor** is a base in the Brønsted–Lowry acid–base theory.

A **proton donor** is an acid in the Brønsted–Lowry acid–base theory.

Proton number (See **atomic number**.)

Protons are fundamental particles carrying the basic unit of positive electric charge and found in the nuclei of all atoms.

Pyrometallurgy is the traditional approach to extractive metallurgy that uses dry solid materials heated to high temperatures. (See also **extractive metallurgy** and **hydrometallurgy**.)

Qualitative cation analysis is a laboratory method, based on a variety of solution equilibrium concepts, for determining the presence or absence of certain cations in a sample.

A **quantum** refers to a discrete unit of energy that is the smallest quantity by which the energy of a system can change.

In quantum theory, **quantum numbers** are used to characterize the state of a system. The state of a hydrogen atom, for example, is characterized by specifying the values of the *principal quantum number, n*; the *orbital angular momentum quantum number, l*; the *magnetic quantum number, m_l*; and a *spin quantum number, m_s*. The permitted values for the quantum numbers, and the relationships among them, are obtained by solving the equations of quantum mechanics. The permitted values of these numbers are interrelated.

A **quaternary carbon** is attached to four carbon atoms.

Quaternary structure is the highest order structure that is found in some proteins. It describes how separate polypeptide chains may be assembled into a larger, more complex structure.

Quicklime is a common name for calcium oxide, CaO.

A **racemic mixture** is a mixture containing equal amounts of the enantiomers of an optically active substance.

A **rad** is a quantity of radiation able to deposit 1×10^{-2} J of energy per kilogram of matter.

A **radial wave function,** $R(r)$, is the part of a wave function that depends only on the distance r when the Schrödinger wave equation is expressed in spherical polar coordinates. (See also **angular wave function**.)

Radical (See **free radical**.)

The **radioactive decay law** states that the rate of decay of a radioactive material—the activity, A—is directly proportional to the number of atoms present.

A **radioactive decay series** is a succession of individual steps whereby an initial radioactive isotope (e.g., ^{238}U) is ultimately converted to a stable isotope (e.g., ^{206}Pb).

Radioactivity is a phenomenon in which small particles of matter (α or β particles) and/or electromagnetic radiation (γ rays) are emitted by unstable atomic nuclei.

A **random error** is an error made by the experimenter in performing an experimental technique or measurement, such as the error in estimating a temperature reading on a thermometer.

Raoult's law states that the vapor pressure of a solution component is equal to the product of the vapor pressure of the pure liquid and its mole fraction in solution: $P_A = x_A P_A^*$.

The **rate constant, k,** is the proportionality constant in a rate law that permits the rate of a reaction to be related to the concentrations of the reactants.

A **rate-determining step** in a reaction mechanism is an elementary process that is instrumental in establishing the rate of the overall reaction, usually because it is the slowest step in the mechanism.

The **rate law (rate equation)** for a reaction relates the reaction rate to the concentrations of the reactants. It has the form: rate $= k[A]^m[B]^n \ldots$.

The **rate of reaction** describes how fast reactants are consumed and products are formed, usually expressed as change of concentration per unit time.

Reactants are the substances that enter into a chemical reaction. This term is often applied to *all* the substances involved in a reversible reaction, but it can also be limited to the substances that appear on the *left* side of a chemical equation—the starting substances. (Substances on the *right* side of the equation are usually called products.)

A **reaction intermediate** is a species formed in one elementary reaction in a reaction mechanism and consumed in a subsequent one. As a result, the species does not appear in the equation for the overall reaction.

A **reaction mechanism** is a set of elementary steps or processes by which a reaction is proposed to occur. The mechanism must be consistent with the stoichiometry and rate law of the overall reaction.

A **reaction profile** is a graphical representation of a chemical reaction in terms of the energies of the reactants, activated complex(es), and products.

The **reaction quotient, Q,** is a quotient formed by writing the activities of the products in the numerator and the activities of the reactants in the denominator, with the activity of each reactant or product raised to a power equal to the corresponding coefficient from the balanced equation. In both the numerator and the denominator, the activities (with each raised to the appropriate power) are combined by multiplying them. For example, for the reaction
$2\,H_2S(g) + CH_4(g) \longrightarrow 4\,H_2(g) + CS_2(g)$,
the reaction quotient is
$Q = a^4_{H_2(g)} \times a_{CS_2(g)} / \{a^2_{H_2S(g)} \times a_{CH_4(g)}\}$.

In a **rearrangement reaction,** a molecule is converted into another of its isomeric forms.

Recrystallization (See **fractional crystallization.**)

A **reducing agent (reductant)** makes possible a reduction process by itself becoming *oxidized*.

A **reducing sugar** is one that is able to reduce $Cu^{2+}(aq)$ to red, insoluble Cu_2O. The sugar must have available an aldehyde group, which is oxidized to an acid.

A **reduction** process is one in which electrons are "gained" and the oxidation state of some atom decreases. (Reduction can only occur in combination with oxidation.) (See also **extractive metallurgy**.)

Refining (See **extractive metallurgy**.)

A **rem** is a unit of radiation related to the rad, but taking into account the varying effects on biological matter of different types of radiation of the same energy.

Representative elements (See **main-group elements**.)

A **reserve battery** is a long-term energy storage battery that is designed to provide high power over a relatively short time. Typically, the electrolyte is isolated from the rest of the battery to prevent self-discharge or chemical degradation.

Resonance occurs when two or more plausible Lewis structures can be written for a species. The true structure is a composite or *hybrid* of these different contributing structures.

Reverse osmosis is the passage through a semipermeable membrane of solvent molecules *from a solution into a pure solvent*. It can be achieved by applying to the solution a pressure in excess of its osmotic pressure.

A **reversible process** is one that can be made to reverse direction by just an infinitesimal change in a system property.

Ribonucleic acid (RNA), through its **messenger RNA (mRNA)** and **transfer RNA (tRNA)** forms, is involved in the synthesis of proteins.

Roasting (See **extractive metallurgy**.)

The **root-mean-square speed** is the square root of the average of the squares of the speeds of all the gas molecules in a gaseous sample.

The **R, S system** is used to indicate the arrangement of the four groups bonded to a chiral center and to provide names that distinguish between optical isomers.

A **salt bridge** is a device (a U-tube filled with a salt solution) used to join two half-cells in an electrochemical cell. The salt bridge permits the flow of ions between the two half-cells.

The **salt effect** is that of ions *different* from those directly involved in a solution equilibrium. The salt effect is also known as the diverse or "uncommon" ion effect.

Salts are ionic compounds in which hydrogen atoms of acids are replaced by metal ions. Salts are produced by the neutralization of acids with bases.

Saponification is the hydrolysis of a triglyceride by a strong base. The products are glycerol and a soap.

Saturated hydrocarbon molecules contain only single bonds between carbon atoms.

A **saturated solution** is one that contains the maximum quantity of solute that is normally possible at the given temperature.

The ***s* block** refers to the portion of the periodic table in which the filling of electron orbitals (aufbau process) involves the *s* subshell of the electronic shell of highest principal quantum number.

The **Schrödinger equation** describes the electron in a hydrogen atom as a matter wave. Solutions to the Schrödinger equation are called wave functions.

The **scientific method** refers to the general sequence of activities—observation, experimentation, and the formulation of hypotheses, laws, and theories—that lead to the advancement of scientific knowledge.

The **second law of thermodynamics** relates to the direction of spontaneous change. One statement of the law is that all spontaneous processes produce an increase in the entropy of the universe.

A **secondary battery** produces electricity from a reversible chemical reaction. When electricity is passed through the battery in the reverse direction the battery is recharged.

A **secondary carbon** is attached to two other carbon atoms.

A **secondary color** is the complement of a **primary color**. When light of a primary color and its complement (secondary) color are added, the result is white light. When they are subtracted, the result is an absence of color (black).

The **secondary structure** of a protein describes the structure or shape of a polypeptide chain, for example, a coiled helix.

A **second-order reaction** is one for which the sum of the concentration-factor exponents in the rate equation is 2.

Self-ionization is an acid–base reaction in which one molecule acts as an acid and donates a proton to another molecule of the same kind acting as a base.

The **shielding effect** refers to the effect of inner-shell electrons in shielding or screening outer-shell electrons from the full effects of the nuclear charge. In effect the inner electrons partially reduce the nuclear charge. (See also **effective nuclear charge**.)

A **side reaction** is a reaction that produces an undesired or unexpected product and accompanies a reaction intended to produce something else.

A **sigma (σ) bond** results from the end-to-end overlap of simple or hybridized atomic orbitals along the straight line joining the nuclei of the bonded atoms.

Significant figures are those digits in an experimentally measured quantity that establish the precision with which the quantity is known.

A **silicone** is an organosilicon polymer containing O—Si—O bonds.

Simple distillation (See **distillation**.)

Simultaneous reactions are two or more reactions that occur at the same time.

A **single covalent bond** results from the sharing of *one pair* of electrons between bonded atoms. It is represented by a single dash sign (—).

A **skeletal structure** is an arrangement of atoms in a Lewis structure to correspond to the actual arrangement found by experiment.

Slaked lime is a common name for calcium hydroxide, $Ca(OH)_2$.

Smog is the general term used to refer to a condition in which polluted air reduces visibility, causes stinging eyes and breathing difficulties, and produces additional minor and major health problems. (See also **industrial smog** and **photochemical smog**.)

$\mathbf{S_N1}$ is the designation for a nucleophilic substitution reaction in which the rate-determining step is unimolecular.

$\mathbf{S_N2}$ is the designation for a nucleophilic substitution reaction in which the rate-determining step is bimolecular.

Soaps are the salts of fatty acids, e.g., $RCOO^-Na^+$, where the R group is a hydrocarbon chain containing from 3 to 21 C atoms. Sodium and potassium soaps are the common soaps used as cleansing agents.

Solders are low-melting alloys used for joining wires or pieces of metal. They usually contain metals such as Sn, Pb, Bi, and Cd.

In a **solid**, atoms or molecules are in close contact, often in a highly organized arrangement. A solid has a definite shape and occupies a definite volume. (See also **crystal**.)

The **solubility** of a substance is the concentration of its saturated solution.

The **solubility product constant,** K_{sp}, is the equilibrium constant that describes the formation of a saturated solution of a slightly soluble ionic compound. It is the product of ionic concentrations, with each term raised to an appropriate power.

A **solute** is a solution component that is dissolved in a solvent. A solution may have several solutes, with the solutes generally present in lesser amounts than is the solvent.

Solution (See **homogeneous mixture**.)

The **solvent** is the solution component in which one or more solutes are dissolved. Usually the solvent is present in greater amount than are the solutes and determines the state of matter in which the solution exists.

An ***sp* hybrid orbital** is one of the pair of orbitals formed by the hybridization of one s and one *p* orbital. The angle between the two orbitals is 180°.

An sp^2 **hybrid orbital** is one of the three orbitals formed by the hybridization of one s and two *p* orbitals. The angle between any two of the orbitals is 120°.

An sp^3 **hybrid orbital** is one of the four orbitals formed by the hybridization of one *s* and three *p* orbitals. The angle between any two of the orbitals is the tetrahedral angle—109.5°.

An sp^3d **hybrid orbital** is one of the five orbitals formed by the hybridization of one *s*, three *p*, and one *d* orbital. The five orbitals are directed to the corners of a trigonal bipyramid.

An sp^3d^2 **hybrid orbital** is one of the six orbitals formed by the hybridization of one *s*, three *p*, and two *d* orbitals. The six orbitals are directed to the corners of a regular octahedron.

***spdf* notation** is a method of describing electron configurations in which the numbers of electrons assigned to each orbital are denoted as superscripts. For example, the electron configuration of Cl is $1s^22s^22p^63s^23p^5$.

The **specific heat** of a substance is the quantity of heat required to change the temperature of one gram of the substance by one degree Celsius.

Spectator ions are ionic species that are present in a reaction mixture but do not take part in the reaction. They are usually eliminated from a chemical equation.

The **spectrochemical series** is a ranking of ligand abilities to produce a splitting of the *d* energy level of a central metal ion in a complex ion.

Speed of light, *c*, has a value of 2.99792458×10^8 m s^{-1}.

A **spontaneous (natural) process** is one that is able to take place in a system left to itself. No external action is required to make the process go, although in some cases the process may take a very long time.

Stalactites and **stalagmites** are limestone ($CaCO_3$) formations in limestone caves produced by the slow decomposition of $Ca(HCO_3)_2$(aq).

A **standard cell potential,** $E°$, is the voltage of an electrochemical cell in which all species are in their standard states. (See also **cell potential**.)

Standard conditions of temperature and pressure (STP) refers to a gas maintained at a temperature of exactly 0 °C (273.15 K) and 100 kPa (1 bar).

A **standard electrode potential,** $E°$, is the electric potential that develops on an electrode when the oxidized and reduced forms of some substance are in their *standard* states. Tabulated data are expressed in terms of the reduction process, that is, standard electrode potentials are standard reduction potentials.

The **standard enthalpy of formation,** $\Delta_f H°$, of a substance is the enthalpy change that occurs in the formation of 1 mol of the substance in its standard state from the reference forms of its elements in their standard states. The reference forms of the elements are their most stable forms at the given temperature and 1 bar pressure.

The **standard enthalpy of reaction,** $\Delta_r H°$, is the enthalpy change per mole of reaction for a reaction in which all reactants and products are in their standard states.

Standard Gibbs energy of reaction, $\Delta_r G°$, is the Gibbs energy change per mole of reaction when the reactants and products are all in their standard states. The equation relating standard free energy change to the equilibrium constant is $\Delta_r G° = -RT \ln K$.

The **standard Gibbs energy of formation,** $\Delta_f G°$, is the Gibbs energy change per mole of reaction for the formation of 1 mol of compound from its elements in their most stable forms at 1 bar pressure.

The **standard hydrogen electrode (SHE)** is an electrode at which equilibrium is established between H_3O^+ (aq, a = 1) and H_2 (g, 1 bar) on an inert (Pt) surface. The standard hydrogen electrode is *arbitrarily* assigned an electrode potential of exactly 0 V.

The **standard molar entropy (*S*°)** is the absolute entropy evaluated when one mole of a substance is in its standard state at a particular temperature.

The **standard reaction entropy** is the difference between the standard molar entropies of the reactants and products, with each term weighted by the stoichiometric coefficient.

The **standard state** of a substance refers to that substance when it is maintained at 1 bar pressure and at the temperature of interest. For a gas it is the (hypothetical) pure gas behaving as an ideal gas at 1 bar pressure and the temperature of interest.

Standardization of a solution refers to establishing the exact concentration of the solution, usually through a titration.

A **standing wave** is a wave motion that reflects back on itself in such a way that the wave contains a certain number of points (nodes) that undergo no motion. A common example is the vibration of a plucked guitar string, and a related example is the description of electrons as matter waves.

Steel is a term used to describe iron alloys containing from 0 to 1.5% C together with other key elements, such as V, Cr, Mn, Ni, W, and Mo.

Step-reaction polymerization is a type of polymerization reaction in which monomers are joined together by the elimination of small molecules between them. For example, a H_2O molecule might be eliminated by the reaction of a H atom from one monomer with an —OH group from another.

A **stereocenter** is an asymmetric carbon atom.

In **stereoisomers**, the number and types of atoms and bonds in molecules are the same, but certain atoms are oriented differently in space. *Cis* and *trans* isomerism is one type of stereoisomerism; optical isomerism is another.

Stoichiometric coefficients are the coefficients used to balance an equation.

A **stoichiometric factor** is a conversion factor relating molar amounts of two species in a chemical reaction (i.e., a reactant to a product, one reactant to another, etc.). The numbers used in formulating the factor are stoichiometric coefficients.

The **stoichiometric number** of a reactant or product in a chemical reaction has the same magnitude as the stoichiometric coefficient but is assigned a positive value for a product and a negative value for a reactant. For example, for the reaction $2\,H_2 + O_2 \longrightarrow 2\,H_2O$, the stoichiometric numbers of H_2, O_2, and H_2O are, respectively, –2, –1, and +2.

Stoichiometric proportions refer to relative amounts of reactants that are in the same mole ratio as implied by the balanced equation for a chemical reaction. For example, a mixture of 2 mol H_2 and 1 mol O_2 is in stoichiometric proportions, and a mixture of 1 mol H_2 and 1 mol O_2 is not, for the reaction $2\,H_2 + O_2 \longrightarrow 2\,H_2O$.

Stoichiometry refers to quantitative measurements and relationships involving substances and mixtures of chemical interest.

A **strong acid** is an acid that is completely ionized in aqueous solution.

A **strong base** is a base that is completely ionized in aqueous solution.

A **strong electrolyte** is a substance that is completely ionized in solution.

A **structural formula** for a compound indicates which atoms in a molecule are bonded together, and whether by single, double, or triple bonds.

Structural isomers have the same number and kinds of atoms, but they differ in their structural formulas.

Sublimation is the passage of molecules from the solid to the gaseous state.

A **subshell** refers to a collection of orbitals of the same type. For example, the three 2*p* orbitals constitute the 2*p* subshell.

A **substance** has a constant composition and properties throughout a given sample and from one sample to another. All substances are either elements or compounds.

In a **substitution reaction,** an atom, an ion, or a group in one molecule is replaced by (substituted with) another.

A **substrate** is the substance that is acted upon by an enzyme in an enzyme-catalyzed reaction. The substrate is converted to products, and the enzyme is regenerated.

A **sugar** is a monosaccharide (simple sugar), a disaccharide, or an oligosaccharide containing up to ten monosaccharide units.

The **superoxide** ion has the structure $[:\ddot{\underset{\cdot\cdot}{O}}-\ddot{\underset{\cdot\cdot}{O}}\!\cdot]^-$.

Superphosphate is a mixture of $Ca(H_2PO_4)_2$ and $CaSO_4$ produced by the action of H_2SO_4 on phosphate rock.

A **supersaturated** solution contains more solute than normally expected for a saturated solution, usually prepared from a solution that is saturated at one temperature by changing its temperature to one where supersaturation can occur.

Surface tension is the energy or work required to extend the surface of a liquid.

The **surroundings** represent that portion of the universe with which a system interacts.

A **suspension** is a heterogeneous fluid containing solid particles that are sufficiently large for sedimentation and, unlike colloids, will settle.

Synthesis gas is a mixture of CO(g) and H_2(g), generally made from coal or natural gas, that can be used as a fuel or in the synthesis of organic compounds.

A **system** is the portion of the universe selected for a thermodynamic study. (See also **open**, **closed**, and **isolated** systems.)

A **systematic error** is one that recurs regularly in a series of measurements because of an inherent error in the measuring system (e.g., through faulty calibration of a measuring device).

Temporary hard water (See **hard water**.)

A **terminal atom** is any atom that is bonded to only one other atom in a molecule or polyatomic ion.

A **termolecular process** is an elementary process in a reaction mechanism in which three atoms or molecules must collide simultaneously.

A **ternary compound** is comprised of *three* elements.

A **tertiary carbon** is attached to three other carbon atoms.

The **tertiary structure** of a protein refers to its three-dimensional structure—for example, the twisting and folding of coils.

The **theoretical yield** is the quantity of product *calculated* to result from a chemical reaction. (See also **actual yield** and **percent yield**.)

A **theory** is a model or conceptual framework with which one is able to explain and make further predictions about natural phenomena.

Thermal energy is energy associated with random molecular motion.

The **thermite reaction** is an oxidation–reduction reaction that uses powdered aluminum metal as a reducing agent to reduce a metal oxide, such as Fe_2O_3, to the free metal.

The **thermodynamic equilibrium constant,** ***K***, is an equilibrium constant expression based on activities. In dilute solutions activities can be replaced by the numerical values of molarities and in ideal gases, by the numerical values of partial pressures in bar. The activities of pure solids and liquids are 1.

The **third law of thermodynamics** states that the entropy of a pure perfect crystal is zero at the absolute zero of temperature, 0 K.

The **titrant** is the solution that is added in a controlled fashion through a buret in a titration reaction. (See also **titration**.)

Titration is a procedure for carrying out a chemical reaction between two solutions by the controlled addition (from a buret) of one solution to the other. In a titration a means must be found, as by the use of an indicator, to locate the equivalence point.

A **titration curve** is a graph of solution pH versus volume of titrant. It outlines how pH changes during an acid–base titration, and it can be used to establish such features as the equivalence point of the titration.

A **torr** is a unit of pressure equal to the unit millimeter of mercury.

The **torsional** energy is the energy difference between the eclipsed and staggered forms of ethane.

The term ***trans*** is used to describe geometric isomers in which two groups are attached on opposite sides of a double bond in an organic molecule, or at opposite corners of a square in a square-planar complex, or at positions above and below the central plane of an octahedral complex. (See also **geometric isomerism**.)

Transition elements or **transition metals** are those elements whose atoms feature the filling of a *d* or *f* subshell of an *inner* electronic shell. If the filling of an *f* subshell occurs, the elements are sometimes referred to as inner transition elements.

The **transition state** in a chemical reaction is a state between the reactants and products. (See also **activated complex** and **reaction profile**.)

Triglycerides are esters of glycerol (1,2,3-propanetriol) with long-chain monocarboxylic (fatty) acids.

In a **triple covalent bond**, *three pairs* of electrons are shared between the bonded atoms. It is represented by a triple-dash sign ($\equiv$).

A **triple point** is a condition of temperature and pressure at which three phases of a substance (usually solid, liquid, and vapor) coexist at equilibrium.

Trouton's rule states that at their normal boiling points the entropies of vaporization of many liquids have about the same value: 87 J mol^{-1} K^{-1}.

A **unimolecular** process is an elementary process in a reaction mechanism in which a single molecule, when sufficiently energetic, dissociates.

A **unit cell** is a small collection of atoms, ions, or molecules occupying positions in a crystalline lattice. An entire crystal can be generated by straight-line displacements of the unit cell in the three perpendicular directions.

Unsaturated hydrocarbon molecules contain one or more carbon-to-carbon multiple bonds.

An **unsaturated solution** contains less solute than the solvent is capable of dissolving under the given conditions.

The **valence bond method** treats a covalent bond in terms of the overlap of pure or hybridized atomic orbitals.

Valence electrons are electrons in the electronic shell of highest principal quantum number, that is, electrons in the outermost shell.

The **valence-shell electron-pair repulsion (VSEPR) theory** is a theory used to predict probable shapes of molecules and polyatomic ions based on the mutual repulsions of electron pairs found in the valence shell of the central atom in the structure.

The **van der Waals equation** is an equation of state for nonideal gases. It includes correction terms to account for intermolecular forces of attraction and for the volume occupied by the gas molecules themselves.

The term **van der Waals forces** is used to describe, collectively, intermolecular forces of the London type and interactions between permanent dipoles.

One type of measure of an atomic size are **van der Waals radii.** van der Waals radii are strictly hard sphere radii measured using atomic distances in closest packed crystals.

Vaporization is the passage of molecules from the liquid to the gaseous state.

Vapor pressure is the pressure exerted by a vapor when it is in dynamic equilibrium with its liquid at a fixed temperature.

A **vapor-pressure curve** is a graph of vapor pressure as a function of temperature.

In a **vicinal dihalide**, two halogen atoms are bonded to adjacent carbons.

Viscosity refers to a liquid's resistance to flow. Its magnitude depends on intermolecular forces of attraction and in some cases, on molecular sizes and shapes.

A **volt (V)** is the SI unit for cell voltage. It is defined as 1 joule per coulomb.

A **voltaic (galvanic) cell** is an electrochemical cell in which a *spontaneous* chemical reaction produces electricity.

Water gas is a mixture of CO(g) and H_2(g), together with some of the noncombustible gases CO_2 and N_2, produced by passing steam [H_2O(g)] over heated coke.

A **wave** is a disturbance that transmits energy through a medium.

The **wavelength** is the distance between successive crests or troughs of a wave motion.

Wave mechanics is a form of quantum theory based on the concepts of wave–particle duality, the Heisenberg uncertainty principle, and the treatment of electrons as matter waves. Mathematical solutions of the equations of wave mechanics are known as **wave functions** (ψ).

Wave–particle duality was postulated by de Broglie and states that at times particles of matter have wave-like properties and vice versa. This was demonstrated in the diffraction pattern observed when electrons were directed at a nickel crystal.

A **weak acid** is an acid that is only partially ionized in aqueous solution.

A **weak base** is a base that it only partially ionized in aqueous solution.

A **weak electrolyte** is a substance that is only partially ionized in solution.

Work is a form of energy transfer between a system and its surroundings that can be expressed as a force acting through a distance.

The **work function** is the minimum energy required to extract an electron from the surface of a metal.

A **zero-order reaction** proceeds at a rate that is *independent* of reactant concentrations. The sum of the concentration-factor exponent(s) in the rate equation is equal to *zero*.

The **zero-point energy** is the lowest possible energy in a quantum mechanical system, such as the "particle-in-a-box" energy corresponding to $n = 1$ (page 326).

Zone refining is a purification process in which a rod of material is subjected to successive melting and freezing cycles. Impurities are swept by a moving molten zone to the end of the rod, which is cut off.

A **zwitterion** is a compound (for example, an amino acid or polypeptide) containing both acid and base groups. Zwitterions, at neutral pH, typically have simultaneously positively charged groups (cations) and negatively charge groups (anions).

Appendix

G Answers to Practice Examples and Selected Exercises

CHAPTER 1

Practice Examples 1a. 177 °C **1b.** –26.1 °C **2a.** 1.46 g/mL **2b.** 2.987 cm **3a.** 2.76 g/cm^3 **3b.** The water level will remain unchanged. **4a.** 1.8 kg ethanol **4b.** 0.857 g/mL **5a.** 21.3 **5b.** 1.1 × 10^6 **6a.** 26.7 **6b.** 15.6
Integrative Example A. 14% **B.** 4.239 × 10^{23} Na atoms
Exercises 1. One theory is preferred over another if it can correctly predict a wider range of phenomena and if it has fewer assumptions. **3.** For a given set of conditions, a cause, is expected to produce a certain result or effect. Although these cause-and-effect relationships may be difficult to unravel at times ("God is subtle"), they nevertheless do exist ("He is not malicious"). **5.** The experiment should be carefully set up so as to create a controlled situation in which one can make careful observations after altering the experimental parameters, preferably one at a time. The results must be reproducible (to within experimental error) and, as more and more experiments are conducted, a pattern should begin to emerge, from which a comparison to the current theory can be made. **7a.** Physical **7b.** Chemical **7c.** Chemical **7d.** Physical **9a.** Homogeneous mixture **9b.** Heterogeneous mixture **9c.** Heterogeneous mixture **9d.** Substance **11a.** If a magnet is drawn through the mixture, the iron filings will be attracted to the magnet and the wood will be left behind. **11b.** When the glass-sucrose mixture is mixed with water, the sucrose will dissolve, whereas the glass will not. The water can then be boiled off to produce pure sucrose. **11c.** Olive oil will float to the top of a container and can be separated from water, which is more dense. It would be best to use something with a narrow opening that has the ability to drain off the water layer at the bottom (i.e., buret). **11d.** The gold flakes will settle to the bottom if the mixture is left undisturbed. The water then can be decanted (i.e., carefully poured off). **13a.** 8.950 × 10^3 **13b.** 1.0700 × 10^4 **13c.** 2.40 × 10^{-2} **13d.** 4.7 × 10^{-3} **13e.** 9.383 × 10^2 **13f.** 2.75482 × 10^5 **15a.** 3.4 × 10^4 cm/s **15b.** 6.378 × 10^3 km **15c.** 7.4 × 10^{-11} m **15d.** 4.6 × 10^5 **17a.** exact number **17b.** measured quantity **17c.** measured quantity **17d.** measured quantity **19a.** 3985 **19b.** 422.0 **19c.** 1.860 × 10^5 **19d.** 3.390 × 10^4 **19e.** 6.321 × 10^4 **19f.** 5.047 × 10^{-4} **21a.** 9.3 × 10^{-4} **21b.** 2.2 × 10^{-1} **21c.** 5.4 × 10^{-1} **21d.** 3.058 × 10^1 **23a.** 2.44 × 10^4 **23b.** 1.5 × 10^3 **23c.** 40.0 **23d.** 2.131 × 10^3 **23e.** 4.8 × 10^{-3} **25a.** 186.30 km/h **25b.** 9.95 km/kg **27a.** 127 mL **27b.** 0.0158 L **27c.** 0.981 L **27d.** 2.65 × 10^6 cm^3 **29a.** 174 cm **29b.** 29 m **29c.** 644 g **29d.** 112 kg **29e.** 7.00 dm^3 **29f.** 3.52 × 10^3 mL **31.** 3245 μg **33.** 1.5 m **35a.** 10. s **35b.** 9.8$\underline{3}$ m/s **35c.** 2.5 min **37.** 2.47 acres **39.** 2.2 × 10^3 g/cm^2, 2.2 × 10^4 kg/m^2 **41.** Low: 14 °F, high: 122 °F **43.** A temperature of –465 °F cannot be achieved because it is below absolute zero. **45a.** 35.1 °M **45b.** –59.2 °M **47.** 0.958 g/mL **49.** 0.790 g/mL **51.** 0.629 kg acetone **53.** 1.0$\underline{7}$ kg fertilizer **55.** 1.04 × 10^4 g iron **57.** iron bar < aluminum foil < water **59.** 1.82 × 10^{-2} mm **61.** 0.9 L blood **63.** 9.2% A, 28.9% B, 48.7% C, 11% D, 3% F **65.** 1.10 × 10^3 g sucrose
Integrative and Advanced Exercises 71. 5.5 × 10^{16} tons **72.** 38.8 m **74a.** 5 mg/1m^2 • 1h **74b.** 4.1 × 10^5 h **76a.** −101.$\underline{25}$ °C **76b.** 160. °C **76c.** −19.1 °C **76d.** 335 °C **80.** 11 g/mL **81.** 1.98 × 10^4 kg sodium hypochlorite **84.** 0.25 g/cm^3 **85a.** 0.867 g/cm^3 **85b.** 1.02 g/cm^3 **85c.** 0.869 g/cm^3 **85d.** %N = 58.4 **86.** 76.4 m **87.** 28.9 g ethanol **88.** 0.119 mm **89.** 2.1 glasses of wine
Feature Problems 94. The information provided in sketches (a) and (d) allows calculation of the density. Densityof the plastic material = 1.12 g/cm^3.
Self-Assessment Exercises 99. (c) experiment **100.** The answer is (e), a natural law. **101.** (c) an element **102.** The answer is (a). **103.** The answer is (c). **104.** (a) should have 3 significant figures and the answer shows (4), (b) should have 3 significant figures and the answer shows 2, (c) should have 2 significant figures and the answer shows 3, (d) 2.83 has the correct number of significant figures in the answer **105.** The answer is (d) and (f). **106.** The answer is (d). **107.** The answer is (b). **109a.** 998.2 g/L **109b.** 998.2 kg/m^3 **109c.** 9.982 × 10^{11} kg/km^3 **110.** Student A is more accurate; student B is more precise. **111.** The answer is (b). **112.** (e), (a), (c), (b), (d), listed in order of increasing significant figures. **113.** The answer is (d). **114.** 0.114 mm **115.** No. **116.** (a) Physical change, (b) Physical change, (c) Chemical change, (d) Chemical change, (e) Physical change

CHAPTER 2

Practice Examples 1a. 0.529 g magnesium nitride **1b.** 6.85 g magnesium **2a.** 1.20 g Mg, 0.80 g oxygen. **2b.** 16.61 g of MgO (product) and 3.39 g of unreacted O_2. **3a.** $^{109}_{47}$Ag **3b.** $^{116}_{50}Sn^{2+}$, $^{117}_{50}Sn^{2+}$, $^{118}_{50}Sn^{2+}$, $^{119}_{50}Sn^{2+}$, and $^{120}_{50}Sn^{2+}$ are all possible answers. **4a.** 16.830885 **4b.** $^{158}_{64}$Gd, 157.9241 u, 9.87340. **5a.** Boron-11 **5b.** In-115 **6a.** 28.08 u **6b.** 7.5% lithium-6, 92.5% lithium-7 **7a.** Li^+, S^{2-}, Ra^{2+}, F^-, I^-, Al^{3+}. **7b.** Na is a main-group metal in group 1(1A). Re is a transition metal in group 7(7B). S is a main-group nonmetal in group 16(6A) I is a main-group nonmetal in group 17(7A). Kr is a nonmetal in group 18(8A). Mg is a main-group metal in group 2(2A). U is an inner transition metal, an actinide. Si is a main-group metalloid in group 14(4A). B is a metalloid in group 13(3A). A1 is a main-group metal in group 13(3A). As is a main-group metalloid in group 15(5A). **8a.** 248 g Cu **8b.** 1.58 × 10^{22} ^{206}Pb atoms **9a.** 3.46 × 10^{21} Pb atoms **9b.** 62.5%
Integrative Example A. 9.14 × 10^5 atoms of ^{63}Cu **B.** 2.4477 × 10^6 servings, No.
Exercises 1. The observations cited do not necessarily violate the law of conservation of mass. The oxide formed when iron rusts is a solid and remains with the solid iron, increasing the mass of the solid by an amount equal to the mass of the oxygen that has combined. The oxide formed when a match burns is a gas and will not remain with the solid product (the ash); the mass of the ash thus is less than that of the match. We would have to collect all reactants and all products and weigh them to determine if the law of conservation of mass is obeyed or violated. **3.** 0.268 g oxygen **5.** Compare the mass before reaction (initial) with that after reaction (final). These data support the law of conservation of mass. **7a.** 0.397 **7b.** 0.659 **7c.** 60.3% **9.** The two samples of sodium chloride have the same %Na. **11.** 0.422 g sulfur **13.** For a given mass of sulfur, the mass of oxygen in the second compound (SO_3) relative to the mass of oxygen in the first compound (SO_2) is in a ratio of 3:2. These results are entirely consistent with the Law of Multiple Proportions because the same two elements, sulfur and oxygen in this case, have reacted together to give two different compounds that have masses of oxygen that are in the ratio of small positive integers for a fixed amount of sulfur. **15a.** These results are consistent with the Law of Multiple Proportions because the masses of hydrogen in the three compounds end up in a ratio of small whole numbers when the mass of nitrogen in all three compounds is normalized to a simple value (1.000 g here). **15b.** Compound A might be N_2H_6 and compound C, N_2H_4. **17.** ≈ 11% **19.** Calculate the charge on each drop, express each in terms of 10^{-19} C. Finally, express each in terms of e = 1.6 × 10^{-19} C. The values are consistent with the charge that Millikan found for that of the electron, and he could have inferred the correct charge from these data, since they are all multiples of e. **21a.** Determine the ratio of the mass of a hydrogen atom to that of an electron. Use the mass of a proton plus that of an electron for the mass of a hydrogen atom. **21b.** Use data in Table 2-1 to compare the mass-to-charge ratios those for the proton (a hydrogen ion, H^+) and the electron. The hydrogen ion is the lightest positive ion available. The mass-to-charge ratio for a positive particle is considerably larger than that for an electron. **23a.** $^{60}_{27}$Co **23b.** $^{32}_{15}$P **23c.** $^{59}_{26}$Fe **23d.** $^{226}_{88}$Ra

25.

Name	Symbol	Number of Protons	Number of Electrons	Number of Neutrons	Mass Number
Sodium	$^{23}_{11}Na$	11	11	12	23
Silicon	$^{28}_{14}Si$	14	14[a]	14	28
Rubidium	$^{85}_{37}Rb$	37	37[a]	48	85
Potassium	$^{40}_{19}K$	19	19	21	40
Arsenic[a]	$^{75}_{33}As$	33[a]	33	42	75
Neon	$^{20}_{10}Ne^{2+}$	10	8	10	20
Bromine[b]	$^{80}_{35}Br$	35	35	45	80
Lead[b]	$^{208}_{82}Pb$	82	82	126	208

[a]This result assumes that a neutral atom is involved.
[b]Insufficient data. Does not characterize a specific nuclide; several possibilities exist. The minimum information needed is the atomic number (or some way to obtain it, such as from the name or the symbol of the element involved), the number of electrons (or some way to obtain it, such as the charge on the species), and the mass number (or the number of neutrons).

27a. 46 protons, 46 electrons, 62 neutrons **27b.** 8.9919908 **29.** 106.936 u **31a.** 2.914071 **31b.** 2.165216 **31c.** 18.50146 **33a.** $^{35}_{17}Cl^-$ **33b.** $^{60}_{27}Co^{3+}$ **33c.** $^{124}_{50}Sn^{2+}$ **35.** calcium **37.** 53 protons, 54 electrons, 70 neutrons. **39.** 95 protons, 95 electrons, 146 neutrons. **41.** There are no chlorine atoms that have a mass of 35.45 u. **43.** 24.31 u **45.** 108.9 u **47.** 40.962 u

49a.

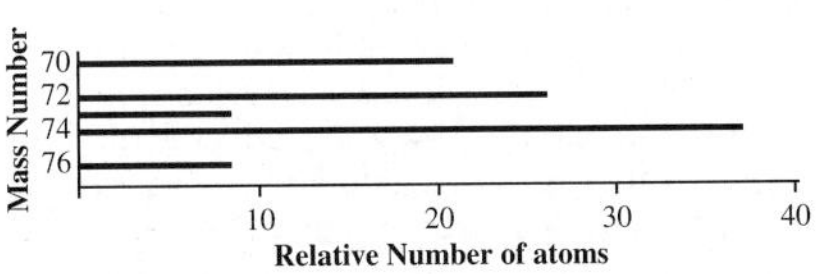

49b. Average atomic mass of germanium = 72.6. The result is only approximately correct because the isotopic masses are given to only two significant figures. **51a.** Ge **51b.** Other elements in group 16(6A) are similar to S: O, Se, and Te. Most of the elements in the periodic table are unlike S, but particularly metals such as Na, K, and Rb. **51c.** Rb **51d.** At **53a.** 118 **53b.** 119 **55a.** 9.51×10^{24} atoms Fe **55b.** 2.81×10^{20} atoms Ag **55c.** 5.1×10^{13} atoms Na **57a.** 6.347 mol Zn **57b.** 1.707×10^{27} atoms Cr **57c.** 3.3×10^{-10} g Au **57d.** $\dfrac{3.1547 \times 10^{-23}\text{g F}}{1 \text{ atom F}}$ **59.** 2.4×10^{22} Cu atoms **61.** 8.7×10^{18} atoms ^{204}Pb **63a.** $1.4\underline{5} \times 10^{-6}$ mol Pb/L **63b.** 8.7×10^{14} Pb atoms/mL **65.** Answer (b).

Integrative and Advanced Exercises
70. 3×10^{15} g/cm^3 **72a.** ^{234}Th **72b.** Ti^{2+}. There is not enough information to determine the mass number. **72c.** $^{110}Sn^{2+}$ **74.** ^{210}Po **76.** 0.36% **77.** 200.6 u **82.** 2.3×10^{21} Si atoms **86.** 50.1% Sn, 32.0% Pb, 17.9% Cd. **87.** 102 cm^3

Feature Problems 89. The sensitivity of Lavoisier's balance can be as little as 0.01 grain, which seems to be the limit of the readability of the balance; alternatively, it can be as large as 3.12 grains, which assumes that all of the error in the experiment is due to the (in)sensitivity of the balance. Convert 0.01 grains and 3.12 grains to grams. Lavoisier's results conform closely to the law of conservation of mass. The difference between reactant mass and product mass (in grams) falls between the maximum error of a common modern laboratory balance and the minimum error of a good quality analytical balance. **92.** 2.43×10^5 km^3 **93.** 159 ppm Rb

Self-Assessment Exercises 97. The element is boron **98.** The answer is (b). **99.** The answer is (d). **100.** 14.0 g/mol X **101.** The answer is (c). **102.** The answer is (a). **103.** The answer is (d). **104.** The answer is (e) **105.** The answer is (d) **106.** The answer is (e) **107.** $^{35}_{17}Cl^+$ **108.** The answer is (d). **109a.** Group 18 **109b.** Group 17 **109c.** Group 13 and Group 1 **109d.** Group 18 **110.** (d) and (f) **111.** The answer is (c). **112.** The answer is (d). **113.** The answer is (b). **114.** The answer is (d) **115.** Fe_2O_3 and Fe_3O_4 **116.** ^{86}Sr = 9.5%, ^{87}Sr = 7.3%. The reason for the imprecision is the low number of significant figures for ^{84}Sr. **117.** 1.2×10^{14} atoms of Au

CHAPTER 3

Practice Examples 1a. 4.0×10^1 g $MgCl_2$ **1b.** 8.12×10^{15} NO_3^- ions, 2.44×10^{16} O **2a.** 3.69×10^{21} Au atoms **2b.** The vapor will be detectable. **3a.** 18.9 g Br **3b.** 175 mL $C_2HBrClF_3$ **4a.** 23.681% C, 3.180% H, 13.81% N, 18.32% P, 41.008% O. **4b.** Both (b) and (e) have the same empirical formula, that is, CH_2O. These two molecules have the same percent oxygen by mass. **5a.** The empirical formula of the compound is $C_3H_7O_3$. The molecular formula is $C_6H_{14}O_6$. **5b.** The empirical formula of the compound is $C_{13}H_{16}O_8Cl_{12}$. The molecular formula is $C_{13}H_{16}O_8Cl_{12}$. **6a.** $C_7H_{14}O_2$ **6b.** C_4H_4S **7a.** For an atom of a free element, the oxidation state is 0; each Cr has an O.S. = +6; each chlorine has an O.S. = +1; each oxygen has an O.S. = –1/2. **7b.** Each S has an O.S. = +2; each Hg has O.S. = +1; the O.S. of Mn is +7; the O.S. of C is 0. **8a.** Li_2O, SnF_2, Li_3N. **8b.** Al_2S_3, Mg_3N_2, V_2O_3. **9a.** cesium iodide, calcium fluoride, iron(II) oxide, chromium(III) chloride. **9b.** calcium hydride, copper(I) chloride, silver(I) sulfide, mercury(I) chloride. **10a.** sulfur hexafluoride, nitrous acid, calcium bicarbonate or calcium hydrogen carbonate, iron(II) sulfate. **10b.** ammonium nitrate, phosphorus trichloride, hypobromous acid, silver(I) perchlorate, iron(III) sulfate. **11a.** BF_3, $K_2Cr_2O_7$, H_2SO_4, $CaCl_2$. **11b.** $Al(NO_3)_3$, P_4O_{10}, $Cr(OH)_3$, HIO_3. **12a.** (a) Not isomers; (b) Molecules are isomers. **12b.** (a) Molecules are isomers; (b) Not isomers. **13a.** (a) alkane; (b) chloroalkane; (c) carboxylic acid; (d) alkene. **13b.** (a) alcohol; (b) carboxylic acid; (c) chloro carboxylic acid; (d) bromoalkene. **14a.** (a) propan-2-ol; (b) 1-iodopropane; (c) 3-methylbutanoic acid; (d) propene. **14b.** (a) 2-chloropropane; (b) 1,4-dichlorobutane; (c) 2-methyl propanoic acid. **15a.** (a) $CH_3(CH_2)_3CH_3$; (b) CH_3CO_2H; (c) $ICH_2(CH_2)_6CH_3$; (d) $CH_2(OH)(CH_2)_3CH_3$.

15b.
(a) 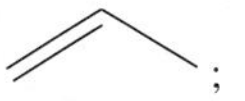 ;
(b) 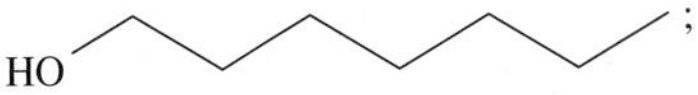

;
(c) 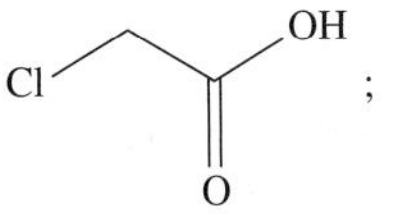

;
(d) 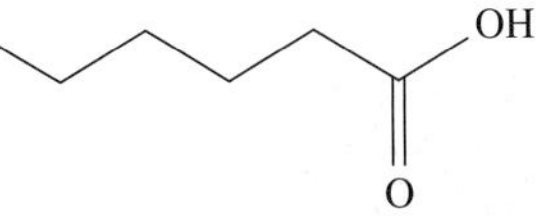

Integrative Example A. The compound is magnesium phosphate, $Mg_3(PO_4)_2$. The compound is $Mg_3(PO_4)_2 \cdot 5H_2O$. **B.** The compound is $Cu(ClO_4)_2 \cdot 6H_2O$. The oxidation state of Cu is +2 and Cl is +7.
Exercises 1a. H_2O_2 **1b.** CH_3CH_2Cl **1c.** P_4O_{10} **1d.** $CH_3CH(OH)CH_3$ **1e.** HCO_2H

3. (b) CH_3CH_2Cl 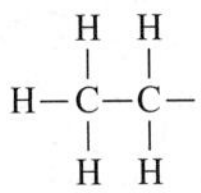

(d) $CH_3CH(OH)CH_3$ 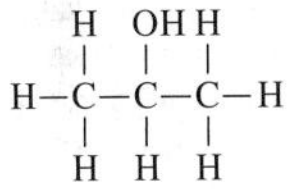

(e) HCO_2H 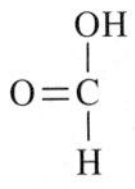

5a. 21 atoms **5b.** 1.29×10^{22} atoms **5c.** 2.195×10^{25} F atoms **7a.** 149.208 u/$C_5H_{11}NO_2S$ molecule **7b.** There are 11 moles of H atoms in each mole of $C_5H_{11}NO_2S$ molecules. **7c.** 60.055 g C **7d.** 2.73×10^{25} C atoms **9.** The greatest number of N atoms is present in 50.0 g N_2O, answer (a). **11a.** 1.25 mol N_2O_4 **11b.** 0.587 mol N atoms **11c.** 0.206 mol N atoms **13.** 3×10^{22} Fe atoms **15a.** False **15b.** True **15c.** False **15d.** False **17a.** 41 atoms **17b.** 0.800 H atoms/O atom **17c.** 1.026 **17d.** Uranium **17e.** 15.1 g of $Cu(UO_2)_2(PO_4)_2 \cdot 8H_2O$ **19.** 15.369 %H **21.** 15.88 %H **23.** 74.046 %C, 7.4570 %H, 8.635 % N, 9.8631 %O. **25a.** 64.06 %Pb **25b.** 45.495 % Fe **25c.** 2.7202 %Mg **27.** Oxide with the largest %Cr will have the largest number of moles of Cr per mole of oxygen. Arranged in order of increasing %Cr: $CrO_3 < CrO_2 < Cr_2O_3 < CrO$. **29.** SO_3 (40.05% S) and S_2O (80.0% S). **31.** $C_4H_{10}O_3$ **33a.** $C_{19}H_{16}O_4$ **33b.** $C_9H_{10}N_4O_2S_2$ **35.** $C_{14}H_{10}$ **37.** $C_{16}H_{10}N_2O_2$ **39.** titanium **41.** 894 u **43a.** 90.49 %C, 9.495 % H. **43b.** C_4H_5 **43c.** C_8H_{10} **45.** CH_4N **47.** $C_{10}H_8$. The compound with the largest number of moles of C per mole of the compound will produce the largest amount of CO_2 and, thus, also the largest mass of CO_2. **49.** 3.710 g CO_2, 1.898 g H_2O. **51a.** C = –4 in CH_4 **51b.** S = +4 in SF_4 **51c.** O = –1 in Na_2O_2 **51d.** C = 0 in $C_2H_3O_2^-$ **51e.** Fe = +6 in FeO_4^{2-} **53.** Cr_2O_3, CrO_2, CrO_3 **55a.** O = +2 in OF_2 **55b.** O = +1 in O_2F_2 **55c.** O = $\frac{-1}{2}$ in CsO_2 **55d.** O = –1 in BaO_2 **57a.** strontium oxide **57b.** zinc sulfide **57c.** potassium chromate **57d.** cesium sulfate **57e.** chromium(III) oxide **57f.** iron(III) sulfate

57g. magnesium hydrogen carbonate or magnesium bicarbonate **57h.** ammonium hydrogen phosphate **57i.** calcium hydrogen sulfite **57j.** copper(II) hydroxide **57k.** nitric acid **57l.** potassium perchlorate **57m.** bromic acid **57n.** phosphorous acid **59a.** carbon disulfide **59b.** silicon tetrafluoride **59c.** chlorine pentafluoride **59d.** dinitrogen pentoxide **59e.** sulfur hexafluoride **59f.** diiodine hexachloride **61a.** $Al_2(SO_4)_3$ **61b.** $(NH_4)_2Cr_2O_7$ **61c.** SiF_4 **61d.** Fe_2O_3 **61e.** C_3S_2 **61f.** $Co(NO_3)_2$ **61g.** $Sr(NO_2)_2$ **61h.** HBr(aq) **61i.** HIO_3 **61j.** PCl_2F_3 **63a.** $TiCl_4$ **63b.** $Fe_2(SO_4)_3$ **63c.** Cl_2O_7 **63d.** $S_2O_8{}^{2-}$ **65a.** chlorous acid **65b.** sulfurous acid **65c.** hydroselenic acid **65d.** nitrous acid **67a.** oxygen difluoride, not ionic. **67b.** xenon difluoride, not ionic. **67c.** copper(II) sulfite, ionic. **67d.** ammonium hydrogen phosphate, ionic. **69.** $MgCl_2 \cdot 6\,H_2O$ **71.** 22.3 g $CuSO_4$ **73.** $CuSiF_6 \cdot 6\,H_2O$ **75.** Answer is (b), butan-2-ol. **77.** Molecules (a), (b), (c), and (d) are structural isomers. **79a.** $CH_3(CH_2)_5CH_3$ **79b.** $CH_3CH_2CO_2H$ **79c.** $CH_3CH_2CH_2CH(CH_3)CH_2OH$ **79e.** CH_3CH_2F **81a.** methanol; CH_3OH; Molecular mass = 32.04 u. **81b.** 2-chlorohexane; $CH_3(CH_2)_3CHClCH_3$; Molecular mass = 120.6 u. **81c.** pentanoic acid; $CH_3(CH_2)_3CO_2H$; Molecular mass = 102.1 u. **81d.** 2-methylpropan-1-ol; $CH_3CH(CH_3)CH_2OH$; Molecular mass = 74.12 u.

Integrative and Advanced Exercises

83. 1.24×10^{23} Li–6 atoms **86.** An additional 2.500 kg mass is required. **89.** It is not possible to have less than 1 molecule of S_8. 426 yoctograms. **91.** CH_4 **93.** % H_2SO_4 = 85.0 %, % H_2O = 15.0 %. **95.** 0.05 ppb **96.** 4.0 mg Kr **99.** 26.9 g/mol **103.** $ZnSO_4 \cdot 7\,H_2O$ **106.** $C_{12}H_8Cl_6O$ **110.** I_2Cl_6

Feature Problems **111.** The empirical formula is CuI **112a.** % P = 4.37%, % K = 4.15%. **112b.** (i) 60.6% P_2O_5; (ii) 53.7% P_2O_5. **112c.** A "17.7-44.6-10.5" fertilizer. **112d.** It is impossible to make a "5-10-5" fertilizer if the only fertilizing components are $(NH_4)_2HPO_4$ and KCl. **114a.** $4.7 \times 10^3\ m^2$ **114b.** 2.5 nm **114c.** 5.8×10^{23} molecules per mole of oleic acid

Self-Assessment Exercises **119.** The answer is (c). **120.** The answer is (b). **121.** The answer is (d). **122.** The answer is (a). **123.** 1.29×10^9 Fe atoms/red blood cell **124.** The answer is (c). **125.** The answer is (c). **126.** The answer is (c). **127.** The answer is (a). **128.** Strontium bicarbonate **129.** The answer is (b). **130.** Molar mass of Li_3P is 51.79 g mol^{-1}. **131.** The answer is (d). **132.** The answer is (d). **133.** $Na_2SO_3 \cdot 7\,H_2O$ **134a.** % Cu = 57.46% **134b.** 719.5 g **135.** $C_8H_9NO_2$ **136a.** 75.71% C, 8.79% H, 15.5% O **136b.** $C_{13}H_{18}O_2$

CHAPTER 4

Practice Examples

1a. $4\,HgS(s) + 4\,CaO(s) \longrightarrow 3\,CaS(s) + CaSO_4(s) + 4\,Hg(l)$

1b. $2\,C_7H_6O_2S(l) + 17\,O_2(g) \longrightarrow 14\,CO_2(g) + 6\,H_2O(l) + 2\,SO_2(g)$

2a. 2.64 mol O_2 **2b.** 8.63 mol Ag **3a.** 5.29 g Mg_3N_2 **3b.** 126 g H_2 **4a.** 0.710 g NH_3 **4b.** 3.50 g $O_2(g)$ **5a.** 3.34 cm^3 alloy **5b.** 0.79 g Cu **6a.** 0.4 mg $H_2(g)$ **6b.** 0.15 g CO_2 **7a.** 0.307 M **7b.** 0.524 M **8a.** 115 g $NaNO_3$ **8b.** 50.9 g $Na_2SO_4 \cdot 10\,H_2O$ **9a.** 0.0675 M **9b.** 0.122 M **10a.** 18.1 mL **10b.** 4.96 g Ag_2CrO_4 **11a.** 936 g PCl_3 **11b.** 1.86 kg $POCl_3$ **12a.** 3.8 g P_4 **12b.** 57 g O_2 remaining **13a.** (a) 30.0 g CH_2O; (b) 25.7 g CH_2O (g); (c) 85.6 % yield. **13b.** 93.7% **14a.** 41.8 g CO_2 **14b.** 69.0 g impure $C_6H_{11}OH$ **15a.** 2.47×10^3 g HNO_3 **15b.** Approximately 56 g of SiO_2 and 30 g of KNO_3 are needed. **16a.** 83 wt% **16b.** 19.47 **17a.** 0.125 mol **17b.** 25.0%

Integrative Example **A.** 78.3% yield **B.** 83.3% Zn

Exercises **1a.** $2\,SO_3 \longrightarrow 2\,SO_2 + O_2$

1b. $Cl_2O_7 + H_2O \longrightarrow 2\,HClO_4$

1c. $3\,NO_2 + H_2O \longrightarrow 2\,HNO_3 + NO$

1d. $PCl_3 + 3\,H_2O \longrightarrow H_3PO_3 + 3\,HCl$

3a. $3\,PbO + 2\,NH_3 \longrightarrow 3\,Pb + N_2 + 3\,H_2O$

3b. $2\,FeSO_4 \longrightarrow Fe_2O_3 + 2\,SO_2 + {}^1\!/_2\,O_2$ or $4\,FeSO_4 \longrightarrow 2\,Fe_2O_3 + 4\,SO_2 + O_2$

3c. $6\,S_2Cl_2 + 16\,NH_3 \longrightarrow N_4S_4 + 12\,NH_4Cl + S_8$

3d. $2\,C_3H_7CHOHCH(C_2H_5)CH_2OH + 23\,O_2 \longrightarrow 16\,CO_2 + 18\,H_2O$

5a. $2\,Mg(s) + O_2(g) \longrightarrow 2\,MgO(s)$

5b. $2\,NO(g) + O_2(g) \rightarrow 2\,NO_2(g)$

5c. $2\,C_2H_6(g) + 7\,O_2(g) \longrightarrow 4\,CO_2(g) + 6\,H_2O(l)$

5d. $Ag_2SO_4(aq) + BaI_2(aq) \longrightarrow BaSO_4(s) + 2\,AgI(s)$

7a. $2\,C_4H_{10}(g) + 13\,O_2(g) \longrightarrow 8\,CO_2(g) + 10\,H_2O(l)$ **7b.** $2\,CH_3CH(OH)CH_3(l) + 9\,O_2(g) \longrightarrow 6\,CO_2(g) + 8\,H_2O(l)$

7c. $CH_3CH(OH)COOH(s) + 3\,O_2(g) \longrightarrow 3\,CO_2(g) + 3\,H_2O(l)$

9a. $NH_4NO_3(s) \xrightarrow{\Delta} N_2O(g) + 2\,H_2O(g)$

9b. $Na_2CO_3(aq) + 2\,HCl(aq) \longrightarrow 2\,NaCl(aq) + H_2O(l) + CO_2(g)$

9c. $2\,CH_4(g) + 2\,NH_3(g) + 3\,O_2(g) \longrightarrow 2\,HCN(g) + 6\,H_2O(g)$

11. $2\,N_2H_4(g) + N_2O_4(g) \longrightarrow 4\,H_2O(g) + 3\,N_2(g)$

13. $2\,Cr(s) + 3\,O_2(g) \longrightarrow 2\,CrO_3(s)$

15. 785 g **17a.** 0.402 mol O_2 **17b.** 128 g $KClO_3$ **17c.** 43.9 g KCl **19.** 96.0 g Ag_2CO_3 **21a.** 12.2 g H_2 **21b.** 48.1 g H_2O **21c.** 284 g CaH_2 **23.** 79.7% Fe_2O_3 **25.** % by mass $B_{10}H_{14}$ = 25.8 **27.** 1.03 g H_2 **29.** Al produces the largest amount of H_2 per gram of metal. **31a.** 0.408 M **31b.** 0.154 M **31c.** 1.53 M **33a.** 1.753 M **33b.** 0.320 M $CO(NH_2)_2$ **33c.** 0.206 M **35a.** 4.73 g

35b. 44.1mL CH_3OH **37a.** $4.7\,\dfrac{\text{mmol } C_6H_{12}O_6}{\text{L}}$

37b. Molarity $= 4.7 \times 10^{-3}\,\dfrac{\text{mol } C_6H_{12}O_6}{\text{L}}$

39. Solution (d) is a 0.500 M KCl solution. **41.** The 46% by mass sucrose solution is the more concentrated. **43.** 0.0820 M **45.** 0.236 M **47.** The ratio of the volume of the volumetric flask to that of the pipet would be 20:1. We could use a 100.0 mL flask and a 5.00 mL pipet, a 1000.0 mL flask and a 50.00 mL pipet, or a 500.0 mL flask and a 25.00 mL pipet. There are many combinations that could be used. **49a.** 0.177 g Na_2S **49b.** 0.562 g Ag_2S

51. 59.4 mL K_2CrO_4 **53.** $8.46 \times 10^{-3}\,\dfrac{\text{mol } HNO_3}{\text{L}}$

55a. 0.0693 mol $AlCl_3$ **55b.** 2.91 M $AlCl_3$ **57.** 12.8 g Ag_2CrO_4 **59.** 0.624g Na **61.** 0.2649 M **63.** 3.00 moles of NO (g) **65.** HNO_3 (aq) is the limiting reactant, it will be completely consumed, leaving some Cu unreacted. **67.** 143 g Na_2CS_3 **69a.** 10.5 g NH_3 **69b.** 10.1 g excess $Ca(OH)_2$ **71.** 211.51 g $CrSO_4$ **73a.** 1.80 mol CCl_4 **73b.** 1.55 mol CCl_2F_2 **73c.** 86.1% yield **75.** 87.6% **81.** A main criterion for choosing a synthesis reaction is how economically it can be run. In the analysis of a compound, on the other hand, it is essential that all of the material present be detected. Therefore, a 100% yield is required; none of the material present in the sample can be lost during the analysis. **84.** 1.22×10^3 g CO_2 **86.** 0.0386 g C_2H_6 **87.** 1.34 kg $AgNO_3$ per kg of I_2 produced.

89a. $SiO_2(s) + 2\,C(s) \xrightarrow{\Delta} Si(s) + 2\,CO(g)$
$Si(s) + 2\,Cl_2(g) \longrightarrow SiCl_4(l)$
$SiCl_4(l) + 2\,H_2(g) \longrightarrow Si(s, \text{ultrapure}) + 4\,HCl(g)$

89b. 885 g C, 5.05×10^3 g Cl_2, 144 g H_2. **91.** 73.33 %

Integrative and Advanced Exercises

93a. $CaCO_3(s) \xrightarrow{\Delta} CaO(s) + CO_2(g)$

93b. $2\,ZnS(s) + 3\,O_2(g) \xrightarrow{\Delta} 2\,ZnO(s) + 2\,SO_2(g)$

93c. $C_3H_8(g) + 3\,H_2O(g) \longrightarrow 3\,CO(g) + 7\,H_2(g)$

93d. $4\,SO_2(g) + 2\,Na_2S(aq) + Na_2CO_3(aq) \longrightarrow CO_2(g) + 3\,Na_2S_2O_3(aq)$

95. 1.96×10^4 g LiOH **96.** 68.0% $CaCO_3$ **98.** $3\,FeS + 5\,O_2 \longrightarrow Fe_3O_4 + 3\,SO_2$ **99.** 8.88 M CH_3CH_2OH **101.** 143 mL **105.** 0.2 cm^2 **106.** 0.365 g Zn **113a.** 0.01628 **113b.** 0.270 mol **116.** 9.2% (C_2H_5) O, 90.8% CH_3CH_2OH. **117.** 16.95 % **118a.** $Cu_5(CrO_4)_2(OH)_6$

118b. $5\,CuSO_4(aq) + 2\,K_2CrO_4(aq) + 6\,H_2O(l)$
$\downarrow$
$Cu_5(CrO_4)_2(OH)_6(s) + 2\,K_2SO_4(aq) + 3\,H_2SO_4(aq)$

119. $C_3H_4O_4(l) + 2\,O_2(g) \longrightarrow 3\,CO_2(g) + 2\,H_2O(l)$ **120.** 5.2 g Fe, mass of excess Fe_2O_3 = 2.1 g. **121.** 0.850 g $AgNO_3$ **122.** The balanced equation for the reaction is: $S_8(s) + 4\,Cl_2(g) \longrightarrow 4\,S_2I_2(l)$. Both "a" and "b" are consistent with the stoichiometry of this equation. **123.** 1.25 g $C_3N_3(OH)_3$ **125.** We must use all of the most concentrated solution and dilute this solution down using the next most concentrated solution. A total of 374 mL may be prepared this way. **129a.** $2\,C_3H_6(g) + 2\,NH_3(g) + 3\,O_2(g) \longrightarrow 2\,C_3H_3N(l) + 6\,H_2O(l)$ **129b.** 555 kg NH_3 **130.** % atom economy for reaction 1 = 17%, for reaction 2 = 63%

Self-Assessment Exercises **138.** The answer is (d). **139.** The answer is (d). **140.** The answer is (a). **141.** The answer is (a). **142.** The answer is (c). **143.** 66.7 mL **144.** The answer is (d). **145.** The answer is (b). **146.** The answer is (d). **147.** Fe_2O_3 is the limiting reagent and 12.6 g of Fe are produced by this reaction **148.** 43.9 g **149a.** $2\,C_8H_{18} + 25\,O_2 \longrightarrow 16\,CO_2 + 18\,H_2O$ **149b.** $2\,C_8H_{18} + 25\,O_2 \longrightarrow 12\,CO_2 + 4\,CO + 18\,H_2O$ **150.** Dolomite is the compound. **151.** The answer is (b). **152.** The answer is (c). **153.** All of the reactions have an AE of 10%.

CHAPTER 5

Practice Examples **1a.** 0.540 M Cl^- **1b.** (a) 7.9×10^{-5} M F^-; (b) 3.1 kg CaF_2 **2a.** (a) $Al^{3+}(aq) + 3\,OH^-(aq) \longrightarrow Al(OH)_3(s)$ (b) No reaction occurs. (c) $Pb^{2+}(aq) + 2\,I^-(aq) \longrightarrow PbI_2(s)$ **2b.** (a) $Al^{3+}(aq) + PO_4{}^{3-}(aq) \longrightarrow AlPO_4(s)$ (b) $Ba^{2+}(aq) + SO_4{}^{2-}(aq) \longrightarrow BaSO_4(s)$ (c) $Pb^{2+}(aq) + CO_3{}^{2-}(aq) \longrightarrow PbCO_3(s)$ **3a.** The acid and base react to form a salt solution of ammonium propionate. $NH_3(aq) + CH_3CH_2COOH(aq) \longrightarrow NH_4{}^+(aq) + CH_3CH_2COO^-(aq)$

3b. $CaCO_3(s) + 2\,CH_3CH_2COOH(aq) \longrightarrow CO_2(g) + H_2O(l) + Ca^{2+}(aq) + 2\,CH_3CH_2COO^-(aq)$

4a. (a) It is not an oxidation–reduction reaction; (b) This is an oxidation–reduction reaction.

4b. Vanadium is oxidized, manganese is reduced.

5a. *Ox.* $\{Al(s) \rightarrow Al^{3+}(aq) + 3\,e^-\} \times 2$

Red. $\{2\,H^+(aq) + 2\,e^- \rightarrow H_2(g)\} \times 3$

Net $2\,Al(s) + 6\,H^+(aq) \rightarrow 2\,Al^{3+}(aq) + 3\,H_2(g)$

5b. *Ox.* $2\,Br^-(aq) \rightarrow Br_2(1) + 2\,e^-$

Red. $Cl_2(g) + 2\,e^- \rightarrow 2\,Cl^-(aq)$

Net $2\,Br^-(aq) + Cl_2(g) \rightarrow Br_2(1) + 2\,Cl^-(aq)$

6a. $MnO_4{}^-(aq) + 8\,H^+(aq) + 5\,Fe^{2+}(aq) \longrightarrow Mn^{2+}(aq) + 4\,H_2O(l) + 5\,Fe^{3+}(aq)$

6b. $3\,UO^{2+}(aq) + Cr_2O_7{}^{2-}(aq) + 8\,H^+(aq) \longrightarrow 3\,UO_2{}^{2+}(aq) + 2\,Cr^{3+}(aq) + 4\,H_2O(l)$

7a. $S(s) + 2\,OH^-(aq) + 2\,OCl^-(aq) \longrightarrow SO_3{}^{2-}(aq) + H_2O(l) + 2\,Cl^-(aq)$ **7b.** $2\,MnO_4{}^-(aq) + 3\,SO_3{}^{2-}(aq) + H_2O(l) \longrightarrow 2\,MnO_2(s) +$

$3\ SO_4^{2-}(aq) + 2\ OH^-(aq)$ **8a.** Since the oxidation state of H is 0 in $H_2(g)$ and is +1 in both $NH_3(g)$ and $H_2O(g)$, hydrogen is oxidized. A substance that is oxidized is called a reducing agent. The oxidation state of the element N decreases during this reaction, meaning that $NO_2(g)$ is reduced. The substance that is reduced is called the oxidizing agent. **8b.** Au has been oxidized and, thus, Au(s) (oxidization state = 0), is the reducing agent. O has been reduced and thus, $O_2(g)$ (oxidation state = 0) is the oxidizing agent. **9a.** 0.1019 M **9b.** 0.130 M **10a.** 65.4% Fe **10b.** 0.03129 M **Integrative Example A.** 49.89% **B.** 1.32% **Exercises 1a.** Weak electrolyte **1b.** Strong electrolyte **1c.** Strong electrolyte **1d.** Nonelectrolyte **1e.** Strong electrolyte **3.** HCl is practically 100% dissociated into ions. The apparatus should light up brightly. A solution of both HCl and $HC_2H_3O_2$ will yield similar results. **5a.** Barium bromide: strong electrolyte **5b.** Propionic acid: weak electrolyte **5c.** Ammonia: weak electrolyte **7a.** 0.238 M K^+ **7b.** 0.334 M NO_3^- **7c.** 0.166 M Al^{3+} **7d.** 0.627 M Na^+ **9.** 3.04×10^{-3} M OH^- **11a.** 3.54×10^{-4} M Ca^{2+} **11b.** 8.39×10^{-3} M K^+ **11b.** 8.39×10^{-3} M K^+ **11c.** 3.44×10^{-3} M Zn^{2+} **13.** The solution containing 8.1 mg K^+ per mL gives the largest K^+ of the three solutions. **15.** 4.3×10^2 mg MgI_2 **17.** 0.732 M **19a.** $Pb^{2+}(aq) + 2\ Br^-(aq) \longrightarrow PbBr_2(s)$ **19b.** No reaction occurs (all are spectator ions). **19c.** $Fe^{3+}(aq) + 3\ OH^-(aq) \longrightarrow Fe(OH)_3(s)$ **21a.** No reaction occurs. **21b.** $Cu^{2+}(aq) + CO_3^{2-}(aq) \rightarrow CuCO_3(s)$ **21c.** $3\ Cu^{2+}(aq) + 2\ PO_4^{3-}(aq) \rightarrow Cu_3(PO_4)_2(s)$ **23a.** Add K_2SO_4(aq); $BaSO_4$(s) will form and $MgSO_4$ will not precipitate. **23b.** H_2O (l); Na_2CO_3(s) dissolves, but $MgCO_3$(s) will not dissolve (appreciably). **23c.** Add KCl(aq); AgCl(s) will form, while $Cu(NO_3)_2$(s) will dissolve. **25a.** $Sr(NO_3)_2(aq) + K_2SO_4(aq)$: $Sr^{2+}(aq) + SO_4^{2-}(aq) \rightarrow SrSO_4(s)$ **25b.** $Mg(NO_3)_2(aq) + NaOH(aq)$: $Mg^{2+}(aq) + 2\ OH^-(aq) \rightarrow Mg(OH)_2(s)$ **25c.** $BaCl_2(aq) + K_2SO_4(aq)$: $Ba^{2+}(aq) + SO^{2-}(aq) \rightarrow BaSO_4(s)$ **27a.** $OH^-(aq) + CH_3CH_2COOH(aq) \rightarrow H_2O(l) + CH_3CH_2COO^-(aq)$ **27b.** No reaction occurs. This is the physical mixing of two acids. **27c.** $FeS(s) + 2\ H^+(aq) \rightarrow H_2S(g) + Fe^{2+}(aq)$ **27d.** $HCO_3^-(aq) + H^+(aq) \rightarrow$ "$H_2CO_3(aq)$" $\rightarrow H_2O(l) + CO_2(g)$ **27e.** $Mg(s) + 2\ H^+(aq) \rightarrow Mg^{2+}(aq) + H_2(g)$ **29.** As a salt: $NaHSO_4(aq) \rightarrow Na^+(aq) + HSO_4^-(aq)$
As an acid: $HSO_4^-(aq) + OH^-(aq) \rightarrow H_2O(l) + SO_4^{2-}(aq)$

31. Use (b), NH_3(aq). NH_3 affords the OH^- ions necessary to form $Mg(OH)_2$(s). **33a.** The O.S. of H is +1, that of O is –2, that of C is +4, and that of Mg is +2 on each side of this equation. This is not a redox equation. **33b.** The O.S. of Cl is 0 on the left and –1 on the right side of this equation. The O.S. of Br is –1 on the left and 0 on the right side of this equation. This is a redox reaction. **33c.** The O.S. of Ag is 0 on the left and +1 on the right side of this equation. The O.S. of N is +5 on the left and +4 on the right side of this equation. This is a redox reaction. **33d.** On both sides of the equation the O.S. of O is −2, that of Ag is +1, and that of Cr is +6. Thus, this is not a redox equation. **35a.** $2\ SO_3^{2-}(aq) + 6\ H^+(aq) + 4\ e^- \rightarrow S_2O_3^{2-}(aq) + 3\ H_2O(l)$ **35b.** $2\ NO_3^-(aq) + 10\ H^+(aq) + 8\ e^- \rightarrow N_2O(g) + 5\ H_2O(l)$ **35c.** $Al(s) + 4\ OH^-(aq) \rightarrow Al(OH)_4^-(aq) + 3\ e^-$ **37a.** $10\ I^-(aq) + 2\ MnO_4^-(aq) + 16\ H^+(aq) \rightarrow 5\ I_2(s) + 2\ Mn^{2+}(aq) + 8\ H_2O(l)$ **37b.** $3\ N_2H_4(l) + 2\ BrO_3^-(aq) \rightarrow 3\ N_2(g) + 2\ Br^-(aq) + 6\ H_2O(l)$ **37c.** $Fe^{2+}(aq) + VO_4^{3-}(aq) + 6\ H^+(aq) \rightarrow Fe^{3+}(aq) + VO^{2+}(aq) + 3\ H_2O(l)$ **37d.** $3\ UO^{2+}(aq) + 2\ NO_3^-(aq) + 2\ H^+(aq) \rightarrow 3\ UO_2^{2+}(aq) + 2\ NO(g) + H_2O(l)$

39a. $2\ MnO_2(s) + ClO_3^-(aq) + 2\ OH^-(aq) \rightarrow 2MnO_4^-(aq) + Cl^-(aq) + H_2O(l)$ **39b.** $2\ Fe(OH)_3(s) + 3\ OCl^-(aq) + 4\ OH^-(aq) \rightarrow 2FeO_4^{2-}(aq) + 3\ Cl^-(aq) + 5\ H_2O(l)$ **39c.** $6\ ClO_2(aq) + 6\ OH^-(aq) \rightarrow 5ClO_3^-(aq) + Cl^-(aq) + 3\ H_2O$ **39d.** $3\ Ag(s) + CrO_4^{2-} + 4\ H_2O(l) \rightarrow 3\ Ag^+(aq) + Cr(OH)_3(s) + 5\ OH^-$ **41a.** $3\ Cl_2(g) + 6\ OH^-(aq) \rightarrow 5\ Cl^-(aq) + ClO_3^-(aq) + 3\ H_2O(l)$ **41b.** $2\ S_2O_4^{2-}(aq) + H_2O(l) \rightarrow 2\ HSO_3^-(aq) + S_2O_3^{2-}(aq)$ **43a.** $5\ NO_2^-(aq) + 2\ MnO_4^-(aq) + 6\ H^+(aq) \rightarrow 5\ NO_3^-(aq) + 2\ Mn^{2+}(aq) + 3\ H_2O(l)$ **43b.** $3\ Mn^{2+}(aq) + 2\ MnO_4^-(aq) + 4\ OH^-(aq) \rightarrow 5\ MnO_2(s) + 2\ H_2O(l)$ **43c.** $Cr_2O_7^{2-}(aq) + 8\ H^+(aq) + 3\ C_2H_5OH \rightarrow 2\ Cr^{3+}(aq) + 7\ H_2O(l) + 3\ CH_3CHO$ **45a.** $CH_4(g) + 4\ NO(g) \rightarrow 2\ N_2(g) + CO_2(g) + 2\ H_2O(g)$ **45b.** $16\ H_2S(g) + 8\ SO_2(g) \rightarrow 3\ S_8(s) + 16\ H_2O(g)$ **45c.** $10\ NH_3(g) + 3\ Cl_2O(g) \rightarrow 6\ NH_4Cl(s) + 2\ N_2(g) + 3\ H_2O(l)$ **47a.** SO_3^{2-}(aq) is the reducing agent; MnO_4^-(aq) is the oxidizing agent. **47b.** H_2(g) is the reducing agent; NO_2(g) is the oxidizing agent. **47c.** $[Fe(CN)_6]^{4-}$(aq) is the reducing agent; H_2O_2(aq) is the oxidizing agent. **49.** 13.3 mL NaOH (aq) soln **51.** 3.546 mL KOH solution **53.** 0.1230 M NaOH **55.** 0.077 M NaOH **57.** Acidic **59.** 34 mL base **61.** The answer is (d). **63.** 1.968×10^{-2} M **65.** 53.23% Fe **67.** 37.0 g $Na_2C_2O_4$
Integrative and Advanced Exercises 71. $3\ Ca^{2+}(aq) + 2\ HPO_4^{2-}(aq) \longrightarrow Ca_3(PO_4)_2(s) + 2\ H^+(aq)$ **74.** 108 ppm Mg **75.** 0.0874 L **80a.** $4\ FeS_2(s) + 15\ O_2(g) + 2\ H_2O(l) \rightarrow 4\ Fe^{3+}(aq) + 8\ SO_4^{2-}(aq) + 4\ H^+(aq)$ **80b.** 23.4 g $CaCO_3$ **83.** 44.6 g Cl_2 **85.** 5.0×10^2 g ClO_2(g) **88a.** 45.9 g **88b.** 1.00 L **89.** % $Mg(OH)_2$ = 21.6; %$Al(OH)_3$ = 78.4. **91.** 0.4346
93a. $CaO(s) + H_2O(l) \rightarrow Ca^{2+}(aq) + 2\ OH^-(aq)$
$H_2PO_4^-(aq) + 2\ OH^-(aq) \rightarrow PO_4^{3-}(aq) + 2\ H_2O(l)$
$HPO_4^-(aq) + OH^-(aq) \rightarrow PO_4^{3-}(aq) + H_2O(l)$
$5\ Ca^{2+}(aq) + 3\ PO_4^{3-}(aq) + OH^-(aq) \rightarrow Ca_5(PO_4)_3OH(s)$

93b. 0.302 kg
Feature Problems 94. $x = 1.03$ **95.** 91.0% MnO_2 **97.** Before the breath test: 8×10^{-4} M; After the breath test 3×10^{-4} mol/L.
Self-Assessment Exercises 102. The answer is (b). **103.** The answer is (d). **104.** The answer is (c). **105.** The answer is (a). **106.** Unknown is $(NH_4)_2SO_4$ **107.** $2\ I^- + Pb^{2+} \rightarrow PbI_2(s)$ **108.** $CO_3^{2-} + 2\ H^+ \rightarrow H_2O(l) + CO_2(g)$ **109a.** $3\ Zn^{2+} + PO_4^{3-} \rightarrow Zn_3(PO_4)_2(s)$ **109b.** $Cu^{2+} + 2\ OH^- \rightarrow Cu(OH)_2(s)$ **109c.** $Ni^{2+} + CO_3^{2-} \rightarrow NiCO_3(s)$ **110a.** Species oxidized: N in NO **110b.** Species reduced: O_2 **110c.** Oxidizing agent: O_2 **110d.** Reducing agent: NO **110e.** Gains electrons: O_2 **110f.** Loses electrons: NO **111a.** $Cl_2(aq) + 2\ OH^-(aq) \rightarrow ClO^-(aq) + Cl^-(aq) + H_2O(l)$ **111b.** $5\ C_2O_4^{2-}(aq) + 2\ MnO_4^-(aq) + 16\ H^+(aq) \rightarrow 10\ CO_2(g) + 2\ Mn^{2+}(aq) + 8\ H_2O(l)$ **112.** The answer is (b). **113.** The answer is (d). **114.** The answer is (a). **115.** The answer is (c). **116.** The answer is (b). **117.** The answer is (d). **118a.** False **118b.** True **118c.** False **118d.** False **118e.** True **119a.** No **119b.** Yes **119c.** Yes **119d.** No

CHAPTER 6

Practice Examples 1a. 760. mmHg **1b.** 1.13 g/cm^3 **2a.** 756.0 mmHg **2b.** 93 mm glycerol **3a.** 139 Torr **3b.** 1.35 kg. It is not necessary to add a mass with the same cross sectional area. **4a.** 24.4 L NH_3 **4b.** 464 K **5a.** 2.11mol He **5b.** 5.59×10^{14} molecules N_2 **6a.** 2.11mL **6b.** 0.378 g O_2 **7a.** 86.5 g/mol **7b.** 30.0 g/mol. This answer is in good agreement with the molar mass of NO. **8a.** 0.159 g/L. When compared to the density of air under the same conditions (1.16 g/L, based on the "average molar mass of air"=28.8 g/mol) the density of He is only about one seventh as much. Thus, helium is less dense ("lighter") than air. **8b.** 32.0 g/mol **9a.** 35.6 g NaN_3 **9b.** 0.619 g Na(l) **10a.** 1.25 L O_2(g) **10b.** 150. L NH_3 **11a.** 13 bar **11b.** 5.5 atm **12a.** 0.0348 atm H_2O(g), 2.47 atm CO_2(g). **12b.** N_2 = 584 mmHg, O_2 pressure = 157 mmHg, CO_2 = 0.27 mmHg, Ar = 7.0 mmHg. **13a.** 0.00278 mol HCl **13b.** 88.4 **14a.** NH_3(g), 661m/s. **14b.** 76.75 K **15a.** 2.1×10^{-4} mol O_2 **15b.** 28.2 s **16a.** 1.90×10^2 g/mol **16b.** 52.3 s **17a.** Cl_2(g) **17b.** Cl_2(g)
Integrative Example A. C_3H_4. There are three possible Lewis structures. **B.** $C_4H_9NO_3$
Exercises 1a. 0.968 atm **1b.** 0.766 atm **1c.** 1.17 atm **1d.** 2.22 atm **3.** 11.4 m benzene **5.** 710 mmHg **7.** 1.03 kg cm^{-2} **9a.** 52.8 L **9b.** 7.27 L **11.** 225 K **13.** 50.6 atm **15.** 255 K **17.** 0.132 g Ar **19a.** 41.3 mg PH_3 **19b.** 7.32×10^{20} molecules **21.** At the higher elevation of the mountains, the atmospheric pressure is lower than at the beach. However, the bag is virtually leak proof; no gas escapes. Thus, the gas inside the bag expands in the lower pressure until the bag is filled to near bursting. **23.** 4.30 L **25.** 4.89 g **27.** 53.2 L **29.** 702 g Kr **31.** 2.1×10^{-11} Pa **33a.** 24.4 L mol^{-1} **33b.** 31.3 L mol^{-1} **35.** 104 g/mol **37.** SF_4 **39.** 55.8 g/mol. The formula is C_4H_8. **41.** 15.56 L/mol **43a.** 1.18 g/L air **43b.** The density of CO_2 is 1.80 g/L CO_2. Since this density is greater than that of air, the balloon will not rise in air. **45.** P_4 **47.** 378 L O_2 **49.** 3.1×10^7 L SO_2 **51.** 10.9% $KClO_3$ **53.** 73.7 L H_2 **55.** 5.59 L gas **57.** 2.06×10^3 g Ne **59.** The answer is (d). **61a.** 842 mmHg **61b.** $P_{benzene}$ = 89.5 mmHg, P_{Ar} = 752 mmHg. **63.** Situation (b) best represents the resulting mixture, as the volume has increased by 50%. **65.** 4.0 atm **67.** 2.37 L H_2(g) **69.** 751 mmHg **71.** 17.9 **73.** 326 m/s **75.** 7.83 g/mol or 7.83 u. **77.** 1.51×10^3 K **79.** 6.17×10^{-21} J **81.** 0.00473 mol NO_2 **83a.** 1.07 **83b.** 1.05 **83c.** 0.978 **83d.** 1.004 **85.** 9.80 h **87a.** P_{ideal} =15.5 bar, P_{vdw} =14.4 bar, P_{ideal} is off by 1.1 bar or +7.6%. **87b.** P_{ideal} =19.7 bar, P_{vdw} = 18.6 bar, P_{ideal} is off by 1.1 bar or 5.9%. **87c.** P_{ideal} = 28.0 bar, P_{vdw} = 27.1 bar, P_{ideal} is off by 0.9 bar or +3.3%. **89.** 3.95×10^7 pm^3/ He atom
Integrative and Advanced Exercises 94. Flask 2 n =0.391, Flask 1 n = 0.609. **97.** 2.25 atm **98.** 542 L **99.** 0.39 atm **101.** 153 mmHg **102.** 19.9% He **105.** 23 L **108.** 2.57×10^4 L **109.** 31 mmHg **111a.** 3.42% H_2O by volume **111b.** 3.42% by number **111c.** 1.95 % H_2O by mass **112a.** 0.40 g/L **112b.** 2.9 atm **114.** χ_{O_3} = 8.05×10^{-4} **115.** 19°C **118a.** 482 m/s **118b.** F_u =1.92×10^{-3} **119.** 13.6 km
121a.

$$0 = V^3 - n\left(\frac{RT + bP}{P}\right)V^2 + \left(\frac{n^2a}{P}\right)V - \frac{n^3ab}{P} = 0$$

121b. 7.37 L
Feature Problems 126. The atomic mass of X is 16 u which corresponds to the element oxygen. The number of atoms of X (oxygen) in each compound is: Nitryl Fluoride = 2 atoms of O, Nitrosyl Fluoride = 1 atom of O, Thionyl Fluoride = 1 atom of O, Sulfuryl Fluoride = 2 atoms of O. **127a.** The N_2(g) extracted from liquid air has some Ar(g) mixed in. **127b.** Because of the presence of Ar(g) [39.95 g/mol], the N_2(g) [28.01 g/mol] from liquid air will have a greater density than N_2(g) from nitrogen compounds. **127c.** Magnesium will react with molecular nitrogen but not with Ar. Thus, magnesium reacts with all the nitrogen in the mixture, but leaves the relatively inert Ar(g) unreacted. **127d.** The densities differ by 0.50%. **129.** Just over 30 km.

Self-Assessment Exercises **133.** The answer is (d). **134.** The answer is (c). **135a.** 8.90 bar **135b.** 2.95 bar **135c.** 5.95 bar **136.** 546 K **137.** The answer is (d). **138.** The answer is (b). **139a.** He has greater average molecular speed **139b.** He strikes the container walls more frequently **139c.** The gases exert equal pressure **140.** The answer is (d). **141.** The answer is (c). **142a.** False **142b.** True **142c.** False **142d.** True **142e.** False **143.** The answer is (c). **144.** The answer is (a). **145.** The answer is (b). **146.** 1.3 L of CO remain. **148.** The answer is (c). **149a.** Ne has higher *a* and *b* values. **149b.** C_3H_8 has higher *a* and *b* values. **149c.** Cl_2 has higher *a* and *b* values. **150.** The height, h is inversely proportional to D. That is, the larger the diameter of the tube, the shorter the height of the liquid. **151.** C_4H_{10} **152.** For every single mark representing CO_2, we need 2060 marks for N_2, 553 marks for O_2, and 25 for Ar.

CHAPTER 7

Practice Examples **1a.** 32.7 kJ **1b.** 4.89 kJ **2a.** 3.0×10^2 g **2b.** 37.9 °C **3a.** -3.83×10^3 kJ/mol **3b.** 6.26 kJ/°C **4a.** –65. kJ/mol **4b.** 33.4 °C **5a.** 114 J **5b.** +14.6 kJ **6a.** $+1.70 \times 10^2$ J **6b.** 179 J of work is done by the system to the surroundings. **7a.** 60.6 g $C_{12}H_{22}O_{11}$ **7b.** 0.15 kJ heat evolved **8a.** +3.31 kJ **8b.** 1.72 kg H_2O **9a.** –124 kJ/mol **9b.** –2006 kJ/mol **10a.** 6 C (graphite) $+\frac{13}{2} H_2(g) + O_2(g) + \frac{1}{2} N_2(g) \rightarrow C_6H_{13}O_2N(s)$ **10b.** The specified reaction is twice the reverse of the formation reaction, and its enthalpy change is minus two times the enthalpy of formation of NH_3 (g): $-2 \times (-46.11\text{ kJ}) = +92.22$ kJ/mol **11a.** –1367 kJ/mol **11b.** -2.5×10^3 kJ/mole of mixture **12a.** –1273 kJ/mol $C_6H_{12}O_6(s)$ **12b.** –184 kJ/mol $CH_3OCH_3(g)$ **13a.** –112.3 kJ/mol AgI(s) formed **13b.** –505.8 kJ/mol $Ag_2CO_3(s)$ formed.

Integrative Example **A.** –125.5 kJ/mol **B.** The contents of the vessel after completion of the reaction are 74 g of $Ca(OH)_2$ and 68 g of H_2O.

Exercises **1a.** $+2.9 \times 10^2$ kJ **1b.** –177 kJ **3a.** 0.385 J g^{-1} °C^{-1} for Zn **3b.** 0.13 J g^{-1} °C^{-1} for Pt **3c.** 0.905 J g^{-1} °C^{-1} for Al **5.** 545 °C **7.** 24.0 °C **9.** 2.3×10^2 J mol^{-1} °C^{-1} **11.** 1.21 J/K **13.** 2.49×10^5 kJ of heat evolved. **15a.** 65.59 kJ **15b.** 3.65×10^3 kJ **15c.** 1.45×10^3 kJ **17a.** 504 kg CH_4 **17b.** -6.20×10^5 kJ of heat energy **17c.** 2.90×10^3 L H_2O **19.** 1.36×10^3 kJ heat **21a.** -5×10^1 kJ/mol **21b.** The ΔT here is known to just one significant figure (0.9 °C). Doubling the amount of KOH should give a temperature change known to two significant figures (1.6 °C) and using twenty times the mass of KOH should give a temperature change known to three significant figures (16.0 °C). **23.** 4.2×10^2 g NH_4Cl **25.** –56 kJ/mol **27.** 2.7×10^2 kJ heat evolved. **29.** 39.6 g **31.** 1.<u>83</u> g H_2O **33.** 0.111 L **35.** 4.98 kJ/ °C **37a.** -2.34×10^3 kJ/mol $C_5H_{10}O_5$ **37b.** $C_5H_{10}O_5(g) + 5 O_2(g) \rightarrow 5 CO_2(g) + 5 H_2O(l)$ $\Delta H = -2.34 \times 10^3$ kJ **39a.** Endothermic **39b.** +18 kJ/mol **41.** 7.72 kJ/°C **43.** 5.23 °C **45.** 15 g **47a.** –3.4 L atm **47b.** -3.5×10^2 J **47c.** –83 cal **49.** When the Ne(g) sample expands into an evacuated vessel it does not push aside any matter, hence no work is done. **51a.** Work is done on the system by the surroundings (compression). **51b.** No pressure-volume work is done. **51c.** Work is done on the system by the surroundings (compression). **53.** 3.21 L **55a.** 0 **55b.** –562 J **55c.** 0.08 kJ **57a.** Yes **57b.** Yes **57c.** The temperature of the gas stays the same if the process is isothermal. **57d.** ΔU for the gas must equal zero by definition (temperature is not changing). **59.** This situation is impossible. An ideal gas expanding isothermally means that $\Delta U = 0 = q + w$, or $w = -q$, not $w = -2q$. **61.** –545 J **63.** According to the First Law of Thermodynamics, the answer is (c). **65a.** –2008 kJ/mol **65b.** –2012 kJ/mol **67.** +30.74 kJ/mol **69.** $\Delta H° = -290.$ kJ/mol **71.** $\Delta H° = -217.5$ kJ/mol **73.** $\Delta H° = -747.5$ kJ/mol **75.** $\Delta H° = -35.8$ kJ/mol **77.** $\Delta H° = -120.$ kJ/mol **79a.** –55.7 kJ/mol **79b.** –1124 kJ/mol **81.** –206.0 kJ/mol **83.** +202.4 kJ/mol **85.** –1366.7 kJ/mol **87.** –102.9 kJ/mol **89.** –55 kJ/mol **91.** 2.40×10^6 kJ/mol **93.** –424 kJ/mol = $\Delta_f H°$ [HCOOH (s)]

Integrative and Advanced Exercises **97.** 3.5 °C. This large a temperature rise is unlikely, as some of the kinetic energy will be converted into forms other than heat, such as sound and the fracturing of the object along with the surface it strikes. In addition, some heat energy would be transferred to the surface. **98.** -1.95×10^3 kJ/mol $C_6H_8O_7$ **104.** The sewage gas produces more heat per liter at STP than than does coal gas. **108.** 303 g $C_6H_{12}O_6$ **110.** The gas mixture contains 85.1% CH_4 and 14.9% C_2H_6, both by volume. **113.** $\Delta H = -1.45 \times 10^3$ kJ/mol, $3 O_2(g) + C_2H_6O(g) \rightarrow 2 CO_2(g) + 3 H_2O(l)$. **114.** 69 seconds **115.** –12 J

Feature Problems **122.** Methanl produces the least amount of CO_2 per unit of energy generated **125a.** The equivalence point occurs with 45.0 mL of 1.00 M NaOH(aq) [45.0 mmol NaOH] added and 15.0 mL of 1.00 M citric acid [15.0 mmol citric acid].

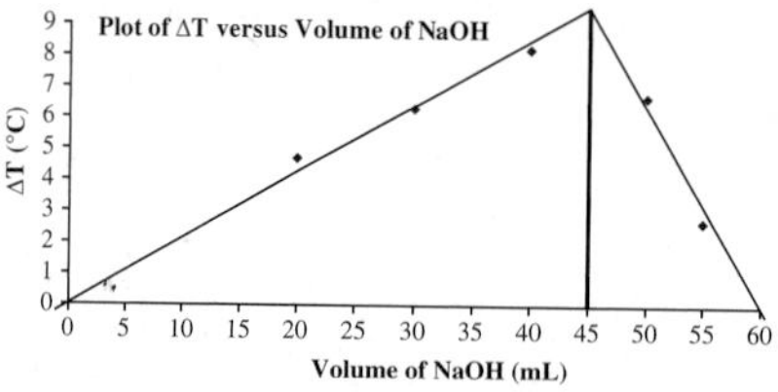

125b. Heat is a product of the reaction, as are chemical species (products). Products are maximized at the exact stoichiometric proportions. Since each reaction mixture has the same volume, and thus about the same mass to heat, the temperature also is a maximum at this point. **125c.** $H_3C_6H_5O_7(s) + 3 OH^-(aq) \rightarrow 3H_2O(l) + C_6H_5O_7^{3-}(aq)$ **128a.** –14<u>7</u> J **128b.**

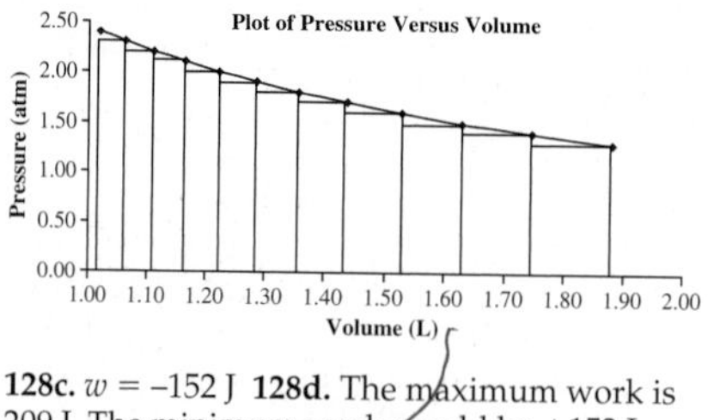

128c. $w = -152$ J **128d.** The maximum work is 209 J. The minimum work would be +152 J. **128e.** $\Delta U = 0$. Because $\Delta U = q + w = 0$, $q = -w$. This means that –209 J corresponds to the maximum work of compression, and –152 J corresponds to the minimum work of compression. **128f.** $q/T = nR \ln V_f/V_i$. Yes, q/T is a state function.

Self-Assessment Exercises **133.** The answer is (b). **134.** The answer is (c). **135.** The answer is (d). **136.** The answer is (a). **137.** The answer is (b). **138a.** $Sn(s) + Cl_2(g) \rightarrow SnCl_2(s)$ **138b.** $7 C(gr) + 3 H_2(g) + O_2(g) \rightarrow C_6H_5COOH(s)$ **138c.** $C(gr) + 1/2 O_2(g) + Cl_2(g) \rightarrow COCl_2(g)$ **139a.** –103.3 kJ mol^{-1} **139b.** –802.0 kJ mol^{-1} **140.** Tf = 29.53 °C **141.** The answer is (b). **142.** The answer is (a). **143.** The answer is (a). **144.** 52.7 °C **145a.** $2 N_2 + O_2 \rightarrow 2 N_2O$ **145b.** $S + O_2 + Cl_2 \rightarrow SO_2Cl_2$ **145c.** $2 CH_3CH_2COOH + 7 O_2 \rightarrow 6 CO_2 + 6 H_2O$ **146.** –218 kJ/mol **147.** Enthalpy of formation for elements (even molecular ones, such as O_2 or Cl_2) is by convention set to 0. While it is possible for the enthalpy of formation of a compound to be near zero, it is unlikely. **148.** From a theoretical standpoint, one can have a situation where the $\Delta U < 0$, but there is enough work done on the system that makes $\Delta H > 0$. **149.** A gas stove works by combustion of a flammable fuel. Once shut off, the heat source instantly disappears. An electric stove works by the principle of heat conduction. Even after the electricity is shut off to the heating coil, it takes time for the coil to cool because of its heat capacity, and therefore it continues to supply heat to the pot. **150.** The answer is (a). **151.** The answer is (b).

CHAPTER 8

Practice Examples **1a.** 4.34×10^{14} Hz **1b.** 3.28 m **2a.** 520 kJ/mol **2b.** 4.612×10^{14} Hz (orange), 6.662×10^{14} Hz (indigo). **3a.** Yes, This is E_9 for n = 9. **3b.** n = 7 **4a.** 486.2 nm **4b.** 121.6 nm (1216 angstroms) **5a.** 80.13 nm **5b.** Be^{3+} **6a.** 1.21×10^{-43} m **6b.** 3.96×10^4 m/s **7a.** 6.4×10^{-43} m **7b.** 1.3 m s^{-1} **8a.** 0.167 out of 1, or 16.7%. **8b.** 100 and 200 pm. **9a.** 0.52 nm **9b.** 150. pm **10a.** Yes **10b.** ℓ = 1 and 2 **11a.** 3*p* **11b.** n = 3; l = 0, 1, 2; m_l = –2, –1, 0, 1, 2 **12a.** (3,2,–2,1) m_s = 1 is incorrect. The values of m_s can only be + ½ or –½. (3,1,–2, ½) m_ℓ = –2 is incorrect. The values of m_ℓ can be +1,0,+1 when ℓ = 1 (3,0,0, ½) All quantum numbers are allowed. (2,3,0, ½) ℓ = 3 is incorrect. The value for ℓ can not be larger than n. (1,0,0, – ½) All quantum numbers are allowed. (2, –1, –1, ½) ℓ = –1 is incorrect. The value for ℓ can not be negative. **12b.** (2,1,1,0) m_s = 0 is incorrect. The values of m_s can only be + ½ or – ½. (1,1,0, ½) ℓ = 1 is incorrect. The value for ℓ is 0 when n=1. (3, –1,1, ½) ℓ = –1 is incorrect. The value for ℓ can not be negative. (0,0,0, – ½) n = 0 is incorrect. The value for n can not be zero. (2,1,2, ½) m_ℓ = 2 is incorrect. The values of m_ℓ can be +1,0,+1 when ℓ = 1. **13a.** (a) and (c) are equivalent. **13b.** excited state of a neutral species **14a.** Ti **14b.** $1s^2 2s^2 2p^6 3s^2 3p^6 3d^{10} 4s^2 4p^6 4d^{10} 5s^2 5p^5$. Each iodine atom has ten 3*d* electrons and one unpaired 5*p* electron.

15a. [Ar] 3*d* [⥮|⥮|↿|↿|↿] 4*s* [⥮]

15b. [Xe] 4*f* [⥮|⥮|⥮|⥮|⥮|⥮|⥮] 5*d* [⥮|⥮|⥮|⥮|⥮] 6*s* [⥮] 6*p* [↿|↿|↿]

16a. (a) Tin is in the 5th period, hence, five electronic shells are filled or partially filled; (b) The 3*p* subshell was filled with Ar; there are six 3*p* electrons in an atom of Sn; (c) The electron configuration of Sn is [Kr] $4d^{10} 5s^2 5p^2$. There are no 5*d* electrons; (d) Both of the 5*p* electrons are unpaired, thus there are two unpaired electrons in a Sn atom. **16b.** (a) The 3*d* subshell was filled at Zn, thus each Y atom has ten 3*d* electrons; (b) Ge is in the 4*p* row; each germanium atom has two 4*p* electrons; (c) We would expect each Au atom to have ten

5*d* electrons and one 6*s* electron. Thus each Au atom should have one unpaired electron.
Integrative Example A. (a) 73.14 pm; (b) 17.7 K. **B.** The possible combinations are $1s \rightarrow np \rightarrow nd$, for example, $1s \rightarrow 3p \rightarrow 5d$. The frequencies of these transitions are calculated as follows:

$1s \rightarrow 3p$: $\nu = 2.92 \times 10^{15}$ Hz
$3p \rightarrow 5d$: $\nu = 2.34 \times 10^{14}$ Hz

The emission spectrum will have lines representing $5d \rightarrow 4p$, $5d \rightarrow 3p$, $5d \rightarrow 2p$, $4p \rightarrow 3s$, $4p \rightarrow 2s$, $4p \rightarrow 1s$, $3p \rightarrow 2s$, $3p \rightarrow 1s$, and $2p \rightarrow 1s$. The difference between the sodium atoms is that the positions of the lines will be shifted to higher frequencies by 11^2.
Exercises 1. 4.68 nm **3a.** True **3b.** False **3c.** False **3d.** True **5.** Choice (c). **7.** 8.3 min **9a.** 4.90×10^{-18} J/photon **9b.** 78.6 kJ/mol **11a.** 3.46×10^{-19} J/photon **11b.** 2.08×10^5 J/mol **13.** 2.1×10^{-5} nm radiation, has the greatest energy per photon. 4.1×10^3 nm has the least amount of energy per photon. **15.** UV radiation **17a.** 6.60×10^{-19} J/photon **17b.** Indium will display the photoelectric effect when exposed to ultraviolet light. It will not display the photoelectric effect when exposed to infrared light. **19a.** $6.9050 \times 10^{14}\ s^{-1}$ **19b.** 397.11 nm **19c.** n = 10 **21.** -1.816×10^{-19} J, $2.740 \times 10^{14}\ s^{-1}$ **23.** $n = 7.9 \approx 8$. **25.** 656.46 nm, 486.26 nm, 434.16 nm, 410.28 nm. **27.** -6.053×10^{-18} J **29a.** $1.384 \times 10^{14}\ s^{-1}$ **29b.** 2166 nm **29c.** Infrared radiation. **31a.** -3.405×10^{-20} J **31b.** $n = 0.2952$. Since n is not an integer, ther is no energy level at -2.500×10^{-19} J. **31c.** $IE_{(\text{from } n = 6)} = 6.053 \times 10^{-20}$ J **33.** $n = 2$ **35a.** Line A is for the transition n = 3 → n = 1, while Line B is for the transition n = 4 → n = 1. **35b.** Hydrogen atom **37a.** Line A is for the transition n = 5 → n = 2, while Line B is for the transition n = 6 → n = 2. **37b.** Be^{3+} cation **39.** Electrons **41.** 9.79×10^{-35} m. The diameter of a nucleus approximates 10^{-15} m, which is far larger than the baseball's wavelength. **43.** The uncertainty relation $\Delta r\ \Delta p \geq h/(4\pi)$ indicates that it is not possible to specify exact values for both the radial position and the momentum along the radial direction. For an electron moving in a circular orbit, the radius is exactly known and Also, because the distance between the electron and nucleus is constant, the electron has no velocity along the radial direction. Therefore, the momentum along the radial direction is exactly zero. The specification of exact values for both r and p violates the Heisenberg uncertainty principle. Alternatively: Because r and p are exactly known, then the corresponding uncertainties are $\Delta r = 0$ and $\Delta p = 0$. However, the product of Δr and Δp cannot be zero; it must be greater than or equal to $h/(4\pi)$. **45.** $\sim 1 \times 10^{-13}$m **47.** $1.4 \times 10^7\ m\ s^{-1}$ **49.** $\lambda = 17$ cm **51.** 0.55 nm **53.** n = 4.0 **55.** Bohr orbits, as originally proposed, are circular, while orbitals can be spherical, or shaped like two tear drops or two squashed spheres, or shaped like four tear drops meeting at their points. Bohr orbits are planar pathways, while orbitals are three-dimensional regions of space in which there is a high probability of finding electrons. The electron in a Bohr orbit has a definite trajectory. Its position and velocity are known at all times. The electron in an orbital, however, does not have a well-known position or velocity. Orbits and orbitals are similar in that the radius of a Bohr orbit is comparable to the average distance of the electron from the nucleus in the corresponding wave mechanical orbital. **57.** Answer (c) is the only one that is correct. **59a.** 5*p* **59b.** 4*d* **59c.** 2*s* **61a.** 1 electron **61b.** 2 electrons **61c.** 10 electrons **61d.** 32 electrons **61e.** 5 electrons **63.** 106 pm **65.** The angular part of the $2p_y$ wave function is $Y(\theta\phi)_{py} = \sqrt{\frac{3}{4\pi}}\ \sin\theta\ \sin\phi$. For all points in the *xz* plane $\phi = 0$, and since the sine of 0° is zero, this means that the entire *xz* plane is a node. Thus, the probability of finding a $2p_y$ electron in the *xz* plane is zero.
67.

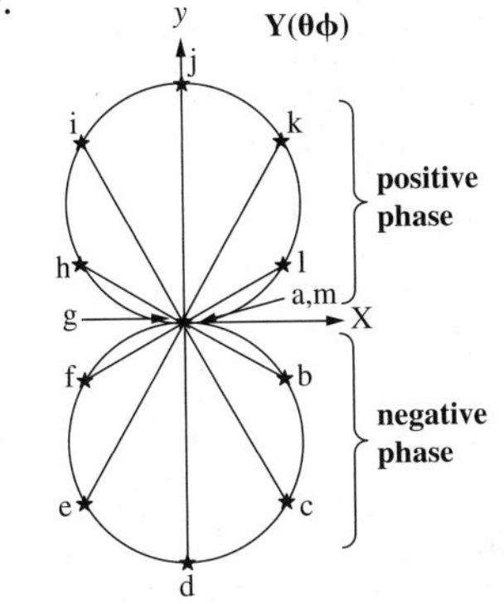

69. A plot of radial probability distribution versus r/a_o for a H_{1s} orbital shows a maximum at 1.0 (that is, $r = a_o$ or r = 53 pm). **71a.** 3*p* **71b.** 3*d* **71c.** 6*f* **73.** $5d_{xyz}$ **75a.** three **75b.** two **75c.** zero **75d.** fourteen **75e.** two **75f.** five **75g.** 32 **77.** Configuration (b) is correct for phosphorus. **79a.** 3 **79b.** ten **79c.** two **79d.** two **79e.** fourteen **81a.** Pb: [Xe] $4f^{14}\,5d^{10}\,6s^2\,6p^2$ **81b.** 114: [Rn] $5f^{14}\,6d^{10}\,7s^2\,7p^2$ **83a.** Excited state **83b.** Excited state **83c.** Ground state **83d.** Excited state **85a.** [Xe]$6s^2 4f^{14} 5d^{10}$ **85b.** [Ar]$4s^2$ **85c.** [Xe]$6s^2 4f^{14} 5d^{10} 6p^4$ **85d.** [Kr]$5s^2 4d^{10} 5p^2$ **85e.** [Xe]$6s^2 4f^{14} 5d^3$ **85f.** [Kr]$5s^2 4d^{10} 5p^5$ **87a.** rutherfordium **87b.** carbon **87c.** vanadium **87d.** tellurium **87e.** Not an element.
Integrative and Advanced Exercises
91a. Because the shortest wavelength of visible light is 390 nm, the photoelectric effect for mercury cannot be obtained with visible light. **91b.** 2.02×10^{-19} J **91c.** $6.66 \times 10^5\ m\ s^{-1}$ **92.** $1.0 \times 10^{20}\ \frac{\text{photons}}{\text{sec}}$ **93.** Green **96.** $m = 3$ and $n = 4$. **101.** 4.8×10^{-3} J **102.** 7×10^2 photons/s **105.** 66.2 m/s
Feature Problems 114. Plot ν on the vertical axis and $1/n^2$ on the horizontal axis. The slope is $b = -3.2881 \times 10^{15}$ Hz and the intercept is 8.2203×10^{14} Hz. **116a.** 121.5 nm and 102.6 nm. **116b.** 97.2 nm, 486.1 nm, 1875 nm, 102.5 nm, 656.3 nm, 121.5 nm. **116c.** The number of lines observed in the two spectra is not the same.
Self-Assessment Exercises 124. Atomic orbitals of multielectron atoms resemble those of the H atom in having both angular and radial nodes. They differ in that subshell energy levels are not degenerate and their radial wave functions no longer conform to the expressions in Table 8.1. **125.** Effective nuclear charge is the amount of positive charge from the nucleus that the valence shell of the electrons actually experiences. This amount is less than the actual nuclear charge, because electrons in other shells shield the full effect. **126.** The p_x, p_y and p_z orbitals are triply degenerate (the are the same energy), and they have the same shape. Their difference lies in their orientation with respect to the arbitrarily assigned *x*, *y*, and *z* axes of the atom. **127.** The difference between the 2*p* and 3*p* orbitals is that the 2*p* orbital has only one node (n = 2-1) which is angular, whereas the 3*p* orbital has two nodes, angular and radial **128.** The answer is (a). **129a.** Velocity of the electromagnetic radiation is fixed at the speed of light in a vacuum. **129b.** Wavelength is inversely proportional to frequency. **129c.** Energy is directly proportional to frequency.

CHAPTER 9

Practice Examples 1a. S **1b.** Ca **2a.** V^{3+} (64 pm) < Ti^{2+} (86 pm) < Ca^{2+} (100 pm) < Sr^{2} + (113 pm) < Br^- (196 pm) **2b.** As **3a.** K < Mg < S < Cl **3b.** Sb **4a.** Cl and Al are paramagnetic; K^+, O^{2-}, and Zn are diamagnetic. **4b.** Cr^{2+}
Integrative Example A. Based on rough approximations of the trends of data, the properties of francium can be approximated. Melting point: 22 °C, density: 2.75 g/cc, atomic radius: 4.25 Å.

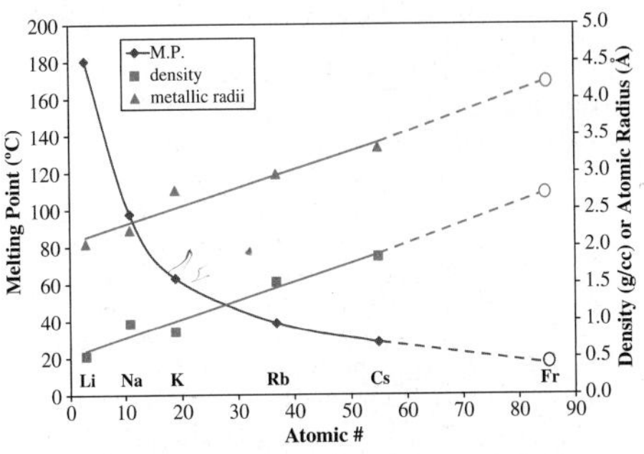

B. Element 168 should be a solid since the trend in boiling point and melting point would put the boiling point temperature above 298 K. The electronic configuration is [Unk] $8s^2\ 5g^{18}\ 6f^{14}\ 7d^{10}\ 8p^6$, where Unk represents element 118.

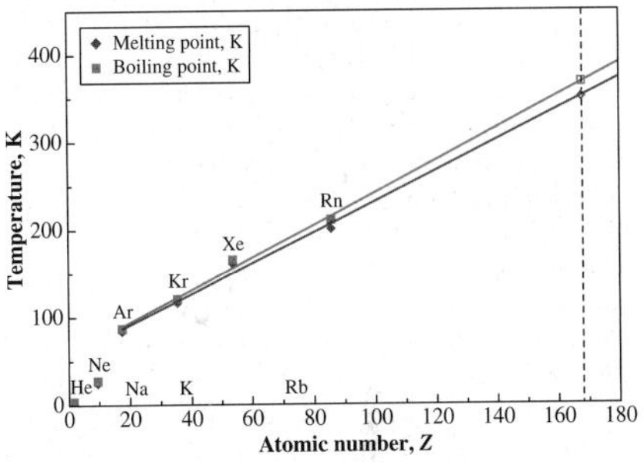

Exercises 1. 14 g/cm³ to 16 g/cm³ **3.** A plot of density versus atomic number clearly shows that density is a periodic property for these two periods of main group elements. It rises, falls a bit, rises again, and falls back to the axis, in both cases. **5.** Mendeleev arranged elements in the periodic table in order of increasing atomic weight. Atomic masses with non-integral values are permissible. Hence, there always is room for an added element between two elements that already are present in the table. On the other hand, Moseley arranged elements in order of increasing atomic number. Only integral (whole number) values of atomic number are permitted. Thus, when elements with all possible integral values in a certain range have been discovered, no new elements are possible in that range. **7a.** 118 **7b.** 119 **7c.** $A_{119} \approx 298$ u and $A_{118} \approx 295$ u. **9a.** Te **9b.** K **9c.** Cs **9d.** N **9e.** P **9f.** Au **11.** Sizes of atoms do not simply increase with atomic number is because electrons often are added successively to the same subshell. These electrons do not fully screen each other from the nuclear charge (they do not effectively get between each other and the nucleus). Consequently, as each electron is added to a subshell and the nuclear charge increases by one unit, all of the electrons in this subshell are drawn more closely into the nucleus, because of the ineffective shielding.

13a. B **13b.** Te **17.** Fe^{2+} and Co^{3+}, Sc^{3+} and Ca^{2+}, F^- and Al^{3+}, Zn^{2+} and Cu^+. **19.** Ions can be isoelectronic without having noble-gas electron configurations. **21.** Cs < Sr < As < S < F **23.** For an ionization potential, negatively charged electron is being separated from a positively charged cation, a process that must always require energy, because unlike charges attract each other. **25.** 9951.5 kJ/mol **27.** –456 kJ, Exothermic **29.** In the case of Na^+, the electron is being removed from a species that is left with a +2 charge, while in the case of Ne, the electron is being removed from a species with a +1 charge. The more highly charged the resulting species, the more difficult it is to produce it by removing an electron. **31a.** Al < Si < S < Cl **31b.** Al < Si < S < Cl **31c.** Cl < S < Si < Al **33.** Fe^{2+} **35a.** Diamagnetic **35b.** Paramagnetic **35c.** Diamagnetic **35d.** Diamagnetic **35e.** Paramagnetic **35f.** Diamagnetic **35g.** Paramagnetic **37.** All atoms with an odd number of electrons must be paramagnetic. There is no way to pair all of the electrons up if there is an odd number of electrons. Many atoms with an even number of electrons are diamagnetic, but some are paramagnetic. **39a.** Elements that one would expect to exhibit the photoelectric effect with visible light should be ones that have a small value of their first ionization energy. Based on Figure 9-9, the alkali metals have the lowest first ionization potentials of these. Cs, Rb, and K are three suitable metals. Metals that would not exhibit the photoelectric effect with visible light are those that have high values of their first ionization energy. Again from Figure 9-9, Zn, Cd, and Hg seem to be three metals that would not exhibit the photoelectric effect with visible light. **39b.** Rn **39c.** +600 kJ/mol **39d.** 5.7 g/cm^3 **41.** O < N < Be < S < K **43.** Cl^+ < Cl < Cl^- **45a.** Statement 1 **45b.** Statement 1 **45c.** Statement 3 **45d.** Statement 4 **45e.** Statement 2 **45f.** Statement 2 **47.** Ga^{4+} and Ge^{5+}

Integrative and Advanced Exercises
49. Both (b) and (e) are compatible with the sketch. **51.** "Gruppe V" or "Gruppe VI." **56a.** A **56b.** B **56c.** A **56d.** B **57.** Since (0.3734)(382) = 143 u, if we subtract 143 from 382 we get 239 u, which is very close to the correct value of 238 u. First we convert to J/g °C: 0.0276 × (4.184 J/cal) = 0.1155 J/g °C, so: 0.1155 = 0.011440 + (23.967/atomic mass); solving for atomic mass yields a value of 230 u, which is within about 3% of the correct value.

60. $E_1 = E_1 \times N_A = \dfrac{8.716 \times 10^{-18}}{1\text{ atom}} \times \dfrac{6.022 \times 10^{23}}{1\text{ mol}} \times \dfrac{1\text{kJ}}{1000\text{J}} = 5249\text{ kJ/mol}$

$E_{i,1}$ for F (1681) ≈ $E_{i,2}$ for Ba (965) < $E_{i,3}$ for Sc (2389) < $E_{i,2}$ for Na (4562) < $E_{i,3}$ for Mg (7733)

Feature Problems **66.** A = 2.30×10^{15} Hz and b = 0.969. The value of A (calculated in this question) is close to the Rydberg frequency ($3.2881 \times 10^{15}\,s^{-1}$), so it is probably the equivalent term in the Moseley equation. The constant b, which is close to unity, could represent the number of electrons left in the K shell after one K-shell electron has been ejected by a cathode ray. Thus, one can think of b as representing the screening afforded by the remaining electron in the K-shell.
67a. [Ne]$3p^1$ = 293 kJ mol^{-1}, [Ne]$4s^1$ = 188 kJ mol^{-1}, [Ne]$3d^1$ = 147 kJ mol^{-1}, [Ne]$4p^1$, = 134 kJ mol^{-1}.
67b. [Ne]$3p^1$: Z_{eff} = 1.42, [Ne]$4s^1$: Z_{eff} = 1.51, [Ne]$3d^1$: Z_{eff} = 1.00, [Ne]$4p^1$: Z_{eff} = 1.28.
67c. [Ne]$3p^1$: $\bar{r}_{3p}$ = 466 pm, [Ne]$4s^1$: $\bar{r}_{4s}$ = 823 pm, [Ne]$3d^1$: $\bar{r}_{3d}$ = 555 pm, [Ne]$4p^1$: $\bar{r}_{4p}$ = 950 pm.
67d. The more deeply an orbital penetrates (i.e., the closer the orbital is to the nucleus), the greater is the effective nuclear charge felt by the electrons in that orbital. It follows then that the 4s orbital will experience the greatest effective nuclear charge and that the Z_{eff} values for the 3p and 4p orbitals should be larger than the Z_{eff} for the 3d orbital.
Self-Assessment Exercises **73.** The answer is (b). **74.** The answer is (a). **75.** The answer is (a). **76.** The answer is (b). **77.** The answer is (a). **78.** The answer is (d). **79.** Cs^+: [Xe], or [Kr] $5s^24d^{10}5p^6$, Cs^{2+}: [Kr] $5s^24d^{10}5p^5$
80. The first ionization energy of Mg is higher than Na because, in the case of Mg, an electron from a filled subshell is being removed, whereas in Na, the removing of one electron leads to the highly stable electron configuration of Ne. The second ionization of Mg is lower than Na, because in the case Mg, the electron configuration of Ar is achieved, whereas in Na, the stable [Ar] configuration is being lost by removing an electron from the filled subshell. **81a.** As **81b.** F^- **81c.** Cl^- **81d.** Carbon **81e.** Carbon **82.** The trends would generally follow higher first ionization energy values for a fuller subshell. The exception is the case of S and P, where P has a slightly higher value. This is because there is a slight energy advantage to having a half-filled subshell with an electron in each of p_x, p_y and p_z orbitals as in the case of P, whereas the S has one extra electron. **83.** The pairs are Ar/Ca, Co/Ni, Te/I and Th/Pa. **84a.** protons = 50 **84b.** neutrons = 69 **84c.** 4d electrons = 10 **84d.** 3s electrons = 2 **84e.** 5p electrons = 2 **84f.** valence shell electrons = 4 **85a.** F **85b.** Sc **85c.** Si **86a.** C **86b.** Rb **86c.** At **87a.** Ba **87b.** S **87c.** Bi **88.** Rb > Ca > Sc > Fe > Te > Br > O > F **89a.** False **89b.** True **89c.** True **89d.** True **89e.** True **90a.** False **90b.** False **90c.** True **90d.** True **91.** These ionization energies are the reverse of electron affinities. **92.** Ionization energy generally increases with Z for a given period. **93.** Mg would have the more positive electron affinity, because Mg has the electron configuration [Ne] 3s2. Therefore, the added electron would need to occupy the higher energy level 3p orbitals, compared to Na (electron configuration = [Ne] 3s1) which would has a partially filled 3s orbital that can accommodate another electron. **94.** Addition of an electron is exothermic because the existing electrons in the atom do not completely shield the incoming electron from the nuclear charge. Therefore, the incoming electron is attracted by the nucleus. The reverse process is easier to think about. Given a valence electron in an atom, energy (such as heat) must be expended to remove that electron. **95.** The answer is (a) atomic radius. Metallic elements tend to occur earlier in a period than nonmetals, and have larger radii. Compared to nonmetal atoms, metal atoms have low ionization energies and low electron affinities. **96.** The answer is (c) Ge. Looking a the Periodic Table, it is seen that Ge falls on the boundary line between metals and non-metals. **97.** The answer is (a) Mg2+. When Mg loses two electrons, it has the same electron configuration as Ne. However, because it has two more protons in its nucleus, the effective charge experienced by the electrons is greater, which decreases the radius. **98.** The answer is (c) the outer-shell electron with the highest orbital quantum number. When an atom is ionized, it loses its electron(s) in the valence shell, which are the outer-most shell, highest energy electrons. The electrons in the valence shell with highest orbital quantum number have the highest energy. Hence, they are the first to be removed when the atom ionizes. **99.** The answer is (c) 4s. The electron configuration of Fe is $1s^22s^22p^63s^23p^63d^64s^2$, whereas the electron configuration for Fe^{2+} is $1s^22s^22p^63s^23p^63d^6$.

CHAPTER 10

Practice Examples

1a. ·Mg·, ·Ge·, K·, :Ne:

1b. ·SN·, [:Br:]$^-$, [·Tl·]$^+$, [:S:]$^{2-}$

2a. (a) [Na]$^+$[:S:]$^{2-}$[Na]$^+$;

(b) [Mg]$^{2+}$[:N:]$^{3-}$[Mg]$^{2+}$[:N:]$^{3-}$[Mg]$^{2+}$

2b. (a) [:I:]$^-$[Ca]$^{2+}$[:I:]$^-$;

(b) [Ba]$^{2+}$[:S:]$^{2-}$; (c) [Li]$^+$[:O:]$^{2-}$[Li]$^+$

3a. 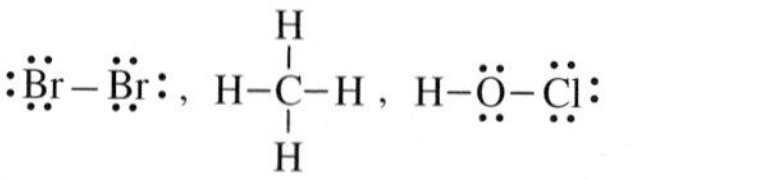

3b.

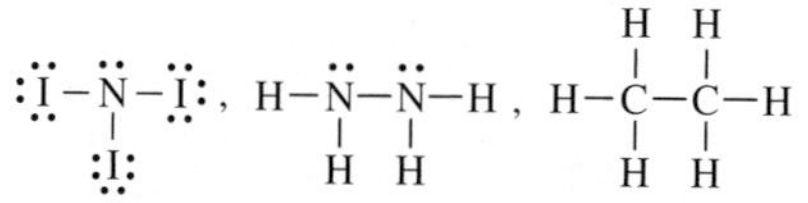

4a. N—H and P—Cl bonds are the most polar of the four bonds cited. **4b.** The P—O bond is the most polar of the four bonds cited. **5a.** The electrostatic potential map that corresponds to IF is the one with the most red in it. **5b.** The electrostatic potential map that corresponds to CH_3OH is the one with the most red in it.

6a. (a) :S=C=S:; (b) H—C≡N:;
(c) :O=C(—Cl:)$_2$;

6b. (a) 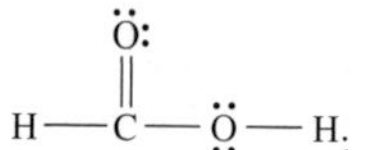

(b) 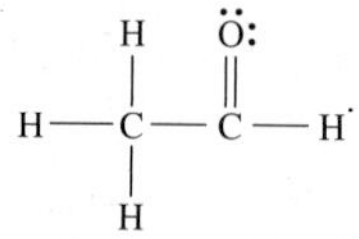

7a. (a) 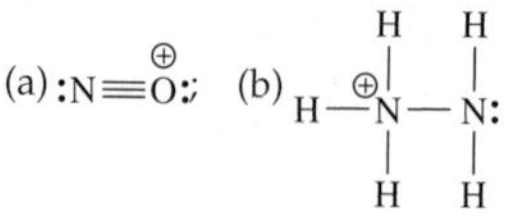

(c) 

7b. (a) 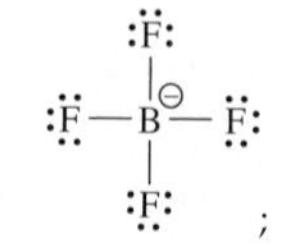

(b) 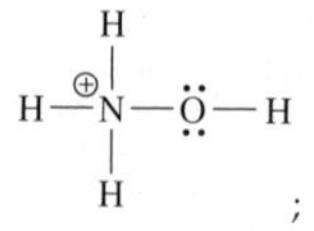

(c) 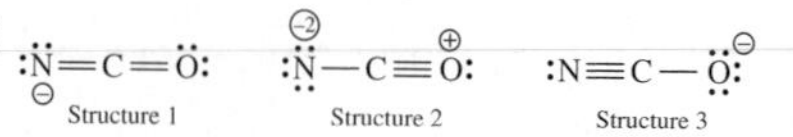

Structure 1 Structure 2 Structure 3

8a. :N̈=Ö—C̈l: is much poorer than the one derived in Example 10-8 because it has a positive formal charge on oxygen, which is the most electronegative atom in the molecule.

8b.

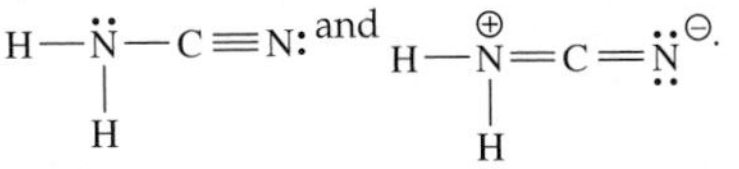

The first structure is the better of the two structures because it has no formal charges.

9a.

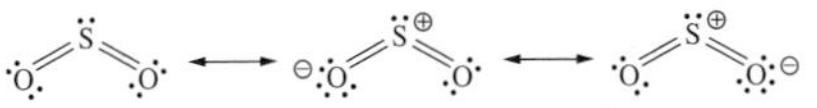

9b.

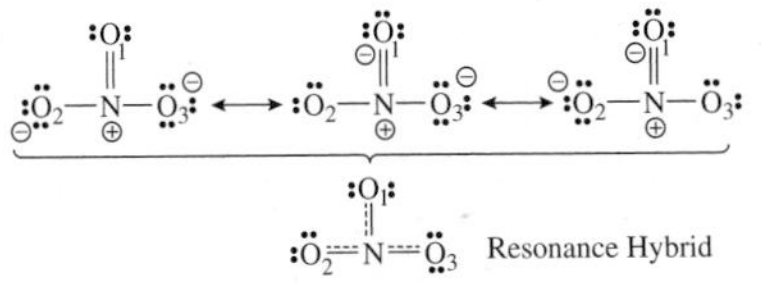

10a. Trigonal pyramidal **10b.** Tetrahedral
11a. Linear **11b.** Linear

12a.

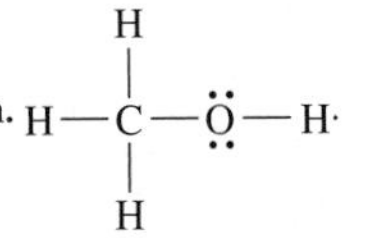

The H—C—H bond angles are ~ 109.5°, as are the H—C—O bond angles. Around the O there are two bonding pairs of electrons and two lone pairs, resulting in a tetrahedral electron-group geometry and a bent molecular shape around the O atom, with a C—O—H bond angle of slightly less than 109.5°. **12b.** The H—N—H bond angle and the H—N—C bond angles are almost the tetrahedral angle of 109.5°, made a bit smaller by the lone pair. The H—C—N angles, the H—C— H angle and the H—C—C angles all are very close to 109.5°. The C—O—H bond angle is made somewhat smaller than 109.5° by the presence of two lone pairs on O. Three electron groups surround the right-hand C, making its electron-group and molecular geometries trigonal planar. The O—C—O bond angle and the O—C—C bond angles all are very close to 120°. **13a.** In H_2O_2, the molecular geometry around each O atom is bent; the bond moments do not cancel. H_2O_2 is polar.
13b. PCl_5 as the only nonpolar species; it is a highly symmetrical molecule in which individual bond dipoles cancel out. **14a.** From Table 10.2, the length of a C—H bond is 110 pm. The length of a C—Br bond is not given in the table. A reasonable value is the average of the C—C and Br—Br bond lengths = 195pm.
14b. :S̈—C≡N:, linear. **15a.** –486 kJ/mol
15b. -4×10^1 kJ/mol NH_3 **16a.** Exothermic
16b. Endothermic
Integrative Example A. P—Cl = –165.5 kJ/mol. Since the geometries of the two molecules differ, the orbital overlap between P and the surrounding Cl atoms will be different and therefore the P—Cl bonds in these two compounds will also be slightly different.

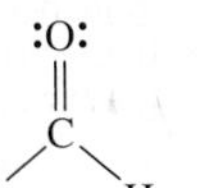

B. (a) Formamide: H_2N–C(=O)–H, bond energy = 2233 kJ/mol;

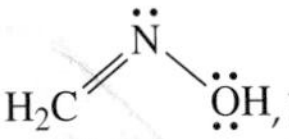

Formaldoxime: H_2C=N–ÖH, bond energy = 2129 kJ/mol. Since BE of formamide is greater than that of formaldoxime, it is more stable, and its conversion endothermic.
(b)

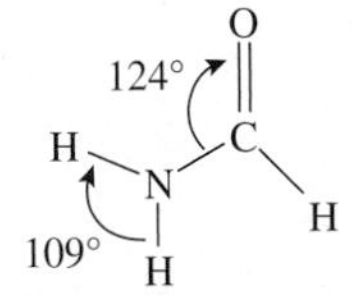

Exercises 1a. :K̈r: **1b.** ·Ġe· **1c.** ·Ṅ: **1d.** ·Ġa
1e. ·Ȧs: **1f.** Rb. **3a.** :F̈—C̈l: **3b.** :Ï—Ï:

3c.

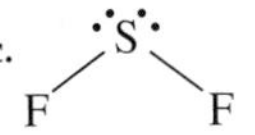

3d.

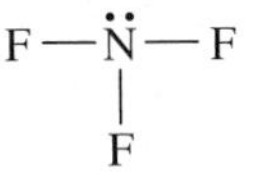

3e.

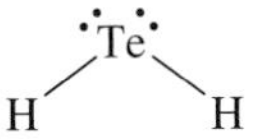

5a. $Cs^{\oplus}$:B̈r:$^{\ominus}$

5b.

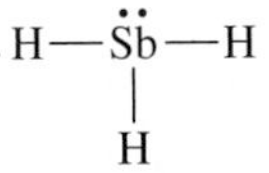

5c.

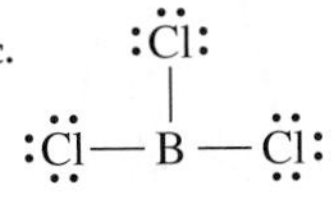

5d. $Cs^{\oplus}$:C̈l:$^{\ominus}$ **5e.** $Li^{\oplus}$:Ö:$^{2-}$ $Li^{\oplus}$

5f. :Ï—C̈l: **7.** NO_2, BF_3, SF_6. **9a.** Has two bonds (4 electrons) to the second hydrogen, and only six electrons around the nitrogen.
9b. It is improperly written as a covalent Lewis structure. CaO is actually an ionic compound.
11. The answer is (c). The flaws with the other answers are as follows: (a) C does not have an octet of electrons; (b) Neither C has an octet of electrons and the total number of valence electrons is incorrect; (c) The total number of valence electrons is incorrect.

13a. $[:\ddot{Cl}:]^-$ $[Ca]^{2+}$ $[:\ddot{Cl}:]^-$

13b. $[Ba]^{2+}$ $[:\ddot{S}:]^{2-}$ **13c.** $[Li]^+$ $[:\ddot{O}:]^{2-}$ $[Li]^+$

13d. $[Na]^+$ $[:\ddot{F}:]^-$ **15a.** $[Li]^+$ $[:\ddot{S}:]^{2-}$ $[Li]^+$

15b. $[Na]^+$ $[:\ddot{F}:]^-$ **15c.** $[:\ddot{I}:]^-$ $[Ca]^{2+}$ $[:\ddot{I}:]^-$

15d. $[:\ddot{Cl}:]^-$
$[:\ddot{Cl}:]^-$ $[Sc]^{3+}$ $[:\ddot{Cl}:]^-$

17a.	H—	—C≡	≡C
	0	0	–1
17b.	=O	—O(×2)	C
	–0	–1	0
17c.	—H(×7)	side C(×2)	central C
	0	0	+1

19. For instance, there are cases where atoms of the same type with the same oxidation state have different formal charges, such as oxygen in ozone, O_3. Another is that formal charges are used to decide between alternative Lewis structures, while oxidation state is used in balancing equations and naming compounds. Also, the oxidation state in a compound is invariant, while the formal charge can change. The most significant difference, though, is that whereas the oxidation state of an element in its compounds is usually not zero, its formal charge usually is. **21a.** +1 **21b.** –1 **21c.** 0
21d. –2 **21e.** 0 **23.** Based on formal charge rules alone, we must conclude that structures (A) and (B) are equally plausible.

25a.

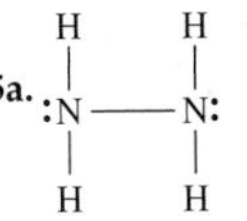

25b. H—Ö—C̈l=Ö

25c.

25d. H—Ö—Ö=H

25e.

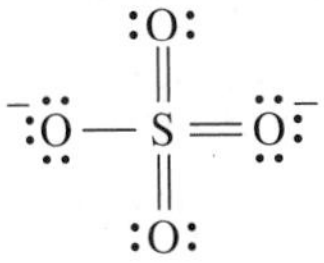

27a.

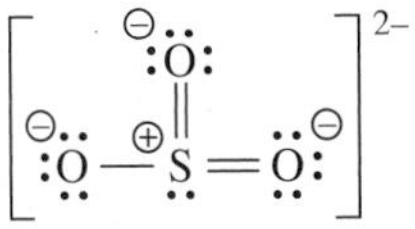

27b. $^{\ominus}$:Ö—N̈=Ö ⟷ Ö=N̈—Ö:$^{\ominus}$

27c.

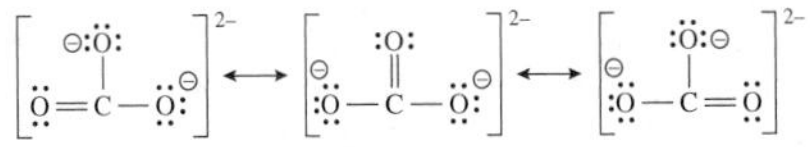

27d. H—Ö=Ö:$^{\ominus}$

29.

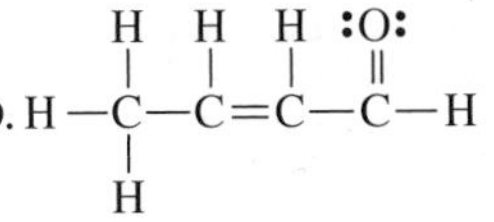

31a.

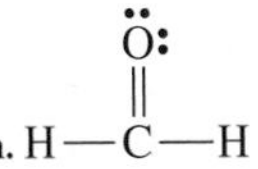

31b.

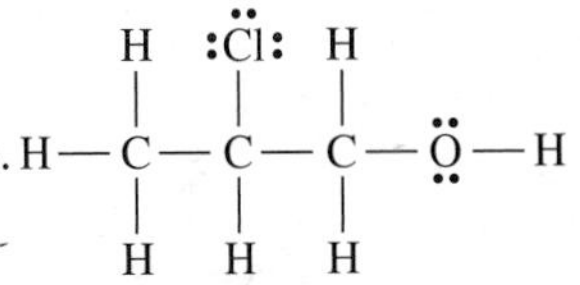

33a.

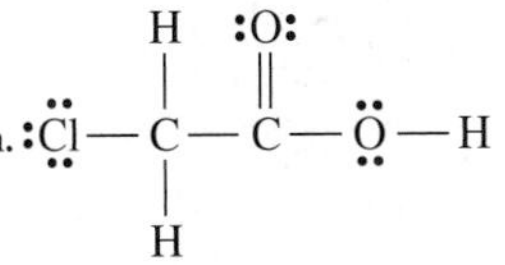

33b.

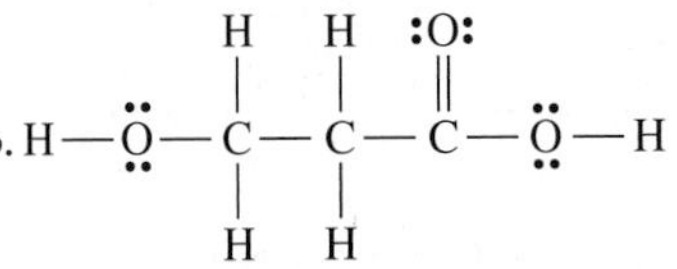

35a. Group 16,

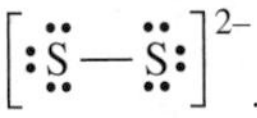

35b. Group 16 except oxygen,

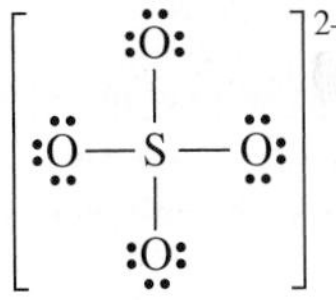

35c. Group 17 except fluorine,

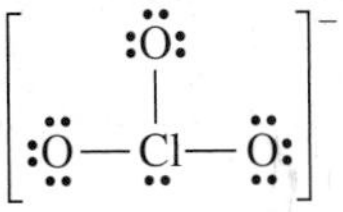

35d. Group 13,

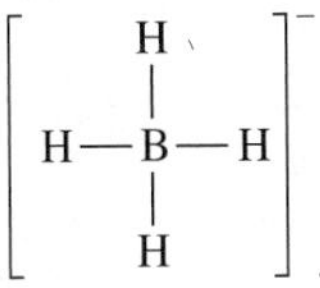

37. C—H < Br—H < F—H < Na—Cl < K—F **39a.** 4% **39b.** 5% **39c.** 60% **39d.** 33% **43.** $F_2C{=}O$ is represented on the left, while $H_2C{=}O$ is represented on the right. **45.** The molecular formulas for the compounds are SF_4 and SiF_4. The electrostatic potential map on the right is for SiF_4.

47. 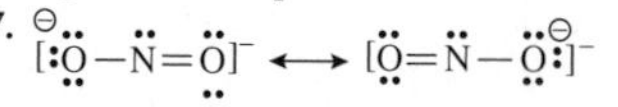

49. The molecule seems best represented as a resonance hybrid of (a) and (b).

51a.

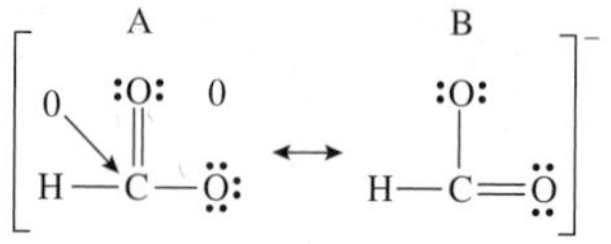

51b.

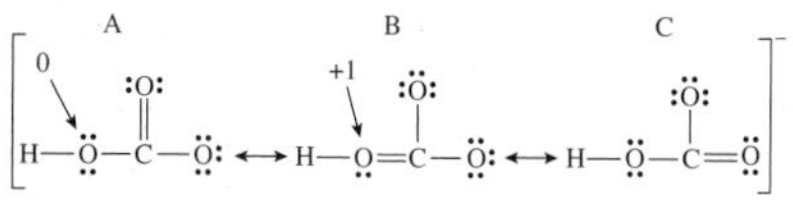

51c.

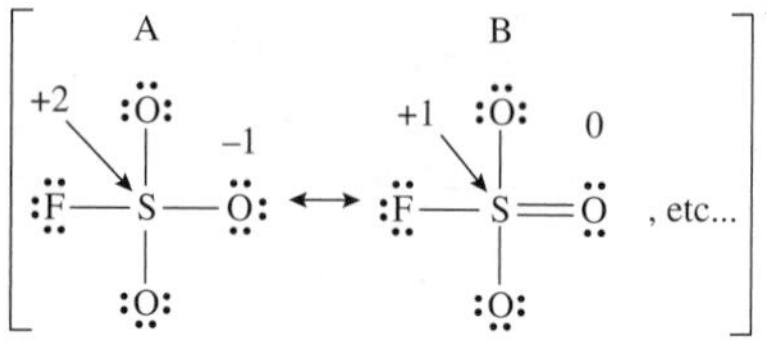

51d.

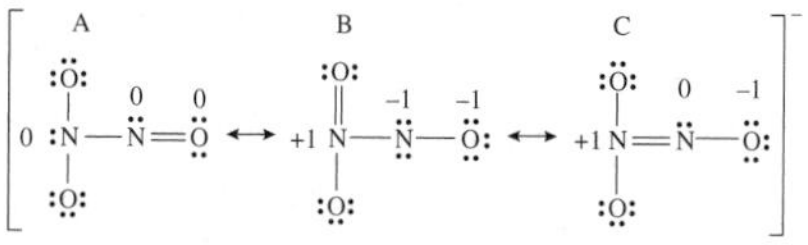

53a. 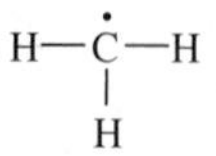 **53b.** 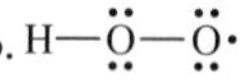

53c. 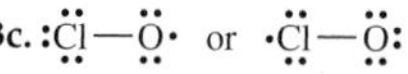

55a. Diamagnetic **55b.** Paramagnetic **55c.** Paramagnetic **55d.** Diamagnetic **55e.** Diamagnetic **55f.** Paramagnetic **57.** SF_4 and ICl_3 require expanded octets. The others do not. **59a.** Linear **59b.** Linear **59c.** Tetrahedral **59d.** Trigonal Planar **59e.** Bent **61a.** Bent **61b.** Planar **61c.** Linear **61d.** Octahedral **61e.** Tetrahedral

63. CO_3^{2-} **65a.** 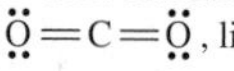, linear., linear.

65b. 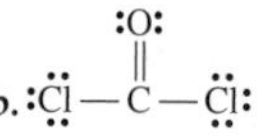, trigonal planar.

65c. 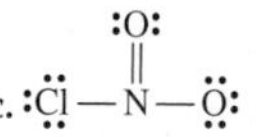 **67a.** Tetrahedral **67b.** Tetrahedral **67c.** Octahedral **67d.** Linear **69.** Tetrahedral **71.** Looking at the structures, the molecular angle/shape depends on the number of valence electron pairs on the central atom. The more pairs there are, the more acute the angle becomes.

73a. 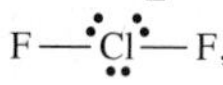, trigonal bipyramidal, linear.

73b. 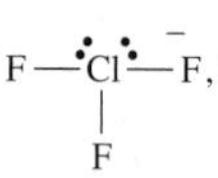, trigonal bipyramidal, t-shaped.

73c. 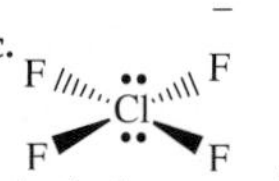, octahedral, square planar.

73d. 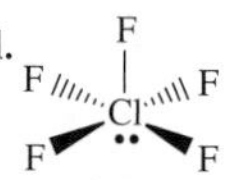, octahedral, square pyramidal.

75.

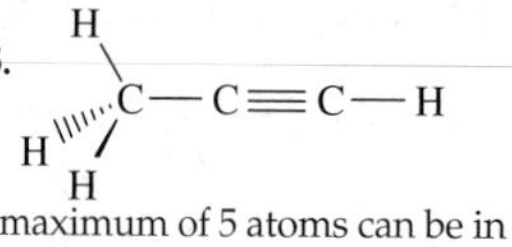

A maximum of 5 atoms can be in the same plane

77.

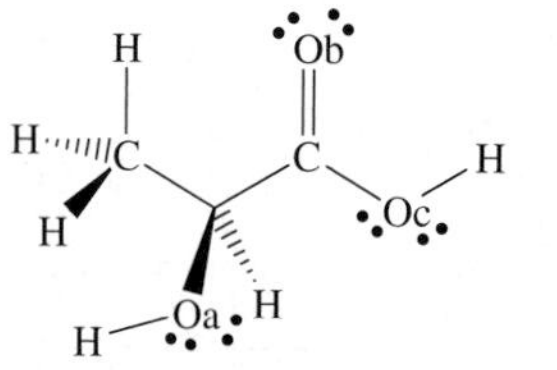

all angles ~ 109.5 except O_c—C—C, O_b=C—C, and O_b=C—O_c which are 120°

79.

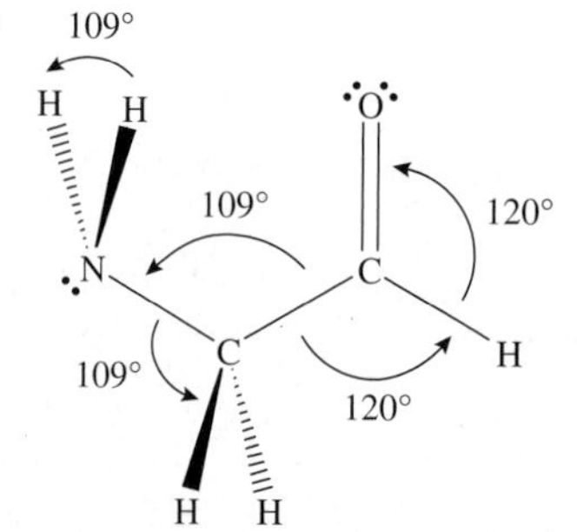

81a. Bent, polar. **81b.** Trigonal pyramidal, polar. **81c.** Bent, polar. **81d.** Planar, nonpolar. **81e.** Octahedral, nonpolar. **81f.** Tetrahedral, polar. **83.** It cannot be linear. **85.** Br_2 possess the longest bond. Single bonds are generally longer than multiple bonds. **87a.** 233 pm **87b.** 172 pm **87c.** 149 pm **87d.** 191 pm **89.** 144 pm **91.** Endothermic **93.** −113 kJ/mol

95a. $\Delta_f H° = 39$ kJ/mol **95b.** $\Delta_f H° = 99$ kJ **97.** −166 kJ/mol **99.** 448 kJ/mol

Integrative and Advanced Exercises

101. −744 kJ/mol **104.** 42.0 g/mol, C_3H_6.

106. 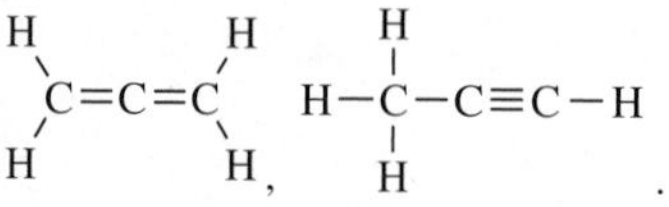

Neither is completely linear.

109. 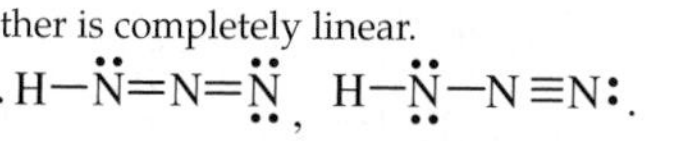

110. Bent **113.** +103 kJ **116.** 307 kJ **120.** F = 0.00199, Cl = 0.00187, Br = 0.00191, I = 0.00192. E_{ea} ~ −260 kJ/mol. E_{ea} for At: ~ −260 kJ/mol.

127.

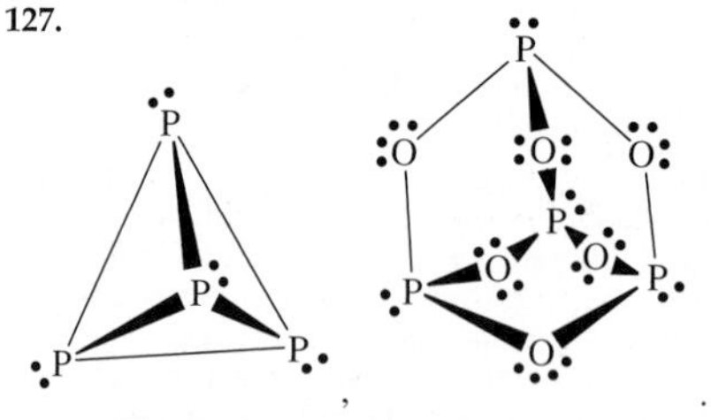

The P—O—P bond is not linear.

128.

	Atom, X				
	H	F	Cl	Br	I
E_i/eV	13.5985	17.423	12.9677	11.8139	10.4513
E_{ea}/eV	0.7542	3.399	3.617	3.365	3.059
D(HX)/eV	4.44	6.39	4.38	3.74	3.07
$D(X_2)$/eV	4.44	2.80	2.468	1.962	1.535
Z_{eff}	1	4.86	5.75	7.25	7.25
r_{cov}/pm	–	71.5	99.4	114.2	133.4

129a. 1.49 D **129b.** Approximately 92°. **129c.** 97.6°

Self-Assessment Exercises

134. The answer is (b). **135.** The answer is (c). **136.** The answer is (a). **137.** The answer is (a). **138.** The answer is (b). **139.** The answer is (c).

140a. 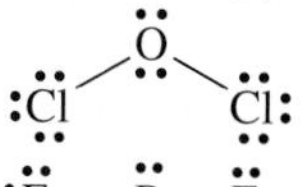

140b. 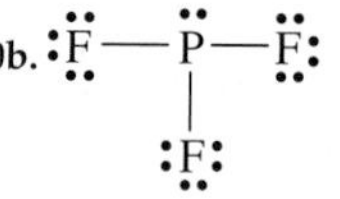

140c. 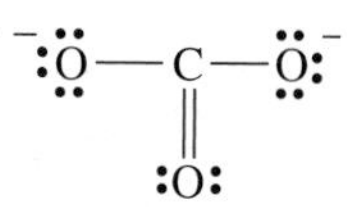

140d. 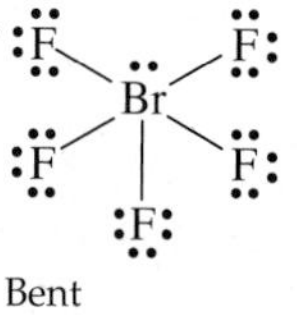

141a. Bent

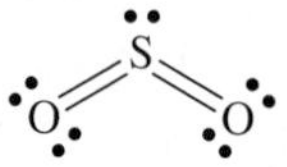

141b. Trigonal pyramidal

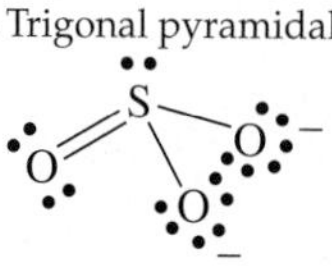

141c. Tetrahedral

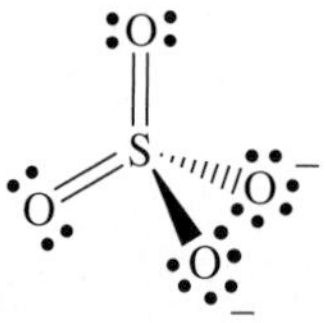

142. The answer is (a).
143. The answer is (d).
144. The answer is (e).
145. The answer is (a).
146.

Electron-Group Geometry	Molecular Shape	
Tetrahedral	Tetrahedral	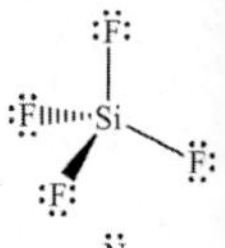
Tetrahedral	Trigonal pyramidal	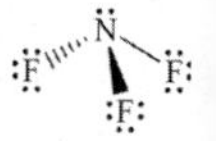
Trigonal pyramidal	Seesaw	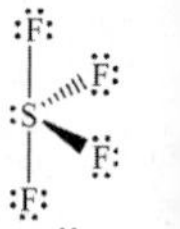
Octahedral	Square pyramidal	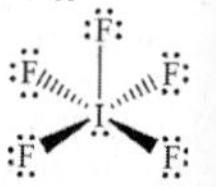

147. CH_4. The more negative the oxidation state of the carbon, the more energy that is generated per gram of that material. **148.** Bi **149.**

Bond	Bond Energy (kJ/mol)	Bond Length (pm)
C—H	414	110
C=O	736	120
C—C	347	154
C—Cl	339	178

150. VSEPR theory is valence shell electron pair repulsion theory. It is based on the premise that electron pairs assume orientations about an atom to minimize electron pair repulsions.
151. Since there are 4 electron pairs around the central atom, the way to maximize the distance between them is to set up a tetrahedral electron group geometry. However, since there are only three atoms bonding to the central atom, the molecular geometry is trigonal pyramidal. **152.** A pyramidal geometry is observed when an atom has one lone pair and is bonded to three other atoms (AX_3E). A bent geometry is observed when an atom has two lone pairs and is bonded to two other atoms (AX_2E_2). For both, the bond angles will be approximately (usually smaller than) 109°.

153.

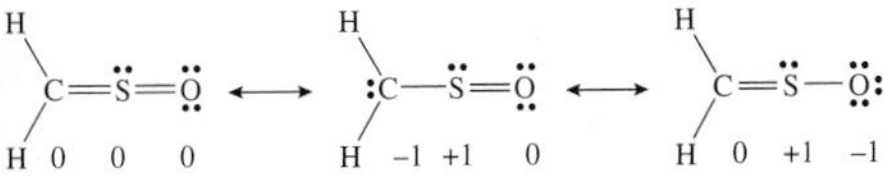

CHAPTER 11

Practice Examples 1a. There are three half-filled 2*p* orbitals on N, and one half-filled 5*p* orbital on I. Each half-filled 2*p* orbital from N will overlap with one half-filled 5*p* orbital of an I. Thus, there will be three N—I bonds. The I atoms will be oriented in the same direction as the three 2*p* orbitals of N: toward the *x*−, *y* −, and *z*-directions of a Cartesian coordinate system. Thus, the I—N—I angles will be approximately 90° (probably larger because the I atoms will repel each other). The three I atoms will lie in the same plane at the points of a triangle, with the N atom centered above them. The molecule is trigonal pyramidal.
1b. There are three half-filled orbitals on N and one half-filled orbital on each H. There will be three N—H bonds, with bond angles of approximately 90°. The molecule is trigonal pyramid. (We obtain the same molecular shape if N is assumed to be sp^3 hybridized, but bond angles are closer to109.5°, the tetrahedral bond angle.) VSEPR theory begins with the Lewis structure and notes that there are three bond pairs and one lone pair attached to N. This produces a tetrahedral electron pair geometry and a trigonal pyramidal molecular shape with bond angles a bit less than the tetrahedral angle of 109.5° because of the lone pair. Since VSEPR theory makes a prediction closer to the experimental bond angle of 107°, it seems more appropriate in this case. **2a.** Bent molecular geometry, sp^3 hybridization. **2b.** See-saw molecular geometry, sp^3d hybridization.
3a. Each central atom is surrounded by four electron pairs, requiring sp^3 hybridization. The carbon atoms have tetrahedral geometry, the oxygen atom is bent. **3b.** The left-most C and the right-most O are surrounded by four electron pairs, and thus require sp^3 hybridization. The central carbon is surrounded by three electron groups and is sp^2 hybridized. **4a.** There are four bond pairs around the left-hand C, requiring sp^3 hybridization. Three of the bonds that form are C—H sigma bonds resulting from the overlap of a half-filled sp^3 hybrid orbital on C with a half-filled 1*s* orbital on H. The other C has two attached electron groups, utilizing *sp* hybridization. The two C atoms join with a sigma bond: overlap of sp^2 on the left- hand C with *sp* on the right-hand C. The three bonds between C and N consist of a sigma bond (*sp* on C with *sp* on N), and two pi bonds (2*p* on C with 2*p* on N).
4b. structure(1) structure(2)

$:N_a{\equiv}N_b^{\oplus}{-}\ddot{\underset{..}{O}}: \longleftrightarrow \ :\underset{..}{N_a}{=}N_b^{\oplus}{=}\ddot{O}$

In both structures, the central N is attached to two other atoms, and possesses no lone pairs. The geometry of the molecule thus is linear and the hybridization on this central N is *sp*. In structure (1) the N≡N bond results from the overlap of three pairs of half-filled orbitals: (1) sp_x on N_b with $2p_x$ on N_a forming a σ bond, (2) $2p_y$ on N_b with $2p_y$ on N_a forming a π bond, and (3) $2p_z$ on N_b with $2p_z$ on N_a also forming a π bond. The N—O bond is a coordinate covalent bond, and forms by the overlap of the full sp_x orbital on N_b with the empty 2*p* orbital on O. In structure (2) the N=O bond results from the overlap of two pairs of half-filled orbitals: (1) sp_y on N_b with $2p_y$ on O forming a σ bond and (2) $2p_z$ on N_b with $2p_z$ on O forming a π bond. The N=N σ bond is a coordinate covalent bond, and is formed by two overlaps: (1) the overlap of the full *spy* orbital on N_b with the empty $2p_y$ orbital on N_a to form a σ bond, and (2) the overlap of the half-filled $2p_x$ orbital on N_b with the half-filled 2*px* orbital on N_a to form a π bond. Based on formal charge arguments, structure (1) is preferred, because the negative formal charge is on the more electronegative atom, O. **5a.** 53 kJ/mol Li_2^+
5b. The bond order in H_2^- is ½. We would expect the ion H_2^- to be stable, with a bond strength about half that of a hydrogen molecule.

6a.

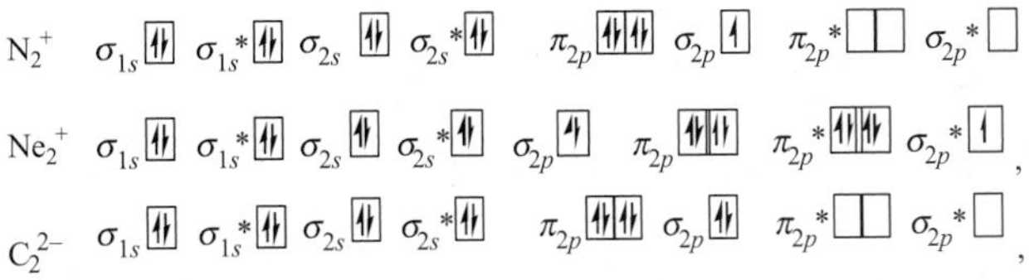

N_2^+: bond order = 2.5.
Ne_2^+: bond order = 0.5.
C_2^{2-}: bond order = 3.0.

6b. The bond length increases as the bond order decreases. Longer bonds are weaker bonds.

7a. CN^+

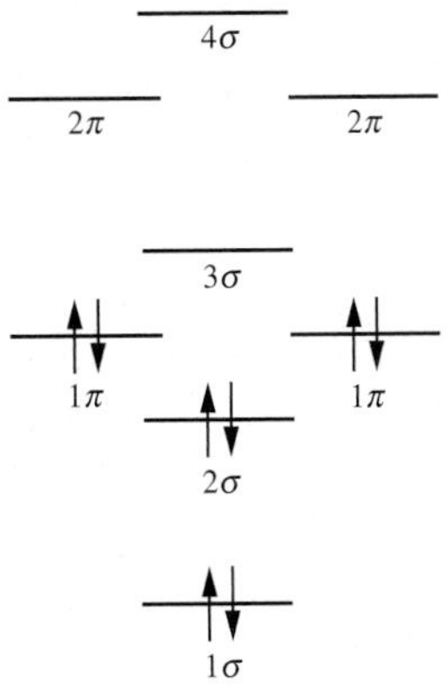

, bond order = 2.0.

7b. BN

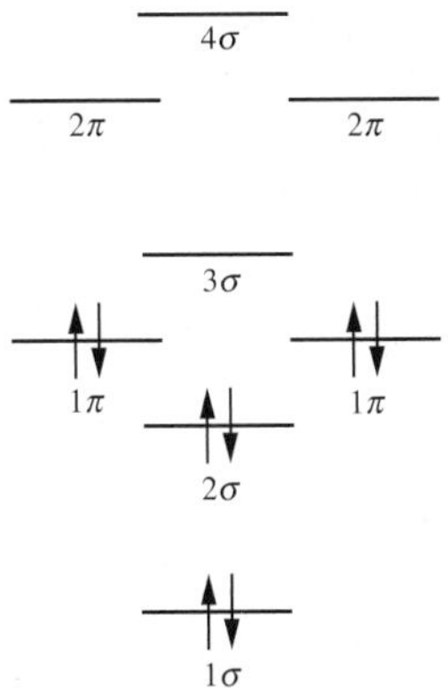

, bond order = 2.0.

8a.

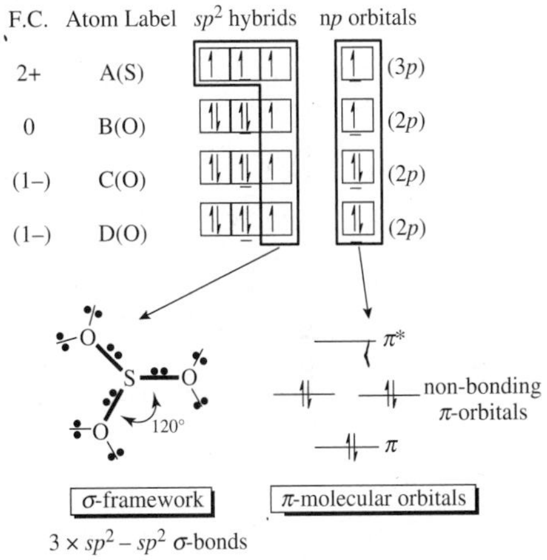

8b.

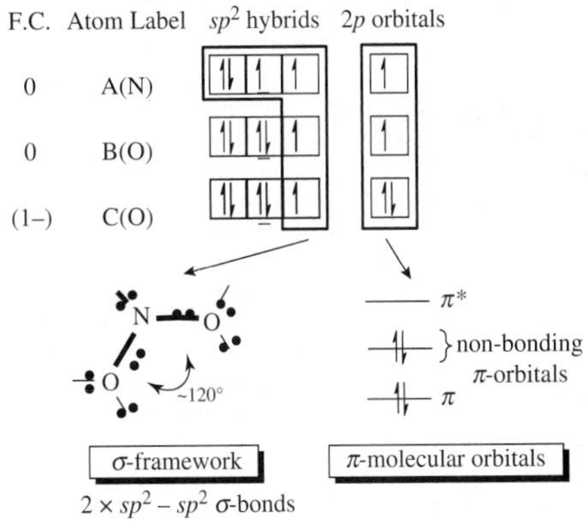

Integrative Example

A. (a)

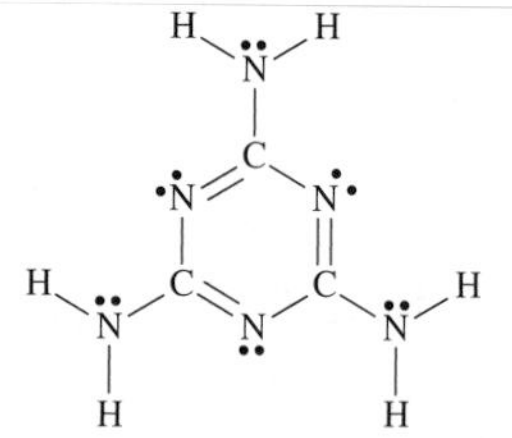

(b) *σ-system*

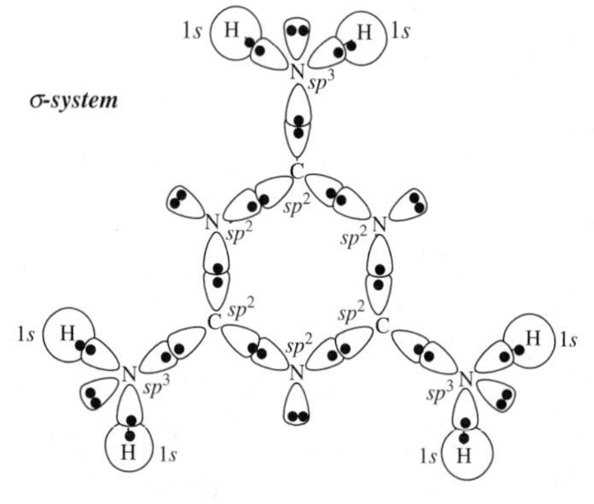

π-system

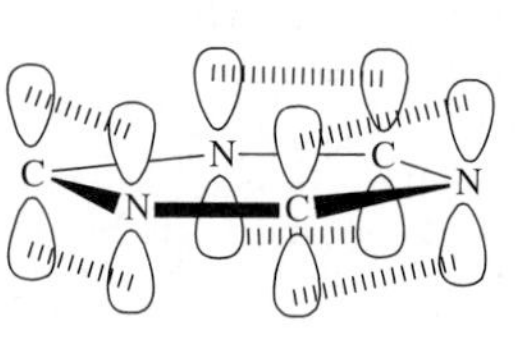

(c)

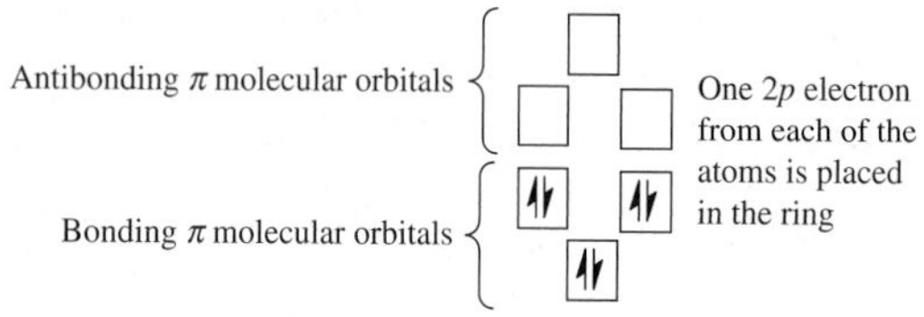

B. (a)

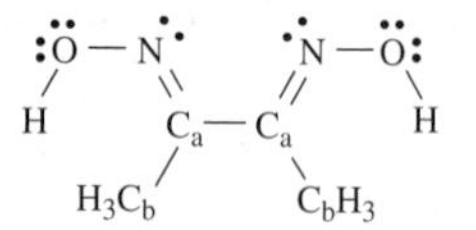

B. (b) C_b—H: σ H(1*s*) – $C_b(sp^3)$
C_a=N: σ $C_a(sp^2)$ – N(sp^2), π: $C_a(2p)$ – N(2*p*)
N—O: σ N(sp^2) – O(sp^3)
C_a—C_a: σ $C_a(sp^2)$ – $C_a(sp^2)$
C_a—C_b: σ $C_b(sp^3)$ – $C_a(sp^2)$
O—H: σ H(1s) – O(sp^3)

Exercises 1. First,valence-bond theory clearly distinguishes between sigma and pi bonds. In valence-bond theory, it is clear that a sigma bond must be stronger than a pi bond, for the orbitals overlap more effectively in a sigma bond (end-to-end) than they do in a pi bond (side-to-side). *Second*, molecular geometries are more directly obtained in valence-bond theory than in Lewis theory. *Third*, Lewis theory does not explain hindered rotation about double bonds. **3a.** Lewis theory does not describe the shape of the water molecule. **3b.** In valence-bond theory using simple atomic orbitals, each H—O bond results from the overlap of a 1s orbital on H with a 2*p* orbital on O. The angle between 2*p* orbitals is 90° so this method initially predicts a 90° bond angle. **3c.** In VSEPR theory the H_2O molecule is categorized as being of the AX_2E_2 type. The lone pairs repel each other more than do the bond pairs, explaining the smaller than 109.5° tetrahedral bond angle. **3d.** In valence-bond theory using hybrid orbitals, each H—O bond results from the overlap of a 1*s* orbital on H with an sp^3 orbital on O. The angle between sp^3 orbitals is 109.5°. **5.** The central atom is sp^2 hybridized in SO_2, CO_3^{2-}, and NO_2^-. **7a.** C is the central atom. The molecule is linear and C is *sp* hybridized. **7b.** N is the central atom. The molecule is trigonal planar around N which is sp^2 hybridized. **7c.** Cl is the central atom. The electron-group geometry around Cl is tetrahedral, indicating that Cl is sp^3 hybridized. **7d.** B is the central atom. The electron-group geometry is tetrahedral, indicating that B is sp^3 hybridized. **9.** The hybridization is sp^3d. Each of the three sigma bonds are formed by the overlap of a 2*p* orbital on F with one of the half-filled dsp^3 orbitals on Cl. The two lone pairs occupy dsp^3 orbitals. **11a.** sp^3d^2. **11b.** *sp* **11c.** sp^3. **11d.** sp^2. **11e.** sp^3d. **13a.** Planar molecule. The hybridization on C is sp^2. **13b.** Linear molecule. The hybridization for each C is *sp*. **13c.** Neither linear nor planar. The shape around the left-hand C is tetrahedral and that C has sp^3 hybridization. **13d.** Linear molecule. The hybridization for C is *sp*. **15a.** H—C≡N:. The H—C bond is a σ bond, and the C ≡ N bond is composed of 1 σ and π bonds.
15b. :N≡C—C≡N: . The C—C bond is a σ bond, and each C ≡ N bond is composed of 1 σ and 2 π bonds.

15c.

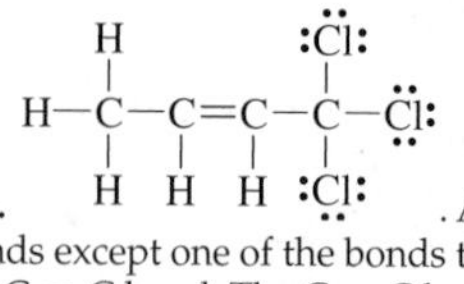

. All bonds are σ bonds except one of the bonds that comprise the C ≡ C bond. The C ≡ C bond is composed of one σ and one π bond. **15d.** H—Ö—N̈=Ö. All single bonds in this structure are σ bonds. The double bond is composed of one σ and one π bond. **17a.** The geometry at C is tetrahedral; C is sp^3 hybridized. Cl—C—Cl bond angles are 109.5°. Each C—Cl bond is represented by σ: C(sp^3)^1— Cl(3*p*)^1 **17b.** The e$^-$ group geometry around N is triangular planar, and N is sp^2 hybridized. The O—N—C bond angle is about 120°. The bonds are:
σ: O($2p_y$)^1— N(sp^2)^1 σ: N(sp^2)^1—Cl($3p_z$)^1
π: O($2p_z$)^1— N($2p_z$)^1.

17c. H—$\ddot{O}_a$—N̈=$\ddot{O}_b$. The geometry of O_a is tetrahedral, O_a is sp^3 hybridized and the H—O_a—N bond angle is (at least close to) 109.5°. The e$^-$ group geometry at N is trigonal, N is sp^2 hybridized, and the O_a—N—O_b bond angle is 120°. The four bonds are represented as follows. σ: H $(1s)^1$ — O_a $(sp^3)^1$
σ : O_a $(sp^3)^1$—N$(sp^2)^1$ σ : N$(sp^2)^1$—$O_b(2p_y)^1$
π: N $(2p_z)^1$—$O_b(2p_z)^1$.

17d. The e$^-$ group geometry around C is trigonal planar; all bond angles around C are 120°, and the hybridization of C is sp^2. The four bonds in the molecules are:
2 × σ : Cl $(3p_x)^1$ —C$(sp^2)^1$ σ : O $(2p_y)^1$ —C $(sp^2)^1$ π : O $(2p_z)^1$ —C $(2p_z)^1$.

19.

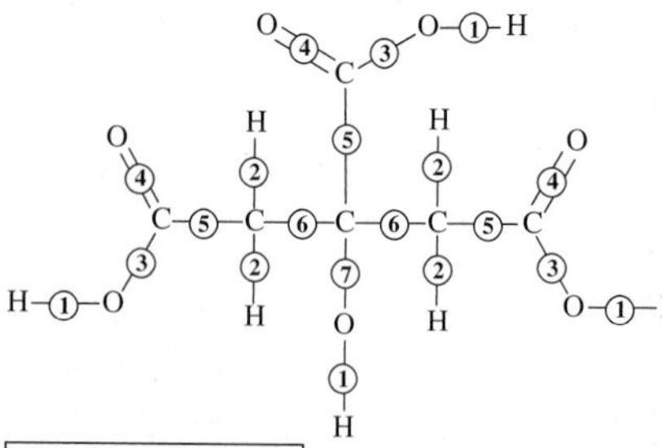

① σ(O(sp^3) - H(1*s*))
② σ(C(sp^3) - H(1*s*))
③ σ(C(sp^2) - O(sp^3))
④ σ(C(sp^2) - O(2*p*))
\+
π(C(2*p*) - O(2*p*))
⑤ σ(C(sp^3) - C(sp^2))
⑥ σ(C(sp^3) - C(sp^3))
⑦ σ(C(sp^3) - O(sp^3))

21a. Central S is sp^2 hybridized and the terrminal O and S atoms are unhybridized

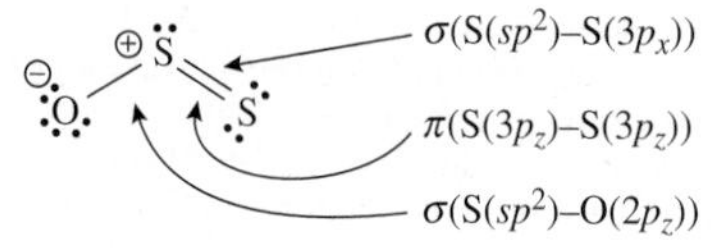

21b. H_a—C_a≡C_b—C_c=Ö (with H_b on C_c)

23. σ : $C_a(sp)^1$—$H_a(1s)^1$ σ : $C_a(sp)^1$—$C_b(sp)^1$
σ : C_b $(sp)^1$— $C_c(sp^2)^1$
σ : $C_c(sp^2)^1$— $H_b(1s)^1$ σ : $C_c(sp^2)^1$—O$(2p_y)^1$
π : $C_a(2p_y)^1$—$C_b(2p_y)^1$
π : $C_a(2p_z)^1$—$C_b(2p_z)^1$ π : $C_c(2p_z)^1$—O$(2p_z)^1$

25.

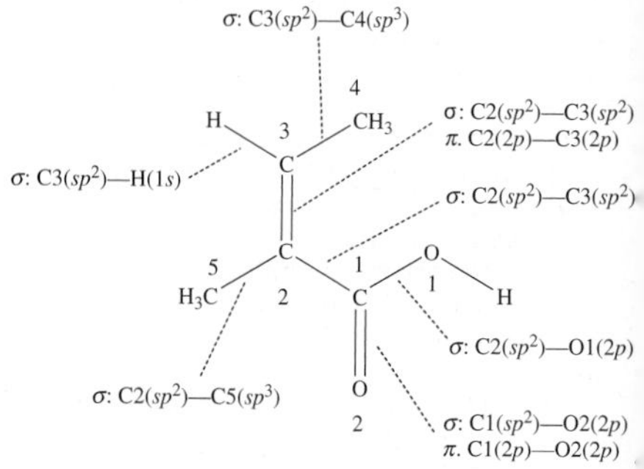

There are 8 atoms that are on the same plane (O1, O2, C1-5, H att. To C3). Furthermore, depending on the angle of rotation of the —CH_3 groups (C4 and C5), two H atoms can also be added to this total.

27. The valence-bond method describes a covalent bond as the result of the overlap of atomic orbitals. Molecular orbital theory describes a bond as a region of space in a molecule where there is a good chance of finding the bonding electrons. The molecular orbital bond does not have to be created from atomic orbitals (although it often is) and the orientations of atomic orbitals do not have to be manipulated to obtain the correct geometric shape. There is little concept of the relative energies of bonding in valence-bond theory. In molecular orbital theory, bonds are ordered energetically. **29.** N_2^- bond order = 2.5 (stable), N_2^{2-} bond order = 2 (stable). **31.** In order to have a bond order higher than three, there would have to be a region in a molecular orbital diagram where four bonding orbitals occur together in order of increasing energy, with no intervening antibonding orbitals. No such region exists in any of the molecular orbital diagrams in Figure 11-25. **33a.** σ_{1s} **33b.** σ_{2s} **33c.** σ^*_{1s} **33d.** σ_{2p}
35a. $1\sigma^2 2\sigma^2 1\pi^4 3\sigma^2 2\pi^1$ **35b.** $1\sigma^2 2\sigma^2 1\pi^4 3\sigma^2$
35c. $1\sigma^2 2\sigma^2 1\pi^4 3\sigma^2$ **35d.** $1\sigma^2 2\sigma^2 1\pi^4 3\sigma^1$
35e. $1\sigma^2 2\sigma^2 1\pi^4 3\sigma^2$ **35f.** $1\sigma^2 2\sigma^2 1\pi^4$
35g. $1\sigma^2 2\sigma^2 1\pi^4$
37a.

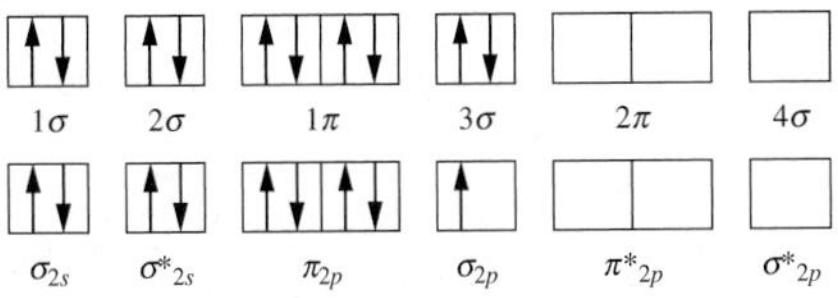

37b. Bond order is 3 for NO^+, 2.5 for N_2^+. **37c.** NO^+ is diamagnetic (all paired electrons), N_2^+ is paramagnetic. **37d.** N_2^+ has the greater bond length, because there is less electron density between the two nuclei.
39.

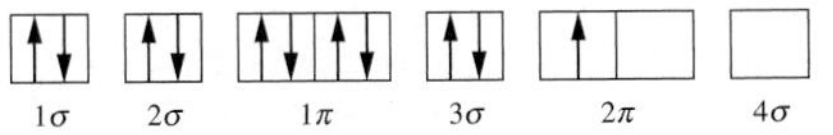

The bond length in CF^+ would be shorter. **41.** The π bonding requires that all six C—C π bonds must be equivalent. This can be achieved by creating six π molecular orbitals—three bonding and three antibonding—into which the 6 π electrons are placed. This creates a single delocalized structure for the C_6H_6 molecule. **43a.** Delocalized molecular orbitals required. **43b.** Delocalized molecular orbitals required. **43c.** Delocalized molecular orbitals not required.

Integrative and Advanced Exercises

46. $[Ne]_{3s}$ The half-filled 3s orbital on each Na overlaps with another to form a σ covalent bond. There are 22 electrons in Na2. These electrons are distributed in the molecular orbitals as follows. σ_{1s} σ_{1s}^* σ_{2s} σ_{2s}^*

σ_{2p} σ_{2p} σ_{2p}^* σ_{2p}^* σ_{3s}

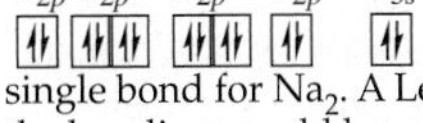

Again, we predict a single bond for Na_2. A Lewis-theory picture of the bonding would have the two Lewis symbols for two Na atoms uniting to form a bond: N· ·N ⟶ Na—Na Thus, the bonding in Na_2 is very much like that in H_2.
48a.

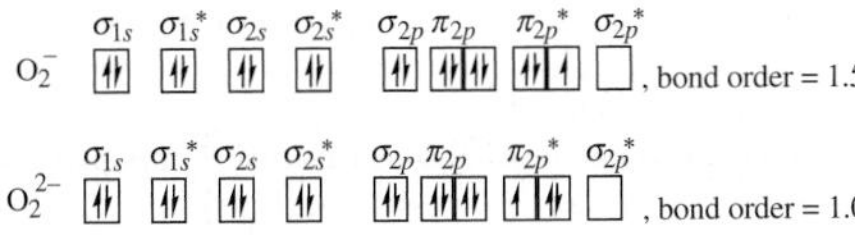

48b. O_2^- [:Ö—Ö·]$^-$, O_2^{2-} [:Ö—Ö:]$^{2-}$

52.

σ : F(2p)—O_a (sp^3) σ : O_a (sp^3)—N(sp^2)
σ : N(sp^2)—O$(2p_y)$ π : N$(2p_z)$—O$(2p_z)$

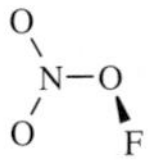

54a. For ClF_3 the electron group geometry is trigonal bipyramidal and the molecule is T-shaped. For AsF_5 the electron group geometry and molecular shape are trigonal bipyramidal. For ClF_2^+ the electron group geometry is tetrahedral and the ion is bent in shape. For AsF_6^- the electron group geometry and the shape of this ion are octahedral. **54b.** Trigonal bipyramidal electron group geometry is associated with sp^3d hybridization. The Cl in ClF_3 and As in AsF_5 both have sp^3d hybridization. Tetrahedral electron group geometry is associated with sp^3 hybridization. Thus Cl in ClF_2^+ has sp^3 hybridization. Octahedral electron group geometry is associated with sp^3d^2 hybridization, which is the hybridization adopted by As in AsF_6^-. **56.** Suppose two He atoms in the excited state $1s^1 2s^1$ unite to form an He_2 molecule. One possible configuration is $\sigma_{1s}^2\sigma_{1s}^{*0}\sigma_{2s}^2\sigma_{2s}^{0*}$. The bond order would be (4-0)/2 = 2. **58.** The orbital diagrams for C and N are as follows. C[He] sp^2 □□□ 2p □, N[He] sp^2 □□□ 2p □ antibonding π molecular orbitals bonding π molecular orbitals

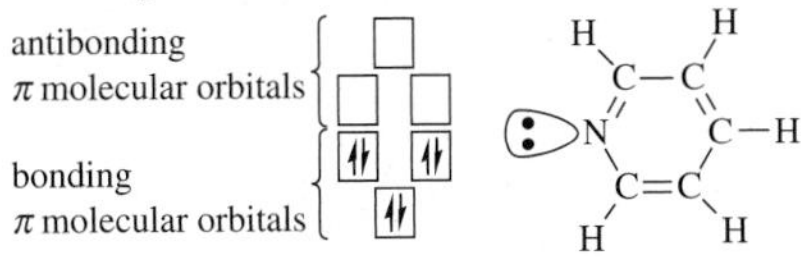

Number of π-bonds = 3
This π-bonding scheme produces three π-bonds, which is identical to the number predicted by Lewis theory.

61.

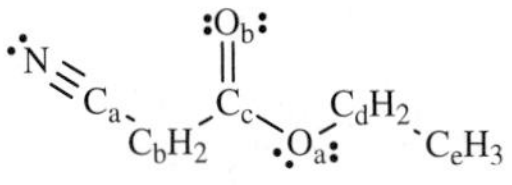

C_b—H, C_d—H, C_e—H: σ H(1s) – C(sp^3)
(all tetrahedral carbon uses sp^3 hybrid orbitals)
C_c=O_b: σ C_c (sp^2) –O_b (2p or sp^2), π :
C_c (2p) – O_b (2p)
C_c—O_a: σ C_c (sp^2) –O_a (2p or sp^3)
C_d—O_a: σ C_d (sp^3) –O_a (2p or sp^3)
C_a ≡ N: σ C_a (sp) –N(sp)
Two mutually perpendicular π -bonds:
C(2p) – N(2p)
C_a—C_b: σ $C_b(sp^3)$ –$C_a(sp)$
C_d—C_e: σ $C_d(sp^3)$ –$C_e(sp^3)$
C_c—C_b: σ $C_b(sp^3)$ –$C_c(sp^2)$
63a. 4.00 watts **63b.** 8.9 amps
68.

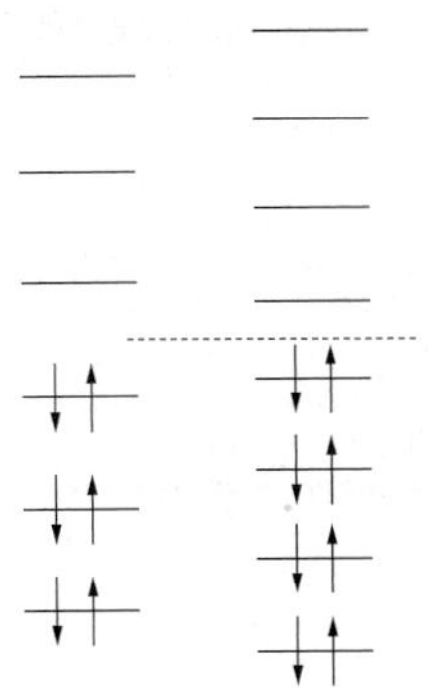

The wavelength corresponding to the excitation of the 1,3,5,7-octatetraene will be longer.

Feature Problems

70a. −205.4 kJ mol^{-1} **70b.** −117.9 kJ mol^{-1}
70c. −148.3 kJ mol^{-1}
70d. $\Delta H^\circ_{atomization}$ = 5358 kJ mol^{-1} (per mole of C_6H_6), Resonance energy = −168 kJ mol^{-1}.
76.

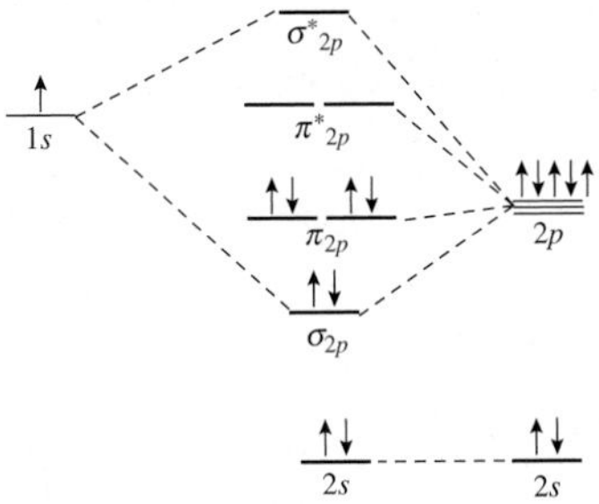

Self-Assessment Exercises

80. The answer is (c). **81.** The answer is (c). **82.** The answer is (a). **83.** The answer is (b). **84.** The answer is (c). **85.** The answer is (d). **86.** The valence-bond method using pure *s* and *p* orbitals incorrectly predicts a trigonal pyramidal shape with 90° F—B—F bond angles. **87.** BrF_5 has six constituents around it; five are fluorine atoms, and the sixth is a lone pair. Therefore, the hybridization is sp^3d^2, but the geometry is square pyramidal. **88.** There are (a) 6 σ and (b) 2 π bonds. **89.** All three are paramagnetic. B_2^- has 3 bonding electrons, so the B−B bond is the strongest. **90.** The answer is (c). **91.** C_2^+ **92.** C_2 has the greater bond energy. **93.** (a) and (c) have resonance structures; (b) is a Lewis structure.

94.

The hybridization of NO_2^- is sp_2. The hybridization of NO_2^+ is *sp*. **95.** The answer is (a). **96.** The answer is (e). **97.** The answer is (d). **98.** The answer is (b). **99a.** There are 5 σ bonds and 3 π bonds. **99b.** C^a is *sp* hybridized, C^b is sp^2 hybridized and O is sp^3 hybridized. **99c.** The lone pairs on the nitrogen are in one of the *sp* hybrid orbitals on N. Both lone pairs on the sulfur are in two of the sp^2 hybrid orbitals on S. The lone pairs on the oxygen are in two of the sp^3 hybrid orbitals on O. **99d.** N—C^a—C^b; C^a is *sp* hybridized, so the ideal bond angle is 180° C^a—C^b=S; C^b is sp^2 hybridized, so the ideal bond angle is 120°. C^a—C^b—O; C^b is sp^2 hybridized, so the ideal bond angle is 120°. C^b—O—H; O is sp^3 hybridized, so the ideal bond angle is 109.5°.

CHAPTER 12

Practice Examples 1a. CH_3CN is polar and thus has the strongest intermolecular forces and should have the highest boiling point. **1b.** $(CH_3)_3CH < CH_3CH_2CH_2CH_3 < SO_3 < C_8H_{18} < C_6H_5CHO$ **2a.** 0.923 kJ **2b.** −1025 J **3a.** 151 Torr **3b.** ≈ 1.7 g/L **4a.** Vapor only **4b.** 0.0435 g H_2O vapor, 0.089 g liquid water. **5a.** 121 mmHg **5b.** 42.9 mmHg **6a.** He, Ne, O_2, O_3, Cl_2, $(CH_3)_2CO$. **6b.** The magnitude of the enthalpy of vaporization is strongly related to the strength of intermolecular forces: the stronger these forces, the more endothermic the vaporization process. The first three substances all are nonpolar and, therefore, their only intermolecular forces are London forces, whose strength primarily depends on molar mass. The substances are arranged in order of increasing molar mass: H_2 = 2.0 g/mol, CH_4 = 16.0 g/mol, C_6H_6 = 78.1 g/mol, and also in order of increasing heat of vaporization. The last substance has a molar mass of 61.0 g/mol, which would produce intermolecular forces smaller than those of C_6H_6 if CH_3NO_2 were nonpolar. But the molecule is definitely polar. Thus, the strong dipole–dipole forces developed between CH_3NO_2 molecules make the enthalpy of vaporization for CH_3NO_2 larger than that for C_6H_6, which is, of course, essentially non-polar. **7a.** Moving from point R to P we begin with $H_2O(g)$ at high temperature (>100 °C). When the temperature reaches the point on the vaporization curve, OC, water condenses at constant temperature (100 °C). Once all of the water is in the liquid state, the temperature drops. When the temperature reaches the point on the fusion curve, OD, ice begins to form at constant temperature (0 °C). Once all of the water has been converted to $H_2O(s)$, the temperature of the sample decreases slightly until point P is reached. Since solids are not very compressible, very little change occurs until the pressure reaches the point on the fusion curve OD. Here, melting begins. A significant decrease in the volume occurs (≈10%) as ice is converted to liquid water. After melting, additional pressure produces very little change in volume because liquids are not very compressible.

7b.

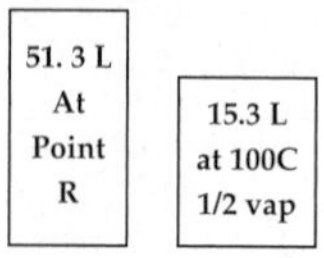

8a. LowerMelting point higher than CaO: MgO, Ga_2O_3, Ca_3N_2, Mg_3N_2. **8b.** NaI
9a. 524 pm **9b.** 6.628×10^7 pm^3

10a. 0.86 g/cm^3 **10b.** $6.035 \times 10^{23} \dfrac{\text{atoms Al}}{\text{mol Al}}$

11a. 402 pm **11b.** 2.21 g/cm^3 **12a.** −669.2 kJ/mol **12b.** −764 kJ/mol

Integrative Example A. At 25.0 °C, the vapor pressure of water is 23.8 mmHg. The vapor pressure for isooctane (using the Clausius-Clapeyron equation) is 43.1 mm Hg, higher than H2O's vapor pressure. **B.** $E_{ea,2} = +881$ kJ.

Exercises 1a. London forces between HCl molecules are expected to be relatively weak; Hydrogen bonding is weak; No hydrogen bonds; Dipole–dipole interactions should be relatively strong. **1b.** Neither hydrogen bonds nor dipole–dipole attractions can occur; London forces are more important. **1c.** No hydrogen bonds; London forces are as strong; Dipole–dipole interactions are important. **1d.** London forces are not very important; Hydrogen bonding is most important interaction between HF molecules. **1e.** H bonds are not important; There are no dipole-dipole interactions; London forces are quite weak.
3. (c) < (b) < (d) < (a). **5.** CH_3OH
7. Three water molecules.
9.

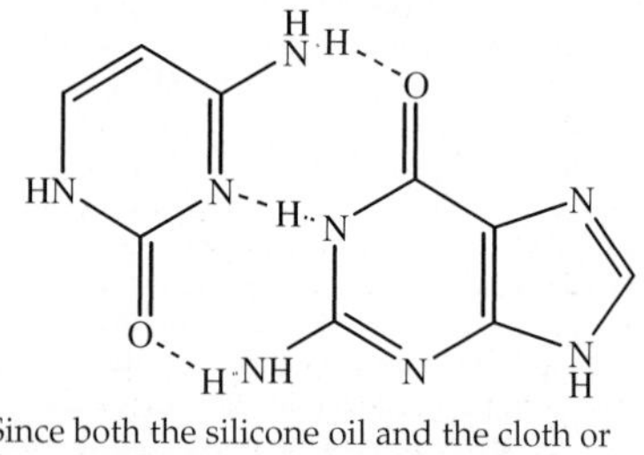

11. Since both the silicone oil and the cloth or leather are composed of relatively nonpolar molecules, they attract each other. Water, on the other hand is polar and adheres very poorly to the silicone oil (actually, the water is repelled by the oil), much more poorly, in fact, than it adheres to the cloth or leather. This is because the oil is more nonpolar than is the cloth or the leather. Thus, water is repelled from the silicone-treated cloth or leather.

13. Molasses, like honey, is a very viscous liquid (high resistance to flow). The coldest temperatures are generally in January (in the northern hemisphere). Viscosity generally increases as the temperature decreases. Hence, molasses at low temperature is a very slow flowing liquid. Thus there is indeed a scientific basis for the expression "slower than molasses in January." **15.** $CH_3CH_2OCH_2CH_3 < CH_3OH < HOCH_2CH_2OH$ **17.** The intermolecular interactions in butanol are dominated by H-bonding, which is much stronger than the London dispersion forces dominant in pentane. **19.** The process of evaporation is endothermic, meaning it requires energy. If evaporation occurs from an uninsulated container, this energy is obtained from the surroundings, through the walls of the container. However, if the evaporation occurs from an insulated container, the only source of the needed energy is the liquid that is evaporating. Therefore, the temperature of the liquid will decrease as the liquid evaporates.
21. 8.88 L C_6H_6 (l) **23.** 40.5 kJ **25.** 221 L methane
27a. 45 mmHg **27b.** 110° C **29.** 226 mmHg
31a. In order to vaporize water in the outer container, heat must be applied (i.e., vaporization is an endothermic process). When this vapor (steam) condenses on the outside walls of the inner container, that same heat is liberated. Thus condensation is an exothermic process. **31b.** 100.00 °C **33.** 120.7 °C
35. 0.907 mmHg
37.

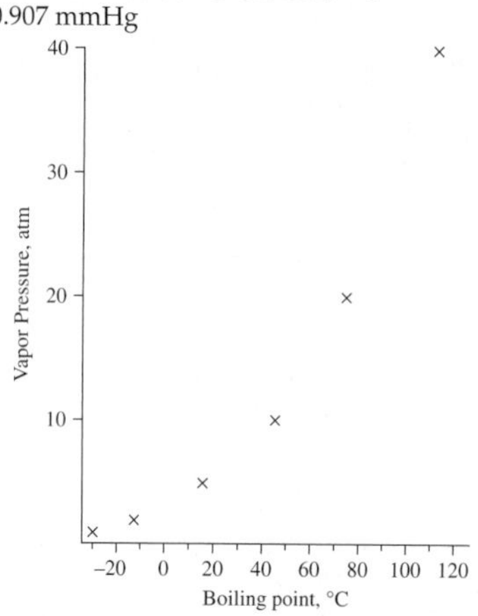

, 6.5 atm.

39. 49.7 kJ/mol **41.** 33 °C **43.** 0.0981 atm
45. Substances that can exist as a liquid at room temperature (about 20 °C) have critical temperature above 20 °C, 293 K. Of the substances listed in Table 12.5, CO_2, HCl, NH_3, SO_2, and H_2O can exist in liquid form at 20 °C.
47a. −776 kJ **47b.** 2.5×10^4 kJ **49.** Liquid and gas. **51a.** The upper-right region of the phase diagram is the liquid region, while the lower-right region is the region of gas. **51b.** As the phase diagram shows, the lowest pressure at which liquid exists is at the triple point pressure, namely, 43 atm. 1.00 atm is far below 43 atm. Thus, liquid cannot exist at this pressure, and solid sublimes to gas instead. **51c.** As we move from point *A* to point *B* by lowering the pressure, initially nothing happens. At a certain pressure, the solid liquefies. The pressure continues to drop, with the entire sample being liquid while it does, until another, lower pressure is reached. At this lower pressure the entire sample vaporizes. The pressure then continues to drop, with the gas becoming less dense as the pressure falls, until point *B* is reached. **53a.** 0.0418 atm **53b.** 0.110 atm
53c. 0.117 atm **55a.** No **55b.** No **55c.** Yes
55d. Possible, yes. **55e.** No **57a.** 0.22 kg
57b. 27 g steam **59.** The liquid in the can is supercooled. When the can is opened, gas bubbles released from the carbonated beverage serve as sites for the formation of ice crystals. The condition of supercooling is destroyed and the liquid reverts to the solid phase. An alternative explanation follows. The process of the gas coming out of solution is endothermic (heat is required). The required heat is taken from the cooled liquid, causing it to freeze.
61. Diamond
63a.

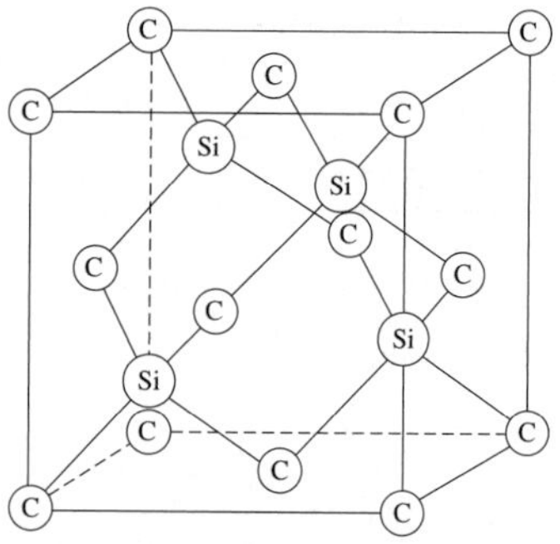

63b. sp^2 hybridization for each atom. The half-filled sp^2 hybrid orbitals of the boron and nitrogen atoms overlap to form the σ bonding structure, and a hexagonal array of atoms. The $2p_z$ orbitals then overlap to form the π bonding orbitals. **65.** We expect forces in ionic compounds to increase as the sizes of ions become smaller and as ionic charges become greater. As the forces between ions become stronger, a higher temperature is required to melt the crystal. In the series of compounds NaF, NaCl, NaBr, and NaI, the anions are progressively larger, and thus the ionic forces become weaker. We expect the melting points to decrease in this series from NaF to NaI. **67.** NaF **69.** In each layer of a closest packing arrangement of spheres, there are six spheres surrounding and touching any given sphere. A second similar layer then is placed on top of this first layer so that its spheres fit into the indentations in the layer below. The two different closest packing arrangements arise from two different ways of placing the third layer on top of these two, with its spheres fitting into the indentations of the layer below. In one case, one can look down into these indentations and see a sphere of the bottom (first) layer. In the other case, no first layer sphere is visible through the indentation.
71a.

, the two top unit cells in the diagram.
71b. The unit cell has one light-colored square per unit cell, 2 circles per unit cell, 1 diamond per unit cell. **71c.** The small square outlined near the center of the figure drawn in part (a).
73. 19.25 g/cm^3 **75a.** 335 pm **75b.** 9.23 g cm^{-3}
75c. 15.45° **77.** 2 Si atoms per unit cell.
79. CaF_2: There are eight Ca^+ ions on the corners for a total of one corner ion per unit cell. There are six Ca^{2+} ions on the faces for a total of three face ions per unit cell. This gives a total of four Ca^{2+} ions per unit cell. There are eight F^- ions, each wholly contained within the unit cell. The ratio of Ca^{2+} ions to F^- ions is 4 Ca^{2+} ions per 8 F^- ions: Ca_4F_8 or CaF_2. TiO_2. There are eight Ti^{4+} ions on the corners for a total of one Ti^{4+} corner ion per unit cell.

There is one Ti^{4+} ion in the center, wholly contained within the unit cell. Thus, there are a total of two Ti^{4+} ions per unit cell. There are four O^{2-} ions on the faces of the unit cell, each shared between two unit cells, for a total of two face atoms per unit cells. There are two O^{2-} ions totally contained within the unit cell. This gives a total of four O^{2-} ions per unit cell. The ratio of Ti^{4+} ions to O^{2-} ions is 2 Ti^{4+} ions per 4 O^{2-} ion: Ti_2O_4 or TiO_2. **81a.** The coordination number of Mg^{2+} is 6 and that of O^{2-} is 6 also. **81b.** Four formula units per unit cell of MgO. **81c.** Length = 424 pm, volume = 7.62×10^{-23} cm^3. **81d.** 3.51 g/cm^3 **83a.** Face centered cubic **83b.** Face centered cubic **83c.** Face centered cubic **85.** Lattice energies of a series such as LiCl(s), NaCl(s), KCl(s), RbCl(s), and CsCl(s) will vary approximately with the size of the cation. A smaller cation will produce a more exothermic lattice energy. Thus, the lattice energy for LiCl(s) should be the most exothermic and CsCl(s) the least in this series, with NaCl(s) falling in the middle of the series. **87.** −645 kJ/mol. $MgCl_2$ is much more stable than MgCl, since considerably more energy is released when it forms.

Integrative and Advanced Exercises
91. In many instances, the substance in the tank is not present as a gas only, but as a liquid in equilibrium with its vapor. As gas is released from the tank, the liquid will vaporize to replace it, maintaining a pressure in the tank equal to the vapor pressure of the substance at the temperature at which the cylinder is stored. This will continue until all of the liquid vaporizes, after which only gas will be present. The remaining gas will be quickly consumed, and hence the reason for the warning. However, the situation of gas in equilibrium with liquid only applies to substances that have a critical temperature above room temperature (20 °C or 293 K). **93.** At 100 °C, the quantity of heat needed is 2257 J/g H_2O. Thus, less heat is needed to vaporize 1.000 g of H_2O at the higher temperature of 100. °C. This makes sense, for at the higher temperature the molecules of the liquid already are in rapid motion. Some of this energy of motion or vibration will contribute to the energy needed to break the cohesive forces and vaporize the molecules. **96.** The final condition is a point on the vapor pressure curve at 30.0 °C. **99a.** 31 atm **99b.** −0.25 °C **100.** 87 °C **103.** 54.3% dimer. We would expect the % dimer to decrease with temperature. **104.** 1.84 kg N_2

107.

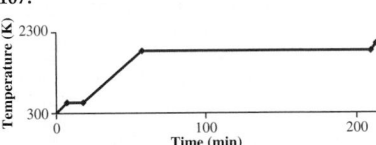

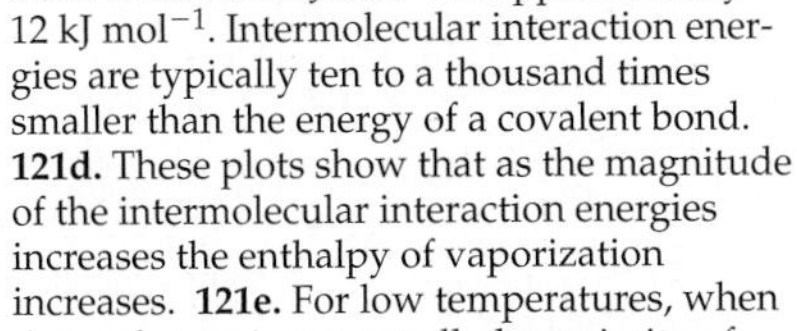

109. 16.0°

Feature Problems 121a. $E = -3.44$ kJ mol^{-1} **121b.** % dipole = 82%; % induced = 4.3%; % dispersion = 13.6% **121c.** The values range from tenths of a kJ mol^{-1} to approximately 12 kJ mol^{-1}. Intermolecular interaction energies are typically ten to a thousand times smaller than the energy of a covalent bond. **121d.** These plots show that as the magnitude of the intermolecular interaction energies increases the enthalpy of vaporization increases. **121e.** For low temperatures, when thermal energies are small, the majority of molecules will have their dipoles aligned head-to-tail. However, at very high temperatures, the molecules will be able to "jiggle" out of alignment, with their dipoles more randomly oriented. **122.** 0.021 J/m^2

123a. $$\ln\left(\frac{P_2}{P_1}\right) = \frac{(15{,}971)}{R}\left(\frac{1}{T_1} - \frac{1}{T_2}\right) + \frac{(14.55)}{R}\ln\left(\frac{T_2}{T_1}\right) - \frac{(0.160)}{R}\left(T_2 - T_1\right)$$

123b. 169 K

Self-Assessment Exercises 131. The answer is (e). **132.** The answer is (b). **133.** The answer is (c). **134.** The answer is (c). **135.** The answer is (d). **136.** $T_2 = 267.17$ K or −6.0 °C **137.** The answer is (b). **138.** The anwer is (a). **139.** The answers are (d) and (f). **140a.** $C_{10}H_{22}$ **140b.** $(CH_3)_2O$ **140c.** CH_3CH_2OH **141.** O_3 is out of place. The correct order of boiling points based on molar masses is: $N_2 < F_2 < Ar < O_3 < Cl_2$. **142.** $Ne < C_3H_8 < CH_3CH_2OH < CH_2OHCHOHCH_2OH < KI < K_2SO_4 < MgO$ **143.** Refer to the photograph on page 538 of water boiling under a reduced pressure. If the vapor is evacuated fast enough, to supply the required ΔH_{vap}, the water may cool to 0 °C and ice may begin to form. **144.** If there is too little benzene(l) in the sealed tube in Figure 12-22 initially, the liquid will all be converted to benzene(g) before T_c is reached. If too much is present initially, the liquid will expand and cause the benzene(l) to condense, and therefore only benzene(l) will be present at the time T_c is reached. **145a.** All the CCl_4 will be in the vapor phase. **145b.** 666 kJ **146a.** 362 pm **146b.** 4.74×10^7 pm^3 **146c.** 4 atoms/unit cell. **146d.** 74% **146e.** 4.221×10^{-22} g **146f.** 8.91 g/cm^3 **147.** The answer is (a). **148.** The answer is (d). **149.** A network covalent solid will have a higher melting point, because it takes much more energy to overcome the covalent bonds in the solid (such as, for example, diamond) than to overcome ionic interactions. **150.** MgO would have the highest melting point **151.** 2.13 Å **152.** The answer is (c).

CHAPTER 13

Practice Examples 1a. 6 microstates
1b. part (b)
2a. (a) $\Delta n_{gas} = -1; \Delta_r S < 0$ and (b) $\Delta n_{gas} = +1; \Delta_r S > 0$
2b. (a) uncertain because there is no gas involved and (b) $\Delta_r S > 0$
3a. $\Delta_{vap}S^o = 83.0$ J mol^{-1} K^{-1}
3b. $\Delta_{tr}H^o = 402$ J mol^{-1} **4a.** $\Delta S = 1.12$ J K^{-1}
4b. $\Delta S = 5.76$ J K^{-1} **5a.** $\Delta_r S^o - 99.4$ J mol^{-1} K^{-1}
5b. $S^o(N_2O_3(g)) = 312.4$ J mol^{-1} K^{-1}
6a. (a) spontaneous at low temperature and non spontaneous at high temperatures and (b) non-spontaneous at all temperatures.
6b. (a) reaction is spontaneous at high temperatures and (b) reaction is spontaneous at low temperatures and non-spontaneous at high ones. **7a.** $\Delta_r G^o = -1484$ kJ
7b. $\Delta_r G^o = -70.48$ kJ mol^{-1}
8a. $P_{NH_3} = 5.0 \times 10^3$ bar **8b.** $Q > 5.81 \times 10^5$
9a. (a) $K = \dfrac{P_{SiCl_4}}{P^2_{Cl_2}} = K_p$ and

(b) $K = \dfrac{[HOCl][H^+][Cl^-]}{P_{Cl_2}}$

9b. $K = \dfrac{[Pb^{2+}]^3 P^2_{NO}}{[NO_3^-]^2[H^+]^8}$

10a. $K = 8.5 \times 10^{-17} = K_{sp}$

10b. $K = 4 \times 10^{-5}$, the reaction proceeds in the forward direction only to a very limited extent. **11a.** $T = 607$ K **11b.** (a) $K = 1.5 \times 10^7$ and (b) $K = 1.7 \times 10^5$ **12a.** $T = 1240$ K
12b. $K = 5 \times 10^9$

Integrative Example A. 1.42 kJ mol^{-1}
B. 167 J K^{-1} mol^{-1}
Exercises 1a. Decrease **1b.** Increase
1c. Increase **3.** $W = 10^{4.111 \times 10^{24}}$ **5a.** Increase **5b.** Increase **5c.** Increase **7.** The first law of thermodynamics states that energy is neither created nor destroyed (thus, "The energy of the universe is constant"). A consequence of the second law of thermodynamics is that entropy of the universe increases for all spontaneous, that is, naturally occurring, processes (and therefore, "the entropy of the universe increases toward a maximum"). **9a.** Increase **9b.** Decrease **9c.** Uncertain **9d.** Decrease **11a.** Negative **11b.** Positive **11c.** Positive **11d.** Uncertain **11e.** Negative **13a.** 5.76 J K^{-1} **13b.** 3.10 J K^{-1} **15a.** $\Delta_{vap}H^o = 44.0$ kJ mol^{-1} and $\Delta_{vap}S^o = 118.9$ J mol^{-1} K^{-1}
15b. Hydrogen bonding is more disrupted in water at 100 °C than at 25 °C, and hence, there is not as much energy needed to convert liquid to vapor (thus $\Delta_{vap}H^o$ has a smaller value at 100 °C). The reason why $\Delta_{vap}S^o$ has a larger value at 25 °C than at 100 °C has to do with dispersion. A vapor at 1 atm pressure (the case at both temperatures) has about the same entropy. On the other hand, liquid water is more disordered (better able to disperse energy) at higher temperatures since more of the hydrogen bonds are disrupted by thermal motion. **17.** $C_6H_5CH_3$ **19.** 354 K **21.** Answer (b) is correct **23a.** Case 2 **23b.** No prediction **23c.** Case 3 **25.** First of all, the process is clearly spontaneous, and therefore $\Delta G < 0$. In addition, the gases are more dispersed when they are at a lower pressure and therefore $\Delta S > 0$. We also conclude that $\Delta H = 0$ because the gases are ideal and thus there are no forces of attraction or repulsion between them. **27a.** An exothermic reaction may not occur spontaneously if, at the same time, the system becomes more ordered (concentrated) that is, $\Delta S^o < 0$. This is particularly true at high temperature, where $T\Delta < S^o$ term dominates the ΔG^o expression. **27b.** A reaction in which $\Delta S^o > 0$ need not be spontaneous if that process is endothermic. This is particularly true at low temperatures, where ΔH^o term dominates the ΔG^o expression. **29.** −285 J mol^{-1} K^{-1}
31a. −506 kJ mol^{-1} **31b.** The reaction proceeds spontaneously in the forward direction.
33a. −1.6 kJ mol^{-1}, reaches equilibrium
33b. −474.3 kJ mol^{-1}, goes to completion
33c. −574.2 kJ mol^{-1}, goes to completion
35a. −3202 kJ mol^{-1} **35b.** −3177 kJ mol^{-1}
37a. −0.0545 kJ mol^{-1} K^{-1} **37b.** −616 kJ mol^{-1}
37c. −600 kJ mol^{-1} **39a.** $K = K_p = K_c(RT)^{-1}$
39b. $K = K_p = K_c$ **39c.** $K = K_p = K_c(RT)^2$
41. 18.9 kJ mol^{-1} **43.** 8×10^{-13} **45a.** 2×10^{13} **45b.** 4×10^5 **45c.** 8×10^4 **47.** −104 kJ mol^{-1}, reaction is spontaneous in the forward direction. **49.** −1.46 kJ mol^{-1}, reaction is spontaneous in the forward direction. **51.** Equation (b) is incorrect. The $T\Delta S^o$ term should be subtracted from the ΔH^o term, not added to it.
53a. 0.659 **53b.** 3.467 kJ mol^{-1} **53c.** The reaction will proceed to the right. **55a.** −23.4 kJ mol^{-1}
55b. 69 kJ mol^{-1} **55c.** 5.38 kJ mol^{-1}
55d. 28.8 kJ mol^{-1} **57.** −204.6 kJ mol^{-1}
59a. 3.0×10^{-21} atm **59b.** High temperature, which favors the products. In addition, the molecular oxygen was removed immediately after it was formed, which causes the equilibrium to shift to the right continuously.
61a. 215.2 J K^{-1}mol^{-1} **61b.** 91 kJ mol^{-1}
61c. 27 kJ mol^{-1} **61d.** 2×10^{-5} **63.** 1.36×10^3 K
65. 6.0×10^{-4} **67.** 650 K **69a.** 0.014 **69b.** 329 K
71. −206.1 kJ mol^{-1} **73a.** 152.1 kJ mol^{-1}
73b. −362.3 kJ mol^{-1}. The reaction is spontaneous
75. 1.7 kJ mol^{-1}. The reaction is not spontaneous.

Integrative and Advanced Exercises
77. For a spontaneous adiabatic process only, ΔS_{surr} is equal to zero. Only for a reversible, adiabatic process will we have $\Delta S = \Delta S_{surr} = \Delta S_{univ} = 0$. **82a.** True **82b.** False **82c.** False **82d.** True **83.** 0.210 mol Br_2, 0.210 mol Cl_2, 0.580 mol BrCl. **87.** 2.01×10^3 K **90.** 297 K **93.** 647 K or 374 °C. **94.** 0.0067 K^{-1} **98.** The increase in translation and rotation on going from solid to liquid is much less than on going from liquid to gas.
Feature Problems 100a. 8.556 kJ mol^{-1} **100b.** 0.0317 **100c.** 23.8 mmHg **100d.** 23.8 mmHg **101a.** When we combine two reactions and obtain the overall value of $\Delta_r G°$, we subtract the value on the plot of the reaction that becomes a reduction from the value on the plot of the reaction that is an oxidation. Thus, to reduce ZnO with elemental Mg, we subtract the values on the line labeled " $2\ Zn + O_2 \rightarrow 2\ ZnO$ " from those on the line labeled " $2\ Mg + O_2 \rightarrow 2\ MgO$". The result for the overall $\Delta G°$ will always be negative because every point on the "zinc" line is above the corresponding point on the "magnesium" line. **101b.** In contrast, the "carbon" line is only below the "zinc" line at temperatures above about 1000 °C. Thus, only at these elevated temperatures can ZnO be reduced by carbon. **101c.** The decomposition of zinc oxide to its elements is the reverse of the plotted reaction, the value of $\Delta_r G°$ for the decomposition becomes negative, and the reaction becomes spontaneous, where the value of $\Delta_r G°$ for the plotted reaction becomes positive. This occurs above about 1850 °C. **101d.** The "carbon" line has a negative slope, indicating that carbon monoxide becomes more stable as temperature rises. The point where CO(g) would become less stable than 2 C(s) and O_2(g) looks to be below −1000 °C (by extrapolating the line to lower temperatures). Based on this plot, it is not possible to decompose CO(g) to C(s) and O_2 (g) in a spontaneous reaction.

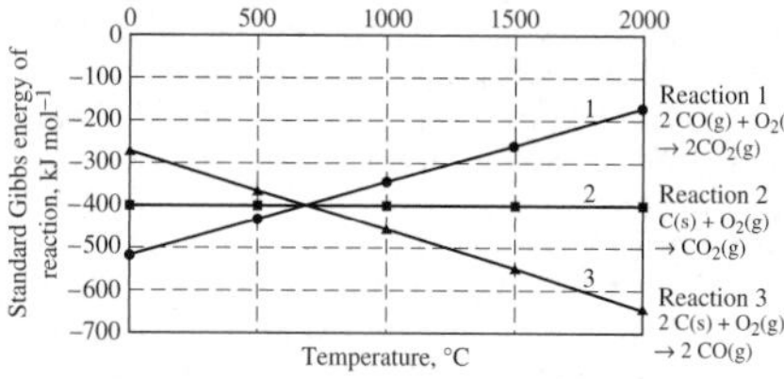

All three lines are straight-line plots of $\Delta_r G°$ vs. T following the equation $\Delta_r G° = \Delta_r H° - T\Delta_r S°$. The general equation for a straight line is given below with the slightly modified Gibbs energy equation as a reference: $\Delta_r G° = -\Delta_r S°\ T + \Delta_r H°$ (here $\Delta_r H°$ assumed constant) $y = mx + b$ ($m = -\Delta_r S°$ = slope of the line). Thus, the slope of each line multiplied by minus one is equal to the $\Delta_r S°$ for the oxide formation reaction. It is hardly surprising, therefore, that the slopes for these lines differ so markedly because these three reactions have quite different $\Delta_r S°$ values ($\Delta_r S°$ for Reaction 1 = −173 J K^{-1}, $\Delta_r S°$ for Reaction 2 = 2.86 J K^{-1}, $\Delta_r S°$ for Reaction 3 = 178.8 J K^{-1}) **101f.** Carbon would appear to be an excellent reducing agent, therefore, because it will reduce virtually any metal oxide to its corresponding metal as long as the temperature chosen for the reaction is higher than the threshold temperature. **104a.** 5.8 J $K^{-1}mol^{-1}$ **104b.** 3.4 J $K^{-1}mol^{-1}$
Self-Assessment Exercises 108. The answer is (d). **109.** The answer is (c). **110.** The answer is (a). **111.** The answer is (b). **112.** The answers are (a) and (d). **113a.** No reaction is expected. **113b.** Forward reaction should occur. **113c.** Forward reaction should occur. **114a.** Entropy change must be accessed for the system and its surroundings ($\Delta_r S_{univ}$), not just for the system alone. **114b.** Equilibrium constant can be calculated from $\Delta_r G°$ ($\Delta_r G° = -RT \ln K$), and K permits equilibrium calculations for nonstandard conditions. **115a.** 57 °C **115b.** 2.8 kJ mol^{-1} **115c.** Consistent with T_{bp}=57 °C. **116a.** 208.6 J $K^{-1}mol^{-1}$, forward reaction is exothermic. **116b.** −186.1 kJ mol^{-1} **116c.** 4.1×10^{32} **116d.** The reaction is spontaneous at all temperatures. **117.** Graph (a) **118.** Carbon dioxide is a gas at room temperature. The melting point of carbon dioxide is expected to be very low. At room temperature and normal atmospheric pressure this process is spontaneous. The entropy of the universe if positive.

CHAPTER 14

Practice Examples 1a. 16.2% ethanol by mass **1b.** (a) 7.29 M ≈ 7 M; (b) 5.28 m ≈ 5 m. **2a.** 0.03593 **2b.** (a) 0.3038 M; (b) 0.3246 m; (c) 0.005815. **3a.** Oxalic acid should be the most readily soluble in water because it is polar and can form hydrogen bonds. **3b.** I_2 should dissolve well in CCl_4 by simple mixing. **4a.** Suggestion 1: 31g NH_4Cl crystallized; Suggestion 2: 46 g of NH_4Cl crystallized. **4b.** KNO_3 (47%) > $KClO_4$ (38%) > K_2SO_4 (21%) **5a.** 0.717 atm O_2 pressure **5b.** 6.328 atm **6a.** P_{hex} = 112 mmHg; P_{pen} = 127 mmHg; P_{tot} = 239 mmHg. **6b.** P_{tol} = 13.0 mmHg; P_{benz} = 51.5 mmHg; P_{tot} = 64.5 mmHg. **7a.** $y_{hexane} = 0.469$, $y_{pentane} = 0.531$. **7b.** $y_t = 0.202$, $y_b = 0.798$. **8a.** 0.857 atm **8b.** $8.2\underline{4} \times 10^{-3}$g **9a.** 8.60×10^4 g/mol **9b.** 0.0105 atm **10a.** (a) 0.122 m; (b) 377 g/mol; (c) $C_{17}H_{20}O_6N_4$. **10b.** Lower than 760.0 mmHg. **11a.** 3.89 atm **11b.** 0.560 mL conc soln
Integrative Example A. 29.3% H_2O, χ_{phenol} = 0.161. **B.** (a) 0.48 mmHg; (b) 0.42 mmHg.
Exercises 1. NH_2OH(s) should be the most soluble. **3.** Salicyl alcohol probably is moderately soluble in both benzene and water. The reason for this assertion is that salicyl alcohol contains a benzene ring, which would make it soluble in benzene, and also can use its −OH groups to hydrogen bond to water molecules. **5.** (c) Formic acid and (f) propylene glycol are soluble in water. (b) Benzoic acid and (d) 1- Butanol are only slightly soluble in water. (a) Iodoform and (e) chlorobenzene are insoluble in water. **7.** KF is probably the most water soluble, based on smallest lattice energy. **9.** 53.7 g NaBr/100 g solution. **11.** 1.83×10^4 L sol'n **13.** For water, the mass in grams and the volume in mL are about equal; the density of water is close to 1.0 g/mL. For ethanol, on the other hand, the density is about 0.8 g/mL. As long as the final solution volume after mixing is close to the sum of the volumes for the two pure liquids, the percent by volume of ethanol will have to be larger than its percent by mass. This would not necessarily be true of other ethanol solutions. It would only be true in those cases where the density of the other component is greater than the density of ethanol. **15.** 21.6 g $HC_2H_3O_2$ **17.** 4.80×10^{-4} M **19.** 1.85 M **21.** 113 mL conc. soln **23.** 0.00654 M **25.** 0.410 m **27.** 54.8 g I_2 **29.** 4.19 M, 5.26 m. **31a.** $\chi_{C_7H_{16}} = 0.187$, $\chi_{C_8H_{18}} = 0.427$, $\chi_{C_9H_{20}} = 0.386$. **31b.** 18.7 mol% C_7H_{16}, 42.7 mol% C_8H_{18}, 38.6 mol% C_9H_{20}. **33a.** 0.00204 **33b.** 0.0101 **35.** 207 mL glycerol **37.** 4.36×10^{16} atoms, $\chi_{Pb} = 1.303 \times 10^{-6}$ **39.** 8.66 m **41a.** Unsaturated **41b.** 3.0 g $KClO_4$ **43.** 4.48×10^{-3} M **45.** 4×10^2 g CH_4 (natural gas) **47.** 1.40×10^{-5} M **49.** Because of the low density of molecules in the gaseous state, the solution volume remains essentially constant as a gas dissolves in a liquid. Changes in concentrations in the solution result from changes in the number of dissolved gas molecules. This number is directly proportional to the mass of dissolved gas. **51.** P_b = 40.6 mmHg, P_t = 16.3 mmHg, P_{tot} = 56.9 mmHg **53.** $P_{sol'n}$ = 23.2 mmHg **55.** $y_e = 0.68$, $y_s = 0.32$. **57.** 1.29×10^3 mmHg **59.** 3.2×10^4 g/mol **61.** Both the flowers and the cucumber contain ionic solutions (plant sap), but both of these solutions are less concentrated than the salt solution. Thus, the solution in the plant material moves across the semipermeable membrane in an attempt to dilute the salt solution, leaving behind wilted flowers and shriveled pickles (wilted/shriveled plants have less water in their tissues). **63.** ≈ 22.4 L solvent **65.** 2.8×10^5 g/mol **67.** 21 atm **69.** 1.2×10^2 g/mol **71a.** 20.°C/m **71b.** Cyclohexane is the better solvent for freezing-point depression determinations of molar mass, because a less concentrated solution will still give a substantial freezing-point depression. **73.** $C_6H_4N_2O_4$ **75.** C_4H_4S **77.** $12\underline{0}$ g NaCl, No. **79.** 1.04 **81a.** −0.186 °C **81b.** −0.372 °C **81c.** −0.372 °C **81d.** −0.558 °C **81e.** −0.372 °C **81f.** −0.186 °C **81g.** < −0.186 °C **83.** The combination of NH_3 (aq) with CH_3COOH (aq), results in the formation of NH_4CH_3COO (aq), which is a solution of the ions NH_4^+ and CH_3COO^-. This solution of ions or strong electrolytes conducts a current very well. **85.** The answer is (d).
Integrative and Advanced Exercises
88. 682 g H_2O **92.** 4.5 kg H_2O **95.** 1×10^1 % palmitic acid **97.** $\chi_{urea} = \chi_{sucrose} = 0.00161$, mole fraction of water = 0.99839. **100a.** 1.56×10^{-3} m^2 **100b.** 8.34×10^{-7} m^2 **105.** Volume % N_2 = 65.38%, volume % O_2 = 34.62%. **107.** 1.3×10^2 g $CuSO_4 \cdot 5H_2O$ **109.** 0.0125 g/L, 0.73 atm. **111.** 5.96 L
Feature Problems 114a. $\chi_{HCl} > 0.50$ **114b.** The composition of HCl(aq) changes as the solution boils in an open container because the vapor has a different composition than does the liquid. Thus, the component with the lower boiling point is depleted as the solution boils. The boiling point of the remaining solution must change as the vapor escapes due to changing composition. **114c.** The azeotrope occurs at the maximum of the curve: at χ_{HCl} = 0.12 and a boiling temperature of 110 °C. **114d.** $\chi_{HCl} = 0.112$ **115a.** 90.04% **115b.** $CaCl_2 \cdot 6\ H_2O$ deliquesces if the relative humidity is over 32%. Thus, $CaCl_2 \cdot 6\ H_2O$ will deliquesce. **115c.** If the substance in the bottom of the desiccator has high water solubility, its saturated solution will have a low χ_{water}, which in turn will produce a low relative humidity. **116a.** Using 0.92% NaCl (aq) and i = 2.0 gives ΔT_f = −0.60 °C, fair agreement with −0.52 °C. **116b.** ΔT_f = −0.60 °C, close to the defined freezing point of −0.52 °C.
Self-Assessment Exercises 120. The answer is (b). **121.** The answer is (d). **122.** The answer is (a). **123.** The answer is (d). **124.** The answer is (c). **125.** Unsaturated **126a.** 0.16 M Na^+ **126b.** 0.32 M **126c.** 8.1 atm **126d.** −0.60 °C **127a.** 7.80 M $C_3H_8O_3$ **127b.** 24.4 M H_2O **127c.** 34.0 m H_2O **127d.** 0.242 **127e.** 75.8% **128.** 1−b, 2−d, 3−d, 4−a. **130.** The answer is (b). **131.** The magnitude of $\Delta H_{lattice}$ is larger than the sum of the $\Delta H_{hydration}$ of the individual ions. **132.** The answer is (a). **133.** The answer is (b). **134.** The answer is (a). **135.** The answer is (e). **136.** The answer is (c).

CHAPTER 15

Practice Examples 1a. $[Cu^{2+}] = \dfrac{x}{1.22}$ **1b.** ≈ 0.0098 M **2a.** 8.3×10^2 **2b.** 6.9×10^{-5}

3a. $K_c = \dfrac{[Ca^{2+}]^5[HPO_4^{2-}]^3}{[H^+]^4}$ **3b.** $K_p = \dfrac{\{P(H_2)\}^4}{\{P(H_2O)\}^4}$
4a. 1.7×10^{-6} **4b.** 95 **5a.** There will be greater quantities of reactants, and smaller quantities of products than there were initially. **5b.** The net reaction will proceed to the right. **6a.** The equilibrium system will shift right. **6b.** (a) Adding extra CaO(s) will have no direct effect on the position of equilibrium; (b) The reaction will shift left; (c) Addition of solid $CaCO_3$ will not have an effect upon the position of equilibrium. **7a.** Decreasing the cylinder volume would have the initial effect of doubling both $[N_2O_4]$ and $[NO_2]$ In order to reestablish equilibrium, some NO_2 will then be converted into N_2O_4. Note, however, that the NO_2 concentration will still ultimately end up being higher than it was prior to pressurization. **7b.** A change in overall volume or total gas pressure will have no effect on the position of equilibrium. **8a.** Greater **8b.** Greater **9a.** 2.3×10^{-4} **9b.** 25.2 g N_2O_4 **10a.** $K_c = 1.5 \times 10^{-4}$, $K_p = 3.7 \times 10^{-3}$. **10b.** $K_p \approx 0.022$ **11a.** 0.481 bar **11b.** 0.716 bar **12a.** 0.262 mol HI **12b.** (a) The amount of N_2O_4 will decrease; (b) 0.0118 mol. **13a.** $[Ag^+] = [Fe^{2+}] = 0.488$ M, $[Fe^{3+}] = 1.20 - 0.488 = 0.71$ M. **13b.** $[V^{3+}] = [Cr^{2+}] = 0.0057$ M, $[V^{2+}] = [Cr^{3+}] = 0.154$ M.
Integrative Example A. $[F6P]_{eq} = 1.113 \times 10^{-6}$. F6P will decrease with an increase in temperature. **B.** (a) 8.47 L; (b) ~ 8.47 L.
Exercises 1a. $2\,COF_2\,(g) \rightleftharpoons CO_2(g) + CF_4\,(g)$, $K_c = \dfrac{[CO_2][CF_4]}{[COF_2]^2}$

1b. $Cu\,(s) + 2\,Ag^+\,(aq) \rightleftharpoons Cu^{2+}\,(aq) + 2\,Ag\,(s)$, $K_c = \dfrac{[Cu^{2+}]}{[Ag^+]^2}$

1c. $S_2O_8^{2-}\,(aq) + 2\,Fe^{2+}\,(aq) \rightleftharpoons 2\,SO_4^{2-}(aq) + 2\,Fe^{3+}\,(aq)$, $K_c = \dfrac{[SO_4^{2-}]^2[Fe^{3+}]^2}{[S_2O_8^{2-}][Fe^{2+}]^2}$

3a. $K_c = \dfrac{[NO_2]^2}{[NO]^2[O_2]}$ **3b.** $K_c = \dfrac{[zn^{2+}]}{[Ag^+]^2}$

3c. $K_c = \dfrac{[OH^-]^2}{[CO_3^{2-}]}$ **5a.** $K_c = \dfrac{[HF]}{[H_2]^{1/2}[F_2]^{1/2}}$

5b. $K_c = \dfrac{[NH_3]^2}{[N_2][H_2]^3}$ **5c.** $K_c = \dfrac{[N_2O]^2}{[N_2]^2[O_2]}$

5d. $K_c = \dfrac{1}{[Cl_2]^{1/2}[F_2]^{3/2}}$ **7a.** 0.0012

7b. 5.63×10^5 **7c.** 0.429 **9.** $K_p = 0.0313$, $K_c = 1.28 \times 10^{-3}$. **11.** 9.7×10^{-16} **13.** 5×10^{14}

15. $K = a(H_2CO_3)/a(CO_2)$, $K = \dfrac{[H_2CO_3]/c°}{P_{CO_2}/P°}$

17. 0.10$\underline{6}$ **19a.** 26.3 **19b.** 0.613 **21.** 9.1×10^{-18} M **23.** 0.516
25a. $NH_3(g) + \frac{7}{4}O_2(g) \rightleftharpoons NO_2(g) + \frac{3}{2}H_2O(g)$
25b. 4.03×10^{19}

27a. $K_c = \dfrac{[CO][H_2O]}{[CO_2][H_2]} = \dfrac{\dfrac{n\{CO\}}{V} \times \dfrac{n\{H_2O\}}{V}}{\dfrac{n\{CO_2\}}{V} \times \dfrac{n\{H_2\}}{V}}$

Since V is present in both the denominator and the numerator, it can be stricken from the expression. **27b.** $K_c = K_p = 0.659$. **29.** $Q_c = 0.82 < K_c = 100$. The mixture described cannot be maintained indefinitely. **31a.** The mixture is not at equilibrium. **31b.** The reaction will proceed to the right. **33.** Amount H_2 = amount I_2 = 0.033 mol. Amount HI = 0.234 mol HI.

35. 4.0×10^{-4} mol Cl_2
37a. $n_{PCl_5} = 0.478$ mol PCl_5,
$n_{PCl_3} = 0.623$ mol PCl_3,
$n_{Cl_2} = x = 0.0725$ mol Cl_2.
37b. Amount PCl_3 = 0.20 mol = amount Cl_2. Amount PCl_5 = 0.41 mol. **39a.** The reaction will shift to the left. **39b.** 26 g C_2H_5OH, 35 g CH_3CO_2H, 32 g $CH_3CO_2C_2H_5$, 68 g H_2O. **41.** 5.58 atm **45.** Sketch (b) is the best representation because it contains numbers of SO_2Cl_2, SO_2, and Cl_2 molecules that are consistent with the K_c for the reaction.

47. $K = \dfrac{[\text{aconitate}]}{[\text{citrate}]}$, $Q = \dfrac{4.0 \times 10^{-5}}{(0.00128)}$

= 0.031. Since $Q = K$, the reaction is at equilibrium. **49.** 749.0 mmHg **51.** 0.529 atm O_2, $P_{total} = 0.601$ atm total. **53a.** 30.1$\underline{4}$ atm **53b.** 32.46 atm **55.** Continuous removal of the product has the effect of decreasing the concentration of the products below their equilibrium values. Thus, the equilibrium system is disturbed by removing the products and the system will attempt to re-establish the equilibrium by shifting toward the right, that is, to generate more products. **57a.** Decreases **57b.** Increases **57c.** No change **57d.** No change **59a.** More NO(g) formed. **59b.** Rate increases. **61.** $P\{N_2(g)\}$ will have decreased when equilibrium is re-established. **63a.** Shifts right, toward products. **63b.** No shift, no change in equilibrium position. **63c.** Shifts right, towards products. **65a.** Hb:O_2 is reduced. **65b.** No effect. **65c.** Hb:O_2 increases. **67.** The pressure on N_2O_4 will initially increase as the crystal melts and then vaporizes, but over time the new concentration decreases as the equilibrium is shifted toward NO_2. **69.** While the amount of calcium carbonate will decrease, its concentration will remain the same because it is a solid.
Integrative and Advanced Exercises
71. $[Cu^{2+}] = 5.4 \times 10^{-16}$ M **73.** $[CH_3COOH]_{eq} = 7.5 \times 10^{-6}$, the fraction reacted is 99.9925%. The reaction essentially goes to completion **77.** $[Fe^{2+}] = 0.111$ M, $[Ag^+] = 0.15$ M, $[Fe^{3+}] = 0.049$ M. **78.** 2.79 atm **80.** 2.1×10^{-3} **82.** 12.0 **85.** 0.177 **86.** 87.1 g/mol **89.** 3.34×10^{-4} **91.** 0.085 moles $CH_4(g)$, 0.070 moles $H_2O(g)$, 0.12 mol CO_2, 0.16 mol H_2. **95.** 24.6 atm
Feature Problems 98. 99.999% of the reactants remain. **99.** 4.0 **100.** ~ 0.015 **101.** 4.80×10^{-8} **103.** C(or) = 6×10^{-4} M, C(aq) = 4×10^{-4} M.
Self-Assessment Exercises 107. The answer is (c). **108.** The answer is (d). **109.** The answer is (a). **110.** The answer is (b). **111.** The answer is (a). **112.** The answer is (c). **113.** 1.9 **114a.** More Cl_2 is produced. **114b.** Less Cl_2 is made. **114c.** Less Cl_2 is made. **114d.** No change. **114e.** Less Cl_2 is made. **115.** SO_2 (g) will be less than SO_2 (aq). **116.** There will be much more product than reactant. **117a.** 0.0578 moles **117b.** 0.232 moles

CHAPTER 16

Practice Examples 1a. (a) Forward direction: HF is the acid and H_2O is the base. Reverse direction: F^- is the base and H_3O^+ is the acid. (b) Forward direction: HSO_4^- is the acid and NH_3 is the base. Reverse direction: SO_4^{2-} is the base and NH_4^+ is the acid. (c) Forward direction: HCl is the acid and CH_3COO^- is the base. Reverse direction: Cl^- is the base and CH_3COOH is the acid.
1b.

$HNO_2(aq) + H_2O(l) \rightleftharpoons NO_2^-(aq) + H_3O^+(aq)$;
$HCO_3^-(aq) + H_2O(l) \rightleftharpoons CO_3^{2-}(aq) + H_3O^+(aq)$;
$PO_4^{3-}(aq) + H_2O(l) \rightleftharpoons HPO_4^{2-}(aq) + OH^-(aq)$;
$HCO_3^-(aq) + H_2O(l) \rightleftharpoons H_2CO_3(aq) \rightarrow CO_2 \cdot H_2O(aq) + OH^-(aq)$

2a. $[H_3O^+] = 1.4 \times 10^{-3}$, $[OH^-] = 7.1 \times 10^{-12} \times$ M.
2b. $[H_3O^+] = 3.16 \times 10^{-6}$, $[OH^-] = 3.16 \times 10^{-9}$, 0.1% of the H_3O^+ produced by water self-ionization **3a.** 99.95% **3b.** 0.17%, pH = 4.77 **4a.** 17% **4b.** 1.4×10^{-4} **5a.** $[I^-] = [H_3O^+] = 0.0025$ M. $[OH^-] = 4.0 \times 10^{-12}$ M.
5b. 1.670 **6a.** 11.519 **6b.** 13.739 **7a.** 2.9×10^{-8} **7b.** 2.6×10^{-6} **8a.** 1.8 **8b.** 2.66 **9a.** 2.29 **9b.** 11.28 **10a.** $[H_3O^+] = [HOOCCH_2COO^-] = 3.7 \times 10^{-2}$ M. $[^-OOCCH_2COO^-] = 2.0 \times 10^{-6}$ M.
10b. $K_{a_1} = 5.3 \times 10^{-2}$, $K_{a_2} = 5.3 \times 10^{-5}$.
11a. $[SO_4^{2-}] = 0.010$M, $[H_3O^+] = 0.21$ M, $[HSO_4^-] = 0.19$ M.
11b. $[HSO_4^-] = 0.014$M, $[H_3O^+] = 0.026$ M, $[SO_4^{2-}] = 0.0060$ M.
12a. (a) acidic; (b) neutral; (c) basic.
12b. $H_2PO_4^-(aq) + H_2O(l) \rightleftharpoons H_3O^+(aq) + HPO_4^{2-}(aq)$ $K_{a_2} = 6.3 \times 10^{-8}$
$H_2PO_4^-(aq) + H_2O(l) \rightleftharpoons OH^-(aq) + H_3PO_4(aq)$ $K_b = 1.4 \times 10^{-12}$
13a. Codeine hydrochloride **13b.** Basic **14a.** 8.08 **14b.** 3.6×10^{-3} M **15a.** $HClO_4$ and CH_2FCOOH are the stronger acids in the pairs. **15b.** H_2SO_3 and CCl_2FCH_2COOH are the stronger acids in the pairs. **16a.** (a) BF_3 is a Lewis acid and NH_3 is a Lewis base. (b) H_2O is a Lewis base and Cr^{3+} is a Lewis acid. **16b.** Hydroxide ion and the chloride ion are the Lewis bases. $Al(OH)_3$ and $SnCl_4$ are the Lewis acids.
Integrative Example A. pH = 5.68 ≈ 5.7 **B.** (a) Determine values of m and n in the formula $EO_m(OH)_n$. HOCl or Cl(OH) has $m = 0$ and $n = 1$. We expect $K_a \approx 10^{-7}$ or $pK_a \approx 7$, which is in good agreement with the accepted $pK_a = 7.52$. HOClO or ClO(OH) has $m = 1$ and $n = 1$. We expect $K_a \approx 10^{-2}$ or $pK_a \approx 2$, in good agreement with the $pK_a = 1.92$. $HOClO_2$ or $ClO_2(OH)$ has $m = 2$ and $n = 1$. We expect K_a to be large and in good agreement with the accepted value of $pK_a = -3$, $K_a = 10^{-pKa} = 10^3$. $HOClO_3$ or $ClO_3(OH)$ has $m = 3$ and $n = 1$. We expect K_a to be very large and in good agreement with the accepted $K_a = -8$, $pK_a = -8$, $K_a = 10^{-pKa} = 10^8$ which turns out to be the case. (b) $K_a = 10^{-2}$. (c)

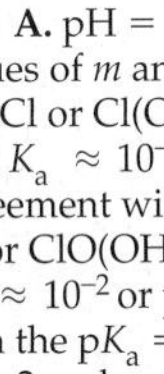

Exercises 1a. Acid **1b.** Base **1c.** Base **1d.** Acid **1e.** Acid
3a.
$HOBr(aq) + H_2O(l) \rightleftharpoons H_3O^+(aq) + OBr^-(aq)$
acid base acid base
3b.
$HSO_4^-(aq) + H_2O(l) \rightleftharpoons H_3O^+(aq) + SO_4^{2-}(aq)$
acid base acid base
3c. $HS^-(aq) + H_2O(l) \rightleftharpoons H_2S(aq) + OH^-(aq)$
base acid acid base
3d.
$C_6H_5NH_3^+(aq) + OH^-(aq) \rightleftharpoons C_6H_5NH_2(aq) + H_2O(l)$
acid base base acid
5. Answer (b). **7a.** Forward **7b.** Reverse **7c.** Reverse **9a.** $[H_3O^+] = 0.00165$ M, $[OH^-] = 6.1 \times 10^{-12}$ M.
9b. $[OH^-] = 0.0087$ M, $[H_3O^+] = 1.1 \times 10^{-12}$ M.
9c. $[OH^-] = 0.00426$ M, $[H_3O^+] = 2.3 \times 10^{-12}$ M.
9d. $[H_3O^+] = 5.8 \times 10^{-4}$ M, $[OH^-] = 1.7 \times 10^{-11}$ M.
11. $[H_3O^+] = 4.0 \times 10^{-14}$ M, pH = 13.40.

13. 1.96 × 10^{-3} M **15.** 8.5 mL **17.** 53.9 mL acid **19.** 10.20 **21.** $[H_3O^+]$ = 0.0098 M, pH = 2.01. **23a.** 0.0030 M **23b.** 2.62 **25.** 2.7 × 10^{-3} **27.** 1.4 g C_6H_5COOH **29.** $[H_3O^+]$ = 0.073 M, pH = 1.14, pOH = 12.86, $[OH^-]$ = 1.4 × 10^{-13} M. **31.** 11.04 **33.** 0.063 g vinegar **35a.** 11.23 **35b.** 34 mg of NaOH **37.** Sketch (b) **39a.** 0.0053 **39b.** 0.53% **41.** 0.0098 M **43.** We would not expect these ionizations to be correct because the calculated degree of ionization is based on the assumption that the $[CH_3COOH]_{initial}$ ≈ $[CH_3COOH]_{initial}$ - $[CH_3COOH]_{equil.}$, which is invalid at the 13 and 42 percent levels of ionization seen here. **45.** Because H_3PO_4 is a weak acid, there is little HPO_4^{2-} (produced in the 2nd ionization) compared to the H_3O^+ (produced in the 1st ionization). In turn, there is little PO_4^{3-} (produced in the 3rd ionization) compared to the HPO_4^{2-}, and very little compared to the H_3O^+. **47a.** $[H_3O^+] = 8.7 \times 10^{-5}$ M, $[HS^-]$ = 8.7 × 10^{-5} M, $[S^{2-}] = 1 \times 10^{-19}$ M. **47b.** $[H_3O^+] = 2.2 \times 10^{-5}$ M, $[HS^-]$ = 2.2 × 10^{-5} M, $[S^{2-}] = 1 \times 10^{-19}$ M. **47c.** $[H_3O^+] = 9.5 \times 10^{-7}$ M, $[HS^-]$ = 9.5 × 10^{-7} M, $[S^{2-}] = 1 \times 10^{-19}$ M. **49a.** $[SO_4^{2-}]$ = 0.011 M, $[HSO_4^-]$ = 0.74 M, $[H_3O^+]$ = 0.76 M. **49b.** $[SO_4^{2-}]$ = 0.0087 M, $[HSO_4^-]$ = 0.066 M, $[H_3O^+]$ = 0.084 M. **49c.** $[SO_4^{2-}] = 6.6 \times 10^{-4}$ M, $[HSO_4^-]$ = 9 × 10^{-5} M, $[H_3O^+] = 1.41 \times 10^{-3}$ M. **51a.** First ionization:

$$C_{20}H_{24}O_2N_2 + H_2O \rightleftharpoons C_{20}H_{24}O_2N_2H^+ + OH^- \qquad pK_{b_1} = 6.0$$

Second ionization:

$$C_{20}H_{24}O_2N_2H^+ + H_2O \rightleftharpoons C_{20}H_{24}O_2N_2H_2^{2+} + OH^- \quad pK_{b_2} = 9.8$$

53. 9.6 **65a.** $HClO_3$ **65b.** HNO_2 **65c.** H_3PO_4 **67a.** HI **67b.** HOClO **67c.** $H_3CCH_2CCl_2COOH$ **69.** The largest K_b belongs to (c) $CH_3CH_2CH_2NH_2$. The smallest K_b is that of (a) o-chloroaniline. **71a.** acid is CO_2 and base is H_2O.

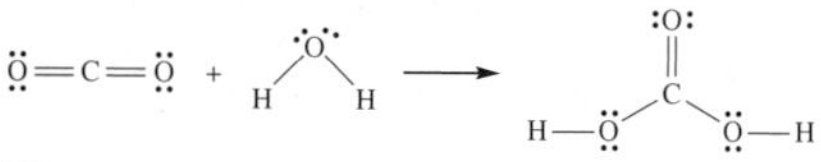

71b.

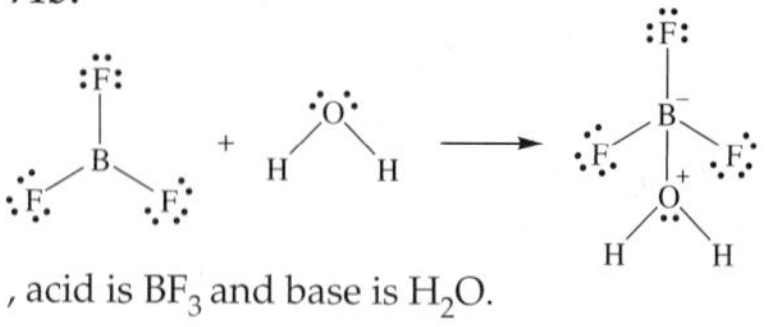

, acid is BF_3 and base is H_2O.

71c.

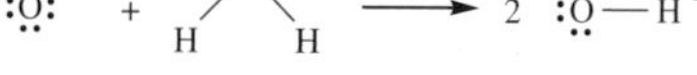

, acid is H_2O and base is O^{2-}.

71d.

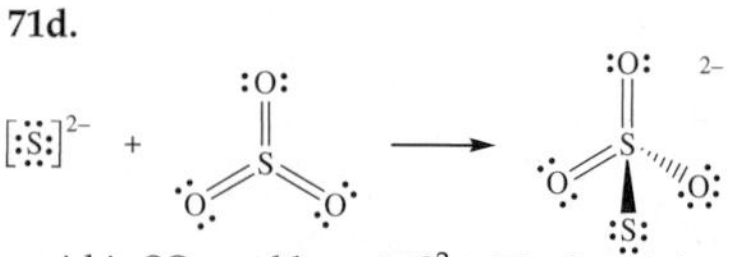

, acid is SO_3 and base is S^{2-}. **73a.** Lewis base **73b.** Lewis acid **73c.** Lewis acid

75a.

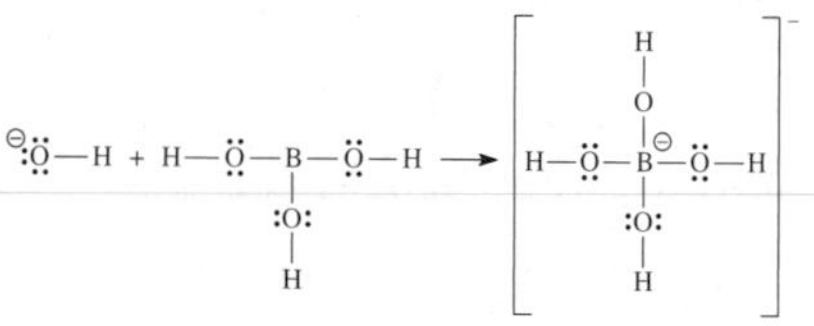

, base is OH^-, acid is $B(OH)_3$.

75b.

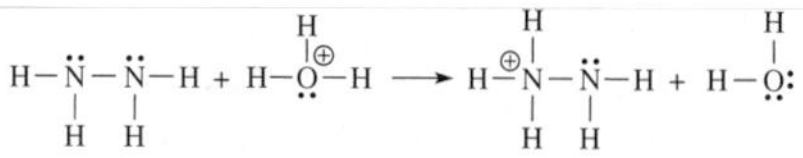

The actual Lewis acid is H^+, which is supplied by H_3O^+

75c.

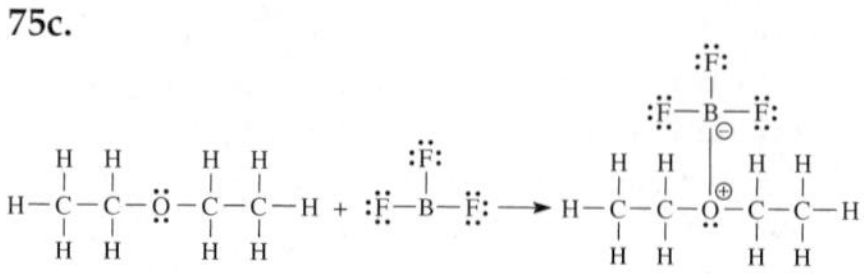

Base: electron pair donor Acid: electron pair acceptor

77.

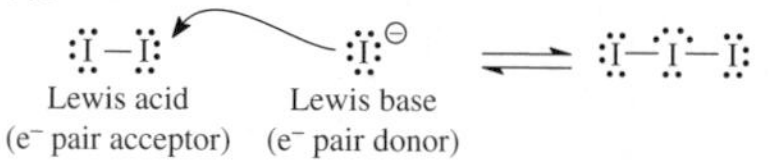

79.

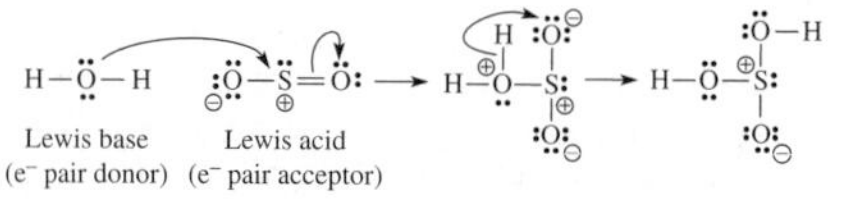

Integrative and Advanced Exercises

81a. Base **81b.** Base **81c.** Acid or Base **81d.** Acid **82.** 2.10 × 10^{-3} M **83a.** Not matched. **83b.** pH 4.6 matched. **83c.** Not matched. **83d.** Not matched. **83e.** pH 10.8 matched. **83f.** Not matched. **83g.** Not matched. **83h.** pH 2.1 matched. **83i.** Not matched. **87.** 5 × 10^{-5} **88.** 0.096 g NH_4Cl. Other solutes don't provide acidic solutions or lead to masses or volumes that are very difficult to measure. **94.** 9.73 g CH_3COOH **95.** 0.16 **98.**

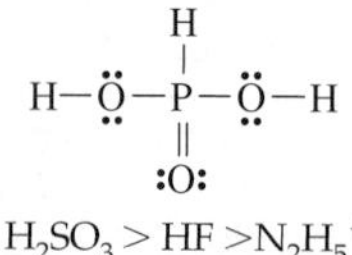

99a. $H_2SO_3 > HF > N_2H_5^+ > CH_3NH_3^+ > H_2O$ **99b.** $OH^- > CH_3NH_2 > N_2H_4 > F^- > HSO_3^-$ **99c.** (i) to the right and (ii) to the left. **100a.** ~ 1 M **100b.** 10^{-16} M **103a.** $CH_3OH < C_6H_5OH < CH_3COOH$ **103b.** HCCH < HCN < HOCN < HOClO

Feature Problems **107a.** 2.15 **107b.** 11.98 **107c.** 9.23

Self-Assessment Exercises **111.** The answer is (a). **112.** The answer is (c). **113.** The answer is (d). **114.** The answer is (e). **115.** The answer is (c). **116.** The answer is (b). **117.** 2.70 **118.** 37.2 mL **119.** 8.58 **120.** 0.10 M HI < 0.05 M H_2SO_4 < 0.05 M $HC_2H_2ClO_2$ < 0.05 M $HC_2H_3O_2$ < 0.05 M NH_4Cl 1.0 M NaBr < 0.05 M $KC_2H_3O_2$ < 0.05 M NH_3 < 0.06 M NaOH < 0.05 M $Ba(OH)_2$ **121.** The solution is basic. $[H_3O^+]$ = 2.14 × 10^{-12} M. $[NH_3]$ = 1.2 M. **122.** The answer is (a). **123.** The answer is (b). **124.** The answer is (d). **125.** 8.2

CHAPTER 17

Practice Examples **1a.** $[H_3O^+]$ = 0.018 M, [HF] = 0.482 M. $[H_3O^+]$ = 0.103 M, [HF] = 0.497 M. **1b.** 30 drops **2a.** $[H_3O^+] = 1.2 \times 10^{-4}$ M, [HCOO] = 0.150 M. **2b.** 15 g $NaCH_3COO$ **3a.** The hydronium ion and the acetate ion react to form acetic acid. All that is necessary to form a buffer is to have approximately equal amounts of a weak acid and its conjugate base together in solution. This will be achieved if we add an amount of HCl equal to approximately half the original amount of acetate ion. **3b.** HCl dissociates essentially completely in water and serves as a source of hydronium ion. This reacts with ammonia to form ammonium ion. Because a buffer contains approximately equal amounts of a weak base (NH_3) and its conjugate acid (NH_4^+), to prepare a buffer we simply add an amount of HCl equal to approximately half the amount of NH_3(aq) initially present. **4a.** 3.94 **4b.** 4.51 **5a.** 21 g **5b.** pH = 5.09 ≈ 5.1 **6a.** (a) 3.53; (b) 3.53; (c) 3.55. **6b.** 1.3 mL **7a.** (a) 0.824; (b) 1.238; (c) 7.00; (d) 11.785. **7b.** (a) 12.21; (b) 11.79; (c) 7.00. **8a.** (a) 2.02; (b) 2.70; (c) 3.18; (d) 8.08. **8b.** (a) 11.15; (b) 9.74; (c) 9.26; (d) 5.20. **9a.** 12.18 **9b.** 10.45

Integrative Example **A.** –0.283 °C **B.** 80 g/mol, $K_b = 1 \times 10^{-6}$.

Exercises **1a.** 0.0892M **1b.** 1.1 × 10^{-13} M **1c.** 4.0 × 10^{-5} M **1d.** 0.0892 M **3a.** 1.05 **3b.** No change. The pH changes are not the same because there is an equilibrium system to be shifted in the first solution, whereas there is no equilibrium, just a change in total ionic strength, for the second solution. **5a.** 0.035 M **5b.** 4.0 × 10^{-4} M **5c.** 3.9 × 10^{-5} M **7.** 0.77 M **9a.** 4.64 **9b.** 9.68 **11a.** Not a buffer. **11b.** Not a buffer. **11c.** Buffer **11d.** Not a buffer. **11e.** Buffer **11f.** Not a buffer. **13.** 8.88 **15a.** 4.8 g $(NH_4)_2SO_4$ **15b.** 0.41 g $(NH_4)_2SO_4$ **17.** 9.16 **19.** Use HHCOO and NaHCOO to prepare a buffer with pH = 3.50. V_{acid} = 100. mL, V_{salt} = 58 mL. **21a.** pH = 3.89 to pH = 5.89. **21b.** 0.100 mol of acid or base per liter of buffer solution. **23a.** 3.48 **23b.** 3.52 **23c.** 1.23

25a. $$\text{pH} = pK_a + \log\frac{[NH_3]}{[NH_4^+]} = 9.26 + \log\frac{0.0720\text{ M}}{0.128\text{ M}} = 9.01 \approx 9.00$$

25b. The Henderson-Hasselbalch equation depends on the assumption that: $[NH_3] \gg 1.8 \times 10^{-5}$ M $\ll [NH_4^+]$. If the solution is diluted to 1.00 L, $[NH_3] = 7.20 \times 10^{-3}$ M, and $[NH_4^+] = 1.28 \times 10^{-2}$ M. These concentrations are consistent with the assumption. However, if the solution is diluted to 1000. L, $[NH_3] = 7.2 \times 10^{-6}$ M, and $[NH_3]$ = 1.28 × 10^{-5} M, and these two concentrations are not consistent with the assumption. Thus, in 1000. L of solution, the given quantities of NH_3 and NH_4^+ will not produce a solution with pH = 9.00. With sufficient dilution, the solution will become indistinguishable from pure water (i.e., its pH will equal 7.00). **25c.** 8.99 **25d.** 1.1mL 1.00 M HCl **27a.** Bromphenol blue changes color in acidic solution. Bromcresol green changes color in acidic solution. Bromthymol changes color in neutral solution. 2,4-Dinitrophenol changes color in acidic solution. Chlorophenol red changes color in acidic solution. Thymolphthalein changes color in basic solution. **27b.** If bromcresol green is green, the pH is probably about pH = 4.7. If chlorophenol red is orange, the pH is probably about pH = 6.0. **29a.** In an acid–base titration, the pH of the solution changes sharply at a definite pH that is known prior to titration. Determining the pH of a solution is more difficult because the pH of the solution is not known precisely in advance. Since each indicator only serves to fix the pH over a quite small region, often less than 2.0 pH units, several indicators—carefully chosen to span the entire range of 14 pH units—must be employed to narrow the pH to ± pH unit or possibly lower. **29b.** An indicator is, after all, a weak acid. Its addition to a solution will affect the acidity of that solution. Thus, one adds only enough indicator to show a color change and not enough to affect solution acidity. **31a.** Red **31b.** Yellow **31c.** Yellow **31d.** Yellow **31e.** Red **31f.** Yellow **33.** The approximate pK_{HIn} = 3.84. This is a relatively good indicator. **35.** 0.1508 M **37.** 11.63 **39a.** 1.346 **39b.** 2.034 **41a.** 2.87 **41b.** 3.51 **43.** The volume of titrant needed to reach the equivalence point is determined by the volume and concentration of the HA solution and the concentration of the NaOH solution, irrespective of whether HA is a strong acid or a weak acid. The pH at the equivalence point, however, is determined by the nature of the salt NaA, and especially by the nature of A^-. If HA is a strong acid, then A^- is a very weak base that will not hydrolyze in water. Consequently, for the titration of a strong acid with NaOH, the solution will be neutral at the equivalence point. If HA is a weak acid, then A^- is a weak base that will hydrolyze to produce OH^- ions. Therefore, for the titration of a weak acid with NaOH, the solution will be basic at the equivalence point.

45a.

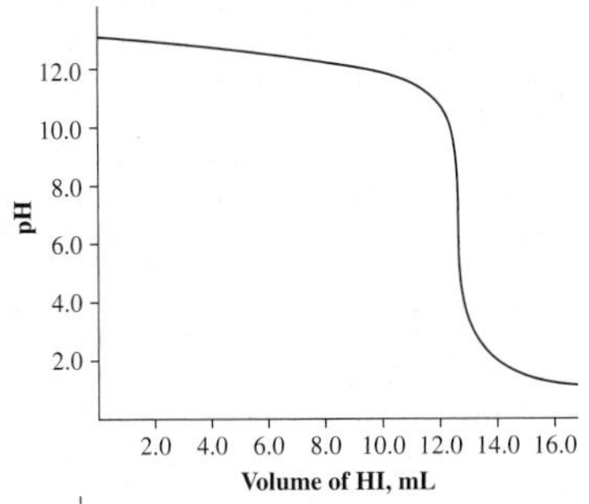

45b.

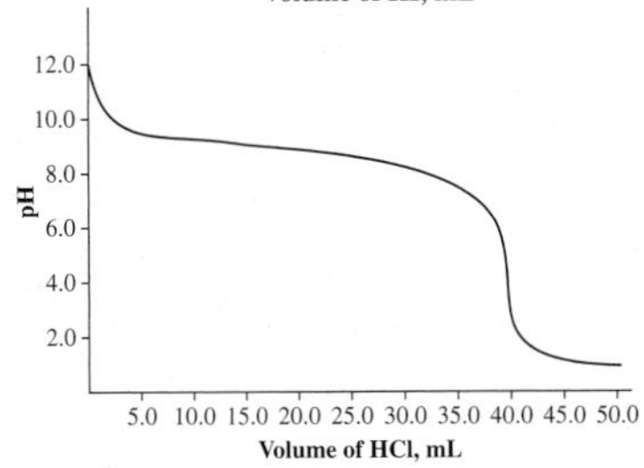

47a. 11.9 mL acid
47b. 17.4 mL acid
47c. 17.5 mL acid
49a.

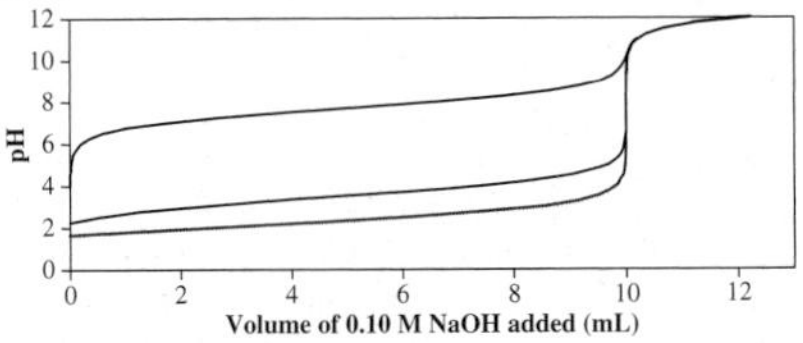

bromthymol blue
49b. See sketch in 49a. Thymol blue. **49c.** See sketch in 49a. Alizarin yellow.
51.

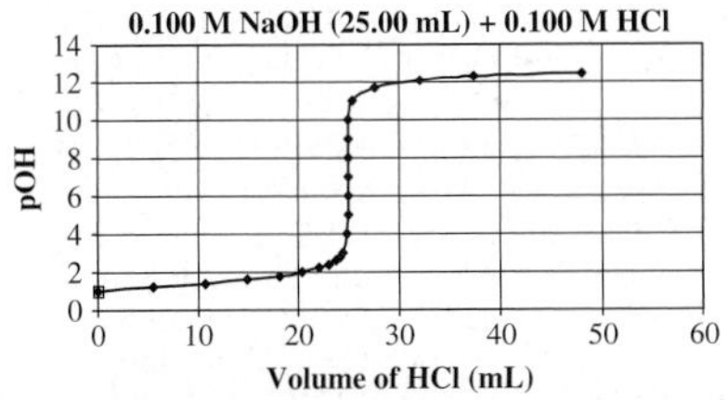

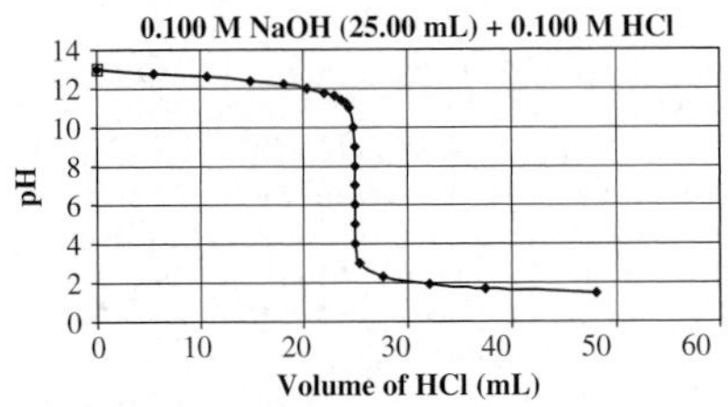

53. Basic. This alkalinity is created by the hydrolysis of the sulfide ion, the anion of a very weak acid.
55a.

$H_3PO_4(aq) + CO_3^{2-}(aq) \rightleftharpoons H_2PO_4^-(aq) + HCO_3^-(aq)$
$H_2PO_4^-(aq) + CO_3^{2-}(aq) \rightleftharpoons HPO_4^{2-}(aq) + HCO_3^-(aq)$
$HPO_4^{2-}(aq) + OH^-(aq) \rightleftharpoons PO_4^{3-}(aq) + H_2O(1)$

55b. CO_3^{2-} is not a strong enough base to remove the third proton from H_3PO_4.
57. $K_{a_1} = 1.6 \times 10^{-3}, K_{a_1} = 1.0 \times 10^{-8}$.
59a. 0.0038 M **59b.** 0.49 M **61a.** No **61b.** No **61c.** Yes **61d.** No **61e.** Yes **61f.** No

Integrative and Advanced Exercises
63a. $HSO_4^-(aq) + OH^-(aq) \longrightarrow SO_4^{2-}(aq) + H_2O(l)$
63b. 2.8% **63c.** bromthymol blue or phenol red. **64a.** 27.8 mL **64b.** 3.3 mL **64c.** 250 mL **66a.** No **66b.** 18% **69a.** $pH = pK_a + \log(f/(1-f))$ **69b.** 9.56 **70a.** 1.6 **70b.** 6.3 g KH_2PO_4, 26 g $Na_2HPO_4 \cdot 12\,H_2O$. **71.** ~3 drops NH_3 **73a.** Equation (1): $K = 1.00 \times 10^{14}$. Equation (2): $K = 1.8 \times 10^9$. **73b.** The extremely large size of each equilibrium constant indicates that each reaction goes essentially to completion.
77a.

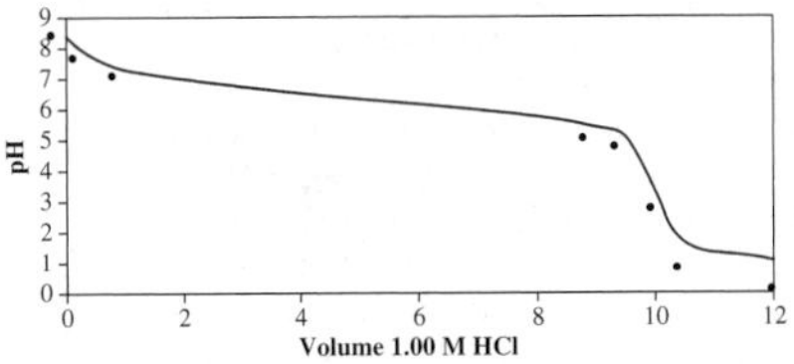

77b.

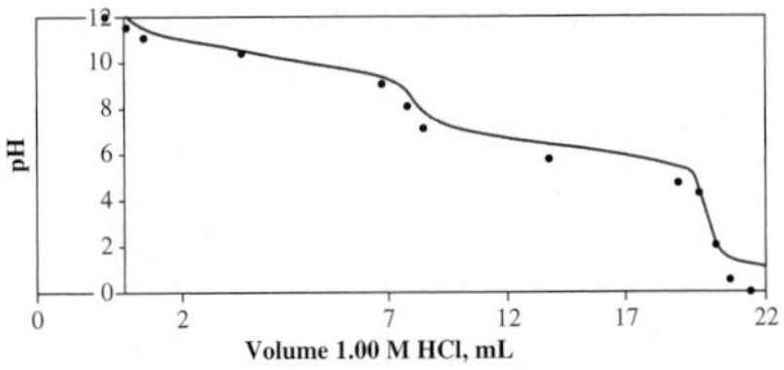

77c. 119 mL **77d.** 189 mL **77e.** 8.3%
81a. A buffer solution is able to react with small amounts of added acid or base. When strong acid is added, it reacts with formate ion.
$HCOO^-(aq) + H_3O^+(aq) \longrightarrow HHCOO(aq) + H_2O$
Added strong base reacts with acetic acid.
$HCH_3COO(aq) + OH^-(aq) \longrightarrow H_2O(1) + CH_3COO^-(aq)$
Therefore neither added strong acid nor added strong base alters the pH of the solution very much. Mixtures of this type are referred to as buffer solutions.
81b. 4.40 **81c.** 3.8 **82a.** Bromthymol blue or phenol red. **82b.** The $H_2PO_4^-/HPO_4^{2-}$ system. **82c.** 5.4 **83.** 12.17 **84.** 8.2

89a. $[H_3O^+] = \sqrt{\dfrac{K_1[CO_2(g)][Ca^{2+}]}{K_2 \cdot K_3 \cdot K_4}}$

89b. 7.57 **91a.** 4.6×10^{-3} **91b.** 4.78
91c.

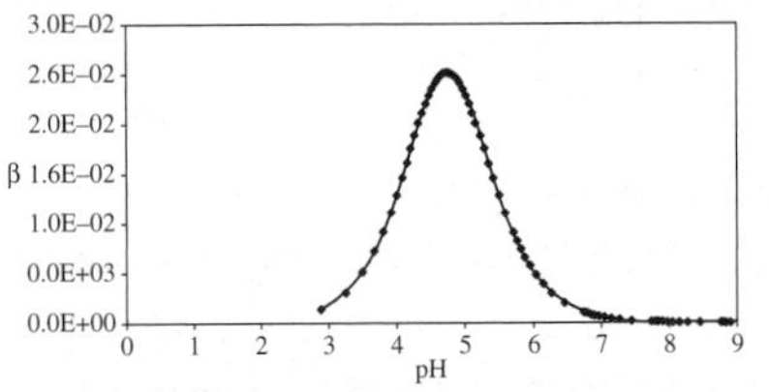

Feature Problems 92a. The two curves cross the point at which half of the total acetate is present as acetic acid and half is present as acetate ion. This is the half equivalence point in a titration, where $pH = pK_a = 4.74$.
92b.

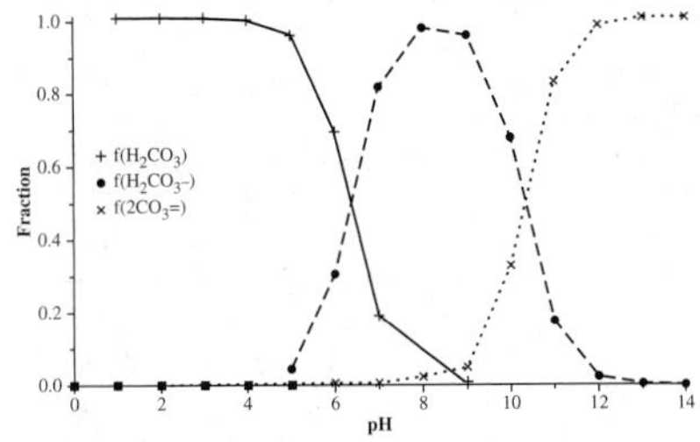

92c.

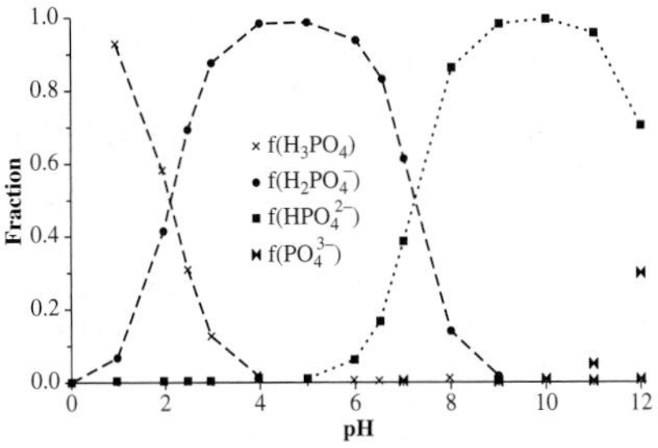

Self-Assessment Exercises
98a. $HHCOO + OH^- \rightarrow HCOO^- + H_2O$, $HCOO^- + H_3O^+ \rightarrow HHCOO + H_2$
98b. $C_6H_5NH_3^+ + OH^- \rightarrow C_6H_5NH_2 + H_2O$, $C_6H_5NH_2 + H_3O^+ \rightarrow C_6H_5NH_3^+ + H_2O$
98c. $H_2PO_4^- + OH^- \rightarrow HPO_4^{2-} + H_2O$, $HPO_4^{2-} + H_3O^+ \rightarrow H_2PO_4^- + H_2O$
99a. The pH at the equivalence point is 7. Use bromthymol blue.

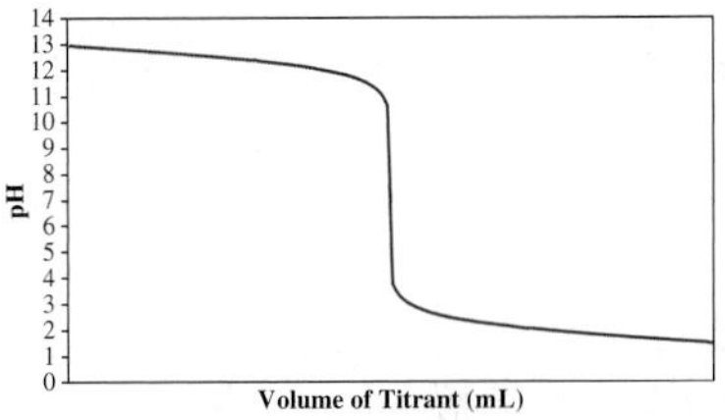

99b. The pH at the equivalence point is ~ 5.3 for a 0.1 M solution. Use methyl red.

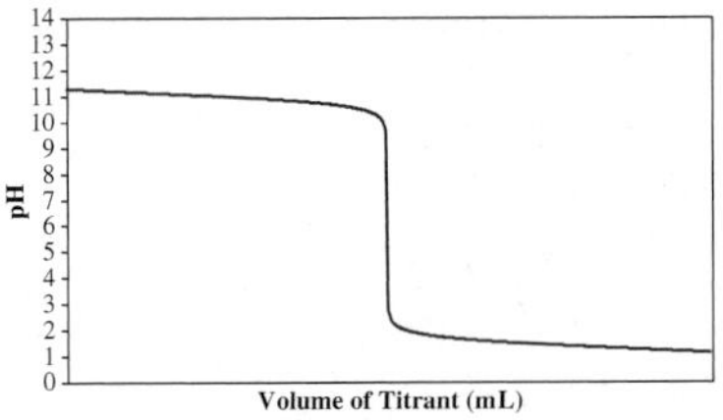

99c. The pH at the equivalence point is ~ 8.7 for a 0.1 M solution. Use phenolphthalein.

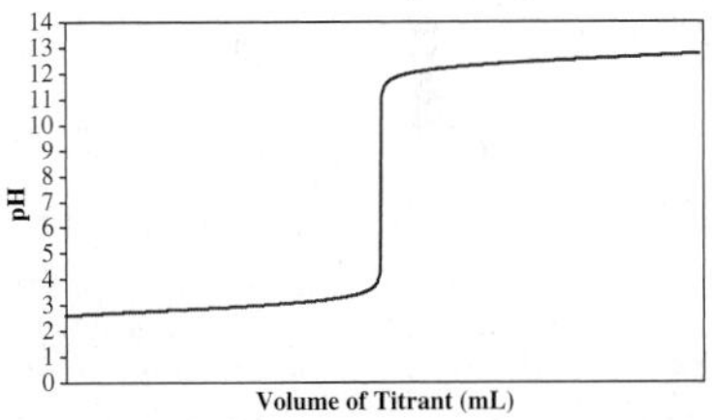

99d. The pH for the first equivalence point ($NaH_2PO_4^-$ to $Na_2HPO_4^{2-}$) for a 0.1 M solution is right around ~ 7. Use bromthymol blue.

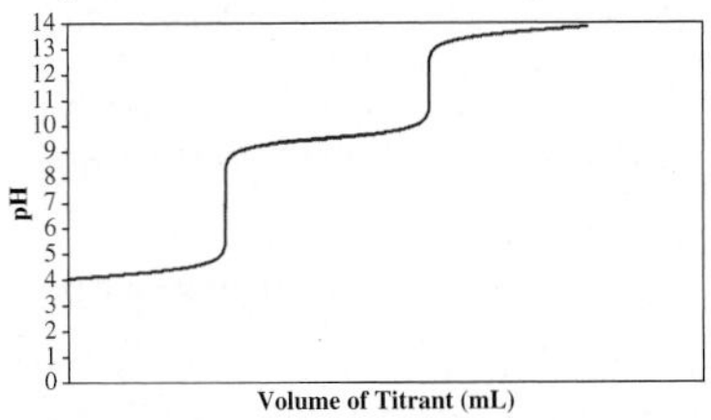

100a. 3.12 **100b.** 4.20 **100c.** 8.00 **100d.** 11.1 **101.** The answer is (c). **102.** The answer is (d). **103.** The answer is (b). **104.** The answer is (b). **105.** 10.32 **106.** 8.75 **107.** The answer is (a). **108.** The answer is (b). **109.** The answer is (b). **110a.** pH > 7 **110b.** pH < 7 **110c.** pH = 7

CHAPTER 18

Practice Examples 1a. (a) $K_{sp} = [Mg^{2+}][CO_3^{2-}]$; (b) $K_{sp} = [Ag^+]^3[PO_4^{3-}]$.
1b. (a) $CaHPO_4(s) \rightleftharpoons Ca^{2+}(aq) + HPO_4^{2-}(aq)$; (b) $K_{sp} = [Ca^{2+}][HPO_4^{2-}]$.

2a. 3×10^{-7} **2b.** 1.9×10^{-9} **3a.** 3.7×10^{-8} M **3b.** 0.55 mg **4a.** 1.3×10^{-4} M **4b.** 9.8×10^{-21} M **5a.** Yes **5b.** 10 drops **6a.** No **6b.** 0.02 M **7a.** 0.12% **7b.** Chromate. $[Ba^{2+}] = 5.5 \times 10^{-7}$ M. **8a.** No **8b.** Yes **9a.** 0.074 M **9b.** 8.7×10^{-3} M **10a.** (a) $Cu^{2+}(aq) + 2\,OH^-(aq) \rightarrow Cu(OH)_2(s)$; (b) $Cu(OH)_2(s) + 4\,NH_3(aq) \rightleftharpoons [Cu(NH_3)_4]^{2+}(aq) + 2\,OH^-(aq)$; (c) $Cu(OH)_2(s) + 4\,NH_3(aq) \rightleftharpoons [Cu(NH_3)_4]^{2+}(aq) + 2\,OH^-(aq)$, $Cu(OH)_2(s) \rightleftharpoons Cu^{2+}(aq) + 2\,OH^-(aq)$. **10b.** (a) $Zn^{2+}(aq) + 4\,NH_3(aq) \rightleftharpoons [Zn(NH_3)_4]^{2+}(aq)$; (b) $[Zn(NH_3)_4]^{2+}(aq) + 4H_3O^+(aq) \rightleftharpoons [Zn(H_2O)_4]^{2+}(aq) + 4NH_4^+(aq)$; (c) $[Zn(H_2O)_4]^{2+}(aq) + 2OH^-(aq) \rightleftharpoons Zn(OH)_2(s) + 4H_2O(l)$; (d) $Zn(OH)_2(s) + 2OH^-(aq) \rightleftharpoons [Zn(OH)_4]^{2-}(aq)$. **11a.** Yes **11b.** No **12a.** 0.58 M **12b.** 0.20 M **13a.** 4×10^{-6} M **13b.** In Example 18-13 the expression for the solubility, *s*, of a silver halide in an aqueous ammonia solution, where $[NH_3]$ is the concentration of aqueous ammonia, was given by:

$\sqrt{K_{sp} \times K_f} = \dfrac{s}{[NH_3] - 2s}$. For all scenarios, the $[NH_3]$ stays fixed at 0.1000 M and K_f is always 1.6×10^7. We see that *s* will decrease as does K_{sp}. **14a.** For FeS, $Q_{spa} = 0.022 < 6 \times 10^2 = K_{spa}$, for Ag_2S, $Q_{spa} = 1.1 \times 10^{-4} > 6 \times 10^{-30} = K_{spa}$. **14b.** 2.7

Integrative Example **A.** 87×10^6 g $Ca(NO_3)_2$ **B.** Ag_2SO_4. No precipitate is remaining.

Exercises **1a.** $K_{sp} = [Ag^+]^2[SO_4^{2-}]$
1b. $K_{sp} = [Ra^{2+}][IO_3^-]^2$
1c. $K_{sp} = [Ni^{2+}]^3[PO_4^{3-}]^2$
1d. $K_{sp} = [PuO_2^{2+}][CO_3^{2-}]$
3a. $K_{sp} = [Cr^{3+}][F^-]^3 = 6.6 \times 10^{-11}$
3b. $K_{sp} = [Au^{3+}]^2[C_2O_4^{2-}]^3 = 1 \times 10^{-10}$
3c. $K_{sp} = [Cd^{2+}]^3[PO_4^{3-}]^2 = 2.1 \times 10^{-33}$
3d. $K_{sp} = [Sr^{2+}][F^-]^2 = 2.5 \times 10^{-9}$

5. $AgI < AgCN < AgIO_3 < Ag_2SO_4 < AgNO_2$. **7.** Yes **9.** 11 mg **11.** 4.5×10^{-3} M **13.** 5.30×10^{-5} **15a.** 1.7×10^{-4} M **15b.** 7.3×10^{-6} M **15c.** 1.4×10^{-8} M **17.** The presence of KI in a solution produces a significant $[I^-]$ in the solution. Not as much AgI can dissolve in such a solution as in pure water, since the ion product, $[Ag^+][I^-]$, cannot exceed the value of K_{sp}. In similar fashion, $AgNO_3$ produces a significant $[Ag^+]$ in solution, again influencing the value of the ion product; not as much AgI can dissolve as in pure water. **19.** 1.5×10^{-5} **21.** 0.079 M **23.** The solubility of Ag_2CrO_4 cannot be lowered to 5.0×10^{-8} M with CrO_4^{2-} as the common ion. The solubility can be lowered to 5.0×10^{-8} M by carefully adding $Ag^+(aq)$. **25.** 27 ppm **27.** Yes **29.** pH > 8.32 **31a.** Yes **31b.** Yes **31c.** No **33.** CaC_2O_4 should precipitate. **35.** 0.052% **37.** 2.5%. $[Cl^-] = 0.16$ M. **39a.** No **39b.** Yes **41a.** AgI **41b.** 2.7×10^{-4} M **41c.** $[Ag^+] = 3.1 \times 10^{-13}$ M. **41d.** Yes **43a.** AgI(s) **43b.** $[I^-] = 4.7 \times 10^{-9}$ M. **43c.** Yes **45.** $MgCO_3(s)$, FeS (s), $Ca(OH)_2(s)$. **47.** 9.60 **49a.** 3.41 **49b.** 0.96 g **51.** Lead(II) ion forms a complex ion with chloride ion. It forms no such complex ion with nitrate ion. The formation of this complex ion decreases the concentrations of free $Pb^{2+}(aq)$ and free $Cl^-(aq)$. Thus, $PbCl_2$ will dissolve in the HCl(aq) up until the value of the solubility product is exceeded. **53.** 2.0×10^{30} **55.** Precipitation of AgI(s) should occur. **57.** 1.4×10^{-6} g **59.** If the $[H_3O^+]$ is maintained just a bit higher than 1.8×10^{-5} M, FeS will precipitate and $Mn^{2+}(aq)$ will remain in solution. Separation will be complete. **61a.** $Q_{spa} = 1.7 \times 10^7 < 3 \times 10^7 = K_{spa}$ for MnS. Precipitation will not occur. **61b.** $[C_2H_3O_2^-] = 0.21$ M. **63.** The most important consequence would be the absence of a valid test for the presence or absence of Pb^{2+}. In addition, if NH_3 is added first, $PbCl_2$ may form $Pb(OH)_2$. If $Pb(OH)_2$ does form, it will be present with Hg_2Cl_2 in the solid, although $Pb(OH)_2$ will not darken with added NH_3. Presence of Ag^+ may be falsely concluded. **65.** (a) and (c) are the valid conclusions.

Integrative and Advanced Exercises **67.** 68% **69.** 38% **71.** 0.012 M **74.** 2 g MnS/L **77a.** Yes **77b.** No **77c.** Yes **80.** 1.2×10^{-16} **82.** 1.6×10^{-3} M **83.** 6.5×10^{-4} M **85a.** $Ag_2SO_4(s) + Ba^{2+}(aq) + 2\,Cl^-(aq) \longrightarrow BaSO_4(s) + 2\,AgCl(s)$ **85b.** 0.875 g $BaSO_4$, 1.07 g AgCl, $Ag_2SO_4(s)$ = 0.63 g, mass dissolved $Ag_2SO_4 = 0.702$ g.

Feature Problems **87.** 3.0×10^{-4} **89a.** 1.7×10^{-4} M **89b.** 1.7×10^{-4} M **89c.** 2.8×10^{-5} M **89d.** 0.059 M **89e.** 1.4×10^{-9} M

Self-Assessment Exercises **93.** The answer is (d). **94.** The answer is (a). **95.** The answer is (c). **96.** The answer is (b). **97.** The answers are (c) and (d). **98.** The answer is (a). **99.** The answer is (c). **100a.** More soluble in basic solution. **100b.** More soluble in acidic solution. **100c.** More soluble in acidic solution. **100d.** Solubility is independent of pH. **100e.** Solubility is independent of pH. **100f.** More soluble in acidic solution. **101.** The answer is NH_3. **102.** Yes **103.** The answer is (b). **104.** The answer is (d). **105.** The answer is (b). **106.** It will decrease the amount of precipitate. **107.** No precipitate will form. **108.** $CuCO_3$ is soluble in $NH_3(aq)$ and CuS(s) is not.

CHAPTER 19

Practice Examples
1a. $Sc(s) + 3\,Ag^+(aq) \rightarrow Sc^{3+}(aq) + 3\,Ag(s)$
1b. Anode: $Al(s) \rightarrow Al^{3+}(aq) + 3\,e^-$;
Cathode: $Ag^+(aq) + e^- \rightarrow Ag(s)$;
Overall:
$Al(s) + 3\,Ag^+(aq) \rightarrow Al^{3+}(aq) + 3\,Ag(s)$;

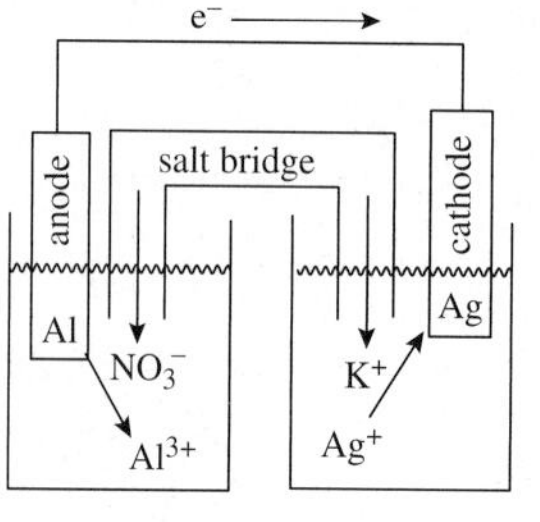

2a. Anode: $Sn(s) \rightarrow Sn^{2+}(aq) + 2\,e^-$;
Cathode: $Ag^+(aq) + 1\,e^- \rightarrow Ag(s)$;
Overall:
$Sn(s) + 2\,Ag^+(aq) \rightarrow Sn^{2+}(aq) + 2\,Ag(s)$.
2b. Anode: $In(s) \rightarrow In^{3+}(aq) + 3\,e^-$;
Cathode: $Cd^{2+}(aq) + 2\,e^- \rightarrow Cd(s)$;
Overall:
$2\,In(s) + 3\,Cd^{2+}(aq) \rightarrow 2\,In^{3+}(aq) + 3\,Cd(s)$.
3a. +0.587 V **3b.** +0.74 V **4a.** −0.48 V **4b.** −0.424V **5a.** −1587 kJ **5b.** +1.229 V **6a.** Silver ion is one example. $E^\circ_{cell} = +0.460$ V. **6b.** The method is not feasible because another reaction occurs that has a more positive cell potential, i.e., Na(s) reacts with water to form H_2(g) and NaOH(aq). **7a.** Not be a feasible method. **7b.** The standard cell potential for reaction (2) is more positive than that for situation (1). Thus, reaction (2) should occur preferentially. Also, if $Sn^{4+}(aq)$ is formed, it should react with Sn(s) to form $Sn^{2+}(aq)$ spontaneously. **8a.** The size of the equilibrium constant indicates that this reaction indeed will go to essentially 100% to completion. **8b.** The equilibrium constant's small size indicates that this reaction will not go to completion. **9a.** +1.815 V **9b.** +0.017 V **10a.** Yes **10b.** 0.0150 **11a.** +0.23 V **11b.** 6.9×10^{-9} **12a.** I_2 (s) and H_2 (g). **12b.** Silver metal Ag^+ (aq). **13a.** 1.89 amperes **13b.** 5.23 h

Integrative Example **A.** 0.137 V **B.** (a) Oxidation: $Al(s) + 4\,OH^-(aq) \rightarrow [Al(OH)_4]^- + 3\,e^-$, Reduction: $O_2(g) + 2\,H_2O(l) + 4\,e^- \rightarrow 4\,OH^-(aq)$, Overall: $4\,Al(s) + 4\,OH^-(aq) + 3\,O_2(g) + 6\,H_2O(l) \rightarrow 4[Al(OH)_4]^-(aq)$. (b) −2.329 V; (c) −1303 kJ/mol; (d) 13.4 g.

Exercises **1a.** 0.340 V < $E°$ < 0.800 V. **1b.** −0.440 V < $E°$ < 0.000 V. **3.** +0.755 V **5.** −2.31V **7a.** The largest positive cell potential will be obtained for the reaction involving the oxidation of Al(s) to $Al^{3+}(aq)$ and the reduction of $Ag^+(aq)$ to Ag(s), $E^\circ_{cell} = 2.476$ V. **7b.** The cell with the smallest positive cell potential will be obtained for the reaction involving the oxidation of Zn(s) to $Zn^{2+}(aq)$ (anode) and the reduction of $Cu^{+2}(aq)$ to Cu(s) (cathode), $E^\circ_{cell} = 1.103$ V. **9a.** Ni^{2+} **9b.** Cd **11a.** Spontaneous **11b.** Nonspontaneous **11c.** Nonspontaneous **11d.** Spontaneous **13a.** Significant extent. **13b.** Significant extent. **13c.** Large extent. **13d.** Significant extent. **13e.** Not to a significant extent. **15a.** Yes **15b.** Yes **15c.** No **17a.** $3\,Ag(s) + NO_3^-(aq) + 4\,H^+(aq) \rightarrow 3\,Ag^+(aq) + NO(g) + 2\,H_2O(l)$ **17b.** $Zn(s) + 2\,H^+(aq) \rightarrow Zn^{2+}(aq) + H(g)$ **17c.** Does not react. **19a.** $2\,Al(s) + 3\,Sn^{2+}(aq) \rightarrow 2\,Al^{3+}(aq) + 3\,Sn(s)$, $E^\circ_{cell} = +1.539$ V. **19b.** $Fe^{2+}(aq) + Ag^+(aq) \rightarrow Fe^{3+}(aq) + Ag(s)$, $E^\circ_{cell} = +0.029$ V. **19c.** $3\,Cr(s) + 2\,Au^{3+}(aq) \rightarrow 3\,Cr^{2+}(aq) + 2\,Au(s)$, $E^\circ_{cell} = 2.42$ V.
19d. $H_2O(l) \rightarrow H^+(aq) + OH^-(aq)$, $E^\circ_{cell} = -0.828$ V **23a.** $E^\circ_{cell} = -0.029$ V, nonsponatneous. **23b.** $E^\circ_{cell} = +0.291$V, spontaneous. **23c.** $E^\circ_{cell} = -0.435$ V, nonspontaneous. **23d.** $E^\circ_{cell} = -0.435$ V, nonspontaneous. **25a.** -1.165×10^3 kJ **25b.** −268 kJ **25c.** -3.1×10^2 kJ **27a.** −0.100 V **27b.** 48.24 kJ mol^{-1} **27c.** 3.5×10^{-9} **27d.** Since *K* is very small the reaction will not go to completion. **29a.** No **29b.** To the left towards reactants. **31.** 1.591V **33.** 0.923 V **35.** 5×10^{-6} M **37a.** 1.249 V **37b.** +0.638 V **39a.** The reactions for which *E* depends on pH are those that contain either H^+ (aq) or OH^- (aq) in the balanced half-equation. **39b.** H^+ (aq)will inevitably be on the left side of the reduction of an oxoanion because reduction is accompanied by not only a decrease in oxidation state, but also by the loss of oxygen atoms. These oxygen atoms appear on the right-hand side as H_2O molecules. The hydrogens that are added to the right-hand side with the water molecules are then balanced with H^+ (aq) on the left-hand side. **39c.** If a half-reaction with H^+ (aq) ions present is transferred to basic solution, it may be rebalanced by adding to each side OH^- (aq) ions equal in number to the H^+ (aq) originally present. This results in $H_2O(l)$ on the side that had H^+ (aq) ions (the left side in this case) and OH^- (aq) ions on the other side (the right side.) **41a.** 6×10^{-38} M **41b.** The reaction goes to completion. **43a.** +0.818 V **43b.** In the standard half-cell, [OH] = 1.00 M, while in the anode half-cell in the case at hand, [OH] = 0.65 M. The forward reaction (dilution of H^+) should be even more spontaneous, (i.e., a more positive voltage will be created), with 1.00 M KOH than with 0.65 M KOH. We expect that E°_{cell} (1.000 M NaOH) should be a little larger than E°_{cell} (0.65 M NaOH), which, is in fact, the case. **45.** 0.119 V **47a.** 0.039 V **47b.** Decrease

47c. 0.025 V **47d.** 0.237 M **47e.** $[Sn^{2+}] = 0.50$ M, $[Pb^{2+}] = 0.18$ M. **49.** $E^\circ_{cell} = -0.03$ V, therefore the reaction will not be spontaneous under standard conditions. $Cl_2(g)$ could be obtained by driving it to the product side with an external voltage or, since E°_{cell} is only slightly negative, by removing products as they are formed and replenishing reactants as they are consumed. **51a.** Cell diagram: Cr (s) | Cr^{2+} (aq), Cr^{3+} (aq)||Fe^{2+} (aq), Fe^{3+} (aq) | Fe (s). **51b.** +1.195 V **53a.** 1.229 V **53b.** +1.992 V **53c.** +2.891 V **55.** Calculate the quantity of charge transferred when 1.00 g of metal is consumed in each cell. The aluminum-air cell has the highest value (1.07×10^4 C). **57.** Cell diagram: Li(s),Li^+(aq)|KOH(satd)|MnO_2(s),Mn(Oh)$_3$(s)|, $E^\circ_{cell} = 2.84$ V. **59a.** Oxidation of iron metal to Fe^{2+}(aq) should be enhanced in the body of the nail (blue precipitate), and hydroxide ion should be produced in the vicinity of the copper wire (pink color), which serves as the cathode. **59b.** A scratch tears the iron and exposes "fresh" metal, which makes it is more susceptible to corrosion. We expect blue precipitate in the vicinity of the scratch. **59c.** Zinc should protect the iron nail from corrosion. There should be almost no blue precipitate; the zinc corrodes instead. The pink color of OH should continue to form. **61.** One way to retard oxidation of the metal would be to convert it into a cathode. Once transformed into a cathode, the metal would develop a positive charge and no longer release electrons (or oxidize). This change in polarity can be accomplished by hooking up the metal to an inert electrode in the ground and then applying a voltage across the two metals in such a way that the inert electrode becomes the anode and the metal that needs protecting becomes the cathode. This way, any oxidation that occurs will take place at the negatively charged inert electrode rather than the positively charged metal electrode. **63a.** 3.3 g Zn **63b.** 0.90 g Al **63c.** 11 g Ag **63d.** 2.9 g Ni **65a.** Requires electrolysis with an applied voltage greater than +1.229V. **65b.** Spontaneous **65c.** Requires electrolysis with an applied voltage greater than +0.236V. **65d.** Requires electrolysis with an applied voltage greater than +0.183V. **67a.** H_2 (g) and O_2 (g). **67b.** $2\,H_2O(l) \rightarrow 2\,H_2\,(g) + O_2\,(g)$, $E^\circ_{cell} = -1.229$ V. **69a.** 1.62 g Zn **69b.** 20.2 min **71.** 1079 C **71b.** 0.7642 A **73a.** Copper **73b.** 37%

Integrative and Advanced Exercises

75. −0.255 V **77a.** 9.38×10^6 kJ **77b.** 2.61×10^3 kWh **80.** 6.9×10^{-4} M **82.** + 127.3 kJ **87.** 1.146 V **90.** 0.051% **92.** The standard hydrogen electrode is the anode. E = 0.223 V. **96a.** Aluminum **96b.** $2\,Al(s) + 3\,Ag_2S(s) \longrightarrow 6\,Ag(s) + 2\,Al^{3+}\,(aq) + 3\,S^{2-}\,(aq)$ **96c.** The dissolved $NaHCO_3$(s) serves as an electrolyte. It would also enhance the electrical contact between the two objects. **96d.** There are several chemicals involved: Al, H_2O, and $NaHCO_3$. Although the aluminum plate will be consumed very slowly because silver tarnish is very thin, it will, nonetheless, eventually erode away. **99.** $Q = 1.61$, $\Delta H^o = -26.8$ kJ, $\Delta S^o = 189.2\ \mathrm{J\,K^{-1}}$, K (at 25 °C) $= 3.77 \times 10^{14}$, K (at 50 °C) $= 1.59 \times 10^{14}$. **100.** 2.26 V

Feature Problems 102a. Equation 1: Na (s) → Na (amalgam, 0.206%); Equation 2: 2 Na (amalgam, 0.206 %) + 2 H^+ (aq, 1M) → 2 Na^+ (1 M) + H_2 (g,1 bar). **102b.** $\Delta G_1 = 8.156 \times 10^4$ J, $\Delta G_2 = -36.033 \times 10^4$ J. **102c.** 2 Na(s)+ 2 H^+(aq) → 2 Na^+ (1 M)+ H_2 (g, 1 atm), $\Delta_r G^\circ = -52.345 \times 10^4$ J. **102d.** $E^\circ_{cell} = 2.713$ V, E° {Na^+ (1 M)/Na (s)}= −2.713 V **104a.** 2.66×10^{-13} F **104b.** 2.26×10^{-14} C **104c.** 1.41×10^5 K^+ ions **104d.** 9.3×10^{11} ions **104e.** 1.5×10^{-7} (~0.000015%) **105.** Reactions with a positive cell potential are reactions for which $\Delta_r G^\circ < 0$, or reactions for which $K > 1$. $\Delta_r S^\circ$, $\Delta_r H^\circ$ and $\Delta_r U^\circ$ cannot be used alone to determine whether a particular electrochemical reaction will have a positive or negative value.

106. $E^\circ_{X^+/X} < 0$ V, $E^\circ_{X^+/X} > E^\circ_{Y^+/Y}$.

107. Proceed similarly to the solution for 20.100.

Self-Assessment Exercises 114a. False **114b.** False **114c.** True **114d.** True **114e.** True **114f.** False **114g.** True **115.** The answer is (b). **116.** The answer is (d). **117.** The answer is (c). **118.** The answer is (a). **119.** The answer is (d). **120.** The answer is (a). **121.** Zn(s) | Zn^{2+}(1M) || H^+(1M), NO_3^-(1M) | NO(g,1atm) | Pt(s), $E^\circ_{cell} = 1.719$ V. **122.** 1.82 **123a.** No **123b.** A net reaction occurs to the left. **124a.** Fe(s) + Cu^{2+}(1 M)→Fe^{2+}(1 M) + Cu(s), $E^\circ_{cell} = -0.780$ V, electron flow from B to A **124b.** Sn^{2+}(1 M) + 2 Ag^+(1 M) → Sn^{4+}(aq) + 2 Ag(s), $E^\circ_{cell} = +0.646$ V, electron flow from A to B. **124c.** Zn(s) + Fe^{2+}(0.0010 M) → Zn^{2+}(0.10 M) + Fe(s), $E^\circ_{cell} = +0.264$ V, electron flow from A to B. **125a.** Cl_2(g) at anode and Cu(s) at cathode. **125b.** O_2(g) at anode and H_2(g) and OH^-(aq) at cathode. **125c.** Cl_2(g) at anode and Ba(l) at cathode. **125d.** O_2(g) at anode and H_2(g) and OH^-(aq) at cathode.

CHAPTER 20

Practice Examples 1a. $4.46 \times 10^{-5}\ \mathrm{M\,s^{-1}}$ **1b.** $1.62 \times 10^{-4}\ \mathrm{M\,s^{-1}}$ **2a.** (a) $3.3 \times 10^{-4}\ \mathrm{M\,s^{-1}}$; (b) 0.37 M. **2b.** 2.17 M. This value differs from the value of 2.15 M determined in Example 20-2b because it used the initial rate of reaction, which is a bit faster than the average rate over the first 400 seconds. **3a.** First-order **3b.** $3.9 \times 10^{-7}\ \mathrm{M\,min^{-1}}$ **4a.** $4.4 \times 10^{-2}\ \mathrm{M^{-2}\,s^{-1}}$ **4b.** $2.4 \times 10^{-7}\ \mathrm{M\,min^{-1}}$ **5a.** 1.0 M **5b.** Substitute the provided values into text Equation 20.13. The two k values agree within the limits of the experimental error and thus, the reaction is first-order in $[H_2O_2]$. **6a.** 64.2% **6b.** 25.1 min **7a.** 272 mmHg **7b.** (a) 1.9×10^{-7} mmHg; (b) 1.56×10^3 mmHg. **8a.** First-order, $k = 6.93 \times 10^{-3}\ \mathrm{s^{-1}}$. **8b.** Zero-order, $k = 9.28 \times 10^{-3}$ M/min. **9a.** 43 s **9b.** 311 K **10a.** The first step is the slow step. Rate of reaction $= k_1[NO_2]^2$. **10b.** (1) The steps of the mechanism add to produce the overall reaction. (2) The predicted rate law, Rate $= (k_1k_3/k_2)[NO_2][F_2]$, agrees with the experimental rate law.

Integrative Example A. (a) 44 kJ/mol; (b) 20. h. **B.** The rate of reaction is: Rate = k2[A*]

$$= \frac{k_2k_1[A]^2}{k_{-1}[A] + k_2}.$$ At low pressures ([A] ~ 0 and hence $k_2 \gg k_{-1}[A]$), the denominator becomes ~ k_2 and the rate law is

$$\text{Rate} = \frac{k_2k_1[A]^2}{k_2} = k_1[A]^2.$$ Second-order with respect to [A]. At high pressures ([A] is large and $k_{-1}[A] \gg K_2$), the denominator becomes ~ $k_{-1}[A]$ and the rate law is

$$\text{Rate} = \frac{k_2k_1[A]^2}{k_{-2}[A]} = \frac{k_2k_1[A]}{k_{-1}}.$$ First-order with respect to [A].

Exercises 1a. $3.1 \times 10^{-4}\ \mathrm{Ms^{-1}}$ **1b.** $3.1 \times 10^{-4}\ \mathrm{Ms^{-1}}$ **1c.** $9.3 \times 10^{-4}\ \mathrm{Ms^{-1}}$ **3.** $1.0 \times 10^{-3}\ \mathrm{M\,s^{-1}}$ **5a.** 0.575 M **5b.** 5.4 min **7a.** 3.52×10^{-5} M/s **7b.** 0.357 M **7c.** 4.5×10^2 s **9a.** 3000. mmHg **9b.** 1400. mmHg **11a.** First-order with respect to A, second-order with respect to B. **11b.** Third-order overall. **11c.** $0.102\ \mathrm{M^{-2}\,s^{-1}}$ **13.** Rate $= k\,[NO]^2[Cl_2]$, $k = 5.70\ \mathrm{M^2\,s^1}$. **15a.** True **15b.** False **17a.** 3.13% **17b.** 0.00193 M/s **19a.** 19 min **19b.** 0.18 g A **21.** 20.6 min **23a.** 218 min **23b.** 2.3 L CO_2 **25a.** The virtual constancy of the rate constant throughout the time of the reaction confirms that the reaction is first-order. **25b.** $1.86 \times 10^{-3}\ \mathrm{s^{-1}}$ **25c.** 0.148 M **27a.** Set II **27b.** Set I **27c.** Set III **29.** 70 s **31a.** $0.010\ \mathrm{M\,s^{-1}}$ **31b.** $0.0048\ \mathrm{M\,s^{-1}}$ **31c.** $0.0034\ \mathrm{M\,s^{-1}}$ **33.** 1.39×10^{-4} M/min **35.** The reaction is second-order in HI at 700 K. Rate $= k[HI]^2$. $k = 0.00118\ \mathrm{M^{-1}s^{-1}}$. **37a.** Zero-order **37b.** 72 s **39a.** +0.022 M/min, +0.089 M/min. **39b.** Second-order **41.** Rate $= k\,[A^2]$ and $k = 0.020\ \mathrm{L\,mol^{-1}min^{-1}}$. **43.** A zero-order reaction has a half life that varies proportionally to $[A]_0$, therefore, increasing $[A]_0$ increases the half-life for the reaction. A second-order reaction's half-life varies inversely proportional to $[A]_0$, that is, as $[A]_0$ increases, the half-life decreases. The reason for the difference is that a zero-order reaction has a constant rate of reaction (independent of $[A]_0$). In a second-order reaction, the rate of reaction increases as the square of the $[A]_0$. **45a.** The rate of a reaction depends on at least two factors other than the frequency of collisions. The first of these is whether each collision possesses sufficient energy to get over the energy barrier to products. The second factor is whether the molecules in a given collision are properly oriented for a successful reaction. **45b.** Although the collision frequency increases relatively slowly with temperature, the fraction of those collisions that have sufficient energy to overcome the activation energy increases much more rapidly. Therefore, the rate of reaction will increase dramatically with temperature. **45c.** The addition of a catalyst has the net effect of decreasing the activation energy of the overall reaction, by enabling an alternative mechanism. The lower activation energy of the alternative mechanism means that a larger fraction of molecules have sufficient energy to react. Thus the rate increases, even though the temperature does not. **47a.** 63 kJ/mol **47b.**

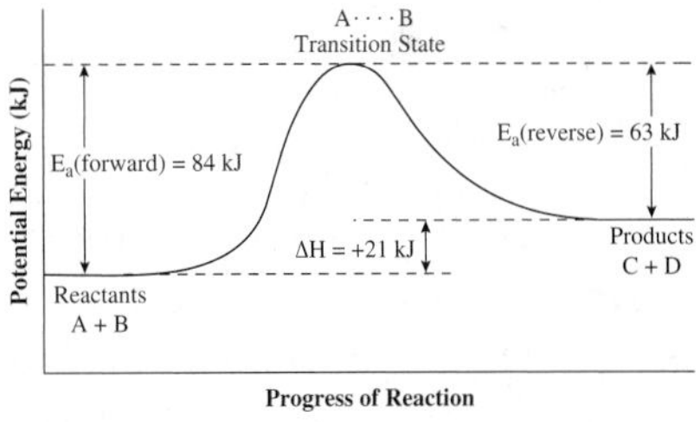

49a. Two **49b.** Three **49c.** Step 3 **49d.** Step 1 **49e.** Endothermic **49f.** Exothermic **51.** 160 kJ/mol **53a.**

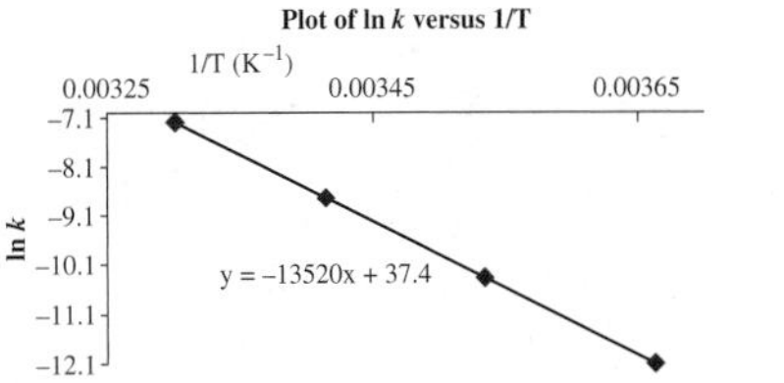

53b. $112\ \mathrm{kJ\,mol^{-1}}$ **53c.** 2.3×10^2 s **55a.** 34.8 kJ/mol **55b.** 61 °C **57a.** 51 kJ/mol **57b.** Since the activation energy for the depicted reaction is 209 kJ/mol, we would not expect this reaction to follow the rule of thumb. **59a.** Although a catalyst is *recovered unchanged from the reaction mixture*, it does "take part in the reaction." Some catalysts actually slow down the rate of a reaction. Usually, however, these negative catalysts are called inhibitors. **59b.** The function of a catalyst is to *change the mechanism of a reaction*. The new mechanism is one that has a different (lower) activation energy (and frequently a different A value), than the original reaction. **61.** Both platinum and an enzyme have a metal center that acts as the active site. Generally speaking, platinum is not dissolved in the reaction solution (heterogeneous), whereas enzymes are generally soluble in the reaction media (homogeneous). Platinum is rather nonspecific, catalyzing many different reactions. An enzyme is quite specific, usually catalyzing only one reaction rather than all reactions of a given class. **63.** An excess of substrate must be present. **65.** The molecularity is equal to the order of the overall reaction only if the elementary process in question is the slowest and, thus, the rate-determining step of the overall reaction. In addition, the elementary process in question should be the only elementary step that influences the rate of the reaction. **67.** The second step, which is slow, is: $H_2 + N_2O_2 \rightarrow H_2O + N_2O$. The rate of this rate-determining step is: Rate $= k_2[H_2][N_2O_2]$. Substituting for $[N_2O_2]$ gives

Rate $= \dfrac{k_2k_1}{k_{-1}}[H_2][NO]^2$. The reaction is first-order in $[H_2]$ and second-order in [NO]. This result conforms to the experimentally determined reaction order.

69. $Cl_2(g) + NO(g) \underset{k_{-1}}{\overset{k_1}{\rightleftharpoons}} NOCl(g) + Cl(g)$

$Cl(g) + NO(g) \underset{k_{-1}}{\overset{k_1}{\rightleftharpoons}} NOCl(g)$

71. $\dfrac{d(S_1:S_2)}{dt} = k_2(S_1:S_2)^* = \dfrac{k_2 \cdot k_1[S_1][S_2]}{k_{-1} + k_2}$

Integrative and Advanced Exercises

74a. First-order **74b.** 0.29 min^{-1} **74c.** 0.103 M/min **74d.** 0.064 M/min **74e.** 0.29_4 M/min. **75.** 1.86 M **78.** {Units of k} $= M_{1-0}\,s^{-1}$ **82.** 1.9×10^2 min

87. $\text{rate}_{\text{overall}} = k_2[CHCl_3]\left(\dfrac{k_1}{k_{-1}}[Cl_2(g)]\right)^{1/2}$, $k = 0.015$.

90. $1.067 \times 10^{-6}\,M^{-1}s^{-1}$ **91.** $0.0275\,s^{-1}$ **94a.** First step. **94b.** No **94c.** Yes **95a.** Both reactions are first-order. **95b.** The second step. **95c.** 0.9235 M **95d.** 0.731 M N_2O

Feature Problems

96a.

Time, min	0	3	6	9	12	15	18	21	24	27	30	∞
V_{N2}, mL	0	10.8	19.3	26.3	32.4	37.3	41.3	44.3	46.5	48.4	50.4	58.3
$[C_6H_5N_2Cl]$, mM	71	58	47	39	32	26	21	17	14	12	10	0

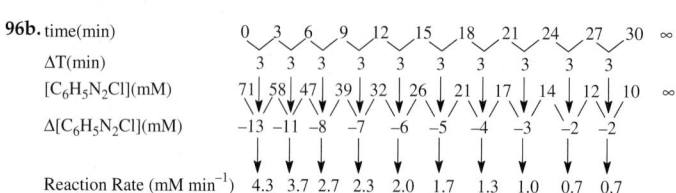

96b.

time(min)	0	3	6	9	12	15	18	21	24	27	30	∞
ΔT(min)	3	3	3	3	3	3	3	3	3	3		
$[C_6H_5N_2Cl]$(mM)	71	58	47	39	32	26	21	17	14	12	10	∞
$\Delta[C_6H_5N_2Cl]$(mM)	−13	−11	−8	−7	−6	−5	−4	−3	−2	−2		
Reaction Rate (mM min^{-1})	4.3	3.7	2.7	2.3	2.0	1.7	1.3	1.0	0.7	0.7		

96c.

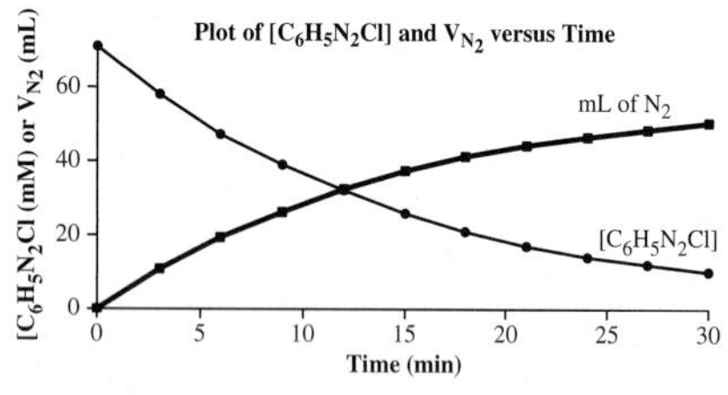

96d. 1.1×10^{-3} M min^{-1} **96e.** 5.5×10^{-3} M min^{-1} **96f.** Rate $= k[C_6H_5N_2Cl]$, $k_{avg} =$ 0.071 min^{-1}. **96g.** Calculated half-life = 9.8 min, estimated half-life = 10.5 min. **96h.** About 20 minutes.

96i.

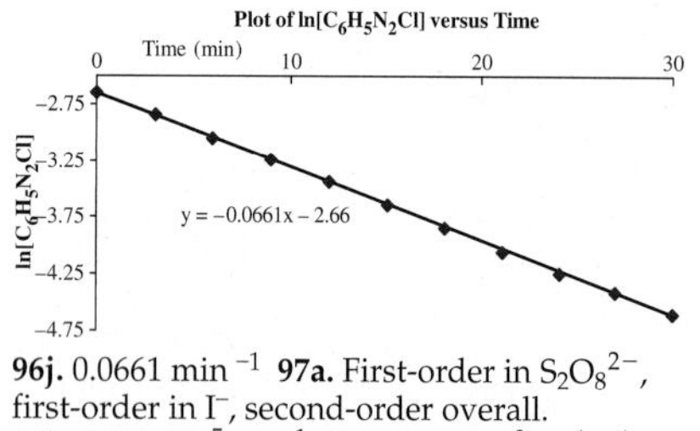

96j. 0.0661 min^{-1} **97a.** First-order in $S_2O_8^{2-}$, first-order in I^-, second-order overall. **97b.** $3.7 \times 10^{-5}\,M\,s^{-1}$ **97c.** $6.2 \times 10^{-3}\,M^{-1}\,s^{-1}$ **97d.** $0.0014\,M^{-1}\,s^{-1}$, $0.0030\,M^{-1}\,s^{-1}$, $0.0062\,M^{-1}\,s^{-1}$, $0.012\,M^{-1}s^{-1}$. **97e.** 51 kJ/mol **97f.** The mechanism is consistent with the stoichiometry. The rate of the slow step of the mechanism is $\text{Rate}_1 = k_1[S_2O_8^{2-}]^1[I^-]^1$. This is exactly the same as the experimental rate law. It is reasonable that the first step be slow since it involves two negatively charged species coming together. We know that like charges repel, and thus this should not be an easy or rapid process.

Self-Assessment Exercises **101.** The answer is (c). **102.** The answers are (b) and (e). **103.** The answer is (a). **104.** The answer is (d). **105.** The answer is (b). **106.** The answer is (c). **107.** $3.0 \times 10^{-3}\,M \cdot s^{-1}$ **108a.** 60.0 min **108b.** 45.0 min **109.** The reaction is second-order, because the half-life doubles with each successive half-life period. **110a.** 0.024 M/min **110b.** 2.312 M **110c.** 1.33 M **111a.** $\dfrac{d[C]}{dt} = k_1[A][B]$. This agrees with the observed reaction rate law. **111b.** Adding the two reactions given, we still get the same overall stoichiometry as part (a). However, with the given proposed reaction mechanisms, the rate law for the product(s) is given as follows: $\dfrac{d[C]}{dt} = \dfrac{k_2k_1[A][B]^2}{k_{-1} + k_2[A]}$. This does not agree with the observed reaction rate law. **112.** The answer is (b). **113.** The answer is (a). **114.** The answer is (c).

CHAPTER 21

Practice Examples

1a. $2\,NaCl(aq) + 2\,H_2O\,(l) \xrightarrow{\text{electrolysis}} 2\,NaOH\,(aq) + H_2\,(g) + Cl_2\,(g)$, $2\,NaOH\,(aq) + 3\,NO_2\,(g) \rightarrow 2\,NaNO_3\,(aq) + NO\,(g) + H_2O(l)$. **1b.** $2\,NaCl(aq) + 2\,H_2O\,(l) \xrightarrow{\text{electrolysis}} 2\,NaOH\,(aq) + H_2\,(g) + Cl_2\,(g)$, $2\,NaOH\,(aq) + SO_2\,(g) \rightarrow Na_2SO_3\,(aq) + H_2O(l)$, $Na_2SO_3\,(aq) + S(s) \xrightarrow{\text{boil}} Na_2S_2O_3\,(aq)$. **2a.** Compound X is most likely Na_2O and compound Y is Na_2O_2. $NaNO_2\,(s) + 3\,Na(s) \rightarrow 2\,Na_2O(s) + \frac{1}{2}N_2(g)$, $Na_2O(s) + \frac{1}{2}O_2\,(g) \rightarrow Na_2O_2(s)$. **2b.** Compound X is likely calcium carbide and compound Y is calcium cyanamide. $Ca(s) + 2\,C(s) \rightarrow CaC_2\,(s)$, $CaC_2\,(s) + N_2\,(g) \rightarrow CaCN_2\,(s) + C(s)$. The structure of the Y anion is linear: $\ddot{N}{=}C{=}\ddot{N}$

3a. $Na_2B_4O_7 \cdot 10\,H_2O(s) + H_2SO_4\,(l) \rightarrow 4\,B(OH)_3\,(s) + Na_2SO_4\,(s) + 5\,H_2O(l)$, $2\,B(OH)_3\,(s) \xrightarrow{\Delta} B_2O_3\,(s) + 3\,H_2O(g)$, $2\,B_2O_3\,(s) + 3\,C\,(s) + 6\,Cl_2\,(g) \xrightarrow{\Delta} 4\,BCl_3\,(g) + 3\,CO_2\,(g)$, $4\,BCl_3\,(g) + 3\,LiAlH_4\,(s) \rightarrow 2\,B_2H_6\,(g) + 3\,LiCl(s) + 3\,AlCl_3\,(s)$. **3b.** $Na_2B_4O_7 \cdot 10\,H_2O(s) + H_2SO_4(aq) \rightarrow 4\,B(OH)_3(s) + Na_2SO_4(aq) + 5\,H_2O(l)$, $2\,B(OH)_3(s) \xrightarrow{\Delta} B_2O_3(s) + 3\,H_2O(l)$, $B_2O_3(s) + 3\,CaF_2(s) + 3\,H_2SO_4(l) \xrightarrow{\Delta} 2\,BF_3(g) + 3\,CaSO_4(S) + 3\,H_2O(g)$.

Integrative Example **A.** $NaCN(aq) \rightarrow Na^+\,(aq) + CN^-\,(aq)$, $Al(NO_3)_3(aq) \rightarrow Al^{3+}\,(aq) + 3\,NO_3^-\,(aq)$. The product $[Al^{3+}][OH^-]^3 = 4.0 \times 10^{-9}$ is much greater than K_{sp} for $Al(OH)_3$ (1.3×10^{-33}) and therefore precipitation will occur. **B.** When $BeCl_2 \cdot 4\,H_2O$ is heated, it decomposes to $Be(OH)_2(s)$, $H_2O(g)$, and HCl(g), as discussed in Section 21-3. $BeCl \cdot 4\,H_2O$ comprises $[Be(H_2O)4]^{2+}$ and Cl^- ions. Because of the high polarizing power of Be^{2+}, it is difficult to remove the coordinated water molecules by heating the solid and the acidity of the coordinated H_2O molecules is enhanced. When $BeCl_2 \cdot 4\,H_2O$ is dissolved in water, $[Be(H_2O)4]^{2+}$ ions react with water, and $[Be(H_2O)_3(OH)]^+$ and H_3O^+ ions are produced. Hence, a solution of $BeCl_2 \cdot 4\,H_2O$ is expected to be acidic. When $CaCl_2 \cdot 6\,H_2O$ is heated, $CaCl_2(s)$ and $H_2O(g)$ are produced. Because the charge density and polarizing power of Ca^{2+} is much less than that of Be^{2+}, it is much easier to drive off the coordinated water molecules by heating the solid. When $CaCl_2 \cdot 6\,H_2O$ is dissolved in water, $Ca^{2+}(aq)$ and $Cl^-(aq)$ are produced. However, the charge density of the Ca^{2+} ion is too low to affect the acidity of water molecules in the hydration sphere of the Ca^{2+} ion. Hence, a solution of $CaCl_2 \cdot 6\,H_2O$ has a neutral pH.

Exercises **1a.** $2\,Cs(s) + Cl_2\,(g) \rightarrow 2\,CsCl(s)$ **1b.** $2\,Na(s) + O_2\,(g) \rightarrow Na_2O_2\,(s)$ **1c.** $Li_2CO_3\,(s) \longrightarrow Li_2O\,(s) + CO_2\,(g)$ **1d.** $Na_2SO_4\,(s) + 4\,C\,(s) \rightarrow Na_2S\,(s) + 4\,CO\,(g)$ **1e.** $K\,(s) + O_2\,(g) \rightarrow KO_2\,(s)$ **3.** Both LiCl and KCl are soluble in water, but Li_3PO_4 is not very soluble. Hence the addition of K_3PO_4 (aq) to a solution of the white solid will produce a precipitate if the white solid is LiCl, but no precipitate if the white solid is KCl. The best method is a flame test; lithium gives a red color to a flame, while the potassium flame test is violet. **5.** Based on molar solubilities: $MgCO_3 < 2\,Li_2CO_3 < Na_2CO_3$. **7a.** 11.2 **7b.** As long as KCl is in excess and the volume of the solution is nearly constant, the solution pH only depends on the number of electrons transferred. **9a.** 71.5% yield **9b.** NH_3 is simply used during the Solvay process to produce the proper conditions for the desired reactions. Any net consumption of NH_3 is the result of unavoidable losses during production.

11. $K = 1.87 \times 10^{-25}$. $Na_2O_2(s)$ is thermodynamically stable with respect to $Na_2O(s)$ and $O_2(g)$ at 298 K.
13.

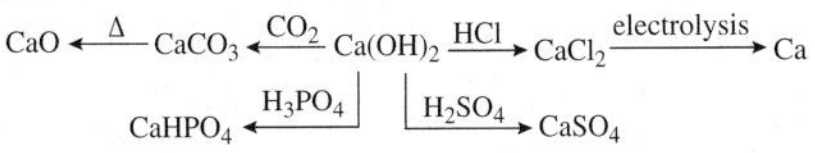

$Ca(OH)_2\ (s) + 2\ HCl\ (aq) \rightarrow CaCl_2\ (aq) + 2\ H_2O(l)$
$CaCl_2\ (l) \xrightarrow{\Delta,\text{electrolysis}} Ca\ (l) + Cl_2\ (g)$
$Ca(OH)_2\ (s) + CO_2\ (g) \rightarrow CaCO_3\ (s) + H_2O\ (g)$
$CaCO_3\ (s) \xrightarrow{\Delta} CaO\ (s) + CO_2\ (g)$
$Ca(OH)_2\ (s) + H_2SO_4\ (aq) \rightarrow CaSO_4\ (s) + 2\ H_2O(l)$
15. $Mg^{2+} + 2\ Cl^-(aq) \rightarrow Mg(s) + Cl_2(g)$ (Overall reaction). Charge is conserved.
17a. $BeF_2\ (s) + Mg(s) \xrightarrow{\Delta} Be(s) + MgF_2\ (s)$
17b. $Ba\ (s) + Br_2\ (l) \rightarrow BaBr_2\ (s)$
17c. $UO_2\ (s) + 2\ Ca\ (s) \rightarrow U\ (s) + 2\ CaO\ (s)$
17d. $MgCO_3 \cdot CaCO_3\ (s) \xrightarrow{\Delta} MgO\ (s) + CaO\ (s) + 2\ CO_2\ (g)$
17e. $2\ H_3PO_4\ (aq) + 3\ CaO(s) \rightarrow Ca_3(PO_4)_2\ (s) + 3\ H_2O\ (l)$
19a. Equilibrium lies slightly to the left.
19b. Equilibrium lies to the left.
19c. Equilibrium lies to the right.
21. The SO_4^{2-} ion is a large polarizable ion. A cation with a high polarizing power will polarize the SO_4^{2-} ion and kinetically assist the decomposition to SO_3. Because Be^{2+} has the largest charge density of the group 2 cations, we expect Be^{2+} to be the most polarizing and thus, $BeSO_4$ will be the least stable with respect to decomposition. **23a.** B_4H_{10} contains 22 valence electrons or 11 pairs. Ten of these pairs could be allocated to form 10 B—H bonds, leaving only one pair to bond the four B atoms together. Clearly an electron deficient situation. **23b.** Two
23c.

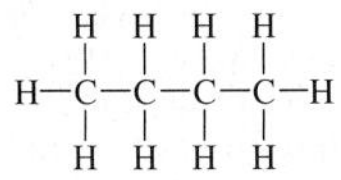

25a. $2\ BBr_3(l) + 3\ H_2(g) \rightarrow 2\ B(s) + 6\ HBr(g)$
25b. $B_2O_3(s) + 3\ C(s) \xrightarrow{\Delta} 3\ CO(g) + 2\ B(s)$, $2\ B(s) + 3\ F_2(g) \xrightarrow{\Delta} 2\ BF_3(g)$
25c. $2\ B(s) + 3\ N_2O(g) \xrightarrow{\Delta} 3\ N_2(g) + B_2O_3(s)$
27a. $2\ Al(s) + 6\ HCl(aq) \rightarrow 2\ AlCl_3\ (aq) + 3\ H_2\ (g)$
27b. $2\ NaOH\ (aq) + 2\ Al\ (s) + 6\ H_2O\ (l) \rightarrow 2\ Na^+\ (aq) + 2\,[Al(OH)_4]^-\ (aq) + 3\ H_2\ (g)$
27c. $2\ Al\ (s) + 3\ SO_4^{2-}\ (aq) + 12\ H^+\ (aq) \longrightarrow 2\ Al^{3+}\ (aq) + 3\ SO_2\ (aq) + 6\ H_2O(l)$
29. $Al^{3+}\ (aq) + 3\ HCO_3^-\ (aq) \longrightarrow Al(OH)_3\ (s) + 3\ CO_2\ (g)$
31. Aluminum and its oxide are soluble in both acid and base. Al(s) is resistant to corrosion only over the pH range 4.5 to 8.5. Thus, aluminum is inert only when the medium to which it is exposed is neither highly acidic nor highly basic. **33.** $[Al(OH)_4]^-\ (aq) + CO_2(aq) \rightarrow Al(OH)_3\ (s) + HCO_3^-\ (aq)$. HCl(aq), being a strong acid, can't be used because it will dissolve the $Al(OH)_3(s)$.
35. $2\ KOH(aq) + 2\ Al(s) + 6\ H_2O(l) \rightarrow 2\ K[Al(OH)_4](aq) + 3\ H_2(g)$, $2\ K[Al(OH)_4](aq) + 4\ H_2SO_4(aq) \rightarrow K_2SO_4(aq) + Al_2(SO_4)_3(aq) + 8\ H_2O(l)$, —crystallize $\rightarrow 2\ KAl(SO_4)_2(s)$.
37. The chemical reaction for the disproportionation of $FB(BF_2)_2$ is
$3\ FB(BF_2)_2 \rightarrow BF_3 + B_8F_{12}$
The structure of B_8F_{12} is

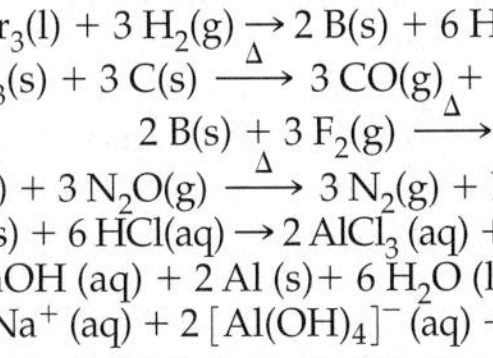

39. In the sense that diamonds react imperceptibly slowly at room temperature (either with oxygen to form carbon dioxide, or in its transformation to the more stable graphite), it is essentially true that "diamonds last forever." However, at elevated temperatures, diamond will burn to form $CO_2(g)$ and thus the statement is false. Eventually, the conversion to graphite occurs.
41a. $3\ SiO_2(s) + 4\ Al(s) \xrightarrow{\Delta} 2\ Al_2O_3(s) + 3\ Si(s)$
41b. $K_2CO_3(s) + SiO_2(s) \xrightarrow{\Delta} CO_2(g) + K_2SiO_3(s)$
41c. $Al_4C_3(s) + 12\ H_2O(l) \rightarrow 3\ CH_4(g) + 4\ Al(OH)_3(s)$
43. A silane is a silicon-hydrogen compound, with the general formula $Si_n\,H_{2n+2}$. A silanol is a compound in which one or more of the hydrogens of silane is replaced by an —OH group. Silicones are produced when silanols condense into chains, with the elimination of a water molecule between every two silanol molecules. **45.** $2\ CH_4\ (g) + S_8\ (g) \longrightarrow 2\ CS_2\ (g) + 4\ H_2S\ (g)$, $CS_2\ (g) + 3\ Cl_2\ (g) \longrightarrow CCl_4\ (l) + S_2Cl_2\ (l)$, $4\ CS_2\ (g) + 8\ S_2Cl_2\ (g) \longrightarrow 4\ CCl_4\ (l) + 3\ S_8\ (s)$. **47.** The formula is $KAl_2(OH)_2(AlSi_3O_{10})$. Since they are not segregated into O_2 units in the formula, all of the oxygen atoms in the mineral must be in the –2 oxidation state. Potassium is obviously in the +1 oxidation state, as are the hydrogen atoms in the hydroxyl groups. Up to this point, we have −24 from the twelve oxygen atoms and +3 from the potassium and hydrogen atoms for a net number of −21 for the oxidation state. We still have three aluminum atoms and three silicon atoms to account for. In oxygen-rich salts such as mica, we would expect that the silicon and the aluminum atoms would be in their highest possible oxidation states, namely +4 and +3, respectively. Since the salt is neutral, the oxidation numbers for the silicon and aluminum atoms must add up to +21. This is precisely the total that is obtained if the silicon and aluminum atoms are in their highest possible oxidation states: $(3 \times (+3) + 3 \times (+4) = +21)$. Consequently, the empirical formula for white muscovite is consistent with the expected oxidation state for each element present.
49a. $PbO\ (s) + 2\ HNO_3\ (aq) \longrightarrow Pb\ (NO_3)_2\ (s) + H_2O(l)$
49b. $SnCO_3\ (s) \xrightarrow{\Delta} SnO(s) + CO_2\ (g)$
49c. $PbO(s) + C(s) \xrightarrow{\Delta} Pb(l) + CO(g)$
49d. $2\ Fe^{3+}\ (aq) + Sn^{2+}\ (aq) \longrightarrow 2\ Fe^{2+}\ (aq) + Sn^{4+}\ (aq)$
49e. $PbS\ (s) + 2\ O_2\ (g) \longrightarrow PbSO_4\ (s)$
51a. The reaction will go to completion.
51b. The reaction is not spontaneous.
51c. The reaction is not spontaneous. **53.** $SnCl_2$
Integrative and Advanced Exercises.
58. $:\ddot{I}{=}\ddot{I}{=}\ddot{I}:$. The Li^+ ion has a high charge density and significant polarizing power. The Li^+ ion will polarize the I_3^- to a significant extent and presumably assists the decomposition of I_3^- to I_2 and I^-. **59.** $K = P_{O_2} = 8.8 \times 10^4$.
62a. 0.350 g CaO **62b.** If the mass of product formed from the starting Ca differs from the 0.350 g CaO predicted, then the product could be a mixture of calcium nitride and calcium oxide. This scenario is not implausible since molecular nitrogen is readily available in the atmosphere. **62c.** The product is 40% by mass CaO. **63a.** Since Li has a higher boiling point (1347 °C) than K (773.9 °C), a reaction similar to (21.4) is not a feasible way of producing Li metal from LiCl. **63b.** Cs has a lower boiling point (678.5 °C) than K (773.9 °C), and thus a reaction similar to (21.4) is a feasible method of producing Cs metal from CsCl. **64a.** −852 kJ
64b. $4\ Al(s) + 3\ MnO_2(s) \rightarrow 2\ Al_2O_3(s) + 3\ Mn(s)$, $\Delta H = -1792$ kJ.

64c. $2\ Al(s) + 3\ MgO(s) \rightarrow Al_2O_3(s) + 3\ Mg(s)$, $\Delta H = +129$ kJ. **66.** $Mg(OH)_2$ should precipitate. **67a.** 81.4 ppm Ca^{2+} **67b.** 48.7 g CaO
67c. Use the K_{sp} expression for $CaCO_3$ to determine the needed $[CO_3^{2-}]$. The carbonate ion concentration can readily be achieved by adding solid Na_2CO_3. **67d.** 2.2×10^2 g Na_2CO_3
68a. 2.68×10^5 g Al **68b.** 1.3 metric tons of coal. **69.** −1.605 V. If the oxidation of C(s) to $CO_2(g)$ did not occur, then the cell reaction would just be the reverse of the formation reaction of Al_2O_3 with $n = 6\ e^-$. $E° = \Delta E° = -2.626$ V.
70. 19 g $Pb(NO_3)_2$ **73a.** 7.4×10^{-3} M
73b. 6.4×10^{-4} M **73c.** 2.0×10^{-3} M
76. We would expect MgO(s) to have a larger value of lattice energy than MgS(s) because of the smaller interionic distance in MgO(s). Lattice energy of MgO = −3789 kJ, Lattice energy of MgS = −3215 kJ. **77.** The empirical formula is M_3C_{60}, $n = 3$.
Feature Problems 78a. $\Delta H° = -277\ \text{kJ} \approx \Delta G°$, $E°_{red} = -2.87$ V. **78b.** $E°_{red} = -3.03$ V.
Self-Assessment Exercises 83. The answer is (c). **84.** The answer is (f). **85a.** $BCl_3NH_3(g)$ (an adduct) **85b.** $KO_2(s)$ **85c.** $Li_2O(s)$
85d. $Ba(OH)_2(aq)$ and $H_2O_2(aq)$. **86.** The thermite reaction is evidence that aluminum will readily extract oxygen from Fe_2O_3. Aluminum can be used for making products that last and for structural purposes, because aluminum develops a coating of Al_2O_3 that protects the metal beneath it. **87.** The answer is (b).
88a. $Li_2CO_3\ (s) \xrightarrow{\Delta} Li_2O(s) + CO_2(g)$
88b. $CaCO_3\ (s) + 2\ HCl(aq) \longrightarrow CaCl_2\ (aq) + H_2O(l) + CO_2\ (g)$
88c. $2\ Al(s) + 2\ NaOH(aq) + 6\ H_2O(l) \longrightarrow 2\ Na[Al(OH)_4](aq) + 2\ H_2(g)$
88d. $BaO(s) + H_2O(l) \longrightarrow Ba(OH)_2$ (aq,limited solubility)
88e. $2\ Na_2O_2\ (s) + 2\ CO_2\ (g) \longrightarrow 2\ Na_2CO_3\ (s) + O_2\ (g)$
89a. $MgCO_3\ (s) + 2\ HCl(aq) \longrightarrow MgCl_2\ (aq) + H_2O(l) + CO_2\ (g)$
89b. $2\ Na(s) + 2\ H_2O(l) \longrightarrow 2\ NaOH(aq) + H_2\ (g)$ followed by the reaction in Exercise 82(c).
89c. $2\ NaCl(s) + H_2SO_4$ (concd.,aq) $\xrightarrow{\Delta} Na_2SO_4(s) + 2\ HCl(g)$
90a. $K_2CO_3\ (aq) + Ba(OH)_2\ (aq) \longrightarrow BaCO_3\ (s) + 2\ KOH(aq)$
90b. $Mg(HCO_3)_2\ (aq) \xrightarrow{\Delta} MgCO_3\ (s) + H_2O(l) + CO_2\ (g)$
90c. $SnO(s) + C(s) \xrightarrow{\Delta} Sn(l) + CO(g)$
90d. $CaF_2\ (s) + H_2SO_4$ (concd. aq) $\xrightarrow{\Delta} CaSO_4\ (s) + 2\ HF(g)$
90e. $NaHCO_3\ (s) + HCl(aq) \longrightarrow NaCl(aq) + H_2O(l) + CO_2\ (g)$
90f. $PbO_2\ (s) + 4\ HBr(aq) \longrightarrow PbBr_2\ (s) + Br_2\ (l) + 2\ H_2O(l)$
90g. $SiF_4\ (g) + 4\ Na(l) \longrightarrow Si(s) + 4\ NaF(s)$
91. $CaSO_4 \cdot 2\ H_2O(s) + (NH_4)_2CO_3\ (aq) \longrightarrow CaCO_3\ (s) + (NH_4)_2SO_4\ (aq) + 2\ H_2O(l)$
92a. $2\ B(OH)_3\ (s) \longrightarrow B_2O_3\ (s) + 3\ H_2O(g)$
92b. No reaction.
92c. $CaSO_4 \cdot 2\ H_2O(s) \xrightarrow{\Delta} CaSO_4 \cdot \frac{1}{2}\ H_2O(s) + \frac{3}{2}\ H_2O(g)$
93a. $Pb(NO_3)_2\ (aq) + 2\ NaHCO_3\ (aq) \longrightarrow PbCO_3\ (s) + 2\ NaNO_3\ (aq) + H_2O(l) + CO_2\ (g)$
93b. $Li_2O(s) + (NH_4)_2CO_3\ (aq) \longrightarrow Li_2CO_3\ (s) + 2\ NH_3\ (g) + H_2O(l)$
93c. $H_2SO_4\ (aq) + BaO_2\ (aq) \longrightarrow H_2O_2\ (aq) + BaSO_4\ (s)$
93d. $2\ PbO(s) + Ca(OCl)_2\ (aq) \longrightarrow CaCl_2\ (aq) + 2\ PbO_2\ (s)$
94a. $CaCO_3(s)$ **94b.** $CaSO_4 \cdot 2\ H_2O(s)$
94c. $BaSO_4(s)$ in water. **94d.** $Al_2O_3(s)$ with Fe^{3+} and Ti^{4+} replacing some Al^{3+} in the crystal structure. **95.** $1.27 \times 10^3\ m^3$

CHAPTER 22

Practice Examples **1a.** 0.622 V **1b.** 1.453 V **2a.** 1.7×10^{-48} **2b.** $K_P = 4 \times 10^{-34}$. 4×10^{-15} % decomposed.

Integrative Example **A.** $E°$ for the disproportionation reaction is negative. The disproportionation of NO_2^- to NO_3^- and NO is not spontaneous. **B.** $E°_{cell}$ for the reaction is positive and therefore disproportionation of HNO_2 to NO_3^- and NO is spontaneous under standard conditions.

Exercises **1.** LiF, BeF_2, BF_3, CF_4, NF_3, OF_2. LiF is an ionic compound, BeF_2 is a network covalent compound, and the others are molecular covalent compounds. **3.** Bi_2O_3 is most basic, P_4O_6 is least basic (and is actually acidic), and Sb_4O_6 is amphoteric. **5.** 7.8×10^5 L air **7a.** Trigonal pyramidal **7b.** Tetrahedral **7c.** Square pyramidal **9.** $3\ XeF_4(aq) + 6\ H_2O(l) \rightarrow 2\ Xe(g) + 3/2\ O_2(g) + 12\ HF(g) + XeO_3(s)$ **11.** The bond energy of the noble gas fluorine is too small to offset the energy required to break the F—F bond. **13.** Iodide ion is slowly oxidized to iodine, which is yellow-brown in aqueous solution, by oxygen in the air: $4\ I^-(aq) + O_2(g) + 4\ H^+(aq) \longrightarrow 2\ I_2(aq) + 2\ H_2O(l)$. **15.** Displacement reactions involve one element displacing another element from solution. The element that dissolves in the solution is more "active" than the element supplanted from solution. Within the halogen group, the activity decreases from top to bottom. Thus, each halogen is able to displace the members of the group below it, but not those above it. The only halogen with sufficient oxidizing power to displace $O_2(g)$ from water is $F_2(g)$. None of the halogens reacts with water to form $H_2(g)$. **17a.** 2×10^6 kg F_2 **17b.** Since there is no chemical oxidizing agent that is stronger than $F_2(g)$, this method of displacement would not work for $F_2(g)$. **19.** $6\ Cl_2(aq) + 6\ H_2O(l) \rightarrow 2\ ClO_3^-(aq) + 12\ H^+(aq) + 10\ Cl^-(aq)$, $E°_{cell} = -0.095$ V. Since the final cell voltage is negative, the disproportionation reaction will not occur spontaneously under standard conditions. **21a.** T-shaped **21b.** Square pyramidal **21c.** Square planar **23a.** 0.085 V **23b.** $x = 7.65 \times 10^{-1}$ M

25a. $2\ HgO(s) \xrightarrow{\Delta} 2\ Hg(l) + O_2(g)$

25b. $2\ KClO_4(s) \xrightarrow{\Delta} 2\ KClO_3(s) + O_2(g)$ **27.** CO_2 **29.** 3×10^{-5} mmHg **31.** The electrolysis reaction is $2\ H_2O(l) \xrightarrow{\text{electrolysis}} 2\ H_2(g) + O_2(g)$. In this reaction, 2 moles of $H_2(g)$ are produced for each mole of $O_2(g)$. By the law of combining volumes, we would expect the volume of hydrogen to be twice the volume of oxygen produced. **33.** 5.89 **35.** 302 kJ/mol **37a.** H_2S, while polar, forms only weak hydrogen bonds. H_2O forms much stronger hydrogen bonds, leading to a higher boiling point. **37b.** All electrons are paired in O_3, producing a diamagnetic molecule. **39a.** $K_{eq} = 3.2 \times 10^{41}$, the reaction goes to completion. **39b.** $K_{eq} = 2.1 \times 10^{-65}$, the reaction does not even come close to going to completion. **39c.** $K_{eq} = 9.0 \times 10^2$, the reaction goes to completion. **39d.** $K_{eq} = 4.8 \times 10^{-7}$, the reaction does not even come close to going to completion. **41.** 30 L **43a.** magnesium sulfide **43b.** hydrogen sulfide **43c.** calcium hydrogen sulfite **43d.** sodium thiosulfate **43e.** tetrasulfur tetranitride

45a. $FeS(s) + 2\ HCl(aq) \rightarrow FeCl_2(aq) + H_2S(aq)$ MnS(s), ZnS(s), etc. also are possible.

45b. $CaSO_3(s) + 2\ HCl(aq) \rightarrow CaCl_2(aq) + H_2O(l) + SO_2(g)$

45c. $SO_2(aq) + MnO_2(s) \rightarrow Mn^{2+}(aq) + SO_4^{2-}(aq)$

45d. $S_2O_3^{2-}(aq) + 2\ H^+(aq) \rightarrow S(s) + SO_2(g) + H_2O(l)$

47. The decomposition of thiosulfate ion is more highly favored in an acidic solution. If the white solid is Na_2SO_4, there will be no reaction with strong acids such as HCl. By contrast, if the white solid is $Na_2S_2O_3$, $SO_2(g)$ will be liberated and a pale yellow precipitate of S(s,rhombic) will form upon addition of HCl(aq). $S_2O_3^{2-}(aq) + 2\ H^+(aq) \rightarrow S(s) + SO_2(g) + H_2O(l)$. **49.** 1.1 **51.** 5.70% **53a.** +4 **53b.** +4 **53c.** +6 **53d.** +1 **53e.** +3

55a. $N_2(g) + 3\ H_2(g) \rightleftharpoons 2\ NH_3(g)$

55b. $4\ NH_3(g) + 5\ O_2(g) \xrightarrow{850\ °C,\ Pt} 4\ NO(g) + 6\ H_2O(g)$

55c. $2\ NO(g) + O_2(g) \rightarrow 2\ NO_2(g)$, $3\ NO_2(g) + H_2O(l) \rightarrow 2\ HNO_3(aq) + NO(g)$.

57. $2\ Na_2CO_3(aq) + 4\ NO(g) + O_2(g) \rightarrow 4\ NaNO_2(aq) + 2\ CO_2(g)$

59. 6×10^9 kg **61a.** $2\ NO_2(g) \rightleftharpoons N_2O_4(g)$ **61b.** $HNO_2(aq) + N_2H_5^+(aq) \rightarrow HN_3(aq) + 2\ H_2O(l) + H^+(aq)$, $HN_3(aq) + HNO_2(aq) \rightarrow N_2(g) + H_2O(l) + N_2O(g)$. **61c.** $H_3PO_4(aq) + 2\ NH_3(aq) \rightarrow (NH_4)_2HPO_4(aq)$

63a. $CH_3(NH)NH_2$ has 20 valence electrons.

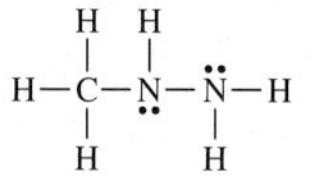

63b. $(CH_3)_2NNH_2$ has 26 valence electrons.

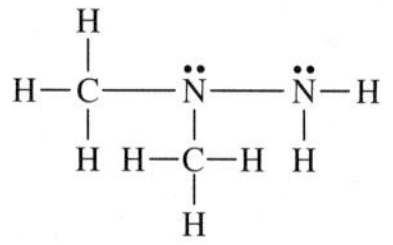

63c. N_2O_4 has 34 valence electrons.

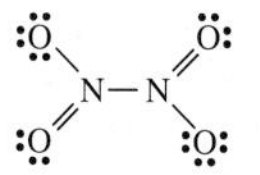

63d. H_3PO_4 has 32 valence electrons.

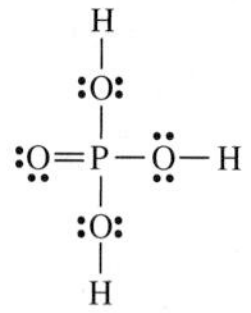

63e. $ClNO_2$ has 24 valence electrons.

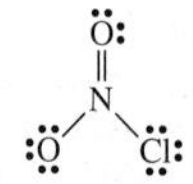

65a. hydrogen phosphate ion **65b.** calcium pyrophosphate or calcium diphosphate **65c.** tetrapolyphosphoric acid **67.** 1.031 V **69a.** Nitrogen atom cannot bond to five fluorine atoms because, as a second-row element, it cannot accommodate more than four electron pairs. **69b.** The NF_3 molecule is trigonal pyramidal. The lone pair on the N atom causes the F—N—F bond angle to decrease from the ideal tetrahedral bond angle of 109° to 102.5°. **71.** $CH_4(g)$, $\Delta_{comb}H° = -890.3$ kJ; $C_2H_6(g)$, $\Delta_{comb}H° = -1559.7$ kJ; $C_3H_8(g)$, $\Delta_{comb}H° = -2219.9$ kJ; $C_4H_{10}(g)$, $\Delta_{comb}H° = -2877.4$ kJ. **73a.** $2\ Al(s) + 6\ HCl(aq) \rightarrow 2\ AlCl_3(aq) + 3\ H_2(g)$ **73b.** $3\ CO(g) + 7\ H_2(g) \rightarrow C_3H_8(g) + 3\ H_2O(g)$ **73c.** $MnO_2(s) + 2\ H_2(g) \xrightarrow{\Delta} Mn(s) + 2\ H_2O(g)$ **75a.** $CaH_2(s)$ produces the most $H_2(g)$. **75b.** CaH_2 produces the greatest amount of H_2 per gram of solid. **77.** Since CH_4 has the highest hydrogen atom to non-hydrogen atom ratio, this molecule has the greatest mass percent hydrogen. **79.** NH_2^- has 8 valence electrons. The first four MO's are fully occupied. Because the $2p_x$ orbital on N is strongly bonding in the bent configuration, the energy of the NH_2^- will be much lower for the bent configuration.

Integrative and Advanced Exercises

81. In step (1), oxygen is converted to H_2O by the reaction $H_2(g) + 1/2\ O_2(g) \rightarrow H_2O(l)$. Step (2) ensures that unreacted hydrogen from step (1) is also converted to $H_2O(l)$. The dehydrated zeolite has a very strong affinity for water molecules and thus, $H_2O(l)$ is removed from the gas mixture. **84.** $NaNO_2$ decolorizes acidic solution of $KMnO_4$ according to the following balanced chemical equation: $2\ KMnO_4 + 3\ H_2SO_4 + 5\ NaNO_2 \rightarrow 5\ NaNO_3 + K_2SO_4 + 2\ MnSO_4 + 3\ H_2O$. $NaNO_3$ does not react with acidic solution of $KMnO_4$. **85.** 0.1716 M **86.** 3.0×10^{-20} J/atom **89.** $2\ NH_4ClO_4(s) \rightarrow N_2(g) + 4\ H_2O(g) + Cl_2(g) + 2\ O_2(g)$ **92.** 9.23 g/cm^3 **94a.** 339 kJ/mol **94b.** The radiation is in the near ultraviolet. $\nu = 8.50 \times 10^{14}\ s^{-1}$, $\lambda = 353$ nm.

95a. $\text{mass P} = 1.000\ \text{g}\ P_4O_{10} \times \frac{1\ \text{mol}\ P_4O_{10}}{283.89\ \text{g}\ P_4O_{10}} \times \frac{4\ \text{mol P}}{1\ \text{mol}\ P_4O_{10}} \times \frac{30.974\ \text{g P}}{1\ \text{mol P}} = 0.436\ \text{g P}$. Thus, multiplying the mass of P_4O_{10} by 0.436 will give the mass (or mass percent) of P.

$$\text{mass}\ Ca_3(PO_4)_2 = 1.000\ \text{g}\ P_4O_{10} \times \frac{1\ \text{mol}\ P_4O_{10}}{283.89\ \text{g}\ P_4O_{10}} \times \frac{4\ \text{mol P}}{1\ \text{mol}\ P_4O_{10}} \times \frac{1\ \text{mol}\ Ca_3(PO_4)_2}{2\ \text{mol P}} \times \frac{310.18\ \text{g}\ Ca_3(PO_4)_2}{1\ \text{mol}\ Ca_3(PO_4)_2} = 2.185\ \text{g}\ Ca_3(PO_4)_2$$

Thus, multiplying the mass of P_4O_{10} (283.88) by 2.185 will give the mass (or mass %) of BPL. **95b.** A %BPL greater than 100% means that the material has a larger %P than does pure $Ca_3(PO_4)_2$. **95c.** 92.4% BPL **99.** 15 g H_2SO_4, 37 g Cl_2. **100.** Solving for $[H^+]$ yields a value of 8×10^{-5} M which corresponds to a pH of 4.1. The solution is still acidic. **102.** Oxidations: $XeF_4(g) + 3\ H_2O \longrightarrow XeO_3(g) + 4\ HF(aq) + 2\ H^+(aq) + 2\ e^-$, $2\ H_2O(l) \longrightarrow O_2(g) + 4\ H^+(aq) + 4\ e^-$. Reduction: $XeF_4(g) + 4\ e^- \longrightarrow Xe(g) + 4\ F^-(aq)$. Net: $3\ XeF_4(g) + 6\ H_2O(l) \longrightarrow 2\ XeO_3(g) + 12\ HF(aq) + 3/2\ O_2(g) + 2\ Xe(g)$ **104.** $[HOCl(aq)] = [H^+(aq)] = [Cl^-(aq)] = 0.030$ M and $[Cl_2(aq)] = 0.060$ M. **105.** 0.70 V **107.** Both molecules are V shaped, with a lone pair of electrons on the central atom. For O_3, the structure is a hybrid of two equivalent structures. In each of these structures, the central atom has a formal charge of +1. The oxygen–oxygen bond order is between 1 and 2. Although many resonance structures can be drawn for SO_2, in the most important structure, the formal charge on the S atom is zero and the sulfur–oxygen bonds are double bonds. **109.** $\Delta_f H° = 639$ kJ mol^{-1}. The formation reaction is very endothermic and thus energetically unfavorable. This result supports the observation that Xe(g) does not react directly with $O_2(g)$ to form $XeO_3(g)$. Because the reaction converts 2.5 moles of gas into 1 mole of gas, we expect $\Delta_f S° < 0$; thus, the reaction is also entropically unfavorable.

Feature Problems **110.** Demonstrate that the three reactions result in the decomposition of water as the net reaction: $2\ H_2O \rightarrow 2\ H_2 + O_2$. First balance each equation and then consider how to manipulate and add the equations together to get the overall equation. **112.** $2\ H^+(aq) + ClO_3^-(aq) + 1\ e^- \rightarrow ClO_2(g) + H_2O(l)$. The standard voltage for this half-reaction is given in Appendix D. $H^+(aq) + ClO_2(aq) + 1\ e^- \rightarrow HClO_2(g)$, $E° = 1.187$ V.

Self-Assessment Exercises **120.** The answer is (b). **121.** The answer is (d). **122.** The answer is (c). **123.** The answer is (a). **124.** The answer is (d). **125.** The answer is (b). **126.** The answer is (a) and (b). **127a.** $Cl_2(g) + 2\ NaOH(aq) \rightarrow NaCl(aq) + NaOCl(aq) + H_2O(l)$ **127b.** $2\ NaI(s) + 2\ H_2SO_4(\text{concd aq}) \rightarrow Na_2SO_4(a) + 2\ H_2O(g) + SO_2(g) + I_2(g)$ **127c.** $Cl_2(g) + 2\ KI_3(aq) \rightarrow 2\ KCl(aq) + 3\ I_2(s)$

127d. $3\ NaBr(s) + H_3PO_4(\text{concd aq}) \rightarrow Na_3PO_4 + 3\ HBr(g)$ **127e.** $5\ HSO_3^-(aq) + 2\ MnO_4^-(aq) + H^+(aq) \rightarrow 5\ SO_4^{2-}(aq) + 2\ Mn^{2+}(aq) + 3\ H_2O(l)$
128a. $2\ KClO_3(s) \rightarrow 2\ KCl(s) + 3\ O_2(g)$. A catalyst such as MnO_2 is required. Electrolysis of H_2O is an alternate method. **128b.** $3\ Cu(s) + 8\ H^+(aq) + 2\ NO_3^-(aq) \rightarrow 3\ Cu^{2+}(aq) + 4\ H_2O(l) + 2\ NO(g)$. Difficult to control reaction so that NO(g) is the only reduction product. **128c.** $Zn(s) + 2\ H^+(aq) \rightarrow Zn^{2+}(aq) + H_2(g)$. A number of other metals can be used. **128d.** $NH_4Cl(aq) + NaOH(aq) \rightarrow NaCl(aq) + H_2O(l) + NH_3(g)$. Other ammonium salts and other bases work as well. **128e.** $CaCO_3(s) + 2\ HCl(aq) \rightarrow CaCl_2(aq) + H_2O(l) + CO_2(g)$. Other carbonates and acids can be used.
129a. $LiH(s) + H_2O(l) \rightarrow LiOH(aq) + H_2(g)$ **129b.** $C(s) + H_2O(g) \rightarrow CO(g) + H_2(g)$ **129c.** $3\ NO_2(g) + H_2O(l) \rightarrow 2\ HNO_3(aq) + NO(g)$ **130.** $3\ Cl_2(g) + I^-(aq) + 3\ H_2O(l) \rightarrow 6\ Cl^-(aq) + IO_3^-(aq) + 6\ H^+(aq)$, $Cl_2(g) + 2\ Br^-(aq) \rightarrow 2\ Cl^-(aq) + Br_2(aq)$. **131.** 5.0 kg^3 of seawater. **132a.** AgAt **132b.** sodium perxenate **132c.** MgPo **132d.** tellurous acid **132e.** K_2SeSO_3 **132f.** potassium perastatate **133.** 0.38 V **134a.** SO_3 **134b.** SO_2 **134c.** Cl_2O_7 **134d.** I_2O_5 **135a.** False **135b.** True **135c.** True

CHAPTER 23

Practice Examples 1a. (a) $2\ Cu_2S(s) + 3\ O_2(g) \rightarrow 2\ Cu_2O(s) + 2\ SO_2(g)$; (b) $WO_3(s) + 3\ H_2(g) \rightarrow W(s) + 3\ H_2O(g)$; (c) $2\ HgO(s) \xrightarrow{\Delta} 2\ Hg(l) + O_2(g)$. **1b.** (a) $3\ Si(s) + 2\ Cr_2O_3(s) \xrightarrow{\Delta} 3\ SiO_2(s) + 4\ Cr(s)$; (b) $2\ Co(OH)_3(s) \xrightarrow{\Delta,\ Air} Co_2O_3(s) + 3\ H_2O(g)$; (c) $Mn^{2+}(aq) + 2\ H_2O(l) \rightarrow MnO_2(s) + 4\ H^+(aq) + 2\ e^-$. **2a.** $NO_3^-(aq) + 3\ V^{3+}(aq) + H_2O(l) \rightarrow NO(g) + 3\ VO^{2+}(aq) + 2\ H^+(aq)$. $E^\circ_{cell} = +0.619V$ **2b.** $-E^\circ$ for the couple must be > -0.041 V and $< +1.13$ V. Fe(s), $Cr^{2+}(aq)$, and Zn(s) will do the job.

Integrative Example

A. $2\ V^{2+} + PtCl_6^{2-} \rightarrow 2\ V^{3+} + PtCl_4^{2-} + 2\ Cl^-$. The formation of $PtCl_6^{2-}$ is favored by using a very low concentration of V^{2+} and very high concentrations of V^{3+}, $PtCl_4^{2-}$, and Cl^-. In practical terms, an external voltage source would be required to make a significant amount of $PtCl_6^{2-}$ from V^{3+}, $PtCl_4^{2-}$, and Cl^-.
B. The disproportionation reaction is $3\ Ti^{2+} \rightarrow 2\ Ti^{3+} + Ti$ and $E^\circ = -1.261$ V. Because $E^\circ < 0$ the reaction is not spontaneous under standard conditions.

Exercises

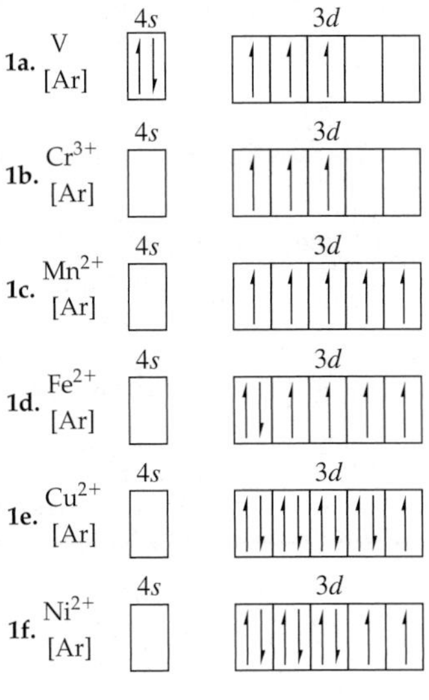

3. A given main-group metal typically displays just one oxidation state, usually equal to its family number in the periodic table. Main group metals do not form a wide variety of complex ions, with Al^{3+}, Sn^{2+}, Sn^{4+}, and Pb^{2+} being major exceptions. On the other hand, most transition metal ions form an extensive variety of complex ions. Most compounds of main group metals are colorless; exceptions occur when the anion is colored. On the other hand, many of the compounds of transition metal cations are colored, due to d-d electron transitions. Virtually every main-group metal cation has no unpaired electrons and hence is diamagnetic. On the other hand, many transition metals cations have one or more unpaired electrons and therefore are paramagnetic.
5. When an electron and a proton are added to a main group element to create the element of next highest atomic number, it is the electron that influences the radius. The electron is added to the outermost shell (n value), which is farthest from the nucleus. Moreover, the added electron is well shielded from the nucleus and hence it is only weakly attracted to the nucleus. Thus, this electron billows out and, as a result, has a major influence on the size of the atom. However, when an electron and a proton are added to a transition metal atom to create the atom of next highest atomic number, the electron it is not added to the outermost shell. The electron is added to the d-orbitals, which are one principal quantum number lower than the outermost shell. Also, the electron in the same subshell as the added electron offers little shielding. Thus it has small effect on the size of the atom. **7.** Manganese. One possible explanation might be its $3d^5 4s^2$ electron configuration. Removing one electron produces an electron configuration ($3d^5 4s^1$) with two half-filled subshells, removing two produces one with a half-filled and an empty subshell. Then there is no point of semistability until the remaining five *d* electrons are removed. **9.** The greater ease of forming lanthanide cations compared to forming transition metal cations, is due to the larger size of lanthanide atoms. The valence (outer shell) electrons of these larger atoms are further from the nucleus, less strongly attracted to the positive charge of the nucleus in diffuse f-orbitals that are do not penetrate effectively and are very effectively shielded by the core electrons. As a result, they are removed much more readily.
11a. $TiCl_4(g) + 4\ Na(l) \xrightarrow{\Delta} Ti(s) + 4\ NaCl(l)$
11b. $Cr_2O_3(s) + 2\ Al(s) \xrightarrow{\Delta} 2\ Cr(l) + Al_2O_3(s)$
11c. $Ag(s) + HCl(aq) \rightarrow$ no reaction
11d. $K_2Cr_2O_7(aq) + 2\ KOH(aq) \rightarrow 2\ K_2CrO_4(aq) + H_2O(l)$
11e. $MnO_2(s) + 2\ C(s) \xrightarrow{\Delta} Mn(l) + 2\ CO(g)$
13a. $Sc(OH)_3(s) + 3\ H^+(aq) \rightarrow Sc^{3+}(aq) + 3\ H_2O(l)$
13b. $3\ Fe^{2+}(aq) + MnO_4^-(aq) + 2\ H_2O(l) \rightarrow 3\ Fe^{3+}(aq) + MnO_2(s) + 4\ OH^-$
13c. $2\ KOH(l) + TiO_2(s) \xrightarrow{\Delta} K_2TiO_3(s) + H_2O(g)$
13d. $Cu(s) + 2\ H_2SO_4(\text{conc, aq}) \rightarrow CuSO_4(aq) + SO_2(g) + 2\ H_2O(l)$
15a. $FeS(s) + 2\ HCl(aq) \rightarrow FeCl_2(aq) + H_2S(g)$
$4\ Fe^{2+}(aq) + O_2(g) + 4\ H^+(aq) \rightarrow 4\ Fe^{3+}(aq) + 2\ H_2O(l)$
$Fe^{3+}(aq) + 3\ OH^-(aq) \rightarrow Fe(OH)_3(s)$
15b. $BaCO_3(s) + 2\ HCl(aq) \rightarrow BaCl_2(aq) + H_2O(l) + CO_2(g)$
$2\ BaCl_2(aq) + K_2Cr_2O_7(aq) + 2\ NaOH(aq) \rightarrow 2\ BaCrO_4(s) + 2\ KCl(aq) + 2\ NaCl(aq) + H_2O(l)$
17. $HgS(s) + O_2(g) \xrightarrow{\Delta} Hg(l) + SO_2(g)$,
$4\ HgS(s) + 4\ CaO(s) \longrightarrow 4\ Hg(l) + 3\ CaS(s) + CaSO_4(s)$.

19. The graph should be similar to that for $2\ Mg(s) + O_2(g) \rightarrow 2\ MgO(s)$. We expect a positive slope with slight changes in the slope after the melting point (839 °C) and boiling point (1484 °C), mainly owing to changes in entropy. The plot will be below the ΔG° line for $2\ Mg(s) + O_2(g) \rightarrow 2\ MgO(s)$ at all temperatures.
21a. $CuO(S) + 2\ H^+(aq) + e^- \rightarrow Cu^+(aq) + H_2O$
21b. $FeO(s) + 2\ H^+(aq) \rightarrow Fe^{3+}(aq) + H_2O + e^-$
23a. $E^\circ_{cell} = -0.704$ V. Does not occur to a significant extent. **23b.** $E^\circ_{cell} = +0.229$ V. Does occur to a significant extent. **23c.** $E^\circ_{cell} = +0.54$ V. Does occur to a significant extent. **25.** $-E^\circ$ for the couple must be > -0.337 V and $< +0.255$ V. Pb(s), Sn(s), $H_2(g)$ will do the job. **27.** Table D-4 contains the following data: Cr^{3+}/Cr^{2+} reduction potential = −0.424 V, $Cr_2O_7^{2-}/Cr^{3+}$ reduction potential = 1.33 V and Cr^{2+}/Cr reduction potential = −0.90 V. The two unknown potentials are $Cr_2O_7^{2-}/Cr^{2+}$, E° = 0.892 V and Cr^{3+}/Cr, E° = − 0.74 V.
29. $Cr_2O_7^{2-}(aq) + H_2O(l) \rightleftharpoons 2\ CrO_4^{2-}(aq) + 2\ H^+(aq)$,
$Pb^{2+}(aq) + CrO_4^{2-}(aq) \rightleftharpoons PbCrO_4(s)$.
31. $3\ Zn(s) + Cr_2O_7^{2-}(aq) + 14\ H^+(aq) \rightarrow 3\ Zn^{2+}(aq) + 2\ Cr^{3+}(aq) + 7\ H_2O(l)$
$Zn(s) + 2\ Cr^{3+}(aq) \rightarrow Zn^{2+}(aq) + 2\ Cr^{2+}(aq)$
$4\ Cr^{2+}(aq) + O_2(g) + 4\ H^+(aq) \rightarrow 4\ Cr^{3+}(aq) + 2\ H_2O(l)$
33a. 0.074 M **33b.** 6.4×10^{-6} M **35.** 1.10 g Cr
37. Dichromate ion is the prevalent species in acidic solution. Oxoanions are better oxidizing agents in acidic solution because increasing the concentration of hydrogen ion favors formation of product. The half-equation is: $Cr_2O_7^{2-}(aq) + 14\ H^+(aq) + 6\ e^- \rightarrow 2\ Cr^{3+}(aq) + 7\ H_2O(l)$. Notice also that CrO_4^{2-} is smaller than is $Cr_2O_7^{2-}$, giving it a higher lattice energy in its compounds, which makes these compounds harder to dissolve. **39.** $E = 0.24$ V. Spontaneous under these conditions.
41. $Fe^{3+}(aq) + K_4[\overset{[II]}{Fe}(CN)_6](aq) \rightarrow K\overset{[III]}{Fe}[\overset{[II]}{Fe}(CN)_6](s) + 3K^+(aq)$
43a. $Cu^{2+}(aq) + H_2(g) \rightarrow Cu(s) + 2\ H^+(aq)$
43b. $Au^+(aq) + Fe^{2+}(aq) \rightarrow Au(s) + Fe^{3+}(aq)$
43c. $2\ Cu^{2+}(aq) + SO_2(g) + 2\ H_2O(l) \rightarrow 2\ Cu^+(aq) + SO_4^{2-}(aq) + 4\ H^+(aq)$ **45a.** No **45b.** Yes **47.** 0.912 V **49a.** 0.02 **49b.** 0.02 atm
51. For ZnO, λ = 413 nm (violet light). For CdS, λ = 479 nm (blue light). The blue light absorbed by CdS is subtracted from the white light incident on the surface of the solid. The remaining reflected light is yellow in this case. When the violet light is subtracted from the white light incident on the ZnO surface, the reflected light appears white.

Advanced and Integrative Exercises

53. First, Ag^+, the oxidation product, forms a very insoluble chloride, AgCl, which probably adheres to the surface of the metal and prevents further reaction. $AuCl_3$ is not noted as being an insoluble chloride. Second, gold(III) forms a very stable complex ion with chloride ion, $[AuCl_4]^-$. This complex ion is much more stable than the corresponding dichloroargentate ion, $[AgCl_2]^-$. **57a.** $Mo(CO)_6$ **57b.** $Os(CO)_5$ **57c.** $Re(CO)_5^-$ **57d.** All of the metal carbonyls have zero dipole moments and very weak intermolecular forces. Therefore one would expect these compounds to be liquids or low melting solids at room temperature. Since chromium carbonyls have more electrons than the other compounds it will be more polarizable and therefore have the strongest dispersion force leading to it being a solid while the other metal carbonyls are liquids. **57e.** This compound would be an ionic, salt-like material consisting of Na^+ and $V(CO)_6^-$ ions.

61a. Dissolution: $AgO(s) + 2\,H^+(aq) \rightarrow Ag^{2+}(aq) + H_2O(l)$
Oxidation: $2\,H_2O(l) \rightarrow O_2(g) + 4\,H^+(aq) + 4\,e^-$
Reduction: $Ag^{2+}(aq) + e^- \rightarrow Ag^+(aq)$
61b. $E^\circ_{cell} = +0.75$ V. Spontaneous. **64.** 85.0% **65a.** More of the 0.1000 M MnO_4^- solution would be required. **65b.** 29.40 mL **67.** % Mn = 3.286%, % Cr = 28.2%. **68a.** $PdC_8H_{14}O_4N_4$ **68b.** 3.80% **71a.** $[Hg—Hg]^{+2}$
71b.

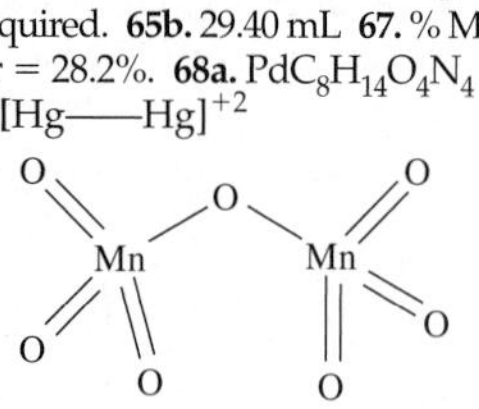

71c.

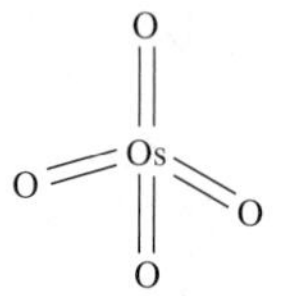

73. The empirical formula is NiTi. The % Ti by mass is 45%.
Feature Problems 74a. If $\Delta n_{gas} = 0$, then $\Delta_r S^\circ \sim 0$ and $\Delta_r G^\circ$ is essentially independent of temperature ($C(s) + O_2(g) \rightarrow CO_2(g)$). If $\Delta n_{gas} > 0$, then $\Delta_r S^\circ > 0$ and $\Delta_r G^\circ$ will become more negative with increasing temperature, hence the graph has a negative slope ($2\,C(s) + O_2(g) \rightarrow 2\,CO\,(g)$). If $\Delta n_{gas} < 0$, then $\Delta_r S^\circ < 0$ and $\Delta_r G^\circ$ will become more positive with increasing temperature, hence the graph has a positive slope ($2\,CO(g) + O_2(g) \rightarrow 2\,CO_2\,(g)$). **74b.** the plot of $\Delta_r G^\circ$ for the *net* reaction as a function of temperature will be a straight line with a slope of $-[\{\Delta_r S^\circ\,(\text{Rxn b})\} - \{\Delta_r S^\circ\,(\text{Rxn c})\}]$ (in kJ/K/mol) and a y-intercept of $[\{\Delta_r H^\circ\,(\text{Rxn b})\} - \{\Delta_r H^\circ\,(\text{Rxn c})\}]$ (in kJ mol^{-1}). the plot of $\Delta_r G^\circ$ vs. T for the reaction will follow the equation $y = -0.176\,x + 172.5$. $P_{CO} = 5.7$ atm.
Self-Assessment Exercises 79a. Impure iron (95% Fe). **79b.** Iron-manganese alloy. **79c.** $Fe(CrO_2)_2$ **79d.** An alloy of Zn and Cu with small amounts of Sn, Pb, and Fe. **79e.** Mixture of HCl(aq) and HNO_3(aq). **79f.** Impure Cu(s) containing SO_2(g). **79g.** Iron alloy with varying quantities of metals such as Cr, Mn, and Ni, and a small and carefully controlled percentage of carbon. **80.** The answers are (c), (f), and (g). **81.** The answer is (b). **82.** The answer is (d). **83.** The answer is (a). **84.** The answer is (d). **85.** The answers are (c) and (e). **86a.** $CrO_3(s)$ **86b.** potassium manganate **86c.** chromium carbonyl **86d.** $BaCr_2O_7$ **86e.** lanthanum(III) sulfate nonahydrate **86f.** $Au(CN)_3 \cdot 3\,H_2O$
87a. $2\,Fe_2S_3\,(s) + 3\,O_2\,(g) + 6\,H_2O(l) \rightarrow 4\,Fe(OH)_3\,(s) + 6S(s)$ **87b.** $2\,Mn^{2+}\,(aq) + 8\,H_2O(l) + 5\,S_2O_8^{2-}\,(aq) \rightarrow 2\,MnO_4^-(aq) + 16\,H^+\,(aq) + 10\,SO_4^{2-}(aq)$ **87c.** $4\,Ag(s) + 8\,CN^-\,(aq) + O_2\,(g) + 2\,H_2O(l) \rightarrow 4\,[Ag(CN_2]^-\,(aq) + 4\,OH^-\,(aq)$ **88.** Atoms of Zn, Cd and Hg have configurations of $4s^2\,3d^{10}$, $5s^2\,4d^{10}$, and $6s^2\,4f^{14}\,5d^{10}$, respectively. One the ns^2 electrons participate in bonding, and in this regard Zn, Cd, and Hg resemble the group 2 elements.
89. HNO_3 is the oxidizing agent for oxidizing the metal to Au^{3+} but Au^{3+} must be stabilized in solution. In aqua regia, Cl^- ions combine with Au^{3+} to form $[AuCl_4]^-$, which is stable in solution. **90.** The Fe^{3+} ion forms a complex ion, $Fe(H_2O)_6^{3+}$, in aqueous solution. The complex behaves as a weak monoprotic acid in solution.

CHAPTER 24

Practice Examples 1a. C.N. = 5, O.S. = +2. **1b.** $[Fe(CN)_6]^{3-}$ **2a.** $K_2[PtCl_6]$
2b. pentaamminethiocyanato-*S*-cobalt(III) chloride
3a.

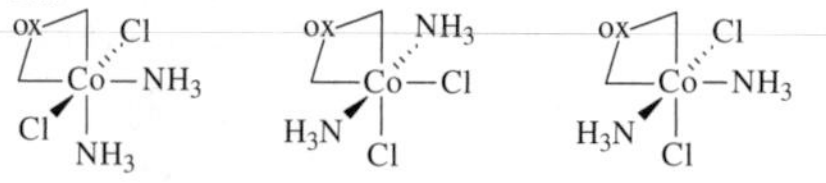

3b.

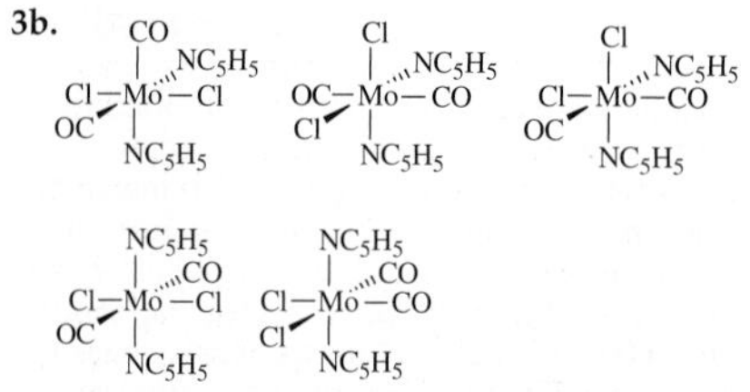

4a. Three **4b.** There are three unpaired electrons in each case. **5a.** $[Co(CN)_4]^{2-}$ must be tetrahedral (3 unpaired electrons) and not octahedral (1 unpaired electron) because the magnetic behavior of a tetrahedral arrangement would agree with the experimental observations (3 unpaired electrons). **5b.** The complex ion must be paramagnetic to the extent of one unpaired electron, regardless of the geometry the ligands adopt around the central metal ion.
6a. We are certain that $[Co(H_2O)_6]^{2+}$ is octahedral with a moderate field ligand. Tetrahedral $[CoCl_4]^{2-}$ has a weak field ligand. The relative values of ligand field splitting for the same ligand are $\Delta_t = 0.44\,\Delta_o$. Thus, $[Co(H_2O)_6]^{2+}$ absorbs light of higher energy, blue or green light, leaving a light pink as the complementary color we observe. $[CoCl_4]^{2-}$ absorbs lower energy red light, leaving blue light to pass through and be seen. **6b.** $K_4\,[Fe(CN)_6] \cdot 3\,H_2O$ should appear yellow. $[Fe(H_2O)_6](NO_3)_2$ should be green.
Integrative Example A. (a) *trans*-chloridobis (ethylenediamine) nitrito-*N*-cobalt(III) nitrite (no reaction with $AgNO_3$)

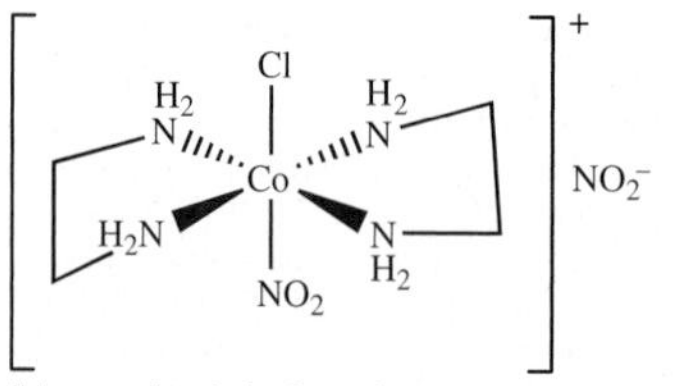

(b) *trans*-bis(ethylenediamine)dinitrito-*N*-cobalt(III) chloride (reacts with $AgNO_3$)

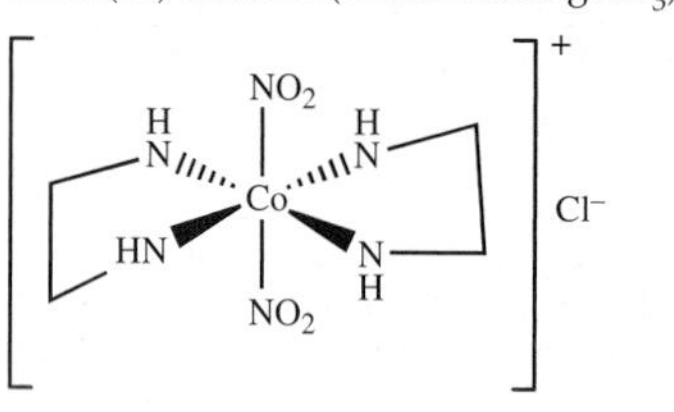

(c) *cis*-bis(ethylenediamine)dinitrito-*N*-cobalt(III) chloride (reacts with $AgNO_3$ and en)

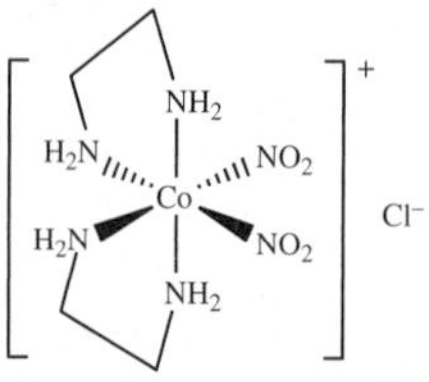

B. $[PtCl(NH_3)_5][Cl]_3$,

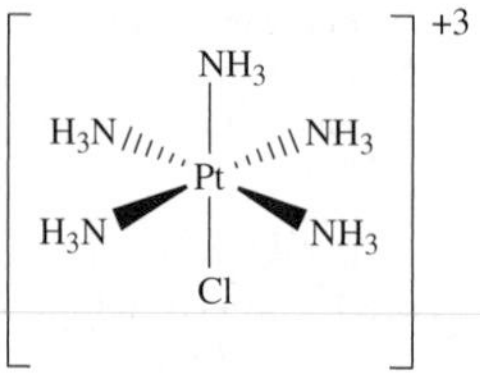

Exercises 1a. $[CrCl_4(NH_3)_2]^-$, diamminetetrachlorochromate(III) ion. **1b.** $[Fe(CN)_6]^{3-}$, hexacyanoferrate(III) ion. **1c.** $[Cr(en)_3]_2[Ni(CN)_4]_3$, tris(ethylenediamine)chromium(III) tetracyanonickelate(II) ion. **3a.** diamminesilver(I) chloride **3b.** tetraamminediaquacopper(II) sulfate **3c.** dichlorido(1,2-ethanediamine) platinum(II) **3d.** pentaaquabromidochromium (III) ion **3e.** rubidium tetrafluoridoargentate (III) **3f.** sodium pentacyanidooxidonitrogenferrate(III)
5a.

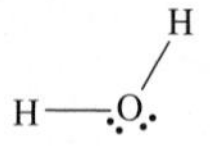

5b.

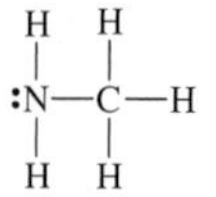

5c.

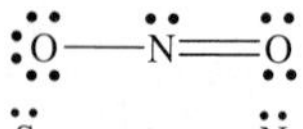

5d.

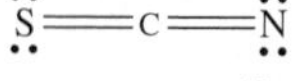

7a.

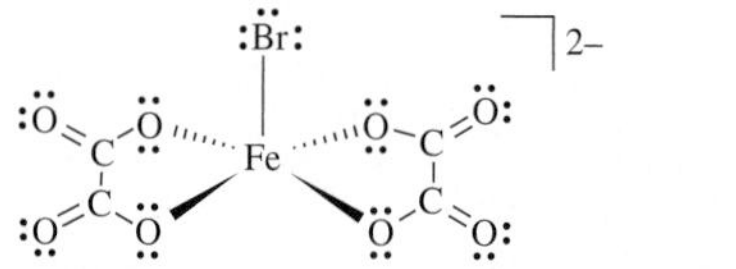

7b.

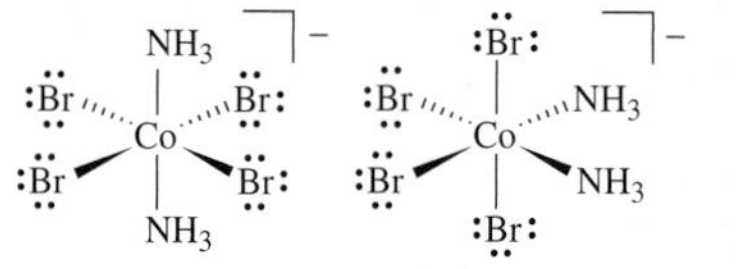

trans *cis*

7c.

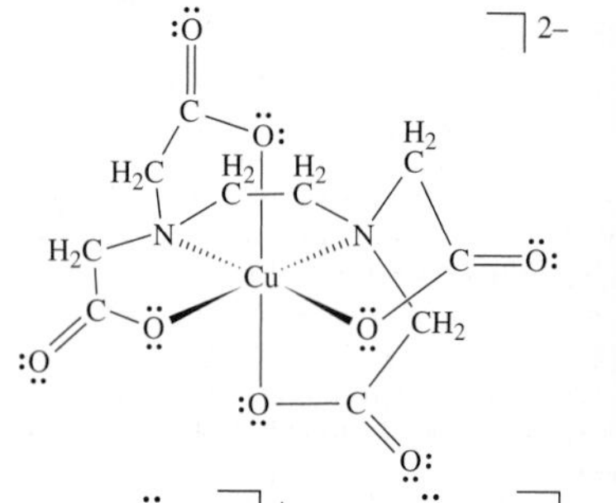

7d.

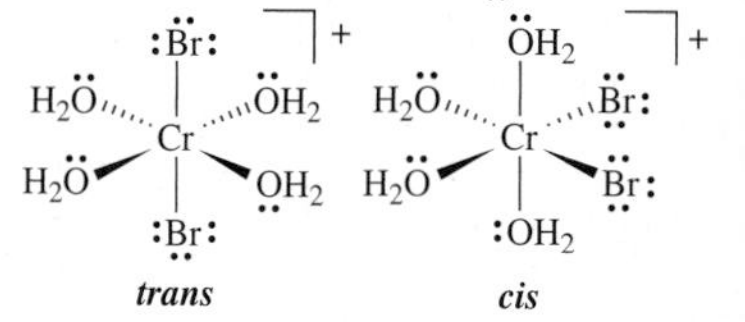

trans *cis*

7e.

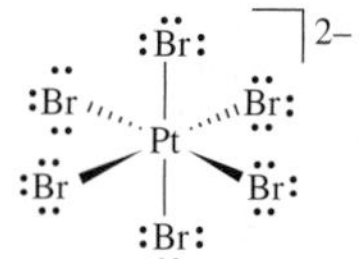

9a.

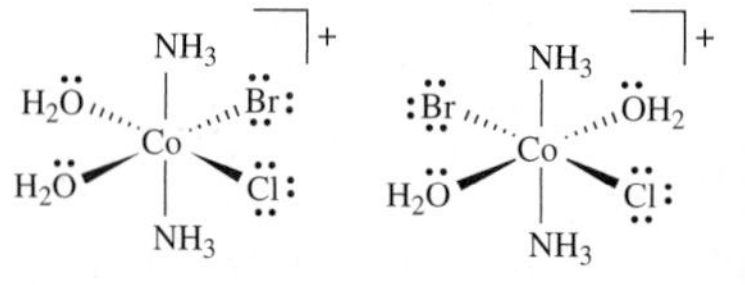

trans-NH_3 *cis*-H_2O *trans*-NH_3 *trans*-H_2O

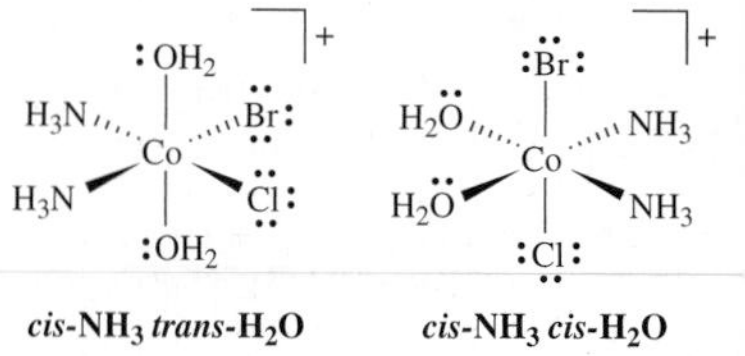

cis-NH_3 *trans*-H_2O *cis*-NH_3 *cis*-H_2O

9b.

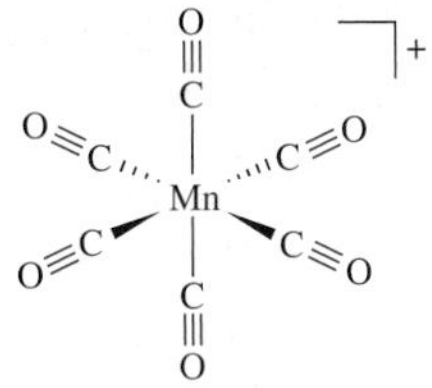

9c.

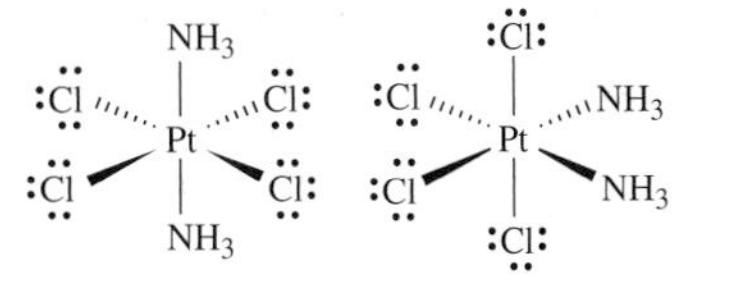

trans *cis*

9d.

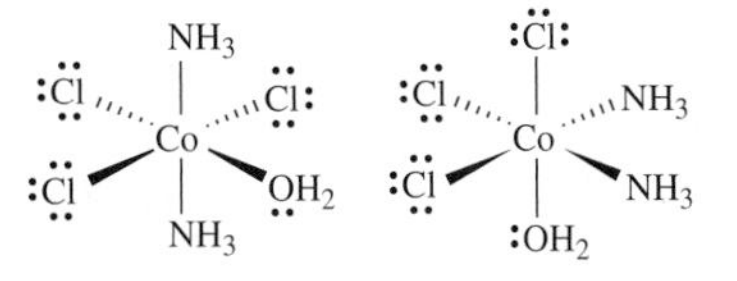

trans *cis*

11a. Tetrahedral structures do not display *cis-trans* isomerism. **11b.** Square planar structures can show *cis-trans* isomerism. **11c.** Linear structures do not display *cis-trans* isomerism. **13a.** Three **13b.** Yes

15.

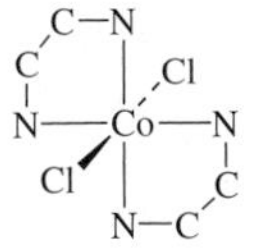

trans-isomer

The *cis*-dichloridobis(ethylenediamine)cobalt (III) ion is optically active. The *trans*-isomer is not optically active. **17.** In crystal field theory, the five *d* orbitals of a central transition metal ion are split into two (or more) groups of different energies. The energy spacing between these groups often corresponds to the energy of a photon of visible light. Thus, the transition-metal complex will absorb light with energy corresponding to this spacing. If white light is incident on the complex ion, the light remaining after absorption will be missing some of its components. Thus, light of certain wavelengths (corresponding to the energies absorbed) will no longer be present in the formerly white light. The resulting light is colored.

19.

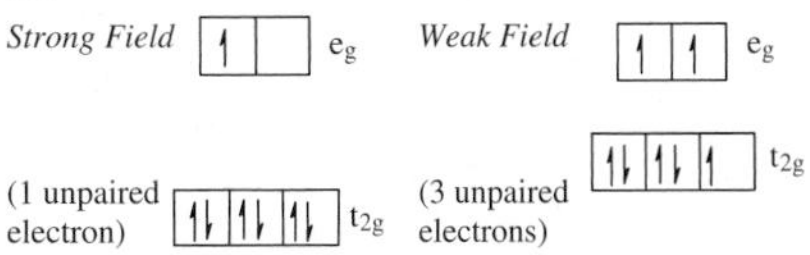

21a. The Mo complex is paramagnetic. The Co complex is diamagnetic. **21b.** Three **23.** The octahedral and tetrahedral configurations have the same number of unpaired electrons. The magnetic behavior cannot be used to determine whether the ammine complex of nickel(II) is octahedral or tetrahedral. **25a.** $Zn(OH)_2(s) + 4\,NH_3(aq) \rightleftharpoons [Zn(NH_3)_4]^{2+}(aq) + 2OH^-(aq)$ **25b.** $Cu^{2+}(aq) + 2\,OH^-(aq) \rightleftharpoons Cu(OH)_2(s)$, $Cu(OH)_2(s) + 4\,NH_3(aq) \rightleftharpoons [Cu(NH_3)_4]^{2+}$(aq, dark blue) $+ 2\,OH^-(aq)$, $[Cu(NH_3)_4]^{2+}(aq) + 4\,H_3O^+(aq) \rightleftharpoons [Cu(H_2O)_4]^{2+}(aq) + 4\,NH_4^+(aq)$. **27.** $[Co(en)_3]^{3+}$ should have the largest overall K_f value. **29.** First: $[Fe(H_2O)_6]^{3+}(aq) + en(aq) \rightleftharpoons [Fe(en)(H_2O)_4]^{3+}(aq) + 2\,H_2O(l)$ Second: $[Fe(en)(H_2O)_4]^{3+}(aq) + en(aq) \rightleftharpoons [Fe(en)_2(H_2O)_2]^{3+}(aq) + 2\,H_2O(l)$ Third: $[Fe(en)_2(H_2O)_2]^{3+}(aq) + en(aq) \rightleftharpoons [Fe(en)_3]^{3+}(aq) + 2\,H_2O(l)$ $K_f = 5.0 \times 10^9 = \beta_3$ **31a.** Aluminum(III) forms a stable (and soluble) hydroxo complex but not a stable ammine complex. **31b.** Although zinc(II) forms a soluble stable ammine complex ion, its formation constant is not sufficiently large to dissolve highly insoluble ZnS. However, it is sufficiently large to dissolve the moderately insoluble $ZnCO_3$. **31c.** Chloride ion forms a stable complex ion with silver(I) ion, that dissolves the AgCl(s) that formed when $[Cl^-]$ is low. **33.** $[Al(H_2O)_6]^{3+}(aq)$ **35a.** $K_{overall} = K_{sp} \times K_f = 8.5$. With a reasonably high $[S_2O_3{}^{2-}]$, this reaction will go essentially to completion. **35b.** NH_3 (aq) cannot be used in the fixing of photographic film because of the relatively small value of K_f for $[Ag(NH_3)_2]^+$ (aq), $K_f = 1.6 \times 10^7$. This would produce a value of $K = 8.0 \times 10^{-6}$ which is nowhere large enough to drive the reaction to completion. **37.** To make the *cis* isomer, ligands that show a strong tendency for directing incoming ligands to positions that are *trans* to themselves must be used. I^- has a stronger tendency than does Cl^- or NH_3 for directing incoming ligands to the *trans* positions, and so it is beneficial to convert $K_2[PtCl_4]$ to $K_2[PtI_4]$ before replacing ligands around Pt with NH_3 molecules.

Integrative and Advanced Exercises

39a. tetraamminecopper(II) ion, $[Cu(NH_3)_4]^{2+}$. **39b.** tetraamminedichloridocobalt(III) chloride, $[CoCl_2(NH_3)_4]Cl$. **39c.** hexachloridoplatinate(IV) ion, $[PtCl_6]^{2-}$. **39d.** sodium tetrachloridocuprate(II), $Na_2[CuCl_4]$, **39e.** potassium pentachloridoantimonate(III), $K_2[SbCl_5]$. **40.** $[Pt(NH_3)_4][PtCl_4]$, tetraammineplatinum(II) tetrachloridoplatinate(II). **45.** The successive acid ionizations of a complex ion such as $[Fe(OH)_6]^{3+}$ are more nearly equal in magnitude than those of an uncharged polyprotic acid such as H_3PO_4 principally because the complex ion has a positive charge. The second proton is leaving a species which has one fewer positive charge but which is nonetheless positively charged. Since positive charges repel each other, successive ionizations should not show a great decrease in the magnitude of their ionization constants. **47a.** 2.02 **47b.** 9.0×10^{-4} M **47c.** $[H_3O^+] = 90.$ M. An impossibly high concentration of H_3O^+ would be required. **51.** Total $[Cl^-] = 0.52$ M. **52a.** $4\,Co^{3+}(aq) + 2\,H_2O(l) \rightarrow 4\,Co^{2+}(aq) + 4\,H^+(aq) + O_2(g)$, $E^\circ_{cell} = +0.59$ V. **52b.** $[Co^{3+}] = 2.2 \times 10^{-28}$ M. **52c.** $E = -0.142$ V. The negative cell potential indicates that the reaction indeed does not occur. **54.** $E = +0.99$ V **57.** The Cr^{3+} ion would have 3 unpaired electrons, each residing in a 3*d* orbital and would be sp^3d^2 hybridized. The hybrid orbitals would be hybrids of 4*s*, 4*p*, and 3*d* (or 4*d*) orbitals. Each Cr–NH_3 coordinate covalent bond is a σ bond formed when a lone pair in an sp^3 orbital on N is directed toward an empty sp^3d^2 orbital on Cr^{3+}. The number of unpaired electrons predicted by valence bond theory would be the same as the number of unpaired electrons predicted by crystal field theory. **62.** The coordination compound is face-centered cubic, K^+ occupies tetrahedral holes, while $PtCl_6{}^{2-}$ occupies octahedral holes. **63.** In order from 0 NH_3 ligands to 6 NH_3 ligands: $K_2[PtCl_6]$, $K[PtCl_5(NH_3)]$, $PtCl_4(NH_3)_2$, $[PtCl_3(NH_3)_3]Cl$, $[PtCl_2(NH_3)_4]Cl_2$, $[PtCl(NH_3)_5]Cl_3$, $[Pt(NH_3)_6]Cl_4$.

Feature Problems 64a. A trigonal prismatic structure predicts three geometric isomers for $[CoCl_2(NH_3)_4]^+$, which is one more than the actual number of geometric isomers found for this complex ion. **64b.** Barring any unusual stretching of the ethylenediamine ligand, a trigonal prismatic structure cannot account for the optical isomerism that was observed for $[Co(en)_3]^{3+}$.

Self-Assessment Exercises 70. The answer is (d). **71.** The answer is (e). **72.** The answer is (b). **73.** The answer is (a). **74.** The answer is (d). **75.** The answer is (c). **76.** The answer is (b). **77a.** pentaamminebromidocobalt(III)sulfate, no isomerism. **77b.** hexaamminechromium(III)hexacyanidocobaltate(III), no isomerism. **77c.** sodiumhexanitrito-N-cobaltate(III), no isomerism **77d.** tris(ethylenediamine)cobalt(III)chloride, two optical isomers. **78a.** $[Ag(CN)_2]^-$ **78b.** $[Pt(NH_3)_3(NO_2)]^+$ **78c.** $[CoCl(en)_2(H_2O)]^{2+}$ **78d.** $K_4[Cr(CN)_6]$

79a.

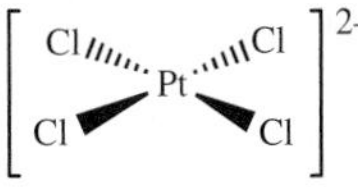

79b.

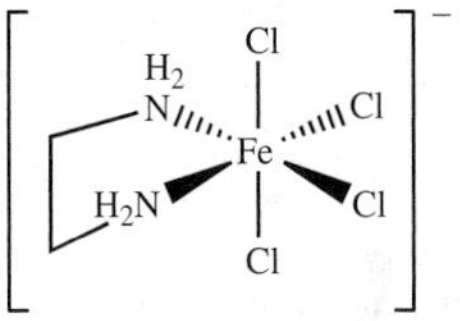

79c.

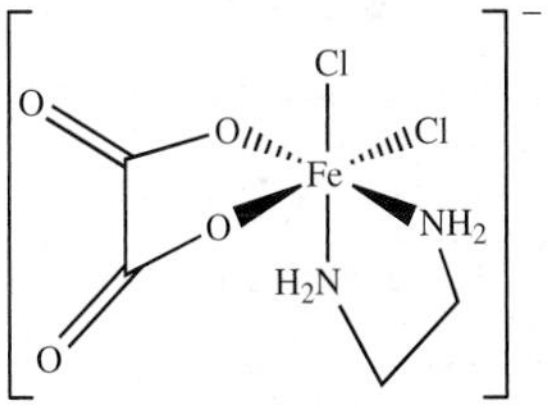

79d.

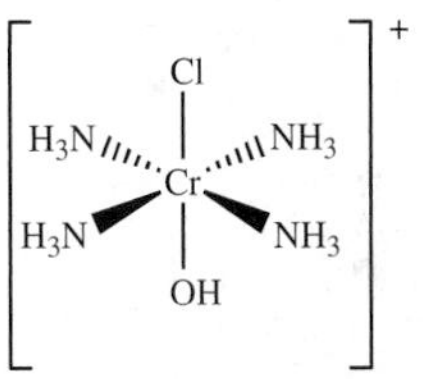

80a. One, **80b.** Two **80c.** Two **80d.** Two **81a.** Coordination isomerism **81b.** Linkage isomerism **81c.** No isomerism **81d.** Geometric isomerism **81e.** Geometric isomerism **82a.** Geometric isomerism (*cis* and *trans*) and optical isomerism in the *cis* isomer. **82b.** Geometric isomerism (*cis* and *trans*), optical isomerism in the *cis* isomer, and linkage isomerism in the thiocyanate ligand. **82c.** No isomers. **82d.** No isomers. **82e.** Two geometric isomers, one with the tridentate ligand occupying meridional positions and the other with the tridentate ligand occupying facial positions. **83.** $[Co(en)_3]^{3+}(aq)$ is yellow and $[Co(H_2O)_6]^{3+}(aq)$ is blue.

CHAPTER 25

Practice Examples

1a. ${}^{241}_{94}Pu \rightarrow {}^{241}_{95}Am + {}^{0}_{-1}\beta$ **1b.** ${}^{58}_{29}Cu \rightarrow {}^{58}_{28}Ni + {}^{0}_{+1}\beta$ **2a.** ${}^{139}_{57}La + {}^{12}_{6}C \rightarrow {}^{147}_{63}Eu + 4{}^{1}_{0}n$ **2b.** ${}^{121}_{51}Sb + {}^{4}_{2}He \rightarrow {}^{124}_{53}I + {}^{1}_{0}n$, ${}^{124}_{53}I \rightarrow {}^{0}_{+1}\beta + {}^{124}_{52}Te$. **3a.** (a) $9.98 \times 10^{-7}\,s^{-1}$; (b) 9.40×10^{12} disintegrations/second; (c) 2.36×10^{18} atoms;

(d) 2.36×10^{12} dis/s. **3b.** 75.8 d **4a.** 4.7 years $\times 10^3$ years **4b.** 13 dis/min (per gram of C) **5a.** 2.544 MeV **5b.** 0.006001 u **6a.** (a) Stable; (b) Radioactive; (c) Radioactive. **6b.** Positron emission by ${}^{17}F$ to produce ${}^{17}O$ and β^- emission by ${}^{22}F$ to produce ${}^{22}Ne$.

Integrative Example A. 0.14 **B.** (a) Yes; (b) 4.67×10^{40}; (c) Yes; (d) Possibly yes.

Exercises

1a. ${}^{55}_{26}\mathrm{Fe} \rightarrow {}^{55}_{27}\mathrm{Co} + {}^{0}_{-1}\mathrm{e}$ **1b.** ${}^{238}_{92}\mathrm{U} \rightarrow {}^{234}_{90}\mathrm{Th} + {}^{4}_{2}\mathrm{He}$

1c. ${}^{222}_{86}\mathrm{U} \rightarrow {}^{218}_{84}\mathrm{Po} + {}^{4}_{2}\mathrm{He}$; ${}^{218}_{84}\mathrm{Po} \rightarrow {}^{214}_{82}\mathrm{Pb} + {}^{4}_{2}\mathrm{He}$

1d. ${}^{64}_{29}\mathrm{Cu} \rightarrow {}^{64}_{30}\mathrm{Zn} + {}^{0}_{-1}\mathrm{e}$; ${}^{64}_{30}\mathrm{Zn} + {}^{64}_{31}\mathrm{Ga} + {}^{0}_{-1}\mathrm{e}$

3. ${}^{14}_{6}\mathrm{C} \rightarrow {}^{14}_{7}\mathrm{N} + {}^{0}_{-1}\mathrm{e}$

5.

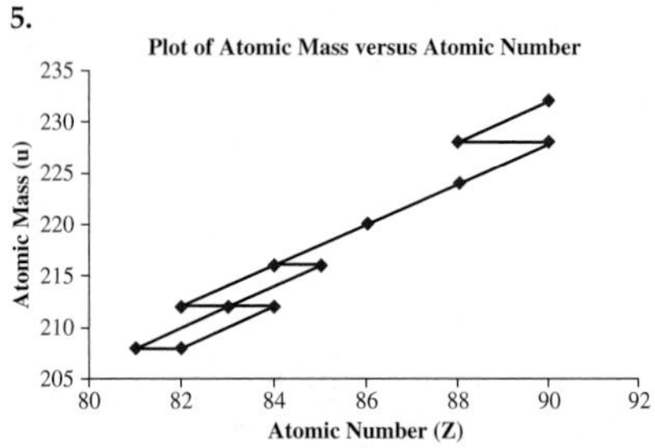

7. In Figure 25–2, only the following mass numbers are represented: 206, 210, 214, 218, 222, 226, 230, 234, and 238. The mass numbers are separated from each other by 4 units. The first of them, 206, equals $(4 \times 51) + 2$, that is $4n + 2$, where $n = 51$. **9a.** ${}^{160}_{74}\mathrm{W} \rightarrow {}^{156}_{72}\mathrm{Hf} + {}^{4}_{2}\mathrm{He}$

9b. ${}^{38}_{17}\mathrm{Cl} \rightarrow {}^{38}_{18}\mathrm{Ar} + {}^{0}_{-1}\beta$ **9c.** ${}^{214}_{83}\mathrm{Bi} \rightarrow {}^{214}_{84}\mathrm{Po} + {}^{0}_{-1}\beta$

9d. ${}^{32}_{17}\mathrm{Cl} \rightarrow {}^{32}_{16}\mathrm{S} + {}^{0}_{+1}\beta$ **11a.** ${}^{7}_{3}\mathrm{Li} + {}^{1}_{1}\mathrm{H} \rightarrow {}^{8}_{4}\mathrm{Be} + \gamma$

11b. ${}^{9}_{4}\mathrm{Be} + {}^{2}_{1}\mathrm{H} \rightarrow {}^{10}_{5}\mathrm{B} + {}^{1}_{0}\mathrm{n}$

11c. ${}^{14}_{7}\mathrm{N} + {}^{1}_{0}\mathrm{n} \rightarrow {}^{14}_{6}\mathrm{C} + {}^{1}_{1}\mathrm{H}$

13. ${}^{238}_{92}\mathrm{U} + {}^{2}_{1}\mathrm{H} \rightarrow {}^{238}_{93}\mathrm{Np} + 2{}^{1}_{0}\mathrm{n}$;

${}^{238}_{93}\mathrm{Np} \rightarrow 5{}^{4}_{2}\mathrm{He} + {}^{218}_{83}\mathrm{Bi}$.

15. ${}^{48}_{20}\mathrm{Ca} + {}^{249}_{98}\mathrm{Cf} \rightarrow {}^{249}_{118}\mathrm{Unk} + {}^{1}_{0}n + {}^{1}_{0}n + {}^{1}_{0}n$

17. ${}^{58}_{26}\mathrm{Fe} + {}^{244}_{94}\mathrm{Pu} \rightarrow {}^{302}_{120}\mathrm{Unk}$ **19a.** ${}^{214}_{84}\mathrm{Po}$ **19b.** ${}^{32}_{15}\mathrm{P}$

19c. ${}^{222}_{86}\mathrm{Rn}$, ${}^{13}_{8}\mathrm{O}$, ${}^{28}_{12}\mathrm{Mg}$, ${}^{80}_{35}\mathrm{Br}$, and ${}^{214}_{84}\mathrm{Po}$.

21. 1.90×10^8 atoms of ${}^{101}_{45}\mathrm{RH}$

23. 87.5 h **25.** 142 days **27.** The object is a bit more than 3000 years old, and thus is probably not from the pyramid era, which occurred about 3000 B.C. **29.** 0.12 **31.** 3.03×10^9 years **33a.** 5.42×10^{-9} J **33b.** 3727 MeV **35.** 7.805 MeV/nucleon **37.** 4.06 MeV **39.** 7.26×10^3 neutrons **41a.** ${}^{20}Ne$ **41b.** ${}^{18}O$ **41c.** ${}^{7}Li$ **43a.** ${}^{29}P$ should decay by β^+ emission. ${}^{33}P$ should decay by β^- emission. **43b.** ${}^{120}I$ should decay by β^+ emission. ${}^{134}I$ should decay by β^- emission. **45.** A "doubly magic" nuclide is one in which the atomic number is a magic number (2, 8, 20, 28, 50, 82, 114) and the number of neutrons also is a magic number (2, 8, 20, 28, 50, 82, 126, 184). An example is ${}^{16}O$. **47.** 1.37 mg **49.** The rem takes into account the quantity of biological damage done by a given dosage of radiation. The rad is the dosage that places 0.010 J of energy into each kilogram of irradiated matter. For living tissue, the rem provides a good idea of how much tissue damage a certain kind and quantity of radiation damage will do. But for nonliving materials, the rad is usually preferred. **51.** One reason why ${}^{90}Sr$ is hazardous is because strontium is in the same family of the periodic table as calcium, and hence often reacts in a similar fashion to calcium. The most likely place for calcium to be incorporated into the body is in bones, where it resides for a long time. Strontium is expected to behave in a similar fashion. Thus, it will be retained in the body for a long time. **53.** Mix a small amount of tritium with the H_2 (g) and detect where the radioactivity appears with a Geiger counter. **55.** The recovered sample will be radioactive. When NaCl(s) and $NaNO_3$ (s) are dissolved in solution, the ions are free to move throughout the solution. A given anion does not remain associated with a particular cation. Thus, all the anions and cations are shuffled and some of the radioactive ${}^{24}Na$ will end up in the crystallized $NaNO_3$. **59.** 2.9×10^3 metric tons **62.** 7.3×10^{-6} g ${}^{90}Sr$ **63.** 29 years **67.** 2.0×10^2 Ci **68.** 1.2×10^7 kJ

Feature Problems

72.

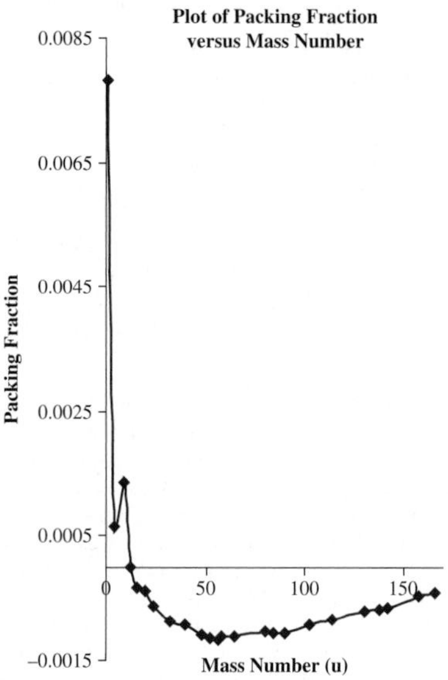

The graph and Fig. 25-6 are almost exactly the inverse of one another, with the maxima of one being the minima of the other. **73a.** The rate of decay depends on both the half-life and the number of radioactive atoms present. In the early stages of the decay chain, the larger number of radium-226, atoms multiplied by the very small decay constant is still larger than the product of the very small number of radon-222 atoms and its much larger decay constant. Only after some time has elapsed, does the rate of decay of radon-222 approach the rate at which it is formed from radium-226 and the amount of radon-222 reaches a maximum. Beyond this point, the rate of decay of radon-222 exceeds its rate of formation.

73b. $\dfrac{dD}{dt} = \lambda_p P - \lambda_d D = \lambda_p N_0 e^{-\lambda_p t} - \lambda_d D$

73c. 1 year **74a.** 87.5 u **74b.** 267.44 ppm **74c.** 2.95% **74d.** 2.11×10^9 years

Self-Assessment Exericises 78. The answer is (c). **79.** The answer is (b). **80.** The answer is (d). **81.** The answer is (c). **82.** The answer is (c). **83.** The answer is (d). **84.** The answer is (d).

85a. ${}^{214}_{88}\mathrm{Ra} \rightarrow {}^{210}_{86}\mathrm{Rn} + {}^{4}_{2}\mathrm{He}$

85b. ${}^{205}_{85}\mathrm{At} \rightarrow {}^{205}_{84}\mathrm{Po} + {}^{0}_{+1}\mathrm{b}$

85c. ${}^{212}_{87}\mathrm{Fr} + \mathrm{e}^- \rightarrow {}^{212}_{86}\mathrm{Rn}$

85d. ${}^{2}_{1}\mathrm{H} + {}^{2}_{1}\mathrm{H} \rightarrow {}^{3}_{2}\mathrm{He} + {}^{1}_{0}\mathrm{n}$

85e. ${}^{241}_{95}\mathrm{Am} + {}^{4}_{2}\mathrm{He} \rightarrow {}^{243}_{97}\mathrm{Bk} + 2{}^{1}_{0}\mathrm{n}$

85f. ${}^{232}_{90}\mathrm{Th} + {}^{4}_{2}\mathrm{He} \rightarrow {}^{232}_{92}\mathrm{U} + 4{}^{1}_{0}\mathrm{n}$ **86.** 76 days **87a.** 174 days **87b.** 287 days **87c.** 375 days **88.** The answer is (b). **89.** The answer is (d). **90.** The answer is (c). **91.** The answer is (a). **92.** The answer is (b).

CHAPTER 26

Practice Examples

1a.

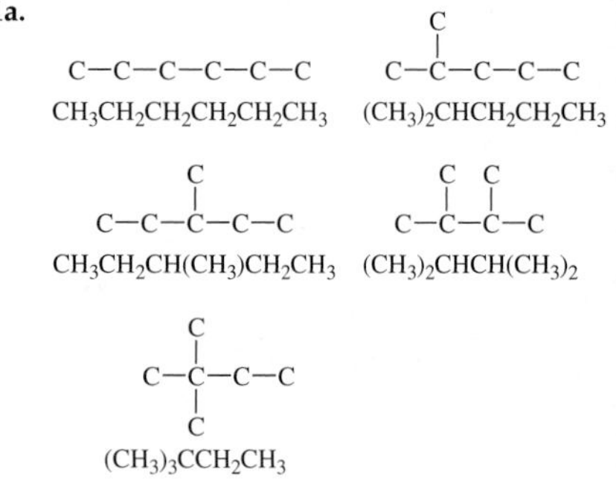

1b.

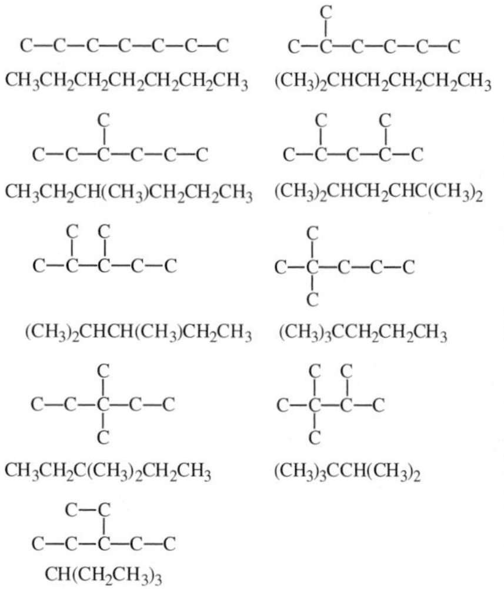

2a. 3,6,6-trimethyloctane **2b.** 3,6-dimethyloctane

3a.

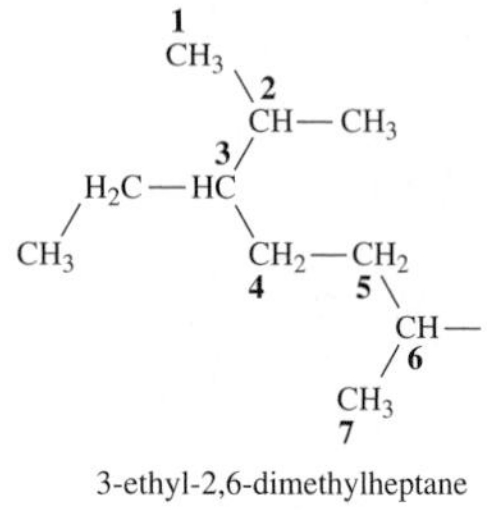

3-ethyl-2,6-dimethylheptane

3b.

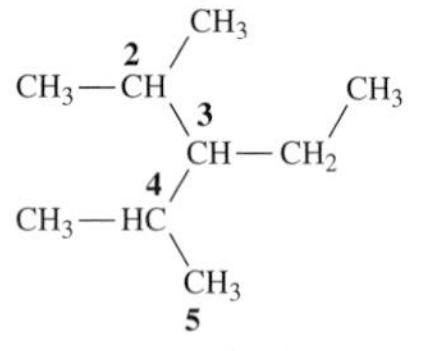

3-ethyl-2,4-dimethylpentane

4a.

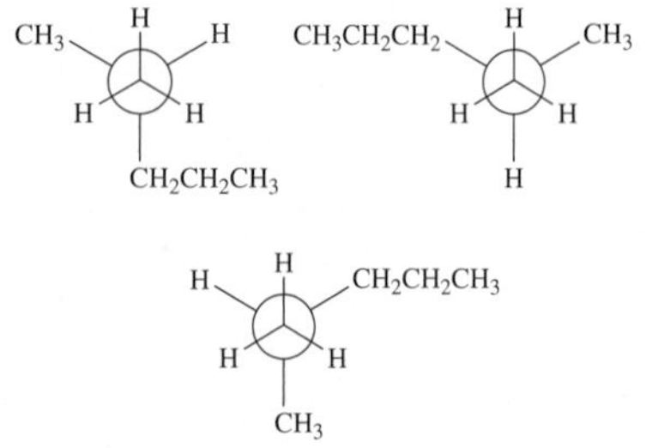

All conformations have the same energy.

4b.

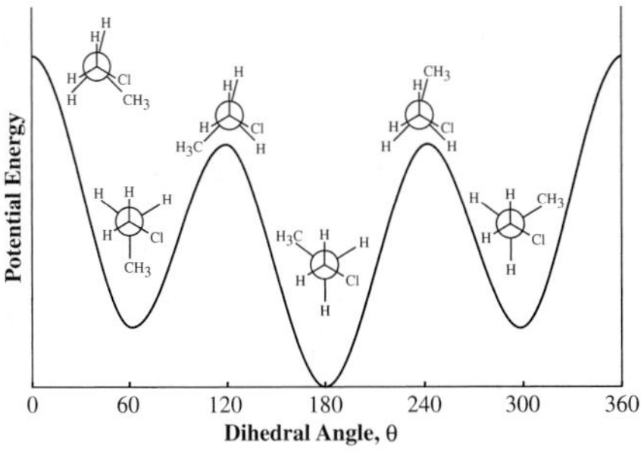

5a.

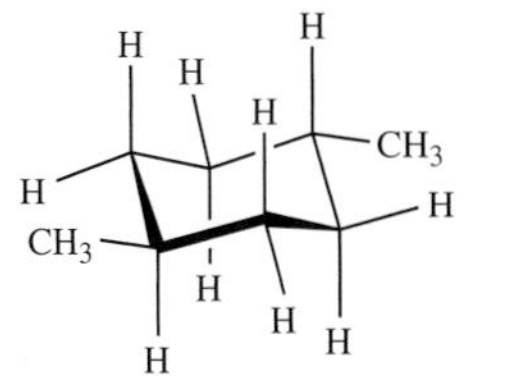

5b.

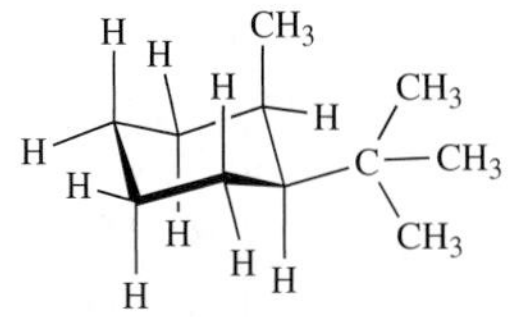

6a. (a) Achiral; (b) Chiral; (c) Chiral.
6b. (a) Achiral; (b) Chiral; (c) Chiral.
7a.

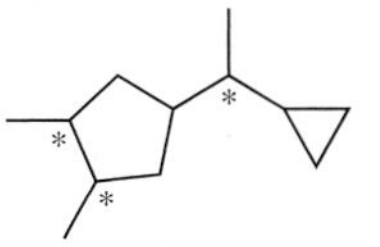

7b.

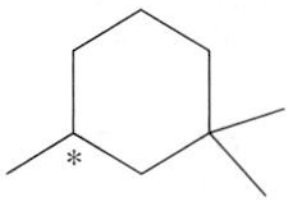

8a. (a) R configuration; (b) R configuration; (c) R configuration. **8b.** (a) Enantiomers; (b) Enantiomers. **9a.** (a) (E) stereoisomer; (b) (Z) stereoisomer; (c) (Z) stereoisomer.

9b. (a) (Z) stereoisomer

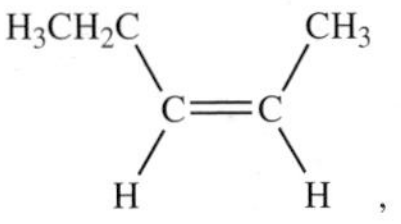

(E) stereoisomer

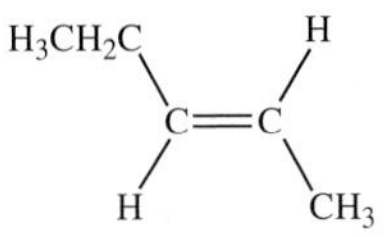

(b) (Z) stereoisomer

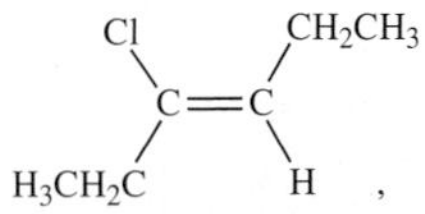

(E) stereoisomer

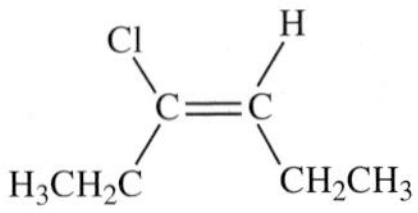

(c) (Z) stereoisomer

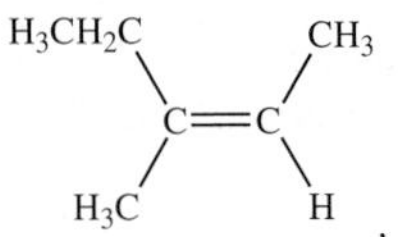

(E) stereoisomer

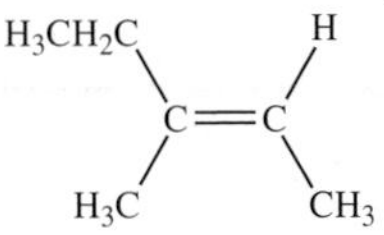

10a.

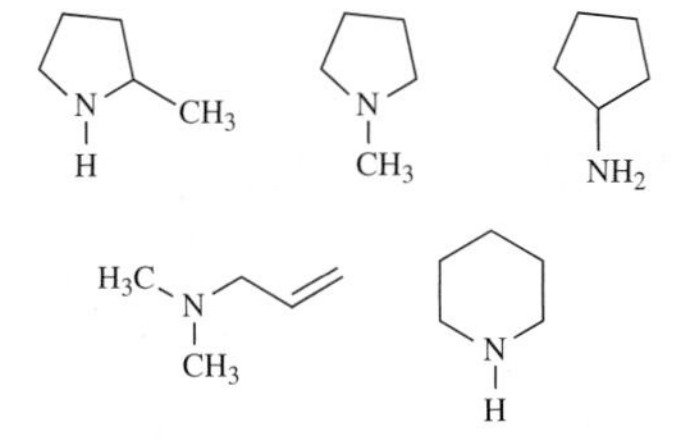

10b.

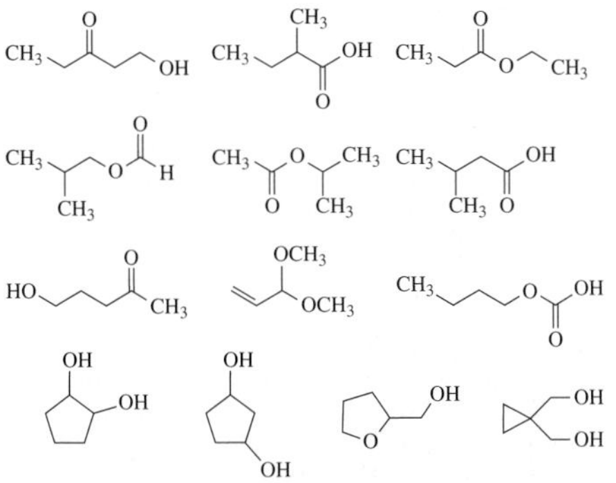

Integrative Example A. CH_3-CH_2-CH_2-OH (Compound A), CH_3-CH_2-CH_2-COOH (Compound B), CH_3-CH_2-CH_2- COO-CH_2-CH_3 (Compound C). **B.** Compounds (a) and (b) are both ethers. Ethers are relatively unreactive and the ether linkage is stable in the presence of most oxidizing and reducing agents, as well as dilute acids and alkalis. Compound (c) is unsaturated secondary alcohol. It will decolorize a Br_2/CCl_4 solution. Compound (d), on the other hand is an aldehyde. On treatment with chromic acid it will be oxidized to carboxylic acid.

Exercises

1a.

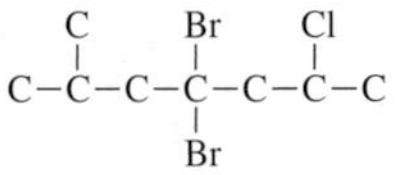

1b.

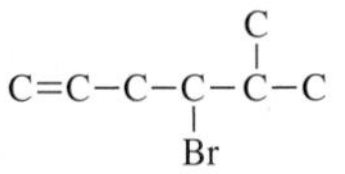

1c.

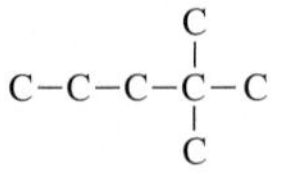

1d.

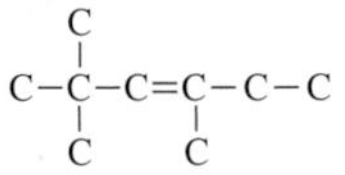

3a.

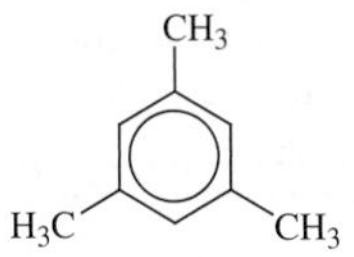

3b.

3c.

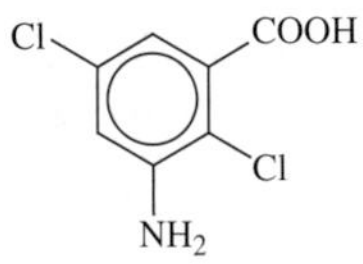

5a.

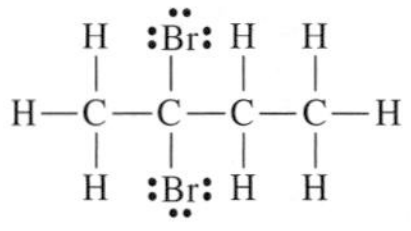

5b.

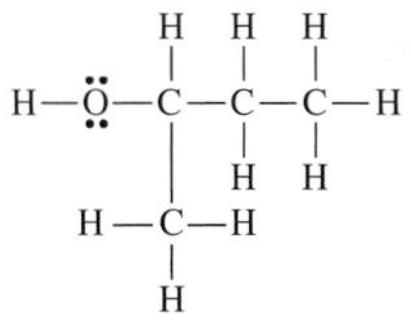

5c.

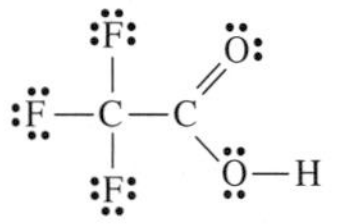

5d.

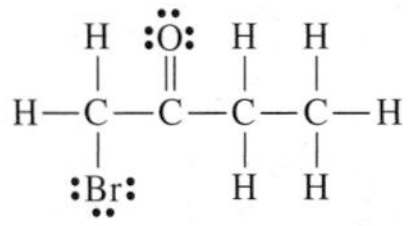

7a. Each carbon atom is sp^3 hybridized. All of the C – H bonds in the structure are sigma bonds, between the 1*s* orbital of H and the sp^3 orbital of C. The C–C bond is between sp^3 orbitals on each C atom.

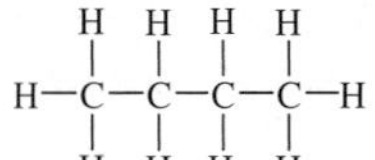

7b. Both carbon atoms are sp^2 hybridized. All of the C – H bonds in the structure are sigma bonds between the 1*s* orbital of H and the sp^2 orbital of C. The C–Cl bond is between the sp^2 orbital on C and the 3*p* orbital on Cl. The C = C double bond is composed of a sigma bond between the sp^2 orbitals on each C atom and a pi bond between the $2p_z$ orbitals on the two C atoms.

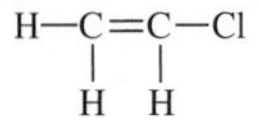

7c. The left-most C atom is sp^3 hybridized, and the C–H bonds to that C atom are between the sp^3 orbitals on C and the 1*s* orbital on H. The other two C atoms are *sp* hybridized. The right-hand C – H bond is between the *sp* orbital on C and the 1*s* orbital on H. The C≡C triple bond is composed of one sigma bond formed by overlap of *sp* orbitals, one from each C atom, and two pi bonds, each formed by the overlap of two 2*p* orbitals, one from each C atom (that is a $2p_y$—$2p_y$ overlap and a $2p_z$—$2p_z$ overlap).

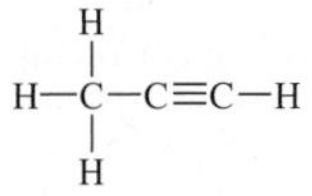

9a. Identical **9b.** Constitutional isomers **9c.** No relationship **9d.** Constitutional isomers **9e.** Stereoisomers

11.

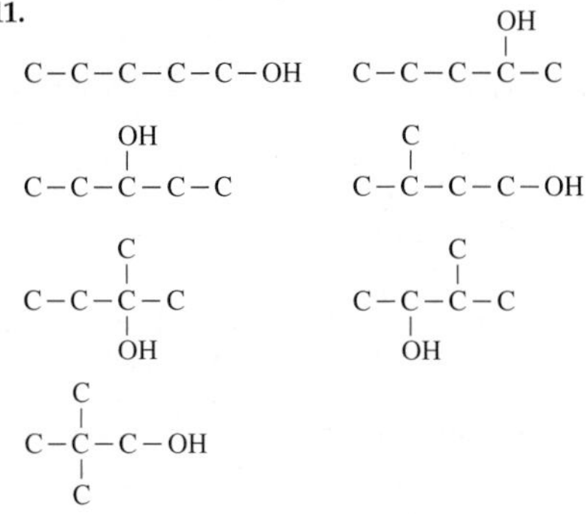

13a. Carbon 2 is chiral. **13b.** There are no chiral carbon atoms. **13c.** Carbon 2 is chiral.

15a.

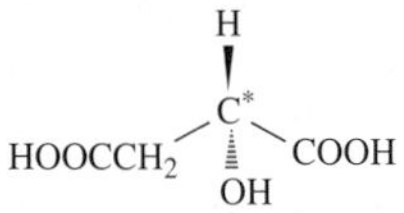

15b. There are no chiral carbon atoms.

15c.

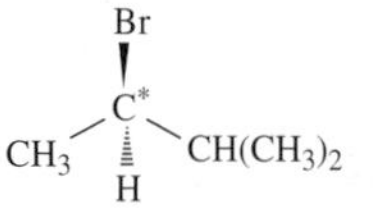

17a. Alkyl halide **17b.** Aldehyde **17c.** Ketone **17d.** alcohol, specifically hydroxyphenol **19a.** The hydroxyl group. **19b.** An aldehyde has a carbonyl group with a hydrogen and a carbon group attached. In a ketone, the carbonyl group is attached to two carbons. **19c.** The essential difference is the presence of the —OH group in acetic acid. **21a.** Carboxylic acid, hydroxyl group and disubstituted aromatic ring. **21b.** Ester group **21c.** Ketone and carboxylic acid group. **21d.** Aldehyde, hydroxyl group, and disubstituted aromatic ring.

23.

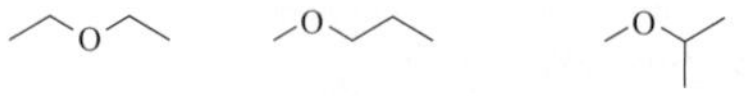

ethoxyethane — diethyl ether; 1-methoxypropane — methyl propyl ether; 2-methoxypropane — isopropyl methyl ether

25.

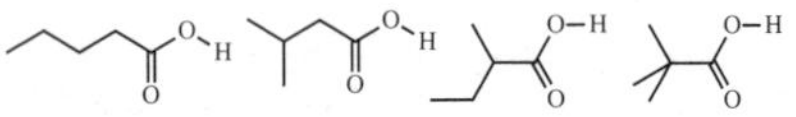

27.

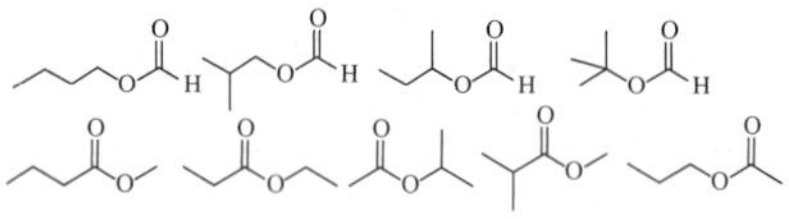

29.

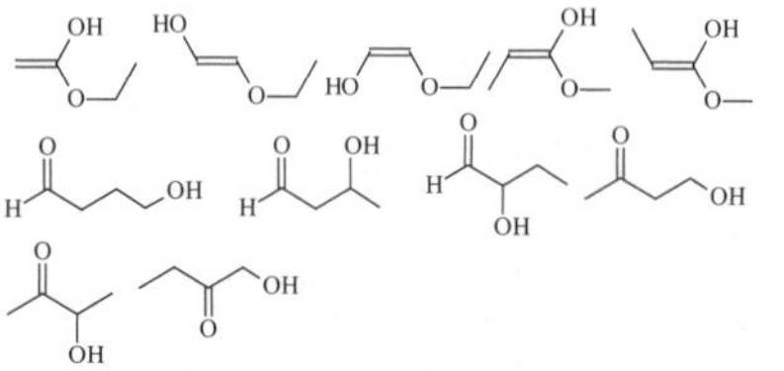

31a. 2,4,6-trimethyloctane **31b.** 4-ethyl-2-methylheptane **31c.** 2-chloro-2-methylpropane **31d.** 3-chloro-7-ethyl-4-methylnonane **33a.** 2,2-dimethylbutane **33b.** 2-methylpropene **33c.** 1,2-dimethylcyclopropane **33d.** 4-methylpent-2-yne **33e.** 3,4-dimethylhexane **33f.** 3,4-dimethyl-2-propylpent-1-ene **35a.** Insufficient **35b.** Sufficient **35c.** Insufficient **35d.** Insufficient **35e.** Sufficient **35f.** Insufficient

37a.

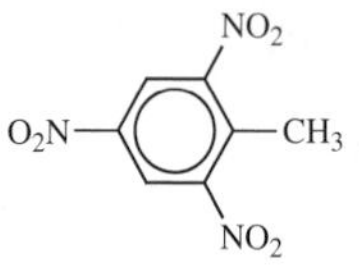

37b.

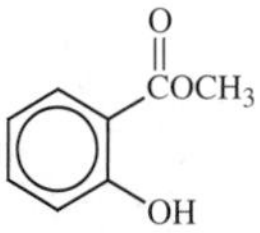

37c. $HOOCCH_2C(OH)(COOH)CH_2COOH$ **39a.** diethylamine **39b.** *p*-nitroaniline **39c.** *N*-ethylcyclopentanamine **39d.** *N*-ethyl-*N*-methylethanamine

41a. 2-methylbutane

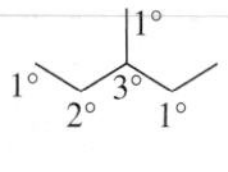

41b. 2,2-dimethylpropane

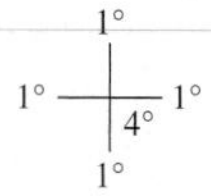

43. The Anti conformation is lowest in energy.

45.

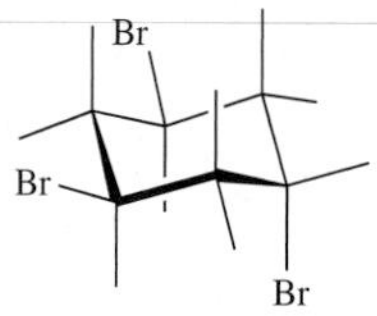

47a.

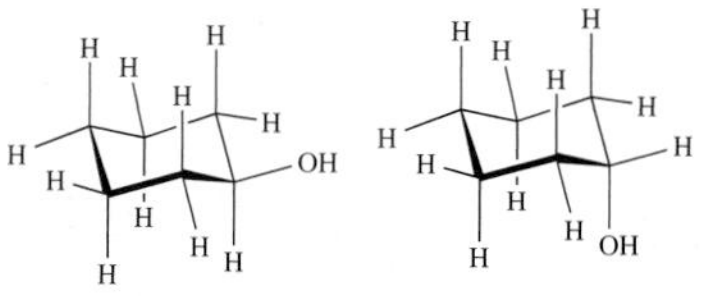

more stable

47b.

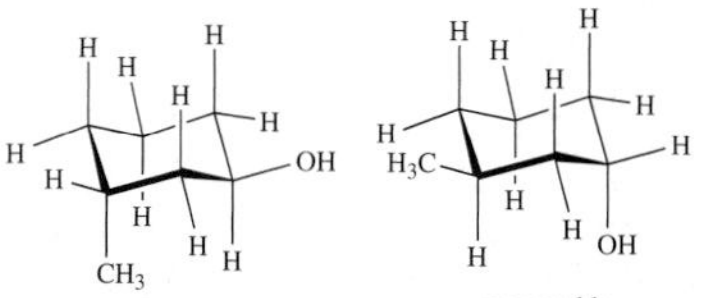

more stable

49. In the case of ethene there are only two carbon atoms between which there can be a double bond. In the case of propene, there can be a double bond only between the central carbon atom and a terminal carbon atom. The case of butene is different since 1-butene is distinct from 2- butene. **51a.** (E) configuration **51b.** (E) configuration **51c.** (Z) configuration **51d.** (E) configuration **51e.** (Z) configuration

53a. 2-chlorobut-2-ene

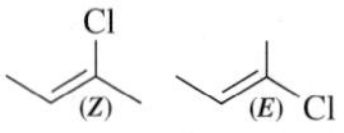

53b. 3-methylpent-2-ene

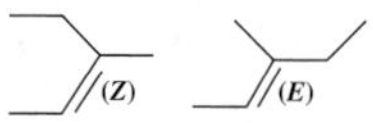

55a.

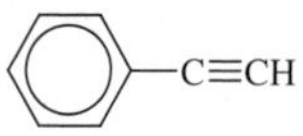

55b.

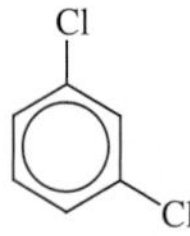

55c. 1,3,5-trihydroxybenzene

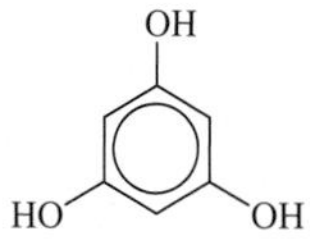

57. 1,2-dichloropropane 1,3-dichloropropane

2,2-dichloropropane 1,1-dichloropropane

59a. Identical **59b.** Identical **59c.** Isomers **59d.** Enantiomers **59e.** Identical **59f.** Identical **61a.** (S)-3-bromo-2-methylpentane **61b.** (S)-1,2-dibromopentane **61c.** (R)-3-(bromomethyl)-1-chloropentan-3-ol **61d.** (S)-1-bromopropan-2-ol **63a.** (Z)-pent-2-ene **63b.** (E)-1-chloro-2-methylbut-1-ene **63c.** (E)-4-(chloromethyl)-3,7-dimethyloct-3-ene **63d.** (Z)-5-bromo-3-(bromomethyl)-2-methylpent-2-enal

65a.

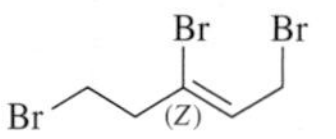

(*Z*)-1,3,5-tribromopent-2-ene

65b.

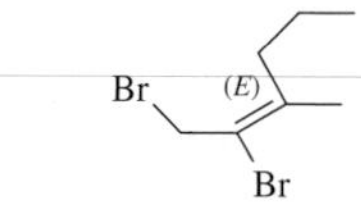

(*E*)-1,2-dibromo-3-methylhex-2-ene

65c.

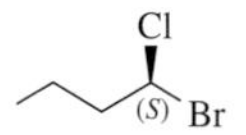

(*S*)-1-bromo-1-chlorobutane

65d.

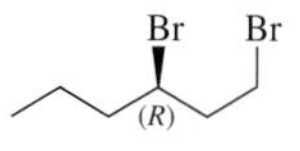

(*R*)-1,3-dibromohexane

65e.

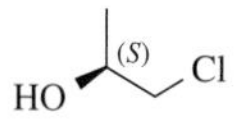

(*S*)-1-chloropropan-2-ol

67a. $C_4H_{11}N$ has zero elements of unsaturation. **67b.** C_4H_6O has two elements of unsaturation. **67c.** $C_9H_{15}ClO$ has two elements of unsaturation. **69.** There are two elements of unsaturation in the molecule (one π bond and one ring structure). The molecular formula is $C_8H_{15}NO$. **71.** Compound 1 will decolorize bromine water (b):

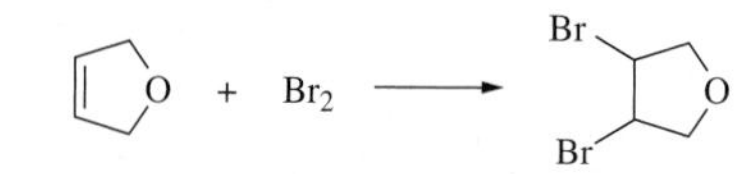

Compound 2 is easily oxidized (a). It will also react with sodium metal to liberate hydrogen gas:

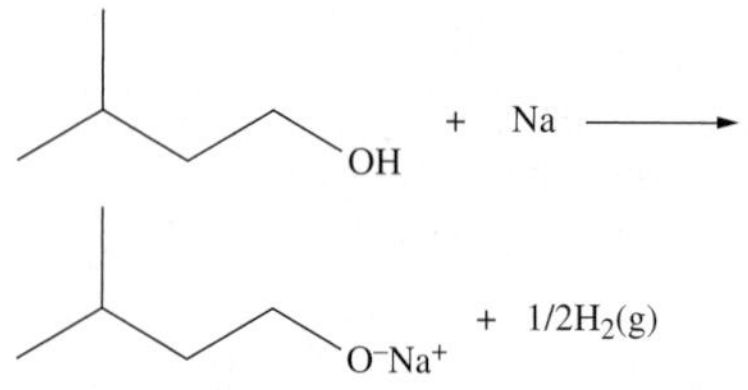

Compound 3 reacts with $Na_2CO_3(aq)$ to generate $CO_2(g)$ (c):

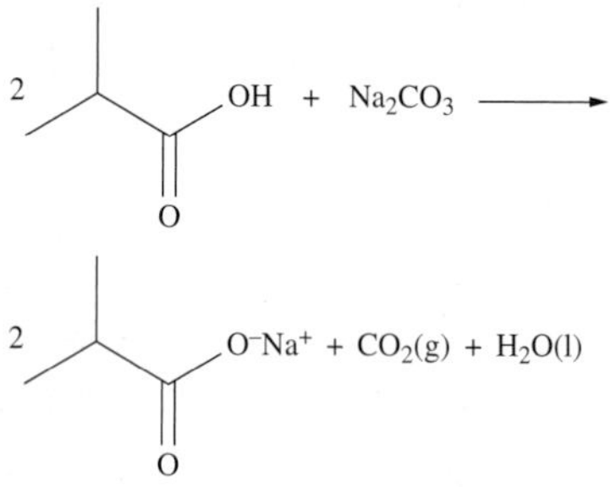

73.

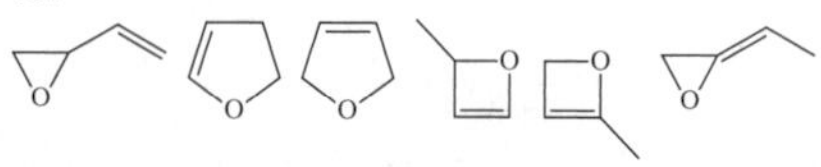

Integrative and Advanced Exercises

75a. cycloocta-1,5-diene

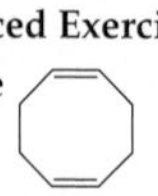

75b. 3,7,11-trimethyldodeca-2,6,10-trien-1-ol $HOCH_2CH{=}C(CH_3)CH_2CH_2CH{=}C(CH_3)CH_2CH_2CH = C(CH_3)_2$ **75c.** 2,6-dimethylhept-5-enal $OHCCH(CH_3)CH_2CH_2CH = C(CH_3)_2$

77.

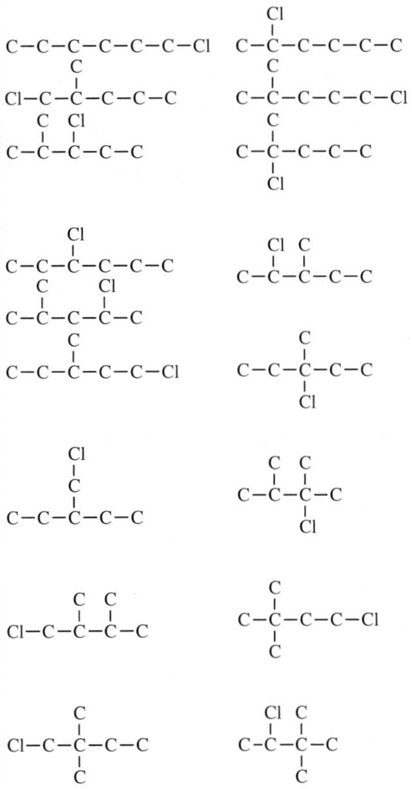

79a. $C_6H_5NO_2 + 7\,H^+ + 2\,Fe \longrightarrow C_6H_5NH_3^+ + 2\,H_2O + 2\,Fe^{3+}$
79b. $3\,C_6H_5CH_2OH + 2\,Cr_2O_7^{2-} + 16\,H^+ \longrightarrow 3\,C_6H_5COOH + 4\,Cr^{3+} + 11\,H_2O$
79c. $3\,CH_3CH = CH_2 + 2\,MnO_4^- + 4\,H_2O \longrightarrow 3\,CH_3CHOHCH_2OH + 2\,MnO_2 + 2\,OH^-$ **81.** $C_4H_8N_2O_2$ **84.** The unknown must be butan-1-ol or butyraldehyde. The identity of the unknown can be established by treatment with a carboxylic acid.
85.

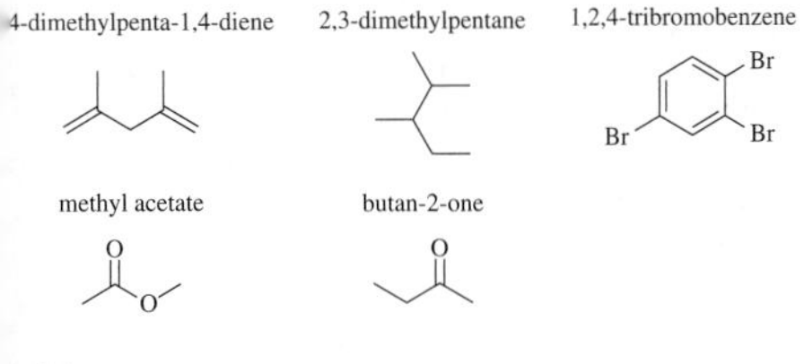

87.

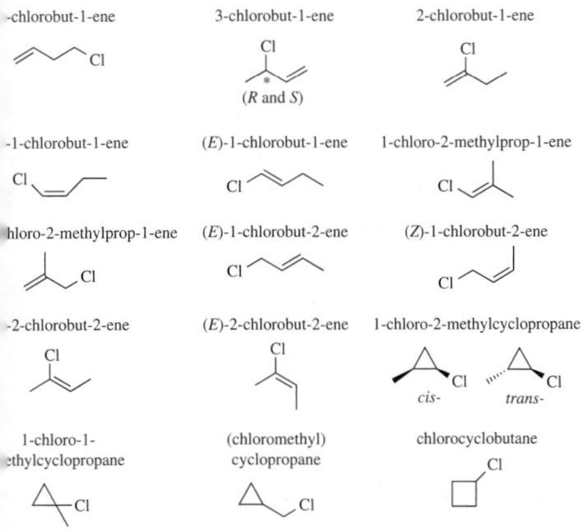

89a. Ester, amine(tertiary), arene.
89b. $C_1 = sp^2, C_2 = sp^3, C_3 = sp^3, C_4 = sp^3$, $N = sp^3$. **89c.** Carbons 2 and 4 are chiral.
91a. alcohol, amine(secondary), arene
91b. $C_1 = sp^2, C_2 = sp^3, C_3 = sp^3, C_4 = sp^3$, $N = sp^3$ **91c.** Carbons 3 and 4 are chiral.
91d. $pK_b = 4.87$
93a.

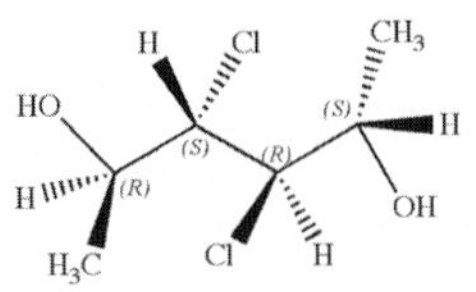

93b.

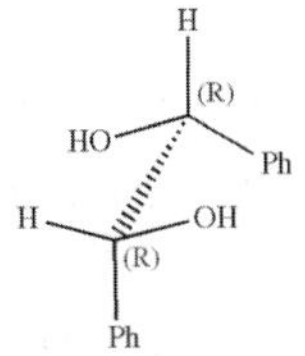

95. Cholesterol has eight chiral centers. The configuration of the carbon atom bonded to the OH group is S and the configuration of the double bond is Z.
Self-Assessment Exercises 101. The answer is (c). **102.** The answer is (b). **103.** The answer is (e). **108.** The answer is (c). **109.** C_4H_8
110a. C_6H_{12} **110b.** C_3H_7OH **110c.** C_6H_5COOH
111a.

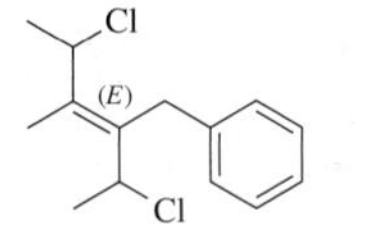

111b.

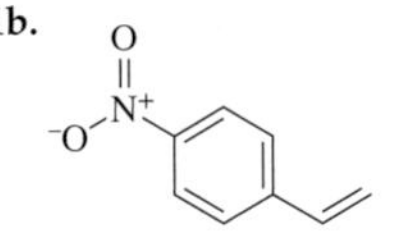

111c.

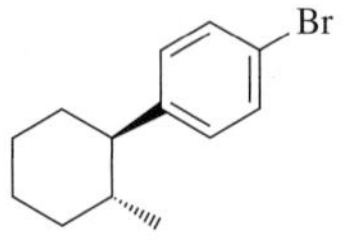

CHAPTER 27

Practice Examples 1a. (a) substitution (b) rearrangement (c) addition
1b. (a) elimination (b) substitution (c) substitution
2a. (a)

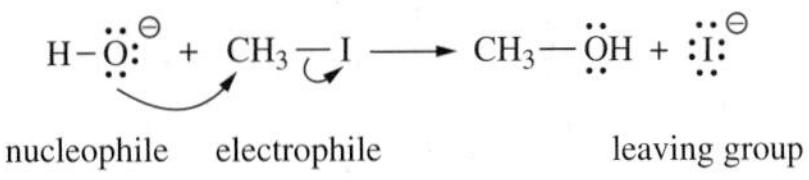

(b)

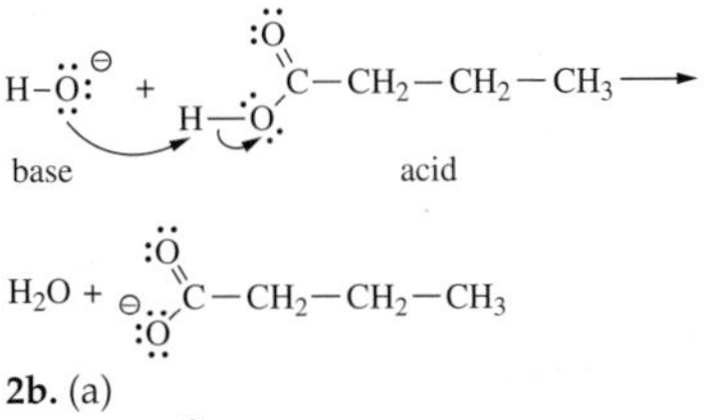

2b. (a)

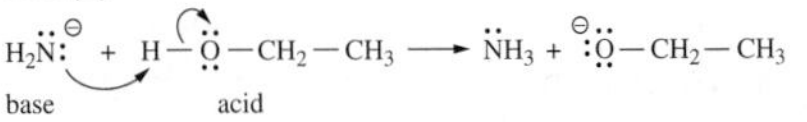

(b)

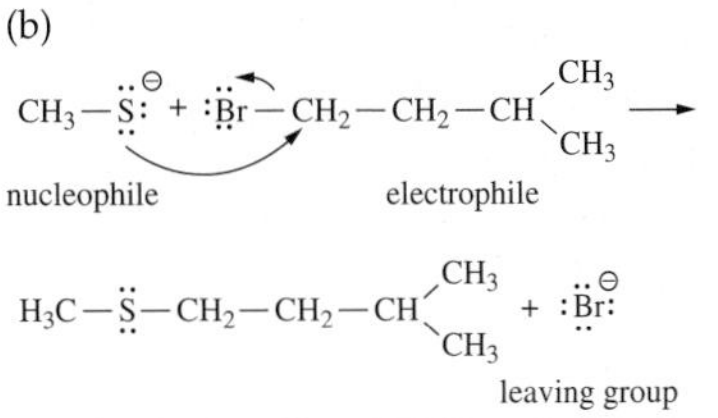

3a. (a) $CH_3C{\equiv}C^- + CH_3Br \xrightarrow{S_N2} CH_3C{\equiv}CCH_3 + Br^-$
(b) no reaction is expected
(c) $CH_3NH_2 + (CH_3)_3CCl \xrightarrow{S_N1} CH_3\overset{+}{N}H_2C(CH_3)_3 + Cl^-$
3b. (a) S_N2 mechanism

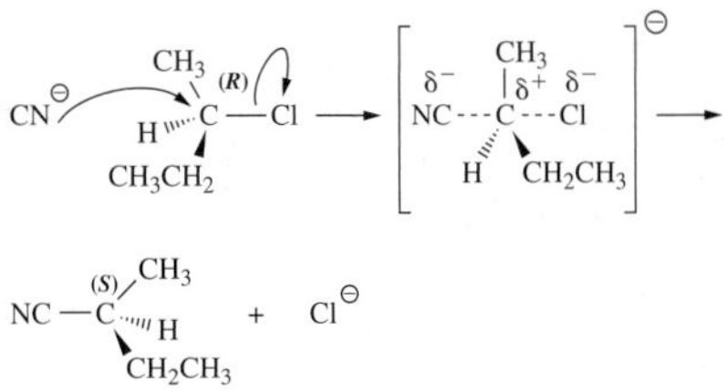

(b) S_N1 mechanism

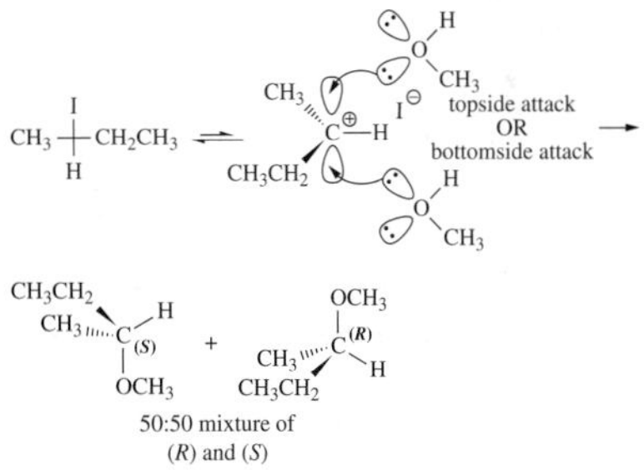

4a. S_N2 mechanism

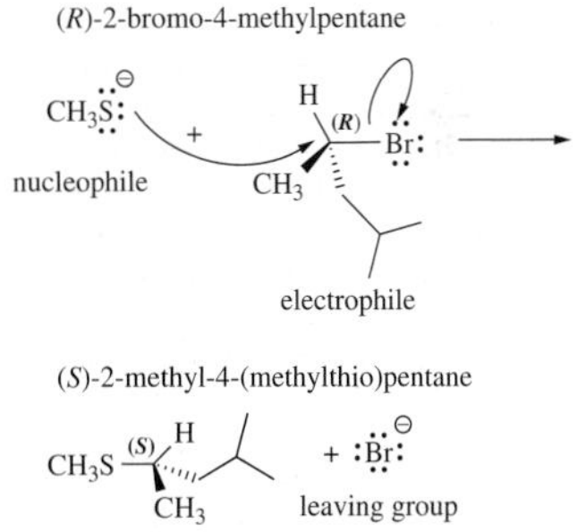

4b. S_N1 mechanism

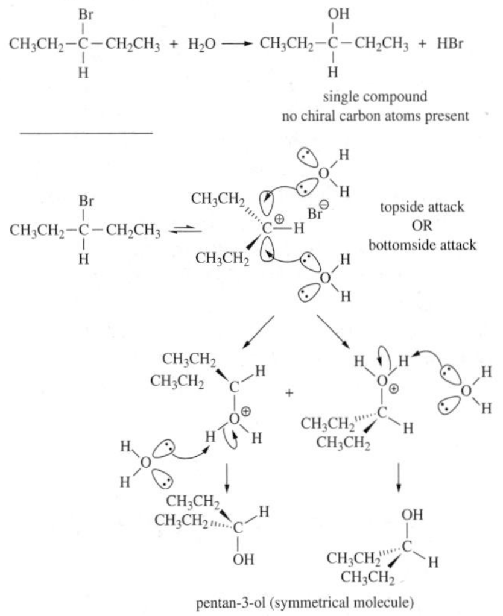

5a. This is an S_N2 reaction leading to the alkyl sulfide product.
5b.

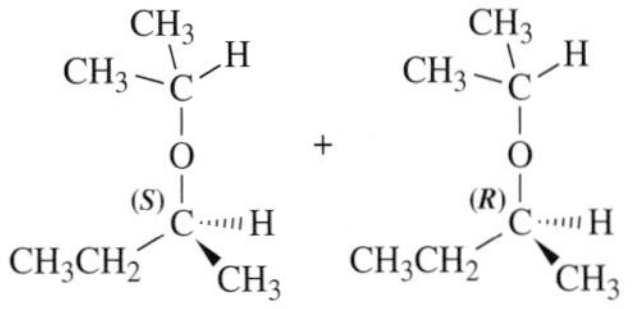

S_N1 (major product)
+
$CH_3{-}CH{=}CH{-}CH_3$
E1 (minor product)

6a. (a)

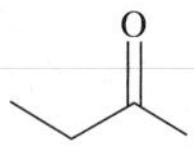

(b)

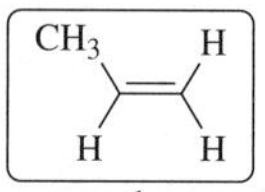

prop-1-ene

6b. (a) The reactions proceed via a S_N2 mechanism.

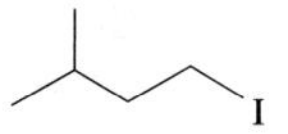

(b)

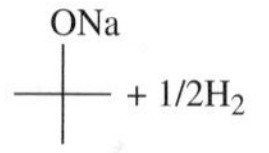

7a. 3-nitro-benzaldehyde

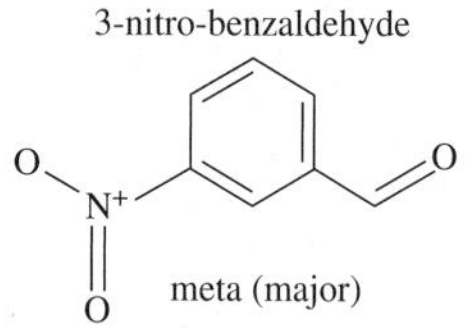

meta (major)

7b.

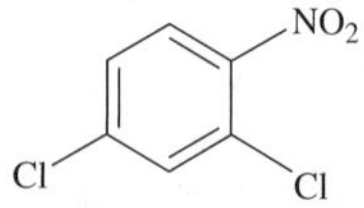

Integrative Example A. Alkanes are relatively unreactive. However, they will undergo bromination reaction in the presence of light. Such reaction will generate bromo-cyclohexane in the first step. In thepresence of a base, elimination will occur resulting in the formation of cyclohexene. Subsequent repetition of these two steps will genererate the desired product, 1,3- cyclohexadiene.

B.

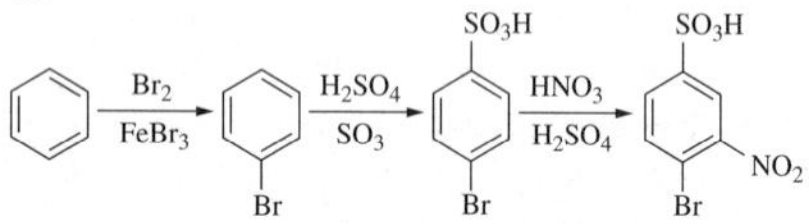

Exercises 1a. Nucleophilic substitution is a fundamental class of substitution reaction in which an "electron rich" nucleophile selectively bonds with or attacks the positive or partially positive charge of an atom attached to a group or atom called the leaving group; the positive or partially positive atom is referred to as an electrophile.

$$\text{R-Br} + \text{OH}^- \rightarrow \text{R-OH} + \text{Br}^-$$

1b. Electrophilic substitution reactions are chemical reactions in which an electrophile displaces another group,

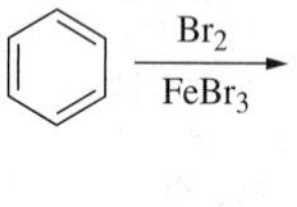

1c. In an addition reaction, a molecule adds across a double or triple bond in another molecule.

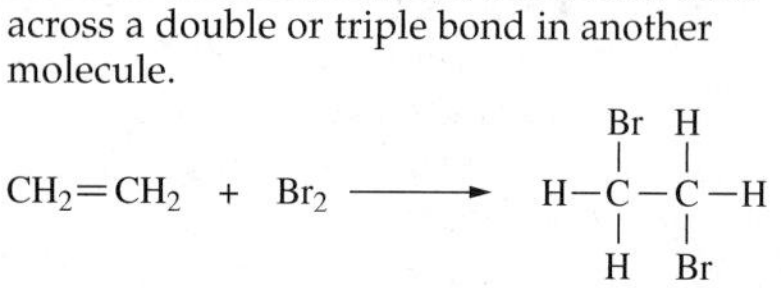

1d. In an elimination reaction, atoms or groups that are bonded to adjacent atoms are eliminated as a small molecule.

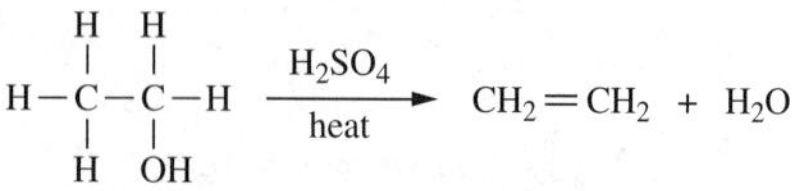

1e. When an organic compound undergoes a rearrangement reaction, the carbon skeleton of a molecule is rearranged.

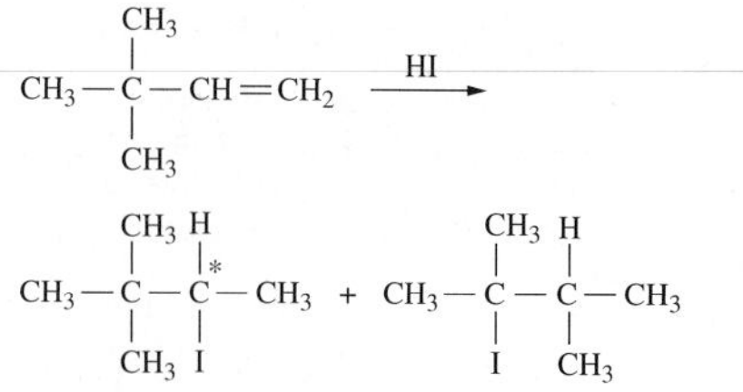

minor product (*S* and *R*) major product (*S* and *R*)

chiral carbon atom indicated by*

3a. substitution reaction **3b.** addition reaction **3c.** substitution reaction **5a.** addition reaction

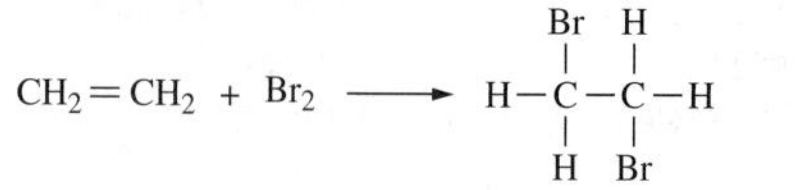

5b. elimination reaction

5c. substitution reaction

7a.

7b.

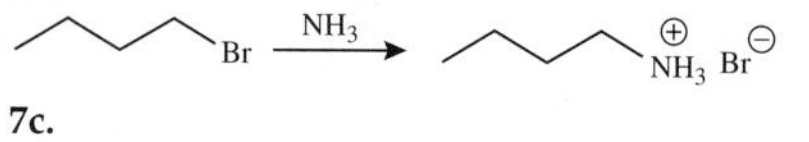

7c.

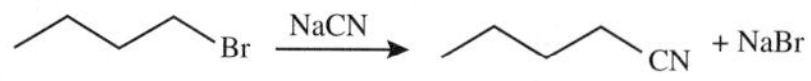

7d.

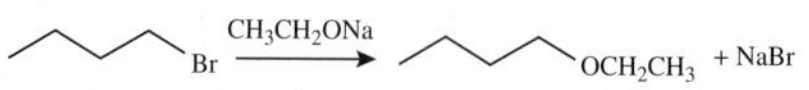

9a. rate=k[$CH_3CH_2CH_2CH_2Br$][NaOH]

9b.

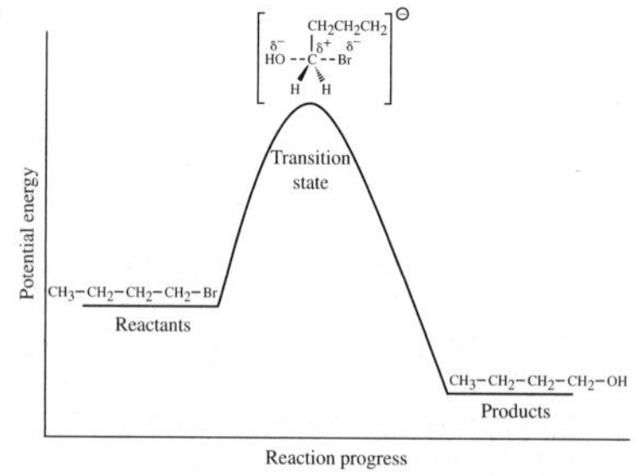

9c. Doubling the concentration of n-butyl bromide will also double the rate. **9d.** Decreasing the concentration of NaOH by a factor of two will decrease the rate by a factor of 2.

11.

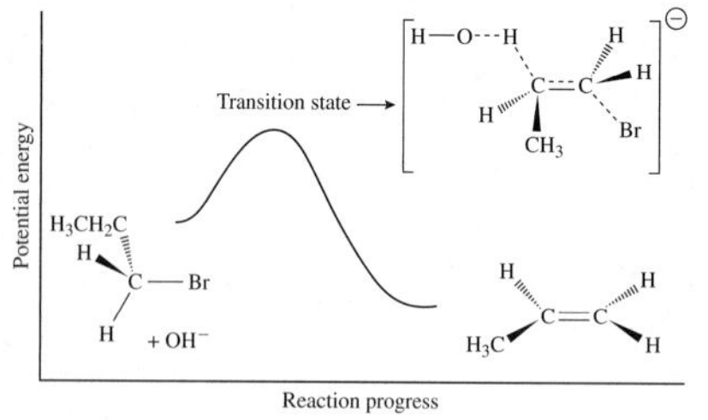

The rate will double.
The rate will decrease by a factor of 2.

13a. The equilibrium favors the formation of the products. $CH_3CH_2O^- + CH_3CH_2CH_2I \rightleftharpoons CH_3CH_2CH_2OCH_2CH_3 + NaI$
($CH_3CH_2O^-$: nucleophile; $CH_3CH_2CH_2I$: electrophile; NaI: leaving group)

13b. The equilibrium favors the formation of the products. $CH_3CH_2NH_3^+ + KI \rightleftharpoons NH_3 + CH_3CH_2I + K^+$
($CH_3CH_2NH_3^+$: electrophile; KI: nucleophile; NH_3: leaving group)

15. Molecule (a) reacts faster in the S_N2 reaction than does molecule (b) because there is less steric hindrance. In molecule (b), the steric hindrance is associated with a methyl group being positioned in the axial position.

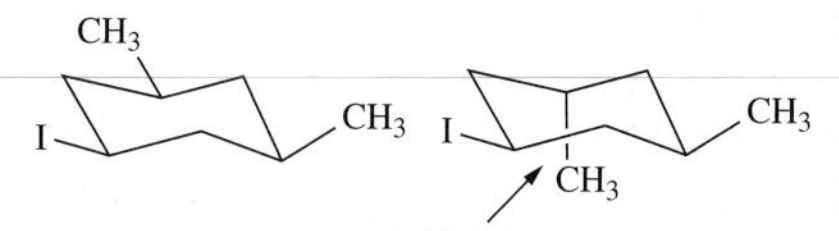

steric hindrance

17. The reaction most likely proceeded via an S_N1 mechanism.

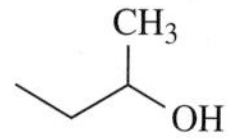

(*S* and *R*)

19. The reaction proceeded via an S_N1 mechanism.

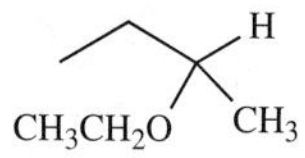

(*S* and *R*)

21a. E_2 mechanism **21b.** S_N2 mechanism

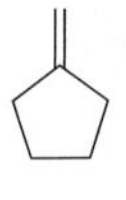

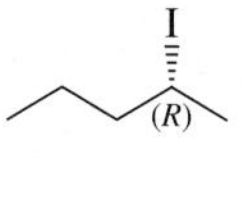

21c. E2 mechanism

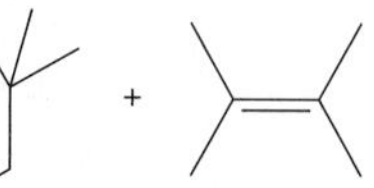

23a. $(CH_3)_2CHCH_2Br$ **23b.** $(CH_3)_3CO^-K^+$ **23c.** $(CH_3)_2CO$ **23d.** There is no reaction.

25.

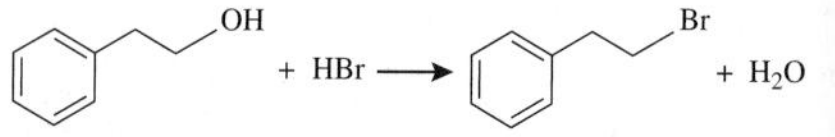

27. If no rearrangement occurs the expected product would be 4,4-dimethyl-pent -2-ene.

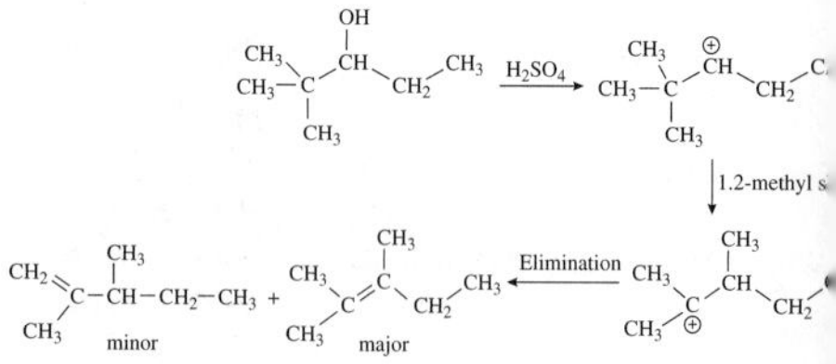

29a. $CH_3—CH_2—CH_3$

29b. $CH_3—CH{=}CH—CH_3$ E/Z major + $CH_3—CH_2—CH{=}CH_2$ minor

31a.

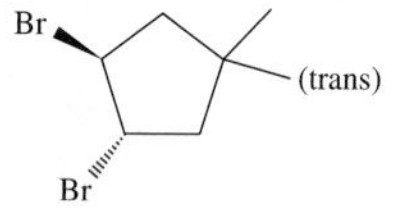

31b.

(*R*)-2-iodobutane (*S*)-2-iodobutane

31c.

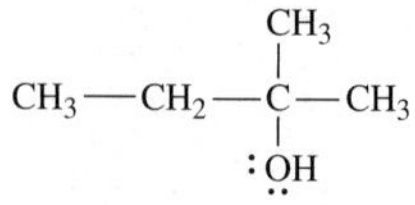

33a.

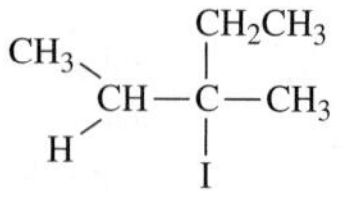

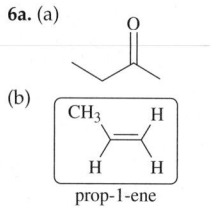

33b.
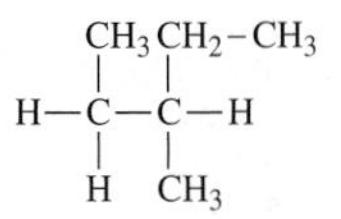

33c.
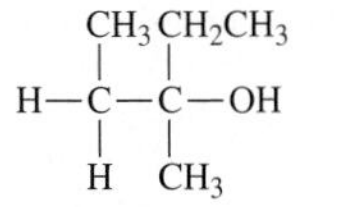

33d.
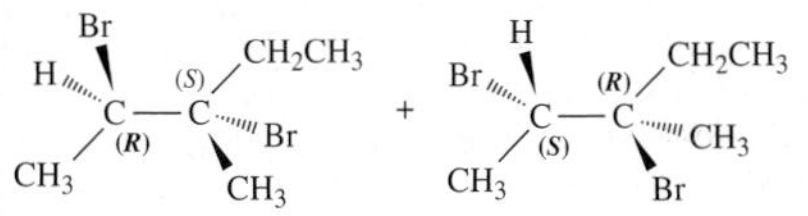

35.
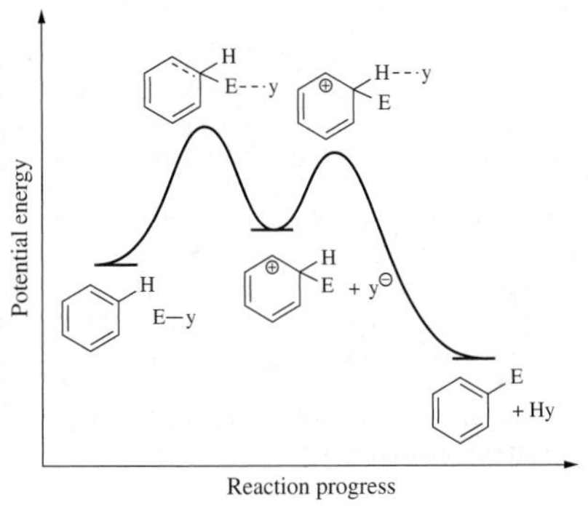

37.
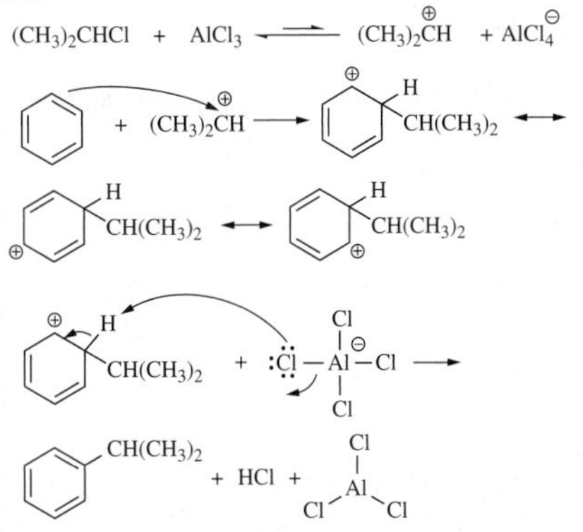

39a.
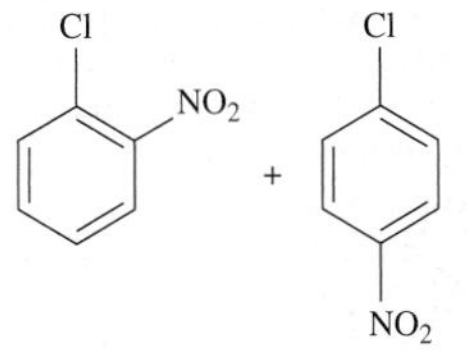

39b.
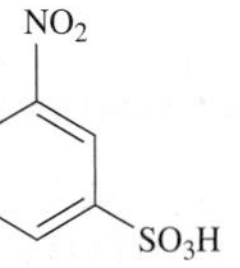

39c.
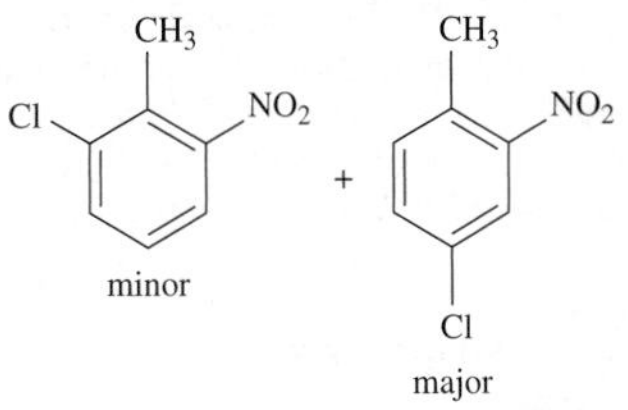

41.
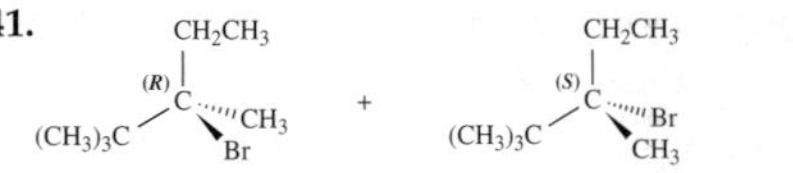
(R)-3-bromo-2,2,3-trimethylpentane (S)-3-bromo-2,2,3-trimethylpentane

43a.
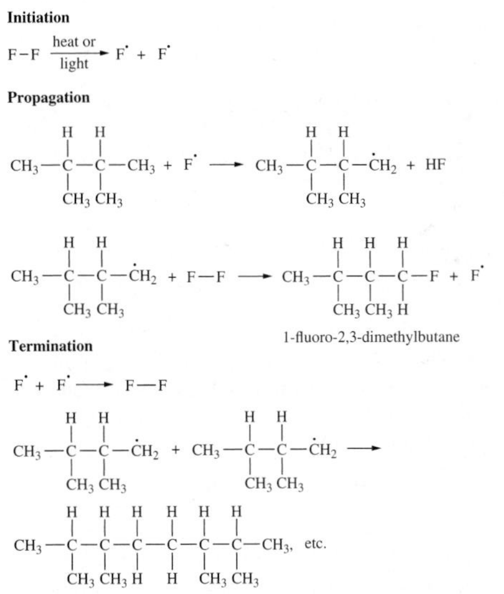

43b. Whereas the flourine radical is so reactive, its attack on alkane is purely statistical. There are six times as many 1° hydrogens and 2° hydrogens in the reactant molecule. **45.** During the polymerization reaction, polymers of different chain-lengths are formed. As a result, we can only speak of average molecular weight.

47.
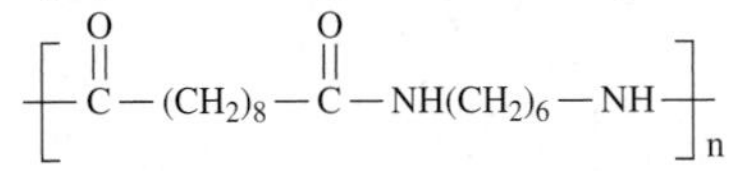

49.
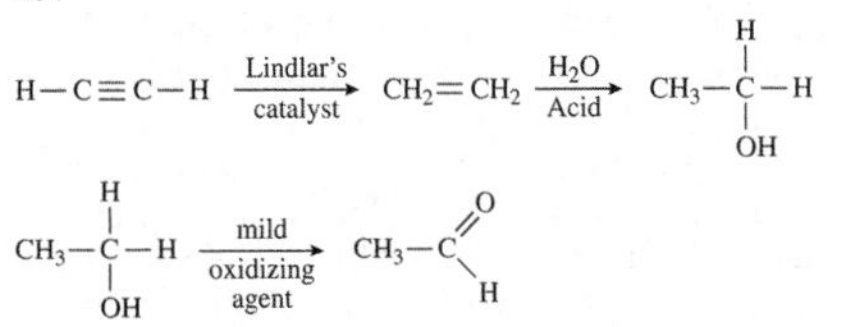

51.
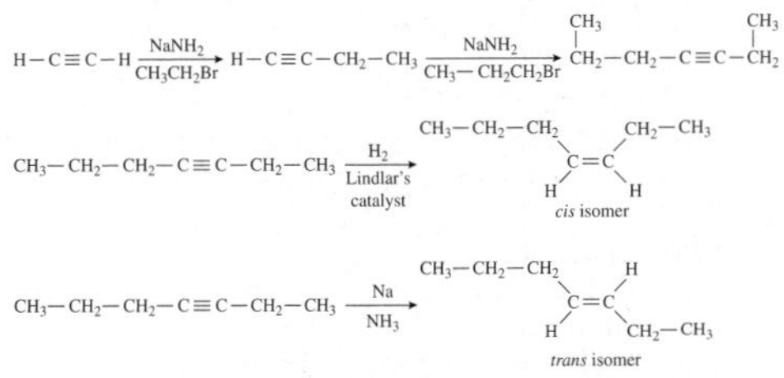

53.
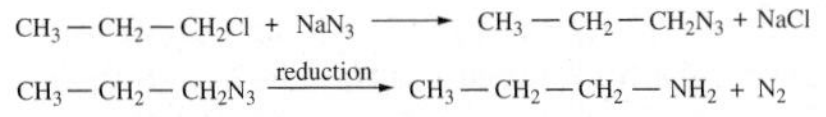

Integrative and Advanced Exercises

55a.
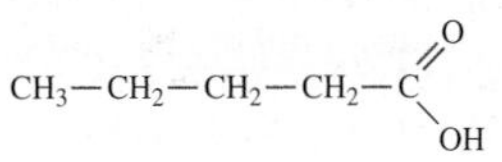

55b.
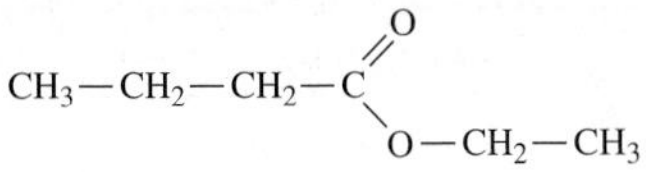

55c.
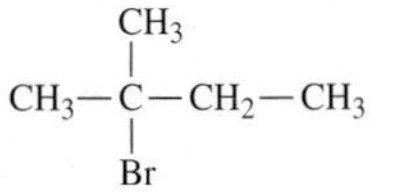

57a.

$$CH_3-CH=CH_2 \xrightarrow[Pt]{H_2} CH_3-CH_2-CH_3$$

57b.
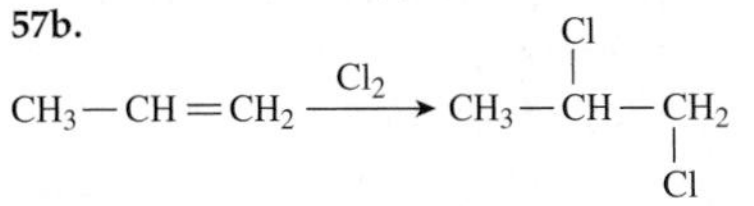

57c.
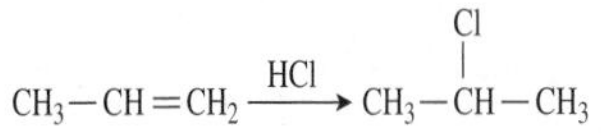

57d.
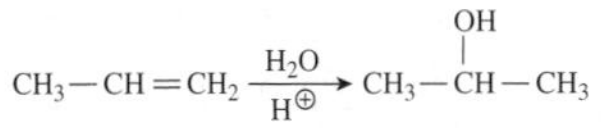

59a. Compound (3) reacts with dilute HCl.
59b. Compound (1) hydrolyzes.
59c. Compound (2) neutralizes dilute NaOH.
61a. $CH_3CH_2NH_3^{\oplus}Cl^{\ominus}$
61b. $(CH_3)_3\overset{\oplus}{N}H_3\ Br^{\ominus}$
61c. no reaction
61d. $CH_3CH_2NH_2 + H_2O$
63. butan-2-one.
65a. Compound (1) and (3).

65b. Compound (2).
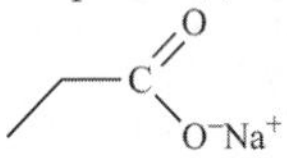

65c. Compound (2).
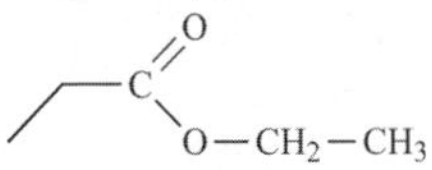

65d. Compound (1) and (3).
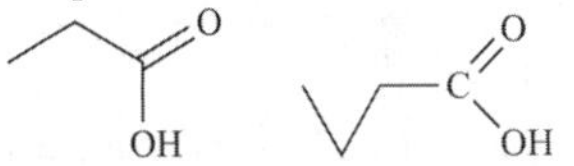

65e. Compound (2) and (3).
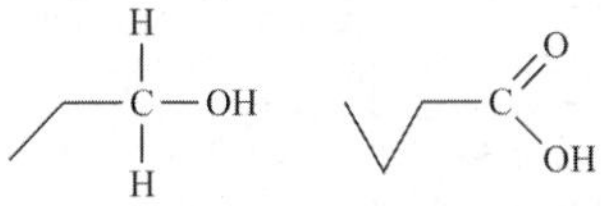

67.
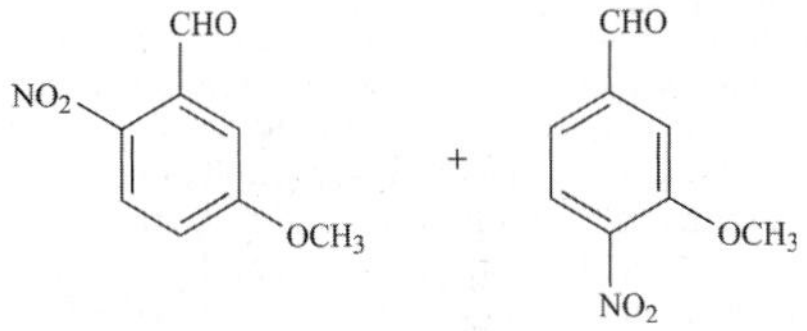

69.
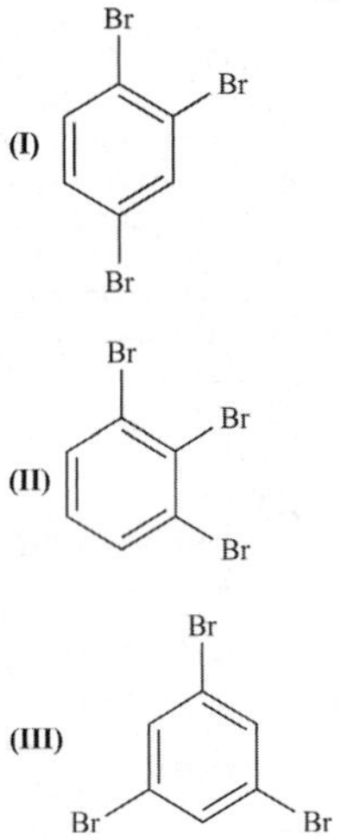

71.

1-chloro-2-methylbutane	31%
1-chloro-3-methylbutane	16%
2-chloro-3-methylbutane	31%
2-chloro-2-methylbutane	21%

73. In the presence of strong acid (HI), the oxygen atom of the —OH group is protonated, forming $CH_3CH_2-\overset{\oplus}{O}H_2$ H_2O is a weaker base than OH^- and is therefore a good leaving group.

75.

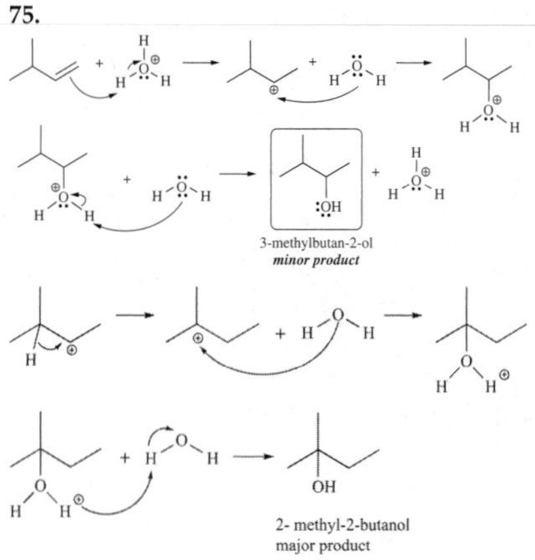

77.

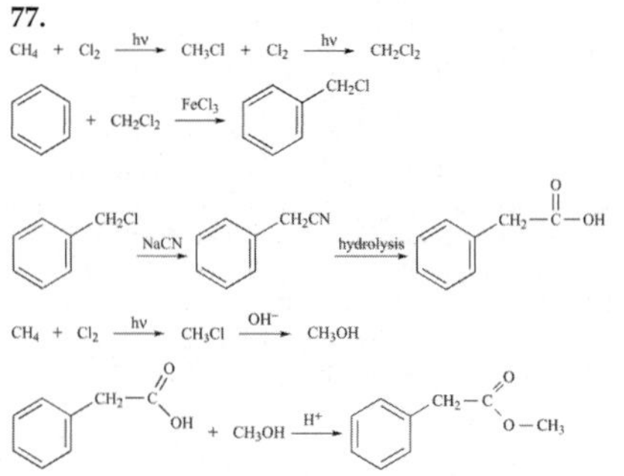

Self-Assessment Exercises 79a. Nucleophilic substitution corresponds to a substitution (either S_N1 or S_N2) for aliphatic compounds. Electrophilic aromatic substitution is typical for aromatic compounds (an atom is replaced by an electrophile) **79b.** An addition reaction is the opposite of an elimination reaction. In an addition reaction, two or more atoms (molecules) combine to form a larger one. **79c.** S_N1 reaction involves the formation of carbocation. S_N2 reaction, on the other hand, is a one-step process in which bond breaking and bond making occur simultaneously at a carbon atom with a suitable leaving group. **79d.** E1 reactions are unimolecular elimination reactions that proceed via carbocation intermediates. E2 reactions are bimolecular, one-step reactions that require an antiperiplanar conformation at the time of π-bond formation and β-bond breaking and do not involve carbocations. **81a.** CN^- **81b.** CH_3I **83.** Ether formation proceeds via an S_N2 mechanism. The combination of $(CH_3)_2CHONa$ and CH_3I will be better for S_N2 (backside attack) because there is less crowding of the carbon in CH_3I than in $(CH_3)_2CHI$. The first reaction will favor E2 mechanism.

85a.

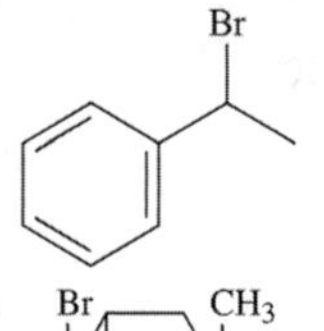

85b.

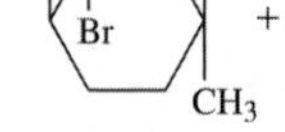

+ enantiomer

(racemic product)

85c.

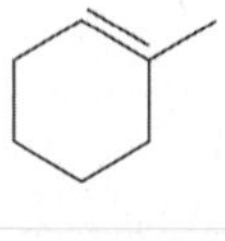

85d.

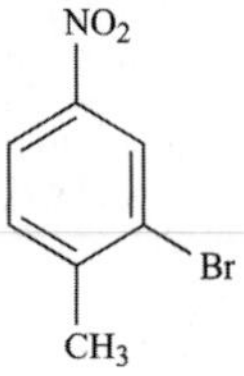

85e.

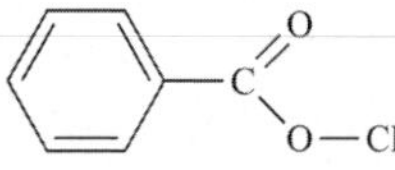

CHAPTER 28

Practice Examples 1a. dithreonylmethionine **1b.**

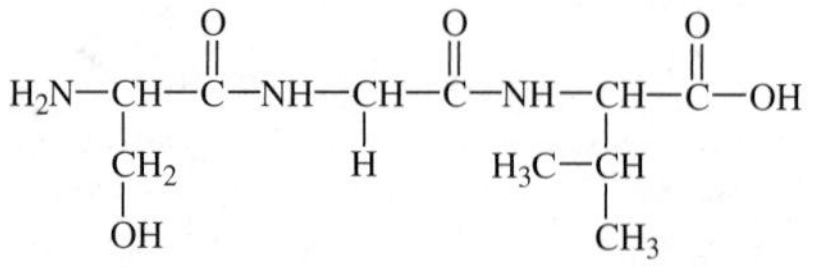

2a. Pentapeptide sequence: Gly-Cys-Val-Phe-Tyr. **2b.** Hexapeptide sequence: Ser-Gly-Gly-Ala-Val-Trp.

Integrative Example A. 0.300 m

B. (a)

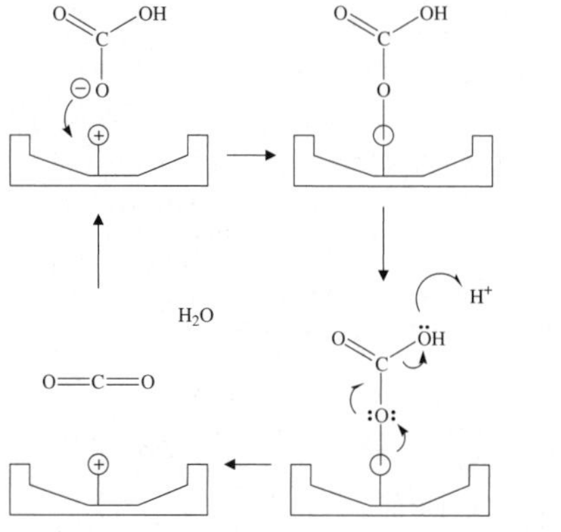

(b) $K_M = 0.01406$ M, $k_2 = 4.8 \times 10^5\ s^{-1}$.

Exercises 1a. $3 \times 10^2\ H_3O^+$ ions **1b.** $1.2 \times 10^5\ K^+$ ions **3.** 5×10^6 protein molecules **5a.** glyceryl palmitolinolenolaurate or glyceryl palmitoeleosterolaurate. **5b.** glyceryl trioleate or triolein. **5c.** myristate **7a.** Both are triglycerides or glycerol esters. Trilaurin is saturated and a fat. Trinolein is unsaturated and an oil. **7b.** Soaps are salts of fatty acids. Phospholipids are derived from glycerols, fatty acids, phosphoric acid, and a nitrogen containing base. Soaps and phospholipids have hydrophilic heads and hydrophobic tails. **9.** Polyunsaturated fatty acids are characterized by a large number of C=C double bonds in their hydrocarbon chain. Stearic acid has no C=C double bonds and therefore is not unsaturated, let alone polyunsaturated. Eleostearic acid has three C=C double bonds and thus is polyunsaturated. Polyunsaturated fatty acids are recommended in dietary programs since saturated fats are linked to a high incidence of heart disease. Of the lipids listed in Table 28-2, safflower oil has the highest percentage of unsaturated fatty acids. **11.** $CH_2OHCHOHCH_2OH$ and $NaOOC(CH_2)_{14}CH_3$. **13a.** L-Glucose, aldohexose. **13b.** D-Erythrulose, ketotetrose.

15a.

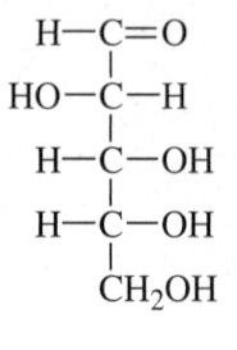

15b.

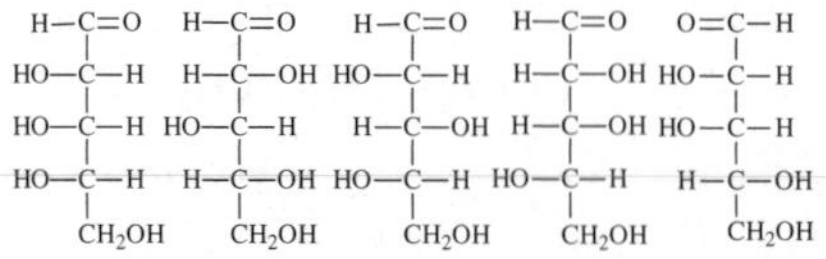

17a. A dextrorotatory compound rotates the plane of polarized light to the right, namely clockwise. **17b.** A levorotatory compound is one that rotates the plane of polarized light to the left, namely counterclockwise. **17c.** A racemic mixture has equal amounts of an optically active compound and its enantiomer. Since these two compounds rotate polarized light by the same amount but in opposite directions, such a mixture does not exhibit a net rotation of the plane of polarized light. **17d.** In organic nomenclature, this designation is given to a chiral carbon atom. The stereogenic center has an *R*-configuration if a curved arrow from the group of highest priority through to the one of lowest priority is drawn in a clockwise direction. **19.** A reducing sugar has a sufficient amount of the straight-chain form present in equilibrium with its cyclic form such that the sugar will reduce Cu^{2+} (aq) to insoluble, red Cu_2O(s). Only free aldehyde groups are able to reduce the copper(II) ion down to copper(I). 0.397 g of Cu_2O should be produced. **21.** Diastereomers **23a.** Enantiomers: *S*-configuration (leftmost structure), *R*-config. (rightmost structure). **23b.** Different molecules: different formulas **23c.** Diastereomers: *S,R* configuration (leftmost structure–top to bottom), *S,S*-configuration (rightmost structure). **23d.** Diastereomers: *R,R*-configuration (leftmost structure–top to bottom), *R, S*-configuation (rightmost structure–top to bottom).

25a.

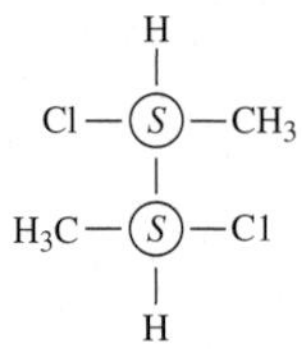

25b.

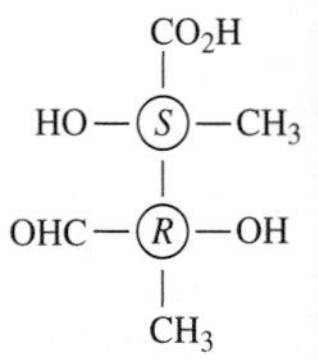

25c.

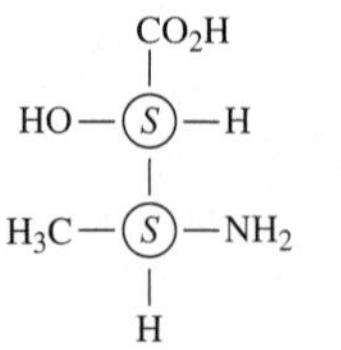

25d.

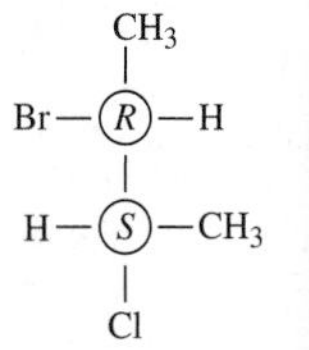

27a. An α -amino acid has an amine group ($-NH_2$) bonded to the same carbon as the carboxyl group ($-COOH$). For example, glycine (H_2NCH_2COOH) is the simplest α -amino acid. **27b.** A zwitterion is a form of an amino acid in which the amine group is protonated $(-NH_3^+)$ and the carboxyl group is deprotonated ($-\ COO^-$). For instance, the zwitterion form of glycine is $^+H_3\ NCH_2\ COO^-$. **27c.** The pH at which the zwitterion form of an amino acid predominates in solution is known as the isoelectric point. The isoelectric point of glycine is $pI = 6.03$. **27d.** The peptide bond is the bond that forms between the carbonyl group of one amino acid and the amine group of another, with the elimination of a water molecule between them. **27e.** Tertiary structure describes how a coiled protein chain further interacts with itself to wrap into a cluster through a combination of salt linkages, hydrogen bonding, and disulfide linkages, to name a few.

29a.

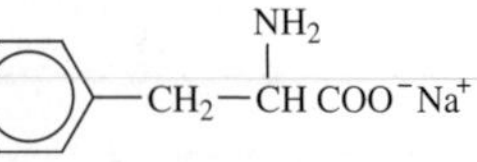

29b.

29c.

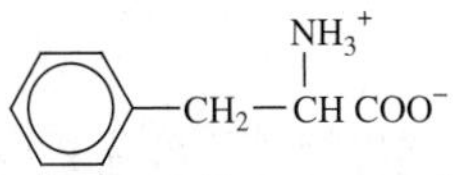

31a.

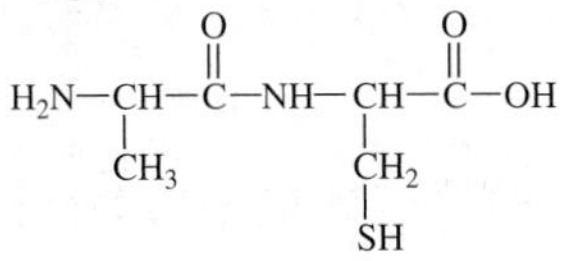

31b.

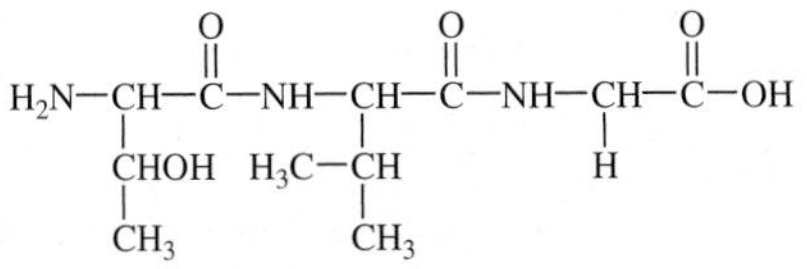

33. Proline will not migrate very effectively under these conditions. Lysine will migrate toward the negatively charged cathode. Aspartic acid will migrate toward the positively charged anode. **35a.** $^{+}H_3NCH(CHOHCH_3)COOH$ **35b.** $^{+}H_3NCH(CHOHCH_3)COO^-$ **35c.** $H_2NCH(CHOHCH_3)COO^-$ **37a.** There are 6 combinations: Lys-Ser-Ala, Lys-Ala-Ser, Ser-Lys-Ala, Ser-Ala-Lys, Ala-Ser-Lys, Ala-Lys-Ser. **37b.** There are 6 combinations: Ala-Ser-Ala-Ser, Ala-Ala-Ser-Ser, Ala-Ser-Ser-Ala, Ser-Ser-Ala-Ala, Ser-Ala-Ser-Ala, Ser-Ala-Ala-Ser. **39a.** Ala-Ser-Gly-Val-Thr-Leu. **39b.** alanyl-seryl-glycyl-valyl-threonyl-leucine, or alanylserylglycylvalylthreonylleucine. **41.** The *primary* structure of an amino acid is the sequence of amino acids in the chain of the polypeptide. The *secondary* structure describes how the protein chain is folded, coiled, or convoluted. The *tertiary* structure of a protein refers to how different parts of the molecules interact with each other to maintain the overall shape of the protein macromolecule. Quaternary structure refers to how two or more protein molecules pack together into a larger protein complex. Not all proteins have a *quaternary* structure since many proteins have only one polypeptide chain.

43.

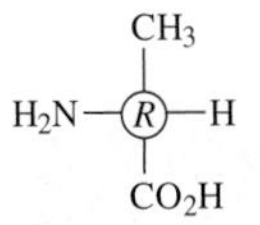

R-alanine

45.

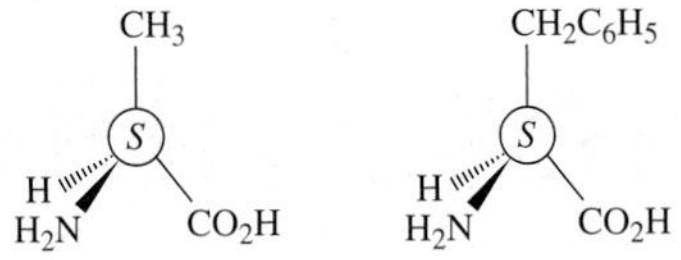

S-alanine *S*-phenylalanine

47. Two major types of nucleic acids are DNA (deoxyribonucleic acid), and RNA (ribonucleic acid). Both of them contain phosphate groups. These phosphate groups alternate with sugars to form the backbone of the molecule. The sugars are deoxyribose in the case of DNA and ribose in the case of RNA. Attached to each sugar is a purine or a pyrimidine base. The purine bases are adenine and guanine. One pyrimidine base is cytosine. In the case of RNA the other pyrimidine base is uracil, while for DNA the other pyrimidine base is thymine. **49.** The complementary sequence to AGC is TCG.

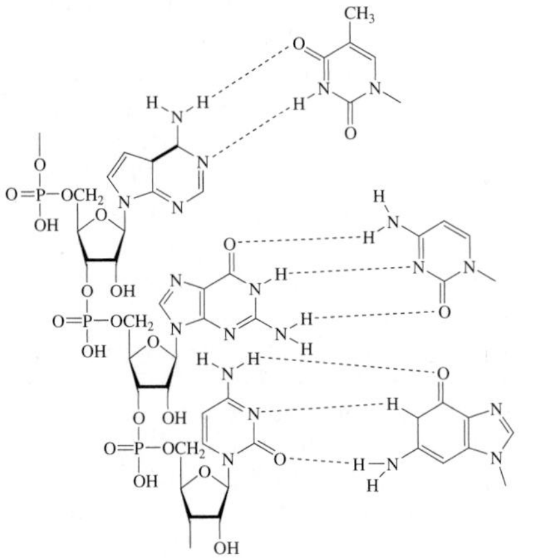

Integrative and Advanced Exercises
52. 2.9×10^3 g/mol. This is a minimum value because it is assumed that 1 Ag^+ ion is all that is necessary to denature each protein molecule. **56.** 9.71 **58.** Nonapeptide: Arg-Pro-Pro-Gly-Phe-Ser-Pro-Phe-Arg. **59.** One example for the required sequence would be AGA CCA CAA CGA. The redundancy enables the organism to produce a given amino acid in several ways in case of transcription error. **62.** Met-Ser-Met-Met-Gly. **65a.** 9.3×10^{-5} **65b.** 1.8×10^5

65c. $\frac{[B]}{[A]} = 2.25$

Feature Problems 66a. Saponification value = 189. Iodine number = 86.0. **66b.** Saponification value = 180. Iodine number = 81.6. **66c.** The iodine number for safflower oil may range between 150 to 144. The saponification value may range between 211 and 179.
Self-Assessment Exercises 72. The answer is (b). **73.** The answer is (d). **74.** The answer is (a). **75.** The answer is (d). **76.** The answer is (e). **77.** The answer is (c). **78.** The answer is (b). **79.** 129 g **80.** RNA chain. **81.** The answer is (e). **82.** The answer is (b). **83.** The answer is (a). **84.** The answer is (d). **85.** The answer is (c). **86.** The answer is (d). **87.** The answer is (b). **88.** Gly-Cys-Val-Phe-Tyr.

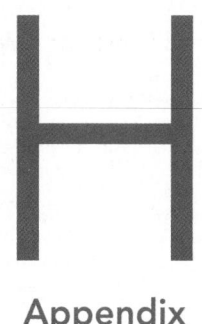

Appendix H Answers to Concept Assessment Questions

Note: Your answers may differ slightly from those given here, depending on the number of steps used to solve a problem and whether any intermediate results were rounded off.

CHAPTER 1

Concept Assessment 1-1. No, an experimental result contrary to that predicted by a hypothesis is reason to reject a hypothesis, not its proof. **1-2.** The product (mg) is the same for the unknown and the "weights" of known mass, regardless of the value of g. The measured mass with a two-pan balance is the same on the moon as on Earth. The single pan electronic balance measures *weight*, which is converted to a "mass" reading. With such a balance calibrated on Earth, the "mass" will appear less when measured on the moon. **1-3.** To find the *one* temperature, substitute $t(°C) = t(°F)$ into the equation $t(°F) = \frac{9}{5}t(°C) + 32$. Solve for $t(°F)$ to obtain the value -40 °F; thus -40 °F = -40 °C. **1-4.** The volume of wood is $1000\ g/(0.68\ g\ cm^{-3}) = 1470\ cm^3$. The wood displaces its own mass of water—1000 g, which has a volume of 1000 cm^3. The fraction of wood under water is $1000/1470 = 0.68$. **1-5.** Yes to both questions. A measuring instrument might yield precise readings but be incorrectly calibrated—measurements might agree with one another but their average might not agree with the actual value. Measurements with an imprecise instrument might differ widely from the actual value, yet their average might, by chance, agree with the actual value. **1-6.** The relationship 1 in = 2.54 cm is a definition—an *exact* quantity. A more precise relationship between meters and inches is $1\ m = 100\ cm \times (1\ in/2.54\ cm) = 39.370079\ in$.

CHAPTER 2

Concept Assessment 2-1. Helmont assumed, incorrectly, that the tree interacted with its surroundings only through the soil, sunlight, and watering. The tree also interacted with the atmosphere, that is, with carbon dioxide gas (see photosynthesis, page 278).
2-2. The *final* mass, magnesium bromide plus unreacted bromine, equals the *initial* mass: 4.15 g + 82.6 g = 86.8 g, but without knowing the mass of unreacted bromine we cannot deduce the mass of magnesium bromide.
2-3. The discovery of cathode rays (electrons) refuted the idea that atoms are indivisible. The discovery of isotopes refuted the idea that all atoms of an element are alike in mass. The idea that atoms combine in simple numerical ratios remains valid. **2-4.** The exception is the protium atom, ${}^1_1H$, the most abundant isotope of hydrogen. It consists of a lone proton as the atomic nucleus. **2-5.** The weighted average atomic mass of 51.996 u, almost exactly 52 u, suggests that chromium might exist exclusively as ${}^{52}_{24}Cr$ atoms. Another possibility (which, in fact, is the case) is that chromium exists as a mixture of isotopes whose weighted average atomic mass is 51.996 u. For zinc, we should conclude two or more isotopes. The weighted average atomic mass, 65.38 u, is too far from 65 u to suggest a single isotopic mass. **2-6.** If all naturally occurring Au atoms have the mass $196.97g\ Au/N_A$, they must all be ${}^{197}_{79}Au$, with no isotopes. If no Ag atom has the mass $107.87g\ Ag/N_A$, there must be two or more naturally occurring isotopes. (In fact there are two: ${}^{107}_{47}Ag$ and ${}^{109}_{47}Ag$.)

CHAPTER 3

Concept Assessment 3-1. For the *molecular* formula count the atoms in the condensed structural formula given: $C_4H_6O_2$. The *empirical* formula has the same ratio but with the smallest possible integers: $C_2H_3O_2$. Visualize the *structural* formula in terms of that of butane in Figure 3-2(a). Replace the $—CH_3$ groups at the two ends of the butane structure with —COOH groups (see structural formula of acetic acid in Figure 3-1). For the simplest *line-angle* formula, draw a line to represent $H_2C—CH_2$ and at an angle at each end of the line attach a —COOH group. **3-2.** Compare the other four quantities to **(d)**—the 20.000 g brass weight.
(a) 0.50 mol O_2 = 16.00 g; this might be the smallest mass but cannot be the largest.
(b) 2.0×10^{23} Cu atoms = 1/3 mol Cu ≈ 21 g; this is now the largest mass. **(c)** 1.0×10^{24} H_2O molecules is more than 1.5 mol H_2O > 27 g; this is now the largest mass. **(e)** the mass of 1.0 mol Ne = 20. g. Conclusion: greatest mass, **(c)**; smallest mass, **(a)**. **3-3.** C contributes the greatest number of atoms—13—and Cl contributes the greatest mass. A Cl atom, of which there are six, has nearly three times the mass of a C atom. **3-4.** Obtain relative numbers of atoms in the formula by multiplying the molar mass by the mass fractions of the elements. In Example 3-5, for example, mol C = $0.6258 \times 230\ g \times (1\ mol\ C/12.011\ g) = 11.98$ mol C. Similarly, obtain 21.97 mol H and 3.995 mol H. Thus, the molecular formula is $C_{12}H_{22}O_4$ and the empirical formula is $C_6H_{11}O_2$. **3-5** In each combustion, 1 mol CO_2 forms for every mole C and 1 mol H_2O for every 2 mol H. Determine these amounts for each combustion:
(a) 5 mol CO_2 and 2 mol H_2O; **(b)** 1.25 mol CO_2 and 2.5 mol H_2O; **(c)** 1 mol CO_2 and 1.5 mol H_2O; **(d)** 6 mol CO_2 and 3 mol H_2O. Thus, C_6H_5OH [response **(d)**] produces a greatest number of moles (and thus mass) of both CO_2 and H_2O. **3-6.** The O.S. of N in NH_3 is -3. The O.S. of N is higher in H_2NNH_2; it is -2. Note also that the molar mass of N_2H_4 is 32 g/mol. **3-7.** No. The greatest mass ratio of H_2O to CO_2 is found in CH_4, which has the highest percent H of all hydrocarbons; and its combustion produces only 2 mol H_2O (36 g) to 1 mol CO_2 (44 g).

CHAPTER 4

Concept Assessment 4-1. (a) The product is $O_2(g)$, not $O(g)$. **(b)** The product is $O_2(g)$ exclusively, not a mixture of $O(g)$ and $O_2(g)$. **(c)** The product is KCl(s), not KClO(s).
4-2 (a) incorrect (3 mol S per *2 mol* H_2S) **(b)** incorrect (stoichiometric coefficients refer to moles, not grams) **(c)** correct (1 mol H_2O/1 mol H_2S is the same as 2 mol H_2O/2 mol H_2S) **(d)** correct (2 of every 3 mole of S on the left are in 2 mol H_2S, yielding 3 mol S on the right) **(e)** incorrect (3 moles of reactants yield 5 moles of product) **(f)** correct (no atoms can be created or destroyed in the reaction) **4-3.** The reaction producing the greatest mass of $O_2(g)$ per gram of reactant is the one having the reactant of lowest molar mass. Clearly that reactant is $NH_4NO_3(s)$ in reaction **(a)**. **4-4 (a)** Tripling the solution volume reduces the molarity to 1/3 of its initial value: 0.050 M NaCl. **(b)** Reducing the volume from 250.0 mL to 200.0 mL increases the molarity by 5/4, that is, to 1.000 M $C_{12}H_{22}O_{11}$. **(c)** The molarity of the first solution is reduced to 1/3, to 0.0900 M KCl; that of the second is reduced to 2/3, to 0.0900 M KCl. The total molarity of the final solution is 0.180 M KCl. **4-5.** The balanced equation is $4\ NH_3(g) + 5\ O_2(g) \longrightarrow 4\ NO(g) + 6\ H_2O(l)$. Starting with 1.0 mol each of NH_3 and O_2, the limiting reactant is O_2. The amounts of products are 0.8 mol NO and 1.2 mol H_2O. The only true statement is **(d)** all the $O_2(g)$ is consumed. **4-6.** The answer must be consistent with the following facts.
(1) The factor 0.90 must appear twice in the setup (which can include $0.90 \times 0.90 = 0.81$); **(2)** CH_3Cl is an intermediate and does not

enter in; **(3)** the only molar masses needed are those of CH_4 (16 g/mol) and CH_2Cl_2 (85 g/mol). The correct response is **(a)**.

CHAPTER 5

Concept Assessment 5-1 (1). (e) 0.025 M $RbNO_3$, the only strong electrolyte of the group. **(2). (a)** a strong electrolyte with a total ion concentration of 0.024 M. **5-2.** Predicted as soluble based on guidelines in Table 5.1: **(a), (c), (g)**. Predicted as insoluble based on Table 5.1: **(b), (e), (h), (i)**. Inconclusive based on Table 5.1: **(d)** Li_2CO_3 might be soluble (Li^+ is a group 1 cation), but also might be one of the exceptions referred to in Table 5.1. **(f)** No data are available in Table 5.1 concerning permanagnates, so the solubility of $Mg(MnO_4)_2$ is uncertain. **5-3.** $H_2O(l)$ is suitable for $K_2CO_3(s)$ and $ZnSO_4(s)$, and HCl(aq) is suitable for CaO(s) and $BaCO_3(s)$. $H_2SO_4(aq)$ would not be suitable for CaO(s) and $BaCO_3(s)$ because $CaSO_4(s)$ and $BaSO_4(s)$ might precipitate. **5-4.** No reaction would occur in **(a)**. Two reduction half-reactions and no oxidation half-reaction. Reaction could occur under appropriate conditions in **(b)** because $Cl_2(g)$ undergoes both oxidation and reduction. **5-5.** This can occur as seen, for example, in the reverse of the disproportionation reaction directly above on this page. **5-6.** An inaccurate statement. An oxidizing agent is necessary to oxidize $Cl^-(aq)$ to $Cl_2(g)$, but neither HCl(aq) or NaOH(aq) is an oxidizing agent. **5-7.** A neutral solution is formed. The HCl and KOH solutions provide equal amounts (3.11×1^{-3} mol) of H^+ and OH^- ions, which neutralize each other.

CHAPTER 6

Concept Assessment 6-1. A water siphon passes over a "hump" from a pool of water at a higher level to a receiver at a lower level. After the siphon is initially filled with water, the pressure of the atmosphere pushes water over the hump, beyond which the water flows freely. In a suction pump, air pressure pushes water up a partially evacuated pipe. **6-2. (a) 6-3. (a) 6-4.** The direct proportionality of V and T must be based on an absolute temperature scale. While a change from 100 K to 200 K produces a doubling of V, a change from 100 °C to 200 °C produces only a 27% increase: $[(200 + 273)/(100 + 273)] = 1.27$. **6-5. (a)** 0.667 L $SO_2(g)$/1.00 L $O_2(g)$, because the actual P and T are immaterial as long as the two gases are compared at the same t and P. **(b)** The 0.667 L $SO_2(g)$ has to be adjusted for the an increase in T (by the factor 298 K/273 K) and a decrease in P (by the factor 100 kPa/98.0 kPa), leading to $V = 0.743$ L $SO_2(g)$. **6-6.** The correct responses—**(b)** and **(e)**—follow from basic ideas. Dalton's law of partial pressures dictates that P_{He} is not affected by any other gases present, and addition of 0.50 mol H_2 (1.0 g) will increase the total mass of gas by 1.0 g, independent of anything else that may happen. By simple estimates the other three statements can be shown to be false. **6-7.** He(g) at 1000 K has twice the u_{rms} as at 250 K (change T to $4T$ in equation 6.20). At 250 K, u_{rms} of $H_2(g)$ exceeds that of He(g) by the factor $\sqrt{2}$ (change M to $\frac{1}{2}M$ in equation 6.20). The two-fold increase in the first case exceeds the $\sqrt{2}$-fold increase in the second. He(g) at 1000 K has a greater u_{rms} than $H_2(g)$ at 250 K. **6-8.** The correct responses are **(a)** and **(c)**. The average kinetic energy of gas molecules depends only on T, and to two significant figures the mass of 0.50 mol He is the same as that of 1.0 mol H_2. **6-9.** Rearrange equation (6.14) to the form $R = MP/dT$. Substitute molar masses, $P = 1$ atm (exactly), $T = 293.2$ K, and the density data. Solve for three values of R and see how closely they conform to the ideal gas constant $R = 0.08206$ L atm mol^{-1} K^{-1}. Increasing adherence to ideal gas behavior: $OF_2(R = 0.0815) < NO(R = 0.08194) < O_2(R = 0.08200)$.

CHAPTER 7

Concept Assessment 7-1. Dynamite exploding in an underground cavern is a close approximation to an isolated system. Titration of an acid with a base is an open system. A steam-filled cylinder in a steam engine with all valves closed constitutes a closed system. **7-2.** Basic principle: law of conservation of energy. Assumptions: no heat loss to surroundings, *d* and *sp. ht.* of $H_2O(l)$ independent of T. Because the mass of hot water is twice that of the colder water, the initial temperature difference of 60.00 °C is divided into a 40.00 °C warming of the cold water and a 20.00 °C cooling of the hot water; final $T = 50.00$ °C. **7-3.** The ΔT of a fixed mass of substance is inversely proportional to its specific heat capacity; thus the object with the smaller ΔT has the greater specific heat capacity. The second question requires us to recognize the difference in enthalpy of transition for the solid and liquid form of water. The enthalpy of fusion for ice is less than the enthalpy of vaporization for the liquid, meaning that the amount of heat required to vaporize water is greater than that for ice. **7-4.** This is accomplished by adding measured amounts of reactants for a reaction in which the heat of reaction is known. **7-5.** This is a closed system. Since the pressure dropped while the volume remained constant then the temperature must have decreased. The internal energy of the system decreased. Therefore the energy transferred across the boundary was in the form of heat. The direction of energy transfer was from the system to the surroundings. **7-6.** The balloon feels warm because the dissolution of $NH_3(g)$ in $H_2O(l)$ is exothermic, $q < 0$. The balloon shrinks because the atmosphere (surroundings) does work on the system, $w > 0$. **7-7.** In the bottom row T is uniform throughout the object while in the top row the object is hotter at the edges than in its interior. Heating in the top row is irreversible; the process is far from equilibrium. The bottom row represents reversible heating; removal of just a tiny amount of heat can change heating to cooling. **7-8.** Enthalpy is a function of state. When a process returns a system to its initial state H returns to its initial value, meaning that $\Delta H = 0$. **7-9.** The enthalpy change in forming 1 mol $C_2H_2(g)$ from its elements is represented by the top line; that for forming 1 mol $C_2H_4(g)$ is the next line down. Both of these lines are above the broken line representing $\Delta_rH = 0$. The formation of 1 mol C_2H_6 has $\Delta_rH < 0$ and is the first line below the broken line. $\Delta_rH°$ for the reaction of interest is represented by the distance between the first and third lines. **7-10.** Yes, this can be done. The additional data needed are heat capacities as a function of T. The procedure is outlined in Figure 7-16.

CHAPTER 8

Concept Assessment 8-1. The wavelength of red light is about 700 nm (see Fig. 8-3). Since frequency and wavelength are reciprocally related, doubling ν halves λ. The frequency-doubled light will have a wavelength about 350 nm—light in the near ultraviolet not visible to the human eye. **8-2.** The threshold wavelength is 91.2 nm. Using 70.0 nm light as compared to 80.0 nm light produces more energetic electrons. Each photon produces one electron, and the number of electrons produced depends on the intensity (number of photons) of the light, provided its wavelength is less than the threshold wavelength. **8-3.** By studying emission spectra from the collision, scientists hope to identify the elements present in the comet and also on Jupiter's surface. **8-4.** The transition $n = 1$ to $n = 4$ corresponds to the greatest ΔE, but it involves *absorption* of a photon not emission. Photons are emitted in the other two transitions, with the transition $n = 4 \longrightarrow n = 2$ corresponding to the greater ΔE and hence shortest wavelength. **8-5.** If the wavelengths are the same then the momenta are the same (equation 8.10). The speed of the proton will have to be 1/2000th of the speed of the electron (that is, $m_p \times u_p = m_e \times u_e$, and $u_p = u_e \times m_e/m_p = u_e \times 1/2000$). **8-6.** The state $n = 2$ has a peak $\frac{1}{4}$ of the length from either end of the box, corresponding to the greatest probability of the particle being at those points. **8-7.** An orbital with three angular nodes has $\ell = 3$; it is an f orbital. One radial node makes for a total of four nodes, and since the total number of nodes is $n - 1$, n must be 5. The orbital is a $5f$. **8-8.** The compound is arsenic. The ground state is $4s^23d^{10}4p^3$ with all three arrows pointing in the same direction, one in each box. The anion is adding one more arrow to one of the $4p$ boxes pointing in the opposite direction.

CHAPTER 9

Concept Assessment 9-1. The ground-state electron configuration of Si ($Z = 14$) is $1s^22s^22p^6\,3s^23p^2$. If the ten core electrons are perfectly effective at screening the outer electrons, they will cancel out ten units of charge. Assuming the outer (valence) electrons do not screen each other, then $Z_{eff} = 14 - 10 = +4$ for the valence electrons. **9-2.** Z_{eff} increases and atomic radius decreases with increasing Z. The blue axis represents Z_{eff} and the blue line, Z_{eff} as a function of Z. The red axis represents atomic radius and the red line, atomic radius as a function of Z. **9-3. (a)** B (at the top of group 13)

(b) Cl (at the right end of the third period) **(c)** P^{3-} in period 3, group 15 (strong electron repulsions in an anion of high negative charge) **(d)** Tl (at the bottom of group 13) **9-4. (a)** C (smallest group-14 atom, at the top of the group) **(b)** Kr (noble gas element in group 18) **(c)** Se (lower ionization energy than Br based on the expected trend; lower than As for the same reason as in the P/S comparison on page 395) **9-5. (a)** group 17 (the smallest atoms in their periods) **(b)** group 2 (a filled *ns* subshell and essentially no affinity for an additional electron) **(c)** group 18 (noble gases have all shells and subshells closed) **9-6 (a)** scandium (Sc^{3+} has noble gas electron configuration) **(b)** tellurium (Te^{2-} has a noble gas electron configuration) **(c)** manganese (Mn^{2+} has the electron configuration [Ar]$3d^5$) **9-7** Cations are smaller than the atoms from which they are derived. Therefore, an Al atom is larger than an Al^{3+} ion and has a greater polarizability.

CHAPTER 10

Concept Assessment 10-1. The first and last symbols are acceptable; each has six dots with two unpaired. The unacceptable symbols have seven and five dots. **10-2.** The bonds are all covalent, with one being coordinate covalent. **10-3.** groups 14, 15, and 16 (for example, the elements C, N, O, P, and S) **10-4. (a)** Br **(b)** Be **(c)** P **10-5.** If covalent bonds between atoms involve equal contributions from all the bonded atoms, there are no formal charges. Where coordinate covalent bonds are formed there will be formal charges. A polyatomic ion must have at least one atom with a formal charge, consistent with the charge on the ion. **10-6.** We can draw two possible Lewis structures. One structure has no formal charges. In the other structure, one of the O atoms (the one bonded to H) has a formal charge of +1 and the other has a formal charge of −1. The structure with formal charges is considered to be unimportant and so we never represent the structure of CH_3CO_2H as a resonance hybrid. **10-7.** The structure of the SO_2 molecule is

best represented as 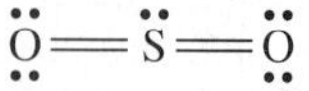, and

so the sulfur–oxygen bonds are best thought of as double bonds. **10-8.** ICl_2^- is a *linear* anion with five electrons pairs around the I atom (AX_2E_3). ICl_2^+ is a *bent* cation with four electron pairs around the I atom (AX_2E_2). The difference of one pair of electrons produces a completely different electron-group geometry and geometric shape. **10-9.**

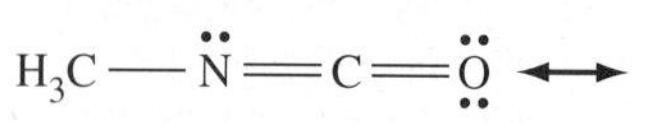

Most satisfactory

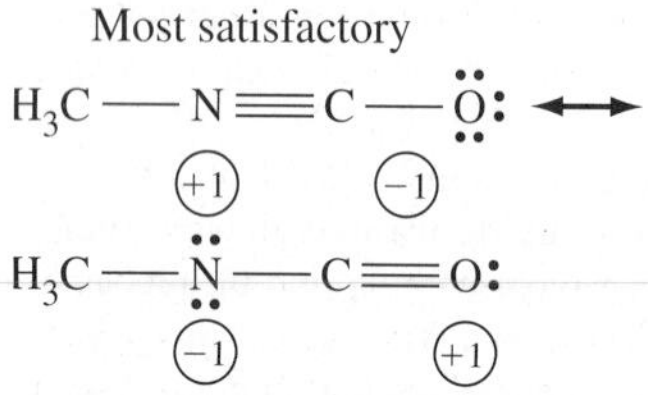

Least satisfactory

10-10. In NH_3 the lone pair of electrons on the N atom pulls electron density away from the H atoms, creating a large resultant dipole moment. In NF_3 the highly electronegative F atoms pull electron density away from the N atom, producing highly polar N—F bonds that counteract the effect of the nitrogen lone pair and a greatly reduced resultant dipole moment. **10-11.** Both the linear NO_2^+ cation and bent NO_2^- anion exhibit resonance that involves double bond character in the N—O bonds. However, electrons in the NO_2^+ cation will be more tightly held by the center of positive charge, leading to shorter N-to-O bond lengths in NO_2^+ than in NO_2^-.

CHAPTER 11

Concept Assessment 11-1. The cation CH_3^+ is isoelectronic with BH_3 and has three pairs of electrons around the C atom; we expect sp^2 hybridization. In the anion CH_3^- there are four electron pairs suggesting sp^3 hybridization, as in CH_4. **11-2.** The sp^3d^2 hybridization scheme corresponds to six electron pairs around a central atom. Similar to PF_5 in the period above it, we expect the compound AsF_5. Now imagine adding F^- to AsF_5 to create $[AsF_6]^-$, which has 6 electron pairs around the As atom and requires sp^3d^2 hybridization. **11-3.** Five bonding electron groups can be accommodated by sp^3d hybridization, but the distribution would be trigonal bipyramidal. What is needed is sp^3d^2 hybridization in the species AX_5E. The lone-pair electrons are directed to a corner of an octahedron, and the remaining five positions determine the molecular geometry—square pyramidal. **11-4.** To complete the octets of the N atoms, they must retain a lone pair of electrons, form a double bond between themselves, and a single bond to a H atom. The hybridization of the N atoms is sp^2. **11-5.** The H_2^+ ion is formed by removal of one electron from H_2—a larger energy requirement than promoting a σ_{1s} electron to the σ_{1s}^* MO in the excited state of H_2. On the other hand, the bond order in H_2^+ is 0.5 and 0 in the excited state of H_2. H_2^+ is a stable species and the excited state of H_2 is not. **11-6.** No. For example, the double bond in C_2 is made up of two π bonds and no σ bond (see Figure 11-26). **11-7.** The molecule NeO is isoelectronic with F_2 and should have a bond order of 1. We expect it to be stable, but it has never been observed. **11-8.** In HCO_2^- three atoms provide *p* orbitals for π bonding, just as in ozone. In the NO_3^- anion four atoms provide *p* orbitals. The delocalized π bonding in HCO_2^- is different than in NO_3^-.

CHAPTER 12

Concept Assessment 12-1. The intermolecular interactions are both London dispersion forces and hydrogen bonds. In substances with small molecules, hydrogen bonding usually dominates. **12-2.** Because the ball drops faster through it, the 10W oil is less viscous than the 40W oil. Viscosity is inversely proportional to *T*, and the lower weight oil (10W) is preferred for low-temperature use (where higher weight oils might solidify). In hot desert regions higher weight oils (40W) are preferred because light-weight oils might become so mobile as to lose their lubricating properties. The strengths of intermolecular forces are directly related to viscosity, and hence the higher viscosity 40W oil has the stronger intermolecular forces. **12-3.** Because of the different elevation (and barometric pressure) between landlocked, mountainous Switzerland and sea-level Manhattan Island, the lower boiling temperature results in a longer cooking time. **12-4.** Hydrogen bonding occurs in NH_3 but not in N_2, resulting in stronger intermolecular attractions and consequently lower vapor pressures, a higher boiling point, and a higher critical temperature in NH_3 than in N_2. **12-5.** The greater number of electrons (much larger molar mass) in CCl_4 causes the intermolecular attractions (London dispersion forces) to outweigh the effect of the polar bonds in CH_3Cl. **12-6.** Dew forms in the condensation of $H_2O(g)$ to $H_2O(l)$ and frost from the deposition of $H_2O(g)$ as $H_2O(s)$. Both processes are exothermic and give off heat to the surroundings—more heat in frost formation, because $\Delta H(\text{deposition}) = \Delta H(\text{condensation}) + \Delta H(\text{freezing})$. **12-7.** Wet books are placed in a cold, evacuated chamber. Moisture in the books freezes and the ice that is formed sublimes to $H_2O(g)$. This process avoids heating and involves a minimum of handling of the damaged books. **12-8.** According to the Bragg equation, $n\lambda = 2d \sin\theta$, if the extra distance traveled by the diffracted wave ($2d \sin\theta$) is to remain the same when *n* is doubled, the wavelength of the wave must be halved, so that $2n(\lambda/2) = n\lambda$. The required multiple is $1/2$. **12-9.** The fcc unit cell contains four C_{60} molecules. The unit cell has four octahedral and eight tetrahedral holes occupied by 12 K atoms. The formula based on the unit cell is $K_{12}(C_{60})_4$, and the molecular formula is $K_3(C_{60})$.

CHAPTER 13

Concept Assessment 13-1. No, spontaneous and nonspontaneous refer to the thermodynamics of a process, whereas instantaneous refers to the rate (or kinetics) of a process. A nonspontaneous process will not occur without external intervention, and a spontaneous reaction is not necessarily fast; it can occur very slowly. **13-2.** Doubling the volume available to the gas in Figure 13-2 is equivalent to doubling the length of the box from *L* to 2*L* in Figure 13-1b. The expansion of the gas in Figure 13-2 seems driven by a tendency to fill all the available volume. The gas expansion can also be explained, however, as the tendency of the system energy to be distributed among the greater number of available energy levels in the 2*L* box compared to the *L* box. **13-3.** Since $\Delta S_{surr} = -\Delta H_{sys}/T$, highly exothermic processes ($\Delta H_{sys} < 0$) cause a large increase in the entropy of the surroundings. Provided ΔS_{sys} is not too negative, the large value of ΔS_{surr} makes $\Delta S_{univ} > 0$, and the process spontaneous. **13-4.** $H_2O(l, 1\text{ atm}) \rightleftharpoons H_2O(g, 1\text{ atm})$ is a process in which $\Delta H > 0$ and $\Delta S > 0$, Because the normal boiling point of water is 100 °C, we know that $\Delta_{vap}G = 0$ and $\Delta_{vap}H = T\Delta_{vap}S$ at this temperature.

Assuming that $\Delta_{vap}H$ and $\Delta_{vap}S$ have constant values, raising the temperature above 100 °C causes the value of $T\Delta_{vap}S$ to become greater than that of $\Delta_{vap}H$. **13-5.** $\Delta_r G° = 326.4$ kJ mol^{-1} means that the Gibbs energy change for the system is 326.4 kJ when 3 mol O_2 is converted into 2 mol O_3. If 1.75 mol O_2 reacts, then the Gibbs energy change for the system is (326.4 kJ/3 mol O_2) × (1.75 mol O_2) = 190.4 kJ.

CHAPTER 14

Concept Assessment 14-1. 3.011×10^{23}. **14-2.** Concentrations are independent of temperature if based solely on mass or temperature-independent properties related to mass, specifically, mass percent, molality, mole fraction, and mole percent. Concentrations based on volumes—volume percent and molarity—are temperature dependent. **14-3.** HCl is undissociated in $C_6H_6(l)$, and the concentration of HCl in $C_6H_6(l)$ should closely follow $P_{HCl(g)}$ above the solution. On the other hand, HCl(g) reacts with $H_2O(l)$ to produce $H_3O^+(aq)$ and $Cl^-(aq)$. The relationship between $P_{HCl(g)}$ and the aqueous concentrations of ions is more complex. **14-4.** Start with $(P_A^* - P_A)/P_A^* = x_B$. Note that $(P_A^* - P_A)/P_A^* = 1 - (P_A/P_A^*)$. According to Raoult's law, $P_A = x_A P_A^*$, which means that $P_A/P_A^* = x_A$. Thus, we arrive at the true statement that $1 - x_A = x_B$. **14-5.** This will happen if both components in an ideal solution have the same vapor pressure. The vapor pressure of the solution will be independent of the solution composition and the line will be parallel to the composition axis. The likelihood of this happening is not very great, although it might be found at one particular temperature where the vapor pressure curves of two liquids cross. **14-6.** The similarities are that two different solutions are involved, water is transported from the more dilute to the more concentrated solution, and the process continues until the two solutions have the same concentration. The chief difference is that water is transported via the vapor phase in Figure 14-20a and through a semipermeable membrane in Figure 14-21. **14-7.** If Figure 14-23 were based on water rather than some other solvent, the two fusion curves would have *negative* rather than positive slopes. However, there would still be a freezing-point depression and a boiling-point elevation. **14-8.** When enough NaCl is present to depress the freezing point to −21 °C, the NaCl(aq) is saturated. Any solute added beyond this point remains as undissolved NaCl(s) and can have no further effect on the freezing point of the solution.

CHAPTER 15

15-1. $$Q = \frac{a_{Cu^+(aq)}a_{H_2(g)}}{a_{Cu(s)}a^2_{H^+(aq)}} = \frac{[Cu^{2+}]P_{H_2}}{[H^+]^2} \times \frac{c°}{P°}$$

15-2. Into the expression $K = [B]/[A]$, substitute $[B] = 54 - [A]$ and the given value of K; solve for [A] and [B]. If $K = 0.02$, [A] (open circles) = 53 and [B] (filled circles) = 1. If $K = 0.5$, [A] = 36 and [B] = 18. If $K = 1$, [A] = [B] = 27. **15-3.** If $K > 1$ for the 2nd reaction, K for the 1st reaction will be the larger of the two, but if $K < 1$ for the 2nd reaction, K for the 1st reaction will be the smaller of the two. **15-4.** Reverse given equation (invert its K value). To that equation add $CH_4(g) + H_2O(g) \rightleftharpoons CO(g) + 3\,H_2(g)$; CO(g) cancels and the overall equation is the one we seek; its K is the ratio of the other two K values. **15-5.** The balanced equation is sufficient to determine the outcome of a reaction that goes to completion. Otherwise, the value of K is required as well. **15-6. (a)** incorrect: would require that CO(g) and $H_2O(g)$ be completely consumed—impossible with $K_p = 10.0$. **(b)** incorrect: would be violation of the law of conservation of mass. **(c)** incorrect; would require the consumption of some $CO_2(g)$, but the direction of net change must be in the forward reaction. **(d)** correct: an outcome that would result from a net change in the forward direction. **(e)** incorrect: sufficient data are given to calculate the composition of the equilibrium mixture. **15-7.** Even though the pressure increases because of the addition of an inert gas, the reaction will shift to the right since the volume of the reaction vessel decreased. **15-8. (a)** True—more $H_2(g)$ will form at the expense of the $H_2S(g)$ and $CH_4(g)$. **(b)** False—an inert gas has no effect on a constant-volume equilibrium condition. **(c)** True—K changes with T and so does the composition of the equilibrium mixture. **(d)** Uncertain—the partial pressures of $H_2S(g)$ and $CH_4(g)$ will rise because a net reaction occurs to the left, but the increase in partial pressures of $CS_2(g)$ and $H_2(g)$ caused by forcing these two gases into a smaller volume will be at least partly offset by the equilibrium shift to the left. **15-9.** Equilibrium shifts in the forward direction, the endothermic reaction. Student B, by holding the beaker, stimulates heat flow into the reaction mixture, probably achieving a higher yield of product.

CHAPTER 16

Concept Assessment 16-1. (a) is a conjugate acid/base pair; HCO_3^- can transfer a proton to a base (e.g., OH^-) yielding CO_3^{2-}, and CO_3^{2-} can react with an acid (e.g., H_3O^+) to reform HCO_3^-. **(b)** is not a conjugate acid/base pair; SO_4^{2-} can be produced from HSO_3^- only through an oxidation process not in an acid–base reaction. **(c)** is not a conjugate acid/base pair; it is a pair of unrelated acids. **(d)** yes; **(e)** no. **16-2.** In this solution, $[H_3O^+] = 10^{-7.00}$ M and $[OH^-] = K_w/[H_3O^+] = 2.88 \times 10^{-14}/1.0 \times 10^{-7} = 2.88 \times 10^{-7}$ M. Because $[OH^-] > [H_3O^+]$, the solution is basic. **16-3.** A concentrated solution of a weak acid may often have a lower pH than a dilute solution of a strong one. For example, the pH of 0.10 M $HC_2H_3O_2$—calculated as pH = 2.89 in Example 16-8—is lower than the pH of 0.0010 M HCl. **16-4.** The bottle labeled $K_a = 7.2 \times 10^{-4}$ contains the more acidic solution. The bottle labeled $K_a = 1.9 \times 10^{-5}$ has the acid with the larger pK_a. **16-5. (a)** The material balance equation is 0.010 M = $[HSO_4^-] + [SO_4^{2-}]$. The charge balance equation is $[H_3O^+] = [HSO_4^-] + 2\,[SO_4^{2-}] + [OH^-]$. **(b)** The material balance equation is 0.025 M = $[NH_3] + [NH_4^+]$. The charge balance equation is $[NH_4^+] + [H_3O^+] = [OH^-]$. **16-6.** The relevant equations are $^+NH_3CH_2CH_2NH_3^+(aq) + H_2O(l) \rightleftharpoons H_3O^+(aq) + NH_2CH_2CH_2NH_3^+(aq)$ pK_1 = 6.85 and $NH_2CH_2CH_2NH_3^+(aq) + H_2O(l)$ $H_2O(l) \rightleftharpoons H_3O^+(aq) + NH_2CH_2CH_2NH_2(aq)$ pK_2 = 9.92. From equation (16.20), the values of the base ionization constants are pK_{b_1} = 14.00–9.92 = 4.08 and pK_{b_2} = 14.00–6.85 = 7.15.

The base ionization reactions are

$NH_2CH_2CH_2NH_2(aq) + H_2O(l) \rightleftharpoons NH_2CH_2CH_2NH_3^+(aq) + OH^-(aq)$ pK_{b_1}=4.08 and $NH_2CH_2CH_2NH_3^+(aq) + H_2O(l) \rightleftharpoons {}^+NH_3CH_2CH_2NH_3^+(aq) + OH^-(aq)$ pK_{b_2}=7.15

$^+NH_3CH(CH_3)COOH + H_2O \rightleftharpoons NH_3CH(CH_3)COO^- + H_3O^+$ pK_a = 2.34

$^+NH_3CH(CH_3)COO^- + H_2O \rightleftharpoons NH_2CH(CH_3)COO^- + H_3O^+$ pK_a = 9.87

pK_{b_1} = 14.00–2.63 = 11.37

pK_{b_2} = 14.00–9.87 = 4.13

$NH_2CH(CH_3)COO^- + H_2O \rightleftharpoons {}^+NH_2CH(CH_3)COO^- + OH^-$ pK_{b_1} = 4.13

$^+NH_3CH(CH_3)COO^- + H_2O \rightleftharpoons {}^+NH_3CH(CH_3)COOH + OH^-$ pK_{b_1} = 11.37.

16-7. Consider HCO_3^-, which can act as an acid: $HCO_3^-(aq) + H_2O(l) \rightleftharpoons H_3O^+(aq) + CO_3^{2-}(aq)$, $K_a K_{a2} = 4.2 \times 10^{-11}$ or as a base: $HCO_3^-(aq) + H_2O(l) \rightleftharpoons H_2CO_3^-(aq) + OH^-(aq)$ $K_b = K_w/K_{a1} = 1.00 \times 10^{-14}/4.4 \times 10^{-7} = 2.3 \times 10^{-8}$. Because K_b is much greater than K_a, $NaHCO_3(aq)$ is basic. The CO_3^{2-} ion can act only as a weak base: $CO_3^{2-}(aq) + H_2O(l) \rightleftharpoons HCO_3^-(aq) + OH^-(aq)$. For CO_3^{2-}, $K_b = K_w/K_a(HCO_3^-) = 1.0 \times 10^{-14}/4.2 \times 10^{-11} = 2.4 \times 10^{-4}$. Thus, $Na_2CO_3(aq)$ is basic. **16-8.** Since equilibrium favors the formation of the weaker base, we need to identify bases that have an ionization constant, K_b, significantly larger than that of CH_3COO^- (say, 100 times larger). For CH_3COO^-, $K_b = K_w/K_a(CH_3COOH) = 1.0 \times 10^{-14}/1.8 \times 10^{-5} = 5.6 \times 10^{-10}$. Of the bases in Table 16.4, only NH_3, $CH_3CH_2NH_2$, and $(CH_3CH_2)_2NH$ have K_b values significantly larger than 5.6×10^{-10}. **16-9.** We should expect pK_a for *ortho*-chlorophenol to be smaller than for phenol because of the electron-withdrawing effect of the Cl atom. (Its pK_a is 8.55 compared with 10.00 for phenol.) **16-10.** Picture three Br atoms joined by single bonds to a Fe(III) atom on which there is also a lone pair of electrons: Br_3Fe: Now imagine that a Br_2 molecule dissociates into Br^+ and Br^- ions. The electron-deficient Br^+, a Lewis acid, attaches to the lone-pair electrons of Fe(III), a Lewis base, forming $[FeBr_4]^+$. The final product is $[FeBr_4]^+Br^-$.

CHAPTER 17

Concept Assessment 17-1. (a) no; NH_4Cl lowers the pH through the common-ion effect **(b)** yes, but only slightly. Diethylamine is a somewhat stronger base than NH_3, but only a rather small amount is being added. **(c)** no;

HCl is a strong acid that will neutralize some of the NH_3, producing an aqueous solution of NH_3 and NH_4Cl. **(d)** no; since the added $NH_3(aq)$ is more dilute than the 0.10 M $NH_3(aq)$ the overall solution will be 0.075 M $NH_3(aq)$. **(e)** yes; $Ca(OH)_2(s)$ is a strong base. **17-2.** benzoic acid/benzoate with a 1:2 ratio. **17-3. (a)** yellow; a low pH **(b)** yellow; a CH_3COOH/CH_3COO^- buffer solution is formed but its pH is about 5 **(c)** yellow; the buffer completely neutralizes the small amount of added OH^- **(d)** red; the buffer capacity is exceeded and the solution becomes basic. **17-4.** Choice **(c)** is the correct one; 0.60 mol $NaCH_3COO$, converts all the HCl to CH_3COOH, producing a $CH_3COOH/NaCH_3COO$ buffer solution of pH $\approx$ 4. Choices **(a)** and **(b)** have essentially no effect on the pH, and while choice **(d)** would neutralize 80% of the acid, the amount of strong acid remaining would still produce a pH $\approx$ 1. **17-5.** This is the titration of a weak base that ionizes in two stages. The titration curve would begin at a moderately high pH; the pH would drop during the titration, and there would be two equivalence points. In general, the curve would resemble that in Figure 17-12, but flipped from bottom to top. The two buffer regions and pH = pK_b values would be in segments of the curve between the two equivalence points. **17-6. (a)** six species: K^+, H_3O^+, I^-, CH_3COO^-, OH^-, CH_3COOH **(b)** most abundant, K^+ (the spectator ion in highest concentration); 2nd most abundant, CH_3COO^- (produced in the neutralization of 3/4 of the CH_3COOH) **(c)** least abundant, OH^- (the final solution is acidic, so $[OH^-] < 10^{-7}$ M); 2nd least abundant, H_3O^+ (final solution a buffer with pH $\approx$ 5)

CHAPTER 18

Concept Assessment 18-1. Because of the large excess of $MgF_2(s)$ the solution would remain saturated even when the volume of solution is doubled; $[Mg^{2+}]$ remains constant. **18-2.** CaF_2 and AgCl are insoluble with $K_{sp} = 5.3 \times 10^{-9}$ and $K_{sp} = 1.8 \times 10^{-}10$, respectively. Since K_{sp} values for $CaCl_2$ and AgF are not found in tables, they can be assumed to be soluble compounds. **18-3.** It is largely immaterial. The only requirement is that the $AgNO_3(aq)$ be concentrated enough to bring about the precipitation without unduly diluting the solution from which the precipitation occurs. **18-4.** It will affect the solubility of CaF_2 more. The formation of F^- is derived from the weak acid HF, and so it will undergo hydrolysis whose equilibrium can be changed on the addition of an acid or base. **18-5. (a)** At pH = 10.00, $[OH^-]=1.0 \times 10^{-4}$ M and $[Mg^{2+}] = 1.8 \times 10^{-3}$ M, an entirely plausible result. **(b)** At pH = 5.00, $[OH^-] = 1 \times 10^{-9}$ M and $[Mg^{2+}] = 1.8 \times 10^{7}$ M, an impossible result. The situation here is that at pH = 5.00 the solution is not one of $Mg(OH)_2$. It is $MgCl_2(aq)$, and $[Mg^{2+}]$ in this solution depends on the solubility of $MgCl_2$. **18-6.** As expected, with $[Cl^-] = 0.0039$ M, the molar solubility of AgCl(s) is less than in pure water because of the common-ion effect. At higher concentrations of $Cl^-(aq)$, AgCl(s) becomes more soluble because of complex ion formation: $AgCl(s) + Cl^-(aq) \longrightarrow [AgCl_2]^-(aq)$. **18-7.** $Pb^{2+}(aq)$ is present because of the formation of the yellow $PbCrO_4(s)$; $Hg_2{}^{2+}(aq)$ is absent because of the negative test for that ion; the presence of $Ag^+(aq)$ is uncertain because it really wasn't tested for. **18-8.** We must consider the formation of $Pb(OH)_3{}^-$, $K_f = 3.8 \times 10^{14}$, in the solution. **18-9.** Will not work because both ions would precipitate: $(NH_4)_2CO_3$, $H_2S(aq)$, and $NaOH(aq)$; will not work because neither ion would precipitate: $HNO_3(aq)$ and $NH_3(aq)$. HCl(aq) will work because $CuCl_2$ is water soluble and AgCl is not.

CHAPTER 19

Concept Assessment 19-1. At the anode, Zn(s) is oxidized to $Zn^{2+}(aq)$ and, to preserve charge balance, $NO_3{}^-(aq)$ migrates in from the salt bridge. At the cathode, $Cu^{2+}(aq)$ is reduced to Cu(s) and, to preserve charge balance, $K^+(aq)$ migrates in from the salt bridge. **19-2.** No changes in mass at the inert Pt(s) electrodes; a gain in mass at the Cu(s) electrode through the half reaction $Cu^{2+}(aq) + 2\,e^- \longrightarrow Cu(s)$; and a loss in mass at the Zn(s) electrode through the half reaction $Zn(s) \longrightarrow Zn^{2+}(aq) + 2\,e^-$. **19-3.** Standard-state conditions for $ClO_4{}^-(aq)$ and $H^+(aq)$ are $a \approx 1$ M; for $Cl_2(g)$, $a = 1$ bar; for $H_2O(l)$, $a = 1$. **19-4.** The cell with $E° > 0$ proceeds toward the formation of more products. A net reaction also occurs in the case where $E° < 0$, but in the reverse direction; the concentrations on the left side of the equation increase and those on the right decrease, until equilibrium is reached. **19-5.** $E_{cell} = E°_{cell}$ if all reactants and products are in their standard states, but also for any set of concentrations where $Q = 1$ in equation (20.18). **19-6.** The precipitate is $PbSO_4(s)$, thereby reducing $[Pb^{2+}]$ in the anode compartment and increasing the value of E_{cell}, so that $E_{cell} > E°_{cell}$. **19-7.** The cell diagram $Pt(s)|Cl_2(g, 1\text{ bar})|\ Cl^-(0.50\text{ M})\|$ $Cl^-(0.10\text{ M})|Cl_2(g, 1\text{ atm})|Pt(s)$ has the net cell reaction: $0.50\text{ M } Cl^-(aq) \longrightarrow 0.10\text{ M } Cl^-(aq)$ and $E_{cell} = -0.0592\text{ V} \times \log(0.10/0.50) = 0.041$ V. **19-8.** In a calomel electrode, reduction potential depends on the chloride potential. Therefore, the standard reduction potential for a calomel electrode has a different chloride concentration from the saturated calomel electrode. **19-9.** Dry cells and lead-acid cells "run down" as the concentrations of reactants and products eventually reach their equilibrium values, where Δ_rG and E_{cell} both become 0. This does not happen in a fuel cell because fuel is continuously added. **19-10.** Both Al and Zn can be used because they are more active than Fe; Ni and Cu are less active and cannot be used.

CHAPTER 20

Concept Assessment 20-1. In the reaction $N_2(g) + 3\,H_2(g) \longrightarrow 2\,NH_3(g)$, compared to the rate of disappearance of N_2, the rate of disappearance of H_2 is three times as great and the rate of formation of NH_3 is twice as great. **20-2.** If the initial and instantaneous rates of reaction are initially equal and remain so throughout a reaction, the concentration-time graph must be a straight line with a negative slope, as seen in Figure 20-3. **20-3.** If the reaction were first order the initial rate would double and if second order, quadruple, thus 1 < order (m) < 2. More precisely, $2^m = 2.83$ and $m = 1.50$ (solve this equation for m: $m \log 2 = \log 2.83$). **20-4.** All four graphs can be plotted on the same sheet of paper. For example, take $[A]_0 = 3.0$ M and $k = 0.20\ s^{-1}$ as the larger of two rate constants and $k = 0.10\ s^{-1}$ as the smaller. Both concentration vs. time graphs will resemble Figure 20-5, but after starting at the same point, $[A]_0 = 3.0$ M, the one with the larger k has a shorter half-life and falls off more rapidly than the other. The two plots of ln k vs. t are straight lines, both starting at $\ln [A]_0 = \ln 3.0 = 1.10$, and having negative slopes, the steeper slope for the larger value of k. **20-5. (a)** If the plot was linear, the reaction would be zero order; **(b)** Look at successive half-lives. If $t_{1/2}$ is constant, the reaction is first order; **(c)** Look at successive half-lives. If $t_{1/2}$ doubles each time, the reaction is second-order. **20-6 (a)** This condition can exist. The reaction is exothermic (similar to Figure 20-10). **(b)** This condition can exist, and the reaction is endothermic (imagine flipping the reaction profile in Figure 20-10 from left to right). **(c)** This condition cannot exist; E_a for an exothermic cannot be less than ΔH. **(d)** This condition can exist; its only distinction is that there is no heat of reaction. **(e)** This condition cannot exist; E_a cannot be negative. **20-7.** Consider this equation from Figure 20-12: $E_a = R \times (-\text{slope of } \ln k \text{ vs. } 1/T)$. The greater the value of E_a, the greater the slope of the graph and the more rapidly the rate of reaction changes with temperature. **20-8.** The means are not the same. The increase in reaction rate caused by the presence of a catalyst is most likely because of a different reaction mechanism that lowers the reaction barrier. The increase in the rate of reaction caused by an increase in temperature is due to more molecules with kinetic energy greater than the barrier; more collisions occur per unit time.

CHAPTER 21

Concept Assessment 21-1. AlF_3 will have the higher melting point. **21-2.** The Na^+, K^+, Rb^+ and Cs^+ ions have relatively low charge densities and are better able to stabilize large, polyatomic anions such as $NO_2{}^-$. Because the Li^+ ion has a very high charge density and high polarizing power, it may kinetically assist the decomposition of polyatomic anions, such as $N_3{}^-$ and $N_2{}^-$, to smaller anions such as O_2^-. Balanced chemical equations for the reactions are: $MNO_3(s) \xrightarrow{\Delta} MNO_2(s) + \frac{1}{2}O_2(g)$ (M = Na, K, Rb, Cs); $2\,LiNO_3(s) \xrightarrow{\Delta} Li_2O(s) + 2\,NO_2(g) + \frac{1}{2}O_2(g)$. **21-3.** The hybridization of Be changes from sp to sp^2 to sp^3 on going from $BeCl_2$ to $(BeCl_2)_2$ and $(BeCl_2)_n$ and the geometry around the Be atom changes from linear to trigonal planar to tetrahedral. **21-4.** A plausible reaction is that Mg(s) is oxidized to MgO(s) and $CO_2(g)$ is reduced to C(s), that is, $2\,Mg(s) + CO_2(g) \longrightarrow 2\,MgO(s) + C(s)$. **21-5.** For all but beryllium,

heating the group 2 carbonates (MCO_3) yields the corresponding oxide (MCO); see equation (21.11). One method for preparing BeO is to burn Be(s) in O_2. **21-6.** The small size of Be^{2+} precludes the possibility of six H_2O molecules coordinating to the central Be^{2+} ion, whereas this happens readily with the Mg^{2+} ion.
21-7. Lithium and magnesium form ions, Li^+ and Mg^{2+}, that have high charge densities and are strongly polarizing. Presumably, when these ions are formed, their charge densities are large enough to stabilize the N^{3-} anion. Balanced chemical equations for the reactions are: $3\ Li(s) + \frac{1}{2}\ N_2(g) \rightarrow Li_3N(s)$; $3\ Mg(s) + N_2(g) \rightarrow Mg_3N_2(s)$. **21-8.** The structure of $B_2H_2(CH_3)_4$ is analogous to diborane (Fig. 21-11a) with two bridging H atoms and four methyl groups ($—CH_3$) in the terminal positions. **21-9.** AlF_3 is an electron-deficient compound and forms $[AlF_4]^-$ in the presence of F^- from KF. BF_3 is a stronger Lewis acid than AlF_3 and abstracts a F^- ion from $[AlF_4]^-$. The result: $[AlF_4]^- + BF_3 \longrightarrow AlF_3(s) + [BF_4]^-$.
21-10. Determine $\Delta H°$ for each reaction using equation (7.21), the only difference is replacing 1 mol CO_2 in equation (21.29) by 1 mol CO in equation (21.30). The difference in the heat of reaction is -110.5 kJ/mol CO(g) $-$ [-393.5 kJ/mol $CO_2(g)$] = 283 kJ. Reaction (21.30) liberates 238 kJ less heat than reaction (21.29). **21-11.** Refer to pages 1018 and 1019 and imagine a chain of tetrahedra starting with SiO_4^{4-} and increasing in increments of SiO_3^{2-}, yielding SiO_4^{4-}, $Si_2O_7^{6-}$, $Si_3O_{10}^{8-}$ $Si_4O_{13}^{10-}$, $Si_5O_{16}^{12-}$, $Si_6O_{19}^{14-}$. Now bend the six-unit silicate chain into a hexagonal ring (similar to that in the third structure on page 1018) by eliminating one O^{2-} between the ends of the chain. The result is the anion $Si_6O_{18}^{12-}$. In beryl, 3 Be^{2+} and 2 Al^{3+} provide the necessary 12 units of positive charge.
21-12. Dissolved O_2 (g) in Sn^{2+}(aq) is able to oxidize Sn^{2+} to Sn^{4+}. The following spontaneous reaction reduces tin(IV) to tin(II): $Sn^{4+}(aq) + Sn(s) \longrightarrow 2\ Sn^{2+}(aq)$; $E°_{cell} = 0.154\ V - (-0.137\ V) = 0.017$ V. Thus, as long as an excess of Sn(s) is present, the Sn^{2+}(aq) can be maintained with little or no Sn^{4+}(aq) present.

CHAPTER 22

Concept Assessment 22-1. The larger EN difference and shorter bond length between Xe and F, compared to Xe and Cl, makes XeF_2 a more stable molecule than $XeCl_2$. **22-2.** The ions do not have the same shape. ICl_2^+ (VSEPR notation, AX_2E_2) has a tetrahedral electron-group geometry and a bent molecular shape. ICl_2^- (VSEPR notation, AX_2E_3) has a trigonal-bipyramidal electron-group geometry, and a linear molecular shape. **22-3.** $CuSO_4$(aq) yields Cu(s) at the Pt cathode and O_2(g) at the Pt anode, and NaI(aq) yields I_2 at the anode and H_2(g) at the cathode. The other three solutions—H_2SO_4(aq), NaOH(aq), and KNO_3(aq)—all yield H_2(g) at the cathode and O_2(g) at the anode. **22-4.** O_3 and O_3^- are both V shaped. The Lewis structures for these species suggest that the central O atom is sp^2 hybridized in O_3 and sp^3 hybridized in O_3^-; thus, the ideal bond angles are 120° for O_3 and 109° in O_3^-. Experiment shows that the O—O—O bond angles in these two molecular species are much closer: 117° in O_3(g) and 114° in KO_3(*s*). The experimental results suggest that valence bond theory not entirely satisfactory for describing the bonding O_3 and O_3^-, and a molecular orbital approach is more appropriate. (According to molecular orbital theory, the extra electron in O_3^- occupies an antibonding orbital. As a result, the oxygen–oxygen bonds in O_3^- are slightly longer than in O_3, but the bond angle is not significantly affected.) **22-5.** The Cl in OCl_2 shows more positive character (blue color) than does F in OF_2 because the electronegativity of Cl is considerably less than that of F.
22-6. The condensed structural formulas for phosphoric acid and phosphorous acid are $OP(OH)_3$ and $HPO(OH)_2$, respectively.
22-7. The Br atom is larger than the Cl atom. A central P atom can accommodate only four Br atoms, as in the tetrahedral PBr_4^+ ion. But the central P atom can accommodate either four Cl atoms, as in the tetrahedral PCl_4^+ ion, or six, as in the octahedral PCl_6^- ion.

CHAPTER 23

Concept Assessment 23-1. Both involve reducing metallic compounds to the free metal, often from the same ores. Pyrometallurgy employs high temperatures, yields impure metals that must be refined, and generates gaseous emissions and solid wastes. Hydrometallurgy involves leaching desired metal ions into an aqueous solution, followed by chemical or electrolytic reduction to the metal. Lower temperatures are used, gaseous emissions are largely eliminated, but liquid waste solutions are generated. **23-2.** $Fe_2O_3(s) + 3\ H_2(g) \xrightarrow{\Delta} 2\ Fe(s) + 3\ H_2O(g)$ **23-3.** The $[Cr_3O_{10}]^{2-}$ anion consists of three tetrahedral structures arranged around the backbone Cr—O—Cr—O—Cr. The central Cr atom of the backbone is bonded to two additional O atoms and the Cr atoms at the ends to three other O atoms for a total of 10 O atoms. The O. S. of Cr is +6 and that of O is −2 (accounting for the 2− charge on the anion). Similar anions are polysilicate (Fig. 21-31) and polyphosphate (Fig. 22-22). **23-4.** The CO groups in $Fe(CO)_5$ are neutral molecules; the sum of the O.S. of the C and O is 0, and the O.S. of Fe is also 0. **23-5.** Au ($Z = 79$, group 11) has the electron configuration $[Xe]4f^{14}5d^{10}6s^1$; the electrons lost in forming Au^{3+} are the 6s and two of the 5*d*, resulting in $[Xe]4f^{14}5d^8$. **23-6.** The five pairs of electrons around the Cd^{2+} central ion in $[CdCl_5]^{3-}$ is consistent with trigonal bipyramidal molecular geometry (see Table 10-1).

CHAPTER 24

Concept Assessment 24-1. Only one possibility for a six-coordinate complex—$[AlCl_3(H_2O)_3]$—a non-electrolyte that does not conduct electricity and yields no precipitate with $AgNO_3$(aq). Three possible four-coordinate complexes: $[AlCl(H_2O)_3]Cl_2$, $[AlCl_2(H_2O)_2]Cl \cdot H_2O$, and $[AlCl_3(H_2O)] \cdot 2\ H_2O$. The first two of these three can be differentiated by conductivity measurements; the first is the better conductor. The third cannot be distinguished from the six-coordinate complex by Werner's method.
24-2. Three N atoms can donate an electron pair, so the ligand is tridentate. **24-3.** The formula of the coordination compound is $K_3[FeBr_2Cl_2(OH)_2]$ and the name is potassium dibromidodichloridodihydroxidoferrate(III).
24-4. As seen from the models in the margin, substitution of a fourth Cl^- in the *mer*-isomer leads to all four Cl^- ligands in the same plane, but this same isomer cannot be obtained by substituting a fourth Cl^- in the *fac*-isomer.
24-5. Structures **(a)** and **(d)** are identical and are geometric isomers of **(b)**, **(c)**, and **(e)**. Structures **(e)** and **(c)** are enantiomers, and **(b)** and **(e)** are identical. **24-6 (i)** 4 **(ii)** 3 **(iii)** 5 **(iv)** 2 **(v)** 1 **24-7.** Both ligands form chelates, but $[Cr(EDTA)]^-$ does so in a single step, while $[Cr(en)_3]^{3+}$ requires a succession of three steps. The β_1 for $[Cr(EDTA)]^-$ is much larger than β_1 or β_2 of $[Cr(en)_3]^{3+}$, but the cumulative formation constant, β_3 (or K_f), of $[Cr(en)_3]^{3+}$ should be similar to β_1 (or K_f) of $[Cr(EDTA)]^-$.

CHAPTER 25

Concept Assessment 25-1. Radioactive decay that changes Z produces a different element. This includes α, β^-, β^+ emissions and electron capture. If there is no change in Z (emission of γ rays) the element remains the same.
25-2. Francium, a radioactive element, is produced in the decay schemes of heavier elements and found only in conjunction with other decay products, not in natural sources of the alkali metals that Fr resembles.
25-3. Radioactive nuclides with very long half-lives have a very low activity; those with very short half-lives have a high activity but they do not persist long. Those with intermediate half-lives may persist in the environment for a significant period of time at a high activity, making them potentially the most hazardous.
25-4. Refer to Table 25.2 and Figure 25-7 and focus on magic numbers and the relative thickness and width of the belt of stability for a fixed proton number and a fixed neutron number. The greatest thickness comes at the magic number, $Z = 50$. (Tin has 10 stable isotopes.) The greatest width comes at the magic number, $N = 82$. (There are 7 stable nuclides with 82 n.) **25-5.** Refer to Figure 25-7. The point representing ^{44}Ca falls in the belt of stability; ^{44}Ca is a stable nuclide. ^{57}Cu falls below the $N = Z$ line and in the region of β^+ emission, while ^{100}Zr falls above the belt of stability and decays by β^- emission. ^{235}U falls above and beyond the belt of stability and decays by α-particle emission.

CHAPTER 26

Concept Assessment 26-1. All structures based on F atoms at two of the vertices of a tetrahedron and H atoms at the remaining two are superimposable. Only one molecule has the formula CH_2F_2. Two possibilities exist for four atoms at the corners of a square—two F atoms on one side (*cis*) or on opposite corners (*trans*). **26-2.** No. A quaternary carbon atom is bonded to four other carbon atoms.
26-3. The preferred IUPAC name is pentan-2-ol. However, the name 2-pentanol is often used.
26-4. The generic formula of an alkane is C_nH_{2n+2}, and for an alkyl halide where an X atom replaces one H atom, $C_nH_{2n+1}X$. If we limit the series to straight-chain alkanes with X as a terminal atom, we have $H(CH_2)_nX$. That is, $n = 1$, HCH_2X or CH_3X; $n = 2$, HCH_2CH_2X or CH_3CH_2X; $n = 3$, $HCH_2CH_2CH_2X$ or $CH_3CH_2CH_2X$; and so on. **26-5.** Yes.
26-6.

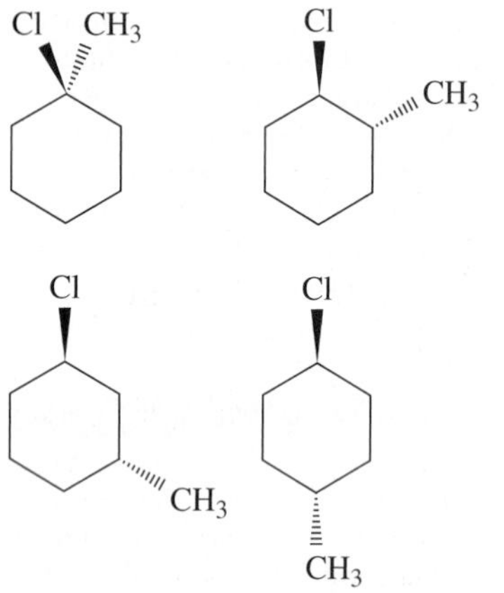

26-7. The conformer with the methyl group in the axial position has higher energy, and so it would release more energy, as heat, if burned.
26-8. The lower energy conformation is

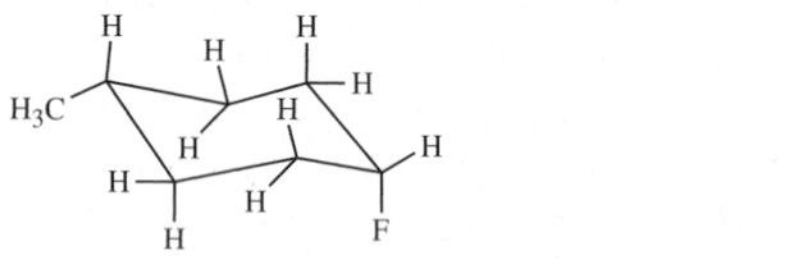

The larger group, $-CH_3$, is in the equatorial position. **26-9.** There are two chiral centers, at the 2nd and 3rd C atoms (to which Br atoms are attached), and four stereoisomers. To show this, sketch a dashed line-wedge structure with the two Br atoms on the same side of the molecule; next to it sketch its nonsuperimposable mirror image. Sketch another structure with the two Br atoms on opposite sides of the molecule and its mirror image, making a total of four stereoisomers.
26-10.

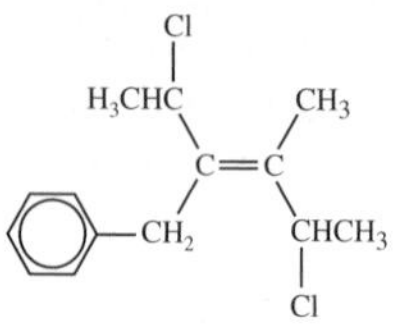

26-11. In butyl acohol —OH attaches to either of the indicated C atoms in the skeleton C—C*—C*—C; in both cases the molecule is the same and there is only one *sec*-butyl alcohol. In pentyl alcohol, —OH attaches to one of the indicated C atoms in C—C*—C#—C*—C. Here, the two structures are the same if attachment is at a C* atom, but a different molecule results if attachment is to the C# atom. The name "*sec*-pentyl alcohol" is inadequate.
26-12.

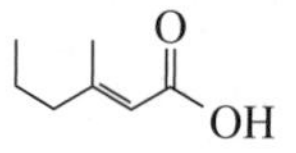

26-13. No, because $C_3H_6O_2$ has only one element of unsaturation. A dialdehyde has two π bonds, and thus two elements of unsaturation.

CHAPTER 27

27-1. CH_3CN is aprotic; NH_3 is protic; $(CH_3)_3N$ is aprotic; $HCONH_2$ is protic; CH_3COCH_3 is aprotic. **27-2.** The other product is the substitution product, $(CH_3)_3C—OCH_2CH_3$, an ether.
Because the substrate is a 3° haloalkane (which disfavors backside attack) and the solvent is polar protic (and stabilizes a carbocation), the ether is formed by the S_N1 mechanism.
27-3.

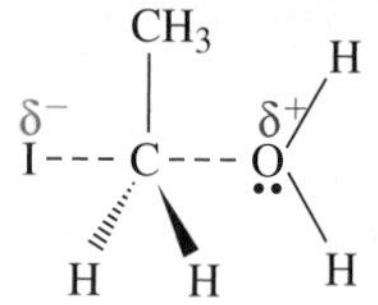

27-4.

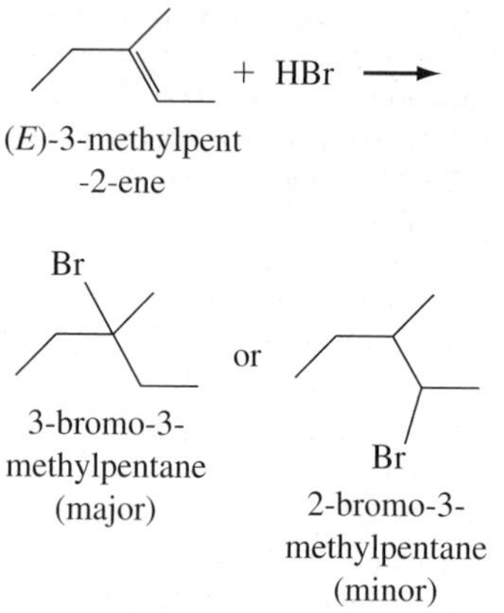

27-5. A mixture of enantiomers would be obtained because hydration proceeds through a carbocation. The carbocation can be attacked from above or below by a water molecule, and so both the (*R*) and (*S*) configurations of butan-2-ol will be produced. **27-6.** The product will be *trans*-1,2-dibromocyclopentane. If the bromonium ion is formed with the Br atom situated above the plane of the ring, then Br^- will attack from below the plane of the ring because it must attack from the backside. Therefore, the *trans* isomer is obtained:

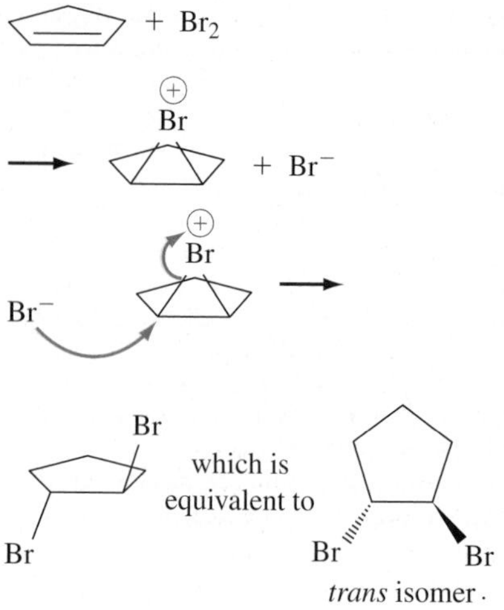

27-7. Once formed, CH_3Cl can react with a Cl to form a $H_2ClC\bullet$ radical. A $H_2ClC\bullet$ radical can then react with a $H_3C\bullet$ radical to form $ClCH_2CH_3$. **27-8.** $(CH_3)_2CBrCH_2CH_3$.

CHAPTER 28

Concept Assessment 28-1. The products are 1 mol glycerol and 1 mol each of sodium palmitate, sodium oleate, and sodium linoleate.
28-2. The mirror image of a (+) enantiomer is the (−) enantiomer, so the mirror image of D-(+)-glucose is L-(−)-glucose. There can be no D-(−)-glucose. **28-3.** This polypeptide structure is shown in Example 28-1. Ionization occurs only at the N-terminal and C-terminal ends of the chain and nowhere else along the chain. Since the N-terminal and C-terminal amino acids are at a pH more than one unit above their isoelectric points, the only significant ionization is at the C-terminal amino acid, which would be present as a 1− anion, making the net charge on the tripeptide also 1−. **28-4.** Helix formation in a protein (Fig. 28-12) requires close proximity of carbonyl and amide groups and formation of hydrogen bonds between them, occurring regularly over an entire macromolecule. In the polysaccharides (Fig. 28-9), O atoms and —OH groups could conceivably form hydrogen bonds between them, but more randomly and not in the very tight helical fashion seen in proteins.

Index

The letters following a page reference indicate: a note (*n*); a figure (*f*); a table (*t*). The letters preceding a page reference indicate: the appendices (A); the online chapter, Chapter 28 (W).

C

D

E

I

J

K

L

M

P

Q

R

S

U

V

X

Y

Z

Selected Physical Constants*

Acceleration due to gravity	g	9.80665 m s^{-2}
Speed of light (in vacuum)	c	$2.99792458 \times 10^8 \text{ m s}^{-1}$
Gas constant	R	$0.0820573 \text{ atm L mol}^{-1}\text{ K}^{-1}$
		$0.083144598 \text{ bar L mol}^{-1}\text{ K}^{-1}$
		$8.3144598 \text{ J mol}^{-1}\text{ K}^{-1}$
Electron charge	e^-	$-1.6021766208 \times 10^{-19}\text{ C}$
Electron rest mass	m_e	$9.10938356 \times 10^{-31}\text{ kg}$
Planck's constant	h	$6.626070040 \times 10^{-34}\text{ J s}$
Faraday constant	F	$9.648533289 \times 10^4\text{ C mol}^{-1}$
Avogadro constant	N_A	$6.022140857 \times 10^{23}\text{ mol}^{-1}$
Boltzmann constant	kB	$1.38064852 \times 10^{-23}\text{ J K}^{-1}$

*Committee on Data for Science and Technology (CODATA) Recommended Values of the Fundamental Physics Constants: 2014 (http://physics.nist.gov/cuu/Constants/index.html)

Some Common Conversion Factors

Length

1 meter (m) = 39.37007874 inches (in.)

1 in. = 2.54 centimeters (cm) (exact)

Mass

1 kilogram (kg) = 2.2046226 pounds (lb)

1 lb = 453.59237 grams (g)

Volume

1 liter (L) = 1000 mL = 1000 cm^3 (exact)

1 L = 1.056688 quart (qt)

1 gallon (gal) = 3.785412 L

Pressure

1 atmosphere (atm) = 101.325 kilopascals (kPa) (exact)
= 1.01325 bar (exact)
= 760 Torr (exact)

1 Torr ≈ 1 millimeter of mercury (mmHg)

Energy

1 joule (J) = 1 N m = 1 kg $\text{m}^2\text{ s}^{-2}$

1 calorie (cal) = 4.184 J (exact)

1 kPa L = 1 J

1 bar L = 100 J

1 atm L = 101.325 J (exact)

1 electronvolt (eV) = $1.602176565 \times 10^{-19}$ J

1 eV/atom = 96.485 kJ mol^{-1}

1 kilowatt hour (kWh) = 3600 kJ (exact)

Mass-energy equivalence:

1 unified atomic mass unit (u)
= $1.660538921 \times 10^{-27}$ kg
= 931.494061 MeV

Force

1 newton (N) = 1 kg m s^{-2}

Some Useful Geometric Formulas

Perimeter of a rectangle = $2l + 2w$

Circumference of a circle = $2\pi r$

Area of a rectangle = $l \times w$

Area of a triangle = $\frac{1}{2}$(base × height)

Area of a circle = πr^2

Area of a sphere = $4\pi r^2$

Volume of a parallelepiped = $l \times w \times h$

Volume of a sphere = $\frac{4}{3}\pi r^3$

Volume of a cylinder or prism = (area of base) × height

$\pi \approx 3.14159$